Primer on the Metabolic Bone Diseases and Disorders of Mineral Metabolism

Seventh Edition

An Official Publication of the American Society for Bone and Mineral Research

NHS Staff Libr
To re this book, please phone
afflibraries@nhs.net.
library c

Primer on the Metabolic Bone Diseases and Disorders of Mineral Metabolism

Seventh Edition

EDITOR-IN-CHIEF

Clifford J. Rosen, M.D.
Maine Medical Center
Research Institute
Scarborough, Maine

SENIOR ASSOCIATE EDITORS

Juliet E. Compston, M.D., FRCP
University of Cambridge
School of Clinical Medicine
Cambridge, United Kingdom

Jane B. Lian, Ph.D.
University of Massachusetts
Medical School
Worcester, Massachusetts

ASSOCIATE EDITORS

Roger Bouillon, M.D., Ph.D.
Katholieke Universiteit Leuven
Laboratory for Experimental Medicine and Endocrinology
Leuven, Belgium

Pierre D. Delmas, M.D., Ph.D.
Hopital Edouard Herriot
Lyon, France

Marie Demay, M.D.
Massachusetts General Hospital and Harvard Medical School
Boston, Massachusetts

Theresa A. Guise, M.D.
University of Virginia
Department of Internal Medicine
Charlottesville, Virginia

Suzanne M. Jan de Beur, M.D.
Johns Hopkins University
Department of Medicine
Baltimore, Maryland

Richard W. Keen, M.D., Ph.D.
Royal National Orthopaedic Hospital
Stanmore, United Kingdom

Laurie Kay McCauley, D.D.S., Ph.D.
University of Michigan, School of Dentistry, Department of Periodontics and Oral Medicine
Ann Arbor, Michigan

Paul D. Miller, M.D., FACP
Colorado Center for Bone Research
Lakewood, Colorado

Socrates Papapoulos, M.D.
Leiden University Medical Center
Department of Endocrinology and Metabolic Diseases
Leiden, The Netherlands

Vicki Rosen, Ph.D.
Harvard School of Dental Medicine
Department of Developmental Biology
Boston, Massachusetts

Ego Seeman, M.D., FRACP
Heidelberg Redatriation Hospital
Department of Endocrinology
West Heidelberg, Australia

Rajesh V. Thakker, M.D., FRCP
Oxford University
Nuffield Department of Clinical Medicine
Oxford, United Kingdom

Published by the American Society for Bone and Mineral Research
Washington, D.C.

Cover Designer: David Spratte, Third Eye Studio
Printer: The Sheridan Press

©2008 by American Society for Bone and Mineral Research. Published by:

American Society for Bone and Mineral Research
2025 M Street, NW, Suite 800
Washington, DC 20036-3309
www.asbmr.org

All rights reserved. This book is protected by copyright. No part of this book may be reproduced in any form or by any means, including photocopying, or utilized by any information storage and retrieval system without written permission from the copyright owner, except for brief quotations embodied in critical articles and reviews. Materials appearing in this book prepared by individuals as part of their official duties as U.S. government employees are not covered by the above-mentioned copyright.

Printed in the USA

Library of Congress Cataloging-in-Publication Data

Primer on the metabolic bone diseases and disorders of mineral metabolism. — 7th ed. / editor-in-chief, Clifford J. Rosen ; senior associate editors, Juliet E. Compston, Jane B. Lian ; associate editors, Roger Bouillon . . . [et al.].

p. ; cm.

Includes bibliographical references and index.

ISBN 978-0-9778882-1-4 (pbk. : alk. paper)

1. Bones—Metabolism—Disorders. 2. Mineral metabolism—Disorders. 3. Bones—Diseases. I. Rosen, Clifford J. II. American Society for Bone and Mineral Research.

[DNLM: 1. Bone Diseases, Metabolic. 2. Bone and Bones—metabolism. 3. Minerals—metabolism. WE 250 P953 2009]

RC931.M45P75 2009
616.7′16—dc22

2008043881

Care has been taken to confirm the accuracy of the information presented and to describe generally accepted practices. However, the authors, editors and publisher are not responsible for errors or omissions or for any consequences from application of the information in this book and make no warranty, expressed or implied, with respect to the currency, completeness, or accuracy of the contents of the publication. Application of this information in a particular situation remains the professional responsibility of the practitioner. The authors have sought to ensure that drug selection and dosage set forth in this text are in accordance with current recommendations and practice at the time of publication. However, in view of ongoing research, changes in government regulations, and the constant flow of information relating to drug therapy and drug reactions, the reader is urged to check the package insert for each drug for any change in indications and dosage and for added warnings and precautions. This is particularly important when the recommended agent is a new or infrequently employed drug. Some drugs and medical devices presented in this publication have Food and Drug Administration (FDA) clearance for limited use in restricted research settings. It is the responsibility of the health care provider to ascertain the FDA (or the provider's own country's equivalent agency) status of each drug or device planned for use in their clinical practice.

Contents

Please also see the following website on Chondrodysplasias: http://www.csmc.edu/10784.html

Please also see the ASBMR ONJ Task Force Report at: http://www.jbmronline.org/doi/full/10.1359/jbmr.0707onj?prevSearch=authorsfield%3A%28shane%29 and The Jackson Laboratory website at www.jax.org

Section XI. Appendix
(Section Editor: Roger Bouillon)

Please visit the Seventh Edition *Primer* website at www.asbmrprimer.org for the expanded and updated Appendix.

Contributing Authors

Silvano Adami, M.D.
John Adams, M.D.
Judith Adams, MBBS, FRCR
Yasemin Alanay, M.D.
Andrew Arnold, M.D.
Emilio Arteaga-Solis, M.D., Ph.D.
John R. Asplin, M.D.
Laura K. Bachrach, M.D.
Shona Bass, Ph.D.
Wesley G. Beamer, Ph.D.
Paolo Bianco, M.D.
Daniel Bikle, M.D., Ph.D.
John P. Bilezikian, M.D.
Heike A. Bischoff-Ferrari, M.D., MPH
Nick Bishop, MRCP, M.D.
Ronald Blasberg, M.D.
Amy E. Bobrowski, M.D.
Jean-Jacques Body, M.D., Ph.D.
Lynda F. Bonewald, Ph.D.
Steven Boonen, M.D., Ph.D.
Adele L. Boskey, Ph.D.
Roger Bouillon, M.D., Ph.D.
Mary L. Bouxsein, Ph.D.
Alan Boyde, Ph.D., BDS, M.D.
Nathalie Bravenboer, Ph.D.
Edward M. Brown, M.D.
Øyvind S. Bruland, M.D., Ph.D.
David A. Bushinsky, M.D.
Thomas O. Carpenter, M.D.
Jacqueline R. Center, Ph.D.
Noe L. Charbonneau
Edward Chow, MBBS, MSc, FRCPC
Sylvia Christakos, Ph.D.
Blaine A. Christiansen, Ph.D.
Yong-Hee Chun, Ph.D.
Gregory A. Clines, M.D., Ph.D.
Denis R. Clohisy, M.D.
Jackie A. Clowes, M.D., Ph.D.
Adi Cohen , M.D., MHS
Michael T. Collins, M.D.
Juliet Compston, M.D., FRCP
Gary J. R. Cook, MBBS, M.Sc.
Cyrus Cooper, D.M., FRCP, MedSci
Felicia Cosman, M.D.
Gilbert J. Cote, Ph.D.
Bess Dawson-Hughes, M.D.
Pierre D. Delmas, M.D., Ph.D.
Linda L. Demer, M.D., Ph.D.
David Dempster, Ph.D.
Elaine Dennison, Ph.D., MBBS
Matthew T. Drake, M.D., Ph.D.
Gaele Ducher, Ph.D.
Richard Eastell, M.D., FRCP
Peter R. Ebeling, M.D., FRACP
Michael J. Econs, M.D.
Paul C. Edwards, D.D.S.
Thomas A. Einhorn, M.D.
John A. Eisman, MBBS, Ph.D.
Florent Elefteriou, Ph.D.
Jessica Ellerman, M.D.
Murray J. Favus, M.D.
Ignac Fogelman, M.D.
Peter A. Friedman, Ph.D.
Robert F. Gagel, M.D.
Claus C. Glüer, Ph.D.
Gopinath Gnanasegaran, MBBS, M.Sc., M.D.
Deborah T. Gold, Ph.D.
Steven R. Goldring, M.D.
David Goltzman, M.D.
David J. J. de Gorter, Ph.D.
Susan Greenspan, M.D.
Theresa A. Guise, M.D.
Neveen A. T. Hamdy, M.D., MRCP
Nicholas Harvey, MBBChir
Robert P. Heaney, M.D.
Angela C. Hirbe, B.A.
Amanda Hird, B.S.
Janet M. Hock, BDS, Ph.D.
Steven P. Hodak, M.D.
Ingrid A. Holm, M.D., MPH
Mara J. Horwitz, M.D.
Keith A. Hruska, M.D.
Suzanne M. Jan de Beur, M.D.
Graeme Jones, M.D., FRACP
Sheila J. Jones, Ph.D.
Stefan Judex, Ph.D.
Harald Jüppner, M.D.
John A. Kanis, M.D.
Frederick S. Kaplan, M.D.
Magnus K. Karlsson, M.D., Ph.D.
Marcel Karperien, Ph.D.
Gerard Karsenty, M.D., Ph.D.
Richard W. Keen, M.D., Ph.D.
Sundeep Khosla, M.D.
Douglas P. Kiel, M.D., MPH
Keith L. Kirkwood, D.D.S., Ph.D.
Michael Kleerekoper, M.D.
Gordon L. Klein, M.D., MPH
John Klingensmith, Ph.D.
Stephen J. Knohl, M.D.
Scott Kominsky, Ph.D.
Christopher S. Kovacs, M.D., FRCP
Carola Krause, M.Sc.
Paul H. Krebsbach, DDS, Ph.D.
Henry M. Kronenberg, M.D.
Craig B. Langman, M.D.
Ching C. Lau, M.D.
Michael A. Levine, M.D., FAAP, FACP

Robert Lindsay, MBChB, Ph.D., FRCP
Paul Lips, M.D., Ph.D.
David G. Little, MBBS, FRACS
Karen Lyons, Ph.D.
Sharmila Majumdar, Ph.D.
David A. Mankoff, M.D., Ph.D.
Joan C. Marini, M.D., Ph.D.
T. John Martin, M.D., DSc
Stephen J. Marx, M.D.
Suresh Mathew, M.D.
Laurie K. McCauley, D.D.S., Ph.D.
Wassim M. McHayleh, M.D.
Elise F. Morgan, Ph.D.
Elizabeth A. Morgan, M.D.
Gregory R. Mundy, M.D.
Dorothy A. Nelson, Ph.D.
Kong Wah Ng, M.D.
Robert A. Nissenson, Ph.D.
Shane A. Norris, Ph.D.
Regis J. O'Keefe, M.D.
Robert N. Ono
Eric S. Orwoll, M.D.
Socrates E. Papapoulos, M.D.
Flavia Pirih, D.D.S., Ph.D.
Lawrence G. Raisz, M.D.
Manoj Ramachandran, BSc, MBBS, MRCS, FRCS
Francesco Ramirez, Ph.D.
Frank Rauch, M.D.
Andrew Ravanelli, B.S.
Robert R. Recker, M.D.
Ian R. Reid, M.D., MBChB
Mara Riminucci, M.D.
David L. Rimoin, M.D., Ph.D.
René Rizzoli, M.D.
Pamela Gehron Robey, Ph.D.
G. David Roodman, M.D., Ph.D.
Clifford J. Rosen, M.D.
F. Patrick Ross, Ph.D.
Clinton Rubin, Ph.D.
Janet Rubin, M.D.
Mishaela R. Rubin, M.D.
Robert K. Rude, M.D.
Mary D. Ruppe, M.D.
Mary Ann C. Sabino, D.D.S., Ph.D.
Lynn Y. Sakai, Ph.D.
Steven J. Scheinman, M.D.
Gerhard Sengle, Ph.D.
Inna Serganova, Ph.D.
Elizabeth Shane, M.D.
Nicholas Shaw, MBChB, FRCPCH
Andrew T. Shields, M.D.
Dolores Shoback, M.D.
Eileen M. Shore, Ph.D.
Shonni J. Silverberg, M.D.
James P. Simmer, Ph.D.
Ethel S. Siris, M.D.
Julie A. Sterling, Ph.D.
Andrew F. Stewart, M.D.
Pawel Szulc, M.D., Ph.D.
Peter ten Dijke, Ph.D.
Rajesh V. Thakker, M.D., FRCP
Anna N. A. Tosteson, Sc.D.
Dwight A. Towler, M.D., Ph.D.
Nathaniel S. Treister, DMD, DMSc
Özge Uluçkan, B.S.
Natasja M. van Schoor, Ph.D.
Dirk Vanderschueren, M.D., Ph.D.
Katrien Venken, Ph.D.
Jean Wactawski-Wende, Ph.D.
Connie M. Weaver, Ph.D.
Kristy Weber, M.D.
Katherine Weilbaecher, M.D.
Robert S. Weinstein, M.D.
Kenneth E. White, Ph.D.
Michael P. Whyte, M.D.
Tania Winzenberg, Ph.D., MBBS
Sook-Bin Woo, DMD
John J. Wysolmerski, M.D.
Yingzi Yang, Ph.D.
Michael J. Zuscik, Ph.D.

Preface to the Seventh Edition

The mission of the ASBMR *Primer on the Metabolic Bone Diseases and Disorders of Mineral Metabolism* is to present core knowledge in this field in a comprehensive yet concise fashion. In the two years since publication of the Sixth Edition, substantial new information has advanced our understanding of bone biology, mineral homeostasis, hormonal regulation, and the associated disease states.

The Seventh Edition reflects these rapid and exciting advances in the basic sciences and translational research. Each of the major sections now includes an overview which provides the reader with an opportunity to understand the scope and implications of the chapters contained within it. Under the guidance of Jane Lian and Juliet Compston, we added several new chapters, and expanded the text by nearly one third. We increased our international representation on the editorial board and among authors. In the Seventh Edition, we added a new section on animal models that includes everything from mouse genetics to phenotyping and structural correlates. We expanded the section related to malignancy and bone under the guidance of Dr. Theresa Guise and included sections on the management of bone pain from cancer and the treatment of metastatic bone disease. Genetics is now covered more extensively, for mouse and humans, and methods for measuring bone mass have been expanded to keep pace with recent technology. Dr. Laurie McCauley spearheaded an expansion of the oral and maxillofacial section to include new concepts, and expand on a discussion of osteonecrosis of the jaw. The appendix is now much larger and includes not only reference data, but Web-based links to multiple sites for ease of use. This appendix will appear on the *Primer* website (www.asbmrprimer.org) and is free to access. In addition, Dr. Bouillon did an outstanding job updating approved drugs for metabolic bone diseases for both national and international members. In total, some chapters have been consolidated, others expanded, but all continue the role of the *Primer* as an essential resource for house staff, graduate students, fellows, and junior faculty. We believe the 7th edition will prove even more valuable for the many scientists, trainees, and physician practitioners who are intrigued and challenged by the complexities of the bone, its regulatory hormones, and the related diseases.

The Seventh Edition of the *Primer* is the first without Dr. Murray Favus as Editor in Chief. We continue to owe him a debt of gratitude for his previous successes in modeling the *Primer* as the most complete and up to date source of information in the field. This edition also represented a complete change in the editorial board, with two new associate editors, Jane Lian and Juliet Compston, as well as new assistant editors: Drs. Vicki Rosen, Socrates Papapoulos, Pierre Delmas, Raj Thakker, Paul Miller, Laurie McCauley, Richard Keene, Ego Seeman, Marie Demay, Roger Bouillon, Theresa Guise, and Suzanne Jan de Beur. As such, this edition is a remarkable accomplishment and a testimony to the hard work, collegiality, and fellowship of this editorial board. It is even more remarkable considering the increase in breadth and depth that was undertaken in this publication. Our initial goal was to expand the number of chapters by nearly one third, and enhance the appendix, both as a reference component and as a gateway for Web-based references.

The Seventh Edition of the *Primer* is the product of a tripartite collaboration involving an experienced editorial board, skilled authors and an enthusiastic and supportive ASBMR editorial staff. The ASBMR staff, seasoned by the initial experience of self-publication with the Fifth Edition, has improved the publication process in many ways that benefited the authors and editors. The JBMR staff under the leadership of Matt Kilby must also be credited with expanding the distribution of the *Primer* and delivering it into the hands of so many more students, trainees, scientists, and physician practitioners than we ever envisioned. Matt's tireless efforts on the part of the *Primer* are greatly appreciated.

The evolution of the *Primer* is an ongoing process, and just as we complete the Seventh Edition we look forward to continuing this work in the Eighth Edition.

Of course, it is the contributing authors who have made the *Primer* a valuable source of information in the bone field. The editorial board is deeply indebted to the many authors who over the years have articulated in print the state of their areas of expertise and to the new authors who have given so much of their valuable time to this endeavor. This impressive group has successfully taken on the challenge of organizing the ever-expanding scientific information into a coherent presentation.

Clifford J. Rosen, M.D.
Maine Medical Center Research Institute
Scarborough, Maine

The untimely death of Dr. Pierre Delmas, one of the associate editors of the *Primer,* is a tragic loss to the Society, to the editors of this book, and to all who work in the field of metabolic bone diseases. His commitment to research, teaching, and promoting public health will be valued for years to come. The Seventh Edition of the *Primer* is dedicated to his legacy.

Preface for Primer

The American Society for Bone and Mineral Research is proud to present the Seventh Edition of the *Primer on the Metabolic Bone Disease and Disorders of Mineral Metabolism*. This book marks the first edition published under its new Editor-in-Chief, Dr. Clifford Rosen, and many notable changes have been made to the text and format as a result of this transition.

A major objective of Dr. Rosen and the new editorial team was to increase international coverage with the Seventh Edition and the accomplishment of this goal has exceeded all expectations. The editorial board and the list of contributing authors closely match the international depth of the ASBMR.

Along with these changes, Dr. Roger Bouillon was tasked with creating an international appendix, which can be seen on the *Primer* Website (www.asbmrprimer.org). Among the tables is a formulary of drugs used in the treatment of bone disorders and their availability throughout the world. The appendix will be located online only to allow extra room for broad coverage of topics in the book, as well as to make this information freely accessible online.

The Seventh Edition of our *Primer* upholds the tradition of excellence set by its predecessors while carving its own niche into our science and our lives. We commend the efforts of everyone involved.

Barbara E. Kream, Ph.D.
President
American Society for Bone and Mineral Research

SECTION I

Molecular Cellular and Genetic Determinants of Bone Structure and Formation (Section Editor: Vicki Rosen)

© 2008 American Society for Bone and Mineral Research

Chapter 1. Skeletal Morphogenesis and Embryonic Development

Yingzi Yang

Developmental Genetics Section, Genetic Disease Research Branch, National Human Genome Research Institute, Bethesda, Maryland

INTRODUCTION

Formation of the skeletal system is one of the hallmarks that distinguish vertebrate animals from invertebrate ones. In higher vertebrates (i.e., birds and mammals), the skeletal system contains mainly cartilage and bone that are mesoderm-derived tissues and formed by chondrocytes and osteoblasts, respectively, during embryogenesis. A common mesenchymal progenitor cell also referred as the osteochondral progenitor gives rise to both chondrocytes and osteoblasts. Skeletal development starts from mesenchymal condensation, during which mesenchymal progenitor cells aggregate at future skeletal locations. Because mesenchymal cells in different parts of the embryo come from different cell lineages, the locations of initial skeletal formation determine which of the three mesenchymal cell lineages contribute to the future skeleton. Neural crest cells from the branchial arches contribute to the craniofacial bone, the sclerotome compartment of the somites gives rise to most axial skeletons, and lateral plate mesoderm form the limb mesenchyme, from which limb skeletons are derived (Fig. 1). Ossification is one of the most critical processes in bone development, and this process is controlled by two major mechanisms: intramembranous and endochondral ossification. Osteochondral progenitors differentiate into osteoblasts to form the membranous bone during intramembranous ossification, whereas during endochondral ossification, osteochondral progenitors differentiate into chondrocytes instead to form a cartilage template of the future bone. The location of each skeletal element also determines its ossification mechanism and unique anatomic properties such as the shape and size. Importantly, the positional identity of each skeletal element is acquired early in embryonic development even before mesenchymal condensation through a process called pattern formation.

Cell–cell communication that coordinates cell proliferation and differentiation plays a critical role in pattern formation. Patterning of the early skeletal system is controlled by several major signaling pathways that also regulate other pattern formation processes. These signaling pathways are mediated by Wnts, Hedgehogs (Hhs), Bone morphogenetic proteins (BMPs), fibroblast growth factors (FGFs), and Notch/Delta. Later in skeletal development in a different cell context, these signaling pathways also control cell fate determination, proliferation, and maturation in the skeleton.

EARLY SKELETAL PATTERNING

In the craniofacial region, neural crest cells are major sources of cells establishing the craniofacial skeleton.[1] It is the temporal- and spatial-dependent reciprocal signaling between and among the neural crest cells and the epithelial cells (surface ectoderm, neural ectoderm or endodermal cells) that ultimately establish the pattern of craniofacial skeleton formed by neural crest cells.[2] Patterning of the axial skeleton can be traced back to the formation of somites, which are segmented mesodermal structure on either side of the neural tube and the underlying notochord. Somites give rise to axial skeleton, striated muscle, and dorsal dermis.[3–6] The repetitive and left-right symmetrical patterning of axial skeleton is controlled by a molecular oscillator or the segmentation clock and gradients of signaling molecules that act in the presomitic mesoderm (PSM) (Fig. 2A). The segmentation clock is operated by cyclic expression of genes, most of which are components of the Notch and canonical Wnt signaling pathways (Fig. 2B).[7,8]

The Notch signaling pathway mediates short-range communication between contacting cells.[9] The majority of cyclic genes are downstream targets of the Notch signaling pathway and code for Hairy/Enhancer of split (Hes) family members, Lunatic fringe (Lfng), and the Notch ligand Delta. The canonical Wnt signaling pathway mediates long-range signaling across several cell diameters. On activation of canonical Wnt signaling, β-catenin is stabilized and translocates to the nucleus where it binds Lef/Tcf factors and activates expression of downstream genes including *Axin2* and *Nkd1*. *Axin2* and *Nkd1* expression oscillates out of phase with Notch signaling components in the mouse PSM (Fig. 2B).[10,11] The Notch and Wnt pathways interact to certain extent within the mechanism of the segmentation clock.

The FGF and retinoic acid (RA) signaling also control somatogenesis by regulating the competence of PSM cells to undergo segmentation. They form two opposing and functionally antagonistic gradients within the PSM (Fig. 2A).[12] RA signaling has an additional role in maintaining left-right bilateral symmetry of somites by interacting and coordinating with the signaling pathways establishing left-right asymmetry of the body axis and the segmentation clock of the somites.[13–16]

The functional significance of segmentation clock in human skeletal development is highlighted by congenital axial skeletal diseases. Abnormal vertebral segmentation (AVS) in humans is a relatively common malformation. For instance. mutations in the NOTCH signaling components cause at least two human disorders, spondylocostal dysostosis (SCD, 277300, 608681, and 609813) and Alagille syndrome (AGS, OMIM 118450 and 610205); both exhibit vertebral column defects.

The formed somite is also patterned along the dorsal-ventral axis by cell signaling from the surface ectoderm, neural tube and the notochord (Fig. 1). Ventralizing signals such as Sonic hedgehog (Shh) from the notochord and ventral neural tube is required to induce sclerotome formation on the ventral side,[17,18] whereas Wnt signaling from the surface ectoderm and dorsal neural tube is required for the formation of dermomyotome on the dorsal side of the somite (Fig. 1).[19,20] The sclerotome gives rise to the axial skeleton and the ribs. In the mouse mutant that lacks *Shh* function, vertebral column and posterior ribs failed to form. The paired domain transcription factor *Pax1* is expressed in the sclerotome, and *Shh* is required to regulate its expression.[21–23] However, axial skeletal phenotypes in *Pax1* mutant mice[24] were far less severe than those in the *Shh* mutants.

Key words: bone, cartilage, embryonic development, chondrocyte, osteoblast, cell signaling, skeletal development, pattern formation PTH-related peptide, Ihh, Wnt, fibroblast growth factor, bone morphogenetic protein, retinoic acid, Notch, Sox9, Runx2, ossification mechanism, intramembranous ossification, endochondral ossification, chondrocyte proliferation, chondrocyte hypertrophy, somatogenesis, segmentation clock

The author states that she has no conflicts of interest.

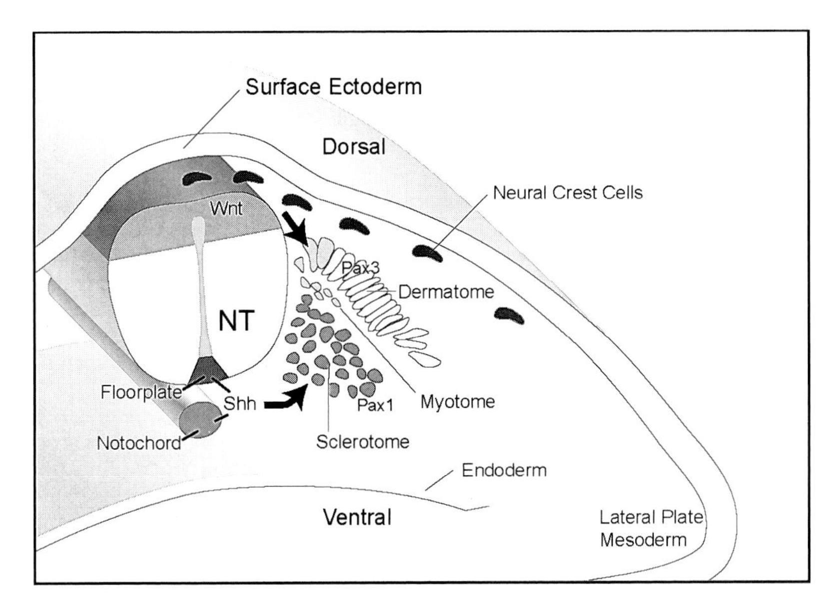

FIG. 1. Cell lineage contribution of chondrocytes and osteoblasts. Neural crest cells are born at the junction of dorsal neural tube and surface ectoderm. In the craniofacial region, neural crest cells from the branchial arches differentiate into chondrocytes and osteoblasts. In the trunk, axial skeletal cells are derived from the ventral somite compartment, sclerotome. Shh secreted from the notochord and floor plate of the neural tube induces the formation of sclerotome which expresses *Pax1*. Wnts produced in the dorsal neural tube inhibits sclerotome formation and induces dermomyotome that expresses *Pax3*. Cells from the lateral plate mesoderm will form the limb mesenchyme, from which limb skeletons are derived.

Limb skeletons are patterned along the proximal-distal (P-D, shoulder to digit tip), anterior-posterior (A-P, thumb to little finger), and dorsal-ventral (D-V, back of the hand to palm) axis (Fig. 3). Along the P-D axis, the limb skeletons form three major segments: humerus or femur at the proximal end, radius and ulna or tibia and fibula in the middle, and carpal/tarsal, metacarpal/metatarsal and digits in the distal end. Along the A-P axis, the radius and ulna have distinct morphological features; so does each of the five digits. Patterning along the D-V limb axis also results in characteristic skeletal shapes and structures. For instance, the sesamoid processes are located ventrally whereas the knee patella forms on the dorsal side of the knee. The 3D limb patterning events are regulated by three signaling centers in the early limb primordium called the limb bud before mesenchymal condensation.

The apical ectoderm ridge (AER), a thickened epithelial structure formed at the distal tip of the limb bud, is the signaling center that directs P-D limb outgrowth (Fig. 3). Canonical Wnt signaling activated by Wnt3 indices AER formation.[25] *Fgf* family members that are expressed in the AER, mainly *Fgf8* and *Fgf4*, are necessary and sufficient to mediate the function of AER.[26–28] *Fgf10* expressed in the presumptive limb mesoderm is required for limb initiation and it also controls limb outgrowth by maintaining *Fgf8* expression in the AER.[29–31]

The second signaling center is the zone of polarizing activity (ZPA), which is a group of mesenchymal cells located at the posterior distal limb margin and immediately adjacent to the AER (Fig. 3B). When ZPA tissue is grafted to the anterior limb bud under the AER, it leads to digit duplication in mirror image of the endogenous ones.[32] *Shh* is expressed in the ZPA and is both necessary and sufficient to mediate ZPA activity in patterning digit identity along the A-P axis.[33,34] However, the A-P axis of the limb is established before Shh signaling. This pre-Shh A-P limb patterning is controlled by combined activities of Gli3, Alx4, and basic helix-loop-helix (bHLH) transcription factors dHand and Twist1. The Gli3 repressor form (Gli3R) and Alx4 establish the anterior limb territory by re-

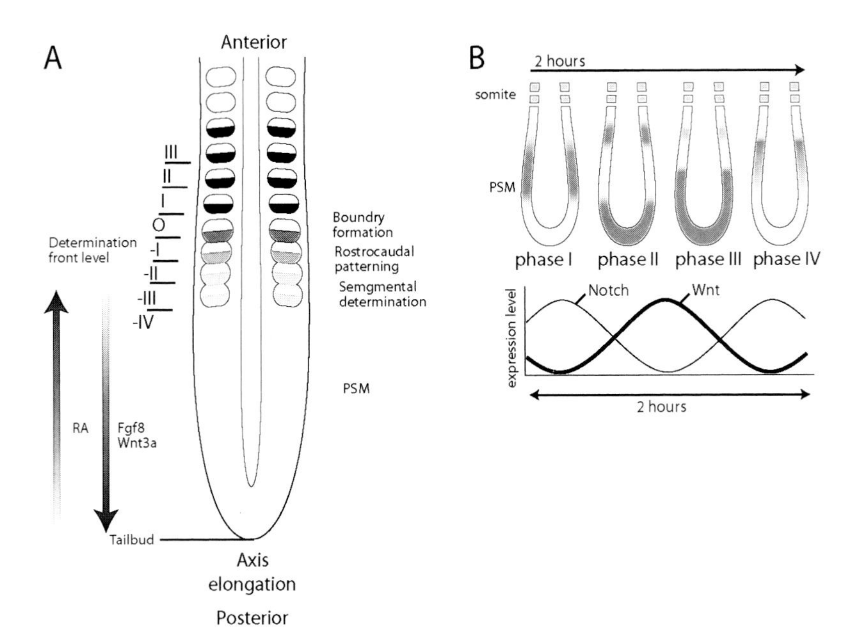

FIG. 2. Periodic and left-right symmetrical somite formation is controlled by signaling gradients and oscillations. (A) Somites form from the presomitic mesoderm (PSM) on either side of the neural tube in an anterior to posterior (A-P) wave. Each segment of the somite is also patterned along the A-P axis. Retinoic acid signaling controls the synchronization of somite formation on the left and right side of the neural tube. The most recent visible somite is marked by "0," whereas the region in the anterior PSM that is already determined to form somites is marked by a determination front that is determined by Fgf8 and Wnt3a gradients. This Fgf signaling gradient is antagonized by an opposing gradient of retinoic acid. (B) Periodic somite formation (one pair of somite/2 h) is controlled by a segmentation clock, the molecular nature of which is oscillated expression of signaling components in the Notch and Wnt pathway. Notch signaling oscillates out of phase with Wnt signaling.

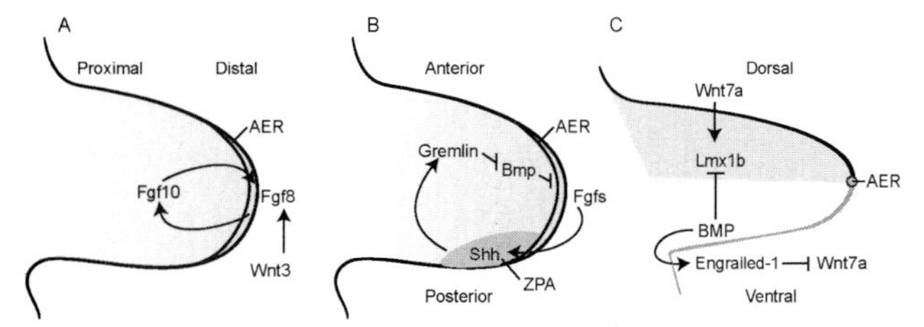

FIG. 3. Limb patterning and growth along the proximal-distal (P-D), anterior-posterior (A-P), and dorsal-ventral (D-V) axes are controlled by signaling interactions and feedback loops. (A) A signaling feedback loop between Fgf10 in the limb mesoderm and Fgf8 in the AER is required to direct P-D limb outgrowth. Wnt3 is required for AER formation. (B) Shh in the ZPA controls A-P limb patterning. A-P and P-D limb patterning and growth are also coordinated through a feedback loop between *Shh* and *Fgfs* expressed in the AER. Fgf signaling from the AER is required for *Shh* expression. *Shh* also maintains AER integrity by regulating *Gremlin* expression. Gremlin is a secreted antagonist of Bmp signaling which promotes AER degeneration. (C) D-V patterning of the limb is determined by Wnt7a and Bmp signaling through regulating the expression of *Lmx1b* in the limb mesenchyme.

stricting *dHand* expression to the posterior limb,[35] which in turn activates Shh expression.[36,37] In addition, the activity of dHand in the posterior limb is antagonized by Twist1 through a dHand-Twist1 heterodimer.[38] Mutations in the human *TWIST1* gene cause Saethre-Chotzen syndrome (SCS; OMIM 101400), one of the most commonly inherited craniosynostosis conditions. The hallmarks of this syndrome are premature fusion of the calvarial bones and limb abnormalities. Mutations in GLI3 gene also cause limb malformations including Greig cephalopolysyndactyly syndrome (GCPS; OMIM 175700) and Pallister-Hall syndrome (PHS; OMIM 146510).

The third signaling center is the non-AER limb ectoderm that covers the limb bud. It sets up the D-V polarity of not only the ectoderm but also the underlying mesoderm of the limb (Fig. 3C).[39,40] Wnt and BMP signaling are required to control D-V limb polarity of both the limb ectoderm and mesoderm. *Wnt7a* is expressed specifically in the dorsal limb ectoderm, and it activates the expression of *Lmx1b*, which encodes a dorsal-specific LIM homeobox transcription factor that determines the dorsal identity.[41,42] *Wnt7a* expression in the ventral ectoderm is suppressed by *En-1*, which encodes a transcription factor that is expressed specifically in the ventral ectoderm.[43] The BMP signaling pathway is also ventralizing in the early limb (Fig. 3C). The function of BMP signaling in the early limb ectoderm is upstream of *En-1* in controlling D-V limb polarity.[44] However, in the mouse limb bud mesoderm, BMPs also have *En-1*–independent ventralization activity by inhibiting *Lmx1b* expression.[45]

Limb development is a coordinated 3D event. Indeed, the three signaling centers interact with each other through interactions of the mediating signaling molecules. First, there is a positive feedback loop between *Shh* and *Fgfs* expressed in the AER, which connects A-P limb patterning with P-D limb outgrowth (Fig. 3B).[46–48] Second, the dorsalizing signal Wnt7a is required for maintaining the expression of *Shh* that patterns the A-P axis.[49,50]

EMBRYONIC CARTILAGE AND BONE FORMATION

The early patterning events determine where and when the mesenchymal cells condense. After that, osteochondral progenitors in the condensation form either chondrocytes or osteoblasts. Sox9 and Runx2, master transcription factors that are required for the determination of chondrocyte and osteoblast cell fates, respectively,[51–54] are both expressed in osteochondral progenitor cells, but *Sox9* expression precedes that of *Runx2* in the mesenchymal condensation in the limb.[55] Early *Sox9*-expressing cells give rise to both chondrocytes and osteoblasts regardless of ossification mechanisms.[56] In addition, loss of *Sox9* function in the limb leads to loss of mesenchymal condensation and *Runx2* expression.[55] Coexpression of *Sox9* and *Runx2* is terminated on chondrocyte and osteoblast differentiation. *Sox9* and *Runx2* expression are quickly segregated into chondrocytes and osteoblasts, respectively. Understanding the mechanism controlling this expression segregation of *Sox9* and *Runx2* in specific cell lineages is fundamental to elucidate the regulation of not only chondrocyte and osteoblast differentiation but also the determination of ossification mechanism. It is clear that cell–cell signaling, particularly those mediated by Wnts and Indian hedgehog (Ihh), are required for cell fate determination of chondrocytes and osteoblasts by controlling the expression of *Sox9* and *Runx2*.

Active canonical Wnt signaling is detected in the developing calvarium and perichondrium where osteoblasts differentiate through either intramembranous or endochondral ossification. Indeed, enhanced canonical Wnt signaling enhanced bone formation and *Runx2* expression, but inhibited chondrocyte differentiation and *Sox9* expression.[57–59] Conversely, removal of *β-catenin* in osteochondral progenitor cells resulted in ectopic chondrocyte differentiation at the expense of osteoblasts during both intramembranous and endochondral ossification.[59–61] Therefore, the mesenchymal progenitor cells in the condensation are at least bipotential in their final cell fate determination. During intramembranous ossification, Wnt signaling in the condensation is higher, which promotes osteoblast differentiation while inhibiting chondrocyte differentiation. During endochondral ossification, however, Wnt signaling in the condensation is kept low such that only chondrocytes can differentiate. Later, when Wnt signaling is upregulated in the periphery of the cartilage, osteoblasts will differentiate. Thus, by manipulating Wnt signaling, mesenchymal progenitor cells, and perhaps even mesenchymal stem cells, can be directed to form only chondrocytes, which is needed in repairing cartilage damage in osteoarthritis, or only form osteoblasts, which will lead to new therapeutic strategies to treat osteoporosis. These studies also provide new insights to tissue engineering that aims to fabricate cartilage or bone in vitro using mesenchymal progenitor cells or stem cells.

Ihh signaling is required for osteoblast differentiation by activating *Runx2* expression only during endochondral bone formation.[62,63] *Ihh* is expressed in newly differentiated chondrocytes, and Ihh signaling does not seem to affect chondrocyte differentiation from mesenchymal progenitors. However, when Hh signaling is inactivated in the perichondrium cells, they ectopically form chondrocytes and express *Sox9* at the expense of *Runx2*. This is similar to what has been observed in the Osterix (*Osx*) mutant embryos, except that, in the $Osx^{-/-}$ embryos, ectopic chondrocytes express both *Sox9* and *Runx2* because *Osx* acts downstream of *Runx2* in osteoblast differentiation.[64] It is still not clear what controls *Ihh*-independent *Runx2* expression during intramembranous ossification. One likely scenario is that the function of *Ihh* is compensated by *Shh* during intramembranous ossification in the developing calvarium.

Both the canonical Wnt and Ihh signaling pathways are required for endochondral bone formation. It is important to understand their genetic epistasis. All vertebrate Hhs including Shh and Ihh signal through the same pathway. Two multipass transmembrane proteins Patched1 (Ptch1) and Smoothened (Smo) receive Hh signaling on the cell membrane. Hh signal-

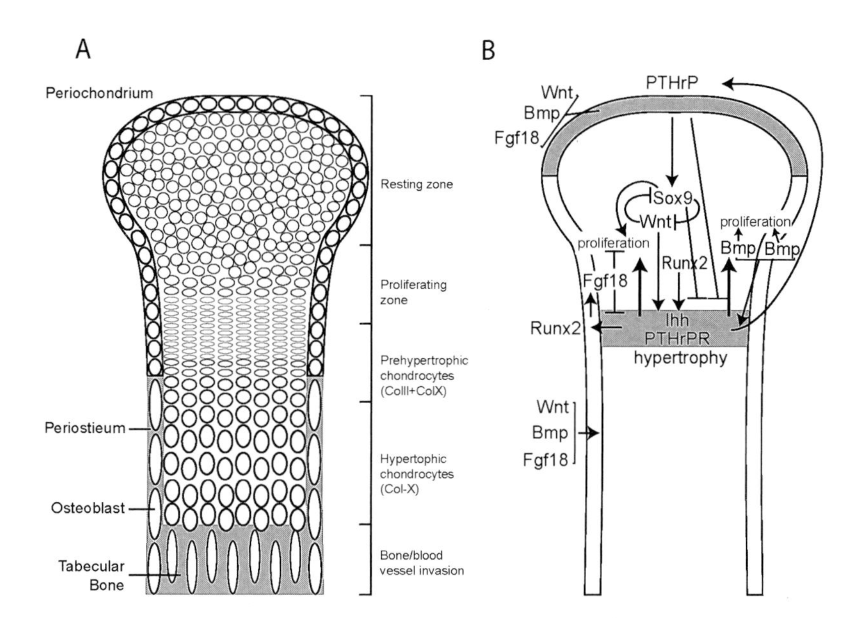

FIG. 4. Chondrocyte proliferation and hypertrophy are tightly controlled by signaling pathways and transcription factors. (A) Schematic drawing of a developing long bone cartilage. Chondrocytes with different properties of proliferation have different morphologies and are located in distinct locations along the longitudinal axis. See text for details. (B) Molecular regulation of chondrocyte proliferation and hypertrophy. Ihh, PTHrP, Wnt, Fgf, and Bmp are major signaling pathways that control chondrocyte proliferation and hypertrophy. A negative feedback loop between Ihh and PTHrP is fundamental in regulating the pace of chondrocyte hypertrophy. Transcription factors Sox9 and Runx2 act inside the cell to integrate signals from different pathways. See text for details.

ing is transduced to intracellular components by Smo, which is inhibited by Ptch1 in the absence of Hh ligands. Smo is activated on Hh binding to Ptch1. This event triggers a cascade that leads to upregulated expression of Hh target genes including *Ptch1*, *Gli1*, and *Hip1*. Therefore, removal of *Ptch1* leads to cell autonomous and ligand-independent upregulation of Hh signaling, whereas loss of *Smo* results in inhibition of Hh signaling. By generating double conditional mutant mice with the use of floxed allele of *Ptch1* and β-*catenin*, Hh signaling is cell autonomously activated, whereas β-*catenin* is cell autonomously inactivated at the same time in the same cells in the developing lone bones.[65] In this double mutant, β-*catenin* is required downstream of not just *Ihh*, but also *Osx* in promoting osteoblast maturation.[65] In contrast, Ihh signaling is not required after *Osx* expression for osteoblast differentiation.[66] The sequential actions of Hh and Wnt signaling in osteoblast differentiation and maturation suggest that Hh and Wnt signaling need to be manipulated at distinct stages during fracture repair and tissue engineering.

The BMP family of secreted growth factors belong to the TGF superfamily. BMPs were identified as secreted proteins that have the ability to promote ectopic cartilage and bone formation.[67] Unlike Ihh and Wnt signaling, BMP signaling promotes the differentiation of both osteoblast and chondrocyte differentiation from mesenchymal progenitors. The mechanisms underlying these unique activities of BMPs have been under intense investigation in the last two decades. During the course, understanding BMP action in chondrogenesis and osteogenesis has benefited greatly from molecular studies of BMP signal transduction.[68] Reducing BMP signaling by removing BMP receptors leads to impaired chondrocyte and osteoblast differentiation and maturation.[69]

FGF ligands and FGF receptors (FGFRs) are both expressed in the developing skeletal system, and the significant role of FGF signaling in skeletal development was first identified by the discovery that achondroplasia (ACH; OMIM 100800), the most common form of skeletal dwarfism in humans, was caused by a missense mutation in FGFR3. Later, hypochondroplasia (HCH; OMIM 146000), a milder form of dwarfism, and thanatophoric dysplasia (TD; OMIM 187600 and 187601), a more severe form of dwarfism, were also found to result from mutations in FGFR3. FGFR3 signaling acts to regulate the proliferation and hypertrophy of the differentiated chondrocytes. However, the function of FGF signaling in mesenchymal condensation and chondrocyte differentiation from progenitors remains to be elucidated because complete genetic inactivation of FGF signaling in mesenchymal condensation has not been achieved. Nevertheless, it is clear that FGF signaling acts in the mesenchymal condensation to control osteoblast differentiation during intramembranous bone formation. Mutations in FGFR 1, 2, and 3 cause craniosynostosis (premature fusion of the cranial sutures). The craniosynostosis syndromes involving FGFR1, 2, and 3 mutations include Apert syndrome (AS; OMIM 101200), Beare-Stevenson cutis gyrata (OMIM 123790), Crouzon syndrome (CS; OMIM 123500), Pfeiffer syndrome (PS; OMIM 101600), Jackson-Weiss syndrome (JWS; OMIM 123150), Muenke syndrome (MS; OMIM 602849), crouzonodermoskeletal syndrome (OMIM 134934), and osteoglophonic dysplasia (OGD; OMIM 166250), a disease characterized by craniosynostosis, a prominent supraorbital ridge, and a depressed nasal bridge, as well as the rhizomelic dwarfism and nonossifying bone lesions. All these mutations are autosomal dominant, and many of them are activating mutations of FGFRs. FGF signaling can promote or inhibit osteoblast proliferation and differentiation depending on the cell context. It does so either directly or through interacting with the Wnt and BMP signaling pathways.

CHONDROCYTE PROLIFERATION AND DIFFERENTIATION IN THE DEVELOPING CARTILAGE

During endochondral bone formation, chondrocytes differentiated from the mesenchymal condensation form the cartilage, which provides a growth template for the future bone. Differentiated chondrocytes inside the cartilage undergo tightly controlled progressive proliferation and hypertrophy, which is required for endochondral bone formation. In the developing cartilage of the long bone, chondrocytes with different properties of proliferation and differentiation are located in distinct zones along the longitudinal axis, and such organization is required for long bone elongation (Fig. 4A). Proliferating chondrocytes express *ColII*, whereas hypertrophic chondrocytes express *ColX*. The chondrocytes that already exit cell cycle, but have not yet become hypertrophic, are prehypertrophic chondrocytes. These cells and early hypertrophic chondrocytes express *Ihh*, which is a master regulator of endochondral bone development by coupling chondrocyte hypertrophy with osteoblast differentiation as Ihh is produced by prehypertrophic chondrocytes and signals to the adjacent perichondrium to induce osteoblast differentiation.[65]

Ihh$^{-/-}$ mice have striking skeletal development defects. Apart from lack of endochondral bone formation, all cartilage elements are small because of a marked decrease in chondrocyte proliferation.[65,70] Ihh also controls the pace of chondrocyte hypertrophy by activating the expression of PTH-related peptide (PTHrP) in articular cartilage and periarticular cells.[62,71] PTHrP acts on the same G protein–coupled receptor used by PTH. This PTH/PTHrP receptor (PPR) is expressed at much higher levels by prehypertrophic and early hypertrophic chondrocytes. PTHrP signaling is required to inhibit precocious chondrocyte hypertrophy primarily by keeping proliferating chondrocytes in the proliferating pool.[72,73] PTHrP signaling is also required to mediate the activity of Ihh in regulating chondrocyte hypertrophy but not proliferation.[74] Therefore, Ihh and PTHrP form a negative feedback loop to control the decision of chondrocytes to leave the proliferating pool and become hypertrophy (Fig. 4B). In this model, PTHrP, secreted from cells at the ends of cartilage, acts on proliferating chondrocytes to keep them proliferating. When chondrocytes are further away from the end of cartilage as a result of cartilage elongation, they are no longer sufficiently stimulated by PTHrP. These cells exit cell cycle and become Ihh-producing prehypertrophic chondrocytes. Ihh stimulates *PTHrP* expression at the ends of cartilage to slow down cell cycle exit of proliferating chondrocytes. This model is supported by experiments using chimeric mouse embryos.[75] Clones of *PPR*$^{-/-}$ chondrocytes differentiate into hypertrophic chondrocytes and produce Ihh within the wildtype proliferating chondrocyte domain. This ectopic *Ihh* expression leads to ectopic osteoblast differentiation in the perichondrium, an upregulation of *PTHrP* expression, and a consequent lengthening of the columns of wildtype proliferating chondrocytes. These studies show that the lengths of proliferating columns, hence the growth potential of cartilages, are critically determined by the Ihh–PTHrP negative feedback loop. Indeed, it has been found that mutations in IHH in humans cause brachydactyly type A1 (OMIM 112500), which exhibit shortened digits phalanges and short body statue.[76] This feedback loop may also act to assure that cells at the hypertrophic front exit cell cycle and undergo hypertrophy at the same time.[72]

Several Wnt ligands including Wnt5a, Wnt5b, Wnt4, and Wnt9a are expressed in the cartilage and perichondrium of mouse embryos.[58,77] Wnt4 and Wnt9a signal through the canonical Wnt pathway, whereas Wnt5a and Wnt5b signal through the β-catenin–independent (noncanonical) pathways to regulate chondrocyte proliferation and hypertrophy. Although the mechanisms vary, in the absence of either canonical or noncanonical Wnt signaling, chondrocyte proliferation is altered, and chondrocyte hypertrophy is delayed.[59,77,78] Furthermore, both the canonical and noncanonical Wnt pathways act in parallel with Ihh signaling to regulate chondrocyte proliferation and differentiation.[65,77] Wnt and Ihh signaling may both regulate common downstream targets such as Sox9 (see below).[77,78]

Many FGF ligands and receptors are expressed in the developing cartilage. The likely functional redundancy of the FGF ligand and receptor members in cartilage development is a challenging issue that needs to be addressed to cleanly dissect the function of FGF signaling. FGFR3 signaling negatively regulates chondrocyte proliferation and hypertrophy.[79–86] Furthermore, FGFR3 signaling does so in part by directly signaling in chondrocytes[87,88] to activate the Janus kinase signal transducer and activator of transcription-1 (Jak–Stat1) and the MAPK pathways.[89] FGFR3 signaling also interacts with the Ihh/PTHrP/BMP signaling pathways.[81,90,91] The proliferation-inhibitory role of Fgf signaling in chondrocytes is not unique to FGFR3. When expressed in growth plate chondrocytes in vivo, both FGFR1 and FGFR3 kinase domains seem to have similar activities.[92]

Because *Fgf18*$^{-/-}$ mice show an increase in chondrocyte proliferation that closely resembles the cartilage phenotypes of *Fgfr3*$^{-/-}$ mice, FGF18 is likely a physiological ligand for FGFR3 during skeletal development. However, the phenotype of the *Fgf18*$^{-/-}$ mouse is more severe than that of the *Fgfr3*$^{-/-}$ mice, suggesting that Fgf18 may also signal through Fgfr1 in hypertrophic chondrocytes and Fgfr2 and -1 in the perichondrium. Mice conditionally lacking *Fgfr2* develop skeletal dwarfism with decreased BMD.[93,94] Osteoblasts also express *Fgfr3*, and mice lacking *Fgfr3* have decreased BMD and develop osteopenia.[95,96] Thus, in osteoblasts, FGF signaling positively regulates bone growth by promoting osteoblast proliferation. Interestingly, mice lacking *Fgf2* also show osteopenia, although much later in development than in *Fgfr2*-deficient mice,[97] suggesting that FGF2 may be a homeostatic factor that replaces the developmental growth factor, FGF18, in adult bones. It is still not clear which FGFR (1, 2, or 3) is actually responding to FGF2/18 in osteoblasts.

Like the other major signaling pathways mentioned above, BMP signaling also acts during later stages of cartilage development. Both in vitro limb explant experiments and in vivo genetic studies showed that BMP signaling promotes chondrocyte proliferation and *Ihh* expression. Addition of BMPs to bone explants increases proliferation of chondrocytes, whereas Noggin blocks chondrocyte proliferation.[91,98] In addition, conditional removal of both *BmpRIA* and *BmpRIB* in differentiated chondrocytes leads to reduced chondrocyte proliferation and *Ihh* expression. BMP signaling also regulates chondrocyte hypertrophy because removal of *BmpRIA* in chondrocytes leads to an expanded hypertrophic zone caused by accelerated chondrocyte hypertrophy and delayed terminal maturation of hypertrophic chondrocytes.[99] Given the function of Ihh in promoting chondrocyte proliferation and controlling the pace of chondrocyte hypertrophy, BMP signaling regulates chondrocyte proliferation and hypertrophy at least in part through *Ihh* expression.

BMP signaling also interacts with FGF signaling through mutual antagonism. In limb explant cultures, BMP and FGF signaling pathways have opposing functions in the growth plate.[91] Furthermore, comparison of cartilage phenotypes of BMP and FGF signaling mutants indicate that these two signaling pathways antagonize each other in regulating chondrocyte proliferation and hypertrophy.[99] The molecular mechanisms of BMP/FGF antagonism requires further study.

Chondrocytes in the developing cartilage transduce distinct signals including Ihh, PTHrP, Wnts, FGFs, and BMPs at the same time. A complete understanding of the molecular regulation of cartilage development will not be achieved without deciphering how these signaling pathways interact with each other and coordinately control common downstream effectors inside the cell. Sox9 and Runx2 are two critical transcription factors that can integrate various signaling inputs in controlling chondrocyte proliferation and differentiation. When *Sox9* was removed from differentiated chondrocytes, chondrocyte proliferation, expression of matrix genes, and the Ihh–PTHrP signaling components were reduced in the cartilage.[55] These phenotypes are very similar to those mutant mice lacking both *Sox5* and *Sox6*, two other Sox-family members that themselves require *Sox9* for expression. *Sox5* and *Sox6* cooperate with *Sox9* in maintaining chondrocyte phenotypes and regulating chondrocyte specific gene expression.[100] Haploinsufficiency of SOX9 protein in humans causes camptomelic dysplasia (CD; OMIM 114290), and defects of CD patients are recapitulated in *Sox9*$^{+/-}$ mice that show cartilage hypoplasia and a perinatal lethal osteochondrodysplasia.[101] Chondrocyte hypertrophy is accelerated in the *Sox9*$^{+/-}$ cartilage but delayed in *Sox9*-overexpressing cartilage.[78,101] Sox9 acts in both PTHrP and Wnt signaling pathways to control chondrocyte hypertrophy. PTHrP signaling in chondrocyte activate protein kinase A (PKA), which promotes Sox9 transcription activity by phosphorylating it.[102] In addition, Sox9 inhibits the canonical Wnt signaling activity by promoting degradation of β-catenin, a central mediator of the canonical Wnt pathway.[78] Thus, Sox9 is a master transcription factor that act in many critical stages of chondrocyte proliferation and differentiation as a central node inside pre-chondrocytes and chondrocytes to receive and integrate multiple signaling inputs.

Runx2 is expressed in prehypertrophic and hypertrophic chondrocytes. It is also highly expressed in perichondrial cells and in osteoblasts. The significant role of *Runx2* in skeletal development is first shown by the striking phenotypes of *Runx2*$^{-/-}$ mice. These mutant mice have no osteoblast differentiation at all.[52,103] Mutations in human *RUNX2* cause cleidocranial dysplasia (CCD; OMIM 119600), an autosomal-dominant condition characterized by hypoplasia/aplasia of clavicles, patent fontanelles, supernumerary teeth, short stature, and other changes in skeletal patterning and growth.[104] Runx2 is the earliest known transcription factor that is required for osteoblast differentiation from mesenchymal progenitors. *Runx2* also controls chondrocyte proliferation and hypertrophy. Chondrocyte hypertrophy is significantly delayed, and *Ihh* expression is reduced in *Runx2*$^{-/-}$ mice, whereas *Runx2* overexpression in the cartilage results in accelerated chondrocyte hypertrophy.[105,106] Furthermore, removing both *Runx2* and *Runx3* completely blocks chondrocyte hypertrophy and *Ihh* expression in mice, indicating Runx transcription factors control *Ihh* expression.[107] Thus, *Runx2* can also be viewed as a master controlling transcription factor and a central node through which other signaling pathways are integrated in coordinate chondrocyte proliferation and hypertrophy. In chondrocytes, *Runx2* acts in the Ihh–PTHrP pathway to regulate cartilage growth by controlling the expression of *Ihh*. However, this will not be its only function because *Runx2* upregulation leads to accelerated chondrocyte hypertrophy, whereas *Ihh* upregulation leads to delayed chondrocyte hypertrophy. One of *Runx2*'s *Ihh*-independent activities is to act in the perichondrium to inhibit chondrocyte proliferation and hypertrophy indirectly by regulating *Fgf18* expression.[108] Interestingly, this role of Runx2 in the perichondrium is antagonist to its role in chondrocytes. Furthermore, histone deacetylases 4 (HDAC4), which modulates cell growth and differentiation by governing chromatin structure and repressing the activity of specific transcription factors, regulates chondrocyte hypertrophy and endochondral bone formation by interacting with and inhibiting the activity of Runx2.[109]

REGULATION OF CHONDROCYTE SURVIVAL

Cartilage is an avascular tissue that develops under a hypoxia condition because chondrocytes, particularly the ones in the middle of the cartilage, do not have access to vascular oxygen delivery.[110] As in other hypoxia conditions, a major mediator of hypoxic response in the developing cartilage is the transcription factor hypoxia–inducible factor 1 (Hif-1) and its oxygen-sensitive component Hif-1α. Removal of *Hif-1α* in cartilage results in chondrocyte cell death in the interior of the growth plate. A downstream target of Hif-1 in regulating chondrocyte hypoxic response is *Vegf*.[111] The extensive cell death seen in the cartilage of mice lacking *Vegfa* has a striking similarity to that observed in mice in which *Hif-1α* is removed in the developing cartilage. In addition, *Vegf* expression in the growth plate is reduced in the absence of *Hif-1α*.[110] However, *Vegf* expression is also regulated through *Hif-1α*–independent mechanisms, possibly as a result of alternative response to hypoxia in the *Hif-1α* cartilage-specific mutant. In this mutant, upregulation of *Vegf* expression and ectopic angiogenesis are observed in chondrocytes surrounding areas of extensive cell death.[110]

CONCLUSIONS

Being able to use autologous cells and tissues to repair damaged bone and cartilage during injury and diseases has obvious advantages and will be a big step forward in regenerative medicine. With the possibility of reprogramming somatic cells into embryonic stem cells (ESCs), one may be able to fabricate cartilage or bone using one's own reprogrammed ESCs. Because bone formation is a process that has been perfected by nature in embryos during vertebrate evolution, understanding the underlying molecular mechanism of cartilage and bone formation in embryonic development will allow us to learn the strategy in achieving directed differentiation of chondrocytes and osteoblasts from ESCs. In addition, such knowledge will significantly promote consistent cartilage or bone repair in vivo or grow functional cartilage or bone in vitro.

ACKNOWLEDGMENTS

This work was supported by the NIH Intramural Research Program. The author thanks Darryl Leja and Julia Fekes for help in preparing the graphics.

REFERENCES

1. Santagati F, Rijli FM 2003 Cranial neural crest and the building of the vertebrate head. Nat Rev Neurosci **4**:806–818.
2. Helms JA, Cordero D, Tapadia MD 2005 New insights into craniofacial morphogenesis. Development **132**:851–861.
3. Christ B, Huang R, Scaal M 2004 Formation and differentiation of the avian sclerotome. Anat Embryol (Berl) **208**:333–350.
4. Gossler A, Hrabe de Angelis M 1998 Somitogenesis. Curr Top Dev Biol **38**:225–287.
5. Hirsinger E, Jouve C, Dubrulle J, Pourquie O 2000 Somite formation and patterning. Int Rev Cytol **198**:1–65.
6. Scaal M, Christ B 2004 Formation and differentiation of the avian dermomyotome. Anat Embryol (Berl) **208**:411–424.
7. Aulehla A, Pourquie O 2006 On periodicity and directionality of somatogenesis. Anat Embryol (Berl) **211**(Suppl 1):3–8.
8. Dequeant ML, Glynn E, Gaudenz K, Wahl M, Chen J, Mushegian A, Pourquie O 2006 A complex oscillating network of signaling genes underlies the mouse segmentation clock. Science **314**:1595–1598.
9. Ilagan MX, Kopan R 2007 SnapShot: Notch signaling pathway. Cell **128**:1246.
10. Aulehla A, Wehrle C, Brand-Saberi B, Kemler R, Gossler A, Kanzler B, Herrmann BG 2003 Wnt3a plays a major role in the segmentation clock controlling somatogenesis. Dev Cell **4**:395–406.
11. Ishikawa A, Kitajima S, Takahashi Y, Kokubo H, Kanno J, Inoue T, Saga Y 2004 Mouse Nkd1, a Wnt antagonist, exhibits oscillatory gene expression in the PSM under the control of Notch signaling. Mech Dev **121**:1443–1453.
12. Moreno TA, Kintner C 2004 Regulation of segmental patterning by retinoic acid signaling during Xenopus somatogenesis. Dev Cell **6**:205–218.
13. Diez del Corral R, Olivera-Martinez I, Goriely A, Gale E, Maden M, Storey K 2003 Opposing FGF and retinoid pathways control ventral neural pattern, neuronal differentiation, and segmentation during body axis extension. Neuron **40**:65–79.
14. Vermot J, Gallego Llamas J, Fraulob V, Niederreither K, Chambon P, Dolle P 2005 Retinoic acid controls the bilateral symmetry of somite formation in the mouse embryo. Science **308**:563–566.

15. Vermot J, Pourquie O 2005 Retinoic acid coordinates somatogenesis and left-right patterning in vertebrate embryos. Nature **435:**215–220.
16. Sirbu IO, Duester G 2006 Retinoic-acid signalling in node ectoderm and posterior neural plate directs left-right patterning of somitic mesoderm. Nat Cell Biol **8:**271–277.
17. Fan CM, Tessier-Lavigne M 1994 Patterning of mammalian somites by surface ectoderm and notochord: Evidence for sclerotome induction by a hedgehog homolog. Cell **79:**1175–1186.
18. Johnson RL, Laufer E, Riddle RD, Tabin C 1994 Ectopic expression of Sonic hedgehog alters dorsal-ventral patterning of somites. Cell **79:**1165–1173.
19. Fan CM, Lee CS, Tessier-Lavigne M 1997 A role for WNT proteins in induction of dermomyotome. Dev Biol **191:**160–165.
20. Capdevila J, Tabin C, Johnson RL 1998 Control of dorsoventral somite patterning by Wnt-1 and beta-catenin. Dev Biol **193:**182–194.
21. Brand-Saberi B, Ebensperger C, Wilting J, Balling R, Christ B 1993 The ventralizing effect of the notochord on somite differentiation in chick embryos. Anat Embryol (Berl) **188:**239–245.
22. Koseki H, Wallin J, Wilting J, Mizutani Y, Kispert A, Ebensperger C, Herrmann BG, Christ B, Balling R 1993 A role for Pax-1 as a mediator of notochordal signals during the dorsoventral specification of vertebrae. Development **119:**649–660.
23. Chiang C, Litingtung Y, Lee E, Young KE, Corden JL, Westphal H, Beachy PA 1996 Cyclopia and defective axial patterning in mice lacking Sonic hedgehog gene function. Nature **383:**407–413.
24. Wallin J, Wilting J, Koseki H, Fritsch R, Christ B, Balling R 1994 The role of Pax-1 in axial skeleton development. Development **120:**1109–1121.
25. Barrow JR, Thomas KR, Boussadia-Zahui O, Moore R, Kemler R, Capecchi MR, McMahon AP 2003 Ectodermal Wnt3/beta-catenin signaling is required for the establishment and maintenance of the apical ectodermal ridge. Genes Dev **17:**394–409.
26. Niswander L, Tickle C, Vogel A, Booth I, Martin GR 1993 FGF-4 replaces the apical ectodermal ridge and directs outgrowth and patterning of the limb. Cell **75:**579–587.
27. Crossley PH, Minowada G, MacArthur CA, Martin GR 1996 Roles for FGF8 in the induction, initiation, and maintenance of chick limb development. Cell **84:**127–136.
28. Sun X, Mariani FV, Martin GR 2002 Functions of FGF signalling from the apical ectodermal ridge in limb development. Nature **418:**501–508.
29. Ohuchi H, Nakagawa T, Yamamoto A, Araga A, Ohata T, Ishimaru Y, Yoshioka H, Kuwana T, Nohno T, Yamasaki M, Itoh N, Noji S 1997 The mesenchymal factor, FGF10, initiates and maintains the outgrowth of the chick limb bud through interaction with FGF8, an apical ectodermal factor. Development **124:**2235–2244.
30. Sekine K, Ohuchi H, Fujiwara M, Yamasaki M, Yoshizawa T, Sato T, Yagishita N, Matsui D, Koga Y, Itoh N, Kato S 1999 Fgf10 is essential for limb and lung formation. Nat Genet **21:**138–141.
31. Min H, Danilenko DM, Scully SA, Bolon B, Ring BD, Tarpley JE, DeRose M, Simonet WS 1998 Fgf-10 is required for both limb and lung development and exhibits striking functional similarity to Drosophila branchless. Genes Dev **12:**3156–3161.
32. Saunders JWJ, Gasseling MT 1968 Ectoderm-mesenchymal interaction in the origin of wing symmetry. In: Fleischmajer R, Billingham RE (eds.) Epithelia-Mesenchymal Interactions. Williams and Wilkins, Baltimore, MD, USA, pp. 78–97.
33. Riddle RD, Johnson RL, Laufer E, Tabin C 1993 Sonic hedgehog mediates the polarizing activity of the ZPA. Cell **75:**1401–1416.
34. Chan DC, Laufer E, Tabin C, Leder P 1995 Polydactylous limbs in Strong's Luxoid mice result from ectopic polarizing activity. Development **121:**1971–1978.
35. te Welscher P, Fernandez-Teran M, Ros MA, Zeller R 2002 Mutual genetic antagonism involving GLI3 and dHAND prepatterns the vertebrate limb bud mesenchyme prior to SHH signaling. Genes Dev **16:**421–426.
36. Charite J, McFadden DG, Olson EN 2000 The bHLH transcription factor dHAND controls Sonic hedgehog expression and establishment of the zone of polarizing activity during limb development. Development **127:**2461–2470.
37. Fernandez-Teran M, Piedra ME, Kathiriya IS, Srivastava D, Rodriguez-Rey JC, Ros MA 2000 Role of dHAND in the anterior-posterior polarization of the limb bud: Implications for the Sonic hedgehog pathway. Development **127:**2133–2142.
38. Firulli BA, Krawchuk D, Centonze VE, Vargesson N, Virshup DM, Conway SJ, Cserjesi P, Laufer E, Firulli AB 2005 Altered Twist1 and Hand2 dimerization is associated with Saethre-Chotzen syndrome and limb abnormalities. Nat Genet **37:**373–381.
39. Tickle C 2003 Patterning systems–from one end of the limb to the other. Dev Cell **4:**449–458.
40. Niswander L 2002 Interplay between the molecular signals that control vertebrate limb development. Int J Dev Biol **46:**877–881.
41. Riddle RD, Ensini M, Nelson C, Tsuchida T, Jessell TM, Tabin C 1995 Induction of the LIM homeobox gene Lmx1 by WNT7a establishes dorsoventral pattern in the vertebrate limb. Cell **83:**631–640.
42. Parr BA, Shea MJ, Vassileva G, McMahon AP 1993 Mouse Wnt genes exhibit discrete domains of expression in the early embryonic CNS and limb buds. Development **119:**247–261.
43. Loomis CA, Harris E, Michaud J, Wurst W, Hanks M, Joyner AL 1996 The mouse Engrailed-1 gene and ventral limb patterning. Nature **382:**360–363.
44. Lallemand Y, Nicola MA, Ramos C, Bach A, Cloment CS, Robert B 2005 Analysis of Msx1; Msx2 double mutants reveals multiple roles for Msx genes in limb development. Development **132:**3003–3014.
45. Ovchinnikov DA, Selever J, Wang Y, Chen YT, Mishina Y, Martin JF, Behringer RR 2006 BMP receptor type IA in limb bud mesenchyme regulates distal outgrowth and patterning. Dev Biol **295:**103–115.
46. Khokha MK, Hsu D, Brunet LJ, Dionne MS, Harland RM 2003 Gremlin is the BMP antagonist required for maintenance of Shh and Fgf signals during limb patterning. Nat Genet **34:**303–307.
47. Niswander L, Jeffrey S, Martin GR, Tickle C 1994 A positive feedback loop coordinates growth and patterning in the vertebrate limb. Nature **371:**609–612.
48. Laufer E, Nelson CE, Johnson RL, Morgan BA, Tabin C 1994 Sonic hedgehog and Fgf-4 act through a signaling cascade and feedback loop to integrate growth and patterning of the developing limb bud. Cell **79:**993–1003.
49. Parr BA, McMahon AP 1995 Dorsalizing signal Wnt-7a required for normal polarity of D-V and A-P axes of mouse limb. Nature **374:**350–353.
50. Yang Y, Niswander L 1995 Interaction between the signaling molecules WNT7a and SHH during vertebrate limb development: Dorsal signals regulate anteroposterior patterning. Cell **80:**939–947.
51. Komori T, Yagi H, Nomura S, Yamaguchi A, Sasaki K, Deguchi K, Shimizu Y, Bronson RT, Gao YH, Inada M, Sato M, Okamoto R, Kitamura Y, Yoshiki S, Kishimoto T 1997 Targeted disruption of Cbfa1 results in a complete lack of bone formation owing to maturational arrest of osteoblasts. Cell **89:**755–764.
52. Otto F, Thornell AP, Crompton T, Denzel A, Gilmour KC, Rosewell IR, Stamp GW, Beddington RS, Mundlos S, Olsen BR, Selby PB, Owen MJ 1997 Cbfa1, a candidate gene for cleidocranial dysplasia syndrome, is essential for osteoblast differentiation and bone development. Cell **89:**765–771.
53. Ducy P, Zhang R, Geoffroy V, Ridall AL, Karsenty G 1997 Osf2/Cbfa1: A transcriptional activator of osteoblast differentiation. Cell **89:**747–754.
54. Bi W, Deng JM, Zhang Z, Behringer RR, de Crombrugghe B 1999 Sox9 is required for cartilage formation. Nat Genet **22:**85–89.
55. Akiyama H, Chaboissier MC, Martin JF, Schedl A, de Crombrugghe B 2002 The transcription factor Sox9 has essential roles in successive steps of the chondrocyte differentiation pathway and is required for expression of Sox5 and Sox6. Genes Dev **16:**2813–2828.
56. Akiyama H, Kim JE, Nakashima K, Balmes G, Iwai N, Deng JM, Zhang Z, Martin JF, Behringer RR, Nakamura T, de Crombrugghe B 2005 Osteo-chondroprogenitor cells are derived from Sox9 expressing precursors. Proc Natl Acad Sci USA **102:**14665–14670.
57. Hartmann C, Tabin CJ 2000 Dual roles of Wnt signaling during chondrogenesis in the chicken limb. Development **127:**3141–3159.
58. Guo X, Day TF, Jiang X, Garrett-Beal L, Topol L, Yang Y 2004 Wnt/beta-catenin signaling is sufficient and necessary for synovial joint formation. Genes Dev **18:**2404–2417.
59. Day TF, Guo X, Garrett-Beal L, Yang Y 2005 Wnt/beta-catenin

signaling in mesenchymal progenitors controls osteoblast and chondrocyte differentiation during vertebrate skeletogenesis. Dev Cell **8:**739–750.
60. Hill TP, Spater D, Taketo MM, Birchmeier W, Hartmann C 2005 Canonical Wnt/beta-catenin signaling prevents osteoblasts from differentiating into chondrocytes. Dev Cell **8:**727–738.
61. Hu H, Hilton MJ, Tu X, Yu K, Ornitz DM, Long F 2005 Sequential roles of Hedgehog and Wnt signaling in osteoblast development. Development **132:**49–60.
62. St-Jacques B, Hammerschmidt M, McMahon AP 1999 Indian hedgehog signaling regulates proliferation and differentiation of chondrocytes and is essential for bone formation. Genes Dev **13:**2072–2086.
63. Long F, Chung UI, Ohba S, McMahon J, Kronenberg HM, McMahon AP 2004 Ihh signaling is directly required for the osteoblast lineage in the endochondral skeleton. Development **131:**1309–1318.
64. Nakashima K, Zhou X, Kunkel G, Zhang Z, Deng JM, Behringer RR, de Crombrugghe B 2002 The novel zinc finger-containing transcription factor osterix is required for osteoblast differentiation and bone formation. Cell **108:**17–29.
65. Mak KK, Chen MH, Day TF, Chuang PT, Yang Y 2006 Wnt/beta-catenin signaling interacts differentially with Ihh signaling in controlling endochondral bone and synovial joint formation. Development **133:**3695–3707.
66. Rodda SJ, McMahon AP 2006 Distinct roles for Hedgehog and canonical Wnt signaling in specification, differentiation and maintenance of osteoblast progenitors. Development **133:**3231–3244.
67. Wozney JM 1989 Bone morphogenetic proteins. Prog Growth Factor Res **1:**267–280.
68. Derynck R, Zhang YE 2003 Smad-dependent and Smad-independent pathways in TGF-beta family signalling. Nature **425:**577–584.
69. Yoon BS, Ovchinnikov DA, Yoshii I, Mishina Y, Behringer RR, Lyons KM 2005 Bmpr1a and Bmpr1b have overlapping functions and are essential for chondrogenesis in vivo. Proc Natl Acad Sci USA **102:**5062–5067.
70. Long F, Zhang XM, Karp S, Yang Y, McMahon AP 2001 Genetic manipulation of hedgehog signaling in the endochondral skeleton reveals a direct role in the regulation of chondrocyte proliferation. Development **128:**5099–5108.
71. Vortkamp A, Lee K, Lanske B, Segre GV, Kronenberg HM, Tabin CJ 1996 Regulation of rate of cartilage differentiation by Indian hedgehog and PTH-related protein. Science **273:**613–622.
72. Karaplis AC, Luz A, Glowacki J, Bronson RT, Tybulewicz VL, Kronenberg HM, Mulligan RC 1994 Lethal skeletal dysplasia from targeted disruption of the parathyroid hormone-related peptide gene. Genes Dev **8:**277–289.
73. Lanske B, Karaplis AC, Lee K, Luz A, Vortkamp A, Pirro A, Karperien M, Defize LH, Ho C, Mulligan RC, Abou-Samra AB, Juppner H, Segre GV, Kronenberg HM 1996 PTH/PTHrP receptor in early development and Indian hedgehog-regulated bone growth. Science **273:**663–666.
74. Karp SJ, Schipani E, St-Jacques B, Hunzelman J, Kronenberg H, McMahon AP 2000 Indian hedgehog coordinates endochondral bone growth and morphogenesis via parathyroid hormone related-protein-dependent and -independent pathways. Development **127:**543–548.
75. Chung UI, Schipani E, McMahon AP, Kronenberg HM 2001 Indian hedgehog couples chondrogenesis to osteogenesis in endochondral bone development. J Clin Invest **107:**295–304.
76. Gao B, Guo J, She C, Shu A, Yang M, Tan Z, Yang X, Guo S, Feng G, He L 2001 Mutations in IHH, encoding Indian hedgehog, cause brachydactyly type A-1. Nat Genet **28:**386–388.
77. Yang Y, Topol L, Lee H, Wu J 2003 Wnt5a and Wnt5b exhibit distinct activities in coordinating chondrocyte proliferation and differentiation. Development **130:**1003–1015.
78. Akiyama H, Lyons JP, Mori-Akiyama Y, Yang X, Zhang R, Zhang Z, Deng JM, Taketo MM, Nakamura T, Behringer RR, McCrea PD, De Crombrugghe B 2004 Interactions between Sox9 and {beta}-catenin control chondrocyte differentiation. Genes Dev **18:**1072–1087.
79. Deng C, Wynshaw-Boris A, Zhou F, Kuo A, Leder P 1996 Fibroblast growth factor receptor 3 is a negative regulator of bone growth. Cell **84:**911–921.
80. Colvin JS, Bohne BA, Harding GW, McEwen DG, Ornitz DM 1996 Skeletal overgrowth and deafness in mice lacking fibroblast growth factor receptor 3. Nat Genet **12:**390–397.
81. Naski MC, Colvin JS, Coffin JD, Ornitz DM 1998 Repression of hedgehog signaling and BMP4 expression in growth plate cartilage by fibroblast growth factor receptor 3. Development **125:**4977–4988.
82. Chen L, Adar R, Yang X, Monsonego EO, Li C, Hauschka PV, Yayon A, Deng CX 1999 Gly369Cys mutation in mouse FGFR3 causes achondroplasia by affecting both chondrogenesis and osteogenesis. J Clin Invest **104:**1517–1525.
83. Chen L, Li C, Qiao W, Xu X, Deng C 2001 A Ser(365)→Cys mutation of fibroblast growth factor receptor 3 in mouse downregulates Ihh/PTHrP signals and causes severe achondroplasia. Hum Mol Genet **10:**457–465.
84. Li C, Chen L, Iwata T, Kitagawa M, Fu XY, Deng CX 1999 A Lys644Glu substitution in fibroblast growth factor receptor 3 (FGFR3) causes dwarfism in mice by activation of STATs and ink4 cell cycle inhibitors. Hum Mol Genet **8:**35–44.
85. Iwata T, Chen L, Li C, Ovchinnikov DA, Behringer RR, Francomano CA, Deng CX 2000 A neonatal lethal mutation in FGFR3 uncouples proliferation and differentiation of growth plate chondrocytes in embryos. Hum Mol Genet **9:**1603–1613.
86. Wang Y, Spatz MK, Kannan K, Hayk H, Avivi A, Gorivodsky M, Pines M, Yayon A, Lonai P, Givol D 1999 A mouse model for achondroplasia produced by targeting fibroblast growth factor receptor 3. Proc Natl Acad Sci USA **96:**4455–4460.
87. Dailey L, Laplantine E, Priore R, Basilico C 2003 A network of transcriptional and signaling events is activated by FGF to induce chondrocyte growth arrest and differentiation. J Cell Biol **161:**1053–1066.
88. Henderson JE, Naski MC, Aarts MM, Wang D, Cheng L, Goltzman D, Ornitz DM 2000 Expression of FGFR3 with the G380R achondroplasia mutation inhibits proliferation and maturation of CFK2 chondrocytic cells. J Bone Miner Res **15:**155–165.
89. Raucci A, Laplantine E, Mansukhani A, Basilico C 2004 Activation of the ERK1/2 and p38 mitogen-activated protein kinase pathways mediates fibroblast growth factor-induced growth arrest of chondrocytes. J Biol Chem **279:**1747–1756.
90. Kronenberg HM 2003 Developmental regulation of the growth plate. Nature **423:**332–336.
91. Minina E, Kreschel C, Naski MC, Ornitz DM, Vortkamp A 2002 Interaction of FGF, Ihh/Pthlh, and BMP signaling integrates chondrocyte proliferation and hypertrophic differentiation. Dev Cell **3:**439–449.
92. Wang Q, Green RP, Zhao G, Ornitz DM 2001 Differential regulation of endochondral bone growth and joint development by FGFR1 and FGFR3 tyrosine kinase domains. Development **128:**3867–3876.
93. Eswarakumar VP, Monsonego-Ornan E, Pines M, Antonopoulou I, Morriss-Kay GM, Lonai P 2002 The IIIc alternative of Fgfr2 is a positive regulator of bone formation. Development **129:**3783–3793.
94. Yu K, Xu J, Liu Z, Sosic D, Shao J, Olson EN, Towler DA, Ornitz DM 2003 Conditional inactivation of FGF receptor 2 reveals an essential role for FGF signaling in the regulation of osteoblast function and bone growth. Development **130:**3063–3074.
95. Valverde-Franco G, Liu H, Davidson D, Chai S, Valderrama-Carvajal H, Goltzman D, Ornitz DM, Henderson JE 2004 Defective bone mineralization and osteopenia in young adult FGFR3–/– mice. Hum Mol Genet **13:**271–284.
96. Xiao L, Naganawa T, Obugunde E, Gronowicz G, Ornitz DM, Coffin JD, Hurley MM 2004 Stat1 controls postnatal bone formation by regulating fibroblast growth factor signaling in osteoblasts. J Biol Chem **279:**27743–27752.
97. Montero A, Okada Y, Tomita M, Ito M, Tsurukami H, Nakamura T, Doetschman T, Coffin JD, Hurley MM 2000 Disruption of the fibroblast growth factor-2 gene results in decreased bone mass and bone formation. J Clin Invest **105:**1085–1093.
98. Minina E, Wenzel HM, Kreschel C, Karp S, Gaffield W, McMahon AP, Vortkamp A 2001 BMP and Ihh/PTHrP signaling interact to coordinate chondrocyte proliferation and differentiation. Development **128:**4523–4534.
99. Yoon BS, Pogue R, Ovchinnikov DA, Yoshii I, Mishina Y, Behringer RR, Lyons KM 2006 BMPs regulate multiple aspects of growth-plate chondrogenesis through opposing actions on FGF pathways. Development **133:**4667–4678.
100. Smits P, Li P, Mandel J, Zhang Z, Deng JM, Behringer RR, de

Crombrugghe B, Lefebvre V 2001 The transcription factors L-Sox5 and Sox6 are essential for cartilage formation. Dev Cell **1**:277–290.
101. Bi W, Huang W, Whitworth DJ, Deng JM, Zhang Z, Behringer RR, de Crombrugghe B 2001 Haploinsufficiency of Sox9 results in defective cartilage primordia and premature skeletal mineralization. Proc Natl Acad Sci USA **98**:6698–6703.
102. Huang W, Chung UI, Kronenberg HM, de Crombrugghe B 2001 The chondrogenic transcription factor Sox9 is a target of signaling by the parathyroid hormone-related peptide in the growth plate of endochondral bones. Proc Natl Acad Sci USA **98**:160–165.
103. Kato M, Patel MS, Levasseur R, Lobov I, Chang BH, Glass DA II, Hartmann C, Li L, Hwang TH, Brayton CF, Lang RA, Karsenty G, Chan L 2002 Cbfa1-independent decrease in osteoblast proliferation, osteopenia, and persistent embryonic eye vascularization in mice deficient in Lrp5, a Wnt coreceptor. J Cell Biol **157**:303–314.
104. Mundlos S, Otto F, Mundlos C, Mulliken JB, Aylsworth AS, Albright S, Lindhout D, Cole WG, Henn W, Knoll JH, Owen MJ, Mertelsmann R, Zabel BU, Olsen BR 1997 Mutations involving the transcription factor CBFA1 cause cleidocranial dysplasia. Cell **89**:773–779.
105. Kim IS, Otto F, Zabel B, Mundlos S 1999 Regulation of chondrocyte differentiation by Cbfa1. Mech Dev **80**:159–170.
106. Takeda S, Bonnamy JP, Owen MJ, Ducy P, Karsenty G 2001 Continuous expression of Cbfa1 in nonhypertrophic chondrocytes uncovers its ability to induce hypertrophic chondrocyte differentiation and partially rescues Cbfa1-deficient mice. Genes Dev **15**:467–481.
107. Yoshida CA, Yamamoto H, Fujita T, Furuichi T, Ito K, Inoue K, Yamana K, Zanma A, Takada K, Ito Y, Komori T 2004 Runx2 and Runx3 are essential for chondrocyte maturation, and Runx2 regulates limb growth through induction of Indian hedgehog. Genes Dev **18**:952–963.
108. Hinoi E, Bialek P, Chen YT, Rached MT, Groner Y, Behringer RR, Ornitz DM, Karsenty G 2006 Runx2 inhibits chondrocyte proliferation and hypertrophy through its expression in the perichondrium. Genes Dev **20**:2937–2942.
109. Vega RB, Matsuda K, Oh J, Barbosa AC, Yang X, Meadows E, McAnally J, Pomajzl C, Shelton JM, Richardson JA, Karsenty G, Olson EN 2004 Histone deacetylase 4 controls chondrocyte hypertrophy during skeletogenesis. Cell **119**:555–566.
110. Schipani E, Ryan HE, Didrickson S, Kobayashi T, Knight M, Johnson RS 2001 Hypoxia in cartilage: HIF-1alpha is essential for chondrocyte growth arrest and survival. Genes Dev **15**:2865–2876.
111. Zelzer E, Mamluk R, Ferrara N, Johnson RS, Schipani E, Olsen BR 2004 VEGFA is necessary for chondrocyte survival during bone development. Development **131**:2161–2171.

Chapter 2. Signal Transduction Cascades Controlling Osteoblast Differentiation

Carola Krause,[1] David J. J. de Gorter,[1] Marcel Karperien,[2] and Peter ten Dijke[1]

[1]Department of Molecular Cell Biology, University Medical Centre, Leiden, The Netherlands; [2]Department of Tissue Regeneration, University of Twente/Faculty Science and Technology Institute for Biomedical Technology, Enschede, The Netherlands

INTRODUCTION

Mesenchymal stem cells are pluripotent cells located in bone marrow, muscles, and fat that can differentiate into a variety of tissues, including bone, cartilage, muscle, and fat.[1,2] Differentiation toward these lineages is controlled by a multitude of cytokines, which regulate the expression of cell lineage–specific sets of transcription factors. Among the cytokines involved in osteoblast differentiation are the Hedgehogs, bone morphogenetic proteins (BMPs), TGF-β, PTH, and WNTs. The signal transduction cascades initiated by these cytokines and their effect on osteoblast differentiation will be discussed in this chapter. Osteoblasts and chondrocytes are thought to differentiate from a common mesenchymal precursor, the osteo-chondrogenic precursor (Fig. 1). The osteoblastic differentiation process can be divided into several stages, including proliferation, extracellular matrix deposition, matrix maturation, and mineralization.[3] To study osteoblast differentiation, the expression level of distinct differentiation markers are used, including alkaline phosphatase (ALP), type I collagen (Col1), bone sialoprotein (BSP), osteopontin (OPN), and osteocalcin (OC). Whereas ALP is used as an early marker, OC is considered to be a late marker for osteoblast differentiation.

RUNX2 AND OSTERIX TRANSCRIPTION FACTORS

An essential event in osteoblast differentiation, and a point of convergence of many signal transduction pathways involved, is activation of the transcription factor Runx2 (also known as Cbfa1) (Fig. 2). Runx2 is a master switch for osteoblast differentiation. This is shown by the fact that Runx2-deficient mice completely lack osteoblasts, fail to form hypertrophic chondrocytes, and produce a cartilaginous skeleton that is completely devoid of mineralized matrix.[4] In humans, heterozygous insertions, deletions, and nonsense mutations leading to translational stop codons in the DNA-binding domain or in the C-terminal transactivating region of the *Runx2* gene underlie the rare skeletal disorder cleidocranial dysplasia (CCD). CCD is characterized by defective development of the cranial bones and the complete or partial absence of the collar bones, emphasizing the importance of Runx2 in bone formation.[5] By interacting with many transcriptional activators and repressors and other co-regulatory proteins, Runx2 can either positively or negatively regulate expression of a variety of osteoblast-specific genes including Col1, ALP, OPN, osteonectin (ON), and OC (Fig. 2).[6–9]

Runx2 also regulates expression of the zinc finger–containing transcription factor Osterix. The promoter of the *Osx* (*Sp7*) gene (which encodes Osterix) contains a consensus Runx2-binding sequence, which suggests that Osterix is a direct Runx2 target.[10] Whereas Runx2 expression is not affected in $Osx^{-/-}$ mice, Osterix expression is lost in Runx2-

The authors state that they have no conflicts of interest.

Key words: osteoblast, bone formation, WNT, bone morphogenetic protein, Hedgehog, PTH, PTH-related peptide, TGF-β, fibroblast growth factor, IGF, Notch, Runx2, Osterix, nuclear factor for activated T cells, β-catenin, Smad

© 2008 American Society for Bone and Mineral Research

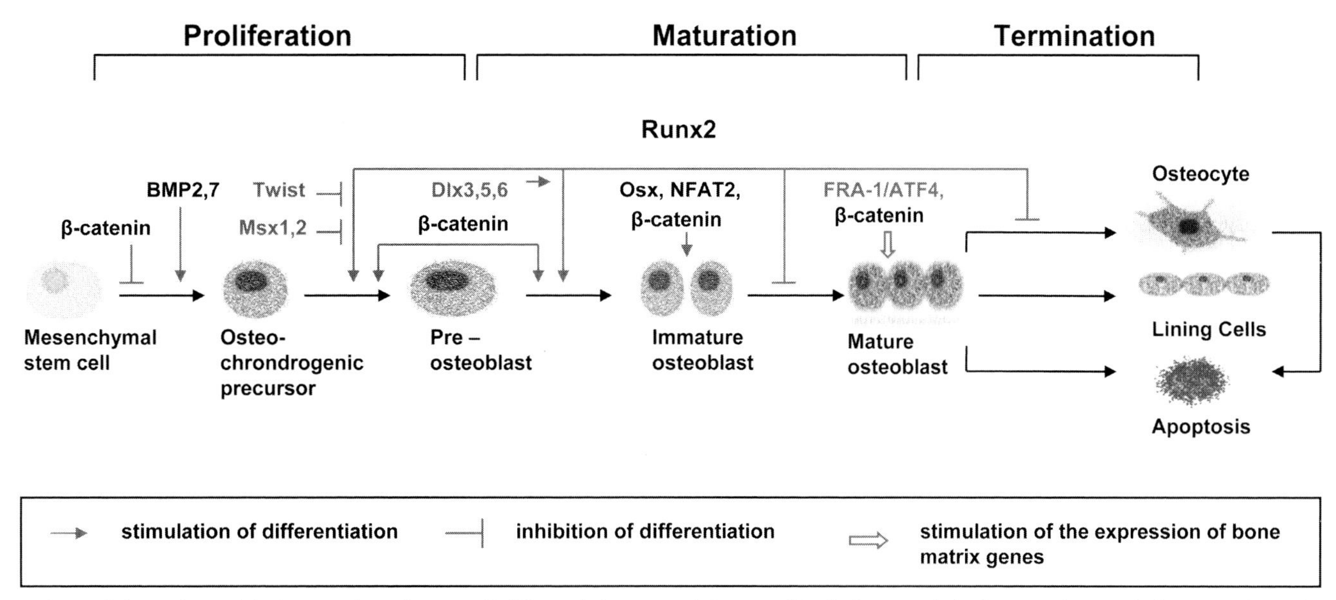

FIG. 1. Schematic model of mesenchymal stem cell differentiation toward the osteoblastic linage and the impact of transcriptional regulators in this process. ATF4, activating transcription factor-4; Dlx, Distalles homeobox; FRA, Fos-related antigen; Osx, Osterix; Runx2, Runt-related transcription factor2.

deficient mice.[11] Similar to mice deficient in Runx2, *Osx*$^{-/-}$ mice lack osteoblasts, showing the requirement of this transcription factor in bone formation. Osterix can interact with nuclear factor for activated T cells 2 (NFAT2), which cooperates with Osterix in controlling transcription of target genes such as *OC*, *OPN*, *ON*, and *Col1*.[11,12] Because nuclear localization of NFAT transcription factors is regulated by the Ca^{2+}-calcineurin pathway, signaling pathways that modulate intracellular Ca^{2+} levels can potentially control Osterix-mediated osteoblast differentiation through NFAT activation (Fig. 3).

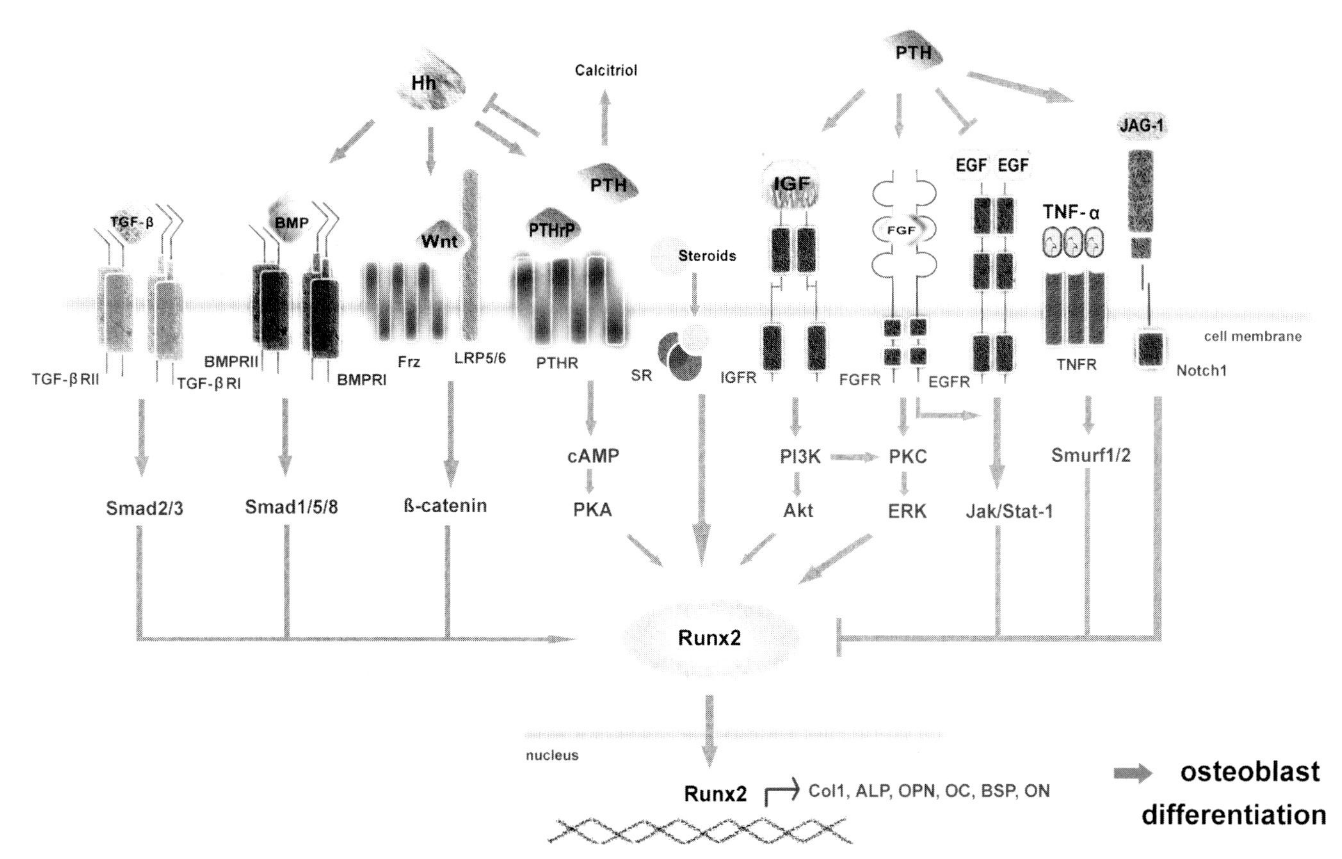

FIG. 2. Schematic model of signaling pathways involved in Runx2-mediated osteoblast differentiation. Col1, collagen type 1; BSP, bone sialoprotein; OPN, osteopontin; ON, osteonectin; OC, osteocalcin; ALP, alkaline phosphatase.

© 2008 American Society for Bone and Mineral Research

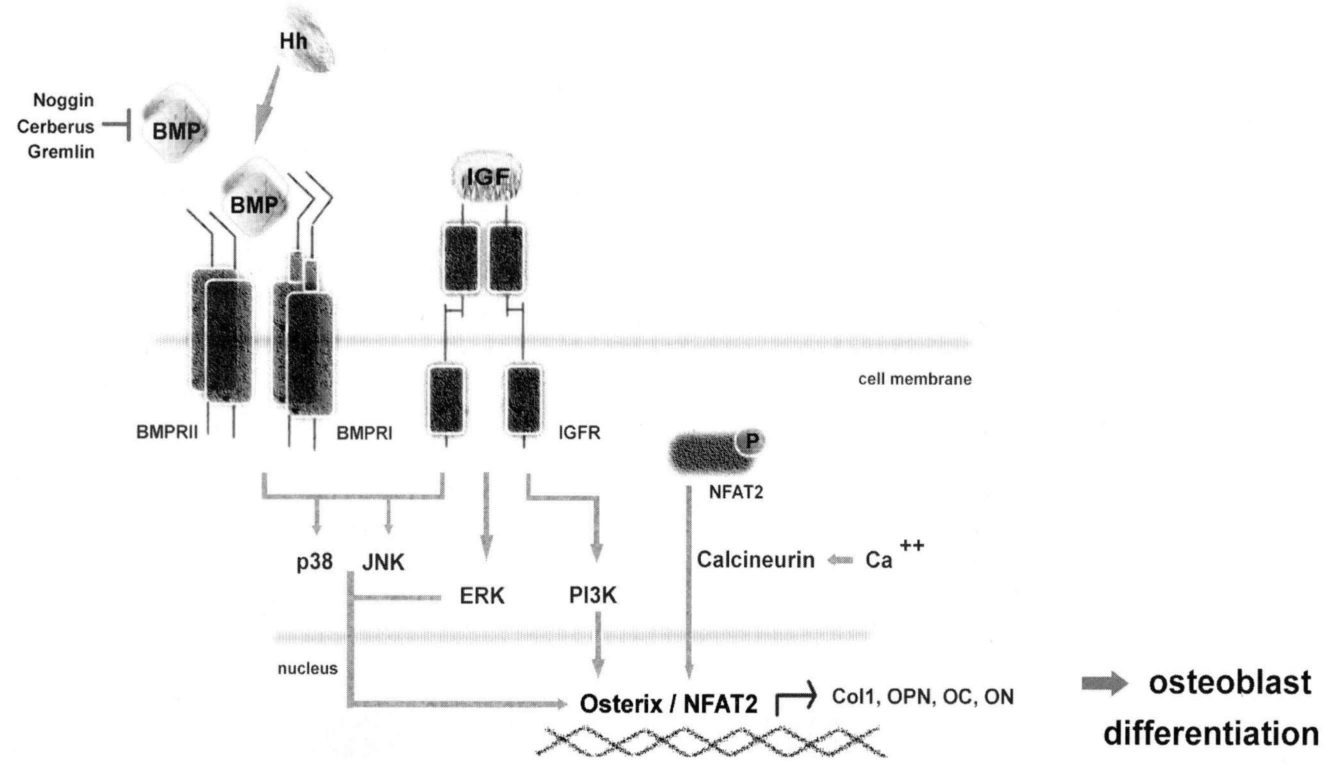

FIG. 3. Schematic model of signaling pathways involved in Osterix/NFAT2-mediated osteoblast differentiation. Col1, collagen type 1; OPN, osteopontin; ON, osteonectin; OC, osteocalcin.

Other transcription factors that are involved in osteoblast differentiation are homeobox proteins such as Msx2, Dlx-3, Dlx-5, Dlx-6, and members of the activator protein 1 (AP-1) family such as Fos, Fra, and ATF4. However, deficiency of these genes does not result in complete loss of osteoblasts like in *Runx2*$^{-/-}$ mice and *Osx*$^{-/-}$ mice, pointing at a facilitory role in osteoblastogenesis.

BMP SIGNALING

BMPs belong to the TGF-β superfamily and were originally identified as the active components in bone extracts capable of inducing bone formation at ectopic sites.[13] BMPs are expressed in skeletal tissue and are required for skeletal development and maintenance of adult bone homeostasis and play an important role in fracture healing.[14,15] Conditional knockout mice deficient in BMP ligands or Smad proteins, the intracellular mediators of BMP signaling, in bone display skeletal defects.[14,15] In addition, several naturally occurring mutations in BMPs or their receptors cause inherited disorders, including fibrodysplasia ossificans progressiva (FOP),[15,16] in which bone is progressively formed at ectopic sites. Thus, the BMP signaling pathway fulfills a key role in skeletal development and bone remodeling.

BMPs bind as dimers to type I and type II serine/threonine receptor kinases, forming an oligomeric complex (Fig. 2). On oligomerization, the constitutively active type II receptors phosphorylate and consequently activate the type I receptors. Subsequently, the activated type I receptors phosphorylate BMP receptor–regulated Smads, Smad1, -5, and -8, at their extreme C termini. The receptor-regulated Smads associate with the Co-Smad, Smad4, and translocate into the nucleus, where they together with other transcription factors bind promoters of target genes and control their expression (Fig. 2).[17,18] For example, Runx2 interacts with Smad1 and -5 and cooperates in controlling BMP-induced osteoblast-specific gene expression and osteogenic differentiation.[19,20] Interestingly, a nonsense mutation found in a CCD patient results in expression of a truncated Runx2 mutant which displayed impaired Smad1 interaction and inhibited BMP-induced ALP activity.[21] Moreover, BMP signaling induces expression of both BMPs and Runx2, thereby generating a positive feedback loop.[20,22] In addition, BMPs control expression of Id (inhibitor of differentiation or inhibitor of DNA binding) proteins.[23,24] Id proteins are inhibitors of basic helix-loop-helix proteins that inhibit osteoblast differentiation. Indeed, BMP-induced bone formation in vivo was found to be suppressed in Id1/Id3 heterozygous knockout mice.[25] Furthermore, BMP2 was found to induce expression of Osterix, which besides Runx2 also seems to be mediated by the p38 and c-Jun N-terminal kinase (JNK) MAP kinase (Fig. 3).[11,26,27]

TGF-β SIGNALING

TGF-β is implicated in the control of proliferation, migration, differentiation, and survival of many different cell types. TGF-β is one of the most abundant cytokines in bone matrix and plays a major role in development and maintenance of the skeleton, affecting both cartilage and bone metabolism.[28] Interestingly, TGF-β can have both positive and negative effects on bone formation depending on the context and concentration.

TGF-β signals through a similar mechanism as the related BMPs. However, on binding to its specific receptors, TGF-β induces activation of Smad2 and -3.[17,18] Smad3 overexpression in mouse osteoblastic MC3T3-E1 cells enhanced the levels of bone matrix proteins, ALP activity, and mineralization.[29] As is the case for Smad1 and -5, Runx2 also interacts with

© 2008 American Society for Bone and Mineral Research

Smad3 and cooperates in regulating TGF-β–induced transcription (Fig. 2).[30] This interaction requires a functional Runx2 C-terminal domain, because a truncated Runx2 mutant derived from a CCD patient is unable to interact with Smad3.[21,30,31] The effects of TGF-β/Smad3 signaling on the function of Runx2 depend on the cell type and promoter context.[30]

WNT SIGNALING

Bone diseases characterized by low or high bone mass caused by decreased or increased osteoblast activity have been associated with loss or gain of function mutations, respectively, in the *low-density lipoprotein receptor-related protein (LRP)5* gene, which encodes a WNT co-receptor (Fig. 2).[32–34] In addition, mutations in the coding region or in regulatory elements of *SOST*, the gene encoding the osteocyte-derived WNT antagonist sclerostin, underlie the rare high bone mass disorders sclerosteosis and Van Buchem disease, respectively.[35] These findings show the importance of WNT signaling in controlling bone formation.

WNTs are secreted glycoproteins that transduce their signals through 7-transmembrane spanning receptors of the frizzled family and LRP5 and -6 to β-catenin (Fig. 2).[36] In the absence of WNT ligand, β-catenin forms a complex with adenomatous polyposis coli (APC), axin, glycogen synthase kinase 3 (GSK3), and casein kinase I (CK1). This complex facilitates phosphorylation and proteasomal degradation of β-catenin. In the presence of WNT ligand, this complex dissociates, and β-catenin accumulates and translocates into the nucleus, where it initiates the transcription of target genes through complex formation with TCF/Lef1 transcription factors.[36] Conditional deletion of β-catenin led osteo-chondrogenic progenitor cells to differentiate into chondrocytes instead of osteoblasts during both intramembranous and endochondral ossification, whereas ectopic WNT signaling enhanced osteoblast differentiation,[37–39] indicating that WNT signaling drives differentiation of osteo-chondrogenic progenitor cells toward the osteoblast lineage (Fig. 1).

HEDGEHOG SIGNALING

Hedgehogs (Hh), of which there are three in mammals (i.e., Sonic, Indian, and Desert hedgehog) are critically important for development. In the endochondral skeleton, Indian hedgehog (Ihh) was found to be indispensable for osteoblast development, because mice deficient in Ihh completely lack osteoblasts in bones formed by endochondral ossification.[40,41]

Cellular responses to the Hh signal are controlled by two transmembrane proteins, the tumor suppressor 12-transmembrane protein Patched-1 (Ptch) and the 7-transmembrane receptor and oncoprotein Smoothened (Smo). The latter has homology to G protein–coupled receptors and transduces the Hh signal. In the absence of Hh, Ptch maintains Smo in an inactive state. With the binding of Hh, Ptch inhibition of Smo is released and intracellular signaling is initiated.[42] The transcriptional response to Hh signaling is mediated by three closely related zinc finger transcription factors termed Gli proteins, Gli1, Gli2, and Gli3, each with different roles and distinct set of target genes. Gli2 functions mainly as a transcriptional activator. In the absence of Hh, Gli3 is processed into a repressor of transcription. In the presence of Hh, however, full-length Gli3 translocates into the nucleus, which has transcriptional activation properties. Gli1 acts only as transcriptional activator and is induced by Hh signals.[42] Gli3 has a pivotal role in limb bud development and regulation of digit number and identity.[43,44] Ihh regulates osteoblast differentiation through a Gli2-mediated increase in expression and function of Runx2.[45]

PTH SIGNALING

PTH and PTH-related peptide (PTHrP) can have both anabolic and catabolic effects on bone. Whereas intermittent PTH administration induces bone formation, continuous treatment with PTH leads to bone loss.[46] The critical role of PTHrP in bone development is evident from loss of function of these genes and their receptors in mice and humans. Mice lacking PTHrP die perinatally, likely from respiratory failure caused by abnormalities in endochondral bone development.[47] A less severe phenotype is found for mice deficient in PTH, which are viable and display a slightly expanded hypertrophic zone.[48] Transgenic mice that overexpress PTHrP under the collagen type II promoter develop shortened limbs caused by delayed mineralization and chondrocyte maturation in the growth plate.[49] PTHrP signal through a 7-transmembrane G protein–coupled cell surface receptor (PTHR1). PTHR1 knockout mice show growth plate abnormalities caused by premature chondrocyte maturation.[50] In humans, loss of function in PTH1R has been linked to Blomstrand lethal osteochondrodysplasia.[51] These patients also suffer from advanced skeletal maturation and premature ossification of the skeleton. On ligand binding to PTHR1, activation of several intracellular signaling pathways can be activated, including the cAMP/protein kinase A (PKA) and protein kinase C (PKC) pathways. Different mechanisms have been suggested to explain the anabolic and catabolic effects of PTH; PTH may have diverse effects on the proliferation, commitment, differentiation, or apoptosis of the osteoblasts. Effects of PTH and PTHrP seem highly context dependent; action varies with cell type, stage of cell differentiation, dosage, and exposure time. For example, the bone anabolic effects of PTH involve an increase in expression and PKA-dependent activity of Runx2.[52] On the other hand, PTH has also shown to repress Runx2 and Osterix expression in osteoblasts.[53]

IGF-1 SIGNALING

IGF-1 is secreted by skeletal cells and is considered to be an auto- or paracrine regulator of osteoblastic cell function.[54] IGF-1–deficient mice were found to develop a smaller, but more compact bone structure.[55] IGF-1 signals through the IGF1 receptor, which like many other receptor tyrosine kinases activates the phosphatidylinositol 3-kinase (PI3K)/Akt and Ras/ERK MAP kinase pathways. Interestingly, Akt1/Akt2 double-knockout mice show a phenotype that is resembling that of mice deficient in the IGF-1 receptor, which includes impaired bone development.[56] Moreover, osteoblastic differentiation by forced expression of Runx2 was found to be inhibited by co-expression of dominant negative Akt or by treatment with IGF-1 antibodies or the pharmacological PI3K-inhibitor LY294002.[57] These data suggest that IGF-1–induced PI3K activation is involved in osteoblast differentiation. In addition, IGF-1 was found to induce expression of Osterix, which occurs in an extracellular-regulated kinase (ERK)-, p38-, and JNK-dependent manner (Fig. 3).[27] Through these mechanisms, IGF-1 can stimulate osteoblast differentiation.

FGF SIGNALING

Fibroblast growth factors (FGFs) are important regulators of endochondral and intramembranous bone formation, development, and apoptosis, affecting both chondrogenesis and

© 2008 American Society for Bone and Mineral Research

osteogenesis.[58] Many human craniosynostosis disorders have been linked to activating mutations in FGF receptors (FGFRs).[58] Disruption of FGFR2 signaling in skeletal tissues results in skeletal dwarfism and decreased BMD caused by disruption of the proliferation of osteoprogenitors and the anabolic function of mature osteoblasts, whereas osteoblast differentiation was found not to be affected.[59,60] FGFs are known to induce proliferation of immature osteoblasts through the Ras/ERK MAPK pathway.[61] In addition, FGF signaling was shown to stimulate expression, DNA binding and transcriptional activities of Runx2, mainly in a PKC-dependent manner (Fig. 2).[62]

NOTCH SIGNALING

Notch proteins are transmembrane receptors that control cell-fate decisions and inhibit osteoblastic differentiation.[63–65] Binding of the transmembrane Notch ligands Delta, Serrate, and Lag2 to Notch receptors induces cleavage of the Notch extracellular domain near the transmembrane region.[63] The resulting membrane-associated Notch is cleaved by Presenilin, generating the intracellular domain of Notch (NICD), which translocates into the nucleus. Here the NCID forms a complex with members of the CSL family (C promoter-binding factor 1 [CBF1], Suppressor of Hairless [Su(H)], and longevity assurance gene-1 [LAG-1]) of DNA-binding proteins, which recruits coactivators (CoA) to drive transcription of target genes.[63] In osteoblasts, Notch stimulates expression of Hey1, which was found to inhibit Runx2 transcriptional activity and mineralization.[64]

CONCLUDING REMARKS

Besides that regulation of Runx2 activity is a point of convergence of many of the signal transduction routes discussed in this chapter, there is also a high degree of cross-talk between these pathways. For example, apart from its C-terminal phosphorylation by BMP type I receptors, Smad1 can be phosphorylated by the ERK, p38, and JNK MAP kinases and subsequently by GSK-3, which results in cytoplasmic retention and increased proteasomal degradation of Smad1.[66,67] In this manner, FGF and WNT signaling can control the duration of BMP signaling.[66–68] On the other hand, β-catenin-TCF/Lef1 can interact with Smad1 and -3 proteins to cooperate in transcription of target genes.[69,70] TGF-β can inhibit BMP2-induced transcription and osteoblast differentiation,[71] and Notch 1 overexpression inhibits osteoblast differentiation by suppressing WNT signaling.[65] Gli2 mediates BMP2 expression in osteoblasts in response to Hh signaling,[72] and Hh signaling is required for accurate β-catenin–mediated Wnt signaling in osteoblasts.[39] TNFα inhibits BMP2-induced Smad activation through induction of NF-κB[73] and in part through SAPK/JNK signaling.[74] Thus, the combined action of the signal transduction pathways induced by bone promoting cytokines determines commitment of mesenchymal stem cells toward the osteoblast lineage and the efficiency of bone formation.

ACKNOWLEDGMENTS

Research on osteoblast differentiation in our laboratory is supported by a grant from the Dutch Organization for Scientific Research (918.66.606).

REFERENCES

1. Caplan AI, Bruder SP 2001 Mesenchymal stem cells: Building blocks for molecular medicine in the 21st century. Trends Mol Med **7**:259–264.
2. Jiang Y, Jahagirdar BN, Reinhardt RL, Schwartz RE, Keene CD, Ortiz-Gonzalez XR, Reyes M, Lenvik T, Lund T, Blackstad M, Du J, Aldrich S, Lisberg A, Low WC, Largaespada DA, Verfaillie CM 2002 Pluripotency of mesenchymal stem cells derived from adult marrow. Nature **418**:41–49.
3. Stein GS, Lian JB 1993 Molecular mechanisms mediating proliferation/differentiation interrelationships during progressive development of the osteoblast phenotype. Endocr Rev **14**:424–442.
4. Otto F, Thornell AP, Crompton T, Denzel A, Gilmour KC, Rosewell IR, Stamp GW, Beddington RS, Mundlos S, Olsen BR, Selby PB, Owen MJ 1997 Cbfa1, a candidate gene for cleidocranial dysplasia syndrome, is essential for osteoblast differentiation and bone development. Cell **89**:765–771.
5. Mundlos S 1999 Cleidocranial dysplasia: Clinical and molecular genetics. J Med Genet **36**:177–182.
6. Lian JB, Javed A, Zaidi SK, Lengner C, Montecino M, van Wijnen AJ, Stein JL, Stein GS 2004 Regulatory controls for osteoblast growth and differentiation: Role of Runx/Cbfa/AML factors. Crit Rev Eukaryot Gene Expr **14**:1–41.
7. Schroeder TM, Jensen ED, Westendorf JJ 2005 Runx2: A master organizer of gene transcription in developing and maturing osteoblasts. Birth Defects Res C Embryo Today **75**:213–225.
8. Harada H, Tagashira S, Fujiwara M, Ogawa S, Katsumata T, Yamaguchi A, Komori T, Nakatsuka M 1999 Cbfa1 isoforms exert functional differences in osteoblast differentiation. J Biol Chem **274**:6972–6978.
9. Kern B, Shen J, Starbuck M, Karsenty G 2001 Cbfa1 contributes to the osteoblast-specific expression of type I collagen genes. J Biol Chem **276**:7101–7107.
10. Nishio Y, Dong Y, Paris M, O'Keefe RJ, Schwarz EM, Drissi H 2006 Runx2-mediated regulation of the zinc finger Osterix/Sp7 gene. Gene **372**:62–70.
11. Nakashima K, Zhou X, Kunkel G, Zhang Z, Deng JM, Behringer RR, de Crombrugghe B 2002 The novel zinc finger-containing transcription factor osterix is required for osteoblast differentiation and bone formation. Cell **108**:17–29.
12. Koga T, Matsui Y, Asagiri M, Kodama T, de Crombrugghe B, Nakashima K, Takayanagi H 2005 NFAT and Osterix cooperatively regulate bone formation. Nat Med **11**:880–885.
13. Urist MR 1965 Bone: Formation by autoinduction. Science **150**:893–899.
14. Chen D, Zhao M, Mundy GR 2004 Bone morphogenetic proteins. Growth Factors **22**:233–241.
15. Gazzerro E, Canalis E 2006 Bone morphogenetic proteins and their antagonists. Rev Endocr Metab Disord **7**:51–65.
16. Shore EM, Xu M, Feldman GJ, Fenstermacher DA, Cho TJ, Choi IH, Connor JM, Delai P, Glaser DL, LeMerrer M, Morhart R, Rogers JG, Smith R, Triffitt JT, Urtizberea JA, Zasloff M, Brown MA, Kaplan FS 2006 A recurrent mutation in the BMP type I receptor ACVR1 causes inherited and sporadic fibrodysplasia ossificans progressiva. Nat Genet **38**:525–527.
17. Feng XH, Derynck R 2005 Specificity and versatility in TGF-β signaling through Smads. Annu Rev Cell Dev Biol **21**:659–693.
18. Massague J, Seoane J, Wotton D 2005 Smad transcription factors. Genes Dev **19**:2783–2810.
19. Javed A, Bae JS, Afzal F, Gutierrez S, Pratap J, Zaidi SK, Lou Y, van Wijnen AJ, Stein JL, Stein GS, Lian JB 2008 Structural coupling of Smad and Runx2 for execution of the BMP2 osteogenic signal. J Biol Chem **14**:14–24.
20. Lee KS, Kim HJ, Li QL, Chi XZ, Ueta C, Komori T, Wozney JM, Kim EG, Choi JY, Ryoo HM, Bae SC 2000 Runx2 is a common target of transforming growth factor β1 and bone morphogenetic protein 2, and cooperation between Runx2 and Smad5 induces osteoblast-specific gene expression in the pluripotent mesenchymal precursor cell line C2C12. Mol Cell Biol **20**:8783–8792.
21. Zhang YW, Yasui N, Ito K, Huang G, Fujii M, Hanai J, Nogami H, Ochi T, Miyazono K, Ito Y 2000 A RUNX2/PEBP2alpha A/CBFA1 mutation displaying impaired transactivation and Smad interaction in cleidocranial dysplasia. Proc Natl Acad Sci USA **97**:10549–10554.
22. Pereira RC, Rydziel S, Canalis E 2000 Bone morphogenetic protein-4 regulates its own expression in cultured osteoblasts. J Cell Physiol **182**:239–246.
23. Ogata T, Wozney JM, Benezra R, Noda M 1993 Bone morphogenetic protein 2 transiently enhances expression of a gene, Id (inhibitor of differentiation), encoding a helix-loop-helix molecule in osteoblast-like cells. Proc Natl Acad Sci USA **90**:9219–9222.
24. Korchynskyi O, ten Dijke P 2002 Identification and functional

© 2008 American Society for Bone and Mineral Research

characterization of distinct critically important bone morphogenetic protein-specific response elements in the Id1 promoter. J Biol Chem **277**:4883–4891.
25. Maeda Y, Tsuji K, Nifuji A, Noda M 2004 Inhibitory helix-loop-helix transcription factors Id1/Id3 promote bone formation in vivo. J Cell Biochem **93**:337–344.
26. Celil AB, Hollinger JO, Campbell PG 2005 Osx transcriptional regulation is mediated by additional pathways to BMP2/Smad signaling. J Cell Biochem **95**:518–528.
27. Celil AB, Campbell PG 2005 BMP2 and insulin-like growth factor-I mediate Osterix (Osx) expression in human mesenchymal stem cells via the MAPK and protein kinase D signaling pathways. J Biol Chem **280**:31353–31359.
28. Janssens K, ten Dijke P, Janssens S, Van Hul W 2005 Transforming growth factor-β1 to the bone. Endocr Rev **26**:743–774.
29. Sowa H, Kaji H, Yamaguchi T, Sugimoto T, Chihara K 2002 Smad3 promotes alkaline phosphatase activity and mineralization of osteoblastic MC3T3-E1 cells. J Bone Miner Res **17**:1190–1199.
30. Alliston T, Choy L, Ducy P, Karsenty G, Derynck R 2001 TGF-β-induced repression of CBFA1 by Smad3 decreases cbfa1 and osteocalcin expression and inhibits osteoblast differentiation. EMBO J **20**:2254–2272.
31. Hanai J, Chen LF, Kanno T, Ohtani-Fujita N, Kim WY, Guo WH, Imamura T, Ishidou Y, Fukuchi M, Shi MJ, Stavnezer J, Kawabata M, Miyazono K, Ito Y 1999 Interaction and functional cooperation of PEBP2/CBF with Smads. Synergistic induction of the immunoglobulin germline Calpha promoter. J Biol Chem **274**:31577–31582.
32. Boyden LM, Mao J, Belsky J, Mitzner L, Farhi A, Mitnick MA, Wu D, Insogna K, Lifton RP 2002 High bone density due to a mutation in LDL-receptor-related protein 5. N Engl J Med **346**:1513–1521.
33. Little RD, Carulli JP, Del Mastro RG, Dupuis J, Osborne M, Folz C, Manning SP, Swain PM, Zhao SC, Eustace B, Lappe MM, Spitzer L, Zweier S, Braunschweiger K, Benchekroun Y, Hu X, Adair R, Chee L, FitzGerald MG, Tulig C, Caruso A, Tzellas N, Bawa A, Franklin B, McGuire S, Nogues X, Gong G, Allen KM, Anisowicz A, Morales AJ, Lomedico PT, Recker SM, Van Eerdewegh P, Recker RR, Johnson ML 2002 A mutation in the LDL receptor-related protein 5 gene results in the autosomal dominant high-bone-mass trait. Am J Hum Genet **70**:11–19.
34. Van Wesenbeeck L, Cleiren E, Gram J, Beals RK, Benichou O, Scopelliti D, Key L, Renton T, Bartels C, Gong Y, Warman ML, De Vernejoul MC, Bollerslev J, Van Hul W 2003 Six novel missense mutations in the LDL receptor-related protein 5 (LRP5) gene in different conditions with an increased bone density. Am J Hum Genet **72**:763–771.
35. van Bezooijen RL, ten Dijke P, Papapoulos SE, Lowik CW 2005 SOST/sclerostin, an osteocyte-derived negative regulator of bone formation. Cytokine Growth Factor Rev **16**:319–327.
36. Clevers H 2006 Wnt/β-catenin signaling in development and disease. Cell **127**:469–480.
37. Day TF, Guo X, Garrett-Beal L, Yang Y 2005 Wnt/β-catenin signaling in mesenchymal progenitors controls osteoblast and chondrocyte differentiation during vertebrate skeletogenesis. Dev Cell **8**:739–750.
38. Hill TP, Spater D, Taketo MM, Birchmeier W, Hartmann C 2005 Canonical Wnt/β-catenin signaling prevents osteoblasts from differentiating into chondrocytes. Dev Cell **8**:727–738.
39. Hu H, Hilton MJ, Tu X, Yu K, Ornitz DM, Long F 2005 Sequential roles of Hedgehog and Wnt signaling in osteoblast development. Development **132**:49–60.
40. St Jacques B, Hammerschmidt M, McMahon AP 1999 Indian hedgehog signaling regulates proliferation and differentiation of chondrocytes and is essential for bone formation. Genes Dev **13**:2072–2086.
41. Long F, Chung UI, Ohba S, McMahon J, Kronenberg HM, McMahon AP 2004 Ihh signaling is directly required for the osteoblast lineage in the endochondral skeleton. Development **131**:1309–1318.
42. Hooper JE, Scott MP 2005 Communicating with Hedgehogs. Nat Rev Mol Cell Biol **6**:306–317.
43. Hui CC, Joyner AL 1993 A mouse model of greig cephalopolysyndactyly syndrome: The extra-toesJ mutation contains an intragenic deletion of the Gli3 gene. Nat Genet **3**:241–246.
44. Litingtung Y, Dahn RD, Li Y, Fallon JF, Chiang C 2002 Shh and Gli3 are dispensable for limb skeleton formation but regulate digit number and identity. Nature **418**:979–983.
45. Shimoyama A, Wada M, Ikeda F, Hata K, Matsubara T, Nifuji A, Noda M, Amano K, Yamaguchi A, Nishimura R, Yoneda T 2007 Ihh/Gli2 signaling promotes osteoblast differentiation by regulating Runx2 expression and function. Mol Biol Cell **18**:2411–2418.
46. Rubin MR, Bilezikian JP 2003 New anabolic therapies in osteoporosis. Endocrinol Metab Clin North Am **32**:285–307.
47. Karaplis AC, Luz A, Glowacki J, Bronson RT, Tybulewicz VL, Kronenberg HM, Mulligan RC 1994 Lethal skeletal dysplasia from targeted disruption of the parathyroid hormone-related peptide gene. Genes Dev **8**:277–289.
48. Miao D, He B, Karaplis AC, Goltzman D 2002 Parathyroid hormone is essential for normal fetal bone formation. J Clin Invest **109**:1173–1182.
49. Weir EC, Philbrick WM, Amling M, Neff LA, Baron R, Broadus AE 1996 Targeted overexpression of parathyroid hormone-related peptide in chondrocytes causes chondrodysplasia and delayed endochondral bone formation. Proc Natl Acad Sci USA **93**:10240–10245.
50. Lanske B, Karaplis AC, Lee K, Luz A, Vortkamp A, Pirro A, Karperien M, Defize LH, Ho C, Mulligan RC, Abou-Samra AB, Juppner H, Segre GV, Kronenberg HM 1996 PTH/PTHrP receptor in early development and Indian hedgehog-regulated bone growth. Science **273**:663–666.
51. Zhang P, Jobert AS, Couvineau A, Silve C 1998 A homozygous inactivating mutation in the parathyroid hormone/parathyroid hormone-related peptide receptor causing Blomstrand chondrodysplasia. J Clin Endocrinol Metab **83**:3365–3368.
52. Krishnan V, Moore TL, Ma YL, Helvering LM, Frolik CA, Valasek KM, Ducy P, Geiser AG 2003 Parathyroid hormone bone anabolic action requires Cbfa1/Runx2-dependent signaling. Mol Endocrinol **17**:423–435.
53. van der Horst G, Farih-Sips H, Lowik CW, Karperien M 2005 Multiple mechanisms are involved in inhibition of osteoblast differentiation by PTHrP and PTH in KS483 Cells. J Bone Miner Res **20**:2233–2244.
54. Canalis E, McCarthy TL, Centrella M 1991 Growth factors and cytokines in bone cell metabolism. Annu Rev Med **42**:17–24.
55. Bikle D, Majumdar S, Laib A, Powell-Braxton L, Rosen C, Beamer W, Nauman E, Leary C, Halloran B 2001 The skeletal structure of insulin-like growth factor I-deficient mice. J Bone Miner Res **16**:2320–2329.
56. Peng XD, Xu PZ, Chen ML, Hahn-Windgassen A, Skeen J, Jacobs J, Sundararajan D, Chen WS, Crawford SE, Coleman KG, Hay N 2003 Dwarfism, impaired skin development, skeletal muscle atrophy, delayed bone development, and impeded adipogenesis in mice lacking Akt1 and Akt2. Genes Dev **17**:1352–1365.
57. Fujita T, Azuma Y, Fukuyama R, Hattori Y, Yoshida C, Koida M, Ogita K, Komori T 2004 Runx2 induces osteoblast and chondrocyte differentiation and enhances their migration by coupling with PI3K-Akt signaling. J Cell Biol **166**:85–95.
58. Ornitz DM 2005 FGF signaling in the developing endochondral skeleton. Cytokine Growth Factor Rev **16**:205–213.
59. Eswarakumar VP, Monsonego-Ornan E, Pines M, Antonopoulou I, Morriss-Kay GM, Lonai P 2002 The IIIc alternative of Fgfr2 is a positive regulator of bone formation. Development **129**:3783–3793.
60. Yu K, Xu J, Liu Z, Sosic D, Shao J, Olson EN, Towler DA, Ornitz DM 2003 Conditional inactivation of FGF receptor 2 reveals an essential role for FGF signaling in the regulation of osteoblast function and bone growth. Development **130**:3063–3074.
61. Franceschi RT, Xiao G 2003 Regulation of the osteoblast-specific transcription factor, Runx2: Responsiveness to multiple signal transduction pathways. J Cell Biochem **88**:446–454.
62. Kim HJ, Kim JH, Bae SC, Choi JY, Kim HJ, Ryoo HM 2003 The protein kinase C pathway plays a central role in the fibroblast growth factor-stimulated expression and transactivation activity of Runx2. J Biol Chem **278**:319–326.
63. Ehebauer M, Hayward P, Martinez-Arias A 2006 Notch signaling pathway. Sci STKE **2006**:cm7.
64. Zamurovic N, Cappellen D, Rohner D, Susa M 2004 Coordinated activation of notch, Wnt, and transforming growth factor-β signaling pathways in bone morphogenic protein 2-induced osteogenesis. Notch target gene Hey1 inhibits mineralization and Runx2 transcriptional activity. J Biol Chem **279**:37704–37715.
65. Deregowski V, Gazzerro E, Priest L, Rydziel S, Canalis E 2006 Notch 1 overexpression inhibits osteoblastogenesis by suppressing Wnt/β-catenin but not bone morphogenetic protein signaling. J Biol Chem **281**:6203–6210.

© 2008 American Society for Bone and Mineral Research

66. Fuentealba LC, Eivers E, Ikeda A, Hurtado C, Kuroda H, Pera EM, De Robertis EM 2007 Integrating patterning signals: Wnt/GSK3 regulates the duration of the BMP/Smad1 signal. Cell **131:**980–993.
67. Sapkota G, Alarcon C, Spagnoli FM, Brivanlou AH, Massague J 2007 Balancing BMP signaling through integrated inputs into the Smad1 linker. Mol Cell **25:**441–454.
68. Nakayama K, Tamura Y, Suzawa M, Harada S, Fukumoto S, Kato M, Miyazono K, Rodan GA, Takeuchi Y, Fujita T 2003 Receptor tyrosine kinases inhibit bone morphogenetic protein-Smad responsive promoter activity and differentiation of murine MC3T3-E1 osteoblast-like cells. J Bone Miner Res **18:**827–835.
69. Hu MC, Rosenblum ND 2005 Smad1, β-catenin and Tcf4 associate in a molecular complex with the Myc promoter in dysplastic renal tissue and cooperate to control Myc transcription. Development **132:**215–225.
70. Labbe E, Letamendia A, Attisano L 2000 Association of Smads with lymphoid enhancer binding factor 1/T cell-specific factor mediates cooperative signaling by the transforming growth factor-β and wnt pathways. Proc Natl Acad Sci USA **97:**8358–8363.
71. Spinella-Jaegle S, Roman-Roman S, Faucheu C, Dunn FW, Kawai S, Gallea S, Stiot V, Blanchet AM, Courtois B, Baron R, Rawadi G 2001 Opposite effects of bone morphogenetic protein-2 and transforming growth factor-β1 on osteoblast differentiation. Bone **29:**323–330.
72. Zhao M, Qiao M, Harris SE, Chen D, Oyajobi BO, Mundy GR 2006 The zinc finger transcription factor Gli2 mediates bone morphogenetic protein 2 expression in osteoblasts in response to hedgehog signaling. Mol Cell Biol **26:**6197–6208.
73. Li Y, Li A, Strait K, Zhang H, Nanes MS, Weitzmann MN 2007 Endogenous TNFalpha lowers maximum peak bone mass and inhibits osteoblastic Smad activation through NF-kappaB. J Bone Miner Res **22:**646–655.
74. Mukai T, Otsuka F, Otani H, Yamashita M, Takasugi K, Inagaki K, Yamamura M, Makino H 2007 TNF-alpha inhibits BMP-induced osteoblast differentiation through activating SAPK/JNK signaling. Biochem Biophys Res Commun **356:**1004–1010.

Chapter 3. Osteoclast Biology and Bone Resorption

F. Patrick Ross

Department of Pathology and Immunology, Washington University School of Medicine, St. Louis, Missouri

CELL BIOLOGY OF THE OSTEOCLAST

Pathological bone loss, regardless of etiology, invariably represents an increase in the rate at which the skeleton is degraded by osteoclasts relative to its formation by osteoblasts. Thus, prevention of conditions such as osteoporosis requires an understanding of the molecular mechanisms of bone resorption.

The osteoclast, the exclusive bone resorptive cell (Fig. 1), is a member of the monocyte/macrophage family and a polykaryon that can be generated in vitro from mononuclear phagocyte precursors resident in a number of tissues.[1] There is, however, general agreement that the principal physiological osteoclast precursor is the bone marrow macrophage. Two cytokines are essential and sufficient for basal osteoclastogenesis, the first being RANKL[1,2] and the second being macrophage-colony stimulating factor (M-CSF), also designated CSF-1.[3] These two proteins, which exist as both membrane-bound and soluble forms (the former is secreted by activated T cells),[4] are produced by marrow stromal cells and their derivative osteoblasts, and thus physiological recruitment of osteoclasts from their mononuclear precursors requires the presence of these nonhematopoietic, bone-residing cells.[1] RANKL, a member of the TNF superfamily, is the key osteoclastogenic cytokine, because osteoclast formation requires its presence or its priming of precursor cells. M-CSF contributes to the proliferation, survival, and differentiation of osteoclast precursors, as well as the survival and cytoskeletal rearrangement required for efficient bone resorption (Fig. 2; a brief summary of the integrated signaling pathways for each osteoclastic regulator discussed in this review is provided later in this review [Fig. 6]). The discovery of RANKL was preceded by identification of its physiological inhibitor osteoprotegerin (OPG), to which it binds with high affinity.[5] In contrast, M-CSF is a moiety long known to regulate the broader biology of myeloid cells, including osteoclasts[3] (see Fig. 6).

Our understanding of how osteoclasts resorb bone derives from two major sources: biochemical and genetic.[2] The unique osteoclastogenic properties of RANKL permit generation of pure populations of osteoclasts in culture and hence the performance of meaningful biochemical and molecular experiments that provide insights into the molecular mechanisms by which osteoclasts resorb bone. Further evidence has come from our capacity to generate mice lacking specific genes, plus the positional cloning of genetic abnormalities in people with abnormal osteoclast function. Key to the resorptive event is the capacity of the osteoclast to form a microenvironment between itself and the underlying bone matrix (Fig. 3A). This compartment, which is isolated from the general extracellular space, is acidified by an electrogenic proton pump (H^+-ATPase) and a Cl^- channel to a pH of ~4.5.[6] The acidified milieu mobilizes the mineralized component of bone, exposing its organic matrix, consisting largely of type 1 collagen that is subsequently degraded by the lysosomal enzyme cathepsin K. The critical role that the proton pump, Cl^- channel, and cathepsin K play in osteoclast action is underscored by the fact that diminished function of each results in a human disease of excess bone mass, namely osteopetrosis or pyknodysostosis.[2,6] Degraded protein fragments are endocytosed and transported in undefined vesicles to the basolateral surface of the cell, where they are discharged into the surrounding intracellular fluid.[7,8] It is also likely that retraction of an osteoclast from the resorptive pit results in release of products of digestion.

The above model of bone degradation clearly depends on physical intimacy between the osteoclast and bone matrix, a role provided by integrins. Integrins are αβ heterodimers with long extracellular and single transmembrane domains.[9] In most instances, the integrin cytoplasmic region is relatively short, consisting of 40–70 amino acids. Integrins are the principal cell/matrix attachment molecules and they mediate osteoclast/bone recognition. Members of the β1 family of integrins, which recognize collagen, fibronectin, and laminin, are present on osteoclasts, but αvβ3 is the principal integrin mediating bone resorption.[10] This heterodimer, like all members

The author states that he has no conflicts of interest.

© 2008 American Society for Bone and Mineral Research

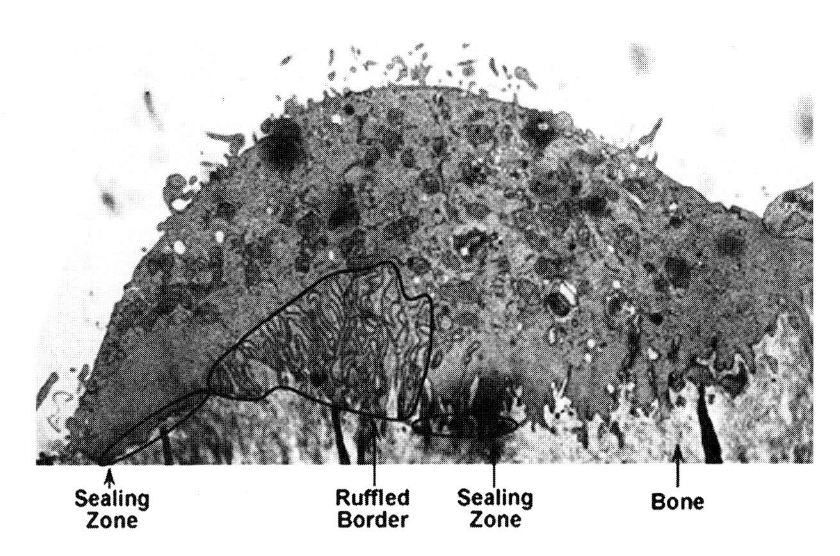

FIG. 1. The osteoclast as a resorptive cell. Transmission electron microscopy of a multinucleated primary rat osteoclast on bone. Note the extensive ruffled border, close apposition of the cell to bone and the partially degraded matrix between the sealing zones. Courtesy of H. Zhao.

of the αv integrin family, recognizes the amino acid motif Arg-Gly-Asp (RGD), which is present in a variety of bone-residing proteins such as osteopontin and bone sialoprotein. Thus, osteoclasts attach to and spread on these substrates in an RGD-dependent manner and, most importantly, competitive ligands arrest bone resorption in vivo. Proof of the pivotal role that αvβ3 has in the resorptive process came with the generation of the β3 integrin knockout mouse, which develops a progressive increase in bone mass because of osteoclast dysfunction.[33] Based on a combination of these in vitro and in vivo observations, small molecule inhibitors of osteoclast function that target αvβ3 have been developed.[11]

Bone resorption also requires a polarization event in which the osteoclast delivers effector molecules like HCl and cathepsin K into the resorptive microenvironment. Osteoclasts are characterized by a unique cytoskeleton, which mediates the resorptive process. Specifically, when the cell contacts bone, it generates two polarized structures, which enable it to degrade skeletal tissue. In the first instance, a subset of acidified vesicles containing specific cargo, including cathepsin K and other matrix metalloproteases (MMPs), are transported, probably through microtubules and actin, to the bone-apposed plasma membrane,[12] to which they fuse in a manner not currently understood, but which may involve PLEKHM1.[13] Insertion of these vesicles into the plasmalemma results in formation of a villous structure, unique to the osteoclast, called the ruffled membrane. This resorptive organelle contains the abundant H^+ transporting machinery to create the acidified microenvironment, whereas the accompanying exocytosis serves as the means by which cathepsin K is secreted (Fig. 3B).

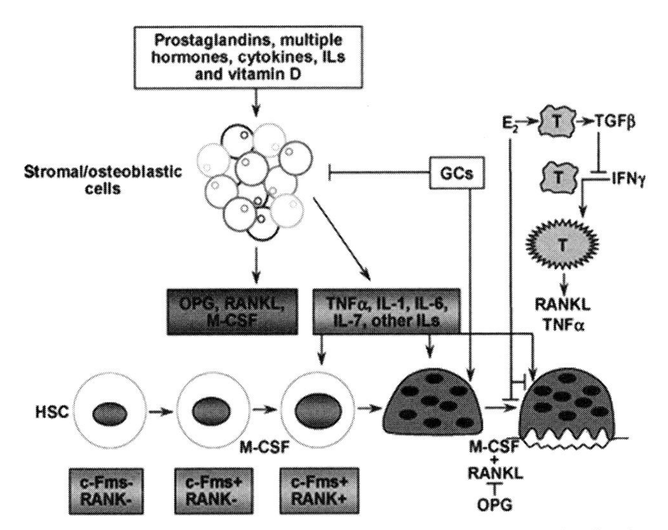

FIG. 2. Role of cytokines, hormones, steroids, and prostaglandins in osteoclast formation. Under the influence of other cytokines (data not shown), hematopoietic stem cells (HSCs) commit to the myeloid lineage, express c-Fms and RANK, the receptors for M-CSF and RANKL, respectively, and differentiate into osteoclasts. Mesenchymal cells in the marrow respond to a range of stimuli, secreting a mixture of pro- and anti-osteoclastogenic proteins, the latter primarily OPG. Glucocorticoids suppress bone resorption indirectly but possibly also target osteoclasts and/or their precursors. Estrogen, by a complex mechanism, inhibits activation of T cells, decreasing their secretion of RANKL and TNF-α; the sex steroid also inhibits osteoblast and osteoclast differentiation and lifespan. A key factor regulating bone resorption is the RANKL/OPG ratio.

In addition to inducing ruffled membrane formation, contact with bone also prompts the osteoclast to polarize its fibrillar actin into a circular structure known as the "actin ring." A separate "sealing zone" surrounds and isolates the acidified resorptive microenvironment in the active cell, but its composition is almost completely unknown. The actin ring, like the ruffled membrane, is a hallmark of the degradative capacity of the osteoclast, because structural abnormalities of either occur in conditions of arrested resorption.[14] In most cells, such as fibroblasts, matrix attachment prompts formation of stable structures known as focal adhesions that contain both integrins and a host of signaling and cytoskeletal molecules, which mediate contact and formation of actin stress fibers. In keeping with the substitution of the actin ring for stress fibers in osteoclasts, these cells form podosomes instead of focal adhesions. Podosomes, which in resorbing osteoclasts are present in the actin ring, consist of an actin core surrounded by αvβ3 and associated cytoskeletal proteins.

The integrin β3 subunit knockout mouse serves as an important tool for determining the role of αvβ3 in the capacity of the osteoclast to resorb bone. Failure to express αvβ3 results in a dramatic osteoclast phenotype, particularly regarding the actin cytoskeleton. The $\beta3^{-/-}$ osteoclast forms abnormal ruffled membranes in vivo and, whether generated in vitro or directly isolated from bone, the mutant cells fail to spread when plated on immobilized RGD ligand or mineralized matrix in physiological amounts of RANKL and M-CSF. Confirming their attenuated resorptive activity, $\beta3^{-/-}$ osteoclasts generate fewer and shallower resorptive lacunae on dentin slices than do their

© 2008 American Society for Bone and Mineral Research

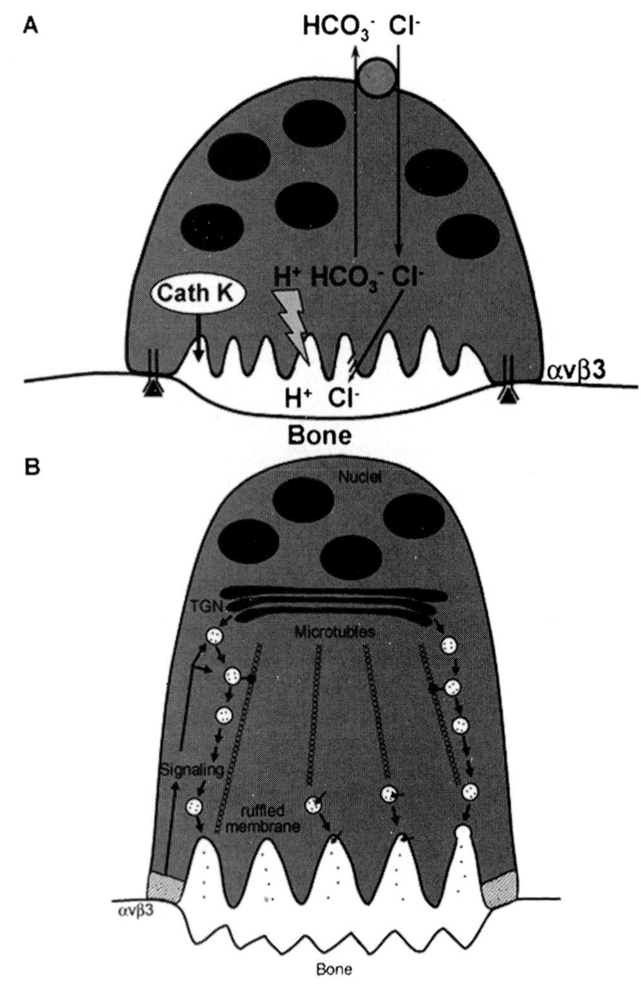

FIG. 3. Mechanism of osteoclastic bone resorption. (A) The osteoclast adheres to bone through the integrin αvβ3, creating a sealing zone, into which is secreted hydrochloric acid and acidic proteases such as cathepsin K, MMP9, and MMP13. The acid is generated by the combined actions of a vacuolar H^+ ATPase; it coupled chloride channel and a basolateral chloride-bicarbonate exchanger. Carbonic anhydrase converts CO_2 into H^+ and HCO^{3-} (data not shown). (B) Integrin engagement results in signals that target acidifying vesicles (♦ = proton pump complex) containing specific cargo (black dots) to the bone-apposed face of the cell. Fusion of these vesicles with the plasma membrane generates a polarized cell capable of secreting the acid and proteases required for bone resorption.

wildtype counterparts. In keeping with attenuated bone resorption in vivo, $\beta 3^{-/-}$ mice are substantially hypocalcemic.[6]

INTEGRIN SIGNALING

Whereas integrins were viewed initially as merely cell attachment molecules, it is now apparent that their capacity to transmit signals to and from the cell interior is equally important, an event that requires that the integrin convert from a default low affinity state to one in which its capacity to bind matrix is significantly enhanced. The process, termed activation, arises from either integrin ligation of their multivalent ligands or indirectly by growth factor signaling.[15]

αvβ3 is absent from osteoclast precursors, but their differentiation under the action of RANKL results in marked upregulation of this heterodimer. The capacity of integrins to transmit intracellular signals to the cytoskeleton heightened interest in the cytoplasmic molecules mediating these events in osteoclasts and αvβ3 signaling in this context is reasonably well understood. The initial signaling event involves the proto-oncogene c-src, which, acting as a kinase and an adaptor protein, regulates formation of lamellipodia and disassembly of podosomes, indicating that c-src controls formation of resorptive organelles of the cell, such as the ruffled membrane, and also arrests migration on the bone surface. There is continuing debate surrounding the molecules which link c-src to the cytoskeleton, one proposal being that the focal adhesion kinase family member Pyk2, acting in concert with c-Cbl, a proto-oncogene and ubiquitin ligase.[16] A second strong candidate is Syk, a nonreceptor tyrosine kinase that is recruited to the active conformation of αvβ3 in osteoclasts in a c-src–dependent manner,[17] where it targets Vav3,[10] a member of the large family of guanine nucleotide exchange factors (GEFs) that convert Rho GTPases from their inactive GDP to their active GTP conformation.

SMALL GTPASES

The Rho family of GTPases is central to remodeling of the actin cytoskeleton in many cell types,[18] and as such plays a central role in osteoclastic bone resorption. On attachment to bone, Rho and Rac bind GTP and translocate to the cytoskeleton. Whereas both small GTPases impact the actin cytoskeleton, Rac and Rho exert distinctive effects. Rho signaling mediates formation of the actin ring and a constitutively active form of the GTPase stimulates podosome formation, osteoclast motility, and bone resorption, whereas dominant negative Rho arrests these events.[19] Rac stimulation in osteoclast precursors prompts appearance of lamellipodia, thus forming the migratory front of the cell to which αvβ3 moves when activated.[20] In sum, it is likely that Rho's effect is principally on cell adhesion, whereas Rac mediates the cytoskeleton's migratory machinery. Importantly, absence of Vav3 blunts Rac but not Rho activity in the osteoclast.[21]

FACTORS REGULATING OSTEOCLAST FORMATION AND/OR FUNCTION

Proteins

In addition to the two key osteoclastogenic cytokines M-CSF and RANKL, a number of other proteins play important roles in osteoclast biology, either in physiological and/or patho-physiological circumstances.

As discussed earlier, OPG, a high-affinity ligand for RANKL that acts as a soluble inhibitor of RANKL, is secreted by cells of mesenchymal origin, both basally and in response to other regulatory signals, including cytokines and bone-targeting steroids.[5] Pro-inflammatory cytokines suppress OPG expression while simultaneously enhancing that of RANKL, with the net effect being a marked increase in osteoclast formation and function. Genetic deletion of OPG in both mice and humans leads to profound osteoporosis,[22] whereas overexpression of the molecule under the control of a hepatic promoter results in severe osteopetrosis.[23] Together, these observations indicate that skeletal and perhaps circulating OPG modulates the bone resorptive activity of RANKL and helps to explain the increased bone loss in clinical situations accompanied by increased levels of TNF-α, interleukin (IL)-1, PTH, or PTH-related protein (PTHrP). Serum PTH levels are increased in hyperparathyroidism of whatever etiology, whereas PTHrP is secreted by metastatic lung and breast carcinoma.[24,25] TNF antibodies or a soluble TNF receptor–IgG fusion protein potently suppress the bone loss in disorders of inflammatory osteolysis such as rheumatoid arthritis.[26]

© 2008 American Society for Bone and Mineral Research

The molecular basis of this observation seems to be that the inflammatory cytokine synergizes with RANKL in a unique manner, most likely because RANKL and TNF each activate a number of key downstream effector pathways, leading to nuclear localization of a range of osteoclastogenic transcription factors (see Fig. 6). Recent evidence suggests a new paradigm linking TNF, IL-1, and the natural secreted inhibitor for the latter cytokine, IL-1 receptor antagonist, which blocks IL-1 function. Specifically, it seems that, at least in murine osteoclasts and their precursors, many of the effects of TNF are mediated through its stimulation of IL-1, which in turn increases expression and secretion of IL-1ra, a set of events that represent a complex control pathway. The significance of IL receptor antagonist is shown by the fact an IgG fusion protein containing the active component of this molecule has been developed and enhances the ability of anti-TNF-α antibodies to decrease bone loss in rheumatoid arthritis.[27]

Elegant studies suggest that interferon γ (IFNγ) is an important suppressor of osteoclast formation and function.[28] Nevertheless, these findings seem to be in conflict with other in vivo observations, including the report that IFNγ treatment of children with osteopetrosis ameliorates the disease[29] and the fact that a number of in vivo studies indicate that IFNγ stimulates bone resorption.[30] This conundrum highlights the importance of discriminating between in vitro culture experiments using single cytokines and results in vivo. Many additional studies have implicated a range of other cytokines in the regulation of the osteoclast. These include a range of interleukins, GM-CSF, IFNβ, stromal cell–derived factor 1 (SDF-1), macrophage inflammatory protein 1 (MIPα), and monocyte chemoattractant protein 1 (MCP-1),[31–34] but at this time the results are either contradictory, as for GM-CSF in the murine versus human systems, or lack direct proof in humans. Future studies are likely to clarify the currently confusing data set. Finally, interactions between immune receptors such as DNAX activating protein of 12 kDa (DAP12) and FC receptor γ (FcRγ), present on osteoclasts and their precursors, and their ligands on cells of the stromal and myeloid/lymphoid lineages are important for transmission of RANK-derived signals.[28] IL-17 is a product of Th17 cells, a recently identified T-cell subset that is generated from uncommitted precursors under the influence of TGF-β, IL-23, and IL-6.[32,33]

Small Molecules

1,25-dihydroxyvitamin D has all the characteristics of a steroid hormone, including a high-affinity nuclear receptor that binds as a heterodimer with the retinoid X receptor to regulate transcription of a set of specific target genes. This active form of vitamin D, generated by successive hydroxylation in the liver and kidney, is a well-established stimulator of bone resorption when present at supraphysiological levels. Studies over many years have indicated that this steroid hormone increases mesenchymal cell transcription of the *RANKL* gene, whereas diminishing that of OPG.[5] Separately, 1,25-dihydroxyvitamin D suppresses synthesis of the pro-osteoclastogenic hormone PTH[34] and enhances calcium uptake from the gut. Taken together, the two latter effects would seem to be antiresorptive, but many studies in humans indicate the net osteolytic action resulting from high levels of this steroid hormone, suggesting that its ability to stimulate osteoclast function overrides any bone anabolic actions.

Loss of estrogen (E_2), most often seen in the context of menopause, is a major reason for the development of significant bone loss in aging. Interestingly, it is now clear that estrogen is the main sex steroid regulating bone mass in both men and women.[35] The mechanisms by which estrogen mediates its osteolytic effects are still incompletely understood, but significant advances have been made over the last decade. The original hypothesis, now considered to only part of the explanation, is that decreased serum E_2 led to increased production, by circulating macrophages, of osteoclastogenic cytokines such as IL-6, TNF, and IL-1. These molecules act on stromal cells and osteoclast precursors to enhance bone resorption by regulating expression of pro- (RANKL, M-CSF) and anti- (OPG) osteoclastogenic cytokines (in the case of mesenchymal cells) and by synergizing with RANKL itself (in the case of myeloid osteoclast precursors; see Fig. 2). However, the understanding that lymphocytes play a key role in mediating several aspects of bone biology has led to a growing realization that the cellular and molecular targets for E_2 are more widespread than previously believed. A model proposes that E_2 impacts the resorptive component of bone turnover (the steroid has separate effects on osteoblasts), at least in part by modulating production by T cells of RANKL and TNF.[30] This effect is itself indirect, with E_2 suppressing antigen presentation by dendritic cells and macrophages by enhancing expression by the same cells of TGFβ. Antigen presentation activates T cells, thereby enhancing their production of RANKL and TNF. As discussed previously, the first molecule is the key osteoclastogenic cytokine, whereas the second potentiates RANKL action and stimulates production by stromal cells of M-CSF and RANKL. This newly discovered interface between T cells and bone resorption also clarifies aspects of inflammatory osteolysis. Finally, some studies indicate that E_2 modulates signaling in pre-osteoclasts and that, acting through reactive oxygen species, it increases the lifespan and/or function of mature osteoclasts.[36]

Both endogenous glucocorticoids and their synthetic analogs, which have been and continue to be a major mainstay of immunosuppressive therapy, are members of a third steroid hormone family having a major impact on bone biology.[37] One consequence of their chronic mode of administration is severe osteoporosis arising from decreased bone formation and resorption with the latter absolutely decreased (low turnover osteoporosis). The majority of the evidence focuses on the osteoblast as the prime target with the steroid increasing apoptosis of these bone-forming cells. However, numerous human studies document a rapid initial decrease in bone resorption, suggesting that the osteoclast and/or its precursors may also be targets. The molecular basis for this latter finding is unclear. However, because osteoblasts are a requisite part of the resorptive cycle, one consequence of their long-term diminution could be decreased osteoclast formation and/or function secondary to lower levels of RANKL and/or M-CSF production. Alternatively, glucocorticoids have been shown to decrease osteoclast apoptosis.[38]

A wide range of clinical information shows that excess prostaglandins stimulate bone loss, but once again, the cellular basis has not been established. Prostaglandins target stromal and osteoblastic cells, stimulating expression of RANKL and suppressing that of OPG.[39] This increase in the RANKL/OPG ratio, seen in a variety of human studies, is sufficient of itself to explain the clinical findings of increased osteoclastic activity. However, highlighting again the dilemma of interpreting in vitro studies there have been a number of studies in which prostaglandins regulate osteoclastogenesis per se in murine cell culture.

Phosphoinositides play distinct and important roles in organization of the osteoclast cytoskeleton.[40] Binding of M-CSF or RANKL to their cognate receptors, c-Fms and RANK, or activation of αvβ3, recruits phosphoinositol-3-kinase (PI3K) to the plasma membrane, where it converts membrane-bound

© 2008 American Society for Bone and Mineral Research

phosphatidylinositol 4,5-bisphosphate into phosphatidylinositol 3,4,5-trisphosphate (Fig. 4). The latter compound is recognized by specific motifs in a wide range of cytoskeletally active proteins,[41] and thus PI3K plays a central role in organizing the cytoskeleton of the osteoclast, including its ruffled membrane. Akt is a downstream target of PI3K and plays an important role in osteoclast function, particularly by mediating RANKL and/or M-CSF–stimulated proliferation and/or survival.[40]

Cell–Cell Interactions in Bone Marrow

Recent evidence has indicated that a number of additional cell types are important for osteoclast biology in a variety of situations (Fig. 5). First, as discussed previously, T cells play a key role in estrogen deficiency bone loss but also are important in a range of inflammatory diseases, most notably rheumatoid arthritis[42] and periodontal disease,[43] where the Th17 subset likely secretes TNF and IL-17, a newly described osteoclastogenic cytokine. Given that both osteoclast precursors and the various lymphocyte subsets, such as T, B, and NK cells, arise from the same stem cell, it is not surprising that some of the same receptors and ligands that mediate the immune process also govern the maturation of osteoclast precursors and the capacity of the mature cell to degrade bone. This interface has given rise to the new discipline of osteo-immunology, which promises to provide important and exciting findings in the future.

Second, whereas it is well established that mesenchymal cells are major mediators of cytokine and prostaglandin action on osteoclasts, it has become clear recently that cells of the same lineage, residing on cortical and trabecular bone, are the site of a hematopoietic stem cell (HSC) niche.[44] Specifically, HSCs reside close to osteoblasts as a result of multiple interactions involving receptors and ligands on both cells types.[45] Furthermore, the mesenchymally derived cells secrete both membrane-bound and soluble factors that contribute to survival and proliferation of multipotent osteoclast precursors, as well as molecules that influence osteoclast formation and function. Both committed osteoblasts and the numerous stromal cells in bone marrow produce a range of proteins both basally

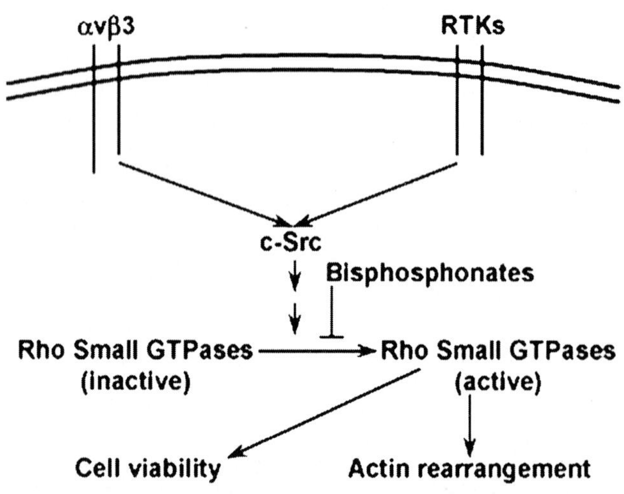

FIG. 4. Regulation and role of small GTPases in osteoclasts. Signals from αvβ3 and/or receptor tyrosine kinases (RTKs) activate small GTPases of the Rho family in a c-src–dependent manner. Bisphosphonates, the potent antiresorptive drugs, block addition of hydrophobic moieties onto the GTPases, preventing their membrane targeting and activation. The active GTPases also regulate cell viability and thus bisphosphonates induce osteoclast death.[51]

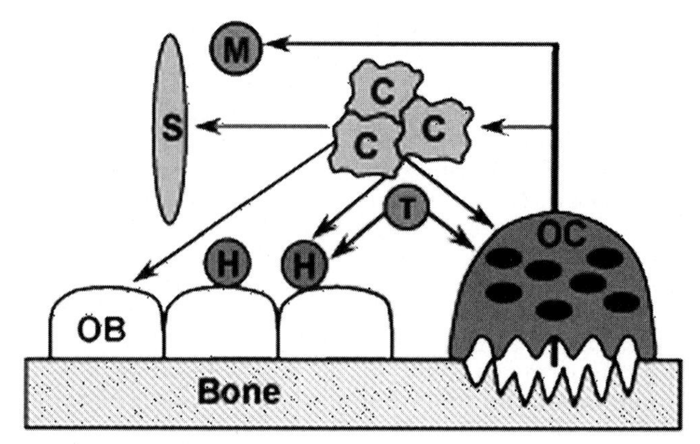

FIG. 5. Cell–cell interactions in bone marrow. Hematopoietic stem cells (H), the precursors of both T cells (T) and osteoclasts (OC), reside in a stem cell niche provided by osteoblasts (OB), which, together with stromal cells (S), derive from mesenchymal stem cells (M). Bone degradation results in release of matrix-associated growth factors (thick vertical line), which stimulate mesenchymal cells and thus bone formation. This "coupling" is an essential consequence of osteoclast activity.[28] After activation, T cells secrete molecules that stimulate osteoclastogenesis and function. Cancer cells (C) release cytokines that activate bone resorption; in turn, matrix-derived factors stimulate cancer cell proliferation, the so-called "vicious cycle."

and in response to hormones and growth factors, resulting in modulation of the capacity of HSCs to become functional osteoclasts.

Third, cancer cells facilitate their infiltration into the marrow cavity by stimulating osteoclast formation and function. An initial stimulus is PTHrP generation by lung and breast cancer cells,[24,25,46] thus enhancing mesenchymal production of RANKL and M-CSF, whereas decreasing that of OPG and possibly chemotactic factors. The resulting increase in matrix dissolution releases bone-residing cytokines and growth factors that, feeding back on the cancer cells, increase their growth and/or survival. This loop has been termed "the vicious cycle."[24] Multiple myeloma seems to use a different but related strategy, namely secretion of MIPα and MCP-1, both of which are chemotactic and proliferative for osteoclast precursors.[47,48] The latter compound has been reported to be secreted by osteoclasts in response to RANKL and enhances osteoclast formation.[2] It seems likely further future studies experiments will uncover additional molecules mediating bone loss in metastatic disease.

Intracellular Signaling Pathways

The discussions above have not described in detail the intracellular signals by which osteoclasts are formed or those by which they degrade bone. The final major section of this review lays out the important pathways involved. Briefly, three major protein classes are involved, adaptors, kinases, and transcription factors (Fig. 6), with one significant exception, RANKL-induced release of Ca^{2+}, a pathway that activates the calmodulin-dependent phosphatase calcineurin. NFAT1c is a major substrate for this enzyme, resulting in its nuclear translocation and subsequent activation of osteoclast-specific genes. Importantly, the potent immunosuppressive drugs FK506 and cyclosporine inhibit calcineurin activity and therefore may target the osteoclast.[49]

The multiplicity of adaptors that link the various receptors to downstream signals precludes providing a meaningful summary, and so we summarize only the modulatory effects of kinases and transcription factors, which together regulate receptor-driven proliferation and/or survival of precursors. Thus,

© 2008 American Society for Bone and Mineral Research

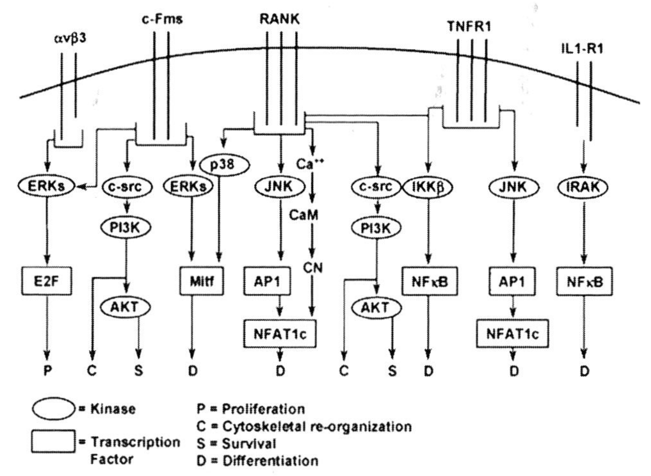

FIG. 6. Osteoclast signaling pathways. Summary of the major receptors, downstream kinases, and effector transcription factors that regulate osteoclast formation and function. Proliferation (P) of precursors is driven chiefly through ERKs and their downstream cyclin targets and E2F; maximal activation of this pathway requires combined signals from c-Fms and the integrin αvβ3. As expected, the cytoskeleton (C) is independent of nuclear control but depends on a series of kinases and their cytoskeletal-regulating targets, whereas differentiation (D) is regulated largely by controlling gene expression. The calcium/calmodulin (CaM)/calcineurin (CN) axis enhances nuclear translocation of NFAT1c, the most distal transcription factor characterized to date. See Refs. 2, 3, 10, 28, 40, and 53–56 for details.

proliferation is mediated by αvβ3 and c-Fms,[10,50] reorganization of the cytoskeleton by αvβ3, c-Fms, and RANK,[2,10] differentiation of mature osteoclasts from myeloid progenitors by c-Fms, RANK, TNFR1, and IL-1R1,[2,50,51] and their function by RANK, TNFR1, and IL-1R1.[52,53] Not shown is the fact that multiple other cytokines and growth factors, targeting the same or other less prominent pathways, or acting indirectly, probably contribute to overall control of bone resorption.[48]

Human Genetics

The text above might suggest that numerous mutations in many genes linked to the osteoclast are likely to have been discovered in humans. In fact, few such genetic changes have been defined, with >50% of those reported being in patients with osteopetrosis caused by defects in the chloride channel that modulates osteoclast acid secretion (Fig. 3). Rare reports link deficiencies in RANK, the proton pump, or carbonic anhydrase II to osteopetrosis, whereas decreased cathepsin K function leads to pyknodysostosis. In contrast, RANK activation manifests as osteolytic bone disease, whereas OPG deficiency leads to a severe form of high turnover osteoporosis.

REFERENCES

1. Suda T, Takahashi N, Udagawa N, Jimi E, Gillespie MT, Martin TJ 1999 Modulation of osteoclast differentiation and function by the new members of the tumor necrosis factor receptor and ligand families. Endocr Rev **20:**345–357.
2. Boyle WJ, Simonet WS, Lacey DL 2003 Osteoclast differentiation and activation. Nature **423:**337–342.
3. Pixley FJ, Stanley ER 2004 CSF-1 regulation of the wandering macrophage: Complexity in action. Trends Cell Biol **14:**628–638.
4. Weitzmann MN, Cenci S, Rifas L, Brown C, Pacifici R 2000 IL-7 stimulates osteoclast formation by upregulating the T-cell production of soluble osteoclastogenic cytokines. Blood **96:**1873–1878.
5. Kostenuik PJ, Shalhoub V 2001 Osteoprotegerin: A physiological and pharmacological inhibitor of bone resorption. Curr Pharm Des **7:**613–635.
6. Teitelbaum SL, Ross FP 2003 Genetic regulation of osteoclast development and function. Nat Rev Genet **4:**638–649.
7. Salo J, Lehenkari P, Mulari M, Metsikko K, Vaananen HK 1997 Removal of osteoclast bone resorption products by transcytosis. Science **276:**270–273.
8. Stenbeck G, Horton MA 2004 Endocytic trafficking in actively resorbing osteoclasts. J Cell Sci **117:**827–836.
9. Hynes RO 2002 Integrins: Bidirectional, allosteric signaling machines. Cell **110:**673–687.
10. Ross FP, Teitelbaum SL 2005 αvβ3 and macrophage colony-stimulating factor: Partners in osteoclast biology. Immunol Rev **208:**88–105.
11. Teitelbaum SL 2005 Osteoporosis and integrins. J Clin Endocrinol Metab **90:**2466–2468.
12. Teitelbaum SL, Abu-Amer Y, Ross FP 1995 Molecular mechanisms of bone resorption. J Cell Biochem **59:**1–10.
13. Van Wesenbeeck L, Odgren PR, Coxon FP, Frattini A, Moens P, Perdu B, MacKay CA, Van Hul E, Timmermans JP, Vanhoenacker F, Jacobs R, Peruzzi B, Teti A, Helfrich MH, Rogers MJ, Villa A, Van Hul W 2007 Involvement of PLEKHM1 in osteoclastic vesicular transport and osteopetrosis in incisors absent rats and humans. J Clin Invest **117:**919–930.
14. Vaananen HK, Zhao H, Mulari M, Halleen JM 2000 The cell biology of osteoclast function. J Cell Sci **113:**377–381.
15. Schwartz MA, Ginsberg MH 2002 Networks and crosstalk: Integrin signalling spreads. Nat Cell Biol **4:**E65–E68.
16. Horne WC, Sanjay A, Bruzzaniti A, Baron R 2005 The role(s) of Src kinase and Cbl proteins in the regulation of osteoclast differentiation and function. Immunol Rev **208:**106–125.
17. Zou W, Kitaura H, Reeve J, Long F, Tybulewicz VLJ, Shattil SJ, Ginsberg MH, Ross FP, Teitelbaum SL 2007 Syk, c-Src, the αvβ3 integrin, and ITAM immunoreceptors, in concert, regulate osteoclastic bone resorption. J Cell Biol **176:**877–888.
18. Jaffe AB, Hall A 2005 Rho GTPases: Biochemistry and biology. Annu Rev Cell Dev Biol **21:**247–269.
19. Chellaiah MA 2005 Regulation of actin ring formation by rho GTPases in osteoclasts. J Biol Chem **280:**32930–32943.
20. Fukuda A, Hikita A, Wakeyama H, Akiyama T, Oda H, Nakamura K, Tanaka S 2005 Regulation of osteoclast apoptosis and motility by small GTPase binding protein Rac1. J Bone Miner Res **20:**2245–2253.
21. Faccio R, Teitelbaum SL, Fujikawa K, Chappel JC, Zallone A, Tybulewicz VL, Ross FP, Swat W 2005 Vav3 regulates osteoclast function and bone mass. Nat Med **11:**284–290.
22. Whyte MP, Obrecht SE, Finnegan PM, Jones JL, Podgornik MN, McAlister WH, Mumm S 2002 Osteoprotegerin deficiency and juvenile Paget's disease. N Engl J Med **347:**175–184.
23. Simonet WS, Lacey DL, Dunstan CR, Kelley M, Chang MS, Luthy R, Nguyen HQ, Wooden S, Bennett L, Boone T, Shimamoto G, DeRose M, Elliott R, Colombero A, Tan HL, Trail G, Sullivan J, Davy E, Bucay N, Renshaw-Gegg L, Hughes TM, Hill D, Pattison W, Campbell P, Sander S, Van G, Tarpley J, Derby J, Lee R, Boyle WJ 1997 Osteoprotegerin: A novel secreted protein involved in the regulation of bone density. Cell **89:**309–319.
24. Clines GA, Guise TA 2005 Hypercalcaemia of malignancy and basic research on mechanisms responsible for osteolytic and osteoblastic metastasis to bone. Endocr Relat Cancer **12:**549–583.
25. Martin TJ 2002 Manipulating the environment of cancer cells in bone: A novel therapeutic approach. J Clin Invest **110:**1399–1401.
26. Zwerina J, Redlich K, Schett G, Smolen JS 2005 Pathogenesis of rheumatoid arthritis: Targeting cytokines. Ann NY Acad Sci **1051:**716–729.
27. Zwerina J, Hayer S, Tohidast-Akrad M, Bergmeister H, Redlich K, Feige U, Dunstan C, Kollias G, Steiner G, Smolen J, Schett G 2004 Single and combined inhibition of tumor necrosis factor, interleukin-1, and RANKL pathways in tumor necrosis factor-induced arthritis: Effects on synovial inflammation, bone erosion, and cartilage destruction. Arthritis Rheum **50:**277–290.
28. Takayanagi H 2005 Mechanistic insight into osteoclast differentiation in osteoimmunology. J Mol Med **83:**170–179.
29. Key LL, Rodriguiz RM, Willi SM, Wright NM, Hatcher HC, Eyre DR, Cure JK, Griffin PP, Ries WL 1995 Long-term treatment of osteopetrosis with recombinant human interferon gamma. N Engl J Med **332:**1594–1599.
30. Cenci S, Toraldo G, Weitzmann MN, Roggia C, Gao Y, Qian WP,

© 2008 American Society for Bone and Mineral Research

Sierra O, Pacifici R 2003 Estrogen deficiency induces bone loss by increasing T cell proliferation and lifespan through IFN-γ-induced class II transactivator. Proc Natl Acad Sci USA **100:** 10405–10410.
31. Kim MS, Day CJ, Selinger CI, Magno CL, Stephens SRJ, Morrison NA 2006 MCP-1-induced human osteoclast-like cells are tartrate-resistant acid phosphatase, NFATc1, and calcitonin receptor-positive but require receptor activator of NFκB ligand for bone resorption. J Biol Chem **281:**1274–1285.
32. Stockinger B, Veldhoen M 2007 Differentiation and function of Th17 T cells. Curr Opin Immunol **19:**281–286.
33. Udagawa N 2003 The mechanism of osteoclast differentiation from macrophages: Possible roles of T lymphocytes in osteoclastogenesis. J Bone Miner Metab **21:**337–343.
34. Goltzman D, Miao D, Panda DK, Hendy GN 2004 Effects of calcium and of the Vitamin D system on skeletal and calcium homeostasis: Lessons from genetic models. J Steroid Biochem Mol Biol **89–90:**485–489.
35. Syed F, Khosla S 2005 Mechanisms of sex steroid effects on bone. Biochem Biophys Res Commun **328:**688–696.
36. Eastell R 2005 Role of oestrogen in the regulation of bone turnover at the menarche. J Endocrinol **185:**223–234.
37. Canalis E, Bilezikian JP, Angeli A, Giustina A 2004 Perspectives on glucocorticoid-induced osteoporosis. Bone **34:**593–598.
38. Weinstein RS, Chen J-R, Powers CC, Stewart SA, Landes RD, Bellido T, Jilka RL, Parfitt AM, Manolagas SC 2002 Promotion of osteoclast survival and antagonism of bisphosphonate-induced osteoclast apoptosis by glucocorticoids. J Clin Invest **109:**1041–1048.
39. Kobayashi T, Narumiya S 2002 Function of prostanoid receptors: Studies on knockout mice. Prostaglandins Other Lipids Mediat **68–69:**557–573.
40. Golden LH, Insogna KL 2004 The expanding role of PI3-kinase in bone. Bone **34:**3–12.
41. DiNitto JP, Cronin TC, Lambright DG 2003 Membrane recognition and targeting by lipid-binding domains. Sci STKE **2003:**re16.
42. Nakashima T, Wada T, Penninger JM 2003 RANKL and RANK as novel therapeutic targets for arthritis. Curr Opin Rheumatol **15:**280–287.
43. Taubman MA, Valverde P, Han X, Kawai T 2005 Immune response: The key to bone resorption in periodontal disease. J Periodontol **76:**2033–2041.
44. Suda T, Arai F, Hirao A 2005 Hematopoietic stem cells and their niche. Trends Immunol **26:**426–433.
45. Taichman RS 2005 Blood and bone: Two tissues whose fates are intertwined to create the hematopoietic stem-cell niche. Blood **105:**2631–2639.
46. Bendre M, Gaddy D, Nicholas RW, Suva LJ 2003 Breast cancer metastasis to bone: It is not all about PTHrP. Clin Orthop **415**(Suppl)**:**S39–S45.
47. Hata H 2005 Bone lesions and macrophage inflammatory protein-1 alpha (MIP-1a) in human multiple myeloma. Leuk Lymphoma **46:**967–972.
48. Kim MS, Day CJ, Morrison NA 2005 MCP-1 is induced by receptor activator of nuclear factor κB ligand, promotes human osteoclast fusion, and rescues granulocyte macrophage colony-stimulating factor suppression of osteoclast formation. J Biol Chem **280:**16163–16169.
49. Seales EC, Micoli KJ, McDonald JM 2006 Calmodulin is a critical regulator of osteoclastic differentiation, function, and survival. J Cell Biochem **97:**45–55.
50. Ross FP 2006 M-CSF, c-Fms and signaling in osteoclasts and their precursors. Ann NY Acad Sci **1068:**110–116.
51. Rogers MJ 2004 From molds and macrophages to mevalonate: A decade of progress in understanding the molecular mode of action of bisphosphonates. Calcif Tissue Int **75:**451–461.
52. Blair HC, Robinson LJ, Zaidi M 2005 Osteoclast signalling pathways. Biochem Biophys Res Commun **328:**728–738.
53. Feng X 2005 Regulatory roles and molecular signaling of TNF family members in osteoclasts. Gene **350:**1–13.
54. Hershey CL, Fisher DE 2004 Mitf and Tfe3: Members of a b-HLH-ZIP transcription factor family essential for osteoclast development and function. Bone **34:**689–696.
55. Lee ZH, Kim H-H 2003 Signal transduction by receptor activator of nuclear factor kappa B in osteoclasts. Biochem Biophys Res Commun **305:**211–214.
56. Wagner EF, Eferl R 2005 Fos/AP-1 proteins in bone and the immune system. Immunol Rev **208:**126–140.

Chapter 4. Osteocytes

Lynda F. Bonewald

Department of Oral Biology, University of Missouri at Kansas City School of Dentistry, Kansas City, Missouri

INTRODUCTION

In the adult skeleton, osteocytes make up >90–95% of all bone cells compared with 4–6% osteoblasts and ~1–2% osteoclasts. These cells are regularly dispersed throughout the mineralized matrix, connected to each other and cells on the bone surface through dendritic processes generally radiating toward the bone surface and the blood supply. The dendritic processes travel through the bone in tiny canals called canaliculi (250–300 nm), whereas the cell body is encased in a lacuna (15–20 μm; Figs. 1 and 2). Osteocytes are thought to function as a network of sensory cells mediating the effects of mechanical loading through this extensive lacuno-canalicular network. Not only do these cells communicate with each other and with cells on the bone surface, but their dendritic processes extend past the bone surface into the bone marrow. Osteocytes have long been thought to respond to mechanical strain to send signals of resorption or formation, and evidence is accumulating to show that this is a major function of these cells. Recently, it has been shown that osteocytes have another important function, to regulate phosphate homeostasis; therefore, the osteocyte network may also function as an endocrine gland. Defective osteocyte function may play a role in a number of bone diseases, especially glucocorticoid-induced bone fragility and osteoporosis in the adult, aging skeleton.

OSTEOCYTE ONTOGENY

Osteoprogenitor cells reside in the bone marrow before differentiating into plump, polygonal osteoblasts on the bone surface.[1,2] By an unknown mechanism, some of these cells are destined to become osteocytes, whereas some become lining cells and some undergo programmed cell death known as apoptosis.[3] Osteoblasts, osteoid–osteocytes, and osteocytes may play distinct roles in the initiation and regulation of mineralization of bone matrix, but Bordier et al.[4] first proposed that osteoid-osteocytes are major regulators of this process. Oste-

Dr. Bonewald has received graduate student support from and has consulted for Procter & Gamble. She also holds a patent on MLO cell lines.

Key words: osteocytes, mechanical load, phosphate metabolism, apoptosis, bone disease

© 2008 American Society for Bone and Mineral Research

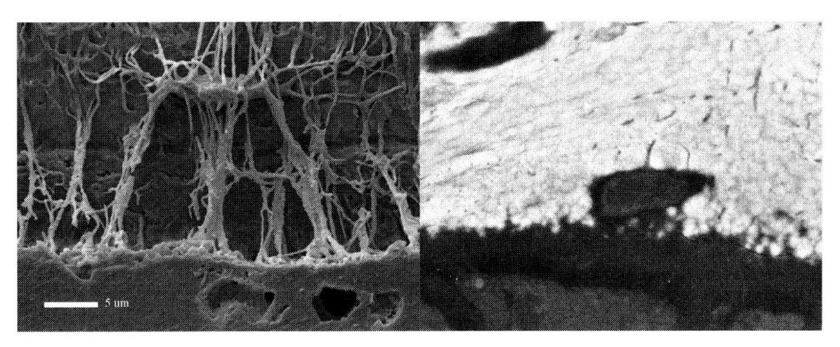

FIG. 1. The embedding osteocyte retains its connectivity with cells on the bone surface. The image on the right is of acid-etched plastic embedded murine cortical bone. With this technique, resin fills the lacuno-canalicular system, osteoid, and marrow, but cannot penetrate mineral. Mild acid is used to remove the mineral leaving behind a resin cast relief. Note the canaliculi connecting the lacunae with the bone surface at the bottom of the image. The image on the right is a from transmission electron microscopy showing a fully embedded osteocyte and an osteoid-osteocyte becoming surrounded by mineral (white). The osteoid is black and the osteoblasts are at the bottom of the image.

oid–osteocytes actively make matrix while simultaneously calcifying this matrix. The osteoblast cell body reduces in size ~30% at the osteoid–osteocyte stage, whereas cytoplasmic processes are forming and are ~70% with complete maturation of the osteocyte (Fig. 1).

Whereas numerous markers for osteoblasts have been identified such as cbfa1, osterix, alkaline phosphatase, and collagen type 1, few markers have been available for osteocytes until recently.[2] In 1996, the markers described for osteocytes were limited to low or no alkaline phosphatase, high casein kinase II, high osteocalcin protein expression, and high CD44. Osteocyte markers such as E11/gp38, phosphate-regulating neutral endopeptidase on chromosome X (Phex), dentin matrix protein 1 (DMP1), and sclerostin have recently been identified (Table 1). Some of these markers are overlapping in expression with osteoblasts, but some have been identified for specific stages of differentiation. Promoters for specific markers have been used to drive green fluorescent protein (GFP) to follow osteoblast to osteocyte differentiation in vivo. Collagen type 1–GFP is strongly expressed in both osteoblasts and osteocytes, osteocalcin–GFP is expressed in a few osteoblastic cells lining the endosteal bone surface and in scattered osteocytes, and the osteocyte-selective promoter, the 8-kb DMP1 driving GFP showed exclusive expression in osteocytes.[5] The actin-bundling proteins, villin, α-actinin, and fimbrin, were shown to be markers for osteocytes with strong signals of fimbrin at branching points in dendrites.[6] It is likely that these actin reorganizing proteins play a role in osteocyte cell body movement within its lacuna and the retraction and extension of dendritic processes.[7]

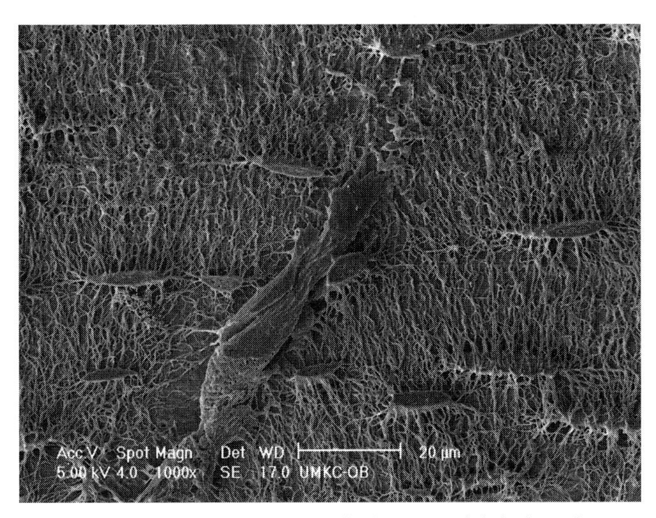

FIG. 2. The osteocyte lacuno-canalicular network is intimately associated with the blood vessel network in the bone matrix. The white marker points to an osteocyte lacunae intimately associated with the blood vessel.

OSTEOCYTES AS ORCHESTRATORS OF BONE (RE)MODELING

Considerable evidence is mounting that osteocytes can conduct and control both bone resorption and bone formation. Some of the earliest supporting data for the theory that osteocytes can send signals of bone resorption were observations that isolated avian osteocytes can support osteoclast formation and activation in the absence of any osteotropic factors[8] as can the osteocyte-like cell line, MLO-Y4.[9] It was suggested that expression of RANKL along exposed osteocyte dendritic processes provides a potential means for osteocytes within bone to stimulate osteoclast precursors at the bone surface.

One of the major means by which osteocytes may support osteoclast activation and formation is through their death. Osteocyte apoptosis can occur at sites of microdamage, and it is proposed that dying osteocytes are targeted for removal by osteoclasts. The expression of anti-apoptotic and pro-apoptotic molecules in osteocytes surrounding microcracks was mapped, and it was found that pro-apoptotic molecules are elevated in osteocytes immediately at the microcrack locus, whereas anti-apoptotic molecules are expressed 1–2 mm from the microcrack.[10] Therefore, those osteocytes that do not undergo apoptosis are prevented from doing so by protective mechanisms, whereas those destined for removal by osteoclasts undergo apoptosis. Targeted ablation of osteocytes was performed using the 10-kb Dmp1 promoter to drive the diphtheria toxin receptor in mice.[11] Injection of a single dose of diphtheria toxin eliminated ~70% of osteocytes in cortical bone in these mice, leading to dramatic osteoclast activation. Therefore, viable osteocytes are necessary to prevent osteoclast activation and maintain bone mass (Fig. 3).

The osteocyte-like cell line MLO-Y4 not only supports osteoclast formation but also osteoblast differentiation,[12] and surprisingly, mesenchymal stem cell differentiation.[13] It is most likely (but remains to be proven) that primary osteocytes can perform all three functions, therefore possessing the unique capacity to regulate all phases of bone remodeling.

OSTEOCYTE CELL DEATH AND APOPTOSIS

It has been proposed that the purpose and function of osteocytes is to die, thereby releasing signals of resorption. Osteocyte cell death can occur in association with pathological conditions, such as osteoporosis and osteoarthritis, leading to increased skeletal fragility.[14] Such fragility is considered to be caused by loss of the ability to sense microdamage and/or signal repair. Several conditions have been shown to result in osteocyte cell death such as oxygen deprivation as occurs during immobilization, withdrawal of estrogen, and glucocorticoid treatment.[14] TNFα and interleukin-1 (IL-1) have been re-

© 2008 American Society for Bone and Mineral Research

TABLE 1. OSTEOCYTE MARKERS

Marker	Expression	Function
E11/gp38	Early, embedding cell[37]	Dendrite formation?[37]
CD44	More highly expressed in osteocytes compared with osteoblasts[38]	Hyaluronic acid receptor associated with E11 and linked to cytoskeleton[39]
Fimbrin	All osteocytes[5]	Dendrite branching?
Phex	Early and late osteocytes[40,41]	Phosphate metabolism[42]
OF45/MEPE	Late osteoblast through osteocytes[43]	Inhibitor of bone formation[43]/regulator of phosphate metabolism[44]
DMP1	Early and mature osteocytes[45]	Phosphate metabolism and mineralization[34]
Sclerostin	Late embedded osteocyte[46]	Inhibitor of bone formation[47]
FGF23	Early and mature osteocytes[35]	Induces hypophosphatemia[42]

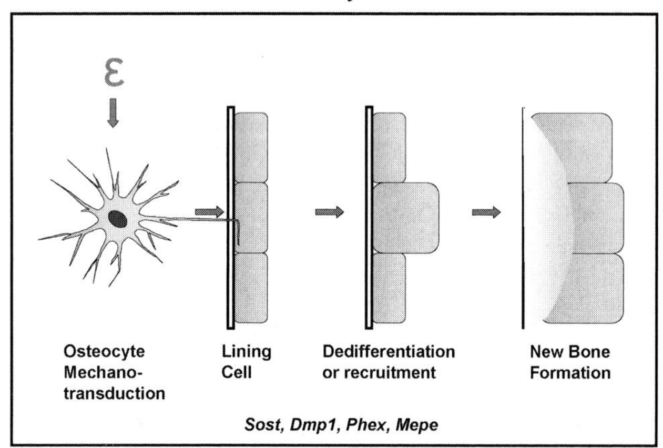

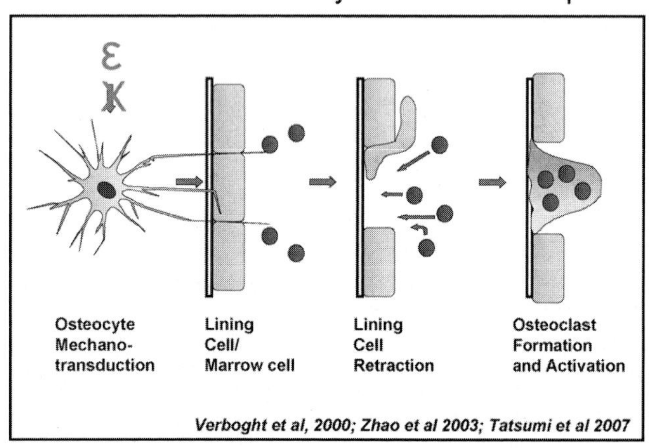

FIG. 3. Osteocytes as orchestrators of bone (re)modeling. Osteocytes play a role in bone formation and mineralization as promoters of mineralization (such as Dmp1 and Phex) and inhibitors of mineralization and bone formation (such as Sost/sclerostin and MEPE/OF45) are highly expressed in osteocytes (top). These supporters and inhibitors of bone formation and mineralization are most likely exquisitely balanced to maintain equilibrium to maintain bone mass. Osteocytes also appear to play a major role in the regulation of osteoclasts, by both inhibiting and activating osteoclastic resorption. It has recently been shown that with loading, the osteocytes send signals inhibiting osteoclast activation (bottom).[11] In contrast, compromised, hypoxic, apoptotic or dying osteocytes especially with unloading appear to send unknown signals to osteoclasts/preosteoclasts on the bone surface to initiate resorption. Therefore, osteocytes within the bone regulate bone formation and mineralization and inhibit osteoclastic resorption, whereas having the capacity to also send signals of osteoclast activation under specific conditions.

ported to increase with estrogen deficiency and also induce osteocyte apoptosis.[2]

Several agents have been found to reduce or inhibit osteoblast and osteocyte apoptosis. These include estrogen, selective estrogen receptor modulators, bisphosphonates, calcitonin, CD40 ligand, calbindin-D28k, monocyte chemotactic proteins MCP1 and 3, and recently mechanical loading through the release of prostaglandin.[2] Osteocyte viability clearly plays a significant role in the maintenance of bone homeostasis and integrity. However, whereas blocking osteocyte apoptosis may improve diseases such as bone loss caused by aging or glucocorticoid therapy, osteocyte apoptosis may be essential for normal damage repair and skeletal replacement. Any agents that block this process may exacerbate conditions in which repair is required. The death processes and consequently the resorption signals sent by dying osteocytes in an aging or glucocorticoid-treated skeleton may be distinct from those in a normal, healthy skeleton in response to microdamage. It will be important to identify and characterize these differences.

OSTEOCYTE MODIFICATION OF THEIR MICROENVIRONMENT

Almost 100 yr ago, it was proposed that osteocytes may resorb their lacunar wall under particular conditions.[15] The term "osteolytic osteolysis" was initially used to describe the enlarged lacunae in patients with hyperparathyroidism[16] and later in immobilized rats[17]. Osteolytic osteolysis has a negative connotation because it was confused with osteoclastic bone resorption. When resorption pits similar to those observed with osteoclasts were not observed with primary avian osteocytes seeded onto dentin slices, it was concluded that osteocytes cannot remove mineralized matrix. Removal of mineral by osteocytes would not be detectable using this approach because these cells are within a lacuna and do not form the characteristic sealed osteoclast resorption lacuna that rapidly decalcifies bone. In contrast to lacunar enlargement by the embedded, mature osteocyte, enlarged lacunae with renal osteodystrophy may be caused by defective mineralization during embedding of the osteoid–osteocyte during bone formation.[2]

In addition to enlargement of the lacunae, changes can take place in the perilacunar matrix. The term "osteocyte halos" was used to describe perilacunar demineralization in rickets[18] and later to describe periosteocytic lesions in X-linked hypophosphatemic rickets.[19] Glucocorticoids in addition to having effects on osteocyte apoptosis seem to also cause osteocytes to not only enlarge their lacunae but to also remove mineral from the perilacunar space thereby generating "halos" of hypomineralized bone.[20] Glucocorticoids may therefore alter or com-

© 2008 American Society for Bone and Mineral Research

promise the metabolism and function of the osteocyte and not just induce cell death.

The capacity to deposit or remove mineral from lacunae and canaliculi has important implications with regard to (1) mineral homeostasis, (2) magnitude of fluid shear stress applied to the cell, and (3) mechanical properties of bone. The surface area of the osteocyte lacuno-canalicular system is several order of magnitudes greater than the bone surface area; therefore, removal of only a few angstroms of mineral would have significant effects on circulating, systemic ion levels. Enlargement of the lacunae and canaliculi would reduce bone fluid flow shear stress, thereby reducing mechanical loading on the osteocyte. As holes in a material act as stress concentrators, enlargement of lacunae would enhance this effect in bone. Therefore, changes in lacunar size and matrix properties could have dramatic effects on bone properties and quality in addition to osteocyte function.

Over three decades ago, it was suggested that the osteocyte not only has matrix destroying capability but also can form new matrix.[21] Osteocyte lacunae were shown to uptake tetracycline, called "periosteocytic perilacunar tetracycline labeling," indicating the ability to calcify or form bone. Therefore, the osteocyte may be capable of both adding and removing mineral from its surroundings.

MECHANOSENSATION AND TRANSDUCTION

Mechanical strain is required for postnatal but not for prenatal skeletal development and maintenance. The postnatal and adult skeleton is able to continually adapt to mechanical loading by the process of adaptive remodeling where new bone is added to withstand increased amounts of loading and bone is removed in response to unloading or disuse. The parameters for inducing bone formation or bone resorption in vivo are fairly well known and well characterized. Frequency, intensity, and timing of loading are all important parameters. Bone mass is influenced by peak applied strain[22] and bone formation rate is related to loading rate.[23] When rest periods are inserted, the loaded bone shows increased bone formation rates compared with bone subjected to a single bout of mechanical loading and improved bone structure and strength is greatest if loading is applied in shorter versus longer increments.[24]

The major challenge in the field of mechanotransduction has been to translate these well-characterized in vivo parameters of mechanical loading to in vitro cell culture models. Theoretical models and experimental studies suggest that flow of bone fluid is driven by extravascular pressure as well as applied cyclic mechanical loading of osteocytes.[25] Mechanical forces applied to bone cause fluid flow through the canaliculi surrounding the osteocyte inducing shear stress and deformation of the cell membrane. It has also been proposed that mechanical information is relayed by cilia, a flagellar-like structure found on every cell.[26,27] Osteocytes may use a combination of means to sense mechanical strain.[28] Theoretical modeling predicts osteocyte wall shear stresses resulting from peak physiologic loads in vivo to be in the range of 8–30 dynes/cm^2,[25] but this has not been confirmed in vivo. It will be a significant advance in the field to be able to actually measure bone fluid flow within the lacuno-canalicular system.

ROLE OF GAP JUNCTIONS AND HEMICHANNELS IN OSTEOCYTE COMMUNICATION

A means by which osteocytes communicate intracellularly is through gap junctions, transmembrane channels that connect the cytoplasm of two adjacent cells, through which molecules with molecular weights <1 kDa can pass. Gap junction channels are formed by members of a family of proteins known as connexins, and Cx43 is the major connexin in bone cells. Much of mechanotransduction in bone is thought to be mediated through gap junctions.

Primary osteocytes and MLO-Y4 osteocyte-like cells[29,30] express large amounts of Cx43, suggesting that Cx43 has another function in addition to being a component of gap junctions. Recently, it has been shown that connexins can form and function as unapposed halves of gap junction channels called hemichannels. Hemichannels directly serve as the pathway for the exit of intracellular PGE_2 in osteocytes induced by fluid flow shear stress[31] and function as essential transducers of the anti-apoptotic effects of bisphosphonates.[32] Hemichannels are now one of several types of openings or channels to the extracellular bone fluid that also includes channels such as calcium, ion, voltage, stretch-activated channels, and others.[33] Therefore, gap junctions at the tip of dendrites seem to mediate a form of intracellular communication, and hemichannels along the dendrite (and perhaps the cell body) seem to mediate a form of extracellular communication between osteocytes.

POTENTIAL ROLE OF OSTEOCYTES IN BONE DISEASE

Osteoid osteocytes play a role in phosphate homeostasis. Once the osteoblast begins to embed in osteoid, molecules such as Dmp1, Phex, and Mepe are elevated (Table 1). Autosomal recessive hypophosphatemic rickets in patients is caused by mutations in Dmp1.[34] Dmp1-null mice have a similar phenotype to hyp mice carrying a Pex mutation, that of osteomalacia and rickets caused by elevated fibroblast growth factor 23 (FGF23) levels in osteocytes.[34,35] The osteocyte lacuno-canalicular system should be viewed as an endocrine organ regulating phosphate metabolism. The unraveling of the interactions of these molecules should lead to insight into diseases of hyper- and hypophosphatemia (see chapter on Phex/FGF23).

The connectivity and structure of the osteocyte lacuno-canalicular system may play a role in bone disease. Osteocyte dendricity may change depending on orientation and with static and dynamic bone formation and has been shown to be disrupted in bone disease.[36] In osteoporotic bone, there is disorientation of the canaliculi and a marked decrease in connectivity, which increases in severity. In contrast, in osteoarthritic bone, a decrease in connectivity is observed, but orientation is intact. In osteomalacic bone, the osteocytes seem viable with high connectivity, but the processes are distorted and the network is chaotic.[36] Variability in complexity and number of dendrites and canaliculi could have a dramatic effect on osteocyte function and viability and on the mechanical properties of bone.

Osteocyte cell death may be responsible for some forms of osteonecrosis. Osteonecrosis is "dead" bone containing empty osteocyte lacunae that does not remodel but can remain in the bone for years. As reviewed above, viable osteocytes are necessary to send signals of (re)modeling. Early proposed mechanisms responsible for osteonecrosis include the mechanical theory, where osteoporosis and the accumulation of unhealed trabecular microcracks result in fatigue fractures; the vascular theory, where ischemia is caused by microscopic fat emboli; and a newer theory of osteocyte apoptosis, where agents induce osteocyte cell death resulting in dead bone that does not remodel.[14] Osteocyte health, compromised status, viability,

© 2008 American Society for Bone and Mineral Research

and capacity to regulate its own death most likely play a highly significant role in the maintenance and integrity of bone. Bone loss in osteoporosis may be caused in part by pathological and not physiological osteocyte cell death.[3] It will be important to develop therapeutics that maintain both osteocyte viability and physiological osteocyte cell death that leads to normal bone repair.

In conclusion, it is most likely that osteocytes use undiscovered specific molecules to regulate bone (re)modeling. With the dramatic increases or maintenance of bone mass being observed with neutralizing antibody to sclerostin, an osteocyte-specific marker, greater effort is being made to identify additional markers and to unravel the mysteries surrounding osteocyte function. It is also likely that new functions will be discovered for these cells, making them a target of investigation, not only to understand basic bone physiology, but also to understand and treat bone disease.

ACKNOWLEDGMENTS

The author's work in osteocyte biology is supported by the National Institutes of Health AR-46798.

REFERENCES

1. Franz-Odendaal TA, Hall BK, Witten PE 2006 Buried alive: How osteoblasts become osteocytes. Dev Dyn **235:**176–190.
2. Bonewald L 2007 Osteocytes. In: Marcus DF, Nelson D, Rosen C (eds.) Osteoporosis, 3rd ed., vol. 1. Elsevier, New York, NY, USA, pp. 169–190.
3. Manolagas SC 2000 Birth and death of bone cells: Basic regulatory mechanisms and implications for the pathogenesis and treatment of osteoporosis. Endocr Rev **21:**115–137.
4. Bordier PJ, Miravet L, Ryckerwaert A, Rasmussen H 1977 Morphological and morphometrical characteristics of the mineralization front. A vitamin D regulated sequence of bone remodeling. In: Meunie PJB (ed.) Bone Histomorphometry. Armour Montagu, Paris, France, pp. 335–354.
5. Kalajzic I, Braut A, Guo D, Jiang X, Kronenberg MS, Mina M, Harris MA, Harris SE, Rowe DW 2004 Dentin matrix protein 1 expression during osteoblastic differentiation, generation of an osteocyte GFP-transgene. Bone **35:**74–82.
6. Tanaka-Kamioka K, Kamioka H, Ris H, Lim SS 1998 Osteocyte shape is dependent on actin filaments and osteocyte processes are unique actin-rich projections. J Bone Miner Res **13:**1555–1568.
7. Veno P, Nicolella DP, Sivakumar P, Kalajzic I, Rowe D, Harris SE, Bonewald L, Dallas SL 2006 Live imaging of osteocytes within their lacunae reveals cell body and dendrite motions. J Bone Miner Res **21:**S1;538.
8. Tanaka K, Yamaguchi Y, Hakeda Y 1995 Isolated chick osteocytes stimulate formation and bone-resorbing activity of osteoclast-like cells. J Bone Miner Metab **13:**61–70.
9. Zhao S, Zhang YK, Harris S, Ahuja SS, Bonewald LF 2002 MLO-Y4 osteocyte-like cells support osteoclast formation and activation. J Bone Miner Res **17:**2068–2079.
10. Verborgt O, Tatton NA, Majeska RJ, Schaffler MB 2002 Spatial distribution of Bax and Bcl-2 in osteocytes after bone fatigue: Complementary roles in bone remodeling regulation? J Bone Miner Res **17:**907–914.
11. Tatsumi S, Ishii K, Amizuka N, Li M, Kobayashi T, Kohno K, Ito M, Takeshita S, Ikeda K 2007 Targeted ablation of osteocytes induces osteoporosis with defective mechanotransduction. Cell Metab **5:**464–475.
12. Heino TJ, Hentunen TA, Vaananen HK 2002 Osteocytes inhibit osteoclastic bone resorption through transforming growth factor-beta: Enhancement by estrogen. J Cell Biochem **85:**185–197.
13. Heino TJ, Hentunen TA, Vaananen HK 2004 Conditioned medium from osteocytes stimulates the proliferation of bone marrow mesenchymal stem cells and their differentiation into osteoblasts. Exp Cell Res **294:**458–468.
14. Weinstein RS, Nicholas RW, Manolagas SC 2000 Apoptosis of osteocytes in glucocorticoid-induced osteonecrosis of the hip. J Clin Endocrinol Metab **85:**2907–2912.
15. Recklinghausen FV 1910 Untersuchungen uber rachitis and osteomalacia. Fischer, Jena, Germany.
16. Belanger LF 1969 Osteocytic osteolysis. Calcif Tissue Res **4:**1–12.
17. Kremlien B, Manegold C, Ritz E, Bommer J 1976 The influence of immobilization on osteocyte morphology:osteocyte differential count and electron microscopic studies. Virchows Arch A Pathol Anat Histol **370:**55–68.
18. Heuck F 1970 Comparative investigations of function of osteocytes in bone resorption. Calcif Tissue Res (Suppl):148–149.
19. Marie PJ, Glorieux FH 1983 Relation between hypomineralized periosteocytic lesions and bone mineralization in vitamin D-resistant rickets. Calcif Tissue Int **35:**443–448.
20. Lane NE, Yao W, Balooch M, Nalla RK, Balooch G, Habelitz S, Kinney JH, Bonewald LF 2006 Glucocorticoid-treated mice have localized changes in trabecular bone material properties and osteocyte lacunar size that are not observed in placebo-treated or estrogen-deficient mice. J Bone Miner Res **21:**466–476.
21. Baud CA, Dupont DH 1962 The fine structure of the osteocyte in the adult compact bone. In: Breese SSJ (ed.) Electron Microscopy, vol. 2. Academic Press, New York, NY, USA, pp. QQ–10.
22. Rubin C 1984 Skeletal strain and the functional significance of bone architecture. Calcif Tissue Int **36:**S11–S18.
23. Turner CH, Forwood MR, Otter MW 1994 Mechanotransduction in bone: Do bone cells act as sensors of fluid flow? FASEB J **8:**875–878.
24. Robling AG, Hinant FM, Burr DB, Turner CH 2002 Shorter, more frequent mechanical loading sessions enhance bone mass. Med Sci Sports Exerc **34:**196–202.
25. Weinbaum S, Cowin SC, Zeng Y 1994 A model for the excitation of osteocytes by mechanical loading-induced bone fluid shear stresses. J Biomech **27:**339–360.
26. Xiao Z, Zhang S, Mahlios J, Zhou G, Magenheimer BS, Guo D, Dallas SL, Maser R, Calvet JP, Bonewald L, Quarles LD 2006 Cilia-like structures and polycystin-1 in osteoblasts/osteocytes and associated abnormalities in skeletogenesis and Runx2 expression. J Biol Chem **281:**30884–30895.
27. Malone AM, Anderson CT, Tummala P, Kwon RY, Johnston TR, Stearns T, Jacobs CR 2007 Primary cilia mediate mechanosensing in bone cells by a calcium-independent mechanism. Proc Natl Acad Sci USA **104:**13325–13330.
28. Bonewald LF 2006 Mechanosensation and transduction in osteocytes. Bonekey Osteovision **3:**7–15.
29. Doty SB 1981 Morphological evidence of gap junctions between bone cells. Calcif Tissue Int **33:**509–512.
30. Kato Y, Windle JJ, Koop BA, Mundy GR, Bonewald LF 1997 Establishment of an osteocyte-like cell line, MLO-Y4. J Bone Miner Res **12:**2014–2023.
31. Cherian PP, Siller-Jackson AJ, Gu S, Wang X, Bonewald LF, Sprague E, Jiang JX 2005 Mechanical strain opens connexin 43 hemichannels in osteocytes: A novel mechanism for the release of prostaglandin. Mol Biol Cell **16:**3100–3106.
32. Plotkin LI, Manolagas SC, Bellido T 2002 Transduction of cell survival signals by connexin-43 hemichannels. J Biol Chem **277:**8648–8657.
33. Klein-Nulend J, Bonewald LF 2008 The osteocyte. In: Bilezikian JP, Raisz LG (eds.) Principles of Bone Biology, vol. 1. Academic Press, San Diego, CA, USA, pp. 151–172.
34. Feng JQ, Ward LM, Liu S, Lu Y, Xie Y, Yuan B, Yu X, Rauch F, Davis SI, Zhang S, Rios H, Drezner MK, Quarles LD, Bonewald LF, White KE 2006 Loss of DMP1 causes rickets and osteomalacia and identifies a role for osteocytes in mineral metabolism. Nat Genet **38:**1310–1315.
35. Liu S, Lu Y, Xie Y, Zhou J, Quarles LD, Bonewald L, Feng JQ 2006 Elevated levels of FGF23 in dentin matrix protein 1 (DMP1) null mice potentially explain phenotypic similarities to Hyp mice. J Bone Miner Res **21:**S1;551.
36. Knothe Tate ML, Adamson JR, Tami AE, Bauer TW 2004 The osteocyte. Int J Biochem Cell Biol **36:**1–8.
37. Zhang K, Barragan-Adjemian C, Ye L, Kotha S, Dallas M, Lu Y, Zhao S, Harris M, Harris SE, Feng JQ, Bonewald LF 2006 E11/gp38 selective expression in osteocytes: Regulation by mechanical strain and role in dendrite elongation. Mol Cell Biol **26:**4539–4552.
38. Hughes DE, Salter DM, Simpson R 1994 CD44 expression in human bone: A novel marker of osteocytic differentiation. J Bone Miner Res **9:**39–44.
39. Ohizumi I, Harada N, Taniguchi K, Tsutsumi Y, Nakagawa S, Kaiho S, Mayumi T 2000 Association of CD44 with OTS-8 in tumor vascular endothelial cells. Biochim Biophys Acta **1497:**197–203.

© 2008 American Society for Bone and Mineral Research

40. Westbroek I, De Rooij KE, Nijweide PJ 2002 Osteocyte-specific monoclonal antibody MAb OB7.3 is directed against Phex protein. J Bone Miner Res **17**:845–853.
41. Ruchon AF, Tenenhouse HS, Marcinkiewicz M, Siegfried G, Aubin JE, DesGroseillers L, Crine P, Boileau G 2000 Developmental expression and tissue distribution of Phex protein: Effect of the Hyp mutation and relationship to bone markers. J Bone Miner Res **15**:1440–1450.
42. The HYP Consortium 1995 A gene (PEX) with homologies to endopeptidases is mutated in patients with X-linked hypophosphatemic rickets. The HYP Consortium. Nat Genet **11**:130–136.
43. Gowen LC, Petersen DN, Mansolf AL, Qi H, Stock JL, Tkalcevic GT, Simmons HA, Crawford DT, Chidsey-Frink KL, Ke HZ, McNeish JD, Brown TA 2003 Targeted disruption of the osteoblast/osteocyte factor 45 gene (OF45) results in increased bone formation and bone mass. J Biol Chem **278**:1998–2007.
44. Rowe PS, Kumagai Y, Gutierrez G, Garrett IR, Blacher R, Rosen D, Cundy J, Navvab S, Chen D, Drezner MK, Quarles LD, Mundy GR 2004 MEPE has the properties of an osteoblastic phosphatonin and minhibin. Bone **34**:303–319.
45. Toyosawa S, Shintani S, Fujiwara T, Ooshima T, Sato A, Ijuhin N, Komori T 2001 Dentin matrix protein 1 is predominantly expressed in chicken and rat osteocytes but not in osteoblasts. J Bone Miner Res **16**:2017–2026.
46. Poole KE, van Bezooijen RL, Loveridge N, Hamersma H, Papapoulos SE, Lowik CW, Reeve J 2005 Sclerostin is a delayed secreted product of osteocytes that inhibits bone formation. FASEB J **19**:1842–1844.
47. Balemans W, Ebeling M, Patel N, Van Hul E, Olson P, Dioszegi M, Lacza C, Wuyts W, Van Den Ende J, Willems P, Paes-Alves AF, Hill S, Bueno M, Ramos FJ, Tacconi P, Dikkers FG, Stratakis C, Lindpaintner K, Vickery B, Foernzler D, Van Hul W 2001 Increased bone density in sclerosteosis is due to the deficiency of a novel secreted protein (SOST). Hum Mol Genet **10**:537–543.

Chapter 5. Connective Tissue Pathways That Regulate Growth Factors

Gerhard Sengle, Noe L. Charbonneau, Robert N. Ono, and Lynn Y. Sakai

Department of Biochemistry and Molecular Biology, Oregon Health and Science University and Shriners Hospital for Children, Portland, Oregon

INTRODUCTION

Different connective tissues perform different physiological functions. To perform these functions, connective tissue cells secrete distinct sets of extracellular matrix (ECM) proteins that are arranged within the individual connective tissue in discrete patterns. The relative abundance of ECM proteins within a tissue and the histological patterns in which these proteins are organized endow connective tissues with their specific developmental, physiological, and homeostatic properties. For example, in bone, type I collagen is the most abundant ECM protein constituent, and type I collagen fibers are organized in long, thick bundles that are consistent with the mechanical properties required of bone.

The contributions of ECM proteins such as collagens and proteoglycans to the mechanical properties of cartilage and bone are better understood today than the contributions of many of the minor constituents of the ECM. However, recent knowledge of the function of a fibril-forming ECM protein named fibrillin[1] indicates important roles for fibrillin microfibrils in the development, growth, and maintenance of skeletal elements. Recent studies of fibrillin microfibrils have created a new, and still emerging, paradigm for understanding the extracellular regulation of growth factor signaling. This chapter summarizes current knowledge about fibrillin microfibrils, their roles in the regulation of growth factor signaling, their molecular partners in the connective tissue, and their relevance to skeletal biology.

THE FIBRILLINOPATHIES

The importance of fibrillin to the skeleton was first appreciated when a mutation in the gene for fibrillin-1 (FBN1) was identified as the cause of the Marfan syndrome.[2] Individuals with Marfan syndrome (OMIM 154700) display major disease features in the skeleton: tall stature and arachnodactyly, scoliosis and chest deformities, joint hypermobility and muscle wasting, pes planus, and craniofacial abnormalities, including a highly arched palate. Skeletal features of Marfan syndrome are thought to be largely caused by the overgrowth of long bones. Multiple features in other organs (cardiovascular, ocular, skin, lung, and central nervous system) also characterize Marfan syndrome, reflecting the ubiquitous tissue distribution of fibrillin-1 and the importance of fibrillin-1 to the affected tissues.

Mutations in fibrillin-2 cause Beals syndrome or congenital contractural arachnodactyly (CCA).[3] Features of CCA (OMIM 121050) include contractures of the small and large joints, crumpled ears, and arachnodactyly. The more limited nature of disease features caused by mutations in fibrillin-2 is thought to reflect the low to null expression levels of FBN2 mRNA in postnatal tissues and the compensatory high levels of FBN1 mRNA in postnatal tissues.

Mutations in FBN1 also cause autosomal dominant Weill-Marchesani syndrome.[4] Skeletal features of Weill-Marchesani syndrome are the opposite of those found in the Marfan syndrome. Individuals with Weill-Marchesani syndrome (OMIM 608328) display short stature, brachydactyly, hypermuscularity, and joint stiffness. Whereas these skeletal features are the opposite of Marfan, ectopia lentis (resulting from a weakness in the suspensory ligament of the lens) is typical of both syndromes.

FIBRILLIN MICROFIBRILS

Fibrillin was first identified as the major protein component of small (10 nm) diameter microfibrils that are ubiquitous in the extracellular space.[1] At the ultrastructural level, fibrillin microfibrils can be distinguished from collagen fibers by their uniform small diameter and a characteristic beaded or hollow

The authors state that they have no conflicts of interest.

Key words: connective tissue, fibrillin microfibrils, growth factors, latent TGFβ binding proteins, bone morphogenetic protein, TGFβ

© 2008 American Society for Bone and Mineral Research

appearance with no periodic banding pattern. Fibrillin microfibrils are often found as bundles of microfibrils and are always present in elastic fibers.

In developing cartilages, fibrillin microfibrils in the perichondrium ring the cartilage matrix and also extend into the cartilage matrix in a chicken-wire pattern. In aging cartilages, fibrillin microfibrils close to the chondrocyte aggregate laterally to form banded thick fibers that can be distinguished from amianthoid fibers.[5] In demineralized bone, fibrillin microfibrils are found in the cement lines, Haversian canals, osteocyte lacunae, canaliculi, and within fibrous structures containing type III collagen.[6]

The protein components of microfibrils include fibrillin-1,[1] fibrillin-2,[7] and fibrillin-3.[8] Interestingly, the expression levels of FBN2 and FBN3 mRNA are highly restricted to fetal development, whereas fibrillin-1 can be found from the time of gastrulation[9] through postnatal life. Microfibrils can be heteropolymers or homopolymers of fibrillins.[10] Gene knockout experiments in mice suggest that fibrillin-2 function is especially important in the interdigital space, because mice null for fibrillin-2 develop syndactyly.[11] Mice null for fibrillin-1 develop early postnatal aortic aneurysm and die, indicating that fibrillin-1 function is required in the aorta after the second week of life.[12] Analyses of skeletal phenotypes in fibrillin-1 homozygous-null mice have not been reported.

The fibrillins and the latent TGFβ binding proteins (LTBPs) form a family of structurally related proteins. Each of the three fibrillins is a modular protein composed of tandemly repeated epidermal growth factor (EGF)-like domains of the calcium-binding type (cbEGF). These stretches of cbEGF domains are interspersed by domains that contain eight cysteines (8-cys) and "hybrid" domains that seem to be similar to both the 8-cys and the cbEGF domains.[13] In addition, each of the fibrillins is flanked by N-terminal and C-terminal domains, and each contains a special region that is either proline-rich, glycine-rich, or proline- and glycine-rich.[8] Whereas the fibrillins are homologous in primary structure, overall domain structure, and size, LTBPs vary in size and in primary structure but retain an overall similarity in domain organization. LTBPs are composed of the same types of modular domains as the fibrillins. One of the three 8-cys domains present in LTBP-1 was shown to bind covalently to LAP, the latency-associated propeptide of TGFβ1 and to facilitate the secretion of latent TGFβ complexes.[14] Hence, 8-cys domains are also referred to as TGFβ binding (TB) domains. LTBP-1 was immunolocalized to fibrillin microfibrils in perichondrium and osteoblast cultures,[15,16] and direct binding of the C-terminal regions of LTBP-1 and -4 to fibrillin-1 was shown.[17]

Other proteins associated with fibrillin microfibrils include elastin, the fibulins,[18,19] microfibril-associated glycoproteins (MAGPs),[20] perlecan,[21] versican,[22] decorin,[23] and biglycan.[24] These associated proteins establish connective tissue pathways that extend from the fibrillin microfibril networks to basement membranes (through perlecan interactions) and from proteoglycan shells around the microfibrils to hyaluronan and to collagen. These connective tissue pathways integrate fibrillin microfibril networks into the histological patterns of specific organs (cartilage and bone compared with muscle and skin) and into the mechanical and physiological properties of specific organs. In addition, these connective tissue pathways may also serve to integrate growth factor signaling into the mechanical and physiological functions of specific organs.

REGULATION OF GROWTH FACTORS BY FIBRILLIN MICROFIBRILS

A working model for the extracellular regulation of the large latent TGFβ complex by fibrillin microfibrils was proposed, based on the demonstration that LTBPs interact directly with fibrillin-1.[17] According to this model, LTBPs target latent TGFβ complexes to the extracellular matrix, where the large LTBP/TGFβ complex is stabilized through C-terminal interactions of LTBPs with fibrillin[17] and N-terminal interactions with other matrix components, possibly fibronectin.[14] Mutant fibrillin-1 (e.g., in Marfan syndrome) or absent fibrillin-1 (in fibrillin-1–deficient mice) was expected to destabilize LTBP/TGFβ complexes, leading either to activation or to loss of TGFβ signaling. This hypothesis was tested in mouse models of deficient and mutant fibrillin-1, and activation of TGFβ signaling was found in the lung,[25] mitral valve,[26] and aorta,[27] indicating that fibrillin microfibrils are negative regulators of TGFβ signaling.

Additional support that activation of TGFβ signaling underlies aortic dissection and aneurysm, the major life-threatening feature of Marfan syndrome, comes from genetic evidence in humans. Mutations in the receptors for TGFβ cause a Marfan-related disorder[28] that has been named Loeys-Dietz syndrome.[29,30] The major phenotypic features of Loeys-Dietz syndrome (OMIM 609192) include aortic aneurysm and dissection, hypertelorism, bifid uvula and/or cleft palate, arterial tortuosity, and sometimes craniosynostosis and mental retardation. Dolichostenomelia, a feature shared with Marfan syndrome, was reported in <20% of individuals with Loeys-Dietz syndrome.[30]

All together, the genetic evidence in humans and mice show that TGFβ signaling and fibrillin-1 share common pathways, at least in the aorta. However, skeletal features in Loeys-Dietz syndrome are not major compared with those found in Marfan syndrome, suggesting that additional signaling mechanisms may be modulating skeletal phenotypes in these two genetic disorders. In fibrillin-1 homozygous mutant mice, analyses of skeletal phenotypes have not been reported and may be precluded by early postnatal death.[12,31,32] However, in heterozygous C1039G fibrillin-1 mutant mice and in a hypomorphic fibrillin-1–deficient mouse,[33] skeletal features (mainly kyphosis and overgrowth of the ribs) were noted,[32] but it has not been reported whether activation of TGFβ signaling is associated with these phenotypes. The tight skin (tsk) mouse, which harbors a large in-frame duplication within the *Fbn1* gene,[34] shows skin fibrosis and excessive long bone growth, but it is unknown whether TGFβ signaling is associated with tsk skeletal phenotypes.

Another related mechanism that may be involved in the pathogenesis of phenotypic features in Marfan syndrome is the interaction between fibrillin and bone morphogenetic proteins (BMPs). Fibrillin-1 binds to the propeptide of BMP-7, and antibodies specific for BMP-7 propeptide and growth factor can be immunolocalized to fibrillin microfibrils.[35] BMP-7 propeptide and the BMP-7 propeptide/growth factor complex bind to fibrillin microfibrils isolated from tissues.[36] Studies of additional members of the TGFβ superfamily indicate that the propeptides of BMP-2, -4, -5, and -10 and growth and differentiation factor (GDF)-5 also interact with fibrillin and that these growth factor complexes may use this interaction as a mechanism for proper and effective positioning of growth factors in the extracellular space.[37] The specificity of these interactions is supported by the findings that propeptides of TGFβs bind to LTBPs but not to fibrillins,[38] that the propeptide of BMP-7 binds to fibrillin but not to LTBP-1,[35] and that the propeptide of myostatin (GDF-8) does not bind to fibrillin[37] but may interact with LTBP-3.[39]

In vivo evidence that fibrillin-2 affects BMP-7 signaling and limb patterning was shown when fibrillin-2 was knocked out in mice.[11] Fibrillin-2–null mice were born with bilateral syndactyly of the forelimbs and hindlimbs. Complete fusion or close

© 2008 American Society for Bone and Mineral Research

apposition of the cartilage elements of digits 2 and 3 or 2, 3, and 4 were observed, along with fewer apoptotic signals in the interdigital spaces. Expression of BMP-4 and responsiveness to BMP-loaded beads were retained in the fibrillin-2–null autopods, showing that loss of fibrillin-2 in the interdigital space was sufficient to dysregulate the BMP signaling required for proper formation of the cartilage elements and regression of interdigital tissue. Furthermore, compound heterozygous mice ($Fbn2^{+/-}$;$BMP\text{-}7^{+/-}$) were both polydactylous and syndactylous (features of the single nulls but not of the single heterozygous mice), indicating that both genes are in the same pathway that controls digit formation.

Fibrillin-1 and fibrillin-2 protein localization is nonoverlapping in the interdigital space of the developing autopod: fibrillin-1 is located mostly at the boundaries or edges of the cartilage elements, whereas fibrillin-2 forms abundant rays extending through the interdigital space to connect the cartilage elements to the ectoderm.[40] These tissue-specific differences in the fibrillins may explain why BMP signaling is dysregulated in the developing autopod but apparently not in other tissues, where it is presumed that fibrillin-1 compensates for the loss of fibrillin-2.

MOLECULAR MECHANISMS ORCHESTRATED ON A MICROFIBRIL SCAFFOLD

In the TGFβ complex, the propeptides of TGFβs confer latency to the growth factor dimer, and the propeptides are disulfide-linked to LTBPs. Therefore, to initiate TGFβ signaling, the TGFβ growth factor dimer must first be released from this large latent complex. BMPs, on the other hand, are not believed to require activation. BMP-9 propeptide/growth factor complex has been shown to be as active as BMP-9 growth factor, using in vitro cell-based assays.[41] Similarly, BMP-7 propeptide/growth factor complex is as active as BMP-7 growth factor.[42] In solution, type II BMP receptors can outcompete BMP-7 propeptides for interaction with the BMP-7 growth factor dimer and displace the propeptides from the propeptide/growth factor complex, indicating that, unlike TGFβ propeptides, BMP-7 propeptide does not block receptor binding and does not confer latency to the growth factor.[42]

Although these growth factors are similar in structure, extracellular mechanisms that control signaling are clearly different. Whereas TGFβ requires activation by proteolytic or dissociative mechanisms and also requires cofactors such as thrombospondin, endoglin, or betaglycan to facilitate signaling, BMPs can activate signaling simply by binding to its receptors, and instead of activators, BMPs require inhibitors like noggin (and several others) to control the amount of extracellular presentation. The role of the matrix as a physical scaffold important in regulating extracellular signaling is only now becoming appreciated. However, the mechanisms involved in matrix regulation of extracellular signaling are largely unknown.

Direct interactions of fibrillins with BMPs and LTBPs with TGFβs have been documented. Nevertheless, it is not clearly understood whether interactions between these proteins, and with the microfibril scaffold composed by these proteins, sequester and store growth factors in the connective tissue or target and concentrate growth factors for immediate use. It is not known how other interactions (e.g., fibrillin/fibulin interactions) mediated by the microfibril scaffold affect these matrix/growth factor interactions. However, because both TGFβ and BMP signals are associated with the same microfibril scaffold, it may be predicted that the matrix scaffold physically integrates these disparate and sometimes opposing, signals. Because perlecan binds to fibrillin,[21] it is also possible that the microfibril scaffold helps to integrate fibroblast growth factor (FGF) signals with BMP and TGFβ signals. The roles of heparan sulfate proteoglycans in the regulation of FGF signaling and skeletal biology have been extensively reviewed.[43,44]

Knockouts of LTBP-2, LTBP-3, LTBP-1L, and an LTBP-4 hypomorphic mouse have been described. Only LTBP-3–null mice showed skeletal phenotypes (craniofacial abnormalities, kyphosis, osteosclerosis of the long bones and vertebrae, and osteopetrosis). Many of the phenotypes found in these mutant mice were reported to be consistent with loss of TGFβ activity, indicating that LTBPs are required in specific tissues for appropriate TGFβ function. However, none of these mouse models phenocopy TGFβ-null mice, suggesting functional redundancies among the four LTBPs and/or additional tissue-specific functions of the individual LTBPs.[14]

An interesting study of LTBP-4–deficient lung fibroblasts showed that BMP-4 signaling is enhanced and that expression of gremlin, an inhibitor of BMP signaling, was decreased.[45] These findings, which were confirmed in lung tissue, were within the context of loss of LTBP-4 and consequent impaired activation of TGFβ. This study directs attention to the unknown mechanisms by which BMP and TGFβ signals may be integrated across the microfibril scaffold.

Many genes encoding other proteins that interact with fibrillin microfibrils have also been knocked out in mice. Mice null for each of the five fibulin genes have been reported. Skeletal phenotypes have not been reported. However, disturbed TGFβ signaling was found associated with aneurysm formation in a fibulin-4 hypomorphic mouse.[46] In humans, mutations in fibulin-4 can result in skeletal features (arachnodactyly) as well as aneurysm and cutis laxa,[47] suggesting that defects in fibulin function may perturb pathways common to fibrillin microfibrils.

Biochemical results have also identified proteoglycans as potential interactors on the microfibril scaffold. Versican,[22] perlecan,[21] and decorin[23] bind to fibrillin, and biglycan binds to elastic fiber components.[24] Perlecan-null mice have defective cartilage formation, possibly caused by both structural effects and dysregulated FGF signaling, and die in the perinatal period.[48] Decorin binds to the TGFβ growth factor,[49] and biglycan modulates BMP-4 signaling to osteoblasts.[50] More studies are needed to understand how these associated proteoglycans may cooperate with the microfibrillar scaffold in the extracellular regulation of growth factor signaling.

SUMMARY

Human genetic disorders have pointed to important functional roles of fibrillins in skeletal biology. Fibrillin-1 controls long bone growth, and mutations in fibrillin-1 can lead either to tall stature and arachnodactyly (in Marfan syndrome) or to short stature and brachydactyly (in Weill-Marchesani syndrome). Fibrillins also control joint function, because mutations in fibrillin-1 or fibrillin-2 can result in either joint hypermobility, joint contractures, or stiff joints. Recent advances have shown that mutations in fibrillins result in dysregulated TGFβ and BMP signaling and have led to the concept that the microfibrillar scaffold, together with its extended partners in specific connective tissue pathways, regulate TGFβ and BMP signaling.

According to this concept of extracellular regulation by the microfibrillar scaffold, growth factor signaling is "fine tuned,"[40] and resulting perturbations in signaling can lead to a broad range of pathologies. These ranges of pathologies may bear only limited resemblance to phenotypes resulting from complete loss of function mouse models of growth factor sig-

© 2008 American Society for Bone and Mineral Research

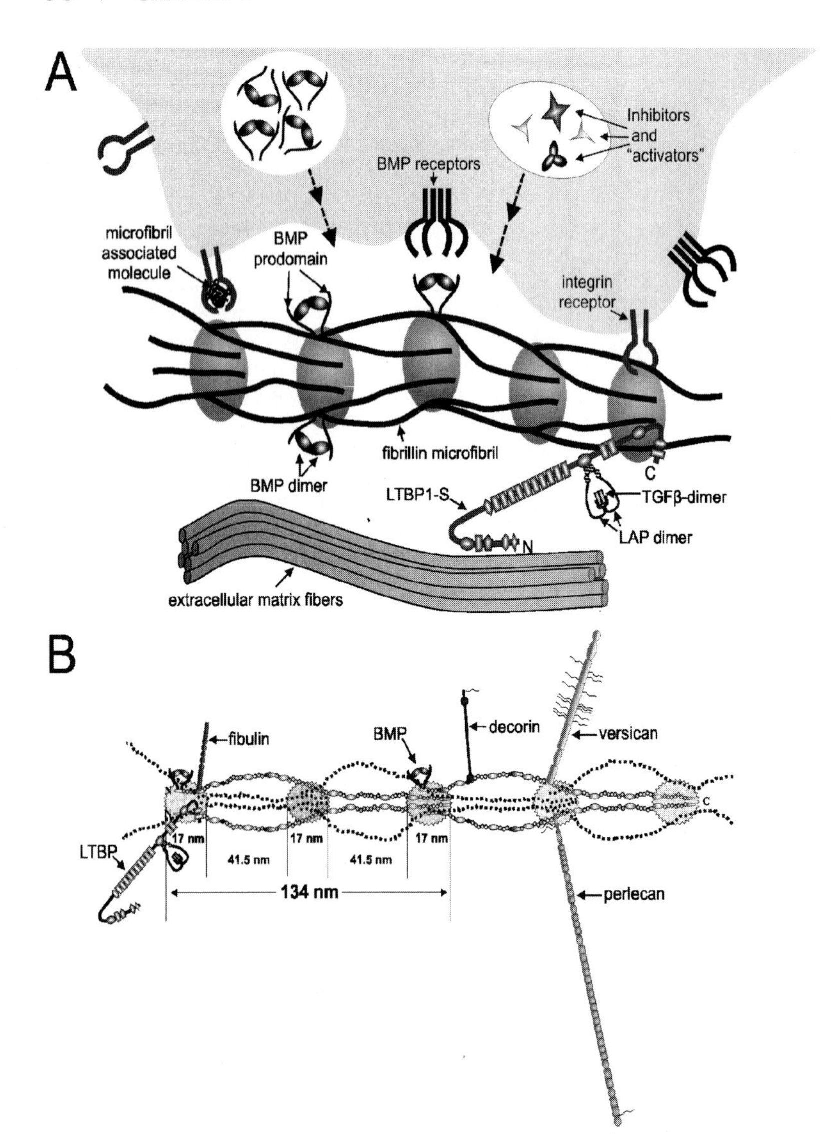

FIG. 1. (A) Fibrillin microfibrils provide a scaffold to which growth factor complexes are targeted and sequestered. Extracellular regulation of targeted growth factor complexes is accomplished through direct interactions between cells, which receive positional information from integrin interactions with microfibril components, and growth factors and through indirect interactions mediated by cellular release of inhibitors and activators of growth factors. (B) A schematic cartoon depicts fibrillin molecules arranged within a single microfibril.[36] Associated proteins (LTBPs, fibulins, BMP complexes, decorin, versican, and perlecan) are positioned along the microfibril according to published interaction data. Because fibulins and proteoglycans interact with other extracellular molecules, fibrillin microfibrils are part of networks that form tissue-specific connective tissue pathways integrating growth factor signaling within an individual connective tissue framework.

naling. Fibrillin-1 homozygous null mice have thus far not elucidated the skeletal phenotypes manifested in the human FBN1 genetic disorders. More sophisticated mouse models are needed to dissect the mechanisms by which the microfibril scaffold regulates growth factor signaling and by which human pathologies are produced.

A model for the extracellular regulation of TGFβ and BMP signals is depicted in Fig. 1A. Cells secrete the large latent TGFβ complex and BMP complexes. Through interactions with LTBPs, which target the latent TGFβ complex to the matrix, and with the propeptides of BMP complexes, fibrillins position and concentrate growth factor signals. Cells may activate or inhibit growth factor signals as they come into contact with the microfibril scaffold. These choices may be regulated by positional information contained in the microfibrils themselves or in associated proteins, some of which are shown in Fig. 1B. In addition, the associated proteins may themselves regulate growth factor signaling by competitive mechanisms (e.g., LTBPs and fibulins may compete for the same binding site on fibrillin) or by differential or sequential interactions with the growth factors themselves. In this manner, the microfibril scaffold may work as a well-oiled machine[51] to regulate growth factor signaling.

© 2008 American Society for Bone and Mineral Research

REFERENCES

1. Sakai LY, Keene DR, Engvall E 1986 Fibrillin, a new 350kD glycoprotein, is a component of extracellular microfibrils. J Cell Biol **103:**2499–2509.
2. Dietz HC, Cutting GR, Pyeritz RE, Maslen CL, Sakai LY, Corson GM, Puffenberger EG, Hamosh A, Nanthakumar EJ, Curristin SM, Stetten G, Meyers DA, Francomano CA 1991 Marfan syndrome caused by a recurrent de novo missense mutation in the fibrillin gene. Nature **352:**337–339.
3. Gupta PA, Putnam EA, Carmical SG, Kaitila I, Steinmann B, Child A, Danesino C, Metcalfe K, Berry SA, Chen E, Delorme CV, Thong MK, Ades LC, Milewicz DM 2002 Ten novel FBN2 mutations in congenital contractural arachnodactyly: Delineation of the molecular pathogenesis and clinical phenotype. Hum Mutat **19:**39–48.
4. Faivre L, Gorlin RJ, Wirtz MK, Godfrey M, Dagoneau N, Samples JR, Le Merrer M, Collod-Beroud G, Boileau C, Munnich A, Cormier-Daire V 2003 In frame fibrillin-1 gene deletion in autosomal dominant Weill-Marchesani syndrome. J Med Genet **40:**34–36.
5. Keene DR, Jordan CD, Reinhardt DP, Ridgway CC, Ono RN, Corson GM, Fairhurst M, Sussman MD, Memoli VA, Sakai LY 1997 Fibrillin-1 in human cartilage: Developmental expression and formation of special banded fibers. J Histochem Cytochem **45:**1069–1082.

6. Keene DR, Sakai LY, Burgeson RE 1991 Human bone contains type III collagen, type VI collagen, and fibrillin. J Histochem Cytochem **39:**59–69.
7. Zhang H, Hu W, Ramirez F 1995 Developmental expression of fibrillin genes suggests heterogeneity of extracellular microfibrils. J Cell Biol **129:**1165–1176.
8. Corson GM, Charbonneau NL, Keene DR, Sakai LY 2004 Differential expression of fibrillin-3 adds to microfibril variety in human and avian, but not rodent, connective tissues. Genomics **83:**461–472.
9. Gallagher BC, Sakai LY, Little CD 1993 Fibrillin delineates the primary axis of the early avian embryo. Dev Dyn **196:**70–78.
10. Charbonneau NL, Dzamba BJ, Ono RN, Keene DR, Corson GM, Reinhardt DP, Sakai LY 2003 Fibrillins can co-assemble in fibrils, but fibrillin fibril composition displays cell-specific differences. J Biol Chem **278:**2740–2749.
11. Arteaga-Solis E, Gayraud B, Lee SY, Shum L, Sakai L, Ramirez F 2001 Regulation of limb patterning by extracellular microfibrils. J Cell Biol **154:**275–281.
12. Carta L, Pereira L, Arteaga-Solis E, Lee-Arteaga SY, Lenart B, Starcher B, Merkel CA, Sukoyan M, Kerkis A, Hazeki N, Keene DR, Sakai LY, Ramirez F 2006 Fibrillins 1 and 2 perform partially overlapping functions during aortic development. J Biol Chem **281:**8016–8023.
13. Corson GM, Chalberg SC, Dietz HC, Charbonneau NL, Sakai LY 1993 Fibrillin binds calcium and is coded by cDNAs that reveal a multidomain structure and alternatively spliced exons at the 5′ end. Genomics **17:**476–484.
14. Rifkin DB 2005 Latent transforming growth factor-β (TGF-β) binding proteins: Orchestrators of TGF-β availability. J Biol Chem **280:**7409–7412.
15. Dallas SL, Miyazono K 1995 Skerry, T.M., Mundy, G.R., Bonewald, L.F.: Dual role for the latent transforming growth factor-beta binding protein in storage of latent TGF-beta in the extracellular matrix and as a structural matrix protein. J Cell Biol **131:**539–549.
16. Dallas SL, Keene DR, Bruder SP, Saharinen J, Sakai LY, Mundy GR, Bonewald LF 2000 Role of the latent transforming growth factor beta binding protein 1 in fibrillin-containing microfibrils in bone cells in vitro and in vivo. J Bone Miner Res **15:**68–81.
17. Isogai Z, Ono RN, Ushiro S, Keene DR, Chen Y, Mazzieri R, Charbonneau NL, Reinhardt DP, Rifkin DB, Sakai LY 2003 Latent transforming growth factor β-binding protein 1 interacts with fibrillin and is a microfibril-associated protein. J Biol Chem **278:**2750–2757.
18. Reinhardt DP, Sasaki T, Dzamba BJ, Keene DR, Chu M-L, Göhring W, Timpl R, Sakai LY 1996 Fibrillin-1 and fibulin-2 interact and are colocalized in some tissues. J Biol Chem **271:**19489–19496.
19. El Hallous E, Sasaki T, Hubmacher D, Getie M, Tiedemann K, Brinckmann J, Bätge B, Davis EC, Reinhardt DP 2007 Fibrillin-1 interactions with fibulins depend on the first hybrid domain and provide an adaptor function to tropoelastin. J Biol Chem **282:**8935–8946.
20. Gibson MA, Hughes JL, Fanning JC, Cleary EG 1986 The major antigen of elastin-associated microfibrils is a 31-kDa glycoprotein. J Biol Chem **261:**11429–11436.
21. Tiedemann K, Sasaki T, Gustafsson E, Göhring W, Bätge B, Notbohm H, Timpl R, Wedel T, Schlötzer-Schrehardt U, Reinhardt DP 2005 Microfibrils at basement membrane zones interact with perlecan via fibrillin-1. J Biol Chem **280:**11404–11412.
22. Isogai Z, Aspberg A, Keene DR, Ono RN, Reinhardt DP, Sakai LY 2002 Versican interacts with fibrillin-1 and links extracellular microfibrils to other connective tissue networks. J Biol Chem **277:**4565–4572.
23. Trask BC, Trask TM, Broekelmann T, Mecham RP 2000 The microfibrillar proteins MAGP-1 and fibrillin-1 form a ternary complex with the chondroitin sulfate proteoglycan decorin. Mol Biol Cell **11:**1499–1507.
24. Reinboth B, Hanssen E, Cleary EG, Gibson MA 2002 Molecular interactions of biglycan and decorin with elastic fiber components: Biglycan forms a ternary complex with tropoelastin and microfibril-associated glycoprotein 1. J Biol Chem **277:**3950–3957.
25. Neptune ER, Frischmeyer PA, Arking DE, Myers L, Bunton TE, Gayraud B, Ramirez F, Sakai LY, Dietz HC 2003 Dysregulation of TGFβ activation contributes to pathogenesis in Marfan syndrome. Nat Genet **33:**407–411.
26. Ng CM, Cheng A, Myers LA, Martinez-Murillo F, Jie C, Bedja D, Gabrielson KL, Hausladen JM, Mecham RP, Judge DP, Dietz HC 2004 TGF-beta-dependent pathogenesis of mitral valve prolapse in a mouse model of Marfan syndrome. J Clin Invest **114:**1586–1592.
27. Habashi JP, Judge DP, Holm TM, Cohn RD, Loeys BL, Cooper TK, Myers L, Klein EC, Liu G, Calvi C, Podowski M, Neptune ER, Halushka MK, Bedja D, Gabrielson K, Rifkin DB, Carta L, Ramirez F, Huso DL, Dietz HC 2006 Losartan, an AT1 antagonist, prevents aortic aneurysm in a mouse model of Marfan syndrome. Science **312:**117–121.
28. Mizuguchi T, Collod-Beroud G, Akiyama T, Abifadel M, Harada N, Morisaki T, Allard D, Varret M, Claustres M, Morisaki H, Ihara M, Kinoshita A, Yoshiura K, Junien C, Kajii T, Jondeau G, Ohta T, Kishino T, Furukawa Y, Nakamura Y, Niikawa N, Boileau C, Matsumoto N 2004 Heterozygous TGFBR2 mutations in Marfan syndrome. Nat Genet **36:**855–860.
29. Loeys BL, Chen J, Neptune ER, Judge DP, Podowski M, Holm T, Meyers L, Leitch CC, Katsanis N, Sharifi N, Xu FL, Myers LA, Spevak PJ, Cameron DE, De Backer J, Hellemans J, Chen Y, Davis EC, Webb CL, Kress W, Coucke P, Rifkin DB, De Paepe AM, Dietz HC 2005 A syndrome of altered cardiovascular, craniofacial, neurocognitive and skeletal development caused by mutations in TGFBR1 or TGFBR2. Nat Genet **37:**275–281.
30. Loeys BL, Schwarze U, Holm T, Callewaert BL, Thomas GH, Pannu H, De Backer JF, Oswald GL, Symoens S, Manouvrier S, Roberts AE, Faravelli F, Greco MA, Pyeritz RE, Milewicz DM, Coucke PJ, Cameron DE, Braverman AC, Byers PH, De Paepe AM, Dietz HC 2006 Aneurysm syndromes caused by mutations in the TGF-beta receptor. N Engl J Med **355:**788–798.
31. Pereira L, Andrikopoulos K, Tian J, Lee SY, Keene DR, Ono R, Reinhardt DP, Sakai LY, Jensen-Biery N, Bunton T, Dietz HC, Ramirez F 1997 Targeting of the gene encoding fibrillin-1 recapitulates the vascular aspect of Marfan syndrome. Nat Genet **17:**218–222.
32. Judge DP, Biery NJ, Keene DR, Geubtner J, Myers L, Huso DL, Sakai LY, Dietz HC 2004 Evidence for a critical contribution of haploinsufficiency in the complex pathogenesis of Marfan syndrome. J Clin Invest **114:**172–181.
33. Pereira L, Lee SY, Gayraud B, Andrikopoulos K, Shapiro SD, Bunton T, Biery NJ, Dietz HC, Sakai LY, Ramirez F 1999 Pathogenetic sequence for aneurysm revealed in mice underexpressing fibrillin-1. Proc Natl Acad Sci USA **96:**3819–3823.
34. Siracusa LD, McGrath R, Ma Q, Moskow JJ, Manne J, Christner PJ, Buchberg AM, Jimenez SA 1996 A tandem duplication within the fibrillin 1 gene is associated with the mouse tight skin mutation. Genome Res **6:**300–313.
35. Gregory KE, Ono RN, Charbonneau NL, Kuo C-L, Keene DR, Bächinger HP, Sakai LY 2005 The prodomain of BMP-7 targets the BMP-7 complex to the extracellular matrix. J Biol Chem **280:**27970–27980.
36. Kuo CL, Isogai Z, Keene DR, Hazeki N, Ono RN, Sengle G, Bächinger HP, Sakai LY 2007 Effects of fibrillin-1 degradation on microfibril ultrastructure. J Biol Chem **282:**4007–4020.
37. Sengle G, Charbonneau NL, Ono RN, Sasaki T, Alvarez J, Keene DR, Bächinger HP, Sakai LY 2008 Targeting of BMP growth factor complexes to fibrillin. J Biol Chem **283:**13874–13888.
38. Saharinen J, Keski-Oja J 2000 Specific sequence motif of 8-cys repeats of TGF-beta binding proteins, LTBPs, reates a hydrophobic interaction site for binding of small latent TGF-beta. Mol Biol Cell **11:**2691–2704.
39. Anderson SB, Goldberg AL, Whitman M 2008 Identification of a novel pool of extracellular pro-myostatin in skeletal muscle. J Biol Chem **283:**7027–7035.
40. Charbonneau NL, Ono RN, Corson GM, Keene DR, Sakai LY 2004 Fine tuning of growth factor signals depends on fibrillin microfibril networks. Birth Defects Res Part C Embryo Today **72:**37–50.
41. Brown MA, Zhao Q, Baker KA, Naik C, Chen C, Pukac L, Singh M, Tsareva T, Parice Y, Mahoney A, Roschke V, Sanyal I, Choe S 2005 Crystal structure of BMP-9 and functional interactions with pro-region and receptors. J Biol Chem **280:**25111–25118.
42. Sengle G, Ono RN, Lyons KM, Bächinger HP, Sakai LY 2008 A new model for growth factor activation: Type II receptors compete with the prodomain for BMP-7. J Molec Biol (in press).
43. Jackson RA, Nurcombe V, Cool SM 2006 Coordinated fibroblast growth factor and heparan sulfate regulation of osteogenesis. Gene **379:**79–91.

© 2008 American Society for Bone and Mineral Research

44. DeCarlo AA, Whitelock JM 2006 The role of heparan sulfate and perlecan in bone-regenerative procedures. J Dent Res **85:**122–132.
45. Koli K, Wempe F, Sterner-Kock A, Kantola A, Komor M, Hofmann WK, von Melchner H, Keski-Oja J 2004 Disruption of LTBP-4 function reduces TGF-beta activation and enhances BMP-4 signaling in the lung. J Cell Biol **167:**123–133.
46. Hanada K, Vermeij M, Garinis GA, de Waard MC, Kunen MG, Myers L, Maas A, Duncker DJ, Meijers C, Dietz HC, Kanaar R, Essers J 2007 Perturbations of vascular homeostasis and aortic valve abnormalities in fibulin-4 deficient mice. Circ Res **100:**738–746.
47. Dasouki M, Markova D, Garola R, Sasaki T, Charbonneau NL, Sakai LY, Chu ML 2007 Compound heterozygous mutations in fibulin-4 causing neonatal lethal pulmonary artery occlusion, aortic aneurysm, arachnodactyly, and mild cutis laxa. Am J Med Genet A **143:**2635–2641.
48. Arikawa-Hirasawa E, Watanabe H, Takami H, Hassell JR, Yamada Y 1999 Perlecan is essential for cartilage and cephalic development. Nat Genet **23:**354–358.
49. Yamaguchi Y, Mann DM, Ruoslahti E 1990 Negative regulation of transforming growth factor-beta by the proteoglycan decorin. Nature **346:**281–284.
50. Chen X-D, Fisher LW, Gehron-Robey P, Young MF 2004 The small leucine-rich proteoglycan biglycan modulates BMP-4 induced osteoblast differentiation. FASEB J **18:**948–958.
51. Engel J 2006 Molecular machines in the matrix? Matrix Biol **25:**200–201.

Chapter 6. The Composition of Bone

Pamela Gehron Robey[1] and Adele L. Boskey[2]

[1]Craniofacial and Skeletal Diseases Branch, National Institute of Dental and Craniofacial Research, National Institutes of Health, Department of Health and Human Services, Bethesda, Maryland; [2]Research Division, Hospital for Special Surgery, and Department of Biochemistry and Graduate Field of Physiology, Biophysics and Systems Biology Cornell University Medical and Graduate Medical Schools, New York, New York

INTRODUCTION

Bone composes the largest proportion of the body's connective tissue mass. Unlike most other connective tissue matrices, bone matrix is physiologically mineralized and is unique in that it is constantly regenerated throughout life as a consequence of bone turnover. Bone as an organ is made up of the cartilaginous joints, the calcified cartilage in the growth plate (in developing individuals), the marrow space, and the cortical and cancellous mineralized structures. Bone as a tissue consists of the mineralized and nonmineralized (osteoid) components of the cortical and cancellous regions of long and flat bones. There are three cell types in bone: (1) the bone-forming osteoblasts, which when engulfed in mineral become (2) osteocytes, and (3) the bone-destroying osteoclasts. Each of these cells communicates with one another by either direct cell contact or through signaling molecules and respond to each other. The detailed properties of these cells have been discussed in numerous publications; see for example the excellent chapter by Lian and Stein.(1) This chapter focuses on the extracellular matrix, which is synthesized primarily by osteoblasts but also contains proteins adsorbed from the circulation.

The preponderance of bone is extracellular matrix (ECM). Information on the gene and protein structure and potential function of bone ECM constituents has exploded during the last two decades. This information has been described in great detail in several recent reviews,(2) to which the reader is referred for specific references, which are too numerous to be listed adequately here. This chapter summarizes the composition of bone and the salient features of the classes of bone matrix proteins. The tables list specific details for the individual ECM components.

THE COMPOSITE

Bone is a composite material whose extracellular matrix consists of mineral, collagen, water, noncollagenous proteins, and lipids in decreasing proportion (depending on age, species, and site). These components have both mechanical and metabolic functions. Understanding of some of the biological functions of these components has come from mouse models and analyses of healthy and diseased human tissues and from cell culture studies.

The Mineral

The mineral phase of bone is a nano-crystalline highly substituted analog of the naturally occurring mineral, hydroxylapatite $[Ca_{10}(PO_4)_6(OH)_2]$. The major substituents are carbonate, magnesium, and acid phosphate, along with other trace elements the content of which depends on diet and environment. Although the precise chemical nature of the initial mineral formed has been debated,(3) it is well accepted that the "biomineral" present in the bones during development is apatitic. The physical and chemical properties of this mineral have been determined by a variety of techniques including chemical analyses, X-ray diffraction, vibrational spectroscopy, energy dispersive electron analysis, nuclear magnetic resonance, small angle scattering, and transmission and atomic force microscopy.(4)

The functions of the mineral are to strengthen the collagen composite, providing more mechanical resistance to the tissue, and also to serve as a source of calcium, phosphate, and magnesium ions for mineral homeostasis. For physicochemical reasons, usually it is the smallest mineral crystals that are lost during remodeling; thus, in osteoporosis, it is not surprising that the larger more perfect crystals persist within the matrix,(5) contributing to the brittle nature of osteoporotic bone, whereas where remodeling is impaired, as in osteopetrosis, the mineral crystals remain small relative to age-matched controls.(5)

Collagen

The basic building block of the bone matrix fiber network is

The authors state that they have no conflicts of interest.

Key words: mineral, hydroxyapatite, collagen, noncollagenous proteins, extracellular matrix

© 2008 American Society for Bone and Mineral Research

TABLE 1. CHARACTERISTICS OF COLLAGEN-RELATED GENES AND PROTEINS FOUND IN BONE MATRIX

Protein/gene	Function	Disease/animal model/phenotype
Type I – *17q21.23, 17q21.3–22* [α1(I)$_2\alpha$2(I)] α1(I)$_3$]	Serves as scaffolding, binds and orients other proteins that nucleate hydroxyapatite deposition	Human mutations—osteogenesis imperfecta (OMIM 166210; 166200; 610854; 259420; 166220); mouse models: oim mouse; mov 14 mouse; brittle mouse—bones mechanically weak; mineral crystals small; some mineral outside collagen
Type X – *6q21–22.3* [α1(I)$_3$]	Present in hypertrophic cartilage of the growth plate, but does not appear to regulate matrix mineralization	Human mutations - Schmid metaphyseal chondrodysplasia (OMIM120110) knockout mouse—no apparent skeletal phenotype,
Type III – *2q24.3–31* [α1(III)]$_3$ Type V – *9q34.2–34.3;2q24.3–31, 9q34.2–34.3*[α1(V)$_2\alpha$2(V)] [α1(V) α2(V) α3(V)]	Present in bone in trace amounts, may regulate collagen fibril diameter, their paucity in bone may explain the large diameter size of bone collagen fibrils	Human mutations in type III—different forms of Ehlers-Danlos syndrome (OMIM 130050); Mutations in type V α1 α2 (OMIM 120215; 120190); mouse model has disrupted fibril arrangement

type I collagen, which is a triple-helical molecule containing two identical α1(I) chains and a structurally similar, but genetically different, α2(I) chain.[6] Collagen α chains are characterized by a Gly-X-Y repeating triplet (where X is usually proline, and Y is often hydroxyproline) and by several post-translational modifications including (1) hydroxylation of certain lysyl or prolyl residues, (2) glycosylation of the hydroxylysine with glucose or galactose residues or both, (3) addition of mannose at the propeptide termini, and (4) formation of intra- and intermolecular covalent cross-links that differ from those found in soft connective tissues. Measurement of these bone-derived collagen cross-links in urine has proven to be good measures of bone resorption.[7] Bone matrix proper consists predominantly of type I collagen; however, trace amounts of type III, V, and fibril-associated collagens (Table 1) may be present during certain stages of bone formation and may regulate collagen fibril diameter.

NONCOLLAGENOUS PROTEINS

Noncollagenous proteins (NCPs) compose 10–15% of the total bone protein content. These proteins are multifunctional, having roles in organizing the extracellular matrix, coordinating cell–matrix and mineral–matrix interactions, and regulating the mineralization process. Knowledge of these functions have come from studies of the isolated proteins in solution, from analyses of mice in which the proteins are ablated (knocked out [KO]), or overexpressed, characterization of human diseases in which these proteins have mutations, and studies using appropriate cell cultures. The tables summarizing the gene and protein structures, and the functions of these proteins are associated with each discussion of the protein families.

Serum-Derived Proteins

Approximately one fourth of the total NCP content is exogenously derived (Table 2). This fraction is largely composed of serum-derived proteins, such as albumin and α_2-HS-glycoprotein, which are acidic in character and bind to bone matrix because of their affinity for hydroxyapatite. Although these proteins are not endogenously synthesized, they may exert effects on matrix mineralization and bone cell proliferation. For example, α_2-HS-glycoprotein, which is the human analog of fetuin, when ablated in mice causes ectopic calcification,[8] suggesting the protein is a mineralization inhibitor. The remainder of the exogenous fraction is composed of growth factors and a large variety of other molecules present in trace amounts, which influence local bone cell activity.[1,3] On a mole-to-mole basis, bone-forming cells synthesize and secrete as many molecules of NCP as of collagen. These molecules can be classified into four general (and sometimes overlapping) groups: (1) proteoglycans, (2) glycosylated proteins, (3) glycosylated proteins with potential cell attachment activities, and (4) γ-carboxylated (gla) proteins. The physiological roles for individual bone protein constituents are not well defined; however, they may participate not only in regulating the deposition of mineral but also in the control of osteoblastic and osteoclastic metabolism.

Proteoglycans

Proteoglycans are macromolecules that contain acidic polysaccharide side chains (glycosaminoglycans) attached to a central core protein, and bone matrix contains several members of this family (Table 3).[9] During initial stages of bone formation, the large chondroitin sulfate proteoglycan, versican, and the glycosaminoglycan, hyaluronan (which is not attached to a protein core), are produced and may delineate areas that will become bone. With continued osteogenesis, versican is replaced by two small chondroitin sulfate proteoglycans, decorin and biglycan, composed of tandem repeats of a leucine-rich

TABLE 2. GENE AND PROTEIN CHARACTERISTICS OF SERUM PROTEINS FOUND IN BONE MATRIX

Protein/gene	Function	Disease/animal model/phenotype
Albumin–*2q11–13*—69 kDa, nonglycosylated, one sulfhydryl, 17 disulfide bonds, high affinity hydrophobic binding pocket	Inhibits hydroxyapatite crystal growth	
α2HS glycoprotein–*3q27–29*—precursor protein of fetuin, cleaved to form A and B chains that are disulfide linked, Ala-Ala and ProPro repeat sequences, N-linked oligosaccharides, cystatin-like domains	Promotes endocytosis, has opsonic properties, chemoattractant for monocytic cells, bovine analog (fetuin) is a growth factor; inhibits calcification	Knockout mouse—adult ectopic calcification

© 2008 American Society for Bone and Mineral Research

TABLE 3. GENE AND PROTEIN CHARACTERISTICS: GLYCOSAMINOGLYCAN-CONTAINING MOLECULES IN BONE

Protein/gene	*Function*	*Disease/animal model/phenotype*
Aggrecan–*15q26.1*—~2.5 × 10^6 intact protein, ~180–370,000 core, ~100 CS chains of 25 kDa, and some KS chains of similar size, G1, G2, and G3 globular domains with hyaluronan binding sites, EGF and CRP-like sequences	Matrix organization, retention of water and ions, resilience to mechanical forces	Human mutation: spondyloepiphyseal dysplasia (OMIM 155760; 608361); mouse models: brachymorphic mouse, accelerated growth plate calcification; cartilage matrix deficiency mouse- shortened stature; nanomelic chick (mutation)—abnormal bone shape
Versican *(PG-100)–5q12–14*—~1 × 10^6 intact protein, ~360 kDa core, ~12 CS chains of 45 kDa, G1 and G3 globular domains with hyaluronan binding sites, EGF and CRP-like sequences	Regulates chondrogenesis; may "capture" space that is destined to become bone	Human mutation: Wagner syndrome (an ocular disorder) (OMIM 143200)
Decorin *(class 1 LRR)–12q13.2*—~130 kDa intact protein, ~38–45 kDa core with 10 leucine-rich repeat sequences, 1 CS chain of 40 kDa	Binds to collagen and may regulate fibril diameter, binds to TGF-β and may modulate activity, inhibits cell attachment to fibronectin	Mouse knockout—no apparent skeletal phenotype although collagen fibrils are abnormal, DCN/BGN double knockout—progeroid form of Ehler's-Danlos syndrome
Biglycan *(class 1 LRR)–Xq27*—~270 kDa intact protein, ~38–45 kDa core protein with 12 leucine-rich repeat sequences, 2 CS chains of 40 kDa	Binds to collagen, TGF-β, and other growth factors; pericellular environment, a genetic determinant of peak bone mass	Knockout mouse, osteopenia; thin bones, decreased mineral content, increased crystal size; short stature
Asporin *(class 1 LRR)–9q21.3*—67 kDa, most likely no GAG chains	Regulates collagen structure	Human polymorphism associated with osteoarthritis (OMIM 608135)
Fibromodulin *(class 2 LRR)–1q32*—59 kDa intact protein, 42 kDa core protein, one N-linked KS chain	Binds to collagen, may regulate fibril formation, binds to TGF-β	Fibromodulin/biglycan double knockout mice—joint laxity and formation of supernumery sesamoid bones
Osteoadherin *(class 2 LRR)*—85 kDa intact protein, 47 kDa core protein, RGD sequence	May mediate cell attachment	
Lumican *(class 2 LRR)–12q21.3-q22*—70–80 kDa intact protein, 37 kDa core protein	Binds to collagen, may regulate fibril formation	Lumican/fibromodulin double knockout mouse—has ectopic calcification and a variant of Ehler's Danlos syndrome (OMIM 130000)
Perlecan–*1p36.1*—five domain heparan sulfate proteoglycan, core protein 400 kDa	Interacts with matrix components to regulate cell signaling; Cephalic development	Transgenic mice with mutated perlecan: Schwartz-Jampel syndrome (OMIM 142461)—mice have impaired mineralization and misshapen skeletons and joint abnormalities. Knockout mice have phenotype resembling thanatophoric dysplasia (TD) type I
Glypican–*Xq26*—lipid-linked heparan sulfate proteoglycan, 14 conserved cysteine residues	Regulates BMP-SMAD signaling; Regulates cell development	Human mutation: Simpson-Golabi-Behmel syndrome (OMIM 300037); knockout mouse has delayed endochondral ossification and impaired osteoclast development
Osteoglycin/Mimecan *(class 3 LRR)–9q22*—299 aa precursor, 105 aa mature protein, no GAG in bone, keratan sulfate in other tissues.	Binds to TGF-β Regulates collagen fibrillogenesis	
Hyaluronan–multigene complex—multiple proteins associated outside of the cell, structure unknown	May work with versican molecule to capture space destined to become bone	

repeat sequence. Decorin has been implicated in the regulation of collagen fibrillogenesis and is distributed predominantly in the ECM space of connective tissues and in bone, whereas biglycan tends to be found in pericellular locales. A heparan sulfate proteoglycan, perlecan is involved in limb patterning, and is found surrounding chondrocytes in the growth plate, whereas the heparan sulfate proteoglycans called glypicans also affect skeletal growth. In addition, in bone, there are other small leucine-rich proteoglycans (SLRPs), including osteoglycin (mimecan), osteoadherin, lumican, asporin, and fibromodulin.[10] Although their exact physiological functions are not known, these proteoglycans are assumed to be important for the integrity of most connective tissue matrices. Deletion of the *biglycan* gene, for example, leads to a significant decrease in the development of trabecular bone, indicating that it is a positive regulator of bone formation.[2] Other functions might arise from the ability of these proteoglycans to bind and modulate the activity of the growth factors in the extracellular space, thereby influencing cell proliferation and differentiation.[1]

Glycosylated Proteins

Glycosylated proteins with diverse functions abound in

© 2008 American Society for Bone and Mineral Research

TABLE 4. GENE AND PROTEIN CHARACTERISTICS OF GLYCOPROTEINS IN BONE MATRIX

Protein/gene	*Function*	*Disease/animal models/phenotype*
Alkaline phosphatase (bone-liver-kidney isozyme)–*1p34–36.1*—two identical subunits of ~80 kDa, disulfide bonded, tissue-specific post-translational modifications	Potential Ca^{2+} carrier, hydrolyzes inhibitors of mineral deposition such as pyrophosphates, increases local phosphate concentration	Human mutations hypophosphatasia (OMIM 171760) (decreased activity), TNAP knockout mouse—growth impaired; decreased mineralization
Osteonectin–*5q31.3-q32*—~35–45 kDa, intramolecular disulfide bonds, α helical amino terminus with multiple low affinity Ca^{2+} binding sites, two EF hand high affinity Ca^{2+} sites, ovomucoid homology, glycosylated, phosphorylated, tissue specific modifications	Regulates collagen organization; May mediate deposition of hydroxyapatite, binds to growth factors, may influence cell cycle, positive regulator of bone formation	Knockout mouse—severe osteopenia, decreased trabecular connectivity; decreased mineral content; increased crystal size
Tetranectin–*3p22-p21.3*—21 kDa protein composed of four identical subunits of 5.8 kDa, sequence homologies with a sialoprotein receptor and G3 domain of aggrecan	Binds to plasminogen, may regulate matrix mineralization	Knockout mouse—no long bone phenotype, spinal deformity, increased mineralization in implant model
Tenascin-C–*9q33*—hexameric structure, six identical chains of 320kDA, Cys-rich, EGF-like repeats, FN type III repeats	Interferes with cell–FN interactions	Knockout mouse—no apparent skeletal phenotype
Tenascin-X–*6p21.3*—hexameric with five N-linked glycosylation sites and multiple EFG and 40 fibronectin type III repeats	Regulates cell–matrix interactions	Human mutation—Ehlers Danlos II phenotype (OMIM 600985)
Secreted phosphoprotein 24–*2q37*—24 kDa secreted phosphoprotein, shares sequence homology with members of the cystatin family of thiol protease inhibitors	Associates with regulators of mineralization in serum, may regulate thiol proteases in bone	

bone. One of the hallmarks of bone formation is the synthesis of high levels of alkaline phosphatase (Table 4). This glycoprotein enzyme, primarily bound to the cell surface through a phosphoinositol linkage, can be cleaved from the cell surface and found within mineralized matrix. The function of alkaline phosphatase in bone cell biology has been the matter of much speculation and remains undefined. Mice lacking tissue non-specific alkaline phosphatase have impaired mineralization, showing the importance of this enzyme for mineral deposition.[11]

TABLE 5. GENE AND PROTEIN CHARACTERISTICS OF SIBLINGS

Protein/gene	*Function*	*Disease/animal models/phenotype*
Osteopontin–*4q21*—44–75 kDa, polyaspartyl stretches, no disulfide bonds, glycosylated, phosphorylated, RGD located 2/3 from the N-terminal	Binds to cells, may regulate mineralization, may regulate proliferation, inhibits nitric oxide synthase, may regulate resistance to viral infection	Knockout mouse—decreased crystal size; increased mineral content; not subject to osteoclastic remodeling
Bone sialoprotein–*4q21*—46–75 kDa, polyglutamyl stretches, no disulfide bonds, 50% carbohydrate, tyrosine-sulfated, RGD near the C terminus	Binds to cells, may initiate mineralization	Knockout mouse—no published data on phenotype
DMP-1–4q21–513 amino acids predicted; serine-rich, acidic, RGD 2/3 from N terminus	Regulator of biomineralization; regulates osteocyte function	Human mutation—dentinogenesis imperfecta and hypophosphatemia (OMIM 600980); knockout mouse—undermineralized with craniofacial and growth plate abnormalities and defective osteocyte function
Dentin sialophosphoprotein–*4q21.3*–gene produces three proteins, dentin sialoprotein, dentin phosphophorin, and dentin glycoprotein. All have RGD sites; dentin phosphophorin is highly phosphorylated	Regulation of biomineralization	Human mutations in dentinal dysplasias and dentinogenesis imperfecta; no bone disease (OMIM 125485); knockout mouse has thinner bones at 9 mo, no significant other bone phenotype, and severe dentin abnormalities
MEPE–*4q21.1*–525 amino acids, two N-glycosylation motifs, a glycosaminoglycan-attachment site, an RGD cell-attachment motif, and phosphorylation motifs	Regulation of biomineralization; regulation of PHEX (phosphaturic hormone) activity	Human—association with oncogenic osteomalacia; knockout mouse—increased bone mass and resistance to ovariectomy-induced bone loss
Other SIBLINGs: enamelin		

© 2008 American Society for Bone and Mineral Research

TABLE 6. GENE AND PROTEIN CHARACTERISTICS OF OTHER RGD-CONTAINING GLYCOPROTEINS

Protein/gene	*Function*	*Disease/animal models*
Thrombospondins *(1–4, COMP)–15Q-1, 6q27, 1q21–24, 5q13, 19p13.1*—450 kDa molecules, three identical disulfide linked subunits of ~150–180 kDa, homologies to fibrinogen, properdin, EGF, collagen, von Willebrand, *P. falciparum,* and calmodulin, RGD at the C-terminal globular domain	Cell attachment (but usually not spreading), binds to heparin, platelets, types I and V collagens, thrombin, fibrinogen, laminin, plasminogen and plasminogen activator inhibitor, histidine rich glycoprotein	Human mutation in COMP—pseudoachondroplasia (OMIM 600310); TSP-2 knockout mouse—large collagen fibrils, thickened bones; spinal deformities
Fibronectin-*2q34*—400 kDa with two nonidentical subunits of ~200 kDa, composed of type I, II, and III repeats, RGD in the 11th type III repeat 2/3 from N terminus	Binds to cells, fibrin heparin, gelatin, collagen	Knockout mouse—lethal before skeletal development
Vitronectin-*17q11*—70 kDa, RGD close to N terminus, homology to somatomedin B, rich in cysteines, sulfated, phosphorylated	Cell attachment protein, binds to collagen, plasminogen and plasminogen activator inhibitor, and to heparin	
Fibrillin 1 and 2-*15q21.1, 5q23-q31*—350 kDa, EGF-like domains, RGD, cysteine motifs	May regulate elastic fiber formation	Human fibrillin 1 mutations (OMIM 134797): Marfan's syndrome, human fibrillin 2 mutations (OMIM 121050)—congenital contractural arachnodactyly

The most abundant NCP produced by bone cells is osteonectin, a phosphorylated glycoprotein accounting for ~2% of the total protein of developing bone in most animal species. Osteonectin is transiently produced in nonbone tissues that are rapidly proliferating, remodeling, or undergoing profound changes in tissue architecture and is also found constitutively in certain types of epithelial cells, cells associated with the skeleton, and in platelets. Its function(s) in bone may be multiple, with potential association with osteoblast growth, proliferation, or both, as well as with matrix mineralization.[2,12] A mouse deficient in osteonectin exhibits a defect in bone formation.[13] Tetranectin, tenascin, and secreted phosphoprotein 24 also have been found in bone matrix, but their precise functions are not yet known.

Small Integrin-Binding Ligand, N-Glycosylated Protein, and Other Glycoproteins With Cell Attachment Activity

All connective tissue cells interact with their extracellular environment in response to stimuli that direct or coordinate (or both) specific cell functions, such as migration, proliferation, and differentiation (Tables 5 and 6). These particular interactions involve cell attachment through transient or stable focal adhesions to extracellular macromolecules, which are mediated by cell surface receptors that subsequently transduce intracellular signals. Bone cells synthesize at least 12 proteins that may mediate cell attachment: members of the small integrin-binding ligand, N-glycosylated protein (SIBLING) family (osteopontin, bone sialoprotein, dentin matrix protein-1, dentin sialophosphoprotein, and matrix extracellular phosphoprotein), type I collagen, fibronectin, thrombospondin(s) (predominantly TSP-2 with lower levels of TSP1, 3, and 4 and COMP), vitronectin, fibrillin, BAG-75, and osteoadherin (which is also a proteoglycan). Many of these proteins are phosphorylated and/or sulfated, and all contain RGD (Arg-Gly-Asn), the cell attachment consensus sequence that binds to the integrin class of cell surface molecules. However, in some cases, cell attachment seems to be independent of RGD, indicating the presence of other sequences or mechanisms of cell attachment. Thrombospondin(s), fibronectin, vitronectin, fibrillin, and osteopontin are expressed in many tissues. Whereas certain types of epithelial cells synthesize bone sialoprotein, it is highly enriched in bone and is expressed by hypertrophic chondrocytes, osteoblasts, osteocytes, and osteo-

TABLE 7. GENE AND PROTEIN CHARACTERISTICS OF GAMMA-CARBOXY GLUTAMIC ACID–CONTAINING PROTEINS IN BONE MATRIX

Protein/gene	*Function*	*Disease/animal model/phenotype*
Matrix Gla protein-*12p13.1*—15 kDa, five gla residues, one disulfide bridge, phosphoserine residues	May function in cartilage metabolism, a negative regulator of mineralization	Human mutations—Keutel syndrome (OMIM 245150)—excessive cartilage calcification; knockout mouse—excessive cartilage calcification
Osteocalcin-*1q25–31*—5 kDa, one disulfide bridge, gla residues located in α helical region	May regulate activity of osteoclasts and their precursors, may mark the turning point between bone formation and resorption, suggested to be a hormone	Knockout mouse—osteopetrotic, thickened bones, decreased crystal size, increased mineral content
Protein S-*3p11-q11.2*—72 kDa	Primarily a liver product, but may be made by osteogenic cells	Human mutations—(OMIM 076080) protein deficiency with osteopenia

© 2008 American Society for Bone and Mineral Research

TABLE 8. EFFECTS OF BONE MATRIX MOLECULES ON MINERALIZATION IN VITRO

Promote or support apatite formation	*Inhibit mineralization*	*Dual function (nucleate and inhibit)*	*No published effect*
Type I collagen	Aggrecan	Biglycan	Decorin
Proteolipid (matrix vesicle nucleational core	α2-HS glycoprotein	Osteonectin	Lumican
BAG-75	Matrix gla protein (MGP)	Fibronectin	Mimecan
Alkaline phosphatase	Osteocalcin	Bone sialoprotein Osteopontin	Tetranectin
		MEPE	Osteoadherin
			Thrombospondin

clasts. In bone, the expression of bone sialoprotein correlates with the appearance of mineral.[14] In solution, it can function as an hydroxyapatite nucleator,[2] is found in association with bone acidic glycoprotein-75 in mineralization foci,[15] and it is upregulated during mineralization in culture.[16] Both osteopontin and bone sialoprotein are known to anchor osteoclasts to bone, and in addition to supporting cell attachment, bind Ca^{2+} with extremely high affinity through polyacidic amino acid sequences. Each SIBLING protein regulates hydroxyapatite formation in solution, and their knockouts have phenotypes that can be correlated with these in vitro functions.[2] It is not immediately clear why there are such a plethora of RGD-containing proteins in bone; however, the pattern of expression varies from one RGD protein to another, as does the pattern of the different integrins that bind to these proteins. This variability indicates that cell–matrix interactions change as a function of maturational stage, suggesting that they also may play a role in osteoblastic maturation.[2] Their post-translational modifications also vary, suggesting that these modifications may also determine their in situ functions.[17]

Gla-Containing Proteins

Three bone-matrix NCPs, matrix gla protein (MGP), osteocalcin (bone gla-protein, BGP), both of which are made endogenously, and protein S (made primarily in the liver but also made by osteogenic cells) are post-translationally modified by the action of vitamin K-dependent γ-carboxylases (Table 7).[2] The di-carboxylic glutamyl (gla) residues enhance calcium binding. MGP is found in many connective tissues, whereas osteocalcin is more bone specific. The physiological roles of both these proteins are unclear; both may function in the control of mineral deposition and remodeling, but osteocalcin has recently been reported to be a hormone involved in regulation of glucose metabolism.[18] MGP-deficient mice develop calcification in extraskeletal sites such as the aorta,[19] implying it is an inhibitor of mineralization. Expression of MGP in blood vessels of the MGP-deficient mice prevents calcification, whereas expression in osteoblasts prevents mineralization.[20] Osteocalcin-deficient mice are reported to have increased BMD compared with normal,[21] but with age, the mineral properties did not show the changes that occurred in age-matched controls, suggesting a role for osteoclast in osteoclast recruitment.[22] In human bone, osteocalcin is concentrated in osteocytes, and its release may be a signal in the bone turnover cascade. Osteocalcin measurements in serum have proved valuable as a marker of bone turnover in metabolic disease states.[7]

OTHER COMPONENTS

The sections above summarize the major components of bone ECM, but there are other minor components that affect the properties of the tissue. For example, there are numerous enzymes that are important for processing the extracellular matrix components. Some of these are cell associated, some are found in the ECM. Readers are referred to other reviews[3,17,23,24] for more details. Growth factors sequestered on NCPs regulate cell–matrix interaction and cell function.[1]

Water accounts for ~10% of the weight of bone, depending on species and bone age. Water is important for cell and matrix nutrition, for control of ion flux, and for maintenance of the collagen structure, because type I collagen contains the bulk of the tissue water. As the bone matrix increases in mineral content, the water content decreases by ~30%,[25] and hence, water content can be used as a surrogate for mechanical strength.

Lipids make up <2% of the dry weight of bone; however, they have some significant effects on bone properties.[26] This is illustrated by a few recent examples including the neutral sphingomyelinase-deficient mouse that has a dwarfed phenotype,[27] the *fro/fro* mice that mimic severe osteogenesis imperfecta and have a chemically induced mutation in sphingomyelinase,[28] the caveolin-deficient mice that have altered mechanical properties,[29] and the report that phospholipase D is involved in the initial formation of bone during embryogenesis.[30]

Each of the components in the organic matrix of bone influences the mechanism of mineral deposition. Some promote mineralization, some inhibit the formation and/or growth of mineral crystals, and some are multifunctional, promoting in some cases and inhibiting in others. The known effects on hydroxyapatite formation in solution for each of the components discussed in this chapter are summarized in Table 8.

REFERENCES

1. Lian JB, Stein GS 2006 The cells of bone. In: Seibel MJ, Robins SJ, Bilezikian JP (eds.) Dynamics of Bone and Cartilage Metabolism. Academic Press, San Diego, CA, USA, pp. 221–258.
2. Zhu W, Robey PG, Boskey AL 2007 The regulatory role of matrix proteins in mineralization of bone. In: Marcus R, Feldman D, Nelson D, Rosen C (eds.) Osteoporosis, 3rd ed., vol. 1. Academic Press, San Diego, CA, USA, pp. 191–240.
3. Grynpas MD, Omelon S 2007 Transient precursor strategy or very small biological apatite crystals? Bone **41:**162–164.
4. Boskey AL 2006 Organic and inorganic matrices. In: Wnek G, Bowlin GL (eds.), Encyclopedia of Biomaterials and Biomedical Engineering. Dekker Encyclopedias, Taylor & Francis Books, London, UK, pp. 1–15
5. Boskey A 2007 Osteoporosis and osteopetrosis. In: Baeuerlein E, Behrens P, Epple M (eds.) Biomineralization in Medicine, vol 3. Wiley, New York, NY, USA, pp. 59–75.
6. Rossert J, de Crombrugghe B 2002 Type I collagen: Structure, synthesis and regulation. In: Bilezikian JP, Raisz LA, Rodan GA (eds.) Principles of Bone Biology, 2nd ed., vol. 1. Academic Press, San Diego, CA, USA, pp. 189–210.
7. Pagani F, Francucci CM, Moro L 2005 Markers of bone turnover: Biochemical and clinical perspectives. J Endocrinol Invest **28:**8–13.
8. Schafer C, Heiss A, Schwarz A, Westenfeld R, Ketteler M, Floege J, Muller-Esterl W, Schinke T, Jahnen-Dechent W 2003 The se-

© 2008 American Society for Bone and Mineral Research

rum protein alpha 2-Heremans-Schmid glycoprotein/fetuin-A is a systemically acting inhibitor of ectopic calcification. J Clin Invest **112**:357–366.
9. Robey PG 2002 Bone proteoglycans and glycoproteins. In: Bilezikian JP, Raisz LA, Rodan GA (eds.) Principles of Bone Biology. Academic Press, San Diego, CA, USA, pp. 225–238.
10. Young MF, Bi Y, Ameye L, Xu T, Wadhwa S, Heegaard A, Kilts T, Chen XD 2006 Small leucine-rich proteoglycans in the aging skeleton. J Musculoskelet Neuronal Interact. **6**:364–365.
11. Anderson HC, Sipe JB, Hessle L, Dhanyamraju R, Atti E, Camacho NP, Millan JL 2004 Impaired calcification around matrix vesicles of growth plate and bone in alkaline phosphatase-deficient mice. Am J Pathol **164**:841–847.
12. Brekken RA, Sage EH 2001 SPARC, a matricellular protein: At the crossroads of cell-matrix communication. Matrix Biol **19**:816–827.
13. Delany AM, Amling M, Priemel M, Howe C, Baron R, Canalis E 2000 Osteopenia and decreased bone formation in osteonectin-deficient mice. J Clin Invest **105**:915–923.
14. Paz J, Wade K, Kiyoshima T, Sodek J, Tang J, Tu Q, Yamauchi M, Chen J 2005 Tissue- and bone cell-specific expression of bone sialoprotein is directed by a 9.0 kb promoter in transgenic mice. Matrix Biol **24**:341–352.
15. Huffman NT, Keightley JA, Chaoying C, Midura RJ, Lovitch D, Veno PA, Dallas SL, Gorski JP 2007 Association of specific proteolytic processing of bone sialoprotein and bone acidic glycoprotein-75 with mineralization within biomineralization foci. J Biol Chem **282**:26002–26013.
16. Gordon JA, Tye CE, Sampaio AV, Underhill TM, Hunter GK, Goldberg HA 2007 Bone sialoprotein expression enhances osteoblast differentiation and matrix mineralization in vitro. Bone **41**:462–473.
17. Qin C, Baba O, Butler WT 2004 Post-translational modifications of sibling proteins and their roles in osteogenesis and dentinogenesis. Crit Rev Oral Biol Med **15**:126–136.
18. Lee NK, Sowa H, Hinoi E, Ferron M, Ahn JD, Confavreux C, Dacquin R, Mee PJ, McKee MD, Jung DY, Zhang Z, Kim JK, Mauvais-Jarvis F, Ducy P, Karsenty G 2007 Endocrine regulation of energy metabolism by the skeleton. Cell **130**:456–469.
19. Luo G, Ducy P, McKee MD, Pinero GJ, Loyer E, Behringer RR, Karsenty G 1997 Spontaneous calcification of arteries and cartilage in mice lacking matrix GLA protein. Nature **386**:78–81.
20. Murshed M, Schinke T, McKee MD, Karsenty G 2004 Extracellular matrix mineralization is regulated locally; different roles of two gla-containing proteins. J Cell Biol **165**:625–630.
21. Ducy P, Desbois C, Boyce B, Pinero G, Story B, Dunstan C, Smith E, Bonadio J, Goldstein S, Gundberg C, Bradley A, Karsenty G 1996 Increased bone formation in osteocalcin-deficient mice. Nature **382**:448–452.
22. Boskey AL, Gadaleta S, Gundberg C, Doty SB, Ducy P, Karsenty G 1998 Fourier transform infrared microspectroscopic analysis of bones of osteocalcin-deficient mice provides insight into the function of osteocalcin. Bone **23**:187–196.
23. Trackman PC 2005 Diverse biological functions of extracellular collagen processing enzymes. J Cell Biochem **96**:927–937.
24. Ge G, Greenspan DS 2006 Developmental roles of the BMP1/TLD metalloproteinases. Birth Defects Res C Embryo Today **78**:47–68.
25. Lees S 2003 Mineralization of type I collagen. Biophys J **85**:204–207.
26. Goldberg M, Boskey AL 1996 Lipids and biomineralizations. Prog Histochem Cytochem **31**:1–187.
27. Stoffel W, Jenke B, Block B, Zumbansen M, Koebke J 2005 Neutral sphingomyelinase 2 (smpd3) in the control of postnatal growth and development. Proc Natl Acad Sci USA **102**:4554–4559.
28. Aubin I, Adams CP, Opsahl S, Septier D, Bishop CE, Auge N, Salvayre R, Negre-Salvayre A, Goldberg M, Guenet JL, Poirier C 2005 A deletion in the gene encoding sphingomyelin phosphodiesterase 3 (Smpd3) results in osteogenesis and dentinogenesis imperfecta in the mouse. Nat Genet **37**:803–805.
29. Rubin J, Schwartz Z, Boyan BD, Fan X, Case N, Sen B, Drab M, Smith D, Aleman M, Wong KL, Yao H, Jo H, Gross TS 2007 Caveolin-1 knockout mice have increased bone size and stiffness. J Bone Miner Res **22**:1408–1418.
30. Gregory P, Kraemer E, Zurcher G, Gentinetta R, Rohrbach V, Brodbeck U, Andres AC, Ziemiecki A, Butikofer P 2005 GPI-specific phospholipase D (GPI-PLD) is expressed during mouse development and is localized to the extracellular matrix of the developing mouse skeleton. Bone **37**:139–147.

Chapter 7. Assessment of Bone Mass and Microarchitecture in Rodents

Blaine A. Christiansen and Mary L. Bouxsein

Orthopaedic Biomechanics Laboratory, Beth Israel Deaconess Medical Center and Harvard Medical School, Boston, Massachusetts

INTRODUCTION

Rodent models are important research tools for studying the musculoskeletal system. Imaging techniques can be used to characterize the skeletal effects of aging or disease, as well as changes in skeletal morphology caused by dietary, genetic, pharmacologic, or mechanical interventions.

Several imaging modalities are available for the assessment of skeletal morphology in animal models (Table 1). Some of these techniques, such as radiographs and peripheral DXA (pDXA) provide relatively inexpensive and fast assessments of bone mass and gross morphology in vivo but have poor resolution and are limited to planar images. Until recently, quantitative histological techniques were the standard for assessing trabecular and cortical bone architecture. Although histological analyses provide unique information on cellularity and dynamic indices of bone remodeling, they have limitations with respect to assessment of bone microarchitecture because structural parameters are derived from stereologic analysis of a few 2D sections, assuming that the underlying structure is plate-like.[1] In comparison, high-resolution 3D imaging techniques, such as μCT, directly measure bone microarchitecture without relying on stereologic models.

In this chapter, we review the imaging techniques commonly used to assess bone mass and microarchitecture in rodents, paying particular attention to their advantages and disadvantages, as well as technical challenges associated with each technique. Although in vivo molecular, or functional, imaging provides a powerful tool for studies in skeletal biology, it is beyond the scope of this chapter and interested readers are directed to other reviews for further information.[2]

RADIOGRAPHS

Although often overlooked, whole body radiographs are an important tool for evaluating gross skeletal morphology in vivo

The authors state that they have no conflicts of interest.

© 2008 American Society for Bone and Mineral Research

TABLE 1. SUMMARY OF SKELETAL IMAGING MODALITIES

Imaging modality	*Approximate resolution*	*2D or 3D*	*In vivo*	*Benefits and typical uses*
Planar radiography	10 lp/mm	2D	x	Rapid, inexpensive visualization of skeletal morphology
Peripheral DXA	180 μm	2D	x	Rapid, highly reproducible measures of bone mass and body composition
Peripheral QCT	70 μm	3D	x	Assessment of bone geometry, bone mass, and vBMD
MRI	40 μm	3D	x	Not typically used for imaging bone in rodents, effective for soft tissue
Ex vivo μCT	6–90 μm	3D		High-resolution imaging of bone microarchitecture and vBMD
In vivo μCT	10–150 μm	3D	x	High-resolution in vivo imaging of bone microarchitecture and vBMD
Synchrotron radiation CT	<1 nm	3D	x	Extremely high-resolution imaging of bone microarchitecture or ultrastructure, highly accurate vBMD measurement

and ex vivo (Fig. 1). Radiographs are produced by the summation of attenuation along a single scan direction. The advantage to planar radiographs is the rapid, relatively inexpensive visualization of skeletal morphology; however, they are limited to qualitative 2D evaluations.

PERIPHERAL DXA

pDXA is commonly used to measure BMC (g), areal BMD (g/cm^2), and body composition (% fat, lean tissue mass) of small animals both in vivo and ex vivo (Fig. 1). DXA imaging uses a compact X-ray source to expose an animal to two X-ray beams with different energy levels. The ratio of attenuation of the high- and low-energy beams allows the separation of bone from soft tissue, as well as lean tissue from fat. A typical voxel size for pDXA measurement in mice and rats is 0.18 × 0.18 mm. In vivo pDXA measurements have been widely used to show the skeletal changes after estrogen deficiency,[3] dietary and pharmacologic interventions,[4,5] and mouse strain–related differences in bone mass and body composition.[6,7]

Generally, pDXA provides highly reproducible measures of bone mass and body composition, with precision errors for whole body measurements <2%,[8,9] with relatively short scan times (<5 min). Precision of bone measurements at individual skeletal sites, such as the lumbar spine or distal femur, are worse (up to 6%) because of the challenges in consistently identifying the region of interest. Additionally, measurements of body composition suffer from poor accuracy, with general underestimation of lean tissue mass and overestimation fat

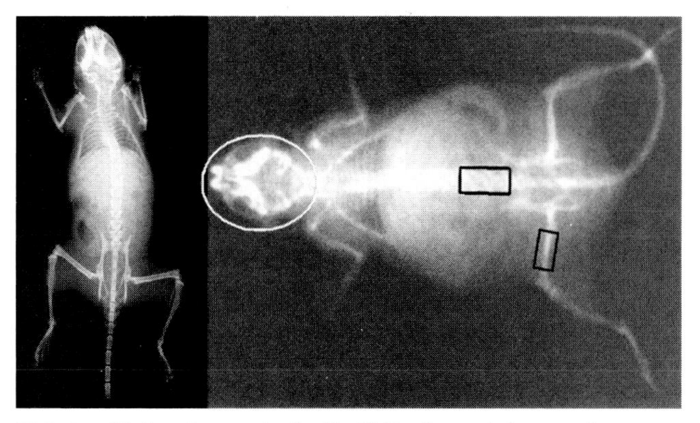

FIG. 1. 2D imaging methods. (Left) Radiograph image of a mouse skeleton showing the capabilities of radiographs to assess skeletal morphology. Image from Faxitron X-Ray, Wheeling, IL (http://www.faxitron.com/). (Right) Image of a mouse scanned with peripheral DXA. The head is typically excluded from measurements of body composition or BMC. The boxes represent areas of interest for analyzing BMC or BMD at the lumbar spine or the femoral diaphysis.

mass.[8,10,11] These errors can largely be eliminated with careful calibration of the DXA system to body fat content measured independently, such as by carcass analysis,[10,11] although few investigators take the time to do this. pDXA has relatively limited spatial resolution, assesses areal (i.e., 2D) BMD rather than volumetric BMD, and cannot distinguish cortical and trabecular bone compartments.

PERIPHERAL QCT

Peripheral QCT (pQCT) is used for 3D assessment of bone geometry, BMC, and volumetric BMD (vBMD) in small animal studies, both in vivo and ex vivo. Commercially available pQCT scanners achieve an in-plane voxel size of ~70 μm. The ability to measure bone mass and morphology in vivo makes pQCT useful for longitudinal assessment of the skeletal response to aging, disease, and/or interventions. The resolution of commercially available pQCT scanners does not allow for effective imaging of trabecular architecture, but they can be used to analyze compartment specific changes in vBMD.

Many studies have used pQCT in rats to monitor changes in cortical bone geometry along with cortical and trabecular vBMD after pharmacologic[12,13] or mechanical[14] interventions and to monitor fracture healing.[15] Importantly, pQCT measures of cross-sectional moment of inertia and cortical BMD provide an index of mechanical strength for rat long bones.[16,17] pQCT has also been used to quantify bone density and geometry in mice.[18,19] It has been suggested that this imaging method yields satisfactory in vivo precision and accuracy in skeletal characterization of mice[20]; however, the voxel size of pQCT (~70 μm) is relatively large compared with the cortical thickness of a mouse femur, typically 100–300 μm. This may introduce errors caused by partial volume averaging, in which a CT voxel samples a region containing two materials with different densities (bone and soft tissue). Because of partial volume averaging, mouse cortical bone could appear thicker and less dense than it actually is. This has been confirmed in mouse femora as well as similarly sized aluminum tubes, where an object thickness to voxel size ratio of 9:1 or less was associated with errors of at least 15% in density measurements.[21] This issue can be partially addressed with the careful selection of an analysis threshold that yields physiologically accurate results; however, because of these concerns, the accuracy of pQCT for assessment of some bone parameters in mice is questionable.

MAGNETIC RESONANCE IMAGING

A few studies have used high-resolution MRI to assess trabecular and cortical bone in small animals,[22–24] although it is more widely used for assessment of soft tissues. Bone mineral

© 2008 American Society for Bone and Mineral Research

lacks free protons and generates no MR signal, whereas soft tissues contain abundant free protons and give a strong signal. This contrast allows for segmentation and quantification of trabecular architecture, yielding results that are highly correlated with those obtained from 2D histology.[22] However, resolution concerns (typical voxel size is 39–137 μm) make this technology more suitable for use in humans or large animal models.[25]

MICROCOMPUTED TOMOGRAPHY

μCT has become the gold standard for the evaluation of bone morphology and microarchitecture of animal models ex vivo. μCT uses X-ray attenuation data taken from multiple viewing angles to reconstruct a 3D representation of the specimen characterizing the spatial distribution of material density (Fig. 2). Current μCT scanners achieve a nominal voxel size of ~6–10 μm on a side, which is sufficient for studying structures such as mouse trabeculae that have widths of ~30–50 μm.[26] There are several advantages to using μCT for assessment of bone mass and morphology in excised specimens: (1) it allows for direct 3D measurement of trabecular morphology such as trabecular thickness and separation, rather than inferring these values based on 2D stereological models,[27–29] as is done with standard histological evaluations; (2) compared with 2D histology, a much larger volume of interest is analyzed; and (3) measurements can be performed with a much faster throughput than typical histologic analyses of undecalcified bone specimens, which can take a week or more for embedding and sectioning. Moreover, assessment of bone morphology by μCT scanning is nondestructive, so samples can subsequently be used for other assays, such as histology or mechanical testing. Finally, μCT scans can provide an estimate of bone tissue mineralization on a voxel-by-voxel basis by comparing material attenuation to that of known standards, although this must be done with care given the constraints of the polychromatic X-ray source.[30] This voxel-based mineral density data, along with the high-resolution morphology of the bone determined from the μCT scan, can be used to create specimen-specific finite element (FE) models of bone.

However, in addition to these numerous advantages of μCT, there are still several issues associated with this imaging technique that must be considered. First, μCT analyses do not provide information about the cellular composition, bone remodeling rates, or mRNA and protein expression patterns in the tissue. Thus, 2D histological analyses are often used to complement structural analyses obtained by μCT. Furthermore, it is important to note that the results obtained from μCT are strongly dependent on a number of technical issues associated with the analysis, including (1) the scan resolution (voxel size), (2) the segmentation algorithm and threshold used to delineate soft tissue from bone; (3) the skeletal site(s) of interest, and (4) identification of the volumes of interest within those sites.

Accordingly, there are many technical considerations that must be addressed to ensure that analysis using μCT is as physiologically accurate as possible. Differences in voxel size have little effect on the evaluation of structures with high relative thickness (i.e., 100–200 μm), such as cortical bone or trabeculae in humans or large animal models. However, when analyzing smaller structures such as mouse or rat trabeculae with approximate dimensions of 30–60 μm, voxel size can have significant effects on the results (Fig. 3).[31] Scanning with low resolution (large voxel size) relative to the size of the structure of interest can cause an underestimation of BMD and overestimation of object thickness; therefore, the highest available scan resolution should be used for analysis of trabecular bone in small animal models.

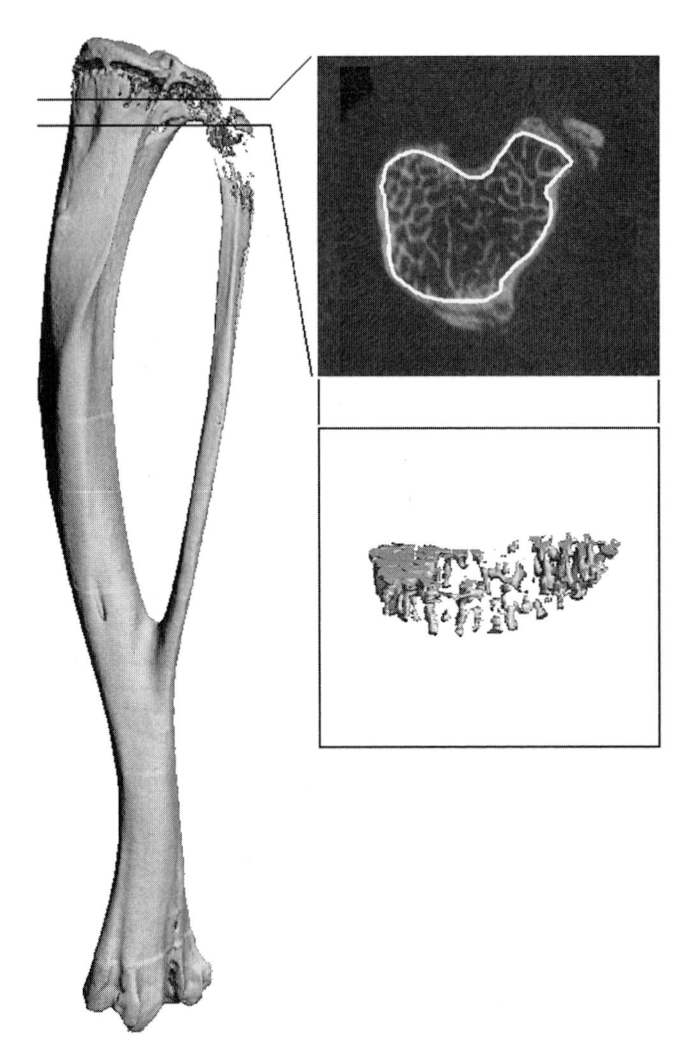

FIG. 2. 3D reconstruction of a mouse tibia scanned by μCT showing a region of interest for analysis of proximal tibial trabecular bone. A volume of trabecular bone is analyzed, excluding the surrounding cortical shell.

Similarly, the segmentation algorithm and threshold used to delineate bone from soft tissue can have significant effects on the results. The higher the threshold selected, the fewer voxels will be segmented as bone (Fig. 4). The goal of selecting a threshold should therefore be to obtain results that are physiologically accurate (i.e., similar to those that would be obtained using 2D histology). However, this "physiologically accurate" threshold value depends on many factors, including the scan resolution, the bone volume fraction (BV/TV), and the mineral density of the bone in the volume of interest. There is no consensus as to whether a global threshold (i.e., the same value for all specimens) versus specimen-specific threshold provides the most accurate results.[32] The threshold value for analysis should be carefully selected for each application and a standardized approach should be used to ensure comparability across groups.

Skeletal sites of interest must also be carefully selected depending on the specific research question. For trabecular bone analysis of mice or rats, the sites most often studied are the proximal tibia, distal femur, and vertebral body. It is also often desirable to analyze more than one skeletal site, because heterogeneity of skeletal sites has been reported.[33,34] For samples that will be mechanically tested, quantifying cortical

© 2008 American Society for Bone and Mineral Research

bone (of the diaphysis for long bones) is critical, because the cortical bone bears the majority of the load during mechanical testing. Similarly, selecting volumes of interest at a particular skeletal site is an important issue. In mice especially, trabecular bone in the metaphyseal region is largely confined to a few millimeters adjacent to the metaphyseal growth plate. Extending a volume of interest beyond this region would include more "empty space," thereby reducing the mean values for BV/TV at that skeletal site and possibly masking relevant differences between study groups (Fig. 5).

μCT has been used for a wide range of studies of bone mass and bone morphology, including the analysis of growth and development,[35] skeletal phenotypes in genetically altered mice, and animal models of disease states such as postmenopausal osteoporosis and renal osteodystrophy. Additionally, μCT has been used to assess the effects of pharmacologic interventions,[36] as well as mechanical loading[37] and unloading.[34] Furthermore, μCT has been used to image macrocracks in cortical bone (Fig. 6),[38] evaluate fracture healing,[39] and has also been used in combination with a perfused contrast agent to assess 3D vascular architecture.[40]

IN VIVO μCT

In vivo μCT is a relatively new imaging technique for studying bone morphology. This technique provides the high resolution of μCT while allowing for longitudinal studies of morphological changes (Fig. 7). In vivo μCT is an ideal strategy for tracking bone changes that occur on a time scale of weeks or months, such as bone loss associated with disuse or ovariectomy or increased bone mass caused by pharmacological or mechanical intervention. By registering 3D images against images from previous time points, it is possible to determine the precise locations of bone formation or resorption.[41,42] Altogether, the ability to perform longitudinal assessments of bone microstructure has the potential to reduce the number of animals needed in a given study and provide novel information about skeletal development, adaptation, repair, and response to disease or therapeutic interventions.

In vivo μCT has been used to follow the rapid trabecular bone loss in rats in the weeks immediately after ovariectomy[42,43] and was able to discern significant "irreversible" reductions in trabecular connectivity after only 2 wk.[43] In vivo μCT has also been used to study the effect of zoledronic acid treatment on bone in ovariectomized rats,[44] as well as changes in the epiphyseal bone of rats after either surgical destabilization of the knee[45] or intra-articular injection of monoiodoacetate (MIA), an inhibitor of glycolysis that promotes loss of articular cartilage.[46]

Despite the clear advantages of in vivo μCT, there are also several concerns about this method of imaging. First, there are concerns about the amount of ionizing radiation delivered during the in vivo μCT scan, particularly when animals are scanned multiple times throughout an experimental period. This radiation may introduce unwanted effects on the tissues or processes of interest, or on the animals in general. Young, growing animals and proliferative biologic processes, such as fracture healing or tumor growth, may be particularly susceptible to radiation exposure. The radiation exposure reported by Waarsing et al.,[41] 0.4 Gy for a single 20-min μCT scan (10 μm voxel size) of a rat hindlimb, is not predicted to have

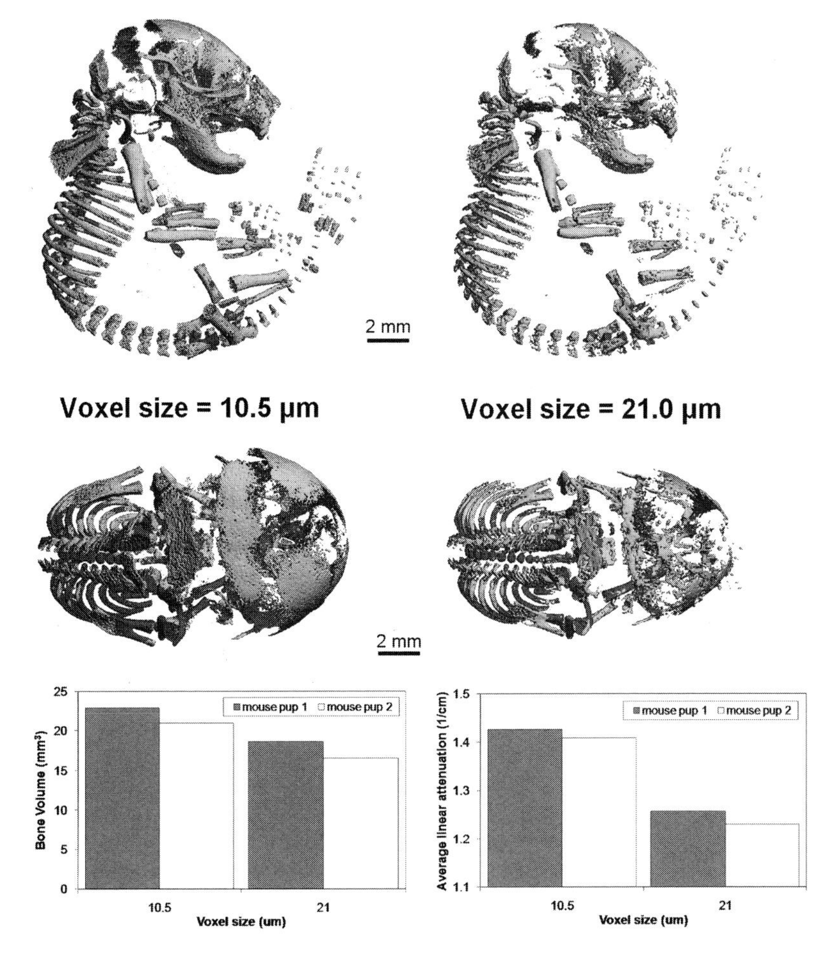

FIG. 3. μCT images of representative 2-day-old C57BL/6 mouse pups scanned using voxel sizes of either 10.5 or 21.0 μm. Bar plots show the sensitivity of measurements of bone volume and average mineral density to variations in voxel size. (Reprinted with permission of Wiley-Liss, a subsidiary of John Wiley & Sons, from Guldberg RE, Lin AS, Coleman R, Robertson G, Duvall C 2004 Microcomputed tomography imaging of skeletal development and growth. Birth Defects Res C Embryo Today **72**:250–255.[55])

© 2008 American Society for Bone and Mineral Research

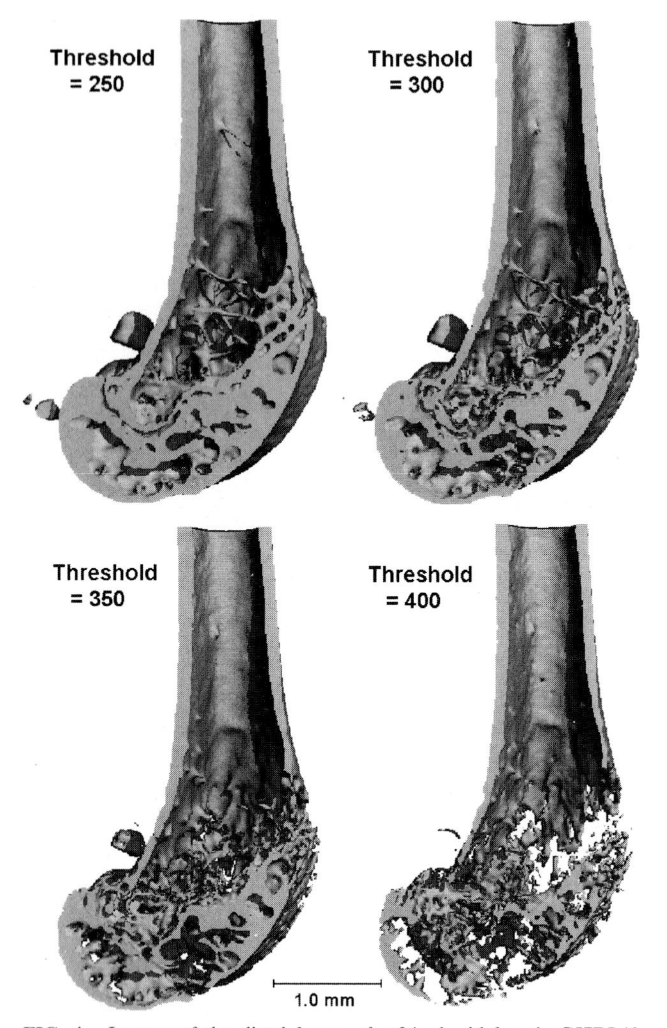

FIG. 4. Images of the distal femur of a 24-wk-old female C57BL/6 mouse showing the sensitivity of μCT analysis to selected threshold values.

significant deleterious effects on bone cells,[47] but the effect of multiple exposures needs additional study. Recently, Klinck et al.[48] performed weekly in vivo μCT scans on the proximal tibia of OVX and SHAM-OVX rats beginning at 12 wk of age. They found no observable effects of radiation in the animals; however, trabecular bone volume was decreased by 8–20% in the irradiated limbs compared with the contralateral nonirradiated limbs. These observations confirm that additional studies are needed to determine the potential effects of repeated in vivo μCT scans and provide strong rationale for the inclusion of measurements an internal nonirradiated control limb in the study design.

A second potential limitation relates to possible movement artifacts that may be introduced because of the animal's breathing. This is a negligible concern when scanning peripheral limbs, but this should be taken into account when scanning the axial skeleton. The final issue associated with in vivo imaging is concern over the ability to accurately and precisely monitor changes in individual bone structures over time. Although theoretically possible and highly intriguing, accomplishing this requires accurate reregistration of images acquired at different time points, an area of active research.[42]

SYNCHROTRON RADIATION CT

In synchrotron radiation (SR)-based μCT (or nano-CT), the polychromatic X-ray source used for standard desktop μCT imaging systems is replaced by a high photon flux monochromatic X-ray beam that is extracted from a synchrotron source. The advantages of this approach relative to standard μCT include (1) the use of an X-ray beam with just one energy, eliminating beam-hardening artifacts and allowing for accurate assessment of tissue mineral density[49]; (2) increased spatial resolution (~1 μm or below); and (3) very high signal to noise ratio.[26,50] The high spatial resolution associated with SR-μCT affords extremely precise assessment of trabecular bone architecture[26] and may be particularly useful for assessment of small-scale bone structures in young animals.[51,52] SR-μCT has also recently been used to study genetic variations in the mineral density and ultrastructural properties of murine cortical bone, including the vascular canals and osteocyte lacunae (Fig. 8).[53,54] Although studies with SR-μCT are generally

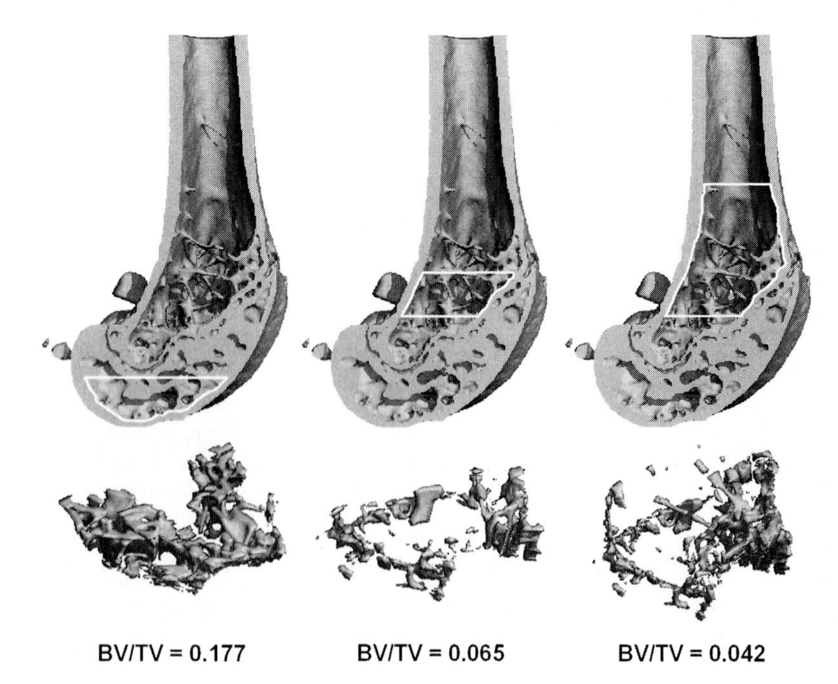

FIG. 5. Volumes of interest at the distal femur of a 24-wk-old female C57BL/6 mouse (threshold = 250). Evaluation of epiphyseal trabecular bone (left) will typically yield a much higher BV/TV than evaluation of metaphyseal trabecular bone. Likewise, a volume of interest limited to the metaphyseal trabecular bone immediately adjacent to the growth plate (center) will yield a higher BV/TV than a larger volume of interest that includes a large amount of "empty space" with few trabeculae (right).

© 2008 American Society for Bone and Mineral Research

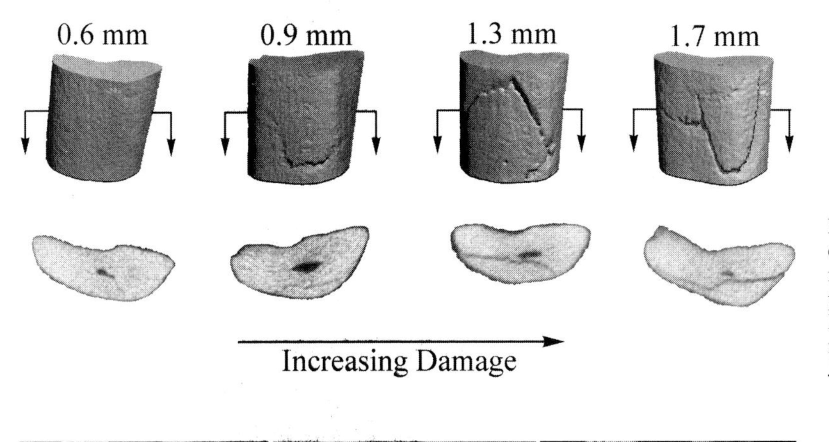

FIG. 6. μCT images showing damage sustained by cyclic compression of a rat ulna. Crack extent, length, and number increased with increasing levels of fatigue displacement. This figure was published in Uthgenannt BA, Silva MJ 2007 Use of the rat forelimb compression model to create discrete levels of bone damage in vivo. J Biomech **40:**317–324,[38] Copyright Elsevier.

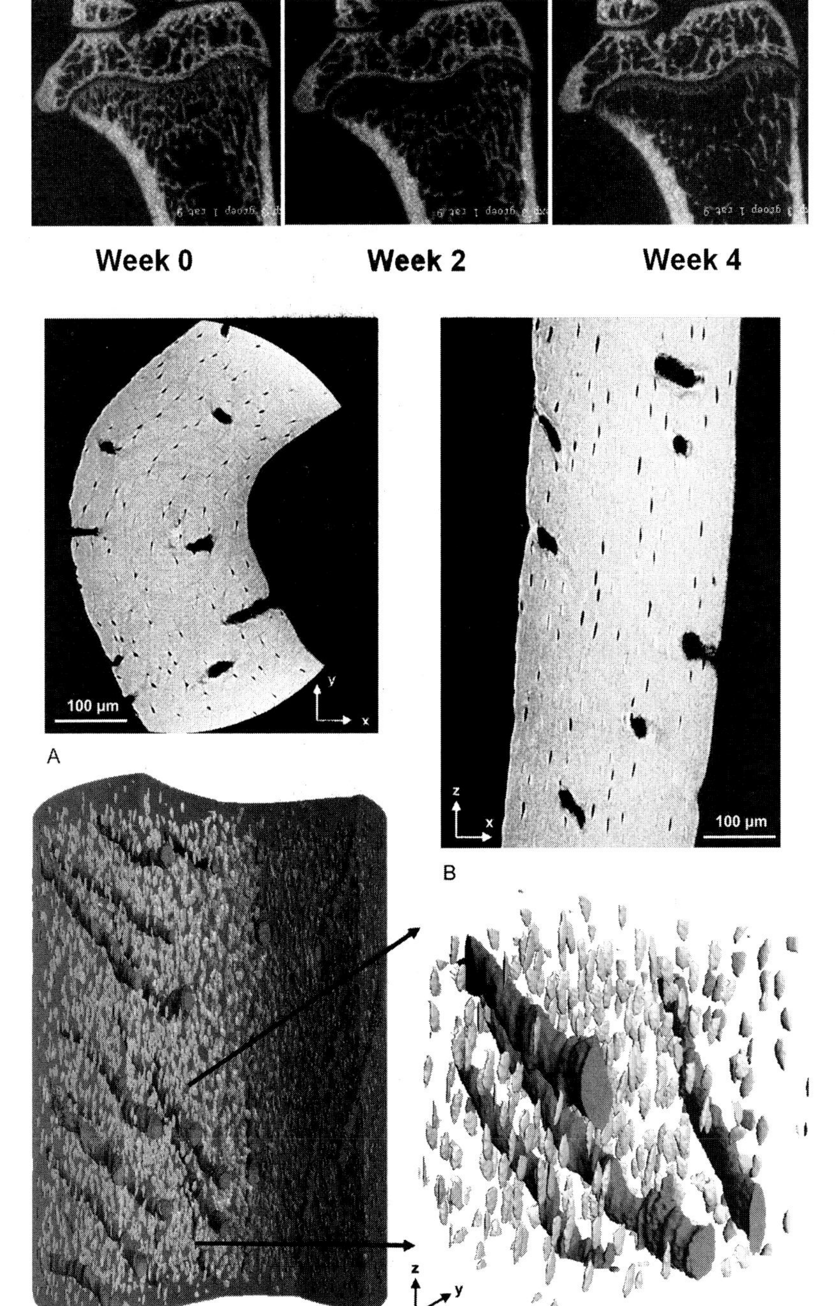

FIG. 7. Images showing typical trabecular bone loss at the proximal tibia of a 30-wk-old Wistar rat after OVX, assessed by in vivo μCT at weeks 0, 2, and 4. Images courtesy of J.E.M. Brouwers, Eindhoven University of Technology.

FIG. 8. Cortical bone from the femoral mid-diaphysis of mice measured with Synchrotron Radiation Nano-Computed Tomography (SR-nCT) at 700-nm nominal resolution. The top row shows the lateral cortical mid-diaphysis of a mouse femur in a transversal (A) and sagittal (B) view. The bottom row is a reconstruction of the canal network and osteocyte lacunae within the same lateral cortical bone. Figure originally published in Schneider et al. 2007.[54] Reproduced with permission.

© 2008 American Society for Bone and Mineral Research

performed on excised specimens, Kinney et al.[50] used SR-μCT in vivo in the rat proximal tibia to show the early deterioration in trabecular architecture after estrogen deficiency.

Altogether, SR-μCT offers extremely high-resolution imaging of microarchitecture and mineral density in excised bone specimens. The disadvantages of the technique are its limited availability (i.e., requires access to synchrotron source), relatively small volume of tissue examined, and technical expertise needed to acquire and analyze the measurements.

CONCLUSIONS

Assessment of skeletal mass and morphology in rodents by nondestructive imaging is an important component of current studies aimed at improving our understanding of musculoskeletal development, growth, adaptation, and disease. Currently, several different imaging modalities aimed at whole animal, organ, tissue, and cellular levels are available and will be increasingly used to study the hierarchy of bone structure.

ACKNOWLEDGMENTS

The authors acknowledge funding from NIH (AR053986 and AR049265). Dr. Christiansen is supported by a NIA-funded Training Grant (AG023480). The authors thank David Panus for assistance with μCT figures.

REFERENCES

1. Parfitt A, Drezner M, Glorieux F, Kanis J, Recker R 1987 Bone histomorphometry: Standardization of nomenclature, symbols and units. J Bone Miner Res **2:**595–610.
2. Mayer-Kuckuk P, Boskey AL 2006 Molecular imaging promotes progress in orthopedic research. Bone **39:**965–977.
3. Binkley N, Dahl DB, Engelke J, Kawahara-Baccus T, Krueger D, Colman RJ 2003 Bone loss detection in rats using a mouse densitometer. J Bone Miner Res **18:**370–375.
4. Brochmann EJ, Duarte ME, Zaidi HA, Murray SS 2003 Effects of dietary restriction on total body, femoral, and vertebral bone in SENCAR, C57BL/6, and DBA/2 mice. Metabolism **52:**1265–1273.
5. Iida-Klein A, Hughes C, Lu SS, Moreno A, Shen V, Dempster DW, Cosman F, Lindsay R 2006 Effects of cyclic versus daily hPTH(1-34) regimens on bone strength in association with BMD, biochemical markers, and bone structure in mice. J Bone Miner Res **21:**274–282.
6. Masinde GL, Li X, Gu W, Wergedal J, Mohan S, Baylink DJ 2002 Quantitative trait loci for bone density in mice: The genes determining total skeletal density and femur density show little overlap in F2 mice. Calcif Tissue Int **71:**421–428.
7. Reed DR, Bachmanov AA, Tordoff MG 2007 Forty mouse strain survey of body composition. Physiol Behav **91:**593–600.
8. Nagy TR, Clair AL 2000 Precision and accuracy of dual-energy X-ray absorptiometry for determining in vivo body composition of mice. Obes Res **8:**392–398.
9. Kolta S, De Vernejoul MC, Meneton P, Fechtenbaum J, Roux C 2003 Bone mineral measurements in mice: Comparison of two devices. J Clin Densitom **6:**251–258.
10. Brommage R 2003 Validation and calibration of DEXA body composition in mice. Am J Physiol Endocrinol Metab **285:**E454–E459.
11. Johnston SL, Peacock WL, Bell LM, Lonchampt M, Speakman JR 2005 PIXImus DXA with different software needs individual calibration to accurately predict fat mass. Obes Res **13:**1558–1565.
12. Gasser JA, Ingold P, Venturiere A, Shen V, Green JR 2008 Long-term protective effects of zoledronic acid on cancellous and cortical bone in the ovariectomized rat. J Bone Miner Res **23:**544–551.
13. Armamento-Villareal R, Sheikh S, Nawaz A, Napoli N, Mueller C, Halstead LR, Brodt MD, Silva MJ, Galbiati E, Caruso PL, Civelli M, Civitelli R 2005 A new selective estrogen receptor modulator, CHF 4227.01, preserves bone mass and microarchitecture in ovariectomized rats. J Bone Miner Res **20:**2178–2188.
14. Silva MJ, Touhey DC 2007 Bone formation after damaging in vivo fatigue loading results in recovery of whole-bone monotonic strength and increased fatigue life. J Orthop Res **25:**252–261.
15. McCann RM, Colleary G, Geddis C, Clarke SA, Jordan GR, Dickson GR, Marsh D 2008 Effect of osteoporosis on bone mineral density and fracture repair in a rat femoral fracture model. J Orthop Res **26:**384–393.
16. Ferretti JL, Capozza RF, Zanchetta JR 1996 Mechanical validation of a tomographic (pQCT) index for noninvasive estimation of rat femur bending strength. Bone **18:**97–102.
17. Ferretti JL, Cointry GR, Capozza RF, Capiglioni R, Chiappe MA 2001 Analysis of biomechanical effects on bone and on the muscle-bone interactions in small animal models. J Musculoskelet Neuronal Interact **1:**263–274.
18. Beamer WG, Donahue LR, Rosen CJ, Baylink DJ 1996 Genetic variability in adult bone density among inbred strains of mice. Bone **18:**397–403.
19. Breen SA, Loveday BE, Millest AJ, Waterton JC 1998 Stimulation and inhibition of bone formation: Use of peripheral quantitative computed tomography in the mouse in vivo. Lab Anim **32:**467–476.
20. Schmidt C, Priemel M, Kohler T, Weusten A, Muller R, Amling M, Eckstein F 2003 Precision and accuracy of peripheral quantitative computed tomography (pQCT) in the mouse skeleton compared with histology and microcomputed tomography (microCT). J Bone Miner Res **18:**1486–1496.
21. Brodt MD, Pelz GB, Taniguchi J, Silva MJ 2003 Accuracy of peripheral quantitative computed tomography (pQCT) for assessing area and density of mouse cortical bone. Calcif Tissue Int **73:**411–418.
22. Weber MH, Sharp JC, Latta P, Sramek M, Hassard HT, Orr FW 2005 Magnetic resonance imaging of trabecular and cortical bone in mice: Comparison of high resolution in vivo and ex vivo MR images with corresponding histology. Eur J Radiol **53:**96–102.
23. Gardner JR, Hess CP, Webb AG, Tsika RW, Dawson MJ, Gulani V 2001 Magnetic resonance microscopy of morphological alterations in mouse trabecular bone structure under conditions of simulated microgravity. Magn Reson Med **45:**1122–1125.
24. Kapadia RD, Stroup GB, Badger AM, Koller B, Levin JM, Coatney RW, Dodds RA, Liang X, Lark MW, Gowen M 1998 Applications of micro-CT and MR microscopy to study pre-clinical models of osteoporosis and osteoarthritis. Technol Health Care **6:**361–372.
25. Jiang Y, Zhao J, White DL, Genant HK 2000 Micro CT and micro MR imaging of 3D architecture of animal skeleton. J Musculoskel Neuron Interact **1:**45–51.
26. Martin-Badosa E, Amblard D, Nuzzo S, Elmoutaouakkil A, Vico L, Peyrin F 2003 Excised bone structures in mice: Imaging at three-dimensional synchrotron radiation micro CT. Radiology **229:**921–928.
27. Hildebrand T, Ruegsegger P 1997 A new method for the model-independent assessment of thickness in three-dimensional images. J Microsc **185:**67–75.
28. Laib A, Hildebrand T, Hauselmann HJ, Ruegsegger P 1997 Ridge number density: A new parameter for in vivo bone structure analysis. Bone **21:**541–546.
29. Hildebrand T, Ruegsegger P 1997 Quantification of Bone Microarchitecture with the Structure Model Index. Comput Methods Biomech Biomed Engin **1:**15–23.
30. Fajardo R, Cory E, Patel N, Nazarian A, Snyder B, Bouxsein M 2007 Specimen size and porosity can introduce error into μCT-based tissue mineral density measurements. Trans Orthop Res Soc **32:**166.
31. Muller R, Koller B, Hildebrand T, Laib A, Gianolini S, Ruegsegger P 1996 Resolution dependency of microstructural properties of cancellous bone based on three-dimensional mu-tomography. Technol Health Care **4:**113–119.
32. Rajagopalan S, Lu L, Yaszemski MJ, Robb RA 2005 Optimal segmentation of microcomputed tomographic images of porous tissue-engineering scaffolds. J Biomed Mater Res A **75:**877–887.
33. Glatt V, Canalis E, Stadmeyer L, Bouxsein ML 2007 Age-related changes in trabecular architecture differ in female and male C57BL/6J mice. J Bone Miner Res **22:**1197–1207.
34. Squire M, Donahue LR, Rubin C, Judex S 2004 Genetic variations that regulate bone morphology in the male mouse skeleton do not define its susceptibility to mechanical unloading. Bone **35:**1353–1360.
35. Hankenson KD, Hormuzdi SG, Meganck JA, Bornstein P 2005 Mice with a disruption of the thrombospondin 3 gene differ in

© 2008 American Society for Bone and Mineral Research

geometric and biomechanical properties of bone and have accelerated development of the femoral head. Mol Cell Biol **25:**5599–5606.
36. von Stechow D, Zurakowski D, Pettit AR, Muller R, Gronowicz G, Chorev M, Otu H, Libermann T, Alexander JM 2004 Differential transcriptional effects of PTH and estrogen during anabolic bone formation. J Cell Biochem **93:**476–490.
37. Christiansen BA, Silva MJ 2006 The effect of varying magnitudes of whole-body vibration on several skeletal sites in mice. Ann Biomed Eng **34:**1149–1156.
38. Uthgenannt BA, Silva MJ 2007 Use of the rat forelimb compression model to create discrete levels of bone damage in vivo. J Biomech **40:**317–324.
39. Gardner MJ, Ricciardi BF, Wright TM, Bostrom MP, van der Meulen MC 2008 Pause insertions during cyclic in vivo loading affect bone healing. Clin Orthop Relat Res **466:**1232–1238.
40. Bolland BJ, Kanczler JM, Dunlop DG, Oreffo RO 2008 Development of in vivo muCT evaluation of neovascularisation in tissue engineered bone constructs. Bone **43:**195–202.
41. Waarsing JH, Day JS, van der Linden JC, Ederveen AG, Spanjers C, De Clerck N, Sasov A, Verhaar JA, Weinans H 2004 Detecting and tracking local changes in the tibiae of individual rats: A novel method to analyse longitudinal in vivo micro-CT data. Bone **34:**163–169.
42. Boyd SK, Davison P, Muller R, Gasser JA 2006 Monitoring individual morphological changes over time in ovariectomized rats by in vivo micro-computed tomography. Bone **39:**854–862.
43. Campbell GM, Buie HR, Boyd SK 2008 Signs of irreversible architectural changes occur early in the development of experimental osteoporosis as assessed by in vivo micro-CT. Osteoporos Int (in press).
44. Brouwers JE, Lambers FM, Gasser JA, van Rietbergen B, Huiskes R 2008 Bone degeneration and recovery after early and late bisphosphonate treatment of ovariectomized wistar rats assessed by in vivo micro-computed tomography. Calcif Tissue Int **82:**202–211.
45. McErlain DD, Appleton CT, Litchfield RB, Pitelka V, Henry JL, Bernier SM, Beier F, Holdsworth DW 2008 Study of subchondral bone adaptations in a rodent surgical model of OA using in vivo micro-computed tomography. Osteoarthritis Cartilage **16:**458–469.
46. Morenko BJ, Bove SE, Chen L, Guzman RE, Juneau P, Bocan TM, Peter GK, Arora R, Kilgore KS 2004 In vivo micro computed tomography of subchondral bone in the rat after intra-articular administration of monosodium iodoacetate. Contemp Top Lab Anim Sci **43:**39–43.
47. Dare A, Hachisu R, Yamaguchi A, Yokose S, Yoshiki S, Okano T 1997 Effects of ionizing radiation on proliferation and differentiation of osteoblast-like cells. J Dent Res **76:**658–664.
48. Klinck RJ, Campbell GM, Boyd SK 2008 Radiation effects on bone architecture in mice and rats resulting from in vivo micro-computed tomography scanning. Med Eng Phys **30:**888–895.
49. Nuzzo S, Lafage-Proust MH, Martin-Badosa E, Boivin G, Thomas T, Alexandre C, Peyrin F 2002 Synchrotron radiation microtomography allows the analysis of three-dimensional microarchitecture and degree of mineralization of human iliac crest biopsy specimens: Effects of etidronate treatment. J Bone Miner Res **17:**1372–1382.
50. Kinney JH, Ryaby JT, Haupt DL, Lane NE 1998 Three-dimensional in vivo morphometry of trabecular bone in the OVX rat model of osteoporosis. Technol Health Care **6:**339–350.
51. Burghardt AJ, Wang Y, Elalieh H, Thibault X, Bikle D, Peyrin F, Majumdar S 2007 Evaluation of fetal bone structure and mineralization in IGF-I deficient mice using synchrotron radiation microtomography and Fourier transform infrared spectroscopy. Bone **40:**160–168.
52. Matsumoto T, Yoshino M, Asano T, Uesugi K, Todoh M, Tanaka M 2006 Monochromatic synchrotron radiation muCT reveals disuse-mediated canal network rarefaction in cortical bone of growing rat tibiae. J Appl Physiol **100:**274–280.
53. Raum K, Hofmann T, Leguerney I, Saied A, Peyrin F, Vico L, Laugier P 2007 Variations of microstructure, mineral density and tissue elasticity in B6/C3H mice. Bone **41:**1017–1024.
54. Schneider P, Stauber M, Voide R, Stampanoni M, Donahue LR, Muller R 2007 Ultrastructural properties in cortical bone vary greatly in two inbred strains of mice as assessed by synchrotron light based micro- and nano-CT. J Bone Miner Res **22:**1557–1570.
55. Guldberg RE, Lin AS, Coleman R, Robertson G, Duvall C 2004 Microcomputed tomography imaging of skeletal development and growth. Birth Defects Res C Embryo Today Rev **72:**250–259.

Chapter 8. Animal Models: Genetic Manipulation

Karen Lyons

Orthopaedic Hospital/UCLA Department of Orthopaedic Surgery, University of California, Los Angeles, California

INTRODUCTION

Genetically manipulated mice have contributed enormously to our identification of genes controlling skeletal development and to the elucidation of their mechanisms of action. Techniques are available to examine the effects of loss-of-function, gain-of-function, and altered structure of gene products. The ability to introduce defined mutations has facilitated the production of animal models of human diseases, cell lineage studies, examination of tissue-specific functions, and dissection of distinct gene functions at specific stages of differentiation within a single cell lineage.

OVEREXPRESSION OF TARGET GENES

The first widely used approach to study gene function in vivo was to produce transgenic mice that overexpress target genes. This requires the full-length coding sequence (cDNA) of a gene to be cloned downstream of a promoter. There are several promoters that have been well characterized and used to drive gene expression in skeletal tissues.

Chondrocytes

The most widely used cartilage-specific promoter is derived from the mouse pro αI(II) collagen gene (*Col2a1*). This promoter drives high levels of expression beginning at the condensation stage, and in the sclerotomal compartment. The *Col11a2* promoter has also been used to overexpress genes in chondrocytes,[1] although some of these promoters also drive expression in perichondrium and osteoblasts.[2] Overexpression in pre-chondrogenic limb mesenchyme has been achieved using the *Prx1* promoter.[3] Promoters that drive high levels of expression in hypertrophic chondrocytes have not been described. Chicken and mouse *Col10a1* promoters allow transgene expression in hypertrophic chondrocytes at low to moderate levels; however, expression is not seen in all hypertrophic

The author states that she has no conflicts of interest.

Key words: knockout, transgenic, embryonic stem cells, gene targeting

© 2008 American Society for Bone and Mineral Research

chondrocytes and is weak.[4] The recent development of a *Col10a1*-based BAC construct that permits robust expression of β-galactosidase (LacZ) in hypertrophic chondrocytes suggests that this vector may permit high levels of gene expression in these cells.[5]

Osteoblasts

Several promoters allow overexpression of genes in osteoblasts. The most frequently used is a 2.3-kb fragment from the rat or mouse *Col1a1* proximal promoter (*2.3Col1a1*). Strong and specific activity is seen in fetal and adult mature osteoblasts and osteocytes.[6] The 3.6-kb proximal *Col1a1* promoter drives strong gene expression at an earlier stage of differentiation (pre-osteoblasts), but it is also expressed in nonosseous tissues.[6] The 1.7-kb mouse osteocalcin (OC) promoter has been used to express genes in mature osteoblasts. However, this promoter is expressed in a low percentage of osteoblasts and at a relatively low level. Consistent with this, the 1.7-kb OC promoter is unable to drive Cre recombinase expression at effective levels.[7] On the other hand, 3.5- to 3.9-kb human OC promoter fragments drive osteoblast-specific expression in a large proportion of mature osteoblasts and osteocytes.[8] Osteocyte-specific promoters have not been widely used, although osteocyte-specific expression has been achieved using the DMP1 promoter.[9]

Tendon/ligament

Tendon patterning and differentiation are seldom studied genetically because of the lack of tissue-specific markers. *Scleraxis* (*Scx*) encodes a transcription factor expressed in all developing tendons and ligaments and their progenitors. The development of transgenic mouse lines harboring alkaline phosphatase (ALP) and green fluorescent protein (GFP) expressed under the control of the *Scx* locus will facilitate isolation of tendon cells and phenotypic analysis of these tissues.[10]

Osteoclasts

There are a variety of promoters that drive high levels of gene expression in osteoclasts and their progenitors. These include *CD11b*, expressed in monocytes, macrophages, and along the osteoclast differentiation pathway from monoculeated progenitor cells and into mature osteoclasts,[11] and *TRACP*, expressed in mature osteoclasts and their precursors.[12]

Advantages/Disadvantages of Overexpression Approaches

The major advantages of the transgenic approach are that it is straightforward and inexpensive, high levels of gene expression can be achieved, and transgenic mice often show an obvious phenotype. Furthermore, transgenic strains in which marker genes such as *LacZ*, *GFP*, and/or *ALP* are expressed allow easy visualization of specific cell types in vivo, and in some cases, permit isolation of specific cell types with a resolution not possible using other methods.[10] Other methods to detect specific lineages are based on site-specific recombination systems (discussed below).

A major caveat of the transgenic approach is that overexpression models often yield nonphysiological levels of protein expression, and this can confound interpretation of the role of the normal gene. Modifications of transgenic approaches can overcome some of the uncertainty caused by activation of pathways to nonphysiological levels. These include the use of transgenes encoding dominant negative variants or natural antagonists. These approaches lead to loss of function and thus target the pathways in their normal physiological context. Also, the site of transgene integration can have consequences on tissue specificity and levels of expression. This can be exploited to examine dose-dependent effects, but care must be taken to assess not only levels of expression but also sites of transgene expression. Moreover, even well-characterized promoters can be expressed at low levels in nontarget tissues, so rigorous analysis should include examination of expression in a range of tissues.

Overexpression of genes that have profound effects in skeletal tissues may confer embryonic lethality, precluding the establishment of stable transgenic lines. By the same token, the transgenic mouse lines that can be established overexpress genes only to an extent that is compatible with survival. Several bigenic systems have been devised to address this issue. One uses the tetracycline (tet) responsive transactivator (rtTA).[13] This bigenic system permits tissue-specific and tet-responsive gene expression. A transactivator (tTA) whose activity is modified by tet or the tet analog doxycline (dox) is expressed under the control of a tissue-specific or ubiquitous promoter. The second component is a strain expressing the gene of interest under the control of the operator sequences of the tet operon (tetO). Depending on whether the transactivator is induced (tTA; tet-on) or repressed (rtTA; tet-off) by dox, expression of target genes can be induced or repressed. A second bigenic strategy uses the Gal4/UAS system. This system uses one transgenic strain expressing the GAL4 transactivator under a tissue-specific or inducible promoter and a second transgenic strain expressing the gene of interest under the control of the UAS sequence, which requires GAL4 binding for activity.[14] These systems have been used to develop stable transgenic lines that permit activation of genes whose overexpression in cartilage and bone would confer lethality.[15,16]

GENE TARGETING

The most powerful and widely used technique for genetic manipulation is gene targeting in mouse embryonic stem (ES) cells (Fig. 1). Briefly, a targeting construct contains a portion of the gene of interest, along with a modification that renders the gene inactive. This is usually accomplished by inserting a drug resistance gene into an exon. The construct is introduced into ES cells, and the drug resistance cassette is used to select colonies that have incorporated the DNA. A fraction of the clones will have integrated the construct by homologous recombination. These are identified by Southern blot analysis.

The ability to eliminate defined genes is possible in the mouse because of the unique properties of ES cells. These cells can be genetically manipulated in culture, yet retain the ability to colonize the germline when injected into a mouse blastocyst. Once incorporated, they can give rise to germ cells, permitting the establishment of mouse strains carrying the modified gene. ES cells from the 129 strain were the first to be derived and are the most frequently used. However, 129 mice are poor breeders and exhibit abnormal immunological characteristics.[17] At present, the only widely available alternatives to 129-derived lines are C57Bl/6 lines. These are less robust, but obviate the need for repeated backcrosses. Newer C57Bl/6 ES lines under development are reported to exhibit efficient germline transmission.[18] It is important to bear in mind that C57BL/6, 129, and outbred strains have different BMD profiles[19] that must be taken into consideration when interpreting skeletal phenotypes.

The most time-consuming step in gene targeting is the introduction of modified ES cells into the germline. This is most commonly done by blastocyst injection, resulting in an F_0 ani-

© 2008 American Society for Bone and Mineral Research

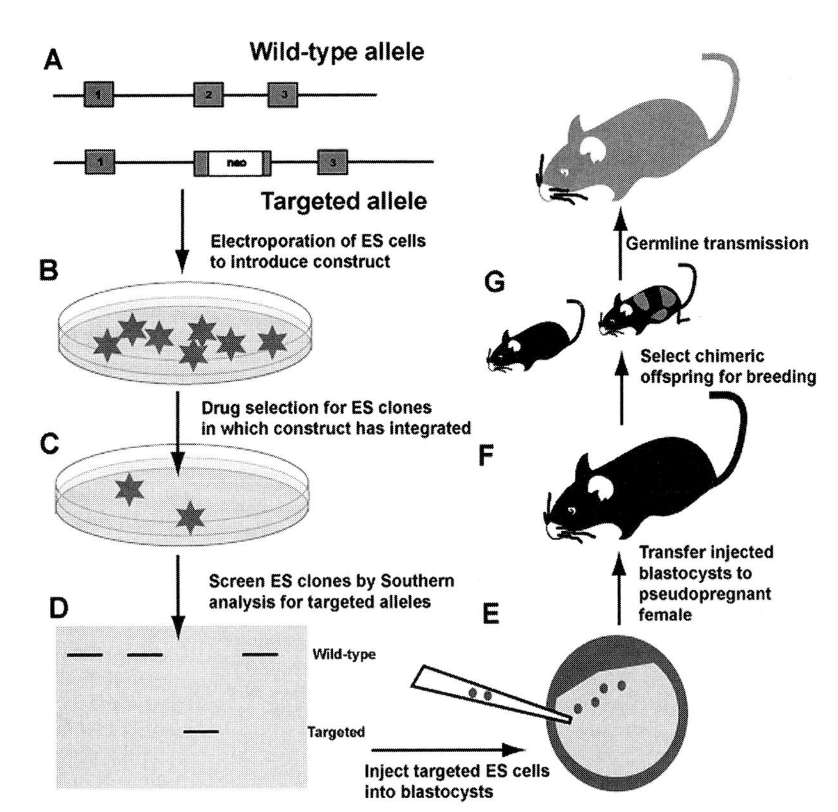

FIG. 1. Steps involved in generation of a knockout mouse model. (A) A targeting construct consisting of a neomycin-positive selection cassette (neo) inserted into or in place of an exon essential for the normal function of the target gene is generated. (B) The construct is introduced into ES cells (derived from 129 strain mice, agouti coat color) by electroporation. (C) Drug selection is used to select for colonies in which the construct has integrated stably. (D) Homologous recombinants are distinguished from colonies in which the targeted construct has integrated randomly by Southern blot analysis. (E) Correctly targeted ES cells are microinjected into a blastocyst derived from a host of a different genetic background (e.g., C57BL/6, black coat color) and transplanted into a pseudopregnant female. (F) Injected blastocysts are transplanted into a host pseudopregant female. (G) Resulting chimeric offspring (mixed coat color) are backcrossed to a mouse of the same genetic background as the host blastocyst (e.g., C57BL/6) to obtain agouti offspring that are derived from the 129 strain ES cells and therefore are heterozygous for the targeted allele.

mal that is partially derived from the modified ES cells. These chimeric mice are bred to obtain F_1 mice that are heterozygous for the defined mutation. A second technique is tetraploid complementation. Mice that are almost completely ES cell–derived are obtained in the F_0 generation, but these exhibit poor viability and growth abnormalities, precluding the use of this technique to study gene function in most aspects of skeletal biology. A recent report from researchers at Regeneron indicates that the recently developed VelociMouse method, using laser-assisted injection of ES cells into eight-cell embryos, yields F_0 mice that are germline competent and almost entirely ES cell derived.[20]

Advantages/Disadvantages of Gene Targeting

Global knockout mouse models are more reproducible than transgenic overexpression models. Phenotypes caused by loss of function provide direct insight into the physiological roles of the ablated gene product. Moreover, novel actions of target genes can emerge because, unlike transgenic models, global knockout models are not limited to a particular tissue or system.

A disadvantage of the global knockout approach is that deletion of genes essential for early development will result in early lethality. On the other hand, many knockout strains do not exhibit an obvious phenotype because of functional redundancy. In this case, creation of double or even triple knockouts may be necessary. Another consideration is that global knockouts usually contain a modified allele in which the selectable cassette used to screen the ES colonies is retained in the locus of interest. On occasion, this leads to effects on neighboring genes. These effects can be shown by comparing phenotypes of mice carrying null alleles in which the selection cassette is left in place with those in which it has been removed. This is discussed below in the context of tissue-specific knockouts.

TISSUE-SPECIFIC AND INDUCIBLE KNOCKOUTS TO STUDY SKELETAL BIOLOGY

The ability to achieve site-specific recombination has revolutionized genetic analysis of skeletal cell function. Tissue-specific recombination circumvents early lethality associated with global knockouts, and tools are available that allow researchers to ablate genes in skeletal cells at specific stages of commitment and differentiation.

Several methods can be used to achieve tissue-specific recombination; these rely on site-specific recombinases derived from bacteriophage (Cre) or yeast (Flp).[13] Cre and Flp recombine DNA at specific target sites, known as loxP and FRT, respectively. Depending on the orientation of the sites, the recombinase catalyzes excision or inversion of DNA flanked by the sites. Two mouse lines are required. For the Cre-loxP system, these are the "floxed" strain, in which the region of the gene targeted for deletion is flanked by loxP sites, and a second transgenic mouse line in which the Cre recombinase is expressed is under the control of an inducible and/or tissue-specific promoter. The floxed mouse strain is produced by homologous recombination, whereas the Cre transgenic mouse line is produced by standard transgenic approaches. In mice carrying both the floxed gene and the Cre transgene, Cre deletes the sequence flanked by loxP sites. The loxP sites are usually placed in introns and generally do not interfere with the normal function of the gene. As such, the floxed target gene functions normally and is deleted only in tissues where Cre is expressed.

Most studies have used constitutively active forms of Cre. However, ligand-regulated forms enable temporal control of gene deletion. The most popular strategy uses fusions of Cre to a ligand binding domain from a mutant estrogen receptor (ER).[21] The ER domain recognizes the synthetic estrogen antagonist 4-OH tamoxifen (T), but is insensitive to endog-

© 2008 American Society for Bone and Mineral Research

enous β-estradiol. In the absence of T, the Cre-ER(T) fusion protein is retained in the cytoplasm and is thus inactive. Binding of T to the ER domain induces a conformational change that permits the fusion protein to enter the nucleus and catalyze recombination.

A second system for achieving temporal control uses the tetracycline (tet) responsive transactivator (rtTA) (reviewed in Ref. 13 and discussed above). Using rtTA, expression of Cre has been detected within hours of dox administration in most tissues. This system has not been widely used to investigate gene function in skeletal tissues, and will not be reviewed further here.

Uncondensed Mesenchyme, Mesenchymal Condensations, and Neural Crest

Prx1-Cre drives expression in early uncondensed limb and head mesenchyme.[22] Dermo1-Cre expresses Cre in mesenchymal condensations.[23] Sox9 is expressed in mesenchymal condensations.[24] A Sox9-Cre knock-in (see below) strain drives Cre-mediated excision in precursors of osteoblasts and chondrocytes in these condensations.[25] The Wnt1-Cre transgene drives expression in migrating neural crest and can be used to ablate genes in all chondrocytes and osteoblasts derived from this source.[26]

Cartilage

The most widely used driver for gene ablation in cartilage is Col2-Cre.[27] Whereas the activity of this promoter seems restricted to chondrocytes in the majority of studies, there have been reports of expression in the perichondrium. A recent study indicated that there may be a brief window during chondrogenesis when Col2-Cre is expressed in the perichondrium.[28] Because Cre expression in perichondrium ablates gene expression in a fraction of osteoblasts in addition to chondrocytes, controls need to be performed to determine the extent to which Cre is expressed in the perichondrium in specific experiments. To date, no Cre lines exhibit sufficient robustness to permit gene deletion in hypertrophic chondrocytes.

The development of tools that permit inducible recombination in postnatal cartilage has been challenging. Several groups have generated inducible cartilage-specific Cre transgenic lines using Cre-ER(T) fusion proteins under the control of the Col2a1 promoter. These strains permit ablation in articular chondrocytes[28] if administered within 2 wk after birth. A chondrocyte-specific doxycycline inducible system has also been developed.[29] Although not all of these lines have been tested for efficiency of recombination in adults, in the one study examining this, the efficiency was very low.[28] Hence, the development of lines that allow ablation in adult articular cartilage remains an important goal.

Osteoblasts

The transcription factor *Osterix* (*Osx1*) is expressed in osteoblast precursors.[30] A Cre-GFP fusion protein has been inserted into the *Osx1* locus on a BAC transgene.[31] This construct allows targeting of osteoblast precursors. Several transgenic lines in which Cre is expressed under the control of the *Col1a1* promoter permit excision of genes in osteoblasts. A 3.6-kb *Col1a1* promoter drives high levels of Cre expression in osteoblasts but also targets tendon and fibrous cells types in the suture, skin, and several organs.[6] *Col1-Cre* lines (2.3 kb) show more restricted expression to mature osteoblasts.[6,7] Osteocalcin (OC)-Cre drives excision in mature osteoblasts but is not activated until just before birth.[8] Few inducible Cre transgenic strains have been developed for bone. An inducible 2.3 Col1-CreER(T) strain that permits tamoxifen-inducible Cre expression in osteoblasts has been developed.[32]

Osteoclasts

Several Cre strains permit ablation of myeloid cells. These include LysMcre mice, in which Cre has been introduced into the *M Lysozyme* locus,[33] and a strain in which the *CD11b* promoter drives Cre expression in macrophages and osteoclasts.[11] Strains permitting Cre-mediated recombination in mature osteoclasts include a Cre knock-in into the *Cathepsin K (Ctsk)* locus[34] and transgenic lines expressing Cre under the control of the *TRACPC* and *Ctsk* promoters.[35]

Advantages/Disadvantages of Inducible Knockouts

The most significant advantage of the Cre-loxP system is its flexibility. A significant feature of the Cre-loxP system is its potential for exploration of gene function in multiple tissues and at multiple time points. There are some caveats to bear in mind when using Cre-loxP systems. It can be difficult to find a promoter that drives Cre expression with sufficient activity to result in complete excision of the target gene. Cre transgenic strains based on identical promoters, but generated in different laboratories, can exhibit different tissue specificities and efficiencies of recombination. For this reason, every study should include control experiments to examine Cre expression in the context of the floxed line of interest. Floxed lines also vary with respect to the kinetics of Cre-mediated recombination. This must be born in mind when attempting to compare phenotypes caused by excision of different genes using a given Cre line.

The presence of the drug selection cassette can have a major effect on expression levels of the target gene in control mice (floxed mice in the absence of the Cre transgene). Prominent examples include a floxed *Fgf8* strain, in which retention of the neo cassette leads to a hypomorphic (reduced function) phenotype,[36] and the *scleraxis* global knockout, in which the presence of the neo cassette led to embryonic lethality by day 9.5 of gestation.[37] In striking contrast, mice homozygous for a *scleraxis* null allele in which the neo cassette was removed are viable as adults.[38] Another important consideration is that in some cases, Cre itself may confer a phenotype because of toxicity. This is especially true if the Cre is expressed as a fusion with GFP.[39]

Finally, with respect to inducible Cre models, the inducer may have a significant impact on the phenotype. Both doxycycline and tamoxifen can have profound effects on cartilage, bone, and osteoclasts independently of target gene deletion.[40,41] In the case of tamoxifen, even the low doses used to catalyze Cre-ER(T)–mediated excision have effects on bone.[42] Thus, a control group such as Cre-negative mice treated with the inducer must be included to examine the impact of this variable to the mutant phenotype under study.

KNOCK-IN MODELS

Homologous recombination in ES cells can also be used to generate knock-in models. In these, a locus of interest is modified such that it encodes an altered gene product. The major applications of this technology have been to generate mouse models of human genetic diseases, to dissect the functions of specific domains in a gene of interest, and to assess the effects of specific signaling pathways. A modification of knock-in technology is commonly used to perform lineage tracing studies.

Advantages/Disadvantages of Knock-in Approaches

The power of knock-in models is that they can be used to

© 2008 American Society for Bone and Mineral Research

determine the effects of subtle changes in protein structure or function. The generation of knock-in models requires care during the design of the targeting construct. Because the consequences of the knock-in are compared with wildtype or other genetically altered strains, it is essential to show that the only change in the targeted locus is the introduced knock-in mutation, and that other modifications, such as an intronic neo cassette, do not alter gene expression.

LINEAGE TRACING AND ACTIVITY REPORTERS

Genetically modified mice have permitted determination of cell lineage relationships and relative contributions of cells from various sources to a given organ with unprecedented resolution. These studies rely on strains that carry a floxed reporter gene, such as LacZ or GFP. For example, the R26R strain carries a floxed LacZ cassette introduced into the ROSA26 locus.[43] When these mice are bred to a Cre-expressing strain, all cells in which Cre is expressed, and all of their descendants, express LacZ. R26R mice are used to test the specificity and efficiency of Cre-expressing transgenic strains, although use of this strain in bone is limited by the fact that osteoblasts express endogenous LacZ. Major insights have been shown using R26R to study osteoprogenitors. For example, Wnt1-Cre;R26R mice unexpectedly showed that frontal bones are derived from the neural crest, but parietal bones are derived from the mesoderm.[44] Other reporters used in lineage tracing include Z/AP and Z/EG mice (which express a LacZ fusion protein in cells in which no Cre is present, and either ALP or GFP in Cre-expressing cells.[45]

Several signaling pathway activity reporter strains have been developed. These allow visual assessment of tissues in which these pathways are active. TCF/LEF reporter mice contain a *lacZ* gene downstream of consensus TCF/LEF binding sites.[46] These mice were used to show that canonical Wnt signaling is required for differentiation of mesenchymal cells into osteoblasts or chondrocytes during fracture repair.[47] A related TCF/LEF reporter strain, TOPGAL, has been used to track β-catenin activity in endochondral bone formation.[48] Tools are also available to monitor BMP pathway activity in vivo. Mice carrying a transgene in which lacZ expression is controlled by a promoter element activated by Smad 1 or 5 can be used to monitor BMP pathway activity in vivo.[49]

FUNCTIONAL GENOMICS

Functional knockout data are available for nearly 4000 of the estimated 25,000 mouse genes. However, most knockout mice have been generated in individual laboratories, and <900 are in public repositories. As a first approach, 251 knockout strains generated by commercial enterprises, Deltagen and Lexicon, were acquired by the NIH. These have been deposited as cryopreserved embryos (http://www.informatics.jax.org/external/ko/). Mutant Mouse Regional Resource Centers (http://www.mmrrc.org) permitted the acquisition and storage of another 320 strains developed in individual laboratories. In 2003, an international effort to mutagenize all protein-coding genes in the mouse was initiated as a cost-effective and systematic mechanism to annotate the mouse genome and facilitate strain distribution.

A second major effort is the use of gene trapping, a high-throughput mutagenesis strategy.

Several excellent reviews discuss gene trapping.[50] Briefly, one of the most widely used strategies involves the production of random insertional mutations in ES cells using vectors that contain a promoterless selection/reporter gene such as β-geo (the β-galactosidase and neomycin resistance fusion gene), flanked upstream by a splice acceptor site and downstream by a poly(A) tail. Insertion of this vector downstream of a promoter disrupts transcription by leading to the generation of a fusion transcript. Gene trap ES cell lines are then used to generate knockout mice.[50]

Advantages/Disadvantages of Gene Trap Approaches

The major advantages of gene trap approaches is that they enable large-scale functional genomics and make cell lines available to all interested scientists for a nominal fee. The International Gene Trap Consortium (http://www.genetrap.org) website provides centralized access to all publicly available annotated gene trap cell lines and permits searches by gene name, chromosomal location, sequence, or biological pathway. At last report, >49,000 cell lines were available, representing ~9000 mouse genes.[51] Researchers can order the ES cells and generate mice through blastocyst injection. It is important to bear in mind that it is the generation of the mice through blastocyst injection, and the subsequent phenotypic analysis, that is the most time-consuming and costly aspect of genetic research.

A potential disadvantage of the gene trap approach is that the fusion protein that is generated may retain partial function or reduced levels of wildtype transcript may still be generated. On the other hand, hypomorphic alleles arising from gene trap insertions can provide invaluable insights into normal gene function if they are used as a part of a genetic series when mice carrying a true null allele are available.

GENERAL CONSIDERATIONS

With all of the successes in genetic manipulation, the real bottleneck is phenotyping. Phenotype is dependent on genetic and environmental factors. As discussed, inbred strains vary considerably in their peak BMD.[19] Moreover, animal housing conditions, number of animals per cage, light cycles, and food intake can have a significant impact on metabolic parameters, hormone levels, and BMD.[52,53] Thus, even genetically identical mice can have different phenotypes when maintained in different facilities. It is important to assess phenotypes at different ages, because effects present at early stages may be compensated for later on. An example of this can be seen in matrix metalloproteinase (MMP)-9–deficient mice.[54] These exhibit a prenatal expansion of the hypertrophic zone because of aberrant angiogenesis, but are normal within a few weeks of birth because of compensatory remodeling. Similarly, some phenotypes manifest only in late stages or when a metabolic stress is applied such as ovariectomy.

Caution must be taken when extrapolating findings in mouse models to functions in humans, because important differences exist between the two species. This is particularly true for endocrine systems. When mouse models are used to study metabolic bone diseases, biomechanical loading on bones is clearly different in mice and humans. Moreover, linear growth in humans ceases after epiphyseal closure, whereas in mice, the growth plate does not fuse. Nonetheless, the similarities outweigh the differences by far, and genetic models are likely to play an increasingly prominent role in every aspect of research in skeletal biology.

REFERENCES

1. Horiki M, Imamura T, Okamoto M, Hayashi M, Murai J, Myoui A, Ochi T, Miyazono K, Yoshikawa H, Tsumaki N 2004 Smad6/Smurf1 overexpression in cartilage delays chondrocyte hypertrophy and causes dwarfism with osteopenia. J Cell Biol **165:**433–445.
2. Li SW, Arita M, Kopen GC, Phinney DG, Prockop DJ 1998 A 1,064 bp fragment from the promoter region of the Col11a2 gene

© 2008 American Society for Bone and Mineral Research

drives lacZ expression not only in cartilage but also in osteoblasts adjacent to regions undergoing both endochondral and intramembranous ossification in mouse embryos. Matrix Biol **17**:213–221.
3. Martin JF, Olson EN 2000 Identification of a prx1 limb enhancer. Genesis **26**:225–229.
4. Campbell MR, Gress CJ, Appleman EH, Jacenko O 2004 Chicken collagen X regulatory sequences restrict transgene expression to hypertrophic cartilage in mice. Am J Pathol **164**:487–499.
5. Gebhard S, Hattori T, Bauer E, Bosl MR, Schlund B, Poschl E, Adam N, de Crombrugghe B, von der Mark K 2007 BAC constructs in transgenic reporter mouse lines control efficient and specific LacZ expression in hypertrophic chondrocytes under the complete Col10a1 promoter. Histochem Cell Biol **127**:183–194.
6. Liu F, Woitge HW, Braut A, Kronenberg MS, Lichtler AC, Mina M, Kream BE 2004 Expression and activity of osteoblast-targeted Cre recombinase transgenes in murine skeletal tissues. Int J Dev Biol **48**:645–653.
7. Dacquin R, Starbuck M, Schinke T, Karsenty G 2002 Mouse alpha1(I)-collagen promoter is the best known promoter to drive efficient Cre recombinase expression in osteoblasts. Dev Dyn **224**:245–251.
8. Zhang M, Xuan S, Bouxsein ML, von Stechow D, Akeno N, Faugere MC, Malluche H, Zhao G, Rosen CJ, Efstratiadis A, Clemens TL 2002 Osteoblast-specific knockout of the insulin-like growth factor (IGF) receptor gene reveals an essential role of IGF signaling in bone matrix mineralization. J Biol Chem **277**:44005–44012.
9. Lu Y, Xie Y, Zhang S, Dusevich V, Bonewald LF, Feng JQ 2007 DMP1-targeted Cre expression in odontoblasts and osteocytes. J Dent Res **86**:320–325.
10. Pryce BA, Brent AE, Murchison ND, Tabin CJ, Schweitzer R 2007 Generation of transgenic tendon reporters, ScxGFP and ScxAP, using regulatory elements of the scleraxis gene. Dev Dyn **236**:1677–1682.
11. Ferron M, Vacher J 2005 Targeted expression of Cre recombinase in macrophages and osteoclasts in transgenic mice. Genesis **41**:138–145.
12. Reddy SV, Hundley JE, Windle JJ, Alcantara O, Linn R, Leach RJ, Boldt DH, Roodman GD 1995 Characterization of the mouse tartrate-resistant acid phosphatase (TRAP) gene promoter. J Bone Miner Res **10**:601–606.
13. Branda CS, Dymecki SM 2004 Talking about a revolution: The impact of site-specific recombinases on genetic analyses in mice. Dev Cell **6**:7–28.
14. Ornitz DM, Moreadith RW, Leder P 1991 Binary system for regulating transgene expression in mice: Targeting int-2 gene expression with yeast GAL4/UAS control elements. Proc Natl Acad Sci USA **88**:698–702.
15. Liu Z, Shi W, Ji X, Sun C, Jee WS, Wu Y, Mao Z, Nagy TR, Li Q, Cao X 2004 Molecules mimicking Smad1 interacting with Hox stimulate bone formation. J Biol Chem **279**:11313–11319.
16. Kobayashi T, Lyons KM, McMahon AP, Kronenberg HM 2005 BMP signaling stimulates cellular differentiation at multiple steps during cartilage development. Proc Natl Acad Sci USA **102**:18023–18027.
17. McVicar DW, Winkler-Pickett R, Taylor LS, Makrigiannis A, Bennett M, Anderson SK, Ortaldo JR 2002 Aberrant DAP12 signaling in the 129 strain of mice: Implications for the analysis of gene-targeted mice. J Immunol **169**:1721–1728.
18. Keskintepe L, Norris K, Pacholczyk G, Dederscheck SM, Eroglu A 2007 Derivation and comparison of C57BL/6 embryonic stem cells to a widely used 129 embryonic stem cell line. Transgenic Res **16**:751–758.
19. Rosen CJ, Beamer WG, Donahue LR 2001 Defining the genetics of osteoporosis: Using the mouse to understand man. Osteoporos Int **12**:803–810.
20. Poueymirou WT, Auerbach W, Frendewey D, Hickey JF, Escaravage JM, Esau L, Dore AT, Stevens S, Adams NC, Dominguez MG, Gale NW, Yancopoulos GD, DeChiara TM, Valenzuela DM 2007 F0 generation mice fully derived from gene-targeted embryonic stem cells allowing immediate phenotypic analyses. Nat Biotechnol **25**:91–99.
21. Feil R, Brocard J, Mascrez B, LeMeur M, Metzger D, Chambon P 1996 Ligand-activated site-specific recombination in mice. Proc Natl Acad Sci USA **93**:10887–10890.
22. Logan M, Martin JF, Nagy A, Lobe C, Olson EN, Tabin CJ 2002 Expression of Cre Recombinase in the developing mouse limb bud driven by a Prxl enhancer. Genesis **33**:77–80.
23. Yu K, Xu J, Liu Z, Sosic D, Shao J, Olson EN, Towler DA, Ornitz DM 2003 Conditional inactivation of FGF receptor 2 reveals an essential role for FGF signaling in the regulation of osteoblast function and bone growth. Development **130**:3063–3074.
24. Akiyama H, Chaboissier MC, Martin JF, Schedl A, de Crombrugghe B 2002 The transcription factor Sox9 has essential roles in successive steps of the chondrocyte differentiation pathway and is required for expression of Sox5 and Sox6. Genes Dev **16**:2813–2828.
25. Akiyama H, Kim JE, Nakashima K, Balmes G, Iwai N, Deng JM, Zhang Z, Martin JF, Behringer RR, Nakamura T, de Crombrugghe B 2005 Osteo-chondroprogenitor cells are derived from Sox9 expressing precursors. Proc Natl Acad Sci USA **102**:14665–14670.
26. Chai Y, Jiang X, Ito Y, Bringas P Jr, Han J, Rowitch DH, Soriano P, McMahon AP, Sucov H 2000 Fate of the mammalian cranial neural crest during tooth and mandibular morphogenesis. Development **127**:1671–1679.
27. Ovchinnikov DA, Deng JM, Ogunrinu G, Behringer RR 2000 Col2a1-directed expression of Cre recombinase in differentiating chondrocytes in transgenic mice. Genesis **26**:145–146.
28. Nakamura E, Nguyen MT, Mackem S 2006 Kinetics of tamoxifen-regulated Cre activity in mice using a cartilage-specific CreER(T) to assay temporal activity windows along the proximodistal limb skeleton. Dev Dyn **235**:2603–2612.
29. Grover J, Roughley PJ 2006 Generation of a transgenic mouse in which Cre recombinase is expressed under control of the type II collagen promoter and doxycycline administration. Matrix Biol **25**:158–165.
30. Nakashima K, Zhou X, Kunkel G, Zhang Z, Deng JM, Behringer RR, de Crombrugghe B 2002 The novel zinc finger-containing transcription factor osterix is required for osteoblast differentiation and bone formation. Cell **108**:17–29.
31. Rodda SJ, McMahon AP 2006 Distinct roles for Hedgehog and canonical Wnt signaling in specification, differentiation and maintenance of osteoblast progenitors. Development **133**:3231–3244.
32. Kim JE, Nakashima K, de Crombrugghe B 2004 Transgenic mice expressing a ligand-inducible cre recombinase in osteoblasts and odontoblasts: A new tool to examine physiology and disease of postnatal bone and tooth. Am J Pathol **165**:1875–1882.
33. Clausen BE, Burkhardt C, Reith W, Renkawitz R, Forster I 1999 Conditional gene targeting in macrophages and granulocytes using LysMcre mice. Transgenic Res **8**:265–277.
34. Nakamura T, Imai Y, Matsumoto T, Sato S, Takeuchi K, Igarashi K, Harada Y, Azuma Y, Krust A, Yamamoto Y, Nishina H, Takeda S, Takayanagi H, Metzger D, Kanno J, Takaoka K, Martin TJ, Chambon P, Kato S 2007 Estrogen prevents bone loss via estrogen receptor alpha and induction of Fas ligand in osteoclasts. Cell **130**:811–823.
35. Chiu WS, McManus JF, Notini AJ, Cassady AI, Zajac JD, Davey RA 2004 Transgenic mice that express Cre recombinase in osteoclasts. Genesis **39**:178–185.
36. Meyers EN, Lewandoski M, Martin GR 1998 An Fgf8 mutant allelic series generated by Cre- and Flp-mediated recombination. Nat Genet **18**:136–141.
37. Brown D, Wagner D, Li X, Richardson JA, Olson EN 1999 Dual role of the basic helix-loop-helix transcription factor scleraxis in mesoderm formation and chondrogenesis during mouse embryogenesis. Development **126**:4317–4329.
38. Murchison ND, Price BA, Conner DA, Keene DR, Olson EN, Tabin CJ, Schweitzer R 2007 Regulation of tendon differentiation by scleraxis distinguishes force-transmitting tendons from muscle-anchoring tendons. Development **134**:2697–2708.
39. Huang WY, Aramburu J, Douglas PS, Izumo S 2000 Transgenic expression of green fluorescence protein can cause dilated cardiomyopathy. Nat Med **6**:482–483.
40. Ryan ME, Greenwald RA, Golub LM 1996 Potential of tetracyclines to modify cartilage breakdown in osteoarthritis. Curr Opin Rheumatol **8**:238–247.
41. Bettany JT, Peet NM, Wolowacz RG, Skerry TM, Grabowski PS 2000 Tetracyclines induce apoptosis in osteoclasts. Bone **27**:75–80.
42. Starnes LM, Downey CM, Boyd SK, Jirik FR 2007 Increased bone mass in male and female mice following tamoxifen administration. Genesis **45**:229–235.
43. Soriano P 1999 Generalized lacZ expression with the ROSA26 Cre reporter strain. Nat Genet **21**:70–71.
44. Jiang X, Iseki S, Maxson RE, Sucov HM, Morriss-Kay GM 2002 Tissue origins and interactions in the mammalian skull vault. Dev Biol **241**:106–116.

© 2008 American Society for Bone and Mineral Research

45. Lobe CG, Koop KE, Kreppner W, Lomeli H, Gertsenstein M, Nagy A 1999 Z/AP, a double reporter for cre-mediated recombination. Dev Biol **208**:281–292.
46. Cheon SS, Cheah AY, Turley S, Nadesan P, Poon R, Clevers H, Alman BA 2002 beta-Catenin stabilization dysregulates mesenchymal cell proliferation, motility, and invasiveness and causes aggressive fibromatosis and hyperplastic cutaneous wounds. Proc Natl Acad Sci USA **99**:6973–6978.
47. Chen Y, Whetstone HC, Lin AC, Nadesan P, Wei Q, Poon R, Alman BA 2007 Beta-catenin signaling plays a disparate role in different phases of fracture repair: Implications for therapy to improve bone healing. PLoS Med **4**:e249.
48. Day TF, Guo X, Garrett-Beal L, Yang Y 2005 Wnt/beta-catenin signaling in mesenchymal progenitors controls osteoblast and chondrocyte differentiation during vertebrate skeletogenesis. Dev Cell **8**:739–750.
49. Monteiro RM, de Sousa Lopes SM, Korchynskyi O, ten Dijke P, Mummery CL 2004 Spatio-temporal activation of Smad1 and Smad5 in vivo: Monitoring transcriptional activity of Smad proteins. J Cell Sci **117**:4653–4663.
50. Stanford WL, Epp T, Reid T, Rossant J 2006 Gene trapping in embryonic stem cells. Methods Enzymol **420**:136–162.
51. Consortium IMK 2007 A mouse for all reasons. Cell **128**:9–13.
52. Nagy TR, Krzywanski D, Li J, Meleth S, Desmond R 2002 Effect of group vs. single housing on phenotypic variance in C57BL/6J mice. Obes Res **10**:412–415.
53. Champy MF, Selloum M, Piard L, Zeitler V, Caradec C, Chambon P, Auwerx J 2004 Mouse functional genomics requires standardization of mouse handling and housing conditions. Mamm Genome **15**:768–783.
54. Vu TH, Shipley JM, Bergers G, Berger JE, Helms JA, Hanahan D, Shapiro SD, Senior RM, Werb Z 1998 MMP-9/gelatinase B is a key regulator of growth plate angiogenesis and apoptosis of hypertrophic chondrocytes. Cell **93**:411–422.

Chapter 9. Animal Models: Allelic Determinants for BMD

Wesley G. Beamer[1] and Clifford J. Rosen[2]

[1]*The Jackson Laboratory, Bar Harbor, Maine;* [2]*Maine Medical Center Research Institute, Scarborough, Maine*

INTRODUCTION

Human studies of twins, three-generational families, and siblings have shown the strong role for heritable factors in BMD. In particular, estimates of genetic heritability of BMD by analyses of monozygotic versus dizygotic twin pairs have pointed that 60–80% of phenotypic variation is caused by genetics. Density of the spine, hip, and forearm are readily measurable in the clinical setting, and BMD is widely recognized as a risk factor for osteoporotic fracture. The importance of BMD to biomechanical strength of bone has driven major efforts to define those genetic factors. Such efforts have been supported by wide availability of advanced technology for accurate and precise measurement of BMD (DXA, QCT), sophisticated statistical analyses (numerous excellent software programs), molecular methods for mapping genes by genotyping of DNA for simple sequence length polymorphisms (SSLPs; variable lengths of nucleotide repeats) or single nucleotide polymorphisms (SNPs), and computer-based publically available genetic data sets (www.ensembl.org/index.html). It is abundantly clear from such work that genetic regulation of the skeleton is profoundly complex with cascades of genes underlying fundamental bone traits.

Numerous studies have been presented over the past decade associating X-ray absorptiometric BMD data with chromosomal regions across the human genome. These studies are driven by phenotype rather than by specific candidate gene(s) drawn from bone biology. Subject groups for genome-wide analyses have varied in numbers, familial relationship, sex, age, menopausal status, skeletal sites, and ethnicity. The common method of reporting relationships between phenotype and genomic location is the LOD score (i.e., the odds favoring linkage of BMD with chromosomal genotype obtained from the logarithm-likelihood ratio of the probability of observed relationship::probability of no relationship). Although a LOD score is not equivalent to a p value, LOD scores of 2.1 and 3.0 do correspond to p values of 10^{-3} and 10^{-4}, respectively. Using either association or linkage experimental designs, investigators have reported statistically suggestive (LOD 1.9+) or significant (LOD 3.3+) scores relating BMD of spine, hip, or other bone site with every chromosome. The association of a phenotype (BMD) to a chromosomal region is commonly termed a QTL and requires further work to identify an actual regulatory gene for the phenotype of interest. Associations for BMD to some genomic regions have been replicated among several studies (i.e., 1p, 1q, 3p, 4p, 11q), whereas other chromosomal associations (i.e., 5, 14, 17, 21) may be more uniquely related to specific features of the population studied or even artifactual in nature.

Meta-analyses of reports from genome-wide searches for human BMD QTLs have been assembled to overcome issues of low statistical power, differences in measurement devices, and weak linkages. The meta-analysis of Lee et al.[(1)] covered nine individual publications presenting association p values and concluded that pooled evidence supported eight BMD QTLs on seven different human chromosomes (Chrs 1pter-p36, 3p22-p14, 6p21-q15, 10p14-q11, 16pter-p12, 20pter-p12, 22q12–22pter). In an alternative approach using actual spine and hip data from nine studies, Ioannidis et al.[(2)] found that pooled evidence supported lumbar spine BMD regulation on Chrs 1p13-q23, 12q24-qter, 3p25-p22, 11p12-q13, 1q32-q42, and 18p11-q12, whereas femoral neck BMD was associated with Chrs 9q21-q31, 9q31-q33, 17p12-q21, 14q13-q24, and 5q14-q23. Altogether, Ioannidis et al. concluded the data supported 12 BMD QTLs on nine chromosomes. Between the two meta-analyses, 14 chromosomes were associated with BMD; however, agreement was limited to Chrs 1, 3, and 18. Clearly, specific genetic regions are critical to BMD, with both sex-limited and site-limited aspects. Nevertheless, evidence for two candidate genes attracted Ioannidis et al. attention: PTH receptor1, *PTHR1* (3p; mouse Chr 9@110.566 Mbp) for lumbar spine and collagen1α1, *COL1A1*, Sp1 binding site (17p; mouse Chr11@94.752 Mbp) for femoral neck. The human data are encouraging and in need of further study.

The physical sizes of the chromosomal regions putatively carrying bone regulatory QTLs are daunting and likely to contain dozens of genes. Narrowing the intervals for candidate gene selection has proven very challenging for a variety of

The authors state that they have no conflicts of interest.

© 2008 American Society for Bone and Mineral Research

reasons, not the least of which is the richness of alleles found at most genetic loci in humans. Nevertheless, the rapid expansion of databases for SNP differences across the human genome has allowed marked refinement of genotyping 3′ and 5′ regulatory elements, as well as within exons and introns of genes, increasing the chance of correctly identifying genes that may affect QTL regulation. Patterns of linked, statistically significant SNPs lend a measure of confidence that a true regulatory sequence for a phenotype may be in hand. The presence of known bone disease genes within a QTL region offers one route for pursuing gene identification studies. In the absence of a phenotype-altering allele for the human gene, the investigator is confronted with the issue of how to confirm and test a causal role for that QTL-related gene. Animal models can provide considerable support through comparative genomic relationships and experimental methods not possible in humans.

ANIMAL MODELS

The impetus to identify fundamental BMD regulatory genes has been addressed by both phenotypic and genetic methods. In this synopsis, a genetic approach is described that does not depend on selecting a candidate gene, which has been prejudged to be important to bone based on established research (i.e., estrogen, dietary fat, vitamin D receptors). Thus, in parallel with genome-wide screens of human populations for genetic regulation of BMD, numerous investigators have adopted inbred strains of laboratory animals as surrogate test systems to locate QTLs for BMD in the genomes of experimental systems. These animal models offer advantages of controlled matings, defined genotypes of subjects, control over environmental factors, economics of testing and maintenance, as well as treatment systems not available for human studies. Genome-wide analyses of populations that are segregating genetic alleles for bone traits—or any trait—depend on (1) accuracy and precision of phenotypic measurements, (2) numbers of subjects, and (3) quality and marker density of genetic maps. Experimental models meet these characteristics and can capitalize on the rapidly developing databases describing orthologous gene linkage relationships among vertebrate species. Accordingly, QTLs located in an experimental species can be used to predict the possible locations of genes with analogous regulatory action in humans.

INBRED AND RECOMBINANT INBRED STRAINS OF MICE

To date, the most widely used species for genome-wide genetic analyses of BMD are inbred strains of laboratory mice. A great deal of basic information may be found in the scientific literature on their physiological systems and genetics. Just over a decade ago, pQCT was applied to ex vivo bones from nine different inbred strain females, showing that peak volumetric BMD (vBMD) was achieved around 4 mo of age and that the inbred strains differed markedly in vBMD values of femurs, vertebrae, and phalanges.[3] Classic genetic analyses were undertaken by crossing low vBMD C57BL/6J (B6) with high vBMD C3H/HeJ (C3H) mice and then intercrossing female and male B6C3F1 mice to obtain (B6C3H)F_2 females for phenotyping bones ex vivo at 4 mo. The F_2 mice, of course, are segregating alleles that govern the differences in BMD observed in the progenitor strains. The resultant distributions of femoral and lumbar densities from F_2 mice were gaussian in form, indicating the participation of many genes in the regulation of peak vBMD for both bone sites. Statistical analyses showed 11 significant QTLs for femoral and lumbar 5 vBMD, with both similarities and differences in QTLs for each bone site. In 9 of 11 QTLs, the allele effects were additive with the *c3h* allele dominant to the *b6* allele. A more limited study was performed with (B6 × CAST)F_2 females,[4] with femoral phenotyping and genotyping limited to the mice in the extreme tails of the density distribution, as suggested by Lander and Botstein.[5] Again, evidence was found for four significant and four suggestive QTLs participating in the regulation of femoral vBMD. Importantly, distal Chr 1 carried QTLs in both F_2 female progenies.

A substantial group of other researchers have conducted analyses of F_2 progeny from matings between more than a dozen inbred mouse strains. Measurement systems used have included DXA for aBMD, pQCT for vBMD, X-ray films for femoral mid-diaphyseal cortical thickness, and μCT for microstructure data. The net result is that one or more BMD QTLs have been associated with all 19 autosomes and Chr X in laboratory mice. A brief summary of these QTL analyses is found in Table 1. In essence, with the experimental control afforded by studies of laboratory mice, it has been amply shown that QTLs for the phenotype of BMD are richly represented across the genome, as has been reported for humans.

A variation on the inbred mouse theme is present in the numerous sets of "new" strains that have been derived from inbreeding many pairs of F_2 mice generated from progenitor inbred strains, such as B6 and C3H. These new strains are designated BXH.[6] Every recombinant inbred (RI) strain carries one half of it genetic material from each progenitor strain, and every member of an RI strain set carries a different assortment of genetic alleles compared with every other member of that set. RI sets have been genotyped (some more than other) for polymorphisms on every chromosome that also differed in the progenitor strains, providing remarkable genetic mapping systems[7] (tables of genetic data for 25 RI sets are found at www.informatics.jax.org/searches/riset_form.shtml). RI sets such as BXH[8] and BXD[9,10] have shown that QTLs for vBMD and microstructure of appendicular (femurs, tibias) and axial (lumbar vertebrae) cortical and trabecular bone are present on numerous chromosomes. Not surprisingly, many of the regions identified in F_2 analyses also were found in RI strain sets. The replication of findings for QTL locations by both RI sets and F_2 analyses has been encouraging to bone geneticists. However, the same problem of large chromosomal regions—noted for human population studies above—has also been true for mice. QTL regions typically extend over major physical distances, encompass chromosomal features such as low or high meiotic recombination activities, gene deserts, and yet QTL regions contain up to a thousand or more genes. Multiple F_2 crosses can be statistically combined for those chromosomes that appear to carry common alleles.[11] This can result in a marked reduction in the physical size of the QTL and sharpen the application of other mapping methods, such as Haplotype Association Mapping,[12] to the task of identifying candidate genes from F_2 data.

CONGENIC STRAINS

One experimental attack on the BMD QTLs distributed across the mouse genome, except for Chr Y, has used a recombination approach through additional matings to isolate a specific QTL on a given chromosome. Congenic strains can confirm the existence of a QTL, quantify the QTL effect within a constant genetic background, and improve the genetic map location of the QTL. Congenic strains are made by repeated backcrossing (for 10 generations) of a genetically marked chromosomal region from a donor strain to a recipient strain. At N10, the marked donor QTL region has replaced the

© 2008 American Society for Bone and Mineral Research

TABLE 1. SUMMARY OF QTLS FOR BMD FOUND IN F_2 AND RI PROGENIES IN FEMALE MICE FROM 12 REPORTS

Chromosome	*Mouse cross*	*Approximate chromosome location*	*Metric and bone site*	*Human homologous region*	*Reference*
Chr 1	BXD RI	Middle	Total body	2q33-q35	(30)
	B6xC3H	Middle	L_5 trab BV/TV	2q33-q35	(31)
	B6x129	Middle	Total body	18q21	(32)
	B6xC3H	Distal	Femur; L5 vBMD	1q24-q25	(16)
	B6xC3H	Distal	L_5 trab BV/TV	1q24-q25	(31)
	B6xCAST	Distal	Femur	1q21-q24	(4)
	B6x129	Distal	Spine	1q22	(32)
	B6xDBA/2	Distal	Total body	1q41-q42	(9)
	MRLxSJL	Distal	Femur, total body	1q41-q42	(33)
	MRLxSJL	Distal	Femur, total body	1q32	(34)
Chr 2	SAMP6xAKR/J	Proximal	Spine	10p15-p11	(35)
	BXD RI	Middle	Total body	2q33-q36	(30)
	B6xDBA/2	Middle	Total body	2q31-q2	(9)
	BXD RI	Middle	Total body	18q21; 2q13–31	(30)
	MRLxSJL	Middle	Total body	11p14/15q15	(34)
	MRLxSJL	Distal	Total body	20q11-q13	(34)
	B6xC3H	Distal	Femur	20q11-q13	(16)
Chr 3	NZBxRF	Proximal	Femur	8q13-q22	(36)
Chr 4	MRLxSJL	Middle	Total body	9q21-q34	(34)
	B6xC3H	Middle	L_5 trab BV/TV	1p32-p31	(31)
	B6xC3H	Distal	Femur, L_5 vBMD	1p34-p33	(16)
	MRLxSJL	Distal	Femur	1p36-p34	(34)
	B6xDBA/2	Distal	Total body	1p36	(9)
Chr 5	B6xCAST	Middle	Femur	4q11-q13	(4)
	B6xDBA/2	Middle	Tibial trab BV/TV	4q12-q13	(10)
Chr 6	B6x129	Middle	Total body, spine	2p11-p13	(32)
	B6xC3H	Middle	Femur	3p25-p24	(16)
Chr 7	SAMP6xAKR/J	Proximal	Spine	19q12-q13	(35)
	BXD RI	Proximal	Total body	19q12-q13/11p14-p15	(30)
	BXD RI	Middle	Total body	15q11-q26	(30)
	NZBxRF	Middle	Femur	15q11-q13	(36)
	B6x129	Middle	Spine	11q13-q14	(32)
	B6xC3H	Distal	L_5 vBMD	10q26	(16)
Chr 8	B6xC3H	Middle	L_5 trab BV/TV	19p13	(31)
	BXD RI	Middle	Total body	16q22-q23	(30)
Chr 9	MRLxSJL	Proximal	Total body	11q23-q24	(34)
	B6xC3H	"	L_5 trab BV/TV	11q23-q24	(31)
	B6xC3H	Middle	L_5 trab BV/TV	15q21-q23	(31)
	MRLxSJL	Middle	Femur	6p12 /15q21-q22	(34)
	B6xC3H	Middle	L_5 vBMD	6q12-q15	(16)
Chr 10	B6xC3H	113.78	L_5 trab BV/TV	12q14-q15	(31)
	NZBxRF	117.09	Femur	12q14-q15	(36)
	B6x129	121.60	Total body	12q14-q15	(32)
Chr 11	SAMP6xAKR/J	89.79	Total body	17q21-q24	(35)
	B6xDBA/2	55.63	Total body	5q31-q35	(9)
	B6xC3H	63.24	Femur, L_5 vBMD	17p13-p11	(16)
	NZBxRF	70.05	Femur	17p13-p12	(36)
	MRLxSJL	83.65	Total body	17q11-q23	(34)
	SAMP6xSAMP2	88.79	Femur	17q21-q23	(12)
	BXD RI	119.08	Total body	17q24-q25	(30)
Chr 12	B6xC3H	Proximal	Femur, L_5 vBMD	2p24-p23	(16)
	B6xC3H	Proximal	L_5 trab BV/TV	2p24-p23	(31)
	B6xC3H	Middle	L_5 trab BV/TV	14q23-q24	(31)
	B6xC3H	Middle	L_5 trab BV/TV	14q23-q24	(31)
	MRLxSJL	Middle	Femur	14q21-q24	(33,34)
	B6xC3H	Distal	L_5 trab BV/TV	14q32	(31)
Chr 13	SAMP6xSAMP2	Proximal	Femur	6p23-p21/7p15-p13	(12)
	B6xCAST	Proximal	Femur	6p23-p21/7p15-p13	(4)
	B6xC3H	Middle	L_5 trab BV/TV	6p24-p23	(31)
	B6xC3H	Middle	Femur	5q23-q35	(16)

continued

© 2008 American Society for Bone and Mineral Research

TABLE 1. SUMMARY OF QTLS FOR BMD FOUND IN F_2 AND RI PROGENIES IN FEMALE MICE FROM 12 REPORTS (CONTINUED)

Chromosome	*Mouse cross*	*Approximate chromosome location*	*Metric and bone site*	*Human homologous region*	*Reference*
Chr 14	MRLxSJL	Proximal	Femur	Unknown	(34)
B6xC3H	Middle	Femur, L_5 vBMD	8p21-p11	(16)	
	B6xC3H	Femur, L_5 vBMD	L_5 trab BV/TV	8p21-p11	(31)
	MRLxSJL	Distal	Total body	13q14-q21	(34)
Chr 15	MRLxSJL	Proximal	Total body	8q22-q23	(34)
	B6xCAST	Middle	Femur	8q24	(4)
Chr 16	B6xC3H	Middle	Femur	3q13-q29	(16)
	SAMP6xAKR/J	Middle	Spine	3q12-q13	(35)
	BXD RI	Middle	Total body	3q12–13	(30)
Chr 17	B6xC3H	Proximal	L_5 trab BV/TV	6q25-q27	(31)
	MRLxSJL	Middle	Femur	19p13	(34)
	MRLxSJL	Middle	Femur	6p21-p12	(33)
Chr 18	B6xC3H	Middle	Femur, L_5 vBMD	5q22-q23	(16)
	MRLxSJL	Middle	Femur	18p11-q21	(34)
	NZBxRF	Distal	Femur	18q12-q23	(36)
Chr 19	MRLxSJL	Middle	Femur	10q11-q23	(33,34)

Map position in mouse genome given as approximations, because mouse genetic maps for several chromosomes are being recalculated; human homologus regions drawn from ±5 Mbp of published best position.

L_5, lumbar 5 vertebra; BV/TV, trabecular volume fraction.Chr 1

recipient's region and can be studied in the absence of mixtures of alleles for other QTL discovered in the initial F_2 cross.

Following up on the initial description of two QTLs associated with the mouse senile osteoporosis model, senescence accelerated mouse P6 (SAMP6), Shimizu et al.[13] reported on the Chr 13 congenic, designated P2.P6-*Pbd2* (recipient.donor-locus), that carries the *Pbd2* (peak bone density2) QTL. The P2.P6-*Pbd2* congenic has lower BMD than the P2 (normal BMD) recipient strain. The authors noted that *Bmp6* was within the 15-cM QTL interval in the P2.P6-*Pbd2* congenic, and, based on a CAG repeat polymorphism within exon 1, proposed *Bmp6* was a candidate gene with the SAMP6 allele responsible for the disordered BMD phenotype. Although the *Bmp6* knockout mouse has a developmental bone phenotype (delayed ossification of the sternum), it is not a close approximation of the skeletal phenotype of the SAMP6.[14] In a subsequent report, a reciprocal congenic, P6.P2-13-*Pbd2*, involving a 2.8-Mbp part of the original QTL, was reported with the SAMP2 QTL allele acting to increase BMD of the SAMP6 mouse.[15] Rather than *Bmp6* as the candidate, the authors report several lines of evidence supporting Secreted frizzled-related protein 4 (*Sfrp4)* as a more promising candidate gene. The evidence included (1) quantitative RT-PCR of SAMP6 calvarial osteoblasts showed a 40-fold increase in expression of sFRP4 above that of the P6.P2-13-*Pbd2* congenic, (2) studies with luciferase indicated the SAMP6 promoter for *Sfrp4* was more active than that of the congenic, and (3) reduced SAMP6 osteoblast proliferation and histomorphometry showing reduced SAMP6 osteoblast numbers compared with the P6.P2-13-*Pbd2* bone. The authors proposed that the *Sfrp4* gene is a negative regulator of peak BMD in the SAMP6 model; any role for *bmp6* was genetically excluded by this new congenic strain.

The complex relationship existing among phenotypes is shown by IGF-I. A genome-wide analysis of $(B6C3H)F_2$ progeny for QTL supporting serum IGF-I levels yielded four QTLs, one of which was on Chr 10 and contained the *Igf1* gene.[16] Nested (overlapping chromosomal intervals) congenic strains were developed for mapping the Chr 10 QTL interval and defining the bone phenotype.[17] One congenic subline, B6.C3H-10-4 carrying *c3h* alleles at *Igf1* increased serum IGF-I along with bone phenotypes of increased femoral mineral, trabecular connectivity, and trabecular number. Although the 18.3-Mbp interval established by the nested congenic sublines contains an estimated 148 genes, as well as *Igf1*, this growth factor locus remains an attractive candidate contributing to the variance observed in the serum IGF-I levels and bone phenotypes of the original $(B6C3H)F_2$ population of mice.

The distal region of Chr 1 in the laboratory mouse was initially shown to harbor regulation for vBMD by genome wide analyses using $(B6CAST)F_2$ and then $(B6C3H)F_2$ female progenies.[4,18] Nested sets of congenic strains were developed for this region of Chr 1 using B6 as the recipient and either CAST/EiJ or C3H/HeJ as the donor strain.[19] In two reports on B6.CAST-1 congenic mice, Edderkaoui et al.[20,21] showed that QTL alleles in this region exert positive and negative effects on BMD, as does sex. Importantly, a very attractive candidate gene, Duffy antigen (*dfy*), now renamed *Darc* (Duffy antigen receptor for chemokines), located at 175.168 Mbp, was proposed as a BMD QTL gene.[22] This gene is highly expressed in CNS structures and to a lesser extent in bone of young B6 mice (Novartis Research Foundation and gene expression database, http://symatlas.gnf.SymAtlas/). This receptor modulates actions of inflammatory cytokines through a proposed switching between homodimeric and heterodimeric forms in the erythrocyte membrane,[23] a feature that points to the possibility that candidate gene expression affecting bone density can involve other than traditional bone cells. Continuing pursuit of the distal Chr 1 with the B6.C3H-1 congenic sublines, Beamer et al.[24] found that this BMD QTL region contained three distinct QTLs (1–3), with QTL1 affecting cortical and trabecular bone only in females, whereas all three linked QTLs affect only trabecular bone in males. QTL1 is ~0.14 Mbp and contains up to three genes related to the P200 Interferon inducible cluster. Interestingly, the *Darc* gene seems to not be within any of these three QTLs associated with *C3H* alleles, suggesting that the initial distal Chr 1 BMD QTL represents a cluster of bone genes. Finally, this distal mouse Chr 1 region shares linkage homology with human 1q21-q24, a region reported to carry density regulation.[25–27]

In a preliminary report, Klein et al.[28] reported that a B6.DBA-4 congenic carrying a QTL was shown to increase

© 2008 American Society for Bone and Mineral Research

BMD and included a substantial portion of the middle of Chr 4. A candidate gene, alkaline phosphatase2 (*Akp2*), was proposed, and evidence of increased serum ALP and histomorphometric mineralizing surface, mineral apposition rates, and bone formation rates were shown in the congenic strain. Sequencing showed an amino acid difference (B6 = leu; DBA/2 = Pro), further supporting a functional role for an *Akp2* variant as a candidate gene for the DMD QTL in mice.

OTHER SPECIES

The laboratory rat (both inbred and outbred strains) has a long history of contribution to experimental biology of every sort. As the genomics era expanded to species in addition to humans and mice, the physical map of the rat genome (NCBI Ensembl Build 36) has fostered genetic studies of BMD regulation. A genome-wide study of (Fisher 344 × Lewis)F_2 progeny found significant BMD QTLs on Chrs 1, 2, 8, and 10 regulating aBMD and vBMD of the lumbar spine and femur.[29] Similar studies by Rubin et al.[30] of F_2 progeny from a cross between white leghorn × red jungle fowl yielded aBMD and pQCT data on chicken femurs that showed significant QTLs for female BMD on Chrs 1 and 2. In both reports, additional QTL regions with suggestive linkage were indicated. Significant QTL regions in rat and chicken models share genetic linkage homology with humans and mice, expanding the opportunities for insights to bone regulation through comparative biologic investigation.

SUMMARY AND CONCLUSIONS

The experimental models being applied to genetic and functional analyses of BMD regulation can provide the resolution and experimental platforms to facilitate BMD QTL gene identification. As can be surmised from Table 1, analyses of progenies from diverse inbred strains have yielded a wealth of BMD regulatory regions throughout the mouse genome. Without doubt, many of these BMD QTLs will participate in other bone traits (size, strength, bone loss, etc.), as may be determined from a quick perusal of bone literature. Thus far, congenic strains have narrowed QTL regions and provided rational evidence for five candidate genes. As research continues to add to that evidence, the bone community will have greater confidence whether a QTL and a candidate gene are one and the same.[31] In turn, homologous linkage of specific genes on defined chromosomes in experimental models and in humans will move investigators and clinicians toward a better appreciation for the complexity of BMD regulation and the relationship between BMD and biomechanical property of bone strength.

ACKNOWLEDGMENTS

The writing of this chapter was supported by AR043618 (WGB) and AR053853 (CJR).

REFERENCES

1. Lee Y, Rho Y, Choi S, Ji J, Song G 2006 Meta-analysis of genome-wide linkage studies for bone mineral density. J Hum Genet **51:**480–486.
2. Ioannidis JPA, Ng MY, Sham PC, Zintzaras E, Lewis CM, Deng H-W, Econs MJ, Karasik D, Devoto M, Kammerer CM, Spector T, Andrew T, Cupples LA, Duncan EL, Foroud T, Kiel DP, Koller DP, Langdahl B, Mitchell AG, Peacock M, Recker R, Shen H, Sol-Church K, Spotila LD, Uitterlinden AG, Wilson SG, Kung AWC, Ralston SH 2007 Meta-analysis of genome wide scans provides evidence for sex- and site specific regulation of bone mass. J Bone Miner Res **22:**173–183.
3. Beamer WG, Donahue LR, Rosen CJ, Baylink DJ 1996 Genetic variability in adult bone density among inbred strains of mice. Bone **18:**397–403.
4. Beamer WG, Shultz KL, Churchill GA, Frankel WN, Baylink DJ, Rosen CJ, Donahue LR 1999 Quantitative trait loci for bone density in C57BL/6J and CAST/EiJ inbred mice. Mamm Genome **10:**1043–1049.
5. Lander ES, Botstein D 1989 Mapping Mendelian factors underlying quantittive traits using RFLP linkage maps. Genetics **121:**185–199.
6. Bailey DW 1981 Recombinant inbred strains and bilineal congenic strains. In: Foster HL, Fox JG (eds.) The Mouse in Biomedical Research. Academic Press, New York, NY, USA, pp. 223–240.
7. Eppig JT, Bult CJ, Kadin JA, Richardson JE, Blake JA, Group MGD 2005 The Mouse Genome Database (MGD): From genes to mice—a community resource for mouse biology. Nucleic Acids Res **33:**D471–D475.
8. Turner CH, Hsieh Y-F, Mueller R, Bouxsein ML, Rosen CJ, McCrann ME, Donahue LR, Beamer WG 2001 Variation in bone biomechanical properties, microstructure, and density in BXH recombinant inbred mice. J Bone Miner Res **16:**206–213.
9. Klein R, Carlos A, Vartanian K, Chambers V, Turner R, Phillips T, Belknap J, Orwoll E 2001 Confirmation and fine mapping of chromosomal regions influencing peak bone mass in mice. J Bone Miner Res **16:**1953–1961.
10. Bower AL, Lang DH, Vogler GP, Vandenbergh DJ, Blizzard DA, Stout JT, McClearn GE, Sharkey NA 2006 QTL Analysis of trabecular bone in BXD F2 and RI mice. J Bone Miner Res **21:**1267–1275.
11. Park Y, Clifford R, Buetow K, Hunter K 2003 Multiple cross and inbred strain haplotype mapping of complex-trait candidate genes. Genome Res **13:**118–121.
12. DiPetrillo K, Wang X, Stylianou IM, Paigen B 2005 Bioinformatics toolbox for narrowing rodent quantitative trait loci. Trends Genet **21:**683–692.
13. Shimizu M, Higuchi K, Kasai S, Tsuboyama T, Matsushita M, Matsumura T, Okudaira S, Mori M, Koizumi A, Nakamura T, Hosokawa M 2002 A congenic mouse and candidate gene at the Chr 13 locus regulating bone density. Mamm Genome **13:**335–340.
14. Shimizu M, Higuchi K, Bennett B, Xia C, Tsuboyama T, Kasai S, Chiba T, Fujisawa H, Kogishi K, Kitado H, Kimoto M, Takeda N, Matsuchita M, Okumura H, Serikawa T, Nakamura T, Johnson TE, Hosokawa M 1999 Identification of peak bone mass QTL in a spontaneously osteoporotic mouse strain. Mamm Genome **10:**81–87.
15. Nakanishi R, Shimizu M, Mori M, Akiyama H, Okudaira S, Otsuki B, Hashimoto B, Higuchi K, Hosokawa M, Tsuboyama T, Nakamura T 2006 Secreted frizzle-related protein 4 is a negative regulator of peak BMD in SAMP6 mice. J Bone Miner Res **21:**1713–1721.
16. Rosen CJ, Churchill GA, Donahue LR, Shultz KL, Burgess JK, Powell DR, Beamer WG 2000 Mapping quantitative trait loci for serum insulin-like growth factor-I levels in mice. Bone **27:**521–528.
17. Delahunty KM, Koczon-Jaremko B, Adamo ML, Horton LG, Lorenzo J, Donahue LR, Ackert-Bicknell CL, Kream BE, Beamer WG, Rosen CJ 2006 Congenic mice provide in vivo evidence for a genetic locus that modulates serum insulin-like growth factor-I and bone acquisition. Endocrinology **147:**3915–3923.
18. Beamer WG, Shultz KL, Donahue LR, Churchill GA, Sen S, Wergedal JE, Baylink DJ, Rosen CJ 2001 Quantitative trait loci for femoral and vertebral bone mineral density in C57BL/6J and C3H/HeJ inbred strains of mice. J Bone Miner Res **16:**1195–1206.
19. Shultz KL, Donahue LR, Bouxsein ML, Baylink DJ, Rosen CJ, Beamer WG 2003 Congenic strains of mice for verification and genetic decomposition of quantitative trait loci for femoral bone mineral density. J Bone Miner Res **18:**175–185.
20. Edderkaoui B, Baylink DJ, Beamer WG, Wergedal JE, Dunn N, Shultz KL, Mohan S 2006 Multiple genetic loci from CAST/EiJ Chromosome 1 affect vBMD either positively or negatively in a C57BL/6J background. J Bone Miner Res **21:**97–104.
21. Edderkaoui B, Baylink DJ, Beamer WG, Shultz KL, Wergedal JE, Mohan S 2007 Genetic regulation of femoral bone mineral density: Complexity of sex effect in Chromosome 1 revealed by congenic cublines of mice. Bone **41:**340–345.
22. Edderkaoui B, Baylink DJ, Beamer WG, Wergedal JE, Porte R, Chaudhuri A, Mohan S 2007 Identification of mouse duffy antigen receptor for chemokines (Darc) as a BMD QTL gene. Genome Res **17:**577–585.
23. Chakera A, Seeber RM, John AE, Eidne KA, Greaves DR 2008

© 2008 American Society for Bone and Mineral Research

The Duffy Antigen/Receptor for Chemokines (DARC) exists in an oligomeric form in living cells and functionally antagonises CCR5 signalling through hetero-oligomerization. Mol Pharmacol (in press).
24. Beamer WG, Shultz KL, Ackert-Bicknell CL, Horton LG, Delahunty KM, Coombs HF III, Donahue LR, Canalis E, Rosen CJ 2007 Genetic dissection of mouse distal Chromosome 1 reveals three linked BMD QTLs with sex-dependent regulation of bone phenotypes. J Bone Miner Res **22:**1187–1196.
25. Koller DL, Econs MJ, Morin PA, Christian JC, Hui SL, Parry P, Curran M, Rodriguez LA, Conneally PM, Joslyn G, Peacock M, Johnston CC, Foroud T 2000 Genome screen for QTLs contributing to normal variation in bone mineral density and osteoporosis. J Clin Endocrinol Metab **85:**3116–3120.
26. Peacock M, Koller D, Lai D, Hui S, Foroud T, Econs MJ 2005 Sex-specific quantitative trait loci contribute to normal variation in bone structure at the proximal femur in men. Bone **37:**467–473.
27. Wilson S, Reed P, Bansal A, Chiano M, Lindersson M, Langdown M, Prince R, Thompson E, Bailey M, Kleyn P, Sambrook P, Shi M, Spector T 2003 Comparison of genome screens for two independent cohorts provides replication of suggestive linkage of bone mineral density to 3p21 and 1p36. Am J Hum Genet **72:**144–155.
28. Klein R, Carlos A, Kansagor J, Olson D, Wagoner W, Larson E, Dinulescu D, Munsey T, Vanek C, Madison D, Lundblad J, Belknap J, Orwoll E 2005 Identification of *Akp2* as a gene that regulates peak bone density in mice. J Bone Miner Res **20:**S1;S9.
29. Koller DL, Alam I, Liu L, Fishburn T, Carr LG, Econs MJ, Foroud T, Turner CH 2005 Genome screen for bone mineral density phenotypes in Fisher 344 and Lewis rat strains. Mamm Genome **16:**578–586.
30. Rubin CJ, Brandstrom H, Wright D, Kerje S, Gunnarsson U, Schutz K, Fredriksson R, Jensen P, Andersson L, Ohlsson C, Malllmin H, Larsson S, Kindmark A 2007 Quantitative trait loci for BMD and bone strength in an intercross between domestic and wildtype chickens. J Bone Miner Res **22:**375–384.
31. Abiola O, Angel JM, Avner P, Bachmanov AA, Belknap JK, Bennett B, Blankenhorn EP, Blizard DA, Bolivar V, Brockmann GA, Buck KJ, Bureau JF, Casley WL, Chesler EJ, Cheverud JM, Churchill GA, Cook M, Crabbe JC, Crusio WE, Darvasi A, de Haan G, Dermant P, Doerge RW, Elliot RW, Farber CR, Flaherty L, Flint J, Gershenfeld H, Gibson JP, Gu J, Gu W, Himmelbauer H, Hitzemann R, Hsu HC, Hunter K, Iraqi FF, Jansen RC, Johnson TE, Jones BC, Kempermann G, Lammert F, Lu L, Manly KF, Matthews DB, Medrano JF, Mehrabian M, Mittlemann G, Mock BA, Mogil JS, Montagutelli X, Morahan G, Mountz JD, Nagase H, Nowakowski RS, O'Hara BF, Osadchuk AV, Paigen B, Palmer AA, Peirce JL, Pomp D, Rosemann M, Rosen GD, Schalkwyk LC, Seltzer Z, Settle S, Shimomura K, Shou S, Sikela JM, Siracusa LD, Spearow JL, Teuscher C, Threadgill DW, Toth LA, Toye AA, Vadasz C, Van Zant G, Wakeland E, Williams RW, Zhang HG, Zou F, Complex Trait Consortium 2003 The nature and identification of quantitative trait loci: A community's view. Nat Rev Genet **4:**911–916.
32. Ishimori N, Li R, Walsh KA, Korstanje R, Rollins J, Petkov P, Pletcher MT, Wiltshire T, Donahue LR, Rosen CJ, Beamer WG, Churchill GA, Paigen BJ 2006 Quantitative trait loci that determine BMD in C57BL/6J and 129S1/SvImJ inbred mice. J Bone Miner Res **21:**105–112.
33. Li XM, Masinde GL, Gu W, Wergedal JE, Mohan S, Baylink DJ 2002 Genetic dissection of femur breaking strength in a large population (MRL/MpJ × SJL/J) of F2 mice: Single QTL effects, epistasis, and pleiotropy. Genomics **79:**734–740.
34. Masinde GL, Li XM, Gu W, Wergedal JE, Mohan S, Baylink DJ 2002 Quantitative trait loci for bone density in mice: The genes determining total skeletal density and femur density show little overlap in F2 mice. Calcif Tissue Int **71:**421–428.
35. Benes H, Weinstein RS, Zheng W, Thaden JJ, Jilka RL, Manolagos SC, Smookler Reis RJ 2000 Chromosomal mapping of osteopenia-associated quantitative trait loci using closely related mouse strains. J Bone Miner Res **15:**626–633.
36. Wergedal JE, Ackert-Bicknell CL, Tsaih S-W, Sheng MH-C, Li R, Mohan S, Beamer WG, Churchill GA, Baylink DJ 2006 Femur mechanical properties in the F2 progeny of an NZB/B1NJ × RF/J cross are regulated predominantly by genetic loci that regulate bone geometry. J Bone Miner Res **22:**1256–1266.

Chapter 10. Neuronal Regulation of Bone Remodeling

Gerard Karsenty[1] and Florent Elefteriou[2]

[1]Department of Genetics and Development and Department of Medicine, Columbia University Medical Center, New York, New York;
[2]Department of Medicine/Clinical Pharmacology, Vanderbilt University Medical Center, Center for Bone Biology, Nashville, Tennessee

INTRODUCTION

All homeostatic functions in vertebrates are regulated by hypothalamic inputs. Bone remodeling, the process whereby bone is constantly renewed during adulthood, is a prototypical homeostatic function (the destruction of bone by osteoclasts followed by de novo bone formation by osteoblasts) that allows repair of micro- and macro-damages and maintenance of bone mass and of the structural properties of the skeleton. Throughout life, this process, driven by the action of osteoclasts and osteoblasts, is regulated through a number of different mechanisms for the skeleton to respond and to adapt to the physiological and mechanical demand. The discovery of a central regulation of bone remodeling during the past few years is in agreement with its homeostatic nature and sheds a new light on the complexity of the mechanisms regulating bone mass. This chapter summarizes the major findings showing that bone remodeling is an integral part of a complex homeostatic control system regulating bone remodeling along with some aspects of energy metabolism and reproduction.

CENTRAL NERVOUS SYSTEM REGULATES BONE REMODELING

The fact that obese patients are protected from osteoporosis and that osteoporosis typically follows gonadal deficiency first suggested the existence of a common mechanism regulating reproduction, body weight, and bone remodeling.[1] Because reproduction and body weight are regulated by the central nervous system (CNS), this hypothesis implied from its inception the existence of a central control of bone remodeling. Again the homeostatic nature of bone remodeling adds conceptual credence to this hypothesis. The first experimental evidence of a functional regulation of bone remodeling by the CNS in vivo was provided by the analysis of mice that lack leptin (the *ob/ob* mice), an adipocyte-derived hormone that appeared during evolution with the skeleton and whose function includes the regulation of food intake and of energy expenditure. Leptin carries out these functions after binding on its receptors, ObRb, located on hypothalamic neurons. Ge-

The authors state that they have no conflicts of interest.

Key words: leptin, sympathetic nervous system, β2-adrenergic receptor, neuropeptide Y, neuromedin U, bone remodeling

© 2008 American Society for Bone and Mineral Research

netic evidence in mice and humans have identified α-melanocyte-stimulating hormone (α-MSH) as the main, if not the only, mediator of leptin anorexigenic function.[2,3] The discovery that absence of leptin in *ob/ob* mice or of its signaling receptor *ObRb* in *db/db* mice results in high bone mass caused by an increase in bone formation and, despite their hypogonadism, indicated that leptin is a negative regulator of bone formation in vivo.[1] This observation was extended to humans shortly after.[4] The demonstration that leptin uses a central relay for regulating bone mass was provided first in mice and then in rats and sheep.[1,5,6] Infusion of low doses of leptin infused intracerebroventricularly significantly decreases trabecular bone mass in *ob/ob* mice. This experiment is particularly telling because of its genetic nature. Indeed, what has been done was to reintroduce, in the brain only, leptin in an animal genetically deprived of it. At no point could leptin be detected in the blood of these animals and yet their high bone mass was completely corrected. Because this intracerebroventricular infusion fully corrected the high bone mass phenotype of *ob/ob* mice, this experiment not only established the existence of leptin-dependent central regulation of bone formation but also showed that leptin had no other mechanism of regulating bone formation in vivo; otherwise, the rescue would never have been complete.

The increased BMD and decreased fracture risk associated with obesity is counterintuitive based on this pathway, where high leptin levels should reduce bone mass. One possible explanation for this observation invokes the notion of leptin resistance, in which leptin levels above a critical threshold create a state of resistance to the hormone. This explanation has been used to explain why obesity is associated with high leptin serum levels and may hold true to explain the increased bone mass in obese patients who may be in a state of low leptin responsiveness centrally.

Follow-up studies performed in wildtype (WT) or *ob/ob* mice in which specific neuronal populations have been lesioned and followed by leptin intracerebroventricular infusions defined a population of ventro-medial hypothalamic (VMH) neurons constituting the major center responsive to leptin for its regulation of bone formation.[7] Indeed, destruction of VMH neurons (that highly express *ObRb*) recapitulated the high bone formation/ high bone mass phenotype of *ob/ob,* mice and leptin intracerebroventricular infusion could not rescue the high bone mass phenotype of VMH-lesioned *ob/ob* mice, thereby showing that leptin requires VMH neurons to inhibit bone formation. Subsequent work showed that the neural mediator of this leptin function was the sympathetic nervous system (SNS).

SYMPATHETIC NERVOUS SYSTEM, DOWNSTREAM OF HYPOTHALAMIC NEURONS, REGULATES BONE FORMATION

The nature of the downstream mechanism whereby hypothalamic neurons regulate the activity of distant cells like osteoblasts and osteoclasts has been determined by the use of in vivo mouse mutant models and pharmacological approaches. A number of observations first pointed to the existence of a neuronal pathway rather than a humoral one: first, *ob/ob* mice have a high bone mass phenotype accompanied by a low sympathetic tone, which may contribute to it[1]; second, electrical stimulation of hypothalamic VMH neurons[8] and leptin injection in the VMH area results in enhanced sympathetic tone[9]; and third, bones are innervated, as evidenced by immunological reactivity for various neuropeptides[10] and linked by nerves to the CNS as evidenced by retrograde neuronal labeling.[11] On the other hand, primary osteoblasts extracted from mouse calvaria express a functional β2-adrenergic receptor (β2AR) but none of the other postsynaptic adrenergic receptors,[7] suggesting that osteoblasts are receptive to sympathetic tone.

Analysis of mice with autonomic dysfunction confirmed the existence of a neuronal pathway between hypothalamic neurons and osteoblasts. Mice lacking dopamine β-hydroxylase (Dbh), the enzyme generating norepinephrine (NE), display a late onset increase in bone mass, indicating that the sympathetic nervous system inhibits bone formation.[7] Supporting this result, mice and rats treated with the nonselective β-adrenergic blocker propranolol exhibit increased bone mass, whereas mice treated with the nonselective β-agonist isoproterenol or the β2AR selective agonists clenbuterol or salbutamol exhibit low bone mass.[7,12] Furthermore, mice lacking *Adrβ2* (the gene encoding β2AR), as opposed to *ob/ob* mice, VMH-lesioned mice, and $Dbh^{-/-}$ mice, have a normal endocrine status but display an increase in bone formation and a decrease in bone resorption, leading to a high bone mass phenotype.[13] Similarly, mice lacking adenylyl cyclase 5, a downstream mediator of β2AR signaling, also display positive changes in bone mass and biomechanical properties.[14] Importantly, leptin intracerebroventricular infusion completely failed to decrease bone mass in the β2AR-deficient mice, showing not only that the SNS through β2AR mediates leptin regulation of bone formation but also that there is no other mediator of this function.

The increased number of osteoblasts and increased bone formation rate observed in β2AR-deficient mice suggested that the autonomic nervous system inhibits osteoblast proliferation and function. Surprisingly, studies of mice lacking critical determinants of the clock machinery, including *Per1* and *-2–* or *Cry1* and *-2–*deficient mice, showed that peripheral clock genes mediate the effect of the autonomic nervous system on osteoblast proliferation.[15] As opposed to what was initially suspected, no evidence was found that central master clock genes work upstream of leptin to inhibit bone formation. Instead, these studies indicated that osteoblasts express peripheral subordinate clock genes that mediate leptin-dependent sympathetic inhibition of bone formation by suppressing the expression of *G1 cyclin*s and their proliferation. Further studies showed that osteoblastic clock genes are under the control of β2AR signaling and uncovered that leptin and sympathetic signaling exert a countervailing and stimulatory effect on osteoblast proliferation through the activator protein (AP)-1 family of transcription factors.[15] These results characterizing the involvement of clock genes in the regulation of bone remodeling are in agreement with the known daily variation in bone marrow cell proliferation, collagen synthesis, and turnover markers.[16,17]

Although osteoclasts express β2AR,[18] the effect of the autonomic nervous system on osteoclasts differentiation was shown to be indirect and mediated by osteoblasts through β2AR and stimulation of the expression of the osteoclast-differentiating factor RANKL.[13] The transcription factor ATF4, involved in osteoblast differentiation,[19] was identified as a target of β2AR signaling. ATF4 is indeed phosphorylated by protein kinase A (PKA) after β2AR stimulation and directly binds to the *Rankl* promoter to activate *Rankl* transcription[13] (Fig. 1).

Is the leptin-dependent anti-osteogenic function of the autonomic nervous system on bone remodeling conserved between mice and human? What has been established is that leptin is a negative independent predictor of BMD in adult.[20] Likewise, the majority of the available retrospective studies suggest a protective effect of β-blockers on fracture risk, in agreement with mouse and rat data.[21–26] Some reports, how-

© 2008 American Society for Bone and Mineral Research

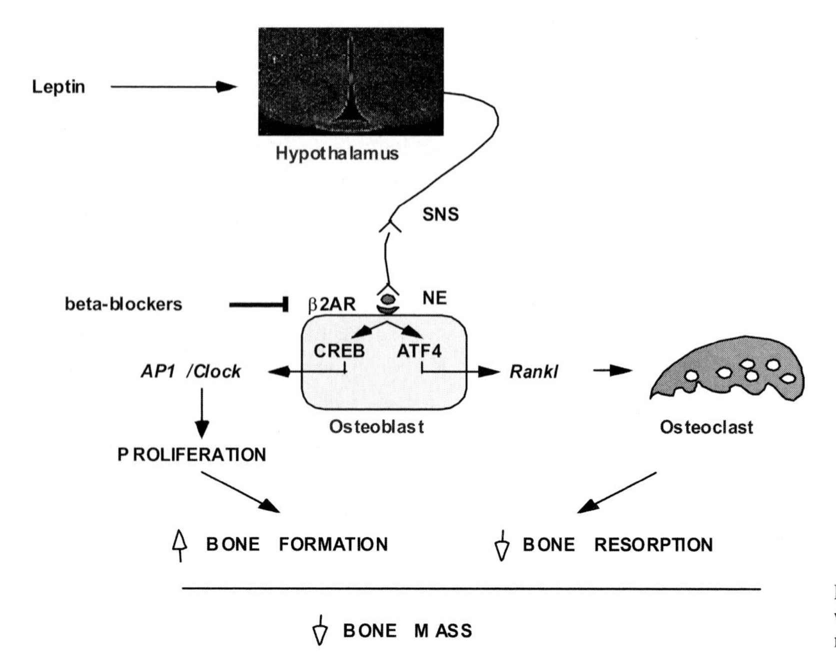

FIG. 1. Representation of the neural pathway by which leptin decreases bone mass. SNS, sympathetic nervous system; NE, norepinephrine.

ever, did not find any significant relationship between β-blocker users and fracture risk.[27,28] Therefore, long-term prospective randomized studies will be needed to show unequivocally a potential protective effect of β-blockers on fracture risk in humans.

REGULATION OF BONE REMODELING BY CART, NEUROMEDIN U, AND THE Y2 RECEPTOR

Cellular histomorphometric analyses have shown that absence of leptin in *ob/ob* mice leads to an increase in osteoblast proliferation and function at the origin of their increase in bone formation and also to an increase in osteoclast number. This was assumed to be secondary to the hypogonadism caused by the absence of this hormone. However, a serendipitous finding showed that the increase in osteoclastogenesis and bone resorption observed in *ob/ob* mice is not caused mainly by gonadal failure. β2AR-deficient mice not only have an increase in bone formation but also display an unexpected decrease in bone resorption parameters whose molecular bases have been elucidated.[13] The surprise came when β2AR-deficient mice were subjected to gonadectomy. In term of bone remodeling, these mice should be the phenocopy of the *ob/ob* mice: they lack both sympathetic tone and gonadal functions. This procedure that increases osteoclast number and decreases bone mass in WT mice did not have this effect in β2AR-deficient mice, indicating that gonadal failure–induced bone loss is blunted by the absence of sympathetic tone but also that the increase in bone resorption observed in *ob/ob* mice is not mainly caused by their hypogonadism; otherwise, β2AR-deficient mice would have a similar increase in bone resorption parameters. In short, these results indicated that leptin and the CNS regulates both bone formation and bone resorption.

Using as an approach a screening for genes whose expression is regulated by leptin but that do not regulate appetite or reproduction led to the discovery that *Cart* (cocaine and amphetamine-regulated transcript) mediates the leptin inhibition of bone resorption.[13] CART is a neuropeptide broadly expressed in the CNS, including hypothalamic neurons, as well as in peripheral tissues such as the pancreas.[29] The importance of CART in regulating bone remodeling and the hypothalamic nature of this regulation is supported by animal models characterized by low or high hypothalamic *Cart* expression and significant bone phenotypes. Low *Cart* expression in *ob/ob* mice accompanies the increased bone resorption observed in these mice, whereas increased hypothalamic *Cart* expression in obese and hyperleptinemic *Mc4r*-deficient mice correlates with their low bone resorption and high bone mass.[13] Furthermore, lack of one *Cart* copy in *Mc4r*-deficient mice rescues their resorption phenotype, indicating that *Cart* increased hypothalamic level in *Mc4r*-deficient mice causes their high bone mass phenotype.[30] Importantly, this CART-mediated regulatory loop of bone resorption is conserved in humans, because individuals lacking *MC4R* have increased CART serum levels and increased BMD associated with decreased bone resorption.[13,30,31] The molecular mode of action of CART on bone resorption is not yet defined.

Neuropeptide U (NMU) is another neuropeptide whose expression is regulated by leptin and that has been recently shown to be involved in the regulation of bone remodeling. NMU is expressed in hypothalamic neurons as well as in the small intestine, and its functions include the regulation of appetite and sympathetic activation,[32] as shown by the obesity of NMU-deficient mice.[33] These latter mice display a high bone mass phenotype caused by an increase in bone formation.[34] The fact that the receptors for NMU, NMU1R, and NMU2R are not detectable in bone, the absence of effect of NMU treatment on osteoblast differentiation in vitro, the expression of NMU2R in hypothalamic neurons, and most importantly, the rescue of the high bone mass of NMU-deficient mice by NMU intracerebroventricular infusion all concur to show that NMU acts through a central relay. That NMU intracerebroventricular infusion could decrease the high bone mass of leptin-deficient mice indicates that NMU acts downstream of leptin to regulate bone formation. Most interestingly, NMU-deficient mice are resistant to the anti-osteogenic effect of leptin and adrenergic agonists. Furthermore, osteoblast number paradoxically increased in NMU-deficient mice treated by leptin intracerebroventricularly, as observed in *Clock*-deficient mice, suggesting that NMU regulates Clock gene's function in osteoblasts (see below). In support of this

© 2008 American Society for Bone and Mineral Research

hypothesis, expression of *Per* genes was downregulated in NMU-deficient bones compared with WT bones.

Neuropeptide Y (NPY) is a well-known target of leptin in the hypothalamus, with the potential to act through at least five Y-receptors (Y1, Y2, Y4, Y5, and Y6) that differ in their distribution in the CNS and the periphery. All of these receptors are expressed in the hypothalamus, and several respond to other ligands, including peptide YY and pancreatic polypeptide (PP). Strong expression of NPY is found in the hypothalamic area, where NPY fibers project from the arcuate nucleus. Expression of NPY is increased in the hypothalamus of *Ob/Ob* mice. Strongly supporting a role of NPY receptor signaling in the regulation of bone formation, mice lacking the Y2 receptor ($Y2^{-/-}$ mice) display an increase in trabecular bone mass that can be reproduced by hypothalamic-specific deletion of *Y2*, indicating that Y2 signaling in the hypothalamus inhibits bone formation.[35] Interestingly, inactivation of both *Y2* and *Y4* receptors led to a further increase in bone mass compared with *Y2* alone, which was accompanied by reduced serum leptin level, suggesting a *Y4*-mediated additional effect of leptin deficiency on the $Y2^{-/-}$ bone phenotype.[36] Thus, Y2 receptor signaling clearly regulates, through a hypothalamic relay, the bone formation arm of the bone remodeling process. Whether NPY is the ligand for Y2 receptor for this function on bone mass and whether the role of NPY on bone formation is restricted to long bones is still a question, because a bone phenotype was observed in NPY-deficient mice in one study but not in another one.[37,38] The effects of Y2 antagonists on bone remains to be tested.

A number of additional molecules have been proposed to be involved, through a central relay, in the regulation of bone remodeling, but in vivo and mechanistic studies are still lacking to firmly show this putative role. These include NO signaling[39] and the central cannabinoid system.[40–44] The endothelial isoform of NO synthase (eNOS) is constitutively expressed in bone, whereas inducible NO synthase (iNOS) is expressed under inflammatory conditions and acts as a mediator of cytokine actions on bone. Both isoforms regulate bone cell function in a cell autonomous manner.[45] In contrast, the neuronal form of NO synthase (nNOS) is not expressed in adult bone under normal conditions[46] but is highly expressed in the CNS. The high bone mass phenotype of nNOS-deficient mice and the high expression of nNOS in hypothalamic neurons suggests that hypothalamic NO signaling may regulate bone mass by a central mechanism.[39] On the other hand, the cannabinoid type 1 (CB1) receptor, mostly known for its involvement in psychotropic, analgesic, and orexigenic processes, is also expressed in the CNS and peripheral nervous system (PNS), and mice lacking *Cb1* exhibit a bone phenotype, which is dependent on strain background differences. Interestingly, CB1 signaling in peripheral neurons inhibits NE release[47] and may therefore regulate the action of the autonomic nervous system on bone.

PERSPECTIVES

The notion that there is a leptin-dependent central control of bone remodeling is not disputed anymore. Likewise, the role of additional neuropeptide in regulating bone remodeling becomes increasingly evident.[10] Among all of them, cGRP (a-calcitonin gene-related peptide) is the sensory neuropeptide whose function in bone homeostasis has been the best shown. Both in vitro and in vivo studies have led to the conclusion that cGRP is an anabolic factor that acts on osteoblasts and stimulates their proliferation and function.[48,49] In vitro data also suggest that cGRP may be involved in modulating the proresorptive effect of the autonomic nervous system by interfering with the action of RANKL.[18] Similarly, in vitro data suggest an inhibitory effect of NPY on isoproterenol-induced osteoclast formation.[50] What has been untouched yet and lies as a major challenge ahead of us is to identify and delineate the neuronal circuitry involved in the central control of bone mass, as well as to characterize the feedback loops regulating these CNS centers. Likewise, if we exclude the SNS, the molecular mode of action of neuropeptides that have been proposed to regulate bone mass remains to be established.

REFERENCES

1. Ducy P, Amling M, Takeda S, Priemel M, Schilling AF, Beil T, Shen J, Vinson C, Rueger JM, Karsenty G 2000 Leptin inhibits bone formation through a hypothalamic relay: A central control of bone mass. Cell **100:**197–207.
2. Zigman JM, Elmquist JK 2003 Minireview: From anorexia to obesity–the yin and yang of body weight control. Endocrinology **144:**3749–3756.
3. Yeo GS, Farooqi IS, Aminian S, Halsall DJ, Stanhope RG, O'Rahilly S 1998 A frameshift mutation in MC4R associated with dominantly inherited human obesity. Nat Genet **20:**111–112.
4. Elefteriou F, Takeda S, Ebihara K, Magre J, Patano N, Kim CA, Ogawa Y, Liu X, Ware SM, Craigen WJ, Robert JJ, Vinson C, Nakao K, Capeau J, Karsenty G 2004 Serum leptin level is a regulator of bone mass. Proc Natl Acad Sci USA **101:**3258–3263.
5. Guidobono F, Pagani F, Sibilia V, Netti C, Lattuada N, Rapetti D, Mrak E, Villa I, Cavani F, Bertoni L, Palumbo C, Ferretti M, Marotti G, Rubinacci A 2006 Different skeletal regional response to continuous brain infusion of leptin in the rat. Peptides **27:**1426–1433.
6. Pogoda P, Egermann M, Schnell JC, Priemel M, Schilling AF, Alini M, Schinke T, Rueger JM, Schneider E, Clarke I, Amling M 2006 Leptin inhibits bone formation not only in rodents, but also in sheep. J Bone Miner Res **21:**1591–1599.
7. Takeda S, Elefteriou F, Levasseur R, Liu X, Zhao L, Parker KL, Armstrong D, Ducy P, Karsenty G 2002 Leptin regulates bone formation via the sympathetic nervous system. Cell **111:**305–317.
8. Perkins MN, Rothwell NJ, Stock MJ, Stone TW 1981 Activation of brown adipose tissue thermogenesis by the ventromedial hypothalamus. Nature **289:**401–402.
9. Satoh N, Ogawa Y, Katsuura G, Numata Y, Tsuji T, Hayase M, Ebihara K, Masuzaki H, Hosoda K, Yoshimasa Y, Nakao K 1999 Sympathetic activation of leptin via the ventromedial hypothalamus: Leptin-induced increase in catecholamine secretion. Diabetes **48:**1787–1793.
10. Elefteriou F 2005 Neuronal signaling and the regulation of bone remodeling. Cell Mol Life Sci **62:**2339–2349.
11. Denes A, Boldogkoi Z, Uhereczky G, Hornyak A, Rusvai M, Palkovits M, Kovacs KJ 2005 Central autonomic control of the bone marrow: Multisynaptic tract tracing by recombinant pseudorabies virus. Neuroscience **134:**947–963.
12. Bonnet N, Brunet-Imbault B, Arlettaz A, Horcajada MN, Collomp K, Benhamou CL, Courteix D 2005 Alteration of trabecular bone under chronic beta2 agonists treatment. Med Sci Sports Exerc **37:**1493–1501.
13. Elefteriou F, Ahn JD, Takeda S, Starbuck M, Yang X, Liu X, Kondo H, Richards WG, Bannon TW, Noda M, Clement K, Vaisse C, Karsenty G 2005 Leptin regulation of bone resorption by the sympathetic nervous system and CART. Nature **434:**514–520.
14. Yan L, Vatner DE, O'Connor JP, Ivessa A, Ge H, Chen W, Hirotani S, Ishikawa Y, Sadoshima J, Vatner SF 2007 Type 5 adenylyl cyclase disruption increases longevity and protects against stress. Cell **130:**247–258.
15. Fu L, Patel MS, Bradley A, Wagner EF, Karsenty G 2005 The molecular clock mediates leptin-regulated bone formation. Cell **122:**803–815.
16. Simmons DJ, Nichols G Jr 1966 Diurnal periodicity in the metabolic activity of bone tissue. Am J Physiol **210:**411–418.
17. Gundberg CM, Markowitz ME, Mizruchi M, Rosen JF 1985 Osteocalcin in human serum: A circadian rhythm. J Clin Endocrinol Metab **60:**736–739.
18. Arai M, Nagasawa T, Koshihara Y, Yamamoto S, Togari A 2003 Effects of beta-adrenergic agonists on bone-resorbing activity in human osteoclast-like cells. Biochim Biophys Acta **1640:**137–142.
19. Yang X, Matsuda K, Bialek P, Jacquot S, Masuoka HC, Schinke T, Li L, Brancorsini S, Sassone-Corsi P, Townes TM, Hanauer A,

© 2008 American Society for Bone and Mineral Research

Karsenty G 2004 ATF4 is a substrate of RSK2 and an essential regulator of osteoblast biology; implication for Coffin-Lowry Syndrome. Cell **117:**387–398.
20. Lorentzon M, Landin K, Mellstrom D, Ohlsson C 2006 Leptin is a negative independent predictor of areal BMD and cortical bone size in young adult Swedish men. J Bone Miner Res **21:**1871–1878.
21. Pasco J, Henry M, Sanders K, Kotowicz M, Seeman E, Nicholson G 2003 Beta-adrenergic blockers reduce the risk of fracture partly by increasing bone mineral density: Geelong osteoporosis study. J Bone Miner Res **19:**19–24.
22. Schlienger RG, Kraenzlin ME, Jick SS, Meier CR 2004 Use of beta-blockers and risk of fractures. JAMA **292:**1326–1332.
23. Rejnmark L, Vestergaard P, Kassem M, Christoffersen BR, Kolthoff N, Brixen K, Mosekilde L 2004 Fracture risk in perimenopausal women treated with beta-blockers. Calcif Tissue Int **75:**365–372.
24. Pasco JA, Henry MJ, Sanders KM, Kotowicz MA, Seeman E, Nicholson GC 2004 Beta-adrenergic blockers reduce the risk of fracture partly by increasing bone mineral density: Geelong Osteoporosis Study. J Bone Miner Res **19:**19–24.
25. Wiens M, Etminan M, Gill SS, Takkouche B 2006 Effects of antihypertensive drug treatments on fracture outcomes: A meta-analysis of observational studies. J Intern Med **260:**350–362.
26. Rejnmark L, Vestergaard P, Mosekilde L 2006 Treatment with beta-blockers, ACE inhibitors, and calcium-channel blockers is associated with a reduced fracture risk: A nationwide case-control study. J Hypertens **24:**581–589.
27. Reid IR, Gamble GD, Grey AB, Black DM, Ensrud KE, Browner WS, Bauer DC 2005 beta-Blocker use, BMD, and fractures in the study of osteoporotic fractures. J Bone Miner Res **20:**613–618.
28. Levasseur R, Marcelli C, Sabatier JP, Dargent-Molina P, Breart G 2005 Beta-blocker use, bone mineral density, and fracture risk in older women: Results from the Epidemiologie de l'Osteoporose prospective study. J Am Geriatr Soc **53:**550–552.
29. Kristensen P, Judge ME, Thim L, Ribel U, Christjansen KN, Wulff BS, Clausen JT, Jensen PB, Madsen OD, Vrang N, Larsen PJ, Hastrup S 1998 Hypothalamic CART is a new anorectic peptide regulated by leptin. Nature **393:**72–76.
30. Ahn JD, Dubern B, Lubrano-Berthelier C, Clement K, Karsenty G 2006 Cart overexpression is the only identifiable cause of high bone mass in melanocortin 4 receptor deficiency. Endocrinology **147:**3196–3202.
31. Orwoll B, Bouxsein ML, Marks DL, Cone RD, Klein RF 2004 Increased bone mass and strength in melanocortin-4 receptor-deficient mice ORS/AAOS Presentations 2003, 71st Annual Meeting of the AAOS, San Francisco, CA, USA, March 2003.
32. Brighton PJ, Szekeres PG, Willars GB 2004 Neuromedin U and its receptors: Structure, function, and physiological roles. Pharmacol Rev **56:**231–248.
33. Hanada R, Teranishi H, Pearson JT, Kurokawa M, Hosoda H, Fukushima N, Fukue Y, Serino R, Fujihara H, Ueta Y, Ikawa M, Okabe M, Murakami N, Shirai M, Yoshimatsu H, Kangawa K, Kojima M 2004 Neuromedin U has a novel anorexigenic effect independent of the leptin signaling pathway. Nat Med **10:**1067–1073.
34. Sato S, Hanada R, Kimura A, Abe T, Matsumoto T, Iwasaki M, Inose H, Ida T, Mieda M, Takeuchi Y, Fukumoto S, Fujita T, Kato S, Kangawa K, Kojima M, Shinomiya KI, Takeda S 2007 Central control of bone remodeling by neuromedin U. Nat Med **13:**1234–1240.
35. Baldock PA, Sainsbury A, Couzens M, Enriquez RF, Thomas GP, Gardiner EM, Herzog H 2002 Hypothalamic Y2 receptors regulate bone formation. J Clin Invest **109:**915–921.
36. Sainsbury A, Baldock PA, Schwarzer C, Ueno N, Enriquez RF, Couzens M, Inui A, Herzog H, Gardiner EM 2003 Synergistic effects of Y2 and Y4 receptors on adiposity and bone mass revealed in double knockout mice. Mol Cell Biol **23:**5225–5233.
37. Baldock PA, Allison S, Sainsbury A, Enriquez RF, Gardiner EM, Herzog H, Eisman JA 2006 Hypothalamic neuropeptide Y exerts a negative effect on cortical bone formation. 28th Annual Meeting of the American Society for Bone and Mineral Research, September 2006, Philadelphia, PA, USA.
38. Elefteriou F, Takeda S, Liu X, Armstrong D, Karsenty G 2003 Monosodium glutamate-sensitive hypothalamic neurons contribute to the control of bone mass. Endocrinology **144:**3842–3847.
39. van't Hof RJ, Macphee J, Libouban H, Helfrich MH, Ralston SH 2004 Regulation of bone mass and bone turnover by neuronal nitric oxide synthase. Endocrinology **145:**5068–5074.
40. Lutz B 2002 Molecular biology of cannabinoid receptors. Prostaglandins Leukot Essent Fatty Acids **66:**123–142.
41. Idris AI, van 't Hof RJ, Greig IR, Ridge SA, Baker D, Ross RA, Ralston SH 2005 Regulation of bone mass, bone loss and osteoclast activity by cannabinoid receptors. Nat Med **11:**774–779.
42. Tam J, Ofek O, Fride E, Ledent C, Gabet Y, Muller R, Zimmer A, Mackie K, Mechoulam R, Shohami E, Bab I 2006 Involvement of neuronal cannabinoid receptor CB1 in regulation of bone mass and bone remodeling. Mol Pharmacol **70:**786–792.
43. Tam J, Alexandrovich A, Di Marzo V, Petrosino S, Trembovler V, Zimmer A, Ledent C, Mackie K, Mechoulam R, Shohami E, Bab I 2006 CB1, but not CB2 cannabinoid receptor mediates stimulation of bone formation induced by traumatic brain injury. 28th Annual Meeting of the American Society for Bone and Mineral Research, September 2006, Philadelphia, PA, USA.
44. Devoto M, Shimoya K, Caminis J, Ott J, Tenenhouse A, Whyte MP, Sereda L, Hall S, Considine E, Williams CJ, Tromp G, Kuivaniemi H, Ala-Kokko L, Prockop DJ, Spotila LD 1998 First-stage autosomal genome screen in extended pedigrees suggests genes predisposing to low bone mineral density on chromosomes 1p, 2p and 4q. Eur J Hum Genet **6:**151–157.
45. van't Hof RJ, Ralston SH 2001 Nitric oxide and bone. Immunology **103:**255–261.
46. Helfrich MH, Evans DE, Grabowski PS, Pollock JS, Ohshima H, Ralston SH 1997 Expression of nitric oxide synthase isoforms in bone and bone cell cultures. J Bone Miner Res **12:**1108–1115.
47. Ishac EJ, Jiang L, Lake KD, Varga K, Abood ME, Kunos G 1996 Inhibition of exocytotic noradrenaline release by presynaptic cannabinoid CB1 receptors on peripheral sympathetic nerves. Br J Pharmacol **118:**2023–2028.
48. Ballica R, Valentijn K, Khachatryan A, Guerder S, Kapadia S, Gundberg C, Gilligan J, Flavell RA, Vignery A 1999 Targeted expression of calcitonin gene-related peptide to osteoblasts increases bone density in mice. J Bone Miner Res **14:**1067–1074.
49. Schinke T, Liese S, Priemel M, Haberland M, Schilling AF, Catala-Lehnen P, Blicharski D, Rueger JM, Gagel RF, Emeson RB, Amling M 2004 Decreased bone formation and osteopenia in mice lacking alpha-calcitonin gene-related peptide. J Bone Miner Res **19:**2049–2056.
50. Amano S, Arai M, Goto S, Togari A 2007 Inhibitory effect of NPY on isoprenaline-induced osteoclastogenesis in mouse bone marrow cells. Biochim Biophys Acta **1770:**966–973.

© 2008 American Society for Bone and Mineral Research

Chapter 11. Skeletal Healing

Michael J. Zuscik and Regis J. O'Keefe

Center for Musculoskeletal Research and Department of Orthopaedics, University of Rochester Medical Center, Rochester, New York

IMPORTANCE OF SKELETAL FRACTURE REPAIR IN VERTEBRATES

The process of skeletal repair is essential for (1) resolution of orthopedic trauma that has caused bony disjunction and (2) healing of surgical interventions that are intended to create bony injury with the aim of inducing a repair response or a therapy. In general, the tissue/cellular/genetic processes that are required for healing and their molecular control are conserved in different skeletal elements. This chapter provides a concise overview of the skeletal healing process at the cellular and molecular level and a discussion of a few key situations that complicate healing. The chapter concludes with a discussion of the current therapeutic paradigms that are aimed at accelerating healing or alleviating failure to heal (i.e., nonunion).

PROCESS OF SKELETAL HEALING

Cellular Contribution to Fracture Repair

As mentioned, in humans, bone repair is a process essential for reconstitution of skeletal integrity after trauma or skeletal surgery. The schematic in Fig. 1, which presents a pictorial representation of the unique morphogenesis of reparative tissue during the phases of bone fracture healing, provides a benchmark for the description of the healing process that follows. This process begins immediately after fracture and involves both intramembraneous and endochondral ossification.[1,2] Overall, healing events that occur in various vertebrate species are similar, except that relative to humans the pace of repair is generally accelerated in smaller animals/rodents. The trauma that induces the fracture initially results in bleeding and formation of a hematoma at the injury site. Hematoma-related infiltration of inflammatory cells and release of cytokines and growth factors as well as the level of mechanical instability at the injury site direct the recruitment and proliferation of various local mesenchymal stem cell (MSC) progenitor populations.[3] A key participant is the periosteal progenitor cell that responds to the inflammation by entering the osteogenic or chondrogenic lineage. Endochondral bone formation always takes place closest to the fracture site where the oxygen tension is low and vascularity is disrupted. Intramembraneous bone formation, on the other hand, always occurs distal to the disjunction where intact vasculature remains present. As mentioned, the mechanical stability of the fractured bone markedly affects the fate of the progenitor cells, with stabilized fractures healing with virtually no evidence of cartilage, whereas nonstabilized fractures produce abundant cartilage at the fracture site.[4]

Given that the periosteum represents a primary source of MSCs that contribute to bone repair, understanding its structure/function is critical for dissecting the tissue and cellular dynamics of the healing process. Overall, periosteum is a vascularized connective tissue that covers the outer surface of cortical bone. It can be separated into two distinct layers: an outer layer that contains fibroblasts and Sharpey's fibers (which facilitate connection to the underlying cortical bone) and an inner layer referred to as the cambium, which contains multipotent mesenchymal stem cells and osteoprogenitor cells that contribute to normal bone growth, healing, and regeneration.[5,6] It is known that the cambium layer in children is much thicker and better vascularized than adults, facilitating faster healing.

Once periosteal MSCs have committed to the chondrogenic or osteogenic lineage, chondrocyte and osteoblast differentiation takes place. Directly overlying the site of the fracture, the ends of the original bone have decreased perfusion caused by disrupted vascularity and necrosis occurs. In this central hypoxic region, MSCs differentiate into chondrocytes, and endochondral bone formation is initiated. This is consistent with the concept that hypoxia is a critical inducer of chondrogenesis.[7] The tissue that forms as the cell population expands is referred to as the callus, and differentiation of MSCs into chondrocytes occurs directionally within the callus with the process initiating in the most central avascular region. Whereas these centrally positioned MSCs persist in the callus area directly overlying the fracture site, chondrocytes that differentiate radially recapitulate the maturation process that occurs in the growth plate, including phases of proliferation, hypertrophy, and terminal differentiation. The calcified cartilage, which acts a template for primary bone formation, is populated by the most terminally differentiated hypertrophic chondrocytes that are contributing to the mineralization of the tissue. Ultimately, terminally differentiated chondrocytes residing in the calcified cartilage matrix undergo apoptosis. Distal to the fracture site and flanking the chondrocytes undergoing endochondral ossification is the location where intramembranous ossification occurs. This process, which proceeds in the zone of injury where blood supply has been better preserved, is characterized by the differentiation of periosteal cells into osteoblasts that directly lay down new mineral without a cartilage intermediate. As mentioned, the better the fracture is fixed (i.e., less instability there is), the greater the ratio of intramembraneous to endochondral ossification in the overall healing process.

Fractures are considered healed when bone stability has been restored by the formation of new bone that bridges the area of fracture. However, this initial woven bone matrix is replaced by organized lamellar bone through a remodeling process that is the critical final step in achieving an anatomically correct skeletal element. This process is governed by osteoclasts, which are present even at earlier stages of healing, but become dominant in this final stage. Similarly, the initial cortical bone is remodeled and replaced at the fracture site, where necrosis occurs secondary to loss of vascularization caused by the injury.

Gene Expression During Bone Repair

Given that the bone repair process is dependent on a combination of endochondral and intramembraneous ossification followed by osteoclast remodeling, the genetic profile of the healing tissue is stage dependent and reflective of the differentiation of these cells. Because the endochondral healing process recapitulates the events that occur during skeletal development, it is no surprise that the genetic profile partially reflects the profile seen in the growth plate chondrocyte hy-

The authors state that they have no conflicts of interest.

Key words: fracture healing, endochondral ossification, intramembraneous bone formation, periosteum, stem cell, chondrocyte, osteoblast

© 2008 American Society for Bone and Mineral Research

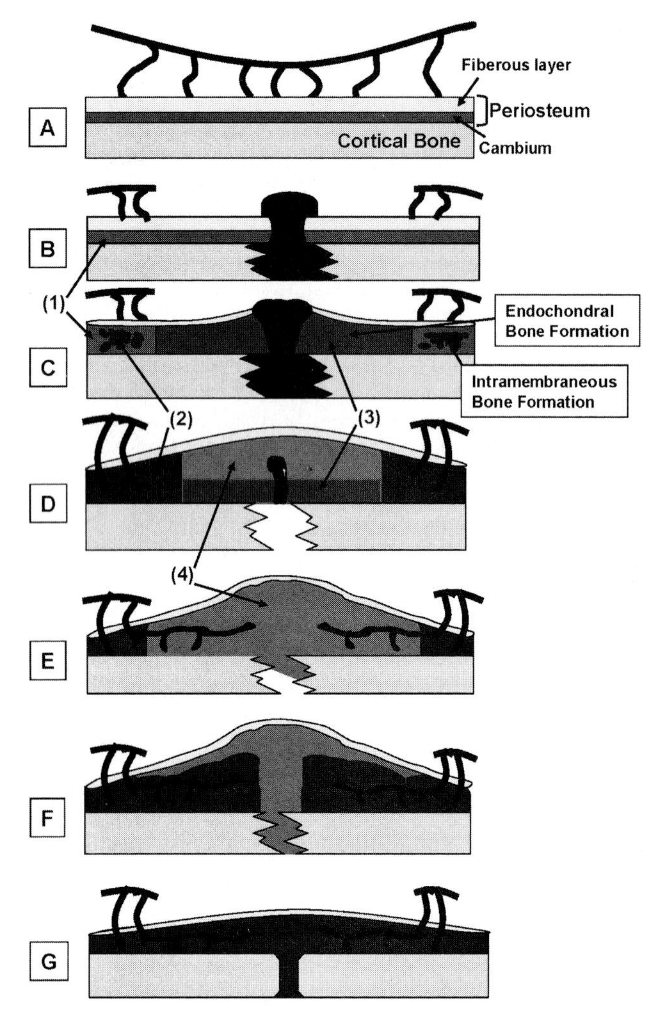

FIG. 1. Tissue morphogenesis during bone repair. (A) Periosteum is a well-microvascularized tissue (vessels in black) consisting of an outer fibrous layer and an inner cambium layer. The cambium layer contains abundant stem/progenitor cells that can differentiate into bone and cartilage. (B) After fracture or osteotomy, blood supply is disrupted at the defect, and a blood clot (hematoma) forms near the disjunction. (C) Progenitor cells residing in the periosteum are recruited to differentiate into osteoblasts to facilitate intramembraneous bone formation where intact blood supply is preserved and chondrocytes to facilitate endochondral bone formation adjacent to the fracture where the tissue is hypoxic. In this panel, osteogenic tissue is labeled (1) with newly mineralized tissue labeled (2). Tissues supporting chondrogenesis are labeled (3). (D) Intramembraneous bone formation proceeds with robust matrix mineralization (1) where blood supply is present distal to the fracture site. Endochondral bone formation proceeds simultaneously with chondrogenic tissue supporting a growing population of chondrocytes that comprise the hypertrophic cartilage which is labeled (4). (E) Cartilage tissue continues to mature, ultimately encompassing the callus nearest the fracture site. Revascularization of the callus also ensues. (F) Chondrocytes in the hypertrophic cartilage undergo terminal differentiation and the matrix is progressively mineralized expanding the portion of the callus that is comprised of woven bone. (G) The remodeling process proceeds with osteoclasts and osteoblasts facilitating the conversion of woven bone into lamellar bone, culminating in the re-creation of the appropriate anatomic shape.

pertrophic program. Overlying this is the genetic profile of osteoblast differentiation that occurs during intramembraneous bone formation. Regarding the endochondral process, mesenchymal cell condensation coincides with the expression of early markers of cartilage formation that include *Sox-9* and *type II collagen*.[8] As chondrocyte differentiation ensues, there is a significant increase in cell volume that is associated with the expression of hypertrophy-associated genes that include *type X collagen*, *MMP9* and *13*, and *osteocalcin*,[1–3] as well *indian hedgehog*.[9] Terminally mature chondrocytes contribute to revascularization through expression of vascular endothelial growth factor[10] and also may initiate the remodeling process through the induction of osteoclast formation/activity by expression of RANKL.[11] Regarding the intramembraneous process, markers of osteoblast differentiation are detected including type I collagen, osteopontin, and osteocalcin. Osteoblasts also contribute to callus revascularization by producing vascular endothelial growth factor.[12] The differentiation process in these cells is driven by the expression of Runx2,[13] a transcription factor required for mineralization. By establishing the temporal and regional gene expression pattern during the healing process, a benchmark has emerged that facilitates monitoring healing rate that may be delayed or accelerated depending on comorbidities (e.g., aging or diabetes) or therapeutic intervention (e.g., bone morphogenetic protein [BMP]-2 treatment), respectively.

Molecular Control of Fracture Healing

The molecular signaling pathways involved in the initiation and tissue morphogenesis that occurs during fracture repair are only superficially understood. Although animals and humans have only very limited capacity to regenerate damaged tissues, it has long been suspected that postnatal bone repair involving endochondral ossification recapitulates some of the essential pathways/factors in limb development.[14] Regarding the developmental process, the most notable regulators are the BMPs that belong to the TGF-β superfamily, Hedgehog and Wnt proteins, fibroblast growth factors (FGFs), and insulin-like growth factors (IGFs). Relative to fracture repair, BMP-2 expression has been observed in early periosteal callus just a few days after cortical bone fracture. Most recently, Tsuji et al.[15] showed that elimination of BMP-2 in the limb disrupts the initiation of postnatal fracture healing, establishing the essential role of BMP-2 in bone repair. In fact, their use has been approved in a number of bone healing situations (discussed below).[16] Evidence has also emerged indicating that Wnt/β-catenin signaling is important in driving osteoblast differentiation in the fracture callus,[17] implicating this pathway as an important participant in the callus mineralization process. Hedgehogs are likely important during the cartilage differentiation phase of healing, where Indian Hedgehog expression is the highest in a rat model of healing.[9] Similarly FGFs[18] and IGFs[19] have also been implicated to play a role during skeletal healing. Overall, however, studies are yet to be undertaken to fully characterize the role of these pathways and factors in the adult fracture healing process.

In addition to cell differentiation processes that recapitulate limb development, genes that are involved in injury and inflammatory responses during bone repair have been shown to play key roles in endochondral bone repair.[20,21] For example, during the endochondral healing phase, a turnover of mineralized cartilage occurs that sets the stage for primary bone formation. This initial remodeling process coincides with the upregulation of macrophage colony-stimulating factor, RANKL, osteoprotegerin, and TNF-α, suggesting that these factors are important in the transition from cartilage to bone.[22,23] During a second remodeling phase, which occurs during the conversion of woven bone into lamellar bone, interleukin (IL)-1 and -6 expression is upregulated, implicating these factors in the recruitment of osteoclasts that are critical for this final remodeling step.[2] Supporting this idea, IL-6

© 2008 American Society for Bone and Mineral Research

knockout mice display delayed callus mineralization and maturation during repair of femoral osteotomy.[24] Thus, there is a clear involvement of pro-inflammatory mediators in the bone repair process.

Several studies have also shown that cyclooxygenase activity is involved in normal bone metabolism and have suggested that nonsteroidal anti-inflammatory drugs (NSAIDs) have a negative impact on bone repair.[25,26] The action of aspirin and other NSAIDs is through inhibition of cyclooxygenase (COX), an enzyme that catalyzes the formation of prostaglandins and thromboxanes from arachidonic acid.[27] The most compelling data implicating COX function during skeletal healing comes from genetic models that show a critical role for COX-2 in the process. Whereas COX-2$^{-/-}$ mice develop normally, bone repair is impaired in adult knockout mice after fracture.[28] Defective healing in these model occurs at the early inflammatory phase and persists into the reparative phase of healing. Histology of fracture callus derived from COX-2$^{-/-}$ mice shows delayed chondrogenesis and persistent mesenchyme at the fracture site. The defective bone healing phenotype coincides with the early induction of COX-2 expression, further showing the requirement of this enzyme in early chondrogenesis in skeletal repair.

Critical to successful bone repair is the revascularization of injured tissues to provide oxygen, facilitate nutrient/metabolic waste management, and deliver a population of precursor cells of hematopoietic origin that may have the potential to contribute to healing. As mentioned previously, support for angiogenesis during the repair process is thought to be modulated by vascular endothelial growth factors (VEGFs) and their cognate receptors VEGFR1 and VEGFR2. It has been shown that exogenous administration of VEGF during mouse femur fracture healing enhances vascular ingrowth into the callus and accelerates repair by promoting bony bridging.[29] This has been borne out in allograft bone healing, where VEGF gene therapy likewise enhanced the healing process.[30] Whereas it has been suggested that osteoblasts are the primary regulators of angiogenesis during healing because of their production of VEGF in response to BMPs,[31] a contribution of VEGF from hypertrophic cartilage in the callus may also occur.[11]

CONDITIONS THAT IMPAIR FRACTURE HEALING AND THERAPEUTIC MODALITIES

The normal progression of the fracture healing process can be significantly compromised by a number of physiologic, pathologic, and environmental factors including aging, diabetes, and cigarette smoking. Clinical data provide evidence for this, and basic research has begun to show the details of the underlying biological basis in some cases. Below is a brief discussion of three of the most important conditions that are documented to impair the process of skeletal healing.

Aging

Whereas it has been known for >30 yr that the rate of fracture healing in reduced with aging, minimal progress has been made toward understanding the mechanisms involved. Studies have suggested that the rate of bone repair is progressively reduced with aging in the pediatric population.[32] Furthermore, numerous studies document that development of nonunion in the aging population is a significant clinical problem.[33–35] Several mechanisms have been proposed to explain reduced/delayed fracture healing in the elderly. Whereas the healing effect(s) caused by altered temporal expression of genes is a matter of conjecture, changes in progenitor cell populations during aging may be important. A reduced number/responsiveness of progenitors,[36] enhanced adipogenic potential at the expense of chondrogenesis and osteogenesis,[37] or altered competency to support osteoclastogenesis at various stages of healing[38] may be involved. There has also been an association between aging and a decrement in endothelial cells and the factors/pathways that modulate them,[39,40] suggesting that impaired blood vessel formation in elderly patients could also affect healing. Ongoing effort aims to address the relative mechanistic contribution of these and other processes to impaired skeletal healing in the elderly.

Diabetes

Documented clinical findings established that fracture healing is impaired in patients with diabetes.[41] Consistent with this, animal models of diabetes show impaired healing evidenced by reduced mesenchymal cell proliferation in the early callus, reduced matrix deposition (collagen), and reduced biomechanical properties in the healed fracture.[42,43] Whereas it is not known if the impaired healing is the result of hypoinsulinemia or hyperglycemia/formation of advanced glycation endproducts (AGEs), insulin treatment to normalize blood glucose in a diabetic rat bone explant model has been shown to reverse the deficit in healing.[44] Interestingly, during diabetic bone healing, osteoblast expression of RAGE (cell surface receptors for AGE) is enhanced,[45] possibly facilitating the effects of AGE on reduced bone repair.[45] Last, in a diabetic rat fracture model, local intramedullary delivery of insulin to the fracture site, which does not provide systemic management of glucose, reversed the healing deficit at both early (mesenchymal cell proliferation and chondrogenesis) and late (mineralization and biomechanical strength) time points.[46] This supports a novel hypothesis predicting that there is a direct anabolic effect of insulin on cells at the fracture site.

Cigarette Smoking

Clinically, smoking has been shown to have a negative impact on skeletal healing after long bone fracture[47] and spinal fusion surgery.[48] Little is known about the mechanism underlying these deficits in skeletal repair, with mesenchymal cell condensation and the process of chondrogenesis hypothesized to be important targets of cigarette smoke. This hypothesis is supported by a recent study suggesting that exposure of mice to second-hand smoke causes delayed chondrogenesis in tibial fracture.[49] Full characterization of the healing process in smokers or animal models of smoke exposure is necessary, and work to identify which components of cigarette smoke are responsible for its effect(s) is important if the underlying molecular mechanisms are to be elucidated.

MOLECULAR THERAPIES TO ENHANCE BONE HEALING

Currently, the only FDA-approved molecular therapy for bone healing is BMP-2. As mentioned previously, the underlying rationale for its therapeutic potential is based on the finding that elimination of BMP-2 in the limb disrupts the initiation of postnatal fracture healing,[15] establishing its essential role in the process. Thus, it is no surprise that a number of animal studies have identified a positive effect of BMP-2 or activation of its signaling pathway on various bone healing situations. As its use has gained attention, clinical data have emerged that supports the use of BMP-2 in clinical situations. For example, recombinant BMP-2 delivered in a collagen sponge with cancellous autograft aids in the healing of tibial diaphyseal fractures.[50,51] Spine fusion patients were also found to have better neck disability and arm pain scores 24 mo

© 2008 American Society for Bone and Mineral Research

postoperatively when INFUSE bone graft (collagen sponge impregnated with BMP-2) was used.[52] Despite the positive outcomes of these and other studies, it should be noted that the clinical and cost effectiveness of BMP-2 in skeletal repair situations remains an open debate.[53]

Because PTH is an approved treatment for enhancing bone mass in osteoporosis patients, its off-label use as a candidate therapy for fracture healing in patients with nonunion has been proposed. Recent studies in human and animals showed compelling evidence of positive actions of PTH on bone fracture healing and repair.[54–56] Whereas the mechanism underlying the ability of PTH to enhance healing remains to be determined, recent findings suggest that PTH increases chondrogenesis and mesenchymal cell proliferation in rat femur fractures.[57] This is supported by the finding that SOX-9 is markedly upregulated by PTH at the early stage of fracture healing.[57] These recent and novel advances suggest that PTH treatment may represent an important emerging molecular therapy to accelerate healing or alleviate development of nonunion.

REFERENCES

1. Einhorn TA 1998 The cell and molecular biology of fracture healing. Clin Orthop **355:**S7–S21.
2. Gerstenfeld LC, Cullinane DM, Barnes GL, Graves DT, Einhorn TA 2003 Fracture healing as a post-natal developmental process: Molecular, spatial, and temporal aspects of its regulation. J Cell Biochem **88:**873–884.
3. Le AX, Miclau T, Hu D, Helms JA 2001 Molecular aspects of healing in stabilized and non-stabilized fractures. J Orthop Res **19:**78–84.
4. Thompson Z, Miclau T, Hu D, Helms JA 2002 A model for intramembranous ossification during fracture healing. J Orthop Res **20:**1091–1098.
5. Augustin G, Antabak A, Davila S 2007 The periosteum Part 1: Anatomy, histology and molecular biology. Injury **38:**1115–1130.
6. Orwoll ES 2003 Toward an expanded understanding of the role of the periosteum in skeletal health. J Bone Miner Res **18:**949–954.
7. Schipani E 2005 Hypoxia and HIF-1 alpha in chondrogenesis. Semin Cell Dev Biol **16:**539–546.
8. Uusitalo H, Salminen H, Vuorio E 2001 Activation of chondrogenesis in response to injury in normal and transgenic mice with cartilage collagen mutations. Osteoarthritis Cartilage **9**(Suppl A):S174–S179.
9. Murakami S, Noda M 2000 Expression of Indian hedgehog during fracture healing in adult rat femora. Calcif Tissue Int **66:**272–276.
10. Pufe T, Wildemann B, Petersen W, Mentlein R, Raschke M, Schmidmaier G 2002 Quantitative measurement of the splice variants 120 and 164 of the angiogenic peptide vascular endothelial growth factor in the time flow of fracture healing: A study in the rat. Cell Tissue Res **309:**387–392.
11. Gerber HP, Vu TH, Ryan AM, Kowalski J, Werb Z, Ferrara N 1999 VEGF couples hypertrophic cartilage remodeling, ossification and angiogenesis during endochondral bone formation. Nat Med **5:**623–628.
12. Athanasopoulos AN, Schneider D, Keiper T, Alt V, Pendurthi UR, Liegibel UM, Sommer U, Nawroth PP, Kasperk C, Chavakis T 2007 Vascular endothelial growth factor (VEGF)-induced upregulation of CCN1 in osteoblasts mediates proangiogenic activities in endothelial cells and promotes fracture healing. J Biol Chem **282:**26746–26753.
13. Kawahata H, Kikkawa T, Higashibata Y, Sakuma T, Huening M, Sato M, Sugimoto M, Kuriyama K, Terai K, Kitamura Y, Nomura S 2003 Enhanced expression of Runx2/PEBP2alphaA/CBFA1/AML3 during fracture healing. J Orthop Sci **8:**102–108.
14. Vortkamp A, Pathi S, Peretti GM, Caruso EM, Zaleske DJ, Tabin C 1998 Recapitulation of signals regulating embryonic bone formation during postnatal growth and in fracture repair. Mech Dev **71:**65–76.
15. Tsuji K, Bandyopadhyay A, Harfe BD, Cox K, Kakar S, Gerstenfeld L, Einhorn T, Tabin CJ, Rosen V 2006 BMP2 activity, although dispensable for bone formation, is required for the initiation of fracture healing. Nat Genet **38:**1424–1429.
16. De Biase P, Capanna R 2005 Clinical applications of BMPs. Injury **36**(Suppl 3):S43–S46.
17. Chen Y, Whetstone HC, Lin AC, Nadesan P, Wei Q, Poon R, Alman BA 2007 Beta-catenin signaling plays a disparate role in different phases of fracture repair: Implications for therapy to improve bone healing. PLoS Med **4:**e249.
18. Szczesny G 2002 Molecular aspects of bone healing and remodeling. Pol J Pathol **53:**145–153.
19. Weiss S, Henle P, Bidlingmaier M, Moghaddam A, Kasten P, Zimmermann G 2007 Systemic response of the GH/IGF-I axis in timely versus delayed fracture healing. Growth Horm IGF Res **18:**205–212.
20. Lehmann W, Edgar CM, Wang K, Cho TJ, Barnes GL, Kakar S, Graves DT, Rueger JM, Gerstenfeld LC, Einhorn TA 2005 Tumor necrosis factor alpha (TNF-alpha) coordinately regulates the expression of specific matrix metalloproteinases (MMPS) and angiogenic factors during fracture healing. Bone **36:**300–310.
21. Baldik Y, Diwan AD, Appleyard RC, Fang ZM, Wang Y, Murrell GA 2005 Deletion of iNOS gene impairs mouse fracture healing. Bone **37:**32–36.
22. Kimble RB, Bain S, Pacifici R 1997 The functional block of TNF but not of IL-6 prevents bone loss in ovariectomized mice. J Bone Miner Res **12:**935–941.
23. Kon T, Cho TJ, Aizawa T, Yamazaki M, Nooh N, Graves D, Gerstenfeld LC, Einhorn TA 2001 Expression of osteoprotegerin, receptor activator of NF-kappaB ligand (osteoprotegerin ligand) and related proinflammatory cytokines during fracture healing. J Bone Miner Res **16:**1004–1014.
24. Yang X, Ricciardi BF, Hernandez-Soria A, Shi Y, Pleshko CN, Bostrom MP 2007 Callus mineralization and maturation are delayed during fracture healing in interleukin-6 knockout mice. Bone **41:**928–936.
25. Ho ML, Chang JK, Wang GJ 1998 Effects of ketorolac on bone repair: A radiographic study in modeled demineralized bone matrix grafted rabbits. Pharmacology **57:**148–159.
26. Sudmann E, Hagen T 1976 Indomethacin-induced delayed fracture healing. Arch Orthop Unfallchir **85:**151–154.
27. Vane JR 1971 Inhibition of prostaglandin synthesis as a mechanism of action for aspirin-like drugs. Nat New Biol **231:**232–235.
28. Zhang X, Schwarz EM, Young DA, Puzas JE, Rosier RN, O'Keefe RJ 2002 Cyclooxygenase-2 regulates mesenchymal cell differentiation into the osteoblast lineage and is critically involved in bone repair. J Clin Invest **109:**1405–1415.
29. Street J, Bao M, deGuzman L, Bunting S, Peale FV Jr, Ferrara N, Steinmetz H, Hoeffel J, Cleland JL, Daugherty A, van Bruggen N, Redmond HP, Carano RA, Filvaroff EH 2002 Vascular endothelial growth factor stimulates bone repair by promoting angiogenesis and bone turnover. Proc Natl Acad Sci USA **99:**9656–9661.
30. Ito H, Koefoed M, Tiyapatanaputi P, Gromov K, Goater JJ, Carmouche J, Zhang X, Rubery PT, Rabinowitz J, Samulski RJ, Nakamura T, Soballe K, O'Keefe RJ, Boyce BF, Schwarz EM 2005 Remodeling of cortical bone allografts mediated by adherent rAAV-RANKL and VEGF gene therapy. Nat Med **11:**291–297.
31. Deckers MM, van Bezooijen RL, van Der Horst G, Hoogendam J, van Der Bent C, Papapoulos SE, Lowik CW 2002 Bone morphogenetic proteins stimulate angiogenesis through osteoblast-derived vascular endothelial growth factor A. Endocrinology **143:**1545–1553.
32. Skak SV, Jensen TT 1988 Femoral shaft fracture in 265 children. Log-normal correlation with age of speed of healing. Acta Orthop Scand **59:**704–707.
33. Nieminen S, Nurmi M, Satokari K 1981 Healing of femoral neck fractures; influence of fracture reduction and age. Ann Chir Gynaecol **70:**26–31.
34. Nilsson BE, Edwards P 1969 Age and fracture healing: A statistical analysis of 418 cases of tibial shaft fractures. Geriatrics **24:**112–117.
35. Hee HT, Wong HP, Low YP, Myers L 2001 Predictors of outcome of floating knee injuries in adults: 89 patients followed for 2–12 years. Acta Orthop Scand **72:**385–394.
36. Gruber R, Koch H, Doll BA, Tegtmeier F, Einhorn TA, Hollinger JO 2006 Fracture healing in the elderly patient. Exp Gerontol **41:**1080–1093.
37. Akune T, Ohba S, Kamekura S, Yamaguchi M, Chung UI, Kubota N, Terauchi Y, Harada Y, Azuma Y, Nakamura K, Kadowaki T, Kawaguchi H 2004 PPARgamma insufficiency enhances osteogenesis through osteoblast formation from bone marrow progenitors. J Clin Invest **113:**846–855.
38. Cao JJ, Wronski TJ, Iwaniec U, Phleger L, Kurimoto P, Bou-

© 2008 American Society for Bone and Mineral Research

dignon B, Halloran BP 2005 Aging increases stromal/osteoblastic cell-induced osteoclastogenesis and alters the osteoclast precursor pool in the mouse. J Bone Miner Res **20:**1659–1668.
39. Brandes RP, Fleming I, Busse R 2005 Endothelial aging. Cardiovasc Res **66:**286–294.
40. Edelberg JM, Reed MJ 2003 Aging and angiogenesis. Front Biosci **8:**s1199–s1209.
41. Loder RT 1988 The influence of diabetes mellitus on the healing of closed fractures. Clin Orthop Relat Res 210–216.
42. Beam HA, Parsons JR, Lin SS 2002 The effects of blood glucose control upon fracture healing in the BB Wistar rat with diabetes mellitus. J Orthop Res **20:**1210–1216.
43. Funk JR, Hale JE, Carmines D, Gooch HL, Hurwitz SR 2000 Biomechanical evaluation of early fracture healing in normal and diabetic rats. J Orthop Res **18:**126–132.
44. Hough S, Avioli LV, Bergfeld MA, Fallon MD, Slatopolsky E, Teitelbaum SL 1981 Correction of abnormal bone and mineral metabolism in chronic streptozotocin-induced diabetes mellitus in the rat by insulin therapy. Endocrinology **108:**2228–2234.
45. Santana RB, Xu L, Chase HB, Amar S, Graves DT, Trackman PC 2003 A role for advanced glycation end products in diminished bone healing in type 1 diabetes. Diabetes **52:**1502–1510.
46. Gandhi A, Beam HA, O'Connor JP, Parsons JR, Lin SS 2005 The effects of local insulin delivery on diabetic fracture healing. Bone **37:**482–490.
47. Schmitz MA, Finnegan M, Natarajan R, Champine J 1999 Effect of smoking on tibial shaft fracture healing. Clin Orthop Relat Res 184–200.
48. Hadley MN, Reddy SV 1997 Smoking and the human vertebral column: A review of the impact of cigarette use on vertebral bone metabolism and spinal fusion. Neurosurgery **41:**116–124.
49. El-Zawawy HB, Gill CS, Wright RW, Sandell LJ 2006 Smoking delays chondrogenesis in a mouse model of closed tibial fracture healing. J Orthop Res **24:**2150–2158.
50. Jones AL, Bucholz RW, Bosse MJ, Mirza SK, Lyon TR, Webb LX, Pollak AN, Golden JD, Valentin-Opran A 2006 Recombinant human BMP-2 and allograft compared with autogenous bone graft for reconstruction of diaphyseal tibial fractures with cortical defects. A randomized, controlled trial. J Bone Joint Surg Am **88:**1431–1441.
51. Swiontkowski MF, Aro HT, Donell S, Esterhai JL, Goulet J, Jones A, Kregor PJ, Nordsletten L, Paiement G, Patel A 2006 Recombinant human bone morphogenetic protein-2 in open tibial fractures. A subgroup analysis of data combined from two prospective randomized studies. J Bone Joint Surg Am **88:**1258–1265.
52. Baskin DS, Ryan P, Sonntag V, Westmark R, Widmayer MA 2003 A prospective, randomized, controlled cervical fusion study using recombinant human bone morphogenetic protein-2 with the CORNERSTONE-SR allograft ring and the ATLANTIS anterior cervical plate. Spine **28:**1219–1224.
53. Garrison KR, Donell S, Ryder J, Shemilt I, Mugford M, Harvey I, Song F 2007 Clinical effectiveness and cost-effectiveness of bone morphogenetic proteins in the non-healing of fractures and spinal fusion: A systematic review. Health Technol Assess **11:**1–50, iii–iv.
54. Andreassen TT, Willick GE, Morley P, Whitfield JF 2004 Treatment with parathyroid hormone hPTH(1-34), hPTH(1-31), and monocyclic hPTH(1-31) enhances fracture strength and callus amount after withdrawal fracture strength and callus mechanical quality continue to increase. Calcif Tissue Int **74:**351–356.
55. Alkhiary YM, Gerstenfeld LC, Krall E, Westmore M, Sato M, Mitlak BH, Einhorn TA 2005 Enhancement of experimental fracture-healing by systemic administration of recombinant human parathyroid hormone (PTH 1-34). J Bone Joint Surg Am **87:**731–741.
56. Komatsubara S, Mori S, Mashiba T, Nonaka K, Seki A, Akiyama T, Miyamoto K, Cao Y, Manabe T, Norimatsu H 2005 Human parathyroid hormone (1-34) accelerates the fracture healing process of woven to lamellar bone replacement and new cortical shell formation in rat femora. Bone **36:**678–687.
57. Nakazawa T, Nakajima A, Shiomi K, Moriya H, Einhorn TA, Yamazaki M 2005 Effects of low-dose, intermittent treatment with recombinant human parathyroid hormone (1-34) on chondrogenesis in a model of experimental fracture healing. Bone **37:**711–719.

Chapter 12. Biomechanics of Fracture Healing

Elise F. Morgan[1,2] and Thomas A. Einhorn[2]

[1]Department of Aerospace and Mechanical Engineering, Boston University, Boston, Massachusetts; [2]Department of Orthopaedic Surgery, Boston University Medical Center, Boston, Massachusetts

INTRODUCTION

Fracture healing involves a dynamic interplay of biological processes that, when properly executed, restore form and function to the injured bone. Given the importance of restoring bone mechanical function, this chapter presents a biomechanical description of fracture healing, with emphasis on methods of assessing the extent of healing and on the role of the local mechanical environment. Fracture healing is often classified as either primary or secondary fracture healing, where the former is characterized by direct cortical reconstruction and the latter involves substantial periosteal callus formation. The techniques for assessing healing that are presented in this chapter apply equally well to primary and secondary healing; however, the overviews of the biomechanical stages of fracture healing and the mechanobiology of fracture healing are largely specific to secondary healing. We also note that this chapter does not include a discussion of the biomechanics of fracture fixation, because this topic has been extensively reviewed elsewhere.[1]

The authors state that they have no conflicts of interest.

BIOMECHANICAL ASSESSMENT OF FRACTURE HEALING

In the laboratory setting, the mechanical properties of a healing bone are commonly assessed by mechanical tests that load the bone in torsion or in three-point bending. Loading the bone in tension (to pull apart the two halves of the bone) or compression (to compact the fractured ends against one another) is less common. The choice of the type of test is dictated by technical as well as physiological considerations. For example, bending and torsion are logical choices when studying fracture healing in long bones, because these bones experience bending and torsional moments in vivo. However, whereas torsion tests subject every cross-section of the callus to the same torque, three-point bending creates a nonuniform bending moment throughout the callus. As a result, failure of the callus during a three-point bend test does not necessarily occur at the weakest cross-section of the callus.

Regardless of the type of mechanical test, the outcome measures that can be obtained are the strength, stiffness, rigidity, and toughness of the healing bone (Fig. 1). For torsion tests, an additional parameter, twist to failure, can be used as a measure

© 2008 American Society for Bone and Mineral Research

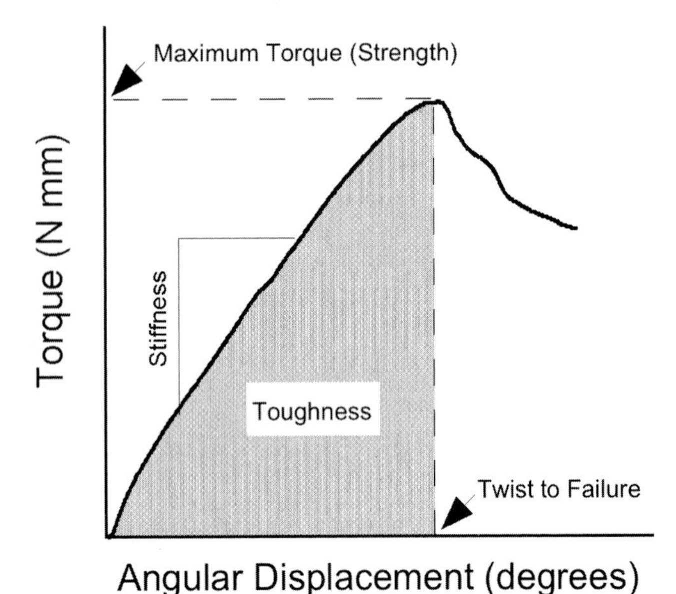

FIG. 1. Representative torque-twist curve for a mouse tibia 21 days after fracture. The curve is annotated to show definitions of basic biomechanical parameters. Torsional rigidity is computed by multiplying the torsional stiffness by the gage length. Analogous definitions hold for bending tests.

of the ductility of the callus. Although strength, a measure of to the force or moment that causes failure, can only be measured once for a given callus, it is possible to obtain more than one measure of stiffness and rigidity. Multistage testing protocols have been reported that apply nondestructive loads to the callus in planes or in loading modes that are different from those used for the stage of the test in which the callus is loaded to failure. With these protocols, it is possible to quantify the bending stiffness in multiple planes[2] or the torsional as well as compressive stiffness.[3]

The mechanical properties shown in Fig. 1 are structural, rather than material, properties. Material properties describe the intrinsic mechanical behavior of a particular type of material (tissue), such as woven bone, fibrocartilage, or granulation tissue. The structural properties of a fracture callus depend on the material properties of the individual callus tissues as well as the spatial arrangement of the tissues and the overall geometry of the callus. Whereas it is possible to use measurements of callus geometry together with those of structural properties to gain some insight into callus tissue material properties,[4] true measurement of these material properties requires direct testing of individual callus tissues.[5]

BIOMECHANICAL STAGES OF FRACTURE HEALING

White et al.[6] used the results of torsion tests performed on healing rabbit tibias at multiple time points (Fig. 2) to define four biomechanical stages of secondary fracture healing. Stage 1 is characterized by extremely low callus stiffness and strength, and failure during the torsion test occurs at the original fracture line. Stage 2 corresponds to a notable increase in callus stiffness and, to a lesser extent, strength. However, it is not until stage 3 that failure during the torsion test occurs at least partly outside of the original fracture line. This stage is also characterized by an increase in callus strength from stage 2. Finally, in stage 4, failure during the torsion test occurs in the intact bone rather than through the original fracture line. It is important to note that, although fracture healing is commonly described in terms of four biological phases (inflammation, soft callus formation, bony callus formation, and remodeling), these phases do not map onto the four biomechanical stages in a one-to-one manner. Stage 1 does correspond to the inflammatory phase, yet stage 2 encompasses the soft callus phase as well as the first part of the bony callus phase. It is the occurrence of bony bridging of the fracture line that is responsible for the increase in stiffness observed in stage 2. The transition from stage 3 to stage 4 roughly corresponds to the start of the remodeling phase.

If the bony callus is sufficiently large, the rigidity and strength of the callus during stage 3 can exceed that of the intact bone. Even though the callus tissues at this stage are not as rigid or as strong as those of well-mineralized lamellar bone, the larger cross-sectional area and moments of inertia of the callus compared with the intact bone can overcompensate for the inferior material properties. However, while robust, the callus at this point in the healing progression is also mechanically inefficient. Through remodeling, the callus is able to retain sufficient mechanical integrity with less mass.

Results of several recent studies further illustrate the biomechanical consequences of individual biological phases of healing. For example, intermittent PTH(1-34) treatment has been shown to increase callus strength,[7,8] primarily as a result of enhanced chondrogenesis.[9] However, whereas PTH treatment leads to an increase in callus size, a slight decrease in the fraction of the callus that is comprised of mineralized tissue was observed,[9] suggesting that the mechanical enhancement results purely from modulation of callus geometry (Fig. 3A). With respect to the later stages of healing, bisphosphonate treatment has been shown to enhance callus strength through inhibition of callus remodeling, resulting in a larger callus and larger proportion of mineralized tissue (Fig. 3B).[10,11]

NONINVASIVE ASSESSMENT OF FRACTURE HEALING

Whereas mechanical tests provide the gold standard mea-

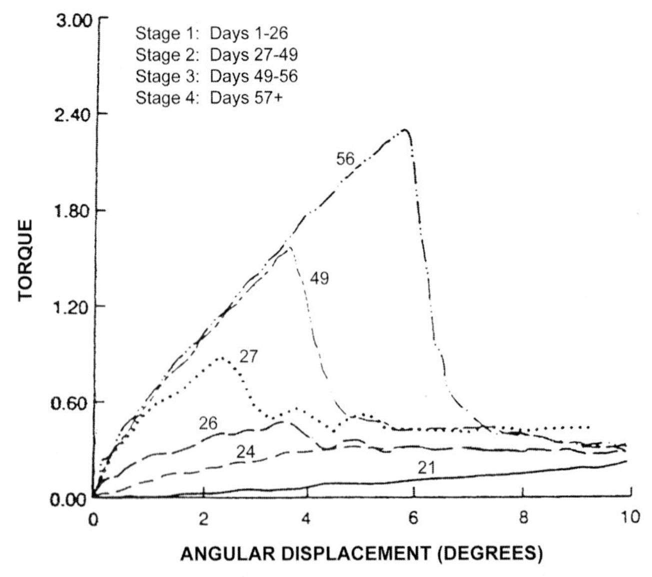

FIG. 2. Torque-twist curves for healing rabbit tibias at various time points (in days) after fracture (after White et al.[6]). Comparison of these curves to that in Fig. 1 shows species differences in the rate of healing. Healing progresses more rapidly in smaller animals (e.g., mouse vs. rabbit), and 21 days postfracture in the mouse corresponds to stage 2. The larger size of the bone and callus in larger animals also leads to greater values of maximum torque, stiffness, and toughness.

© 2008 American Society for Bone and Mineral Research

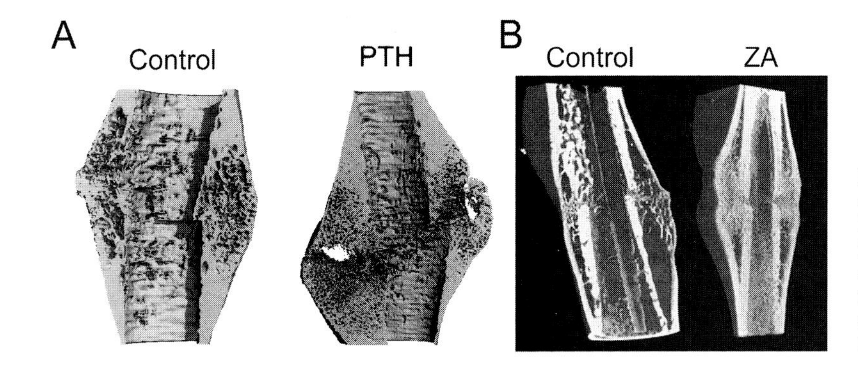

FIG. 3. (A) Longitudinal cut-away views of 3D μCT reconstructions of representative saline- (control) and PTH-treated murine fracture calluses at 14 days after fracture (data from Ref. 9). (B) Longitudinal cross-sections of the fracture callus and cortex at 6 wk after fracture in rats treated with saline (control) and zolendronic acid (ZA) beginning 2 wk after fracture (from Ref. 10; images not to scale).

sures of healing in laboratory studies of fracture healing, clinical assessment of healing requires noninvasive methods. Multiple noninvasive approaches to measuring callus stiffness, whether in axial loading or bending, have been reported, and the clinical feasibility of several has been shown. Typically, these measurements rely on measuring the displacement across the fracture gap or the pin-to-pin displacement under a known force or bending moment.[12–14] If an external fixator is present, it is necessary to consider only the fraction of the applied load that is borne by the callus as opposed to the fixator. From these approaches, quantitative criteria for healing have been put forth. For example, it has been proposed that a fracture can be considered healed when the bending stiffness (the ratio of the applied bending moment to the angular displacement) exceeds 15 Nm/deg,[12] and that in distraction osteogenesis, external fixation can be removed when the fraction of the axial force borne by the fixator is <10%.[15]

Other noninvasive methods of assessing healing provide surrogate, rather than direct, measures of callus mechanical properties and include acoustic emission,[16] ultrasound,[17–19] and CT imaging. Direct comparisons of CT and standard radiographic analyses have indicated that the former can yield comparable or better predictions of callus compressive strength[20] and torsional strength and stiffness[21,22] and more definitive diagnoses of healing progression[23] and of nonunions.[24] However, no consensus currently exists as to which CT-derived measures, or combinations of measures, best predict callus strength and stiffness.

Importantly, the vast majority of noninvasive approaches to monitoring healing focus on callus stiffness and not callus strength. Whereas noninvasive measures of stiffness may provide valuable information about the healing process, a method to evaluate strength would be more clinically meaningful because it would theoretically provide information regarding the ability bear weight and carry loads. In this respect, acoustic methods may pose a considerable advantage, because analysis of ultrasonic wave propagation across the fracture gap can be used to detect bony bridging of the gap. Another viable approach is CT-based finite element analysis, in which CT images are used to construct a finite element model of the callus. This approach was recently shown for estimating callus stiffness[25]; however, it is important to note that this approach requires either knowledge of or assumptions about the elastic and failure properties of the callus tissues.

MECHANOBIOLOGY OF FRACTURE HEALING

Fracture healing is one of the most frequently used scenarios for studies of the effects the local mechanical environment on skeletal tissue differentiation. Mechanical loading of a fracture callus occurs most commonly as a consequence of weight bearing; however, dynamization, or applied micromotion, of the fracture gap has also been studied. Results of these studies have shown that the effects of loading depend heavily on the mode,[26–29] rate,[30,31] and magnitude of loading,[32,33] as well as gap size.[32] Application of cyclic compressive displacements can enhance healing through increased callus formation and more rapid ossification and bridging.[34,35] This effect was found to be greatest for an intermediate strain rate (40 mm/s) as opposed to fast (400 mm/s) or slow (2 mm/s) rates.[36] However, the benefits of applied cyclic compressive displacements seem to be limited to displacements that induce an interfragmentary strain (defined as the ratio of applied displacement to the gap size) of 7% or less.[26,32,37,38]

Tensile displacements applied across the fracture gap can also promote bone formation. This is widely seen in distraction osteogenesis, a clinical procedure for lengthening bones that involves creating an osteotomy and subsequently applying repeated tensile displacements across the gap. Interestingly, in contrast to the effects of cyclic compressive loading, bone formation in distraction osteogenesis occurs primarily by intramembranous ossification. The effect of shear or transverse movement at the fracture site is controversial, with some studies reporting enhanced bone healing[28,29] and others reporting increased fibrous tissue formation and delayed bony healing.[27,39] A series of studies investigating the use of shear movement and also a bending motion to an osteotomy gap reported that these two types of applied motion result in the formation of cartilage rather than bone within the gap.[40,41]

In parallel with some of the earlier experimental studies summarized above, Perren[42] and Perren and Cordey[43] proposed the interfragmentary strain theory, which states that only tissue that is capable of withstanding the present value of interfragmentary strain can form in the fracture gap. This theory is consistent with observations that granulation tissue forms initially in the gap, followed by cartilage and then bone. The successive formation of each type of tissue further reduces the interfragmentary strain that occurs as a result of the applied load and allows a stiffer tissue to form next.

The interfragmentary strain theory presents an oversimplified description of the mechanical environment within the fracture gap in that it uses one scalar (interfragmentary strain) to describe a multiaxial strain field that varies as a function of position within the gap. More recent theories of the mechanobiology of skeletal tissue differentiation have sought to account for this complexity by considering the distributions of local mechanical stimuli present throughout the fracture gap (Fig. 4). Carter et al.[44] have proposed that different combinations of hydrostatic pressure and tensile strain promote formation of different skeletal tissues, whereas Claes and Heigle[45] have postulated that these two stimuli regulate intramembranous versus endochondral ossification. Prendergast and colleagues[46,47] have instead proposed that the two key

© 2008 American Society for Bone and Mineral Research

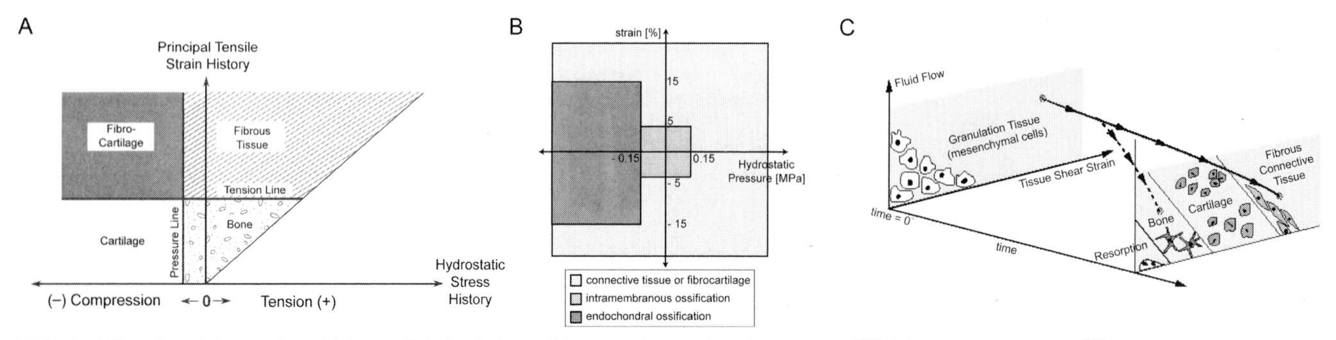

FIG. 4. Theories of the mechanobiology of skeletal tissue differentiation by (A) Carter et al.,[44] (B) Claes and Heigle,[45] and (C) Prendergast et al.[47]

stimuli are shear strain and fluid flow. Direct comparison of these theories' predictions to histological analyses of bone healing suggests that the most accurate predictions are those based on shear strain and fluid flow.[48] However, each of these theories is unable to predict certain histological features of the fracture healing process,[46,48] indicating that the definitive role of the local mechanical environment in modulating healing has yet to be elucidated fully.

SUMMARY

An essential outcome in fracture healing is restoration of sufficient mechanical integrity to allow weight bearing and activities of daily living. Thus, biomechanical analyses of fracture healing are critical for thorough assessment of the repair process. At present, the biomechanical progression of secondary fracture healing is well characterized, and standardized in vitro methods of quantifying the extent of healing have been established. Noninvasive methods of measuring the regain of bone stiffness have also been reported; however, noninvasive methods of measuring the regain of strength are still under development. Studies to date on the effects of mechanical factors indicate that it is possible to augment healing by mechanical loading, and the growing body of literature in this area suggests that further enhancements in healing may be possible. Thus, an understanding of the biomechanics of fracture healing can be applied not only to the assessment of healing but also to development of new repair strategies.

REFERENCES

1. Chao EYS, Aro HT 1997 Biomechanics of fracture fixation. In: Mow VC, Hayes WC (eds.) Basic Orthopaedic Biomechanics. Lippincott-Raven, Philadelphia, PA, USA, pp. 317–352.
2. Foux A, Black RC, Uhthoff HK 1990 Quantitative measures for fracture healing: An in-vitro biomechanical study. J Biomech Eng **112:**401–406.
3. Tsiridis E, Morgan EF, Bancroft JM, Song M, Kain M, Gerstenfeld L, Einhorn TA, Bouxsein ML, Tornetta P III 2007 Effects of OP-1 and PTH in a new experimental model for the study of metaphyseal bone healing. J Orthop Res **25:**1193–1203.
4. Ulrich-Vinther M, Andreassen TT 2005 Osteoprotegerin treatment impairs remodeling and apparent material properties of callus tissue without influencing structural fracture strength. Calcif Tissue Int **76:**280–286.
5. Leong PL, Morgan EF 2008 Measurement of fracture callus material properties via nanoindentation. Acta Biomaterialia **4:**1569–1575.
6. White AA III, Panjabi MM, Southwick WO 1977 The four biomechanical stages of fracture repair. J Bone Joint Surg Am **59:**188–192.
7. Alkhiary YM, Gerstenfeld LC, Krall E, Sato M, Westmore M, Mitlak B, Einhorn TA 2004 Parathyroid Hormone (1-24; Teriparitide) Enhances Experimental Fracture Healing. Orthopaedic Research Society, San Francisco, CA, USA.
8. Andreassen TT, Ejersted C, Oxlund H 1999 Intermittent parathyroid hormone (1-34) treatment increases callus formation and mechanical strength of healing rat fractures. J Bone Miner Res **14:**960–968.
9. Kakar S, Einhorn TA, Vora S, Miara LJ, Hon G, Wigner NA, Toben D, Jacobsen KA, Al-Sebaei MO, Song M, Trackman PC, Morgan EF, Gerstenfeld LC, Barnes GL 2007 Enhanced Chondrogenesis and Wnt-signaling in Parathyroid Hormone Treated Fractures. J Bone Miner Res **22:**1903–1912.
10. Amanat N, McDonald M, Godfrey C, Bilston L, Little D 2007 Optimal timing of a single dose of zoledronic acid to increase strength in rat fracture repair. J Bone Miner Res **22:**867–876.
11. Little DG, McDonald M, Bransford R, Godfrey CB, Amanat N 2005 Manipulation of the anabolic and catabolic responses with OP-1 and zoledronic acid in a rat critical defect model. J Bone Miner Res **20:**2044–2052.
12. Richardson JB, Cunningham JL, Goodship AE, O'Connor BT, Kenwright J 1994 Measuring stiffness can define healing of tibial fractures. J Bone Joint Surg Br **76:**389–394.
13. Hente R, Cordey J, Perren SM 2003 In vivo measurement of bending stiffness in fracture healing. Biomed Eng Online **2:**8.
14. Ogrodnik PJ, Moorcroft CI, Thomas PB 2001 A fracture movement monitoring system to aid in the assessment of fracture healing in humans. Proc Inst Mech Eng [H] **215:**405–414.
15. Aarnes GT, Steen H, Ludvigsen P, Waanders NA, Huiskes R, Goldstein SA 2005 In vivo assessment of regenerate axial stiffness in distraction osteogenesis. J Orthop Res **23:**494–498.
16. Watanabe Y, Takai S, Arai Y, Yoshino N, Hirasawa Y 2001 Prediction of mechanical properties of healing fractures using acoustic emission. J Orthop Res **19:**548–553.
17. Gerlanc M, Haddad D, Hyatt GW, Langloh JT, St Hilaire P 1975 Ultrasonic study of normal and fractured bone. Clin Orthop Relat Res 111:175–180.
18. Glinkowski W, Gorecki A 2006 Clinical experiences with ultrasonometric measurement of fracture healing. Technol Health Care **14:**321–333.
19. Brown SA, Mayor MB 1976 Ultrasonic assessment of early callus formation. Biomed Eng **11:**124–127,136.
20. Jamsa T, Koivukangas A, Kippo K, Hannuniemi R, Jalovaara P, Tuukkanen J 2000 Comparison of radiographic and pQCT analyses of healing rat tibial fractures. Calcif Tissue Int **66:**288–291.
21. Augat P, Merk J, Genant HK, Claes L 1997 Quantitative assessment of experimental fracture repair by peripheral computed tomography. Calcif Tissue Int **60:**194–199.
22. den Boer FC, Bramer JA, Patka P, Bakker FC, Barentsen RH, Feilzer AJ, de Lange ES, Haarman HJ 1998 Quantification of fracture healing with three-dimensional computed tomography. Arch Orthop Trauma Surg **117:**345–350.
23. Grigoryan M, Lynch JA, Fierlinger AL, Guermazi A, Fan B, MacLean DB, MacLean A, Genant HK 2003 Quantitative and qualitative assessment of closed fracture healing using computed tomography and conventional radiography. Acad Radiol **10:**1267–1273.
24. Kuhlman JE, Fishman EK, Magid D, Scott WW Jr, Brooker AF, Siegelman SS 1988 Fracture nonunion: CT assessment with multiplanar reconstruction. Radiology **167:**483–488.
25. Shefelbine SJ, Simon U, Claes L, Gold A, Gabet Y, Bab I, Muller

© 2008 American Society for Bone and Mineral Research

R, Augat P 2005 Prediction of fracture callus mechanical properties using micro-CT images and voxel-based finite element analysis. Bone **36:**480–488.
26. Augat P, Merk J, Wolf S, Claes L 2001 Mechanical stimulation by external application of cyclic tensile strains does not effectively enhance bone healing. J Orthop Trauma **15:**54–60.
27. Schell H, Epari DR, Kassi JP, Bragulla H, Bail HJ, Duda GN 2005 The course of bone healing is influenced by the initial shear fixation stability. J Orthop Res **23:**1022–1028.
28. Bishop NE, van Rhijn M, Tami I, Corveleijn R, Schneider E, Ito K 2006 Shear does not necessarily inhibit bone healing. Clin Orthop Relat Res **443:**307–314.
29. Park SH, O'Connor K, McKellop H, Sarmiento A 1998 The influence of active shear or compressive motion on fracture-healing. J Bone Joint Surg Am **80:**868–878.
30. Wolf S, Augat P, Eckert-Hubner K, Laule A, Krischak GD, Claes LE 2001 Effects of high-frequency, low-magnitude mechanical stimulus on bone healing. Clin Orthop **385:**192–198.
31. Goodship AE, Cunningham JL, Kenwright J 1998 Strain rate and timing of stimulation in mechanical modulation of fracture healing. Clin Orthop **355**(Suppl)**:**S105–S115.
32. Claes L, Augat P, Suger G, Wilke HJ 1997 Influence of size and stability of the osteotomy gap on the success of fracture healing. J Orthop Res **15:**577–584.
33. Claes L, Eckert-Hubner K, Augat P 2002 The effect of mechanical stability on local vascularization and tissue differentiation in callus healing. J Orthop Res **20:**1099–1105.
34. Goodship AE, Kenwright J 1985 The influence of induced micromovement upon the healing of experimental tibial fractures. J Bone Joint Surg Br **67:**650–655.
35. Claes LE, Wilke HJ, Augat P, Rubenacker S, Margevicius KJ 1995 Effect of dynamization on gap healing of diaphyseal fractures under external fixation. Clin Biomech (Bristol, Avon) **10:**227–234.
36. Goodship AE, Watkins PE, Rigby HS, Kenwright J 1993 The role of fixator frame stiffness in the control of fracture healing. An experimental study. J Biomech **26:**1027–1035.
37. Augat P, Margevicius K, Simon J, Wolf S, Suger G, Claes L 1998 Local tissue properties in bone healing: Influence of size and stability of the osteotomy gap. J Orthop Res **16:**475–481.
38. Claes LE, Heigele CA, Neidlinger-Wilke C, Kaspar D, Seidl W, Margevicius KJ, Augat P 1998 Effects of mechanical factors on the fracture healing process. Clin Orthop **355**(Suppl)**:**S132–S147.
39. Augat P, Burger J, Schorlemmer S, Henke T, Peraus M, Claes L 2003 Shear movement at the fracture site delays healing in a diaphyseal fracture model. J Orthop Res **21:**1011–1017.
40. Cullinane DM, Fredrick A, Eisenberg SR, Pacicca D, Elman MV, Lee C, Salisbury K, Gerstenfeld LC, Einhorn TA 2002 Induction of a neoarthrosis by precisely controlled motion in an experimental mid-femoral defect. J Orthop Res **20:**579–586.
41. Cullinane DM, Salisbury KT, Alkhiary Y, Eisenberg S, Gerstenfeld L, Einhorn TA 2003 Effects of the local mechanical environment on vertebrate tissue differentiation during repair: Does repair recapitulate development? J Exp Biol **206:**2459–2471.
42. Perren SM 1979 Physical and biological aspecs of fracture healing with special reference to internal fixation. Clin Orthop **138:**175–180.
43. Perren SM, Cordey J 1980 The concept of interfragmentary strain. In: Uhthoff HK (ed.) Current Concepts of Internal Fixation of Fractures. Springer, Berlin, Germany, pp. 63–77.
44. Carter DR, Beaupre GS, Giori NJ, Helms JA 1998 Mechanobiology of skeletal regeneration. Clin Orthop **355**(Suppl)**:**S41–S55.
45. Claes LE, Heigele CA 1999 Magnitudes of local stress and strain along bony surfaces predict the course and type of fracture healing. J Biomech **32:**255–266.
46. Lacroix D, Prendergast PJ 2002 A mechano-regulation model for tissue differentiation during fracture healing: Analysis of gap size and loading. J Biomech **35:**1163–1171.
47. Prendergast PJ, Huiskes R, Soballe K 1997 ESB Research Award 1996. Biophysical stimuli on cells during tissue differentiation at implant interfaces. J Biomech **30:**539–548.
48. Isaksson H, Wilson W, van Donkelaar CC, Huiskes R, Ito K 2006 Comparison of biophysical stimuli for mechano-regulation of tissue differentiation during fracture healing. J Biomech **39:**1507–1516.

© 2008 American Society for Bone and Mineral Research

SECTION II

Skeletal Physiology
(Section Editor: Ego Seeman)

Please also see the following websites: Tanner Staging: https://courses.stu.qmul.ac.uk/SMD/kb/humandevelopment/stage1/learninglandscapefolders/growthpubertyetc/topic8picture2.htm; http://www.ncbi.nlm.nih.gov/bookshelf/br.fcgi?book = endocrin&partid = 29&rendertype = box&id = A1059; Growth Charts: http://www.chartsgraphsdiagrams.com/HealthCharts

© 2008 American Society for Bone and Mineral Research

Chapter 13. Fetal and Neonatal Bone Development

Frank Rauch

Genetics Unit, Shriners Hospital for Children, Montreal, Quebec, Canada

INTRODUCTION

Bone development during the fetal and neonatal periods is characterized by extremely rapid growth. In the 7 mo of fetal life, body length increases by ~45 cm. After a brief deceleration at around the time of birth, longitudinal growth resumes at a rate that is about twice as fast as at the peak of the pubertal growth spurt.[1] Bone in the earliest phases of development is exposed to conditions that differ markedly from those of subsequent periods of the life cycle. In particular, the unique nutritional, hormonal, and mechanical environment sets the fetal and neonatal skeleton apart from later developmental stages.

Studying fetal and neonatal bone development in human is difficult for a variety of ethical and methodological reasons. However, many widely used animal models fail to replicate key characteristics of the human feto-placental unit or of the postnatal adaptation process. It is therefore often unclear whether findings in animal models can be applied to the human situation. Where possible, this brief overview therefore focuses on observations that were made in humans.

FETAL BONE DEVELOPMENT

By the start of fetal life (ninth week after conception), skeletal patterning is complete, and the basic shapes of all bones are established. The skeletal soft tissue templates undergo ossification and growth, resulting in a dramatic increase in size but only small changes in shape. For example, the femur increases 2.3-fold in length and 2.1-fold in mid-diaphyseal width during the second half of pregnancy, indicating that dimensional relationships change little.[2,3] Concomitant with the increase in bone size, the internal structure is modified. The average thickness of trabeculae in the proximal femoral metaphysis increases by about one third during the second half of gestation.[4] Cortical thickness at the femoral midshaft approximately doubles during the same period.[2] Together, these changes in external and internal bone structure lead to a rapid rise in skeletal mass, especially during the third trimester of gestation. At the end of pregnancy, the human skeleton weighs ~100 g, of which 65 g is mineral matter, including ~30 g of calcium.[5]

Transfer of Mineral From Maternal Blood to the Fetal Skeleton

The organic and mineral raw materials for building the skeletal structure need to be extracted from the maternal circulation through the placenta. Insufficient transfer of organic compounds results in global intrauterine growth retardation affecting all tissues, whereas lack of mineral primarily results in skeletal abnormalities.

Calcium and phosphorus are carried across the placenta by active processes.[6] Transplacental transport of calcium is thought to be a one-way process, because fetal-to-maternal flow of calcium is assumed to be <1% of the maternal-to-fetal flow.[6] The serum levels of both minerals are higher in the fetus than in the mother, which is probably necessary to maintain the high mineralization rates in the fetal skeleton.[6] Both calcium and phosphorus are essential components of the bone mineral phase. However, whereas a number of situations are known where insufficient calcium supply becomes a limiting factor for skeletal development, no conditions have been reported where lack of phosphorus leads to skeletal abnormalities in the fetus. Maybe as a consequence of this, the regulators of fetal phosphorus homeostasis have not been studied in detail.

The main determinant of fetal calcium levels is likely to be PTH-related protein (PTHrP), which also seems to regulate the fetal–placental calcium transport.[6] PTHrP is produced in many fetal tissues, but the predominant source of circulating PTHrP seems to be the placenta. PTH and 1,25 dihydroxyvitamin D concentrations are low in fetal serum, in accordance with high calcium levels. 1,25 dihydroxyvitamin D does not seem to be a major player in fetal mineral homeostasis, because neither 25-hydroxyvitamin D-1-α-hydroxylase deficiency nor absence of vitamin D receptors lead to clinically recognizable skeletal problems in the fetus or newborn.[6]

Mechanical Aspects

Mechanical forces have an essential role in healthy fetal bone development. As in later life, the largest forces acting on the skeleton are caused by muscle action. When the loads imposed by muscle action are diminished, fetal bone development is impaired. For example, mice lacking striated muscles have abnormal skeletal shape.[7] Pharmacological inhibition of muscle contraction during fetal development decreases periosteal expansion in rats.[8] Mechanical stimulation also seems to be important for the creation of secondary ossification centers in the epiphyses of long bones.[8] The importance of the fetal muscle–bone interaction is also obvious in newborns with muscular hypotonia of intrauterine onset. The diameter of long bones is decreased in such disorders and fractures often occur at birth.[8]

NEWBORN BONE DEVELOPMENT

The newborn period is defined as the time from birth until the age of 28 days. However, changes in the skeleton occur somewhat slower than in many other organ systems. Therefore, this overview also takes the first few postnatal months into account.

The immediate postnatal period is characterized by rapid adaptational changes in many organ systems. The skeleton has to adapt to the sudden interruption of placental supply of nutrition and hormones, as well as to the transition from a mechanical environment where movements occur against the resistance of the uterine wall to an environment with unrestricted movement.[9]

After the placental supply of calcium is cut off, the neonate becomes dependent on intestinal sources for obtaining the raw materials for skeletal growth. For the first few days of life, milk intake is typically low and body weight decreases. However, after the age of 5–10 days, oral intake usually becomes sufficient to resume steady weight gain and growth.

Neonatal serum calcium levels decrease within hours after birth, leading to a secondary increase in PTH levels and subsequently of 1,25 dihydroxyvitamin D.[6] PTHrP, crucial for maintaining fetal calcium levels, loses its dominant role in calcium homeostasis.

The author states that he has no conflicts of interest.

Key words: bone development, calcium, newborn, parathyroid hormone, vitamin D

© 2008 American Society for Bone and Mineral Research

Some bone marker studies suggest that bone formation decreases in the first few days after birth, whereas bone resorption increases.[10,11] Once the early adaptational hurdles have been cleared and the phase of regular and rapid growth sets in, markers of both bone formation and resorption increase.

During the first few months of life, the volumetric density of long bone shafts such as the femoral diaphysis decreases by ~30%.[5,12] This is mostly because of an expansion in marrow cavity size relative to outer bone size (Fig. 1). Apart from this macroscopic change in bone geometry, there is also a postnatal decrease in cortical bone density that contributes to the overall decline in whole bone density (Fig. 1). These changes have been classically called "physiologic osteoporosis of infancy."[12] However, the most important criterion for the presence of osteoporosis—increased bone fragility—is absent.

It should also be noted that the decrease in volumetric density on the whole bone level does not mean that bone mass is lost. Rather, it reflects a redistribution of bone tissue from the endocortical and intracortical surfaces to the periosteal surface.[12] In addition, new net bone formation proceeds rapidly at the periosteal surface. Consequently, bone mineral mass increases considerably during the first months of life (Fig. 1). The determinants of these postnatal changes of the skeleton are not clear but are likely to involve both hormonal and mechanical factors.[9]

DISORDERS OF FETAL AND NEONATAL BONE DEVELOPMENT

Bone Dysplasias

This is the largest group of disorders affecting fetal and neonatal bone. They are typically caused by mutations in genes that encode structural components of the organic bone matrix or that are otherwise important for the function of bone cells. A discussion of this group of disorders is beyond the scope of this overview.

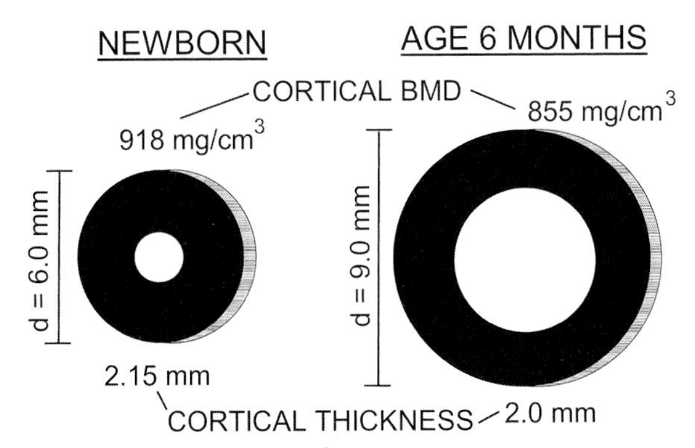

FIG. 1. A model of bone development at the femoral midshaft from birth to 6 mo of age. Adapted from Reference 12. At birth, the external bone diameter, d, is ~6.0 mm, and cortical thickness is 2.15 mm. Thus, ~90% of the total bone cross-sectional area is made up of cortical bone, and only 10% is taken up by the marrow cavity. Six months later, the external bone diameter has increased to 9.0 mm, and cortical thickness has decreased to 2.0 mm. Now, cortical bone takes up only 70% of the cross-section, whereas the marrow cavity represents 30% of the total bone cross-sectional area. In addition, cortical BMD has decreased by 7%. Together these changes represent a drop in total volumetric BMD of ~30%. At the same time, the total mass of mineral in this hypothetical bone slice of 2 mm thickness has increased by 58% because of the increase in bone size.

Maternal Hypocalcemia

At the peak of fetal skeletal development, ~250 mg of calcium are incorporated into bone matrix per day. When the supply of calcium to the skeleton does not match the demand imposed by the prevailing growth rate, a mineralization disorder arises. This can occur when the mother is chronically hypocalcemic (e.g., because of hypoparathyroidism, pseudohypoparathyroidism, chronic renal failure, or severe vitamin D deficiency). The fetus or newborn may present with severe skeletal abnormalities such as widening of the growth plates, cupping of the metaphyses, thin cortices, bone deformities, and fractures, as well as elevated serum levels of PTH. These are the typical features of calcipenic rickets, but in the neonatal literature, such cases have confusingly been reported as fetal/neonatal/congenital hyperparathyroidism.[6] The severity of this disorder in fetal calcium homeostasis can vary widely, ranging from a lack of clinical manifestations to death in utero.

Maternal Hypercalcemia

Conditions that lead to maternal hypercalcemia, such as maternal primary hyperparathyroidism or familial hypocalciuric hypercalcemia, can lead to suppression of the fetal parathyroid glands, probably because of increased calcium flux across the placenta to the fetus.[6] This reversible suppression of parathyroid glands can lead to "late-onset" (after the third day of life) hypocalcemia in the newborn but does not lead to obvious skeletal abnormalities.

Calcium Receptor Defects

When the fetus is heterozygous for an inactivating calcium receptor mutation and the mother is normocalcemic, fetal primary hyperparathyroidism may result.[13] Skeletal finding can include periosteal erosions, metaphyseal fraying, and metaphyseal fractures. Both the hyperparathyroidism and the bone disorder are usually self-limiting, and such patients will go on to have familial hypocalciuric hypercalcemia, a benign disorder. In contrast, if the fetus is homozygous for an inactivating calcium receptor mutation, neonatal severe hyperparathyroidism can ensue. This is a very serious condition with often extremely high serum calcium levels and marked signs of hyperparathyroid bone disease, which requires parathyroidectomy.[14]

Maternal Vitamin D Deficiency

As mentioned earlier, severe cases of maternal vitamin D deficiency may cause hypocalcemia in the mother and can result in a picture of calcipenic rickets in the newborn.[15] This, however, is a very rare occurrence. Most babies of vitamin D–deficient mothers are asymptomatic at birth but can present with hypocalcemic symptoms (seizures, neuromuscular irritability, stridor, apnea) after the third day of life or with rickets (typically after the age of 2 mo).[16,17] Epidemiological studies suggest that vitamin D deficiency during pregnancy may have long lasting effects on bone development even in children who do not develop symptoms.[18]

Metabolic Bone Disease of Prematurity

Both osteomalacia/rickets (i.e., an inadequate amount of mineral relative to the amount of bone tissue) and osteopenia (i.e., an inadequate amount of bone tissue) can occur after premature birth. When premature babies are fed regular human milk, the supply of both calcium and phosphorus is low, but the critical factor leading to osteomalacia/rickets is the lack of phosphorus.[19] Serum phosphate levels decrease, and there

© 2008 American Society for Bone and Mineral Research

is not enough substrate for incorporation into the organic bone matrix. Osteopenia results from diminished synthesis and/or increased resorption of organic bone matrix. This can be caused by severe systemic disease, a drug side effect, or lack of mechanical stimulation (disuse osteoporosis).

Fortunately, bone disease of prematurity seems to be a self-limiting disease. The fracture risk is increased up to the age of 2 yr but not later on.[20] In any case, bone disorders in premature babies are much less common now than in the 1980s, when extremely high fracture rates were reported.[20] Both improved nutritional management and shorter postnatal immobilization periods (because of considerably decreased ventilation times) may have contributed to this result.

REFERENCES

1. Bogin B 1998 Patterns of human growth. In: Ulijaszek SJ, Johnston FE, Preece MA (eds.) The Cambridge Encyclopedia of Human Growth and Development. Cambridge University Press, Cambridge, UK, pp. 90–96.
2. Rodriguez JI, Palacios J, Rodriguez S 1992 Transverse bone growth and cortical bone mass in the human prenatal period. Biol Neonate **62:**23–31.
3. Chitty LS, Altman DG 2002 Charts of fetal size: Limb bones. BJOG **109:**919–929.
4. Salle BL, Rauch F, Travers R, Bouvier R, Glorieux FH 2002 Human fetal bone development: histomorphometric evaluation of the proximal femoral metaphysis. Bone **30:**823–828.
5. Trotter M, Hixon BB 1974 Sequential changes in weight, density, and percentage ash weight of human skeletons from an early fetal period through old age. Anat Rec **179:**1–18.
6. Kovacs CS, Kronenberg HM 1997 Maternal-fetal calcium and bone metabolism during pregnancy, puerperium, and lactation. Endocr Rev **18:**832–872.
7. Gomez C, David V, Peet NM, Vico L, Chenu C, Malaval L, Skerry TM 2007 Absence of mechanical loading in utero influences bone mass and architecture but not innervation in Myod-Myf5-deficient mice. J Anat **210:**259–271.
8. Nowlan NC, Murphy P, Prendergast PJ 2007 Mechanobiology of embryonic limb development. Ann N Y Acad Sci **1101:**389–411.
9. Rauch F, Schoenau E 2001 The developing bone: Slave or master of its cells and molecules? Pediatr Res **50:**309–314.
10. Hogler W, Schmid A, Raber G, Solder E, Eibl G, Heinz-Erian P, Kapelari K 2003 Perinatal bone turnover in term human neonates and the influence of maternal smoking. Pediatr Res **53:**817–822.
11. Schoenau E, Rauch F 2003 Biochemical markers of bone metabolism. In: Glorieux FH, Pettifor J, Jueppner H (eds.) Pediatric Bone. Academic Press, San Diego, CA, USA, pp. 339–357.
12. Rauch F, Schoenau E 2001 Changes in bone density during childhood and adolescence: An approach based on bone's biological organization. J Bone Miner Res **16:**597–604.
13. Bai M, Pearce SH, Kifor O, Trivedi S, Stauffer UG, Thakker RV, Brown EM, Steinmann B 1997 In vivo and in vitro characterization of neonatal hyperparathyroidism resulting from a de novo, heterozygous mutation in the Ca2+-sensing receptor gene: Normal maternal calcium homeostasis as a cause of secondary hyperparathyroidism in familial benign hypocalciuric hypercalcemia. J Clin Invest **99:**88–96.
14. Waller S, Kurzawinski T, Spitz L, Thakker R, Cranston T, Pearce S, Cheetham T, Vanthoff WG 2004 Neonatal severe hyperparathyroidism: Genotype/phenotype correlation and the use of pamidronate as rescue therapy. Eur J Pediatr **163:**589–594.
15. Park W, Paust H, Kaufmann HJ, Offermann G 1987 Osteomalacia of the mother--rickets of the newborn. Eur J Pediatr **146:**292–293.
16. Specker B 2004 Vitamin D requirements during pregnancy. Am J Clin Nutr **80:**1740S–1747S.
17. Ladhani S, Srinivasan L, Buchanan C, Allgrove J 2004 Presentation of vitamin D deficiency. Arch Dis Child **89:**781–784.
18. Javaid MK, Crozier SR, Harvey NC, Gale CR, Dennison EM, Boucher BJ, Arden NK, Godfrey KM, Cooper C 2006 Maternal vitamin D status during pregnancy and childhood bone mass at age 9 years: A longitudinal study. Lancet **367:**36–43.
19. Bishop N 1989 Bone disease in preterm infants. Arch Dis Child **64:**1403–1409.
20. Bishop N, Sprigg A, Dalton A 2007 Unexplained fractures in infancy: Looking for fragile bones. Arch Dis Child **92:**251–256.

Chapter 14. Skeletal Development in Childhood and Adolescence

Laura K. Bachrach

Division of Pediatric Endocrinology, Stanford University School of Medicine, Stanford, California

INTRODUCTION

Childhood and adolescence are periods of enormous skeletal growth when >90% of adult bone mass is acquired.[1] This process involves changes in the size, shape, and material properties of bone—parameters that determine skeletal strength.[2–4] To add to the complexity, skeletal changes vary in tempo and magnitude within different regions of the skeleton and within the trabecular and cortical compartments. Although heritable factors (including sex and race) determine 60–80% of the variability in skeletal development, modifiable factors related to lifestyle and illness can influence bone growth, modeling, and remodeling.[1,5]

Despite the proliferation of pediatric bone research over the past three decades, many uncertainties remain. How much of the gain in bone mass can be attributed to increased bone size and how much to gains in volumetric density? Do sex and racial differences in the rates of stress fractures and osteoporosis have their roots in childhood[2,3]? If so, do they reflect differences in bone mass, size, or distribution and at what stage of development do they appear? What factor(s) predict bone fragility and fractures in childhood[6]? How much can modifiable variables such as calcium intake and activity influence bone size, shape, and density?

Details regarding skeletal development in childhood have been obscured by limitations of densitometry. Until recently, pediatric bone studies have relied principally on measures of BMC and areal BMD (aBMD) measured by DXA. These data are challenging to interpret in growing subjects.[2,5,7] Only re-

Dr. Bachrach has served on the Data Safety Monitoring Board for Novartis and as a consultant to Roche and Takeda Pharmaceuticals.

Key words: puberty, bone mineral accrual, peak bone mass, sex, race, pediatrics, fractures, bone geometry, growth, pQCT, DXA, QCT, MRI, periosteum, endosteum, BMD, cortical thickness

© 2008 American Society for Bone and Mineral Research

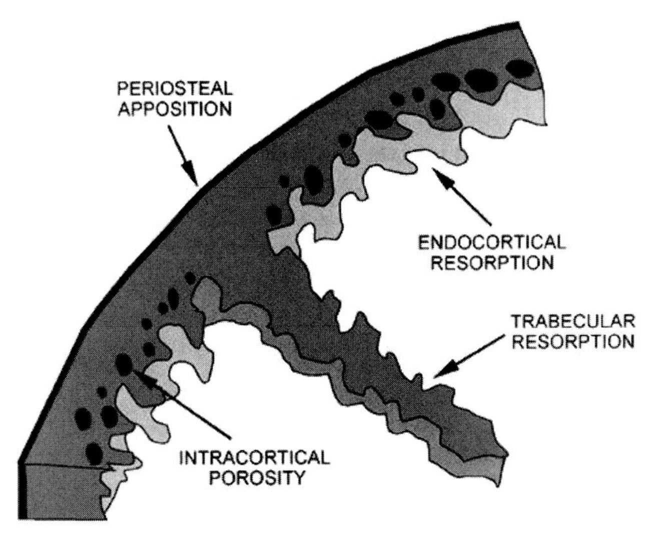

FIG. 1. The mineralized bone mass of the skeleton is defined externally by its periosteal envelope and internally by the endocortical, intracortical, and trabecular components of its endosteal envelope. (Reproduced with permission from the American Society for Bone and Mineral Research from Duan Y, Beck TJ, Wang X-F, Seeman E 2003 Structural and biomechanical basis of sexual dimorphism in femoral neck fragility has its origins in growth and aging. J Bone Miner Res **18**:1767.)

cently have alternative methods been used to explore the evolution of bone size, geometry, and changes within bone compartments.

Understanding the changes in bone mass, size, geometry, and material properties in childhood and adolescence is not simply an academic exercise. Rather, this knowledge is essential to unraveling the determinants of bone fragility. This chapter will provide a broad overview of the skeletal changes during childhood and adolescence with a particular emphasis on the peripubertal period and the emergence of sex and racial differences in parameters that determine bone strength.

EVALUATING SKELETAL DEVELOPMENT

Changes in bone size can be captured accurately by measuring height, sitting height, and other anthropometric parameters. Assessing changes at all levels of bone organization is considerably more challenging.[4] Figure 1 depicts the skeletal processes that are active at the outer (periosteal) and inner (endosteal) surfaces and within the trabecular and cortical compartments during childhood and adolescence. The material property of bone ($BMD_{material}$) is determined by the extent of mineralization of the organic matrix of bone exclusive of marrow spaces, canals, and lacunae. Newly acquired bone deposited during bone remodeling has less density because the matrix contains less mineral and more water; density increases as mineral is added over the next 6 mo during secondary mineralization.[4] $BMD_{material}$ can be assessed only by bone biopsy or by measuring the ash weight of dried bone. $BMD_{compartment}$ reflects the mineral mass per unit volume within the trabecular or cortical regions of the bone, inclusive of marrow spaces, canals, and lacunae. This parameter will increase if $BMD_{material}$ increases or if there are changes in the number or thickness of trabeculae. $BMD_{compartment}$ can be assessed noninvasively through use of QCT or pQCT. BMD_{total} is defined as the mineral mass within the entire bone or region of bone bounded by the periosteum. Gains in BMD_{total} occur with increases in $BMD_{material}$, relative increases in the amount of denser cortical bone, or relative decreases in the trabecular compartment of bone. BMD_{total} can be assessed with DXA, QCT, and pQCT.

Although bone densitometry is discussed in detail elsewhere in this *Primer*, it is important to underscore the technical challenges of each method as a cautionary note when interpreting pediatric bone mineral research to date.[5,7,8] None of these noninvasive methods measures the true "density" of bones, defined as the material properties of the bone. Despite references throughout the literature to BMD, these devices provide measures of apparent density contained within the area or volume of bone in the region of interest. DXA is widely available, precise, and safe for use in children. DXA provides a measure of BMC, bone area, and areal BMD (aBMD; BMC within the projected area of bone measured). Because aBMD does not adjust for the depth of bone, values are affected by size and will be greater in larger individuals. This is a critical limitation given the marked variability in bone and body size in growing children. DXA cannot distinguish trabecular from cortical compartments nor does the technique provide direct measurements of bone geometry. Several approaches have been proposed to address these limitations. BMC has been adjusted for height or for estimates of bone volume derived from bone area measurements.[5] Given the strong association between muscle and bone development, other investigators have proposed correcting BMC for lean body mass.[9] 2D DXA measurements have also been manipulated to estimate bone geometry, based on assumptions concerning the 3D shape of bone.[10] To date, there is no consensus among pediatric bone researchers about the optimal method(s) to address the limitations of DXA.[11]

QCT provides 3D assessments of bone, from which apparent volumetric BMD (vBMD, g/cm^3) can be determined directly in the axial and appendicular skeleton.[8] QCT can distinguish between trabecular and cortical bone and accurately measures bone size and geometry. pQCT offers the advantages of QCT with minimal radiation exposure.[8] Total and cortical cross-sectional bone area, cortical thickness, and vBMD can be measured in the long bones of the arm or leg. Bone dimensions and density are used to calculate estimates of bone strength (stress-strain index). pQCT has several limitations in growing children.[12,13] In regions of the bone such as distal forearm where the cortical compartment is very narrow, pQCT may not be able to distinguish trabecular from cortical bone. There is also considerable variability in trabecular density along the length of the metaphysis; altering the position of the slice of tibial bone studied by only 1 mm resulted in a 17% difference in vBMD.[13] This makes it critical to standardize measurement sites or to rely on mean density from multiple slices for both cross-sectional and longitudinal studies.

Densitometry data acquired using DXA, QCT, and pQCT are not interchangeable because of technical differences between the methodologies. Although measurements of spine BMC using DXA and QCT are highly correlated, there is little correlation between aBMD by DXA and vBMD measured using QCT.[14] This discrepancy reflects the influence of bone size on DXA aBMD measurements. When comparing DXA and pQCT measurements, whole body bone area and BMC corrected for height (by DXA) are the strongest correlates of bone geometry and strength by pQCT.[15]

MRI has been used to study 3D changes in bone geometry in the growing child. Estimates of midfemur total bone volume derived from DXA measurements differ from those obtained using MRI, likely because long bones are irregular in shape are not cylindrical, as assumed in DXA models.[16] Observations of skeletal geometry based on DXA calculations must be interpreted with this caveat in mind.

© 2008 American Society for Bone and Mineral Research

Data from bone densitometry, histomorphometry, and in vivo testing of tissue material properties assess components that contribute to skeletal strength. However, whole bone strength is more than the sum of its parts.[17] Tests of material properties, bone geometry, or mass correlate with but may not predict the resistance of whole bone to fracture. At best, they are measurable surrogates that provide some insight into the complexity of skeletal development.

BONE GROWTH AND BONE MASS

Changes in bone size, geometry, and mineral accrual follow a similar but not identical pattern through childhood and adolescence. Skeletal development proceeds at different rates in the axial and appendicular skeleton. Through the prepuberty period, bone growth and mineral accrual are faster in the legs and arms than in the spine. Spine growth accelerates in early and midpuberty, and growth at both axial and appendicular sites slows at late puberty.[18,19] These observations have clinical implications because chronic disease in childhood or adolescence will have varying effects on bone growth or mineral accrual depending on the timing of illness.

Maximal bone growth precedes mineral acquisition throughout childhood and early adolescence. Girls at 7 yr of age have reached 80% of their adult stature but have acquired only 40% of their expected adult bone mass.[18] Seven-year-old boys have achieved 70% of their adult height but only 35% of their expected peak bone mass.[19] Because peak rates of mineral accrual lag 8 mo behind peak height velocity, bone is relatively undermineralized during peripuberty, contributing to the increased incidence of fractures.[20] There are sex and racial differences in the timing of puberty and in the magnitude of skeletal growth.[2,21,22] Girls have earlier onset of puberty than boys[21]; blacks mature earlier than whites.[22] Men achieve a final height that averages 13 cm (5 in) greater than that of women.

The gains in bone mass reported using DXA must be interpreted in light of these differences in skeletal growth and maturity. Between 6 and 16 yr of age, total body BMC increases 2.5- and 3-fold in girls and boys, respectively; aBMD at the spine nearly doubles.[23] Increasing bone size accounts for much of the observed gains in both BMC and BMD, although gains in volumetric BMD occur at some skeletal sites during puberty as discussed below.[2,22,24]

CHANGES IN MATERIAL PROPERTY OF BONE ($BMD_{material}$)

Insight into how the material properties of bone evolve through childhood and adolescence is limited because $BMD_{material}$ cannot be assessed by noninvasive techniques. Analysis of transiliac crest bone biopsies from healthy 1- to 22-yr-old youths showed a decline in metabolic activity with age.[25] Rates of bone remodeling decreased within both cortical and trabecular compartments, although bone formation per bone surface remained higher in cortical bone. Within the cortical bone compartment, there was more remodeling activity in the inner than in the outer region. These observations suggest that bone remodeling activity is controlled locally within bone. Regional differences in bone porosity may influence the biomechanical strength of bone.[25]

The bone histomorphometry studies cited above lacked the power to detect potential sex differences in $BMD_{material}$. However, one study in young adults (17–46 yr old) found no differences in tissue-level quality of cortical bone.[26] Tibias were measured and bone samples extracted from the midcortex to test tissue-level mechanical properties. Females had narrower tibias (for body size) than males, but there were no sex differences in tissue stiffness, strength, toughness, and damageability by in vitro tests.

CHANGES IN VOLUMETRIC BMD

Although bone size and mass increase throughout the skeleton during childhood and adolescence, gains in volumetric BMD are much more limited. Increases in apparent volumetric BMD (BMAD) using DXA have been noted at the spine during adolescence.[22,27] Between 9 and 20 yr of age, spine BMAD increased only 20%, whereas aBMD (which reflects bone growth and density) increased by 55%.[22] Estimates of volumetric BMD did not increase with age in the femoral neck and midshaft in some studies[22,27] but increased in the femoral midshaft during late puberty in others.[18,19]

Apparent volumetric BMD measured directly by QCT seems to increase during adolescence only in the axial skeleton.[24] Gains in both cortical and trabecular volumetric BMD (vBMD) were observed to an equal extent in boys and girls.[24] Increases in trabecular vBMD have been attributed to gains in thickness of trabeculae rather than to increases of trabecular number or $BMD_{material}$.[2] Cortical bone vBMD in the long bones (midshaft of the femur and radius) as measured by QCT did not change throughout childhood and adolescence.[28,29]

In contrast to QCT findings, studies using pQCT have observed gains in cortical vBMD of the long bones during puberty.[12,30–32] At the radial diaphysis, boys developed much larger bones and acquired more bone mass than girls but vBMD increased more in girls. Between 6 and 40 yr of age, gains occurred in cross-sectional area (116% versus 50%) and BMC (140% versus 111%) in males and females, respectively, whereas vBMD increased by 48% in females and 23% in males.[12] When subjects were matched for developmental stage and cortical width, sex differences became apparent only in late puberty, when girls had 3–4% greater vBMD at this site than boys.[30] In a similar manner, sex differences in the distal radius were increased in late puberty. Total and cortical vBMD at this site increased between age 15 and adulthood in both males and females; trabecular vBMD in the distal forearm increased only in males, increasing the sex difference for this parameter between prepuberty and young adulthood.[31] At the tibia, girls had a greater cortical density than boys from pre- and early puberty to late puberty.[32,33] These finding suggest there is less intracortical remodeling in girls than in boys, resulting in less porous cortical bone. Discrepant results obtained using QCT and pQCT likely reflect methodological differences between the techniques or variability in the specific subregion of bone studied.

CHANGES IN BONE GEOMETRY

As long bones increase in length, they increase in width as new bone is added to the outer periosteal surface.[3,34] Resorption occurs at the inner endocortical surface of bone, which allows the inner medullary (marrow) cavity of bone to expand. Cortical thickness is determined by the net changes occurring at the periosteal and endosteal surfaces of bone. Even without an increase in cortical thickness, the distribution of bone mass further away from the neutral axis of bone increases the resistance of bone to bending or torsion.[35]

More than three decades ago, Garn[34] observed sex differences in bone geometry of the second metacarpals using radiogrametry. Bones in males had larger diameters, whereas in females, bones were narrower with smaller medullary cavities. These changes were postulated to reflect the differential ef-

© 2008 American Society for Bone and Mineral Research

fects of sex steroids on the direction of bone growth and geometry. Subsequent studies using DXA, QCT, and pQCT have both confirmed and refuted this model of sex-specific bone growth. Contraction of the medullary cavity in females during puberty has not been observed consistently. Furthermore, several studies have observed sex differences in bone geometry before puberty, suggestive that determinants other than sex steroids are programming bone size and density.

Using DXA models of bone structure, sex differences in femoral bone geometry emerge during puberty. The width of cortical bone (adjusted for height) was similar in both sexes but the cortex in males was distributed further away from the neutral axis of bone. Total cortical width increased at puberty in girls through small increases in periosteal diameter and further additions at the endocortical surface.[18]

In contrast, QCT studies of bone geometry found sex differences in the axial but not the appendicular skeleton.[36] The vertebrae of females had smaller cross-sectional areas than males matched for height and weight, whereas vertebral heights and volumetric density were similar. No sex differences were seen in either the total cross-sectional area or the cortical bone area of the femoral midshaft, a site whose geometry is shaped by weight and biomechanical forces.

Sex differences in tibial bone geometry have been reported in pQCT studies of prepubertal and pubertal subjects.[33,37] Boys exhibited greater periosteal expansion, bone diameter, total bone area, and cortical area than girls. Gains in these geometric parameters contributed to greater calculated indices of bone strength. In all subjects, muscle cross-sectional area was a key correlate of tibial bone strength and geometry, supporting the hypothesis that bones adapt to the muscle forces applied to them.[38] No increased endosteal apposition in the tibia was observed even in late puberty.[37] This is in contrast to studies of the upper extremity. Using pQCT of the radius,[12,39] girls had a reduction in endocortical diameter during mid to late puberty, indicating increased endocortical bone apposition. Higher estrogen concentrations have been associated with smaller medullary cavities, greater cortical thickness and increased total vBMD in pubertal girls.[39]

An MRI study of the femoral shaft in healthy 6–25 yr olds found that total bone area, cortical bone area, and medullary area increased significantly between prepuberty and adulthood. These parameters were significantly greater in males than in females at all ages. After correcting for bone size and body weight, the sex difference persisted for all parameters except medullary area.[16] This study, like the QCT study, concluded that the pattern of periosteal and endocortical expansion during growth of the femur was similar in both sexes with expansion of the medullary space during growth. Differences in findings may be attributed to technical challenges of estimating bone geometry from 2D DXA images.[16]

It remains to be determined whether these divergent results reflect different skeletal adaptations at weight-bearing or non–weight-bearing sites or technical differences and limitations in the densitometry techniques used.

ETHNIC DIFFERENCES IN BONE ACQUISITION

Racial and ethnic differences in rates of osteoporosis have their roots in childhood. Studies using DXA have reported that blacks had greater BMC or aBMD than nonblacks (whites, Hispanics, and Asians) at some skeletal sites.[22,40] These differences have been detected before puberty in some but not all studies; differences increase in late puberty as blacks continue to gain more bone than nonblacks. These differences persisted after adjusting for bone size with estimates of vBMD.[22]

Studies using QCT indicate that healthy black youths have significantly greater trabecular vBMD in the spine than whites, although vertebral bone size was similar.[28] These differences emerged at Tanner stage IV and V as blacks gained more vBMD than whites. In contrast, in the appendicular skeleton, racial differences were related to bone size. In the midshaft of the femur, blacks had greater total cross-sectional area than whites but similar cortical vBMD and cortical bone area.

Several mechanisms have been proposed to account for racial differences in bone mass and geometry.[40] Earlier pubertal maturation and larger bone size among blacks account for some differences. However, BMC remains greater in blacks after adjusting for bone size, Tanner stage, and age.[40] Black-white differences in body composition, bone turnover, and efficiency in calcium metabolism may account for residual differences.[40,41]

Differences among nonblack groups (Hispanics, Asians, whites) studied by DXA have been much more subtle and largely attributed to differences in bone size.[22] A study using pQCT found that pre- and early pubertal whites had significantly greater cortical area of the tibia, whereas Asian girls had significantly greater cortical density.[33]

HERITABLE AND MODIFIABLE INFLUENCES ON SKELETAL DEVELOPMENT

Beyond sex and racial differences in skeletal development, there is considerable variability among individuals in bone size, geometry, and mass. Heritable factors account for an estimated 60–80% of this variability, influencing rates of bone gained (peak bone mass) or subsequent rates of bone loss.[1] A number of candidate genes have been linked to bone.[42] These genetic factors influence skeletal development through effects on bone size, efficiency of calcium absorption, the response to biomechanical stimuli, and other mechanisms.

Genetic potential can be reached only when diet, physical activity, and hormone production are adequate. A number of chronic childhood illnesses and the medications used to treat them can compromise bone growth or mineral accrual.[1,5] Chronic undernutrition, immobilization, inflammation, sex steroid deficiency, and glucocorticoid excess are key skeletal risk factors. As with studies of healthy youth, much of the research on childhood osteoporosis has been based on DXA measurements of BMC and aBMD. Failure to consider delayed bone growth and maturity results in overestimation of apparent deficits.[5] For example, young patients (4–26 yr of age) with Crohn disease were found to have significantly reduced whole body for BMC for age and race.[43] Adjusting for the confounding effects of short stature and delayed puberty reduced the differences between patients and controls; correcting for reduced lean body mass in Crohn patients eliminated the remaining differences.[43] Finding that bone mass is appropriate for reduced bone size, and lean body mass does not indicate that bone strength is adequate. In fact, increased fracture rates have been observed in Crohn disease[44] and other chronic disorders such as anorexia nervosa[45] and muscular dystrophy.[46]

More controversy surrounds the potential to enhance skeletal development through lifestyle. Skeletal loading during weight-bearing activity is perhaps the most effective stimulus for skeletal development.[38] Some, but not all, observation studies have shown that bone mineral accrual (as measured by DXA) is greater in more active than sedentary youth.[47] Variability in results likely reflects the different instruments used to quantify activity, errors in self-reporting, and the influence of selection bias favoring participation of larger, stronger youths in sport. More convincing evidence for the skeletal effects of

© 2008 American Society for Bone and Mineral Research

activity comes from studies of elite racket sports players. The diameter and total and cortical bone area of the dominant radius (measured by pQCT) are significantly greater than in the nondominant, "control" arm of these athletes.[48] Some controlled intervention trials have shown increases in BMC, BMD, or bone size in response to increased high-impact activity.[47,49]

Calcium is a key nutrient for bone health. A number of observational studies have reported a positive association between dietary calcium intake and bone mass as measured by DXA.[50] The risk of osteoporotic fracture is increased in older women who consumed little calcium through childhood and early adulthood.[51] Milk avoidance has also been linked to childhood fractures.[52] These data suggest that bone quality or quantity is compromised in a calcium-deficient state. Establishing criteria for calcium sufficiency has been challenging.[50] A meta-analysis of pediatric calcium supplementation trials in healthy youths concluded that the skeletal benefits were small, increasing BMD by an estimated 1.7% and only in the arm.[53] Unlike activity, calcium seems to influence primarily bone mass (and not geometry) at cortical sites by reducing bone remodeling.[54]

There is evidence for interactions between activity and other modifiable variables. Several studies have suggested that the skeletal response to activity varies by pubertal stage. Sensitivity of the bone "mechanostat" to loading is increased in the presence of estrogen, potentially making early and midpuberty an optimal time for intervention.[47] The ratios of vBMD and cross-sectional area of cortical bone to muscle cross-sectional area are greater in pubertal girls than boys, suggestive that calcium in packed in at puberty in girls in response to loading.[12,30,31] By late puberty, bone growth is complete, making changes in bone geometry less likely. Some randomized controlled trials have observed an interaction (synergism) between activity and calcium such that a combination of increased calcium intake and activity had greater effects than either alone. Increases in BMC at loaded sites (femur, tibia-fibula) were 2–3% greater in boys given additional calcium fortified foods and moderate impact activity than subjects in the placebo or no-exercise groups.[55] The clinical implications of these small gains with diet and activity are uncertain, and the duration of benefit remains controversial. Lifestyle effects on bone are discussed in more detail elsewhere in this *Primer*.

SKELETAL DEVELOPMENT AND BONE FRAGILITY

Peak bone mass reached by early adulthood is a key determinant of osteoporotic risk in older adults.[1] Recent studies indicate that bone size and mass are similarly linked to the risk of childhood fracture. Forearm fractures occur commonly during the peripubertal period (ages 8–12 in girls and 11–14 in boys).[56] As in older adults, individuals with a history of fracture have significantly lower BMC, aBMD, and estimated vBMD measured by DXA than controls.[57] Bone size also contributes to fracture risk. Both cross-sectional and prospective studies have found that lower total body BMC and bone area for height were the best predictors of fracture.[6,58]

SUMMARY

Changes in bone size, mass, and distribution of mass before and during puberty evolve at varying rates throughout the skeleton. Development is largely controlled by heritable factors resulting in sex and racial differences. Modifiable factors can compromise bone growth and mineral accrual or potentially result in modest gains. Use of QCT, pQCT, and MRI to study skeletal development in children has helped to distinguish changes in bone size, geometry, and volumetric density from increases in mass. At the same time, this research has produced inconsistent findings depending on the skeletal site examined and the technology used. A few observations can be made with certainty, however. Males develop bigger bones than females both in length and width, providing them with a biomechanical advantage. Females may acquire greater cortical bone density than males in some regions, creating calcium reserves for subsequent demands of pregnancy and lactation. Differences in the development of the axial and appendicular skeleton and variability in the geometry of loaded and unloaded sites provide evidence for local control of bone development. The mechanism(s) behind sex, race, and other genetic controls of skeletal mass and shape remain to be determined. However, observation of differences in before puberty suggests that genetic programming can begin early in life. Further work is needed to enhance understanding of all the skeletal parameters that provide the underpinning of bone strength. This research is essential to identifying the genetic and lifestyle determinants of optimal skeletal development—knowledge needed to combat the increasing prevalence of fracture in both adults and children.[59]

REFERENCES

1. Heaney RP, Abrams S, Dawson-Hughes B, Looker A, Marcus R, Matkovic V, Weaver C 2000 Peak bone mass. Osteoporos Int **11**:985–1009.
2. Seeman E 1997 From density to structure: Growing up and growing old on the surfaces of bone. J Bone Miner Res **12**:509–521.
3. Duan Y, Beck TJ, Wang X-F, Seeman E 2003 Structural and biomechanical basis of sexual dimorphism in femoral neck fragility has its origins in growth and aging. J Bone Miner Res **18**:1766–1774.
4. Rauch F, Schoenau E 2001 Changes in bone density during childhood and adolescence: An approach based on bones biological organization. J Bone Miner Res **16**:1547–1555.
5. Bachrach LK 2005 Osteoporosis and measurement of bone mass in children and adolescents. Endocrinol Metab Clin North Am **34**:521–535.
6. Clark EM, Ness AR, Bishop NJ, Tobias JH 2006 Association between bone mass and fractures in children a prospective cohort study. J Bone Miner Res **21**:1489–1495.
7. Leonard MB, Shults J, Elliott DM, Stallings VA, Zemel BS 2004 Interpretation of whole body dual energy x-ray absorptiometry measures in children: Comparison with peripheral quantitative computed tomography. Bone **34**:1044–1052.
8. Specker BL, Schoenau E 2005 Quantitiative bone analysis in children: Current methods and recommendations. J Pediatr **146**:726–731.
9. Crabtree NJ, Kibirige MS, Fordham JN, Banks LM, Muntoni F, Chinn D, Boivin CM, Shaw NJ 2004 The relationship between lean body mass and bone mineral content in paediatric health and disease. Bone **35**:965–972.
10. Beck TJ 2007 Extending DXA beyond bone mineral density: Understanding hip structure analysis. Curr Osteoporos Rep **5**:49–55.
11. Bachrach LK 2007 Osteoporosis in children: Still a diagnostic challenge. J Clin Endocrinol Metab **92**:2030–2032.
12. Neu CM, Rauch F, Manz F, Schoenau E 2001 Modeling of cross-sectional bone size, mass and geometry at the proximal radius: A study of normal bone development using peripheral quantitative computed tomography. Osteoporos Int **12**:538–547.
13. Lee DC, Gilsanz V, Wren TAL 2007 Limitations of peripheral quantitative computed tomography metaphyseal bone density measurements. J Clin Endocrinol Metab **92**:4248–4253.
14. Wren TA, Liu X, Pitukcheewanont P, Gilsanz V 2005 Bone acquisition in healthy children and adolescents: Comparisons of dual-energy x-ray absorptiometry and computed tomography measures. J Clin Endocrinol Metab **90**:1925–1928.
15. Leonard MB, Shults J, Elliott DM, Stallings VA, Zemel BS 2004 Interpretation of whole body dual energy x-ray absorptiometry measures in children: Comparison with peripheral quantitative computed tomography. Bone **34**:1044–1052.

© 2008 American Society for Bone and Mineral Research

16. Hogler W, Blimkie CJR, Cowell CT, Kemp AF, Briody J, Wiebe P, Farpour-Lambert N, Duncan CS 2003 Woodhead HJ. A comparison of bone geometry and cortical density at the mid-femur between prepuberty and young adulthood using magnetic resonance imaging. Bone **33:**771–778.
17. Van der Meulen MCH, Jepsen KJ, Mikic B 2001 Understanding bone strength: Size isn't everything. Bone **29:**101–104.
18. Bass S, Delmas PD, Pearce G, Hendrich E, Tabensky A, Seeman E 1999 The differing tempo of growth in bone size, mass and density in girls is region-specific. J Clin Invest **104:**795–804.
19. Bradney M, Karlsson MK, Duan Y, Stuckey S, Bass S, Seeman E 2000 Heterogeneity in the growth of the axial and appendicular skeleton in boys: Implications for the pathogenesis of bone fragility in men. J Bone Miner Res **15:**1871–1878.
20. Bailey DA, McKay HA, Mirwald RL, Crocker PRE, Faulkner RA 1999 A six-year longitudinal study of the relationship of physical activity to bone mineral accrual in growing children: The University of Saskatchewan Bone Mineral Accrual Study. J Bone Miner Res **14:**1672–1679.
21. Bonjour JP, Theintz G, Buchs B, Slosman D, Rizzoli R 1991 Critical years and stages of puberty for spinal and femoral bone mass accumulation during adolescence. J Clin Endocrinol Metab **73:**555–563.
22. Bachrach LK, Hastie T, Wang M-C, Narasimhan B, Marcus R 1999 Bone mineral acquisition in healthy Asian, Hispanic, Black and Caucasian youth. A longitudinal study. J Clin Endocrinol Metab **84:**4702–4712.
23. Kalkwarf HJ 2007 Zemel BS, Gilsanz V, Lappe JM, Horlick M, Oberfield S, Mahboubi S, Fan B, Frederick MM, Winer K, Shepherd JA. The Bone Mineral Density in Childhood Study: Bone mineral content and density according to age, sex, and race. J Clin Endocrinol Metab **92:**2087–2099.
24. Gilsanz V, Boechat MI, Rote TF, Loro ML, Sayre JW, Goodman WG 1994 Gender differences in vertebral body sizes in children and adolescents. Radiology **190:**673–677.
25. Rauch F, Travers R, Glorieux FH 2007 Intracortical remodeling during human bone development: A histomorphometric study. Bone **40:**274–280.
26. Tommasini SM, Nasser P, Jepsen KJ 2007 Sexual dimorphism affects tibia size and shape but not tissue-level mechanical properties. Bone **40:**498–505.
27. Lu PW, Cowell CT, Lloyd-Jones SA, Briody JN, Howman-Giles R 1996 Volumetric bone mineral density in normal subjects, aged 5–27 year. J Clin Endocrinol Metab **81:**1586–1590.
28. Gilsanz V, Skaggs DL, Kovanlikaya A, Sayre J, Loro ML, Kaufman F, Korenman SG 1998 Differential effect of race on the axial and appendicular skeletons of children. J Clin Endocrinol Metab **83:**1420–1427.
29. Skaggs DL, Loro ML, Pitukcheewanont P, Tolo V, Gilsanz V 2001 Increased body weight and decreased radial cross-sectional dimensions in girls with forearm fractures. J Bone Miner Res **16:**1337–1342.
30. Schoenau E, Neu CM, Rauch F, Manz F 2002 Gender-specific pubertal changes in volumetric cortical bone mineral density at the proximal radius. Bone **31:**110–113.
31. Neu CM, Manz F, Rauch F, Merkel A, Schoenau E 2001 Bone densities and bone size at the distal radius in healthy children and adolescents: A study using peripheral quantitative computed tomography. Bone **28:**227–232.
32. Kontulainen SA, Macdonald HM, McKay HA 2006 Change in cortical bone density and its distribution differs between boys and girls during puberty. J Clin Endocrinol Metab **91:**2555–2561.
33. Macdonald H, Kontulainen S, Petit M, Janssen P, McKay H 2006 Bone strength and its determinants in pre-and early pubertal boys and girls. Bone **39:**598–608.
34. Garn SM 1972 The course of bone gain and the phases of bone loss. Orthop Clin North Am **3:**503–520.
35. Turner CH, Burr DB 1993 Basic biochemical measurements of bone: A tutorial. Bone **14:**595–608.
36. Gilsanz V, Kovanlikaya A, Costin G, Rose TF, Sayre J, Kaufman F 1997 Differential effect of gender on the size of the bones in the axial and appendicular skeletons. J Clin Endocrinol Metab **82:**1603–1607.
37. Kontulainen SA, Macdonald HM, Khan KK, McKay HA 2005 Examining bone surfaces across puberty: A 20 month pQCT trial. J Bone Miner Res **20:**1202–1207.
38. Frost HM 1987 Bone "mass" and the "mechanostat": A proposal. Anat Rec **219:**1–9.
39. Wang Q, Nicholson PHF, Suuriniemi M, Lyytikainen A, Helkala E, Alen M, Suominen H, Cheng S 2004 Relationship of sex hormones to bone geometric properties and mineral density in early pubertal girls. J Clin Endocrinol Metab **89:**1698–1703.
40. Hui SL, Dimeglio LA, Longcope C, Peacock M, McClintock R, Perkins AJ, Johnston R 2003 CC. Differences in bone mass between black and white American children: Attributable to body build, sex hormone levels, or bone turnover? J Clin Endocrinol Metab **88:**642–649.
41. Braun M, Palacios C, Wigeertz K, Jackman LA, Bryant RJ, McCabe LD, Martin BR, McCabe GP, Peacock M, Weaver CM 2007 Racial differences in skeletal calcium retention in adolescent girls on a range of controlled calcium intakes. Am J Clin Nutr **85:**1657–1663.
42. Liu Y-L, Shen J, Xiao P, Xiong D-H, Li L-H, Recker RR, Deng H-W 2004 Molecular genetic studies of gene identification for osteoporosis: A 2004 update. J Bone Miner Res **21:**1511–1535.
43. Burnham JM, Shults J, Semeao E, Foster B, Zemel BS, Stallings VA, Leonard MB 2004 Whole body BMC in pediatric Crohn disease: Independent effects of altered growth, maturation and body composition. J Bone Miner Res **19:**1961–1968.
44. van Staa TP, Cooper C, Brusse LS, Leufkens H, Javaid MK, Arden NK 2003 Inflammatory bowel disease and the risk of fracture. Gastroenterology **125:**1591–1597.
45. Vestergaard P, Emborg C, Stoving RK, Hagen C, Mosekilde L, Brixen K 2003 Patients with eating disorders. A high-risk group for fractures. Orthop Nurs **22:**325–331.
46. Larson CM, Henderson RC 2000 Bone mineral density and fractures in boys with Duchenne muscular dystrophy. J Pediatr Orthop **20:**71–74.
47. MacKelvie KJ, Khan KM, McKay HA 2002 Is there a critical period for bone response to weight-bearing exercise in children and adolescents? A systematic review. Br J Sports Med **36:**250–257.
48. Kontulainen S, Sievanen H, Kannus P, Pasanen M, Vuori I 2002 Effect of long-term impact-loading on mass, size, and estimated strength of humerus and radius of female racquet sports players: A peripheral quantitative computed tomography study between young and old starters and controls. J Bone Miner Res **17:**2281–2289.
49. Hind K, Burrows M 2006 Weight-bearing exercise and bone mineral accrual in children and adolescents: A review of controlled trials. Bone **40:**14–27.
50. Wosje KS, Specker BL 2000 Role of calcium in bone health during childhood. Nutr Rev **58:**253–268.
51. Kalkwarf HJ, Khoury HC, Lanphear BP 2003 Milk intake during childhood and adolescence, adult bone density, and osteoporotic fractures in US women. Am J Clin Nutr **77:**257–265.
52. Goulding A, Rockell JE, Black RE, Grant AM, Jones IE, Williams SM 2004 Children who avoid drinking cow's milk are at increased risk for prepubertal bone fractures. J Am Diet Assoc **104:**250–253.
53. Winzenberg T, Shaw K, Fryer J, Jones G 2006 Effects of calcium supplementation on bone density in healthy children: Meta-analysis of randomized controlled trials. BMJ **14:**333:775.
54. Heaney RP 2001 The bone remodeling transient: Interpreting interventions involving bone-related nutrients. Nutr Rev **39:**327–334.
55. Bass SL, Naughton G, Saxon L, Iuliano-Burns S, Daly R, Briganti EM, Hume C, Nowson C 2007 Exercise and calcium combined results in greater osteogenic effect than either alone: A blinded randomized placebo-controlled trial in boys. J Bone Miner Res **22:**458–464.
56. Cooper C, Dennison EM, Leufkens HG, Bishop N, van Staa TP 2004 Epidemiology of childhood fractures in Britain: A study using the General Practice Research Database. J Bone Miner Res **19:**1976–1981.
57. Goulding A, Grant AM, Williams SM 2005 Bone and body composition of children and adolescents with repeated forearm fractures. J Bone Miner Res **20:**2090–2096.
58. Jones G, Ma D, Cameron F 2006 Bone density interpretation and relevance in Caucasian children aged 9–17 years of age: Insights from a population-based fracture study. J Clin Densitom **9:**202–209.
59. Khosla S, Melton LJ III, Dekutoski MB, Achenbach SJ, Oberg AL, Riggs BL 2003 Incidence of childhood distal forearm fractures over 30 years: A population-based study. JAMA **290:**1479–1485.

© 2008 American Society for Bone and Mineral Research

Chapter 15. Ethnic Differences in Bone Acquisition

Shane A. Norris[1] and Dorothy A. Nelson[2]

[1]*MRC Mineral Metabolism Research Unit, University of the Witwatersrand, Johannesburg, South Africa;* [2]*Internal Medicine, Wayne State University, Detroit, Michigan*

INTRODUCTION

It is known that there are population differences in the risk of osteoporosis and fractures. Research conducted in ethnically diverse populations is particularly valuable for making inferences about ethnic differences in hip fracture incidence. Considerable interest has been placed on defining the determinants of physiological variations in skeletal growth and development to better understand and predict skeletal health in adults. Studies have identified a number of ethnic differences in the acquisition of bone mass, dietary calcium intake, and calcium homeostasis that may be related to the differences noted among adults and later fracture risk.

There is significant variation in fracture risk across geographical regions and ethnic groups. A cross-national study in five countries between 1990 and 1992 found that the greatest incidence of hip fracture rates occurred in Iceland, intermediate rates in Hungary and Hong Kong, and the lowest rates in Beijing.[(1)] Another cross-national study found that the highest hip fracture rates from hospital admissions were reported for European and North American countries, and the lowest fractures rates were observed in South America countries.[(2)] Hip fracture incidence among black South Africans was reported to be exceptionally low in a study conducted between 1950 and 1964[(3)]; there are no recent studies in African populations except in Morocco and Cameroon, where again, very low hip fracture incidence was reported.[(4)] Recent trend data indicate that there is an increase in hip fracture incidence in Denmark, static rates in Scandinavia and Asia, and a decline in age-adjusted incidence rates in the United States.[(5,6)] A noteworthy review by Maggi et al.[(7)] highlighted that hip fracture incidence increases with age in all ethnic groups, but the increase occurs earlier in non-Hispanic white populations than in black, Asian, and Hispanic populations. It has been assumed that fracture risk in adulthood in any population reflects lifelong genetic and environmental influences and especially those affecting the accumulation of peak bone mass (PBM).

ETHNIC DIFFERENCES IN BONE ACQUISITION, MASS, AND GEOMETRY

Most reports of ethnic differences in bone mass during childhood, based on absorptiometric methods, have indicated a higher bone mass among blacks compared with whites.[(8–10)] This has led to the generalization that black population groups have greater bone mass and strength compared with white population groups and have been given as an explanation for the lower fracture rates observed in black population groups. However, not all U.S. studies have observed ethnic differences in BMC and BMD in children.[(11,12)] Furthermore, studies of black South Africans have shown that their bone mass at most sites does not exceed that of age-matched white South Africans.[(13,14)] After correcting for differences in body size, black South African children have similar BMD at the midshaft radius and lumbar spine to their white peers, but femoral neck BMD is increased in the former. In a comparative study between white, black, and mixed ancestral children in South Africa and white and black children in the United States, the results showed similar patterns of whole body BMC between black and white children in South Africa and those in the United States, with the black children having significantly higher bone mineral content, after adjusting for body size differences. Furthermore, children of mixed ancestral origin had greater whole body bone mass than both white and African children in South Africa and the United States (Fig. 1).[(15)] However, black children in The Gambia have significantly less bone mass compared with British children.[(16)] These data describe the difficulty in generalizing about an ethnic group, when obviously ethnic gradations in bone mass exist.

A separate example can be found in a study of white and black children that included a large subgroup of Chaldean, an Iraqi ethnic group, who considered themselves "white." The Chaldean children's whole body bone mass was significantly greater than non-Chaldean white children and was not different from other participants who considered themselves black.[(17)]

Pediatric studies using CT have indicated that, regardless of sex, ethnicity has significant and differential effects on the density and the size of the bones in the axial and appendicular skeletons.[(18)] In the axial skeleton, the density of cancellous bone in the vertebral bodies is greater in black than in U.S. white adolescents. This difference first becomes apparent during late stages of puberty and persists throughout life.[(19)] The magnitude of the increase from prepubertal to postpubertal values is substantially greater in black than in white subjects (34% versus 11%, respectively).[(18)] The cross-sectional areas of the vertebral bodies, however, do not differ between black and white children.[(18)] Thus, theoretically, the structural basis for the lower vertebral bone strength and the greater incidence of fractures in the axial skeleton of white subjects resides in their lower cancellous BMD. In contrast, in the appendicular skeleton, ethnicity influences the cross-sectional areas of the femora, but not the cortical bone area nor the material density of cortical bone.[(18)] Although values for femoral cross-sectional area increase with height, weight, and other anthropometric parameters in all children, this measurement is substantially greater in black children.[(18)] Because the same amount of cortical bone placed further from the center of the bone results in greater bone strength, the skeletal advantage for blacks in the appendicular skeleton is likely the consequence of the greater cross-sectional size of the bones.[(20)] Data from Asian and Hispanic youth suggest that their bone mass is similar to that of whites but much is lower than that of black children.[(21)] Differences in bone and body size account for much of the observed ethnic differences in BMD among non-Hispanic, Hispanic, and Asian children.[(21)]

ETHNIC DIFFERENCES IN DIETARY CALCIUM AND CALCIUM HOMEOSTASIS

Habitual dietary calcium intakes vary considerably across different ethnic groups, with some populations consuming between 300 and 400 mg/d, whereas others may consume three times that. The earliest data suggesting an influence of dietary calcium on PBM come from a study of two Croatian populations with substantially different calcium intakes.[(22)] The differences seen in bone mass were present at 30 yr of age, suggesting that the effects of dietary calcium probably occurred during growth rather than in adulthood. Moreover, some epi-

The authors state that they have no conflicts of interest.

© 2008 American Society for Bone and Mineral Research

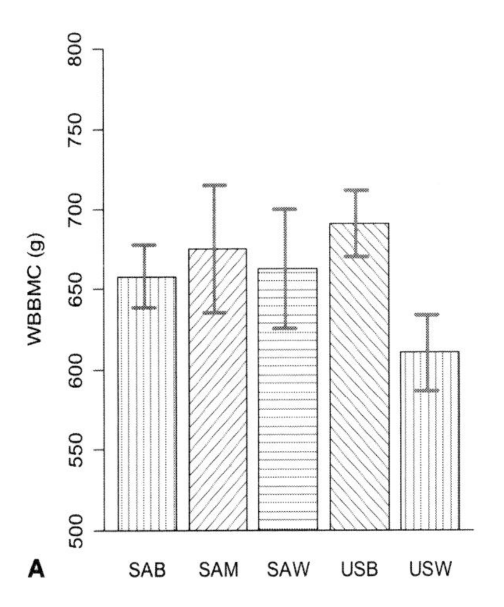

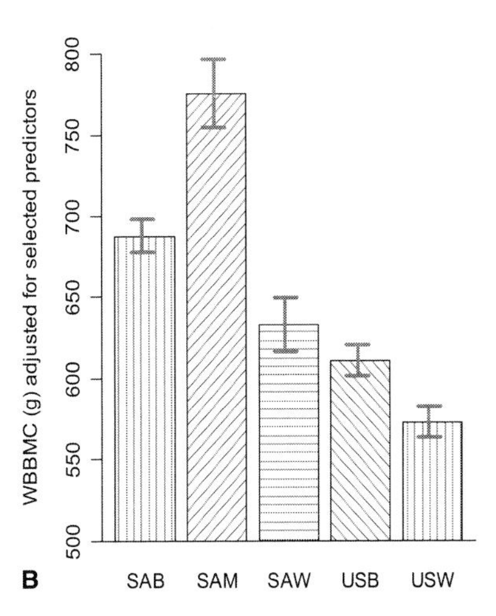

FIG. 1. Whole body BMC (A) unadjusted means and (B) adjusted means for selected predictors, with simultaneous 95% CIs (SAB, South African black; SAM, South African mixed ancestral origin; SAW, South African white; USB, black American; USW, white American). T-bars representing the CIs will overlap if the differences in means are not significant (Reproduced with permission from the American Society for Bone and Mineral Research from Micklesfield L, Norris SA, Nelson DA, Lambert EV, van der Merwe L, Pettifor JM 2007 Comparisons of body size, composition and whole body bone mass between North American and South African children. J Bone Miner Res **22:**1869–1877.)

demiological studies have shown an increased prevalence of osteoporosis in regions where dietary calcium intake is extremely low.[23]

The most convincing evidence that calcium consumption influences rates of bone mineral accrual rates comes from controlled supplementation trials in young healthy subjects. These studies showed that subjects given additional calcium for 1–3 yr have greater gains than did controls.[24–27] Although bone size increased as a result of added dietary calcium in two studies,[24,27] the response to calcium supplementation varied by skeletal site, pretreatment dietary calcium intake, and pubertal development. Greater bone mineral gains have been generally reported at cortical skeletal sites in prepubertal subjects and in girls whose habitual dietary intake was <850 mg/d.[27] The magnitude of gains in BMC or BMD in most studies was modest (<5%), and the majority of studies have not shown lasting effects once the supplements are withdrawn. Furthermore, Winzenberg et al.[28] showed through a meta-analysis of randomized controlled trails that after the cessation of calcium supplementaion, the modest effect on BMD dissappeared except at the upper limb, but it is unlikely that this persisting effect would reduce the risk of fracture in later life.

Few studies have addressed the effect of calcium supplementation in children other than whites living in developed countries. A study conducted on 10-yr-old Chinese girls found that a 2-yr fortified school milk supplementation trial improved bone mineral acquisition; however, 3 yr after supplement withdrawal, there were no long-lasting effects on bone mass.[29] Dibba et al.[30] have studied the effect of calcium supplements in a group of prepubertal black children living in The Gambia, whose calcium intakes averaged ~300 mg/d before supplementation. During supplementation, BMC and BMD at the midshaft and distal radius increased, and osteocalcin levels fell in the treatment group, but no effect on radial size or height was detected. At follow-up 12–24 mo later, the calcium-supplemented group still showed a small advantage in BMC at the midshaft of the radius.[31]

Abrams et al.[32] studied aspects of calcium metabolism in African-American, Mexican-American, and white children. They found higher PTH concentrations in Mexican-American girls compare with white girls, despite lack of vitamin D deficiency, although ethnic differences in 25(OH)D and PTH concentrations did not significantly affect calcium absorption, excretion, or bone calcium kinetics. In studies comparing black and white girls, calcium absorption was greater in both pre- and postmenarcheal blacks, who also had greater calcium deposition and marginally lower urinary calcium excretion.[33]

SUMMARY

The literature on ethnic variation in hip fracture incidence confirms that many factors contribute to skeletal health. Skeletal mass is accrued throughout childhood and adolescence, and many ethnic differences seem to be present throughout growth. Interethnic studies in bone mass acquisition are important to help elucidate possible etiologies in the early pathogenesis of osteoporosis and the development of interventions that may ameliorate the manifestations of the disease in later life. Particularly, exploring the environmental and cultural milieu during bone mass acquisition and its impact on dietary exposures, calcium metabolism, bone mass, and geometry is needed understand the broad range of diversity in different ethnic groups and differing risk profiles.

REFERENCES

1. Schwartz AW, Kelsey JL, Maggi S, Tuttleman M, Ho SC, Jonsson PV, Poor G, Sisson de Castro JA, Xu L, Matkin CC, Nelson LM, Heyse SP 1999 International variation in the incidence of hip fractures: Cross-national project on osteoporosis for the World Health Organization Program for Research on Aging. Osteoporos Int **9:**242–253.
2. Bacon WE, Maggi S, Looker A, Harris T, Nair CR, Giaconi J, Honkanen R, Ho SC, Peffers KA, Torring O, Gass R, Gonzalez N 1996 International comparison of hip fracture rates in 1988–89. Osteoporos Int **6:**69–75.
3. Solomon L 1968 Osteoporosis and fracture of the femoral neck in the South African Bantu. J Bone Joint Surg Br **50:**2–11.
4. El Maghraoui A, Koumba BA, Jroundi I, Achemlal L, Bezza A, Tazi MA 2005 Epidemiology of hip fractures in 2002 in Rabat, Morocco. Osteoporos Int **16:**597–602.
5. Melton LJ III, Atkinson EJ, Madhok R 1996 Downturn in hip fracture incidence. Public Health Rep **111:**146–150.
6. Giversen IM 2006 Time trend of age-adjusted incidence rates of first hip fractures: A register-based study among older people in Viborg County, Denmark, 1987-1997. Osteoporos Int **17:**552-564.
7. Maggi S, Kelsey JL, Litvak L, Heyse SP 1991 Incidence of hip fractures in the elderly: A cross-national analysis. Osteoporos Int **1:**232–241.
8. Nelson DA, Simpson PM, Johnson CC, Barondness DA, Kleerekoper M 1997 The accumulation of whole body skeletal mass in third- and fourth-grade children: Effects of age, gender, ethnicity and body composition. Bone Miner **20:**73–78.
9. Li JY, Specker BL, Ho ML, Tsang RC 1989 Bone mineral content in black and white children 1 to 6 years of age. Early appearance of race and sex differences. Am J Dis Child **143:**1346–1349.
10. Thomas KA, Cook SD, Bennett JT, Whitecloud TSHI, Rice JC 1991 Femoral neck and lumbar spine mineral densities in a normal population 3-20 years of age. J Pediatr Orthop **11:**48–58.
11. Southard RN, Morris JD, Mahan JD, Hayes JR, Torch MA, Som-

© 2008 American Society for Bone and Mineral Research

mer A, Zipf WB 1991 Bone mass in healthy children: Measurements with quantitative DXA. Radiology **179:**735–738.
12. Bachrach LK, Hastie T, Wang MC, Balasubramanian N, Marcus R 1999 Bone mineral acquisition in healthy asian, hispanic, black, and caucasian youth: A longitudinal study. J Clin Endocrinol Metab **84:**4702–4712.
13. Solomon L 1979 Bone density in ageing Caucasian and African populations. Lancet **2:**1326–1330.
14. Daniels ED, Pettifor JM, Schnitzler CM, Moodley GP, Zachen D 1997 Differences in mineral homeostasis, volumetric bone mass and femoral neck axis length in black and white South African women. Osteoporos Int **7:**105–112.
15. Micklesfield L, Norris SA, Nelson DA, Lambert EV, van der Merwe L, Pettifor JM 2007 Comparisons of body size, composition and whole body bone mass between North American and South African children. J Bone Miner Res **22:**1869–1877.
16. Prentice A, Laskey MA, Shaw J, Cole TJ, Fraser DR 1990 Bone mineral content of Gambian and British children aged 0-36 months. Bone Miner **10:**211–224.
17. Nelson DA, Barondess DA 1997 Whole body bone, fat and lean mass in children: Comparison of three ethnic groups. Am J Phys Anthropol **103:**157–162.
18. Gilsanz V, Skaggs DL, Kovanlikaya A, Sayre J, Loro ML, Kaufman F, Korenman SG 1998 Differential effect of race on the axial and appendicular skeletons of children. J Clin Endocrinol Metab **83:**1420–1427.
19. Kleerekoper M, Nelson DA, Flynn MJ, Pawluszka AS, Jacobsen G, Peterson EL 1994 Comparison of radiographic absorptiometry with dual energy x-ray absorptiometry and quantitative computed tomography in normal older white and black women. J Bone Miner Res **9:**1745–1750.
20. Van der Meulen MCH, Beaupre GS, Carter DR 1993 Mechanobiologic influences in long bone cross-sectional growth. Bone **14:**635–642.
21. Bachrach LK, Hastie T, Wang MC, Balasubramanian N, Marcus R 1999 Bone mineral acquisition in healthy asian, hispanic, black, and caucasian youth: A longitudinal study. J Clin Endocrinol Metab **84:**4702–4712.
22. Matkovic V, Kostial K, Simonovic I, Buzinz R, Brodarec A, Nordin BE 1979 Bone status and fracture rates in two regions of Yugoslavia. Am J Clin Nutr **32:**540–549.
23. Heaney RP 1992 Calcium in the prevention and treatment of osteoporosis. J Intern Med **231:**169–180.
24. Lee WTK, Leung SSF, Leung DM, Cheng JC 1996 A follow-up study on the effects of calcium-supplement withdrawal and puberty on bone acquisition in children. Am J Clin Nutr **64:**71–77.
25. Lloyd T, Andon MB, Rollings N, Martel JK, Landis JR, Demers LM, Eggli DF, Kieselhorst K, Kulin HE 1993 Calcium supplementation and bone mineral density in adolescent girls. JAMA **270:**841–844.
26. Johnston CC Jr, Miller JZ, Slemenda CW, Reister TK, Hui S, Christian JC, Peacock M 1992 Calcium supplementation and increases in bone mineral density in children. N Engl J Med **327:**82–87.
27. Bonjour JP, Carrie AL, Ferrari S, Clavien H, Slosman D, Theintz G, Rizzoli R 1997 Calcium-enriched foods and bone mass growth in prepubertal girls: A randomized double blind, placebo-controlled trial. J Clin Invest **99:**1287–1294.
28. Winzenberg T, Shaw K, Fryer J, Jones G 2006 Effects of calcium supplementation on bone density in healthy children: Meta-analysis of randomized controlled trials. BMJ **333:**775–780.
29. Zhu K, Zhang Q, Foo LH, Trube A, Ma G, Hu X, Du X, Cowell CT, Fraser DR, Greenfield H 2006 Growth, bone mass, and vitamin D status of Chinese adolescent girls 3y after withdrawal of milk supplementation. Am J Clin Nutr **83:**714–721.
30. Dibba B, Prentice A, Ceesay M, Stirling DM, Cole TJ, Poskitt EM 2000 Effect of calcium supplementation on bone mineral accretion in Gambian children accustomed to a low-calcium diet. Am J Clin Nutr **71:**544–549.
31. Dibba B, Prentice A, Ceesay M, Mendy M, Darboe S, Stirling DM, Cole TJ, Poskitt EM 2002 Bone mineral contents and plasma osteocalcin concentrations of Gambian children 12 and 24 mo after the withdrawal of a calcium supplement. Am J Clin Nutr **76:**681–686.
32. Abrams SA, Copeland KC, Gunn SK, Stuff JE, Clarke LL, Ellis KJ 1999 Calcium absorption and kinetics are similar in 7- and 8-year-old Mexican-American and Caucasian girls despite hormonal differences. J Nutr **129:**666–671.
33. Abrams SA, O'Brien KO, Liang LK, Stuff JE 1995 Differences in calcium absorption and kinetics between black and white girls aged 5-16 years. J Bone Miner Res **10:**829–833.

Chapter 16. Calcium and Other Nutrients During Growth

Tania Winzenberg and Graeme Jones

Menzies Research Institute, Hobart, Tasmania, Australia

INTRODUCTION

BMD in later life is a function of peak bone mass and the rate of subsequent bone loss.[1] Childhood is potentially an important time to intervene, because modeling suggests that a 10% increase in peak bone mass will delay the onset of osteoporosis by 13 yr.[2] In addition, low BMD in childhood is a risk factor for childhood fractures,[3] suggesting that optimizing age-appropriate bone mass could also have a more immediate preventive effect on fracture rates in children. This chapter reviews key nutritional influences on childhood bone development.

CALCIUM

It is widely accepted that an adequate calcium intake in childhood is important for bone development, although the results of observational and intervention studies are mixed.[4] In case-control studies, low calcium/dairy intake has been found to be associated with increased fracture risk in 11- to 13-yr-old boys, but this result has not been confirmed in other groups.[5–7] Low calcium/dairy intake has been found to be associated with recurrent fracture in both sexes.[8,9]

High levels of calcium intake for children are recommended in many developed countries (350–800 mg/d for children and 800–1300 mg/d for adolescents).[10] In a growing rat model, calcium intake showed threshold behavior where, below a given level of calcium intake, skeletal calcium accumulation was related to intake, but above this level, skeletal accumulation remained constant. Modeling of data from calcium balance studies in 348 children[11] suggested that there is a calcium threshold in children, varying with age to up to 1730 mg in 9–17 yr olds. A similar threshold at ~1300 mg was described in girls 12–15 yr of age.[12] However, the relationship between

The authors state that they have no conflicts of interest.

Key words: children, diet, nutrition, calcium, milk, dairy, vitamin D, sodium, salt, fruit and vegetables, pregnancy, breast feeding, carbonated beverages, cola, soft drink

© 2008 American Society for Bone and Mineral Research

short-term calcium balance studies and achieving bone outcome improvements from longer-term calcium supplementation is open to question. In a meta-analysis of randomized controlled trials (RCTs),[13,14] bone outcomes were no different above or below calcium intakes of 1400 mg/d, casting doubt on the clinical relevance of the balance studies' results.

This meta-analysis also[13,14] found that calcium supplementation had no effect on BMD at the femoral neck or lumbar spine. Whereas supplementation had a small effect on total body BMC, this did not persist once supplementation ceased. There was a small persistent effect on upper limb BMD, equivalent to a 1.7 percentage point greater increase in BMD in the supplemented compared with the control group, which might reduce the absolute risk of fracture at the peak childhood fracture incidence by, at most, 0.2% per annum (p.a.). Furthermore, there was no evidence to suggest that increasing the duration of supplementation led to increasing effects or that the effect size varied with baseline calcium intakes, down to a level of <600 mg/d. Thus, the small increase in BMD at the upper limb from increasing intake from an average 700 to 1200 mg/d is unlikely to result in a clinically significant decrease in fracture risk. Because the meta-analysis only included placebo-controlled trials, some RCTs of calcium supplementation using dairy products were not included.[15–21] However, qualitatively, the results of these studies were not dissimilar, mainly showing no effect[15] or only small to moderate short-term effects,[17–20] which did not persist after supplementation ceased.[18,21] In the only study with a larger effect, the intervention group had substantially higher levels of vitamin D intake,[16] making it unclear how much of the effect was caused by calcium and how much was caused by vitamin D.

VITAMIN D

Vitamin D deficiency is common in children, especially in late adolescence.[22–25] There is observational evidence that mild vitamin D deficiency (<50 nM) may affect bone mineralization. However, the effectiveness of vitamin D supplementation for improving bone development and peak bone mass in children remains uncertain, because RCTs of vitamin D supplementation alone in healthy children with BMD outcomes give inconsistent results, possibly because of variation in compliance, doses given, and baseline vitamin D levels.[20,26–28] In positive studies, effect sizes were 1.3% over 2 yr for total body (TB) BMC,[20] 2% for lumbar spine (LS) and femoral BMC in 1 yr,[28] and 5% for total hip BMC in 1 yr.[27] These results suggest that vitamin D supplementation could deliver improvements in bone health and childhood fracture incidence that are of clinical and public health significance. However, no studies have been powered for fracture, and it is not known if effects accumulate with ongoing supplementation.

FRUIT AND VEGETABLES

Fruit and vegetable intake is postulated to have effects on bone through a number of mechanisms. These include the induction of a mild metabolic alkalosis, vitamin K, vitamin C, antioxidants, and phytoestrogens, although phytoestrogens alone have little effect on bone turnover in children.[29] Cross-sectional studies provide data supporting a positive relationship between fruit and vegetable intake and bone outcomes in children. In 8-yr-old children,[30] urinary potassium was positively associated with both fruit and vegetable intake and BMD, whereas girls at Tanner stage 2[31] consuming at least three servings of fruit and vegetable per day had higher bone area, lower urinary calcium excretion, and lower PTH levels than those consuming more than three servings per day, although there were no differences in BMD, urinary deoxypyridinoline, or serum osteocalcin between these groups. In another study, 12-yr-old girls consuming high amounts of fruit had higher heel BMD than moderate fruit consumers.[32] In adolescent boys and girls,[33] significant positive associations were observed between spine size-adjusted BMC (SA-BMC) and fruit intake and, in boys, between femoral neck SA-BMC and fruit intake.

Over 7 yr, fruit and vegetable intake was an independent predictor of TBBMC in boys but not girls.[34] In children 10–15 yr of age,[35] stiffness index (SI) over 1 yr measured by quantitative ultrasound was positively associated with fruit, vegetable, and soybean intake. Girls increasing fruit intake had a 4.7% greater increase in SI than those who did not, girls increasing vegetable intake had a 3.6% greater increase, and boys increasing vegetable intake had a 2.4% greater increase. Fruit and vegetable intake has been increased in children by RCT interventions, with increases in the range of 0.3–0.99 servings per day.[36] Further research is needed to confirm if bone health is changed by clinically significant amounts by such increases.

DIET IN PREGNANCY

Nutritional influences on childhood bone development may begin in utero, and because of in utero programming, such influences may affect both early skeletal development and the acquisition of bone mass throughout childhood. Studies examining this are sparse. In an exploratory study, maternal dietary intake of magnesium, phosphorus, potassium, and protein during the third trimester of pregnancy has been shown to be positively associated, and maternal fat intake to be negatively associated, with BMD in their children at 8 yr of age.[37] Another study[38] examining associations between maternal diet at 32 wk gestation and BMC and BMD at age 9 yr found that maternal magnesium intake was positively associated with total body BMC and BMD, but not when adjusted for child's height, and that maternal intake of potassium was positively associated with spinal BMC and BMD, until adjusted for child's weight. Maternal folate intake was positively associated with spinal BMC adjusted for bone area (BA) after adjusting for both weight and height of children. The effect sizes in this study were smaller than those observed in other studies.[37] Maternal serum 25-hydroxyvitamin D in late pregnancy was positively associated with whole body and LS BMC in the same children at age 9.[39]

RCTs examining childhood bone outcomes from supplementation interventions in pregnancy are lacking. However, maternal vitamin D supplementation in pregnancy resulted in lower bone-specific alkaline phosphatase levels and smaller fontanelle size (suggesting improved skull ossification)[40] in infants and in lower cord serum alkaline phosphatase[41] and greater crown-heel length[41] in neonates. Zinc supplementation in pregnancy in a poor area in a developing country resulted in increased fetal femur diaphysis length.[42] In a retrospective cohort study, maternal use of vitamin D supplements was associated with increased BMD at the distal radius and femoral neck, although not at the lumbar spine.[43] Although limited, these data provide support for further research into nutritional interventions in pregnancy.

BREAST-FEEDING

An RCT in preterm infants showed that early exposure to breast milk for as little as 4 wk resulted in higher bone mass in later life, postulated to be caused by changes in bone cell programming from such exposure.[44] Generally, studies show that

© 2008 American Society for Bone and Mineral Research

human milk–fed infants have lower bone accretion compared with formula-fed infants, possibly because of low vitamin D content and decreasing phosphorus content of human milk with continued lactation.[45] However, data on the long-term effects of breast feeding on bone health in children born at term suggest that this initial lower bone accretion is temporary, and catch-up growth occurs later in childhood. This includes data from an RCT of infant feeding comparing two different formulas and breast feeding, in which initial differences in BMC accretion did not persist past 12 mo of age,[46] as well as longitudinal observational data. In 8-yr-old children born at term,[47] breast-fed children had higher femoral neck, LS, and TB BMD compared with bottle-fed children, and the effect was most marked in children breast fed for >3 mo. In 7- to 9-year-old children born at term, being breast fed was not associated with broadband ultrasound attenuation (BUA) or speed of sound (SOS), but in breast-fed children, duration of breast feeding was positively associated with metacarpal diameter.[48]

Other observational studies with bone measures at younger ages[49,50] did not show associations between breast feeding and BMD. However, in a retrospective study, premenopausal women who had been breast fed for >3 mo had greater cortical thickness at the radius and a trend toward greater cortical area and cortical BMC at the radius, but not at other sites.[51] Importantly, breast feeding was protective for childhood fractures in a longitudinal study of prepubertal children[52] and in a case control study of children 4–15 yr of age,[9] although this was not observed in a longitudinal study of fracture risk from birth to 18 yr.[53]

SALT

Urinary sodium excretion has been shown to be associated with urinary calcium excretion in girls,[54–56] although not with an acute sodium chloride load.[56] Despite this, in the few studies assessing bone outcomes in children, urinary sodium excretion has not in turn been shown to be associated with BMD,[30,55] although dietary sodium intake was associated with size-adjusted bone area but not BMC in a cross-sectional study of 10-yr-old girls.[57] Urinary sodium has also been shown to be associated with a high bone turnover state in adolescent boys (unpublished data). Whether high dietary sodium intake adversely affects other bone outcomes in children is uncertain. Initially, more longitudinal studies are needed to determine whether sodium intake does in fact have a clinically important effect on bone in children.

SOFT DRINKS AND MILK AVOIDANCE

Carbonated beverage consumption has been linked with decreased BMD in girls but not boys.[58,59] In both sexes, carbonated beverage intake is associated with increased fracture risk. Low milk intake and a higher consumption of carbonated beverages were independent fracture risk factors in children with recurrent fractures.[9] Other studies have reported increased fracture risk with higher cola intake but not noncola carbonated beverage intake.[60,61] It is unclear if this effect is caused by milk replacement. In one study,[7] the association between cola drinks and fracture persisted after adjustment for milk intake but not after adjustment for television, computer, and video watching, suggesting the latter mediates the effect on fracture risk.

Milk avoidance also seems to have deleterious effects on children's bone. Prepubertal children who avoid milk have lower total body BMC and areal BMD,[62] as well as an increased risk of childhood fracture.[63] The effects of low milk consumption in childhood may extend into adult life, with low childhood milk consumption being shown to be associated with lower BMD[64] and higher risk of fracture in adult life in women.[65]

In conclusion, there is increasing evidence linking a number of nutritional factors with children's bone development. Calcium supplementation has been studied to the greatest extent, but its effects are of limited public health significance. This makes the exploration of other nutritional approaches of key importance.

ACKNOWLEDGMENTS

TW receives a NHMRC GP Training Fellowship and GJ receives a NHMRC Practitioner Fellowship, which supported this work.

REFERENCES

1. Hansen MA, Overgaard K, Riis BJ, Christiansen C 1991 Role of peak bone mass and bone loss in postmenopausal osteoporosis: 12 year study. BMJ **303:**961–964.
2. Hernandez CJ, Beaupre GS, Carter DR 2003 A theoretical analysis of the relative influences of peak BMD, age-related bone loss and menopause on the development of osteoporosis. Osteoporos Int **14:**843–847.
3. Clark EM, Tobias JH, Ness AR 2006 Association between bone density and fractures in children: A systematic review and meta-analysis. Pediatrics **117:**e291–e297.
4. Lanou AJ, Berkow SE, Barnard ND 2005 Calcium, dairy products, and bone health in children and young adults: A reevaluation of the evidence. Pediatrics **115:**736–743.
5. Goulding A, Jones IE, Taylor RW, Williams SM, Manning PJ 2001 Bone mineral density and body composition in boys with distal forearm fractures: A dual-energy x-ray absorptiometry study. J Pediatr **139:**509–515.
6. Goulding A, Cannan R, Williams SM, Gold EJ, Taylor RW, Lewis-Barned NJ 1998 Bone mineral density in girls with forearm fractures. J Bone Miner Res **13:**143–148.
7. Ma D, Jones G 2004 Soft drink and milk consumption, physical activity, bone mass, and upper limb fractures in children: A population-based case-control study. Calcif Tissue Int **75:**286–291.
8. Goulding A, Grant AM, Williams SM 2005 Bone and body composition of children and adolescents with repeated forearm fractures. J Bone Miner Res **20:**2090–2096.
9. Manias K, McCabe D, Bishop N 2006 Fractures and recurrent fractures in children; varying effects of environmental factors as well as bone size and mass. Bone **39:**652–657.
10. FAO/WHO 2001 FAO/WHO Expert Consultation on Human Vitamin and Mineral Requirements. World Health Organization, Geneva, Switzerland.
11. Matkovic V, Heaney RP 1992 Calcium balance during human growth: Evidence for threshold behavior. Am J Clin Nutr **55:**992–996.
12. Jackman LA, Millane SS, Martin BR, Wood OB, McCabe GP, Peacock M, Weaver CM 1997 Calcium retention in relation to calcium intake and postmenarcheal age in adolescent females. Am J Clin Nutr **66:**327–333.
13. Winzenberg TM, Shaw K, Fryer J, Jones G 2006 Calcium supplementation for improving bone mineral density in children. Cochrane Database Syst Rev **2006:**CD005119.
14. Winzenberg T, Shaw K, Fryer J, Jones G 2006 Effects of calcium supplementation on bone density in healthy children: Meta-analysis of randomised controlled trials. BMJ **333:**775.
15. Lau EMC, Lee WTK, Leung S, Cheng J 1992 Milk supplementation—a feasible and effective way to enhance bone gain for Chinese adolescents in Hong Kong? J Appl Nutr **44:**16–21.
16. Chan GM, Hoffman K, McMurry M 1995 Effects of dairy products on bone and body composition in pubertal girls. J Pediatr **126:**551–556.
17. Cadogan J, Eastell R, Jones N, Barker ME 1997 Milk intake and bone mineral acquisition in adolescent girls: Randomised, controlled intervention trial. BMJ **315:**1255–1260.
18. Merrilees MJ, Smart EJ, Gilchrist NL, Frampton C, Turner JG,

© 2008 American Society for Bone and Mineral Research

Hooke E, March RL, Maguire P 2000 Effects of diary food supplements on bone mineral density in teenage girls. Eur J Nutr **39:**256–262.
19. Lau EM, Lynn H, Chan YH, Lau W, Woo J 2004 Benefits of milk powder supplementation on bone accretion in Chinese children. Osteoporos Int **15:**654–658.
20. Du X, Zhu K, Trube A, Zhang Q, Ma G, Hu X, Fraser DR, Greenfield H 2004 School-milk intervention trial enhances growth and bone mineral accretion in Chinese girls aged 10-12 years in Beijing. Br J Nutr **92:**159–168.
21. Zhu K, Zhang Q, Foo LH, Trube A, Ma G, Hu X, Du X, Cowell CT, Fraser DR, Greenfield H 2006 Growth, bone mass, and vitamin D status of Chinese adolescent girls 3 y after withdrawal of milk supplementation. Am J Clin Nutr **83:**714–721.
22. Looker AC, Dawson-Hughes B, Calvo MS, Gunter EW, Sahyoun NR 2002 Serum 25-hydroxyvitamin D status of adolescents and adults in two seasonal subpopulations from NHANES III. Bone **30:**771–777.
23. Jones G, Blizzard C, Riley MD, Parameswaran V, Greenaway TM, Dwyer T 1999 Vitamin D levels in prepubertal children in southern Tasmania: Prevalence and determinants. Eur J Clin Nutr **53:**824–829.
24. Jones G, Dwyer T, Hynes KL, Parameswaran V, Greenaway TM 2005 Vitamin D insufficiency in adolescent males in Southern Tasmania: Prevalence, determinants, and relationship to bone turnover markers. Osteoporos Int **16:**636–641.
25. Rockell JE, Skeaff CM, Williams SM, Green TJ 2006 Serum 25-hydroxyvitamin D concentrations of New Zealanders aged 15 years and older. Osteoporos Int **17:**1382–1389.
26. El-Hajj Fuleihan G, Nabulsi M, Tamim H, Maalouf J, Salamoun M, Choucair M, Veith R 2004 Impact of vitamin D supplementation on musculoskeletal parameters in adolescents: A randomised trial. J Bone Miner Res **19:**S13.
27. El-Hajj Fuleihan G, Nabulsi M, Tamim H, Maalouf J, Salamoun M, Khalife H, Choucair M, Arabi A, Vieth R 2006 Effect of vitamin D replacement on musculoskeletal parameters in school children: A randomized controlled trial. J Clin Endocrinol Metab **91:**405–412.
28. Viljakainen HT, Natri AM, Karkkainen M, Huttunen MM, Palssa A, Jakobsen J, Cashman KD, Molgaard C, Lamberg-Allardt C 2006 A positive dose-response effect of vitamin D supplementation on site-specific bone mineral augmentation in adolescent girls: A double-blinded randomized placebo-controlled 1-year intervention. J Bone Miner Res **21:**836–844.
29. Jones G, Dwyer T, Hynes K, Dalais FS, Parameswaran V, Greenaway TM 2003 A randomized controlled trial of phytoestrogen supplementation, growth and bone turnover in adolescent males. Eur J Clin Nutr **57:**324–327.
30. Jones G, Riley MD, Whiting S 2001 Association between urinary potassium, urinary sodium, current diet, and bone density in prepubertal children. Am J Clin Nutr **73:**839–844.
31. Tylavsky FA, Holliday K, Danish R, Womack C, Norwood J, Carbone L 2004 Fruit and vegetable intakes are an independent predictor of bone size in early pubertal children. Am J Clin Nutr **79:**311–317.
32. McGartland CP, Robson PJ, Murray LJ, Cran GW, Savage MJ, Watkins DC, Rooney M, Boreham C 2004 Fruit and vegetable consumption and bone mineral density: The Northern Ireland Young Hearts Project. Am J Clin Nutr **80:**1019–1023.
33. Prynne CJ, Mishra GD, O'Connell MA, Muniz G, Laskey MA, Yan L, Prentice A, Ginty F 2006 Fruit and vegetable intakes and bone mineral status: A cross sectional study in 5 age and sex cohorts. Am J Clin Nutr **83:**1420–1428.
34. Vatanparast H, Baxter-Jones A, Faulkner RA, Bailey DA, Whiting SJ 2005 Positive effects of vegetable and fruit consumption and calcium intake on bone mineral accrual in boys during growth from childhood to adolescence: The University of Saskatchewan Pediatric Bone Mineral Accrual Study. Am J Clin Nutr **82:**700–706.
35. Hirota T, Kusu T, Hirota K 2005 Improvement of nutrition stimulates bone mineral gain in Japanese school children and adolescents. Osteoporos Int **16:**1057–1064.
36. Knai C, Pomerleau J, Lock K, McKee M 2006 Getting children to eat more fruit and vegetables: A systematic review. Prev Med **42:**85–95.
37. Jones G, Riley MD, Dwyer T 2000 Maternal diet during pregnancy is associated with bone mineral density in children: A longitudinal study. Eur J Clin Nutr **54:**749–756.
38. Tobias JH, Steer CD, Emmett PM, Tonkin RJ, Cooper C, Ness AR 2005 Bone mass in childhood is related to maternal diet in pregnancy. Osteoporos Int **16:**1731–1741.
39. Javaid MK, Crozier SR, Harvey NC, Gale CR, Dennison EM, Boucher BJ, Arden NK, Godfrey KM, Cooper C 2006 Maternal vitamin D status during pregnancy and childhood bone mass at age 9 years: A longitudinal study. Lancet **367:**36–43.
40. Brooke OG, Brown IR, Bone CD, Carter ND, Cleeve HJ, Maxwell JD, Robinson VP, Winder SM 1980 Vitamin D supplements in pregnant Asian women: Effects on calcium status and fetal growth. BMJ **280:**751–754.
41. Marya RK, Rathee S, Dua V, Sangwan K 1988 Effect of vitamin D supplementation during pregnancy on foetal growth. Indian J Med Res **88:**488–492.
42. Merialdi M, Caulfield LE, Zavaleta N, Figueroa A, Costigan KA, Dominici F, Dipietro JA 2004 Randomized controlled trial of prenatal zinc supplementation and fetal bone growth. Am J Clin Nutr **79:**826–830.
43. Zamora SA, Rizzoli R, Belli DC, Slosman DO, Bonjour JP 1999 Vitamin D supplementation during infancy is associated with higher bone mineral mass in prepubertal girls. J Clin Endocrinol Metab **84:**4541–4544.
44. Morley R, Lucas A 1994 Influence of early diet on outcome in preterm infants. Acta Paediatr Suppl **405:**123–126.
45. Specker B 2004 Nutrition influences bone development from infancy through toddler years. J Nutr **134:**691S–695S.
46. Specker BL, Beck A, Kalkwarf H, Ho M 1997 Randomized trial of varying mineral intake on total body bone mineral accretion during the first year of life. Pediatrics **99:**E12.
47. Jones G, Riley M, Dwyer T 2000 Breast feeding in early life and bone mass in prepubertal children: A longitudinal study. Osteoporos Int **11:**146–152.
48. Micklesfield L, Levitt N, Dhansay M, Norris S, van der Merwe L, Lambert E 2006 Maternal and early life influences on calcaneal ultrasound parameters and metacarpal morphometry in 7- to 9-year-old children. J Bone Miner Metab **24:**235–242.
49. Kurl S, Heinonen K, Jurvelin JS, Lansimies E 2002 Lumbar bone mineral content and density measured using a Lunar DPX densitometer in healthy full-term infants during the first year of life. Clin Physiol Funct Imaging **22:**222–225.
50. Young RJ, Antonson DL, Ferguson PW, Murray ND, Merkel K, Moore TE 2005 Neonatal and infant feeding: Effect on bone density at 4 years. J Pediatr Gastroenterol Nutr **41:**88–93.
51. Laskey MA, de Bono S, Smith EC, Prentice A 2007 Influence of birth weight and early diet on peripheral bone in premenopausal Cambridge women: A pQCT study. J Musculoskelet Neuronal Interact **7:**83.
52. Ma DQ, Jones G 2002 Clinical risk factors but not bone density are associated with prevalent fractures in prepubertal children. J Paediatr Child Health **38:**497–500.
53. Jones IE, Williams SM, Goulding A 2004 Associations of birth weight and length, childhood size, and smoking with bone fractures during growth: Evidence from a birth cohort study. Am J Epidemiol **159:**343–350.
54. O'Brien KO, Abrams SA, Stuff JE, Liang LK, Welch TR 1996 Variables related to urinary calcium excretion in young girls. J Pediatr Gastroenterol Nutr **23:**8–12.
55. Matkovic V, Ilich JZ, Andon MB, Hsieh LC, Tzagournis MA, Lagger BJ, Goel PK 1995 Urinary calcium, sodium, and bone mass of young females. Am J Clin Nutr **62:**417–425.
56. Duff TL, Whiting SJ 1998 Calciuric effects of short-term dietary loading of protein, sodium chloride and potassium citrate in prepubescent girls. J Am Coll Nutr **17:**148–154.
57. Hoppe C, Molgaard C, Michaelsen KF 2000 Bone size and bone mass in 10-year-old Danish children: Effect of current diet. Osteoporos Int **11:**1024–1030.
58. Whiting S, Heaky A, Psiuk S, Mirwald R, Kowalski K, Bailey DA 2001 Relationship between carbonated and other low nutrient dense beverages and bone mineral content of adolescents. Nutr Res **21:**1107–1115.
59. McGartland C, Robson PJ, Murray L, Cran G, Savage MJ, Watkins D, Rooney M, Boreham C 2003 Carbonated soft drink consumption and bone mineral density in adolescence: The Northern Ireland Young Hearts project. J Bone Miner Res **18:**1563–1569.
60. Wyshak G, Frisch RE 1994 Carbonated beverages, dietary calcium, the dietary calcium/phosphorus ratio, and bone fractures in girls and boys. J Adolesc Health **15:**210–215.

© 2008 American Society for Bone and Mineral Research

61. Petridou E, Karpathios T, Dessypris N, Simou E, Trichopoulos D 1997 The role of dairy products and non alcoholic beverages in bone fractures among schoolage children. Scand J Soc Med **25**:119–125.
62. Black RE, Williams SM, Jones IE, Goulding A 2002 Children who avoid drinking cow milk have low dietary calcium intakes and poor bone health. Am J Clin Nutr **76**:675–680.
63. Goulding A, Rockell JE, Black RE, Grant AM, Jones IE, Williams SM 2004 Children who avoid drinking cow's milk are at increased risk for prepubertal bone fractures. J Am Diet Assoc **104**:250–253.
64. Vatanparast H, Whiting SJ 2004 Early milk intake, later bone health: Results from using the milk history questionnaire. Nutr Rev **62**:256–260.
65. Kalkwarf HJ, Khoury JC, Lanphear BP 2003 Milk intake during childhood and adolescence, adult bone density, and osteoporotic fractures in US women. Am J Clin Nutr **77**:257–265.

Chapter 17. Growing a Healthy Skeleton: The Importance of Mechanical Loading

Gaele Ducher,[1] Shona Bass,[1] and Magnus K. Karlsson[2]

[1]Centre for Physical Activity and Nutrition Research, Deakin University, Burwood, Victoria, Australia; [2]Clinical and Molecular Osteoporosis Research Unit, Department of Clinical Sciences, Lund University, Department of Orthopaedics, Malmö University Hospital, Malmö, Sweden

INTRODUCTION

Paradoxically, during human growth, the only constant is change: change in stature, mass, proportionality, body composition, and shape. These changes at the somatic level induce changes in biomechanical conditions, which coupled with changing patterns of physical activity, provide the growing skeleton with a continually varying mechanical strain environment. This chapter reviews the existing evidence regarding the site-, sex-, and maturity-specific adaptations of bone to loading and identifies the gaps in our knowledge to develop efficient exercise prescription for long-term bone health.

ADAPTATIVE RESPONSE OF BONE TO LOADING

It has been estimated from natural experiments that mechanical use could determine over 40% of postnatal bone strength.[1] The skeleton responds to the mechanical stimuli by initiating or inhibiting bone modeling and remodeling to keep typical peak strains within a safe physiological range. This feedback regulation of bone strength—best known as the mechanostat[2]—depends on the characteristics of the loading. The results of animal studies have convincingly shown that the rate of bone formation is enhanced when loads are applied dynamically.[3] High magnitude, high frequency, and unusually distributed strains are also key features of osteogenic stimuli. The mechanical load needed to stimulate osteogenesis decreases as the strain magnitude and frequency increases.[4,5] Bone cell mechanosensitivity becomes saturated after a few loading cycles[6] but recovers after rest; thus, separating loading into short bouts with periods of rest optimizes the response to loading.[7–9]

In an attempt to put these findings into practice, the osteogenic potential of exercise was estimated according to the magnitude and frequency of loading and the periods of recovery between of the exercise bouts.[10] Overall, the loading characteristics associated with an improvement in bone strength are very specific, making the general prescription of exercise (for cardiovascular health or weight management) unsuitable for skeletal health.

The authors state that they have no conflicts of interest.

EFFECTS OF EXERCISE ON BONE STRENGTH DURING GROWTH

Exercise has long been shown to enhance bone mineral accrual during growth. In young female gymnasts, areal BMD has been shown to increase 30–85% more rapidly than in sedentary children.[11] Young tennis players display marked side-to-side difference in bone mass (10–15%) in comparison with age-matched controls (<5%).[12,13]

Exercise during growth is important because of the associated changes in bone geometry that translate to greater increases in bone strength than an increase in bone mass alone[12,14,15] (Fig. 1). 3D techniques such as the pQCT and MRI have identified periosteal expansion (bone laid down on the periosteal surface) at loaded sites in young athletes. Bone size was ~10% greater when comparing the upper limbs of young prepubertal gymnasts and normally active children[16,17] or the playing and nonplaying arms of young prepubertal tennis players.[12,18] No such increase in bone size has been detected in the lower limbs. Alternatively at this site, bone may be preferentially laid down on the endosteal surface: for instance, despite there being no difference in bone size, cortical cross-sectional area was 5–12% greater in the lower limbs of young runners or gymnasts compared with controls.[17,19,20] The osteogenic response in the upper and lower limbs is site specific.[17,21] It is not possible to determine with confidence, however, whether the observed differences are the result of different loading histories (weight bearing versus non–weight bearing), different thresholds for osteogenesis, or different load magnitudes relative to bone size and strength (Fig. 2).

A less documented mechanism to increase bone strength is the redistribution of the bone mass to the areas submitted to high mechanical strains. Bone shape changes to cope with loading, as reported in animals[7,22–24] and in humans,[25–27] without necessarily marked increases in bone mass and bone size. The complexity of the response to loading is also illus-

Key words: exercise, loading, physical activity, growth and development, puberty, bone mass, bone geometry, bone size, bone strength, DXA, pQCT, MRI

Electronic Databases: 1. Pubmed—www.ncbi.nlm.nih.gov/sites/entrez?. 2.Sportdiscus—http://www.sirc.ca/products/sportdiscus.cfm. 3. Embase—http://www.info.embase.com/. 4. Web of Knowledge—http://www.webofknowledge.com/.

© 2008 American Society for Bone and Mineral Research

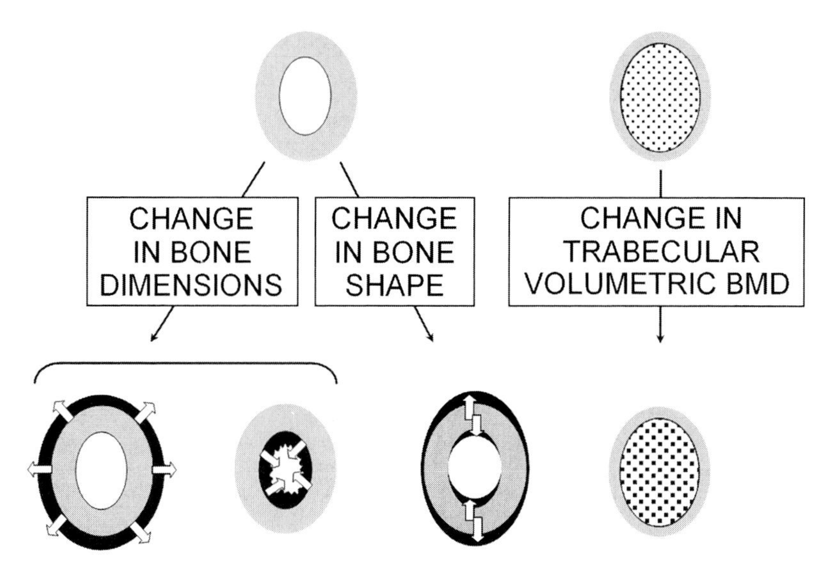

FIG. 1. Potential changes in bone mass and shape that underpin the exercise-induced increase in bone strength in children and adolescents. The different mechanisms depicted are not mutually exclusive and in many instances are combined. Changes in bone dimensions and bone shape are the preferential mechanisms in long bone shafts in response to exercise during growth. In long bone ends that are rich in trabecular bone, the increase in bone size is limited, and thus, exercise alternatively promotes an increase in trabecular volumetric density (trabecular vBMD).

trated by the heterogeneity of the geometrical adaptations along the length of a bone, in terms of magnitude and modality of the response (periosteal and/or endosteal apposition).[12,14,18,27]

The data in elite athletes provide us with a model of what is possible rather than probable in physically active children. Controlled trials in normally active children have shown that exercise intervention is associated with much lower skeletal benefits.[28] The mean increases in bone parameters over 6 mo reach 1–5% in prepubertal and early pubertal children (Tanner stage 1–3), and <2% in pubertal children (Tanner stages 4 and 5). However, such benefits should not be underestimated because animal studies showed that a small increase in bone mass (<15%) can generate a >2-fold increase in bone strength.[7,29]

Intervention studies in pre- and peripubertal girls and boys have been mostly of short duration (8–12 mo),[30–38] except a few studies with a longer follow-up[39–44]; most were controlled but not all were randomized. These interventions resulted in a 1.3–5% greater increase in BMD at the legs. Findings were equivocal at the spine, with studies that reported an increase in BMC or areal BMD,[30–33,40] whereas other studies did not.[34,35,37,38] Daily sessions of shorter bouts of exercise can also enhance bone mass accrual.[45,46] However, bone structural parameters were not markedly affected by such programs,[45] and the increase in bone strength did not reach statistical significance, except in prepubertal boys.[46]

Designing an exercise program for specific populations with limited physical abilities is even more challenging than in normally active children. Low-magnitude vibrations (<100 microstrains) applied at a high frequency (10–90 Hz), which correspond to the stimulus induced by muscular contraction,[47] can be an alternative to high magnitude loads. In children with limited mobility associated with a disability, trabecular volumetric density increased after 6 mo of a standing program with low-magnitude vibrations.[48] Further experiments are warranted to confirm the skeletal benefits of low-magnitude vibrations on trabecular bone and their potential effects on cortical bone.[49]

The osteogenic benefits achieved from exercise during growth are also maturity and sex dependent.[50] Exercise interventions seem to be most effective when initiated during pre- or early puberty.[51] It is thought that exercise may preferentially affect the surface of bone that is undergoing apposition during growth.[52] Accordingly, the prepubertal skeleton shows the capacity to respond to loading by adding more bone on the periosteal surface than would normally occur through growth-induced periosteal apposition.[12,16,18] Several studies also mentioned exercise-induced endosteal apposition in pre-

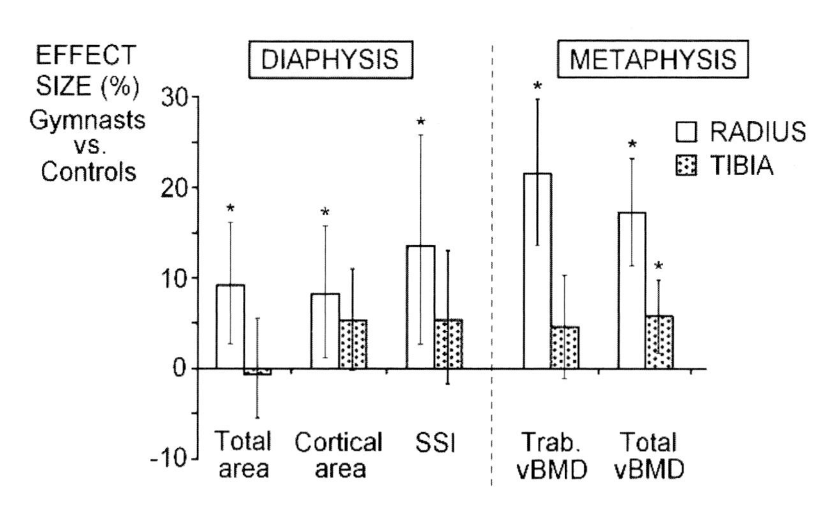

FIG. 2. Osteogenic response to exercise in the upper vs. lower limbs. The comparison of bone parameters (as measured by pQCT) between prepubertal gymnasts and normally active children showed that the between-group difference amounted to 8–21% at the radius, whereas it reached only 0–5% at the tibia. The observed differences in the osteogenic response between the upper and lower limbs may be caused by different loading histories (i.e., weight bearing vs. non–weight bearing), different thresholds for osteogenesis, or different magnitudes of loading relative to bone size and strength. vBMD, volumetric BMD (mg/mm^3); Trab, trabecular; SSI, stress-strain index (mm^3, which is a measure of diaphyseal bone bending and torsional strength). Bars indicate 95% CIs. *Between-group difference at $p < 0.05$–0.0001. (Adapted with permission from Elsevier from Ward KA, Roberts SA, Adams JE, Mughal MZ 2005 Bone geometry and density in the skeleton of prepubertal gymnasts and school children. Bone **36:**1012–1018.)

© 2008 American Society for Bone and Mineral Research

pubertal boys,[17,18,30] whereas such a response was not reported in prepubertal girls observed in the same conditions.[12,17]

The onset of puberty is associated with marked changes in sex steroids in both boys and girls. Whereas androgens stimulate periosteal apposition, a minimum level of estrogens also seems essential for pubertal periosteal bone expansion.[53–55] The rise in testosterone levels in peripubertal boys—in association with low levels of estrogens—drives periosteal expansion.[18] In contrast, periosteal expansion slows down in peripubertal girls as high levels of estrogens alternatively promote endosteal apposition, as shown in competitive female tennis players.[12] Apposition of bone on the periosteal (outer) surface rather than the endosteal (inner) surface is a more effective means of increasing the bending and torsional strength.[56] The enlargement of bone cross-section in response to loading has been reported to increase from pre- to peripuberty in male but not in female tennis players.[12,18] Consequently, the window of opportunity to strengthen the skeleton may be shorter in girls than in boys.

MAINTENANCE OF THE SKELETAL BENEFITS AFTER EXERCISE CESSATION

The exercise-induced skeletal benefits obtained during growth cannot be considered clinically important unless they are maintained into late adulthood, when fractures occur. Intuitively, this would seem unlikely because the mechanostat theory predicts a decrease in bone strength if exercise is reduced or ceased in adulthood.[57] However, >7 yr after an exercise intervention in prepubertal children, the skeletal benefits (i.e., increase in bone mass) were partly maintained.[58] Even 8 yr after the end of their career, former elite gymnasts 18–35 yr of age still had stronger bones than controls (polar stress strain index +10–25%) despite experiencing similar levels of exercise than the control group since retirement.[59] Former national-level tennis players showed significant side-to-side differences in bone size, cortical area, and bone strength index 1.5–3 yr after retirement.[14] Because the mature skeleton is thought to lose bone mass essentially through remodeling on the endosteal envelope and to a much lower extent on the periosteal envelope,[60] the structural adaptations obtained during growth may be better preserved.[61] However, the aforementioned studies included retired athletes who were younger than 40 yr of age. The limited data in 70- to 80-yr-old retired athletes suggests that the exercise-induced benefits in areal BMD may be eroded in those who have substantially decreased their training volumes,[62] whereas structural benefits may persist[63] because former male older athletes were shown to have fewer fractures than matched controls.[64]

CONCLUSIONS AND PERSPECTIVES

Childhood and adolescence is a critical time when the skeleton is sculptured to best suit its long-term functional needs. Recent research has focused on the structural adaptations underpinning the improvement in bone strength in response to loading, such as changes in bone size, cortical thickness, and volumetric BMD. Future developments include changes in bone shape and also bone microarchitecture. Improving exercise prescription also requires identifying the window of opportunity to improve bone strength in girls and boys and investigating whether the magnitude of the skeletal benefits is different between sexes during different stages of maturation. There is weak evidence supporting the notion that the skeletal gains obtained during growth are maintained despite exercise cessation.[61,65] Further studies are needed to determine (1) the minimum threshold of exercise during growth that is necessary to obtain a clinically significant increase in bone strength[28] and (2) the minimum threshold of exercise during adulthood that is needed to maintain the benefits (gained during growth) and prevent osteoporosis. This is a pivotal area of research that underpins future decisions regarding the role of exercise during growth for improved bone health in the aged.

REFERENCES

1. Frost HM, Schoenau E 2000 The "muscle-bone unit" in children and adolescents: A 2000 overview. J Pediatr Endocrinol Metab **13:**571–590.
2. Frost HM 1987 Bone mass and the mechanostat: A proposal. Anat Rec **219:**1–9.
3. Turner CH 1998 Three rules for bone adaptation to mechanical stimuli. Bone **23:**399–407.
4. Hsieh YF, Turner CH 2001 Effects of loading frequency on mechanically induced bone formation. J Bone Miner Res **16:**918–924.
5. Cullen DM, Smith RT, Akhter MP 2001 Bone-loading response varies with strain magnitude and cycle number. J Appl Physiol **91:**1971–1976.
6. Rubin CT, Lanyon LE 1984 Regulation of bone formation by applied dynamic loads. J Bone Joint Surg Am **66:**397–402.
7. Robling AG, Hinant FM, Burr DB, Turner CH 2002 Improved bone structure and strength after long-term mechanical loading is greatest if loading is separated into short bouts. J Bone Miner Res **17:**1545–1554.
8. Robling AG, Hinant FM, Burr DB, Turner CH 2002 Shorter, more frequent mechanical loading sessions enhance bone mass. Med Sci Sports Exerc **34:**196–202.
9. Srinivasan S, Weimer DA, Agans SC, Bain SD, Gross TS 2002 Low-magnitude mechanical loading becomes osteogenic when rest is inserted between each load cycle. J Bone Miner Res **17:**1613–1620.
10. Turner CH, Robling AG 2003 Designing exercise regimens to increase bone strength. Exerc Sport Sci Rev **31:**45–50.
11. Bass S, Pearce G, Bradney M, Hendrich E, Delmas PD, Harding A, Seeman E 1998 Exercise before puberty may confer residual benefits in bone density in adulthood: Studies in active prepubertal and retired female gymnasts. J Bone Miner Res **13:**500–507.
12. Bass SL, Saxon L, Daly RM, Turner CH, Robling AG, Seeman E, Stuckey S 2002 The effect of mechanical loading on the size and shape of bone in pre-, peri-, and postpubertal girls: A study in tennis players. J Bone Miner Res **17:**2274–2280.
13. Kannus P, Haapasalo H, Sankelo M, Sievänen H, Pasanen M, Heinonen A, Oja P, Vuori I 1995 Effect of starting age of physical activity on bone mass in the dominant arm of tennis and squash players. Ann Intern Med **123:**27–31.
14. Haapasalo H, Kontulainen S, Sievänen H, Kannus P, Järvinen M, Vuori I 2000 Exercise-induced bone gain is due to enlargement in bone size without a change in volumetric bone density: A peripheral quantitative computed tomography study of the upper arms of male tennis players. Bone **27:**351–357.
15. Kontulainen S, Sievänen H, Kannus P, Pasanen M, Vuori I 2002 Effect of long-term impact-loading on mass, size, and estimated strength of humerus and radius of female racquet-sports players: A peripheral quantitative computed tomography study between young and old starters and controls. J Bone Miner Res **17:**2281–2289.
16. Dyson K, Blimkie CJ, Davison KS, Webber CE, Adachi JD 1997 Gymnastic training and bone density in pre-adolescent females. Med Sci Sports Exerc **29:**443–450.
17. Ward KA, Roberts SA, Adams JE, Mughal MZ 2005 Bone geometry and density in the skeleton of pre-pubertal gymnasts and school children. Bone **36:**1012–1018.
18. Ducher G, Black J, Saxon L, Daly R, Turner CH, Bass S 2007 Exercise-induced changes in the macro-architectural parameters of cortical bone as boys progress through puberty: A large window of opportunity? J Bone Miner Res **22:**S1;S486.
19. Greene DA, Naughton GA, Briody JN, Kemp A, Woodhead H, Corrigan L 2005 Bone strength index in adolescent girls: Does physical activity make a difference? Br J Sports Med **39:**622–627.
20. Duncan CS, Blimkie CJ, Kemp A, Higgs W, Cowell CT, Woodhead H, Briody JN, Howman-Giles R 2002 Mid-femur geometry

© 2008 American Society for Bone and Mineral Research

and biomechanical properties in 15- to 18-yr-old female athletes. Med Sci Sports Exerc **34:**673–681.

21. Heinonen A, Sievanen H, Kannus P, Oja P, Vuori I 2002 Site-specific skeletal response to long-term weight-training seems to be attributable to principal loading modality: A pQCT study of female weightlifters. Calcif Tissue Int **70:**469–474.
22. Warden SJ, Fuchs RK, Castillo AB, Nelson IR, Turner CH 2007 Exercise when young provides lifelong benefits to bone structure and strength. J Bone Miner Res **22:**251–259.
23. Hiney KM, Nielsen BD, Rosenstein D, Orth MW, Marks BP 2004 High-intensity exercise of short duration alters bovine bone density and shape. J Anim Sci **82:**1612–1620.
24. Carlson KJ, Judex S 2007 Increased non-linear locomotion alters diaphyseal bone shape. J Exp Biol **210:**3117–3125.
25. Cheng S, Sipilä S, Taaffe DR, Puolakka J, Suominen H 2002 Change in bone mass distribution induced by hormone replacement therapy and high-impact physical exercise in postmenopausal women. Bone **31:**126–135.
26. Macdonald HM, Cooper DML, Kontulainen S, McKay HA 2007 A school-based physical activity intervention positively affects changes in Imax in pre- and early pubertal boys. J Bone Miner Res **22:**S1;S12.
27. Jones HH, Priest JD, Hayes WC, Tichenor CC, Nagel DA 1977 Humeral hypertrophy in response to exercise. J Bone Joint Surg Am **59:**204–208.
28. Hind K, Burrows M 2007 Weight-bearing exercise and bone mineral accrual in children and adolescents: A review of controlled trials. Bone **40:**14–27.
29. Warden SJ, Hurst JA, Sanders MS, Turner CH, Burr DB, Li J 2005 Bone adaptation to a mechanical loading program significantly increases skeletal fatigue resistance. J Bone Miner Res **20:**809–816.
30. Bradney M, Pearce G, Naughton G, Sullivan C, Bass S, Beck T, Carlson J, Seeman E 1998 Moderate exercise during growth in prepubertal boys: Changes in bone mass, size, volumetric density, and bone strength: A controlled prospective study. J Bone Miner Res **13:**1814–1821.
31. Morris FL, Naughton G, Gibbs JL, Carlson JS, Wark JD 1997 Prospective ten-month exercise intervention in premenarcheal girls: Positive effects on bone and lean mass. J Bone Miner Res **12:**1453–1462.
32. Fuchs RK, Bauer JJ, Snow CM 2001 Jumping improves hip and lumbar spine bone mass in prepubescent children: A randomized controlled trial. J Bone Miner Res **16:**148–156.
33. Heinonen A, Sievänen H, Kannus P, Oja P, Pasanen M, Vuori I 2000 High-impact exercise and bones of growing girls: A 9-month controlled trial. Osteoporos Int **11:**1010–1017.
34. McKay HA, Petit MA, Schutz RW, Prior JC, Barr SI, Khan KM 2000 Augmented trochanteric bone mineral density after modified physical education classes: A randomized school-based exercise intervention study in prepubescent and early pubescent children. J Pediatr **136:**156–162.
35. Petit MA, McKay HA, MacKelvie KJ, Heinonen A, Kahn KM, Beck TJ 2002 A randomized school-based jumping intervention confers site and maturity-specific benefits on bone structural properties in girls: A hip structural analysis study. J Bone Miner Res **17:**363–372.
36. Van Langendonck L, Claessens AL, Vlietinck R, Derom C, Beunen G 2003 Influence of weight-bearing exercises on bone acquisition in prepubertal monozygotic female twins: A randomized controlled prospective study. Calcif Tissue Int **72:**666–674.
37. Iuliano-Burns S, Saxon L, Naughton G, Gibbons K, Bass S 2003 Regional specificity of exercise and calcium during skeletal growth in girls: A randomized controlled trial. J Bone Miner Res **18:**156–162.
38. Bass SL, Naughton G, Saxon L, Iuliano-Burns S, Daly R, Briganti EM, Hume C, Nowson C 2007 Exercise and calcium combined results in a greater osteogenic effect than either factor alone: A blinded randomised placebo-controlled trial in boys. J Bone Miner Res **22:**458–464.
39. Nichols DL, Sanborn CF, Love AM 2001 Resistance training and bone mineral density in adolescent females. J Pediatr **139:**494–500.
40. MacKelvie KJ, Khan KM, Petit MA, Janssen PA, McKay HA 2003 A school-based exercise intervention elicits substantial bone health benefits: A 2-year randomized controlled trials in girls. Pediatrics **112:**e447–e452.
41. MacKelvie KJ, Petit MA, Khan KM, Beck TJ, McKay HA 2004 Bone mass and structure are enhanced following a 2-year randomized controlled trial of exercise in prepubertal boys. Bone **34:**755–764.
42. Linden C, Ahlborg HG, Besjakov J, Gardsell P, Karlsson MK 2006 A school curriculum-based exercise program increases bone mineral accrual and bone size in prepubertal girls: Two-year data from the pediatric osteoporosis prevention (POP) study. J Bone Miner Res **21:**829–835.
43. Linden C, Stenevi S, Gärdsell P, Karlsson M 2006 A five year school curriculum based exercise program in girls during early adolescence is associated with a large bone size and a thick cortical shell—pQCT data from the Prospective Pediatric Osteoporosis Prevention Study (POP-Study). J Bone Miner Res **21:**S1;S38.
44. Linden C, Gärdsell P, Ahlborg H, Karlsson M 2005 A school curriculum based exercise program increase bone mineral accrual in boys and girls during early adolescence—Four years data from the POP study (Pediatric Osteoporosis Prevention study)—a prospective controlled intervention study in 221 children. J Bone Miner Res **20:**S1;S4.
45. McKay HA, MacLean L, Petit M, MacKelvie-O'Brien K, Janssen P, Beck T, Khan KM 2005 "Bounce at the bell": A novel program of short bouts of exercise improves proximal femur bone mass in early pubertal children. Br J Sports Med **39:**521–526.
46. Macdonald HM, Kontulainen SA, Khan KM, McKay HA 2007 Is a school-based physical activity intervention effective for increasing tibial bone strength in boys and girls? J Bone Miner Res **22:**434–446.
47. Huang R, Rubin C, McLeod K 1999 Changes in postural muscle dynamics as a function of age. J Gerontol A Biol Sci Med Sci **54:**B353–B357.
48. Ward K, Alsop C, Caulton J, Rubin C, Adams J, Mughal Z 2004 Low magnitude mechanical loading is osteogenic in children with disabling conditions. J Bone Miner Res **19:**360–369.
49. Gilsanz V, Wren TA, Sanchez M, Dorey F, Judex S, Rubin C 2006 Low-level, high-frequency mechanical signals enhance musculoskeletal development of young women with low BMD. J Bone Miner Res **21:**1464–1474.
50. Daly RM 2007 The effect of exercise on bone mass and structural geometry during growth. In: Daly RM, Petit MA (eds.) Optimizing Bone Mass and Strength. The Role of Physical Activity and Nutrition During Growth, vol. 51. Karger, Basel, Switzerland, pp. 33–49.
51. Hughes JM, Novotny SA, Wetzsteon RJ, Petit M 2007 Lessons learned from school-based skeletal loading intervention trials: Putting research into practice. In: Daly RM, Petit MA (eds.) Optimizing Bone Mass and Strength. The Role of Physical Activity and Nutrition During Growth, vol. 51. Karger, Basel, Switzerland, p. 162.
52. Ruff CB, Walker A, Trinkaus E 1994 Postcranial robusticity in *Homo.* III: Ontogeny. Am J Phys Anthropol **93:**35–54.
53. Vanderschueren D, Venken K, Ophoff J, Bouillon R, Boonen S 2006 Clinical review: Sex steroids and the periosteum–reconsidering the roles of androgens and estrogens in periosteal expansion. J Clin Endocrinol Metab **91:**378–382.
54. Rochira V, Zirilli L, Madeo B, Aranda C, Caffagni G, Fabre B, Montangero VE, Roldan EJA, Maffei L, Carani C 2007 Skeletal effects of long-term estrogen and testosterone replacement treatment in a man with congenital aromatase deficiency: Evidences of a priming effect of estrogen for sex steroids action on bone. Bone **40:**1662–1668.
55. Bouillon R, Bex M, Vanderschueren D, Boonen S 2004 Estrogens are essential for male pubertal periosteal bone expansion. J Clin Endocrinol Metab **89:**6025–6029.
56. Turner CH, Burr DB 1993 Basic biomechanical measurements of bone: A tutorial. Bone **14:**595–608.
57. Gafni RI, Baron J 2007 Childhood bone mass acquisition and peak bone mass may not be important determinants of bone mass in late adulthood. Pediatrics **119**(Suppl 2)**:**S131–S136.
58. Gunter KB, Baxter-Jones A, Mirwald R, Almstedt HC, Durski S, Fuller-Hayes AA, Snow CM 2007 Jump starting skeletal health: Bone increases from jumping exercise persist seven years post intervention. J Bone Miner Res **22:**S1;S506.
59. Eser P, Hill B, Ducher G, Bass S 2007 The long-term macro-architectural benefits of high impact exercise during growth: A study of retired gymnasts. J Bone Miner Res **22;**S1;S486.
60. Riggs BL, Melton LJ III, Robb RA, Camp JJ, Atkinson EJ, Peterson JM, Rouleau PA, McCollough CH, Bouxsein ML, Khosla S 2004 Population-based study of age and sex differences in bone

© 2008 American Society for Bone and Mineral Research

volumetric density, size, geometry, and structure at different skeletal sites. J Bone Miner Res **19:**1945–1954.
61. Karlsson M, Bass S, Seeman E 2001 The evidence that exercise during growth or adulthood reduces the risk of fragility fractures is weak. Best Pract Res Clin Rheumatol **15:**429–450.
62. Karlsson MK, Linden C, Karlsson C, Johnell O, Obrant K, Seeman E 2000 Exercise during growth and bone mineral density and fractures in old age. Lancet **355:**469–470.
63. Karlsson MK, Ahlborg H, Obrant K, Nyquist F, Lindberg H, Karlsson C 2002 Exercise during growth and young adulthood is associated with reduced fracture risk in old ages. J Bone Miner Res **17:**S1;S297.
64. Nordstrom A, Karlsson C, Nyquist F, Olsson T, Nordstrom P, Karlsson M 2005 Bone loss and fracture risk after reduced physical activity. J Bone Miner Res **20:**202–207.
65. Ducher G, Bass SL 2007 Exercise during growth: Compelling evidence for the primary prevention of osteoporosis? BoneKEy **4:**171–180.

Chapter 18. Pregnancy and Lactation

Christopher S. Kovacs[1] and Henry M. Kronenberg[2]

[1]*Faculty of Medicine–Endocrinology, Health Sciences Centre, Memorial University of Newfoundland, St. John's, Newfoundland, Canada;* [2]*Endocrine Unit, Massachusetts General Hospital and Harvard Medical School, Boston, Massachusetts*

INTRODUCTION

Normal pregnancy places a demand on the calcium homeostatic mechanisms of the human female, because the fetus and placenta draw calcium from the maternal circulation to mineralize the fetal skeleton. Similar demands are placed on the lactating woman to supply sufficient calcium to breast milk and enable continued skeletal growth in a nursing infant. Despite a similar magnitude of calcium demand presented to pregnant and lactating women, the adjustments made in each of these reproductive periods differ significantly (Fig. 1). These hormone-mediated adjustments normally satisfy the daily calcium needs of the fetus and infant without long-term consequences to the maternal skeleton. Detailed references on this subject are available in several comprehensive reviews.[1–3]

PREGNANCY

The developing fetal skeleton accretes ~30 g of calcium by term and ~80% of it during the third trimester when the fetal skeleton is rapidly mineralizing. This calcium demand seems to be largely met by a doubling of maternal intestinal calcium absorption, mediated by 1,25-dihydroxyvitamin D [calcitriol or 1,25(OH)D] and other factors.

Mineral Ions and Calcitropic Hormones

Normal pregnancy results in characteristic alterations in serum chemistries and calciotropic hormones.[1] The total serum calcium falls early in pregnancy because of a fall in serum albumin, but the ionized calcium (the physiologically important fraction) remains constant. Serum phosphate levels remain normal.

In studies of women from North America and Europe, the serum PTH level, when measured with two-site "intact" assays, falls to the low-normal range (i.e., 10–30% of the mean nonpregnant value) during the first trimester but increases steadily to the midnormal range by term. In contrast, the PTH value did not suppress in studies of women from Asia and Gambia and may reflect the lower calcium and vitamin D intakes in those populations. Total 1,25(OH)D levels double early in pregnancy and maintain this increase until term; free 1,25(OH)D levels are increased from the third trimester and possibly earlier. The rise in 1,25(OH)D may be largely independent of changes in PTH, because PTH levels are typically low at the time of the increase in 1,25(OH)D. Maternal kidneys likely account for most, if not all, of the rise in 1,25(OH)D during pregnancy, although the decidua, placenta, and fetal kidneys may contribute a small amount.[1] The renal 1-α-hydroxylase is upregulated in response to factors such as PTH-related protein (PTHrP), estradiol, prolactin, and placental lactogen. Serum calcitonin levels are also increased during pregnancy.

PTHrP levels are increased during pregnancy, as determined by assays that detect fragments encompassing amino acids 1–86. Because PTHrP is produced by many tissues in the fetus and mother, it is not clear which source(s) contribute to the rise detected in the maternal circulation. PTHrP may contribute to the elevation in 1,25(OH)D and suppression of PTH during pregnancy. PTHrP may have other roles such as regulating placental calcium transport in the fetus.[1,4] Also, PTHrP may have a role in protecting the maternal skeleton during pregnancy, because the carboxyl-terminal portion of PTHrP (osteostatin) has been shown to inhibit osteoclastic bone resorption.[5]

Pregnancy induces significant changes in other hormones, including sex steroids, prolactin, placental lactogen, and IGF-1. Each of these may have direct or indirect effects on calcium and bone metabolism during pregnancy, but these issues have been largely unexplored.

Intestinal Absorption of Calcium

Intestinal absorption of calcium is doubled during pregnancy from as early as 12 wk of gestation; this seems to be a major maternal adaptation to meet the fetal need for calcium. This increase may be largely the result of a 1,25(OH)D-mediated increase in intestinal calbindin$_{9K}$-D and other proteins, although it is notable that intestinal calcium absorption doubles well before the rise in free 1,25(OH)D levels in humans and other mammals. Prolactin, placental lactogen, and other factors may also stimulate intestinal calcium absorption. Enhancing calcium absorption early in pregnancy may allow the maternal skeleton to store calcium in advance of the peak fetal demands that occur later in pregnancy.

Key words: pregnancy, lactation, weaning, calcium, magnesium, phosphorus, parathyroid hormone, parathyroid hormone–related protein, calcitonin, vitamin D, calcitriol, estradiol, hyperparathyroidism, hypoparathyroidism, familial hypocalciuric, hypercalcemia, pseudohypoparathyroidism, vitamin D deficiency/insufficiency, bone markers, BMD

The authors state that they have no conflicts of interest.

© 2008 American Society for Bone and Mineral Research

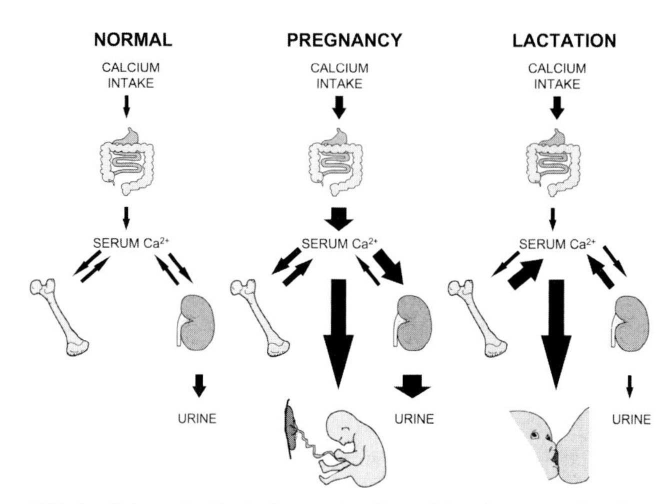

FIG. 1. Schematic illustration contrasting calcium homeostasis in human pregnancy and lactation compared with normal. The thickness of arrows indicates a relative increase or decrease with respect to the normal and nonpregnant state. Although not shown, the serum (total) calcium is decreased during pregnancy, whereas the ionized calcium remains normal during both pregnancy and lactation. (Adapted with permission from The Endocrine Society © 1997: Kovacs CS, Kronenberg HM 1997 Maternal-fetal calcium and bone metabolism during pregnancy, puerperium and lactation. Endocr Rev **18:**832–872.[24])

Renal Handling of Calcium

The 24-h urine calcium excretion is increased as early as the 12th week of gestation and may exceed the normal range. The elevated calcitonin levels of pregnancy might also promote renal calcium excretion. Because fasting urine calcium values are normal or low, the increase in 24-h urine calcium likely reflects the increased intestinal absorption of calcium (absorptive hypercalciuria).

Skeletal Calcium Metabolism

Animal models indicate that histomorphometric parameters of bone turnover are increased during pregnancy and that BMC may increase; however, comparable histomorphometric data are not available for human pregnancy. In one study,[6] 15 women who electively terminated a pregnancy in the first trimester (8–10 wk) had bone biopsy evidence of increased bone resorption, including increased resorption surface and increased numbers of resorption cavities. These findings were not present in biopsies obtained from nonpregnant controls or in biopsies obtained at term from 13 women who had elective C-sections.

Most human studies of skeletal calcium metabolism in pregnancy have examined changes in serum markers of bone formation and urine markers of bone resorption. These studies are fraught with a number of confounding variables, including lack of prepregnancy baseline values; effects of hemodilution in pregnancy on serum markers; increased glomerular filtration rate (GFR); altered creatinine excretion; placental, uterine, and fetal contribution to the levels of markers in blood; degradation and clearance by the placenta; and lack of diurnally timed or fasted specimens. Given these limitations, many studies have reported that urinary markers of bone resorption (24-h collection) are increased from early to midpregnancy (including deoxypyridinoline, pyridinoline, and hydroxyproline). Conversely, serum markers of bone formation (generally not corrected for hemodilution or increased GFR) are often decreased from prepregnancy or nonpregnant values in early or midpregnancy, rising to normal or above before term (including osteocalcin, procollagen I carboxypeptides, and bone-specific alkaline phosphatase). Total alkaline phosphatase rises early in pregnancy largely because of contributions from the placental fraction; it is not a useful marker of bone formation in pregnancy.

Based on the scant bone biopsy data and the measurements of bone markers (with aforementioned confounding factors), one may cautiously conclude that bone resorption is increased in pregnancy, from as early as the 10th week of gestation. There is comparatively little maternal–fetal calcium transfer occurring at this stage of pregnancy compared with the peak rate of calcium transfer in the third trimester. One might have anticipated that markers of bone resorption would increase particularly in the third trimester, but no marked increase is seen at that time.

Changes in skeletal calcium content have been assessed through the use of sequential areal BMD (aBMD) studies during pregnancy. Because of concerns about fetal radiation exposure, few such studies have been done. Such studies are confounded by the changes in body composition, weight, and skeletal volumes during normal pregnancy that can lead to artifactual changes in the aBMD reading obtained. Using single and/or dual-photon absorptiometry, several prospective studies did not find a significant change in cortical or trabecular aBMD during pregnancy.[1] Several recent studies have used DXA before conception (range 1–8 mo prior, but not always stated) and after delivery (range 1–6 wk postpartum).[2] Most studies involved 16 or fewer subjects. One study found no change in lumbar spine aBMD measurements obtained preconception and within 1–2 wk after delivery, whereas the other studies reported decreases of 4–5% in lumbar aBMD with the postpartum measurement taken between 1 and 6 wk after delivery. Because the puerperium is associated with BMD losses of 1–3%/mo (see LACTATION), it is possible that obtaining the second measurement 2–6 wk after delivery contributed to the bone loss documented in many of the studies. Other longitudinal studies have found a progressive decrease during pregnancy in indices thought to correlate with volumetric BMD, as determined by ultrasonographic measurements at another peripheral site, the os calcis. None of all the aforementioned studies can address the question as to whether skeletal calcium content is increased early in pregnancy in advance of the third trimester. Further studies, with larger numbers of patients, will be needed to clarify the extent of bone loss during pregnancy.

It seems certain that any acute changes in bone metabolism during pregnancy do not cause long-term changes in skeletal calcium content or strength. Numerous studies of osteoporotic or osteopenic women have failed to find a significant association of parity with BMD or fracture risk.[1,7]

Osteoporosis in Pregnancy

Occasionally, a woman may present with fragility fractures and low BMD during or shortly after pregnancy; the possibility that the woman had low BMD before pregnancy cannot be excluded. Some women may experience excessive resorption of calcium from the skeleton because of changes in mineral metabolism induced by pregnancy and other factors such as low dietary calcium intake and vitamin D insufficiency. The apparently increased rate of bone resorption in pregnancy may contribute to fracture risk, because a high rate of bone turnover is an independent risk factor for fragility fractures outside of pregnancy. Therefore, fragility fractures in pregnancy or the puerperium may be a consequence of preexisting low BMD and increased bone resorption, among other possible factors.

© 2008 American Society for Bone and Mineral Research

During lactation, additional changes in mineral metabolism occur that may further increase fracture risk in some women (see below).

Focal, transient osteoporosis of the hip is a rare, self-limited form of pregnancy-associated osteoporosis. It is probably not a manifestation of altered calciotropic hormone levels or mineral balance during pregnancy but rather might be a consequence of local factors. These patients present with unilateral or bilateral hip pain, limp, and/or hip fracture in the third trimester. There is objective evidence of reduced BMD of the symptomatic femoral head and neck, which has been shown by MRI to be the consequence of increased water content of the femoral head and the marrow cavity; a joint effusion may also be present. The symptoms and the radiological appearance usually resolve within 2–6 mo postpartum.

Primary Hyperparathyroidism

Although probably a rare condition (there are no data available on its prevalence), primary hyperparathyroidism in pregnancy has been associated in the literature with an alarming rate of adverse outcomes in the fetus and neonate, including a 30% rate of spontaneous abortion or stillbirth. The adverse postnatal outcomes are thought to result from suppression of the fetal and neonatal parathyroid glands; this suppression may occasionally be prolonged after birth for months. To prevent these adverse outcomes, surgical correction of primary hyperparathyroidism during the second trimester has been almost universally recommended. Several case series have found elective surgery to be well tolerated and to dramatically reduce the rate of adverse events compared with the earlier cases reported in the literature. Many of the women in those early cases had a relatively severe form of primary hyperparathyroidism that is not often seen today (symptomatic, with nephrocalcinosis and renal insufficiency). Although mild, asymptomatic primary hyperparathyroidism during pregnancy has been followed conservatively with successful outcomes, complications continue to occur, so that, in the absence of definitive data, surgery during the second trimester remains the most common recommendation.[8]

Familial Hypocalciuric Hypercalcemia

Although familial hypocalciuric hypercalcemia (FHH) has not been reported to adversely affect the mother during pregnancy, maternal hypercalcemia has caused fetal and neonatal parathyroid suppression with subsequent tetany.

Hypoparathyroidism and Pseudohypoparathyroidism

Early in pregnancy, hypoparathyroid women may have fewer hypocalcemic symptoms and require less supplemental calcium. This is consistent with a limited role for PTH in the pregnant woman and suggests that an increase in 1,25(OH)D and/or increased intestinal calcium absorption will occur in the absence of PTH. However, it is clear from other case reports that some pregnant hypoparathyroid women may require increased calcitriol replacement to avoid worsening hypocalcemia. It is important to maintain a normal ionized calcium level in pregnant women because maternal hypocalcemia caused by hypoparathyroidism has been associated with the development of intrauterine fetal hyperparathyroidism and fetal death. The ionized calcium rather than the total calcium should be followed because of the fall of serum albumin during pregnancy. Late in pregnancy, hypercalcemia may occur in hypoparathyroid women unless the calcitriol dose is substantially reduced or discontinued. This effect may be mediated by the increasing levels of PTHrP in the maternal circulation in late pregnancy.

In limited case reports of pseudohypoparathyroidism, pregnancy has been noted to normalize the serum calcium level, reduce the PTH level by one half, and increase the 1,25(OH)D level 2- to 3-fold.[9] The mechanism by which these changes occur despite pseudohypoparathyroidism remains unclear.

Vitamin D Deficiency and Insufficiency

There are no comprehensive studies of the effects of vitamin D deficiency or insufficiency on human pregnancy, but the available data from small clinical trials of vitamin D supplementation, observational studies, and case reports suggest that, consistent with animal studies, vitamin D deficiency is not associated with any worsening of maternal calcium homeostasis and that the fetus will have a normal serum calcium and fully mineralized skeleton at term.[10] At least in animal studies, maternal intestinal calcium absorption upregulates during pregnancy despite absence of vitamin D.

LACTATION

The typical daily loss of calcium in breast milk has been estimated to range from 280 to 400 mg, although daily losses as great as 1000 mg calcium have been reported. A temporary demineralization of the skeleton seems to be the main mechanism by which lactating humans meet these calcium requirements. This demineralization does not seem to be mediated by PTH or 1,25(OH)D but may be mediated by PTHrP in the setting of a fall in estrogen levels.

Mineral Ions and Calcitropic Hormones

The mean ionized calcium level of exclusively lactating women is increased, although it remains within the normal range. Serum phosphate levels are also higher during lactation, and the level may exceed the normal range. Because reabsorption of phosphate by the kidneys seems to be increased, the increased serum phosphate levels may, therefore, reflect the combined effects of increased flux of phosphate into the blood from diet and from skeletal resorption in the setting of decreased renal phosphate excretion.

Intact PTH, as determined by a two-site IRMA assay, has been found to be reduced 50% or more during the first several months of lactation. It rises to normal at weaning but may rise above normal postweaning. In contrast to the high 1,25(OH)D levels of pregnancy, maternal free and bound 1,25(OH)D levels fall to normal within days of parturition and remain there throughout lactation. Calcitonin levels fall to normal after the first 6 wk postpartum. Mice lacking the calcitonin gene lose twice the normal amount of BMC during lactation, which indicates that physiological levels of calcitonin may protect the maternal skeleton from excessive resorption during this time period.[11] Whether calcitonin plays a similar role in human physiology is unknown.

PTHrP levels, as measured by two-site IRMA assays, are significantly higher in lactating women than in nonpregnant controls. The source of PTHrP may be the breast, because PTHrP has been detected in breast milk at concentrations exceeding 10,000 times the level found in the blood of patients with hypercalcemia of malignancy or normal human controls. Furthermore, lactating mice with the *PTHrP* gene ablated only from mammary tissue have lower blood levels of PTHrP than control lactating mice.[12] Studies in animals suggest that PTHrP may regulate mammary development and blood flow and the calcium content of milk. In addition, PTHrP reaching the maternal circulation from the lactating breast may cause resorption of calcium from the maternal skeleton, renal tubular reabsorption of calcium, and (indirectly) suppression of

© 2008 American Society for Bone and Mineral Research

PTH. In support of this hypothesis, deletion of the *PTHrP* gene from mammary tissue at the onset of lactation resulted in more modest losses of BMC during lactation in mice.[12] In humans, PTHrP levels correlate with the amount of BMD lost, negatively with PTH levels, and positively with the ionized calcium levels of lactating women.[13–15] Furthermore, observations in aparathyroid women provide evidence of the impact of PTHrP in calcium homeostasis during lactation (see below).

Intestinal Absorption of Calcium

Intestinal calcium absorption decreases to the nonpregnant rate from the increased rate of pregnancy. This corresponds to the fall in 1,25(OH)D levels to normal.

Renal Handling of Calcium

In humans, the glomerular filtration rate falls during lactation, and the renal excretion of calcium is typically reduced to levels as low as 50 mg/24 h. This suggests that the tubular reabsorption of calcium must be increased, perhaps by the actions of PTHrP.

Skeletal Calcium Metabolism

Histomorphometric data from animals consistently show increased bone turnover during lactation, and losses of 35% or more of bone mineral are achieved during 2–3 wk of normal lactation in the rat.[1] Comparative histomorphometric data are lacking for humans, and in place of that, serum markers of bone formation and urinary markers of bone resorption have been assessed in numerous cross-sectional and prospective studies of lactation. Some of the confounding factors discussed with respect to pregnancy apply to the use of these markers in lactating women. In this instance, the GFR is reduced, and the intravascular volume is contracted. Urinary markers of bone resorption (24-h collection) have been reported to be elevated 2- to 3-fold during lactation and are higher than the levels attained in the third trimester. Serum markers of bone formation (not adjusted for hemoconcentration or reduced GFR) are generally high during lactation and increased over the levels attained during the third trimester. Total alkaline phosphatase falls immediately postpartum because of loss of the placental fraction, but may still remain above normal because of the elevation in the bone-specific fraction. Despite the confounding variables, these findings suggest that bone turnover is significantly increased during lactation.

Serial measurements of aBMD during lactation (by SPA, DPA, or DXA) have shown a fall of 3–10.0% in BMC after 2–6 mo of lactation at trabecular sites (lumbar spine, hip, femur, and distal radius), with smaller losses at cortical sites and whole body.[1,7] These aBMD changes are in accordance with studies in rats, mice, and primates in which the skeletal resorption has been shown to occur largely at trabecular and to a lesser degree at endocortical surfaces. The loss occurs at a peak rate of 1–3%/mo, far exceeding the rate of 1–3%/yr that can occur in women with postmenopausal osteoporosis who are considered to be losing bone rapidly. Loss of mineral from the maternal skeleton seems to be a normal consequence of lactation and may not be preventable by raising the calcium intake above the recommended dietary allowance. Several studies have shown that calcium supplementation does not significantly reduce the amount of bone lost during lactation.[16–19] Not surprisingly, the lactational decrease in BMD correlates with the amount of calcium lost in the breast milk.[20]

The mechanisms controlling the rapid loss of skeletal calcium content are not well understood. The reduced estrogen

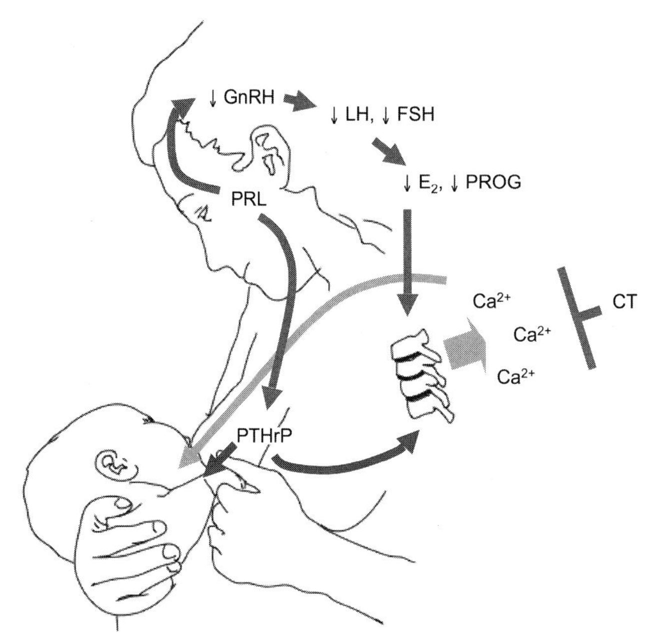

FIG. 2. The breast is a central regulator of skeletal demineralization during lactation. Suckling and prolactin both inhibit the hypothalamic gonadotropin-releasing hormone (GnRH) pulse center, which in turn suppresses the gonadotropins (luteinizing hormone [LH] and follicle-stimulating hormone [FSH]), leading to low levels of the ovarian sex steroids (estradiol and progesterone). PTHrP production and release from the breast is controlled by several factors, including suckling, prolactin, and the calcium receptor. PTHrP enters the bloodstream and combines with systemically low estradiol levels to markedly upregulate bone resorption. Increased bone resorption releases calcium and phosphate into the blood stream, which reaches the breast ducts and is actively pumped into the breast milk. PTHrP also passes into milk at high concentrations, but whether swallowed PTHrP plays a role in regulating calcium physiology of the neonate is unknown. Calcitonin (CT) may inhibit skeletal responsiveness to PTHrP and low estradiol. (Reprinted with permission from Springer Science and Business Media ©2005: Kovacs CS 2005 Calcium and bone metabolism during pregnancy and lactation. J Mammary Gland Biol Neoplasia **10**:105–118.)

levels of lactation are clearly important but are unlikely to be the sole explanation. To estimate the effects of estrogen deficiency during lactation, it is worth noting the alterations in calcium and bone metabolism occur in reproductive-age women who have estrogen deficiency induced by gonadotropin-releasing hormone (GnRH) agonist therapy for endometriosis and other conditions. Six months of acute estrogen deficiency induced by GnRH agonist therapy leads to 1–4% losses in trabecular (but not cortical) aBMD, increased urinary calcium excretion, and suppression of 1,25(OH)D and PTH.[1] In lactation, women are not as estrogen deficient but lose more aBMD (at both trabecular and cortical sites), have normal (as opposed to low) 1,25(OH)D levels, and have reduced (as opposed to increased) urinary calcium excretion. The difference between isolated estrogen deficiency and lactation is caused by the effects of PTHrP, which complements the effects of estrogen withdrawal in lactation. Stimulated in part by suckling and high prolactin levels, the effects of PTHrP and estrogen deficiency combine to coordinate the marked skeletal resorption that occurs during lactation (Fig. 2).

The BMD losses of lactation seem to be substantially reversed during weaning.[1,7,17] This corresponds to a gain in BMD of 0.5–2%/mo in the woman who has weaned her infant. The mechanism for this restoration of BMD is uncertain and

© 2008 American Society for Bone and Mineral Research

largely unexplored. In the long term, the consequences of lactation-induced depletion of bone mineral seem clinically unimportant. The vast majority of epidemiological studies of pre- and postmenopausal women have found no adverse effect of a history of lactation on peak bone mass, BMD, or hip fracture risk.

Osteoporosis of Lactation

Rarely, a woman will suffer a fragility fracture during lactation, and osteoporotic readings will be confirmed by DXA. Like osteoporosis in pregnancy, this may represent a coincidental, unrelated disease; the woman may have had low BMD before conception. Alternatively, some cases might represent an exacerbation of the normal degree of skeletal demineralization that occurs during lactation and a continuum from changes in BMD and bone turnover that may have occurred during pregnancy. For example, excessive PTHrP release from the lactating breast into the maternal circulation could conceivably cause excessive bone resorption, osteoporosis, and fractures. PTHrP levels were high in one case of lactational osteoporosis and were found to remain elevated for months after weaning.[21]

Hypoparathyroidism and Pseudohypoparathyroidism

Levels of calcitriol and calcium supplementation required for treatment of hypoparathyroid women fall early in the postpartum period, especially if the woman breastfeeds, and hypercalcemia may occur if the calcitriol dose is not substantially reduced.[22] As observed in one case, this is consistent with PTHrP reaching the maternal circulation in amounts sufficient to allow stimulation of 1,25(OH)D synthesis and maintenance of normal (or slightly increased) maternal serum calcium.[23]

The management of pseudohypoparathyroidism has been less well documented. Because these patients are likely resistant to the renal actions of PTHrP and the placental sources of 1,25(OH)D are lost at parturition, the calcitriol requirements might well increase and may require further adjustments during lactation.

Vitamin D Deficiency and Insufficiency

The available data from small clinical trials, observational studies, and case reports indicate that lactation proceeds normally regardless of vitamin D status, and breast milk calcium content is unaffected by vitamin D deficiency or supplementation.[10] This is likely because the maternal calcium homeostasis is dominated by skeletal resorption induced by estrogen deficiency and PTHrP. It is the neonate who will suffer the consequences of being born of a vitamin D–deficient mother, especially if exclusively breast fed, because both vitamin D and 25-hydroxyvitamin D penetrate poorly into breast milk.

IMPLICATIONS

In both pregnancy and lactation, novel regulatory systems specific to these settings complement the usual regulators of calcium homeostasis. The fetal calcium demand is met in large part by intestinal calcium absorption, which more than doubles from early in pregnancy, an adaptation that may not be fully explained by the increase in 1,25(OH)D levels. In comparison, skeletal calcium resorption is a dominant mechanism by which calcium is supplied to the breast milk, whereas renal calcium conservation is also apparent. These changes during lactation seem to be driven by PTHrP in association with estrogen deficiency rather than PTH and vitamin D. Consistent with this, treatment of lactating women with calcium supplements has little or no impact on bone loss.

These observations indicate that the maternal adaptations to pregnancy and lactation have evolved differently over time, such that dietary calcium absorption dominates in pregnancy, whereas lactation programs an obligatory but temporary skeletal calcium loss that is completely restored after weaning. The rapidity of calcium regain by the skeleton of the lactating woman occurs through a mechanism that is not understood. A full elucidation of the mechanism of bone restoration after lactation might lead to the development of novel approaches to the treatment of osteoporosis and other metabolic bone diseases. Finally, whereas it is apparent that some women will experience fragility fractures as a consequence of pregnancy or lactation, the vast majority of women can be assured that the changes in calcium and bone metabolism during pregnancy and lactation are normal, healthy, and without adverse consequences in the long term.

REFERENCES

1. Kovacs CS, Kronenberg HM 1997 Maternal-fetal calcium and bone metabolism during pregnancy, puerperium and lactation. Endocr Rev **18:**832–872.
2. Kovacs CS, El-Hajj Fuleihan G 2006 Calcium and bone disorders during pregnancy and lactation. Endocrinol Metab Clin North Am **35:**21–51.
3. Wysolmerski JJ 2007 Conversations between breast and bone: Physiological bone loss during lactation as evolutionary template for osteolysis in breast cancer and pathological bone loss after menopause. BoneKEy **4:**209–225.
4. Kovacs CS, Lanske B, Hunzelman JL, Guo J, Karaplis AC, Kronenberg HM 1996 Parathyroid hormone-related peptide (PTHrP) regulates fetal-placental calcium transport through a receptor distinct from the PTH/PTHrP receptor. Proc Natl Acad Sci USA **93:**15233–15238.
5. Cornish J, Callon KE, Nicholson GC, Reid IR 1997 Parathyroid hormone-related protein-(107-139) inhibits bone resorption in vivo. Endocrinology **138:**1299–1304.
6. Purdie DW, Aaron JE, Selby PL 1988 Bone histology and mineral homeostasis in human pregnancy. Br J Obstet Gynaecol **95:**849–854.
7. Sowers M 1996 Pregnancy and lactation as risk factors for subsequent bone loss and osteoporosis. J Bone Miner Res **11:**1052–1060.
8. Schnatz PF, Curry SL 2002 Primary hyperparathyroidism in pregnancy: Evidence-based management. Obstet Gynecol Surv **57:**365–376.
9. Breslau NA, Zerwekh JE 1986 Relationship of estrogen and pregnancy to calcium homeostasis in pseudohypoparathyroidism. J Clin Endocrinol Metab **62:**45–51.
10. Kovacs CS 2008 The role of vitamin D in pregnancy and lactation: Maternal, fetal and neonatal outcomes from clinical and animal studies. Am J Clin Nutr (in press).
11. Woodrow JP, Sharpe CJ, Fudge NJ, Hoff AO, Gagel RF, Kovacs CS 2006 Calcitonin plays a critical role in regulating skeletal mineral metabolism during lactation. Endocrinology **147:**4010–4021.
12. VanHouten JN, Dann P, Stewart AF, Watson CJ, Pollak M, Karaplis AC, Wysolmerski JJ 2003 Mammary-specific deletion of parathyroid hormone-related protein preserves bone mass during lactation. J Clin Invest **112:**1429–1436.
13. Kovacs CS, Chik CL 1995 Hyperprolactinemia caused by lactation and pituitary adenomas is associated with altered serum calcium, phosphate, parathyroid hormone (PTH), and PTH-related peptide levels. J Clin Endocrinol Metab **80:**3036–3042.
14. Dobnig H, Kainer F, Stepan V, Winter R, Lipp R, Schaffer M, Kahr A, Nocnik S, Patterer G, Leb G 1995 Elevated parathyroid hormone-related peptide levels after human gestation: Relationship to changes in bone and mineral metabolism. J Clin Endocrinol Metab **80:**3699–3707.
15. Sowers MF, Hollis BW, Shapiro B, Randolph J, Janney CA, Zhang D, Schork A, Crutchfield M, Stanczyk F, Russell-Aulet M 1996 Elevated parathyroid hormone-related peptide associated with lactation and bone density loss. JAMA **276:**549–554.
16. Kolthoff N, Eiken P, Kristensen B, Nielsen SP 1998 Bone mineral

© 2008 American Society for Bone and Mineral Research

changes during pregnancy and lactation: A longitudinal cohort study. Clin Sci (Colch) **94**:405–412.
17. Polatti F, Capuzzo E, Viazzo F, Colleoni R, Klersy C 1999 Bone mineral changes during and after lactation. Obstet Gynecol **94**:52–56.
18. Kalkwarf HJ, Specker BL, Bianchi DC, Ranz J, Ho M 1997 The effect of calcium supplementation on bone density during lactation and after weaning. N Engl J Med **337**:523–528.
19. Cross NA, Hillman LS, Allen SH, Krause GF 1995 Changes in bone mineral density and markers of bone remodeling during lactation and postweaning in women consuming high amounts of calcium. J Bone Miner Res **10**:1312–1320.
20. Laskey MA, Prentice A, Hanratty LA, Jarjou LM, Dibba B, Beavan SR, Cole TJ 1998 Bone changes after 3 mo of lactation: Influence of calcium intake, breast-milk output, and vitamin D-receptor genotype. Am J Clin Nutr **67**:685–692.
21. Reid IR, Wattie DJ, Evans MC, Budayr AA 1992 Post-pregnancy osteoporosis associated with hypercalcaemia. Clin Endocrinol (Oxf) **37**:298–303.
22. Caplan RH, Beguin EA 1990 Hypercalcemia in a calcitriol-treated hypoparathyroid woman during lactation. Obstet Gynecol **76**:485–489.
23. Mather KJ, Chik CL, Corenblum B 1999 Maintenance of serum calcium by parathyroid hormone-related peptide during lactation in a hypoparathyroid patient. J Clin Endocrinol Metab **84**:424–427.
24. Kovacs CS, Kronenberg HM 1997 Maternal-fetal calcium and bone metabolism during pregnancy, puerperium and lactation. Endocr Rev **18**:832–872.

Chapter 19. Menopause

Ian R. Reid

Department of Medicine, University of Auckland, Auckland, New Zealand

INTRODUCTION

Menopause refers to the cessation of menstruation, which occurs at ~48–50 yr of age in healthy women. The decline in ovarian hormone production is gradual and starts several years before the last period. Changes in bone mass and calcium metabolism are evident during this perimenopausal transition. Estrogen is the ovarian product that has the greatest impact on mineral metabolism, although both progesterone and ovarian androgens may have some influence. Menopause ushers in a period of bone loss that extends until the end of life and that is the central contributor to the development of osteoporotic fractures in older women.

EFFECTS ON BONE

Before the menopause, there is virtually no bone loss in most regions of the skeleton,[1,2] and fracture rates are stable. The most obvious effect of menopause on bone is an increase in the incidence of fractures—in the forearms and vertebrae, this is clearly apparent within the first postmenopausal decade. It is attributable to the rapid decline in bone mass that occurs in the perimenopausal years. Bone loss is more marked in trabecular than in cortical bone because the former has a greater surface area over which bone resorption can take place (e.g., loss of 1.4%/yr at the hip compared with 1.6%/yr at the spine in a recent study[3]). Thus, the fractures that occur early in menopause are in trabecular-rich regions of the skeleton such as the distal forearm and vertebrae. The loss of bone and increase in fracture rates are preventable with estrogen replacement.

The perimenopausal increase in bone loss is driven by increased bone resorption.[4] Bone biopsies in normal postmenopausal women show an increase in the proportion of bone surfaces at which resorption is taking place and an increase in the depth of resorption pits. These changes follow from an increase in the activation frequency of remodeling units and a prolongation of their resorptive phase. Indices of bone resorption are twice the levels found in premenopausal women, whereas markers of bone formation are only ~50% above premenopausal levels,[5] leading to negative bone balance. The resulting loss of bone leads to the perforation and loss of trabeculae,[6] and endosteal resorption, cortical thinning, and increased porosity in cortical bone.[7] The changes in histomorphometric indices and biochemical markers can be returned to premenopausal levels with estrogen replacement therapy.

The changes in bone turnover that accompany menopause are in part accounted for by the direct actions of estrogen on bone cells. Estrogen receptors are present in both osteoblasts and osteoclasts. Estrogen promotes the development of osteoblasts in preference to adipocytes from their common precursor cell,[8] increases osteoblast proliferation,[9] and increases production of a number of osteoblast proteins (e.g., IGF-1, type I procollagen, TGF-β, and bone morphogenetic protein-6). Thus, estrogen tends to have an anabolic effect on the isolated osteoblast, which is complemented by its inhibition of apoptosis in osteocytes[10] and osteoblasts.[11] In vivo, however, the initiation of estrogen replacement therapy is usually associated with a reduction in osteoblast numbers and activity.[12] This is accounted for by the tight coupling of osteoblast activity to that of osteoclasts and the overriding effect of estrogen to reduce osteoclastic bone resorption. However, there is now evidence that high concentrations of estrogen increase some histomorphometric indices of osteoblast activity (e.g., mean wall thickness) in humans, possibly by increasing osteoblast synthesis of growth factors.[13]

Estrogen's suppression of osteoclast activity is contributed to by increased osteoclast apoptosis,[14] reduced osteoblast/stromal cell production of RANKL, and increased production of osteoprotegerin.[15] These direct effects are buttressed by estrogen action on bone marrow stromal and mononuclear cells, which produce cytokines, such as interleukin-1 (IL-1), IL-6, and TNF-α, potent stimulators of osteoclast recruitment and/or activity.[16] Estrogen also regulates T-cell production of TNF-α through changes in IL-7[17] and interferon-γ.[18] Estrogen decreases production of each of these cytokines[19] and modulates levels of IL-1 receptors.[20] Bone loss after ovariectomy is reduced by blockers of these cytokines, and blockade

The author states that he has no conflicts of interest.

Key words: bone mass, BMD, estrogen, progesterone, osteoporosis, fractures, histomorphometry, bone formation, bone resorption, cytokines, growth factors, calcium balance, calcium absorption, renal tubular reabsorption of calcium, parathyroid hormone, serum calcium fractions, acid-base

© 2008 American Society for Bone and Mineral Research

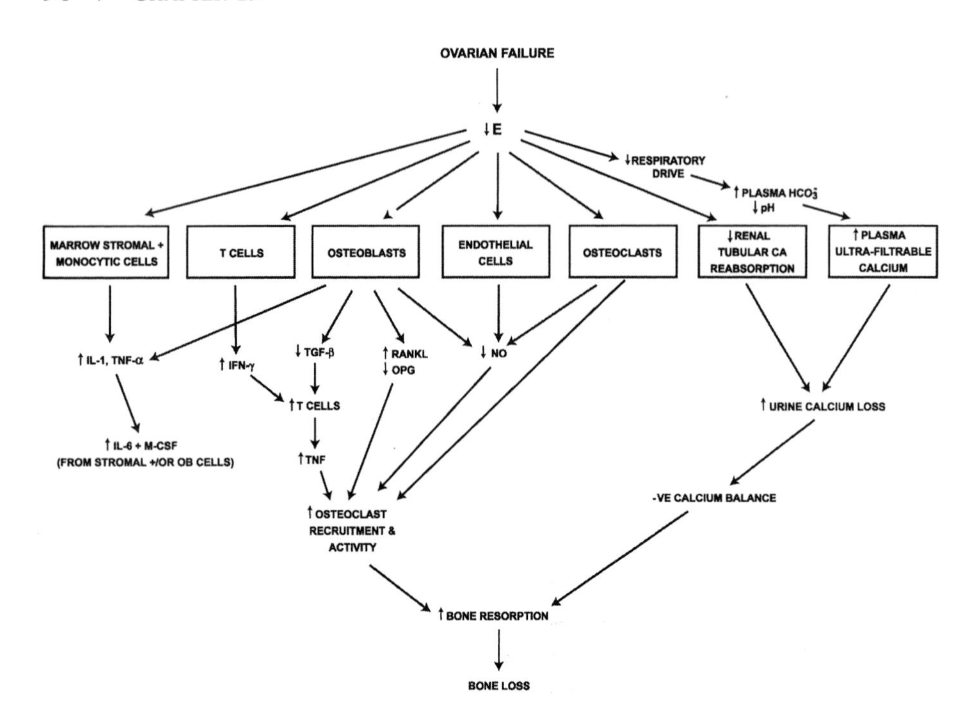

FIG. 1. The potential pathways by which menopause leads to bone loss. For simplicity, it does not show a contribution from loss of any anabolic effect of estrogen on the osteoblast. The fall in ovarian production of androgens and progesterone also contributes to some of these changes.

of TNF-α and IL-1 has been shown to reduce bone resorption in postmenopausal women.[21] IL-1 and TNF-α may act in part by regulating stromal cell production of IL-6 and macrophage-colony stimulating factor.[22] Estrogen's reduction of bone resorption is also contributed to by its causing increased levels of TGF-β, which inhibits osteoclast differentiation through regulation of T-cell production of TNF-α.[23] These cytokine interactions have recently been reviewed in detail.[24] Estrogen also has antioxidant effects, decreasing reactive oxygen species in mouse bone,[25] thereby reducing osteocyte apoptosis,[26] and contributing to reduced bone resorption through increased levels of NO.[27] Estrogen's effects on osteoclasts and bone marrow cytokines may also be supported by its regulation of other systemic bone active factors, such as growth hormone[28] and follicle-stimulating hormone (FSH).[29]

EFFECTS ON CALCIUM METABOLISM

The bone loss that follows menopause is accompanied by negative changes in external calcium balance, which are approximately equally contributed to by decreases in intestinal calcium absorption and by increases in urinary calcium loss.[30] Menopause is associated with reduced circulating concentrations of total, but not free, 1,25-dihydroxyvitamin D [$1,25(OH)_2D$], implying that its main effect is on vitamin D–binding protein. However, intestinal mucosal cells contain estrogen receptors and respond directly to 17β-estradiol with enhanced calcium transport,[31] probably through regulation of the epithelial calcium channel CaT1,[32] suggesting estrogen's effects are independent of vitamin D.

In the kidney, it is clear that tubular reabsorption of calcium is higher in the presence of estrogen.[33,34] One study[34] found higher PTH concentrations in the presence of estrogen and inferred that this was the mechanism of the renal calcium conservation. However, higher PTH levels have not been the finding in a number of other studies. Thus, it is likely that estrogen directly modulates renal tubular calcium absorption through its own receptor in the kidney, as suggested by in vitro studies of renal tubule cells that show a stimulatory effect of 17β-estradiol on calcium membrane transport.[35]

The changes in the handling of calcium by the gut and kidney could each be a cause of postmenopausal bone loss or they could represent homeostatic responses to it. If the former was the case, PTH concentrations would be elevated in postmenopausal women to maintain plasma calcium concentrations in the face of intestinal and renal losses. This, in turn, would cause bone loss. If, on the other hand, bone loss was the primary event, suppression of PTH would be expected, leading to secondary declines in intestinal and renal calcium absorption. The effect of menopause on PTH concentrations has been addressed many times without any consistent pattern emerging. This suggests that estrogen has direct effects on bone, kidney, and gut, and the opposing effects of these actions on PTH secretion leads to inconsistent changes in PTH concentrations. Furthermore, estrogen itself may directly modulate PTH secretion.[36,37]

There are small but consistently demonstrable effects of menopause on circulating concentrations of calcium. Total calcium is 0.05 mM higher after menopause.[33,38] This is partly attributable to a contraction of the plasma volume and resulting increase in albumin concentrations that occurs in the absence of estrogen[39,40] and partly to an increase in plasma bicarbonate, which leads to an increase in the complexed fraction of plasma calcium. The higher bicarbonate levels of postmenopausal women are attributable to a respiratory acidosis that results from the loss of the respiratory stimulatory effects of progesterone on the central nervous system, an action that is potentiated by estrogen.[41,42] Despite changes in protein-bound and complexed calcium fractions, ionized calcium concentrations are usually found to be the same in pre- and postmenopausal women.

SUMMARY

The effects of menopause on skeletal physiology are summarized in Fig 1. The major effect is an increase in bone turnover, which is dominantly an increase in bone resorption. This results in bone loss that may be contributed to by reductions in both intestinal and renal tubular absorption of calcium. Bone loss persists throughout the entire postmenopausal period and results in a high risk of fractures in those women whose peak bone mass was in the lower part of the normal range.

© 2008 American Society for Bone and Mineral Research

REFERENCES

1. Chapurlat RD, Garnero P, Sornay-Rendu E, Arlot ME, Claustrat B, Delmas PD 2000 Longitudinal study of bone loss in pre- and perimenopausal women: Evidence for bone loss in perimenopausal women. Osteoporos Int **11**:493–498.
2. Sowers MR, Jannausch M, McConnell D, Little R, Greendale GA, Finkelstein JS, Neer RM, Johnston J, Ettinger B 2006 Hormone predictors of bone mineral density changes during the menopausal transition. J Clin Endocrinol Metab **91**:1261–1267.
3. Macdonald HM, New SA, Campbell MK, Reid DM 2005 Influence of weight and weight change on bone loss in perimenopausal and early postmenopausal Scottish women. Osteoporos Int **16**:163–171.
4. Heaney RP, Recker RR, Saville PD 1978 Menopausal changes in bone remodeling. J Lab Clin Med **92**:964–970.
5. Garnero P, Sornayrendu E, Chapuy MC, Delmas PD 1996 Increased bone turnover in late postmenopausal women is a major determinant of osteoporosis. J Bone Miner Res **11**:337–349.
6. Akhter MP, Lappe JM, Davies KM, Recker RR 2007 Transmenopausal changes in the trabecular bone structure. Bone **41**:111–116.
7. Cooper DM, Thomas CD, Clement JG, Turinsky AL, Sensen CW, Hallgrimsson 2007 Age-dependent change in the 3D structure of cortical porosity at the human femoral midshaft. Bone **40**:957–965.
8. Okazaki R, Inoue D, Shibata M, Saika M, Kido S, Ooka H, Tomiyama H, Sakamoto Y, Matsumoto T 2002 Estrogen promotes early osteoblast differentiation and inhibits adipocyte differentiation in mouse bone marrow stromal cell lines that express estrogen receptor (ER) alpha or beta. Endocrinology **143**:2349–2356.
9. Fujita M, Urano T, Horie K, Ikeda K, Tsukui T, Fukuoka H, Tsutsumi O, Ouchi Y, Inoue S 2002 Estrogen activates cyclin-dependent kinases 4 and 6 through induction of cyclin D in rat primary osteoblasts. Biochem Biophys Res Commun **299**:222–228.
10. Tomkinson A, Reeve J, Shaw RW, Noble BS 1997 The death of osteocytes via apoptosis accompanies estrogen withdrawal in human bone. J Clin Endocrinol Metab **82**:3128–3135.
11. Gohel A, McCarthy MB, Gronowicz G 1999 Estrogen prevents glucocorticoid-induced apoptosis in osteoblasts in vivo and in vitro. Endocrinology **140**:5339–5347.
12. Vedi S, Compston JE 1996 The effects of long-term hormone replacement therapy on bone remodeling in postmenopausal women. Bone **19**:535–539.
13. Bord S, Beavan S, Ireland D, Horner A, Compston JE 2001 Mechanisms by which high-dose estrogen therapy produces anabolic skeletal effects in postmenopausal women: Role of locally produced growth factors. Bone **29**:216–222.
14. Kameda T, Mano H, Yuasa T, Mori Y, Miyazawa K, Shiokawa M, Nakamaru Y, Hiroi E, Hiura K, Kameda A, Yang NN, Hakeda Y, Kumegawa M 1997 Estrogen inhibits bone resorption by directly inducing apoptosis of the bone-resorbing osteoclasts. J Exp Med **186**:489–495.
15. Syed F, Khosla S 2005 Mechanisms of sex steroid effects on bone. Biochem Biophys Res Commun **328**:688–696.
16. Manolagas SC, Jilka RL 1995 Mechanisms of disease: Bone marrow, cytokines, and bone remodeling - Emerging insights into the pathophysiology of osteoporosis. N Engl J Med **332**:305–311.
17. Ryan MR, Shepherd R, Leavey JK, Gao YH, Grassi F, Schnell FJ, Qian WP, Kersh GJ, Weitzmann MN, Pacifici R 2005 An IL-7-dependent rebound in thymic T cell output contributes to the bone loss induced by estrogen deficiency. Proc Natl Acad Sci USA **102**:16735–16740.
18. Cenci S, Toraldo G, Weitzmann MN, Roggia C, Gao YH, Qian WP, Sierra O, Pacifici R 2003 Estrogen deficiency induces bone loss by increasing T cell proliferation and lifespan through IFN-gamma-induced class II transactivator. Proc Natl Acad Sci USA **100**:10405–10410.
19. Rogers A, Eastell R 1998 Effects of estrogen therapy of postmenopausal women on cytokines measured in peripheral blood. J Bone Miner Res **13**:1577–1586.
20. Sunyer T, Lewis J, Collin-Osdoby P, Osdoby P 1999 Estrogen's bone-protective effects may involve differential IL-1 receptor regulation in human osteoclast-like cells. J Clin Invest **103**:1409–1418.
21. Charatcharoenwitthaya N, Khosla S, Atkinson EJ, McCready LK, Riggs BL 2007 Effect of blockade of TNF-alpha and interleukin-1 action on bone resorption in early postmenopausal women. J Bone Miner Res **22**:724–729.
22. Kimble RB, Srivastava S, Ross FP, Matayoshi A, Pacifici R 1996 Estrogen deficiency increases the ability of stromal cells to support murine osteoclastogenesis via an interleukin-1-and tumor necrosis factor-mediated stimulation of macrophage colony-stimulating factor production. J Biol Chem **271**:28890–28897.
23. Gao YH, Qian WP, Dark K, Toraldo G, Lin ASP, Guldberg RE, Flavell RA, Weitzmann MN, Pacifici R 2004 Estrogen prevents bone loss through transforming growth factor beta signaling in T cells. Proc Natl Acad Sci USA **101**:16618–16623.
24. Weitzmann MN, Pacifici R 2006 Estrogen deficiency and bone loss: An inflammatory tale. J Clin Invest **116**:1186–1194.
25. Almeida M, Han L, Martin-Millan M, Plotkin LI, Stewart SA, Roberson PK, Kousteni S, O'Brien CA, Bellido T, Parfitt AM, Weinstein RS, Jilka RL, Manolagas SC 2007 Skeletal involution by age-associated oxidative stress and its acceleration by loss of sex steroids. J Biol Chem **282**:27285–27297.
26. Mann V, Huber C, Kogianni G, Collins F, Noble B 2007 The antioxidant effect of estrogen and selective estrogen receptor modulators in the inhibition of osteocyte apoptosis in vitro. Bone **40**:674–684.
27. Ralston SH 1997 The Michael-Mason-Prize essay 1997—nitric oxide and bone—what a gas. Br J Rheumatol **36**:831–838.
28. Friend KE, Hartman ML, Pezzoli SS, Clasey JL, Thorner MO 1996 Both oral and transdermal estrogen increase growth hormone release in postmenopausal women—a clinical research center study. J Clin Endocrinol Metab **81**:2250–2256.
29. Sun L, Peng Y, Sharrow AC, Iqbal J, Zhang Z, Papachristou DJ, Zaidi S, Zhu LL, Yaroslavskiy BB, Zhou H, Zallone A, Sairam MR, Kumar TR, Bo W, Braun J, Cardoso-Landa L, Schaffler MB, Moonga BS, Blair HC, Zaidi M 2006 FSH directly regulates bone mass. Cell **125**:247–260.
30. Heaney RP, Recker RR, Saville PD 1978 Menopausal changes in calcium balance performance. J Lab Clin Med **92**:953–963.
31. Arjandi BH, Salih MA, Herbert DC, Sims SH, Kalu DN 1993 Evidence for estrogen receptor-linked calcium transport in the intestine. Bone Miner **21**:63–74.
32. Van Cromphaut SJ, Rummens K, Stockmans I, Van Herck E, Dijcks FA, Ederveen A, Carmeliet P, Verhaeghe J, Bouillon R, Carmeliet G 2003 Intestinal calcium transporter genes are upregulated by estrogens and the reproductive cycle through vitamin D receptor-independent mechanisms. J Bone Miner Res **18**:1725–1736.
33. Nordin BEC, Wlshart JM, Clifton PM, McArthur R, Scopacasa F, Need AG, Morris HA, O'Loughlin PD, Horowitz M 2004 A longitudinal study of bone-related biochemical changes at the menopause. Clin Endocrinol (Oxf) **61**:123–130.
34. McKane WR, Khosla S, Burritt MF, Kao PC, Wilson DM, Ory SJ, Riggs BL 1995 Mechanism of renal calcium conservation with estrogen replacement therapy in women in early postmenopause—A clinical research center study. J Clin Endocrinol Metab **80**:3458–3464.
35. Dick IM, Liu J, Glendenning P, Prince RL 2003 Estrogen and androgen regulation of plasma membrane calcium pump activity in immortalized distal tubule kidney cells. Mol Cell Endocrinol **212**:11–18.
36. Duarte B, Hargis GK, Kukreja SC 1988 Effects of estradiol and progesterone on parathyroid hormone secretion from human parathyroid tissue. J Clin Endocrinol Metab **66**:584–587.
37. Greenberg C, Kukreja SC, Bowser EN, Hargis GK, Henderson WJ, Williams GA 1987 Parathyroid hormone secretion: Effect of estradiol and progesterone. Metabolism **36**:151–154.
38. Sokoll LJ, Dawson-Hughes B 1989 Effect of menopause and aging on serum total and ionized calcium and protein concentrations. Calcif Tissue Int **44**:181–185.
39. Aitken JM, Lindsay R, Hart DM 1974 The redistribution of body sodium in women on long-term estrogen therapy. Clin Sci Mol Med **47**:179–187.
40. Minkoff JR, Young G, Grant B, Marcus R 1986 Interactions of medroxyprogesterone acetate with estrogen on the calcium-parathyroid axis in post-menopausal women. Maturitas **8**:35–45.
41. Bayliss DA, Millhorn DE 1992 Central neural mechanisms of progesterone action: Application to the respiratory system. J Appl Physiol **73**:393–404.
42. Orr-Walker BJ, Horne AM, Evans MC, Grey AB, Murray MAF, McNeil AR, Reid IR 1999 Hormone replacement therapy causes a respiratory alkalosis in normal postmenopausal women. J Clin Endocrinol Metab **84**:1997–2001.

© 2008 American Society for Bone and Mineral Research

Chapter 20. Age-Related Bone Loss

Douglas P. Kiel,[1] Clifford J. Rosen,[2] and David Dempster[3,4]

[1]*Institute for Aging Research, Hebrew SeniorLife, and Harvard Medical School, Boston, Massachusetts;* [2]*Maine Medical Center Research Institute, Portland, Maine;* [3]*Regional Bone Center, Helen Hayes Hospital, West Haverstraw, New York;* [4]*Department of Pathology, College of Physicians and Surgeons of Columbia University, New York, New York*

INTRODUCTION

Age-related fractures are the most common manifestation of osteoporosis and are responsible for the greatest proportion of the morbidity and mortality from this disease. Over the next quarter century, as the population ages, fracture prevalence will also rise. Biochemical, biomechanical, and nonskeletal factors contribute to fragility fractures in the elderly. In this overview, we will focus on the quantitative and qualitative changes in the skeleton and the nonskeletal pathways that contribute to osteoporotic fractures in older individuals.

AGE-RELATED CHANGES IN BONE QUALITY AND QUANTITY THAT CONTRIBUTE TO FRACTURE RISK

Numerous studies have documented a progressive reduction in BMD at nearly every skeletal site with aging[1]; however, fracture risk also climbs with age, independent of BMD.[2] Therefore, other skeletal and nonskeletal factors must contribute to overall fracture risk. Recent advances in imaging technology and the availability of more longitudinal studies showed significant changes in trabecular microarchitecture that can be linked to bone strength and ultimately fracture risk in the elderly.[3] Other qualitative factors that are influenced by age include the degree of mineralization, microcrack number and frequency, anisotropy, skeletal geometry, matrix changes such as advanced glycation endproducts,[4,5] and the periosteal response to trabecular bone loss. The latter is particularly intriguing because the loss of trabecular elements may be accompanied by an increase in the cortical shell diameter.[6] This characteristic may be more apparent in men than women and may serve to protect the skeleton during active bone loss.[7] Finally, recent attention has focused on the role of marrow fat in the bone marrow compartment because adipogenesis at this site increases with aging,[8] possibly at the expense of bone. These age-related changes can be visualized by MRI and could have structural consequences, although neither the function nor the fate of marrow adipocytes is known.[9]

Currently, it is difficult to clinically measure qualitative characteristics of bone; however, newer techniques such as high-resolution pQCT is able to assess microarchitecture, which may itself contribute independently to fracture risk. Also, risk factors such as age and previous fracture capture some of these qualitative determinants of fracture risk. In contrast to the limited ability to measure qualitative changes in bone, quantifying BMD and loss of BMD can be easily assessed by DXA measurements. BMD changes with aging contribute to the risk of future fracture. For example, over a lifespan, women lose ~42% of their spinal and 58% of their femoral areal BMD.[1] The early postmenopausal years are associated with accelerated loss of bone from the cancellous compartment, followed by more gradual but sustained loss of both cancellous and cortical bone in both sexes. The rapid postmenopausal bone loss is attributed to an increase in resorption depth resulting in perforation and, ultimately, loss of trabecular plates.[10,11] The deleterious effect of increased erosion depth within each remodeling unit is exacerbated by an increase in remodeling activation frequency during this period.[12] The slower, age-related bone loss is caused by a decrease in osteon wall thickness,[13] with erosion depth remaining constant in cancellous bone[14] and increased on the endocortex.[15] Surprisingly, rates of bone loss in the eighth and ninth decades of life may be comparable to or even exceed those found in the immediate peri- and postmenopausal period of some women.[16,17] This is because of uncoupling in the bone remodeling cycle of older individuals, resulting in a marked increase in bone resorption but no change or a decrease in bone formation.[16,18] The latter scenario is particularly intriguing because aging is associated with a significant increase in stromal cell differentiation into the fat lineage and greater marrow adiposity, which may be associated with fewer stromal cells committed to the osteoblast lineage.

Alterations in bone turnover can be detected by biochemical markers of bone remodeling that include bone resorption indices (e.g., urinary and serum breakdown products of type I collagen) and bone formation markers (e.g., osteocalcin, procollagen peptide, bone-specific alkaline phosphatase). In general, bone turnover markers are significantly higher in older rather than younger postmenopausal women, and these indices are inversely related to BMD.[19] For example in the EPIDOS trial of elderly European females, the highest levels of osteocalcin, N-telopeptide, C-telopeptide, and bone-specific alkaline phosphatase were noted for those in the lowest tertile of femoral BMD.[20] Also, increased bone resorption indices were associated with a greater fracture risk independent of BMD.[20] For those women in EPIDOS with low BMD and a high bone resorption rate, there was a nearly 5-fold greater risk of a hip fracture. Similar findings have been noted in other cohorts composed of elderly individuals.[21]

In contrast to consistently high bone resorption indices, bone formation markers in older persons are more variable. Serum osteocalcin levels are high in elderly individuals and likely represent an increase in osteoblast number with increased bone formation rate at the tissue level coupled to the increase in bone resorption. As previously mentioned, this increased formation rate is unable to parallel the increase in bone resorption at the level of the bone remodeling unit.[20] On the other hand, bone-specific alkaline phosphatase and procollagen peptide levels have been reported to be high, normal, or low in elderly men and women.[22] Bone histomorphometric indices in elders are also quite variable, and the number of tetracycline-labeled biopsies in normal subjects is extremely limited.[23] Morphologically, the age-associated increase in the ratio of marrow adipocytes to hematopoietic/stromal tissue may contribute to the reduced bone formation rate at the level of the bone remodeling unit.[24] Thus, although there is strong evidence for an age-associated rise in bone resorption, changes

Dr. Kiel has received grants or contracts for research from Amgen, Merck, Novartis, and Hologic. He has served on the speakers bureau for Merck, Lilly, Roche, GlaxoSmithKline, and Novartis and has consulted for Amgen, Merck, Novartis, Lilly, Roche, GlaxoSmithKline, Hologic, and Procter & Gamble. All other authors state that they have no conflicts of interest.

Key words: aging, osteoporosis, treatment, epidemiology, hormones and receptors, nutrition, falls, fracture

© 2008 American Society for Bone and Mineral Research

in bone formation are inconsistent. Notwithstanding the limitations of biochemical markers, it is generally assumed there is uncoupling of the remodeling unit that leads to bone loss, altered skeletal architecture, and an increased propensity to fractures.

FACTORS THAT CONTRIBUTE TO AGE-RELATED BONE LOSS

Nutritional Factors

Increased bone resorption in older individuals can be attributed to a number of factors including calcium and/or vitamin D deficiency. Both are more common in older persons and are caused by a number of conditions, including dietary changes, lack of sunlight exposure, malabsorption, and the use of certain drugs. It is estimated that upward of 80% of elderly postmenopausal women may have vitamin D deficiency as defined by a 25-hydroxyvitamin D of <20 ng/ml (50 nM).[25,26] Low vitamin D concentrations could result not only in reduced bone mass but also altered muscle function[27,28] and an increased risk for falls.[29] Clinical trials show that the administration of vitamin D, in doses of ≥800 IU/d, can prevent fractures.[30–32] The National Academy of Science recommends a minimal daily requirement for calcium intake in people >65 years of age to be 1500 mg/d, and for vitamin D, to be 600 IU/d.[33] It is likely that this 600 IU/d recommendation for vitamin D will be increased to at least 800 IU/d in the coming years, especially because vitamin D supplementation also may reduce falls,[29] risk of malignancy,[34] and even prevent flu-like illnesses,[35] all common in older persons.

A balanced diet that includes micro- and macronutrients is essential for the overall health of the older individual. Besides calcium and vitamin D, other nutritional factors may play a role in age-related osteoporosis. Although total protein intake seems to be beneficial to the skeleton, there is still a debate as to whether high animal protein intake creates an acidic environment in bone, leading to loss of calcium from the skeleton.[36–38] Protein/calorie malnutrition stimulates bone resorption and impairs bone formation both directly and through other mechanisms such as reduced serum IGF-I.[39] Vitamin K deficiency may contribute to an increased risk of osteoporotic fractures, possibly through effects on the carboxylation of bone proteins such as osteocalcin,[40] although clinical trials have not shown a benefit on skeletal health in the short term. Other micronutrients such as the B vitamins and vitamin C have been linked to osteoporotic risk in some but not all studies. At least one observational study suggested that increased vitamin A intake might be associated with low BMD.[41] A case-control study showed a significant relationship between excessive intake of vitamin A and age-related fractures.[42]

Hormonal Factors

Estrogen deficiency has long been recognized as a major cause of bone loss in the first decade after menopause. More recently, investigators have identified a strong relationship between endogenous estrogen and bone mass in elderly men and women. In one prospective study, Slemenda et al.[43] noted that both estrogens and androgens were independent predictors of bone loss in older postmenopausal women. In both the Rancho Bernardo cohort and the Framingham Cohort, estradiol levels were strongly related to BMD at the spine, hip, and forearm.[44,45] Akhter et al.[46] have recently reported the results from the first transmenopausal bone biopsy study. Bone microarchitecture was assessed in paired biopsies taken before and 12 months after the last menstrual period. A significant deterioration in most structural parameters was observed, and trabecular connectivity density was negatively correlated with activation frequency, confirming the concept that postmenopausal bone loss is caused by increased bone turnover. Estrogen deficiency has also been shown to increase the depth of resorption cavities.[10]

Men also suffer from age-related bone loss, and evidence suggests that absolute estrogen levels, rather than testosterone concentrations, are essential for maintenance of BMD. In the Rancho Bernardo cohort, serum estradiol levels in elderly men correlated closely with bone mass at several sites,[45] and low estradiol levels in men are associated with increased risk of hip fracture.[47] Recently, Falahati et al.[48] showed that small amounts of estradiol were essential for preventing bone resorption in men, in part by upregulating osteoprotegerin (OPG).[49] Endogenous testosterone also plays a role in regulating bone turnover possibly more on the formation side than in respect to resorption. Most short-term testosterone treatment trials in older men show significant increases in BMD, although the dose and long-term risk-benefit ratio of testosterone replacement is not known. Serum testosterone levels decline with age at a rate of ~1.2%/yr, whereas sex hormone binding globulin (SHBG) levels rise. Men treated with androgen antagonists or gonadotropin agonists for prostate cancer metastases rapidly lose bone mass and may be at high risk for subsequent osteoporotic fractures.[50] Overall, it seems likely that both androgens and estrogens are important in the elderly man. Whether changes in male hormone levels are causally related to age-related bone loss in men will have to await large-scale prospective studies.

Changes in other circulating factors may be related to bone loss with aging. For example, growth hormone (GH) secretion declines 14% per decade and is the principle cause for low serum IGF-I concentrations in both elderly men and women.[51] Serum IGF-I concentrations in some studies but not others are directly related to BMD, and in one study, low IGF-I was an independent predictor of hip fracture.[52] Similarly, the adrenal androgens, DHEA and DHEA-S, also decline precipitously with age and are 10–20% of young adult serum levels.[53] OPG levels are also lower in older persons than in the younger postmenopausal woman, but whether this can lead to increased bone loss has yet to be shown.[54] Aging is also associated with a generalized cytokinemia, including greater serum levels of interleukin (IL)-6 and TNF, as well as C-reactive protein (CRP). Whether these changes are a function of other disease processes, and therefore may contribute to bone loss in the elderly, has not been established.

Heritable and Environmental Factors

Age-related bone loss can be dramatic in some individuals, and this decline cannot be attributed solely to hormonal or nutritional factors. Several investigators have hypothesized that there is genetic programming that, when triggered by environmental factors, may lead to bone loss. Some animal models, but not others, have shown a heritable component to age-related bone loss.[55] Recently, Bouxsein et al.[56] showed that, after ovariectomy, inbred strains of mice lost bone at very different rates. This would suggest that genetic programming may be operative in determining the rate of bone loss with estrogen deprivation. In humans, the multiplicity of environmental factors makes the determination of fracture heritability complicated, although recent publications also suggest a genetic component.[57] On the other hand, environmental factors, including inactivity and loss of muscle mass, smoking, vitamin D deficiency, and medications such as glucocorticoids and anticonvulsants, all may contribute to an excessive rate of bone loss in some elderly.

© 2008 American Society for Bone and Mineral Research

FRACTURES AND FALLS IN THE ELDERLY

In elderly individuals, the increased risk of fracture is caused by decreased bone strength, as reflected by BMD, and other factors including propensity to fall, inability to correct a postural imbalance, characteristics of the faller such as height and muscle activity, the orientation of the fall, adequacy of local tissue shock absorbers, and characteristics of the impact surface. The resistance of a skeletal structure to failure (i.e., fracture) depends on the geometry of the bone, the material properties of the calcified tissue, and the location and direction of the loads to which the bone is subjected (i.e., during a fall or other activities). Most of the energy from a fall dissipates before actual injury, and yet the residual force at impact remains two orders of magnitude greater than the energy required to fracture elderly femurs. This would suggest that a simple fall on the hip is easily capable of fracturing the proximal femur.

Falls in older people are rarely from a single cause. Falls usually occur when a threat to the normal homeostatic mechanisms that maintain postural stability is superimposed on underlying age-related declines in balance, strength, sensory function, and cardiovascular function. In some cases, this may involve an acute illness such as a fever or infection, an environmental stress such as a newly initiated drug, or an unsafe walking surface. Regardless of the nature of the stress, an elderly person may not be able to compensate because of either age-related declines in function or severe chronic disease. It is unlikely for an extrinsic stress to completely explain the circumstances of a fall. Older persons, by virtue of their age alone, experience declines in physiologic function, have greater numbers of chronic diseases, acute illnesses, and hospitalizations, and use multiple medications. Superimposed on these age-related characteristics, challenges to postural control may have a greater impact in aged persons according to their risk-taking behavior and opportunity to fall. Thus, those individuals who are completely immobile may not be at risk for falling despite multiple predisposing factors. On the other hand, persons who are either vigorous or only slightly frail may be at higher risk compared with individuals in between those extremes, caused in part by more risk taking and inability to compensate for postural changes. Despite the importance of falls, BMD still remains a major predictor of fracture risk,[2] regardless of age.[58] In general, both BMD and nonskeletal fracture risk factors may be combined to estimate absolute risk of fracture. Chronological age itself is a potent independent risk factor for fracture for any given measurement of BMD.

APPROACH TO FRACTURE PREVENTION IN THE ELDERLY PATIENT

Because older persons have lower BMD to start, are continuing to lose bone, and are in the age group most likely to fracture, interventions would be expected to be most cost-effective when initiated in these individuals. The interventions can be divided into two groups: (1) those that reduce the applied load to the skeleton (fall prevention, passive protective systems) and (2) those that preserve or increase BMD.

Interventions That Reduce the Applied Load

Interventions to prevent falls must be predicated on an assessment of fall risk. This should include a history of falls because a history of falls is the single most important risk factor for a subsequent fall. If that history is positive, additional information can be obtained surrounding the events of the fall, because this information may identify important factors for targeting risk factor modification strategies. The physical assessment of fall risk should include orthostatic vital sign measurement, a test of visual acuity, hearing, cardiac exam, extremity exam, and a test of the postural stability system as a whole using any of several recently developed assessment tools such as the "Get Up and Go" test[59,60] or the Short Portable Physical Performance Battery.[61] Because some of the unfavorable outcomes of major fractures, such as hip fractures, are highly dependent on the premorbid status of an older patient, fracture prevention efforts should include a thorough assessment of underlying disability and frailty of the older person, because these factors influence long-term outcomes.[62]

Pooled results from several fall intervention studies suggest that an intervention in which older people are assessed by a health professional trained to identify intrinsic and environmental risk factors is likely to reduce the fall rate (OR = 0.79; 95% CI = 0.65–0.96).[63] Because falls to the side that impact on the hip are the primary determinant of hip fracture,[64] protective trochanteric padding devices have been developed. Over the past decade, results from efficacy trials of external hip protectors conducted outside the United States have been conflicting. Recent meta-analyses concluded that the effectiveness of hip protectors in an institutional setting was uncertain.[65]

Interventions That Preserve or Increase BMD

In addition to the attention to adequate basic nutritional factors, the use of therapeutic agents in the treatment of osteoporosis may be synergistic with fall prevention in the elderly person. Because older persons are at the greatest risk of fracture, and because fracture reduction has been shown for the bisphosphonates,[66–70] estrogen therapy,[71,72] nasal calcitonin,[73] risedronate,[67] strontium ranelate,[74] and PTH,[75] potentially fewer elderly persons would have to be treated for less duration to prevent fractures than a younger population at lower risk of fracture.

REFERENCES

1. Melton LJ III, Ilstrup DM, Beckenbaugh RD, Riggs BL 1982 Hip fracture recurrence A population-based study. Clin Orthop **167:**131–138.
2. Cummings SR, Nevitt MC, Browner WS, Stone K, Fox KM, Ensrud KE, Cauley J, Black D, Vogt TM 1995 Risk factors for hip fracture in white women. Study of Osteoporotic Fractures Research Group. N Engl J Med **332:**767–773.
3. Bouxsein ML 2003 Bone quality: Where do we go from here? Osteoporos Int **14**(Suppl 5)**:**118–127.
4. Saito M, Fujii K, Soshi S, Tanaka T 2006 Reductions in degree of mineralization and enzymatic collagen cross-links and increases in glycation-induced pentosidine in the femoral neck cortex in cases of femoral neck fracture. Osteoporos Int **17:**986–995.
5. Viguet-Carrin S, Roux JP, Arlot ME, Merabet Z, Leeming DJ, Byrjalsen I, Delmas PD, Bouxsein ML 2006 Contribution of the advanced glycation end product pentosidine and of maturation of type I collagen to compressive biomechanical properties of human lumbar vertebrae. Bone **39:**1073–1079.
6. Ahlborg HG, Johnell O, Turner CH, Rannevik G, Karlsson MK 2003 Bone loss and bone size after menopause. N Engl J Med **349:**327–334.
7. Seeman E 2007 The periosteum–a surface for all seasons. Osteoporos Int **18:**123–128.
8. Justesen J, Stenderup K, Ebbesen EN, Mosekilde L, Steiniche T, Kassem M 2001 Adipocyte tissue volume in bone marrow is increased with aging and in patients with osteoporosis. Biogerontology **2:**165–171.
9. Gomberg BR, Saha PK, Song HK, Hwang SN, Wehrli FW 2000 Topological analysis of trabecular bone MR images. IEEE Trans Med Imaging **19:**166–174.
10. Eriksen EF, Langdahl B, Vesterby A, Rungby J, Kassem M 1999 Hormone replacement therapy prevents osteoclastic hyperactivity: A histomorphometric study in early postmenopausal women. J Bone Miner Res **14:**1217–1221.

© 2008 American Society for Bone and Mineral Research

11. Weinstein RS, Hutson MS 1987 Decreased trabecular width and increased trabecular spacing contribute to bone loss with aging. Bone **8**:137–142.
12. Recker R, Lappe J, Davies KM, Heaney R 2004 Bone remodeling increases substantially in the years after menopause and remains increased in older osteoporosis patients. J Bone Miner Res **19**:1628–1633.
13. Lips P, Courpron P, Meunier PJ 1978 Mean wall thickness of trabecular bone packets in the human iliac crest: Changes with age. Calcif Tissue Res **26**:13–17.
14. Croucher PI, Garrahan NJ, Mellish RW, Compston JE 1991 Age-related changes in resorption cavity characteristics in human trabecular bone. Osteoporos Int **1**:257–261.
15. Parfitt AM 1984 Age-related structural changes in trabecular and cortical bone: Cellular mechanisms and biomechanical consequences. Calcif Tissue Int **36**(Suppl 1):S123–S128.
16. Ensrud KE, Palermo L, Black DM, Cauley J, Jergas M, Orwoll E, Nevitt MC, Fox KM, Cummings SR 1995 Hip and calcaneal bone loss increase with advancing age: Longitudinal results from the study of osteoporotic fractures. J Bone Miner Res **10**:1778–1787.
17. Hannan MT, Felson DT, Dawson-Hughes B, Tucker KL, Cupples LA, Wilson PW, Kiel DP 2000 Risk factors for longitudinal bone loss in elderly men and women: The Framingham Osteoporosis Study. J Bone Miner Res **15**:710–720.
18. Ross PD, Knowlton W 1998 Rapid bone loss is associated with increased levels of biochemical markers. J Bone Miner Res **13**:297–302.
19. Dresner-Pollak R, Parker RA, Poku M, Thompson J, Seibel MJ, Greenspan SL 1996 Biochemical markers of bone turnover reflect femoral bone loss in elderly women. Calcif Tissue Int **59**:328–333.
20. Garnero P, Hausherr E, Chapuy MC, Marcelli C, Grandjean H, Muller C, Cormier C, Breart G, Meunier PJ, Delmas PD 1996 Markers of bone resorption predict hip fracture in elderly women: The EPIDOS Prospective Study. J Bone Miner Res **11**:1531–1538.
21. Garnero P, Sornay-Rendu E, Claustrat B, Delmas PD 2000 Biochemical markers of bone turnover, endogenous hormones and the risk of fractures in postmenopausal women: The OFELY study. J Bone Miner Res **15**:1526–1536.
22. Bollen AM, Kiyak HA, Eyre DR 1997 Longitudinal evaluation of a bone resorption marker in elderly subjects. Osteoporos Int **7**:544–549.
23. Recker RR, Kimmel DB, Parfitt AM, Davies KM, Keshawarz N, Hinders S 1988 Static and tetracycline-based bone histomorphometric data from 34 normal postmenopausal females. J Bone Miner Res **3**:133–144.
24. Verma S, Rajaratnam JH, Denton J, Hoyland JA, Byers RJ 2002 Adipocytic proportion of bone marrow is inversely related to bone formation in osteoporosis. J Clin Pathol **55**:693–698.
25. Heaney RP 2000 Vitamin D: How much do we need, and how much is too much? Osteoporos Int **11**:553–555.
26. Rosen CJ 2005 Clinical practice. Postmenopausal osteoporosis. N Engl J Med **353**:595–603.
27. Bischoff-Ferrari HA, Dietrich T, Orav EJ, Hu FB, Zhang Y, Karlson EW, Dawson-Hughes B 2004 Higher 25-hydroxyvitamin D concentrations are associated with better lower-extremity function in both active and inactive persons aged > or =60 y. Am J Clin Nutr **80**:752–758.
28. Holick MF 2007 Vitamin D deficiency. N Engl J Med **357**:266–281.
29. Bischoff-Ferrari HA, Dawson-Hughes B, Willett WC, Staehelin HB, Bazemore MG, Zee RY, Wong JB 2004 Effect of vitamin D on falls: A meta-analysis. JAMA **291**:1999–2006.
30. Bischoff-Ferrari HA, Willett WC, Wong JB, Giovannucci E, Dietrich T, Dawson-Hughes B 2005 Fracture prevention with vitamin D supplementation: A meta-analysis of randomized controlled trials. JAMA **293**:2257–2264.
31. Dawson-Hughes B, Harris SS, Krall EA, Dallal GE 1997 Effect of calcium and vitamin D supplementation on bone density in men and women 65 years of age or older. N Engl J Med **337**:670–676.
32. Recker RR, Hinders S, Davies KM, Heaney RP, Stegman MR, Lappe JM, Kimmel DB 1996 Correcting calcium nutritional deficiency prevents spine fractures in elderly women. J Bone Miner Res **11**:1961–1966.
33. NIH 2000 Osteoporosis Prevention, Diagnosis and Therapy. NIH Consensus Development Conference, vol. 17. NIH, Washington, DC.
34. Lappe JM, Travers-Gustafson D, Davies KM, Recker RR, Heaney RP 2007 Vitamin D and calcium supplementation reduces cancer risk: Results of a randomized trial. Am J Clin Nutr **85**:1586–1591.
35. Aloia JF, Li-Ng M 2007 Correspondence. Epidemiol Infect **135**:1095–1098.
36. Dawson-Hughes B, Harris SS 2002 Calcium intake influences the association of protein intake with rates of bone loss in elderly men and women. Am J Clin Nutr **75**:773–779.
37. Hannan MT, Tucker KL, Dawson-Hughes B, Cupples LA, Felson DT, Kiel DP 2000 Effect of dietary protein on bone loss in elderly men and women: The Framingham Osteoporosis Study. J Bone Miner Res **15**:2504–2512.
38. Tucker KL, Chen H, Hannan MT, Cupples LA, Wilson PW, Felson D, Kiel DP 2002 Bone mineral density and dietary patterns in older adults: The Framingham Osteoporosis Study. Am J Clin Nutr **76**:245–252.
39. Schurch MA, Rissoli R, Slosman D, Vadas L, Vergnaud P, Bonjour JP 1998 Protein supplements increase serum insulin-like growth factor-I levels and attenuate proximal femur bone loss in patients with recent hip fracture a randomized, double-blind, placebo-controlled trial. Ann Intern Med **128**:801–809.
40. McKeown NM, Jacques PF, Gundberg CM, Peterson JW, Tucker KL, Kiel DP, Wilson PW, Booth SL 2002 Dietary and nondietary determinants of vitamin K biochemical measures in men and women. J Nutr **132**:1329–1334.
41. Promislow JH, Goodman-Gruen D, Slymen DJ, Barrett-Connor E 2002 Retinol intake and bone mineral density in the elderly: The Rancho Bernardo Study. J Bone Miner Res **17**:1349–1358.
42. Michaelsson K, Lithell H, Vessby B, Melhus H 2003 Serum retinol levels and the risk of fracture. N Engl J Med **348**:287–294.
43. Slemenda C, Hui SL, Longcope C, Johnston CC 1987 Sex steroids and bone mass: A study of changes about the time of menopause. J Clin Invest **80**:1261–1269.
44. Amin S, Zhang Y, Sawin CT, Evans SR, Hannan MT, Kiel DP, Wilson PW, Felson DT 2000 Association of hypogonadism and estradiol levels with bone mineral density in elderly men from the Framingham study. Ann Intern Med **133**:951–963.
45. Greendale GA, Edelstein S, Barrett-Connor E 1997 Endogenous sex steroids and bone mineral density in older women and men: The Rancho Bernardo Study. J Bone Miner Res **12**:1833–1843.
46. Akhter MP, Lappe JM, Davies KM, Recker RR 2007 Transmenopausal changes in the trabecular bone structure. Bone **41**:111–116.
47. Amin S, Zhang Y, Felson DT, Sawin CT, Hannan MT, Wilson PW, Kiel DP 2006 Estradiol, testosterone, and the risk for hip fractures in elderly men from the Framingham Study. Am J Med **119**:426–433.
48. Falahati-Nini A, Riggs BL, Atkinson EJ, O'Fallon WM, Eastell R, Khosla S 2000 Relative contributions of testosterone and estrogen in regulating bone resorption and formation in normal elderly men. J Clin Invest **106**:1553–1560.
49. Khosla S, Atkinson EJ, Dunstan CR, O'Fallon WM 2002 Effect of estrogen versus testosterone on circulating osteoprotegerin and other cytokine levels in normal elderly men. J Clin Endocrinol Metab **87**:1550–1554.
50. Smith MR, Finkelstein JS, McGovern FJ, Zietman AL, Fallon MA, Schoenfeld DA, Kantoff PW 2002 Changes in body composition during androgen deprivation therapy for prostate cancer. J Clin Endocrinol Metab **87**:599–603.
51. Rosen CJ, Donahue LR, Hunter SJ 1994 Insulin-like growth factors and bone: The osteoporosis connection. Proc Soc Exp Biol Med **206**:83–102.
52. Bauer DC, Rosen CJ, Cauley J, Cummings SR 1997 Low serum IGF-I but not IGFBP-3 predicts hip and spine fractures. Bone **20**(Suppl):561.
53. Barrett-Connor E, Kritz-Silverstein D, Edelstein SL 1993 A prospective study of dehydroepiandrosterone sulfate (DHEAS) and bone mineral density in older men and women. Am J Epidemiol **137**:201–206.
54. Stern A, Laughlin GA, Bergstrom J, Barrett-Connor E 2007 The sex-specific association of serum osteoprotegerin and receptor activator of nuclear factor kappaB legend with bone mineral density in older adults: The Rancho Bernardo study. Eur J Endocrinol **156**:555–562.
55. Halloran BP, Ferguson VL, Simske SJ, Burghardt A, Venton LL, Majumdar S 2002 Changes in bone structure and mass with advancing age in the male C57BL/6J mouse. J Bone Miner Res **17**:1044–1050.
56. Bouxsein ML, Myers KS, Shultz KL, Donahue LR, Rosen CJ, Beamer WG 2005 Ovariectomy-induced bone loss varies among inbred strains of mice. J Bone Miner Res **20**:1085–1092.
57. Deng HW, Mahaney MC, Williams JT, Li J, Conway T, Davies

© 2008 American Society for Bone and Mineral Research

KM, Li JL, Deng H, Recker RR 2002 Relevance of the genes for bone mass variation to susceptibility to osteoporotic fractures and its implications to gene search for complex human diseases. Genet Epidemiol **22:**12–25.
58. Nevitt MC, Johnell O, Black DM, Ensrud K, Genant HK, Cummings SR 1994 Bone mineral density predicts non-spine fractures in very elderly women. Study of Osteoporotic Fractures Research Group. Osteoporos Int **4:**325–331.
59. Mathias A, Nayak USL, Isaacs B 1986 Balance in elderly patients: The "get-up and go" test. Arch Phys Med Rehabil **67:**387–389.
60. Tinetti ME 1986 Performance-oriented assessment of mobility problems in elderly patients. J Am Geriatr Soc **34:**119–126.
61. Guralnik JM, Ferrucci L, Pieper CF, Leveille SG, Markides KS, Ostir GV, Studenski S, Berkman LF, Wallace RB 2000 Lower extremity function and subsequent disability: Consistency across studies, predictive models, and value of gait speed alone compared with the short physical performance battery. J Gerontol A Biol Sci Med Sci **55:**M221–M231.
62. Leibson CL, Tosteson AN, Gabriel SE, Ransom JE, Melton LJ 2002 Mortality, disability, and nursing home use for persons with and without hip fracture: A population-based study. J Am Geriatr Soc **50:**1644–1650.
63. Gillespie LD, Gillespie WJ, Robertson MC, Lamb SE, Cumming RG, Rowe BH 2003 Interventions for preventing falls in elderly people. Cochrane Database Syst Rev **4:**CD000340.
64. Greenspan SL, Myers E, Kiel DP, Parker RA, Hayes WC, Resnick NM 1998 Fall direction, bone mineral density, and function: Risk factors for hip fracture in frail nursing home elderly. Am J Med **104:**539–545.
65. Parker MJ, Gillespie WJ, Gillespie LD 2006 Effectiveness of hip protectors for preventing hip fractures in elderly people: Systematic review. BMJ **332:**571–574.
66. Black DM, Cummings SR, Karpf DB, Cauley JA, Thompson DE, Nevitt MC, Bauer DC, Genant HK, Haskell WL, Marcus R, Ott SM, Torner JC, Quandt SA, Reiss TF, Ensrud KE 1996 Randomised trial of effect of alendronate on risk of fracture in women with existing vertebral fractures. Lancet **348:**1535–1541.
67. McClung MR, Geusens P, Miller PD, Zippel H, Bensen WG, Roux C, Adami S, Fogelman I, Diamond T, Eastell R, Meunier PJ, Reginster JY 2001 Effect of risedronate on the risk of hip fracture in elderly women. Hip Intervention Program Study Group. N Engl J Med **344:**333–340.
68. Chesnut IC, Skag A, Christiansen C, Recker R, Stakkestad JA, Hoiseth A, Felsenberg D, Huss H, Gilbride J, Schimmer RC, Delmas PD 2004 Effects of oral ibandronate administered daily or intermittently on fracture risk in postmenopausal osteoporosis. J Bone Miner Res **19:**1241–1249.
69. Black DM, Delmas PD, Eastell R, Reid IR, Boonen S, Cauley JA, Cosman F, Lakatos P, Leung PC, Man Z, Mautalen C, Mesenbrink P, Hu H, Caminis J, Tong K, Rosario-Jansen T, Krasnow J, Hue TF, Sellmeyer D, Eriksen EF, Cummings SR 2007 Once-yearly zoledronic acid for treatment of postmenopausal osteoporosis. N Engl J Med **356:**1809–1822.
70. Lyles KW, Colon-Emeric CS, Magaziner JS, Adachi JD, Pieper CF, Mautalen C, Hyldstrup L, Recknor C, Nordsletten L, Moore KA, Lavecchia C, Zhang J, Mesenbrink P, Hodgson PK, Abrams K, Orloff JJ, Horowitz Z, Eriksen EF, Boonen S 2007 Zoledronic Acid and Clinical Fractures and Mortality after Hip Fracture. N Engl J Med **357:**1799–1809.
71. Cauley JA, Robbins J, Chen Z, Cummings SR, Jackson RD, LaCroix AZ, LeBoff M, Lewis CE, McGowan J, Neuner J, Pettinger M, Stefanick ML, Wactawski-Wende J, Watts NB 2003 Effects of estrogen plus progestin on risk of fracture and bone mineral density: The Women's Health Initiative randomized trial. JAMA **290:**1729–1738.
72. Rossouw JE, Anderson GL, Prentice RL, LaCroix AZ, Kooperberg C, Stefanick ML, Jackson RD, Beresford SA, Howard BV, Johnson KC, Kotchen JM, Ockene J 2002 Risks and benefits of estrogen plus progestin in healthy postmenopausal women: Principal results From the Women's Health Initiative randomized controlled trial. JAMA **288:**321–333.
73. Chesnut CH III, Silverman S, Andriano K, Genant H, Gimona A, Harris S, Kiel D, LeBoff M, Maricic M, Miller P, Moniz C, Peacock M, Richardson P, Watts N, Baylink D 2000 A randomized trial of nasal spray salmon calcitonin in postmenopausal women with established osteoporosis: The prevent recurrence of osteoporotic fractures study. PROOF Study Group. Am J Med **109:**267–276.
74. Seeman E, Vellas B, Benhamou C, Aquino JP, Semler J, Kaufman JM, Hoszowski K, Varela AR, Fiore C, Brixen K, Reginster JY, Boonen S 2006 Strontium ranelate reduces the risk of vertebral and nonvertebral fractures in women eighty years of age and older. J Bone Miner Res **21:**1113–1120.
75. Neer RM, Arnaud CD, Zanchetta JR, Prince R, Gaich GA, Reginster JY, Hodsman AB, Eriksen EF, Ish-Shalom S, Genant HK, Wang O, Mitlak BH 2001 Effect of parathyroid hormone (1-34) on fractures and bone mineral density in postmenopausal women with osteoporosis. N Engl J Med **344:**1434–1441.

© 2008 American Society for Bone and Mineral Research

SECTION III

Mineral Homeostasis
(Section Editor: Ego Seeman)

© 2008 American Society for Bone and Mineral Research

Chapter 21. Regulation of Calcium and Magnesium

Murray J. Favus[1] and David Goltzman[2]

[1]Department of Medicine, University of Chicago, Chicago, Illinois; [2]Center for Advanced Bone and Periodontal Research, McGill University, Montreal, Quebec, Canada

CALCIUM

Distribution

Total Body Distribution. In adults, the body contains ~1000 g of Ca, of which 99% is located in the mineral phase of bone as the hydroxyapatite crystal $[Ca_{10}(PO_4)_6(OH)_2]$. The crystal plays a key role in the mechanical weight-bearing properties of bone and serves as a ready source of Ca to support a number of Ca-dependent biological systems and to maintain blood ionized Ca within the normal range. The remaining 1% of total body Ca is located in the blood, extracellular fluid, and soft tissues. In serum, total Ca is 10^{-3} M and is the most frequent measurement of serum Ca levels. Of the total Ca, the ionized fraction (50%) is the biologically functional portion of total Ca and can be measured clinically; 40% of the total is bound to albumin in a pH-dependent manner; and the remaining 10% exists as a complex of either citrate or PO_4 ions.

Cell Levels. Cytosol Ca is ~10^{-6} M, which creates a 1000-fold gradient across the plasma membrane (extracellular fluid [ECF] Ca is 10^{-3} M) that favors Ca entry into the cell. There is also an electrical charge across the plasma membrane of ~50 mV with the cell interior negative. Thus, the chemical and electrical gradients across the plasma membrane favor Ca entry, which the cell must defend against to preserve cell viability. Ca-induced cell death is largely prevented by several mechanisms including extrusion of Ca from the cell by ATP-dependent energy driven Ca pumps and Ca channels; Na-Ca exchangers; and binding of intracellular Ca by proteins located in the cytosol, endoplasmic reticulum (ER), and mitochondria. Ca binding to ER and mitochondrial sites buffer intracellular Ca and can be mobilized to maintain cytosol Ca levels and to create pulsatile peaks of Ca to mediate membrane receptor signaling that regulate a variety of biological systems.

Blood Levels. Ca in the blood is normally transported partly bound to plasma proteins (~45%), notably albumin, partly bound to small anions such as phosphate and citrate (~10%), and partly in the free or ionized state (~45%).[1] Although only the ionized Ca is available to move into cells and activate cellular processes, most clinical laboratories report total serum Ca concentrations. Concentrations of total Ca in normal serum generally range between 8.5 and 10.5 mg/dl (2.12–2.62 mM), and levels above this are considered to be hypercalcemic. The normal range of ionized Ca is 4.65–5.25 mg/dl (1.16–1.31 mM). When protein concentrations, and especially albumin concentrations, fluctuate, total Ca levels may vary, whereas the ionized Ca may remain relatively stable. Dehydration or hemoconcentration during venipuncture may elevate serum albumin and falsely elevate total serum Ca. Such elevations in total Ca, when albumin levels are increased, can be "corrected" by subtracting 0.8 mg/dl from the total Ca for every 1.0 g/dl by which the serum albumin concentration is >4 g/dl. Conversely, when albumin levels are low, total Ca can be corrected by adding 0.8 mg/dl for every 1.0 g/dl by which the albumin is <4 g/dl. Even in the presence of a normal serum albumin, changes in blood pH can alter the equilibrium constant of the albumin–Ca^{2+} complex, with acidosis reducing the binding and alkalosis enhancing it. Consequently major shifts in serum protein or pH requires direct measurement of the ionized Ca level to determine the physiologic serum calcium level.

Mineral Homeostasis

The ECF concentration of calcium is tightly maintained within a rather narrow range because of the importance of the Ca ion to numerous cellular functions including cell division, cell adhesion and plasma membrane integrity, protein secretion, muscle contraction, neuronal excitability, glycogen metabolism, and coagulation.

The skeleton, the gut, and the kidney each plays a major role in assuring Ca homeostasis. Overall, in a typical individual, if 1000 mg of Ca is ingested in the diet per day, ~200 mg will be absorbed. Approximately 10 g of Ca will be filtered daily through the kidney, and most will be reabsorbed, with ~200 mg being excreted in the urine. The normal 24-h excretion of Ca may, however, vary between 100 and 300 mg/d (2.5–7.5 mmol/d). The skeleton, a storage site of ~1 kg of Ca, is the major Ca reservoir in the body. Ordinarily, as a result of normal bone turnover, ~500 mg of Ca is released from bone per day, and the equivalent amount is accreted per day (Fig. 1).

Tight regulation of the ECF calcium concentration is maintained through the action of Ca-sensitive cells that modulate the production of hormones.[2–5] These hormones act on specific cells in bone, gut, and kidney, which can respond by altering fluxes of Ca to maintain ECF Ca. Thus, a reduction in ECF Ca stimulates release of PTH from the parathyroid glands in the neck. This hormone can act to enhance bone resorption and liberate both Ca and phosphate from the skeleton. PTH can also enhance Ca reabsorption in the kidney while at the same time inhibit phosphate reabsorption producing phosphaturia. Hypocalcemia and PTH itself can both stimulate the conversion of the inert metabolite of vitamin D, 25-hydroxyvitamin D_3 [$25(OH)D_3$] to the active moiety 1,25-dihydroxyvitamin D_3 [$1,25(OH)_2D_3$],[6] which in turn will enhance intestinal Ca absorption, and to a lesser extent, renal phosphate reabsorption. The net effect of the mobilization of Ca from bone, the increased absorption of Ca from the gut, and the increased reabsorption of filtered Ca along the nephron is to restore the ECF Ca to normal and to inhibit further production of PTH and $1,25(OH)_2D_3$. The opposite sequence of events [i.e., diminished PTH and $1,25(OH)_2D_3$ secretion], along with stimulation of renal Ca sensing receptor (CaSR), occurs when the ECF Ca is raised above the normal range. The effect of suppressing the release of PTH and $1,25(OH)_2D_3$ and stimulating CaSR diminishes skeletal Ca release, decreases intestinal Ca absorption and renal Ca reabsorption, and restores the elevated ECF Ca to normal.

PTH and $1,25(OH)_2D_3$ Actions on Target Tissues

Intestinal Ca Transport. Net intestinal Ca absorption can be determined by the external balance technique in which a diet of known composition with a known amount of Ca is ingested, and urine Ca excretion and fecal Ca loss are measured. Negative absorption occurs when net absorption declines to ~200 mg Ca/d (5.0 mmol). The portion of dietary Ca absorbed varies with age and amount of Ca ingested and may vary from 20% to 60%. Rates of net Ca absorption are high in growing chil-

The authors state that they have no conflicts of interest.

© 2008 American Society for Bone and Mineral Research

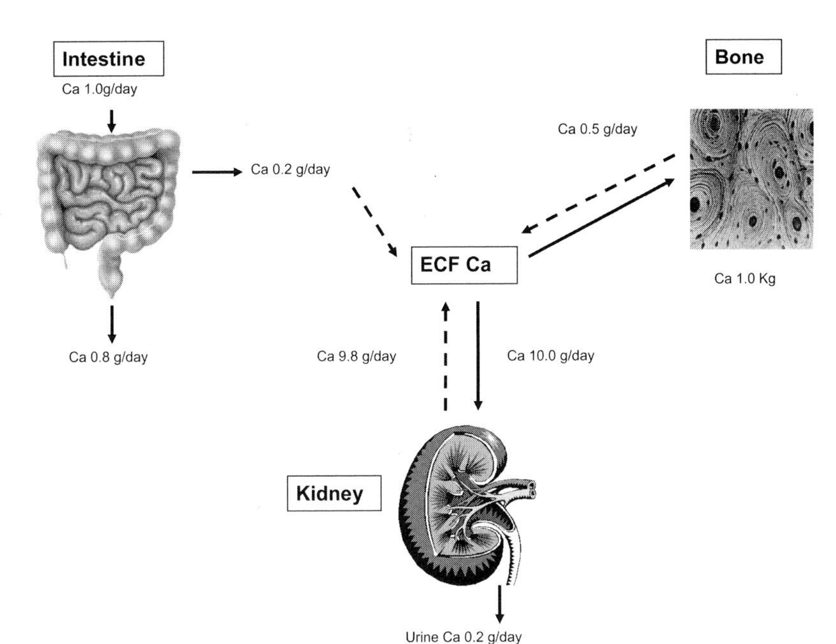

FIG. 1. Calcium balance. On average, in a typical adult, ~1 g of elemental calcium (Ca^{+2}) is ingested per day. Of this, ~200 mg/d will be absorbed and 800 mg/d excreted. Approximately 1 kg of Ca is stored in bone and ~500 mg/d is released by resorption or deposited during bone formation. Of the 10 g of Ca filtered through the kidney per day, only ~200 mg or less appears in the urine, the remainder being reabsorbed.

dren; during grow spurts in adolescence; and during pregnancy and lactation. The efficiency of Ca absorption increases during prolonged dietary Ca restriction to absorb the greatest portion of that ingested. Net absorption declines with age in men and women, and so increased Ca intake is required to compensate for the lower absorption rate. Fecal Ca losses vary between 100 and 200 mg/d (2.5–5.0 mmol). Fecal Ca is composed of unabsorbed dietary Ca and Ca contained in intestinal, pancreatic, and biliary secretions. Secreted Ca is not regulated by hormones or serum Ca.

Because of the large surface area of the duodenum and jejunum, 90% of absorbed Ca occurs in these regions. Increased Ca requirements stimulate expression of the epithelial cellular Ca active transport system in duodenum, ileum, and throughout the colon sufficient to increase fractional Ca absorption from 20–45% in older men and women to 55–70% in children and young adults.$1{,}25(OH)_2D_3$ increases the efficiency of the small intestine and colon to absorb dietary Ca. Active Ca absorption accounts for absorption of 10–15% of a dietary load.[7] Active transcellular intestinal absorption involves three sequential cellular steps: a rate-limiting step involving transfer of luminal Ca into the intestinal cell through the epithelial Ca channel, expression of TRPV6, a channel-associated protein, annexin2, and calbindin-D9K, and to a lesser extent, the basolateral extrusion system PMCA1b.[8,9] Reductions in dietary Ca intake can increase PTH secretion and $1{,}25(OH)_2D_3$ production, which can enhance fractional Ca absorption and compensates for the dietary reduction.

Intestinal epithelial Ca transport includes both an energy-dependent, cell-mediated saturable active process that is largely regulated by $1{,}25(OH)_2D_3$, and a passive, diffusional paracellular path of absorption that is driven by transepithelial electrochemical gradients. The cell-mediated pathway involving the TRPV6 Ca channel is saturable with a Kt (1/2 maximal transport) of 1.0 mM. Passive diffusion increases linearly with luminal Ca concentration and is not regulated by $1{,}25(OH)_2D_3$. In adults fed a diet low in Ca, enhanced $1{,}25(OH)_2D_3$ production increases the efficiency of absorption through an increase in saturable Ca transport. During high dietary Ca intake absorption, $1{,}25(OH)_2D_3$ is suppressed and passive paracellular transport accounts for most all absorption. Causes of increased and decreased intestinal Ca absorption are listed in Table 1.

Renal Ca Handling. The kidney plays a central role in ensuring Ca balance, and PTH has a major role in fine-tuning this renal function[10–12] by stimulating both renal Ca reabsorption (proximal tubule) and excretion (distal nephron). Multiple influences of Ca handling are listed in Table 2. Descriptions of the molecular actions of PTH on the kidney are found elsewhere in the *Primer*. PTH has little effect on modulating Ca fluxes in the proximal tubule where 65% of the filtered Ca is reabsorbed, coupled to the bulk transport of solutes such as sodium and water.[11] In this nephron region, PTH can also stimulate the $25(OH)D_3$-1α hydroxylase [1α(OH)ase], leading to increased synthesis of $1{,}25(OH)_2D_3$.[13] A reduction in ECF Ca can itself stimulate $1{,}25(OH)_2D_3$ production but whether this occurs through the CaSR is presently unknown. Finally PTH can also inhibit Na and $HC0_3^-$ reabsorption in the proximal tubule by inhibiting the apical type 3 Na^+/H^+ exchanger[14] and the basolateral Na^+/K^+-ATPase[15] by inhibiting apical Na^+/Pi^- cotransport.

About 20% of filtered Ca is reabsorbed in the cortical thick ascending limb of the loop of Henle (CTAL) and 15% is reabsorbed in the distal convoluted tubule (DCT). At both sites, PTH binds to the PTH receptor (PTHR)[16,17] and enhances Ca reabsorption. In the CTAL, at least, this seems to occur by

TABLE 1. CONDITIONS THAT INCREASE OR DECREASE INTESTINAL CA ABSORPTION

Increased Ca absorption	*Decreased Ca absorption*
Increased renal $1{,}25(OH)_2D_3$ production	Decreased renal $1{,}25(OH)_2D_3$ production
Growth	Vitamin D deficiency
Pregnancy	Chronic renal insufficiency
Lactation	Hypoparathyroidism
Primary hyperparathyroidism	Vitamin D–dependent rickets type 1
Idiopathic hypercalciuria	Aging
Increased extrarenal $1{,}25(OH)_2D_3$ production	Normal $1{,}25(OH)_2D_3$ production
Sarcoid and other granulomatous diseases	Glucocorticoid excess
B-cell lymphoma	Hyperthyroidism

© 2008 American Society for Bone and Mineral Research

TABLE 2. HORMONES AND CONDITIONS THAT REGULATE URINE CA AND MG EXCRETION

Hormones/conditions	*Calcium*	*Magnesium*
Hypercalcemia	I	D
Hypocalcemia	D	I
Hypermagnesemia	—	I
Hypomagnesemia	D	D
Renal insufficiency	D	D
Tubular reabsorption		
Increased		
ECF volume contraction	I	I
Hypocalcemia	I	I
Thiazide diuretics	I	—
Phosphate administration	I	I
Metabolic alkalosis	I	I
Parathyroid hormone	I	I
Parathyroid hormone related peptide	I	I
Familial hypocalciuric hypercalcemia	I	—
Decreased		
ECF volume expansion	D	D
Hypercalcemia	D	D
Phosphate deprivation	D	D
Metabolic acidosis	D	—
Loop diuretics	D	D
Cyclosporin A	D	D
Autosomal dominant hypocalcemia	D	—
Dent's disease	D	—
Bartter's syndrome	D	—
Gittelman's syndrome	—	D

D, decreased GFR or tubule reabsorption; I, increased GFR or tubule reabsorption; —, either modest effects are present or that no specific information is available.

increasing the activity of the Na/K/2 Cl co-transporter that drives NaCl reabsorption and stimulates paracellular Ca and Mg reabsorption.[18] The CaSR is also resident in the CTAL,[19] where increased ECF Ca activates phospholipase A2, thereby reducing the activity of the Na/K/2Cl co-transporter and of an apical K channel and diminishing paracellular Ca reabsorption. Consequently, a raised ECF Ca antagonizes the effect of PTH in this nephron segment and ECF Ca can in fact participate in this way in the regulation of its own homeostasis. Inhibition of NaCl reabsorption and loss of NaCl in the urine may contribute to the volume depletion observed in severe hypercalcemia. ECF Ca may therefore act in a manner analogous to "loop" diuretics such as furosemide.

In the DCT, PTH can also influence[8] luminal Ca transfer into the renal tubule cell through the transient receptor potential channel (TRPV5), translocation of Ca across the cell from apical to basolateral surface involving proteins such as calbindin-D28K, and finally active extrusion of Ca from the cell into the blood through an Na^+/Ca exchanger, designated NCX1. PTH markedly stimulates Ca reabsorption in the DCT primarily by augmenting NCX1 activity through a cyclic AMP-mediated mechanism.

Bone Resorption and Ca Release. In bone, the PTHR is localized on cells of the osteoblast phenotype that are of mesenchymal origin[20] but not on osteoclasts that are of hematogenous origin. The major physiologic role of PTH seems to be to maintain normal Ca homeostasis by enhancing osteoclastic bone resorption and liberating Ca into the ECF. Bone formation and resorption are discussed in detail elsewhere in the *Primer*.

Mediators of Bone Remodeling. Calcitropic hormones PTH, PTH-related peptide (PTHrP), and $1,25(OH)_2D_3$ initiate osteoclastic bone resorption and increase the activation frequency of bone remodeling. Physiologic control of bone turnover can be disrupted by an excess of each of these calcitropic hormones, resulting in altered ECF Ca homeostasis and hypercalcemia. The molecular basis for physiologic and pathologic states of bone turnover is detailed elsewhere in the *Primer*.

Regulation of Hormone Production and Actions on Ca Homeostasis

PTH Production. A major regulator of parathyroid gland secretion of PTH is ECF Ca. The relationship between ECF Ca and PTH secretion is governed by a steep inverse sigmoidal curve which is characterized by a maximal secretory rate at low ECF Ca, a midpoint or "set point," which is the level of ECF Ca, which half-maximally suppresses PTH, and a minimal secretory rate at high ECF Ca.[21,22] The parathyroid glands detect ECF Ca through a CaSR.[23] Sustained hypocalcemia can eventually lead to parathyroid cell proliferation[24] and an increased total secretory capacity of the parathyroid gland. $1,25(OH)_2D_3$ reduces PTH synthesis and parathyroid cell proliferation.[25] Molecular events in PTH secretion and CaSR function are found elsewhere in the *Primer*.

Vitamin D Production and Actions. The renal production of $1,25(OH)_2D_3$ is stimulated by hypocalcemia, hypophosphatemia and elevated PTH levels. The renal $1\alpha(OH)$ase is also potently inhibited by $1,25(OH)_2D_3$ as part of a negative feedback loop. The molecular details of the vitamin D metabolic pathway are described elsewhere in the *Primer*.

Vitamin D is essential for normal mineralization of bone that may be caused by an indirect effect by enhancing intestinal calcium and phosphate absorption and maintaining these ions within a range that facilitates hydroxyapatite deposition in bone matrix. A major indirect function of $1,25(OH)_2D_3$ on bone seems to be to enhance mobilization of Ca stores when dietary Ca is insufficient to maintain a normal ECF Ca.[26] As with PTH,[27] $1,25(OH)_2D_3$ enhances osteoclastic bone resorption by binding to receptors in the pre-osteoblastic stromal cell and stimulating the RANK/RANK system to enhance the proliferation, differentiation, and activation of the osteoclastic system from its monocytic precursors.[28] Endogenous and exogenous $1,25(OH)_2D_3$ have also been reported to have an anabolic role in vivo.[29,30] $1,25(OH)_2D_3$ has a direct effect on renal Ca handling through stimulation of CaSR. It remains controversial whether $1,25(OH)_2D_3$ plays a direct role in enhancing tubular Ca reabsorption.

MAGNESIUM

Total Body Distribution

There is ~1.04 mol (25 g) of Mg in the adult, of which ~66% is within the skeleton, 33% is intracellular, and 1% is in the ECF including blood.[1,2] Mg content of the hydroxyapatite crystal in bone varies widely and is mainly on the surface of bone where a portion is in equilibrium with ECF Mg. Mg is the most abundant divalent cation within cells, where it is found at a concentration of ~5×10^{-4} M in the cytosol. In the cells, it serves as a co-factor and regulates a number of essential biological systems.[1] The concentration of Mg in the ECF approaches that of the intracellular environment. Both intracellular and ECF are tightly regulated by factors that are poorly understood.

© 2008 American Society for Bone and Mineral Research

Cellular Content

Ionic cytosolic Mg accounts for 5–10% of total cellular Mg. Cytosol Mg is regulated by binding to intracellular organelles, of which 60% is within mitochondria where it participates in phosphate transport and ATP utilization. Control of intracellular Mg is poorly understood.

Homeostasis

Of the total serum Mg, 70% is either ionic or complexed, and the remaining 30% is protein bound.[1,2] Blood levels are not as tightly regulated as Ca but fluctuate with influx and efflux across the ECF with changes in intestinal Mg absorption, net renal Mg reabsorption, and influx and efflux across bone. Blood ionic Mg regulates PTH secretion, but the potency is less than that of Ca.

Intestinal Absorption

Mg is a requirement for bone health; however, unlike Ca, Mg is found in all food groups and is especially rich in foods of cellular origin. Therefore, Mg deficiency caused by inadequate intake does not occur in the absence of severe defects in intestinal or renal function. Net intestinal Mg absorption is in direct proportion to dietary Mg ingestion. Under conditions of stable Mg intake, external Mg balance studies show that when Mg intake is >28 mg (2 mmol), Mg absorption exceeds Mg secretion, and Mg balance becomes positive. The efficiency of Mg absorption is 35–40% over the range of usual intakes (168–720 mg/d or 7–30 mmol). Net Mg absorption also varies with dietary constituents such as phosphate, which forms insoluble complexes with Mg and thereby reduces Mg absorption. In contrast to its actions on Ca and P absorption, $1,25(OH)_2D_3$ does not stimulate Mg absorption. There is no correlation between serum $1,25(OH)_2D_3$ levels and net Mg absorption.[31]

Absorptive and secretory Mg fluxes across both small intestine and colon are largely voltage dependent, indicating the presence of a large paracellular pathway of Mg transport that is driven by luminal Mg concentrations. The Mg ion channel TRPM6 has been identified in the apical membrane of intestinal brush border epithelial cells that seems to play an important role in Mg homeostasis.[31] Whether TRPM6 is regulated by PTH or $1,25(OH)_2D_3$ has yet to be determined.

Renal Handling

Ultrafiltrable Mg is 70% of the total serum Mg (ionized plus complexed). Based on the urine Mg excretion (~24 mmol/24 h), ~95% of the filtered load of Mg undergoes tubular reabsorption before the final urine is formed. A small fraction of reabsorbed Mg (15%) occurs along the proximal tubule, whereas ~70% of filtered Mg is reabsorbed along the cortical TALH.[18,32] Mg ion may also stimulate basolateral membrane CaR, which decreases renal Mg reabsorption. DCT Mg reabsorption is through a transcellular transport process and accounts for ~10% of Mg reabsorption.

Mg reabsorption is highly regulated, with a number of factors that may increase or decrease renal tubule Mg reabsorption (Table 2). Because there is little distal tubule Mg reabsorption, ECF volume expansion decreases Mg reabsorption and increases urine Mg excretion. Hypermagnesemia increases urine Mg excretion at least in part through an activation of CaR.[19] In contrast, hypomagnesemia increases TALH Mg reabsorption and decreases urine Mg excretion. Loop diuretics increase urine Mg excretion, and thiazide diuretic agents have a minimal effect of Mg transport (Table 2). The Mg ion channel TRPM6 is found in the apical membrane of the renal distal convoluted tubule and may be involved in Mg homeostasis in both the kidney and intestine.

REFERENCES

1. Walser M 1961 Ion association: VI. Interactions between calcium, magnesium, inorganic phosphate, citrate, and protein in normal human plasma. J Clin Invest **40:**723–735.
2. Parfitt AM, Kleerekoper M 1980 Clinical disorders of calcium, phosphorus and magnesium metabolism. In: Maxwell MH, Kleeman CR (eds.) Clinical Disorders of Fluid and Electrolyte Metabolism, 3rd ed. McGraw-Hill, New York, NY, USA, pp. 947–1151.
3. Stewart AF, Broadus AE 1987 Mineral metabolism. In: Felig P, Baxter ID, Broadus AE, Frohman LA (eds.) Endocrinology and Metabolism, 2nd ed. McGraw-Hill, New York, NY, USA, pp. 1317–1453.
4. Bringhurst FR, Demay MB, Kronenberg HM 1988 Hormones and disorders of mineral metabolism. In: Wilson JD, Foster DW, Kronenberg HM, Larsen PR (eds.) Williams Textbook of Endocrinology, 9th ed. Saunders, Philadelphia, PA, USA, pp. 1155–1209.
5. Brown EM 2001 Physiology of calcium homeostasis. In: Bilezikian JP, Marcus R, Levine MA (eds.) The Parathyroids: Basic and Clinical Concepts, 2nd ed. Academic Press, San Diego, CA, USA, pp. 167–181.
6. Fraser DR, Kodicek E 1973 Regulation of 25-hydroxycholecalciferol-1-hydroxylase activity in kidney by parathyroid hormone. Nat New Biol **241:**163–166.
7. Favus MF 1992 Intestinal absorption of calcium, magnesium and phosphorus. In: Coe FL, Favus MJ (eds.) Disorders of Bone and Mineral Metabolism. Raven, New York, NY, USA, pp. 57–81.
8. Hoenderop JGJ, Nilius B, Bindels RJM 2005 Calcium absorption across epithelia. Physiol Rev **85:**373–422.
9. Van de Graaf SFJ, Boullart I, Hoenderop JGJ, Bindels RJM 2004 Regulation of the epithelial Ca^{2+} channels TRPV5 and TRPV6 by 1α,25-dihydroxy Vitamin D3 and dietary Ca^{2+}. J Steroid Biochem Mol Biol **89–90:**303–308.
10. Friedman PA, Gesek FA 1995 Cellular calcium transport in renal epithelia: Measurement, mechanisms, and regulation. Physiol Rev **75:**429–471.
11. Nordin BE, Peacock M 1969 Role of kidney in regulation of plasma-calcium. Lancet **2:**1280–1283.
12. Rouse D, Suki WN 1990 Renal control of extracellular calcium. Kidney Int **38:**700–708.
13. Brenza HL, Kimmel-Jehan C, Jehan F, Shinki T, Wakino S, Anazawa H, Suda T, DeLuca HF 1998 Parathyroid hormone activation of the 25-hydroxyvitamin D3-1a-hydroxylase gene promoter. Proc Natl Acad Sci USA **95:**1387–1391.
14. Azarani A, Goltzman D, Orlowski J 1995 Parathyroid hormone and parathyroid hormone-related peptide inhibit the apical Na+/H+ exchanger NHE-3 isoform in renal cells (OK) via a dual signaling cascade involving protein kinase A and C. J Biol Chem **270:**20004–20010.
15. Derrickson BH, Mandel LJ 1997 Parathyroid hormone inhibits Na(+)-K(+)-ATPase through Gq/G11 and the calcium-independent phospholipase A2. Am J Physiol **272:**F781–F788.
16. Juppner H, Abou-Samra AB, Freeman MW, Kong XF, Schipani E, Richards J, Kolakowski LF Jr, Hock J, Potts JT Jr, Kronenberg HM, Segre GVA 1991 G protein-linked receptor for parathyroid hormone and parathyroid hormone-related peptide. Science **254:**1024–1026.
17. Abou-Samra AB, Juppner H, Force T, Freeman MW, Kong XF, Schipani E, Urena P, Richards J, Bonventre JV, Potts JT Jr, Kronenberg HM, Segre GV 1992 Expression cloning of a common receptor for parathyroid hormone and parathyroid hormone-related peptide from rat osteoblast-like cells: A single receptor stimulates intracellular accumulation of both cAMP and inositol triphosphates and increases intracellular free calcium. Proc Natl Acad Sci USA **89:**2732–2736.
18. De Rouffignac C, Quamme GA 1994 Renal magnesium handling and its hormonal control. Physiol Rev **74:**305–322.
19. Hebert SC 1996 Extracellular calcium-sensing receptor: Implications for calcium and magnesium handling in the kidney. Kidney Int **50:**2129–2139.
20. Rouleau MF, Mitchell J, Goltzman D 1988 In vivo distribution of parathyroid hormone receptors in bone: Evidence that a predominant osseous target cell is not the mature osteoblast. Endocrinology **123:**187–191.

© 2008 American Society for Bone and Mineral Research

21. Potts JT Jr, Juppner H 1997 Parathyroid hormone and parathyroid hormone-related peptide in calcium homeostasis, bone metabolism, and bone development: The proteins, their genes, and receptors. In: Avioli LV, Krane SM (eds.) Metabolic Bone Disease, 3rd ed. Academic Press, New York, NY, USA, pp. 51–84.
22. Grant FD, Conlin PR, Brown EM 1990 Rate and concentration dependence of parathyroid hormone dynamics during stepwise changes in serum ionized calcium in normal humans. J Clin Endocrinol Metab **71**:370–378.
23. Brown EM, Gamba G, Riccardi D, Lombardi M, Butters R, Kifor O, Sun A, Hediger MA, Lytton J, Hebert SC 1993 Cloning and characterization of an extracellular Ca(2+)-sensing receptor from bovine parathyroid. Nature **366**:575–580.
24. Kremer R, Bolivar I, Goltzman D, Hendy GN 1989 Influence of calcium and 1,25-dihydroxycholecalciferol on proliferation and proto-oncogene expression in primary cultures of bovine parathyroid cells. Endocrinology **125**:935–941.
25. Goltzman D, Miao D, Panda DK, Hendy GN 2004 Effects of calcium and of the vitamin D system on skeletal and calcium homeostasis: Lessons from genetic models. J Steroid Biochem Mol Biol **89–90**:485–489.
26. Li YC, Pirro AE, Amling M, Delling G, Baron R, Bronson R, Demay MB 1997 Targeted ablation of the vitamin D receptor: An animal model of vitamin D-dependent rickets type II with alopecia. Proc Natl Acad Sci USA **94**:9831–9835.
27. Lee SK, Lorenzo JA 1999 Parathyroid hormone stimulates TRANCE and inhibits osteoprotegerin messenger ribonucleic acid expression in murine bone marrow cultures: Correlation with osteoclast-like cell formation. Endocrinology **140**:3552–3561.
28. Takahashi N, Udagawa N, Takami M, Suda T 2002 Cells of bone: Osteoclast generation. In: Bilezikian JP, Raisz LG, Rodan GA (eds.) Principles of bone biology, 2nd ed. Academic Press, San Diego, CA, USA, pp. 109–126.
29. Panda DK, Miao D, Bolivar I, Li J, Huo R, Hendy GN, Goltzman D 2004 Inactivation of the 25-dihydroxyvitamin D-1alpha-hydroxylase and vitamin D receptor demonstrates independent effects of calcium and vitamin D on skeletal and mineral homeostasis. J Biol Chem **279**:16754–16766.
30. Xue Y, Karaplis AC, Hendy GN, Goltzman D, Miao D 2006 Exogenous 1,25-dihydroxyvitamin D3 exerts a skeletal anabolic effect and improves mineral ion homeostasis in mice which are homozygous for both the 1{alpha} hydroxylase and parathyroid hormone null alleles. Endocrinology **147**:4801–4810.
31. Schmulen AC, Leman M, Pak CY, Zerwekh J, Morawski S, Fordtran JS, Vergne-Marini P 1980 Effect of $1,25(OH)_2D_3$ on jejunal absorption of magnesium in patients with chronic renal disease. Am J Physiol **238**:G349–G351.
32. Yu ASL 2004 Renal transport of calcium, magnesium, and phosphate. In: Brenner BM (ed.) The Kidney, 7th ed. Saunders, Philadelphia, PA, USA, pp. 535–572.

Chapter 22. Fetal Calcium Metabolism

Christopher S. Kovacs

Faculty of Medicine–Endocrinology, Health Sciences Centre, Memorial University of Newfoundland, St. John's, Newfoundland, Canada

INTRODUCTION

Much of normal mineral and bone homeostasis in the adult can be explained by the interactions of PTH, 1,25-dihydroxyvitamin D or calcitriol (1,25-D), calcitonin, and the sex steroids. In contrast, comparatively little is known about how mineral and bone homeostasis is regulated in the fetus. Because of obvious limitations in studying human fetuses, human regulation of fetal mineral homeostasis must be largely inferred from studies in animals, and some observations in animals may not apply to humans. This chapter briefly reviews existing human and animal data; for more information and references, the reader is referred to two comprehensive reviews on the subject.[1,2]

Fetal mineral metabolism has been adapted to maintain an extracellular level of calcium (and other minerals) that is physiologically appropriate for fetal tissues and to provide sufficient calcium (and other minerals) to fully mineralize the skeleton before birth. Mineralization occurs rapidly in late gestation, such that a human accretes 80% of the required 30 g of calcium in the third trimester, whereas a rat accretes 95% of the required 12.5 mg of calcium in the last 5 days of its 3-wk gestation.

MINERALS IONS AND CALCIOTROPIC HORMONES

A consistent finding among human and other mammalian fetuses is a total and ionized calcium concentration that is significantly higher than the maternal level during late gestation. Similarly, serum phosphate is significantly elevated, and serum magnesium is minimally elevated above the maternal concentration. The physiological importance of these elevated levels is not known; complete mineralization of the skeleton and survival to term have been noted in genetically manipulated mice in which the fetal blood calcium is not raised above the maternal level. The normal increase in the fetal calcium level is robustly maintained despite maternal hypocalcemia from a variety of causes. For example, adult humans and mice with nonfunctional vitamin D receptors (VDRs) are hypocalcemic, but *Vdr*-null fetuses have normal serum calcium concentrations.[3]

Calcitropic hormone levels are also maintained at levels that differ from the adult. These differences seem to reflect the relatively different roles that these hormones play in the fetus and are not an artifact of altered metabolism or clearance of these hormones. Intact PTH levels are much lower than maternal PTH levels near the end of gestation, but it is unknown whether fetal PTH levels are low throughout gestation after the formation of the parathyroids or only in late gestation. Despite its low level, PTH is important for fetal development because fetal mice lacking parathyroids or PTH are hypocalcemic and have undermineralized skeletons.[4–6] Circulating 1,25-D levels are also lower than the maternal level in late gestation and seem to be largely if not completely derived from fetal sources. The low circulating levels of 1,25-D in the fetus may be a response to high serum phosphate and low PTH. 1,25-D may also be relatively unimportant for fetal mineral

---

The author states that he has no conflicts of interest.

Key words: fetus, pregnancy, calcium, magnesium, phosphorus, PTH, PTH-related protein, calcitonin, vitamin D, calcitriol, estradiol, hyperparathyroidism, hypoparathyroidism, familial hypocalciuric hypercalcemia, vitamin D deficiency/insufficiency, placenta, placental calcium transport, calcium receptor, rickets

© 2008 American Society for Bone and Mineral Research

homeostasis, because several vitamin D deficiency models, and *Vdr*-null mice, have normal serum mineral concentrations and fully mineralized skeletons at term.[3] Fetal calcitonin levels are higher than maternal levels and are thought to reflect increased synthesis of the hormone. Apart from responding appropriately to changes in the serum calcium concentration, there is little evidence of an essential role for calcitonin in fetal mineral homeostasis.[7]

At term, cord blood levels of PTH-related protein (PTHrP) are up to 15-fold higher than simultaneous PTH levels; this hormone and paracrine/autocrine factor is produced in many tissues and plays multiple roles during embryonic and fetal development (see chapter on PTH-related protein). The absence of PTHrP (in the *Pthrp*-null fetal mouse) leads to abnormalities of chondrocyte differentiation and skeletal development,[8] modest hypocalcemia,[9] and reduced placental calcium transfer (see below). Such *Pthrp*-null fetuses have increased PTH levels[5] but still remain modestly hypocalcemic, indicating that PTH does not make up for lack of PTHrP in maintaining a normal calcium concentration in the fetal circulation. The converse is also true, that PTHrP cannot make up for absence of PTH, given the aforementioned hypocalcemia in fetuses lacking either parathyroids or PTH.

The role (if any) of the sex steroids in fetal skeletal development and mineral accretion is unknown, largely because the relevant analyses have not been performed in the relevant mouse models, and corresponding human data are absent. Estrogen receptor α and β knockout mice have been shown to have altered skeletal metabolism that develops postnatally, but the fetal skeleton has not been examined in detail. Similarly, postnatal skeletal roles of RANK, RANKL, and osteoprotegerin have been shown in relevant knockout mice, but the role that this system plays in fetal mineral metabolism is not yet known.

FETAL PARATHYROIDS

Intact parathyroid glands are required for maintenance of normal fetal calcium, magnesium, and phosphate levels; lack of parathyroids causes the fetal blood calcium to fall below the maternal level in mice, whereas lack of either PTH or PTHrP causes the fetal calcium to fall to the maternal level. Fetal parathyroids and PTH are also required for normal accretion of mineral by the skeleton and may be required for regulation of placental mineral transfer, as discussed below. Studies in fetal lambs have indicated that the fetal parathyroids may contribute to mineral homeostasis by producing both PTH and PTHrP, whereas detailed study of rats indicated that the fetal parathyroids produce only PTH. Whether human fetal parathyroids produce PTH alone, or PTH and PTHrP together, is unclear.

CALCIUM-SENSING RECEPTOR

The parathyroid calcium-sensing receptor (CaSR) regulates the serum calcium level in adults by inhibiting PTH, but it does not seem to set the high serum calcium level of fetuses. Instead, the CaSR may be responsible for the suppression of PTH in response to the elevated fetal blood calcium. Inactivating mutations of the CaSR (*Casr*-null fetuses) increased the serum calcium, PTH, 1,25-D, and bone turnover of fetuses and resulted in a lower skeletal calcium content by term.[10] The CaSR is also expressed within placenta, as shown in humans and mice, and this may indicate a role for the CaSR in the regulation of placental mineral transfer. *Casr*-null fetuses have a reduced rate of placental calcium transfer, but whether this is a direct consequence of the loss of placental CaSR is not known.[10]

FETAL KIDNEYS AND AMNIOTIC FLUID

Fetal kidneys partly regulate calcium homeostasis by adjusting the relative reabsorption and excretion of calcium, magnesium, and phosphate in response to the filtered load and other factors, such as PTHrP and PTH. The fetal kidneys also synthesize 1,25-D, but because *Vdr* null fetal mice displayed normal calcium homeostasis, placental calcium transfer, and skeletal mineral content, it seems likely that renal production of 1,25-D is relatively unimportant.

Renal calcium handling in fetal life may have minimal importance because calcium excreted by the kidneys is not permanently lost. Fetal urine is the major source of fluid and solute in amniotic fluid, and through fetal swallowing of amniotic fluid, the excreted calcium is made available again to the fetus.

PLACENTAL MINERAL ION TRANSPORT

As noted above, the bulk of placental calcium and other mineral transfer occurs late in gestation at a rapid rate. Active transport of calcium, magnesium, and phosphate across the placenta is necessary for the fetal requirement to be met; only placental calcium transfer has been studied in detail. Analogous to calcium transfer across the intestinal mucosa, it has been theorized that calcium diffuses into calcium-transporting cells through maternal-facing basement membranes, is carried across these cells by calcium binding proteins, and is actively extruded at the fetal-facing basement membranes by Ca^{2+}-ATPase.

Data from animal models indicate that a normal rate of maternal-to-fetal calcium transfer can usually be maintained despite the presence of maternal hypocalcemia or maternal hormone deficiencies such as aparathyroidism, vitamin D deficiency, and absence of the VDR. Whether the same is true for human pregnancies is less certain (see below). A "normal" rate of maternal–fetal calcium transfer does not necessarily imply that the fetus is unaffected by maternal hypocalcemia. Instead, it is an indication of the resilience of the fetal–placental unit to be able to extract the required amount of calcium from a hypocalcemic maternal circulation.

Fetal regulation of placental calcium transfer has been studied in a number of different animal models. Thyroparathyroidectomy in fetal lambs results in a reduced rate of placental calcium transfer, suggesting that the parathyroids are required for this process.[11] In contrast, mice lacking parathyroids as a consequence of ablation of the *Hoxa3* gene have a normal rate of placental calcium transfer.[4] The discrepancy between these findings in lambs and mice may be caused by whether the parathyroids are an important source of PTHrP in the circulation, as discussed above. Studies in fetal lambs and in *Pthrp*-null fetal mice are in agreement that PTHrP, and in particular midmolecular forms of PTHrP, stimulate placental calcium transfer.[9,12,13] There is preliminary evidence that PTH may also stimulate placental calcium transfer,[14] whereas calcitonin or 1,25-D are not required.[3,7]

FETAL SKELETON

A complete cartilaginous skeleton with digits and intact joints is present by the eighth week of gestation in humans. Primary ossification centers form in the vertebrae and long bones between the 8th to 12th weeks, but it is not until the third trimester that the bulk of mineralization occurs. At the 34th week of gestation, secondary ossification centers form in the femurs, but otherwise most epiphyses are cartilaginous at birth, with secondary ossification centers appearing in other bones in the neonate and child.[15]

The skeleton must undergo substantial growth and be sufficiently mineralized by the end of gestation to support the organism, but, as in the adult, the fetal skeleton participates in the regulation of mineral homeostasis. Calcium accreted by the fetal skeleton can be subsequently resorbed to help maintain the concentration of calcium in the blood. Functioning fetal parathyroid glands are needed for normal skeletal mineral accretion, and both hypoparathyroidism (thyroparathyroidectomized fetal lambs and aparathyroid fetal mice) and hyperparathyroidism (including *Casr*-null fetal mice) reduce the net amount of skeletal mineral accreted by term.

Further comparative study of fetal mice lacking parathyroids, PTH, or PTHrP has clarified the relative and interlocking roles of PTH and PTHrP in the regulation of the development and mineralization of the fetal skeleton. PTHrP produced locally in the growth plate directs the development of the cartilaginous scaffold that is later broken down and transformed into endochondral bone,[16] whereas PTH controls the mineralization of bone through its contribution to maintaining the fetal blood calcium and magnesium.[5] In the absence of PTHrP, a severe chondrodysplasia results,[8] but the fetal skeleton is fully mineralized.[5] In the absence of parathyroids or PTH, endochondral bone forms normally but is significantly undermineralized.[5,6,14] The blood calcium and magnesium were significantly reduced in fetuses lacking parathyroids or PTH, and this may explain why lack of PTH impaired skeletal mineralization. That is, by reducing the amount of mineral presented to the skeletal surface and to osteoblasts, lack of PTH thereby impaired mineral accretion by the skeleton.

FETAL RESPONSE TO MATERNAL HYPERPARATHYROIDISM

In humans, maternal primary hyperparathyroidism has been associated with adverse fetal outcomes, including spontaneous abortion and stillbirth, which are thought to result from suppression of the fetal parathyroid glands.[17] Because PTH cannot cross the placenta, fetal parathyroid suppression may result from increased calcium flux across the placenta to the fetus, facilitated by maternal hypercalcemia. Similar suppression of fetal parathyroids occurs when the mother has hypercalcemia caused by familial hypocalciuric hypercalcemia. Chronic elevation of the maternal serum calcium in mice results in suppression of the fetal PTH level,[10] but fetal outcome is not notably affected by this.

FETAL RESPONSE TO MATERNAL HYPOPARATHYROIDISM

Maternal hypoparathyroidism during human pregnancy can cause fetal hyperparathyroidism. This is characterized by fetal parathyroid gland hyperplasia, generalized skeletal demineralization, subperiosteal bone resorption, bowing of the long bones, osteitis fibrosa cystica, rib, and limb fractures, low birth weight, spontaneous abortion, stillbirth, and neonatal death. Similar skeletal findings have been reported in the fetuses and neonates of women with pseudohypoparathyroidism, renal tubular acidosis, and chronic renal failure. These changes in human skeletons differ from what has been found in animal models of maternal hypocalcemia, in which the fetal skeleton and the blood calcium is generally normal.

FETAL RESPONSE TO MATERNAL VITAMIN D DEFICIENCY

25-hydroxyvitamin D (25-D) readily crosses the placenta, such that cord blood levels of 25-D are within 20% of the maternal value. This means that if the mother is vitamin D deficient or insufficient, the fetus will be as well. As mentioned earlier, animal models of vitamin D deficiency, and *Vdr*-null mice, indicate that fetal calcium metabolism and skeletal mineral content will be normal at term despite maternal hypocalcemia and vitamin D deficiency. The limited human data are consistent with this, indicating that calcium homeostasis and skeletal mineral content in a human fetus may be unaffected by vitamin D deficiency, but that the neonate will be at risk for hypocalcemia, and the infant will later be at risk for the development of rickets.[18] The difference between maternal hypoparathyroidism (which causes fetal hyperparathyroidism in humans and animal models) and maternal vitamin D deficiency (which has little or no effect on fetal calcium homeostasis) may be explained by the effect of secondary hyperparathyroidism in the mother, which makes the hypocalcemia more modest compared with hypoparathyroidism.

Fetal vitamin D deficiency or insufficiency may not be completely benign, given evidence that vitamin D insufficiency in utero or postnatally may increase the risk of islet cell antibodies and type 1 diabetes.[19–21] Recent evidence has also associated vitamin D deficiency or insufficiency in the adult with cancer, multiple sclerosis, and other chronic diseases; however, as yet, no study has specifically examined whether vitamin D status during fetal development affects the risk of developing any of these diseases.

INTEGRATED FETAL CALCIUM HOMEOSTASIS

The evidence discussed in the preceding sections suggests the following summary models.

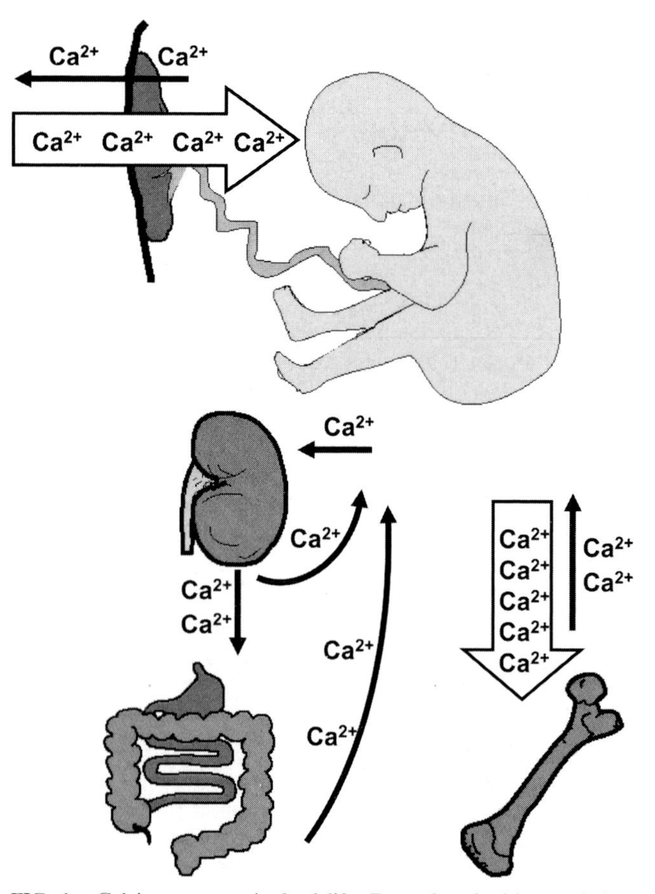

FIG. 1. Calcium sources in fetal life. Reproduced with permission from Pediatric Bone: Biology and Diseases, vol. 1, issue 1, Glorieux FH, Pettifor JM, Jüppner H, Fetal Mineral Homeostasis, pp. 271–302; Copyright Elsevier 2003.

© 2008 American Society for Bone and Mineral Research

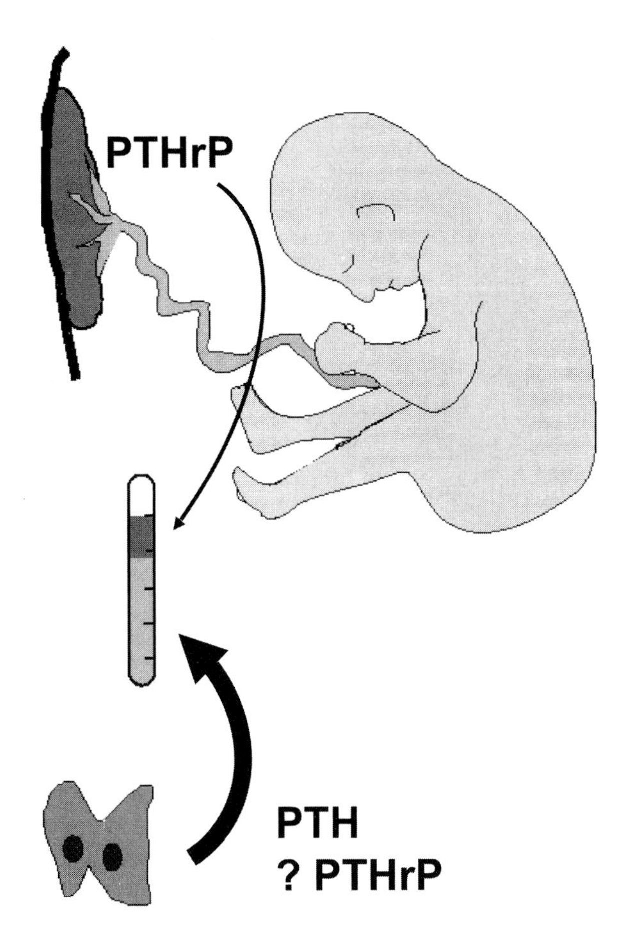

FIG. 2. Fetal blood calcium regulation. PTH and PTHrP both contribute to blood calcium regulation, with blood calcium represented schematically as a thermometer (light gray, contribution of PTH; dark gray, contribution of PTHrP). In the absence of PTHrP, the blood calcium falls to the maternal level. In the absence of parathyroids (*Hoxa3* null), the blood calcium falls well below the maternal calcium concentration, whereas in the absence of PTH (*Pth* null), the blood calcium equals the maternal level. In the absence of both PTHrP and PTH (*Hoxa3/Pthrp* double mutant), the blood calcium falls even further than in the absence of the parathyroids. Reproduced with permission from Pediatric Bone: Biology and Diseases, vol. 1, issue 1, Glorieux FH, Pettifor JM, Jüppner H, Fetal Mineral Homeostasis, pp. 271–302; Copyright Elsevier 2003.

Calcium Sources

The main flux of calcium and other minerals is across the placenta and into fetal bone, but calcium is also made available to the fetal circulation through several routes (Fig. 1). The kidneys reabsorb calcium; calcium excreted by the kidneys into the urine and amniotic fluid may be swallowed and reabsorbed; calcium is also resorbed from the developing skeleton. Some calcium returns to the maternal circulation (backflux). The maternal skeleton is a potential source of mineral, and it may be compromised in mineral deficiency states to provide to the fetus.

Blood Calcium Regulation

The fetal blood calcium is set at a level higher than the maternal level through the actions of PTHrP and PTH acting in concert (among other potential factors; Fig. 2). The CaSR suppresses PTH in response to the high calcium level, but the low level of PTH is required for maintaining the blood calcium and facilitating mineral accretion by the skeleton. 1,25-D synthesis and secretion are, in turn, suppressed by low PTH and high blood calcium and phosphate. The parathyroids may play a central role by producing PTH and PTHrP or may produce PTH alone, whereas PTHrP is produced by the placenta and other fetal tissues.

PTH and PTHrP, both present in the fetal circulation, independently and additively regulate fetal blood calcium. Neither hormone can make up for absence of the other: if one is missing, the blood calcium is reduced, and if both are missing, the blood calcium is reduced even further. PTH may contribute to the blood calcium through actions on the PTH/PTHrP (PTH1) receptor in classic target tissues (kidney, bone), whereas PTHrP may contribute through placental calcium transfer and actions on the PTH1 receptor and other receptors.

The normal elevation of the fetal blood calcium above the maternal calcium concentration was historically taken as proof that placental calcium transfer is an active process. However, the fetal blood calcium level is not simply determined by the rate of placental calcium transfer, because placental calcium transfer is normal in aparathyroid mice and increased in mice lacking the PTH1 receptor, but both phenotypes have significantly reduced blood calcium levels.[4,9] Also, *Casr*-null fetuses

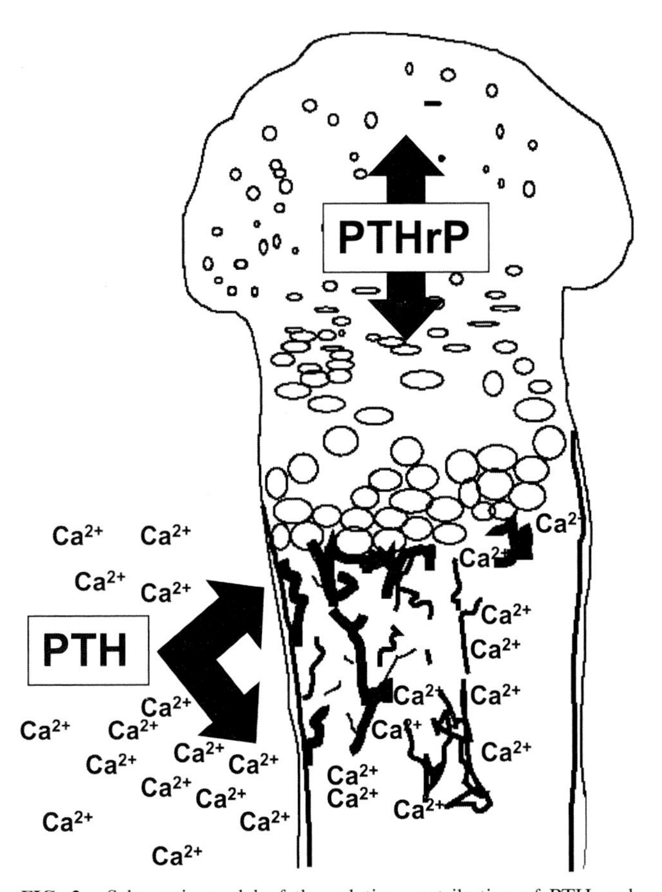

FIG. 3. Schematic model of the relative contribution of PTH and PTHrP to endochondral bone formation and skeletal mineralization. PTHrP is produced within the cartilaginous growth plate and directs the development of this scaffold that will later be broken down and replaced by bone. PTH reaches the skeleton systemically from the parathyroids and directs the accretion of mineral by the developing bone matrix. Reproduced with permission from Pediatric Bone: Biology and Diseases, vol. 1, issue 1, Glorieux FH, Pettifor JM, Jüppner H, Fetal Mineral Homeostasis, pp. 271–302; Copyright Elsevier 2003.

© 2008 American Society for Bone and Mineral Research

have reduced placental calcium transfer but markedly increased blood calcium levels.[10]

Placental Calcium Transfer

Placental calcium transfer is regulated by PTHrP but apparently not by PTH, and the placenta (and possibly the parathyroids) is likely an important source of PTHrP.

Skeletal Mineralization

PTH and PTHrP have separate roles with respect to skeletal development and mineralization (Fig. 3). PTH normally acts systemically to direct the mineralization of the bone matrix by maintaining the blood calcium at the adult level, and possibly by direct actions on osteoblasts within the bone matrix. In contrast, PTHrP acts both locally within the growth plate to direct endochondral bone development and outside of bone to affect skeletal development and mineralization by contributing to the regulation of the blood calcium and placental calcium transfer. PTH may have the more dominant effect to maintain skeletal mineral accretion.

REFERENCES

1. Kovacs CS 2003 Fetal mineral homeostasis. In: Glorieux FH, Pettifor JM, Jüppner H (eds.) Pediatric Bone: Biology and Diseases. Academic Press, San Diego, CA, USA, pp. 271–302.
2. Kovacs CS, Kronenberg HM 1997 Maternal-fetal calcium and bone metabolism during pregnancy, puerperium and lactation. Endocr Rev **18**:832–872.
3. Kovacs CS, Woodland ML, Fudge NJ, Friel JK 2005 The vitamin D receptor is not required for fetal mineral homeostasis or for the regulation of placental calcium transfer. Am J Physiol Endocrinol Metab **289**:E133–E144.
4. Kovacs CS, Manley NR, Moseley JM, Martin TJ, Kronenberg HM 2001 Fetal parathyroids are not required to maintain placental calcium transport. J Clin Invest **107**:1007–1015.
5. Kovacs CS, Chafe LL, Fudge NJ, Friel JK, Manley NR 2001 PTH regulates fetal blood calcium and skeletal mineralization independently of PTHrP. Endocrinology **142**:4983–4993.
6. Noseworthy CS, Fudge NJ, Karsenty G, Karaplis AC, Kovacs CS 2004 Parathyroid hormone (PTH) regulates placental calcium transfer independently of PTH-related protein (PTHrP). J Bone Miner Res **19**:S1;SA523.
7. McDonald KR, Fudge NJ, Woodrow JP, Friel JK, Hoff AO, Gagel RF, Kovacs CS 2004 Ablation of calcitonin/calcitonin gene related peptide-α impairs fetal magnesium but not calcium homeostasis. Am J Physiol Endocrinol Metab **287**:E218–E226.
8. Karaplis AC, Luz A, Glowacki J, Bronson RT, Tybulewicz VL, Kronenberg HM, Mulligan RC 1994 Lethal skeletal dysplasia from targeted disruption of the parathyroid hormone-related peptide gene. Genes Dev **8**:277–289.
9. Kovacs CS, Lanske B, Hunzelman JL, Guo J, Karaplis AC, Kronenberg HM 1996 Parathyroid hormone-related peptide (PTHrP) regulates fetal-placental calcium transport through a receptor distinct from the PTH/PTHrP receptor. Proc Natl Acad Sci USA **93**:15233–15238.
10. Kovacs CS, Ho-Pao CL, Hunzelman JL, Lanske B, Fox J, Seidman JG, Seidman CE, Kronenberg HM 1998 Regulation of murine fetal-placental calcium metabolism by the calcium-sensing receptor. J Clin Invest **101**:2812–2820.
11. Care AD, Caple IW, Abbas SK, Pickard DW 1986 The effect of fetal thyroparathyroidectomy on the transport of calcium across the ovine placenta to the fetus. Placenta **7**:417–424.
12. Care AD, Abbas SK, Pickard DW, Barri M, Drinkhill M, Findlay JB, White IR, Caple IW 1990 Stimulation of ovine placental transport of calcium and magnesium by mid-molecule fragments of human parathyroid hormone-related protein. Exp Physiol **75**:605–608.
13. Rodda CP, Kubota M, Heath JA, Ebeling PR, Moseley JM, Care AD, Caple IW, Martin TJ 1988 Evidence for a novel parathyroid hormone-related protein in fetal lamb parathyroid glands and sheep placenta: Comparisons with a similar protein implicated in humoral hypercalcaemia of malignancy. J Endocrinol **117**:261–271.
14. Noseworthy CS, Karsenty G, Karaplis AC, Kovacs CS 2007 PTH upregulates placental calcium transfer in response to fetal hypocalcemia while PTHrP does not. J Bone Miner Res **22**;S1;S396.
15. Moore KL, Persaud TVN 1998 The Developing Human, 6th ed. WB Saunders, Philadelphia, PA, USA.
16. Karsenty G 2001 Chondrogenesis just ain't what it used to be. J Clin Invest **107**:405–407.
17. Schnatz PF, Curry SL 2002 Primary hyperparathyroidism in pregnancy: Evidence-based management. Obstet Gynecol Surv **57**:365–376.
18. Kovacs CS 2008 Vitamin D in pregnancy and lactation: Maternal, fetal and neonatal outcomes from clinical and animal studies. Am J Clin Nutr (in press).
19. Stene LC, Ulriksen J, Magnus P, Joner G 2000 Use of cod liver oil during pregnancy associated with lower risk of Type I diabetes in the offspring. Diabetologia **43**:1093–1098.
20. Chiu KC, Chu A, Go VL, Saad MF 2004 Hypovitaminosis D is associated with insulin resistance and beta cell dysfunction. Am J Clin Nutr **79**:820–825.
21. Hypponen E, Laara E, Reunanen A, Jarvelin MR, Virtanen SM 2001 Intake of vitamin D and risk of type 1 diabetes: A birth-cohort study. Lancet **358**:1500–1503.

Chapter 23. Fibroblast Growth Factor-23

Kenneth E. White[1] and Michael J. Econs[1,2]

[1]Department of Medical and Molecular Genetics, Indiana University School of Medicine, Indianapolis, Indiana; [2]Department of Medicine, Indiana University School of Medicine, Indianapolis, Indiana

INTRODUCTION

Disorders of phosphate (Pi) homeostasis, in concert with powerful in vitro and in vivo studies, have shown that fibroblast growth factor-23 (FGF23) is central to the control of renal Pi and vitamin D homeostasis. Although the molecular mechanisms are unique to each disorder, elevated FGF23 is associated with syndromes manifested by hypophosphatemia with paradoxically low or normal $1{,}25(OH)_2$ vitamin D [$1{,}25(OH)_2D$] and include: autosomal dominant hypophosphatemic rickets (ADHR), X-linked hypophosphatemic rickets (XLH), tumor-induced osteomalacia (TIO), and autosomal recessive hypophosphatemic rickets (ARHR). Heritable disorders of hyperphosphatemia and often elevated $1{,}25(OH)_2D$, such as tumoral calcinosis (TC), are associated with reduced

Dr. White receives royalties from licensing FGF23 to Kirin Pharmaceuticals and receives funding from Kirin Pharmaceuticals to study FGF23. Dr. Econs has received royalties and a research grant from Kirin Pharmaceuticals and has a consultant agreement with Kirin Pharmaceuticals but has not received any money for consulting.

Key words: fibroblast growth factor-23, phosphate, phosphate metabolism, vitamin D, hypophosphatemia, hyperphosphatemia, Klotho, fibroblast growth factor receptor

© 2008 American Society for Bone and Mineral Research

FGF23 activity. These collective findings have provided unique insight into the activity of FGF23 on renal Pi and vitamin D metabolism.

THE *FGF23* GENE AND PROTEIN

The *FGF23* gene resides on human chromosome 12p13 (mouse chromosome 6), is comprised of three coding exons, and contains an open reading frame of 251 residues.[1] The tissue with the highest FGF23 expression is bone. FGF23 mRNA is observed in osteoblasts, osteocytes, flattened bone-lining cells, and osteoprogenitor cells.[2] Quantitative PCR showed that FGF23 mRNA was most highly expressed in long bone, followed by thymus, brain, and heart.[3]

Western blot analyses showed that wildtype FGF23 is secreted as a full-length 32-kDa species, as well as cleavage products of 12 and 20 kDa.[3–5] Cleavage of FGF23 occurs within a subtilisin-like proprotein convertase (SPC) proteolytic site ($_{176}RXXR_{179}/S_{180}$) that separates the conserved FGF-like N-terminal domain from the variable C-terminal tail.

FGF23 ACTIVITY

FGF23 has overlapping function with PTH to reduce renal Pi reabsorption but has opposite effects on $1,25(OH)_2D$. The two primary transport proteins responsible for Pi reabsorption in the kidney are the type II sodium-phosphate co-transporters NPT2a and NPT2c, expressed in the apical membrane of the proximal tubule. FGF23 delivery leads to renal Pi wasting through the downregulation of both Npt2a and Npt2c.[6]

Normally, hypophosphatemia is a strong positive stimulator for increasing serum $1,25(OH)_2D$. However, patients with TIO, ADHR, XLH, and ARHR manifest hypophosphatemia with paradoxically low or inappropriately normal $1,25(OH)_2D$. In mice, the expression of the 1α(OH)ase enzyme and the catabolic 24(OH)ase are decreased and increased, respectively, when the animals are exposed to FGF23.[4] Thus, the effects of FGF23 on the renal vitamin D metabolic enzymes is most likely responsible for the reductions in $1,25(OH)_2D$ in the setting of persistent hypophosphatemia in ADHR, XLH, TIO, and ARHR patients.

REGULATION OF FGF23 IN VIVO

In humans, dietary Pi supplementation increased FGF23, whereas Pi restriction and the addition of Pi binders suppressed serum FGF23,[7] indicating that FGF23 plays a role in maintenance of Pi homeostasis. In animal studies, the FGF23 response to serum Pi has been more dramatic than in the human studies. Mice given high and low Pi diets produce the expected correlations between FGF23 and dietary Pi intake.[8]

Vitamin D has important regulatory effects on FGF23. In mice, injections of 20–200 ng $1,25(OH)_2D$ led to dose-dependent increases in serum FGF23.[9] These changes in FGF23 occurred before changes in serum Pi, indicating that FGF23 may be directly regulated by vitamin D. Physiologically, this would be consistent with results examining the role of FGF23 in vitamin D metabolism. FGF23 has been shown to downregulate the 1α(OH)ase mRNA[6,9]; thus, as $1,25(OH)_2D$ rises in the blood as a product of 1α(OH)ase activity, vitamin D stimulates FGF23, which completes the feedback loop and downregulates 1α(OH)ase expression.

FGF23 RECEPTORS

FGF23 is a member of a unique class of FGFs including FGF19 and FGF21 that are endocrine as opposed to paracrine/autocrine factors. FGF23 requires the co-receptor Klotho (KL) for bioactivity. *KL*-null mice have severe calcifications and markedly elevated serum Pi,[10] which parallels *Fgf23*-null mice[11,12] and that of TC patients. However, both the *KL*-null and *Fgf23*-null mice have more extreme phenotypes than that observed in patients. Importantly, these defects in the *KL*-null and *Fgf23*-null mice can be ameliorated with a low Pi diet to reduce serum Pi.[13] In parallel with *Fgf23*-null mice, *KL*-null mice have increased Npt2a in the proximal tubule,[14] indicating that the hyperphosphatemia is secondary to increased renal reabsorption of Pi.

KL is produced as two isoforms caused by alternative splicing of the same gene. Membrane bound KL (mKL) is a 130-kDa single-pass transmembrane protein characterized by a large extracellular domain and a very short (10 residue) intracellular domain that does not possess signaling capabilities.[15] The secreted form of KL (sKL) is ~80 kDa and is spliced within exon 3 to result in a KL protein that does not contain the transmembrane domain and is thus secreted into the circulation.[15]

The most likely mechanism for FGF23 signaling through KL is the recruitment of canonical FGF receptors (FGFRs) to form heteromeric complexes. One group has identified a specific complex between FGFR1c and KL.[16] In contrast, others showed that multiple FGFRs (FGFR1c, FGFR3c, and FGFR4) can interact with KL and FGF23 and signal through mitogen-activated protein kinase (MAPK) cascades.[17] Importantly, within the kidney, KL localizes to the distal tubule[14]; however, FGF23 mediates its effects on NPT2a, NPT2c, and vitamin D within the proximal tubule.[4,6] Therefore, the mechanisms underlying a local distal convoluted tubule–proximal tubule (DCT–PT) axis in the kidney after FGF23 delivery are unclear.

SERUM ASSAYS

FGF23 is measured in the circulation using several assays. One extensively used assay is a C-terminal FGF23 ELISA with both the capture and detection antibodies binding C-terminal to the $FGF23_{176}RXXR_{179}/S$ cleavage site.[18] This assay thus recognizes full-length FGF23 and C-terminal proteolytic fragments. The C-terminal assay is quantified relative to standards comprised of FGF23-conditioned media from stable cell lines. The normal mean for this assay is 55 ± 50 Reference Units (RU)/ml, and the upper limit of normal is 150 RU/ml. In a study with a large number of controls and TIO patients, this ELISA was used to test the levels of FGF23 in TIO and XLH[18] and showed that serum FGF23 is detectable in normal individuals. The mean FGF23 was >10-fold elevated in TIO patients and rapidly fell after tumor resection. Importantly, many XLH patients (13 of 21) had elevated FGF23 compared with controls,[18] and in those with "normal" FGF23, these levels may be "inappropriately normal" in the setting of hypophosphatemia.

An intact FGF23 ELISA assay has been developed that uses conformation-specific monoclonal antibodies that span the $_{176}RXXR_{179}/S_{180}$ SPC site and thus recognize N- and C-terminal portions of FGF23.[19] This assay detects a mean circulating concentration of 29 pg/ml in normal individuals. The published upper limit of normal is 54 pg/ml.[19] The results of these two assays generally agree with regard to the relative ranges of FGF23 concentrations in XLH and in TIO patients and that FGF23 is elevated in most XLH patients. Based on limited data from two TIO patients undergoing resection, the half-life of FGF23 is between 20 and 50 min.[20,21]

© 2008 American Society for Bone and Mineral Research

FGF23-ASSOCIATED SYNDROMES

Disorders Associated With Increased FGF23 Bioactivity

Autosomal Dominant Hypophosphatemic Rickets. Importantly, ADHR (OMIM 193100) is distinguished from other hereditary hypophosphatemias by having either early or delayed onset with variable expressivity.[22] The ADHR mutations replace arginine (R) residues at positions 176 or 179 with glutamine (Q) or tryptophan (W) within the FGF23 subtilisin-like proprotein convertase (SPC) cleavage site, $_{176}RXXR_{179}/S_{180}$[1,4,23] (Table 1b). After insertion of the ADHR mutations into wildtype FGF23, FGF23 secreted from mammalian cells was primarily a full-length (32 kDa), active polypeptide, as opposed to the 32-kDa and cleavage products typically observed for wildtype FGF23 expression.[5]

Tumor-Induced Osteomalacia. TIO is an acquired disorder of isolated renal Pi wasting that is associated with tumors. TIO patients present with similar biochemistries as patients with ADHR,[24] and osteomalacia is seen on bone biopsy. Clinical symptoms include muscle weakness, fatigue, and bone pain.[24] Insufficiency fractures are common, and proximal muscle weakness can become severe.[24] FGF23 is elevated in patients with TIO,[18,19] and tumors that cause TIO have a dramatic overexpression of FGF23 mRNA.[23] Surgical resection of the tumor results in rapid decreases in serum FGF23.[18]

X-Linked Hypophosphatemic Rickets. XLH (OMIM 307800) is caused by inactivating mutations in *PHEX* (phosphate-regulating gene with homologies to endopeptidases on the X chromosome).[25] PHEX is a member of the M13 family of membrane-bound metalloproteases and shows the highest expression in bone cells such as osteoblasts, osteocytes, and odontoblasts in teeth.[26]

Reports have established that FGF23 is elevated in many XLH patients.[18,19] Although it was initially thought that PHEX might cleave FGF23, this is not the case.[3] Instead, FGF23 mRNA expression is markedly increased in *Hyp* mice (mouse model of XLH) bone.[3,8] The elevated FGF23 mRNA levels indicate that the increase in serum FGF23 in XLH is caused by overproduction by skeletal cells, as opposed to a decreased rate of FGF23 degradation by cell surface proteases after secretion into the circulation. Although the interactions between FGF23 and PHEX are most likely indirect (see ARHR below), the encoded proteins have overlapping expression in bone.[2,3,26] At present, the PHEX substrate is unknown.

Autosomal Recessive Hypophosphatemic Rickets. Dentin matrix protein-1 (DMP1), a member of the small integrin-binding ligand, N-linked glycoprotein (SIBLING) family, is highly expressed in osteocytes. Both *Dmp1*-null mice and patients with ARHR (OMIM 241520) manifest rickets and osteomalacia with isolated renal Pi wasting associated with elevated FGF23. Mutational analyses showed that one ARHR family carried a mutation that ablated the DMP1 start codon, and a second family exhibited a deletion in the DMP1 C terminus.[27] Mutations have also been identified in DMP1 splicing sites, which likely result in nonfunctional protein.[28] Mechanistic studies using the *Dmp1*-null mouse showed that loss of DMP1 causes defective osteocyte maturation, leading to elevated FGF23 expression and pathological changes in bone mineralization.[27] Importantly, *Dmp1*-null mice are biochemical phenocopies of the *Hyp* mouse, and patients with ARHR and XLH (as well as the *Dmp1*-null and *Hyp* mice) share a unique bone histology characterized by distinctive periosteocytic lesions.[27] Thus, these findings suggest that PHEX may also have a role in osteocyte maturation in a parallel pathway to DMP1 that leads to overexpression of FGF23.

Other Heritable Disorders Involving Elevated FGF23. In addition to the disorders described above, FGF23 is also upregulated in several bone dysplasias that manifest documented isolated renal Pi wasting. These disorders include McCune Albright Syndrome (OMIM 174800),[2] caused by activating mutations in G_s; opsismodysplasia (OMIM 258480)[29]; osteoglophonic dysplasia (OMIM 166250),[30] which is caused by activating mutations in FGFR1; and epidermal nevus syndrome (ENS).[31]

Disorders Associated With Reduced FGF23 Bioactivity

Familial Tumoral Calcinosis. TC (OMIM 211900) is an autosomal recessive disorder characterized by dental abnormalities and soft tissue periarticular and vascular calcification.[32] Biochemical abnormalities include hyperphosphatemia, increased percent tubular reabsorption of phosphate (TRP), and inappropriately normal or elevated $1{,}25(OH)_2D$. Calcium and PTH are usually within the normal ranges, although PTH may be suppressed. Hyperostosis-hyperphosphatemia syndrome

TABLE 1. SUMMARY OF HERITABLE AND ACQUIRED DISORDERS INVOLVING FGF23

Disorder	*Mutated gene*	*Mutation consequence*	*Relationship to FGF23*	*Effect on serum Pi*	*Effect on serum 1,25D*	*Intact FGF23 ELISA concentration.(Kainos)*	*C-terminal FGF23 ELISA concentration(Immutopics)*
ADHR	*FGF23*	Gain of function	Stabilize full-length, active FGF23	↓	↔	↔ or ↑	↔ or ↑
XLH	*PHEX*	Loss of function	Increased FGF23 production in osteocytes	↓	↔	↔ or ↑	↔ or ↑
ARHR	*DMP1*	Loss of function	Increased FGF23 production in osteocytes	↓	↔	↔ or ↑	↔ or ↑
TIO	—	—	FGF23over-produced by tumor	↓	↔	↔ or ↑	↔ or ↑
TC/HHS	*FGF23* or *GALNT3*	Loss of function	Destabilize full-length, active FGF23	↑	↔ or ↑	↓	↑
TC	*KL*	Loss of function	Decreased FGF23-dependent signaling	↑	↔ or ↑	↑	↑

© 2008 American Society for Bone and Mineral Research

(HHS) is a rare metabolic disorder characterized by a biochemical profile that is identical to TC, with localized hyperostosis.[33,34]

TC/HHS Caused by GALNT3 Mutations. The first gene identified for heritable TC was UDP-*N*-acetyl-α-D-galactosamine: polypeptide *N*-acetylgalactosaminyl transferase-3 (*GALNT3*).[35] GALNT3 is expressed in the Golgi and initiates *O*-linked glycosylation of nascent proteins. These TC patients were originally reported to manifest serum FGF23 levels ~30-fold above the normal mean when assessed with the C-terminal FGF23 ELISA.[35] Importantly, it was subsequently shown that the TC patients did indeed have elevated C-terminal FGF23; however, the same individuals had low FGF23 when measured with the intact FGF23 ELISA (Table 1).[36] These findings were confirmed by showing that loss of GALNT3 resulted in the production of nonfunctional FGF23 protein caused by intracellular degradation.[37] FGF23 is *O*-glycosylated on specific residues within the $_{176}RH\underline{T}R_{179}/S_{180}$ site (at threonine 178); thus, the lack of glycosylation at this residue is thought to destabilize intact active FGF23.[37]

HHS was also found to be caused by inactivating mutations in *GALNT3*,[33] and these patients also manifest inappropriate C-terminal to intact FGF23 ELISA values (Table 1).[33,34] Indeed, some of the HHS mutations are the same as those that result in TC,[34] indicating that genetic background may influence disease phenotype and/or that TC and HHS may represent a spectrum of severity within the same disease.

TC Caused by FGF23 Mutations. TC can also be caused by recessive, inactivating mutations in the *FGF23* gene.[38–40] These mutations have all been missense mutations (S71G, M96T, S129F) within the FGF23 N-terminal FGF-like domain. The TC alterations destabilize FGF23, as supported by the findings that the TC patients with *FGF23* mutations have the same FGF23 ELISA pattern as GALNT3-TC patients (i.e., markedly elevated C-terminal concentrations, in concert with low intact values)[38,39] and the fact that these mutants are cleaved before cellular secretion.[38–40] Thus, the common denominator in GALNT3-TC and FGF23-TC is the lack of production of intact FGF23. This lack of intact FGF23 results in elevation of serum Pi through increased renal reabsorption, which in turn results in elevated secretion of nonfunctional FGF23 fragments through a positive feedback cycle.

TC Caused by KL Mutations. KL is a co-receptor for FGF23 and was therefore tested as a candidate gene for TC in a 13-yr-old girl with hypothesized end organ defects in renal FGF23 bioactivity. This patient manifested hyperphosphatemia, hypercalcemia, elevated PTH, elevated intact and C-terminal FGF23[41] (~100- to 550-fold elevation of the normal means), and ectopic calcifications in the heel and brain. She had normal pubertal development, and her disease paralleled *KL*-null mice with regard to ectopic calcifications and dramatic elevation of circulating FGF23.[16] This patient had a novel recessive mutation in a highly conserved residue (Histidine193Arginine [H193R]) in the extracellular domain of KL (KL1 domain). Mutant KL expression was markedly reduced compared with that of wildtype KL, which resulted in a striking reduction in the ability of KL to mediate FGF23-dependent signaling.[41] Thus, an inactivating H193R KL mutation results in a TC phenotype and shows that KL is required for FGF23 bioactivity.

Chronic Kidney Disease. FGF23 has been measured in patients with chronic kidney disease (CKD), because these patients are hyperphosphatemic. FGF23 is elevated in CKD, but it is unknown as to whether this represents a compensatory response to the hyperphosphatemia or is in part caused by the lack of renal FGF23 clearance. One report has shown that higher FGF23 levels are a predictor of increased progression of renal disease in patients with nondiabetic CKD.[42] However, the pathophysiological significance of these findings remains to be elucidated.

REFERENCES

1. Consortium ADHR 2000 Autosomal dominant hypophosphataemic rickets is associated with mutations in FGF23. Nat Genet **26:**345–348.
2. Riminucci M, Collins MT, Fedarko NS, Cherman N, Corsi A, White KE, Waguespack S, Gupta A, Hannon T, Econs MJ, Bianco P, Gehron Robey P 2003 FGF-23 in fibrous dysplasia of bone and its relationship to renal phosphate wasting. J Clin Invest **112:**683–692.
3. Liu S, Guo R, Simpson LG, Xiao ZS, Burnham CE, Quarles LD 2003 Regulation of fibroblastic growth factor 23 expression but not degradation by PHEX. J Biol Chem **278:**37419–37426.
4. Shimada T, Mizutani S, Muto T, Yoneya T, Hino R, Takeda S, Takeuchi Y, Fujita T, Fukumoto S, Yamashita T 2001 Cloning and characterization of FGF23 as a causative factor of tumor-induced osteomalacia. Proc Natl Acad Sci USA **98:**6500–6505.
5. White KE, Carn G, Lorenz-Depiereux B, Benet-Pages A, Strom TM, Econs MJ 2001 Autosomal-dominant hypophosphatemic rickets (ADHR) mutations stabilize FGF-23. Kidney Int **60:**2079–2086.
6. Larsson T, Marsell R, Schipani E, Ohlsson C, Ljunggren O, Tenenhouse HS, Juppner H, Jonsson KB 2004 Transgenic mice expressing fibroblast growth factor 23 under the control of the alpha1(I) collagen promoter exhibit growth retardation, osteomalacia, and disturbed phosphate homeostasis. Endocrinology **145:**3087–3094.
7. Burnett SM, Gunawardene SC, Bringhurst FR, Juppner H, Lee H, Finkelstein JS 2006 Regulation of C-terminal and intact FGF-23 by dietary phosphate in men and women. J Bone Miner Res **21:**1187–1196.
8. Perwad F, Azam N, Zhang MY, Yamashita T, Tenenhouse HS, Portale AA 2005 Dietary and serum phosphorus regulate fibroblast growth factor 23 expression and 1,25-dihydroxyvitamin D metabolism in mice. Endocrinology **146:**5358–5364.
9. Shimada T, Hasegawa H, Yamazaki Y, Muto T, Hino R, Takeuchi Y, Fujita T, Nakahara K, Fukumoto S, Yamashita T 2004 FGF-23 is a potent regulator of vitamin D metabolism and phosphate homeostasis. J Bone Miner Res **19:**429–435.
10. Tsujikawa H, Kurotaki Y, Fujimori T, Fukuda K, Nabeshima Y 2003 Klotho, a gene related to a syndrome resembling human premature aging, functions in a negative regulatory circuit of vitamin D endocrine system. Mol Endocrinol **17:**2393–2403.
11. Shimada T, Kakitani M, Yamazaki Y, Hasegawa H, Takeuchi Y, Fujita T, Fukumoto S, Tomizuka K, Yamashita T 2004 Targeted ablation of Fgf23 demonstrates an essential physiological role of FGF23 in phosphate and vitamin D metabolism. J Clin Invest **113:**561–568.
12. Sitara D, Razzaque MS, Hesse M, Yoganathan S, Taguchi T, Erben RG, Jap H, Lanske B 2004 Homozygous ablation of fibroblast growth factor-23 results in hyperphosphatemia and impaired skeletogenesis, and reverses hypophosphatemia in Phex-deficient mice. Matrix Biol **23:**421–432.
13. Segawa H, Yamanaka S, Ohno Y, Onitsuka A, Shiozawa K, Aranami F, Furutani J, Tomoe Y, Ito M, Kuwahata M, Tatsumi S, Imura A, Nabeshima Y, Miyamoto KI 2006 Correlation between hyperphosphatemia and type II Na/Pi cotransporter activity in klotho mice. Am J Physiol Renal Physiol **292:**F769–F779.
14. Li SA, Watanabe M, Yamada H, Nagai A, Kinuta M, Takei K 2004 Immunohistochemical localization of Klotho protein in brain, kidney, and reproductive organs of mice. Cell Struct Funct **29:**91–99.
15. Matsumura Y, Aizawa H, Shiraki-Iida T, Nagai R, Kuro-o M, Nabeshima Y 1998 Identification of the human klotho gene and its two transcripts encoding membrane and secreted klotho protein. Biochem Biophys Res Commun **242:**626–630.
16. Urakawa I, Yamazaki Y, Shimada T, Iijima K, Hasegawa H, Okawa K, Fujita T, Fukumoto S, Yamashita T 2006 Klotho converts canonical FGF receptor into a specific receptor for FGF23. Nature **444:**770–774.

© 2008 American Society for Bone and Mineral Research

17. Kurosu H, Ogawa Y, Miyoshi M, Yamamoto M, Nandi A, Rosenblatt KP, Baum MG, Schiavi S, Hu MC, Moe OW, Kuro-o M 2006 Regulation of fibroblast growth factor-23 signaling by klotho. J Biol Chem **281:**6120–6123.
18. Jonsson KB, Zahradnik R, Larsson T, White KE, Sugimoto T, Imanishi Y, Yamamoto T, Hampson G, Koshiyama H, Ljunggren O, Oba K, Yang IM, Miyauchi A, Econs MJ, Lavigne J, Juppner H 2003 Fibroblast growth factor 23 in oncogenic osteomalacia and X-linked hypophosphatemia. N Engl J Med **348:**1656–1663.
19. Yamazaki Y, Okazaki R, Shibata M, Hasegawa Y, Satoh K, Tajima T, Takeuchi Y, Fujita T, Nakahara K, Yamashita T, Fukumoto S 2002 Increased circulatory level of biologically active full-length FGF-23 in patients with hypophosphatemic rickets/osteomalacia. J Clin Endocrinol Metab **87:**4957–4960.
20. Khosravi A, Cutler CM, Kelly MH, Chang R, Royal RE, Sherry RM, Wodajo FM, Fedarko NS, Collins MT 2007 Determination of the elimination half-life of fibroblast growth factor-23. J Clin Endocrinol Metab **92:**2374–2377.
21. Takeuchi Y, Suzuki H, Ogura S, Imai R, Yamazaki Y, Yamashita T, Miyamoto Y, Okazaki H, Nakamura K, Nakahara K, Fukumoto S, Fujita T 2004 Venous sampling for fibroblast growth factor-23 confirms preoperative diagnosis of tumor-induced osteomalacia. J Clin Endocrinol Metab **89:**3979–3982.
22. Econs MJ, McEnery PT 1997 Autosomal dominant hypophosphatemic rickets/osteomalacia: Clinical characterization of a novel renal phosphate-wasting disorder. J Clin Endocrinol Metab **82:**674–681.
23. White KE, Jonsson KB, Carn G, Hampson G, Spector TD, Mannstadt M, Lorenz-Depiereux B, Miyauchi A, Yang IM, Ljunggren O, Meitinger T, Strom TM, Juppner H, Econs MJ 2001 The autosomal dominant hypophosphatemic rickets (ADHR) gene is a secreted polypeptide overexpressed by tumors that cause phosphate wasting. J Clin Endocrinol Metab **86:**497–500.
24. Ryan EA, Reiss E 1984 Oncogenous osteomalacia. Review of the world literature of 42 cases and report of two new cases. Am J Med **77:**501–512.
25. Hyp Consortium 1995 A gene (PEX) with homologies to endopeptidases is mutated in patients with X-linked hypophosphatemic rickets. The HYP Consortium. Nat Genet **11:**130–136.
26. Beck L, Soumounou Y, Martel J, Krishnamurthy G, Gauthier C, Goodyer CG, Tenenhouse HS 1997 Pex/PEX tissue distribution and evidence for a deletion in the 3′ region of the Pex gene in X-linked hypophosphatemic mice. J Clin Invest **99:**1200–1209.
27. Feng JQ, Ward LM, Liu S, Lu Y, Xie Y, Yuan B, Yu X, Rauch F, Davis SI, Zhang S, Rios H, Drezner MK, Quarles LD, Bonewald LF, White KE 2006 Loss of DMP1 causes rickets and osteomalacia and identifies a role for osteocytes in mineral metabolism. Nat Genet **38:**1310–1315.
28. Lorenz-Depiereux B, Bastepe M, Benet-Pages A, Amyere M, Wagenstaller J, Muller-Barth U, Badenhoop K, Kaiser SM, Rittmaster RS, Shlossberg AH, Olivares JL, Loris C, Ramos FJ, Glorieux F, Vikkula M, Juppner H, Strom TM 2006 DMP1 mutations in autosomal recessive hypophosphatemia implicate a bone matrix protein in the regulation of phosphate homeostasis. Nat Genet **38:**1248–1250.
29. Zeger MD, Adkins D, Fordham LA, White KE, Schoenau E, Rauch F, Loechner KJ 2007 Hypophosphatemic rickets in opsismodysplasia. J Pediatr Endocrinol Metab **20:**79–86.
30. White KE, Cabral JM, Davis SI, Fishburn T, Evans WE, Ichikawa S, Fields J, Yu X, Shaw NJ, McLellan NJ, McKeown C, Fitzpatrick D, Yu K, Ornitz DM, Econs MJ 2005 Mutations that cause osteoglophonic dysplasia define novel roles for FGFR1 in bone elongation. Am J Hum Genet **76:**361–367.
31. Hoffman WH, Jueppner HW, Deyoung BR, O'Dorisio MS, Given KS 2005 Elevated fibroblast growth factor-23 in hypophosphatemic linear nevus sebaceous syndrome. Am J Med Genet A **134:**233–236.
32. Prince MJ, Schaeffer PC, Goldsmith RS, Chausmer AB 1982 Hyperphosphatemic tumoral calcinosis: Association with elevation of serum 1,25-dihydroxycholecalciferol concentrations. Ann Intern Med **96:**586–591.
33. Frishberg Y, Topaz O, Bergman R, Behar D, Fisher D, Gordon D, Richard G, Sprecher E 2005 Identification of a recurrent mutation in GALNT3 demonstrates that hyperostosis-hyperphosphatemia syndrome and familial tumoral calcinosis are allelic disorders. J Mol Med **83:**33–38.
34. Ichikawa S, Guigonis V, Imel EA, Courouble M, Heissat S, Henley JD, Sorenson AH, Petit B, Lienhardt A, Econs MJ 2007 Novel GALNT3 mutations causing hyperostosis-hyperphosphatemia syndrome result in low intact fibroblast growth factor 23 concentrations. J Clin Endocrinol Metab **92:**1943–1947.
35. Topaz O, Shurman DL, Bergman R, Indelman M, Ratajczak P, Mizrachi M, Khamaysi Z, Behar D, Petronius D, Friedman V, Zelikovic I, Raimer S, Metzker A, Richard G, Sprecher E 2004 Mutations in GALNT3, encoding a protein involved in O-linked glycosylation, cause familial tumoral calcinosis. Nat Genet **36:**579–581.
36. Garringer HJ, Fisher C, Larsson TE, Davis SI, Koller DL, Cullen MJ, Draman MS, Conlon N, Jain A, Fedarko NS, Dasgupta B, White KE 2006 The role of mutant UDP-N-acetyl-alpha-D-galactosamine-polypeptide N-acetylgalactosaminyltransferase 3 in regulating serum intact fibroblast growth factor 23 and matrix extracellular phosphoglycoprotein in heritable tumoral calcinosis. J Clin Endocrinol Metab **91:**4037–4042.
37. Frishberg Y, Ito N, Rinat C, Yamazaki Y, Feinstein S, Urakawa I, Navon-Elkan P, Becker-Cohen R, Yamashita T, Araya K, Igarashi T, Fujita T, Fukumoto S 2007 Hyperostosis-hyperphosphatemia syndrome: A congenital disorder of O-glycosylation associated with augmented processing of fibroblast growth factor 23. J Bone Miner Res **22:**235–242.
38. Benet-Pages A, Orlik P, Strom TM, Lorenz-Depiereux B 2005 An FGF23 missense mutation causes familial tumoral calcinosis with hyperphosphatemia. Hum Mol Genet **14:**385–390.
39. Larsson T, Yu X, Davis SI, Draman MS, Mooney SD, Cullen MJ, White KE 2005 A novel recessive mutation in fibroblast growth factor-23 causes familial tumoral calcinosis. J Clin Endocrinol Metab **90:**2424–2427.
40. Araya K, Fukumoto S, Backenroth R, Takeuchi Y, Nakayama K, Ito N, Yoshii N, Yamazaki Y, Yamashita T, Silver J, Igarashi T, Fujita T 2005 A novel mutation in fibroblast growth factor 23 gene as a cause of tumoral calcinosis. J Clin Endocrinol Metab **90:**5523–5527.
41. Ichikawa S, Imel EA, Kreiter ML, Yu X, Mackenzie DS, Sorenson AH, Goetz R, Mohammadi M, White KE, Econs MJ 2007 A homozygous missense mutation in human KLOTHO causes severe tumoral calcinosis. J Clin Invest **117:**2684–2691.
42. Fliser D, Kollerits B, Neyer U, Ankerst DP, Lhotta K, Lingenhel A, Ritz E, Kronenberg F, Kuen E, Konig P, Kraatz G, Mann JF, Muller GA, Kohler H, Riegler P 2007 Fibroblast growth factor 23 (FGF23) predicts progression of chronic kidney disease: The Mild to Moderate Kidney Disease (MMKD) Study. J Am Soc Nephrol **18:**2600–2608.

© 2008 American Society for Bone and Mineral Research

Chapter 24. Gonadal Steroids

Katrien Venken, Steven Boonen, Roger Bouillon, and Dirk Vanderschueren

Bone Research Unit, Laboratory for Experimental Medicine and Endocrinology, Department of Experimental Medicine, Katholieke Universiteit Leuven, Leuven, Belgium

INTRODUCTION

Bone and calcium metabolism is regulated actively by a number of calciotropic hormones [PTH, 1,25(OH)$_2$ vitamin D, and calcitonin], whereas other hormones (e.g., thyroid and growth hormone) including sex steroid hormones have a more gradual yet highly important regulatory role. Indeed, gonadal (sex) steroids regulate skeletal maturation and preservation in both men and women, as already recognized in the 1940s by Albright and Reifenstein.(1) In the late 1980s, sex steroid receptors were discovered in bone cells.(2–4) Our understanding of sex steroid receptor activation and subsequent translation into biological skeletal actions is, however, still incomplete. First, the metabolism of gonadal steroids is complex; androgens, for instance, can be converted into estrogens by the P450 aromatase enzyme in bone and stimulate both estrogen and androgen receptors. Hence, sex steroids may have not only endocrine but also paracrine and autocrine skeletal effects. Moreover, circulating sex steroid concentrations do not necessarily reflect their biological activity. Because of strong binding to sex hormone binding globulin, only small fractions of serum sex steroids are available for skeletal metabolism and action. Finally, gonadal steroids have both genomic and nongenomic effects in bone and nonbone cells. These cellular events ultimately translate into important biological actions on bone mineral acquisition and maintenance. This chapter will give an overview of our current understanding of gonadal steroid metabolism, receptor activation, and their most relevant cellular and biological actions on bone, with special emphasis on the well-established biological action of sex steroids (androgens and estrogens), because the physiological relevance of progesterone action on bone is less clear.

METABOLISM OF GONADAL STEROIDS

Androgens, estrogens, and progesterone are known as gonadal (sex) steroids. Androgens are C-19 steroids secreted from the testes in men and from the adrenals in both men and women. The most important circulating androgen in men is testosterone (T), of which ~95% is secreted by the testes. In serum, the majority of T is bound to sex hormone–binding globulin (SHBG; 50–60%, with high affinity) and albumin (40–50%, with low affinity), whereas only 1–2% of T remains free.(5) The albumin-bound T and the free T represent the fractions available for biological action and are therefore referred to as the bioavailable T (bioT). Unbound T diffuses passively through the cell membranes into the target cell, where it binds to the specific androgen receptor (AR; Fig. 1).(6) In peripheral tissues, T can be irreversibly converted to the more potent 5α-dihydrotestosterone (DHT) by the enzyme 5α-reductase.(7) In addition, T has the unique feature that it can be converted into 17β-estradiol (E$_2$) by the P450 aromatase enzyme, even within bone, and subsequently can exert its effects through estrogen receptor α (ERα) or β (ERβ) (Fig. 1).(8) In addition, the adrenal cortex and, to a lesser extent, the gonads secrete large amounts of dehydroepiandrosterone (DHEA), DHEA sulfate (DHEA-S), and androstenedione (Fig. 1). Although only weakly androgenic, they are an important source of substrate for the extragonadal synthesis of potent sex steroids (i.e., these C-19 androgens can be metabolized to estrone [E$_1$] by the aromatase enzyme or to T by steroid sulfatase, 17β-hydroxysteroid dehydrogenase [17β-HSD] and/or 3β-HSD; Fig. 1).(9,10) Thus, depending on the relative activity of P450 aromatase, 5α-reductase, 17β-HSD, 3β-HSD, and steroid sulfatase, T and C-19 androgens may either predominantly activate the AR or the ERs. Because these enzymes are all expressed in bone tissue, local metabolism of androgens in bone may be of significant physiological importance. Thus, sex steroids are in a large part synthesized locally in peripheral tissues, providing individual target tissues with the means to adjust synthesis and metabolism of sex steroids to their local requirements. In men, only ~20% of E$_2$ is directly secreted by the testes. The other 80% of E$_2$ in men is derived from aromatization of T and androstenedione in peripheral tissues, mainly in adipose tissue.(11)

In women, the steroidogenic pathways are essentially the same as in men. More than 95% of E$_2$ and most of E$_1$ in premenopausal women is secreted from the ovaries, which express large amounts of P450 aromatase enzyme (and therefore secrete only low amounts of T). Also, 20–40% of E$_2$ in women is bound to SHBG. The ovaries also secrete progesterone, with high levels during the luteal phase of the menstrual cycle and during pregnancy. Moreover, progesterone serves as a precursor for all other gonadal steroid hormones.(9) Ovarian E$_2$ production ceases after menopause in women. By that time, circulating E$_2$ and E$_1$ concentrations in postmenopausal women are solely derived from the conversion of adrenal androgens into estrogens, mainly within fat. Estrogen concentrations in postmenopausal women are generally lower than in men of the same age. In contrast to women, total T concentrations decrease only marginally (~1%/yr) during aging in men, whereas total E$_2$ concentrations remain constant. However, SHBG increases (+120%) markedly in aging men, resulting in important reductions of bioavailable T and E$_2$ in elderly men.(12)

RECEPTORS, GENOMIC, AND NONGENOMIC EFFECTS OF GONADAL STEROIDS

Gonadal Steroid Receptors

The AR, ERα, ERβ, and progesterone receptor (PR) are all gonadal steroid receptors that belong to the same nuclear receptor superfamily; i.e., a family of ligand-activated transcription factors.(13) They are composed of three independent but interacting functional domains; the NH$_2$-terminal or A/B domain, the C or DNA-binding domain (DBD), and the D/E/F or ligand-binding domain (Fig. 2).(14–16) The sequence homology in the DBD is high among the sex steroid receptors. The N-terminal domain of these receptors encodes a ligand-independent activation function (AF-1), a region of the receptor involved in protein–protein interactions and transcriptional activation of target gene expression. The COOH-terminal region, or ligand-binding domain (LBD), mediates ligand bind-

Key words: sex steroid receptors, androgens, estrogens, progesterone, androgen receptor, estrogen receptor α, estrogen receptor β, progesterone receptor, genomic sex steroid action, nongenomic sex steroid action, osteoporosis, periosteum, endosteum, trabecular bone

The authors state that they have no conflicts of interest.

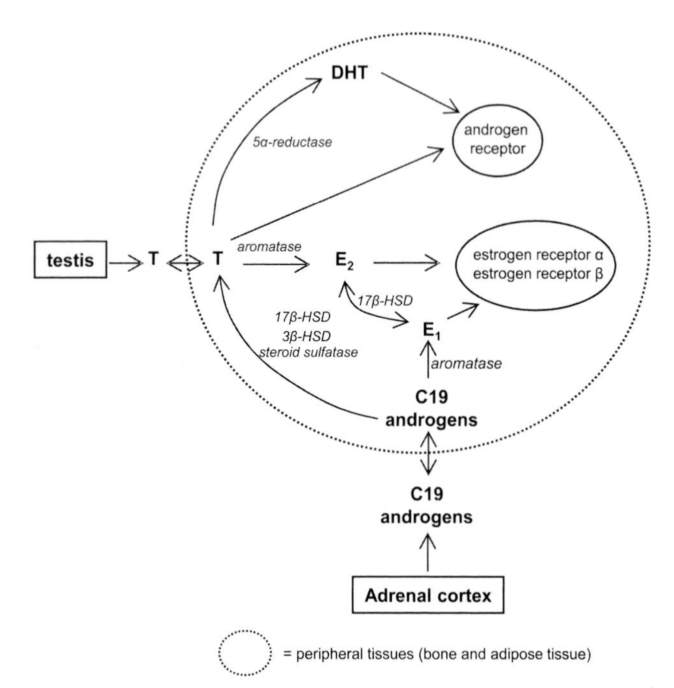

FIG. 1. Overview of the metabolism and action of sex steroids in men. HSD, hydroxysteroid dehydrogenase.

ing, receptor dimerization, nuclear translocation, and transcription of target gene expression (see below; Fig. 2).

Genomic Sex Steroid Signaling

In the absence of hormone, the receptor is kept in an inactive state and sequestered in a multiprotein complex by means of chaperone molecules such as heat shock proteins (Hsp) Hsp70 and Hsp90. Hormone binding to the receptor induces an activating conformational change within the receptor, resulting in dissociation of the chaperone proteins, translocation of the receptor to the nucleus, and increased phosphorylation (Fig. 3). The hormone-bound receptor dimerizes with another hormone-bound receptor, and this dimer binds with high affinity to specific DNA sequences (i.e., estrogen, androgen, or progesterone responsive elements [ERE, ARE, or PRE, respectively]) located within the regulatory region of the target gene.(17) The DNA-bound receptor contacts the general transcription apparatus either directly or indirectly through co-regulatory proteins that act coordinated to influence the transcription of the target gene.(18) These co-regulatory proteins may include either factors enhancing transactivation (co-activators such sex steroid co-activator 1, 2, and 3) or factors reducing transactivation (co-repressors). The process of gene modulation through this classical mechanism takes at least 30–45 min, and the time required to produce significant levels of protein is in the order of hours.(19)

The above-mentioned mechanism provides an explanation for the regulation of genes with a functional responsive element sequence within the promoter region. Reports of hormone-bound receptor activation of genes without a responsive element sequence led to the discovery that sex steroid receptors can also modulate gene expression at alternative regulatory DNA sequences such as activating protein-1 (AP-1)(20–22) and specificity protein-1 (SP-1).(23) In this context, the sex steroid receptor is recruited to the specific promoter complex where it interacts with other DNA-bound transcription factors such as c-*jun*, c-*fos*, and other co-activator proteins.

Nongenomic Sex Steroid Signaling

The observation that a variety of cell types respond to sex steroids within seconds or minutes makes a classical genomic signaling unlikely and suggests that sex steroids may elicit nongenomic effects, possibly through cell surface receptors linked to intracellular signal transduction proteins(24,25) (Fig. 3). Indeed, binding sites for estrogens,(26) androgens,(27) and progesterone(28) have been identified on plasma membranes. The membrane-initiated steroid signaling results in activation of conventional second messenger signal transduction cascades, including activation of protein kinase A (PKA), protein kinase C (PKC), cellular tyrosine kinases, mitogen-activated protein kinases (MAPKs), phosphatidylinositol 3-kinase (PI3K), and Akt- and Src/Shc/ERK signaling(29,30) (Fig. 3).

In the context of nongenomic sex steroid signaling, it was previously shown that androgens and estrogens can transmit anti-apoptotic effects on osteoblasts in vitro with similar efficiency through either AR, ERα, or ERβ.(31) These effects are mediated by Src/Shc/ERK signaling and seem sex nonspecific.(32) Interestingly, the synthetic compound 4-estren-3α,17β-diol (estren) was described to increase bone mass in ovariectomized female and orchidectomized male mice through nongenomic signaling.(33) Moreover, these positive effects on bone occur without adverse effects on reproductive organs. However, later it was clearly shown that estren is able to increase seminal vesicles and uterine weight,(34,35) with even evidence for transcriptional (i.e., genomic) activity through the AR.(35–37) In addition, in vivo data showed that no cross-reactivity exists between ARs and ERs and their ligands.(38) Although it has become clear that the mechanisms of sex steroids and their receptors are complex and that a wide diversity of cellular pathways may be involved, their significance with respect to bone metabolism has not been clarified yet. Moreover, in contrast to the abundance of data on estrogen and androgen action on bone, little is known about the role and biological relevance of progesterone on bone.(4) In this regard, the next section will focus entirely on the biological effects of androgens and estrogens on bone tissue during growth and maintenance.

BIOLOGICAL EFFECTS OF SEX STEROIDS

Men are better protected against osteoporosis because, compared with women, they gain more bone during puberty and lose less during aging. The explanation for this sex difference seems to be 2-fold. First, the traditional concept of sexual dimorphic bone growth considered "male hormones" (androgens) as stimulatory and "female hormones" (estrogens) as inhibitory for bone growth, in particular for periosteal (i.e., radial) bone growth. Second, men lose less bone than women during aging because they do not have an equivalent of menopause, a phase when ovarian production of "female hormones" ceases. More recently, it became clear that androgens and estrogens cannot be considered as male or female hormones, respectively. As will be pointed out in the following sections,

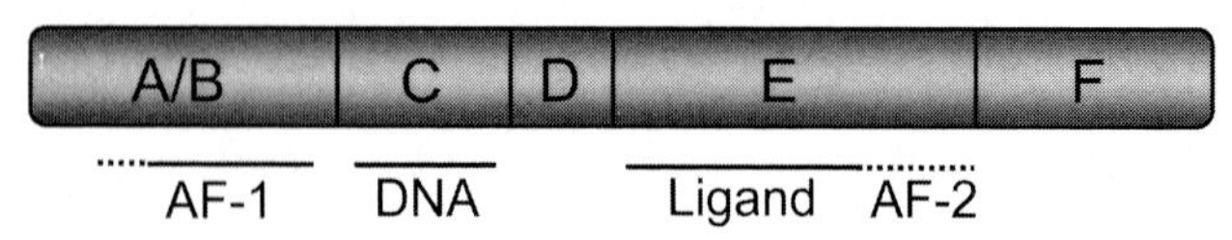

FIG. 2. Diagrammatic representation of the domain structure of nuclear receptors. The A/B domain at the NH_2 terminus contains the AF-1 site where other transcription factors interact. The C/D domain contains the two-zinc finger structure that binds to DNA, and the E/F domain contains the ligand binding pocket and the AF-2 domain that directly contacts co-activator peptides.

© 2008 American Society for Bone and Mineral Research

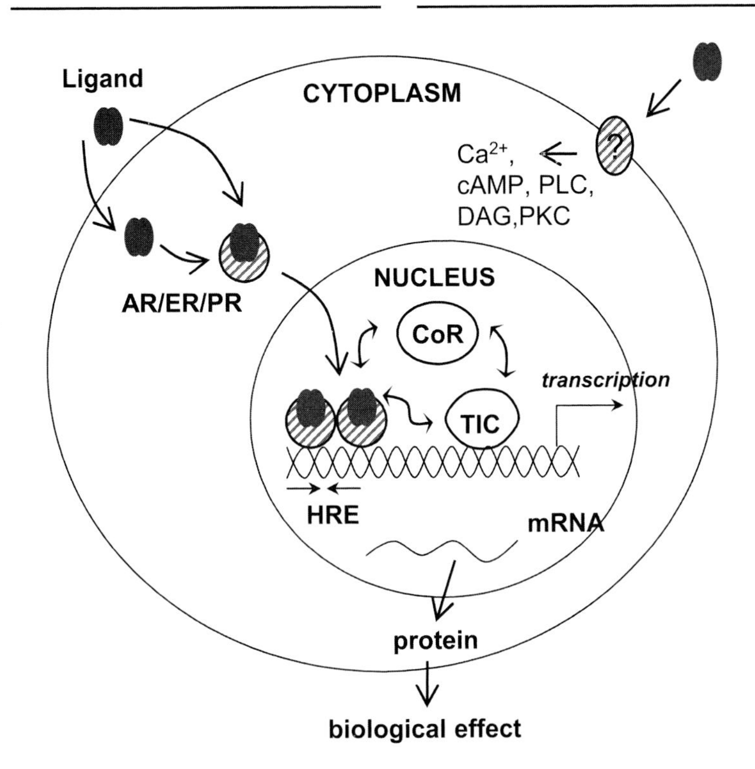

FIG. 3. Genomic and nongenomic gonadal steroid signaling. Genomic action: the ligand binds to its specific receptor, which induces a conformational change. The receptor translocates to the nucleus and undergoes dimerization. The receptor dimer binds to specific DNA sequences, so-called hormone response elements. The DNA bound receptor contacts the general transcription initiation complex (TIC) either directly or indirectly through co-regulatory proteins (CoR). Nongenomic action: the ligand binds a, yet undefined, plasma membrane receptor, which results in the activation of second messenger signal transduction pathways including Ca^{2+}, (PLC), cAMP, phospholipase C, diacylglycerol (DAG), protein kinase C (PKC). ER, estrogen receptor; AR, androgen receptor; PR, progesterone receptor; HRE, hormone response element.

androgens and estrogens and their respective receptors have specific actions on the different bone envelopes during bone growth and maintenance in both sexes.

EFFECTS OF SEX STEROIDS ON BONE GROWTH

Periosteal Bone Formation

The production of sex steroids at the start of puberty is inextricably linked with the increase in bone accretion during this period. As they enter puberty, boys and girls experience an increase in bone size resulting from an enlargement of the outer diameter of bone and a concomitant widening of the medullary diameter.[39] These changes in bone geometry occur in both sexes, but they occur to a substantially greater extent in boys than in girls. The greater male bone size results from an enhanced periosteal bone formation.[40] In androgen-deficient growing male rats, periosteal bone formation is significantly impaired, whereas it is increased in estrogen-deficient female rats, resulting in the traditional view of stimulatory androgens in males versus inhibitory estrogens in females on periosteal growth.[41] In males, direct AR-mediated androgen action on periosteal bone formation was shown in studies of mice and rats with either a natural mutation or genetic manipulation of the gene encoding the AR.[42–44] Stimulatory actions of androgens on periosteal bone have also been shown in female rats treated with an anti-androgen[45] and in hirsute women with increased bone mass (corrected for bone size) after higher (biological) androgen activity.[46] Moreover, cortical bone mass in female mice is increased by ERβ disruption, indicating that ERβ may inhibit periosteal bone expansion in females.[47,48] The finding that AR, ERα, and ERβ are expressed in human, rat, and mouse osteoblasts, osteoclasts, and osteocytes further supports the notion that sex steroids can act on the skeleton through direct stimulation of their receptor.[3]

Because the observations that men suffering from either estrogen deficiency (because of a mutation in the aromatase gene) or resistance (secondary to a mutation in the *ERα* gene) have delayed skeletal development and impaired bone mineral gain,[49–51] the crucial role of estrogens in male skeletal growth has received much attention. In addition, the observation that estrogen treatment of an adolescent aromatase–deficient boy with already (supra)normal T levels results in a significant increased bone size of the distal radius challenged the traditional view of stimulatory androgens versus inhibitory estrogens on periosteal bone formation.[51] This observation indicated that periosteal bone formation in males may not be solely dependent on androgen action, but also, at least in part, on estrogen action. In this context, targeted deletion of the gene for either aromatase or ERα, but not ERβ, in male mice results in decreased cortical bone area and cortical thickness. Also, young growing male rats treated with an aromatase inhibitor show a reduced cortical bone growth. However, the findings in these mice and rats may be confounded by decreased serum IGF-I levels, indicating that estrogen action may be rather indirect through modulation of the growth hormone (GH)–IGF-I axis. In conclusion, it seems that both AR and ERα are required in males to obtain optimal periosteal bone expansion.

Endocortical Bone Formation and Resorption

Medullary bone expansion during growth results from greater endocortical bone resorption than formation. Overall, net endocortical bone resorption is less than periosteal bone formation, leading to a radial bone expansion and cortical thickening during growth.[39] Medullary bone expansion during growth is limited in females compared with males (i.e., endocortical contraction occurs in both female rodents and in women [relative to male] at the end of puberty, at least at some skeletal sites).[52] This suggests that estrogens are involved in bone mineral accumulation at the endocortical bone surface.[53]

© 2008 American Society for Bone and Mineral Research

Enchondral Bone Formation and Epiphyseal Closure

Sex steroids stimulate longitudinal bone growth at start of puberty. This stimulatory action on enchondral bone formation depends on ERα activation.[3,54] The latter effect may, at least partly, be attributed to indirect activation of the GH–IGF-I axis by estrogens.[55] An AR-dependent effect on enchondral bone formation is rather uncertain as bone length in AR-disrupted mice and rats is unchanged.[3] At the end of puberty, ERα activation induces epiphyseal growth plate closure in both sexes, as shown by the absence of epiphyseal fusion in men with aromatase deficiency and ERα disruption.[49,50]

BONE MINERAL MAINTENANCE

Gonadal Steroid Deficiency and Replacement

Sex steroid deficiency in both sexes results in high bone turnover (i.e., osteoclast and osteoblast numbers and activity both increase, but the former exceeds the latter with a loss of bone mass and strength). As a result, hypogonadal men, similar to postmenopausal women, develop osteoporosis. Androgens, similar to estrogen replacement in postmenopausal women, prevent bone loss in hypogonadal men.[56] According to studies in transgenic mice, gonadal steroids are able to maintain trabecular bone mass by both AR and ERα activation in both sexes.[3] Androgens therefore have a "dual mode of action." However, under physiological circumstances, AR signaling seems to be the most important for the maintenance of trabecular bone mass in males, whereas ERα signaling is the dominant pathway in females.[44,47]

Molecular Mechanism of the Antiresorptive Action of Sex Steroids

Sex steroids downregulate osteoclast precursors (osteoclastogenesis) in bone marrow. This action indirectly relates to the regulation of the production of multiple cytokines by different cell types in the bone marrow microenvironment.[57] Antiresorptive effects of estrogen are mediated by the downregulation of cytokines that are involved in the regulation of osteoclast formation, such as interleukin-1 (IL-1) and TNF-α produced by monocytes and IL-6, granulocyte macrophage-colony stimulating factor (GM-CSF), and macrophage-colony stimulating factor (M-CSF) produced by stromal cells and osteoblasts[58] (Fig. 4). Moreover, TGF-β mediates the stimulatory effect of E_2 on the induction of osteoclast apoptosis in vitro and in vivo, and it directly acts on the osteoclast to decrease its activity[59] (Fig. 4). E_2 can also suppress osteoclastogenesis through enhanced production of osteoprotegerin (OPG) and decreased RANKL by osteoblasts.[60] Interestingly, a direct ERα-mediated anti-apoptotic effect on osteoclasts has been described in female osteoclast-specific ERα knockout mice.[61] Androgens also decrease osteoclast formation and resorption through increased production of OPG by osteoblasts.[60,62] In addition, androgens—at least in vitro—may also directly reduce osteoclastogenesis independently of osteoblasts.[63,64] In vitro data also indicate that sex steroids stimulate proliferation and differentiation of osteoblasts and inhibit osteoblasts apoptosis.[2,57]

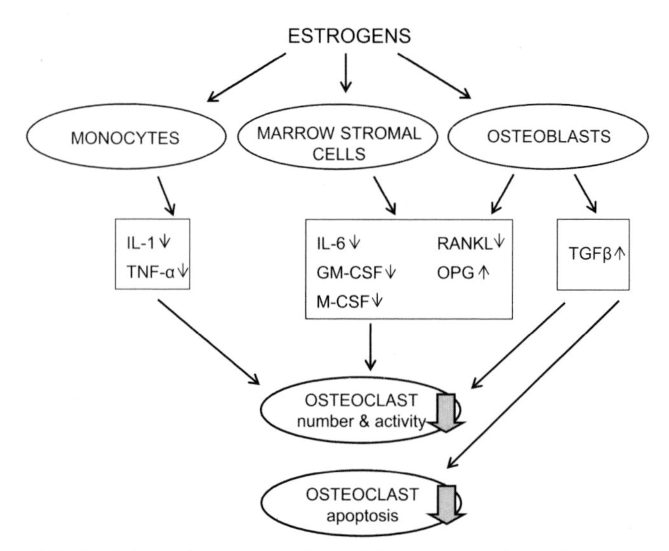

FIG. 4. Schematic representation of the antiresorptive action of estrogen on bone. Estrogen downregulates the production in bone of cytokines involved in bone resorption, such as IL-1, TNF-α, IL-6, GM-CSF, and M-CSF. Downregulation of these cytokines decreases the number and activity of osteoclasts. Estrogen also decreases TGF-β, resulting in a decreased osteoclast number and activity and increased osteoclast apoptosis. Estrogen increases OPG, the decoy receptor of RANKL in binding to its receptor RANK on osteoclasts.

Age-Related Bone Loss and Sex Steroids

Age-related trabecular bone loss in men is primarily caused by thinning of trabeculae, whereas trabecular bone loss in postmenopausal women is typically characterized by a disruption of trabeculae. The resulting loss of connectivity has a much more profound impact on bone strength than trabecular thinning, leading to similar decreases in trabecular bone volume.[65] The acceleration of trabecular bone loss in women after menopause (as well as after ovariectomy) is caused by the high bone turnover because of estrogen deficiency. Likewise, the same type of trabecular bone loss is observed in hypogonadal men.

In both sexes, age-related cortical bone loss is caused by reduced periosteal bone formation and, at the same time, increased endocortical bone resorption. In elderly men and women, ongoing periosteal bone formation is unable to offset endosteal bone resorption, resulting in cortical thinning.[66,67] In aged rodents, it seems that periosteal bone is less responsive to androgens than estrogens, because androgen deficiency and replacement hardly affects periosteal bone in aging rodents,[68,69] whereas estrogens increase cortical thickness through endocortical contraction/apposition.[38] To what extent periosteal bone formation and endocortical bone resorption in the elderly is also related to age-related changes in gonadal steroid concentrations remains unclear.

Overall, the gonadal androgen T or the adrenal androgen DHEAS only show weak or no correlations with BMD in elderly men, whereas serum E_2—in particular bioavailable E_2—is positively associated with BMD, rates of bone loss, and fracture occurrence.[12,70,71] In addition, E_2 is negatively correlated with markers of bone resorption.[12] Interestingly, bone loss seems associated with a polymorphism in the aromatase gene in elderly men, even independently of circulating E_2 concentrations.[72] Hence, men with higher local estrogen synthesis may be protected against bone loss. Therefore, aromatization of androgens into estrogens and subsequent activation of ER may have an important role in male skeletal homeostasis during aging. Whether there exist a serum threshold concentrations for E_2 below which bone loss increases in both men and women remains a matter of debate.[73] In addition, intracrine or paracrine actions of estrogens may be more relevant for skeletal homeostasis in the elderly than previously anticipated. In postmenopausal women, aromatase inhibitors

© 2008 American Society for Bone and Mineral Research

cause bone loss and increased fracture risk, showing the importance of estrogen synthesis in these women with already low circulating estrogen concentrations.[74]

It is unknown whether the frequently observed subnormal concentrations of bioavailable T also contribute to bone loss in elderly men. In fact, intervention trials have not yet been able to prove a benefit of T replacement in elderly men with partial androgen deficiency.[75,76] Therefore, androgen replacement with the specific purpose of promoting bone health and preventing osteoporosis cannot be recommended in this patient group until extensive clinical data on long-term efficacy and safety become available.

Selective Estrogen and Androgen Modulators

Estrogen replacement for osteoporosis in women is restricted as a consequence of unwanted extraskeletal effects at sites such as the uterus, breast, and the cardiovascular system.[77,78] Partial estrogen receptor agonists having agonistic estrogen-like effects on bone and antagonistic effects on reproductive tissues would be of interest for the treatment of postmenopausal bone loss. The so-called first-generation selective ER modulator (SERM) tamoxifen, initially used to treat and prevent breast cancer, showed a positive estrogen-like effect in the rodent and human skeleton[79] but an unfavorable long-term safety profile on the uterus.[80] Raloxifene, a second-generation SERM, lacks the uterotropic activity associated with tamoxifen and therefore represents an improved agonist/antagonist profile.[79] As a results, raloxifene has been approved as a drug for the prevention and treatment of postmenopausal osteoporosis, and clinical studies confirmed that raloxifene effectively reduced fracture risk, without adverse effects in reproductive tissues.[81,82] The recognition of SERMs as agents able to elicit estrogenic effects in a tissue-specific manner remains a major challenge to the pharmaceutical industry to continue to develop new ER ligands that retain the beneficial effects in tissues such as bone, brain, and the cardiovascular system but that lack the mitogenic and perhaps carcinogenic actions in the breast and uterus. SERMs are thought to exert different biological actions in different tissues compared with full agonists, such as E_2, through distinct conformational changes in the ER on binding to ligand. These different conformations, in turn, result in tissue specificity, because those mechanisms involved in ER-dependent stimulation of transcription are thought to vary according to tissue and promoter context.[83] However, it has not been possible yet to define a single unifying mechanism to explain why SERMs preferentially activate transcription in certain tissues but not others. The successful clinical application of SERMs stimulated a great interest in the discovery and development of selective AR modulators (SARMs). One uniform characteristic of these compounds is that they are not substrates for aromatase and 5α-reductase.[84] Unlike the agonistic–antagonistic properties of SERMs in target tissues, these AR ligands are known to act as full agonists in anabolic organs (bone and muscle) but as partial agonists in androgenic tissues (prostate and seminal vesicles). However, (preclinical), data supporting the bone selectivity of these agents are still rather limited.[84] The goal of SARMs is to reproduce the beneficial effects of androgens in men and women without undesirable effects such as prostate stimulation or acne. The discovery of SARMs not only provides a potentially significant therapeutic advance for androgen replacement therapy but also provides model compounds to further study the molecular mechanism of AR action.

REFERENCES

1. Albright F, Reifenstein EC 1948 Metabolic bone disease: Osteoporosis. In: Albright F, Reifenstein EC (eds.) The Parathyroid Glands and Metabolic Bone Disease. Williams and Wilkins, Baltimore, MD, USA, pp. 145–204.
2. Riggs BL, Khosla S, Melton LJ III 2002 Sex steroids and the construction and conservation of the adult skeleton. Endocr Rev **23**:279–302.
3. Vanderschueren D, Vandenput L, Boonen S, Lindberg MK, Bouillon R, Ohlsson C 2004 Androgens and bone. Endocr Rev **25**:389–425.
4. Prior JC 1990 Progesterone as a bone-trophic hormone. Endocr Rev **11**:386–398.
5. Vermeulen A, Verdonk L 1968 Studies on the binding of testosterone to human plasma. Steroids **11**:609–635.
6. Giorgi EP, Stein WD 1981 The transport of steroids into animal cells in culture. Endocrinology **108**:688–697.
7. Wilson JD 2001 The role of 5alpha-reduction in steroid hormone physiology. Reprod Fertil Dev **13**:673–678.
8. Simpson ER, Clyne C, Rubin G, Boon WC, Robertson K, Britt K, Speed C, Jones M 2002 Aromatase–a brief overview. Annu Rev Physiol **64**:93–127.
9. Longcope C 1998 Metabolism of estrogens and progestins. In: Fraser IS Jr, Lobo RA, Whitehead MI (eds.) Estrogens and Progestens in Clinical Practice. Churchill Livingstone, New York, NY, USA, pp. 89–94.
10. Griffin HG, Wilson JD 1994 Disorders of the testes and male reproductive tract. In: Wilson JD, Foster DW (eds.) Textbook of Endocrinology. Saunders, Philadelphia, PA, USA, pp. 799–852.
11. Kaufman JM, Vermeulen A 2005 The decline of androgen levels in elderly men and its clinical and therapeutic implications. Endocr Rev **26**:833–876.
12. Khosla S, Melton LJ, Atkinson EJ, O-Fallon WM, Klee GG, Riggs BL 1998 Relationship of serum sex steroid levels and bone turnover markers with bone mineral density in men: A key role for bio-available estrogen. J Clin Endocrinol Metab **83**:2266–2275.
13. Mangelsdorf DJ, Thummel C, Beato M, Herrlich P, Schutz G, Umesono K, Blumberg B, Kastner P, Mark M, Chambon P, Evans RM 1995 The nuclear receptor superfamily: The second decade. Cell **83**:835–839.
14. Nilsson S, Makela S, Treuter E, Tujague M, Thomsen J, Andersson G, Enmark E, Pettersson K, Warner M, Gustafsson JA 2001 Mechanisms of estrogen action. Physiol Rev **81**:1535–1565.
15. Quigley CA, De Bellis A, Marschke KB, el-Awady MK, Wilson EM, French FS 1995 Androgen receptor defects: Historical, clinical, and molecular perspectives. Endocr Rev **16**:271–321.
16. Savouret JF, Chauchereau A, Misrahi M, Lescop P, Mantel A, Bailly A, Milgrom E 1994 The progesterone receptor. Biological effects of progestins and antiprogestins. Hum Reprod **9**(Suppl 1):7–11.
17. Tsai MJ, O'Malley BW 1994 Molecular mechanisms of action of steroid/thyroid receptor superfamily members. Annu Rev Biochem **63**:451–486.
18. McKenna NJ, Lanz RB, O'Malley BW 1999 Nuclear receptor coregulators: Cellular and molecular biology. Endocr Rev **20**:321–344.
19. Shang Y, Myers M, Brown M 2002 Formation of the androgen receptor transcription complex. Mol Cell **9**:601–610.
20. Kushner PJ, Agard DA, Greene GL, Scanlan TS, Shiau AK, Uht RM, Webb P 2000 Estrogen receptor pathways to AP-1. J Steroid Biochem Mol Biol **74**:311–317.
21. Whitmarsh AJ, Davis RJ 1996 Transcription factor AP-1 regulation by mitogen-activated protein kinase signal transduction pathways. J Mol Med **74**:589–607.
22. Wiren KM, Toombs AR, Zhang XW 2004 Androgen inhibition of MAP kinase pathway and Elk-1 activation in proliferating osteoblasts. J Mol Endocrinol **32**:209–226.
23. Safe S 2001 Transcriptional activation of genes by 17 beta-estradiol through estrogen receptor-Sp1 interactions. Vitam Horm **62**:231–252.
24. Falkenstein E, Tillmann HC, Christ M, Feuring M, Wehling M 2000 Multiple actions of steroid hormones–a focus on rapid, nongenomic effects. Pharmacol Rev **52**:513–556.
25. Losel RM, Falkenstein E, Feuring M, Schultz A, Tillmann HC, Rossol-Haseroth K, Wehling M 2003 Nongenomic steroid action: Controversies, questions, and answers. Physiol Rev **83**:965–1016.

© 2008 American Society for Bone and Mineral Research

26. Pietras RJ, Szego CM 1977 Specific binding sites for oestrogen at the outer surfaces of isolated endometrial cells. Nature **265:**69–72.
27. Konoplya EF, Popoff EH 1992 Identification of the classical androgen receptor in male rat liver and prostate cell plasma membranes. Int J Biochem **24:**1979–1983.
28. Wasserman WJ, Pinto LH, O'Connor CM, Smith LD 1980 Progesterone induces a rapid increase in $[Ca^{2+}]_{in}$ of Xenopus laevis oocytes. Proc Natl Acad Sci USA **77:**1534–1536.
29. Song RX, McPherson RA, Adam L, Bao Y, Shupnik M, Kumar R, Santen RJ 2002 Linkage of rapid estrogen action to MAPK activation by ERalpha-Shc association and Shc pathway activation. Mol Endocrinol **16:**116–127.
30. Migliaccio A, Castoria G, Di Domenico M, de Falco A, Bilancio A, Lombardi M, Barone MV, Ametrano D, Zannini MS, Abbondanza C, Auricchio F 2000 Steroid-induced androgen receptor-oestradiol receptor beta-Src complex triggers prostate cancer cell proliferation. EMBO J **19:**5406–5417.
31. Kousteni S, Bellido T, Plotkin LI, O'Brien CA, Bodenner DL, Han L, Han K, DiGregorio GB, Katzenellenbogen JA, Katzenellenbogen BS, Roberson PK, Weinstein RS, Jilka RL, Manolagas SC 2001 Nongenotropic, sex-nonspecific signaling through the estrogen or androgen receptors: Dissociation from transcriptional activity. Cell **104:**719–730.
32. Kousteni S, Han L, Chen JR, Almeida M, Plotkin LI, Bellido T, Manolagas SC 2003 Kinase-mediated regulation of common transcription factors accounts for the bone-protective effects of sex steroids. J Clin Invest **111:**1651–1664.
33. Kousteni S, Chen JR, Bellido T, Han L, Ali AA, O'Brien CA, Plotkin L, Fu Q, Mancino AT, Wen Y, Vertino AM, Powers CC, Stewart SA, Ebert R, Parfitt AM, Weinstein RS, Jilka RL, Manolagas SC 2002 Reversal of bone loss in mice by nongenotropic signaling of sex steroids. Science **298:**843–846.
34. Moverare S, Dahllund J, Andersson N, Islander U, Carlsten H, Gustafsson JA, Nilsson S, Ohlsson C 2003 Estren is a selective estrogen receptor modulator with transcriptional activity. Mol Pharmacol **64:**1428–1433.
35. Windahl SH, Galien R, Chiusaroli R, Clement-Lacroix P, Morvan F, Lepescheux L, Nique F, Horne WC, Resche-Rigon M, Baron R 2006 Bone protection by estrens occurs through non-tissue-selective activation of the androgen receptor. J Clin Invest **116:**2500–2509.
36. Centrella M, McCarthy TL, Chang WZ, Labaree DC, Hochberg RB 2004 Estren (4-estren-3alpha,17beta-diol) is a prohormone that regulates both androgenic and estrogenic transcriptional effects through the androgen receptor. Mol Endocrinol **18:**1120–1130.
37. Krishnan V, Bullock HA, Yaden BC, Liu M, Barr RJ, Montrose-Rafizadeh C, Chen K, Dodge JA, Bryant HU 2005 The nongenotropic synthetic ligand 4-estren-3alpha17beta-diol is a high-affinity genotropic androgen receptor agonist. Mol Pharmacol **67:**744–748.
38. Moverare S, Venken K, Eriksson AL, Andersson N, Skrtic S, Wergedal J, Mohan S, Salmon P, Bouillon R, Gustafsson JA, Vanderschueren D, Ohlsson C 2003 Differential effects on bone of estrogen receptor alpha and androgen receptor activation in orchidectomized adult male mice. Proc Natl Acad Sci USA **100:**13573–13578.
39. Seeman E 1997 From density to structure: Growing up and growing old on the surfaces of bone. J Bone Miner Res **12:**509–521.
40. Seeman E 2001 Clinical review 137: Sexual dimorphism in skeletal size, density, and strength. J Clin Endocrinol Metab **86:**4576–4584.
41. Turner RT, Wakley GK, Hannon KS 1990 Differential effects of androgens on cortical bone histomorphometry in gonadectomized male and female rats. J Orthop Res **8:**612–617.
42. Vandenput L, Swinnen JV, Boonen S, Van Herck E, Erben RG, Bouillon R, Vanderschueren D 2004 Role of the androgen receptor in skeletal homeostasis: The androgen-resistant testicular feminized male mouse model. J Bone Miner Res **19:**1462–1470.
43. Kawano H, Sato T, Yamada T, Matsumoto T, Sekine K, Watanabe T, Nakamura T, Fukuda T, Yoshimura K, Yoshizawa T, Aihara K, Yamamoto Y, Nakamichi Y, Metzger D, Chambon P, Nakamura K, Kawaguchi H, Kato S 2003 Suppressive function of androgen receptor in bone resorption. Proc Natl Acad Sci USA **100:**9416–9421.
44. Venken K, De Gendt K, Boonen S, Ophoff J, Bouillon R, Swinnen JV, Verhoeven G, Vanderschueren D 2006 Relative impact of androgen and estrogen receptor activation in the effects of androgens on trabecular and cortical bone in growing male mice: A study in the androgen receptor knock-out mouse model. J Bone Miner Res **21:**576–585.
45. Lea C, Kendall N, Flanagan AM 1996 Casodex (a nonsteroidal antiandrogen) reduces cancellous, endosteal, and periosteal bone formation in estrogen-replete female rats. Calcif Tissue Int **58:**268–272.
46. Dagogo-Jack S, al-Ali N, Qurttom M 1997 Augmentation of bone mineral density in hirsute women. J Clin Endocrinol Metab **82:**2821–2825.
47. Windahl SH, Vidal O, Andersson G, Gustafsson JA, Ohlsson C 1999 Increased cortical bone mineral content but unchanged trabecular bone mineral density in female ERbeta(–/–) mice. J Clin Invest **104:**895–901.
48. Windahl SH, Hollberg K, Vidal O, Gustafsson JA, Ohlsson C, Andersson G 2001 Female estrogen receptor beta–/– mice are partially protected against age-related trabecular bone loss. J Bone Miner Res **16:**1388–1398.
49. Smith EP, Boyd J, Frank GR, Takahashi H, Cohen RM, Specker B, Williams TC, Lubahn DB, Korach KS 1994 Estrogen resistance caused by a mutation in the estrogen-receptor gene in a man. N Engl J Med **331:**1056–1061.
50. Gennari L, Nuti R, Bilezikian JP 2004 Aromatase activity and bone homeostasis in men. J Clin Endocrinol Metab **89:**5898–5907.
51. Bouillon R, Bex M, Vanderschueren D, Boonen S 2004 Estrogens are essential for male pubertal periosteal bone expansion. J Clin Endocrinol Metab **89:**6025–6029.
52. Bass S, Delmas PD, Pearce G, Hendrich E, Tabensky A, Seeman E 1999 The differing tempo of growth in bone size, mass, and density in girls is region-specific. J Clin Invest **104:**795–804.
53. Jarvinen TL, Kannus P, Sievanen H 2003 Estrogen and bone–a reproductive and locomotive perspective. J Bone Miner Res **18:**1921–1931.
54. Vanderschueren D, Venken K, Bouillon R 2004 Animal models for gender-based skeletal differences. In: Legato M (ed.) Principles of Gender-Specific Medicine. Elsevier Academic Press, San Diego, CA, USA, pp. 1043–1051.
55. Juul A 2001 The effects of oestrogens on linear bone growth. Hum Reprod Update **7:**303–313.
56. Behre HM, Kliesch S, Leifke E, Link TM, Nieschlag E 1997 Long-term effect of testosterone therapy on bone mineral density in hypogonadal men. J Clin Endocrinol Metab **82:**2386–2390.
57. Manolagas SC 2000 Birth and death of bone cells: Basic regulatory mechanisms and implications for the pathogenesis and treatment of osteoporosis. Endocr Rev **21:**115–137.
58. Riggs BL 2000 The mechanisms of estrogen regulation of bone resorption. J Clin Invest **106:**1203–1204.
59. Hughes DE, Dai A, Tiffee JC, Li HH, Mundy GR, Boyce BF 1996 Estrogen promotes apoptosis of murine osteoclasts mediated by TGF-β. Nat Med **2:**1132–1136.
60. Michael H, Harkonen PL, Vaananen HK, Hentunen TA 2005 Estrogen and testosterone use different cellular pathways to inhibit osteoclastogenesis and bone resorption. J Bone Miner Res **20:**2224–2232.
61. Nakamura T, Imai Y, Matsumoto T, Sato S, Takeuchi K, Igarashi K, Harada Y, Azuma Y, Krust A, Yamamoto Y, Nishina H, Takeda S, Takayanagi H, Metzger D, Kanno J, Takaoka K, Martin TJ, Chambon P, Kato S 2007 Estrogen prevents bone loss via estrogen receptor alpha and induction of Fas ligand in osteoclasts. Cell **130:**811–823.
62. Chen Q, Kaji H, Kanatani M, Sugimoto T, Chihara K 2004 Testosterone increases osteoprotegerin mRNA expression in mouse osteoblast cells. Horm Metab Res **36:**674–678.
63. Bellido T, Jilka RL, Boyce BF, Girasole G, Broxmeyer H, Dalrymple SA, Murray R, Manolagas SC 1995 Regulation of interleukin-6, osteoclastogenesis, and bone mass by androgens. The role of the androgen receptor. J Clin Invest **95:**2886–2895.
64. Huber DM, Bendixen AC, Pathrose P, Srivastava S, Dienger KM, Shevde NK, Pike JW 2001 Androgens suppress osteoclast formation induced by RANKL and macrophage-colony stimulating factor. Endocrinology **142:**3800–3808.
65. Riggs BL, Parfitt AM 2005 Drugs used to treat osteoporosis: The critical need for a uniform nomenclature based on their action on bone remodeling. J Bone Miner Res **20:**177–184.
66. Seeman E 2002 Pathogenesis of bone fragility in women and men. Lancet **359:**1841–1850.
67. Seeman E 2003 Periosteal bone formation—a neglected determinant of bone strength. N Engl J Med **349:**320–323.

© 2008 American Society for Bone and Mineral Research

68. Vandenput L, Boonen S, Van Herck E, Swinnen JV, Bouillon R, Vanderschueren D 2002 Evidence from the aged orchidectomized male rat model that 17beta-estradiol is a more effective bone-sparing and anabolic agent than 5alpha-dihydrotestosterone. J Bone Miner Res **17:**2080–2086.
69. Venken K, Boonen S, Van Herck E, Vandenput L, Kumar N, Sitruk-Ware R, Sundaram K, Bouillon R, Vanderschueren D 2005 Bone and muscle protective potential of the prostate-sparing synthetic androgen 7alpha-methyl-19-nortestosterone: Evidence from the aged orchidectomized male rat model. Bone **36:**663–670.
70. Slemenda CW, Longcope C, Zhou L, Hui SL, Peacock M, Johnston CC 1997 Sex steroids and bone mass in older men. Positive associations with serum estrogens and negative associations with androgens. J Clin Invest **100:**1755–1759.
71. Greendale G, Edelstein S, Barrett-Connor E 1997 Endogenous sex steroids and bone mineral density in older women and men. The Rancho Bernardo Study. J Bone Miner Res **12:**1833–1841.
72. Van Pottelbergh I, Goemaere S, Kaufman JM 2003 Bioavailable estradiol and an aromatase gene polymorphism are determinants of bone mineral density changes in men over 70 years of age. J Clin Endocrinol Metab **88:**3075–3081.
73. Riggs BL, Khosla S, Melton LJ III 1998 A unitary model for involutional osteoporosis: Estrogen deficiency causes both type I and type II osteoporosis in postmenopausal women and contributes to bone loss in aging men. J Bone Miner Res **13:**763–773.
74. Eastell R 2007 Aromatase inhibitors and bone. J Steroid Biochem Mol Biol **106:**157–161.
75. Snyder PJ, Peachey H, Hannoush P, Berlin JA, Loh L, Holmes JH, Dlewati A, Staley J, Santanna J, Kapoor SC, Attie MF, Haddad JG Jr, Strom BL 1999 Effect of testosterone treatment on bone mineral density in men over 65 years of age. J Clin Endocrinol Metab **84:**1966–1972.
76. Kenny AM, Prestwood KM, Marcello KM, Raisz LG 2000 Determinants of bone density in healthy older men with low testosterone levels. J Gerontol A Biol Sci Med Sci **55A:**M492–M497.
77. Rossouw JE, Anderson GL, Prentice RL, LaCroix AZ, Kooperberg C, Stefanick ML, Jackson RD, Beresford SA, Howard BV, Johnson KC, Kotchen JM, Ockene J 2002 Risks and benefits of estrogen plus progestin in healthy postmenopausal women: Principal results From the Women's Health Initiative randomized controlled trial. JAMA **288:**321–333.
78. Anderson GL, Limacher M, Assaf AR, Bassford T, Beresford SA, Black H, Bonds D, Brunner R, Brzyski R, Caan B, Chlebowski R, Curb D, Gass M, Hays J, Heiss G, Hendrix S, Howard BV, Hsia J, Hubbell A, Jackson R, Johnson KC, Judd H, Kotchen JM, Kuller L, LaCroix AZ, Lane D, Langer RD, Lasser N, Lewis CE, Manson J, Margolis K, Ockene J, O'Sullivan MJ, Phillips L, Prentice RL, Ritenbaugh C, Robbins J, Rossouw JE, Sarto G, Stefanick ML, Van Horn L, Wactawski-Wende J, Wallace R, Wassertheil-Smoller S 2004 Effects of conjugated equine estrogen in postmenopausal women with hysterectomy: The Women's Health Initiative randomized controlled trial. JAMA **291:**1701–1712.
79. Smith CL, O'Malley BW 2004 Coregulator function: A key to understanding tissue specificity of selective receptor modulators. Endocr Rev **25:**45–71.
80. Anonymous 1998 Tamoxifen for early breast cancer: An overview of the randomised trials. Early Breast Cancer Trialists' Collaborative Group. Lancet **351:**1451–1467.
81. Johnell O, Kanis JA, Black DM, Balogh A, Poor G, Sarkar S, Zhou C, Pavo I 2004 Associations between baseline risk factors and vertebral fracture risk in the Multiple Outcomes of Raloxifene Evaluation (MORE) Study. J Bone Miner Res **19:**764–772.
82. Seeman E, Crans GG, Diez-Perez A, Pinette KV, Delmas PD 2006 Anti-vertebral fracture efficacy of raloxifene: A meta-analysis. Osteoporos Int **17:**313–316.
83. McDonnell DP, Clemm DL, Hermann T, Goldman ME, Pike JW 1995 Analysis of estrogen receptor function in vitro reveals three distinct classes of antiestrogens. Mol Endocrinol **9:**659–669.
84. Chen J, Kim J, Dalton JT 2005 Discovery and therapeutic promise of selective androgen receptor modulators. Mol Interv **5:**173–188.

Chapter 25. Parathyroid Hormone

Robert A. Nissenson[1] and Harald Jüppner[2]

[1]Endocrine Research Unit, VA Medical Center, Departments of Medicine and Physiology, University of California, San Francisco, California; [2]Endocrine Unit and Pediatric Nephrology Unit, Departments of Medicine and Pediatrics, Harvard Medical School, Massachusetts General Hospital, Boston, Massachusetts

INTRODUCTION

The parathyroid glands first appear during evolution with the movement of animals from an aquatic environment to a terrestrial environment deficient in calcium. Maintenance of adequate levels of plasma ionized calcium (1.0–1.3 mM) is required for normal neuromuscular function, bone mineralization, and many other physiological processes. Chief cells in the parathyroid gland secrete PTH in response to very small decrements in blood ionized calcium to maintain the normocalcemic state. As discussed later, PTH accomplishes this task by promoting bone resorption and releasing calcium from the skeletal reservoir; by inducing renal conservation of calcium and excretion of phosphate; and by indirectly enhancing intestinal calcium absorption by increasing the renal production of the active vitamin D metabolite $1,25(OH)_2$ vitamin D. Serum ionized calcium and $1,25(OH)_2$ vitamin D produce feedback inhibition of the secretion of PTH, whereas serum phosphate increases PTH secretion. The interplay between serum calcium, PTH, $1,25(OH)_2$ vitamin D, and phosphate permit serum ionized calcium levels to be maintained within very narrow limits over a wide range of dietary calcium intake. This chapter summarizes our current understanding of the biology of PTH secretion and action. A nice historical perspective on this field has recently been published.[1]

The authors state that they have no conflicts of interest.

THE PTH POLYPEPTIDE FAMILY

Mammalian PTH is a single-chain 84 amino acid polypeptide hormone that is expressed almost exclusively in the parathyroid gland, with lesser expression in the rodent hypothalamus and thymus. The appearance of PTH precedes that of the parathyroid gland in evolution, because two forms of PTH have been detected in teleosts that lack discrete parathyroid glands.[2,3] The physiological roles of PTH in fish have not yet been definitively established, although it has suggested based on sites of expression that PTH may play a role in the development of the teleost neural system, cartilage, and bone.[4] Teleost PTHs display significant amino acid homology to mammalian PTHs in the 1-34 sequence of the mature peptides (i.e., the region required for binding to and activating the G-protein–coupled PTH1 receptor; Fig. 1). Teleosts display widespread expression of a PTH1 receptor that presumably mediates the physiological actions of PTH.

© 2008 American Society for Bone and Mineral Research

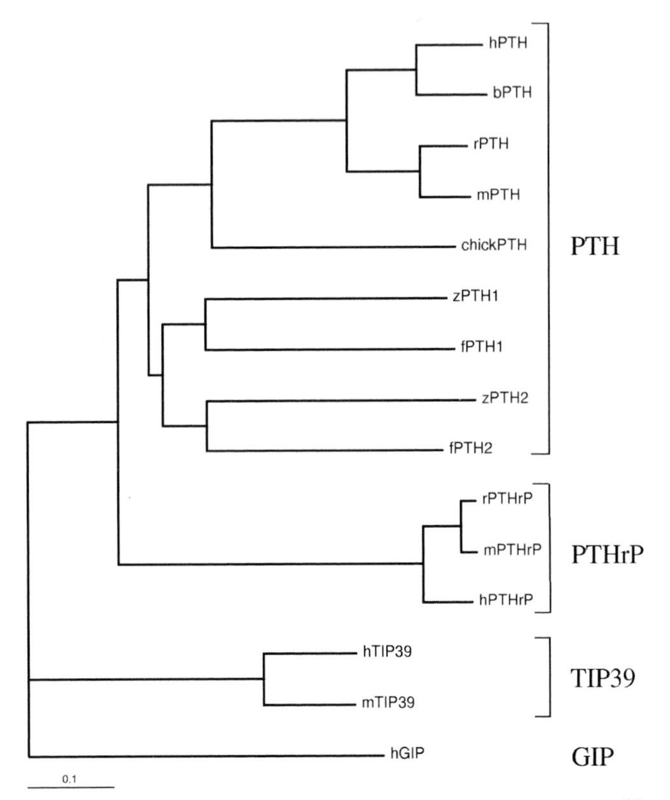

FIG. 1. Phylogenetic analysis of PTH and related polypeptides.[3] Copyright 2004, The Endocrine Society.

PTH-related peptide (PTHrP) displays sequence homology with PTH that is limited to the amino-terminal 1-34 region of both peptides.[5] PTHrP was originally identified as the humoral mediator of hypercalcemia of malignancy[5–7] and is now known to play a number of important physiological roles (e.g., control of endochondral bone development, smooth muscle tone, and morphogenesis of the mammary gland).[8] The *PTHrP* gene has structural similarity to that of PTH, and the genes are presumed to be derived from a common ancestral gene. This gene duplication event occurred at least five hundred million years ago, as teleosts are known to express homologs of mammalian PTHrP.[9] PTH and PTHrP also display homology to TIP39 (tuberoinfundibular peptide of 39 amino acids),[10] a factor that is expressed in the brain and testes and acts through the G-protein–coupled PTH2 receptor.[11] Recently, TIP39 expression in the testis has been shown to be essential for sperm development.[12] Homology amongst PTH, PTHrP, and TIP39 is reflected in the organization of the genes encoding these polypeptides (Fig. 2).

PTH SYNTHESIS AND SECRETION

There is a single mammalian PTH gene that in humans is present on the short arm of chromosome 11. The primary translation product is the precursor molecule prepro-PTH that includes a 25 amino acid pre sequence, a 6 amino acid pro sequence, and an 84 amino acid mature PTH sequence.[13] The pre sequence functions as a signal sequence that directs the nascent polypeptide to the machinery that transports it across the membrane of the endoplasmic reticulum (ER), where the pre sequence is cleaved. The function of the pro sequence is not as clearly defined, but it seems to be required for efficient ER transport of the polypeptide and may play a role in subsequent events such as protein folding.[14] The pro sequence seems to be cleaved by the protease furin, producing the mature 1-84 PTH polypeptide. Once produced and packaged into secretory vesicles with the parathyroid chief cell, PTH(1-84) is subject to alternative fates. The mature hormone can be secreted through a classical exocytotic mechanism or it may be cleaved by calcium-sensitive proteases present within secretory vesicles, resulting in the production and secretion of fragments of PTH(1-84) that lack the amino-terminal domain and are thus inactive with respect to responses mediated by the PTH1 receptor.[15] Cleavage of circulating PTH(1-84) to carboxyl-fragments can also occur in peripheral tissues such as liver and kidney.[16] Historically, cleavage of PTH(1-84) has been viewed as a mechanism for biological inactivation of the hormone, but there is suggestive evidence that carboxyl-terminal fragments of PTH may display unique biological properties.[17]

REGULATION OF PTH SECRETION BY EXTRACELLULAR CALCIUM

The major physiological function of the parathyroid glands is to act as a "calciostat," sensing the prevailing blood ionized calcium level and adjusting the secretion of PTH accordingly (Fig. 3). The relationship between ionized calcium and PTH secretion is a steep sigmoidal one, allowing significant changes in PTH secretion in response to very small changes in plasma ionized calcium. The midpoint of this curve ("set-point") is a reflection of the sensitivity of the parathyroid gland to suppression by extracellular calcium.

Alteration in plasma ionized calcium affects the secretion of PTH(1-84) by multiple mechanisms. Short-term increases in extracellular ionized calcium produce increased levels of intracellular free calcium in the parathyroid cell, resulting in activation of calcium-sensitive proteases in secretory vesicles. As a result, there is increased cleavage of PTH(1-84) into carboxyl-

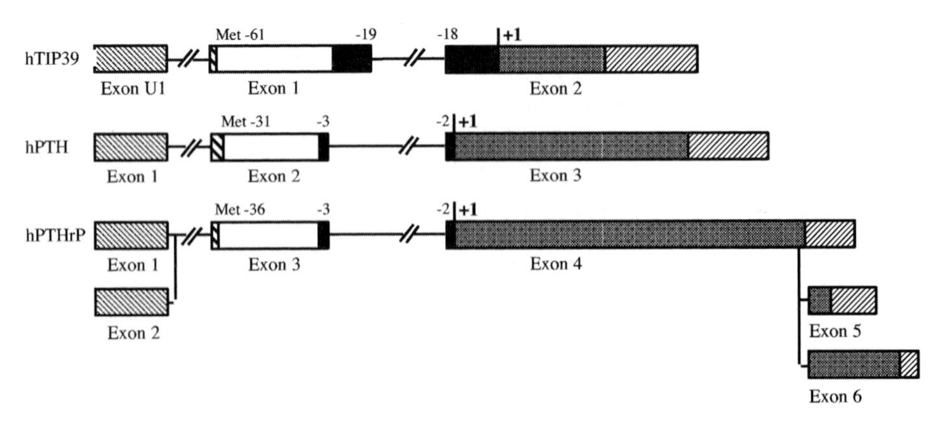

FIG. 2. Diagrammatic structures of the genes encoding human TIP39, PTH, and PTHrP. Boxed areas represent exons (the 5′ end of exon U1 in the *TIP39* gene is not known). White boxes denote pre sequences, black boxes are pro sequences, gray stippled boxes are mature protein sequences, and striped boxes are noncoding regions. The small striped boxes preceding the white boxes denote untranslated exonic sequences. The positions of the initiator methionines are also indicated. +1 represents the relative position of the beginning of the secreted protein.[42] Copyright 2002, The Endocrine Society.

© 2008 American Society for Bone and Mineral Research

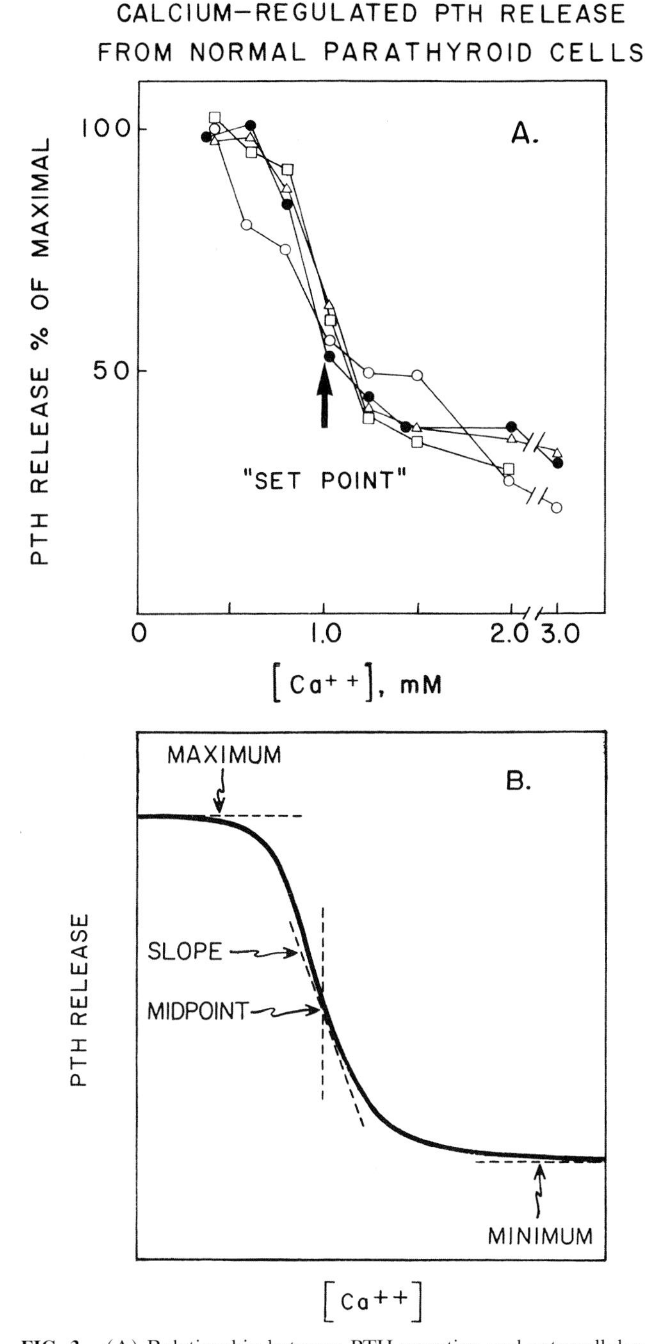

FIG. 3. (A) Relationship between PTH secretion and extracellular calcium in normal human parathyroid cells. Dispersed parathyroid cells were incubated with indicated levels of calcium, and PTH was determined by radioimmunoassay. [Reproduced with permission from Brown EM 1980 Set-point for calcium: Its role in normal and abnormal parathyroid secretion. In: Cohn DV, Talmage RV, Matthews JL (eds.) Hormonal Control of Calcium Metabolism, Proceedings of the Seventh International Conference on Calcium Regulating Hormones. Excerpta Medica, Amsterdam, The Netherlands, pp. 35–43]. (B) The four parameters describing the inverse sigmoidal relationship between the extracellular calcium concentration and PTH release in vivo and in vitro: maximal secretory rate; slope of the curve at the midpoint; midpoint or set-point of the curve (the level of calcium producing 50% of the maximal decrease in secretory rate; minimal secretory rate.[43] Copyright 1983, The Endocrine Society.

terminal fragments. Increased extracellular calcium also inhibits the release of stored PTH from secretory vesicles, although the molecular details of this regulation are not well defined. Long-term changes in plasma ionized calcium (e.g., chronic dietary calcium deficiency) result in alteration in the expression of the PTH gene and in the number of PTH-secreting parathyroid cells.

Extracellular calcium is sensed at the surface of the parathyroid cell, and this leads to suppression of PTH secretion. Calcium sensing is initiated by the binding of calcium to a "sensing" receptor (CaR) present at high levels on the plasma membrane of parathyroid cells.[18] Unlike intracellular calcium-binding proteins, which have an affinity for free calcium in the nanomolar range (consistent with intracellular levels of free calcium), CaR is presumed to bind free calcium with an affinity in the millimolar range. CaR is a member of the G-protein–coupled receptor superfamily. It contains calcium binding elements in its extracellular domain and signaling determinants in its cytoplasmic regions. Calcium binding to CaR triggers activation of the G-proteins Gq and (to a lesser extent) Gi, resulting in stimulation of phospholipase C and inhibition of adenylyl cyclase, respectively.[19,20] This results in an increase in intracellular calcium and a decrease in cyclic AMP levels in parathyroid cells. By mechanisms that are not fully defined, activation of these signaling pathways suppresses the synthesis and secretion of PTH. When blood ionized calcium falls, there is less signaling by the CaRs on the parathyroid cell and PTH secretion consequently increases. The essential role of the CaR can best be seen in humans bearing loss-of-function mutations in the *CaR* gene. In the heterozygous state, such mutations result in familial hypocalciuric hypercalcemia (FHH), characterized by inappropriately high levels of PTH secretion in the face of hypercalcemia.[21,22] These individuals are quantitatively resistant to the suppressive effect of calcium of PTH secretion because of the reduced number of parathyroid CaRs. In the homozygous state, patients display a severe increase in PTH secretion with life-threatening hypercalcemia (neonatal severe primary hyperparathyroidism). Mice with homozygous and heterozygous disruption of the *CaR* gene display similar phenotypes.[23] Of interest, deletion in mice of the function of Gq and Gi, the G-proteins associated with CaR, results in neonatal severe primary hyperparathyroidism, confirming the role of these G-proteins in CaR signaling.[24] Point mutations in the CaR that produce constitutive signaling have also been described, and these are associated with autosomal dominant hypocalcemia in humans.[25,26]

Limited information is available on the mechanisms by which CaR signaling suppresses PTH gene expression. Some studies suggest that transcription of the PTH gene is negatively regulated by calcium.[27] More recently, attention has been focused on the post-transcriptional effect of calcium to reduce the stability of PTH mRNA. Treatment of parathyroid cells with high calcium results in the binding of a protein factor(s) to the untranslated regions of the *PTH* gene and to destabilization of PTH mRNA.[28] This effect seems to be mediated by CaR-dependent increases in the level of intracellular free calcium.[29] The identity of the relevant factor(s) remains to be established.

Under normal physiological conditions, there is minimal proliferation of parathyroid cells. However, chronic hypocalcemia elicits an increase in both the size and number of parathyroid cells.[30] Normalization of serum calcium suppresses the hyperplasia through signaling by CaR, although the molecular details have not yet been worked out.

It is of considerable interest that CaR is expressed in a number of tissues outside of the parathyroid gland including kidney, C-cells of the thyroid gland, gut, bone, cartilage, and many

© 2008 American Society for Bone and Mineral Research

others.[25] An important function of CaR in the kidney is to signal the inhibition of calcium reabsorption in the cortical thick ascending limb. This allows plasma calcium to regulate renal calcium excretion independently of PTH, and a reduction in this signaling contributes to the hypercalcemia and hypocalciuria seen in patients with FHH. Although the physiological role of the CaR in other peripheral tissues is not well understood, recent studies with conditional knockout models suggests that the expression of CaR in chondrocytes and osteoblasts is essential for normal endochondral bone development (W. Chang, C. Tu, D. Bikle, and D. Shoback, personal communication).

Several pharmacological agents that interact with CaR have been developed, and these are effective in altering the ability of the CaR to signal.[26,31] So-called calcimimetic drugs bind to transmembrane regions in the CaR and increase the receptor's sensitivity to extracellular calcium. This results in an increase in receptor signaling and thus suppression of PTH secretion. Calcimimetic drugs have clinical utility in the medical management of hyperparathyroidism. Calcilytic drugs act as pharmacological antagonists of the CaR, reducing the receptor's sensitivity to calcium thus increasing the secretion of PTH.[32]

REGULATION OF PTH SECRETION BY 1,25$(OH)_2$ VITAMIN D

For many years, it has been known that vitamin D deficiency is linked to excessive production of PTH. This is because of reduced suppression of PTH secretion by extracellular calcium and by 1,25$(OH)_2$ vitamin D. This frequently occurs in the setting of chronic renal failure where 1,25$(OH)_2$ vitamin D production is diminished, serum calcium is reduced, and phosphate levels are increased. As described below, hyperphosphatemia has an independent effect to increase the secretion of PTH.

The suppression of PTH secretion by 1,25$(OH)_2$ vitamin D results from the inhibition of transcription of the PTH gene.[33] This seems to involve 1,25$(OH)_2$ vitamin D–induced binding of the vitamin D receptor to negative regulatory elements in the PTH gene promoter[34] and 1,25$(OH)_2$ vitamin D–induced association of the vitamin D receptor with a transcriptional repressor.[35] 1,25$(OH)_2$ vitamin D and calcium act coordinately to suppress expression of the PTH gene and to inhibit parathyroid cell proliferation.

REGULATION OF PTH SECRETION BY PLASMA PHOSPHATE, α-KLOTHO, AND FIBROBLAST GROWTH FACTOR 23

It has long been known that hyperphosphatemia (as in chronic renal failure) is associated with parathyroid hyperplasia and hyperparathyroidism. This effect of hyperphosphatemia is in part caused by the binding of plasma phosphate to free calcium, which lowers blood ionized calcium, thus stimulating PTH synthesis, secretion, and parathyroid cell number.[36] However, serum phosphate also seems to directly affect the parathyroid gland, increasing PTH synthesis by promoting the stability of PTH mRNA.[28]

In response to hyperphosphatemia, fibroblast growth factor 23 (FGF23) is secreted by osteocytes/osteoblasts and acts on the kidney to inhibit renal phosphate reabsorption.[37] This requires the binding of FGF23 to cognate renal FGF receptors, an interaction that requires a co-factor, namely the transmembrane protein α-Klotho.[38] Interestingly, α-Klotho is also expressed in parathyroid cells, where it may promote PTH secretion through the maintenance of plasma membrane Na^+/K^+ ATPase activity.[39] It has been suggested that α-Klotho serves as a mediator of the effects of plasma phosphate on PTH secretion,[40] although direct evidence is not currently available. Further complicating matters is the observation that FGF23 acts directly on the parathyroid gland to inhibit PTH secretion.[41] The integrated action of phosphate, FGF23, and α-Klotho on the parathyroid gland almost certainly constitutes a new and important physiological mechanism for the control of PTH secretion.

REFERENCES

1. Potts JT 2005 Parathyroid hormone: Past and present. J Endocrinol **187:**311–325.
2. Hogan BM, Danks JA, Layton JE, Hall NE, Heath JK, Lieschke GJ 2005 Duplicate zebrafish pth genes are expressed along the lateral line and in the central nervous system during embryogenesis. Endocrinology **146:**547–551.
3. Gensure RC, Ponugoti B, Gunes Y, Papasani MR, Lanske B, Bastepe M, Rubin DA, Juppner H 2004 Identification and characterization of two parathyroid hormone-like molecules in zebrafish. Endocrinology **145:**1634–1639.
4. Guerreiro PM, Renfro JL, Power DM, Canario AV 2007 The parathyroid hormone family of peptides: Structure, tissue distribution, regulation, and potential functional roles in calcium and phosphate balance in fish. Am J Physiol Regul Integr Comp Physiol **292:**R679–R696.
5. Strewler GJ, Stern PH, Jacobs JW, Eveloff J, Klein RF, Leung SC, Rosenblatt M, Nissenson RA 1987 Parathyroid hormonelike protein from human renal carcinoma cells. Structural and functional homology with parathyroid hormone. J Clin Invest **80:**1803–1807.
6. Suva LJ, Winslow GA, Wettenhall RE, Hammonds RG, Moseley JM, Diefenbach-Jagger H, Rodda CP, Kemp BE, Rodriguez H, Chen EY, Hudson PJ, Martin TJ, Wood WI 1987 A parathyroid hormone-related protein implicated in malignant hypercalcemia: Cloning and expression. Science **237:**893–896.
7. Broadus AE, Mangin M, Ikeda K, Insogna KL, Weir EC, Burtis WJ, Stewart AF 1988 Humoral hypercalcemia of cancer. Identification of a novel parathyroid hormone-like peptide. N Engl J Med **319:**556–563.
8. Gensure RC, Gardella TJ, Juppner H 2005 Parathyroid hormone and parathyroid hormone-related peptide, and their receptors. Biochem Biophys Res Commun **328:**666–678.
9. Abbink W, Flik G 2007 Parathyroid hormone-related protein in teleost fish. Gen Comp Endocrinol **152:**243–251.
10. Usdin TB, Hoare SR, Wang T, Mezey E, Kowalak JA 1999 TIP39: A new neuropeptide and PTH2-receptor agonist from hypothalamus. Nat Neurosci **2:**941–943.
11. Usdin TB, Dobolyi A, Ueda H, Palkovits M 2003 Emerging functions for tuberoinfundibular peptide of 39 residues. Trends Endocrinol Metab **14:**14–19.
12. Usdin TB, Paciga M, Riordan T, Kuo J, Parmelee A, Petukova G, Camerini-Otero RD, Mezey E 2008 Tuberoinfundibular peptide of 39 residues is required for germ cell development. Endocrinology **149:**4292–4300.
13. Kemper B, Habener JF, Mulligan RC, Potts JT Jr, Rich A 1974 Pre-proparathyroid hormone: A direct translation product of parathyroid messenger RNA. Proc Natl Acad Sci USA **71:**3731–3735.
14. Wiren KM, Ivashkiv L, Ma P, Freeman MW, Potts JT Jr, Kronenberg HM 1989 Mutations in signal sequence cleavage domain of preproparathyroid hormone alter protein translocation, signal sequence cleavage, and membrane-binding properties. Mol Endocrinol **3:**240–250.
15. Habener JF, Kemper B, Potts JT Jr 1975 Calcium-dependent intracellular degradation of parathyroid hormone: A possible mechanism for the regulation of hormone stores. Endocrinology **97:**431–441.
16. D'Amour P 2006 Circulating PTH molecular forms: What we know and what we don't. Kidney Int Suppl 102:S29–S33.
17. Murray TM, Rao LG, Divieti P, Bringhurst FR 2005 Parathyroid hormone secretion and action: Evidence for discrete receptors for the carboxyl-terminal region and related biological actions of carboxyl-terminal ligands. Endocr Rev **26:**78–113.
18. Brown EM, Gamba G, Riccardi D, Lombardi M, Butters R, Kifor O, Sun A, Hediger MA, Lytton J, Hebert SC 1993 Cloning and characterization of an extracellular Ca(2+)-sensing receptor from bovine parathyroid. Nature **366:**575–580.

© 2008 American Society for Bone and Mineral Research

19. Brown EM, MacLeod RJ 2001 Extracellular calcium sensing and extracellular calcium signaling. Physiol Rev **81**:239–297.
20. Chang W, Chen TH, Pratt S, Shoback D 2000 Amino acids in the second and third intracellular loops of the parathyroid Ca2+-sensing receptor mediate efficient coupling to phospholipase C. J Biol Chem **275**:19955–19963.
21. Pearce SH, Williamson C, Kifor O, Bai M, Coulthard MG, Davies M, Lewis-Barned N, McCredie D, Powell H, Kendall-Taylor P, Brown EM, Thakker RV 1996 A familial syndrome of hypocalcemia with hypercalciuria due to mutations in the calcium-sensing receptor. N Engl J Med **335**:1115–1122.
22. Pollak MR, Seidman CE, Brown EM 1996 Three inherited disorders of calcium sensing. Medicine (Baltimore) **75**:115–123.
23. Ho C, Conner DA, Pollak MR, Ladd DJ, Kifor O, Warren HB, Brown EM, Seidman JG, Seidman CE 1995 A mouse model of human familial hypocalciuric hypercalcemia and neonatal severe hyperparathyroidism. Nat Genet **11**:389–394.
24. Wettschureck N, Lee E, Libutti SK, Offermanns S, Robey PG, Spiegel AM 2007 Parathyroid-specific double knockout of Gq and G11 alpha-subunits leads to a phenotype resembling germline knockout of the extracellular Ca2+ -sensing receptor. Mol Endocrinol **21**:274–280.
25. Egbuna OI, Brown EM 2008 Hypercalcaemic and hypocalcaemic conditions due to calcium-sensing receptor mutations. Best Pract Res Clin Rheumatol **22**:129–148.
26. Hu J, Spiegel AM 2007 Structure and function of the human calcium-sensing receptor: Insights from natural and engineered mutations and allosteric modulators. J Cell Mol Med **11**:908–922.
27. Russell J, Sherwood LM 1987 The effects of 1,25-dihydroxyvitamin D3 and high calcium on transcription of the pre-proparathyroid hormone gene are direct. Trans Assoc Am Physicians **100**:256–262.
28. Moallem E, Kilav R, Silver J, Naveh-Many T 1998 RNA-Protein binding and post-transcriptional regulation of parathyroid hormone gene expression by calcium and phosphate. J Biol Chem **273**:5253–5259.
29. Ritter CS, Pande S, Krits I, Slatopolsky E, Brown AJ 2008 Destabilization of parathyroid hormone mRNA by extracellular Ca2+ and the calcimimetic R-568 in parathyroid cells: Role of cytosolic Ca and requirement for gene transcription. J Mol Endocrinol **40**:13–21.
30. Cozzolino M, Brancaccio D, Gallieni M, Galassi A, Slatopolsky E, Dusso A 2005 Pathogenesis of parathyroid hyperplasia in renal failure. J Nephrol **18**:5–8.
31. Steddon SJ, Cunningham J 2005 Calcimimetics and calcilytics—fooling the calcium receptor. Lancet **365**:2237–2239.
32. Nemeth EF, Delmar EG, Heaton WL, Miller MA, Lambert LD, Conklin RL, Gowen M, Gleason JG, Bhatnagar PK, Fox J 2001 Calcilytic compounds: Potent and selective Ca2+ receptor antagonists that stimulate secretion of parathyroid hormone. J Pharmacol Exp Ther **299**:323–331.
33. Silver J, Russell J, Sherwood LM 1985 Regulation by vitamin D metabolites of messenger ribonucleic acid for preproparathyroid hormone in isolated bovine parathyroid cells. Proc Natl Acad Sci USA **82**:4270–4273.
34. Okazaki T, Igarashi T, Kronenberg HM 1988 5′-flanking region of the parathyroid hormone gene mediates negative regulation by 1,25-(OH)2 vitamin D3. J Biol Chem **263**:2203–2208.
35. Kim MS, Fujiki R, Murayama A, Kitagawa H, Yamaoka K, Yamamoto Y, Mihara M, Takeyama K, Kato S 2007 1Alpha,25(OH)2D3-induced transrepression by vitamin D receptor through E-box-type elements in the human parathyroid hormone gene promoter. Mol Endocrinol **21**:334–342.
36. Naveh-Many T, Rahamimov R, Livni N, Silver J 1995 Parathyroid cell proliferation in normal and chronic renal failure rats. The effects of calcium, phosphate, and vitamin D. J Clin Invest **96**:1786–1793.
37. Fukumoto S 2008 Physiological regulation and disorders of phosphate metabolism–pivotal role of fibroblast growth factor 23. Intern Med **47**:337–343.
38. Urakawa I, Yamazaki Y, Shimada T, Iijima K, Hasegawa H, Okawa K, Fujita T, Fukumoto S, Yamashita T 2006 Klotho converts canonical FGF receptor into a specific receptor for FGF23. Nature **444**:770–774.
39. Imura A, Tsuji Y, Murata M, Maeda R, Kubota K, Iwano A, Obuse C, Togashi K, Tominaga M, Kita N, Tomiyama K, Iijima J, Nabeshima Y, Fujioka M, Asato R, Tanaka S, Kojima K, Ito J, Nozaki K, Hashimoto N, Ito T, Nishio T, Uchiyama T, Fujimori T, Nabeshima Y 2007 alpha-Klotho as a regulator of calcium homeostasis. Science **316**:1615–1618.
40. Brownstein CA, Adler F, Nelson-Williams C, Iijima J, Li P, Imura A, Nabeshima Y, Reyes-Mugica M, Carpenter TO, Lifton RP 2008 A translocation causing increased alpha-klotho level results in hypophosphatemic rickets and hyperparathyroidism. Proc Natl Acad Sci USA **105**:3455–3460.
41. Ben-Dov IZ, Galitzer H, Lavi-Moshayoff V, Goetz R, Kuro-o M, Mohammadi M, Sirkis R, Naveh-Many T, Silver J 2007 The parathyroid is a target organ for FGF23 in rats. J Clin Invest **117**:4003–4008.
42. John MR, Arai M, Rubin DA, Jonsson KB, Juppner H 2002 Identification and characterization of the murine and human gene encoding the tuberoinfundibular peptide of 39 residues. Endocrinology **143**:1047–1057.
43. Brown EM 1983 Four-parameter model of the sigmoidal relationship between parathyroid hormone release and extracellular calcium concentration in normal and abnormal parathyroid tissue. J Clin Endocrinol Metab **56**:572–581.

Chapter 26. Parathyroid Hormone–Related Protein

John J. Wysolmerski

Section of Endocrinology and Metabolism, Department of Internal Medicine, Yale University School of Medicine, New Haven, Connecticut

INTRODUCTION

In a 1941 case report in the *New England Journal of Medicine*, Fuller Albright first postulated that tumors associated with hypercalcemia might elaborate a PTH-like humor.[1] Work in the 1980s and 1990s subsequently led to the biochemical characterization of a specific syndrome of humoral hypercalcemia of malignancy (HHM) and the fulfillment of Albright's predictions by the isolation of PTH-related protein (PTHrP) and the characterization of its gene.[2–6] We now understand that PTHrP and PTH are related molecules that can both stimulate the same type I PTH/PTHrP receptor (PTH1R).[7–9] PTHrP usually serves a local autocrine, paracrine, or intracrine role and normally does not circulate. However, in patients with HHM, PTHrP does reach the circulation and mimics the systemic actions of PTH. Another chapter will discuss malignancy-associated hypercalcemia in detail. This chapter will outline the normal physiology of PTHrP.

The author states that he has no conflicts of interest.

Key words: parathyroid hormone–related protein, bone, nuclear transport, cartilage, mammary gland, vascular smooth muscle

© 2008 American Society for Bone and Mineral Research

PTHrP GENE AND THE *PTH/PTHrP* GENE FAMILY

PTHrP is encoded by a single-copy gene located on the short arm of chromosome 12. The human gene consists of eight exons and at least three promoters.[3,6,8] Alternative splicing at the 3′ end of the gene gives rise to three different classes of mRNA coding for specific translation products of 139, 141, or 173 amino acids. The physiological significance of these different PTHrP transcripts remains unclear, and in rodents and lower vertebrates such as birds and fish, the gene has a much simpler structure.[7,8,10] PTHrP mRNA has been found in almost every organ at some time during its development or functioning. Many different hormones and growth factors regulate the transcription and/or stability of PTHrP mRNA. As with PTH, the calcium-sensing receptor (CaR) has been found to regulate PTHrP gene expression in many cells.[11,12] Another common theme is the observation that mRNA levels are induced by mechanical deformation.[8] The reader is referred to other reviews for a comprehensive discussion of the sites and regulation of PTHrP expression.[7–9]

The *PTHrP* and *PTH* genes share structural elements and sequence homology, suggesting that they are related genes.[3,5,6,8] The exon/intron organization of that portion of both genes encoding the prepro sequences and the initial portion of the mature peptides is identical. Furthermore, there is high sequence homology at the amino-terminal portion of both genes such that the peptides share 8 of the first 13 amino acids and a high degree of predicted secondary structure over the next 21 amino acids. These common sequences allow both peptides to bind and activate the same PTH1R, which ultimately explains the ability of PTHrP to cause hypercalcemia in HHM.[4] The above-mentioned structural similarities together with the location of the two genes on related chromosomes in the human genome (short arm of chromosome 11 for PTH; short arm of chromosome 12 for PTHrP) indicate that the two genes arose from a common ancestor through a process of gene duplication. The recent demonstration of two *PTH* genes and two *PTHrP* genes in several species of fish suggests that these genes split from their common origin before the radiation of the fishes during evolution.[10] Furthermore, fish contain a separate gene, *PTH-L*, that is intermediate between the *PTHrP* and *PTH* genes in its characteristics and may represent their original ancestor.[10] Thus, PTHrP is a member of an ancient family of PTH peptides that seems to be larger and more diverse in lower vertebrates than in mammals.

PTHrP IS A POLYHORMONE

Similar to the *proopiomelanocortin* (*POMC*) gene, the primary translation product of PTHrP can undergo a variety of post-translational processing events to give rise to an overlapping series of biological peptides.[8] The prepro sequences from amino acids –36 to 1 direct the nascent protein into the endoplasmic reticulum so that it can enter the secretory pathway, after which they are removed. The primary sequence of PTHP contains clusters of basic amino acids that act as recognition sequences for processing enzymes responsible for generating different PTHrP peptides in a cell type–specific manner. The details of PTHrP processing and the biological significance of the different PTHrP peptides are not entirely clear, but several specific secreted forms of PTHrP have been defined. First, PTHrP(1-36) is secreted from several cell types.[8,13] In addition, longer forms of amino-terminal containing PTHrP are secreted from keratinocytes and mammary epithelial cells and circulate in patients with cancer and during lactation.[14–16] The amino terminus is necessary for interaction with the PTH1R. The secretion of midregion peptides including amino acids 38–94, 38–95, and 38–101 has also been described.[8,17] The biology of these specific secretory forms is unclear, but the midregion of PTHrP stimulates placental calcium transport and modulates renal bicarbonate handling, and this portion of the molecule contains nuclear localization signals (see below).[18–20] Finally, C-terminal fragments consisting of amino acids 107–138 and 109–138 have been described. These peptides have been suggested to inhibit osteoclast function and stimulate osteoblast proliferation.[8,19]

PTHrP RECEPTORS

The amino terminus of PTHrP binds to and activates the PTH1R, a prototypical, seven transmembrane-containing, G protein–coupled receptor (GPCR), which is a member of class B of the large family of GPCRs.[7,21] As with the *PTHrP* gene, *PTH1R* is one of several related PTH receptor genes that most likely arose through gene duplication events. Although PTH can bind to other receptors in this family, PTHrP can only interact with the PTH1R. This receptor has been described to couple to both $G_{\alpha s}$ and $G_{\alpha q11}$ and signal through the cAMP and protein kinase A pathway and through the generation of inositol phosphates, diacylglycerol, and intracellular calcium transients.[7,21] The vast majority of studies in vitro suggest that this receptor binds PTHrP and PTH with equal affinity and that both peptides generate identical biological effects. This is also true when amino-terminal fragments of PTH and PTHrP are infused into animals.[7,8] However, the human PTH1R may respond differently to PTH and PTHrP. Human subjects subjected to continuous infusion of the two peptides for 72 h were found to be become hypercalcemic, with lower doses of PTH(1-34) than PTHrP(1-36).[22] In these same studies, PTHrP was also much weaker than PTH at stimulating the renal 1-α-hydroxylase enzyme producing 1,25-dihydroxyvitamin D. This may be explained by physical differences in the binding of the two peptides to different conformational states of the receptor, so that the duration of cAMP production is shorter for PTHrP(1-36) than for PTH(1-34).[23]

The existence of biological actions for midregion and C-terminal peptides of PTHrP implies the possibility of additional receptors for these forms of PTHrP. However, no such receptors have been identified to date.

NUCLEAR PTHrP

Immunohistochemical studies have localized PTHrP to the nucleus of many different cell types.[19,24] There are several potential mechanisms by which PTHrP can avoid secretion and remain in the cell.[19,24] Once in the cytoplasm, PTHrP seems to shuttle into and out of the nucleus in a regulated fashion. This is dependent on a specific nuclear localization sequence (NLS) located between amino acids 84 and 93 and requires binding to microtubules and a specific shuttle protein known as importin β1, which allows PTHrP to transit the nuclear pore.[19,24] Nuclear export is facilitated by a related shuttle protein known as CRM1 and likely requires a different recognition sequence in the C-terminal region of the peptide.[19] The regulation of nuclear trafficking of PTHrP is not fully understood, but phosphorylation at Thr^{85} by the cell cycle–regulated, cyclin-dependent kinase, $p34^{cdc2}$, seems to regulate nuclear import in a cell cycle–dependent fashion.[19] The function of nuclear PTHrP remains obscure, but it has been described to bind RNA and in some cells PTHrP localizes to the nucleolus. This has led to the suggestion that it may be involved in regulating RNA trafficking, ribosomal dynamics, and/or protein translation.[19,24] In cell lines, nuclear PTHrP has been implicated in the regulation of proliferation and/or

© 2008 American Society for Bone and Mineral Research

apoptosis. Whatever the exact function(s) of nuclear PTHrP, it is likely to be important, because replacement of the endogenous mouse PTHrP gene with a mutant version encoding a protein that cannot enter the nucleus is lethal.[25]

PHYSIOLOGICAL FUNCTIONS OF PTHrP

PTHrP has been found in at least some cells of almost all organs. Like many growth factors or cytokines, a variety of effects have been ascribed to PTHrP. The reader is referred to more comprehensive reviews for a complete discussion of these findings.[8,9] What follows is a brief outline of areas where PTHrP has been rigorously documented to have physiological effects in intact organisms.

Skeleton

Biological functions of PTHrP have been discovered through the study of genetically altered mice, the first of which involved disruption of the *PTHrP* gene by homologous recombination.[26] Lack of PTHrP results in alterations in chondrocyte differentiation in the growth plates of long bones and in costal cartilage that lead to short-limbed dwarfism and a shield chest that interferes with breathing and causes perinatal death. Disrupting the *PTH1R* gene generates a similar phenotype, and overexpressing PTHrP or a constitutively active PTH1R within growth plate chondrocytes in transgenic mice produces the opposite effect.[27–29] These animal models have documented that amino-terminal PTHrP acts through the PTH1R to coordinate the rate of chondrocyte differentiation to maintain the orderly growth of long bones during development.[30] As shown in Fig. 1, the growth plate consists of columns of proliferating and differentiating chondrocytes that progressively enlarge to prehypertrophic and hypertrophic chondrocytes, which secrete matrix and undergo apoptosis to form a calcified scaffold that is remodeled into bone in the primary spongiosum. PTHrP is secreted primarily by immature chondrocytes at the top of the columns in response to another molecule known as Indian Hedgehog (Ihh) produced by differentiating hypertrophic chondrocytes. PTHrP, in turn, acts on its receptor located on proliferating and prehypertrophic cells to slow their rate of differentiation into hypertrophic cells. In this manner, Ihh and PTHrP act in a local negative feedback loop to regulate the rate of chondrocyte differentiation (Fig. 1).[30]

PTHrP is also produced in other cartilaginous sites such as the perichondrium that surrounds the costal cartilage and the subarticular chondrocyte population immediately subjacent to the hyaline cartilage lining the joint space.[31,32] In both of these sites, PTHrP seems to prevent hypertrophic differentiation of chondrocytes and the inappropriate encroachment of bone into these structures.[31,32] It is the failure of this function that leads to the shield chest noted in the PTHrP knockout mice.

In addition to its functions in cartilage, PTHrP has important anabolic functions in bone. Heterozygous PTHrP-null mice are normal at birth but develop trabecular osteopenia with age.[33] In addition, selective deletion of the *PTHrP* gene from osteoblasts results in a decreased bone mass, reduced bone formation and mineral apposition, and a reduction in the formation and survival of osteoblasts.[34] These data suggest that PTHrP acts as an important local anabolic factor in the skeleton. Osteoblast cell lines in culture produce PTHrP and its production in vitro can be stimulated by mechanical deformation, raising the possibility that it may be involved in mediating the anabolic response to skeletal loading. However, despite the clear phenotype in the osteoblast-specific PTHrP-knockout mice, there is disagreement over the PTHrP-expressing population(s) of osteoblasts in the skeleton and even if the gene is normally expressed within these cells.[31,32] Nonetheless, the osteopenia in these animal models suggests the intriguing possibility that intermittent PTH treatment invokes an anabolic response in bone by mimicking the natural functions of local PTHrP in the skeleton.

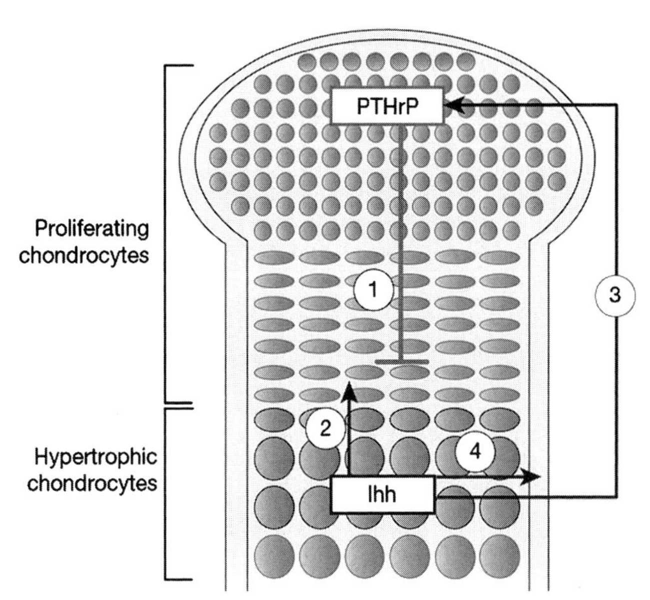

FIG. 1. PTHrP and Ihh act as part of a negative feedback loop regulating chondrocyte proliferation and differentiation. The chondrocyte differentiation program proceeds from undifferentiated chondrocytes at the end of the bone, to proliferative chondrocytes within the columns, and to prehypertrophic and terminally differentiated hypertrophic chondrocytes nearest the primary spongiosum. PTHrP is made by undifferentiated and proliferating chondrocytes at the ends of long bones. It acts through PTH1R on proliferating and prehypertrophic chondrocytes to delay their differentiation, maintain their proliferation and delay the production of Ihh, which is made by hypertrophic cells (1). Ihh, in turn, increases the rate of chondrocyte proliferation (2) and stimulates the production of PTHrP at the ends of the bone (3). Ihh also acts on perichondrial cells to generate osteoblasts of the bone collar (4). (Reprinted with permission from Macmillan Publishers: Kronenberg HM 2003 Developmental regulation of the growth plate. Nature **423**:332–336.)

Mammary Gland

Not long after its discovery, PTHrP mRNA was found to be expressed in the lactating breast, and PTHrP protein was measured in high concentrations in milk.[35,36] It is now known that PTHrP has important functions during breast development, is involved in regulating systemic calcium metabolism during lactation, and contributes to the pathophysiology of breast cancer.

Like other epidermal appendages, the mammary gland initially forms as a bud-like invagination of epidermal cells that grow down into a developing fatty stroma as a branching tube that becomes the mammary duct system. These processes are regulated by a series of sequential and reciprocal interactions between the epithelial cells in the bud and ducts and adjacent mesenchymal cells in the stroma.[37] In mice, as soon as the mammary bud begins to form, epithelial cells produce PTHrP, which interacts with the PTH1R expressed on surrounding mesenchymal cells. This interaction is necessary for proper differentiation of the dense mammary mesenchyme that surrounds the embryonic mammary bud so that these mesenchy-

© 2008 American Society for Bone and Mineral Research

mal cells can maintain the mammary fate of the epithelial cells, initiate outgrowth of the duct system, and stimulate the formation of the specialized epidermis that comprises the nipple.[38] PTHrP- or PTH1R-knockout mice lack mammary glands because loss of PTHrP signaling interrupts the vital cross-talk between epithelial and mesenchymal cells (Fig. 2). The formation of the breast in human fetuses is similar to the formation of the fetal mammary gland in mice, and PTHrP is necessary for the formation of breast epithelium in humans as well.[39]

During puberty, the mammary duct system expands to fill out the surrounding fatty stroma to form the mature virgin breast. This phase of mammary development is governed by systemic hormones, which induce the formation of terminal end buds (TEBs) at the tips of the prepubertal ducts. These bulbous structures serve as the sites of active proliferation, differentiation, and stromal remodeling as the ducts lengthen and branch. During puberty, PTHrP seems to interact with stromal cells surrounding the TEBs to regulate the growth of the mammary ducts in response to estrogen.[40,41]

PTHrP is also made by breast epithelial cells during lactation, and large quantities are secreted into milk.[16,35] Although its function in milk is unclear, PTHrP is secreted from the lactating breast into the circulation, where it participates in the regulation of systemic calcium metabolism. Milk production requires a great deal of calcium, an important source of which is the maternal skeleton. Elevated rates of bone resorption and rapid bone loss are well documented in both nursing women and rodents.[42] During lactation, elevated levels of PTHrP correlate with bone loss in humans, and circulating levels of PTHrP correlate directly with rates of bone resorption and inversely with bone mass in mice.[15,43] In addition, mammary-specific disruption of the *PTHrP* gene during lactation reduces circulating PTHrP levels, lowers bone turnover, and preserves bone mass, showing that the lactating breast secretes PTHrP into the circulation to increase bone resorption.[16] The lactating breast also expresses the CaR, which signals to suppress PTHrP secretion in response to increases in calcium delivery to the breast.[12] These interactions define a classical endocrine negative feedback loop, whereby mammary cells secrete PTHrP to mobilize calcium from the bone. Calcium, in turn, feeds back to inhibit further PTHrP secretion from the breast. Therefore, during lactation, the breast and bone engage in a conversation, which leads to the mobilization of skeletal calcium stores to ensure a steady supply of calcium for milk production. Interestingly, fish PTHrP fulfills a similar function to mobilize calcium stored in scales to be used during egg production.[10] Thus, this reproductive function of PTHrP is ancient, and the actions of PTHrP during lactation may represent one of the principal evolutionary pressures that resulted in PTHrP and PTH retaining the use of the same PTH1R.

Placenta

During pregnancy, calcium must be actively transported across the placenta from mother to fetus. Furthermore, the responsible placental pump maintains a higher calcium concentration in the fetus compared with the mother, so that calcium must be transported against a gradient.[42] In PTHrP$^{-/-}$ mice, this gradient is lost, and PTHrP-deficient fetuses are relatively hypocalcemic, suggesting that fetal PTHrP is important in mediating placental calcium transport from the mother.[20] The source of the PTHrP is likely the placenta itself and placental production of PTHrP has been shown to be regulated by the CaR.[44,45] Interestingly, experiments in sheep and mice have shown that it is midregion PTHrP, not the amino-terminal portion, that is responsible for placental calcium transport.[18,20]

Smooth Muscle and the Cardiovascular System

PTHrP is expressed in many different smooth muscle cell beds.[8] In these sites, mechanical deformation seems to increase the expression of PTHrP, which, in turn, acts in an autocrine or paracrine fashion through the PTH1R to relax the muscle cell or structure that has been stretched.[8,9,46–48] In accommodative structures such as the stomach, bladder, or uterus, this feedback loop may be important in allowing for gradual filling. In the vasculature, PTHrP is induced by vasoconstrictive agents and stretch itself and acts as a vasodilator to

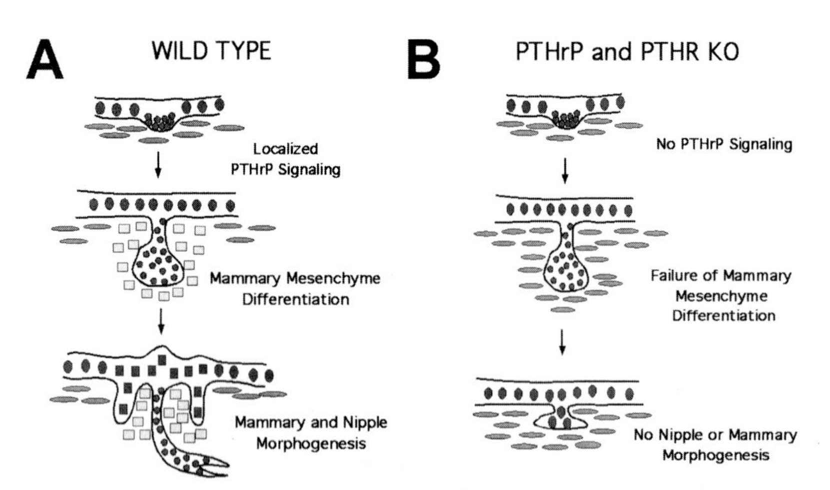

FIG. 2. PTHrP regulates mesenchymal cell fate during embryonic mammary development. (A) During normal mammary development, PTHrP is secreted by epithelial cells within the forming mammary bud (dark circles) and interacts with the immature dermal mesenchyme (gray ovals) to induce formation of the dense mammary mesenchyme (light squares). These cells, in response to PTHrP, maintain the fate of the mammary epithelial cells, initiate branching morphogenesis, and induce the formation of the specialized nipple skin (dark squares). (B) In PTHrP- or PTH1R-knockout embryos, the mammary bud forms, but the mammary mesenchyme does not. As a result, the mammary epithelial cells revert to an epidermal fate (dark ovals), morphogenesis fails, and the nipple never forms. (Adapted with permission from Company of Biologists: Foley J, Dann P, Hong J, Cosgrove J, Dreyer BE, Rimm D, Dunbar, ME, Philbrick WM, Wysolmerski JJ 2001 Parathyroid hormone-related protein maintains mammary epithelial fate and triggers nipple skin differentiation during embryonic breast development. Development **128**:513–525.)

© 2008 American Society for Bone and Mineral Research

resistance vessels. Given these actions, PTHrP may act as a local modulator of blood flow.[49]

PTHrP regulates the proliferation of vascular smooth muscle cells. Secreted amino-terminal PTHrP can inhibit the proliferation of vascular smooth muscle cells by acting through the PTH1R on the cell surface. However, the midregion and C-terminal portions of PTHrP, acting in the nucleus, stimulate the proliferation of these cells. This latter effect results from destabilization of a cell cycle regulatory protein known as $p27^{kip1}$, which leads to progression through the G_1/S checkpoint.[50,51] This pathway seems to be active during development, because the rate of proliferation of smooth muscle cells in the aorta of $PTHrP^{-/-}$ embryos is reduced.[51] Furthermore, PTHrP expression is upregulated by vascular damage after balloon angioplasty and in atherosclerotic lesions in rodents and in humans. Several studies have suggested that PTHrP plays an important role in the response of these cells to injury and may contribute to the pathophysiologic development of a neo-intima after angioplasty.[52,53]

PTHrP has been found in cardiomyocytes and co-localizes with atrial natriuretic peptide in granules within atrial cells in the rat heart.[54] In certain genetic backgrounds, the absence of the PTH1R causes widespread cardiomyocyte death during midgestation in developing mice.[55] However, this phenotype is not seen in $PTHrP^{-/-}$ mice. Thus, it is not clear if this lesion reflects the actions of a different ligand for the receptor or alternatively an imbalance between the actions of secreted and intracellular forms of PTHrP as a result of loss of the receptor. In isolated hearts, PTHrP has both positive ionotropic and chronotropic effects and may affect coronary blood flow.[56]

Teeth

Developing teeth become surrounded by alveolar bone but must maintain a cavity or crypt that is free from bone to allow for proper morphogenesis. After teeth are formed, they must erupt through the roof of the dental crypt to emerge into the oral cavity. The process of tooth eruption relies on geographically uncoupled bone turnover in which osteoclasts form over the crown of the tooth to resorb the overlying bone and osteoblasts at the base of the tooth propel it upward out of the crypt. In the absence of PTHrP, teeth develop but they do not erupt. Normally, just before the onset of eruption, PTHrP is produced by epithelial, stellate reticulum cells and signals to stromal, dental follicle cells to drive the formation of osteoclasts above the crypt. In the absence of PTHrP, these osteoclasts do not appear, eruption fails to occur, and the surrounding bone encroaches leading to impaction of the tooth.[57–59]

Pancreatic Islets

PTHrP is expressed by all four neuroendocrine cell types within the pancreatic islets.[60] In β cells, it is stored within secretory granules and is co-released with insulin.[61] Pancreatic islets express the PTH1R and β cells respond to PTHrP by activating phospholipase C and increasing intracellular calcium.[62] Overexpression of PTHrP in β cells leads to an increased β-cell mass, hyperinsulinemia, and hypoglycemia, because of the combination of increased β-cell proliferation, increased insulin production, and inhibition of β-cell apoptosis.[62,63] It is not clear how these actions of PTHrP relate to normal islet physiology, especially because there are no obvious defects in islet development in the $PTHrP^{-/-}$ mice. Nonetheless, the death of these mice at birth has precluded any examination of potential roles of PTHrP in the physiology of islets.

Central and Peripheral Nervous Systems

PTHrP and the PTH1R are both widely expressed within the brain, including regions of the cortex, the cerebellum, and the hippocampus, hypothalamus, and pituitary.[64,65] In cultured hippocampal neurons, PTHrP has been shown to be secreted in response to calcium influx through L-type calcium channels on depolarization. In turn, PTHrP can act on the PTH1R on these same neurons to dampen L-type channel activity, giving rise to the idea that PTHrP acts in an autocrine/paracrine short feedback loop to protect neurons from damage caused by prolonged or repeated depolarization, so-called "excitotoxicity."[66] The $PTHrP^{-/-}$ mice have been rescued from neonatal death by the reintroduction of transgenic PTHrP into chondrocytes.[67] Consistent with the role of PTHrP in protecting from excitotoxicity, these "PTHrP-rescue" mice were shown to be much more sensitive to kainate-induced seizures.[66]

In addition to neurons, PTHrP has also been shown to be expressed in glia and astrocytes.[68–70] Interestingly, the *PTHrP* gene is expressed in fetal and malignant glial cells but not in mature glia in the adult brain.[70] However, its expression can be induced in reactive glia in an injury model in rats.[69] Consistent with the enhanced expression of PTHrP in less differentiated glial cells, in humans, the level of PTHrP expression in glial tumors was shown to correlate with more aggressive behavior of the tumor and was predictive of a poor outcome for the patient.[71] In a similar fashion, a recent report showed upregulation of PTHrP in dedifferentiated Schwann cells after crush injury in the peripheral nervous system (sciatic nerve).[72] PTHrP was shown to inhibit the differentiation of these cells, suggesting that it contributed to maintaining the dedifferentiated state necessary for nerve regeneration.

CONCLUSIONS

PTHrP was discovered as the cause of the clinical syndrome of HHM. It is evolutionarily and functionally related to PTH and shares the same PTH1R. The common use of this receptor, in turn, allows PTHrP to act as a hormone mimicking the actions of PTH during reproduction, a function preserved from fish to mammals. Although the conservation of these relationships through evolution allows PTHrP to cause hypercalcemia when it is secreted into the circulation by tumors, we have come to understand that PTHrP is generally a locally produced and locally acting growth factor that participates in normal development and physiology at many diverse sites. The power of mouse genetics has provided the tools to begin to catalogue the biological functions of PTHrP. However, much remains to be accomplished to understand its functions fully.

ACKNOWLEDGMENTS

This work was supported by Grants DK077565, DK55501, and DK069542 from the NIH. The author thanks Drs. Arthur Broadus, William Philbrick, Carolyn Macica, Joshua Van-Houten, Andrew Karaplis, and Rupangi Vasavada for valuable conversations during the preparation of this chapter; and Drs. Arthur Broadus, Robert Nissenson, and Gordon Strewler, the authors of previous versions of this chapter in earlier editions of the Primer, for help in galvanizing my efforts for this edition.

REFERENCES

1. Mallory TB 1941 Case records of the Massachusetts General Hospital. Case #27461. N Engl J Med **225:**789–791.
2. Burtis WJ, Wu T, Bunch C, Wysolmerski JJ, Insogna KL, Weir EC, Broadus AE, Stewart AF 1987 Identification of a novel

© 2008 American Society for Bone and Mineral Research

17,000-dalton parathyroid hormone-like adenylate cyclase-stimulating protein from a tumor associated with humoral hypercalcemia of malignancy. J Biol Chem **262:**7151–7156.
3. Mangin M, Webb AC, Dreyer BE, Posillico JT, Ikeda K, Weir EC, Stewart AF, Bander NH, Milstone LM, Barton DE, Francke U, Broadus AE 1988 An identification of a cDNA encoding a parathyroid hormone-like peptide from a human tumor asociated with humoral hypercalcemia of malignancy. Proc Natl Acad Sci USA **85:**597–601.
4. Stewart AF, Horst RL, Deftos LJ, Cadman EC, Lang R, Broadus AE 1980 Biochemical evaluation of patients with cancer-associated hypercalcemia: Evidence for humoral and nonhumoral groups. N Engl J Med **303:**1377–1383.
5. Strewler GJ, Stern PH, Jacobs JW, Evelott J, Klein RF, Leung SC, Rosenblatt M, Nissenson RA 1987 Parathyroid hormone-like protein from human renal carcinoma cells. Structural and functional homology with parathyroid hormone. J Clin Invest **80:**1803–1807.
6. Suva LJ, Winslow GA, Wettenhall RE, Hammonds RG, Moseley JM, Diefenbach-Jagger H, Rodda CP, Kemp BE, Rodriguez H, Chen EY 1987 A parathyroid hormone-related protein implicated in malignant hypercalcemia: Cloning and expression. Science **237:**893–896.
7. Gensure RC, Gardella TJ, Juppner H 2005 Parathyroid hormone and parathyroid hormone-related peptide, and their receptors. Biochem Biophys Res Commun **328:**666–678.
8. Philbrick WM, Wysolmerski JJ, Galbraith S, Holt EH, Orloff JJ, Yang KH, Vasavada R, Weir EC, Broadus AE, Stewart AF 1996 Defining the roles of parathyroid hormone-related protein in normal physiology. Physiol Rev **76:**127–173.
9. Strewler GJ 2000 The physiology of parathyroid hormone-related protein. N Engl J Med **342:**177–185.
10. Guerreiro PM, Renfro JL, Power DM, Canario AV 2007 The parathyroid hormone family of peptides: Structure, tissue distribution, regulation, and potential functional roles in calcium and phosphate balance in fish. Am J Physiol Regul Integr Comp Physiol **292:**R679–R696.
11. Chattopadhyay N 2006 Effects of calcium-sensing receptor on the secretion of parathyroid hormone-related peptide and its impact on humoral hypercalcemia of malignancy. Am J Physiol Endocrinol Metab **290:**E761–E770.
12. VanHouten J, Dann P, McGeoch G, Brown EM, Krapcho K, Neville M, Wysolmerski JJ 2004 The calcium-sensing receptor regulates mammary gland parathyroid hormone-related protein production and calcium transport. J Clin Invest **113:**598–608.
13. Orloff JJ, Reddy D, de Papp AE, Yang KH, Soifer NE, Stewart AF 1994 Parathyroid hormone-related protein as a prohormone: Posttranslational processing and receptor interactions. Endocr Rev **15:**40–60.
14. Burtis WJ, Brady TG, Orloff JJ, Ersbak JB, Warrell RP Jr, Olson BR, Wu TL, Mitnick ME, Broadus AE, Stewart AF 1990 Immunochemical characterization of circulating parathyroid hormone-related protein in patients with humoral hypercalcemia of cancer. N Engl J Med **322:**1106–1112.
15. Sowers MF, Hollis BW, Shapiro B, Randolph J, Janney CA, Zhang D, Schork A, Crutchfield M, Stanczyk F, Russell-Aulet M 1996 Elevated parathyroid hormone-related peptide associated with lactation and bone density loss. JAMA **276:**549–554.
16. VanHouten JN, Dann P, Stewart AF, Watson CJ, Pollak M, Karaplis AC, Wysolmerski JJ 2003 Mammary-specific deletion of parathyroid hormone-related protein preserves bone mass during lactation. J Clin Invest **112:**1429–1436.
17. Soifer NE, Dee KE, Insogna KL, Burtis WJ, Matovcik LM, Wu TL, Milstone LM, Broadus AE, Philbrick WM, Stewart AF 1992 Parathyroid hormone-related protein. Evidence for secretion of a novel mid-region fragment by three different cell types. J Biol Chem **267:**18236–18243.
18. Care AD, Abbas SL, Pickard DW, Barri M, Drinkhill M, Findley JBC, White IR, Caple IW 1990 Stimulation of ovine placental transport of calcium and magnesium by mid-molecule fragments of human parathyroid hormone-related protein. Exp Physiol **75:**605–608.
19. Jans DA, Thomas RJ, Gillespie MT 2003 Parathyroid hormone-related protein (PTHrP): A nucleocytoplasmic shuttling protein with distinct paracrine and intracrine roles. Vitam Horm **66:**345–384.
20. Kovacs CS, Lanske B, Hunzelman JL, Guo J, Karaplis AC, Kronenberg HM 1996 Parathyroid hormone-related peptide (PTHrP) regulates fetal-placental calcium transport through a receptor distinct from the PTH/PTHrP receptor. Proc Natl Acad Sci USA **93:**15233–15238.
21. Jupnner H, Abou-Samra AB, Freeman M, Kong XF, Schipani E, Richards J, Kolakowski LF Jr, Hock J, Potts JT Jr, Kronenberg HM, Segre GV 1991 A G protein-linked receptor for parathyroid hormone and parathyroid hormone-related peptide. Science **254:**1024–1026.
22. Horwitz MJ, Tedesco MB, Sereika SM, Syed MA, Garcia-Ocana A, Bisello A, Hollis BW, Rosen CJ, Wysolmerski JJ, Dann P, Gundberg C, Stewart AF 2005 Continuous PTH and PTHrP infusion causes suppression of bone formation and discordant effects on 1,25(OH)2 vitamin D. J Bone Miner Res **20:**1792–1803.
23. Dean T, Vilardaga JP, Potts JT Jr, Gardella TJ 2008 Altered selectivity of parathyroid hormone (PTH) and PTH-related protein for distinct conformations of the PTH/PTHrP receptor. Mol Endocrinol **22:**156–166.
24. Fiaschi-Taesch NM, Stewart AF 2003 Minireview: Parathyroid hormone-related protein as an intracrine factor–trafficking mechanisms and functional consequences. Endocrinology **144:**407–411.
25. Miao D, Su H, He B, Gao J, Xia Q, Goltzman D, Karaplis AC 2005 Deletion of the mid- and carboxyl regions of PTHrP produces growth retardation and early senescence in mice. J Bone Miner Res **20:**S14.
26. Karaplis AC, Luz A, Glowacki J, Bronson RT, Tybulewicz VL, Kronenberg HM, Mulligan RC 1994 Lethal skeletal dysplasia from targeted disruption of the parathyroid hormone-related peptide gene. Genes Dev **8:**277–289.
27. Lanske B, Karaplis AC, Lee K, Luz A, Vortkamp A, Pirro A, Karperien M, Defize LH, Ho C, Mulligan RC, Abou-Samra A-B, Jueppner H, Segre GV, Kronenberg HM 1996 PTH/PTHrP receptor in early development and Indian hedgehog-regulated bone growth. Science **273:**663–666.
28. Schipani E, Lanske B, Hunzelman JL, Luz A, Kovacs CS, Lee K, Pirro A, Kronenberg HM, Jueppner H 1997 Targeted expression of constitutively active receptors for parathyroid hormone and parathyroid hormone-related peptide. Proc Natl Acad Sci USA **94:**13689–13694.
29. Weir EC, Philbrick WM, Amling M, Niff LA, Baron R, Broadus AE 1996 Targeted overexpression of parathyroid hormone-related peptide in chondrodysplasia and delayed endochondrial bone formation. Proc Natl Acad Sci USA **93:**10240–10245.
30. Kronenberg HM 2006 PTHrP and skeletal development. Ann N Y Acad Sci **1068:**1–13.
31. Chen X, Macica C, Nasiri A, Judex S, Broadus AE 2007 Mechanical regulation of PTHrP expression in entheses. Bone **41:**752–759.
32. Chen X, Macica CM, Dreyer BE, Hammond VE, Hens JR, Philbrick WM, Broadus AE 2006 Initial characterization of PTH-related protein gene-driven lacZ expression in the mouse. J Bone Miner Res **21:**113–123.
33. Amizuka N, Karaplis AC, Henderson JE, Warshawsky H, Lipman ML, Matsuki Y, Ejiri S, Tanaka M, Izumi N, Ozawa H, Goltzman D 1996 Haploinsufficiency of parathyroid hormone-related peptide (PTHrP) results in abnormal postnatal bone development. Dev Biol **175:**166–176.
34. Miao D, He B, Jiang Y, Kobayashi T, Soroceanu MA, Zhao J, Su H, Tong X, Amizuka N, Gupta A, Genant HK, Kronenberg HM, Goltzman D, Karaplis AC 2005 Osteoblast-derived PTHrP is a potent endogenous bone anabolic agent that modifies the therapeutic efficacy of administered PTH 1-34. J Clin Invest **115:**2402–2411.
35. Budayr AA, Halloran BR, King JC, Diep D, Nissenson RA 1989 High levels of a parathyroid hormone-like protein in milk. Proc Natl Acad Sci USA **86:**7183–7185
36. Thiede MA, Rodan GA 1988 Expression of a calcium-mobilizing parathyroid hormone-like peptide in lactating mammary tissue. Science **242:**278–280.
37. Robinson GW 2007 Cooperation of signalling pathways in embryonic mammary gland development. Nat Rev Genet **8:**963–972.
38. Hens JR, Wysolmerski JJ 2005 Key stages of mammary gland development: Molecular mechanisms involved in the formation of the embryonic mammary gland. Breast Cancer Res **7:**220–224.
39. Wysolmerski JJ, Cormier S, Philbrick WM, Dann P, Zhang JP, Roume J, Delezoide AL, Silve C 2001 Absence of functional type 1 parathyroid hormone (PTH)/PTH-related protein recpetors in humans is associated with abnormal breast development and tooth impaction. J Clin Endocrinol Metab **86:**1788–1794.
40. Dunbar ME, Dann P, Brown CW, Van Houton J, Dreyer B, Phil-

© 2008 American Society for Bone and Mineral Research

brick WP, Wysolmerski JJ 2001 Temporally regulated overexpression of parathyroid hormone-related protein in the mammary gland reveals distinct fetal and pubertal phenotypes. J Endocrinol **171**:403–416.
41. Dunbar ME, Young P, Zhang JP, McCaughern-Carucci JF, Lanske B, Orloff JJ, Karaplis AC, Cunha GR, Wysolmerski JJ 1998 Stromal cells are critical targets in the regulation of mammary ductal morphogenesis by parathyroid hormone-related protein (PTHrP). Dev Biol **203**:75–89.
42. Kovacs CS 2001 Calcium and bone metabolism in pregnancy and lactation. J Clin Endocrinol Metab **86**:2344–2348.
43. VanHouten JN, Wysolmerski JJ 2003 Low estrogen and high parathyroid hormone-related peptide levels contribute to accelerated bone resorption and bone loss in lactating mice. Endocrinology **144**:5521–5529.
44. Hellman P, Ridefelt P, Juhlin C, Akerstrom G, Rastad J, Gylfe E 1992 Parathyroid-like regulation of parathyroid-hormone-related protein release and cytoplasmic calcium in cytotrophoblast cells of human placenta. Arch Biochem Biophys **293**:174–180.
45. Kovacs CS, Ho C, Seidman CE, Seidman JG, Kronenberg HM 1996 Parathyroid calcium sensing receptor regulates fetal blood calcium and fetal-maternal calcium gradient independently of the maternal calcium levels. J Bone Miner Res **22**:S121.
46. Thiede MA, Daifotis AG, Weir EC, Brines ML, Burtis WJ, Ikede, Dreyer BE, Garfield RE, Broadus AE 1990 Intrauterine occupancy controls expression of the parathyroid hormone-related peptide gene in pre-term rat myometrium. Proc Natl Acad Sci USA **87**:6969–6973
47. Thiede MA, Harm SC, McKee RL, Grasser WA, Duong LT, Leach RM Jr 1991 Expression of the parathyroid hormone-related protein gene in the avian oviduct: Potential role as a local modulator of vascular smooth muscle tension and shell gland motility during the egg-laying cycle. Endocrinology **129**:1958–1966.
48. Yamamoto M, Harm SC, Grasser WA, Thiede MA 1992 Parathyroid hormone-related protein in the rat urinary bladder: A smooth muscle relaxant produced locally in response to mechanical stretch. Proc Natl Acad Sci USA **89**:5326–5330.
49. Massfelder T, Helwig JJ 1999 Parathyroid hormone-related protein in cardiovascular development and blood pressure regulation. Endocrinology **140**:1507–1510.
50. Fiaschi-Taesch N, Sicari BM, Ubriani K, Bigatel T, Takane KK, Cozar-Castellano I, Bisello A, Law B, Stewart AF 2006 Cellular mechanism through which parathyroid hormone-related protein induces proliferation in arterial smooth muscle cells: Definition of an arterial smooth muscle PTHrP/p27kip1 pathway. Circ Res **99**:933–942.
51. Massfelder T, Dann P, Wu TL, Vasavada R, Helwig JJ, Stewart AF 1997 Opposing mitogenic and anti-mitogenic actions of parathyroid hormone-related protein in vascular smooth muscle cells: A critical role for nuclear targeting. Proc Natl Acad Sci USA **94**:13630–13635.
52. Fiaschi-Taesch N, Takane KK, Masters S, Lopez-Talavera JC, Stewart AF 2004 Parathyroid-hormone-related protein as a regulator of pRb and the cell cycle in arterial smooth muscle. Circulation **110**:177–185.
53. Ishikawa M, Akishita M, Kozaki K, Toba K, Namiki A, Yamaguchi T, Orimo H, Ouchi Y 2000 Expression of parathyroid hormone-related protein in human and experimental atherosclerotic lesions: Functional role in arterial intimal thickening. Atherosclerosis **152**:97–105.
54. Deftos LJ, Burton DW, Brandt DW 1993 Parathyroid hormone-like protein is a secretory product of atrial myocytes. J Clin Invest **92**:727–735.
55. Qian J, Colbert MC, Witte D, Kuan CY, Gruenstein E, Osinska H, Lanske B, Kronenberg HM, Clemens TL 2003 Midgestational lethality in mice lacking the parathyroid hormone (PTH)/PTH-related peptide receptor is associated with abrupt cardiomyocyte death. Endocrinology **144**:1053–1061.
56. Halapas A, Tenta R, Pantos C, Cokkinos DV, Koutsilieris M 2003 Parathyroid hormone-related peptide and cardiovascular system. In Vivo **17**:425–432.
57. Boabaid F, Berry JE, Koh AJ, Somerman MJ, McCauley LK 2004 The role of parathyroid hormone-related protein in the regulation of osteoclastogenesis by cementoblasts. J Periodontol **75**:1247–1254.
58. Calvi LM, Shin HI, Knight MC, Weber JM, Young MF, Giovannetti A, Schipani E 2004 Constitutively active PTH/PTHrP receptor in odontoblasts alters odontoblast and ameloblast function and maturation. Mech Dev **121**:397–408.
59. Philbrick WM, Dreyer BE, Nakchbandi IA, Karaplis AC 1998 Parathyroid hormone-related protein is required for tooth eruption. Proc Natl Acad Sci USA **95**:11846–11851.
60. Asa SL, Henderson J, Goltzman D, Drucker DJ 1990 Parathyroid hormone-like peptide in normal and neoplastic human endocrine tissues. J Clin Endocrinol Metab **71**:1112–1118.
61. Plawner LL, Philbrick WM, Burtis WJ, Broadus AE, Stewart AF 1995 Cell type-specific secretion of parathyroid hormone-related protein via the regulated versus the constitutive secretory pathway. J Biol Chem **270**:14078–14084.
62. Vasavada RC, Wang L, Fujinaka Y, Takane KK, Rosa TC, Mellado-Gil JM, Friedman PA, Garcia-Ocana A 2007 Protein kinase C-zeta activation markedly enhances beta-cell proliferation: An essential role in growth factor mediated beta-cell mitogenesis. Diabetes **56**:2732–2743.
63. Vasavada RC, Cavaliere C, D'Ercole AJ, Dann P, Burtis WJ, Madlener AL, Zawalich K, Zawalich W, Philbrick W, Stewart AF 1996 Overexpression of parathyroid hormone-related protein in the pancreatic islets of transgenic mice causes islet hyperplasia, hyperinsulinemia, and hypoglycemia. J Biol Chem **271**:1200–1208.
64. Weaver DR, Deeds JD, Lee K, Segre GV 1995 Localization of parathyroid hormone-related peptide (PTHrP) and PTH/PTHrP receptor mRNAs in rat brain. Brain Res Mol Brain Res **28**:296–310.
65. Weir EC, Brines ML, Ikeda K, Burtis WJ, Broadus AE, Robbins RJ 1990 Parathyroid hormone-related peptide gene is expressed in the mammalian central nervous system. Proc Natl Acad Sci USA **87**:108–112.
66. Chatterjee O, Nakchbandi IA, Philbrick WM, Dreyer BE, Zhang JP, Kaczmarek LK, Brines ML, Broadus AE 2002 Endogenous parathyroid hormone-related protein functions as a neuroprotective agent. Brain Res **930**:58–66.
67. Wysolmerski JJ, Philbrick WM, Dunbar ME, Lanske B, Kronenberg HM, Karaplis AC, Broadus AE 1998 rescue of the parathyroid hormone-related protein knockout mouse demonstrates that parathyroid hormone-related protein is essential for mammary gland development. Development **125**:1285–1294.
68. Chattopadhyay N, Evliyaoglu C, Heese O, Carroll R, Sanders J, Black P, Brown EM 2000 Regulation of secretion of PTHrP by Ca(2+)-sensing receptor in human astrocytes, astrocytomas, and meningiomas. Am J Physiol Cell Physiol **279**:C691–C699.
69. Funk JL, Trout CR, Wei H, Stafford G, Reichlin S 2001 Parathyroid hormone-related protein (PTHrP) induction in reactive astrocytes following brain injury: A possible mediator of CNS inflammation. Brain Res **915**:195–209.
70. Shankar PP, Wei H, Davee SM, Funk JL 2000 Parathyroid hormone-related protein is expressed by transformed and fetal human astrocytes and inhibits cell proliferation. Brain Res **868**:230–240.
71. Pardo FS, Lien WW, Fox HS, Efird JT, Aguilera JA, Burton DW, Deftos LJ 2004 Parathyroid hormone-related protein expression is correlated with clinical course in patients with glial tumors. Cancer **101**:2622–2628.
72. Macica CM, Liang G, Lankford KL, Broadus AE 2006 Induction of parathyroid hormone-related peptide following peripheral nerve injury: Role as a modulator of Schwann cell phenotype. Glia **53**:637–648.

© 2008 American Society for Bone and Mineral Research

Chapter 27. Ca^{2+}-Sensing Receptor

Edward M. Brown

Division of Endocrinology, Diabetes and Hypertension, Brigham and Women's Hospital and Harvard Medical School, Boston, Massachusetts

INTRODUCTION

Complex terrestrial organisms, including humans, maintain a virtually constant level of their extracellular ionized calcium (Ca^{2+}_o) concentration, with a normal range of 1.1–1.3 mM.[1,2] This provides a constant supply of calcium ions for their numerous extracellular roles, i.e., acting as a co-factor for clotting factors, adhesion molecules, and numerous other proteins, modulating neuronal excitability, and providing a source of calcium for its intracellular functions.[1,2] In addition, salts of calcium and phosphate provide the mineral phase of the skeleton, which protects vital organs and facilitates locomotion and other movements. Bone also serves as a nearly inexhaustible supply of calcium and phosphate when dietary sources are insufficient for the body's requirements.[2]

The resting level of the cytosolic calcium concentration (Ca^{2+}_i), in contrast, is ~100 nM, nearly 10,000-fold lower than that of Ca^{2+}_o.[3] Ligand-induced activation of cell surface receptors that initiate influx of Ca^{2+} and/or its release from intracellular stores can elevate Ca^{2+}_i 10- to 100-fold.[3] Changes in Ca^{2+}_i serve a key intracellular second messenger function, regulating processes as diverse as cellular motility, differentiation, proliferation, and apoptosis, as well as muscular contraction and hormonal secretion.[3] All intracellular Ca^{2+} ultimately derives from Ca^{2+} in the extracellular fluids (ECFs). Therefore, maintaining a nearly constant level of Ca^{2+}_o ensures that calcium is available for its host of intracellular roles.

The "guardian" of the near constancy of Ca^{2+}_o in mammals is a homeostatic system made up of the parathyroid glands, calcitonin (CT)-secreting C-cells, kidney, bone, and intestine, as detailed elsewhere in this volume.[1,2] Key components of this homeostatic mechanism are several types of cells that can "sense" small perturbations in Ca^{2+}_o from its normal value and respond so as to return Ca^{2+}_o to normal.[4] Calcium ions cross the cell membrane through a variety of ion channels and transporters,[3] but the mechanism underlying Ca^{2+}_o sensing was unknown for many years. This chapter describes the properties and functions of a G protein–coupled receptor (GPCR), the Ca^{2+}_o-sensing receptor (CaSR), which plays a central role in Ca^{2+}_o homeostasis by virtue of its capacity to sense Ca^{2+}_o. It is the principal mechanism in parathyroid cells, C-cells, and several nephron segments in the kidney for measuring the level of Ca^{2+}_o. As such, it can serve as the body's "thermostat for Ca^{2+}_o" or "calciostat" through its capacity to modulate the functions of those cell types just enumerated that participate in Ca^{2+}_o homeostasis. Additional details on the roles of these cells and tissues and those of intestine and bone in mineral ion homeostasis can be found elsewhere in this text.

ISOLATION AND PROPERTIES OF THE CaSR

Expression cloning in *Xenopus laevis* oocytes enabled isolation of a full-length, functional clone of the bovine parathyroid CaSR.[5] Conventional, hybridization-based approaches permitted the cloning of cDNAs encoding CaSRs from human parathyroid and kidney, rat kidney, brain (namely striatum),[6] and C-cells, rabbit kidney, and chicken parathyroid.[4] All exhibit highly similar amino acid sequences (≥84% identical) and predicted structures, and they represent tissue and species homologs of the same ancestral *CaSR* gene (so-called orthologs). Similar genes have been isolated from evolutionarily more distant species, such as salmon and dogfish shark,[7] which exhibit ~60–70% identity to the human CaSR. Thus, the CaSR originated before the migration of vertebrates from the oceans onto dry land and is thought to regulate processes that maintain stability of Ca^{2+}_o in bony fish and elasmobranchs (e.g., sharks). In fact, the CaSR possesses the capacity to sense changes not only in Ca^{2+}_o but also in salinity[7] and to permit appropriate adaptations of ion transport systems so as to maintain ionic homeostasis during the migration of salmon from fresh to saltwater and vice versa.

Figure 1 shows the predicted topology of the CaSR protein. Its three principal structural domains include (1) a large, ~600 amino acid extracellular amino-terminal domain (ECD), (2) a seven membrane-spanning motif characteristic of the superfamily of GPCRs, and (3) a carboxyl-terminal (C-) tail of ~200 amino acids. Key Ca^{2+}_o-binding sites reside within the CaSR's ECD,[8] although an engineered CaSR lacking the entire ECD still exhibits some responsiveness to polyvalent cations.[9] Thus, there also may be binding sites for Ca^{2+}_o within the CaSR's transmembrane domains (TMDs) and/or the extracellular loops linking the transmembrane helices. The steep slope of the relationship between Ca^{2+}_o and the activation of the CaSR suggests "positive cooperativity" caused by the presence of several Ca^{2+}_o-binding sites, thereby ensuring that the receptor responds over the narrow physiological range of Ca^{2+}_o regulating PTH secretion.[1]

The cell surface, biologically active form of the CaSR is a dimer,[10] with the monomers linked by disulfide bonds within their ECDs involving cysteines 129 and 131.[11] The receptor is heavily glycosylated and, whereas the glycosylation is important for cell surface expression, it is not essential for biological activity per se.[9] Within its intracellular loops and C-tail, the human CaSR harbors five predicted protein kinase C (PKC) sites. Activation of PKC diminishes CaSR-mediated stimulation of phospholipase C (PLC), primarily through phosphorylation of a single, key PKC site in the C-tail of the CaSR at threonine T888.[12] This PKC-induced phosphorylation of the C-tail confers negative feedback regulation of CaSR-mediated stimulation of PLC.

On the plasma membrane, the CaSR in parathyroid resides in caveolae, which are flask-shaped invaginations of the plasma membrane.[13] In this locale, the CaSR binds to caveolin-1, a key cholesterol-binding, structural protein of caveolae that interacts with multiple other proteins, including signaling proteins (e.g., G proteins).[14] The CaSR's C-tail interacts with filamin-A, an actin-binding protein, which, like caveolin-1, acts as a scaffold for multiple proteins.[15] The CaSR's ability to activate mitogen-activated protein kinases (MAPK), i.e., extracellular signal-regulated kinase 1 and 2 (ERK1/2) (see next section) is dependent, in part, on its binding to filamin-A.[15] The CaSR is notable for the limited extent to which it is desensitizes on repetitive exposure to concentrations of Ca^{2+}_o

Dr. Brown receives royalties related to the sale of Sensipar from NPS Pharmaceuticals.

Key words: Ca^{2+}-sensing receptor, parathyroid, C-cell, kidney, intestine, PTH, calcitonin, bone, cartilage, G protein–coupled receptor, signal transduction

© 2008 American Society for Bone and Mineral Research

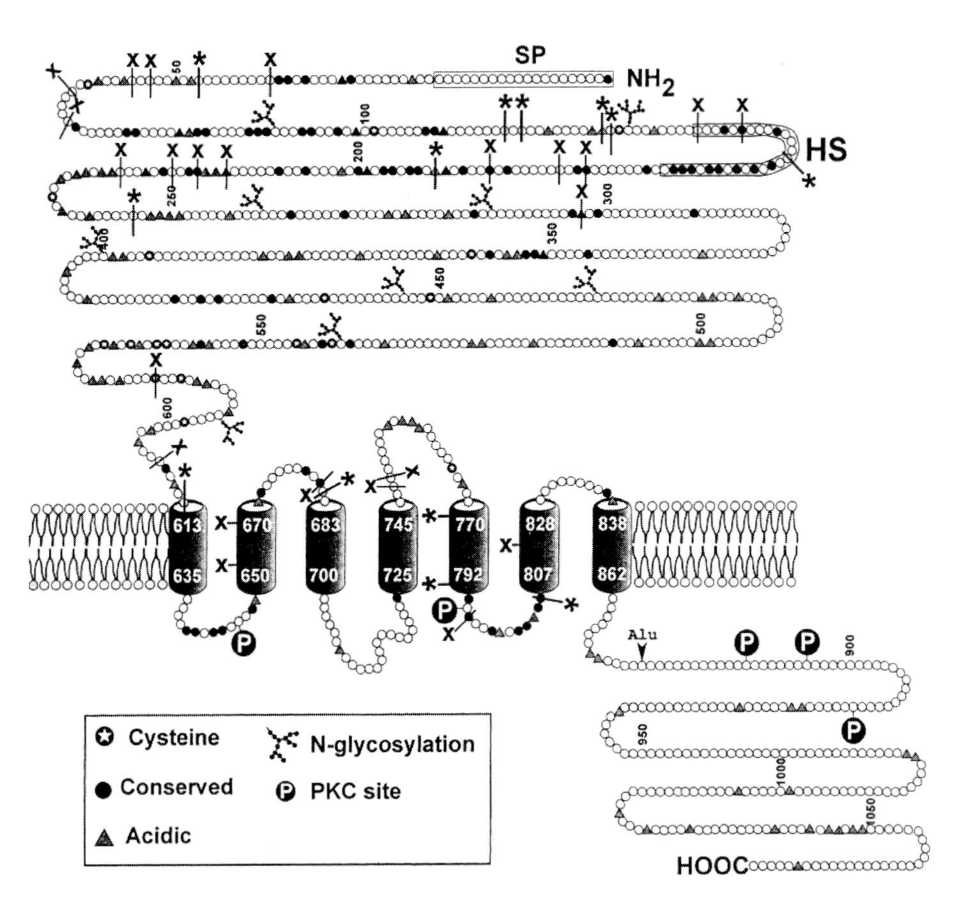

FIG. 1. Predicted structure of the human CaSR (see text for details). SP, signal peptide; HS, hydrophobic segment. Xs show examples of sites of naturally occurring inactivating mutations, and asterisks indicate locations of activating mutations. (Reproduced in modified form with permission from Elsevier from Brown EM, Bai M, Pollak M 1999 In: Avioli L, Krane SM (eds.) Metabolic Bone Disease and Clinically Related Diseases. Academic Press, San Diego, CA, USA, pp. 479–499. Copyright Elsevier 1999.)

that activate it—a property that may be essential for it to function as an effective calciostat in parathyroid and perhaps other cells. Its resistance to desensitization depends, in part, on its being tethered to the cytoskeleton by filamin-A.[16]

INTRACELLULAR SIGNALING BY THE CaSR

Activating the CaSR stimulates phospholipases C, A_2 (PLA_2), and D (PLD) in bovine parathyroid cells.[4] The high $Ca^{2+}{}_o$-induced, transient elevation in $Ca^{2+}{}_i$ in bovine parathyroid cells is the result of PLC activation and consequent IP_3-mediated release of intracellular Ca^{2+} stores.[4] High $Ca^{2+}{}_o$ also produces sustained increases in $Ca^{2+}{}_i$ through incompletely defined influx pathway(s) for Ca^{2+} that probably include activation of one or more Ca^{2+}-permeable, nonselective cation channels of the transient receptor potential (TRP) family. High $Ca^{2+}{}_o$ decreases agonist-evoked cAMP accumulation in parathyroid cells[4] by inhibiting adenylate cyclase through the inhibitory G protein, G_i.[17] High $Ca^{2+}{}_o$-elicited diminution of cAMP in other cells types, however, can result from high $Ca^{2+}{}_o$-induced inhibition of a Ca^{2+}-inhibitable isoform of adenylate cyclase or increased degradation of cAMP through activation of phosphodiesterase.[18] The CaSR also activates MAPKs, including ERK1/2, p38 MAPK, and c-Jun amino-terminal (JNK) through both PKC- and tyrosine kinase–dependent pathways.[19]

CaSR GENE AND ITS REGULATION

The human *CaSR* gene is on the long arm of chromosome 3 in band 3q13.3–21.[20] The rat and mouse *CaSR* genes reside on chromosomes 11 and 16, respectively.[20] The *CaSR* gene has seven exons. The first exon is comprised of upstream untranslated sequences, and the second contains the translational start site. The next four exons encode the remainder of the ECD, and the seventh exon encodes the rest of the CaSR from its first TMD to the C terminus.[21]

$1,25(OH)_2D_3$ increases CaSR expression in certain cell types, including parathyroid and kidney.[22] High $Ca^{2+}{}_o$ likewise enhances CaSR expression in the parathyroid gland.[23] The high $Ca^{2+}{}_o$- and $1,25(OH)_2D_3$-evoked increases in CaSR expression may contribute to the known inhibition of parathyroid function by these two factors. Interleukin-1β increases the level of CaSR mRNA modestly in parathyroid gland slices, which may contribute to the associated reduction in PTH secretion.[24]

There are several situations in which CaSR expression decreases. Calf parathyroid cells show a rapid, 80–85% reduction in CaSR expression after being placed in culture,[25] and this is likely to be a major factor contributing to the associated loss of high $Ca^{2+}{}_o$-evoked inhibition of PTH secretion. The expression of the CaSR in rat kidney decreases in a model of chronic renal insufficiency induced by subtotal nephrectomy,[26] perhaps secondary to the concomitant lowering of circulating $1,25(OH)_2D_3$ levels. Reduced CaSR expression might contribute to the hypocalciuria observed in human renal insufficiency, because reduced renal CaSR expression and/or activity enhances tubular reabsorption of Ca^{2+} in humans with reduced CaSR activity in the kidney as a consequence of inactivating mutations of the receptor,[27,28] causing relative or absolute hypocalciuria.

ROLES OF THE CaSR IN TISSUES MAINTAINING $Ca^{2+}{}_o$ HOMEOSTASIS

Parathyroid

The parathyroid glands express arguably the highest levels

of CaSR mRNA and protein in the body.[29] Abundant evidence supports the CaSR's key role as the mediator of the high Ca^{2+}_o-induced inhibition of PTH secretion. Through this action (e.g., CaR-mediated stimulation of PTH secretion in response to hypocalcemia), the CaSR in the parathyroid provides a "floor," which the Ca^{2+}_o homeostatic system uses to vigorously defend against hypocalcemia. As just noted, the decrease in CaSR expression in bovine parathyroid cells maintained in culture is associated with a marked loss of high Ca^{2+}_o-evoked suppression of PTH release.[4] Moreover, persons with familial hypocalciuric hypercalcemia (FHH), who are heterozygous for inactivating mutations of the *CaSR* gene,[28] or mice heterozygous for targeted disruption of the *CaSR* gene,[30] exhibit mild to moderate right-shifts in their set-points for Ca^{2+}_o-regulated PTH release (the level of Ca^{2+}_o half-maximally suppressing PTH release). Furthermore, humans and mice homozygous for loss of the CaSR exhibit much more severely impaired high Ca^{2+}_o-induced inhibition of PTH release.[28,30] Thus, the degree of "Ca^{2+}_o resistance" of the parathyroid is inversely related to the number of normally functioning CaSR alleles. Finally, calcimimetics, allosteric activators of the CaSR, acutely inhibit PTH secretion in vivo and in vitro,[31] further proving the CaSR's key role in regulating PTH secretion.

Despite 30 yr or more of study, the most important intracellular signaling pathway(s) by which the CaSR suppresses PTH release are poorly defined. Recent evidence, however, showed that activating $G_{q/11}$ is essential, because mice with knockout of both of these G proteins have severe hyperparathyroidism similar to that present in mice homozygous for knockout of the CaSR.[32] Through poorly defined mechanism(s), the CaSR eventually induces polymerization of the actin-based cytoskeleton, which may represent a physical barrier to the secretion of PTH-containing secretory vesicles.[33]

Another CaSR-regulated process in parathyroid cells is PTH gene expression, because the calcimimetic CaSR activator, NPS R-568, decreases the elevated level of preproPTH mRNA in rats with experimentally induced secondary hyperparathyroidism.[34] This effect is primarily post-transcriptional, occurring through modulation of the binding of the transacting factor, AUF1, to preproPTH mRNA. Finally, the CaSR directly or indirectly suppresses parathyroid cellular proliferation, because individuals homozygous for inactivating CaSR mutations[28] or homozygous CaSR knockout mice[30] exhibit marked parathyroid cellular hyperplasia. Furthermore, treating rats with experimentally induced renal insufficiency with a calcimimetic mitigates the parathyroid hyperplasia that would otherwise take place in this setting,[35] providing further documentation that CaSR activation suppresses parathyroid cellular proliferation.

C-Cells

In contrast to the effect of Ca^{2+}_o on PTH secretion, increasing the level of Ca^{2+}_o enhances CT secretion—a response conforming to the usual positive relationship between Ca^{2+} and stimulation of exocytosis observed in most other hormone-secreting cells. Recent studies have shown that human, rat, and rabbit C cells express the CaSR,[4,36] and the CaSR cloned from a rat C-cell tumor cell line is identical to that expressed in rat kidney. Available data, albeit limited, indicate that the CaSR is the mediator of high Ca^{2+}_o-stimulated CT secretion; namely, mice heterozygous for knockout of the *CaSR* gene exhibit a rightward shift in their relationship between serum Ca^{2+} concentration and blood CT levels.[37] In contrast to the role of the CaR in parathyroid, the receptor in the C-cell serves more as a "ceiling" to defend against hypercalcemia.

Kidney

In rat kidney, the CaSR resides along nearly the entire nephron; the highest levels of expression of the CaSR protein are at the basolateral surface of the cells of the cortical thick ascending limb (CTAL).[38] This nephron segment has a key role in PTH-regulated divalent cation reabsorption.[39] The CaSR also has a basolateral localization in the distal convoluted tubule (DCT), where PTH likewise stimulates Ca^{2+} reabsorption. The CaSR also resides at the base of the microvilli of the proximal tubular brush border, on the basolateral surface of the epithelial cells of the medullary thick ascending limb (MTAL),[38] and on the luminal surface of the inner medullary collecting duct (IMCD).[40] These latter three nephron segments do not participate directly in the regulation of renal Ca^{2+}_o reabsorption by Ca^{2+}_o-regulating hormones, such as PTH and $1,25(OH)_2D_3$, but are sites where the CaSR could potentially modulate the handling of other solutes and/or water (see below).

In the proximal tubule, the CaSR suppresses PTH-induced phosphaturia[41] and enhances vitamin D receptor expression.[42] The latter could conceivably participate in the direct, high Ca^{2+}_o-elicited lowering of circulating $1,25(OH)_2D_3$ levels, because activation of the VDR reduces production of $1,25(OH)_2D_3$ and increases production of $24,25(OH)_2D_3$. The CaSR's location on the basolateral surface in the CTAL supports its role as the mediator of the known inhibitory action of high peritubular but not luminal Ca^{2+}_o on Ca^{2+} and Mg^{2+} reabsorption in perfused tubular segments from this region of the nephron.[41] The CaSR suppresses PTH-stimulated divalent cation reabsorption in the CTAL by acting in a "lasix-like" manner to inhibit the overall activity of the Na-K-2Cl cotransporter. This co-transporter contributes to the generation of the lumen-positive, transepithelial potential gradient driving passive paracellular reabsorption of ~50% of NaCl and most of the Ca^{2+} and Mg^{2+} in this nephron segment.[43] Others, however, have observed discrepant findings, including high Ca^{2+}_o-induced inhibition of Ca^{2+} transport in CTAL without any associated decrease in NaCl or Mg^{2+} transport, as well as inhibitory actions of raising peritubular Ca^{2+} not only on paracellular but also transepithelial Ca^{2+} transport.[41] Additional studies are needed to resolve these discrepancies. In any event, the documentation of direct, CaSR-mediated inhibition of Ca^{2+} reabsorption in CTAL proves that hypercalcemia-induced hypercalciuria has two distinct CaSR-mediated components: (1) inhibition of PTH release, which secondarily reduces Ca^{2+} reabsorption, and (2) direct suppression of tubular transport of Ca^{2+} in the CTAL. The direct inhibitory action of the CaSR on renal tubular Ca^{2+} reabsorption, like CaR-stimulated CT secretion, represents a "ceiling" that defends against hypercalcemia resulting from a calcium load.

It is not currently known whether the CaSR modulates PTH-enhanced Ca^{2+} reabsorption in DCT, which could potentially occur through inhibition of PTH-stimulated cAMP accumulation. However, extracellular Ca^{2+} modulates the expression of key, vitamin D–inducible genes involved in transcellular Ca^{2+} transport in this segment of the nephron, specifically, the apical uptake channel TRPV5, calbindin D_{28K}, and the basolateral calcium pump, PMCA1b, and sodium-calcium exchanger, NCX1.[44] These changes occur both indirectly, through high Ca^{2+}-evoked alterations in PTH release and, consequently, $1,25(OH)_2D_3$ production, as well as by seemingly direct renal actions of Ca^{2+}_o that are independent of $1,25(OH)_2D_3$. The contribution of these actions of Ca^{2+}_o to regulating Ca^{2+} transport in DCT under normal circumstances is not presently known.

© 2008 American Society for Bone and Mineral Research

Intestine

The intestine plays an important role in maintaining $Ca^{2+}{}_o$ homeostasis because of its capacity for regulated absorption of dietary Ca^{2+} through the action of $1,25(OH)_2D_3$.[2] The duodenum is the predominant site for $1,25(OH)_2D_3$-mediated, transcellular intestinal Ca^{2+} absorption. The first step in this process is the entry of Ca^{2+} through the Ca^{2+}-permeable channel, TRPV6, down a favorable electrochemical gradient, specifically a ~10,000-fold higher concentration of extracellular relative to intracellular Ca^{2+} and an interior of the cell that is negative relative to the extracellular fluid.[45] Calcium ions then diffuse from the apical to the basolateral cell surface, facilitated by the vitamin D–dependent Ca^{2+}-binding protein, calbindin D_{9K}. The last step is extrusion of Ca^{2+} across the basolateral membrane of the cell by the Ca^{2+}-ATPase and/or the Na^+-Ca^{2+} exchanger.[44] At higher luminal Ca^{2+} concentrations, a passive paracellular pathway of Ca^{2+} absorption becomes dominant. The ileum and jejunum absorb less Ca^{2+} by the transcellular route than does the duodenum, but the human colon absorbs substantial amounts of Ca^{2+} by vitamin D–dependent and –independent pathways, particularly in the cecum,[46] where TRPV6 is present at levels comparable to those in the duodenum.

The CaSR is expressed throughout rat intestine; the greatest expression levels are on the basal surface of the small intestinal epithelial cells, the large and small intestinal crypts, and in the enteric nervous system.[47] Does the intestinal CaSR contribute to systemic $Ca^{2+}{}_o$ homeostasis? $Ca^{2+}{}_o$ modulates several intestinal functions. Hypercalcemia reduces the absorption of dietary Ca^{2+}.[48] Recent studies have also documented apparently direct actions of dietary and/or blood Ca^{2+} on the expression of TRPV6, calbindin D_{9K}, and PMCA1b in mice that lack the *1-hydroxylase* gene. The alterations in the levels of these genes are of uncertain physiological relevance but suggest that the gastrointestinal (GI) tract per se has the ability to sense $Ca^{2+}{}_o$.[44] The CaSR in the enteric nervous system, which regulates secretomotor functions of the GI tract, could potentially contribute to the known capacity of hypo- and hypercalcemia to enhance and decrease, respectively, GI motility.[2] Recent studies have shown that activation of the CaSR in the colon markedly reduces fluid secretion, potentially affording a novel approach to the treatment of diarrheal states (e.g., through the use of a calcimimetic agent).[18]

Bone and Cartilage

The level of $Ca^{2+}{}_o$ within the microenvironment of bone probably fluctuates substantially during osteoclastic bone resorption and osteoblastic bone formation.[4] In fact, $Ca^{2+}{}_o$ underneath resorbing osteoclasts can be as high as 8–40 mM.[49] Moreover, $Ca^{2+}{}_o$ has several actions on the function of bone cells in vitro of potential physiological relevance, although it is currently unknown whether these actions of $Ca^{2+}{}_o$ take place in vivo. For example, high $Ca^{2+}{}_o$ stimulates the proliferation and chemotaxis of pre-osteoblasts, potentially enhancing their availability at sites of recent bone resorption, promotes their differentiation to mature osteoblasts, and increases their capacity to mineralize bone proteins in vitro.[4,50] In addition, raising $Ca^{2+}{}_o$ inhibits both the formation and activity of osteoclasts in vitro.[51] If these actions of $Ca^{2+}{}_o$ on bone cells take place in vivo, increases in $Ca^{2+}{}_o$ could promote net transfer of Ca^{2+} into bone by enhancing bone formation and inhibiting bone resorption.

Some investigators have failed to detect CaSR expression in osteoblast-like and osteoclast-like cells and have suggested the presence of another $Ca^{2+}{}_o$-sensing mechanism in osteoblasts, such as the basic amino acid–sensing, family C receptor, GPRC6A.[52] Other studies have provided strong evidence, however, for the CaSR's presence in various cells originating from bone and bone marrow. These CaSR-expressing cells are made up of cells from both the osteoclast and osteoblast lineages when examined in situ in bone sections, as well as in osteoblast-like cell lines.[53,54] Very recent data presented in abstract form have shown that knockout of the CaSR exclusively in immature murine osteoblasts in vivo results in a runted phenotype and death after several weeks, strongly supporting a key biological role for the CaSR in immature osteoblasts.[55] With regard to cells of the osteoclast lineage, pre-osteoclast-like cells generated in vitro show CaSR expression, and osteoclasts derived from rabbit or mouse bone also express the receptor.[56] In sections from murine, rat, and bovine bones, however, only a minority of multinucleated osteoclasts expressed CaSR mRNA and protein.[54]

Although chondrocytes, the cells that form cartilage, do not participate directly in $Ca^{2+}{}_o$ homeostasis, they play a key role in skeletal development and growth by affording a cartilaginous template of the future skeleton that is gradually replaced by actual bone. Moreover, the growth plate is the site where chondrocytes are critical participants in the longitudinal growth of bones that continues until the skeleton is fully mature at the end of the pubertal growth spurt. The availability of Ca^{2+} is important for promoting normal growth and differentiation of chondrocytes and resultant skeletal growth in vivo.[57] Moreover, some cartilage cells in intact bone express CaSR mRNA and protein, including the hypertrophic chondrocytes of the growth plate, which are key participants in the growth of long bones.[54] Raising $Ca^{2+}{}_o$ dose-dependently reduces the levels of the mRNAs that encode important cartilaginous proteins, including aggrecan, the α_1 chains of types II and X collagen, and alkaline phosphatase, in a chondrocytic cell line (RCJ3.1C5.18 cells).[58] These actions are likely CaSR mediated, because the use of a CaSR antisense oligonucleotide to lower CaSR expression reversed the action of high $Ca^{2+}{}_o$ on aggrecan mRNA expression.[54] Recent data presented in abstract form have documented that selective knockout of the CaSR in chondrocytes of mice results in an embryonic lethal phenotype, confirming an essential role of the CaSR in chondrogenesis.[59]

CaSR INTEGRATES $Ca^{2+}{}_o$ AND WATER METABOLISM

In addition to its central role in $Ca^{2+}{}_o$ homeostasis, a growing body of evidence has shown the CaSR's presence in several cells and tissues where it seems to promote integration of seemingly unrelated homeostatic systems, as exemplified by the CaSR's apparent ability to integrate mineral and water metabolism.[43] A long-recognized clinical observation is that some hypercalcemic patients manifest a decrease in urinary concentrating capacity, which in some cases progresses to frank nephrogenic diabetes insipidus.[60] Moreover, hypercalciuria, even without hypercalcemia, can decrease urinary concentrating ability, as shown by the development of enuresis in some hypercalciuric children, which resolves after the hypercalciuria is corrected by restriction of dietary calcium.[61] The CaSR is present in two nephron segments, MTAL and IMCD, that are important participants in the urinary concentrating mechanism,[38,40] providing clues into how it might contribute to the hypercalcemia-evoked impairment of renal concentrating capacity. Elevating $Ca^{2+}{}_o$, most likely by stimulating CaSRs present on the apical membrane of the epithelial cells of the IMCD, reversibly diminishes vasopressin-activated water flow by 30–40% in perfused IMCD tubules.[40] Moreover, inducing chronic hypercalcemia in rats with vitamin D de-

© 2008 American Society for Bone and Mineral Research

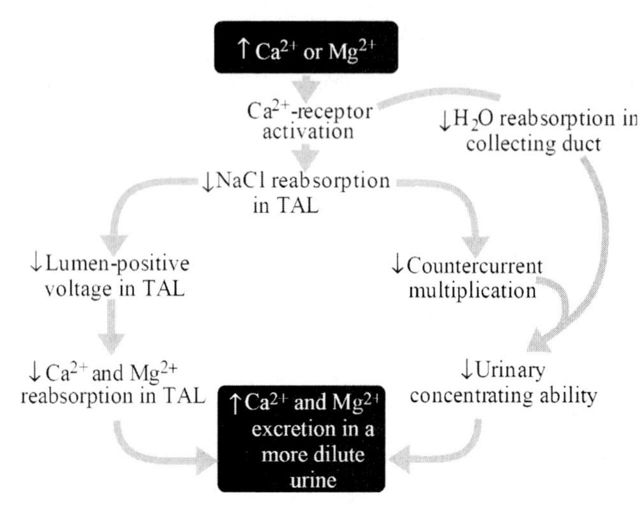

FIG. 2. Mechanisms through which the CaR reduces maximal urinary-concentrating ability (see text for additional details). Elevated concentrations of Ca^{2+} in the blood activate the CaSR in the CTAL, which inhibits the reabsorption of Ca^{2+}, thereby increasing the concentration of Ca^{2+} in the tubular fluid. Concomitant activation of the CaSR in the MTAL reduces transport of NaCl into the medullary interstitium, and, as a consequence, decreases maximal urinary concentrating ability by inhibiting countercurrent multiplication. If the level of Ca^{2+} in the tubular fluid remains high when it reaches the IMCD, activation of the CaSR on the apical membrane of the tubular epithelial cells directly inhibits vasopressin-stimulated reabsorption of water, further decreasing the concentration of Ca^{2+} in the final urine, and presumably the risk of forming Ca^{2+}-containing kidney stones. (Reproduced with permission from Brown EM, Hebert SC 1997 Novel insights into the physiology and pathophysiology of Ca^{2+} homeostasis from the cloning of an extracellular Ca^{2+}-sensing receptor. Reg Peptide Lett **VII:**43–47.)

creases expression of the aquaporin-2 (AQP-2) water channel protein (but not its mRNA), which is inserted in the apical membrane of the IMCD in response to vasopressin to promote water reabsorption in the setting of dehydration.[62] This reduced expression of the AQP-2 protein would further decrease vasopressin-elicited water reabsorption. Furthermore, high $Ca^{2+}{}_o$-elicited, CaSR-mediated diminution in NaCl reabsorption in the MTAL,[43,63] by reducing the medullary countercurrent gradient, would further decrease maximal urinary-concentrating power during hypercalcemia (Fig. 2).

Is the impaired renal water handling in hypercalcemic patients of any physiological importance? It may provide a mechanism that integrates the handling of divalent cations and water by the kidney, thereby allowing appropriate "trade-offs" under specific physiological situations in how the kidney coordinates calcium and water metabolism.[43] For example, when a systemic Ca^{2+} load must be gotten rid of, the resultant CaSR-induced decrease in PTH release as well as the direct inhibition of Ca^{2+} reabsorption in the distal tubule caused by increases in $Ca^{2+}{}_o$ promote calciuria. The resultant elevation in luminal $Ca^{2+}{}_o$ in IMCD, particularly in a dehydrated individual, could potentially promote the formation of Ca^{2+}-containing renal stones if it were it not for the accompanying, CaSR-mediated diminution of maximal urinary ability. Furthermore, abundant CaSRs are present in the subfornical organ (SFO),[64] an important hypothalamic thirst center, which could potentially mediate a physiologically appropriate stimulation of drinking if there were any overt elevation in $Ca^{2+}{}_o$. Enhanced water intake could forestall any dehydration that could otherwise result from the renal loss of free water resulting from the concomitant, CaSR-induced reduction in urinary concentrating ability (Fig. 2).

OTHER AGONISTS AND MODULATORS OF THE CaSR

Several divalent ($Ca^{2+}{}_o$, $Mg^{2+}{}_o$, and $Sr^{2+}{}_o$) and trivalent cations (i.e., $Gd^{3+}{}_o$), and even organic polycations, such as spermine, activate the CaSR (Fig. 3).[4,8] They all probably bind to the CaSR's ECD.[8] Only a few of these polycationic CaSR activators, however, reside in biological fluids at concentrations that are capable of stimulating the CaSR. $Mg^{2+}{}_o$ and spermine, in addition to $Ca^{2+}{}_o$, have the potential to be physiologically relevant CaSR activators. In particular microenvironments (e.g., the lumen of the GI tract and CNS), the levels of spermine are high enough to activate the CaSR even at levels of $Ca^{2+}{}_o$ insufficient to do so by themselves.[65]

Does the CaSR also serve as a $Mg^{2+}{}_o$ sensor? Some evidence supporting the CaSR's role in sensing and "setting" $Mg^{2+}{}_o$ derives from the experiments-in-nature that initially proved the CaSR's role as a central element in $Ca^{2+}{}_o$ homeostasis. Namely, individuals with hypercalcemia because of heterozygous inactivating CaSR mutations (e.g., the syndrome FHH) have serum Mg^{2+} levels in the high normal or mildly elevated range.[66] Conversely, persons who have activating CaSR mutations can manifest mild hypomagnesemia.[67] Therefore, inactivating and activating mutations of the CaSR reset not only $Ca^{2+}{}_o$ but also $Mg^{2+}{}_o$. The regulation of $Mg^{2+}{}_o$ homeostasis by the CaSR may occur, in part, in the parathyroid gland, where hypermagnesemia suppresses PTH release, or in the CTAL of the kidney, where elevated $Mg^{2+}{}_o$ reduces the reabsorption of not only Mg^{2+} but also Ca^{2+}.[39]

In addition to the polycationic CaSR agonists noted above, novel "calcimimetic," allosteric activators of the CaSR are currently being used clinically. These are small hydrophobic molecules, which activate the CaSR by increasing its apparent affinity for $Ca^{2+}{}_o$ as a result of interacting with its TMDs rather than at or near the $Ca^{2+}{}_o$-binding sites in the ECD (Fig. 3).[68] Calcimimetics are called "modulators" rather than "agonists" because they only activate the CaSR in the presence of $Ca^{2+}{}_o$. Polycationic CaSR agonists (e.g., $Ca^{2+}{}_o$ or spermine), in contrast, activate the receptor even in the absence of $Ca^{2+}{}_o$.[68] Calcimimetics are approved for use in patients with stage 5 chronic kidney disease (i.e., with glomerular filtration rate [GFR] < 15 ml/min, who are generally on dialysis), as well as those with parathyroid cancer. In patients on dialysis, calcimimetics afford an effective way to decrease PTH secretion through their direct action on the CaSR, whereas at the same time modestly reducing serum phosphorus and calcium-phosphate product, perhaps through a reduction in bone turnover caused by the decrease in circulating PTH levels.[69] About 60% of patients with parathyroid cancer exhibit clinically significant reductions in serum calcium concentration when treated with a calcimimetic,[70] thereby prolonging the time during which hypercalcemia can be controlled medically in this difficult-to-treat disease. CaSR antagonists, called "calcilytics," are also in clinical trials as a possible treatment of osteoporosis. Because intermittently administering exogenous PTH subcutaneously can increase BMD substantially, administration of a short-acting calcilytic by mouth might accomplish this same goal by evoking a "pulse" of endogenous PTH secretion.[71]

Amino acids are another type of endogenous CaSR modulator (Fig. 3).[72] Stimulation of the CaSR by amino acids, particularly the aromatic amino acids, phenylalanine, tyrosine, histidine, and tryptophan, only takes place when $Ca^{2+}{}_o$ is ≥ 1 mM. Although the physiological implications of amino acid–induced activation of the CaSR are not clear, this action may help to explain several long recognized but poorly understood findings that seem to link $Ca^{2+}{}_o$ and protein metabolism. For

© 2008 American Society for Bone and Mineral Research

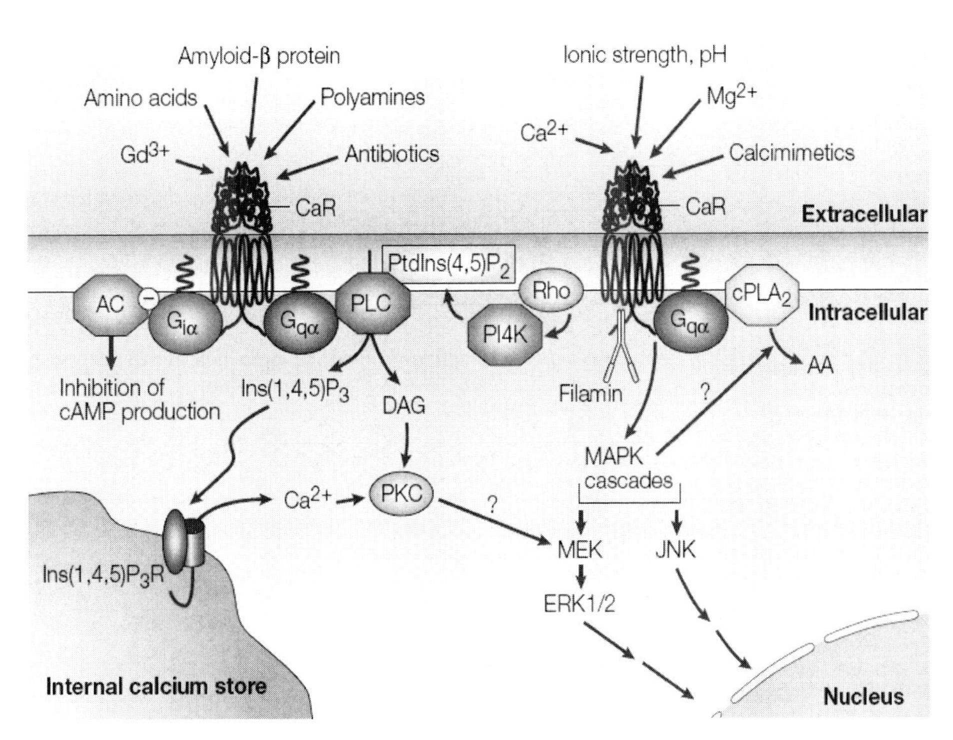

FIG. 3. Multiple agonists and other factors modulating the activity of the CaSR as well as the numerous intracellular signaling pathways through which the receptor can regulate cellular function. Ca^{2+}, Mg^{2+}, aminoglycoside antibiotics, spermine and other polyamines, and amyloid β peptides are examples of polycationic agonists of the CaSR. Aromatic amino acids and calcimimetics—drugs that activate the receptor and are used to control hyperparathyroidism—are allosteric modulators of the CaSR. The former bind to a site near a putative calcium-binding site in the receptor's ECD, whereas the latter bind to the CaSR's transmembrane domains; both increase its apparent affinity for polycationic agonists. AA, arachidonic acid; AC, adenylate cyclase; cAMP, cyclic AMP; $cPLA_2$, cytosolic phospholipase A_2; DAG, diacylglycerol; ERK, extracellular signal-regulated kinase; $G_{\alpha i}$ and $G_{\alpha q}$, α subunits of the i- and q-type heterotrimeric G-proteins, respectively, $Ins(1,4,5)P_3$, inositol 1,4,5-trisphosphate; $Ins(1,4,5)P_3R$, 1,4,5-trisphosphate receptor; JNK, Jun N-terminal kinase; MAPK, mitogen-activated protein kinase; MEK, MAPK kinase; PI4K, phosphatidylinositol 4-kinase; PKC, protein kinase C; PLC, phospholipase C; $PtdIns(4,5)P_2$, phoasphatidylinositol-4,5-bisphosphate. (Reprinted with permission from Macmillan Publishers from Hofer A, Brown EM 2003 Extracellular calcium sensing and signaling. Nat Rev Cell Mol Biol **4:**530–538.)

instance, ingestion of a diet high in protein nearly doubles the rate of urinary calcium excretion compared with that observed on a low protein intake, possibly by activating the CaSR in the CTAL. Furthermore, decreasing dietary protein almost doubles both serum PTH and $1,25(OH)_2D_3$ levels in normal women.[73] It is possible, but not yet proven, that these latter changes in PTH and $1,25(OH)_2D_3$ result from the associated decrease in the circulating concentrations of amino acids, despite little, if any, alteration in $Ca^{2+}{}_o$, presumably by modulating the activity the CaSR in parathyroid gland and kidney. The CaSR in the GI tract, where acute ingestion of Ca^{2+} and/or aromatic amino acids stimulate gastrin, gastric acid, and cholecystokinin secretion,[74,75] might be an especially suitable location for sensing dietary mineral ions and protein, which are often ingested together (i.e., as in milk). Additional studies are necessary, therefore, to study whether the CaSR, in fact, serves as a physiologically significant amino acid receptor, thereby acting not only as a $Ca^{2+}{}_o$-sensing receptor but likewise as a more generalized "nutrient sensor" in the GI tract and elsewhere.

REFERENCES

1. Brown EM 1991 Extracellular Ca^{2+} sensing, regulation of parathyroid cell function, and role of Ca^{2+} and other ions as extracellular (first) messengers. Physiol Rev **71:**371–411.
2. Bringhurst FR, Demay MB, Kronenberg HM 1998 Hormones and disorders of mineral metabolism. In: Wilson JD, Foster DW, Kronenberg HM, Larsen PR (eds.) Williams Textbook of Endocrinology, 9th ed. W.B. Saunders, Philadelphia, PA, USA, pp. 1155–1209.
3. Berridge MJ, Bootman MD, Roderick HL 2003 Calcium signalling: Dynamics, homeostasis and remodelling. Nat Rev Mol Cell Biol **4:**517–529.
4. Brown EM, MacLeod RJ 2001 Extracellular calcium sensing and extracellular calcium signaling. Physiol Rev **81:**239–297.
5. Brown EM, Gamba G, Riccardi D, Lombardi M, Butters R, Kifor O, Sun A, Hediger MA, Lytton J, Hebert SC 1993 Cloning and characterization of an extracellular $Ca^{(2+)}$-sensing receptor from bovine parathyroid. Nature **366:**575–580.
6. Ruat M, Molliver ME, Snowman AM, Snyder SH 1995 Calcium sensing receptor: Molecular cloning in rat and localization to nerve terminals. Proc Natl Acad Sci USA **92:**3161–3165.
7. Nearing J, Betka M, Quinn S, Hentschel H, Elger M, Baum M, Bai M, Chattopadyhay N, Brown EM, Hebert SC, Harris HW 2002 Polyvalent cation receptor proteins (CaRs) are salinity sensors in fish. Proc Natl Acad Sci USA **99:**9231–9236.
8. Brauner-Osborne H, Jensen AA, Sheppard PO, O'Hara P, Krogsgaard-Larsen P 1999 The agonist-binding domain of the calcium-sensing receptor is located at the amino-terminal domain. J Biol Chem **274:**18382–18386.
9. Hu J, Spiegel AM 2003 Naturally occurring mutations in the extracellular Ca^{2+}-sensing receptor: Implications for its structure and function. Trends Endocrinol Metab **14:**282–288.
10. Ward DT, Brown EM, Harris HW 1998 Disulfide bonds in the extracellular calcium-polyvalent cation-sensing receptor correlate with dimer formation and its response to divalent cations in vitro. J Biol Chem **273:**14476–14483.
11. Ray K, Hauschild BC, Steinbach PJ, Goldsmith PK, Hauache O, Spiegel AM 1999 Identification of the cysteine residues in the amino-terminal extracellular domain of the human $Ca^{(2+)}$ receptor critical for dimerization. Implications for function of monomeric $Ca^{(2+)}$ receptor. J Biol Chem **274:**27642–27650.
12. Davies SL, Ozawa A, McCormick WD, Dvorak MM, Ward DT 2007 Protein kinase C-mediated phosphorylation of the calcium-sensing receptor is stimulated by receptor activation and attenuated by calyculin-sensitive phosphatase activity. J Biol Chem **282:**15048–15056.
13. Kifor O, Diaz R, Butters R, Kifor I, Brown EM 1998 The calcium-

© 2008 American Society for Bone and Mineral Research

sensing receptor is localized in caveolin-rich plasma membrane domains of bovine parathyroid cells. J Biol Chem **273:**21708–21713.

14. Williams TM, Lisanti MP 2004 The Caveolin genes: From cell biology to medicine. Ann Med **36:**584–595.
15. Awata H, Huang C, Handlogten ME, Miller RT 2001 Interaction of the calcium-sensing receptor and filamin, a potential scaffolding protein. J Biol Chem **276:**34871–34879.
16. Zhang M, Breitwieser GE 2005 High affinity interaction with filamin A protects against calcium-sensing receptor degradation. J Biol Chem **280:**11140–11146.
17. Gerbino A, Ruder WC, Curci S, Pozzan T, Zaccolo M, Hofer AM 2005 Termination of cAMP signals by Ca^{2+} and G_s via extracellular Ca^{2+} sensors: A link to intracellular Ca^{2+} oscillations. J Cell Biol **171:**303–312.
18. Geibel J, Sritharan K, Geibel R, Geibel P, Persing JS, Seeger A, Roepke TK, Deichstetter M, Prinz C, Cheng SX, Martin D, Hebert SC 2006 Calcium-sensing receptor abrogates secretagogue-induced increases in intestinal net fluid secretion by enhancing cyclic nucleotide destruction. Proc Natl Acad Sci USA **103:**9390–9397.
19. McNeil SE, Hobson SA, Nipper V, Rodland KD 1998 Functional calcium-sensing receptors in rat fibroblasts are required for activation of SRC kinase and mitogen-activated protein kinase in response to extracellular calcium. J Biol Chem **273:**1114–1120.
20. Janicic N, Soliman E, Pausova Z, Seldin MF, Riviere M, Szpirer J, Szpirer C, Hendy GN 1995 Mapping of the calcium-sensing receptor gene (CASR) to human chromosome 3q13.3-21 by fluorescence in situ hybridization, and localization to rat chromosome 11 and mouse chromosome 16. Mamm Genome **6:**798–801.
21. Pearce SH, Trump D, Wooding C, Besser GM, Chew SL, Grant DB, Heath DA, Hughes IA, Paterson CR, Whyte MP, Thakker RV 1995 Calcium-sensing receptor mutations in familial benign hypercalcemia and neonatal hyperparathyroidism. J Clin Invest **96:**2683–2692.
22. Canaff L, Hendy GN 2002 Human calcium-sensing receptor gene. Vitamin D response elements in promoters P1 and P2 confer transcriptional responsiveness to 1,25-dihydroxyvitamin D. J Biol Chem **277:**30337–30350.
23. Mizobuchi M, Hatamura I, Ogata H, Saji F, Uda S, Shiizaki K, Sakaguchi T, Negi S, Kinugasa E, Koshikawa S, Akizawa T 2004 Calcimimetic compound upregulates decreased calcium-sensing receptor expression level in parathyroid glands of rats with chronic renal insufficiency. J Am Soc Nephrol **15:**2579–2587.
24. Nielsen PK, Rasmussen AK, Butters R, Feldt-Rasmussen U, Bendtzen K, Diaz R, Brown EM, Olgaard K 1997 Inhibition of PTH secretion by interleukin-1 beta in bovine parathyroid glands in vitro is associated with an up-regulation of the calcium- sensing receptor mRNA. Biochem Biophys Res Commun **238:**880–885.
25. Brown AJ, Zhong M, Ritter C, Brown EM, Slatopolsky E 1995 Loss of calcium responsiveness in cultured bovine parathyroid cells is associated with decreased calcium receptor expression. Biochem Biophys Res Commun **212:**861–867.
26. Mathias R, Nguyen H, Zhang M, Portale A 1998 Expression of the renal calcium-sensing receptor is reduced in rats with experimental chronic renal insufficiency. J Am Soc Nephrol **9:**2067–2074.
27. Hauache OM 2001 Extracellular calcium-sensing receptor: Structural and functional features and association with diseases. Braz J Med Biol Res **34:**577–584.
28. Thakker RV 2004 Diseases associated with the extracellular calcium-sensing receptor. Cell Calcium **35:**275–282.
29. Kifor O, Moore FD Jr, Wang P, Goldstein M, Vassilev P, Kifor I, Hebert SC, Brown EM 1996 Reduced immunostaining for the extracellular Ca^{2+}-sensing receptor in primary and uremic secondary hyperparathyroidism. J Clin Endocrinol Metab **81:**1598–1606.
30. Ho C, Conner DA, Pollak MR, Ladd DJ, Kifor O, Warren HB, Brown EM, Seidman JG, Seidman CE 1995 A mouse model of human familial hypocalciuric hypercalcemia and neonatal severe hyperparathyroidism. Nat Genet **11:**389–394.
31. Nemeth EF, Steffey ME, Hammerland LG, Hung BC, Van Wagenen BC, DelMar EG, Balandrin MF 1998 Calcimimetics with potent and selective activity on the parathyroid calcium receptor. Proc Natl Acad Sci USA **95:**4040–4045.
32. Wettschureck N, Lee E, Libutti SK, Offermanns S, Robey PG, Spiegel AM 2007 Parathyroid-specific double knockout of G_q and G_{11} alpha-subunits leads to a phenotype resembling germline knockout of the extracellular Ca^{2+}-sensing receptor. Mol Endocrinol **21:**274–280.
33. Quinn SJ, Kifor O, Kifor I, Butters RR Jr, Brown EM 2007 Role of the cytoskeleton in extracellular calcium-regulated PTH release. Biochem Biophys Res Commun **354:**8–13.
34. Levi R, Ben-Dov IZ, Lavi-Moshayoff V, Dinur M, Martin D, Naveh-Many T, Silver J 2006 Increased parathyroid hormone gene expression in secondary hyperparathyroidism of experimental uremia is reversed by calcimimetics: Correlation with posttranslational modification of the trans acting factor AUF1. J Am Soc Nephrol **17:**107–112.
35. Colloton M, Shatzen E, Miller G, Stehman-Breen C, Wada M, Lacey D, Martin D 2005 Cinacalcet HCl attenuates parathyroid hyperplasia in a rat model of secondary hyperparathyroidism. Kidney Int **67:**467–476.
36. Freichel M, Zink-Lorenz A, Holloschi A, Hafner M, Flockerzi V, Raue F 1996 Expression of a calcium-sensing receptor in a human medullary thyroid carcinoma cell line and its contribution to calcitonin secretion. Endocrinology **137:**3842–3848.
37. Fudge NJ, Kovacs CS 2004 Physiological studies in heterozygous calcium sensing receptor (CaSR) gene-ablated mice confirm that the CaSR regulates calcitonin release in vivo. BMC Physiol **4:**5.
38. Riccardi D, Hall AE, Chattopadhyay N, Xu JZ, Brown EM, Hebert SC 1998 Localization of the extracellular Ca^{2+}/polyvalent cation-sensing protein in rat kidney. Am J Physiol **274:**F611–F622.
39. de Rouffignac C, Quamme G 1994 Renal magnesium handling and its hormonal control. Physiol Rev **74:**305–322.
40. Sands JM, Naruse M, Baum M, Jo I, Hebert SC, Brown EM, Harris HW 1997 Apical extracellular calcium/polyvalent cation-sensing receptor regulates vasopressin-elicited water permeability in rat kidney inner medullary collecting duct. J Clin Invest **99:**1399–1405.
41. Ba J, Friedman PA 2004 Calcium-sensing receptor regulation of renal mineral ion transport. Cell Calcium **35:**229–237.
42. Maiti A, Beckman MJ 2007 Extracellular calcium is a direct effecter of VDR levels in proximal tubule epithelial cells that counter-balances effects of PTH on renal Vitamin D metabolism. J Steroid Biochem Mol Biol **103:**504–508.
43. Hebert SC, Brown EM, Harris HW 1997 Role of the $Ca^{(2+)}$-sensing receptor in divalent mineral ion homeostasis. J Exp Biol **200:**295–302.
44. Thebault S, Hoenderop JG, Bindels RJ 2006 Epithelial Ca^{2+} and Mg^{2+} channels in kidney disease. Adv Chronic Kidney Dis **13:**110–117.
45. Hoenderop JG, van der Kemp AW, Hartog A, van de Graaf SF, van Os CH, Willems PH, Bindels RJ 1999 Molecular identification of the apical Ca^{2+} channel in 1, 25- dihydroxyvitamin D_3-responsive epithelia. J Biol Chem **274:**8375–8378.
46. Favus MJ 1992 Intestinal absorption of calcium, magnesium and phosphorus. In: Coe FL, Favus MJ (eds.) Disorders of Bone and Mineral Metabolism. Raven Press, New York, NY, USA, pp. 57–81.
47. Chattopadhyay N, Cheng I, Rogers K, Riccardi D, Hall A, Diaz R, Hebert SC, Soybel DI, Brown EM 1998 Identification and localization of extracellular $Ca^{(2+)}$-sensing receptor in rat intestine. Am J Physiol **274:**G122–G130.
48. Krishnamra N, Angkanaporn K, Deenoi T 1994 Comparison of calcium absorptive and secretory capacities of segments of intact or functionally resected intestine during normo-, hypo-, and hypercalcemia. Can J Physiol Pharmacol **72:**764–770.
49. Silver IA, Murrils RJ, Etherington DJ 1988 Microlectrode studies on the acid microenvironment beneath adherent macrophages and osteoclasts. Exp Cell Res **175:**266–276.
50. Quarles LD 1997 Cation-sensing receptors in bone: A novel paradigm for regulating bone remodeling? J Bone Miner Res **12:**1971–1974.
51. Zaidi M, Adebanjo OA, Moonga BS, Sun L, Huang CL 1999 Emerging insights into the role of calcium ions in osteoclast regulation. J Bone Miner Res **14:**669–674.
52. Pi M, Faber P, Ekema G, Jackson PD, Ting A, Wang N, Fontilla-Poole M, Mays RW, Brunden KR, Harrington JJ, Quarles LD 2005 Identification of a novel extracellular cation-sensing G-protein-coupled receptor. J Biol Chem **280:**40201–40209.
53. Dvorak MM, Siddiqua A, Ward DT, Carter DH, Dallas SL, Nemeth EF, Riccardi D 2004 Physiological changes in extracellular calcium concentration directly control osteoblast function in the absence of calciotropic hormones. Proc Natl Acad Sci USA **101:**5140–5145.
54. Chang W, Tu C, Chen T-H, Komuves L, Oda Y, Pratt S, Miller S,

© 2008 American Society for Bone and Mineral Research

Shoback D 1999 Expression and signal transduction of calcium-sensing receptors in cartilage and bone. Endocrinology **140:**5883–5893.
55. Chang W, Tu C, Chen T, Liu B, Elalieh H, Dvorak M, Clemens T, Kream B, Halloran B, Bikle D, Shoback D 2007 Conditional knockouts in early and mature osteoblasts reveals a critical role for Ca2+ receptors in bone development. J Bone Mineral Res **22:**S79.
56. Mentaverri R, Yano S, Chattopadhyay N, Petit L, Kifor O, Kamel S, Terwilliger EF, Brazier M, Brown EM 2006 The calcium sensing receptor is directly involved in both osteoclast differentiation and apoptosis. FASEB J **20:**2562–2564.
57. Jacenko O, Tuan RS 1995 Chondrogenic potential of chick embryonic calvaria: I. Low calcium permits cartilage differentiation. Dev Dyn **202:**13–26.
58. Chang W, Tu C, Bajra R, Komuves L, Miller S, Strewler G, Shoback D 1999 Calcium sensing in cultured chondrogenic RCJ3.1C5.18 cells. Endocrinology **140:**1911–1919.
59. Tu C, Elalieh H, Chen T, Liu B, Hamilton M, Shoback D, Bikle D, Chang W 2007 Expression of Ca^{2+} receptors in cartilage is essential for embryonic skeletal develoment in vivo. J Bone Miner Res **22:**S1;S50.
60. Gill JJ, Bartter F 1961 On the impairment of renal concentrating ability in prolonged hypercalcemia and hypercalciuria in man. J Clin Invest **40:**716–722.
61. Valenti G, Laera A, Gouraud S, Pace G, Aceto G, Penza R, Selvaggi FP, Svelto M 2002 Low-calcium diet in hypercalciuric enuretic children restores AQP2 excretion and improves clinical symptoms. Am J Physiol Renal Physiol **283:**F895–F903.
62. Sands JM, Flores FX, Kato A, Baum MA, Brown EM, Ward DT, Hebert SC, Harris HW 1998 Vasopressin-elicited water and urea permeabilities are altered in IMCD in hypercalcemic rats. Am J Physiol **274:**F978–F985.
63. Wang W, Lu M, Balazy M, Hebert SC 1997 Phospholipase A2 is involved in mediating the effect of extracellular Ca^{2+} on apical K^+ channels in rat TAL. Am J Physiol **273:**F421–F429.
64. Rogers KV, Dunn CK, Hebert SC, Brown EM 1997 Localization of calcium receptor mRNA in the adult rat central nervous system by in situ hybridization. Brain Res **744:**47–56.
65. Quinn SJ, Ye CP, Diaz R, Kifor O, Bai M, Vassilev P, Brown E 1997 The Ca^{2+}-sensing receptor: A target for polyamines. Am J Physiol **273:**C1315–C1323.
66. Strewler GJ 1994 Familial benign hypocalciuric hypercalcemia–from the clinic to the calcium sensor. West J Med **160:**579–580.
67. Pearce SH, Williamson C, Kifor O, Bai M, Coulthard MG, Davies M, Lewis-Barned N, McCredie D, Powell H, Kendall-Taylor P, Brown EM, Thakker RV 1996 A familial syndrome of hypocalcemia with hypercalciuria due to mutations in the calcium-sensing receptor. N Engl J Med **335:**1115–1122.
68. Nemeth EF, Fox J 1999 Calcimimetic compounds: A direct approach to controlling plasma levels of parathyroid hormone in hyperparathyroidism. Trends Endocrinol Metab **10:**66–71.
69. Block GA, Martin KJ, de Francisco AL, Turner SA, Avram MM, Suranyi MG, Hercz G, Cunningham J, Abu-Alfa AK, Messa P, Coyne DW, Locatelli F, Cohen RM, Evenepoel P, Moe SM, Fournier A, Braun J, McCary LC, Zani VJ, Olson KA, Drueke TB, Goodman WG 2004 Cinacalcet for secondary hyperparathyroidism in patients receiving hemodialysis. N Engl J Med **350:**1516–1525.
70. Silverberg SJ, Rubin MR, Faiman C, Peacock M, Shoback DM, Smallridge RC, Schwanauer LE, Olson KA, Klassen P, Bilezikian JP 2007 Cinacalcet hydrochloride reduces the serum calcium concentration in inoperable parathyroid carcinoma. J Clin Endocrinol Metab **92:**3803–3808.
71. Gowen M, Stroup GB, Dodds RA, James IE, Votta BJ, Smith BR, Bhatnagar PK, Lago AM, Callahan JF, DelMar EG, Miller MA, Nemeth EF, Fox J 2000 Antagonizing the parathyroid calcium receptor stimulates parathyroid hormone secretion and bone formation in osteopenic rats. J Clin Invest **105:**1595–1604.
72. Conigrave AD, Quinn SJ, Brown EM 2000 L-amino acid sensing by the extracellular Ca^{2+}-sensing receptor. Proc Natl Acad Sci USA **97:**4814–4819.
73. Kerstetter JE, Caseria DM, Mitnick ME, Ellison AF, Gay LF, Liskov TA, Carpenter TO, Insogna KL 1997 Increased circulating concentrations of parathyroid hormone in healthy, young women consuming a protein-restricted diet. Am J Clin Nutr **66:**1188–1196.
74. Itami A, Kato M, Komoto I, Doi R, Hosotani R, Shimada Y, Imamura M 2001 Human gastrinoma cells express calcium-sensing receptor. Life Sci **70:**119–129.
75. Mangel AW, Prpic V, Wong H, Basavappa S, Hurst LJ, Scott L, Garman RL, Hayes JS, Sharara AI, Snow ND, Walsh JH, Liddle RA 1995 Phenylalanine-stimulated secretion of cholecystokinin is calcium dependent. Am J Physiol **268:**G90–G94.

Chapter 28. Vitamin D: Production, Metabolism, Mechanism of Action, and Clinical Requirements

Daniel Bikle,[1] John Adams,[2] and Sylvia Christakos[3]

[1]*Veteran Affairs Medical Center and Department of Medicine and Dermatology, University of California, San Francisco, California;* [2]*Orthopaedic Hospital, University of California, Los Angeles, California;* [3]*Department of Biochemistry and Molecular Biology, New Jersey Medical School, Newark, New Jersey*

VITAMIN D_3 PRODUCTION

Vitamin D_3 is produced from 7-dehydrocholesterol (7-DHC; Fig. 1). Although irradiation of 7-DHC was known to produce pre-D_3 (which subsequently undergoes a temperature-dependent rearrangement of the triene structure to form D_3, lumisterol, and tachysterol), the physiologic regulation of this pathway was not well understood until the studies of Holick et al.[1–3] They showed that the formation of pre-D_3 under the influence of solar or UVB irradiation (maximal effective wavelength between 290 and 310 nm) is relatively rapid and reaches a maximum within hours. Both the degree of epidermal pigmentation and the intensity of exposure correlate with the time needed to achieve this maximal concentration of pre-D_3 but do not alter the maximal level achieved. Although pre-D_3 levels reach a maximum level, the biologically inactive lumisterol and tachysterol accumulate with continued UV exposure. Thus, short exposure to sunlight would be expected to lead to a prolonged production of D_3 in the exposed skin because of the relatively slow thermal conversion of pre-D_3 to D_3 and of lumisterol to pre-D_3. Prolonged exposure to sunlight would not produce toxic amounts of D_3 because of the photoconversion of pre-D_3 to lumisterol and tachysterol. Melanin in the epidermis, by absorbing UV irradiation, can also reduce the effectiveness of sunlight in producing D_3 in the skin. Sunlight exposure increases melanin production and therefore provides

The authors state that they have no conflicts of interest.

Key words: vitamin D, bone, intestine, kidney, osteomalacia, immune function, keratinocytes, vitamin D metabolism, vitamin D mechanism of action, cancer, malabsorption

© 2008 American Society for Bone and Mineral Research

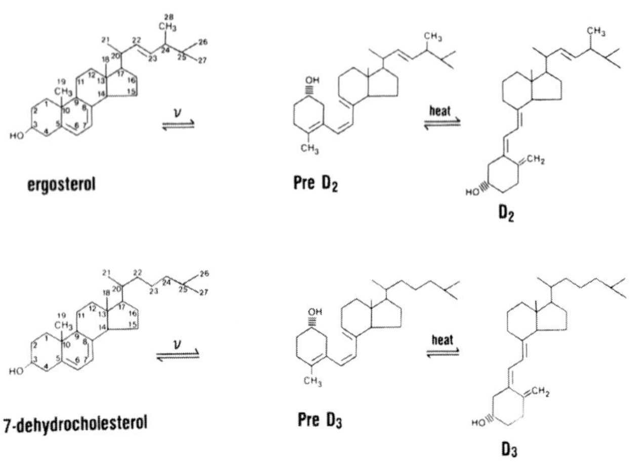

FIG. 1. The photolysis of ergosterol and 7-dehydrocholesterol to vitamin D_2 (ergocalciferol) and vitamin D_3 (cholecalciferol), respectively. An intermediate is formed after photolysis, which undergoes a thermal-activated isomerization to the final form of vitamin D. The rotation of the A-ring puts the 3β-hydroxyl group into a different orientation with respect to the plane of the A-ring during production of vitamin D.

another mechanism by which excess D_3 production can be prevented. As just noted, the intensity of UV irradiation is also important for D_3 production and is dependent on latitude; In Edmonton, Canada (52° N), very little D_3 is produced in exposed skin from mid-October to mid-April, whereas in San Juan (18° N) the skin is able to produce D_3 all year long.[4] Clothing and sunscreen effectively prevent D_3 production in the covered areas.

VITAMIN D METABOLISM

Vitamin D, which by itself is biologically inert, must be ferried into the circulation bound to the serum vitamin D–binding protein (DBP) to be metabolically converted to the prohormone, 25-hydroxyvitamin D [25(OH)D; Fig. 2]. There are a number of cytochrome P450 enzymes capable of converting vitamin D to 25(OH)D. These enzymes are principally found in the liver, exhibit a high capacity for substrate vitamin D, and release 25(OH)D back into the circulation and not into the bile. As such, serum 25(OH)D is the most reliable indicator of whether too little or too much vitamin D is entering the host.[5]

25(OH)D is biologically inert unless present in intoxicating concentrations in the blood because of the ingestion of large amounts of vitamin D. Otherwise, it must be converted to $1,25(OH)_2D$, the specific, naturally occurring ligand for the vitamin D receptor (VDR) through CYP27B1 hydroxylase (Fig. 2). The 1-hydroxylase is a heme-containing, inner mitochondrial membrane-embedded, cytochrome P450 mixed function oxidase requiring molecular oxygen and a source of electrons for biological activity. Although the proximal renal tubular epithelial cell is the richest source of 1-hydroxylase and responsible for generating the relatively large amounts of $1,25(OH)_2D$ that are needed to achieve the endocrine functions of the hormone in mineral ion homeostasis, this enzyme is also encountered in a number of extrarenal sites, including immune cells and a variety of normal and malignant epithelia,[6] where it functions to provide $1,25(OH)_2D$ for intracrine or paracrine access to the VDR in these and neighboring cells. As discussed below, the VDR has an extraordinarily broad distribution among human tissues. There are four major recognized means of regulating the 1-hydroxylase: (1) controlling the availability of substrate 25(OH)D to the enzyme; (2) controlling the amount of CYP27B1 hydroxylase expressed; (3) altering the activity of the enzyme by co-factor availability; and (4) controlling the amount and activity of the alternatively spliced, catabolic CYP24 hydroxylase.

For the kidney, CYP27B1 hydroxylase substrate is provided by the endocytic internalization of filtered, megalin-bound DBP carrying 25(OH)D into the proximal tubular cell from the urinary side of that cell. Regulation of CYP27B1 in the proximal nephron is principally controlled at the level of transcription, with circulating PTH and fibroblast growth factor 23 (FGF-23) being the major stimulator and inhibitor of *CYP27B1* gene expression, respectively (see below). In the kidney, the CYP24 hydroxylase, also a mitochondrial P450, serves not only to limit the amount of $1,25(OH)_2D$ leaving the kidney for distant target tissues by accelerating its catabolism to $1,24,25(OH)_3D$ but also by shunting available substrate 25(OH)D away from 1-hydroxylase. In both cases, the 24-hydroxylated products are biologically inert and degraded by the same enzyme to side chain–cleaved, water-soluble catabolites. The *CYP24 hydroxylase* gene is under stringent transcriptional control by $1,25(OH)_2D$ itself, providing a robust means of proximate, negative feedback regulation of the amount of $1,25(OH)_2D$ made in and released from the kidney.[7] By comparison, the activity of some of the extrarenal, intracrine/paracrine-acting 1-hydroxylase, like that which occurs in disease-activated macrophages, seems to be primarily governed by the availability of extracellular substrate 25(OH)D to the enzyme. It is postulated that this is caused by the expression of an amino-terminally truncated splice variant of the *CYP24* gene that cannot be transported into mitochondria.[8,9] The result is generation of a noncatalytically active enzyme, albeit one that can serve as a cytoplasmic reservoir for the CYP24 substrates, $1,25(OH)_2D$ and 25(OH)D. Also contrary to renal 1-hydroxylase, the extrarenal CYP27B1 hydroxylase, at least in macrophages, (1) is immune to control by either PTH or FGF-23 (receptors for these molecules are not expressed to any degree in inflammatory cells), (2) is susceptible to induction by Toll-like receptors (TLR) ligands shed by microbial agents, and (3) can be upregulated by nontraditional electron

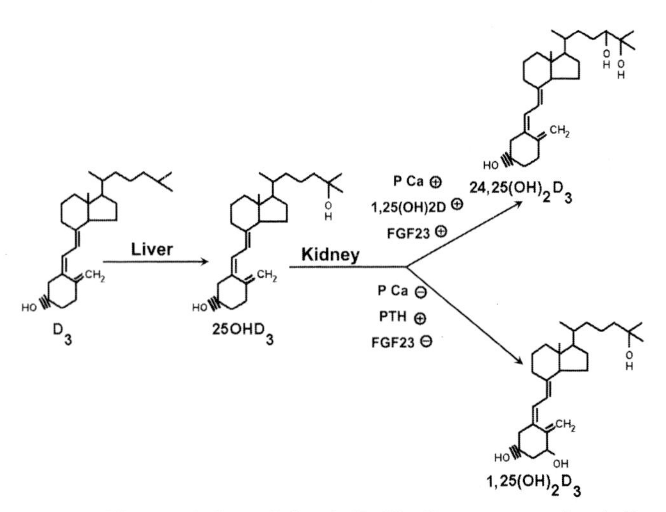

FIG. 2. The metabolism of vitamin D. The liver converts vitamin D to 25OHD. The kidney converts 25OHD to $1,25(OH)_2D_3$ and $24,25(OH)_2D_3$. Control of metabolism is exerted primarily at the level of the kidney, where low serum phosphorus, low serum calcium, low FGF23, and high parathyroid hormone (PTH) levels favor production of $1,25(OH)_2D_3$, whereas high serum phosphorus, calcium, FGF23, and $1,25(OH)_2D_3$ and low PTH favor $24,25(OH)_2D_3$ production.

© 2008 American Society for Bone and Mineral Research

donors like NO.[6] CYP27B1 hydroxylase in keratinocytes, on the other hand, shares features of both the renal and macrophage 1-hydroxylase in that it is associated with very active CYP24 hydroxylase, which limits the levels of $1,25(OH)_2D$ in the cell, is stimulated by cytokines such as TNF-α[10] and IFN-γ[11] but not by c-AMP, and is induced by TLR2 activation.[12]

TRANSPORT OF VITAMIN D IN THE BLOOD

For the hormone $1,25(OH)_2D$ to reach any of its target tissues, with the exception of the skin where it can be both produced and act locally as just described, vitamin D must be able to escape its synthetic site in the skin or it absorption site in the gut and be transported to tissues expressing one of the vitamin D 25-hydroxylase genes. From there, 25(OH)D must travel to tissue sites expressing the *CYP27B1 hydroxylase* gene, and synthesized $1,25(OH)_2D$ must be able to gain access to target tissues containing cells expressing the VDR for the genomic actions of the sterol hormone to be realized. The serum DBP, a member of the albumin family of proteins, is the specific chaperone for vitamin D and its metabolites in the serum.[13] It has a high capacity (<5% saturated with vitamin D metabolites in humans) and is bound with high affinity (nM range) by vitamin D, particularly the 25-hydroxylated metabolites 25(OH)D, $24,25(OH)_2D$, and $1,25(OH)_2D$.[14] DBP is produced mainly in the liver and is freely filterable across the glomerulus into the urine. DBP has a serum half-life of 2.5–3.0 days, indicating that it must be largely reclaimed from the urine once filtered. Reclamation is achieved by DBP being bound by the endocytic, low-density lipoprotein (LDL)-like co-receptor molecules megalin and cubulin embedded in the plasma membrane of the proximal renal tubular epithelial cell, with eventual transcellular transport and return to the circulation through intracellular DBP (IDBP) chaperones in the heat protein-70 family.[15] No human, DBP-null homozygote has yet been described, suggesting that, unlike the DBP-null mice that are both viable and fertile,[16] such a human genotype would be embryonically lethal.

INTERNALIZATION OF VITAMIN D METABOLITES

Once bound to DBP and shuttled to sites of metabolism, action, and/or catabolism, vitamin D metabolites must gain access to the interior of their target cell and arrive safely at their intracellular destination (i.e., nucleus for transaction through the VDR, inner mitochondrial membrane for access to the CYP27B and CYP24 hydroxylases). Although possible, it seems unlikely that simple diffusion of the sterol metabolite off the serum DBP and simple diffusion through the plasma membrane to a specific intracellular destination, the so-called "free hormone" hypothesis, can account for the required specificity for targeted metabolite delivery. Current observations suggest that, similar to that which occurs in the kidney, there exists a plasma membrane–anchored receptor for DBP that is endocytically internalized with intracellular chaperones moving the metabolite(s) to specific intracellular destinations (e.g., the CYP27B1 and VDR).[17]

MECHANISM OF ACTION

The mechanism of action of the active form of vitamin D, $1,25(OH)_2D_3$, is similar to that of other steroid hormones. The intracellular mediator of $1,25(OH)_2D_3$ function is the VDR. $1,25(OH)_2D_3$ binds stereospecifically to VDR, which is a high-affinity, low-capacity intracellular receptor that has extensive homology with other members of the superfamily of nuclear receptors including receptors of steroid and thyroid hormones. VDR functions as a heterodimer with the retinoid X receptor (RXR) for activation for vitamin D target genes. Once formed, the $1,25(OH)_2D_3$-VDR-RXR heterodimeric complex interacts with specific DNA sequences (vitamin D response elements [VDRE]) within the promoter of target genes, resulting in either activation or repression of transcription.[18–21] In general, for activation of transcription, the VDRE consensus consists of two direct repeats of the hexanucleotide sequence GGGTGA separated by three nucleotide pairs. The mechanisms involved in VDR-mediated transcription after binding of the $1,25(OH)_2D_3$-VDR-RXR heterodimeric complex to DNA are now beginning to be defined. TFIIB, several TATA binding protein associated factors (TAFs), as well as the p160 co-activators known also as steroid receptor activator-1, -2, and -3 (SRC-1, SRC-2, and SRC-3), which have histone acetylase (HAT) activity, have been reported to be involved in VDR-mediated transcription. In addition to acetylation, methylation also occurs on core histones. Recent studies have indicated that cooperativity between histone methyltransferases and p160 co-activators may also play a fundamental role in VDR-mediated transcriptional activation.[22] VDR-mediated transcription is also mediated by the co-activator complex DRIP (VDR interacting protein). This complex does not have HAT activity but rather functions, at least in part, through recruitment of RNA polymerase II. It has been suggested that the SRC/CREB-binding protein (CBP) co-activator complex is recruited first for chromatin remodeling followed by the recruitment of the transcription machinery by the DRIP complex.[20,23] In addition, a number of promoter-specific transcription factors including YY1 and CCAAT enhancer binding proteins β and δ have been reported to modulate VDR-mediated transcription.[24–26] It has been suggested that cell- and promoter-specific functions of VDR may be mediated through differential recruitment of coactivators.

OVERVIEW OF VITAMIN D REGULATION OF CALCIUM AND PHOSPHATE METABOLISM

The classic actions of $1,25(OH)_2D_3$ involve its regulation of calcium and phosphate flux across three target tissues: bone, gut, and kidney. The mechanisms by which $1,25(OH)_2D_3$ operates in these tissues will be described in more detail below. However, the receptor for $1,25(OH)_2D_3$ (VDR) is widespread and not limited to these classic target tissues. Indeed, the list of tissues not containing the VDR is probably shorter than the list of tissues that contain the VDR. Furthermore, as discussed previously, a number of these tissues express CYP27B1 and therefore can make their own $1,25(OH)_2D_3$. The biological significance of these observations is found in the large number of nonclassical actions of vitamin D including effects on cellular proliferation and differentiation, regulation of hormone secretion, and immune modulation. Examples of these actions will be discussed below. In at least the classic actions of vitamin D, $1,25(OH)_2D_3$ acts in concert with two peptide hormones, PTH and FGF-23 (Fig. 3). In each case, feedforward and feedback loops are operative. PTH is the major stimulator of $1,25(OH)_2D_3$ production in the kidney. $1,25(OH)_2D_3$ in turn suppresses PTH production directly by a transcriptional mechanism and indirectly by increasing serum calcium levels. Calcium acts through the calcium receptor (CaR) in the parathyroid gland to suppress PTH release. $1,25(OH)_2D_3$ increases the levels of CaR in the parathyroid gland just as calcium increases the $1,25(OH)_2D_3$ receptor (VDR) in the parathyroid gland, further enhancing the negative influence of calcium and $1,25(OH)_2D_3$ on PTH secretion. FGF-23, on the other hand, inhibits $1,25(OH)_2D_3$ production by the kidney while increasing the expression of CYP24, whereas $1,25(OH)_2D_3$ stimulates

© 2008 American Society for Bone and Mineral Research

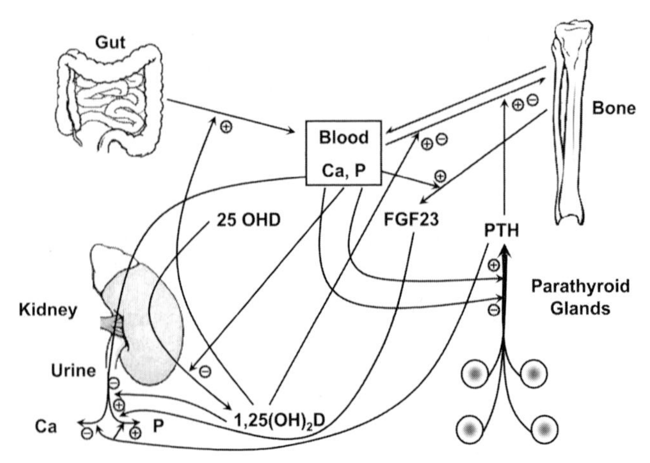

FIG. 3. $1,25(OH)_2D_3$ interacts with other hormones, in particular FGF23 and PTH, to regulate calcium and phosphate homeostasis. As noted in the legend to figure 2, FGF23 inhibits whereas PTH stimulates $1,25(OH)_2D_3$ production by the kidney. In turn $1,25(OH)_2D_3$ inhibits PTH production but stimulates that of FGF23. Calcium and phosphate in turn regulate FGF23, PTH, and so $1,25(OH)_2D_3$ indirectly.

FGF-23 production. Dietary phosphate also regulates FGF-23 levels [high phosphate stimulates, an effect independent of $1,25(OH)_2D_3$ as shown by phosphate regulation of FGF-23 in the VDR-null mice]. Whether phosphate has its own receptor like calcium is unclear. FGF-23 expression is found in a number of tissues including the parathyroid gland, but its greatest expression is in osteocytes, bone-lining cells, and active osteoblasts. Thus, in considering the mechanisms of action of vitamin D and its active metabolite $1,25(OH)_2D_3$ in vivo, the roles of PTH and FGF-23 must also be considered.

CLASSIC TARGET TISSUES

Bone

Whether $1,25(OH)_2D_3$ acts directly on bone or whether the anti-rachitic effects of $1,25(OH)_2D_3$ are indirect and are caused by $1,25(OH)_2D_3$ stimulation of intestinal calcium and phosphorus absorption resulting in increased incorporation of calcium and phosphorus into bone has been a matter of debate. VDR ablated mice (VDR knockout mice) develop secondary hyperparathyroidism, hypocalcemia, and rickets after weaning.[27,28] However, when VDR KO mice are fed a rescue diet containing high levels of calcium, phosphorus, and lactose, serum ionized calcium and PTH levels are normalized, and rickets and osteomalacia are prevented, suggesting that a major effect of $1,25(OH)_2D_3$ is the provision of calcium and phosphate to bone from the intestine rather than a direct action on bone.[29] In vitro studies, however, support a direct effect of $1,25(OH)_2D_3$ on bone.[30] $1,25(OH)_2D_3$ can stimulate the formation of bone-resorbing osteoclasts.[30] However, VDR is not present in osteoclasts but rather in osteoprogenitor cells, osteoblast precursors, and mature osteoblasts. Stimulation of osteoclast formation by $1,25(OH)_2D_3$ involves upregulation by $1,25(OH)_2D_3$ in osteoblastic cells of RANKL (osteoclast differentiating factor) and requires cell to cell contact between osteoblastic cells and osteoclast precursors.[31] Osteoclastogenesis inhibitory factor/osteoprotegerin, which is a decoy receptor for RANKL and antagonizes RANKL function thus blocking osteoclastogenesis, is downregulated by $1,25(OH)_2D_3$.[31] $1,25(OH)_2D_3$ has also been reported to stimulate the production in osteoblasts of the calcium-binding proteins osteocalcin and osteopontin.[32,33] Runx2, a transcriptional regulator of osteoblast differentiation, is also regulated by $1,25(OH)_2D_3$.[34] Transgenic mice overexpressing VDR in osteoblastic cells have increased bone formation, further indicating direct effects of $1,25(OH)_2D_3$ on bone.[35] Thus, effects of $1,25(OH)_2D_3$ on bone are diverse and can affect formation or resorption.

Intestine

When the demand for calcium increases from a diet deficient in calcium, from growth, or from pregnancy or lactation, the synthesis of $1,25(OH)_2D_3$ is increased, stimulating the rate of calcium absorption. In VDR KO mice, a major defect is in intestinal calcium absorption, suggesting that a principal action of $1,25(OH)_2D_3$ to maintain calcium homeostasis is increased intestinal calcium absorption.[36–38] It is thought that intestinal calcium absorption is comprised of two different modes of calcium transport: the saturable process that is mainly transcellular and a diffusional mode that is nonsaturating, requires a lumenal free calcium concentration >2–6 mM and is paracellular (the movement of calcium is across tight junctions and intracellular spaces and is directly related to the concentration of calcium in the intestinal lumen). The saturable component of intestinal calcium absorption is observed predominantly in the duodenum. The diffusional, nonsaturable process is observed all along the intestine (duodenum, jejunum, and ileum). $1,25(OH)_2D_3$ has been reported to affect both the transcellular and the paracellular path.[37,38] The transcellular process is comprised of three $1,25(OH)_2D_3$-regulated steps: the entry of calcium across the brush border membrane, intracellular diffusion, and the energy requiring extrusion of calcium across the basolateral membrane.[37,38] It is thought that the calcium-binding protein, calbindin, which is induced by $1,25(OH)_2D_3$ in the intestine, acts to facilitate the diffusion of calcium through the cell interior toward the basolateral membrane. Recent studies in which calbindin-D_{9k}–null mutant mice showed no change in $1,25(OH)_2D_3$-mediated intestinal calcium absorption and in serum calcium levels compared with wildtype mice[36,39] provide evidence that calbindin alone is not responsible for $1,25(OH)_2D_3$-mediated intestinal calcium absorption.

$1,25(OH)_2D_3$ also affects calcium extrusion from the enterocyte. The plasma membrane calcium pump (PMCA) has been reported to be stimulated by $1,25(OH)_2D_3$, suggesting that intestinal calcium absorption may involve a direct effect of $1,25(OH)_2D_3$ on calcium pump expression.[38]

The rate of calcium entry into the enterocyte is also increased by $1,25(OH)_2D_3$. Recently a calcium selective channel, TRPV6, which is co-localized with calbindin and is induced by $1,25(OH)_2D_3$, was cloned from rat duodenum.[40,41] It has been suggested that TRPV6 plays a key role in vitamin D–dependent calcium entry into the enterocyte.

In addition to intestinal calcium absorption, $1,25(OH)_2D_3$ also results in enhanced intestinal phosphorus absorption. Although the mechanisms involved have been a matter of debate, it has been suggested that $1,25(OH)_2D_3$ stimulates the active transport of phosphorus.[42]

Kidney

A third target tissue involved in $1,25(OH)_2D_3$-mediated mineral homeostasis is the kidney. $1,25(OH)_2D_3$ has been reported to enhance the actions of PTH on calcium transport in the distal tubule, at least in part, by increasing PTH receptor mRNA and binding activity in distal tubule cells.[43] $1,25(OH)_2D_3$ also induces the synthesis of the calbindins in the distal tubules.[37] It has been suggested that calbindin-D_{28k}

© 2008 American Society for Bone and Mineral Research

stimulates the high-affinity system in the distal luminal membrane, and calbindin-D_{9k} enhances the ATP-dependent calcium transport of the basolateral membrane.[37] Similar to studies in the intestine, an apical calcium channel, TRPV5, which is co-localized with the calbindins and induced by $1{,}25(OH)_2D_3$, has been identified in the distal convoluted tubule and distal connecting tubules.[40] Calbindin-D_{28k} was reported to associate directly with TRPV5 and to control TRPV5-mediated calcium influx.[44] Thus, $1{,}25(OH)_2D_3$ affects calcium transport in the distal tubule by enhancing the action of PTH and by inducing TRPV5 and the calbindins. Another important effect of $1{,}25(OH)_2D_3$ in the kidney is the inhibition of the $25(OH)D_3$ 1 α-hydroxylase enzyme (CYP27B1) and the induction of the 24-hydroxylase enzyme (CYP24).[45] Besides calcium transport in the distal nephron and modulation of the $25(OH)D_3$ hydroxylases, effects of $1{,}25(OH)_2D_3$ on phosphate reabsorption in the proximal tubule have been suggested. Vitamin D has been reported to increase or decrease renal phosphate reabsorption depending on the parathyroid status and on experimental conditions.

NONCLASSICAL TARGET TISSUES

Parathyroid Glands

The parathyroid glands are an important target of $1{,}25(OH)_2D_3$. As discussed previously, $1{,}25(OH)_2D_3$ inhibits the synthesis and secretion of PTH and prevents the proliferation of the parathyroid gland to maintain normal parathyroid status.[46,47] It has also been shown that $1{,}25(OH)_2D_3$ upregulates calcium sensing receptor (CaSR) transcription,[48] suggesting that $1{,}25(OH)_2D_3$ sensitizes the parathyroid gland to calcium inhibition.

Pancreas

The pancreas was one of the first nonclassical target tissues in which receptors for $1{,}25(OH)_2D_3$ were identified.[49] Although $1{,}25(OH)_2D_3$ has been reported to play a role in insulin secretion, the exact mechanisms remain unclear. Autoradiographic data and immunocytochemical studies have localized VDR and calbindin-D_{28k},respectively, in pancreatic β cells.[50,51] Studies using calbindin-D_{28k}–null mutant mice have indicated that calbindin-D_{28k}, by regulating intracellular calcium, can modulate depolarization-stimulated insulin release.[52] In addition to modulating insulin release, calbindin-D_{28k}, by buffering calcium, can protect against cytokine-mediated destruction of β cells.[53] These findings have important therapeutic implications for type 1 diabetes and the prevention of cytokine destruction of pancreatic β cells, as well as type 2 diabetes and the potentiation of insulin secretion.

IMMUNOBIOLOGY OF VITAMIN D

Nonclassical regulation of immune responses by $1{,}25(OH)_2D$ was first reported 25 yr ago with the discovery of the presence of the VDR in activated human inflammatory cells[54] and the ability of disease-activated macrophages to make $1{,}25(OH)_2D$.[55] Recent studies have shown that $1{,}25(OH)_2D$ regulates both innate and adaptive immunity, but in opposite directions, namely promoting the former while repressing the latter.

Vitamin D and Innate Immunity

Innate immunity encompasses the ability of the host immune system to recognize and respond to an offending antigen. In 1986, Rook et al.[56] described studies using cultured human macrophages in which they showed that $1{,}25(OH)_2D$ can inhibit the growth of *Mycobacterium tuberculosis*. Although this seminal report was widely cited, it is only in the last 3 yr that more comprehensive appraisals of the antimicrobial effects of vitamin D metabolites have been published. In silico screening of the human genome showed the presence of a VDRE in the promoter of the human gene for cathelicidin, whose product LL37 is an antimicrobial peptide capable of killing bacteria.[57] Subsequent studies confirmed the ability of $1{,}25(OH)_2D$[58] and its precursor 25(OH)D[59] to induce expression of cathelicidin in cells of the monocyte/macrophage and epidermal lineage, respectively, highlighting the potential for intracrine/autocrine induction of antimicrobial responses in cells that also express the 25(OH)D-activating enzyme, CYP27B1. Although detectable in many cell types, functionally significant expression and activity of CYP27B1 seem to be dependent on cell-specific stimulation of a broad spectrum of immune surveillance proteins, the TLRs. The TLRs are an extended family of host, noncatalytic transmembrane pattern-recognition receptors (PRRs) that interact with specific pathogen-associated membrane patterns (PAMPs) shed by infectious agents and trigger the innate immune response in the host.[60]

In this regard, Liu et al.[61] recently used DNA array to characterize changes in gene expression after activation of the human macrophage TLR2/1 dimer by one of the PAMPs for *M. tuberculosis.* Macrophages, but not dendritic cells, thus treated showed increased expression of both *CYP27B1* and *VDR* genes and gene products and showed intracrine induction of the antimicrobial cathelicidin gene with subsequent mycobacterial killing in response to 25(OH)D and $1{,}25(OH)_2D$. In fact, microbial killing was more efficiently achieved with the prohormone 25(OH)D than with $1{,}25(OH)_2D$ at similar extracellular concentrations, indicating that the robustness of the human innate response to microbial challenge is dependent on the serum 25(OH)D status of the host. This concept was confirmed in these studies by the ability of 25(OH)D-sufficient serum to rescue a deficient, cathelicidin-driven antimicrobial response in human macrophages conditioned in vitamin D–deficient serum. A similar vitamin D–directed antimicrobial-generating capacity has been recently observed in wounded skin,[12] suggesting that TLR-driven expression of cathecidin, requiring the intracellular synthesis and genomic action of $1{,}25(OH)_2D$, is a common response feature to infectious agent invasion. Although not yet proven in a clinical setting, it is possible that increasing the serum 25(OH)D level of the host to the normal range (>30 ng/ml) will augment the effectiveness of the innate immune response to such commonly encountered microbial agents as *M. tuberculosis* and HIV known to trip macrophage TLR signaling pathways. Reinforcing these events is the ability of locally generated $1{,}25(OH)_2D$ to escape the confines of the cell in which it is made to act on neighboring VDR-expressing monocytes to promote their maturation to mature macrophages,[62] thus acting as a feedforward signal to further enhance the innate immune response.

Vitamin D and the Adaptive Immune Response

The adaptive immune response is generally defined by T and B lymphocytes and their ability to produce cytokines and immunoglobulins, respectively, to specifically combat the source of the antigen presented to them by cells (i.e., macrophages, dendritic cells) of the innate immune response. As previously noted,[54] the presence of VDR in activated, but not resting, human T and B lymphocytes was the first observation implicating these cells as targets for the noncalciotropic responses to $1{,}25(OH)_2D$. Contrary to the role of locally produced $1{,}25(OH)_2D$ to promote the innate immune response,

© 2008 American Society for Bone and Mineral Research

the hormone exerts a generalized dampening effect on lymphocyte function. With respect to B cells, $1,25(OH)_2D$ suppresses proliferation and immunoglobulin production and retards the differentiation of B-lymphocyte precursors to mature plasma cells. With regard to T cells, $1,25(OH)_2D$, acting through the VDR, inhibits the proliferation of uncommitted T_H (helper) cells. This results in diminished numbers of T_H capable of maturing to IFNγ-elaborating, macrophage-stimulating T_{H1} cells and, to a lesser extent, to interleukin (IL)-4–, IL-5–, and IL-13–producing, B cell–activating T_{H2} cells. On the other hand, the hormone promotes the proliferation of immunosuppressive regulatory T cells, so called T_{REG}s,[63] and promotes their accumulation at sites of inflammation by stimulating expression of the T-cell homing molecule, CCL22, by dendritic cells[64] in the local inflammatory microenvironment. In fact, it is this generalized ability of $1,25(OH)_2D$ to quell the adaptive immune response, which has prompted the use of the hormone and its analogs in the adjuvant treatment of inflammatory autoimmune and neoplastic disorders.

In summary, the collective, concerted action of $1,25(OH)_2D$ is to promote the host's response to an invading pathogen while simultaneously acting to limit what might be an overzealous immune response to that pathogen, representative of the process of tolerance. Once again, a good example is that of infection with the intracellular pathogen *M. tuberculosis.* In this case, the pathogen evokes an exceptionally robust innate immune response that is fueled by the endogenous production of $1,25(OH)_2D$ by macrophages and dendritic cells, which also express the CYP27B1, at the site of host invasion. If substantial amounts of $1,25(OH)_2D$ escape the confines of the macrophage or the dendritic cell, the immunostimulation of VDR-expressing, activated lymphocytes in that environment is quelled by $1,25(OH)_2D$. If the innate immune response is extreme and enough $1,25(OH)_2D$ finds its way into the general circulation, as may occur in human granuloma-forming disorders such as sarcoidosis and tuberculosis, an endocrine effect of the hormone, most notably hypercalciuria and hypercalcemia, can be observed.

KERATINOCYTE FUNCTION IN EPIDERMIS AND HAIR FOLLICLES

$1,25(OH)_2D$-Regulated Epidermal Differentiation

$1,25(OH)_2D_3$ is likely to be an autocrine or paracrine factor for epidermal differentiation because it is produced in the keratinocyte by the same enzyme, CYP27B1, as found in the kidney, but under normal circumstances, keratinocyte production of $1,25(OH)_2D_3$ does not seem to contribute to circulating levels.[65] The receptors for and the production of $1,25(OH)_2D_3$ decrease with differentiation. Stimulation of differentiation is accompanied by the rise in mRNA and protein levels of involucrin and transglutaminase,[66] as well as the late differentiation markers filaggrin and loricrin.[67] The mechanisms by which $1,25(OH)_2D_3$ alter keratinocyte differentiation are multiple and include induction of the calcium receptor enhancing the effects of calcium on differentiation and induction of the phospholipase C family, which provide second messengers such as diacyl glycerol and inositol trisphosphate to the differentiation process. Although the most striking feature of the VDR-null mouse is the development of alopecia (also found in many but not all patients with mutations in the VDR), these mice also exhibit a defect in epidermal differentiation as shown by reduced levels of involucrin and loricrin, loss of keratohyalin granules, loss of the calcium gradient, and disruption of lamellar body production and secretion resulting in defective barrier function. Furthermore, both VDR and $1,25(OH)_2D_3$ production are required for normal antimicrobial peptide expression in response to epidermal injury.[12]

VDR Regulation of Hair Follicle Cycling

As noted above, alopecia is a well-known part of the phenotype of many patients with mutations in their VDR.[68] Vitamin D deficiency or lack of CYP27B1 per se is not associated with alopecia, and the alopecia can be rescued with mutants of VDR that fail to bind $1,25(OH)_2D_3$ or its co-activators.[69] Recent attention has been paid to both hairless (Hr), a putative transcription factor capable of binding the VDR and suppressing at least its ligand-dependent transcriptional activity, and β-catenin, which like Hr binds to VDR and may regulate its transcriptional activity (or vice versa). Hr mutations in both mice and humans and transcriptionally inactivating β-catenin mutations in mice result in phenocopies of the VDR-null animal with regard to the morphologic changes observed in hair cycling. In these models, the abnormality leading to alopecia develops during catagen at the end of the developmental cycle, precluding the re-initiation of anagen. Both VDR-null mice and those with disrupted Wnt signaling lose stem cells from the bulge, perhaps as a result of the loss of the interaction of the bulge with the dermal papilla.[70,71] Thus, although the mechanism by which VDR controls hair follicle cycling is not established, hair follicle cycling represents the best example by which VDR regulates a physiologic process that is independent of its ligand, $1,25(OH)_2D_3$, and therefore points to a novel mechanism of action for this transcriptional regulator.

NUTRITIONAL CONSIDERATIONS

Defining Vitamin D Sufficiency

Serum 25(OH)D levels provide a useful surrogate for assessing vitamin D status, because the conversion of vitamin D to 25(OH)D is less well controlled (i.e., primarily substrate dependent) than the subsequent conversion of 25(OH)D to $1,25(OH)_2D$. $1,25(OH)_2D$ levels, unlike 25(OH)D levels, are well maintained until the extremes of vitamin D deficiency because of secondary hyperparathyroidism and therefore do not provide a useful index for assessing vitamin D deficiency, at least in the initial stages. Historically, vitamin D sufficiency was defined as the level of 25(OH)D sufficient to prevent rickets in children and osteomalacia in adults. Levels of 25(OH)D <5 ng/ml (or 12 nM) are associated with a high prevalence of rickets or osteomalacia, and current "normal" levels of 25(OH)D are often stated to include levels as low as 15 ng/ml. However, there is a growing consensus that these lower limits of normal are too low. Although there is currently no consensus on the optimal levels, most experts define vitamin D deficiency as levels of 25(OH)D <30 ng/ml.[5] With this definition of vitamin D deficiency, a very large proportion of the population in both the developed and developing world are vitamin D deficient.

Impact on the Musculoskeletal System

This rethinking of the definition of vitamin D sufficiency comes from the appreciation that vitamin D affects a large number of physiologic functions in addition to bone mineralization. 25(OH)D levels are inversely proportional to PTH levels such that PTH levels increase at levels of 25(OH)D <30–40 ng/ml. Intestinal calcium transport increases 45–65% when the 25(OH)D levels are increased from 20 to 32 ng/ml. Large epidemiologic surveys showed a positive correlation between 25(OH)D levels and BMD, with no evidence for a plateau <30 ng/ml, and vitamin D and calcium supplementation

© 2008 American Society for Bone and Mineral Research

showed improvement in BMD in older individuals. Similarly a positive association between 25(OH)D levels and muscle function (e.g., walking speed, sit-to-stand) has been shown, even over the interval of 20–38 ng/ml, although the correlation is strongest at lower levels. Vitamin D supplementation with at least 800 IU improved lower extremity function, decreased body sway, and reduced falls. Most importantly, adequate levels of vitamin D and calcium supplementation prevent fractures.[72,73]

Impact Beyond the Musculoskeletal System

The impact of vitamin D extends beyond the musculoskeletal system and the regulation of calcium homeostasis. Vitamin D deficiency is a well-known accompaniment of various infectious diseases such as tuberculosis, and $1,25(OH)_2D_3$ has long been recognized to potentiate the killing of mycobacteria by monocytes. The nutritional aspect of these observations has recently been illuminated by the observation that the monocyte, when activated by mycobacterial lipopeptides, expresses CYP27B1, producing $1,25(OH)_2D_3$ from circulating 25(OH)D, and in turn inducing cathelicidin, an antimicrobial peptide that enhances killing of the *Mycobacterium*. Inadequate 25(OH)D levels abort this process.[61] Vitamin D deficiency and/or living at higher latitudes (with less sunlight) is associated with a number of autoimmune diseases including type 1 diabetes mellitus, multiple sclerosis, and Crohn's disease.[74] 25(OH)D levels are also inversely associated with type 2 diabetes mellitus and metabolic syndrome, and some studies have shown that vitamin D and calcium supplementation may prevent the progression to diabetes mellitus in individuals with glucose intolerance. Improvements in both insulin secretion and action have been observed. The potential to prevent certain cancers may be the most compelling reason for adequate vitamin D nutrition. A large body of epidemiologic data exists documenting the inverse correlation of 25(OH)D levels, latitude, and/or vitamin D intake with cancer incidence.[5] Although numerous types of cancers show reduction,[75] most attention has been paid to breast, colon, and prostate. A prospective 4-yr trial with 1100 IU vitamin D and 1400–1500 mg calcium showed a 77% reduction in cancers after the first year of study,[76] including a reduction in both breast and colon cancers. In this study, vitamin D supplementation raised the 25(OH)D levels from a mean of 28.8 to 38.4 ng/ml, with no changes in the placebo or calcium-only arms of the study.

Vitamin D Treatment Strategies

Adequate sunlight exposure is the most cost-effective means of obtaining vitamin D. Whole body exposure to sunlight has been calculated to provide the equivalent of 10,000 IU vitamin D_3.[77] A 0.5 minimal erythema dose of sunlight (i.e., one half the dose required to produce a slight reddening of the skin) or UVB radiation to the arms and legs, which can be achieved in 5–10 min on a bright summer day, has been calculated to be the equivalent of 3000 IU vitamin D_3.[5] However, concerns regarding the association between sunlight and skin cancer and/or aging have limited this approach, perhaps to the extreme, although it remains a viable option for those unable or unwilling to benefit from oral supplementation. Current recommendations for daily vitamin D supplementation (200 IU for children and young adults, 400 IU for adults 51–70 yr of age, and 600 IU for adults >71 yr of age) are too low and do not maintain 25(OH)D at the desired level for many individuals. Studies have shown that for every 100 IU vitamin D_3 supplementation administered, the 25(OH)D levels rise by 0.5–1 ng/ml.[77,78] Seven hundred to 800 IU seems to be the lower limit of vitamin D supplementation required to prevent fractures and falls. Unfortified food contains little vitamin D, with the exception of wild salmon and other fish products such as cod liver oil. Milk and other fortified beverages typically contain 100 IU/8-oz serving. Vitamin D_2 is substantially less potent than vitamin D_3, in part because it is more rapidly cleared. Therefore, if vitamin D_2 is used, it needs to be given at least weekly. Toxicity caused by vitamin D supplementation has not been observed at doses <10,000 IU/d.[79]

REFERENCES

1. Holick MF, McLaughlin JA, Clark MB, Doppelt SH 1981 Factors that influence the cutaneous photosynthesis of previtamin D3. Science **211**:590–593.
2. Holick MF, McLaughlin JA, Clark MB, Holick SA, Potts JT, Anderson RR, Blank IH, Parrish JA 1980 Photosynthesis of previtamin D3 in human and the physiologic consequences. Science 210:203–205.
3. Holick MF, Richtand NM, McNeill SC, Holick SA, Frommer JE, Henley JW, Potts JT 1979 Isolation and identification of previtamin D3 from the skin of exposed to ultraviolet irradiation. Biochemistry **18**:1003–1008.
4. Webb AR, Kline L, Holick MF 1988 Influence of season and latitude on the cutaneous synthesis of vitamin D3: Exposure to winter sunlight in Boston and Edmonton will not promote vitamin D3 synthesis in human skin. J Clin Endocrinol Metab **67**:373–378.
5. Holick MF 2007 Vitamin D deficiency. N Engl J Med **357**:266–281.
6. Hewison M, Burke F, Evans KN, Lammas DA, Sansom DM, Liu P, Modlin RL, Adams JS 2007 Extra-renal 25-hydroxyvitamin D3-1alpha-hydroxylase in human health and disease. J Steroid Biochem Mol Biol **103**:316–321.
7. Zierold C, Darwish HM, DeLuca HF 1995 Two vitamin D response elements function in the rat 1,25-dihydroxyvitamin D 24-hydroxylase promoter. J Biol Chem **270**:1675–1678.
8. Ren S, Nguyen L, Wu S, Encinas C, Adams JS, Hewison M 2005 Alternative splicing of vitamin D-24-hydroxylase: A novel mechanism for the regulation of extrarenal 1,25-dihydroxyvitamin D synthesis. J Biol Chem **280**:20604–20611.
9. Wu S, Ren S, Nguyen L, Adams JS, Hewison M 2007 Splice variants of the CYP27b1 gene and the regulation of 1,25-dihydroxyvitamin D3 production. Endocrinology **148**:3410–3418.
10. Bikle DD, Pillai S, Gee E, Hincenbergs M 1989 Regulation of 1,25-dihydroxyvitamin D production in human keratinocytes by interferon-gamma. Endocrinology **124**:655–660.
11. Bikle DD, Pillai S, Gee E, Hincenbergs M 1991 Tumor necrosis factor-alpha regulation of 1,25-dihydroxyvitamin D production by human keratinocytes. Endocrinology **129**:33–38.
12. Schauber J, Dorschner RA, Coda AB, Buchau AS, Liu PT, Kiken D, Helfrich YR, Kang S, Elalieh HZ, Steinmeyer A, Zugel U, Bikle DD, Modlin RL, Gallo RL 2007 Injury enhances TLR2 function and antimicrobial peptide expression through a vitamin D-dependent mechanism. J Clin Invest **117**:803–811.
13. Cooke NE, Haddad JG 1989 Vitamin D binding protein (Gc-globulin). Endocr Rev **10**:294–307.
14. Liang CJ, Cooke NE 2005 Vitamin D-binding protein In: Pike JW, Glorieux FH, Feldman D (eds.) Vitamin D, 2nd ed. Academic Press, San Diego, CA, USA, pp. 117–134.
15. Willnow TE, Nykjaer A 2005 Endocytic pathways for 25-(OH) vitamin D3 In: Pike JW, Glorieux FH, Feldman D (eds.) Vitamin D, 2nd ed. Academic Press, San Diego, CA, USA, pp. 153–163
16. Safadi FF, Thornton P, Magiera H, Hollis BW, Gentile M, Haddad JG, Liebhaber SA, Cooke NE 1999 Osteopathy and resistance to vitamin D toxicity in mice null for vitamin D binding protein. J Clin Invest **103**:239–251.
17. Adams JS 2005 "Bound" to work: The free hormone hypothesis revisited. Cell **122**:647–649.
18. Christakos S, Dhawan P, Liu Y, Peng X, Porta A 2003 New insights into the mechanisms of vitamin D action. J Cell Biochem **88**:695–705.
19. DeLuca HF 2004 Overview of general physiologic features and functions of vitamin D. Am J Clin Nutr **80**:1689S–1696S.
20. Rachez C, Freedman LP 2000 Mechanisms of gene regulation by vitamin D(3) receptor: A network of coactivator interactions. Gene **246**:9–21.
21. Sutton AL, MacDonald PN 2003 Vitamin D: More than a "bone-a-fide" hormone. Mol Endocrinol **17**:777–791.

© 2008 American Society for Bone and Mineral Research

22. Zhong Y, Christakos S 2007 Novel Mechanisms of gene regulation by vitamin D (3) receptor: A network of coactivator interactions. Gene **246:**9–21.
23. Christakos S, Dhawan P, Peng X, Obukhov AG, Nowycky MC, Benn BS, Zhong Y, Liu Y, Shen Q 2007 New insights into the function and regulation of vitamin D target proteins. J Steroid Biochem Mol Biol **103:**405–410.
24. Dhawan P, Peng X, Sutton AL, MacDonald PN, Croniger CM, Trautwein C, Centrella M, McCarthy TL, Christakos S 2005 Functional cooperation between CCAAT/enhancer-binding proteins and the vitamin D receptor in regulation of 25-hydroxyvitamin D3 24-hydroxylase. Mol Cell Biol **25:**472–487.
25. Guo B, Aslam F, van Wijnen AJ, Roberts SG, Frenkel B, Green MR, DeLuca H, Lian JB, Stein GS, Stein JL 1997 YY1 regulates vitamin D receptor/retinoid X receptor mediated transactivation of the vitamin D responsive osteocalcin gene. Proc Natl Acad Sci USA **94:**121–126.
26. Raval-Pandya M, Dhawan P, Barletta F, Christakos S 2001 YY1 represses vitamin D receptor-mediated 25-hydroxyvitamin D(3)24-hydroxylase transcription: Relief of repression by CREB-binding protein. Mol Endocrinol **15:**1035–1046.
27. Li YC, Pirro AE, Amling M, Delling G, Baron R, Bronson R, Demay MB 1997 Targeted ablation of the vitamin D receptor: An animal model of vitamin D-dependent rickets type II with alopecia. Proc Natl Acad Sci USA **94:**9831–9835.
28. Yoshizawa T, Handa Y, Uematsu Y, Takeda S, Sekine K, Yoshihara Y, Kawakami T, Arioka K, Sato H, Uchiyama Y, Masushige S, Fukamizu A, Matsumoto T, Kato S 1997 Mice lacking the vitamin D receptor exhibit impaired bone formation, uterine hypoplasia and growth retardation after weaning. Nat Genet **16:**391–396.
29. Amling M, Priemel M, Holzmann T, Chapin K, Rueger JM, Baron R, Demay MB 1999 Rescue of the skeletal phenotype of vitamin D receptor-ablated mice in the setting of normal mineral ion homeostasis: Formal histomorphometric and biomechanical analyses. Endocrinology **140:**4982–4987.
30. Raisz LG, Trummel CL, Holick MF, DeLuca HF 1972 1,25-dihydroxycholecalciferol: A potent stimulator of bone resorption in tissue culture. Science **175:**768–769.
31. Yasuda H, Shima N, Nakagawa N, Yamaguchi K, Kinosaki M, Mochizuki S, Tomoyasu A, Yano K, Goto M, Murakami A, Tsuda E, Morinaga T, Higashio K, Udagawa N, Takahashi N, Suda T 1998 Osteoclast differentiation factor is a ligand for osteoprotegerin/osteoclastogenesis-inhibitory factor and is identical to TRANCE/RANKL. Proc Natl Acad Sci USA **95:**3597–3602.
32. Prince CW, Butler WT 1987 1,25-Dihydroxyvitamin D3 regulates the biosynthesis of osteopontin, a bone-derived cell attachment protein, in clonal osteoblast-like osteosarcoma cells. Coll Relat Res **7:**305–313.
33. Price PA, Baukol SA 1980 1,25 Dihydroxyvitamin D3 increases synthesis of the vitamin K-dependent bone protein by osteosarcoma cells. J Biol Chem **255:**11660–11663.
34. Drissi H, Pouliot A, Koolloos C, Stein JL, Lian JB, Stein GS, van Wijnen AJ 2002 1,25-(OH)2-vitamin D3 suppresses the bone-related Runx2/Cbfa1 gene promoter. Exp Cell Res **274:**323–333.
35. Gardiner EM, Baldock PA, Thomas GP, Sims NA, Henderson NK, Hollis B, White CP, Sunn KL, Morrison NA, Walsh WR, Eisman JA 2000 Increased formation and decreased resorption of bone in mice with elevated vitamin D receptor in mature cells of the osteoblastic lineage. FASEB J **14:**1908–1916.
36. Akhter S, Kutuzova GD, Christakos S, DeLuca HF 2007 Calbindin D9k is not required for 1,25-dihydroxyvitamin D3-mediated Ca2+ absorption in small intestine. Arch Biochem Biophys **460:**227–232.
37. Raval-Pandya M, Porta AR, Christakos S 1998 Mechanism of action of 1,25 dihydroxyvitamin D_3 on intestinal calcium absorption and renal calcium transportation. In: Holick MF (ed.) Vitamin D. Physiology, Molecular Biology and Clinical Applications. Humana Press, Totowa, NJ, USA, pp. 163–173.
38. Wasserman RH, Fullmer CS 1995 Vitamin D and intestinal calcium transport: Facts, speculations and hypotheses. J Nutr **125:**1971S–1979S.
39. Kutuzova GD, Akhter S, Christakos S, Vanhooke J, Kimmel-Jehan C, Deluca HF 2006 Calbindin D(9k) knockout mice are indistinguishable from wild-type mice in phenotype and serum calcium level. Proc Natl Acad Sci USA **103:**12377–12381.
40. Hoenderop JG, Nilius B, Bindels RJ 2003 Epithelial calcium channels: From identification to function and regulation. Pflugers Arch **446:**304–308.
41. Peng JB, Chen XZ, Berger UV, Vassilev PM, Tsukaguchi H, Brown EM, Hediger MA 1999 Molecular cloning and characterization of a channel-like transporter mediating intestinal calcium absorption. J Biol Chem **274:**22739–22746.
42. Williams KB, DeLuca HF 2007 Characterization of intestinal phosphate absorption using a novel in vivo method. Am J Physiol Endocrinol Metab **292:**E1917–E1921.
43. Sneddon WB, Barry EL, Coutermarsh BA, Gesek FA, Liu F, Friedman PA 1998 Regulation of renal parathyroid hormone receptor expression by 1, 25-dihydroxyvitamin D3 and retinoic acid. Cell Physiol Biochem **8:**261–277.
44. Lambers TT, Weidema AF, Nilius B, Hoenderop JG, Bindels RJ 2004 Regulation of the mouse epithelial Ca2(+) channel TRPV6 by the Ca(2+)-sensor calmodulin. J Biol Chem **279:**28855–28861.
45. Omdahl JL, Bobrovnikova EA, Choe S, Dwivedi PP, May BK 2001 Overview of regulatory cytochrome P450 enzymes of the vitamin D pathway. Steroids **66:**381–389.
46. Demay MB, Kiernan MS, DeLuca HF, Kronenberg HM 1992 Sequences in the human parathyroid hormone gene that bind the 1,25- dihydroxyvitamin D3 receptor and mediate transcriptional repression in response to 1,25-dihydroxyvitamin D3. Proc Natl Acad Sci USA **89:**8097–8101.
47. Martin KJ, Gonzalez EA 2004 Vitamin D analogs: Actions and role in the treatment of secondary hyperparathyroidism. Semin Nephrol **24:**456–459.
48. Canaff L, Hendy GN 2002 Human calcium-sensing receptor gene. Vitamin D response elements in promoters P1 and P2 confer transcriptional responsiveness to 1,25-dihydroxyvitamin D. J Biol Chem **277:**30337–30350.
49. Christakos S, Norman AW 1979 Studies on the mode of action of calciferol. XVIII. Evidence for a specific high affinity binding protein for 1,25 dihydroxyvitamin D3 in chick kidney and pancreas. Biochem Biophys Res Commun **89:**56–63.
50. Clark SA, Stumpf WE, Sar M, DeLuca HF, Tanaka Y 1980 Target cells for 1,25 dihydroxyvitamin D3 in the pancreas. Cell Tissue Res **209:**515–520.
51. Morrissey RL, Bucci TJ, Richard B, Empson N, Lufkin EG 1975 Calcium-binding protein: Its cellular localization in jejunum, kidney and pancreas. Proc Soc Exp Biol Med **149:**56–60.
52. Sooy K, Schermerhorn T, Noda M, Surana M, Rhoten WB, Meyer M, Fleischer N, Sharp GW, Christakos S 1999 Calbindin-D(28k) controls [Ca(2+)](i) and insulin release. Evidence obtained from calbindin-d(28k) knockout mice and beta cell lines. J Biol Chem **274:**34343–34349.
53. Rabinovitch A, Suarez-Pinzon WL, Sooy K, Strynadka K, Christakos S 2001 Expression of calbindin-D(28k) in a pancreatic islet beta-cell line protects against cytokine-induced apoptosis and necrosis. Endocrinology **142:**3649–3655.
54. Provvedini DM, Tsoukas CD, Deftos LJ, Manolagas SC 1983 1,25-dihydroxyvitamin D3 receptors in human leukocytes. Science **221:**1181–1183.
55. Adams JS, Sharma OP, Gacad MA, Singer FR 1983 Metabolism of 25-hydroxyvitamin D3 by cultured pulmonary alveolar macrophages in sarcoidosis. J Clin Invest **72:**1856–1860.
56. Rook GA, Steele J, Fraher L, Barker S, Karmali R, O'Riordan J, Stanford J 1986 Vitamin D3, gamma interferon, and control of proliferation of Mycobacterium tuberculosis by human monocytes. Immunology **57:**159–163.
57. Wang TT, Nestel FP, Bourdeau V, Nagai Y, Wang Q, Liao J, Tavera-Mendoza L, Lin R, Hanrahan JW, Mader S, White JH 2004 Cutting edge: 1,25-dihydroxyvitamin D3 is a direct inducer of antimicrobial peptide gene expression. J Immunol **173:**2909–2912.
58. Gombart AF, Borregaard N, Koeffler HP 2005 Human cathelicidin antimicrobial peptide (CAMP) gene is a direct target of the vitamin D receptor and is strongly up-regulated in myeloid cells by 1,25-dihydroxyvitamin D3. FASEB J **19:**1067–1077.
59. Weber G, Heilborn JD, Chamorro Jimenez CI, Hammarsjo A, Torma H, Stahle M 2005 Vitamin D induces the antimicrobial protein hCAP18 in human skin. J Invest Dermatol **124:**1080–1082.
60. Medzhitov R 2007 Recognition of microorganisms and activation of the immune response. Nature **449:**819–826.
61. Liu PT, Stenger S, Li H, Wenzel L, Tan BH, Krutzik SR, Ochoa MT, Schauber J, Wu K, Meinken C, Kamen DL, Wagner M, Bals R, Steinmeyer A, Zugel U, Gallo RL, Eisenberg D, Hewison M, Hollis BW, Adams JS, Bloom BR, Modlin RL 2006 Toll-like re-

© 2008 American Society for Bone and Mineral Research

ceptor triggering of a vitamin D-mediated human antimicrobial response. Science **311**:1770–1773.

62. Kreutz M, Andreesen R, Krause SW, Szabo A, Ritz E, Reichel H 1993 1,25-dihydroxyvitamin D3 production and vitamin D3 receptor expression are developmentally regulated during differentiation of human monocytes into macrophages. Blood **82**:1300–1307.
63. Penna G, Adorini L 2000 1 Alpha,25-dihydroxyvitamin D3 inhibits differentiation, maturation, activation, and survival of dendritic cells leading to impaired alloreactive T cell activation. J Immunol **164**:2405–2411.
64. Penna G, Amuchastegui S, Giarratana N, Daniel KC, Vulcano M, Sozzani S, Adorini L 2007 1,25-Dihydroxyvitamin D3 selectively modulates tolerogenic properties in myeloid but not plasmacytoid dendritic cells. J Immunol **178**:145–153.
65. Bikle DD, Nemanic MK, Whitney JO, Elias PW 1986 Neonatal human foreskin keratinocytes produce 1,25-dihydroxyvitamin D3. Biochemistry **25**:1545–1548.
66. Su MJ, Bikle DD, Mancianti ML, Pillai S 1994 1,25-Dihydroxyvitamin D3 potentiates the keratinocyte response to calcium. J Biol Chem **269**:14723–14729.
67. Hawker NP, Pennypacker SD, Chang SM, Bikle DD 2007 Regulation of Human Epidermal Keratinocyte Differentiation by the Vitamin D Receptor and its Coactivators DRIP205, SRC2, and SRC3. J Invest Dermatol **127**:874–880.
68. Malloy PJ, Pike JW, Feldman D 1999 The vitamin D receptor and the syndrome of hereditary 1,25-dihydroxyvitamin D-resistant rickets. Endocr Rev **20**:156–188.
69. Skorija K, Cox M, Sisk JM, Dowd DR, MacDonald PN, Thompson CC, Demay MB 2005 Ligand-independent actions of the vitamin D receptor maintain hair follicle homeostasis. Mol Endocrinol **19**:855–862.
70. Bikle DD, Elalieh H, Chang S, Xie Z, Sundberg JP 2006 Development and progression of alopecia in the vitamin D receptor null mouse. J Cell Physiol **207**:340–353.
71. Cianferotti L, Cox M, Skorija K, Demay MB 2007 Vitamin D receptor is essential for normal keratinocyte stem cell function. Proc Natl Acad Sci USA **104**:9428–9433.
72. Bischoff-Ferrari HA, Willett WC, Wong JB, Giovannucci E, Dietrich T, Dawson-Hughes B 2005 Fracture prevention with vitamin D supplementation: A meta-analysis of randomized controlled trials. JAMA **293**:2257–2264.
73. Chapuy MC, Arlot ME, Duboeuf F, Brun J, Crouzet B, Arnaud S, Delmas PD, Meunier PJ 1992 Vitamin D3 and calcium to prevent hip fractures in the elderly women. N Engl J Med **327**:1637–1642.
74. Ponsonby AL, McMichael A, van der Mei I 2002 Ultraviolet radiation and autoimmune disease: Insights from epidemiological research. Toxicology **181-182**:71–78.
75. Boscoe FP, Schymura MJ 2006 Solar ultraviolet-B exposure and cancer incidence and mortality in the United States, 1993-2002. BMC Cancer **6**:264, pp. 1–9.
76. Lappe JM, Travers-Gustafson D, Davies KM, Recker RR, Heaney RP 2007 Vitamin D and calcium supplementation reduces cancer risk: Results of a randomized trial. Am J Clin Nutr **85**:1586–1591.
77. Vieth R 1999 Vitamin D supplementation, 25-hydroxyvitamin D concentrations, and safety. Am J Clin Nutr **69**:842–856.
78. Heaney RP, Davies KM, Chen TC, Holick MF, Barger-Lux MJ 2003 Human serum 25-hydroxycholecalciferol response to extended oral dosing with cholecalciferol. Am J Clin Nutr **77**:204–210.
79. Hathcock JN, Shao A, Vieth R, Heaney R 2007 Risk assessment for vitamin D. Am J Clin Nutr **85**:6–18.

© 2008 American Society for Bone and Mineral Research

SECTION IV

Investigation of Metabolic Bone Diseases (Section Editor: Pierre Delmas)

© 2008 American Society for Bone and Mineral Research

Chapter 29. DXA in Adults and Children

Judith Adams[1] and Nick Bishop[2]

[1]*Manchester Royal Infirmary, Imaging Science and Biomedical Engineering, University of Manchester, Manchester, United Kingdom;* [2]*Academic Unit of Child Health, University of Sheffield, Sheffield Children's Hospital, Sheffield, United Kingdom*

INTRODUCTION

The first DXA scanners were introduced in the late 1980s, and DXA is now the most widely used and available bone densitometry technique[(1,2)]; there are ~27,000 central DXA scanners worldwide. Dual-energy X-ray beams are required to correct BMD measurements for overlying soft tissue and are produced by a variety of techniques by different manufacturers, and the energies are selected to optimize separation of mineralized and soft tissue components of the skeletal site analyzed.[(3)] If a single-energy photon beam is used, this is applicable only to peripheral skeletal sites that have to be placed in a water bath to correct for overlying soft tissues.

Technical Aspects and Developments

Original DXA scanners used a pencil X-ray beam with a single detector, scanning in a rectilinear fashion across the anatomical site (scans times: 5–10 min; spatial resolution = 1 mm). Technical developments have taken place over recent years,[(4,5)] which include a fan-beam X-ray source and a bank of detectors that enable faster scanning (1 min/site) with improved image quality and spatial resolution (0.5 mm).

Scanning Sites and Measurements Provided

DXA can be applied to sites of the skeleton at which osteoporotic fractures occur; in the central skeleton, this includes the lumbar spine (L_1–L_4; Fig. 1A) and proximal femur (total hip, femoral neck, trochanter, and Ward's area; Fig. 1B). DXA can also be applied to peripheral skeletal sites (forearm and calcaneus), using either full-sized or dedicated peripheral DXA scanners. Central DXA measures of lumbar spine, femoral neck, and total hip are currently used as the "gold standard" for the clinical diagnosis of osteoporosis by bone densitometry (Figs. 1A and 1B).

DXA X-ray attenuation values are converted to BMC (g); bone area (BA; cm) is calculated by summing the pixels within the bone edges (software algorithms detect the bone edges). "Areal" BMD (BMD_a; g/cm^2) is calculated by dividing BMC/BA.[(3–5)] Because the DXA image is a 2D image of a 3D object, the depth of bones is not taken into account. This fact results in DXA being size-dependent, one of its significant limitations, particular in children in whom the bones change in size and shape during growth, and in patients whose disease might result in small stature or slender bones.

DXA provides BMD_a of integral (cortical and trabecular) bone. The cortical/trabecular ratios vary in different sites (50/50 postero-anterior [PA] lumbar spine; 10/90 lateral lumbar spine; 60/40 total hip; 80/20 total body; 5/95 calcaneus; 95/5 distal radius; 40/60 ultra-distal radius [depending on site of region of interest]). Because of the different composition of bones and rates of change in various skeletal sites, measurements in different sites in the same individual will not give the same BMD results. Correlations between BMD measurements made in the same patient vary between $r = 0.4$ and $r = 0.9$; it is not possible to predict from a DXA BMD measurement made in one site what the BMD will be in another site using DXA or other bone densitometry methods.[(6)] In research studies, BMD measurements in different anatomical sites and by various bone density methods (DXA, QCT, and quantitative ultrasound [QUS]) may be complementary.

With appropriate software, whole body DXA scanning can be performed, from which is extracted whole body and regional BMC (g) and whole body and regional body composition (lean [muscle] and fat mass; Fig. 2).[(2,3)]

Precision

Precision measures the reproducibility of a bone densitometry technique and is usually expressed as a CV or standardized CV (takes into account range of measurements).[(7,8)] Precision for DXA total hip and lumbar spine DXA = 1–2%; femoral neck and trochanter CV = 2.5%; and Ward's area = 2.5–5% (site not applied in clinical practice). In peripheral sites, CV = 1% in the distal forearm, 2.5% in the ultra-distal forearm, and 1.4% in the calcaneus. Measurement sites generally used in clinical diagnosis (in contrast to research) are mean BMD for lumbar spine (L_1–L_4), femoral neck, and total hip[(8)]; in hyperparathyroidism (primary or secondary), a forearm measurement is relevant, because cortical bone can be lost preferentially from this site. Precision is optimized by using the minimum number of expert, highly motivated, and well-trained technical staff; it is not ideal to have a large number of staff who rotate through different departments and perform BMD scanning only infrequently.

Accuracy

Accuracy is how close the BMD measured by densitometry is to the actual calcium content of the bone (ash weight). The accuracy of DXA lies in the region of 10–15%; the inaccuracies are related to marrow fat and DXA taking soft tissue as a reference.[(3,9)] DXA makes some assumptions about body composition and soft tissues, so inaccuracies may occur with excessively under- or overweight subjects and with large changes in weight between scans, but exactly how this can be corrected for in adults is uncertain.

Sensitivity

Sensitivity is the ability of the measurement to discriminate between patients with and without fractures and to measure small changes with time and/or treatment. A statistically significant change in BMD is calculated as 2.77 multiplied by precision (CV) at the site of measurement to provide least significant change (LSC).[(10)] LSC in the spine = 5.3%; total hip = 5.0% and femoral neck = 6.9%. Because changes in BMD are generally small, it is essential in an individual patient to leave an adequate time interval between DXA measures, usually 18–24 mo.

Fracture Prediction

Because whether a patient develops a fracture depends on factors in addition to BMD (age, whether the patient falls, the nature of the fall, and the patient's response to the fall), it is impossible for BMD techniques to completely discriminate be-

Dr. Bishop has consulted for Procter & Gamble, Novartis, Roche, and Alliance for Better Bone Health. Dr. Adams states that she has no conflicts of interest.

 © 2008 American Society for Bone and Mineral Research

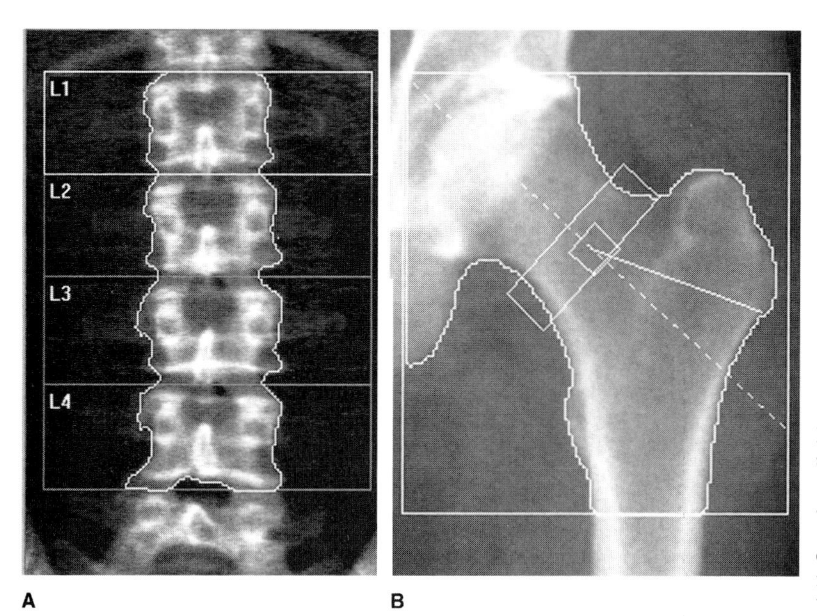

FIG. 1. (A) DXA of normal lumbar spine L_1–L_4. Results are generally expressed as a mean "areal" density (BMD_a; g/cm^2) for all four vertebrae. (B) DXA of left hip; although BMD_a is provided in a number of different sites (femoral neck, oblong box; Ward's area, small box; trochanter; and total hip), for clinical diagnosis of osteoporosis (WHO T-score at or below −2.5), femoral neck and total hip are used.

tween those with and without fractures. However, the lower the DXA BMD, the more at risk the patient is of suffering a fracture.[11] DXA BMD measurements made in any skeletal site (central and peripheral) are predictive of fracture, with the risk of fracture increased in individuals with the lower BMD. The relative risk (RR) of fracture per 1 SD decrease in BMD below the age-adjusted mean varies between 1.4 and 2.6.[11] This reduction in BMD in predicting fracture is as good as a rise of 1 SD in blood pressure is in predicting stroke and a 1 SD rise in cholesterol is in predicting myocardial infarction. Site-specific measurements are best in predicting fracture in that particular anatomical place.

Radiation Doses

DXA involves very low radiation doses (pDXA: calcaneus, 0.03 μSv; forearm, 0.5 μSv; spine, 2–4 μSv; femur, 2–5 μSv; whole body, 1–3 μSv) that are similar to those of natural background radiation (2400 μSv/yr; ~7 μSv/d).[12,13]

Clinical Applications

There has been much debate concerning the appropriate use of bone densitometry, particularly in population screening in women at menopause, and the cost-effectiveness of such a program has not been established. However, there is consensus that central DXA bone densitometry is the "gold standard" measurement to make in appropriate individuals to assess fracture risk and make the diagnosis of osteoporosis in terms of bone densitometry.[1,14] Bone densitometry has high specificity but low sensitivity. Selection of patients who would most appropriately be referred for DXA bone densitometry is based on a case-finding strategy in those who have had a fragility (low-trauma) fracture or have other strong risk factors (estrogen/testosterone deficiency, primary hypogonadism, height loss, low trauma fractures, radiographic osteopenia, oral glucocorticoid therapy [prednisolone > 7.5 mg/d for >3 mo], rheumatoid arthritis, and other secondary causes of osteoporosis). There may be national differences in such referral guidelines based on local reimbursement policies and economic constraints.

Artefacts

These can cause inaccuracies in DXA and are most common in the lumbar spine in the elderly population. All the calcium in the path of the X-ray beam will contribute to measured BMD. If there is degenerative disc disease with osteophytes, osteoarthritis with hyperostosis of facet joints, or vertebral fracture (same amount of calcium before fracture contained in

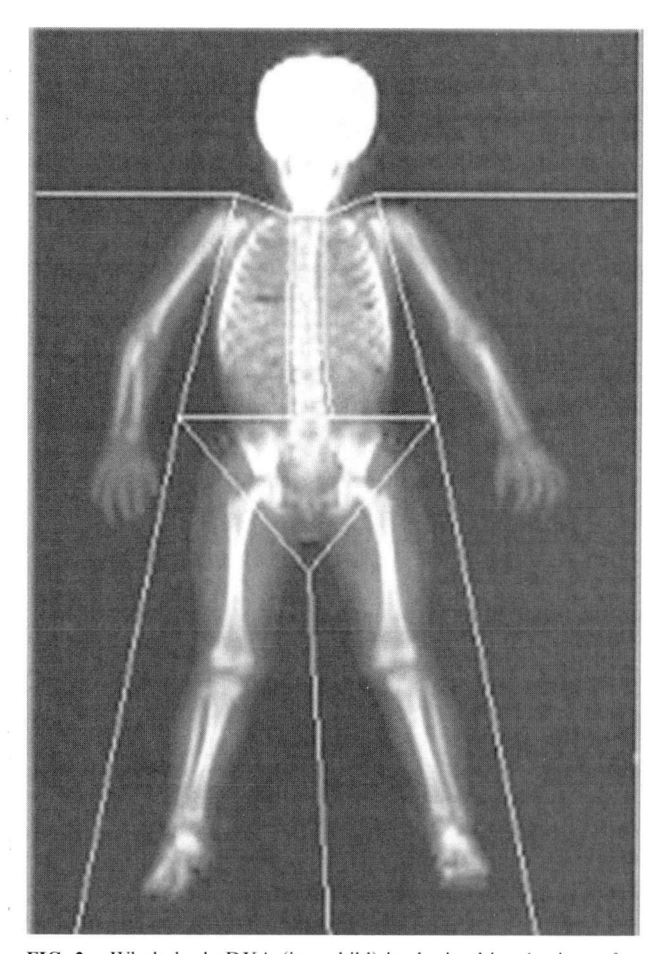

FIG. 2. Whole body DXA (in a child) is obtained in ~1 min on fan beam scanners and provides total and regional information about the skeleton (less head) and body composition (lean muscle and fat mass).

© 2008 American Society for Bone and Mineral Research

TABLE 1. CAUSES OF ARTEFACTS RESULTING IN ERRORS IN DXA BMD[a]

Overestimation of BMD
Spinal degeneration and hyperostosis (osteophytes)
Vertebral fracture
Extraneous calcification (lymph nodes, aortic calcification)
Overlying metal (navel rings, surgical rods/plates, Myodil)
Sclerotic metastases
Vertebral hemangioma
Ankylosing spondylitis with paravertebral ossification
Poor positioning of femoral neck (inadequate internal rotation)
Excessive body weight
Strontium ranelate therapy
Underestimation of BMD
Laminectomy
Lytic metastases
Low body weight
Other
DXA will not differentiate low calcium content in bone being caused by osteoporosis (reduced amount of bone; quantitative abnormality) and osteomalacia (reduced calcium/osteoid ratio; qualitative abnormality).

vertebra which is reduced in area) present, BMD will be falsely elevated.[1–3] Other etiologies can also cause false elevation or underestimation of BMD measured by DXA (Table 1; Figs. 3A and 3B); it is essential that DXA images be scrutinized for such artefacts. Vertebra affected significantly by artefact should be excluded from analysis; however, for clinical diagnosis, a minimum of two vertebrae must be available for analysis. Strontium ranelate is a relatively new oral therapy for osteoporosis; some of the apparent increase in BMD is artefactual, related to the high atomic number of strontium that accumulates in bone and contributes to X-ray attenuation; methods to correct for this have been proposed.[15] It is important to note that DXA will not differentiate low calcium content in bone being caused by osteoporosis (reduced amount of bone; quantitative abnormality) and osteomalacia (reduced calcium/osteoid ratio; qualitative abnormality).

To overcome the problems of degenerative disc disease and hyperostosis in PA DXA of the spine, lateral DXA has been developed. On scanners with "C" arms, lateral scanning can be performed with the patient remaining in the supine position; otherwise, the patient has to be repositioned in the lateral decubitus position, which limits its clinical practicality and precision (CV = 2.8–5.9% in lateral decubitus position, 1.6–2% in the supine position). Because L_1 and L_2 may have ribs superimposed and L_4 is overlain by the iliac crest, L_3 may be the only vertebra that can be analyzed in lateral DXA. Therefore, although lateral DXA may be a more sensitive predictor of vertebral fracture than PA spinal DXA, its limited precision and impracticality means that it is not often performed in clinical practice.

Because of these artefacts on PA DXA of lumbar spine, it has been suggested that, in the elderly population (>65 yr of age), only the proximal femur (femoral neck, total hip) should be scanned. However, monitoring change is performed optimally in the lumbar spine if there are no artefacts.[10]

Interpretation

When a BMD measurement has been made, this has to be interpreted as normal or abnormal and a report must be formulated that will be of assistance to the referring clinician.[16] For this, it is essential that age-, sex-, and ethnically matched reference data are available. The scanner manufacturer supplies such normal reference databases. These databases are predominantly, but not exclusively, drawn from a white American-based population. There is a paucity of appropriate reference ranges for children and certain ethnic minorities (e.g., Asians). A patient's results can be interpreted in terms of the SD from the mean of either sex-matched peak bone mass (PBM; T-score) or age-matched BMD (Z-score).

The World Health Organization (WHO) has defined osteoporosis in terms of bone densitometry as a T-score at or below −2.5 in the lumbar spine, femoral neck, and total femur.[17] The definition applies to DXA of lumbar spine, proximal femur (femoral neck), and the distal third of the forearm but not to other techniques (e.g., QCT, QUS) or other anatomical sites (e.g., calcaneus),[18] nor is it yet confirmed to be applicable to younger women and men. Until PBM has been reached (i.e., in children and young adults up to ~20 yr), interpretation can be made only by comparison with age-matched mean (Z-score).[19–22] Variations in the mean and SD of reference ranges may alter the number of patients identified as osteoporotic; in DXA of the hip, use of the NHANES reference database is preferred for whites.[23]

Although there is consensus on the definition of osteoporosis in bone densitometry (WHO T-score below −2.5), there is as yet no consensus on levels of BMD that justify therapeutic interventions that are cost-effective. This is perhaps not surprising, because it is the individual patient that is being treated and not the BMD result, and age is a strong independent predictor of fracture. Other factors (e.g., age, previous low trauma fracture over age 50, parental hip fracture, glucocorticoid therapy, rheumatoid arthritis, smoking, alcohol consumption) contribute to prediction of fracture risk. The WHO has recently published algorithms for calculating 10-yr fracture risk for patients from such clinical risk factors that will make the decision to instigate therapy more cost-effective.[24]

Monitoring Change in BMD

Whether DXA should be used to monitor change in BMD with time and therapy is contentious and varies between different countries and their guidelines. In calculating change over time, the absolute BMD values (g/cm^2) should be used; statistically significant change is 2.77 × CV%. In longitudinal studies in an individual patient, an adequate interval of time (18–24 mo) between measures is required to show significant change, unless large changes in BMD are anticipated (e.g., after organ transplantation together with large doses of glucocorticoids).[10]

Different manufacturers use different edge detection algorithms and analyze different regions of interest (ROIs) for analysis in the hip, so results from different scanners are not interchangeable. In longitudinal studies, it is vital to use the same scanner and software program. With technical developments, it may become necessary to replace a scanner. To cross-calibrate between the old and new scanner, scanning patients (~100, with a spread of BMD from high to low) and phantoms (scanner manufacturer or European Spine Phantom) will generally allow the required calculations to be made.[25,26]

Research Application

There are other scanning capabilities on central DXA scanners, which currently remain as research applications. These include whole body scanning for total and regional BMC, lean muscle, and fat mass (Fig. 2) in adults, children, and neonates.[27] Software programs are available for measuring BMD around prostheses after hip and knee arthroplasty, in bone

© 2008 American Society for Bone and Mineral Research

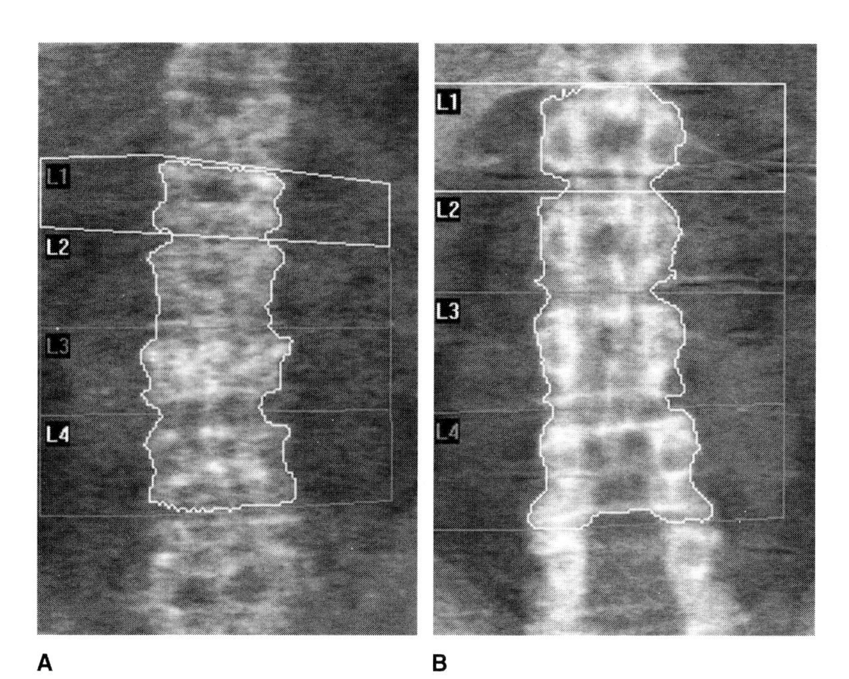

FIG. 3. Artefacts on DXA. (A) Lumbar spine vertebral fracture of L_1 and L_3. BMD at these levels will be falsely elevated (same BMC as nonfractured vertebra but in a smaller projected area, giving higher apparent BMD). (B) Laminectomy L_4. Removal of the laminae and spinous process will falsely reduce the BMD of this vertebra. DXA images must be carefully scrutinized for such artefacts, and affected vertebra should be excluded from analysis. A minimum of two vertebrae are required for classification/interpretation.

specimens and small animals, and application in novel anatomical sites (hand, mandible).[(27–29)] Although there may be no specific commercial software program available for scanning such sites, programs for scanning conventional sites (e.g., forearm) can be used, with analysis being performed by hand-placed ROIs; these would not be as precise as automated ROIs.

Anatomical measurements can also be derived from DXA. The hip axis length (HAL) has been found to be predictive of hip fracture; normal HAL in postmenopausal women is 10.5 ± 0.62 cm; HAL of 11.0 cm increased hip fracture 2-fold, and HAL of 11.5 cm increased the risk 4-fold.[(30)] HAL is the distance from the inner margin of the bony pelvis to the lateral border of the femur along a line drawn through the midline of the femoral neck and parallel to its margins. There will be some magnification of the hip geometry with fan-beam scanners, so that corrections have to be applied. Structural components of the proximal femur (size and shape), in combination with BMD, have been found to improve fracture risk prediction.[(31)] Hip structure analysis (HSA) has been introduced to extract geometric strength information from hip DXA, by making mathematical calculations of the distribution of calcium in DXA cross-sections in the femoral neck, trochanteric region, and proximal femoral shaft[(32)] and has been applied retrospectively in several large research studies. Its application in clinical practice is still to be defined.

Vertebral Fracture Assessment

Lateral views of thoracic and lumbar spine (T_4–L_4) can be obtained with fan-beam scanners, using dual- or single-energy scanning, with the patient either in the supine ("C" arm scanners) or lateral decubitus position (Fig. 4). From these, a visual assessment for vertebral fractures or morphometric assessment of vertebral shape can be made.[(33,34)] Such morphometric analysis has the potential for automation by application of computer analysis techniques (e.g., active shape and appearance models) that may potentially make them more practical for use in a clinical setting.[(35)] Vertebral fracture assessment (VFA) has several advantages over conventional radiography: lower dose of ionizing radiation (12 μSv single energy; 42 μSv) and avoiding problems of the divergent X-ray beam of radiography that can distort vertebral shape causing apparent biconcavity of endplates—DXA uses a lateral scan projection method, with simultaneous movement of X-ray source and detectors along the spine, so the X-ray beam is always parallel to the vertebral endplates and avoids the "bean can" artefact of vertebral endplates caused by the parallax effect on radiographs. Guidelines suggest that elderly patients (women >70 yr; men >80 yr) who have low BMD with historical height loss of 4 cm (1.6 in) in women (6 cm; 2.4 in in men) or prospective height loss of >2 cm (0.8 in) in women (3 cm; 1.2 in in men) and other risk factors should undergo vertebral fracture assessment using lateral vertebral assessment (LVA) or instantaneous vertebral assessment (IVA) by DXA to detect vertebral fractures.[(21)] The method has been shown to be satisfactory for excluding vertebral fractures being present.[(36)] However, more scientific studies are required to confirm the exact clinical role of this alternative technology to conventional spinal radiography for the identification of vertebral fractures.

DXA IN CHILDREN

The purpose of any assessment of bone size or mass in children is to provide data relevant to the current, or future, health of the skeleton. DXA is the most widely used quantitative bone imaging technique in pediatric practice, but many aspects of its use, and the interpretation of the data obtained, remain contentious.[(19–22,37,38)]

DXA provides estimates of bone size in two dimensions and bone mass within that envelope, the value of bone mass adjusted for size being BMD_a (g/cm^2). To overcome this limitation to some extent, a "calculated volumetric BMD" (bone mineral apparent density [BMAD; g/cm^3]) can be made in the spine and femoral neck.[(39–41)] Although in adult practice there is general acceptance that DXA BMD below certain absolute values are associated with increased fracture risk, there is no such absolute threshold for children. The fact that this is reflected in manufacturer-provided reference values for BMD_a, which increase in a similar way to height and weight during childhood and adolescence, clearly indicate that DXA is not measuring true volumetric BMD, but some composite measure

© 2008 American Society for Bone and Mineral Research

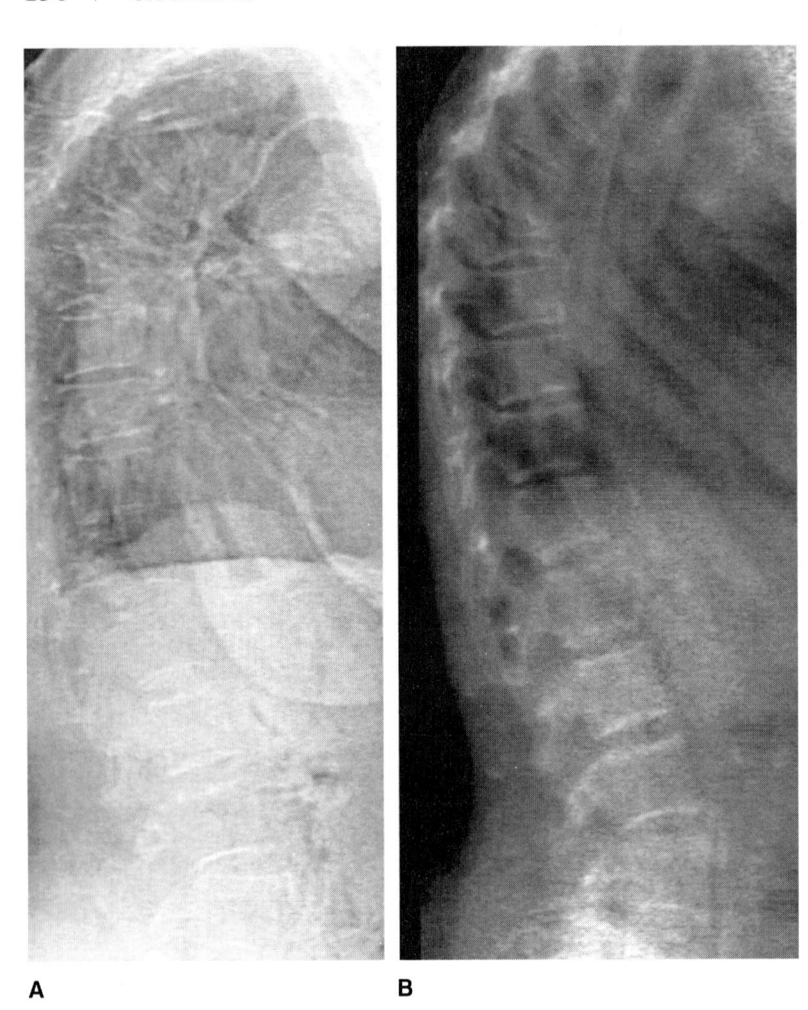

FIG. 4. VFA can be made on DXA scanners on (A) single-energy and (B) dual-energy images (grade 3 severe crush fracture of T_8) at much lower radiations doses (1/100th to 1/50th) than in conventional spinal radiographs; some of the thoracic vertebrae are better visualized on dual, than on single, energy images. Six-point morphometry (MXA) and visual assessment of the vertebrae can be made. Aortic calcification can also be visualized and graded.

of bone size and mass. This is not necessarily a disadvantage because bone size, especially in the tubular bones that children are most prone to fracture, is an important predictor of bone strength.

Advantages of DXA in children are short scan time, low radiation dose, and general widespread availability. DXA measures of bone mass in healthy children are predictive of fracture risk both at the measurement site (in the forearm) and elsewhere; total body less head (TBLH; Fig. 2) BMC, adjusted for weight, height, and bone area, is the measure that has been found, in a prospective cohort study at age 9.9 yr, to be most strongly associated with fracture risk over the following 2 yr.[42] However, there are no data for the predictive value of DXA at other ages in apparently healthy children and no similar data for children with bone disease.

Children are a difficult group to study, because as they grow, their bone mass should increase. They also fracture frequently—by the end of teenage years, up to one half of all boys and one third of girls will have sustained a fracture.[43,44] A single fracture in an otherwise healthy child should not therefore require study of skeletal health. Most requests for DXA in children will be in the groups thought to be at increased risk of fracture. These include children with primary bone diseases such as osteogenesis imperfecta and idiopathic juvenile osteoporosis; chronic immobilization (cerebral palsy and Duchenne dystrophy); inflammatory conditions (Crohn's disease, cystic fibrosis, and juvenile idiopathic arthritis); endocrine disorders such as anorexia nervosa and Cushing's syndrome (but not Turner's syndrome); after chemotherapy or organ transplantation; and thalassemia major. Other requests come when a child has recurrent fractures in the absence of an obvious underlying predisposition.

In children, when should a DXA scan be performed and should it be repeated? Although there are many studies using DXA to show efficacy of a specific intervention in children, little has been published about the practical application of DXA in the clinical setting. It can be reasonably assumed that a DXA would be only part of an assessment of bone health and would be undertaken either because of a perceived increase in fracture risk or because of bone disease.

Measurement sites for DXA in children are typically the lumbar spine (L_1–L_4) and total body, where precision is similar to that achieved in adults. Forearm and proximal femur have been used in some studies, as has the lateral distal femur where deformity and contracture preclude use of DXA in the normal measurement sites. Normative reference data are available for spine, femoral neck, total body, and lateral distal femur.[20,45]

T-scores must not be used in children who have not yet reached PBM; measurements are usually reported as sex specific and in relation to age (Z-score; at or below −2.0 being reduced in children who are normal in size for their age). The diagnosis of osteoporosis in children should not be made on the basis of densitometric criteria alone. Terminology such as "low BMD for chronological age" may be used if the Z-score is below −2.0; terms such as osteoporosis and osteopenia should not be used.[19,21] Adjustments have been made in research studies in healthy children to account for the assumed shape of vertebrae (cylindrical, cuboidal) and the femoral

© 2008 American Society for Bone and Mineral Research

neck,[39–41] and some reference data using these have been published[45] or the expected relationship of bone size and mass with body size or elements of body composition such as lean body mass.[46–49] Some of the current generation of scanners provide methods of adjusting for body size at both the spine and total body, but there is a lack of information currently as to which adjustment (if any) should be used in clinical practice. In some studies, adjustment for bone age or pubertal status has been undertaken, but estimation of bone age is a specialized technique, and there is more than one method of assessment in common use. There is a lack of data for ethnic subgroups. It is unclear from the published data whether any one of the proposed adjustments is optimal in terms of either fracture prediction or skeletal health assessment.

In diseases in childhood in which fracture risk is increased, initiation of DXA BMD should be at the discretion of the treating clinician, and the monitoring interval should reflect the severity of the disease. There is little evidence to support monitoring at intervals of <6 mo in any setting. Although there is increased risk of fracture for children receiving glucocorticoid therapy, it is difficult to disentangle the effects of these from those of the underlying bone disease. Currently it does not seem appropriate that DXA should be any more frequent in children receiving such therapy.

Interpretation of DXA results depends on the clinical context. A diagnosis of osteoporosis should not be made on the basis of a bone mass measurement in isolation. If a child presents with recurrent fractures but no clinically apparent underlying disease and bone mass is within the expected range, reassurance can be offered. If there is an underlying problem, further monitoring and additional imaging may be required. Such a clinical setting would be a child with apparently mild osteogenesis imperfecta with bone mass in the normal range who may have occult fractures of thoracic vertebrae. Low bone mass in the presence of vertebral, or recurrent, fractures resulting in loss of independent mobility and chronic bone pain should prompt evaluation of the need for active intervention.

CONCLUSIONS

DXA offers a precise technique, with acceptable accuracy, for measuring BMD in central and peripheral skeletal sites in adults and children, using very low doses of radiation (similar to levels of natural background radiation). DXA is currently regarded as the "gold standard" for BMD measurements for the diagnosis of osteoporosis in adults. However, there are some important limitations (BMD_a, size dependency, measurement of integral [cortical and trabecular] bone), of which users and operators need to be aware. Good precision depends on scanners being operated by skilled and appropriately trained staff and quality assurance protocols being in place. DXA can be used in adults to diagnose osteoporosis using the WHO threshold (T-score = –2.5 or below in lumbar spine L_1–L_4, femoral neck, and total hip), predict fractures, contribute to decisions on patient management and therapeutic intervention, and perhaps also monitor change in BMD (except in the forearm). There is potential for visual and morphometric assessment of vertebral fractures from lateral DXA images. There are increasing, and varied, applications of DXA in research studies in novel sites and in both humans and animals. There are particular issues for DXA in children in whom the size dependency is a limitation, and to date, there is no consensus on whether size correction should be applied and which method is optimum.

REFERENCES

1. Blake GM, Fogelman I 2007 The role of DXA bone density scans in the diagnosis and treatment of osteoporosis. Postgrad Med J **83:**509–517.
2. Adams JE 2008 Dual energy X-ray absorptiometry. In: Grampp S (ed.) Radiology of Osteoporosis, 2nd ed. Springer-Verlag, Berlin, Germany, pp. 105–124.
3. Blake GM, Wahner HW, Fogelman I (eds.) 1999 The Evaluation of Osteoporosis: Dual Energy X-ray Absorptiometry and Ultrasound in Clinical Practice. Martin Dunitz, London, UK.
4. Engelke K, Gluer CC 2006 Quality and performance measures in bone densitometry: Part 1: Errors and diagnosis. Osteoporos Int **17:**1283–1292.
5. Fogelman I, Blake GM 2005 Bone densitometry; and update. Lancet **366:**2068–2070.
6. Grampp S, Genant HK, Mathur A, Lang P, Jergas M, Takada M, Gluer CC, Lu Y, Chavez M 1997 Comparisons of non-invasive bone mineral measurements in assessing age-related loss, fracture discrimination and diagnostic classification. J Bone Miner Res **12:**697–711.
7. Gluer CC, Blake G, Blunt BA, Jergas M, Genant HK 1995 Accurate assessment of precision errors: How to measure the reproducibility of bone densitometry techniques. Osteoporos Int **5:**262–270.
8. Kanis JA, Gluer C for the Committee of the Scientific Advisors, International Osteoporosis Foundation 2000 An update in the diagnosis and assessment of osteoporosis with densitometry. Osteoporos Int **11:**192–202.
9. Blake GM, Fogelman I 2008 How important are BMD accuracy errors for the clinical interpretation of DXA scans? J Bone Miner Res **23:**457–462.
10. Gluer CC 1999 Monitoring skeletal changes by radiological techniques. J Bone Miner Res **14:**1952–1962.
11. Marshall D, Johnell O, Wedel H 1996 Meta-analysis of how well measures of bone density predict occurrence of osteoporotic fractures. BMJ **312:**1254–1259.
12. Kalender WA 1992 Effective dose values in bone mineral measurements by photon absorptiometry and computed tomography. Osteoporos Int **2:**82–87.
13. Blake GM, Naeem M, Boutros M 2006 Comparison of effective dose to children and adults from dual energy X-ray absorptiometry examinations. Bone **38:**935–942.
14. Compston J 2005 Guidelines for the management of osteoporosis: The present and the future. Osteoporos Int **16:**1173–1176.
15. Blake GM, Fogelman I 2007 The correction of BMD measurements for bone strontium content. J Clin Densitom **10:**259–265.
16. Miller P, Bonnick SL, Rosen CJ 1996 Consensus of an international panel on the clinical utility of bone mass measurements in the detection of low bone mass in the adult population. Calcif Tissue Int **58:**207–214.
17. World Health Organization Study Group 1994 Assessment of Fracture Risk and Its Application to Screening for Postmenopausal Osteoporosis. World Health Organization, Geneva, Switzerland.
18. Miller P 2000 Controversies in bone mineral density diagnostic classification. Calcif Tissue Int **66:**317–319.
19. National Osteoporosis Society 2004 A Practical Guide to Bone Densitometry in Children. National Osteoporosis Society, Camerton, UK.
20. Faulkner RA, Bailey DA, Drinkwater DT, Wilkinson AA, Houston CS, McKay HA 1993 Regional and total body bone mineral content, bone mineral density, and total body tissue composition in children 8–16 years of age. Calcif Tissue Int **53:**7–12.
21. International Society for Clinical Densitometry 2005 ISCD. Available at www.ISCD.org. Accessed March 30, 2008.
22. Sawyer AJ, Bachrach LK, Fung EB (eds.) 2006 Bone Densitometry in Growing Patients; Guidelines for Clinical Practice. Humana Press, Totowa, NJ, USA.
23. Looker AC, Wahner HW, Dunn WL, Calvo MS, Harris TB, Heyse SP, Johnston CC Jr, Lindsay R 1998 Updated data on proximal femur bone mineral levels of US adults. Osteoporos Int **8:**468–489.
24. World Health Organisation (WHO) Fracture Risk Assessment Tool (FRAX) Available at http://www.shef.ac.uk/FRAX/. Accessed March 28, 2008.
25. Genant HK, Grampp S, Glueer CC, Faulkner KG, Jergas M, Engelke K, Hagiwara S, van Kuijk C 1994 Universal standardisation

© 2008 American Society for Bone and Mineral Research

for the dual X-ray absorptiometry: Patient and phantom cross-calibration results. J Bone Miner Res **9:**1503–1514.
26. Kalender WA, Felsenberg D, Genant HK, Dequeker J, Reeve J 1995 The European Spine Phantom: A tool for standardisation and quality control in spinal bone mineral measurement by DXA and QCT. Eur J Radiol **20:**83–92.
27. Khoo WW 2000 Body composition measurements during infancy. Ann N Y Acad Sci **904:**383–392.
28. Drage NA, Palmer RM, Blake GM, Wilson R, Crane F, Fogelman I 2007 A comparison of bone mineral density in the spine, hip and jaw of edentulous subjects. Clin Oral Implants Res **18:**496–500.
29. Haugeberg G, Emery P 2005 Value of dual-energy X-ray absorptiometry as a diagnostic tool in early rheumatoid arthritis. Rheum Dis Clin North Am **31:**715–728.
30. Faulkner KG, McClung M, Cummings SR 1994 Automated evaluation of hip axis length for predicting hip fracture. J Bone Miner Res **9:**1065–1070.
31. Gregory JS, Stewart A, Undrill PE, Reid DM, Aspden RM 2005 Bone shape, structure and density as determinants of osteoporotic hip fracture: A pilot study investigating the combination of risk fracture. Invest Radiol **40:**591–597.
32. Beck TJ 2007 Extending DXA beyond bone mineral density: Understanding hip structure analysis. Curr Osteoporos Rep **5:**49–55.
33. Link TM, Guglielmi G, van Kuijk C, Adams JE 2005 Radiologic assessment of osteoporotic fracture: Diagnostic and prognostic implications. Eur Radiol **158:**1521–1532.
34. Ferrar L, Jiang G, Adams J, Eastell R 2005 Identification of vertebral fractures: An update. Osteoporos Int **16:**717–728.
35. Roberts MG, Cootes TF, Pacheco EM, Adams JE 2007 Quantitative fracture detection on Dual Energy X-ray Absorptiometry (DXA) images using shape and appearance models. Acad Radiol **14:**1166–1178.
36. Rea JA, Li J, Blake GM, Steiger P, Genant HK, Fogelman I 2000 Visual assessment of vertebral deformity by X-ray absorptiometry: A highly predictive method to exclude vertebral deformity. Osteoporos Int **11:**660–668.
37. van Rijn RR, Van DS, I, Link TM, Grampp S, Guglielmi G, Imhof H, Gluer C, Adams JE, van Kuijk C 2003 Bone densitometry in children: A critical appraisal. Eur Radiol **13:**700–710.
38. Crabtree NJ, Leonard MB, Zemel BS 2006 Dual energy X-ray absorptiometry. In: Sawyer AJ, Bachrach LK, Fung EB (eds.) Bone Densitometry in Growing Patients; Guidelines for Clinical Practice. Humana Press, pp. 41–58.
39. Carter DR, Bouxsein ML, Marcus R 1992 New approaches for interpreting projected bone densitometry data. J Bone Miner Res **7:**137–145.
40. Kroger H, Kotaniemi A, Vainio P, Alhava E 1992 Bone densitometry of the spine and femur in children by dual-energy x-ray absorptiometry. Bone Miner **17:**75–85.
41. Lu PW, Cowell CT, Lloyd-Jones SA, Briody J, Howman-Giles R 1996 Volumetric bone mineral density in normal subjects, aged 5–27 years. J Clin Endocrinol Metab **81:**1586–1590.
42. Clark EM, Ness AR, Bishop NJ, Tobias JH 2006 Association between bone mass and fractures in children: A prospective cohort study. J Bone Miner Res **21:**1489–1495.
43. Cooper C, Dennison EM, Leufkens HG, Bishop N, van Staa TP 2004 Epidemiology of childhood fractures in Britain: A study using the general practice research database. J Bone Miner Res **19:**1976–1981.
44. Jones IE, Williams SM, Dow N, Goulding A 2002 How many children remain fracture-free during growth? a longitudinal study of children and adolescents participating in the Dunedin Multidisciplinary Health and Development Study. Osteoporos Int **13:**990–995.
45. Ward KA, Ashby RL, Roberts SA, Adams JE, Mughal MZ 2007 UK reference data for the Hologic QDR Discovery dual energy X-ray absorptiometry scanner in healthy children aged 6–17 years. Arch Dis Child **92:**53–59.
46. Prentice A, Parsons TJ, Cole TJ 1994 Uncritical use of bone mineral density in absorptiometry may lead to size-related artifacts in the identification of bone mineral determinants. Am J Clin Nutr **60:**837–842.
47. Warner JT, Cowan FJ, Dunstan FD, Evand WD, Webb DK, Gregory JW 1998 Measured and predicted bone mineral content in healthy boys and girls aged 6–18 years: Adjustment for body size and puberty. Acta Paediatr **87:**244–249.
48. Molgaard C, Thomsen BL, Prentice A, Cole TJ, Michaelsen KF 1997 Whole body bone mineral content in healthy children and adolescents. Arch Dis Child **76:**9–15.
49. Crabtree NJ, Kibirige MS, Fordham J, Banks LM, Muntoni F, Chinn D, Boivin CM, Shaw NJ 2004 The relationship between lean body mass and bone mineral content in paediatric health and disease. Bone **35:**965–972.

Chapter 30. Quantitative Computed Tomography in Children and Adults

Claus C. Glüer

Medizinische Physik, Klinik für Diagnostische Radiologie, Universitätsklinikum Schleswig-Holstein, Kiel, Germany

INTRODUCTION

BMD is one of the strongest predictors of fracture risk, but today's standard densitometric method, DXA, has limitations, particularly for characterizing changes in bone status in subjects on treatment. QCT is a potential alternative for the future. Technical and clinical considerations are outlined in this chapter.

METHODOLOGY

X-ray–based CT provides 3D morphological and compositional information. On the reconstructed CT images, tissue contrast is predominantly determined by X-ray absorption of its heavier elements (e.g., calcium). Image gray values are expressed as CT number measured in Hounsfield units (HUs), but for QCT, a reference phantom of known composition is scanned together with the patient that allows conversion of HUs into calcium-hydroxyapatite (Ca-HAP) equivalent BMD. For earlier measurements that were calibrated to K_2HPO_4, a correction factor needs to be applied to express results in Ca-HAP, an issue important, for example, for comparison with published reference ranges. Unlike DXA, BMD measured by QCT reflects volumetric BMD and not an areal projection.

Single-energy QCT was initially applied to the radius by Rüegsegger et al.[1] followed by measurements at the spine by Genant and Boyd.[2] Dual-energy QCT approaches improve

The author states that he has no conflict of interest.

Key words: BMD, absorptiometry, QCT, pQCT, calibration, reference data, radiation exposure, error sources, T-scores, Z-scores, Tanner stage, mineralization, fracture risk, diagnosis, monitoring, bone geometry, microstructure, finite element analysis, quality assurance

© 2008 American Society for Bone and Mineral Research

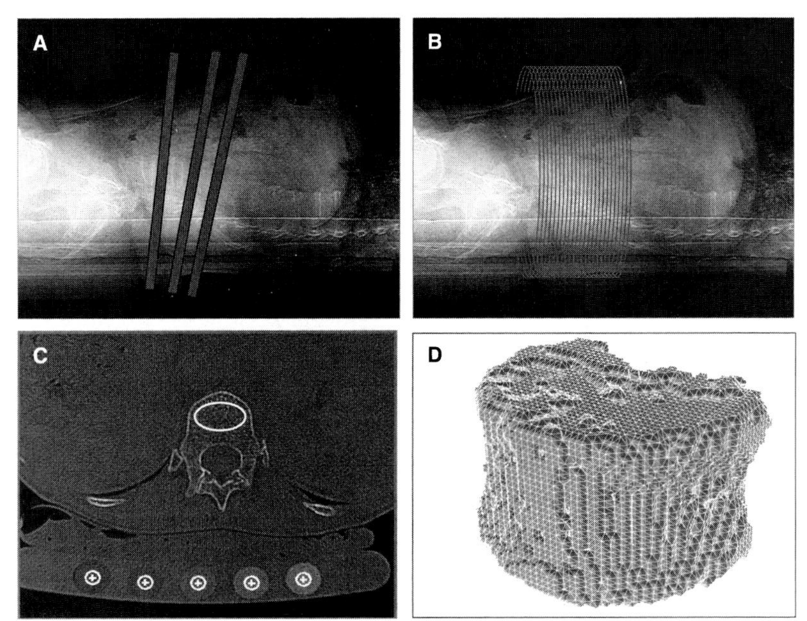

FIG. 1. Illustration of QCT procedure: localizer image used for either tilted single slice placement at three adjacent vertebrae (A) or spiral volumetric multislice acquisition covering the same three vertebrae (B); typical single slice technique image showing elliptical trabecular region of interest evaluated and calibration phantom underneath patient (C); stack of images segmented along the cortical rim depicting the volume of interest evaluated in the multi slice approach (D).

the accuracy of the measurement but at the expense of poorer precision and increased radiation exposure, and these approaches have not achieved a significant clinical role. For a QCT measurement, first a lateral planar overview measurement of the region to be scanned is generated (Figs. 1A and 1B), which allows placement of the measurement regions at the anatomically correct positions. Until the 1990s, the measurement region consisted of multiple individual and spatially separated CT slices, each ~5–10 mm thick (single-slice QCT; Fig. 1A). Today, spiral CT approaches yield a complete 3D dataset of the volume of interest (multislice QCT; Fig. 1B), consisting of 10–100 s of CT slices. Data acquisition is fast, with scan times on the order of 1–10 s. The spatial resolution has improved substantially, approaching 100–200 μm for human studies in vivo at peripheral measurement sites.[3] At central measurement sites, in-plane resolution is poorer by a factor of 2, and slice thickness is 300–500 μm.

Spinal QCT

For single slice approaches L_1–L_3 or T_{12}–L_3 are measured, and a minimum of two vertebrae should be evaluated. At 80 kVp, 140 mAs, and 8–10 mm slice thickness located at the center of the vertebrae (Fig. 1C), good precision at low radiation exposure and lower fat errors can be achieved.[4] With the more recent multislice approach (Fig. 1D) L_1–L_3 are measured, typically at 120 kVp, 50 mAs, a slice thickness of 1–2 mm, and pitch = 1. Trabecular bone is evaluated in an ellipse in the anterior part of the vertebral body or in larger regions encompassing most of the trabecular volume. The trabecular region shows high responsiveness early after menopause, better than other approaches. For research, the cortical rim can also be evaluated, but it is only with advancing age that the cortical width seen on QCT images is more than an artifact of partial volume effects.

Hip QCT

In recent years, QCT hip analysis software has been marketed by that is based on multislice data acquisition with 120 kVp, 100 mAs, and 3 mm slice thickness (Mindways Software, Austin, TX, USA). The software allows both calculation of trabecular and cortical compartments in the femoral neck and trochanter and provides estimated DXA results (CTXA) calculated from their QCT data. Few publications are available thus far.

pQCT has mostly been carried out at the distal forearm. Using small-angle fan beams, typical single-slice pQCT X-ray tube settings are 45–60 kVp and 140–400 mAs and 1–3 mm slice thickness in plane pixel sizes of 100–300 μm can be achieved. BMD is measured at the distal radius (4% or radius length, mostly trabecular) and at a more proximal location (15–65% radius length, cortical). Publications are few, and therefore the clinical use remains unclear. Recently, a multislice high-resolution pQCT approach featuring an isotropic voxel size and slice thickness of 80 μm, sufficient to depict microstructural information, has been commercialized.

Radiation Exposure

The level of desired image resolution, the location of the measurement region (proximity to radiation-sensitive organs), and the size of the measurement region determine the level of radiation exposure. By variation of the X-ray energy, radiation levels can be optimized, recognizing that the choice of the energy also affects accuracy (low tube voltage settings of ~80 kVp lead to lower radiation levels and to smaller fat-related accuracy errors than measurements at 120 kVp). For earlier single-slice QCT approaches of T_{12}–L_2, low radiation levels down to 50–100 μSv have been reported.[5] For current multislice approaches, the radiation exposure of QCT may vary substantially, reaching >1 mSv for measurements at the proximal femur or for high-resolution measurements at the spine. The radiation exposure for pQCT measurements is small: ~3-μSv effective dose for radial high-resolution pQCT.[3] Further improvements of spatial resolution are, however, restricted by high local skin radiation exposure.

CLINICAL APPLICATION

Reference Data

Reference ranges have been published for single-slice spinal QCT for a European and a North American population. For

© 2008 American Society for Bone and Mineral Research

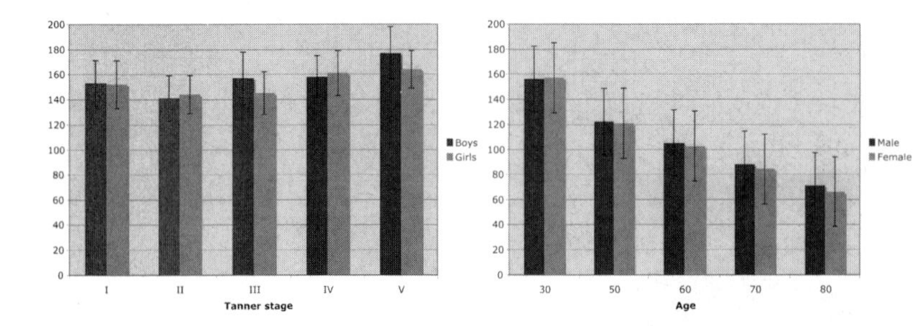

FIG. 2. Spinal trabecular BMD reference data measured by single slice QCT as function of sexual development for boys and girls (left[14]) or age in adult women (right[6]). Data obtained in single energy mode at 80–85 kVp.

the two largest studies, QCT methods differed with regard to scanner, calibration standard, kVp, and slice thickness, and results depended on the model fit selected, but still BMD results deviated by less than one half a population SD.[6] Assuming that a dual-energy approach provides the most accurate data, volumetric BMD of a central slice remains fairly constant between age 20 and 40 at ~140 mg/ml and decreases to ~60 mg/ml at age 80, with a population SD of ~25 mg/ml.[6] Men have the same volumetric BMD at young ages, a similar population SD, and an only slightly slower decrease to a level of ~70 mg/ml at age 80.[6] Single-energy approaches yield BMD data ~10–15% higher, depending on the kVp setting (Fig. 2, right[6]). Multislice spinal QCT BMD levels should be somewhat higher because the more dense regions closer to the endplates are included in the volume evaluated—no reference data have been published yet.

For pQCT, reference ranges have been published for the single-slice approach.[7] Single and multislice approaches have been cross-calibrated using the European Forearm Phantom,[8] and resulting reference data have been published.[9] For the new high-resolution multislice device, no reference data have been published yet.

Diagnosis of BMD Status in Adults

The diagnostic threshold given by the WHO of a T-score of −2.5 cannot be applied to QCT. Efforts have been made to define equivalent diagnostic thresholds for QCT, either based on equivalent fracture risk or equivalent fracture prevalence. QCT decreases faster with age than DXA. Therefore, diagnostic thresholds for QCT would have to be lower than −2.5, and they cannot be constant: they will decrease with age. For example, a level of 72 mg/ml for trabecular bone of the vertebrae,[10] equivalent to a QCT T-score of approximately −3.2, may be appropriate at age 50, but it is not low enough for older subjects. For QCT techniques using protocols that differ from those used for the above-mentioned published reference data, correction factors need to be applied. Additionally, for multislice approaches, the differences between the single and multislice volumes need to be accounted for.

The correlations between spinal trabecular BMD by QCT (or spinal DXA-based BMD) and radial trabecular BMD by pQCT are weak ($r = 0.4$[11]). Therefore, it is not possible to use pQCT to estimate BMD status at the spine. Because the diagnostic relevance of an abnormally low radial BMD is difficult to judge, pQCT results should not be used for diagnosis.

Diagnosis of BMD Status in Children

DXA as a projectional technique has many limitations for the assessment of bone status in children,[12] most importantly the impact of bone size, but also variable geometry and changing tissue composition (trabecular versus cortical bone and red versus yellow marrow). QCT approaches may thus be of interest, but the radiation exposure is higher than for DXA. QCT is less affected by size, but it is more sensitive to bone marrow changes.[13] The stage of sexual development seems to be the strongest predictor of volumetric BMD[14] and, therefore, it is useful to use reference data based on Tanner stages (Fig. 2, left[14]). Within the five groups of Tanner stage, no differences between boys and girls were observed.[14] Deviations from normal could be derived by calculating Tanner Z-scores (i.e., expressing the deviation in multiples of SD of the normal BMD variability within a given Tanner stage). T-scores are not meaningful for children. Unlike DXA, QCT allows differentiation between small bones and low BMD. If chronological age and bone age (e.g., according to Greulich-Peyl or Tanner-Whitehouse scores) differ, even QCT may not provide the complete answer, because it is not possible to distinguish low bone organ density from low tissue mineralization (e.g., in rickets).

Peripheral measurements have less of a radiation exposure problem. Moreover, pQCT of the forearm shows important relationships between bone growth and mechanical or hormonal factors impacting during adolescence. However, the many different variables generated from pQCT may confuse a clinical user. For single-slice pQCT, reference curves for trabecular or cortical bone at the 4% measurement site may be a good choice.[15,16] Similar to spinal volumetric BMD and unlike spinal areal BMD results of DXA, volumetric BMD does not change in girls and boys between ages 6 and 15 and later only increases for boys. However, pQCT results may also be biased: partial volume effects need to be taken in consideration, particularly at younger ages and for cortical measures. Moreover, discordance of spinal and peripheral BMD developments would affect the role of pQCT in children.

The indications for bone densitometry by DXA in children have been addressed in a position paper by the British National Osteoporosis Society.[12] QCT and pQCT can be used instead of DXA to reduce the impact of error sources, but for spinal QCT, this comes at the expense of higher radiation exposure. To date, QCT remains a tool for research—a very powerful tool.

Fracture Risk Assessment

For spinal QCT, there is only one old prospective fracture study in which QCT was compared with DXA. However, in a large number of cross-sectional studies, spinal QCT in general discriminated fracture patients at least as well as DXA and in some studies was significantly stronger (Table 1). In one study on discrimination of hip fractures, spinal multislice QCT showed poor performance, indicating that femoral QCT approaches are required for this task.

For pQCT, no prospective studies have been published. The cross-sectional studies for the single-slice approach are difficult to interpret because the authors usually tested a large number of pQCT variables and thus it is unclear whether the observed association (Table 1) would hold in a retest in independent samples. For multislice pQCT, only data from one Japanese

© 2008 American Society for Bone and Mineral Research

TABLE 1. AGE-ADJUSTED GRADIENTS OF RISK PER SD POPULATION VARIATION FOR QCT AND PQCT

Technique	*Measurement site*	*Variable*	*Gradient of risk vertebral fractures*	*Gradient of risk hip fractures*
QCT, single slice	Spine	Trabecular BMD	2.3[36]*	
			3.7[37]†	
			2.0[38]	
			2.2[39]	
			2.9[40]	
			2.9[41]	
			3.4[42]	
QCT, multislice	Spine	Trabecular BMD	1.9[18]†	1.2[43]
pQCT, single slice	Ultradistal radius (4% site)	Trabecular BMD	1.2[44]‡	2.4[44]
			1.3[41]‡	1.5[46]
			1.5[45]	1.9[47]
				1.6[45]
		Cortical area	1.9[48]	
pQCT, multislice	Distal radius	Trabecular BMD	1.6[38]	
	Radius metaphysis	Cortical area fraction	2.0[38]	
	Radius diaphysis	Integral BMD	1.8[38]	

To partially adjust for subject selection bias, they have been standardized to DXA of the spine for vertebral fractures and DXA of the femur for hip fractures according to the formula $OR_{adjusted} = OR_{unadjusted} \times$ [OR(DXA from meta analysis)/OR(DXA in that population)]. OR(DXA from meta analysis) is 1.9 for vertebral fractures and 2.6 for hip fractures.[35]

* Only one prospective study included and risk ratio specified; for all other cross-sectional studies gradient of risk is given as OR.

† Unadjusted for DXA ORs because of discrepant DXA hip and/or spine data.

‡Values are not significant at $p < 0.05$.

group on an older device is available. For high-resolution pQCT, two recent studies showed fracture discrimination (for densitometric and microstructural variables) independent of DXA for various structural variables,[3,17] but another study on vertebral fracture reported insignificant discrimination.[18]

Clinical Interpretation

For spinal single-slice QCT, the risk of fracture at a given T-score level is smaller than for DXA because a larger fraction of the T-score reflects age-associated risk, leaving less to BMD-associated risk. Therefore, for a given level of risk, the T-score of QCT has to be lower than the DXA T-score and that difference increases with age. Given the poor performance of spinal QCT in predicting hip fracture and the increasing prevalence of hip fractures with advancing age, the differences between risk-equivalent T-scores of DXA and QCT increase further at older age. Evidence how to use QCT for treatment decisions is limited. A comparison of patient data with appropriate reference ranges may provide valuable insight, but intervention thresholds would have to be lower for QCT than for DXA if expressed as T-scores. From the perspective of evidence-based medicine, a valid DXA scan will in general still be preferable for making treatment decisions. For the future, the development of risk-based intervention criteria for QCT is indicated: instead of using T-scores, a patient's risk could better be estimated by correcting the average age-associated risk by the QCT-associated risk derived from Z-scores and the technique's standardized gradient of risk (Table 1).[19] However, this requires more data on QCT-associated fracture risk. Evidence for risk assessment in men is very limited, and thus no recommendation can be made.

Monitoring Changes in Bone Status

A necessary requirement for monitoring is good longitudinal sensitivity (i.e., that the technique is able to detect the changes early on). Longitudinal sensitivity is defined as the ratio of response rate divided by (long-term) precision errors.[20] Compared with DXA, precision errors of spinal CT approaches are somewhat higher. Response rates are about twice as large as for DXA for antiresorptive drugs[21] and three times as large during early osteoanabolic treatment.[22,23] It is not clear yet to what extent QCT changes reflect antifracture efficacy of drugs, but improvements in finite element–derived bone strength under treatment were larger than the DXA changes seen in the same patients.[23]

Data for single-slice pQCT of the radius are limited, but response rates seem to be smaller. Longitudinal sensitivity of pQCT is poorer than that of spinal DXA.[24] Potentially, multislice high-resolution pQCT techniques may show improved performance.

PERSPECTIVES

In addition to a more comprehensive 3D evaluation of volumetric BMD in different bone compartments, three areas of research point to innovative extension beyond the current possibilities.

Bone geometry can be accurately assessed by 3D QCT approaches, and this is of interest to understand bone growth, to investigate geometric strength under specific loading conditions, to study the impact of treatment, and to assess error sources of projectional techniques. Because of the complex 3D shape and the relevance as a fracture site, QCT measurements of femur are of greatest importance. Geometric measures, like section modulus or buckling ratio, allow a direct assessment of mechanical competence under bending or buckling forces. For long bones, geometric strength indices have been developed that show improved correlation with breaking strength in specimen studies.[25–27] Also sex-dependent growth-associated geometric changes can be studied and provide valuable insight in the adaptive processes of bone modeling and remodeling.[28]

Microstructural assessments by high-resolution CT (HRCT) allow visualization of trabecular microstructure and some aspects of cortical porosity. Best image quality at 80-μm voxel size can be achieved at peripheral measurement sites. Here structural information can be extracted,[3,17,29] but it remains to be shown whether this improves vertebral or hip fracture

© 2008 American Society for Bone and Mineral Research

risk assessment.[18] For HRCT measurements at the spine, image quality is not as good, with 160- to 200-μm voxel size in plane and a slice thickness of ~200–300 μm, but this is sufficient to depict and quantify trabecular separation. Improved fracture discrimination[30] and responsiveness to treatment[22] have been reported.

QCT-based finite element analysis (FEA), a powerful analysis tool well established in engineering sciences, allows modeling of the mechanical competence of whole bones under specific loading conditions.[31–33] Initial results have shown that additional information about treatment effects can be obtained.[23] The method can also be used to identify weak regions that are most likely to fracture under loads. FEA also provides a direct estimate of the forces a specific bone can withstand, and it is possible to derive indices of fracture risk[34] that may provide independent information about fracture risk.[18]

SUMMARY

The renewed interest in QCT has two primary reasons: the deficiencies of DXA in monitoring treatment and the prospect that QCT may not only do better but also yield direct measures of strength (and, addressing the combination of these two aspects, variations in strength induced by treatment). Already today, QCT is a valuable clinically applicable tool for the assessment of bone status. However, prospective studies on risk assessment and clinical guidelines for appropriate use are lacking. In a clinical setting, a diagnostic QCT-based evaluation currently can only be carried out if the scanner, calibration method, and the protocol used are well characterized compared with published references data. Recent developments, however, document that the potential for further advances in QCT, specifically for clinical research, is very substantial.

ACKNOWLEDGMENTS

Valuable input by Judith Adams, Keenan Brown, Christian Graeff, and Andres Laib is gratefully acknowledged.

REFERENCES

1. Rüegsegger P, Elsasser U, Anliker M, Gnehm H, Kind H, Prader A 1976 Quantification of bone mineralization using computed tomography. Radiology **121:**93–97.
2. Genant HK, Boyd D 1977 Quantitative bone mineral analysis using dual energy computed tomography. Invest Radiol **12:**545–551.
3. Boutroy S, Bouxsein ML, Munoz F, Delmas PD 2005 In vivo assessment of trabecular bone microarchitecture by high-resolution peripheral quantitative computed tomography. J Clin Endocrinol Metab **90:**6508–6515.
4. Cann CE 1981 Low-dose CT scanning for quantitative spinal mineral analysis. Radiology **140:**813–815.
5. Kalender WA 1992 Effective dose values in bone mineral measurements by photon absorptiometry and computed tomography. Osteoporos Int **2:**82–87.
6. Kalender WA, Felsenberg D, Louis O, Lopez P, Klotz E, Osteaux M, Fraga J 1989 Reference values for trabecular and cortical vertebral bone density in single and dual-energy quantitative computed tomography. Eur J Radiol **9:**75–80.
7. Butz S, Wuster C, Scheidt-Nave C, Gotz M, Ziegler R 1994 Forearm BMD as measured by peripheral quantitative computed tomography (pQCT) in a German reference population. Osteoporos Int **4:**179–184.
8. Pearson J, Ruegsegger P, Dequeker J, Henley M, Bright J, Reeve J, Kalender W, Felsenberg D, Laval-Jeantet AM, Adams JE, Birkenhager JC, Fischer M, Geusens P, Hesch RD, Hyldstrup L, Jaeger P, Jonson R, Kroger H, van Lingen A, Mitchell A, Reiners C, Schneider P 1994 European semi-anthropomorphic phantom for the cross-calibration of peripheral bone densitometers: Assessment of precision accuracy and stability. Bone Miner **27:**109–120.
9. Reeve J, Kroger H, Nijs J, Pearson J, Felsenberg D, Reiners C, Schneider P, Mitchell A, Ruegsegger P, Zander C, Fischer M, Bright J, Henley M, Lunt M, Dequeker J 1996 Radial cortical and trabecular bone densities of men and women standardized with the European Forearm Phantom. Calcif Tissue Int **58:**135–143.
10. Lafferty FW, Rowland DY 1996 Correlations of dual-energy X-ray absorptiometry, quantitative computed tomography, and single photon absorptiometry with spinal and non-spinal fractures. Osteoporos Int **6:**407–415.
11. Grampp S, Genant HK, Mathur A, Lang P, Jergas M, Takada M, Glüer CC, Lu Y, Chavez M 1997 Comparisons of noninvasive bone mineral measurements in assessing age-related loss, fracture discrimination, and diagnostic classification. J Bone Miner Res **12:**697–711.
12. National Osteoporosis Society 2004 A Practical Guide to Bone Densitometry in Children. National Osteoporosis Society, Bath, UK.
13. Glüer CC, Genant HK 1989 Impact of marrow fat on accuracy of quantitative CT. J Comput Assist Tomogr **13:**1023–1035.
14. Gilsanz V, Boechat MI, Roe TF, Loro ML, Sayre JW, Goodman WG 1994 Gender differences in vertebral body sizes in children and adolescents. Radiology **190:**673–677.
15. Neu CM, Manz F, Rauch F, Merkel A, Schoenau E 2001 Bone densities and bone size at the distal radius in healthy children and adolescents: A study using peripheral quantitative computed tomography. Bone **28:**227–232.
16. Rauch F, Schoenau E 2005 Peripheral quantitative computed tomography of the distal radius in young subjects—new reference data and interpretation of results. J Musculoskelet Neuronal Interact **5:**119–126.
17. Sornay-Rendu E, Boutroy S, Munoz F, Delmas PD 2007 Alterations of cortical and trabecular architecture are associated with fractures in postmenopausal women, partially independent of decreased BMD measured by DXA: The OFELY study. J Bone Miner Res **22:**425–433.
18. Melton LJ III, Riggs BL, Keaveny TM, Achenbach SJ, Hoffmann PF, Camp JJ, Rouleau PA, Bouxsein ML, Amin S, Atkinson EJ, Robb RA, Khosla S 2007 Structural determinants of vertebral fracture risk. J Bone Miner Res **22:**1885–1892.
19. Dachverband Osteologie e V 2006 Osteoporose-Leitlinie. Schattauer, Stuttgart, Germany.
20. Glüer CC 1999 Monitoring skeletal changes by radiological techniques. J Bone Miner Res **14:**1952–1962.
21. Black DM, Greenspan SL, Ensrud KE, Palermo L, McGowan JA, Lang TF, Garnero P, Bouxsein ML, Bilezikian JP, Rosen CJ 2003 The effects of parathyroid hormone and alendronate alone or in combination in postmenopausal osteoporosis. N Engl J Med **349:**1207–1215.
22. Graeff C, Timm W, Nickelsen TN, Farrerons J, Marin F, Barker C, Glüer CC 2007 Monitoring teriparatide-associated changes in vertebral microstructure by high-resolution CT in vivo: Results from the EUROFORS study. J Bone Miner Res **22:**1426–1433.
23. Keaveny TM, Donley DW, Hoffmann PF, Mitlak BH, Glass EV, San Martin JA 2007 Effects of teriparatide and alendronate on vertebral strength as assessed by finite element modeling of QCT scans in women with osteoporosis. J Bone Miner Res **22:**149–157.
24. Schneider PF, Fischer M, Allolio B, Felsenberg D, Schroder U, Semler J, Ittner JR 1999 Alendronate increases bone density and bone strength at the distal radius in postmenopausal women. J Bone Miner Res **14:**1387–1393.
25. Müller ME, Webber CE, Bouxsein ML 2003 Predicting the failure load of the distal radius. Osteoporos Int **14:**345–352.
26. Ferretti JL, Capozza RF, Zanchetta JR 1996 Mechanical validation of a tomographic (pQCT) index for noninvasive estimation of rat femur bending strength. Bone **18:**97–102.
27. Siu WS, Qin L, Leung KS 2003 pQCT bone strength index may serve as a better predictor than bone mineral density for long bone breaking strength. J Bone Miner Metab **21:**316–322.
28. Seeman E 1997 Osteoporosis in men. Baillieres Clin Rheumatol **11:**613–629.
29. MacNeil JA, Boyd SK 2007 Accuracy of high-resolution peripheral quantitative computed tomography for measurement of bone quality. Med Eng Phys **29:**1096–1105.
30. Ito M, Ikeda K, Nishiguchi M, Shindo H, Uetani M, Hosoi T, Orimo H 2005 Multi-detector row CT imaging of vertebral microstructure for evaluation of fracture risk. J Bone Miner Res **20:**1828–1836.
31. van Rietbergen B 2001 Micro-FE analyses of bone: State of the art. Adv Exp Med Biol **496:**21–30.

© 2008 American Society for Bone and Mineral Research

32. Chevalier Y, Pahr D, Allmer H, Charlebois M, Zysset P 2007 Validation of a voxel-based FE method for prediction of the uniaxial apparent modulus of human trabecular bone using macroscopic mechanical tests and nanoindentation. J Biomech **40:**3333–3340.
33. Crawford RP, Cann CE, Keaveny TM 2003 Finite element models predict in vitro vertebral body compressive strength better than quantitative computed tomography. Bone **33:**744–750.
34. Bouxsein ML, Melton LJ III, Riggs BL, Muller J, Atkinson EJ, Oberg AL, Robb RA, Camp JJ, Rouleau PA, McCollough CH, Khosla S 2006 Age- and sex-specific differences in the factor of risk for vertebral fracture: A population-based study using QCT. J Bone Miner Res **21:**1475–1482.
35. Marshall D, Johnell O, Wedel H 1996 Meta-analysis of how well measures of bone mineral density predict occurrence of osteoporotic fractures. BMJ **312:**1254–1259.
36. Ross PD, Genant HK, Davis JW, Miller PD, Wasnich RD 1993 Predicting vertebral fracture incidence from prevalent fractures and bone density among non-black, osteoporotic women. Osteoporos Int **3:**120–126.
37. Yu W, Glüer C-C, Grampp S, Jergas M, Fuerst T, Wu CY, Lu Y, Fan B, Genant HK 1995 Spinal bone mineral assessment in postmenopausal women: A comparison between dual x-ray absorptiometry and quantitative computed tomography. Osteoporos Int **5:**433–439.
38. Tsurusaki K, Ito M, Hayashi K 2000 Differential effects of menopause and metabolic disease on trabecular and cortical bone assessed by peripheral quantitative computed tomography (pQCT). Br J Radiol **73:**14–22.
39. Bergot C, Laval-Jeantet AM, Hutchinson K, Dautraix I, Caulin F, Genant HK 2001 A comparison of spinal quantitative computed tomography with dual energy X-ray absorptiometry in European women with vertebral and nonvertebral fractures. Calcif Tissue Int **68:**74–82.
40. Guglielmi G, Cammisa M, De Serio A, Scillitani A, Chiodini I, Carnevale V, Fusilli S 1999 Phalangeal US velocity discriminates between normal and vertebrally fractured subjects. Eur Radiol **9:**1632–1637.
41. Grampp S, Genant HK, Mathur A, Lang P, Jergas M, Takada M, Glüer C-C, Lu Y, Chavez M 1997 Comparisons of noninvasive bone mineral measurements in assessing age-related loss, fracture discrimintion, and diagnostic classification. J Bone Miner Res **12:**697–711.
42. Duboeuf F, Jergas M, Schott AM, Wu CY, Gluer CC, Genant HK 1995 A comparison of bone densitometry measurements of the central skeleton in post-menopausal women with and without vertebral fracture. Br J Radiol **68:**747–753.
43. Lang TF, Augat P, Lane NE, Genant HK 1998 Trochanteric hip fracture: Strong association with spinal trabecular bone mineral density measured with quantitative CT. Radiology **209:**525–530.
44. Formica CA, Nieves JW, Cosman F, Garrett P, Lindsay R 1998 Comparative assessment of bone mineral measurements using dual X-ray absorptiometry and peripheral quantitative computed tomography. Osteoporos Int **8:**460–467.
45. Clowes JA, Eastell R, Peel NF 2005 The discriminative ability of peripheral and axial bone measurements to identify proximal femoral, vertebral, distal forearm and proximal humeral fractures: A case control study. Osteoporos Int **16:**1794–1802.
46. Augat P, Fan B, Lane NE, Lang TF, LeHir P, Lu Y, Uffmann M, Genant HK 1998 Assessment of bone mineral at appendicular sites in females with fractures of the proximal femur. Bone **22:**395–402.
47. Majumdar S, Link TM, Augat P, Lin JC, Newitt D, Lane NE, Genant HK 1999 Trabecular bone architecture in the distal radius using magnetic resonance imaging in subjects with fractures of the proximal femur. Osteoporos Int **10:**231–239.
48. Grampp S, Lang P, Jergas M, Gluer CC, Mathur A, Engelke K, Genant HK 1995 Assessment of the skeletal status by peripheral quantitative computed tomography of the forearm: Short-term precision in vivo and comparison to dual X-ray absorptiometry. J Bone Miner Res **10:**1566–1576.

Chapter 31. Magnetic Resonance Imaging of Bone

Sharmila Majumdar

Musculo-Skeletal and Quantitative Imaging Research Group, Department of Radiology, University of California, San Francisco, California

INTRODUCTION

3D imaging techniques that show bone structure are emerging as important contenders for defining bone quality, at least partially. Techniques such as μCT have recently been developed and provide high-resolution images of the trabecular and cortical architecture. This method is routinely used in specimen evaluation and has recently been extended to in vivo animal and human extremity imaging. Another recent development in the assessment of trabecular bone structure is the use of MRI, a nonionizing technique that makes it possible to obtain noninvasive bone biopsies at multiple anatomic sites.

MRI OF BONE

Trabecular bone consists of a network of rod-like elements interconnected by plate-like elements, immersed in bone marrow composed partly of water and partly of fat. Magnetic susceptibility of trabecular bone is substantially different from that of bone marrow. This gives rise to susceptibility gradients at the bone–marrow interface. Magnetic inhomogeneity arising from these susceptibility gradients depends on the static magnetic field strength, number of bone–bone marrow interfaces, and the size of individual trabeculae.[1–3] These effects cause dephasing of spins and signal decay at a rate known as T2*. In a voxel partly occupied by bone and partly by marrow, the static inhomogeneity induced intravoxel dephasing of spins leads to signal cancellation within the voxel. T2* methods have been used to quantify trabecular bone, and these measures have been related to bone strength, osteoporotic status, and therapeutic response.[4]

Beside the tissue composition, the small dimensions of the trabecular elements (~100 μm) require very high imaging resolutions. The suitability of an MRI method (acquisition and analysis) for depicting bone microstructure depends on its ability to yield images with high enough signal in a reasonable acquisition time and its ability to derive trabecular structural measurements from the images accurately and reproducibly. The three competing factors to be considered in high-resolution MRI (HR-MRI) are signal-to-noise ratio (SNR), spatial resolution, and imaging time. Spatial resolution and SNR are both directly related to imaging time but are inversely related to each other. Recent technique developments in tra-

Dr. Majumdar has received research grants from Merck and GlaxoSmithKline.

© 2008 American Society for Bone and Mineral Research

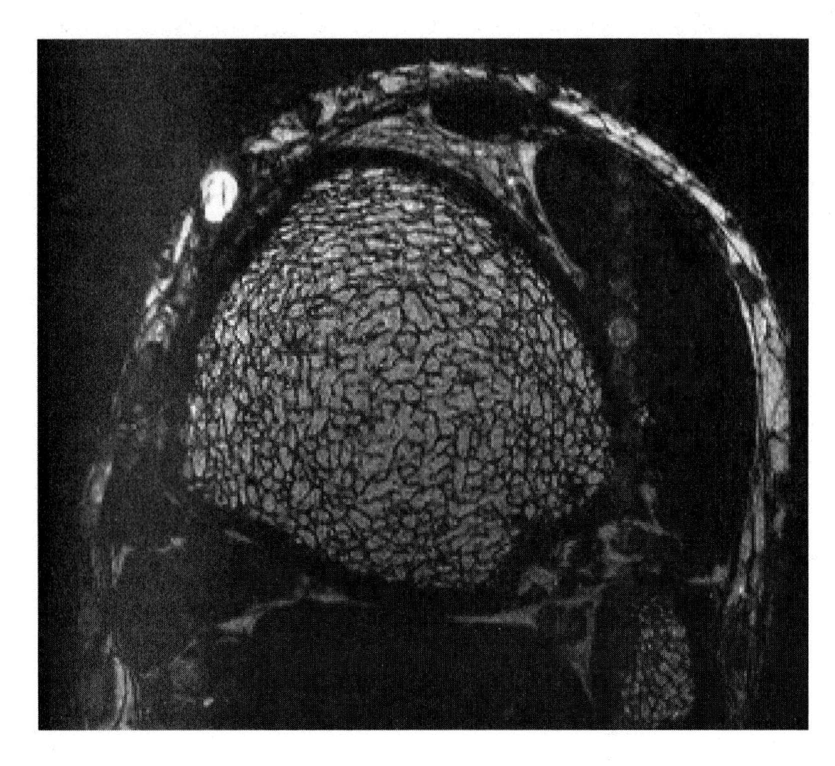

FIG. 1. Distal tibia imaged at 3 T depicting the trabecular bone as a dark network against bright bone marrow.

becular bone MRI technique reflect all these considerations and have been aiming for increasing SNR and accelerating total acquisition times.

MR pulse sequences can be broadly classified into spin-echo and gradient echo sequences. Ideally, 3D spin echo (SE) sequences are better suited for imaging of trabecular bone microarchitecture than gradient echo (GE)-based sequences because they are less sensitive to the thickening of trabeculae caused by susceptibility differences. However, GE sequences can be used with short repetition time (TR) because of their higher SNR efficiency and can thus acquire a 3D volume in shorter scan time and avoid patient motion artifacts.[5,6] 3D-SE type pulse sequences with variable flip angle like rapid SE excitation (RASEE),[7,8] large-angle spin-echo imaging,[9] and subsequently fast 3D large-angle spin-echo imaging (FLASE),[10] and a new fully balanced steady-state 3D-spin-echo pulse sequence have also been developed.[11] The choice of pulse sequence for trabecular bone imaging is still a topic of active research. Availability of the sequences at multiple centers, their robustness, and total imaging time versus the anatomical coverage are typical considerations.

SNR is linearly proportional to the static magnetic field strength, perhaps making 3 T preferable over 1.5-T magnets. Phan et al.[12] imaged the trabecular microarchitecture in 40 cadaveric calcaneus specimens at 1.5 and 3 T and compared them with μCT as the gold standard and found that correlations between trabecular structural parameters derived from 3-T MRI and μCT were significantly higher ($p < 0.05$) than correlations between structural parameters obtained from 1.5-T MRI and μCT. Preliminary experiments conducted on a 7-T GE Signa scanner yielded a 2-fold increase in SNR for HR-MRI of trabecular bone.[13,14] Figure 1 shows an example of a high-resolution MR image obtained at 3 T, where the bone marrow is bright, and trabecular bone is depicted as dark striations. Image such as these can be analyzed to derive structural measures of microarchitecture.

The most common structural measures analogous to quantitative histomorphometry derived from MR images include app. BV/TV, app.Tb.N, app.Tb.Sp, and app. Tb.Th[15,16] and require the images to be subdivided into a bone and marrow component or binarized. Because the MR images are not acquired at true microscopic resolutions, Majumdar et al.[17] described these measures derived from MR images as "apparent" measurements, which, although obtained in the limited-resolution regimen, are highly correlated to the "true" structure. Binarization of an MR image is not a trivial task; because of partial volume effects, multiple techniques have been developed that operate directly on the grayscale image. Recognizing the fuzzy nature of the images caused by partial volume effects, Saha and Wehrli[18] applied a fuzzy distance transform (FDT) technique for computing trabecular thickness and observed an improved robustness in the computation against loss of resolution. Digital topological analysis techniques have also been applied to quantify the number of surface and curved edges, junctions, and interiors in the trabecular network.[19]

RELATIONSHIP OF MR-DERIVED STRUCTURE MEASURES TO BONE STRENGTH, FRACTURE, AND OSTEOPOROTIC STATUS: RESPONSE TO THERAPY

Several studies relating the measures of trabecular structure obtained using MRI to measures of bone strength in vitro have been conducted.[20–23] A relationship between whole bone strength and bone structure measures have been shown in radii and in the proximal femur.[24,25]

High-resolution MR images of the distal radius were obtained at 1.5 T in premenopausal normal, postmenopausal normal, and postmenopausal osteoporotic women.[26] Significant differences were evident in spinal BMD, radial trabecular BMD, trabecular bone volume fraction, trabecular spacing, and trabecular number between the postmenopausal nonfracture and the postmenopausal osteoporotic subjects. Trabecular

© 2008 American Society for Bone and Mineral Research

spacing and trabecular number showed moderate correlation with radial trabecular BMD but correlated poorly with radial cortical BMC.

Distance transformation techniques were applied to the 3D image of the distal radius of postmenopausal patients, and structural indices such as app.Tb.N, app.Tb.Th, and app.Tb.Sp were determined without model assumptions.[27] A new metric index, the apparent intraindividual distribution of separations (app.Tb.Sp.SD), was introduced. It was found that app.Tb.Sp.SD discriminates fracture subjects from nonfracture patients as well as DXA measurements of the radius and the spine but not as well as DXA of the hip. MR-derived measures of trabecular bone architecture in the distal radius[28] and calcaneus[29] were obtained in 20 subjects with hip fractures and 19 age-matched postmenopausal controls, in addition to BMD measures at the hip DXA and the distal radius (pQCT). Measures of app. Tb.Sp and app. Tb.N in the distal radius showed significant ($p < 0.05$) differences between the two groups, as did hip BMD measures. However, radial trabecular BMD measures showed only a marginal difference ($p = 0.05$). In the calcaneus, significant differences between both patient groups were obtained using morphological parameters.

Sagittal MR images of the calcaneus were obtained in 50 men (26 patients with osteoporosis and 24 age-matched healthy control subjects).[30] Structural parameters, especially connectivity parameters, showed significant differences between control subjects and patients ($p < 0.05$).

In addition, in vivo images have also been combined with microfinite element analysis in a limited set of subjects. Newitt et al.[31] studied subjects in two groups: postmenopausal women with normal BMD ($n = 22$; mean age, 58 ± 7 yr) and postmenopausal women with spine or femur BMD −1 SD to −2.5 SD below young normal ($n = 37$; mean age, 62 ± 11 yr). Anisotropy of trabecular bone microarchitecture, as measured by the ratios of the MIL values (MIL1/MIL3, etc.) and the anisotropy in elastic modulus (E1/E3, etc.) were greater in the osteopenic group.

Ninety-one postmenopausal osteoporotic women were followed for 2 yr ($n = 46$ for nasal spray calcitonin, $n = 45$ for placebo).[32] MRI measurements of trabecular structure were obtained at the distal radius and calcaneus in addition to DXA-BMD at the spine/hip/wrist/calcaneus (obtained yearly). MRI assessment of trabecular microarchitecture at individual regions of the distal radius showed preservation (no significant loss), in the treated group compared with significant deterioration in the placebo control group.

Trabecular bone structure of the tibia was studied in 10 men with severe, untreated hypogonadism and age- and race-matched eugonadal men. Two composite topological indices were determined: the ratio of surface voxels (representing plates) to curve voxels (representing rods), which is higher when architecture is more intact, and the erosion index, a ratio of parameters expected to increase on architectural deterioration to those expected to decrease, which is higher when deterioration is greater. The surface/curve ratio was 36% lower ($p = 0.004$) and the erosion index was 36% higher ($p = 0.003$) in the hypogonadal men than in the eugonadal men.[33] In contrast, BMD of the spine and hip was not significantly different between the two groups. After 24 mo of testosterone treatment, BMD of the spine increased 7.4% ($p < 0.001$) and of the total hip increased 3.8% ($p = 0.008$). Architectural parameters assessed by MRI also changed: the surface-to-curve ratio increased 11% ($p = 0.004$) and the topological erosion index decreased 7.5% ($p = 0.004$).[34]

Until recently, in vivo MRI of trabecular microarchitecture was limited to peripheral sites such as the distal tibia and femur, radius, and calcaneus because of SNR limitations. However, the main sites of osteoporotic fractures are nonperipheral regions such as the vertebral bodies (spine) and the proximal femur (hip). High-resolution MRI has only recently been applied to the proximal femur[35] by using SNR efficient sequences, high magnetic field strength (3 T), and phased array coils.

Although significant work has been done using MRI to measure trabecular bone structure, there is little work on the macroarchitectural geometry of the cortex, which may play an equally important role for bone strength. Recently, Gomberg et al.[36] studied cortical shell geometry of the femur, further expanding the potential role of MR in characterizing bone.

Imaging trabecular and cortical microarchitecture, characterizing the features of trabecular and cortical bone, has been an area of fertile and ongoing research. Beyond relating microarchitecture to the biomechanical properties of bone in specimens, advances have been made to extend these measures in vivo in human subjects. In this context, relationships between age, fracture status, and even post-therapeutic response have been studied. New advances in MR (nonionizing, peripheral sites, calcaneus, and femur) are ongoing and evolving at a rapid pace, and with the establishment of robust analysis methodologies and normative databases, have the potential for further clinical use in the coming years.

REFERENCES

1. Majumdar S 1991 Quantitative study of the susceptibility difference between trabecular bone and bone marrow: Computer simulations. Magn Reson Med **22:**101–110.
2. Weisskoff RM, Zuo CS, Boxerman JL, Rosen BR 1994 Microscopic susceptibility variation and transverse relaxation: Theory and experiment. Magn Reson Med **31:**601–610.
3. Ford JC, Wehrli FW, Chung HW 1993 Magnetic field distribution in models of trabecular bone. Magn Reson Med **30:**373–379.
4. Link TM, Majumdar S, Augat P, Lin JC, Newitt D, Lane NE, Genant HK 1998 Proximal femur: Assessment for osteoporosis with T2* decay characteristics at MR imaging. Radiology **209:**531–536.
5. Majumdar S, Link TM, Augat P, Lin JC, Newitt D, Lane NE, Genant HK 1999 Trabecular bone architecture in the distal radius using magnetic resonance imaging in subjects with fractures of the proximal femur. Magnetic Resonance Science Center and Osteoporosis and Arthritis Research Group. Osteoporos Int **10:**231–239.
6. Newitt DC, van Rietbergen B, Majumdar S 2002 Processing and analysis of in vivo high-resolution MR images of trabecular bone for longitudinal studies: Reproducibility of structural measures and micro-finite element analysis derived mechanical properties. Osteoporos Int **13:**278–287.
7. Jara H, Wehrli FW, Chung H, Ford JC 1993 High-resolution variable flip angle 3D MR imaging of trabecular microstructure in vivo. Magn Reson Med **29:**528–539.
8. Bogdan AR, Joseph PM 1990 RASEE: A rapid spin-echo pulse sequence. Magn Reson Imaging **8:**13–19.
9. DiIorio G, Brown JJ, Borrello JA, Perman WH, Shu HH 1995 Large angle spin-echo imaging. Magn Reson Imaging **13:**39–44.
10. Ma J, Wehrli FW, Song HK 1996 Fast 3D large-angle spin-echo imaging 3D FLASE. Magn Reson Med **35:**903–910.
11. Krug R, Han ET, Banerjee S, Majumdar S 2006 Fully balanced steady-state 3D-spin-echo (bSSSE) imaging at 3 Tesla. Magn Reson Med **56:**1033–1040.
12. Phan CM, Matsuura M, Bauer JS, Dunn TC, Newitt D, Lochmueller EM, Eckstein F, Majumdar S, Link TM 2006 Trabecular bone structure of the calcaneus: Comparison of MR imaging at 3.0 and 1.5 T with micro-CT as the standard of reference. Radiology **239:**488–496.
13. Zuo J, Bolbos R, Hammond K, Li X, Majumdar S 2008 Reproducibility of the quantitative assessment of cartilage morphology and trabecular bone structure with magnetic resonance imaging at 7 T. Magn Reson Imaging **26:**560–566.
14. Krug R, Carballido-Gamio J, Banerjee S, Stahl R, Carvajal L, Xu D, Vigneron D, Kelley DA, Link TM, Majumdar S 2007 In vivo

© 2008 American Society for Bone and Mineral Research

bone and cartilage MRI using fully-balanced steady-state free-precession at 7 tesla. Magn Reson Med **58:**1294–1298.
15. Parfitt AM 1983 Assessment of trabecular bone status. Henry Ford Hosp Med J **31:**196–198.
16. Parfitt AM, Mathews CH, Villanueva AR, Kleerekoper M, Frame B, Rao DS 1983 Relationships between surface, volume, and thickness of iliac trabecular bone in aging and in osteoporosis. Implications for the microanatomic and cellular mechanisms of bone loss. J Clin Invest **72:**1396–1409.
17. Majumdar S, Newitt D, Mathur A, Osman D, Gies A, Chiu E, Lotz J, Kinney J, Genant H 1996 Magnetic resonance imaging of trabecular bone structure in the distal radius: Relationship with X-ray tomographic microscopy and biomechanics. Osteoporos Int **6:**376–385.
18. Saha PK, Wehrli FW 2004 Measurement of trabecular bone thickness in the limited resolution regime of in vivo MRI by fuzzy distance transform. IEEE Trans Med Imaging **23:**53–62.
19. Gomberg BR, Saha PK, Song HK, Hwang SN, Wehrli FW 2000 Topological analysis of trabecular bone MR images. IEEE Trans Med Imaging **19:**166–174.
20. Majumdar S, Newitt D, Mathur A, Osman D, Gies A, Chiu E, Lotz J, Kinney J, Genant H 1996 Magnetic resonance imaging of trabecular bone structure in the distal radius: Relationship with X-ray tomographic microscopy and biomechanics. Osteoporos Int **6:**376–385.
21. Hwang SN, Wehrli FW, Williams JL 1997 Probability-based structural parameters from three-dimensional nuclear magnetic resonance images as predictors of trabecular bone strength. Med Phys **24:**1255–1261.
22. Pothuaud L, Laib A, Levitz P, Benhamou CL, Majumdar S 2002 Three-dimensional-line skeleton graph analysis of high-resolution magnetic resonance images: A validation study from 34-microm-resolution microcomputed tomography. J Bone Miner Res **17:**1883–1895.
23. Majumdar S, Kothari M, Augat P, Newitt DC, Link TM, Lin JC, Lang T, Lu Y, Genant HK 1998 High-resolution magnetic resonance imaging: Three-deimensional trabecular bone architecture and biomechanical properties. Bone **22:**445–454.
24. Link TM, Vieth V, Langenberg R, Meier N, Lotter A, Newitt D, Majumdar S 2003 Structure analysis of high resolution magnetic resonance imaging of the proximal femur: In vitro correlation with biomechanical strength and BMD. Calcif Tissue Int **72:**156–165.
25. Link TM, Bauer J, Kollstedt A, Stumpf I, Hudelmaier M, Settles M, Majumdar S, Lochmuller EM, Eckstein F 2004 Trabecular bone structure of the distal radius, the calcaneus, and the spine: Which site predicts fracture status of the spine best? Invest Radiol **39:**487–497.
26. Majumdar S, Genant H, Grampp S, Newitt D, Truong V, Lin J, Mathur A 1997 Correlation of trabecular bone structure with age, bone mineral density and osteoporotic status: In vivo studies in the distal radius using high resolution magnetic resonance imaging. J Bone Miner Res **12:**111–118.
27. Laib A, Newitt DC, Lu Y, Majumdar S 2002 New model-independent measures of trabecular bone structure applied to in vivo high-resolution MR images. Osteoporos Int **13:**130–136.
28. Majumdar S, Link T, Augat P, Lin JC, Newitt D, Lane NE, Genant HK 1999 Trabecular bone architecture in the distal radius using MR imaging in subjects with fractures of the proximal femur. Osteoporos Int **10:**231–239.
29. Link TM, Majumdar S, Augat P, Lin JC, Newitt D, Lu Y, Lane NE, Genant HK 1998 In vivo high resolution MRI of the calcaneus: Differences in trabecular structure in osteoporosis patients. J Bone Miner Res **13:**1175–1182.
30. Boutry N, Cortet B, Dubois P, Marchandise X, Cotten A 2003 Trabecular bone structure of the calcaneus: Preliminary in vivo MR imaging assessment in men with osteoporosis. Radiology **227:**708–717.
31. Newitt DC, Majumdar S, van Rietbergen B, von Ingersleben G, Harris ST, Genant HK, Chesnut C, Garnero P, MacDonald B 2002 In vivo assessment of architecture and micro-finite element analysis derived indices of mechanical properties of trabecular bone in the radius. Osteoporos Int **13:**6–17.
32. Chesnut CH III, Majumdar S, Newitt DC, Shields A, Van Pelt J, Laschansky E, Azria M, Kriegman A, Olson M, Eriksen EF, Mindeholm L 2005 Effects of salmon calcitonin on trabecular microarchitecture as determined by magnetic resonance imaging: Results from the QUEST Study*. J Bone Miner Res **20:**1548–1561.
33. Benito M, Gomberg B, Wehrli FW, Weening RH, Zemel B, Wright AC, Song HK, Cucchiara A, Snyder PJ 2003 Deterioration of trabecular architecture in hypogonadal men. J Clin Endocrinol Metab **88:**1497–1502.
34. Benito M, Vasilic B, Wehrli FW, Bunker B, Wald M, Gomberg B, Wright AC, Zemel B, Cucchiara A, Snyder PJ 2005 Effect of testosterone replacement on trabecular architecture in hypogonadal men. J Bone Miner Res **20:**1785–1791.
35. Krug R, Banerjee S, Han ET, Newitt DC, Link TM, Majumdar S 2005 Feasibility of in vivo structural analysis of high-resolution magnetic resonance images of the proximal femur. Osteoporos Int **16:**1307–1314.
36. Gomberg BR, Saha PK, Wehrli FW 2005 Method for cortical bone structural analysis from magnetic resonance images. Acad Radiol **12:**1320–1332.

Chapter 32. Radionuclide Scintigraphy in Metabolic Bone Disease

Gopinath Gnanasegaran,[1] Gary J. R. Cook,[2] and Ignac Fogelman[1]

[1]*Department of Nuclear Medicine, Guys & St. Thomas' Hospital NHS Foundation Trust, London, United Kingdom;* [2]*Department of Nuclear Medicine and PET, Royal Marsden Hospital, London, United Kingdom*

INTRODUCTION

Radionuclide bone imaging remains the most widely used method for detection of benign and metastatic involvement of the skeleton. Current γ cameras systems are able to perform high-resolution imaging in short scan times (whole body acquisitions or localized views of the skeleton). More recently, single photon emission CT (SPECT) imaging has become widely available and is becoming routine in nuclear medicine, leading to improved sensitivity and specificity for lesion detection. Currently hybrid SPECT/CT and positron emission tomography (PET)/CT scanners are the newest additions to the diagnostic armamentarium.

Bisphosphonate compounds such as methylene diphosphonate (MDP), labeled with ^{99m}Tc, are the most commonly used radiopharmaceuticals for bone scintigraphy. Their exact mechanism of localization in bone is not fully understood, but it is probable that they are adsorbed onto the surface of hydroxyapatite crystals. The degree of accumulation in bone is dependent on local blood flow but is influenced more strongly

The authors state that they have no conflicts of interest.

Key words: radionuclide, scintigraphy, metabolic, bone

© 2008 American Society for Bone and Mineral Research

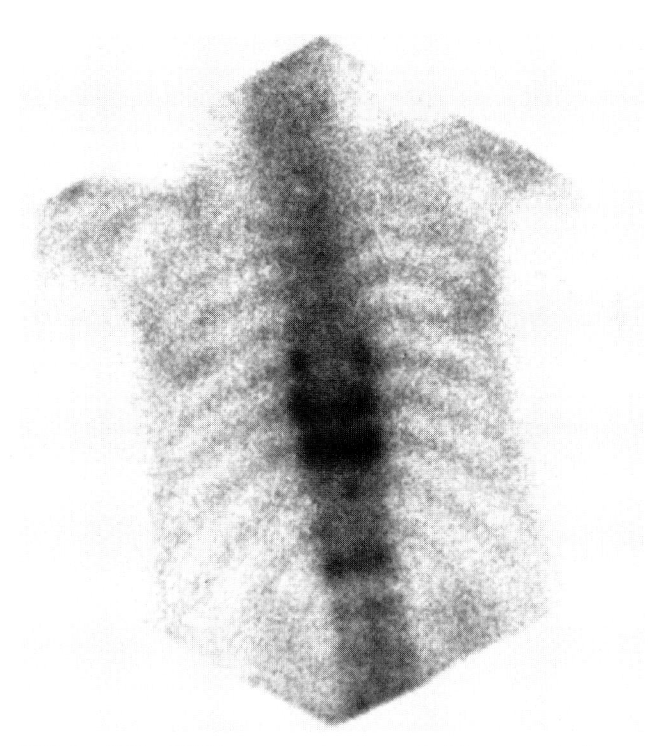

FIG. 1. ^{99m}Tc-MDP bone scan showing typical linear uptake in vertebral fractures. The different intensity suggests they occurred at different times.

by the degree of osteoblastic activity and hence bone formation. In general, pathologic processes that involve bone result in increased local bone turnover, with both osteoblast and osteoclast activity being increased.

BONE METASTASES

Radionuclide bone imaging is routinely used for detecting bone metastases in patients with prostate and breast cancer. In general, the success of imaging bone lesions depends on the osteoblastic response, which accompanies bone destruction by a metastasis.[1] However, rarely with aggressive metastases where bone is unable to mount sufficient osteoblastic response, there may be false negatives, and this is most commonly seen in multiple myeloma. The radionuclide imaging of bone metastases is extensively reviewed by Gnanasegaran et al.[2] and Van der Wall and Clark[3] and is beyond the scope of this chapter.

OSTEOPOROSIS

The bone scan has no role in the diagnosis of osteoporosis per se but is most often used in established osteoporosis to diagnose vertebral or other fractures (e.g., sacral, pelvis, or rib) and has a valuable role in evaluation and management of patients with back pain.[4]

The characteristic appearance of a vertebral fracture is of intense, linearly increased tracer uptake at the affected level (Fig. 1). Although the bone scan may become positive immediately after a fracture, this is site specific, and in the spine, can take up to 2 wk to become abnormal.[5] The changes diminish gradually over a period of time, with the scan normalizing between 3 and 18 mo after the incident, the average being between 9 and 12 mo.[5,6] The degree of intensity is also useful in assessing the age of fractures.

In general, if a patient is complaining of back pain with multiple vertebral fractures on radiographs but has a normal bone scan, this essentially excludes recent fracture as a cause of patient's symptoms. In these clinical scenarios, other causes of pain should be considered and may well be identified on the bone scan (e.g., facetal joint disease).

In general, vertebral fracture is defined on the basis of morphometry,[7] but morphometric abnormalities are not specific to fracture and, for example, may be caused by congenital vertebral anomalies. The bone scan may therefore have a role in deciding whether a morphometric abnormality is related to a fracture, provided that it is acquired within several months of the start of symptoms.

The radionuclide bone scan, being highly sensitive, is useful in identifying unsuspected osteoporotic fractures at other sites such as ribs, pelvis, and hip. It also has an important role in assessing suspected fractures where radiography is unhelpful, either because of poor sensitivity related to the anatomic site of the fracture (e.g., sacrum) or because adequate views are not obtainable because of the patient's discomfort or mobility.[4]

A radionuclide bone scan also may be valuable in patients in whom back pain persists for longer than one would expect after vertebral fracture (e.g., because of additional unsuspected vertebral fractures). In addition, we are increasingly becoming aware that osteoporotic patients with chronic back pain may have unsuspected abnormalities affecting the facet joints.[4,8] It is not known whether this is related to physical disruption of the joint at the time of vertebral collapse or is caused by subsequent secondary degenerative or inflammatory changes. To identify abnormalities in the facet joints, SPECT imaging is essential.

PAGET'S DISEASE

The radionuclide bone scan is invaluable in Paget's disease, both for diagnosis and to define the extent of skeletal involvement. It provides a simple way to evaluate the whole skeleton with greater sensitivity than radiographic skeletal surveys.[9] Characteristically, affected long bones show intensely increased activity, which starts at the end of a bone and spreads either proximally or distally, often showing a "V" or "flame-shaped" leading edge (Fig. 2). Another clue that a scintigraphic abnormality is caused by Paget's disease rather than other focal skeletal pathology is that a whole bone is often involved, and this is most often evident in the pelvis, scapula, and vertebrae. The common differential diagnosis is fibrous dysplasia. However, lack of preservation of bony outline in fibrous dysplasia differentiates it from Paget's disease.

The role of the bone scan with regard to treatment is not well defined. A radionuclide bone scan can be obtained ~3–6 mo after therapy, and it is important to note that pagetic lesions may often respond in a heterogeneous manner, even in individual patients.[10] After intravenous bisphosphonate therapy, some bones may completely normalize, whereas the majority show some improvement, and a small proportion remains unchanged.

The bone scan may provide a sensitive means to assess persistent metabolically active disease or to identify reactivation of disease, influencing decisions on the timing of further treatment, although there is no evidence to support such an approach. It is important to be aware that the bone scan appearances can be unusual and indeed bizarre after successful bisphosphonate treatment, with resultant heterogeneous uptake sometimes mimicking metastatic disease, and hence,

© 2008 American Society for Bone and Mineral Research

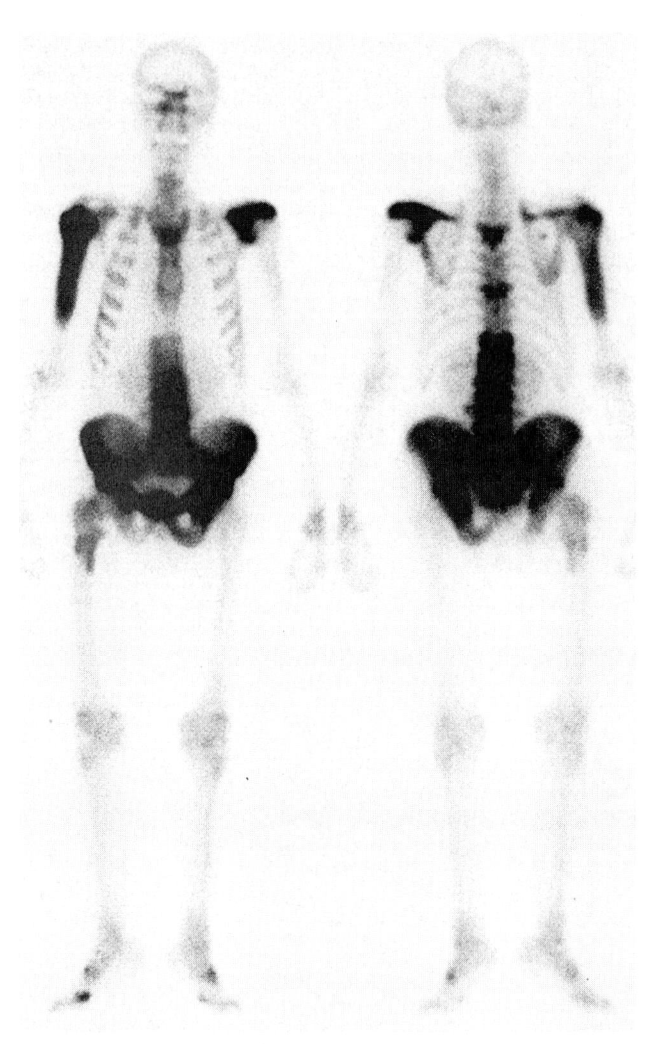

FIG. 2. ^{99m}Tc-MDP anterior and posterior whole body bone scan showing intense activity within a number of bones, with features typical of Paget's disease.

knowledge of the relevant history and previous treatments is essential for correct interpretation.

Several quantitative methods have been reported to evaluate the absolute skeletal uptake of tracer, but none have been found to be of enough value to be used in routine clinical practice.[11]

The radionuclide bone scan may occasionally identify complications of Paget's disease. Although osteosarcoma complicating Paget's disease is very rare (<1%),[12] clues that sarcomatous change may have occurred include a change to heterogeneous and irregular uptake within an area of bone, perhaps with some photon-deficient areas corresponding to bone destruction. However, the bone scan may be misleading in the event of fracture or sarcomatous change or both. Thus, in general, the radionuclide bone scan is not reliable in the diagnosis of skeletal complications of Paget's disease, and complementary radiological correlation is necessary.

In general, 18F-fluorodeoxyglucose (18F-FDG) PET can differentiate benign from malignant tissue in many tumors, and in principle, could be a useful tool in distinguishing the benign changes of Paget's disease from associated osteosarcomas.[13] However, it is important to be aware that some FDG uptake may be seen in patients with more active Paget's disease.[13] 18F-fluoride PET is reported to be useful to measure the activity of Paget's disease of bone and as a promising noninvasive tool to monitor the therapeutic efficacy of bisphosphonate regimens in Paget's disease[14] but essentially provides the same information as a conventional bone scan.

HYPERPARATHYROIDISM

Most cases of primary hyperparathyroidism are asymptomatic and are unlikely to be associated with changes on bone scintigraphy. The diagnosis is a biochemical one, and the bone scan therefore has no routine role in diagnosis. Bone scans are often used to help differentiate the causes of hypercalcemia, in particular, hyperparathyroidism versus malignancy, and so typical features of metabolic bone disorders may be recognized. A bone scan may show a number of features in hyperparathyroidism, but the most important is the generalized increased uptake throughout the skeleton, identified as increased contrast between bone and soft tissues. Indeed, renal activity normally clearly seen on a bone scan may not be evident. This appearance has been termed a "superscan" because of the apparent high-quality images. Other typical features that have been described include a prominent calvarium and mandible, beading of the costochondral junctions (rosary beads), and prominent uptake in the sternum (the "tie" sign), sometimes with horizontal lines ("striped-tie" sign).

Severe forms of hyperparathyroidism may be associated with ectopic calcification, which may lead to uptake of bone radiopharmaceuticals into the soft tissue.

Nuclear medicine is the most frequently used modality for localizing abnormal parathyroid glands before surgery. In general, it is performed using dual-phase 99mTc-sestamibi imaging (early and delayed imaging at 15 min and 2–3 h, respectively). 99mTc-sestamibi is the current radionuclide tracer of choice for radionuclide localization of parathyroid adenomas. However, it is less useful in patients with secondary hyperparathyroidism, where the predominant lesion is often hyperplasia.

The most frequent cause of false-positive results in 99mTc-sestamibi parathyroid imaging is the solid thyroid nodule (either solitary or multinodular gland), thyroid carcinoma, lymphoma, and other causes of lymphadenopathy.[15] False-negative results may arise from failure of 99mTc-sestamibi imaging to identify some smaller parathyroid lesions, which relates to system resolution and to the amount of tracer uptake by the parathyroid tissue.[15] The addition of 99mTc-sestamibi SPECT to radionuclide scintigraphy is reported to increase the sensitivity and accuracy. More recently, 99mTc-sestamibi SPECT/CT has been used in the localization of adenomas. However, it has been reported to offer no significant additive value over conventional SPECT for identifying a normally located parathyroid adenoma, but may assist in localizing ectopic parathyroid adenomas.[16]

PET imaging using [11C] methionine is reported to correctly localize abnormal parathyroid glands in patients with recurrent hyperparathyroidism in whom conventional nuclear medicine techniques had failed.[17] However, its role in routine practice is debatable because short-lived tracers (cyclotron needed) will not be available in most centers.

RENAL OSTEODYSTROPHY

Renal osteodystrophy is caused by a combination of several metabolic bone disorders as a consequence of chronic renal dysfunction and often shows the most severe cases seen on the bone scan. It may be made up of osteoporosis, osteomalacia, adynamic bone, and secondary hyperparathyroidism in varying

© 2008 American Society for Bone and Mineral Research

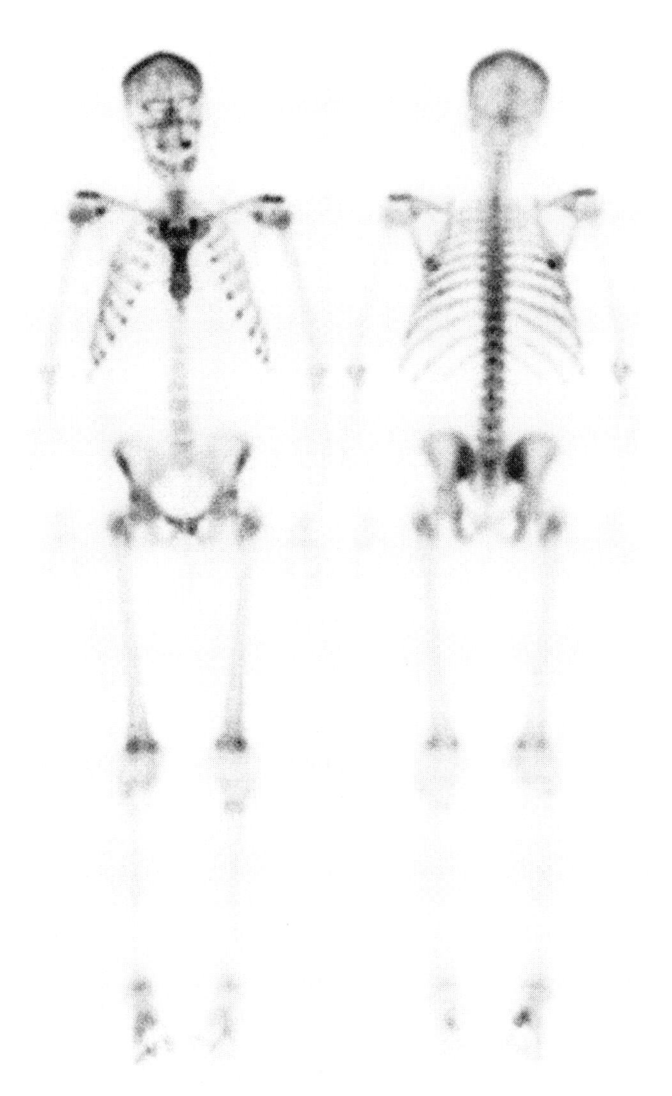

FIG. 3. Whole body bone scan showing enhanced uptake throughout the skeleton (increased bone/soft tissue ratio) in a patient with chronic renal failure and biochemistry consistent with osteomalalcia.

degrees. The commonest bone scan appearance is a superscan reflecting hyperparathyroidism (Fig. 3). A clue in differentiating this type of scintigraphic pattern from others is that there may be a lack of bladder activity.

OSTEOMALACIA

Patients with osteomalacia usually show similar features on a bone scan as described in hyperparathyroidism, although in the early stages of the disease, they may appear normal.[18] Tracer avidity may reflect diffuse uptake in osteoid, although more probably it is caused by the secondary hyperparathyroidism that is present. The presence of focal lesions may represent pseudofractures or true fractures. Although osteomalacia may be suspected from the biochemistry, it is a histologic diagnosis. However, the typical bone scan features may well be helpful in supporting the diagnosis. The detection of pseudofractures with this technique is more sensitive than that with radiography.[19]

CONCLUSION

The radioisotope bone scan remains a valuable method in the evaluation of malignant and metabolic bone diseases, providing functional and potentially quantitative information on bone metabolism. Recent advances in hybrid anatomical/functional imaging will consolidate the roles of radioisotope imaging in skeletal disease.

REFERENCES

1. O'Sullivan JM, Cook GJ 2002 A review of the efficacy of bone scanning in prostate and breast cancer. Q J Nucl Med **46:**152–159.
2. Gnanasegaran G, Cook GJ, Fogelman I 2007 Musculoskeletal system. In: Biersack H-J, Freeman LM (eds.) Clinical Nuclear Medicine. Springer-Verlag, Berlin, Germany, pp. 241–262.
3. Van der Wall H, Clarke S 2004 The evaluation of malignanacy: Metastatic bone disease. In: Ell PJ, Gambhir SS (eds.) Nuclear Medicine in Clinical Diagnosis and Treatment, 3rd ed., vol. 1. Churchill Livingstone, Philadelphia, PA, USA, pp. 641–655.
4. Cook GJR, Hannaford E, Lee M, Clarke SEM, Fogelman I 2002 The value of bone scintigraphy in the evaluation of osteoporotic patients with back pain. Scand J Rheumatol **31:**245–248.
5. Spitz J, Lauer I, Tittel K, Wiegand H 1993 Scintimetric evaluation of remodeling after bone fractures in man. J Nucl Med **34:**1403–1409.
6. Fogelman I, Carr D 1980 A comparison of bone scanning and radiology in the evaluation of patients with metabolic bone disease. Clin Radiol **31:**321–326.
7. Eastell R, Cedel SL, Wahner HW, Riggs BL, Melton LJ 1991 Classification of vertebral fractures. J Bone Miner Res **6:**207–215.
8. Ryan PJ, Evans PA, Gibson T, Fogelman I 1992 Osteoporosis and chronic back pain: A study with single photon emission computed bone scintigraphy. J Bone Miner Res **7:**1455–1459.
9. Fogelman I, Carr D 1980 A comparison of bone scanning and radiology in the assessment of patients with symptomatic Paget's disease. Eur J Nucl Med **5:**417–421.
10. Ryan PJ, Gibson T, Fogelman I 1992 Bone scintigraphy following pamidronate therapy for Paget's disease of bone. J Nucl Med **33:**1589–1593.
11. Fogelman I, Bessent RG, Gordon D 1981 A critical assessment of bone scan quantitation (bone to soft tissue ratios) in the diagnosis of metabolic bone disease. Eur J Nucl Med **6:**93–97.
12. Boutin RD, Spitz DJ, Newman JS, Lenchik L, Steinbach LS 1998 Complications in Paget disease at MR imaging. Radiology **209:**641–651.
13. Cook GJ, Maisey MN, Fogelman I 1997 Fluorine-18-FDG PET in Paget's disease of bone. J Nucl Med **7:**1495–1497.
14. Installe J, Nzeusseu A, Bol A, Depresseux G, Devogelaer JP, Lonneux M 2005 (18)F-fluoride PET for monitoring therapeutic response in Paget's disease of bone. J Nucl Med **6:**1650–1658.
15. Palestro CJ, Tomas MB, Tronco GG 2005 Radionuclide Imaging of the Parathyroid Glands. Semin Nucl Med **5:**266–276.
16. Gayed IW, Kim EE, Broussard WF, Evans D, Lee J, Broemeling LD, Ochoa BB, Moxley DM, Erwin WD, Podoloff DA 2005 The value of 99mTc-sestamibi SPECT/CT over conventional SPECT in the evaluation of parathyroid adenomas or hyperplasia. J Nucl Med **46:**248–252.
17. Beggs AD, Hain SF 2005 Localization of parathyroid adenomas using 11C-methionine positron emission tomography. Nucl Med Commun **26:**133–136.
18. Fogelman I, McKillop JH, Bessent RG, Boyle IT, Turner JG, Greig WR 1978 The role of bone scanning in osteomalacia. J Nucl Med **19:**245–248.
19. Fogelman I, McKillop JH, Greig WR, Boyle IT 1977 Pseudofractures of the ribs detected by bone scanning. J Nucl Med **18:**1236–1237.

© 2008 American Society for Bone and Mineral Research

Chapter 33. Assessment of Fracture Risk

John A. Kanis

WHO Collaborating Centre for Metabolic Bone Diseases, University of Sheffield Medical School, Sheffield, United Kingdom

INTRODUCTION

The increasing prevalence and awareness of osteoporosis, together with the development of treatments of proven efficacy, will increase the demand for management of patients with osteoporosis. This in turn will require widespread facilities for the assessment of osteoporosis. Measurements of bone mineral are a central component of any provision, because osteoporosis is defined in terms of BMD and microarchitectural deterioration of bone tissue. There are no satisfactory clinical tools available to assess bone quality independently of BMD, so that for practical purposes, the assessment of osteoporosis depends on the measurement of skeletal mass, as assessed by measurements of BMD.(1)

The clinical significance of osteoporosis is the fractures that arise with their attendant morbidity and mortality. Although bone mass is an important component of the risk of fracture, other abnormalities occur in the skeleton that contribute to fragility. In addition, a variety of nonskeletal factors, such as the liability to fall and force of impact, contribute to fracture risk. Because BMD forms but only component of fracture risk, accurate assessment of fracture risk should ideally take into account other readily measured indices of fracture risk, particularly those that add information to that provided by BMD.

ASSESSMENT OF RISK WITH BMD

The use of bone mass measurements for prognosis depends on accuracy. Accuracy in this context is the ability of the measurement to predict fracture. In general, all absorptiometric techniques have high specificity but low sensitivity, which vary with the cut-off chosen to designate high risk. Many cross-sectional prospective population studies indicate that the risk for fracture increases by a factor of 1.5–3.0 for each SD decrease in BMD (Table 1).(2) The ability of BMD to predict fracture is comparable to the use of blood pressure to predict stroke and significantly better than serum cholesterol to predict myocardial infarction.(2) Accuracy is improved by site specific measurements (Table 1), so that for hip fractures, the risk might ideally be measured at the hip. In the immediate postmenopausal population, measurements at any site (hip, spine, and wrist) predict any osteoporotic fracture equally well with a gradient of risk of ~1.5 per SD decrease in BMD.

The highest gradient of risk is found at the hip to predict hip fracture, where the gradient of risk is 2.6. Thus, an individual with a T-score of –3 SD at the hip would have a 2.6^3, or greater than 15-fold, higher risk than an individual with a T-score of 0 SD. In contrast, the same T-score at the spine would yield a much lower risk estimate—~4-fold increased (1.6^3). This emphasizes the importance of accuracy or gradient of risk in the categorization of fracture risk.

Despite these performance characteristics, it should be recognized that, just because BMD is normal, there is no guarantee that a fracture will not occur—only that the risk is decreased. Conversely, if BMD is in the osteoporotic range, fractures are more likely, but not invariable. At the age of 50 yr, the proportion of women with osteoporosis who will fracture their hip, spine, forearm, or proximal humerus in the next 10 yr (i.e., positive predictive value) is ~45%. The detection rate for these fractures (sensitivity) is, however, low, and 96% of such fractures would occur in women without osteoporosis.(3) The low sensitivity is one of the reasons why widespread population-based screening is not widely recommended in women at the time of the menopause.

AGE

The performance characteristics of the test can, however, be improved by the concurrent consideration of risk factors that operate independently of BMD. Perhaps the best example is age. The same T-score with the same technique at any one site has a different significance at different ages. For any BMD, fracture risk is much higher in the elderly than in the young.(4,5) This is because age contributes to risk independently of BMD. In addition, the performance characteristics of BMD vary with age. For example, at the age of 50 yr, hip fracture risk increased 3.7-fold per SD decrease in femoral neck BMD, whereas at the age of 80 yr, the gradient of risk is 2.3.(6) The impact of age on hip fracture probability is shown in Table 2. Thus, the consideration of age and BMD together increases the range of risk that can be identified.

There are, however, a large number of additional risk factors that provide information on fracture risk independently of both age and BMD.

OTHER CLINICAL RISK FACTORS

A large number of additional risk factors for fracture have been identified. For the purposes of risk assessment, interest lies in those factors that contribute significantly to fracture risk over and above that provided by BMD measurements or age.(7) A caveat is that some risk factors are not amenable to particular treatments, so that the relationship between absolute probability of fracture and reversible risk is important. Liability to falls is an appropriate example where the risk of fracture is high, but treatment with agents affecting bone metabolism may have little effect on risk.

Many risk factors for osteoporotic fracture have been identified.(7) In general, risk factor scores show relatively poor specificity and sensitivity in predicting either BMD or fracture risk.(8–10) Moreover, some risk factors vary in importance according to age. For example, risk factors for falling, such as visual impairment, reduced mobility, and treatment with sedatives, are more strongly predictive of fracture in the elderly than in younger individuals.(11)

Over the past few years, a series of meta-analyses has been undertaken to identify clinical risk factors that could be used in case finding strategies with or without the use of BMD. These are summarized in Table 3 with their predictive value for hip fracture risk.(12)

(1) Low body mass index (BMI). A low BMI is a significant risk factor for hip fracture. Thus, the risk is nearly 2-fold increased comparing individuals with a BMI of 25 and 20 kg/m^2 (Table 3). It is important to note that comparison of 25 versus 30 kg/m^2 is not associated with a halving of risk

Key words: osteoporosis, bone densitometry, probability, fracture, clinical risk fractures, sensitivity, gradient of fracture risk, life expectancy, family history of fracture, rheumatoid arthritis, secondary osteoporosis, glucocorticoids, smoking, alcohol

The author states that he has no conflicts of interest.

© 2008 American Society for Bone and Mineral Research

Table 1. Age-Adjusted Relative Increase in Risk of Fracture (With 95% CIs) in Women for Every 1 SD Decrease in BMD (Absorptiometry) Below the Mean Value for Age (From 2)

Site of measurement	*Forearm fracture*	*Hip fracture*	*Vertebral fracture*	*All fractures*
Distal radius	1.7 (1.4–2.0)	1.8 (1.4–2.2)	1.7 (1.4–2.1)	1.4 (1.3–1.6)
Femoral neck	1.4 (1.4–1.6)	2.6 (2.0–3.5)	1.8 (1.1–2.7)	1.6 (1.4–1.8)
Lumbar spine	1.5 (1.3–1.8)	1.6 (1.2–2.2)	2.3 (1.9–2.8)	1.5 (1.4–1.7)

(Adapted with permission from the BMJ Publishing Group from Marshall D, Johnell O, Wedel H 1996 Meta-analysis of how well measures of bone mineral density predict occurrence of osteoporotic fractures. Br Med J **312:**1254–1259.)

(i.e., leanness is a risk factor rather than obesity being a protective factor). It is also important to note that the value of BMI in predicting fractures is very much diminished when adjusted for BMD.

(2) Many studies indicate that history of fragility fracture is an important risk factor for further fracture.[13] Fracture risk is approximately doubled in the presence of a prior fracture (Table 3). The increase in risk is even more marked for a vertebral fracture after a previous spine fracture. The risks are in part independent of BMD. In general, adjustment for BMD would decrease the relative risk by 10–20% (see Table 3).

(3) A family history of fragility fractures is a significant risk factor that is largely independent of BMD. A family history of hip fracture is a stronger risk factor than a family history of other osteoporotic fractures and is independent of BMD.

(4) Smoking is a risk factor that is in part dependent on BMD.

(5) Glucocorticoids are an important cause of osteoporosis and fractures.[14] The fracture risk conferred by the use of glucocorticoids is, however, not solely dependent on bone loss, and BMD independent risks have been identified.

(6) Alcohol. The relationship between alcohol intake and fracture risk is dose dependent. Where alcohol intake is on average two units or less daily, there is no increase in risk. Indeed, some studies suggest that BMD and fracture risk may be reduced. Intakes of three or more units daily are associated with a dose-dependent increase in risk.

(7) Rheumatoid arthritis. There are many secondary causes of osteoporosis (e.g., inflammatory bowel disease, endocrine disorders), but in most instances, it is uncertain to what extent this is dependent on low BMD or other risk factors such as the use of glucocorticosteroids. In contrast, rheumatoid arthritis causes a fracture risk independently of BMD and the use of glucocorticoids.

BIOCHEMICAL ASSESSMENT OF FRACTURE RISK

Bone markers are increased after the menopause, and in several studies, the rate of bone loss varies according to the marker value.[15] Thus, a potential clinical application of biochemical indices of skeletal metabolism is in assessing fracture risk. Prospective studies have shown an association of osteoporotic fracture with indices of bone turnover independent of BMD in women at the time of menopause and elderly women.[16]

INTEGRATING RISK FACTORS

The multiplicity of these risk factors poses problems in the units of risk to be used. The T-score becomes of little value in that different T-score thresholds for treatment would be needed for each combination of risk factors. Although the use of relative risks is feasible, the metric of risk best suited for clinicians is the absolute risk (or probability) of fracture.

FRACTURE PROBABILITY

The absolute risk of fracture depends on age and life expectancy, as well as the current relative risk. In general, remaining lifetime risk of fracture decreases with age, especially after the age of 70 yr or so, because the risk of death with age outstrips the increasing incidence of fracture with age. Estimates of lifetime risk are of value in considering the burden of osteoporosis in the community and the effects of intervention strategies. For several reasons, they are less relevant for assessing risk of individuals in whom treatment might be envisaged[5] so that the International Osteoporosis Foundation and the World Health Organization recommend that risk of fracture should be expressed as a short-term absolute risk (i.e., probability over a 10-yr interval).[17] The period of 10 yr covers the likely duration of treatment and the benefits that may continue once treatment is stopped.

The major advantage of using absolute fracture probability is that it standardizes the output from the multiple techniques

Table 2. Ten-Year Probability of Hip Fracture (%) in Men and Women From Sweden According to Age and T-Score for BMD at the Femoral Neck

	*T-score (SD units) with fixed gradient of risk**						*Variable gradients of risk†*					
Age (yr)	*+1*	*0*	*−1*	*−2*	*−3*	*−4*	*+1*	*0*	*−1*	*−2*	*−3*	*−4*
Men												
50	0.1	0.2	0.6	1.9	5.5	15.4	0.1	0.2	0.8	2.6	8.6	26.6
60	0.1	0.3	0.8	2.2	6.0	16.0	0.1	0.4	0.9	2.5	6.7	17.1
70	0.2	0.6	1.8	4.8	12.9	31.6	0.5	1.2	2.5	5.4	11.4	23.0
80	0.4	1.1	2.9	7.7	19.2	41.7	1.8	3.2	5.7	10.0	17.2	28.5
Women												
50	0.1	0.2	0.4	1.1	2.8	7.0	0.0	0.1	0.3	0.9	3.2	10.7
60	0.1	0.4	1.0	2.7	7.1	17.9	0.1	0.3	0.8	2.3	6.7	18.9
70	0.2	0.7	1.9	5.3	14.1	34.6	0.3	0.8	2.1	5.2	12.8	29.4
80	0.3	0.8	2.4	6.8	18.6	43.6	1.1	2.3	4.8	9.9	19.8	36.9

*Assumes a fixed gradient of risk (2.6/SD).

†Assumes a gradient of risk that varies with age.

(Reproduced with permission from the American Society for Bone and Mineral Research from Johnell O, Kanis JA, Oden A, Johansson H, De Laet C, Delmas P, Eisman JA, Fujiwara S, Kroger H, Mellstrom D, Meunier PJ, Melton LJ, O'Neill T, Pols H, Reeve J, Silman A, Tenenhouse A 2007 Productive value of BMD for hip and other fractures. J Bone Miner Res **22:**774.)

© 2008 American Society for Bone and Mineral Research

TABLE 3. RISK RATIO (RR) FOR OSTEOPOROTIC FRACTURE AND 95% CIS ASSOCIATED WITH RISK FACTORS ADJUSTED FOR AGE, WITH AND WITHOUT ADJUSTMENT FOR BMD[11]

	Without BMD		*With BMD*	
Risk indicator	*RR*	*95% CI*	*RR*	*95% CI*
Body mass index				
20 vs. 25 kg/m^2	1.27	1.16–1.38	1.02	0.92–1.13
30 vs. 25 kg/m^2	0.89	0.81–0.98	0.96	0.86–1.08
Prior fracture after 50 yr	1.86	1.72–2.01	1.76	1.60–1.93
Parental history of hip fracture	1.54	1.25–1.88	1.54	1.25–1.88
Current smoking	1.29	1.17–1.43	1.13	1.00–1.25
Ever use of systemic corticosteroids	1.65	1.42–1.90	1.66	1.42–1.92
Alcohol intake >2 units daily	1.38	1.16–1.65	1.36	1.13–1.63
Rheumatoid arthritis	1.56	1.20–2.02	1.47	1.12–1.92

(Reprinted with permission from Elsevier from Kanis JA, McCloskey EV 1996 Evaluation of the risk of hip fracture. Bone **18**:127–132.)

and sites used for assessment. The estimated probability will of course depend on the performance characteristics (gradient of risk) provided by any technique at any one site. Moreover, it also permits the presence or absence of risk factors other than BMD to be incorporated as a single metric. This is important because, as mentioned, there are many risk factors that give information over and above that provided by BMD and age.

The general relationship between relative risk and 10-yr probability of hip fracture is shown in Table 4.[3] For example, a woman at the age of 60 yr has on average a 10-yr probability of hip fracture of 2.4%. In the presence of a prior fragility fracture, this risk is increased ~2-fold, and the probability increases to 4.8%. The integration of risk factors is not new and has been successfully applied in the management of coronary heart disease.[18]

TABLE 4. TEN-YEAR PROBABILITY OF FRACTURE IN MEN AND WOMEN FROM SWEDEN ACCORDING TO AGE AND THE RISK (RR) RELATIVE TO THE AVERAGE POPULATION[3]

	Age (yr)			
RR	*50*	*60*	*70*	*80*
Hip fracture				
Men				
1	0.84	1.26	3.68	9.53
2	1.68	2.50	7.21	17.89
3	2.51	3.73	10.59	25.26
4	3.33	4.94	13.83	31.75
Women				
1	0.57	2.40	7.87	18.00
2	1.14	4.75	15.1	32.0
3	1.71	7.04	21.7	42.9
4	2.27	9.27	27.7	51.6
Hip, clinical spine, humeral, or Colles' fracture				
Men				
1	3.3	4.7	7.0	12.6
2	6.5	9.1	13.5	23.1
3	9.6	13.3	19.4	13.9
4	12.6	17.3	24.9	39.3
Women				
1	5.8	9.6	16.1	21.5
2	11.3	18.2	29.4	37.4
3	16.5	26.0	40.0	49.2
4	21.4	33.1	49.5	58.1

(Adapted with permission from Elsevier from Kanis JA, Johnell O, Oden A, De Laet C, Jonsson B, Dawson A 2002 Ten-year risk of osteoporotic fracture and the effect of risk factors on screening strategies. Bone **30**:252.)

Algorithms that integrate the weight of clinical risk factors for fracture risk, with or without information on BMD, have been developed by the WHO Collaborating Center for Metabolic Bone Diseases at Sheffield, UK.[19] The risk factors used are given in Table 5. The FRAX tool (www.shef.ac.uk/FRAX) computes the 10-yr probability of hip fracture or a major osteoporotic fracture. A major osteoporotic fracture is a clinical spine, hip, forearm, and humerus fracture. Probabilities can be computed for the index countries (including Japan, China, the United States, the United Kingdom, Sweden, Turkey, France, and Spain). Where a country is not represented (because of the lack of epidemiological data), a surrogate may be chosen.

Where computer access is limited, paper charts can be downloaded that give fracture probabilities for each index country (www.shef.ac.uk/FRAX) according to the number of clinical risk factors. A specimen chart is given in Table 6. For example, a woman from the United Kingdom who is 60 yr of age with a T-score of −2.0 SD at the femoral neck with a prior forearm fracture and rheumatoid arthritis (i.e., two clinical risk factors) has a 10-yr fracture probability of 16% (10–24%). The range is not a confidence interval but, because the weight of different risk factors varies, is a true range.

The assessment takes no account of prior treatment nor of dose responses for several risk factors. For example, two prior fractures carry a much higher risk than a single prior fracture. Dose responses are also evident for glucocorticoid use. A prior clinical vertebral fracture carries an ~2-fold higher risk than other prior fractures. Because it is not possible to model all such scenarios with the FRAX algorithm, these limitations should temper clinical judgment. A further limitation is that the FRAX algorithm uses T-scores for femoral neck BMD.

TABLE 5. CLINICAL RISK FACTORS USED FOR THE ASSESSMENT OF FRACTURE PROBABILITY[19]

Age
Sex
Low body mass index
Previous fragility fracture, particularly of the hip, wrist and spine including morphometric vertebral fracture
Parental history of hip fracture
History of fragility fracture
Glucocorticoid treatment (≥5 mg prednisolone daily for 3 mo or more)
Current smoking
Alcohol intake ≥3 units daily
Rheumatoid arthritis
Other secondary causes of osteoporosis
- Untreated hypogonadism in men and women (e.g., premature menopause, bilateral oophorectomy or orchidectomy, anorexia nervosa, chemotherapy for breast cancer, hypopituitarism)
- Inflammatory bowel disease (e.g., Crohn's disease and ulcerative colitis). It should be noted that the risk is in part dependent on the use of glucocorticoids, but an independent risk remains after adjustment for glucocorticoid exposure.
- Prolonged immobility (e.g., spinal cord injury, Parkinson's disease, stroke, muscular dystrophy, ankylosing spondylitis)
- Organ transplantation
- Type I diabetes
- Thyroid disorders (e.g., untreated hyperthyroidism, overtreated hypothyroidism)
- Chronic obstructive pulmonary disease

© 2008 American Society for Bone and Mineral Research

TABLE 6. FRAX TABLE FOR THE 10-yr PROBABILITY (%) OF A MAJOR OSTEOPOROTIC FRACTURE (CLINICAL SPINE, HIP, FOREARM, AND HUMERUS FRACTURE) ACCORDING TO BMD, THE NUMBER OF CLINICAL RISK FACTORS (CRFs) FOR WOMEN 60 yr OF AGE IN THE UNITED KINGDOM

Number of CRFs	BMD T-score (femoral neck)					
	−4.0	*−3.0*	*−2.0*	*−1.0*	*0*	*1.0*
0	23	12	7.7	5.5	4.6	4.1
1	32 (29–37)	18 (15–21)	11 (8.2–14)	8.0 (5.5–11)	6.8 (4.5–9.5)	6.0 (3.9–8.4)
2	44 (38–54)	25 (19–34)	16 (10–24)	12 (6.7–18)	9.8 (5.4–16)	8.6 (4.6–14)
3	58 (48–68)	35 (25–49)	23 (14–36)	16 (8.7–28)	14 (6.9–25)	12 (5.9–22)
4	71 (59–78)	46 (35–59)	31 (22–44)	22 (14–35)	19 (11–31)	17 (9.4–28)

The range is not a confidence interval but, because the weight of different risk factors varies, is a true range.
©World Health Organization Collaborating Center for Metabolic Bone Diseases, University of Sheffield, Sheffield, UK.

CONCLUSION

The diagnosis of osteoporosis centers on the assessment of BMD at the hip using DXA. However, other sites and validated techniques can be used for fracture prediction. Several clinical risk factors contribute to fracture risk independently of BMD. These include age, prior fragility fracture, smoking, excess alcohol use, a family history of hip fracture, rheumatoid arthritis, and the use of oral corticosteroids. The use of these risk factors in conjunction with BMD improves sensitivity of fracture prediction without adverse effects on specificity.

REFERENCES

1. World Health Organization 1994 Assessment of Fracture Risk and Its Application to Screening for Postmenopausal Osteoporosis. World Health Organization, Geneva, Switzerland.
2. Marshall D, Johnell O, Wedel H 1996 Meta-analysis of how well measures of bone mineral density predict occurrence of osteoporotic fractures. BMJ **312:**1254–1259.
3. Kanis JA, Johnell O, Oden A, De Laet C, Jonsson B, Dawson A 2001 Ten year risk of osteoporotic fracture and the effect of risk factors on screening strategies. Bone **30:**251–258.
4. Hui SL, Slemenda CW, Johnston CC 1988 Age and bone mass as predictors of fracture in a prospective study. J Clin Invest **81:**1804–1809.
5. Kanis JA, Johnell O, Oden A, Dawson A, De Laet C, Jonsson B 2001 Ten year probabilities of osteoporotic fractures according to BMD and diagnostic thresholds. Osteoporos Int **12:**989–995.
6. Johnell O, Kanis JA, Oden A, Johansson H, De Laet C, Delmas P, Eisman JA, Fujiwara S, Kroger H, Mellstrom D, Meunier PJ, Melton LJ III, O'Neill T, Pols H, Reeve J, Silman A, Tenenhouse A 2005 Predictive value of bone mineral density for hip and other fractures. J Bone Miner Res **20:**1185–1194.
7. Kanis JA 2002 Diagnosis of osteoporosis and assessment of fracture risk. Lancet **359:**1929–1936.
8. Cummings SR, Nevitt MC, Browner WS, Stone K, Fox KM, Ensrud KE, Cauley J, Black D, Vogt TM 1995 Risk factors for hip fracture in white women. N Engl J Med **332:**767–773.
9. Ribot C, Pouilles JM, Bonneu M, Tremollieres F 1992 Assessment of the risk of postmenopausal osteoporosis using clinical risk factors. Clin Endocrinol (Oxf) **36:**225–228.
10. Poor G, Atkinson EJ, O'Fallon WM, Melton LJ III 1995 Predictors of hip fractures in elderly men. J Bone Miner Res **10:**1900–1907.
11. Kanis JA, McCloskey EV 1996 Evaluation of the risk of hip fracture. Bone **18**(Suppl 3)**:**127–132.
12. Kanis JA, Johnell O 2005 Requirements for DXA for the management of osteoporosis in Europe. Osteoporos Int **36:**22–32.
13. Klotzbuecher CM, Ross PD, Landsman PB, Abbot TA, Berger M 2000 Patients with prior fractures have increased risk of future fractures: A summary of the literature and statistical synthesis. J Bone Miner Res **15:**721–727.
14. Van Staa TP, Leufkens HGM, Cooper C 2002 The epidemiology of corticosteroid-induced osteoporosis: A meta-analysis. Osteoporos Int **13:**777–787.
15. Delmas PD Ed. 2000 The use of biochemical markers of bone turnover in the management of post-menopausal osteoporosis. Osteoporos Int **11**(Suppl 6)**:**S1–S76.
16. Johnell O, Oden A, DeLaet C, Garnero P, Delmas PD, Kanis JA 2002 Biochemical indices of bone turnover and the assessment of fracture probability. Osteoporos Int **13:**523–526.
17. Kanis JA, Black D, Cooper C, Dargent P, Dawson-Hughes B, De Laet C, Delmas P, Eisman J, Johnell O, Jonsson B, Melton L, Oden A, Papapoulos S, Pols H, Rizzoli R, Silman A, Tenenhouse A, International Osteoporosis Foundation 2002 A new approach to the development of assessment guidelines for osteoporosis. Osteoporos Int **13:**527–536.
18. Dyslipidaemia Advisory Group on behalf of the Scientific Committee of the National Heart Foundation of New Zealand 1996 National Heart Foundation Clinical Guidelines for the assessment and management of dyslipidaemia. N Z Med J **109:**224–232.
19. Kanis JA on behalf of the World Health Organization Scientific Group 2008 Assessment of Osteoporosis at the Primary Health Care Level. WHO Collaborating Centre, University of Sheffield, Sheffield, UK.

© 2008 American Society for Bone and Mineral Research

Chapter 34. Biochemical Markers of Bone Turnover in Osteoporosis

Pawel Szulc and Pierre D. Delmas

INSERM Research Unit 831 and Université de Lyon, Lyon, France

INTRODUCTION

Bone metabolism is characterized by two opposite activities coupled at a basic multicellular unit (BMU). During bone resorption, dissolution of bone mineral and catabolism of bone matrix by osteoclasts results in the formation of resorptive cavity and the release of bone matrix components. Then, during bone formation, osteoblasts synthesize bone matrix that fills in the resorption cavity and undergoes mineralization. There are two groups of biochemical bone turnover markers (BTMs). Bone formation is assessed by osteocalcin (OC), bone-specific alkaline phosphatase (BALP) and N- and C-terminal propeptides of type I procollagen (P1NP and P1CP). Bone resorption is assessed by N- and C-terminal cross-linking telopeptides of type I collagen (NTX-I and CTX-I), C-terminal cross-linking telopeptides of type I collagen generated by metalloproteinase (CTX-MMP, ICTP), deoxypyridinoline (DPD), hydroxylysine glycosides, or isoform 5b of TRACP (TRACP5b).

GENERAL LIMITATIONS OF BTMS AND THEIR ANALYTICAL AND PREANALYTICAL VARIABILITY

The analytical variability (assessed by the interassay and intra-assay coefficients of variation) depends on the BTM and the measurement method. The preanalytical variability has a strong effect on the BTM levels. It comprises a large number of factors (Table 1), which may coexist in one person (e.g., in a patient with rheumatoid arthritis, BTMs depend on the disease, corticotherapy, and limited mobility). Collection of serum should be performed in standardized conditions, preferably in the fasting state in the morning. For urinary collection, the choice between a spot (usually second morning void) and a 24-h collection is a trade-off between the biological interest and practical reliability. The 24-h collection reflects the overall bone metabolism, whereas the spot collection may be performed in a controlled way.

SELECTED DETERMINANTS OF THE PREANALYTICAL VARIABILITY

Circadian rhythm has a strong impact on the variability of BTMs, especially bone resorption markers that peak in the second half of the night and have their nadir in the afternoon.[1,2] The amplitude is higher for CTX-I than for other BTMs. Food intake has a strong effect on bone resorption. This postprandial decrease in the serum CTX-I is most probably mediated by glucagon-like peptide 2, the synthesis of which is stimulated by food intake.[3]

In the elderly, bone metabolism is strongly influenced by the vitamin D and calcium status.[4,5] BTM levels are increased mainly in the institutionalized and home-bound vitamin D–deficient elderly who have lower 25-hydroxycholecalciferol [25(OH)D] concentrations and higher PTH concentrations than the ambulatory ones. 25(OH)D is lower and PTH and BTM levels are higher during winter.

BTM levels are increased in patients with bone metastases.[6,7] ICTP and α-α-CTX-I (nonisomerized form of CTX-I) seem to be the most sensitive markers of bone involvement.[6–8] Their levels are positively associated with the spread of bone metastases, the progression of the disease, and skeletal-related events. During an anticancer treatment, poor decrease in the BTM levels predicts the progression of bone disease and death.[9,10]

BTM levels are influenced by a recent fracture. During the first hours after fracture, OC decreases because of high cortisol secretion (stress).[11,12] Then, bone formation and resorption increase reflecting the healing of the fracture. The BTM levels are increased mainly for 4 mo after fracture, and then they decrease for up to 1 yr.

Endogenous and exogenous corticosteroids increase the risk of osteoporosis. They inhibit bone formation. The decrease in the OC level is most rapid followed by a delayed and milder decrease in P1CP and P1NP.[13] Bone resorption can increase; however, data are less consistent. Low-dose prednisone (5 mg/d) decreased bone formation but not bone resorption.[14] Inhaled corticosteroids induce a dose- and drug-dependent decrease in the OC concentration without significant effect on other BTMs.[15,16]

Inhibitors of aromatase (used in the treatment of breast cancer) reduce the residual secretion of estrogens leading to an acceleration of bone turnover and bone loss.[17,18] This increase is not observed in the case of the concomitant treatment with bisphosphonates or tamoxifen.

CHANGES IN BTM LEVELS AFTER MENOPAUSE

In young adults, quantity of bone formed at every BMU is equal to the quantity of bone removed by resorption. After menopause, BTMs increase rapidly.[19,20] Bone formation increases to fill in the higher number of resorption cavities, which increases serum levels of bone formation markers. Because the quantity of bone formed is lower than the quantity of bone resorbed, there is a net bone loss at the BMU level. The increased number of BMUs is the principal determinant of the postmenopausal BTM levels and BMD.[19–21]

ASSOCIATION BETWEEN BTM LEVELS AND RATE OF BONE LOSS

In some studies, baseline BTM levels are correlated with the subsequent bone loss,[22] which suggests that the bone turnover rate determines the subsequent bone loss. However, for a given BTM level, there is a large scatter of individual values of bone loss,[23] and BTM cannot be used for prediction of the accelerated bone loss at the individual level.

ASSOCIATION BETWEEN BONE TURNOVER RATE AND RISK OF FRACTURE

Increased BTM levels predict fragility fractures independently of age, BMD, and prior fracture[24–27] in prospective cohort and case-control studies. This association has been

The authors state that they have no conflicts of interest.

Key words: bone turnover, osteocalcin, C-terminal cross-linking telopeptides of type I collagen, deoxypyridinoline, osteoporosis

© 2008 American Society for Bone and Mineral Research

TABLE 1. DETERMINANTS OF THE PREANALYTICAL VARIABILITY OF BONE TURNOVER

Controllable determinants
- Circadian variation
- Menstrual variation
- Seasonal variation
- Fasting and food intake (serum β-CTX-I)
- Exercise and physical activity

Determinants that cannot be easily modified
- Age
- Sex
- Menopausal status
- Vitamin D deficit and secondary hyperparathyroidism
- Diseases characterized by an acceleration of bone turnover
  - Primary hyperparathyroidism
  - Thyrotoxicosis
  - Acromegaly
  - Paget's disease
  - Bone metastases
- Diseases characterized by a dissociation of bone turnover
  - Cushing's disease
  - Multiple myeloma
- Diseases characterized by a low bone turnover
  - Hypothyroidism
  - Hypoparathyroidism
  - Hypopituitarism
  - Growth hormone deficit
- Renal impairment (depending on the stage)
- Recent fracture
- Chronic diseases associated with limited mobility
  - Stroke
  - Hemiplegia
  - Dementia
  - Alzheimer's disease
- Medications
  - Oral corticosteroids
  - Inhaled corticosteroids (only osteocalcin)
  - Aromatase inhibitors (anti-aromatases)
  - Gonadoliberin agonists
  - Anti-epileptic drugs
  - Thiazolidinediones
  - Heparin
  - Vitamin K antagonists

found in postmenopausal and elderly women, but not in the frail elderly where incident falls were the strongest predictor of fracture. BTM levels are predictive of all fractures, vertebral fractures, hip fracture, and multiple fractures. The fractures are predicted mainly by the bone resorption markers and BALP but not by less specific BTMs (total alkaline phosphatase, hydroxyproline). Analysis of three major predictors (i.e., BMD T-score < −2.5, BTM level > 2 SD above the mean in premenopausal women, and prior fracture) showed that each of them contributed to the prediction of hip fracture. (Fig. 1).[28] Increased urinary CTX-I level remained a significant predictor of hip fracture after adjustment for heel broadband ultrasound attenuation. Osteopenic women with high BTM levels have a risk of fracture similar to that of osteoporotic women, whereas those with normal BTM levels have a low fracture risk similar to that of women with normal BMD. BTM measurements may help to identify women who will benefit the best from the anti-osteoporotic treatment, a way to improve the cost effectiveness of treatment.[29,30]

High bone turnover is associated with lower BMD, faster bone loss, and poor bone microarchitecture both in the trabecular compartment (trabecular perforations, loss of trabeculae, poor trabecular connectivity) and in the cortical compartment (cortical thinning, increased porosity).[31,32] Thus, women with high bone turnover experience more severe decrease in bone strength. However, they may also have experienced a higher bone loss and greater structural weakening of bone since achievement of their peak bone mass. Consequently, remaining bone may sustain higher stress, leading to a more rapid fatigue of bone tissue and the deterioration of its material mechanical properties.

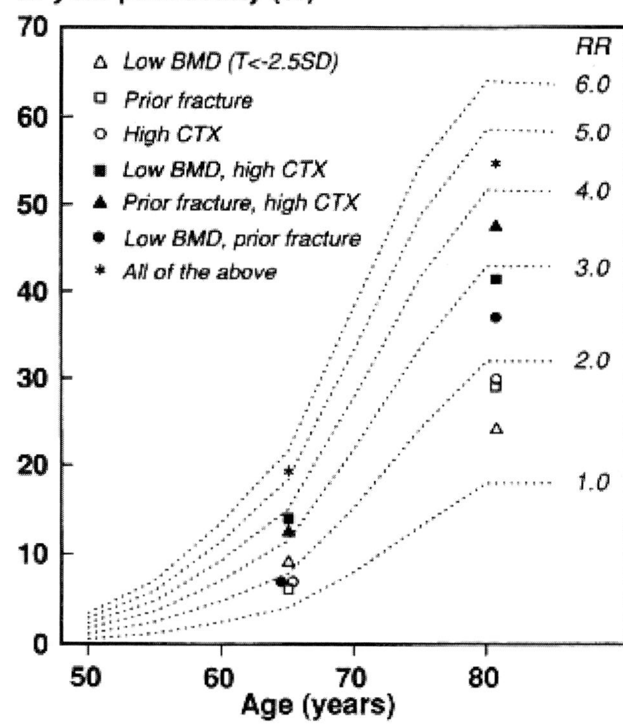

FIG. 1. Ten-year probability of hip fracture in Swedish women according to age and relative risk. The symbols show the effect of risk factors on fracture probability derived from women 65 (OFELY study) or 80 yr of age (EPIDOS study). Data from the OFELY study are derived from information on all fractures. The following threshold values were used for the risk factors: low BMD, T-score < −2.5; high CTX, above premenopausal values. (Reproduced with kind permission of Springer Science and Business Media from Johnell O, Oden A, de Laet C, Garnero P, Delmas PD, Kanis JA 2002 Biochemical indices of bone turnover and the assessment of fracture probability. Osteoporos Int **13**:523–526.)

Resorption cavities trigger stress risers, leading to the local weakening of the trabecula.[33] High bone turnover is associated with a higher fraction of recently formed partly mineralized bone, which may have suboptimal mechanical resistance. Shorter periods between metabolic cycles leave less time for the post-translational modifications of bone matrix proteins (cross-linking, β-isomerization of collagen type I).[34] Low isomerization of type I collagen, assessed by the α/β ratio of urinary CTX, was associated with a higher risk of fracture independently of other predictors, including the bone turnover rate.[35] Thus, the maturation of bone matrix may play a role in the skeletal fragility.

BIOCHEMICAL BTMs AND ANTI-OSTEOPOROTIC TREATMENT

In pharmaceutical trials, BTMs provide information on the metabolic effect of drugs on bone turnover, help to establish

© 2008 American Society for Bone and Mineral Research

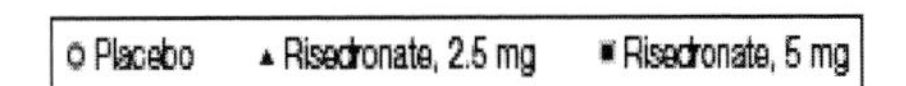

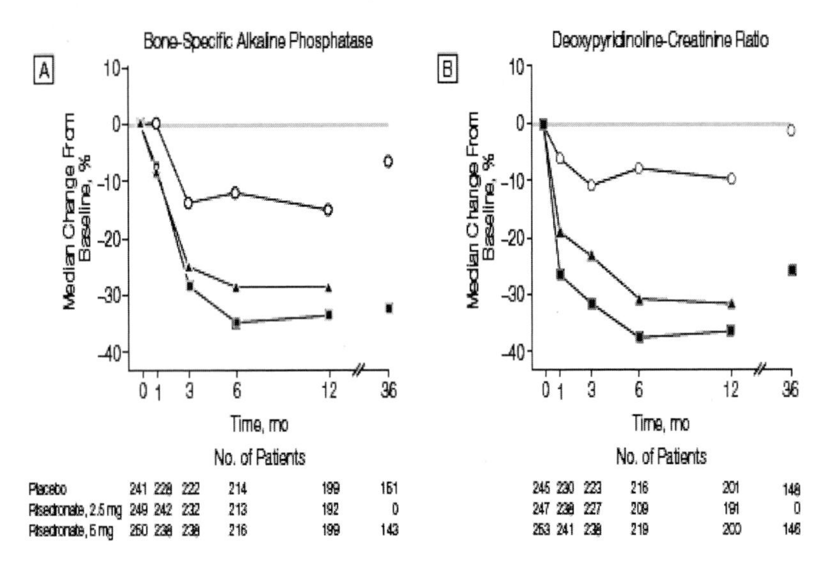

FIG. 2. Median percentage change from baseline in bone alkaline phosphatase and deoxypyridinoline-creatinine ratio in postmenopausal osteoporotic women treated with oral risedronate. (Reproduced with permission from the American Medical Association from Harris ST, Watts NB, Genant HK, McKeever CD, Hangartner T, Keller M, Chesnut CH III, Brown J, Eriksen EF, Hoseyni S, Axelrod DW, Miller PD 1999 Effects of risedronate treatment on vertebral and nonvertebral fractures in women with postmenopausal osteoporosis. A randomized controlled trial. JAMA **282:**1344–1352.) Copyright © 1999 American Medical Association.

the adequate dose, and predict the increase in BMD and the decrease in the fracture risk. Their use may improve compliance to treatment.

METABOLIC EFFECT OF ANTI-OSTEOPOROTIC TREATMENT

Changes in BTM levels depend on the mechanism of action of the drugs. Inhibition of bone resorption by antiresorptive drugs results in a rapid decrease in bone resorption (Fig. 2).[36] As bone formation continues in BMUs activated before treatment, bone formation is stable and then decreases when osteoblasts fill in the lower number of BMUs formed during treatment. BMD increases rapidly during the early period, when bone resorption is reduced and bone formation is still at the baseline level. Changes in BTMs during antiresorptive therapy depend on the route of administration and dose of the drug, degree of inhibition of bone resorption, and the cellular mechanism of action of the drug. For instance, intravenous bisphosphonates decrease BTM levels faster than when orally administered.[37]

During bone formation-stimulating treatment, recombinant human PTH(1-34) (teriparatide) induced a rapid increase in bone formation (especially P1NP) followed by an increase in bone resorption.[38,39] BMD increases rapidly during the early phase and mainly in trabecular bone. Strontium ranelate slightly increases BALP and slightly lowers serum CTX-I at the beginning of the therapy,[40] and then both plateau throughout treatment.

DOSE-FINDING STUDIES

BTMs help to establish the appropriate dose of anti-osteoporotic drugs because the treatment-induced changes in BTM are more rapid compared with BMD. Oral antiresorptive drugs induce a dose-dependent decrease in bone resorption (3 mo) and bone formation (6 mo), followed by an increase in BMD (1 yr). The higher the dose, the lower the steady-state BTM level and the higher the increase in BMD. Such trends are observed for transdermal 17β-estradiol, raloxifene, alendronate, risedronate, and ibandronate.

The first dose of intermittent treatment with subcutaneous monoclonal anti-RANKL antibody (denosumab) or intravenous bisphosphonates induces a very rapid dose-dependent decrease in bone resorption.[41,42] The higher the dose of the drug, the longer the period of the decreased bone resorption, the lower the levels of bone resorption markers before injection of the drug, and the higher the increase in BMD.

During treatment with bone formation-stimulating PTH, dose-dependent increases in BMD and BTM levels are also observed, especially for lumbar spine BMD and bone formation markers.[43]

DECREASE IN BTM LEVELS AND ANTIFRACTURE EFFICACY OF ANTIRESORPTIVE TREATMENT

Changes in BMD induced by antiresorptive therapy are weakly associated with their antifracture efficacy. For any change in BMD, fracture risk is lower in the treated than in the placebo group.[44] Thus, BMD is a poor surrogate measure of the antifracture efficacy of this treatment. The early decrease in BTM levels (6–12 mo) is associated with the long-term antifracture efficacy (2–3 yr) of the antiresorptive agent.[45–47] For a given decrease in BTM levels and for a given BTM level during treatment, the incidence of vertebral fracture was similar in the active treatment and placebo groups.

BTM LEVELS AND ANTIFRACTURE EFFICACY OF BONE FORMATION-STIMULATING TREATMENT

Teriparatide-induced early increase in BTM levels is correlated positively with the subsequent increase in BMD, especially with the increase in trabecular volumetric BMD, probably because there are more BMUs in the trabecular bone.[38]

CHANGES IN BONE TURNOVER AFTER DISCONTINUATION OF ANTI-OSTEOPOROTIC TREATMENT

Hormone replacement therapy (HRT) is active during its administration. Discontinuation of HRT results in a rapid increase in BTM levels to the pretreatment level and is followed by a decrease in BMD and an increase in fracture risk.[48–50]

Bisphosphonates have a strong affinity to bone, are accu-

© 2008 American Society for Bone and Mineral Research

mulated in bone, and are not metabolized. After withdrawal of short-term treatment (ibandronate, 9–12 mo), BTMs increased rapidly. The lower the cumulative dose, the sooner the BTMs returned to baseline.[51] After withdrawal of alendronate administered for several years, BTM increased moderately and BMD decreased more slowly.[52–54]

Withdrawal of PTH(1-84) after 1 yr of treatment was followed by a return of BTM levels to baseline values and a decrease in volumetric BMD of trabecular bone.[55]

COMBINATION THERAPY AND BTMs

Three designs combining antiresorptive and PTH have been studied in postmenopausal women: both drugs administered jointly, antiresorptive treatment followed by PTH, and PTH followed by antiresorptive treatment.

Alendronate and PTH(1-84) administered jointly rapidly decreased bone resorption (serum CTX-I) but less than alendronate alone and temporarily increased bone formation (P1NP, BALP) but less that PTH(1-84) alone.[56] Then, bone formation decreased and remained slightly below baseline levels. The time course of BTM levels during this therapy was more strongly determined by alendronate, which is consistent with the similar changes in BMD in the combination therapy group and in the group receiving alendronate alone.

The effect of PTH treatment on BTM levels after antiresorptive treatment depends on the degree of inhibition of bone turnover. After a strong suppression of bone resorption by alendronate, the increase in BTM levels induced by teriparatide was delayed and smaller than after raloxifene therapy.[57] In the women treated with risedronate, BTMs were higher, and teriparatide induced a greater increase in BTMs than in those treated with alendronate.[58]

Alendronate administered after PTH(1-84) induces a marked decrease in BTM levels that are indistinguishable from those in women treated with alendronate alone.[55,59] This strong inhibition of bone turnover may prevent the resorption of the bone synthesized under PTH(1-84) treatment, which results in an additional increase in BMD.

ASSOCIATION BETWEEN BTM LEVELS AND ADHERENCE WITH ANTIRESORPTIVE TREATMENT

Low compliance is a serious problem during anti-osteoporotic treatment, leading to an increased risk of fracture.[60] BTM change reflects the degree of compliance to risedronate treatment in postmenopausal osteoporotic women.[61] The better the compliance, the greater the average decrease in bone turnover. Moreover, measurement of BTM levels may also improve the persistence with antiresorptive treatment (e.g., risk of treatment discontinuation was lower in women who received positive information corresponding to a >30% decrease in NTX-I urinary excretion).[62,63]

BTMs IN MEN

In boys, the growth spurt starts later and lasts longer than in girls. Therefore, young men enter the phase of consolidation (formation of peak BMD after growth arrest) later than women. Men have wider bones even after adjustment for body size. At the age of 20–25 yr, men have BTM levels higher than women because men have more active bone turnover in longer and wider bones. Then, BTMs decrease and attain their lowest levels between 50 and 60 yr of age.[64–66] After the age of 60, bone formation remains stable or increases slightly. Bone resorption increases progressively after the age of 60. However, bone resorption markers, which have produced meaningful data in women, do not necessarily reflect the status of bone resorption in elderly men.

Men with high bone turnover have lower BMD; thus, age-related bone loss in men results at least in part from increased bone resorption. Elderly men with high BTM levels have a faster subsequent bone loss; however, this association is weak.[67–69] In a nested case-control study, a high level of CTX-MMP (ICTP) was associated with an increased risk of incident clinical fractures.[70] Recent prospective large cohort studies showed that BTMs do not predict osteoporotic fractures in elderly men.[67,68]

EFFECT OF ANTI-OSTEOPOROTIC TREATMENT ON BTM IN MEN

Testosterone replacement therapy (TRT) inhibits bone turnover in hypogonadism if the normal bioT concentration has been achieved.[71,72] During TRT, bone resorption decreases promptly, but decrease in urinary excretion per milligram creatinine may be partly related to the increase in muscle mass. Bone formation increases during the first 6 mo of TRT (direct stimulatory effect), levels off, and finally decreases, reflecting the general slowdown of bone turnover.

The equivalent daily doses of bisphosphonates (alendronate 10 mg, risedronate 5 mg) similarly decreased BTMs in men with low BMD and in elderly men after stroke.[73,74] In men, teriparatide increased bone formation (P1NP) after 1 mo and bone resorption after 3 mo of treatment.[75] In growth hormone (GH)-deficient men, recombinant human GH accelerated bone turnover. BTMs attained their peak values after 6–12 mo and then decreased progressively.[76,77]

CONCLUSIONS

BTMs improve our understanding of the relationship between bone turnover, BMD, bone fragility, and the effect of anti-osteoporotic treatment (metabolic effect, antifracture efficacy). Data on BTMs show that the rate of bone turnover (spontaneous or modified by the therapy) is an important determinant of bone fragility in postmenopausal and elderly women. Preliminary data suggest that the use of BTMs may improve the cost-effectiveness of the anti-osteoporotic treatment. From a clinical point of view, measurement of BTMs may help to identify postmenopausal women at high risk of fracture and may improve persistence with antiresorptive treatment. However, practical guidelines for the use of BTMs in the clinical management of postmenopausal osteoporosis are still lacking.

REFERENCES

1. Qvist P, Christgau S, Pedersen BJ, Schlemmer A, Christiansen C 2002 Circadian variation in the serum concentration of C-terminal telopeptide of type I collagen (serum CTx): Effects of gender, age, menopausal status, posture, daylight, serum cortisol, and fasting. Bone **31**:57–61.
2. Bollen AM, Martin MD, Leroux BG, Eyre DR 1995 Circadian variation in urinary excretion of bone collagen cross-links. J Bone Miner Res **10**:1885–1890.
3. Henriksen DB, Alexandersen P, Bjarnason NH, Vilsboll T, Hartmann B, Henriksen EE, Byrjalsen I, Krarup T, Holst JJ, Christiansen C 2003 Role of gastrointestinal hormones in postprandial reduction of bone resorption. J Bone Miner Res **18**:2180–2189.
4. Chapuy MC, Schott AM, Garnero P, Hans D, Delmas PD, Meunier PJ 1996 Healthy elderly French women living at home have secondary hyperparathyroidism and high bone turnover in winter. J Clin Endocrinol Metab **81**:1129–1133.

© 2008 American Society for Bone and Mineral Research

5. Theiler R, Stähelin HB, Kräzlin M, Tyndall A, Bischoff HA 1999 High bone turnover in the elderly. Arch Phys Med Rehabil **80:**485–489.
6. Voorzanger-Rousselot N, Juillet F, Mareau E, Zimmermann J, Kalebic T, Garnero P 2006 Association of 12 serum biochemical markers of angiogenesis, tumour invasion and bone turnover with bone metastases from breast cancer: A cross-secional and longitudinal evaluation. Br J Cancer **95:**506–514.
7. Leeming DJ, Koizumi M, Byrjalsen I, Li B, Qvist P, Tanko LB 2006 The relative use of eight collagenous and noncollagenous markers fo diagnosis of skeletal metastases in breast, prostate, or lung cancer patients. Cancer Epidemiol Biomarkers Prev **15:**32–38.
8. Leeming DJ, Delling G, Koizumi M, Henriksen K, Karsdal MA, Li B, Qvist P, Tanko LB, Byrjalsen I 2006 Alpha CTX as a biomarkers of skeletal invasion of breast cancer: Immunolocalization and the load dependency of urinary excretion. Cancer Epidemiol Biomarkers Prev **15:**1392–1395.
9. Brown JE, Cook RJ, Major P, Lipton A, Saad F, Smith M, Lee KA, Zheng M, Hei YJ, Coleman RE 2005 Bone turnover markers as predictors of skeletal complications in prostate cancer, lung cancer, and other solid tumors. J Natl Cancer Inst **97:**59–69.
10. Coleman RE, Major P, Lipton A, Brown JE, Lee KA, Smith M, Saad F, Zheng M, Hei YJ, Seaman J, Cook R 2005 Predictive value of bone resorption and formation markers in cancer patients with bone metastases receiving the bisphosphonate zoledronic acid. J Clin Oncol **23:**4925–4935.
11. Ivaska KK, Gerdhem P, Akesson K, Garnero P, Obrant KJ 2007 Effect of fracture on bone turnover markers: A longitudinal study comparing marker levels before and after injury in 113 elderly women. J Bone Miner Res **22:**1155–1164.
12. Stoffel K, Engler H, Kuster M, Riesen W 2007 Changes in biochemical markers after lower limb fractures. Clin Chem **53:**131–134.
13. Dovio A, Perazzolo L, Osella G, Ventura M, Termine A, Milano E, Bertolotto A, Angeli A 2004 Immediate fall of bone formation and transient increase of bone resorption in the course of high-dose, short-term glucocorticoid therapy in youn patients with multiple sclerosis. J Clin Endocrinol Metab **89:**4923–4928.
14. Ton FJ, Gunawardene SC, Lee H, Neer RM 2005 Effects of low-dose prednisone on bone metabolism. J Bone Miner Res **20:**464–470.
15. Richy F, Bousquet J, Ehrlich GE, Meunier PJ, Israel E, Morii H, Devogelaer JP, Peel N, Haim M, Bruyere O, Reginster JY 2003 Inhaled corticosteroids effects on bone in asthmatic and COPD patients: A quantitative systematic review. Osteoporos Int **14:**179–190.
16. Martin RJ, Szefler SJ, Chinchilli VM, Kraft M, Dolovich M, Boushey HA, Cherniack RM, Craig TJ, Drazen JM, Fagan JK, Fahy JV, Fish JE, Ford JG, Isreal E, Kunselman SJ, Lazarus SC, Lemanske RF Jr, Peters SP, Sorkness CA 2002 Systemic effect comparisons of six inhaled corticosteroid preparations. Am J Respir Crit Care Med **165:**1377–1383.
17. Brufsky A, Harker WG, Beck JT, Carroll R, Tan-Chiu E, Seidler C, Hohneker J, Lacerna L, Petrone S, Perez EA 2007 Zoledronic acid inhibits adjuvant letrozole-induced bone loss in postmenopausal women with early breast cancer. J Clin Oncol **25:**829–836.
18. Confavreux CB, Fontana A, Guastalla JP, Munoz F, Brun J, Delmas PD 2007 Estrogen-dependent increase in bone turnover and bone loss in postmenopausal women with breast cancer treated with anastrozole. Prevention with bisphosphonates. Bone **41:**346–352.
19. Garnero P, Sornay-Rendu E, Chapuy MC, Delmas PD 1996 Increased bone turnover in late postmenopausal women is a major determinant of osteoporosis. J Bone Miner Res **11:**337–349.
20. Garnero P, Mulleman D, Munoz F, Sornay-Rendu E, Delmas PD 2003 Long-term variability of markers of bone turnover in postmenopausal women and implications for their clinical use: The OFELY study. J Bone Miner Res **18:**1789–1794.
21. Schneider DL, Barrett-Connor EL 1997 Urinary N-telopeptide levels discriminate normal, osteopenic, and osteoporotic bone mineral density. Arch Intern Med **157:**1241–1245.
22. Stepan J 2000 Prediction of bone loss in premenopausal women. Osteoporos Int **11**(Suppl 6):S45–S54.
23. Rogers A, Hannon RA, Eastell R 2000 Biochemical markers as predictors of rates of bone loss after menopause. J Bone Miner Res **15:**1398–1404.
24. Garnero P, Sornay-Rendu E, Claustrat B, Delmas PD 2000 Biochemical markers of bone turnover, endogenous hormones and the risk of fractures in postmenopausal women: The OFELY study. J Bone Miner Res **15:**1526–1536.
25. Gerdhem P, Ivaska KK, Alatalo SL, Halleen JM, Hellman J, Isaksson A, Pettersson K, Väänänen HK, Akesson K, Obrant KJ 2004 Biochemical markers of bone metabolism and prediction of fracture in elderly women. J Bone Miner Res **19:**386–393.
26. Sornay-Rendu E, Munoz F, Garnero P, Duboeuf F, Delmas PD 2005 Identification of osteopenic women at high risk of fracture: The OFELY study. J Bone Miner Res **20:**1813–1819.
27. Garnero P, Hausher E, Chapuy MC, Marcelli C, Grandjean H, Muller C, Cormier C, Bréart G, Meunier PJ, Delmas PD 1996 Markers of bone resorption predict hip fracture in elderly women: The Epidos prospective study. J Bone Miner Res **11:**1531–1538.
28. Johnell O, Oden A, de Laet C, Garnero P, Delmas PD, Kanis JA 2002 Biochemical indices of bone turnover and the assessment of fracture probability. Osteoporos Int **13:**523–526.
29. Delmas PD, Licata AA, Reginster JY, Crans GG, Chen P, Misurski DA, Wagman RB, Mitlak BH 2006 Fracture risk reduction during treatment with teriparatide is independent of pretreatment bone turnover. Bone **39:**237–243.
30. Bauer DC, Garnero P, Hochberg MC, Santora A, Delmas PD, Exing SK, Black DM 2006 Pretreatment levels of bone turnover and the antifracture efficacy of alendronate: The Fracture Intervention Trial. J Bone Miner Res **21:**292–299.
31. Garnero P, Sornay-Rendu E, Duboeuf F, Delmas PD 1999 Markers of bone turnover predict postmenopausal forearm bone loss over 4 years: The Ofely Study. J Bone Miner Res **14:**1614–1621.
32. Seeman E, Delmas PD 2006 Bone quality: The material and structural basis of bone strength and fragility. N Engl J Med **354:**2250–2261.
33. Dempster DW 2000 The contribution of trabecular architecture to calcellous bone quality. J Bone Miner Res **15:**20–23.
34. Viguet-Carin S, Roux JP, Arlot ME, Merabet Z, Leeming DJ, Byrjalsen I, Delmas PD, Bouxsein ML 2006 Contribution of the advanced glycation end product pentosidine and of maturation of type I collagen to compressive biomechanical properties of human lumbar vertebrae. Bone **39:**1073–1079.
35. Garnero P, Cloos P, Sornay-Rendu E, Qvist P, Delmas PD 2002 Type I collagen racemization and isomerization and the risk of fracture in postmenopausal women: The OFELY prospective study. J Bone Miner Res **17:**826–833.
36. Liberman UM, Weiss SR, Broll J, Minne HW, Quan H, Bell NH, Rodriguez-Portales J, Downs RW Jr, Dequeker J, Favus M 1995 Effect of oral alendronate on bone mineral density and the incidence of fractures in postmenopausal osteoporosis. The Alendronate Phase III Osteoporosis Treatment Study Group. N Engl J Med **333:**1437–1443.
37. Thiébaud D, Burckhardt P, Kriegbaum H, Huss H, Milder H, Juttmann JR, Schöter KH 1997 Three monthly intravenous injections of ibandronate in the treatment of postmenopausal osteoporosis. Am J Med **103:**298–307.
38. Chen P, Satterwhite JH, Licata AA, Lewiecki EM, Sipos AA, Misurski DM, Wagman RB 2005 Early changes in biochemical markers of bone formation predict BMD response to teriparatide in postmenopausal women with osteoporosis. J Bone Miner Res **20:**962–970.
39. Eastell R, McCloskey EV, Glover S, Rogers A, Garnero P, Lowery J, Belleli R, Wright TM, John MR 2007 Rapid and robust biochemical response to teriparatide therapy for osteoporosis. J Bone Miner Res **22:**S1;S322.
40. Meunier PJ, Roux C, Seeman E, Ortolani S, Badurski JE, Spector TD, Cannata J, Balogh A, Lemmel EM, Pors-Nielsen S, Rizzoli R, Genant HK, Reginster JY 2004 The effects of strontium ranelate on the risk of vertebral fracture in women with postmenopausal osteoporosis. N Engl J Med **350:**459–468.
41. McClung MR, Lewiecki EM, Cohen SB, Bolognese MA, Woodson GC, Moffett AH, Peacock M, Miller PD, Lederman SN, Chesnut CH, Lain D, Kivitz AJ, Holloway DL, Zhang C, Peterson MC, Bekker PJ 2006 Denosumab in postmenopausal women with low bone mineral density. N Engl J Med **354:**821–831.
42. Reid IR, Brown JP, Burckhardt P, Horowitz Z, Richardson P, Trechsel U, Widmer A, Devogelaer JP, Kaufman JM, Jaeger P, Body JJ, Meunier PJ 2002 Intravenous zoledronic acid in postmenopausal women with low bone mineral density. N Engl J Med **346:**653–661.
43. Hodsman AB, Hanley DA, Ettinger MP, Bolognese MA, Fox J, Metcalfe AJ, Lindsay R 2003 Efficacy and safety of human para-

© 2008 American Society for Bone and Mineral Research

thyroid hormone-(1-84) inincreasing bone mineral density in postmenopausal osteoporosis. J Clin Endocrinol Metab **88:**5212–5220.
44. Sarkar S, Mitlak BH, Wong M, Stock JL, Black DM, Harper KD 2002 Relationships between bone mienral density and incident vertebral fracture risk with raloxifene therapy. J Bone Miner Res **17:**1–10.
45. Bauer DC, Black DM, Garnero P, Hochberg M, Ott S, Orloff J, Thompson DE, Ewing SK, Delmas PD 2004 Change in bone turnover and hip, non-spine, and vertebral fracture in alendronate-treated women: The Fracture Intervention Trial. J Bone Miner Res **19:**1250–1258.
46. Reginster JY, Sarkar S, Zegels B, Henrotin Y, Bruyere O, Agnusdei D, Collette J 2004 Reduction in PINP, a marker of bone metabolism, with raloxifene treatment and its relationship with vertebral fracture risk. Bone **34:**344–351.
47. Eastell R, Hannon RA, Garnero P, Campbell MJ, Delmas PD 2007 Relationship of early changes in bone resorption to the reduction in fracture risk with risedronate: Review of statistical analysis. J Bone Miner Res **22:**1656–1660.
48. Sornay-Rendu E, Garnero P, Munoz F, Duboeuf F, Delmas PD 2003 Effect of withdrawal of hormone replacement therapy on bone mass and bone turnover: The OFELY study. Bone **33:**159–166.
49. Greenspan SL, Emkey RD, Bone HG, Weiss SR, Bell NH, Downs RW Jr, McKeever C, Miller SS, Davidson M, Bolognese MA, Mulloy AL, Heyden N, Wu M, Kaur A, Lombardi A 2002 Significant differential effects of alendronate, estrogen, or combination therapy on the rate of bone loss after discontinuation of treatment of postmenopausal osteoporosis. A randomized, double-blind, placebo-controlled trial. Ann Intern Med **137:**875–883.
50. Cauley JA, Seeley DG, Ensrud K, Ettinger B, Black DM, Cummings SR 1995 - Estrogen replacement therapy and fractures in older women. Study of Osteoporotic Fractures Research Group. Ann Intern Med **122:**9–16.
51. Ravn P, Christensen JO, Baumann M, Clemmensen B 1998 Changes in biochemical markers and bone mass after withdrawal of ibandronate treatment: Prediction of bone mass changes during treatment. Bone **22:**559–564.
52. Bone HG, Hosking D, Devolelaer JP, Tucci JR, Emkey RD, Tonino RP, Rodriguez-Portales JA, Downs RW, Gupta J, Santora AC, Liberman UA 2004 Ten years' experience with alendronate for osteoporosis in postmenopausal women. N Engl J Med **350:**1189–1199.
53. Tonino RP, Meunier PJ, Emkey R, Rodriguez-Portales JA, Menkes CJ, Wasnich RD, Bone HG, Santora AC, Wu M, Desai R, Ross PD 2000 Skeletal benefits of alendronate: 7-year treatment of postmenopausal osteoporotic women. J Clin Endocrinol Metab **85:**3109–3115.
54. Black DM, Schwartz AV, Ensrud KE, Cauley JA, Levis S, Quandt SA, Satterfield S, Wallace RB, Bauer DC, Palermo L, Wehren LE, Lombardi A, Santora AC, Cummings SR 2006 Effects of continuing or stopping alendronate after 5 years of treatment. The Fracture Intervention Trial Long-term Extension (FLEX): A randomized trial. JAMA **296:**2927–2938.
55. Black DM, Bilezikian JP, Ensrud KE, Greenspan SL, Palermo L, Hue T, Lang TF, McGowan JA, Rosen CJ 2005 One year alendronate after one year of parathyroid hormone (1-84) for osteoporosis. N Engl J Med **353:**555–565.
56. Black DM, Greenspan SL, Ensrud KE, Palermo L, McGowan JA, Lang TF, Garnero P, Bouxsein ML, Bilezikian JP, Rosen CJ 2003 The effects of parathyroid hormone and alendronate alone or in comination in postmenopausal osteoporosis. N Engl J Med **349:**1207–1215.
57. Ettinger B, San Martin J, Crans G, Pavo I 2004 Differential effects of teriparatide on BMD after treatment with raloxifene or alendronate. J Bone Miner Res **19:**745–751.
58. Delmas PD, Watts N, Miller P, Cahall D, Bilezikian J, Lindsay R 2007 Bone turnover markers demonstrate greater earlier responsiveness to teriparatide following treatment with risedronate compared with alendronate: The OPTAMISE study. J Bone Miner Res **22:**S1;S27.
59. Rittmaster RS, Bolognese M, Ettinger MP, Hanley DA, Hodsman AB, Kendler DL, Rosen CJ 2000 Enhancement of bone mass in osteoporotic women with parathyroid hormone followed by alendronate. J Clin Endocrinol Metab **85:**2129–2134.
60. Caro JJ, Ishak KJ, Huybrechts KF, Raggio G, Naujoks C 2004 The impact of compliance with osteoporosis therapy on fracture rates in actual practice. Osteoporos Int **15:**1003–1008.
61. Eastell R, Garnero P, Vrijens B, van de Lengerijt L, Pols HAP, Ringe JD, Roux C, Watts NB, Cahall D, Delmas PD 2003 Influence of patient compliance with risedronate therapy on bone turnover marker and bone mineral density response: The IMPACT study. Calcif Tissue Int **72:**408.
62. Clowes JA, Peel NFA, Eastell R 2004 The impact of monitoring on adherence and persistence with anti-resorptive treatment for postmenopausal osteoporosis: A randomized controlled trial. J Clin Endocrinol Metab **89:**1117–1123.
63. Delmas PD, Vrijens B, Eastell R, Roux C, Pols HAP, Ringe JD, Grauer A, Cahall D, Watts NB 2007 Effect of monitoring bone turnover markers on peristence with risedronate treatment of postmenopausal osteoporosis. J Clin Endocrinol Metab **92:**1296–1304.
64. Szulc P, Garnero P, Munoz F, Marchand F, Delmas PD 2001 Cross-sectional evaluation of bone metabolism in men. J Bone Miner Res **16:**1642–1650.
65. Fatayerji D, Eastell R 1999 Age-related changes in bone turnover in men. J Bone Miner Res **14:**1203–1210.
66. Khosla S, Melton LJ III, Atkinson EJ, O'Fallon WM, Klee GG, Riggs BL 1998 Relationship of serum sex steroid levels and bone turnover markers with bone mineral density in men and women: A key role for bioavailable estrogen. J Clin Endocrinol Metab **83:**2266–2274.
67. Szulc P, Montella A, Delmas PD 2008 High bone turnover is associated with accelerated bone loss but not with increased fracture risk in men aged 50 and over – the prospective MINOS study. Ann Rheum Dis 67:1249–1255.
68. Bauer DC, Garnero P, Harrison SL, Cauley JA, Eastell R, Barrett-Connor E, Orwoll ES 2007 Biochemical markers of bone turnover, hip bone loss and non-spine fracture in men: A prospective study. J Bone Miner Res **22:**S1;S21.
69. Dennison E, Eastell R, Fall CHD, Kellingray S, Wood PJ, Cooper C 1999 Determinants of bone loss in elderly men and women: A prospective population-based study. Osteoporos Int **10:**384–391.
70. Meier C, Nguyen TV, Center JR, Seibel MJ, Eisman JA 2005 Bone resorption and osteoporotic fractures in elderly men: The dubbo osteoporosis epidemiology study. J Bone Miner Res **20:**579–587.
71. Amory JK, Watts NB, Easley KA, Sutton PR, Anawalt BD, Matsumoto AM, Bremner WJ, Tenover L 2004 Exogenous testosterone or testosterone with finasteride increases bone mineral density in older men with low serum testosterone. J Clin Endocrinol Metab **89:**503–510.
72. Wang C, Swerdloff RS, Iranmanesh A, Dobs A, Snyder PJ, Cunningham G, Matsumoto AM, Weber T, Berman N 2001 Effects of transdermal testosterone gel on bone turnover markers and bone mineral density in hypogonadal men. Clin Endocrinol (Oxf) **54:**739–750.
73. Orwoll E, Ettinger M, Weiss S, Miller P, Kendler D, Graham J, Adami S, Weber K, Lorenc R, Pietschmann P 2000 Alendronate treatment of osteoporosis in men. N Engl J Med **343:**604–610.
74. Sato Y, Iwamoto J, Kanoko T, Satoh K 2005 Risedronate sodium therapy for prevention of hip fracture in men 65 years or older after stroke. Arch Intern Med **165:**1743–1748.
75. Orwoll ES, Scheele WH, Paul S, Adami S, Syversen U, Diez-Perez A, Kaufman JM, Clancy AD, Gaich GA 2003 The effect of teriparatide [human parathyroid hormone (1-34)] therapy on bone density in men with osteoporosis. J Bone Miner Res **18:**9–17.
76. Sneppen SB, Hoeck HC, Kollerup G, Sorensen OH, Laurberg P, Feldt-Rasmussen U 2002 Bone mineral content and bone metabolism during physiological GH treatment in GH-deficient adults—an 18-month randomised, placebo-controlled, double blinded trial. Eur J Endocrinol **146:**187–195.
77. Välimäki MJ, Salmela PI, Salmi J, Viikari J, Kataja M, Turunen H, Soppi E 1999 Effects of 42 months of GH treatment on bone mineral density and bone turnover in GH-deficient adults. Eur J Endocrinol **149:**545–554.

© 2008 American Society for Bone and Mineral Research

Chapter 35. Bone Biopsy and Histomorphometry in Clinical Practice

Robert R. Recker

Department of Medicine, Section of Endocrinology, Osteoporosis Research Center, Creighton University Medical Center, Omaha, Nebraska

INTRODUCTION

Histological examination of undecalcified transilial bone biopsy specimens is a valuable and well-established clinical and research tool for studying the etiology, pathogenesis, and treatment of metabolic bone diseases. In this chapter, we will review the underlying organization and function of bone cells; identify a set of basic structural and kinetic histomorphometric variables; outline an approach to interpretation of findings, with examples from a range of metabolic bone diseases; describe techniques for obtaining, processing, and analyzing transilial biopsy specimens; and identify clinical situations in which bone histomorphometry can be useful.

ORGANIZATION AND FUNCTION OF BONE CELLS

Intermediary Organization of the Skeleton

In what he termed the intermediary organization (IO) of the skeleton, Frost[1] described four discrete functions of bone cells: growth, modeling, remodeling, and fracture repair. Although each involves osteoclasts and osteoblasts, the coordinated outcomes differ greatly. Growth elongates the skeleton; modeling shapes it during growth; remodeling removes and replaces bone tissue; and fracture repair heals sites of structural failure.

The remodeling IO, which predominates during adult life, is the focus of this chapter. Coordinated groups of bone cells (i.e., osteoclasts, osteoblasts, osteocytes, and lining cells) comprise the basic multicellular units (BMUs) that carry out bone remodeling. Basic structural units (BSUs) are the packets of new bone that BMUs form.[2] All adult-onset metabolic bone disease involves derangement of the remodeling IO.

Bone Cells

Osteoclasts, large-to-giant cells that are typically multinucleated, resorb bone (both its matrix, or osteoid, and mineral). They excavate shallow pits on the surface of cancellous bone, and they appear at the leading edge of tunnels (cutting cones) in haversian bone. Light microscopy shows an irregular cell shape, foamy, acidophilic cytoplasm, a perimeter zone of attachment to the bone, a ruffled border that appears in the fluid cavity between the cell and the mineralized bone matrix, and positive staining for TRACP.

Osteoblasts form new bone at sites of resorption. They produce the collagenous and noncollagenous constituents of bone matrix and participate in mineralization.[3] Under light microscopy, they appear as plump cells lined up at the surface of unmineralized osteoid. As the site matures, the cells lose their plump appearance.

Osteocytes, derived from osteoblasts, remain at the remodeling site, and are buried as bone formation advances at remodeling sites. They reside individually in small lacunae within the mineralized bone matrix. Their cytoplasmic processes extend through a fine network of narrow canaliculi to form an interconnected network that extends throughout living bone. This network is well situated to monitor the local strain environment and local microdamage and to initiate organized bone cell work in response to changes in strain or microdamage.

Lining cells, also of osteoblast origin, cover cancellous and endocortical bone surfaces. By light microscopy, they appear as elongated, flattened, darkly stained nuclei. The localization and initiation of remodeling probably involves these cells.

Bone Remodeling Process

Remodeling occurs on cancellous and haversian bone surfaces. The first step is activation of osteoclast precursors to form osteoclasts that begin to excavate a cavity. After removal of ~0.05 mm^3 of bone tissue, the site remains quiescent for a short time. Then activation of osteoblast precursors occurs at the site, and the excavation is refilled. The average length of time required to complete the remodeling cycle is ~6 mo[4] and ~4 wk for resorption and the rest for formation.

The healthy bone remodeling system accesses the required building materials within a favorable physiologic milieu to replace fully a packet of aged, microdamaged bone tissue with new, mechanically competent bone. However, overuse can overwhelm the capacity of the system to repair microdamage (the stress fractures that occur in military recruits are an example). The healthy bone remodeling system modifies bone architecture to meet changing mechanical needs. However, the system also promptly reduces the mass of underused bone (the bone loss of extended bedrest, paralysis, or space travel are examples). All bone loss occurs through bone remodeling. The bone remodeling system responds to nutritional and humoral as well as mechanical influences. Among the effects of vitamin D deficiency in adults, for example, is impaired mineralization of bone matrix. Finally, as other chapters describe, bone remodeling involves complex signaling processes between and within bone cells, and metabolic bone diseases of genetic origin involve defects at this level. Figures 1–3 present representative photomicrographs from human transilial biopsy specimens. An extensive atlas has also been published.[5]

BASIC HISTOMORPHOMETRIC VARIABLES

Bone biopsy specimens for histomorphometric examination are ordinarily obtained at the transilial site and shipped to specialized laboratories for processing and microscopic analysis. Later sections of this chapter outline these procedures. Of the dozens of measurements and calculations that have been devised, we provide here descriptions of several frequently used variables. Together they describe a basic set of structural and kinetic features. Nomenclature is as approved by a committee of the American Society of Bone and Mineral Research.[6]

Structural Features

Core width (C.Wi) represents the thickness of the ilium (i.e.,

---

Dr. Recker has received research funding from Merck, Novartis, Procter & Gamble, Roche, and Wyeth.

Key words: bone biopsy, bone histomorphometry, bone remodeling, bone histology, osteoporosis

© 2008 American Society for Bone and Mineral Research

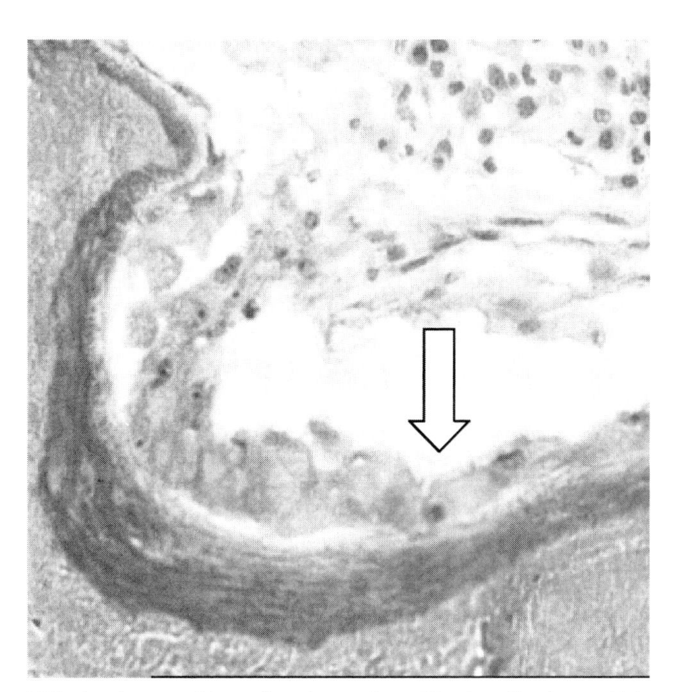

FIG. 1. A normal bone-forming surface. Unmineralized osteoid is covered with plump osteoblasts, as identified by the arrow.

distance between periosteal surfaces, in mm) at the point of biopsy. Cortical width (Ct.Wi) is the combined thickness, in millimeters, of both cortices. Cortical porosity (Ct.Po) is the area of intracortical holes as percent of total cortical area.

Cancellous bone volume (BV/TV) is the percent of total marrow area (including trabeculae) occupied by cancellous bone. Wall thickness (W.Th) is the mean distance in micrometers between resting cancellous surfaces (i.e., surfaces without osteoid or Howship's lacunae) and corresponding cement lines.

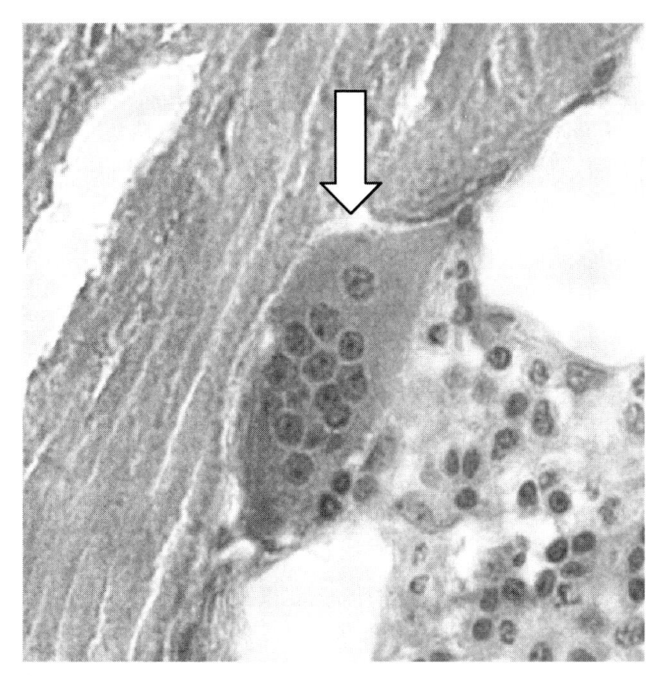

FIG. 2. A normal bone-resorbing surface. The arrow locates a multinucleated osteoclast in a Howship's lacuna.

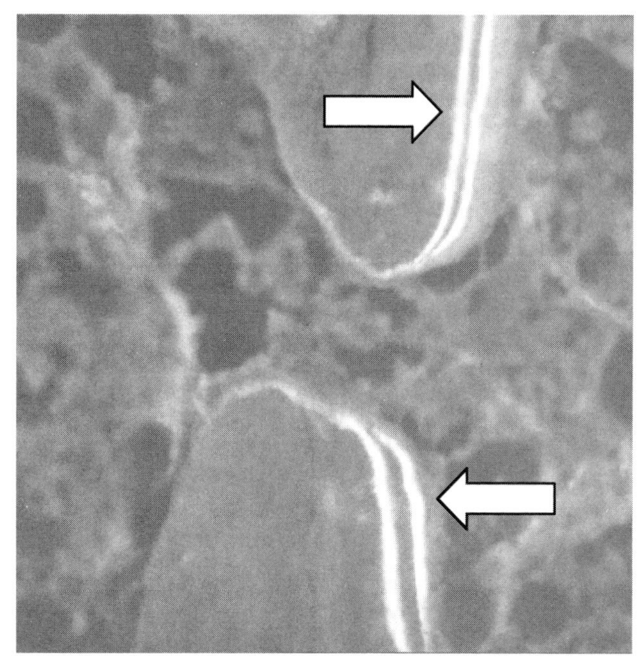

FIG. 3. The arrows identify two mineralizing surfaces with fluorescent double labels.

Trabecular thickness (Tb.Th) is the mean distance across individual trabeculae, in micrometers, and trabecular separation (Tb.Sp) is the mean distance, also in micrometers, between trabeculae. Trabecular number (Tb.N) per millimeter is calculated as (BV/TV)/Tb.Th. These variables can be used to evaluate trabecular connectivity.[7] Other measures of trabecular connectivity include the ratio of nodes to free ends,[8] star volume,[9,10] and trabecular bone pattern factor (TBPf).[11]

Eroded surface (ES/BS) is the percent of cancellous surface occupied by Howship's lacunae, with and without osteoclasts. Osteoblast surface (Ob.S/BS) and osteoclast surface (Oc.S/BS) identify the percent of cancellous surface occupied by osteoblasts and osteoclasts, respectively. Osteoid surface (OS/BS) is the percent of cancellous surface with unmineralized osteoid, with and without osteoblasts. Osteoid thickness (O.Th) is the mean thickness, in micrometers, of the osteoid on cancellous surfaces.

Kinetic Features

A fluorochrome labeling agent, taken orally on a strict schedule before biopsy, deposits a fluorescent double-label at sites of active mineralization and allows rates of change to be determined.[12] Mineralizing surface (MS/BS) is the percent of cancellous surface that is mineralizing and thus labeled. The most accurate version of MS/BS includes surfaces with a double label plus one half of those with a single label.[13] Clear definition of MS/BS is crucial, because it is used to calculate bone formation rates, bone formation periods, and mineralization lag time.

Mineral appositional rate (MAR), is the rate (in μm/d) at which new bone is being added to cancellous surfaces. MAR represents distance between labels at doubly labeled surfaces divided by the marker interval (span in days between the midpoints of each labeling period). This and all measurements of thickness must be corrected for obliquity (i.e., the randomness of the angle between the plane of the section and the plane of the cancellous surface) by use of a scaling factor.[7]

© 2008 American Society for Bone and Mineral Research

Activation frequency (Ac.f) is the probability that a new remodeling cycle will begin at any point on the cancellous bone surface. Bone formation rates (BFR/BV and BFR/BS) are estimates of cancellous bone volume (in $mm^3/mm^3/yr$) and cancellous bone surface (in $mm^3mm^2/mm/yr$), respectively, that are being replaced annually; BFR/BS = Ac.f × W.Th.[14] Formation period (FP) is the time in years required to complete a new cancellous BSU. Mineralization lag time (Mlt) is the interval in days between osteoid formation and mineralization. The most accurate version of Mlt is calculated as O.Th/MAR × MS/OS. Microcrack density (Cr.d.) is the number of microcracks per area of mineralized bone ($\#/mm^2$), and microcrack length (Cr.L) is the average length (in mm) of visualized microcracks.[15] Apoptosis can be quantified as the percent of total osteocytes that are identified as apoptotic using special stains.[16]

INTERPRETATION OF FINDINGS

Reference Data

In 1988, Recker et al.[4] published the results of a study to establish reference values for histomorphometric variables in postmenopausal white women. The 34 healthy subjects were evenly distributed into three age groups: 45–54, 55–64, and 65–74 yr of age. They ranged broadly in age at menopause and in years since menopause at the time of biopsy. A comparative study of 12 blacks and 13 whites, 19–46 yr of age, has also been published.[17]

In 2000, Glorieux et al.[18] reported histomorphometric data from 58 white subjects in each of five age groups: 1.5–6.9, 7.0–10.9, 11.0–13.9, 14.0–16.9, and 17.0–22.9 yr of age. Biopsy specimens were obtained during corrective orthopedic surgeries, but the subjects had been ambulatory and otherwise healthy. The report includes within-subject CVs derived from analysis of adjacent duplicate biopsy specimens in eight subjects.

A recent paper from our center reported Ac.f in several sets of transilial biopsy specimens. In 50 paired transilial biopsy specimens taken during perimenopause and again in early postmenopause, a year after last menses, median values for Ac.f increased from 0.13/yr to 0.24/yr, respectively ($p < 0.001$).[19] Ac.f. was even higher (median, 0.37; $p < 0.01$) in another group of ostensibly normal women[4] who were postmenopausal by an average of 13 yr. Others have published reference databases.[20–23]

Replacement of Normal Marrow Elements

A variety of hematopoietic cells and a varying proportion of fat cells normally occupy the marrow space at the transilial biopsy site. If these normal marrow elements have been displaced by fibrous tissue (osteitis fibrosa), clumps of tumor cells, or sheets of abnormal hematopoietic cells, this change will be obvious to the histomorphometrist. The biopsy preparations described here preserve cellular detail, spatial relationships, and architectural features. However, this approach is unsuitable for hematological diagnosis because of the time that histomorphometry laboratories require to generate a report (typically, ≥4 wk).

Cortical Bone Deficit

Both the angle of the biopsy and site-to-site variation in cortical thickness at the biopsy site influence Ct.Wi. Nevertheless, low BMD at the lumbar spine and/or proximal femur is often reflected in low values for Ct.Wi.[24] Evidence of trabeculation of the cortex (i.e., formation of a transitional zone with characteristic coarse trabeculae) indicates that cortical bone once present in the area adjacent to the marrow space has been lost.[25]

Cancellous Bone Deficit

Low BV/TV indicates a cancellous bone deficit. Generalized trabecular thinning (decreased Tb.Th) and/or complete loss of trabecular elements (poor trabecular connectivity) may contribute to this deficit. The latter finding (e.g., low Tb.N with high Tb.Sp) characterizes bone that is more fragile than its overall mass would suggest.

Altered Bone Remodeling

Ac.f is an indicator of overall level of remodeling activity in cancellous bone. Values for Ac.f correlate with excretion of bone resorption markers ($r = 0.71$, unpublished observations). In biopsy specimens from ostensibly healthy women, it is rare that label cannot be found in cancellous areas in a subject who had followed the fluorochrome labeling protocol. However, a recent paper from our laboratory, cited earlier, reported three cases of no label (i.e., zero Ac.f) among women with untreated postmenopausal osteoporosis.[19]

Abnormal Osteoid Morphology

The characteristic arrangement of osteoid (collagen) fibers in lamellar and woven bone is readily apparent. Woven bone in transilial specimens is generally associated with either Paget's disease or renal osteodystrophy. It can also occur in osteitis fibrosa. In osteogenesis imperfecta, collagen abnormalities may be subtle enough to escape detection.

Accumulation of Unmineralized Osteoid

Parfitt[14] has described the complex relationships between dynamic indices of bone formation and static indices of osteoid accumulation. Increases in OS/BS, O.Th, and Mlt indicate failure of osteoid to mineralize normally. If mineralization is arrested completely, no double label will be seen, and Mlt is unmeasurable.[26]

FINDINGS IN METABOLIC BONE DISEASE

In Table 1, key histomorphometric findings that characterize representative types of metabolic bone disease are identified. For further information, the reader should consult disease-specific chapters in this volume and the current literature.

Postmenopausal Osteoporosis

Osteoporosis in postmenopausal women is characterized by a cortical bone deficit with trabeculation of endocortical bone and a cancellous bone deficit with poor trabecular connectivity. Decreases in Tb.Th are modest, and dynamic measures vary widely.[27,28] Median Ac.f. remains high in specimens from women with postmenopausal osteoporosis, but values vary widely.[19]

Glucocorticoid-Induced Osteoporosis

Early in treatment, Ac.f is increased; later, Ac.f, MAR, and MS/BS are all decreased. In femoral specimens from patients with glucocorticoid-induced osteonecrosis, abundant apoptotic osteocytes and lining cells have been reported.[29] This has led to questions as to how osteocytes function to maintain bone mechanical integrity independent of bone mass and/or bone remodeling.

© 2008 American Society for Bone and Mineral Research

TABLE 1. PATTERNS OF KEY HISTOMORPHOMETRIC FINDINGS THAT CHARACTERIZE SEVERAL TYPES OF METABOLIC BONE DISEASE

	Marrow spaces	*Cortical bone*	*Cancellous bone*	*Bone remodeling*	*Osteoid morphology*	*Osteoid mineralization*
Postmenopausal osteoporosis	—	Cortical bone deficit with endocortical trabeculation	Cancellous bone deficit with poor trabecular connectivity	Ac.f generally increased, but values vary widely	—	—
Glucocorticoid-induced osteoporosis	—	Cortical bone deficit	Cancellous bone deficit	Early, increased Ac.f; later, decreased Ac.f	—	—
Primary hyperparathyroidism	Peritrabecular fibrosis may be seen	Cortical bone deficit, incr. Ct.Po, endocortical trabeculation	Typically unremarkable	Increased Ac.f	Woven bone may be seen	—
Hypogonadism (males and females)	—	Cortical bone deficit	Cancellous bone deficit, sometimes with poor trabecular connectivity	Increased Ac.f	—	—
Hypovitaminosis D osteopathy	Fibrous tissue may be seen	—	—	Early, increased Ac.f	—	Early, increased OS/BS; later, incr. MLT and O.Th. double label may be absent
Hypophosphatemic osteopathy	Fibrous tissue may be seen	—	—	—	—	Increased MLT and O.Th; double label may be absent
Renal osteodystrophy (high turnover type)	Fibrous tissue may be seen	Endocortical trabeculation	Osteoblast, osteocyte, and trabecular abnormalities	Markedly increased remodeling activity	Woven bone may be seen	Increased OS/BS
Renal osteodystrophy (low turnover type)	—	—	—	Markedly decreased remodeling activity	—	Increased OS/BS (osteomalacic type); decreased OS/BS (adynamic type)
Renal osteodystrophy (mixed type)	Fibrous tissue may be seen	—	Variable BV/TV	Patchy remodeling activity	Irregular, woven bone and osteoid may be seen	Increased OS/BS and O.Th

Primary Hyperparathyroidism

Primary hyperparathyroidism leads to a cortical bone deficit, with increased Ct.Po and trabeculation of endocortical bone.[30] Ct.Po correlates positively with fasting serum PTH.[31] BV/TV is generally preserved, and normal cancellous bone architecture is maintained.[32,33] Osteoid with a woven appearance and peritrabecular fibrosis is also seen.[34]

Hypogonadism

Hypogonadism in both women and men increases Ac.f and leads to deficits of both cortical bone and trabecular bone. At low levels of BV/TV and/or Tb.Th, loss of trabecular connectivity occurs.[35]

Hypovitaminosis D Osteopathy

Vitamin D depletion of any etiology leads to hypovitaminosis D osteopathy (HVO). Parfitt[26] describes three stages. In HVOi (pre-osteomalacia), Ac.f and OS/BS are increased, but O.Th is not. Accumulation of unmineralized osteoid characterizes both HVOii and HVOiii (osteomalacia), with Mlt and O.Th clearly increased (i.e., Mlt > 100 days and O.Th > 12.5 μm after correction for obliquity). Some double label can be seen in HVOii, but not in HVOiii. A cortical bone deficit also characterizes advanced HVO; secondary hyperparathyroidism in response to reduced serum ionized calcium is usual, and fibrous tissue in the marrow spaces is frequently seen. Life-long subclinical vitamin D insufficiency may contribute to the development of osteoporosis later in life in both men and women.

Low bone mass and bone disease with osteomalacic features occurs among patients treated with anti-epileptic drugs (AEDs).[36] Hepatic enzyme-inducing AEDs have been most clearly associated with these problems, but the newer AEDs cannot be exonerated at this time.[37]

Hypophosphatemic Osteopathy

Phosphate depletion of any etiology also leads to osteoma-

© 2008 American Society for Bone and Mineral Research

lacia, with histomorphometric findings similar to those of advanced HVO.[26] These cases involve defects in phosphorus metabolism manifest as defects in renal tubular reabsorption of phosphorus. However, most cases are not caused by a primary renal tubular abnormality, but instead, are caused by an abnormality in plasma phosphorus homeostasis.[38] Secondary hyperparathyroidism occurs variably. Transilial biopsy can be quite useful to assess the efficacy of treatment.

Gastrointestinal Bone Disease

Evidence of HVO has been reported in a variety of absorptive and digestive disorders.[39] However, these conditions also may promote deficiency of calcium and other nutrients. Malabsorption is not the only issue. For example, a calcium balance study of asymptomatic patients with celiac disease showed increased endogenous fecal calcium; the gut seemed to "weep" calcium into its lumen.[40] Bone histomorphometry may also reflect the results of treatment (i.e., corticosteroids or surgery). Parfitt[26] describes a histomorphometric profile of low bone turnover, often with evidence of HVO and secondary hyperparathyroidism, that represents the result of multiple insults to bone health in these patients.

Renal Osteodystrophy

At least three patterns of histomorphometric findings have been described among patients with end-stage renal disease (ESRD): high bone turnover with osteitis fibrosa (hyperparathyroid bone disease); low bone turnover (including osteomalacic and adynamic subtypes); and mixed osteodystrophy with high bone turnover, altered bone formation, and accumulation of unmineralized osteoid.[41–44]

At this time, transilial bone biopsy remains a useful "gold standard" on which to base decisions about treatment of bone disease in ESRD.[41] A dramatic example is the evaluation of bone pain and fractures in a chronic dialysis patient with hypercalcemia. If the biopsy shows high bone turnover and osteitis fibrosa, partial parathyroidectomy may be indicated. However, if the biopsy shows little turnover (little or no fluorochrome label), with or without extensive aluminum deposits, parathyroidectomy is contraindicated, and treatment with a chelating agent may be indicated. The same biopsy can also help determine the extent of vitamin D deprivation and indicate the adequacy of vitamin D treatment.

OBTAINING THE SPECIMEN

In this section, we outline the procedures for obtaining bone biopsy specimens, processing them, and carrying out histomorphometric analysis. For greater detail, we recommend another recent publication.[45]

Fluorochrome Labeling

In the clinical setting, tetracyclines are the only suitable fluorochrome labeling agents.[12] Demeclocycline (150 mg, four times daily) or tetracycline hydrochloride (250 mg, four times daily) are commonly used. A schedule of 3 days on, 14 days off, 3 days on, and 5–14 days off before biopsy (abbreviated as 3–14–3–5) produces good results, with a marker interval of 17 days.[13] Tetracyclines must be taken on an empty stomach; thus, oral intake must be avoided for at least 1 h before and after each dose.

Biopsy Procedure

Specimens require use of a trephine with an inner diameter of ≥7.5 mm. The teeth should be sharpened (and reconditioned, if necessary) after every two to three procedures. Transilial bone biopsy is performed in outpatient minor surgery, with the usual procedures (e.g., the surgeon scrubs and uses a cap, mask, gown, and gloves, and the site is prepared and draped) and precautions (e.g., pulse oximetry and blood pressure monitoring). Before the procedure, the patient should be off aspirin for at least 3 days and have had nothing orally for 4 h. If a second biopsy is done on another occasion, it should always be on the side opposite the first; there is thus a practical limit of two transilial biopsy specimens per patient. The gowned patient lies in the supine position on the surgical table, and midazolam (2.5–5 mg) is given through a forearm intravenous catheter.

The biopsy site is ~2 cm posterior to the anterior-superior spine, which is ~2 cm inferior to the iliac crest. The skin, subcutaneous tissues, and periosteum on both sides of the ilium are infiltrated with local anesthetic. The periosteum is accessed by a 2-cm skin incision and blunt dissection. The trephine is inserted and advanced with steady, gentle pressure and a deliberate pace. The specimen—an intact, unfractured core with both cortices and the intervening cancellous bone—is transferred into a 20-ml screw-cap vial containing 70% ethanol. (Note that certain special procedures, presently used in research settings, require unfixed specimens.)

The bony defect is packed with Surgicel. After local pressure to facilitate hemostasis, the wound is closed with three to five stitches and covered by a pressure dressing. Follow-up care is specified clearly (i.e., dressing in place and absolutely dry for 48 h; then a daily shower is allowed; no bathing or strenuous physical activity until suture removal, 1 week after the procedure). The procedure produces localized aching for ~2 days and a small scar at the site.

Patients typically describe feeling something "like a cramp" as the trephine advances, and the bone biopsy procedure described here rarely evokes more than mild discomfort. There is risk of bleeding in some situations (e.g., liver disease, hemodialysis, or medications that compromise hemostasis). Local bruising sometimes occurs, but hematoma is uncommon. In an early survey, physicians who were obtaining transilial biopsy specimens reported adverse events in 0.7% of 9131 biopsy specimens, that is, 22 with hematomas, 17 with pain for >7 days, 11 with transient neuropathy, 6 with wound infection, 2 with fracture, and 1 with osteomyelitis. No cases of death or permanent disability were reported.[46]

SPECIMEN PROCESSING AND ANALYSIS

Specimen Handling and Processing

For routine histomorphometry, the bone biopsy specimen should remain in 70% ethanol for at least 48 h for proper fixation. This solution is suitable for shipping and long-term storage at room temperature. The specimen vials should be filled to capacity with 70% ethanol for shipping, handling, and storage.

Steps in laboratory processing include dehydrating, defatting, embedding, sectioning, mounting, de-plasticizing, staining, and microscopic examination.

After proper trimming, the tissue block is sectioned parallel to the long axis of the biopsy core. Two or more sets of sections are obtained at an intervals of 400 μm, beginning 35–40% into the embedded specimen. Unstained sections 8–10 μm thick are used to examine osteoid morphology and to measure fluorochrome-labeled surfaces. Sections 5–7 μm thick stained with toluidine blue are used to measure wall thickness. Sections 5 μm thick with Goldner's stain[47] are used for other histomorphometric measurements.

© 2008 American Society for Bone and Mineral Research

Table 2. Examples of Clinical Situations in Which Bone Histomorphometry Can Provide Useful Information

1. When there is excessive skeletal fragility in unusual circumstances
2. When a mineralizing defect is suspected
3. To evaluate adherence to treatment in a malabsorption syndrome
4. To characterize the bone lesion in renal osteodystrophy
5. To diagnose and assess response to treatment in vitamin D–resistant osteomalacia and similar disorders
6. When a rare metabolic bone disease is suspected

Reprinted with kind permission from Springer Science and Business Media from Barger-Lux MJ, Recker RR 2005 Towards understanding bone quality: Transilial bone biopsy and bone histomorphometry. Clin Rev Bone Miner Metab **4**:167–176. Fig. 1, ©2007.[49]

Microscopy

The histomorphometric variables are derived from data gathered at the microscope. These data include the width of both cortices and—in defined sectors of cancellous bone—volumes of bone, osteoid, and marrow; total trabecular perimeter; perimeters with features of formation (Fig. 1) or resorption (Fig. 2); thickness of osteoid and osteon walls; and interlabel width. Methods have been described for unbiased sampling of microscopic features.[48]

Our histomorphometry laboratory uses an interactive image analysis system. A digital camera mounted on the microscope presents the microscopic images on-screen, and measurements are made using a mouse. Fluorescent light at a wavelength of 350 nm is used to examine fluorochrome labels (Fig. 3).

INDICATIONS FOR BONE BIOPSY AND HISTOMORPHOMETRY

The purpose of bone histomorphometry in the clinical setting is to gather information (i.e., to establish a diagnosis, clarify a prognosis, or evaluate adherence or response to treatment) on which to base informed clinical decisions. As is the case for every invasive procedure, the risk, discomfort, and expense should be proportionate to the importance of the information to be gained. Given these caveats, the number of clinical indications for this procedure is limited.

Clinicians can manage most metabolic bone diseases, including osteoporosis, without the aid of a bone biopsy. However, there are some situations in which bone biopsy after fluorochrome labeling is appropriate, as outlined in Table 2.

Bone histomorphometry has been, and remains, crucial for assessing the mechanism of action, safety, and efficacy of new bone-active agents. Preclinical animal work includes serial biopsy specimens at multiple skeletal sites, using different colored fluorochrome labels (e.g., calcein or xylenol orange). Testing of every new bone-active treatment should include bone biopsy in at least a subset of human subjects. Trabecular bone histomorphometry provides a method for examining both bone properties and bone physiology. Cortical bone histomorphometry is seldom used because it requires obtaining a rib biopsy, a procedure more risky, expensive, and painful than transilial biopsy. Because of the random orientation of the transilial specimen with regard to the long axis of haversian systems, little information on cortical bone remodeling can be obtained from transilial specimens.

ACKNOWLEDGMENTS

The authors thank Susan Bare and Toni Howard for assistance in describing technical methods and preparing digital photomicrographs.

REFERENCES

1. Frost HM 1986 Intermediary Organization of the Skeleton. CRC Press, Boca Raton, FL, USA.
2. Frost HM 1973 Bone Remodeling and Its Relationship to Metabolic Bone Diseases. Charles C. Thomas, Springfield, IL, USA.
3. Marotti G, Favia A, Zallone AZ 1972 Quantitative analysis on the rate of secondary bone mineralization. Calcif Tiss Res **10**:67–81.
4. Recker RR, Kimmel DB, Parfitt AM, Davies KM, Keshawarz N, Hinders S 1988 Static and tetracycline-based bone histomorphometric data from 34 normal postmenopausal females. J Bone Miner Res **3**:133–144.
5. Malluche HH, Faugere MC 1986 Atlas of Mineralized Bone Histology. Karger, New York, NY, USA.
6. Parfitt AM, Drezner MK, Glorieux FH, Kanis JA, Malluche H, Meunier PJ, Ott SM, Recker RR 1987 Bone histomorphometry: Standardization of nomenclature, symbols, and units. J Bone Miner Res **2**:595–610.
7. Parfitt AM 1983 The physiologic and clinical significance of bone histomorphometric data. In: Recker RR (ed.) Bone Histomorphometry: Techniques and Interpretation. CRC Press, Boca Raton, FL, USA, pp. 143–224.
8. Garrahan NJ, Mellish RWE, Compston JE 1986 A new method for the two-dimensional analysis of bone structure in human iliac crest biopsies. J Microsc **142**:341–349.
9. Vesterby A, Gundersen HJG, Melsen F 1989 Star volume of marrow space and trabeculae of the first lumbar vertebra: Sampling efficiency and biological variation. Bone **10**:7–13.
10. Vesterby A, Gundersen HJG, Melsen F, Mosekilde L 1991 Marrow space star volume in the iliac crest decreases in osteoporotic patients after continuous treatment with fluoride, calcium, and vitamin D2 for five years. Bone **12**:33–37.
11. Hahn M, Vogel M, Pompesius-Kempa M, Delling G 1992 Trabecular bone pattern factor: A new parameter for simple quantification of bone microarchitecture. Bone **13**:327–330.
12. Frost HM 1969 Measurement of human bone formation by means of tetracycline labelling. Can J Biochem Physiol **41**:331–342.
13. Schwartz MP, Recker RR 1982 The label escape error: Determination of the active bone-forming surface in histologic sections of bone measured by tetracycline double labels. Metab Bone Dis Relat Res **4**:237–241.
14. Parfitt AM 2002 Physiologic and pathogenetic significance of bone histomorphometric data. In: Coe FL, Favus M (eds.) Disorders of Bone and Mineral Metabolism, 2nd ed. Lippincott Williams & Wilkins, Philadelphia, PA, USA, pp. 469–485.
15. Chapurlat RD, Arlot M, Burt-Pichat B, Chavassieux P, Roux JP, Portero-Muzy N, Delmas PD 2007 Microcrack frequency and bone remodeling in postmenopausal osteoporotic women on long-term bisphosphonates: A bone biopsy study. J Bone Miner Res **22**:1502–1509.
16. Jilka RL, Weinstein RS, Parfitt AM, Manolagas SC 2007 Perspective: Quantifying osteoblast and osteocyte apoptosis: Challenges and rewards. J Bone Miner Res **22**:1492–1505.
17. Weinstein RS, Bell NH 1988 Diminished rates of bone formation in normal black adults. N Engl J Med **319**:1698–1701.
18. Glorieux FH, Travers R, Taylor A, Bowen JR, Rauch F, Norman M, Parfitt AM 2000 Normative data for iliac bone histomorphometry in growing children. Bone **26**:103–109.
19. Recker R, Lappe J, Davies KM, Heaney R 2004 Bone remodeling increases substantially in the years after menopause and remains increased in older osteoporosis patients. J Bone Miner Res **19**:1628–1633.
20. Parfitt AM, Travers R, Rauch F, Glorieux FH 2000 Structural and cellular changes during bone growth in healthy children. Bone **27**:487–494.
21. Cosman F, Morgan D, Nieves J, Shen V, Luckey M, Dempster D, Lindsay R, Parisien M. 1997 Resistance to bone resorbing effects of PTH in black women.J Bone Miner Res **12**:958–966.
22. Han Z-H, Palnitkar S, Rao DS, Nelson D, Parfitt AM 1997 Effects of ethnicity and age or menopause on the remodeling and turnover of iliac bone: Implications for mechanisms of bone loss. J Bone Miner Res **12**:498–508.
23. Dahl E, Nordal KP, Halse J, Attramadal A 1988 Histomorphometric analysis of normal bone from the iliac crest of Norwegian subjects. Bone Miner **3**:369–377.

© 2008 American Society for Bone and Mineral Research

24. Cosman R, Schnitzer MB, McCann PD, Parisien MV, Dempster DW, Lindsay R 1992 Relationships betwen quantitative histological measurements and noninvasive assessments of bone mass. Bone **13**:237–242.
25. Keshawarz NM, Recker RR 1984 Expansion of the medullary cavity at the expense of cortex in postmenopausal osteoporosis. Metab Bone Dis Relat Res **5**:223–228.
26. Parfitt AM 1998 Osteomalacia and related disorders. In: Avioli LV, Krane SM (eds.) Metabolic Bone Disease and Clinically Related Disorders. Academic Press, Boston, MA, USA, pp. 327–386.
27. Kimmel DB, Recker RR, Gallagher JC, Vaswani AS, Aloia JF 1990 A comparison of iliac bone histomorphometric data in postmenopausal osteoporotic and normal subjects. Bone Miner **11**:217–235.
28. Recker RR, Barger-Lux MJ 2001 Bone remodeling findings in osteoporosis. In: Marcus R, Feldman D, Kelsey J (eds.) Osteoporosis, 2nd ed. Academic Press, San Diego, CA, USA, pp. 59–70.
29. Weinstein RS, Nicholas RW, Manolagas SC 2000 Apoptosis of osteocytes in glucocorticoid-induced osteonecrosis of the hip. J Clin Endocrinol Metab **85**:2907–2912.
30. Ericksen E 2002 Primary hyperparathyroidism: Lessons from bone histomorhometry. J Bone Miner Res **17**:S2;N95–N97.
31. van Doorn L, Lips P, Netelenbos JC, Hackeng WHL 1993 Bone histomorphometry and serum concentrations of intact parathyroid hormone (PTH(1-84)) in patients with primary hyperparathyroidism. Bone Miner **23**:233–242.
32. Parisien M, Mellish RWE, Silverberg SJSE, Lindsay R, Bilezikian JP 1992 Maintenance of cancellous bone connectivity in primary hyperparathyroidism: Trabecular strut analysis. J Bone Miner Res **7**:913–919.
33. Uchiyama T, Tanizawa T, Ito A, Endo N, Takahashi HE 1999 Microstructure of the trabecula and cortex of iliac bone in primary hyperparathyroidism patients determined using histomorphometry and node-strut analysis. J Bone Miner Res **17**:283–288.
34. Monier-Faugere M-C, Langub MC, Malluche HH 1998 Bone biopsies: A modern approach. In: Avioli LV, Krane SM (eds.) Metabolic Bone Disease and Clinically Related Disorders, 3rd ed. Academic Press, San Diego, CA, USA, pp. 237–273.
35. Audran M, Chappard D, Legrand E, Libouban H, Basle MF 2001 Bone microarchitecture and bone fragility in men: DXA and histomorphometry in humans and in the orchidectomized rat model. Calcif Tissue Int **69**:214–217.
36. Pack AM, Morrell MJ 2004 Epilepsy and bone health in adults. Epilepsy Behav **5**:S24–S29.
37. Fitzpatrick LA 2004 Pathophysiology of bone loss in patients receiving anticonvulsant therapy. Epilepsy Behav **5**:S3–S15.
38. Antoniucci DM, Yamashita T, Portale AA 2006 Dietary phosphorus regulates serum fibroblast growth factor-23 concentrations in healthy men. J Clin Endocrinol Metab **91**:3144–3149.
39. Arnala I, Kemppainen T, Kroger H, Janatuinen E, Alhava EM 2001 Bone histomorphometry in celiac disease. Ann Chir Gynaecol **90**:100–104.
40. Ott SM, Tucci JR, Heaney RP, Marx SJ 1997 Hypocalciuria and abnormalities in mineral and skeletal homeostasis in patients with celiac sprue without intestinal symptoms. J Clin Endocrinol Metab **4**:206.
41. Pecovnik BB, Bren A 2000 Bone histomorphometry is still the golden standard for diagnosing renal osteodystrophy. Clin Nephrol **54**:463–469.
42. Parker CR, Blackwell PJ, Freemont AJ, Hosking DJ 2002 Biochemical measurements in the prediction of histologic subtype of renal transplant bone disease in women. Am J Kidney Dis **40**:396.
43. Elder G 2002 Pathophysiology and recent advances in the management of renal osteodystrophy. J Bone Miner Res **17**:2094–2105.
44. Malluche HH, Langub MC, Monier-Faugere MC 1997 Pathogenisis and histology of renal osteodystrophy. J Bone Miner Res **7**:S184–S187.
45. Recker RR, Barger-Lux MJ 2001 Transilial bone biopsy. In: Bilezikian JP, Raisz L, Rodan GA, eds. Principles of Bone Biology, 2nd ed. Academic Press, San Diego, CA, USA, pp. 1625–1634.
46. Rao DS, Matkovic V, Duncan H 1980 Transiliac bone biopsy: Complications and diagnostic value. Henry Ford Hosp Med J **28**:112–118.
47. Goldner J 1938 A modification of the Masson trichrome technique for routine laboratory purposes. Am J Pathol **14**:237–243.
48. Kimmel DB, Jee WSW 1983 Measurements of area, perimeter, and distance: Details of data collection in bone histomorphometry. In: Recker RR (ed.) Bone Histomorphometry: Techniques and Interpretation. CRC Press, Boca Raton, FL, USA, pp. 80–108.
49. Barger-Lux MJ, Recker RR 2005 Towards understanding bone quality: Transilial bone biopsy and bone histomorphometry. Clin Rev Bone Miner Metab **4**:167–176.

Chapter 36. Vertebral Fracture Assessment

Jackie A. Clowes[1] and Richard Eastell[2]

[1]Endocrine Research Unit and Division of Rheumatology, Mayo Clinic, Rochester, Minnesota; [2]University of Sheffield, Sheffield, United Kingdom

INTRODUCTION

The accurate detection of vertebral fracture is essential for risk assessment of individual patients in clinical practice, for determining the prevalence and incidence of osteoporosis in the population, and for evaluating drug efficacy in clinical trials.

Identifying vertebral fractures is particularly important in risk assessment because a prevalent vertebral fracture is the strongest predictor of subsequent vertebral fracture and the strongest predictor of any subsequent osteoporotic fracture.(1) Thus, a vertebral fracture results in a 4.4-fold increase risk of future vertebral fracture in those with a prevalent fracture.(2) Furthermore, the presence of one vertebral fracture predicts both the location and severity of subsequent fractures.(3,4) Subjects with multiple vertebral fractures are more likely to suffer from greater morbidity including reduced lung function, slower gait, and chronic back pain; decreased quality of life including loss of self esteem, loss of independence, and social isolation; and disability.(5,6) There is also an increased risk of death.(7) As a consequence, the National Osteoporosis Foundation recommends patients with vertebral fracture receive drug therapy irrespective of BMD T-scores.(8) Therefore, establishing a history of prior osteoporotic fracture is essential in fracture risk assessment.

Several approaches have been developed and refined to diagnose vertebral fractures. These include morphometric analysis using radiographic standards and, more recently, the documentation, standardization, and quantification of the visual approach used by an expert reader to establish a diagnosis of

The authors state that they have no conflicts of interest.

Key words: vertebral fracture, morphometric analysis, semiquantitative morphometric analysis, vertebral fracture assessment, visual read, algorithm-based qualitative evaluation, definition, non-osteoporotic fracture deformity, normal vertebrae, fracture risk

© 2008 American Society for Bone and Mineral Research

vertebral fractures. Given the increasing availability of newer technologies, it is paramount that technologists and clinicians interpreting these results are aware of the problems and pitfalls in the accurate diagnosis of a vertebral fracture.

Identifying whether a vertebral fracture is present is, however, difficult because the shape of normal vertebrae varies widely from one vertebra to another and between individuals. In addition, errors in projection can be misleading and vertebrae may be abnormal because of non-osteoporotic fracture deformities.

Another difficulty in identifying vertebral fractures is that up to 50% of vertebral fractures are asymptomatic and therefore do not come to the attention of physicians. Many of these asymptomatic vertebral fractures are identified as an incidental finding during other investigations. Even when chest radiographs or vertebral images are obtained, only ~35–50% of all radiographic vertebral fractures are correctly reported.[9,10] Only 19% reached clinical attention and result in the appropriate initiation of osteoporotic therapy.[10] Clear and accurate reporting of vertebral fractures is certainly essential and educational tools are available (www.iofbonehealth.org). In addition, routine screening for vertebral fractures using lumbar and thoracic radiographs is not recommended, particularly in low-risk individuals because of the high radiation dose. More recently, however, the availability of vertebral imaging using DXA means images of near radiographic quality are available at a fraction of the radiation dose.

IMAGING TECHNOLOGIES

Conventional Radiographs

The gold standard in vertebral fracture ascertainment remains evaluation by the expert reader. Traditionally vertebral fractures have been diagnosed using conventional radiographs imaging of the lumbar and thoracic vertebrae. There are, however, significant limitations to this approach, including the fact that the quality of the image obtained has tremendous impact on the ability to detect fractures. It is difficult to acquire images in the presence of scoliosis because the vertebral bodies are projected obliquely making evaluation of the vertebral endplate problematic. In addition, high levels of radiation are required, and even an optimum image has limited resolution and ability to visualize the vertebral endplate. However, the advantage of conventional radiographs is the low cost, wide availability, and convenience.

A standardized protocol should be established to visualize the C_7–S_1 vertebrae with the minimum of parallax caused by off-axis distortion. The anterior-posterior (AP) image is of value to visualize the pedicles, determine the site of osteoporotic fractures, and exclude congenital anomalies (e.g., Cupid's bow). If a patient's management will be altered by follow-up imaging, a lateral image will usually provide sufficient information. If a vertebral fracture is suspected, conventional radiographs are indicated. More sophisticated examinations using MRI or CT are usually only required in the presence of localized pain, focal neurological signs, or symptoms suggesting cord compression or a radiculopathy, or the clinical suspicion of primary or metastatic lesions.[11] Osteoporotic fractures are rare above T_4, and in this setting, it is very important to consider metastasis and, if appropriate, investigate for a primary lesion.

DXA Images

The advantage of using DXA to image vertebrae is the low cost, wide availability (at least in the United States), convenience, and low radiation dose. The effective radiation dose for DXA is ~3 μSv compared with a lateral lumbar spine radiograph dose of ~600 μSv. By comparison, the typical background radiation at sea level is 7 μSv/d. Imaging vertebral fractures using DXA is termed vertebral fracture assessment (VFA). The disadvantage of VFA is the poor image resolution compared with conventional radiographs, CT, or MRI and the increased difficulty in imaging the thoracic spine, especially above T_7. Between 5% and 15% of thoracic vertebrae can be visualized by conventional radiographs and not by DXA imaging.

The sensitivity and specificity of this approach compared with conventional radiographs varies with the approach used to define a vertebral fracture (morphometric, semiquantitative [SQ], or visual identification). Irrespective of the approach used, sensitivity is generally only moderate for mild fractures (Genant grade 1), with results as low as 54%. This is at least in part because of the lower image resolution of this technique. The sensitivity for identifying moderate to severe vertebral fractures (Genant grades 2 and 3) is substantial higher (90–94%). Specificity is high, with results between 94% and 99% compared with conventional radiographs. Morphometric and visual evaluation using DXA images in children may be especially problematic because currently available software cannot detect the vertebrae in most children.[12]

One advantage of DXA imaging is that the scans are not subject to the same degree of projection distortion as conventional radiographs because the X-ray beam is always orthogonal to the spine. This reduces the degree of parallax observed. However, this approach, just like conventional radiographs, still has difficulty imaging the spine in the presence of scoliosis because the vertebral bodies are projected obliquely making evaluation of the vertebral endplate difficult. Imaging requires a fan-beam scanner with the appropriate software. The dual-energy mode reduces the frequency that soft tissues obscure the endplates compared with the single-energy mode, although this is associated with a higher radiation dose. Side-by-side viewing facilitates identification of incident vertebral fractures. Because of the low cost, low radiation dose, and convenience, this approach is beginning to play an increasing role in screening for vertebral fractures in clinical practice.[13] Irrespective of the approach used to define a vertebral fracture, it is recommended to confirm the diagnosis using conventional radiographs.[14]

APPROACHES TO IDENTIFICATION OF VERTEBRAL FRACTURES

Morphometric Analysis

Morphometric analysis uses measurements of vertebral height to define vertebral fractures. A normative database is established against which the vertebrae are compared. There are a number of different morphometric approaches that vary in the criteria by which they define a vertebral fracture and in the reference data used.[15–17] The most widely used approaches to identify prevalent and incident vertebral fractures are the two different algorithms proposed by McCloskey et al.[16] and Eastell et al.[17]

Morphometric analysis has a high sensitivity and moderate to high specificity to discriminate between normal vertebrae and fractured vertebrae. Furthermore, all the morphometric approaches for defining prevalent and incident vertebral fractures are correlated with clinical risk factors for vertebral fracture. However, the approach used can have a significant impact on the prevalence of vertebral fractures identified, varying from 3% to 90%. A loss in vertebral height of 20–25% is usually used to define an incident vertebral fracture. If this definition is used, it results in comparable abilities to predict

© 2008 American Society for Bone and Mineral Research

incident vertebral fractures irrespective of the approach used to define a baseline fracture.[18]

In general, good to excellent agreement (per vertebra, κ = 0.87–0.93; per subject, κ = 0.81–0.91) has been observed between quantitative morphometric approaches to identify vertebral deformities using either DXA or conventional radiographs images.[19] However, there is only moderate agreement (per vertebra, κ = 0.70–0.79; per subject, κ = 0.67–0.75) observed when comparing the same algorithm between DXA and radiographs images.[19] The agreement between qualitative radiological assessment and quantitative morphometric evaluation of conventional radiographs or DXA images was higher in osteoporotic populations with a κ score of 0.86 and 0.71, respectively, compared with a κ score of 0.59 and 0.47 in the reference population.[20]

Deformity identification by DXA has significant limitations compared with conventional radiographs, because 5–15% of vertebrae primarily in the upper thoracic spine cannot be visualized or have very poor image quality. In addition, DXA images are less effective at identifying mild grade 1 deformities with a sensitivity as low as 22%. DXA images (or VFA) are more effective at identifying moderate to severe deformities with a sensitivity of 81.6% for grade 2 deformities.[21] Finally, the precision error of both DXA images and conventional radiographs images are small compared with the 20–25% reduction in vertebral height most commonly used to define incident vertebral fracture. The precision error is less for conventional images compared with DXA images.[21]

Probably the most important limitation of using quantitative morphometric analysis is the failure to differentiate between variation in vertebral shape or size caused by nonvertebral fracture deformities compared with vertebral fractures (discussed below). As a consequence, quantitative morphometric analysis results in a high false-positive rate. This has significant implications in determining the prevalence of vertebral fractures, interpreting clinical trials, and determining the most appropriate therapy for patients.

SQ Analysis

SQ analysis combines measurements of vertebral height with subsequent evaluation of all vertebrae with a short vertebral height by an expert reader. This combined approach enables identification of non-osteoporotic fracture vertebral deformities, which are not identified using morphometric analysis alone. As a consequence, SQ analysis should reduce the number of false-positive results. The most widely used SQ approach is that of Genant et al.[22] The SQ approach is preferable to morphometric analysis alone. Baseline or prevalent vertebral fractures are graded from 0 (normal) to 3 (severe), and incident fractures are defined as an increase of >1 or =1 grade on follow-up radiographs. Genant grade 1 corresponds to a 20–25% reduction in anterior, middle, and/or posterior height, grade 2 corresponds to a 25–40% reduction in any height, and grade 3 corresponds to >40% reduction in any vertebral height.

An SQ approach has been applied to both conventional radiographs and to DXA images. Mild grade 1 SQ vertebral deformities are frequently not associated with low BMD. The interobserver agreement for conventional radiograph or DXA images is similar with a κ score of 0.53 (95% CI, 0.46, 0.60) and 0.51 (95% CI, 0.44, 0.58), respectively. In contrast, a visual read using the ABQ (algorithm-based qualitative approach; see following section) method was associated with low BMD and an interobserver agreement for radiography and DXA images of 0.74 (95% CI, 0.60, 0.87) and 0.65 (95% CI, 0.48, 0.81), respectively. Although the prevalence of radiographic vertebral fractures identified by ABQ and SQ was similar, agreement between ABQ and SQ was only moderate. This approach is currently recommended by the ISCD for diagnosing vertebral fractures with VFA.

Visual Identification by an Expert Reviewer

Vertebral fracture diagnosis by an expert reader is generally considered the gold standard. Despite this, there can be considerable disagreement between expert reviewers and inconsistencies by any one reviewer. One controversy, which applies to all methodologies, is whether mild deformities or short vertebral heights are true fractures, normal variation, or merely a consequence of vertebral remodeling with aging. In some studies, mild deformities were identified in women with moderate or severe fractures and therefore may represent a risk factor for vertebral fractures or be a consequence of osteoporosis.[23] However, the majority of evidence suggests that short vertebral height (SVH) or non-osteoporotic fracture vertebral deformities are not associated with low BMD or prevalent vertebral fractures.[24] This suggests that SVH is caused by either normal variation or a consequence of vertebral remodeling with aging. In contrast, moderate and severe vertebral deformities are strongly associated with low BMD and subsequent risk of all fractures and vertebral fractures and have the strongest clinical correlates with increased back pain and disability.

Visual identification seems to be superior to morphometric and SQ approaches at identifying SVHs or non-osteoporotic fracture vertebral deformities.[24,25] However, even among expert readers, there is disagreement. One systematic approach is to define vertebral fractures based on endplate cracks or breaks as the primary event with the subsequent evaluation of vertebral height.[14] The algorithm-based qualitative (ABQ) approach differs from quantitative morphometric evaluations that focus solely on variation in vertebral height.[15–17] It also differs from SQ morphometric approaches in which the initial step is identifying variation in vertebral height, and these short vertebrae are subsequently evaluated for changes at the endplate and cortex.[22] Initial studies suggest mild vertebral fractures identified using the ABQ method have a stronger association with osteoporosis than mild vertebral fractures identified using SQ approaches.[24,25] Furthermore, because ABQ readers focus on the endplate rather then short vertebral height, there is less disagreement between ABQ readers diagnosing mild fractures.

Visual evaluation comparing conventional radiographs and DXA images suggests a good level of agreement κ = 0.79, sensitivity (91.9%), and negative predictive value (98.0%) for DXA images.[26] It also has excellent negative predictive value and therefore is capable of distinguishing subjects with very low risk of vertebral deformities from those with possible vertebral fractures. However, the use of this technique is limited because of the fact that studies establishing whether the ABQ approach can predict incident fracture in longitudinal studies have not yet been performed. In addition, to date, there are no data on whether this approach can detect a reduction of incident fractures with treatment.

IDENTIFICATION OF VERTEBRAL FRACTURE

Defining a Normal Vertebrae

One difficulty arises because of the wide range of within- and between-individual variation in vertebral size and shape that represents normal variants (Table 1). This is a problem, which applies to the interpretation of vertebral fractures and nonvertebral deformities by all imaging modalities. A normal vertebra is one that does not meet the criteria for a vertebral

© 2008 American Society for Bone and Mineral Research

Table 1. Normal Vertebra and Normal Vertebral Variants

Diagnosis	*X-ray image*	*Description*
Normal vertebrae	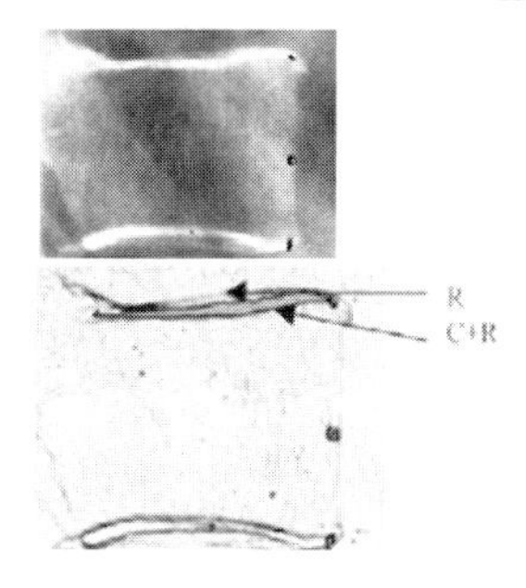	Appearance of vertebral endplates in a normal vertebra. R represents the vertebral ring line. C + R represent the central endplate with in the vertebral ring overlapping the vertebral ring line.
Normal variant—short anterior and middle vertebral height	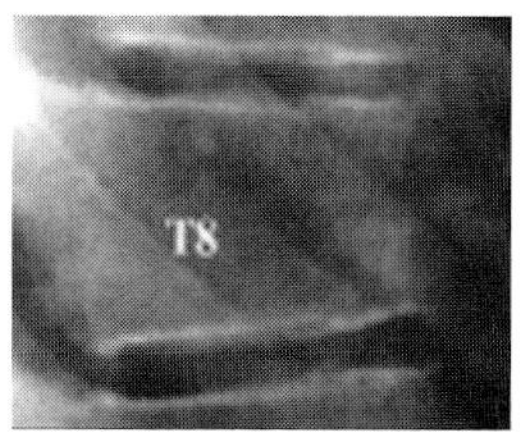	Short anterior vertebral height in T_8. No endplate fracture. The vertebral ring lines in both the superior and inferior endplates are straight and are converging anteriorly.
Normal variant—short anterior vertebral height with degenerative changes anteriorly	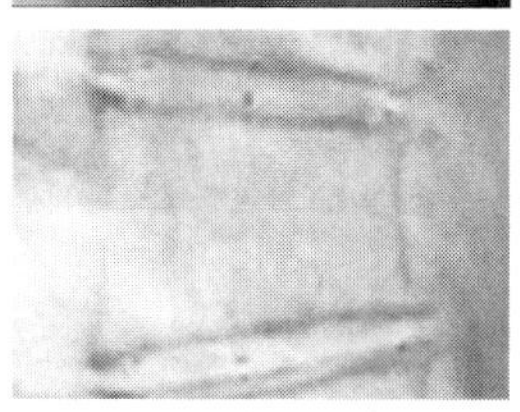	Short anterior vertebral height in T_8. No endplate fracture. The vertebral ring lines in both the superior and inferior endplates are straight and are converging anteriorly. Degenerative changes anteriorly.
Normal variant—short posterior vertebral height	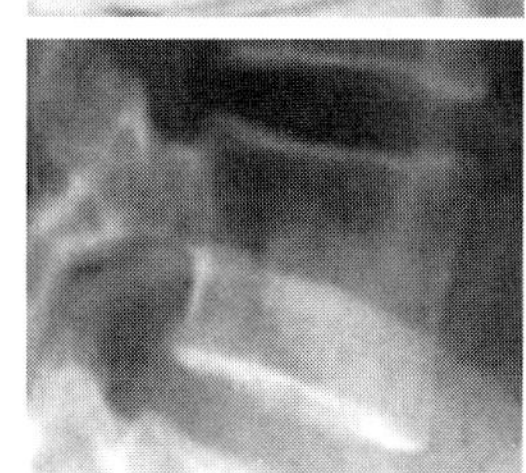	Short posterior height at vertebrae L_4.

fracture or non-osteoporotic fracture deformities. Thoracic vertebrae are generally smaller and may exhibit a small degree of wedging without breaks or buckling of the superior or inferior endplate or the strong outer vertebral ring consisting of the anterior, posterior and lateral cortex of the vertebrae. The height of the vertebrae is frequently similar to the depth (AP) and width (lateral) of the vertebrae. Lumbar vertebrae are larger then thoracic vertebrae, do not typically exhibit wedging in normal vertebrae, and have a greater width and depth compared with their height.

On a true lateral projection, there are two lines seen at the superior and inferior surface of the normal vertebrae (Table 1). One line represents the vertebral ring on one side and the second denser line represents the superimposing of both the central endplate and the opposite vertebral ring.(14) Considerable care needs to be taken if there is an oblique projection of a normal vertebra because it is easy to overlook (or misdiagnose) a vertebral fracture. Oblique projections arise because of the parallax effect of diverging X-rays or poor patient positioning. Developmental abnormalities can result in a short anterior, middle, or posterior vertebral height in either isolated vertebrae or several adjacent vertebrae. There are, however, no associated endplate breaks or cortical breaks (Table 1).

Defining a Vertebral Fracture

It is widely agreed that a vertebral fracture involves altered vertebral shape with loss of parallelism of the vertebral endplates and usually decreased vertebral height in the anterior, mid-, or posterior vertebral height. However, one characteristic feature, which is increasingly recognized as prerequisite to the diagnosis of vertebral fractures, is the presence of breaks in the cortex of the vertebrae. These breaks always occur in the center of either the superior or inferior endplates that are the weakest area of the endplate because it is furthest from the strong outer vertebral ring. As a consequence, the endplate buckles or collapses under pressure because of the intervertebral disc and results in a concave appearance to the superior and/or inferior endplate (Table 2). If a concavity extend beyond the inner border of the vertebral ring, it is unlikely to represent an osteoporotic fracture. More severe fractures result in additional buckling of the anterior, lateral, and, very rarely, the posterior cortical bone, resulting in loss of vertebral height.(14) These changes result in the morphological appearances of wedge and crush fractures, respectively (Table 2).

A vertebral fracture initially involves a crack in the superior or inferior endplate with or without the simultaneous loss of vertebral height. As the severity of the fracture progresses, the

© 2008 American Society for Bone and Mineral Research

Table 2. Vertebral Fractures

Diagnosis	*X-ray image*	*Classical description*
Concave superior endplate fracture	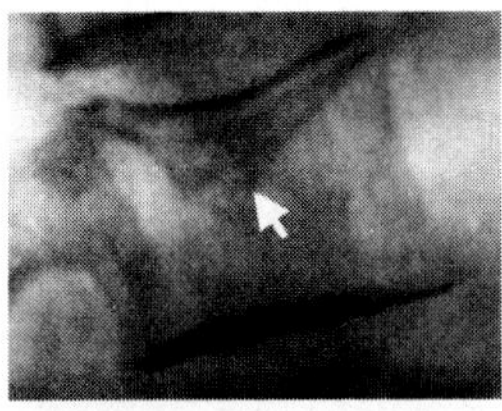	There is a concave line caused by the depression of the central endplate within the vertebral ring (arrow heads) in the inferior endplate of vertebrae L_3. The two vertebral ring lines remain intact (lateral and right).
Inferior endplate fracture with early wedge deformity (fracture)	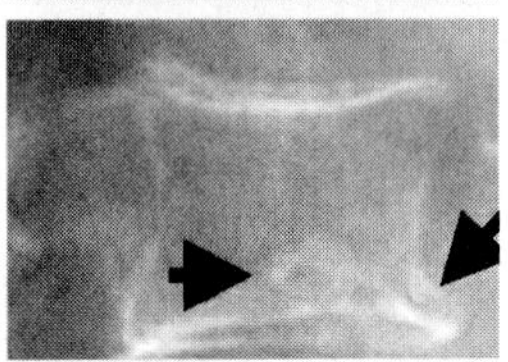	There is a concave line caused by the depression of the central endplate within the vertebral ring (arrow heads) in the inferior endplate of vertebrae L_3. The two vertebral ring lines remain intact (lateral and right). The anterior inferior cortex of L_3 is buckled (arrow) because of a fracture above the inferior vertebral ring.
Wedge deformity (fracture) and endplate fracture	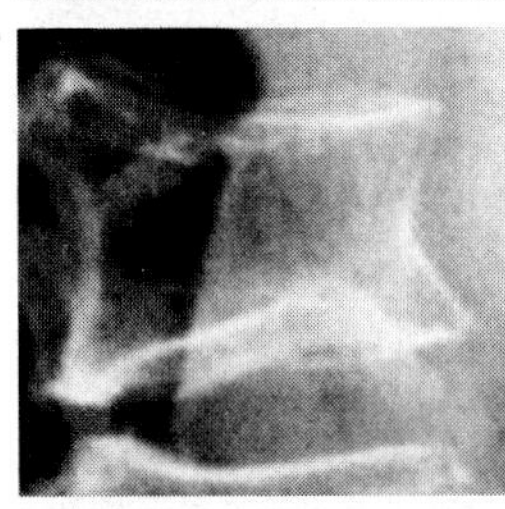	Osteoporotic wedge fracture of inferior endplate. Depression in the central endplate within the vertebral ring. Anterior vertebral ring is displaced. Fracture of the anterior cortex of the vertebral body.
Compression fracture and endplate fracture	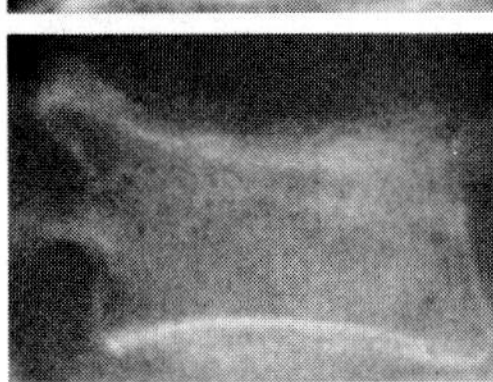	Osteoporotic compression fracture of superior endplate. Depression in the central endplate within the vertebral ring. Anterior and posterior vertebral ring is displaced. Fracture of the anterior and posterior cortex of the vertebral body.

vertebral ring fractures resulting in loss of height and buckling of the anterior, lateral, and occasionally posterior cortex. Highlighting these features is important because there is considerable variation in vertebral shape resulting in osteoporotic and non-osteoporotic deformities that can result in considerable intraobserver error even among expert readers.

Finally, it is important to identify the different characteristics of high trauma burst fractures. There is usually a history of a high trauma injury (e.g., car accident, fall from a significant height) immediately resulting in acute, severe, localized back pain with localized tenderness. Frequently, there is a "burst" appearance with the vertebral body extending beyond the plane of the adjacent vertebrae in the anterior, lateral, and posterior directions with associated loss in vertebral height.

Identifying Non-Osteoporotic Fracture Deformities

Lesions that do not meet the criteria for normal vertebrae or vertebral fractures are termed non-osteoporotic fracture deformities. These include variation in the shape or size of the vertebrae or the vertebral endplate caused by degenerative changes, Scheuermann's disease, Schmorls nodes, congenital deformities, and metastatic lesions. It is important to reinforce that SVH is not always an osteoporotic fracture.[24,25] SVH may be developmental, resulting in a short anterior height. Developmental abnormalities may also have deep step-like endplates without angulations, cracks, or breaks in either the endplate or cortex (Table 3). In addition, the aging skeleton, particularly in women, may develop slight wedging because of remodeling; however, there is no depression or break in the endplate or cortex.

Scheuermann's disease is frequently associated with a short anterior vertebral height in combination with irregularity of the whole of the superior and/or inferior endplates. Scheuermann's disease may appear in isolated vertebrae or several adjacent vertebrae (Table 3). There should be no associated endplate or cortical breaks or fractures. Scheuermann's disease is most commonly observed in the thoracic spine and may be associated with degenerative changes of the anterior cortex. Adolescent Scheuermann's disease is associated with elongated vertebral bodies and, in adulthood, these vertebrae frequently develop degenerative changes.

Abnormalities of the vertebral endplates may also cause difficulties in the differential diagnosis of vertebral fractures. Schmorl's nodes consist of a rounded flask-like break in the superior or inferior endplate in either the AP or lateral view, which very rarely extend beyond 25% of the endplate (Table 3). They are found in 38–75% of the population and are formed by extrusion or herniation of the nuclear material from the intervertebral disk into the vertebral body. It is widely assumed that they are asymptomatic and have no clinical consequence,[27] although more recent studies suggest that at least some lesions may be associated with sudden onset, localized, nonradiating back pain and tenderness.[28] The lesions typically have a well-demarcated cortex with intact cortical margins. Edema may be present in acute lesions. It has been postulated that Schmorl's nodes may progress into, or predispose to, osteoporotic vertebral fractures; however, to date, there is no evidence to support this suggestion (Table 3).

In osteomalacia, there is a uniform or symmetrical concavity of the superior and inferior endplates throughout all the ver-

© 2008 American Society for Bone and Mineral Research

TABLE 3. NON-OSTEOPOROTIC FRACTURE DEFORMITIES

Diagnosis	*X-ray image*	*Classical description*
Scheurmann's disease	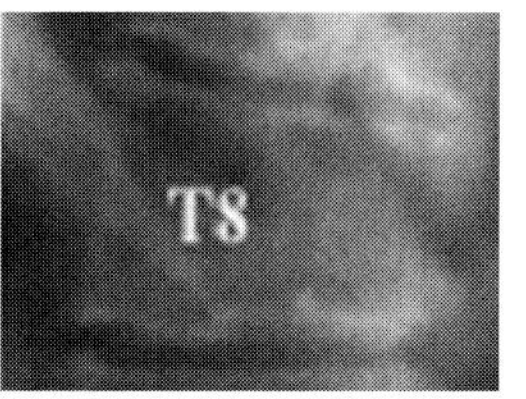	Endplate and vertebral rings are irregular indicating Scheuermann's disease. No fracture of endplate within the vertebral ring. Short vertebral height of T_8. Scheuermanns disease is often seen in association with degenerative changes.
Schmorl's node—degenerative disc disease.	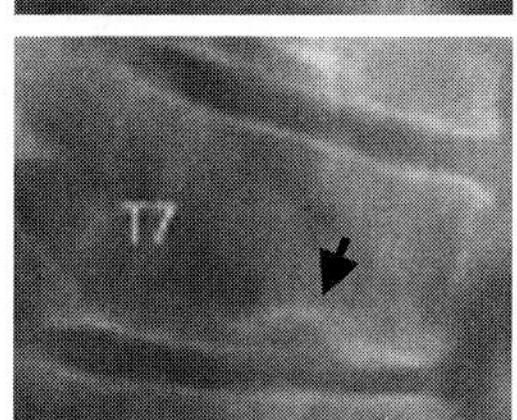	Inferior Schmorl's node (arrow). With short middle and anterior vertebral height in T_7 due to normal variant. Early degenerative changes anteriorly.
Degenerative changes	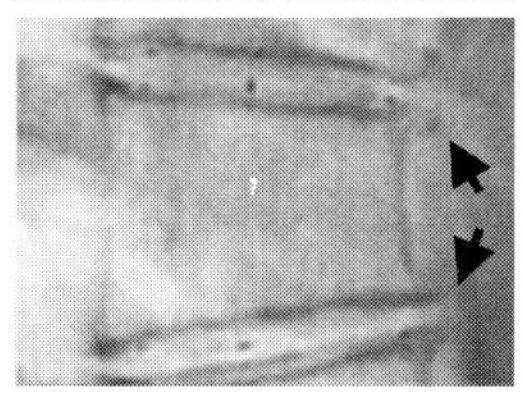	Degenerative changes anteriorly (arrows) in a vertebra with a short anterior vertebral height. No endplate fracture.
Cupid's bow (developmental deformity)	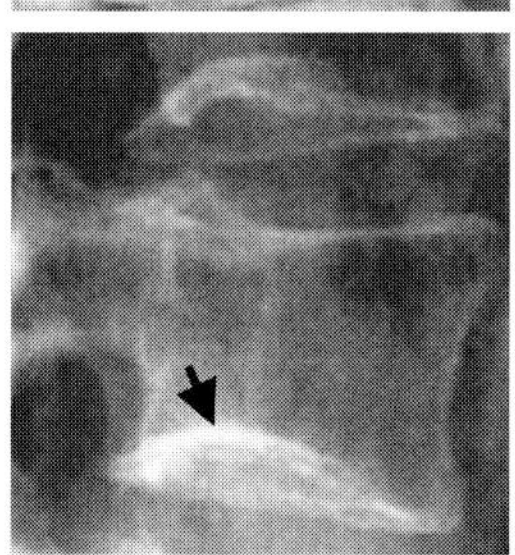	Lateral view lumbar spine with a deeper posterior endplate at L_3 (and L_2 above). AP view lumbar spine shows changes are caused by a "Cupid's bow" (arrows) developmental deformity.
	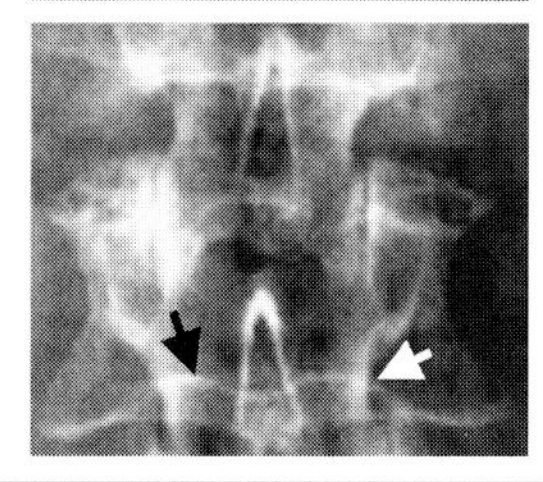	

tebrae. This is frequently associated with generalized thinning and reduced density of all the vertebral cortices and vertebral bodies.

Cupid's bow (Table 3) is a congenital deformity most commonly affecting the inferior end plates of the third, fourth, and fifth lumbar vertebral bodies. It frequently results in paired parasagittal concavities, which are best viewed in the AP projection. When viewed in the lateral projection, the concavities superimpose, lying in the posterior portion of the vertebral body. It is important not to confuse this lesion with a vertebral fracture because these abnormalities are not associated with decreased BMD or increased risk of fracture.[29]

Isolated vertebral fractures above T_4 are uncommon and warrant further evaluation for malignancy. Furthermore, the presence of a heterogeneous density within the vertebral body is highly suspicious of metastatic lesions and warrants further investigation. Additional features of malignancy may include irregular or unusual outlines of the vertebrae caused by destruction of the cortex of the vertebral bodies and pedicles. Identifying whether a collapsed vertebrae is caused by metastasis is particularly difficult. Suspicion should be increased if the clinical history or clinical signs suggest malignancy or there is a normal BMD density scan. Additional imaging may include CT, MRI, and radionucleotide imaging.

CLINICAL RECOMMENDATIONS

Screening for Vertebral Fractures

The current recommendations for using VFA (fracture assessment by DXA imaging) by the ISCD are summarized in Table 4.[13] Ultimately, the decision to obtain spine imaging depends on the availability and cost of different technologies, availability of local expertise, and the cost of primary and secondary fracture prevention. A number of specific clinical scenarios are discussed below.

Although risk factors can provide guidance to identify which patients require screening for osteoporosis, very few subjects would be prepared to initiate long-term therapy to prevent a

© 2008 American Society for Bone and Mineral Research

TABLE 4. ISCD RECOMMENDATIONS FOR SCREENING FOR VERTEBRAL FRACTURES USING VFA

1. When results may influence clinical management
2. If BMD is indicated then consider performing VFA if clinically indicated:
   - Documented height loss >2 cm (0.75 in)
   - Historical height loss of >4 cm (1.5 in) since young adult
   - History of fracture after age 50 yr
   - Commitment to long term oral or parental glucocorticoid therapy
   - History or findings suggestive of vertebral fracture not documented by previous radiographic imaging

fracture without confirmation of a diagnosis of osteoporosis using a DXA scan. If treatment is initiated based on a diagnosis of osteoporosis, it can be argued that there is no added benefit from identifying a vertebral fracture. One exception may be if a physician will alter therapy based on fracture risk assessment. If moderate or severe vertebral fractures or multiple vertebral fractures were identified then this may alter the choice of treatment. For example, an anabolic drug (e.g., PTH) may be used before considering long-term therapy with an anti-catabolic drug (e.g., bisphosphonates). In general, severe or multiple vertebral fractures result in some degree of back pain or disability. In additions, multiple risk factors are usually present. A question that has not yet been addressed is whether an increasing number of fractures or type of fracture influences patient's compliance. It is reasonable, however, to consider a full evaluation of subsequent fracture risk if compliance is a concern.

If a patient has osteopenia and a fragility fracture (at any site), the majority of physicians would intervene with therapy. If a patient has osteopenia and no previous history of fragility fractures, spine imaging is indicated because if a vertebral fracture is identified, treatment should be considered. Patients with osteopenia and vertebral fractures have been shown to benefit from treatment with raloxifene, risedronate, and alendronate. Therefore, it is reasonable to screen all patients with osteopenia using VFA if it will alter the management of the patient. In one study, 15.9% of patients 60–69 yr of age and 44.9% of those >70 yr of age had a previously undiagnosed vertebral fracture on VFA.[30]

If a subject is on long-term corticosteroids and has a history of an osteoporotic fracture (at any site), the majority of physicians would consider therapy irrespective of whether spine imaging showed a vertebral fracture. Therefore, additional imaging is probably not warranted unless a comprehensive risk assessment is required. Patients on long-term corticosteroids without a prior history of osteoporotic fracture warrant spine imaging, particularly if there is a history of back pain or height loss or age is >65 yr.

Risk Factors for Vertebral Fractures

Clinical risk factors that independently predict prevalent vertebral fractures are height loss, past nonvertebral osteoporotic fracture, history of back pain, and age. Furthermore, the Margolis back pain score and rib-pelvis distance is associated with the presence of multiple vertebral fractures.[31] One of the simplest and most effective risk factors for prevalent vertebral fractures is a measured height loss of >2 cm (0.75 in) or a recalled height loss of >4 cm (1.5 in). Thus, a measured height loss of >2 cm had a 35% sensitivity and 94% specificity for new vertebral fractures.[32] A recalled height loss of >4 cm is associated with a 3-fold increase in the risk of a identifying a prevalent vertebral fracture.[33] Therefore, spine imaging is probably indicated because identification of a prevalent vertebral fracture will prompt therapy.

SUMMARY

As technologies develop further, it may well be possible to incorporate biomechanical properties of bone such as bone size, shape, and quality, including microarchitectural characteristics. Finally, it is clearly established that the recognition of vertebral fractures is poor,[9] and yet effective therapies are available. It is essential to accurately and reproducibly identify vertebral fractures. Approaches have been developed and refined to diagnose vertebral fractures. The gold standard remains a visual read by an expert reader, and this has been standardized and applied to newer technologies. These approaches have enhanced a physician's ability to perform a comprehensive risk evaluation by identifying vertebral fractures. Routine screening for vertebral fractures in patients evaluated for osteoporosis is not yet recommended for routine clinical practice. However, as techniques and approaches to interpretation improve further, it is conceivable that vertebral fracture evaluation will become an integral part of fracture risk assessment in patients.

ACKNOWLEDGMENTS

We thank Drs. Lynn Ferrar and Guirong Jiang for providing many of the images used in this manuscript.

REFERENCES

1. Black DM, Arden NK, Palermo L, Pearson J, Cummings SR 1999 Prevalent vertebral deformities predict hip fractures and new vertebral deformities but not wrist fractures. Study of Osteoporotic Fractures Research Group. J Bone Miner Res **14**:821–828.
2. Klotzbuecher CM, Ross PD, Landsman PB, Abbott TA, Berger M 2000 Patients with prior fractures have an increased risk of future fractures: A summary of the literature and statistical synthesis. J Bone Miner Res **15**:721–739.
3. Reeve J, Lunt M, Felsenberg D, Silman AJ, Scheidt-Nave C, Poor G, Gennari C, Weber K, Lorenc R, Masaryk P, Cannata JB, Dequeker J, Reid DM, Pols HA, Benevolenskaya LI, Stepan JJ, Miazgowski T, Bhalla A, Bruges AJ, Eastell R, Lopes-Vaz A, Lyritis G, Jajic I, Woolf AD, Banzer D, Reisinger W, Todd CJ, Felsch B, Havelka S, Hoszowski K, Janott J, Johnell O, Raspe HH, Yershova OB, Kanis JA, Armbrecht G, Finn JD, Gowin W, O'Neill TW 2003 Determinants of the size of incident vertebral deformities in European men and women in the sixth to ninth decades of age: The European Prospective Osteoporosis Study (EPOS). J Bone Miner Res **18**:1664–1673.
4. Lunt M, O'Neill TW, Felsenberg D, Reeve J, Kanis JA, Cooper C, Silman AJ 2003 Characteristics of a prevalent vertebral deformity predict subsequent vertebral fracture: Results from the European Prospective Osteoporosis Study (EPOS). Bone **33**:505–513.
5. O'Neill TW, Cockerill W, Matthis C, Raspe HH, Lunt M, Cooper C, Banzer D, Cannata JB, Naves M, Felsch B, Felsenberg D, Janott J, Johnell O, Kanis JA, Kragl G, Lopes VA, Lyritis G, Masaryk P, Poor G, Reid DM, Reisinger W, Scheidt-Nave C, Stepan JJ, Todd CJ, Woolf AD, Reeve J, Silman AJ 2004 Back pain, disability, and radiographic vertebral fracture in European women: A prospective study. Osteoporos Int **15**:760–765.
6. Ettinger B, Black DM, Nevitt MC, Rundle AC, Cauley JA, Cummings SR, Genant HK 1992 Contribution of vertebral deformities to chronic back pain and disability. The Study of Osteoporotic Fractures Research Group. J Bone Miner Res **7**:449–456.
7. Kado DM, Browner WS, Palermo L, Nevitt MC, Genant HK, Cummings SR 1999 Vertebral fractures and mortality in older women: A prospective study. Study of Osteoporotic Fractures Research Group. Arch Intern Med **159**:1215–1220.
8. National Osteoporosis Foundation 1999 Physician's Guide to Prevention and Treatment of Osteoporosis. National Osteoporosis Foundation, Washington, DC, USA.
9. Delmas PD, Vande LL, Watts NB, Eastell R, Genant H, Grauer

© 2008 American Society for Bone and Mineral Research

A, Cahall DL 2005 Underdiagnosis of vertebral fractures is a worldwide problem: The IMPACT study. J Bone Miner Res **20**:557–563.
10. Gehlbach SH, Bigelow C, Heimisdottir M, May S, Walker M, Kirkwood JR 2000 Recognition of vertebral fracture in a clinical setting. Osteoporos Int **11**:577–582.
11. Lentle BC, Brown JP, Khan A, Leslie WD, Levesque J, Lyons DJ, Siminoski K, Tarulli G, Josse RG, Hodsman A 2007 Recognizing and reporting vertebral fractures: Reducing the risk of future osteoporotic fractures. Can Assoc Radiol J **58**:27–36.
12. Mayranpaa MK, Helenius I, Valta H, Mayranpaa MI, Toiviainen-Salo S, Makitie O 2007 Bone densitometry in the diagnosis of vertebral fractures in children: Accuracy of vertebral fracture assessment. Bone **41**:353–359.
13. Vokes T, Bachman D, Baim S, Binkley N, Broy S, Ferrar L, Lewiecki EM, Richmond B, Schousboe J 2006 Vertebral fracture assessment: The 2005 ISCD Official Positions. J Clin Densitom **9**:37–46.
14. Jiang G, Eastell R, Barrington NA, Ferrar L 2004 Comparison of methods for the visual identification of prevalent vertebral fracture in osteoporosis. Osteoporos Int **15**:887–896.
15. Smith-Bindman R, Cummings SR, Steiger P, Genant HK 1991 A comparison of morphometric definitions of vertebral fracture. J Bone Miner Res **6**:25–34.
16. McCloskey EV, Spector TD, Eyres KS, Fern ED, O'Rourke N, Vasikaran S, Kanis JA 1993 The assessment of vertebral deformity: A method for use in population studies and clinical trials. Osteoporos Int **3**:138–147.
17. Eastell R, Cedel SL, Wahner HW, Riggs BL, Melton LJ 1991 Classification of vertebral fractures. J Bone Miner Res **6**:207–215.
18. Melton LJ III, Wenger DE, Atkinson EJ, Achenbach SJ, Berquist TH, Riggs BL, Jiang G, Eastell R 2006 Influence of baseline deformity definition on subsequent vertebral fracture risk in postmenopausal women. Osteoporos Int **17**:978–985.
19. Rea JA, Chen MB, Li J, Blake GM, Steiger P, Genant HK, Fogelman I 2000 Morphometric X-ray absorptiometry and morphometric radiography of the spine: A comparison of prevalent vertebral deformity identification. J Bone Miner Res **15**:564–574.
20. Ferrar L, Jiang G, Barrington NA, Eastell R 2000 Identification of vertebral deformities in women: Comparison of radiological assessment and quantitative morphometry using morphometric radiography and morphometric X-ray absorptiometry. J Bone Miner Res **15**:575–585.
21. Rea JA, Chen MB, Li J, Marsh E, Fan B, Blake GM, Steiger P, Smith IG, Genant HK, Fogelman I 2001 Vertebral morphometry: A comparison of long-term precision of morphometric X-ray absorptiometry and morphometric radiography in normal and osteoporotic subjects. Osteoporos Int **12**:158–166.
22. Genant HK, Jergas M, Palermo L, Nevitt M, Valentin RS, Black D, Cummings SR 1996 Comparison of semiquantitative visual and quantitative morphometric assessment of prevalent and incident vertebral fractures in osteoporosis The Study of Osteoporotic Fractures Research Group. J Bone Miner Res **11**:984–996.
23. Sone T, Tomomitsu T, Miyake M, Takeda N, Fukunaga M 1997 Age-related changes in vertebral height ratios and vertebral fracture. Osteoporos Int **7**:113–118.
24. Ferrar L, Jiang G, Cawthon PM, San Valentin R, Fullman R, Lambert L, Cummings SR, Black DM, Orwoll E, Barrett-Connor E, Ensrud K, Fink HA, Eastell R 2007 Identification of vertebral fracture and non-osteoporotic short vertebral height in men: The MrOS study. J Bone Miner Res **22**:1434–1441.
25. Ferrar L, Jiang G, Armbrecht G, Reid DM, Roux C, Gluer CC, Felsenberg D, Eastell R 2007 Is short vertebral height always an osteoporotic fracture? The Osteoporosis and Ultrasound Study (OPUS). Bone **41**:5–12.
26. Rea JA, Li J, Blake GM, Steiger P, Genant HK, Fogelman I 2000 Visual assessment of vertebral deformity by X-ray absorptiometry: A highly predictive method to exclude vertebral deformity. Osteoporos Int **11**:660–668.
27. Resnick D, Niwayama G 1978 Intravertebral disk herniations: Cartilaginous (Schmorl's) nodes. Radiology **126**:57–65.
28. Wagner AL, Murtagh FR, Arrington JA, Stallworth D 2000 Relationship of Schmorl's nodes to vertebral body endplate fractures and acute endplate disk extrusions. AJNR Am J Neuroradiol **21**:276–281.
29. Chan KK, Sartoris DJ, Haghighi P, Sledge P, Barrett-Connor E, Trudell DT, Resnick D 1997 Cupid's bow contour of the vertebral body: Evaluation of pathogenesis with bone densitometry and imaging-histopathologic correlation. Radiology **202**:253–256.
30. Schousboe JT, Ensrud KE, Nyman JA, Kane RL, Melton LJ III 2006 Cost-effectiveness of vertebral fracture assessment to detect prevalent vertebral deformity and select postmenopausal women with a femoral neck T-score>-2.5 for alendronate therapy: A modeling study. J Clin Densitom **9**:133–143.
31. Tobias JH, Hutchinson AP, Hunt LP, McCloskey EV, Stone MD, Martin JC, Thompson PW, Palferman TG, Bhalla AK 2007 Use of clinical risk factors to identify postmenopausal women with vertebral fractures. Osteoporos Int **18**:35–43.
32. Siminoski K, Jiang G, Adachi JD, Hanley DA, Cline G, Ioannidis G, Hodsman A, Josse RG, Kendler D, Olszynski WP, Ste Marie LG, Eastell R 2005 Accuracy of height loss during prospective monitoring for detection of incident vertebral fractures. Osteoporos Int **16**:403–410.
33. Siminoski K, Warshawski RS, Jen H, Lee K 2006 The accuracy of historical height loss for the detection of vertebral fractures in postmenopausal women. Osteoporos Int **17**:290–296.

Chapter 37. Molecular Diagnosis of Bone and Mineral Disorders

Robert F. Gagel and Gilbert J. Cote

Department of Endocrine Neoplasia and Hormonal Disorders, University of Texas M.D. Anderson Cancer Center, Houston, Texas

INTRODUCTION

With the complete sequencing of the human genome, we are moving from a period of progressive discovery to one of where genetic analysis has become a common tool for the study of the molecular basis and clinical diagnosis of human disease. The past 20 yr have seen exponential growth in the identification of genetic abnormalities that cause bone and mineral disorders. There are now specific genetic forms of: osteoporosis and osteosclerosis; multiple causes of hypocalcemia, hypercalcemia, hypophosphatemia, hyperphosphatemia; and numerous examples of hereditary bone dysplasia. These observations have enriched our understanding of the hormonal and signal transduction pathways involved in bone formation, bone remodeling, and mineral homeostasis and have provided new therapeutic targets for treatment of a variety of bone disorders.

Diagnostic use of this type of information has quickly made its way into the clinical practice of medicine. For example, within 3 yr after the description of missense mutations in the *RET* proto-oncogene in multiple endocrine neoplasia, type 2 (MEN2), genetic testing for these mutations has replaced prior

Dr. Gagel has consulted for Eli Lilly, Novartis, and Merck. Dr. Cote states that he has no conflicts of interest.

© 2008 American Society for Bone and Mineral Research

nongenetic approaches. In addition, mutational analysis of the *MEN1* gene (multiple endocrine neoplasia type 1), *RET* gene (MEN2), *HPRT2* gene (hyperparathyroidism-jaw tumor syndrome gene, which is involved in some examples of familial isolated hyperparathyroidism), and the *CASR* gene (calcium sensing receptor) can be important for evaluation of hypercalcemia. This rapid acquisition of new information and its application to disease management underscores the importance of acquiring a fundamental knowledge of testing strategies and an understanding of the power and limitations of current approaches to genetic testing.

The single most important resource for up-to-date information related to specific genetic syndromes is provided by Online Mendelian Inheritance in Man (OMIM), available to all physicians without charge on the World Wide Web (http://www.ncbi.nlm.nih.gov/omim).[1] This concise but complete reference is an excellent starting point for genetic information relating to bone and mineral disorders and provides an intuitive, searchable textual database that is updated on a regular basis. For each genetic disorder, identifiable by a specific OMIM number, a detailed and well-referenced review discussing the mapping and identification of the causative gene, the spectrum of clinical presentation and management, molecular genetics, and specific animals models is available. Additional links provide a clinical synopsis, whether genetically related disorders exist, and a detailed description of the causative gene, including associated mutations. Because OMIM is incorporated into the National Center for Biotechnology Information (NCBI) website (www.ncbi.nlm.nih.gov), direct links to PubMed citations, genetic sequence, and expression profiles are easily accessed. Other databases provide specific information on the availability of genetic testing and individual gene mutations. The most widely used resource for information about genetic testing is available at www.genetests.org, a site that provides a searchable database of research and clinical sources for genetic testing and a variety of educational materials, including gene reviews and PowerPoint presentations.[2] Detailed information regarding the association of specific gene mutations with various disorders can be found at the Human Gene Mutation Database (www.hgmd.org).[3] This site provides mutation data on >2000 genes and has direct links to several external websites including OMIM.

In the past 5–10 yr, genetic testing for diagnosis has largely moved from individual research laboratories to large clinical laboratories that have been government certified to follow the Clinical Laboratory Improvement Amendments (CLIA; www.CMS.hhs.gov/clia) and mandated practices of quality assurance.[4] The CLIA program, originally put in place to ensure quality laboratory testing, has incorporated specific standards for genetic testing. With the widespread availability of genetic testing for use in the clinical setting, the goal of this chapter is to provide a basic overview of the primary techniques used to identify mutations and to briefly discuss the interpretations of test results.

GENERAL SCREENING TO DETECT MUTATIONS

This summary will focus only on the application of direct DNA sequencing for the detection of specific nucleotide mutations, small insertions, or small deletions (typically <50 base pairs). Whereas it is possible to use sequencing to precisely map the DNA breakpoints of large insertions, deletions, or rearrangements, standard PCR approaches typically fail to identify these defects. For genetic disorders where DNA breakage is common, clinical laboratories will frequently supplement DNA sequence analysis with karyotype analysis, Southern blotting, or other newer approaches.[5] Indeed, continuing refinement of array based comparative hybridization technology may 1 day lead to the replacement of standard karyotype analysis.[6] This is a particularly important consideration when interpreting negative results (discussed below).

Over the years, several different strategies have been developed for identification of specific mutations. Cost, reproducibility, and specificity have typically been the driving factors in application of various methods to clinical testing. A benefit of the human genome project has been the continual reduction in the costs associated with direct DNA sequencing along with an accompanying increase in reliability. Direct DNA sequencing remains the "gold standard" for mutation detection. This has led most commercial laboratories to adopt automated PCR and DNA sequencing as the technique of choice. However, for genetic disorders with a limited and defined set of mutations, an array of equally reliable methodologies that allow high-throughput analysis of DNA polymorphisms is also finding its way into clinical mutation screening applications. Many use capillary electrophoresis, high-performance liquid chromatography (HPLC), or mass spectrometry approaches to separate DNA fragments or techniques that rely on fluorescence-based imaging, such as DNA melting curve analysis or enzymatic detection, of mismatches to define mutations.[5–11] Major factors for deciding which technique to use include the size of the DNA sequence to be examined, the spectrum of mutations that cause the disease, and access to newer equipment and technologies. For example, several disorders are known to result from mutations affecting collagen. Collagen genes are large and comprised of numerous exons (>50), making it impractical to sequence the many exons of the many collagen genes without the application of automation.

The remainder of this chapter will focus on a single technique, standard automated DNA sequencing, and interpretation of genetic results. We recently entered an era where several high-throughput approaches are just becoming available for full genome sequencing.[12] As we move into the future, issues regarding the practical, ethical, and clinical use of genetic testing will continue to evolve as decisions based on sequence aberrants found throughout the whole genome.[12,13]

DIRECT DNA SEQUENCING

Direct DNA sequencing of PCR products is universally accepted as the most specific method for detection of genetic mutations.[5–10] Direct DNA sequencing of a PCR product derived from genomic DNA permits analysis of both alleles or copies of the gene and the identification of new or unreported mutations. The disadvantages of this method include its complexity, the requirement for expensive equipment, and the potential difficulties associated with analyzing >500–750 nucleotides in a single sequencing reaction. The process has been largely computer automated with an assembly line of robots coupling DNA isolation from blood, PCR amplification, sequencing, and output of results. Large clinical laboratories, with the means to invest in equipment, routinely use direct sequencing because it is more practical and cost effective, even for large genes. The methodology is based on the standard Sanger dideoxysequencing method.[14] PCR products are first typically enzymatically treated to remove unused oligonucleotide primers and to strip the deoxynucleotide triphosphates (dNTPs) of their phosphates. This step is required so that a new oligonucleotide primer can be added to direct the site-specific initiation of the sequencing reaction by DNA polymerase, similar to the annealing and elongation steps of PCR. Unlike PCR, however, the dNTP mixture includes fluorescently labeled dideoxynucleotides, which randomly terminate elongation when they become incorporated into the newly syn-

© 2008 American Society for Bone and Mineral Research

thesized DNA strand (called "chain termination"). Each of the four dideoxynucleotides is labeled with a different color fluorescent label, which serves to identify the specific nucleotide ending the DNA chain. The mixture of end-labeled DNA fragments are separated based on their size by capillary electrophoresis. As individual end-labeled DNA fragments, which differ in size by a single nucleotide, pass by a laser detector, the fluorescent nucleotide is read. The DNA sequence is represented as a four-colored chromatograph with separate peaks representing individual nucleotides (Fig. 1). Single nucleotide mutations/polymorphisms are denoted by the presence of two nucleotides (peaks) migrating with the same apparent molecular weight. Insertions and deletions have a much more distinctive chromatograph pattern because the sequence frameshift results in an overlapping pattern of peaks from the point of the genetic defect (Fig. 1).

SOURCES OF ERROR IN GENETIC TESTING

It is important that clinicians be aware of the frequency and nature of genetic testing errors.[8,9,15] Both false positive and negative results clearly have the potential to adversely impact outcome of patient care. Sample mix-up, especially in the setting of family screening where many family members share a common last name, may occur in up to 5% of analyses. These errors may occur at the time of blood drawing, during subsequent analysis or recording, or even at the clinical testing laboratory. A second potential source of error is contamination by DNA from individuals who harbor a disease-causing mutation. The funneling of large numbers of samples to a few laboratories for analysis of a single disease further increases the chance of contamination. The extreme sensitivity of PCR analysis makes it possible that a positive result could occur as a result of airborne contamination of a reaction tube. This has become less of a concern at larger clinical laboratories that have taken specific measures to prevent sample mix-up and contamination through the use of automation and computerized coding.[9] A third source of error is the failure to amplify both alleles, thereby resulting in the possibility of a false-negative result because only the normal allele is included in the analysis. The most common explanation for amplification failure is a random polymorphism (DNA sequence change) acting to reduce oligonucleotide primer hybridization during the PCR reaction. Other causes for amplification failure include genetic defects involving insertion, deletion, or rearrangement of DNA sequence at the site of amplification. To help rule out these possibilities, it is always helpful to note whether an immediate family member or close relative has tested positive using the same PCR oligonucleotide primer set. Finally, despite the robust nature of PCR and DNA sequencing, the methodologies themselves are not error free.[9] Therefore, it is important to note whether DNA sequencing was performed on both the sense and antisense strands to safeguard against possible errors in sequencing.

Even under ideal conditions, mutation analysis errors will to occur. Government-mandated standards and procedures in individual laboratories have reduced but not eliminated these errors completely.[4,9] If genetic testing is to be used as the sole determinant for decision making in disease management, it is important for the clinician to be aware of the possibility of error and to take steps to minimize the impact on patient care. Many clinical testing laboratories and GeneTests.org provide estimates of the frequency of new mutation detection, the potential for large deletions or rearrangements, and the inherent error rate associated with the test results. Therefore, if not provided, the clinician should always request the gene-specific sensitivity and specificity that the clinical laboratory routinely

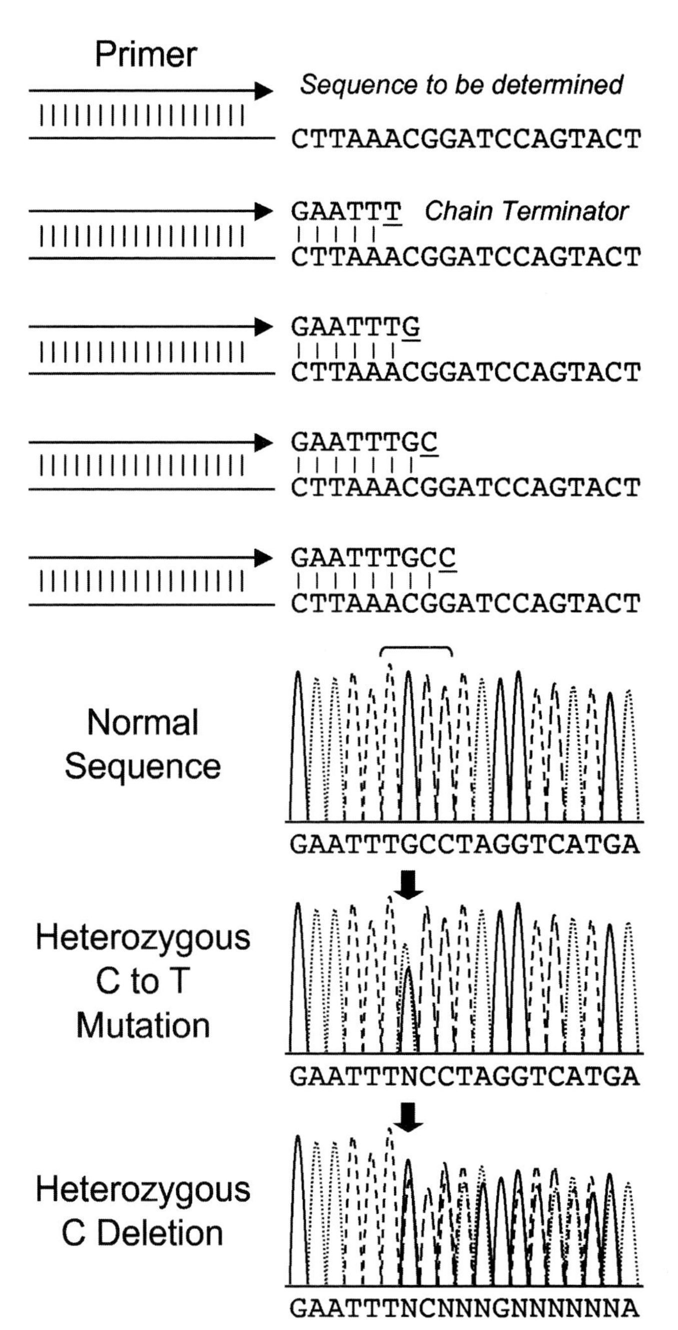

FIG. 1. DNA sequencing. Similar to the first step of PCR, an oligonucleotide primer is hybridized to a DNA target adjacent to the sequence to be determined and serves as a template for extension by a DNA polymerase. Unlike, PCR a precise mixture normal nucleotides with fluorescent dideoxynucleotides (underlined) causes the growth of the newly synthesized DNA strand to terminate selectively at each position. The mixture of newly synthesized DNA fragments is then denatured, separated based on sized by capillary electrophoresis and the specific end-labeled nucleotide detected by a laser scanner. The chromatograph simulates a representitive sequencing output with the typical black (G), red (T), green (A), and blue (C) colors replaced by dotted and dashed lines. The bracketed region highlights the position of the four terminated DNA fragments shown above. Additional chromatographs provide examples of a heterozygous point mutation and a deletion. The arrow indicates the defect.

observes. One simple approach that will eliminate the majority of these errors is to repeat each analysis, whether positive or negative, in a different laboratory on an independently ob-

© 2008 American Society for Bone and Mineral Research

tained sample. This approach will eliminate most sample mix-up, DNA contamination, and technical errors. Sending the sample to a separate laboratory that uses a different primer set for PCR amplification will also reduce the likelihood of a single allele amplification error.

INTEGRATION OF GENETIC INFORMATION INTO CLINICAL MANAGEMENT

Genetic testing has several important clinical uses. Identification of a specific disease-causing mutation may clarify and simplify patient management. For example, the identification of a mutation of the calcium sensing receptor causative for familial hypercalcemic hypocalciuria in an individual with an atypical clinical presentation may prevent unnecessary parathyroid surgery. For other disorders, such as multiple endocrine neoplasia type 2, the identification of a specific mutation may lead to a specific action (thyroidectomy) in a child.[8,15] However, genetic testing also imposes additional ethical and legal duties on the physician to ensure that the patient is properly counseled before testing and additionally when receiving their test results. Unlike most clinical testing, obtaining a blood sample for genetic testing requires informed patient consent. Unfortunately, a simple discussion of the utility of testing in diagnosis and treatment is insufficient. The patient needs to be informed about the limitations of testing, genetic testing options, alternative tests, inheritance and risk models, genetic health risks to family members, and a wide variety of insurance and legal issues. In addition, the fear of genetic discrimination sometimes becomes a deterrent, preventing patients from seeking genetic testing. The complexity of these many issues, including state to state differences in the laws regarding the use of genetic information, make it is advisable to involve a genetic counselor when possible (www.nsgc.org). In addition to dealing with issues surrounding consent and disclosure of test results, a genetic counselor frequently can help the patient with many of the burdens associated with having a genetic disorder. For example, when caring for a patient with a hereditary cancer syndrome, the clinician has an ethical duty to warn the patient about genetic health risks to his or her family members. Genetic counselors can typically write a "family letter" to be offered to their family members as an educational tool and to relieve the patient from the burden of remembering a lot of technical information.

In other situations, the benefits of genetic screening may be more ambiguous. The identification of a mutation in an individual with a severe and fatal form of osteogenesis imperfecta may not alter therapy for the patient; however, detection of the mutation may make prenatal genetic screening possible. Identification and categorization of these mutations are also important because gene therapy strategies, especially for single gene defects, are evolving rapidly. The discovery that mutation of a single gene can be associated with multiple disease phenotypes has also led to a rethinking of how skeletal disorders should be classified. Major classifications that originally emerged based primarily either on biochemical evidence or radiographic evidence[16,17] now include knowledge of genetic defect as a major classifier.[18] Each classification group serves as a unique entry point in facilitating patient diagnosis and treatment. Included in the Appendix of this *Primer* is a table providing an extensive listing of serum mineral and skeletal disorders and their specific genetic defects. Information in this table was primarily obtained by a keyword search of the OMIM website (discussed above). We have grouped the disorders according to serum mineral findings and followed this list with disorders grouped according to a recent International Nosology and Classification of Constitutive Disorders of Bone consensus, which is a combination of morphological findings and molecular defects.[18]

REFERENCES

1. McKusick-Nathans Institute of Genetic Medicine, Johns Hopkins University and National Center for Biotechnology Information, National Library of Medicine 2007 Online Mendelian Inheritance in Man OMIM (TM). Available at http://www.ncbi.nlm.nih.gov/omim/. Accessed October 15, 2007.
2. GeneTests Medical Genetics Information Resource (database online). Available at http://www.genetests.org. Accessed October 15, 2007.
3. Stenson PD, Ball EV, Mort M, Phillips AD, Shiel JA, Thomas NS, Abeysinghe S, Krawczak M, Cooper DN 2003 The Human Gene Mutation Database (HGMD®): 2003 update. Hum Mutat **21:**577–581.
4. Williams LO, Cole EC, Lubin IM, Iglesias NI, Jordan RL, Elliot LE 2003 Quality assurance in human molecular genetics testing, status and recommendations. Arch Pathol Lab Med **127:**1353–1358.
5. Sellner LN, Taylor GR 2004 MLPA and MAPH: New techniques for detection of gene deletions. Hum Mutat **23:**413–419.
6. Aradhya S, Cherry AM 2007 Array-based comparative genomic hybridization: Clinical contexts for targeted and whole-genome designs. Genet Med **9:**553–559.
7. Green ED, Klapholz S, Birren B 1998 Genome Analysis: A Laboratory Manual. Detecting Genes, vol. 2. Cold Spring Harbor Laboratory Press, New York, NY, USA.
8. Hoff AO, Cote GJ, Gagel RF 2002 Laboratory evaluation and screening of genetic endocrine diseases. In: L. Martini (ed.) Modern Endocrinology. Genetics in Endocrinology. Lippincott Williams & Wilkins, Philadelphia, PA, USA, pp. 189–220.
9. Strom CM 2005 Mutation detection, interpretation, and applications in the clinical laboratory setting. Mutat Res **573:**160–167.
10. Tomita N, Oto M 2004 Molecular genetic diagnosis of familial tumors. Int J Clin Oncol **9:**246–256.
11. Tost J, Gut IG 2005 Genotyping single nucleotide polymorphisms by MALDI mass spectrometry in clinical applications. Clin Biochem **38:**335–350.
12. Ropers HH 2007 New perspectives for the elucidation of genetic disorders. Am J Hum Genet **81:**199–207.
13. Wolf SM, Kahn JP 2007 Working group on genetic testing in disability insurance. Genetic testing and the future of disability insurance: Ethics, law & policy. J Law Med Ethics **35:**6–32.
14. Sanger F, Nicklen S, Coulson AR 1977 DNA sequencing with chain-terminating inhibitors. Proc Natl Acad Sci USA **74:**5463–5467.
15. Gagel RF, Cote GJ 2002 The role of the *RET* proto-oncogene in multiple endocrine neoplasia, type 2. In: Bilezikian JP, Raisz LG, Rodan GA (eds.) Principles of Bone Biology, 2nd ed. Academic Press, New York, NY, USA, pp. 1067–1078.
16. Hall CM 2002 International nosology and classification of constitutional disorders of bone. Am J Med Genet **113:**65–77.
17. Superti-Furga A, Bonafe L, Rimoin DL 2001 Molecular-pathogenetic classification of genetic disorders of the skeleton. Am J Med Genet **106:**282–293.
18. Superti-Furga A, Unger S 2007 Nosology and classification of genetic skeletal disorders: 2006 revision. Am J Med Genet **143:**1–18.

© 2008 American Society for Bone and Mineral Research

SECTION V

Osteoporosis
(Section Editor: Paul Miller and Socrates Papapoulos)

Please also see the following websites: http://www.nice.org.uk and www.thecochranelibrary.com (requires subscription)

© 2008 American Society for Bone and Mineral Research

Chapter 38. Epidemiology of Osteoporotic Fractures

Nicholas Harvey, Elaine Dennison, and Cyrus Cooper

Department of Rheumatology, The MRC Epidemiology Resource Centre, University of Southampton, Southampton General Hospital, Southampton, United Kingdom

INTRODUCTION

Osteoporosis is a skeletal disease characterized by low bone mass and microarchitectural deterioration of bone tissue with a consequent increase in bone fragility and susceptibility to fracture.[1] The term osteoporosis was first introduced in France and Germany during the last century. It means "porous bone" and initially implied a histological diagnosis, but was later refined to mean bone that was normally mineralized but reduced in quantity. Historically, osteoporosis has been difficult to define: a focus on BMD may not encompass all the risk factors for fracture, whereas a fracture-based definition will not enable identification of at risk populations. In 1994, the World Health Organization (WHO)[2] convened to resolve this issue, defining osteoporosis in terms of BMD and previous fracture. Thus, the WHO definition does not take into account microarchitectural changes that may weaken bone independently of any effect on BMD.

More recently, there has been a move toward assessment of individualized 5- or 10-yr absolute risk.[3] This has the advantage of incorporating risk factors that are partly independent of BMD, such as age and previous fracture, and thus allows decisions regarding commencement of therapy to be made more readily. Osteoporotic fracture has a huge impact economically, in addition to its effect on health. Osteoporotic fractures cost the United States approximately $17.9 billion/yr, with the cost in the United Kingdom being £1.7 billion (Table 1 summarizes fracture impact for a Western population).[4] Hip fractures contribute most to these figures.

FRACTURE EPIDEMIOLOGY

Incidence and Prevalence

The 2004 report from the U.S. Surgeon General highlighted the enormous burden of osteoporosis-related fractures.[7] An estimated 10 million Americans >50 yr of age have osteoporosis, and there are ~1.5 million fragility fractures each year. Another 34 million Americans are at risk for the disease. A study of British fracture occurrence indicated that population risk is similar in the United Kingdom.[5] Thus, one in two women 50 yr of age will have an osteoporotic fracture in their remaining lifetime; the figure for men is one in five.

Fracture incidence in the community is bimodal, showing peaks in youth and in the very elderly. In young people, fractures of long bones predominate, usually after substantial trauma, and are more frequent in boys than girls. In this group, the question of bone strength rarely arises, although there are now data suggesting that this may not be entirely irrelevant as a risk factor.[8] Over the age of 35 yr, fracture incidence in women climbs steeply, so that rates become twice those in men. Before studies ascertaining vertebral deformities radiographically, rather than by clinical presentation, this peak was thought to be mainly caused by hip and distal forearm fracture, but as Fig. 1 shows, vertebral fracture has now been shown to make a significant contribution.

Hip Fracture. In most populations, hip fracture incidence increases exponentially with age (Fig. 1). Above 50 yr of age, there is a female to male incidence ratio of around two to one.[9] Overall, ~98% of hip fractures occur among people ⩾35 yr of age, and 80% occur in women (because there are more elderly women than men). Worldwide, there were an estimated 1.66 million hip fractures in 1990[10]: ~1.19 million in women and 463,000 in men. The majority occur after a fall from standing height or less; 90% occur in people >50 yr of age, and 80% are in women.[11] Recent work has characterized the age- and sex-specific incidence in the UK population, using the General Practice Research Database (GPRD; which includes 6% of the UK population). Thus, the lifetime risk of hip fracture for 50 yr olds in the United Kingdom is 11.4% and 3.1% for women and men, respectively.[5] Most of this increased risk is accrued in old age, such that a 50-yr-old woman's 10-yr risk of hip fracture is 0.3%, rising to 8.7% when she is 80 yr old.[5] The corresponding figures for men are 0.2% and 2.9%, respectively. Hip fractures are seasonal, with an increase in winter in temperate countries, but their mainly indoor occurrence would imply that this increase is not caused by slipping on icy pavements: possible causes may include slowed neuromuscular reflexes and lower light in winter weather. The direction of falling is important, and a fall directly onto the hip (sideways) is more likely to cause a fracture than falling forward.[12]

Incidence rates vary substantially from one population to another, and incidence is usually greater in whites than nonwhites, although there are differences within populations of a given sex or race. In Europe, hip fracture rates vary 7-fold between countries.[13] These findings suggest an important role for environmental factors in the etiology of hip fracture, but factors studied thus far, such as smoking, alcohol consumption, activity levels, obesity, and migration status, have not explained these trends.

Vertebral Fracture. Recent data from the European Vertebral Osteoporosis Study (EVOS) have shown that the age-standardized population prevalence across Europe was 12.2% for men and 12.0% for women 50–79 yr of age.[14] Figure 2 shows the prevalence by age and sex in this population. Historically, it was believed that vertebral fractures were more common in men than women, but the EVOS data suggest that this is not the case at younger ages: the prevalence of deformities in 50–60 yr olds is similar, if not higher in men, possibly because of a greater incidence of trauma.[14] The majority of vertebral fractures in elderly women occur through normal activities such as lifting rather than through falling.

Many vertebral fractures are asymptomatic, and there is disagreement about the radiographic definition of deformities in those patients who do present. Thus, in studies using radiographic screening of populations, the incidence of all vertebral deformities has been estimated to be three times that of hip fracture, with only one third of these coming to medical attention.[15] Data from EVOS have allowed accurate assessment of radiographically determined vertebral fractures in a large

The authors state that they have no conflicts of interest.

Key words: osteoporosis, epidemiology, fractures, risk factors, hip, wrist, vertebral, early life determinants, vitamin D, childhood, adulthood, mortality, morbidity

© 2008 American Society for Bone and Mineral Research

TABLE 1. IMPACT OF OSTEOPOROSIS-RELATED FRACTURES[4]

	Hip	*Spine*	*Wrist*
Lifetime risk (%)			
Women	14	28	13
Men	3	6	2
Cases per year	70,000	120,000	50,000
Hospitalization (%)	100	2–10	5
Relative survival	0.83	0.82	1.00

Costs: all sites combined ~£1.7 billion.

population. At age 75–79 yr, the incidence of vertebral fractures so-defined was 13.6 per 1000 person-years for men and 29.3 per 1000 person-years for women.[6] This compares with 0.2 per 1000 person-years for men and 9.8 per 1000 person-years in 75–84 yr olds, where the fractures were defined by clinical presentation in an earlier study from Rochester, MN, USA.[15] The overall age-standardized incidence in EVOS was 10.7 per 1000 person-years in women and 5.7 per 1000 person-years in men.

Figure 3 shows the incidence rates of morphometrically (by radiograph) defined vertebral fracture in EVOS compared with rates derived from other population-based radiographic studies. It shows that the heterogeneity in vertebral fracture incidence is markedly lower than the geographic variation in age- and sex-adjusted rates of hip fracture. The figure contrasts incidence rates for vertebral fractures defined radiographically with those identified by clinical diagnosis or hospitalization for the fracture. It clearly shows the enormous shortfall in clinical recognition and hospital identification of this important osteoporotic fracture.

Distal Forearm Fracture. Wrist fractures show a different pattern of occurrence to hip and vertebral fractures. There is an increase in incidence in white women between the ages of 45 and 60 yr, followed by a plateau.[16] This may relate to altered neuromuscular reflexes with aging, and as a result, a tendency to fall sideways or backward and thus not to break the fall with an outstretched arm. Most wrist fractures occur in women, and 50% occur in women >65 yr old. Data from the GPRD show that a woman's lifetime risk of wrist fracture at 50 yr old is 16.6%, falling to 10.4% at 70 yr. The incidence in men is low and does not rise much with aging (lifetime risk, 2.9% at age 50 yr and 1.4% at age 70 yr).[5]

Clustering of Fractures in Individuals

Epidemiological studies have suggested that patients with different types of fragility fractures are at increased risk of developing other types of fracture. For example, the presence of a previous vertebral deformity leads to a 7- to 10-fold increase in the risk of subsequent vertebral deformities.[17] This is a comparable level of increased risk to that seen for individuals who have sustained one hip fracture to sustain a second. Furthermore, data from Rochester, MN, USA, suggest that the risk of a hip fracture is increased 1.4-fold in women and 2.7-fold in men after the occurrence of a distal forearm fracture.[18] The corresponding figures for subsequent vertebral fracture are 5.2 and 10.7. Data from EVOS have shown that prevalent vertebral deformity predicts incident hip fracture with a rate ratio of 2.8–4.5, and this increases with the number of vertebral deformities.[19] The number and morphometry of baseline vertebral deformities also predict incident vertebral fracture.[20] The incidence of new vertebral fracture within a year of an incident vertebral fracture is 19.2%,[21] and in Rochester, the cumulative incidence of any fracture 10 yr after baseline event was 70%. These data emphasize the importance of prompt therapeutic action on discovering vertebral deformities.

Time Trends and Future Projections

Life expectancy is increasing around the globe, and the number of elderly individuals is rising in every geographic region. The world population is expected to rise from the current 323 million individuals ≥65 yr of age to 1555 million by 2050. These demographic changes alone can be expected to increase the number of hip fractures occurring among people ≥35 yr of age worldwide: the incidence is estimated to rise from 1.66 million in 1990 to 6.26 million in 2050. Assuming a constant age-specific rate of fracture, as the number of over 65s increases from 32 million in 1990 to 69 million in 2050, the number of hip fractures in the United States will increase 3-fold.[22] In the United Kingdom, the number of hip fractures may increase from 46,000 in 1985 to 117,000 in 2016.[23] However, in the developed world, recent studies from Switzerland and Finland suggest that the age-adjusted incidence of hip fracture has declined over the last decade.[24,25] The reason for these changes may be an increase in obesity or better screening and treatment for osteoporosis and might potentially partly offset the impact of the projected increase in the elderly population.

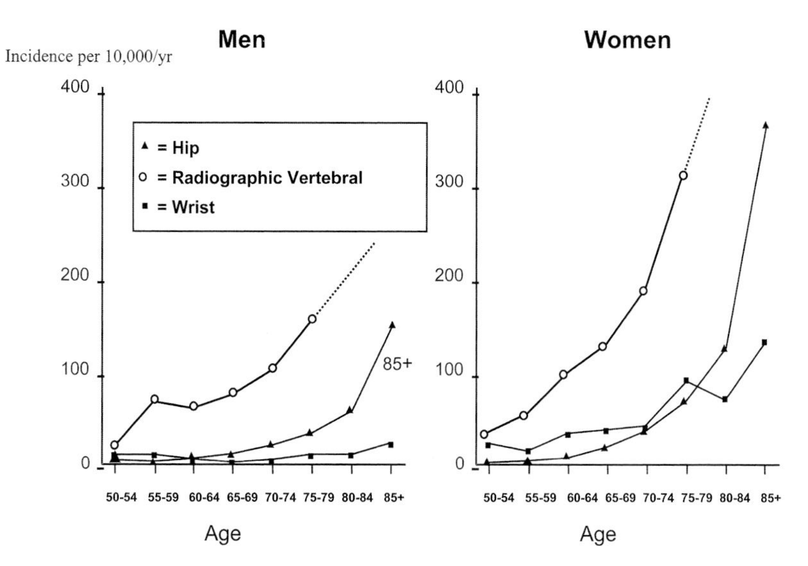

FIG. 1. Radiographic vertebral, hip, and wrist fracture incidence by age and sex.[5,6]

© 2008 American Society for Bone and Mineral Research

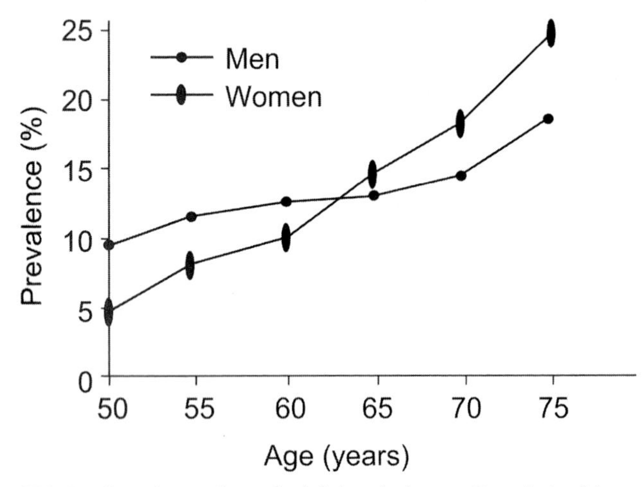

FIG. 2. Prevalence of vertebral deformity by sex. Data derived from the European Vertebral Osteoporosis Survey. (Reproduced with permission from the American Society for Bone and Mineral Research: J Bone Miner Res 1996:**11**:1011–1019.)

This decrease in age-adjusted incidence has not been recorded in the third world, and thus the increase in the elderly population, together with the adoption of westernized lifestyles, is likely to ensure an increase in the worldwide burden of osteoporotic fractures in future generations. This increase is likely to be uneven across the globe, with the increase in the elderly population in Latin America and Asia, potentially leading to a shift in the geographical distribution of hip fractures, with only one quarter occurring in Europe and North America.(22)

Geography

There is variation in the incidence of hip fracture within populations of a given race and sex.(13,26,27) Thus, age-adjusted hip fracture incidence rates are higher among white residents of Scandinavia than comparable subjects in the United States or Oceania. Within Europe the range of variation is ~11-fold.(26) These differences were not explained by variation in activity levels, smoking, obesity, alcohol consumption, or migration status.(13) The EVOS study showed a 3-fold difference in the prevalence of vertebral deformities between countries, with the highest rates in Scandinavia. The prevalence range between centers was 7.5–19.8% for men and 6.2–20.7% for women. The differences were not as great as those seen for hip fracture in Europe, and some of the differences could be explained by levels of physical activity and body mass index.(14)

MORTALITY AND MORBIDITY

Mortality

Mortality patterns have been studied for the three most frequent osteoporotic fractures. In Rochester, MN, USA, survival rates 5 yr after hip and vertebral fractures were found to be ~80% of those expected for men and women of similar age without fractures.(28)

Hip Fracture. Hip fracture mortality is higher in men than women, increases with age,(28) and is greater for those with co-existing illnesses and poor prefracture functional status. There are ~31,000 excess deaths within 6 mo of the ~300,000 hip fractures that occur annually in the United States. About 8% of men and 3% of women >50 yr of age die while hospitalized for their fracture. In the United Kingdom, 12-mo survival after hip fracture for men is 63.3% versus 90.0% expected, and for women, 74.9% versus 91.1% expected.(5) The risk of death is greatest immediately after the fracture and decreases gradually over time. The cause of death is not usually directly attributable to the fracture itself but to other chronic diseases, which led both to the fracture and to the reduced life expectancy.

Vertebral Fracture. In contrast to this pattern, vertebral fractures are associated with increased mortality well beyond a year after fracture.(28) Again, it seems that it is co-morbid conditions that are responsible. The impairment of survival after vertebral fracture also markedly worsens as time from diagnosis of the fracture increases. This is in contrast to the pattern of survival for hip fractures. In the UK GPRD study, the observed survival in women 12 mo after vertebral fracture was 86.5% versus 93.6% expected. At 5 years, survival was 56.5% observed and 69.9% expected.(5)

Morbidity

In the United States, 7% of survivors of all types of fracture have some degree of permanent disability and 8% require long-term nursing home care. Overall, a 50-yr-old white American woman has a 13% chance of experiencing functional decline after any fracture.(29)

Hip Fracture. As with mortality, hip fractures contribute most to osteoporosis-associated disability. Patients are prone to developing acute complications such as pressure sores, bronchopneumonia, and urinary tract infections. Perhaps the most important long-term outcome is impairment of the ability to walk. Fifty percent of those ambulatory before the fracture are unable to walk independently afterward. Age is an important determinant of outcome, with 14% of 50- to 55-yr-old hip fracture victims being discharged to nursing homes versus 55% of those >90 yr old.(29)

Vertebral Fracture. Despite only a minority of vertebral fractures coming to clinical attention, they account for 52,000 hospital admissions in the United States and 2188 in England and Wales each year in patients ⩾45 yr of age. The major clinical consequences of vertebral fracture are back pain, kyphosis, and height loss. Quality of life (QUALEFFO) scores decrease as the number of vertebral fractures increases.(30)

Distal Forearm Fracture. Wrist fractures do not seem to increase mortality.(5) Although wrist fractures may impact on some activities such as writing or meal preparation, overall, few patients are completely disabled, despite over one half reporting only fair to poor function at 6 mo.(29)

Low Bone Mass in Children

There has been considerably less study of fractures in childhood. Most evidence comes from two large European studies that describe the epidemiology of fractures in childhood.(31–33) In Malmo, Sweden, the overall incidence of fracture was 212 per 10,000 girls and 257 per 10,000 boys, with 27% of girls and 42% of boys sustaining a fracture between birth and 16 yr of age. Fractures of the distal radius occurred most commonly, followed by fractures of the phalanges of the hand.(32,33) A follow-up study in Malmo between 1993 and 1994 found the incidence of fracture had decreased by almost 10% since the original study.(34) A similar pattern was found in the UK GPRD.(31) The overall incidence of fracture was 133.1 per

© 2008 American Society for Bone and Mineral Research

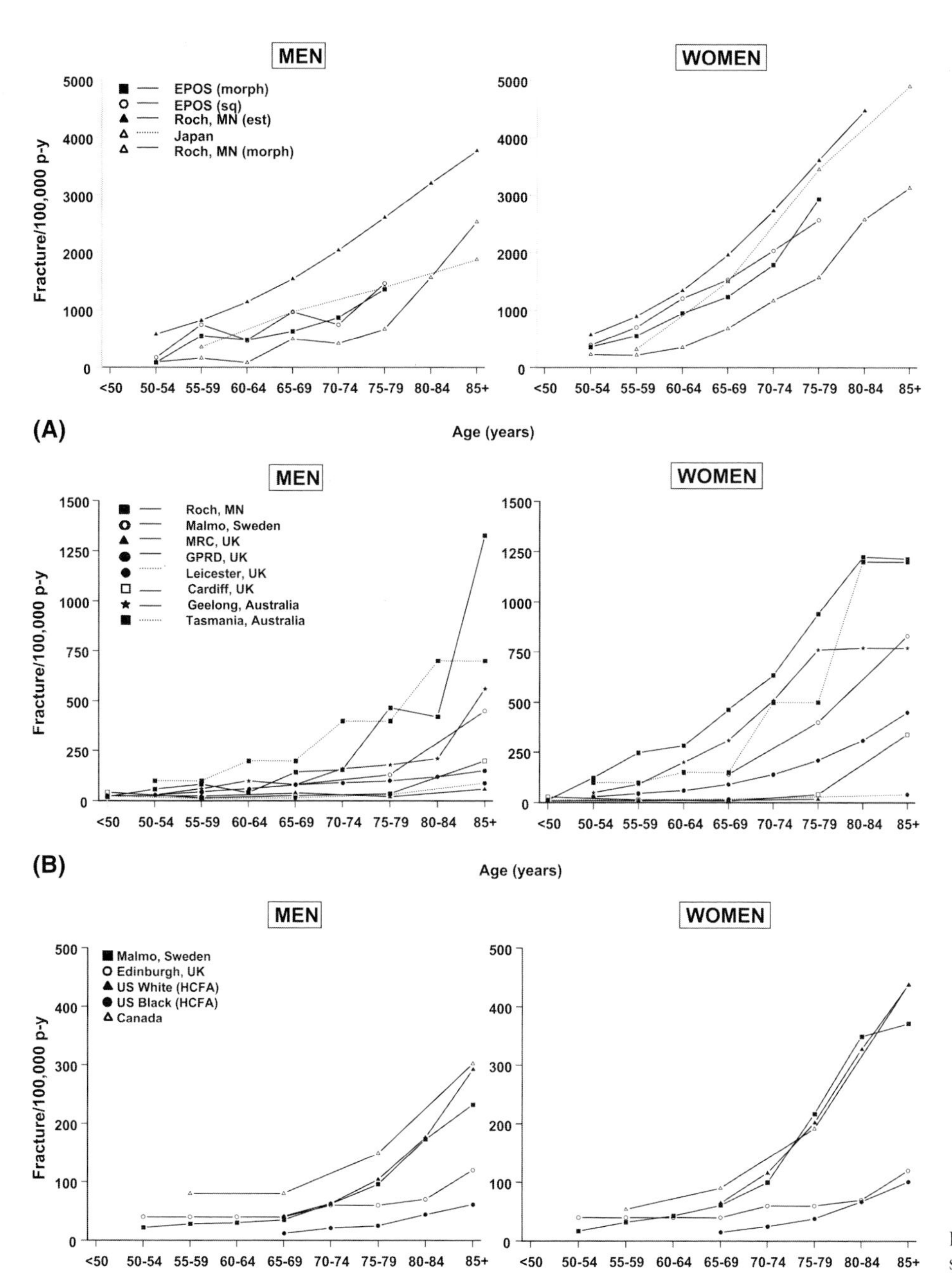

FIG. 3. Incidence of vertebral fracture in several populations, defined morphometrically (A), clinically (B), and by hospitalization for fracture (C).

10,000 children, with fractures being more common in boys than girls, with an incidence of 161.6 per 10,000 and 102.9 per 10,000, respectively. Again, the most common fracture site in both sexes was the radius/ulna, with a total of 39.3 per 10,000 per year. That these fractures are not solely caused by trauma is suggested by several recent studies documenting lower areal and volumetric BMD in children with distal forearm fractures than age- and sex-matched controls.[35,36]

Early Life Influences on Adult Fragility Fracture

The importance of bone mineral accrual in childhood and achievement of adequate peak bone mass (PBM) in early adulthood has been emphasized in recent work, showing that PBM is a major determinant osteoporosis risk in later life.[37] Over the last 20 yr, evidence has accrued that the early environment may have long-term influences on future bone health. This phenomenon of "developmental plasticity," whereby a single genotype may lead to different phenotypes, dependent on the prevailing environmental milieu, is well established in the natural world. There is a growing body of epidemiological evidence that a poor intrauterine environment leads to lower bone mass in adult life, both in the third and sixth/seventh decades.[38–40] Additionally, work in Finland has shown an association between poor infant and childhood growth and increased risk of hip fracture seven decades later.[41,42] Physiological studies have implicated the PTH/vitamin D axis in mechanisms underlying this phenomenon, such that mothers who are deficient in vitamin D in late pregnancy have children with decreased bone mass in childhood.[43] This novel area of research may lead ultimately to innovative strategies to im-

© 2008 American Society for Bone and Mineral Research

prove bone health in children, with a subsequent reduction in the burden of osteoporotic fracture in future generations.

CONCLUSION

Osteoporosis is therefore a disease that has a huge effect on public health. The impact of osteoporotic fracture is massive, not just for individuals, but for the health service, economy, and population as a whole. The characterization of some of the risk factors for inadequate PBM, involutional loss, and fracture, coupled with new pharmacologic therapies, means that we are now in a position to develop novel preventative and therapeutic strategies, both for the entire population and those at highest risk.

ACKNOWLEDGMENTS

We thank the Medical Research Council (UK), Arthritis Research Campaign (UK), National Osteoporosis Society (UK), International Osteoporosis Foundation, and European Union Network in Male Osteoporosis for funding this work.

REFERENCES

1. Anonymous 1993 Consensus development conference: Diagnosis, prophylaxis and treatment of osteoporosis. Am J Med **941:**646–650.
2. World Health Organization Study Group 1994 Assessment of Fracture Risk and Its Application to Screening for Postmenopausal Osteoporosis. World Health Organization, Geneva, Switzerland.
3. Kanis JA, Johnell O, Oden A, Dawson A, De Laet C, Jonsson B 2001 Ten year probabilities of osteoporotic fractures according to BMD and diagnostic thresholds. Osteoporos Int **12:**989–995.
4. Department of Health 1994. Advisory Group on Osteoporosis. Department of Health, London, UK.
5. van Staa TP, Dennison EM, Leufkens HG, Cooper C 2001 Epidemiology of fractures in England and Wales. Bone **29:**517–522.
6. The European Prospective Osteoporosis Study (EPOS) Group 2002 Incidence of vertebral fracture in Europe: Results from the European Prospective Osteoporosis Study (EPOS). J Bone Miner Res **17:**716–724.
7. U.S. Department of Health and Human Services 2004 Bone Health and Osteoporosis: A Report of the Surgeon General. U.S. Department of Health and Human Services, Rockville, MD, USA.
8. Goulding A, Jones IE, Taylor RW, Manning PJ, Williams SM 2000 More broken bones: A 4-year double cohort study of young girls with and without distal forearm fractures. J Bone Miner Res **15:**2011–2018.
9. Melton LJ 1988 Epidemiology of fractures. In: Riggs BL, Melton LJ (eds.) Osteoporosis: Etiology, Diagnosis and Management. Raven Press, New York, NY, USA, pp.133–154.
10. Cooper C, Melton LJ 1992 Epidemiology of osteoporosis. Trends Endocrinol Metab **314:**224–229.
11. Gallagher JC, Melton LJ, Riggs BL, Bergstrath E 1980 Epidemiology of fractures of the proximal femur in Rochester, Minnesota. Clin Orthop **150:**163–171.
12. Nevitt MC, Cummings SR 1993 Type of fall and risk of hip and wrist fractures: The study of osteoporotic fractures. The Study of Osteoporotic Fractures Research Group. J Am Geriatr Soc **41:**1226–1234.
13. Johnell O, Gullberg B, Allander E, Kanis JA 1992 The apparent incidence of hip fracture in Europe: A study of national register sources. MEDOS Study Group. Osteoporos Int **2:**298–302.
14. O'Neill TW, Felsenberg D, Varlow J, Cooper C, Kanis JA, Silman AJ 1996 The prevalence of vertebral deformity in European men and women: The European Vertebral Osteoporosis Study. J Bone Miner Res **11:**1010–1018.
15. Cooper C, Atkinson EJ, O'Fallon WM, Melton LJ 1992 Incidence of clinically diagnosed vertebral fractures: A population-based study in Rochester, Minnesota, 1985-1989. J Bone Miner Res **7:**221–227.
16. Melton LJ, Cooper C 2001 Magnitude and impact of osteoporosis and fractures. In: Marcus R, Feldman D, Kelsey J (eds.) Osteoporosis, 2nd ed. Academic Press, San Diego, CA, USA, pp. 557–567.
17. Ross PD, Davis JW, Epstein RS, Wasnich RD 1991 Pre-existing fractures and bone mass predict vertebral fracture incidence in women. Ann Intern Med **114:**919–923.
18. Cuddihy MT, Gabriel SE, Crowson CS, O'Fallon WM, Melton LJ 1999 Forearm fractures as predictors of subsequent osteoporotic fractures. Osteoporos Int **9:**469–475.
19. Ismail AA, Cockerill W, Cooper C, Finn JD, Abendroth K, Parisi G, Banzer D, Benevolenskaya LI, Bhalla AK, Armas JB, Cannata JB, Delmas PD, Dequeker J, Dilsen G, Eastell R, Ershova O, Falch JA, Felsch B, Havelka S, Hoszowski K, Jajic I, Kragl U, Johnell O, Lopez VA, Lorenc R, Lyritis G, Marchand F, Masaryk P, Matthis C, Miazgowski T, Pols HA, Poor G, Rapado A, Raspe HH, Reid DM, Reisinger W, Janott J, Scheidt-Nave C, Stepan J, Todd C, Weber K, Woolf AD, Ambrecht G, Gowin W, Felsenberg D, Lunt M, Kanis JA, Reeve J, Silman AJ, O'Neill TW 2001 Prevalent vertebral deformity predicts incident hip though not distal forearm fracture: Results from the European Prospective Osteoporosis Study. Osteoporos Int **12:**85–90.
20. Lunt M, O'Neill TW, Felsenberg D, Reeve J, Kanis JA, Cooper C, Silman AJ 2003 Characteristics of a prevalent vertebral deformity predict subsequent vertebral fracture: Results from the European Prospective Osteoporosis Study (EPOS). Bone **33:**505–513.
21. Lindsay R, Silverman SL, Cooper C, Hanley DA, Barton I, Broy SB, Licata A, Benhamou L, Geusens P, Flowers K, Stracke H, Seeman E 2001 Risk of new vertebral fracture in the year following a fracture. JAMA **285:**320–323.
22. Cooper C, Campion G, Melton LJ 1992 Hip fractures in the elderly: A world-wide projection. Osteoporos Int **2:**285–289.
23. Royal College of Physicians 1989 Fractured neck of femur: Prevention and management. Summary and report of the Royal College of Physicians. J R Coll Physicians Lond **23:**8–12.
24. Chevalley T, Guilley E, Herrmann FR, Hoffmeyer P, Rapin CH, Rizzoli R 2007 Incidence of hip fracture over a 10-year period (1991-2000): Reversal of a secular trend. Bone **40:**1284–1289.
25. Kannus P, Niemi S, Parkkari J, Palvanen M, Vuori I, Jarvinen M 2006 Nationwide decline in incidence of hip fracture. J Bone Miner Res **21:**1836–1838.
26. Elffors I, Allander E, Kanis JA, Gullberg B, Johnell O, Dequeker J, Dilsen G, Gennari C, Lopes Vaz AA, Lyritis G 1994 The variable incidence of hip fracture in southern Europe: The MEDOS Study. Osteoporos Int **4:**253–263.
27. Nagant DD, Devogelaer JP 1988 Increase in the incidence of hip fractures and of the ratio of trochanteric to cervical hip fractures in Belgium. Calcif Tissue Int **42:**201–203.
28. Cooper C, Atkinson EJ, Jacobsen SJ, O'Fallon WM, Melton LJ 1993 Population-based study of survival after osteoporotic fractures. Am J Epidemiol **137:**1001–1005.
29. Chrischilles EA, Butler CD, Davis CS, Wallace RB 1991 A model of lifetime osteoporosis impact. Arch Intern Med **151:**2026–2032.
30. Oleksik A, Lips P, Dawson A, Minshall ME, Shen W, Cooper C, Kanis J 2000 Health-related quality of life in postmenopausal women with low BMD with or without prevalent vertebral fractures. J Bone Miner Res **15:**1384–1392.
31. Cooper C, Dennison EM, Leufkens HG, Bishop N, van Staa TP 2004 Epidemiology of childhood fractures in Britain: A study using the General Practice Research Database. J Bone Miner Res **19:**1976–1981.
32. Landin LA 1997 Epidemiology of children's fractures. J Pediatr Orthop B **6:**79–83.
33. Landin LA 1983 Fracture patterns in children. Analysis of 8,682 fractures with special reference to incidence, etiology and secular changes in a Swedish urban population 1950-1979. Acta Orthop Scand Suppl **202:**1–109.
34. Tiderius CJ, Landin L, Duppe H 1999 Decreasing incidence of fractures in children: An epidemiological analysis of 1,673 fractures in Malmo, Sweden, 1993–1994. Acta Orthop Scand **70:**622–626.
35. Jones IE, Taylor RW, Williams SM, Manning PJ, Goulding A 2002 Four-year gain in bone mineral in girls with and without past forearm fractures: A DXA study (dual energy X-ray absorptiometry). J Bone Miner Res **17:**1065–1072.
36. Clark EM, Ness AR, Bishop NJ, Tobias JH 2006 Association between bone mass and fractures in children: A prospective cohort study. J Bone Miner Res **21:**1489–1495.
37. Hernandez CJ, Beaupre GS, Carter DR 2003 A theoretical analysis of the relative influences of peak BMD, age-related bone loss and menopause on the development of osteoporosis. Osteoporos Int **14:**843–847.
38. Cooper C, Cawley M, Bhalla A, Egger P, Ring F, Morton L,

© 2008 American Society for Bone and Mineral Research

Barker D 1995 Childhood growth, physical activity, and peak bone mass in women. J Bone Miner Res **10**:940–947.

39. Gale CR, Martyn CN, Kellingray S, Eastell R, Cooper C 2001 Intrauterine programming of adult body composition. J Clin Endocrinol Metab **86**:267–272.
40. Cooper C, Fall C, Egger P, Hobbs R, Eastell R, Barker D 1997 Growth in infancy and bone mass in later life. Ann Rheum Dis **56**:17–21.
41. Javaid MK, Eriksson JG, Valimaki MJ, Forsen T, Osmond C, Barker DJ, Cooper C 2005 Growth in infancy and childhood predicts hip fracture risk in late adulthood. Bone **36**(Suppl 1):S38.
42. Cooper C, Eriksson JG, Forsen T, Osmond C, Tuomilehto J, Barker DJ 2001 Maternal height, childhood growth and risk of hip fracture in later life: A longitudinal study. Osteoporos Int **12**:623–629.
43. Javaid MK, Crozier SR, Harvey NC, Gale CR, Dennison EM, Boucher BJ, Arden NK, Godfrey KM, Cooper C 2006 Maternal vitamin D status during pregnancy and childhood bone mass at age 9 years: A longitudinal study. Lancet **367**:36–43.

Chapter 39. Overview of Pathogenesis

Lawrence G. Raisz

New England Musculoskeletal Institute, Department of Medicine, University of Connecticut Health Center, Farmington, Connecticut

INTRODUCTION

Osteoporosis has become a major medical problem in the last half century, largely as the result of increased longevity and a changing lifestyle.[1] Osteoporosis has been defined as "a skeletal disorder characterized by comprised bone strength predisposing to an increased risk of fracture."[2] However, this definition could include many skeletal disorders. This chapter will focus on primary osteoporosis, which in itself is heterogeneous and presumably involves multiple pathogenetic mechanisms. This has led to use of terms such as postmenopausal, senile osteoporosis, and idiopathic osteoporosis. The first two overlap clinically, whereas the third is used for younger men and premenopausal women with osteoporotic fragility fractures without identifiable secondary causes. However, primary osteoporosis in older individuals is still essentially idiopathic; we do not know the precise pathways that lead to skeletal fragility. Because the clinical consequence of osteoporosis is an increased risk of fragility fractures, we must consider all of the factors that increase fracture risk as pathogenetic mechanisms. Clinically the time-course, distribution of fractures, and changes in bone mass, architecture, and turnover are quite variable in osteoporosis, which is consistent with the many different pathogenetic mechanisms that have been implicated.

BASIC PATHOGENETIC MECHANISMS

The major pathways that can result in skeletal fragility and increase fracture risk are outlined in Table 1. We assume that failure to achieve not only optimal peak bone mass but also optimal bone strength in the late teens or early 20s will contribute to skeletal fragility later in life, although the direct proof of this would require longitudinal studies of 30- to 50-yr duration. The fact that there are genetic determinants of fracture risk that are both dependent on and independent of bone mass supports this concept.[3,4] In any case, the major determinants of peak bone mass and strength are genetic. The recent finding of a high bone mass phenotype caused by activating mutations of the Wnt signaling pathway or loss of the Wnt inhibitor sclerostin has raised an interesting question.[5,6] Why did humans not evolve generally to have such a high bone mass and decreased fracture risk? It is likely that, in evolution, a lighter skeleton and greater mobility was favored for survival of the species, whereas age-related bone loss and skeletal fragility would have little effect. The importance of the lipoprotein receptor–related protein-5 (LRP-5) pathway is reinforced by the finding that inactivating mutations of the *LRP-5* gene can result in early onset of osteoporosis.[7,8]

A large number of genetic polymorphisms that can affect bone mass have been identified.[9] Some polymorphisms can affect fracture risk independent of BMD, such as a polymorphism of the $\alpha 1$ collagen gene, located in the first intron, which could alter in transcriptional control rather than protein sequence.[9] Other studies suggest that fracture risk can be affected independent of bone mass. The finding that high homocysteine levels can increase fracture risk, although not seen in all studies,[10–12] could represent an acquired effect on bone fragility, affecting collagen cross-links.[13] Nonenzymatic glycation of collagen that occurs with age and is accelerated in diabetes could also affect both matrix and mineral quality.[14,15] These findings indicate that the original concept of osteoporosis as a disorder in which the mineral and matrix are chemically normal but simply decreased in amount and defective in architecture is not correct. Subtle changes in collagen structure or cross-linking and changes in the mineral could affect skeletal fragility. Of course, genes can also affect fracture risk by their effects on macroarchitecture of the skeleton or on other factors such as neuromuscular function and the risk of falling.

In addition to the important impact of genetics, lifestyle factors can contribute further to the optimal or the inadequate development of the skeleton, as well as to subsequent maintenance of skeletal strength.[16] Epidemiologic evidence for this is abundant. Calcium, vitamin D, and physical activity are probably the most important determinants of peak bone mass and strength.[17] However, total nutrition, protein intake, and micronutrients such as vitamins B_6, B_{12}, and folic acid, which may be related to homocysteine, and vitamin K may be important.[18,19] There are also complex interactions between the gene and the environment,[20] as well as among different nutrients.[21] Negative influences on the skeleton that can affect bone mass and fracture risk include immobilization, smoking, and excessive alcohol intake.[22]

Because peak bone mass has a Guassian distribution, values 1 or 2 SD below the mean will occur in 14% and 2% of young men and women, respectively; however, they rarely have fragility fractures. Progressively increasing fracture risk occurs with loss of bone mass, and deterioration of microarchitecture associated with estrogen deficiency and aging results. There is evidence that some bone loss begins in the 20s, particularly for

Dr. Raisz has served as a consultant for Novartis, Servier International, Procter & Gamble, Akros Pharma, Pfizer, and Omeros.

© 2008 American Society for Bone and Mineral Research

Table 1. Pathogenesis of Osteoporosis

Failure to achieve optimal peak bone mass and strength	*Accelerated bone loss caused by increased resorption*	*Inadequate formation response during remodeling*	*Increased falls*
Genetics	Estrogen efficiency	Impaired cell renewal	Neural and muscular impairment
Nutrition	Calcium and vitamin D deficiency	Decreased growth factors	Drugs
Lifestyle	Secondary hyperparathyroidism	Cytokines	Environmental hazards
	Cytokines		

trabecular bone, but is then accelerated.[23] This acceleration requires an increase in osteoclastic activity. Whereas the precise mechanism of accelerated resorption is not known, estrogen deficiency is clearly important in both men and women.[24] This has been attributed to effects of estrogen directly on the hematopoietic precursors of osteoclasts, on the osteoblast–osteoclast interaction that regulates bone resorption and on other hematopoietic cells such as lymphocytes that may produce cytokines influencing this response.[25–27] Fracture risk is increased at the lowest levels of estrogen,[28,29] and quite low doses can prevent bone loss.[30,31] Androgens may play an additional role both directly on bone resorption and indirectly through their effects on muscle.[27,32] Many other hormones, neural mediators, and cytokines have been implicated in increased bone resorption of osteoporosis.[33–38] There are limited data on these fractures in humans.[39] Oxidative stress may be involved in bone loss of estrogen deficiency, possibly by stimulating TNFα production.[40,41] Much of the evidence derives from the use of inhibitors, and it is important to recognize that, whereas the inhibited factor may be playing a role in the complex concert that regulates osteoclast formation and activity, the specific agonists that is being regulated by estrogen or which changes with aging may be a different player in that concert.[36]

A second mechanism for increased bone resorption is deficiency of calcium and vitamin D, leading to secondary hyperparathyroidism.[17] This is particularly important in the elderly with low intakes of calcium, poor sun exposure, and lack of vitamin D supplementation. However, an increased level of PTH associated with low circulating levels of 25 hydroxyvitamin D is quite common at all ages and in all populations studied. Vitamin D deficiency can also impair physical performance and the risk of falling.[42,43] Both calcium and vitamin D are need to reverse this.[44]

Increased bone resorption alone does not necessarily lead to skeletal fragility. High rates of bone remodeling in adolescence are associated with a gain in bone mass and strength. Thus, by definition, there must be a relative defect in the formation response to accelerated bone remodeling in osteoporosis. Moreover, based on histomorphometric data, it is likely that, in some osteoporotic patients, the major defect is not accelerated resorption but decreased formation.[45,46] The mechanisms of this defect in formation are not well understood. There is a progressive decrease in the size of packets of new bone that are formed during trabecular remodeling that begins in the 20s.[47] Age-related changes in a number of growth factors have been described, particularly insulin-like growth factor-1 (IGF-1).[48] Some cytokines, such as TNFα, may inhibit bone formation and stimulate bone resorption, although studies on this are inconsistent.[49,50] There maybe accelerated cell death of osteoblasts and osteocytes or impairment of osteocyte signaling.[51,52] Estrogen and androgen deficiency could also play a role in bone formation through Wnt signaling.[53] Lack of mechanical stimulation caused by decreased physical activity contributes to impaired bone formation through multiple pathways.[54–56]

Finally, the risk of nonvertebral fractures depends on the frequency and type of falls. Impaired vision or poor visual correction will increase the risk of falls, as will a large number of drugs that impair balance or cause postural hypotension.[57]

NEW CONCEPTS

Although the mechanisms listed above have focused largely on the skeleton or on neuromuscular function that directly affect the skeleton, there is emerging evidence that the skeleton might be involved in more global changes in cell and organ function or metabolic regulation. For example, the increase in adipogenesis that occurs with aging may also take place in the marrow, so that more adipocytes and fewer osteoblasts are formed from precursor cells. The recent finding of bone loss with thiazolidinediones therapy supports this.[58] The skeleton is richly innervated and feedback mechanisms from the CNS involving leptin and neuropeptides can affect bone.[59] This may involve interactions with energy metabolism.[60] Age-related changes in skeletal vasculature could also contribute to bone loss.[61] Although none of these pathways has yet been convincingly shown to affect skeletal fragility in humans, further studies may show them to be relevant, and this could alter our methods of diagnosing, preventing, and treating osteoporosis.

CONCLUSION

The emergence of osteoporosis and fragility fractures as a major cause of morbidity and mortality has it roots both in evolution and in human progress. Humans evolved with a relatively light skeleton that enhanced mobility but was prone to traumatic fracture, with little concern for the effects of aging on the skeleton. With progress to an industrialized society and a remarkable increase in longevity, skeletal fragility became common, and fractures occurred with minimal or no trauma. As we learn more and more about the mechanisms that have produced this state, it should be possible to reduce the burden of fractures through interventions that help to achieve optimal peak bone mass, reduce excessive resorption, enhance formation, and reduce the risk of falling.

REFERENCES

1. Raisz LG 2005 Pathogenesis of osteoporosis: Concepts, conflicts, and prospects. J Clin Invest **115:**3318–3325.
2. NIH 2001 Consensus Conference: Osteoporosis prevention, diagnosis, and therapy. JAMA **285:**785–795.
3. Bollerslev J, Wilson SG, Dick IM, Islam FM, Ueland T, Palmer L, Devine A, Prince RL 2005 LRP5 gene polymorphisms predict bone mass and incident fractures in elderly Australian women. Bone **36:**599–606.
4. Ioannidis JP, Ralston SH, Bennett ST, Brandi ML, Grinberg D, Karassa FB, Langdahl B, van Meurs JB, Mosekilde L, Scollen S, Albagha OM, Bustamante M, Carey AH, Dunning AM, Enjuanes A, van Leeuwen JP, Mavilia C, Masi L, McGuigan FE, Nogues X, Pols HA, Reid DM, Schuit SC, Sherlock RE, Uitterlinden AG 2004 Differential genetic effects of ESR1 gene polymorphisms on osteoporosis outcomes. JAMA **292:**2105–2114.

© 2008 American Society for Bone and Mineral Research

5. Balemans W, Ebeling M, Patel N, Van Hul E, Olson P, Dioszegi M, Lacza C, Wuyts W, Van Den Ende J, Willems P, Paes-Alves AF, Hill S, Bueno M, Ramos FJ, Tacconi P, Dikkers FG, Stratakis C, Lindpaintner K, Vickery B, Foernzler D, Van Hul W 2001 Increased bone density in sclerosteosis is due to the deficiency of a novel secreted protein (SOST). Hum Mol Genet **10:**537–543.
6. Little RD, Carulli JP, Del Mastro RG, Dupuis J, Osborne M, Folz C, Manning SP, Swain PM, Zhao SC, Eustace B, Lappe MM, Spitzer L, Zweier S, Braunschweiger K, Benchekroun Y, Hu X, Adair R, Chee L, FitzGerald MG, Tulig C, Caruso A, Tzellas N, Bawa A, Franklin B, McGuire S, Nogues X, Gong G, Allen KM, Anisowicz A, Morales AJ, Lomedico PT, Recker SM, Van Eerdewegh P, Recker RR, Johnson ML 2002 A mutation in the LDL receptor-related protein 5 gene results in the autosomal dominant high-bone-mass trait. Am J Hum Genet **70:**11–19.
7. Gong Y, Slee RB, Fukai N, Rawadi G, Roman-Roman S, Reginato AM, Wang H, Cundy T, Glorieux FH, Lev D, Zacharin M, Oexle K, Marcelino J, Suwairi W, Heeger S, Sabatakos G, Apte S, Adkins WN, Allgrove J, Arslan-Kirchner M, Batch JA, Beighton P, Black GC, Boles RG, Boon LM, Borrone C, Brunner HG, Carle GF, Dallapiccola B, De Paepe A, Floege B, Halfhide ML, Hall B, Hennekam RC, Hirose T, Jans A, Juppner H, Kim CA, Keppler-Noreuil K, Kohlschuetter A, LaCombe D, Lambert M, Lemyre E, Letteboer T, Peltonen L, Ramesar RS, Romanengo M, Somer H, Steichen-Gersdorf E, Steinmann B, Sullivan B, Superti-Furga A, Swoboda W, van den Boogaard MJ, Van Hul W, Vikkula M, Votruba M, Zabel B, Garcia T, Baron R, Olsen BR, Warman ML 2001 LDL receptor-related protein 5 (LRP5) affects bone accrual and eye development. Cell **107:**513–523.
8. Hartikka H, Makitie O, Mannikko M, Doria AS, Daneman A, Cole WG, Ala-Kokko L, Sochett EB 2005 Heterozygous mutations in the LDL receptor-related protein 5 (LRP5) gene are associated with primary osteoporosis in children. J Bone Miner Res **20:**783–789.
9. Ralston SH 2007 Genetics of osteoporosis. Proc Nutr Soc **66:**158–165.
10. McLean RR, Jacques PF, Selhub J, Tucker KL, Samelson EJ, Broe KE, Hannan MT, Cupples LA, Kiel DP 2004 Homocysteine as a predictive factor for hip fracture in older persons. N Engl J Med **350:**2042–2049.
11. Perier MA, Gineyts E, Munoz F, Sornay-Rendu E, Delmas PD 2007 Homocysteine and fracture risk in postmenopausal women: The OFELY study. Osteoporos Int **18:**1329–1336.
12. van Meurs JB, Dhonukshe-Rutten RA, Pluijm SM, van der Klift M, de Jonge R, Lindemans J, de Groot LC, Hofman A, Witteman JC, van Leeuwen JP, Breteler MM, Lips P, Pols HA, Uitterlinden AG 2004 Homocysteine levels and the risk of osteoporotic fracture. N Engl J Med **350:**2033–2041.
13. Herrmann M, Widmann T, Herrmann W 2006 Re: "Elevated serum homocysteine and McKusick's hypothesis of a disturbed collagen cross-linking: What do we really know?" Bone **39:**1385–1286.
14. Saito M, Fujii K, Soshi S, Tanaka T 2006 Reductions in degree of mineralization and enzymatic collagen cross-links and increases in glycation-induced pentosidine in the femoral neck cortex in cases of femoral neck fracture. Osteoporos Int **17:**986–995.
15. Tang SY, Zeenath U, Vashishth D 2007 Effects of non-enzymatic glycation on cancellous bone fragility. Bone **40:**1144–1151.
16. Lock CA, Lecouturier J, Mason JM, Dickinson HO 2006 Lifestyle interventions to prevent osteoporotic fractures: A systematic review. Osteoporos Int **17:**20–28.
17. Lips P 2001 Vitamin D deficiency and secondary hyperparathyroidism in the elderly: Consequences for bone loss and fractures and therapeutic implications. Endocr Rev **22:**477–501.
18. Kaneki M, Hosoi T, Ouchi Y, Orimo H 2006 Pleiotropic actions of vitamin K: Protector of bone health and beyond? Nutrition **22:**845–852.
19. Baines M, Kredan MB, Usher J, Davison A, Higgins G, Taylor W, West C, Fraser WD, Ranganath LR 2007 The association of homocysteine and its determinants MTHFR genotype, folate, vitamin B12 and vitamin B6 with bone mineral density in postmenopausal British women. Bone **40:**730–736.
20. Ferrari S, Manen D, Bonjour JP, Slosman D, Rizzoli R 1999 Bone mineral mass and calcium and phosphate metabolism in young men: Relationships with vitamin D receptor allelic polymorphisms. J Clin Endocrinol Metab **84:**2043–2048.
21. Chevalley T, Bonjour JP, Ferrari S, Rizzoli R 2008 High protein intake enhances the postive impact of physical activity on bone mineral content in pre-pubertal boys. J Bone Miner Res **23:**131–142.
22. Wong PK, Christie JJ, Wark JD 2007 The effects of smoking on bone health. Clin Sci (Lond) **113:**233–241.
23. Khosla S, Riggs BL, Atkinson EJ, Oberg AL, McDaniel LJ, Holets M, Peterson JM, Melton LJ III 2006 Effects of sex and age on bone microstructure at the ultradistal radius: A population-based non-invasive in vivo assessment. J Bone Miner Res **21:**124–131.
24. Falahati-Nini A, Riggs BL, Atkinson EJ, O'Fallon WM, Eastell R, Khosla S 2000 Relative contributions of testosterone and estrogen in regulating bone resorption and formation in normal elderly men. J Clin Invest **106:**1553–1560.
25. Taxel P, Kaneko H, Lee SK, Aguila HL, Raisz LG, Lorenzo JA 2008 Estradiol rapidly inhibits osteoclastogenesis and RANKL expression in bone marrow cultures in postmenopausal women: A pilot study. Osteoporos Int **19:**193–199.
26. Pacifici R 2007 Estrogen deficiency, T cells and bone loss. Cell Immunol (in press).
27. Syed F, Khosla S 2005 Mechanisms of sex steroid effects on bone. Biochem Biophys Res Commun **328:**688–696.
28. Amin S, Zhang Y, Felson DT, Sawin CT, Hannan MT, Wilson PW, Kiel DP 2006 Estradiol, testosterone, and the risk for hip fractures in elderly men from the Framingham Study. Am J Med **119:**426–433.
29. Cummings SR, Browner WS, Bauer D, Stone K, Ensrud K, Jamal S, Ettinger B 1998 Endogenous hormones and the risk of hip and vertebral fractures among older women. Study of Osteoporotic Fractures Research Group. N Engl J Med **339:**733–738.
30. Prestwood KM, Kenny AM, Kleppinger A, Kulldorff M 2003 Ultralow-dose micronized 17beta-estradiol and bone density and bone metabolism in older women: A randomized controlled trial. JAMA **290:**1042–1048.
31. Huang A, Ettinger B, Vittinghoff E, Ensrud KE, Johnson KC, Cummings SR 2007 Endogenous estrogen levels and the effects of ultra low-dose transdermal estradiol therapy on bone turnover and bone density in postmenopausal women. J Bone Miner Res **22:**1791–1797.
32. Kohn FM 2006 Testosterone and body functions. Aging Male **9:**183–188.
33. Ammann P, Rizzoli R, Bonjour JP, Bourrin S, Meyer JM, Vassalli P, Garcia I 1997 Transgenic mice expressing soluble tumor necrosis factor-receptor are protected against bone loss caused by estrogen deficiency. J Clin Invest **99:**1699–1703.
34. Gao Y, Qian WP, Dark K, Toraldo G, Lin AS, Guldberg RE, Flavell RA, Weitzmann MN, Pacifici R 2004 Estrogen prevents bone loss through transforming growth factor beta signaling in T cells. Proc Natl Acad Sci USA **101:**16618–16623.
35. Horwitz MCL 2002 IL-10, IL-4, the LIF/IL-6 family, and additional cytokines. In: Bilezikian JP, Raisz LG, Rodan GA (eds.) Principles of Bone Biology. Academic Press, San Diego, CA, USA, pp. 961–977.
36. Kawaguchi H, Pilbeam CC, Vargas SJ, Morse EE, Lorenzo JA, Raisz LG 1995 Ovariectomy enhances and estrogen replacement inhibits the activity of bone marrow factors that stimulate prostaglandin production in cultured mouse calvariae. J Clin Invest **96:**539–548.
37. Pierroz DD, Bouxsein ML, Rizzoli R, Ferrari SL 2006 Combined treatment with a beta-blocker and intermittent PTH improves bone mass and microarchitecture in ovariectomized mice. Bone **39:**260–267.
38. Weitzmann MN, Cenci S, Rifas L, Haug J, Dipersio J, Pacifici R 2001 T cell activation induces human osteoclast formation via receptor activator of nuclear factor kappaB ligand-dependent and -independent mechanisms. J Bone Miner Res **16:**328–337.
39. Charatcharoenwitthaya N, Khosla S, Atkinson EJ, McCready LK, Riggs BL 2007 Effect of blockade of TNF-alpha and interleukin-1 action on bone resorption in early postmenopausal women. J Bone Miner Res **22:**724–729.
40. Jagger CJ, Lean JM, Davies JT, Chambers TJ 2005 Tumor necrosis factor-alpha mediates osteopenia caused by depletion of antioxidants. Endocrinology **146:**113–118.
41. Lean JM, Jagger CJ, Kirstein B, Fuller K, Chambers TJ 2005 Hydrogen peroxide is essential for estrogen-deficiency bone loss and osteoclast formation. Endocrinology **146:**728–735.
42. Bischoff-Ferrari HA, Dawson-Hughes B, Willett WC, Staehelin HB, Bazemore MG, Zee RY, Wong JB 2004 Effect of vitamin D on falls: A meta-analysis. JAMA **291:**1999–2006.

© 2008 American Society for Bone and Mineral Research

43. Wicherts IS, van Schoor NM, Boeke AJ, Visser M, Deeg DJ, Smit J, Knol DL, Lips P 2007 Vitamin D status predicts physical performance and its decline in older persons. J Clin Endocrinol Metab **92:**2058–2065.
44. Boonen S, Lips P, Bouillon R, Bischoff-Ferrari HA, Vanderschueren D, Haentjens P 2007 Need for additional calcium to reduce the risk of hip fracture with vitamin d supplementation: Evidence from a comparative metaanalysis of randomized controlled trials. J Clin Endocrinol Metab **92:**1415–1423.
45. Eriksen EF, Hodgson SF, Eastell R, Cedel SL, O'Fallon WM, Riggs BL 1990 Cancellous bone remodeling in type I (postmenopausal) osteoporosis: Quantitative assessment of rates of formation, resorption, and bone loss at tissue and cellular levels. J Bone Miner Res **5:**311–319.
46. Parfitt AM, Villanueva AR, Foldes J, Rao DS 1995 Relations between histologic indices of bone formation: Implications for the pathogenesis of spinal osteoporosis. J Bone Miner Res **10:**466–473.
47. Lips P, Courpron P, Meunier PJ 1978 Mean wall thickness of trabecular bone packets in the human iliac crest: Changes with age. Calcif Tissue Res **26:**13–17.
48. Rosen CJ 2004 Insulin-like growth factor I and bone mineral density: Experience from animal models and human observational studies. Best Pract Res Clin Endocrinol Metab **18:**423–435.
49. Iqbal J, Sun L, Kumar TR, Blair HC, Zaidi M 2006 Follicle-stimulating hormone stimulates TNF production from immune cells to enhance osteoblast and osteoclast formation. Proc Natl Acad Sci USA **103:**14925–14930.
50. Zhou FH, Foster BK, Zhou XF, Cowin AJ, Xian CJ 2006 TNF-alpha mediates p38 MAP kinase activation and negatively regulates bone formation at the injured growth plate in rats. J Bone Miner Res **21:**1075–1088.
51. Bonewald L 2007 Osteocytes as dynamic, multifunctional cells. Ann NY Acad Sci **1116:**281–290.
52. Jilka R, Weinstein RS, Parfitt AM, Manolagas SC 2007 Quantifying osteoblast and osteocyte aptoptosis: Challenges and rewards. J Bone Miner Res **22:**1492–1501.
53. Armstrong VJ, Muzylak M, Sunters A, Zaman G, Saxon LK, Price JS, Lanyon LE 2007 Wnt/beta-catenin signaling is a component of osteoblastic bone cell early responses to load-bearing and requires estrogen receptor alpha. J Biol Chem **282:**20715–20727.
54. Kapur S, Baylink DJ, Lau KH 2003 Fluid flow shear stress stimulates human osteoblast proliferation and differentiation through multiple interacting and competing signal transduction pathways. Bone **32:**241–251.
55. Triplett JW, O'Riley R, Tekulve K, Norvell SM, Pavalko FM 2007 Mechanical loading by fluid shear stress enhances IGF-1 receptor signaling in osteoblasts in a PKCzeta-dependent manner. Mol Cell Biomech **4:**13–25.
56. Wadhwa S, Godwin SL, Peterson DR, Epstein MA, Raisz LG, Pilbeam CC 2002 Fluid flow induction of cyclo-oxygenase 2 gene expression in osteoblasts is dependent on an extracellular signal-regulated kinase signaling pathway. J Bone Miner Res **17:**266–274.
57. Tinetti ME, Gordon C, Sogolow E, Lapin P, Bradley EH 2006 Fall-risk evaluation and management: Challenges in adopting geriatric care practices. Gerontologist **46:**717–725.
58. Lazarenko OP, Rzonca SO, Hogue WR, Swain FL, Suva LJ, Lecka-Czernik B 2007 Rosiglitazone induces decreases in bone mass and strength that are reminiscent of aged bone. Endocrinology **148:**2669–2680.
59. Karsenty G 2006 Convergence between bone and energy homeostases: Leptin regulation of bone mass. Cell Metab **4:**341–348.
60. Lee NK, Sowa H, Hinoi E, Ferron M, Ahn JD, Confavreux C, Dacquin R, Mee PJ, McKee MD, Jung DY, Zhang Z, Kim JK, Mauvais-Jarvis F, Ducy P, Karsenty G 2007 Endocrine regulation of energy metabolism by the skeleton. Cell **130:**456–469.
61. Prisby RD, Ramsey MW, Behnke BJ, Dominguez JM II, Donato AJ, Allen MR, Delp MD 2007 Aging reduces skeletal blood flow, endothelium-dependent vasodilation, and NO bioavailability in rats. J Bone Miner Res **22:**1280–1288.

Chapter 40. Nutrition and Osteoporosis

Connie M. Weaver[1] and Robert P. Heaney[2]

[1]Department of Foods and Nutrition, Purdue University, West Lafayette, Indiana; and [2]Creighton University, Omaha, Nebraska

INTRODUCTION

Nutrients are essential to the viability of all cells including those in bone. However, it is the whole diet rather than individual nutrients that determines many factors that influence bone, including nutrition adequacy of all essential nutrients, the presence or absence of inhibitors to absorption and utilization of individual nutrients, the energy available for growth and maintenance of bone and adiposity, and acid-base balance. Diet and other lifestyle choices around the world lead to shortages of some nutrients that are particularly important to bone (i.e., calcium, protein, and vitamin D). Our ability to accurately link and quantify the role of individual nutrients or whole diet in building and maintaining bone is handicapped by methodological limitations in assessing dietary intakes and the time lag for seeing consequences of diet on bone. Although nutrition is an important component for treating those with osteoporosis as addressed in another chapter, the more important role of diet is preventive. The cumulative effect of diet over the life span influences development of peak bone mass and its subsequent maintenance. Osteoporosis has been called a pediatric disorder because adult peak bone mass is largely determined during childhood.

The authors state that they have no conflicts of interest.

ROLE OF DIET IN BUILDING PEAK BONE MASS

Rapid skeletal growth occurs in infancy and adolescence. During growth, there is a high demand for nutrients. For bone mineral matrix formation, calcium, phosphorus, and magnesium are particularly important. Vitamin D status is important for active calcium absorption across the gut. Many nutrients are important for collagen synthesis, including protein, copper, zinc, and iron. Meeting nutrient needs is easier during infancy through breast feeding or carefully developed infant formulas. With the exception of its vitamin D content, the nutrient profile of breast milk is relatively constant and nearly independent of the diet of mothers.(1) In contrast, the pubertal growth spurt occurs at a life stage where diet becomes increasingly influenced by peers. This is an extremely important period for development of peak bone mass. During the 4 yr surrounding peak bone mass accretion, ~40% of peak bone mass is acquired. Peak bone mass velocity determined from a longitudinal study in white boys and girls was 409 g/d in boys and 325 g/d in girls.(2) During this period, controlled feeding studies on a range of calcium intakes in black and white girls showed that calcium intake explained 12.3% of the variance in skeletal calcium retention compared with the 13.7% explained by race, whereas a measure of sexual maturity explained an additional

© 2008 American Society for Bone and Mineral Research

4%.[3] That fraction caused by calcium intake emphasizes the importance of nutrition at this life stage.

In addition to providing raw materials for growth, diet can alter regulators of growth that affect bone accretion. In a randomized, controlled trial of a pint of milk a day in early pubertal girls, serum IGF-1 increased, which was thought to be partly casual for the increase in BMD in the intervention group relative to the control group.[4] Diet can alter timing of menarche, perhaps through modulating growth hormones. This was shown in a study of girls 7.9 yr of age who were randomized to products fortified with dairy minerals and followed for 16 yr until approximate peak mass had been achieved.[5] Girls who had received the mineral complex, although only for 1 yr, achieved menarche almost 5 mo earlier on average than the control group. Earlier menarche with longer exposure to estrogen resulted in greater bone accretion at six skeletal sites.

The concern for low peak bone mass is risk of fracture, especially later in life. Fracture risk is also of concern in childhood, particularly during the period of relatively low BMD when bone consolidation lags behind growth.[6] Fracture incidence in childhood has dramatically increased over the last three decades.[7]

The persistence of effects of short-term dietary interventions throughout life is controversial. The advantage in bone gain with supplementation of a calcium rich source during randomized, controlled trials largely disappeared on follow-up after cessation of the intervention in some trials,[8] but not others.[9] Additionally, the possibility of catch-up growth has been raised in a randomized controlled calcium supplementation trial from prepuberty to peak bone mass.[10] Although final BMD was not different on average for total body or radius, total hip BMD showed no catch-up growth, nor did taller women.[10,11] Evidence of life-long consequences of diet early in life comes from those with eating disorders who fail to recover BMD.[12]

ROLE OF NUTRITION IN MAINTAINING BONE MASS

Bone mass is ultimately determined by the genetic program as modified by current and past mechanical loading and limited or permitted by nutrition. The genetic potential cannot be reached or maintained if dietary calcium intake and absorption are insufficient. The aggregate total of bone resorptive activity is controlled systemically by PTH, which in turn responds to the demands of extracellular fluid calcium ion homeostasis and not to the structural need for bone mass.

Calcium is the principal cation of bone mineral. Bone constitutes a very large nutrient reserve for calcium, which, over the course of evolution, acquired a secondary, structural function that is responsible for its importance for osteoporosis. Bone strength varies as the approximate second power of bone structural density. Accordingly, any decrease in bone mass produces a corresponding decrease in bone strength. Whereas reserves are designed to be used in times of need, such use would normally be temporary. Sustained, unbalanced withdrawals deplete the reserves and thereby reduce bone strength. The principal skeletal role of calcium once peak bone mass has been achieved is to offset obligatory losses of calcium through sweat, desquamated skin, and excreta.

Whenever absorbed calcium intake is insufficient to meet the demands of growth and/or the drain of cutaneous and excretory losses, resorption will be stimulated and bone mass will be reduced as the body scavenges the calcium released in bone resorption.

In addition to depleting or limiting bone mass, low calcium intakes directly cause fragility through this PTH-stimulated increase in bone remodeling. Resorption pits on trabeculae cause applied loads to shift to adjacent bone, leading to increased strain locally. In this way excessive remodeling is itself a fragility factor, altogether apart from its effect on bone mass. When adequate calcium is absorbed, PTH-stimulated remodeling decreases immediately,[13] and with it, fragility. Fracture rate responses in the major treatment trials show the predicted prompt reduction in fracture risk.[14,15]

DIETARY PATTERNS

Recommended food patterns around the world attempt to meet national nutrient requirements and to promote health and reduce risk of disease. The food group most associated with bone health is the dairy group. This food group provides between 20% and 75% of recommended calcium, as well as high-quality protein, phosphorus, magnesium, and potassium. Recommended intakes of dairy products range from one serving or less per day in some Asian countries to three servings a day for most people in North America. Intakes on average are much less than the recommended levels for many populations. Dairy intakes in much of the world have been low persistently, possibly related to high incidence of lactose maldigestion. Milk consumption has declined over the last half century in the United States, concurrent with increased consumption of soft drinks.[16] Adequacy of milk intake has been associated with adequacy of a number of nutrients in children including calcium, potassium, magnesium, zinc, iron, riboflavin, vitamin A, folate, and vitamin D.[17] Alternative sources to replace this whole package of nutrients are not typically consumed in sufficient amounts to replace milk.[18] Milk consumption within various cultures has been positively associated with bone health. High dairy-consuming regions have better bone measures than low dairy-consuming regions in Yugoslavia[19] and China.[20] Milk avoiders have higher risk of fracture than milk drinking counterparts in children[21] and adults.[22] Retrospective studies show that milk drinking in childhood is inversely associated with risk of hip fracture later in life.[23]

Two other food groups have also been associated with bone health by affecting acid-base balance—fruits and vegetables positively and meats, fish, and poultry negatively. Sulfur-rich amino acids from the meat groups favor an acidic ash that increases calciuria, whereas fruits and vegetables favor an alkaline ash largely because of potassium content, which has a protective effect against calcium loss in the urine. However, recently studies have challenged this paradigm for both of these food groups. Some vegetables and herbs have the ability to decrease bone resorption, but the effect is independent of the alkaline load or potassium content.[24] Dietary potassium reduces calciuria, but it is offset by decreased calcium absorption so there is no change in calcium balance.[25] The calciuria induced by protein does not lead to increased bone resorption or affect the overall calcium economy.[26] In fact, protein intake negatively predicts age-related bone loss and protein supplements decrease fracture rates in the elderly.[27]

Dietary salt is the largest predictor of urinary calcium excretion.[28] Sodium and calcium share transport proteins in the kidney. In adolescence, less sodium, and consequently less calcium, is excreted by black compared with white girls, presumably because of racial differences in renal transport.[29]

CONCLUSIONS

Bone health rests on a combination of mechanical loading and adequate intakes of a broad array of macro- and micronutrients. Most important of these are calcium, vitamin D, and protein. Most diets, inadequate in one key nutrient, will be inadequate in several. Optimal protection of bone requires a

© 2008 American Society for Bone and Mineral Research

diet rich in all the essential nutrients. Mononutrient supplementation regimens will often be inadequate to ensure optimal nutritional protection of bone health.

REFERENCES

1. Prentice A, Laskey A, Jarjon LMA 1999 Lactation and bone development: Implications for calcium requirement of infants and lactating mothers. In: Bonjour JP, Tsang RC (eds). Nutrition and Bone Development. Lippincott-Raven, Philadelphia, PA, USA, pp. 127–145.
2. Bailey DA, McKay HA, Mirwald RL, Crocker PRE, Faulkner RA 1999 A six-year longitudinal study of the relationship of physical activity to bone mineral accrual in growing children: The University of Saskatchewan bone mineral accrual study. J Bone Miner Res **14:**1672–1679.
3. Braun M, Palacios C, Wigertz K, Jackman LA, Bryant RJ, McCabe LD, Martin BR, McCabe GP, Peacock M, Weaver CM 2007 Racial differences in skeletal calcium retention in adolescent girls on a range of controlled calcium intakes. Am J Clin Nutr **85:**1657–1663.
4. Cadogan J, Eastell R, Jones N, Barker ME 1997 Milk intake and bone mineral acquisition in adolescent girls: Randomized, controlled intervention trial. BMJ **315:**1255–1260.
5. Chevalley T, Rizzoli R, Hans D, Ferrari S, Bonjour J-P 2005 Interaction between calcium intake and menarcheal age on bone mass gain: An eight-year follow-up study from prepuberty to postmenarche. J Clin Endocrinol Metab **90:**44–51.
6. Bailey DA, Martin AD, McKay AA, Whiting S, Miriwald R 2000 Calcium accretion in girls and boys during puberty: A longitudinal analysis. J Bone Miner Res **15:**2245–2250.
7. Khosla S, Melton LJ III, Dekutoski MB, Achenbach SJ, Oberg AL, Riggs BL 2003 Incidence of childhood distal forearm fractures over 30 years. JAMA **290:**1479–1485.
8. Lee WTK, Leung SSF, Leung DMY, Cheng JCY 1996 A follow-up study on the effect of calcium-supplement withdrawal and puberty on bone acquisition of children. Am J Clin Nutr **64:**71–77.
9. Ghatge KD, Lambert HL, Barker ME, Eastell R 2001 Bone mineral gain following calcium supplementation in teenage girls is preserved two years after withdrawal of the supplement. J Bone Miner Res **16:**S173.
10. Matkovic V, Goel PK, Badenhop-Stevens NE, Landoll JD, Li B, Ilich JZ, Skugor M, Nagode L, Mobley SL, Ha EJ, Hangartner T, Clairmont A 2005 Calcium supplementation and bone mineral density in females from childhood to young adulthood: A randomized controlled trial. Am J Clin Nutr **81:**175–188.
11. Matkovic V, Landoll JD, Badenhop-Stevens NE, Ha Y-Y, Crnevic-Orlic Z, Li B, Goel P 2004 Nutrition influences skeletal development from childhood to adulthood: A study of hip, spine, and forearm in adolescent females. J Nutr **134:**701S–705S.
12. Biller BMK, Caughlin JF, Sake V, Schoenfeld D, Spratt DI, Klitanski A 1991 Osteopenia in women with hypothalamic amenorrhea: A prospective study. Obstet Gynecol **78:**996–1001.
13. Wastney ME, Martin BR, Peacock M, Smith D, Jiang XY, Jackman LA, Weaver CM 2000 Changes in calcium kinetics in adolescent girls induced by high calcium intake. J Clin Endocrinol Metab **85:**4470–4475.
14. Chapuy MC, Arlot ME, Duboeuf F, Brun J, Crouzet B, Arnaud S, Delmas PD, Meunier PJ 1992 Vitamin D3 and calcium to prevent hip fractures in elderly women. N Engl J Med **327:**1637–1642.
15. Dawson-Hughes B, Harris SS, Krall EA, Dallal GE 1997 Effect of calcium and vitamin D supplementation on bone density in men and women 65 years of age or older. N Engl J Med **337:**670–676.
16. U.S. Department of Health and Human Services and U.S. Department of Agriculture 2005 Dietary Guidelines for Americans, 2005, 6th ed. US Government Printing Office, Washington, DC, USA.
17. Ballow C, Kuester S, Gillespie C 2000 Beverage choices affect adequacy of children's nutrient intakes. Arch Pediatr Adolesc Med **154:**1148–1152.
18. Gao X, Wilde PE, Lichtenstein AH, Tucker KL 2006 Meeting adequate intake for dietary calcium without dairy foods in adolescents aged 9 to 18 years (National Health and Nutrition Examination Survey 2001-2002). J Am Diet Assoc **106:**1759–1765.
19. Matkovic V, Kostial K, Siminovic I, Buzina R, Brodarec A, Nordin BEC 1979 Bone status and fracture rates in two regions of Yugoslavia. Am J Clin Nutr **32:**540–549.
20. Hu J-F, Zhao X-H, Jia J-B, Parpia B, Campbell TC 1993 Dietary calcium and bone density among middle-aged and elderly women in China. Am J Clin Nutr **58:**219–227.
21. Goulding A, Rockell JE, Black RE, Grant AM, Jones IE, Williams SM 2004 Children who avoid drinking cow's milk are at increased risk for prepubertal bone fractures. J Am Diet Assoc **104:**250–253.
22. Honkanen R, Kroger H, Alhava E, Turpeinen P, Tuppurainen M, Suarikoski S 1997 Lactose intolerance associated with fractures in weight-bearing bones in Finnish women aged 38-57 years. Bone **21:**473–477.
23. Kalkwarf HJ, Khoury JC, Lanphear BP 2003 Milk intake during childhood and adolescence, adult bone density, and osteoporotic fractures in US women. Am J Clin Nutr **77:**257–265.
24. Muhlbauer RC, Lozano A, Reiuli A 2002 Onion and a mixture of vegetables, salads, and herbs affect bone resorption in the rat by a mechanism independent of their base exceeds. J Bone Miner Res **17:**1230–1236.
25. Rafferty K, Davies KM, Heaney RP 2005 Potassium intake and the calcium economy. J Am Coll Nutr **24:**99–106.
26. Kerstetter JE, O'Brien KO, Caseria DM, Wall DE, Insogna KL 2005 The impact of dietary protein on calcium absorption and kinetic measures of bone turnover in women. J Clin Endocrinol Metab **90:**26–31.
27. Bonjour JP 2005 Dietary protein: An essential nutrient for bone health. J Am Coll Nutr **24:**5265–5365.
28. Nordin BE, Need AG, Morris HA, Horowitz M 1993 The nature and significance of the relationship between urinary sodium and urinary calcium in women. J Nutr **123:**1615–1622.
29. Wigertz K, Palacios C, Jackman LA, Martin BR, McCabe LD, McCabe GP, Peacock M, Pratt JH, Weaver CM 2005 Racial differences in calcium retention in response to dietary salt in adolescent girls. Am J Clin Nutr **81:**845–850.

Chapter 41. Role of Sex Steroids in the Pathogenesis of Osteoporosis

Matthew T. Drake and Sundeep Khosla

Department of Internal Medicine, Division of Endocrinology, Diabetes, Metabolism, and Nutrition, Mayo Clinic, Rochester, Minnesota

INTRODUCTION

Significant bone loss occurs with aging in both men and women, leading to alterations in skeletal microarchitecture and an increased incidence of fractures.[1] Much work over the past two decades has significantly increased our understanding of the role that sex steroids (primarily estrogen and testosterone) play in the development and progression of osteoporosis in both females and males.

The authors state that they have no conflicts of interest.

CHANGES IN BONE MASS AND STRUCTURE WITH AGING

The composite DXA data shown in Fig. 1 demonstrates that, at menopause, women undergo rapid trabecular bone loss.[2]

Key words: estrogen, testosterone, menopause, osteoporosis, DXA, volumetric BMD

Key References: 1. http://edrv.endojournals.org/cgi/content/full/23/3/279. 2. http://www.jbmronline.org/doi/full/10.1359/jbmr.1997.12.4.509. 3. http://edrv.endojournals.org/cgi/content/full/21/2/115.

© 2008 American Society for Bone and Mineral Research

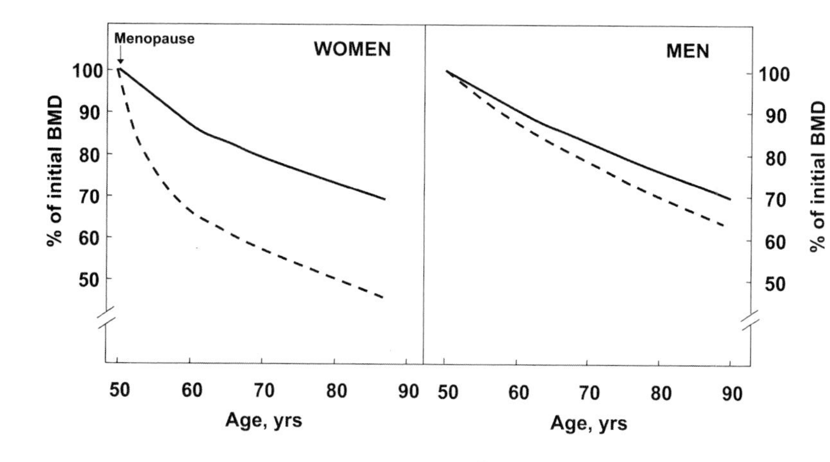

FIG. 1. Patterns of age-related bone loss in women and men. Dashed lines, trabecular bone; solid lines, cortical bone. The figure is based on multiple cross-sectional and longitudinal studies using DXA. (Reprinted with permission from Elsevier from Khosla S, Riggs BL 2005 Pathophysiology of age-related bone loss and osteoporosis. Endocrinol Metab Clin North Am **34:**1015–1030, Copyright 2005.)

While somewhat variable in length, this period of accelerated bone loss generally extends for 5–10 yr, with loss of ~20–30% of trabecular bone, but only 5–10% of cortical bone. Approximately 8–10 yr after menopause, a second slow and continuous phase of bone loss, in which cortical and trabecular bone losses occur at approximately equal rates, becomes predominant. This second phase extends throughout the remaining female lifespan. Comparatively, from middle life, men show slow progressive trabecular and cortical bone loss that is nearly equivalent to the latter phase seen in postmenopausal women. However, because men do not undergo the equivalent of menopause, a parallel early accelerated phase of bone loss does not occur, and thus overall loss of both trabecular and cortical bone is lessened.

More recent work has challenged the prevailing notion that there is relative maintenance of skeletal integrity in both sexes in the period between completion of puberty and middle age. In comparison with DXA, QCT permits measurement of volumetric BMD (vBMD) and articulates a more precise separation of trabecular and cortical components than the areal BMDs provided by DXA. Cross-sectional studies using QCT show that in the spine (composed primarily of trabecular bone), large decreases (~55% in women and ~45% in men) in vBMD occur, beginning in the third decade (Fig. 2).[3] In contrast, cortical vBMD measured at the distal radius in the same cohort shows little change in either sex until midlife. Thereafter, approximately linear declines in cortical bone occur in both sexes, although cumulative decreases are greater in women (28%) than men (18%, $p < 0.01$), reflecting the period of rapid bone loss in early menopausal women. Multivariate analysis in this cohort showed that the changes in trabecular microarchitecture at the wrist were most closely associated with reductions in sex steroid levels in those subjects >60 yr of age.[4,5] These cross-sectional finding have been confirmed by longitudinal studies in which vBMD was followed at both the distal radius and tibia, showing substantial loss of trabecular bone beginning shortly after the completion of puberty in both men and women, an age range during which sex steroid levels are defined as normal.[6] The relative contribution of bone loss during these years to the future development of skeletal fragility remains to be determined.

In addition to changes in bone mass that occur in both sexes during the aging process, changes in bone cross-sectional area also occur in various skeletal sites. Importantly, despite a net decrease in cortical area and thickness caused by endocortical resorption, outward cortical displacement resulting from ongoing periosteal apposition has been shown by DXA to occur in both men and women, leading to an increase in bone strength for bending stresses and partially offsetting the decrease in bone strength resulting from cortical thinning.[7]

Together, these changes in bone quantity and structure in aging women and men lead to changes in the annual incidence of osteoporotic fractures (Fig. 3).[2] Distal forearm fractures rise markedly in women around the menopause and plateau ~15 yr after menopause. Similarly, vertebral fracture rates also begin to rise with the onset on menopause, but in contrast to wrist fractures, continue to increase for the remainder of the female lifespan. Hip fracture rates in women initially parallel those of vertebral fractures, but rise markedly later in life. In contrast, men never have a significant incidence of distal forearm fractures. Cross-sectional studies of the wrist in both sexes using pQCT show that, whereas women undergo both trabecular loss and increased trabecular spacing over the lifespan, men begin young adult life with relatively thicker trabeculae and primarily sustain thinning rather than loss of trabeculae with aging.[8] Thus, in addition to having larger bones, elderly

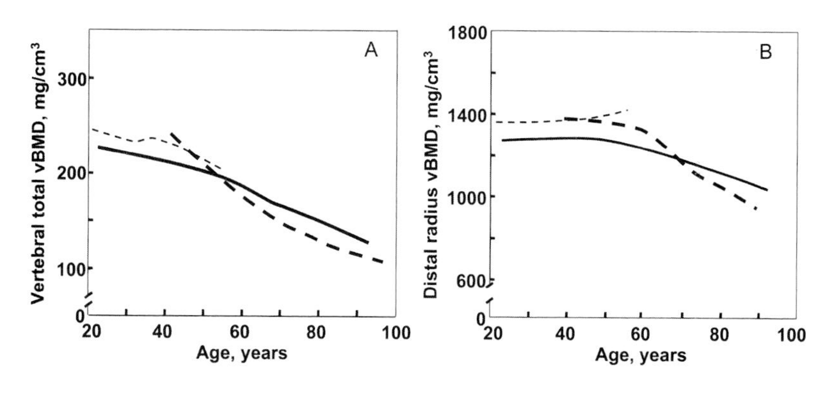

FIG. 2. (A) Values for vBMD (mg/cm³) of the total vertebral body in a population sample of Rochester, MN, women and men between the ages of 20 and 97 yr. Thin-dashed line, premenopausal women; thick-dashed line, postmenopausal women; solid line, men. (B) Values for cortical vBMD at the distal radius in the same cohort. Line coding as in A. All changes with age were significant ($p < 0.05$). (Adapted with permission of the American Society for Bone and Mineral Research from Riggs BL, Melton LJI, Robb RA, Camp JJ, Atkinson EJ, Peterson JM, Rouleau PA, McCollough CH, Bouxsein ML, Khosla S 2004 Population-based study of age and sex differences in bone volumetric density, size, geometry, and structure at different skeletal sites. J Bone Miner Res **19:**1945–1954.)

© 2008 American Society for Bone and Mineral Research

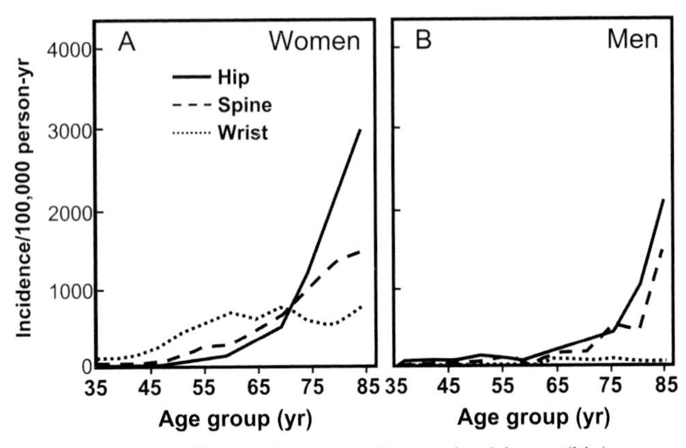

FIG. 3. Age-specific incidence rates for proximal femur (hip), vertebral (spine), and distal forearm (wrist) fractures in Rochester, MN, women (A) and men (B). (Adapted with permission from Elsevier from Cooper C, Melton LJ 1992 Epidemiology of osteoporosis. Trends Endocrinol Metab **3**:224–229, Copyright 1992.)

men have comparatively more trabeculae at the distal forearm than women, likely contributing to the rarity of wrist fractures in aged men. The incidence of both vertebral and hip fractures in men is roughly commensurate with that in females, although delayed by about one decade relative to women, again likely reflecting the lack of a male menopause and the associated rapid skeletal loss seen during this period in women.

SEX STEROIDS AND BONE LOSS IN WOMEN

The relationship between diminishing estrogen levels in women caused by ovarian failure and the development of postmenopausal osteoporosis has been recognized for over six decades.[9] Relative to premenopausal levels, during the menopausal transition, serum estradiol (E2) levels decrease by 85–90%, and serum estrone (E1; a 4-fold weaker estrogen) decrease by 65–75%.[10] Temporally associated with this decline are changes in bone formation and resorption rates. Before menopause, rates of bone formation and resorption are approximately equivalent in the coupled process of bone remodeling. With the onset of menopause, however, there is an increase in the basic multicellular unit (BMU) activation frequency rate, an extension of the resorption period,[11] and a shortening of the formation period.[12] Accordingly, as assessed by biochemical markers, bone resorption at the menopause increases by 90%, whereas bone formation increases by only 45%.[13] This imbalance leads to the accelerated phase of bone loss described above and an efflux of skeletal-derived calcium to the extracellular fluid. Compensatory mechanisms including increased renal calcium clearance,[14] decreased intestinal calcium absorption,[15] and partial suppression of PTH secretion,[16] however, all serve to limit the development of hypercalcemia, but contribute to the establishment of a negative total body calcium balance derived from skeletal losses. Importantly, these compensatory effects seem directly related to estrogen deficiency, because estrogen repletion, at least in early menopause, leads to preservation of both renal calcium reclamation and intestinal calcium absorption.[17]

The effects of estrogen on modulating skeletal metabolism at a cellular and molecular level are the subject of active study. Estrogen plays a central role in osteoblast biology, promoting the differentiation of bone marrow stromal cells toward the osteoblast lineage, increasing the differentiation of preosteoblasts to osteoblasts, and limiting apoptosis of both osteoblasts and osteocytes.[18,19] In addition, estrogen increases the osteoblastic production of growth factors (IGF-1[20] and TGF-β[21]) and synthesis of procollagen.[20] Furthermore, estrogen limits osteoblast apoptosis.[22] Direct evidence that pharmacologic dosages of estrogen can provide anabolic effects in postmenopausal women comes from histomorphometric studies of iliac crest biopsies from subjects receiving prolonged treatment with percutaneous estrogen, in whom a 61% increase in trabecular bone volume and a 12% increase in trabecular wall thickness relative to baseline was found after 6 yr of continuous treatment.[23] Accordingly, impaired bone formation caused by estrogen deficiency becomes apparent soon after the onset of menopause.

Estrogen also plays a central role in regulating osteoclast biology. As shown in both in vitro and in vivo studies, estrogen suppresses production of RANKL, the central molecule in osteoclast development, from bone marrow stromal/osteoblast precursor cells, T cells, and B cells.[24] Furthermore, estrogen increases production of the soluble RANKL decoy receptor osteoprotegerin (OPG) by osteoblast lineage cells.[25] Thus, through modulation of RANKL and OPG levels, estrogen limits the exposure of osteoclast lineage cells (which express RANK) to RANKL, effectively regulating osteoclast development. With diminishing estrogen levels, as occur in menopausal women, the relative ratios of RANKL and OPG are altered, leading to increased osteoclast development and activity. Additional estrogen-suppressible cytokines produced by osteoblasts and bone marrow mononuclear cells also seem to play central roles in mediating bone resorption and include interleukin-1 (IL-1), IL-6,[26] TNF-α,[27] macrophage colony stimulating factor (M-CSF),[28] and prostaglandins.[29] In support of this, a recent study showed that, in early postmenopausal women induced to undergo acute estrogen withdrawal, pharmacologic blockade of either IL-1 or TNF-α activity was able to partially blunt the resultant rise in bone resorption markers.[30]

In addition to modulating osteoclast development, estrogen can both directly and indirectly promote apoptosis of both osteoclast lineage cells and mature osteoclasts. Thus, estrogen induces direct apoptosis of osteoclast lineage cells by leading to a reduction in c-*jun* activity, thereby limiting activator protein-1 (AP-1)-dependent transcription.[31,32] Alternatively, estrogen can induce osteoblastic cell production of TGF-β, indirectly leading to osteoclast apoptosis.[11] The importance of direct effects of estrogen on osteoclast apoptosis was recently shown in studies using mice with selective deletion of estrogen receptor α in osteoclasts.[33] With estrogen deficiency, these pro-apoptotic effects on osteoclasts are lost.

Like estrogen, testosterone has a primary effect on bone to limit resorption, although at least part of this effect seems to derive from aromatization of testosterone to estrogen.[34] In vitro, testosterone can both stimulate osteoblast proliferation[35] (albeit weakly) and limit osteoblast apoptosis.[12] Whereas testosterone likely plays a role in increasing bone formation (and perhaps periosteal apposition) in women, at present there are very little data on the role of testosterone in the maintenance of skeletal integrity in postmenopausal women.

SEX STEROIDS AND BONE LOSS IN MEN

Relative to postmenopausal women, elderly men lose one half as much bone and sustain one third as many fragility fractures.[16] Although men do not undergo the hormonal equivalent of menopause, work over the past decade has delineated substantial changes in biologically available sex steroid levels over the male lifespan largely related to a >2-fold age-related rise in sex hormone binding globulin (SHBG) levels.[36] Whereas circulating sex steroids bound to SHBG have limited capacity to reach targets tissue, free (1–3% of total) and albu-

© 2008 American Society for Bone and Mineral Research

TABLE 1. SPEARMAN CORRELATION COEFFICIENTS RELATING RATES OF CHANGE IN BMD AT THE RADIUS AND ULNA TO SERUM SEX STEROID LEVELS AMONG A SAMPLE OF ROCHESTER, MN, MEN STRATIFIED BY AGE

	Young		*Middle-aged*		*Elderly*	
	Radius	*Ulna*	*Radius*	*Ulna*	*Radius*	*Ulna*
T	−0.02	−0.19	−0.18	−0.25*	0.13	0.14
E_2	0.33†	0.22*	0.03	0.07	0.21*	0.18*
E_1	0.35‡	0.34†	0.17	0.23*	0.16	0.14
Bio T	0.13	−0.04	0.07	0.01	0.23†	0.27†
Bio E_2	0.30†	0.20	0.14	0.21*	0.29†	0.33‡

Reproduced from Khosla S, Melton LJ, Atkinson EJ, O'Fallon WM 2001 Relationship of serum sex steroid levels to longitudinal changes in bone density in young versus elderly men. J Clin Endocrinol Metab **86**:3555–3561. Copyright 2001, The Endocrine Society.

* $p < 0.05$.
† $p < 0.01$.
‡ $p < 0.001$.
T, testosterone; E_2, estradiol; E_1, estrone; Bio, bioavailable.

min-associated (35–55% of total) sex steroids have ready biological availability. As a result of these changes in SHBG levels, bioavailable estrogen and testosterone levels decline by 47% and 64%, respectively, over the male lifespan, as shown in a cross-sectional study of 346 men between the ages of 23 and 90 in Rochester, MN.[36]

Although testosterone is the major sex steroid in men, available evidence from both cross-sectional[37–42] and longitudinal[43] studies showed that male BMD at various sites correlates better with serum bioavailable estradiol levels than with testosterone levels. In a longitudinal study, young men (age, 22–39 yr) were compared with older men (age, 60–90 yr) over a 4-yr interval to distinguish the effects of sex steroid levels on the final stages of skeletal maturation versus age-related bone loss.[43] Whereas BMD in the distal forearm (reflecting principally cortical bone changes) of young men increased by 0.42–0.43%/yr, distal forearm BMD in older men decreased by 0.49–0.66%/yr. As shown in Table 1, both the increase in BMD in younger men and the decrease in BMD in older men were more closely associated with bioavailable estradiol levels than with testosterone levels. Perhaps most intriguingly, in older men, there seemed to be a threshold level of bioavailable estradiol of ~40 pM (11 pg/ml) below which the rate of bone loss was most clearly associated with estradiol levels. Interestingly, vBMD analysis showed that this threshold effect seems to be more pronounced at cortical sites compared with trabecular sites.[4] Above this level, there was no firm relationship between the rate of bone loss and bioavailable estradiol levels. Similar findings by other investigators have been supportive of this threshold effect.[44]

Although suggestive, the association studies described above do not provide direct evidence for a causal role of estrogen in maintenance of the aging male skeleton. To address the relative roles of both estrogen and testosterone in skeletal maintenance in elderly men, Falahati-Nini et al.[34] used pharmacologic suppression of endogenous estrogen and testosterone production (through treatment with both a gonadotropin releasing hormone [GnRH] agonist and an aromatase inhibitor) in elderly men. Physiologic estrogen (E) and testosterone (T) levels were maintained by the placement of topical estrogen and testosterone patches, and baseline markers of bone resorption and formation were obtained. Subjects underwent randomization to one of four groups: group A (−T, −E) discontinued both patches; group B (−T, +E) discontinued only the testosterone patch; group C (+T, −E) discontinued only the estrogen patch; and group D (+T, +E) continued both patches. Because suppression of endogenous sex steroid production was maintained for the entire 3-wk period, changes in bone metabolism resulting from either estrogen or testosterone could be distinguished.

As seen in Fig. 4A, the significant increases in bone resorption seen in group A (−T, −E) were completely prevented by treatment with both testosterone and estrogen (group D). Whereas estrogen alone was almost completely able to prevent the rise in bone resorption (group B), testosterone alone was much less potent (group C). In comparison, the marked decreases in bone formation seen with dual sex steroid deficiency in group A were completely prevented by continuation of estrogen and testosterone (group D). Interestingly, serum osteocalcin levels were only slightly diminished with either estrogen or testosterone alone, whereas levels of serum amino-terminal propeptide of type I collagen (P1NP) were sustained with estrogen (group B) but not testosterone. In summary, these results are consistent with a dominant role for estrogen in the maintenance of skeletal integrity in aging men.

As noted above, bioavailable testosterone levels decrease to an even greater extent than bioavailable estrogen levels in aging men. Despite this decline, however, the role of testosterone in mediation of age-related bone loss in men is less clear than that of estrogen. As seen in Fig. 4, testosterone does have slight antiresorptive effects and also plays a role in mediating bone formation. In a recent study by Nair et al.,[45] a statistically significant increase in BMD at the femoral neck was seen in a cohort of elderly males who received low-dose testosterone replacement for 2 yr. Interestingly, however, no increase in BMD was seen at any other site (anterior-posterior lumbar spine, total hip, and ultradistal radius) examined. In addition,

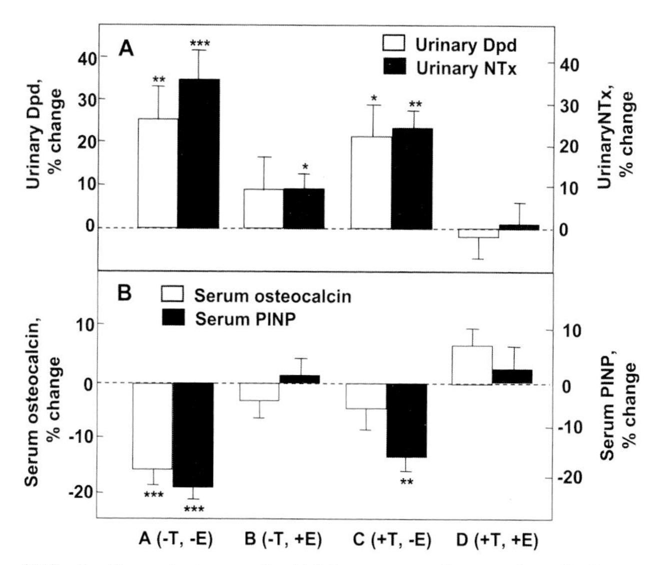

FIG. 4. Percent changes in (A) bone resorption markers (urinary deoxypyridinoline [Dpd] and N-telopeptide of type I collagen [NTX]) and (B) bone formation markers (serum osteocalcin and N-terminal extension peptide of type I collagen [PINP]) in a group of elderly men (mean age, 68 yr) made acutely hypogonadal and treated with an aromatase inhibitor (group A), estrogen alone (group B), testosterone alone (group C), or both estrogen and testosterone (group D). Significance for change from baseline: *$p < 0.05$; **$p < 0.01$; ***$p < 0.001$. (Adapted with permission from American Society for Clinical Investigation from Falahati-Nini A, Riggs BL, Atkinson EJ, O'Fallon WM, Eastell R, Khosla S 2000 Relative contributions of testosterone and estrogen in regulating bone resorption and formation in normal elderly men. J Clin Invest **106**:1553–1560.[34])

© 2008 American Society for Bone and Mineral Research

whether the effects of testosterone replacement on BMD were direct effects of testosterone or mediated through aromatization to estrogen is unclear. Finally, testosterone likely also plays a role in cortical appositional growth, although this have been most pronounced in studies of rodents[46] and may play a less important role in human biology.

OSTEOPOROSIS AND NON–SEX STEROID HORMONE CHANGES WITH AGING

In addition to changes in sex steroid levels that occur with aging in both men and women, sex steroid–independent factors including reductions in the production of growth factors important for osteoblast differentiation and function also occur. Thus, with aging, both the frequency and amplitude of growth hormone secretion is diminished,[47] leading to decreased hepatic production of both IGF-1 and IGF-2, an effect that may contribute to decreased bone formation with aging.[48,49] Additionally, aging is associated with increased levels of the IGF inhibitory binding protein, IGFBP-2, which also correlates inversely with bone mass in elderly subjects.[50] Finally, it is likely that there are intrinsic changes in osteoblast and perhaps osteoclast lineage cells as a consequence of the aging process.[51] These changes, which are probably independent of changes in sex steroids or other hormonal factors, are the subject of ongoing studies in animal models and in humans.

REFERENCES

1. Riggs BL, Khosla S, Melton LJ 2002 Sex steroids and the construction and conservation of the adult skeleton. Endocr Rev **23:**279–302.
2. Khosla S, Riggs BL 2005 Pathophysiology of age-related bone loss and osteoporosis. Endocrinol Metab Clin North Am **34:**1015–1030.
3. Riggs BL, Melton LJI, Robb RA, Camp JJ, Atkinson EJ, Peterson JM, Rouleau PA, McCollough CH, Bouxsein ML, Khosla S 2004 Population-based study of age and sex differences in bone volumetric density, size, geometry, and structure at different skeletal sites. J Bone Miner Res **19:**1945–1954.
4. Khosla S, Riggs BL, Robb RA, Camp JJ, Achenback SJ, Oberg AL, Rouleau PA, Melton LJ III 2005 Relationship of volumetric bone density and structural parameters at different skeletal sites to sex steroid levels in women. J Clin Endocrinol Metab **90:**5096–5103.
5. Khosla S, Melton LJ III, Robb RA, Camp JJ, Atkinson EJ, Oberg AL, Rouleau PA, Riggs BL 2005 Relationship of volumetric BMD and structural parameters at different skeletal sites to sex steroid levels in men. J Bone Miner Res **20:**730–740.
6. Riggs BL, Melton LJ III, Oberg AL, Atkinson EJ, Khosla S 2005 Substantial trabecular bone loss occurs in young adult women and men: A population-based longitudinal study. J Bone Miner Res **20:**S1;S4.
7. Seeman E 1997 From density to structure: Growing up and growing old on the surfaces of bone. J Bone Miner Res **12:**509–521.
8. Khosla S, Riggs BL, Atkinson EJ, Oberg AL, McDaniel LJ, Holets M, Peterson JM, Melton LJ III 2006 Effects of sex and age on bone microstructure at the ultradistal radius: A population-based noninvasive in vivo assessment. J Bone Miner Res **21:**124–131.
9. Albright F, Smith PH, Richardson AM 1941 Postmenopausal osteoporosis. JAMA **116:**2465–2474.
10. Khosla SK, Atkinson EJ, Melton LJ III, Riggs BL 1997 Effects of age and estrogen status on serum parathyroid hormone levels and biochemical markers of bone turnover in women: A population-based study. J Clin Endocrinol Metab **82:**1522–1527.
11. Hughes DE, Dai A, Tiffee JC, Li HH, Mundy GR, Boyce BF 1996 Estrogen promotes apoptosis of murine osteoclasts mediated by TGF-beta. Nat Med **2:**1132–1136.
12. Manolagas SC 2000 Birth and death of bone cells: Basic regulatory mechanisms and implications for the pathogenesis and treatment of osteoporosis. Endocr Rev **21:**115–137.
13. Garnero P, Sornay-Rendu E, Chapuy M, Delmas PD 1996 Increased bone turnover in late postmenopausal women is a major determinant of osteoporosis. J Bone Miner Res **11:**337–349.
14. Young MM, Nordin BEC 1967 Effects of natural and artificial menopause on plasma and urinary calcium and phosphorus. Lancet **2:**118–120.
15. Gennari C, Agnusdei D, Nardi P, Civitelli R 1990 Estrogen preserves a normal intestinal responsiveness to 1,25-dihydroxyvitamin D3 in oophorectomized women. J Clin Endocrinol Metab **71:**1288–1293.
16. Riggs BL, Khosla S, Melton LJI 1998 A unitary model for involutional osteoporosis: Estrogen deficiency causes both type I and type II osteoporosis in postmenopausal women and contributes to bone loss in aging men. J Bone Miner Res **13:**763–773.
17. McKane R, Khosla S, Burritt M, Kao P, Wilson DM, Riggs BL 1994 Mechanism of renal calcium conservation with estrogen replacement therapy in perimenopausal women. J Bone Miner Res **9:**S1;388.
18. Chow J, Tobias JH, Colston KW, Chambers TJ 1992 Estrogen maintains trabecular bone volume in rats not only by suppression of bone resorption but also by stimulation of bone formation. J Clin Invest **89:**74–78.
19. Qu Q, Perala-Heape M, Kapanen A, Dahllund J, Salo J, Vaananen HK, Harkonen P 1998 Estrogen enhances differentiation of osteoblasts in mouse bone marrow culture. Bone **22:**201–209.
20. Ernst M, Heath JK, Rodan GA 1989 Estradiol effects on proliferation, messenger ribonucleic acid for collagen and insulin-like growth factor-I, and parathyroid hormone-stimulated adenylate cyclase activity in osteoblastic cells from calvariae and long bones. Endocrinology **125:**825–833.
21. Oursler MJ, Cortese C, Keeting PE, Anderson MA, Bonde SK, Riggs BL, Spelsberg TC 1991 Modulation of transforming growth factor-beta production in normal human osteoblast-like cells by 17beta-estradiol and parathyroid hormone. Endocrinology **129:**3313–3320.
22. Gohel A, McCarthy M-B, Gronowicz G 1999 Estrogen prevents glucocorticoid-induced apoptosis in osteoblasts in vivo and in vitro. Endocrinology **140:**5339–5347.
23. Khastgir G, Studd J, Holland N, Alaghband-Zadeh J, Fox S, Chow J 2001 Anabolic effect of estrogen replacement on bone in postmenopausal women with osteoporosis: Histomorphometric evidence in a longitudinal study. J Clin Endocrinol Metab **86:**289–295.
24. Eghbali-Fatourechi G, Khosla S, Sanyal A, Boyle WJ, Lacey DL, Riggs BL 2003 Role of RANK ligand in mediating increased bone resorption in early postmenopausal women. J Clin Invest **111:**1221–1230.
25. Hofbauer LC, Khosla S, Dunstan CR, Lacey DL, Spelsberg TC, Riggs BL 1999 Estrogen stimulates gene expression and protein production of osteoprotegerin in human osteoblastic cells. Endocrinology **140:**4367–4370.
26. Jilka RL, Hangoc G, Girasole G, Passeri G, Williams DC, Abrams JS, Boyce B, Broxmeyer H, Manolagas SC 1992 Increased osteoclast development after estrogen loss: Mediation by interleukin-6. Science **257:**88–91.
27. Ammann P, Rizzoli R, Bonjour J, Bourrin S, Meyer J, Vassalli P, Garcia I 1997 Transgenic mice expressing soluble tumor necrosis factor-receptor are protected against bone loss caused by estrogen deficiency. J Clin Invest **99:**1699–1703.
28. Tanaka S, Takahashi N, Udagawa N, Tamura T, Akatsu T, Stanley ER, Kurokawa T, Suda T 1993 Macrophage colony-stimulating factor is indispensable for both proliferation and differentiation of osteoclast progenitors. J Clin Invest **91:**257–263.
29. Kawaguchi H, Pilbeam CC, Vargas SJ, Morse EE, Lorenzo JA, Raisz LG 1995 Ovariectomy enhances and estrogen replacement inhibits the activity of bone marrow factors that stimulate prostaglandin production in cultured mouse calvariae. J Clin Invest **96:**539–548.
30. Charatcharoenwitthaya N, Khosla S, Atkinson EJ, McCready LK, Riggs BL 2007 Effect of blockade of TNF-α and interleukin-1 action on bone resorption in early postmenopausal women. J Bone Miner Res **22:**724–729.
31. Shevde NK, Bendixen AC, Dienger KM, Pike JW 2000 Estrogens suppress RANK ligand-induced osteoclast differentiation via a stromal cell independent mechanism involving c-Jun repression. Proc Natl Acad Sci USA **97:**7829–7834.
32. Srivastava S, Toraldo G, Weitzmann MN, Cenci S, Ross FP, Pacifici R 2001 Estrogen decreases osteoclast formation by down-regulating receptor activator of NF-kB ligand (RANKL)-induced JNK activation. J Biol Chem **276:**8836–8840.
33. Nakamura T, Imai Y, Matsumoto T, Sato S, Takeuchi K, Igarashi

© 2008 American Society for Bone and Mineral Research

K, Harada Y, Azuma Y, Krust A, Yamamoto Y, Nishina H, Takeda S, Takayanagi H, Metzger D, Kanno J, Takaoka K, Martin TJ, Chambon P, Kato S 2007 Estrogen prevents bone loos via estrogen receptor alpha and induction of fas ligand in osteoclasts. Cell **130:**811–823.
34. Falahati-Nini A, Riggs BL, Atkinson EJ, O'Fallon WM, Eastell R, Khosla S 2000 Relative contributions of testosterone and estrogen in regulating bone resorption and formation in normal elderly men. J Clin Invest **106:**1553–1560.
35. Kasperk CH, Wergedal JE, Farley JR, Linkhart TA, Turner RT, Baylink DJ 1989 Androgens directly stimulate proliferation of bone cells in vitro. Endocrinology **124:**1576–1578.
36. Khosla S, Melton LJI, Atkinson EJ, O'Fallon WM, Klee GG, Riggs BL 1998 Relationship of serum sex steroid levels and bone turnover markers with bone mineral density in men and women: A key role for bioavailable estrogen. J Clin Endocrinol Metab **83:**2266–2274.
37. Slemenda CW, Longcope C, Zhou L, Hui SL, Peacock M, Johnston C 1997 Sex steroids and bone mass in older men: Positive associations with serum estrogens and negative associations with androgens. J Clin Invest **100:**1755–1759.
38. Greendale GA, Edelstein S, Barrett-Connor E 1997 Endogenous sex steroids and bone mineral density in older women and men: The Rancho Bernardo study. J Bone Miner Res **12:**1833–1843.
39. Center JR, Nguyen TV, Sambrook PN, Eisman JA 1999 Hormonal and biochemica parameters in the determination of osteoporosis in elderly men. J Clin Endocrinol Metab **84:**3626–3635.
40. van den Beld AW, de Jong FH, Grobbee DE, Pols HAP, Lamberts SWJ 2000 Measures of bioavailable serum testosterone and estradiol and their relationships with muscle strength, bone density, and body composition in elderly men. J Clin Endocrinol Metab **85:**3276–3282.
41. Amin S, Zhang Y, Sawin CT, Evans SR, Hannan MT, Kiel DP, Wilson PWF, Felson DT 2000 Association of hypogonadism and estradiol levels with bone mineral density in elderly men from the Framingham study. Ann Intern Med **133:**951–963.
42. Szulc P, Munoz F, Claustrat B, Garnero P, Marchand F 2001 Bioavailable estradiol may be an important determinant of osteoporosis in men: The MINOS study. J Clin Endocrinol Metab **86:**192–199.
43. Khosla S, Melton LJ, Atkinson EJ, O'Fallon WM 2001 Relationship of serum sex steroid levels to longitudinal changes in bone density in young versus elderly men. J Clin Endocrinol Metab **86:**3555–3561.
44. Gennari L, Merlotti D, Martini G, Gonnelli S, Franci B, Campagna S, Lucani B, Canto ND, Valenti R, Gennari C, Nuti R 2003 Longitudinal association between sex hormone levels, bone loss, and bone turnover in elderly men. J Clin Endocrinol Metab **88:**5327–5333.
45. Nair KS, Rizza RA, O'Brien P, Dhatariya K, Short KR, Nehra A, Vittone JL, Klee GG, Basu A, Basu R, Cobelli C, Toffolo G, Dalla Man C, Tindall DJ, Melton LJ III, Smith GE, Khosla S, Jensen MD 2006 DHEA in elderly women and DHEA or testosterone in elderly men. N Engl J Med **355:**1647–1659.
46. Wakley GK, Shutte DE, Hannon KS, Turner RT 1991 The effects of castration and androgen replacement therapy on bone: A histomorphometric study in the rat. J Bone Miner Res **6:**325–330.
47. Ho KY, Evans WS, Blizzard RM, Veldhuis JD, Merriam GR, Samojlik E, Furlanetto R, Rogol AD, Kaiser DL, Thorner MO 1987 Effects of sex and age on the 24-hour profile of growth hormone secretion in man: Importance of endogenous estradiol concentrations. J Clin Endocrinol Metab **64:**51–58.
48. Bennett A, Wahner HW, Riggs BL, Hintz RL 1984 Insulin-like growth factors I and II, aging and bone density in women. J Clin Endocrinol Metab **59:**701–704.
49. Boonen S, Mohan S, Dequeker J, Aerssens J, Vanderschueren D, Verbeke G, Broos P, Bouillon R, Baylink DJ 1999 Down-regulation of the serum stimulatory components of the insulin-like growth factor (IGF) system (IGF-I, IGF-II, IGF binding protein [BP]-3, and IGFBP-5) in age-related (type II) femoral neck osteoporosis. J Bone Miner Res **14:**2150–2158.
50. Amin S, Riggs BL, Atkinson EJ, Oberg AL, Melton LJ III, Khosla S 2004 A potentially deleterious role of IGFBP-2 on bone density in aging men and women. J Bone Miner Res **19:**1075–1083.
51. Moerman EJ, Teng K, Lipschitz DA, Lecka-Czernik B 2004 Aging activates adipogenic and suppresses osteogenic programs in mesenchymal marrow stroma/stem cells: The role of PPAR-gamma2 transcription factor and TGF-beta/BMP signaling pathways. Aging Cell **3:**379–389.

Chapter 42. Genetics of Osteoporosis

Jacqueline R. Center and John A. Eisman

Bone and Mineral Research Program, Garvan Institute of Medical Research, and St. Vincent's Hospital Sydney, New South Wales, Australia

INTRODUCTION

Genetic factors play an important role in many of the measures used to assess osteoporosis and the risk of fracture. Initial studies of monozygotic (identical) and dizygotic (nonidentical) twins and subsequently family studies showed a high heritability, up to 60–80% of various measures of bone structure, and a clear, albeit modest, heritability of fracture risk.

The demonstration of high heritability has led to a large body of studies aimed at identifying the genes responsible involving humans and animals. The types of studies can be broadly divided into a candidate gene approach or genome-wide search strategy using either linkage or association studies. The candidate gene approach involves directly testing variation in genes known to be involved in bone biology for their role in osteoporosis and fracture risk. The genome-wide search strategy involves systematically screening all genes using DNA markers uniformly distributed throughout the entire genome. Regions identified by genome scanning with polymorphic markers typically span 5–10 cM, and their localization has to be progressively refined until a gene can be identified. Association studies are either population-based or case-control studies relating a polymorphism of a certain candidate gene to the desired phenotype. Association studies are also suitable for a genome-wide analysis where identified single nucleotide polymorphisms (SNPs) are in linkage disequilibrium (LD) with the phenotype studied. Linkage studies relate the inheritance of genetic markers to the inheritance of phenotypes within families.

These studies have not always been very successful: Some of the reasons for this lie in the complexity of the disease itself and some relate to methodological and statistical issues related to the types of studies. Osteoporosis is now widely accepted as being multifactorial with several genes involved, each having a small to moderate effect on various parameters affecting bone

Dr. Eisman has served as a consultant for Amgen, deCodo, Eli Lilly, GE-Lunar, Merck, Novartis, Roche-GlaxoSmithKline, Sanofi-Aventis, and Servier, has served on Scientific Advisory Boards for Amgen, Eli Lilly, Merck, Novartis, Roche-GlaxoSmithKline, Sanofi-Aventis, Servier, and Wyeth and has received research funding from Amgen, Eli Lilly, Merck, Novartis, Roche-GSK, Sanofi-Aventis, and Servier. Dr. Center states that she has no conflicts of interest.

Key words: genetics, osteoporosis, fracture

© 2008 American Society for Bone and Mineral Research

physiology and risk of fracture. Gene–gene interactions and gene–environment interactions potentially increase this complexity.

Linkage studies have been very successful in identifying rare monogenic diseases and, while they have been applied to complex diseases such as osteoporosis, they suffer from low statistical power when the gene effect is only modest and inherently are unable to determine the size of the genetic effect. In contrast, association studies are easier to perform, and the large numbers required for enough power to show a small effect may be easier to obtain; however, to date, these studies in osteoporosis have often resulted in inconsistent results because of too small sample sizes, problems with population stratification, and variations in phenotype classification.

Nevertheless, the genetic studies performed over the last few decades have provided valuable insights into the pathophysiology of osteoporosis. Recent resources such as the Hapmap project and the decreasing costs of genome wide scanning have resulted in consortiums pooling much larger populations for genome-wide association studies for multifactorial diseases that may overcome some of the statistical problems that have plagued reproducibility from earlier studies in osteoporosis. The challenge facing the researcher is whether genetic information obtained can be used in a clinically relevant manner that will enhance either the prediction of fracture or response to treatment after readily measurable clinical and historical factors have been taken into account.

This chapter will not catalog the wide range of genes and loci that have now been associated with bone biology and pathology but will rather set out the background data showing the genetic association with bone biology, followed by a few examples of the more well-studied bone-related genes based on both candidate gene and genome-wide approaches. It will also touch on some of the statistical issues, the emerging field of pharmacogenomics, the clinical use of genetic information, and ideas on the way forward.

GENETIC CONTRIBUTION TO OSTEOPOROSIS

The possibility of osteoporosis being a genetic disease arose after a number of heritability twin and family studies showed that genetic factors contributed 60–80% to the heritability of BMD, the most widely used surrogate marker of bone fragility in the peripheral and axial skeleton.[1–8] Relatives of individuals with osteoporotic fractures were also found to have lower BMD than their age- and sex-adjusted reference values.[9,10] Other markers of fracture risk including bone turnover, quantitative ultrasound, and bone geometry were also found to be highly heritable, although the heritability differed between cortical and trabecular bone sites and between various geometric bone measures.[4,11–18]

A number of analyses suggest that different loci or sets of genes regulate bone size, shape, and density at either the lumbar spine and hip and distinct from fracture susceptibility per se.[19–21] There may also be differences in the regulation of BMD between younger and older individuals[19] and between men and women. For example, in the *LRP5* gene, polymorphisms have been associated with bone mass in men but not necessarily so in women.[22–24] However, as outlined below, apparent differences in replicability for all genetic studies may be related to differences in sample sizes, populations, and other confounding issues.

Fracture-based studies are fewer than those based on surrogate markers but favor a genetic basis for the fracture event. A family history of fracture has been shown to be a risk for fracture, somewhat surprisingly independent of osteoporosis.[25,26] Wrist fractures were shown to have a heritability of ~25% in a large cohort of U.S. women and their first-degree relatives[27] and almost 50% in a UK study of >6500 twins.[28] Again, the risk seemed to be mainly independent of BMD. Osteoporotic fractures were also shown to be heritable (h^2 = 0.27 for any fracture and 0.48 for hip fracture) in the large Swedish twin registry, although this decreased with age.[29] Perhaps consistent with the diminution of genetic effect with age, another study of Finnish twins showed virtually no genetic association with fracture in their elderly women.[30] Nevertheless, the modest genetic contribution of fracture, at least in younger people, in candidate gene studies also seems to be essentially independent of BMD.[31–33]

These studies show the multifactorial genetic inheritance of fracture risk and that, although both BMD and other surrogate markers may be inherited to a large degree, the genes affecting these surrogate markers are likely to be different from other genes affecting fracture risk, related in part to other individual and environmental influences.[34]

CANDIDATE GENE APPROACHES

The vitamin D receptor (*VDR*) gene was the first candidate gene studied in relation to osteoporosis. Multiple studies followed the first initial reports of an association between bone turnover and BMD with polymorphic markers (*Bsm*, *Apa*, and *Taq*) in the *VDR* gene.[35,36] The findings have not been replicable in all populations, but a recent meta-analysis concluded that there was a modest effect on spine BMD between the high-risk genotype of the *Bsm* polymorphism (BB) compared with the other genotypes.[37] However, other meta-analyses have not shown any clear association between *VDR* polymorphisms and osteoporosis with the *Bsm*, *Apa*, *Taq*, *Fok1*, or *cdX2* promoter polymorphisms.[38,39] A modest association with fracture has been reported in a number of studies,[31,32,39–41] but not related to an effect by BMD.

A number of other genetic loci, predominantly identified on the basis of their role in calcium metabolism, have since been identified as being associated with osteoporosis and fracture risk with some of the more widely studied ones including the collagen type Iα1 gene (*COL1A1*), lipoprotein receptor–related protein 5 (*LRP5*), estrogen receptor α (*ER*α), bone morphogenic proteins (BMPs), sclerostin (SOST), and CBFA1 or Runx2.[42]

Perhaps the most robust of these is the Sp1 polymorphic site in the promoter of the *COL1A1* gene,[43] which has been shown to be related both to BMD and fracture risk, particularly vertebral fracture risk, again largely independent of BMD.[44–46] Collagen type 1 is the major protein in bone, and mutation of one of the collagen 1 genes results in the clinical disease of osteogenesis imperfecta with associated high risk of fracture at a young age and low BMD. Interestingly, complete failure to produce one copy of either collagen chain generally results in less severe disease than production of an abnormal peptide that can not form the normal heterotrimer of two α1 and one α2 peptide chains. Unbalanced expression of COL1A1 and COL1A2 mRNA and thus unbalanced levels of the collagen type α1 (I) and α2 (I) gene products may be the explanation for the effects on BMD and fracture reported with the polymorphic Sp1binding site.[44]

In all of these studies, the effect size of any of these genes is small, and osteoporosis is now widely accepted as being multifactorial with several genes being involved, each having a small to moderate effect on various parameters affecting bone physiology and risk of fracture. The inherent limitation of the candidate gene approach, in that only previously identified genes can be studied, needs to be appreciated.

© 2008 American Society for Bone and Mineral Research

GENOME-WIDE APPROACHES

Human Studies

The whole gene search strategy, with no a priori assumption of underlying mechanism, has resulted in some limited successes in the search for osteoporosis genes. Perhaps the most remarkable is the *LRP5* gene discovery. Linkage analysis of data from a family with osteoporosis-pseudoglioma syndrome (a disorder characterized by low bone mass) localized the abnormality to the same region of chromosome 11 as that found by other researchers of a high bone mass family using a genome-wide linkage analysis.(47–49) This led to the discovery of the role of the Wnt signaling pathway in bone biology. Following the identification of the LRP5 pathway, there have been several studies supporting an association between SNPs or polymorphisms in this gene and osteoporosis risk in the general population.(22,50–55)

Another successful example of the genome-wide approach is the identification of the *BMP2* gene in a highly resourced study using a combined linkage and association strategy. A genome-wide scan of >1000 individuals from 207 nuclear families of Icelandic ancestry linked the *BMP2* gene to a combined phenotype of low BMD and fracture. Subsequent LD association analysis showed a contribution of this gene to osteoporosis and fracture risk in a Danish population at large.(56) However, variations in the *BMP2* gene were not associated with variations in BMD alone (i.e., not the previously used BMD plus fracture phenotype) in another population of Americans, emphasizing the importance of categorizing the phenotype.(57)

An interesting example of a genome-wide approach is that of an association study using subjects selected from the extremes of the population (extreme-truncate selection). Theoretically, this approach may have increased power to detect a genetic association compared with the more commonly used unselected populations and could therefore minimize genotyping costs by needing a smaller sample size. In a study of 344 subjects selected for having either high or low BMD, polymorphisms of the Wnt gene locus were found to be robustly associated with variations in BMD, showing the efficiency of this study design.(58)

Animal Studies

The genome-wide scan approach using linkage studies has been applied to rodent (predominately mouse) studies attempting to identify genes underlying complex traits such as BMD. These essentially depend on identification of strains for high or low bone mass at a particular site. These are cross-bred and selectively backcrossed on to one parental strain until the phenotype of one strain has effectively been transferred onto a nearly pure background of the alternate mouse strain. Genetic analysis using genome-wide searches can allow the identification of the transferred locus relatively straightforwardly. However, even these studies are complicated by the presence of competing genetic effectors in the parental strains (i.e., each strain carries some loci that dispose to both higher and lower bone mass as well as divergence in other structural measures).(59–62) Although a large number of quantitative trait loci (QTLs) that regulate BMD have been identified by these genetic mapping studies, identifying the genetic region responsible within this region is extremely time consuming, and only one gene to date, the *ALOX15* gene, has been identified from mouse crosses.(63) Identification of any particular locus can be compared with human studies by synteny and by comparison of genes identified from the various animal genomes. Indeed, polymorphisms in the human *ALOX12* gene, functionally similar to the mouse *ALOX15* gene, have been found to be associated with BMD in two separate populations.(64,65)

These animal crossing studies, as for family-based human studies, suffer from one major and self-evident but often ignored limitation: Such analyses, even if fully successful, can only identify genetic loci that differ among those selected individuals or families. An example of this relates to variants of the alkaline phosphatase gene associated with peak bone mass in one set of studies but not in another. This was simply because the two strains used in one study were invariant at the site in question.(66,67) This limitation means that lack of reproducibility of a study between strains of animals or between human populations may reflect inherent differences between the samples/strains studied rather than lack of effect of a gene variant per se.

STATISTICAL ISSUES

Despite some of the successes reported above, a large majority of the association studies on candidate genes that have been performed over the last decade have not been able to reliably reproduce the findings of the initial study. Similarly, across the fewer genome-wide studies that have been performed, although a large number of regions have been identified as being associated with osteoporosis, these are predominantly across differing regions. There are a number of statistical issues, accounting for much of this lack of reproducibility.

Lack of power is probably the most important factor affecting association and linkage studies, although some of the issues related to lack of power differ across these two study designs. Small sample size has been a major problem for many genetic studies. Because the effect size of individual genes is small (RR = ~1.5), sample sizes in the 1000s are generally required. The resources required for this kind of study are large, and meta-analyses can provide some solutions in this regard.(37,45,68) However, issues such as positive publication bias, differing phenotypes, and ascertainment and population stratification, as well as the likelihood of the association itself, also need to be considered for sound meta-analytical approaches.(69) Meta-analysis of individual data are more time consuming but may be more informative in relation to some of these issues.

For association studies in particular, the degree of LD between a polymorphism and the phenotype in question can vary across the genome and in different populations.(69) Thus, testing only one or some of the polymorphic markers may result in different results in different populations. Population stratification is also an issue affecting particularly association studies that can result in both false-positive and false-negative outcomes.(70) Cases and controls should be selected from a comparable genetic background and as much as possible with similar environmental influences(71) to decrease the heterogeneity between them. Possible solutions to this are genotyping ancestry informative markers, using the transmission disequilibrium test (TDT)(72) for family-based association studies or other proposed statistical methods for unrelated association studies.(73,74)

Multiple comparisons performed on the same data set increase the type 1 error, or false-positive rate, and need to be accounted for. Although beyond the scope of this chapter, the standard Bonferroni approach may be too conservative because the various genetic markers are not necessarily independent of each other.(71) This is important for the design and interpretation of both candidate gene and genome-wide link-

© 2008 American Society for Bone and Mineral Research

age and association studies and, although various methods have been proposed to deal with this issue, more robust methods are needed.

Clearly, phenotype classification is also important in complex diseases such as osteoporosis as previously alluded to, with different genetic loci probably responsible for different components of phenotypes, although which phenotype to address remains an issue. Other issues such as errors in genotyping and gene–gene and gene–environment interactions also need to be considered and may be responsible for some of the reproducibility problems reported.

PHARMACOGENOMICS

Pharmacogenomics, the use of genetic information to predict the outcome of response to drug treatment, is still in its infancy. As with other complex diseases, there is some variability in the therapeutic response of drugs to treat osteoporosis, but limited tools are available to predict which individuals will respond best to therapy. However, genetic factors may mediate that response.

Polymorphisms of the VDR were shown to be related to gut calcium absorption and BMD response, depending on the level of dietary calcium intake, with the bb genotype having the greater response in several[(75–77)] but not all studies.[(78–80)] Further evidence for the interaction between *VDR* polymorphisms and antiresorptive drug effect on BMD was shown in a series of studies by Palomba et al.,[(81–83)] whereby the increase in BMD was different in those women on hormone replacement therapy (HRT) and alendronate from those on raloxifene depending on the *Bsm* polymorphism of the VDR.

The ERα (*XbaI* and *PvuII* polymorphisms) have also been associated with response to HRT in several studies. XX homozygosity was associated with an increase in spine BMD in white women,[(84)] whereas presence of the *P* allele was associated with a higher BMD in Thai women[(85)] and lower fracture risk in a Finnish study.[(86)] In the latter study, the polymorphism was only associated with fracture risk in those women on HRT but not in the nontreated group. These gene–drug interactions may also confound some of the association studies described above.[(87)]

CLINICAL USE OF GENETIC INFORMATION

The slow progress in identifying the genetic basis of osteoporosis highlights an important clinical point about complex diseases and the use of identifying the genetic variants. As previously mentioned, linkage studies have been successful in identifying rare monogenic diseases where the gene effect is clearly large. However, this is not the situation for a common disease such as osteoporosis. Indeed, for a gene to have a large effect, its frequency must be small (as in rare monogenic diseases), and conversely, smaller effect variants can be common. Thus, one may question the use of identifying any such gene with a relatively modest effect size (<2). However, in a simple model of gene–gene effect (either additive or multiplicative) and taking into account the genotype prevalence, the risk ratios of each gene, Yang et al.[(88)] estimated that, for common genetic variants (genotype frequency ≥ 10%), even if the disease risk associated with each gene was relatively weak (≤1.5), only a limited number of genes would be needed to explain a substantial (>30%) population attributable fraction of the disease. This genetic information has great potential to help inform decisions about osteoporosis risk and even possibly treatment options.

THE WAY FORWARD

It may now be possible to overcome the problems of traditional candidate gene association and linkage studies using large genome-wide association studies or more focused studies using individuals with extreme phenotypes. This is now possible but not simple. It is estimated that there are >10 million common SNPs across the genome, accounting for ~90% of the variation among individuals.[(89)]

The advances in genetic possibilities have resulted from (1) the Human Genome Project, (2) extensive SNP databases, (3) the International HapMap resource that has documented genomic variation and LD patterns in four population samples, (4) advances in genotyping techniques with decreasing costs of commercial chips with >100,000 SNPs now available, (5) international consortiums pooling large, well-characterized data sets, and (6) more powerful statistical analysis techniques including prediction of the tagging SNPs (i.e., the subset of SNPs that can capture the proportion of the variants that are in strong LD) and nontags,[(89)] therefore reducing the number of SNPs needed to be genotyped, methods for analyzing interactions between unlinked loci,[(90)] and population genetics models for imputing data from untyped SNPs.[(91,92)]

Examples of some progress in genome-wide association studies can be found in other complex diseases such as diabetes and celiac and heart disease.[(93–96)] Most recently, the Wellcome Trust Case Control Consortium in the United Kingdom examined ~2000 individuals for seven major diseases and ~3000 controls and identified 24 independent significant association signals. However, even with this large number of subjects, power was adequate only for some common variants of relatively large effect.[(97)] In osteoporosis research, the GENOMOS and deCODE studies[(98,99)] have been the most successful in this regard with collaboration and pooling of several large data sets from different European populations. However, it is to be expected that in osteoporosis, as in other complex diseases, specific causal genetic variants will have small effects, with risk ratios generally <2. Although this in itself is not critical because the combination of relatively few of the risk genes may account for a significant population proportion of the disease, it does raise the issue of dealing with the interpretation of weak gene–disease associations and distinguishing the real from the false positive. In this context, the open, worldwide collaboration set up as the Human Genome Epidemiology Network (HuGENet) is to be applauded, with one of its tasks being to develop a systematic approach for assessing the evidence for gene–disease association.[(100)] Replication studies, critical for evaluation of initial findings, will be facilitated by the availability of large public databases such as the database of Genotypes and Phenotypes (dbGaP) developed by the NIH's National Library of Medicine.[(101)]

In summary, genetic factors modify determinants of bone strength, including density, architecture, and turnover, as well as fracture risk per se, all most likely through different sets of genes. The complexity of the interactions between an individual's unique genetic profile and gene–gene, gene–environment, and gene–drug interactions has meant that vast data sets with highly advanced statistical techniques are needed. Advances in genetic tools and information available mean that large genome-wide association studies are now possible. In the future, not only might it be possible to unravel the genetic code behind a complex disease such as osteoporosis, but individualizing treatment, through genetic profiling, may become a possibility, thus using a more rationale basis for treatment decisions and pharmacological advice.

REFERENCES

1. Debunker J, Nijs J, Verstraeten A, Geusens P, Gevers G 1987 Genetic determinants of bone mineral content at the spine and radius: A twin study. Bone **8**:207–209.

© 2008 American Society for Bone and Mineral Research

2. Pocock NA, Eisman JA, Hopper JL, Yeates MG, Sambrook PN, Eberl S 1987 Genetic determinants of bone mass in adults. A twin study. J Clin Invest **80**:706–710.
3. Smith DM, Nance WE, Kang KW, Christian JC, Johnston CC Jr 1973 Genetic factors in determining bone mass. J Clin Invest **52**:2800–2808.
4. Deng HW, Stegman MR, Davies KM, Conway T, Recker RR 1999 Genetic determination of variation and covariation of peak bone mass at the hip and spine. J Clin Densitom **2**:251–263.
5. Keen RW, Hart DJ, Arden NK, Doyle DV, Spector TD 1999 Family history of appendicular fracture and risk of osteoporosis: A population-based study. Osteoporos Int **10**:161–166.
6. Danielson ME, Cauley JA, Baker CE, Newman AB, Dorman JS, Towers JD, Kuller LH 1999 Familial resemblance of bone mineral density (BMD) and calcaneal ultrasound attenuation: The BMD in mothers and daughters study. J Bone Miner Res **14**:102–110.
7. Cohen-Solal ME, Baudoin C, Omouri M, Kuntz D, De Vernejoul MC 1998 Bone mass in middle-aged osteoporotic men and their relatives: Familial effect. J Bone Miner Res **13**:1909–1914.
8. Sowers MR, Burns TL, Wallace RB 1986 Familial resemblance of bone mass in adult women. Genet Epidemiol **3**:85–93.
9. Seeman E, Hopper JL, Bach LA, Cooper ME, Parkinson E, McKay J, Jerums G 1989 Reduced bone mass in daughters of women with osteoporosis. N Engl J Med **320**:554–558.
10. Evans RA, Marel GM, Lancaster EK, Kos S, Evans M, Wong SY 1988 Bone mass is low in relatives of osteoporotic patients. Ann Intern Med **109**:870–873.
11. Flicker L, Faulkner KG, Hopper JL, Green RM, Kaymacki B, Nowson CA, Young D, Wark JD 1996 Determinants of hip axis length in women aged 10–89 years: A twin study. Bone **18**:41–45.
12. Livshits G, Yakovenko K, Kobyliansky E 2003 Quantitative genetic study of radiographic hand bone size and geometry. Bone **32**:191–198.
13. Slemenda CW, Turner CH, Peacock M, Christian JC, Sorbel J, Hui SL, Johnston CC 1996 The genetics of proximal femur geometry, distribution of bone mass and bone mineral density. Osteoporos Int **6**:178–182.
14. Lee M, Czerwinski SA, Choh AC, Towne B, Demerath EW, Chumlea WC, Sun SS, Siervogel RM 2004 Heritability of calcaneal quantitative ultrasound measures in healthy adults from the Fels Longitudinal Study. Bone **35**:1157–1163.
15. Drozdzowska B, Pluskiewicz W 2001 Quantitative ultrasound at the calcaneus in premenopausal women and their postmenopausal mothers. Bone **29**:79–83.
16. Arden NK, Baker J, Hogg C, Baan K, Spector TD 1996 The heritability of bone mineral density, ultrasound of the calcaneus and hip axis length: A study of postmenopausal twins. J Bone Miner Res **11**:530–534.
17. Nguyen TV, Center JR, Eisman JA 2004 Bone mineral density-independent association of quantitative ultrasound measurements and fracture risk in women. Osteoporos Int **15**:942–947.
18. Nguyen TV, Blangero J, Eisman JA 2000 Genetic epidemiological approaches to the search for osteoporosis genes. J Bone Miner Res **15**:392–401.
19. Ralston SH, Galwey N, MacKay I, Albagha OM, Cardon L, Compston JE, Cooper C, Duncan E, Keen R, Langdahl B, McLellan A, O'Riordan J, Pols HA, Reid DM, Uitterlinden AG, Wass J, Bennett ST 2005 Loci for regulation of bone mineral density in men and women identified by genome wide linkage scan: The FAMOS study. Hum Mol Genet **14**:943–951.
20. Peacock M, Koller DL, Fishburn T, Krishnan S, Lai D, Hui S, Johnston CC, Foroud T, Econs MJ 2005 Sex-specific and non-sex-specific quantitative trait loci contribute to normal variation in bone mineral density in men. J Clin Endocrinol Metab **90**:3060–3066.
21. Turner CH, Sun Q, Schriefer J, Pitner N, Price R, Bouxsein ML, Rosen CJ, Donahue LR, Shultz KL, Beamer WG 2003 Congenic mice reveal sex-specific genetic regulation of femoral structure and strength. Calcif Tissue Int **73**:297–303.
22. Ferrari SL, Deutsch S, Choudhury U, Chevalley T, Bonjour JP, Dermitzakis ET, Rizzoli R, Antonarakis SE 2004 Polymorphisms in the low-density lipoprotein receptor-related protein 5 (LRP5) gene are associated with variation in vertebral bone mass, vertebral bone size, and stature in whites. Am J Hum Genet **74**:866–875.
23. Koh JM, Jung MH, Hong JS, Park HJ, Chang JS, Shin HD, Kim SY, Kim GS 2004 Association between bone mineral density and LDL receptor-related protein 5 gene polymorphisms in young Korean men. J Korean Med Sci **19**:407–412.
24. van Meurs JB, Rivadeneira F, Jhamai M, Hugens W, Hofman A, van Leeuwen JP, Pols HA, Uitterlinden AG 2006 Common genetic variation of the low-density lipoprotein receptor-related protein 5 and 6 genes determines fracture risk in elderly white men. J Bone Miner Res **21**:141–150.
25. Kanis JA, Johansson H, Oden A, Johnell O, De Laet C, Eisman JA, McCloskey EV, Mellstrom D, Melton LJ III, Pols HA, Reeve J, Silman AJ, Tenenhouse A 2004 A family history of fracture and fracture risk: A meta-analysis. Bone **35**:1029–1037.
26. Cummings SR, Nevitt MC, Browner WS, Stone K, Fox KM, Ensrud KE, Cauley J, Black D, Vogt TM 1995 Risk factors for hip fracture in white women. Study of Osteoporotic Fractures Research Group. N Engl J Med **332**:767–773.
27. Deng HW, Chen WM, Recker S, Stegman MR, Li JL, Davies KM, Zhou Y, Deng H, Heaney R, Recker RR 2000 Genetic determination of Colles' fracture and differential bone mass in women with and without Colles' fracture. J Bone Miner Res **15**:1243–1252.
28. Andrew T, Antioniades L, Scurrah KJ, Macgregor AJ, Spector TD 2005 Risk of wrist fracture in women is heritable and is influenced by genes that are largely independent of those influencing BMD. J Bone Miner Res **20**:67–74.
29. Michaelsson K, Melhus H, Ferm H, Ahlbom A, Pedersen NL 2005 Genetic liability to fractures in the elderly. Arch Intern Med **165**:1825–1830.
30. Kannus P, Palvanen M, Kaprio J, Parkkari J, Koskenvuo M 1999 Genetic factors and osteoporotic fractures in elderly people: Prospective 25 year follow up of a nationwide cohort of elderly Finnish twins. BMJ **319**:1334–1337.
31. Nguyen TV, Esteban LM, White CP, Grant SF, Center JR, Gardiner EM, Eisman JA 2005 Contribution of the collagen I alpha1 and vitamin D receptor genes to the risk of hip fracture in elderly women. J Clin Endocrinol Metab **90**:6575–6579.
32. Garnero P, Munoz F, Borel O, Sornay-Rendu E, Delmas PD 2005 Vitamin D receptor gene polymorphisms are associated with the risk of fractures in postmenopausal women, independently of bone mineral density. J Clin Endocrinol Metab **90**:4829–4835.
33. van Meurs JB, Schuit SC, Weel AE, van der Klift M, Bergink AP, Arp PP, Colin EM, Fang Y, Hofman A, van Duijn CM, van Leeuwen JP, Pols HA, Uitterlinden AG 2003 Association of 5′ estrogen receptor alpha gene polymorphisms with bone mineral density, vertebral bone area and fracture risk. Hum Mol Genet **12**:1745–1754.
34. Lei SF, Jiang H, Deng FY, Deng HW 2007 Searching for genes underlying susceptibility to osteoporotic fracture: Current progress and future prospect. Osteoporos Int **18**:1157–1175.
35. Morrison NA, Qi JC, Tokita A, Kelly PJ, Crofts L, Nguyen TV, Sambrook PN, Eisman JA 1994 Prediction of bone density from vitamin D receptor alleles. Nature **367**:284–287.
36. Morrison NA, Yeoman R, Kelly PJ, Eisman JA 1992 Contribution of trans-acting factor alleles to normal physiological variability: Vitamin D receptor gene polymorphism and circulating osteocalcin. Proc Natl Acad Sci USA **89**:6665–6669.
37. Thakkinstian A, D'Este C, Eisman J, Nguyen T, Attia J 2004 Meta-analysis of molecular association studies: Vitamin D receptor gene polymorphisms and BMD as a case study. J Bone Miner Res **19**:419–428.
38. Zintzaras E, Rodopoulou P, Koukoulis GN 2006 BsmI, TaqI, ApaI and FokI polymorphisms in the vitamin D receptor (VDR) gene and the risk of osteoporosis: A meta-analysis. Dis Markers **22**:317–326.
39. Uitterlinden AG, Ralston SH, Brandi ML, Carey AH, Grinberg D, Langdahl BL, Lips P, Lorenc R, Obermayer-Pietsch B, Reeve J, Reid DM, Amedei A, Bassiti A, Bustamante M, Husted LB, Diez-Perez A, Dobnig H, Dunning AM, Enjuanes A, Fahrleitner-Pammer A, Fang Y, Karczmarewicz E, Kruk M, van Leeuwen JP, Mavilia C, van Meurs JB, Mangion J, McGuigan FE, Pols HA, Renner W, Rivadeneira F, van Schoor NM, Scollen S, Sherlock RE, Ioannidis JP 2006 The association between common vitamin D receptor gene variations and osteoporosis: A participant-level meta-analysis. Ann Intern Med **145**:255–264.
40. Fang Y, van Meurs JB, Bergink AP, Hofman A, van Duijn CM, van Leeuwen JP, Pols HA, Uitterlinden AG 2003 Cdx-2 polymorphism in the promoter region of the human vitamin D receptor gene determines susceptibility to fracture in the elderly. J Bone Miner Res **18**:1632–1641.

© 2008 American Society for Bone and Mineral Research

41. Fang Y, van Meurs JB, d'Alesio A, Jhamai M, Zhao H, Rivadeneira F, Hofman A, van Leeuwen JP, Jehan F, Pols HA, Uitterlinden AG 2005 Promoter and 3′-untranslated-region haplotypes in the vitamin d receptor gene predispose to osteoporotic fracture: The rotterdam study. Am J Hum Genet **77:**807–823.
42. Ralston SH, de Crombrugghe B 2006 Genetic regulation of bone mass and susceptibility to osteoporosis. Genes Dev **20:**2492–2506.
43. Grant SF, Reid DM, Blake G, Herd R, Fogelman I, Ralston SH 1996 Reduced bone density and osteoporosis associated with a polymorphic Sp1 binding site in the collagen type I alpha 1 gene. Nat Genet **14:**203–205.
44. Mann V, Hobson EE, Li B, Stewart TL, Grant SF, Robins SP, Aspden RM, Ralston SH 2001 A COL1A1 Sp1 binding site polymorphism predisposes to osteoporotic fracture by affecting bone density and quality. J Clin Invest **107:**899–907.
45. Mann V, Ralston SH 2003 Meta-analysis of COL1A1 Sp1 polymorphism in relation to bone mineral density and osteoporotic fracture. Bone **32:**711–717.
46. Ralston SH, Uitterlinden AG, Brandi ML, Balcells S, Langdahl BL, Lips P, Lorenc R, Obermayer-Pietsch B, Scollen S, Bustamante M, Husted LB, Carey AH, Diez-Perez A, Dunning AM, Falchetti A, Karczmarewicz E, Kruk M, van Leeuwen JP, van Meurs JB, Mangion J, McGuigan FE, Mellibovsky L, del Monte F, Pols HA, Reeve J, Reid DM, Renner W, Rivadeneira F, van Schoor NM, Sherlock RE, Ioannidis JP 2006 Large-scale evidence for the effect of the COLIA1 Sp1 polymorphism on osteoporosis outcomes: The GENOMOS study. PLoS Med **3:**e90.
47. Gong Y, Slee RB, Fukai N, Rawadi G, Roman-Roman S, Reginato AM, Wang H, Cundy T, Glorieux FH, Lev D, Zacharin M, Oexle K, Marcelino J, Suwairi W, Heeger S, Sabatakos G, Apte S, Adkins WN, Allgrove J, Arslan-Kirchner M, Batch JA, Beighton P, Black GC, Boles RG, Boon LM, Borrone C, Brunner HG, Carle GF, Dallapiccola B, De Paepe A, Floege B, Halfhide ML, Hall B, Hennekam RC, Hirose T, Jans A, Juppner H, Kim CA, Keppler-Noreuil K, Kohlschuetter A, LaCombe D, Lambert M, Lemyre E, Letteboer T, Peltonen L, Ramesar RS, Romanengo M, Somer H, Steichen-Gersdorf E, Steinmann B, Sullivan B, Superti-Furga A, Swoboda W, van den Boogaard MJ, Van Hul W, Vikkula M, Votruba M, Zabel B, Garcia T, Baron R, Olsen BR, Warman ML 2001 LDL receptor-related protein 5 (LRP5) affects bone accrual and eye development. Cell **107:**513–523.
48. Boyden LM, Mao J, Belsky J, Mitzner L, Farhi A, Mitnick MA, Wu D, Insogna K, Lifton RP 2002 High bone density due to a mutation in LDL-receptor-related protein 5. N Engl J Med **346:**1513–1521.
49. Little RD, Carulli JP, Del Mastro RG, Dupuis J, Osborne M, Folz C, Manning SP, Swain PM, Zhao SC, Eustace B, Lappe MM, Spitzer L, Zweier S, Braunschweiger K, Benchekroun Y, Hu X, Adair R, Chee L, FitzGerald MG, Tulig C, Caruso A, Tzellas N, Bawa A, Franklin B, McGuire S, Nogues X, Gong G, Allen KM, Anisowicz A, Morales AJ, Lomedico PT, Recker SM, Van Eerdewegh P, Recker RR, Johnson ML 2002 A mutation in the LDL receptor-related protein 5 gene results in the autosomal dominant high-bone-mass trait. Am J Hum Genet **70:**11–19.
50. Ferrari SL, Rizzoli R 2005 Gene variants for osteoporosis and their pleiotropic effects in aging. Mol Aspects Med **26:**145–167.
51. Bollerslev J, Wilson SG, Dick IM, Islam FM, Ueland T, Palmer L, Devine A, Prince RL 2005 LRP5 gene polymorphisms predict bone mass and incident fractures in elderly Australian women. Bone **36:**599–606.
52. Koay MA, Brown MA 2005 Genetic disorders of the LRP5-Wnt signalling pathway affecting the skeleton. Trends Mol Med **11:**129–137.
53. Koller DL, Ichikawa S, Johnson ML, Lai D, Xuei X, Edenberg HJ, Conneally PM, Hui SL, Johnston CC, Peacock M, Foroud T, Econs MJ 2005 Contribution of the LRP5 gene to normal variation in peak BMD in women. J Bone Miner Res **20:**75–80.
54. Lev D, Binson I, Foldes AJ, Watemberg N, Lerman-Sagie T 2003 Decreased bone density in carriers and patients of an Israeli family with the osteoporosis-pseudoglioma syndrome. Isr Med Assoc J **5:**419–421.
55. Mizuguchi T, Furuta I, Watanabe Y, Tsukamoto K, Tomita H, Tsujihata M, Ohta T, Kishino T, Matsumoto N, Minakami H, Niikawa N, Yoshiura K 2004 LRP5, low-density-lipoprotein-receptor-related protein 5, is a determinant for bone mineral density. J Hum Genet **49:**80–86.
56. Styrkarsdottir U, Cazier JB, Kong A, Rolfsson O, Larsen H, Bjarnadottir E, Johannsdottir VD, Sigurdardottir MS, Bagger Y, Christiansen C, Reynisdottir I, Grant SF, Jonasson K, Frigge ML, Gulcher JR, Sigurdsson G, Stefansson K 2003 Linkage of osteoporosis to chromosome 20p12 and association to BMP2. PLoS Biol **1:**E69.
57. Ichikawa S, Johnson ML, Koller DL, Lai D, Xuei X, Edenberg HJ, Hui SL, Foroud TM, Peacock M, Econs MJ 2006 Polymorphisms in the bone morphogenetic protein 2 (BMP2) gene do not affect bone mineral density in white men or women. Osteoporos Int **17:**587–592.
58. Sims A-M, Shepard N, Carter N, Doan T, Dowling A, Duncan EL, Eisman JA, Jones G, Nicholson G, Prince R, Seeman E, Thomas G, Wass JA, Brown MA 2008 Genetic analyses in a sample of individuals with high or low bone density demonstrates association with multiple Wnt pathway genes. J Bone Miner Res **23:**499–506.
59. Klein RF 2002 Genetic regulation of bone mineral density in mice. J Musculoskelet Neuronal Interact **2:**232–236.
60. Klein RF, Mitchell SR, Phillips TJ, Belknap JK, Orwoll ES 1998 Quantitative trait loci affecting peak bone mineral density in mice. J Bone Miner Res **13:**1648–1656.
61. Bouxsein ML, Uchiyama T, Rosen CJ, Shultz KL, Donahue LR, Turner CH, Sen S, Churchill GA, Muller R, Beamer WG 2004 Mapping quantitative trait loci for vertebral trabecular bone volume fraction and microarchitecture in mice. J Bone Miner Res **19:**587–599.
62. Jepsen KJ, Akkus OJ, Majeska RJ, Nadeau JH 2003 Hierarchical relationship between bone traits and mechanical properties in inbred mice. Mamm Genome **14:**97–104.
63. Klein RF, Allard J, Avnur Z, Nikolcheva T, Rotstein D, Carlos AS, Shea M, Waters RV, Belknap JK, Peltz G, Orwoll ES 2004 Regulation of bone mass in mice by the lipoxygenase gene Alox15. Science **303:**229–232.
64. Ichikawa S, Koller DL, Johnson ML, Lai D, Xuei X, Edenberg HJ, Klein RF, Orwoll ES, Hui SL, Foroud TM, Peacock M, Econs MJ 2006 Human ALOX12, but not ALOX15, is associated with BMD in white men and women. J Bone Miner Res **21:**556–564.
65. Mullin BH, Spector TD, Curtis CC, Ong GN, Hart DJ, Hakim AJ, Worthy T, Wilson SG 2007 Polymorphisms in ALOX12, but not ALOX15, are significantly associated with BMD in postmenopausal women. Calcif Tissue Int **81:**10–17.
66. Foreman JE, Blizard DA, Gerhard G, Mack HA, Lang DH, Van Nimwegen KL, Vogler GP, Stout JT, Shihabi ZK, Griffith JW, Lakoski JM, McClearn GE, Vandenbergh DJ 2005 Serum alkaline phosphatase activity is regulated by a chromosomal region containing the alkaline phosphatase 2 gene (Akp2) in C57BL/6J and DBA/2J mice. Physiol Genomics **23:**295–303.
67. Klein R, Carlos A, Kansagor J, Olson D, Wagoner W, Ea L, Dinulescu D, Munsey T, Vanek C, Madisin D, Lundblad J, Belknap J, Orwoll E 2005 Identification of Akp2 as a gene that regualtes peak bone mass in mice. J Bone Miner Res **20:**S1;S9.
68. Ioannidis JP, Ntzani EE, Trikalinos TA, Contopoulos-Ioannidis DG 2001 Replication validity of genetic association studies. Nat Genet **29:**306–309.
69. Shen H, Liu Y, Liu P, Recker RR, Deng HW 2005 Nonreplication in genetic studies of complex diseases–lessons learned from studies of osteoporosis and tentative remedies. J Bone Miner Res **20:**365–376.
70. Marchini J, Cardon LR, Phillips MS, Donnelly P 2004 The effects of human population structure on large genetic association studies. Nat Genet **36:**512–517.
71. Chanock SJ, Manolio T, Boehnke M, Boerwinkle E, Hunter DJ, Thomas G, Hirschhorn JN, Abecasis G, Altshuler D, Bailey-Wilson JE, Brooks LD, Cardon LR, Daly M, Donnelly P, Fraumeni JF Jr, Freimer NB, Gerhard DS, Gunter C, Guttmacher AE, Guyer MS, Harris EL, Hoh J, Hoover R, Kong CA, Merikangas KR, Morton CC, Palmer LJ, Phimister EG, Rice JP, Roberts J, Rotimi C, Tucker MA, Vogan KJ, Wacholder S, Wijsman EM, Winn DM, Collins FS 2007 Replicating genotype-phenotype associations. Nature **447:**655–660.
72. Lewis CM 2002 Genetic association studies: Design, analysis and interpretation. Brief Bioinform **3:**146–153.
73. Devlin B, Roeder K 1999 Genomic control for association studies. Biometrics **55:**997–1004.
74. Pritchard JK, Rosenberg NA 1999 Use of unlinked genetic markers to detect population stratification in association studies. Am J Hum Genet **65:**220–228.
75. Kiel DP, Myers RH, Cupples LA, Kong XF, Zhu XH, Ordovas J, Schaefer EJ, Felson DT, Rush D, Wilson PW, Eisman JA, Holick

© 2008 American Society for Bone and Mineral Research

MF 1997 The BsmI vitamin D receptor restriction fragment length polymorphism (bb) influences the effect of calcium intake on bone mineral density. J Bone Miner Res **12**:1049–1057.
76. Dawson-Hughes B, Harris SS, Finneran S 1995 Calcium absorption on high and low calcium intakes in relation to vitamin D receptor genotype. J Clin Endocrinol Metab **80**:3657–3661.
77. Wishart JM, Horowitz M, Need AG, Scopacasa F, Morris HA, Clifton PM, Nordin BE 1997 Relations between calcium intake, calcitriol, polymorphisms of the vitamin D receptor gene, and calcium absorption in premenopausal women. Am J Clin Nutr **65**:798–802.
78. Francis RM, Harrington F, Turner E, Papiha SS, Datta HK 1997 Vitamin D receptor gene polymorphism in men and its effect on bone density and calcium absorption. Clin Endocrinol (Oxf) **46**:83–86.
79. Kinyamu HK, Gallagher JC, Knezetic JA, DeLuca HF, Prahl JM, Lanspa SJ 1997 Effect of vitamin D receptor genotypes on calcium absorption, duodenal vitamin D receptor concentration, and serum 1,25 dihydroxyvitamin D levels in normal women. Calcif Tissue Int **60**:491–495.
80. Rauch F, Radermacher A, Danz A, Schiedermaier U, Golucke A, Michalk D, Schonau E 1997 Vitamin D receptor genotypes and changes of bone density in physically active German women with high calcium intake. Exp Clin Endocrinol Diabetes **105**:103–108.
81. Palomba S, Numis FG, Mossetti G, Rendina D, Vuotto P, Russo T, Zullo F, Nappi C, Nunziata V 2003 Effectiveness of alendronate treatment in postmenopausal women with osteoporosis: Relationship with BsmI vitamin D receptor genotypes. Clin Endocrinol (Oxf) **58**:365–371.
82. Palomba S, Numis FG, Mossetti G, Rendina D, Vuotto P, Russo T, Zullo F, Nappi C, Nunziata V 2003 Raloxifene administration in post-menopausal women with osteoporosis: Effect of different BsmI vitamin D receptor genotypes. Hum Reprod **18**:192–198.
83. Palomba S, Orio F Jr, Russo T, Falbo A, Tolino A, Manguso F, Nunziata V, Mastrantonio P, Lombardi G, Zullo F 2005 BsmI vitamin D receptor genotypes influence the efficacy of antiresorptive treatments in postmenopausal osteoporotic women. A 1-year multicenter, randomized and controlled trial. Osteoporos Int **16**:943–952.
84. Deng HW, Li J, Li JL, Johnson M, Gong G, Davis KM, Recker RR 1998 Change of bone mass in postmenopausal Caucasian women with and without hormone replacement therapy is associated with vitamin D receptor and estrogen receptor genotypes. Hum Genet **103**:576–585.
85. Ongphiphadhanakul B, Chanprasertyothin S, Payatikul P, Tung SS, Piaseu N, Chailurkit L, Chansirikarn S, Puavilai G, Rajatanavin R 2000 Oestrogen-receptor-alpha gene polymorphism affects response in bone mineral density to oestrogen in post-menopausal women. Clin Endocrinol (Oxf) **52**:581–585.
86. Salmen T, Heikkinen AM, Mahonen A, Kroger H, Komulainen M, Saarikoski S, Honkanen R, Maenpaa PH 2000 The protective effect of hormone-replacement therapy on fracture risk is modulated by estrogen receptor alpha genotype in early postmenopausal women. J Bone Miner Res **15**:2479–2486.
87. Xiong DH, Long JR, Recker RR, Deng HW 2003 Pharmacogenomic approaches to osteoporosis. Pharmacogenomics J **3**:261–263.
88. Yang Q, Khoury MJ, Friedman J, Little J, Flanders WD 2005 How many genes underlie the occurrence of common complex diseases in the population? Int J Epidemiol **34**:1129–1137.
89. Eyheramendy S, Marchini J, McVean G, Myers S, Donnelly P 2007 A model-based approach to capture genetic variation for future association studies. Genome Res **17**:88–95.
90. Marchini J, Donnelly P, Cardon LR 2005 Genome-wide strategies for detecting multiple loci that influence complex diseases. Nat Genet **37**:413–417.
91. Marchini J, Howie B, Myers S, McVean G, Donnelly P 2007 A new multipoint method for genome-wide association studies by imputation of genotypes. Nat Genet **39**:906–913.
92. Pe'er I, de Bakker PI, Maller J, Yelensky R, Altshuler D, Daly MJ 2006 Evaluating and improving power in whole-genome association studies using fixed marker sets. Nat Genet **38**:663–667.
93. Scott LJ, Mohlke KL, Bonnycastle LL, Willer CJ, Li Y, Duren WL, Erdos MR, Stringham HM, Chines PS, Jackson AU, Prokunina-Olsson L, Ding CJ, Swift AJ, Narisu N, Hu T, Pruim R, Xiao R, Li XY, Conneely KN, Riebow NL, Sprau AG, Tong M, White PP, Hetrick KN, Barnhart MW, Bark CW, Goldstein JL, Watkins L, Xiang F, Saramies J, Buchanan TA, Watanabe RM, Valle TT, Kinnunen L, Abecasis GR, Pugh EW, Doheny KF, Bergman RN, Tuomilehto J, Collins FS, Boehnke M 2007 A genome-wide association study of type 2 diabetes in Finns detects multiple susceptibility variants. Science **316**:1341–1345.
94. Zeggini E, Weedon MN, Lindgren CM, Frayling TM, Elliott KS, Lango H, Timpson NJ, Perry JR, Rayner NW, Freathy RM, Barrett JC, Shields B, Morris AP, Ellard S, Groves CJ, Harries LW, Marchini JL, Owen KR, Knight B, Cardon LR, Walker M, Hitman GA, Morris AD, Doney AS, McCarthy MI, Hattersley AT 2007 Replication of genome-wide association signals in UK samples reveals risk loci for type 2 diabetes. Science **316**:1336–1341.
95. Helgadottir A, Thorleifsson G, Manolescu A, Gretarsdottir S, Blondal T, Jonasdottir A, Jonasdottir A, Sigurdsson A, Baker A, Palsson A, Masson G, Gudbjartsson DF, Magnusson KP, Andersen K, Levey AI, Backman VM, Matthiasdottir S, Jonsdottir T, Palsson S, Einarsdottir H, Gunnarsdottir S, Gylfason A, Vaccarino V, Hooper WC, Reilly MP, Granger CB, Austin H, Rader DJ, Shah SH, Quyyumi AA, Gulcher JR, Thorgeirsson G, Thorsteinsdottir U, Kong A, Stefansson K 2007 A common variant on chromosome 9p21 affects the risk of myocardial infarction. Science **316**:1491–1493.
96. van Heel DA, Franke L, Hunt KA, Gwilliam R, Zhernakova A, Inouye M, Wapenaar MC, Barnardo MC, Bethel G, Holmes GK, Feighery C, Jewell D, Kelleher D, Kumar P, Travis S, Walters JR, Sanders DS, Howdle P, Swift J, Playford RJ, McLaren WM, Mearin ML, Mulder CJ, McManus R, McGinnis R, Cardon LR, Deloukas P, Wijmenga C 2007 A genome-wide association study for celiac disease identifies risk variants in the region harboring IL2 and IL21. Nat Genet **39**:827–829.
97. Wellcome Trust Case Control Consortium 2007 Genome-wide association study of 14,000 cases of seven common diseases and 3,000 shared controls. Nature **447**:661–678.
98. Styrkarsdottir U, Halldorsson BV, Gretarsdottir S, Gudbjartsson DF, Walters GB, Ingvarsson T, Jonsdottir T, Saemundsdottir J, Center JR, Nguyen TV, Bagger Y, Gulcher JR, Eisman JA, Christiansen C, Sigurdsson G, Kong A, Thorsteinsdottir U, Stefansson K 2008 Multiple Genetic Loci for Bone Mineral Density and Fractures. N Engl J Med. **358**:2355–2365.
99. Richards JB, Rivadeneira F, Inouye M, Pastinen TM, Soranzo N, Wilson SG, Andrew T, Falchi M, Gwilliam R, Ahmadi KR, Valdes AM, Arp P, Whittaker P, Verlaan DJ, Jhamai M, Kumanduri V, Moorhouse M, van Meurs JB, Hofman A, Pols HA, Hart D, Zhai G, Kato BS, Mullin BH, Zhang F, Deloukas P, Uitterlinden AG, Spector TD 2008 Bone mineral density, osteoporosis, and osteoporotic fractures: A genome-wide association study. Lancet **371**:1505–1512.
100. Khoury MJ, Little J, Gwinn M, Ioannidis JP 2007 On the synthesis and interpretation of consistent but weak gene-disease associations in the era of genome-wide association studies. Int J Epidemiol **36**:439–445.
101. Campbell H, Manolio T 2007 Commentary: Rare alleles, modest genetic effects and the need for collaboration. Int J Epidemiol **36**:445–448.

© 2008 American Society for Bone and Mineral Research

Chapter 43. Overview of Osteoporosis Treatment

Michael Kleerekoper

St. Joseph Mercy Hospital, Ann Arbor, Michigan; Department of Medicine, Wayne State University School of Medicine, Detroit, Michigan

Osteoporosis is a systemic skeletal disorder that is manifest clinically after a fracture has occurred. The goal of clinical care is to prevent, to the extent possible, adverse health outcomes such as fracture, and the tools to detect individuals at increased risk of fracture have been detailed in preceding chapters. The following section in the *Primer* details the interventions available and necessary to minimize risk of fracture, and this chapter provides an overview of that section.

An osteoporosis-related fragility fracture is arbitrarily defined as a fall after trauma equal to or less than a fall from a standing height. Regrettably, this definition is more often than not ignored when patients present to Emergency Departments with a clinically apparent fracture, and the majority of such fractures go unrecognized as osteoporosis. The end result of this neglect is further fracture as this is abundant evidence that "the best predictor of tomorrow's fracture is yesterday's fracture." Equally importantly, there is gathering evidence that even fractures resulting from major trauma may herald future fragility fractures.

The only osteoporosis-related fracture that is not always associated with an acute event, including a fall, is a vertebral fracture. All others result from a fall or similar event. The chapter on "Prevention of Falls" details the many circumstances that lead to an increased risk of falling. Many of these are age related, coincident with the time that most bone loss has already occurred and probably beyond the likelihood of even very effective osteoporosis therapies to prevent the fracture. As the population ages, there are insufficient health care professionals to provide the one-on-one fall prevention check-up and teaching that is needed. However, there are a number of reader friendly guides that clinicians should make available to all elderly patients, particularly those that are identified as being frail. There are also a number of factors that the clinician can address directly with the patient. Checking for proximal muscle weakness only requires asking the patient to stand from a sitting position without use of the arms. Such patients need to be checked for reversible causes of a proximal myopathy (e.g., vitamin D deficiency) but must also be prescribed a walking aid (cane or walker) or possibly be wheelchair restricted. Failing vision increases the likelihood of an accidental fall, and regular ophthalmologic examination is critical. Patients with known diseases likely to interfere with gait and/or balance (cerebro-vascular accident, Parkinson's disease, vertigo from any cause) also need special attention. The biggest problem the clinician faces in this at-risk (of falling) population is judicious and appropriate use of medication that may adversely affect patient functioning.

The ideal preventive therapy for an osteoporosis-related fracture is the optimization of peak bone mass and the prevention of any bone loss once peak bone mass has been attained, The chapter on "Calcium and Vitamin D" details the current information concerning optimal requirements for calcium and vitamin D for maintaining bone health and those conditions that might impede the attainment of these optimum requirements. This an area of nutrition and health that has undergone considerable scrutiny and change over the past decade, but there is general agreement that much of the world's population, even in developed countries, are not yet meeting these requirements. This has a negative impact on maintaining skeletal health and, equally importantly, on maintaining the muscle strength necessary to minimize the risk of falls. In addition to inadequate intake, there are a number of diseases and medications that impede absorption of calcium. Often overlooked in this regard is the potential for a high-sodium diet to increase urinary calcium losses with resultant secondary hyperparathyroidism and increased bone loss. Also overlooked too often in clinical practice is the recognition that subjects in the placebo arms of all controlled clinical trials of therapies for osteoporosis received calcium and/or vitamin supplements. Both calcium and vitamin D, alone or in combination, have been shown in several controlled clinical trials to improve BMD and, in some studies, reduce fracture occurrence. One unresolved issue is the discussion concerning the different merits of cholecalciferol (vitamin D_3) and ergocalciferol (vitamin D_2)—the bulk of the data suggesting that there is little to choose between these preparations.

The importance of muscle strength is emphasized in each of the chapters discussed above and is amplified in the chapter "Exercise and the Prevention of Osteoporosis." Rather than focus on the clinical impact of preservation of muscle mass and strength, this chapter provides important insight to the molecular mechanisms whereby mechanical effects impact stromal cells, osteoblasts, and osteocytes at the cellular and molecular level. This has already been transferred to the clinical arena with a number of studies addressing the effect of low impact forces on skeletal metabolism. That this cellular and molecular understanding might translate into pharmacologic therapy does not seem to be too far into the future.

Clearly the greatest advance in the field of osteoporosis treatment in the last 10–15 yr has been the development of highly effective pharmacologic therapy. Unlike other therapies that are approved for treatment of conditions leading to an adverse health outcome (acute myocardial infraction, cerebrovascular accident) before regulatory approval, the osteoporosis therapies have to show that they have antifracture effectiveness (i.e., they significantly reduce the number of fractures compared with placebo) in randomized double-blind controlled clinical trials. This higher standard has obvious advantages for prescribing with respect to safety and efficacy. The downside is that the requirement for a fracture outcome makes it prohibitively expensive to conduct active comparator trials, so there is little information that indicates one therapy is better than any other.

These agents fall into one of two categories: those that inhibit bone resorption (antiresorptive) and those that stimulate bone formation (anabolic). The earliest recognized antiresorptive therapy was estrogen ± progestin, and it remains a very effective agent for prevention of early postmenopausal bone loss. Its use for osteoporosis has been largely supplanted by other agents because more attention has been paid to possible adverse health outcomes when offered for the first time to women who are >10 yr postmenopause. The benefits of estrogen for management of the early postmenopause are still evident, including preservation of bone mass, but regulatory authorities have strongly cautioned against the use of estrogen solely for the prevention and treatment of osteoporosis. The next of the antiresorptive drugs to appear was calcitonin, which is still widely used because of its each of use and excellent

Dr. Kleerekoper is Chief Medical Officer at Micro MRI Inc. He is also a consultant to Roche and has received honoraria from Roche and GlaxoSmithKline.

© 2008 American Society for Bone and Mineral Research

safety profile. Whereas head-to-head studies against newer therapies are lacking, it does seem that calcitonin is not as effective as other options at improving BMD or reducing fracture risk. There are reports that calcitonin has an analgesic effect when started early after the occurrence of a clinically apparent vertebral fracture. Although not in the order that it became available, raloxifene should be discussed here because it has hormonal-like action that puts it in the same class category as estrogen and calcitonin. The mechanism of action of these drugs is detailed in the corresponding chapters but, in brief, they differ from other antiresorptive drugs in that their effect lasts only as long as the therapy is continued. Interruption of therapy rapidly erases the protective effect on bone mass, but it is unclear how long antifracture effectiveness may last once therapy is discontinued.

The other group of antiresorptive drugs is the bisphosphonates that effect bone remodeling by actions quite different from those drugs just described above. In general, they have a long half-life in the skeleton, allowing dosing schedules ranging from daily to yearly, depending on the specific bisphosphonate. This long in-vivo half-life preserves any increase in bone mass for prolonged periods of time, 1 yr or longer, after therapy is discontinued. In general, the effect on BMD is greater than with raloxifene and calcitonin, but there is a limited relationship between the effect on bone mass and the antifracture effectiveness of therapy. Intuitively, one might have expected that the several dosing options would have improved patient compliance with long-term therapy, but unfortunately, that is not the case—an important unsolved issue in clinical medicine. The bisphosphonates are very potent inhibitors of bone remodeling, and there has been intense speculation that prolonged use might ultimately produce a so-called "frozen bone" state in which the bone may become more brittle. The evidence to support this is limited at present, but there have been recent reports of unusual and unexpected stress fractures in long-term bisphosphonate users. This possibility is discussed in detail in the chapter on bisphosphonates and will require vigilance as the duration of therapy increases.

More recently, drugs that directly stimulate bone formation, teriparatide and strontium ranelate, have been approved for treatment of osteoporosis. Experience with these drugs is not as extensive as with the antiresorptive, but to date, there seems to be no major concerns about safety. The effectiveness has been clearly shown in controlled clinical trials. The trials of teriparatide were stopped prematurely because osteosarcoma had been found in rats treated with teriparatide from birth. To date, there is no evidence to suggest that this happens in humans but, as a precaution, use of this drug is not recommended in patients who might be at risk for osteosarcoma (e.g., Paget's disease of bone, previous irradiation therapy to the skeleton). Experience with strontium is less than with teriparatide, and the mechanism of action is quite different, as detailed in the appropriate chapter. The clinical trials data indicate that improvements in bone mass with these anabolic drugs are greater than with antiresorptives, but antifracture effectiveness has not been fully compared. In general, use of anabolic agents is reserved for those patients who have already sustained an osteoporosis-related fracture, although the antifracture effectiveness of antiresorptives has been shown in patients who have sustained a fracture before therapy.

The process of bone remodeling begins with osteoblast-mediated bone resorption, which is rapid, ~10 days, followed by a slower period, ~3 mo, of bone formation to refill the resorption cavity. With aging, this process slows down, particularly on the formation side. Given that there are no currently available agents that work independently on both aspects of this remodeling cycle, it is logical to consider the concept of combining antiresorptive and anabolic therapies, either sequentially or concurrently. Another approach would be to begin with antiresorptive therapy in patients with shown increased resorption and anabolic agents in patients that show decreased bone resorption. This was attempted unsuccessfully several years ago before the current complement of drugs was available, and the results did not support this speculation. Disappointingly, to date, this has also been the case with the newer targeted therapies, but clinical trials directly addressing this concept are in process. For now, it can be concluded that both antiresorptive and anabolic agents have shown antifracture effectiveness independent of the pretreatment remodeling characteristics.

Early chapters in the *Primer* detail the developing understanding of the molecular and cellular regulation of osteoclast and osteoblast function and their interactions. As new information is uncovered, therapies targeted at specific aspects of this work are being identified, and some are already in clinical trials. Data on phase 1 and phase 2 trials for some of these have already been published, and the underlying hypothesis for each compound has to date proven correct in these trials. Whether these result in clinical improvement over the existing excellent therapies remains to be shown.

No drug for any medical condition can be expected to be effective if it is either not taken regularly or not taken properly by the patient for whom it is prescribed. Regrettably, compliance remains an issue for the available osteoporosis therapies. This is surprising given their ease of use for most people. Even the option of once weekly or once monthly bisphosphonate therapy has not made much in-road into this vexed clinical problem. More recently, bisphosphonates are available as a once every 3 mo intravenous injection (ibandronate) or annual intravenous infusion (zoledronic acid). It is too early to tell whether this will improve patient compliance. A major issue seems to be the long time interval between initiating therapy and showing improvement in BMD. For some patients, there is no apparent change in BMD even after 2 yr. Even in patients who respond with a brisk increase in BMD, the wait is long because lack of medical insurance coverage reimbursement for the repeat test forces most patients to wait at least 12 mo before getting any feedback that therapy is working. Biochemical markers of bone remodeling do provide feedback concerning effectiveness as early as 3 mo into therapy, sometimes earlier, but this is not widely applied, and the published data do not seem to shown any improvement in compliance.

The end result of undertreatment or no treatment at all is addressed in the final two chapters of this section. Patients with osteoporosis that remain undiagnosed and untreated (i.e., the majority of older adults at risk for fracture), end up in the hands of skilled orthopedic surgeons with good early outcomes for an incident fracture. The most devastating of these fractures, the hip fracture, still results in an unacceptable level of early mortality, and even as the early mortality is declining, restoration of full prefracture function happens in <50% of hip fracture patients. Equally importantly, the occurrence of any osteoporosis-related fracture significantly increases the risk of future such fractures. Failure to recognize a distal forearm (Colles') fracture, the earliest such fracture, is not generally recognized as osteoporosis, yet it clearly increases that patient's risk of a future wrist, spine, and hip fracture and should alert all clinicians to initiate therapy to minimize that risk.

The challenges of therapy for osteoporosis have been met, and improved options are being developed. Costs associated with management after a fracture has occurred are also decreasing, and the outcomes are improving. However, the cost of care for osteoporosis continues to increase.

© 2008 American Society for Bone and Mineral Research

Chapter 44. Prevention of Falls

Heike A. Bischoff-Ferrari

Department of Rheumatology and Institute of Physical Medicine and Rehabilitation, University Hospital Zurich, Zurich, Switzerland; Jean Mayer USDA Human Nutrition Research Center on Aging, Tufts Boston, Massachusetts

INTRODUCTION

More than 90% of fractures occur after a fall, and fall rates increase with age.[1] Thus, critical for the understanding and prevention of fractures at later age is their close relationship with muscle weakness[2] and falling.[3,4] In fact, antiresorptive treatment alone may not reduce fractures among individuals ≥80 yr of age in the presence of nonskeletal risk factors for fractures despite an improvement in bone metabolism.[5]

This chapter will review the epidemiology of falls and their importance in regard to fracture risk among older individuals. Finally, fall prevention strategies are reviewed based on data from randomized controlled trials.

EPIDEMIOLOGY AND COST OF FALLS

Thirty percent of those ≥65 yr of age and 40–50% of those ≥80 yr of age report having had a fall over the past year.[1,6] Serious injuries occur with 10–15% of falls, resulting in fractures in 5% and hip fracture in 1–2%.[3] As an independent determinant of functional decline,[7] falls lead to 40% of all nursing home admissions.[8] Recurrent fallers may have close to 4-fold increased odds of sustaining a fall-related fracture compared with individuals with a single fall.[9]

Because of the increasing proportion of older individuals, annual costs from all fall-related injuries in the United States in persons ≥65 yr of age were projected to increase from $20.3 billion in 1994 to $32.4 billion in 2020, including medical, rehabilitation, hospital costs, and the costs of morbidity and mortality.[10]

FALL DEFINITION AND ASSESSMENT

Buchner et al.[11] created a useful fall definition for the common database of the FICSIT (Frailty and Injuries: Cooperative Studies of Intervention Techniques) trials. Falls were defined as "unintentionally coming to rest on the ground, floor, or other lower level." Coming to rest against furniture or a wall was not considered a fall.[11] Challenging for their assessment is that falls tend to be forgotten if not associated with significant injury,[12] requiring short periods of follow-up. Thus, high-quality fall assessment in older persons requires a prospective ascertainment of falls and their circumstances, ideally in short periods of time periods (<3 mo).[12] Fall reports may be performed by postcards, phone calls, or diary/calendar, although the usefulness and comprehensiveness of different ascertainment methods have not been compared directly. Furthermore, fall assessment has not been standardized across randomized controlled trials or large epidemiologic data sets,[13] which prevented falls from being included in the WHO criteria for hip fracture risk prediction, despite falls being the primary risk factor for hip fracture.[13]

FALL MECHANICS AND RISK OF FRACTURE

Mechanistically, the circumstances[14] and the direction[15] of a fall determine the type of fracture, whereas BMD and factors that attenuate a fall, such as better strength or better padding, critically determine whether a fracture will take place when the faller lands on a certain bone.[16] Moreover, falling may affect BMD through increased immobility from self-restriction of activities.[17] It is well known that falls may lead to psychological trauma known as fear of falling.[18] After their first fall, ~30% of persons develop fear of falling,[17] resulting in self-restriction of activities[17] and decreased quality of life. Figure 1 shows the fall-fracture construct that describes the complexity of osteoporosis prevention introduced by nonskeletal risk factors for fractures among older individuals.

Consistent with the understanding that factors unrelated to bone are at play in fracture epidemiology, the circumstances of different fractures are strikingly different. Hip fractures tend to occur in less active individuals falling indoors from a standing height with little forward momentum, and they tend to fall sideways or straight down on their hip.[16] On the other hand, other nonvertebral fractures, such as distal forearm or humerus fractures, tend to occur among more active older individuals who are more likely to be outdoors and have a greater forward momentum when they fall.[19]

Supporting the notion of bone not to be seen in isolation, fracture risk caused by falling is increased among individuals with osteoarthritis of the weight-bearing joints despite having increased BMD compared with controls.[20] One prospective study found that prevalent knee pain caused by osteoarthritis increased the risk of falling by 26% and the risk of hip fracture 2-fold.[21]

Some studies indicate that falls caused by snow and ice may play an important role in seasonality of fractures.[22,23] One cause of the increased fracture risk in winter compared with summer may be that older persons are more likely to slip and fall during periods of snow and ice.[24] On the other hand, hip fractures, which mostly occur indoors,[25] may be less affected by snow and ice with a smaller increase in winter versus summer fracture rates compared with distal forearm, humerus, and ankle fractures.[26]

RISK FACTORS FOR FALLS

Falls are a hallmark of age and becoming frail, and falls are often heralded by the onset of gait instability, visual impairment or its correction by multifocal glasses, drug treatment with antidepressants, anticonvulsants/barbiturates, or benzodiazepines, weakness, cognitive impairment, vitamin D deficiency, poor mental health, home hazards, or a combination of several risk factors.[27,28] The seemingly inseparable relationship of falls to worsening health status and the complexity of factors involved in falling has led to pessimism on the part of physicians when faced with falling, especially recurrent falling. However, there is a growing body of literature that should encourage the standardized assessment of falls and application of fall prevention strategies for fracture prevention.

FALL PREVENTION STRATEGIES

Fall prevention by risk factor reduction has been tested in a number of approaches. Multifactorial approaches, such as medical and occupational therapy assessment or adjustment in

The author state that she has no conflicts of interest.

Key words: falls, vitamin D, meta-analyses, fear of falling, fractures, risk factors, muscle weakness, exercise, multifactorial interventions, elderly, public health, economics

© 2008 American Society for Bone and Mineral Research

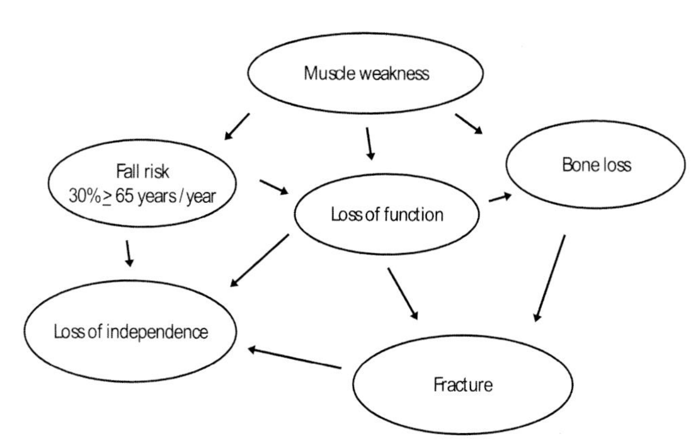

FIG. 1. Nonskeletal fall-fracture construct.

medications, behavioral instructions, and exercise programs, as shown in the PROFET-trial (Prevention of Falls in the Elderly Trial)[29] and FICSIT trials (Cooperative Studies of Intervention Techniques),[30] as well as single intervention strategies, such as Tai Chi balance training[31] and exercise,[30]reduced falls by 25–50%. Multifactorial approaches may be especially useful in high-risk populations for falls, such as older individuals in care institutions,[32] whereas less evidence is available for exercise benefits in very frail older individuals.[32] Significant limitations of multifactorial interventions and exercise programs are their cost, high implementation time, and limited long-term adherence. The latter may explain the lack of data regarding fracture prevention endpoints with exercise, Tai Chi, or multifactorial interventions.

Tai Chi has been successful in reducing falls among healthy older individuals[31,33] and physically inactive community-dwelling older individuals,[34] whereas frail older individuals[35] and fallers[33] may not benefit as much. Furthermore, Tai Chi may not improve BMD,[36] and fracture prevention has not been explored as an endpoint with Tai Chi intervention programs.

FALL PREVENTION STRATEGIES WITH EVIDENCE FOR FRACTURE REDUCTION

Two interventions among older individuals resulted in both fall and fracture reduction. One is cataract surgery, with limited evidence from one trial. Three hundred six women >70 yr of age, with cataracts, were randomized to expedited (~4 weeks) or routine (12-mo wait) surgery. Over a 12-mo follow-up, the rate of falling was reduced by 34% in the expedited group (rate ratio, 0.66; 95% CI, 0.45–0.96) accompanied by a significantly lower number persons with a new fracture (p = 0.04).[37]

With evidence from several trials, supplementation with vitamin D may reduce the risk of falls[38] and fractures[39] in older individuals. An important risk factor for falls is muscle weakness, which is a prominent feature of the clinical syndrome of vitamin D deficiency and may plausibly mediate fracture risk through an increased susceptibility to falls. The vitamin D receptor (VDR) is expressed in human muscle tissue,[40] and vitamin D bound to its nuclear receptor in muscle tissue may lead to de novo protein synthesis,[41,42] followed by a relative increase in the diameter and number of type II muscle fibers.[42]

Summarizing five high-quality double-blind randomized controlled trials (RCTs; n = 1237) a meta-analysis published in 2004 found that vitamin D reduced the risk of falling by 22% (pooled corrected OR = 0.78; 95% CI = 0.64, 0.92) compared with calcium or placebo.[38] This risk reduction was independent of the type of vitamin D, duration of therapy, and sex. For the two trials with 259 subjects using 800 IU of cholecalciferol per day over 2–3 mo, the corrected pooled OR was 0.65 (95% CI = 0.40, 1.00),[38] whereas 400 IU was insufficient in reducing falls.[19] The importance of dose of vitamin D in regard to antifall efficacy was confirmed by one more recent double-blind RCT among 124 nursing home residents receiving 200, 400, 600, or 800 IU vitamin D compared with placebo over a 5-mo period.[43] Participants in the 800 IU group had a 72% lower rate of falls than those taking placebo or a lower dose of vitamin D (rate ratio = 0.28; 95% CI = 0.11–0.75).[43]

Long-term supplementation over 3 yr with 700 IU D_3 plus 500 mg calcium among community-dwelling older women reduced the odds of falling by 46% (OR = 0.54; 95% CI = 0.30–0.97).[44] Similarly, long-term supplementation among institutionalized individuals with ergocalciferol for 2 yr, initially 10,000 IU given once weekly and then 1000 IU daily, reduced the incident rate ratio for falling by 26% (RR = 0.73, 95% CI = 0.57–0.95).[45] A 49% fall reduction among older individuals with a history of a fall was shown in the most recent 12-mo trial with 1000 IU ergocalciferol compared with placebo (OR = 0.61; 95% CI = 0.37–0.99).[46]

All evidence considered, the reduction of falls with oral or injectable vitamin D with or without calcium was 11% (pooled OR = 0.89; 95% CI = 0.80–0.99) in a 2007 meta-analysis,[47] even after the inclusion of open design trials[48,49] and trials with <60% adherence.[48,50] Finally, it is important to note that vitamin D may address each component of the fall-fracture construct as shown in Fig. 1, including strength,[51] balance,[52] lower extremity function,[53,54] falling,[43,44] BMD,[55,56] risk of hip and nonvertebral fractures,[39] and the risk of nursing home admission.[57]

CONCLUSIONS

Fall risk reduction is a significant component of fracture prevention at older age, and the public health impact of falls is significant. Falls can be reduced by a number of interventions, with vitamin D offering efficacy on several levels of the fall-fracture construct at later ages (Fig. 1). To study falls and the fall-fracture risk profile from different interventions and cohort studies better, fall definition and ascertainment need to be standardized.

REFERENCES

1. Tinetti ME 1988 Risk factors for falls among elderly persons living in the community. N Engl J Med **319:**1701–1707.
2. Cummings SR, Nevitt MC, Browner WS, Stone K, Fox KM, Ensrud KE, Cauley J, Black D, Vogt TM 1995 Risk factors for hip fracture in white women. Study of Osteoporotic Fractures Research Group. N Engl J Med **332:**767–773.
3. Centers for Disease Control and Prevention 2006 Fatalities and injuries from falls among older adults–United States, 1993-2003 and 2001-2005. MMWR Morb Mortal Wkly Rep **55:**1221–1224.
4. Schwartz AV, Nevitt MC, Brown BW Jr, Kelsey JL 2005 Increased falling as a risk factor for fracture among older women: The study of osteoporotic fractures. Am J Epidemiol **161:**180–185.
5. McClung MR, Geusens P, Miller PD, Zippel H, Bensen WG, Roux C, Adami S, Fogelman I, Diamond T, Eastell R, Meunier P, Reginster JY 2001 Effect of risedronate on the risk of hip fracture in elderly women. Hip Intervention Program Study Group. N Engl J Med **344:**333–340.
6. Campbell AJ, Reinken J, Allan BC, Martinez GS 1981 Falls in old age: A study of frequency and related clinical factors. Age Ageing **10:**264–270.
7. Tinetti ME, Williams CS 1998 The effect of falls and fall injuries on functioning in community-dwelling older persons. J Gerontol A Biol Sci Med Sci **53:**M112–M119.

© 2008 American Society for Bone and Mineral Research

8. Tinetti ME, Williams CS 1997 Falls, injuries due to falls, and the risk of admission to a nursing home. N Engl J Med **337**:1279–1284.
9. Pluijm SM, Smit JH, Tromp EA, Stel VS, Deeg DJ, Bouter LM, Lips P 2006. A risk profile for identifying community-dwelling elderly with a high risk of recurrent falling: Results of a 3-year prospective study. Osteoporos Int. **17**(3):417-25.
10. Englander F, Hodson TJ, Terregrossa RA 1996 Economic dimensions of slip and fall injuries. J Forensic Sci **41**:733–746.
11. Buchner DM, Hornbrook MC, Kutner NG, Tinetti ME, Ory MG, Mulrow CD, Schechtman KB, Gerety MB, Fiatorone MA, Wolf SL 1993 Development of the common data base for the FICSIT trials. J Am Geriatr Soc **41**:297–308.
12. Cummings SR, Nevitt MC, Kidd S 1988 Forgetting falls. The limited accuracy of recall of falls in the elderly. J Am Geriatr Soc **36**:613–616.
13. Kanis JA, Borgstrom F, De Laet C, Johansson H, Johnell O, Jonsson B, Oden A, Zethraeus N, Pfelger P, Khaltaev N 2005 Assessment of fracture risk. Osteoporos Int **16**:581–589.
14. Cummings SR, Nevitt MC 1994 Non-skeletal determinants of fractures: The potential importance of the mechanics of falls. Study of Osteoporotic Fractures Research Group. Osteoporos Int **4**(Suppl 1):67–70.
15. Nguyen ND, Frost SA, Center JR, Eisman JA, Nguyen TV 2007 Development of a nomogram for individualizing hip fracture risk in men and women. Osteoporos Int **17**:17.
16. Nevitt MC, Cummings SR 1993 Type of fall and risk of hip and wrist fractures: The study of osteoporotic fractures. The Study of Osteoporotic Fractures Research Group. J Am Geriatr Soc **41**:1226–1234.
17. Vellas BJ, Wayne SJ, Romero LJ, Baumgartner RN, Garry PJ 1997 Fear of falling and restriction of mobility in elderly fallers. Age Ageing **26**:189–193.
18. Arfken CL, Lach HW, Birge SJ, Miller JP 1994 The prevalence and correlates of fear of falling in elderly persons living in the community. Am J Public Health **84**:565–570.
19. Graafmans WC, Ooms ME, Hofstee HM, Bezemer PD, Bouter LM, Lips P 1996 Falls in the elderly: A prospective study of risk factors and risk profiles. Am J Epidemiol **143**:1129–1136.
20. Arden NK, Nevitt MC, Lane NE, Gore LR, Hochberg MC, Scott JC, Pressman AR, Cummings SR 1999 Osteoarthritis and risk of falls, rates of bone loss, and osteoporotic fractures. Study of Osteoporotic Fractures Research Group. Arthritis Rheum **42**:1378–1385.
21. Arden NK, Crozier S, Smith H, Anderson F, Edwards C, Raphael H, Cooper C 2006 Knee pain, knee osteoarthritis, and the risk of fracture. Arthritis Rheum **55**:610–615.
22. Bulajic-Kopjar M 2000 Seasonal variations in incidence of fractures among elderly people. Inj Prev **6**:16–19.
23. Hemenway D, Colditz GA 1990 The effect of climate on fractures and deaths due to falls among white women. Accid Anal Prev **22**:59–65.
24. Ralis ZA 1981 Epidemic of fractures during period of snow and ice. Br Med J (Clin Res Ed) **282**:603–605.
25. Carter SE, Campbell EM, Sanson-Fisher RW, Gillespie WJ 2000 Accidents in older people living at home: A community-based study assessing prevalence, type, location and injuries. Aust N Z J Public Health **24**:633–636.
26. Bischoff-Ferrari HA, Orav JE, Barrett JA, Baron JA 2007 Effect of seasonality and weather on fracture risk in individuals 65 years and older. Osteoporos Int **24**:1225–1233.
27. Tinetti ME, Inouye SK, Gill TM, Doucette JT 1995 Shared risk factors for falls, incontinence, and functional dependence. Unifying the approach to geriatric syndromes. JAMA **273**:1348–1353.
28. Mowe M, Haug E, Bohmer T 1999 Low serum calcidiol concentration in older adults with reduced muscular function. J Am Geriatr Soc **47**:220–226.
29. Close J, Ellis M, Hooper R, Glucksman E, Jackson S, Swift C 1999 Prevention of falls in the elderly trial (PROFET): A randomized controlled trial. Lancet **353**:93–97.
30. Province MA, Hadley EC, Hornbrook MC, Lipsitz LA, Miller JP, Mulrow CD, Ory MG, Sattin RW, Tinetti ME, Wolf SL 1995 The effects of exercise on falls in elderly patients. A preplanned meta-analysis of the FICSIT Trials. Frailty and Injuries: Cooperative Studies of Intervention Techniques. JAMA **273**:1341–1347.
31. Wolf SL, Barnhart HX, Kutner NG, McNeely E, Coogler C, Xu T 1996 Reducing frailty and falls in older persons: An investigation of Tai Chi and computerized balance training. Atlanta FICSIT Group. Frailty and Injuries: Cooperative Studies of Intervention Techniques. J Am Geriatr Soc **44**:489–497.
32. Oliver D, Connelly JB, Victor CR, Shaw FE, Whitehead A, Genc Y, Vanoli A, Martin FC, Gosney MA 2007 Strategies to prevent falls and fractures in hospitals and care homes and effect of cognitive impairment: Systematic review and meta-analyses. BMJ **334**:82.
33. Voukelatos A, Cumming RG, Lord SR, Rissel C 2007 A randomized, controlled trial of tai chi for the prevention of falls: The Central Sydney tai chi trial. J Am Geriatr Soc **55**:1185–1191.
34. Li F, Harmer P, Fisher KJ, McAuley E, Chaumeton N, Eckstrom E, Wilson NL 2005 Tai Chi and fall reductions in older adults: A randomized controlled trial. J Gerontol A Biol Sci Med Sci **60**:187–194.
35. Wolf SL, Sattin RW, Kutner M, O'Grady M, Greenspan AI, Gregor RJ 2003 Intense tai chi exercise training and fall occurrences in older, transitionally frail adults: A randomized, controlled trial. J Am Geriatr Soc **51**:1693–1701.
36. Lee MS, Pittler MH, Shin BC, Ernst E 2008 Tai chi for osteoporosis: A systematic review. Osteoporos Int **19**:139–146.
37. Harwood RH, Foss AJ, Osborn F, Gregson RM, Zaman A, Masud T 2005 Falls and health status in elderly women following first eye cataract surgery: A randomised controlled trial. Br J Ophthalmol **89**:53–59.
38. Bischoff-Ferrari HA, Dawson-Hughes B, Willett CW, Staehelin HB, Bazemore MG, Zee RY, Wong JB 2004 Effect of vitamin D on falls: A meta-analysis. JAMA **291**:1999–2006.
39. Bischoff-Ferrari HA, Willett WC, Wong JB, Giovannucci E, Dietrich T, Dawson-Hughes B 2005 Fracture prevention with vitamin D supplementation: A meta-analysis of randomized controlled trials. JAMA **293**:2257–2264.
40. Bischoff-Ferrari HA, Borchers M, Gudat F, Durmuller U, Stahelin HB, Dick W 2004 Vitamin D receptor expression in human muscle tissue decreases with age. J Bone Miner Res **19**:265–269.
41. Boland R 1986 Role of vitamin D in skeletal muscle function. Endocr Rev **7**:434–447.
42. Sorensen OH, Lund B, Saltin B, Andersen RB, Hjorth L, Melsen F, Moseklide L 1979 Myopathy in bone loss of ageing: Improvement by treatment with 1 alpha-hydroxycholecalciferol and calcium. Clin Sci (Colch) **56**:157–161.
43. Broe KE, Chen TC, Weinberg J, Bischoff-Ferrari HA, Holick MF, Kiel DP 2007 A higher dose of vitamin d reduces the risk of falls in nursing home residents: A randomized, multiple-dose study. J Am Geriatr Soc **55**:234–239.
44. Bischoff-Ferrari HA, Orav EJ, Dawson-Hughes B 2006 Effect of cholecalciferol plus calcium on falling in ambulatory older men and women: A 3-year randomized controlled trial. Arch Intern Med **166**:424–430.
45. Flicker L, MacInnis RJ, Stein MS, Scherer SC, Mead KE, Nowson CA, Thomas J, Lowndes C, Hopper JL, Wark JD 2005 Should older people in residential care receive vitamin D to prevent falls? Results of a randomized trial. J Am Geriatr Soc **53**:1881–1888.
46. Prince RL, Austin N, Devine A, Dick IM, Bruce D, Zhu K 2008 Effects of ergocalciferol added to calcium on the risk of falls in elderly high-risk women. Arch Intern Med **168**:103–108.
47. Cranny A, Horsley T, O'Donnell S, Weiler H, Puil L, Ooi D, Atkinson S, Ward L, Moher D 2007 Effectiveness and safety of vitamin D in relation to bone health. Available at http://www.ahrqgov/clinic/tp/vitadtp.htm. Accessed August 2007.
48. Porthouse J, Cockayne S, King C, Saxon L, Steele E, Aspray T, Baverstock M, Birks Y, Dumville J, Francis R, Iglesias C, Puffer S, Sutcliffe A, Watt I, Togerson DJ 2005 Randomised controlled trial of calcium and supplementation with cholecalciferol (vitamin D3) for prevention of fractures in primary care. BMJ **330**:1003.
49. Harwood RH, Sahota O, Gaynor K, Masud T, Hosking DJ 2004 A randomised, controlled comparison of different calcium and vitamin D supplementation regimens in elderly women after hip fracture: The Nottingham Neck of Femur (NONOF) Study. Age Ageing **33**:45–51.
50. Grant AM, Avenell A, Campbell MK, McDonald AM, MacLennan GS, McPherson GC, Anderson FH, Cooper C, Francis RM, Donaldson C, Gillespie WJ, Robinson CM, Togerson DJ, Wallace WA 2005 Oral vitamin D3 and calcium for secondary prevention of low-trauma fractures in elderly people (Randomised Evaluation of Calcium Or vitamin D, RECORD): A randomised placebo-controlled trial. Lancet **365**:1621–1628.
51. Bischoff HA, Stahelin HB, Dick W, Akos R, Knecht M, Salis C, Nebiker M, Theler R, Pfeifer M, Begereow B, Lew RA, Conzel-

© 2008 American Society for Bone and Mineral Research

mann M 2003 Effects of vitamin D and calcium supplementation on falls: A randomized controlled trial. J Bone Miner Res **18:**343–351.
52. Pfeifer M, Begerow B, Minne HW, Abrams C, Nachtigall D, Hansen C 2000 Effects of a short-term vitamin D and calcium supplementation on body sway and secondary hyperparathyroidism in elderly women. J Bone Miner Res **15:**1113–1118.
53. Bischoff-Ferrari HA, Dietrich T, Orav EJ, Hu FB, Zhang Y, Karlson EW, Dawson-Hughes B 2004 Higher 25-hydroxyvitamin D concentrations are associated with better lower-extremity function in both active and inactive persons aged ≥60 y. Am J Clin Nutr **80:**752–758.
54. Wicherts IS, van Schoor NM, Boeke AJ, Visser M, Deeg DJ, Smit J, Knol DL, Lips P 2007 Vitamin D status predicts physical performance and its decline in older persons. J Clin Endocrinol Metab **6:**2058–2065.
55. Dawson-Hughes B, Harris SS, Krall EA, Dallal GE 1997 Effect of calcium and vitamin D supplementation on bone density in men and women 65 years of age or older. N Engl J Med **337:**670–676.
56. Bischoff-Ferrari HA, Dietrich T, Orav EJ, Dawson-Hughes B 2004 Positive association between 25-hydroxy vitamin d levels and bone mineral density: A population-based study of younger and older adults. Am J Med **116:**634–639.
57. Visser M, Deeg DJ, Puts MT, Seidell JC, Lips P 2006 Low serum concentrations of 25-hydroxyvitamin D in older persons and the risk of nursing home admission. Am J Clin Nutr. **84:**616-622.

Chapter 45. Orthopedic Surgical Principles of Fracture Management

Manoj Ramachandran[1] and David G. Little[2]

[1]*Barts and the London NHS Trust, Whitechapel, London, United Kingdom;* [2]*Paediatrics and Child Health, University of Sydney, Orthopaedic Research and Biotechnology, The Children's Hospital at Westmead, Westmead, New South Wales, Australia*

INTRODUCTION

A broad range of fractures and associated injuries present to the orthopedic or trauma surgeon for management. The principle that fracture immobilization supports both alignment and union of the fracture, while minimizing discomfort, has been known since antiquity. The energy of the injury, associated soft tissue injury, and displacement of the fracture usually guide intervention.

The initial management of fractures consists of realignment of the broken limb segment and immobilization of the fractured extremity once the initial assessment, evaluation, and management of any life-threatening injury are completed. The aim of fracture treatment is to obtain union of the fracture in the most anatomical position compatible with maximal functional recovery of the extremity or the spine with minimal complications. This is accomplished by obtaining and subsequently maintaining a reduction of the fracture with an immobilization technique that allows the fracture to heal and, at the same time, provides the patient with functional aftercare. Either nonoperative or surgical means may be used. Any surgical technique, if chosen, should minimize additional soft tissue and bone injury, which can delay fracture healing.

In open fractures, in addition to the treatment aims outlined above, the prevention of infection is vital.[1] This is achieved by urgent wound irrigation and debridement (with serial irrigations and débridements every 24–48 h until the wounds are clean and closed), antibiotic administration, and tetanus vaccination. If soft tissue coverage over the injury is inadequate, soft tissue transfers or free flaps are performed when the wound is clean, and the fracture is definitively treated.

TREATMENT PRINCIPLES

Fracture management can be divided into nonoperative and operative techniques. Nonoperative technique consists of a closed reduction (required if the fracture is significantly displaced or angulated), achieved by applying traction to the long axis of the injured limb and reversing the mechanism of injury. This is followed by a period of immobilization with casting. Casts are made from fiberglass or plaster of Paris. Complications of casts can include the development of pressure ulcers, thermal burns during plaster hardening, and joint stiffness. Traction (skin or skeletal) is rarely used nowadays for definitive fracture management.

If the fracture cannot be reduced (e.g., because of soft tissue interposition), surgical intervention may be needed. Indications for surgery include failed nonoperative management, unstable fractures that cannot be adequately maintained in a reduced position, displaced intra-articular fractures (>2 mm), impending pathologic fractures, unstable or complicated open fractures, fractures in growth areas in skeletally immature individuals that have increased risk for growth arrest, and nonunions or malunions that have failed to respond to nonoperative treatment.[2] Contraindications to internal fixation include active infection (local or systemic), soft tissues that compromise the overlying fracture or the surgical approach (e.g., burns and previous surgical scars), medical conditions that contraindicate surgery or anesthesia (e.g., recent myocardial infarction), and cases in which amputation would be more appropriate (e.g., severe neurovascular injury).

SURGICAL OPTIONS

The treatment goals for surgical fracture management, as outlined by the Association for the Study of Internal Fixation (ASIF), are anatomic reduction of the fracture fragments (for the correction of length, angulation, and rotation and for intraarticular fractures for the restoration of the joint surface), stable internal fixation to cope with physiological biomechanical demands, preservation of blood supply, and active, pain-free mobilization of adjacent muscles and joints.[3] The objectives of open reduction and internal fixation include adequate exposure and reduction of the fracture, followed by maintenance of the reduction by stabilization using one or a combination of the methods below.

Kirschner Wires

Kirschner wires, or K-wires, placed percutaneously or

Dr. Little has received royalties from Novartis. Dr. Ramachandran states that he has no conflicts of interest.

© 2008 American Society for Bone and Mineral Research

through a mini-open approach, are commonly used for temporary and definitive treatment of fractures. However, they have poor resistance to torque and bending forces and rely on friction with the bone for maintenance of reductions. Therefore, when they are used as the sole form of fixation, casting or splinting is used in conjunction. K-wires are commonly used as adjunctive fixation for screws or plates and screws that involve fractures around joints.

Plates and Screws

Plates and screws are commonly used in the management of articular fractures because they provide strength and stability to neutralize the forces on the injured limb for functional postoperative aftercare. Plate designs vary, depending on the anatomic region and size of the bone the plate is used for. All plates should be applied with minimal stripping of the soft tissue. Five main plate designs exist:

1. Buttress (antiglide) plates counteract the compression and shear forces that commonly occur with fractures that involve the metaphysis and epiphysis. These plates are commonly used with interfragmentary screw fixation and require anatomical contouring to achieve stable fixation.
2. Compression plates counteract bending, shear, and torsional forces by providing compression across the fracture site by the eccentrically loaded holes in the plate. Compression plates are commonly used in the long bones, especially the fibula, radius, and ulna, and in nonunion or malunion surgery.
3. Neutralization plates are used in combination with interfragmentary screw fixation. The interfragmentary compression (lag) screws provide compression at the fracture site. This plate function neutralizes bending, shear, and torsional forces on the lag screw fixation, as well as increases the stability of the construct. These plates are commonly used for fractures involving the fibula, radius, ulna, and humerus.
4. Bridge plates are useful in the management of multifragmented diaphyseal and metaphyseal fractures. Indirect reduction techniques are preferred in bridge plating without disrupting the soft tissue attachments to the bone fragments.
5. Tension band plate technique converts tension forces into compressive forces, thereby providing stability (e.g., in oblique olecranon fractures).

More recently, locking plates have been introduced. A locking plate acts like an internal fixator. There is no need to anatomically contour the plate onto the bone, thus reducing bone necrosis and allowing for a minimally invasive technique. Locking screws directly anchor and lock onto the plate, thereby providing angular and axial stability. These screws are incapable of toggling, sliding, or becoming dislodged, thus reducing the possibility of a secondary loss of reduction, as well as eliminating the possibility of intraoperative overtightening of the screws. The locking plate is indicated for osteoporotic fractures, for short and metaphyseal segment fractures, and for bridging comminuted areas. These plates are also appropriate for metaphyseal areas where subsidence may occur or prostheses are involved.

The technique of minimally invasive percutaneous plate osteosynthesis (MIPPO) with indirect reduction is becoming increasingly popular. This involves the use of anatomically preshaped plates and instrumentation to safely and effectively insert the plate percutaneously or through limited incisions.[4] Advantages of MIPPO may include faster bone healing, reduced infection rate, decreased need for bone grafting, less postoperative pain, faster rehabilitation, and more aesthetic results. Some disadvantages include difficulty with indirect reduction, increased radiation exposure caused by more intraoperative radiographs being used, malunion, pseudoarthrosis through diastases, and delayed union with flexible fixation in simple fractures.

Intramedullary Nails

These nails operate like an internal splint that shares the load with the bone and can be flexible or rigid, locked or unlocked, and reamed or unreamed. Locked intramedullary nails provide relative stability to maintain bone alignment and length and to limit rotation. Ideally, the intramedullary nail allows for compressive forces at the fracture site, which stimulates bone healing.

Intramedullary nails are commonly used for femoral and tibial diaphyseal fractures and, occasionally, humeral diaphyseal fractures. The advantages of intramedullary nails include minimally invasive procedures, early postoperative ambulation, and early range of motion being permitted in adjacent joints. Reaming may also increase union rates possibly by providing the equivalent of a bone graft to the fractured region.

External Fixation

External fixation provides fracture stabilization at a distance from the fracture site, without interfering with the soft tissue structures that are near the fracture. This technique not only provides stability for the extremity and maintains bone length, alignment, and rotation without requiring casting, but it also allows for inspection of the soft tissue structures that are vital for fracture healing. Indications for external fixation (temporarily or as definitive care) are as follows[5]:

1. Open fractures that have significant soft tissue disruption (e.g., type II or III open fractures)
2. Soft tissue injury (e.g., burns)
3. Pelvic fractures
4. Severely comminuted and unstable fractures
5. Fractures associated with bony defects
6. Limb-lengthening and bone transport procedures
7. Fractures associated with infection or nonunion

Complications of external fixation include pin tract infection, pin loosening or breakage, interference with joint motion, neurovascular damage during pin placement, malalignment caused by poor placement of the fixator, delayed union, and malunion. Modern external fixators allow for postoperative adjustment to effect more anatomical reduction of the fracture in the weeks after intervention.

COMMON FRACTURES IN OSTEOPOROTIC PATIENTS

Vertebral compression fractures, Colles' (distal radius) fractures, hip fractures, and other peripheral (nonvertebral) fractures all occur in patients with osteoporosis. These fractures may indicate the need for treatment of osteoporosis for secondary fracture prevention.

Osteoporotic fractures are usually low-energy injuries. Some fractures such as fatigue fractures of the pelvis have no displacement. Colles' fractures and hip fractures are usually displaced and require reduction and fixation. Colles' fractures are sometimes held in a cast, but may require wire fixation or low-profile plating to maintain reduction. Colles' fractures nearly always heal, but significant malunion can interfere with function. Intertrochanteric hip fractures are usually fixed with sliding hip screw devices that allow compression of the fracture fragments on weight bearing. Subcapital neck of femur frac-

© 2008 American Society for Bone and Mineral Research

tures require joint arthroplasty of some form in the majority of cases because of the high incidence of nonunion and avascular necrosis. In recent times, the development of locking plate technology has improved the surgeon's ability to internally fix fractures in osteoporotic bone, but further research is needed, because in many cases, optimal fixation cannot be achieved. It is a principle of management for all osteoporotic fractures to institute load bearing and functional tasks as quickly as possible to minimize loss of function or mobility.

INTERVENTION FOR VERTEBRAL FRACTURES IN OSTEOPOROTIC PATIENTS

Acute vertebral compression fractures can be painful and lead to disability, whereas multiple "silent" compression fractures lead to kyphosis and loss of height. Whereas in many clinically apparent fractures the pain settles in a few weeks, it has been estimated that one third of fractures can remain chronically painful. Whereas most vertebral compression fractures are managed nonoperatively, there has been an increase in intervention for painful acute fracture to minimize morbidity. These techniques are known as vertebroplasty and kyphoplasty.

In vertebroplasty, a percutaneous approach to the vertebral body is made under fluoroscopic control, either through or adjacent to the pedicles. Bone cement, usually polymethylmethacrylate (PMMA), is injected into the fracture under pressure by a cannula while the cement is in a fluid state. This acutely stabilizes the fracture and results in an immediate reduction of pain that is significant enough in many cases to allow immediate return to activities of daily living. Pain reduction is thought to be from stabilization of the fracture, although heat necrosis of nerve endings from the exothermic setting of the cement has also been suggested. Vertebroplasty makes little difference to the spinal alignment because the injection of cement usually does little to change the wedging deformity of the fracture. Kyphoplasty is designed to address these limitations. In this technique, cannulae are placed usually bilaterally to allow the introduction of balloon tamps, which are expanded with radio-opaque saline. Some elevation of the end plate and thus correction of deformity can be achieved by this method. The balloon expansion creates a void into which cement is injected at a slightly more viscous state than in vertebroplasty. The literature available suggests that, whereas correction of vertebral morphology is achievable, overall spinal balance is usually not affected because alterations in shape can be accommodated by disc spaces and deformity at other levels.[6]

There are no direct comparisons of the two techniques, and both are effective in relieving pain in a few days in the majority of individuals. One nonrandomized study showed decreases in pain, rapid return to function, and decreased hospital stay in vertebroplasty versus conservatively treated patients.[7] A systematic review has shown good evidence for a therapeutic effect and some superiority for kyphoplasty over nonoperative treatment.[8] There is a suggestion from this meta-analysis that kyphoplasty may be associated with fewer cement leakage events than vertebroplasty. However, randomized studies are still lacking. Serious complications include neurological sequelae and have been reported to run at ~1%.

The morbidity of the intervention itself for vertebroplasty and kyphoplasty are minor compared with the alternative of open surgery for internal fixation, giving the techniques a favorable risk to harm ratio. It is clear that not all the estimated 700,000 vertebral fractures occurring per year in the United States could or should be treated by such vertebral augmentation techniques. It is important that proper patient selection is used, because only acute painful lesions are likely to be corrected by these techniques.

REFERENCES

1. Gustilo RB, Merkow RL, Templeman D 1990 The management of open fractures. J Bone Joint Surg Am **72**:299–304.
2. Canale ST 2003 Campbell's Operative Orthopaedics, 10th ed. Mosby-Year Book, St. Louis, MO, USA.
3. Ruedi TP, Buckley R, Moran C, eds. 2007 AO Principles of Fracture Management, 2nd ed. Thieme Medical Publishers, New York, NY, USA.
4. Krettek C, Schandelmaier P, Miclau T, Tscherne H 1997 Minimally invasive percutaneous plate osteosynthesis (MIPPO) using the DCS in proximal and distal femoral fractures. Injury **28**(Suppl 1):A20–A30.
5. Bucholz RW, Heckman JD, Court-Brown C, Tornetta P III, Koval KJ, Wirth MA, eds. 2005 Rockwood & Green's Fractures in Adults, 6th ed. Lippincott Williams & Wilkins, Philadelphia, PA, USA.
6. Pradhan BB, Bae HW, Kropf MA, Patel VV, Delamarter RB 2006 Kyphoplasty reduction of osteoporotic vertebral compression fractures: Correction of local kyphosis versus overall sagittal alignment. Spine **31**:435–441.
7. Diamond TH, Bryant C, Browne L, Clark WA 2006 Clinical outcomes after acute osteoporotic vertebral fractures: A 2-year non-randomised trial comparing percutaneous vertebroplasty with conservative therapy. Med J Aust **184**:113–117.
8. Taylor RS, Fritzell P, Taylor RJ 2007 Balloon kyphoplasty in the management of vertebral compression fractures: An updated systematic review and meta-analysis. Eur Spine J **16**:1085–1100.

Chapter 46. Exercise and the Prevention of Osteoporosis

Clinton Rubin,[1] Janet Rubin,[2] and Stefan Judex[1]

[1]*Department of Biomedical Engineering, State University of New York, Stony Brook, New York;* [2]*Department of Medicine, University of North Carolina School of Medicine, Chapel Hill, North Carolina*

INTRODUCTION

Osteopenia, a condition of diminished bone mass, becomes osteoporosis when mechanical demands exceed the ability of the skeletal structure to support them. Consequences of bone loss are exacerbated by an age-related decrease in muscle strength and postural stability, markedly increasing the risk of falling and fracture. Mechanical signals generated by exercise can mitigate bone loss as well as help preserve the musculoskeletal "system."

Dr. Clinton Rubin has consulted for and owns stock in Juvent Medical, Inc. All other authors state that they have no conflicts of interest.

Key words: mechanical, osteoblast, osteocyte, osteoporosis, exercise, strain, adaptation, tissue, adipocyte, obesity

© 2008 American Society for Bone and Mineral Research

Bone structure is enhanced by exercise and compromised by disuse.[1] Cross-sectional studies illustrate the sensitivity of bone morphology to physical extremes. Astronauts subject to microgravity lose up to 2% of hip BMD each month,[2] whereas professional tennis players possess 35% more bone in the dominant arm.[3] A broader benefit can be seen in long-trained professionals involved in a variety of demanding activities, including soccer players, weight lifters, speed skaters, and gymnasts.[4,5]

Several prospectively designed trials emphasize that functional loading result in increased bone mass. Intense exercise in young army recruits[6] stimulated large increases in BMD, whereas a 10-mo, high-impact strength building regimen in children significantly increased femoral neck BMD.[7] A number of longitudinal exercise studies, however, have reported only modest increases in bone mass. For example, a 1-yr high-resistance strength training study in young women significantly increased muscle strength but failed to influence bone mass.[8] The range of results and the complex mechanical milieu generated by exercise suggests that some components of the load-bearing regimen are more influential than others.

MECHANICAL FACTORS REGULATING BONE CELL RESPONSE

Skeletal loads and bending moments resolve into strain in the bone tissue. The strain levels actually "experienced" by bone cells in vivo is unclear, and may be as much as 10 times that experienced by the matrix.[9] Bone cells also experience interstitial fluid flow during and dynamic pressure changes during mechanical loading.[10] Functional loading also induces pressure in the intramedullary cavity, shear forces through canaliculi, and dynamic electric fields as interstitial fluid flows past charged bone crystals. Certainly, the complex loading environment of the skeleton generates a diverse range of mechanical signals that are ultimately inseparable, but biologically may differentially influence tissue, cell, and molecular activity.

Animal models show that bone remodeling is sensitive to changes in strain magnitude,[11] the number of loading cycles,[12] the distribution of the loading,[13] and the rate of strain.[14] Importantly, the load signal must be dynamic (time-varying) as static loads are ignored by the skeleton,[15] whereas the anabolic potential increases when rest periods are inserted between the mechanical events.[16] Even extremely low-magnitude bone strains, three orders of magnitude below peak strains generated during strenuous activity, when induced at a high frequencies similar to the spectral content of muscle contractibility,[17] are anabolic to bone tissue,[18] enhancing not only bone mass but bone quality and strength (Fig. 1).

Physiologic levels of strain reduce osteocyte apoptosis, suggesting matrix deformation is critical even to the survival of these cells.[19,20] Too much strain, however, induces matrix microdamage, exacerbating death of adjacent cells.[21] Factors other than matrix strain also cause an adaptive cellular response (e.g., acceleration, independent of direct loading can be anabolic to bone), suggesting that cells act as "accelerometers" as well as responding to deformation,[22] emphasizing that by-products of the strain signal, such as shear stress or strain-generated potentials, may be critical to regulating the biology of the adaptive response. However, how do the cells sense these mechanical signals?

MECHANICALLY RESPONSIVE BONE CELLS

The sensitivity of bone cells to mechanical signals, including stromal cells, osteoblasts, and osteocytes, has been well documented,[23] but it is difficult to designate a critically responsive cell. Whereas the osteoblast is critical for the adaptive response, the osteocyte, representing 95% of adult skeletal cells, may prove key to bone tissue plasticity.[24] The antenna-like 3D morphology of this osteocyte syncytium, interconnected by regulated gap-junctional connexins,[25] is ideally configured to perceive and even amplify biophysical stimuli.[26] The connectivity of this network deteriorates markedly with age and may contribute to the progressive loss of sensitivity of bone to chemical and physical signals.[27]

Marrow stromal cells change proliferation and gene expression in response to mechanical stimulation,[28] and through mechanical regulation of RANKL expression[29] also affect osteoclast number and function. The osteoclast also responds directly to mechanical signals limiting bone resorption.[30] Other cells present in bone, such as shear-sensitive endothelial cells in the penetrating vasculature, likely contribute to the adaptive response by producing NO,[31] which has plesiomorphic effects on the skeleton.[32]

The role of mesenchymal stem cells (MSCs) in regulating the adaptive response to mechanical signals has also been proposed.[33] Anabolic low-magnitude mechanical signals bias the fate of MSCs, driving them toward a musculoskeletal lineage while suppressing adipogenesis.[34] Considering the interdependence of fat and bone tissue,[35] it may be that exercise-based prevention of obesity and osteoporosis could be achieved through regulation of MSC lineage rather than nec-

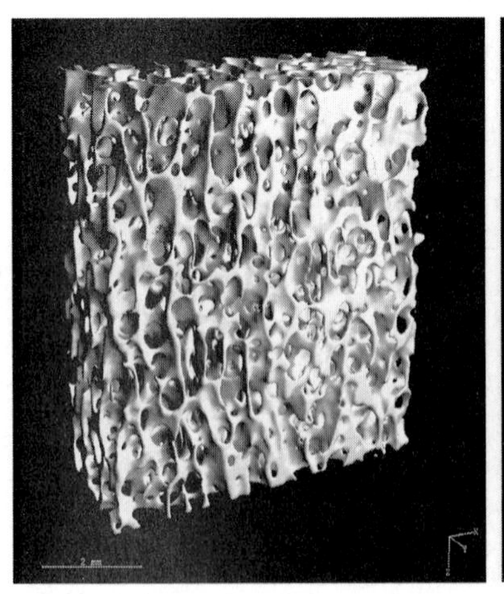

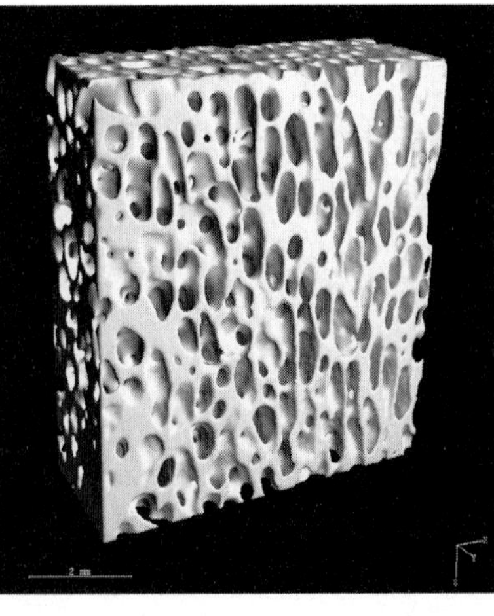

FIG. 1. μCT of the distal femur of adult (8 yr) sheep, comparing a control animal (left) to an animal subject to 20 min/d of 30 Hz (cycles per second) of a low-level (0.3*g*) mechanical vibration for 1 yr.[55] The large increase in trabecular BMD results in enhanced bone strength, achieved with tissue strains three orders of magnitude below those that cause damage to the tissue. These data suggest that specific mechanical parameters may represent a nonpharmacologic basis for the treatment of osteoporosis.[52]

© 2008 American Society for Bone and Mineral Research

essarily the resident cell population in fat (adipocytes) or bone (osteocytes).

TRANSDUCING MECHANICAL SIGNALS INTO CELLULAR RESPONSE

When bone strains of 3000 microstrain, realized during strenuous activity,[36] are resolved to the level of the cell, these deformations are on the order of Angstroms, requiring an exquisitely sensitive receptive system. Furthermore, cell mechanoreceptors must either be in contact with the outside, through the cell membrane and its attachment to substrate, or the mechanoreceptor must be able to sense byproducts of load, such as fluid shear on the apical membrane. Whereas in sensory organs there are examples of channels that are regulated by movement of mechanosensory bristles[37] or by tension waves,[38] a unified model of proximal events inducing intracellular signal transduction in nonsensory tissues does not yet exist. Multiple candidates for mechanoreceptors exist.

Ion channel activity in osteoblasts stimulated by stretch/strain of the membrane[39] or by PTH[40] have been associated with bone cell activation. Patch-clamp techniques show at least three classes of mechanosensitive ion channels.[41] In limb bone cultures, gadolinium chloride, which blocks some stretch/shear-sensitive cation channels, blocked load-related increases in PGI2 and NO.[42]

Membrane deformation and shear across the membrane, as well as pressure transients, are transmitted to the cytoskeleton and ultimately to the cell–matrix adhesion proteins that anchor the cell in place.[43] Both membrane spanning integrins, which couple the cell to its extracellular environment, and a large number adhesion-associated linker proteins are potential molecular mechanotransducers. The architecture of the cytoskeleton with its microfilamentous and microtubular network linking adhesion receptors to the cell nucleus may also play a role in perceiving small deformations and directly informing the nucleus.[44]

Cells possess a complex organizational structure that supports compartmentalization of signals within an equally complex plasma membrane that contains several phases of liquid-ordered and liquid-disordered lipid.[45] The organized lipid rafts may sense mechanical signals. In endothelial cells, shear stress causes signaling molecules to translocate to caveolar lipid rafts, and if caveolae are disassembled, both proximal and downstream mechanical signals, including MAPK activation, are abrogated.[46]

With the multiplicity of mechanical signals presented to the cell, it is likely that no single mechanosensor or receptor mechanism perceives the mechanical environment (Fig. 2). At the very least, multiple mechanosensors interact to integrate both mechanical and chemical information from the microenvironment.

Because the distal responses to mechanical factors are similar to those elicited by ligand-receptor pairing and result in changes in gene expression, mechanotransduction must eventually end up using similar intracellular signaling cascades. Mechanical forces have been shown to activate every type of signal transduction cascade, including cAMP,[47] IP3 and intracellular calcium,[28,48] guanine regulatory proteins,[49] and MAPK,[50] among others.[51]

TRANSLATING MECHANICAL SIGNALS TO THE CLINIC

With growing evidence that low magnitude mechanical signals can be anabolic to bone,[18] two clinical studies are summarized here indicating that these signals are osteogenic in the young skeleton. In the first study, the ability of these low-level signals to improve bone mass was examined in children with disabling conditions.[52] Children were randomized to stand on an actively vibrating (0.3*g*, 90 Hz) or placebo device for 10 min/d. Over a 6-mo trial, proximal tibial volumetric trabecular BMD (vTBMD) in children on active devices increased by 6.3%, whereas vTBMD decreased by 11.9% on placebo devices ($p = 0.0033$). In the second study, a 12-mo trial was conducted in 48 young women with low BMD and at least one skeletal fracture.[53] Subjects were randomly assigned either into a daily, low-magnitude whole body vibration group (10 min/d, 30 Hz, 0.3*g*) or control. Intention-to-treat data indicated that cancellous bone in the lumbar vertebrae and cortical bone in the femoral midshaft of the experimental group increased by 2.0% ($p = 0.06$) and 2.3% ($p = 0.04$), respectively, compared with controls. Importantly, cross-sectional area of paraspinous musculature was 4.9% greater ($p = 0.002$) in the experimental group versus controls.

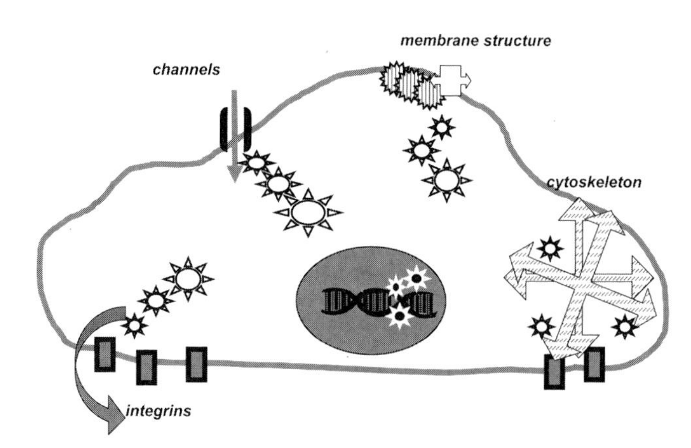

FIG. 2. Skeletal loading generates deformation of the hard tissue with strain across the cell's substrate, pressure in the intramedullary cavity and within the cortices with transient pressure waves, shear forces through canaliculi that cause drag over cells, and dynamic electric fields as interstitial fluid flows past charged bone crystals. Some composite of these physical signals interacts with the morphology of the cell through interactions such as potentiated by the membrane-matrix or membrane-nucleus structures or distortion of the membrane itself to regulate transcriptional activity of the cell.

SUMMARY

Because bone geometry and material properties vary between individuals, there are also genome-specific sensitivities to mechanical loading that may help explain the variability in exercise-based trials.[54] However, evidence in animals and humans indicates that exercise in general, and mechanical signals in particular, are both anabolic and anti-catabolic to bone tissue and benefit both bone quantity and quality. The challenge remains to identify those parameters within the complex mechanical milieu induced by exercise that are critical to driving anabolism in bone and that represent the basis of a nondrug strategy to control BMD. Whereas pharmaceutical interventions can prevent osteoporosis, exercise, in its many varied forms, is self-targeting, endogenous to bone tissue, and autoregulated, causing site-specific positive adaptation in bone mass and structure.

REFERENCES

1. Frost HM 1987 Bone "mass" and the "mechanostat": A proposal. Anat Rec **219:**1–9.
2. Lang T, LeBlanc A, Evans H, Lu Y, Genant H, Yu A 2004 Cortical and trabecular bone mineral loss from the spine and hip in long-duration spaceflight. J Bone Miner Res **19:**1006–1012.

© 2008 American Society for Bone and Mineral Research

3. Jones HH, Priest JD, Hayes WC, Tichenor CC, Nagel DA 1977 Humeral hypertrophy in response to exercise. J Bone Joint Surg Am **59:**204–208.
4. Heinonen A, Oja P, Kannus P, Sievanen H, Haapasalo H, Manttari A, Vuori I 1995 Bone mineral density in female athletes representing sports with different loading characteristics of the skeleton. Bone **17:**197–203.
5. Snow-Harter C, Whalen R, Myburgh K, Arnaud S, Marcus R 1992 Bone mineral density, muscle strength, and recreational exercise in men. J Bone Miner Res **7:**1291–1296.
6. Leichter I, Simkin A, Margulies JY, Bivas A, Steinberg R, Giladi M, Milgrom C 1989 Gain in mass density of bone following strenuous physical activity. J Orthop Res **7:**86–90.
7. McKay HA, MacLean L, Petit M, MacKelvie-O'Brien K, Janssen P, Beck T, Khan KM 2005 "Bounce at the Bell": A novel program of short bouts of exercise improves proximal femur bone mass in early pubertal children. Br J Sports Med **39:**521–526.
8. Heinonen A, Sievanen H, Kannus P, Oja P, Vuori I 1996 Effects of unilateral strength training and detraining on bone mineral mass and estimated mechanical characteristics of the upper limb bones in young women. J Bone Miner Res **11:**490–501.
9. Nicolella DP, Moravits DE, Gale AM, Bonewald LF, Lankford J 2006 Osteocyte lacunae tissue strain in cortical bone. J Biomech **39:**1735–1743.
10. Piekarski K, Munro M 1977 Transport mechanism operating between blood supply and osteocytes in long bones. Nature **269:**80–82.
11. Rubin CT, Lanyon LE 1985 Regulation of bone mass by mechanical strain magnitude. Calcif Tissue Int **37:**411–417.
12. Rubin CT, Lanyon LE 1984 Regulation of bone formation by applied dynamic loads. J Bone Joint Surg Am **66:**397–402.
13. Lanyon LE, Goodship AE, Pye CJ, MacFie JH 1982 Mechanically adaptive bone remodelling. J Biomech **15:**141–154.
14. O'Connor JA, Lanyon LE, MacFie H 1982 The influence of strain rate on adaptive bone remodelling. J Biomech **15:**767–781.
15. Lanyon LE, Rubin CT 1984 Static vs dynamic loads as an influence on bone remodelling. J Biomech **17:**897–905.
16. Srinivasan S, Weimer DA, Agans SC, Bain SD, Gross TS 2002 Low-magnitude mechanical loading becomes osteogenic when rest is inserted between each load cycle. J Bone Miner Res **17:**1613–1620.
17. Huang RP, Rubin CT, McLeod KJ 1999 Changes in postural muscle dynamics as a function of age. J Gerontol A Biol Sci Med Sci **54:**B352–B357.
18. Rubin C, Turner AS, Bain S, Mallinckrodt C, McLeod K 2001 Anabolism: Low mechanical signals strengthen long bones. Nature **412:**603–604.
19. Noble BS, Peet N, Stevens HY, Brabbs A, Mosley JR, Reilly GC, Reeve J, Skerry TM, Lanyon LE 2003 Mechanical loading: Biphasic osteocyte survival and targeting of osteoclasts for bone destruction in rat cortical bone. Am J Physiol Cell Physiol **284:**C934–C943.
20. Gross TS, Akeno N, Clemens TL, Komarova S, Srinivasan S, Weimer DA, Mayorov S 2001 Selected Contribution: Osteocytes upregulate HIF-1alpha in response to acute disuse and oxygen deprivation. J Appl Physiol **90:**2514–2519.
21. Verborgt O, Gibson GJ, Schaffler MB 2000 Loss of osteocyte integrity in association with microdamage and bone remodeling after fatigue in vivo. J Bone Miner Res **15:**60–67.
22. Garman R, Rubin C, Judex S 2007 Small oscillatory accelerations, independent of matrix deformations, increase osteoblast activity and enhance bone morphology. PLoS ONE **2:**e653.
23. Rubin J, Rubin C, Jacobs CR 2006 Molecular pathways mediating mechanical signaling in bone. Gene **367:**1–16.
24. Cowin SC, Weinbaum S 1998 Strain amplification in the bone mechanosensory system. Am J Med Sci **316:**184–188.
25. Yellowley CE, Li Z, Zhou Z, Jacobs CR, Donahue HJ 2000 Functional gap junctions between osteocytic and osteoblastic cells. J Bone Miner Res **15:**209–217.
26. Han Y, Cowin SC, Schaffler MB, Weinbaum S 2004 Mechanotransduction and strain amplification in osteocyte cell processes. Proc Natl Acad Sci USA **101:**16689–16694.
27. Rubin CT, Bain SD, McLeod KJ 1992 Suppression of the osteogenic response in the aging skeleton. Calcif Tissue Int **50:**306–313.
28. Li YJ, Batra NN, You L, Meier SC, Coe IA, Yellowley CE, Jacobs CR 2004 Oscillatory fluid flow affects human marrow stromal cell proliferation and differentiation. J Orthop Res **22:**1283–1289.
29. Rubin J, Fan X, Biskobing DM, Taylor WR, Rubin CT 1999 Osteoclastogenesis is repressed by mechanical strain in an in vitro model. J Orthop Res **17:**639–645.
30. Wiltink A, Nijweide PJ, Scheenen WJ, Ypey DL, Van Duijn B 1995 Cell membrane stretch in osteoclasts triggers a self-reinforcing Ca2+ entry pathway. Pflugers Arch **429:**663–671.
31. Davis ME, Cai H, Drummond GR, Harrison DG 2001 Shear stress regulates endothelial nitric oxide synthase expression through c-Src by divergent signaling pathways. Circ Res **89:**1073–1080.
32. Fan X, Roy E, Zhu L, Murphy TC, Ackert-Bicknell C, Hart CM, Rosen C, Nanes MS, Rubin J 2004 Nitric oxide regulates receptor activator of nuclear factor-kappaB ligand and osteoprotegerin expression in bone marrow stromal cells. Endocrinology **145:**751–759.
33. David V, Martin A, Lafage-Proust MH, Malaval L, Peyroche S, Jones DB, Vico L, Guignandon A 2007 Mechanical loading down-regulates peroxisome proliferator-activated receptor gamma in bone marrow stromal cells and favors osteoblastogenesis at the expense of adipogenesis. Endocrinology **148:**2553–2562.
34. Rubin CT, Capilla E, Luu YK, Busa B, Crawford H, Nolan DJ, Mittal V, Rosen CJ, Pessin JE, Judex S 2007 Adipogenesis is inhibited by brief, daily exposure to high-frequency, extremely low-magnitude mechanical signals. Proc Natl Acad Sci USA **104:**17879–17884.
35. Rosen CJ, Bouxsein ML 2006 Mechanisms of disease: Is osteoporosis the obesity of bone? Nat Clin Pract Rheumatol **2:**35–43.
36. Rubin CT, Lanyon LE 1984 Dynamic strain similarity in vertebrates; an alternative to allometric limb bone scaling. J Theor Biol **107:**321–327.
37. Sukharev S, Corey DP 2004 Mechanosensitive channels: Multiplicity of families and gating paradigms. Sci STKE **2004:**re4.
38. Morris CE 1990 Mechanosensitive ion channels. J Membr Biol **113:**93–107.
39. Duncan RL, Hruska KA, Misler S 1992 Parathyroid hormone activation of stretch-activated cation channels in osteosarcoma cells (UMR-106.01). FEBS Lett **307:**219–223.
40. Ferrier J, Ward A, Kanehisa J, Heersche JN 1986 Electrophysiological responses of osteoclasts to hormones. J Cell Physiol **128:**23–26.
41. Davidson RM, Tatakis DW, Auerbach AL 1990 Multiple forms of mechanosensitive ion channels in osteoblast-like cells. Pflugers Arch **416:**646–651.
42. Rawlinson SC, Pitsillides AA, Lanyon LE 1996 Involvement of different ion channels in osteoblasts' and osteocytes' early responses to mechanical strain. Bone **19:**609–614.
43. Katsumi A, Orr AW, Tzima E, Schwartz MA 2004 Integrins in mechanotransduction. J Biol Chem **279:**12001–12004.
44. Ingber DE 2005 Mechanical control of tissue growth: Function follows form. Proc Natl Acad Sci USA **102:**11571–11572.
45. Simons K, Toomre D 2000 Lipid rafts and signal transduction. Nat Rev Mol Cell Biol **1:**31–39.
46. Rizzo V, Sung A, Oh P, Schnitzer JE 1998 Rapid mechanotransduction in situ at the luminal cell surface of vascular endothelium and its caveolae. J Biol Chem **273:**26323–26329.
47. Lavandero S, Cartagena G, Guarda E, Corbalan R, Godoy I, Sapag-Hagar M, Jalil JE 1993 Changes in cyclic AMP dependent protein kinase and active stiffness in the rat volume overload model of heart hypertrophy. Cardiovasc Res **27:**1634–1638.
48. Dassouli A, Sulpice JC, Roux S, Crozatier B 1993 Stretch-induced inositol trisphosphate and tetrakisphosphate production in rat cardiomyocytes. J Mol Cell Cardiol **25:**973–982.
49. Gudi S, Huvar I, White CR, McKnight NL, Dusserre N, Boss GR, Frangos JA 2003 Rapid activation of Ras by fluid flow is mediated by Galpha(q) and Gbetagamma subunits of heterotrimeric G proteins in human endothelial cells. Arterioscler Thromb Vasc Biol **23:**994–1000.
50. Rubin J, Murphy TC, Fan X, Goldschmidt M, Taylor WR 2002 Activation of extracellular signal-regulated kinase is involved in mechanical strain inhibition of RANKL expression in bone stromal cells. J Bone Miner Res **17:**1452–1460.
51. Judex S, Zhong N, Squire ME, Ye K, Donahue LR, Hadjiargyrou M, Rubin CT 2005 Mechanical modulation of molecular signals which regulate anabolic and catabolic activity in bone tissue. J Cell Biochem **94:**982–994.

© 2008 American Society for Bone and Mineral Research

52. Ward K, Alsop C, Caulton J, Rubin C, Adams J, Mughal Z 2004 Low magnitude mechanical loading is osteogenic in children with disabling conditions. J Bone Miner Res **19:**360–369.
53. Gilsanz V, Wren TA, Sanchez M, Dorey F, Judex S, Rubin C 2006 Low-Level, High-Frequency Mechanical Signals Enhance Musculoskeletal Development of Young Women With Low BMD. J Bone Miner Res **21:**1464–1474.
54. Judex S, Donahue LR, Rubin CT 2002 Genetic predisposition to osteoporosis is paralleled by an enhanced sensitivity to signals anabolic to the skeleton. FASEB J **16:**1280–1282.
55. Rubin C, Turner AS, Muller R, Mittra E, McLeod K, Lin W, Qin YX 2002 Quantity and quality of trabecular bone in the femur are enhanced by a strongly anabolic, noninvasive mechanical intervention. J Bone Miner Res **17:**349–357.

Chapter 47. Calcium and Vitamin D

Bess Dawson-Hughes

Bone Metabolism Laboratory, Jean Mayer USDA Human Nutrition Research Center on Aging at Tufts University, Boston, Massachusetts

INTRODUCTION

Calcium is required for the bone formation phase of bone remodeling. Typically, ~5 nmol (200 mg) of calcium is removed from the adult skeleton and replaced each day. To supply this amount, one would need to consume ~600 mg of calcium, because calcium is not very efficiently absorbed. Calcium also affects bone mass through its impact on the remodeling rate. An inadequate intake of calcium results in reduced calcium absorption, a lower circulating ionized calcium concentration, and an increased secretion of PTH, a potent bone-resorbing agent. A high remodeling rate leads to bone loss; it is also an independent risk factor for fracture. Dietary calcium at sufficiently high levels, usually 1000 mg/d or more, lowers the bone remodeling rate by ~10–20% in older men and women, and the degree of suppression seems to be dose related.[1] The reduction in remodeling rate accounts for the increase in BMD that occurs in the first 12–18 mo of treatment with calcium.

With aging, there is a decline in calcium absorption efficiency in men and women. This may be related to loss of intestinal vitamin D receptors or resistance of these receptors to the action of $1,25(OH)_2D$. Diet composition, season, and race also influence calcium absorption efficiency.

Vitamin D is acquired from the diet and from skin synthesis on exposure to UVB rays. The best clinical indicator of vitamin D status is the serum 25-hydroxyvitamin D [25(OH)D] level. Serum 25(OH)D levels are lower in individuals using sunscreens and in those with more pigmented skin. Season is an important determinant of vitamin D levels. In much of the temperate zone, skin synthesis of vitamin D does not occur during the winter. Consequently, 25(OH)D levels fall in the winter and early spring. Serum PTH levels vary inversely with 25(OH)D levels. These cyclic changes are not benign. Bone loss is greater in the winter/spring when 25(OH)D levels are lowest (and PTH levels are highest) than in the summer/fall when 25(OH)D levels are highest (and PTH levels are lowest).

Serum 25(OH)D levels decline with aging for several reasons. There is less efficient skin synthesis of vitamin D with aging as a result of an age-related decline in the amount of 7-dehydrocholesterol, the precursor to vitamin D, in the epidermal layer of skin.[2] Also, older individuals as a group spend less time outdoors. There does not seem to be an impairment in the intestinal absorption of vitamin D with aging.[3]

IMPACT ON BMD

Calcium and vitamin D support bone growth in children and adolescents and lower rates of bone loss in adults and the elderly. A meta-analysis of 15 trials found that calcium alone in adults caused positive mean percentage BMD changes from baseline of 1.7% at lumbar spine, 1.6% at the hip, and 1.9% at the distal radius.[4] In one trial, the effects of calcium from food (milk powder) and supplement sources on changes in BMD in older postmenopausal women were compared and found to be similar.[5]

Higher serum 25(OH)D levels have been associated with higher BMD of the hip in young and older adult men and women in the National Health and Nutrition Evaluation Survey III (NHANES III).[6] This association was present at 25(OH)D levels up through the upper end of the reference range (92 nM or 36.8 ng/ml). Supplementation with vitamin D also reduces rates of bone loss in older adults.[7] To sustain the reduced turnover rate and higher bone mass induced by increased calcium and vitamin D intakes, the higher intakes need to be maintained.

IMPACT ON MUSCLE STRENGTH AND FALLING

In NHANES III women ≥60 yr of age, higher 25(OH)D levels were associated with improved lower extremity function (faster walking and sit-to-stand speeds).[8] Similarly, in 1234 elderly men and women living in the Netherlands, concentrations of 25(OH)D <50 nM (or 20 ng/ml) were associated with reduced physical performance.[9] The mechanism(s) by which vitamin D influences muscle performance/strength are not well established but are likely to involve the nuclear vitamin D receptors known to be present in muscle. Supplementation with 1000 IU of vitamin D_2 daily over a 2-yr period, compared with placebo, significantly increased the diameter of fast twitch type II muscle fibers in older women who had had a recent stroke.[10]

A meta-analysis of five randomized placebo-controlled vitamin D intervention trials showed that supplementation lowered risk of falling by 22%.[11] The trials in which the effect was most apparent used doses of 20 μg (800 IU)/d. This observation was supported by a recent placebo-controlled trial in 625 assisted-living residents, mean age 83.4 yr, in which treatment for 2 yr with 1200 IU/d of vitamin D_2 significantly reduced risk of all falls (0.73; 95% CI 0.57–0.95).[12] Not all studies have detected an effect of vitamin D on falls,[13] perhaps in this case because of poor compliance. The impact of vitamin D on falls is likely to be mediated by its effect on muscle strength; vitamin D may also influence balance, but this has not been established.

The author states that she has no conflicts of interest.

Key words: calcium, viatmin D, falls, fractures

© 2008 American Society for Bone and Mineral Research

IMPACT ON FRACTURE RATES

Several small studies have examined the impact of calcium on fracture rates. The Shea meta-analysis of these studies[4] found that calcium alone (versus placebo) tended to lower risk of vertebral fractures (RR, 0.77; 95% CI, 0.54–1.09) but not nonvertebral fractures (RR, 0.86; 95% CI, 0.43–1.72). The studies in this analysis range from 18 mo to 4 yr in duration. The recent RECORD trial in older people with a prior fracture showed that supplementation with 1000 mg/d of calcium did not lower fracture risk over 45 mo.[11] Poor compliance may have contributed. At 24 mo into the 60-mo study, only 54.5% of subjects were still taking pills.

The effect of supplemental vitamin D on fracture incidence has been examined in several randomized controlled trials. A recent meta-analysis of randomized controlled trials showed that supplementation with vitamin D lowers risk of hip fracture by 26% and any nonvertebral fracture by 23%.[14] The trials using doses of 17.5–20 μg (700–800 IU)/d were positive and those using 10 μg (400 IU)/d were neutral. Serum 25(OH)D levels were measured in most of these trials, and fracture risk reduction was inversely related to the serum 25(OH)D level achieved with supplementation.[14]

Several large trials published more recently have shown no significant effect of vitamin D on fracture risk.[13,15,16] In a secondary prevention study, the RECORD trial, 20 μg (800 IU)/d of vitamin D with and without 1000 mg of supplemental calcium over a 5-yr period did not lower fracture risk over a 45-mo period in people ≥70 yr of age.[13] A small, nonrandom subset of participants had 25(OH)D levels measured, and the mean value during supplementation was 62 nM (24.8 ng/ml), which is in the range of other negative studies. This relatively low value is a reflection of poor compliance. As indicated earlier, only 54.5% of the overall sample was taking the supplements at 24 mo.

An open trial of 20 μg (800 IU) of vitamin D plus 1000 mg of calcium versus no treatment for an average of 25 mo in 3314 women ≥70 yr of age had a null result. Serum 25(OH)D levels were not measured, but compliance may have been a factor in this trial also. Only 60% of subjects were still taking any pills at 12 mo. Finally, the Women's Health Initiative identified no significant effect of vitamin D (10 μg or 400 IU) together with 1000 mg of calcium daily on fracture rates in the group as a whole, although there was hip fracture risk reduction in the subset who were older and in those who were not taking their own supplements throughout the trial.[16] A subset of the women in this trial had a mean 25(OH)D level of 59 nM (or 23.6 ng/ml) after 1 yr on the supplements.

From the available evidence, it seems that a serum 25(OH)D level of 75 nM (30 ng/ml) or above is needed to lower risk of fracture. The reduced risk of fracture seems to result from effects of vitamin D on both muscle and bone metabolism. In many of the positive trials, calcium was given along with vitamin D; based on this and other evidence,[17,18] it is prudent to replete both calcium and vitamin D.

ROLE IN PHARMACOTHERAPY

In recent randomized, controlled trials testing the antifracture efficacy of the antiresorptive therapies alendronate, risedronate, raloxifene, and calcitonin and the anabolic drug, PTH(1-34), calcium and vitamin D have been given to both the control and intervention groups. This allows one to define the impact of these drugs in calcium- and vitamin D–replete patients and to conclude that any efficacy of the drugs is beyond that associated with calcium and vitamin D alone. Based on the evidence that follows, however, one can not conclude that these drugs would have the same efficacy in calcium- and vitamin D–deficient patients. In a comparative analysis of the impact of estrogen on BMD in early post menopausal women who did and did not take calcium supplements, Nieves et al.[19] found that the BMD gains at the spine, hip, and forearm were several-fold greater in the women who increased their calcium intakes than in those who took estrogen without added calcium. From this, it seems that calcium enables estrogen to be more effective in building BMD. In the Mediterranean Osteoporosis Study (MEDOS) in southern Europe,[20] use of nasal calcitonin was associated with a nonsignificant decrease in vertebral fracture risk (RR = 0.78), as was use of calcium alone (RR = 0.82). Use of calcitonin and calcium together, however, was associated with a significant reduction in vertebral fracture risk (RR = 0.63), suggesting that the effects of calcium and this antiresorptive therapy are additive. Little information is available on a potential interaction of other osteoporosis treatments with calcium intake. One can certainly infer that an adequate calcium intake is essential for an optimal response to treatment with the bone-building drug, PTH(1-34). Little direct evidence is available of an interaction of vitamin D with pharmacotherapy, but because vitamin D works in concert with calcium, adequate vitamin D status is very likely to be an important component of the therapy.

INTAKE REQUIREMENTS

Calcium intake recommendations vary enormously worldwide. Recommendations by the U.S. National Academy of Sciences (NAS) are among the highest. The NAS recommended intakes of calcium are as follows: ages 1–3 yr, 500 mg/d; 4–8 yr, 800 mg/d; 9–18 yr, 1300 mg/d; 19–50 yr, 1000 mg/d; 51+ yr, 1200 mg/d.[21] Lower calcium intakes would likely be adequate for populations with lower intakes of salt and protein.

Among women in the United States, fewer than 1 in 10 up to age 70 yr and fewer than 1 in 100 over that age meet the calcium requirement through their diets. Among males, no more than 25% in any age group has an adequate calcium intake from the diet. Without major dietary changes, most of the American population will need to rely on fortified foods and supplements to meet calcium requirements. Calcium from calcium carbonate, the most commonly used supplement, is better absorbed when taken with a meal.[22,23] Absorption from all supplements is more efficient in doses of 500 mg or less.[24] Thus, individuals requiring >500 mg/d from supplements should take it in divided doses. The Safe Upper Limit for calcium set by the NAS is 2500 mg/d.[21]

The vitamin D intake recommendations of the NAS, made in 1997, are as follows: for adult men and women up to age 50 yr, 5 μg/d (200 IU); 51–70 yr, 10 μg/d (400 IU); 71+ yr, 15 μg/d (600 IU).[21] These recommendations are based on the amount of vitamin D needed to maximally suppress PTH secretion. However, for reasons that are not entirely clear, variability in this endpoint is large across study populations. Several studies have placed the 25(OH)D level needed for maximal PTH suppression in the range of 75–110 nM (30–44 ng/ml), whereas another places it as low as 25 nM (10 ng/ml).[25] The currently recommended vitamin D intake of 600 IU/d (15 μg) for men and women ≥71 yr of age is not adequate to bring most of the elderly population to 25(OH)D levels of 75–80 nM (30–32 ng/ml), the level apparently needed to lower fracture risk.

The increase in 25(OH)D with supplementation is inversely related to the starting level. At low starting levels, 1 μg (40 IU) of vitamin D will increase serum 25(OH)D by 1.2 nM (0.48 ng/ml); at a higher starting level of 70 nM (28 ng/ml), the increase from this dose would be only about 0.7 nM (0.28 ng/ml).[26,27] The average older man and woman will need an

© 2008 American Society for Bone and Mineral Research

intake of at least 20–25 μg/d (800–1000 IU) to reach a serum 25(OH)D level of 75 nM (30 ng/ml). The NAS has placed the safe upper limit for vitamin D at 50 μg/d (2000 IU).[21]

Vitamin D is available in two forms: the plant-derived ergocalciferol (D_2) and cholecalciferol (D_3), which is of animal origin. For years, these forms were considered to be equipotent in humans. Some studies have suggested that vitamin D_3 increases serum 25(OH)D level more efficiently than vitamin D_2.[28,29] Moreover, vitamin D_2 is not accurately measured in all 25(OH)D assays.[30] For these reasons, vitamin D_3, when available, is the preferred form for clinical use.

In conclusion, adequate intakes of calcium and vitamin D are essential preventative measures and essential components of any therapeutic regimen for osteoporosis. Many men and women will need supplements to meet the intake requirements. Current evidence suggests that a 25(OH)D level of 75 nM (30 ng/ml) or higher along with an adequate calcium intake is needed to lower fracture risk. The average older person will need 20–25 μg/d (800–1000 IU) of vitamin D_3 to reach this level.

ACKNOWLEDGMENTS

Any opinions, findings, conclusions, or recommendations expressed in this chapter are those of the author and do not necessarily reflect the view of the U.S. Department of Agriculture.

REFERENCES

1. Elders PJ, Netelenbos JC, Lips P, van Ginkel FC, Khoe E, Leeuwenkamp OR, Hackeng WH, van der Stelt PF 1991 Calcium supplementation reduces vertebral bone loss in perimenopausal women: A controlled trial in 248 women between 46 and 55 years of age. J Clin Endocrinol Metab **73:**533–540.
2. MacLaughlin J, Holick MF 1985 Aging decreases the capacity of human skin to produce vitamin D3. J Clin Invest **76:**1536–1538.
3. Harris SS, Dawson-Hughes B 2002 Plasma vitamin D and 25(OH)D responses of young and old men to supplementation with vitamin D3. J Am Coll Nutr **21:**357–362.
4. Shea B, Wells G, Cranney A, Zytaruk N, Robinson V, Griffith L, Ortiz Z, Peterson J, Adachi J, Tugwell P, Guyatt G 2002 VII. Meta-analysis of calcium supplementation for the prevention of postmenopausal osteoporosis. Endocr Rev **23:**552–559.
5. Prince R, Devine A, Dick I, Criddle A, Kerr D, Kent R, Price R, Randell A 1995 The effects of calcium supplementation (milk powder or tablets) and exercise on bone density in postmenopausal women. J Bone Miner Res **10:**1068–1075.
6. Bischoff-Ferrari HA, Dietrich T, Orav EJ, Dawson-Hughes B 2004 Positive association between 25-hydroxy vitamin D levels and bone mineral density: A population-based study of younger and older adults. Am J Med **116:**634–639.
7. Ooms ME, Roos JC, Bezemer PD, van der Vijgh WJ, Bouter LM, Lips P 1995 Prevention of bone loss by vitamin D supplementation in elderly women: A randomized double-blind trial. J Clin Endocrinol Metab **80:**1052–1058.
8. Bischoff-Ferrari HA, Dietrich T, Orav EJ, Zhang Y, Karlson EW, Dawson-Hughes B 2004 Higher 25-hydroxyvitamin D levels are associated with better lower extremity function in both active and inactive adults 60+ years of age. Am J Clin Nutr **80:**752–758.
9. Wicherts IS, van Schoor NM, Boeke AJ, Visser M, Deeg DJ, Smit J, Knol DL, Lips P 2007 Vitamin D status predicts physical performance and its decline in older persons. J Clin Endocrinol Metab **92:**2058–2065.
10. Sato Y, Iwamoto J, Kanoko T, Satoh K 2005 Low-dose vitamin D prevents muscular atrophy and reduces falls and hip fractures in women after stroke: A randomized controlled trial. Cerebrovasc Dis **20:**187–192.
11. Bischoff-Ferrari HA, Dawson-Hughes B, Willett WC, Staehelin HB, Bazemore MG, Zee RY, Wong JB 2004 Effect of vitamin D on falls: A meta-analysis. JAMA **291:**1999–2006.
12. Flicker L, MacInnis RJ, Stein MS, Scherer SC, Mead KE, Nowson CA, Thomas J, Lowndes C, Hopper JL, Wark JD 2005 Should older people in residential care receive vitamin D to prevent falls? Results of a randomized trial. J Am Geriatr Soc **53:**1881–1888.
13. Grant AM, Avenell A, Campbell MK, McDonald AM, MacLennan GS, McPherson GC, Anderson FH, Cooper C, Francis RM, Donaldson C, Gillespie WJ, Robinson CM, Torgerson DJ, Wallace WA, Group RT 2005 Oral vitamin D3 and calcium for secondary prevention of low-trauma fractures in elderly people (Randomised Evaluation of Calcium Or vitamin D, RECORD): A randomised placebo-controlled trial. Lancet **365:**1621–1628.
14. Bischoff-Ferrari HA, Willett WC, Wong JB, Giovannucci E, Dietrich T, Dawson-Hughes B 2005 Fracture prevention by vitamin D supplementation: A meta-analysis of randomized controlled trials. JAMA **293:**2257–2264.
15. Porthouse J, Cockayne S, King C, Saxon L, Steele E, Asprey T, Baverstock M, Birks Y, Dumville J, Francis R, Iglesias C, Puffer S, Sutcliffe A, Watt I, Torgerson DJ 2005 Randomized controlled trial of calcium and supplementation with cholecalciferol (vitamin D3) for prevention of fractures in primary care. BMJ **330:**1003–1008.
16. Jackson RD, LaCroix AZ, Gass M, Wallace RB, Robbins J, Lewis CE, Bassford T, Beresford SA, Black HR, Blanchette P, Bonds DE, Brunner RL, Brzyski RG, Caan B, Cauley JA, Chlebowski RT, Cummings SR, Granek I, Hays J, Heiss G, Hendrix SL, Howard BV, Hsia J, Hubbell FA, Johnson KC, Judd H, Kotchen JM, Kuller LH, Langer RD, Lasser NL, Limacher MC, Ludlam S, Manson JE, Margolis KL, McGowan J, Ockene JK, O'Sullivan MJ, Phillips L, Prentice RL, Sarto GE, Stefanick ML, Van Horn L, Wactawski-Wende J, Whitlock E, Anderson GL, Assaf AR, Barad D Women's Health Initiative I 2006 Calcium plus vitamin D supplementation and the risk of fractures. N Engl J Med **354:**669–683.
17. Boonen S, Lips P, Bouillon R, Bischoff-Ferrari HA, Vanderschueren D, Haentjens P 2007 Need for additional calcium to reduce the risk of hip fracture with vitamin d supplementation: Evidence from a comparative metaanalysis of randomized controlled trials. J Clin Endocrinol Metab **92:**1415–1423.
18. Tang BM, Eslick GD, Nowson C, Smith C, Bensoussan A 2007 Use of calcium or calcium in combination with vitamin D supplementation to prevent fractures and bone loss in people aged 50 years and older: A meta-analysis. Lancet **370:**657–666.
19. Nieves JW, Komar L, Cosman F, Lindsay R 1998 Calcium potentiates the effect of estrogen and calcitonin on bone mass: Review and analysis. Am J Clin Nutr **67:**18–24.
20. Johnell O, Gullberg B, Kanis JA, Allander E, Elffors L, Dequeker J, Dilsen G, Gennari C, Lopes VA, Lyritis G 1995 Risk factors for hip fracture in European women: The MEDOS Study. Mediterranean Osteoporosis Study. J Bone Miner Res **10:**1802–1815.
21. Standing Committee on the Scientific Evaluation of Dietary Reference Intakes FaNBIoM 1997 Dietary Reference Intakes for Calcium, Phosphorus, Magnesium, Vitamin D and Fluoride. National Academy Press, Washington, DC, USA.
22. Heaney RP, Smith KT, Recker RR, Hinders SM 1989 Meal effects on calcium absorption. Am J Clin Nutr **49:**372–376.
23. Recker RR 1985 Calcium absorption and achlorhydria. N Engl J Med **313:**70–73.
24. Harvey JA, Zobitz MM, Pak CY 1988 Dose dependency of calcium absorption: A comparison of calcium carbonate and calcium citrate. J Bone Miner Res **3:**253–258.
25. Dawson-Hughes B, Heaney RP, Holick MF, Lips P, Meunier PJ, Vieth R 2005 Estimates of optimal vitamin D status. Osteoporos Int **16:**713–716.
26. Vieth R, Ladak Y, Walfish PG 2003 Age-related changes in the 25-hydroxyvitamin D versus parathyroid hormone relationship suggest a different reason why older adults require more vitamin D. J Clin Endocrinol Metab **88:**185–191.
27. Heaney RP, Davies KM, Chen TC, Holick MF, Barger-Lux MJ 2003 Human serum 25-hydroxycholecalciferol response to extended oral dosing with cholecalciferol. Am J Clin Nutr **77:**204–210.
28. Trang HM, Cole DE, Rubin LA, Pierratos A, Siu S, Vieth R 1998 Evidence that vitamin D3 increases serum 25-hydroxyvitamin D more efficiently than does vitamin D2. Am J Clin Nutr **68:**854–858.
29. Armas LA, Hollis BW, Heaney RP 2004 Vitamin D2 is much less effective than vitamin D3 in humans. J Clin Endocrinol Metab **89:**5387–5391.
30. Binkley N, Krueger D, Cowgill CS, Plum L, Lake E, Hansen KE, DeLuca HF, Drezner MK 2004 Assay variation confounds the diagnosis of hypovitaminosis D: A call for standardization. J Clin Endocrinol Metab **89:**3152–3157.

© 2008 American Society for Bone and Mineral Research

Chapter 48. Estrogens and SERMs

Robert Lindsay

Helen Hayes Hospital, West Haverstraw, New York

INTRODUCTION

More than 60 yr have passed since Fuller Albright established the relationship between estrogen deficiency and osteoporosis and began experimenting with estrogens as treatment. In 1947 Albright showed that estrogen intervention reversed the negative calcium balance seen in osteoporosis and proposed that estrogen increased the rate of bone formation. Henneman and Wallach, in a review of Albright's patients, showed that those on estrogen did not lose height, a surrogate for vertebral fracture, whereas those untreated individuals with osteoporosis continued to have sporadic height reduction. The first placebo-controlled clinical trials of estrogen intervention were initiated in the late 1960s showing clearly that estrogen intervention prevented bone loss and suggested prevention of vertebral fractures. Since then, a large number of clinical trials, mostly with BMD as the primary outcome, have been completed including a few with fracture outcomes. For many years estrogens were assumed to have several beneficial effects on the health of postmenopausal women including cardiovascular and CNS benefits. However, the Women's Health Initiative (WHI), the largest estrogen therapy/hormone therapy (ET/HT) study ever performed cast doubt on many of these benefits, but confirmed the role of ET/HT in prevention of osteoporosis and clinical fractures. Those studies involving 26,000 women between 50 and 79 yr provided robust data the ET/HT can reduce the risk of all fractures including fractures of the hip. However, other more controversial issues have been raised by WHI, including the failure to reduce cardiovascular risk, and increased breast cancer risk (HT only). Consequently, clinicians are gradually coming to terms with how best to include these agents in the management of osteoporosis in their postmenopausal patients.

PATHOGENESIS

Sex steroids play important roles in skeletal homeostasis throughout life, influencing growth and development. Estrogens are responsible for epiphyseal closure in both sexes and for the maintenance of skeletal mass in healthy premenopausal women and probably among aging males also. Estrogen deficiency (anorexia, athletic induced amenorrhea, hyperprolactinemia, the use of depot medroxyprogesterone acetate [DMPA] as a contraceptive, etc.) in women at any age is associated with loss of bone mass. However, by virtue of its ubiquitous nature, loss of ovarian estrogen supply at the time of menopause is a major pathogenetic factor in the development of low osteoporosis-related fractures among aging women. Biochemical data indicate that bone remodeling increases across menopause, with a more rapid rise in resorption, the initial process in remodeling, and a slower rise in bone formation. Histomorphometric data are somewhat sparse but generally agree that bone remodeling increases after menopause, although some data do not support that view. Recent data confirmed the rapid deterioration in cancellous architecture that occurs in time periods as short as 1 yr in the early years after menopause. The mechanisms include increased activation of new remodeling cycles and increased recruitment and more prolonged survival of osteoclast teams. The combination of these effects may result in larger resorption cavities increased risk of trabecular penetration, more points of stress, and a higher likelihood of microfractures. Finally, there may also be a failure to synthesize enough new tissue to repair the damage created by resorption, although the data are sparse on this point. The greater architectural damage, with loss of trabeculae with increased spacing between trabeculae, that occurs in the skeletons of postmenopausal women supports the conclusion that the dominant effect of estrogen deficiency are both increased bone resorption and remodeling. The duration of rapid bone loss is not well established and may be quite heterogeneous, and in susceptible individuals, may continue into old age.

Some recent data support the concept that much of the influence of estrogen deficiency and conversely estrogen action on bone is mediated by modulation of one of several local cytokines, of which the RANK/RANKL system may be crucial. Also important may be modulation of interleukin secretion (IL-1,IL-6) TNF-α, lymphotoxin, macrophage-colony stimulating factor (M-CSF), and granulocyte macrophage-colony stimulating factor (GM-CSF). Estrogen may also stimulate secretion of TGF-β, which inhibits bone resorption and stimulates bone formation. Estrogen also stimulates the secretion of BMP-6 in human osteoblast cell lines. Although there are many pathways affected by loss of estrogen, the net effect seems to be an increase in activation of new remodeling sites and increased avidity of the osteoclast population.

ESTROGEN INTERVENTION

Albright originally showed that estrogens reversed the negative calcium balance that he had found in postmenopausal women with osteoporosis. Since the early 1970s, overwhelming data from controlled clinical trials have shown that estrogen intervention prevents bone loss among estrogen-deficient women, reducing bone remodeling to premenopausal levels. The vast majority of these studies were performed using doses of estrogen considered "replacement" doses (i.e., 0.625 mg conjugated equine estrogen [CEE]) or the equivalent, thought to be the "lowest effective dose." Three relatively small studies in this early period were able to show a fracture benefit from estrogen intervention. One study involved transdermal estrogen intervention in patients with osteoporosis and showed ~50% reduction in the risk of vertebral fracture over 1 and 2 yr of observation. The other two studies involved oral estrogen and were double-blind placebo controlled over 9–10 yr of observation. One study showed a reduction in radiological vertebral fracture rate (the primary outcome of the modern clinical trials in this field) of ~70%. The third study showed a reduction in clinical fractures. In addition, a significant literature from epidemiological studies over the past 20 yr also suggested that estrogens reduce the risk of hip and other nonvertebral fractures. The concern has been raised that this is a healthy user effect—that older women who used hormones had more health-seeking behaviors than those who did not.

Two recent European clinical trials have shown that intervention with estrogen and progestin in the early years after menopause reduce the risk of Colles' fracture and other clinical nonvertebral fractures. In the United States, two separate NIH-supported efforts evaluated the effects of estrogen with or without a progestin on bone. The first of these studies evaluated women close to menopause, the so-called PEPI study, and

---

Dr. Lindsay has served as a consultant for Proctor & Gamble, NPS, Sanofi-Aventis, Roche-GlaxoSmithKline, Novartis, and Wyeth.

 © 2008 American Society for Bone and Mineral Research

had BMD as the primary outcome. That study of some 800 postmenopausal women replicated other published clinical trials showing that 0.625 mg of CEE per day over 3 yr prevented bone loss and that the addition of medroxyprogesterone acetate (MPA) or micronized progesterone had no measurable effect on this surrogate outcome. Furthermore, the study confirmed that >90% of women who took estrogen in the then standard dose did not lose bone and that when treatment was stopped bone loss was reinstated. The more recently completed Women's Health Initiative (WHI) had a more clinically relevant outcome—namely the effect of hormones (CEE, or CEE + MPA) on fractures in a fairly healthy, although somewhat overweight, population of postmenopausal women. This study has resulted in a number of publications that have raised interesting questions, but the fracture data are solid, and in agreement with all we know about estrogen deficiency, bone loss, and fracture risk. WHI consisted of two cohorts of women (50 and 79 yr). The larger cohort of ~16,000 asymptomatic, relatively healthy, naturally postmenopausal women were enrolled into a controlled trial of CEE (0.625 mg/d) plus MPA (5 mg/d) for an intended 8 yr of observation. This study was halted after a median observation period of 5.2 yr because the Data Safety Monitoring Board perceived excess harm with hormone replacement therapy (HRT) use, based primarily on a global index. The second study involved ~10,000 hysterectomized women enrolled in a study of CEE (0.625 mg/d). This study was discontinued after a median observation period of 6.2 yr because of an increased risk of stroke. Fractures were a secondary outcome of both studies, but unlike other osteoporosis studies, patients were not recruited based on BMD or prior fractures. BMD was measured on a subpopulation of ~10% in the combination study, and in that sample, ~9% had osteoporosis by BMD (T-score below −2.5). In both studies, there were a large number of clinical fractures, documented by radiographs and confirmed centrally. In each study, overall fracture risk was reduced by hormonal intervention. It is of no small interest that in this relatively young, low-risk population, the risk of hip fracture was also reduced by ~30% ($p < 0.05$ using nominal confidence limits). Thus, as a result of these two studies, there is now robust evidence that estrogen in a standard dose reduces the risk of fractures among postmenopausal women. The WHI fracture data confirmed the previous smaller controlled clinical trials, the epidemiological data, and the large number of positive BMD studies on estrogens. WHI is also in agreement with the known pathophysiology of osteoporosis in postmenopausal women. A reasonable conclusion can now be drawn that estrogen intervention, with or without a progestin, reduces the risk of fractures among postmenopausal women.

MECHANISM OF FRACTURE RISK REDUCTION

Estrogens are among a group of agents that reduce remodeling rate in the skeleton. These include bisphosphonates, calcitonin, selective estrogen receptor modulators (SERMs; raloxifene), and calcium and vitamin D. Calcium and vitamin D seem to have little effect on bone loss in the early years after menopause when estrogen deficiency is the principal driver. However, at both younger and older ages they are important, in the former to ensure adequate skeletal growth and in the latter to treat the secondary hyperparathyroidism that occurs with increased age as absorption efficiency declines. In the past, it has been assumed that these agents reduce fracture risk by increasing BMD measured by DXA. Recently, that conclusion has been challenged. There is in fact little relationship between the change in BMD and fracture risk reduction, and it seems that much of the effect is related to greater accumulation of mineral into already existed bone tissue. Post hoc analyses have suggested that <20% of the fracture effect can be accounted for by the BMD change. Additionally, the fracture effects begin to appear fairly soon after drug treatment has begun (within a few months) and are temporally related to the decline in bone remodeling that is seen. This has led to the suggestion that this may not be coincidence, but in fact cause and effect, and that it is the decline in remodeling that leads to fracture risk decline. In some large clinical trials, a relationship between reduction in remodeling and fracture risk reduction has been shown.

EFFECTS OF LOWER DOSES OF HORMONE THERAPY

During the past year, low-dose forms of hormone therapy, lower than those previously considered to be standard, have been introduced in the United States. Whereas fracture studies have not been performed at these lower doses, there is some evidence to support the idea that these doses will have fracture benefits. Recent data have suggested that even the relatively small amount of estrogen produced by postmenopausal women have skeletal effects. In a population-based study of women >65 yr of age, Cummings et al. showed that women with estradiol levels between 10 and 15 pg/ml had higher bone mass that those with estradiol levels <5 pg/ml. In a logical extension of these data, it was shown that women with the lowest estradiol levels had the highest risk of vertebral and hip fractures and that this risk further increased in women with high levels of sex hormone–binding globulin, which would be expected to lower the biologically active estradiol further.

Evaluation of low-dose estrogen administration (0.3 mg CEE and 12.5 μg patch) has also shown biological effects on bone. In a multidose study, we showed reductions in bone remodeling at all doses of conjugated equine estrogens administered with and without the addition of a progestin. Biochemical markers of remodeling were only slightly less suppressed at 0.3 mg/d CEE than at 0.625 mg/d, with little noted effect of the progestin (MPA). In this study, bone mass increased at all doses, and >90% of individuals responded to treatment by either maintaining or gaining bone mass at spine or hip. Other studies of relatively low doses of hormone therapy confirmed these findings. A recent study of what has been called an ultra-low dose (transdermal delivery of 0.0125 mg estradiol) again confirms reduction in remodeling and preservation of bone mass. These doses of estrogen all provide increases in circulating estrogens that would at least raise circulating estradiol levels to those suggested to be biologically active in untreated older women. CEE will of course provide other estrogens in addition. The reductions in turnover are similar to those seen with other antiresorptive agents, suggesting that fracture studies might yield positive results, if indeed as argued above, it is the reduction in bone remodeling rate that is the principal mechanism of the antifracture efficacy.

EFFECTS OF DISCONTINUING HORMONES

It is clear that discontinuing hormone therapy results in bone loss, which begins immediately on stopping therapy. However, what is not clear is whether there is accelerated rate of loss comparable to that seen after ovariectomy. Some data suggest that there is an immediate increase in risk of fracture after stopping, whereas other studies suggest that more prolonged reduction in risk occurs even when treatment is discontinued. In general, the impression from all the data available is that, to obtain the fracture benefit, continued therapy is necessary.

© 2008 American Society for Bone and Mineral Research

CLINICAL USE OF HORMONE THERAPY FOR OSTEOPOROSIS

For many years, therapy with estrogen, with or without a progestin, was the only pharmacologic approach to osteoporosis. Today several agents are approved for the prevention or treatment of osteoporosis, including bisphosphonates, calcitonin, and a SERM. Since the publication of WHI, several organizations have provided guidance to physicians about the use of hormone therapy in the postmenopausal population. In general, these statements recommend that hormones be used for treatment of menopausal symptoms and for the shortest period of time compatible with obtaining prolonged symptom relief. Hormones, it is argued, should only be used for osteoporosis after all other treatments have been considered and when all the risks and benefits are carefully explained to the patient (surely we do that with all medicines?). However, hormones now have more substantial and robust fracture data in a non-osteoporosis population than any other agent. In the early postmenopausal population, and especially after hysterectomy, when estrogen alone can be used, the risk benefit equation for the use of hormones may be quite different than that seen in the WHI population overall. Detailed analyses of WHI may inform clinicians about such differences and there may indeed be populations for whom hormones can still be recommended as first-line treatment. The risk benefit equation might be considerably different for estrogens other than CEE and MPA and indeed for the lower doses that seem to have beneficial effects on bone remodeling. Clearly those issues require further study. Like any good study, WHI raises more questions than it provides answers. Unfortunately, for clinicians and their patients, these questions will be difficult to address in appropriate clinical studies.

SELECTIVE ESTROGEN RECEPTOR MODULATORS

The concept that synthesized molecules could interact with the estrogen receptors and produce estrogen-like effects at least in some tissues is attractive. To date only one of these molecules, raloxifene, has come to market, and several have failed clinical trials. At present two SERMs are completing phase III studies (bazodoxifene and lasofoxifene) and may obtain FDA approval soon.

The spectrum of effects that these agents have varies from compound to compound. In general, all act as estrogen antagonists on the brain and breast and agonists on bone. The effects on the uterus can be antagonistic in the presence of estrogen (bazodoxifene) or neutral in the absence of estrogen (raloxifene).

The effects on bone turnover and BMD seem in general to be somewhat less than those seen with full doses of estrogen or the bisphosphonates. Raloxifene reduces the occurrence of vertebral fractures but has not shown efficacy against nonvertebral fractures. Consequently, raloxifene was most often prescribed for women with low bone mass in the spine but with a more normal BMD in the hip. Because raloxifene can cause an increase in menopausal symptoms, it is generally restricted to asymptomatic women. The recent approval for prevention of breast cancer seems likely to increase its use, particularly among women in their 50s and early 60s. It is assumed that raloxifene acts on the skeleton by estrogen receptor (ER)α and effects the same pathways as estrogens. However, the somewhat lesser effects on turnover and BMD suggest that there may be a difference. Whether other SERMs are consistent with the raloxifene effect or can produce effects on bone more like estrogens is not yet clear. Raloxifene is prescribed as a 60 mg/d tablet and in addition to causing menopausal symptoms in some women is also associated with an increase in the risk of thrombo-embolic disease like estrogens. Two SERMs are currently approaching possible approval by the FDA, bazodoxifene and lasofoxifene, and their position in our therapeutic armamentarium will need to be assessed when full details of their phase III results are available.

One interesting development is the use of the SERM in conjunction with estrogen. Phase II data in which bazodoxifene is used along with low-dose CEE suggests a positive effect on bone with uterine protection, but the effects on cardiovascular disease, the brain, and breast cancer are unknown.

CONCLUSIONS

The recent data from WHI clearly support the efficacy of hormone therapy for prevention of fractures among postmenopausal and hysterectomized women. These data confirm the large body of information in the literature supporting the concept that estrogen deficiency is a major pathogenetic factor for osteoporosis among aging women and that intervention with estrogen can reduce the impact of all fractures. For clinicians, the challenge will be to determine how to incorporate the data from WHI into their practice to insure that those who need therapy can be adequately treated, whereas those who might be at risk for osteoporosis, but in whom hormones are contraindicated, receive appropriate guidance. SERMs are gradually emerging as an option for osteoporosis prevention and perhaps treatment, particularly because at least one (raloxifene) has positive effects on ER-positive breast cancer risk. The combination of a SERM with conjugated estrogens for osteoporosis prevention may be an interesting addition.

SELECTED READINGS

Cauley JA, Robbins J, Chen Z, Cummings SR, Jackson RD, LaCroix AZ, LeBoff M, Lewis CE, McGowan J, Neuner J, Pettinger M, Stefacik ML, Wactawski-Wende J, Watts NB, Womens Health Initiative Investigators 2003 Effects of estrogen plus progesterone on the risk of fractures and bone mineral density. JAMA **290:**1729–1738.

Cauley JA, Seeley DG, Ensrud K, Ettinger B, Black D, Cummings SR 1995 Estrogen replacement therapy and fractures in older women. Ann Intern Med **122:**9–16.

Dane C, Dane B, Cetin A, Erginbas M 2007 Comparison of the effects of raloxifene and low-dose hormone replacement therapy on bone mineral density and bone turnover in the treatment of postmenopausal osteoporosis. Gynecol Endocrinol **23:**398–403.

Eastell R, Barton I, Hannon RA, Chines A, Garnero P, Delmas PD 2003 Relationship of early changes in bone resorption to the reduction in fracture risk with risedronate. J Bone Miner Res **18:**1051–1056.

Gambacciani M, Cappagli B, Ciaponi M, Pepe A, Vacca F, Genazzani AR 2008 Ultra low-dose hormone replacement therapy and bone protection in postmenopausal women. Maturitas **59:**2–6.

Greendale GA, Espeland ME, Slone S, Marcus R, Barrett-Connor E, PEPI Safety Followup Study. 2002 Bone mass response to discontinuation or long term use of hormone replacement therapy. Arch Intern Med **162:**665–672.

Huang AJ, Ettinger B, Vittinghoff E, Ensrud KE, Johnson KC, Cummings SR 2007 Endogenous estrogen levels and the effects of ultra-low-dose transdermal estradiol therapy on bone turnover and BMD in postmenopausal women. J Bone Miner Res **22:**1791–1797.

Rossouw JE, Anderson GL, Prentice RL, LaCroix AZ, Kooperberg C, Stafanick ML, Jackson RD, Beresford SA, Howard BV, Johnson KC, Kotchen JM, Ockene J 2002 Risks and benefits of estrogen plus progestin in health postmenopausal women. Results from the Women's Health Initiative randomized controlled trial. JAMA **288:**321–333.

Sarkar S, Mitlak B, Wong M, Stock JL, Black DM, Harper KD 2002 Relationships between bone mineral density and incident vertebral fracture risk with raloxifene therapy. J Bone Miner Res **17:**1–10.

Writing Group for the PEPI 1996 Effects of hormone therapy on bone mineral density: Results from the postmenopausal estrogen/piogestin interventions. JAMA **276:**1389–1396.

© 2008 American Society for Bone and Mineral Research

Chapter 49. Bisphosphonates for Postmenopausal Osteoporosis

Socrates E. Papapoulos

Department of Endocrinology and Metabolic Diseases, Leiden University Medical Center, Leiden, The Netherlands

INTRODUCTION

Bisphosphonates (BPs) are synthetic compounds that have high affinity for calcium crystals, concentrate selectively in the skeleton, and decrease bone resorption. The first BP was synthesized in the 19th century, but their relevance to medicine was recognized in the 1960s and were first given to patients with osteoporosis in the early 1970s. Currently alendronate, ibandronate, risedronate, and zoledronate are approved for the treatment of osteoporosis, whereas clodronate, etidronate, and pamidronate are available in some countries.

PHARMACOLOGY

BPs are synthetic analogs of inorganic pyrophosphate in which the oxygen atom that connects the two phosphates is replaced by a carbon (Fig. 1). This substitution renders BPs resistant to biological degradation and suitable for clinical use. BPs have two additional side chains (R1 and R2) that allow the synthesis of a large number of analogs with different pharmacological properties (Fig. 1). A hydroxyl substitution at R1 enhances the affinity of BPs for calcium crystals, whereas the presence of a nitrogen atom in R2 enhances their potency and determines their mechanism of action. The whole molecule is responsible for the action of BPs on bone resorption and probably also for their affinity for bone mineral.[1,2]

The intestinal absorption of BPs is poor (<1%) and decreases further in the presence of food or calcium that binds them. Oral BPs should be given in the fasting state 30–60 min before meals with water. BPs are cleared rapidly from the circulation; ~50% of the administered dose concentrates in the skeleton, primarily at active remodeling sites, whereas the rest is excreted unmetabolized in urine. Skeletal uptake depends on the rate of bone turnover, renal function, and on the structure of BPs.[3] The capacity of the skeleton to retain BP is large, and saturation of binding sites with the doses used in the treatment of osteoporosis is unlikely even if these are given for a very long time. At the bone surface, BPs inhibit bone resorption, and they are subsequently embedded in bone where they remain for long and are pharmacologically inactive. The elimination of BPs from the body is multiexponential, and this should be considered when calculating half-lives; the calculated terminal half-life of elimination from the skeleton can be as long as 10 yr, and pamidronate has been detected in urines of patients for up to 8 yr after discontinuation of treatment.

The decrease of bone resorption by BPs is followed by a slower decrease in the rate of bone formation, because of the coupling of the two processes, so that a new steady state at a lower rate of bone turnover is reached 3–6 mo after the start of treatment. This level of bone turnover remains constant during the whole period of treatment, showing that the accumulation of BP in the skeleton is not associated with a cumulative effect on bone remodeling. In addition to decreasing the rate of bone turnover to premenopausal levels, BPs preserve or may improve trabecular and cortical architecture, improve the hypomineralization of osteoporotic bone, increase areal BMD, and may also have a favorable effect on osteocytes. The relevant clinical outcome of these actions is the decrease of the risk of fractures (Fig. 2).

At the cellular level, BPs inhibit the activity of osteoclasts.[1,4] BPs bound to bone hydroxyapatite are released in the acidic environment of the resorption lacunae under the osteoclasts and are taken up by them. BPs without a nitrogen atom in their molecule (Fig. 1) incorporate into ATP and generate metabolites that induce osteoclast apoptosis. Nitrogen-containing BPs (N-BPs) induce changes in the cytoskeleton of osteoclasts, leading to their inactivation and potentially apoptosis. This action is mainly the result of inhibition of farnesyl pyrophosphate synthase (FPPS), an enzyme of the mevalonate biosynthetic pathway. FPPS is responsible for the formation of isoprenoid metabolites required for the prenylation of small GTPases that are important for cytoskeletal integrity and function of osteoclasts. There is a close relation between the degree of inhibition of FPPS and the antiresorptive potencies of N-BPs. In addition, the inhibition of FPPS by N-BPs leads to accumulation of IPP, a metabolite immediately upstream of FPPS, which reacts with AMP leading to the production of a new metabolite which induces osteoclast apoptosis.

ANTIFRACTURE EFFICACY

All BPs given daily in adequate doses significantly reduce the risk of vertebral fractures by 35–65% (Fig. 3).[5–11] Also illustrated in Fig. 3 are the large differences in the incidence of fractures among placebo-treated patients. Results, therefore, of different clinical trials should not be used to compare efficacy of individual BPs. For that, head-to head studies are needed, but these are not available. The overall efficacy and consistency of daily BPs to reduce the risk of vertebral fractures has been shown by meta-analyses of randomized control trials (RCTs) for alendronate and risedronate.[12–14] In studies in which radiographs were taken annually (e.g., VERT study with risedronate), the effect of the BP in reducing the risk of vertebral fractures was already evident after 1 yr, showing rapid protection of skeletal integrity. This was also shown for moderate and severe vertebral fractures with ibandronate[15] and for clinical vertebral fractures with alendronate.[16] A posthoc analysis of risedronate trials reported a significant reduction in clinical vertebral fractures as early as 6 mo after the start of treatment.[17]

The efficacy of daily oral BPs in reducing the risk of nonvertebral fractures was explored in a number of RCTs. It should be noted that definitions and adjudication procedures of nonvertebral fractures were different among clinical trials. A recent meta-analysis of the Cochrane Collaboration reported an overall reduction of the risk of nonvertebral fractures in women with osteoporosis of 23% (RR, 0.77; 95% CI, 0.74–0.94) with alendronate and 20% (RR, 0.80; 95% CI, 0.72–0.90) with risedronate.[13,14] The corresponding risk reductions

Dr. Papapoulos has received research support and/or honoraria from Amgen, Eli-Lilly, Merck & Co, Novartis, Procter & Gamble, Roche/GSK, and Servier.

Key words: bisphosphonates, alendronate, clodronate, ibandronate, pamidronate, risedronate, zoledronate, vertebral fractures, nonvertebral fractures

© 2008 American Society for Bone and Mineral Research

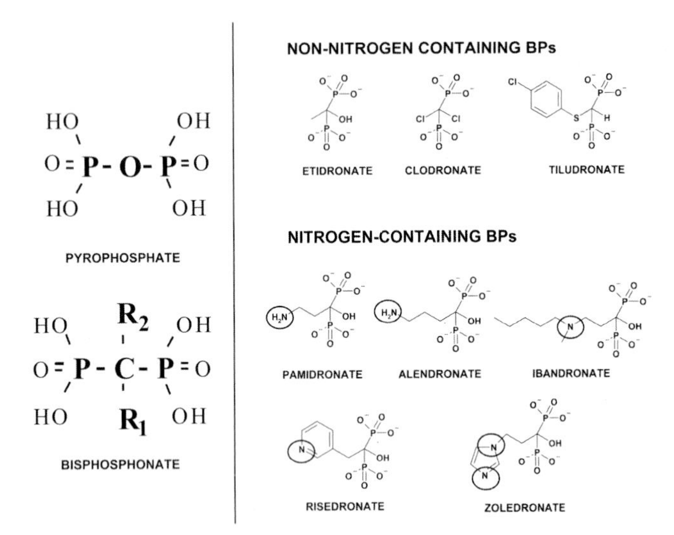

FIG. 1. (Left) Structure of pyrophosphate and geminal bisphosphonates. (Right) Structures of clinically used bisphosphonates (acid forms are depicted).

for hip fractures were 53% (RR, 0.47; 95% CI, 0.26–0.85) with alendronate and 26% (RR, 0.74; 95% CI, 0.59–0.94%) with risedronate. These estimates are in agreement with earlier published meta-analyses.[18,19] With daily ibandronate, a reduction (69%) in the risk of nonvertebral fractures was reported in a population at high risk (femoral neck BMD < −3.0) by posthoc analysis.[10] As with vertebral fractures, the effect of BPs on nonvertebral fractures occurs early after the start of treatment.

Daily administration of BPs, although highly efficacious, is inconvenient and may also be associated with gastrointestinal adverse effects. These reduce adherence to treatment and can diminish the therapeutic response.[20,21] To overcome both these problems, once-weekly formulations, the sum of seven daily doses, have been developed for alendronate and risedronate and were shown to significantly improve patient adherence to treatment while sustaining the same pharmacodynamic response as daily treatment.[22,23] Thus, daily and weekly BPs are pharmacologically equivalent and should be considered as continuous administration, whereas the term intermittent or cyclical administration should be reserved for treatments with drug-free intervals >2 wk.[3]

INTERMITTENT ADMINISTRATION OF BPs

Results of early attempts to give BPs intermittently to patients with osteoporosis were equivocal, but a meta-analysis of studies with cyclical etidronate showed a significant reduction in the risk of vertebral but not of nonvertebral fractures.[24] The efficacy of intermittent administration on N-BPs was explored in studies with ibandronate, which indicated that dose and dosing intervals are important determinants of the response to intermittent BP therapy, which in turn depends on the safety and tolerability of the administered dose.[25] In the BONE study,[10] in which women with postmenopausal osteoporosis received orally daily or intermittent ibandronate that provided equivalent total systemic annual doses, there was a significant reduction of the incidence of new vertebral fractures by 62% and 50%, respectively, after 3 yr. However, the magnitude of decrease of bone resorption and the subsequent gains in BMD were lower with the intermittent regimen compared with daily administration, suggesting that higher total doses should be given. This was tested with oral ibandronate 150 mg once-monthly and intravenous injections of 3 mg every 3 mo, which induced significantly higher increases in BMD compared with the oral daily dose.[26,27] Consistent with these data, these higher doses significantly reduced the risk of nonvertebral fractures by 38% compared with the daily oral regimen. Recently, an once-monthly oral preparation of risedronate has also become available.

The efficacy of intermittent administration of zoledronate, the most potent N-BP, in reducing the risk of osteoporotic fractures was examined in the HORIZON trial in which postmenopausal women with osteoporosis were randomized to receive 15-min infusions of zoledronate 5 mg or placebo once yearly.[28] Compared with placebo, zoledronate reduced the incidence of vertebral fractures by 70% and that of hip fractures by 41% after 3 yr. The risk of nonvertebral fractures was also significantly reduced by 25%. The effect of zoledronate on vertebral fractures was significant already at 1 yr. In a second controlled study,[29] zoledronate infusions given within 90 days after surgical repair of a hip fracture decreased significantly the rate of new clinical fractures by 35% and improved patient survival (28% reduction in deaths from any cause).

LONG-TERM EFFECTS ON BONE FRAGILITY

Skeletal fragility on long-term BP therapy has been examined in extensions of three clinical trials for 7–10 yr.[30–32] None of these extension studies was specifically designed to

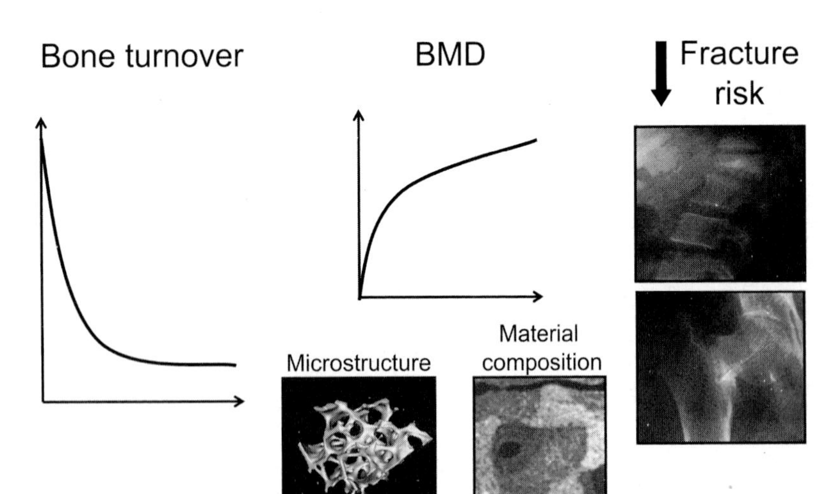

FIG. 2. Schematic presentation of effects of bisphosphonates on bone metabolism and strength in osteoporosis.

© 2008 American Society for Bone and Mineral Research

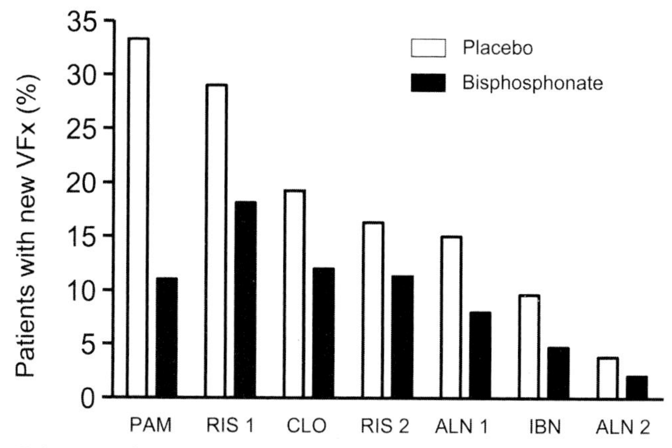

FIG. 3. Incidence of fractures in patients with osteoporosis treated with daily oral placebo (open bars) or bisphosphonate (closed bars). PAM, pamidronate[5]; RIS 1, risedronate (VERT multinational study)[9]; CLO, clodronate[11]; RIS 2, risedronate (VERT North America study)[8]; ALN 1, alendronate (FIT 1 study)[6]; IBN, ibandronate (BONE study)[10]; ALN 2, alendronate (FIT 2 study).[7]

assess antifracture efficacy, but rather safety and efficacy on surrogate endpoints, and the consistency of the effect of BPs over longer periods were evaluated. In all three studies, the incidence of nonvertebral fractures was constant with time. In the extension of the FIT trial (FLEX), patients who received alendronate for, on average, 5 yr were randomized to placebo, alendronate 5 mg/d, or alendronate 10 mg/d and were followed for another 5 yr.[32] At the end of the 10-yr observation period, the incidence of nonvertebral and hip fractures in the ALN/PBO group was similar to that of the ALN/ALN groups. Interestingly women who entered the extension with osteoporosis, as assessed by BMD, and continued treatment with alendronate showed a significant reduction in the risk of nonvertebral fractures during the 5-yr extension. In addition, the incidence of clinical vertebral fractures was lower in the ALN/ALN groups compared with the ALN/PBO group (2% versus 5%). These results suggest that alendronate treatment should be continued in patients at high risk, whereas discontinuation of treatment after 5 yr may be considered in patients at low risk for fractures.

SPECIAL ISSUES RELATED TO TREATMENT OF OSTEOPOROSIS WITH BPs

Resolution of the Effect on Bone Remodeling

The speed of reversal of the effect of BPs on bone after cessation of treatment is slow and different from that of other anti-osteoporotic treatments, such as estrogens. This is probably because of the slow release from the BP from the skeleton, which may not be enough to offer full skeletal protection but may be sufficient to slow down the rate of bone remodeling. The rate of reversal of the effect may be different among BPs depending on their pharmacological properties, particularly their bone binding affinity, but no head-to-head studies have addressed this question yet.

Excessive Suppression of Bone Remodeling

There have been concerns that the decrease of bone remodeling by BPs may compromise bone integrity, if this decrease is excessive, leading to increased bone fragility. In all clinical studies of BPs reported thus far, there have been no safety concerns, but there have been a few animal and human case studies that raised questions concerning the long-term safety of BPs on bone.

Numerous studies in different animal models with N-BPs given at a wide range of doses and time intervals have consistently shown preservation or improvement of bone strength. In only one study with high doses of clodronate given to healthy dogs has an increase in fracture incidence been reported. Earlier reports of potential compromise of the biomechanical competence of bone caused by increases in microdamage accumulation in bone biopsies of healthy dogs treated with high BP doses were not substantiated by later animal and human studies.[33,34]

Odvina et al.[35] reported nine patients treated with alendronate for 1–8 yr who developed atraumatic nonspinal fractures at skeletal sites that are not typically associated with osteoporosis and attributed these to excessive suppression of bone remodeling, histologically evaluated, by alendronate. In another case series, Goh et al.[36] reported that 9 of 13 patients with low energy subtrochanteric fractures had been treated with alendronate between 2.5 and 5 yr. The authors concluded that these fractures were caused by excessive suppression of bone turnover by alendronate, although no histological or biochemical data were provided. Since then, more cases of such fractures on BP treatment have been described, but documentation of a potential pathogenetic association has been poor.[37]

In human controlled studies of osteoporosis, bone biopsies taken between 6 mo and 10 yr of treatment with different BPs showed no evidence of excessive suppression of bone remodeling, the incidence of nonvertebral fractures was not increased with long-term therapy, and bone turnover markers increased after cessation of treatment, indicating metabolically active bone. In addition, an analysis of the FIT data showed that higher decreases of bone turnover were associated with larger decreases in the incidence of nonvertebral and hip fractures,[38] a finding supported by the above-mentioned analysis of the ibandronate studies. Moreover, in studies of patients treated with BPs followed by treatment with teriparatide, significant increases in bone markers have been reported, indicating that bisphosphonate-treated bone can readily respond to stimuli.[39] This conclusion is further supported by a study of zoledronate treatment of patients previously treated with alendronate that showed that alendronate-treated bone reacted normally to an acute bisphosphonate load, as provided by zoledronate, indicating that metabolic activity was preserved.[40]

The bulk of evidence indicates that the decrease of bone turnover by BPs, at the doses used in the treatment of postmenopausal osteoporosis, protects skeletal integrity. However, potential intrinsic or external risk factors that may contribute to increased bone fragility in selected patients treated with BPs warrant further study.

Osteonecrosis of the Jaw

Osteonecrosis of the jaw (ONJ) is defined as exposed bone in the mandible, maxilla, or both that persists for at least 8 wk in the absence of previous irradiation or metastases in the jaw. It has been reported mainly in patients with malignant diseases receiving high intravenous doses of BPs, the background incidence in the population and its pathogenesis are not known, and a causal relation with BPs has not been established. In patients with osteoporosis treated with BPs, ONJ is rare, and an incidence between 1:10,000 and <1:100,000 patient-years has been estimated.[41,42] In the clinical trials of yearly infusions of zoledronate, two adjudicated cases of ONJ were re-

© 2008 American Society for Bone and Mineral Research

ported among 9892 patients with osteoporosis: one in the placebo-treated group and one in the zoledronate-treated group.[28,29]

ADVERSE EFFECTS

Considering the different molecules, doses, routes of administration, and multiple indications for their use, BPs are safe compounds. Specific adverse effects related to the use of BPs in osteoporosis include gastrointestinal toxicity associated with the oral, particularly daily, use of N-BPs and symptoms related to an acute phase reaction, mainly after the first exposure to intravenous N-BPs. Renal toxicity is not a concern if the instructions for intravenous administration are followed. Recently, a significant increase in the incidence of atrial fibrillation, reported as serious adverse event, was observed in patients receiving zoledronate compared with those receiving placebo,[28] and a similar, nonsignificant trend was observed in the FIT study with alendronate. A biological explanation for this effect is not apparent, and further analyses of clinical trials with ibandronate and risedronate and a new study with zoledronate did not confirm such an association.

GENERAL CONCLUSIONS

BPs, because of their efficacy, safety, and ease of administration, are generally accepted as first-line therapy for osteoporosis. Selection of a BP for the treatment of an individual patient should be based on review of efficacy data, risk profile of the BP, and values and preferences of the patient. Despite progress in our understanding of the antifracture actions of BPs and their long-term effects on bone, there are still questions that remain to be addressed. These include potential, clinically relevant, differences among BPs, optimal selection of patients for treatment, and use of BPs in combination with bone-forming agents.

REFERENCES

1. Russell RGG, Watts NB, Ebetino FH, Rogers MJ 2008 Machanisms of action of bisphosphonates: Similarities and differences and their potential influence on clinical efficacy. Osteoporos Int **19:**733–759.
2. Papapoulos SE 2006 Bisphosphonate actions: Physical chemistry revisited. Bone **38:**613–616.
3. Cremers SC, Pillai G, Papapoulos SE 2005 Pharmacokinetics/pharmacodynamics of bisphosphonates: Use for optimisation of intermittent therapy for osteoporosis. Clin Pharmacokinet **44:**551–570.
4. Rogers MJ 2004 From molds and macrophages to mevalonate: A decade of progress in understanding the molecular mode of action of bisphsposphonates. Calcif Tissue Int **75:**451–461.
5. Brumsen C, Papapoulos SE, Lips P, Geelhoed-Duijvestijn PHLM, Hamdy NAT, Landman JO, McCloskey EV, Netelenbos JC, Pauwels EKJ, Roos JC, Valentijn RM, Zwinderman AH 2002 Daily oral pamidronate in women and men with osteoporosis: A 3-year randomized placebo-controlled clinical trial with a 2-year open extension. J Bone Miner Res **17:**1057–1064.
6. Black DM, Cummings SR, Karpf DB, Cauley JA, Thompson DE, Nevitt MC, Bauer DC, Genant HK, Haskell WL, Marcus R, Ott SM, Torner JC, Quandt SA, Reiss TF, Ensrud KE 1996 Randomised trial of effect of alendronate on risk of fracture in women with existing vertebral fractures. Fracture Intervention Trial Research Group. Lancet **348:**1535–1541.
7. Cummings SR, Black DM, Thompson DE, Applegate WB, Barrett-Connor E, Musliner TA, Palermo L, Prineas R, Rubin SM, Scott JC, Vogt T, Wallace R, Yates AJ, LaCroix AZ 1998 Effect of alendronate on risk of fracture in women with low bone density but without vertebral fractures: Results from the Fracture Intervention Trial. JAMA **280:**2077–2082.
8. Harris ST, Watts NB, Genant HK, McKeever CD, Hangartner T, Keller M, Chesnut CH III, Brown J, Eriksen EF, Hoseyni MS, Axelrod DW, Miller PD 1999 Effects of risedronate treatment on vertebral and nonvertebral fractures in women with postmenopausal osteoporosis: A randomized controlled trial. Vertebral Efficacy With Risedronate Therapy (VERT) Study Group. JAMA **282:**1344–1352.
9. Reginster J, Minne HW, Sorensen OH, Hooper M, Roux C, Brandi ML, Lund B, Ethgen D, Pack S, Roumagnac I, Eastell R 2000 Randomized trial of the effects of risedronate on vertebral fractures in women with established postmenopausal osteoporosis. Vertebral Efficacy with Risedronate Therapy (VERT) Study Group. Osteoporos Int **11:**83–91.
10. Chesnut CH, Ettinger MP, Miller PD, Baylink DJ, Emkey R, Harris ST, Wasnich RD, Watts NB, Schimmer RC, Recker RR 2004 Effects of oral ibandronate administered daily or intermittently on fracture risk in postmenopausal osteoporosis. J Bone Miner Res **19:**1241–1249.
11. McCloskey E, Selby P, Davies M, Robinson J, Francis RM, Adams J, Kayan K, Beneton M, Jalava T, Pylkkänen L, Kenraali J, Aropuu S, Kanis JA 2004 Clodronate reduces vertebral fracture risk in women with postmenopausal or secondary osteoporosis: Results of a double-blind, placebo-controlled 3-year study. J Bone Miner Res **19:**728–736.
12. Cranney A, Guyatt G, Griffith L, Wells G, Tugwell P, Rosen C; Osteoporosis Methodology Group and The Osteoporosis Research Advisory Group 2002 Meta-analyses of therapies for postmenopausal osteoporosis. IX: Summary of meta-analyses of therapies for postmenopausal osteoporosis. Endocr Rev **23:**570–578.
13. Wells G, Cranney A, Peterson J, Boucher M, Shea B, Robinson V, Coyle D, Tugwell P 2008 Risedronate for the primary and secondary prevention of osteoporotic fractures in postmenopausal women. Cochrane Database Syst Rev **23:**CD004523.
14. Wells GA, Cranney A, Peterson J, Boucher M, Shea B, Robinson V, Coyle D, Tugwell P 2008 Alendronate for the primary and secondary prevention of osteoporotic fractures in postmenopausal women. Cochrane Database Syst Rev **23:**CD001155.
15. Felsenberg D, Miller P, Armbrecht G, Wilson K, Schimmer RC, Papapoulos SE 2005 Oral ibandronate significantly reduces the risk of vertebral fractures of greater severity after 1, 2, and 3 years in postmenopausal women with osteoporosis. Bone **37:**651–654.
16. Black DM, Thompson DE, Bauer DC, Ensrud K, Musliner T, Hochberg MC, Nevitt MC, Suryawanshi S, Cummings SR; Fracture Intervention Trial 2000 Fracture risk reduction with alendronate in women with osteoporosis: The Fracture Intervention Trial. FIT Research Group. J Clin Endocrinol Metab **85:**4118–4124.
17. Roux C, Seeman E, Eastell R, Adachi J, Jackson RD, Felsenberg D, Songcharoen S, Rizzoli R, Di Munno O, Horlait S, Valent D, Watts NB 2004 Efficacy of risedronate on clinical vertebral fractures within six months. Curr Med Res Opin **20:**433–439.
18. Papapoulos SE, Quandt SA, Liberman UA, Hochberg MC, Thompson DE 2005 Meta-analysis of the efficacy of alendronate for the prevention of hip fractures in postmenopausal women. Osteoporos Int **16:**468–474.
19. Nguyen ND, Eisman JA, Nguyen TV 2006 Anti-hip fracture efficacy of biophosphonates: A Bayesian analysis of clinical trials. J Bone Miner Res **21:**340–349.
20. Caro JJ, Ishak KJ, Huybrechts KF, Raggio G, Naujoks C 2004 The impact of compliance with osteoporosis therapy on fracture rates in actual practice. Osteoporos Int **15:**1003–1008.
21. Siris ES, Harris ST, Rosen CJ, Barr CE, Arvesen JN, Abbott TA, Silverman S 2006 Adherence to bisphosphonate therapy and fracture rates in osteoporotic women: Relationship to vertebral and nonvertebral fractures from 2 US claims databases. Mayo Clin Proc **81:**1013–1022.
22. Schnitzer T, Bone HG, Crepaldi G, Adami S, McClung M, Kiel D, Felsenberg D, Recker RR, Tonino RP, Roux C, Pinchera A, Foldes AJ, Greenspan SL, Levine MA, Emkey R, Santora AC II, Kaur A, Thompson DE, Yates J, Orloff JJ 2000 Therapeutic equivalence of alendronate 70 mg once-weekly and alendronate 10 mg daily in the treatment of osteoporosis. Alendronate Once-Weekly Study Group. Aging (Milano) **12:**1–12.
23. Brown JP, Kendler DL, McClung MR, Emkey RD, Adachi JD, Bolognese MA, Li Z, Balske A, Lindsay R 2002 The efficacy and tolerability of risedronate once a week for the treatment of postmenopausal osteoporosis. Calcif Tissue Int **71:**103–111.
24. Adachi JD, Rizzoli R, Boonen S, Li Z, Meredith MP, Chesnut CH III 2005 Vertebral fracture risk reduction with risedronate in postmenopausal women with osteoporosis: A meta-analysis of individual patient data. Aging Clin Exp Res **17:**150–156.
25. Papapoulos SE, Schimmer RC 2007 Changes in bone remodelling

© 2008 American Society for Bone and Mineral Research

and antifracture efficacy of intermittent bisphosphonate therapy: Implications from clinical studies with ibandronate. Ann Rheum Dis **66**:853–858.
26. Reginster JY, Adami S, Lakatos P, Greenwald M, Stepan JJ, Silverman SL, Christiansen C, Rowell L, Mairon N, Bonvoisin B, Drezner MK, Emkey R, Felsenberg D, Cooper C, Delmas PD, Miller PD 2006 Efficacy and tolerability of once-monthly oral ibandronate in postmenopausal osteoporosis: 2 year results from the MOBILE study. Ann Rheum Dis **65**:654–661.
27. Eisman JA, Civitelli R, Adami S, Czerwinski E, Recknor C, Prince R, Reginster JY, Zaidi M, Felsenberg D, Hughes C, Mairon N, Masanauskaite D, Reid DM, Delmas PD, Recker RR 2008 Efficacy and tolerability of intravenous ibandronate injections in postmenopausal osteoporosis: 2-year results from the DIVA study. J Rheumatol **35**:488–497.
28. Black DM, Delmas PD, Eastell R, Reid IR, Boonen S, Cauley JA, Cosman F, Lakatos P, Leung PC, Man Z, Mautalen C, Mesenbrink P, Hu H, Caminis J, Tong K, Rosario-Jansen T, Krasnow J, Hue TF, Sellmeyer D, Eriksen EF, Cummings SR; HORIZON Pivotal Fracture Trial 2007 Once-yearly zoledronic acid for treatment of postmenopausal osteoporosis. N Engl J Med **356**:1809–1822.
29. Lyles KW, Colón-Emeric CS, Magaziner JS, Adachi JD, Pieper CF, Mautalen C, Hyldstrup L, Recknor C, Nordsletten L, Moore KA, Lavecchia C, Zhang J, Mesenbrink P, Hodgson PK, Abrams K, Orloff JJ, Horowitz Z, Eriksen EF, Boonen S; HORIZON Recurrent Fracture Trial 2007 Zoledronic acid and clinical fractures and mortality after hip fracture. N Engl J Med **357**:1799–1809.
30. Bone HG, Hosking D, Devogelaer JP, Tucci JR, Emkey RD, Tonino RP, Rodriguez-Portales JA, Downs RW, Gupta J, Santora AC, Liberman UA 2004 Alendronate phase III osteoporosis treatment study group. N Engl J Med **350**:1189–1199.
31. Mellström DD, Sörensen OH, Goemaere S, Roux C, Johnson TD, Chines AA 2004 Seven years of treatment with risedronate in women with postmenopausal osteoporosis. Calcif Tissue Int **75**:462–468.
32. Black DM, Schwartz AV, Ensrud KE, Cauley JA, Levis S, Quandt SA, Satterfield S, Wallace RB, Bauer DC, Palermo L, Wehren LE, Lombardi A, Santora AC, Cummings SR; FLEX Research Group 2006 Effects of continuing or stopping alendronate after 5 years of treatment: The Fracture Intervention Trial Long-term Extension (FLEX): A randomized trial. JAMA **296**:2927–2938.
33. Allen MR, Iwata K, Phipps R, Burr DB 2006 Alterations in canine vertebral bone turnover, microdamage accumulation, and biomechanical properties following 1-year treatment with clinical treatment doses of risedronate or alendronate. Bone **39**:872–879.
34. Chapurlat RD, Arlot M, Burt-Pichat B, Chavassieux P, Roux JP, Portero-Muzy N, Delmas PD, Chapurlat RD, Arlot M, Burt-Pichat B, Chavassieux P, Roux J-P, Portero-Muzy N, Delmas PD 2007 Microcrack frequency and bone remodeling in postmenopausal osteoporotic women on long-term bisphosphonates: A bone biopsy study. J Bone Miner Res **22**:1502–1509.
35. Odvina CV, Zerwekh JE, Rao DS, Maalouf N, Gottschalk FA, Pak CY 2005 Severely suppressed bone turnover: A potential complication of alendronate therapy. J Clin Endocrinol Metab **90**:1294–1301.
36. Goh SK, Yang KY, Koh JS, Wong MK, Chua SY, Chua DT, Howe TS 2007 Subtrochanteric insufficiency fractures in patients on alendronate therapy: A caution. J Bone Joint Surg Br **89**:349–353.
37. Kwek EB, Goh SK, Koh JS, Png MA, Howe TS 2008 An emerging pattern of subtrochanteric stress fractures: A long-term complication of alendronate therapy? Injury **39**:224–231.
38. Bauer DC, Black DM, Garnero P, Hochberg M, Ott S, Orloff J, Thompson DE, Ewing SK, Delmas PD; Fracture Intervention Trial Study Group 2004 Change in bone turnover and hip, non-spine, and vertebral fracture in alendronate-treated women: The fracture intervention trial. J Bone Miner Res **19**:1250–1258.
39. Obermaier-Pietsch CB, Marin F, McCloskey E, Hadji P, Simoes ME, Barker C, Oertel H, Nickelsen TN, Boonen S 2007 Effects of prior antiresorptive therapy on markers of bone formation after 6 months of teriparatide treatment in postmenopausal women with osteoporosis: Results from the Eurofors trial. Calcif Tissue Int **80**(Suppl 1):S137.
40. McClung M, Recker R, Miller P, Fiske D, Minkoff J, Kriegman A, Zhou W, Adera M, Davis J 2007 Intravenous zoledronic acid 5 mg in the treatment of postmenopausal women with low bone density previously treated with alendronate. Bone **41**:122–128.
41. Khosla S, Burr D, Cauley J, Dempster DW, Ebeling PR, Felsenberg D, Gagel RF, Gilsanz V, Guise T, Koka S, McCauley LK, McGowan J, McKee MD, Mohla S, Pendrys DG, Raisz LG, Ruggiero SL, Shafer DM, Shum L, Silverman SL, Van Poznak CH, Watts N, Woo SB, Shane E; American Society for Bone and Mineral Research 2007 Bisphosphonate-associated osteonecrosis of the jaw: Report of a task force of the American Society for Bone and Mineral Research. J Bone Miner Res **22**:1479–1491.
42. Rizzoli R, Burlet N, Cahall D, Delmas PD, Eriksen EF, Felsenberg D, Grbic J, Jontell M, Landesberg R, Laslop A, Wollenhaupt M, Papapoulos S, Sezer O, Sprafka M, Reginster JY 2008 Osteonecrosis of the jaw and bisphosphonate treatment for osteoporosis. Bone **42**:841–847.

Chapter 50. Strontium Ranelate in the Prevention of Osteoporotic Fractures

René Rizzoli

Division of Bone Diseases [World Health Organization Collaborating Center for Osteoporosis Prevention], Department of Rehabilitation and Geriatrics, Geneva University Hospital and Faculty of Medicine, Geneva, Switzerland

INTRODUCTION

Strontium (Sr) is an alkaline earth divalent cation, which is a trace element in the human body, representing 0.00044% of the body mass.[1] With an atomic weight of 87.6, it is more than twice as heavy as calcium. A normal diet contains between 2 and 4 mg of Sr per day, mostly from vegetables and cereals. Sr competes with calcium for intestinal absorption, which is ~20% of the intake, with a ratio of Sr to calcium intestinal absorption of ~0.6–0.7.[2] Sr is eliminated through the kidney and through gastrointestinal tract secretion. Although Sr and calcium share a common transport system in the renal tubule, the clearance of the former seems to be higher than that of the latter. In bone, Sr is mainly adsorbed onto the crystal surface, with less than one atom of Sr replacing one calcium atom in the hydroxyapatite crystal,[3] without altering the lattice structure. Sr is primarily found in newly deposited bone tissue. After stopping treatment, >50% of bone strontium disappears within 10 wk.[3] Strontium ranelate {5[bis(carboxymethyl)amino]-2-carboxy-4-cyano-3-thiophenacetic acid distrontium salt}, which is approved for the treatment of osteoporosis in Europe and in

Dr. Rizzoli has served on the speakers board and participated on the advisory board for Servier, Novartis, Roche, Nycomed, Merck, and Amgen.

© 2008 American Society for Bone and Mineral Research

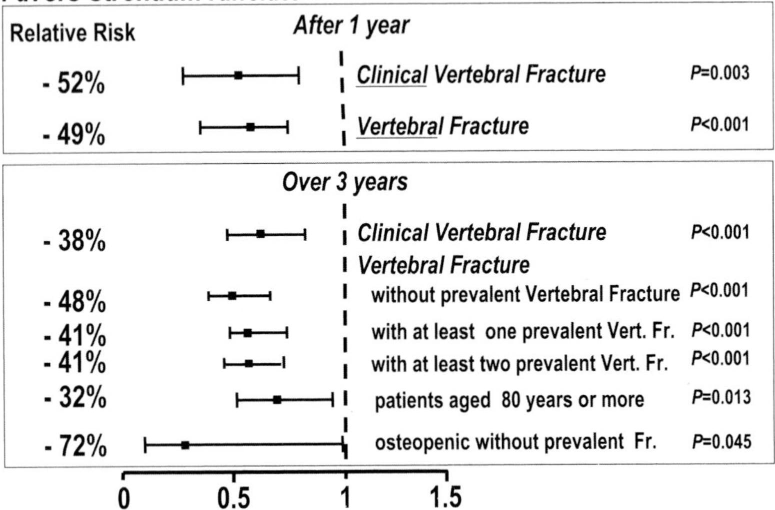

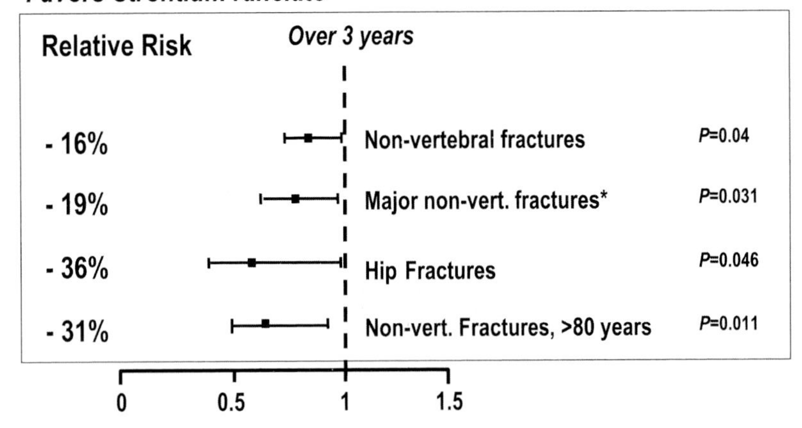

FIG. 1. Effects of strontium ranelate on (A) vertebral and (B) nonvertebral fracture risk. Relative risks and 95% CIs. *Includes humerus, pelvis-sacrum, ribs-sternum, hip, clavicle, and wrist.

many countries worldwide, is made up of an organic anion (ranelate) and two stable Sr cations. Less than 3% ranelate is absorbed in the intestine. Ranelate is not metabolized and is mainly excreted through the kidney. The therapeutic dose of 2 g/d of strontium ranelate provides 8 mmol of Sr; 8 mmol of calcium represent 320 mg.

STRONTIUM RANELATE ANTIFRACTURE EFFICACY

Strontium ranelate was studied in a phase 3 program consisting of two studies. The Spinal Osteoporosis Therapeutic Intervention (SOTI) study aimed at assessing the effect of 2 g daily of strontium ranelate on the risk of vertebral fractures, and the Treatment of Peripheral Osteoporosis (TROPOS) trial aimed at evaluating the effect of strontium ranelate on nonvertebral fractures. All patients included in these two studies had previously participated in a run-in study, the Fracture International Run-in Strontium Ranelate Trials (FIRST), aimed at normalizing the calcium and vitamin D status of all patients before entry into either SOTI or TROPOS trials. Patients received a calcium/vitamin D supplement throughout the studies, which varied from 500 to 1000 mg of calcium and from 400 to 800 IU of vitamin D_3. Among >9000 postmenopausal women with osteoporosis who took part in the FIRST study, 1649 patients, with a mean age of 70 yr, were included in SOTI,[4] and 5091 patients, with a mean age of 77 yr, were included in TROPOS.[5]

In the SOTI study, treatment with strontium ranelate for 3 yr was associated with a 41% reduction in relative risk of experiencing a new vertebral fracture, as assessed by semiquantitative evaluation (Fig. 1). The relative risk of experiencing a new vertebral fracture was already significantly reduced by 49% in the strontium ranelate group by the end of the first year of treatment.

In the TROPOS study, which was designed to assess strontium ranelate efficacy in preventing nonvertebral fractures in postmenopausal women with osteoporosis, ambulatory postmenopausal women were eligible if they had femoral neck BMD corresponding to a T-score < −2.5 and were older than 74 yr of age or were between 70 and 74 yr of age with one additional risk factor for fracture. In the 5091 patients recruited, strontium ranelate was associated with a 16% relative risk reduction in all nonvertebral fractures over a 3-yr follow-up period ($p = 0.04$) and with a 19% reduction in risk of major nonvertebral osteoporotic fractures ($p = 0.031$). In a high-risk fracture subgroup (women older than 74 yr and with femoral-

© 2008 American Society for Bone and Mineral Research

neck BMD T-score < –2.4 SD, using the Third National Health and Nutrition Examination Survey [NHANES III] reference range), treatment was associated with a 36% reduction in risk of hip fracture ($p = 0.046$). Yearly vertebral X-rays were performed in 3640 patients, in whom there was a reduction in the relative risk of new vertebral fracture of 39% over 3 yr in the strontium ranelate group ($p < 0.001$) and of 45% ($p < 0.001$) at the first year of treatment, indicating a highly consistent effect between both trials. In these 3640 patients, 66.4% had no prevalent vertebral fracture at inclusion. The risk of experiencing a first vertebral fracture in these patients was reduced by 45% ($p < 0.001$). In the subgroup of patients with at least one prevalent fracture ($n = 1224$), the risk of experiencing a first vertebral fracture was reduced by 32% ($p < 0.001$). An extension of the phase 3 trials up to 5 yr has shown a sustained antifracture efficacy with –24% and –15% reduction in vertebral and nonvertebral fractures, respectively, after 5 yr of placebo-controlled study.[6]

Data from SOTI and TROPOS were pooled to address several prespecified research questions. The efficacy of strontium ranelate was specifically shown in elderly patients (age ≥ 80 yr) with a significant reduction in both vertebral (–32% and –31%) and nonvertebral (–31% and –26%) risks of fracture over 3[7] and 5 yr, respectively. Between the age of 50 and 65 yr of age ($n = 353$), strontium ranelate decreased the risk of vertebral fracture by 47% ($p = 0.006$). In the 446 patients with a lumbar spine osteopenia, and without prevalent fracture, strontium ranelate reduced vertebral fracture risk by 59% ($p = 0.039$).[8] The antifracture efficacy of strontium ranelate could be shown whatever the severity of osteoporosis (baseline BMD, number of prevalent fracture, family history, or body mass index)[9] or the level of bone remodeling.

SAFETY OF STRONTIUM RANELATE

In general, strontium ranelate was well tolerated, without major adverse events. During clinical trials, the most common side effects were nausea and diarrhea (~7% versus 5% in the placebo group over 5 yr), headache, and skin irritation. All were mild and transient and did not lead to withdrawal from the studies. By pooling SOTI and TROPOS trial data, a significant increase in the risk of venous-thrombosis embolism event was found (relative risk: 1.42; $p = 0.036$). However, no coagulation abnormality has been detected in relation with strontium ranelate treatment. Recently, a few cases of DRESS syndrome (drug rash with eosinophilia and systemic symptoms) were recorded.

QUALITY OF LIFE AND COST-EFFECTIVENESS

Using a validated instrument specifically developed to assess quality of life in patients with osteoporosis, in the patients followed in the SOTI trial, strontium ranelate produced beneficial effects on both physical and emotional measures, together with a significant increase in the number of patients free of back pain by the first year and over 3 yr of treatment.[10]

Adapting a Markov model to the conditions of SOTI and TROPOS trials, with Swedish costs and epidemiological data, strontium ranelate treatment was shown to be cost-effective and even cost-saving in patients older than 74 yr of age with a T-score < –2.4 or in patients older than 80 yr of age.[11]

RECENT DEVELOPMENTS

Based on preclinical data suggesting that strontium ranelate may influence chondrocyte metabolism, a subgroup of 2617 postmenopausal women from TROPOS study had measurement of urinary type II collagen degradation products. Strontium ranelate was associated with a 10–20% lower value compared with placebo.[12] Spine osteoarthritis progression (osteophytes, disc space narrowing, and sclerosis) was evaluated in 1105 osteoporotic women form SOTI and TROPOS. A 42% reduction ($p = 0.0005$) reduction in the proportion of patients with radiological worsening was observed over 3 yr.[13]

MECHANISMS OF ANTIFRACTURE EFFICACY

In the treated group of SOTI, lumbar spine BMD increased from baseline by 12.7% at the lumbar spine, 7.2% at the femoral neck, and 8.6% at the total hip. BMD adjusted for strontium content[14] increased by 6.8% from baseline at the lumbar spine after 3 yr compared with a decrease of 1.3% in the placebo group.[4] In a pooled analysis of both trials, changes in femoral neck BMD were predictive of the reduction of vertebral fracture risk.[15] Each percent increase in femoral neck BMD at either 1 or 3 yr was associated with a 3% decrease in vertebral fracture risk, with changes in BMD explaining 76% of the variance of fracture risk reduction.

In both trials, bone-specific alkaline phosphatase increased in the strontium ranelate group by ~8%, whereas serum type I collagen C-telopeptide cross-links decreased as well by 12%, indicating a mild decreased bone resorption and maintained, or even increased, bone formation. This pattern is quite different from what is observed with inhibitors of bone resorption, where markers of bone formation are decreased in a commensurate way to the changes of the markers of bone resorption.

Comparing 49 transiliac bone biopsies collected in strontium ranelate–treated patients to 92 biopsies obtained either at baseline or in the placebo group, higher mineral apposition rate in cancellous bone (+9%, $p = 0.019$) and osteoblast perimeters (+38%, $p = 0.047$) were found.[16] Submitting 20 and 21 biopsies obtained at 3 yr of strontium ranelate treatment and placebo, respectively, to μCT, higher cortical thickness (+18%, $p = 0.008$) and trabecular number (+14%, $p = 0.008$), together with lower trabecular separation (–22%, $p = 0.01$), without any modification in cortical porosity, were shown.[16] This can explain the decrease in vertebral and nonvertebral, including the hip, fracture risk observed in strontium ranelate–treated patients.

A large series of preclinical studies have shown that strontium ranelate treatment increases bone ultimate strength, mainly through an increase in plastic energy,[17] improves intrinsic bone tissue quality, increasing elastic modulus, hardness, and dissipated energy, as assessed by nanoindentation,[18] reduces bone loss in various experimental models,[19,20] and does not impair the process of fracture healing.[21] A direct inhibition of strontium on osteoclast differentiation and activity has been reported.[22,23] Strontium stimulates proliferation and differentiation of cells of the osteoblast lineage, through mechanisms possibly involving a calcium-sensing receptor for the former[24] and prostaglandins for the latter.[25,26]

REFERENCES

1. Pors Nielsen S 2004 The biological role of strontium. Bone **35**:583–588.
2. Marcus CS, Lengemann FW 1962 Absorption of Ca45 and Sr85 from solid and liquid food at various levels of the alimentary tract of the rat. J Nutr **77**:155–160.
3. Farlay D, Boivin G, Panczer G, Lalande A, Meunier PJ 2005 Long-term strontium ranelate administration in monkeys preserves characteristics of bone mineral crystals and degree of mineralization of bone. J Bone Miner Res **20**:1569–1578.
4. Meunier PJ, Roux C, Seeman E, Ortolani S, Badurski JE, Spector TD, Cannata J, Balogh A, Lemmel EM, Pors-Nielsen S, Rizzoli R,

© 2008 American Society for Bone and Mineral Research

Genant HK, Reginster JY 2004 The effects of strontium ranelate on the risk of vertebral fracture in women with postmenopausal osteoporosis. N Engl J Med **350:**459–468.
5. Reginster JY, Seeman E, De Vernejoul MC, Adami S, Compston J, Phenekos C, Devogelaer JP, Curiel MD, Sawicki A, Goemaere S, Sorensen OH, Felsenberg D, Meunier PJ 2005 Strontium ranelate reduces the risk of nonvertebral fractures in postmenopausal women with osteoporosis: Treatment of Peripheral Osteoporosis (TROPOS) study. J Clin Endocrinol Metab **90:**2816–2822.
6. Reginster JY, Meunier PJ, Roux C 2006 Strontium ranelate: An antiosteoporotic treatment demonstrated vertebral and nonvertebral antifracture efficacy over 5 years in postmenopausal osteoporotic women. Osteoporos Int **17**(Suppl 2)**:**S14.
7. Seeman E, Vellas B, Benhamou C, Aquino JP, Semler J, Kaufman JM, Hoszowski K, Varela AR, Fiore C, Brixen K, Reginster JY, Boonen S 2006 Strontium ranelate reduces the risk of vertebral and nonvertebral fractures in women eighty years of age and older. J Bone Miner Res **21:**1113–1120.
8. Seeman E, Devogelaer JP, Lorenc RS, Spector T, Brixen K, Balogh A, Stucki G, Reginster JY 2008 Strontium ranelate reduces the risk of vertebral fractures in patients with osteopenia. J Bone Miner Res **23:**433–438.
9. Roux C, Reginster JY, Fechtenbaum J, Kolta S, Sawicki A, Tulassay Z, Luisetto G, Padrino JM, Doyle D, Prince R, Fardellone P, Sorensen OH, Meunier PJ 2006 Vertebral fracture risk reduction with strontium ranelate in women with postmenopausal osteoporosis is independent of baseline risk factors. J Bone Miner Res **21:**536–542.
10. Marquis P, Roux C, de la Loge C, Diaz-Curiel M, Cormier C, Isaia G, Badurski J, Wark J, Meunier PJ 2008 Strontium ranelate prevents quality of life impairment in post-menopausal women with established vertebral osteoporosis. Osteoporos Int **19:**503–510.
11. Borgstrom F, Jonsson B, Ström O, Kanis JA 2006 An economic evaluation of strontium ranelate in the treatment of osteoporosis in a Swedish setting: based on the results of the SOTI and TROPOS trials. Osteoporos Int **17:**1781–1793.
12. Alexandersen P, Karsdal MA, Qvist P, Reginster JY, Christiansen C 2007 Strontium ranelate reduces the urinary level of cartilage degradation biomarker CTX-II in postmenopausal women. Bone **40:**218–222.
13. Bruyere O, Delferriere D, Roux C, Wark JD, Spector T, Devogelaer JP, Brixen K, Adami S, Fechtenbaum J, Kolta S, Reginster JY 2008 Effects of strontium ranelate on spinal osteoarthritis progression. Ann Rheum Dis **67:**335–339.
14. Blake GM, Fogelman I 2005 Long-term effect of strontium ranelate treatment on BMD. J Bone Miner Res **20:**1901–1904.
15. Bruyere O, Roux C, Detilleux J, Slosman DO, Spector TD, Fardellone P, Brixen K, Devogelaer JP, Diaz-Curiel M, Albanese C, Kaufman JM, Pors-Nielsen S, Reginster JY 2007 Relationship between bone mineral density changes and fracture risk reduction in patients treated with strontium ranelate. J Clin Endocrinol Metab **92:**3076–3081.
16. Arlot ME, Jiang Y, Genant HK, Zhao J, Burt-Pichat B, Roux JP, Delmas PD, Meunier PJ 2008 Histomorphometric and μCT analysis of bone biopsies from postmenopausal osteoporotic women treated with strontium ranelate. J Bone Miner Res **23:**215–222.
17. Ammann P, Shen V, Robin B, Mauras Y, Bonjour JP, Rizzoli R 2004 Strontium ranelate improves bone resistance by increasing bone mass and improving architecture in intact female rats. J Bone Miner Res **19:**2012–2020.
18. Ammann P, Badoud I, Barraud S, Dayer R, Rizzoli R 2007 Strontium ranelate treatment improves trabecular and cortical intrinsic bone tissue quality, a determinant of bone strength. J Bone Miner Res **22:**1419–1425.
19. Marie PJ 2005 Strontium as therapy for osteoporosis. Curr Opin Pharmacol **5:**633–636.
20. Marie PJ, Hott M, Modrowski D, De Pollak C, Guillemain J, Deloffre P, Tsouderos Y 1993 An uncoupling agent containing strontium prevents bone loss by depressing bone resorption and maintaining bone formation in estrogen-deficient rats. J Bone Miner Res **8:**607–615.
21. Cebesoy O, Tutar E, Kose KC, Baltaci Y, Bagci C 2007 Effect of strontium ranelate on fracture healing in rat tibia. Joint Bone Spine **74:**590–593.
22. Baron R, Tsouderos Y 2002 In vitro effects of S12911-2 on osteoclast function and bone marrow macrophage differentiation. Eur J Pharmacol **450:**11–17.
23. Takahashi N, Sasaki T, Tsouderos Y, Suda T 2003 S 12911-2 inhibits osteoclastic bone resorption in vitro. J Bone Miner Res **18:**1082–1087.
24. Chattopadhyay N, Quinn SJ, Kifor O, Ye C, Brown EM 2007 The calcium-sensing receptor (CaR) is involved in strontium ranelate-induced osteoblast proliferation. Biochem Pharmacol **74:**438–447.
25. Choudhary S, Halbout P, Alander C, Raisz L, Pilbeam C 2007 Strontium ranelate promotes osteoblastic differentiation and mineralization of murine bone marrow stromal cells: Involvement of prostaglandins. J Bone Miner Res **22:**1002–1010.
26. Pi M, Quarles LD 2004 A novel ation-sensing mechanism in osteoblasts is a molecular target for strontium. J Bone Miner Res **19:**862–869.

Chapter 51. Parathyroid Hormone Treatment for Osteoporosis

Felicia Cosman[1,2] and Susan Greenspan[3]

[1]Department of Clinical Medicine, Columbia College of Physicians and Surgeons, Columbia University, New York, New York; [2]Clinical Research Center Helen Hayes Hospital, West Haverstraw, New York, New York; [3]Osteoporosis Prevention and Treatment Center, University of Pittsburgh, Pittsburgh, Pennsylvania

INTRODUCTION

As a result of its unique mechanism of action, PTH, the only approved anabolic therapy for bone, produces larger increments in bone mass (particularly in the spine), than those seen with antiresorptive therapies. PTH treatment first stimulates bone formation and subsequently stimulates both bone resorption and formation, whereas the balance remains positive for formation, even in this latter phase of PTH activity.[1–3] The growth of new bone with PTH permits restoration of bone microarchitecture, including improved trabecular connectivity and enhanced cortical thickness.[4,5] Bone formation may also be induced on the outer periosteal surface,[6–8] possibly affecting bone size and geometry, with additional beneficial effects on bone strength.[6–12]

This chapter reviews the clinical trial data using PTH as both monotherapy and in combination and sequential regimens with antiresorptive agents in women and in men and briefly overviews trials in a few special populations. PTH will be referred to as teriparatide (TPD) when it is the recombinant

Dr. Cosman has consulted for Eli Lilly and Merck. Dr. Greenspan has consulted for Proctor & Gamble and has received research grants from Proctor & Gamble, Alliance for Better Bone Health, and Eli Lilly.

© 2008 American Society for Bone and Mineral Research

human PTH(1-34) fragment produced by Lilly (Indianapolis, IN, USA) or the hPTH(1-34) produced by biochemical synthetic methods (Bachem, Torrance, CA, USA), and PTH(1-84) as the intact human recombinant molecule developed by NPS Pharmaceuticals (Salt Lake City, UT, USA), approved and available in Europe. PTH without other designation denotes either of the compounds.

CANDIDATES FOR ANABOLIC THERAPY

Good candidates for PTH are women and men who are at high risk of future osteoporosis-related fractures, including those with vertebral compression fractures (clinical or radiographic), other osteoporosis-related fractures with BMD in the osteoporosis range, or very low BMD even in the absence of fractures (T-score < –3). PTH should also be recommended for individuals who have been on prior antiresorptive agents, who have had a suboptimal response to treatment, defined as incident fractures or active bone loss during therapy, or who have persistent osteoporosis despite therapy. Individuals who might be at elevated risk for osteosarcoma, such as those with a history of Paget's disease, bone irradiation, unexplained elevations in alkaline phosphatase, adults with open epiphyses, and children, should not receive PTH treatment. Furthermore, people with metastatic bone cancer, primary bone cancer, myeloma, hyperparathyroidism, and hypercalcemia should not receive PTH. The PTH treatment course is recommended to be 18–24 mo.

POSTMENOPAUSAL OSTEOPOROSIS

TPD as Monotherapy

The largest study of TPD action was that of Neer et al.[13] in 1637 postmenopausal women with prevalent vertebral fractures, of average age 70 who were randomized to receive TPD, 20 or 40 μg, or placebo by daily subcutaneous injection. After a median treatment period of 19 mo, TPD increased spine BMD by 9.7% (20-μg dose) and 13.7% (400-μg dose) and hip and total body BMD to a lesser extent. A small decrease in radius BMD was seen (significant at the higher dose). Vertebral fracture risk reductions were 65% and 69%, respectively, with an absolute v risk of 4% in the high-dose group (19/434) and 5% in the low-dose group (22/444) versus 14% in the placebo group (64/448). There was also a reduction in the incidence of new or worsening back pain in both TPD groups. In patients with incident vertebral fractures, height loss was reduced (mean 0.21 cm lost in TPD compared with 1.11 cm in placebo). Incident nonvertebral fractures were reduced by 40% (6% incidence in TPD versus 10% in placebo) and by 50% for those defined as fragility fractures (no differences between TPD groups). Despite the small decrease in radius BMD, there was an apparent reduction in wrist fracture occurrence in TPD-treated women (although too small a number to statistically evaluate). There were also numerically fewer hip fractures in TPD-treated patients, although again, there were too few to evaluate statistically.

Although transient increases in serum calcium were common when measured within 6 h of TPD injection, sustained increases (confirmed with at least one subsequent measurement) were seen in only 3% of patients assigned to the 20-μg group and 11% of those assigned to the 40-μg group. There were no significant differences between TPD or placebo groups with respect to deaths, hospitalizations, cardiovascular disorders, renal stones, or gout, despite an average increase in 24-h urine calcium of 40 mg/d and an increase in serum uric acid of up to 25%. Animal studies have shown that administration of high-dose TPD to rodents is associated with osteogenic sarcoma, dependent on dose and duration of administration.[14,15] The relevance of this finding to humans is unclear. In patients with endogenous hyperparathyroidism or parathyroid cancer, there is no evidence of an increased risk of osteogenic sarcoma. Furthermore, there has only been one case of osteosarcoma in all of the clinical trials and postmarketing experience, which currently exceeds >600,000 patients.[16] Furthermore, in Neer et al.,[13] new cancer diagnoses occurred in fewer women assigned to TPD (2% versus 4% in placebo), differences that were or approached significance ($p = 0.03$ for 20 μg and 0.07 for 40 μg). Possible side effects of TPD are dizziness and leg cramps, redness and irritation at injection sites, headache, nausea, arthralgias, myalgias, lethargy, and weakness. The higher dose produced more side effects and withdrawals. TPD-induced BMD changes in the trial of Neer et al. were not dependent on patient age, baseline BMD, or prior fracture history,[17] but were related to baseline biochemical bone turnover indices.[18] Furthermore, early PTH-induced changes in bone turnover markers (at 1 and 3 mo) were predictive of ultimate change in spine BMD and bone structure.[18,19]

Two smaller studies have evaluated surrogate endpoints comparing TPD to alendronate.[1,12] In the first,[12] where 146 women were randomized to receive TPD (40 μg/d) versus alendronate (10 mg/d), spine BMD increased 15% in the TPD group versus 6% in the alendronate group after 1 yr. Although there were fewer fractures in the TPD than the alendronate group at the end of 14 mo (3/73 versus 10/73), several of the fractures were minor (toe fractures).

McClung et al.[1] studied 203 postmenopausal women with osteoporosis randomized to receive TPD (20 μg/d) or alendronate for 18 mo. Biochemical turnover increased substantially in the TPD group (formation earlier than resorption) and decreased substantially in the alendronate group (resorption earlier than formation). In TPD-treated women, markers peaked within 6 mo, suggesting developing resistance, as has been seen in other TPD trials.[13,20,21] Spine BMD by DXA increased 10.3% in TPD-treated women versus 5.5% in alendronate-treated women. Volumetric spine BMD by QCT in a subset of women increased 19% in the TPD group versus 3.8% in the alendronate group. Femoral neck BMD by DXA increased similarly in both groups, although by QCT, cortical volumetric femoral neck BMD increased 7.7% in the alendronate group and decreased 1.2% in the TPD group. The spine BMD change correlated with the procollagen type 1 amino-terminal propeptide (P1NP) increment in the TPD group and with the P1NP decrement in the alendronate group ($r = 0.53$ and –0.51, respectively). Clinical fracture incidence was similar in the TPD (nine fractures) and alendronate (eight fractures) groups, but no radiographs were done to evaluate vertebral fractures. Moderate or severe back pain was reported significantly less often in women assigned to TPD versus alendronate (15% versus 33%, $p = 0.003$).

PTH(1-84) as Monotherapy

One study using the full intact PTH(1-84) molecule was performed in 217 women of mean age 64.5 yr.[22] Patients were randomly assigned to receive placebo or one of three different PTH(1-84) doses (50, 75, or 100 μg). There was a dose-dependent increase in spine BMD; however, there was no increase in hip or total body BMD. Further studies involving PTH(1-84) have now been completed. The TOP (Treatment of Osteoporosis) trial was an 18-mo, randomized, double-blind, study of 2532 postmenopausal women with osteoporosis randomized to 100 mg of recombinant PTH(1-84) or placebo by daily subcutaneous injection.[23] To be enrolled, those ≥55

© 2008 American Society for Bone and Mineral Research

needed spine or hip BMD ≤ –2.5, or ≤ –2 with a prevalent vertebral fracture. Those 45–54 yr of age required a T-score ≤ –3 or ≤ –2.5 with a fracture. Mean age was 64 yr, and 19% of subjects had a prevalent vertebral fracture. Average change in spine BMD was 7% in PTH(1-84)–treated subjects compared with those on placebo. In the per protocol adherent population (n = 1870), new or worsened vertebral fracture incidence was 3.4% in placebo and 1.4% in PTH(1-84) (relative risk reduction 58%), with reductions in both those with and without prevalent vertebral fracture. Nonvertebral fracture incidence was not reduced. The incidence of hypercalcemia was significantly higher in PTH(1-84)–treated women (28.3% versus 4.5% in placebo).[23] PTH(1-84) therapy is currently available in Europe. There have been no head to head trials comparing PTH(1-84) with TPD.

PTH and Antiresorptive Combination/Sequential Therapy

The rationale supporting the concept that PTH and antiresorptive agents could produce synergistic or at least additive benefits on BMD and on bone strength is as follows: PTH stimulates bone growth, restores microarchitecture, and seems to expand bone size, whereas antiresorptives decrease stress risers and cortical porosity and increase bone mineralization. Despite this, studies on combination therapy in previously untreated women and men have produced results suggesting no clear benefit to starting PTH(1-84) and alendronate together[24] or starting TPD after a brief course of alendronate.[25,26] Data also suggest that prior long-term alendronate treatment might blunt (but by no means eliminate) the magnitude of BMD accrual induced by TPD.[27–29] These clinical situations (previously untreated versus long-term bisphosphonate-treated individuals) must be distinguished when reviewing the clinical trial data. Furthermore, in patients on prior antiresorptive agents, emerging evidence suggests that there are differences in biochemical and BMD responses when patients are switched to TPD compared with when TPD is added to ongoing antiresorptive therapy.[30]

Combination Therapy in Treatment Naïve Women: PTH(1-84) and Alendronate

Black et al.[24] examined the potential additive or synergistic effect of co-administering PTH(1-84) with alendronate versus each agent alone in 238 previously treatment-naïve patients. BMD of the anteroposterior (AP) spine by DXA increased similarly in the PTH(1-84) alone and combination groups (6.3% and 6.1%, respectively). Total hip BMD (by DXA) increased in the combination group (1.9%) but not with PTH(1-84) alone (0.3%). Radial BMD decreased more in PTH(1-84) alone (3.4%) than in the combination group (1.1%). QCT measured increase in the lumbar spine and total hip were similar between the PTH(1-84) and combination groups, but trabecular spine BMD increased more with PTH(1-84) alone (25.5%) than with the combination (12.6%). In contrast, QCT-assessed cortical BMD decreased in the hip (1.7%) with PTH(1-84) alone but was unchanged in the combination group. Cortical volume of the femoral neck of the hip (but not the total hip) increased significantly in PTH(1-84)–treated versus combination-treated women. These data provide no clear evidence of synergistic or additive effects when PTH(1-84) is combined with alendronate versus PTH(1-84) alone in previously untreated women, although there was no evidence of blunting of PTH(1-84) effect by DXA. Evidence of a blunted effect was only apparent by QCT. Which of these outcomes (DXA versus QCT, trabecular versus whole bone) assessed in this trial is the most highly predictive of bone strength is unknown. There were only a small number of fractures in this study, and no differences among groups were seen. Incident vertebral fractures were not reported.

Preliminary data from a study in women pretreated with alendronate for a 6-mo period also suggested that spine BMD gain by DXA, and even more impressively by QCT, was lower in those given TPD with continued alendronate compared with those given TPD alone.[25] Based on this and the trial outlined above, at the current time, there is insufficient evidence to recommend using PTH with a bisphosphonate in treatment naïve patients.

Combination Therapy in Treatment Naïve Women: TPD and Raloxifene

Deal et al.[31] randomized 137 postmenopausal treatment-naive women to receive TPD or TPD plus raloxifene. The bone formation marker P1NP rose similarly in the two groups, whereas the bone resorption marker carboxyterminal telopeptide of collagen 1 (CTX) increased more in the TPD plus raloxifene versus TPD alone group. Spine BMD increments were similar in the two groups, whereas hip BMD increased more in the TPD plus raloxifene group.

PTH Therapy in Women on Established Bisphosphonate or Raloxifene Treatment

Patients maintained and stabilized on long-term antiresorptive treatment are a distinct, but clinically very important, population, because many of these patients have fractures or do not achieve a BMD above osteoporotic range and thus might benefit from anabolic therapy. Two studies have evaluated BMD and biochemical outcomes in women pretreated with long-term alendronate.[28,29] Cosman et al.[28] randomized 126 women, average age 68 yr, who had been on alendronate for at least 1 yr (average, 3.2 yr), to continue alendronate and to receive daily TPD, cyclic TPD (given in a 3 mo on/3 mo off regimen), or alendronate alone. TPD stimulated increments in bone formation rapidly and resorption markers more slowly, with percentage changes for bone formation substantially higher than for bone resorption. This difference was magnified in the cyclic group. Spine BMD rose 6.1% in the daily TPD group and 5.4% in the cyclic TPD group ($p < 0.001$ for each TPD group, no group difference), whereas BMD was unchanged in the alendronate alone group. Although not powered for fracture outcome, a trend toward decreased vertebral fracture occurrence was seen in the TPD groups (one in the daily group, two in the cyclic group, and four in the alendronate alone group had new or worsening vertebral deformities on radiograph, p = 0.2 group difference). Cyclic TPD administration might take advantage of the early phase of TPD action, characterized by more pure stimulation of bone formation, and avoid the latter phase of TPD action, characterized by stimulation of both formation and resorption. The clinical use of this approach deserves further study. Moreover, this study clearly showed that, in patients on long-term alendronate, TPD can increase spine BMD substantially.

In a separate study, Cosman et al.[32] evaluated postmenopausal women on raloxifene for at least 1 yr (n = 42) with persistent osteoporosis and randomized them to stay on raloxifene alone or to receive raloxifene plus TPD. The TPD plus raloxifene group had an increment of ~10% in the lumbar spine and 3% in the total hip, whereas those randomized to the raloxifene alone group had no BMD change. Increases in both biochemical turnover markers at 3 mo correlated with increases in spine BMD at 1 yr.

© 2008 American Society for Bone and Mineral Research

In an observational study where TPD was given to women after cessation of long-term alendronate or raloxifene,[29] bone turnover markers increased, as did spine BMD, but these increases were somewhat delayed and of lower magnitude in patients pretreated with alendronate compared with those in patients pretreated with raloxifene. A transient reduction in hip BMD was seen at 6 mo in the group previously on alendronate, but this reversed in the latter 12 mo of administration. In contrast to the protocol above, where alendronate was continued during TPD treatment,[28] in this study,[29] antiresorptive agents were discontinued when TPD was initiated.

A recent study has now assessed the impact of continuing the antiresorptive agent while treating with TPD compared with stopping the antiresorptive agent and switching to TPD in women treated with prior antiresorptive agents for at least 1 yr.[30] Patients on either alendronate or raloxifene were separately randomized to continue or stop their antiresorptive when TPD was initiated. Although an anabolic response was seen both biochemically and densitometrically in all groups of patients, 6-mo results show biochemical turnover markers increasing more in those randomized to the switch design and BMD increasing more in those assigned to the add design. The group where prior alendronate was stopped showed slight hip BMD loss at 6 mo, whereas the group where alendronate was discontinued did not show this loss. Final 18-mo results from this study are pending.

In an observational study of women previously treated with risedronate ($n = 146$) or alendronate ($n = 146$) for >2 yr and then switched to TPD for 12 mo (OPTAMISE), women pretreated with risedronate showed a greater biochemical and BMD response at the spine and hip and trabecular spine by QCT compared with those previously on alendronate.[33] Biochemical responses showed increments in bone resorption already within 1 mo in both groups of patients, an outcome not seen within the first month in treatment-naïve patients treated with TPD.[1,13] Furthermore, increases in bone resorption at 1 mo are not seen in patients on prior antiresorptive therapy when the antiresorptive agent is continued during administration of TPD.[20,28,34]

These results suggest that not all combinations of anabolic and antiresorptive therapies should be avoided, particularly after alendronate and raloxifene. Furthermore, it is unclear to what degree prior bisphosphonates blunt the anabolic response to TPD and how blunting should best be assessed (biochemically versus densitometrically). Whether the frequency of bisphosphonate administration plays a role in modifying the response to TPD and whether intravenous bisphosphonates, especially those administered once yearly, will produce different responses to TPD administration is unknown.

PTH and Hormone Therapy

In 52 women with osteoporosis, average age 60 yr, treated with hormone therapy (HT),[20,34] daily TPD produced rapid increases in markers of bone formation and delayed increases in markers of bone resorption.[34] This period of time, where augmentation of bone formation exceeds stimulation of bone resorption, has been referred to as the anabolic window and may represent the most efficient bone building opportunity with TPD. Furthermore, bone turnover levels remained elevated for only 18–24 mo, after which marker levels decreased.[20] The mechanism of this apparent resistance to TPD has still not been determined. BMD increased by ~14% over 3 yr in women receiving TPD + HT, with evidence of the most rapid rise in BMD within the first 6 mo. Total body and hip BMD increased by 4% in patients on TPD + HT. Although the study was not powered to assess fracture occurrence, after 3 yr of treatment, vertebral deformity occurrence was significantly reduced in patients receiving TPD + HT compared with HT alone.[20]

Another study of similar design performed in women who had previously been treated with HT showed BMD increments by DXA in the TPD group of 30% in the lumbar spine and 12% in the femoral neck versus placebo.[35] No fracture data were presented from this trial, and the data have never been published in a peer-reviewed journal. A third study was performed in 247 women, where one subgroup had been on prior HT (as in the previously discussed two trials), and a second subgroup consisted of treatment-naïve women about to receive HT for the first time.[36] In the former group, there were BMD increments of ~11% in the spine and 3% in the total hip in women randomized to teriparatide (40 μg/d). In the women receiving de novo HT, there were increases caused by HT itself (4% in the spine and 2% in the total hip) and larger increases in the group receiving HT with TPD (16% in the spine and 6% in the hip). The increases from TPD seemed additive to those of HT, although not synergistic.

PTH TREATMENT OF MEN

In a small study, men with idiopathic osteoporosis were randomized to receive TPD or placebo.[21] Biochemical markers of bone turnover increased rapidly with TPD administration and spine BMD rose ~12%, with a plateau between 12 and 18 mo. In the femoral neck and total hip, BMD increased 5% and 4%, respectively, and radius BMD did not change significantly.

A subsequent multicenter trial of TPD[37] was performed in 437 men (mean age, 49 yr) with primary idiopathic or hypogonadal osteoporosis. Subjects were randomized to TPD 20 or 40 μg daily or placebo. After ~1 yr, spine BMD rose 5.4% and 8.5% in the 20 and 40 μg groups, respectively, with no change in the placebo group. There were also dose-dependent increases in BMD at hip sites and total body. Of the original enrollees, 355 men participated in an observational follow-up study. Lateral spine radiographs repeated after ~18 mo of follow-up (including use of antiresorptive therapy in a substantial proportion of the men) showed a 50% reduction in vertebral fracture risk in those men initially assigned to TPD compared with those who had received placebo ($p = 0.07$).[38]

In a third study, 83 men with osteoporosis were assigned to TPD at 40 μg/d, alendronate alone, or TPD after 6 mo of alendronate pretreatment (with ongoing alendronate).[26] A substantial proportion of men in both TPD groups required dose adjustment (by 25–50%) because of hypercalcemia or side effects. After a total of 24 mo of TPD administration, spine BMD increased most in the TPD alone group (18.1%) compared with that in the combination group (14.8%) or alendronate alone (7.9%). Similar trends were seen for the lateral spine and femoral neck, but for the total hip and total body, increases were similar in the three treatment groups. In contrast, in the radius, BMD decreased in the TPD alone group with slight increases in the other groups. Spine trabecular BMD on QCT increased 48% with TPD alone, 17% with the combination, and 3% with alendronate alone.

PTH IN SPECIAL POPULATIONS

Glucocorticoid-Treated Patients

PTH could conceivably be a preferred treatment for glucocorticoid osteoporosis, because some of the major pathophysiologic skeletal problems with glucocorticoid administration are reduced osteoblast function and lifespan, both of which might be counteracted by PTH. Women with a variety of rheumatologic conditions on glucocorticoids and being treated with HT

© 2008 American Society for Bone and Mineral Research

were randomized to TPD + HT or continued HT alone.[39] TPD resulted in a 12% increase in spine BMD by DXA and a smaller increase in femoral neck BMD. No fracture results were reported.

In an 18-mo active comparator trial of TPD versus alendronate for the treatment of glucocorticoid-induced osteoporosis, patients treated with TPD had BMD increases of 7.2% at the spine and 3.8% at the total hip, both significantly greater than the changes of 3.4% at the spine and 2.4% at the hip seen with alendronate therapy.[40] Furthermore, fewer new vertebral fractures occurred in the teriparatide group compared with the alendronate group (0.6% versus 6.1%, $p = 0.004$). There were no differences in nonvertebral fractures between the groups.

Premenopausal Women

In premenopausal women with endometriosis being treated with a gonadotropin-releasing hormone (GnRH) analog to induce acute estrogen deficiency,[41] women randomized to TPD maintained bone mass in the hip and increased bone mass in the lumbar spine, especially in the lateral spine, compared with bone loss at all sites in women receiving placebo over a 6- to 12-mo treatment period.

PERSISTENCE OF EFFECT

A series of observational studies suggested that BMD is lost in individuals who do not take antiresorptive agents after cessation of TPD or PTH(1-84), whereas antiresorptive therapy can maintain PTH-induced gains or even provide further increments in BMD after a course of PTH.[20,38,39,42–44] Black et al.[45] have now provided clinical trial confirmation of this observation. Subjects originally randomized to 1 yr of treatment with PTH(1-84) were subsequently randomized to receive alendronate or placebo for an additional year. Over 2 yr, women who received alendronate after PTH(1-84) had significant increases in spine BMD of 12.1% compared with 4.1% in women in the PTH(1-84) followed by placebo. Trabecular BMD at the spine assessed by QCT showed a 31% increase in women on PTH(1-84) followed by alendronate compared with 14% in those assigned to PTH(1-84) followed by placebo. BMD at the femoral neck and total hip was increased above baseline in all groups except those receiving PTH, followed by placebo. These data suggest that, after 1 yr of PTH treatment, the gains in bone mass are preserved or further improved with alendronate but lost in patients not on antiresorptive therapy.

RECHALLENGE WITH PTH

Women originally randomized to daily or cyclic TPD in addition to ongoing alendronate were followed for a year after TPD was discontinued.[46] BMD remained stable in these women during this year. A second 15-mo course of TPD was given to those volunteers who still had osteoporosis. The rechallenge with TPD produced similar biochemical and BMD changes to those seen during the first course of therapy.[46]

CONCLUSION

PTH is a unique approach to osteoporosis treatment. Because of the underlying effects it produces on the microarchitecture, macroarchitecture, and mass of bone, PTH may be able to ensure more long-term protection against fracture occurrence than antiresorptive agents alone; however, data proving this principle are lacking. Antiresorptive agents are clearly needed after PTH to maintain PTH-induced gains. There are still many unanswered questions concerning PTH therapy, including the optimal duration and regimen of therapy and the mechanism underlying resistance to PTH effect after 18 mo. Different PTH peptides and alternative forms of delivery (oral, nasal, inhaled, transdermal) are currently under study.

REFERENCES

1. McClung MR, San Martin J, Miller PD, Civitelli R, Bandiera F, Omizo M, Donley DW, Dalsky GP, Eriksen EF 2005 Opposite bone remodeling effects of teriparatide and alendronate in increasing bone mass. Arch Intern Med **165:**1762–1768.
2. Arlot M, Meunier PJ, Boivin G, Haddock L, Tamayo J, Correa-Rotter R, Jasqui S, Donley DW, Dalsky GP, Martin JS, Eriksen EF 2005 Differential effects of teriparatide and alendronate on bone remodeling in postmenopausal women assessed by histomorphometric parameters. J Bone Miner Res **20:**1244–1253.
3. Lindsay R, Cosman F, Zhou H, Bostrom MP, Shen V, Cruz JD, Nieves JW, Dempster DW 2006 A novel tetracycline labeling schedule for longitudinal evaluation of the short-term effects of anabolic therapy with a single iliac crest bone biopsy: Early actions of teriparatide. J Bone Miner Res **21:**366–373.
4. Jiang Y, Zhao JJ, Mitlak BH, Wang O, Genant HK, Eriksen EF 2003 Recombinant human parathyroid hormone (1-34) [teriparatide] improves both cortical and cancellous bone structure. J Bone Miner Res **18:**1932–1941.
5. Dempster DW, Cosman F, Kurland ES, Zhou H, Nieves JW, Woelfert L, Shane E, Plavetic K, Muller R, Bilezikian JP, Lindsay R 2001 Effects of daily treatment with parathyroid hormone on bone microarchitecture and turnover in patients with osteoporosis: A paired biopsy study. J Bone Miner Res **16:**1846–1853.
6. Parfitt AM 2002 PTH and periosteal bone expansion. J Bone Miner Res **17:**1741–1743.
7. Burr D 2005 Does early PTH treatment compromise bone strength? The balance between remodeling, porosity, bone mineral, and bone size. Curr Osteoporo Rep **3:**19–24.
8. Lindsay R, Zhou H, Cosman F, Nieves JW, Dempster DW, Hodsman AB 2007 Effects of a one-month treatment with parathyroid hormone (1-34) on bone formation on cancellous, endocortical and periosteal surfaces of the human ilium. J Bone Miner Res **22:**495–502.
9. Rehman Q, Lang TF, Arnaud CD, Modin GW, Lane NE 2003 Daily treatment with parathyroid hormone is associated with an increase in vertebral cross-sectional area in postmenopausal women with glucocorticoid-induced osteoporosis. Osteoporos Int **14:**77–81.
10. Zanchetta JR, Bogado C, Ferretti JL, Wang O, Wilson MG, Sato M, Gaich GA, Dalsky GP, Myers SL 2003 Effects of teriparatide [recombinant parathyroid hormone (1–34)] on cortical bone in postmenopausal women with osteoporosis. J Bone Miner Res **18:**539–543.
11. Uusi-Rasi K, Semanick LM, Zanchetta JR, Bogado CE, Eriksen EF, Sato M, Beck TJ 2005 Effects of teriparatide [rhPTH (1–34)] treatment on structural geometry of the proximal femur in elderly osteoporotic women. Bone **36:**948–958.
12. Body J-J, Gaich GA, Scheele WH, Kulkami PM, Miller PD, Peretz A, Dore RK, Correa-Rotter R, Papaoiannou A, Cumming DC, Hodsman AB 2002 A randomized double-blind trial to compare the efficacy of teriparatide [recombinant human parathyroid hormone (1–34)] with alendronate in postmenopausal women with osteoporosis. J Clin Endocrinol Metab **87:**4528–4535.
13. Neer RM, Arnaud CD, Zanchetta JR, Prince R, Gaich GA, Reginster JY, Hodsman AB, Eriksen EF, Ish-Shalom S, Genant HK, Wang O, Mitlak BH 2001 Effect of parathyroid hormone (1–34) on fractures and bone mineral density in postmenopausal women with osteoporosis. N Engl J Med **344:**1434–1441.
14. Vahle JL, Sato M, Long GG, Young JK, Francis PC, Engelhardt JA, Westmore MS, Linda Y, Nold JB 2002 Skeletal changes in rats given daily subcutaneous injections of rhPTH(1-34) for 2 years and relevance to human safety. Toxicol Pathol **30:**312–321.
15. Vahle JL, Long GG, Sandusky G, Westmore M, Ma YL, Sato M 2004 Bone neoplasms in F344 rats given teriparatide [rhPTH (1-34)] are dependent on duration of treatment and dose. Toxicol Pathol **32:**426–438.
16. Harper KD, Krege JH, Marcus R, Mitlak BH 2007 Osteosarcoma and teriparatide? J Bone Miner Res **22:**334.
17. Marcus R, Wang O, Satterwhite J, Mitlak B 2003 The skeletal response to teriparatide is largely independent of age, initial bone mineral density, and prevalent vertebral fractures in postmenopausal women with osteoporosis. J Bone Miner Res **18:**18–23.

© 2008 American Society for Bone and Mineral Research

18. Chen P, Satterwhite JH, Licata AA, Lewiecki EM, Sipos AS, Misurski DM, Qwagman RB 2005 Early changes in biochemical markers of bone formation predict BMD response to teriparatide in postmenopausal women with osteoporosis. J Bone Miner Res **20:**962–970.
19. Dobnig H, Sipos A, Jiang Y, Fahrleitner-Pammer A, Ste-Marie LG, Gallagher JC, Pavo I, Wang J, Eriksen EF 2005 Early changes in biochemical markers of bone formation correlate with improvements in bone structure during teriparatide therapy. J Clin Endocrinol Metab **90:**3970–3977.
20. Cosman F, Nieves J, Woelfert L, Formica C, Gordon S, Shen V, Lindsay R 2001 Parathyroid hormone added to established hormone therapy: Effects on vertebral fracture and maintenance of bone mass after parathyroid hormone withdrawal. J Bone Miner Res **16:**925–931.
21. Kurland ES, Cosman F, McMahon DJ, Rosen CJ, Lindsay R, Bilezikian JP 2000 Parathyroid hormone as a therapy for idiopathic osteoporosis in men: Effects on bone mineral density and bone markers. J Clin Endocrinol Metab **85:**3069–3076.
22. Hodsman AB, Hanley DA, Ettinger MP, Bolognese MA, Fox J, Metcalfe AJ, Lindsay R 2003 Efficacy and safety of human parathyroid hormone-(1-84) in increasing bone mineral density in postmenopausal osteoporosis. J Clin Endocrinol Metab **88:**5212–5220.
23. Greenspan SL, Bone HG, Ettinger MP, Hanley DA, Lindsay R, Zanchetta JR, Blosch CM, Mathisen AL, Morris SA, Marriott TB 2007 Effect of recombinant human parathyroid hormone (1-84) on vertebral fracture and bone mineral density in postmenopausal women with osteoporosis. Ann Intern Med **146:**326–339.
24. Black DM, Greenspan S, Ensrud KE, Palermo L, McGowan JA, Lang TF, Garnero P, Bouxsein ML, Bilezikian JP, Rosen CJ 2003 The effects of parathyroid hormone and alendronate alone or in combination in postmenopausal osteoporosis. N Engl J Med **349:**1207–1215.
25. Neer RM, Hayes A, Rao A, Finkelstein J. 2002 Effects of parathyroid hormone, alendronate, or both on bone density in osteoporotic postmenopausal women. J Bone Miner Res **17:**S1;S039–S135.
26. Finkelstein JS, Hayes A, Hunzelman JL, Wyland JJ, Neer RM 2003 The effects of parathyroid hormone, alendronate, or both in men with osteoporosis. N Engl J Med **349:**1216–1226.
27. Cosman F, Nieves J, Woelfert L, Shen V, Lindsay R 1998 Alendronate does not block the anabolic effect of PTH in postmenopausal osteoporotic women. J Bone Miner Res **13:**1051–1055.
28. Cosman F, Nieves J, Zion M, Woelfert L, Luckey M, Lindsay R 2005 Daily and Cyclic parathyroid hormone in women receiving alendronate. N Engl J Med **353:**566–575.
29. Ettinger B, San Martin J, Crans G, Pavo I 2004 Differential effects of teriparatide on BMD after treatment with raloxifene or alendronate. J Bone Miner Res **19:**745–751.
30. Cosman F, Wermers RA, Recknor C, Mauck KF, Xie L, Glass EV, Krege JH. 2007 Efficacy of adding teriparatide versus switching to teriparatide in postmenopausal women with osteoporosis previously treated with raloxifene or alendronate. J Bone Miner Res **22:**S1;O423.
31. Deal C, Omizo M, Schwartz EN, Eriksen EF, Cantor P, Wang J, Glass EV, Myers SL, Krege JH 2005 Combination teriparatide and raloxifene therapy for postmenopausal osteoporosis: Results from a 6-month double-blind placebo-controlled trial. J Bone Miner Res **20:**1905–1911.
32. Cosman F, Nieves J, Zion M, Barbuto N, Lindsay R 2008 Effect of prior and ongoing raloxifene therapy on response to PTH and maintenance of BMD after PTH therapy. Osteoporosis Int **19:**529–535.
33. Miller P, Watts N, Lindsay R, Meeves S, Lang T, Delmas PD, Bilezikian JP 2007 Patients previously treated with risedronate demonstrate greater responsiveness to teriparatide than those previously with alendronate: The OPTAMISE study. J Bone Miner Res **22:**S1;S26.
34. Lindsay R, Nieves J, Formica C, Henneman E, Woelfert L, Shen V, Dempster D, Cosman F 1997 Randomised controlled study of effect of parathyroid hormone on vertebral-bone mass and fracture incidence among postmenopausal women on oestrogen with osteoporosis. Lancet **350:**550–555.
35. Roe EB, Sanchez SD, del Puerto GA, Pierini E, Bacchetti P, Cann CE, Arnaud CD 1999 Parathyroid hormone 1–34 (hPTH 1–34) and estrogen produce dramatic bone density increases in postmenopausal osteoporosis- results from a placebo-controlled randomized trial. J Bone Miner Res **12:**S1;S137.
36. Ste-Marie LG, Schwartz SL, Hossain A, Desaiah D, Gaich GA 2006 Effect of teriparatide [rh PTH(1-34)] on BMD when given to postmenopausal women receiving hormone replacement therapy. J Bone Miner Res **21:**283–291.
37. Orwoll ES, Scheele WH, Paul S, Adami S, Syresen V, Diez-Perez A, Kaufman JM, Clancy AD, Gaich GA 2003 The effects of teriparatide [human parathyroid hormone (1–34)] therapy on bone density in men with osteoporosis. J Bone Miner Res **18:**9–17.
38. Kaufman JM, Orwoll E, Goemaere S, San Martin J, Hossain A, Dalsky GP, Lindsay R, Mitlak BH 2005 Teriparatide effects on vertebral fractures and bone mineral density in men with osteoporosis: Treatment and discontinuation of therapy. Osteoporos Int **16:**510–516.
39. Lane NE, Sanchez S, Modin GW, Genant HK, Pierini E, Arnaud CD 2000 Bone mass continues to increase at the hip after parathyroid hormone treatment is discontinued in glucocorticoid-induced osteoporosis: Results of a randomized controlled clinical trial. J Bone Miner Res **15:**944–951.
40. Saag KS, Shane E, Boonen S, Marin F, Donley DW, Taylor KA, Dalsky GP, Marcus R 2007 Teriparatide or alendronate in glucocorticoid-induced osteoporosis. N Engl J Med **357:**2028–2039.
41. Finkelstein JS, Klibanski A, Arnold AL, Toth TL, Hornstein MD, Neer RM 1998 Prevention of estrogen deficiency-related bone loss with human parathyroid hormone-(1–34): A randomized, controlled trial. JAMA **280:**1067–1073.
42. Lindsay R, Scheele WH, Neer R, Pohl G, Adami S, Mautlaen C, Reginster JY, Stepan JJ, Myers SL, Mitlak BH 2004 Sustained vertebral fracture risk reduction after withdrawal of teriparatide in postmenopausal women with osteoporosis. Arch Intern Med **164:**2024–2030.
43. Kurland ES, Heller SL, Diamond B, McMahon DJ, Cosman F, Bilezikian JP 2004 The importance of bisphosphonate therapy in maintaining bone mass in men after therapy with teriparatide. Osteoporos Int **15:**992–997.
44. Rittmaster RS, Bolognese M, Ettinger MP, Hanley DA, Hodsman AB, Kendler DL, Rosen CJ 2000 Enhancement of bone mass in osteoporotic women with parathyroid hormone followed by alendronate. J Clin Endocrinol Metab **85:**2129–2134.
45. Black DM, Bilezikian JP, Ensrud KE, Greenspan SL, Palermo L, Hue T, Lang TF, McGowan JA, Rosen CJ 2005 One year of alendronate after one year of parathyroid hormone (1-84) for osteoporosis. N Engl J Med **353:**555–565.
46. Cosman F, Nieves J, Zion M, Barbuto N, Lindsay R 2005 Effects of PTH(1-34) rechallenge 1 year after the first PTH course in patients on long-term alendronate. J Bone Miner Res **20:**S1;S21.

© 2008 American Society for Bone and Mineral Research

Chapter 52. Calcitonin

Silvano Adami

Rheumatology Unit, University of Verona, Verona, Italy

INTRODUCTION

Calcitonin (CT)is a 32 amino acid peptide secreted by the C cells of the thyroid gland in response to an increase in serum calcium. CT interacts with a specific G-protein–coupled receptor expressed on the osteoclast surfaces on their final stages of differentiation. This interaction is followed by flattening of the ruffled borders and withdrawal from sites of active bone resorption, thus causing inhibition of bone resorption. In vitro, continuous exposure to CT is associated with loss of the inhibitory effect of CT on osteoclasts, called "escape," but the pharmacological relevance of this observation is uncertain.[1,2] CT also has receptors on renal tubuli where it causes natriuresis through an effect on the Na/H exchanger, which causes an increase in calcium excretion.[3]

The exact physiological role of CT in calcium homeostasis and skeletal metabolism in humans is not known, because, in patients who underwent total thyroidectomy and had undetectable serum levels of the hormone, specific symptoms have never been reported.

CT as a therapeutic agent was introduced for the first time in the early 1970s for the treatment of Paget's disease of bone and then for hypercalcemic conditions. It was registered for the treatment of postmenopausal osteoporosis in the 1980s when the synthetic hormone, cheaper and purer than the extractable one, became available. Even though both human and eel CT have been also developed and commercialized, salmon CT has been by far the most extensively studied because of its greater potency.

EFFECTS OF CALCITONIN ON POSTMENOPAUSAL OSTEOPOROSIS

The first trials for the treatment of postmenopausal osteoporosis were carried out with subcutaneous CT, most often 100 IU/d. With today's standards, these studies would be considered at best exploratory, but the lack of any alternative to hormone replacement therapy convinced the U.S. Food and Drug Administration to approve the use of injectable CT for the treatment of osteoporosis in 1984 and the nasal spray formulation a few years later.

The poor tolerability and compliance to the subcutaneous formulation led to the development of nasal spray salmon CT. The bioavailability of this formulation has never been ascertained in a convincing manner. In early studies, adjuvant in the form of bile salts was added to promote the mucosal absorption, but these were removed after further development because of their irritating action.

One phase 2 study[4] and 10 small clinical pilot studies were inconclusive regarding the ideal dose to be tested in a phase 3 study. Thus, three daily doses (100, 200, and 400 IU) were tested in the pivotal trial PROOF[5] over 5 yr, including 1255 osteoporotic women, of whom 78% had prevalent vertebral fractures.

None of the doses were associated with significant changes in femoral neck BMD, and the values were not reported. Lumbar spine BMD rose significantly compared with placebo in the three doses. The results relative to placebo are summarized in Table 1.

Dr. Adami has been on Speakers' Bureaus for Eli Lilly and Roche/GlaxoSmithKline and has been on the Advisory Boards for Amgen, Eli Lilly, Merck, Novartis, Proctor & Gamble, and Wyeth.

Despite the minor changes in lumbar spine BMD, the PROOF study showed vertebral fracture efficacy of nasal CT after 5 yr of treatment. The relative risk (RR) of developing a new vertebral fracture in the 200-IU group compared with placebo was 0.67 (95% CI: 0.47 to –0.97); however, the RR reduction was not significant for the 100- and 400-IU groups. The proportion of completers in this study was rather poor, ranging from 32% to 41%. This justified a posthoc analysis among participants who completed the first 3 yr of treatment. The results were similar to those observed in the main analysis (Table 1), with a similar significant reduction in vertebral fracture risk in the 200-IU group.

Some decreases in nonvertebral fracture rate were observed in the three active arms, and the changes were statistically significant for the 100-IU group.

The main limitation of this study remains the high discontinuation rate, which was not uniform across the investigational centers. Also unexplained is the lack of dose response for reduction of fracture risk, even though a dose–response relationship was appreciable for both BMD and serum C-telopeptide (CTX) changes. It was postulated that CT therapy may be associated with improvements in bone microstructure that are not detected by BMD. To address this problem, a 2-yr randomized placebo-controlled study with 200 U of nasal CT including ~30 patients per group was carried out.[6] In this study, no significant changes in BMD as measured by DXA were observed. However, in women on nasal CT, trabecular microarchitecture as determined by high-resolution MRI exhibited a preservation of trabecular number and spacing and bone volume/total volume compared with the significant worsening observed in the placebo group. Combined μCT/histomorphometric analysis of iliac crest bone biopsies did not show significant differences between treatment and placebo groups. The results of this study suggested that even a minor suppression of bone turnover as that obtained with CT therapy may lead to a complete block of the age-related deterioration of the microarchitecture of trabecular bone. However, it also seems likely that to obtain improvements in the cortical bony structure, a stronger inhibition of bone turnover is warranted.

EFFECTS OF CALCITONIN ON OTHER TYPES OF OSTEOPOROSIS

Nasal spray CT has been tested in men with idiopathic osteoporosis in only one small study of 1 yr. The large BMD changes observed at the spine BMD (+7.1%) with no changes at the hip need to be confirmed.[7]

CT has been tested in a number of other forms of osteoporosis secondary to glucocorticoid administration, prolonged immobilization, rheumatoid arthritis, multiple myeloma, orchidectomy, and algodystrophy.[8] The doses, either subcutaneously or by nasal spray, were variable. Unfortunately, none of these small studies met the modern scientific standards, and their design was almost invariably rather poor.

CALCITONIN THERAPY AND PAIN

In a number of studies, an analgesic effect with both injectable and nasal CT has been reported in a variety of clinical models.[9] A rapid resolution of symptoms has been reported in patients with recent vertebral fractures. The main limitations of these studies were the level of blindness of the scien-

© 2008 American Society for Bone and Mineral Research

Table 1. Lumbar Spine BMD Changes Relative to Placebo (as Derived From a Figure From Ref. 5) and RR Reduction (RRR) of Fracture (FX) Risk According to Daily Doses of Nasal Spray Salmon Calcitonin

Spray calcitonin dose	Year 1	Year 3		Year 5		
	BMD changes	*BMD changes*	*Percent RRR vertebral FX*	*BMD changes*	*Percent RRR vertebral FX*	*Percent RRR nonvertebral FX*
100 IU/d	1.3%*	0.6%	9	0.6%	15	36*
200 IU/d	1.0%*	0.6%	34*	0.7%	33	12
400 IU/d	1.0%*	1.1%*	29	1.1%	16	19

* Statistically significant at $p > 0.05$ or less.

tists caused by the typical side effects of CT particularly when given parenterally, the comparability of the small groups for the expected wide variability of clinical symptoms, and the inclusion of patients with moderate clinical symptoms that tend to attenuate spontaneously with time.

Reduction of pain has been also reported in patients with chronic pain, such as the pain associated with spinal deformity caused by previous vertebral fractures, Paget's disease, and bony metastasis.[8] In these latter cases, analgesia may have involved a decrease in bone resorption activity.

However, positive results have been also reported in patients with migraine, phantom limb pain syndrome, and lumbar spinal stenosis,[8] conditions that are not associated with metabolic abnormalities of bone remodeling. This suggested that the analgesic action of CT is not mediated by its effect on bone metabolism. Several hypotheses have been put forward including CT binding on the central nervous system, stimulation of β-endorphin synthesis, and an effect on cyclo-oxygenase activity. The results have been often inconsistent or even contradictory, and the mechanisms by which CT may induce pain relief remain obscure.

CLINICAL USE IN OSTEOPOROSIS

CT is indicated in several countries for the treatment of osteoporosis in women at least 5 yr after menopause. After the introduction of other drugs with more solid and consistent antifracture efficacy, its use is most often limited to patients who cannot tolerate the new agents or with less severe forms of osteoporosis. Nasal CT administration may be associated with rhinitis and parenteral CT with facial flushing and/or nausea, but the safety profile is excellent. This remains the main reason for its maintenance in the armamentarium for the management of postmenopausal osteoporosis and even for further development. Thus, in the near future, an oral formulation of salmon CT may became available at the completion of a pivotal phase 3 trial, which followed a phase 2 study reporting that 1 mg/d of oral CT is associated with changes in bone markers somewhat greater than those seen with 200 U nasal spray CT.[10]

REFERENCES

1. Azria M, Copp SH, Zanelli JM 1995 25 years of salmon calcitonin: From synthesis to therapeutic use. Calcif Tissue Int **57:**405–408.
2. Hoff AO, Cote GJ, Gagel RF 2002 Calcitonin. In: Marcus R, Feldman D, Kelsey J (eds.) Osteoporosis. Academic Press, San Diego, CA, USA, pp. 247–255.
3. Chakraborty M, Chatterjee D, Gorelick FS, Baron R 1994 Cell cycle-dependent and kinase –specific regulation of the apical Na/H exchanger and the Na,K-ATPase in the kidney cell line LLC-PK1 by Calcitonin. Proc Natl Acad Sci USA **91:**2115–2119.
4. Overgaard K, Hansen MA, Jensen SB, Christiansen C 1992 Effect of calcitonin given intranasally on bone mass and fracture rates in established osteoporosis. A dose response study. BMJ **305:**556–561.
5. Chesnut CH, Silverman S, Andriano K, Genant H, Gimona A, Harris S, Kiel D, LeBoff M, Maricic M, Miller P, Moniz C, Peacock M, Richardson P, Watts N, Baylink D, for the PROOF Study Group 2000 A randomized trial of nasal spray calcitonin in postmenopausal women with established osteoporosis. The Prevent Recurrence of Osteoporotic Fractures Study. Am J Med **109:**267–276.
6. Chesnut CH, Majumdar S, Newitt DC, Shields A, Van Pelt J, Laschansky E, Azria M, Kriegman A, Olson M, Erickson EF, Mindeholm L 2005 Effects of salmon calcitonin on trabecular microarchitecture as determined by magnetic resonance imaging:results from the QUEST study. J Bone Miner Res **20:**1548–1561.
7. Trovas GP, Lyritis GP, Galanos A, Raptou P, Constantelou E 2002 A randomized trial of nasal spray calcitonin in men with idiopathic osteoporosis: Effects on bone mineral density and bone markers. J Bone Miner Res **17:**521–527.
8. Civitelli R 2002 Calcitonin for treatmet of osteoporosis. In: Marcus R, Feldman D, Kelsey J (eds.) Osteoporosis. Academic Press, San Diego, CA, USA, pp. 651–673.
9. Silverman SL, Azria M 2002 The analgesic role of calcitonin following osteoporotic fracture. Osteoporos Int **13:**858–867.
10. Tanko LB, Bagger YZ, Alexanderson P, Devogelaer JP, Reginster JY, Chick R, Olson M, Benmammar H, Mindeholm L, Azria M, Christiansen C 2004 Safety and efficacy of a novel salmon calcitonin (sCT) technology based oral formulation in healthy postmenopausal women: Acute and 3 month effects on biomarkers of bone turnover. J Bone Miner Res **19:**1531–1538.

© 2008 American Society for Bone and Mineral Research

Chapter 53. Combination Anabolic and Antiresorptive Therapy for Osteoporosis

John P. Bilezikian

Department of Medicine, College of Physicians and Surgeons, Columbia University, New York, New York

INTRODUCTION

Combination therapy can be viewed in three contexts: antiresorptive therapy followed by anabolic therapy; simultaneous combined antiresorptive and anabolic therapy; and antiresorptive therapy after anabolic therapy.

ANTIRESORPTIVE THERAPY FOLLOWED BY ANABOLIC THERAPY

Many patients who are treated for osteoporosis with anabolic therapy [teriparatide (PTH1-34) in the United States; teriparatide or PTH(1-84) in Europe and other countries] have previously been treated with bisphosphonates or other antiresorptives. This approach can be considered in two respects: "add-on" therapy in which PTH therapy is added onto the antiresorptive therapy that continues or "switch" therapy in which the antiresorptive is stopped when the anabolic is started.

Add-On Therapy

Lindsay et al.[1] treated postmenopausal women, previously given estrogen for at least 1 yr, with teriparatide. Increases in vertebral BMD began with no delay and increased in a linear fashion during the entire 3-yr study. Similar results were obtained by Lane et al.[2] in a study of postmenopausal women with glucocorticoid-induced osteoporosis treated with estrogen followed by teriparatide and estrogen. When Cosman et al.[3] used the bisphosphonate alendronate as the antecedent antiresorptive, the addition of teriparatide was also associated with prompt increases in BMD. The same BMD gains were seen whether teriparatide was used in a 3-mo cyclical fashion or continuously while alendronate was also continued.

Switch Therapy

The study of Ettinger et al.[4] was designed to switch from antiresorptive therapy to teriparatide after 28 mo of therapy with either raloxifene or alendronate. In subjects who had been on raloxifene, the switch to teriparatide was associated with rapid increases in BMD, whereas subjects who had been on alendronate experienced a delay in gains at the lumbar spine and, in fact, a transient reduction of BMD in the hip. The data were interpreted to suggest that the delay after alendronate was caused by its more powerful effects to reduce bone resorption than raloxifene and that the greater potency of alendronate, in this regard, retarded the initial actions of teriparatide to increase bone turnover and, thus, BMD. To support this idea, the response to teriparatide has been shown to be a function of the level of baseline bone turnover in subjects not previously treated with any therapy for osteoporosis: the higher the level of turnover, the more robust the densitometric response to teriparatide.[5] Some support for this hypothesis also comes from a trial in which patients who had previously been treated with either risedronate or alendronate were switched to teriparatide.[6] Individuals who were first treated with risedronate had bone turnover marker levels that were higher than those first treated with alendronate and responded more exuberantly to teriparatide with increases in BMD.

CONCURRENT USE OF ANABOLIC AND ANTIRESORPTIVE THERAPY

It is attractive to consider simultaneous combination therapy with an antiresorptive and PTH as potentially more beneficial than monotherapy with either class of therapeutic given that their mechanisms of action are quite different from each other. If bone resorption is being inhibited (antiresorptive) while bone formation is being stimulated (anabolic), combination therapy might give better results than therapy with either agent alone. Despite the intuitive appeal of this reasoning, important data to the contrary have been provided by Black et al.[7] in the PaTH study and by Finkelstein et al.[8] These two investigators independently conducted trials using a form of PTH alone, alendronate alone, or the combination of PTH and alendronate. Black et al. studied postmenopausal women with 100 μg of PTH(1-84). The study of Finkelstein et al. involved men treated with 40 μg of teriparatide. Both studies used DXA and QCT to measure areal or volumetric BMD, respectively. With either DXA or QCT, monotherapy with PTH exceeded densitometric gains with combination therapy or alendronate alone at the lumbar spine. Measurement of trabecular bone by QCT, in fact, showed that combination therapy was associated with substantially smaller increases in BMD than therapy with PTH alone, equivalent to the changes seen with alendronate alone. Bone turnover markers followed the expected course for anabolic (increases) or antiresorptive (decreases) therapy alone. However, for combination therapy, bone markers followed the course of alendronate, not PTH therapy, with reductions in bone formation and bone resorption markers. This suggests that the impaired response to combination therapy, in comparison with PTH alone, might be caused by the dominating effects of alendronate on bone remodeling dynamics when both drugs are used together.

The results of these two combination therapy studies led to the concept that an antiresorptive agent that did not impair the anabolic actions of teriparatide to increase bone formation while mitigating its affects on bone resorptive might in fact be a means by which combination therapy would be more beneficial than monotherapy. Deal et al.[9] addressed this point by studying the effects of raloxifene as the antiresorptive agent in combination with teriparatide in a "proof of concept" 6-mo clinical trial. As hypothesized, Deal et al.[9] showed that the combination of teriparatide and raloxifene had greater densitometric effects than monotherapy with teriparatide in postmenopausal osteoporosis. As expected, moreover, raloxifene did not impair the actions of teriparatide to stimulate bone formation in contrast to the PaTH study in which alendronate did impair teriparatide's anabolic actions. Bone formation markers increased to the same extent in the teriparatide only group and in the teriparatide and raloxifene groups. In contrast, when raloxifene was present along with teriparatide,

Dr. Bilezikian has served as a consultant for Merck.

Key words: osteoporosis, anabolic therapy, antiresorptive therapy, combination therapy

© 2008 American Society for Bone and Mineral Research

bone resorption markers were significantly lower than the teriparatide-only group. The change in total hip BMD was significantly greater in subjects treated with both teriparatide and raloxifene than with teriparatide alone. The effect of raloxifene, a less potent antiresorptive than alendronate, seems to allow teriparatide to stimulate bone formation, unimpeded, but does impair the ability of teriparatide to stimulate bone resorption.

ANTIRESORPTIVES AFTER ANABOLIC THERAPY

Teriparatide and PTH(1-84) are approved for a treatment period of 18–24 mo. A major question follows about the consequences of discontinuing therapy after this relatively short period of time. Some concerns relate to the fact that new bone matrix is not fully mineralized after PTH therapy. Therefore, this new bone matrix could be at risk for resorption if a period of consolidation with an antiresorptive is not used.[10] Published data addressing this concern were initially based on observational trials. These studies, using either bisphosphonate[11,12] or estrogen[1,13] therapy after PTH, suggested that antiresorptive treatment may be necessary to maintain densitometric gains achieved during PTH administration. With a stronger experimental design, the PaTH study has provided prospective data in a rigorously controlled, blinded fashion to address this issue.[14] Postmenopausal women who had received PTH(1-84) for 12 mo were randomly assigned to an additional 12 mo of therapy with 10 mg of alendronate daily or placebo. In subjects who received alendronate, there was a further 4.9% gain in lumbar spine BMD, whereas those who received placebo experienced a substantial decline. By QCT analysis, the net increase over 24 mo in lumbar spine BMD among those treated with alendronate after PTH(1-84) was 30%. In those who received placebo after PTH(1-84), the net change in BMD was only 13%. There were similar, dramatic differences in hip BMD when those who followed PTH with alendronate were compared with those who were treated with placebo after PTH (13% versus 5%). The results of this study establish the importance of following PTH or teriparatide therapy with an antiresorptive with regard to BMD.

Data on this approach to maintain fracture protection comes only from observational trials. In a 30-mo observational cohort following the pivotal clinical trial of teriparatide,[15] subjects were given the option of switching to a bisphosphonate or not taking any further medications after teriparatide.[16] A majority (60%) were treated with antiresorptive therapy after PTH discontinuation. As noted from the PaTH trial, gains in BMD were maintained in those who chose to begin antiresorptive therapy immediately after teriparatide. Reductions in BMD were progressive throughout the 30-mo observational period in subjects who elected not to follow teriparatide with any therapy. In a group that did not begin antiresorptive therapy until 6 mo after teriparatide discontinuation, major reductions in BMD were seen during these first 6 mo but no further reductions were observed after antiresorptive therapy was instituted.[16] Despite these densitometric data, the effect of previous therapy with teriparatide and/or subsequent therapy with a bisphosphonate on fracture prevention persisted for as long as 31 mo after teriparatide discontinuation. Nonvertebral fragility fractures were reported by proportionately fewer women previously treated with PTH (followed with or without a bisphosphonate) compared with those treated with placebo (with or without a bisphosphonate; $p < 0.03$). In a logistic regression model, bisphosphonate use for 12 mo or longer was said to add little to overall risk reduction of new vertebral fractures in this posttreatment period. However, it is hard to be sure of this conclusion because the data were not actually separately analyzed into those who did or did not follow teriparatide treatment with an antiresorptive. One might anticipate a residual but transient protection against fracture after PTH treatment without follow-up antiresorptive therapy, which could wane over time. Additional studies are needed to address fracture outcomes specifically. However, based particularly on the PaTH trial, the importance of following PTH or teriparatide therapy with an antiresorptive agent to maintain increases in bone mass is clear.

CONCLUSIONS

With the availability of antiresorptives and anabolic therapy for the treatment of osteoporosis, we are beginning to understand how best to use them in combination and in sequence. Further studies are needed to elucidate more clearly differences among the antiresorptives vis a vis the three aspects of combination and sequential therapy summarized in this report. However, the information available at this time gives guidance into ways in which simultaneous or sequential therapy with antiresorptives can be used to maximal advantage.

REFERENCES

1. Lindsay R, Nieves J, Formica C, Henneman E, Woelfert L, Shen V, Dempster D, Cosman F 1997 Randomised controlled study of effect of parathyroid hormone on vertebral-bone mass and fracture incidence among postmenopausal women on oestrogen with osteoporosis. Lancet **23:**550–555.
2. Lane NE, Sanchez S, Modin GW, Genant HK, Pierini E, Arnaud CD 1998 Parathyroid hormone treatment can reverse corticosteroid-induced osteoporosis. Results of a randomized controlled clinical trial. J Clin Invest **102:**1627–1633.
3. Cosman FJ, Nieves M, Zion L, Woelfert M, Luckey M, Lindsay R 2005 Daily and cyclic parathyroid hormone in women receiving alendroante. N Engl J Med **353:**566–575.
4. Ettinger BJ, San Martin G, Crans G, Pavo I 2004 Differential effects of teriparatide on BMD after treatment with raloxifene or alendronate. J Bone Miner Res **19:**745–751.
5. Kurland ES, Cosman F, McMahon DJ, Rosen DJ, Lindsay R, Bilezikian JP 2000 Parathyroid hormone as a therapy for idiopathic osteoporosis in men: Effects on bone mineral density and bone markers. J Clin Endocrinol Metab **85:**3069–3076.
6. Miller P, Lindsay R, Watts N, Meeves S, Lang T, Delmas P, Bilezikian J 2007 Patients previously treated with risedronate demonstrate greater responsiveness to teriparatide than those previously treated with alendronate: The Optamise trial. J Bone Miner Res **22:**S1;S26.
7. Black DM, Greenspan SL, Ensrud KE, Palermo L, McGowan JA, Lang TF, Garnero P, Bouxsein ML, Bilezikian JP, Rosen CJ 2003 The effects of parathyroid hormone and alendronate alone or in combination in postmenopausal osteoporosis. N Engl J Med **349:**1207–1215.
8. Finkelstein JS, Hayes A, Hunzelman JL, Wyland JJ, Lee H, Neer RM 2003 The effects of parathyroid hormone, alendronate, or both in men with osteoporosis. N Engl J Med **349:**1216–1226.
9. Deal CM, Omizo EN, Schwartz EF, Eriksen EF, Cantor P, Wang J, Glass EV, Myers SL, Krege JH 2005 Combination teriparatide and raloxifene therapy for postmenopausal osteoporosis: Results from a 6-month double-blind placebo-controlled trial. J Bone Miner Res **20:**1905–1911.
10. Misof BM, Roschger P, Cosman F, Kurland ES, Tesch W, Messmer P, Dempster DW, Nieves J, Shane E, Fratzl P, Klaushofer K, Bilezikian J, Lindsay R 2003 Effects of intermittent parathyroid hormone administration on bone mineralization density in iliac crest biopsies from patients with osteoporosis: A paired study before and after treatment. J Clin Endocrinol Metab **88:**1150–1156.
11. Lindsay R, Scheele WH, Neer R, Pohl G, Adami S, Mautalen C, Reginster JY, Stepan JJ, Myers SL, Mitlak BH 2004 Sustained vertebral fracture risk reduction after withdrawal of teriparatide in postmenopausal women with osteoporosis. Arch Intern Med **164:**2024–2030.

© 2008 American Society for Bone and Mineral Research

12. Kurland ES, Heller SL, Diamond B, McMahon DJ, Cosman F, Bilezikian JP 2004 The importance of bisphosphonate therapy in maintaining bone mass in men after therapy with teriparatide [human parathyroid hormone(1-34)]. Osteoporos Int **15:**992–997.
13. Lane NE, Sanchez S, Modin GW, Genant HK, Pierini E, Arnaud CD 2000 Bone mass continues to increase at the hip after parathyroid hormone treatment is discontinued in glucocorticoid-induced osteoporosis: Results of a randomized controlled clinical trial. J Bone Miner Res **15:**944–951.
14. Black DM, Bilezikian JP, Ensrud KE, Greenspan SL, Palermo L, Hue T, Lang TF, McGowan JA, Rosen CJ 2005 One year of alendronate after one year of parathyroid hormone (1-84) for osteoporosis. N Engl J Med **353:**555–565.
15. Neer RM, Arnaud CD, Zanchetta JR, Prince R, Gaich GA, Reginster JY, Hodsman AB, Eriksen EF, Ish-Shalom S, Genant HK, Wang O, Mitlak BH 2001 Effect of parathyroid hormone (1-34) on fractures and bone mineral density in postmenopausal women with osteoporosis. N Engl J Med **344:**1434–1441.
16. Prince R, Sipos A, Hossain A, Syversen U, Ish-Shalom S, Marcinowska E, Halse J, Lindsay R, Dalsky GP, Mitlak BH 2005 Sustained nonvertebral fragility fracture risk reduction after discontinuation of teriparatide treatment. J Bone Miner Res **20:**1507–1513.

Chapter 54. Compliance and Persistence With Osteoporosis Medications

Deborah T. Gold

Departments of Psychiatry and Behavioral Sciences, Sociology, and Psychology and Neuroscience Center for the Study of Aging and Human Development, Duke University Medical Center, Durham, North Carolina

INTRODUCTION

The last decade has been one of tremendous progress in the prevention, diagnosis, management, and treatment of osteoporosis, and multiple medications for the prevention and/or treatment of osteoporosis have been approved by the U.S. Food and Drug Administration (FDA). Despite this, the 2004 Surgeon General's Report on Bone Health and Osteoporosis suggests that American bone health status is in jeopardy and will continue to worsen because of population aging.[1] How has this happened in the 21st century? Given that we have excellent tools for prevention and treatment of bone loss, why are so many people compromised by this disease? Former U.S. Surgeon General C. Everett Koop said, "Drugs don't work in patients who don't take them," and patients with osteoporosis have poor rates of taking their medications as prescribed. Studies on medication-taking behavior in osteoporosis show that patients are not following recommended courses of treatment.

DEFINITION OF TERMS

An impediment to understanding why patients are not taking medications is the inconsistency in use of the terms compliance, persistence, and adherence. To standardize the terms, the International Society for Pharmacoeconomic and Outcomes Research (ISPOR) has proposed the following. Compliance is following the timing, dose, and frequency of medication as prescribed.[2] Persistence is continuing to take medication for the prescribed length of time.[3] Adherence is a synonym for compliance.[4,5] Several recent articles urge researchers to adopt this terminology,[6–8] but confusion remains. For example, one recent editorial urges use of the ISPOR definitions, yet its authors use these terms incorrectly.[9] Although implementing a uniform terminology will be difficult, it is essential for progress in this area.

Another crucial step in standardizing these terms includes measurement issues for compliance and persistence. To this point, the measurement of compliance differs if data are prospective or retrospective. Prospectively, compliance is measured over time and is reported as a proportion (e.g., 80% compliant).[5] Retrospective compliance is measured by calculating the medication possession ratio (MPR). This is calculated as the number of days' supply of medication received divided by the length of the follow-up period and is dependent on refill information.[10]

Persistence is measured by determining the number of days for which the patient had medication available (retrospective) or the number of days the patient continued to take medication (prospective).[5] In certain studies, dichotomizing persistence (as persistent or nonpersistent) may be more useful if there has been a predefined period of time being described.[11]

Studies have found that compliance and persistence with osteoporosis medications are poor.[12–15] Reasons for this vary by individual and medication. They include but are not limited to side effects,[16,17] perception that physicians view osteoporosis as unimportant,[18] convenience and tolerability of dosing regimen,[19] that benefits of medications are not easily perceived by patients,[20] and cost.[21]

THE DOSING CHALLENGE

Some researchers who did early work on compliance and adherence with osteoporosis medications called for changes in medication (specifically the bisphosphonates) that would require less frequent dosing.[12] The rigid routine of oral bisphosphonates—taken on an empty stomach, with 6–8 oz of water with a 30-min upright wait before eating or drinking anything else—discourages many patients from complying, especially if they have any gastric side effects. Weekly dosing is now available for two bisphosphonates (alendronate and risedronate). Most research shows that weekly dosing of bisphosphonates leads to better compliance than daily dosing.[22–24] The availability of the weekly dosing interval has offered health care professionals one means by which they can encourage compliant and persistent behavior.

Bisphosphonates are the current drugs of choice for osteoporosis prevention and treatment in the United States. Alendronate has daily (10 mg/d) and weekly (70 mg/wk) dosing.

The author states that she has no conflicts of interest.

Key words: osteoporosis, medication compliance, persistence, bisphosphonate, dosing, fracture risk

© 2008 American Society for Bone and Mineral Research

Risedronate has daily (5 mg/d) and weekly (35 mg/wk) dosing plus a monthly option (75 mg/d, 2 consecutive d/mo). Ibandronate has a monthly oral formulation (150 mg/mo) and an injectable version (3 mg/every 3 mo). A 2.5-mg daily dose was developed but is not available. The newest bisphosphonate is zoledronic acid, given as a 5-mg infusion once yearly.[25]

The other FDA-approved drugs for osteoporosis do not have the same dosing requirements as the bisphosphonates. They are raloxifene (oral, 60 mg/d), calcitonin (nasal spray, 200 iu/d; injection, variable dose), estrogen (varying forms such as oral or patch; varying doses), and teriparatide (20 μg/d, SC; maximum duration is 24 mo).[25] However, only teriparatide has been shown to have significantly better persistence than the other medications.[26]

BENEFITS OF COMPLIANCE AND PERSISTENCE WITH OSTEOPOROSIS MEDICATIONS

Several studies using large health services databases have documented the relationship between medication compliance and persistence and reduced fracture risk. McCombs et al.[27] found that patients who complied for 1 yr (<25%) significantly reduced their risk of hip ($p < 0.01$) and vertebral fracture ($p < 0.05$). Caro et al.[28] used Canadian health services databases to find that patients who were highly compliant for their entire follow-up period had a 25% reduction in fractures ($p < 0.0001$). Finally, Huybrechts et al.[29] found in a managed care database that ~25% of women remained compliant for the entire study mean follow-up = 1.7 yr). Poor compliers had a 31% higher risk of fractures ($p < 0.0001$); in multivariate analyses, poor compliers still had a 17% increased risk of fractures.

The studies discussed above examined women using any osteoporosis drug. Three additional studies examined the relationship between compliance with bisphosphonates and fracture risk. Siris et al.[30] found that refill compliant and persistent patients (20%) had significant relative risk reductions of 20–45% for all fractures. Using data from the Netherlands, van den Boogaard et al.[31] found that 1-yr patients persistent with bisphosphonate therapy showed significant fracture reduction (26%); 2-yr persistent patients reduced fractures by 32%. Finally, Gold et al.[10] found that persistent patients were 26% less likely to fracture than nonpersistent patients ($p = 0.045$). The message from these studies is clear: taking bisphosphonates for osteoporosis does reduce fracture rates.

INTERVENTIONS TO IMPROVE COMPLIANCE AND PERSISTENCE

Nearly every study of compliance and persistence referenced here has the same refrain: the poor compliance and persistence with osteoporosis medications must be improved. Some call for more extensive research on these issues. Others demand community or physician-based interventions to teach patients more about the importance of complying with prescribed osteoporosis medicines.

Only a few studies have actually tried to implement compliance interventions. Perhaps the most widely cited is that of Clowes et al.,[32] in which 75 postmenopausal women were randomized to one of three arms: (1) no monitoring, (2) nurse monitoring, or (3) bone marker monitoring. The nurse-monitored group improved compliance 57% ($p = 0.04$) compared with the no monitoring group. Bone marker data did not improve compliance. Another intervention by Delmas et al.[33] randomized 2382 women to groups with reinforcement or no reinforcement. At 1 yr, persistence was high and similar for both groups ($p = 0.160$). In the reinforcement group, persistence depended on the bone marker message. When marker results were positive, persistence improved significantly; when negative, persistence was stable or lower.

FUTURE OF OSTEOPOROSIS COMPLIANCE AND PERSISTENCE ISSUES

Although the 2004 Surgeon General's Report talks about a current population whose bone health is in jeopardy, there is positive news as well. Substantial improvements can occur if we can implement the tools that we already have. The most optimistic intervention findings are those reported by Clowes et al.[32] that showed that monitoring by a health care professional (in this case, a nurse) improved compliance by 57% over no monitoring. Furthermore, despite not having been powered for persistence analyses, the monitored group showed a 25% increase in persistence over those not monitored. Interestingly, it was the contact with the health care professional—and not bone marker monitoring—that reinforced correct patient medication behavior.

There is some precedent for establishing a national program focused on one disease. The National Diabetes Education Program is supported by the NIH, the Centers for Disease Control and Prevention (CDC), and >200 public and private organizations.[34] Its mission is to reduce the morbidity and mortality associated with diabetes through education and research. The American Society of Diabetes Educators is focused on improved quality of life as well as self-management of people with diabetes.[35] Similar programs, with support from the NIH, CDC, and other organizations focused on osteoporosis (e.g., the National Osteoporosis Foundation) could establish and sustain a similar program to educate and support patients and ultimately reduce the disease burden of osteoporosis and improve quality of life for its victims.

REFERENCES

1. U.S. Department of Health and Human Services 2004 Bone Health and Osteoporosis: A Report of the Surgeon General. U.S. Department of Health and Human Services, Office of the Surgeon General, Rockville, MD, USA.
2. Steiner JF, Prochazka AV 1997 The assessment of refill compliance using pharmacy records. J Clin Epidemiol **50:**105–116.
3. Emkey RD, Ettinger M 2006 Improving compliance and persistence with bisphosphonate therapy for osteoporosis. Am J Med **119**(Suppl 1)**:**S18–S24.
4. ISPOR. Medication Compliance and Persistence Special Interest Group Accomplishments.Available at: http://www.ispor.org/sigs/MCP_accomplishments.asp#definition. Accessed January 5, 2008.
5. Cramer JA, Roy A, Burrell A, Fairchild CJ, Fuldeore MJ, Ollendorf DA, Wong PK 2008 Medication compliance and persistence: Terminology and definitions. Value Health.
6. Cramer JA, Gold DT, Silverman SL, Lewiecki EM 2007 A systematic review of persistence and compliance with bisphosphonates for osteoporosis. Osteoporos Int **18:**1023–1031.
7. Seeman E, Compston J, Adachi J, Brandi ML, Cooper C, Dawson-Hughes B, Jonsson B, Pols H, Cramer JA 2007 Non-compliance: The Achilles' heel of anti-fracture efficacy. Osteoporos Int **18:**711–719.
8. Silverman SL, Gold DT, Cramer JA 2007 Reduced fracture rates observed only in patients with proper persistence and compliance with bisphosphonate therapies. South Med J **100:**1214–1218.
9. Lekkerkerker F, Kanis JA, Alsayed N, Bouvenot G, Burlet N, Cahall D, Chines A, Delmas P, Dreiser RL, Ethgen D, Hughes N, Kaufman JM, Korte S, Kreutz G, Laslop A, Mitlak B, Rabenda V, Rizzoli R, Santora A, Schimmer R, Tsouderos Y, Viethel P, Reginster JY 2007 Group for the Respect of Ethics and Excellence in Science (GREES) Adherence to treatment of osteoporosis: A need for study. Osteoporos Int **18:**1311–1317.
10. Gold DT, Martin BC, Frytak JR, Amonkar MM, Cosman F 2007 A claims database analysis of persistence with alendronate therapy and fracture risk in post-menopausal women with osteoporosis. Curr Med Res Opin **23:**585–594.
11. Cramer JA, Amonkar MM, Hebborn A, Altman R 2005 Compli-

© 2008 American Society for Bone and Mineral Research

ance and persistence with bisphosphonate dosing regimens among women with postmenopausal osteoporosis. Curr Med Res Opin **21:**1453–1460.
12. Solomon DH, Avorn J, Katz JN, Finkelstein JS, Arnold M, Polinski JM, Brookhart MA 2005 Compliance with osteoporosis medications. Arch Intern Med **165:**2414–2419.
13. Rossini M, Bianchi G, Di Munno O, Giannini S, Minisola S, Sinigaglia L, Adami S 2006 Treatment of Osteoporosis in clinical Practice (TOP) Study Group. Determinants of adherence to osteoporosis treatment in clinical practice. Osteoporos Int **17:**914–921.
14. Segal E, Tamir A, Ish-Shalom S 2003 Compliance of osteoporotic patients with different treatment regimens. Isr Med Assoc J **5:**859–862.
15. Gold DT, Safi W, Trinh H 2006 Patient preference and adherence: Comparative US studies between two bisphosphonates, weekly risedronate and monthly ibandronate. Curr Med Res Opin **22:**2383–2391.
16. Kamatari M, Koto S, Ozawa N, Urao C, Suzuki Y, Akasaka E, Yanagimoto K, Sakota K 2007 Factors affecting long-term compliance of osteoporotic patients with bisphosphonate treatment and QOL assessment in actual practice: Alendronate and risedronate. J Bone Miner Metab **25:**302–309.
17. Ettinger B, Pressman A, Schein J, Chan J, Silver P, Connolly N 1998 Alendronate use among 812 women: Prevalence of gastrointestinal complaints, noncompliance with patient instructions, and discontinuation. J Manag Care Pharm **4:**488–492.
18. Hamdy RC 2006 Osteoporosis: We are neglecting our own? South Med J **99:**447–448.
19. Downey TW, Foltz SH, Boccuzzi SJ, Omar MA, Kahler KH 2006 Adherence and persistence associated with the pharmacologic treatment of osteoporosis in a managed care setting. South Med J **99:**570–575.
20. Lespessailles E 2007 A forgotten challenge when treating osteoporosis: Getting patients to take their meds. Joint Bone Spine **74:**7–8.
21. Badamgarav E, Fitzpatrick LA 2006 A new look at osteoporosis outcomes: The influence of treatment, compliance, persistence, and adherence. Mayo Clin Proc **81:**1009–1012.
22. Simon JA, Lewiecki EM, Smith ME, Petruschke RA, Wang L, Palmisano JJ 2002 Patient preference for once-weekly alendronate 70 mg versus once-daily alendronate 10 mg: A multicenter, randomized, open-label, crossover study. Clin Ther **24:**1871–1886.
23. Schnitzer T, Bone HG, Crepaldi G, Adami S, McClung M, Kiel D, Felsenberg D, Recker RR, Tonino RP, Roux C, Pinchera A, Foldes AJ, Greenspan SL, Levine MA, Emkey R, Santora AC II, Kaur A, Thompson DE, Yates J, Orloff JJ 2000 Therapeutic equivalence of alendronate 70 mg once-weekly and alendronate 10 mg daily in the treatment of osteoporosis. Alendronate Once-Weekly Study Group. Aging Clin Exp Res **12:**1–12.
24. Claxton AJ, Cramer J, Pierce C 2001 A systematic review of the associations between dose regimens and medication compliance. Clin Ther **23:**1296–1310.
25. National Osteoporosis Foundation. Medications to Prevent and Treat Osteoporosis. Available at: http://www.nof.org/patientinfo/medications.htm/. Accessed December 10, 2007.
26. Arden NK, Earl S, Fisher DJ, Cooper C, Carruthers S, Goater M 2006 Persistence with teriparatide in patients with osteoporosis: The UK experience. Osteoporos Int **17:**1626–1629.
27. McCombs JS, Thiebaud P, McLaughlin-Miley C, Shi J 2004 Compliance with drug therapies for the treatment and prevention of osteoporosis. Maturitas **48:**271–287.
28. Caro JJ, Ishak KJ, Huybrechts KF, Raggio G, Naujoks C 2004 The impact of compliance with osteoporosis therapy on fracture rates in actual practice. Osteoporos Int **15:**1003–1008.
29. Huybrechts KF, Ishak KJ, Caro JJ 2006 Assessment of compliance with osteoporosis treatment and its consequences in a managed care population. Bone **38:**922–928.
30. Siris ES, Harris ST, Rosen CJ, Barr CE, Arvesen JN, Abbott TA, Silverman S 2006 Adherence to bisphosphonate therapy and fracture rates in osteoporotic women: Relationship to vertebral and nonvertebral fractures from 2 US claims databases. Mayo Clin Proc **81:**1013–1022.
31. van den Boogaard CH, Breekveldt-Postma NS, Borggreve SE, Goettsch WG, Herings RM 2006 Persistent bisphosphonate use and the risk of osteoporotic fractures in clinical practice: A database analysis study. Curr Med Res Opin **22:**1757–1764.
32. Clowes JA, Peel NF, Eastell R 2004 The impact of monitoring on adherence and persistence with antiresorptive treatment for postmenopausal osteoporosis: A randomized controlled trial. J Clin Endocrinol Metab **89:**1117–1123.
33. Delmas PD, Vrijens B, Eastell R, Roux C, Pols HA, Ringe JD, Grauer A, Cahall D, Watts NB 2007 Improving Measurements of Persistence on Actonel Treatment (IMPACT) Investigators. Effect of monitoring bone turnover markers on persistence with risedronate treatment of postmenopausal osteoporosis. J Clin Endocrinol Metab **92:**1296–1304.
34. National Diabetes Education Program. Available at http://ndep.nih.gov/. Accessed January 3, 2008.
35. American Association of Diabetes Educators. Available at www.diabeteseducator.org/. Accessed December 22, 2007.

Chapter 55. Cost-Effectiveness of Osteoporosis Treatment

Anna N. A. Tosteson

Multidisciplinary Clinical Research Center in Musculoskeletal Diseases, Dartmouth Medical School, The Dartmouth Institute for Health Policy and Clinical Practice, Dartmouth College, Lebanon, New Hampshire

INTRODUCTION

Osteoporosis affects a large proportion of the elderly population and is associated with fractures that are costly in both human and economic terms.[1] In 2005, the U.S. population sustained an estimated 2 million incident fractures at a cost of $16.9 billion.[2] With annual fracture-related expenditures projected to increase to $25.3 billion by 2025, there is widespread recognition that growing elderly populations and constrained health care budgets will continue to challenge health care systems to find cost-effective approaches to osteoporosis care. Cost-effectiveness analysis is a form of economic evaluation that estimates the value of an intervention by weighing the expected net increase in cost of an intervention against its expected net gain in health.[3] The rationale for cost-effectiveness analysis is that when health care resources are limited, expenditures should be planned to maximize health outcomes within available resources. The cost-effectiveness of new treatments relative to current care standards is one attribute that policy-makers may consider when making formulary coverage decisions. In this chapter, the methodology of cost-effectiveness analysis is described, recent developments in the cost-effectiveness of osteoporosis care are discussed, and key findings are highlighted.

Dr. Tosteson has consulted for Amgen and Procter & Gamble.

© 2008 American Society for Bone and Mineral Research

OVERVIEW OF METHODS FOR COST-EFFECTIVENESS ANALYSIS

Cost-Effectiveness Ratio

The incremental cost-effectiveness ratio (ICER), which estimates expected cost per unit of health gained, is the primary outcome measure used to characterize value in cost-effectiveness studies. Consider two alternative treatments, A and B, where the average cost of A is higher than the average cost of B; the ICER is defined as follows:

$$\text{ICER} = \frac{(\text{Cost}_A - \text{Cost}_B)}{(\text{Effectiveness}_A - \text{Effectiveness}_B)}$$

Using this definition, the value of each more costly intervention is judged relative to the improvement in health that it provides over and above health outcomes associated with the less costly alternative.

Choice of Comparator. When assessing cost-effectiveness of a new osteoporosis intervention, the standard of care that is used as the basis for comparison (i.e., the comparator) may have a marked impact on the intervention's estimated value. Cost-effectiveness analyses of osteoporosis prevention conducted before 2002, when Women's Health Initiative findings were published,[4] typically included hormone therapy as a comparator. Choice of comparator today depends on whether treatment is being considered for a man or woman and whether the individual has established osteoporosis. Unless cost-effectiveness is measured relative to a reasonable alternative, the estimated ICER may not provide a meaningful estimate of an intervention's value. Whereas ICERs computed relative to "no intervention" have meaning for the minority of patients who have no other viable treatment option, for the majority of patients in whom less costly treatments are possible, they are potentially misleading estimates of value. In general, when new interventions are compared with "no intervention" (technically an average rather than an ICER), they will have more favorable value than when compared with active treatment comparators.

Model-Based Analyses. Estimating an ICER typically requires mathematical modeling to project expected health and cost implications of alternative treatments over a longer time horizon than can be observed in any clinical trial[5] and/or to expand the treatments and/or population subgroups considered. Most analyses use Markov state-transition models,[6] which are comprised of a discrete number of health states, each with an associated cost and health state value (i.e., health utility), along with annual probabilities of transition among the health states. Other modeling methods that detail the biological processes related to bone health have also been proposed.[7]

Estimating the Cost of Osteoporosis Treatment

To estimate the net difference in cost of a new treatment relative to a comparator several types of direct medical costs should be considered (Table 1). The cost of medical care in future years of life may also be included. Against these costs, potential savings that may accrue because of fracture prevention are considered and include the cost of acute fracture care, rehabilitation services (if required), and costs of ongoing fracture-related disability (if present). Differences in the cost of providing health care from country to country make generalizability of cost-effectiveness findings across countries challenging.

TABLE 1. COMPONENTS OF DIRECT MEDICAL COSTS TO CONSIDER WHEN ASSESSING THE COST-EFFECTIVENESS OF OSTEOPOROSIS TREATMENT

Cost component
Medication
Acquisition
Health care services for routine monitoring
Health care services for treatment side-effects/sequelae
Fracture
Acute care services
Rehabilitation services
Ongoing disability services
Extended life years
Health care services

Indirect costs of an illness are those that are associated with a loss in productivity caused by morbidity and mortality. Such costs may be incurred by the individual who sustains a fracture and/or by the care givers. However, there is scant evidence available to address the latter. The human capital approach, which values productivity changes based on lost earnings,[8] has been applied to assess the cost of fractures in some U.S. cost-of-illness studies,[9–11] but to date, such costs have not been included in cost-effectiveness analyses of osteoporosis treatment. Some argue that productivity costs are adequately reflected in the denominator of the cost-effectiveness ratio when quality-adjusted life years (QALYs) are used to measure effectiveness.[3]

Whether each of these potential costs/savings are included in the analysis depends on the perspective that is taken. For informing public policy decision makers, the societal perspective is generally most desirable. The marked impact that perspective may have on the cost-effectiveness of osteoporosis treatment under a health care system such as the United States where different payors are responsible for health care at different ages is shown by an example that underscores the disparity between who pays for the prevention and who realizes potential long-term savings. Consider 5-yr treatment of high-risk 55-yr-old women from two perspectives: (1) a private insurer who pays for health care services up until age 65 and (2) a government insurer who pays after age 65 (i.e., Medicare in the United States). For the private insurer who pays for the costs of the treatment and monitoring, but will realize limited savings because of fractures averted, treatment does not seem cost-effective. In contrast, the government payor only benefits because fractures are averted and sees treatment as cost-saving. This simplistic example suggests that optimal decisions for public health require a broad perspective that considers the full time horizon of costs and benefits.

Estimating the Effectiveness of Osteoporosis Treatment

QALYs. The recommended measure for assessing the effectiveness of health interventions is the QALY,[3] which takes both length of life and quality of life into account. Use of QALYs facilitates comparisons of economic value across disease areas (e.g., interventions to control diabetes can be compared with osteoporosis treatments). Cost-effectiveness studies that report ICERs as cost per QALY gained are often referred to as cost-utility analyses, because to estimate QALYs, health state values or "utilities" that reflect preferences for various health states, are used.

Whereas QALYs have the potential to incorporate the intangible fracture-related costs of pain and suffering, data on

© 2008 American Society for Bone and Mineral Research

health state values for fracture-related health outcomes are required. Evidence on the impact of fractures on QALYs was summarized in two reviews,[12,13] with a more recent report addressing the impact of fractures in Sweden.[14] The absolute QALY losses associated with fracture vary based on who is asked (e.g., patient who sustained a vertebral fracture versus patient imagining a vertebral fracture) and how they are asked (e.g., visual analog scale, time tradeoff), yet published studies consistently report health state values for fracture-related outcomes that are significantly below ideal health. Although a growing literature addresses preference-based measures of health in osteoporosis, many cost-effectiveness studies continue to rely on expert opinion regarding the quality-of-life impact of fractures both initially and in the long term.[15–17]

When evaluating the value of osteoporosis treatment, it is important to consider the potential adverse impact that treatment side effects may have on estimates of quality-adjusted life expectancy. The potential for side effects to offset quality-of-life gains caused by fracture prevention was first highlighted in studies of the role of hormone therapy in osteoporosis prevention.[18,19]

Number of Fractures Prevented. The value of osteoporosis treatment is sometimes reported in disease-specific terms as number of fractures prevented, which is problematic for two reasons. First, some osteoporosis interventions, such as raloxifene, have extraskeletal health effects that go unaccounted for when value is reported in terms of cost per fracture prevented. Second, inherent differences in human and economic costs of different fracture types (e.g., wrist versus hip) make interpreting cost per fracture prevented challenging. To address this, analysts sometimes report costs in terms of specific fracture types (e.g., cost per hip fracture prevented or per vertebral fracture prevented) or in "hip fracture equivalent units."[20]

COST-EFFECTIVENESS OF OSTEOPOROSIS TREATMENT

Cost-Effectiveness Analysis and Clinical Practice Guidelines

As constrained health care budgets are increasingly felt, guideline developers recognize that costs cannot be entirely ignored.[21] One approach to setting treatment thresholds that has seen growing application in the osteoporosis literature is to identify the absolute fracture risk at which the cost per QALY gained falls below a "willingness-to-pay" per QALY gained threshold.[15–17,20,22,23] This approach was used by the National Osteoporosis Foundation (NOF) to identify a 10-yr absolute hip fracture risk at which treatment cost $60,000 per QALY gained or lower for treatment relative to no intervention.[23] Whereas intervention thresholds defined by absolute 10-yr fracture risk may previously have been viewed as somewhat academic, the recent release of the WHO fracture risk assessment tool, FRAX, facilitates such risk predictions for previously untreated populations.[24] A report from the NOF guide committee provides insight into specific clinical factors that meet the intervention thresholds (3% for 10-yr hip fracture risk or 20% for hip, wrist, spine, and shoulder fracture risks combined) based on an adaptation of FRAX for the U.S. population.[25,26]

Cost-Effectiveness of Osteoporosis Treatment

Fracture risk, treatment cost, the impact of fractures on health-related quality of life, and the durability of treatment[27] all influence the value of osteoporosis treatment. A recent U.S. analysis of an unspecified treatment that reduces fracture incidence by 35% relative to no intervention[23] showed the marked improvement in cost-effectiveness that average-risk women attain with advancing age because of their higher absolute fracture risk as shown in Fig. 1. For example, a treatment costing $900 per year costs in excess of $580,000 per QALY gained for a 50-yr-old woman whose 10-yr hip fracture risk is 0.7% compared with only $4,000 per QALY gained for an 80-yr-old white woman whose 10-yr hip fracture risk is 13.5%.

Before 1993, most osteoporosis cost-effectiveness studies assessed the value of hormone therapy.[28,29] More recently, several reviews and technology assessments have addressed the value of other osteoporosis treatments with evaluation of bisphosphonates being a frequent focus.[29–33] Studies also address the value of calcium and vitamin D,[34] raloxifene,[35–39] teriparatide,[40,41] calcitonin,[42] and strontium ranelate,[43,44] as well as the cost-effectiveness of selective treatment strategies[45,46] and special populations.[47]

Bisphosphonate treatment is generally cost-effective when used in moderately high-risk populations such as women over age 65 with osteoporosis.[31] Whereas treatment of elderly populations with calcium and vitamin D is potentially cost saving,[34] studies of raloxifene differ in their findings depending on the comparators included and the side effects considered.[35–38] Discrepant findings have also been reported for teriparatide use among women at high risk of fracture.[40,41]

SUMMARY

Because of the size of the elderly population that is at risk for complications of osteoporosis-related fractures, it is imperative that cost-effective approaches to osteoporosis management be identified. Whereas a growing literature addresses the value of specific treatments for various population subgroups, clinical practice guidelines identify cost-effective intervention thresholds on the basis of absolute 10-yr fracture risk. For the U.S. population, cost-effective treatment intervention thresholds of 3% or greater 10-yr hip fracture risk or 20% or greater hip, wrist, clinical spine, and shoulder fracture risk combined have been recommended.[26] The advent of risk assessment tools for predicting 10-yr fracture risk[24] should facilitate efficient targeting of therapy to those individuals who stand to benefit most from osteoporosis treatment.

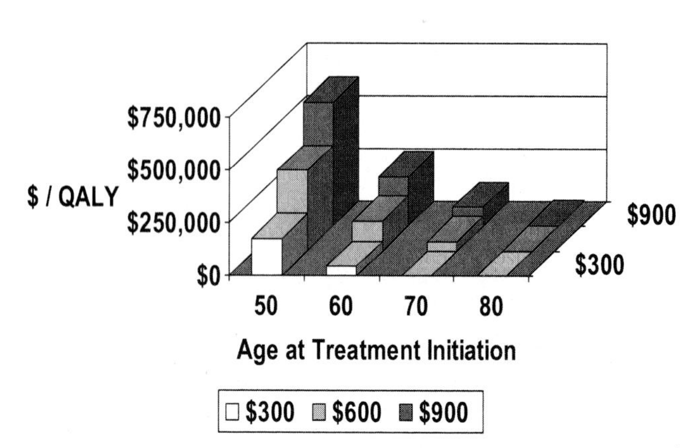

FIG. 1. The impact that annual treatment cost ($300, $600, or $900) has on cost per QALY gained at different ages of treatment initiation when treatment reduces fracture incidence by 35% and 5-yr losses in health-related quality of life after fracture are modeled.[23]

© 2008 American Society for Bone and Mineral Research

ACKNOWLEDGMENTS

This work was supported by the National Institute of Arthritis and Musculoskeletal and Skin Diseases (P60-AR048094) and the National Institute on Aging (R01-AG12262).

REFERENCES

1. Department of Health and Human Services 2004 Bone Health and Osteoporosis: A Report of the Surgeon General. Department of Health and Human Services, Rockville, MD, USA.
2. Burge R, Dawson-Hughes B, Solomon DH, Wong JB, King A, Tosteson A 2007 Incidence and economic burden of osteoporosis-related fractures in the United States, 2005-2025. J Bone Miner Res **22**:465–475.
3. Gold M, Siegel J, Russell L, Weinstein M 1996 Cost-Effectiveness in Health and Medicine. Oxford University Press, New York, NY, USA.
4. Rossouw JE, Anderson GL, Prentice RL, LaCroix AZ, Kooperberg C, Stefanick ML, Jackson RD, Beresford SA, Howard BV, Johnson KC, Kotchen JM, Ockene J 2002 Risks and benefits of estrogen plus progestin in healthy postmenopausal women: Principal results From the Women's Health Initiative randomized controlled trial. JAMA **288**:321–333.
5. Tosteson AN, Jönsson B, Grima DT, O'Brien BJ, Black DM, Adachi JD 2001 Challenges for model-based economic evaluations of postmenopausal osteoporosis interventions. Osteoporos Int **12**:849–857.
6. Sonnenberg FA, Beck JR 1993 Markov models in medical decision making: A practical guide. Med Decis Making **13**:322–338.
7. Vanness DJ, Tosteson AN, Gabriel SE, Melton LJ III 2005 The need for microsimulation to evaluate osteoporosis interventions. Osteoporos Int **16**:353–358.
8. Hodgson T, Meiners M 1982 Cost-of-illness medthodology: A guide to current practices and procedures. Milbank Mem Fund Q **60**:429–462.
9. Holbrook T, Grazier K, Kelsey J, Sauffer R 1984 The Frequency of Occurrence, Impact, and Cost of Musculoskeletal Conditions in the United States. American Academy of Orthopaedic Surgeons, Rosemont, IL, USA.
10. Praemer A, Furner S, Rice D 1992 Musculoskeletal Conditions in the United States. American Academy of Orthopaedic Surgeons, Rosemont, IL, USA.
11. Praemer A, Furner S, Rice D 1999 Musculoskeletal Conditions in the United States. American Academy of Orthopaedic Surgeons, Rosemont, IL, USA.
12. Brazier JE, Green C, Kanis JA 2002 A systematic review of health state utility values for osteoporosis-related conditions. Osteoporos Int **13**:768–776.
13. Tosteson AN, Hammond CS 2002 Quality-of-life assessment in osteoporosis: Health-status and preference-based measures. Pharmacoeconomics **20**:289–303.
14. Borgström F, Zethraeus N, Johnell O, Lidgren L, Ponzer S, Svensson O, Abdon P, Ornstein E, Lunsjo K, Thorngren KG, Sernbo I, Rehnberg C, Jönsson B 2006 Costs and quality of life associated with osteoporosis-related fractures in Sweden. Osteoporos Int **17**:637–650.
15. Kanis JA, Borgström F, Zethraeus N, Johnell O, Oden A, Jönsson B 2005 Intervention thresholds for osteoporosis in the UK. Bone **36**:22–32.
16. Kanis JA, Johnell O, Oden A, Borgström F, Johansson H, De Laet C, Jönsson B 2005 Intervention thresholds for osteoporosis in men and women: A study based on data from Sweden. Osteoporos Int **16**:6–14.
17. Kanis JA, Johnell O, Oden A, De Laet C, Oglesby A, Jönsson B 2002 Intervention thresholds for osteoporosis. Bone **31**:26–31.
18. Weinstein MC 1980 Estrogen use in postmenopausal women–costs, risks, and benefits. N Engl J Med **303**:308–316.
19. Weinstein MC, Schiff I 1983 Cost-effectiveness of hormone replacement therapy in the menopause. Obstet Gynecol Surv **38**:445–455.
20. Kanis JA, Oden A, Johnell O, Jönsson B, de Laet C, Dawson A 2001 The burden of osteoporotic fractures: A method for setting intervention thresholds. Osteoporos Int **12**:417–427.
21. Guyatt G, Baumann M, Pauker S, Halperin J, Maurer J, Owens DK, Tosteson AN, Carlin B, Gutterman D, Prins M, Lewis SZ, Schunemann H 2006 Addressing resource allocation issues in recommendations from clinical practice guideline panels: Suggestions from an American College of Chest Physicians task force. Chest **129**:182–187.
22. Borgström F, Johnell O, Kanis JA, Jönsson B, Rehnberg C 2006 At what hip fracture risk is it cost-effective to treat? International intervention thresholds for the treatment of osteoporosis. Osteoporos Int **17**:1459–1471.
23. Tosteson AN, Melton LJ III, Dawson-Hughes B, Baim S, Favus MJ, Khosla S, Lindsay RL 2008 Cost-effective osteoporosis treatment thresholds: The United States perspective. Osteoporos Int **19**:437–447.
24. Kanis JA, Johnell O, Oden A, Johansson H, McCloskey E 2008 FRAX and the assessment of fracture probability in men and women from the UK. Osteoporos Int **19**:385–397.
25. Dawson-Hughes B, Tosteson AN, Melton LJ III, Baim S, Favus MJ, Khosla S, Lindsay RL 2008 Implications of absolute fracture risk assessment for osteoporosis practice guidelines in the USA. Osteoporos Int **19**:449–458.
26. National Osteoporosis Foundation 2008 Clinician's Guide to Prevention and Treatment of Osteoporosis. National Osteoporosis Foundation, Washington, DC, USA.
27. Jönsson B, Kanis J, Dawson A, Oden A, Johnell O 1999 Effect and offset of effect of treatments for hip fracture on health outcomes. Osteoporos Int **10**:193–199.
28. Anonymous 1998 Osteoporosis: Review of the evidence for prevention, diagnosis and treatment and cost-effectiveness analysis. Executive summary. Osteoporos Int **8**(Suppl 4):S3–S6.
29. Fleurence RL, Iglesias CP, Torgerson DJ 2006 Economic evaluations of interventions for the prevention and treatment of osteoporosis: A structured review of the literature. Osteoporos Int **17**:29–40.
30. Zethraeus N, Borgström F, Strom O, Kanis JA, Jönsson B 2007 Cost-effectiveness of the treatment and prevention of osteoporosis—a review of the literature and a reference model. Osteoporos Int **18**:9–23.
31. Fleurence RL, Iglesias CP, Johnson JM 2007 The cost effectiveness of bisphosphonates for the prevention and treatment of osteoporosis: A structured review of the literature. Pharmacoeconomics **25**:913–933.
32. Urdahl H, Manca A, Sculpher MJ 2006 Assessing generalisability in model-based economic evaluation studies: A structured review in osteoporosis. Pharmacoeconomics **24**:1181–1197.
33. Stevenson M, Jones ML, De Nigris E, Brewer N, Davis S, Oakley J 2005 A systematic review and economic evaluation of alendronate, etidronate, risedronate, raloxifene and teriparatide for the prevention and treatment of postmenopausal osteoporosis. Health Technol Assess **9**:1–160.
34. Torgerson DJ, Kanis JA 1995 Cost-effectiveness of preventing hip fractures in the elderly population using vitamin D and calcium. QJM **88**:135–139.
35. Goeree R, Blackhouse G, Adachi J 2006 Cost-effectiveness of alternative treatments for women with osteoporosis in Canada. Curr Med Res Opin **22**:1425–1436.
36. Mobley LR, Hoerger TJ, Wittenborn JS, Galuska DA, Rao JK 2006 Cost-effectiveness of osteoporosis screening and treatment with hormone replacement therapy, raloxifene, or alendronate. Med Decis Making **26**:194–206.
37. Kanis JA, Borgström F, Johnell O, Oden A, Sykes D, Jönsson B 2005 Cost-effectiveness of raloxifene in the UK: An economic evaluation based on the MORE study. Osteoporos Int **16**:15–25.
38. Borgström F, Johnell O, Kanis JA, Oden A, Sykes D, Jönsson B 2004 Cost effectiveness of raloxifene in the treatment of osteoporosis in Sweden: An economic evaluation based on the MORE study. Pharmacoeconomics **22**:1153–1165.
39. Armstrong K, Chen TM, Albert D, Randall TC, Schwartz JS 2001 Cost-effectiveness of raloxifene and hormone replacement therapy in postmenopausal women: Impact of breast cancer risk. Obstet Gynecol **98**:996–1003.
40. Lundkvist J, Johnell O, Cooper C, Sykes D 2006 Economic evaluation of parathyroid hormone (PTH) in the treatment of osteoporosis in postmenopausal women. Osteoporos Int **17**:201–211.
41. Liu H, Michaud K, Nayak S, Karpf DB, Owens DK, Garber AM 2006 The cost-effectiveness of therapy with teriparatide and alendronate in women with severe osteoporosis. Arch Intern Med **166**:1209–1217.
42. Coyle D, Cranney A, Lee KM, Welch V, Tugwell P 2001 Cost effectiveness of nasal calcitonin in postmenopausal women: Use of

© 2008 American Society for Bone and Mineral Research

Cochrane Collaboration methods for meta-analysis within economic evaluation. Pharmacoeconomics **19:**565–575.
43. Borgström F, Jönsson B, Strom O, Kanis JA 2006 An economic evaluation of strontium ranelate in the treatment of osteoporosis in a Swedish setting: Based on the results of the SOTI and TROPOS trials. Osteoporos Int **17:**1781–1793.
44. Stevenson M, Davis S, Lloyd-Jones M, Beverley C 2007 The clinical effectiveness and cost-effectiveness of strontium ranelate for the prevention of osteoporotic fragility fractures in postmenopausal women. Health Technol Assess **11:**1–134.
45. Schousboe JT, Ensrud KE, Nyman JA, Melton LJ III, Kane RL 2005 Universal bone densitometry screening combined with alendronate therapy for those diagnosed with osteoporosis is highly cost-effective for elderly women. J Am Geriatr Soc **53:**1697–1704.
46. Schousboe JT, Taylor BC, Fink HA, Kane RL, Cummings SR, Orwoll ES, Melton LJ III, Bauer DC, Ensrud KE 2007 Cost-effectiveness of bone densitometry followed by treatment of osteoporosis in older men. JAMA **298:**629–637.
47. Kanis JA, Stevenson M, McCloskey EV, Davis S, Lloyd-Jones M 2007 Glucocorticoid-induced osteoporosis: A systematic review and cost-utility analysis. Health Technol Assess **11:**1–231.

Chapter 56. Future Therapies for Osteoporosis

Kong Wah Ng[1,2,3] and T. John Martin[2,3]

[1]*Department of Endocrinology and Diabetes, St. Vincent's Health, Melbourne, Australia;* [2]*St. Vincent's Institute, Melbourne, Australia;* [3]*University of Melbourne Department of Medicine, Victoria, Australia*

INTRODUCTION

As understanding has increased of the metabolic pathways governing the formation and function of osteoblasts and osteoclasts, new insights have identified specific points of intervention that are guiding the development of the next generation of therapies for osteoporosis. The clinical outcomes of these new therapies may be predicted because the actions of the selected targets are known, and in some cases, preclinical evidence is fulfilling those predictions. The aims are to develop therapies to improve the fracture risk reduction if possible, to avoid the possibility of long-term effects on bone structure, and to find drugs whose effects reverse with cessation of therapy and drugs that stimulate bone formation or that inhibit resorption without inhibiting bone formation at the same time. This chapter will discuss several candidate molecules that have been identified with the potential to act as antiresorptive or anabolic agents. Studies with these molecules are still either in preclinical or in early investigational stages without fracture data. Nonetheless, preliminary results hold the promise that at least some of these new therapies may develop into effective means of treating and preventing osteoporosis.

ANTIRESORPTIVES

In the last 10 yr, several bisphosphonates and a selective estrogen receptor modulator (SERM) have been shown in careful, thorough clinical trials to reduce fracture incidence in osteoporosis by 30–50%.(1) This is the starting point. Although it might reasonably be asked whether we need other antiresorptives, the real and potential limitations of existing therapies are sufficient to warrant the continued search for new approaches. Newer therapies may be better suited for particular indications or provide greater efficacy, safety, or convenience. There may be a limit to the safe reduction in fracture risk that can be achieved with resorption inhibitors, and it remains to be seen whether that limit can be reached simply by effecting more powerful inhibition of bone resorption.(2) The candidate resorption inhibitors to be considered below can be divided into those whose predominant action is to inhibit osteoclast formation, those that inhibit osteoclast activity, and a potentially interesting new class that might inhibit resorption without concomitant inhibition of bone formation.

The authors state that they have no conflicts of interest.

© 2008 American Society for Bone and Mineral Research

Inhibition of Osteoclast Formation

Osteoblasts express a membrane protein, RANKL, regulated by osteotropic hormones, PTH, and calcitriol, as well as cytokines such as interleukin-6. RANKL plays a critical role in osteoclast differentiation, activation, and survival. The binding of RANKL to its receptor, RANK, expressed in mononuclear hematopoietic precursors, initiates the processes that ultimately lead to the formation of multinucleate osteoclasts. Osteoprotegerin (OPG) acts as a decoy receptor for RANKL to suppress osteoclast formation. Studies in genetically altered mice have clearly established the essential physiological roles of RANKL and OPG in controlling osteoclast formation and activity and showed a pathway obviously rich in targets for pharmaceutical development. Denosumab (Amgen, Thousand Oaks, CA, USA) is a fully human monoclonal antibody that binds with high affinity and specificity to RANKL to inhibit its action. In a phase II study, subcutaneous injections of denosumab to postmenopausal women with low BMD every 3 or 6 mo for 12 mo significantly decreased bone resorption and increased BMD at the lumbar spine, hip, and distal third of radius.(3) Bone resorption marker indices decreased rapidly after denosumab injection, as did markers of bone formation, reflecting the coupling of formation to resorption. This proof of concept that substantial bone resorption inhibition can be achieved by neutralizing RANKL shows an exceptionally prolonged and powerful action, which is being investigated in a phase III study that uses subcutaneous injection of denosumab every 6 mo.

Inhibition of Osteoclast Action

Cathepsin K Inhibitors. Cathepsin K is selectively expressed in osteoclasts and is the predominant cysteine protease in these cells. It accumulates in lysosomal vesicles and is localized at the ruffled border in actively resorbing osteoclasts, discharging into the acidified, sealed resorption space beneath the osteoclast when the lysosomal vesicles fuse with the cell membrane. Defects in the gene encoding cathepsin K are linked to the clinical condition pycnodysostosis (OMIM 265800), an autosomal recessive dysplasia characterized by skeletal defects in-

Key words: cathepsin K inhibition, integrin receptor inhibition, chloride-7 channel inhibition, Wnt signaling pathway, sclerostin, DKK-1, glycogen synthase kinase

cluding dense, brittle bones, short stature, and poor bone remodeling.[4] Similarly, the deletion of the cathepsin K gene in mice results in osteopetrosis.[5] The substantial body of evidence indicating that cathepsin K plays an important role in bone resorption makes it a target for inhibition of resorption.

Peptide inhibitors designed to inhibit cysteine proteases by binding at the substrate site to mimic a cathepsin K–substrate complex are in clinical development.[6,7] Preclinical data have shown that cathepsin K inhibitors act as antiresorptive agents that prevent bone loss while allowing bone formation to continue. A 12-mo study with an orally bioavailable specific inhibitor of human cathepsin K, balicatib (compound AAE581; Novartis, Cambridge, MA, USA), in postmenopausal women showed an increase in lumbar spine and hip BMD associated with statistically significant decrease in markers of bone resorption but not in those of bone formation.[8] Other potent cathepsin K inhibitors currently under study include SB-462795 (relacatib; GlaxoSmithKline, Research Triangle Park, NC, USA) and CRA-013783 (Merck, West Point, PA, USA).

Other New Approaches to Resorption Inhibition. A relatively selective osteoclast-specific structure, which seems to play a rate-limiting role in osteoclast activity, is the $\alpha v\beta 3$ integrin receptor, which is produced in osteoclasts and in budding blood vessels and leukocytes. Treatment of rats with the disintegrins, echistatin or kistrin, which bind with high affinity to $\alpha v\beta 3$, inhibits bone resorption stimulated by PTH or estrogen deficiency. Moreover, small molecular weight compounds that mimic the tripeptide RGD sequence, recognized by the integrin, were shown to have similar effects.[9] One such compound, L-000845704 (Merck) was used in a 12-mo, randomized, double-blind placebo-controlled study in 227 women with low BMD. Treatment was associated with significantly increased lumbar spine BMD and decreases in markers of bone resorption and formation.[10]

Bone Resorption Inhibitors That Do Not Inhibit Bone Formation

With existing marketed resorption inhibitory drugs, the coupling of bone formation to resorption is illustrated by the decrease in bone formation that accompanies resorption inhibition (e.g., with bisphosphonates, estrogens and SERMs). This seems very likely to be the case also with the anti-RANKL, denosumab. Will it be possible to develop drugs that inhibit bone resorption without inhibiting formation—in other words uncoupling bone formation from resorption? There are some indications that it might be possible. Cathepsin K inhibition is one of these, as mentioned above.

An essential component of osteoclastic bone resorption is acidification of the resorption lacuna, which reduces the pH to ~4 to dissolve bone mineral. Passive transport of chloride through chloride channels preserves electroneutrality in the course of the acidification process, and preventing chloride transport will lead to a rapid hyperpolarization of the membrane, preventing further secretion of protons, thus resulting in an inhibition of further bone resorption. A vacuolar H^+ATPase in the osteoclast membrane plays a key role in this process by mediating the active transport of protons. Inhibitors of this enzyme, for example, bafilomycin, have been shown to inhibit osteoclastic bone resorption in vitro and in vivo. A relatively osteoclast-selective H^+ATPase inhibitor has been shown to inhibit ovariectomy-induced bone loss in rats.[11] In mice deficient for either *c-src*[12] or the chloride-7 channel (*CLCN7*),[13] bone resorption is inhibited without any inhibition of the rate or extent of formation. In each of these mouse mutations, osteoclast numbers are maintained, but the osteoclasts are unable to resorb bone. This is the case also in human subjects with inactivating mutations either of *CLCN7* (OMIM 166600 and OMIM 25900)[14] or the vacuolar H^+ATPase.[15]

One possibility is that osteoclasts are able to generate a factor (or factors) that can contribute to bone formation, despite the fact that they do not resorb bone.[16–18] Early data with an orally delivered CLCN7 inhibitor showed that it inhibited bone loss in the ovariectomized rat without inhibiting bone formation.[13] A target in a related category is Src, a protein tyrosine kinase. *Src*$^{-/-}$ mice formed increased numbers of osteoclasts, but the cells failed to resorb bone because they did not form ruffled borders.[19] In a phase I study in healthy males, AZD0530, a highly selective, dual-specific, orally available inhibitor of Src and Abl kinases was shown to suppress bone resorption markers reversibly with variable changes in bone formation markers.[20]

It is possible that such inhibitors of resorption, at the moment cathepsin K, CLCN7, vacuolar H^+ATPase, and Src, might conceivably provide a new class of resorption inhibitory drug that does not inhibit bone formation. If they are safe and at least as effective in fracture reduction as other inhibitors, they could provide a real advance. For example, they might more effectively be combined with anabolic therapy than those resorption inhibitors (e.g., bisphosphonates, and likely, anti-RANKL) that lead to inhibition of bone formation.

ANABOLIC AGENTS

Antiresorptive agents do not reconstruct the skeleton, but until recently, no therapeutic approach was available to restore bone once it had been lost. That situation has changed with the development of PTH as a highly effective anabolic therapy for the skeleton, despite its better known action as a resorptive hormone. The anabolic effectiveness of PTH requires that it be administered intermittently, and this has been achieved with the use of daily injections that rapidly achieve a peak level in blood that is not maintained.[21]

Recent research has shed new light on the control of bone formation, and the effect of PTH treatment is such that it is important to learn from the action of PTH and to determine whether any of the newly recognized pathways play any part in PTH action. Such studies are at a very early stage, but some potential targets are emerging for development of anabolic agents. There is particular interest in the possibility of modulating the activity of components of the Wnt canonical signaling pathway to produce a net anabolic effect. The first link between Wnt signaling and human bone disease came from observations that inactivating mutations in the low-density lipoprotein receptor–related protein 5 (*LRP5*) cause osteoporosis-pseudoglioma syndrome (OPPG, OMIM 259770), characterized by severely decreased bone mass.[22] Conversely, a syndrome of high bone mass was found to be caused by a gain-of-function mutation of *LRP5* (OMIM 601884).[23] These genetic syndromes were reproduced with the appropriate genetic manipulations in mice.[24,25]

Wnt Signaling Targets

The Wnt/β-catenin signaling pathway offers several targets that may be suitable for pharmacological intervention at a number of specific points. These include extracellular agonists and the points of interaction of antagonists, especially the secreted frizzled-related proteins (SFRPs), dickkopf (DKK) proteins, and sclerostin, as well as regulation within the cell of glycogen synthase kinase -3β (GSK-3β), the enzyme that plays a crucial role in determining availability of β-catenin for the transcriptional effects that are essential for Wnt signaling (see

© 2008 American Society for Bone and Mineral Research

http://www.stanford.edu/%7Ernusse/wntwindow.html). The primary aim of these interventions is to increase Wnt/β-catenin canonical signaling to increase bone mass. Initial success in animal models has been reported with the inhibition of DKK-1, GSK-3, and sclerostin. Another potential target is SFRP-1. Deletion of *Sfrp-1* enhances osteoblast proliferation, differentiation, and function, whereas it suppresses osteoblast and osteocyte apoptosis.[26]

Inhibition of DKK-1 Action. LRP5 can form a ternary complex with DKK-1 and Kremen (a receptor for DKK), which triggers rapid internalization and depletion of LRP5, leading to inhibition of the canonical Wnt signaling pathway. Inhibition of interactions between DKK-1 and LRP5 would release LRP5 to activate the Wnt pathway. Genetic studies with mice lacking a single allele of DKK-1 showed a markedly increased trabecular bone volume and elevated trabecular bone formation rate,[27] whereas transgenic overexpression of DKK-1 under the control of the *Col1A1* promoter caused severe osteopenia.[28] The production of DKK-1 by multiple myeloma cells has been invoked as a contributing factor to the reduced bone formation in the lytic bone lesions of myeloma.[29] A study using antibodies raised against DKK-1 in the treatment of a mouse model of multiple myeloma showed increased numbers of osteoblasts, reduced number of osteoclasts, and reduced myeloma burden in the antibody-treated mice.[30] Initial drug development attempts in osteoporosis also use monoclonal ani-DKK-1 reagents, but the possibility of directing small molecules to prevent the DKK-1–LRP5 interaction is being explored.

Inhibition of GSK-3. Inhibition of GSK-3 would prevent the phosphorylation of β-catenin, leading to stabilization of β-catenin independently of Wnt interactions with the receptor complex. Mice treated with lithium chloride as a GSK-3 inhibitor showed increased bone formation and bone mass.[31] Treatment of ovariectomized rats with an orally active dual GSK α/β inhibitor, LY603281-31-8, for 2 mo resulted in an increase in the number of trabeculae and connectivity, as well as trabecular area and thickness. BMD at cancellous and cortical sites was increased, and this was associated with increased strength. The increased bone formation shown on histomorphometric analysis was associated with increased expression of mRNA markers of the osteoblast phenotype, such as bone sialoprotein, type 1 collagen, osteocalcin, alkaline phosphate, and runx-2.[32] When the same drug was compared with PTH in a cDNA microarray study, ovariectomy in 6-mo-old rats resulted in decreased markers of osteogenesis and chondrogenesis, as well as increased adipogenesis, both of which were reversed by PTH and the GSK-3 α/β inhibitor.[33]

Inhibition of Sclerostin. Sclerostin, the protein product of *SOST*, is produced in bone by osteocytes and is a circulating inhibitor of the Wnt-signaling pathway that achieves this by binding to LRP5 and LRP6.[34] High BMD in Van Buchem's disease is caused by an inactivating mutation in the *SOST* gene (OMIM 607363).[35] Inhibition of production or action of sclerostin resulting in enhanced Wnt canonical signaling would be predicted to lead to increased bone mass. Indeed, in preclinical studies reported in abstract form at the time of writing, monoclonal antibody against sclerostin has been shown to promote bone formation rapidly in monkeys and ovariectomized rats. Considerable increases in bone formation rates and in amounts of trabecular bone took place rapidly without increases in resorption parameters.[36,37] Sclerostin is produced in bone virtually exclusively in osteocytes. Of particular interest is the fact that PTH rapidly reduces sclerostin mRNA and protein production by osteoblasts in vitro and in bone in vivo,[38,39] raising the possibility that transient reduction of sclerostin output by osteocytes in response to intermittent PTH could mediate enhanced osteoblast differentiation and bone formation[40] and reduced osteoblast apoptosis.[41] Such a mechanism would offer real possibilities as a drug target, and the mechanism of this inhibition is all the more interesting, with the recent finding[42] that the cyclic AMP–mediated effect of PTH to diminish sclerostin production operates through a long range enhancer, MEF2, the discovery of which came from the pursuit of the nature of the van Buchem's disease mutation. There may be small molecule approaches amenable to sclerostin regulation, in addition to antibody neutralization of its activity.

Safety and Specificity

Any new therapy emerging from manipulation of the Wnt canonical signaling pathway will need to ensure first that it is safe and second that its action can be targeted specifically to bone. Wnt proteins are critical signaling proteins involved in developmental biology, with roles in early axis specification, brain patterning, intestinal development, and limb development. In adults, Wnt proteins play a vital role in tissue maintenance, with aberrations in Wnt signaling leading to diseases such as adenomatous polyposis.[43] Inhibition of GSK-3 results in increased cyclin D1, cyclin E, and c-Myc, and overexpression of these cell cycle regulators has been linked to tumor formation.[44] All relevant possibilities of side effects of enhanced Wnt signaling need to be kept in mind throughout preclinical studies.

CONCLUSION

The approach of targeting specific peptides that are known to play important roles in osteoblast formation or osteoclast action have yielded several promising candidates with the potential to become effective antiresorptive or anabolic agents to treat osteoporosis. Exciting possibilities of new anabolic therapies are evident, and the benefit of this approach is clear from the satisfactory increase in bone formation that is achieved with PTH treatment. The early evidence that it might be possible to develop resorption inhibitors that can uncouple bone resorption from bone formation is exciting, potentially offering an advantage over currently available antiresorptive agents. Nonetheless, all these predictions are based largely on preclinical data, and it is hoped that properly conducted clinical trials in the coming years will see the emergence of new therapies that are effective, durable, and safe, and at an affordable cost.

REFERENCES

1. Delmas PD 2002 Treatment of postmenopausal osteoporosis. Lancet **359:**2018–2026.
2. Martin TJ, Seeman E 2007 New mechanisms and targets in the treatment of bone fragility. Clin Sci **112:**77–91.
3. McClung MR, Lewiecki EM, Cohen SB, Bolognese MA, Woodson GC, Moffett AH, Peacock M, Miller PD, Lederman SN, Chesnut CH, Lain D, Kivitz AJ, Holloway DL, Zhang C, Peterson MC, Bekker PJ, AMG 162 Bone Loss Study Group 2006 Denosumab in postmenopausal women with low bone mineral density. N Engl J Med **354:**821–831.
4. Gelb BD, Shi GP, Chapman HA, Desnick RJ 1996 Pycnodysostosis, a lysosomal disease caused by cathepsin K deficiency. Science **273:**1236–1238.
5. Saftig P, Hunziker E, Wehmeyer O, Jones S, Boyde A, Rommerskirch W, Moritz JD, Schu P, von Figura K 1998 Impaired osteoclastic bone resorption leads to osteopetrosis in cathepsin-K-deficient mice. Proc Natl Acad Sci USA **95:**13453–13458.

© 2008 American Society for Bone and Mineral Research

6. Yamashita DS, Dodds RA 2000 Cathepsin K and the design of inhibitors of cathepsin K. Curr Pharm Des **6:**1–24.
7. Tavares FX, Boncek V, Deaton DN, Hassell AM, Long ST, Miller AB, Payne AA, Miller LR, Shewchuk LM, Wells-Knecht K, Willard DH, Wright LL, Zhou HQ 2004 Design of potent, selective, and orally bioavailable inhibitors of cysteine protease cathepsin K. J Med Chem **47:**588–599.
8. Adami S, Supronik J, Hala T, Brown JP, Garnero P, Haemmerle S, Ortmann CE, Bouisset F, Trechsel U 2006 Effect of One Year Treatment with the Cathepsin-K Inhibitor, Balicatib, on Bone Mineral Density (BMD) in Postmenopausal Women with Osteopenia/Osteoporosis. J Bone Miner Res **21:**S1;S24.
9. Coleman PJ, Brashear KM, Askew BC, Hutchinson JH, McVean CA, Duong le T, Feuston BP, Fernandez-Metzler C, Gentile MA, Hartman GD, Kimmel DB, Leu CT, Lipfert L, Merkle K, Pennypacker B, Prueksaritanont T, Rodan GA, Wesolowski GA, Rodan SB, Duggan ME 2004 Non peptide alphavebeta 3 antagonists. Part 11; discovery and preclinical evaluation of potent alphavebeta 3 antagonists for the prevention and treatment of osteoporosis. J Med Chem **47:**4829–4837.
10. Murphy MG, Cerchio K, Stoch SA, Gottesdiener K, Wu M, Recker R, for the L-000845704 Study Group 2005 Effect of L-000845704, an αvβ3 integrin antagonist, on markers of bone turnover and bone mineral density in postmenopausal osteoporotic women. J Clin Endocrinol Metab **90:**2022–2028.
11. Visentin L, Dodds RA, Valente M, Misiano P, Bradbeer JN, Oneta S, Liang X, Gowen M, Farina C 2000 A selective inhibitor of the osteoclastic V-H (+)–ATPase prevents bone loss in both thyroparathyroidechomized and ovariectomized rats. J Clin Invest **106:**309–318.
12. Marzia M, Sims NA, Voit S, Migliaccio S, Taranta A, Bernadini S, Faraggiana T, Yoneda T, Mundy GR, Boyce BF, Baron R, Teti A 2000 Decreased c-src expression enhances osteoblastic differentiation and bone formation. J Cell Biol **151:**311–320.
13. Schaller S, Henriksen K, Sorensen MG, Karsdal MA 2005 The role of chloride channels in osteoclasts: ClC-7 as a target for osteoporosis treatment. Drug News Perspect **18:**489–495.
14. Brockstedt H, Bollerslev J, Melsen F, Mosekilde L 1996 Cortical bone remodelling in autosomal dominant osteopetrosis: A study of two different phenotypes. Bone **18:**67–72.
15. Del Fattore A, Peruzzi B, Rucci N, Recchia I, Cappariello A, Longo M, Fortunati D, Ballanti F, Jacobini M, Luciani M, Devito R, Pinto R, Caniglia M, Lanino E, Mesina C, Cesaro C, Letizia C, Bianchini G, Fryssira H, Grabowski P, Shaw N, Bishop N, Hughes D, Kapur R, Datta H, Taranta A, Fornari D, Migliaccio S, Teti A 2005 Clinical, genetic, and cellular analysis of 49 osteopetrotic patients: Implications for diagnosis and treatment. J Med Genet **43:**315–325.
16. Martin TJ, Sims NA 2005 Osteoclast-derived activity in the coupling of bone formation to resorption. Trends Mol Med **11:**76–81.
17. Karsdal M, Martin TJ, Bollerslev J, Christiansen C, Henriksen K 2007 Are non-resorbing osteoclasts sources of bone anabolic activity? J Bone Miner Res **22:**487–494.
18. Lee S-H, Rho J, Jeong D, Sul J-Y, Kim T, Kim N, Kang J-S, Miyamoto T, Suda T, Lee S-K, Pignolo RJ, Koczon-Jaremko B, Lorenzo J, Choi Y 2007 v-ATPase V0 subunit d2-deficient mice exhibit impaired osteoclast fusion and increased bone formation. Nat Med **12:**1401–1409.
19. Boyce BF, Xing L, Yao Z, Shakespeare WC, Wang Y, Metcalf CA III, Sundaramoorthi R, Dalgarno DC, Iuliucci JD, Sayer TK 2006 Future anti-catabolic therapeutic targets in bone disease. Ann NY Acad Sci **1068:**447–457.
20. Hannon RA, Clack G, Swaisland A, Churchman C, Finkelman RD, Eastell R 2005 The effect of AZD0530, a highly selective Src inhibitor, on bone turnover in healthy males. J Bone Miner Res **20:**S1;S372.
21. Frolik CA, Black EC, Cailn RL, Sutterwhite JH, Brown-Augsburger PL, Sato M, Hick JM 2003 Anabolic and catabolic bone effects of human parathyroid hormone (1-34) are predicted by duration of hormone exposure. Bone **33:**372–379.
22. Gong Y, Slee RB, Fukai N, Rawadi G, Roman-Roman S, Reginato AM, Wang H, Cundy T, Glorieux FH, Lev D, Zacharin M, Oexle K, Marcelino J, Suwairi W, Heeger S, Sabatakos G, Apte S, Adkins WN, Allgrove J, Arslan-Kirchner M, Batch JA, Beighton P, Black GC, Boles RG, Boon LM, Borrone C, Brunner HG, Carle GF, Dallapiccola B, De Paepe A, Floege B, Halfhide ML, Hall B, Hennekam RC, Hirose T, Jans A, Jüppner H, Kim CA, Keppler-Noreuil K, Kohlschuetter A, LaCombe D, Lambert M, Lemyre E, Letteboer T, Peltonen L, Ramesar RS, Romanengo M, Somer H, Steichen-Gersdorf E, Steinmann B, Sullivan B, Superti-Furga A, Swoboda W, van den Boogaard MJ, Van Hul W, Vikkula M, Votruba M, Zabel B, Garcia T, Baron R, Olsen BR, Warman ML, Osteoporosis-Pseudoglioma Syndrome Collaborative Group 2001 LDL receptor-related 5 (LRP5) affects bone accrual and eye development. Cell **107:**513–523.
23. Boyden LM, Mao J, Belsky J, Mitzner L, Farhi A, Mitnick MA, Wu D, Insogna K, Lifton RP 2002 High bone density due to a mutation in LDL-receptor-related protein 5. N Engl J Med **346:**1513–1521.
24. Kato M, Patel MS, Levasseur R, Lobov I, Chang BH, Glass DA, Hartmann C, Li L, Hwang TH, Brayton CF, Lang RA, Karsenty G, Chan L 2002 Cbfa1-independent decrease in osteoblast proliferation, osteopenia, and persistent embryonic eye vascularization in mice deficient in Lrp5, a Wnt coreceptor. J Cell Biol **157:**303–304.
25. Babij P, Zhao W, Small C, Kharode Y, Yaworsky PJ, Bouxsein ML, Reddy PS, Bodine PV, Robinson JA, Bhat B, Marzolf J, Moran RA, Bex F 2003 High bone mass in mice expressing a mutant LRP5 gene. J Bone Miner Res **18:**960–974.
26. Bodine PV, Zhao W, Kharode YP, Bex FJ, Lambert AJ, Goad MB, Gaur T, Stein GS, Lian JB, Komm BS 2004 The Wnt antagonist secreted frizzled-related protein-1 is a negative regulator of trabecular bone formation. Mol Endocrinol **18:**1222–1237.
27. Morvan F, Boulukos K, Clément-Lacroix P, Roman Roman S, Suc-Royer I, Vayssière B, Ammann P, Martin P, Pinho S, Pognonec P, Mollat P, Niehrs C, Baron R, Rawadi G 2006 Deletion of a single allele of the Dkk1 gene leads to an increase in bone formation and bone mass. J Bone Miner Res **21:**934–945.
28. Li J, Sarosi I, Cattley RC, Pretorius J, Asuncion F, Grisanti M, Morony S, Adamu S, Geng Z, Qiu W, Kostenuik P, Lacey DL, Simonet WS, Bolon B, Qian X, Shalhoub V, Omnisky MS, Zhu Ke H, Li X, Richards WG 2006 Dkk1-mediated inhibition of Wnt signaling in bone results in osteopenia. Bone **39:**754–766.
29. Tian E, Zhan F, Walker R, Rasmussen E, Ma Y, Barlogie B, Shaughnessy JD Jr 2003 The role of the Wnt-signaling antagonist DKK-1 in the development of osteolytic lesions in multiple myeloma. N Engl J Med **349:**2483–2494.
30. Yacooby S, Ling W, Zhan F, Walker R, Barlogie B, Shaughnessy JD Jr 2007 Antibody-based inhibition of DKK1 suppresses tumor-induced bone resorption and multiple myeloma growth in vivo. Blood **109:**2106–2111.
31. Clément-Lacroix P, Ai M, Morvan F, Roman-Roman S, Vayssière B, Belleville C, Estrera K, Warman ML, Baron R, Rawadi G 2005 Lrp5-independent activation of Wnt signaling by lithium chloride increases bone formation and bone mass in mice. Proc Natl Acad Sci USA **102:**17406–17411.
32. Kulkarni NH, Onyia JE, Zeng QQ, Tian X, Liu M, Halladay DL, Frolik CA, Engler T, Wei T, Kriauciunas A, Martin TJ, Sato M, Bryant HU, Ma YL 2006 Orally bioavailable GSK-3α/β dual inhibitor increases markers of cellular differentiation in vitro and bone mass in vivo. J Bone Miner Res **21:**910–920.
33. Kulkarni NH, Wei T, Kumar A, Dow ER, Stewart TR, Shou J, N'cho M, Sterchi DL, Gitter BD, Higgs RE, Halladay DL, Engler TA, Martin TJ, Bryant HU, Ma YL, Onyia JE 2007 Changes in osteoblast, chondrocyte, and adipocyte lineages mediate the bone anabolic actions of PTH and small molecule GSK-3 inhibitor. J Cell Biochem **102:**1504–1518.
34. Li X, Zhang Y, Kang H, Liu W, Liu P, Zhang J, Harris SE, Wu D 2005 Sclerostin binds to LRP5/6 and antagonizes canonical Wnt signaling. J Biol Chem **280:**19883–19887.
35. Ott SM 2005 Review Sclerostin and Wnt signaling – the pathway to bone strength. J Clin Endocrinol Metab **90:**6741–6743.
36. Ominsky M, Stouch B, Doellgast G, Gong J, Cao J, Gao Y, Tipton B, Haldankar R, Winters A, Chen Q, Graham K, Zhou L, Hale M, Henry A, Lightwood D, Moore A, Popplewell A, Robinson M, Vlasseros F, Jolette J, Smith SY, Kostenuik PJ, Simonet WS, Lacey DL, Paszty C 2006 Administration of sclerostin monoclonal antibodies to female cynomolgus monkeys results in increased bone formation, bone mineral density and bone strength. J Bone Miner Res **21:**S1;S44.
37. Ominsky MS, Warmington KS, Asuncion FJ, Tan HL, Grisanti MS, Geng Z, Stephens P, Henry A, Lawson A, Lightwood D, Perkins V, Kirby H, Moore A, Robinson M, Li X, Kostenuik PJ, Simonet WS, Lacey DL, Paszty C 2006 Sclerostin monoclonal antibody treatment increases bone strength in aged osteopenic ovariectomized rats. J Bone Miner Res **21:**S1;S44.

© 2008 American Society for Bone and Mineral Research

38. Keller H, Kneissel M 2005 SOST is a target gene for PTH in bone. Bone **37**:148–158.
39. Silvestrini G, Ballanti P, Leopizzi M, Sebastiani M, Berni S, Di Vito M, Bonucci E 2007 Effects of intermittent parathyroid hormone (PTH) administration on SOST mRNA and protein in rat bone. J Mol Histol **38**:261–269.
40. Van Bezooijen RL, ten Dijke P, Papapoulos SE, Löwik CW 2005 SOST/sclerostin, an osteocyte-derived negative regulator of bone formation. Cytokine Growth Factor Rev **16**:319–327.
41. Sutherland MK, Geoghegan JC, Yu C, Winkler DG, Latham JA 2004 Unique regulation of SOST, the sclerosteosis gene, by BMPs and steroid hormones in human osteoblasts. Bone **35**:448–454.
42. Leuptin O, Kramer I, Colette NM, Loots GG, Natt F, Kneissel M, Keller H 2007 Control of the SOST bone enhancer by PTH via MEF2 transcription factors. J Bone Miner Res **22**:S1;S13.
43. Krishnan V, Bryant HU, MacDougald OA 2006 Regulation of bone mass by Wnt signaling. J Clin Invest **116**:1202–1209.
44. Dong J, Peng J, Zhang H, Mondesire WH, Jian W, Mills GB, Hung MC, Meric-Bernstam F 2005 Role of glycogen synthase kinase 3beta in rapamycin-mediated cell cycle regulation and chemosensitivity. Cancer Res **65**:1961–1972.

Chapter 57. Juvenile Osteoporosis

Frank Rauch[1] and Nick Bishop[2]

[1]*Genetics Unit, Shriners Hospital for Children, Montréal, Québec, Canada;* [2]*Sheffield Children's Hospital, University of Sheffield, Sheffield, United Kingdom*

INTRODUCTION

Taken literally, the phrase "juvenile osteoporosis" means "osteoporosis in children and adolescents," and thus does not refer to any particular form of osteoporosis in this age group. However, in the scientific literature and in clinical practice, the term "juvenile osteoporosis" is usually used to refer to idiopathic juvenile osteoporosis (IJO). This chapter therefore discusses IJO rather than the entirety of osteoporotic conditions that may occur in young people.

Osteoporosis in childhood and adolescence is most commonly secondary to a spectrum of diverse conditions, such as osteogenesis imperfecta (OI), prolonged immobilization, and chronic inflammatory disease. Bone loss may be worsened by treatment with anticonvulsants and steroids, respectively, but may also improve as the underlying condition improves. Serious diseases such as leukemia also may temporarily present as osteoporosis. If no underlying cause can be detected, IJO is said to be present.

IJO was first described as a separate entity by Dent and Friedman[(1)] four decades ago. According to the classical description, IJO is a self-limiting disease that develops in a prepubertal, previously healthy child, leads to metaphyseal and vertebral compression fractures, and is characterized radiologically by radiolucent areas in the metaphyses of long bones, dubbed "neo-osseous osteoporosis."[(2)] However, it is clear that there are many children and adolescents who have low bone mass (defined as a body size–adjusted BMC or BMD measured by DXA at the spine or total body that is >2 SD below the mean; i.e., a Z-score ≤ -2) and who sustain fractures after minimal trauma, but whose clinical findings do not correspond to the classical description of Dent and Friedman. Medical science has not yet reserved a particular name for the condition of such patients, but logically, they should be diagnosed with IJO if no other etiology for low bone mass and fractures is found. Thus, it may be useful to distinguish "classical IJO" (for patients whose presentation is similar to the description of Dent and Friedman) from "IJO in the wider sense" (for patients who do not match the description of Dent and Friedman, but nevertheless have unexplained fractures with low bone mass).

Most reviews on the topic state that IJO is an extremely rare disease, because <200 patients have been published under that label. This is probably because few patients present with the classical picture. However, most clinicians who see children and adolescents with fractures could probably list a few of their patients who have osteoporosis without recognizable etiology. In our clinical settings, IJO is ~10 times less common than OI.

The authors state that they have no conflicts of interest.

© 2008 American Society for Bone and Mineral Research

PATHOPHYSIOLOGY

The etiology of IJO is unknown. One study found a normal rise in serum osteocalcin in six IJO patients after calcitriol was administered orally, which was postulated to indicate "normal osteoblast function."[(3)] However, the fact that osteoblasts in this test released normal amounts of osteocalcin into the circulation does not necessarily mean that they also deposited matrix on the bone surface in a normal fashion.

Early histomorphometric reports on IJO were limited to static methods to quantify bone metabolism, described single cases, or did not have adequate control groups.[(4–8)] No conclusive picture emerged from these reports. More recent studies using dynamic histomorphometry showed that IJO was characterized by a markedly reduced activation frequency and therefore low remodeling activity.[(9)] In addition, the amount of bone formed at each remodeling site was abnormally low. No evidence was found for increased bone resorption. Interestingly, the bone formation defect was limited to bone surfaces that were exposed to the bone marrow environment; no abnormalities were detected in intracortical and periosteal surfaces.[(10)] These results suggested that, in IJO, impaired osteoblast performance decreases the ability of cancellous bone to adapt to the increasing mechanical needs during growth. This results in load failure at sites where cancellous bone is essential for stability. The initial trigger of the decrease in osteoblast performance remains nevertheless elusive.

Two recent reports have indicated that heterozygous mutations in the low-density lipoprotein receptor-related protein 5 (LRP5) can result in low bone mass with fractures in some children.[(11,12)] The frequency of such mutations in children with classical IJO as opposed to childhood osteoporosis generally remains to be determined.

CLINICAL FEATURES

Classical IJO typically develops in a prepubertal (mostly between 8 and 12 yr of age), previously healthy child of either

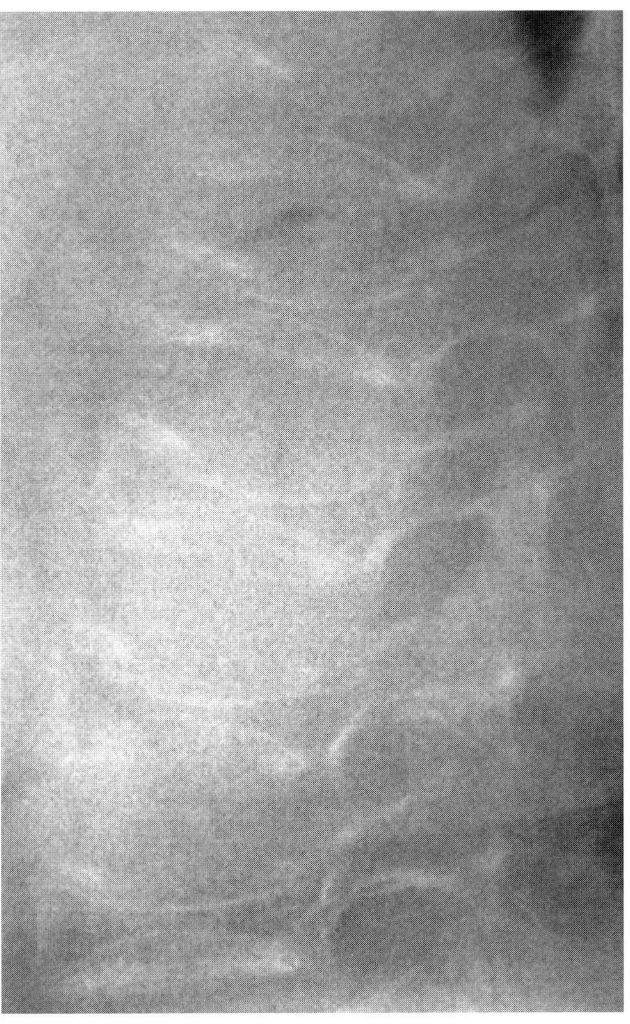

FIG. 1. Lateral lumbar spine radiograph of a 10-yr-old girl with IJO. Compression fractures of all vertebral bodies and severe osteoporosis are evident. At the time of this radiograph, lumbar spine areal BMD Z-score was −4.9 (at a height Z-score of −2.5).

sex.[13] However, a series of 21 children who were presented as having IJO indicated a mean age at onset of 7 yr with a range of 1–13 yr and no sex differences.[14]

Symptoms begin with an insidious onset of pain in the lower back, hips, and feet and difficulty walking.[1] Knee and ankle pain and fractures of the lower extremities may be present, as well as diffuse muscle weakness. Vertebral compression fractures are frequent, resulting in a short back (Fig. 1). Long bone fractures, mostly at metaphyseal sites, may occur. Physical examination may be entirely normal or show thoracolumbar kyphosis or kyphoscoliosis, pigeon chest deformity, loss of height, deformities of the long bones, and limp.

RADIOLOGICAL FEATURES

Children with fully expressed classical IJO present with generalized osteopenia and collapsed or biconcave vertebrae. Disc spaces may be widened asymmetrically because of wedging of the vertebral bodies. Long bones usually have normal diameter and cortical width, unlike the thin, gracile bones of children with OI. The typical radiographic finding in IJO is neo-osseous osteoporosis, a radiolucent band at sites of newly formed metaphyseal bone. This localized metaphyseal weakness can give rise to fractures, often at the distal tibias and adjacent to the knee and hip joints.[7] Nevertheless, "neo-osseous osteoporosis" is not a prerequisite for diagnosing IJO.

BIOCHEMICAL FINDINGS

Biochemical studies of bone and mineral metabolism have not detected any consistent abnormality in children with IJO.[14–16]

Bone Biopsy

Iliac bone biopsies show low trabecular bone volume but largely preserved core width (i.e., a normal outer size of the biopsy specimen) and cortical width.[9,10] Tetracycline double-labeling shows a low extent of mineralizing surface (a sign of decreased remodeling activity) and low mineral apposition rate (a sign of weakness of the individual osteoblast team at a remodeling site). There is no indication of a mineralization defect. Osteoclasts are normal in appearance and number.

Differential Diagnosis

The diagnosis of IJO is made by the exclusion of known etiologies for low bone mass and fractures. The list of conditions that may be associated with bone fragility in children and adolescents is rather long (Table 1). The exclusion of most of these disorders is usually not difficult. The most frequent diagnostic problem facing a clinician is probably to separate IJO from OI type I and adaptational problems during growth.

Table 2 presents the typical distinguishing features between IJO and OI type I. Apart from bone fragility and low bone mass, most patients with OI type I have associated extraskeletal connective tissue signs, such as bluish or grayish sclera,

TABLE 1. FORMS OF OSTEOPOROSIS IN CHILDREN, ACCORDING TO CURRENT LITERATURE

I. Primary
Osteogenesis imperfecta
Idiopathic juvenile osteoporosis
II. Secondary
Endocrine disorders
Cushing syndrome
Thyrotoxicosis
Anorexia nervosa
Inflammatory disorders
Juvenile arthritis
Dermatomyositis
Systemic lupus erythematosis
Inflammatory bowel disease
Cystic fibrosis
Chronic hepatitis
Malabsorption syndromes
Biliary atresia
Inborn errors of metabolism
Homocystinuria
Glycogen storage disease type 1
Immobilization
Cerebral palsy
Duchenne dystrophy
Hematology/oncology
Acute lymphoblastic leukemia
Thalassemia
Severe congenital neutropenia

© 2008 American Society for Bone and Mineral Research

TABLE 2. DIFFERENTIAL DIAGNOSIS BETWEEN IDIOPATHIC JUVENILE OSTEOPOROSIS (IJO) AND OSTEOGENESIS IMPERFECTA (OI) TYPE I

	IJO	*OI type I*
Family history	Negative	Often positive
Onset	Late prepubertal	Birth or soon after
Duration	1–5 yr	Lifelong
Clinical findings	Metaphyseal fractures	Long bone diaphyseal fractures
	No signs of connective tissue Involvement	Blue sclerae, joint hyperlaxity, sometimes abnormal dentition
	Abnormal gait	
Growth rate	Normal	Normal or low
Radiologic findings	Vertebral compression fractures	Vertebral compression fractures
	Long bones: predominantly	"Narrow bones" (low diameter of metaphyseal involvement ["neo-diaphyses osseous osteoporosis"])
	No wormian bones	Wormian bones (skull)
Bone biopsy	Decreased bone turnover	Increased bone turnover
	Normal amount of osteocytes	Hyperosteocytosis
Genetic testing	Patients	Mutations affecting collagen type I in most patients

dentinogenesis imperfecta, joint hyperlaxity, and wormian bones (on skull X-rays). However, the extraskeletal involvement can be absent or too subtle to be clinically recognizable in some OI patients. In this situation, genetic analysis of the genes that code for the two collagen type I α chains (*COL1A1* and *COL1A2*) can be helpful. Mutations affecting a glycine residue in either gene or those leading to a quantitative defect in *COL1A1* expression are diagnostic of OI (this issue was somewhat obfuscated by a case report of two brothers who had glycine mutations in the *COL1A2* gene and therefore suffered from OI but were presented as having IJO[17]). A negative collagen type I mutation analysis does not rule out OI, because false-negative results can occur. LRP5 sequencing may be informative in some cases.

An iliac bone biopsy, preferably after tetracycline double-labeling, may also contribute to clarifying the diagnosis. Microscopically, a "lack of activity" is usually noted in IJO, whereas there is "hypercellularity" in OI. In histomorphometric terms, this translates into low activation frequency and bone surface based remodeling parameters in IJO and an increase in these values in OI. Also, hyperosteocytosis is a common feature in OI, whereas the amount of osteocytes appears to be normal in IJO.

Fractures and low bone mass may also occur in healthy prepubertal children. Indeed, during late prepuberty and early puberty, fracture rates are almost as high as in postmenopausal women.[18–20] Similar to IJO, such fractures frequently involve metaphyseal bone sites, especially the distal radius. This may reflect problems in the adaptation of the skeleton, in particular the metaphyseal cortex, to the increasing mechanical needs during growth.[21] Growing children and adolescents who had a few forearm fractures and who have borderline low areal BMD at the spine are frequently encountered in pediatric bone clinics. The medical literature is silent as to the distinction between IJO and patients who may have an "adaptational bone problem." We propose that classical IJO should only be diagnosed when vertebral compression fractures are present (with or without extremity fractures).

Treatment

There is no treatment with proven benefit to the patient. The effect of any kind of medical intervention is difficult to judge in IJO, because the disease is rare, has a variable course, and usually resolves without treatment. Nevertheless, given the current enthusiasm for pediatric bisphosphonate treatment, many IJO patients probably are receiving treatment with such drugs. A number of case reports have described increasing BMD and clinical improvement after treatment with bisphosphonates was started.[22–24] In any case, attempts at medical therapies should complement rather than replace orthopedic and rehabilitative measures, such as physiotherapy. Review at 6-mo intervals is also warranted in children not receiving bisphosphonates. Changes in the shape of the spine should be monitored carefully, and early referral to a specialist pediatric spine surgeon should be made in any progressive cases.

Prognosis

The disease process is only active in growing children, and spontaneous recovery is the rule after 3–5 yr of evolution.[14] However, in some of the most severe cases reported to date, deformities and severe functional impairment persisted, which left them wheelchair bound with cardiorespiratory abnormalities.[1,14] Preventing such deformities with attendant loss of function should be the focus of attention during the active phase of the disease.

REFERENCES

1. Dent CE, Friedman M 1965 Idiopathic juvenile osteoporosis. Q J Med **34:**177–210.
2. Dent CE 1977 Osteoporosis in childhood. Postgrad Med J **53:**450–457.
3. Bertelloni S, Baroncelli GI, Di Nero G, Saggese G 1992 Idiopathic juvenile osteoporosis: Evidence of normal osteoblast function by 1,25-dihydroxyvitamin D3 stimulation test. Calcif Tissue Int **51:**20–23.
4. Cloutier MD, Hayles AB, Riggs BL, Jowsey J, Bickel WH 1967 Juvenile osteoporosis: Report of a case including a description of some metabolic and microradiographic studies. Pediatrics **40:**649–655.
5. Gooding CA, Ball JH 1969 Idiopathic juvenile osteoporosis. Radiology **93:**1349–1350.
6. Jowsey J, Johnson KA 1972 Juvenile osteoporosis: Bone findings in seven patients. J Pediatr **81:**511–517.
7. Smith R 1980 Idiopathic osteoporosis in the young. J Bone Joint Surg Br **62-B:**417–427.
8. Evans RA, Dunstan CR, Hills E 1983 Bone metabolism in idiopathic juvenile osteoporosis: A case report. Calcif Tissue Int **35:** 5–8.
9. Rauch F, Travers R, Norman ME, Taylor A, Parfitt AM, Glorieux FH 2000 Deficient bone formation in idiopathic juvenile osteoporosis: A histomorphometric study of cancellous iliac bone. J Bone Miner Res **15:**957–963.

© 2008 American Society for Bone and Mineral Research

10. Rauch F, Travers R, Norman ME, Taylor A, Parfitt AM, Glorieux FH 2002 The bone formation defect in idiopathic juvenile osteoporosis is surface-specific. Bone **31**:85–89.
11. Toomes C, Bottomley HM, Jackson RM, Towns KV, Scott S, Mackey DA, Craig JE, Jiang L, Yang Z, Trembath R, Woodruff G, Gregory-Evans CY, Gregory-Evans K, Parker MJ, Black GC, Downey LM, Zhang K, Inglehearn CF 2004 Mutations in LRP5 or FZD4 underlie the common familial exudative vitreoretinopathy locus on chromosome 11q. Am J Hum Genet **74**:721–730.
12. Hartikka H, Makitie O, Mannikko M, Doria AS, Daneman A, Cole WG, Ala-Kokko L, Sochett EB 2005 Heterozygous mutations in the LDL receptor-related protein 5 (LRP5) gene are associated with primary osteoporosis in children. J Bone Miner Res **20**:783–789.
13. Teotia M, Teotia SP, Singh RK 1979 Idiopathic juvenile osteoporosis. Am J Dis Child **133**:894–900.
14. Smith R 1995 Idiopathic juvenile osteoporosis: Experience of twenty-one patients. Br J Rheumatol **34**:68–77.
15. Saggese G, Bertelloni S, Baroncelli GI, Perri G, Calderazzi A 1991 Mineral metabolism and calcitriol therapy in idiopathic juvenile osteoporosis. Am J Dis Child **145**:457–462.
16. Saggese G, Bertelloni S, Baroncelli GI, Di Nero G 1992 Serum levels of carboxyterminal propeptide of type I procollagen in healthy children from 1st year of life to adulthood and in metabolic bone diseases. Eur J Pediatr **151**:764–768.
17. Dawson PA, Kelly TE, Marini JC 1999 Extension of phenotype associated with structural mutations in type I collagen: Siblings with juvenile osteoporosis have an alpha2(I)Gly436–> Arg substitution. J Bone Miner Res **14**:449–455.
18. Landin LA 1997 Epidemiology of children's fractures. J Pediatr Orthop B **6**:79–83.
19. Cooper C, Dennison EM, Leufkens HG, Bishop N, van Staa TP 2004 Epidemiology of childhood fractures in Britain: A study using the general practice research database. J Bone Miner Res **19**:1976–1981.
20. Khosla S, Melton LJ III, Dekutoski MB, Achenbach SJ, Oberg AL, Riggs BL 2003 Incidence of childhood distal forearm fractures over 30 years: A population-based study. JAMA **290**:1479–1485.
21. Rauch F, Neu C, Manz F, Schoenau E 2001 The development of metaphyseal cortex–implications for distal radius fractures during growth. J Bone Miner Res **16**:1547–1555.
22. Hoekman K, Papapoulos SE, Peters AC, Bijvoet OL 1985 Characteristics and bisphosphonate treatment of a patient with juvenile osteoporosis. J Clin Endocrinol Metab **61**:952–956.
23. Brumsen C, Hamdy NA, Papapoulos SE 1997 Long-term effects of bisphosphonates on the growing skeleton. Studies of young patients with severe osteoporosis. Medicine (Baltimore) **76**:266–283.
24. Kauffman RP, Overton TH, Shiflett M, Jennings JC 2001 Osteoporosis in children and adolescent girls: Case report of idiopathic juvenile osteoporosis and review of the literature. Obstet Gynecol Surv **56**:492–504.

Chapter 58. Glucocorticoid-Induced Osteoporosis

Robert S. Weinstein

Division of Endocrinology and Metabolism, Center for Osteoporosis and Metabolic Bone Diseases, Department of Internal Medicine, and the Central Arkansas Veterans Healthcare System, University of Arkansas for Medical Sciences, Little Rock, Arkansas

INTRODUCTION

Glucocorticoid-induced osteoporosis (GIO) is the second most common form of osteoporosis and is the most common iatrogenic form of osteoporosis.[1,2] Fractures may occur in 30–50% of patients receiving chronic glucocorticoid therapy and many are asymptomatic, possible because of glucocorticoid-induced analgesia.[3,4] However, even asymptomatic prevalent fractures are important because they further reduce vital capacity in patients with chronic lung disease and increase the risk of subsequent fractures independently of BMD.[1,3] In addition to GIO, glucocorticoid administration causes the most frequent and devastating form of osteonecrosis.[5] The adverse effects of glucocorticoids on the skeleton are primarily directly on the osteoblasts, osteocytes, and osteoclasts, decreasing the production of both osteoblasts and osteoclasts and increasing the apoptosis of osteoblasts and osteocytes while prolonging the lifespan of osteoclasts.[6–10] Thus, bisphosphonates are effective in the preservation of BMD and prevention of fractures in GIO but less so than in postmenopausal osteoporosis. Consequently, anabolic therapy with intermittent injections of PTH, a regimen that decreases apoptosis of osteoblasts and osteocytes, seems quite promising.

CLINICAL FEATURES

Physicians have long been perplexed by the occasional patient who becomes Cushingoid when treated with relatively small amounts of glucocorticoids, whereas other patients seem to be remarkably resistant. Another puzzle is that glucocorticoid-induced fractures may sometimes occur without a Cushingoid habitus. The explanation may be that the sensitivity to exogenous glucocorticoids in different tissues is mediated by inherited gradations in the local activity of the 11β-hydroxysteroid dehydrogenase (11β-HSD) system. This remarkable system is a natural prereceptor modulator of corticosteroid action. Two isoenzymes, 11β-HSD1 and 11β-HSD2, catalyze the interconversion of hormonally active glucocorticoids (such as cortisol or prednisolone) and inactive glucocorticoids (such as cortisone or prednisone). The 11β-HSD1 enzyme is an activator and the 11β-HSD2 enzyme is an inactivator. The ability of any glucocorticoid to bind to the glucocorticoid receptor (GR) depends on the presence of a hydroxyl group at C-11. Therefore, any tissue expressing 11β-HSDs can regulate the exposure of the resident cells to active glucocorticoids. Increased fractures caused by GIO in the elderly may be caused by increases in 11β-HSD1 that occur with aging and glucocorticoid administration.[11] In addition, glucocorticoid sensitivity has also been postulated to result from polymorphisms in the *GR* gene. Another explanation for the individual susceptibility to fractures is the underlying condition that requires glucocorticoid therapy.[2,12–14] Fractures may be more common with high dose glucocorticoids and other immunosuppressive drugs used in transplant recipients,[15] inflammatory bowel disease and the accompanying malabsorption,[16–18] severe rheumatoid arthritis and the relative

Dr. Weinstein has served on the Merck Consultants Board.

Key words: corticosteroid osteoporosis, glucocorticoid-induced osteoporosis, 11β-hydroxysteroid dehydrogenase, bone cell apoptosis, fractures, osteoblasts, osteocytes, osteoclasts, histomorphometry, BMD, osteonecrosis, cancellous bone, cortical bone, side effects, kyphoplasty

© 2008 American Society for Bone and Mineral Research

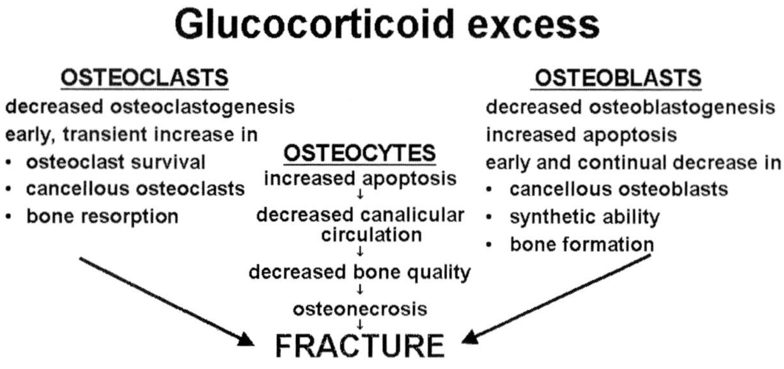

Glucocorticoid excess

OSTEOCLASTS
decreased osteoclastogenesis
early, transient increase in
• osteoclast survival
• cancellous osteoclasts
• bone resorption

OSTEOCYTES
increased apoptosis
↓
decreased canalicular circulation
↓
decreased bone quality
↓
osteonecrosis
↓
FRACTURE

OSTEOBLASTS
decreased osteoblastogenesis
increased apoptosis
early and continual decrease in
• cancellous osteoblasts
• synthetic ability
• bone formation

FIG. 1. Diagram showing the direct effects of glucocorticoid excess on bone cells and the pathway to fractures and osteonecrosis.

immobilization in addition to the systemic inflammation,[14] and chronic pulmonary disease and the rib fractures provoked by fits of coughing[19] than with myasthenia gravis[20] or multiple sclerosis.[21] However, the absence of severe underlying disease or the presence of youth does not convey protection from glucocorticoid-induced bone disease. Osteoporosis and osteonecrosis have been reported in young and older patients receiving depot, topical, or nasal glucocorticoids for dermatitis, rhinitis, or hay fever and even with modest over-replacement for pituitary or adrenal insufficiency or congenital adrenal hyperplasia.[22–31]

Bone loss in GIO is biphasic, with a reduction in BMD of 6–12% within the first year, followed by a slower annual loss of ~3%; however, there is more to consider than just reduced BMD.[32] Bone quality is also an issue.[33] Importantly, the relative risk of fracture increases rapidly and escalates by as much as 75% within the first 3 mo after initiation of steroid therapy. This effect is offset shortly after discontinuation of therapy, which suggests that glucocorticoids do more than just decrease bone mass.[34] Consistent with these concepts, glucocorticoid-treated patients with vertebral fractures are ~10 yr younger and have higher BMD but suffer twice the risk of fracture than patients with vertebral fractures caused by postmenopausal osteoporosis.[18,34] GIO is diffuse, affecting both cortical and cancellous bone, but there is a distinct predilection for fractures in regions of the skeleton with abundant cancellous bone such as the ribs, lumbar spine, and proximal femur. There may be no safe dose because an increase in vertebral fractures is noted with as little as 2.5 mg/d of prednisolone.[2,3,34] In addition, replacing oral with inhaled glucocorticoids, alternate day regimens, and intermittent therapy has failed to prevent GIO.[35–39]

BONE HISTOMORPHOMETRY

Histomorphometric studies in patients receiving long-term glucocorticoid treatment consistently have shown reduced numbers of osteoblasts on cancellous bone and diminished wall width, a measure of the work performed by these cells.[1,6,40,41] The decreased osteoblasts are caused by glucocorticoid-induced reductions in the production of new osteoblast precursors and premature apoptosis of mature, matrix-secreting osteoblasts.[6] Therefore, it should not be surprising that a decrease in the bone formation rate is an expected histological finding in glucocorticoid-treated patients.[1] Inadequate numbers of osteoblasts are also an important cause of the reduction in cancellous bone area and decrease in trabecular width, a result of incomplete cavity repair during bone remodeling.[1,6,40] With glucocorticoid excess, cancellous bone area is often <12% (normal, 21.6 ± 4.5 SD)[42] and correlates with the decreased wall width.[41] Cortical bone at the iliac crest shows increased porosity, whereas cortical width ranges from clearly subnormal to within normal limits.[43]

Some clinical histomorphometric studies of GIO have reported moderate increases in the erosion perimeter, but others showed no significant change. However, increased erosion perimeter may occur merely because of a delay in bone formation.[1,6] When carefully measured, osteoclast numbers in patients receiving chronic glucocorticoid treatment are within the normal range or just slightly above normal.[1,6,40–42]

DIRECT EFFECTS OF GLUCOCORTICOIDS ON BONE CELLS

Osteoblasts and Osteocytes

Glucocorticoids affect every tissue in the body, and thus, the negative impact of excess glucocorticoids on the skeleton could be caused by direct effects on bone cells or indirect effects on extraskeletal tissues or through chemical mediators. Using the 11β-HSD system as a tool to determine the contribution of direct effects of glucocorticoids on osteoblastic/osteocytic cells in vivo, O'Brien et al.[7] overexpressed 11β-HSD2 in mice using the murine osteocalcin gene 2 (OG2) promoter, which is active only in mature osteoblasts and osteocytes. This strategy shields osteoblasts and osteocytes from glucocorticoids before the drugs reach the GR. Under control of this promoter, the transgene did not affect normal bone development or turnover as shown by identical bone density, strength, and histomorphometry in adult transgenic and wildtype animals. The absence of a baseline phenotype facilitated interpretation of a glucocorticoid challenge. Administration of excess glucocorticoids induced equivalent bone loss in wildtype and transgenic mice. This was as expected because cancellous osteoclasts were unaffected by this transgene. However, mice harboring the OG2–11β-HSD2 transgene were protected from glucocorticoid-induced apoptosis of osteoblasts and osteocytes. Prevention of osteoblast and osteocyte apoptosis, in turn, resulted in the preservation of cancellous osteoblast numbers and osteoid production, thereby preventing the expected decrease in bone formation caused by administration of excess glucocorticoids. More strikingly, bone strength was preserved in the transgenic animals despite a loss of BMD, suggesting that the maintenance of osteocyte viability independently contributed to bone strength possibility because of maintenance of the bone circulation through the canalicular system. These results showed that the adverse affects of glucocorticoids on bone formation are primarily the result of direct actions on osteoblasts, because they are abrogated by deflecting the actions of the hormones on this cell type (Fig. 1).

Osteoclasts

Glucocorticoid administration results in a rapid loss of BMD

© 2008 American Society for Bone and Mineral Research

caused by an imbalance in osteoblast and osteoclast numbers. Whereas glucocorticoid administration dramatically reduces the production of both osteoblast and osteoclast precursors as measured in ex vivo bone marrow cell cultures,[6,10] surprisingly, cancellous osteoclast number does not decrease as does the number of osteoblasts, because of the ability of glucocorticoids to promote osteoclast lifespan, an effect mediated by the glucocorticoid receptor as shown by its blockade with RU 486, a potent glucocorticoid receptor antagonist.[10] This effect is powerful enough that it is not preventable by bisphosphonates, possibly accounting for the reduced ability of bisphosphonates to protect BMD in GIO compared with postmenopausal or male osteoporosis.[44–47]

To determine whether glucocorticoids act directiy on osteoclasts in vivo to promote their lifespan and whether this contributes to the rapid loss of bone with glucocorticoid, the 11β-HSD system was again used by overexpressing 11β-HSD2 in mice using the murine TRACP promoter.[8] When challenged with prednisolone, there were equivalent increases in cancellous osteoblast apoptosis and equivalent decreases in osteoblast number, osteoid perimeter, and rate of bone formation in wildtype and transgenic mice. In contrast, glucocorticoids stimulated expression of the mRNA for the calcitonin receptor, an osteoclast product, in wildtype but not transgenic mice. Consistent with the previous findings that glucocorticoids decrease osteoclast precursors and prolong the lifespan of mature osteoclasts, glucocorticoids decreased cancellous osteoclast number in the transgenic mice but not in the wildtype mice. In accordance with this decrease in osteoclast number, the loss of BMD observed in wildtype mice was prevented in transgenic mice. These results show that the early, rapid loss of bone caused by glucocorticoid excess results primarily from direct actions on osteoclasts (Fig. 1). Furthermore, deflection of glucocorticoid action in osteoclasts did not prevent the expected glucocorticoid-induced decrease in osteoblast number or increase in the prevalence of osteoblast apoptosis.

INDIRECT EFFECTS OF GLUCOCORTICOIDS ON BONE CELLS

Lack of a Role of PTH and Hypogonadism in GIO

Evidence firmly indicates that there is "no longer unambiguous, compelling support" for elevated levels of PTH in acute or chronic administration of glucocorticoids.[48–52] Bone densitometry studies also suggest that the preferential loss of appendicular cortical bone with preservation of vertebral cancellous bone typical of PTH excess is not present in GIO in which a major loss of lumbar spinal density occurs before the appendicular skeleton is involved. Moreover, bone histomorphometry shows that the increased osteoid, osteoblasts, bone formation rate, and osteoclasts typical of increased PTH levels are absent in GIO. In summary, the evidence has clearly shifted away from the notion of secondary hyperparathyroidism in GIO. Further support of this conclusion is that, today, the benefit of daily injections of PTH in GIO are impressive.[48,53]

Decreased production of sex steroids is another means by which glucocorticoid excess might stimulate bone resorption. However, loss of gonadal function in either sex stimulates the production of osteoblasts and osteoclasts in the bone marrow, resulting in an increase in cancellous osteoblasts, osteoclasts, and bone turnover—changes that are quite distinct from those found in GIO. Actually, the increases in osteoblast and osteoclast precursors, incidence of new remodeling cycles at sites on cancellous bone, and rate of bone remodeling that occur after the loss of sex steroids are abrogated by glucocorticoid excess.[54] Even low-dose prednisone decreases bone turnover in estrogen-deficient postmenopausal women.[55] Further evidence against a role for sex steroid deficiency in the pathogenesis of GIO was supplied by Pearce et al.,[49] who found a 4.6% decrease in spinal BMD after 6 mo of treatment to decrease anti-sperm antibody formation in infertile men with 50 mg/d of prednisolone, despite the maintenance of a normal testosterone sex hormone–binding globulin ratio and restoration of fertility. Additional strong evidence against the concept that hypogonadism is universal in glucocorticoid-treated patients is provided by registries containing records of thousands of successful pregnancies in woman receiving prednisone to prevent rejection of renal transplants.[56,57] Furthermore, as many as one half of patients with Cushing's disease are amenorrheic, and fracture prevalence (~76%) is identical in both amenorrheic and amenorrheic woman.[58]

Indirect Effects on Bone Cell Progenitors

Although the consensus favors the contention that the vast majority of GIO results primarily from direct effects on bone cells,[59] evidence suggests that some of the actions of glucocorticoid excess on the production of osteoblasts may be indirect, mediated by enhanced expression of dickkopf-1, an antagonist of the Wnt signaling pathway important for bone formation.[1,59,60] Glucocorticoid administration also directly suppresses bone morphogenetic proteins and runt-related transcription factor 2, factors needed to induce osteoblast differentiation, giving the coup de grace to osteoblastogenesis.

Some of the actions of glucocorticoid excess on osteoclasts may also be indirect. An early and transient increase in bone resorption caused by glucocorticoid treatment may be explained as follows: in vivo, glucocorticoids downregulate the mRNA for osteoprotegerin (OPG), a soluble decoy receptor for RANKL, which is essential for the support of osteoclasts. Meanwhile, RANKL mRNA increases in preosteoblastic cells.[61,62] Decreased levels of OPG would allow RANKL to increase osteoclastogenesis by unopposed binding to its specific receptor, RANK, on the surface of hematopoietic osteoclast progenitor cells. The resultant decrease in the OPG/RANKL ratio would contribute to an increase in the number of cancellous osteoclasts on bone and the early, rapid bone loss typical of GIO. Nonetheless, the transient nature of this early increase may be partly explained by the glucocorticoid-induced reduction in osteoblastogenesis,[54,59] eventually resulting in a decline in osteoclast-supporting marrow stromal preosteoblasts. In support of this contention, in vitro studies suggest that glucocorticoid-induced downregulation of osteoprotegerin is only transient.[62]

OSTEONECROSIS

A devastating accompaniment of long-term glucocorticoid therapy is osteonecrosis, which causes collapse of the femoral neck or proximal humerus in as many as 25% of patients who receive high-dose or long-term therapy.[5] However, the name, osteonecrosis, may be misleading, because it has not been shown that, in GIO, the bone cells die by necrosis. Indeed, the cell swelling and inflammatory responses that characterize necrosis in soft tissues usually do not occur in glucocorticoid-induced osteonecrosis. The disorder has been attributed to fat emboli, microvascular tamponade of the blood vessels of the femoral head by marrow fat or fluid retention, and poorly mending fatigue fractures. However, abundant apoptotic osteocytes have been identified in sections of whole femoral heads obtained during total hip replacement for glucocorticoid-induced osteonecrosis, whereas apoptotic bone cells were absent from femoral specimens removed because of traumatic or sickle cell osteonecrosis, suggesting that the so-called glu-

© 2008 American Society for Bone and Mineral Research

cocorticoid-induced osteonecrosis actually is osteocyte apoptosis (Fig. 1).[63,64] Glucocorticoid-induced osteocyte apoptosis, a cumulative and irreparable defect, could uniquely disrupt the mechanosensory function of the osteocyte network and thus start the inexorable sequence of events leading to collapse of the femoral head. Glucocorticoid-induced osteocyte apoptosis would also explain the correlation between total steroid dose and the incidence of osteonecrosis and its occurrence after glucocorticoid administration has ceased.[5,65,66]

MANAGEMENT OF GIO

Responsibilities of Glucocorticoid Therapy

Under some circumstances (vasculitis, rheumatoid arthritis, lupus, status asthmaticus, inflammatory bowel disease), high-dose glucocorticoid therapy is almost obligatory, and the clinician always hopes that the course of treatment will be brief but the days turn to weeks, the weeks to months, and the months to years often without addressing the complications of these drugs. It is the responsibility of physicians who prescribe glucocorticoids to educate their patients about these side effects and complications including osteoporosis and osteonecrosis, cataract and glaucoma, hypokalemia, hyperglycemia, hypertension, hyperlipidemia, weight gain, fluid retention, easy bruisability, susceptibility to infection, impaired healing, myopathy, adrenal insufficiency, and the steroid withdrawal syndrome. All patients receiving long-term glucocorticoid therapy should carry a steroid therapy card or medication identification jewelry. Malpractice suits for failure to document disclosure of the skeletal complications to patients are not rare.[5] Despite this, the bone complications are ignored by >50% of the specialists who prescribe glucocorticoids.[44]

General Measures

Laboratory tests are indicated to exclude vitamin D deficiency, renal insufficiency with secondary hyperparathyroidism, malabsorption, and male hypogonadism. Prevalent fractures, vertebral morphological assessment, spinal radiographs, or changes in height measurements using a stadiometer may help identify patients at risk for further fractures. However, disparity between bone quantity and quality in GIO makes BMD or ultrasound measurements inadequate for identifying patients at risk for fractures, although serial measurement may still be useful after intervention.[3,67] Biochemical markers of bone metabolism are also inadequate in GIO.[2,52] Part of the explanation is that, after long-term glucocorticoid therapy, bone is in a low turnover state characterized by decreased osteoblast number and bone formation rate with low-normal to normal numbers of osteoclasts, histomorphometric features that prevent clearly distinguishing biochemical markers of bone formation, or resorption in GIO from normal values. Prevention is far more effective than treatment of established GIO.

Specific Treatments

The cornerstones of treatment for GIO are the aminobisphosphonates (alendronate or risedronate), which have been shown to have antifracture efficacy in patients 17–85 yr of age.[68,69] Aminobisphosphonates should be prescribed for all patients if glucocorticoid therapy is planned for >3 mo or multiple courses are envisioned, as well as in already established GIO, especially when there is a history of prior fracture, but attempts to target only high-risk individuals may not be possible because of the glucocorticoid-induced discrepancy between bone quantity and quality. If oral therapy cannot be tolerated or esophageal disease is present, intravenous therapy with pamidronate or ibandronate should be instituted.[70,71] Fever during an acute-phase reaction to intravenous aminobisphosphonate therapy may aggravate multiple sclerosis. Calcium alone or with active vitamin D analogs may not preserve BMD or prevent fractures.[72] These analogs, calcitonin, and estrogens should be avoided.[2,73] The duration of therapy should be as long as glucocorticoids are prescribed because the skeletal protection afforded by the aminobisphosphonates undergoes rapid waning if they are discontinued while glucocorticoids are continued.[74] However, even after glucocorticoids are discontinued, patients with very low BMD may still require treatment with an aminobisphosphonate. Teratogenic effects of aminobisphosphonate administration have not been well studied, so the treatment must be carefully considered in women of childbearing age.

GIO is prominently a disease of reduced bone formation, and therefore, anabolic therapy with intermittent administration of PTH is particularly effective.[53] This approach would prevent glucocorticoid-induced decrements in the number of osteoblasts and bone formation rate by averting osteoblast apoptosis.[75] However, the anabolic effect may be less than that shown in patients with postmenopausal osteoporosis because of the attenuating influence of concurrent glucocorticoid therapy.[76]

It should be noted that, in a patient with GIO, incapacitating adjacent vertebral fractures have been reported days after kyphoplasty, suggesting that great caution should be exercised in recommending the procedure in patients receiving glucocorticoids.[77]

REFERENCES

1. Weinstein RS 2001 Glucocorticoid-induced osteoporosis. Rev Endocr Metab Disord **2:**65–73.
2. Canalis E, Mazziotti G, Giustina A, Bilezikian JP 2007 Glucocorticoid-induced osteoporosis: Pathophysiology and therapy. Osteoporosis Int **18:**1319–1328.
3. Angeli A, Guglielmi G, Dovio A, Capelli G, de Feo D, Giannini S, Giorgino R, Moro L, Giustina A 2006 High prevalence of asymptomatic vertebral fractures in post-menopausal women receiving chronic glucocorticoid therapy: A cross-sectional outpatient study. Bone **39:**253–259.
4. Salerno A, Hermann R 2006 Efficacy and safety of steroid use for postoperative pain relief. J Bone Joint Surg Am **88:**1361–1372.
5. Mankin HF 1992 Nontraumatic necrosis of bone (osteonecrosis). N Engl J Med **326:**1473–1479.
6. Weinstein RS, Jilka RL, Parfitt AM, Manolagas SC 1998 Inhibition of osteoblastogenesis and promotion of apoptosis of osteoblasts and osteocytes by glucocorticoids: Potential mechanisms of the deleterious effects on bone. J Clin Invest **102:**274–282.
7. O'Brien CA, Jia D, Plotkin LI, Bellido T, Powers CC, Stewart SA, Manolagas SC, Weinstein RS 2004 Glucocorticoids act directly on osteoblasts and osteocytes to induce their apoptosis and reduce bone formation and strength. Endocrinology **145:**1835–1841.
8. Jia D, O'Brien CA, Stewart SA, Manolagas SC, Weinstein RS 2006 Glucocorticoids act directly on osteoclasts to increase their lifespan and reduce bone density. Endocrinology **147:**5592–5599.
9. Sambrook PN, Hughes DR, Nelsen AE, Robinson BG, Mason RS 2003 Osteocyte viability with glucocorticoid treatment: Relation to histomorphometry. Ann Rheum Dis **62:**1215–1217.
10. Weinstein RS, Chen JR, Powers CC, Stewart SA, Landes RD, Bellido T, Jilka RL, Parfitt AM, Manolagas SC 2002 Promotion of osteoclast survival and antagonism of bisphosphonate-induced osteoclast apoptosis by glucocorticoids. J Clin Invest **109:**1041–1048.
11. Cooper MS, Rabbitt EH, Goddard PE, Bartlett WA, Hewison M, Stewart PM 2002 Osteoblastic 11β-hydroxysteroid dehydrogenase type 1 activity increases with age and glucocorticoid exposure. J Bone Miner Res **17:**979–986.
12. Vestergaard P, Rejnmark L, Mosekilde L 2006 Methotrexate, azathioprine, cyclosporine, and risk of fractures. Calcif Tissue Int **79:**69–75.

© 2008 American Society for Bone and Mineral Research

13. Weinstein RS, Manolagas SC 2005 Apoptosis in glucocorticoid-induced bone disease. Curr Opin Endocrinol Diabetes **12:**219–223.
14. Saag KG, Koehnke R, Caldwell JR, Brasington R, Burmeister LF, Zimmerman B, Kohler JA, Furst DE 1994 Low dose long-term corticosteroid therapy in rheumatoid arthritis: An analysis of serious adverse events. Am J Med **96:**115–123.
15. Maalouf NM, Shane E 2005 Osteoporosis after solid organ transplantation. J Clin Endocrinol Metab **90:**2456–2465.
16. Bernstein CN, Blanchard JF, Leslie W, Wajda A, Yu BN 2000 The incidence of fractures among patients with inflammatory bowel disease. Ann Intern Med **133:**795–799.
17. Jahnsen J, Falch JA, Mowinckel P, Aadland E 2002 Vitamin D status, parathyroid hormone and bone mineral density in patients with inflammatory bowel disease. Scand J Gastroenterol **37:**192–199.
18. Stockbrugger RW, Schoon EJ, Bollani S, Mills PR, Israeli E, Landgraf L, Felsenberg D, Ljunghall S, Nygard G, Persson T, Graffner H, Bianchi Porro G, Ferguson A 2002 Discordance between the degree of osteopenia and the prevalence of spontaneous vertebral fractures in Crohn's disease. Aliment Pharmacol Ther **16:**1519–1527.
19. Jorgensen NR, Schwarz P, Holme I, Hendriksen BM, Petersen LJ, Becker V 2007 The prevalence of osteoporosis in patients with chronic obstructive pulmonary disease: A cross sectional study. Respir Med **101:**177–185.
20. Wakata N, Nemoto H, Sugimoto H, Nomoto N, Konno S, Hayashi N, Arak Y, Nakazato A 2004 Bone density in myasthenia gravis patients receiving long-term prednisolone therapy. Clin Neurol Neurosurg **106:**139–141.
21. Khachanova NV, Demina TL, Smirnov AV, Gusev EI 2006 Risk factors of osteoporosis in women with multiple sclerosis. Zh Nevrol Psikhiatr Im S S Korsakova **3:**56–63.
22. Nathan AW, Rose GL 1979 Fatal iatrogenic Cushing's syndrome. Lancet **I:**207.
23. Nasser SMS, Ewan PW 2001 Depot corticosteroid treatment for hay fever causing avascular necrosis of both hips. BMJ **322:**1589–1591.
24. McLean CJ, Lobo RFJ, Brazier DJ 1995 Cataracts, glaucoma, and femoral avascular necrosis caused by topical corticosteroid ointment. Lancet **345:**330.
25. Champion PK 1974 Cushing's syndrome secondary to abuse of dexamethasone nasal spray. Arch Intern Med **134:**750–751.
26. Licata AA 2005 Systemic effects of fluticasone nasal spray: Report of two cases. Endocr Pract **11:**194–196.
27. Kubo T, Kojima A, Yamazoe S, Ueshima K, Yamamoto T, Hirasawa Y 2001 Osteonecrosis of the femoral head that developed after long-term topical steroid application. J Orthop Sci **6:**92–94.
28. Williams PL, Corbett M 1983 Avascular necrosis of bone complicating corticosteroid replacement therapy. Ann Rheum Dis **42:**276–279.
29. Snyder S 1984 Avascular necrosis and corticosteroids. Ann Intern Med **100:**770.
30. Zelissen PMJ, Croughs RJM, van Rijk PP, Raymakers JA 1994 Effect of glucocorticoid replacement therapy on bone mineral density in patients with Addison disease. Ann Intern Med **120:**207–210.
31. King JA, Wisniewski AB, Bankowski BJ, Carson KA, Zacur HA, Migeon CJ 2006 Long-term corticosteroid replacement and bone mineral density in adult women with classical congenital adrenal hyperplasia. J Clin Endocrinol Metab **91:**865–869.
32. LoCascio V, Bonucci E, Imbimbo B, Ballanti P, Adami S, Milani S, Tartarotti D, DellaRocca C 1990 Bone loss in response to long-term glucocorticoid therapy. Bone Miner **8:**39–51.
33. Weinstein RS 2000 Perspective: True strength. J Bone Miner Res **15:**621–625.
34. van Staa TP, Laan RF, Barton IP, Cohen S, Reid DM, Cooper C 2003 Bone density threshold and other predictors of vertebral fracture in patients receiving oral glucocorticoid therapy. Arthrit Rheum **48:**3224–3229.
35. van Staa TP, Leufkins H, Cooper C 2001 Use of inhaled glucocorticoids and risk of fractures. J Bone Miner Res **16:**581–588.
36. Gluck OS, Murphy WA, Hahn TJ, Hahn B 1981 Bone loss in adults receiving alternate day glucocorticoid therapy, A comparison with daily therapy. Arthrit Rheum **24:**892–898.
37. Samaras K, Pett S, Gowers A, McMurchie M, Cooper DA 2005 Iatrogenic Cushing's syndrome with osteoporosis and secondary adrenal failure in human immunodeficiency virus-infected patients receiving inhaled corticosteroids and ritonavir-boosted protease inhibitors: Six cases. J Clin Endocrinol Metab **90:**4394–4398.
38. Sosa M, Saavedra P, Valero C, Guanabens N, Nogues X, del Pino-Montes J, Mosquera J, Alegre J, Gomez-Alonso C, Munoz-Torres M, Quesada M, Perez-Cano R, Jodar E, Torrijos A, Lozano-Tonkin C, Diaz-Curiel M 2006 Inhaled steroids do not decrease bone mineral density but increase risk of fractures: Data from the GIUMO study group. J Clin Densitom **9:**154–158.
39. de Vries F, Bracke M, Leufkens HGM, Lammers JWJ, Cooper C, van Staa TP 2007 Fracture risk with intermittent high-dose oral glucocorticoid therapy. Arthrit Rheum **56:**208–214.
40. Dempster DW 1989 Bone histomorphometry in glucocorticoid-induced osteoporosis. J Bone Miner Res **4:**137–141.
41. Dempster DW, Arlot MA, Meunier PJ 1983 Mean wall thickness and formation periods of trabecular bone packets in corticosteroid-induced osteoporosis. Calcif Tissue Int **35:**410–417.
42. Weinstein RS, Bell NH 1988 Diminished Rate of Bone Formation in Normal Black Adults. N Engl J Med **319:**1689–1701.
43. Vedi S, Elkin SL, Compston JE 2005 A histomorphometric study of cortical bone of the iliac crest in patients treated with glucocorticoids. Calcif Tissue Int **77:**79–83.
44. Curtis JR, Westfall AO, Allison JJ, Becker A, Casebeer L, Freeman A, Spettell CM, Weissman NW, Wilke S, Saag KG 2005 Longitudinal patterns in the prevention of osteoporosis in glucocorticoid-treated patients. Arthrit Rheum. **52:**2485–2494.
45. Adachi JD, Saag KG, Delmas PD, Liberman UA, Emkey RD, Seeman E, Lane NE, Kaufman JM, Poubelle PE, Hawkins F, Correa-Rotter R, Menkes CJ, Rodriguez-Portales JA, Schnitzer TJ, Block JA, Wing J, McIlwain HH, Westhovens R, Brown J, Melo-Gomes JA, Gruber BL, Yanover MJ, Leite MO, Siminoski KG, Nevitt MC, Sharp JT, Malice MP, Dumortier T, Czachur M, Carofano W, Daifotis A 2001 Two-year effects of alendronate on bone mineral density and vertebral fracture in patients receiving glucocorticoids: A randomized, double-blind, placebo-controlled extension trial. Arthritis Rheum **44:**202–211.
46. Liberman UA, Weiss SR, Bröll J, Minne HW, Quan H, Bell NH, Rodriguez-Portales J, Downs RW, Dequeker J, Favus M 1995 Effect of oral alendronate on bone mineral density and the incidence of fractures in postmenopausal osteoporosis. The Alendronate Phase III Osteoporosis Treatment Study Group. N Engl J Med **333:**1437–1443.
47. Orwoll E, Ettinger M, Weiss S, Miller P, Kendler D, Graham J, Adami S, Weber K, Lorenc R, Pietschmann P, Vandormael K, Lombardi A 2000 Alendronate for the treatment of osteoporosis in men. N Engl J Med **343:**604–610.
48. Rubin MA, Bilezikian JP 2002 The role of parathyroid hormone in the pathogenesis of glucocorticoid-induced osteoporosis: A reexamination of the evidence. J Clin Endocrinol Metab **87:**4033–4041.
49. Pearce G, Tabensky DA, Delmas PD, Gordon Baker HW, Seeman E 1998 Corticosteroid-induced bone loss in men. J Clin Endocrinol Metab **83:**801–806.
50. Hattersley AT, Meeran K, Burrin J, Hill P, Shiner R, Ibbertson HK 1994 The effect of long-and short-term corticosteroids on plasma calcitonin and parathyroid hormone levels. Calcif Tissue Int **54:**198–202.
51. Pas-Pacheco E, Fuleihan GEH, LeBoff MS 1995 Intact parathyroid hormone levels are not elevated in glucocorticoid-treated subjects. J Bone Miner Res **10:**1713–1718.
52. Jacobs JW, de Nijs RN, Lems WF, Geusens PP, Laan RF, Huisman AM, Algra A, Buskens E, Hofbauer LC, Oostveen AC, Bruyn GA, Dijkmans BA, Bijlsma JW 2007 Prevention of glucocorticoid induced osteoporosis with alendronate or alfacalcidol: Relations of change in bone mineral density, bone markers, and calcium homeostasis. J Rheumatol **34:**1051–1057.
53. Saag KG, Shane E, Boonen S, Martin F, Donley DW, Taylor KA, Dalsky GP, Marcus R 2007 Teriparatide or alendronate in glucocorticoid-induced osteoporosis. N Engl J Med **357:**2028–2039.
54. Weinstein RS, Jia D, Powers CC, Stewart SA, Jilka RL, Parfitt AM, Manolagas SC 2004 The skeletal effects of glucocorticoid excess override those of orchidectomy in mice. Endocrinology **145:**1980–1987.
55. Ton FN, Gunawardene SC, Lee H, Neer RM 2005 Effects of low-dose prednisone on bone metabolism. J Bone Miner Res **20:**464–470.
56. Rizzoni G, Ehrich JH, Broyer M, Brunner FP, Brynger H, Fassbinder W, Geerlings W, Selwood NH, Tufveson G, Wing AJ 1992

© 2008 American Society for Bone and Mineral Research

Successful pregnancies in woman on renal replacement therapy: Report from the EDTA registry. Nephrol Dial Transplant **7**:279–287.
57. European Renal Association-European Dialysis and Transplantation Association IV 2002 Section 10: Pregnancy in renal transplant recipients. Guidelines. Nephrol Dial Transplant **Suppl 4**:50–55.
58. Tauchmanovà L, Pivonello R, Di Somma C, Rossi R, De Martino MC, Camera L, Klain M, Salvatore M, Lombardi G, Colao A 2006 Bone demineralization and vertebral fractures in endogenous cortisol excess: Role of disease etiology and gonadal status. J Clin Endocrinol Metab **91**:1779–1784.
59. Canalis E, Bilezikian JP, Angeli A, Giustina A 2004 Perspectives on glucocorticoid-induced osteoporosis. Bone **34**:593–598.
60. Ohnaka K, Taniguchi H, Kawate H, Nawata H, Takayanagi R 2004 Glucocorticoid enhances the expression of dickkopf-1 in human osteoblasts: Novel mechanism of glucocorticoid-induced osteoporosis. Biochem Biophys Res Commun **318**:259–264.
61. Hofbauer LC, Gori F, Riggs BL, Lacey DL, Dunstan CR, Spelsberg TC, Khosla S 1999 Stimulation of osteoprotegerin ligand and inhibition of osteoprotegerin production by glucocorticoids in human osteoblastic lineage cells: Potential paracrine mechanisms of glucocorticoid-induced osteoporosis. Endocrinology **140**:4382–4389.
62. Vidal NO, Brandstrom H, Jonsson KB, Ohlsson C 1998 Osteoprotegerin mRNA is expressed in primary human osteoblast-like cells: Down-regulation by glucocorticoids. J Endocrinol **159**:191–195.
63. Calder JDF, Buttery L, Revell PA, Pearse M, Polak JM 2004 Apoptosis—a significant cause of bone cell death in osteonecrosis of the femoral head. J Bone Joint Surg Br **86**:1209–1213.
64. Weinstein RS, Nicholas RW, Manolagas SC 2000 Apoptosis of osteocytes in glucocorticoid-induced osteonecrosis of the hip. J Clin Endocrinol Metab **85**:2907–2912.
65. Zizic TM, Marcoux C, Hungerford DS, Dansereau J-V, Stevens MB 1985 Corticosteroid therapy associated with ischemic necrosis of bone in systemic lupus erythematosus. Am J Med **79**:596–604.
66. Felson DT, Anderson JJ 1987 Across-study evaluation of association between steroid dose and bolus steroids and avascular necrosis of bone. Lancet **I**:902–905.
67. Weinstein RS 2007 Is long-term glucocorticoid therapy associated with a high prevalence of asymptomatic vertebral fractures in postmenopausal women? Nat Clin Pract Endocrinol Metab **3**:86–87.
68. Adachi JD, Saag KG, Delmas PD, Liberman UA, Emkey RD, Seeman E, Lane NE, Kaufman JM, Poubelle PE, Hawkins F, Correa-Rotter R, Menkes CJ, Rodriguez-Portales JA, Schnitzer TJ, Block JA, Wing J, McIlwain HH, Westhovens R, Brown J, Melo-Gomes JA, Gruber BL, Yanover MJ, Leite MO, Siminoski KG, Nevitt MC, Sharp JT, Malice MP, Dumortier T, Czachur M, Carofano W, Daifotis A 2001 Two-year effects of alendronate on bone mineral density and vertebral fracture in patients receiving glucocorticoids: A randomized, double-blind, placebo-controlled extension trial. Arthrit Rheum **44**:202–211.
69. Reid DM, Hughes RA, Laan RF, Sacco-Gibson NA, Wenderoth DH, Adami S, Eusebio RA, Devogelaer JP 2000 Efficacy and safety of daily risedronate in the treatment of corticosteroid-induced osteoporosis in men and women: A randomized trial. J Bone Miner Res **15**:1006–1013.
70. Boutsen Y, Jamart J, Esselinckx W, Stoffel M, Devogelaer JP 1997 Primary prevention of glucocorticoid-induced osteoporosis with intermittent intravenous pamidronate: A randomized trial. Calcif Tissue Int **61**:266–271.
71. Ringe JD, Dorst A, Faber H, Ibach K, Sorenson F 2003 Intermittent intravenous ibandronate injections reduce vertebral fracture risk in corticosteroid-induced osteoporosis: Results from a long-term comparative study. Osteoporos Int **14**:801–807.
72. de Nijs RN, Jacobs JW, Lems WF, Laan RF, Algra A, Huisman AM, Buskens E, de Laet CE, Oostveen AC, Geusens PP, Bruyn GA, Dijkmans BA, Bijlsma JW 2006 Alendronate or alfacalcidol in glucocorticoid-induced osteoporosis. N Engl J Med **355**:675–684.
73. Cranney A, Welch V, Adachi J, Homik J, Shea B, Suarez-Almazor ME, Tugwell P, Wells G 2005 Calcitonin for preventing and treating corticosteroid induced osteoporosis (Review). Cochrane Library **I**:1–31.
74. Emkey R, Delmas PD, Goemaere S, Liberman UA, Poubelle PE, Daifotis AG, Verbruggen N, Lombardi A, Czachur M 2003 Changes in bone mineral density following discontinuation or continuation of alendronate therapy in glucocorticoid-treated patients: A retrospective, observational study. Arthrit Rheum **48**:1102–1108.
75. Jilka RL, Weinstein RS, Bellido T, Roberson P, Parfitt AM, Manolagas SC 1999 Increased bone formation by prevention of osteoblast apoptosis with PTH. J Clin Invest **104**:439–446.
76. Oxlund H, Ortoft G, Thomsen JS, Danielsen CC, Ejersted C, Andreassen TT 2006 The anabolic effect of PTH on bone is attenuated by simultaneous glucocorticoid treatment. Bone **39**:244–252.
77. Donovan MA, Khandji AG, Siris E 2004 Multiple adjacent vertebral fractures after kyphoplasty in a patient with steroid-induced osteoporosis. J Bone Miner Res **19**:712–713.

Chapter 59. Inflammation-Induced Bone Loss in the Rheumatic Diseases

Steven R. Goldring

Departments of Orthopedics and Rheumatology, Hospital for Special Surgery, Weill Medical College of Cornell University, New York, New York

INTRODUCTION

The inflammatory joint diseases include a diverse group of disorders that share in common the presence of inflammatory and destructive changes that adversely affect the structure and function of articular and periarticular tissues. In many of these disorders, the inflammatory processes that target the joint tissues may affect extra-articular tissues and organs, and in addition, there may be generalized effects on systemic bone remodeling. Attention will focus on rheumatoid arthritis (RA), systemic lupus erythematosus (SLE), and the seronegative spondyloarthropathies, which include ankylosing spondylitis, reactive arthritis (formerly designated as Reiter's syndrome), the arthritis of inflammatory bowel disease, juvenile onset-spondyloarthropathy, and psoriatic arthritis. The discussion will be limited to ankylosing spondylitis, which is the prototypical spondyloarthropathy.

SYNOVIAL INFLAMMATION

In RA and SLE, the synovial lining is the initial site of the

The author states that he has no conflicts of interest.

Key words: arthritis, inflammation, bone remodeling, bone resorption, bone formation, cytokines

© 2008 American Society for Bone and Mineral Research

inflammatory process. Under physiological conditions, the synovium forms a thin membrane that lines the surface of the joint cavity and is responsible for generation of the synovial fluid that contributes to joint lubrication and nutrition for the chondrocytes that populate the articular cartilage. In patients with RA and SLE, the synovium becomes a site of an intense immune-mediated inflammatory process that results in synovial proliferation and production of potent inflammatory cytokines and soluble mediator that are responsible for the clinical signs of joint inflammation.[1,2] In patients with RA, this inflammatory process ultimately leads to destruction of the joint tissues. Although the clinical signs of inflammation in SLE and RA are similar, the synovitis associated with SLE does not lead to direct destruction of articular cartilage and bone, suggesting that different inflammatory and immunologic processes are involved in the pathogenesis of the synovial lesion in these two conditions. Of interest, joint deformities do develop in patients with SLE, but these are attributable to alterations in the integrity of periarticular connective tissues rather than destruction of the articular cartilage and bone.[3]

Synovial inflammation also is present in the seronegative spondyloarthropathies and, as in RA, may be associated with focal destruction of periarticular bone. In addition, in conditions such as ankylosing spondylitis (AS) or psoriatic arthritis, the inflammatory process also may involve the entheses, which are sites of periarticular ligament and tendon insertions. In these circumstances, in addition to the peripheral joints, the inflammatory process also may target articulations between the vertebral bodies of the axial skeleton.[1,4–6] The inflammatory process at these sites, although associated with focal bone loss, more commonly is accompanied by enhanced bone formation that may lead to fusion or ankylosis of adjacent vertebrae.

RHEUMATOID ARTHRITIS

RA is a systemic inflammatory disorder characterized by symmetrical polyarthritis. Four major forms of pathologic skeletal remodeling can be observed in this disorder, including focal marginal articular erosions, subchondral bone loss, periarticular osteopenia, and systemic osteoporosis.

Focal Joint Margin and Subchondral Bone Erosions

Focal marginal joint erosions are the radiologic hallmark of RA. Histopathologic examination of these sites of focal bone loss shows the presence of inflamed synovial tissue that has attached to the bone surface forming a mantle or covering referred to as "pannus." The interface between the pannus and adjacent bone is frequently lined by resorption lacunae containing mono- and multinucleated cells with phenotypic features of authentic osteoclasts, thus implicating osteoclasts as the principal cell type responsible for the focal synovial resorptive process.[7–9] Similar sites of focal bone loss are present on the endosteal surface of the subchondral bone adjacent to the marrow space. Resorption lacunae lined by osteoclast-like cells are frequently present at these sites, which conform to regions where the inflamed synovium has penetrated the marrow space and advanced to the subchondral bone surfaces. These areas conform to sites of so-called bone marrow edema visualized by MRI.[10] Importantly, erosion of the subchondral bone contributes to joint destruction by providing access of the inflamed tissue to the deep zones of the articular cartilage, which is subject to destructive activities of the inflamed tissues.

More definitive evidence implicating osteoclasts in the pathogenesis of focal articular bone erosions has come from the use of genetic approaches in which investigators have induced inflammatory arthritis with features of RA in mice lacking the ability to form osteoclasts.[11–13] In these models, the inability to form osteoclasts results in protection from focal articular bone resorption despite the presence of extensive synovial inflammation.

The propensity of the synovial lesion in RA to induce osteoclast-mediated bone resorption can be attributed to the production by cells within the inflamed tissue of a wide variety of products with the capacity to recruit osteoclast precursors and induce their differentiation and activation. These include a spectrum of chemokines, as well as RANKL, interleukin-1 (IL-1), IL-6, IL-11, IL-15, IL-17, monocyte-colony-stimulating factor, TNF-α, prostaglandins, and PTH-related peptide.[2] Among these products, particular attention has focused on RANKL, which is produced by both synovial fibroblasts and T cells within the synovial tissue.[14–16] The critical role of this cytokine in the pathogenesis of focal bone erosions is suggested by the observations that blocking the activity of RANKL in animal models of RA with osteoprotegerin (OPG) results in marked attenuation of articular bone erosions.[15,17,18] The pivotal role of RANKL in the resorptive process is further supported by the results obtained in the genetic models in which deletion of RANKL[19] or disruption of its signaling pathway[13] protects animals from articular bone erosions in models of inflammatory arthritis. More recently, in a preliminary report,[20] blockade of RANKL with Denosumab (Amgen, Thousand Oaks, CA, USA), a monoclonal antibody that blocks RANKL activity, was shown to significantly reduce articular bone erosions in a group of patients with RA, providing further evidence that osteoclasts and osteoclast-mediated bone resorption represent a rational therapeutic target for preventing articular bone destruction in RA. Interestingly, bisphosphonates, which show beneficial effects in protecting from systemic bone loss in RA, have not been effective in reducing focal joint destruction, with the exception of a recent publication in which the investigators used a protocol involving the sequential administration of zoledronic acid.[21] Although there may be limitations with respect to the use of these agents to prevent joint destruction, as discussed below, there is clearly a role for bisphosphonates in treating and preventing systemic bone loss in RA.[22]

Bone Formation in RA

An additional striking feature of the focal marginal and subchondral bone loss in RA is the virtual absence of bone repair. The recent studies of Diarra et al.[23] have provided insights into the mechanism involved in the uncoupling of bone resorption and formation in this form of inflammatory arthritis. They showed that cells in the inflamed RA synovial tissue produced dickkopf-1 (DKK-1), the inhibitor of the wingless (Wnt)-signaling pathway that plays a critical role in osteoblast-mediated bone formation. Synovial fibroblasts, endothelial cells, and chondrocytes were the principal sources of the DKK-1. They furthermore showed that TNF-α was a potent inducer of DKK-1, thus implicating this pro-inflammatory mediator in both bone formation and bone resorption. An additional, somewhat surprising observation was that inhibition of DKK-1 with a blocking antibody produced beneficial effects not only on bone formation but also suppressed osteoclast-mediated bone resorption. The effects on suppression of bone resorption were attributed to downregulation of RANKL production by the inflamed synovium and upregulation of OPG. These observations have clear implications with respect to future therapeutic strategies to prevent bone loss in RA.[24]

Periarticular Osteopenia in RA

Focal marginal bone erosions are the radiographic hallmark

© 2008 American Society for Bone and Mineral Research

of RA, but the earliest skeletal feature of RA is the development of periarticular osteopenia. Of importance, there is evidence that the juxta-artcular bone loss has high predictive value with respect to the subsequent development of marginal joint erosions in the hand.[25] There are few studies correlating the histopathologic changes associated with periarticular osteopenia. Shimizu et al.[26] examined the periarticular bone obtained from a series of patients with RA undergoing joint arthroplasty and observed evidence of both increased bone resorption and formation based on histomorphometric analysis. Examination of the bone marrow in the juxta-articular tissues frequently shows the presence of focal accumulations of inflammatory cells, including lymphocytes and macrophages, and these cells are a likely source of cytokines and related pro-inflammatory mediators that could adversely affect bone remodeling. Immobilization and reduced mechanical loading are additional factors that have been implicated in the pathogenesis of periarticular bone loss.

Generalized Osteoporosis in RA

The final skeletal feature of RA is the presence of generalized osteoporosis. Numerous studies have document that patients with RA have lower BMD and an increased risk of fracture.[27–31] The presence in patients with RA of multiple confounding factors that influence bone remodeling has made it difficult to define in a given patient the underlying pathogenic mechanism responsible for the reduced bone mass. These include the effects of sex, age, nutritional state, level of physical activity, disease duration and severity, and the use of medications such as glucocorticoids that can adversely affect bone remodeling. Lodder et al.[29] evaluated the relationship between bone mass and disease activity in a cohort of patients with RA with low to moderate disease activity and observed that disease activity was a significant contributory factor to systemic bone loss, supporting earlier observations made by several other investigators.

Several different approaches have been used to gain insights into the mechanism responsible for systemic bone loss in RA, including histomorphometric analysis of bone biopsies, measurement of urinary and serum biomarkers of bone remodeling, and the assessment of serum cytokine levels. Earlier studies using histomorphometric analysis suggested that the decrease in bone mass was attributable to depressed bone formation.[32] In contrast, Gough et al.,[33] as well as several other investigators, observed the presence of increased bone resorption based on assessment of urinary markers. Of interest, Garnero et al.[34] observed that a high urinary C-terminal crosslinking telopeptide of type I collagen (CTX-1) level (a marker of bone resorption) predicted risk of radiographic progression independent of rheumatoid factor or erythrocyte sedimentation rate. More recently, several groups of investigators have used the indices of bone remodeling and/or serum cytokine levels to assess the effects of treatment interventions on focal articular and systemic bone loss in RA patients.[35,36] Results indicate that suppression of signs of inflammation and improved functional status are reflected in improvement in the level and pattern of expression of these markers.

The disturbance in systemic bone remodeling in RA has been attributed to the adverse effects of pro-inflammatory cytokines that are released into the circulation from sites of synovial inflammation and act in a manner similar to endocrine hormones to regulate systemic bone remodeling. Although the serum levels of multiple osteoclastogenic cytokines are elevated in RA patients, particular attention has focused on the levels of RANKL and OPG. In a recent publication, Geusens et al.[37] observed that the ratio of circulating OPG/RANKL in early RA predicted subsequent bone destruction. In another study, Vis et al.[35] showed that anti-TNF therapy with infliximab was accompanied by decreased systemic bone loss and that these effects correlated with a fall in serum RANKL levels. There also is evidence that cytokines and mediators released from inflamed joints can adversely affect bone formation. This conclusion is supported by the observations of Diarra et al.,[23] who detected elevated levels of DKK-1, an inhibitor bone formation, in the sera of patients with RA. Of interest, levels of DKK-1 were not increased compared with controls in patients with AS, which is associated with focal increases in periarticular bone formation, as discussed in the following section.

These studies and the related investigations described in the preceding discussion highlight the importance of monitoring patients with RA for evidence of systemic bone loss and for the institution of early therapeutic interventions that have been shown to reduce the long-term risks of fracture and disability.[22] Similar approaches should be considered in patients with SLE and related forms of inflammatory arthritis who also are at risk for the development of systemic osteoporosis and fracture.

ANKYLOSING SPONDYLITIS

As described above, AS is characterized by inflammation in the entheses and the synovial lining of peripheral joints. Examination of the synovial lesion shows many of the same features as the RA synovium, including synovial lining hyperplasia, lymphocytic infiltration, and pannus formation. These lesions are associated with focal marginal joint erosions, although the joint involvement usually is restricted to a few sites, and the pattern is asymmetrical. Unlike RA, the inflammation also may involve the distal interphalangeal joints of the hands and toes.

In contrast to the pattern of articular bone remodeling in RA, in patients with AS, the inflammatory process may be accompanied by evidence of increased bone formation. This is particularly the case at sites of entheseal inflammation, such as ligament and tendon insertion sites, especially in the spine. To study the mechanism responsible for the enhanced bone formation, Braun et al.[38] obtained biopsies from the sacroiliac joints of patients with AS. They noted the presence of dense infiltrates of lymphocytes, similar to the RA synovial lesion; however, unlike the RA synovium, they also detected foci of endochondral ossification. More recently, Lories et al.[39] analyzed synovial tissues from a series of patients with AS or RA and noted the presence of elevated levels of bone morphogenic proteins-2 and -6 in tissues from both patient populations. They speculated that the differential pattern of new bone formation in AS could be related to the skeletal site of the inflammatory process and that at the entheses, which are not affected in RA, these bone growth factors could be responsible for the increased bone formation.

The pattern of enhanced local periarticular bone formation may be reflected in the serum levels of bone remodeling indices. For example, Franck and Ittel[40] examined the serum levels of osteocalcin and alkaline phosphatase in a cohort of patients with psoriatic arthritis and found that levels were elevated compared with controls, including patients with psoriasis without joint involvement. As discussed in the preceding section, Diarra et al.[23] observed that patients with AS did not exhibit elevated DKK-1 serum levels and concluded that the absence of this bone formation inhibitor could account for the tendency of patients with AS to form excessive bone at sites of inflammation.

Despite the tendency of patients with AS to produce exces-

© 2008 American Society for Bone and Mineral Research

sive bone formation at sites of inflammation, many individuals exhibit evidence of spinal osteopenia. This has been attributed to the adverse effects of immobilization that results from spinal ankylosis, although decreased BMD also has been detected in patients even in the absence of boney ankylosis.[41] These authors and others have suggested that, as in the other forms of inflammatory arthritis, the bone loss is related to the adverse effects of joint inflammation on systemic bone remodeling.

ACKNOWLEDGMENTS

This work was supported in part by National Institute of Health Grant NIAMS R01 AR45472.

REFERENCES

1. Schett G 2007 Joint remodelling in inflammatory disease. Ann Rheum Dis **66**(Suppl 3):iii42–iii44.
2. Walsh NC, Crotti TN, Goldring SR, Gravallese EM 2005 Rheumatic diseases: The effects of inflammation on bone. Immunol Rev **208**:228–251.
3. Santiago MB, Galvao V 2008 Jaccoud arthropathy in systemic lupus erythematosus: Analysis of clinical characteristics and review of the literature. Medicine (Baltimore) **87**:37–44.
4. McGonagle D, Benjamin M, Marzo-Ortega H, Emery P 2002 Advances in the understanding of entheseal inflammation. Curr Rheumatol Rep **4**:500–506.
5. Benjamin M, Moriggl B, Brenner E, Emery P, McGonagle D, Redman S 2004 The "enthesis organ" concept: Why enthesopathies may not present as focal insertional disorders. Arthritis Rheum **50**:3306–3313.
6. Benjamin M, McGonagle D 2001 The anatomical basis for disease localisation in seronegative spondyloarthropathy at entheses and related sites. J Anat **199**:503–526.
7. Gravallese EM, Harada Y, Wang JT, Gorn AH, Thornhill TS, Goldring SR 1998 Identification of cell types responsible for bone resorption in rheumatoid arthritis and juvenile rheumatoid arthritis. Am J Pathol **152**:943–951.
8. Bromley M, Woolley DE 1984 Histopathology of the rheumatoid lesion. Identification of cell types at sites of cartilage erosion. Arthritis Rheum **27**:857–863.
9. Bromley M, Woolley DE 1984 Chondroclasts and osteoclasts at subchondral sites of erosion in the rheumatoid joint. Arthritis Rheum **27**:968–975.
10. Jimenez-Boj E, Nobauer-Huhmann I, Hanslik-Schnabel B, Dorotka R, Wanivenhaus AH, Kainberger F, Trattnig S, Axmann R, Tsuji W, Hermann S, Smolen J, Schett G 2007 Bone erosions and bone marrow edema as defined by magnetic resonance imaging reflect true bone marrow inflammation in rheumatoid arthritis. Arthritis Rheum **56**:1118–1124.
11. Redlich K, Hayer S, Ricci R, David J, Tohidast-Akrad M, Kollias G, Steiner G, Smolen JS, Wagner EF, Schett G 2002 Osteoclasts are essential for TNF-alpha-mediated joint destruction. J Clin Invest **110**:1419–1427.
12. Pettit AR, Ji H, von Stechow D, Muller R, Goldring SR, Choi Y, Benoist C, Gravallese EM 2001 TRANCE/RANKL knockout mice are protected from bone erosion in a serum transfer model of arthritis. Am J Pathol **159**:1689–1699.
13. Li P, Schwarz EM, O'Keefe RJ, Ma L, Boyce BF, Xing L 2004 RANK signaling is not required for TNFalpha-mediated increase in CD11(hi) osteoclast precursors but is essential for mature osteoclast formation in TNFalpha-mediated inflammatory arthritis. J Bone Miner Res **19**:207–213.
14. Romas E, Bakharevski O, Hards DK, Kartsogiannis V, Quinn JM, Ryan PF, Martin TJ, Gillrspie MT 2000 Expression of osteoclast differentiation factor at sites of bone erosion in collagen-induced arthritis. Arthritis Rheum **43**:821–826.
15. Kong YY, Feige U, Sarosi I, Bolon B, Tafuri A, Morony S, Capparelli C, Li J, Elliot R, McCabe S, Wong T, Campagnuolo G, Moran E, Bogach ER 1999 Activated T cells regulate bone loss and joint destruction in adjuvant arthritis through osteoprotegerin ligand. Nature **402**:304–309.
16. Gravallese EM, Manning C, Tsay A, Naito A, Pan C, Amento E, Goldring SR 2000 Synovial tissue in rheumatoid arthritis is a source of osteoclast differentiation factor. Arthritis Rheum **43**:250–258.
17. Romas E, Gillespie MT, Martin TJ 2002 Involvement of receptor activator of NFkappaB ligand and tumor necrosis factor-alpha in bone destruction in rheumatoid arthritis. Bone **30**:340–346.
18. Redlich K, Hayer S, Maier A, Dunstan C, Tohidast-Akrad M, Lang S, Turk B, Pietschmann P, Woloszczuk W, Haralambous S, Kollias G, Steiner G, Smolen J, Schett G 2002 Tumor necrosis factor-α-mediated joint destruction is inhibited by targeting osteoclasts with osteoprotegerin. Arthritis Rheum **46**:785–792.
19. Pettit AR, Walsh NC, Manning C, Goldring SR, Gravallese EM 2006 RANKL protein is expressed at the pannus-bone interface at sites of articular bone erosion in rheumatoid arthritis. Rheumatology (Oxford) **45**:1068–1076.
20. van der Heide D, Cohen S, Sharp JT, Ory P, Zhou L, Tsuji W 2007 Denosumab inhibits RANKL, reducing progression of the total Sharp score and bone erosions in patients with rheumatoid arthritis. Arthritis Rheum **56**:S299.
21. Jarrett SJ, Conaghan PG, Sloan VS, Papanastasiou P, Ortmann CE, O'Connor PJ, Grainger AJ, Emery P 2006 Preliminary evidence for a structural benefit of the new bisphosphonate zoledronic acid in early rheumatoid arthritis. Arthritis Rheum **54**:1410–1414.
22. Goldring SR, Gravallese EM 2004 Bisphosphonates: Environmental protection for the joint? Arthritis Rheum **50**:2044–2047.
23. Diarra D, Stolina M, Polzer K, Zwerina J, Ominsky MS, Dwyer D, Korb A, Smolen J, Hoffman M, Scheinecker C, van der Heide D, Landewe R, Lacey D, Richards WG 2007 Dickkopf-1 is a master regulator of joint remodeling. Nat Med **13**:156–163.
24. Goldring SR, Goldring MB 2007 Eating bone or adding it: The Wnt pathway decides. Nat Med **13**:133–134.
25. Stewart A, Mackenzie LM, Black AJ, Reid DM 2004 Predicting erosive disease in rheumatoid arthritis. A longitudinal study of changes in bone density using digital X-ray radiogrammetry: A pilot study. Rheumatology (Oxford) **43**:1561–1564.
26. Shimizu S, Shiozawa S, Shiozawa K, Imura S, Fujita T 1985 Quantitative histologic studies on the pathogenesis of periarticular osteoporosis in rheumatoid arthritis. Arthritis Rheum **28**:25–31.
27. Orstavik RE, Haugeberg G, Uhlig T, Mowinckel P, Falch JA, Halse JI, Kvein TK 2005 Incidence of vertebral deformities in 255 female rheumatoid arthritis patients measured by morphometric X-ray absorptiometry. Osteoporos Int **16**:35–42.
28. Orstavik RE, Haugeberg G, Mowinckel P, Hoiseth A, Uhlig T, Falch JA, Halse JI, McCloskey E, Kvein TK 2004 Vertebral deformities in rheumatoid arthritis: A comparison with population-based controls. Arch Intern Med **164**:420–425.
29. Lodder MC, de Jong Z, Kostense PJ, Molenaar ET, Staal K, Voskuyl AE, Hazes JM, Dijkmans BA, Lema WF 2004 Bone mineral density in patients with rheumatoid arthritis: Relation between disease severity and low bone mineral density. Ann Rheum Dis **63**:1576–1580.
30. Haugeberg G, Orstavik RE, Kvien TK 2003 Effects of rheumatoid arthritis on bone. Curr Opin Rheumatol **15**:469–475.
31. Forslind K, Keller C, Svensson B, Hafstrom I 2003 Reduced bone mineral density in early rheumatoid arthritis is associated with radiological joint damage at baseline and after 2 years in women. J Rheumatol **30**:2590–2596.
32. Compston JE, Vedi S, Croucher PI, Garrahan NJ, O'Sullivan MM 1994 Bone turnover in non-steroid treated rheumatoid arthritis. Ann Rheum Dis **53**:163–166.
33. Gough A, Sambrook P, Devlin J, Huissoon A, Njeh C, Robbins S, Nguyen T, Emery P 1998 Osteoclastic activation is the principal mechanism leading to secondary osteoporosis in rheumatoid arthritis. J Rheumatol **25**:1282–1289.
34. Garnero P, Landewe R, Boers M, Verhoeven A, Van Der Linden S, Christgau S, van der Heijde D, Boonen A, Geusens P 2002 Association of baseline levels of markers of bone and cartilage degradation with long-term progression of joint damage in patients with early rheumatoid arthritis: The COBRA study. Arthritis Rheum **46**:2847–2856.
35. Vis M, Havaardsholm EA, Haugeberg G, Uhlig T, Voskuyl AE, van de Stadt RJ, Dijkmans BA, Woolf AD, Kvein TK, Lems WF 2006 Evaluation of bone mineral density, bone metabolism, osteoprotegerin and receptor activator of the NFkappaB ligand serum levels during treatment with infliximab in patients with rheumatoid arthritis. Ann Rheum Dis **65**:1495–1499.
36. Seriolo B, Paolino S, Sulli A, Ferretti V, Cutolo M 2006 Bone metabolism changes during anti-TNF-alpha therapy in patients with active rheumatoid arthritis. Ann N Y Acad Sci **1069**:420–427.
37. Geusens PP, Landewe RB, Garnero P, Chen D, Dunstan CR,

© 2008 American Society for Bone and Mineral Research

Lems WF, Stinissen P, vander Heijde DM, van der Linden S, Boers M 2006 The ratio of circulating osteoprotegerin to RANKL in early rheumatoid arthritis predicts later joint destruction. Arthritis Rheum **54:**1772–1777.
38. Braun J, Bollow M, Neure L, Seipelt E, Seyrekbasan F, Herbst H, Eggens U, Distler A, Sieper J 1995 Use of immunohistologic and in situ hybridization techniques in the examination of sacroiliac joint biopsy specimens from patients with ankylosing spondylitis. Arthritis Rheum **38;**499–505.
39. Lories RJ, Derese I, Ceuppens JL, Luyten FP 2003 Bone morphogenetic proteins 2 and 6, expressed in arthritic synovium, are regulated by proinflammatory cytokines and differentially modulate fibroblast-like synoviocyte apoptosis. Arthritis Rheum **48:**2807–2818.
40. Franck H, Ittel T 2000 Serum osteocalcin levels in patients with psoriatic arthritis: An extended report. Rheumatol Int **19:**161–164.
41. Will R, Palmer R, Bhalla AK, Ring F, Calin A 1989 Osteoporosis in early ankylosing spondylitis: A primary pathological event? Lancet **2:**1483–1485.

Chapter 60. Osteoporosis: Other Secondary Causes

Neveen A. T. Hamdy

Department of Endocrinology and Metabolic Diseases, Leiden University Medical Center, Leiden, The Netherlands

INTRODUCTION

Bone loss is an inevitable consequence of aging, starting some years before menopause, accelerating after its onset, and continuing throughout life in both men and women. A very large number of heterogeneous causes, collectively grouped as "secondary causes of osteoporosis" may also lead to bone loss through a number of mechanisms, independently of age or estrogen deficiency. A secondary cause for osteoporosis can be found in about two thirds of men, in more than one half of premenopausal women, and in about one fifth of postmenopausal women.[1] Secondary causes of osteoporosis are legion, ranging from easily identifiable specific disease states such as systemic inflammatory disorders, malignancy, endocrinopathies, and use of medication, particularly glucocorticoids, to more "occult" conditions such as vitamin D deficiency, hypercalciuria, or hyperparathyroidism. These latter causes of osteoporosis can only be diagnosed by a high degree of suspicion and confirmed by appropriate investigations and may be identified as underlying cause of unexpected bone loss in up to 63% of patients.[2–4] A large number of these secondary osteoporoses are individually discussed elsewhere in the *Primer*. This chapter focuses on osteoporosis associated with systemic inflammatory diseases, diabetes mellitus, and mastocytosis.

OSTEOPOROSIS ASSOCIATED WITH SYSTEMIC INFLAMMATORY DISORDERS

The RANKL/osteoprotegerin (OPG) ratio is the primary determinant of osteoclastogenesis and therefore of the maintenance of bone mass.[5,6] In inflammatory disorders, T-cell activation leads to increased expression of T cell–derived RANKL,[7,8] and glucocorticoids, often used to control disease activity, decrease osteoblast number and function and inhibit OPG expression.[9] In these inflammatory disorders, underlying disease activity alters the RANKL/OPG ratio, and this is further exacerbated by glucocorticoid use, the combined effect of both potentially leading to significant bone loss.

Inflammatory arthritis

Rheumatoid arthritis, discussed elsewhere in the *Primer*, represents the prototype of a systemic inflammatory disorder, in which inflammation triggers the increased expression of RANKL from activated T cells and from synovial fibroblasts, which is not matched by an increase in OPG, resulting in local bone loss: joint erosions, periarticular bone loss, and systemic osteoporosis.[10]

Inflammatory Bowel Diseases

In Crohn's disease, the pathophysiology of osteoporosis is multifactorial including the effect of inflammatory cytokines (interleukin [IL]-6, IL-1, TNF-α) mediating disease activity,[11] intestinal malabsorption caused by disease activity or intestinal resection, the use of glucocorticoids, inability to achieve peak bone mass when the disease starts in childhood, malnutrition, immobilization, low body mass index (BMI), smoking, and hypogonadism.[12] However, patients with Crohn's disease are relatively young, and the exact relationship between the host of factors potentially deleterious to the skeleton and increased risk for osteoporosis and fractures is unclear. Opinion also remains divided on the prevalence of fractures[13–19] and on bone loss in the long term.[13,20–24] The question of whether all patients with Crohn's disease should be treated with bone protective agents or whether these agents should be restricted to patients at increased risk for osteoporosis remains unsolved. Ileum resection has been identified as the single most significant risk factor for osteoporosis, followed by age, which is of relevance in predicting overall lifetime risk as Crohn's disease peaks in the second and third decade, with osteoporosis potentially becoming clinically significant only as patients grow older.[13] Maintaining a vitamin D–replete status prevents bone loss, and the judicious use of corticosteroids may counteract the deleterious effects of cytokine-driven disease activity on the skeleton.[13] Pharmacodynamic studies suggest that, in patients with reasonably well-controlled disease activity, the nitrogen-containing bisphosphonate alendronate is adequately absorbed from the gut and retained in the skeleton, despite underlying chronic inflammatory gut changes and/or gut resection.[25] Whether this applies during an exacerbation of Crohn's disease is not clear, because acute gut inflammation may be potentially associated with either decreased or indeed increased absorption of an orally administered bisphosphonate.

The author has reported no conflicts of interest.

Key words: T-cell activation, RANKL/OPG ratio, inflammatory arthritis, inflammatory bowel disease, chronic obstructive pulmonary disease, diabetes mellitus, bone marrow mastocytosis, urine histamine metabolites, premenopausal women, men, adipogenesis, osteogenesis, insulin, peroxisome proliferator activator receptor, screening secondary causes of osteoporosis, Crohn's disease

© 2008 American Society for Bone and Mineral Research

Chronic Obstructive Pulmonary Disease

In chronic obstructive pulmonary disease (COPD), pro-inflammatory cytokines, in particular TNF-α, are the driving force behind the pathophysiology of the disease process.[26] Elevated inflammatory markers reflect not only severity of lung disease but also the likelihood of increased risk for co-morbidities, particularly cardiovascular disease, diabetes, and osteoporosis.[27,28] A high prevalence of osteoporosis was thus observed within the first year after diagnosis among 2699 COPD patients from the UK General Practice Research Database (GPRD).[29] Data from >9500 subjects from the Third National Health and Nutrition Examination Survey conducted in the United States between 1988 and 1994 also showed that airflow obstruction was associated with increased odds of osteoporosis compared with no airflow obstruction (OR, 1.9; 95% CI, 1.4–2.5) and that these odds increased with increased severity of airways obstruction (OR, 2.4; 95% CI, 1.3–4.4; $p < 0.005$).[30] Loss of bone mass seems to be associated with increased excretion of bone collagen protein breakdown products, suggesting a protein catabolic state, which may not only lead to bone loss but also to loss of skeletal muscle mass and function and progressive disability.[31] Continuous users of systemic glucocorticoids are more than twice likely to have one or more vertebral fractures compared with non users.[32] In a large case-controlled study including >100,000 cases from the GPRD, an association between inhaled corticosteroids at daily doses equivalent to >1600 μg beclomethasone and increased fracture risk disappeared after adjustment for disease severity, suggesting that, in COPD, it is disease severity rather than inhaled corticosteroids that increase fracture risk.[33] Factors other than chronic inflammation and corticosteroid use also contribute to bone loss and increased fracture risk in patients with COPD. These include vitamin D deficiency or insufficiency, reduced skeletal muscle mass and strength, immobilization, low BMI and changes in body composition, hypogonadism, reduced levels of IGFs, smoking, increased alcohol intake, and genetic factors.[34] The morbidity associated with vertebral fractures is particularly high in patients with COPD because these have been shown to be associated with restrictive changes in pulmonary function, significant decreases in forced expiratory volume, and up to 9% reduction of predicted vital capacity for each additional thoracic vertebral compression fracture.[35,36]

OSTEOPOROSIS ASSOCIATED WITH DIABETES MELLITUS

The deleterious effects of diabetes mellitus (DM) on the skeleton are multifactorial, and both types 1 and 2 DM are associated with increased fracture risk.[37–40] Data from the Iowa Women's Health study suggested that women with type 1 DM are 12 times more likely to sustain hip fractures than women without DM and that women with type 2 DM have a 1.7-fold increased risk of sustaining hip fractures despite maintaining a normal bone mass.[39] The high prevalence of fractures in type 2 DM is likely to be because of long-term complications such as retinopathy-induced visual impairment and neuropathy-induced decreased balance, resulting in increased risk for falls.[40] The main mechanism of bone loss is decreased bone formation and the osteoporosis of DM is one of low bone turnover.[41,42] Insulin and amylin have an anabolic effect on bone, and their decrease in type 1 DM may lead to impaired bone formation, primarily because of a decrease in IGF-1 concentrations. In vitro studies also showed that sustained exposure to high glucose concentrations results in osteoblast dysfunction and poor metabolic control has a clear negative impact on bone mass. In type 1 DM, decreased peak bone mass also plays a role when the disease manifests itself before skeletal growth is complete. Microvascular complications and decreased mechanical stress caused by neuropathy and/or myopathy contribute to the increased fracture risk at later stages of the disease.[43]

In DM, there is increased bone marrow adiposity, which has also been linked with the osteoporosis of aging, glucocorticoid use, and immobility.[44] Evidence has been mounting for a clear interdependence of adipogenesis and osteogenesis, and several members of the nuclear hormone receptor family control the critical adipogenic and osteogenic steps.[44,45] In the bone marrow microenvironment, the inverse relationship between adipogenic and osteogenic differentiation was shown to be mediated at least in part through cross-talk between the pathways activated by steroid receptors (estrogen, thyroid, corticosteroid, and growth hormone receptors), the peroxisome proliferator activator receptors (PPARs), and other cytokine and paracrine factors. The PPARs play a central role in initiating adipogenesis in bone marrow and other stromal like cells in vitro and in vivo,[46] and their ligands (rosiglitazone and pioglitazone) play a prominent role in the treatment of type 2 DM. These ligands induce adipogenesis and inhibit osteogenesis in vitro, which may explain the recently reported increased incidence of fractures in patients with DM using these agents.[47]

OSTEOPOROSIS ASSOCIATED WITH MASTOCYTOSIS

In all forms of mastocytosis, the proximity of the mast cell to bone remodeling surfaces and the production by this cell of a large number of chemical mediators and cytokines capable of modulating bone turnover translates in skeletal involvement, ranging from severe osteolysis to significant osteosclerosis, with osteoporosis being the most frequently observed pathology.[48,49] Bone loss is exacerbated by the use of glucocorticoids. Clinical manifestations include generalized bony pain, which may be incapacitating, and is often resistant to conventional analgesia, particularly in cases of extensive bone marrow involvement or rapidly progressive disease. Osteoporosis may be associated with systemic manifestations of enhanced mast cell activity such as flushes and gastrointestinal symptoms[49] or may also be the sole presentation of bone marrow mastocytosis,[50–52] in which case the osteoporotic process may be severe and progressive. The diagnosis can only be confidently established by histologic examination of bone marrow biopsies showing the pathognomonic feature of bone marrow infiltration with a large number of morphologically abnormal mast cells, individually or in aggregates of >15 cells.[49] Bone marrow mastocytosis is an important "occult" cause of secondary osteoporosis, shown to be present in up to 9% of men with "idiopathic osteoporosis."[52] Serum tryptase may be normal, and the measurement of the 24-h urine excretion of *N*-methyl histamine represents a valuable noninvasive surrogate to bone marrow biopsies in establishing the diagnosis and evaluating the degree of mast cell load.[52,53]

WHO NEEDS TO BE SCREENED FOR SECONDARY CAUSES FOR OSTEOPOROSIS?

The high prevalence of potentially reversible secondary causes for osteoporosis, which may be identified with a sensitivity of 92% by cost-effective laboratory studies,[54] dictates that the majority of patients with osteoporosis would require at least some laboratory evaluation before start of treatment including a full blood count, serum biochemistry panel, 24-h urine calcium excretion, and 25-hydroxyvitamin D measure-

© 2008 American Society for Bone and Mineral Research

ments. Secondary causes for osteoporosis should be particularly sought in young patients,[55] premenopausal women,[56] men younger than 65 yr of age,[57] in all patients with unexpected or severe osteoporosis, in those with accelerated bone loss, or those experiencing bone loss under treatment with conventional osteoporosis therapy. Further laboratory tests should be requested to confirm or exclude hypogonadism, thyrotoxicosis, celiac disease, hypercortisolism, mastocytosis, and multiple myeloma. If suspicion remains high, or in the case of fragility fractures in the presence of a normal BMD, a double tetracycline–labeled transiliac bone biopsy with bone marrow evaluation may be indicated to establish a mineralization defect, or a bone marrow disorder, particular a nonsecretory myeloma or mastocytosis.

CONCLUSIONS

Secondary causes of osteoporosis are very common, particularly in premenopausal women and men with osteoporosis, while also being the cause of accelerated bone loss in postmenopausal and age-related osteoporosis. In addition to representing significant comorbidity in specific disease entities such as inflammatory disorders, malignant disease, bone marrow disorders, and endocrinopathies, secondary osteoporosis is also commonly associated with often silent disturbances in calcium homeostasis such as vitamin D deficiency, hypercalciuria, malabsorption, and hyperparathyroidism, all of which are readily reversible and easily detectable by standard laboratory testing. The ubiquitous nature of "secondary osteoporosis" suggests that diverse medical disciplines need to better interact to meet some of the challenges presented by osteoporosis as chronic comorbidity of specific disease entities. Screening for secondary causes for osteoporosis should represent an intrinsic part of the optimal management of any patient with osteoporosis.

REFERENCES

1. Painter SE, Kleerekoper M, Camacho PM 2006 Secondary osteoporosis: A review of the recent evidence. Endocr Pract **12**:436–445.
2. Johnson BE, Lucasey B, Robinson RG, Lukert BP 1989 Contributing diagnoses in osteoporosis. The value of a complete medical evaluation. Arch Intern Med **149**:1069–1072.
3. Freitag A, Barzel US 2002 Differential diagnosis of osteoporosis. Gerontology **48**:98–102.
4. Deutschmann HA, Weger M, Weger W, Kotanko P, Deutschmann MJ, Skrabal F 2002 Search for occult secondary osteoporosis: Impact of identified possible risk factors on bone mineral density. J Intern Med **252**:389–397.
5. Boyle WJ, Simonet WS, Lacey DL 2003 Osteoclast differentiation and activation. Nature **423**:337–342.
6. Walsh MC, Kim N, Kadono Y, Rho J, Lee SY, Lorenzo J, Choi Y 2006 Osteoimmunology: Interplay between the immune system and bone metabolism. Annu Rev Immunol **24**:33–36.
7. Teitelbaum SL 2006 Osteoclasts: Culprits in inflammatory osteolysis. Arthritis Res Ther **8**:201.
8. Boyce BF, Schwartz EM, Xing L 2006 Osteoclast precursors: Cytokine stimulated immunomodulators of inflammatory bone disease. Curr Opin Rheumatol **18**:427–432.
9. Hofbauer LC, Gori F, Riggs BL, Lacey DL, Dunstan CR, Spelsberg TC, Khosla S 1999 Stimulation of osteoprotegerin ligand and inhibition of osteoprotegerin production by glucocorticoids in human osteoblastic lineage cells: Potential paracrine mechanisms of glucocorticoid- induced osteoporosis. Endocrinology **140**:4382–4389.
10. Kong YY, Feige U, Sarosi I, Bolon B, Tafuri A, Morony S, Capparelli C, Li J, Elliott R, McCabe S, Wong T, Campagnuolo G, Moran E, Bogoch ER, Van G, Nguyen LT, Ohashi PS, Lacey DL, Fish E, Boyle WJ, Penninger JM 1999 Activated T cells regulate bone loss and joint destruction in adjuvant arthritis through osteoprotegerin ligand. Nature **402**:304–309.
11. Moschen AR, Kaser A, Enrich B, Ludwiczek O, Gabriel M, Obrist P, Wolf AM, Tilg H 2005 The RANKL/OPG system is activated in inflammatory bowel disease and relates to the state of bone loss. Gut **54**:479–487.
12. Compston J 2003 Osteoporosis in inflammatory bowel disease. Gut **52**:63–64.
13. van Hogezand RA, Banffer D, Zwinderman AH, McCloskey EV, Griffoen G, Hamdy NA 2006 Ileum resection is the most predictive factor for osteoporosis in patients with Crohn's disease. Osteoporos Int **17**:535–542.
14. Loftus EV, Crowson CS, Sandborn WJ, Tremaine WJ, O'Fallon WM, Melton LJ III 2002 Long-term fracture risk in patients with Crohn's disease: A population-based study in Olmsted County, Minnesota. Gastroenterology **123**:168–425.
15. Bernstein CN, Blanchard JF, Leslie W, Wajda A, Yu BN 2000 The incidence of fracture among patients with inflammatory bowel disease. A population-based cohort study. Ann Intern Med **133**:795–799.
16. van Staa TP, Cooper C, Brosse LS, Leufkens H, Javaid MK, Arden NK 2003 Inflammatory bowel disease and the risk of fracture. Gastroenterology **125**:1591–1597.
17. Card T, West J, Hubbard R, Logan F 2004 Hip fractures in patients with inflammatory bowel disease and their relationship to corticosteroid use: A population-based study. Gut **53**:251–255.
18. Klaus J, Armbrecht G, Steinkamp M, Bruckel J, Rieber A, Adler G, Reinshagen M, Felsenberg D, von Tirpitz C 2002 High prevalence of osteoporotic vertebral fractures in patients with Crohn's disease. Gut **51**:654–658.
19. Vestergaard P, Mosekilde L 2000 Fracture risk is increased in Crohn's disease, but not ulcerative colitis. Gut **46**:176–181.
20. Vestergaard P, Mosekilde L 2002 Fracture Risk in Patients with Celiac Disease, Crohn's Disease, and Ulcerative Colitis: A nationwide Follow-up Study of 16,416 Patients in Denmark. Am J Epidemiol **156**:1–10.
21. Schulte C, Dignass AU, Mann K, Goebell H 1999 Bone loss in patients with inflammatory bowel disease is less than expected: A follow-up study. Scand J Gastroenterol **34**:696–702.
22. Clements D, Motley RJ, Evans WD, Harries AD, Rhodes J, Coles RJ, Compston JE 1992 Longitudinal study of cortical bone loss in patients with inflammatory bowel disease. Scand J Gastroenterol **27**:1055–1060.
23. Roux C, Abitbol V, Chaussade S, Kolta S, Guillemant S, Dougados M, Amor B, Couturier D 1995 Bone loss in patients with inflammatory bowel disease: A prospective study. Osteoporos Int **5**:156–160.
24. Jahnsen J, Falch JA, Mowinckel AE 2004 Bone mineral density in patients with inflammatory bowel disease: A population-based prospective two-year follow-up study. Scand J Gastroenterol **39**:145–153.
25. Cremers SC, van Hogezand R, Banffer D, den Hartigh J, Vermeij P, Papapoulos SE, Hamdy NA 2005 Absorption of the oral bisphosphonate alendronate in osteoporotic patients with Crohn's disease. Osteoporos Int **16**:1727–1730.
26. Franciosi LG, Page CP, Celli BR, Cazzola M, Walker MJ, Danhof M, Rabe KF, Della Pasqua OE 2006 Markers of disease severity in chronic obstructive pulmonary disease. Pulm Pharmacol Ther **19**:189–199.
27. Gan WQ, Man SF, Senthilselvan A, Sin DD 2004 Association between chronic obstructive pulmonary disease and systemic inflammation: A systematic review and a meta-analysis. Thorax **59**:574–580.
28. Sevenoaks MJ, Stockley RA 2006 Chronic Obstructive Pulmonary Disease, inflammation and co-morbidity: A common inflammatory phenotype? Respir Res **7**:70–78.
29. Soriano JB, Visick GT, Muellerova H, Payvandi N, Hansell AL 2005 Patterns of comorbidities in newly diagnosed COPD and asthma in primary care. Chest **128**:2099–2107.
30. Sin DD, Man JP, Man SF 2003 The risk of osteoporosis in Caucasian men and women with obstructive airways disease. Am J Med **114**:10–14.
31. Bolton CE, Ionexcu AA, Shiels KM, Pettit RJ, Edwards PH, Stone MD, Nixon LS, Evans WD, Griffiths TL, Shale DJ 2004 Associated loss of fat-free mass and bone mineral density in chronic obstructive pulmonary disease. Am J Respir Crit Care Med **170**:1286–1293.
32. McEvoy CE, Ensrud KE, Bender E, Genant HK, Yu W, Griffith JM, Niewoehner DE 1998 Association between corticosteroid use

© 2008 American Society for Bone and Mineral Research

and vertebral fractures in older men with chronic obstructive pulmonary disease. Am J Respir Crit Care Med **157**:704–709.
33. de Vries F, van Staa TP, Bracke MSGM, Cooper C, Leufkens HGM, Lammers J-WJ 2005 Severity of obstructive airway disease and risk of osteoporotic fracture. Eur Respir J **25**:879–884.
34. Ionescu AA, Schoon E 2003 Osteoporosis in chronic obstructive pulmonary disease. Eur Respir J **22**(Suppl 46):S64–S75.
35. Schlaich C, Minne HW, Bruckner T, Wagner G, Gebest HJ, Grunze M, Ziegler R, Leidig-Bruckner G 1998 Reduced pulmonary function in patients with spinal osteoporotic fractures. Osteoporos Int **8**:261–267.
36. Leech JA, Dulberg C, Kellie S, Pattee L, Gay J 1990 Relationship of lung function to severity of osteoporosis in women. Am Rev Respir Dis **141**:68–71.
37. Hofbauer LC, Brueck CC, Singh SK, Dobnig H 2007 Osteoporosis in patients with diabetes mellitus. J Bone Miner Res **22**:1317–1328.
38. Inzerillo AM, Epstein S 2004 Osteoporosis and diabetes mellitus. Rev Endocr Metab Disord **5**:261–268.
39. Nicodemus KK, Folsom AR, Iowa Women's Health Study 2001 Type 1 and type 2 diabetes and incident hip fractures in postmenopausal women. Diabetes Care **24**:1192–1197.
40. de Liefde II, van der Klift M, de Laet CE, van Daele PL, Hofman A, Pols HA 2005 Bone mineral density and fracture risk in type-2 diabetes mellitus: The Rotterdam Study. Osteoporos Int **16**:1713–1720.
41. Bouillon R, Bex M, Van Herck E, Laureys J, Dooms L, Lesaffre E, Ravussin E 1995 Influence of age, sex, and insulin on osteoblast function: Osteoblast dysfunction in diabetes mellitus. J Clin Endocrinol Metab **80**:1194–1202.
42. Goodman WG, Hori MT 1984 Diminished bone formation in experimental diabetes. Relationship to osteoid maturation and mineralization. Diabetes **33**:825–831.
43. Kemink SA, Hermus AR, Swinkels LM, Lutterman JA, Smals AG 2000 Osteopenia in insulin-dependent diabetes mellitus; prevalence and aspects of pathophysiology. J Endocrinol Invest **23**:295–303.
44. Rosen CJ, Bouxsein ML 2006 Mechanisms of disease: Is osteoporosis the obesity of bone? Nat Clin Proct Rheumatol **2**:35–43.
45. Gimble JM, Zvonic S, Floyd ZE, Kassem M, Nuttall ME 2006 Playing with fat and bone. J Cell Biochem **98**:251–266.
46. Botolin S, Faugere MC, Malluche H, Orth M, Meyer R, McCabe LR 2005 Increased bone adiposity and peroxisomal proliferator-activated receptor-gamma2 expression in type I diabetic mice. Endocrinology **146**:3622–3631.
47. Kahn SE, Zinman B, Lachin JM, Haffner SM, Herman WH, Holman RR, Kravitz BG, Yu D, Heise MA, Aftring RP, Viberti G 2008 Diabetes Outcome Progression Trial (ADOPT) Study Group rosiglitazone associated fractures in type 2 diabetes: An analysis from ADOPT. Diabetes Care **31**:845–851.
48. Andrew SM, Freemont AJ 1993 Skeletal mastocytosis. J Clin Pathol **46**:1033–1035.
49. Valent P, Akin C, Escribano L, Fodinger M, Hartmann K, Brockow K, Castells M, Sperr WR, Kluin-Nelemans HC, Hamdy NA, Lortholary O, Robyn J, van Doormaal J, Sotlar K, Hauswirth AW, Arock M, Hermine O, Hellman A, Triggiani M, Niedoszytko M, Schwartz LB, Orfao A, Horny HP, Metcalfe DD 2007 Standards and standardization in mastocytosis: Consensus statements on diagnostics, treatment recommendations and response criteria. Eur J Clin Invest **37**:435–453.
50. Lidor C, Frisch B, Gazit D, Gepstein R, Hallel T, Mekori YA 1990 Osteoporosis as the sole presentation of bone marrow mastocytosis. J Bone Miner Res **5**:871–876.
51. De Gennes C, Kuntz D, de Vernejoul MC 1992 Bone mastocytosis: A report of nine cases with a bone histomorphometric study. Clin Orthop **279**:281–291.
52. Brumsen C, Papapoulos SE, Lentjes EG, Kluin PM, Hamdy NA 2002 A potential role for the mast cell in the pathogenesis of idiopathic osteoporosis in men. Bone **31**:556–561.
53. Oranje AP, Mulder PG, Heide R, Tank B, Riezebos P, van Toorenenbergen AW 2002 Urinary N-methylhistamine as an indicator of bone marrow involvement in mastocytosis. Clin Exp Dermatol **27**:502–506.
54. Tannenbaum C, Clark J, Schwartzman K, Wallenstein S, Lapinski R, Meier D, Luckey M 2002 Yield of laboratory testing to identify secondary contributors to osteoporosis in otherwise healthy women. J Clin Endocrinol Metab **87**:4431–4437.
55. Khosla S, Lufkin EG, Hodgson SF, Fitzpatrick LA, Melton LJ III 1994 Epidemiology and clinical features of osteoporosis in young individuals. Bone **15**:551–555.
56. Peris P, Guanabens N, Martinez de Osaba MJ, Monegal A, Alvarez L, Pons F, Ros I, Cerda D, Munoz-Gomez J 2002 Clinical characteristics and etiologic factors of premenopausal osteoporosis in a group of Spanish women. Semin Arthritis Rheum **32**:64–70.
57. Ebeling PR 1998 Osteoporosis in men. New insights into aetiology, pathogenesis, prevention and management. Drugs Aging **13**:421–434.

Chapter 61. Transplantation Osteoporosis

Peter R. Ebeling

Department of Medicine (RMH/WH), The University of Melbourne, Western Hospital, Footscray, Victoria, Australia

INTRODUCTION

Transplantation is an established therapy for end-stage diseases of the kidney, endocrine pancreas, heart, liver, lung, and intestines and for many hematological disorders. Improved survival rates, because of the addition of calcineurin inhibitors, cyclosporine A, and tacrolimus, to immunosuppressive treatment have been accompanied by a greater awareness of the long-term complications of transplantation such as fractures and osteoporosis.[1,2]

Pretransplantation bone disease and immunosuppressive therapy result in rapid bone loss and increased fracture rates early after transplantation. Patients should be assessed and pretransplantation bone disease and vitamin D deficiency should be treated. Treatment is indicated in the immediate posttransplantation period irrespective of bone mineral density, because further bone loss will occur in the first several months after transplantation. Long-term organ transplant recipients should also have bone mass measurement and treatment of osteoporosis.

Oral and intravenous bisphosphonates are the most promising approach for the management of transplantation osteoporosis. Active vitamin D metabolites may have additional benefits in reducing hyperparathyroidism, particularly after kidney transplantation.

PREEXISTING BONE DISEASE

Chronic Kidney Disease

In chronic kidney disease (CKD)–bone and mineral disease, one or more types of bone disease may be present including osteitis fibrosa cystica as a result of secondary hyperparathyroidism (SHPT), low turnover bone disease (osteomalacia,

Dr. Ebeling has received research grants from Amgen, GlaxoSmithKline, and Servier and has received honoraria from Merck, Eli Lilly, GlaxoSmithKline, and Amgen.

© 2008 American Society for Bone and Mineral Research

adynamic bone disease, or aluminum bone disease), osteoporosis, mixed bone disease, and β_2-microglobulin amyloidosis. In addition, hypogonadism, both in men and women, metabolic acidosis, and certain medications (loop diuretics, heparin, warfarin, glucocorticoids, or immunosuppressive agents) also adversely affect bone health.

Adynamic bone disease needs exclusion before treatment with bisphosphonates, which reduce bone turnover further. It is commonly associated with osteoporosis and occurs early in CKD. On bone histomorphometry, there is a scarcity of bone cells, reduced osteoid thickness, and a low bone formation rate.[3] The factors reducing bone turnover are a low vitamin D system and high phosphate and fibroblast growth factor 23 (FGF23), which override the stimulatory effect of PTH in early CKD. Bone histomorphometry is the best method to assess bone turnover in CKD.

In dialysis patients, fracture risk is increased with older age, female sex, white race,[4] duration of hemodialysis,[5] diabetic nephropathy,[6] peripheral vascular disease,[4] low spine BMD, and low bone turnover states.

Congestive Heart Failure

Low BMD is also common in patients with severe congestive heart failure (CHF).[7] Mild renal insufficiency, vitamin D deficiency, SHPT, increased bone resorption markers, and use of loop diuretics all contribute.

End-Stage Liver Disease

Osteoporosis and fractures commonly accompany chronic liver disease.[1,8] Studies show body mass index (BMI) before liver transplantation (LT), cholestatic liver disease, and older age are important risk factors.[9,10]

Chronic Respiratory Failure

Osteoporosis occurs in up to 61% of patients awaiting lung transplantation. Hypoxia, hypercapnia, smoking, low BMI, chronic obstructive pulmonary disease (COPD), and glucocorticoids all contribute.[11] In cystic fibrosis (CF), there are additional risk factors (pancreatic insufficiency, vitamin D deficiency, calcium malabsorption, hypogonadism, genetic factors, and inactivity).

Candidates for Bone Marrow Transplantation

Hypogonadism secondary to the effects of high-dose chemotherapy, total body irradiation (TBI), and glucocorticoids (GCs) reduce BMD in bone marrow transplantation (BMT) candidates, and women are particularly sensitive. Ovarian insufficiency occurs in the majority,[12,13] although some young, premenarchal women may recover ovarian function. Testosterone levels decline acutely after BMT related to a reduction in luteinizing hormone and then return to normal in most men.[14,15] There may be long-term impairment of spermatogenesis with elevated follicle-stimulating hormone (FSH) occurring in 47% of men.[12,13] In BMT candidates, osteopenia was present in 24% and osteoporosis in 4%.[16]

SKELETAL EFFECTS OF IMMUNOSUPPRESSIVE DRUGS

Glucocorticoids

High doses of GCs are commonly prescribed immediately after transplantation and are weaned rapidly. Doses are increased at the time of rejection episodes. The highest GC-associated rates of bone loss are in the first 3–12 mo after transplant. Trabecular sites are predominantly affected. More recent immunosuppressive regimens have limited GC use.

However, even small doses of GCs are associated with marked increases in fracture risk in epidemiological studies.[17] GCs reduce bone formation by decreasing osteoblast replication and differentiation and increasing apoptosis. Osteoblast genes, including *type I collagen, osteocalcin, insulin-like growth factors, bone matrix proteins,* and *TGF*β, are downregulated, whereas *RANKL* is upregulated. The direct and indirect effects of GCs on bone resorption, although less than effects on bone formation, also contribute to the rapid increase in post-transplant fracture risk. This period is characterized by high bone remodeling and bone resorption rates.

Calcineurin Inhibitors

Cyclosporine (CsA) has markedly reduced rejection episodes and improved post-transplant survival. Although in vivo rodent studies suggest that CsA has independent adverse effects to increase bone turnover,[18] kidney transplant patients receiving CsA without GCs[19,20] do not lose bone. Tacrolimus (FK506), another calcineurin inhibitor (CI), also causes trabecular bone loss in the rat.[18] Both cardiac[21] and liver[22] transplant recipients sustained rapid bone loss with tacrolimus. However, tacrolimus may cause less bone loss in humans than CsA[23,24] and may also protect the skeleton by reducing GC use.

Other Immunosuppressive Agents

Limited information is available regarding the effects of other immunosuppressive drugs on BMD and bone metabolism. However, azathioprine, sirolimus (rapamycin), mycophenolate mofetil, and daclizumab may also protect the skeleton by reducing GC use. In vitro studies suggest rapamycin inhibits osteoblast proliferation and differentiation.[25]

MANAGEMENT OF TRANSPLANTATION OSTEOPOROSIS

Before Organ Transplantation

All candidates for transplantation should have bone densitometry of the hip and spine and spinal X-rays to diagnose prevalent fractures. Any secondary causes of osteoporosis should be identified and treated. Vitamin D deficiency should be corrected and all patients should receive calcium and vitamin D (1000–1500 mg of calcium and at least 800 IU of vitamin D per day). Replacement doses of vitamin D should be selected to achieve a 25(OH)D concentration $\geq$30 ng/ml. Patients with kidney failure should be evaluated, and SHPT should be treated.

After Organ Transplantation

Risk factors for post-transplant bone loss and fractures are shown in Table 1. Bone loss is most rapid immediately after transplantation. Fractures often occur in the first year after transplantation and may affect patients with either low or normal pretransplant BMD. Therefore, the majority of patients may benefit from treatment instituted immediately after transplantation. Patients who present after being transplanted months or years before should also be assessed for treatment.

Most therapeutic trials have focused on the use of vitamin D metabolites and antiresorptive drugs, particularly oral and intravenous bisphosphonates. Hormone therapy with estrogen ± progestin helps protect the skeleton in women receiving liver, lung, and bone marrow transplantation, but does not prevent

© 2008 American Society for Bone and Mineral Research

TABLE 1. RISK FACTORS FOR POST-TRANSPLANT BONE LOSS AND FRACTURES

Contributing factors	*Mechanisms*
Aging Low body mass index Hypogonadism Calcium and vitamin D deficiency Tobacco Alcohol abuse Cholestasis (liver disease) Organ failure (heart, lung, liver, kidney) Pancreatic insufficiency (cystic fibrosis) Physical inactivity	Low pretransplant BMD
High-dose prednisone	Decreased bone formation Direct effect Decreased gonadal function Reduced intestinal and renal calcium transport
Calcineurin inhibitors Cyclosporine or FK506	Increased bone resorption Decreased renal function and $1{,}25(OH)_2D$ Increased PTH secretion Possible direct effect
Calcineurin inhibitor Sirolimus	Decreased bone formation Possible direct effect

bone loss. Because amenorrhea is a common sequela of BMT in premenopausal women, they should receive hormone replacement therapy (HRT). Hypogonadism is common in male cardiac and bone marrow transplant recipients because of chronic illness and hypothalamic-pituitary-adrenal suppression by GCs and CsA. Testosterone levels fall immediately after transplantation and normalize 6–12 mo later.

Kidney Transplantation. CKD–bone and mineral disease improves after transplantation; however, HPT may persist. Bone resorption remains elevated in a substantial proportion of kidney transplant recipients, and there is GC-induced osteoblast dysfunction.[26,27] Cross-sectional studies of patients evaluated several years after kidney transplantation (KT) have reported osteoporosis in 17–49% at the spine, 11–56% at the femoral neck, and 22–52% at the radius.[1] There is a correlation between cumulative GC dose and BMD. Rates of bone loss are greatest in the first 6–18 mo after transplantation and range from 4–9% at the spine to 5–8% at the hip. Increasing time since transplantation is a risk factor for low BMD, and BMD remains low up to 20 yr after transplantation.

In kidney transplant patients, fractures affect appendicular sites (hips, long bones, ankles, feet) more commonly than axial sites (spine and ribs).[28] Women and patients transplanted for diabetic nephropathy are at particularly increased risk of fractures. The majority of fractures after kidney transplantation occur within the first 3 yr; however, fractures continue to increase the longer the post-transplant period.[29]

Prevention and Treatment. Calcium and vitamin D supplementation alone are not effective[30]; however, bisphosphonates reduce bone loss after kidney transplantation.[31] Alendronate, calcitriol, and calcium treatment was also superior to calcitriol and calcium treatment alone, either early (6 mo) or late (5 yr) after kidney transplantation.[32,33] Intermittent calcitriol during the first 3 mo after renal transplantation was more effective than calcium in preserving total hip BMD over 1 yr.[34]

Renal safety issues and dosing schedules are different for intravenous bisphosphonates. Zoledronic acid (ZA) may cause acute renal failure, with the induction of acute tubular necrosis related to the rate of its infusion rather than the dose. Reduction of the ZA infusion rate to 15 min has not resulted in any cases of increased creatinine levels. The risk of renal failure is also increased by preexisting renal impairment, so the infusion rate should be reduced to one half the recommended rate in patients with glomerular filtration rate (GFR) <30 ml/min or baseline serum creatinine concentration >2 mg/dl.[35] The former measurement is the more accurate. A study comparing two infusions of 4 mg of ZA with placebo at 2 wk and 3 mo after KT showed prevention of early bone loss by ZA.[37]

A large, recent systematic review of 24 trials with 1299 patients showed any treatment of bone disease reduced the risk of fracture by 49% (95% CI, 0.27–0.99) compared with placebo.[37] Bisphosphonates and active vitamin D analogs had beneficial effects on the BMD at the spine and femoral neck. Bisphosphonates were better at preventing bone loss compared with vitamin D analogs. An unexpected finding was a reduction in the risk of graft rejection associated with bisphosphonate therapy. The authors concluded a trial comparing oral or parenteral bisphosphonates with calcitriol starting at the time of transplant and using fractures as a primary end-point is now needed.

Kidney-Pancreas Transplantation. Severe osteoporosis commonly complicates kidney-pancreas transplantation (SPK) in recipients with type 1 diabetes, with up to 58% having femoral neck osteoporosis, whereas vertebral or nonvertebral fractures were documented in 45%.[1] Other retrospective studies have documented a fracture prevalence of 26–49% up to 8.3 yr after SPK.[38] A prospective study addressed osteoporosis and SHPT in SPK. Hyperparathyroidism was common, affecting 68% of SPK recipients. Fractures were related to low pretransplant femoral neck BMD.[39]

Lung Transplantation. The prevalence of osteoporosis is as high as 73% in lung transplantation (LT) recipients. Fracture rates are also high during the first year, ranging from 18% to 37%. Bone turnover is also increased.[40] Repeated doses of intravenous pamidronate prevented lumbar spine (LS) and femoral neck (FN) bone loss in LT recipients.[41,42]

Cardiac Transplantation. Osteoporosis is common in long-term cardiac transplantation (CT) recipients, affecting 28% at the spine and 20% at the FN. The most rapid rate of bone loss occurs in the first year. Spinal BMD declines by 6–10% during the first 6 mo, with little decrease thereafter. In some studies, there has been partial recovery of spinal BMD in later years. Femoral neck BMD falls by 6–11% in the first year and stabilizes thereafter in most cases. BMD declines at the largely cortical proximal radius site over the second and third years, perhaps reflecting post-transplant SHPT. Vitamin D deficiency and testosterone deficiency (in men) are associated with more severe bone loss. Testosterone levels fall immediately after CT and normalize after 6–12 mo. Some studies have found correlations between GC dose and bone loss. Vertebral fracture prevalence rates range between 22% and 35% in long-term CT recipients. Vertebral fracture incidence ranges from 33% to 36% during the first 1–3 yr after CT.[43,44]

© 2008 American Society for Bone and Mineral Research

Prevention and Treatment

Vitamin D and Calcitriol. Calcium and vitamin D alone do not prevent bone loss after CT. However, 25(OH)D (calcidiol) has been associated with significant increases in spinal BMD 18 mo after CT.[1]

A double-blind study of CT or LT recipients compared placebo and or calcitriol (0.5–0.75 μg/d) and calcium for either 12 or 24 mo after transplantation.[45] Although spinal bone loss at 2 yr was similar in groups, bone loss from the FN was prevented. Another study compared rates of bone loss in patients randomized to receive calcitriol (0.5 μg/d) or two cycles of etidronate during the first 6 mo after CT or LT and followed for an additional 12 mo.[46] Significant and comparable bone loss (3–8%) occurred at the spine and FN in both treatment groups but was less than in historical controls.

Other studies observed that CT recipients randomized to either alphacalcidol or cyclic etidronate sustained considerable bone loss at the spine and FN during the first year after transplantation.[1] Another study of calcitriol[47] found no protective benefit. Thus, data regarding calcitriol and prevention of post-CT bone loss are inconsistent. Monitoring of serum and urine calcium levels is also needed.

Bisphosphonates. An open-label study of a single intravenous dose of pamidronate (60 mg) followed by four cycles of etidronate (400 mg every 3 mo) and daily low-dose calcitriol (0.25 μg) prevented spinal and FN bone loss and reduced fracture rates in heart transplant recipients compared with historical controls.[48]

In a 1-yr trial, in which 149 patients were randomized immediately after heart transplantation to receive either alendronate (10 mg/d) or calcitriol (0.25 μg twice daily), bone loss at the spine and hip was prevented by both regimens compared with a prospectively recruited nonrandomized reference group of 27 patients who received only calcium and vitamin D.[49] A 1-yr follow-up of the same group assessed the effects of treatment withdrawal on BMD and bone turnover markers.[50] Whereas BMD did not change in either the former alendronate or calcitriol group, bone resorption increased in the calcitriol group. This suggests that antiresorptive therapy may be discontinued 1 yr after transplant in CT recipients without rapid bone loss. However, these patients still require observation to ensure that BMD remains stable in the long term.

Exercise. Resistance exercise significantly improved lumbar spine BMD after LT[51] and heart[52] transplantation when used alone and in combination with alendronate. However, these small studies used highly variable lateral BMD measurements.

Liver Transplantation. Bone loss and fracture rates after liver transplantation are highest in the first 6–12 mo. Spine BMD declines by 2–24% during the first year in early studies, but bone loss has been in more recent studies. Fracture rates range from 24% to 65%, and the ribs and vertebrae are the most common sites. Women with primary biliary cirrhosis seem to be at greatest risk. Older age and pretransplant spinal and femoral neck BMD predicted post-transplantation fractures in one recent prospective study[8] and pretransplant vertebral fractures predicted post-transplant vertebral fractures in two recent prospective studies.[1,53]

Prevention and Treatment. A randomized trial of intravenous ibandronate in liver[54] transplant recipients found a significant protective effect on BMD at 1 yr. However, in a randomized trial of either a single dose of intravenous pamidronate administered 1–3 mo before liver transplantation or no treatment, pamidronate had no effect on BMD or fractures.[53] A recent randomized, double-blind trial studied 62 adults having liver transplantation who received treatment with either infusions of 4 mg ZA or saline within 7 days of transplantation and again at 1, 3, 6, and 9 mo after transplantation.[55] ZA significantly prevented bone loss from the LS, FN, and total hip by 3.8–4.7%, and differences were greatest 3 mo after transplant, but only remained significant at the total hip at 12 mo. Vitamin D deficiency should be corrected before giving ZA after liver transplantation. Alendronate also prevents bone loss after liver transplantation. An uncontrolled, prospective study of 136 patients with end-stage liver diseases showed alendronate prevented bone loss in patients with osteopenia and led to an increase in BMD at the spine and FN in patients with osteoporosis.[8] Another study of 59 patients having liver transplantation used historical controls and showed alendronate treatment led to significant increases in spinal, FN, and total hip BMD at 12 mo, being higher than in historical controls.[56]

Small Bowel Transplantation. Small bowel transplantation (SBT) is being increasingly used for severe inflammatory bowel disease. It may also include concomitant liver, pancreas, and stomach transplantation. In a cross-sectional study of 81 patients who had SBT 2.2 yr previously, BMD at the spine, total hip, and FN was reduced by ~0.8 SD compared with age- and sex-matched controls with small bowel disease, and in a longitudinal study of 9 patients, bone loss occurred at the spine (2.6%), total hip, and FN (by ~15%) 1.3 yr after SBT.[57]

Bone Marrow Transplantation. Bone marrow or stem cell transplantation (BMT) is the treatment of choice for patients with many hematological malignancies, the majority of whom will survive for many years. Up to 29% and 52% of survivors have osteopenia at the spine and FN, respectively.[1] The pathogenesis of post-BMT osteoporosis is complex and related to effects of treatment and effects on the bone marrow stromal cell (MSC) compartment.[58,59] Similar to solid organ transplantation, bone resorption increases, whereas bone formation decreases,[1,60] resulting in early, rapid bone loss. In addition to osteoporosis, osteomalacia and avascular necrosis may occur.

Many groups have shown that dramatic bone loss from the proximal femur occurs within the first 12 mo of allogeneic BMT.[1,61,62] Spinal bone loss is less. Most studies suggest that little additional bone loss occurs after this time. Studies of long-term survivors of BMT have shown that losses from the proximal femur are not regained.[63] After autologous BMT, bone loss from the proximal femur occurs early but is less.[64]

Contributing factors to bone loss after BMT include cumulative GC exposure. Bone loss has been related to duration of CsA exposure[61] and may also be a direct effect of graft versus host disease (GVHD) itself on bone cells. Abnormal cellular or cytokine-mediated bone marrow function may affect bone turnover and BMD after BMT.[1] Both myeloablative treatment and BMT stimulate the early release of cytokines. BMT also has adverse effects on bone marrow osteoprogenitors. Osteocyte viability is decreased after BMT, and bone marrow stromal cells are damaged by high-dose chemotherapy, TBI, GCs, and CsA; reducing osteoblastic differentiation from osteoprogenitor cells.[65] Colony forming units-fibroblasts (CFU-f) are reduced for up to 12 yr after BMT.[1]

Avascular necrosis develops in 10–20% of allo-BMT survivors, a median of 12 mo after BMT.[59,66] GC treatment of chronic GVHD is the most important risk factor. Avascular necrosis seems to be related to decreased numbers of bone marrow CFU-f colonies in vitro but not to BMD values.[66] Avascular necrosis may thus be facilitated by a deficit in bone MSC regeneration and low osteoblast numbers after BMT.[65]

© 2008 American Society for Bone and Mineral Research

Prevention and Treatment. Risedronate or intravenous ZA given 12 mo after BMT prevents spinal and proximal femoral bone loss.[67,68] ZA effects may be related to improved osteoblast recovery and increased osteoblast numbers after BMT. Two randomized studies recently assessed the effectiveness of intravenous pamidronate in preventing bone loss after BMT. The first studied 99 allogeneic BMT recipients, who were randomized to received calcium and vitamin D daily, hormone therapy with estrogen in women or testosterone in men, or the same treatments plus intravenous 60 mg pamidronate infusions before and 1, 2, 3, 6, and 9 mo after BMT.[60] In the pamidronate group, LS BMD remained stable but decreased significantly in the other group at 12 mo. Total hip BMD and FN BMD decreased by 5.1% and 4.2%, respectively, in the pamidronate group and by 7.8% and 6.2%, respectively, in the other group at 12 mo. Thus, pamidronate reduced bone loss more than in those treated with calcium, vitamin D, and sex steroid replacement alone.

A larger randomized, multicenter open-label 12-mo prospective study compared intravenous pamidronate (90 mg/mo) beginning before conditioning versus no pamidronate.[69] All 116 patients also received calcitriol (0.25 μg/d) and calcium, which were continued for a further year. Pamidronate significantly reduced bone loss at the spine, FN, and total hip at 12 mo. However, BMD of the femoral neck and total hip was still 2.8% and 3.5% lower than baseline, respectively, with pamidronate. Only the BMD benefit at the total hip remained significant between the two groups at 24 mo (Fig. 1). This study also showed benefits of pamidronate therapy were restricted to patients receiving an average daily prednisolone dose >10 mg and cyclosporin therapy for >5 mo within the first 6 mo of alloSCT. Most BMD benefits were lost 12 mo after stopping pamidronate. A small uncontrolled, prospective study of a single 4-mg ZA infusion in allogeneic BMT patients with either osteoporosis or rapid bone loss after allogeneic BMT[70] showed reduced bone loss at the spine and proximal femur in most patients.

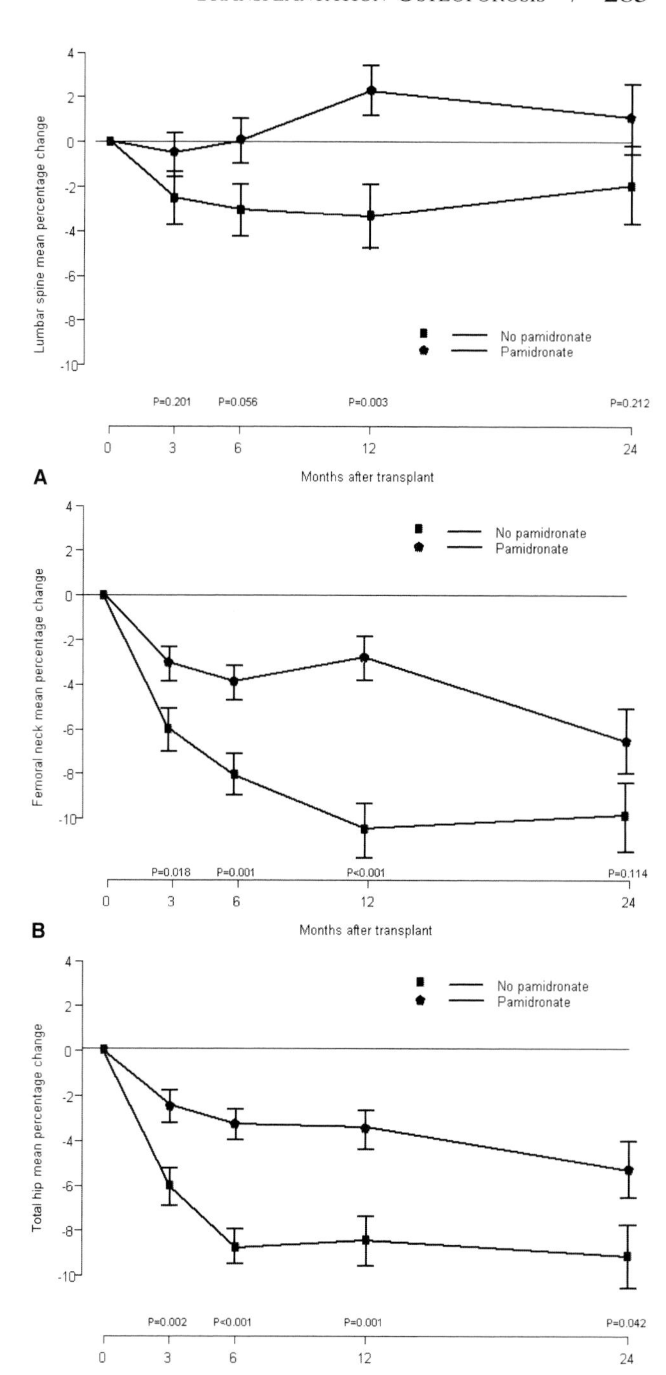

FIG 1. Mean percentage changes in (A) lumbar spine BMD, (B) femoral neck BMD, and (C) total hip BMD (±SE) for each treatment group at each assessment time. The *p* values refer to the outcome of a *t*-test comparing the two treatment groups (no pamidronate vs. pamidronate) at each assessment time. (Reproduced with permission from Grigg A, Shuttleworth P, Reynolds J, Schwarer A, Szer J, Bradstock K, Hui C, Herrmann R, Ebeling P 2006 Pamidronate reduces bone loss after allogeneic stem cell transplantation. J Endocrinol Metab **91**:3835–3843, Copyright 2006, The Endocrine Society.)

SUMMARY AND CONCLUSIONS

Pretransplantation bone disease and immunosuppressive therapy result in a severe form of osteoporosis characterized by rapid bone loss and increased fracture rates, early after transplantation. There is increased bone resorption and decreased bone formation, suggestive of uncoupling of bone turnover. In the late post-transplant period, with weaning of GC doses, bone formation begins to increase and underlying high bone turnover results in osteoporosis. The rates of bone loss and fractures reported in recent studies are lower than those of 10 yr ago; however, they remain too high. Transplant candidates should be assessed, and pretransplantation bone disease should be treated. Preventive therapy initiated in the immediate post-transplantation period is indicated, because further bone loss will occur in the first several months after transplantation. All organ transplant recipients should be considered at risk for post-transplantation bone loss and fractures, because it is impossible to identify patients with the highest fracture risk. Long-term organ transplant recipients should also have bone mass measurement and treatment of osteoporosis.

Bisphosphonates are the most promising approach for the prevention and treatment of transplantation osteoporosis. Active vitamin D metabolites may have additional benefits in reducing hyperparathyroidism, particularly after kidney transplantation. Promising new agents for transplantation osteoporosis include anabolic agents that stimulate bone formation, namely PTH(1-34) or teriparatide, the anti-catabolic drugs, human antibodies to RANKL (denosumab), and cathepsin K inhibitors. PTH(1-34) and other PTH1 receptor agonists may have a specific role after BMT in stimulating MSC differentiation into the osteoblast lineage and reducing adipogenesis.[71,72]

© 2008 American Society for Bone and Mineral Research

Several issues remain regarding the administration of bisphosphonates for transplantation bone disease, including the optimal route of administration and duration of therapy. Treatment may only need to be given for 1 yr after cardiac transplantation, but its optimal duration is less clear after other transplants. It is also uncertain at what level of renal impairment oral bisphosphonates should be avoided, and whether this level is the same for intravenous bisphosphonates. Another special consideration in using bisphosphonates in KT recipients is adynamic bone disease (see above). Large multicenter trials comparing treatment with oral or parenteral bisphosphonates and calcitriol and starting at the time of transplantation are recommended for KT.

Despite some continuing uncertainties, much has been learned about transplantation osteoporosis. Armed with this information, it is critical to act to prevent and treat this disabling disease.

ACKNOWLEDGMENTS

The author thanks Drs. Elizabeth Shane, Adi Cohen, and Stuart Sprague for previous contributions to this work.

REFERENCES

1. Cohen A, Sambrook P, Shane E 2004 Management of bone loss after organ transplantation. J Bone Miner Res **19**:1919–1932.
2. Cohen A, Shane E 2003 Osteoporosis after solid organ and bone marrow transplantation. Osteoporos Int **14**:617–630.
3. Gal-Moscovici A, Sprague SM 2007 Osteoporosis and chronic kidney disease. Semin Dial **20**:423–430.
4. Stehman-Breen CO, Sherrard DJ, Alem AM, Gillen DL, Heckbert SR, Wong CS, Ball A, Weiss NS 2000 Risk factors for hip fracture among patients with end-stage renal disease. Kidney Int **58**:2200–2205.
5. Alem AM, Sherrard DJ, Gillen DL, Weiss NS, Beresford SA, Heckbert SR, Wong C, Stehman-Breen C 2000 Increased risk of hip fracture among patients with end-stage renal disease. Kidney Int **58**:396–399.
6. Ball AM, Gillen DL, Sherrard D, Weiss NS, Emerson SS, Seliger SL, Kestenbaum BR, Stehman-Breen C 2002 Risk of hip fracture among dialysis and renal transplant recipients. JAMA **288**:3014–3018.
7. Shane E, Mancini D, Aaronson K, Silverberg SJ, Seibel MJ, Addesso V, McMahon DJ 1997 Bone mass, vitamin D deficiency and hyperparathyroidism in congestive heart failure. Am J Med **103**:197–207.
8. Monegal A, Navasa M, Guanabens N, Peris P, Pons F, Martinez de Osaba MJ, Ordi J, Rimola A, Rodes J, Munoz-Gomez J 2001 Bone disease after liver transplantation: A long-term prospective study of bone mass changes, hormonal status and histomorphometric characteristics. Osteoporos Int **12**:484–492.
9. Millonig G, Graziadei IW, Eichler D, Pfeiffer KP, Finkenstedt G, Muehllechner P, Koenigsrainer A, Margreiter R, Vogel W 2005 Alendronate in combination with calcium and vitamin D prevents bone loss after orthotopic liver transplantation: A prospective single-center study. Liver Transpl **11**:960–966.
10. Ninkovic M, Love SA, Tom B, Alexander GJ, Compston JE 2001 High prevalence of osteoporosis in patients with chronic liver disease prior to liver transplantation. Calcif Tissue Int **69**:321–326.
11. Tschopp O, Boehler A, Speich R, Weder W, Seifert B, Russi EW, Schmid C 2002 Osteoporosis before lung transplantation: Association with low body mass index, but not with underlying disease. Am J Transplant **2**:167–172.
12. Keilholz U, Max R, Scheibenbogen C, Wuster C, Korbling M, Haas R 1997 Endocrine function and bone metabolism 5 years after autologous bone marrow/blood-derived progenitor cell transplantation. Cancer **79**:1617–1622.
13. Tauchmanova L, Selleri C, Rosa GD, Pagano L, Orio F, Lombardi G, Rotoli B, Colao A 2002 High prevalence of endocrine dysfunction in long-term survivors after allogeneic bone marrow transplantation for hematologic diseases. Cancer **95**:1076–1084.
14. Valimaki M, Kinnunen K, Volin L, Tahtela R, Loyttniemi E, Laitinen K, Makela P, Keto P, Ruutu T 1999 A prospective study of bone loss and turnover after allogeneic bone marrow transplantation: Effect of calcium supplementation with or without calcitonin. Bone Marrow Transplant **23**:355–361.
15. Kananen K, Volin L, Laitinen K, Alfthan H, Ruutu T, Valimaki MJ 2005 Prevention of bone loss after allogeneic stem cell transplantation by calcium, vitamin D, and sex hormone replacement with or without pamidronate. J Clin Endocrinol Metab **90**:3877–3885.
16. Schulte C, Beelen D, Schaefer U, Mann K 2000 Bone loss in long-term survivors after transplantation of hematopoietic stem cells: A prospectiv estudy. Osteoporos Int **11**:344–353.
17. Van Staa TP, Leufkens HG, Abenhaim L, Zhang B, Cooper C 2000 Use of oral corticosteroids and risk of fractures. J Bone Miner Res **15**:993–1000.
18. Epstein S 1996 Post-transplantation bone disease: The role of immunosuppressive agents on the skeleton. J Bone Miner Res **11**:1–7.
19. Ponticelli C, Aroldi A 2001 Osteoporosis after organ transplantation. Lancet **357**:1623.
20. McIntyre HD, Menzies B, Rigby R, Perry-Keene DA, Hawley CM, Hardie IR 1995 Long-term bone loss after renal transplantation: Comparison of immunosuppressive regimens. Clin Transplant **9**:20–24.
21. Stempfle HU, Werner C, Echtler S, Assum T, Meiser B, Angermann CE, Theisen K, Gartner R 1998 Rapid trabecular bone loss after cardiac transplantation using FK506 (tacrolimus)-based immunosuppression. Transplant Proc **30**:1132–1133.
22. Park KM, Hay JE, Lee SG, Lee YJ, Wiesner RH, Porayko MK, Krom RA 1996 Bone loss after orthotopic liver transplantation: FK 506 versus cyclosporine. Transplant Proc **28**:1738–1740.
23. Goffin E, Devogelaer JP, Depresseux G, Squifflet JP, Pirson Y 2001 Osteoporosis after organ transplantation. Lancet **357**:1623.
24. Monegal A, Navasa M, Guanabens N, Peris P, Pons F, Martinez de Osaba MJ, Rimola A, Rodes J, Munoz-Gomez J 2001 Bone mass and mineral metabolism in liver transplant patients treated with FK506 or cyclosporine A. Calcif Tissue Int **68**:83–86.
25. Singha UK, Jiang Y, Yu S, Luo M, Lu Y, Zhang J, Xiao G 2008 Rapamycin inhibits osteoblast proliferation and differentiation in MC3T3-E1 cells and primary mouse bone marrow stromal cells. J Cell Biochem **103**:434–436.
26. Julian BA, Laskow DA, Dubovsky J, Dubovsky EV, Curtis JJ, Quarrles LD 1991 Rapid loss of vertebral bone density after renal transplantation. N Engl J Med **325**:544–550.
27. Monier-Faugere M, Mawad H, Qi Q, Friedler R, Malluche HH 2000 High prevalence of low bone turnover and occurrence of osteomalacia after kidney transplantation. J Am Soc Nephrol **11**:1093–1099.
28. Ramsey-Goldman R, Dunn JE, Dunlop DD, Stuart FP, Abecassis MM, Kaufman DB, Langman CB, Salinger MH, Sprague SM 1999 Increased risk of fracture in patients receiving solid organ transplants. J Bone Miner Res **14**:456–463.
29. Sprague SM, Josephson MA 2004 Bone disease after kidney transplantation. Semin Nephrol **24**:82–90.
30. Wissing KM, Broeders N, Moreno-Reyes R, Gervy C, Stallenberg B, Abramowicz D 2005 A controlled study of vitamin D_3 to prevent bone loss in renal-transplant patients receiving low doses of steroids. Transplantation **79**:108–115.
31. Grotz W, Nagel C, Poeschel D, Cybulla M, Petersen KG, Uhl M, Strey C, Kirste G, Olschewski M, Reichelt A, Rump LC 2001 Effect of ibandronate on bone loss and renal function after kidney transplantation. J Am Soc Nephrol **12**:1530–1537.
32. Kovac D, Lindic J, Kandus A, Bren AF 2001 Prevention of bone loss in kidney graft recipients. Transplant Proc **33**:1144–1145.
33. Giannini S, Dangel A, Carraro G, Nobile M, Rigotti P, Bonfante L, Marchini F, Zaninotto M, Dalle Carbonare L, Sartori L, Crepaldi G 2001 Alendronate prevents further bone loss in renal transplant recipients. J Bone Miner Res **16**:2111–2117.
34. Torres A, Garcia S, Gomez A, Gonzalez A, Barrios Y, Concepcion MT, Hernandez D, Garcia JJ, Checa MD, Lorenzo V, Salido E 2004 Treatment with intermittent calcitriol and calcium reduces bone loss after renal transplantation. Kidney Int **65**:705–712.
35. Miller PD 2005 Treatment of osteoporosis in chronic kidney disease and end-stage renal disease. Curr Osteoporos Rep **3**:5–12.
36. Schwarz CL, Mitterbauer CL, Heinze G, Woloszczuk W, Haas M, Oberbauer R 2004 Nonsustained effect of short-term bisphosphonate therapy on bone turnover three years after renal transplantation. Kidney Int **65**:304–309.
37. Palmer SC, McGregor DO, Strippoli GFM 2007 Interventions for

© 2008 American Society for Bone and Mineral Research

preventing bone disease in kidney transplant recipients. Cochrane Database Syst Rev **18:**CD005015.
38. Chiu MY, Sprague SM, Bruce DS, Woodle ES, Thistlethwaite JR Jr, Josephson MA 1998 Analysis of fracture prevalence in kidney-pancreas allograft recipients. J Am Soc Nephrol **9:**677–683.
39. Smets YFC, De Fijter JW, Ringers J, Lemkes HHPJ, Hamdy NAT 2004 Long-term follow-up study on bone mineral density and fractures after simultaneous pancreas-kidney transplantation. Kidney Int **66:**2070–2076.
40. Shane E, Papadopoulos A, Staron RB, Addesso V, Donovan D, McGregor C, Schulman LL 1999 Bone loss and fracture after lung transplantation. Transplantation **68:**220–227.
41. Aris RM, Lester GE, Renner JB, Winders A, Denene Blackwood A, Lark RK, Ontjes DA 2000 Efficacy of pamidronate for osteoporosis in patients with cystic fibrosis following lung transplantation. Am J Respir Crit Care Med **162:**941–946.
42. Trombetti A, Gerbase MW, Spiliopoulos A, Slosman DO, Nicod LP, Rizzoli R 2000 Bone mineral density in lung-transplant recipients before and after graft: Prevention of lumbar spine post-transplantation-accelerated bone loss by pamidronate. J Heart Lung Transplant **19:**736–743.
43. Shane E, Rivas M, Staron RB, Silverberg SJ, Seibel M, Kuiper J, Mancini D, Addesso V, Michler RE, Factor-Litvak P 1996 Fracture after cardiac transplantation: A prospective longitudinal study. J Clin Endocrinol Metab **81:**1740–1746.
44. Leidig-Bruckner G, Hosch S, Dodidou P, Ritchel D, Conradt C, Klose C, Otto G, Lange R, Theilmann L, Zimmerman R, Pritsch M, Zeigler R 2001 Frequency and predictors of osteoporotic fractures after cardiac or liver transplantation: A follow-up study. Lancet **357:**342–347.
45. Sambrook P, Henderson NK, Keogh A, MacDonald P, Glanville A, Spratt P, Bergin P, Ebeling P, Eisman J 2000 Effect of calcitriol on bone loss after cardiac or lung transplantation. J Bone Miner Res **15:**1818–1824.
46. Henderson K, Eisman J, Keogh A, MacDonald P, Glanville A, Spratt P, Sambrook P 2001 Protective effect of short-tem calcitriol or cyclical etidronate on bone loss after cardiac or lung transplantation. J Bone Miner Res **16:**565–571.
47. Stempfle HU, Werner C, Echtler S, Wehr U, Rambeck WA, Siebert U, Uberfuhr P, Angermann CE, Theisen K, Gartner R 1999 Prevention of osteoporosis after cardiac transplantation: A prospective, longitudinal, randomized, double-blind trial with calcitriol. Transplantation **68:**523–530.
48. Bianda T, Linka A, Junga G, Brunner H, Steinert H, Kiowski W, Schmid C 2000 Prevention of osteoporosis in heart transplant recipients: A comparison of calcitriol with calcitonin and pamidronate. Calcif Tissue Int **67:**116–121.
49. Shane E, Addesso V, Namerow PB, McMahon DJ, Lo SH, Staron RB, Zucker M, Pardi S, Maybaum S, Mancini D 2004 Alendronate versus calcitriol for the prevention of bone loss after cardiac transplantation. N Engl J Med **350:**767–776.
50. Cohen A, Addesso V, McMahon DJ, Staron RB, Namerow P, Maybaum S, Mancini D, Shane E 2006 Discontinuing antiresorptive therapy one year after cardiac transplantation: Effect on bone density and bone turnover. Transplantation **81:**686–691.
51. Mitchell MJ, Baz MA, Fulton MN, Lisor CF, Braith R 2003 Resistance training prevents vertebral osteoporosis in lung transplant recipients. Transplantation **76:**557–562.
52. Braith RW, Magyari PM, Fulton MN, Lisor CF, Vogel SE, Hill JA, Aranda JM Jr 2006 Comparison of calcitonin versus calcitonin and resistance exercise as prophylaxis for osteoporosis in heart transplant recipients. Transplantation **81:**1191–1195.
53. Ninkovic M, Skingle SJ, Bearcroft PW, Bishop N, Alexander CJ, Compston JE 2000 Incidence of vertebral fractures in the first three months after orthotopic liver transplantation. Eur J Gastroenterol Hepatol **12:**931–935.
54. Hommann M, Abendroth K, Lehmann G, Patzer N, Kornberg A, Voigt R, Seifert S, Hein G, Scheele J 2002 Effect of transplantation on bone: Osteoporosis after liver and multivisceral transplantation. Transplant Proc **34:**2296–2298.
55. Crawford BAL, Kam C, Pavlovic J, Byth K, Handelsman DJ, Angus PW, McCaughan GW 2006 Zoledronic acid prevents bone loss after liver transplantation: A randomized, double-blind, placebo-controlled trial. Ann Intern Med **144:**239–248.
56. Karasu Z, Kilic M, Tokat Y 2006 The prevention of bone fractures after liver transplantation: Experience with alendronate treatment. Transplant Proc **38:**1448–1452.
57. Awan KS, Wagner JM, Martin D, Medich DL, Perera S, Abu Elmagd K, Greenspan SL 2007 Bone loss following small bowel transplantation. J Bone Miner Res **22:**S356.
58. Banfi A, Podesta M, Fazzuoli L, Sertoli MR, Venturini M, Santini G, Cancedda R, Quarto R 2001 High-dose chemotherapy shows a dose-dependent toxicity to bone marrow osteoprogenitors: A mechanism for post-bone marrow transplantation osteopenia. Cancer **92:**2419–2428.
59. Lee WY, Cho SW, Oh ES, Oh KW, Lee JM, Yoon KH, Kang MI, Cha BY, Lee KW, Son HY, Kang SK, Kim CC 2002 The effect of bone marrow transplantation on the osteoblastic differentiation of human bone marrow stromal cells. J Clin Endocrinol Metab **87:**329–335.
60. Kananen K, Volin L, Laitinen K, Alfthan H, Ruutu T, Välimäki MJ 2005 Prevention of bone loss after allogeneic stem cell transplantation by calcium, vitamin D, and sex hormone replacement with or without pamidronate. J Clin Endocrinol Metab **90:**3877–3885.
61. Ebeling P, Thomas D, Erbas B, Hopper L, Szer J, Grigg A 1999 Mechanism of bone loss following allogeneic and autologous hematopoeitic stem cell transplantation. J Bone Miner Res **14:**342–350.
62. Ebeling PR 2005 Bone disease after bone marrow transplantation. In: Compston J, Shane E (eds.) Bone Disease of Organ Transplantation. Elsevier, Academic Press, San Diego, CA, USA, pp. 339–352.
63. Lee WY, Kang MI, Baek KH, Oh ES, Oh KW, Lee KW, Kim SW, Kim CC 2002 The skeletal site-differential changes in bone mineral density following bone marrow transplantation: 3-year prospective study. J Korean Med Sci **17:**749–754.
64. Gandhi MK, Lekamwasam S, Inman I, Kaptoge S, Sizer L, Love S, Bearcroft PW, Milligan TP, Price CP, Marcus RE, Compston JE 2003 Significant and persistent loss of bone mineral density in the femoral neck after haematopoietic stem cell transplantation: Long-term follow-up of a prospective study. Br J Haematol **121:**462–468.
65. Ebeling PR 2005 Is defective osteoblast function responsible for bone loss from the proximal femur despite pamidronate therapy? J Clin Endocrinol Metab **90:**4414–4416.
66. Tauchmanova L, De Rosa G, Serio B, Fazioli F, Mainolfi C, Lombardi G, Colao A, Salvatore M, Rotoli B, Selleri C 2003 Avascular necrosis in long-term survivors after allogeneic or autologous stem cell transplantation: A single center experience and a review. Cancer **97:**2453–2461.
67. Tauchmanova L, Selleri C, Esposito M, Di Somma C, Orio F Jr, Bifulco G, Palomba S, Lombardi G, Rotoli B, Colao A 2003 Beneficial treatment with risedronate in long-term survivors after allogeneic stem cell transplantation for hematological malignancies. Osteoporos Int **14:**1013–1019.
68. Tauchmanova L, Ricci P, Serio B, Lombardi G, Colao A, Rotoli B, Selleri C 2005 Short-term zoledronic acid treatment increases bone mineral density and marrow clonogenic fibroblast progenitors after allogeneic stem cell transplantation. J Clin Endocrinol Metab **90:**627–634.
69. Grigg AP, Shuttleworth P, Reynolds J, Schwarer AP, Szer J, Bradstock K 2006 Pamidronate reduces bone loss after allogeneic stem cell transplantation. J Clin Endocrinol Metab **91:**3835–3843.
70. D'Souza AB, Grigg AP, Szer J, Ebeling PR 2006 Zoledronic acid prevents bone loss after allogeneic haemopoietic stem cell transplantation. Intern Med J. **36:**600–603.
71. Rickard DJ, Wang FL, Rodriguez-Rojas AM, Wu Z, Trice WJ, Hoffman SJ, Votta B, Stroup GB, Kumar S, Nuttall ME 2006 Intermittent treatment with parathyroid hormone (PTH) as well as a non-peptide small molecule agonist of the PTH1 receptor inhibits adipocyte differentiation in human bone marrow stromal cells. Bone **39:**1361–1372.
72. Chan GK, Miao D, Deckelbaum R, Bolivar I, Karaplis A, Goltzman D 2003 Parathyroid hormone-related peptide interacts with bone morphogenetic protein 2 to increase osteoblastogenesis and decrease adipogenesis in pluripotent C3H10T mesenchymal cells. Endocrinology **144:**5511–5520.

© 2008 American Society for Bone and Mineral Research

Chapter 62. Osteoporosis in Men

Eric S. Orwoll

Oregon Health Sciences University, Portland, Oregon

INTRODUCTION

Osteoporosis in men is now recognized as an important public health problem, and there is a much greater understanding of the disorder. Effective diagnostic, preventive, and treatment strategies have been developed. Moreover, the study of osteoporosis in men has revealed male-female differences that in turn have fostered a greater understanding of bone biology in general. Nevertheless, there are important pathophysiological and clinical issues that remain unresolved and research continues to be very active.

SKELETAL DEVELOPMENT

Bone mass accumulation in males occurs gradually during childhood and accelerates dramatically during adolescence. Peak bone mass is closely tied to pubertal development, and male–female differences in the skeleton appear during adolescence.[1] The rapid increase in bone mass occurs somewhat later in boys than girls; the majority of the increase has occurred by an average age of 16 yr in girls and age 18 yr in boys. Moreover, whereas trabecular BMD accumulation is similar in boys and girls, boys generally develop thicker cortices and larger bones than do girls, even when adjusted for body size. These differences may provide important biomechanical advantages that could in part underlie the lower fracture risk observed in men later in life. The reasons for these sexual differences in skeletal development are unclear but could be related to differences in sex steroid action (androgens may stimulate periosteal bone formation and bone expansion), growth factor concentrations, and mechanical forces exerted on bone (e.g., by greater muscle action or activity). Sex-specific affects of a variety of genetic loci have been reported in animals and humans, suggesting that the origin of sex differences in skeletal phenotypes is complex. Finally, despite these average sex differences, there is wide variation in bone mass and structure in men after adjustment for body size and considerable overlap with the range of similar measures in women.

EFFECTS OF AGING ON THE SKELETON IN MEN

As in women, aging is associated with large changes in bone mass and architecture in men.[2] Trabecular bone loss (e.g., in the vertebrae and proximal femur) occurs during midlife and accelerates in later life. The magnitude of these changes is similar, but probably slightly less, than those in women. Endocortical bone loss with resulting cortical thinning also takes place in long bones, but that process seems to be accompanied by a concomitant increase in periosteal bone expansion that tends to preserve the breaking strength of bone.[3] Although still uncertain, the increase in periosteal bone formation that occurs in men may be greater than that in women and has been postulated to contribute to the lower fracture risk observed in older men. In general, the pattern of age-related bone loss in men is similar in men and women, but in men, there is no concomitant to the accelerated phase of loss associated with the menopause in women. In the elderly of both sexes, the rate of bone loss accelerates with increasing age.

FRACTURE EPIDEMIOLOGY

Fractures are common in men. The data concerning fractures in men are derived primarily from the study of white populations. In them, the incidence of fracture is bimodal, with a peak of fracture incidence in adolescence and mid-adulthood, a lower incidence between 40 and 60 yr, and a dramatic increase after the age of 70 yr (Fig. 1).[4] The types of fractures sustained in younger and older men are different, with long bone fractures being common in younger men, whereas vertebral and hip fractures predominate in the elderly. These differences suggest that the etiologies of fractures at these two periods of life are distinct. In younger men, trauma may play a larger role, whereas in older men skeletal fragility and fall propensity are likely to be major factors.

The exponential increase in fracture incidence in older men is as dramatic as the similar increase that occurs in women, but it begins 5–10 yr later in life. This delay, combined with the longer life expectancy in women, underlies the greater burden of osteoporotic fractures in women. Nevertheless, the age-adjusted incidence of hip fracture in men is one quarter to one third that in women, and 20–25% of hip fractures occur in men.[2] The consequences of fracture in men are at least as great as in women, and in fact, elderly men seem to be more likely to die and to suffer disability than women after a hip fracture. Older men suffer lower rates of long bone fractures than do women.[5] There is less information concerning vertebral fracture epidemiology in men, but the age-adjusted incidence appears to be high—~50% that in women.[6] In younger men, the prevalence of vertebral fracture is actually greater in men than in women, at least in part the result of higher rates of spinal trauma experienced by men. Although there are inadequate data, the epidemiology of fracture in men seems to be dramatically influenced by both race and geography.[7,8] For instance, black men have a much lower likelihood of fractures than whites, and Asian men have a lower likelihood of suffering hip fracture than whites. Much more information is needed concerning these differences and their causation.

Fractures in men are related to a variety of risk factors. Certainly skeletal fragility makes fracture more likely. This trait is most commonly measured as reduced BMD, but almost certainly has other components (biomechanically important alterations in bone geometry, material properties, etc.). Aging and a previous history of fracture are independently associated with a higher probability of future fracture, and men of lower weight have a higher fracture risk.[2,9] Finally, falling becomes much more common with increasing age in men, and falls are strongly associated with increased fracture risk.[10]

CAUSES OF OSTEOPOROSIS IN MEN

The causation of osteoporosis in men is commonly heterogeneous, and most osteoporotic men have several factors that contribute to the disease. One half to two thirds of men with osteoporosis have secondary osteoporosis—or that associated with other medical conditions, medications, or lifestyle factors that result in bone loss and fragility (Table 1).[2,7] The most important include alcohol abuse, glucocorticoid excess, and

Dr. Orwoll has consulted for Merck, Eli Lilly & Co., and Servier, has received research support from Amgen, Pfizer, Eli Lilly & Co., Novartis, Zelos Therapeutics, Solvay Pharmaceuticals, and Imaging Therapeutics and has received honoraria from Merck.

© 2008 American Society for Bone and Mineral Research

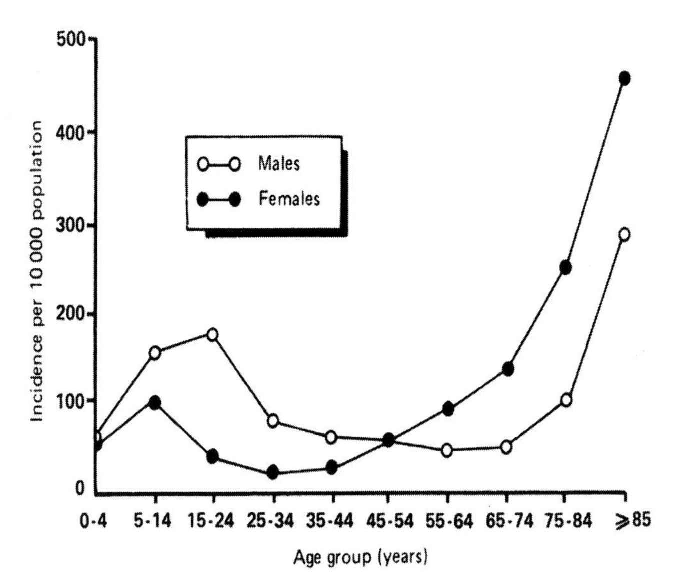

FIG. 1. Average annual fracture incidence rate per 10,000 population in Leicester, UK, by age, group, and sex.

hypogonadism. An important fraction of osteoporotic men, however, have idiopathic disease.

Idiopathic Osteoporosis

Osteoporosis of unknown etiology can present in men of any age[11] but is most dramatic in younger men who are otherwise unlikely to be affected. Several possible etiologies have been considered. Most prominent among them are genetic factors, because BMD and the risk of fracture are highly heritable. The specific genes that may be responsible are uncertain.

Hypogonadism

Sex steroids are clearly important for skeletal health in men, both during growth and the attainment of peak bone mass and the maintenance of bone strength in adults.[11] Hypogonadism is associated with low BMD, the development of hypogonadism results in increased bone remodeling and rapid bone loss (at least in its early phases), and testosterone replacement increases BMD in hypogonadal men. One of the most important causes of severe hypogonadism is androgen deprivation therapy for prostate cancer; in this situation, bone loss is rapid, and the risk of fractures is clearly increased. Gonadal function and sex steroid levels decline with age in men, and it has been postulated that the decline may be an important risk factor age-related bone loss and fracture risk. The strength of this association remains somewhat uncertain.

The relative roles of estrogens and androgens in skeletal physiology in men are uncertain.[12–14] Estrogen is essential for normal bone development in young men, as evidenced by the immature development and low bone mass in men with aromatase deficiency and their reversal with estrogen therapy. Moreover, estrogen is correlated with bone remodeling, BMD, and rate of BMD loss in older men, apparently more strongly than is testosterone. However, testosterone is independently related to indices of bone resorption and formation and may stimulate periosteal bone formation.[15–17] In early reports, low levels of either estradiol or testosterone have been associated with increased fracture risk. The relative roles of estrogen and androgen must be better defined.

EVALUATION OF OSTEOPOROSIS IN MEN

Guidelines for the evaluation of osteoporosis in men are not well validated, but there are several recommendations that can be made confidently.

BMD Measurements

BMD measures are at least as effective in men as in women in predicting the risk of future fractures.[18] In light of the prevalence of osteoporosis and the high incidence of fractures in men, BMD measures are performed too infrequently. Two groups of men clearly would benefit from BMD testing.

- Men >50 yr of age who have suffered a fracture, including those with vertebral deformity. Younger men who suffer low-trauma fractures should also be assessed.
- Men who have known secondary causes of bone loss should have BMD determined. These include men treated with glucocorticoids or other medications associated with osteoporosis, men with hypogonadism of any cause including those treated with androgen deprivation therapy, or men who have alcoholism. Many other risk factors may also prompt BMD measures (Table 1).

Screening BMD measures in older men have been recommended (e.g., >70 yr of age).[19] However, a recent cost-effectiveness analysis suggested that screening may be more appropriate at later ages.[20]

Currently, the presence of reduced BMD in men is commonly quantified with T-scores and using a grading system parallel to that used in women (BMD T-score −1.0 to −2.5 = low bone mass; BMD T-score < −2.5 = osteoporosis). Whether BMD measurements in men should be interpreted using T-scores based on a male-specific reference range or using the same reference range used in women has been controversial. Whereas analyses of large population level data suggest that the BMD–fracture risk association is the same in men and women, other reports from carefully studied cohorts indicate there are differences, and those differences may be influenced by age.[21,22] Furthermore, it would be more appropriate to base treatment decisions on assessments of absolute fracture risk.

TABLE 1. CAUSES OF OSTEOPOROSIS AND BONE LOSS IN MEN

Primary
Aging
Idiopathic
Secondary
Hypogonadism
Glucocorticoid excess
Alcoholism, tobacco abuse
Renal insufficiency
Gastrointestinal, hepatic disorders; malabsorption
Hyperparathyroidism
Hypercalciuria
Anticonvulsants
Thyrotoxicosis
Chronic respiratory disorders
Anemias, hemoglobinopathies
Immobilization
Osteogenesis imperfecta
Homocystinuria
Systemic mastocytosis
Neoplastic diseases and chemotherapy
Rheumatoid arthritis

© 2008 American Society for Bone and Mineral Research

Table 2. Evaluation of Osteoporosis in Men: Laboratory Tests

Serum calcium, phosphorus, creatinine, alkaline phosphatase, liver function tests
Complete blood count (protein electrophoresis in those >50 yr of age)
Serum 25(OH)vitamin D and PTH
Serum testosterone and luteinizing hormone
24-h urine calcium and creatinine
Targeted diagnostic testing in men with signs, symptoms, or other indications of secondary disorders

When an etiology is not apparent after the above, additional testing may be appropriate: thyroid function tests, 24-h urine cortisol, biochemical indices of remodeling, and immunological tests for sprue.

Men who have been selected for androgen deprivation therapy deserve special note because the risk of bone loss and fractures is clearly increased, especially in the first 5 yr after sex steroid deficiency is induced.[23,24] Men starting high-dose glucocorticoid therapy present the same challenges and should be similarly managed. When anti-androgen (or glucocorticoid) therapy is begun, a BMD assessment is appropriate. If it is normal, routine preventative measures are reasonable. A repeat BMD measurement should be done in 1–2 yr. If BMD is reduced at the onset of therapy, more aggressive preventive measures should be considered (e.g., bisphosphonate therapy). In men with osteoporosis even before anti-androgen (or glucocorticoid) therapy is begun, pharmacological approaches to prevent further bone loss or fractures are warranted.

Clinical Evaluation

The clinical evaluation of men found to have low BMD should include a careful history and physical examination designed to identify any factors that may contribute to deficits in bone mass. Attention should be paid to lifestyle factors, nutrition (especially calcium and vitamin D nutrition), activity level, and family history. A history of previous fracture should be identified, and fall risk should be assessed. This information should be used to formulate recommendations for prevention and treatment.

Laboratory Testing

In a man undergoing an evaluation for osteoporosis, laboratory testing is intended to identify correctable causes of bone loss. Appropriate tests are shown in Table 2.

OSTEOPOROSIS PREVENTION IN MEN

The essentials of fracture prevention in men are similar to those in women. In early life, excellent nutrition and exercise seem to have positive effects on bone mass. These principles and the avoidance of lifestyle factors known to be associated with bone loss (Table 1) remain important throughout life. Calcium and vitamin D probably provide beneficial effects on bone mass and fractures in men as in women. Recommendations for both sexes include 1000 mg of calcium for those 30–50 yr of age and 1200 mg for those >50 yr of age, with suggested vitamin D intakes of 1000 IU/d. In those at risk for falls (e.g., with reduced strength, poor balance, previous falls), attempts to increase strength and balance may be beneficial.

TREATMENT OF OSTEOPOROSIS IN MEN

Ensuring adequate calcium and vitamin D intake and appropriate physical activity are essential foundations for preserving and enhancing bone mass in men who have osteoporosis. Secondary causes of osteoporosis should be identified and treated. In addition, there are pharmacological therapies that have been shown to enhance BMD, and in some cases, reduce fracture risk in men. Although the available data are not as extensive as in women, these therapies seem to be as effective in affecting BMD and in reducing fracture risk in men. The treatment indications for these drugs are similar in men and women.

Idiopathic Osteoporosis

Alendronate, risedronate, and PTH are effective in improving BMD,[25,26] regardless of age or gonadal function. Although the trials are relatively small, each is also apparently effective in reducing vertebral fracture risk. The effects of other bisphosphonates have not yet been reported in men, but probably have similar effects.

Glucocorticoid-Induced Osteoporosis

Bisphosphonate therapy (e.g., alendronate, risedronate) is effective in improving BMD and, although the data are not extensive, also probably reduces fracture risk.[27,28]

Hypogonadal Osteoporosis

Bisphosphonates and PTH therapy are effective in increasing BMD in hypogonadal men. Moreover, bisphosphonate treatment can prevent the bone loss that is common after androgen deprivation therapy for prostate cancer. Testosterone replacement therapy results in increases in serum levels of both estradiol and testosterone and improves BMD in men with established hypogonadism,[2] but whether fracture risk is reduced is unknown. In older men with less severe, age-related reductions in gonadal function, the usefulness of testosterone is less certain. Relatively high doses (200 mg of intramuscular testosterone every 2 wk) is associated with an increase in BMD and strength in older men with low testosterone levels,[29,30] but its impact on fracture risk has not been examined. Lower doses (e.g., dermal administration of testosterone) seem to have lesser effects.[31] Once again, the effects of testosterone replacement on fracture risk are uncertain. Moreover, the long-term risks of testosterone therapy in older men are unknown. Therefore, testosterone replacement therapy is appropriate for the management of the hypogonadal syndrome, but the treatment of osteoporosis in a man with low testosterone levels is most confidently undertaken with a bisphosphonate or PTH.

REFERENCES

1. Seeman E 2001 Sexual dimorphism in skeletal size, density, and strength. J Clin Endocrinol Metab **86:**4576–4584.
2. Marcus R, Feldman D, Kelsey J 2001 Osteoporosis, 2nd ed. Academic Press, San Diego, CA, USA.
3. Seeman E 2002 Pathogenesis of bone fragility in women and men. Lancet **359:**1841–1850.
4. Donaldson LJ, Cook A, Thomson RG 1990 Incidence of fractures in a geographically defined population. J Epidemiol Commun Health **44:**241–245.
5. Ismail AA, Pye SR, Cockerill WC, Lunt M, Silman AJ, Reeve J, Banzer D, Benevolenskaya LI, Bhalla A, Armas JB, Cannata JB, Cooper C, Delmas PD, Dequeker J, Dilsen G, Falch JA, Felsch B, Felsenberg D, Finn JD, Gennari C, Hoszowski K, Jajic I, Janott J, Johnell O, Kanis JA, Kragl G, Vaz AL, Lorenc R, Lyritis G, Marchand F, Masaryk P, Matthis C, Miazgowski T, Naves-Diaz M, Pols HAP, Poor G, Rapido A, Raspe HH, Reid DM, Reisinger W, Scheidt-Nave C, Stepan J, Todd C, Weber K, Woolf AD, O'Neill

© 2008 American Society for Bone and Mineral Research

TW 2002 Incidence of limb fracture across Europe: Results from the European prospective osteoporosis study (EPOS). Osteoporos Int **13:**565–571.
6. Group EPOSE 2002 Incidence of vertebral fracture in Europe: Results from the European prospective osteoporosis study (EPOS). J Bone Miner Res **17:**716–724.
7. Amin S, Felson DT 2001 Osteoporosis in men. Rheum Dis Clin North Am **27:**19–47.
8. Schwartz AV, Kelsey JL, Maggi S, Tuttleman M, Ho SC, Jonsson PV, Poor G, Sisson de Castro JA, Xu L, Matkin CC, Nelson LM, Heyse SP 1999 International variation in the incidence of hip fractures: Cross-national project on osteoporosis for the world health organization proram for research aging. Osteoporos Int **9:**242–253.
9. Nguyen TV, Eisman JA, Kelly PJ, Sambrook PN 1996 Risk factors for osteoporotic fractures in elderly men. Am J Epidemiol **144:**258–261.
10. Chan BKS, Marshall LM, Lambert LC, Cauley JA, Ensrud KE, Orwoll ES, Cummings SR 2005 The risk of non-vertebral and hip fracture and prevalent falls in older men: The MrOS Study. J Bone Miner Res **20:**S385.
11. Vanderschueren D, Boonen S, Bouillon R 2000 Osteoporosis and osteoportic fractures in men: A clinical perspective. Baillieres Best Pract Res Clin Endocrinol Metab **14:**299–315.
12. Khosla S, Melton J III 2002 Estrogen and the male skeleton. J Clin Endocrinol Metab **87:**1443–1450.
13. Vanderscheuren D, Boonen S, Bouillon R 1998 Action of androgens versus estrogens in male skeletal homeostasis. Bone **23:**391–394.
14. Orwoll ES 2003 Men, bone and estrogen: Unresolved issues. Osteoporos Int **14:**93–98.
15. Leder BZ, Le Blanc KM, Schoenfeld DA, Eastell R, Finkelstein J 2003 Differential effects of androgens and estrogens on bone turnover in normal men. J Clin Endocrinol Metab **88:**204–210.
16. Falahati-Nini A, Riggs BL, Atkinson EJ, O'Fallon WM, Eastell E, Khosla S 2000 Relative contributions of testosterone and estrogen in regulating bone resorption and formation in normal elderly men. J Clin Invest **106:**1553–1560.
17. Orwoll ES 2001 Androgens: Basic biology and clinical implication. Calcif Tissue Int **69:**185–188.
18. Nguyen ND, Pongchaiyakul C, Center JR, Eisman JA, Nguyen TV 2005 Identification of high-risk individuals for hip fracture: A 14-year prospective study. J Bone Miner Res **20:**1921–1928.
19. Binkley NC, Schmeer P, Wasnich RD, Lenchik L 2002 What are the criteria by which a densitometric diagnosis of osteoporosis can be made in males and non-caucasians? J Clin Densitom **5:**19–27.
20. Schousboe JT, Taylor BC, Fink HA, Kane RL, Cummings SR, Orwoll ES, Melton LJ III, Bauer DC, Ensrud KE 2007 Cost-effectiveness of bone densitometry followed by treatment of osteoporosis in older men. JAMA **298:**629–637.
21. Johnell O, Kanis JA, Oden A, Johansson H, De Laet C, Delmas P, Eisman JA, Fujiwara S, Kroger H, Mellstrom D, Meunier PJ, Melton LJ III, O'Neill T, Pols H, Reeve J, Silman A, Tenenhouse A 2005 Predictive value of BMD for hip and other fractures. J Bone Miner Res **20:**1185–1194.
22. Cummings SR, Cawthon PM, Ensrud KE, Cauley JA, Fink HA, Orwoll ES Osteoporotic Fractures in Men Research G, Study of Osteoporotic Fractures Research G 2006 BMD and risk of hip and nonvertebral fractures in older men: A prospective study and comparison with older women. J Bone Miner Res **21:**1550–1556.
23. Shahinian VB, Kuo YF, Freeman JL, Goodwin JS 2005 Risk of fracture after androgen deprivation for prostate cancer. N Engl J Med **352:**154–164.
24. Nielsen M, Brixen K, Walter S, Andersen J, Eskildsen P, Abrahamsen B 2007 Fracture Risk Is Increased in Danish Men With Prostate Cancer: A Nationwide Register Study. Presented at the 29th Annual Meeting of the American Society for Bone and Mineral Research, September 16–19, 2007, Honolulu, HI, USA.
25. Orwoll E, Ettinger M, Weiss S, Miller P, Kendler D, Graham J, Adami S, Weber K, Lorenc R, Pietschmann P, Vandormael K, Lombardi A 2000 Alendronate for the treatment of osteoporosis in men. N Engl J Med **343:**604–610.
26. Orwoll ES, Scheele WH, Paul S, Adami S, Syversen U, Diez-Perez A, Kaufman JM, Clancy AD, Gaich GA 2003 The effect of teriparatide [human parathyroid hormone (1-34)] therapy on bone density in men with osteoporosis. J Bone Miner Res **18:**9–17.
27. Adachi JD, Bensen WG, Brown J, Hanley D, Hodsman A, Josse R, Kendler DL, Lentle B, Olszynski W, Tenenhouse A, Chines AA 1997 Intermittent etidronate therapy to prevent corticosteroid-induced osteoporosis. N Engl J Med **337:**382–387.
28. Reid DM, Hughes RA, Laan RFJM, Sacco-Gibson NA, Wenderoth DHSA, Eusebio RA, Devogelaer JP 2000 Efficacy and safety of daily residronate in the treatment of corticosteroid-induced osteoporosis in men and women: A randomized trial. J Bone Miner Res **15:**1006–1013.
29. Page ST, Amory JK, Bowman FD, Anawalt BD, Matsumoto AM, Bremner WJ, Tenover JL 2005 Exogenous testosterone (T) alone or with finasteride increases physical performance, grip strength, and lean body mass in older men with low serum T. J Clin Endocrinol Metab **90:**1502–1510.
30. Amory JK, Watts NB, Easley KA, Sutton PR, Anawalt BD, Matsumoto AM, Bremner WJ, Tenover JL 2004 Exogenous testosterone or testosterone with finasteride increases bone mineral density in older men with low serum testosterone. J Clin Endocrinol Metab **89:**503–510.
31. Snyder PJ, Peachey H, Berlin JA, Hannoush P, Haddad G, Dlewati A, Santanna J, Loh L, Lenrow DA, Holmes JH, Kapoor SC, Atkinson LE, Strom BL 2000 Effects of testosterone replacement in hypogonadal men. J Clin Endocrinol Metab **85:**2670–2677.

Chapter 63. Premenopausal Osteoporosis

Adi Cohen and Elizabeth Shane

Division of Endocrinology, Department of Medicine College of Physicians & Surgeons, Columbia University, New York, New York

INTRODUCTION

In this chapter, we will discuss issues specific to the diagnosis, clinical evaluation, and management of premenopausal women who present with low BMD and/or fractures.

PREMENOPAUSAL WOMEN WITH LOW BMD

In postmenopausal women, osteoporosis may be diagnosed before a fracture has occurred by using BMD T-scores. However, in premenopausal women, the World Health Organization (WHO) criteria for diagnosis of osteoporosis and osteopenia do not apply to and generally should not be used to categorize BMD measurements. This is because the predictive relationship between BMD and fractures is much weaker in premenopausal than in postmenopausal women. In premeno-

The authors state that they have no conflicts of interest.

Key words: osteoporosis, premenopausal, BMD, fracture, bisphosphonates, Z-score, anorexia nervosa, secondary causes, glucocorticoid-induced osteoporosis, idiopathic osteoporosis, cancer treatment, pregnancy, lactation, teriparatide

Electronic Databases: 1. BMAD Applet—http://www-stat-class.stanford.edu/pediatric-bones/. 2. Bone Density and Fracture Risk—http://courses.washington.edu/bonephys/opFxRisk.html.

© 2008 American Society for Bone and Mineral Research

pausal women, the incidence and prevalence of fractures is orders of magnitude lower than in postmenopausal women.[1–4] Young women with low BMD, but without other risk factors for fracture (e.g., ongoing rapid bone loss, previous fracture, glucocorticoid therapy), are usually at very low short-term risk of fracture.

Therefore, the International Society for Clinical Densitometry (ISCD) recommends using Z-scores (comparison with an age-matched reference population) to categorize BMD measurements in premenopausal women. Young women with BMD Z-scores below –2.0 should be categorized as having BMD that is "below expected range for age" and those with Z-scores above –2.0 should be categorized as having BMD that is "within the expected range for age."[5] The term "osteoporosis" should be used with restraint in premenopausal women, particularly when the diagnosis is based on BMD alone, when the BMD is at the low end of or just below the normal range, and there is no history of fracture. The term "osteopenia" should not be used in this age group because 15% of a normal population of young women would be expected to have T-scores between –1.0 and –2.49.[6] An erroneous diagnosis of "osteoporosis" or "osteopenia," based only on a low BMD measurement in an otherwise healthy premenopausal woman, may have both emotional and economic long-term consequences.

In premenopausal women, the two most common reasons for low BMD measurements are low peak bone mass and small bone size. When interpreting low BMD measurements in premenopausal women younger than 30 yr of age, one must consider that they may not yet have reached peak bone mass. Alternatively, premenopausal women may have low BMD because their peak bone mass is low for genetic reasons or because of an illness, medication exposure, or life style choice (alcohol, tobacco, low calcium intake, lack of exercise) that negatively affected accrual of bone mass during adolescence. However, even a Z-score below –2.0 may not signify decreased bone strength or increased short-term risk of fracture if BMD is stable and bone microarchitecture, turnover, and mineralization are normal.

Small stature and body size (and therefore bone size) influence DXA measurements, leading to underestimates of true volumetric BMD in petite individuals.

PREMENOPAUSAL WOMEN WITH A HISTORY OF LOW-TRAUMA FRACTURE

The diagnosis of osteoporosis in premenopausal women is most secure when there is a history of low trauma fracture(s). Although the relationship between low BMD and fracture is not nearly as robust in premenopausal as in older women, several studies have shown that young women with low BMD are at higher risk for fractures than young women with normal BMD.[7,8] Premenopausal women with Colles fractures have been found to have significantly lower BMD at the nonfractured radius,[9] lumbar spine, and femoral neck[10] than controls without fractures. In addition, stress fractures in female military recruits and athletes are associated with lower BMD than controls.[8]

In some groups of young women, particularly women on glucocorticoids, fractures occur even though BMD is normal. In a study of 160 women receiving high-dose glucocorticoids, 7 of 16 premenopausal women (44%) had fractures despite normal BMD.[11] Because vertebral fracture prevalence may be as high as 21% in young women with autoimmune disorders on high doses of glucocorticoids,[12] screening for asymptomatic vertebral fractures with radiographs or DXA-based vertebral fracture assessment should be conducted, even when BMD is normal.

Table 1. Secondary Causes of Osteoporosis in Premenopausal Women

Amenorrhea (e.g., anorexia nervosa, pituitary diseases, medications)
Cushing's syndrome
Hyperthyroidism
Primary hyperparathyroidism
Vitamin D, calcium, and/or other nutrient deficiency
• Gastrointestinal malabsorption (celiac disease, inflammatory bowel disease, cystic fibrosis, postoperative states)
• Anorexia nervosa
Rheumatoid arthritis, SLE, other inflammatory conditions
Renal disease
Liver disease
Hypercalciuria
Alcoholism
Connective tissue diseases
• Osteogeneis imperfecta
• Marfan syndrome
• Ehlers Danlos syndrome
Medications
• Glucocorticoids
• Immunosuppressants (e.g., cyclosporine)
• Anti-epileptic drugs (particularly cytochrome P450 inducers such as phenytoin, carbamazepine)
• Cancer chemotherapy
• GnRH agonists (when used to suppress ovulation)
• Heparin
Idiopathic osteoporosis

Several studies have shown that fractures before menopause predict postmenopausal fractures.[2,3,13] In the Study of Osteoporotic Fractures (SOF), women with a history of premenopausal fracture were 35% more likely to fracture during the early postmenopausal years than women without a history of premenopausal fracture.[3] These findings suggest that certain life-long traits, such as fall frequency, neuromuscular protective response to falls, bone mass, or various aspects of bone quality, can affect life-long fracture risk.[2]

SECONDARY CAUSES OF OSTEOPOROSIS IN PREMENOPAUSAL WOMEN

Most premenopausal women with low-trauma fractures or low BMD have an underlying disorder or medication exposure that has interfered with bone mass accrual during adolescence and/or has caused excessive bone loss after reaching peak bone mass. In a population study from Olmstead County, MN, 90% of men and women 20–44 yr of age with osteoporotic fractures were found to have a secondary cause.[14] In contrast, several case series of young women with osteoporosis evaluated in tertiary centers report that only 50% have secondary causes,[15,16] likely reflecting referral bias of more obscure cases to specialists.

Potential secondary causes are listed in Table 1. Many of these are discussed elsewhere in this *Primer*. The main goal of the evaluation of a premenopausal woman with low-trauma fractures or low BMD is to diagnose and treat any correctable secondary cause. Often this can be accomplished by a detailed history and physical examination, although an exhaustive biochemical evaluation may be necessary.

© 2008 American Society for Bone and Mineral Research

OSTEOPOROSIS ASSOCIATED WITH PREGNANCY AND LACTATION

It is not uncommon for premenopausal women with low-trauma fractures to present around the time of pregnancy and lactation. Bone loss and osteoporosis associated with pregnancy and lactation are considered in depth elsewhere in the *Primer*.

IDIOPATHIC OSTEOPOROSIS

Premenopausal women with osteoporosis, in whom no definable cause can be found after a detailed evaluation, are said to have idiopathic osteoporosis (IOP). IOP is predominantly reported in whites, and family history of osteoporosis is common.[14,16,17] Mean age at diagnosis is 35 yr. Multiple fractures of sites rich in cancellous bone (e.g., spine, ribs) may occur over 5–10 yr.[14,16] While the pathogenesis of IOP remains uncertain and is likely heterogeneous, some studies have implicated low IGF-1 and osteoblast dysfunction.[14,18] Subclinical estrogen deficiency and increased bone turnover may be present.[19] Bone biopsy and high-resolution imaging have shown decreased bone formation, increased bone resorption, altered bone microarchitecture, and decreased mechanical competence in comparison with normal control women.[20,21]

EVALUATION OF THE PREMENOPAUSAL WOMAN WITH OSTEOPOROSIS

Premenopausal women with low BMD (Z-score ≤ –2.0) and those with a low-trauma fracture regardless of their BMD should undergo a thorough evaluation for secondary causes of bone loss. Identification of a contributing condition often helps to guide management of the affected individual.

A careful medical history is essential, including information about family history, fractures, kidney stones, amenorrhea or premenopausal estrogen deficiency, timing of recent pregnancies and lactation, dieting and exercise behavior, subtle gastrointestinal symptoms, and medications, including over-the-counter supplements. Physical examination can identify signs of Cushing's syndrome, thyrotoxicosis, or connective tissue disorders (e.g., blue sclerae in some forms of osteogenesis imperfecta). Measurement or estimation of volumetric BMD may be considered in petite individuals.

The laboratory evaluation (Table 2) should be aimed at identifying secondary causes such as hyperthyroidism, hyperparathyroidism, Cushing's syndrome, early menopause, renal or liver disease, celiac disease, malabsorption, and idiopathic hypercalciuria. Bone turnover markers and follow-up BMD testing may help to distinguish those with stably low BMD from those with ongoing bone loss who may be at higher short-term risk of fracture. In premenopausal women with a history of low-trauma fracture(s) and no known secondary cause, bone biopsy may be indicated. In the authors' experience, a bone biopsy may occasionally identify unsuspected causes of bone fragility, such as Gaucher's disease or mastocytosis.

TABLE 2. LABORATORY EVALUATION

Initial laboratory evaluation
Complete blood count
Electrolytes, renal function
Serum calcium, phosphate
Serum albumin, transaminases, total alkaline phosphatase
Serum TSH
Serum 25-hydroxyvitamin D
24-h urine for calcium and creatinine
Additional laboratory evaluation
Estradiol, LH, FSH, prolactin
PTH
1,25-dihydroxyvitamin D
24-h urine for free cortisol
Iron/TIBC, ferritin
Celiac screen
Serum/urine protein electrophoresis
ESR or CRP
Bone turnover markers
Transiliac crest bone biopsy

MANAGEMENT ISSUES

General Measures

For all patients, one should recommend a set of general measures that benefit bone health: adequate weight-bearing exercise,[22,23] nutrition (protein, calories, calcium, vitamin D), and lifestyle modifications (smoking cessation, avoidance of excess alcohol).

Pharmacological therapy is rarely justified for premenopausal women with isolated low BMD and no history of fractures, in whom there is no identifiable secondary cause, particularly if the Z-score is > –3.0. Low BMD in such young woman may be caused by genetic low peak bone mass or to past insults to the growing or adult skeleton (poor nutrition, medications, estrogen deficiency) that are no longer operative. Such young women usually have low short-term risk of fracture. Moreover, Peris et al.[24] recently reported slight BMD improvement and no further fractures in women with unexplained osteoporosis managed with only calcium (total intake of 1500 mg/d), vitamin D (400–800 IU/d), and exercise. BMD should be remeasured after 1 or 2 yr to ascertain that it is stable or to identify the unusual patient with ongoing bone loss.

In women with low BMD or low-trauma fractures and a known secondary cause, address the underlying cause if possible. Women with estrogen deficiency should receive estrogen (unless contraindicated), those with celiac disease should begin a gluten-free diet, those with primary hyperparathyroidism may benefit from parathyroidectomy, and those with idiopathic hypercalciuria may benefit from thiazide diuretics.

In some women, it is not possible to address or alleviate the secondary cause directly. Premenopausal women requiring long-term glucocorticoids and those being treated for breast cancer may require pharmacological therapy to prevent excessive bone loss or fractures. Options for treatment include antiresorptives drugs, such as estrogen or bisphosphonates, or anabolic agents such as teriparatide. Selective estrogen receptor modulators (SERMS), such as raloxifene, should not be used to treat bone loss in menstruating women because they block estrogen action on bone and lead to further bone loss.[25]

Bisphosphonates

The U.S. Food and Drug Administration has approved bisphosphonates only for premenopausal women on glucocorticoids. Because bisphosphonates accumulate in the maternal skeleton, cross the placenta, accumulate in the fetal skeleton,[26] and cause toxic effects in pregnant rats,[27] they should be used with caution in women who may become pregnant. Whereas several reports document normal pregnancies and fetal outcomes in women receiving bisphosphonates,[28–30] mild neonatal abnormalities were reported in two women with osteogenesis imperfecta treated with pamidronate before conception.[31] The potential for fetal abnormalities should be considered when choosing bisphosphonates for a premeno-

© 2008 American Society for Bone and Mineral Research

pausal woman, and bisphosphonates should be reserved for those with fragility fractures or ongoing bone loss caused by glucocorticoids or cancer chemotherapy. The few studies that have rigorously examined bisphosphonates in premenopausal women are summarized below.

Glucocorticoid-Induced Osteoporosis. Bisphosphonates are approved for prevention and treatment of glucocorticoid-induced osteoporosis. However, relatively few premenopausal women participated in the relevant large clinical trials. A few studies have shown protective effects of intermittent cyclical etidronate and oral pamidronate in premenopausal women with autoimmune and connective tissue diseases.[32,33] Guidelines from the American College of Rheumatology recommend bisphosphonates for prevention and treatment of glucocorticoid-induced osteoporosis in premenopausal women taking at least 5 mg of prednisone or equivalent per day.[34] However, because of the potential for harm to the fetus in women who may become pregnant, they also urge great caution in the use of bisphosphonates in premenopausal women.[34]

Bisphosphonate Use for Other Secondary Causes of Osteoporosis. Intravenous and oral bisphosphonates prevent bone loss in premenopausal women experiencing ovarian failure in the setting of treatment for breast cancer.[35–37] Bisphosphonates may lower fracture risk in young individuals with osteogenesis imperfecta. Both alendronate and risedronate have been shown to significantly increase BMD in young women with anorexia.[38,39] Bisphosphonates have also been associated with substantial increases of 11–23% in lumbar spine BMD[28] in women with pregnancy- and lactation-associated osteoporosis. Because BMD is expected to increase postpartum and after weaning in normal women, it is not clear to what extent bisphosphonate use provided an incremental benefit for these patients.

Teriparatide/PTH(1-34)

There are even fewer data on the effects of teriparatide or PTH(1-34) in premenopausal women. In young women treated with the gonadotropin-releasing hormone (GnRH) analog nafarelin for endometriosis, spine BMD declined by 4.9%, whereas those treated with PTH(1-34) 40 μg daily together with nafarelin had an increase of 2.1% ($p < 0.001$).[40] It is not clear whether these results would apply to premenopausal women with normal gonadal status.

A recent study comparing teriparatide and alendronate for glucocorticoid-induced osteoporosis included some premenopausal women. Teriparatide was associated with significantly greater increases in lumbar spine and total hip BMD and resulted in significantly fewer incident vertebral fractures than alendronate.[41] The BMD responses in premenopausal women were similar to those in men and postmenopausal women, but no fractures occurred in either premenopausal group.

Because the long-term effects of teriparatide use in young women are not known, use of this medication should be reserved for those at highest risk for fracture or those who are experiencing recurrent fractures.

SUMMARY AND CONCLUSIONS

Premenopausal women with low-trauma fracture(s) or low BMD (Z-score ≤ −2.0) should have a thorough evaluation for secondary causes of osteoporosis and bone loss. In most, a secondary cause can be found, the most common being glucocorticoid excess, anorexia nervosa, premenopausal estrogen deficiency, and celiac disease. Identification and treatment of the underlying cause should be the focus of management. In women with low BMD, no fractures, and no known secondary cause, a low BMD may not necessarily mean that bone strength is decreased, and pharmacologic therapy is rarely justified. However, women with an ongoing cause of bone loss and those who have had or continue to have low-trauma fractures may require pharmacological intervention with bisphosphonates or teriparatide. Few high-quality clinical trials exist to provide guidance, and there are no data showing that such an intervention actually reduces the risk of future fractures. Although certain bisphosphonates are approved for use in premenopausal women in the setting of ongoing glucocorticoid exposure, these medications should be used with caution in women who may become pregnant in the future.

REFERENCES

1. Thompson PW, Taylor J, Dawson A 2004 The annual incidence and seasonal variation of fractures of the distal radius in men and women over 25 years in Dorset, UK. Injury **35:**462–466.
2. Wu F, Mason B, Horne A, Ames R, Clearwater J, Liu M, Evans MC, Gamble GD, Reid IR 2002 Fractures between the ages of 20 and 50 years increase women's risk of subsequent fractures. Arch Intern Med **162:**33–36.
3. Hosmer WD, Genant HK, Browner WS 2002 Fractures before menopause: A red flag for physicians. Osteoporos Int **13:**337–341.
4. Hui SL, Slemenda CW, Johnston CC Jr 1988 Age and bone mass as predictors of fracture in a prospective study. J Clin Invest **81:**1804–1809.
5. Lewiecki EM 2005 Premenopausal bone health assessment. Curr Rheumatol Rep **7:**46–52.
6. Kanis JA 2002 Diagnosis of osteoporosis and assessment of fracture risk. Lancet **359:**1929–1936.
7. Lauder TD, Dixit S, Pezzin LE, Williams MV, Campbell CS, Davis GD 2000 The relation between stress fractures and bone mineral density: Evidence from active-duty Army women. Arch Phys Med Rehabil **81:**73–79.
8. Lappe J, Davies K, Recker R, Heaney R 2005 Quantitative ultrasound: Use in screening for susceptibility to stress fractures in female army recruits. J Bone Miner Res **20:**571–578.
9. Wigderowitz CA, Cunningham T, Rowley DI, Mole PA, Paterson CR 2003 Peripheral bone mineral density in patients with distal radial fractures. J Bone Joint Surg Br **85:**423–425.
10. Hung LK, Wu HT, Leung PC, Qin L 2005 Low BMD is a risk factor for low-energy Colles' fractures in women before and after menopause. Clin Orthop **435:**219–225.
11. Kumagai S, Kawano S, Atsumi T, Inokuma S, Okada Y, Kanai Y, Kaburaki J, Kameda H, Suwa A, Hagiyama H, Hirohata S, Makino H, Hashimoto H 2005 Vertebral fracture and bone mineral density in women receiving high dose glucocorticoids for treatment of autoimmune diseases. J Rheumatol **32:**863–869.
12. Borba VZ, Matos PG, da Silva Viana PR, Fernandes A, Sato EI, Lazaretti-Castro M 2005 High prevalence of vertebral deformity in premenopausal systemic lupus erythematosus patients. Lupus **14:**529–533.
13. Honkanen R, Tuppurainen M, Kroger H, Alhava E, Puntila E 1997 Associations of early premenopausal fractures with subsequent fractures vary by sites and mechanisms of fractures. Calcif Tissue Int **60:**327–331.
14. Khosla S, Lufkin EG, Hodgson SF, Fitzpatrick LA, Melton LJ III 1994 Epidemiology and clinical features of osteoporosis in young individuals. Bone **15:**551–555.
15. Moreira Kulak CA, Schussheim DH, McMahon DJ, Kurland E, Silverberg SJ, Siris ES, Bilezikian JP, Shane E 2000 Osteoporosis and low bone mass in premenopausal and perimenopausal women. Endocr Pract **6:**296–304.
16. Peris P, Guanabens N, Martinez de Osaba MJ, Monegal A, Alvarez L, Pons F, Ros I, Cerda D, Munoz-Gomez J 2002 Clinical characteristics and etiologic factors of premenopausal osteoporosis in a group of Spanish women. Semin Arthritis Rheum **32:**64–70.
17. Kulak CAM, Schussheim DH, McMahon DJ, Kurland E, Silver-

© 2008 American Society for Bone and Mineral Research

berg SJ, Siris ES, Bilezikian JPES 2000 Osteoporosis and low bone mass in premenopausal and perimenopausal women. Endocr Pract **6**:296–304.
18. Reed BY, Zerwekh JE, Sakhaee K, Breslau NA, Gottschalk F, Pak CY 1995 Serum IGF 1 is low and correlated with osteoblastic surface in idiopathic osteoporosis. J Bone Miner Res **10**:1218–1224.
19. Rubin MR, Schussheim DH, Kulak CA, Kurland ES, Rosen CJ, Bilezikian JP, Shane E 2005 Idiopathic osteoporosis in premenopausal women. Osteoporos Int **16**:526–533.
20. Donovan MA, Dempster D, Zhou H, McMahon DJ, Fleischer J, Shane E 2005 Low bone formation in premenopausal women with idiopathic osteoporosis. J Clin Endocrinol Metab **90**:3331–3336.
21. Cohen A, Recker RR, Guo XE, Zhang XH, Lappe J, Eisenberg HF, McMahon DJ, Shane E 2007 Abnormal trabecular microarchitecture and mechanical competence in premenopausal women with idiopathic osteoporosis (IOP) can be detected by high resolution peripheral quantitative computed tomography (HRpQCT). American Society for Bone and Mineral Research, 29th Annual Meeting, September 16–19, 2007, Honolulu, HI, USA.
22. Wallace BA, Cumming RG 2000 Systematic review of randomized trials of the effect of exercise on bone mass in pre- and postmenopausal women. Calcif Tissue Int **67**:10–18.
23. Mein AL, Briffa NK, Dhaliwal SS, Price RI 2004 Lifestyle influences on 9-year changes in BMD in young women. J Bone Miner Res **19**:1092–1098.
24. Peris P, Monegal A, Martinez MA, Moll C, Pons F, Guanabens N 2007 Bone mineral density evolution in young premenopausal women with idiopathic osteoporosis. Clin Rheumatol **26**:958–961.
25. Powles TJ, Hickish T, Kanis JA, Tidy A, Ashley S 1996 Effect of tamoxifen on bone mineral density measured by dual-energy x-ray absorptiometry in healthy premenopausal and postmenopausal women. J Clin Oncol **14**:78–84.
26. Patlas N, Golomb G, Yaffe P, Pinto T, Breuer E, Ornoy A 1999 Transplacental effects of bisphosphonates on fetal skeletal ossification and mineralization in rats. Teratology **60**:68–73.
27. Minsker DH, Manson JM, Peter CP 1993 Effects of the bisphosphonate, alendronate, on parturition in the rat. Toxicol Appl Pharmacol **121**:217–223.
28. O'Sullivan SM, Grey AB, Singh R, Reid IR 2006 Bisphosphonates in pregnancy and lactation-associated osteoporosis. Osteoporos Int **17**:1008–1012.
29. Biswas PN, Wilton LV, Shakir SA 2003 Pharmacovigilance study of alendronate in England. Osteoporos Int **14**:507–514.
30. Chan B, Zacharin M 2006 Maternal and infant outcome after pamidronate treatment of polyostotic fibrous dysplasia and osteogenesis imperfecta before conception: A report of four cases. J Clin Endocrinol Metab **91**:2017–2020.
31. Munns CF, Rauch F, Ward L, Glorieux FH 2004 Maternal and fetal outcome after long-term pamidronate treatment before conception: A report of two cases. J Bone Miner Res **19**:1742–1745.
32. Nzeusseu Toukap A, Depresseux G, Devogelaer JP, Houssiau FA 2005 Oral pamidronate prevents high-dose glucocorticoid-induced lumbar spine bone loss in premenopausal connective tissue disease (mainly lupus) patients. Lupus **14**:517–520.
33. Nakayamada S, Okada Y, Saito K, Tanaka Y 2004 Etidronate prevents high dose glucocorticoid induced bone loss in premenopausal individuals with systemic autoimmune diseases. J Rheumatol **31**:163–166.
34. American College of Rheumatology Ad Hoc Committee on Glucocorticoid-Induced Osteoporosis 2001 Recommendations for the prevention and treatment of glucocorticoid-induced osteoporosis: 2001 update. Arthritis Rheum **44**:1496–1503.
35. Fuleihan Gel H, Salamoun M, Mourad YA, Chehal A, Salem Z, Mahfoud Z, Shamseddine A 2005 Pamidronate in the prevention of chemotherapy-induced bone loss in premenopausal women with breast cancer: A randomized controlled trial. J Clin Endocrinol Metab **90**:3209–3214.
36. Delmas PD, Balena R, Confravreux E, Hardouin C, Hardy P, Bremond A 1997 Bisphosphonate risedronate prevents bone loss in women with artificial menopause due to chemotherapy of breast cancer: A double-blind, placebo-controlled study. J Clin Oncol **15**:955–962.
37. Gnant MF, Mlineritsch B, Luschin-Ebengreuth G, Grampp S, Kaessmann H, Schmid M, Menzel C, Piswanger-Soelkner JC, Galid A, Mittlboeck M, Hausmaninger H, Jakesz R 2007 Zoledronic acid prevents cancer treatment-induced bone loss in premenopausal women receiving adjuvant endocrine therapy for hormone-responsive breast cancer: A report from the Austrian Breast and Colorectal Cancer Study Group. J Clin Oncol **25**:820–828.
38. Golden NH, Iglesias EA, Jacobson MS, Carey D, Meyer W, Schebendach J, Hertz S, Shenker IR 2005 Alendronate for the treatment of osteopenia in anorexia nervosa: A randomized, double-blind, placebo-controlled trial. J Clin Endocrinol Metab **90**:3179–3185.
39. Miller KK, Grieco KA, Mulder J, Grinspoon S, Mickley D, Yehezkel R, Herzog DB, Klibanski A 2004 Effects of risedronate on bone density in anorexia nervosa. J Clin Endocrinol Metab **89**:3903–3906.
40. Finkelstein JS, Klibanski A, Arnold AL, Toth TL, Hornstein MD, Neer RM 1998 Prevention of estrogen deficiency-related bone loss with human parathyroid hormone-(1-34): A randomized controlled trial. JAMA **280**:1067–1073.
41. Taylor KA, Saag KG, Shane E, Boonen S, Donley DW, Marin F, Warner MR, Dalsky GP, Marcus R 2007 Active comparator trial of teriparatide versus alendronate in the treatment of glucocorticoid-induced osteoporosis The International Society for Clinical Densitometry, 13th Annual Meeting, March 14–17, 2007, Tampa, FL, USA.

Chapter 64. Skeletal Effects of Drugs

Juliet Compston

Department of Medicine, University of Cambridge School of Medicine, Cambridge, United Kingdom

INTRODUCTION

Iatrogenic bone loss, caused by therapies for nonskeletal diseases, is a growing and important cause of osteoporosis (Table 1). Recent additions to the list include aromatase inhibitors, androgen deprivation therapy, thiazolidenediones, and proton pump inhibitors. In contrast, some interventions for nonskeletal diseases may be protective to the skeleton. Osteoporosis associated with glucocorticoids, progestagens, excess thyroid hormone, chemotherapy, and calcineurin inhibitors is described in other sections, and this chapter focuses on the remaining drugs listed in Table 1. Drugs that may have protective skeletal effects are also considered.

ANTI-HORMONAL DRUGS

Androgen Deprivation Therapy

Androgen deprivation therapy (ADT) for carcinoma of the

Dr. Compston has served as a consultant and/or received speaking fees from Novartis, Amgen, Procter & Gamble, Nycomed, Wyeth, Lilly, Servier, Shire, and Roche.

Key words: osteoporosis, fracture, anti-hormone drugs, thiazolidenediones, proton pump inhibitors, anti-epileptic drugs, selective serotonin reuptake inhibitors, heparin

© 2008 American Society for Bone and Mineral Research

TABLE 1. DRUGS ASSOCIATED WITH OSTEOPOROSIS

Glucocorticoids
Antihormonal drugs
Thiazolidenediones
Protein pump inhibitors
Heparin
Anticonvulsants
Calcineurin inhibitors
Thyroxine
Chemotherapy
SSRIs

prostate encompasses a number of options, including bilateral orchidectomy, gonadotrophin-releasing hormone (GnRH) analog therapy, and anti-androgenic agents (cyproterone acetate, flutamide, and bicalutamide). Its use has increased considerably in recent years, and osteoporosis has emerged as a common complication.

ADT is associated with increased rates of bone loss at multiple skeletal sites, and increased fracture rates have also been reported.[1,2] Examination of a large database containing medical records of >50,000 men revealed that, in those treated with ADT for prostate cancer and surviving at least 5 yr after diagnosis, 19.4% had a fracture compared with 12.6% of men not receiving ADT. In men who received at least nine doses of GnRH analogs, the relative risk was 1.45 (95% CI, 1.36–1.56).[3] In another retrospective study, the clinical fracture rate in men treated for prostate cancer with GnRH analogs was 7.91/100 person-years compared with 6.55/100 person-years in men with early prostate cancer who were not treated with GnRH analogs, translating into a relative risk of 1.23 (95% CI, 1.09–1.34)[4]; the largest increase in risk was seen for hip fractures (RR, 1.76; 95% CI, 1.33–2.33).

A number of interventions have been shown to have beneficial effects on BMD in men treated with ADT. These include raloxifene, pamidronate, zoledronic acid, and alendronate. Of these, the treatment effect was greatest with either zoledronate 4 mg, given as an intravenous infusion every 3 mo or as a single infusion,[5,6] or oral alendronate 70 mg once weekly.[7] However, all these studies have been relatively short term (≤1 yr), and effects on fracture risk have not been shown.

Strategies to prevent fracture in men receiving ADT for prostate cancer currently lack a robust evidence base, but it seems reasonable to assess fracture risk, using measurement of BMD and clinical risk factors, in all men at the start of treatment and to repeat this every 1–2 yr in those at moderate risk. In view of the rapid bone loss that occurs, which itself is an independent risk factor for fracture, a case can be made for intervention at a T-score of –2 or lower, although this decision may be modified to some extent by age and other risk factors. Based on current evidence, once weekly oral alendronate or once yearly intravenous zoledronic acid can be regarded as front-line options.

Aromatase Inhibitors

Aromatase inhibitors (AIs) are now regarded as front-line adjuvant therapy in women with estrogen receptor–positive breast cancer. They reduce endogenous oestrogen production by 80–90% by blocking the peripheral conversion of androgens to estrogen and have largely replaced the selective estrogen receptor modulator, tamoxifen, as the preferred treatment option for postmenopausal women. The most commonly used AIs in clinical practice are exemestane, anastrozole, and letrozole; the former is steroidal, whereas the latter two are nonsteroidal.[8] In contrast to tamoxifen, which is bone protective in postmenopausal women (but has the opposite effect in premenopausal women), AIs have adverse effects on BMD and probably also fracture risk.

Interpretation of studies of the effects of AIs on bone are complicated by the use of tamoxifen as a comparator in many studies and also, in some studies, use of tamoxifen before AI therapy. In addition, comparative data on the relative effects of AIs on BMD and fracture rate are currently lacking. Nevertheless, the existing biomarker data indicate that all three AIs increase bone turnover, although some studies indicate a proportionately greater effect of exemstane on formation than resorption.[9] Increased rates of bone loss have also been reported; for example, in the ATAC (Arimidex, Tamoxifen, Alone or in Combination) trial, median rates of bone loss at the spine and hip were 4.1% and 3.9%, respectively, over 2 yr,[10] and in a study of letrozole versus placebo after 5 yr of tamoxifen therapy, losses in the spine and hip over 2 yr were 5.35% and 3.5%, respectively.[11]

Significant increases in fracture rate have been shown in several trials of postmenopausal women with breast cancer, although interpretation of the data is again complicated by the use of tamoxifen as a comparator and treatment with tamoxifen before AI therapy in the majority of studies. In the ATAC study, the incidence of fractures after 5 yr was 11% in women treated with anastrozole and 7.7% in those receiving tamoxifen ($p < 0.001$).[12] Comparison of letrozole and tamoxifen produced similar results, with fracture rates of 8.6% and 5.8%, respectively ($p < 0.001$) at a median follow-up period of 51 mo.[13] However, comparison of letrozole with placebo in women with breast cancer after completion of 5 yr of tamoxifen therapy did not show any significant difference in fracture rate (3.6% in letrozole group versus 2.9% placebo; $p = 0.24$).[14] Finally, in the Intergroup Exemestane Study, in which women who were disease free after 2–3 yr with tamoxifen were randomized to continue tamoxifen or switched to exemestane to complete a total of 5 yr of therapy, a significantly higher fracture rate was seen in the women taking exemestane (7% versus 5%, $p = 0.003$).[15] Collectively, these data indicate that AI therapy is associated with a significantly higher fracture risk than tamoxifen but may not increase fracture risk above that seen in untreated women. This suggests that the difference in fracture risk between women treated with AIs and tamoxifen is at least partly caused by a protective effect of the latter.

Prevention of bone loss associated with AI therapy is currently undergoing evaluation in several studies, and recently, Brufsky et al.[16] have reported that zoledronic acid, 4 mg every 6 mo, was successful in preventing letrozole-induced bone loss over the 1-yr study period. At present, it seems reasonable to advise risk assessment, including BMD measurements, in all postmenopausal women treated with AIs and to perform repeat measurements at 1- to 2-yr intervals in those at moderate risk (based on age, BMD, and other risk factors). As in men on ADT, the risk threshold for intervention may be lower than in untreated women because of the higher rates of bone loss. Based on currently available evidence, bisphosphonates are likely to be effective in preventing bone loss but only zoledronic acid has been evaluated to date and no fracture data are available.

THIAZOLIDINEDIONES

Thiazolidinediones (TZDs) are ligands for peroxisome proliferator-activated receptor γ (PPARγ) and are widely used in the treatment of type 2 diabetes. Activation of PPARγ increases marrow adiposity, increases insulin sensitivity, and sup-

© 2008 American Society for Bone and Mineral Research

presses bone formation. Transgenic mice models have shown that PPARγ deficiency is associated with high bone mass,[17] whereas activation of PPARγ induces bone loss.[18] The mechanisms by which suppression of bone formation occurs have not been fully established but may include inhibition of the Wnt/β-catenin signaling pathway, inhibition of osteoblast differentiation genes including Runx2 and osterix, and suppression of insulin-like growth factor production.[19]

The effects of TZDs in humans are particularly relevant in view of the increased risk of fracture associated with type 2 diabetes. In an observational study, increased rates of bone loss in the lumbar spine, trochanter, and whole body were reported in older diabetic women treated with TZDs.[20] Grey et al.,[21] in a randomized controlled trial in healthy postmenopausal women, showed that administration of the TZD rosiglitazone for 14 wk resulted in significant bone loss in the hip (mean, 1.9% versus 0.2% in the placebo group) and significant suppression of biochemical markers of bone formation. In the recent ADOPT study (A Diabetes Outcome Progression Trial), a randomized controlled trial in 4360 patients with type 2 diabetes, the effects of treatment with rosiglitazone were compared with those of metformin and glibenclamide.[22] In women, there was a statistically significant increase in the incidence of fractures, mainly affecting the foot, hand, and upper arm (9.3% in the rosiglitazone group versus 1.54% and 1.29% in the metformin and glibenclamide groups, respectively). Subsequently, similar findings have been reported in women treated with pioglitazone.[23] Although no significant increase in fracture risk has been shown in men treated with TZDs, a recent study has shown increased rates of bone loss in diabetic men treated with rosiglitazone.[24] Adverse skeletal effects therefore have to be considered when weighing up the risk/benefit balance associated with TZDs, particularly in high-risk individuals. The contribution to fracture of factors other than reduced BMD in this population, including increased risk of falling, merits further study.

ACID-SUPPRESSIVE MEDICATIONS

Increased risk of fracture has been reported in individuals treated with acid-suppressive medications. Grisso et al.[25] reported an association between use of the H_2 receptor blocker, cimetidine, and hip fracture in a case-control study of men with hip fracture (OR, 2.5; 95% CI, 1.4–4.6). Subsequently, in a case-control study from Denmark, proton pump inhibitor (PPI) use within the last year was shown to be associated with a small increase in overall fracture risk (adjusted OR, 1.18; 95% CI, 1.12–1.43), hip fracture risk (adjusted OR, 1.45; 95% CI, 1.28–1.65), and spine fracture risk (adjusted OR, 1.60; 95% CI, 1.25–2.04).[26] Interestingly, in this study, the use of histamine H_2 receptor antagonists within the last year was associated with a significantly reduced risk of fracture (adjusted OR, 0.88; 95% CI, 0.82–0.95), although a significant increase in fracture risk was observed with antacid medications other than PPIs and H_2 receptor blockers. Analysis of data from the Study of Osteoporotic Fractures showed that women taking a PPI or H_2 receptor blocker (grouped together) had a significantly increased risk of nonspine fracture (RH, 1.18; 95% CI, 1.01–1.39), but total hip BMD and rates of bone loss at this site were similar between users and nonusers of these acid-suppressive medications.[27] Most recently, in a nested case-control study from the General Practice Research Database in the UK, a significant increase in hip fracture risk was found in PPI users, which increased significantly with dose and with duration of use.[28]

Existing data thus support a causal association between acid-suppressive medication and fracture, although the limitations of observational studies, particularly the effects of potential but unmeasured confounding factors, have to be recognized, and the discrepant findings with respect to H_2 receptor blockers remain unexplained. The mechanism by which PPIs and possibly other acid-suppressive medications increase fracture risk is unknown. Reduced intestinal calcium absorption resulting from increased gastric pH may contribute, although the relatively small reduction in absorption of calcium carbonate shown in one study of postmenopausal women taking omeprazole seems unlikely to be solely responsible for the development of increased fracture risk within 1 yr of starting therapy.[29] Inhibition of the osteoclastic proton pump would be expected to have beneficial skeletal actions, although it is unknown whether such effects are associated with PPI use in vivo. Finally, data from the Study of Osteoporotic Fractures indicated that the increase in fracture risk associated with PPIs may not be BMD driven. Nevertheless, the association has important clinical implications because PPI use is common in the elderly population and might attenuate the effects of bone protective interventions.

ANTI-EPILEPTIC DRUGS

An association between anti-epileptic drugs (AEDs) and increased fracture risk has been reported in several observational studies, and reduced BMD and increased rates of bone loss have also been reported.[30] The underlying pathogenesis is unclear; vitamin D deficiency, trauma during seizures, increased risk of falling, and co-medications including glucocorticoids may all contribute. In a few patients with severe vitamin D deficiency, osteomalacia or rickets may be present.[31] Currently, there are insufficient data to distinguish between the skeletal effects of specific AED regimens.

Management guidelines to prevent and treat bone disease in AED users have been proposed, although at present, these lack a robust evidence base. Routine prophylaxis of vitamin D deficiency should be considered in high-risk individuals, for example the elderly and institutionalized (higher than normal doses may be required in patients taking some AEDs), and in such cases, calcium supplements should also be given. Routine bone densitometry in all AED users cannot be justified at present, although BMD should be measured in those who present with fracture. Treatment of established osteoporosis in this population has not been specifically evaluated.

SELECTIVE SEROTONIN RECEPTOR UPTAKE INHIBITORS

Selective serotonin receptor uptake inhibitors (SSRIs) are widely prescribed as antidepressants and have been associated with reduced BMD in older men,[32] increased rates of hip bone loss in older women,[33] and increased fracture risk in men and women $\geq$50 yr of age.[34] Serotoninergic pathways are known to exist in bone, and adverse skeletal effects of inhibition of these pathways have been reported in animals, providing a plausible mechanism for the effects observed in humans.[35]

HEPARIN

Long-term heparin therapy, which nowadays is virtually restricted to prophylaxis against thromboembolism in high-risk women during pregnancy, is associated with an increased risk of osteoporosis.[36] Reduced BMD, increased rates of bone loss, and increased fracture risk have all been reported during long-term heparin administration,[37,38] although the mechanisms responsible for bone loss have not been established. The use of low molecular weight heparin and of newer anti-

© 2008 American Society for Bone and Mineral Research

thrombotic agents such as fondaparinux may be associated with fewer adverse skeletal effects. Calcium and vitamin D supplements are often advocated but, in common with other antiresorptive regimens, have not been formally evaluated in this situation.

DRUGS THAT MAY PROTECT AGAINST OSTEOPOROSIS

β-Blockers

In some studies, a significant protective effect of β-blocker therapy on fracture risk has been reported, although this finding has not been universal.[39–41] In a recent case-control study, a significant decrease in hip/femur fracture risk was associated with current β-blocker use, but was only present in patients with a history of using other antihypertensive medications and was not related to cumulative exposure, suggesting that the association may not be causal.[42]

Thiazides

Reduced fracture risk in thiazide users has been reported in several prospective observational studies, and small increases in BMD have also been shown in randomized controlled trials.[40,43–45] Increased renal calcium reabsorption is thought to play a role in these beneficial effects, although because this is transient, other mechanisms are likely to operate.

Statins

Statins inhibit the enzyme 3-hydroxy-3-methyl-glutaryl-coenzyme A reductase in the mevalonate pathway, thus reducing cholesterol biosynthesis but also preventing the prenylation of GTP binding proteins and thus inhibiting osteoclast activity. Beneficial skeletal effects of statins in animals have been shown in vitro and in vivo,[46] but studies in humans have produced conflicting results. Two recent meta-analyses have been conducted: one on BMD effects and the other on fracture.[47,48] The former concluded that statins had small but significant benefits on BMD in the hip, whereas the latter suggested that statin use is likely to reduce hip fractures, although effects on other fracture types require further study.

REFERENCES

1. Body J-J, Bergmann P, Boonen S, Boutsen Y, Devogelaer J-P, Goemaere S, Reginster J-Y, Rozenberg S, Kaufman JM 2007 Management of cancer treatment-induced bone loss in early breast and prostate cancer—a consensus paper of the Belgian Bone Club. Osteoporos Int **18:**1439–1450.
2. Greenspan SL, Coates P, Sereika SM, Nelson JB, Trump DL, Resnick NM 2005 Bone loss after initiation of androgen deprivation therapy in patients with prostate cancer. J Clin Endocrinol Metab **90:**6410–6417.
3. Shahinian VB, Kuo YF, Freeman JL, Goodwin JS 2005 Risk of fracture after androgen deprivation therapy for prostate cancer. N Engl J Med **352:**154–164.
4. Smith MR, Boyce SP, Moyneur E, Duh MS, Raut MK, Brandman J 2006 Risk of clinical fractures after gonadotrophin-releasing hormone agonist therapy for prostate cancer. J Urol **175:**136–139.
5. Smith MR, Eastham J, Gleason DM, Shasha D, Tchekmedyian S, Zinner M 2003 Randomised controlled trial of zoledronic acid to prevent bone loss in men receiving androgen deprivation therapy for non-metastatic prostate cancer. J Urol **169:**2008–2012.
6. Michaelson MD, Kaufman DS, Lee H, McGovern FJ, Kantoff PW, Fallon MA, Finkelstein JS, Smith MR 2007 Randomised controlled trial of annual zoledronic acid to prevent gonadotropin-releasing hormone agonist-induced bone loss in men with prostate cancer. J Clin Oncol **25:**1038–1042.
7. Greenspan SL, Nelson JB, Trump DL, Resnick NM 2007 Effect of once-weekly oral alendronate on bone loss in men receiving androgen deprivation therapy for prostate cancer. Ann Intern Med **146:**416–424.
8. McCloskey E 2006 Effects of third-generation aromatase inhibitors on bone. Eur J Cancer **42:**1044–1051.
9. Goss PE, Hadji P, Subar M, Abreu P, Thomsen T, Banke-Bochita J 2007 Effects of steroidal and non-steroidal aromatase inhibitors on markers of bone turnover in healthy postmenopausal women. Breast Cancer Res **10:**R52.
10. Eastell R, Hannon RA, Cuzick J, Dowsett M, Clack G, Adams JE 2006 Effect of an aromatase inhibitor on BMD and bone turnover markers: 2-year results of the anastrozole, tamoxifen, alone or in combination (ATAC) trial. J Bone Miner Res **21:**1215–1223.
11. Perez EA, Josse RG, Pritchard KI, Ingle JM, Martino S, Findlay BP, Shenkier TN, Tozer RG, Palmer MJ, Shepherd LE, Liu S, Tu D, Goss PE 2006 Effect of letrozole versus placebo on bone mineral density in women with primary breast cancer completing 5 or more years of adjuvant tamoxifen: A comparison study to NCIC CTG MA.17. J Clin Oncol **24:**3629–3635.
12. Howell A, Cuzick J, Baum M, Buzdar A, Dowsett M, Forbes JF, Hoctin-Boes G, Houghton J, Locker GY, Tobias JS 2005 Results of the ATAC (Arimidex, Tamoxifen, Alone or in Combination) trial after completion of 5 years' adjuvant treatment for breast cancer. Lancet **365:**60–62.
13. Coates AS, Keshaviah A, Thurlimann B, Mouridsen H, Mauriac L, Forbes JF, Parisdaens R, Castiglione-Gertsch M, Gelber RD, Colleoni M, Lang I, Del Mastro L, Smith I, Chirgwin J, Nogaret JM, Pienkowski T, Wardley A, Jacobsen EH, Price KN, Goldhirsch A 2007 Five years of letrozole compared with tamoxifen as initial adjuvant therapy for postmenopausal women with endocrine-responsive early breast cancer: Update of study BIG 1-98. J Clin Oncol **25:**486–492.
14. Goss PE, Ingle JN, Martino S, Robert NJ, Muss HB, Picart MJ, Castiglione M, Tu D, Shepherd LE, Pritchard KI, Livingston RB, Davidson NE, Norton L, Perez EA, Abrams JS, Therasse P, Palmer MJ, Pater JL 2003 A randomised trial of letrozole in postmenopausal women after five years of tamoxifen therapy for early-stage breast cancer. N Engl J Med **349:**1793–1802.
15. Coleman RE, Banks LM, Girgis SI, Kilburn LS, Vrdoljak E, Fox J, Cawthorn J, Patel A, Snowdon CF, Hall E, Bliss JM, Coombes RC 2007 Skeletal effects of exemestane on bone mineral density, bone biomarkers, and fracture incidence in postmenopausal women with early breast cancer participating in the Intergroup Exemestane Study (IES): A randomised controlled study. Lancet Oncol **8:**119–127.
16. Brufsky A, Harker WG, Beck JT, Carroll R, Tan-Chiu E, Seidler C, Hohneker J, Lacerna L, Petrone S, Perez EA 2007 Zoledronic acid inhibits adjuvant letrozole-induced bone loss in postmenopausal women with early breast cancer. J Clin Oncol **25:**829–836.
17. Cock TA, Back J, Elefteriou F, Karsenty G, Kastner P, Chan S, Auwerx J 2004 Enhanced bone formation in lipodystrophic PPARgamma(hyp/hyp) mice relocates haematopoiesis to the spleen. EMBO Rep **5:**1007–1012.
18. Ali AA, Weinstein RS, Stewart SA, Parfitt AM, Manolagas SC, Jilka RL 2005 Rosiglitazone causes bone loss in mice by suppressing osteoblast differentiation and bone formation. Endocrinology **146:**1226–1235.
19. Lecka-Czernik B, Ackert-Bicknell C, Adamo ML, Marmolejos V, Churchill GA, Shockley KR, Reid IR, Grey A, Rosen C 2007 Activation of peroxisome proliferator-activated receptor gamma (PPARgamma) by rosiglitazone suppresses components of the insulin-like growth factor regulatory system in vitro and in vivo. Endocrinology **148:**903–911.
20. Schwartz AV, Sellmeyer DE, Vittinghoff E, Palermo L, Lecka-Czernik B, Feingold KR, Strotmeyer ES, Resnick HE, Carbone L, Beamer BA, Park SW, Lane NE, Harris TB, Cummings SR 2006 Thiazolidinedione use and bone loss in older diabetic adults. J Clin Endocrinol Metab **91:**3276–3278.
21. Grey A, Bolland M, Gamble G, Wattie D, Horne A, Davidson J, Reid IR 2007 The peroxisome-proliferator-activated receptor-gamma agonist rosiglitazone decreases bone formation and bone mineral density in healthy postmenopausal women: A randomised controlled trial. J Clin Endocrinol Metab **92:**1305–1310.
22. Kahn SE, Haffner SM, Heise MA, Herman WH, Holman RR, Jones NP, Kravitz BG, Lachin JM, O'Neill MC, Zinman B, Viberti G 2006 Glycaemic durability of rosiglitazone, metformin, or glyburide monotherapy. N Engl J Med **355:**2427–2443.

© 2008 American Society for Bone and Mineral Research

23. Short R 2007 Fracture risk is a class effect of glitazones. BMJ **334:**551.
24. Yaturu S, Bryant B, Jain SK 2007 Thiazolidinedione treatment decreases bone mineral density in type 2 diabetic men. Diabetes Care **30:**1574–1576.
25. Grisso JA, Kelsey JL, O'Brien LA, Miles CG, Sidney S, Maislin G, LaPann K, Moritz D, Peters B 1997 Risk factors for hip fracture in men. Hip Fracture Study Group. Am J Epidemiol **145:**786–793.
26. Vestergaard P, Rejnmark L, Mosekilde L 2006 Proton pump inhibitors, histamine H_2 receptor antagonists, and other antacid medications and the risk of fracture. Calcif Tissue Int **19:**76–83.
27. Yu EW, Shinoff C, Blackwell T, Ensrud K, Hillier T, Bauer DC 2006 Use of acid-suppressive medications and risk of bone loss and fracture in postmenopausal women. J Bone Miner Res **21:**S1;S281.
28. Yang Y-X, Lewis JD, Epstein S, Metz DC 2006 Long-term proton pump inhibitor therapy and risk of hip fracture. JAMA **296:**2947–2953.
29. O'Connell MB, Madden DM, Murray AM, Heaney RP, Kerzner LJ 2005 Effects of proton pump inhibitors on calcium carbonate absorption in women: A randomised crossover trial. Am J Med **120:**778–781.
30. Petty SJ, O'Brien TJ, Wark JD 2007 Anti-epileptic medication and bone health. Osteoporos Int **18:**129–142.
31. Lifshitz F, Maclaren NK 1973 Vitamin D-dependent rickets in institutionalised, mentally retarded children receiving long-term anticonvulsant therapy. J Pediatr **83:**612–620.
32. Haney EM, Chan BK, Diem SJ, Ensrud KE, Cauley JA, Barrett-Connor E, Orwoll E, Bliziotes MM 2007 Association of low bone mineral density with selective serotonin reuptake inhibitors in older men. Arch Intern Med **167:**1246–1251.
33. Diem SJ, Blackwell TL, Stone KL, Yaffe K, Haney EM, Bliziotes MM, Ensrud KE 2007 Use of antidepressants and rates of hip bone loss in older women: The study of osteoporotic fractures. Arch Intern Med **167:**1240–1245.
34. Richards JB, Papaioannou A, Adachi JD, Joseph L, Whitson HE, Prior JC, Goltzman D 2007 Effect of selective serotonin reuptake inhibitors on the risk of fracture. Arch Intern Med **167:**188–194.
35. Warden SJ, Robling AG, Sanders MS, Bliziotes MM, Turner CH 2005 Inhibition of the serotonin (5-hydroxytryptamine) transporter reduces bone accrual during growth. Endocrinology **146:**685–693.
36. deSweit M, Ward P, Fidler A, Horsman A, Katz D, Letsky E, Peacock M, Wise PH 1983 Prolonged heparin therapy in pregnancy causes bone demineralisation. Br J Obstet Gynaecol **90:**1129–1134.
37. Dalhman T 1993 Osteoporotic fractures and the recurrence of thromboembolism during pregnancy and the puerperium in 184 women undergoing thromboprophylaxis with heparin. Am J Obstet Gynecol **168:**1265–1270.
38. Barbour L, Kick S, Steiner J, LoVerde M, Heddleston L, Lear J, Baron A, Barton P 1994 A prospective study of heparin-induced osteoporosis in pregnancy using bone densitometry. Am J Obstet Gynecol **170:**862–869.
39. Wiens M, Etminan M, Gill SS, Takkouche B 2006 Effects of antihypertensive drug treatments on fracture outcomes: A meta-analysis of observational studies. J Intern Med **260:**350–362.
40. Reid IR, Gamble GD, Grey AB, Black DM, Ensrud KE, Browner WS, Bauer DC 2005 Beta-blocker use, BMD, and fractures in the study of osteoporotic fractures. J Bone Miner Res **20:**613–618.
41. Bonnet N, Gadois C, McCloskey E, Lemineur G, Lespressailles E, Courteix D, Benhamou CL 2007 Protective effect of beta blockers in postmenopausal women: Influence on fractures, bone density, micro and macroarchitecture. Bone **40:**1209–1216.
42. de Vries F, Souverein PC, Cooper C, Leufkens HGM, van Staa TP 2007 Use of β-blockers and the risk of hip/femur fracture in the United Kingdom and the Netherlands. Calcif Tissue Int **80:**69–75.
43. La Croix AZ, Wienpahl J, White LR, Wallace RB, Scherr PA, George LK, Cornoni-Huntley J, Ostfield AM 1990 Thiazide diuretic agents and the incidence of hip fracture. N Engl J Med **322:**286–290.
44. La Croix AZ, Ott SM, Ichikawa L, Scholes D, Barlow WE 2000 Low-dose hydrochlorothiazide and preservation of bone mineral density in older adults. A randomised double-blind placebo-controlled trial. Ann Intern Med **133:**516–526.
45. Bolland MJ, Ames RW, Horne AM, Orr-Walker BJ, Gamble GD, Reid IR 2007 The effect of treatment with a thiazide diuretic for 4 years on bone density in normal postmenopausal women. Osteoporos Int **18:**479–486.
46. Mundy G, Garret R, Harris S, Chan JC, Chen D, Rossini G, Boyce B, Zhao M, Gutierrez G 1999 Stimulation of bone formation in vitro and in rodents by statins. Science **286:**1946–1949.
47. Uzzan B, Cohen RM, Nicolas P, Cucherat M, Perret G-Y 2007 Effect of statins on bone mineral density: A meta-analysis of clinical studies. Bone **40:**1581–1587.
48. Nguyen D, Wang CY, Eisman JA, Nguyen TV 2007 On the association between statin and fracture: A Bayesian consideration. Bone **40:**813–820.

Chapter 65. Abnormalities in Bone and Calcium Metabolism After Burns

Gordon L. Klein

Department of Pediatrics, University of Texas Medical Branch at Galveston and Shriners Burns Hospital, Galveston, Texas

INTRODUCTION

After burns of ≥40% total body surface area, children lose 2% of total body BMC by time of discharge 8 wk later, progressing to ~4% at 6 mo postburn. Lumbar spine BMC is lost more acutely, with ~10% being lost in hospital and remaining at this level by 6 mo.[1] There is a decrease in growth velocity for the first year after burn injury, whereas weight fluctuates acutely according to fluid balance but increases as catabolism falls. Lumbar spine BMD Z-scores are skewed to the negative, the majority being −1 to −3, and do not recover for a mean of 5 yr postburn.[2] The risk of postburn fracture is also elevated.[2] To understand why bone loss occurs, it is necessary to examine the changes brought about by the burn injury on bone histomorphometry and calcium homeostasis, followed by an examination of the body's responses to burns and how these responses can help explain the bone loss.

BONE HISTOMORPHOMETRY

Iliac crest bone biopsies taken 2 wk postburn in children[3] and 3 wk postburn in adults[4] show a lack of surface osteoblasts consistent with apoptosis, as well as markedly reduced surface uptake of doxycycline compared with normal controls

The author states that he has no conflicts of interest.

Key words: burns, bone resorption, bone formation, vitamin D deficiency, calcium-sensing receptor upregulation, PTH, recombinant human growth hormone, oxandrolone, bisphosphonates

© 2008 American Society for Bone and Mineral Research

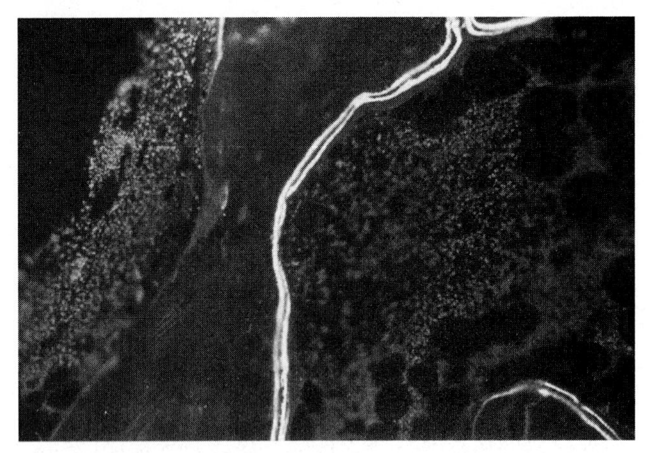

FIG. 1. Iliac crest bone biopsy from a normal subject showing full surface doxycycline uptake and normal double label.

and complete absence of double labels (Figs. 1 and 2). These findings suggest a reduction in osteoblast numbers and function.[3,4] Trabecular width and spacing is normal,[3] in contrast to postmenopausal osteoporosis.

STRESS RESPONSE

The stress response, which begins immediately after burn injury, involves the sustained production of a mean 8-fold increase of endogenous glucocorticoids.[3,5] Analysis of bone using RT-PCR revealed a trend toward decreased glucocorticoid receptor mRNA expression in bone biopsies obtained from burned children 2 wk after the injury, and culture of the marrow stromal cells from those biopsies showed a decrease in the markers of osteoblast differentiation, such as alkaline phosphatase, type I collagen, core-binding factor α 1 (cbfa 1, or runx 2), and bone morphogenetic protein (BMP)-2 compared with normal controls.[5] The reduced number of surface osteoblasts and the diminished osteoblast differentiation is consistent with the mode of action of glucocorticoids on bone. Thus, endogenous glucocorticoid production may be as deleterious to bone as exogenous glucocorticoid administration for medicinal purposes.

DISORDERED CALCIUM HOMEOSTASIS

Another feature of the effect of burn injury on bones is disordered calcium homeostasis. With few exceptions, the majority of children develop sustained hypocalcemia that begins acutely and is of uncertain duration, although sometimes resolving by time of discharge 6–8 wk later. Hypocalcemia is accompanied by hypoparathyroidism and urinary calcium wasting, averaging twice normal.[6] Magnesium depletion, which universally occurs after burn injury, was initially suspected as the cause.[6] However, vigorous attempts at magnesium repletion failed to improve either hypocalcemia or the parathyroid response to it.[7] Studies in a sheep model of burn injury reproduced the acute hypocalcemia and hypomagnesemia. Removal of the parathyroids within 48 h of the burn showed a 50% upregulation of sheep parathyroid calcium-sensing receptor mRNA and increased cell membrane receptor presence on the chief cells of the parathyroids of burned sheep compared with sham-burned controls.[8] Thus, potentially, the human parathyroid calcium-sensing receptor is similarly upregulated, with a consequent reduction of the set point for circulating calcium suppression of PTH production (i.e., low circulating calcium is sufficient to suppress PTH production). This hypocalcemic hypoparathyroidism seen in children postburn[6] leads to urinary calcium wasting and thus to the relative lack of available calcium to replenish bone mass.

SYSTEMIC INFLAMMATORY RESPONSE

Whereas it has not been established how the hypocalcemic, hypercalciuric hypoparathyroidism results from burn injury, a candidate mechanism could be the cytokines produced in abundance by the systemic inflammatory response, which develops within 24 h of the burn injury as a result of skin barrier damage. Specifically, circulating levels of interleukin (IL)-1β and IL-6 are markedly elevated.[3] These cytokines normally stimulate the osteoblast to produce RANKL, which stimulates osteoclastogenesis. However, by 2 wk postburn, there are very few detectable osteoblasts and osteoclasts and low resorption, as determined by urinary deoxypyridinoline measurements.[3] A possible alternative site for the actions of these cytokines would be the parathyroid chief cells. Nielsen et al.[9] and Toribio et al.[10] have found that in vitro incubation of either bovine[9] or equine[10] parathyroid cells with IL-1β resulted in increased calcium receptor mRNA and reduced PTH mRNA and secretion. Moreover, IL-6 incubation was found to reduce PTH secretion.[10] Therefore, one possible pathogenic mechanism for the hypocalcemic, hypercalciuric hypoparathyroidism is by cytokine-stimulated upregulation of the parathyroid calcium-sensing receptor. It should be pointed out that before the disappearance of osteoblasts, both glucocorticoids and inflammatory cytokines can cause a transient increase in bone resorption and thus increased bone loss, but this has not as yet been adequately documented.

VITAMIN D DEFICIENCY

Another abnormality of the burn injury that may contribute to bone loss is progressive vitamin D deficiency. Acutely, although patients receive vitamin D_3 of at least 400 international units (10 μg) daily, serum levels of 25-hydroxyvitamin D [25(OH)D and 1,25-dihydroxyvitamin D ($1,25(OH)_2D$] cannot be adequately measured because both vitamin D binding protein[4] and albumin[3] are low. However, by 6 mo postburn, both serum albumin and total protein are normal. By 14 mo postburn, serum levels of 25(OH)D are low.[11] At 2 yr postburn, $1,25(OH)_2D$ levels are normal. However, in a cross-sectional study, by 7 yr postburn, not only were the vast majority of 25(OH)D levels still low, but one half the measured serum $1,25(OH)_2D$ levels were also low.[12] This finding suggests a progressive vitamin D deficiency after discharge from the hospital. This may be due in part to limited sunlight exposure and lack of vitamin D supplementation, but also to the

FIG. 2. Iliac crest bone biopsy from a burned subject showing patchy surface doxycycline uptake and absent double label.

© 2008 American Society for Bone and Mineral Research

failure of the burn scar and the adjacent normal-appearing skin to convert a normal percentage of the precursor, 7-dehydrocholesterol, to previtamin D_3. This conversion is ~25% that of normal controls.[11] Furthermore, not only the sample from the burn scar but also samples of skin from the adjacent normal-appearing areas were significantly deficient in the precursor, 7-dehydrocholesterol,[11] suggesting that the skin may be biochemically abnormal after a burn injury. Serum levels of 25(OH)D varied directly with BMD of the lumbar spine[12] and inversely with serum PTH concentration.[12] Therefore, inadequate serum 25(OH)D levels may contribute to the failure to recover bone mass postburn.

RESUMPTION OF REMODELING

Bone remodeling resumes in all patients, regardless of treatment, by ~1 yr postburn.[13] However, despite resumption of remodeling, bone mass lost during the first year postburn is not recovered.[2]

TREATMENT

To date, three medications have been used to treat burn-associated bone loss. Recombinant human growth hormone (rhGH) and oxandrolone are both used as anabolic agents. Recombinant hGH is used at a dose of 0.05 mg/kg/d[14] subcutaneously, or oxandrolone is used orally, at a dose of 0.1 mg/kg twice daily.[15] In both settings, the medications produced a significant increase in insulin-like growth factor (IGF)-1 within 6 wk[16] of the burn, in lean body mass at 6 mo after the burn,[14,15] and in total body and lumbar spine BMC at 12 mo.[14,15] Interestingly, lumbar spine BMD was not increased after 1 yr of treatment, suggesting that there was a corresponding increase in bone area, resulting in larger and therefore stronger bones. Although no side effects have been observed with rhGH therapy, hyperglycemia is a potential complication. With oxandrolone, two patients developed clitoral hypertrophy, which disappeared after cessation of the drug.[15] The third medication used has been the bisphosphonate pamidronate. This is given intravenously over 12 h at a dose of 1.5 mg/kg, with a maximum dose of 90 mg, in a 5% dextrose and water solution.[1] It is administered within 10 days postburn, repeated 1 wk later, and not given again.[1] The effects seen are greater on trabecular than cortical bone, because lumbar spine BMC and BMD are preserved at discharge compared with total body BMC, although both are higher than placebo values at 6 mo postburn.[1] Also, 2 yr after the burn injury, lumbar spine BMC and BMD remain higher in the group receiving pamidronate.[13] Although hypocalcemia is a potential side effect of the pamidronate treatment, no more severe hypocalcemia was seen in the patients receiving the two doses of pamidronate than the placebo controls.[1] These treatments, while successful in part, do not have regulatory approval in children.

CONCLUSION

Burn injury produces a variety of bodily responses that adversely affect bone and calcium metabolism. It is likely that there is acute bone resorption as a result of an acute increase in endogenous glucocorticoids and pro-inflammatory cytokine production, which can both stimulate osteoblast production of RANKL. However, although this process could explain some of the acute bone loss, we do not have evidence to document its occurrence at this point. We are making an attempt to better understand these changes, treat them appropriately, and determine in what other situations these same responses may be active to look for heretofore unrecognized bone problems to identify and correct.

ACKNOWLEDGMENTS

This work was supported by funding from the National Institutes of Health Research Grant 1P50 GM 60338, the National Institute for Disability and Rehabilitation Research Grant H133A020102-05, and Shriners Hospitals for Children Grants 8640, 8680, and 15877.

REFERENCES

1. Klein GL, Wimalawansa SJ, Kulkarni G, Sherrard DJ, Sanford AP, Herndon DN 2005 The efficacy of acute administration of pamidronate on the conservation of bone mass following severe burn injury in children: A double-blind, randomized, controlled study. Osteoporos Int **16:**631–635.
2. Klein GL, Herndon DN, Langman CB, Rutan TC, Young WE, Pembleton G, Nusynowitz M, Barnett JL, Broemeling LD, Sailer DE, McCauley RL 1995 Long-term reduction in bone mass after severe burn injury in children. J Pediatr **126:**252–256.
3. Klein GL, Herndon DN, Goodman WG, Langman CB, Phillips WA, Dickson IR, Eastell R, Naylor KE, Maloney NA, Desai M, Benjamin D, Alfrey AC 1995 Histomorphometric and biochemical characterization of bone following acute severe burns in children. Bone **17:**455–460.
4. Klein GL, Herndon DN, Rutan TC, Sherrard DJ, Coburn JW, Langman CB, Thomas ML, Haddad JG Jr, Cooper CW, Miller NL, Alfrey AC 1993 Bone disease in burn patients. J Bone Miner Res **8:**337–345.
5. Klein GL, Bi LX, Sherrard DJ, Beavan SR, Ireland D, Compston JE, Williams WG, Herndon DN 2004 Evidence supporting a role of glucocorticoids in short-term bone loss in burned children. Osteoporos Int **15:**468–474.
6. Klein GL, Nicolai M, Langman CB, Cuneo BF, Sailer DE, Herndon DN 1997 Dysregulation of calcium homeostasis after severe burn injury in children: Possible role of magnesium depletion. J Pediatr **131:**246–251.
7. Klein GL, Langman CB, Herndon DN 2000 Persistent hypoparathyroidism following magnesium repletion in burn-injured children. Pediatr Nephrol **14:**301–304.
8. Murphey ED, Chattopadhyay N, Bai M, Kifor O, Harper D, Traber DL, Hawkins HK, Brown EM, Klein GL 2000 Up-regulation of the parathyroid calcium-sensing receptor after burn injury in sheep: A potential contributory factor to postburn hypocalcemia. Crit Care Med **28:**3885–3890.
9. Nielsen PK, Rasmussen AK, Butters R, Feldt-Rasmussen U, Bendtzen K, Diaz R, Brown EM, Olgaard K 1997 Inhibition of PTH secretion by interleukin-1 beta in bovine parathyroid glands *in vitro* is associated with an up-regulation of the calcium-sensing receptor mRNA. Biochem Biophys Res Commun **238:**880–885.
10. Toribio RE, Kohn CW, Capen CC, Rosol TJ 2003 Parathyroid hormone (PTH) secretion, PTH mRNA, and calcium-sensing receptor mRNA expression in equine parathyroid cells, and effects of interleukin (IL)-1, IL-6 and tumor necrosis factor-alpha on equine parathyroid cell function. J Mol Endocrinol **31:**609–620.
11. Klein GL, Chen TC, Holick MF, Langman CB, Price H, Celis MM, Herndon DN 2004 Synthesis of vitamin D in skin after burns. Lancet **363:**291–292.
12. Klein GL, Langman CB, Herndon DN 2002 Vitamin D depletion following burn injury in children: A possible factor in post-burn osteopenia. J Trauma **52:**346–350.
13. Przkora R, Herndon DN, Sherrard DJ, Chinkes DL, Klein GL 2007 Pamidronate preserves bone mass for at least 2 years following acute administration for pediatric burn injury. Bone **41:**297–302.
14. Hart DW, Herndon DN, Klein G, Lee SB, Celis M, Mohan S, Chinkes DL, Wolf SE 2001 Attenuation of post-traumatic muscle catabolism and osteopenia by long-term growth hormone therapy. Ann Surg **233:**827–834.
15. Murphy KD, Thomas S, Mlcak RP, Chinkes DL, Klein GL, Herndon DN 2004 Effects of long-term oxandrolone administration in severely burned children. Surgery **136:**219–224.
16. Klein GL, Wolf SE, Langman CB, Rosen CJ, Mohan S, Keenan BS, Matin S, Steffen C, Nicolai M, Sailer DE, Herndon DN 1998 Effect of therapy with recombinant human growth hormone on insulin-like growth factor system components and serum levels of biochemical markers of bone formation in children following severe burn injury. J Clin Endocrinol Metab **83:**21–24.

© 2008 American Society for Bone and Mineral Research

SECTION VI

Disorders of Mineral Homeostasis
(Section Editors: Marie Demay and Suzanne Jan de Beur)

Please also see the online appendix at www.asbmrprimer.org

© 2008 American Society for Bone and Mineral Research

Chapter 66. Primary Hyperparathyroidism

Shonni J. Silverberg[1] and John P. Bilezikian[1,2]

[1]*Department of Medicine and Pharmacology, College of Physicians and Surgeons, Columbia University, New York, New York;*
[2]*Department of Medicine, College of Physicians and Surgeons, Columbia University, New York, New York*

INTRODUCTION

Primary hyperparathyroidism is one of the two most common causes of hypercalcemia and thus ranks high as a key diagnostic possibility in anyone with an elevated serum calcium concentration. Primary hyperparathyroidism and malignancy together account for 90% of all hypercalcemic patients. Other potential causes of hypercalcemia are considered after the first two are ruled out or if there is reason to believe that a different cause is likely. The differential diagnosis of hypercalcemia, as well as features of hypercalcemia of malignancy, are considered elsewhere in this *Primer*. This chapter will deal exclusively with the clinical presentation, evaluation, and therapy of primary hyperparathyroidism.

Primary hyperparathyroidism is a relatively common endocrine disease, with an incidence as high as 1 in 500 to 1 in 1000. The high visibility of primary hyperparathyroidism in the population today marks a dramatic change from several generations ago, when it was considered to be a rare disorder. A 4- to 5-fold increase in incidence was noted in the early 1970s, because of the widespread use of the multichannel biochemistry autoanalyzer, which provided serum calcium determinations in patients being evaluated for a set of completely unrelated complaints. Primary hyperparathyroidism occurs at all ages but is most frequent in the sixth decade of life. Women are affected more often than men by a ratio of 3:1. The majority of patients are postmenopausal women, often presenting within the first decade after menopause. When found in children, an unusual event, it is important to consider the possibility that the disease is a harbinger of an endocrinopathy with genetic basis, such as multiple endocrine neoplasia type I or II.

Primary hyperparathyroidism is a hypercalcemic state resulting from excessive secretion of PTH from one or more of the four parathyroid glands. The disease is caused by a benign, solitary adenoma 80% of the time. A parathyroid adenoma is a collection of chief cells surrounded by a rim of normal tissue at the outer perimeter of the gland. In the patient with a parathyroid adenoma, the remaining three parathyroid glands are usually normal. Less commonly (in 15–20% of patients with primary hyperparathyroidism), the disease is caused by hyperplasia of all four parathyroid glands. Four-gland hyperplasia may occur sporadically or in association with multiple endocrine neoplasia type I or II. Parathyroid carcinoma is a very rare form of primary hyperparathyroidism, occurring in <0.5% of patients with hyperparathyroidism. Pathological examination of the malignant tissue might show mitoses, vascular or capsular invasion, and fibrous trabeculae. However, unless gross local or distant metastases are present, the diagnosis of parathyroid cancer is difficult to make microscopically. Specific genetic studies and immunohistochemical analyses (*HRPT2* gene and parafibromin staining) may help to distinguish benign from malignant parathyroid tissue when standard approaches are not clear.

The pathophysiology of primary hyperparathyroidism relates to the loss of normal feedback control of PTH by extracellular calcium. Under virtually all other hypercalcemic conditions, the parathyroid gland is suppressed, and PTH levels are low. In adenomas, the parathyroid cell loses its normal sensitivity to calcium, whereas in primary hyperparathyroidism caused by hyperplasia of the parathyroid glands, the "set-point" for calcium is not changed for a given parathyroid cell; instead, an increase in the number of cells gives rise to hypercalcemia.

The underlying cause of primary hyperparathyroidism is not known. External neck irradiation in childhood, recognized in some patients, is unlikely to be causative in the majority of patients. The molecular basis for primary hyperparathyroidism continues to be elusive. The clonal origin of most parathyroid adenomas suggests a defect at the level of the gene controlling growth of the parathyroid cell or the expression of PTH. Patients with primary hyperparathyroidism have been discovered in whom the *PTH* gene is rearranged to a site adjacent to the *cyclin D1* gene, leading to overexpression of this important cell cycle regulator. Loss of one copy of the MEN1 tumor suppressor gene located on chromosome 11 has also been seen in sporadic parathyroid adenomas. Abnormalities in the p53 tumor suppressor gene have not been described in primary hyperparathyroidism. Other genes under study for a possible role in the development of sporadic parathyroid adenomas are the calcium-sensing receptor gene, the vitamin D receptor gene, and RET. To date, such studies have not been revealing.

SIGNS AND SYMPTOMS

"Classical" primary hyperparathyroidism is associated with skeletal and renal complications. The skeletal disease seen commonly until the latter part of the 20th century, was *osteitis fibrosa cystica*, characterized by subperiosteal resorption of the distal phalanges, tapering of the distal clavicles, a "salt and pepper" appearance of the skull, bone cysts, and brown tumors of the long bones. Overt hyperparathyroid bone disease is now seen in <5% of patients in the United States with primary hyperparathyroidism.

Like the skeleton, the kidney is also less commonly involved in primary hyperparathyroidism than before. The incidence of kidney stones has declined from 33% in the 1960s to 15–20% now. Nephrolithiasis, nevertheless, is still the most common clinical complication of the hyperparathyroid process. Other renal features of primary hyperparathyroidism include diffuse deposition of calcium–phosphate complexes in the parenchyma (nephrocalcinosis). Hypercalciuria (daily calcium excretion of 250 mg for women or 300 mg for men) is seen in up to 30% of patients. In the absence of any other cause, primary hyperparathyroidism may be associated with a reduction in creatinine clearance.

The classic neuromuscular syndrome of primary hyperparathyroidism included a definable myopathy that has virtually disappeared. In its place, however, is a less well-defined syndrome characterized by easy fatigue, a sense of weakness, and a feeling that the aging process is advancing faster than it should. This is sometimes accompanied by an intellectual weariness and a sense that cognitive faculties are less sharp. In some studies, psychodynamic evaluation has seemed to show a

Dr. Silverberg received a research grant from Amgen and currently has research funding from Merck & Co. Dr. Bilezikian is a consultant for Merck.

Key words: primary hyperparathyroidism, hypercalcemia, metabolic bone disease, PTH

© 2008 American Society for Bone and Mineral Research

distinct psychiatric profile. Preliminary results of a recent randomized study of patients with mild hyperparathyroidism suggest that there are distinct psychiatric features of the disease. However, these features are not improved after parathyroidectomy. Thus, there are no clear data supporting surgical intervention for the purpose of improving psychiatric symptoms at this time.

Gastrointestinal manifestations of primary hyperparathyroidism have classically included peptic ulcer disease and pancreatitis. Peptic ulcer disease is not likely to be linked in a pathophysiologic way to primary hyperparathyroidism unless type I multiple endocrine neoplasia is present. Pancreatitis is virtually never seen anymore as a complication of primary hyperparathyroidism because the hypercalcemia tends to be so mild. Like peptic ulcer disease, the association between primary hyperparathyroidism and hypertension is tenuous.

Although there may be an increased incidence of hypertension in primary hyperparathyroidism, it is rarely corrected or improved after successful surgery. In classical primary hyperparathyroidism, cardiovascular features included myocardial, valvular, and vascular calcification, with subsequent increased cardiovascular mortality. Although such overt involvement is not seen in mild primary hyperparathyroidism today, there are data supporting increased vascular stiffness in mild primary hyperparathyroidism, and ongoing studies are investigating the presence of other subtle cardiovascular manifestations. Other potential organ systems that in the past were affected by the hyperparathyroid state are now relegated to being archival curiosities. These include gout and pseudo-gout, anemia, band keratopathy, and loose teeth.

CLINICAL FORMS OF PRIMARY HYPERPARATHYROIDISM

In the United States today, classical primary hyperparathyroidism is rarely seen. Instead, the most common clinical presentation of primary hyperparathyroidism is characterized by asymptomatic hypercalcemia with serum calcium levels within 1 mg/dl above the upper limits of normal. Most patients do not have specific complaints and do not show evidence for any target organ complications. They have often been discovered in the course of a routine multichannel screening test. Rarely, a patient will show serum calcium levels in the life-threatening range, so-called acute primary hyperparathyroidism or parathyroid crisis. These patients are invariably symptomatic of hypercalcemia. Although this is an unusual presentation of primary hyperparathyroidism, it does occur and should always be considered in any patient who presents with acute hypercalcemia of unclear etiology.

The very earliest manifestation of primary hyperparathyroidism may present with isolated elevations in PTH levels, whereas serum calcium is still normal. This constellation may be seen in patients undergoing evaluation for low BMD or in other individuals who are receiving comprehensive screening tests for their skeletal health. Recent data suggest that some of these patients may have an early stage of a more symptomatic form of the disease, in which skeletal abnormalities are more prominent. In such patients, it is important to rule out causes of secondary hyperparathyroidism. It is particularly important to be sure that these individuals do not have vitamin D insufficiency, which can lower serum calcium levels into the normal range. Re-evaluation of such individuals after vitamin D repletion and/or after correction of other secondary causes of hyperparathyroidism is necessary to secure the diagnosis. Unusual clinical presentations of primary hyperparathyroidism include multiple endocrine neoplasia types I and II, familial primary hyperparathyroidism not associated with any other endocrine disorder, familial cystic parathyroid adenomatosis, and neonatal primary hyperparathyroidism.

EVALUATION AND DIAGNOSIS OF PRIMARY HYPERPARATHYROIDISM

The history and the physical examination rarely give any clear indications of primary hyperparathyroidism but are helpful because of the paucity of specific manifestations of the disease. The diagnosis of primary hyperparathyroidism is established by laboratory tests. The biochemical hallmarks of primary hyperparathyroidism are hypercalcemia and elevated levels of PTH. The serum phosphorus tends to be in the lower range of normal. In only approximately one third of patients is the serum phosphorus concentration frankly low. The serum alkaline phosphatase activity may be elevated when bone disease is present. More specific markers of bone formation activity (bone-specific alkaline phosphatase and osteocalcin) and bone resorption activity (urinary deoxypyridinoline and N-telopeptide of collagen) will be above normal when there is active bone involvement but otherwise tend to be in the upper range of normal. The actions of PTH to alter acid–base handling in the kidney will lead, in some patients, to a small increase in the serum chloride concentration and a concomitant decrease in the serum bicarbonate concentration. Urinary calcium excretion is elevated in 30% of patients. The circulating 1,25-dihydroxyvitamin D concentration is elevated in 25% of patients with primary hyperparathyroidism, although it is of little diagnostic value because 1,25-dihydroxyvitamin D levels are increased in other hypercalcemic states, such as sarcoidosis, other granulomatous diseases, and some lymphomas. 25-Hydroxyvitamin D levels tend to be in the lower end of the normal range. With the newer definition of vitamin D insufficiency, most patients with primary hyperparathyroidism have levels that are below that defined cut-point (i.e., <30 ng/ml).

Skeletal X-rays are not cost effective in the evaluation of the patient with primary hyperparathyroidism because the vast majority of patients lack specific radiologic manifestations. On the other hand, bone mineral densitometry has proven to be an essential component of the evaluation because of its great sensitivity to detect early changes in bone mass. Patients with primary hyperparathyroidism tend to show a pattern of bone involvement that preferentially affects the cortical as opposed to the cancellous skeleton (Fig. 1). The typical pattern is a reduction in BMD of the distal third of the forearm, a site

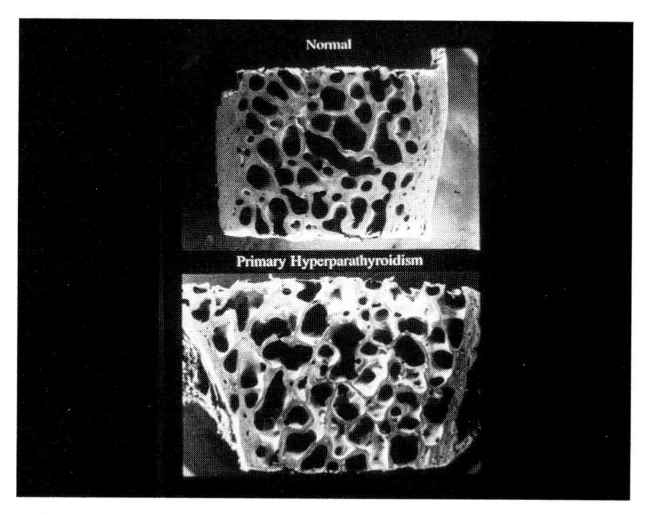

FIG. 1. Scanning electron micrograph shows thin cortices in primary hyperparathyroidism.

© 2008 American Society for Bone and Mineral Research

enriched in cortical bone, and relative preservation of the lumbar spine, a site enriched in cancellous bone. The hip region, best typified by the femoral neck, tends to show values intermediate between the distal radius and the lumbar spine because its composition is a more equal mixture of cortical and cancellous elements. A small subset of patients (15%) present with an atypical BMD profile, characterized by vertebral osteopenia or osteoporosis. Even other patients with primary hyperparathyroidism can show uniform reductions in BMD at all sites. Bone densitometry has become an invaluable aspect of the evaluation of primary hyperparathyroidism because it gives a more accurate assessment of the degree of involvement of the skeleton than any other approach. It is important to underscore the importance of measuring distal forearm BMD in primary hyperparathyroidism. Information obtained from this complete densitometric analysis is used to make recommendations for parathyroid surgery or for conservative medical observation (see following sections).

Measurement of the circulating PTH concentration is the most definitive way to establish the diagnosis of primary hyperparathyroidism. In the presence of hypercalcemia, an elevated level of PTH virtually establishes the diagnosis. A PTH level in the mid or upper end of the normal range in the face of hypercalcemia is also consistent with the diagnosis of primary hyperparathyroidism. The standard assay for measurement of PTH is the immunoradiometric (IRMA) or immunochemiluminometric (ICMA) assays that measures the "intact" molecule. This assay also measures large carboxyterminal fragments [PTH(7–84) is one such fragment] of PTH in addition to full-length PTH(1-84) and can therefore overestimate the amount of bioactive hormone in the serum. A newer assay, specific for PTH(1-84) only, seems to be elevated somewhat more frequently in patients with primary hyperparathyroidism. Although this assay may offer somewhat increased diagnostic sensitivity in cases where the intact IRMA is within the normal range (albeit inappropriately), the intact IRMA currently remains the standard assay in the diagnosis of the disease. The clinical use of PTH measurements in the differential diagnosis of hypercalcemia is a result both of refinements in assay techniques and of the fact that the most common other cause of hypercalcemia, namely hypercalcemia of malignancy, is associated with suppressed levels of hormone. There is no cross-reactivity between PTH and PTH-related peptide (PTHrP; the major causative factor in humoral hypercalcemia of malignancy) in the immunoradiometric assays for PTH. The only hypercalcemic disorders in which the PTH concentration might be elevated are familial hypocalciuric hypercalcemia (FHH) and those related to lithium or thiazide diuretic use. Although in many cases, thiazide diuretics unmask primary hyperparathyroidism by inhibiting calcium excretion, there are cases in which the condition reverses entirely with withdrawal of the drug. Ultimately, the only secure way to make the diagnosis of drug-related primary hyperparathyroidism is to withdraw the medication (if it is safe to do so) and to confirm persistent hypercalcemia and elevated PTH levels 2–3 mo later.

TREATMENT OF PRIMARY HYPERPARATHYROIDISM

Surgery

Primary hyperparathyroidism is cured when the abnormal parathyroid tissue is removed. Whereas it is clear that surgery is appropriate in patients with classical symptoms of primary hyperparathyroidism, there is considerable controversy concerning the need for intervention in patients who have no clear signs or symptoms of their disease. In 2002, a workshop was conducted at the NIH to review the available data on this group of patients. The results of that meeting have led to a revision of the guidelines for management of asymptomatic primary hyperparathyroidism that were first recommended by the 1990 Consensus Development Conference on this subject (Table 1). Patients are always advised to have surgery if they have symptomatic disease, such as overt bone disease or kidney stones, or if they have survived an episode of acute primary hyperparathyroidism with life-threatening hypercalcemia. Asymptomatic patients are now advised to have surgery if the serum calcium is 1 mg/dl above the upper limit of normal. Marked hypercalciuria (400 mg daily excretion) or significantly reduced creatinine clearance (30% more than age- and sex-matched controls) is another general indication for surgery. If bone mass, as determined by bone densitometry, is more than 2.5 SD below young normal control subjects (T-score ≤ 2.5) at any site, surgery should be recommended. Finally, the patient with primary hyperparathyroidism who is <50 yr old is at greater risk for progression of the hyperparathyroid disease process than older patients and should be advised to undergo parathyroidectomy. Adherence to these guidelines for surgery, however, is dependent on both the physician and the patient. Because surgery is an acceptable approach even in patients who do not meet surgical guidelines, some physicians will recommend surgery for all patients with primary hyperparathyroidism; other physicians will not recommend surgery unless clear-cut complications of primary hyperparathyroidism are present. Similarly, some patients cannot tolerate the idea of living with a curable disease and will seek surgery in the absence of the aforementioned guidelines. Other patients, with co-existing medical problems, may not wish to face the risks of surgery, although surgical indications are present.

TABLE 1. GUIDELINES FOR USE IN PATIENTS WITH ASYMPTOMATIC PRIMARY HYPERPARATHYROIDISM

Measure	*Guidelines for surgery*	*Guidelines for follow-up of nonsurgical patients*
Serum calcium	>1 mg/dl above normal	Two times per year
Urinary calcium	>400 mg/d	Do not measure
Renal function		
Creatinine clearance	Reduced by 30%	Do not measure
Serum creatinine		Measure annually
BMD	T-score < −2.5 any site	Annually three sites
Age	<50 yr	

Guidelines describe patients in whom surgical intervention is desirable and criteria for follow-up of nonsurgical patients. It is recommended that all symptomatic patients be sent for parathyroidectomy. (Adapted from NIH Workshop on Asymptomatic Primary Hyperparathyroidism, 2002.)

Parathyroid surgery requires exceptional expertise and experience. The glands are notoriously variable in location, requiring the surgeon's knowledge of typical ectopic sites such as intrathyroidal, retroesophageal, the lateral neck, and the mediastinum. Although the classical approach to parathyroid surgery is to remove the abnormal gland and to ascertain that the other glands are normal by visual inspection, advances in surgery have led to other approaches. In the patient with single gland disease, minimally invasive parathyroidectomy (MIP) is rapidly becoming the procedure of choice. It depends on preoperative localization by the most widely used localization modalities: technetium-99m-sestamibi or ultrasound. The surgeon limits the operative field specifically to the site of the visualized adenoma. MIP also requires capability to measure intraoperative PTH levels. Within a few minutes after resection, an intraoperative PTH level is obtained. If the PTH level falls by 50% and is within the normal range, the adenoma that has

© 2008 American Society for Bone and Mineral Research

been removed is considered to be the only source of abnormal glandular activity, and the operation is terminated. There is some concern about this rule, because if the PTH level falls by >50% but remains frankly elevated, other glandular sources of PTH should be considered. If there is reason to believe that the PTH has not normalized, the operation is converted to a more complete search for all abnormal parathyroid tissue. In the case of multiglandular disease, the approach is to remove all tissue except for a fragment of parathyroid tissue that is left in situ or autotransplanted in the nondominant forearm.

Postoperatively, the patient may experience a brief period of transient hypocalcemia, during which time the normal but suppressed parathyroid glands regain their sensitivity to calcium. This happens within the first few days after surgery but can be prevented in most cases by providing patients with several grams of calcium on a daily basis during the first postoperative week. Prolonged postoperative symptomatic hypocalcemia as a result of rapid deposition of calcium and phosphate into bone ("hungry bone" syndrome) is rarely seen today. Such patients may require parenteral calcium for symptomatic hypocalcemia. Permanent hypoparathyroidism is a potential complication of surgery in those who have had previous neck surgery or who undergo subtotal parathyroidectomy (for multiglandular disease). Another rare complication of parathyroid surgery is unilateral damage to the recurrent laryngeal nerve, which can lead to hoarseness and reduced voice volume.

A number of localization tests are available to define the site of abnormal parathyroid tissue preoperatively. Among the noninvasive tests, scintigraphy, ultrasonography, CT, and MRI are available. Radioisotopic imaging is now performed most commonly with technetium-99m-sestamibi. Sestamibi is taken up both by thyroid and parathyroid tissue but persists in the parathyroid glands. Using ^{123}I along with technetium-99m-sestamibi can provide even greater specificity because uptake of the iodine tracer by the thyroid can be specifically "removed" by computer subtraction technology. Parathyroid localization using scintigraphy offers important localization data that are mandatory before any planned minimally invasive surgery or for repeat parathyroid exploration.

Planar imaging may miss small or ectopic lesions. More precise localization is achieved by using single photon emission CT (SPECT) as an adjunct to scintigraphic scanning. It is important to note that these techniques are not particularly helpful in patients with primary hyperparathyroidism caused by hyperplasia. Invasive localization tests with arteriography and selective venous sampling for PTH in the draining thyroid veins are available when noninvasive studies have not been successful. The value of preoperative localization tests in patients about to undergo parathyroid surgery is controversial. In patients who have not had previous neck surgery, there is little evidence that such tests prevent failed operations or shorten operating time. An experienced parathyroid surgeon will find the abnormal parathyroid gland(s) 95% of the time in the patient who has not had previous neck surgery. Preoperative localization used to be reserved for those who had had previous neck surgery but now it is used routinely in all patients whether they have had previous neck surgery. In patients who are to undergo the MIP procedure, preoperative localization is mandatory. Radioisotopic imaging and ultrasound are best for parathyroid tissue that is located in proximity to the thyroid, whereas CT and MRI approaches are better for ectopically located parathyroid tissue. Arteriography and selective venous studies are reserved for those individuals in whom the noninvasive studies have not been successful.

After successful parathyroid surgery, the patient is cured. Serum biochemistries normalize, and the PTH level returns to normal. In addition, bone mass, as documented by BMD measurements, improves substantially in the first 1–3 yr after surgery. The cumulative increase in bone mass at the lumbar spine and femoral neck can be as high as 12%, a rather impressive improvement. This increase is sustained for at least a decade after parathyroidectomy. It is particularly noteworthy that the lumbar spine, a site where PTH seems to protect from age-related and estrogen deficiency bone loss, is a site of rapid and substantial improvement. Those patients who present with evidence of vertebral osteopenia or osteoporosis sustain an even more impressive improvement in spine BMD after cure and should therefore be routinely referred for surgery, regardless of the severity of their hypercalcemia.

Medical Management

Patients who are not surgical candidates for parathyroidectomy seem to do well when they are managed conservatively. Data on patients with primary hyperparathyroidism followed for up to a decade showed that the biochemical indices of disease. These include the serum calcium, phosphorus, PTH, 25-hydroxyvitamin D, 1,25-dihydroxyvitamin D, and urinary calcium excretion. BMD measures of bone mass also remain remarkably stable over the first decade of observation. More specific markers of bone formation and bone resorption also do not seem to change. However, those patients followed for longer than that period begin to show evidence of bone loss, particularly at the more cortical sites (hip and radius).

Although most patients do have a stable course in the first decade after diagnosis, 25% of patients with asymptomatic primary hyperparathyroidism will have biochemical or bone densitometric evidence of disease progression over a 10-yr period. Second, those under the age of 50 yr have a far higher incidence of progressive disease than do older patients (65% versus 23%). This supports the notion that younger patients should be referred for parathyroidectomy. Finally, today, as in the day of classical primary hyperparathyroidism, patients with symptomatic disease do poorly when observed without surgery. Thus, the data support the safety of observation without surgery only in selected patients with asymptomatic primary hyperparathyroidism.

The longitudinal data on patients who do not have parathyroid surgery also support the need for medical monitoring. In those patients who do not undergo surgery, a set of general medical guidelines is recommended (Table 1). Routine medical follow-up usually includes visits twice yearly with serum calcium determinations. Yearly assessment of serum creatinine and bone densitometry at the spine, hip, and distal one-third site of the forearm is also recommended. Adequate hydration and ambulation are always encouraged. Thiazide diuretics and lithium are to be avoided if possible, because they may lead to worsening hypercalcemia. Dietary intake of calcium should be moderate. There is no good evidence that patients with primary hyperparathyroidism show significant fluctuations of their serum calcium as a function of dietary calcium intake. High calcium intakes (>1 g/d) should be avoided in patients whose 1,25-dihydroxyvitamin D level is elevated. Low calcium diets should also be avoided because they could theoretically lead to further stimulation of PTH secretion.

We still lack an effective and safe therapeutic agent approved for the medical management of primary hyperparathyroidism. Oral phosphate will lower the serum calcium in patients with primary hyperparathyroidism by 0.5–1 mg/dl. Phosphate seems to act by three mechanisms: interference with absorption of dietary calcium, inhibition of bone resorption, and inhibition of renal production of 1,25-dihydroxyvitamin D. Phosphate, however, is not recommended as an approach to management, because of concerns related to ec-

© 2008 American Society for Bone and Mineral Research

topic calcification in soft tissues as a result of increasing the calcium–phosphate product. Moreover, oral phosphate may lead to an undesirable further elevation of PTH levels. Gastrointestinal tolerance is another limiting feature of this approach. In postmenopausal women, estrogen therapy remains an option in those women desiring hormone replacement for treatment of symptoms of menopause. The rationale for estrogen use in primary hyperparathyroidism is based on the known antagonism by estrogen of PTH-mediated bone resorption. Although the serum calcium concentration does tend to decline after estrogen administration (by 0.5 mg/dl), PTH levels and the serum phosphorous concentration do not change. Estrogen replacement may have a salutary effect on BMD in these patients as well. Preliminary data suggest that the selective estrogen receptor modulator, raloxifene, may have a similar effect on serum calcium levels in postmenopausal women with primary hyperparathyroidism. Bisphosphonates have also been considered as a possible medical approach to primary hyperparathyroidism. Alendronate improves vertebral BMD in patients with primary hyperparathyroidism who choose not to have surgery but does not affect the underlying disorder. Finally, a more targeted approach to the medical therapy of primary hyperparathyroidism is to interfere specifically with the production of PTH. Calcimimetic agents that alter the function of the extracellular calcium-sensing receptor offer an exciting new approach to primary hyperparathyroidism. These agents increase the affinity of the parathyroid cell calcium receptor for extracellular calcium, leading to increased intracellular calcium, a subsequent reduction in PTH synthesis and secretion, and ultimately a fall in the serum calcium. One such agent, cinacalcet HCl, is approved for lowering serum calcium in parathyroid cancer. Clinical trials in benign disease have shown normalization of serum calcium for up to 3 yr, but no change in BMD by DXA was documented. The effect of this agent on fracture incidence in patients with primary hyperparathyroidism is unknown.

ACKNOWLEDGMENTS

This work was supported in part by NIH Grants DK32333, DK074457, and DK66329.

SUGGESTED READING

Arnold A, Shattuck TM, Mallya SM, Krebs LJ, Costa J, Gallagher J, Wild Y, Saucier K 2002 Molecular pathogenesis of primary hyperparathyroidism. J Bone Miner Res **17:**N30–N36.

Bilezikian JP, Potts JT, El-Hajj Fuleihan G, Kleerekoper M, Neer R, Peacock M, Rastad J, Silverberg SJ, Udelsman R, Wells SA 2002 Summary statement from a workshop on asymptomatic primary hyperparathyroidism: A perspective for the 21st century. J Bone Miner Res **17:**N2–N11.

Bollerslev J, Jansson S, Mollerup CL, Nordenstrom J, Lundgren E, Torring O, Varhaug JE, Baranowski M, Aanderud S, Franco C, Freyschuss B, Isaksen GA, Ueland T, Rosen T 2007 Medical observation compared to parathyroidectomy for asymptomatic primary hyperparathyroidism: A prospective, randomized trial. J Clin Endocrinol Metab **92:**1687–1692.

Cetani F, Pinchera A, Pardi E, Cianferotti L, Vignali E, Picone A, Miccoli P, Viacava P, Marcocci C 1999 No evidence for mutations in the calcium-sensing receptor gene in sporadic parathyroid adenomas. J Bone Miner Res **14:**878–882.

Dempster DW, Müller R, Zhou H, Kohler T, Shane E, Parisien M, Silverberg SJ, Bilezikian JP 2007 Preserved three-dimensional cancellous bone structure in mild primary hyperparathyroidism. Bone **41:**19–24.

Doppman JL 2001 Preoperative localization of parathyroid tissue in primary hyperparathyroidism. In: Bilezikian JP, Marcus R, Levine MA (eds.) The Parathyroids, 2nd ed. Raven Press, New York, NY, USA, pp. 475–486.

Fitzpatrick LA 2001 Acute primary hyperparathyroidism. In: Bilezikian JP, Marcus R, Levine MA (eds.) The Parathyroids, 2nd ed. Raven Press, New York, NY, USA, pp. 527–360.

Kahn AA, Bilezikian JP, Kung A, Ahmed MM, Dubois SJ, Ho AYY, Schussheim DH, Rubin MR, Shaikh AM, Silverberg SJ, Standish TI, Syed Z, Syed ZA 2004 Alendronate in primary hyperparathyroidism: A double-blind, randomized, placebo-controlled trial. J Clin Endocrinol Metab **89:**3319–3325.

Lowe H, McMahon DJ, Rubin MR, Bilezikian JP, Silverberg SJ 2007 Normocalcemic primary hyperparathyroidism: Further characterization of a new clinical phenotype. J Clin Endocrinol Metab **92:**3001–3005.

Parisien MV, Silverberg SJ, Shane E, de la Cruz L, Lindasy R, Bilezikian JP, Dempster DW 1990 The histomorphometry of bone in primary hyperparathyroidism: Preservation of cancellous bone structure. J Clin Endocrinol Metab **70:**930–938.

Peacock M, Bilezikian JP, Klassen P, Guo MD, Turner SA, Shoback DS 2005 Cinacalcet maintains normocalcemia in patients with primary hyperparathyroidism. J Clin Endocrinol Metab **90:**135–141.

Rubin MR, Lee K, Silverberg SJ 2003 Raloxifene lowers serum calcium and markers of bone turnover in primary hyperparathyroidism. J Clin Endocrinol Metab **88:**1174–1178.

Shattuck TM, Välimäk S, Obara T, Gaz RD, Clark OH, Shoback D, Wierman ME, Tojo K, Robbins CM, Carpten JD, Farnebo LO, Larsson C, Andrew Arnold A 2003 Somatic and germ-line mutations of the *HRPT2* gene in sporadic parathyroid carcinoma. New Engl J Med 349:1722-1729

Silverberg SJ 2002 Non-classical target organs in primary hyperparathyroidism. J Bone Miner Res **17:**N117–N125.

Silverberg SJ, Bilezikian JP 2003 "Incipient" primary hyperparathyroidism: A "forme fruste" of an old disease. J Clin Endocrinol Metab **88:**5348–5352.

Silverberg SJ, Brown I, Bilezikian JP 2002 Youthfulness as a criterion for surgery in primary hyperparathyroidism. Am J Med **113:**681–684.

Silverberg SJ, Brown I, LoGerfo P, Gao P, Cantor T, Bilezikian JP 2003 Clinical utility of an immunoradiometric assay for whole PTH (1– 84) in primary hyperparathyroidism. J Clin Endocrinol Metab **88:**4725–4730.

Silverberg SJ, Locker FG, Bilezikian JP 1996 Vertebral osteopenia: A new indication for surgery in primary hyperparathyroidism. J Clin Endocrinol Metab **81:**4007–4012.

Silverberg SJ, McMahon DJ, Udesky J, Fleischer J, Shane E, Jacobs T, Siris E, Bilezikian JP 2006 Natural history of untreated mild primary hyperparathyroidism: Bone density after 15 years. J Bone Miner Res **21:**S118.

Silverberg SJ, Rubin MR, Faiman C, Peacock M, Shoback DM, Smallridge R, Schwanauer LE, Olson KA, Turner SA, Bilezikian JP 2007 Cinacalcet HCI effectively reduces the serum calcium concentration in parathyroid carcinoma. J Clin Endocrinol Metab **92:**3803–3808.

Silverberg SJ, Shane E, de la Cruz L, Dempster DW, Feldman F, Seldin D, Jacobs TP, Siris ES, Cafferty M, Parisien MV, Lindsay R, Clemens TL, Bilezikian JP 1989 Skeletal disease in primary hyperparathyroidism. J Bone Miner Res **4:**283–291.

Silverberg SJ, Shane E, Dempster DW, Bilezikian JP 1999 Vitamin D deficiency in primary hyperparathyroidism. Am J Med **107:**561–567.

Silverberg SJ, Shane E, Jacobs TP, Siris E, Bilezikian JP 1999 The natural history of treated and untreated asymptomatic primary hyperparathyroidism: A ten year prospective study. N Engl J Med **41:**1249–1255.

Tan M-H, Morrison C, Wang P, Yang X, Haven CJ, Zhang C, Zhao P, Tretiakova MS, Korpi-Hyovalti E, Burgess JR, Soo KC, Cheah W-K, Cao B, Resau J, Morreau H, Teh BT 2004 Loss of parafibromin immunoreactivity is a distinguishing feature of parathyroid carcinoma. Clin Cancer Res **10:**6629–6637.

Turken SA, Cafferty M, Silverberg SJ, de la Cruz L, Cimino C, Lange DJ, Lovelace RE, Bilezikian JP 1989 Neuromuscular involvement in mild, asymptomatic primary hyperparathyroidism. Am J Med **87:**553–557.

Udelsman R 2002 Surgery in primary hyperparathyroidism: The patient without previous neck surgery. J Bone Miner Res **17:**N126–N132.

Wells SA, Doherty GM 2001 The surgical management of primary hyperparathyroidism. In: Bilezikian JP, Marcus R, Levine MA (eds.) The Parathyroids, 2nd ed. Raven Press, New York, NY, USA, pp. 487–498.

Wermers RA, Khosla S, Atkinson EJ, Grant CS, Hodgson SF, O'Fallon WM, Melton LJ 1998 Survival after diagnosis of PHPT: A population based study. Am J Med **104:**115–122.

© 2008 American Society for Bone and Mineral Research

Chapter 67. Non-Parathyroid Hypercalcemia

Mara J. Horwitz, Steven P. Hodak, and Andrew F. Stewart

Division of Endocrinology and Metabolism, The University of Pittsburgh School of Medicine, Pittsburgh, Pennsylvania

PATHOPHYSIOLOGY OF HYPERCALCEMIA

As discussed elsewhere in the *Primer*, the normal total serum calcium concentration of 9.5 mg/dl can be divided into three components: ionized serum calcium (~4.2 mg/dl), serum calcium complexed to anions such as phosphate, sulfate, carbonate, etc., (~0.3 mg/dl), and calcium bound to serum proteins, principally albumin (~4.5 mg/dl). Hypercalcemia is defined as a serum calcium >2 SD above the normal mean in a given laboratory, commonly 10.6 mg/dl for total serum calcium and 1.25 mM for ionized serum calcium. There is no formal grading system for defining the severity of hypercalcemia. In general, however, serum calcium concentrations <12 mg/dl can be considered mild, those between 12 and 14 mg/dl are moderate, and those >14 mg/dl are severe.

The serum calcium concentration is tightly regulated by the flux of ionized serum calcium to and from four physiologic compartments: the skeleton, the intestine, the kidney, and serum binding proteins. Hypercalcemia, therefore, always results from an abnormality in calcium flux between the extracellular fluid (ECF) and one or a combination these compartments. Said another way, hypercalcemia can result only from one of four mechanisms: abnormal binding of calcium to serum proteins or abnormal flux of calcium into extracellular fluid from the gastrointestinal (GI) tract, the skeleton, or the kidney. Combinations of the above three mechanisms are common. Understanding hypercalcemia in this kind of mechanistic construct is critical for accurate diagnosis and is essential for effective treatment of hypercalcemia. For example, hypercalcemia from vitamin D intoxication or milk-alkali syndrome commonly arises from increased GI and renal absorption of calcium and would not be expected to respond to antiresorptive agents such as bisphosphonates. Conversely, humoral hypercalcemia of malignancy results principally from increased skeletal and renal calcium absorption and therefore is not influenced by restricting dietary calcium intake.

CLINICAL SIGNS AND SYMPTOMS OF HYPERCALCEMIA

Hypercalcemia raises the electrical potential difference across cell membranes and increases the depolarization threshold. Clinically, this is manifest as a spectrum of neurological symptoms ranging from mild tiredness to obtundation to coma. There is no precise serum calcium level that leads to impaired neurological function. Instead, the presence or absence and degree of neurological symptoms depends on the abruptness of onset of hypercalcemia, the age and underlying neurological status of the patient, and comorbidities and medications such as narcotics and neuroleptics.

Hypercalcemia acts directly at the nephron to prevent normal reabsorption of water, leading to a functional form of nephrogenic diabetes and polyuria. This may lead to thirst, prerenal azotemia, and significant dehydration, which are common clinical features of hypercalcemia. Hypercalcemia may also cause precipitation of calcium phosphate salts in the renal interstitium (nephrocalcinosis), vasculature, cardiac conduction system, the cornea (visible as so-called "band keratopathy"), and the gastric mucosa. Hypercalcemia may lead to renal failure from obstructive uropathy, nephrolithiasis, nephrocalcinosis, and prerenal causes, including dehydration and a reversible component of hypercalcemia-induced afferent arteriolar vasoconstriction.

Hypercalcemia can also lead to electrocardiographic abnormalities, the most specific of which is a prolonged Q-Tc interval. Hypercalcemia increases the depolarization threshold of skeletal and smooth muscle, making them more refractory to neuronal activation. Resulting decreases in muscle contraction manifest clinically as skeletal muscle weakness and constipation. Nausea, anorexia, vomiting, and flushing are also common. Finally, hypercalcemia may lead to abdominal pain and pancreatitis.

DISORDERS THAT LEAD TO HYPERCALCEMIA

The complete differential diagnosis of hypercalcemia is shown in Table 1. In this chapter, we consider non-parathyroid causes of hypercalcemia. The PTH-dependent family of hypercalcemic disorders including primary and tertiary hyperparathyroidism, their inherited variants, and familial hypocalciuric hypercalcemia, also known as familial benign hypercalcemia, are discussed elsewhere in the *Primer*.

Cancer

Malignancy-associated hypercalcemia (MAHC) accounts for ~90% of hypercalcemia encountered among hospitalized patients. The first patient with MAHC was reported in 1921, immediately after the development of clinical methods to measure serum calcium. MAHC can be subdivided into four mechanistic subtypes: humoral hypercalcemia of malignancy (HHM); local osteolytic hypercalcemia (LOH); $1{,}25(OH)_2$ vitamin D–induced hypercalcemia; and authentic ectopic hyperparathyroidism.

HHM. HHM is the most common form of MAHC, accounting for ~80% of subjects in large series of unselected patients with MAHC. HHM results from the secretion of PTH-related protein (PTHrP) by HHM-associated tumors. Whereas almost any kind of tumor may cause HHM, the most common types are squamous carcinomas of any origin (lung, esophagus, skin, and cervix are common sites) and breast and renal carcinomas.

HHM was first described in the 1940s and 1950s in patients in whom hypercalcemia associated with cancer was corrected after successful removal of the responsible tumor. Investigators deduced the humoral nature of the syndrome from these events, but the responsible "humor" was not identified until 1987, when PTHrP was purified and sequenced, and its cDNA and gene were cloned. We now understand that PTHrP leads to a dramatic uncoupling of bone resorption from formation, by activating osteoclastic bone resorption and suppressing osteoblastic bone formation (Fig. 1B). As a result, enormous net quantities of calcium of up to 700–1000 mg/d leave the skeleton, typically causing marked hypercalcemia. In addition, the anticalciuric effects of PTHrP prevent or restrict effective renal calcium clearance. Finally, HHM is associated with reductions

The authors state that they have no conflicts of interest.

Key words: hypercalcemia, cancer, medications, granulomatous disease, hyperthyroidism, vitamin D toxicity, milk-alkali syndrome, humoral hypercalcemia of malignancy, osteolytic hypercalcimia, immobilization

© 2008 American Society for Bone and Mineral Research

Table 1. Differential Diagnosis of Hypercalcemia

PTH-dependent hypercalcemia*
Cancer
HHM
LOH*
$1,25(OH)_2$ vitamin D and lymphoma/dysgerminoma
Authentic ectopic PTH secretion
Other
Granulomatous disorders
Endocrine disorders
Immobilization
Milk-alkali syndrome
Total parenteral nutrition
Abnormal protein binding
Medications
Vitamin D
Vitamin A
Lithium
PTH
Estrogen/SERMs
Aminophylline and theophylline
Foscarnet
Growth hormone
8-chloro-cyclic AMP
Chronic and acute renal failure
End stage liver disease
Manganese intoxication
Fibrin glue
Hypophosphatemia
Pediatric syndromes*

*This topic is covered in another section of the *Primer*.

in circulating $1,25(OH)_2D$ levels, which in turn limits intestinal calcium absorption. Thus, in pathophysiologic terms, HHM results from enhanced skeletal resorption coupled with an inability to clear calcium through the kidney.

HHM is also associated with a reduction in the renal phosphorus threshold, which results in phosphaturia and hypophosphatemia. The HHM syndrome is also characterized by a marked increase in cyclic AMP excretion by the kidney, termed "nephrogenous cyclic AMP" or "NcAMP." The bone resorption, increased renal tubular calcium resorption, phosphaturia, and increase in NcAMP excretion that characterize HHM reflect interaction of circulating PTHrP with the common PTH/PTHrP receptor in the skeleton and the renal tubule. Surprisingly, HHM syndrome is also associated with paradoxical reductions in $1,25(OH)_2D$ and osteoblastic bone formation (Fig. 1B). These observations stand in striking contrast to primary hyperparathyroidism (HPT), in which $1,25(OH)_2D$ is increased, and both osteoblast and osteoclast activities are increased but remain coupled (Fig. 1A). Why PTH and PTHrP, which act through the identical receptor, should produce directionally opposite physiological effects, remains unexplained despite the initial description of these differences almost 30 yr ago.

Bone scans, bone biopsies, and autopsy show few or no skeletal metastases in patients with HHM. This finding emphasizes the humoral nature of the syndrome and stands in contrast to patients with LOH, described below.

LOH. The tumors that most commonly produce LOH are breast cancer and hematological neoplasms (myeloma, lymphoma, leukemia) in subjects with widespread skeletal involvement. By the 1940s, series of patients with hypercalcemia associated with these malignancies were reported, all of which documented extensive marrow invasion by the tumor. In the 1960s and 1970s, these malignancies were shown to be associated with marked activation of osteoclasts adjacent to the sites of marrow infiltration by the malignancy. Earlier in this era, it was widely believed that hypercalcemia in the majority of patients with MAHC resulted from LOH. However, today, large series have shown that LOH accounts for only ~20% of patients with MAHC. Indeed, evidence suggests that the frequency of MAHC caused by LOH may be decreasing with the widespread use of bisphosphonates to prevent skeletal fractures, metastases, and pain in patients with myeloma and breast cancer.

In the 1970s, several authors began to search for the "osteoclast-activating factors" (OAFs) responsible for LOH. Locally produced osteoclast-activating cytokines are now known to include interleukins-1 and -6, PTHrP, and macrophage inflammatory protein-1α. This area has been reviewed elsewhere in the *Primer*.

Patients with LOH are characterized at bone biopsy or autopsy by extensive skeletal metastases or marrow infiltration (Fig. 1C). Bone scintigraphic scans are generally intensely and widely positive in patients with metastatic disease from solid tumors but may be completely negative despite extensive marrow involvement in patients with multiple myeloma, reflecting a reduction on bone formation.

In mechanistic terms, LOH can be thought of as primarily a resorptive (skeletally derived) form of hypercalcemia in which massive removal of calcium from the skeleton exceeds the normal ability of the kidney to clear calcium. As the dehydration associated with such marked hypercalcemia occurs, the hypercalcemia is exacerbated by a typical decline in renal function as well.

$1,25(OH)_2D$-Induced Hypercalcemia. In the 1970s, reports began to occur of patients with lymphomas in whom hypercalcemia occurred as a result of increased production of $1,25(OH)_2D$. There have been ~60 patients described to date with this syndrome. There is no particular histopathological correlation: all types of lymphomas have been reported to cause this syndrome. In the past several years, the same syndrome has been reported in patients with ovarian dysgerminomas.

The primary pathophysiological abnormality in this syndrome is that the malignant cells or adjacent normal cells overexpress the enzyme 1-α-hydroxylase that converts normal circulating quantities of the precursor 25(OH)D to abnormally elevated circulating concentrations of the active form of vitamin D: $1,25(OH)_2D$. This pathophysiology can be viewed as the malignant counterpart of events that occur in sarcoidosis (see section on Granulomatous Diseases below). Because $1,25(OH)_2D$ activates intestinal calcium absorption, this syndrome is principally an absorptive form of hypercalcemia, although decreased renal clearance of calcium may also develop as a consequence of the dehydration caused by the hypercalcemia.

Authentic Ectopic Hyperparathyroidism. From the 1950s through the 1980s, when the HHM syndrome finally was shown to be caused by PTHrP, the hypercalcemia of HHM was widely ascribed to "ectopic secretion of PTH" by offending neoplasms. With the demonstration in the 1980s that the responsible factor in HHM was PTHrP, not PTH, authentic ectopic secretion of PTH by cancers was believed not to occur. This changed in the 1990s with the description of rare cases in which production of authentic PTH, and not PTHrP, by tumors was shown to cause MAHC. At the time of this writing, ~10 cases have been reported in which convincing evidence

© 2008 American Society for Bone and Mineral Research

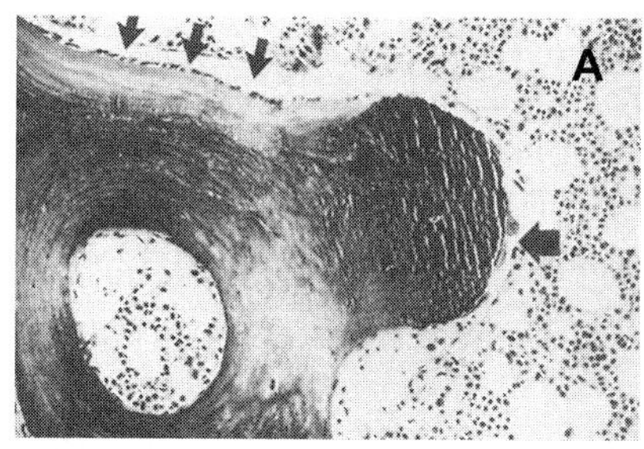

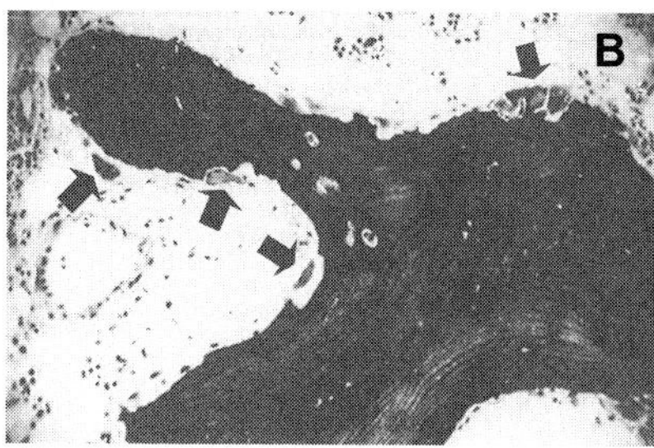

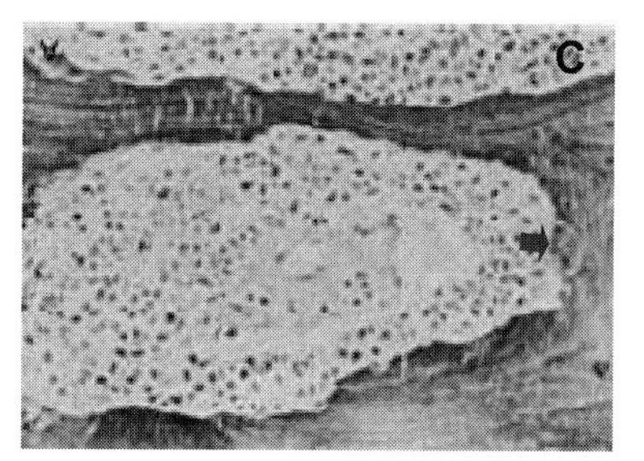

FIG. 1. Comparison of bone histology in a patient with HPT (A), HHM (B), and LOH caused by leukemia (C). In HPT, HHM, and LOH, osteoclastic activity is accelerated (large, thick arrows), although it is higher in HHM than in HPT. In HPT, osteoblastic activity (thin arrows) and osteoid are increased, but both are markedly decreased in HHM and LOH. This uncoupling of formation from resorption in HHM and LOH plays the major rule in causing hypercalcemia. (Reprinted with permission from Stewart AF, Vignery A, Silvergate A, Ravin ND, LiVolsi V, Broadus AE, Baron R 1982 Quantitative bone histomorphometry in humoral hypercalcemia of malignancy: uncoupling of bone cell activity. J Clin Endocrinol Metab **55**:219–227.) Copyright 1982, The Endocrine Society

exists for hypercalcemia resulting from authentic ectopic secretion of PTH from malignant tumors. Thus, authentic ectopic secretion of PTH does exist, but it is very rare.

Other Mechanisms for MAHC. The four categories described above account for >99% of patients with MAHC. Occasionally, however, patients who do not fit any of these categories have been described. For example, there are rare case reports in which none of the above scenarios could be invoked and in which elevated circulating concentrations of prostaglandin E2 may have been responsible.

Granulomatous Diseases

Almost every single disease associated with granuloma formation has been reported to cause hypercalcemia. The most common is sarcoidosis, but tuberculosis (both from *M. bovis* and *Mycobacterium avium*), berylliosis, histoplasmosis, coccidiomycosis, *Pneumocystis*, inflammatory bowel disease, histocytosis X, foreign body granulomas, and granulomatous leprosy have all been associated with the syndrome. In the case of sarcoidosis, ~10% of patients become hypercalcemic and 20% become hypercalciuric during the course of their disease.

The mechanism in sarcoidosis and tuberculosis is inappropriate production of $1,25(OH)_2D$ by the granulomas as a result of increased activity of 1-α-hydroxylase, the enzyme that converts 25(OH)D to its active form, $1,25(OH)_2D$. The reasons for this are unknown, but this results in elevated circulating concentrations of $1,25(OH)_2D$, which in turn lead to intestinal hyperabsorption of calcium, hypercalciuria, and ultimately hypercalcemia.

The syndrome reverses with the eradication of granulomas (e.g., by glucocorticoids or anti-tuberculous medications and by oral or intravenous hydration coupled with lowering dietary intake of vitamin D and calcium). Because sunlight is a source of vitamin D, sunlight exposure should be reduced.

Other Endocrine Disorders

Hyperparathyroidism is the classical endocrine disorder associated with hypercalcemia, but there are four other endocrine disorders that may cause hypercalcemia as well.

Hyperthyroidism has been reported to cause increases in ionized or total serum calcium in up to 50% of affected patients. In general, the hypercalcemia is mild (range, 10.7–11.0 mg/dl), but it has been reported to be as high as 13 mg/dl in rare cases. It is believed to result from increases in osteoclastic bone resorption.

Addisonian Crisis has also been reported to cause hypercalcemia with increases in both ionized as well as total calcium. In general, hypercalcemia is mild and responds to standard therapy for hypoadrenalism (fluid resuscitation and intravenous glucocorticoids). The cause is not known but at least in some cases may be caused by the underlying disorder, such as tuberculosis, that led to the hypoadrenalism. In others, it is possible that the associated volume contraction may have led to factitious increases in total serum calcium through relative hyperalbuminemia. In such cases, although the measured concentration of total serum calcium is elevated, ionized serum calcium remains normal. However, at least in some cases, ionized serum calcium has been reported to have been elevated.

Pheochromocytoma has been associated with hypercalcemia. In some case, the hypercalcemia is caused by primary hyperparathyroidism in the setting of multiple endocrine neoplasia type 2, and hypercalcemia corrects with parathyroid surgery. In some patients, however, hypercalcemia reverses with removal of the pheochromocytoma, and some of these tumors have been shown to secrete PTHrP. In other cases, it has been suggested that catecholamine secretion by the pheochromocytoma is sufficient to activate bone resorption.

The VIP-oma syndrome is caused by vasoactive intestinal polypeptide (VIP) secretion by pancreatic islet or other neuroendocrine tumors and is associated with severe watery diarrhea ("pancreatic cholera"), hypokalemia, and achlorhydria (the WDHA syndrome). Interestingly, 90% of patients with this rare syndrome have been reported to be hypercalcemic, although the mechanism is unknown. VIP has been shown to stimulate osteoclastic bone resorption in vitro, suggesting at least one potential mechanism.

Immobilization in association with another cause of high bone turnover (such as the high turnover associated with youth, hyperparathyroidism, myeloma or breast cancer with bone metastases, and Paget's disease) may cause hypercalcemia. The classic examples are the hypercalcemia that was prevalent in the polio epidemics and that still regularly accompanies paraplegia and quadriplegia. The age dependence of this phenomenon is evident from the observation that hyper-

© 2008 American Society for Bone and Mineral Research

calcemia never occurs in elderly subjects with strokes that result in complete immobilization, yet regularly occurs in children and young adults with similar degrees of immobilization from spinal cord injury or astronauts during space flight.

Immobilization suppresses osteoblastic bone formation and markedly increases osteoclastic bone resorption, leading to complete uncoupling of these two normally tightly coupled processes. The result is massive loss of calcium from the skeleton, with resultant hypercalcemia and reductions in BMD. The process is most effectively reversed by restoration of normal weight-bearing mobility. Alternate options are bisphosphonates and measures to increase renal calcium clearance (hydration and loop diuretics).

Milk-Alkali Syndrome

Originally described in 1949, this syndrome was initially described in patients who developed moderate or severe hypercalcemia when treated with large amounts of milk (several quarts or gallons per day) and absorbable antacids (e.g., baking soda, or sodium bicarbonate) for peptide ulcer disease. Additional features of the syndrome are metabolic alkalosis caused by antacid ingestion and renal failure caused by hypercalcemia.

Contemporary reports occur primarily in patients taking large amounts of calcium carbonate for peptic ulcer or esophageal reflux symptoms. Because calcium is absorbed with only moderate efficiency, normal dietary intake (800–2000 mg/d) does not cause hypercalcemia. In contrast, daily doses in excess of 4000 mg/d can induce hypercalciuria and hypercalcemia in normal adults. Indeed, in many case reports, the daily intake has been in the range of 10–20 g of elemental calcium per day. Because one standard antacid tablet contains ~120–200 mg of elemental calcium, and each packet or roll contains some 10 tablets, it is clear that subjects must consume multiple packages of antacids for this syndrome to develop. The hypercalcemia reverses with hydration and correction of excessive calcium ingestion. Renal damage, however, may be permanent.

Total Parenteral Nutrition

Patients with short bowel syndrome or are otherwise unable to eat normally by mouth, who receive chronic total parenteral nutrition (TPN), have been reported to develop hypercalcemia. In some cases, it was a result of the inclusion of excessive amounts of calcium or vitamin D supplements in early TPN solutions. In other early cases, it seems to have been associated with the use of collagen lysates contaminated with aluminum. Case reports have been rare in recent years, but the etiology of recent cases is uncertain.

Abnormal Protein Binding

Hypercalcemia may be "artifactual" or "factitious" in some settings. In general, this refers to situations in which the total serum calcium is elevated but the ionized serum calcium is normal. In one example, severe dehydration may lead to increases in serum albumin concentration and in the albumin-bound component of total serum calcium. This results in an increase in total but not ionized serum calcium. This can be suspected in the setting of volume contraction with hyperalbuminemia and confirmed by measurement of ionized serum calcium.

An analogous situation has been reported in subjects with multiple myeloma or Waldenstrom's macroglobulinemia, whose monoclonal immunoglobulin specifically recognizes calcium ion. In these cases, patients have displayed severe increases in total serum calcium in the absence of neurologic or EKG abnormalities and in the absence of symptoms or signs of hypercalcemia. In these cases, the ionized serum calcium was found to be normal, as was the urinary calcium excretion, and the patients' immunoglobulins were shown to have an abnormal affinity for serum calcium. Treatment with agents to lower serum calcium such as mithramycin precipitated hypocalcemic seizures, despite a total serum calcium that remained in or above the normal range.

Medications

A number of medications may cause hypercalcemia. Calcium containing antacids are included above in the section on milk-alkali syndrome.

Vitamin D intoxication from standard vitamin D preparations has been reported in association with inappropriate addition by dairies or manufacturers of vitamin D to milk or to infant formula. Vitamin D intoxication may also occur with use of doses of vitamin D in excess of 50,000 units two or three times per week as a treatment for hypoparathyroidism or metabolic bone diseases such as osteoporosis. Vitamin D analogs such as calcitriol [1,25$(OH)_2$D] used in the treatment of hypoparathyroidism, chronic renal failure, and metabolic bone disease may also cause hypercalcemia. The mechanism in all of the above is a combination of increased intestinal calcium absorption and bone resorption induced by vitamin D, together with reductions in renal ability to clear calcium as a result of dehydration.

Vitamin A intoxication can cause hypercalcemia. This may occur through the excessive use of vitamin supplements, or, in Antarctic explorers, as a result of eating sled dog livers. More recently, the use of retinoic acid derivatives for the treatment of dermatologic disorders or as chemotherapy agents has also been associated with induction of hypercalcemia.

Thiazide diuretics such as hydrochlorothiazide or chlorthalidone commonly cause mild hypercalcemia. This has been ascribed to their ability to induce distal renal tubular calcium reabsorption, although it has been reported to occur in anephric subjects on dialysis, suggesting additional mechanisms.

Lithium-Induced Hypercalcemia. Lithium has been reported to cause hypercalcemia in as many as 5% of patients. It has been suggested that lithium may actually induce parathyroid hyperplasia or induce parathyroid adenomas, but it is also possible that the coincidence of lithium use with hyperparathyroidism may represent the simultaneous occurrence of two common clinical syndromes. There are also well-documented patients treated with lithium whose hypercalcemia reversed with cessation of lithium therapy and in whom hypercalcemia therefore was clearly lithium induced. Proposed mechanisms include lithium-induced activation of PTH secretion and lithium-induced stimulation of renal calcium reabsorption, both of which have been documented in vitro or in animals. Whether these are responsible for the hypercalcemia that occurs in humans is not certain.

PTH, both the PTH(1-34) and PTH(1-84) forms used for treatment of osteoporosis, is associated with hypercalcemia in a substantial minority of patients so treated. In general, it is mild and requires little or no treatment or a reduction in the dose of PTH or supplemental calcium, but it can be severe and require discontinuation of PTH therapy.

Other medications that are known to cause hypercalcemia with no known mechanism include estrogens and the selective estrogen receptor modifier (SERM) tamoxifen in women with breast cancer and extensive skeletal metastatic disease; aminophylline and theophylline when used in large, supratherapeutic doses in subjects with bronchospastic disease; Foscarnet, an antiviral agent used in HIV/AIDS; growth hormone

© 2008 American Society for Bone and Mineral Research

treatment in subjects with severe burns and also in subjects with HIV/AIDS; and 8-chloro-cyclic AMP used as an anticancer agent.

Acute and Chronic Renal Failure

The recovery phase from acute renal failure caused by rhabdomyolysis has been associated with hypercalcemia. Typically, this follows an episode of severe hyperphosphatemia and hypocalcemia in the acute, oliguric phase, accompanied by severe secondary hyperparathyroidism. It has been ascribed to residual effects of PTH on bone turnover, as well as release of calcium phosphate precipitated into soft tissues such as skeletal muscle during the early hypocalcemic, hyperphosphatemic phase.

Chronic renal failure and dialysis are associated with hypercalcemia. Frequently, it is associated with the use of calcitriol or other vitamin D analogs used to prevent secondary hyperparathyroidism or with the use of oral calcium binding agents and supplements. Hypercalcemia in this population may also result from tertiary hyperparathyroidism, as discussed elsewhere in the *Primer*.

Hypophosphatemia

Severe dietary phosphate deprivation in rats causes hypophosphatemia associated with hypercalcemia. This has not been documented in humans but merits attention when considering the management of patients with hypercalcemia, because moderate to severe hypophosphatemia commonly accompanies hyperparathyroidism, cancer hypercalcemia, and other disorders that cause hypercalcemia. It is the authors' anecdotal experience that hypercalcemia may be refractory to treatment in the presence of severe hypophosphatemia but responds nicely to appropriate measures once the hypophosphatemia has been corrected.

Miscellaneous

Several other disease and syndromes associated with hypercalcemia include end-stage liver disease in patients awaiting liver transplantation; manganese intoxication in people exposed to well water contaminated with manganese derived from improperly disposed batteries; fibrin glue when used to treat refractory recurrent pneumothorax in children; and other pediatric syndromes that are covered in more detail elsewhere in the *Primer*.

MANAGEMENT

Management of hypercalcemia optimally targets the underlying pathophysiology. For example, removal of a parathyroid adenoma, discontinuing or reducing the dose of an offending medication such as vitamin D or PTH, or eradication of a tumor responsible for hypercalcemia with chemotherapy or surgery would be optimal management strategies. Of course, sometimes these are not possible or therapy must be begun before a definitive diagnosis is made. In these cases, targeting the underlying pathophysiology is most appropriate. Thus, in patients whose hypercalcemia is primarily based on accelerated bone resorption (e.g., LOH, HHM, immobilization), therapy should include agents that block bone resorption, such as the intravenous bisphosphonates zoledronate or pamidronate. In patients whose hypercalcemia is principally GI in origin [e.g., sarcoidosis, milk-alkali syndrome, vitamin D intoxication, $1,25(OH)_2D$-secreting lymphomas], reducing or eliminating oral calcium intake and vitamin D and sunlight may be most appropriate. For those with important renal contributions to hypercalcemia (e.g., dehydration), increasing renal calcium clearance by increasing the glomerular filtration rate (GFR) using saline infusions and blocking renal calcium absorption using loop diuretics such as furosemide are appropriate. Of course, many patients have contributions from several sources (e.g., dehydration plus increased bone resorption in LOH, increased GI calcium absorption plus dehydration in patients with sarcoidosis, or a combination of PTHrP-induced bone resorption plus PTHrP-induced renal calcium retention in HHM), and optimal therapy targets each of these components.

For hyperparathyroidism and its variants, specific therapy is discussed elsewhere in the *Primer*. For patients with cancer, the most effective long-term therapy is tumor eradication. If that is not possible, or while waiting for a response to chemotherapy, aggressive hydration with saline, keeping a careful watch for signs of congestive heart failure, accompanied by a loop diuretic such as furosemide is appropriate. Limiting oral calcium intake is not important in HHM and LOH, because intestinal calcium absorption is already low as a result of the low $1,25(OH)_2D$ concentrations in these patients and because cachexia is a common feature of these patients. On the other hand, in patients with $1,25(OH)_2D$-induced hypercalcemia from lymphoma or dysgerminoma, reducing oral calcium and vitamin D intake is important. Whereas some physicians wait to see the magnitude of the decline in serum calcium induced by hydration and diuresis, the authors' practice, when the serum calcium exceeds 12.0 mg/dl, is to institute antiresorptive therapy with an intravenous bisphosphonate such as zoledronate or pamidronate soon after the discovery of hypercalcemia. Specific doses and regimens have been reviewed in detail recently.

For the granulomatous diseases, correcting the underlying cause is critical, where possible (e.g., tuberculosis). In sarcoid, limiting calcium and vitamin D intake and sun exposure are important, together with oral or parenteral hydration. Glucocorticoid therapy may be necessary to treat the granulomas, to lower intestinal calcium absorption, and to lower $1,25(OH)_2D$ concentrations.

For immobilization-induced hypercalcemia, weight-bearing ambulation is the mainstay of therapy. Often, however, this is not possible because of spinal cord injury or pain. Here, intravenous bisphosphonates are effective and important.

For the remainder of the diagnoses in Table 1, correcting the underlying disorder or withdrawing or reducing the dose of the offending medication corrects the serum calcium.

ACKNOWLEDGMENTS

This work was supported by NIH Grants DK51081 and DK073039.

SUGGESTED READING

Adams JS, Gacad MA 1985 Characterization of 1 hydroxylation of vitamin D_3 sterols by cultured alveolar macrophages from patients with sarcoidosis. J Exp Med **161:**755–765.

Barbour GL, Coburn JW, Slatopolsky E, Norman AW, Horst RL 1981 Hypercalcemia in an anephric patient with sarcoidosis: Evidence for extrarenal generation of 1,25-dihydroxyvitamin D. N Engl J Med **305:**440–443.

Beall DP, Scofield RH 1995 Milk-alkali syndrome associated with calcium carbonate consumption. Medicine **74:**89–96.

Bergstrom WH 1978 Hypercalciuria and hypercalcemia complicating immobilization. Am J Dis Child **132:**553–554.

Burtis WJ, Brady TG, Orloff JJ, Ersbak JB, Warrell RP, Olson BR, Wu TL, Mitnick MA, Broadus AE, Stewart AF 1990 Immunochemical characterization of circulating parathyroid hormone-related protein in patients with humoral hypercalcemia of malignancy. N Engl J Med **322:**1106–1112.

© 2008 American Society for Bone and Mineral Research

Chandra SV, Shukla GS, Srivastava RS 1981 An exploratory study of manganese exposure to welders. Clin Toxicol **18:**407–416.

Chappard D, Minaire P, Privat C, Berard E, Mendoza-Sarmiento J, Tournebise H, Basle MH, Audran W, Rebel A, Picot C 1995 Effects of tiludronate on bone loss in paraplegic patients. J Bone Miner Res **10:**112–118.

Elfatih A, Anderson NR, Fahie-Wilson MN, Gama R 2007 Pseudo-pseudohypercalcaemia, apparent primary hyperparathyroidism and Waldenström's macroglobulinaemia. J Clin Pathol **60:**436–437.

Evans KN, Taylor H, Zehnder D, Kilby MD, Bulmer JN, Shah F, Adams JS, Hewison M 2004 Increased expression of 25-hydroxyvitamin D-1alpha-hydroxylase in dysgerminomas: A novel form of humoral hypercalcemia of malignancy. Am J Pathol **165:**807–813.

Gayet S, Ville E, Durand JM, Mars ME, Morange S, Kaplanski G, Gallais H, Soubeyrand J 1997 Foscarnet-induced hypercalcemia in AIDS. AIDS **11:**1068–1070.

Gerhardt A, Greenberg A, Reilly JJ, Van Thiel DH 1987 Hypercalcema complication of advanced chronic liver disease. Arch Intern Med **147:**274–277.

Ghaferi AA, Chojnacki KA, Long WD, Cameron JL, Yeo CJ 2008 Pancreatic VIPomas: Subject Review and One Institutional Experience. J Gastrointest Surg **12:**382–393.

Gkonos PJ, London R, Hendler ED 1984 Hypercalcemia and elevated 1,25-dihydroxyvitamin D levels in a patient with end stage renal disease and active tuberculosis. N Engl J Med **311:**1683–1685.

Haden ST, Stoll AL, McCormick S, Scott J, Fuleihan GE 1979 Alterations in parathyroid dynamics in lithium-treated subjects. J Clin Endocrinol Metab **82:**2844–2848.

Holick MF, Shao Q, Liu WW, Chen TC 1992 The vitamin D content of fortified milk and infant formula. N Engl J Med **326:**1178–1181.

Horwitz MJ, Tedesco MB, Sereika SM, Syed MA, Garcia-Ocaña A, Bisello A, Hollis BW, Rosen CJ, Wysolmerski JJ, Dann P, Gundberg CM, Stewart AF 2005 Continuous infusion of parathyroid hormone versus parathyroid hormone-related protein in humans: Discordant effects on $1{,}25(OH)_2$ vitamin D and prolonged suppression of bone formation. J Bone Miner Res **20:**1792–1803.

Jacobus CH, Holick MF, Shao Q, Chen TC, Holm IA, Kolodny JM, Fuleihan GE, Seely EW 1992 Hypervitaminosis D. associated with drinking milk. N Engl J Med **326:**1173–1177.

Klein GL, Horst RL, Norman AW, Ament ME, Slatopolsky E, Coburn JW 1981 Reduced serum levels of la, 25-dihydroxyvitamin D during long-term total parenteral nutrition. Ann Intern Med **94:**638–643.

Knox JB, Demling RH, Wilmore DW, Sarraf P, Santos AA 1995 Hypercalcemia associated with the use of human growth hormone in an adult surgical intensive care unit. Arch Surg **130:**442–445.

Llach F, Felsenfeld AJ, Haussler MR 1981 The pathophysiology of altered calcium metabolism in rhabdomyolysis-induced acute renal failure. N Engl J Med **305:**117–123.

McPherson ML, Prince SR, Atamer E, Maxwell DB, Ross-Clunis H, Estep H 1986 Theophylline-induced hypercalcemia. Ann Intern Med **105:**52–54.

Merlini G, Fitzpatrick LA, Siris ES, Bilezikian JP, Birken A, Beychok A, Osserman EF 1984 A human myeloma immunoglobulin G binding four moles of calcium associated with asymptomatic hypercalcemia. J Clin Immunol **4:**185–196.

Miller PD, Bilezikian JP, Diaz-Curiel M, Chen P, Marin F, Krege JH, Wong M, Marcus R 2007 Occurrence of hypercalciuria in patients with osteoporosis treated with teriparatide. J Clin Endocrinol Metab **92:**3535–3541.

Muls E, Bouillon R, Boelaert J, Lamberigts G, Van Imschool S, Daneels P, DeMoor P 1982 Etiology of hypercalcemia in a patient with Addison's disease. Calcif Tissue Int **34:**523–526.

Mune T, Katakami H, Kato Y, Yasuda K, Matsukura S, Miura K 1993 Production and secretion of parathyroid hormone–related protein in pheochromocytoma: Participation of an α-adrenergic mechanism. J Clin Endocrinol Metab **76:**757–762.

Nussbaum SR, Gaz RD, Arnold A 1990 Hypercalcemia and ectopic secretion of PTH by an ovarian carcinoma with rearrangement of the gene for PTH. N Engl J Med **323:**1324–1328.

Orwoll ES 1982 The milk-alkali syndrome: Current concepts. Ann Intern Med **97:**242–248.

Ott SM, Maloney NA, Klein GL, Alfrey AC, Ament ME, Lobourn JW 1983 Aluminum is associated with low bone formation in patients receiving chronic parenteral nutrition. Ann Intern Med **96:**910–914.

Parker MS, Dokoh S, Woolfenden JM, Buchsbaum HW 1984 Hypercalcemia in coccidioidomycosis. Am J Med **76:**341–343.

Porter RH, Cox BG, Heaney D, Hostetter TH, Stinebaugh BJ, Suki WN 1978 Treatment of hypoparathyroid patients with chlorthalidone. N Engl J Med **298:**577.

Roodman GD 2004 Mechanisms of bone metastasis. N Engl J Med **350:**1655–1664.

Rosen HN, Moses AC, Gundberg C, Kung VT, Seyedin SM, Chen T, Holick M, Greenspan SL 1993 Therapy with parenteral pamidronate prevents thyroid hormone-induced bone turnover in humans. J Clin Endocrinol Metab **77:**664–669.

Rosenthal N, Insogna KL, Godsall JW, Smaldone L, Waldron JA, Stewart AF 1985 Elevations in circulating 1,25 dihydroxyvitamin D in three patients with lymphoma-associated hypercalcemia. J Clin Endocrinol Metab **60:**29–33.

Ross DS, Nussbaum SR 1989 Reciprocal changes in parathyroid hormone and thyroid function after radioiodine treatment of hyperthyroidism. J Clin Endocrinol Metab **68:**1216–1219.

Sakoulas G, Tritos NA, Lally M, Wanke C, Hartzband P 1997 Hypercalcemia in an AIDS patient treated with growth hormone. AIDS **11:**1353–1356.

Sarkar S, Hussain N, Herson V 2003 Fibrin Glue for Persistent Pneumothorax in Neonates. J Perinatol **23:**82–84.

Saunders MP, Salisbury AJ, O'Byrne KJ, Long L, Whitehouse RM, Talbot DC, Mawer EB, Harris AL 1997 A novel cyclic adenosine monophosphate analog induces hypercalcemia via production of 1,25-dihydroxyvitamin D in patients with solid tumors. J Clin Endocrinol Metab **83:**4044–4048.

Stewart AF 2005 Hypercalcemia Associated with Cancer. N Engl J Med **352:**373–379.

Stewart AF, Adler M, Byers CM, Segre GV, Broadus AE 1982 Calcium homeostasis in immobilization: An example of resorptive hypercalciuria. N Engl J Med **306:**1136–1140.

Stewart AF, Broadus AE 2006 Malignancy-associated hypercalcemia. In: DeGroot L, Jameson LJ (eds.) Endocrinology, 5th ed. Saunders, Philadelphia, PA, USA, pp. 1555–1565.

Stewart AF, Hoecker J, Segre GV, Mallette LE, Amatruda T, Vignery A 1985 Hypercalcemia in pheochromocytoma: Evidence for a novel mechanism. Ann Intern Med **102:**776–779.

Stewart AF, Horst R, Deftos LJ, Cadman EC, Lang R, Broadus AE 1980 Biochemical evaluation of patients with cancer-associated hypercalcemia: Evidence for humoral and non-humoral groups. N Engl J Med **303:**1377–1383.

Stewart AF, Vignery A, Silvergate A, Ravin ND, LiVolsi V, Broadus AE, Baron R 1982 Quantitative bone histomorphometry in humoral hypercalcemia of malignancy: Uncoupling of bone cell activity. J Clin Endocrinol Metab **55:**219–227.

Valente JD, Elias AN, Weinstein GD 1983 Hypercalcemia associated with oral isotretinoin in the treatment of severe acne. JAMA **250:**1899.

Valentin-Opran A, Eilon G, Saez S, Mundy GR 1985 Estrogens stimulate release of bone-resorbing activity in cultured human breast cancer cells. J Clin Invest **72:**726–731.

Vasikaran SD, Tallis GA, Braund WJ 1994 Secondary hypoadrenalism presenting with hypercalcaemia. Clin Endocrinol (Oxf) **41:**261–265.

Verner JV, Morrison AB 1974 Endocrine pancreatic islet disease with diarrhea. Arch Intern Med **133:**492–500.

Villablanca J, Khan AA, Avramis VI, Seeger RC, Matthay KC, Ramsay NK, Reynolds CP 1995 Phase I trail of 13-*cis*-retinoic acid in children with neuroblastoma following bone marrow transplantation. J Clin Oncol **13:**894–901.

Wermers RA, Kearns AE, Jenkins GD, Melton LJ III 2007 Incidence and clinical spectrum of thiazide-associated hypercalcemia. Am J Med **120:**911.e9–911.e15.

© 2008 American Society for Bone and Mineral Research

Chapter 68. Hypocalcemia: Definition, Etiology, Pathogenesis, Diagnosis, and Management

Dolores Shoback

University of California, San Francisco, California; Endocrine Research Unit, San Francisco, California; Department of Veterans Affairs Medical Center, San Francisco, California

DEFINITION

Hypocalcemia or the presence of a low ionized calcium (Ca^{2+}) concentration, defined as a level that falls below the lower limit of the normal range, usually 1.00–1.25 mM, is a commonly encountered clinical problem with multiple causes. The level of ionized Ca^{2+} is a critical determinant in many vital cellular functions including secretion, skeletal and cardiac muscle contraction, cardiac conduction, blood clotting, and neurotransmission. Approximately 50% of the total serum Ca^{2+} is in the ionized fraction, with the remainder being protein-bound (45–50%), predominantly to albumin, or complexed to circulating anions (<5%). Under most clinical circumstances, total serum Ca^{2+} is the only value that the clinician has when making an initial determination of the state of serum Ca^{2+} homeostasis in the patient. Ionized Ca^{2+} determinations are not routine measurements in most clinical settings. Therefore, the clinician must make the first assessment based on total serum Ca^{2+} levels.

The total serum Ca^{2+} concentration is a reliable indicator of the serum ionized Ca^{2+} concentration under most but not all circumstances. One important common situation where total serum Ca^{2+} poorly reflects the ionized Ca^{2+} concentration is when hypoalbuminemia is present. When serum albumin is depressed, total serum Ca^{2+} often falls to subnormal levels. This can be mistaken for hypocalcemia. Many recommend that a bedside estimation of the corrected serum total Ca^{2+} be done in the hypoalbuminemic patient to determine whether there is real concern for hypocalcemia. This estimation is often done from the following formula: adjusted total Ca^{2+} = measured total Ca^{2+} + [0.8 × (4.0 − measured serum albumin)]. The normal range for the serum total Ca^{2+} is compared with this adjusted or "corrected" total Ca^{2+} to determine whether the adjusted value "corrects up" into the normal range. It is far better, however, when there is any question, to establish that the ionized Ca^{2+} is truly low by making a direct measurement. Estimates of the ionized Ca^{2+} are poor surrogates for actual measurements because, in addition to albumin, disturbances in pH and other circulating substances (e.g., citrate, phosphate, paraproteins) can influence the serum total Ca^{2+}, and these confounding factors are not considered in this estimation. It is imperative that the clinician establishes that the ionized Ca^{2+} concentration is indeed reduced, before an exhaustive workup for an etiology of hypocalcemia is undertaken. Full evaluation may be costly and is unjustified if there is only weak evidence that the serum ionized Ca^{2+} level is subnormal.

ETIOLOGY AND PATHOGENESIS

Numerous etiologies can explain why the ionized Ca^{2+} is low. The disorders can be broadly classified as ones in which there is inadequate PTH or vitamin D production, PTH or vitamin D resistance, or a miscellaneous cause (Table 1). The latter category encompasses a large and diverse spectrum of conditions that the endocrine clinician encounters in the course of practice. It is incumbent on the clinician to be aware of the etiologies, pathogenic mechanisms, intricacies of diagnostic testing including sequencing for mutations and other genetic analysis, and the best approaches to therapies in patients with hypocalcemia.

Hypoparathyroidism is a rare diagnosis in general. In terms of frequency, hypoparathyroidism is most commonly the sequela of parathyroid or thyroid surgery during which damage to, devitalization, and or inadvertent removal of most or all functioning parathyroid tissue have occurred. Perhaps next in frequency is the mild hypocalcemia and hypoparathyroidism caused by constitutively activating mutations of the Ca^{2+}-sensing receptor (CaSR). These mutations lead to the inappropriate suppression of PTH secretion at subnormal serum Ca^{2+} levels. The disorder presents as autosomal dominant hypocalcemia (ADH) in families and may go unrecognized because the hypocalcemia is often mild. The biochemical hallmark of ADH is impressive hypercalciuria, which worsens with attempts to treat the hypocalcemia with Ca^{2+} salts and vitamin D metabolites. Exacerbation of hypercalciuria with nephrocalcinosis and renal failure can result from these efforts. Such renal complications occur because the constitutively active CaSRs in the kidney misperceive prevailing serum Ca^{2+} concentrations as higher than they are, and this enhances renal excretion of Ca^{2+}. It has recently been appreciated that patients can develop antibodies that activate parathyroid and renal CaSRs. This produces an acquired form of hypocalcemia with low PTH and elevated urinary Ca^{2+} levels, thus mimicking the genetic disorder. These rare individuals often have other autoimmune disorders.

Destruction of the parathyroid glands on an immune basis can occur in isolation or as part of the type 1 autoimmune polyglandular syndrome (APS1). This is an autosomal recessive disorder caused by mutations in the autoimmune regulator (*AIRE-1*) gene. APS1 includes mucocutaneous candidiasis and adrenal insufficiency most commonly, as well as other autoimmune manifestations, and typically presents in childhood and adolescence. The parathyroid autoantigen in ~50% of patients with APS1 has recently been identified as NACHT leucine-rich-repeat protein 5 (NALP5), a putative signaling molecule, whose role as yet in parathyroid physiology is unknown. Patients with APS1 and with isolated autoimmune hypoparathyroidism also often generate antibodies against the CaSR, although it is unclear what the role, if any, the CaSR plays in the pathogenesis of the tissue destruction.

As noted above, there are multiple modes of inheritance of hypoparathyroidism, depending on the molecule involved. Autosomal recessive mutations in the gene encoding the transcription factor glial cell missing B (GCMB) are a rare cause of hypoparathyroidism. GCMB is essential for the development of the parathyroid glands. Mutations in a gene near SOX3 on the X chromosome underlie the pathogenesis of X-linked hypoparathyroidism. Another syndrome of HDR (hypoparathyroidism, deafness, renal anomalies) has been clearly shown to be caused by mutations in the transcription factor GATA3. Variable penetrance of the renal anomalies and hearing deficits have been observed. The well-known DiGeorge syndrome, a result of multiple developmental anomalies of the third and fourth branchial pouches, includes a spectrum of hypoparathy-

The author states that she has no conflicts of interest.

TABLE 1. ETIOLOGIES FOR HYPOCALCEMIA

Inadequate PTH production
Hypoparathyroidism
PTH gene mutations
Autosomal recessive (168450.0002)
Autosomal dominant (168450.0001)
X-linked hypoparathyroidism (307700)
Parathyroid gland agenesis
GCMB mutations (603716)
Postsurgical
Autoimmune
Isolated
Polyglandular failure syndrome type 1 (240300 and 607358)
Acquired antibodies that activate the Ca^{2+}-sensing receptor
Postradiation therapy
Secondary to infiltrative processes
Iron overload: hemochromatosis, thalassemia after transfusions
Wilson's disease
Metastatic tumor
Constitutively active Ca^{2+}-sensing receptor mutations (145980)
Magnesium excess
Magnesium deficiency
Syndromes with component of hypoparathyroidism
DiGeorge syndrome (188400)
HDR (hypoparathyroidism, deafness, renal anomalies) syndrome (146255 and 256340)
Blomstrand lethal chondrodysplasia
Kenney-Caffey syndrome (244460)
Sanjad-Sakati syndrome (241410)
Kearns-Sayre syndrome (530000)
Inadequate vitamin D production
Vitamin D deficiency
Nutritional deficiency
Lack of sunlight exposure
Malabsorption
End-stage liver disease and cirrhosis
Chronic kidney disease
PTH resistance
Pseudohypoparathyroidism
Magnesium deficiency
Vitamin D resistance
Pseudovitamin D deficiency rickets (vitamin D–dependent rickets type 1)
Vitamin D–resistant rickets (vitamin D–dependent rickets type 2)
Miscellaneous causes
Substances interfering with the laboratory assay for total Ca^{2+}—certain gadolinium salts in contrast agents given during MRI/MRA, particularly in patients with chronic renal failure
Hyperphosphatemia
Phosphate retention caused by acute or chronic renal failure
Excess phosphate absorption caused by enemas, oral supplements
Massive phosphate release caused by tumor lysis or crush injury
Drugs
Foscarnet
Intravenous bisphosphonate therapy—especially in patients with vitamin D insufficiency or ficiency
Rapid transfusion of large volumes of citrate-containing blood
Acute critical illness–multiple contributing etiologies
"Hungry bone syndrome" or recalcification tetany
Post-thyroidectomy for Grave's disease
Post-parathyroidectomy
Osteoblastic metastases
Acute pancreatitis
Rhabdomyolysis

roidism, thymic aplasia and immunodeficiency, cardiac defects, cleft palate, and abnormal facies. A variety of other very rare genetic syndromes is worth considering when a patient presents with a constellation of features that includes hypoparathyroidism (e.g., Kenney-Caffey, Kearns-Sayre, Sanjad-Sakati, and other syndromes; Table 1). The different forms of PTH resistance syndromes or pseudohypoparathyroidism, their clinical features, and genetic mechanisms are described elsewhere in the *Primer*.

In contrast to the rarity of hypoparathyroidism, vitamin D deficiency and disordered vitamin D metabolism are more common causes of hypocalcemia. Resistance to vitamin D, like that of PTH, remains very rare indeed. Whereas vitamin D deficiency and insufficiency occur in multiple clinical settings (elderly patients, postmenopausal women with fractures, nursing home residents, and so forth), it is much less common to see frankly low ionized Ca^{2+} values in such patients, particularly when 25-hydroxyvitamin D [25(OH)D] levels are just mildly depressed. Generally, low ionized Ca^{2+} values are the result of longstanding, severe vitamin D deficiency and chronically low 25(OH)D levels and along with significant degree of secondary hyperparathyroidism. Nevertheless, it is essential in the evaluation of patients with low ionized Ca^{2+} values that vitamin D inadequacy, disordered vitamin D activation in the liver and kidney, and reduced vitamin D–mediated signaling are considered as contributors to the etiology of the hypocalcemia.

Disorders of magnesium (Mg^{2+}) homeostasis bear mentioning because both Mg^{2+} excess and deficiency can produce generally mild hypocalcemia that is caused by functional (and reversible) hypoparathyroidism. Hypomagnesemia, often of a transient and correctable nature, accompanies a vast number of clinical situations particularly in ill and hospitalized patients (e.g., malnutrition, pancreatitis, chronic alcohol abuse, diarrhea, diuretic and antibiotic therapy, chemotherapy agents such as cisplatin derivatives). In sum, low serum Mg^{2+} levels, seen in conjunction with these clinical entities, require evaluation and often at least short-term therapy. Primary renal Mg^{2+} wasting states such as Gitelman's syndrome, caused by mutations in the renal thiazide-sensitive NaCl co-transporter, are more persistent and require long-term Mg^{2+} and other electrolyte replacement therapy to correct the biochemical parameters and clinical symptoms. Other rare entities that involve primary renal Mg^{2+} wasting include autosomal recessive disorders caused by mutations in the *paracellin-1* gene or in the Na-K ATPase subunit (*FXYD2* gene).

Hypomagnesemia also interferes with PTH action at the target organs, bone, and kidney, particularly PTH receptor–mediated activation of adenylate cyclase through the stimulatory G protein α subunit (α_s). Mg^{2+} is a co-factor for the adenylate cyclase enzyme complex. Hence, chronic hypomagnesemia produces a functional state of PTH resistance. More importantly, the normal physiologic response to hypocalcemia is lacking in the patient with Mg^{2+} depletion. Intact PTH levels are inappropriately low or normal in the presence of hypomagnesemia. Once the Mg^{2+} depletion is corrected, parathyroid function returns to normal.

Hypermagnesemia, in contrast, activates parathyroid CaSRs, thereby suppressing PTH secretion directly. Mg^{2+} levels high enough to ligate the CaSR tend to occur only in patients with chronic kidney disease or in the rare instance when Mg^{2+} is used for tocolytic therapy for preterm labor.

The experienced clinician will recognize that hypocalcemia occurs in the heterogeneous conditions listed in the category "Miscellaneous" (Table 1). In terms of frequency, pancreatitis is the most common disorder and is often associated with a low serum Ca^{2+}. This has been ascribed to the precipitation of

Ca^{2+}-containing salts in the inflamed pancreatic tissue and the presence of excess free fatty acids in the circulation. In many such patients, pancreatitis may progress rapidly with hemorrhage, hypotension, and sepsis as complicating features. Thus, hypocalcemia in patients with pancreatitis often correlates with illness severity.

Acute and chronic hyperphosphatemia can cause low total serum Ca^{2+}. The most common cause for chronic hyperphosphatemia is chronic kidney disease (CKD), which is readily apparent on the initial chemistry panel. Hypocalcemia in CKD has many contributing factors—poor nutrition, low 1,25-dihydroxyvitamin D production, malabsorption, and so forth. Most of these need to be addressed to effectively manage hypocalcemia in patients with CKD. Acute changes in phosphate balance can also lower serum Ca^{2+}. In any situation where large amounts of phosphate are rapidly absorbed into the intravascular compartment, there is the potential for the serum ionized Ca^{2+} to fall, even to symptomatic levels. This can be seen with phosphate-containing enemas and supplements, especially when the latter are given intravenously for the treatment of hypophosphatemia. Also, in the setting of acute tumor lysis caused by cytolytic therapy for high-grade lymphomas, sarcomas, leukemias, and solid tumors, cell breakdown with the rapid release of phosphate from intracellular nucleotides can quickly depress serum ionized Ca^{2+}. The drug foscarnet, given to patients with AIDS or other immunocompromised states to treat viral infections, can also lower both serum Ca^{2+} and Mg^{2+}, sometimes to symptomatic levels. Treating patients who are normocalcemic with intravenous aminobisphosphonates (e.g., zoledronic acid, pamidronate), which block bone resorption dramatically, has the potential to cause a low ionized Ca^{2+}. However, this is relatively infrequent unless concomitant vitamin D deficiency/insufficiency has gone unrecognized.

The "hungry bone syndrome" or recalcification tetany can occur after parathyroidectomy for any form of hyperparathyroidism or after thyroidectomy in patients with hyperthyroidism. Skeletal uptake of Ca^{2+} and phosphate is intense because of the presence of a mineral-depleted bone matrix and the instantaneous removal by surgery of the stimulus for maintaining high rates of bone resorption—either PTH or the thyroid hormones. Depending on the severity of the bone hunger, hypocalcemia and hypophosphatemia can persist for weeks and require large doses of Ca^{2+} and vitamin D metabolites to control. If there has been no permanent damage to the parathyroid glands, intact PTH levels should eventually rise appropriately to supranormal levels. In some cases, however, the viability of the remaining parathyroid glands or the suppression of function of the remaining glands by a previously dominant adenoma may confuse the picture. Careful management and repeated mineral and PTH analyses will usually allow the diagnosis to become clear over time.

The entity of pseudohypocalcemia caused by gadolinium (Gd^{3+})-containing MRI agents has received considerable attention lately because of the frequency of performing MR angiography in general and the use of the procedure in patients with CKD. The clearance of Gd^{3+} in such patients is very prolonged. Total serum Ca^{2+} levels, as measured by standard arsenazo III reagents, in patients with Gd^{3+} contrast agents in the circulation will appear to be low. This is because Gd^{3+} complexes with this Ca^{2+}-sensitive dye and blocks the colorimetric detection of Ca^{2+}. Because ionized Ca^{2+} is measured in a completely different manner, ionized Ca^{2+} levels will be normal in these individuals, and there are no symptoms of hypocalcemia.

Acute and critical illness, often in the intensive care unit (ICU) setting, is frequently accompanied by hypocalcemia, including frankly low ionized Ca^{2+} values. This entity is typically multifactorial with poor nutrition, vitamin D insufficiency, renal dysfunction, acid-base disturbances, cytokines, and other factors contributing. It is prudent to follow the ionized Ca^{2+} values and treat as deemed appropriate based on clinical circumstances.

TABLE 2. SIGNS AND SYMPTOMS OF HYPOCALCEMIA

Symptoms
Paresthesias
Circumoral and acral tingling
Increased neuromuscular irritability
Tetany
Muscle cramping and twitching
Muscle weakness
Abdominal cramping
Laryngospasm
Bronchospasm
Altered central nervous system function
Seizures of all types: grand mal, petit mal, focal
Altered mental status and sensorium
Papilledema, pseudotumor cerebri
Choreoathetoid movements
Depression
Coma
Generalized fatigue
Cataracts
Congestive heart failure
Signs
Chvostek's sign
Trousseau's sign
Prolongation of the QT-c interval
Basal ganglia and other intracerebral calcifications

SIGNS AND SYMPTOMS

Patients with low ionized Ca^{2+} values can present with no symptoms or with significant morbidity (Table 2). Their presentation depends on the severity and chronicity of the disturbance. Chronic hypocalcemia, despite even very low levels of ionized Ca^{2+}, can be asymptomatic, and the only clue is the presence of the positive Chvostek's sign. Neuromuscular irritability is the most frequent cause for symptoms that include tetany, carpopedal spasms, muscle twitching and cramping, circumoral tingling, abdominal cramps, and in severe cases laryngospasm, bronchospasm, seizures, and even coma. Basal ganglia and other intracerebral calcifications can be seen on imaging studies. Ocular findings include cataracts, particularly when there are longstanding elevations in the Ca^{2+}-phosphate product, and pseudotumor cerebri. Longstanding hypocalcemia can also cause congestive heart failure because of cardiomyopathy that reverses with management of the usually very low ionized Ca^{2+} levels. Hypocalcemia is well known for its effects on cardiac conduction that are manifested on the ECG as prolongation of the QT-c interval. In addition, the patient often feels a sense of generalized weakness, fatigue, and depression that often lifts as the mineral disturbance(s) (and vitamin D deficiency, if present) are successfully treated.

DIAGNOSIS: TESTS AND INTERPRETATION

The mainstays of diagnostic testing are determination of the serum ionized Ca^{2+}, total Mg^{2+}, phosphate, intact PTH, and 25(OH)D values. Measuring the latter metabolite is the best way to exclude vitamin D deficiency or insufficiency. Considerable attention has been directed to the reliability of contemporary 25(OH)D assays and clinically relevant cut-points for diagnosing vitamin D deficiency/insufficiency. Intact PTH measured in a reliable assay will readily disclose inappropriately low, low normal, or even undetectable values in hypoparathyroid states and generally normal levels, but inappropriately so in patients with Mg^{2+} depletion. In marked contrast, patients with vitamin D deficiency and pseudohypoparathyroidism will have elevated levels of PTH or secondary hyperparathyroidism. Phosphate levels are low in vitamin D deficiency and elevated or at the high end of the normal range in patients with hypoparathyroidism and pseudohypoparathyroidism—an important distinguishing measurement to make. Accurate determination of 24-h urinary Ca^{2+} or Mg^{2+} excretion, depending on the primary disturbance (hypocalcemia versus hypomagnesemia), can be extremely helpful. Strikingly elevated urinary Ca^{2+} levels in the asymptomatic mildly hypocalcemic patient suggest ADH. Milder elevations are expected in hypoparathyroid states. In contrast, vitamin D deficiency with secondary hyperparathyroidism and osteomalacia classically produces hypocalciuria, in an effort by the kidney, under the influence of PTH, to conserve Ca^{2+} for systemic needs. The presence of significant Mg^{2+} in the urine in a hypomagnesemic patient strongly suggests primary renal Mg^{2+} wasting and not gastrointestinal losses.

MANAGEMENT OF HYPOCALCEMIA: ACUTE AND CHRONIC

The goals of treatment are to alleviate symptoms, heal demineralized bones when osteomalacia is present, maintain an acceptable ionized Ca^{2+} or total serum Ca^{2+}, and avoid hypercalciuria (urine $Ca^{2+} > 300$ mg/24 h), renal dysfunction, stones, and nephrocalcinosis. When clinical circumstances dictate urgent treatment, intravenous Ca^{2+} salts are used. Seizures, severe tetany, laryngospasm, bronchospasm, or altered mental status are strong indicators for intravenous therapy. In contrast, if the patient is minimally symptomatic—despite low numbers—the oral regimen outlined below can be used. The preferred intravenous salt is Ca gluconate (10 ml of a 10% solution = 93 mg elemental Ca^{2+} or 1 ampule). Ten milliliters is infused slowly over 10 min to address symptoms, and this can be repeated once or twice more. Generally, an infusion is begun. Our method is to prepare a drip (10 ampules of 10% Ca gluconate in 1 liter D5W) and infuse it at the rate of 10–100 ml/h to control symptoms and restore the ionized Ca^{2+} to the lower end of the normal range ~1.0 mM. Based on the patient's weight, one calculates the infusion rate to deliver 0.3–1.0 mg elemental Ca^{2+}/kg/h. Once the patient is stabilized, a chronic oral regimen is begun. Infusion rates are tapered down as serum ionized Ca^{2+} reaches the target, symptoms resolve, and oral medications are tolerated. Ionized Ca^{2+} measurements should be used to guide therapy.

Chronic management of hypocalcemia uses oral Ca^{2+} supplements, vitamin D metabolites, and sometimes thiazide diuretics. When Mg^{2+} depletion is the source of hypocalcemia, Mg^{2+} deficits are generally large and poorly reflected by the serum Mg^{2+} level, because Mg^{2+} is predominantly an intracellular cation. Supplementation with Mg^{2+} salts over an extended period of time will usually be needed to replenish total body Mg^{2+} stores. Serum Ca^{2+} and PTH secretory capacity will return to normal.

Ca^{2+} supplements of all types work to treat hypocalcemia. A few general principles are worth emphasizing. It is best to divide up the supplements throughout the day and time their administration to coincide with meals because this will enhance absorption. The most efficient means of supplementation is in the form of Ca carbonate or citrate salts. The former is ~40% and the latter ~21% elemental Ca^{2+} by weight. Generally, 500–1000 mg elemental Ca^{2+} twice or three times a day is a reasonable starting dose and can be escalated upward. This is done based on patient tolerance, compliance, and clinical goals. When Ca^{2+} supplements are insufficient to reach the target serum Ca^{2+}, vitamin D metabolites are prescribed. When renal function is intact, ergocalciferol (vitamin D_2) or cholecalciferol (vitamin D_3) may be used. Doses in the range of 25,000–50,000 units daily (occasionally even more) may be needed to treat hypoparathyroidism, pseudohypoparathyroidism, or vitamin D deficiency in patients with malabsorption. Care must be exercised because these forms of vitamin D have a long tissue half-life (weeks to months) because of long-term storage in fat, and toxicity may be difficult to predict and treat. Calcitriol (1,25-dihydroxyvitamin D; 0.25 or 0.5 μg once or twice daily) is preferred by many clinicians because of its rapid onset and offset of action (1–3 days) and ease of titration, despite its greater costs. Doxercalciferol [$1\alpha(OH)D_2$; initial dose 2.5 μg daily and titrated as needed] is another alternative for treatment, although published data on this vitamin D analog are sparse in hypocalcemic and hypoparathyroid disorders. In patients who develop hypercalciuria on these regimens and or have difficulty achieving the serum Ca^{2+} goal safely, one can take advantage of the Ca^{2+}-retaining actions of thiazide diuretics. Effective doses typically range from hydrochlorothiazide 50–100 mg/d, although lower doses can be tried. Serum Ca^{2+}, phosphorus, and creatinine should be monitored regularly along with 25(OH)D levels if vitamin D therapy is used to avoid toxicity.

SUGGESTED READING

Ali A, Christie PT, Grigorieva IV, Harding B, Van Esch H, Ahmed SF, Bitner-Glindzicz M, Blind E, Bloch C, Christin P, Clayton P, Gecz J, Gilbert-Dussardier B, Guillen-Navarro E, Hackett A, Halac I, Hendy GN, Lalloo F, Mache CJ, Mughal Z, Ong AC, Rinat C, Shaw N, Smithson SF, Tolmie J, Weill J, Nesbit MA, Thakker RV 2007 Functional characterization of GATA3 mutations causing the hypoparathyroidism-deafness-renal (HDR) dysplasia syndrome: Insight into mechanisms of DNA binding by the GATA3 transcription factor. Hum Mol Genet **16:**265–275.

Alimohammadi M, Bjorklund P, Hallgren A, Pontynen N, Szinnai G, Shikama N, Keller MP, Ekwall O, Kinkel SA, Husebye ES, Gustafsson J, Rorsman F, Peltonen L, Betterle C, Perheentupa J, Akerstrom G, Westin G, Scott HS, Hollander GA, Kampe O 2008 Autoimmune polyendocrine syndrome type 1 and NALP5, a parathyroid autoantigen. N Engl J Med **358:**1018–1028.

Bowl MR, Nesbit MA, Harding B, Levy E, Jefferson A, Volpi E, Rizzoti K, Lovell-Badge R, Schlessinger D, Whyte MP, Thakker RV 2005 An interstitial deletion-insertion involving chromosomes 2p25.3 and Xq27.1, near SOX3, causes X-linked recessive hypoparathyroidism. J Clin Invest **115:**2822–2831.

Brasier AR, Nussbaum SR 1988 Hungry bone syndrome: Clinical and biochemical predictors of its occurrence after parathyroid surgery. Am J Med **84:**654–660.

Eisenbarth G, Gottlieb PA 2004 Autoimmune polyendocrine syndromes. N Engl J Med **350:**2068–2079.

Gavalas NG, Kemp EH, Krohn KJ, Brown EM, Watson PF, Weetman AP 2007 The calcium-sensing receptor is a target of autoimmune polyendocrine syndrome type 1. J Clin Endocrinol Metab **92:**2107–2114.

Holick MF 2007 Vitamin D deficiency. N Engl J Med **357:**266–281.

Kifor O, McElduff A, LeBoff MS, Moore FD Jr, Butters R, Gao P, Cantor TL, Kifor I, Brown EM 2004 Activating antibodies to the calcium-sensing receptor in two patients with autoimmune hypoparathyroidism. J Clin Endocrinol Metab **89:**548–556.

Kobrynski LJ, Sullivan KE 2007 Velocardiofacial syndrome, DiGeorge syndrome: The chromosome 22q11.2 deletion syndromes. Lancet **370:**1443–1452.

Lienhardt A, Bai M, Lagarde JP, Rigaud M, Zhang Z, Jiang Y, Kottler ML, Brown EM, Garabedian M 2001 Activating mutations of the calcium-sensing receptor: Management of hypocalcemia. J Clin Endocrinol Metab **86**:5313–5323.

Prince MR, Erel HE, Lent RW, Blumenfeld J, Kent KC, Bush HL, Wang Y 2003 Gadodiamide administration causes spurious hypocalcemia. Radiology **227**:639–646.

Rosen CJ, Brown S 2003 Severe hypocalcemia after intravenous bisphosphonate therapy in occult vitamin D deficiency. N Engl J Med **348**:1503–1504.

Shoback D, Sellmeyer D, Bikle D 2007 Mineral metabolism and metabolic bone disease. In: Shoback D, Gardner D (eds.) Basic and Clinical Endocrinology, 8th ed. Lange Medical Books/McGraw-Hill, New York, NY, USA, pp. 281–345.

Thakker RV 2004 Genetics of endocrine and metabolic disorders: Parathyroid. Rev Endocrinol Metab Dis **5**:37–51.

Thomee C, Schubert SW, Parma J, Le PQ, Hashemolhosseini S, Wegner M, Abramowicz MJ 2005 GCMB mutation in familial isolated hypoparathyroidism with residual secretion of parathyroid hormone. J Clin Endocrinol Metab **90**:2487–2492.

Vivien B, Langeron O, Morell E, Devilliers C, Carli PA, Coriat P, Riou B 2005 Early hypocalcemia in severe trauma. Crit Care Med **33**:1946–1952.

Chapter 69. Disorders of Phosphate Homeostasis

Mary D. Ruppe[1] and Suzanne M. Jan de Beur[2]

[1]Division of Endocrinology, Department of Medicine, University of Texas Health Science Center at Houston, Houston, Texas; [2]Division of Endocrinology and Metabolism, Department of Medicine, The Johns Hopkins School of Medicine, Baltimore, Maryland

INTRODUCTION

Phosphorus is a critical element in skeletal development, bone mineralization, membrane composition, nucleotide structure, and cellular signaling. The physiological control of phosphate homeostasis and the major hormonal regulators of phosphate homeostasis [e.g., fibroblast growth factor 23 (FGF23), 1,25$(OH)_2$D, PTH] have been previously discussed in detail. Serum phosphorus concentration is regulated by diet, hormones, pH, and changes in renal, skeletal, and intestinal function. The focus of this chapter is the molecular basis of human disorders of phosphate homeostasis. In recent years, advances in defining the precise molecular defects in both acquired and inherited hypo and hyperphosphatemic syndromes have catapulted our understanding of phosphate homeostatic mechanisms to a new level.

HYPOPHOSPHATEMIA

Hypophosphatemia is common. In the hospital setting, up to 5% of patients are hypophosphatemic. Among alcoholic patients and those with severe sepsis, up to a 30–50% prevalence has been reported. Hypophosphatemia is defined as an abnormally low concentration of inorganic phosphate in the serum. Moderate hypophosphatemia, a serum concentration between 1 and 2.5 mg/dl, is not typically symptomatic. Severe hypophosphatemia, defined as serum levels <1.0 mg/dl, is frequently symptomatic and requires therapy.

Clinical Consequences

The clinical manifestations of hypophosphatemia are dependent on the severity and chronicity of the phosphorus depletion. Common clinical settings in which severe hypophosphatemia is observed include chronic alcoholism, nutritional repletion in at risk individuals, treatment of diabetic ketoacidosis, and in critical illness.

The symptoms of hypophosphatemia are a direct consequences of intracellular phosphorus depletion: (1) tissue hypoxia caused by reduced 2,3-diphosphoglycerate (2,3-DPG) in the erythrocyte that increases affinity of hemoglobin for oxygen and (2) diminished tissue content of ATP that compromises cellular function. A full list of the clinical manifestations of hypophosphatemia is found in Table 1.

Causes of Hypophosphatemia

The three major mechanisms by which hypophosphatemia can occur are redistribution of phosphorus from extracellular fluid into cells, increased urinary excretion, and decreased intestinal absorption (Table 2). The diagnosis of hypophosphatemia is often evident from the history; if however, the diagnosis remains obscure, measurement of urinary phosphate excretion is indicated. Urinary phosphate excretion can be measured from a 24-h collection or from a random urine specimen by calculation of fractional excretion of filtered phosphate (FEPO4): $FEPO_4 = [UPO_4 \times PCr \times 100]/[PPO_4 \times UCr]$, where U and P are urine and plasma concentrations of phosphate (PO_4) and creatinine (Cr). In the setting of hypophosphatemia, fractional excretion of phosphate >5% or the 24 h urinary phosphate excretion that exceeds 100 mg/d is indicative of renal phosphate wasting.

Intracellular Shifts. Stimulation of glycolysis increases the synthesis of phosphorylated carbohydrates in the skeletal muscle and liver, which is derived from inorganic phosphorus in the extracellular fluid. Consequently, serum phosphate concentrations and urinary excretion fall, and if underlying phosphate depletion is present, severe hypophosphatemia often results. Examples of clinical situations in which this occurs include treatment of diabetic ketoacidosis and refeeding of malnourished patients with anorexia or alcoholism.

Similarly, the fall in carbon dioxide during acute respiratory alkalosis triggers a rise in intracellular pH that stimulates glycolysis through phosphofructokinase. Extreme hyperventilation may result in severe hypophosphatemia (<1.0 mg/dl).

Hungry bone syndrome, a rare complication of hyperparathyroidism after parathyroidectomy, results in rapid reminer-

The authors state that they have no conflicts of interest.

Key words: hypophosphatemia, hyperphosphatemia, fibroblast growth factor 23, KLOTHO, X-linked hypophosphatemic rickets, autosomal dominant hypophosphatemic rickets, tumor-induced osteomalacia, autosomal recessive hypophosphatemic rickets, hereditary hypophosphatemic rickets with hypercalciuria, DMP-1, NaPi IIa, NaPi IIc, tumoral calcinosis, GALNT3

Electronic Databases: 1. OMIM—http://www.ncbi.nlm.nih.gov/sites/entrez?db=omim.

© 2008 American Society for Bone and Mineral Research

TABLE 1. CLINICAL CONSEQUENCES OF HYPOPHOSPHATEMIA

Central nervous system
Irritability, parathesias (early manifestations)
Confusion, seizures, delirium, coma (late manifestations)
Hematopoietic system
Erythrocyte: hemolysis, altered oxygen affinity
Leukocyte: diminished phagocytosis and chemotaxis
Platelets: thrombocytopenia
Muscle
Skeletal muscle dysfunction: proximal myopathy, myalgia, subjective weakness, rhabdomyolysis when acute hypophosphatemia is superimposed on severe phosphorus depletion (e.g., alcoholism)
Smooth muscle dysfunction: ileus, dysphagia
Cardiac: cardiomyopathy, congestive heart failure
Bone and mineral metabolism
Increased bone resorption (early)
Impaired bone mineralization (rickets in children and osteomalacia in adults)
Renal
Decreased glomerular filtration rate
Changes in tubular transport with resultant electrolyte imbalances (hypercalciuria, hypermagnesuria, bicarbonaturia, glycosuria, hypophosphaturia, decreased proximal tubular sodium transport)
Metabolic abnormalities
Diminished gluconeogenesis
Insulin resistance
Increased calcitriol
Hypoparathyroidism
Metabolic acidosis
Reduced hydrogen ion excretion
Suppressed ammonia production
Diminished renal tubular reabsorption of bicarbonate

alization of bone after abrupt cessation of the bone resorption by PTH. Severe hypocalcemia and hypophosphatemia results from movement of calcium and phosphorus from the extracellular space into the bone matrix.

Markedly proliferative tumors such as leukemia in blast crisis and aggressive lymphoma can precipitate hypophosphatemia through phosphorus uptake from the extracellular fluid into rapidly proliferating cells.

Decreased Intestinal Absorption. Dietary phosphate intake usually ranges from 800 to 1500 mg/d, which far exceeds gastrointestinal losses. Eighty percent of dietary phosphorus is absorbed in the small intestine. Calcitriol promotes absorption of dietary phosphorus in the gastrointestinal (GI) tract.

The most common cause of hypophosphatemia from decreased intestinal absorption is vitamin D deficiency or resistance, which is discussed in detail elsewhere in the *Primer*. Other causes include aluminum- and magnesium-containing antacids that bind ingested and secreted phosphate. With high doses and prolonged use, hypophosphatemia, osteomalacia, and myopathy ensue. Because of rapid renal adaption, it is rare for poor intake alone to cause hypophosphatemia. However, if phosphate deprivation is severe and prolonged (<100 mg/d) or coexists with chronic diarrhea, colonic secretion can lead to hypophosphatemia.

Treatment of hypophosphatemia that results from decreased intestinal absorption focuses on repletion of vitamin D, nutritional therapy, or discontinuation of offending medications. Oral phosphorus supplements are indicated only if symptoms are present.

Increased Urinary Excretion. Hypophosphatemia secondary to renal phosphate wasting has a wide differential diagnosis (Table 1). These disorders can result from primary renal transport defects, excess PTH, overproduction of the phosphaturic hormone, FGF23, from normal or dysplastic bone, ectopic production of FGF23 or other phosphaturic proteins from tumors, and overexpression of KLOTHO, a co-factor important for FGF23 signaling (Fig. 1).

TUMOR-INDUCED OSTEOMALACIA

Tumor-induced osteomalacia (TIO), or oncogenic osteomalacia, is an acquired, paraneoplastic syndrome of renal phosphate wasting that resembles genetic forms of hypophosphatemic rickets. First described in 1947, clinical and experimental studies implicate the humoral factor(s) that tumors produce in the profound biochemical and skeletal alterations that characterize TIO.

Clinical and Biochemical Manifestations

Although the preponderance of TIO patients are adults

TABLE 2. CAUSES OF HYPOPHOSPHATEMIA

Decrease intestinal absorption
Vitamin D deficiency or resistance
Nutritional deficiency: low sun exposure, low dietary intake
Malabsorption: celiac disease, Crohn's disease, gastrectomy, bowel resection, pancreatitis, gastric bypass, chronic diarrhea
Chronic liver disease
Chronic renal disease
Increased catabolism: anticonvulsant therapy
Vitamin D receptor defects: vitamin D–dependent rickets type 2
Vitamin D synthetic defects:
1α-hydroxylase (CYP27B1): vitamin D–dependent rickets type 1
25-hydroxylase (CYP27A1)
Nutritional deficiency: alcoholism, anorexia, starvation
Antacids containing aluminum or magnesium
Increased urinary losses
Renal phosphate wasting syndromes
X-linked hypophosphatemic rickets
Autosomal dominant hypophosphatemic rickets
Autosomal recessive hypophosphatemic rickets
Hereditary hypophophatemic rickets with hypercalciuria
Tumor-induced osteomalacia
Fanconi syndrome
Dent's disease
Fibrous dysplasia
Osteoglophonic dysplasia
α KLOTHO excess
Linear nevus sebaceous syndrome
NPT-2a dominant negative mutations
N4ERF1 mutations
Primary and secondary hyperparathyroidism
Osmotic diuresis: diabetic ketoacidosis (DKA)
Medications: calcitonin, diuretics, glucocorticoids, bicarbonate
Acute volume expansion
Intracellular shifts
Increased insulin: refeeding, treatment of DKA, insulin therapy
Hungry bone syndrome
Acute respiratory alkalosis
Tumor consumption: leukemia blast crisis, lymphoma
Sepsis
Sugars: glucose, fructose, glycerol
Recovery from metabolic acidosis: DKA

© 2008 American Society for Bone and Mineral Research

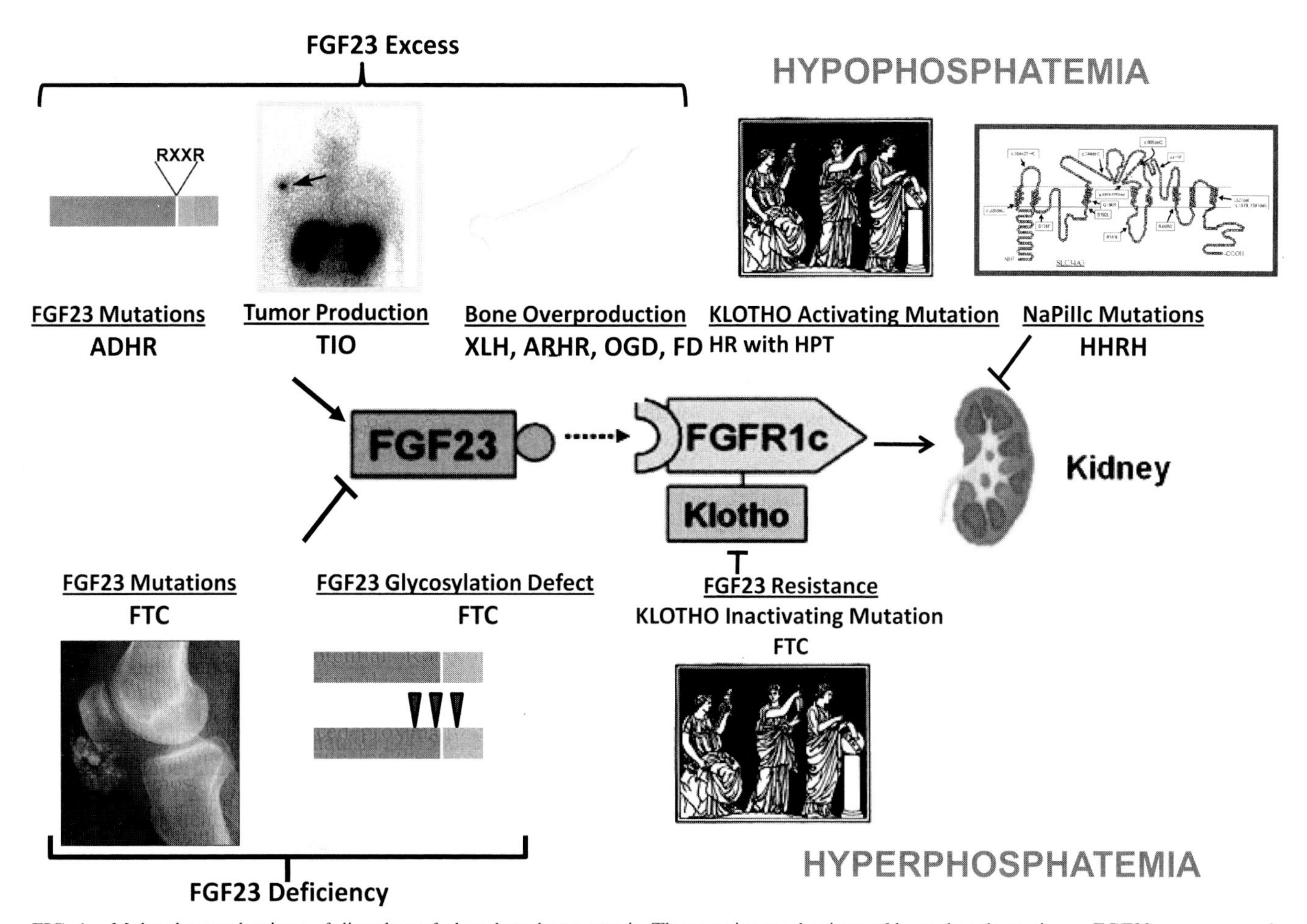

FIG. 1. Molecular mechanisms of disorders of phosphate homeostasis. Three major mechanisms of hypophosphatemia are FGF23 excess caused by ectopic production as in TIO, excess bone production seen in XLH, ARHR, ADHR, FD, and OGD, and mutation in the *FGF23* gene that renders the protein resistant to inactivation. Hypophosphatemia may also be secondary to excess KLOTHO, the co-factor necessary for FGF23 signaling as seen in a patient with hypophosphatemic rickets with hyperparathyroidism. Finally, homozygous inactivating mutations in SLC34A3, which encodes NaPiIIc, or dominant negative mutations in SLC34A2 that encodes NaPiIIa result in phosphate wasting caused by the absence of sodium-phosphate co-transporters. Hyperphosphatemia is caused by FGF23 deficiency, either through inactivating mutations in FGF23, aberrant glycosylation of FGF23 caused by GALNT3 mutations, or FGF23 resistance caused by inactivating KLOTHO mutations.

(usually diagnosed in the sixth decade), this syndrome may present at any age. These patients report long-standing progressive muscle and bone pain. Children with TIO display rachitic features including gait disturbances, growth retardation, and skeletal deformities. The occult nature of TIO delays its recognition, and the average time from onset of symptoms to a correct diagnosis often exceeds 2.5 yr. Once the syndrome is recognized, an average of 5 yr elapses from the time of diagnosis to the identification of the underlying tumor. Until the underlying tumor is identified, other renal phosphate wasting syndromes must be considered. Identification of previously normal serum phosphorus level in an adult patient supports the diagnosis of TIO, although in rare instances patients with autosomal dominant hypophosphatemic rickets (ADHR) can present in adulthood. In situations when inherited hypophosphatemic rickets must be excluded, genetic testing for mutations in the *PHEX* gene and *FGF23* gene is indicated.

The biochemical hallmarks of TIO are low serum concentrations of phosphorus, phosphaturia, secondary to reduced proximal renal tubular phosphate reabsorption, and frankly low or inappropriate normal levels of serum calcitriol that are expected to be elevated in the face of hypophosphatemia (Table 3). Calcium and PTH are typically normal. Bone histomorphometry shows severe osteomalacia with clear evidence of a mineralization defect with increased mineralization lag time and excessive osteoid (Fig. 2). The dual defect of renal phosphate wasting in concert with impaired calcitriol [$1,25(OH)_2D_3$] synthesis results in poor bone mineralization and fractures.

The mesenchymal tumors that are associated with TIO are characteristically slow growing, polymorphous neoplasms with the preponderance being phosphaturic mesenchymal tumor, mixed connective tissue type (PMTMCT; Fig. 2). Characterized by an admixture of spindle cells, osteoclast-like giant cells, prominent blood vessels, cartilage-like matrix, and metaplastic bone, these tumors occur equally in soft tissue and bone. Although typically benign, malignant variants of PMTMCT have been described.

These mesenchymal tumors ectopically express and secrete FGF23 and other phosphaturic proteins. FGF23, a circulating fibroblast growth factor produced by osteocytes and osteoblasts, has two currently known physiologic functions: first, FGF23 promotes internalization of NaPiIIa and NaPiIIc from the brush border membrane and thus reduces reabsorption of urinary phosphorus resulting in hypophosphatemia; second, it diminishes protein expression of the 25 hydroxy-1-α-hydroxylase enzyme that converts vitamin D to its active form, $1,25(OH)_2D$ and disrupts the compensatory increase in

© 2008 American Society for Bone and Mineral Research

TABLE 3. CHARACTERISTICS OF RENAL PHOSPHATE WASTING DISORDERS

Disease (OMIM)	*Defect*	*Biochemical features*	*Pathogenesis*
TIO	Mesenchymal tumor	• Renal phosphate wasting and hypophosphatemia • Inappriopriately low 1,25 $(OH)_2$ D • Absence of hypercalcemia or hyperparathyroidism	Ectopic, unregulated production of FGF23 and other phosphatonins sFRP-4, MEPE, FGF7
Fanconi's	Renal proximal tubular defect that alters absorption of Pi, glucose, amino acids, bicarbonate	Similar to TIO with additional features of glucosuria, amino aciduria, and bicarbonate wasting	Proximal renal tubular damage caused by multiple myeloma, lymphoma, amyloidosis, light chain disease, nephrotic syndrome, drugs, heavy metals
Fibrous dysplasia (174800)	*GNAS* gain of function	Biochemically indistinguishable from TIO	Increased FGF23 production from fibrous dysplastic bone
XLH (307800)	*PHEX* gene loss of function	Biochemically indistinguishable from TIO	Increased FGF23 synthesis from bone
ADHR (193100)	*FGF23* gene gain of function	Biochemically indistinguishable from TIO	Increased circulating intact FGF23 caused by mutations that render it resistant to cleavage
ARHR (241520)	*DMP1* gene loss of function	Biochemically indistinguishable from TIO	Loss of DMP1 causes impaired osteocyte differentiation and increased production of FGF23
HHRH (241530)	*SLC34A3* (NaPiIIc) loss of function	• Renal phosphate wasting • Appropriately elevated 1,25 $(OH)_2$D • Elevated urine calcium and high/normal serum calcium • PTH is normal or suppressed	Loss of function mutations in NaPiIIc results in renal phosphate wasting without defect in 1,25$(OH)_2$ synthesis. Hypophosphatemia stimulates 1,25$(OH)_2$D production with high/normal serum calcium and elevated urine calcium. FGF23 is not elevated.
NaPiIIa mutations (182309)	*NPT-2* gene mutations	• Similar to HHRH	Dominant negative mutations in NaPiIIa results in renal phosphate wasting without defect in 1,25$(OH)_2$ synthesis.
HR and HPT (612089)	α Klotho translocation	Biochemically indistinguishable from TIO with hyperparathyroidism	Increased Klotho, FGF23, and downstream FGF23 signaling
OGD (166250)	*FGFR1* gain of function	Biochemically indistinguishable from TIO	Increased production of FGF23 by dyplastic bone

1,25$(OH)_2$D triggered by hypophosphatemia. Circulating levels of FGF23 are elevated in most patients with TIO. After surgical resection, FGF23 levels plummet. Other secreted proteins such as MEPE (matrix extracellular phosphoglycoprotein), FGF7, and sFRP4 (secreted frizzled related protein 4) are highly expressed in mesenchymal tumors associated with TIO, but the role of each of these "phosphatonins" in the disease process remains obscure.

Treatment

Detection and localization of the culprit tumor in TIO is imperative because complete surgical resection is curative. However, the mesenchymal tumors that cause this syndrome are often small, slow growing, and frequently found in a variety of anatomical locations, including the long bones, the distal extremities, the nasopharynx, the sinuses, and the groin. The

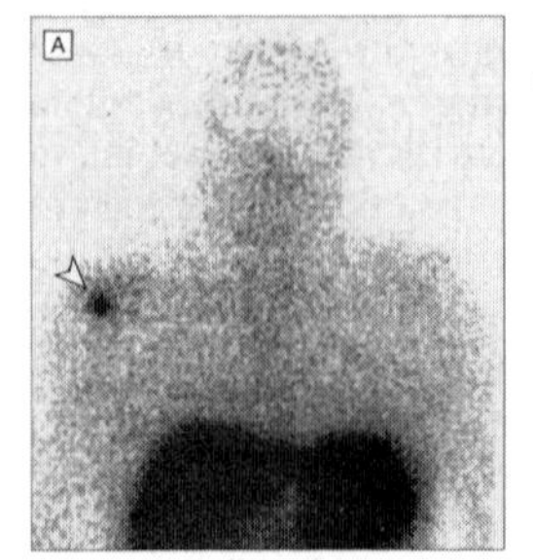

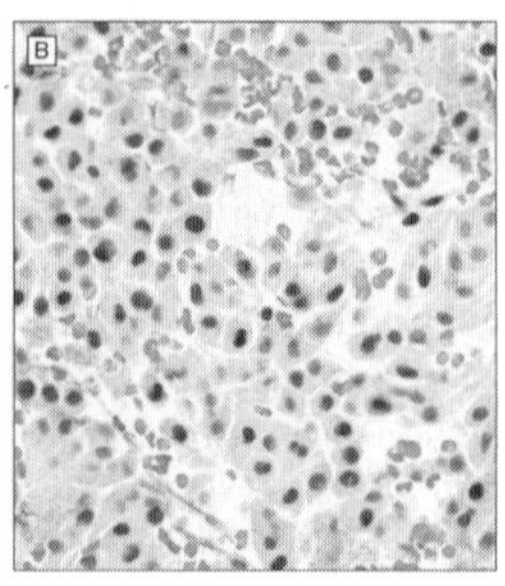

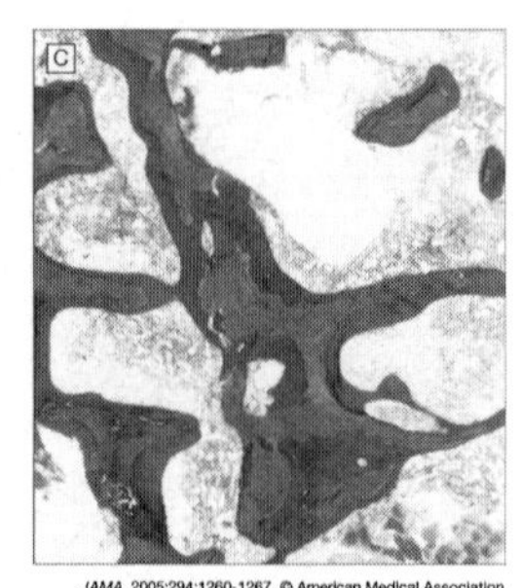

FIG. 2. Radiographic and histologic features in TIO. (A) Octreotide scan showing small mesenchymal tumor in the head of the humerus. (B) Hemiangiopericytoma with with numerous pericytes and vascular channels (H&E strain). (C) Bone biopsy with Goldner stain. Excessive osteoid or unmineralized bone matrix composed mainly of collagen stains pink. Minerlized bone stains blue. This bone biopsy shows severe osteomalacia.

© 2008 American Society for Bone and Mineral Research

size and obscure locations make the tumors difficult to localize with conventional imaging techniques. Because in vitro studies show that many mesenchymal tumors express somatostatin receptors (SSTRs), ^{111}In-pentetreotide scintigraphy (octreotide scan; Fig. 2), a scanning technique that uses a radiolabeled somatostatin analog, has been used to successfully detect and localize these tumors in some patients with TIO. Successful tumor localization has been reported in a few patients with other imaging techniques such as whole body MRI and positron emission tomography. In a single instance, venous sampling for FGF23 was used to confirm an identified mass was the causative tumor in a patient with TIO.

The definitive treatment for TIO is complete resection, which results in rapid correction of the biochemical perturbations and remineralization of bone. However, even after the diagnosis of TIO is made, the tumor often remains obscure or incompletely resected. Therefore, frequently, medical management is necessary. The current practice is to treat TIO with phosphorus supplementation in combination with calcitriol. The phosphorus supplementation serves to replace ongoing renal phosphorus loss, and the calcitriol supplements replace insufficient renal production of 1,25 dihydroxyvitamin D and enhance renal and gastrointestinal phosphorus reabsorption. Generally, patients are treated with phosphorus (2 g/d), in divided doses, and calcitriol (1–3 μg/d). In some cases, administration of calcitriol alone may improve the biochemical abnormalities seen in TIO and heal the osteomalacia. Therapy and dosing should be tailored to improve symptoms, maintain fasting phosphorus in the low normal range, normalize alkaline phosphatase, and control secondary hyperparathyroidism, without inducing hypercalcemia or hypercalciuria. With appropriate treatment, muscle and bone pain will improve, and healing of the osteomalacia will ensue. Monitoring for therapeutic complications of high doses of calcitriol and phosphorus is important to prevent unintended hypercalcemia, nephrocalcinosis, and nephrolithiasis. Although parathyroid autonomy has been reported in only a few cases of TIO, the true incidence is likely higher with prolonged treatment with phosphorus (alone or in combination with calcitriol) because it stimulates parathyroid function that can eventually lead to autonomy. To assess safety and efficacy of therapy, monitoring of serum and urine calcium, renal function, and parathyroid status is recommended at least every 3 mo. There has been a recent report of medically induced hypoparathyroidism through the use cinacalcet in two patients who did not tolerate medical therapy with phosphorus and calcitriol. This therapy allowed a decrease in the phosphorus dose to one that was tolerated. Octreotide in vitro and in vivo has been shown to inhibit secretion of hormones by many neuroendocrine tumors; however, because of the limited and mixed experience with octreotide treatment in TIO, this therapy should be reserved for the most severe cases that are refractory to current medical therapy.

X-LINKED HYPOPHOSPHATEMIC RICKETS

First described by Albright in 1939, X-linked hypophosphatemic rickets (XLH) is characterized by growth retardation, rachitic and osteomalacic bone disease, and dental abscesses. It is the most common disorder of renal phosphate wasting, occurring in ~1:25,000 live births.

Genetics

The X-linked inheritance was first detailed in 1958 by Winters in a study of a North Carolina kindred. It was not until the 1990s that the genetic basis of XLH was elucidated as mutations in *PHEX* (phosphate-regulating gene with homologies to endopeptidases on the X chromosome). To date, >180 mutations have been described (PHEX database: www.PHEXdb.mcgill.ca). The *PHEX* gene codes for a protein of unknown function that is a member of the M13 family of membrane-bound metalloproteases. Subsequent studies have shown that PHEX is present in osteoblasts, osteocytes, and odontoblasts but not in kidney tubules. Two recent studies have evaluated genotype–phenotype correlations in patients with XLH. Those with truncating mutations tended toward more severe disease and family members in more recent generations had milder phenotypes. Neither study showed a phenotype–genotype correlation.

Clinical and Biochemical Manifestations

Before children begin to walk, clinical findings may be limited. The majority of testing that is done on infants is because of a known family history of XLH. After the child is ambulatory, progressive lower extremity bowing may become apparent with a decrease in height velocity. In addition, there may be dental manifestations, including abscessed noncarious teeth, enamel defects, enlarged pulp chambers, and taurodontism. Adults may exhibit bone and joint pain from osteomalacia, pseudofractures, and enthesopathy. Biochemically, the laboratory findings in XLH are indistinguishable from TIO (Table 3), with low serum concentrations of phosphorus, phosphaturia, and frankly low or inappropriate normal levels of serum 1,25$(OH)_2$D. Calcium and PTH are typically normal. This biochemical profile suggests that regulators of phosphate other than parathyroid hormone and vitamin D play a role in the development of XLH

Amassing evidence suggests that FGF23 is key in the pathogenesis of XLH. The endogenous substrate for PHEX remains unknown. Whereas PHEX does not seem to cleave FGF23 directly, serum FGF23 levels are elevated in many XLH patients, and FGF23 expression is increased in the bones of *hyp* mice. These observations suggest that PHEX is involved in downregulation and control of FGF23; however, the precise interplay between FGF23 and PHEX is not currently understood.

The diagnosis of XLH is based on a consistent medical history and physical examination, radiological evidence of rachitic disease, appropriate biochemical findings, and a family history consistent with multigenerational or sporadic occurrence of the disorder. Mutational analysis of the *PHEX* gene is available; however, studies have shown that mutations can only be found in 50–70% of affected individuals.

Treatment and Complications

Treatment is similar for most patients consisting of high dose oral phosphate administered three to five times daily and high-dose calcitriol, the active form of vitamin D. This treatment leads to resolution of radiographic rickets and improved but not normal growth. Subsequent studies have shown that age at initiation of therapy, height at initiation of therapy, and possibly sex of the patient influence peak height attainment. Patients who do not respond to pharmacologic therapy may require surgical intervention to correct the lower extremity deformities.

Because of lower bone turnover and the closure of the epiphyseal plates, the therapeutic requirements drop dramatically as children enter adulthood. In adults, the role of therapy is unclear. As adults, patients may be treated with no medications, a low dose of calcitriol, low doses of phosphorus, or with both calcitriol and phosphorus. It is unknown which patients require long-term treatment and which patients can have their treatment safely discontinued as adults. Symptomatic adults may benefit from therapy with improvement of symptoms and

© 2008 American Society for Bone and Mineral Research

in bone architecture and prevention of osteomalacia. Although data have shown that BMD may be increased at the spine of patients with XLH, this may reflect, in part, calcific enthesopathy, and it is unclear if there is a change in long term fracture rates. Complications of treatment are similar to those in TIO.

AUTOSOMAL DOMINANT HYPOPHOSPHATEMIC RICKETS

Autosomal dominant hypophosphatemic rickets (ADHR) is a rare form of hypophosphatemic rickets with clinical characteristic similar to XLH.

Genetics

Multiple early reports documented an inheritance pattern that included male-to-male transmission. A subset of patients with ADHR has a delayed onset of symptoms, and there is an incomplete penetrance of disease within affected families. Positional mapping, cloning, and sequence analysis established that FGF23 was mutated in ADHR. The mutations in FGF23 occur within a subtilisin proprotein convertase cleavage site leading to impaired cleavage of the full-length active protein, thereby prolonging its activity, mimicking FGF23 excess.

Clinical and Biochemical Manifestations

The clinical and biochemical findings in ADHR are similar to those observed in XLH. Clinically, there can be short stature with lower extremity deformities accompanied by hypophosphatemia, renal phosphate wasting, and inappropriately low levels of calcitriol (Table 3). In contrast to XLH, there are instances of delayed onset, incomplete penetrance, and rarely, resolution of the phosphate wasting. Within the same family, there can be variable presentations with two subgroups of affected individuals described. Those with childhood onset have biochemical and clinical similarities to XLH, whereas those presenting later often lack lower extremity deformities, presumably because of fusion of the growth plate before the development of hypophosphatemia.

FGF23 is central in the pathogenesis of ADHR. Missense mutations in one of two arginine residues at positions 176 or 179 have been identified in affected members of ADHR families. This clustering of missense mutations suggests that they are activating mutations. Furthermore, the mutated arginine residues, located in the consensus proprotein convertase cleavage RXXR motif, prevent inactivation of FGF23 and thus result in prolonged or enhanced FGF23 action.

Treatment

Treatment is similar to that of patients with XLH, consisting of phosphate and calcitriol. As in XLH, patients who do not respond to pharmacologic therapy may require surgical intervention.

OTHER DISORDERS OF RENAL PHOSPHATE WASTING

Hereditary Hypophosphatemic Rickets With Hypercalciuria

Hereditary hypophosphatemic rickets with hypercalciuria (HHRH) is a rare genetic form of hypophosphatemic rickets characterized by hypophosphatemia, renal phosphate wasting, and preserved responsiveness of $1{,}25(OH)_2D_3$ to hypophosphatemia (Table 3). This appropriate increase in calcitriol leads to increased calcium absorption from the gastrointestinal tract and thus to hypercalciuria and nephrolithiasis. The genetic defect in HHRH is loss of function mutations in the gene that encodes NaPiIIc (SLC34A3), one of three subtypes of the type II sodium phosphate co-transporters. HHRH is clinically similar to TIO, with bone pain, osteomalacia, and muscle weakness as prominent features, yet the distinction is easily made with biochemical testing. Both syndromes are characterized by hypophosphatemia because of decreased renal phosphorus reabsorption; however, patients with HHRH exhibit elevated levels of calcitriol and hypercalciuria that distinguish it from TIO, XLH, and ADHR. Treatment consists of phosphate supplements alone.

Autosomal Recessive Hypophosphatemic Rickets

The genetic defect in autosomal recessive hypophosphatemic rickets (ARHR) was identified as loss of function mutations in dentin matrix protein 1 (DMP-1), a matrix protein related to MEPE and a member of the SIBLING (small integrin binding ligand N-linked glycoprotein) family. Interestingly, this protein seems to have two functions: it translocates into the nucleus to regulate gene transcription early in osteocyte proliferation and, likely in response to calcium fluxes, becomes phosphorylated and is exported to the extracellular matrix to facilitate mineralization by hydroxyapatite in a process that requires appropriate cleavage of the full-length protein. Loss of DMP-1 function in ARHR leads to modestly and variably increased serum FGF23, dramatically increased expression of FGF23 in bone, defects in osteocyte maturation, and impaired skeletal mineralization. It seems that the immature osteocytes overproduce FGF23, which acts on the kidney resulting in phosphaturia and impaired calcitriol synthesis. As in XLH, treatment consists of high dose phosphate and calcitriol.

Hypophosphatemic Rickets With Hyperparathyroidism

Brownstein et al. reported a patient with hypophosphatemic rickets, renal phosphate wasting, inappropriately normal $1{,}25(OH)_2D$, and hyperparathyroidism secondary to a genetic translocation resulting in increased levels of α-KLOTHO, the co-factor necessary for FGF23 to bind and activate its receptor (Table 3). Interestingly and somewhat unexpectedly, FGF23 serum levels are also markedly elevated in this disorder. These findings implicate α-KLOTHO in the regulation of serum phosphate, of FGF23 expression, and of parathyroid function.

Fibrous Dysplasia

In polyostotic fibrous dysplasia, FGF23 is overproduced and results in renal phosphate wasting and hypophosphatemia (Table 3). In fibrous dysplasia, bone lesions that replace medullary bone with fibrous tissue, are associated with disorders such as McCune Albright syndrome, a somatic gene defect that leads to hormone-independent activation of G-protein ($Gs\alpha$)-coupled signaling. The etiology of the phosphate abnormality is not clear, but recent evidence has implicated FGF23 as a key component of the phosphate loss in both McCune Albright related and isolated fibrous dysplasia, the degree of fibrous dysplasia correlating with the degree of phosphate wasting.

Osteoglophonic Dysplasia

As in fibrous dysplasia, renal phosphate wasting and a lower than expected calcitriol level have been observed in osteoglophonic dysplasia, a rare form of dwarfism caused by activating mutations in FGF receptor 1 (FGFR1; Table 3). The mechanism of the phosphate wasting is from the high burden of nonossifying lesions often seen in these patients. FGF23 pro-

© 2008 American Society for Bone and Mineral Research

duced by the abnormal bone is likely responsible, because the extent of bone lesions is correlated with FGF23 levels and the degree of phosphate wasting.

Renal Type IIa Sodium-Phosphate Co-Transporter Mutations

A hypophosphatemic syndrome was described with heterozygous, dominant negative, mutations in the renal type IIa sodium-phosphate co-transporter gene (*NPT-2*). Two patients with hypophosphatemia secondary to renal phosphate wasting and osteopenia or nephrolithiasis were described. The prominent symptoms of bone pain and muscle weakness seen in TIO are absent in those with *NPT-2* mutations. Similar to what is seen in HHRH, there is hypercalciuria and elevated calcitriol levels.

In summary, either through mutations of the sodium-phosphate transporters themselves as in HHRH, damage to the proximal renal tubule as in Fanconi's syndrome or through aberrant regulation of FGF23, decreased expression or function of the renal sodium-phosphate co-transporters likely represent the common pathway in renal phosphate wasting observed in these syndromes (Fig. 1)

HYPERPHOSPHATEMIA

Serum inorganic phosphorus levels are generally maintained between 2.5 and 4.5 mg/dl; 90.75–1.45 nM) in adults and between 6 and 7 mg/dl in children <2 yr old.

In steady state, oral phosphate loads of up to 4000 mg/d can be efficiently excreted by the kidneys with minimal rise in serum phosphorus through downregulation of the sodium phosphate co-transporter in the proximal renal tubules. Increased PTH secretion also contributes to increased renal phosphate excretion because excess phosphorus complexes with calcium, resulting in a decreased in ionized calcium, which stimulates PTH secretion. There are four general mechanisms whereby phosphate entry into the extracellular fluid can outstrip the rate of renal excretion: acute exogenous phosphate loads, redistribution of intracellular phosphate to the extracellular space, decreased renal excretion, and pseudohyperphosphatemia caused by interference with analytical detection methods.

Causes of Hyperphosphatemia

Acute Exogenous Phosphate Load. Overzealous administration of exogenous phosphate as intravenous phosphate, phosphate-containing laxatives, Fleet's phosphosoda enemas, and parenteral nutrition may result in hyperphosphatemia (Table 4).

Redistribution of Phosphorus to Extracellular Space. Conditions associated with increased catabolism or tissue destruction often result in hyperphosphatemia. Clinical situations such as rhabdomyolysis, tumor lysis syndrome, crush injuries, severe hyperthermia, systemic infection, fulminant hepatitis, and hemolytic anemia can be complicated by hyperphosphatemia. In tumor lysis syndrome, cytotoxic therapy damages rapidly expanding tumors such as lymphomas and leukemias, releasing intracellular substances with resultant hyperphosphatemia, renal failure, hyperkalemia, hyperuricemia, and hypocalcemia. Metabolic acidosis reduces glycolysis and thus cellular phosphate utilization. In lactic acidosis, this effect is compounded by tissue hypoxia with ATP dephosphorylation to AMP. In diabetic ketoacidosis, reduced glycolysis and insulin deficiency play a role in the hyperphosphatemia seen on initial clinical presentation.

TABLE 4. CAUSES OF HYPERPHOSPHATEMIA

Decrease renal excretion
Renal insufficiency/failure
Hypoparathyroidism
Pseudohypoparathyroidism
Acromegaly
Bisphosphonates (etidronate)
Tumoral Calcinosis
FGF23 inactivating gene mutations
GALNT3 mutations causing aberrant FGF23 glycosylation
KLOTHO inactivating mutations causing FGF23 resistance
Acute phosphate load
Exogenous phosphate: phosphate containing laxatives, Fleet's phosphosoda enemas, intravenous phosphate administration
Redistribution to extracellular space
Tumor lysis
Rhabdomyolysis
Acidosis
Hemolytic anemia
Catabolic states
Increased bone resorption
Pseudohyperphosphatemia
Hyperglobulinemia
Hyperlipidemia
Hemolysis
Hyperbilirubinemia

Pseudohyperphosphatemia. Hyperphosphatemia may be spurious because of interference with analytical detection methods particularly in the setting of hyperglobulinemia (multiple myeloma, Waldenstrom's macroglobulinemia, monoclonal gammopathy), hyperlipidemia (including liposomal amphotericin B), hemolysis, and hyperbilirubinemia.

Decreased Renal Excretion

Renal Failure. The most common cause of hyperphosphatemia is impaired renal excretion caused by renal failure. In stage II and III chronic kidney disease, phosphorus homeostasis is preserved by a progressive reduction in the fraction of filtered phosphate resorbed by the renal tubules because of an increase in FGF23 levels. As renal failure progresses and the number of functional nephrons diminishes, phosphorus balance can no longer be maintained with reductions in tubular reabsorption of phosphate, and hyperphosphatemia ensues. Hyperphosphatemia results in an increased filtered load per nephron and the renal clearance rate is re-established but at a higher serum phosphorus level.

Hypoparathyroidism and Pseudohypoparathyroidism. Either deficient secretion of PTH (hypoparathyroidism) or renal resistance to PTH (PHP) results in increased phosphate reabsorption and leads to hyperphosphatemia and hypocalcemia.

Acromegaly. Growth hormone and insulin-like growth factor I directly stimulates phosphate reabsorption in the kidney, and thus some patients with acromegaly exhibit hyperphosphatemia.

Bisphosphonates (Etidronate). Some bisphosphonates (etidronate, pamidronate, alendronate) are associated with hyperphosphatemia. Mechanisms including cellular phosphate redistribution and decrease renal excretion have been invoked.

© 2008 American Society for Bone and Mineral Research

FAMILIAL TUMORAL CALCINOSIS

Familial tumoral calcinosis (FTC; OMIM 211900) is an inherited disorder notable for progressive deposition of calcium phosphate crystals in periarticular spaces and soft tissues. There are both hyperphosphatemic and normophosphatemic forms (OMIM 610455) of this disorder.

Genetics

Both autosomal recessive and autosomal dominant inheritance has been noted to occur with this disorder. To date, there have been mutations found in four different genes among the families with FTC. In the normophosphatemic form, mutations in sterile α motif domain 9 (SAMD9) have been described. For the hyperphosphatemic form, inactivating mutations have been found in UDP-*N*-acetyl-α-D-galactosamine (GALNT3), FGF23, and KLOTHO (Fig. 1). The mutations in FGF23 lead to FGF23 deficiency, mutations in GALNT3 cause aberrant FGF23 glycosylation that increases its inactivation, whereas the mutations in KLOTHO, a co-factor necessary for FGF23 to bind to the FGF receptor, lead to FGF23 resistance.

Clinical and Biochemical Manifestations

Patients with FTC have heterotopic calcifications that are typically painless and slow growing. The masses can occasionally infiltrate into adjacent structures, including muscle and tendon. The clinical complications are related to masses near and infiltration of skin, marrow, teeth, blood vessels, and nerves. Range of motion is generally not affected unless the masses become large. A variably present feature of the disease is an abnormality in dentition, characterized by short bulbous roots, pulp stones, and radicular dentin deposited in swirls.

Biochemically, along with the hyperphosphatemia, there is increased $1,25(OH)_2D$ with normal calcium and alkaline phosphatase levels. Urinary phosphate reabsorption is frequently supranormal. Radiographs show large aggregates of irregularly dense calcified lobules. The joints are generally not affected.

Treatment

Medical therapy with aluminum hydroxide along with dietary phosphate and calcium deprivation has been reported. Recently, use of the phosphate binder, sevelamer, and the carbonic anhydrase inhibitor, acetazolamide, have been reported to successfully decrease tumor burden. Surgical intervention is generally considered an option in patients when the masses are painful, interfere with function, or are cosmetically unacceptable.

Clinical Manifestations of Hyperphosphatemia

Hypocalcemia and Tetany. The most common and clinically significant consequence of acute, short-term hyperphosphatemia is hypocalcemia and tetany. With rapid elevations in phosphate load, hypocalcemia and tetany can occur with serum phosphate levels as low as 6 mg/dl. In addition to its effect on the calcium phosphate product and the resultant soft tissue deposition, hyperphosphatemia suppresses the renal 1-a-hydroxylase, reducing circulating $1,25\ (OH)_2D$ that further aggravates hypocalcemia by impairing intestinal phosphate absorption. Hypocalcemia that results from renal failure characteristically develops slowly and thus rarely results in tetany. Occasionally, during the acute phase of tumor lysis or rhabdomyolysis, tetany from hyperphosphatemia is observed.

Soft Tissue Calcification. In contrast, consequences of chronic hyperphosphatemia include soft tissue calcification and in the case of renal failure, secondary hyperparathyroidism, and renal osteodystrophy. In the setting of chronic kidney disease, the defenses against mineralization are compromised, and hyperphosphatemia promotes mineral deposition in soft tissues. In addition, hyperphosphatemia stimulates vascular cells to undergo osteogenic differentiation. Calcification of coronary arteries and heart valves is associated with hypertension, congestive heart failure, coronary artery disease, and myocardial infarction. Medial calcification of peripheral arteries associated with hyperphosphatemia may lead to calciphylaxis, a disorder with high mortality and morbidity.

Secondary Hyperparathyroidism and Renal Osteodystrophy. Hyperphosphatemia that results from renal failure plays a critical role in the development of secondary hyperparathyroidism and renal osteodystrophy. This is discussed in detail elsewhere in the *Primer*.

Treatment of Hyperphosphatemia

Treatment of hyperphosphatemia should address the underlying etiology. In the case of acute exogenous phosphorus overload, prompt discontinuation of supplemental phosphate and hydration are indicated. These measures should allow rapid renal excretion and correction of hyperphosphatemia. When transcellular shifts are the cause of hyperphosphatemia (e.g., tumor lysis, rhabdomyolysis), dietary phosphate restriction and diuresis are often successful. In diabetic ketoacidosis (DKA), treatment with insulin and treatment of acidosis reverses the hyperphosphatemia. Phosphate binders such as calcium salts, sevelamer, and lanthanum carbonate, along with dietary restriction, are indicated in renal failure and FTC. Symptomatic hypocalcemia caused by acute severe hyperphosphatemia can be life threatening and should be treated promptly. Phosphate excretion can be increased by saline infusion, but this may exacerbate hypocalcemia. Hemodialysis may be indicated in acute hyperphosphatemia in the setting of renal dysfunction.

ACKNOWLEDGMENTS

This work was supported by the National Institutes of Health [R01 DK073273 (SJ), U54 RR023561 (SJ)].

SUGGESTED READING

Anonymous 1995 A gene (PEX) with homologies to endopeptidases is mutated in patients with X-linked hypophosphatemic rickets. The HYP Consortium. Nat Genet **11**:130–136.

Anonymous 2000 Autosomal dominant hypophosphataemic rickets is associated with mutations in FGF23. The ADHR Consortium. Nat Genet **26**:345–348.

Benet-Pages A, Orlik P, Strom TM, Lorenz-Depiereux B 2005 An FGF23 missense mutation causes familial tumoral calcinosis with hyperphosphatemia. Hum Mol Genet **14**:385–390.

Bergwitz C, Roslin NM, Tieder M, Loredo-Osti JC, Bastepe M, Abu-Zahra H, Frappier D, Burkett K, Carpentr TO, Anderson D, Garabedian M, Sermet I, Fujiwara TM, Morgan K, Tenenhouse HS, Juppner H 2006 SLC34A3 mutations in patients with hereditary hypophosphatemic rickets with hypercalciuria predict a key role for the sodium-phosphate cotransporter NaPi-IIc in maintaining phosphate homeostasis. Am J Hum Genet **78**:179–192.

Berndt T, Craig TA, Bowe AE, Vassiliadis J, Reczek D, Finnegan R, Jan de Beur SM, Schiavi SC, Kumar R 2003 Secreted frizzled-related protein 4 is a potent tumor-derived phosphaturic agent. J Clin Invest **112**:785–794.

Brownstein CA, Adler F, Nelson-Williams C, Iimima J, Li P, Imura A, Nabeshima Y, Reyes-Mugica M, Carpenter TO, Lifton RP 2008 A translocation causing increased α-klotho level results in hypophosphatemic rickets and hyperparathyroidism. Proc Natl Acad Sci USA **105**:3455–3460.

Cai Q, Hodgson SF, Kao PC, Lennon VA, Klee GG Zinsmiester AR,

© 2008 American Society for Bone and Mineral Research

Kumar R 1994 Brief report: Inhibition of renal phosphate transport by a tumor product in a patient with oncogenic osteomalacia. N Engl J Med 330:1645–1649.
Carpenter TO, Ellis BK, Insogna KL, Philbrick WM, Sterpka J, Shimkets R 2005 Fibroblast growth factor 7: An inhibitor of phosphate transport derived from oncogenic osteomalacia-causing tumors. J Clin Endocrinol Metab 90:1012–1020.
Glorieux FH, Scriver CR, Reade TM, Goldman H, Roseborough A 1972 Use of phosphate and vitamin D to prevent dwarfism and rickets in X-linked hypophosphatemia. N Engl J Med 287:481–487.
Ichikawa S, Imel EA, Kreiter ML, Yu X, Mackenzie DS, Sorenson AH, Goetz R, Mohammadi M, White KE, Econs MJ 2007 A homozygous missense mutation in human KLOTHO causes severe tumoral calcinosis. J Clin Invest 117:2684–2691.
Jan de Beur SM 2005 Tumor-induced osteomalacia. JAMA 294:1260–1267.
Jan de Beur SM, Streeten EA, Civelek AC, McCarthy EF, Ruibe L, Marx SJ, Onobrakpeya O, Raisz LG, Watts NB, Sharon M, Levine MA 2002 Localisation of mesenchymal tumours by somatostatin receptor imaging. Lancet 359:761–763.
Jonsson KB, Zahradnik R, Larsson T, White KE, Sugimoto T, Imanishi Y, Yamamoto T, Hampson G, Koshiyama H, LjunggrenO, Oba K, Yang IM, Miyauchi A, Econs MJ, Lavigne J, Juppner H 2003 Fibroblast growth factor 23 in oncogenic osteomalacia and Xlinked hypophosphatemia. N Engl J Med 348:1656–1663.
Kurosu H, Ogawa Y, Miyoshi M. Yamamoto M, Nandi A, Rosenblatt JP, Baum MG, Schiavi S, Hu MC, Moe OW, Kuro-o M 2006 Regulation of fibroblast growth factor-23 signaling by KLOTHO. J Biol Chem 281:6120–6123.
Lorenz-Depiereux B, Bastepe M, Benet-Pages A, Amyere M. Wagenstaller J, Muller-Barth U, Badenhoop K, Kaiser SM, Rittmaster RS, Shlossberg AH, Olivares JL, Loris C, Ramos FJ, Glorieux F, Vikkula M, Juppner H, Strom TM 2006 DMP1 mutations in autosomal recessive hypophosphatemia implicate a bone matrix protein in the regulation of phosphate homeostasis. Nat Genet 38:1248–5.
Lorenz-Depiereux B, Benet-Pages A, Eckstein G, Tenenbaum-Rakover Y, Wagenstaller J, Tiosano D, Gershoni-Baruch R, Albers N, Lichtner P, Schnabel D, Hochberg Z, Strom TM 2006 Hereditary hypophosphatemic rickets with hypercalciuria is caused by mutations in the sodium-phosphate cotransporter gene SLC34A3. Am J Hum Genet 78:193–201.
Prie D, Huart V, Bakouh N, Planelles G, Dellis O, Gerard B, Hulin P, Benque-Blanchet F, Silve C, Grandchamp B, Friedlander G 2002 Nephrolithiasis and osteoporosis associated with hypophosphatemia caused by mutations in the type 2a sodium-phosphate cotransporter. N Engl J Med 347:983–991.
Reid IR, Hardy DC, Murphy WA, Teitelbaum SL, Bergfeld MA, Whyte MP 1989 X-linked hypophosphatemia: A clinical, biochemical, and histopathologic assessment of morbidity in adults. Medicine (Baltimore) 68:336–352.
Riminucci M, Collins MT, Fedarko NS, Cherman N, Corsi A, White KE, Waguespack S, Gupta A, Hannon T, Econs MJ, Bianco P, Gehron Robey P 2003 FGF-23 in fibrous dysplasia of bone and its relationship to renal phosphate wasting. J Clin Invest 112:683–692.
Rowe PS, Kumagai Y, Gutierrez G, Garrett IR, Blacher R, Rosen D, Cundy J, Navvab S, Chen D, Drezner MK, Quarles LD, Mundy GR 2004 MEPE has the properties of an osteoblastic phosphatonin and minhibin. Bone 34:303–319.
Shimada T, Mizutani S, Muto T, Yoneya T, Hino R, Takeda S, Takeuchi Y, Fujita T, Fukumoto S, Yamashita T 2001 Cloning and characterization of FGF23 as a causative factor of tumor-induced osteomalacia. Proc Natl Acad Sci USA 98:6500–6505.
Topaz O, Shurman DL, Bergman R, Indelman M, Ratajczak P, Mizrachi M, Khamaysi Z, Behar D, Petronius D, Friedman V, Zelikovic I, Raimer S, Metzker A, Richard G, Sprecher E 2004 Mutations in GALNT3, encoding a protein involved in O-linked glycosylation, cause familial tumoral calcinosis. Nat Genet 36:579–581.
Urakawa I, Yamazaki Y, Shimada T, Iijima K, Hasegawa H, Okawa K, Fujita T, Fukumoto S, Yamashita T 2006 KLOTHO converts canonical FGF receptor into a specific receptor for FGF23. Nature 444:770–774.
Verge CF, Lam A, Simpson JM, Cowell CT, Howard NJ, Silink M 1991 Effects of therapy in X-linked hypophosphatemic rickets. N Engl J Med 325:1843–1848.

Chapter 70. Magnesium Depletion and Hypermagnesemia

Robert K. Rude

Division of Endocrinology, Keck School of Medicine, University of Southern California, Los Angeles, California

HYPOMAGNESEMIA/MAGNESIUM DEPLETION

Magnesium (Mg) depletion, as determined by low serum Mg levels, is present in 10% of patients admitted to city hospitals and as many as 65% of patients in intensive care units. Hypomagnesemia and/or Mg depletion is usually caused by losses of Mg from either the gastrointestinal tract or the kidney, as outlined in Table 1.

Causes of Mg Depletion

The Mg content of upper intestinal tract fluids is ~1 mEq/liter; therefore, vomiting and nasogastric suction may contribute to Mg depletion. The Mg content of diarrheal fluids is higher (up to 15 mEq/liter), and consequently, Mg depletion is common in acute and chronic diarrhea, regional enteritis, and ulcerative colitis. Malabsorption syndromes may also result in Mg deficiency. Steatorrhea and resection or bypass of the small bowel, particularly the ileum, often results in intestinal Mg malabsorption. Last, acute severe pancreatitis is associated with hypomagnesemia, which may be caused by the clinical problem causing the pancreatitis, such as alcoholism, or to saponification of Mg in necrotic pancreatic fat. A primary defect in intestinal Mg absorption, which presents early in life with hypomagnesemia, hypocalcemia, and seizures, has been described as an autosomal recessive disorder linked to chromosome 9q22 (OMIM 602014). This disorder seems to be caused by mutations in *TRPM6*, which expresses a protein involved with active intestinal Mg transport.

Excessive excretion of Mg into the urine may be the basis of Mg depletion. Renal Mg reabsorption is proportional to tubular fluid flow and to sodium and calcium excretion. Therefore, chronic parenteral fluid therapy, particularly with saline, and volume expansion states, such as primary aldosteronism, and hypercalciuric states may result in Mg depletion. Hypercalcemia decreases renal Mg reabsorption mediated by calcium binding to the calcium-sensing receptor in the thick ascending limb of Henle. Osmotic diuresis caused by glucosuria will re-

Key words: magnesium metabolism, magnesium deficiency, hypomagnesemia, hypermagnesemia, magnesium transport, renal magnesium wasting, hereditary disorders of magnesium transport

The author states that he has no conflicts of interest.

© 2008 American Society for Bone and Mineral Research

Table 1. Causes of Mg Deficiency

1.	Gastrointestinal disorders
	a. Prolonged nasogastric suction/vomiting
	b. Acute and chronic diarrhea
	c. Intestinal and biliary fistulas
	d. Malabsorption syndromes
	e. Extensive bowel resection or bypass
	f. Acute hemorrhagic pancreatitis
	g. Protein-calorie malnutrition
	h. Primary intestinal hypomagnesemia (mutation of *TRPM6*)
2.	Renal loss
	a. Chronic parenteral fluid therapy
	b. Osmotic diuresis (glucose, urea, mannitol)
	c. Hypercalcemia
	d. Alcohol
	e. Diuretics (e.g., furosemide)
	f. Aminoglycosides
	g. Cisplatin
	h. Cyclosporin
	i. Amphotericin B
	j. Pentamidine
	k. Tacrolimus
	l. Proton-pump inhibitors
	m. Metabolic acidosis
	n. Chronic renal disorders with Mg wasting
	o. Primary renal hypomagnesemia (genetic defects in Mg transport)
	p. Mutation of mitochondrial RNA
3.	Endocrine and metabolic disorders
	a. Diabetes mellitus (glycosuria)
	b. Phosphate depletion
	c. Primary hyperparathyroidism (hypercalcemia)
	d. Hypoparathyroidism (hypercalciuria, hypercalcemia caused by overtreatment with vitamin D)
	e. Primary aldosteronism
	f. Hungry bone syndrome
	g. Excessive lactation

sult in urinary Mg wasting. Diabetes mellitus is probably the most common clinical disorder associated with Mg depletion.

An increasing list of drugs causes renal Mg wasting and Mg depletion. The major site of renal Mg reabsorption is at the loop of Henle; therefore, diuretics such as furosemide results in marked Mg wasting. Aminoglycosides cause a reversible renal lesion that results in hypermagnesuria and hypomagnesemia. Similarly, amphotericin B therapy has been reported to result in renal Mg wasting. Other renal Mg-wasting agents include cisplatin, cyclosporin, tacrolimus, pentamidine, and proton pump inhibitors. A rising blood alcohol level is associated with hypermagnesemia and is one factor contributing to Mg depletion in chronic alcoholism. Metabolic acidosis caused by diabetic ketoacidosis, starvation, or alcoholism may also result in renal Mg wasting.

Several renal Mg wasting disorders have been described that may be genetic or sporadic. One form, which is autosomal recessive, results from mutations in the *paracellin-1* gene on chromosome 3q27–29 (OMIM 603959). This disorder is characterized by low serum Mg, hypercalciuria, and nephrocalcinosis. Another autosomal dominant form of isolated renal Mg wasting and hypomagnesemia has been linked to chromosome 11q23 and identified as a mutation on the Na^+,K-ATPase γ-subunit of the *FXYD2* gene (OMIM 154020). Gitelman's syndrome (familial hypokalemia-hypomagnesemia syndrome; OMIM 263800) is an autosomal recessive disorder caused by a genetic defect of the thiazide-sensitive NaCl co-transporter gene on chromosome 16q3. Autosomal dominant hypocalcemia (OMIM 601199), a mutation of the Ca-sensing receptor, also results in hypomagnesemia. Genetic defects of *TRPM6*, *TRPM7*, and *TRPM9*, members of a cation transport family, may also cause renal Mg wasting (OMIM 602014). Recently, a mutation of mitochondrial RNA has been associated with hypomagnesemia and the metabolic syndrome.

Hypomagnesemia may accompany a number of other disorders. Phosphate depletion results in urinary Mg wasting and hypomagnesemia. Hypomagnesemia may also accompany the "hungry bone" syndrome, a phase of rapid bone mineral accretion in subjects with hyperparathyroidism or hyperthyroidism after surgical treatment. Finally, chronic renal tubular, glomerular, or interstitial diseases may be associated with renal Mg wasting.

Manifestations of Magnesium Depletion

Because Mg depletion is usually secondary to another disease process or to a therapeutic agent, the features of the primary disease process may complicate or mask Mg depletion. A high index of suspicion is therefore warranted.

Neuromuscular hyperexcitability may be the presenting complaint. Latent tetany, as elicited by positive Chvostek's and Trousseau's signs, or spontaneous carpal-pedal spasm may be present. Frank generalized seizures may also occur. Although hypocalcemia often contributes to the neurological signs, hypomagnesemia without hypocalcemia has been reported to result in neuromuscular hyperexcitability. Other signs may include vertigo, ataxia, nystagmus, and athetoid and choreiform movements, as well as muscular tremor, fasciculation, wasting, and weakness.

Electrocardiographic abnormalities of Mg depletion in humans include prolonged P-R interval and Q-T interval. Mg depletion may also result in cardiac arrhythmias. Supraventricular arrhythmias have been described. Ventricular premature complexes, ventricular tachycardia, and ventricular fibrillation are more serious complications. A common laboratory feature of Mg depletion is hypokalemia. During Mg depletion there is loss of potassium from the cell with intracellular potassium depletion and an inability of the kidney to conserve potassium. Attempts to replete the potassium deficit with potassium therapy alone are not successful without simultaneous Mg therapy. This biochemical feature may be a contributing cause of the electrocardiologic findings and cardiac arrhythmias discussed above.

Hypocalcemia is a common manifestation of moderate to severe Mg depletion. The hypocalcemia may be a major contributing factor to the increased neuromuscular excitability often present in Mg depleted patients. The pathogenesis of hypocalcemia is multifactorial. In normal subjects, acute changes in the serum Mg concentration will influence PTH secretion in a manner similar to calcium through binding to the calcium-sensing receptor. An acute fall in serum Mg stimulates PTH secretion, whereas hypermagnesemia inhibits PTH secretion. During chronic and severe Mg depletion, however, PTH secretion is impaired. Most patients will have serum PTH concentrations that are undetectable or inappropriately normal for the degree of hypocalcemia. Some patients, however, may have serum PTH levels above the normal range that may reflect early magnesium depletion. Regardless of the basal circulating PTH concentration, an acute injection of Mg stimulates PTH secretion. Impaired PTH secretion therefore seems to be a major factor in hypomagnesemia-induced hypocalcemia. Hypocalcemia in the presence of normal or elevated serum PTH concentrations also suggests end-organ resistance to PTH. Patients with hypocalcemia caused by Mg depletion have

© 2008 American Society for Bone and Mineral Research

TABLE 2. BONE FEATURES AT 6 MO OF LOW MAGNESIUM DIET AS DETERMINED BY μCT*

	Control diet	*Low magnesium diet*
Percent trabecular bone volume (BV/TV)	20.6 ± 9.8 (7)	10.2 ± 5.0 (7)†
Trabecular width (μm)	57.7 ± 12.1 (7)	45.4 ± 7.9 (7)†
Trabecular umber (TbN)	3.4 ± 1.1 (7)	2.1 ± 0.8 (7)†

Data are means ± SD (*n*).

*Reproduced with permission from the American Society for Nutrition from Rude RK, Gruber HE, Norton J, Wei LY, Frausto A, Mills BG 2004 Bone loss induced by dietary magnesium reduction to 10% of nutrition requirement in the rats is associated with increased release of substance P and tumor necrosis factor-a. J Nutr **134:**79–85. Copyright 2004 by American Society for Nutrition. Permission conveyed through Copyright Clearance Center, Inc.

†Significantly different than corresponding control values ($p < 0.01$).

both renal and skeletal resistance to exogenously administered PTH as manifested by subnormal urinary cAMP and phosphate excretion and diminished calcemic response. This renal and skeletal resistance to PTH is reversed after several days of Mg therapy. The basis for the defect in PTH secretion and PTH end-organ resistance is unclear but may be caused by a defect in the adenylate cyclase and/or phospholipase C second messenger systems because they are important in PTH secretion and mediating PTH effects in kidney and bone. Magnesium is necessary for the activity of the G proteins in both enzyme systems. Magnesium is also necessary for substrate formation (MgATP) and is an allosteric activator of adenylate cyclase. Recent data have suggested that Mg deficiency disinhibits Gα subunits and mimics activation of the calcium-sensing receptor.

Clinically, patients with hypocalcemia caused by Mg depletion are resistant not only to PTH but also to calcium and vitamin D therapy. The vitamin D resistance may be caused by impaired metabolism of vitamin D, because serum concentrations of 1,25-dihydroxyvitamin D are low.

Epidemiological studies have suggested that a low Mg diet may be a risk factor for osteoporosis. Dietary Mg intake in the population falls significantly below the recommended daily allowance (RDA).

Mg supplements may also result in an increase in bone mass. Mg deficiency in animal models causes a decrease in bone mass (Table 2) and increase in skeletal fragility. These changes occur even when the diet Mg intake in rats is only 50% of normal. The mechanism(s) may involve abnormalities in PTH and $1,25(OH)_2$ vitamin D formation and/or action and an increase in bone-resorbing inflammatory cytokines.

Diagnosis of Magnesium Depletion

Measurement of the serum Mg concentration is the most commonly used test to assess Mg status. The normal serum Mg concentration ranges from 1.5 to 1.9 mEq/liter (1.8–2.2 mg/dl), and a value <1.5 mEq/liter usually indicates Mg depletion. Mg is principally an intracellular cation, and only ~1% of the body Mg content is in the extracellular fluid compartments. The serum Mg concentration therefore may not reflect the intracellular Mg content. Ion-selective electrodes for Mg are now available; however, different instruments differ in accuracy and may give misleading results in sera with low Mg levels. Because vitamin D and calcium therapy are relatively ineffective in correcting the hypocalcemia, there must be a high index of suspicion for the presence of Mg depletion. Patients with Mg depletion severe enough to result in hypocalcemia are usually significantly hypomagnesemic. However, occasionally, patients may have normal serum Mg concentrations. Mg deficiency in the presence of a normal serum Mg concentration has been shown by measuring intracellular Mg or by whole body retention of infused Mg. Therefore, hypocalcemic patients who are at risk for Mg depletion, but who have normal serum Mg levels, should receive a trial of Mg therapy. The Mg tolerance test (or retention test) is an accurate means of assessing Mg status; correlation with skeletal muscle Mg content and Mg balance have been shown. This test seems to be discriminatory in patients with normal renal function; however, its usefulness may be limited if the patient has a renal Mg wasting disorder or is on a medication that induces renal Mg wasting. A suggested protocol for the Mg tolerance test is shown in Table 3.

Therapy

Patients who present with signs and symptoms of Mg depletion should be treated with Mg. These patients will usually be hypomagnesemic and/or have an abnormal Mg tolerance test. The extent of the total body Mg deficit is impossible to predict, but it may be as high as 200–400 mEq. Under these circumstances, parenteral Mg administration is usually indicated. An effective treatment regimen is the administration of 2 g $MgSO_4 \cdot 7\,H_2O$ (16.2 mEq Mg) as a 50% solution every 8 h intramuscularly. Because these injections are painful, a continuous intravenous infusion of 48 mEq over 24 h may be preferred. Either regimen will usually result in a normal to slightly elevated serum Mg concentration. Despite the fact that PTH secretion increases within minutes after beginning Mg administration, the serum calcium concentration may not return to normal for 3–7 days. This probably reflects slow restoration of intracellular Mg. During this period of therapy, serum Mg concentration may be normal, but the total body deficit may not yet be corrected. Magnesium should be continued until the clinical and biochemical manifestations (hypocalcemia and hypokalemia) of Mg depletion are resolved.

Patients who are hypomagnesemic and have seizures or an acute arrhythmia may be given 8–16 mEq of Mg as an intravenous injection over 5–10 min, followed by 48 mEq/d intravenously. Ongoing Mg losses should be monitored during therapy. If the patient continues to lose Mg from the intestine or kidney, therapy should be continued for a longer duration. Once repletion has been accomplished, patients usually can maintain a normal Mg status on a regular diet. If repletion is accomplished and the patient cannot eat, a maintenance dose of 8 mEq should be given daily. Patients who have chronic Mg loss from the intestine or kidney may require continued oral Mg supplementation. A daily dose of 300–600 mg of elemental

TABLE 3. SUGGESTED PROTOCOL FOR USE OF MAGNESIUM TOLERANCE TEST

I. Collect baseline 24-h urine for magnesium/creatinine ratio*

II. Infuse 0.2 mEq (2.4 mg) elemental magnesium per kilogram lean body weight in 50 ml 5% dextrose over 4 h

III. Collect urine (starting with infusion) for magnesium and creatinine for 24 h

IV. Percentage magnesium retained is calculated by the following formula:

$$\%\text{Mg retained} = \frac{[\text{Postinfusion 24 hr urine Mg} - \text{Preinfusion urine Mg/creatinine} \times \text{Postinfusion urine creatinine}]}{\text{Total elemental Mg infused}} \times 100$$

V. Criteria for Mg deficiency:
>50% retention at 24 h = definite deficiency
>25% retention at 24 h = probable deficiency

*A fasting shorter timed urine (2-h spot) may be used.

© 2008 American Society for Bone and Mineral Research

Mg can be given in divided doses to avoid the cathartic effect of Mg.

Caution should be taken during Mg therapy in patients with any degree of renal failure. If a decrease in glomerular filtration rate exists, the dose of Mg should be halved, and the serum Mg concentration must be monitored daily. If hypermagnesemia ensues, therapy must be stopped.

HYPERMAGNESEMIA

Magnesium intoxication is not a frequently encountered clinical problem, although mild to moderate elevations in the serum Mg concentration may be seen in as many as 12% of hospitalized patients.

Symptomatic hypermagnesemia is virtually always caused by excessive intake or administration of Mg salts. The majority of patients with hypermagnesemia have concomitant renal failure. Hypermagnesemia is usually seen in patients with renal failure who are receiving Mg as an antacid, enema, or infusion. Hypermagnesemia is also seen in acute renal failure in the setting of rhabdomyolysis.

Large amounts of oral Mg have rarely been reported to cause symptomatic hypermagnesemia in patients with normal renal function. The rectal administration of Mg for purgation may result in hypermagnesemia. Mg is a standard form of therapy for pregnancy-induced hypertension (preeclampsia and eclampsia) and may cause Mg intoxication in the mother and in the neonate. Modest elevations in the serum Mg concentration may be seen in familial hypocalcemic hypercalcemia, lithium ingestion, and during volume depletion.

Signs and Symptoms

Neuromuscular symptoms are the most common presenting problem of Mg intoxication. One of the earliest demonstrable effects of hypermagnesemia is the disappearance of the deep tendon reflexes. This is reached at serum Mg concentrations of 4–7 mEq/liter. Depressed respiration and apnea caused by paralysis of the voluntary musculature may be seen at serum Mg concentrations in excess of 8–10 mEq/liter. Somnolence may be observed at levels as low as 3 mEq/liter and above.

Moderate elevations in the serum Mg concentration of 3–5 mEq/liter result in a mild reduction in blood pressure. High concentrations may result in severe symptomatic hypotension. Mg can also be cardiotoxic. At serum Mg concentrations >5 mEq/liter, electrocardiographic findings of prolonged P-R intervals and increased QRS duration and QT interval are seen. Complete heart block, as well as cardiac arrest, may occur at concentrations >15 mEq/liter.

Hypermagnesemia causes a fall in the serum calcium concentration. The hypocalcemia may be related to the suppressive effect of hypermagnesemia on PTH secretion or to hypermagnesemia-induced PTH end-organ resistance. A direct effect of Mg on decreasing the serum calcium is suggested by the observation that hypermagnesemia causes hypocalcemia in hypoparathyroid subjects as well.

Other nonspecific manifestations of Mg intoxication include nausea, vomiting, and cutaneous flushing at serum levels of 3–9 mEq/liter.

Therapy

The possibility of Mg intoxication should be anticipated in any patient receiving Mg, especially if the patient has a reduction in renal function. Mg therapy should merely be discontinued in patients with mild to moderate elevations in the serum Mg level. The kidney will excrete excess Mg, and any symptoms or signs of Mg intoxication will resolve. Patients with severe Mg intoxication may be treated with intravenous calcium. Calcium will antagonize the toxic effects of Mg. This antagonism is immediate but transient. The usual dose is an infusion of 100–200 mg of elemental calcium over 5–10 min. If the patient is in renal failure, peritoneal dialysis or hemodialysis against a low dialysis Mg bath will rapidly and effectively lower the serum Mg concentration.

SUGGESTED READING

Brown EM, MacLeod RJ 2001 Extracellular calcium sensing and extracellular calcium signaling. Physiol Rev **81:**239–297.

Carpenter TO, DeLuca MC, Zhang JH, Bejnerowicz G, Tartamella L, Dziura J, Peterson KF, Belfoy D, Cohen D 2006 A randomized controlled study of effects of dietary magnesium oxide supplementation on bone mineral content in healthy girls. J Clin Endocrinol Metab **91:**4866–4872.

Cholst IN, Steinberg SF, Trooper PJ, Fox HE, Segre GV, Bilezikian JP 1984 The influence of hypermagnesemia on serum calcium and parathyroid hormone levels in human subjects. N Engl J Med **310:**1221–1225.

Fassler CA, Rodriguez RM, Badesch DB, Stone WJ, Marini JJ 1985 Magnesium toxicity as a cause of hypotension and hypoventilation: Occurrence in patients with normal renal function. Arch Intern Med **145:**1604–1606.

Konrad M, Schlingmann KP, Gundermann T 2004 Insights into the molecular nature of magnesium homeostasis Am J Renal Physiol **286:**F599–F605.

Quamme GA 1997 Renal magnesium handling: New insights in understanding old problems. Kidney Int **52:**1180–1195.

Quitterer U, Hoffmann M, Freichel M, Lohse MJ 2001 Paradoxical block of parathormone secretion is mediated by increased activity of G alpha subunits. J Biol Chem **276:**6763–6769.

Romani AMP 2007 Magnesium homeostasis in mammalian cells. Front Biosci **12:**308–331.

Rude RK 1998 Magnesium deficiency: A heterogeneous cause of disease in humans. J Bone Miner Res **13:**749–758.

Rude RK 2001 Magnesium deficiency in parathyroid function. In: Bilezikian JP (ed.) The Parathyroids. Raven Press, New York, NY, USA, pp. 763–777.

Rude RK 2005 Magnesium. In: Coates P, Blackman MR, Cragg G, Levine M, Moss J, White J (eds.) Encyclopedia of Dietary Supplements. Marcel Dekker, New York, NY, USA, pp. 445–455.

Rude RK, Gruber HE 2004 Magnesium deficiency and osteoporosis: Animal and human observations. J Nutr Biochem **15:**710–716.

Rude RK, Gruber HE, Norton HJ, Wei LY, Frausto A, Kilburn J 2005 Dietary magnesium reduction to 25% of nutrient requirement disrupts bone and mineral metabolism in the rat. Bone **37:**211–219.

Rude RK, Gruber HE, Norton HJ, Wei LY, Frausto A, Kilburn J 2006 Nutrient requirement reduction of dietary magnesium by only 50% in the rat disrupts bone and mineral metabolism. Osteoporos Int **17:**1022–1032.

Rude RK, Gruber HE, Norton HJ, Wei LY, Frausto A, Mills BG 2004 Bone loss induced by dietary magnesium reduction to 10% of nutrition requirement in the rats is associated with increased release of substance P and tumor necrosis factor-a. J Nutr **134:**79–85.

Rude RK, Gruber HE, Wei LY, Frausto A 2005 Immunolocalization of RANKL is increased and OPG decreased during dietary magnesium deficiency in the rat. Nutr Metab **2:**24–31.

Rude RK, Shils ME 2006 Magnesium. In: Shils ME (ed.) Modern Nutrition in Health and Disease. Lippincott Williams and Wilkins, Philadelphia, PA, USA, pp. 223–247.

Schlingmann KP, Gudermann T 2005 A critical role to TRPM channel-kinase for human magnesium transport. J Physiol **566:**301–308.

Schlingmann KP, Konrad M, Seyberth HW 2004 Genetics of hereditary disorders of magnesium homeostasis. Pediatr Nephrol **19:**13–25.

Tong GM, Rude RK 2005 Magnesium deficiency in critical illness. J Intensive Care Med **20:**3–17.

Vetter T, Lohse MJ 2002 Magnesium and the parathyroid. Curr Opin Nephrol Hypertens **11:**403–410.

Voets T, Nilius B, Hoefs S, van der Kemp AW, Droogmans G, Bindels RJ, Hoenderop JG 2004 TRPM6 forms the Mg2+ influx channel involved in intestinal and renal Mg2+ absorption. J Biol Chem **279:**19–25.

Wilson FH, Hariri A, Farhi A, Zhao H, Petersen KF, Toka HR, Nelson-Williams C, Raja KM, Kashgarian M, Shulman GI, Scheinman SJ, Lifton RP 2004 A Cluster of Metabolic defects caused by mutation in a mitochrondrial tRNA. Science **306:**1190–1194.

© 2008 American Society for Bone and Mineral Research

Chapter 71. Vitamin D–Related Disorders

Paul Lips,[1,2] Natasja M. van Schoor,[2] and Nathalie Bravenboer[1,3]

[1]*Department of Endocrinology, VU University Medical Center, Amsterdam, The Netherlands;* [2]*EMGO Institute, VU University Medical Center, Amsterdam, The Netherlands;* [3]*Department of Clinical Chemistry, VU University Medical Center, Amsterdam, The Netherlands*

INTRODUCTION

Nutritional vitamin D deficiency causes insufficient mineralization and rickets in the growing child or osteomalacia in the adult, when the epiphyseal lines have closed. This is the classical vitamin D–related disease.[1] The active vitamin D metabolite 1,25-dihydroxyvitamin D [1,25(OH)$_2$D] stimulates the absorption of calcium and phosphate from the gut and makes calcium and phosphate available for mineralization. In mild or moderate vitamin D deficiency, a lower serum calcium concentration causes stimulation of the parathyroid glands (Table 1). The increased serum PTH increases the conversion of 25-hydroxyvitamin D [25(OH)D] to 1,25(OH)$_2$D as a compensatory mechanism. However, the rise of PTH increases bone resorption. In this way, vitamin D deficiency may also cause bone loss and contribute to the pathogenesis of osteoporosis[2] (Fig. 1).

Rickets and osteomalacia are associated with muscular weakness, and recently it has become clear that mild and moderate vitamin D deficiency may be associated with decreased physical performance and falls.[3–5] In vitro and in vivo studies have shown that 1,25(OH)$_2$D has many actions such as stimulating osteoblasts and longitudinal growth, stimulating the development of the immune system, decreasing proliferation and stimulating differentiation of many cell types, stimulating insulin release, and increasing insulin sensitivity.[6–8] During the last years, vitamin D deficiency has been associated with autoimmune diseases such as diabetes mellitus type 1 and multiple sclerosis, infectious diseases such as tuberculosis, diabetes mellitus type 2, and several types of cancer.[9,10] In this chapter, the more classical vitamin D–related disorders will be discussed such as nutritional rickets and osteomalacia, including neuromuscular features and pathways leading to osteoporotic fractures. Rickets and osteomalacia caused by the absence of 1α-hydroxylase and decrease or absent function of the vitamin D receptor will also be discussed.

NUTRITIONAL RICKETS AND OSTEOMALACIA

The classical picture of nutritional rickets was first described in the 17th century by Whistler and Glisson.[2,11] The association between rickets and low sunshine exposure was first recognized in the 19th century, and in the beginning of the 20th century, rickets was experimentally cured by artificial UV light or sunshine.[11] The spectrum of rickets and osteomalacia includes vitamin D–related and other causes (Table 2).

Vitamin D Deficiency: Definition and Threshold

Vitamin D$_3$ is produced by the skin after sunlight exposure and is for a small part available from dietary sources. Fatty fish is the most important dietary source. Fortification of foods (e.g., milk) with vitamin D is practiced in the United States, Sweden, and Ireland. After production in the skin, vitamin D is hydroxylated in the liver to 25(OH)D, which is further hydroxylated in the kidney to 1,25(OH)$_2$D, the active metabolite.

Vitamin D status is usually assessed by measuring the serum concentration of 25(OH)D.[12] There is no consensus about the required level of serum 25(OH)D. It has been defined by some as the concentration of serum 25(OH)D that maximally suppresses serum PTH. In studies among healthy elderly in the United States, France, and The Netherlands, serum PTH was at its lowest point when serum 25(OH)D > 75 nM (30 ng/ml).[13] Another approach is to define thresholds of serum 25(OH)D for different outcomes. Clinical rickets or osteomalacia usually occurs with severe vitamin D deficiency [i.e., very low levels of serum 25(OH)D, often <12.5 nM (5 ng/ml)]. In a round table discussion of experts, the proposed minimally required level of serum 25(OH)D, optimal for fracture prevention, varied between 50 and 75 nM (20–30 ng/ml).[14] According to a review based on epidemiological data, the most advantageous serum concentration of 25(OH)D for BMD, lower extremity function, falls, fractures, and colorectal cancer is >75 nM (30 ng/ml).[15] The evidence from clinical trials to prevent fractures with vitamin D is less clear.[16] A consensus at the last Vitamin D Workshop agreed that serum 25(OH)D should be >50 nM (20 ng/ml).[17] For the purpose of this chapter, vitamin D deficiency is defined as serum 25(OH)D <25 nM (10 ng/ml). Higher levels of serum 25(OH)D between 25 and 50 nM (10–20 ng/ml); i.e., vitamin D insufficiency, are also associated with secondary hyperparathyroidism and muscle weakness.[13] Both vitamin D deficiency and insufficiency should be prevented [i.e., serum 25(OH)D should be > 50 nM]. The staging of vitamin D status is presented in Table 1.

Epidemiology and Risk Groups

Vitamin D deficiency is very common in certain risk groups, such as premature and dysmature children, the elderly, and non-Western immigrants. Elderly people have decreased dermal synthesis, and older nursing home residents who do not come outside frequently are at high risk. Non-Western immigrants migrating to countries at higher latitudes with limited UVB irradiation are at high risk because of more pigmented skin, the habit to stay out of the sun, the wearing of well-covering clothes, and their diet. Non-Western pregnant women and their children are at a very high risk, and serum 25(OH)D was <25 nM (10 ng/ml) in 85% of pregnant women in a Dutch survey.[18,19]

Large differences in vitamin D status exist between various countries of Europe, the Middle East, and Asia.[20] In Europe, a north-south gradient was observed for serum 25(OH)D with higher levels in Scandinavia and lower levels in Southern and Eastern European countries. This points to other determinants (e.g., nutrition, food fortification, and supplement use). Very low serum 25(OH)D levels have been reported in the Middle East in veiled women. In India, China, and Japan, vitamin D deficiency is more common than in Southeast Asia.

In a meta-analysis of randomized controlled trials, it was found that a vitamin D dose of 700–800 IU (17.5–20 µg) daily reduced the relative risk of hip fracture by 26% and any nonvertebral fracture by 23% versus calcium or placebo.[21] How-

Dr. Lips has received grants from Merck, Procter & Gamble, Aventis, and Wyeth. In addition, he has served as a speaker for Merck, Procter & Gamble, and Servier. All other authors state that they have no conflicts of interest.

Key words: vitamin D, PTH, osteomalacia, rickets, nutrition, osteoblasts and stem cells, bone densitometry, bone histomorphometry, bone mineralization, bone scintigraphy

© 2008 American Society for Bone and Mineral Research

TABLE 1. CLASSIFICATION OF VITAMIN D–DEFICIENT STATUS

	25(OH)D [nM (ng/ml)]	$1,25(OH)_2D$ [pM (pg/ml)]	PTH increase	Bone histology
Severe deficiency	<12.5 (<5)	Normal or low	>30%	Incipient or overt osteomalacia
Moderate deficiency	12.5–25 (5–10)	Normal	15–30%	High turnover
Insufficiency	25–50 (10–20)	Normal	5–15%	Normal or high turnover
Replete	>50 (>20)	Normal	—	Normal

ever, since the publication of this meta-analysis, several randomized controlled trials reporting no effect on fracture incidence were reported. In a Cochrane review, only the combination of vitamin D with calcium reduced hip fractures (pooled RR of seven trials = 0.81; 95% CI: 0.68–0.96) and nonvertebral fractures (pooled RR of seven trials = 0.87; 95% CI: 0.78–0.97) in institutionalized elderly.[22] No significant effect on vertebral fractures was observed. The addition of calcium to vitamin D is essential for its preventive effect on fractures.[23]

Recently, it has been shown that mild to moderate vitamin D deficiency may also be associated with reduced muscle strength, impaired physical performance, and falls.[3–5,24] Vitamin D supplementation in combination with calcium may reduce the risk of falls with 22% compared with calcium alone or placebo (pooled OR of five trials = 0.78; 95% CI: 0.64–0.92).[25] The pathway from vitamin D deficiency to falls and fractures is shown in Fig. 1.

Pathogenesis

Whereas 25(OH)D is the major circulating metabolite and store of vitamin D, almost all vitamin D actions are ascribed to the active metabolite $1,25(OH)_2D$. This metabolite behaves as a steroid hormone that functions through a nuclear receptor, the vitamin D receptor (VDR). After binding to the VDR and dimerization with the retinoid X receptor, the complex binds to specific DNA regions. Proteins are formed such as the calcium binding protein or calbindin thereby increasing the active calcium absorption in the gut. The active metabolite also stimulates bone resorption and decreases PTH secretion. This results in a balanced serum calcium concentration.

Studies in transgenic mice showed that direct binding of $1,25(OH)_2D_3$ to the VDR in cells of the osteoblastic lineage results in a catabolic or anabolic effect depending on the maturation of these cells. The predominant action seems enhancement of RANKL and reduction of osteoprotegerin, thereby stimulating osteoclast formation, but stimulation of osteoblastic genes involved in bone formation such as osteocalcin and osteopontin has also been shown.[26,27]

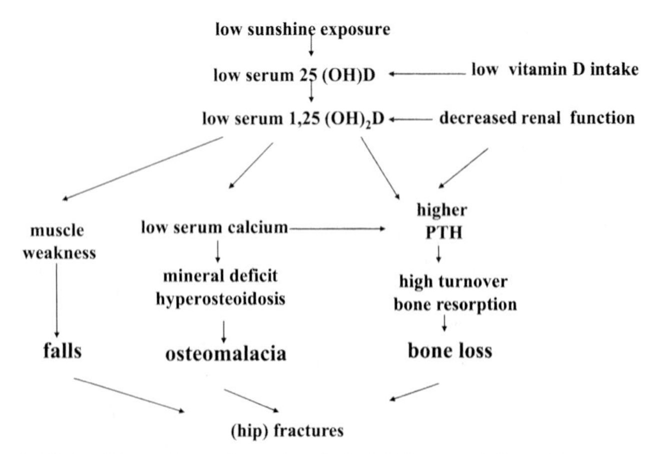

FIG. 1. The pathway from vitamin D deficiency to falls and fractures.

In case of children with vitamin D deficiency, the newly formed bone of the growth plate does not mineralize, and cartilage proliferation is prolonged. The growth plate becomes thick, wide, and irregular. This results in the clinical diagnosis of rickets.[1]

In adults, the newly formed bone matrix, the osteoid, does not mineralize and osteomalacia occurs (Fig. 2). Osteomalacia may be suspected on a clinical basis, especially in case of risk factors, including malabsorption, celiac disease, or severe liver disease (Table 2). Laboratory abnormalities may also point to the diagnosis. Vitamin D also has many nonskeletal effects. The VDR is not only observed in the classical target cells for $1,25(OH)_2D$ but also in the skin, promyelocytes, lymphocytes, colon cells, pituitary cells, and ovarian cells.[28] The function of vitamin D in these tissues has not yet been fully elucidated. The active vitamin D metabolite decreases cell proliferation, and this is applied in practice for the treatment of psoriasis.[11] It is a potent immunomodulator and may play a role in the prevention of autoimmune diseases such as multiple sclerosis and diabetes mellitus type 1.[10]

Clinical Picture

The causes of rickets and osteomalacia are summarized in Table 2. The clinical picture of rickets is characterized by decreased longitudinal growth, widening of the epiphyseal zones, and painful swelling around these.[29] The bowing of the tubular bones is caused by poor mineralization of the growing skeleton. Special features include the rachitic rosary caused by swelling of the cartilage of the ribs. Rachitic children do not grow well and exhibit slowed developmental milestones such as walking. The clinical features of osteomalacia include bone pain, muscular weakness, and difficulty with walking. The muscular weakness is preferential to the proximal muscles around shoulder and pelvic girdle. It is manifest with standing up from a chair or stair climbing. When serum calcium is very low in patients with rickets or osteomalacia, symptoms of low serum calcium level may prevail such as tetany and convulsions. Fractures may occur because the quantity of mineralized bone in osteomalacia is similar to that in osteoporosis.

Laboratory

Vitamin D deficiency is diagnosed by measuring serum 25(OH)D. Rickets and osteomalacia are associated with a serum 25(OH)D <25 nM and often <12.5 nM.[30] Characteristic laboratory findings include a low serum calcium and phosphate concentration or overt hypophosphatemia and an increased level of alkaline phosphatase. The serum concentration of PTH is usually increased (secondary hyperparathyroidism). The serum concentration of $1,25(OH)_2D$ is not helpful because it is normal in most patients with vitamin D deficiency.[12,29] When serum 25(OH)D is very low, the serum $1,25(OH)_2D$ level may be low because of substrate deficiency.[2,31] The 24-h urinary excretion of calcium is low to very low in patients with

© 2008 American Society for Bone and Mineral Research

TABLE 2. CAUSES OF RICKETS AND OSTEOMALACIA

Vitamin D–related rickets/osteomalacia
Nutritional: low sunshine exposure, low dietary intake
Malabsorption: celiac disease, Crohn's disease, gastrectomy, gastric bypass, bowel resection, pancreatitis
Impaired hydroxylation in liver: severe chronic liver disease
Impaired renal function: renal osteodystrophy/osteomalacia
Increased renal loss: nephrotic syndrome
Increased catabolism: anti-convulsant therapy
Inborn errors of metabolism
Nonfunctioning 25-hydroxylase (OMIM 600081)
Absent 1α-hydroxylase: pseudovitamin D deficiency rickets (vitamin D–dependent rickets type 1 OMIM 264700)
Nonfunctioning VDR: hereditary vitamin D resistant rickets (vitamin D–dependent rickets type 2, OMIM 277440)
Hypophosphatemic rickets/osteomalacia: renal phosphate wasting
X-linked hypophosphatemic rickets, OMIM 307800*
Autosomal dominant hypophosphatemic rickets, OMIM 193100*
Hereditary hypophosphatemic rickets with hypercalciuria, OMIM 241530
Oncogenic osteomalacia*
Fanconi syndrome, metabolic acidosis
Calcium deficiency: very low calcium intake in children
Miscellaneous:
Aluminium intoxication
Cadmium intoxication
Etidronate overdose (in Paget's disease)
Hypophosphatasia, OMIM 146300

* Associated with low serum $1{,}25(OH)_2D$.

vitamin D deficiency because of low calcium absorption from the gut and enhanced tubular reabsorption of calcium caused by secondary hyperparathyroidism.

Radiology

Radiological pictures of patients with rickets show decreased mineralization around the epiphysis, unsharp bone margins and less contrast. The number of ossification centers is decreased. Treatment is followed by rapid improvement (Fig. 3).

In osteomalacia, the radiographs show less contrast and are less sharp as if the patient had moved during the X-ray. The classical sign of osteomalacia is the pseudofracture, also named Looser zone. It is a radiolucent line through one cortical plate, often with sclerosis at the margins (Fig. 4). The pseudofractures are visible on bone scintigraphy as hot spots.

The assessment of BMD by DXA may show low values compatible with osteopenia (T-score −1 to −2.5) or osteoporosis (T-score < −2.5). It indicates that the amount of mineralized bone in patients with osteomalacia is similar to that in patients with osteoporosis. The amount of nonmineralized bone (osteoid tissue) may be high, leading to large and fast increases of BMD (up to 50%) when proper therapy has been instituted.

Bone Histology

A transiliac bone biopsy after tetracycline double labeling provides a certain diagnosis of osteomalacia or can exclude it. The biopsy should be included in methylmethacrylate without prior decalcification. Sections should be stained with Goldner

FIG. 2. Osteomalacia in a patient with celiac disease. Goldner stain: mineralized bone is black and osteoid tissue is gray. Besides thick osteoid seams, increased bone resorption by multinucleated osteoclasts (arrows) is visible.

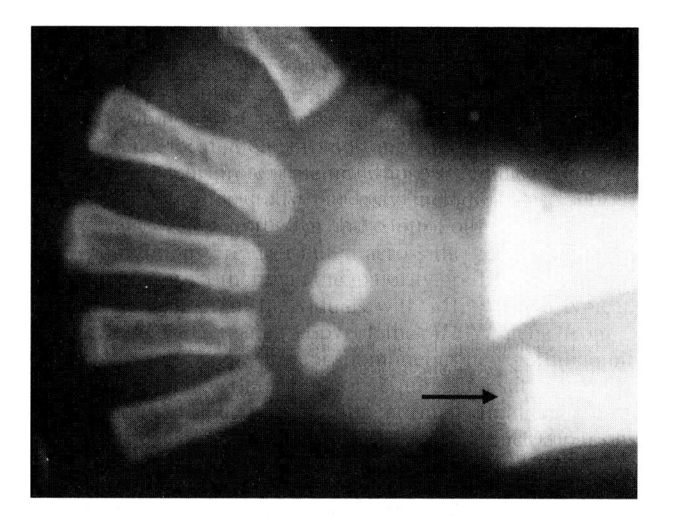

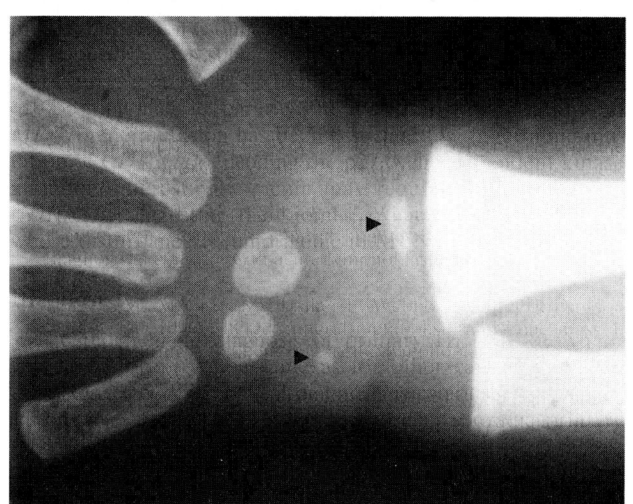

FIG. 3. Rickets in a child 1.5 yr of age. (Top) Florid rickets with typical unsharp concave margin of the ulna (arrow). (Bottom) Healing of rickets and appearance of new ossification centers (arrowheads).

© 2008 American Society for Bone and Mineral Research

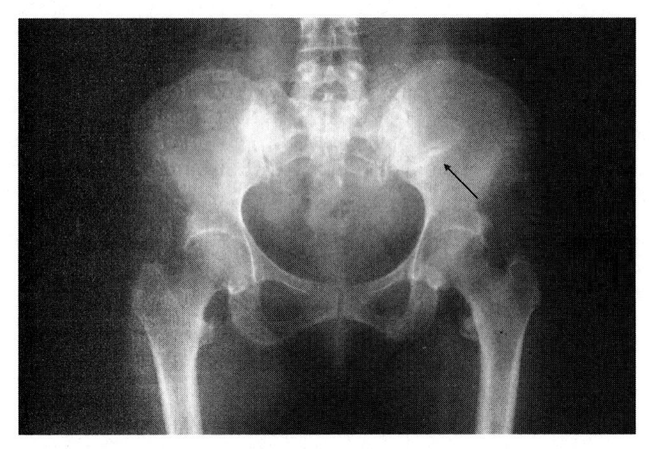

FIG. 4. Pseudofracture (arrow) in the left os ilium in a patient with osteomalacia.

or von Kossa. Unstained sections should be made for tetracycline fluorescence. Typically, the nonmineralized bone matrix (osteoid) comprises >5% of the bone (Fig. 2; osteoid volume >5%, usually >10%). The osteoid surface is extended, >70%, and osteoid seams are thick, >15 μm or more than four lamellae.[1] Bone resorption is usually increased because of secondary hyperparathyroidism. Howship's (resorption) lacunae are visible with many multinucleated osteoclasts (Fig. 2). In frank osteomalacia, double tetracycline labels are not visible, but labeling may appear diffuse and of low intensity. The mineralization index including osteoid thickness, osteoid volume, mineralization rate, and bone formation rate may be helpful.

Treatment

The symptoms of rickets and osteomalacia may disappear with remarkably low doses of vitamin D_3 (e.g., 800–1200 IU/d, 20–30 μg/d) unless intestinal absorption is impaired as in celiac disease.[32] However, a higher initial dose (vitamin D_3 2000 IU/d, 50 μg/d) may be prescribed to bring serum 25(OH)D in the sufficient range (>50 nM or >20 ng/ml) within a few weeks. Alternatively, vitamin D_2 50,000 IU (1250 μg) per week might be prescribed for 4 wk when available. For maintenance, infants and small children may require as little as 400 IU (10 μg) of vitamin D_3 per day. Adults require 800 IU (20 μg) per day as a maintenance dose. Either vitamin D_3 (cholecalciferol) or D_2 (ergocalciferol) may be prescribed, but vitamin D_2 is somewhat less effective.[33] A calcium supplement should always be added. It can be prescribed as calcium carbonate, citrate, or glubionate. A dose of 1000 mg elemental calcium per day in children and 1500–2000 mg/d in adults is necessary. A special problem is rickets or osteomalacia in patients with celiac disease. A gluten-free diet is essential, but higher doses of vitamin D and calcium may be required.

After treatment of severe vitamin D deficiency, serum calcium and phosphate quickly rise into the normal range. The alkaline phosphatase level may initially increase but then falls, within weeks to months to normal levels. At the same time, the urinary excretion of calcium increases to the normal range.

The radiological changes of rickets may heal very rapidly. Ossification centers may become visible within weeks, and healing within a few months is the rule (Fig. 3). Pseudofractures in adults may heal more slowly, and complete disappearance may take >1 yr.

Recently, the required doses of vitamin D and the appropriate level of 25(OH)D have been discussed at international meetings and in scientific papers.[14,15,30] The required dose depends on the baseline serum 25(OH)D level, the desired level, and the dosing interval. When the baseline serum 25(OH)D is ~20 nM, a dose of 400 IU/d leads to a mean serum 25(OH)D of 55 nM and a dose of 800 IU/d to 70 nM after 3 mo.[34] In general, 800 IU/d (20 μg/d) is a sufficient dose to obtain a serum 25(OH)D level >50 nM (20 ng/ml) in most people. When the desired 25(OH)D level is set at 75 nM, much higher vitamin D doses are needed.[35]

Prevention

In Table 3, the recommendations for dietary intake in the United States (http://www.iom.edu/Object.File/Master/7/296/webtablevitamins.pdf), Europe (http://ec.europa.eu/food/fs/sc/scf/out157_en.pdf), and The Netherlands[36] are presented. It should be noted that the required levels of serum 25(OH)D and the daily requirement of vitamin D are under discussion. Some experts believe that the required levels should be higher. In a recent nursing home study, it was shown that an intake of 600 IU (15 μg) daily reduced the percentage of persons having a serum 25(OH)D <50 nM from 100% to 10.9% within 4 mo.[37] A dose of 800 IU (20 μg) should be sufficient in most people. The latter dose is supported by a meta-analysis on fracture prevention.[21]

In the United States, the tolerable upper intake level of vitamin D is 1000 IU (25 μg)/d for children 0–1 yr of age and 2000 IU (50 μg)/d for all other age groups including pregnant and lactating women. The European Commission has defined a tolerable upper intake level of vitamin D of 1000 IU (25 μg)/d until 10 yr of age and 2000 IU (50 μg)/d for all other groups. However, long-term use of lower doses may lead to adverse effects as was shown in the Women's Health Initiative, in which significantly more kidney stones occurred in the group receiving 1000 mg calcium and 400 IU (10 μg) vitamin D daily compared with placebo for a period of 7 yr (HR = 1.17; 95% CI: 1.02–1.34).[16] No significant differences in mortality, cardiovascular disease, cancer, and gastrointestinal symptoms were observed. In a Cochrane review, hypercalcemia was more common when vitamin D or its analogs were given compared with placebo or calcium (pooled RR of 14 trials = 2.38; 95% CI: 1.52–3.71). The risk was particularly high with calcitriol (pooled RR of three trials = 14.94; 95% CI: 2.95–75.61). There was no evidence that vitamin D increased gastrointestinal symptoms or renal disease.

RICKETS CAUSED BY IMPAIRED 1α-HYDROXYLATION OR BY A DEFECTIVE VITAMIN D RECEPTOR

Introduction

The observation that some forms of rickets could not be cured by regular doses of vitamin D led to the discovery of rare inherited abnormalities of vitamin D metabolism or the vitamin D receptor. These inborn errors of metabolism have been confirmed by several mouse knockout models.[38]

Pseudovitamin D Deficiency Rickets (Vitamin D–Dependent Rickets Type I, OMIM 264700)

After the discovery of $1,25(OH)_2D$, it became apparent that some patients with congenital rickets had very low serum concentrations of $1,25(OH)_2D$, which did not increase after vitamin D supplementation.[39] It turned out that these patients did not have 1α-hydroxylase activity. Inactivating mutations of the 1α-hydroxylase gene were identified in affected children[40] The disease is autosomal recessive.

© 2008 American Society for Bone and Mineral Research

TABLE 3. RECOMMENDED DIETARY INTAKE OF VITAMIN D IN MICROGRAMS PER DAY*

Age	*United States*†	*Europe*‡	*The Netherlands*† *(light-colored skin, sunlight exposure)*	*The Netherlands*† *(no sunlight exposure)*
0–3 yr	5	10 (infants§)	5	10
4–50 yr	5	5 (adults)	2.5	5
			5 (51–60 yr)	
51–70 yr	10	10 (>60–65 yr)	7.5 (61–70 yr)	10
>70 yr	15		12.5	15
Pregnant or lactating women	5	10	7.5	10

* Micrograms can be converted to International Units (IU) by multiplying by 40.
† Adequate intake. According to experts, vitamin D_3 20 μg/d (800 IU/d) is needed to increase serum 25(OH)D over 50 nM in 95% of the population.[14,34]
‡ Individual European countries often have more detailed recommendations.
§ Infants from 4 wk onward.

Clinical Features. Patients with pseudovitamin D deficiency rickets present soon after birth with rickets and signs of hypocalcemia, tetany, or convulsions. Laboratory investigations show low serum concentrations of calcium and phosphate and elevated alkaline phosphatase. Serum 25(OH)D is within the normal range, but serum $1,25(OH)_2D$ is very low or undetectable. Radiologic examination and bone biopsy show features indistinguishable from nutritional rickets.

Treatment. The patients should be treated with $1,25(OH)_2D$ (calcitriol 0.5–1 μg/d) or 1α-hydroxyvitamin D (alphacalcidol 0.5–1.5 μg/d) to restore serum $1,25(OH)_2D$ to normal levels. This leads to cure of rickets within a few months with normalization of serum calcium, phosphate, and alkaline phosphatase. Another marker of adequate therapy is serum PTH, which should be within reference limits. Overtreatment results in hypercalciuria, hypercalcemia, urolithiasis, and nephrocalcinosis. Serum calcium, serum creatinine, and 24-h urinary calcium excretion should be checked at regular intervals, in children every month, and when stable every 3 mo, and in adults every 3–6 mo. Patients and parents should be instructed on symptoms of underdosage [i.e., hypocalcemia (numbness, pins and needles, tetany) and overdosage, i.e., hypercalcemia (thirst, polyuria, nausea, headache)]. The requirement of calcitriol or alphacalcidol increases in pregnancy 50–100%, and frequent monitoring is necessary.[41] Therapy should be continued for life, and accidental discontinuation will lead to hypocalcemia within a few days, later followed by signs of rickets or osteomalacia within weeks to months.

Hereditary Vitamin D–Resistant Rickets (Vitamin D–Dependent Rickets Type 2, OMIM 277440)

True resistance to $1,25(OH)_2D$ was discovered because some children with congenital rickets did not respond to treatment with calcitriol.[42] In fact, these children had a high serum $1,25(OH)_2D$, leading to the suspicion of a (post)receptor defect. After cloning of the vitamin D receptor (VDR), mutations have been identified at the DNA binding domain, the $1,25(OH)_2D$ binding domain, and other domains.[43] Hereditary vitamin D–resistant rickets is an autosomal recessive disease.

Clinical Features. Affected children born to normal heterozygote parents present early in life with rickets and signs of hypocalcemia, including tetany and convulsions. The first kindreds had alopecia, but later the disease was recognized in children without alopecia. The degree of hormone resistance (i.e., hypocalcemia) may also vary. Laboratory examination showed low serum calcium and serum phosphate and a high alkaline phosphatase level. Serum PTH is increased and serum $1,25(OH)_2D$ is elevated. The latter is caused by stimulation of renal 1α-hydroxylase by the increased PTH level.

Radiological examination shows the signs of rickets with widened epiphyseal zones and unsharp radiolucent bones. Pseudofractures may be present.

Treatment. The success of treatment is variable and depends on the degree of hormone resistance. When some VDR function is present, a pharmacologic rather than physiologic dose of calcitriol or alphacalcidol can improve calcium absorption and heal the rickets. In case of complete resistance, active vitamin D metabolites are ineffective. Such severely affected individuals can be treated with calcium infusion, which overcomes the defective calcium absorption.[44] Rickets can be cured by calcium infusion, and this confirms that mineralization depends on the presence of adequate calcium and phosphate concentrations rather than on the action of $1,25(OH)_2D$. Because calcium absorption from the intestine also has a passive component by diffusion, which is independent of vitamin D, very high doses of oral calcium can be effective as well.

VITAMIN D INTOXICATION

Vitamin D intoxication is caused by increased ingestion of vitamin D or one of its active metabolites (e.g., alphacalcidol or calcitriol). In case of intoxication by regular vitamin D, the pathogenesis is not completely clear. The serum 25(OH)D concentration is increased over 200 to 1500 nM, but the serum $1,25(OH)_2D$ level usually is normal.[12] Several mechanisms have been proposed. The high amount of 25(OH)D could act on the vitamin D receptor. More probably, the high serum 25(OH)D could compete with $1,25(OH)_2D$ for binding places of vitamin D binding globulin, thereby increasing the bioavailable (free) serum $1,25(OH)_2D$ levels. In addition, extrarenal activation of 25(OH)D by local 1α-hydroxylase in intestine and bone tissue may contribute to the hypercalcemia.

Clinical Features. The patient presents with signs and symptoms of hypercalcemia (i.e., thirst, polyuria, nausea, headache). The cause may be obvious, either supplementation with high doses of vitamin D or a somewhat high dose of active metabolite. Less obvious causes may be an over-the-counter supplement or table sugar to which vitamin D has been added.[45,46] The patient may appear dehydrated and even confused. Laboratory examination shows hypercalcemia, normal or increased

© 2008 American Society for Bone and Mineral Research

serum phosphate, increased serum creatinine or blood urea nitrogen, a very high 25(OH)D level, a normal $1,25(OH)_2D$ level, and suppressed serum PTH.

Treatment. The patient should be rehydrated with saline, and corticosteroids may be given to decrease the formation of $1,25(OH)_2D$. The half-life of 25(OH)D is ~25 days, so complete cure may take some time. Calcium intake should be restricted as long as hypercalcemia is present.

Intoxication With Active Metabolites. Intoxication with alphacalcidol or calcitriol may appear quickly, but, when recognized, also disappears soon with appropriate measures because the half-life is short (~7 h). Treatment should consist of dose reduction or temporary arrest of the active metabolite, rehydration, and restriction of oral calcium intake.

Increased Endogenous Production of 1,25(OH)2D. Endogenous intoxication with calcitriol occurs in the patient with granulomatous disorders or lymphoproliferative diseases when committed macrophages contain 1α-hydroxylase, thus permitting the inappropriate conversion of 25(OH)D into $1,25(OH)_2D$. The latter has been reported from patients with sarcoidosis, tuberculosis, rheumatoid arthritis, Hodgkin's disease, and non-Hodgkin lymphoma.[47–49] Whereas the renal formation of $1,25(OH)_2D$ is tightly regulated by feedback control, the extrarenal production in these diseases is not. The production of $1,25(OH)_2D$ depends on the amount of substrate [i.e., 25(OH)D], and a positive correlation between serum 25(OH)D and $1,25(OH)_2D$ is observed in these disorders, whereas it is not in normal circumstances.[12] The treatment consists of management of the hypercalcemia and hypercalciuria and treatment of the underlying disease. Glucocorticoids decrease the hydroxylation of 25(OH)D into $1,25(OH)_2D$ and may help to correct the hypercalcemia.

REFERENCES

1. Parfitt AM 2005 Vitamin D and the pathogenesis of rickets and osteomalacia. In: Feldman D, Pike JW, Glorieux FH (eds.) Vitamin D, 2nd ed. Elsevier Academic Press, San Diego, CA, USA, pp. 1029–1048.
2. Lips P 2001 Vitamin D deficiency and secondary hyperparathyroidism in the elderly: Consequences for bone loss and fractures and therapeutic implications. Endocr Rev **22**:477–501.
3. Bischoff-Ferrari HA, Dietrich T, Orav EJ, Hu FB, Zhang YQ, Karlson EW, Dawson-Hughes B 2004 Higher 25-hydroxyvitamin D concentrations are associated with better lower-extremity function in both active and inactive persons aged >= 60 y. Am J Clin Nutr **80**:752–758.
4. Wicherts IS, van Schoor NM, Boeke AJ, Visser M, Deeg DJ, Smit J, Knol DL, Lips P 2007 Vitamin d status predicts physical performance and its decline in older persons. J Clin Endocrinol Metab **92**:2058–2065.
5. Snijder MB, van Schoor NM, Pluijm SM, van Dam RM, Visser M, Lips P 2006 Vitamin D status in relation to one-year risk of recurrent falling in older men and women. J Clin Endocrinol Metab **91**:2980–2985.
6. Nagpal S, Na S, Rathnachalam R 2005 Noncalcemic actions of vitamin D receptor ligands. Endocr Rev **26**:662–687.
7. Norman AW 2006 Minireview: Vitamin D receptor: New assignments for an already busy receptor. Endocrinology **147**:5542–5548.
8. Lips P 2006 Vitamin D physiology. Prog Biophys Mol Biol **92**:4–8.
9. Peterlik M, Cross HS 2005 Vitamin D and calcium deficits predispose for multiple chronic diseases. Eur J Clin Invest **35**:290–304.
10. Hypponen E, Laara E, Reunanen A, Jarvelin MR, Virtanen SM 2001 Intake of vitamin D and risk of type 1 diabetes: A birth-cohort study. Lancet **358**:1500–1503.
11. Holick MF 1994 McCollum Award Lecture, 1994: Vitamin D–new horizons for the 21st century. Am J Clin Nutr **60**:619–630.
12. Lips P 2007 Relative value of 25(OH)D and 1,25(OH)(2)D measurements. J Bone Miner Res **22**:1668–1671.
13. Lips P, Pluijm S, Smit JH, van Schoor NM 2005 Vitamin D status and the threshold for secondary hyperparathyroidism in the Longitudinal Aging Study Amsterdam (LASA). Bone **36**:S141–S142.
14. Dawson-Hughes B, Heaney RP, Holick MF, Lips P, Meunier PJ, Vieth R 2005 Estimates of optimal vitamin D status. Osteoporos Int **16**:713–716.
15. Bischoff-Ferrari HA, Giovannucci E, Willett WC, Dietrich T, Dawson-Hughes B 2006 Estimation of optimal serum concentrations of 25-hydroxyvitamin D for multiple health outcomes. Am J Clin Nutr **84**:18–28.
16. Jackson RD, LaCroix AZ, Gass M, Wallace RB, Robbins J, Lewis CE, Bassford T, Beresford SAA, Black HR, Blanchette P, Bonds DE, Brunner RL, Brzyski RG, Caan B, Cauley JA, Chlebowski RT, Cummings SR, Granek I, Hays J, Heiss G, Hendrix SL, Howard BV, Hsia J, Hubbell FA, Johnson KC, Judd H, Kotchen JM, Kuller LH, Langer RD, Lasser NL, Limacher MC, Ludlam S, Manson JE, Margolis KL, McGowan J, Ockene JK, O'Sullivan MJ, Phillips L, Prentice RL, Sarto GE, Stefanick ML, Van Horn L, Wactawski-Wende J, Whitlock E, Anderson GL, Assaf AR, Barad D 2006 Calcium plus vitamin D supplementation and the risk of fractures. N Engl J Med **354**:669–683.
17. Norman AW, Bouillon R, Whiting SJ, Vieth R, Lips P 2007 13th Workshop consensus for vitamin D nutritional guidelines. J Steroid Biochem Mol Biol **103**:204–205.
18. van der Meer IM, Karamali NS, Boeke AJP, Lips P, Middelkoop BJC, Verhoeven I, Wuister D 2006 High prevalence of vitamin D deficiency in pregnant non-Western women in The Hague, Netherlands. Am J Clin Nutr **84**:350–353.
19. Dijkstra SH, van Beek A, Janssen JW, de Vleeschouwer LH, Huysman WA, van den Akker EL 2007 High prevalence of vitamin D deficiency in newborns of high-risk mothers. Arch Dis Child **92**:750–753.
20. Lips P 2007 Vitamin D status and nutrition in Europe and Asia. J Steroid Biochem Mol Biol **103**:620–625.
21. Bischoff-Ferrari HA, Willett WC, Wong JB, Giovannucci E, Dietrich T, Dawson-Hughes B 2005 Fracture prevention with vitamin D supplementation - A meta-analysis of randomized controlled trials. JAMA **293**:2257–2264.
22. Avenell A, Gillespie WJ, Gillespie LD, O'Connell DL 2005 Vitamin D and vitamin D analogues for preventing fractures associated with involutional and post-menopausal osteoporosis. Cochrane Database Syst Rev 3: CD 000227.
23. Boonen S, Lips P, Bouillon R, Bischoff-Ferrari HA, Vanderschueren D, Haentjens P 2007 Need for additional calcium to reduce the risk of hip fracture with vitamin d supplementation: Evidence from a comparative metaanalysis of randomized controlled trials. J Clin Endocrinol Metab **92**:1415–1423.
24. Visser M, Deeg DJH, Lips P 2003 Low vitamin D and high parathyroid hormone levels as determinants of loss of muscle strength and muscle mass (Sarcopenia): The Longitudinal Aging Study Amsterdam. J Clin Endocrinol Metab **88**:5766–5772.
25. Bischoff-Ferrari HA, Dawson-Hughes B, Willett WC, Staehelin HB, Bazemore MG, Zee RY, Wong JB 2004 Effect of vitamin D on falls—a meta-analysis. JAMA **291**:1999–2006.
26. Christakos S, Dhawan P, Peng X, Obukhov AG, Nowycky MC, Benn BS, Zhong Y, Liu Y, Shen Q 2007 New insights into the function and regulation of vitamin D target proteins. J Steroid Biochem Mol Biol **103**:405–410.
27. Goltzman D 2007 Use of genetically modified mice to examine the skeletal anabolic activity of vitamin D. J Steroid Biochem Mol Biol **103**:587–591.
28. Chertow BS, Baker GR, Henry HL, Norman AW 1980 Effects of Vitamin-D Metabolites on Bovine Parathyroid-Hormone Release Invitro. Am J Physiol **238**:E384–E388.
29. Pettifor JM 2005 Vitamin D deficiency and nutritional rickets in children. In: Feldman D, Pike JW, Glorieux FH (eds.) Vitamin D, 2nd ed. Academic Press, San Diego, CA, USA, pp. 1065–1083.
30. Lips P 2004 Which circulating level of 25-hydroxyvitamin D is appropriate? J Steroid Biochem Mol Biol **89-90**:611–614.
31. Bouillon RA, Auwerx JH, Lissens WD, Pelemans WK 1987 Vitamin D status in the elderly: Seasonal substrate deficiency causes 1,25-dihydroxycholecalciferol deficiency. Am J Clin Nutr **45**:755–763.
32. Holick MF 2007 Vitamin D deficiency. N Engl J Med **357**:266–281.
33. Armas LA, Hollis BW, Heaney RP 2004 Vitamin D2 is much less effective than vitamin D3 in humans. J Clin Endocrinol Metab **89**:5387–5391.

© 2008 American Society for Bone and Mineral Research

34. Lips P, Wiersinga A, van Ginkel FC, Jongen MJ, Netelenbos JC, Hackeng WH, Delmas PD, van der Vijgh WJ 1988 The effect of vitamin D supplementation on vitamin D status and parathyroid function in elderly subjects. J Clin Endocrinol Metab **67**:644–650.
35. Heaney RP, Davies KM, Chen TC, Holick MF, Barger-Lux MJ 2003 Human serum 25-hydroxycholecalciferol response to extended oral dosing with cholecalciferol. Am J Clin Nutr **77**:204–210.
36. Health Council of The Netherlands 2000 Dietary Reference Values: Calcium, Vitamin D, Thiamin, Riboflavin, Niacin, Pantothemic Acid, and Biotin. Health Council of The Netherlands, The Hague, The Netherlands.
37. Chel V, Wijnhoven HA, Smit JH, Ooms M, Lips P 2007 Efficacy of different doses and time intervals of oral vitamin D supplementation with or without calcium in elderly nursing home residents. Osteoporos Int **19**:663–671.
38. Panda DK, Miao D, Bolivar I, Li J, Huo R, Hendy GN, Goltzman D 2004 Inactivation of the 25-hydroxyvitamin D 1alpha-hydroxylase and vitamin D receptor demonstrates independent and interdependent effects of calcium and vitamin D on skeletal and mineral homeostasis. J Biol Chem **279**:16754–16766.
39. Fraser D, Kooh SW, Kind HP, Holick MF, Tanaka Y, DeLuca HF 1973 Pathogenesis of hereditary vitamin-D-dependent rickets. An inborn error of vitamin D metabolism involving defective conversion of 25-hydroxyvitamin D to 1 alpha,25-dihydroxyvitamin D. N Engl J Med **289**:817–822.
40. Kitanaka S, Takeyama K, Murayama A, Sato T, Okumura K, Nogami M, Hasegawa Y, Niimi H, Yanagisawa J, Tanaka T, Kato S 1998 Inactivating mutations in the 25-hydroxyvitamin D-3 1 alpha-hydroxylase gene in patients with pseudovitamin D-deficiency rickets. N Engl J Med **338**:653–661.
41. Glorieux FH 1990 Calcitriol treatment in vitamin D-dependent and vitamin D-resistant rickets. Metabolism **39**:10–12.
42. Brooks MH, Bell NH, Love L, Stern PH, Orfei E, Queener SF, Hamstra AJ, DeLuca HF 1978 Vitamin-D-dependent rickets type II. Resistance of target organs to 1,25-dihydroxyvitamin D. N Engl J Med **298**:996–999.
43. Malloy PJ, Hochberg Z, Tiosano D, Pike JW, Hughes MR, Feldman D 1990 The molecular basis of hereditary 1,25-dihydroxyvitamin D3 resistant rickets in seven related families. J Clin Invest **86**:2071–2079.
44. Balsan S, Garabedian M, Larchet M, Gorski AM, Cournot G, Tau C, Bourdeau A, Silve C, Ricour C 1986 Long-term nocturnal calcium infusions can cure rickets and promote normal mineralization in hereditary resistance to 1,25-dihydroxyvitamin D. J Clin Invest **77**:1661–1667.
45. Koutkia P, Chen TC, Holick MF 2001 Vitamin D intoxication associated with an over-the-counter supplement. N Engl J Med **345**:66–67.
46. Vieth R, Pinto TR, Reen BS, Wong MM 2000 Vitamin D poisoning by table sugar. Lancet **359**:672.
47. Barbour GL, Coburn JW, Slatopolsky E, Norman AW, Horst RL 1981 Hypercalcemia in an anephric patient with sarcoidosis: Evidence for extrarenal generation of 1,25-dihydroxyvitamin D. N Engl J Med **305**:440–443.
48. Davies M, Mawer EB, Hayes ME, Lumb GA 1985 Abnormal vitamin D metabolism in Hodgkin's lymphoma. Lancet **1**:1186–1188.
49. Hewison M, Kantorovich V, Liker HR, Van Herle AJ, Cohan P, Zehnder D, Adams JS 2003 Vitamin D-mediated hypercalcemia in lymphoma: Evidence for hormone production by tumor-adjacent macrophages. J Bone Miner Res **18**:579–582.

Chapter 72. Paget's Disease of Bone

Ethel S. Siris[1] and G. David Roodman[2]

[1]*Department of Medicine, Columbia University Medical Center, New York, New York;* [2]*Department of Medicine, University of Pittsburgh, Pittsburgh, Pennsylvania*

INTRODUCTION

Paget's disease of bone is a localized disorder of bone remodeling. The process is initiated by increases in osteoclast-mediated bone resorption, with subsequent compensatory increases in new bone formation, resulting in a disorganized mosaic of woven and lamellar bone at affected skeletal sites. This structural change produces bone that is expanded in size, less compact, more vascular, and more susceptible to deformity or fracture than is normal bone.[1] Clinical signs and symptoms will vary from one patient to the next depending on the number and location of affected skeletal sites, as well as on the degree and extent of the abnormal bone turnover. It is believed that most patients are asymptomatic, but a substantial minority may experience a variety of symptoms, including bone pain, secondary arthritic problems, bone deformity, excessive warmth over bone from hypervascularity, and a variety of neurological complications caused in most instances by compression of neural tissues adjacent to pagetic bone.

ETIOLOGY

Although Paget's disease is the second most common bone disease after osteoporosis, little still is known about its pathogenesis—why it is highly localized, the potential role paramyxoviral infection might play, the basis for the unusual geographic distribution, and the contribution of a genetic component to the disease process.

It is abundantly clear that there is a strong genetic predisposition involved in the pathophysiology of Paget's disease. Paget's disease occurs commonly in families and can be transmitted vertically in an autosomal dominant pattern between generations in an affected family. In patients with Paget's disease described in several clinical series, 15–30% have positive family histories of the disorder.[2–4] Familial aggregation studies in a U.S. population[5] suggest that the risk of a first-degree relative of a pagetic subject developing the condition is seven times greater than is the risk for someone who does not have an affected relative.

Several genetic loci have been linked to familial Paget's disease, and three genes have been identified in these locations. A mutation in the *RANK* gene has been reported,[6] but this mutation is not found in the overwhelming majority of patients with familial Paget's disease.[7]

The most frequent mutations linked to Paget's disease are in a gene on 5q35-QTER, which encodes an ubiquitin binding protein, sequestasome-1 (SQSTM1/p62^{ZIP}).[8] Mutations in SQSTM1 have been found in 30% of patients with familial Paget's disease, with the *P392L* mutation being the most fre-

Dr. Roodman has consulted for Amgen, Novartis, Merck, and Millenium. He has served on the Speakers' Bureau for Novartis. Dr. Siris has consulted for Novartis.

Key words: Paget's disease, bisphosphonates

© 2008 American Society for Bone and Mineral Research

quent.[9] Sequestasome-1 plays an important role in the NF-κB signaling pathway. It binds either TRAF-6 in the interleukin (IL)-1 or RANKL signaling pathway, and RIP-1 in the TNF signaling pathway to activate NF-κB. SQSTM1 mutations are associated with a variable clinical phenotype, including no evidence of Paget's disease in at least one or two individuals, and no gene dose effect seen between heterozygotes and homozygotes. Recent in vitro and in vivo studies have suggested that the *P392L* mutation in $p62^{zip}$ is a predisposing mutation for Paget's disease and does not cause Paget's disease.[10] Human osteoclast precursors transfected with the $p62^{P342L}$ do not form osteoclasts characteristic of Paget's disease, and transgenic mice with the $p62^{P392L}$ mutation targeted to the osteoclast lineage develop progressive osteopenia and not Paget's disease. Furthermore, mice in which the normal *p62* gene has been replaced with $p62^{P392L}$ do not develop Paget's disease.

There is a restricted geographic distribution for the occurrence of Paget's disease. Paget's disease is most common in Europe, North America, Australia, and New Zealand in persons of Anglo-Saxon descent and is extremely uncommon in Asia, Africa, and Scandinavia. There is a remarkable focus of Paget's disease in Lancashire, UK, where 6.3–8.3% of people older than 55 yr of age in several Lancashire towns had radiographs showing Paget's disease.[11]

Some recent studies have reported an apparent decline in the frequency and severity of Paget's disease in both New Zealand and the United Kingdom.[12,13] The basis for this decline is unknown, but the changes are too rapid to be explained by a genetic cause and cannot be explained by migration patterns of persons with a predisposition to Paget's disease.

For >30 yr, studies have suggested that Paget's disease may result from a chronic paramyxoviral infection. This is based on ultrastructural studies by Rebel et al.,[14] who showed that nuclear and, less commonly, cytoplasmic inclusions that were similar to nucleocapsids from paramyxoviruses were present in osteoclasts from Paget's disease patients. Mills et al.[15] also reported that the measles virus nucleocapsid antigen was present in osteoclasts from patients with Paget's disease, but not from patients with other bone diseases. In some specimens, both measles virus and respiratory syncytial virus nucleocapsid proteins were shown by immunocytochemistry on serial sections. Gordon et al.,[16] using in situ hybridization studies, examined specimens from English patients with Paget's disease and found canine distemper virus nucleocapsid protein in 11 of 25 patients. Mee et al.,[17] using highly sensitive in situ PCR techniques, found that osteoclasts from 12 of 12 English patients with Paget's disease expressed canine distemper virus (CDV) nucleocapsid transcripts.

Kurihara et al.[18] have provided evidence for a possible pathophysiologic role for measles virus in the abnormal osteoclast activity in Paget's disease. They transfected the measles virus nucleocapsid gene into normal human osteoclast precursors and showed that the osteoclasts that formed expressed many of the abnormal characteristics of pagetic osteoclasts. However, other workers have been unable to confirm the presence of measles virus or CDV in pagetic osteoclasts[19] so that the role of a chronic paramyxoviral infection in Paget's disease remains controversial. Recently, Kurihara et al.[20] have targeted the measles virus nucleocapsid gene to cells in the osteoclast lineage in transgenic mice, and ~40% of these mice develop localized bone lesions that are similar to lesions seen in patients with Paget's disease. However, these results do not show a cause and effect relationship between measles virus and Paget's disease.

Among the many questions that need to be explained to understand a putative viral or genetic etiology of Paget's disease are as follows. (1) Because paramyxoviral infections such as measles virus occur worldwide, why does Paget's disease have a very restricted geographic distribution? (2) How does the virus persist in osteoclasts in patients who are immunocompetent for such long periods of time, because measles virus infections generally occur in children rather than adults, and Paget's disease is usually diagnosed in elderly patients over the age of 55? (3) How do the virus and the genetic mutations linked to Paget's disease contribute to the development of Paget's disease? (4) What is the explanation for the variable phenotypic presentation of patients with familial Paget's disease, especially that some of these patients who carry the mutated gene do not have Paget's disease even though they are >70 yr of age? (5) Finally, why does expression of the most frequent mutation linked to Paget's disease, $p62^{P392L}$, not induce Paget's disease in animal models? More likely, environmental factors and genetic factors are both required for patients to develop Paget's disease.

PATHOLOGY

Histopathologic Findings in Paget's Disease

The initiating lesion in Paget's disease is an increase in bone resorption. This occurs in association with an abnormality in the osteoclasts found at affected sites. Pagetic osteoclasts are more numerous than normal and contain substantially more nuclei than do normal osteoclasts, with up to 100 nuclei per cell noted by some investigators. In response to the increase in bone resorption, numerous osteoblasts are recruited to pagetic sites where active and rapid new bone formation occurs. It is generally believed that the osteoblasts are intrinsically normal.[21,22]

In the earliest phases of Paget's disease, increased osteoclastic bone resorption dominates, a picture appreciated radiographically by an advancing lytic wedge or "blade-of-grass" lesion in a long bone or by osteoporosis circumscripta, as seen in the skull. At the level of the bone biopsy, the structurally abnormal osteoclasts are abundant. After this, there is a combination of increased resorption and relatively tightly coupled new-bone formation, produced by the large numbers of osteoblasts present at these sites. During this phase, and presumably because of the accelerated nature of the process, the new bone that is made is abnormal. Newly deposited collagen fibers are laid down in a haphazard rather than a linear fashion, creating more primitive woven bone. The woven-bone pattern is not specific for Paget's disease, but it does reflect a high rate of bone turnover. The end product is the so-called mosaic pattern of woven bone plus irregular sections of lamellar bone linked in a disorganized way by numerous cement lines representing the extent of previous areas of bone resorption. The bone marrow becomes infiltrated by excessive fibrous connective tissue and by an increased number of blood vessels, explaining the hypervascular state of the bone. Bone matrix at pagetic sites is usually normally mineralized, and tetracycline labeling shows increased calcification rates. It is not unusual, however, to find areas of pagetic biopsies in which widened osteoid seams are apparent, perhaps reflecting inadequate calcium/phosphorus products in localized areas where rapid bone turnover heightens mineral demands.

In time, the hypercellularity at a locus of affected bone may diminish, leaving the end product of a sclerotic, pagetic mosaic without evidence of active bone turnover. This is so-called burned out Paget's disease. Typically, all phases of the pagetic process can be seen at the same time at different sites in a particular subject. Scanning electron microscopy affords an excellent view of the chaotic architectural changes that occur in pagetic bone and provides the visual imagery that makes com-

© 2008 American Society for Bone and Mineral Research

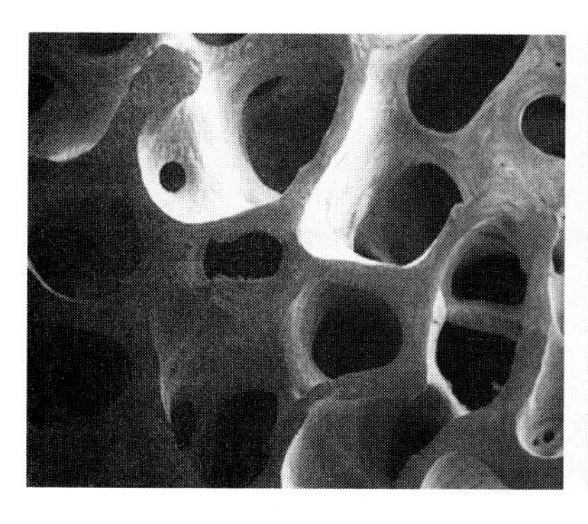

FIG. 1. Scanning electron micrographs with sections of normal bone (left) and pagetic bone (right). Both samples were taken from the iliac crest. The normal bone shows the trabecular plates and marrow spaces to be well preserved, whereas the pagetic bone has completely lost this architectural appearance. Extensive pitting of the pagetic bone is apparent caused by dramatically increased osteoclastic bone resorption. [Photographs courtesy of Dr. David Dempster; reproduced with permission from Lippincott from Siris ES, Canfield RE 1995 Paget's disease of bone. In: Becker KL (ed.) Principles and Practice of Endocrinology and Metabolism, 2nd ed. JB Lippincott, Philadelphia, PA, USA, pp. 585–594 (http://lww.com).]

prehensible the loss of structural integrity. Figure 1 compares the appearances of normal and of pagetic bone using this technique.

BIOCHEMICAL PARAMETERS OF PAGET'S DISEASE

Measurement of biochemical markers of bone turnover are very useful clinically in the assessment of the extent and severity of disease in the untreated state and for monitoring the response to treatment.[23] Increases in the urinary excretion of biomarkers of bone resorption such as collagen cross-links and associated peptides (e.g., N-telopeptide of type 1 collagen [NTX], C-telopeptide of type 1 collagen [CTX], deoxypyridinoline [DPD]) reflect the primary lesion in Paget's disease, the increase in bone resorption. Increases in osteoblastic activity are associated with elevated levels of bone formation markers including serum total alkaline phosphatase, bone-specific alkaline phosphatase, and procollagen type-1 N-terminal propeptide (P1NP). In untreated patients, the values of collagen cross-links and serum alkaline phosphatase rise in proportion to each other, offering a reflection of the preserved coupling between resorption and formation. From the clinical perspective, the degree of elevation of these indices offers an approximation of the extent or severity of the abnormal bone turnover, with higher levels reflecting a more active, ongoing localized metabolic process. Interestingly, the patients with the highest alkaline phosphatase elevations (e.g., >10 times the upper limit of normal) typically have involvement of the skull as at least one site of the disorder. Active monostotic disease (other than skull) may have lower biochemical values than polyostotic disease. Lower values (e.g., <3 times the upper limit of normal) may reflect a lesser extent of involvement (i.e., fewer sites on bone scans or radiographs) or a lesser degree of increased bone turnover at affected skeletal sites. However, mild elevations in a patient with highly localized disease (e.g., the proximal tibia) may be associated with symptoms and clear progression of disease at the affected site over time. Indeed, a so-called "normal" alkaline phosphatase (e.g., a value a slightly less than the upper limit of normal for the assay) may not truly be normal for the pagetic patient. Today many would argue that to be confident that the value is normal (and the disease quiescent), a result in the middle of the normal range is required.

In addition to offering some estimate of the degree of increased bone turnover, measurement of a bone resorption marker and serum alkaline phosphatase is useful in observing the disorder over time and especially for monitoring the effects of treatment. Potent bisphosphonates are capable of normalizing the biochemical markers (i.e., producing a remission of the bone remodeling abnormality) in a majority of patients and bringing the markers to near normal in most others so that the monitoring role has heightened importance in assessing treatment effects. A bone resorption marker such as the N- or C-telopeptide of collagen may become normal in days to a few weeks after bisphosphonate therapy is initiated. At this time, it is simplest and least expensive in the practice setting to monitor serum alkaline phosphatase as the sole biochemical end point, with a baseline measure and subsequent follow-up tests at intervals appropriate for the therapy used (see later). If a patient has concomitant elevations of liver enzymes, a measurement of bone-specific alkaline phosphatase or P1NP may be helpful. Serum osteocalcin, however, is not a useful measurement in Paget's disease.

Serum calcium levels are typically normal in Paget's disease, but they may become elevated in two special situations. First, if a patient with active, usually extensive Paget's disease is immobilized, the loss of the weight-bearing stimulus to new bone formation may transiently uncouple resorption and accretion, so that increasing hypercalciuria and hypercalcemia may occur. This is a relatively infrequent occurrence. Alternatively, when hypercalcemia is discovered in an otherwise healthy, ambulatory patient with Paget's disease, coexistent primary hyperparathyroidism may be the cause. Inasmuch as increased levels of PTH can drive the intrinsic pagetic remodeling abnormality to even higher levels of activity, correction of primary hyperparathyroidism in such cases is indicated. If neither hyperparathyroidism nor immobilization is the cause, the hypercalcemia needs to be investigated further.

Several investigators have commented on the 15–20% prevalence of secondary hyperparathyroidism (associated with normal levels of serum calcium) in Paget's disease, typically seen in patients with very high levels of serum alkaline phosphatase.[24,25] The increase in PTH is believed to reflect the need to increase calcium availability to bone during phases of very active pagetic bone formation, particularly in subjects in whom dietary intake of calcium is inadequate. Secondary hyperparathyroidism and transient decreases in serum calcium also can occur in some patients being treated with potent bisphosphonates such as pamidronate, alendronate, risedronate, or zoledronic acid. This results from the effective and rapid suppression of bone resorption in the setting of ongoing new-bone formation.[26] Later, as restoration of coupling occurs with time, PTH levels fall. The problem can be largely avoided by being certain that such patients are and remain replete in both calcium and vitamin D.

Elevations in serum uric acid and serum citrate have been described in Paget's disease and are of unclear clinical significance.[1] Gout has been noted in this disorder, but it is uncertain whether it is more common in pagetic patients than in nonpagetic subjects. Hypercalciuria may occur in some patients with Paget's disease, presumably because of the increased bone resorption, and kidney stones are occasionally found as a consequence of this abnormality.[1]

© 2008 American Society for Bone and Mineral Research

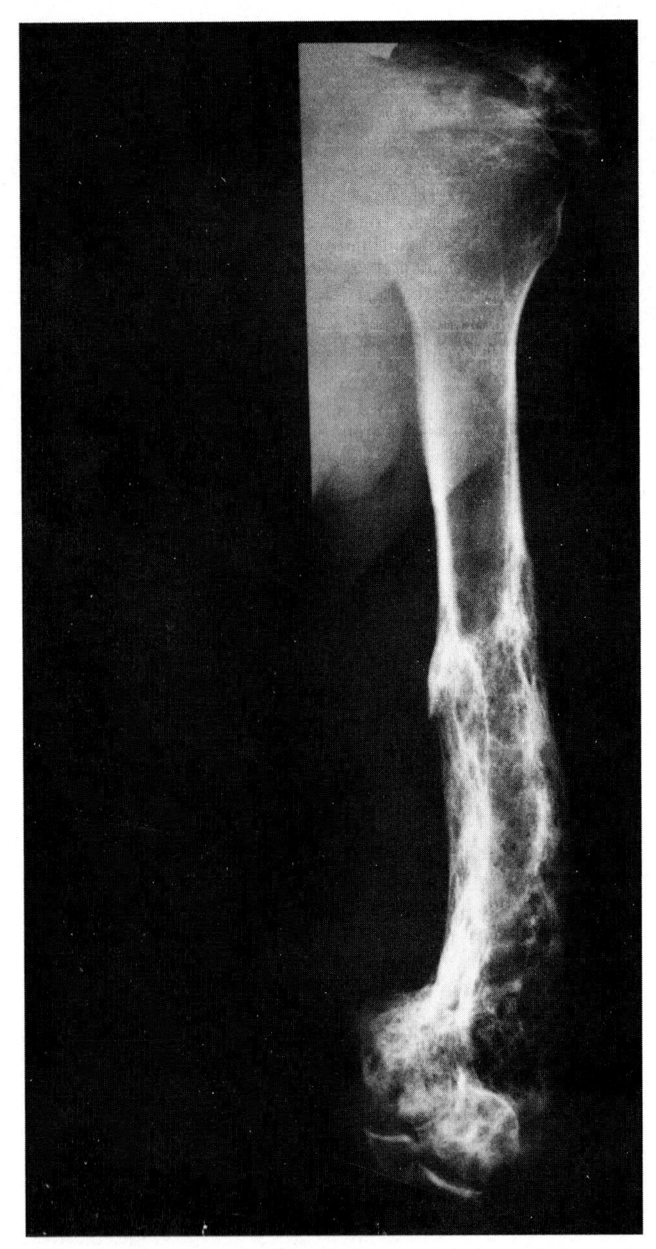

FIG. 2. Radiograph of a humerus showing typical pagetic change in the distal half, with cortical thickening, expansion, and mixed areas of lucency and sclerosis, contrasted with normal bone in the proximal half.

CLINICAL FEATURES

Paget's disease affects both men and women, with most series describing a slight male predominance. It is rarely observed to occur in individuals younger than 25 yr of age, it is thought to develop as a clinical entity after the age of 40 in most instances, and it is most commonly diagnosed in people over the age of 50. In a survey of >800 selected patients in the United States, 600 of whom had symptoms, the average age at diagnosis was 58 yr.[27] It seems likely that many patients have the disorder for a period of time before any diagnosis is made, especially because it is often an incidental finding.

It is important to emphasize the localized nature of Paget's disease. It may be monostotic, affecting only a single bone or portion of a bone (Fig. 2), or may be polyostotic, involving two or more bones. Sites of disease are often asymmetric. A patient might have a pagetic right femur with a normal left, involvement of only one half the pelvis, or involvement of several noncontiguous vertebral bodies. Clinical observation suggests that, in most instances, sites affected with Paget's disease when the diagnosis is made are the only ones that will show pagetic change over time. Although progression of disease within a given bone may occur, the sudden appearance of new sites of involvement years after the initial diagnosis is uncommon. This information can be very reassuring for patients who often worry about extension of the disorder to new areas of the skeleton as they age.

The most common sites of involvement include the pelvis, femur, spine, skull, and tibia. The bones of the upper extremity, as well as the clavicles, scapulae, ribs, and facial bones, are less commonly involved, and the hands and feet are only rarely affected. It is generally believed that most patients with Paget's disease are asymptomatic and that the disorder is most often diagnosed when an elevated serum alkaline phosphatase is noted on routine screening or when a radiograph taken for an unrelated problem shows typical skeletal changes. The development of symptoms or complications of Paget's disease is influenced by the particular areas of involvement, the interrelationship between affected bone and adjacent structures, the extent of metabolic activity, and presence or absence of disease progression within an affected site.

Signs and Symptoms

Bone pain from a site of pagetic involvement, experienced either at rest or with motion, is probably the most common symptom. The direct cause of the pain may be difficult to characterize and requires careful evaluation. Pagetic bone associated with a high turnover state has an increased vascularity, leading to a sensation of warmth of the skin overlying bone (e.g., skull or tibia) that some patients perceive as an unpleasant sensation. Small transverse lucencies along the expanded cortices of involved weight-bearing bones or advancing, lytic, blade-of-grass lesions sometimes cause pain. It is postulated that microfractures frequently occur in pagetic bone and can cause discomfort for a period of days to weeks.

A bowing deformity of the femur or tibia can lead to pain for several possible reasons. A bowed limb is typically shortened, resulting in specific gait abnormalities that can lead to abnormal mechanical stresses. Clinically severe secondary arthritis can occur at joints adjacent to pagetic bone (e.g., the hip, knee, or ankle). The secondary gait problems also may lead to arthritic changes on the contralateral nonpagetic side, particularly at the hip.

Back pain in pagetic patients is another difficult symptom to assess. Nonspecific aches and pains may emanate from enlarged pagetic vertebrae in some instances; vertebral compression fractures also may be seen at pagetic sites. In the lumbar area, spinal stenosis with neural impingement may arise, producing radicular pain and possibly motor impairment. Degenerative changes in the spine may accompany pagetic changes, and it is useful for the clinician to determine which symptoms arise as a consequence of the pagetic process and which result from degenerative disease of nonpagetic vertebrae. Kyphosis may occur, or there may be a forward tilt of the upper back, particularly when a compression fracture or spinal stenosis is present. Treatment options will differ, depending on the basis of the symptoms. When Paget's disease affects the thoracic spine, there may rarely be syndromes of direct spinal cord compression with motor and sensory changes. Several cases of apparent direct cord compression with loss of neural function have now been documented to have resulted from a vascular

© 2008 American Society for Bone and Mineral Research

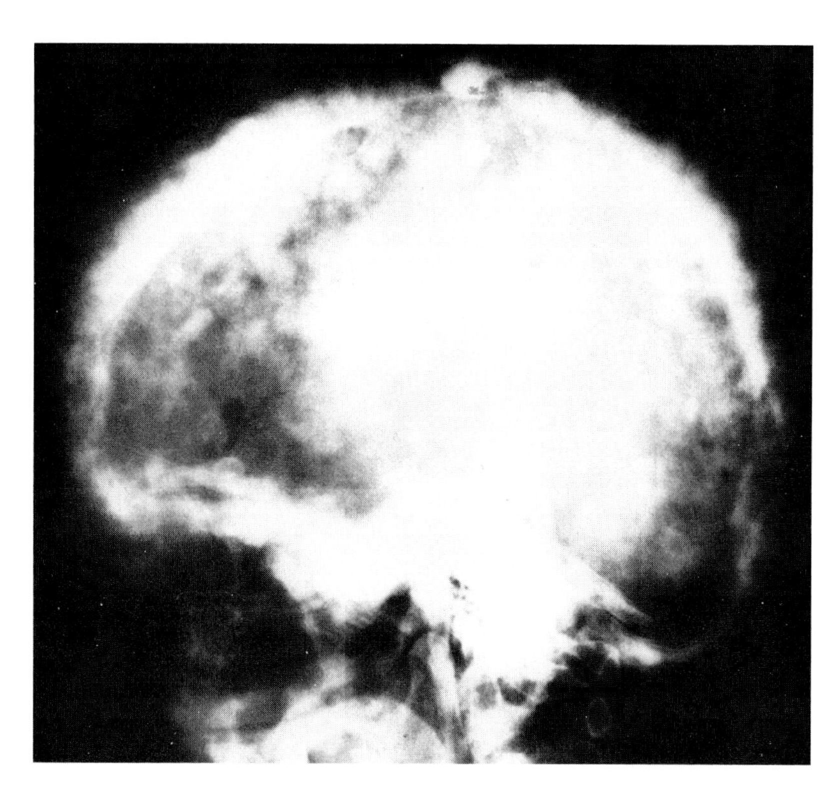

FIG. 3. Typical "cotton-wool" appearance of an enlarged pagetic skull with marked osteoblastic change. The patient had an increase in head size and deafness.

steal syndrome, whereby hypervascular pagetic bone "steals" blood from the neural tissue.[28]

Paget's disease of the skull, shown radiographically in Fig. 3, may be asymptomatic, but common complaints in up to one third of patients with skull involvement may include an increase in head size with or without frontal bossing or deformity, or headache, sometimes described as a band-like tightening around the head. Hearing loss may occur as a result of isolated or combined conductive or neurosensory abnormalities; recent data suggest cochlear damage from pagetic involvement of the temporal bone with loss of BMD in the cochlear capsule is an important component.[29] Cranial nerve palsies (such as in nerves II, VI, and VII) occur rarely. With extensive skull involvement, a softening of the base of the skull may produce platybasia, or flattening, with the development of basilar invagination, so that the odontoid process begins to extend upward as the skull sinks downward on it. This feature can be appreciated by various radiographic measures including skull radiographs and CT or MRI scans. Although many patients with severe skull changes may have radiographic evidence of basilar invagination, a relatively small number develop a very serious complication, such as direct brainstem compression or an obstructive hydrocephalus and increased intracranial pressure caused by blockage of cerebrospinal fluid flow. Pagetic involvement of the facial bones may cause facial deformity, dental problems, and, rarely, narrowing of the airway. Mechanical changes of these types may lead to a nasal intonation when the patient is speaking.

Fracture through pagetic bone is an occasional and serious complication. These fractures may be either traumatic or pathologic, particularly involving long bones with active areas of advancing lytic disease; the most common involve the femoral shaft or subtrochanteric area.[30] The increased vascularity of actively remodeling pagetic bone (i.e., with a moderately increased serum alkaline phosphatase) may lead to substantial blood loss in the presence of fractures caused by trauma. Fractures also may occur in the presence of areas of malignant degeneration, a rare complication of Paget's disease. Far more common are the small fissure fractures along the convex surfaces of bowed lower extremities, which may be asymptomatic, stable, and persistent for years, but sometimes a more extensive transverse lucent area extends medially from the cortex and may lead to a clinical fracture with time. As described later, there are data indicating that blade-of-grass lytic areas and these larger transverse fractures may respond to antipagetic treatment and heal. These types of lesions warrant radiographic follow-up over time. Conversely, the smaller fissure fractures typically do not change with treatment and, in the absence of new pain, rarely require extensive radiographic monitoring. In most cases, fracture through pagetic bone heals normally, although some groups have reported as high as a 10% rate of nonunion.

Neoplastic degeneration of pagetic bone is a relatively rare event, occurring with an incidence of $<1\%$. This abnormality has a grave prognosis, typically manifesting itself as new pain at a pagetic site. The most common site of sarcomatous change seems to be the pelvis, with the femur and humerus next in frequency.[31] The majority of the tumors are classified as osteogenic sarcomas, although both fibrosarcomas and chondrosarcomas are also seen. Typically osteosarcomas are osteolytic, although these lesions involve cells of osteoblastic lineage.[32] Current treatment regimens emphasize maximal resection of tumor mass and chemotherapy and sometimes radiotherapy. Unfortunately, in these typically elderly patients, death from massive local extension of disease or from pulmonary metastases occurs in the majority of cases in 1–3 yr.

Benign giant cell tumors also may occur in bone affected by Paget's disease. These lesions may present clinically as localized masses at the affected site. Radiographic evaluation may show lytic changes. Biopsy shows clusters of large osteoclast-like cells, which some authors believe represent reparative granulomas.[33] These tumors may show a remarkable sensitivity to glucocorticoids, so in many instances, the mass will shrink or even disappear after treatment with prednisone or

© 2008 American Society for Bone and Mineral Research

dexamethasone.[34] Anecdotal evidence also suggests possible shrinkage of benign giant cell tumors in pagetic bone with thalidomide.

DIAGNOSIS

When Paget's disease is suspected, the diagnostic evaluation should include a careful medical history and physical examination. The possibility of a positive family history and a symptom history should be ascertained. Gout, pseudogout, and arthritis are all possible complications of Paget's disease. Rarely, patients with underlying intrinsic heart disease may develop congestive heart failure in the presence of severe Paget's disease. There are also reports suggesting that patients may have an increased incidence of calcific aortic disease.[35] Angioid streaks are seen on funduscopic examination of the eye in some patients with polyostotic Paget's disease. The physical examination also should note the presence or absence of warmth, tenderness, or bone deformity in the skull, spine, pelvis, and extremities, as well as evidence of loss of range of motion at major joints or leg length discrepancy.

Laboratory tests include measurement of serum alkaline phosphatase and in some cases a marker of bone resorption, as described earlier. Radiographic studies (bone scans and conventional radiographs) complete the initial evaluation. Bone biopsy is not usually indicated, because the characteristic radiographic and laboratory findings are diagnostic in most instances.

Bone scans are the most sensitive means of identifying pagetic sites and are most useful for this purpose. Scans are nonspecific, however, and also can be positive in nonpagetic areas that have degenerative changes or, more ominously, may reflect metastatic disease. Plain radiographs of bones noted to be positive on the bone scan provide the most specific information, because the changes noted on the radiograph are usually characteristic to the point of being pathognomonic. Examples of these are shown in Figs. 2 and 3. Enlargement or expansion of bone, cortical thickening, coarsening of trabecular markings, and typical lytic and sclerotic changes may be found. Radiographs also provide data on the status of the joints adjacent to involved sites, identify fissure fractures, indicate the degree to which lytic or sclerotic lesions predominate, and show the presence or absence of deformity or fracture.

Repeated scans or radiographs are usually unnecessary in observing patients over time, unless new symptoms develop or current symptoms become significantly worse. The possibility of an impending fracture or, rarely, of sarcomatous change should be borne in mind in these situations. Although imaging studies such as CT or MRI scans are not usually required in routine cases, a CT scan may be helpful in the assessment of a fracture where radiographs are not sufficient, and MRI scans are quite useful in assessing the possibility of sarcoma, giant cell tumor, or metastatic disease at a site of Paget's disease, in which case discovery of an accompanying soft tissue mass aids in diagnosis. Anecdotal data suggest that positron emission tomography (PET) scans of sclerotic lesions in patients with Paget's disease may help distinguish pagetic lesions from bone metastases, because the former are likely to be minimally to nonmetabolic compared with marked hypermetabolic changes seen with bone metastases.[36]

The characteristic X-ray and clinical features of Paget's disease usually eliminate problems with differential diagnosis. However, an older patient may occasionally present with severe bone pain, elevations of the serum alkaline phosphatase and urinary N-telopeptide, a positive bone scan, and less-than-characteristic radiographic areas of lytic or blastic change. Here the possibility of metastatic disease to bone or some other form of metabolic bone disease (e.g., osteomalacia with secondary hyperparathyroidism) must be considered. Old radiographs and laboratory tests are very helpful in this setting, because normal studies a year earlier would make a diagnosis of Paget's disease less likely. A similar dilemma occurs when someone with known and established Paget's disease develops multiple painful new sites; here, too, the likelihood of metastatic disease must be carefully considered, and bone biopsy for a tissue diagnosis may be indicated.

TREATMENT

Anti-Pagetic Therapy

Specific anti-pagetic therapy consists of those agents capable of suppressing the activity of pagetic osteoclasts. Currently approved agents available by prescription in the United States include six bisphosphonate compounds, orally administered etidronate, tiludronate, alendronate, and risedronate and intravenously administered pamidronate and zoledronic acid, and parenterally administered synthetic salmon calcitonin. Each of these is discussed briefly below. A more detailed review of these agents including more information on dosing regimens, clinical trial results, and side effects was recently published.[37]

Between the mid-1970s when treatments became available for the first time and the mid-1990s, the mainstays of therapy were calcitonin and etidronate. However, these agents should generally be replaced as the first lines of therapy by the newer bisphosphonates, pamidronate, alendronate, risedronate, and zoledronic acid, all progressively more potent than either etidronate or calcitonin, offering the potential for greater disease suppression and frank remission (i.e., normalization of pagetic indices) for prolonged periods of 18 mo or more. Other symptomatic treatments for Paget's disease, including analgesics, anti-inflammatory drugs, use of orthotics or canes, and selected orthopedic and neurosurgical interventions, have important roles in management in many patients.

Two logical indications for treatment of Paget's disease are to relieve symptoms and to prevent future complications. It has been clearly shown that suppression of the pagetic process by any of the available agents can effectively ameliorate certain symptoms in the majority of patients. Symptoms such as bone aches or pain (probably the most common complaints of Paget's disease), excessive warmth over bone, headache caused by skull involvement, low back pain secondary to pagetic vertebral changes, and some syndromes of neural compression (e.g., radiculopathy and some examples of slowly progressive brainstem or spinal cord compression) are the most likely to be relieved. Pain caused by a secondary arthritis from pagetic bone involving the spine, hip, knee, ankle, or shoulder may or may not respond to anti-pagetic treatment. Filling in of osteolytic blade-of-grass lesions in weight-bearing bones has been reported in some treated cases with either calcitonin or bisphosphonates. On the other hand, a bowed extremity or other bone deformity will not change after treatment, and clinical experience indicates that deafness is unlikely to improve, although limited studies suggest that progression of hearing loss may be slowed[38] or even, in one case with pamidronate, reversed.[39]

A second indication for treatment is to prevent the development of late complications in those patients deemed to be at risk, based on their sites of involvement and evidence of active disease, as shown by elevated levels of bone turnover markers. Admittedly, it has not been proven that suppression of pagetic bone turnover will prevent future complications. However, there is a restoration of normal patterns of new bone deposition in biopsy specimens after suppression of pagetic activity. It

© 2008 American Society for Bone and Mineral Research

is also clear that active, untreated disease can continue to undergo a persistent degree of abnormal bone turnover for many years, with the possibility of severe bone deformity over time. Indeed, substantial (e.g., 50%) but incomplete suppression of elevated indices of bone turnover with older therapies has been associated with disease progression[40]; with more potent bisphosphonates such as pamidronate, alendronate, risedronate, and zoledronic acid, however, indices become normal after treatment for extended periods in the majority of patients and approach normal in most of the rest.

Thus, in the view of some investigators, the presence of asymptomatic but active disease (i.e., a serum alkaline phosphatase above normal) at sites where the potential for later problems or complications exists (e.g., weight-bearing bones, areas near major joints, vertebral bodies, extensively involved skull) is an indication for treatment.[37] The need for treatment in this setting may be particularly valid in patients who are younger, for whom many years of co-existence with the disorder are likely. However, even in the elderly, one can justify treatment if a degree of bone deformity is present that might create serious problems in the next few years. Others argue that the evidence does not yet support such use, because it has not been shown in clinical trials that disease suppression reduces progression of deformity.[41]

Although controlled studies are not available to prove efficaciousness in this situation, the use of a potent bisphosphonate before elective surgery on pagetic bone also is recommended.[42] The goal here is to reduce the hypervascularity associated with moderately active disease (e.g., a 3-fold or more elevation in serum alkaline phosphatase) to reduce the amount of blood loss at operation.

Recently, recommendations for the management of Paget's disease have been published as guideline or management documents by consensus panels in the United States,[37] the United Kingdom,[41] and Canada,[43] and the reader is referred to these thoughtful reviews.

Bisphosphonates

The discussion that follows will consider this class of drugs in their ascending order of potency and to some extent in terms of their historical development. It should be emphasized that, whereas any of these medications might be chosen in a specific case, the agents that are considered to be first line at present are the nitrogen-containing bisphosphonates, namely intravenous pamidronate and zoledronic acid and oral alendronate and risedronate.

Etidronate is commercially available as Didronel (Procter & Gamble, Mason, OH, USA) in a 200- or 400-mg tablet. Five milligrams per kilogram per day will provide a 50% lowering of biochemical indices and a reduction in symptoms in the majority of patients.[44] The dose of etidronate is limited by the fact that higher doses that could more effectively reduce the increased bone resorption can also impair mineralization, compelling the use of lower doses given for no longer than 6 mo at a time. The recommended regimen is 5 mg/kg/day (i.e., 400 mg in most patients, taken with a small amount of water midway in a 4-h fast any time of day) for a 6-mo period, followed by at least 6 mo of no treatment. Etidronate is contraindicated in the presence of advancing lytic changes in a weight-bearing bone. Secondary resistance to etidronate with long-term use has been described.

Tiludronate is available as Skelid (Sanofi-Aventis, Bridgewater, NJ, USA) in a 200-mg tablet. The recommended dosage is 400 mg daily for 3 mo (taken with 6–8 oz of water at least 2 h away from food, and the patient should not lie down for the next 30 min) with a 3-mo post-treatment observation period, after which the serum alkaline phosphatase is likely to have reached its nadir. Results with tiludronate (i.e., magnitude of decrease in serum alkaline phosphatase) are similar to slightly better than those with etidronate, but less drug is needed, and the likelihood of a mineralization problem is largely eliminated.[45]

Pamidronate is the first of the more potent, nitrogen-containing bisphosphonates, and its use is associated with a normalization of pagetic indices in the majority of patients.[46] The dose should be individualized to the needs of the specific patient, and there really is no single best dose. The package insert for pamidronate, available as Aredia (Novartis, East Hanover, NJ, USA), recommends three daily infusions of 30 mg each, over a period of 4 h each time, in 500 ml of normal saline or 5% dextrose in water. In clinical practice, this is probably not the best mode of administration. Clinical experience indicates that patients with relatively mild disease may have a substantial reduction of alkaline phosphatase to normal or near normal with a single 60- to 90-mg infusion given over a 2-h period in 300–500 ml of 5% dextrose in water. Patients with more moderate to severe disease (e.g., serum alkaline phosphatase levels >3–4 times normal) may require multiple infusions of 60–90 mg infused as described and given on a once weekly or biweekly basis, primarily based on physician and patient convenience. Total doses in the range of 300–500 mg may be required in some severe cases (serum alkaline phosphatase ~10–20 times normal), given over a number of weeks. Recently, there has been recognition that some patients develop secondary resistance to pamidronate after repeated use.[47]

Alendronate is available as a 40-mg dose of oral Fosamax (Merck, West Point, PA, USA), given for 6 mo. It is taken once daily on arising after an overnight fast with 8 oz of plain water, with the patient remaining upright for the next 30 min. A majority of subjects in clinical trials (>70%) had a normal alkaline phosphatase level after a 6-mo course, and many of these remained in biochemical remission for the next 12–18 mo.[48]

Risedronate is available as Actonel (Procter & Gamble and Sanofi-Aventis) at a daily oral dose of 30 mg for 2 mo. A follow-up measurement of serum alkaline phosphatase 1 mo later is useful; if the value is not yet normal or near normal, a third or fourth month could be offered with a good likelihood of normalcy or near normalcy of indices thereafter. In the pivotal clinical trial, 80% of the patients achieved a normal alkaline phosphatase level 6 mo after initiation of 2 mo of treatment, with a period of subsequent disease suppression (18 mo) similar to that achieved with pamidronate or alendronate.[49] Risedronate is taken in a manner identical to that described earlier for alendronate.

Zoledronic acid (or zoledronate) is sold as Reclast (Novartis) at a dose of 5 mg, administered as a single 15-min intravenous infusion. In a clinical trial comparing the single 5-mg intravenous dose of zoledronate with 2 mo of 30 mg/d oral risedronate, a normal level of serum alkaline phosphatase was achieved by 89% of zoledronate subjects compared with 58% of risedronate subjects.[50] The great potency and ease of use of zoledronate makes it a logical first choice for most patients.

It is important to emphasize the need for full repletion of both calcium and vitamin D before and during treatment with the potent bisphosphonates to avoid hypocalcemia and secondary hyperparathyroidism. Calcium and vitamin D repletion should be maintained thereafter in these patients as a general principle.

Side effects seen with the oral agents, alendronate and risedronate, include upper gastrointestinal symptoms consistent with esophageal irritation in a minority of individuals. The first

© 2008 American Society for Bone and Mineral Research

ever dose of either pamidronate or zoledronate in a patient who has not previously received a nitrogen-containing bisphosphonate can be associated with a flu-like reaction for 1–2 days after treatment with fever, headache, myalgia, and arthralgia, ameliorated by using acetaminophen or a nonsteroidal anti-inflammatory drug (NSAID); this reaction is unlikely to occur with subsequent doses. Finally, relatively rare cases of uveitis or iritis have been described with nitrogen-containing bisphosphonates. In such patients, either etidronate or tiludronate can be given, because these compounds do not contain the nitrogen atom.

Osteonecrosis of the jaw has recently been described as a complication typically after dental extractions in patients receiving relatively high doses of potent bisphosphonates such as zoledronic acid and pamidronate given primarily for management of bone metastases. Seven patients with Paget's disease have also been reported to have had this complication, most of whom were given very high doses for prolonged periods of time outside the usual prescribing guidelines.[51] This topic is discussed in detail elsewhere in the *Primer*.

Calcitonin

The polypeptide hormone, salmon calcitonin, is available therapeutically as a synthetic formulation for parenteral administration. It is sold as Miacalcin (Novartis), injected subcutaneously or intramuscularly. The usual starting dose is 100 U (0.5 ml; the drug is available in a 2-ml vial), generally self-injected subcutaneously, initially on a daily basis. Symptomatic benefit may be apparent in a few weeks, and the biochemical benefit (typically about a 50% reduction from baseline in serum alkaline phosphatase) is usually seen after 3–6 mo of treatment. After this period, many clinicians reduce the dose to 50–100 U every other day or three times weekly. Escape from the efficacy of salmon calcitonin may sometimes occur after a variable period of benefit. The main side effects of parenteral salmon calcitonin include, in a minority of patients, the development of nausea or queasiness, with or without flushing of the skin of the face and ears. It is apparent that the newer bisphosphonates offer both greater effectiveness and ease of use, suggesting that this agent will be used in the future primarily by patients who do not tolerate oral or intravenous bisphosphonate therapy.

Intranasal calcitonin is available as Miacalcin Nasal Spray (Novartis). It seems to have a lower incidence of the side effects described earlier. The optimal dose in Paget's disease with the present formulation is not known, but anecdotal evidence suggests that, in occasional patients with mild disease, the 200-U single spray dose given daily may lower biochemical indices and relieve mild symptoms, such as increased warmth in a pagetic tibia. It is not specifically approved for use in Paget's disease.

Other Therapies

Analgesics such as acetaminophen, aspirin, and NSAIDS may be tried empirically with or without anti-pagetic therapy to relieve pain. Pagetic arthritis (i.e., osteoarthritis caused by deformed pagetic bone at a joint space) may cause periods of pain that are often helped by some of these agents.

Surgery on pagetic bone[52] may be necessary in the setting of established or impending fracture. Elective joint replacement, more complex with Paget's disease than with typical osteoarthritis, is often very successful in relieving refractory pain. Rarely, osteotomy is performed to alter a bowing deformity in the tibia. Neurosurgical intervention is sometimes required in cases of spinal cord compression, spinal stenosis, or basilar invagination with neural compromise. Although medical management may be beneficial and adequate in some instances, all cases of serious neurological compromise require immediate neurological and neurosurgical consultation to allow the appropriate plan of management to be developed. As improved therapies emerge, long-term suppression of pagetic activity may have a preventive role in Paget's disease and, possibly, may obviate the need for surgical management in many cases.

REFERENCES

1. Kanis JA 1998 Pathophysiology and Treatment of Paget's Disease of Bone, 2nd ed. Martin Dunitz, London, UK.
2. Siris ES, Canfield RE, Jacobs TP 1980 Paget's disease of bone. Bull NY Acad Med **56:**285–304.
3. Morales-Piga AA, Rey-Rey JS, Corres-Gonzalez J, Garcia-Sagredo IM, Lopez-Abente G 1995 Frequency and characteristics of familial aggregation of Paget's disease of bone. J Bone Miner Res **10:**663–670.
4. McKusick VA 1972 Heritable Disorders of Connective Tissue, 5th ed. CV Mosby, St. Louis, MO, USA.
5. Siris ES, Ottman R, Flaster E, Kelsey JL 1991 Familial aggregation of Paget's disease of bone. J Bone Miner Res **6:**495–500.
6. Sparks AB, Peterson SN, Bell C, Loftus BJ, Hocking L, Cahill DP, Frassica FJ, Streeten EA, Levine MA, Fraser CM, Adams MD, Broder S, Venter JC, Kinzler KW, Vogelstein B, Ralston SH 2001 Mutation screening of the TNFRSF11A gene encoding receptor activator of NF kappa B (RANK) in familial and sporadic Paget's disease of bone and osteosarcoma. Calcif Tissue Int **68:**151–155.
7. Hocking L, Slee F, Haslam SI, Cundy T, Nicholson G, van Hul W, Ralston SH 2000 Familial Paget's disease of bone: Patterns of inheritance and frequency of linkage to chromosome 18q. Bone **26:**577–580.
8. Laurin N, Brown JP, Morissette J, Raymond V 2002 Recurrent mutation of the gene encoding sequestosome 1 (SQSTM1/p62) in Paget disease of bone. Am J Hum Genet **70:**1582–1588.
9. Hocking LJ, Herbert CA, Nicholls RK, Williams F, Bennett ST, Cundy T, Nicholson GC, Wuyts W, Van Hul W, Ralston SH 2001 Genomewide search in familial Paget disease of bone shows evidence of genetic heterogeneity with candidate loci on chromosomes 2q36, 10p13, and 5q35. Am J Hum Genet **69:**1055–1061.
10. Kurihara N, Hiruma Y, Zhou H, Subler MA, Dempster DW, Singer FR, Reddy SV, Gruber HE, Windle JJ, Roodman GD 2007 Mutation of the Sequestosome 1 (p62) gene increases osteoclastogenesis but does not induce Paget's disease. J Clin Invest **117:**133–142.
11. Barker DJP, Chamberlain AT, Guyer PH, Gardner MJ 1980 Paget's disease of bone: The Lancashire focus. BMJ **280:**1105–1107.
12. Cooper C, Schafheutle K, Dennison E, Kellingray S, Guyer P, Barker D 1999 The epidemiology of Paget's disease in Britain: Is the prevalence decreasing? J Bone Miner Res **14:**192–197.
13. Cundy T, McAnulty K, Wattie D, Gamble G, Rutland M, Ibbertson HK 1997 Evidence for secular changes in Paget's disease. Bone **20:**69–71.
14. Rebel A, Malkani K, Basle M, Bregeon C 1997 Is Paget's disease of bone a viral infection? Calcif Tissue Res **22**(Suppl)**:**283–286.
15. Mills BG, Singer FR, Weiner LP, Suffin SC, Stabile E, Holst P 1984 Evidence for both respiratory syncytial virus and measles virus antigens in the osteoclasts of patients with Paget's disease of bone. Clin Orthop **183:**303–311.
16. Gordon MT, Mee AP, Sharpe PT 1994 Paramyxoviruses in Paget's disease. Semin Arthritis Rheum **23:**232–234.
17. Mee AP, Dixon JA, Hoyland JA, Davies M, Selby PL, Mawer EB 1998 Detection of canine distemper virus in 100% of Paget's disease samples by in situ-reverse transcriptase-polymerase chain reaction. Bone **23:**171–175.
18. Kurihara N, Reddy SV, Menaa C, Anderson D, Roodman GD 2000 Osteoclasts expressing the measles virus nucleocapsid gene display a pagetic phenotype. J Clin Invest **105:**607–614.
19. Ooi CG, Walsh CA, Gallagher JA, Fraser WD 2000 Absence of measles virus and canine distemper virus transcripts in long-term bone marrow cultures from patients with Paget's disease of bone. Bone **27:**417–421.
20. Kurihara N, Zhou H, Reddy SV, Garcia-Palacios V, Subler MA, Dempster DW, Windle JJ, Roodman GD 2006 Expression of

© 2008 American Society for Bone and Mineral Research

measles virus nucleocapsid protein in osteoclasts indures Paget's disease–like bone lesions in mice. J Bone Miner Res **21**:446–455.
21. Rebel A, Basle M, Pouplard A, Malkani K, Filmon R, Lepatezour A 1980 Bone tissue in Paget's disease of bone: Ultrastructure and immunocytology. Arthritis Rheum **23**:1104–1114.
22. Singer FR, Mills BG, Gruber HE, Windle JJ, Roodman GD 2006 Ultrastructure of bone cells in Paget's disease of bone. J Bone Miner Res **21**:S2;P51–P54.
23. Shankar S, Hosking DJ 2006 Biochemical assessment of Paget's disease of bone. J Bone Miner Res **21**:S2;P22–P27.
24. Meunier PJ, Coindre JM, Edouard CM, Arlot ME 1980 Bone histomorphometry in Paget's disease: Quantitative and dynamic analysis of pagetic and non-pagetic bone tissue. Arthritis Rheum **23**:1095–1103.
25. Siris ES, Clemens TP, McMahon D, Gordon AG, Jacobs TP, Canfield RE 1989 Parathyroid function in Paget's disease of bone. J Bone Miner Res **4**:75–79.
26. Siris ES, Canfield RE 1994 The parathyroids and Paget's disease of bone. In: Bilezikian J, Levine M, Marcus R (eds.) The Parathyroids. Raven Press, New York, NY, USA, pp. 823–828.
27. Siris ES 1991 Indications for medical treatment of Paget's disease of bone. In: Singer FR, Wallach S (eds.) Paget's Disease of Bone: Clinical Assessment. Present and Future Therapy. Elsevier, New York, NY, USA, pp. 44–56.
28. Herzberg L, Bayliss E 1980 Spinal cord syndrome due to noncompressive Paget's disease of bone: A spinal artery steal phenomenon reversible with calcitonin. Lancet **2**:13–15.
29. Monsell EM 2004 The mechanism of hearing loss in Paget's disease of bone. Laryngoscope **114**:598–606.
30. Barry HC 1980 Orthopedic aspects of Paget's disease of bone. Arthritis Rheum **23**:1128–1130.
31. Wick MR, Siegal GP, Unni KK, McLeod RA, Greditzer HB 1981 Sarcomas of bone complicating osteitis deformans (Paget's disease). Am J Surg Pathol **5**:47–59.
32. Hansen MF, Seton M, Merchant A 2006 Osteosarcoma in Paget's disease of bone. J Bone Miner Res **21**:S2;P58–P63.
33. Upchurch KS, Simon LS, Schiller AL, Rosenthal DI, Campion EW, Krane SM 1983 Giant cell reparative granulomas of Paget's disease of bone: A unique clinical entity. Ann Intern Med **98**:35–40.
34. Jacobs TP, Michelsen J, Polay J, D'Adamo AC, Canfield RE 1979 Giant cell tumor in Paget's disease of bone: Familial and geographic clustering. Cancer **44**:742–747.
35. Strickenberger SA, Schulman SP, Hutchins GM 1987 Association of Paget's disease of bone with calcific aortic valve disease. Am J Med **82**:953–956.
36. Sundaram M 2006 Imaging of Paget's disease and fibrous dysplasia of bone. J Bone Miner Res **21**:S2;P28–P30.
37. Siris ES, Lyles KW, Singer FR, Meunier PJ 2006 Medical management of Paget's disease of bone: Indications for treatment and review of current therapies. J Bone Miner Res **21**:S2;94–98.
38. El-Sammaa M, Linthicum FH, House HP, House JW 1986 Calcitonin as treatment for hearing loss in Paget's disease. Am J Otol **7**:241–243.
39. Murdin L, Yeoh LH 2005 Hearing loss treated with pamidronate. J R Soc Med **98**:272–274.
40. Meunier PI, Vignot E 1995 Therapeutic strategy in Paget's disease of bone. Bone **17**:489S–491S.
41. Selby PL, Davie MWJ, Ralston SH, Stone MD 2002 Guidelines on the management of Paget's disease of bone. Bone **31**:366–373.
42. Kaplan FS 1999 Surgical management of Paget's disease. J Bone Miner Res **14**:S2;34–38.
43. Drake WM, Kendler DL, Brown JP 2001 Consensus statement on the modern therapy of Paget's disease of bone from a Western Osteoporosis Alliance Symposium. Clin Ther **23**:620–626.
44. Canfield R, Rosner W, Skinner J, McWhorter J, Resnick L, Feldman F, Kammerman S, Ryan K, Kunigonis M, Bohne W 1977 Diphosphonate therapy of Paget's disease of bone. J Clin Endocrinol Metab **44**:96–106.
45. McClung MR, Tou CPK, Goldstein NH, Picot C 1995 Tiludronate therapy for Paget's disease of bone. Bone **17**:493S–496S.
46. Harinck HI, Papapoulos SE, Blanksrna HJ, Moolenaar AJ, Vermeij P, Bijvoet OL 1987 Paget's disease of bone: Early and late responses to three different modes of treatment with aminohydroxypropylidene bisphosphonate (APD). BMJ **295**:1301–1305.
47. Gutteridge DH, Ward LC, Stewart GO, Retallack RW, Will RK, Prince RL, Criddle A, Bhagat CI, Stuckey BG, Price RI, Kent GN, Faulkner DL, Geelhoed E, Gan SK, Vasikaran S 1999 Paget's disease: Acquired resistance to one aminobisphosphonate with retained response to another. J Bone Miner Res **14**:S2;79–84.
48. Siris E, Weinstein RS, Altman R, Conte JM, Favus M, Lombardi A, Lyles K, McIlwain H, Murphy WA Jr, Reda C, Rude R, Seton M, Tiegs R, Thompson D, Tucci JR, Yates AJ, Zimering M 1996 Comparative study of alendronate vs. etidronate for the treatment of Paget's disease of bone. J Clin Endocrinol Metab **81**:961–967.
49. Miller PD, Adachi JD, Brown JP, Khairi RA, Lang R, Licata AA, McClung MR, Ryan WG, Singer FR, Siris ES, Tenenhouse A, Wallach S, Bekker PJ, Axelrod DW 1997 Risedronate vs. etidronate: Durable remission with only two months of 30 mg risedronate. J Bone Miner Res **12**:S269.
50. Reid IR, Miller P, Lyles K, Fraser W, Brown J, Saidi Y, Mesenbrink P, Su G, Pak J, Zelenakas K, Luchi M, Richardson P, Hosking D 2005 A single infusion of zoledronic acid improves remission rates in Paget's disease: A randomized controlled comparison with risedronate. N Engl J Med **353**:898–908.
51. Khosla S, Burr D, Cauley J, Dempster DW, Ebeling PR, Felsenberg D, Gagel RF, Gilsanz V, Guise T, Koka S, McCauley LK, McGowan J, McKee MD, Mohla S, Pendrys DG, Raisz LG, Ruggiero SL, Shafer DM, Shum L, Silverman SL, Van Poznak CH, Watts N, Woo SB, Shane E; American Society for Bone and Mineral Research. 2007 Bisphosphonate-associated osteonecrosis of the jaw: Report of a task force of the American Society for Bone and Mineral Research. J Bone Miner Res **22**:1479–1491.
52. Parvizi J, Klein GR, Sim FH 2006 Surgical management of Paget's disease of bone. J Bone Miner Res **21**:S2;P75–P82.

Chapter 73. Chronic Kidney Disease Mineral Bone Disorder (CKD-MBD)

Keith A. Hruska and Suresh Mathew

Division of Pediatric Nephrology, Department of Pediatrics, Washington University, St. Louis, Missouri

DEFINITION

Chronic kidney disease mineral bone disorder (CKD-MBD), is a term coined by the Kidney Disease Improving Global Outcomes (KDIGO) Foundation[1] to replace the term, renal osteodystrophy (ROD), in recognition of several pathophysiologic discoveries of the 21st century. ROD is now reserved as a term for the various skeletal pathologies, and CKD-MBD is used for the broader syndrome of mineral metabolism, cardiovascular, and skeletal disorders complicating CKD. The first of these pathophysiologic discoveries is that the skeletal remodeling disorders caused by CKD contribute directly to heterotopic mineralization, especially vascular calcification, and the

Dr. Hruska has consulted for Abbott, Shire, Genzyme, and Celgene. Dr. Mathew states that he has no conflicts of interest.

© 2008 American Society for Bone and Mineral Research

disordered mineral metabolism that accompany CKD.[2,3] Second, the disorders in mineral metabolism associated with CKDs are key factors contributing to the excess mortality observed in CKD.[4,5] Third, CKD or renal injury impairs skeletal anabolism decreasing osteoblast function and bone formation rates.[3,6] In short, a multiorgan system has been defined involving the kidney, skeleton, parathyroid glands, and the cardiovasculature that fails in CKD.

PATHOBIOLOGY

The earliest histological abnormalities of bone in CKD-MBD are seen after a relatively mild reduction in the glomerular filtration rate (creatinine clearances between 40 and 70 ml/min; stage 2 CKD).[7] In addition, elevated PTH levels and elevated fibroblast growth factor 23 (FGF23) levels may be observed before detectable changes in serum phosphorus (Pi), calcitriol [1,25(OH)$_2$D$_3$], or calcium (Ca).[8] If hyperparathyroidism is prevented or treated, a low-turnover osteodystrophy, the adynamic bone disorder, is observed, uncovering the effects of kidney injury on the skeleton. By stage 5 CKD (creatinine clearance <15 ml/min), skeletal histological pathology is found in virtually all patients.[9] The increasing incidence of CKD in the United States and across the world, and the role of the CKD-MBD in its high mortality rate, make CKD-MBD a major health issue for Americans and all developed societies.[10,11]

Pathogenesis

Renal injuries produce a loss of skeletal anabolism manifested as a decrease in bone formation rates that derive from osteoblast activity.[6] The loss of anabolism caused by kidney injury occurs in the presence of normal PTH, vitamin D, Ca, and phosphate (PO_4) levels, but it is not usually observed because abnormalities in these factors stimulate an adaptation, PTH secretion, and secondary hyperparathyroidism. The sustained increase in PTH levels produced through adaptation to CKD produce an unwanted disorder of skeletal remodeling a high-turnover osteodystrophy, or osteitis fibrosa (see below).

Pathogenetic Factors in Secondary Hyperparathyroidism

Hyperphosphatemia. As renal injury decreases nephron number, the stimulus to hyperphosphatemia caused by a reduction in filtered phosphate is reversed through PTH- and FGF23-mediated reductions in tubular epithelial phosphate transport. The increase in phosphate excretion per remaining nephron restores phosphate homeostasis at the cost of higher PTH and FGF23 levels and maintains normal phosphate excretion. In stages IV and V CKD, when renal injury is severe enough that the glomerular filtration rate reaches levels of <30% of normal, hyperphosphatemia becomes fixed because of insufficient renal excretion despite high PTH and FGF23 levels.[12] Studies showed that failure of Ca and Pi deposition into the skeleton or excess resorption of the skeleton also contribute to abnormal Ca and Pi levels in CKD.[3,13] Hyperphosphatemia decreases serum Ca through physicochemical binding and suppresses 1α-hydroxylase activity, which results in further lowering of circulating calcitriol levels. Moreover, a direct stimulatory effect of phosphate on parathyroid gland cells, independent of Ca and calcitriol, produces increased secretion and nodular hyperplasia of parathyroid gland cells.[14,15] Finally, hyperphosphatemia is a signaling mechanism for induction of heterotopic mineralization of the vasculature in CKD.[16,17]

Calcitriol Deficiency. As CKD advances, the functioning nephron mass is decreased, leading to decreased calcitriol production by proximal tubular 25-hydroxy cholecalciferol 1α-hydroxylase,[18] and this is reinforced by the increased phosphate load in remaining nephrons and increased FGF23 levels, leading to calcitriol deficiency. The latter in turn decreases intestinal calcium absorption and leads to hypocalcemia. Calcitriol deficiency in cases of advanced kidney failure in turn diminishes tissue levels of vitamin D receptors (VDRs), in particular, the VDR of parathyroid gland cells.[19] Because the chief cell VDR suppresses the expression of prepro-PTH mRNA, lower circulating calcitriol levels together with a low number of vitamin D receptors in patients with CKD result in stimulation of both synthesis and secretion of PTH.[20]

Hypocalcemia. As CKD progresses, hypocalcemia develops because of decreased intestinal Ca absorption. Low blood levels of ionized Ca stimulate PTH secretion, whereas high calcium concentrations suppress it. The action of Ca on parathyroid gland chief cells is mediated through a calcium sensor; a G-protein–coupled plasma membrane receptor (CASR) is expressed in chief cells, kidney tubular epithelia, and widely throughout the body at lower levels.[21,22] The short-term stimulation of PTH secretion induced by low Ca is caused by exocytosis of PTH packaged in granules, and longer-term stimulation results from an increase in the number of cells that secrete PTH. More prolonged hypocalcemia induces changes in intracellular PTH degradation and mobilization of a secondary storage pool. Within days or weeks of the onset of hypocalcemia, prepro-PTH mRNA expression is stimulated. This effect is exerted through a negative calcium response element located in the upstream flanking region of the gene for PTH. Expression of the Ca receptor has been shown to be suppressed by calcitriol deficiency and stimulated by calcitriol administration, suggesting an additional regulatory mechanism of the active vitamin D metabolite on PTH production. The decreased number of calcium-sensing receptors with low circulating calcitriol may, at least in part, explain the relative insensitivity of parathyroid gland cells to Ca in patients undergoing dialysis.

Hyperparathyroidism. All of the mechanisms discussed above result in increased production of PTH and nodular hyperplasia of the parathyroid glands in CKD. The size of the parathyroid glands progressively increases during CKD and in dialyzed patients paralleling serum PTH levels. This increase in gland size is mainly caused by diffuse cellular hyperplasia. Monoclonal chief cell growth also develops, resulting in the formation of nodules. Nodular hyperplastic glands have less vitamin D receptor and Ca-sensing receptors compared with diffusely hyperplastic glands, promoting parathyroid gland resistance to calcitriol and Ca. Sustained elevation in PTH levels, whereas adaptive to maintain osteoblast surfaces, produce an abnormal phenotype of osteoblast function with relatively less type 1 collagen and more RANKL ligand production than anabolic osteoblasts. This leads to a high turnover osteodystrophy, PTH receptor desensitization, and excess bone resorption.

FGF23. FGF23 is the original phosphatonin (phosphate excretion regulating hormone) discovered in studies of autosomal dominant hypophosphatemic rickets and oncogenic osteomalacia.[23,24] FGF23 levels progressively rise during the course of CKD,[25] and the roles of FGF23 in regulating phosphate homeostasis and calcitriol synthesis in CKD are just being elucidated. Furthermore, FGF23 is produced mainly by osteocytes and osteoblasts, and it is a mineralization inhibitor.

© 2008 American Society for Bone and Mineral Research

It is unclear whether FGF23 contributes to the mineralization defects observed in skeletal remodeling in CKD. FGF23 represents a direct bone-kidney connection in the multiorgan system involved in the CKD-MBD.

Hypogonadism. Patients with CKD have various states of gonadal dysfunction. Estrogen and testosterone deficiency significantly contribute to pathogenesis of osteopenia in CKD.

Other Factors. Inflammatory mediators, acidosis, and aluminum are all potentially critical factors in the CKD-MBD that have not been well studied or have been clinically eliminated, and space prevents further description of their role. Some patients with CKD are treated with glucocorticoids, which have important skeletal side effects. Patients maintained on chronic dialysis have retention and deposition of β2-microglobulin in periarticular depots. Additionally, alterations in growth factors and other hormones involved in the regulation of bone remodeling may be disordered in CKD, thus contributing to the development of the CKD-MBD.

Pathology of ROD

ROD is not a uniform disorder. Depending on the relative contribution of the different pathogenic factors discussed above and their treatment, various pathologic patterns of bone remodeling are expressed in CKD.[26]

Predominant Hyperparathyroid Bone Disease, High-Turnover ROD, and Osteitis Fibrosa. Sustained excess PTH results in increased bone turnover. Osteoclasts, osteoblasts, and osteocytes are found in abundance. Disturbed osteoblastic activity results in a disorderly production of collagen, which results in formation of woven bone. Accumulation of fibroblastic osteoprogenitors not in the osteoblastic differentiation program results in collagen deposition (fibrosis) in the peritrabecular and marrow space. The nonmineralized component of bone, osteoid, is increased.

Low-Turnover Bone Disease and Adynamic Bone Disorder. Low-turnover uremic osteodystrophy is the other end of the spectrum of renal osteodystrophy. The histologic hallmark of this group is a profound decrease in bone turnover, because of a low number of active remodeling sites, suppression of bone formation, and low resorption, which is not as decreased as formation. The result is a low-turnover osteopenic condition. The majority of trabecular bone is covered by lining cells, with few osteoclasts and osteoblasts. Bone structure is predominantly lamellar. The extent of mineralizing surfaces is markedly reduced. Usually only a few thin, single tetracycline labels are observed. Two subgroups can be identified in this type of ROD, depending on the cause of events leading to a decline in osteoblast activity: the adynamic bone disorder (ABD) and low-turnover osteomalacia from aluminum intoxication, bisphosphonate administration, or other factors.

Mixed Uremic Osteodystrophy and High-Turnover ROD Plus a Mineralization Defect. Mixed uremic osteodystrophy is caused primarily by hyperparathyroidism and defective mineralization with or without increased bone formation caused by aluminum. These features may coexist in varying degrees in different patients. Increased numbers of heterogeneous remodeling sites can be seen. The number of osteoclasts is usually increased. Because active foci with numerous cells, woven osteoid seams, and peritrabecular fibrosis coexist next to lamellar sites with a more reduced activity, greater production of lamellar or woven osteoid causes an accumulation of osteoid with normal or increased thickness of osteoid seams.

Associated Features

Osteoporosis and Osteosclerosis. A negative balance of bone formation and bone resorption in cortical and cancellous bone is common in ROD. In these cases, bone densitometry will detect osteopenia or osteoporosis. The prevalence of osteoporosis in the population with CKD exceeds the prevalence in the general population.[27,28] Osteoporosis may be observed in CKD before dialysis is required for end-stage kidney failure.[29] When bone turnover is high, as in secondary hyperparathyroidism with osteitis fibrosa, bone resorption rates are in excess of bone formation, and osteopenia progressing to osteoporosis may result. When bone turnover is low, although both bone formation rates and bone resorption may be reduced, resorption is in excess and loss of bone mass occurs. Thus, osteoporosis caused by CKD itself may be observed with either high-turnover[29,30] or low-turnover[31] forms of osteodystrophy. When bone resorption exceeds bone formation rates in CKD, positive phosphorus and Ca balance results in hyperphosphatemia and hypercalcemia without an increase in skeletal mineral deposition, but with a stimulation of heterotopic mineralization, especially of the vasculature. The failure of the skeleton to absorb positive phosphate balance in CKD is an important stimulus to heterotopic mineralization and links the skeleton and osteoporosis in CKD to cardiovascular events and mortality.[32] From the above discussion, it is apparent that four forms of osteoporosis complicate the CKD-MBD: caused by high-turnover ROD, caused by low-turnover ROD, osteoporosis preexistent to renal disease, and osteoporosis caused by gonadal hormone deficiency.

Patients with secondary hyperparathyroidism might have a loss or gain in bone volume depending on bone balance. When the bone balance is positive, osteosclerosis may be observed when osteoblasts are active in depositing new bone (especially woven), thus superseding bone resorption. Cancellous bone volume may be increased along with a loss of cortical bone caused by severe secondary hyperparathyroidism, but this is in part because of deposition of woven immature collagen fibrils instead of lamellar fibrils. Thus, bone strength suffers despite the increase in mass detected by DXA. This is relatively rare in the 21st century because of improved therapy of secondary hyperparathyroidism.

Bone Aluminum, Iron, Lanthanum, and Bisphosphonate Accumulation. During CKD, these substances accumulate in bone at the mineralization front, at the cement lines, or diffusely. The extent of stainable aluminum at the mineralization front correlates with histologic abnormalities in mineralization. Aluminum deposition is most severe in cases of low-turnover osteomalacia. However, it can be observed in all histological forms of ROD. In patients in whom an increased aluminum burden develops, bone mineralization and bone turnover progressively decrease. These abnormalities are reversed with removal of the aluminum. Iron also accumulates at the mineralization front and can cause low-turnover forms of ROD similar to aluminum, although much less is known of skeletal iron intoxication than aluminum. Lanthanum has recently been added as a rare earth ion administered to CKD patients as a phosphate binder. It is poorly absorbed, and its levels in bone are much less than aluminum. Whether it will prove to have long-term toxic effects seems doubtful but is unknown. Five-year data indicate that the levels of skeletal accumulation remain below those with any biological or toxic effects. Lanthanum disappearance from bone deposits is slow but not as slow

© 2008 American Society for Bone and Mineral Research

as bisphosphonate disappearance. Bisphosphonates are drugs used in the treatment of osteoporosis and hypercalcemia. There are increasing instances of bisphosphonate use in patients with CKD, especially for treatment of vascular calcification. However, the nature of the bone remodeling abnormalities in CKD, especially with woven bone formation and mineralization defects, lend a high level of risk to skeletal deposition of a substance that once deposited may not be removed. Such a risk of long-term retention of an active drug inhibiting bone turnover is now being recognized with use of bisphosphonates in osteogenesis imperfecta and the rare side effect of the drugs in osteonecrosis of the jaw.[33]

Clinical Manifestations. Patients with mild to moderate kidney insufficiency are rarely symptomatic due to ROD and its skeletal pathology. However, we must consider vascular calcification a complication of ROD, and ROD as a cause of vascular stiffness. Vascular calcification is a clinically important ROD complication causing an increase in systolic blood pressure, a widening of the pulse pressure, and an increase in pulse wave velocity in CKD. All leading to left ventricular hypertrophy, diastolic dysfunction, and coronary ischemia.

Vascular calcification in CKD is complex, and pathologically is of two types: neointimal and arterial medial. Atherosclerotic neointimal calcification is multifactorial, but it involves activation of an osteoblastic differentiation program in cells of the neointima of atherosclerotic plaques. The skeleton contributes to the pathogenesis of vascular calcification. Signals deriving from the skeleton are direct causes of the vascular mineralization. One such signal is hyperphosphatemia.[32] Diffuse calcification of arterial tunica media is referred to as Mönckeberg's sclerosis. CKD is the most common cause of Mönckeberg's sclerosis, especially when it complicates diabetes mellitus.

Symptoms of ROD related to the skeleton appear in patients with advanced kidney failure. Clinical manifestations are preceded by abnormal Ca, Pi, and PTH levels that prompt the physician to intervene and prevent these complications in many cases. When symptoms related to the skeleton occur, they are usually insidious, subtle, nonspecific, and slowly progressive.

Heterotopic Mineralization, Calciphylaxis, and Tumoral Calcinosis. Heterotopic tissue calcification may occur in the eyes and manifest as band keratopathy in the sclera or induce an inflammatory response known as the red eye syndrome in the conjunctiva. Calcium deposits are also found in the lungs and lead to restrictive lung disease. Deposits in the myocardium might cause arrhythmias, annular calcifications, valvular calcification, or myocardial dysfunction. Most soft tissue calcifications are attributed to the increased Ca phosphate product contributed to by renal osteodystrophy and excess bone resorption. The syndrome of calciphylaxis is characterized by vascular calcification in the tunica media of peripheral arteries. These calcifications induce painful violaceous skin lesions that progress to ischemic necrosis. This syndrome is associated with serious complications and often death.

Tumoral calcinosis is a form of soft tissue calcification that involves the periarticular tissues. Calcium deposits may grow to enormous size and interfere with the function of adjacent joints and organs. Although its pathogenesis is poorly understood, CKD complicated by hyperphosphatemia is the most common association with nonfamilial cases. The recent discoveries of three single gene mutations in FGF23, Klotho, and GALNT3 causing familial tumoral calcinosis shed light on the role of hyperphosphatemia in its pathogenesis.[34–36] That is because FGF23 is a hormone-stimulating excretion,[37] and Klotho is a co-receptor for FGF23 in the proximal tubule of the kidney.[38] The role of GALNT3 in hyperphosphatemia is unknown.

Bone Pain, Fractures, and Skeletal Deformities. Bone pain is usually vague, ill defined, and deep seated. It may be diffuse or localized in the lower part of the back, hips, knees, or legs. Weight bearing and changes in position commonly aggravate it. Bone pain may progress slowly to the degree that patients are completely incapacitated. Bone pain in patients with CKD usually does not cause physical signs; however, local tenderness may be apparent with pressure. Occasionally, pain can occur suddenly at one joint of the lower extremities and mimic acute arthritis or periarthritis not relieved by heat or massage. A sharp chest pain may indicate rib fracture. Spontaneous fractures or fractures after minimal trauma may also occur in vertebrae (crush fractures) and in tubular bones.

Bone pain and bone fractures can be observed in all patients with CKD independent of the underlying histologic bone disease, especially when osteoporosis is present.[27] However, low-turnover osteomalacia and aluminum-related bone disease are associated with the most severe bone pain and the highest incidence of fractures and incapacity.

Skeletal deformities can be observed in children and adults. Most children with CKD have growth retardation, and bone deformities may develop from vitamin D deficiency (rickets) or secondary hyperparathyroidism. In rickets, bowing of the long bones is seen, especially the tibias and femora, with typical genu valgum that becomes more severe with adolescence. Long-standing secondary hyperparathyroidism in children may be responsible for slipped epiphyses secondary to impaired transformation of growth cartilage into regular metaphyseal spongiosa. This complication most commonly affects the hips, becomes obvious in preadolescence, and causes limping but is usually painless. When the radius and ulna are involved, ulnar deviation of the hands and local swelling may occur. In adults, skeletal deformities can be observed in cases of severe osteomalacia or osteoporosis and include lumbar scoliosis, thoracic kyphosis, and recurrent rib fractures.

DIAGNOSIS

ROD has heretofore been characterized pathologically, and the only unequivocal tool for the exact diagnosis is bone biopsy. A new clinical imperative has derived from the discovery of the role of the CKD-MBD in the high mortality rates of CKD, which is forcing a recharacterization of ROD based on noninvasive assessments that can be determined frequently on an ongoing basis to provide more effective noninvasive diagnosis and follow-up. The KDIGO recommendations suggest that ROD be characterized for rates of bone turnover, bone mass, and mineralization. Current practice is to use PTH levels, which have been correlated with bone turnover in biopsy studies to estimate remodeling (Table 1). Bone mass is not routinely assessed in CKD, and bone mineralization disorders are controversial because of the disappearing influence of aluminum.

Abnormalities in serum Ca, phosphorus, and alkaline phosphatase levels indicate ROD but are not helpful when used alone to indicate bone turnover or mineralization. Serum PTH levels are better indicators of bone turnover (Table 1). Careful assessment of the predictive value of intact PTH levels for bone turnover from analysis of biopsy studies shows that all patients with serum PTH levels of 100 pg/ml or less have low bone turnover.[39] Values of serum PTH levels >500 pg/ml are 100% and 95.5% specific for high-bone turnover in patients maintained on hemodialysis and peritoneal dialysis, respec-

© 2008 American Society for Bone and Mineral Research

Table 1. Pathology and Diagnosis of Bone Turnover in CKD

I. Predominant hyperparathyroidism, high-turnover ROD
  a. Intact PTH > 500 pg/ml
  b. Elevated alkaline phosphatase or bone-specific alkaline phosphatase
II. Low-turnover disease
  a. Adynamic bone disorder
    1. Intact PTH < 100 pg/ml
    2. Normal alkaline phosphatase or bone-specific alkaline phosphatase
    3. Low osteocalcin
  b. Osteomalacia
    1. Intact PTH < 100 pg/ml
    2. Normal alkaline phosphatase or bone-specific alkaline phosphatase
    3. Low osteocalcin
    4. Elevated Al^{3+}
III. Mixed uremic osteodystrophy
  a. PTH > 300 pg/ml
  b. Elevated Al^{3+}
V. Unknown
  a. PTH > 100 < 500 pg/ml

tively.[39] For the majority of dialyzed patients, that is, those with serum PTH levels between 65 and 500 pg/ml, bone turnover unfortunately cannot be predicted accurately by the "intact" PTH assay. In addition to serum PTH values, certain risk factors for low bone turnover have been isolated and include peritoneal dialysis, diabetes, advanced age, high Ca content in the dialysate, high doses of Ca-based phosphate binders, aggressive vitamin D therapy, or previous parathyroidectomy. However, in individual patients, discrepancies between risk factors, PTH levels, and bone turnover are frequent (Table 1, unknown); this situation calls for improved noninvasive techniques of measuring bone turnover. Studies are underway testing the hypothesis that modern biomarkers of bone turnover, when added to PTH levels, along with determination of bone mass will more adequately define the osteodystrophy of CKD on an ongoing basis.

Skeletal radiographs are an insensitive measure of bone mass, and DXA, the standard tool for assessment of bone mass and density in postmenopausal and age-related osteoporosis, is insufficient for use in CKD to estimate bone health and guide therapy. DXA measures heterotopic mineralization that is common in CKD-MBD, and DXA does not distinguish between woven and lamellar bone. Thus, it is insensitive to a major cause of bone fragility in CKD and to changes in bone mass. The best means of estimating bone mass on an ongoing basis in CKD-MBD has not been determined, but the increased sensitivity of pCT to detect changes in bone mass makes it a candidate tool that is currently being studied.

PREVENTION AND TREATMENT

Therapeutic intervention in the CKD-MBD should begin in stage 3 CKD (creatinine clearance <60 ml/min/1.73 m^2) when positive phosphate balance and calcitriol deficiency occur.

Control of Serum Phosphorus and Calcium

Dietary phosphorus restriction in CKD is limited because of nutritional needs. Phosphorus is present in most protein-containing food products. The current recommendations for protein intake in stage 5 CKD are at least 1.2 (hemodialysis) and 1.3 g/kg/d (peritoneal dialysis), which provide a minimum of 1 g of phosphorus/d. The available dialytic methods are inefficient in removing phosphorus because of compartmentalization and slow efflux of phosphorus from the exchangeable space. Hemodialysis for 4 h three times a week removes ~3 g of phosphorus/wk in the face of roughly 7 g of intake. Nocturnal dialysis for longer periods and daily dialysis are effective in maintaining normal phosphorus levels. Therefore, the addition of phosphate binders taken with meals is required. Currently used phosphate binders are Ca carbonate, Ca acetate (PhosLo; Fresenuis), sevelamer (Renagel; Genzyme), lanthanum carbonate (Fosrenol; Shire), and others still in development. Calcium citrate should be avoided because it promotes intestinal aluminum absorption. Aluminum-containing phosphate binders, although more potent than Ca salts, should not be used because of the risk of aluminum-related bone disease. Calcium-containing phosphate binders are effective, but increase the serum Ca, sometimes causing hypercalcemia and an increase in Ca × PO_4 products. This may suppress bone formation and stimulate vascular calcification. Non–Ca-containing phosphate binders, sevelamer and lanthanum carbonate, are alternatives to Ca salts. Sevelamer has been shown to be associated with less progression of coronary and aortic calcifications compared with Ca-containing phosphate binders. Tolerability and cost are limiting factors of sevelamer therapy. Lanthanum carbonate is a newer potent and well-tolerated phosphate binder. Lanthanum does not seem to accumulate to toxic levels in bone as aluminum does, and long-term experience with $LaCO_3$ is accumulating to indicate its safety. Ferric citrate is available, but it may share the effects of aluminum on bone remodeling.

Hypocalcemia in CKD may be corrected by control of serum phosphorus and vitamin D treatment. Calcium salt administration between meals as a Ca supplement should be limited to patients with symptomatic hypocalcemia.

Use of Vitamin D and Its Metabolites

Vitamin D deficiency is very common in CKD, and repletion should be the first step. Replacement of the deficient hormone, calcitriol, with active vitamin D analogs may begin in patients with CKD, and it is routine for dialysis patients. Vitamin D analogs are effective in suppressing secondary hyperparathyroidism. In moderate hyperparathyroidism with or without mineralization defects, daily oral administration of calcitriol (0.25–0.5 μg/d of calcitriol) usually decreases serum PTH levels and improves mineralization. It is advisable to start with low doses and increase the daily dose in steps of 0.25 μg to adjust PTH levels to target values after 2 wk of therapy. Episodes of hypercalcemia may occur and can be circumvented by decreasing oral Ca salts if serum phosphate levels permit or lowering the dialysate Ca content. Despite these measures, however, hypercalcemia may persist requiring cessation of the active D analogs. Intravenous treatment regimens have become predominant using two or three times per week doses as high as 5 μg of calcitriol. These measures are effective but the positive response is clearly reduced if the parathyroid glands undergo monoclonal growth transformation and become refractory to the action of calcitriol. The vitamin D analogs 19-Nor-1-α-25-dihydroxyvitamin D_2 (Zemplar; Abbott) and doxercalciferol (Hecterol; Genzyme) for control of secondary hyperparathyroidism have been introduced, and their use has largely displaced calcitriol. They have somewhat different profiles of activity at the various tissues affected by calcitriol.

Use of Calcimimetics

Calcimimetic agents that are allosteric modulators of the Ca-sensing receptor have been introduced for the control of

© 2008 American Society for Bone and Mineral Research

secondary hyperparathyroidism. This class of therapeutic agents suppresses PTH synthesis and secretion while simultaneously lowering serum Ca and phosphorus in CKD. Therapy with cinacalcet (Sensipar; Amgen) seems to increase the proportion of subjects achieving K/DOQI PTH and Ca × PO_4 targets. However, it remains to be seen whether this will translate into improvement of bone health or cardiovascular outcomes.

Parathyroidectomy

Despite treatment, overt secondary hyperparathyroidism develops in some patients and may necessitate parathyroidectomy. Indications for parathyroidectomy include (1): persistent hypercalcemia, (2) persistent hyperphosphatemia and high Ca phosphate product, (3) progressive and symptomatic soft tissue calcification with high bone turnover (including calciphylaxis), (4) severe progressive and symptomatic hyperparathyroidism when rapid reduction in PTH is required and vitamin D pulse therapy has failed, and (5) refractory pruritus. Before parathyroidectomy, histological evidence of severe hyperparathyroidism and absence of aluminum accumulation should be documented by biopsy when possible.

The most frequently used surgical approaches to parathyroidectomy are subtotal parathyroidectomy and total parathyroidectomy with parathyroid autotransplantation. Patients undergoing parathyroidectomy require careful follow-up and meticulous management. Postoperative hypocalcemia should be anticipated and treated with oral and intravenous Ca. The use of calcitriol may minimize the need for large doses of Ca salts; however, it may interfere with successful function of transplanted glands.

Treatment of ABD

At this time, ABD should be managed by measures to increase PTH levels and increase remodeling. These measures include reduction in Ca-containing phosphate binders and/or the dialysate Ca content. Discontinuation of vitamin D analogs and calcimimetics may be necessary. No specific treatment is available for the ABD at present. Thus, preventive measures should be carefully considered because of the morbidity of vascular calcification associated with this form of ROD.

ACKNOWLEDGMENTS

The writing of this chapter was supported by NIH Grants DK070790 and AR41677.

REFERENCES

1. Moe S, Drueke T, Cunningham J, Goodman W, Martin K, Olgaard K, Ott S, Sprague S, Lameire N, Eknoyan G 2006 Definition, evaluation, and classification of renal osteodystrophy: A position statement from Kidney Disease: Improving Global Outcomes (KDIGO). Kidney Int **69:**1945–1953.
2. Mathew S, Lund R, Strebeck F, Tustison KS, Geurs T, Hruska KA 2007 Reversal of the adynamic bone disorder and decreased vascular calcification in chronic kidney disease by sevelamer carbonate therapy. J Am Soc Nephrol **18:**122–130.
3. Davies MR, Lund RJ, Mathew S, Hruska KA 2005 Low turnover osteodystrophy and vascular calcification are amenable to skeletal anabolism in an animal model of chronic kidney disease and the metabolic syndrome. J Am Soc Nephrol **16:**917–928.
4. Stevens LA, Djurdjev O, Cardew S, Cameron EC, Levin A 2004 Calcium, phosphate, and parathyroid hormone levels in combination and as a function of dialysis duration predict mortality: Evidence for the complexity of the association between mineral metabolism and outcomes. J Am Soc Nephrol **15:**770–779.
5. Block GA, Hulbert-Shearon TE, Levin NW, Port FK 1998 Association of serum phosphorus and calcium X phosphate product with mortality risk in chronic hemodialysis patients: A national study. Am J Kidney Dis **31:**607–617.
6. Lund RJ, Davies MR, Brown AJ, Hruska KA 2004 Successful treatment of an adynamic bone disorder with bone morphogenetic protein-7 in a renal ablation model. J Am Soc Nephrol **15:**359–369.
7. Malluche HH, Ritz E, Lange HP 1976 Bone histology in incipient and advanced renal failure. Kidney Int **9:**355–362.
8. Craver L, Marco MP, Martinez I, Rue M, Borras M, Martin ML, Sarro F, Valdivielso JM, Fernandez E 2007 Mineral metabolism parameters throughout chronic kidney disease stages 1-5–achievement of K/DOQI target ranges. Nephrol Dial Transplant **22:**1171–1176.
9. Malluche HH, Faugere MC 1990 Renal bone disease 1990: Challenge for nephrologists. Kidney Int **38:**193–211.
10. Foley RN, Parfrey PS, Sarnak MJ 1998 Clinical epidemiology of cardiovascular disease in chronic renal disease. Am J Kidney Dis **32:**S112–S119.
11. Slinin Y, Foley RN, Collins AJ 2005 Calcium, phosphorus, parathyroid hormone, and cardiovascular disease in hemodialysis patients: The USRDS Waves 1, 3, and 4 Study. J Am Soc Nephrol **16:**1788–1793.
12. Slatopolsky E, Robson AM, Elkan I, Bricker NS 1968 Control of phosphate excretion in uremic man. J Clin Invest **47:**1865–1874.
13. Kurz P, Monier-Faugere M-C, Bognar B, Werner E, Roth P, Vlachojannis J, Malluche HH 1994 Evidence for abnormal calcium homeostasis in patients with adynamic bone disease. Kidney Int **46:**855–861.
14. Moallem E, Kilav R, Silver J, Naveh-Many T 1998 RNA-protein binding and post-transcriptional regulation of parathyroid hormone gene expression by calcium and phosphate. J Biol Chem **273:**5253–5259.
15. Naveh-Many T, Rahamimov R, Livni N, Silver J 1995 Parathyroid cell proliferation in normal and chronic renal failure rats. J Clin Invest **96:**1786–1793.
16. Jono S, McKee MD, Murry CE, Shioi A, Nishizawa Y, Mori K, Morii H, Giachelli CM 2000 Phosphate regulation of vascular smooth muscle cell calcification. Circ Res **87:**e10–e17.
17. Li X, Yang HY, Giachelli CM 2006 Role of the sodium-dependent phosphate cotransporter, Pit-1, in vascular smooth muscle cell calcification. Circ Res **98:**905–912.
18. Goodman WG, Quarles LD 2008 Development and progression of secondary hyperparathyroidism in chronic kidney disease: Lessons from molecular genetics. Kidney Int **74:**276–288.
19. Naveh-Many T, Marx R, Keshet E, Pike JW, Silver J 1990 Regulation of 1,25-dihydroxyvitamin D3 receptor gene expression by 1,25-dihydroxyvitamin D3 in the parathyroid in vivo. J Clin Invest **86:**1968–1975.
20. Silver J, Russell J, Sherwood LM 1985 Regulation by vitamin D metabolites of messenger ribonucleic acid for preproparathyroid hormone in isolated bovine parathyroid cells. Proc Natl Acad Sci USA **82:**4270–4273.
21. Brown EM, Gamba G, Riccardi D, Lombardi M, Butters R, Kifor O, Sun A, Hediger MA, Lytton J, Hebert SC 1993 Cloning and characterization of an extracellular Ca2+-sensing receptor from bovine parathyroid. Nature **366:**575–580.
22. Brown EM, Hebert SC 1995 A cloned Ca^{2+} sensing receptor: A mediator of direct effects of extracellular Ca^{2+} on renal function? J Am Soc Nephrol **6:**1530–1540.
23. The ADHR consortium, Group 1: White KE, Evans WE, O'Riordan JLH, Speer MC, Econs JJ, Groups 2: Lorenz-Depiereux B, Grabowski M, Meitinger T, Strom TM 2000 Autosomal dominant hypophosphatemic rickets is associated with mutations in FGF23. Nat Genet **26:**345–348.
24. White KE, Jonsson KB, Carn G, Hampson G, Spector TD, Mannstadt M, Lorenz-Depiereux B, Miyauchi A, Yang IM, Ljunggren O, Meitinger T, Strom TM, Juppner H, Econs MJ 2001 The autosomal dominant hypophosphatemic rickets (ADHR) gene is a secreted polypeptide overexpressed by tumors that cause phosphate wasting. J Clin Endocrinol Metab **86:**497–500.
25. Larsson T, Nisbeth U, Ljunggren O, Juppner H, Jonsson KB 2003 Circulating concentration of FGF-23 increases as renal function declines in patients with chronic kidney disease, but does not change in response to variation in phosphate intake in healthy volunteers. Kidney Int **64:**2272–2279.
26. Andress DL, Sherrard DJ 2007 The osteodystrophy of chronic renal failure. In: Schrier RW (ed.) Diseases of the Kidney and Urinary Tract, 8th ed. Lippincott Williams & Wilkins, Philadelphia, PA, USA, pp. 2431–2459.

© 2008 American Society for Bone and Mineral Research

27. Alem AM, Sherrard DJ, Gillen DL, Weiss NS, Beresford SA, Heckbert SR, Wong C, Stehman-Breen C 2000 Increased risk of hip fracture among patients with end-stage renal disease. Kidney Int **58:**396–399.
28. Cunningham J, Sprague S, Cannata-Andia J, Coco M, Cohen-Solal M, Fitzpatrick L, Goltzmann D, Lafage-Proust MH, Leonard M, Ott S, Rodriguez M, Stehman-Breen C, Stern P, Weisinger J 2004 Osteoporosis in chronic kidney disease. Am J Kidney Dis **43:**566–571.
29. Rix M, Andreassen H, Eskildsen P, Langdahl B, Olgaard K 1999 Bone mineral density and biochemical markers of bone turnover in patients with predialysis chronic renal failure. Kidney Int **56:**1084–1093.
30. Stehman-Breen C 2001 Bone mineral density measurements in dialysis patients. Semin Dial **14:**228–229.
31. Coco M, Rush H 2000 Increased incidence of hip fractures in dialysis patients with low serum parathyroid hormone. Am J Kidney Dis **36:**1115–1121.
32. Mathew S, Tustison K, Sugatani T, Chaudhary LR, Rifas L, Hruska KA 2008 The mechanism of phosphorus as a cardiovascular risk factor in chronic kidney disease. J Am Soc Nephrol **19:**1092–1105.
33. Khosla S, Burr D, Cauley J, Dempster DW, Ebeling PR, Felsenberg D, Gagel RF, Gilsanz V, Guise T, Koka S, McCauley LK, McGowan J, McKee MD, Mohla S, Pendrys DG, Raisz LG, Ruggiero SL 2007 Bisphosphonate-associated osteonecrosis of the jaw: Report of a task force of the American Society for Bone and Mineral Research. J Bone Miner Res **22:**1479–1491.
34. Ichikawa S, Imel EA, Kreiter ML, Yu X, Mackenzie DS, Sorenson AH, Goetz R, Mohammadi M, White KE, Econs MJ 2007 A homozygous missense mutation in human KLOTHO causes severe tumoral calcinosis. J Clin Invest **117:**2684–2691.
35. Ichikawa S, Lyles KW, Econs MJ 2005 A novel GALNT3 mutation in a pseudoautosomal dominant form of tumoral calcinosis: Evidence that the disorder is autosomal recessive. J Clin Endocrinol Metab **90:**2420–2423.
36. Benet-Pages A, Orlik P, Strom TM, Lorenz-Depiereux B 2005 An FGF23 missense mutation causes familial tumoral calcinosis with hyperphosphatemia. Hum Mol Genet **14:**385–390.
37. Schiavi SC, Kumar R 2004 The phosphatonin pathway: New insights in phosphate homeostasis. Kidney Int **65:**1–14.
38. Kurosu H, Ogawa Y, Miyoshi M, Yamamoto M, Nandi A, Rosenblatt KP, Baum MG, Schiavi S, Hu MC, Moe OW, Kuro-o M 2006 Regulation of fibroblast growth factor-23 signaling by Klotho. J Biol Chem **281:**6120–6123.
39. Eknoyan G, Levin A, Levin NW 2003 K/DOQI clinical practice guidelines on bone metabolism and disease in chronic kidney disease. Am J Kidney Dis **42:**S12–S28;S52–S57.

Chapter 74. Disorders of Mineral Metabolism in Childhood

Thomas O. Carpenter

Department of Pediatrics, Yale University School of Medicine, New Haven, Connecticut

INTRODUCTION

Disorders of mineral homeostasis in children may present differently than in adults. This chapter outlines disorders of mineral metabolism that occur in children, emphasizing the specific features of the age group.

DISORDERS OF CALCIUM HOMEOSTASIS

Hypocalcemia

Clinical Presentation. In the newborn with acute hypocalcemia, jitteriness, hyperacusis, irritability, and limb-jerking may occur, with progression to generalized or focal clonic seizures. Laryngospasm may lead to a misdiagnosis of croup. Atrioventricular heart block occurs in prematures with hypocalcemia, and electrocardiograms should be performed in newborns with significant bradycardia.[1] Apnea, tachycardia, tachypnea, cyanosis, edema, and vomiting have been reported in newborns with hypocalcemia.

Transient Hypocalcemia of the Newborn. Early neonatal hypocalcemia occurs during the first 3 days of life and is seen in prematures, infants of diabetic mothers, and asphyxiated infants. The premature infant has an exaggerated postnatal depression in circulating calcium (Ca), such that total Ca levels may drop below 7.0 mg/dl, but the proportional drop in ionized Ca is less. PTH insufficiency may contribute to early neonatal hypocalcemia in prematures; a delay in the phosphaturic action of PTH and resultant hyperphosphatemia may further decrease serum Ca.

Late neonatal hypocalcemia presents as tetany between 5 and 10 days of life, occurs more frequently in term that in premature infants, and is usually not correlated with birth trauma or asphyxia. Late neonatal hypocalcemia is associated with maternal vitamin D insufficiency. An increased occurrence of late neonatal hypocalcemia in winter has been noted.

Hypocalcemia associated with magnesium (Mg) deficiency may present as late neonatal hypocalcemia. Severe hypomagnesemia (circulating levels of Mg < 0.8 mg/dl) may occur in congenital defects of intestinal Mg absorption or renal tubular reabsorption.[2] Hypocalcemia in this setting may be refractory to therapy unless Mg levels are corrected.

Maternal hyperparathyroidism may result in neonatal hypocalcemia. Serum Pi is often >8 mg/dl and symptoms may be exacerbated by high Pi intake. Maternal hypercalcemia results in increased Ca delivery to the fetus, thereby suppressing parathyroid gland responsivity. As a result, normal Ca levels are not maintained postpartum because of persistent parathyroid gland suppression.

Persistent Hypocalcemia Presenting in Childhood.
Hypoparathyroidism. Persistent hypocalcemia detected in childhood may be caused by congenital hypoparathyroidism. Mutations in genes involved in parathyroid gland development, PTH processing, PTH secretion, PTH structure, and PTH resistance have been identified. The most frequently identified disorder of parathyroid gland developmental is the DiGeorge anomaly (OMIM 188400), which is comprised of hypoparathyroidism, T-cell incompetence caused by a partial or absent thymus, and conotruncal heart defects (tetralogy of Fallot, truncus arteriosus) or aortic arch abnormalities. These structures are derived from the third and fourth pharyngeal

Dr. Carpenter has consulted for Kirin and Enobia and has received a grant from Kirin.

© 2008 American Society for Bone and Mineral Research

pouches and are seen in association with microdeletions of chromosome 22q11.2.[3] Cleft palate and facial dysmorphism may occur. Some present late in childhood because of mild parathyroid defects not apparent in infancy.[4] Deletion of *TBX1*, which encodes a T-box transcription factor, is sufficient to cause the cardiac, parathyroid, thymic, facial, and velopharyngeal features of the syndrome; however, variability in phenotype with similar genetic defects occurs.[5] Other genetic defects result in disrupted parathyroid gland development (146200, 307700; e.g., loss of *TBCE* [241410 and 244460], *GCM2* [603716], or *GATA3* [146255]), abnormal PTH processing and molecular structure (*PTH*, 168450), abnormal PTH secretory dynamics (*CaSR*), and resistance to PTH action (*GNAS*, 103580). Individuals with classic PTH resistance (pseudohypoparathyroidism), caused by loss of function of the Gs α protein, often do not develop clinically evident hypocalcemia until a few years of age.

Acquired hypoparathyroidism in children is most commonly caused by autoimmune destruction of the glands (autoimmune polyendocrinopathy syndrome type 1 [240300]). Manifestations include adrenal insufficiency, mucocutaneous candidiasis, and hypoparathyroidism and are caused by loss of function mutations in the *AIRES* gene, which encodes an autoimmune regulator with features of a transcription factor. Surgery for thyroid disorders may result in inadvertent removal of parathyroid tissue.

Vitamin D–Related Hypocalcemia. Vitamin D deficiency, increasingly reported in North America, is most commonly observed in black infants that are breastfed or have limited dietary intake of dairy products. Older age groups may be affected.[6] Rarely, inherited defects in vitamin D metabolism (mutations in *CYP1b*) or the receptor (*VDR*) cause vitamin D–related hypocalcemia.

Other Causes of Hypocalcemia. Severe hypocalcemia has been induced in children when Pi enema preparations have been administered rectally or orally.[7] The Pi load can result in extreme hyperphosphatemia (up to 20 mg/dl), life-threatening hypocalcemia, and hypomagnesemia. Such preparations should never be administered to infants <2 yr of age. Hyperphosphatemia with rhabdomyolysis has resulted in hypocalcemia. Rotavirus infections may induce malabsorption-related hypocalcemia.[8]

Calcium malnutrition has been considered rare in North America; however, recent trends of decreased dietary intake of dairy products has resulted in an increasing incidence of this diagnosis in infants.[9]

Infantile osteopetrosis may present with hypocalcemia caused by impaired bone resorption. Decreases in ionized Ca occur in infants undergoing exchange transfusions with citrated blood products or receiving lipid infusions. Citrate and fatty acids complex with ionized Ca, reducing the free Ca compartment. Alkalosis secondary to adjustments in ventilatory assistance may provoke a shift from ionized to protein-bound Ca.

Treatment of Hypocalcemia.

In the Neonate. Early neonatal hypocalcemia is usually treated when total serum Ca is <6 mg/dl (1.50 mM; or ionized Ca <3 mg/dl, 0.75 mM) in prematures or <7 mg/dl (1.75 mM) in term infants. Therapy of acute tetany consists of intravenous (never intramuscular) Ca gluconate (10% solution) given slowly (<1 ml/min); 1–3 ml will usually arrest seizure activity. Doses should not exceed 20 mg of elemental Ca/kg body weight and may be repeated up to four times per 24 h. After successful management of acute emergencies, maintenance therapy is achieved by intravenous administration of 20–50 mg of elemental Ca/kg body weight/24 h. Ca glubionate is a commonly used oral supplement (most preparations provide 115 mg of elemental Ca/5 ml). Management of late neonatal tetany should include low-phosphate formula such as Similac PM 60/40, in addition to Ca supplements. Therapy can usually be discontinued after several weeks.

The place of vitamin D in the management of transient hypocalcemia is less clear. A significant portion of intestinal Ca absorption in newborns occurs by facilitated diffusion and is not vitamin D dependent. Thus, vitamin D metabolites may not be as useful for the short-term management of transient hypocalcemia as added Ca. In persistent hypoparathyroidism, calcitriol is used in the long term. Four hundred to 800 units of vitamin D daily will prevent vitamin D deficiency in premature infants. Overt vitamin D–deficient rickets should respond within 4 wk to 1000–2000 units of daily oral vitamin D. Such patients should receive at least 40 mg of elemental Ca/kg body weight/d.

In the Older Child. Ca and active vitamin D metabolites are usually titrated to maintain serum Ca in an asymptomatic range without incurring hypercalciuria. In hypoparathyroidism, a target serum Ca of 7.5–9.0 mg/dl is recommended, because higher serum Ca levels are more likely to result in hypercalciuria. In autosomal dominant hypocalcemia, because of activating mutations in the calcium sensing receptor (CaSR), thiazide diuretics can be helpful if symptomatic hypocalcemia coexists with hypercalciuria. Hypercalciuria is not typically observed in patients with pseudohypoparathyroidism type 1a when serum Ca is maintained in the usual normal range.

Disorders of Hypercalcemia

Infants are usually asymptomatic with mild to moderate hypercalcemia (11.0–12.5 mg/dl). More severe hypercalcemia may lead to failure to thrive, poor feeding, hypotonia, vomiting, seizures, lethargy, polyuria, dehydration, and hypertension. Hypercalcemia is discussed in detail elsewhere in the *Primer*. Syndromes with specific childhood features are described below.

Severe neonatal hyperparathyroidism (SNHP, 239200) presents in the first few days of life. Serum Ca levels may reach 30 mg/dl. Serum Pi is low, and serum PTH is elevated.. Nephrocalcinosis may be present on ultrasonographic examination. SNHP is a rare autosomal recessive disorder caused by homozygous CaSR loss-of-function mutations,[10] occurring in families with familial hypocalciuric hypercalcemia (FHH, 145980); SNHP is life threatening and usually requires emergency extirpation of the parathyroid glands.

In Williams syndrome (194050) hypercalcemia may occur in the neonatal period. Growth failure, a characteristic facies, cardiovascular abnormalities (usually supravalvular aortic stenosis or peripheral pulmonic stenosis), delayed psychomotor development, and selective mental deficiency may be present. A deletion of the elastin gene is often found in Williams syndrome. Hypercalcemia usually subsides spontaneously by 1 yr of age but may rarely persist for longer. Treatment has traditionally consisted of a vitamin D–free, low Ca diet and in severe settings, corticosteroids; pamidronate can be used, with fewer potential complications than glucocorticoids.[11]

Subcutaneous fat necrosis is a self-limited disorder of infancy presenting with hypercalcemia and erythematous or violaceous skin. Affected areas contain mononuclear cell infiltrates, often coexistent with calcification. Pamidronate is useful when significant hypercalcemia is unresponsive to dietary Ca and vitamin D restriction.

In vitamin D intoxication, increased circulating 25(OH)D

© 2008 American Society for Bone and Mineral Research

Table 1. Normative Values for Serum Phosphate by Age[28]

Age (yr)	Mean	2.5th percentile	97.5th percentile
0–0.5	6.7 (2.15)	5.8 (1.88)	7.5 (2.42)
2	5.6 (1.81)	4.4 (1.43)	6.8 (2.20)
4	5.5 (1.77)	4.3 (1.38)	6.7 (2.15)
6	5.3 (1.72)	4.1 (1.33)	6.5 (2.11)
8	5.2 (1.67)	4.0 (1.29)	6.4 (2.06)
10	5.1 (1.63)	3.8 (1.24)	6.2 (2.01)
12	4.9 (1.58)	3.7 (1.19)	6.1 (1.97)
14	4.7 (1.53)	3.6 (1.15)	6.0 (1.92)
16	4.6 (1.49)	3.4 (1.10)	5.8 (1.88)
20	4.3 (1.39)	3.1 (1.01)	5.5 (1.78)
Adult	3.6 (1.15)	2.7 (0.87)	4.4 (1.41)

Values are shown as mg/dl and mM in parentheses.

levels occur, but $1,25(OH)_2D$ levels are usually low. Vitamin A intoxication may result in bone pain, hypercalcemia, headache, pseudotumor cerebri, and an exfoliative erythematous rash. Alopecia and ear discharge may be present. Hypercalcemia is mediated by increased bone resorption. To establish the diagnosis of vitamin A intoxication, serum retinyl ester levels should be determined.[12]

Other conditions in children in which hypercalcemia may be manifest include Down syndrome, skeletal dysplasias (such as Jansen's, 156400), hypophosphatasia (241500, 241510), and osteogenesis imperfecta (120150 and others). Endogenous overproduction of $1,25(OH)_2D$ occurs in granulomatous diseases, such as cat-scratch disease. Other causes include those commonly encountered in adults: immobilization, malignancy, and acquired hyperparathyroidism.

Treatment of Hypercalcemia. Management of acute hypercalcemia consists of administration of intravenous saline. Furosemide (1 mg/kg) is frequently given intravenously at 6- to 8-h intervals. Bisphosphonates for unremitting hypercalcemia in children has become widely accepted.[11] Pamidronate has been highly successful in the management of hypercalcemia.

DISORDERS OF PHOSPHATE HOMEOSTASIS

Disorders of Hypophosphatemia

Serum Pi is relatively higher in young children compared with adults. Unfortunately, lapses in diagnosis of childhood hypophosphatemia occur because this clinical difference is not always recognized (Table 1).

Etiology of Hypophosphatemia in Children. Hypophosphatemia may result from decreased Pi supply, excessive renal losses, or intracellular/extracellular compartmental movement of Pi. "Supply" problems result from dietary deficiency or limited intestinal absorption of Pi. Reduced dietary intake occurs in breastfed premature infants, because human milk is relatively low in Pi content. Fortifiers have been developed to restore mineral content to human milk; these may result in hypercalcemia, so monitoring may be indicated when used. Rickets caused by inadequate dietary Pi can be treated with 20–25 mg of elemental phosphorus/kg body weight/d, given orally in three to four divided doses.

Hypophosphatemia secondary to renal losses is encountered clinically in the setting of several primary Pi wasting disorders, of which X-linked hypophosphatemia (XLH, 307800) is the most common.[13] XLH may be suspected in the setting of a family history, but affected adult members may never have been correctly diagnosed. XLH typically presents in the second or third year of life, with progressive leg bowing. Progression of bowing over 6 mo in a child older than 18 mo of age requires further study. Children may be incorrectly diagnosed with other disorders (typically metaphyseal dysplasias). The delay in a correct diagnosis of XLH may result in the child missing early medical therapy, which has beneficial effects on growth and leg alignment. Circulating fibroblast growth factor 23 (FGF23) is elevated in XLH and mediates renal Pi wasting. Additionally, FGF23 decreases 25(OH)D 1-hydroxylase (*CYP27B1*) message synthesis, limiting $1,25(OH)_2D$ production, despite hypophosphatemia. Treatment consists of the administration of Pi in conjunction with $1,25(OH)_2D_3$ (calcitriol). Doses of these agents vary and require adjustment based on clinical monitoring. The range of doses for both Pi and calcitriol on a per weight basis are extremely wide; doses for calcitriol are usually 20–50 ng/kg/d, usually given in two divided doses, and phosphorus, 0.25–2 g/d in three to five divided doses. Because of the propensity to develop autonomous hyperparathyroidism with this therapy, and because of the concern of potential vitamin D intoxication, monitoring is suggested at 3- to 4-mo intervals in treated children. Serum Ca, Pi, and alkaline phosphatase and urinary Ca and creatinine excretion should be determined. Circulating PTH should be measured at least twice yearly. Accurate height measurements and assessment of the bow defect should be performed at all visits. Radiographs of the epiphyses of the distal femur and proximal tibia are obtained every 2 yr or more frequently if bow deformities fail to correct or if progressive skeletal disease is grossly evident. Therapy in adulthood is variably recommended, dependent on symptomatology. Pi should never be given as monotherapy in XLH, but always with calcitriol.

Soft tissue calcification of the renal medullary pyramids (nephrocalcinosis) may occur secondary to the mineral load that this treatment provides and can be detected with sonograms. Most compliant patients show nephrocalcinosis within 3–4 yr of beginning therapy, but significant clinical sequelae in patients with mild nephrocalcinosis are not generally observed. Calcification of the entheses is described in the third decade and is not thought to be associated with treatment.

Several other hypophosphatemic disorders are described in which elevated FGF23 occurs, including tumor-induced osteomalacia (caused by neoplastic overproduction of FGF23), autosomal dominant hypophosphatemic rickets (193100) caused by mutations that disrupt proteolytic cleavage of FGF23),[14] and in fibrous dysplasia/McCune Albright syndrome (174800).[15] Most recently, autosomal recessive hypophosphatemic rickets (ARHR, 241520, caused by mutations in dentin matrix protein 1 [DMP1]) has been described.[16] In that PHEX, FGF23, and DMP1 are products of the osteocyte, a central role for this cell in regulation of mineralization has been speculated. This group of disorders is discussed elsewhere in the *Primer*.

Renal Pi losses occur in hereditary hypophosphatemic rickets with hypercalciuria (HHRH, 241530) secondary to mutations in the renal NaPi (2c) cotransporter (*SLC34A3*).[17] In contrast to XLH and other FGF23-mediated disorders, circulating $1,25(OH)_2D$ is elevated in HHRH; hypercalciuria is common and renal stones may occur. Osteoporosis may develop. HHRH is treated with oral phosphate without vitamin D metabolites. Generalized tubular dysfunction (Fanconi syndrome, 134600) may occur in conditions such as cystinosis, Lowe's syndrome, and Wilson's disease. Finally, hypophosphatemia may occur in Dent's disease (300009), an X-linked recessive disorder caused by mutations in *CLCN5*, which encodes a renal tubular chloride channel.

© 2008 American Society for Bone and Mineral Research

Intracellular/Extracellular Shifts. Acute hypophosphatemia may occur with acute movement from the extracellular space to intracellular compartments. This is typically seen with correction of diabetic ketoacidosis when insulin-induced intracellular Pi uptake can acutely decrease serum Pi. Another such setting is in the refeeding syndrome, as with nutritional rehabilitation of anorexia nervosa. Pi levels reach a nadir within a week of refeeding; slow oral feeds minimize the severity of this phenomenon. Targeting a minimum 4-day weight gain between 0.36 and 0.55 kg has been recommended to reduce complications of refeeding in anorexic adolescents.[18]

Hyperphosphatemia

Increases in serum Pi above age-appropriate ranges are uncommon in children with normal renal function. Acute elevations in circulating Pi are often accompanied by a reciprocal decrease in serum Ca, precipitating tetany and seizures. When serum Pi and/or Ca concentrations are sufficiently elevated to result in chronic elevations in the Ca × Pi product (in children, >60 mg^2/dl^2 is often quoted as undesirable), there is a risk of soft tissue calcification, involving blood vessels, renal parenchyma, skin, cornea, and joints. Serum Pi may increase with rapid lysis of bulky tumors (tumor lysis syndrome). Hyperphosphatemia is a biochemical hallmark of hypoparathyroidism and pseudohypoparathyroidism, accompanied by hypocalcemia. Hyperphosphatemia is observed in chronic kidney disease, where progressive loss of nephrons results in limited capacity to excrete Pi. A primary disorder of renal Pi excretion, hyperphosphatemic tumoral calcinosis (HTC, 211900), can result from loss-of-function mutations in FGF23[19] or *GALNT3* (which encodes an enzyme which initiates O-glycosylation of FGF23, an important step in its secretion and/or trafficking).[20] HTC has been recently attributed to a loss-of-function mutation in klotho, a factor necessary for FGF23 receptor signaling.[21] As expected from the impaired FGF23 activity, HTC has the converse biochemical phenotype of XLH [hyperphosphatemia caused by an increased tubular maximum threshold for Pi reabsorption and increased circulating $1,25(OH)_2D$].

DISORDERS OF MAGNESIUM

Disorders of Hypomagnesemia

Familial hypomagnesemia with secondary hypocalcemia (602014) is an autosomal recessive disease caused by mutations in the TRPM6 ion channel, resulting in electrolyte abnormalities in the newborn period.[22] This syndrome presents at several weeks of age with tetany or seizures. Hypocalcemia may be refractory to therapy unless Mg levels are corrected. Mutations in a renal tubular paracellular transport protein paracellin, of the claudin family and encoded by *CLDN16*, may also cause hypomagnesemia, hypocalcemia, and hypercalciuria (248250).[23] CLDN19, another member of the claudin family, has been implicated as causal to heritable hypomagnesemia (248190).[24] Gitelman's syndrome (263800) is an autosomal recessive disorder of Mg and potassium wasting with metabolic alkalosis and hypocalciuria, caused by mutations in the gene encoding a thiazide-sensitive Na-Cl cotransporter (*SLC12A3*).[2]

Treatment of Hypomagnesemia. For acute symptomatic hypomagnesemia, Mg sulfate is given intravenously using cardiac monitoring or intramuscular as a 50% solution at a dose of 0.1–0.2 ml/kg. One or two doses may treat transient hypomagnesemia: a dose may be repeated after 12–24 h. Patients with primary defects in Mg metabolism require long-term oral Mg supplements; it is best to give these in several divided doses through the day as to avoid diarrhea. We begin oral supplementation at 5 mg of elemental Mg/kg body weight/d. A variety of salts are available for oral use; we have had limited complications with Mg oxide.

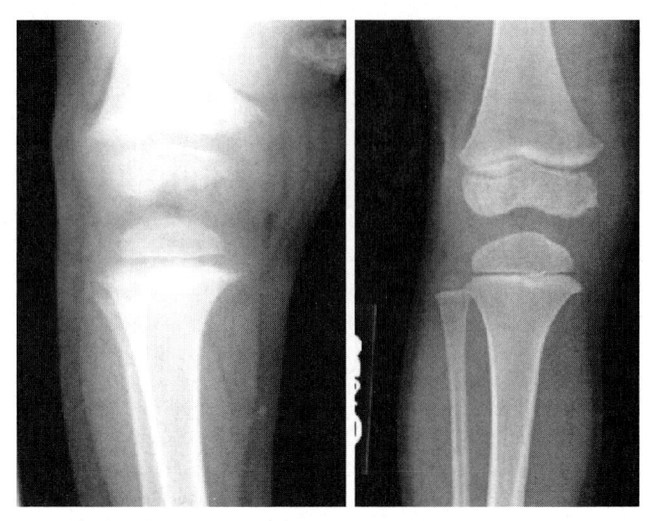

FIG. 1. (Left) Radiograph of the right knee in an infant with vitamin D deficiency rickets. Classic epiphyseal deformities are seen, with metaphyseal flaring and frayed edges of the metaphyseal growth plate junction. (Right) A normal knee is shown for comparison.

Hypermagnesemia

Hypermagnesemia is unusual in pediatrics, but may occur transiently after fetal exposure to maternal Mg infusions used in the management of eclampsia/preeclampsia. Severe hypermagnesemia can result in apnea, respiratory depression, and cardiac arrhythmias. Hypocalcemia may also result from hypermagnesemia.

SKELETAL MANIFIESTATIONS OF DISORDERS OF CALCIUM AND PHOSPHATE

The typical skeletal abnormality in the growing child with a paucity of available Ca or Pi is rickets. The use of the term "rickets" in clinical settings generally refers to the growth plate cartilage abnormalities observed in the long bones. Rickets is manifest by radiographic findings such as widened metaphyses, irregular or "frayed" metaphyseal edges, and "cupped" metaphyseal deformations (Fig. 1). Rickets is usually accompanied by osteomalacia in bone tissue. The histologic correlate of the radiographic findings at the growth plate is the expansion of the hypertrophic zone of chondrocytes.[25] Weight bearing on the undermineralized skeleton results in characteristic bowing. A child with overt rickets may have minimal leg deformity before walking; however, enlarged wrists or costo-chondral junctions (rachitic rosary) are typical. An osteomalacic skull (craniotabes) may be present.

Deficiency, or nutritional rickets refers to restricted vitamin D stores, which may result from limited intake and/or limited sunlight exposure. Dietary Ca deficiency can result in a similar clinical picture, and some children have a mixed deficiency of Ca and vitamin D.[9] Treatment is accomplished by provision of adequate vitamin D, traditionally suggested as 2000 units daily, and providing adequate dietary calcium. Clinical and biochemical findings observed in certain inherited conditions mimic severe nutritional rickets. Mutations in encoding 1α-hydroxylase (*CYP27B1*; 1-α hydroxylase deficiency, or vitamin

© 2008 American Society for Bone and Mineral Research

D–dependent rickets, type 1, 264700), or the vitamin D receptor (*VDR*; hereditary vitamin D resistance or vitamin D–dependent rickets, type 2, 277440) are well described. Mutations in a vitamin D 25-hydroxylase enzyme have been described as well (60081).[26] Therapy with calcitriol is generally recommended for these conditions, although hereditary vitamin D resistance may require exceptionally high dosages of this metabolite and is sometimes completely unresponsive to calcitriol. Provision of intravenous calcium has been useful in such settings, with gradual progression to enteral calcium treatment.[27]

X-linked hypophosphatemia (XLH) is discussed above in the section on hypophosphatemia and elsewhere in the *Primer*.

REFERENCES

1. Stefanaki E, Koropuli M, Stefanaki S, Tsilimigaki A 2005 Atrioventricular block in preterm infants caused by hypocalcemia: A case report and review of the literature. Eur J Obst Gynecol Reprod Biol **120:**115–116.
2. Schlingmann KP, Konrad M, Seyberth HW 2004 Genetics of hereditary disorders of magnesium homeostasis. Pediatr Nephrol **19:**13–25.
3. Webber SA, Hatchwell E, Barber JC, Daubeney PE, Crolla JA, Salmon AP, Keeton BR, Temple IK, Dennis NR 1996 Importance of microdeletions of chromosomal region 22q11 as a cause of selected malformations of the ventricular outflow tracts and aortic arch: A three-year prospective study. J Pediatr **129:**26–32.
4. Sykes KS, Bachrach LK, Siegel-Bartelt J, Ipp M, Kooh SW, Cytrynbaum C 1997 Velocardiofacial syndrome presenting as hypocalcemia in early adolescence. Arch Pediatr Adolesc Med **151:**745–747.
5. Thakker RV 2004 Genetics of endocrine and metabolic disorders: Parathyroid. Rev Endocr Metab Dis **5:**37–51.
6. Gordon CM, DePeter KC, Feldman HA, Grace E, Emans SJ 2004 Prevalence of vitamin D deficiency among healthy adolescents. Arch Pediatr Adolesc Med **158:**531–537.
7. Walton DM, Thomas DC, Aly HZ, Short BL 2000 Morbid hypocalcemia associated with phosphate enema in a six-week-old infant. Pediatrics **106:**E37.
8. Foldenauer A, Vossbeck S, Pohlandt F 1998 Neonatal hypocalcaemia associated with rotavirus diarrhoea. Eur J Pediatr **157:**838–842.
9. DeLucia MC, Mitnick ME, Carpenter TO 2003 Nutritional rickets with normal circulating 25-hydroxyvitamin D: A call for reexamining the role of dietary calcium intake in North American children. J Clin Endocrinol Metab **88:**3539–3545.
10. Pidasheva S, D'Souza-Li L, Canaff L, Cole DE, Hendy GN 2004 CASRdb: Calcium-sensing receptor locus-specific database for mutations causing familial (benign) hypocalciuric hypercalcemia, neonatal severe hyperparathyroidism, and autosomal dominant hypocalcemia. Hum Mutat **24:**107–111.
11. Lteif AN, Zimmerman D 1998 Bisphosphonates for treatment of childhood hypercalcemia. Pediatrics **102:**990–993.
12. Carpenter TO, Pettifor JM, Russell RM, Pitha J, Mobarhan S, Ossip MS, Wainer S, Anast CS 1987 Severe hypervitaminosis A in siblings: Evidence of variable tolerance to retinol intake. J Pediatr **111:**507–512.
13. Holm IA, Econs MJ, Carpenter TO 2003 Familial hypophosphatemia and related disorders. In: Glorieux FH, Juppner H, Pettifor JM (eds.) Pediatric Bone: Biology & Diseases. Academic Press, San Diego, CA, USA, pp. 603–631.
14. White KE, Jonsson KB, Carn G, Hampson G, Spector TD, Mannstadt M, Lorenz-Depiereux B, Miyauchi A, Yang IM, Ljunggren O, Meitinger T, Strom TM, Juppner H, Econs MJ 2001 The autosomal dominant hypophosphatemic rickets (ADHR) gene is a secreted polypeptide overexpressed by tumors that cause phosphate wasting. J Clin Endocrinol Metab **86:**497–500.
15. Riminucci M, Collins MT, Fedarko NS, Cherman N, Corsi A, White KE, Waguespack S, Gupta A, Hannon T, Econs MJ, Bianco P, Gehron Robey P 2003 FGF-23 in fibrous dysplasia of bone and its relationship to renal phosphate wasting. J Clin Invest **112:**683–692.
16. Feng JQ, Ward LM, Liu S, Lu Y, Xie Y, Yuan B, Y6u X, Rauch F, Davis SI, Zhang S, Rios H, Drezner MK, Quarles LD, Bonewald LF, White KE 2006 Loss of DMP1 causes rickets and osteomalacia and identifies a role for osteocytes in mineral metabolism. Nat Genet **38:**1310–1315.
17. Bergwitz C, Roslin NM, Tieder M, Loredo-Osti JC, Bastepe M, Abu-Zahra H, Carpenter TO, Anderson D, Garabedian M, Sermet I, Fujiwara TM, Morgan KN, Tenenhouse HS, Jüppner H 2006 *SLC34A3* mutations in patients with hereditary hypophosphatemic rickets with hypercalciuria (HHRH) predict a key role for the sodium-phosphate cotransporter NaPi-IIc in maintaining phosphate homeostasis and skeletal function. Am J Hum Genet **78:**179–192.
18. Fisher M 2006 Treatment of eating disorders in children, adolescents, and young adults. Pediatr Rev **27:**5–16.
19. Benet-Pagès A, Orlik P, Strom TM, Lorenz-Depiereux B 2005 An FGF23 missense mutation causes familial tumoral calcinosis with hyperphosphatemia. Hum Mol Genet **14:**385–390.
20. Topaz O, Shurman DL, Bergman R, Indelman M, Ratajczak P, Mizrachi M, Khamaysi Z, Behar D, Petronius D, Friedman V, Zelikovic I, Raimer S, Metzker A, Richard G, Sprecher E 2004 Mutations in *GALNT3*, encoding a protein involved in O-linked glycosylation, cause familial tumoral calcinosis. Nat Genet **36:**579–581.
21. Ichikawa S, Imel EA, Kreiter ML, Yu X, Mackenzie DS, Sorenson AH, Goetz R, Mohammadi M, White KE, Econs MJ 2007 A homozygous missense mutation in human KLOTHO causes severe tumoral calcinosis. J Clin Invest **117:**2684–2691.
22. Schlingmann KP, Weber S, Peters M, Niemann Nejsum L, Vitzhum H, Klingel K, Kratz M, Haddad E, Ristoff E, Dinour D, Syrrou M, Nielsen S, Sassen M, Waldegger S, Seyberth HW, Konrad M 2002 Hypomagnesemia with secondary hypocalcemia is caused by mutations in *TRPM6*, a new member of the TRPM gene family. Nat Genet **31:**166–170.
23. Simon DB, Lu Y, Choate KA, Velazquez H, Al-Sabban E, Praga M, Casari G, Bettinelli A, Colussi G, Rodriguez-Soriano J, McCredie D, Milford D, Sanjad S, Lifton RP 1999 Paracellin-1, a renal tight junction protein required for paracellular Mg2-resorption. Science **285:**103–106.
24. Konrad M, Schaller A, Seelow D, Pandey AV, Waldegger S, Lesslauer A, Vitzthum H, Suzuki Y, Luk JM, Becker C, Schlingmann KP, Schmid M, Rodriguez-Soriano J, Ariceta G, Cano F, Enriquez R, Juppner H, Bakkaloglu SA, Hediger MA, Gallati S, Neuhauss SC, Nurnberg P, Weber S 2006 Mutations in the tight-junction gene claudin 19 (CLDN19) are associated with renal magnesium wasting, renal failure, and severe ocular involvement. Am J Hum Genet **79:**949–957.
25. Sabbagh Y, Carpenter TO, Demay MB 2005 Hypophosphatemia leads to rickets by impairing caspase-mediated apoptosis of hypertrophic chondrocytes. Proc Natl Acad Sci USA **102:**9637–9642.
26. Cheng JB, Levine MA, Bell NH, Mangelsdorf DJ, Russell DW 2004 Genetic evidence that the human CYP2R1 enzyme is a key vitamin D 25-hydroxylase. Proc Natl Acad Sci USA **101:**7711–7715.
27. Balsan S, Garabedian M, Larchet M, Gorski AM, Cournot G, Tau C, Bourdeau A, Silve C, Ricour C 1986 Long-term nocturnal calcium infusions can cure rickets and promote normal mineralization in hereditary resistance to 1,25-dihydroxyvitamin D. J Clin Invest **77:**1661–1667.
28. Brodehl J, Gellissen K, Weber HP 1982 Postnatal development of tubular phosphate reabsorption. Clin Nephrol **17:**163–171.

© 2008 American Society for Bone and Mineral Research

Chapter 75. Hypoparathyroidism and Pseudohypoparathyroidism

Mishaela R. Rubin[1] and Michael A. Levine[2]

[1]*Division of Endocrinology, Columbia University College of Physicians and Surgeons, New York, New York;* [2]*Division of Endocrinology and Diabetes, The Children's Hospital of Philadelphia and Department of Pediatrics, University of Pennsylvania School of Medicine, Philadelphia, Pennsylvania*

INTRODUCTION

The term functional hypoparathyroidism refers to a group of metabolic disorders in which hypocalcemia and hyperphosphatemia occur either from a failure of the parathyroid glands to secrete adequate amounts of biologically active PTH or, less commonly, from an inability of PTH to elicit appropriate biological responses in its target tissues. Plasma concentrations of PTH are low or absent in patients with true hypoparathyroidism (HP). In contrast, plasma concentrations of PTH are elevated in patients with pseudohypoparathyroidism (PHP) and reflect the failure of target tissues to respond appropriately to the biological actions of PTH.

There are several important features that distinguish PHP from HP. First, circulating levels of PTH are elevated in patients with PHP (i.e., secondary or adaptive hyperparathyroidism) and low in patients with HP. Second, the fractional excretion of urinary calcium is elevated in HP and low in PHP. In the absence of PTH, active transport of calcium from the glomerular filtrate in the distal renal tubule is decreased. Urinary excretion of calcium is low or normal in HP patients who are hypocalcemic, but as treatment restores the serum calcium level to normal the renal filtered load of calcium increases, and urinary calcium excretion increases pari passu. In contrast, the distal renal tubule is responsive to PTH in patients with PHP, so that urinary calcium excretion remains low relative to the filtered load of calcium (unless PTH levels are suppressed by overtreatment). Third, skeletal responsiveness to PTH is intact in patients with PHP, and thus excessive bone remodeling can occur in patients who have very high levels of PTH. As a corollary, PTH can also induce release of calcium (and phosphate) from skeletal storage pools when circulating levels of $1,25(OH)_2D$ are normal. Fourth, HP is associated with low bone turnover and reduced biochemical markers of bone resorption and formation, whereas bone turnover is normal or elevated in PHP. Protracted low bone turnover in HP leads to bone mass that is greater than age-and sex-matched controls, especially at the lumbar spine. Bone histomorphometry shows that subjects with HP have significantly increased cancellous bone volume, trabecular width and cortical width, and profoundly suppressed dynamic skeletal indices.

CLINICAL MANIFESTATIONS OF FUNCTIONAL HYPOPARATHYROIDISM

The signs and symptoms of functional hypoparathyroidism are principally manifestations of hypocalcemia. Hypocalcemia causes increased neuromuscular irritability, a condition termed tetany. Patients may complain of paresthesias, particularly in the distal extremities and face, as well as muscle cramps. When hypocalcemia is severe, patients may experience laryngospasm or seizures. Clinical signs of latent tetany include the Chvostek sign and Trousseau sign. Other clinical features of chronic hypocalcemia include pseudopapilledema, increased intracranial pressure, and dry, rough skin. Prolonged hypocalcemia and hyperphosphatemia, with an elevated calcium × phosphate product, will lead to posterior subcapsular cataracts and calcification of intracranial structures, which can be detected by CT or MRI. Rarely, calcification of the basal ganglia can cause extrapyramidal neurological dysfunction. Hypocalcemia can cause prolongation of the corrected Q-T interval on EKG and reversible heart failure. Occasionally, patients may be asymptomatic and diagnosed only after low serum calcium is detected by routine blood screening.

DIFFERENTIAL DIAGNOSIS OF FUNCTIONAL HYPOPARATHYROIDISM

The PTH infusion test facilitates the diagnosis of PHP and enables distinction between HP and the several variants of PHP the syndrome (Fig. 1). Patients with HP show a robust increase in urinary excretion of nephrogenous cAMP and phosphate. In contrast, patients with PHP type 1 fail to show an appropriate increase in urinary excretion of both cAMP and phosphate, whereas subjects with the less common type 2 form show a normal increase in urinary cAMP excretion but do not manifest a phosphaturic response (see below). The causes of PHP and HP are presented in Table 1.

SYNDROMES OF PTH RESISTANCE

Pseudohypoparathyroidism Type 1

The blunted nephrogenous cAMP response to PTH in subjects with PHP type 1 is caused by a deficiency of the α subunit of Gs ($G\alpha_s$), the signaling protein that couples PTH1R to stimulation of adenylyl cyclase. Molecular and biochemical studies have provided a basis for distinguishing between two forms of PHP type 1: patients with generalized deficiency of $G\alpha_s$, caused by mutations within exons 1–13 of the *GNAS* gene, are classified as PHP type 1a (OMIM 103580), whereas patients with more restricted deficiency of $G\alpha_s$ caused by mutations that affect imprinting of *GNAS*, are classified as PHP type 1b (OMIM 603233). PHP 1a and PHP 1b also differ in the pattern of hormone resistance and in the expression of additional somatic features. PHP type 1c is likely a variant of PHP 1a (see below).

PHP Type 1a. PHP type 1a (MIM 103580) is the most common variant and most readily recognizable form of PHP. Subjects have short stature, round facies, brachydactyly of hands and/or feet, mild mental retardation, subcutaneous ossifications (a clinical constellation termed Albright hereditary osteodystrophy [AHO]; see below), and obesity (Fig. 2). PHP type 1a results from heterozygous mutations on the maternal allele of the imprinted *GNAS* gene (20q13.2-q13.3) that reduce expression or function of the $G\alpha_s$ protein (Fig. 3). Because $G\alpha_s$ is required for normal transmembrane signal transduction by many hormones and neurotransmitters, subjects with PHP type 1a also have resistance to other hormones (e.g., thyroid stimulating hormone, gonadotropins, calcitonin, and growth hormone releasing hormone) whose target tissues show predominant expression of the maternal *GNAS* allele. Primary

The authors state that they have no conflicts of interest.

© 2008 American Society for Bone and Mineral Research

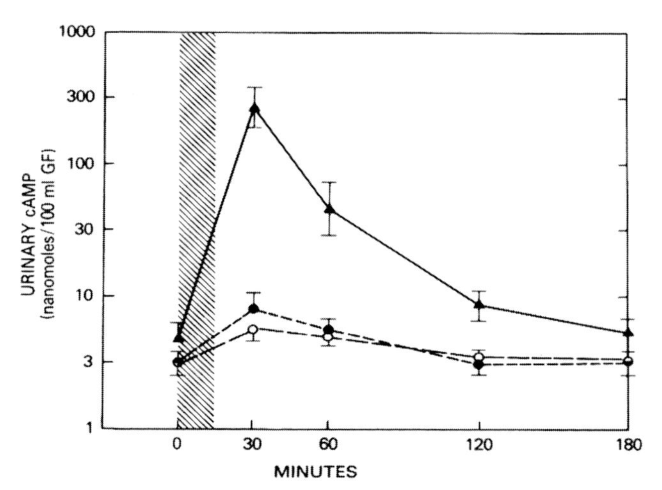

FIG. 1. PTH infusion. cAMP excretion in urine in response to the intravenous administration of bovine parathyroid extract (300 USP units) from 9:00 to 9:15 a.m. The peak response in normals (▲) is 50- to 100-fold times basal; patients with PHP type Ia (●) or PHP type Ib (○) show only a 2- to 5-fold response.

hypothyroidism, without goiter, and growth hormone deficiency are common associated endocrinopathies.

Responsiveness to other hormones (e.g., adrenocorticotropic hormone [ACTH], vasopressin) is normal in tissues in which *GNAS* is not imprinted and both parental alleles are expressed. Subjects with paternally inherited *GNAS* mutations have phenotypic features of AHO without hormonal resistance, a condition termed pseudopseudohypoparathyroidism (pseudo-PHP). Subjects with pseudo-PHP have a normal urinary cAMP response to PTH, which distinguishes them from occasional patients with PHP type 1a who maintain normal serum calcium levels without treatment. It is not unusual to find extended families in which some members will have only AHO (i.e., pseudo-PHP), whereas others will have hormone resistance as well (i.e., PHP type 1a), based on the parental origin of the identical *GNAS* mutation.

The *GNAS* gene is a complex transcriptional unit that derives considerable plasticity through use of alternative first exons, alternative splicing of downstream exons, antisense transcripts, and reciprocal imprinting (Fig. 3). $G\alpha_s$ is encoded by exons 1–13 and is synthesized as a 52- or 45-kDa protein based on the inclusion or exclusion of exon 3, respectively. There seems to be little difference in the signal transducing properties of the two $G\alpha_s$ isoforms. Upstream of exon 1 are three alternative first exons that each splice onto exons 2–13 to create novel transcripts (Fig. 3). These include XL, which is expressed only from the paternal allele and which generates a transcript with overlapping open reading frames that encodes $XL\alpha_s$ and ALEX. The two proteins are interacting co-factors and are specifically expressed in neuroendocrine cells. XLαs is a much larger signaling protein than $G\alpha_s$ (≈78 versus 45–52 kDa) and can interact with receptors for PTH and a variety of other hormones in vitro, but the native receptors that interact with XLαs in vivo are presently unknown. A second alternative promoter encodes the secretory protein Nesp55, which is expressed only from the maternal allele and shares no protein homology with $G\alpha_s$. An exon 1A (associated first exon) transcript is derived only from the paternal allele and does not encode a known protein. These alternative first exons are associated with promoters that contain differentially methylated regions (DMRs) that are methylated on the nonexpressed allele (Fig. 3). In contrast, the promoter for exon 1 is within a CpG island but is unmethylated on both alleles in all tissues. The *cis*-acting elements that control tissue-specific paternal imprinting of $G\alpha_s$ seem to be located within the primary imprint region in exon 1A, because paternal deletion of the exon 1A DMR (differentially methylated region) in mice is associated with increased $G\alpha_s$ expression.

Private mutations have been found in nearly all of the AHO kindreds studied, although a 4-base deletion in exon 7 has been detected in multiple families, and an unusual missense mutation in exon 13 (A366S; see below) has been identified in two unrelated young boys, suggesting that these two regions may be genetic "hot spots." Small deletions or point mutations can be identified in ~80% of AHO patients using PCR-based techniques, and larger genomic rearrangements or uniparental disomy may account for AHO in other patients.

Post-zygotic somatic mutations in the *GNAS* gene that enhance activity of the protein are found in many autonomous

TABLE 1. CLASSIFICATION OF HYPOPARATHYROIDISM

	MIM	*Genetic defect*
Disorders of parathyroid gland formation		
DiGeorge sequence/Catch-22	188400	22q11; *TBX1*
Hypoparathyroidism, sensorineural deafness, and renal dysplasia syndrome	146255	10p; *GATA3*
Hypoparathyroidism-retardation-dysmorphism and Kenny-Caffey syndromes	241410, 244460	1q42–43; *TBCE*
Autosomal isolated hypoparathyroidism	146200	6p23–24; *GCM2*
X-linked hypoparathyroidism	307700	Xq27
Parathyroid gland destruction		
Surgery		
Radiation therapy and infiltration		
Autoimmune polyendocrinopathy-candidiasis-esctodermal dystrophy (APECED)	240300	21q22.3; *AIRE*
Reduced parathyroid gland function		
Autosomal dominant hypocalcemic hypercalciuria (ADHH)	146200	3q13.3–21; *CASR*
PTH gene mutations		11p15 *PTH*
Antibodies to the CASR		
Other causes of hypoparathyroidism		
Mitochondrial disease (see text)		Mitochondrial tRNA
Burns		
Resistance to PTH		
Pseudohypoparathyroidism	103580, 603233, 174800	*GNAS*
Transient pseudohypoparathyroidism of the newborn		
Hypomagnesemia		

© 2008 American Society for Bone and Mineral Research

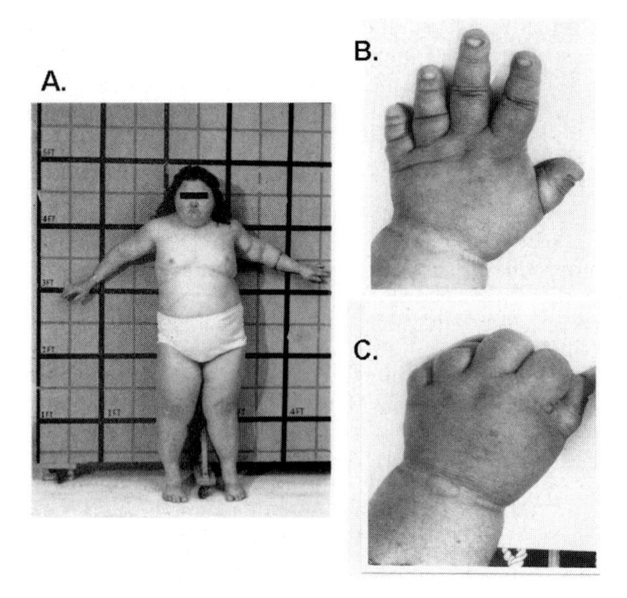

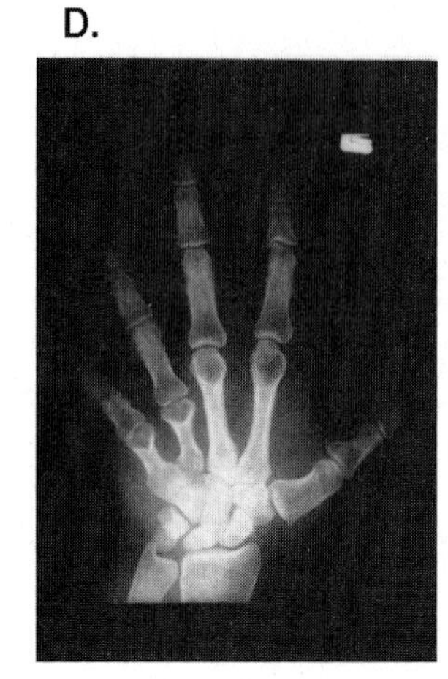

FIG. 2. AHO. Typical features of AHO. The female in the picture shows short stature, sexual immaturity (A), and brachydactyly (B). Note the extreme shortening of digit IV and distal phalanx of digit I (width greater than length) in B and D and the replacement of knuckles by dimples (Archibald sign) in C.

endocrine tumors and affected tissues of patients with McCune-Albright syndrome (MIM 174800). These mutations lead to constitutive activation of adenylyl cyclase and result in proliferation and autonomous hyperfunction of hormonally responsive cells. Clinically significant effects are more likely to ensue when *GNAS* activating mutations occur on the maternally derived allele, which is preferentially expressed in imprinted tissues. The clinical significance of $G\alpha_s$ activity as a determinant of hormone action is further emphasized by the description by of two unrelated males with both precocious puberty and PHP type 1a. These two subjects had identical *GNAS* mutations in exon 13 (A366S) that resulted in a temperature-sensitive form of $G\alpha_s$. This $G\alpha_s$ is constitutively active in the cooler environment of the testis while being rapidly degraded in other tissues at normal body temperature. Thus, different tissues in these two individuals could show hormone resistance (to PTH and thyroid-stimulating hormone [TSH]), hormone responsiveness (to ACTH), or hormone independent activation (to luteinizing hormone [LH]).

PHP Type 1b. Subjects with PHP type 1b (MIM 603233) lack typical features of AHO but may have mild brachydactyly. Levels of $G\alpha_s$ are normal in accessible tissues. PTH resistance is the principal manifestation of hormone resistance, but some patients have slightly elevated serum levels of TSH and normal serum concentrations of thyroid hormones as evidence of associated TSH resistance. Despite renal resistance to PTH, subjects with PHP type 1b who have elevated levels of PTH often manifest skeletal lesions similar to those that occur in patients with hyperparathyroidism.

An epigenetic defect that results in switching of the maternal *GNAS* allele to a paternal pattern of methylation (i.e., paternal epigenotype) is a consistent finding in sporadic and familial PHP type 1b. Causative mutations have been identified in most cases of familial but not sporadic PHP type 1b, including two microdeletions in the *STX16* gene located ~220 kb centromeric of *GNAS* exon 1A and deletions that remove the DMR encompassing exon NESP55 and exons 3 and 4 of the antisense transcript (Fig. 3). In each case, inheritance of a mu-

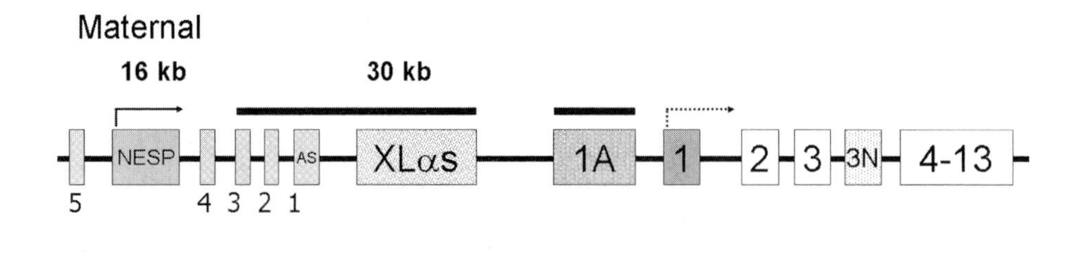

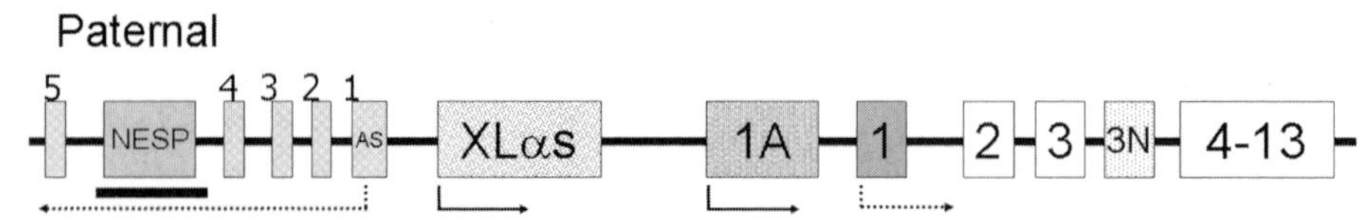

FIG. 3. *GNAS* gene. General organization of the *GNAS* gene complex. The *GNAS* gene complex consists of 13 exons that encode the signaling protein $G\alpha_s$. Upstream of exon 1 are three alternative first exons that are labeled exon 1A, XLαs, and Nesp55; exons 1–5 for the NESP antisense transcript (AS) are also depicted. The three alternative exons are spliced to exons 2–13 to produce unique transcripts. The DMRs are denoted above the respective promoters, and arrows denote the direction of transcription. Nesp55 is transcribed exclusively from the maternal allele; XLαs and exon 1A are transcribed exclusively from the paternal allele. Nesp AS (antisense) and exon 1a transcripts produce noncoding RNA's. $G\alpha_s$ transcripts are biallelically expressed except in a small number of tissues, such as the renal proximal tubules, thyroid, gonads, and pituitary somatotrophs, where expression is preferentially from the maternal allele.

© 2008 American Society for Bone and Mineral Research

tation from a female (or spontaneous mutation of a maternally derived allele) abolishes the maternal *GNAS* epigenotype. It is likely that the conversion of the maternal *GNAS* allele to a "paternal" epigenotype leads to transcriptional silencing of the Gαs promoter in imprinted tissues, with the result that little or no $G\alpha_s$ is expressed from either *GNAS* allele in these tissues.

PHP Type 1c. In rare instances, patients with PHP type 1 and features of AHO show resistance to multiple hormones in the absence of a demonstrable biochemical defect in G_s or G_i. Molecular studies suggest that these patients have *GNAS* mutations that result in functional defects of $G\alpha_s$ that are not apparent in conventional in vitro assays.

AHO. Haploinsufficiency of *GNAS* is associated with an unusual constellation of anomalies that are collectively termed AHO and that include short stature, round faces, brachydactyly, dental defects, and heterotopic ossification of the skin and subcutaneous tissues (Fig. 2). Additional features had been associated with AHO, such as obesity and sensory-neural abnormalities, but these defects seem limited to patients with PHP type 1a who have maternal *GNAS* mutations that cause abnormal Gs_α signaling in the central nervous system. Patients who manifest AHO and have normal hormonal responsiveness caused by inactivating mutations on the paternal *GNAS* allele are considered to have the genetically related disorder pseudo-PHP (see above). Short stature and brachydactyly may be caused in part by premature fusion of epiphyses in tubular and long bones, which implies a requirement of two functional copies of *GNAS* for normal growth plate maturation.

The heterotopic ossifications are the most unique feature of AHO and distinguish true AHO from a variety of clinical phenocopies. The ossifications are not calcifications and are unrelated to the serum levels of calcium and phosphorus. Rather, the ossifications are islands of true membranous bone. These bone islands occur in the absence of a preexisting or associated lesion, as opposed to secondary types of cutaneous ossification that occur by metaplastic reaction to inflammatory, traumatic, and neoplastic processes. Recent studies suggest that $G\alpha_s$ deficiency can induce ectopic expression of Cbfa1/Runx2 in mesenchymal stem cells with consequent differentiation of these cells into osteoblasts.

AHO is a unifying feature of PHP type 1a and pseudo-PHP, and based on the parental origin of the *GNAS* mutation, within a given kindred, affected members of one generation can have only AHO (i.e., pseudo-PHP) or can have hormone resistance as well (i.e., PHP type 1a).

Osteoma Cutis and Progressive Osseous Heteroplasia. Osteoma cutis and progressive osseous heteroplasia (POH) represent alternative manifestations of AHO in which only heterotopic ossification occurs. In osteoma cutis, ectopic ossification is limited to the superficial skin, whereas in POH, heterotopic ossification involves the skin, subcutaneous tissue, muscles, tendons, and ligaments. POH can be disabling because extensive dermal ossification occurs during childhood, followed by widespread ossification of skeletal muscle and deep connective tissue. Nodules and lace-like webs of heterotopic bone extend from the skin into the subcutaneous fat and deep connective tissues and may cross joints, thus leading to stiffness, joint locking, and permanent immobility.

Heterozygous inactivating *GNAS* mutations have been identified in most patients with osteoma cutis and POH, and in each case, the defective allele was paternally inherited. Al though patients with POH lack other features of AHO or PHP, maternal transmission of the defective *GNAS* allele leads to the complete PHP type 1a phenotype in affected children.

PHP Type 2

In subjects with PHP type 2, PTH resistance is characterized by a reduced phosphaturic response to administration of PTH, despite a normal increase in urinary cAMP excretion. These observations suggest that the PTH receptor-adenylyl cyclase complex functions normally to increase nephrogenous cAMP in response to PTH, but that intracellular cAMP is unable to act on downstream targets such as the sodium phosphate transporter. PHP type 2 lacks a clear genetic or familial basis, and a similar clinical and biochemical picture occurs in patients with severe deficiency of vitamin D. Taken together, it is likely that most, if not all, cases of PHP type 2 are actually examples of unsuspected vitamin D deficiency.

Transient PHP of the Newborn

Although most newborns with late-onset hypocalcemia (i.e., onset of hypocalcemia after days 5–7 of life) have low levels of PTH, ~25% of affected babies will have elevated levels of PTH. Twitchiness or seizure is most often the initial clinical sign of hypocalcemia. Hypocalcemia is associated with hyperphosphatemia, which is caused by a high transport maximum of the phosphate/glomerular filtration rate despite elevated PTH levels. Serum levels of magnesium and vitamin D metabolites are typically normal. Intravenous administration of PTH(1-34) produces normal responses of plasma and/or urine cAMP, but the phosphaturic response to the PTH infusion is typically impaired. Affected newborns respond to treatment with calcium and/or 1α hydroxylated metabolites of vitamin D. The condition seems to be transient, with normal serum levels of calcium, phosphorus, and PTH maintained without treatment by age 6 mo. These features are suggestive of delayed maturation of the post-cAMP signaling pathway in the proximal renal tubule.

Circulating Inhibitors as a Cause of PTH Resistance

Past studies had noted an apparent dissociation between circulating levels of immunoreactive and bioactive PTH in patients with PHP type 1, and plasma from many of these patients diminished the biological activity of exogenous PTH in in vitro cytochemical bioassays. Although the basis for this inhibitory effect is unknown, one potential explanation is the accumulation of N-terminally truncated PTH fragments, such as hPTH(7-84), that can inhibit the calcemic and phosphaturic actions of hPTH(1-34) or hPTH(1-84). Circulating levels of PTH(7-84) immunoreactivity are elevated in patients with PHP types 1a and 1b, and the proportion of PTH(7-84)–like fragments to biologically active PTH(1-84) is increased. Calcitriol treatment of patients with PHP type 1 increases the phosphaturic response to PTH (1-34), and one mechanism to explain this observation may be suppression of PTH(7-84)–like fragment levels. Although it is conceivable that circulating hPTH(7-84)–like fragments contribute to PTH resistance in some patients with PHP, it is likely that these circulating antagonists arise as a consequence of sustained secondary hyperparathyroidism and do not have a significant role in the primary pathophysiology of the disorder.

Magnesium Deficiency

Magnesium deficiency can impair parathyroid secretion of PTH and PTH action. In either case, treatment with magnesium will restore parathyroid function and/or PTH responsiveness.

© 2008 American Society for Bone and Mineral Research

SYNDROMES OF DEFICIENT SYNTHESIS OR SECRETION OF PTH

Developmental Disorders

Genetic forms of HP may derive from dysgenesis of the parathyroid glands. The parathyroid glands (PGs) derive from the pharyngeal pouches, which are transient structures that differentiate from the foregut endoderm during embryonic development. The PGs are evolutionarily homologous to pharyngeal gill slits in fish, and development of both of these organs is critically dependent on expression of an evolutionarily conserved hierarchy of gene expression. The relevant genes for PG development include the transcription factors Hoxa3, Pax9, Eya1, GCM2, and Tbx1. In humans, the PGs first appear during the fifth week of gestation; the superior glands derive from the fourth pharyngeal pouches, and the thymus and inferior PGs derive from the third pharyngeal pouches.

DiGeorge Sequence/Catch-22. The DiGeorge sequence (DGS) results from dysembryogenesis of the third and fourth pharyngeal pouches and is associated with hypoplasia of the thymus and parathyroid glands. The term DiGeorge sequence or anomaly is more appropriate than "syndrome" because the constellation of defects does not result from a single cause but rather the failure of an embryological field to develop normally. These patients also often manifest conotruncal cardiac abnormalities, cleft palate, and dysmorphic facies. Hypoparathyroidism is present in up to 60% of patients with DGS. DGS is the leading cause of persistent hypocalcemia of the newborn (~1:10,000), but hypoparathyroidism may resolve during childhood. Thymic defects are associated with impaired T cell–mediated immunity and frequent infections.

Molecular mapping has attributed most (70–80%) cases of DGS to hemizygous microdeletions within a critical 250-kb region of 22q11.21–q11.23. Although many genes are located within this region, the presence in some patients with DGS of point mutations that inactivate the *TBX1* gene suggests that this may be the critically important gene. *TBX1* encodes a T-box transcription factor that is widely expressed in non-neural crest cells, cranial mesenchyme, and pharyngeal pouches, and its distribution certainly corresponds to the clinical phenotype of DGS.

DGS most commonly arises from de novo mutations, but autosomal dominant inheritance can occur. Microdeletions of 22q11 are the most common cause of continuous gene deletion syndromes in humans and are present in ~1:3000 newborns. In addition to DGS, deletions within 22q11 can cause the conotruncal anomaly face syndrome and velocardiofacial syndrome (VCFS). VCFS is typically diagnosed later in childhood, and hypocalcemia has been found to be present in up to 20% of cases. Because of the phenotypic variability of the various overlapping syndromes, these conditions are all included within the acronym "CATCH-22," representing a syndrome of Cardiac abnormality, Abnormal facies, Thymic hypoplasia, Cleft palate, and Hypocalcemia with deletion or chromosome 22q11.

DGS has also been reported to arise in patients with deletions of 10p13, 17p13, and 18q21. Gestational diabetes, as well as exposure to alcohol and other toxins (e.g., retinoids) in the intrauterine stage, can also cause similar phenotypic syndromes.

Hypoparathyroidism, Sensorineural Deafness, and Renal Dysplasia Syndrome. Deletions within two nonoverlapping regions of 10p have been found to contribute to a phenotype similar to DGS, namely the hypoparathyroidism, sensorineural deafness, and renal dysplasia syndrome (HDR, MIM146255). Unlike DGS/CATCH-22, individuals with HDR do not exhibit cardiac, palatal, or immunologic abnormalities. The HDR disorder is caused by haplo-insufficiency of the GATA binding protein-3 (*GATA3*) gene, which is located within a 200-kb critical HDR deletion region on 10p14–10pter and encodes a carboxy-terminal zinc-finger protein essential for DNA binding. It is expressed in the developing vertebrate kidney, otic vesicle, and parathyroids, as well as the central nervous system and organs of T-cell development.

Hypoparathyroidism-Retardation-Dysmorphism and Kenny-Caffey Syndromes. Hypoparathyroidism-retardation-dysmorphism syndrome (HRD; MIM 241410), also known as Sanjad-Sakati (SS) syndrome, is a rare form of autosomal recessive hypoparathyroidism associated with other developmental anomalies. In addition to parathyroid dysgenesis, affected patients have severe growth and mental retardation, microcephaly, microphthalmia, small hands and feet, and abnormal teeth. This disorder is seen almost exclusively in individuals of Arab descent. Kenny-Caffey (KS; MIM 244460) syndrome is an allelic disorder that is characterized by hypoparathyroidism, dwarfism, medullary stenosis of the long bones, and eye abnormalities. Both disorders are caused by mutations in the tubulin-specific chaperone E (*TBCE*) gene on chromosome 1q42–43, which encodes a protein required for folding of α-tubulin and its heterodimerization with β-tubulin, although a second gene locus for this disorder is also probable (KS2).

Isolated Hypoparathyroidism. Hypoparathyroidism may also occur as an isolated condition with either an X-linked or autosomal dominant inhibitor pattern of inheritance. The leading cause of autosomal recessive isolated hypoparathyroidism is inactivation of the *GCM2* (*GCMB*) gene at 6p23–24 (MIM 146200). Mutations of *GCM2* can also cause autosomal dominant hypoparathyroidism. *GCM2* encodes a member of a small family of unique transcription factors that were originally identified in *Drosophila* that lacked glial cells, hence the name *gcm* for "glial cells missing." Two mammalian homologs exist: *GCM1,* which is principally expressed in the placenta and controls placental branching and vasculogenesis, and *GCM2,* which is expressed predominantly if not exclusively in the developing and mature parathyroid gland. Inactivation of *GCM2* at 6p23–24 causes parathyroid aplasia or dysplasia with severe isolated hypoparathyroidism in newborns. In contrast, mutations that reduce the biological activity of the GCM2 protein can cause a milder phenotype. Heterozygous mutation of *GCM2* does not seem to cause any disturbance in parathyroid development or function.

Isolated hypoparathyroidism can also be inherited as an X-linked recessive trait (MIM 307700). Affected males present with infantile hypocalcemic seizures, whereas hemizygous females are unaffected. Autopsy of an affected individual showed complete agenesis of the parathyroid glands as the cause of hypoparathyroidism. Linkage analysis has localized the underlying mutation to a 1.5-Mb region on Xq26–q27, and recent molecular studies have identified a deletion-insertion involving chromosomes Xq27 and 2p25 as the basis for the defect. These findings have also suggested that a gene known as Sry-box 3 (*SOX3*), within a 906-kb X-linked recessive HPT critical region, might play a role in the embryonic development of the parathyroid glands.

Parathyroid Gland Destruction

Surgery. The most common cause of hypoparathyroidism in adults is surgical excision of or damage to the parathyroid glands as a result of total thyroidectomy for thyroid cancer,

© 2008 American Society for Bone and Mineral Research

radical neck dissection for other cancers, or repeated operations for primary hyperparathyroidism. Prolonged hypocalcemia, which may develop immediately or weeks to years after neck surgery, suggests permanent hypoparathyroidism. Postoperative hypoparathyroidism occurs in ~1% of thyroid and parathyroid procedures. In patients with a higher risk of developing permanent hypoparathyroidism, parathyroid tissue may be autotransplanted into the brachioradialis or sternocleidomastoid muscle at the time of parathyroidectomy or cryopreserved for subsequent transplantation as necessary. However, non-autologous transplantation of parathyroid tissue is not a therapeutic option for hypoparathyroidism.

Radiation Therapy and Infiltration. Rarely, hypoparathyroidism has also been described in a small number of patients who receive extensive radiation to the neck and mediastinum. It is also reported in metal overload diseases such as hemochromatosis and Wilson's disease and in neoplastic or granulomatous infiltration of the parathyroid glands. Hypoparathyroidism may be present in as many as 14% of patients with thalassemia who develop iron overload caused by frequent blood transfusion. Hypoparathyroidism has also been observed in association with HIV disease.

Autoimmune Polyendocrinopathy-Candidiasis-Esctodermal Dystrophy. Autoimmune destruction of the parathyroids occurs most commonly in association with the complex of immune-mediated disorders that comprises autoimmune polyendocrinopathy-candidiasis-esctodermal dystrophy (APECED; MIM 240300) syndrome, also known as APS I (autoimmune polyglandular syndrome type I). The genetic etiology has been traced to mutations of the *autoimmune regulator* (*AIRE*) gene on chromosome 21q22.3, which encodes a unique protein with characteristics of a transcription factor. APS I can be either sporadic or autosomal recessive and has been associated with >40 different mutations of the *AIRE* gene. Although the disorder occurs worldwide, it is most prevalent in Finns, Sardinians, and Iranian Jews, and common gene mutations in some populations suggest possible founder effects.

The syndrome's classic triad constitutes the "HAM" complex of hypoparathyroidism, adrenal insufficiency, and mucocutaneous candidiasis. The immune defect may be associated with cytotoxic antibodies that damage or destroy the parathyroid glands. Recently, antibodies that react with the NACHT leucine-rich-repeat protein 5 (NALP5), which is expressed predominantly in the cytoplasm of parathyroid chief cells, have been identified in ~50% of APS 1 patients with HP. An alternative pathophysiological mechanism has been proposed that is based on the presence of circulating antibodies bind and activate the calcium sensing receptor (CaSR), thereby reducing PTH secretion from parathyroid cells. The temporal progression of the HAM complex is quite predictable, with the appearance of mucocutaneous candidiasis and hypoparathyroidism in the first decade of life, followed by primary adrenal insufficiency before 15 yr of age. The candidiasis may affect the skin, nails, and mucous membranes of the mouth and vagina and is often resistant to treatment. Addison's disease can mask the presence of hypoparathyroidism or may manifest only after improvement of the hypoparathyroidism, with a reduced requirement for calcium and vitamin D. By diminishing gastrointestinal absorption of calcium and increasing renal calcium excretion, glucocorticoid therapy for the adrenal insufficiency may exacerbate the hypocalcemia and could cause complications if introduced before the hypoparathyroidism is recognized.

Some patients do not manifest all three primary elements of the HAM complex, whereas other individuals may develop additional endocrinopathies such as hypogonadism, insulin-dependent diabetes, hypothyroidism, and hypophysitis. Nonendocrine components of the disorder that occur frequently include malabsorption, pernicious anemia, vitiligo, alopecia, nail and dental dystrophy, autoimmune hepatitis, and biliary cirrhosis.

Reduced Synthesis or Secretion of PTH

Autosomal Dominant Hypocalcemic Hypercalciuria. Autosomal dominant hypocalcemia (ADHH; also autosomal dominant hypocalcemic hypercalciuria, MIM 146200) most commonly occurs as a result of an activating mutation of the gene encoding the CaSR at 3q13.3–21. Most mutations have been identified in the transmembrane and extracellular domains of the receptor and lower the set point for extracellular calcium sensing. The CaSR is widely expressed, but activating mutations have their most profound effects on calcium-induced signaling in the parathyroid gland and the kidney. The effect of the activating mutation on the parathyroid cell is to reduce PTH secretion and thereby produce a state of functional hypoparathyroidism. In the tubule cells of the thick ascending limb of the loop of Henle, activated CaSRs stimulate calciuresis and increase the fractional excretion of calcium (FeCa), thus producing relative (or absolute) hypercalciuria relative to the filtered load of calcium. Nephrocalcinosis and nephrolithiasis are common complications of vitamin D therapy. Although in most cases, the degree of hypocalcemia and hypercalciuria are mild and well tolerated, in some patients, severe hypocalcemia occurs.

ADHH may be sporadic and has been shown to arise from de novo activating mutations.

PTH Gene Mutations. Rare mutations in the *PTH* gene (11p15.3–p15.1) have been associated with impaired synthesis and secretion of PTH. The human *PTH* gene contains three exons that encode the prepro-PTH hormone. Isolated hypoparathyroidism has been found in a family with a single base substitution in exon 2 of the *PTH* gene, apparently impeding conversion of prepro-PTH to pro-PTH. In another family with autosomal recessive isolated hypoparathyroidism, the entire exon 2 of the *PTH* gene was deleted, preventing generation of a mature secretory peptide. A missense mutation in exon 2 of the prepro-PTH gene has also been found, potentially impairing proteolytic cleavage of the prepro-hormone.

Anti-CaSR Antibodies. Autoimmune hypoparathyroidism had been previously thought to be caused by the binding of cytotoxic autoantibodies to parathyroid cells. However, many patients with late onset primary hypoparathyroidism have circulating antibodies that activate the CaSR, which do not produce irreversible destruction of the parathyroid glands. It seems that there might be a specific autoimmune reaction against the CaSR on parathyroid cells, although detection of antibodies against the receptor appears to be influenced by the assay system used.

Other Causes of Hypoparathyroidism

Mitochondrial Disease. Several syndromes caused by deletions in mitochondrial DNA have been associated with hypoparathyroidism. These include Kearns-Sayre syndrome (encephalomyopathy, ophthalmoplegia, retinitis pigmentosa, heart block), Pearson Marrow-Pancreas syndrome (sideroblastic anemia, neutropenia, thrombocytopenia, pancreatic dysfunction), and maternally inherited diabetes and deafness syndrome. Hypoparathyroidism has also been described in MELAS (Mitochondrial myopathy, Encephalopathy, Lactic

© 2008 American Society for Bone and Mineral Research

Acidosis, and Stroke-like episodes) syndrome, because of point mutations in mitochondrial tRNA. Because renal magnesium wasting is frequently seen in these conditions, a readily reversible form of hypoparathyroidism caused by hypomagnesemia should also be considered. In addition, mutations in the mitochondrial trifunctional protein (MTP), resulting in long-chain 3-hydroxy-acyl-coenzyme A dehydrogenase (LCHAD) deficiency, or combined MTP deficiency, have been associated with hypoparathyroidism in a few unrelated patients. This condition manifests as nonketotic hypoglycemia, cardiomyopathy, hepatic dysfunction, and developmental delay and is associated with maternal fatty liver of pregnancy.

Post Burn. In individuals who have sustained severe burns, there is evidence for upregulation of the CaSR. Lower than normal concentrations of serum calcium suppress PTH secretion, thus leading to hypocalcemia and hypoparathyroidism.

Maternal Hyperparathyroidism and Hypomagnesemia. An infant who is exposed in utero to maternal primary hyperparathyroidism or hypercalcemia can have suppressed parathyroid function and hypocalcemia during the first few weeks of life, but the duration may be up to 1 yr of age. Although therapy may be required acutely, the disorder is usually self-limited. Hypomagnesemia caused by defective intestinal absorption or renal tubular reabsorption of magnesium may impair secretion of PTH and cause hypoparathyroidism. Magnesium replacement will correct the hypoparathyroidism.

DIAGNOSTIC ALGORITHM

PHP should be considered in any patient with functional hypoparathyroidism (i.e., hypocalcemia and hyperphosphatemia) and an elevated plasma concentration of PTH. Hypomagnesemia and severe vitamin D deficiency can produce biochemical features of PTH resistance in some patients, and thus plasma concentrations of magnesium and 25(OH)D must be measured. Unusual initial manifestations of PHP include neonatal hypothyroidism, unexplained cardiac failure, seizures, intracerebral calcification of basal ganglia and frontal lobes, dyskinesia and other movement disorders, and spinal cord compression.

PHP or pseudo-PHP may be suspected in patients who present with somatic features of AHO. However, several aspects of AHO, such as obesity, round face, brachydactyly, and mental retardation, also occur in other congenital disorders (e.g., Prader-Willi syndrome, acrodysostosis, Ullrich-Turner syndrome). An interesting phenocopy of AHO occurs in subjects who have small terminal deletions of chromosome 2q37 [del(2)(q37.3)]. These patients have normal endocrine function and normal $G\alpha_s$ activity.

The classical tests for PHP, the Ellsworth-Howard test and later modifications by Chase, Melson, and Aurbach, involved the administration of 200–300 USP units of purified bovine PTH or parathyroid extract. Although these preparations are no longer available, the synthetic hPTH(1-34) peptide has been approved for human use, and several protocols for its use in the differential diagnosis of hypoparathyroidism have been developed. These protocols are based on intravenous infusion of the peptide, but similar results may be obtained after subcutaneous injection of hPTH(1-34) or hPTH(1-84), albeit requiring administration of higher doses of peptide. The patient should be fasting except for fluids (250 ml of water hourly from 6:00 a.m. to noon). Two control urine specimens are collected before 9:00 a.m. Synthetic human PTH(1-34) peptide (0.625 μg/kg body weight to a maximum of 25 μg for intravenous use and 40 μg for subcutaneous use) is administered at 9:00 a.m. either by subcutaneous injection or intravenous infusion over 15 min, and experimental urine specimens are collected from 9:00 to 9:30, 9:30 to 10:00, 10:00 to 11:00, and 11:00 to 12:00. Blood samples should be obtained at 9:00 a.m. and 11:00 a.m. for measurement of serum creatinine and phosphorous concentrations. Urine samples are analyzed for cAMP, phosphorous, and creatinine concentrations, and results are expressed as nanomoles of cAMP per 100 ml glomerular filtrate (GF) and TmP/GFR (ratio of the renal tubular maximum rate of phosphate reabsorption to the glomerular filtration rate).

Normal subjects usually show a 10- to 20-fold increase in urinary cAMP excretion and a 20–30% decrease in TmP/GFR, whereas patients with PHP type 1 (both type 1a and type 1b), regardless of their serum calcium concentration, will show markedly blunted responses (Fig. 1). Normal children and patients with HP have more robust responses. Thus, this test can distinguish patients with so-called "normocalcemic" PHP (i.e., patients with PTH resistance who are able to maintain normal serum calcium levels without treatment) from subjects with pseudo-PHP (who will have a normal urinary cAMP response to PTH). Measurement of plasma cAMP or plasma 1,25-dihydroxyvitamin D after infusion of hPTH(1-34) may also differentiate PHP type 1 from other causes of hypoparathyroidism.

The diagnosis of PHP type 2 requires exclusion of magnesium depletion or vitamin D deficiency. Documentation of elevated serum PTH and nephrogenous cAMP is a prerequisite for a definitive diagnosis of PHP type 2. These subjects have a normal urinary cAMP response to infusion of PTH but characteristically fail to show a phosphaturic response. Unfortunately, interpretation of the phosphaturic response to PTH is often complicated by random variations in phosphate clearance, and it is sometimes not possible to classify a phosphaturic response as normal or subnormal regardless of the criteria used.

Genetic testing can assist with the diagnosis of HP and PHP. Mutational analysis of the *CaSR*, *AIRE*, and *GNAS* genes is now available as approved tests by several clinical laboratories. In contrast, genetic testing for PHP type 1b and most other forms of genetic hypoparathyroidism is still considered investigational.

THERAPY

Calcium and Vitamin D

The goal of therapy in hypoparathyroidism is to restore serum calcium as close to normal as possible. The main pharmacologic agents available are supplemental calcium and vitamin D preparations. Phosphate binders, to lower serum phosphate, and thiazide diuretics, to decrease urinary calcium excretion, may be useful ancillary agents. The major limitation to restoration of normocalcemia is the development of hypercalciuria, with a resulting risk for nephrolithiasis. With the loss of the renal calcium-conserving effect of PTH, the enhanced calcium absorption of the gut induced by vitamin D therapy results in an increased filtered load of calcium that is readily cleared through the kidney. Consequently, urinary calcium excretion frequently increases in response to vitamin D supplementation well before serum calcium is normalized. It is thus often advisable to maintain a low normal serum calcium concentration to prevent chronic hypercalciuria. Avoidance of hypercalciuria is probably most important for patients with hypercalciuric hypocalcemia caused by activating mutations of the *CaSR* gene. Happily, patients with PHP type 1 rarely develop hypercalciuria as long as PTH levels remain unsuppressed as the distal tubule remains responsive.

© 2008 American Society for Bone and Mineral Research

Approximately 1–2 g (25–50 mg/kg) of elemental calcium per day are recommended to supply adequate calcium and to manage dietary phosphorous intake, and optimal results are achieved when calcium supplements are taken with meals. Patients are unable to convert parent compounds of vitamin D to fully active forms because lack of PTH and hyperphosphatemia both inhibit the renal 1-α hydroxylase enzyme that converts 25-hydroxyvitamin D to 1,25-dihydoxyvitamin D. The 1-α hydroxylated vitamin D metabolites (e.g., calcitriol) are the preferred forms of vitamin D because these drugs circumvent the enzymatic block in vitamin D activation. Because calcitriol has a plasma half-life of only hours, and body stores do not accumulate, calcitriol must be given several times each day. In contrast, alfacalcidol [$1\alpha(OH)D_3$] has a longer half-life and may be given once daily. A variety of parent vitamin D preparations have been used in the past, including vitamin D_3 or D_2; however, typically very large concentrations are required to raise the serum calcium. In this case, body stores of vitamin D can accumulate in massive amounts, increasing the risk of severe and prolonged vitamin D toxicity. Hydrochlorothiazide therapy has been effective in reducing the vitamin D requirement, but potassium supplementation is necessary to offset the thiazide-induced hypokalemia.

PTH Treatment

Preliminary data have suggested that PTH replacement has a salutary clinical effect in HP. In one study, 27 adults with HP were randomized to PTH(1-34) or calcium with vitamin D for 3 yr and were given a mean dose of either PTH(1-34) 36.7 μg twice daily or calcitriol 0.47 μg twice daily. Serum calcium levels were found to be similar in the calcitriol and PTH(1-34) groups, whereas urinary calcium levels were lower with PTH(1-34) than calcitriol. Markers of bone turnover were significantly higher in the PTH group. BMD did not show significant between-group differences. In a different study, PTH(1-84) was administered in an open-label fashion at 100 μg every other day for up to 24 mo to 40 subjects with HP. Serum calcium remained within the normal range, whereas requirements for calcium and calcitriol supplementation fell significantly, by as much as 34% and 38%, respectively. The results of these preliminary studies suggest that treatment with PTH may restore calcium homeostasis in patients with hypoparathyroidism, allowing for serum calcium levels to be maintained while permitting substantial reductions in needs for calcium and vitamin D.

SUGGESTED READING

Ding C, Buckingham B, Levine MA 2001 Familial isolated hypoparathyroidism caused by a mutation in the gene for the transcription factor GCMB. J Clin Invest **108:**1215–1220.

Sherwood LM, Santora AC 1994 Hypoparathyroid states in the differential diagnosis of hypocalcemia. In: Bilezikian JP, Marcus R, Levine MA (eds.) The Parathyroids. Raven Press, New York, NY, USA, pp. 747–752.

Stewart AF 2004 Translational implications of the parathyroid calcium receptor. N Engl J Med **351:**324–326.

Thakker RV 2001 Genetic developments in hypoparathyroidism. Lancet **357:**974–976.

Winer KK, Ko CW, Reynolds JC, Dowdy K, Keil M, Peterson D, Gerber LH, McGarvey C, Cutler GB Jr 2003 Long-term treatment of hypoparathyroidism: A randomized controlled study comparing parathyroid hormone-(1–34) versus calcitriol and calcium. J Clin Endocrinol Metab **88:**4214–4220.

Chapter 76. Familial Hyperparathyroidism (Including MEN, FHH, and HPT-JT)

Andrew Arnold[1] and Stephen J. Marx[2]

[1]*Center for Molecular Medicine and Division of Endocrinology and Metabolism, University of Connecticut School of Medicine, Farmington, Connecticut;* [2]*Genetics and Endocrinology Section, Metabolic Diseases Branch, National Institute of Diabetes and Digestive and Kidney Diseases, National Institutes of Health, Bethesda, Maryland*

INTRODUCTION

Individuals with familial primary hyperparathyroidism (HPT), defined by the combination of hypercalcemia with elevated or nonsuppressed serum PTH, are a small and important subgroup of all cases with primary HPT. Their familial syndromes include multiple endocrine neoplasia (MEN) types 1, 2A, and 4, familial (benign) hypocalciuric hypercalcemia (FHH), neonatal severe hyperparathyroidism (NSHPT), hyperparathyroidism-jaw tumor syndrome (HPT-JT), and familial isolated hyperparathyroidism (FIHP). These syndromes exhibit mendelian inheritance patterns, and the main genes for most have been defined (Table 1). As more knowledge accumulates on genetic contributions to complex phenotypes, additional genetic loci may be identified as contributing, including some loci for less penetrant and more subtle predispositions to primary HPT.

FAMILIAL HYPOCALCIURIC HYPERCALCEMIA

FHH (OMIM 145980; familial benign hypocalciuric hypercalcemia) is an autosomal dominant syndrome with lifelong high penetrance for both hypercalcemia and relative hypocalciuria. The prevalence of FHH is similar to MEN1; either accounts for ~2% of asymptomatic hypercalcemics.

Clinical Expressions

Persons with FHH are usually asymptomatic. Easy fatigue, weakness, thought disturbance, or polydipsia are less common and less severe than in typical primary HPT. Nephrolithiasis or even hypercalciuria are as frequent as in normals. Bone radiographs are usually normal. There is an increased incidence of chondrocalcinosis (usually clinically silent) and of premature vascular calcification. Bone mass and susceptibility to fracture are normal. Hypercalcemia has virtually 100% penetrance at

The authors state that they have no conflicts of interest.

Key words: parathyroid, familial hypocalciuric hypercalcemia, PTH/PTH-related protein, familial hyperparathyroidism

© 2008 American Society for Bone and Mineral Research

TABLE 1. OUTLINE OF SYNDROMES OF FAMILIAL HYPERPARATHYROIDISM WITH EMPHASIS ON MAJOR FEATURES AND FEATURES THAT DISTINGUISH AMONG THE SYNDROMES

Syndrome	*Main gene and mutation*	*Parathyroid gland aspects*	*Aspects outside parathyroid gland*
FHH	*CASR*-	High Ca^{2+} begins at birth Nl gland size; abnl set-point Persist after subtotal PTX Manage without intervention	Relative hypocalciuria
NSHPT	*CASR*=	High Ca^{2+} begins at birth Very large PT glands Ca > 16 mg%; needs total PTX	Relative hypocalciuria
MEN1	*MEN1*-	Begins age 20 Asymm adenomas; Freq post-PTX late recur	Tumors among 20 tissues (Pituit, entero-pancreat, foregut carcinoid, dermis, etc). Easiest to find in case and close relatives.
MEN4	*p27KIP1*-	Like MEN1	Like MEN1 (based on only four cases)
MEN2A	*RET*+	Begins age 30; maximum penetrance of HPT 30%	C-cell cancer that is preventable Pheochromocytoma, may be bilat.
HPT-JT	*HRPT2*-	High Ca^{2+} can occur by age 10, often later 15% PT cancer: microcystic	Benign jaw tumors, renal cysts Uterine tumors
FIHP	(chrom 2p??)	No specific features	None by definition Occult other syndrome may emerge

FHH, familial hypocalciuric hypercalcemias; NSHPT, neonatal severe primary hyperparathyroidism; MEN, multiple endocrine neoplasia (can be type 1, 2A, or 4); HPT-JT, hyperparathyroidism jaw tumor syndrome; FIHP, familial isolated hyperparathyroidism; -, heterozygous inactivation; +, heterozygous activation; =, homozygous inactivation.

all ages. Its onset in infancy is clinically remarkable and can be diagnostically useful; the serum calcium level is similar to that in typical primary HPT with a normal ratio of free to bound calcium. Serum magnesium is typically in the high range of normal or modestly elevated, and serum phosphate is modestly depressed. Creatinine clearance or urinary excretion of calcium is normal, with affected and unaffected family members showing a similar distribution of values.

Parathyroid function, including serum PTH and $1,25(OH)_2D$, is usually normal, with modest elevations in 5–10% of cases. Such "normal" parathyroid function indices in the presence of lifelong hypercalcemia are inappropriate and reflect the primary role for the parathyroids in causing this hypercalcemia. FHH cases often have mild enlargement of the parathyroid glands that may not be recognized at surgery and are evident only by careful measurement. Standard subtotal parathyroidectomy in FHH results in only a very transient lowering of serum calcium, followed by similar hypercalcemia within days.

Pathogenesis/Genetics

Most cases result from heterozygous inactivating mutation of the *CASR* gene. Two FHH kindreds not linked genetically to the *CASR* locus at chromosome 3q, one linked to 19p and the other to 19q, must represent mutation in two other unidentified genes.

There is a selective and mild increase in the parathyroid glandular "set-point" for serum calcium suppression of PTH secretion. This is more striking for the related NSHPT (from mutation in both *CASR* alleles; see below). There also is a disturbed calcium-sensing receptor (Ca-S-R) function, intrinsic to the kidneys in FHH. Normally, the Ca-S-R functions in the kidney to maintain tubular calcium reabsorption in the direction that would correct for an alteration in the receptor-sensed serum calcium. The tubular reabsorption of calcium is normally increased principally by PTH; in FHH, it is high and remains increased even after an intended or unintended total parathyroidectomy. Ca-S-Rs are normally expressed in additional tissues outside of the parathyroids and kidneys, but clinical dysfunction there has not been reported in FHH or NSHPT.

The usual distinctions between typical primary HPT and the primary HPT of FHH can be blurred in some kindreds with *CASR* mutations. One large kindred with a germline missense mutation in *CASR* had a HPT syndrome unlike FHH. In contrast to FHH, these individuals display hypercalciuria, monoclonal parathyroid adenomas, and a benefit from subtotal parathyroidectomy. Several other small families with *CASR* loss-of-function mutations have contained some members with features partly resembling typical primary HPT. FHH can also be caused by antibodies against the Ca-S-R and is associated with other autoimmune features, but this is without *CASR* mutation. This is rare and generally not familial.

Diagnosis

In the presence of hypercalcemia, a normal PTH just like a low urine calcium warns about possible FHH. The diagnosis usually is made from typical clinical features, sometimes including failed parathyroidectomy, in one or more members of a family. The normal urinary calcium in the face of hypercalcemia reflects increased renal tubular reabsorption of calcium (i.e., relative hypocalciuria). Because total urinary calcium excretion depends heavily on glomerular filtration rate (GFR) and collection interval, total calcium excretion is not a practical index to distinguish a case of FHH from typical HPT. The ratio of renal calcium clearance to creatinine clearance

$$\begin{aligned} Ca_{Cl}/Cr_{Cl} &= [Ca_u \times V/Ca_s]/[Cr_u \times V/Cr_s] \\ &= [Ca_u \times Cr_s]/[Cr_u \times Ca_s] \end{aligned}$$

is an empirical index. In hypercalcemic cases, this clearance ratio in FHH averages one third of that in typical primary HPT, and a cut-off value <0.01 (valid units will all cancel out) is helpful for diagnosis.

Family screening for FHH traits can be important to estab-

© 2008 American Society for Bone and Mineral Research

lish familial involvement, to characterize the syndrome, and to diagnose a carrier. Because of high penetrance for hypercalcemia in all FHH carriers, an accurate assignment for each relative at risk can usually be made from one determination of serum calcium (preferably ionized or albumin-adjusted). *CASR* mutation analysis has occasional roles in diagnosis of the syndrome, particularly with an inconclusive clinical evaluation of the family. *CASR* mutation may be undetectable if located outside the tested coding exons (largely explaining 30% falsely "normal" testing).

Management

Despite lifelong hypercalcemia, FHH is compatible with survival into the ninth decade. Because of the generally benign course and lack of response to subtotal parathyroidectomy, few cases should undergo parathyroidectomy. In rare situations, such as relapsing pancreatitis, very high PTH, or serum calcium persistently >14 mg/dl, parathyroidectomy may be indicated. Chronic hypercalcemia in FHH should rarely be treated, and it has been resistant to medications (diuretics, bisphosphonates, phosphates, or estrogens). In theory, drugs that act like Ca^{2+} to stimulate the normal or even the mutated calcium-sensing receptor (calcimimetics) might be effective in FHH. There is no detailed information about this use. Without a positive family history or *CASR* mutation, the management of sporadic hypocalciuric hypercalcemia is challenging; this should generally be managed as typical FHH, unless the features of typical HPT become more prominent.

NEONATAL SEVERE PRIMARY HYPERPARATHYROIDISM

Clinical Expressions

NSHPT (OMIM 239200) is an extremely rare neonatal state of life-threatening, severe hypercalcemia, respiratory distress, and massive enlargement of all parathyroid glands.

Pathogenesis/Genetics

This disorder typically results from homozygous or compound heterozygous *CASR* inactivating mutation. In NSHPT, it is uncertain if the hypercellular parathyroids are polyclonal or are overgrown by a monoclonal component.

Diagnosis

Diagnosis is usually based on the unique clinical features, often combined with parental consanguinity and/or FHH in first-degree relatives.

Management

Urgent total parathyroidectomy can be life-saving.

MULTIPLE ENDOCRINE NEOPLASIA TYPE 1

MEN1 (OMIM 131100) is a rare heritable disorder with an estimated prevalence of 2–3 per 100,000. It is classically defined as tumors in two of the three main tissues (parathyroids, anterior pituitary, and enteropancreatic endocrine cells), although affected persons are known to be predisposed to tumors in many other endocrine and nonendocrine tissues. By extension, familial MEN1 is defined as MEN1 with a first-degree relative showing tumor in at least one of the three main tissues.

Clinical Expressions

Primary HPT is the most penetrant endocrine component of MEN1, occurring in almost all carriers by age 50 and is the initial clinical manifestation of the disorder in most cases. Approximately 2% of all cases of primary HPT are caused by MEN1. Some of the other tumors associated with MEN1 include duodenal gastrinomas, bronchial or thymic carcinoids, gastric enterochromaffin-like tumors, adrenocortical adenomas, lipomas, facial angiofibromas, and truncal collagenomas. Familial isolated HPT, discussed below, can be another expression of *MEN1* mutation. Such families should be followed for development of other tumors of MEN1.

Primary HPT in MEN1 has a number of features different from the common sporadic (nonfamilial) form of HPT. The male to female ratio is ~1.0 in MEN1, in contrast to the female predominance in sporadic HPT. HPT in MEN1 presents ~30 yr earlier, typically in the second to fourth decade of life, and has been found as early as 8 yr of age. Multiple parathyroid gland involvement is typical in MEN1; these tumors may vary widely in size, with an average 10:1 ratio between the largest and smallest tumors. A powerful drive to parathyroid tumorigenesis exists in MEN1, reflected by an impressively high rate of recurrent HPT (~50% after 12 yr) after apparently successful subtotal parathyroidectomy.

Pathogenesis/Genetics

Familial MEN1 follows an autosomal dominant inheritance pattern, and the main genetic basis of the classic syndrome is an inactivating germline mutation of the *MEN1* tumor suppressor gene, located on chromosome band 11q13. *MEN1* encodes the nuclear protein menin, whose specific cellular functions remain under study but may include transcriptional regulation of gene expression and control of genome stability. *MEN1* mutations are distributed across the gene without any apparent pattern, other than the majority predicting truncation or absence of menin. Individuals with MEN1 have typically inherited one inactivated copy of the *MEN1* gene from an affected parent or may have a spontaneous new germline mutation. The outgrowth of a tumor requires the subsequent somatic (acquired) inactivation of the normal, remaining copy of the *MEN1* gene in one clonal precursor cell. Such a parathyroid cell, for example, would be devoid of *MEN1*'s tumor suppressor function, contributing to a selective growth advantage over its neighbors and clonal proliferation. This process of mutational loss of a tumor suppressor is analogous to that for many tumor genes such as *HRPT2, p53, BRCA1,* and *APC*. Very rarely, an MEN1-like phenotype (MEN4; OMIM 610755) that includes parathyroid and pituitary disease without *MEN1* gene mutation has been associated with germline mutation of *CDKN1B,* encoding the p27 cyclin-dependent kinase inhibitor. Interestingly, menin may regulate transcription of *CDKN1B*, suggesting their joint participation in a pathway whose disturbance is tumorigenic in MEN1 target cells.

Diagnosis

The biochemical diagnosis of primary HPT in known or suspected MEN1 is the same as for other causes of typical HPT. Once the biochemical diagnosis is established, the indications for surgery are similar to those in sporadic primary HPT and include symptomatic hypercalcemia, nephrolithiasis, fractures, and/or decreased bone mass, which has been observed in women with MEN1 by 35 yr of age.

Direct sequencing for germline *MEN1* mutations is commercially available, but the indications for such testing remain under discussion. Gene analyses, typically limited to the coding

© 2008 American Society for Bone and Mineral Research

region and near to it, fail to detect *MEN1* mutation in ~30% of typical MEN1 kindreds. Most such kindreds probably have *MEN1*-inactivating mutations, such as deletions or noncoding mutations, that are missed by the current sequencing approach. However, the yield of detectable *MEN1* mutation in cases with a sporadic MEN1 phenotype limited to parathyroid plus pituitary disease is much lower, ~10%, suggesting the existence of other predisposing genes. *CDKN1B* mutation (MEN4) is found rarely in such instances, and still-undetermined gene(s) may contribute. In contrast to the clear clinical importance of testing for *RET* gene mutations in possible MEN2, presymptomatic gene diagnosis has not been established to broadly improve morbidity or mortality in MEN1, and biochemical screening with serum calcium, PTH, etc., provides a non–DNA-based alternative for carrier ascertainment. An uncommon exception is seemingly sporadic gastrinoma; *MEN1* mutation occurs in one quarter of such cases, and its finding can lead to avoidance of abdominal surgery that would otherwise be indicated. Also, *MEN1* gene testing can be diagnostically helpful when clinical diagnosis is inconclusive but a suspicion of MEN1 exists: for example, in a young adult with isolated multigland HPT. For established MEN1 carriers or other family members at risk for developing manifestations of MEN1, screening is recommended because benefit seems likely. The extent to which such screening should extend beyond regular clinical assessments to include routine imaging or biochemical testing can reasonably take cost-effectiveness factors into account. Screening for primary HPT can be accomplished with serum calcium and PTH levels.

Management

Because direct evidence is lacking, opinions differ as to the optimal timing of surgical treatment of HPT in MEN1. Early presymptomatic intervention might, on one hand, lead to better long-term bone health. On the other hand, because of the high rate of recurrent HPT, a policy of deferring surgery might decrease a person's total number of operations and thereby decrease the risk of surgical complications.

The multiplicity of parathyroid tumors in MEN1, the need to identify all abnormal parathyroid glands at surgery, and the inability of imaging tools to reliably detect all tumors collectively diminish the need for preoperative tumor localization studies in unoperated patients. For the same reasons, a suspected or firm preoperative diagnosis of MEN1 argues against performing minimally invasive parathyroidectomy. Even at initial surgery for multiple parathyroid tumors, intraoperative PTH measurement can be helpful. Preoperative tumor imaging/localization is useful before reoperation in patients with recurrent or persistent disease. The initial operation most frequently performed in MEN1 patients is 3.5 gland subtotal parathyroidectomy with transcervical near-total thymectomy. A parathyroid remnant is usually left in situ and may be marked with a clip, but alternatively, the remnant can be autotransplanted to the forearm during an intended complete parathyroidectomy. The efficacy of transcervical thymectomy is unproven but seems reasonable because it may cure incipient thymic carcinoids or prevent their development; in addition, the thymus is a common site for parathyroid tumors in MEN1 patients with recurrent HPT. Involvement of a highly experienced parathyroid surgical team is crucial to optimal outcome.

Management of the pituitary, enteropancreatic, and other tumors of MEN1 outside the parathyroids are beyond the scope of this chapter. It should be emphasized that MEN1-associated malignancies cause fully one third of the deaths in MEN1 cases, and for most of these cancers, no effective prevention or cure currently exists because of their problematic locations.

MULTIPLE ENDOCRINE NEOPLASIA TYPE 2A

MEN2 is subclassified into three major clinical syndromes with mutations in the same gene, *RET*: MEN2A, MEN2B, and familial medullary thyroid cancer (FMTC). Of these, MEN2A (OMIM 171400) is the most common and the only one that manifests HPT.

Clinical Expressions

MEN2A is a heritable predisposition to medullary thyroid or C-cell cancer (MTC), pheochromocytoma, and primary HPT. The respective frequency of these tumors in an adult carrier of MEN2A is >90% for MTC, 40–50% for pheochromocytoma, and 20% for HPT. This lower penetrance of HPT in MEN2A contrasts with the high penetrance found in all other recognized familial hyperparathyroid syndromes. MTC, a potentially lethal manifestation of MEN2A, evolves from preexisting parafollicular C-cell hyperplasia, and its calcitonin production provides a useful marker for monitoring large tumors. Despite the pharmacologic properties of calcitonin, mineral metabolism is generally normal in the setting of metastatic MTC and its often dramatic hypercalcitoninemia. Pheochromocytomas in MEN2A can be unilateral or bilateral. Extraadrenal or malignant pheochromocytomas are under-represented in MEN2A cases compared with cases with sporadic or some other familial forms of pheochromocytoma.

Pathogenesis/Genetics

MEN2A is inherited in an autosomal dominant pattern, with men and women affected in equal proportions, and the responsible genetic defect is germline mutation of the *RET* proto-oncogene on chromosome 10. The RET protein is a receptor tyrosine kinase that normally transduces growth and differentiation signals in developing tissues including those derived from the neural crest. There are both differences and much overlap in the specific *RET* gene mutations underlying MEN2A and FMTC; in contrast, MEN2B is caused by entirely distinct *RET* mutations. Why parathyroid disease fails to develop in the latter syndromes remains unclear. Unlike the numerous and seemingly random different inactivating mutations of *MEN1* that are typical of a tumor suppressor mechanism, *RET* mutations in MEN2A are limited in number, reflecting the need for highly specific gain-of-function changes in selected RET protein domains to activate this oncoprotein. Germline *RET* mutation is detectable in >95% of MEN2A families. *RET* mutation at codon 634 is the most frequent and is highly associated with the expression of HPT in MEN2A.

Diagnosis

HPT in MEN2A is often asymptomatic, and its biochemical diagnosis, as well as indications for surgical treatment, parallel those in sporadic primary HPT. DNA sequencing for germline *RET* mutations is central to clinical management and worthy of emphasis for its role in prevention of MTC. For diagnosis of MEN2A, *RET* sequence testing is superior to immunoassay of basal or stimulated calcitonin, except in the vary rare (5%) MEN2 family without detectable *RET* mutation.

Management

If MEN2A is a possibility, evidence of pheochromocytoma should be sought before parathyroidectomy and, if present, the

© 2008 American Society for Bone and Mineral Research

pheochromocytoma(s) should be removed before parathyroid surgery. Primary HPT in MEN2A is almost always multiglandular, but less than four overtly enlarged glands may be present. Thus, bilateral neck exploration to identify all abnormal glands is advisable in known or suspected MEN2A, with resection of hypercellular parathyroid tissue (up to 3.5 glands) being the most common surgical approach. Asymptomatic enlargement of the parathyroids is commonly found during thyroid surgery in MEN2A, and such incidentally enlarged glands are generally removed. Issues of preoperative tumor localization in unoperated patients are similar to MEN1. In contrast to MEN1, however, recurrent HPT is infrequent after apparently successful resection of enlarged glands, similar to the excellent long-term outcome of surgically treated patients with nonfamilial multiple primary tumors.

DNA diagnosis can guide prophylactic or curative thyroidectomy (i.e., sufficiently early in childhood as to minimize the likelihood that C-cell cancer metastases will have already occurred). Laparoscopic adrenalectomy has greatly improved the management of pheochromocytoma in MEN2A. Incompletely treated postoperative chronic hypocortisolism remains a major cause of morbidity and death after bilateral adrenalectomy.

HYPERPARATHYROIDISM—JAW TUMOR SYNDROME

HPT-JT syndrome (OMIM 145001) is a rare, autosomal dominant predisposition to primary HPT, ossifying or cementifying fibromas of the mandible and maxilla, renal manifestations including cysts, hamartomas, or Wilms tumors, and uterine tumors. In "classical" HPT-JT kindreds, HPT is the most penetrant manifestation at 80% among adults, followed by 30% for ossifying fibromas, and slightly lower for a renal lesion. As mentioned below, familial isolated HPT and apparently sporadic parathyroid carcinoma can represent occult or allelic variant presentations of the genetic defect in HPT-JT.

Clinical Expressions

HPT in HPT-JT may develop as early as the first decade of life. Although all parathyroids are at risk, surgical exploration can show a solitary parathyroid tumor rather than multigland disease, in contrast to typical findings in HPT of MEN1 and MEN2A. Parathyroid neoplasms can be macro- or microcystic, and whereas most tumors are classified as adenomas, the incidence of parathyroid carcinoma (15–20% of HPT) is markedly over-represented in HPT-JT cases. In contrast, parathyroid cancer almost never occurs in MEN1, MEN2, or FHH. Dissemination of parathyroid cancer to the lungs can occur in the early 20s in HPT-JT. After a period of normocalcemia, treated cases may manifest recurrent HPT, and a solitary tumor asynchronously originating in a different parathyroid gland may prove responsible. Ossifying fibromas in HPT-JT are generally benign but still may be large and destructive. Often they are small, asymptomatic, and identified as incidental findings on dental radiographs. They are clearly distinct from the classic, osteoclast-rich "brown tumors" resulting from severe HPT.

Pathogenesis/Genetics

Germline mutation of the *HRPT2* gene (also *CDC73*), located on chromosome arm 1q, causes HPT-JT. The yield of *HRPT2* mutation detection in HPT-JT kindreds is ~60–70%; the remaining kindreds most likely also have *HRPT2* mutations that evade detection because of their location outside the sequenced coding region. Mutations of *HRPT2* cause tumors by inactivating or eliminating its protein product parafibromin, consistent with a classical "two-hit" tumor suppressor mechanism also shown in sporadic parathyroid carcinoma. Parafibromin's normal cellular function may involve regulation of gene expression and chromatin modification. Importantly, many cases with seemingly sporadic presentations of parathyroid carcinoma (OMIM 608266) also harbor germline mutations in *HRPT2*, thus representing newly ascertained HPT-JT or a variant syndrome.

Diagnosis

The biochemical diagnosis of HPT in HPT-JT parallels that in sporadic primary HPT. Recognition of *HRPT2* mutation in classic or variant HPT-JT has opened the door to DNA-based carrier identification in at-risk family members, aimed at preventing or curing parathyroid malignancy.

Management

Management in HPT-JT includes monitoring and surgery to address the predilection to parathyroid malignancy. The finding of biochemical HPT should lead promptly to surgery. All parathyroids should be identified at operation, signs of malignancy sought, and appropriate resection of abnormal glands performed. Because of malignant potential, the consideration of prophylactic total parathyroidectomy (perhaps even for euparathyroid carriers) has been raised as an alternative approach, but is not favored by most in view of difficulties with lifelong management of surgical hypoparathyroidism, the incomplete penetrance of parathyroid cancer in the syndrome, and the plausible idea that close biochemical monitoring for recurrent HPT combined with early surgery will prevent metastatic disease or its progression.

FAMILIAL ISOLATED HYPERPARATHYROIDISM

Clinical Expressions and Diagnosis

FIHP (OMIM 145000) is clinically defined as familial HPT without the extraparathyroid manifestations of another syndromal category. The diagnosis of FIHP can therefore change if features of another syndrome develop.

Pathogenesis/Genetics

FIHP is genetically heterogeneous and can be caused by variant expressions of germline mutations in *MEN1*, *HRPT2*, and *CASR*. The majority of families do not have detectable mutation in any of these three genes. One unidentified gene may be on the short arm of chromosome 2 (OMIM 610071).

Management

Clinical monitoring and management must take into consideration the possibility that additional features of a genetically defined HPT syndrome or a previously unidentified syndrome could emerge or become detectable. For example, the heightened risk of parathyroid carcinoma must be borne in mind in FIHP when the genetic basis is not established and occult *HRPT2* mutation is possible. DNA testing should be considered (e.g., when results might impact on the advisability of, or approach to, parathyroid surgery or to further gene testing in relatives).

OVERLAPPING CONSIDERATIONS AMONG ALL FORMS OF FAMILIAL HPT

Multifocal Parathyroid Gland Hyperfunction

For three hereditary HPT syndromes (MEN1, MEN2A, and HPT-JT), the germline mutation in the parathyroid cell causes

© 2008 American Society for Bone and Mineral Research

susceptibility to postnatal development of mono or oligoclonal parathyroid tumor. For two others (FHH and NSHPT), the phenotype is fully expressed around birth with no postnatal delay. The common and critical feature of the five is that every parathyroid cell carries the germline mutation, which is either sufficient to cause the phenotype in all parathyroid cells or places all parathyroid cells at risk for clonal proliferation/hyperfunction pending the acquisition of additional defects (second hits, etc.).

Detection of Asymptomatic Carriers

Once a syndrome has been diagnosed, testing for the carrier state in asymptomatic relatives should be considered. The concept of the carrier must include disease predisposition even without an identified mutation. Where possible, testing of germline DNA is the gold standard; however, testing of traits that are expressed early and with high penetrance (such as hypercalcemia) is often a major alternative. The possible benefits include guidance of an important intervention and information to the subject, family, and physician. Among all HPT syndromes, only testing for MEN2 can lead to a major intervention, thyroidectomy, of almost certain efficacy in reducing mortality (from medullary thyroid carcinoma). Testing for HPT-JT can lead to management that may lessen mortality associated with parathyroid malignancy; testing in other syndromes is mainly for information to physician and patient and is less urgent.

Specialized Surgery for Multigland Parathyroid Tumors

Many aspects of surgery for parathyroid adenoma require modification when multigland involvement is expected. Efforts are made intraoperatively to know when sufficient pathologic tissue has been removed. Traditionally, identification of all four parathyroid glands is pursued. Rapid measurement of PTH intraoperatively to see a drop from high to low can be useful in ascertaining the presence of overactive parathyroid tissue remaining in vivo.

Efforts to identify all four parathyroid glands and remove several glands result in increased postoperative hypoparathyroidism. To minimize or treat this, in some centers, small fragments of the most normal-appearing parathyroid tissue are autografted to the nondominant forearm. In other centers, because of the reported suboptimal success of such grafts, the surgeon leaves a small remnant of parathyroid in the neck, attached to its own vascular pedicle. In either case, one may cryopreserve fragments of the most normal tissue for possible delayed autograft for late postoperative hypoparathyroidism. The malignant potential of parathyroid tissue in HPT-JT argues against the autografting option there.

Other parathyroid complications are also more frequent after surgery for multigland HPT, including familial HPT, than after surgery for adenoma. These include complications to surrounding tissues, such as recurrent nerve damage, and postoperative persistent HPT. The most obvious reason for the latter is a failure to explore for multigland disease. However, the fact that one or more of four glands can be overactive suggests that the likelihood of just one abnormally located tumor would be about four times as high as in a case with adenoma. True recurrent HPT is a late complication that is almost exclusively limited to familial HPT. True recurrence is defined as HPT after a 3- to 6-mo postoperative period of documented euparathyroidism. It might arise when a small tumorous remnant becomes overactive or when a previously normal but at-risk gland begets a tumor. True recurrence reaches ~50% at 12 yr postoperatively in MEN1. Its frequency has not been measured in other HPT syndromes or in sporadic multigland HPT, but it is probably lower than in MEN1.

ACKNOWLEDGMENTS

This work was supported in part by the intramural program of NIDDK.

SUGGESTED READING

Brandi ML, Gagel RF, Angeli A, Bilezikian JP, Beck-Peccoz P, Bordi C, Conte-Devolx B, Falchetti A, Gheri RG, Libroia A, Lips CJ, Lombardi G, Mannelli M, Pacini F, Ponder BA, Raue F, Skogseid B, Tamburrano G, Thakker RV, Thompson NW, Tomassetti P, Tonelli F, Wells SA Jr, Marx SJ 2001 Guidelines for diagnosis and therapy of MEN type 1 and type 2. J Clin Endocrinol Metab **86:**5658–5671.

Carling T, Szabo E, Bai M, Ridefelt P, Westin G, Gustavsson P, Trivedi S, Hellman P, Brown EM, Dahl N, Rastad J 2000 Familial hypercalcemia and hypercalciuria caused by a novel mutation in the cytoplasmic tail of the calcium receptor. J Clin Endocrinol Metab **85:**2042–2047.

Carpten JD, Robbins CM, Villablanca A, Forsberg L, Presciuttini S, Bailey-Wilson J, Simonds WF, Gillanders EM, Kennedy AM, Chen JD, Agarwal SK, Sood R, Jones MP, Moses TY, Haven C, Petillo D, Leotlela PD, Harding B, Cameron D, Pannett AA, Hoog A, Heath H III, James-Newton LA, Robinson B, Zarbo RJ, Cavaco BM, Wassif W, Perrier ND, Rosen IB, Kristoffersson U, Turnpenny PD, Farnebo LO, Besser GM, Jackson CE, Morreau H, Trent JM, Thakker RV, Marx SJ, Teh BT, Larsson C, Hobbs MR 2002 HRPT2, encoding parafibromin, is mutated in HPT-jaw tumor syndrome. Nat Genet **32:**676–680.

El-Hajj Fuleihan G, Brown EM, Heath H III 2002 Familial benign hypocalciuric hypercalcemia and neonatal primary hyperparathyroidism. In: Bilezikian JP, Raisz LG, Rodan GA (eds.) Principles of Bone Biology, 2nd ed. Academic Press, San Diego, CA, USA, pp. 1031–1045.

Kouvaraki MA, Shapiro SE, Perrier ND, Cote GJ, Gagel RF, Hoff AO, Sherman SI, Lee JE, Evans DB 2005 RET proto-oncogene: A review and update of genotype-phenotype correlations in hereditary medullary thyroid cancer and associated endocrine tumors. Thyroid **15:**531–544.

Marx SJ 2005 Molecular genetics of multiple endocrine neoplasia types 1 and 2. Nat Rev Cancer **5:**367–375.

Marx SJ, Simonds WF, Agarwal SK, Burns AL, Weinstein LS, Cochran C, Skarulis MC, Spiegel AM, Libutti SK, Alexander HR Jr, Chen CC, Chang R, Chandrasekharappa SC, Collins FS 2002 Hyperparathyroidism in hereditary syndromes: Special expressions and special managements. J Bone Miner Res **17:**S1;N37–N43.

Pellegata NS, Quintanilla-Martinez L, Siggelkow H, Samson E, Bink K, Höfler H, Fend F, Graw J, Atkinson MJ 2006 Germ-line mutations in p27Kip1 cause a multiple endocrine neoplasia syndrome in rats and humans. Proc Nat Acad Sci USA **103:**15558–15563.

Shattuck TM, Valimaki S, Obara T, Gaz RD, Clark OH, Shoback D, Wierman ME, Tojo K, Robbins CM, Carpten JD, Farnebo LO, Larsson C, Arnold A 2003 Somatic and germ-line mutations of the *HRPT2* gene in sporadic parathyroid carcinoma. N Engl J Med **349:**1722–1729.

Simonds WF, James-Newton LA, Agarwal SK, Yang B, Skarulis MC, Hendy GN, Marx SJ 2002 Familial isolated hyperparathyroidism: Clinical and genetic characteristics of 36 kindreds. Medicine (Baltimore) **81:**1–26.

Simonds WF, Robbins CM, Agarwal SK, Hendy GN, Carpten JD, Marx SJ 2004 Familial isolated hyperparathyroidism is rarely caused by germline mutation in *HRPT2*, the gene for the hyperparathyroidism-jaw tumor syndrome. J Clin Endocrinol Metab **89:**96–102.

Warner J, Nyholt DR, Busfield F, Epstein M, Burgess J, Stranks S, Hill P, Perry-Keene D, Learoyd D, Robinson B, Teh BT, Prins JB, Cardinal JW 2006 Familial isolated hyperparathyroidism is linked to a 1.7 Mb region on chromosome 2p13.3-14 J Med Genet **43:**e12.

© 2008 American Society for Bone and Mineral Research

SECTION VII

Cancer and Bone
(Section Editor: Theresa Guise)

© 2008 American Society for Bone and Mineral Research

Chapter 77. Overview of Mechanisms in Cancer Metastases to Bone

Gregory A. Clines and Theresa A. Guise

Division of Endocrinology and Metabolism, The University of Virginia, Charlottesville, Virginia

INTRODUCTION

The metastasis of cancer cells to the skeleton is a complication of malignancy that disrupts normal bone homeostasis and remodeling, weakens bone, and causes pathological fractures.[1] New mechanisms of bone metastasis have been implicated that involve all steps of metastasis: migration from the primary tumor into the circulation, invasion into a distant site, and propagation within a hospitable environment. This multistep process requires the cooperation of many genes, each individually involved in every step of metastasis. A comprehensive study to identify gene products that enhance breast cancer metastatic potential to bone was reported using a highly osteolytic subpopulation of the breast cancer cell line MDA-MB-231.[2] Eleven genes were identified that have a >4-fold expression pattern in the highly bone metastatic line compared with parental MDA-MB-231. Four of these gene products, interleukin-11 (IL-11), connective tissue growth factor (CTGF), the chemokine receptor CXCR4, and matrix metalloproteinase 1 (MMP-1), were further analyzed. Overexpression of not a single gene but of a combination of two or more in parental MDA-MB-231 enhanced in vivo osteolytic capacity.

CANCER CELL HOMING AND ADHESION TO BONE

New cellular players have also been implicated in cancer cell homing to bone. Platelets can direct cancer cells to bone by "coating" circulating tumor cells, thereby protecting them from immune surveillance and enhancing the ability to adhere to disrupted vascular endothelium.[3] Cancer cells also promote platelet aggregation and the release of platelet lysophosphatidic acid (LPA).[4] LPA receptors on cancer cells enhance cellular proliferation and stimulate the production of osteolytic factors.[4] The stromal cell–derived factor 1a also has a role of homing cancer cells to bone. This ligand is present in tissues that represent common sites of metastasis, including bone marrow,[5–7] and has an affinity for the cancer cell chemokine receptor CXCR4, another important cancer to bone homing factor.[2] The integrin $\alpha v\beta 3$, which binds the RGD peptide sequence found on a variety of extracellular matrix proteins, is important in tumor cell homing.[8,9] In animal models of bone metastasis, avb3 antagonists prevented invasion of cancer cells into bone.[10] Tumor secretion of osteopontin and bone sialoprotein, small integrin-binding proteins, also support metastasis by promoting tumor cell invasion, extracellular matrix degradation, and survival at metastatic sites.[11]

OSTEOLYTIC BONE DISEASE

Once tumor cells arrive in bone, a "vicious cycle" of reciprocal bone–cancer cellular signals regulate osteolytic metastases. The secretion of PTH-related protein (PTHrP) by breast cancer cells into the metastasis microenvironment stimulates osteoclastic bone resorption and release of transforming growth factor-β (TGFβ) from the mineralized bone matrix, which in turn stimulates the cancer cells to produce even more PTHrP.[12,13] TGF-β regulates other pro-metastatic factors that enhance bone metastasis such as IL-6, -8, and -11, TNF-α, vascular endothelial growth factor, and CXCR4, and as such, may represent a central target for bone metastases therapy.[14–16] RUNX2, a transcription factor that specifies osteoblast fate, is also expressed in breast cancer cells and promotes breast cancer cells to function as osteoblast surrogates to support osteoclast formation.[17] Calcium ions may contribute to osteoclastic bone metastasis. Osteoclastic resorption of bone releases high concentrations of ionized calcium during dissolution of the bone. Calcium activates the calcium-sensing receptor (CaSR) expressed by breast cancer cells[18] that regulates tumor secretion of PTHrP, further increasing the osteolytic response to the tumor cells.[19,20]

OSTEOBLASTIC BONE DISEASE

Endothelin 1 (ET-1) is abundantly secreted by prostate cancers and is a principal factor of the osteoblastic response to metastasis.[21] ET-1 stimulates the osteoblast, through the endothelin A receptor (ETAR), to form pathological new bone.[21] ET-1 activates the WNT signaling pathway by reducing autocrine production of the WNT antagonist DKK1.[22] The WNT signaling pathway is a key osteoblast regulatory pathway critical for normal osteoblast differentiation and function.[23] The formation of osteoblastic bone metastasis is likely dually dependent on the downregulation of microenvironment DKK1 secretion from osteoblasts by tumor-produced ET-1 and from prostate cancer cells themselves.[22,24] Although bone morphogenetic proteins (BMPs) have been implicated in osteoblastic disease, BMP-7 has been shown in some studies to prevent bone metastases. This secreted factor inhibits bone metastasis by antagonizing TGF-β signaling and reversing epithelial-to-mesenchymal transition of cancer cells.[25] Platelet-derived growth factor may promote osteoblastic disease as well as invasive and behavior of prostate epithelium.[26]

A puzzling question has been the role of prostate-specific antigen (PSA) in osteoblastic metastasis. PSA is a serine protease that may indirectly activate the osteoblast by processing factors to an active form: PSA cleaves (1) insulin-like growth factor from its binding protein, (2) processes latent TGF-β to the active form, and (3) cleaves PTHrP into fragments that may activate ETAR.[27–29] PSA may have additional direct effects on the osteoblast.[30–32] Despite convincing in vitro evidence that PSA has biological activity contributing to bone metastasis, in vivo studies showing a clear causal relationship are lacking.[33]

CONCLUSION

Metastatic cancer cells with an affinity to bone flourish within this microenvironment. These cells have the ability not only to proliferate in bone but to coax osteoblasts and osteoclasts to produce factors within the bone microenvironment that further stimulate cancer cell growth. Discovery of the molecular mechanisms that control bone metastasis has translated into therapeutic preclinical and clinical trials. Blockade of

Dr. Guise has consulted for Amgen, has received a research grant from Scios, and has served on speakers bureaus for Amgen, Merck, and Novartis. Dr. Clines reports no conflict of interest.

Key words: bone metastasis, breast cancer, prostate cancer, TGF-β, PTH-related peptide, endothelin-1

© 2008 American Society for Bone and Mineral Research

PTHrP and/or TGF-β to reduce bone metastasis has shown promise in preclinical models.[12,17,34–36] In recent clinical trails, ETAR blockade reduces the progression of osteoblastic lesions and may reduce mortality as well.[37,38] Many unanswered questions remain regarding pathophysiology and treatment. Does accelerated bone turnover facilitate tumor growth in bone, as suggested in animal models?[39,40] How do other cell types, such as immune and stem cells, contribute to tumor growth in bone? What controls tumor dormancy in bone? What is the role of the premetastatic niche? Of angiogenesis? What is the benefit of preemptive bone targeted therapies to prevent bone metastasis and the role of combined therapies? Will inhibitors of RANKL and cathepsin K be as effective as bisphosphonates to treat metastatic bone disease? Will these bone resorption inhibitors also be associated with osteonecrosis of the jaw? What are the most effective radiopharmaceutical therapies and pain therapies?

ACKNOWLEDGMENTS

The authors are supported by grants from the NIH (CA118428, CA069158, DK065837, and DK067333), the Department of Defense, The Prostate Cancer Foundation, and The University of Virginia (Cancer Center, Mellon Institute, Gerald D. Aurbach Endowment).

REFERENCES

1. Clines GA, Guise TA 2005 Hypercalcaemia of malignancy and basic research on mechanisms responsible for osteolytic and osteoblastic metastasis to bone. Endocr Relat Cancer **12**:549–583.
2. Kang Y, Siegel PM, Shu W, Drobnjak M, Kakonen SM, Cordon-Cardo C, Guise TA, Massague AJ 2003 A multigenic program mediating breast cancer metastasis to bone. Cancer Cell **3**:537–549.
3. Palumbo JS, Talmage KE, Massari JV, La Jeunesse CM, Flick MJ, Kombrinck KW, Jirouskova M, Degen JL 2005 Platelets and fibrin(ogen) increase metastatic potential by impeding natural killer cell-mediated elimination of tumor cells. Blood **105**:178–185.
4. Boucharaba A, Serre CM, Gres S, Saulnier-Blache JS, Bordet JC, Guglielmi J, Clezardin P, Peyruchaud O 2004 Platelet-derived lysophosphatidic acid supports the progression of osteolytic bone metastases in breast cancer. J Clin Invest **114**:1714–1725.
5. Muller A, Homey B, Soto H, Ge N, Catron D, Buchanan ME, McClanahan T, Murphy E, Yuan W, Wagner SN, Barrera JL, Mohar A, Verastegui E, Zlotnik A 2001 Involvement of chemokine receptors in breast cancer metastasis. Nature **410**:50–56.
6. Sun YX, Schneider A, Jung Y, Wang J, Dai J, Wang J, Cook K, Osman NI, Koh-Paige AJ, Shim H, Pienta KJ, Keller ET, McCauley LK, Taichman RS 2005 Skeletal localization and neutralization of the SDF-1(CXCL12)/CXCR4 axis blocks prostate cancer metastasis and growth in osseous sites in vivo. J Bone Miner Res **20**:318–329.
7. Wang J, Loberg R, Taichman RS 2006 The pivotal role of CXCL12 (SDF-1)/CXCR4 axis in bone metastasis. Cancer Metastasis Rev **25**:573–587.
8. Sung V, Stubbs JT III, Fisher L, Aaron AD, Thompson EW 1998 Bone sialoprotein supports breast cancer cell adhesion proliferation and migration through differential usage of the alpha(v)beta3 and alpha(v)beta5 integrins. J Cell Physiol **176**:482–494.
9. Felding-Habermann B, O'Toole TE, Smith JW, Fransvea E, Ruggeri ZM, Ginsberg MH, Hughes PE, Pampori N, Shattil SJ, Saven A, Mueller BM 2001 Integrin activation controls metastasis in human breast cancer. Proc Natl Acad Sci USA **98**:1853–1858.
10. Clezardin P, Clement-Lacroix P 2006 Prevention of breast cancer bone metastasis by an integrin alpha v beta 3 antagonist. Cancer Treat Rev **32**(Suppl 3):S18.
11. Bellahcene A, Castronovo V, Ogbureke KU, Fisher LW, Fedarko NS 2008 Small integrin-binding ligand N-linked glycoproteins (SIBLINGs): Multifunctional proteins in cancer. Nat Rev Cancer **8**:212–226.
12. Guise TA, Yin JJ, Taylor SD, Kumagai Y, Dallas M, Boyce BF, Yoneda T, Mundy GR 1996 Evidence for a causal role of parathyroid hormone-related protein in the pathogenesis of human breast cancer-mediated osteolysis. J Clin Invest **98**:1544–1549.
13. Yin JJ, Selander K, Chirgwin JM, Dallas M, Grubbs BG, Wieser R, Massague J, Mundy GR, Guise TA 1999 TGF-beta signaling blockade inhibits PTHrP secretion by breast cancer cells and bone metastases development. J Clin Invest **103**:197–206.
14. Kakonen SM, Kang Y, Carreon MR, Niewolna M, Kakonen RS, Chirgwin JM, Massague J, Guise TA 2002 Breast cancer cell lines selected from bone metastases have greater metastatic capacity and express increased vascular endothelial growth factor (VEGF), interleukin-11 (IL-11), and parathyroid hormone-related protein (PTHrP). J Bone Miner Metab **17**(Suppl 1):M060.
15. de la Mata J, Uy HL, Guise TA, Story B, Boyce BF, Mundy GR, Roodman GD 1995 Interleukin-6 enhances hypercalcemia and bone resorption mediated by parathyroid hormone-related protein in vivo. J Clin Invest **95**:2846–2852.
16. Bendre MS, Gaddy-Kurten D, Mon-Foote T, Akel NS, Skinner RA, Nicholas RW, Suva LJ 2002 Expression of interleukin 8 and not parathyroid hormone-related protein by human breast cancer cells correlates with bone metastasis in vivo. Cancer Res **62**:5571–5579.
17. Javed A, Barnes GL, Pratap J, Antkowiak T, Gerstenfeld LC, van Wijnen AJ, Stein JL, Lian JB, Stein GS 2005 Impaired intranuclear trafficking of Runx2 (AML3/CBFA1) transcription factors in breast cancer cells inhibits osteolysis in vivo. Proc Natl Acad Sci USA **102**:1454–1459.
18. Yamaguchi T, Chattopadhyay N, Brown EM 2000 G protein-coupled extracellular Ca2+ (Ca2+o)-sensing receptor (CaR): Roles in cell signaling and control of diverse cellular functions. Adv Pharmacol **47**:209–253.
19. Buchs N, Manen D, Bonjour JP, Rizzoli R 2000 Calcium stimulates parathyroid hormone-related protein production in Leydig tumor cells through a putative cation-sensing mechanism. Eur J Endocrinol **142**:500–505.
20. Sanders JL, Chattopadhyay N, Kifor O, Yamaguchi T, Butters RR, Brown EM 2000 Extracellular calcium-sensing receptor expression and its potential role in regulating parathyroid hormone-related peptide secretion in human breast cancer cell lines. Endocrinology **141**:4357–4364.
21. Yin JJ, Mohammad KS, Kakonen SM, Harris S, Wu-Wong JR, Wessale JL, Padley RJ, Garrett IR, Chirgwin JM, Guise TA 2003 A causal role for endothelin-1 in the pathogenesis of osteoblastic bone metastases. Proc Natl Acad Sci USA **100**:10954–10959.
22. Clines GA, Mohaddad KS, Bao Y, Stephens O, Suva LJ, Shaughnessy JD, Fox JW, Chirgwin JM, Guise TA 2007 Dickkopf homolog 1 mediates endothelin-1-stimulated new bone formation. Mol Endocrinol **22**:486–498.
23. Westendorf JJ, Kahler RA, Schroeder TM 2004 Wnt signaling in osteoblasts and bone diseases. Gene **341**:19–39.
24. Hall CL, Bafico A, Dai J, Aaronson SA, Keller ET 2005 Prostate cancer cells promote osteoblastic bone metastases through Wnts. Cancer Res **65**:7554–7560.
25. Buijs JT, Rentsch CA, van der Horst G, van Overveld PGM, Schwaninger R, Henriquez NV, Papapoulos SE, Pelger RCM, Vukicevic S, Cecchini MG, Lowik CWGM, van der Pluijm G 2007 BMP7, a putative regulator of epithelial homeostasis in the human prostate, is a potent inhibitor of prostate cancer bone metastasis in vivo. Am J Pathol **171**:1047–1057.
26. Yi B, Williams PJ, Niewolna M, Wang Y, Yoneda T 2002 Tumor-derived platelet-derived growth factor-BB plays a critical role in osteosclerotic bone metastasis in an animal model of human breast cancer. Cancer Res **62**:917–923.
27. Fielder PJ, Rosenfeld RG, Graves HC, Grandbois K, Maack CA, Sawamura S, Ogawa Y, Sommer A, Cohen P 1994 Biochemical analysis of prostate specific antigen-proteolyzed insulin-like growth factor binding protein-3. Growth Regul **4**:164–172.
28. Killian CS, Corral DA, Kawinski E, Constantine RI 1993 Mitogenic response of osteoblast cells to prostate-specific antigen suggests an activation of latent TGF-beta and a proteolytic modulation of cell adhesion receptors. Biochem Biophys Res Commun **192**:940–947.
29. Mohammad KS, Guise TA, Chirgwin JM 2003 PTHrP stimulates new bone formation by molecular mimicry of endothelin-1. J Bone Miner Res **18**:S2;S26.
30. Cohen P, Peehl DM, Graves HC, Rosenfeld RG 1994 Biological effects of prostate specific antigen as an insulin-like growth factor binding protein-3 protease. J Endocrinol **142**:407–415.

© 2008 American Society for Bone and Mineral Research

31. Mundy GR 2002 Metastasis to bone: Causes, consequences and therapeutic opportunities. Nat Rev Cancer **2:**584–593.
32. Roodman GD 2004 Mechanisms of bone metastasis. N Engl J Med **350:**1655–1664.
33. Williams SA, Singh P, Isaacs JT, Denmeade SR 2007 Does PSA play a role as a promoting agent during the initiation and/or progression of prostate cancer? Prostate **67:**312–329.
34. Mohammad KS, Stebbins EG, Niewolna M, Mckenna CR, Walton H, Peng XH, Li G, Murphy A, Chakravarty S, Higgins LS, Wong DH, Guise TA 2006 TGFbeta signaling blockade reduces osteolytic bone metastases and enhances bone mass. Cancer Treat Rev **32**(Suppl 3)**:**S29.
35. Javelaud D, Mohammad KS, McKenna CR, Fournier P, Luciani F, Niewolna M, Andre J, Delmas V, Larue L, Guise TA, Mauviel A 2007 Stable overexpression of Smad7 in human melanoma cells impairs bone metastasis. Cancer Res **67:**2317–2324.
36. Sterling JA, Oyajobi BO, Grubbs B, Padalecki SS, Munoz SA, Gupta A, Story B, Zhao M, Mundy GR 2006 The hedgehog signaling molecule Gli2 induces parathyroid hormone-related peptide expression and osteolysis in metastatic human breast cancer cells. Cancer Res **66:**7548–7553.
37. Nelson JB 2005 Endothelin receptor antagonists. World J Urol **23:**19–27.
38. James ND, Borre M, Zonnenberg B, Beuzeboc T, Morris T, Phung D, Dawson N 2007 ZD4054, a potent, specific endothelin A receptor antagonist, improves overall survival in pain-free or mildly symptomatic patients with hormone-resistant prostate cancer (HRPC) and bone metastases ECCO 14, the European Cancer Conference, Barcelona, Spain, September 23–27, 2007.
39. Schneider A, Kalikin LM, Mattos AC, Keller ET, Allen MJ, Pienta KJ, McCauley LK 2005 Bone turnover mediates preferential localization of prostate cancer in the skeleton. Endocrinology **146:**1727–1736.
40. Padalecki SS, Carreon MR, Grubbs BR, Guise TA 2003 Hypogonadism causes bone loss and increased bone metastases in a model of mixed osteolytic/oteoblastic metastases: Prevention by zoledronic acid. J Bone Miner Res **18:**S2;S36.

Chapter 78. Clinical Imaging and Animal Modeling in Osseous Metastatic Disease

Andrew T. Shields,[1] Inna Serganova,[2] Ronald Blasberg,[2] and David A. Mankoff[1]

[1]*University of Washington Medical Center, Seattle, Washington;* [2]*Memorial Sloan-Kettering Cancer Center, New York, New York*

INTRODUCTION

Bone metastases are a common complication certain cancers (lung, breast, prostate, kidney, and thyroid, among others) and not only confer the potential for significant morbidity, such as hypercalcemia, bone pain, spinal cord compression, and increased risk for pathologic fracture, but also significantly influence staging, treatment, and course of the underlying disease. Imaging has a number of roles in the management of patients at risk for osseous metastases, including staging, assessing extent of disease, monitoring response to therapy, assessing for recurrence, evaluating sites of bone pain, and identifying sites at risk for pathologic fracture.

PATHOPHYSIOLOGY

Bone metastases can be broadly categorized as osteolytic, osteoblastic, or mixed (both osteolytic and osteoblastic), based on their localized effect on density of the underlying bone.

Osteolytic lesions are bone metastases resulting in osteoclastic bone resorption with relatively little osteoblastic response, typified by carcinomas of the kidney, lung, breast, gastrointestinal tract, thyroid, and multiple myeloma. In contrast, with osteoblastic (osteosclerotic) metastases, osteoclastic bone resorption is followed by a relatively vigorous osteoblastic response, resulting in localized net formation of new bone of greater density than the normal bone it replaced—most commonly from prostate cancer in men and uterine or ovarian carcinoma in women. Mixed lesions have both osteoblastic and osteolytic components. Although any primary may produce mixed lesions, the primaries that do so most commonly are breast and lung.[1]

Dr. Mankoff has received research support from Pfizer. All other authors state that they have no conflicts of interest.

CLINICAL IMAGING TECHNIQUES

Plain Film Radiography

Plain film radiography is a map of tissue density. Consequently, osteolytic and osteoblastic lesions, which alter bone tissue density, can be identified by this approach. However, an osteolytic lesion must show resorption of at least 30–50% of normal bone to be consistently detectable by radiography.[2] The use of the whole body plain film survey for patients at risk for metastatic disease has long since been supplanted by other techniques that are both more sensitive and more cost effective.[3] An exception is assessment of multiple myeloma, in which the whole body plain film survey is still used—the osteolytic lesions of myeloma are difficult to detect on radionuclide bone scan but have a characteristic appearance on plain film.[4] Nevertheless, radiography can be a useful adjunct to other imaging techniques. For example, lesions identified on radionuclide bone scan that are concerning for but not diagnostic of metastatic disease may be further characterized by radiographic assessment of the area of concern.[5]

Additionally, plain film radiography may play a role in evaluation of new complaints of bone pain in a patient at risk for osseous metastatic disease.

Because of its high resolution, radiography may be used to assess details such as cortical thickness, because risk of pathologic fracture may be increased if a significant amount of the cortex is destroyed by tumor.[4]

Computed Tomography

As with plain films, CT is a map of tissue density, so osteolytic and osteoblastic lesions may be identified by this approach (Fig. 1A). However, the tomographic nature of CT results in an improved target-to-background ratio relative to plain films, improving sensitivity. Whereas the resolution of CT is lower than that of radiography, it is superior to that of radionuclide bone scans and allows measurement of lesion

© 2008 American Society for Bone and Mineral Research

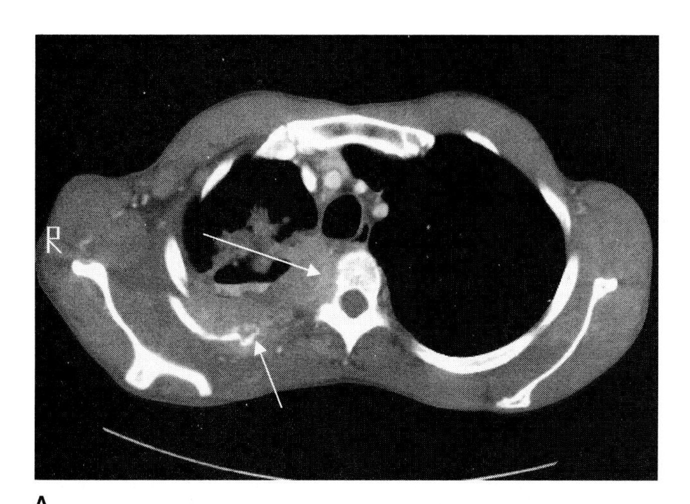

A

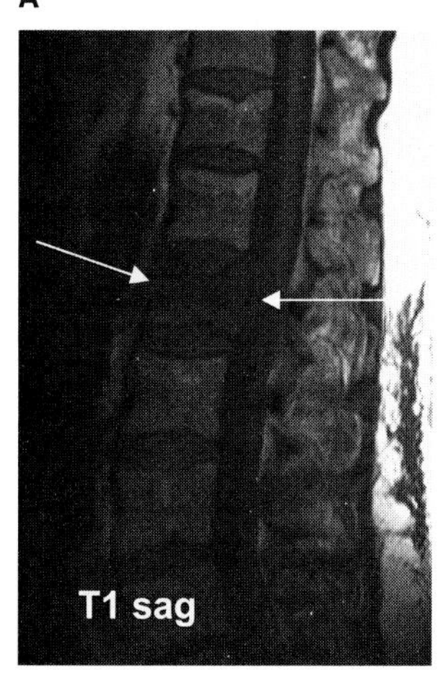

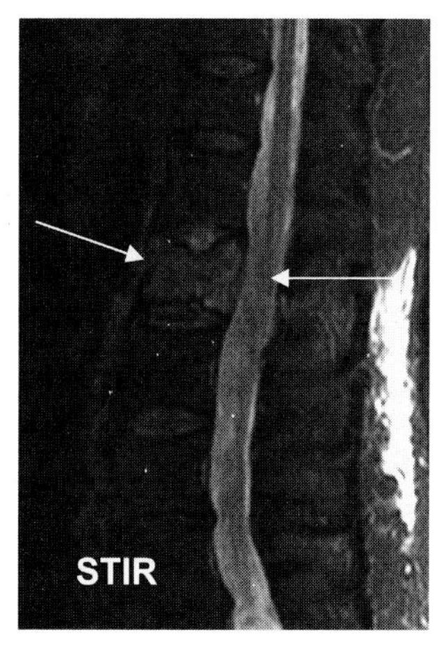

B

FIG. 1. (A) CT showing rib destruction and vertebral invasion by Pancoast tumor (courtesy of Julie Takasugi, MD). (B) MRI of bone metastasis in patient with breast cancer. Note extrinsic compression of thecal sac, which is difficult to image using other modalities (courtesy of Julie Takasugi, MD).

size, assessment of cortical reaction, and evaluation of alterations in adjacent and distant soft tissue.

As with plain films, CT is useful in assessment of new-onset bone pain and for better characterization of lesions deemed suspicious on radionuclide bone scan.

However, it is difficult with CT to perform whole body scanning. Additionally, concerns have been raised regarding the substantial increase in the utilization of CT in recent years and the possibility of increased risks of developing cancer caused by the attendant radiation exposure.[6]

Magnetic Resonance Imaging

MRI, which provides tomographic images related to the concentration and mobility characteristics of hydrogen atoms, is not reliant solely on alterations in tissue density and is thus capable of detecting metastases that have infiltrated the bone marrow before they provoke an osseous response. As a result, MRI is more sensitive than plain films, CT, or radionuclide bone scans in detection of early lesions and marrow-based lesions[7] (Fig. 1B).

The role of MRI in management of osseous metastases has been limited because of, among other factors, cost, availability, and absence of whole body scanning capability. Newly developed MRI scanners with whole body capability may help address this last concern.[8]

Radionuclide Bone Scans

Radionuclide tracer techniques can detect nanomolar—and in some cases, picomolar—concentrations of administered imaging agents, providing the ability to probe tumor biology without disturbing it.[9] Radionuclide bone scintigraphy involves intravenous administration of trace amounts of a radioactive bone-seeking agent (typically technetium-99m methylene diphosphonate [Tc-99m MDP]), followed 3–4 h later by single-photon γ camera imaging. Osteoblastic metastases, which are characterized by locally increased bone formation and blood flow, concentrate greater amounts Tc-99m MDP than does normal adjacent bone and thus appear "hot" on bone scan.

Bone scans do not image the metastasis itself. Rather, they image the reparative response to bone damage done by the metastasis. If the metastasis does not provoke an osseous response, the bone scan will be negative. Osteolytic lesions, which cause a localized disappearance of bone but produce little reparative response, are difficult to detect on bone scan. Therefore, for tumors displaying predominantly osteolytic lesions (such as thyroid, kidney, or multiple myeloma), radionuclide bone scans are relatively insensitive.

Advantages of radionuclide bone scans include high sensitivity to osteoblastic lesions and the capability of whole body imaging. However, because there are many osseous abnormalities that can increase bone formation and blood flow (e.g., fracture, osteomyelitis, postsurgical change, degenerative change, Paget's), bone scans are often nonspecific. Consequently, radionuclide bone scans are frequently used to identify areas of concern, with follow-up utilization of higher-resolution anatomic imaging (e.g., CT or plain film) for better characterization.

Positron Emission Tomography

Positron emission tomography (PET) is a specialized form of radionuclide imaging that offers a wide range of possible radiopharmaceuticals, higher spatial resolution than conventional (single-photon) radionuclide imaging, and, importantly, the ability to accurately quantify regional radiopharmaceutical concentration. The most relevant radiopharmaceuticals for clinical bone imaging are ^{18}F-sodium fluoride and ^{18}F-fluorodeoxyglucose (FDG).

^{18}F-Sodium Fluoride

^{18}F-sodium fluoride, as fluoride ion, is incorporated into newly formed bone as part of the bone mineralization process, behaving in a fashion chemically identical to native, nonradioactive fluoride[10] and in a fashion similar to the bone scanning agent Tc-99m MDP. However, unlike Tc-99m MDP, ^{18}F-fluoride—being a PET tracer—can provide quantitative physiologic information. For example, compartmental models of fluoride kinetics derived from fluoride PET imaging have been used to estimate regional skeletal blood flow[11] and rate of bone mineralization[12] in normal bone,[13–15] osteoporotic bone, treated osteoporotic bone,[16] bone allografts,[17] Paget's disease,[18] and bone metastases.[19]

Although the higher contrast and superior spatial resolution of fluoride PET relative to Tc-99m bone scanning may result in improved sensitivity for detection of metastatic lesions,[20] detecting a greater number of lesions may affect clinical management in only a minority (10% in one study[21]) of patients.

© 2008 American Society for Bone and Mineral Research

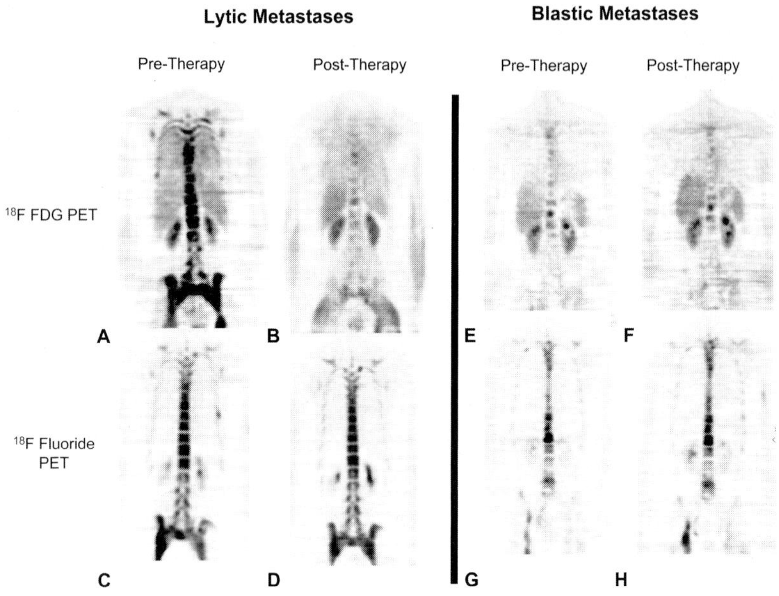

FIG. 2. (Left) Patient with extensive lytic metastases of the spine and pelvis. FDG PET shows hypermetabolic tumor (A) showing near-complete resolution with chemotherapy (B). FDG also identifies a greater volume of lytic metastases in this patient than does fluoride PET. Fluoride PET (C) shows mild improvement (D) but also indicates persistence of increased bone turnover, which is not necessarily reflective of residual viable tumor.. (Right) Patient with blastic metastases of the spine and pelvis. Because blastic metastases increased bone turnover but are not necessarily hypermetabolic, in this case, fluoride PET identifies a greater volume of disease (G) than does FDG (E). Fluoride PET also shows progression of disease (H), which is more difficult to identify with this patient on FDG imaging (F). Clinically, the patient on the left improved, whereas the patient on the right progressed.

Using fluoride PET/CT to measure both bone formation (using the "PET" aspect of PET/CT) and bone structure (using the "CT" aspect) offers an intriguing possibility for both detecting and characterizing bone lesions and looks promising in early studies.[22]

^{18}F-fluoride imaging offers the potential to quantify regional uptake in serial studies as a measure of response to therapy. How best to apply fluoride PET to monitoring therapeutic response of bone metastases remains an active and important area of investigation.

Fluorodeoxyglucose

FDG has proven extremely useful in oncology because the rate of FDG uptake and trapping is a quantitative indicator of the rate of glucose metabolism.[23] However, the rate of glucose metabolism may not be the only factor causing increased FDG uptake in tumors—increased uptake may also be associated with tumor biologic features consistent with dedifferentiation and clinically aggressive behavior.[24,25]

For primarily lytic lesions, such as lung cancer and multiple myeloma, FDG PET is quite accurate and is increasingly used in the routine staging of patients with these diseases in place of the relatively insensitive (for lytic lesions) Tc-99m MDP bone scanning.[26] For cancers with primarily blastic (sclerotic) bone metastases such as prostate cancer, FDG PET is less useful and is not routinely used in clinic practice. For tumors with both lytic and blastic metastases, such as breast cancer, FDG PET is complementary to bone scanning and potentially may detect many of the lesions missed by Tc-99m MDP bone scintigraphy. FDG PET may be particularly helpful for special situations where conventional imaging may miss disease, such as diffuse marrow infiltration.[27]

Whereas ^{18}F-fluoride PET and Tc99m bone scintigraphy reflect the bony reaction to tumor metastases, FDG PET reflects the metabolism of the tumor itself. Consequently, FDG PET may be particularly helpful for monitoring response to therapy, especially for lytic lesions for which bone scintigraphy is not effective in therapeutic monitoring.[28] A recent study of FDG PET/CT in patients with treated bone metastases suggests that the combination of functional and anatomic imaging provided by FDG PET/CT may be particularly helpful in response evaluation.[29] Preliminary studies also suggest that selective use of FDG and fluoride PET may be useful for measuring response in a broad range of bone metastasis phenotypes[19] (Fig. 2).

A number of tracers beyond fluoride and FDG have been tested for cancer imaging,[30] including ^{18}F-fluoroestradiol (FES) for estrogen receptors, ^{11}C-thymidine and ^{18}F-fluorothymidine (FLT) for cellular proliferation, ^{11}C-acetate for lipid biosynthesis, and ^{18}F-choline, which has high uptake in prostate cancer. One intriguing possibility is the use of PET tumor receptor imaging to examine receptor expression in bone metastases, where sampling and tissue analysis can be challenging. Preliminary studies of radiopharmaceuticals specific to tumor neo-vessels have shown promise for imaging tumor angiogenesis[31] and may be particularly helpful in bone metastases, where angiogenesis seems to play an important role.

ANIMAL MODELS FOR IMAGING OSSEOUS METASTATIC DISEASE

The metastatic process involves a series of critical steps as cells from the primary tumor migrate, colonize, and grow. To study these steps in greater detail, various animal imaging strategies have been developed, including fluorescence, bioluminescence, and animal-tailored versions of modalities used in patient care (CT, PET, and MRI).

Fluorescence Imaging

Fluorescence imaging measures light emitted from molecular probes and has been used widely, including for assessment of tumor cell arrival, localization, and colonization. For example, fluorescent imaging techniques have shown that, 1 h after intracardiac injection of breast tumor cells into athymic mice, single tumor cells are found in distal metaphyses; that by 1 wk, these single cells form numerous 2–10 cell foci; that by 2 wk, they form fewer but larger (≥50 cells) foci; and that by 4 wk, they generally form a single large mass.[32]

Fluorescence imaging combined with laser-scanning microscopy has been used in living tissues to shed light on the migration characteristics of metastatic cells escaping from primary tumors. For example, this imaging technique has shown in vivo that certain carcinoma cells migrate under a chemotac-

© 2008 American Society for Bone and Mineral Research

tic signal toward blood vessels, with proximity to blood vessels causing rounded, nonpolarized cells to become highly polarized—suggesting a process that could increase efficiency of cell migration out of the primary tumor mass and into capillaries.[33]

Bioluminescence Imaging

Bioluminescence imaging (BLI) involves the detection of visible light produced during enzymatic (luciferase) oxidation of a specific substrate (luciferin), and has been used to assess the growth of metastases in a number of animal models.[34–36] For example, BLI coupled with fluorescence microscopy has helped elucidate tropism and tumor growth kinetics of individual single cell progenies, showing that some single cell progeny metastasize efficiently to bone, others better to lung, and still others to the adrenal gland, reflecting the organ-specific distribution of breast cancer metastases observed in patients.[37]

Animal Versions of Clinical Modalities—CT, PET, and MRI

High-resolution μCT is a powerful tool for imaging and quantifying bone morphometric and structural changes and can be combined with other techniques to obtain physiologic as well as anatomic information. For example, μCT has been combined in vivo with BLI to assess the kinetics of growth and appearance of bone lesions in intraosseous prostate cancer in mice, showing that tumor growth assessed by BLI was highly correlated with osteolysis assessed by μCT and that μCT can be used to assess events early in the metastatic process.[38]

μPET using ^{18}F-fluoride has been used to assess malignant changes in the mouse skeleton.[39] For example, tumors resulting from tibial injection of a prostate cancer cell line in mice readily show high ^{18}F-fluoride uptake on μPET scans, indicating very high bone turnover, with osteoblastic lesions ~2 mm diameter clearly visible.

Combining small-animal PET/CT and BLI technology has yielded fusion images containing both molecular and anatomic information in a mouse metastatic melanoma model and confirmed co-registration of metastatic sites in PET and CT images of a melanoma cell line a mouse xenograft model.[40]

MRI is commonly used to track tumor size, location, and metastatic burden because of its ability to produce high-resolution tomographic images of entire organ systems without the use of ionizing radiation.[41] MRI contrast agents, such as superparamagnetic iron oxide nanoparticles and micron-sized iron oxide particles,[42] have been used to tag cell populations to allow their detection over time in animals and humans. A multimodality approach—for example, a combination of MRI and BLI, which has been used in the noninvasive assessment of bone precursor cell biology in vivo—may make it possible to track and determine the time course of organ distribution and gene expression of a single metastatic cell.[43]

SUMMARY

Appropriate use of imaging modalities, alone or in combination, can optimize assessment of patients at risk for bone metastasis. Exciting new work in PET and animal modeling deepens our understanding of this common clinical condition and points to potential new directions for diagnosis and treatment of patients with osseous metastatic disease.

REFERENCES

1. Greenspan A, Jundt G, Remagen W 2006 Metastases. In: Differential Diagnosis in Orthopaedic Oncology, 2nd edition, Lippincott Williams & Wilkins, Philadelphia PA, USA, pp.458–480.
2. Bunker SR, Kleiner BD 1988 Bone imaging. In: Harrington KD (ed.) Orthopedic Management of Metastatic Bone Disease. Mosby, St. Louis, MO, USA, pp.35–52.
3. Mall JC, Bekerman C, Hoffer PB, Gottschalk A 1976 A unified radiological approach to the detection of skeletal metastases. Radiology **118:**323–329.
4. Rosenthal DI 1997 Radiologic diagnosis of bone metastases. Cancer **80**(8 Suppl)**:**1595–1607.
5. Helms C 1989 Fundametals of Skeletal Radiology, 1st ed. WB Saunders, Philadelphia, PA, USA.
6. Brenner DJ, Hall EJ 2007 Computed tomography—an increasing source of radiation exposure. N Engl J Med **357:**2277.
7. Taoka T, Mayr NA, Lee HJ, Yuh WTC, Simonson TM, Rezai K, Berbaum KS 2001 Factors influencing visualization of vertebral metastases on MR imaging versus bone scintigraphy. AJR Am J Roentgenol **176:**1525–1530.
8. Schmidt GP, Kramer H, Reiser MF, Glaser C 2007 Whole-body magnetic resonance imaging and positron emission tomography—computed tomography in oncology. Top Magn Reson Imaging **18:**193–202.
9. Tewson T, Krohn K 1998 PET radiopharmeceuticals: State-of-the-art and future prospects. Semin Nucl Med **28:**221–234.
10. Even-Sapir E 2005 Imaging of malignant bone involvement by morphologic, scintigraphic, and hybrid modalities. J Nucl Med **46:**1356–1367.
11. Frost ML, Cook GJ, Blake GM, Marsden PK, Fogelman I 2007 The relationship between regional bone turnover measured using ^{18}F-fluoride positron emission tomography and changes in BMD is equivalent to that seen for biochemical markers of bone turnover. J Clin Densitom **10:**46–54.
12. Hawkins RA, Choi Y, Huang SC, Hoh CK, Dahlbom M, Schiepers C, Satyamurthy N, Barrio JR, Phelps ME 1992 Evaluation of the skeletal kinetics of fluorine-18-fluoride ion with PET. J Nucl Med **33:**633–642.
13. Blake GM, Park-Holohan SJ, Cook GJ, Fogelman I 2001 Quantitative studies of bone with the use of ^{18}F-fluoride and 99mTc-methylene diphosphonate. Semin Nucl Med **31:**28–49.
14. Cook GJ, Lodge MA, Blake GM, Marsden PK, Fogelman I 2000 Differences in skeletal kinetics between vertebral and humeral bone measured by ^{18}F-fluoride positron emission tomography in postmenopausal women. J Bone Miner Res **15:**763–769.
15. Schiepers C, Nuyts J, Bormans G, Dequeker J, Bouillon R, Mortelmans L, Verbruggen A, De Roo M 1997 Fluoride kinetics of the axial skeleton measured in vivo with fluorine-18-fluoride PET. J Nucl Med **38:**1970–1976.
16. Frost ML, Cook GJ, Blake GM, Marsden PK, Benatar NA, Fogelman I 2003 A prospective study of risedronate on regional bone metabolism and blood flow at the lumbar spine measured by ^{18}F-fluoride positron emission tomography. J Bone Miner Res **18:**2215–2222.
17. Brenner W, Vernon C, Conrad EU, Eary JF 2004 Assessment of the metabolic activity of bone grafts with (18)F-fluoride PET. Eur J Nucl Med Mol Imaging **31:**1291–1298.
18. Installe J, Nzeusseu A, Bol A, Depresseux G, Devogelaer JP, Lonneux M 2005 (18)F-fluoride PET for monitoring therapeutic response in Paget's disease of bone. J Nucl Med **46:**1650–1658.
19. Gralow JR, Brenner W, Linden HM, Livingston RB, Ellis GK, Schubert EK, Peterson LM, Muzi MS, Mankoff DA 2005 Changes in tumor metabolism and local bone turnover in patients treated for bone-dominant metastatic breast cancer measured by fluorodeoxyglucose (FDG) and fluoride positron emission tomography (PET). Breast Cancer Res Treat **94**(suppl 1):261.
20. Even-Sapir E, Mishani E, Flusser G, Metser U 2007 ^{18}F-Fluoride positron emission tomography and positron emission tomography/computed tomography. Semin Nucl Med **37:**462–469.
21. Schirrmeister H, Glatting G, Hetzel J, Nüssle K, Arslandemir C, Buck AK, Dziuk K, Gabelmann A, Reske SN, Hetzel M 2001 Prospective evaluation of the clinical value of planar bone scans, SPECT, and (18)F-labeled NaF PET in newly diagnosed lung cancer. J Nucl Med **42:**1800–1804.
22. Even-Sapir E, Metser U, Flusser G, Zuriel L, Kollender Y, Lerman H, Lievshitz G, Ron I, Mishani E 2004 Assessment of malignant skeletal disease: Initial experience with 18F-fluoride PET/CT and comparison between ^{18}F-fluoride PET and ^{18}F-fluoride PET/CT. J Nucl Med **45:**272–278.
23. Mankoff DA, Muzi M, Zabib H 2004 Quantitative analysis of

© 2008 American Society for Bone and Mineral Research

nuclear oncologic images. In: Zabib H (ed.) Qunatitative Analysis of Nuclear Medicine Images. Springer, Hingham, MA, USA, pp. 494–536.
24. Spence AM, Muzi M, Graham MM, O'Sullivan F, Krohn KA, Link JM, Lewellen TK, Lewellen B, Freeman SD, Berger MS, Ojemann GA 1998 Glucose metabolism in human malignant gliomas measured quantitatively with PET, 1-[C-11]glucose and FDG: Analysis of the FDG lumped constant. J Nucl Med **39**:440–448.
25. Bos R, van Der Hoeven JJ, van Der Wall E, van Der Groep P, van Diest PJ, Comans EF, Joshi U, Semenza GL, Hoekstra OS, Lammertsma AA, Molthoff CF 2002 Biologic correlates of (18)fluorodeoxyglucose uptake in human breast cancer measured by positron emission tomography. J Clin Oncol **20**:379–387.
26. Schirrmeister H, Arslandemir C, Glatting G, Mayer-Steinacker R, Bommer M, Dreinhöfer K, Buck A, Hetzel M 2004 Omission of bone scanning according to staging guidelines leads to futile therapy in non-small cell lung cancer. Eur J Nucl Med Mol Imaging **31**:964–968.
27. Aydin A, Yu JQ, Zhuang H, Alavi A 2005 Detection of bone marrow metastases by FDG-PET and missed by bone scintigraphy in widespread melanoma. Clin Nucl Med **30**:606–607.
28. Clamp A, Danson S, Nguyen H, Cole D, Clemons M 2004 Assessment of therapeutic response in patients with metastatic bone disease. Lancet Oncol **5**:607–616.
29. Du Y, Cullum I, Illidge TM, Ell PJ 2007 Fusion of metabolic function and morphology: Sequential [18F]fluorodeoxyglucose positron-emission tomography/computed tomography studies yield new insights into the natural history of bone metastases in breast cancer. J Clin Oncol **25**:3440–3447.
30. Kelloff GJ, Krohn KA, Larson SM, Weissleder R, Mankoff DA, Hoffman JM, Link JM, Guyton KZ, Eckelman WC, Scher HI, O'Shaughnessy J, Cheson BD, Sigman CC, Tatum JL, Mills GQ, Sullivan DC, Woodcock J 2005 The progress and promise of molecular imaging probes in oncologic drug development. Clin Cancer Res **11**:7967–7985.
31. Beer AJ, Haubner R, Sarbia M, Goebel M, Luderschmidt S, Grosu AL, Schnell O, Niemeyer M, Kessler H, Wester HJ, Weber WA, Schwaiger M 2006 Positron emission tomography using [18F]Galacto-RGD identifies the level of integrin alpha(v)beta3 expression in man. Clin Cancer Res **12**:3942–3949.
32. Phadke PA, Mercer RR, Harms JF, Jia Y, Frost AR, Jewell JL, Bussard KM, Nelson S, Moore C, Kappes JC, Gay CV, Mastro AM, Welch DR 2006 Kinetics of metastatic breast cancer cell trafficking in bone. Clin Cancer Res **12**:1431–1440.
33. Wyckoff JB, Jones JG, Condeelis JS, Segall JE 2000 A critical step in metastasis: In vivo analysis of intravasation at the primary tumor. Cancer Res **60**:2504–2511.
34. de Wet JR, Wood KV, DeLuca M, Helinski DR, Subramani S 1987 Firefly luciferase gene: Structure and expression in mammalian cells. Mol Cell Biol **7**:725–737.
35. Contag CH, Spilman SD, Contag PR, Oshiro M, Eames B, Dennery P, Stevenson DK, Benaron DA 1997 Visualizing gene expression in living mammals using a bioluminescent reporter. Photochem Photobiol **66**:523–531.
36. Contag PR, Olomu IN, Stevenson DK, Contag CH 1998 Bioluminescent indicators in living mammals. Nat Med **4**:245–247.
37. Minn AJ, Kang Y, Serganova I, Gupta GP, Giri DD, Doubrovin M, Ponomarev V, Gerald WL, Blasberg R, Massagué J 2005 Distinct organ-specific metastatic potential of individual breast cancer cells and primary tumors. J Clin Invest **115**:44–55.
38. Fritz V, Louis-Plence P, Apparailly F, Noël D, Voide R, Pillon A, Nicolas JC, Müller R, Jorgensen C 2007 Micro-CT combined with bioluminescence imaging: A dynamic approach to detect early tumor-bone interaction in a tumor osteolysis murine model. Bone **40**:1032–1040.
39. Berger F, Lee YP, Loening AM, Chatziioannou A, Freedland SJ, Leahy R, Lieberman JR, Belldegrun AS, Sawyers CL, Gambhir SS 2002 Whole-body skeletal imaging in mice utilizing microPET: Optimization of reproducibility and applications in animal models of bone disease. Eur J Nucl Med Mol Imaging **29**:1225–1236.
40. Deroose CM, De A, Loening AM, Chow PL, Ray P, Chatziioannou AF, Gambhir SS 2007 Multimodality imaging of tumor xenografts and metastases in mice with combined small-animal PET, small-animal CT, and bioluminescence imaging. J Nucl Med **48**:295–303.
41. Weissleder R, Ntziachristos V 2003 Shedding light onto live molecular targets. Nat Med **9**:123–128.
42. Hinds KA, Hill JM, Shapiro EM, Laukkanen MO, Silva AC, Combs CA, Varney TR, Balaban RS, Koretsky AP, Dunbar CE 2003 Highly efficient endosomal labeling of progenitor and stem cells with large magnetic particles allows magnetic resonance imaging of single cells. Blood **102**:867–872.
43. Mayer-Kuckuk P, Gade TP, Buchanan IM, Doubrovin M, Ageyeva L, Bertino JR, Boskey AL, Blasberg RG, Koutcher JA, Banerjee D 2005 High-resolution imaging of bone precursor cells within the intact bone marrow cavity of living mice. Mol Ther **12**:33–41.

Chapter 79. Metastatic Solid Tumors to Bone

Gregory R. Mundy and Julie A. Sterling

Vanderbilt Center for Bone Biology, Vanderbilt University, Nashville, Tennessee

IMPORTANCE OF THE PROBLEM

Metastatic bone disease is a common manifestation of advanced disease in patients with the most frequent malignancies such as breast cancer, prostate cancer, and lung cancer. Eighty percent of patients with osteolytic bone disease have breast or prostate cancer, the most common malignancies of women and men, respectively. There are 250,000 people who die of breast cancer and 100,000 from prostate cancer each year worldwide. In most of these patients, the bulk of the tumor burden is likely present in bone by the time the patients have advanced disease, because the bone marrow provides an enormous reservoir for tumor cells. Approximately 75–95% of patients with advanced prostate or breast cancer have bone metastases, the frequency depending on the sensitivity and precision of the techniques used for detection. The lifetime risk of developing a bone metastasis is similar to that of developing a hip fracture. One important consequence of bone metastasis is that, once the tumor is present in bone, the cancer is no longer curable. This is particularly important because patients are living longer with these common cancers, and the skeletal manifestations associated with metastatic bone disease will likely markedly affect their quality of life.

SPECTRUM OF BONE METASTASES

Bone metastases are most often destructive or osteolytic, although in some, there is a major osteoblastic (or bone form-

The authors state that they have no conflicts of interest.

Key words: osteolytic tumors, osteoblastic tumors, and metastasis to bone

Electronic Databases: 1. PubMed—www.bonekey-ibms.org. 2. BoneKEy—www.ncbi.nlm.nih.gov/sites/entrez.

© 2008 American Society for Bone and Mineral Research

ing) element. For purposes of simplicity, most oncologists classify metastases into osteolytic, osteoblastic, or mixed. In fact, it is probable that bone metastases represent a spectrum of bone cell responses to adjacent tumor cells, rather than distinct entities. In all types of metastases, a major component of the metastatic process is an increase in osteoclastic bone resorption, which is evidenced by marked increases in markers of bone resorption and by beneficial responses of all types of bone metastases to bisphosphonates, drugs that target osteoclastic bone resorption.

CONCEPT OF THE VICIOUS CYCLE

At the metastatic site in bone, a vicious cycle exists between tumor cells and bone-resorbing osteoclasts. This was first shown when bisphosphonates were found to reduce osteoclastic bone resorption at metastatic sites and, in parallel, tumor burden. Later it was found that whenever osteoclastic bone resorption was reduced at metastatic sites, and by whatever mechanism, tumor burden was also decreased. For example, when antisera to PTH-related peptide (PTHrP)[1] or drugs that block PTHrP transcription[2] were used in a mouse model of bone metastases, tumor burden decreased. This is because of a bidirectional interaction between tumor cells and osteoclasts at the metastatic site, so that when osteoclastic bone resorption is inhibited, there is a parallel concomitant decrease in tumor burden.

Later, the molecular mechanisms were determined in specific preclinical models. In some models of human breast cancer, the mediator of osteoclastic bone resorption is PTHrP (through RANKL),[1] although as noted below, other mediators have also been described. The resorbing bone releases TGFβ in active form, which is in turn responsible for stimulating PTHrP expression by the tumor cells.

As research in this area advances, further molecular mechanisms will be added to this basic conceptual framework. It is also now apparent that many host cells in addition to osteoclasts may have important roles in the vicious cycle in the bone metastatic site, including osteoblasts, hematopoietic progenitors, T and B cells, and fibroblasts (Fig. 1).

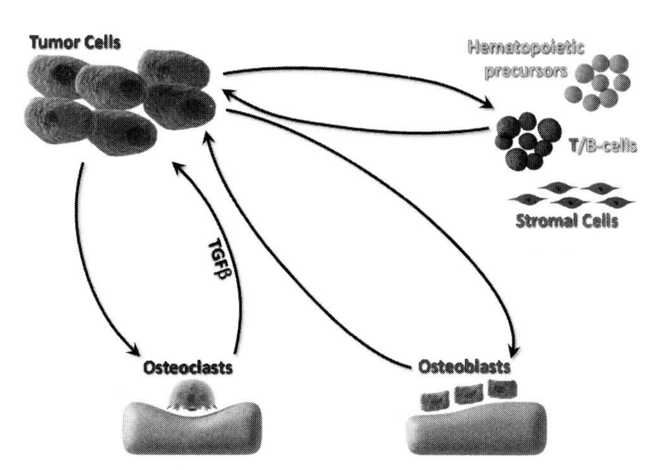

FIG. 1. The vicious cycle of bone destruction once tumor cells metastasize to bone was originally described as an interaction between osteoclasts and tumor cells. As research has progressed, it has become clear that, in addition to interacting with osteoclasts, that tumor cells interact with many cell types in the bone marrow to cause either an osteolytic or osteoblastic response.

CELLULAR MECHANISMS OF OSTEOLYTIC BONE METASTASIS

Most evidence indicates that the cellular mechanism responsible for bone destruction in patients with metastatic bone disease is osteoclastic. Osteoclasts are invariably found at tumor–bone interfaces with detailed searches. The osteoclasts seem to form resorption lacunae in a different manner than they do in other types of bone diseases. Using sophisticated scanning electron microscopic approaches, Boyde et al.[3] have found that osteoclasts cause deep resorption bays resembling inverted ice cream cones, which is different from what happens in diseases such as primary hyperparathyroidism, where the resorption defects are found to be more shallow and which they likens to "shallow licks."

Tumor cells that have become established in the bone microenvironment secrete factors that stimulate osteoclastic bone resorption. In some models of breast cancer metastasis to bone, PTHrP is the major factor used to monitor the ability of a tumor to induce osteolytic metastases.[1] PTHrP is regulated by TGFβ released from the bone matrix as tumor cells degrade the surrounding bone.[4] PTHrP mediates bone resorption by stimulating osteoblast expression of RANKL.[5–7] RANKL binds to and activates its receptor RANK expressed on osteoclasts to mediate bone resorption.

We have recently shown that the Hedgehog signaling pathway controls PTHrP expression in breast cancer cells through its intracellular mediator Gli2, which is also an important mediator of Hedgehog signaling during development.[8] It is not understood why tumor cells express developmental factors such as members of the Hedgehog signaling pathway and PTHrP, but this may represent the common capacity of tumors to express latent or dormant developmental pathways.

Whereas our studies have indicated that PTHrP is the major regulator of osteolytic bone destruction, we have delineated many other molecular mechanisms responsible for osteolytic bone metastasis models using in vivo approaches. For example, when tissue inhibitor of MMP2 (TIMP2) was overexpressed in MDA-231 cells, osteolysis was significantly reduced.[9] Forced expression of E-cadherin in cells that do not normally express this molecule significantly reduces bone destruction.[10] Consistent with these findings, Yang et al.[11] later showed that the transcription factor Twist caused a decrease in E-cadherin expression with activation of mesenchymal markers and less cell–cell adhesion and greater capacity for metastasis. In contrast, suppression of Twist decreased metastatic potential. Finally, we showed that laminin antagonists are effective inhibitors of osteolytic bone metastases in a murine model of bone metastasis caused by melanoma.[12]

Other groups have studied various factors produced by tumor cells that induce osteolysis, although currently the best evidence is for PTHrP. These other factors include IL-8, Runx2, and macrophage colony-stimulating factor (M-CSF). IL-8 has been shown to induce osteolysis and is expressed by certain cancer cells. Several studies have shown a correlation in some human breast tumors between IL-8 expression and increased incidence of bone metastases in patients.[13–15] Some breast cancer cells also secrete M-CSF, a factor that is expressed at higher concentrations in patients with breast and prostate tumor metastases to bone than in other patients and that can stimulate RANKL expression.[16,17] In addition, Runx2, a necessary molecule for normal bone formation and development, is expressed in breast cancer cells, and inhibition of Runx2 expression by tumor cells decreases the capacity of tumor cells to induce osteolysis,[18,19] in part through the capacity of Runx2 to regulate tumor-derived MMP-9.[20] Similarly, Lynch et al.[21] found in PC3 cells after intratibial injection that osteoclast-derived MMP7 was involved in osteolysis caused by PC3 cells and may work at least in part by cleaving

© 2008 American Society for Bone and Mineral Research

RANKL. These studies all show the possible diversity of mechanisms responsible for tumor-induced bone destruction.

OSTEOBLAST IMPAIRMENT IN OSTEOLYTIC METASTASES

Osteoclasts are inhibited specifically by bisphosphonates and denosumab (a RANKL inhibitor). These agents act on osteoclasts and have no direct well-documented effects on osteoblasts. This results in limiting the extent of osteolysis but does not repair the destroyed bone, suggesting that osteoblasts are also inhibited in osteolytic metastases. In the majority of patients, osteoblast activity is diminished and osteoblast differentiation is impaired.[22] The impairment of osteoblast differentiation seen in most patients has remained unexplained, because the accompanying increase in osteoclast activity caused by the presence of the tumor cells would be expected to increase osteoblast differentiation. However, very recent data suggest that the impairment in osteoblast differentiation may be (1) extremely important in influencing tumor burden at that site and (2) is dependent on TGFβ, because maneuvers that impair TGFβ signaling enhance osteoblast differentiation and reduce tumor burden.[23–27] Thus, the state of osteoblast differentiation at the metastatic site seems to be dependent on ambient TGFβ signaling.

OSTEOBLASTIC METASTASES

In some cancers, most notably prostate and ~20% of breast cancer metastases, the metastatic lesion in bone is predominantly osteoblastic. The pathophysiology is still not thoroughly understood. There seems to be a strong resorption component, because bone resorption markers are usually markedly increased, and bisphosphonates reduce the bone pain and other skeletal-related events.

The murine models of osteoblastic metastases are not as well studied as the mechanisms responsible for osteolytic metastases, and there is no currently available ideal model of osteoblastic metastasis caused by prostate cancer. Osteoblastic metastases have been best studied preclinically in murine models of human breast cancer. Yin et al.[28] showed that endothelin-1 (ET-1) was responsible in one murine model of human breast cancer, and Yi et al.[29] showed that platelet-derived growth factor-BB (PDGF-BB) was the mediator in a separate model of breast cancer metastasis to bone. In addition, the Wnt signaling pathway may play an important role in the osteoblastic response and in the transition between the lytic and blastic phases. For example, mixed prostate lesions initially produce the Wnt inhibitor DKK-1, which is decreased as the tumors transition into an osteoblastic phase.[30,31] Furthermore, ET-1 can regulate the expression of DKK-1, suggesting a potentially important relationship between the Wnt signaling pathway and ET-1 expression.[31]

CONCEPT OF THE PREMETASTATIC NICHE

Recently, much interest has been focused on the "premetastatic niche" formed by non-neoplastic host cells at the metastatic site before arrival of the metastasizing cancer cells. The work of Kaplan et al.[32] suggests that the presence of tumor cells in the primary site leads to a series of early host cellular and molecular events in the bone marrow that prepares a niche for the later homing and metastasis of the cancer cells from the primary site. They propose that these host events and the formation of this premetastatic niche determine the site specificity of metastasis. The host cells that have been most extensively studied in the formation of the premetastatic niche include hematopoietic progenitor cells, cancer-associated fibroblasts, and endothelial cells, but also likely include osteoclasts and their progenitors, T and B cells, and osteoblasts.[33] Osteoclast precursors responsible ultimately for the osteolytic lesion are likely derived from cells of the premetastatic niche.

PRECLINICAL MODELS OF BONE METASTASES

The interest in bone metastases over the past 15 yr has led to exploration of a number of preclinical in vivo approaches to understand the mechanisms responsible. The in vivo technique most widely used is the intracardiac injection method, which was pioneered by Arguello et al. [34] and then later by Nakai et al.[12] This has proven to be a reliable approach to identifying mechanisms involved in the metastatic process and for the identification of antagonists or inhibitors of the process. Other approaches have been used less frequently and include orthotopic inoculation[35–37] and spontaneous development of tumors such as the PyMT tumor. Intratibial injections of tumor cells have also been used to mimic the metastatic process.[38] We have some reservations about this approach, but it may be the only useful technique available to answer some specific questions. Our reservations are that it is impossible not to cause injury using this technique,[39] and the subsequent repair process will be superimposed on the effects of the tumor. Second, the tumor cells are inoculated as a bolus, which is very different from what happens during metastasis, where the tumor cells reach the endosteal bone surface by a slower trickle.

CLINICAL MANIFESTATIONS OF BONE METASTASES

The clinical manifestations of bone metastases have been well defined in the past decade from the placebo arms of the bisphosphonate trials.[40] The clinical features include episodes of intractable bone pain, need for radiation therapy or powerful opiate analgesics for relief of pain, spinal cord compression, other nerve compression syndromes, pathologic fractures, and episodes of hypercalcemia. From these studies, it can be concluded that untreated patients with breast cancer will have a major skeletal-related event (such as episodes of severe bone pain or pathologic fracture) every 3–4 mo.[40] This frequency is halved in patients taking powerful nitrogen-containing bisphosphonates such as zoledronic acid or pamidronate.[41]

OSTEOPOROSIS

Osteoporosis will be discussed in detail in a separate section of the *Primer*. However, it is important to remember that cancer is a major risk factor for osteoporosis. In patients with breast cancer and prostate cancer, who make up 80% of all patients with osteolytic bone disease, osteoporosis is very common, both as a consequence of therapy directed at the cancer (in breast cancer, patients are often treated with aromatase inhibitors, selective estrogen receptor modulators (SERMS), corticosteroids, and chemotherapy. Prostate cancer patients are often treated with androgen deprivation therapy, but osteoporosis is also common in treatment in naïve patients. Kanis et al.[42] showed that, in untreated patients, there is a 5-fold increase in vertebral fractures. The mechanisms are unknown, but likely related to molecules produced by the tumor cells that inhibit the Wnt pathway essential for normal bone formation. There are large ongoing clinical trials testing the efficacy of bisphosphonates in treating patients with osteoporosis related to breast cancer, but some of these patients may ulti-

© 2008 American Society for Bone and Mineral Research

mately turn out to do better with anabolic therapy if the cellular mechanism responsible for osteoporosis is inhibition of bone formation.

MANAGEMENT

Diagnostic approaches to metastatic bone disease will not be considered here, and the reader is referred to clinical oncology texts for consideration of this topic.

With respect to specific therapy for bone metastases, the possibilities are radiation therapy or bisphosphonates. Radiation therapy is useful for localized pain in the skeleton, but its use is limited by total dose and to localized sites of bone pain. The mechanism by which it works is unknown.

Bisphosphonates are the mainstay of treatment for metastatic bone disease. Bisphosphonates such as zoledronic acid and pamidronate can be expected to reduce the skeletal-related events associated with cancer by ~50%.[41] These are effective drugs for this situation. These drugs also have markedly reduced the episodes of hypercalcemia associated with malignant disease. However, the nitrogen-containing bisphosphonates, and particularly zoledronic acid, are associated with impaired renal function in patients in whom the drug is given too quickly, and osteonecrosis of the jaw (ONJ) in some patients. ONJ has been a dark cloud overshadowing the whole field over the past few years, but its frequency may decline with the use of lower doses and more prolonged infusions and avoiding bisphosphonate therapy in patients with active dental disease. In patients taking powerful nitrogen-containing bisphosphonates, the frequency is ~5% over 5 yr.[43] The pathophysiology is unknown but may be related to uptake of powerful bisphosphonates at sites of active bone turnover in the mandible in patients with coincident dental disease. It is much less frequent in patients given bisphosphonates for osteoporosis, where lower doses of oral agents are more often used.[44–48]

Although bisphosphonates have been available for more than a decade for patients with breast cancer and its associated bone disease, they are far from ideal therapies. They reduce the frequency of skeletal-related events (such as episodes of severe bone pain and pathologic fracture) in women with breast cancer by only ~50%.[4,41] Although this is better than no specific treatment for bone disease, it is still far from optimal for the following reasons:

(1) Fractures and episodes of bone pain remain a frequent problem, even in patients maximally treated with bisphosphonates.
(2) Osteolytic lesions do not heal with bisphosphonate treatment, except in rare circumstances.
(3) Many patients in remission from disease and treated with bisphosphonates continue to show enhanced bone fragility with continuing episodes of bone pain and fractures. Many patients do not wish to take bisphosphonates because of the recent negative publicity associated with ONJ.[49]

Denosumab, a humanized monoclonal antibody to RANKL, will probably be available for the treatment of metastatic bone disease within the next few years. It is a very powerful and effective osteoclast inhibitor, maybe even more powerful than the bisphosphonates. It will be interesting to see if it produces even greater decreases in skeletal-related events than those found with the bisphosphonates. It may not, because both classes of drugs are powerful resorption inhibitors and cannot be expected to restore bone that has already been lost. Because bone is not spontaneously repaired in these patients when the tumors go into remission or osteoclasts are maximally inhibited, it is conceivable that an anabolic agent will be required to further optimize reduction in skeletal-related events.

Other approaches advancing through the drug development pipeline include other bone resorption inhibitors, potential inhibitors of PTHrP and anti-TGFβ therapeutic approaches. Anti-PTHrP approaches hold promise. Small molecule drugs such as 6-thioguanine have been very effective in the treatment of hypercalcemia and osteolysis in preclinical models.[2] Other possibilities are PTH receptor antagonists, antagonists of PTHrP signal transduction in the cancer cell, and neutralizing antisera to PTHrP. The latter has been very effective in preclinical models.[1,50]

REFERENCES

1. Guise TA, Yin JJ, Taylor SD, Kumagai Y, Dallas M, Boyce BF, Yoneda T, Mundy GR 1996 Evidence for a causal role of parathyroid hormone-related protein in the pathogenesis of human breast cancer-mediated osteolysis. J Clin Invest **98:**1544–1549.
2. Gallwitz WE, Guise TA, Mundy GR 2002 Guanosine nucleotides inhibit different syndromes of PTHrP excess caused by human cancers in vivo. J Clin Invest **110:**1559–1572.
3. Boyde A, Maconnachie E, Reid SA, Delling G, Mundy GR 1986 Scanning electron microscopy in bone pathology: Review of methods, potential and applications. Scan Electron Microsc **Pt 4:**1537–1554.
4. Yin JJ, Selander K, Chirgwin JM, Dallas M, Grubbs BG, Wieser R, Massague J, Mundy GR, Guise TA 1999 TGF-beta signaling blockade inhibits PTHrP secretion by breast cancer cells and bone metastases development. J Clin Invest **103:**197–206.
5. Michigami T, Ihara-Watanabe M, Yamazaki M, Ozono K 2001 Receptor activator of nuclear factor kappaB ligand (RANKL) is a key molecule of osteoclast formation for bone metastasis in a newly developed model of human neuroblastoma. Cancer Res **61:**1637–1644.
6. Martin TJ, Gillespie MT 2001 Receptor activator of nuclear factor kappa B ligand (RANKL): Another link between breast and bone. Trends Endocrinol Metab **12:**2–4.
7. Kitazawa S, Kitazawa R 2002 RANK ligand is a prerequisite for cancer-associated osteolytic lesions. J Pathol **198:**228–236.
8. Sterling JA, Oyajobi BO, Grubbs B, Padalecki SS, Munoz SA, Gupta A, Story B, Zhao M, Mundy GR 2006 The hedgehog signaling molecule Gli2 induces parathyroid hormone-related peptide expression and osteolysis in metastatic human breast cancer cells. Cancer Res **66:**7548–7553.
9. Yoneda T, Sasaki A, Dunstan C, Williams PJ, Bauss F, De Clerck YA, Mundy GR 1997 Inhibition of osteolytic bone metastasis of breast cancer by combined treatment with the bisphosphonate ibandronate and tissue inhibitor of the matrix metalloproteinase-2. J Clin Invest **99:**2509–2517.
10. Mbalaviele G, Dunstan CR, Sasaki A, Williams PJ, Mundy GR, Yoneda T 1996 E-cadherin expression in human breast cancer cells suppresses the development of osteolytic bone metastases in an experimental metastasis model. Cancer Res **56:**4063–4070.
11. Yang J, Mani SA, Donaher JL, Ramaswamy S, Itzykson RA, Come C, Savagner P, Gitelman I, Richardson A, Weinberg RA 2004 Twist, a master regulator of morphogenesis, plays an essential role in tumor metastasis. Cell **117:**927–939.
12. Nakai M, Mundy GR, Williams PJ, Boyce B, Yoneda T 1992 A synthetic antagonist to laminin inhibits the formation of osteolytic metastases by human melanoma cells in nude mice. Cancer Res **52:**5395–5399.
13. Bendre MS, Margulies AG, Walser B, Akel NS, Bhattacharrya S, Skinner RA, Swain F, Ramani V, Mohammad KS, Wessner LL, Martinez A, Guise TA, Chirgwin JM, Gaddy D, Suva LJ 2005 Tumor-derived interleukin-8 stimulates osteolysis independent of the receptor activator of nuclear factor-kappaB ligand pathway. Cancer Res **65:**11001–11009.
14. Bendre MS, Montague DC, Peery T, Akel NS, Gaddy D, Suva LJ 2003 Interleukin-8 stimulation of osteoclastogenesis and bone resorption is a mechanism for the increased osteolysis of metastatic bone disease. Bone **33:**28–37.
15. Bendre M, Gaddy D, Nicholas RW, Suva LJ 2003 Breast cancer metastasis to bone: It is not all about PTHrP. Clin Orthop **415**(Suppl)**:**S39–S45.
16. Mancino AT, Klimberg VS, Yamamoto M, Manolagas SC, Abe E

© 2008 American Society for Bone and Mineral Research

2001 Breast cancer increases osteoclastogenesis by secreting M-CSF and upregulating RANKL in stromal cells. J Surg Res **100:**18–24.
17. McDermott RS, Deneux L, Mosseri V, Vedrenne J, Clough K, Fourquet A, Rodriguez J, Cosset JM, Sastre X, Beuzeboc P, Pouillart P, Scholl SM 2002 Circulating macrophage colony stimulating factor as a marker of tumour progression. Eur Cytokine Netw **13:**121–127.
18. Javed A, Barnes GL, Pratap J, Antkowiak T, Gerstenfeld LC, van Wijnen AJ, Stein JL, Lian JB, Stein GS 2005 Impaired intranuclear trafficking of Runx2 (AML3/CBFA1) transcription factors in breast cancer cells inhibits osteolysis in vivo. Proc Natl Acad Sci USA **102:**1454–1459.
19. Pratap J, Lian JB, Javed A, Barnes GL, van Wijnen AJ, Stein JL, Stein GS 2006 Regulatory roles of Runx2 in metastatic tumor and cancer cell interactions with bone. Cancer Metastasis Rev **25:**589–600.
20. Pratap J, Javed A, Languino LR, van Wijnen AJ, Stein JL, Stein GS, Lian JB 2005 The Runx2 osteogenic transcription factor regulates matrix metalloproteinase 9 in bone metastatic cancer cells and controls cell invasion. Mol Cell Biol **25:**8581–8591.
21. Lynch CC, Hikosaka A, Acuff HB, Martin MD, Kawai N, Singh RK, Vargo-Gogola TC, Begtrup JL, Peterson TE, Fingleton B, Shirai T, Matrisian LM, Futakuchi M 2005 MMP-7 promotes prostate cancer-induced osteolysis via the solubilization of RANKL. Cancer Cell **7:**485–496.
22. Stewart AF, Vignery A, Silverglate A, Ravin ND, LiVolsi V, Broadus AE, Baron R 1982 Quantitative bone histomorphometry in humoral hypercalcemia of malignancy: Uncoupling of bone cell activity. J Clin Endocrinol Metab **55:**219–227.
23. Mohammad KS, Stebbins EG, Niewolna M, McKenna CR, Walton H, Peng XH, Li G, Murphy A, Chakravarty S, Higgins LS, Wong DH, Guise TA 2006 TGF-beta signaling blockade reduces osteolytic bone metastases and enchances bone mass. Cancer Treat Rev **32**(Suppl 3):S29.
24. Guise TA, Stebbins EG, Mohammad KS, Niewolna M, McKenna CR, Mison AP, Li G, Schimmoller F, Murphy A, Chakravarty S 2006 Blocking TBF-beta signaling in breast cancer: Effects on bone and bone metastases. Cancer Treat Rev **32:**S26.
25. Takeuchi K, Abe M, Hiasa M, Kitazoe K, Hashimoto T, Ozaki S, Kido S, Inoue D, Matsumoto T 2006 A TGF-beta receptor kinase inhibitor, restores bone formation which disrupts myeloma-induced microenvironment. Cancer Treat Rev **32**(Suppl 3):S25.
26. Epstein J 2006 Myeloma, bone, and disease control. Cancer Treat Rev **32**(Suppl 3):S23.
27. Matsumoto T, Abe M 2006 Myeloma bone disease: A vicious cycle by interactions between myeloma cells and bone marrow cells. Cancer Treat Rev **32**(Suppl 3):S23.
28. Yin JJ, Mohammad KS, Kakonen SM, Harris S, Wu-Wong JR, Wessale JL, Padley RJ, Garrett IR, Chirgwin JM, Guise TA 2003 A causal role for endothelin-1 in the pathogenesis of osteoblastic bone metastases. Proc Natl Acad Sci USA **100:**10954–10959.
29. Yi B, Williams PJ, Niewolna M, Wang Y, Yoneda T 2002 Tumor-derived platelet-derived growth factor-BB plays a critical role in osteosclerotic bone metastasis in an animal model of human breast cancer. Cancer Res **62:**917–923.
30. Hall CL, Kang S, MacDougald OA, Keller ET 2006 Role of Wnts in prostate cancer bone metastases. J Cell Biochem **97:**661–672.
31. Clines GA, Mohammad KS, Bao Y, Stephens OW, Suva LJ, Shaughnessy JD Jr, Fox JW, Chirgwin JM, Guise TA 2007 Dickkopf homolog 1 mediates endothelin-1-stimulated new bone formation. Mol Endocrinol **21:**486–498.
32. Kaplan RN, Rafii S, Lyden D 2006 Preparing the "soil": The premetastatic niche. Cancer Res **66:**11089–11093.
33. Yang L, DeBusk LM, Fukuda K, Fingleton B, Green-Jarvis B, Shyr Y, Matrisian LM, Carbone DP, Lin PC 2004 Expansion of myeloid immune suppressor Gr+CD11b+ cells in tumor-bearing host directly promotes tumor angiogenesis. Cancer Cell **6:**409–421.
34. Arguello F, Baggs RB, Frantz CW 1988 A murine model of experimental metastasis to bone and bone marrow. Cancer Res **48:**6876–6881.
35. Yoneda T 2000 Cellular and molecular basis of preferential metastasis of breast cancer to bone. J Orthop Sci **5:**75–81.
36. Hiraga T, Williams PJ, Ueda A, Tamura D, Yoneda T 2004 Zoledronic acid inhibits visceral metastases in the 4T1/luc mouse breast cancer model. Clin Cancer Res **10:**4559–4567.
37. Kuperwasser C, Dessain S, Bierbaum BE, Garnet D, Sperandio K, Gauvin GP, Naber SP, Weinberg RA, Rosenblatt M 2005 A mouse model of human breast cancer metastasis to human bone. Cancer Res **65:**6130–6138.
38. Morrissey C, Kostenuik PJ, Brown LG, Vessella RL, Corey E 2007 Host-derived RANKL is responsible for osteolysis in a C4-2 human prostate cancer xenograft model of experimental bone metastases. BMC Cancer **7:**148.
39. Marcelli C, Yates AJ, Mundy GR 1990 In vivo effects of human recombinant transforming growth factor beta on bone turnover in normal mice. J Bone Miner Res **5:**1087–1096.
40. Coleman RE 2001 Metastatic bone disease: Clinical features, pathophysiology and treatment strategies. Cancer Treat Rev **27:**165–176.
41. Hortobagyi GN, Theriault RL, Porter L, Blayney D, Lipton A, Sinoff C, Wheeler H, Simeone JF, Seaman J, Knight RD 1996 Efficacy of pamidronate in reducing skeletal complications in patients with breast cancer and lytic bone metastases. Protocol 19 Aredia Breast Cancer Study Group. N Engl J Med **335:**1785–1791.
42. Kanis JA, McCloskey EV, Powles T, Paterson AH, Ashley S, Spector T 1999 A high incidence of vertebral fracture in women with breast cancer. Br J Cancer **79:**1179–1181.
43. Wilkinson GS, Kuo YF, Freeman JL, Goodwin JS 2007 Intravenous bisphosphonate therapy and inflammatory conditions or surgery of the jaw: A population-based analysis. J Natl Cancer Inst **99:**1016–1024.
44. Marx RE, Stern D 2002 Oral and Maxillofacial Pathology: A Rationale for Treatment. Quintessence Publishing, Hanover Park, IL, USA.
45. American Dental Association Council on Scientific Affairs 2006 Dental management of patients receiving oral bisphosphonate therapy: Expert panel recommendations. J Am Dent Assoc **137:**1144–1150.
46. American Association of Oral and Maxillofacial Surgeons 2007 Position paper on bisphosphonate-related osteonecrosis of the jaws. J Oral Maxillofac Surg **65:**369–376.
47. Ruggiero SL, Fantasia J, Carlson E 2006 Bisphosphonate-related osteonecrosis of the jaw: Background and guidelines for diagnosis, staging and management. Oral Surg Oral Med Oral Pathol Oral Radiol Endod **102:**433–441.
48. Ruggiero SL, Mehrotra B, Rosenberg TJ, Engroff SL 2004 Osteonecrosis of the jaws associated with the use of bisphosphonates: A review of 63 cases. J Oral Maxillofac Surg **62:**527–534.
49. Durie BG, Katz M, Crowley J 2005 Osteonecrosis of the jaw and bisphosphonates. N Engl J Med **353:**99–102.
50. Kukreja SC, Shevrin DH, Wimbiscus SA, Ebeling PR, Danks JA, Rodda CP, Wood WI, Martin TJ 1988 Antibodies to parathyroid hormone-related protein lower serum calcium in athymic mouse models of malignancy-associated hypercalcemia due to human tumors. J Clin Invest **82:**1798–1802.

© 2008 American Society for Bone and Mineral Research

Chapter 80. Hematologic Malignancies and Bone

Wassim M. McHayleh,[1] Jessica Ellerman,[1] and G. David Roodman[1,2]

[1]*Department of Medicine/Hematology-Oncology, University of Pittsburgh Medical Center, Pittsburgh, Pennsylvania;* [2]*VA Pittsburgh Healthcare System, Medicine/Hematology-Oncology, Pittsburgh, Pennsylvania*

INTRODUCTION

Hematologic malignancies can have multiple effects on bone. They can induce hypercalcemia or metastasize to bone and cause severe bone pain and pathologic fractures. Bone involvement can range from very frequent, such as in myeloma, to infrequent as seen in lymphoma or leukemia. Lesions are generally osteolytic, but can be osteoblastic. In both cases, there is dysregulation of the normal bone remodeling process. Increased osteoclastic bone resorption, increased renal tubular calcium reabsorption, and impaired glomerular filtration are all important pathogenic factors that can result in hypercalcemia in patients with hematologic malignancies, with PTH-related peptide (PTHrP) playing an important role in the development of hypercalcemia in many of these cases.

HEMATOLOGIC MALIGNANCIES INVOLVING BONE

Multiple Myeloma

Multiple myeloma (MM) is a clonal malignancy of terminally differentiated plasma cells accumulating in the bone marrow compartment, in association with the monoclonal protein paraprotein in serum and/or urine, decreased normal immunoglobulin (Ig) levels, and lytic bone disease.[1] The identity of the cells responsible for the initiation and maintenance of this disease remains unclear largely because of the difficulty growing MM cells in vitro and in vivo.[2] It has been proposed that these abnormal precursor B cells originate in the lymph nodes and migrate to the bone marrow, which provides a microenvironment conducive to terminal plasma cell differentiation.[3] MM is the second most common and accounts for 10% of all hematologic malignancies. It is the malignancy with the highest prevalence of bone involvement. The annual incidence and prevalence of MM in the United States is ~15,000 and 45,000, respectively.[4] Skeletal manifestations, particularly osteolytic bone lesions, represent the most prominent source of pain and disability in MM. Bone pain occurs in 60–70% of patients, hypercalcemia in 15–20% of patients, and osteolytic lesions and pathological fractures in 60% of myeloma patients.[5,6] Even when complete remission is attained, osteolytic lesions rarely heal. Osteolytic lesions most often include the axial skeleton, skull, and femur; however, diffuse osteopenia is also found.

Bone biopsies have shown that initially MM bone lesions show coupled bone resorption and formation but become uncoupled with advanced disease with suppressed or absent bone formation.[7,8] In MM as in other malignancies, bone destruction is mediated by osteoclasts (OCLs).

OCL activity is increased in MM because MM cells produce or induce osteoclastogenic factors in the bone microenvironment that directly increase OCL formation and activity and decrease production of osteoprotegerin, a soluble decoy receptor for RANKL produced by marrow stromal cells.[9] When myeloma cells adhere to bone marrow stromal cells through binding of surface VLA-4 ($\alpha_4\beta_1$ integrin) to vascular cell adhesion molecule (VCAM)-1 expressed on stromal cells, production of osteoclastogenic cytokines by marrow stromal cells such as RANKL, macrophage-colony stimulating factor (M-CSF), interleukin-11 (IL-11), and IL-6 and by MM cells such as macrophage inflammatory protein-1α (MIP-1α) and IL-3 are increased.[10–13] RANKL increases OCL formation and survival by binding to its receptor RANK on OCL precursor cells and OCLs,[14] whereas MIP-1α acts as a chemotactic factor for OCL precursors and can induce differentiation of OCL progenitors contributing to OCL formation.[15–17] MIP-1α also acts directly on MM cells to promote growth, survival, and migration of MM cells by inducing the activation of multiple signaling pathways crucial for MM cell growth and survival.[18] Furthermore, the bone destructive process releases growth factors from the bone matrix, which increase the growth of MM cells. This results in a "vicious cycle" of bone destruction and results in increased tumor mass (Fig. 1).

Lytic lesions in MM are best visualized by radiographic imaging or MRI. Bone scans severely underestimate the extent of bone disease because bone formation normally associated with osteoclastic bone resorption is severely suppressed in MM.[19] MRI has greater sensitivity than conventional radiography or bone densitometry for detecting bone lesions in myeloma patients.[20]

Hypercalcemia can occur in 15–20% of MM patients because of efflux of calcium into the extracellular fluid.[21] However, bone destruction alone does not explain the development of hypercalcemia in MM patients. Most MM patients with significant bone disease do not develop hypercalcemia. Hypercalcemia is often only seen in the later stages of the disease and in patients with renal dysfunction. Elevated PTHrP production by myeloma cells has been reported,[22] but elevated PTHrP levels are not consistently found in MM patients. Hypercalcemia is most common in patients with greater tumor volume and renal failure regardless of their serum PTHrP levels.[21]

Because of the decreased osteoblast activity in MM, serum alkaline phosphatase and osteocalcin are normal or decreased in patients with myeloma. The mechanisms responsible for the suppressed osteoblast activity are just being identified. Recent studies have focused on the role of Dickkopf-1 (DKK1) as a mediator of the osteoblast suppression in MM.[23] DKK1 is an inhibitor of Wnt signaling and inhibits differentiation of osteoblast precursors and enhances OCL formation and bone resorption. Murine models of myeloma disease treated with anti-DKK1 antibodies show increased bone formation and decreased OCL formation and tumor mass.[24] In addition to DKK1, IL-3, IL-7, and soluble frizzle-related protein-2 have also been reported to be produced by MM cells and inhibit osteoblast differentiation.[25–27]

Rarely, patients with MM can develop osteosclerotic rather than osteolytic lesions.[28] This occurs in patients with POEMS (polyneuropathy, organomegaly, endocrinopathy, monoclonal protein and skin involvement) syndrome. The major problem for these patients is severe neuropathy. The mechanisms responsible for the development of sclerotic lesions are unknown.

Treatment of MM bone disease involves treatment of the underlying malignancy and inhibiting OCL formation and activity. Intravenous bisphosphonates given every 3–4 wk are the current treatment of choice for inhibiting osteoclastic activity

Dr. Roodman has consulted for Amgen, Novartis, Merck, and Millenium and has served on a speakers bureau for Novartis. All other authors report no conflicts of interest.

Key words: hematologic malignancies, bone, multiple myeloma

© 2008 American Society for Bone and Mineral Research

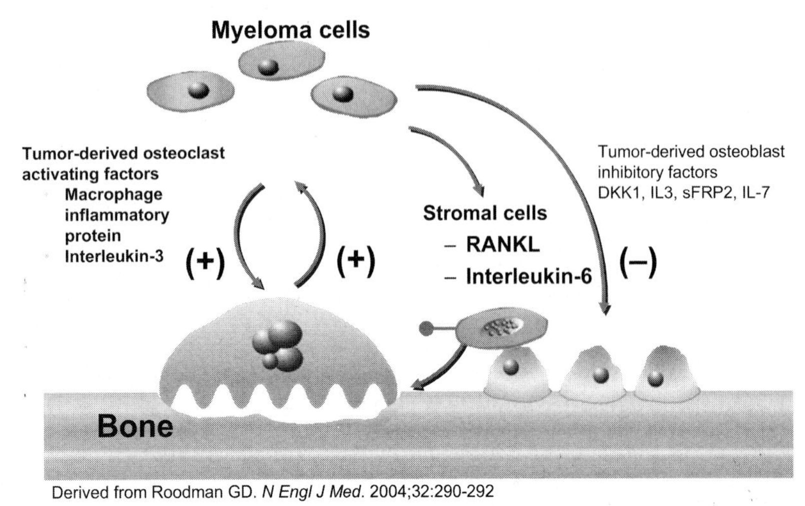

FIG. 1. Vicious cycle of myeloma bone disease. Myeloma cells produce factors that directly or indirectly activate osteoclasts such as MIP-1α and IL-3. In addition, they induce RANKL and IL-6 production by marrow stromal cells to enhance osteoclast formation. The bone destructive process releases growth factors that increase the growth of myeloma cells, further exacerbating the osteolytic process. Myeloma cells also produce DKK1, IL-3, soluble frizzle-related protein-2, and IL-7, which suppress osteoblast differentiation and new bone formation.

in MM. Experimental models of MM suggest that bisphosphonates may induce MM cell apoptosis, although this has not been clearly shown in patients. The major mechanism of action of bisphosphonates is inhibition of OCL activity and reduction of bone resorption without restoration of bone repair. Bisphosphonates alleviate bone pain, treat hypercalcemia, and reduce pathologic fractures. However, side effects of bisphosphonate therapy include renal toxicity and osteonecrosis of the jaw.[29] However, these complications occur in only a small percentage of patients. Bisphosphonate therapy alone has not changed overall survival rates in MM patients unless used in combination with other myeloma therapies.[30] A new therapy in clinical trial for MM bone disease is a human monoclonal antibody denosumab (AMG-12) that binds to RANKL with high affinity and specificity.[31] Denosumab inhibits the RANKL–RANK interaction and mimicking the effects of osteoprotegerin.

Bortezomib (velcade), a proteosome antagonist used to treat MM, is also a potential bone anabolic agent in MM. Bortezomib inhibits the proteasome pathway critical for myeloma survival. Preliminary studies of bortezomib in MM patients indicate increased osteoblast activity with enhanced new bone formation through activation of osteoblasts and inhibition of OCLs. Increased bone formation markers, such as alkaline phosphatase and osteocalcin, have been observed with bortezomib therapy compared with patients treated solely with dexamethasone.[32,33]

Adult T-Cell Leukemia/Lymphoma

Adult T-cell leukemia (ATL) is a malignancy of $CD4^+$ T cells caused by infection with human T-lymphotrophic virus (HTLV-1). It was originally reported in southern Japan, but cases have been reported in the United States as well. ATL cells characteristically produce many chemokines that can affect bone remodeling including IL-1, IL-6, TNF-α, and MIP-1α/MIP-1β.[33,34] Circulating ATL cells rapidly infiltrate into a variety of tissues, mediated by MIP-1α induction of integrin-mediated adhesion to the endothelium and subsequent transmigration.[33] MIP-1α has been implicated in the chemotaxis of monocytes, including OCL progenitor cells and the production of the osteoclastogenic factors IL-6, PTHrP, and RANKL by osteoblasts or stromal cells in ATL.[35,36] IL-1 and PTHrP have both been reported to mediate bone destruction in ATL, with elevated PTHrP levels in these patients and increased concentrations of IL-1 and PTHrP in media conditioned by ATL cells in vitro.[37]

In addition to lytic bone lesions, >70% of ATL patients develop hypercalcemia during the course of their disease.[38] Several mechanisms for the hypercalcemia have been proposed. The HTLV-1 tax protein has been reported transactivate the gene for PTHrP[39] and increase its production, but increased transcription of PTHrP can also occur in a tax gene–independent manner.[40] MIP-1α has also been suggested as a mediator for the hypercalcemia in ATL, by enhancing OCL formation[34] and inducing RANKL expression on ATL cells in an autocrine fashion. RANKL expression was found to be increased on ATL cells in patients with hypercalcemia but not in patients without hypercalcemia.[41] Bone involvement has also been reported in more classical forms of acute lymphoid leukemia and is thought to be mediated by PTHrP production by the malignant cells.[42]

Non-Hodgkin's Lymphoma

Approximately 1–9% of patients with non-Hodgkin's lymphoma (NHL) present with bone involvement. However, 7–25% of all patients with NHL will eventually develop bone findings during the course of their disease. Hypercalcemia is rare in NHL.[43] The most common histologic subtypes of NHL that present with bone manifestations include histiocytic, undifferentiated, and poorly differentiated NHL. Overall, patients with diffuse rather than nodular patterns of involvement more frequently have lytic lesions.[43] Typical radiographic findings include bone destruction with a lytic or moth-bitten pattern. NHL bony involvement has a predilection for the axial skeleton. As in other hematologic malignancies, osteoclastogenic factors produced locally by the tumor or the marrow microenvironment play an important role in the bone destruction. Increased serum levels of PTHrP have been reported to play a role in hypercalcemia in NHL patients.[44]

Hodgkin's Disease

Bone involvement in Hodgkin's disease is uncommon and seldom encountered at diagnosis. Sites of involvement include the spine, pelvis, femur, humerus, ribs, sternum, scapula, and base of the skull.[45,46] However, as with NHL, vertebral and femoral involvements are the most common sites involved.[47] The most frequent presentation is that of a localized, solitary osteoblastic mass,[45] and it is associated with mixed cellularity and nodular sclerosing disease.[46] Bone biopsies will often show fibrosis and a mixed inflammatory infiltrate with rare atypical cells. Radiologic findings include a vertebral sclerotic

© 2008 American Society for Bone and Mineral Research

pattern along with periosteal reaction and hypertrophic pulmonary osteoarthropathy.[47] Like NHL, radiographic patterns cannot predict histologic type or the prognosis of the disease and must be used with clinical staging to predict prognosis.

Bone disease in patients with HD can be lytic, blastic, or mixed. Increased new bone formation by tumor cell stimulation of osteoblast activity occurs at sites of previous osteoclastic activity. Excess production of $1,25(OH)_2$ vitamin D_3 or PTHrP by the lymphoma cells has been reported as the most common cause of hypercalcemia in Hodgkin's disease.[44,48]

REFERENCES

1. Hideshima T, Anderson KC 2002 Molecular mechanisms of novel therapeutic approaches for multiple myeloma. Nat Rev Cancer **2:**927–937.
2. Matsui W, Huff CA, Wang Q, Malehorn MT, Barber J, Tanhehco Y, Smith BD, Civin CI, Jones RJ 2004 Characterization of clonogenic multiple myeloma cells. Blood **103:**2332–2336.
3. Tricot G 2000 New insights into role of microenvironment in multiple myeloma. Lancet **355:**248–250.
4. Singhal S, Mehta J 2006 Multiple myeloma. Clin J Am Soc Nephrol **1:**1322–1330.
5. Callander NS, Roodman GD 2001 Myeloma bone disease. Semin Hematol **38:**276–285.
6. Melton LJ III, Kyle RA, Achenbach SJ, Oberg AL, Rajkumar SV 2005 Fracture risk with multiple myeloma: A population-based study. J Bone Miner Res **20:**487–493.
7. Bataille R, Chappard D, Marcelli C, Dessauw P, Sany J, Baldet P, Alexandre C 1989 Mechanisms of bone destruction in multiple myeloma: The importance of an unbalanced process in determining the severity of lytic bone disease. J Clin Oncol **7:**1909–1914.
8. Taube T, Beneton MN, McCloskey EV, Rogers S, Greaves M, Kanis JA 1992 Abnormal bone remodelling in patients with myelomatosis and normal biochemical indices of bone resorption. Eur J Haematol **49:**192–198.
9. Pearse RN, Sordillo EM, Yaccoby S, Wong BR, Liau DF, Colman N, Michaeli J, Epstein J, Choi Y 2001 Multiple myeloma disrupts the TRANCE/ osteoprotegerin cytokine axis to trigger bone destruction and promote tumor progression. Proc Natl Acad Sci USA **98:**11581–11586.
10. Gunn WG, Conley A, Deininger L, Olson SD, Prockop DJ, Gregory CA 2006 A crosstalk between myeloma cells and marrow stromal cells stimulates production of DKK1 and IL-6: A potential role in the development of lytic bone disease and tumor progression in multiple myeloma. Stem Cells **24:**986–991.
11. Giuliani N, Colla S, Rizzoli V 2004 New insight in the mechanism of osteoclast activation and formation in multiple myeloma: Focus on the receptor activator of NF-kappaB ligand (RANKL). Exp Hematol **32:**685–691.
12. Choi SJ, Cruz JC, Craig F, Chung H, Devlin RD, Roodman GD, Alsina M 2000 Macrophage inflammatory protein 1-alpha is a potential osteoclast stimulatory factor in multiple myeloma. Blood **96:**671–675.
13. Lee JW, Chung HY, Ehrlich LA, Jelinek DF, Callander NS, Roodman GD, Choi SJ 2004 IL-3 expression by myeloma cells increases both osteoclast formation and growth of myeloma cells. Blood **103:**2308–2315.
14. Ehrlich LA, Roodman GD 2005 The role of immune cells and inflammatory cytokines in Paget's disease and multiple myeloma. Immunol Rev **208:**252–266.
15. Abe M, Hiura K, Wilde J, Moriyama K, Hashimoto T, Ozaki S, Wakatsuki S, Kosaka M, Kido S, Inoue D, Matsumoto T 2002 Role for macrophage inflammatory protein (MIP)-1alpha and MIP-1beta in the development of osteolytic lesions in multiple myeloma. Blood **100:**2195–2202.
16. Choi SJ, Oba Y, Gazitt Y, Alsina M, Cruz J, Anderson J, Roodman GD 2001 Antisense inhibition of macrophage inflammatory protein 1-alpha blocks bone destruction in a model of myeloma bone disease. J Clin Invest **108:**1833–1841.
17. Oyajobi BO, Franchin G, Williams PJ, Pulkrabek D, Gupta A, Munoz S, Grubbs B, Zhao M, Chen D, Sherry B, Mundy GR 2003 Dual effects of macrophage inflammatory protein-1alpha on osteolysis and tumor burden in the murine 5TGM1 model of myeloma bone disease. Blood **102:**311–319.
18. Lentzsch S, Gries M, Janz M, Bargou R, Dörken B, Mapara MY 2003 Macrophage inflammatory protein 1-alpha (MIP-1 alpha) triggers migration and signaling cascades mediating survival and proliferation in multiple myeloma (MM) cells. Blood **101:**3568–3573.
19. Huston A, Roodman GD 2007 Myeloma bone disease. In: Anderson K, Ghobrial I (eds.) Multiple Myeloma: Translational and Emerging Therapies (Translational Medicine). Informa, New York, NY, USA, pp. 45–60.
20. Dimopoulos MA, Moulopoulos LA, Datseris I, Weber D, Delasalle K, Gika D, Alexanian R 2000 Imaging of myeloma bone disease—implications for staging, prognosis and follow-up. Acta Oncol **39:**823–827.
21. Oyajobi BO 2007 Multiple myeloma/hypercalcemia. Arthritis Res Ther **9:**S4.
22. Zeimer H, Firkin F, Grill V, Slavin J, Zhou H, Martin TJ 2000 Assessment of cellular expression of parathyroid hormone-related protein mRNA and protein in multiple myeloma. J Pathol **192:**336–341.
23. Tian E, Zhan F, Walker R, Rasmussen E, Ma Y, Barlogie B, Shaughnessy JD Jr 2003 The role of the Wnt-signaling antagonist DKK1 in the development of osteolytic lesions in multiple myeloma. N Engl J Med **349:**2483–2494.
24. Yaccoby S, Ling W, Zhan F, Walker R, Barlogie B, Shaughnessy JD Jr 2007 Antibody-based inhibition of DKK1 suppresses tumor-induced bone resorption and multiple myeloma growth in vivo. Blood **109:**2106–2111.
25. Ehrlich LA, Chung HY, Ghobrial I, Choi SJ, Morandi F, Colla S, Rizzoli V, Roodman GD, Giuliani N 2005 IL-3 is a potential inhibitor of osteoblast differentiation in multiple myeloma. Blood **106:**1407–1414.
26. Giuliani N, Colla S, Morandi F, Lazzaretti M, Sala R, Bonomini S, Grano M, Colucci S, Svaldi M, Rizzoli V 2005 Myeloma cells block RUNX2/CBFA1 activity in human bone marrow osteoblast progenitors and inhibit osteoblast formation and differentiation. Blood **106:**2472–2483.
27. Oshima T, Abe M, Asano J, Hara T, Kitazoe K, Sekimoto E, Tanaka Y, Shibata H, Hashimoto T, Ozaki S, Kido S, Inoue D, Matsumoto T 2005 Myeloma cells suppress bone formation by secreting a soluble Wnt inhibitor, sFRP-2. Blood **106:**3160–3165.
28. Dispenzieri A, Gertz MA 2004 Treatment of POEMS syndrome. Curr Treat Options Oncol **5:**249–257.
29. Lentzsch S, Ehrlich L, Roodman GD 2007 Pathophysiology of multiple myeloma bone disease. Hematol Oncol Clin North Am **21:**1035–1049.
30. Berenson JR, Lichtenstein A, Porter L, Dimopoulos MA, Bordoni R, George S, Lipton A, Keller A, Ballester O, Kovacs MJ, Blacklock HA, Bell R, Simeone J, Reitsma DJ, Heffernan M, Seaman J, Knight RD 1996 Efficacy of pamidronate in reducing skeletal events in patients with advanced multiple myeloma. Myeloma Aredia Study Group. N Engl J Med **334:**488–493.
31. Body JJ, Facon T, Coleman RE, Lipton A, Geurs F, Fan M, Holloway D, Peterson MC, Bekker PJ 2006 A study of the biological receptor activator of nuclear factor-kappaB ligand inhibitor, denosumab, in patients with multiple myeloma or bone metastases from breast cancer. Clin Cancer Res **12:**1221–1228.
32. Zangari M, Esseltine D, Lee CK, Barlogie B, Elice F, Burns MJ, Kang SH, Yaccoby S, Najarian K, Richardson P, Sonneveld P, Tricot G 2005 Response to bortezomib is associated to osteoblastic activation in patients with multiple myeloma. Br J Haematol **131:**71–73.
33. Giuliani N, Morandi F, Tagliaferri S, Lazzaretti M, Bonomini S, Crugnola M, Mancini C, Martella E, Ferrari L, Tabilio A, Rizzoli V 2007 The proteasome inhibitor bortezomib affects osteoblast differentiation in vitro and in vivo in multiple myeloma patients. Blood **110:**334–338.
34. Okada Y, Tsukada J, Nakano K, Tonai S, Mine S, Tanaka Y 2004 Macrophage Inflammatory Protein-1a Induces Hypercalcemia in Adult T-Cell Leukemia. J Bone Miner Res **19:**1105–1110.
35. Tanaka Y, Maruo A, Fujii K, Nomi M, Nakamura T, Eto S, Minami Y 2000 ICAM-1 discriminates functionally different populations of human osteoblasts: Characteristic involvement of cell cycle regulators. J Bone Miner Res **15:**1912–1913.
36. Han JH, Choi SJ, Kurihara N, Koide M, Oba Y, Roodman GD 2001 Macrophage inflammatory protein-1alpha is an osteoclastogenic factor in myeloma that is independent of receptor activator of nuclear factor kappaB ligand. Blood **97:**3349–3353.

© 2008 American Society for Bone and Mineral Research

37. Roodman GD 1997 Mechanisms of bone lesions in multiple myeloma and lymphoma. Cancer **80:**1557–1563.
38. Kiyokawa T, Yamaguchi K, Takeya M, Takahashi K, Watanabe T, Matsumoto T, Lee SY, Takatsuki K 1987 Hypercalcemia and osteoclast proliferation in adult T-cell leukemia. Cancer **59:**1187–1191.
39. Hollsberg P, Hafler DA 1993 Seminars in medicine of the Beth Israel Hospital, Boston. Pathogenesis of diseases induced by human lymphotropic virus type I infection. N Engl J Med **328:**1173–1182.
40. Richard V, Lairmore MD, Green PL, Feuer G, Erbe RS, Albrecht B, D'Souza C, Keller ET, Dai J, Rosol TJ 2001 Humoral hypercalcemia of malignancy: Severe combined immunodeficient/beige mouse model of adult T-cell lymphoma independent of human T-cell lymphotropic virus type-1 tax expression. Am J Pathol **158:**2219–2228.
41. Nosaka K, Miyamoto T, Sakai T, Mitsuya H, Suda T, Matsuoka M 2002 Mechanism of hypercalcemia in adult T-cell leukemia: Overexpression of receptor activator of nuclear factor kappaB ligand on adult T-cell leukemia cells. Blood **99:**634–640.
42. Inukai T, Hirose K, Inaba T, Kurosawa H, Hama A, Inada H, Chin M, Nagatoshi Y, Ohtsuka Y, Oda M, Goto H, Endo M, Morimoto A, Imaizumi M, Kawamura N, Miyajima Y, Ohtake M, Miyaji R, Saito M, Tawa A, Yanai F, Goi K, Nakazawa S, Sugita K 2007 Hypercalcemia in childhood acute lymphoblastic leukemia: Frequent implication of parathyroid hormone-related peptide and E2A-HLF from translocation 17;19. Leukemia **21:**288–296.
43. Pear BL 1974 Skeletal manifestations of the lymphomas and leukemias. Semin Roenigenol **9:**229–239.
44. Firkin F, Seymour JF, Watson AM, Grill V, Martin TJ 1996 Parathyroid hormone-related protein in hypercalcaemia associated with haematological malignancy. Br J Haematol **94:**486–492.
45. Ozdemirli M, Mankin HJ, Aisenberg AC, Harris NL 1996 Hodgkin's disease presenting as a solitary bone tumor. A report of four cases and review of the literature. Cancer **77:**79–88.
46. Borg MF, Chowdhury AD, Bhoopal S, Benjamin CS 1993 Bone involvement in Hodgkin's disease. Australas Radiol **37:**63–66.
47. Franczyk J, Samuels T, Rubenstein J, Srigley J, Morava-Protzner I 1989 Skeletal lymphoma. Can Assoc Radiol J **40:**75–79.
48. Seymour JF, Gagel RF 1993 Calcitrol: The major humoral mediator of hypercalcemia in Hodgkin's disease and non-Hodgkin's lymphomas. Blood **82:**1383–1394.

Chapter 81. Osteogenic Osteosarcoma

Janet M. Hock[1] and Ching C. Lau[2]

[1]*Maine Institute for Human Genetics and Health, Brewer, Maine;* [2]*Baylor College of Medicine, Texas Children's Hospital, Houston, Texas*

INTRODUCTION

Osteogenic osteosarcoma is a primary malignant tumor of bone, derived from primitive bone-forming mesenchymal cells and exhibiting tumor osteoid formation. Although rare, osteosarcoma is the most common malignant bone tumor in adolescents and young adults (Fig. 1). Annual age-adjusted rates are 1/100,000 for men and 0.6/100,000 for women. Differential diagnosis for osteosarcoma includes aneurysmal bone cyst, fibrosarcoma, fibrochondrosarcoma, fibrous dysplasia, giant cell tumors, benign and malignant fibrous histiocytomas of bone, nonossifying fibroma, osteomas, aggressive osteoblastoma, and severe osteitis fibrosa with prolonged high PTH. Osteosarcoma is classified as osteoblastic, chondroblastic, or fibroblastic based on qualitative histology assessment of extracellular matrix. About 14% of osteosarcoma present as lytic lesions with no osteoid matrix. Rare osteosarcoma may occur in organs such as breast or prostate. There are no known risk factors for osteosarcoma. Animal studies and osteosarcoma in older adults suggest osteosarcoma may arise from chronic proliferative dysplasia within bone.

Most osteosarcomas occur in the metaphysis of long bones, such as femur, tibia, or humerus. Because these sites exhibit rapid growth and bone turnover, a popular perception is that deregulation of skeletal growth contributes to the etiology of osteosarcoma. This is not likely. A case-control study of 152 children with osteosarcoma failed to find any significant risk factors for osteosarcoma among demographic characteristics, height of parents, body size, height or weight of study subjects, age at onset of secondary sexual characteristics, occupational exposure of parents, or common maternal risk factors during pregnancy. Osteosarcoma in aged humans occurs in patients with Paget's disease (50%) and postirradiation patients (50%) and likely has a different etiology. Osteosarcoma in 1% of humans with Paget's disease often exhibits loss of heterozygosity at chromosome 18q, at the same locus implicated in familial expansile osteolysis and familial Paget's disease.[1,2] Although *TNFRSF11A* encoding RANK lies in this region, mutations in this gene do not contribute to osteosarcoma.[3] In humans, the risk of osteosarcoma is greatly enhanced by certain chemical carcinogens and radionuclides.

CLINICAL OVERVIEW

After a biopsy for diagnosis, current treatment of osteosarcoma typically uses multiagent preoperative (neoadjuvant) chemotherapy, followed by definitive surgery to resect tumor and postoperative chemotherapy. The resected tumor is assessed for necrosis. Patients whose tumors display ≥90% necrosis have an excellent prognosis and continue to receive chemotherapy. Patients whose tumors display <90% necrosis have a much higher risk of relapse and poor outcome, even after complete resection of the primary tumor.

About 80% of patients with osteosarcoma initially present with nonmetastatic disease. Before the introduction of neoadjuvant chemotherapy, tumor surgery led to survival of only 20–30% of patients. The addition of systemic chemotherapy with surgery increased long-term survival for nonmetastatic disease to 60–70%. The concept of neoadjuvant chemotherapy before definitive surgery was first introduced in the late 1970s to facilitate limb-sparing procedures by reducing tumor size. Although osteosarcoma is typically resistant to radiation therapy, radiation may be attempted for local control when patients decline surgery.

Radical surgical resection of osteosarcoma uses amputation, rotational plasty, or limb-sparing salvage procedure. Amputation is often the method of choice in young children with sub-

The authors state that they have no conflicts of interest.

Key words: osteosarcoma, clinical research, genetics, cytogenetics, osteoblast, human, dog, mouse

© 2008 American Society for Bone and Mineral Research

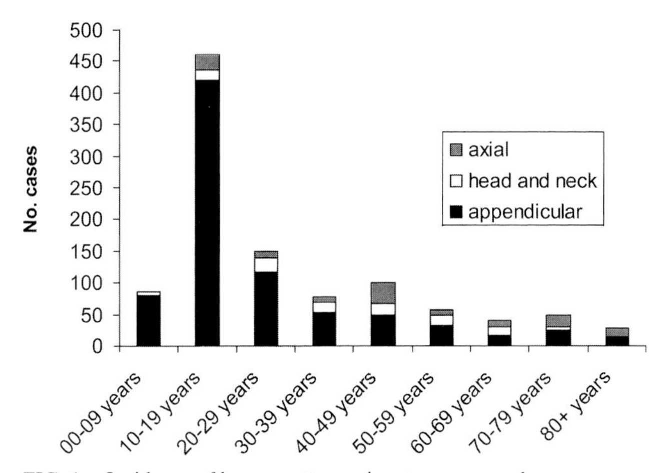

FIG. 1. Incidence of human osteogenic osteosarcoma shown as number of cases per decade of life in the United States. Sixty-six percent (698/1052 cases) of osteosarcoma occurs in the first 30 yr of life; 44% of cases (462/1052 cases) occur between 10 and 19 yr of age. SEER data 2000–2004.

stantial remaining growth potential. Rotational plasty is a modified amputation to preserve the extremity distal to the tumor site. The preserved extremity is left attached to the neurovascular pedicle, advanced proximally, and rotated to allow a distal joint to substitute for the resected joint. This procedure is most often performed in young children for whom en bloc resection of the knee or hip joint is unnecessary. In the former case, the ankle joint serves as the new knee joint. In the latter case, the knee joint serves as a hinge hip joint. Intensified, high-dose postsurgical chemotherapy in patients with a poor response has been unsuccessful. At the present time, osteosarcoma chemotherapeutics include doxorubicin, cisplatin, high-dose methotrexate, ifosfamide, and cyclophosphamide, which are toxic at increased doses. Clinical studies have tried, with mixed results, to minimize the toxicities of high-dose chemotherapy using autologous CD34+ stem cell transplantation.[4]

The histopathology of cancer tissue resected by definitive surgery, after neoadjuvant chemotherapy, is graded as follows, and used to predict prognosis: grade I, no tumor response; grade II, extensive necrosis with >10% viable tumor; grade III, extensive necrosis with scattered foci of <10% viable tumor; grade IV, complete necrosis.

Grades III and IV responses are generally regarded as "favorable responses," whereas grades I and II are "poor (unfavorable) responses." In a Children's Cancer Group (CCG study of 206 patients, patients with >95% tumor necrosis had an 8-yr event-free survival rate of 81% compared with 46% for patients with unfavorable histology.[5] The prognostic value of histological response declines over time. A retrospective study of putative prognostic factors was conducted on 420 patients, 3–63 yr of age, treated with surgery and chemotherapy for stage 2 high-grade osteosarcoma in long bones.[6] The 5-yr metastases-free survival rate was calculated as 61% at the 95% CI. Poor histological response was not a significant predictor of time to metastases by 2 yr after the original diagnosis. Tumor size had no significant prognostic value. Diagnosis of chondroblastic subtype increased in value over time as a predictor of metastases with low probability of survival.

There are currently no prognostic tools that consistently distinguish, at the time of diagnosis, those children with osteosarcoma who will respond to therapy from those who will not respond. The published molecular signatures of chemoresistance[7–9] do not overlap, probably because small sample size and differing therapies introduce significant variability.

STRATEGIES TO BETTER UNDERSTAND OSTEOSARCOMAGENESIS

Cellular Mechanisms

Osteoblasts arise by clonal expansion of bone marrow or periosteal mesenchymal stromal stem cells. One hypothesis holds that osteosarcoma is caused by transformed osteoblasts in which cell cycling has been disrupted. Deregulation of G1-S cell cycle checkpoints may occur because of deletion and epigenetic methylation events affecting genes such as *INK4A, CDK4, and MDM2.* An alternate hypothesis is that catastrophic genomic instability in a stromal stem cell or its immediate progeny leads to rapid clonal expansion, resulting in clusters of malignant osteoprogenitor cells synthesizing a highly abnormal extracellular osteoid-like matrix. Loss of heterozygosity analyses showed that 36/56 human osteosarcomas lost polyhomeotic-like 3 (*PHC-3*); this gene forms part of the polycomb complex hPRC-H that stabilizes cells in G_0, a common state for stem cells.[10]

When implanted as xenografts in immunodeficient mice, few of the available osteoblast-like osteosarcoma cell lines develop lung metastases, unless they have been transformed by oncogenes such as v-fos, K-ras, or v-myc and/or have been passaged several times through animals. A successful alternative has been to use osteosarcoma tissue itself as a xenograft in immunodeficient mice to test new therapies.[11] Human SaOS2 osteosarcoma cell xenografts rarely metastasize to the lungs. After repeated passage of SaOS2 through mice, highly metastatic SaOS-7 were harvested and proved to be a useful model to study novel aerosol therapies targeting suppression of lung metastases.[12–15] The extent to which either osteoclasts, which are hematopoietic in origin, or osteocytes, which are extremely long-lived cells that proliferate vigorously when released in vitro from bone, may contribute to osteosarcomagenesis is not known.

Global Genomic Phenotype (Cytogenetics)

Specific genetic aberration(s) underlying the pathogenesis of osteosarcoma remain uncertain. Unlike other sarcomas, such as synovial sarcoma, alveolar rhabdomyosarcoma, and Ewing's sarcoma, in which there are characteristic recurrent chromosomal translocations, osteosarcomas do not have specific translocations or other molecular genetic abnormalities. Most osteosarcomas display complex numerical and structural chromosomal abnormalities with significant cell-to-cell variation and heterogeneity, suggesting that genomic instability is inherent in osteosarcoma development. Multiple cytogenetic abnormalities occur in osteosarcoma. Early studies used conventional G-banding techniques to show clonal chromosomal and numerical abnormalities. Combining comparative genomic hybridization, spectral karyotyping, and fluorescence in situ hybridization to analyze a panel of 25 osteosarcoma specimens provides a more detailed assessment of the complex cytogenetic aberrations in osteosarcoma. The most frequently detected amplifications were of chromosomal bands 6p12-p21 (28%), 17p11.2 (32%), and 12q13-q14 (8%). Several other recurrent chromosomal losses (2q, 3p, 9, 10p, 12q, 13q, 14q, 15q, 16, 17p, and 18q) and chromosomal gains (Xp, Xq, 5q, 6p, 8q, 17p, and 20q) were identified, as well as several recurrent breakpoint clusters and nonrecurrent reciprocal translocations. Interestingly one of the most frequently seen amplicons in 8q24 in human osteosarcoma harbors the oncogene MYC, which induces osteosarcoma in mice.[16] MYC overexpression can induce genomic instability and defective double strand

© 2008 American Society for Bone and Mineral Research

break DNA repair.[17] This, together with abnormal telomere biology, may underlie the complex genomic instability characteristic of osteosarcoma.

Genetic Phenotypes

Syndromes such as Li-Fraumeni syndrome and hereditary retinoblastoma, in which p53 and Rb, respectively, are nonfunctional, exhibit a higher risk for osteosarcoma.[18] MDM2 and TP53 mutually regulate each other, so polymorphisms or mutations in MDM2 will deregulate p53 to enhance osteosarcoma risk. RB, TP53, and MDM2 deregulation are common to many types of human cancer. In contrast, the RecQ family of helicases, which is required for genome stability, seems to be more specific for the skeleton. Mutations in three of the five known RecQ helicases in humans increase the risk of cancer, including osteosarcoma, skeletal defects, and premature aging. Three autosomal recessive disorders associated with mutations in the *RecQL4* gene: Rothmund-Thomson, RAPADILINO, and Baller-Gerold syndrome result in skeletal dysplasias, including greatly increased risk of osteosarcoma.[19] The function of RecQL4 is poorly understood but seems nonredundant for osteoblast integrity.

Molecular Phenotype

There is no one cluster of mutated genes common to all osteosarcoma. Not surprisingly, given the dominance of osteoblast phenotypes in osteosarcoma, c-*fos* and PTH receptor 1 (PTHR1) expression and c-Jun N terminal kinase (JNK)-activator protein 1 (AP-1) signal transduction are all elevated in osteosarcoma. Although overexpression of c-*fos* and c-myc in bones of mice may induce osteosarcoma and their inactivation decreases or eliminates tumor burden by cell senescence and apoptosis, there is no evidence that a single gene target is an effective therapy for human osteosarcoma. The spectrum of genes identified by gene arrays and the genetic mouse models are comprehensively reviewed by Kansara and Thomas.[20] A recent prospective case-control study of 124 osteosarcoma cases compared with 87 controls screened SNPs for 13 genes in the IGF family.[21] A significant association was found for two SNPs (exon 16, Ex16 + 88G>A; intron 16, IVS16 + 15C>t) in the *IGF2R* gene on chromosome 6q26. Ex16 + 88G>A resides in a region predicted to be a CpG island. A detailed study of 15 of the osteosarcoma cases showed a loss of methylation at that CpG site, but whether this epigenetic change plays a role in sarcomagenesis remains unknown.

A potential etiology of the complex unbalanced karyotypes that characterize osteosarcoma is telomere dysfunction. Telomeres are nucleoprotein structures that cap chromosome ends and serve at least three protective functions: preventing chromosomes from being recognized as damaged DNA, preventing chromosomal end-to-end fusions and recombinations, and accommodating the loss of DNA that occurs with each round of replication. Telomere length is one of the checkpoints that determines when normal human somatic cells stop dividing.[22,23] As cells divide, telomere length gradually decreases to a critical size, at which point senescence is triggered by a p53-dependent process. Human cells may bypass this checkpoint by inactivating the p53 pathways and continue to divide until telomeres become very short; chromosomal instability ensues, triggering apoptosis. Rare cells bypass this second checkpoint by activating mechanisms that lengthen telomeres, a central feature of cancer cells. About 85% of cancers activate an enzyme called telomerase, which lengthens telomeres, and the other 15% of cancers use a recombination-based method called alternative telomere lengthening (ALT).[24,25] At least 50% of osteosarcoma samples are dependent on the ALT mechanism to maintain telomeres.[26–28] The ALT and telomerase mechanisms are different means to the same end but are not equivalent. ALT cell lines have more genomic instability and more translocations than telomerase-positive cell lines.[26] In a mouse model, ALT-dependent tumors had decreased metastatic potential compared with telomerase-dependent tumors.[29] ALT-dependent human osteosarcoma may have different clinical behavior than telomerase-dependent osteosarcoma. In a comprehensive study of telomere maintenance mechanisms in 62 OS cases, a subset lacking both telomerase activity and evidence of ALT correlated with favorable prognosis.[30]

Animal Models of Osteosarcoma

Spontaneous osteogenic osteosarcoma is frequently found in dogs, especially large breeds. Inbred strains, such as the Scottish deerhound, present with an even higher incidence rate of 150/1000 dogs compared with 7/100,000 dogs in general. Human and canine osteosarcomas share common features, such as a predilection for long bone metaphysis; high-grade malignancy; a high rate of metastasis; and the lung as the most common site of metastases. Other useful models include genetically engineered mice and osteosarcoma cell or tissue engraftment into mice. In a study using canine osteosarcoma xenografts in mice, ezrin was linked to metastatic progression.[31] Ezrin is a protein that physically connects the actin cytoskeleton to the cell membrane and seems to provide an early survival advantage to cancer cells that reach the lung.[32] High ezrin expression in spontaneous osteosarcoma in dogs was associated with early development of metastases, suggesting that high ezrin may be a biomarker for poor prognosis. Further studies are needed to determine whether ezrin has a role in human osteosarcoma. The Fas pathway is also inextricably linked to osteosarcoma metastatic mechanisms. Fas ligand, a transmembrane protein that is expressed in certain human organs including the lungs, induces apoptosis by interacting with its receptor Fas on the surface membrane of susceptible cells. Fas is a member of the TNF receptor family. When metastatic osteosarcoma cell lines expressing high or low Fas were injected into nude mice, high Fas-expressing cells caused significantly fewer metastatic lung tumor nodules than low Fas-expressing controls. Increasing Fas expression increased sensitivity of osteosarcoma cells to Fas-induced cell death. Although primary human osteosarcoma often expresses high Fas, lung metastases exhibit little or no Fas expression. Loss of Fas expression may be one mechanism by which osteosarcoma cells evade host resistance mechanisms in the lung, where they most often metastasize. The Fas pathway may serve as a potential target for novel osteosarcoma therapy.

IMMUNE BIOLOGY OF OSTEOSARCOMA

The immune biology of osteosarcoma is not well understood. Suppression of normal immune surveillance may underlie the resistance of osteosarcoma to immune detection and elimination. 501 mice, in which SV40 transgene drives the α-amylase promoter, developed osteosarcoma that metastasized to lungs.[33] SV40 binds and inactivates p53, thereby blocking Rb-mediated pathways in cell cycle progression. Development of osteosarcoma closely followed in time the development of immune tolerance to the SV40 antigen as mice aged.[34] A comparison of osteosarcoma xenografts in immunocompetent and immunodeficient mice showed that metastases were reduced in the presence of interferon-γ secreting T cells, suggesting a role for T cell–based immunosurveillance.[35]

© 2008 American Society for Bone and Mineral Research

SUMMARY OF LIMITATIONS OF CURRENT KNOWLEDGE AND SUGGESTIONS FOR FUTURE WORK

There remains a significant unmet need to improve diagnosis and prognosis of osteosarcoma and disease-free survival. Despite significant gains in genomic information on human osteosarcoma, there has been no improvement in survival rates since the introduction of neoadjuvant chemotherapy in the 1970s. Human syndromes predisposed to osteosarcoma, spontaneous osteosarcoma in dogs and mouse xenograft, and genetically engineered mice have proved useful in providing new insights into metastatic progression. The lack of risk factors and the rarity of this cancer make prevention unlikely. However, new diagnostics and therapies could reduce mortality, extend the long-term survival for humans diagnosed with metastatic osteosarcoma and prevent recurrence or relapse after prolonged dormancy of up to two decades.

REFERENCES

1. Hansen M, Nellissery M, Bhatia P 1999 Common mechanisms of osteosarcoma and Paget's disease. J Bone Miner Res **14:**S2;39–44.
2. Nellissery M, Padalecki SS, Brkanac Z, Singer FR, Roodman GD, Unni KK, Leach RJ, Hansen MF 1998 Evidence for a novel osteosarcoma tumor-suppressor gene in the chromosome 18 region genetically linked with Paget disease of bone. Am J Hum Genet **63:**817–824.
3. Sparks A, Peterson SN, Bell C, Loftus BJ, Hocking L, Cahill DP, Frassica FJ, Streeten EA, Levine MA, Fraser CM, Adams MD, Broder S, Venter JC, Kinzler KW, Vogelstein B, Ralston SH 2001 Mutation screening of the TNFRS11A gene encoding receptor activator of NFkappaB (RANK) in familial and sporadic Paget's Disease of bone and osteosarcoma. Calcif Tissue Int **68:**151–155.
4. Ek E, Choong P 2006 The role of high dose therapy and autologous stem cell transplantation for pediatric bone and soft tissue sarcomas. Expert Rev Anticancer Ther **6:**225–237.
5. Provisor A, Ettinger LJ, Nachman JB, Krailo MD, Makley JT, Yunis EJ, Huvos AG, Betcher DL, Baum ES, Kisker CT, Miser JS 1997 Treatment of nonmetastatic osteosarcoma of the extremity with preoperative and postoperative chemotherapy: A report from the Children's Cancer Group. J Clin Oncol **15:**76–84.
6. Kim M, Cho WH, Song WS, Lee SY, Jeon DG 2007 Time dependency of prognostic factors in patients with stage II osteosarcomas. Clin Orthop **463:**157–165.
7. Man T, Chintagumpala M, Visvanathan J, Shen J, Perlaky L, Hicks J, Johnson M, Davino N, Murray J, Helman L, Meyer W, Triche T, Wong KK, Lau CC 2005 Expression profiles of osteosarcoma that can preict response to chemotherapy. Cancer Res **65:**8142–8150.
8. Mintz M, Sowers R, Brown KM, Hilmer SC, Mazza B, Huvos AG, Meyers PA, Lafleur B, McDonough WS, Henry MM, Ramsey KE, Antonescu CR, Chen W, Healey JH, Daluski A, Berens ME, MacDonald TJ, Gorlick R, Stephan DA 2005 An expression signature classifies chemotherapy-resistant pediatric osteosarcoma. Cancer Res **65:**1748–1754.
9. Ochi KY, Daigo T, Katagiri S, Nagayama T, Tsunoda A, Myoui N, Naka N, Araki I, Kudawara M, Ieguchi Y, Toyama J, Toguchida H, Yoshikawa YN 2004 Prediction of response to neoadjuvant chemotherapy for osteosarcoma by gene-expression profiles. Int J Oncol **24:**647–655.
10. Deshpande A, Akunowicz JD, Reveles XT, Patel BB, Saria EA, Gorlick RG, Naylor SL, Leach RJ, Hansen MF 2007 PHC3, a component of the hPRC-H complex, associates with E2F6 during G0 and is lost in osteosarcoma tumors. Oncogene **26:**1714–1722.
11. Dass C, Eugene T, Choong P 2007 Human xenograft osteosarcoma models with spontaneous metastasis in mice: Clinical relevance and applicability for drug testing. J Cancer Res Clin Oncol **133:**193–198.
12. Duan X, Jia S, Koshkina N, Kleinerman E 2006 Intranasal interleukin-12 gene therapy enhanced the activity of ifosfamide against osteosarcoma lung metastases. Cancer Immunol Immunother **106:**1382–1388.
13. Gordon N, Arndt C, Hawkins D, Doherty D, Inwards C, Munsell M, Stewart J, Koshkina N, Kleinerman E 2005 Fas expression in lung metastasis from osteosarcoma patients. J Pediatr Hematol Oncol **27:**611–615.
14. Jia S, Worth L, Densmore C, Xu B, Duan X, Kleinerman E 2003 Aerosol gene therapy with PEI: IL-12 eradicates osteosarcoma lung metastases. Clin Cancer Res **9:**3462–3468.
15. Jia S, Zhou R, Kleinerman E 2003 Nude mouse lung metastases models of osteosarcoma and Ewing's sarcoma for evaluating new therapeutic strategies. Methods Mol Med **74:**495–505.
16. Felsher D, Bishop J 1999 Reversible tumorigenesis by MYC in hematopoietic lineages. Mol Cell **4:**199–207.
17. Karlsson A, Deb-Basu D, Cherry A, Turner S, Ford J, Felsher DW 2003 Defective double-strand DNA break repair and chromosomal translocations by MYC overexpression. Proc Natl Acad Sci USA **100:**9974–9979.
18. Papachristou D, Papavassiliou A 2007 Osteosarcoma and chondrosarcoma: New signalling pathways as targets for novel therapeutic interventions. Int J Biochem Cell Biol **39:**857–862.
19. Dietschy T, Shevelev I, Stagljar I 2007 The molecular role of the Rothmund-Thomson-, RAPIDILINO-, and Baller-Gerold-gene product, RECQL4: Recent progress (review). Cell Mol Life Sci **64:**796–802.
20. Kansara M, Thomas D 2007 Molecular pathogenesis of osteosarcoma. DNA Cell Biol **26:**1–18.
21. Savage S, Woodson K, Walk E, Modi W, Liao J, Douglass C, Hoover RN, Chanock SJ 2007 Analysis of genes critical for growth regulation identifies Insulin-like Growth Factor 2 Receptor variations with possible functional significance as risk factors for osteosarcoma. Cancer Epidemiol Biomarkers Prev **16:**1667–1674.
22. Harley CB 1991 Telomere loss: Mitotic clock or genetic time bomb? Mutat Res **256:**271–282.
23. Bodnar A, Ouellette M, Frolkis M 1998 Extension of life-span by introduction of telomerase into normal human cells. Science **279:**349–352.
24. Shay J, Bacchetti S 1997 A survey of telomerase activity in human cancer. Eur J Cancer **33:**787–791.
25. Bryan T, Englezou A, Dalla-Pozza L 1997 Evidence for an alternative mechanism for maintaining telomere length in human tumors and tumor-derived cell lines. Nat Med **3:**1271–1274.
26. Scheel C, Schaefer KL, Jauch A 2001 Alternative lengthening of telomeres is associated with chromosomal instability in osteosarcomas. Oncogene **20:**3835–3844.
27. Aue G, Muralidhar B, Schwartz HS 1998 Telomerase activity in skeletal sarcomas. Ann Surg Oncol **5:**627–634.
28. Sangiorgi L, Gobbi GA, Lucarelli E 2001 Presence of telomerase activity in different musculoskeletal tumor histotypes and correlation with aggressiveness. Int J Cancer **95:**156–161.
29. Chang S, Khoo CM, Naylor ML 2003 Telomere-based crisis: Functional differences between telomerase activation and ALT in tumor progression. Genes Dev **17:**88–100.
30. Ulaner G, Huang H-Y, Otero J, Zhao Z, Ben-Porat L, Satagopan JM, Gorlick R, Meyers P, Healey JH, Huvos AG, Hoffman AR, Ladanyi M 2003 Absence of a telomere maintenance mechanism as a favorable prognostic factor in patients with osteosarcoma. Cancer Res **63:**1759–1763.
31. Khanna C, Khan J, Nguyen P, Prehn J, Cayor J, Yeung C, Trepel J, Meltzer P, Helman L 2001 Metastasis-associated differences in gene expression in a murine model of osteosarcoma. Cancer Res **61:**3750–3759.
32. Khanna C, Wan X, Bose S, Cassaday R, Olomu O, Mendoza A, Yeung C, Gorlick R, Hewitt S, Helman L 2004 The membrane cytoskeleton linker ezrin is necessary for osteosarcoma metastases. Nat Med **10:**182–186.
33. Knowles B, McCarrick J, Fox N, Solter D, Damjanov I 1990 Osteosarcomas in transgenic mice expressing the alpha-amylase-SV40 T-antigen hybrid gene. Am J Pathol **137:**259–262.
34. Schell T, Knowles B, Tevethia S 2000 Sequential loss of cytotoxic T lymphocyte responses to Simian virus 40 Large T antigen epitopes in T antigen transgenic mice developing osteosarcoma. Cancer Res **60:**3002–3012.
35. Merchant M, Melchionda F, Sinha M, Khanna C, Helman L, Mackall C 2007 Immune reconsitution prevents metastatic recurrence of murine osteosarcoma. Cancer Immunol Immunother **56:**1037–1046.

© 2008 American Society for Bone and Mineral Research

Chapter 82. Skeletal Complications of Breast and Prostate Cancer Therapies

Özge Uluçkan, Elizabeth A. Morgan, Angela C. Hirbe, and Katherine Weilbaecher

Departments of Medicine and Cellular Biology and Physiology, Division of Oncology, Washington University School of Medicine, St. Louis, Missouri

INTRODUCTION

Breast cancer accounts worldwide for 23% of cancer cases in women, and prostate cancer accounts for 12% of cases in men. Whereas advances in nonsurgical treatment options such as chemotherapy, hormonal therapy, and radiation are improving survival rates in patients with these diseases, these therapies also carry significant side effects. This chapter focuses on one such category of side effects: cancer treatment–induced skeletal complications such as bone loss and fractures, which are a growing cause of morbidity in this patient population.

BREAST CANCER

Adjuvant Hormonal Therapy

In adults, the skeleton undergoes complete turnover every 10 yr. Bone mass maintenance is a balance between the activity of osteoblasts, which form bone, and osteoclasts, which resorb it. Estrogen plays a key regulatory role in this cycle of bone remodeling by mediating effects through the estrogen receptor (ER), present on several cell types in the bone. Estrogen stimulates osteoblasts to produce osteoprotegerin (OPG), a decoy receptor for RANK.[1] OPG blocks the binding of RANKL to RANK on osteoclasts, leading to impaired osteoclast activity and decreased bone resorption. Additionally, estrogen is believed to directly induce apoptosis of bone-resorbing osteoclasts.[2] Thus, in premenopausal women, estrogen both inhibits bone remodeling and suppresses bone resorption, contributing to bone strength. As estrogen levels decline after menopause, this regulation diminishes and bone resorption increases out of proportion to bone formation, leading to a net loss in bone and weakened bony microarchitecture. Despite the persistence of low levels of circulating estrogen in the postmenopausal state (produced by the conversion of peripheral tissue androgens to estrogen by the aromatase enzyme), bone mass can decrease by as much as 3%/yr in the first 5 yr after menopause.[3] The estrogen receptor (ER) is expressed by 70% of breast tumors,[4] and circulating estrogen can promote the growth of ER-positive tumors. Current breast cancer therapies exploit this relationship either by decreasing circulating estrogen levels or by blocking or downregulating the receptor itself. Whereas some of the estrogen-mimicking agents seem to be bone sparing, others that disrupt the estrogen–skeleton axis cause adverse effects on bone remodeling, leading to decreased BMD and an increased risk of osteoporosis and fracture.

Selective Estrogen Receptor Modulators. Tamoxifen is a selective estrogen receptor modulator (SERM) that binds to the ER and acts as an estrogen antagonist in breast tissue. Tamoxifen is routinely used as adjuvant therapy in patients with ER-positive breast cancers and as a breast cancer preventative therapy in high-risk patients.[5] In bone, tamoxifen has both positive and negative effects depending on the menopausal state; premenopausal women taking tamoxifen may experience bone loss, whereas the drug seems to have agonistic effects in postmenopausal women.[6,7] Two trials in postmenopausal women with breast cancer showed statistically significant increases in BMD in the groups receiving tamoxifen versus placebo; however, no differences in fracture incidence were found.[5,6,8] The SERM, raloxifene, which is approved for the prevention of osteoporosis in postmenopausal women, can also be used to decrease the risk of developing invasive breast cancers in postmenopausal women.[9–12] The NSABP study of tamoxifen and raloxifene (STAR) P-2 trial, a trial in postmenopausal women with increased breast cancer risk, showed no difference in fracture incidence between the raloxifene or tamoxifen groups, although this trial did not include comparison with placebo.[13] In summary, SERMs, when used to treat invasive breast cancer or to decrease breast cancer risk, have favorable effects on BMD but have not been shown to prevent fracture incidence.

Aromatase Inhibitors. Postmenopausal women maintain a low level of circulating estrogen because of the aromatization of androgens to estrogen in tissues such as fat and muscle by the cytochrome *P*450 aromatase enzyme. Clinical trials evaluating aromatase inhibitors (AIs) in the adjuvant therapy of breast cancer have shown decreased cancer recurrences and improved disease-free survival in women who received AIs compared with tamoxifen.[8] There are two major classes of AIs: the nonsteroidal reversible inhibitors, such as anastrozole and letrozole, and the steroidal irreversible inhibitors, such as exemestane.[14] Evaluation of skeletal health endpoints in the AI clinical trials showed that both nonsteroidal and steroidal AIs produce significant decreases in BMD and may increase fracture incidence in postmenopausal women with breast cancer.[6,13]

The Arimidex, Tamoxifen, Alone or in Combination (ATAC) trial showed that, at 5 yr, there was a statistically significant increase in fracture rate for women taking the nonsteroidal AI, anastrozole, compared with tamoxifen (11% versus 7.7%, $p < 0.0001$). The Breast International Group (BIG) 1-98 study, which compared letrozole versus tamoxifen as initial adjuvant therapy in postmenopausal women with primary breast cancer, showed a statistically significant increase in bone fractures in the letrozole-treated group over 5 yr (number of fractures = 211 [letrozole] versus 141 [tamoxifen], $p < 0.001$).

Exemestane and its metabolite, 17-hydroxyexemestane, have androgenic properties and have bone-sparing effects in ovariectomized rats.[15] However, in humans, no significant bone-sparing effects have conclusively been shown. A trial comparing the effects of exemestane versus placebo on BMD in 147 women after surgical resection of early breast cancer reported a higher rate of annual BMD loss in the femoral neck, but not spine, in the exemestane group compared with placebo (2.72% versus 1.48%, $p = 0.024$).[6] Comparing the switch to exemestane versus continuation of tamoxifen in postmenopausal women with resected breast cancer showed that there was a decrease in BMD in the lumbar spine (2.7%, $p < 0.0001$) and hip (1.4%, $p < 0.0001$) after 6 mo compared with baseline in women who switched to exemestane; however, the rate of decrease in BMD were not as pronounced by 2 yr after the switch.[16] The Z-FAST trial evaluated the effect on bone loss

The authors state that they have no conflicts of interest.

© 2008 American Society for Bone and Mineral Research

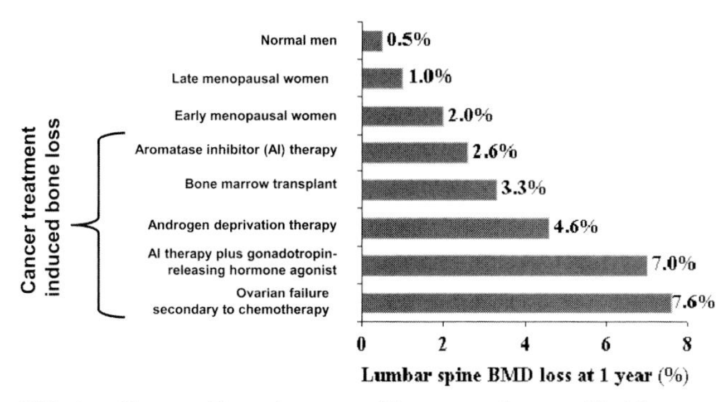

FIG. 1. Extent of bone loss caused by cancer therapy. Healthy men lose bone at a rate of 0.5%/yr beginning at middle age, whereas menopausal women lose bone at a rate of 1–2%/yr. Cancer treatments such as AI therapy, ADT, and chemotherapy accelerate this process, leading to significant bone loss and subsequent skeletal complications.[6,16,18,49,50] (Printed with permission from Postgraduate Institute for Medicine and Interlink Healthcare Communications, joint sponsors of the CME Lecture Series titled *Skeletal Complications Across the Cancer Continuum.* Slide/Lecture Kit release date: June 2005.)

of upfront versus delayed zoledronic acid (ZA) in postmenopausal women on adjuvant letrozole for early breast cancer. BMD was 4.4% higher in the lumbar spine ($p < 0.0001$) and 3.3% higher in the hip ($p < 0.0001$) in patients that received ZA at the time of randomization compared with patients who initiated ZA after 1 yr of letrozole.[6] In summary, clinical trials showed that both classes of AIs result in bone loss (Fig. 1). As such, patients treated with AIs should be monitored carefully for changes in BMD and treated appropriately.

Selective Estrogen Receptor Downregulators. Selective estrogen receptor downregulators (SERDs) downregulate cellular levels of the ER and act as pure ER antagonists without any agonist effects. Fulvestrant is currently the only SERD used clinically and is approved for second-line treatment of advanced and metastatic breast cancer in postmenopausal women. In animal studies, fulvestrant was shown to increase bone turnover; however, the opposite results were obtained when the rats were ovariectomized.[17] Currently, BMD is not an endpoint in clinical trials testing the efficacy of fulvestrant.

Chemotherapy

Chemotherapy can have direct effects on the bone microenvironment, ultimately leading to decreased BMD.[18] Chemotherapy induced ovarian failure, changes in diet, and physical activity, and supportive medications such as dexamethasone and hematopoietic growth factors can impact BMD. BMD scores of postmenopausal women who received adjuvant chemotherapy were ~0.5 SD lower than women who had not received chemotherapy.[18]

The incidence of primary ovarian failure in women receiving breast cancer chemotherapy ranges from 20% to 90% depending on age and treatment regimen.[19–21] Primary ovarian failure leads to a sudden decrease in estrogen production and can result in bone loss[18,22] (Fig. 1). Studies examining the BMD of premenopausal women with breast cancer under chemotherapy found a significant decrease in BMD in women who received chemotherapy and experienced amenorrhea.[13,18] These studies suggest that chemotherapy leads to an increased risk of entering early menopause and, subsequently, an increased risk of bone loss.

Several groups have shown that bisphosphonates are able to reduce bone loss associated with breast cancer chemotherapy. Studies examining the effects of risedronate or clodronate compared with placebo on BMD in premenopausal women receiving chemotherapy, cyclophosphamide methotrexate fluorouracil (CMF) treatment, and/or tamoxifen showed that the change in BMD in the bisphosphonate group was significantly decreased compared with placebo.[18]

In short, chemotherapy in the breast cancer setting has a negative impact on BMD in women undergoing treatment. This could be attributed to direct effects on the bone microenvironment and indirect effects caused by the decrease in estrogen that occurs after chemotherapy-induced primary ovarian failure.

PROSTATE CANCER

Androgen Deprivation Therapy

Prostate cancer patients who are diagnosed with metastatic disease or whose other clinical features suggest a poor chance for cure often receive palliative therapy in the form of androgen deprivation therapy (ADT) as a first-line treatment. ADT includes surgical castration, pharmacologic castration with agents such as luteinizing hormone–releasing hormone agonists (LHRH-a), and anti-androgen therapy with agents such as flutamide, nilutamide, bicalutamide, or cyproterone.[23] One potential complication that can arise from surgical castration or gonadotropin-releasing hormone (GnRH) agonist therapy is a decrease in BMD[24–26] (Fig. 1). ADT causes 2–8% loss of BMD at the lumbar spine and 1.8–6.5% at the hip annually.[26] This is believed to be caused by a decrease in estrogen, because androgens are converted to estrogen by aromatases that are essential to maintain skeletal health.[27] In addition there may be a direct effect of androgens on osteoblast differentiation such that a decrease in androgens would lead to decreased bone formation.[28]

Several retrospective studies have shown this link between ADT and increased bone loss and fracture risk. These studies showed an increased risk of osteoporotic fractures in men with prostate cancer with orchiectomy and those receiving LHRH-a.[6] Vitamin D levels, low calcium diet, the duration of ADT, and alcohol excess were risk factors for osteoporosis and fractures in men with prostate cancer[24,26,29,30]; 19.4% of the men who received ADT suffered at least one fracture compared with 12.6% of men who did not receive ADT ($p < 0.001$) in a recent study.[6] All of these data strongly suggest that ADT is a risk factor for osteoporosis and fracture in men with prostate cancer.

A number of recent studies have suggested that bisphosphonate therapy can reduce this increased risk for bone loss and subsequent fracture in men receiving ADT. Accessing the effect of ZA on BMD during ADT in 106 men with prostate cancer showed that, whereas the men in the control group had a 2.2% decrease in BMD, the men in the ZA group had a 5.6% increase in BMD ($p < 0.001$).[26] A single dose of ZA was shown to be effective in preventing GnRH agonist–induced bone loss in men with nonmetastatic prostate cancer over 1 yr.[31] Currently studies evaluating the effect on fracture reduction by antiresorptive agent ZA and the monoclonal antibody against RANK (AMG162) are underway.[32]

The use of SERMs may also combat bone loss in men receiving ADT. A comparison of BMD scores in 48 men with nonmetastatic prostate cancer who received 12 mo of raloxifene (60 mg/d) or no raloxifene during concurrent treatment with a GnRH agonist showed that men receiving raloxifene demonstrated an increase in total hip BMD of 1.1 ± 0.4%,

© 2008 American Society for Bone and Mineral Research

whereas the no raloxifene group showed a decrease in total hip BMD of 2.6 ± 0.7% ($p < 0.001$).[26]

Antiandrogen compounds, which competitively inhibit activation of the androgen receptor by testosterone, can be used as prostate cancer treatment alone or in combination with other forms of ADT. In men with nonmetastatic prostate cancer, monotherapy with the nonsteroidal anti-androgen bicalutamide provides similar survival rates to castration.[33] The observation that bicalutamide monotherapy significantly increases serum concentrations of testosterone and estradiol compared with baseline[6] suggests that bicalutamide monotherapy may have a bone-protective effect in men with nonmetastatic prostate cancer. Three independent studies showed that BMD is significantly increased in lumbar spine and hip in prostate cancer patients on bicalutamide monotherapy compared with castration or leuprolide treatment, a GnRH agonist.[6,34,35] In summary, it seems that bicalutamide monotherapy may offer skeleto-protective benefits in men with nonmetastatic prostate cancer. An important next step will be to evaluate its effect on fracture rate compared with other forms of ADT.

RADIATION-INDUCED FRACTURES

Breast conservation surgery combined with radiotherapy has become the standard of care for patients with early-stage breast cancer.[36] One potential complication of this treatment is rib fracture after X-ray exposure, although few studies have investigated this phenomenon. Examining the incidence of various radiation-induced complications in 1624 patients with early stage breast cancer treated between 1968 and 1985 showed that the median follow-up time for survivors was 79 mo. The incidence of rib fracture was between 0.4% and 2.2% depending on the type of linear accelerator used.[37] The incidence of radiation-induced complications in 294 women receiving surgery and radiotherapy treatment between 1990 and 1992 was evaluated, and the risk of rib fractures was found to be 0.3%.[36]

Current recommendations for prostate cancer treatment suggest that radiotherapy is an option for nonmetastatic patients with a long life expectancy (>5 yr) or for patients for whom surgery is a contraindication.[38–40] Whereas a possible side effect of radiotherapy for prostate cancer could be pelvic fracture, there are no clinical studies examining this potential risk. Case reports have documented pelvic fracture in men who have received pelvic irradiation[41]; however, fracture data for the newer radiotherapies are not yet available.[42]

OSTEONECROSIS OF THE JAW

Osteonecrosis of the jaw (ONJ) is a clinical condition where bones of the mandible or maxilla are exposed in the oral cavity and has been associated with chronic bisphosphonate use.[43,44] ONJ has been reported in patients receiving chronic oral and intravenous bisphosphonate therapy for bone metastases and less commonly observed when administered for osteoporosis.[16,45] The rate of ONJ in patients receiving monthly intravenous bisphosphonate for bone metastases is in the range of 1–10%.[46]

In patients receiving bisphosphonates, ONJ is often associated with poor dental hygiene or after invasive dental procedures such as tooth extractions and placement of oral implants.[47] Inhibition of the osteoclast by bisphosphonates is hypothesized to disrupt the critical balance between the osteoclast and the osteoblast. In a situation where healing of the bone is necessary, such as after chronic inflammation and infection associated with gum disease, the disruption of the dynamic and coupled processes of bone resorption and formation may contribute to the development of ONJ. Meticulous oral hygiene, antibiotics, and the discontinuation of bisphosphonates are currently recommended for therapy of ONJ. The diagnosis of ONJ is a clinical diagnosis made by physical examination. Biopsy of the affected bone can be associated with worsening of the situation. More studies must be initiated to determine the exact mechanisms and cause of this complication and how it can be prevented and treated.

CONCLUSION

Chemotherapy and hormonal therapies for breast and prostate cancer have the potential to lead to significant bone loss primarily through the disruption of the bone-enhancing properties of estrogen. Current recommendations for avoiding the skeletal complications of cancer therapy include adequate intake of calcium and vitamin D, regular weight-bearing exercise, cessation of smoking, reduction in alcohol intake, and bisphosphonate therapy for osteoporotic patients.[21,48] Patients who are being treated with hormonal therapies are at increased risk for skeletal complications and should have regular BMD monitoring by DXA. The role of antiresorptive, osteoclast inhibitor therapy to prevent cancer therapy–associated bone loss is under active investigation. It is recommended that patients who initiate bisphosphonate therapy receive a thorough oral examination and treatment for dental infections before initiating bisphosphonate therapy.[6,43] In addition to treating the disease, careful monitoring of bone health is now an essential component of the treatment of both breast and prostate cancer.

REFERENCES

1. Hofbauer LC, Hicok KC, Chen D, Khosla S 2002 Regulation of osteoprotegerin production by androgens and anti-androgens in human osteoblastic lineage cells. Eur J Endocrinol **147:**269–273.
2. Hughes DE, Dai A, Tiffee JC, Li HH, Mundy GR, Boyce BF 1996 Estrogen promotes apoptosis of murine osteoclasts mediated by TGF-beta. Nat Med **2:**1132–1136.
3. Riggs BL, Khosla S, Melton LJ III 1998 A unitary model for involutional osteoporosis: Estrogen deficiency causes both type I and type II osteoporosis in postmenopausal women and contributes to bone loss in aging men. J Bone Miner Res **13:**763–773.
4. Harvey JM, Clark GM, Osborne CK, Allred DC 1999 Estrogen receptor status by immunohistochemistry is superior to the ligand-binding assay for predicting response to adjuvant endocrine therapy in breast cancer. J Clin Oncol **17:**1474–1481.
5. Cuzick J, Forbes JF, Sestak I, Cawthorn S, Hamed H, Holli K, Howell A 2007 Long-term results of tamoxifen prophylaxis for breast cancer–96-month follow-up of the randomized IBIS-I trial. J Natl Cancer Inst **99:**272–282.
6. Khan MN, Khan AA 2008 Cancer treatment-related bone loss: A review and synthesis of the literature. Curr Oncol **15**(Suppl 1)**:**S30–S40.
7. Cosman F 2003 Selective estrogen-receptor modulators. Clin Geriatr Med **19:**371–379.
8. Lester J, Coleman R 2005 Bone loss and the aromatase inhibitors. Br J Cancer **93**(Suppl 1)**:**S16–S22.
9. Cummings SR, Duong T, Kenyon E, Cauley JA, Whitehead M, Krueger KA 2002 Serum estradiol level and risk of breast cancer during treatment with raloxifene. JAMA **287:**216–220.
10. Ettinger B, Black DM, Mitlak BH, Knickerbocker RK, Nickelsen T, Genant HK, Christiansen C, Delmas PD, Zanchetta JR, Stakkestad J, Gluer CC, Krueger K, Cohen FJ, Eckert S, Ensrud KE, Avioli LV, Lips P, Cummings SR 1999 Reduction of vertebral fracture risk in postmenopausal women with osteoporosis treated with raloxifene: Results from a 3-year randomized clinical trial. Multiple Outcomes of Raloxifene Evaluation (MORE) Investigators. JAMA **282:**637–645.
11. Lippman ME, Krueger KA, Eckert S, Sashegyi A, Walls EL, Jamal S, Cauley JA, Cummings SR 2001 Indicators of lifetime estrogen exposure: Effect on breast cancer incidence and interaction

© 2008 American Society for Bone and Mineral Research

with raloxifene therapy in the multiple outcomes of raloxifene evaluation study participants. J Clin Oncol **19:**3111–3116.
12. Martino S, Cauley JA, Barrett-Connor E, Powles TJ, Mershon J, Disch D, Secrest RJ, Cummings SR 2004 Continuing outcomes relevant to Evista: Breast cancer incidence in postmenopausal osteoporotic women in a randomized trial of raloxifene. J Natl Cancer Inst **96:**1751–1761.
13. Hadji P 2008 Menopausal symptoms and adjuvant therapy-associated adverse events. Endocr Relat Cancer **15:**73–90.
14. Simpson ER, Dowsett M 2002 Aromatase and its inhibitors: Significance for breast cancer therapy. Recent Prog Horm Res **57:**317–338.
15. Goss PE, Qi S, Josse RG, Pritzker KP, Mendes M, Hu H, Waldman SD, Grynpas MD 2004 The steroidal aromatase inhibitor exemestane prevents bone loss in ovariectomized rats. Bone **34:**384–392.
16. Guise TA 2006 Bone loss and fracture risk associated with cancer therapy. Oncologist **11:**1121–1131.
17. Vanderschueren D, Vandenput L, Boonen S, Lindberg MK, Bouillon R, Ohlsson C 2004 Androgens and bone. Endocr Rev **25:**389–425.
18. Lester J, Dodwell D, McCloskey E, Coleman R 2005 The causes and treatment of bone loss associated with carcinoma of the breast. Cancer Treat Rev **31:**115–142.
19. Bines J, Oleske DM, Cobleigh MA 1996 Ovarian function in premenopausal women treated with adjuvant chemotherapy for breast cancer. J Clin Oncol **14:**1718–1729.
20. Goodwin PJ, Ennis M, Pritchard KI, Trudeau M, Hood N 1999 Risk of menopause during the first year after breast cancer diagnosis. J Clin Oncol **17:**2365–2370.
21. Ramaswamy B, Shapiro CL 2003 Osteopenia and osteoporosis in women with breast cancer. Semin Oncol **30:**763–775.
22. Shapiro CL, Phillips G, Van Poznak CH, Jackson R, Leboff MS, Woodard S, Lemeshow S 2005 Baseline bone mineral density of the total lumbar spine may predict for chemotherapy-induced ovarian failure. Breast Cancer Res Treat **90:**41–46.
23. Loblaw DA, Mendelson DS, Talcott JA, Virgo KS, Somerfield MR, Ben-Josef E, Middleton R, Porterfield H, Sharp SA, Smith TJ, Taplin ME, Vogelzang NJ, Wade JL Jr, Bennett CL, Scher HI 2004 American Society of Clinical Oncology recommendations for the initial hormonal management of androgen-sensitive metastatic, recurrent, or progressive prostate cancer. J Clin Oncol **22:**2927–2941.
24. Krupski TL, Smith MR, Lee WC, Pashos CL, Brandman J, Wang Q, Botteman M, Litwin MS 2004 Natural history of bone complications in men with prostate carcinoma initiating androgen deprivation therapy. Cancer **101:**541–549.
25. Diamond TH, Higano CS, Smith MR, Guise TA, Singer FR 2004 Osteoporosis in men with prostate carcinoma receiving androgen-deprivation therapy: Recommendations for diagnosis and therapies. Cancer **100:**892–899.
26. Holmes-Walker DJ, Woo H, Gurney H, Do VT, Chipps DR 2006 Maintaining bone health in patients with prostate cancer. Med J Aust **184:**176–179.
27. Hansen KA, Tho SP 1998 Androgens and bone health. Semin Reprod Endocrinol **16:**129–134.
28. Vanderschueren D, Bouillon R 1995 Androgens and bone. Calcif Tissue Int **56:**341–346.
29. Diamond TH, Bucci J, Kersley JH, Aslan P, Lynch WB, Bryant C 2004 Osteoporosis and spinal fractures in men with prostate cancer: Risk factors and effects of androgen deprivation therapy. J Urol **172:**529–532.
30. Planas J, Morote J, Orsola A, Salvador C, Trilla E, Cecchini L, Raventos CX 2007 The relationship between daily calcium intake and bone mineral density in men with prostate cancer. BJU Int **99:**812–815.
31. Michaelson MD, Kaufman DS, Lee H, McGovern FJ, Kantoff PW, Fallon MA, Finkelstein JS, Smith MR 2007 Randomized controlled trial of annual zoledronic acid to prevent gonadotropin-releasing hormone agonist-induced bone loss in men with prostate cancer. J Clin Oncol **25:**1038–1042.
32. Smith MR 2006 Treatment-related osteoporosis in men with prostate cancer. Clin Cancer Res **12:**6315s–6319s.
33. Pronzato P, Rondini M 2005 Hormonotherapy of advanced prostate cancer. Ann Oncol **16**(Suppl 4)**:**iv80–iv84.
34. Tyrrell CJ, Blake GM, Iversen P, Kaisary AV, Melezinek I 2003 The non-steroidal antiandrogen, bicalutamide ('Casodex'), may preserve bone mineral density as compared with castration: Results of a preliminary study. World J Urol **21:**37–42.
35. Sieber PR, Keiller DL, Kahnoski RJ, Gallo J, McFadden S 2004 Bicalutamide 150 mg maintains bone mineral density during monotherapy for localized or locally advanced prostate cancer. J Urol **171:**2272–2276.
36. Meric F, Buchholz TA, Mirza NQ, Vlastos G, Ames FC, Ross MI, Pollock RE, Singletary SE, Feig BW, Kuerer HM, Newman LA, Perkins GH, Strom EA, McNeese MD, Hortobagyi GN, Hunt KK 2002 Long-term complications associated with breast-conservation surgery and radiotherapy. Ann Surg Oncol **9:**543–549.
37. Pierce SM, Recht A, Lingos TI, Abner A, Vicini F, Silver B, Herzog A, Harris JR 1992 Long-term radiation complications following conservative surgery (CS) and radiation therapy (RT) in patients with early stage breast cancer. Int J Radiat Oncol Biol Phys **23:**915–923.
38. Aus G, Abbou CC, Bolla M, Heidenreich A, Schmid HP, van Poppel H, Wolff J, Zattoni F 2005 EAU Guidelines on Prostate Cancer. Eur Urol **48:**546–551.
39. Pirtskhalaishvili G, Hrebinko RL, Nelson JB 2001 The treatment of prostate cancer: An overview of current options. Cancer Pract **9:**295–306.
40. Nag S, Beyer D, Friedland J, Grimm P, Nath R 1999 American Brachytherapy Society (ABS) recommendations for transperineal permanent brachytherapy of prostate cancer. Int J Radiat Oncol Biol Phys **44:**789–799.
41. Csuka M, Brewer BJ, Lynch KL, McCarty DJ 1987 Osteonecrosis, fractures, and protrusio acetabuli secondary to x-irradiation therapy for prostatic carcinoma. J Rheumatol **14:**165–170.
42. Stein ME, Boehmer D, Kuten A 2007 Radiation therapy in prostate cancer. Recent Results Cancer Res **175:**179–199.
43. Woo SB, Hande K, Richardson PG 2005 Osteonecrosis of the jaw and bisphosphonates. N Engl J Med **353:**99–102.
44. Purcell PM, Boyd IW 2005 Bisphosphonates and osteonecrosis of the jaw. Med J Aust **182:**417–418.
45. Marx RE 2003 Pamidronate (Aredia) and zoledronate (Zometa) induced avascular necrosis of the jaws: A growing epidemic. J Oral Maxillofac Surg **61:**1115–1117.
46. Khosla S, Burr D, Cauley J, Dempster DW, Ebeling PR, Felsenberg D, Gagel RF, Gilsanz V, Guise T, Koka S, McCauley LK, McGowan J, McKee MD, Mohla S, Pendrys DG, Raisz LG, Ruggiero SL, Shafer DM, Shum L, Silverman SL, Van Poznak CH, Watts N, Woo SB, Shane E 2007 Bisphosphonate-associated osteonecrosis of the jaw: Report of a task force of the American Society for Bone and Mineral Research. J Bone Miner Res **22:**1479–1491.
47. Lugassy G, Shaham R, Nemets A, Ben-Dor D, Nahlieli O 2004 Severe osteomyelitis of the jaw in long-term survivors of multiple myeloma: A new clinical entity. Am J Med **117:**440–441.
48. Saad F, Olsson C, Schulman CC 2004 Skeletal morbidity in men with prostate cancer: Quality-of-life considerations throughout the continuum of care. Eur Urol **46:**731–739.
49. Eastell R, Adams JE, Coleman RE, Howell A, Hannon RA, Cuzick J, Mackey JR, Beckmann MW, Clack G 2008 Effect of anastrozole on bone mineral density: 5-year results from the anastrozole, tamoxifen, alone or in combination trial 18233230. J Clin Oncol **26:**1051–1057.
50. Grant M, Hausmaninger H, Samonigg H, Mlineritsch B, Taucher S, Luschin-Ebengreuth G, Jakesz R 2002 Changes in bone mineral density caused by anastrozole or tamoxifen in combination with goserelin (+/– zoledronate) as adjuvant treatment for hormone receptor-positive premenopausal breast cancer: Results of a randomized multicenter trial. 25th Annual San Antonio Breast Cancer Symposium, December 11, 2002, San Antonio, TX, USA.

© 2008 American Society for Bone and Mineral Research

Chapter 83. Bone Cancer and Pain

Mary Ann C. Sabino[1] and Denis R. Clohisy[2]

[1]*Division of Oral and Maxillofacial Surgery, University of Minnesota School of Dentistry, Minneapolis, Minnesota;* [2]*Department of Orthopedic Surgery and Cancer Center, University of Minnesota School of Medicine, Minneapolis, Minnesota*

EPIDEMIOLOGY OF BONE CANCER PAIN

Bone cancer pain is one of the most difficult pain entities to treat and significantly affects the quality of life for many patients with bone cancer. It can also be frustrating for many clinicians to treat because of its recalcitrant nature. Significant insight into what causes bone cancer pain has been generated in the past decade because of the development of novel animal models. These insights have led to the development of new therapeutic strategies and refinement of older ones to improve our understanding and hope of new mechanistically based therapies in the future for treating this life-changing medical condition.

Pain is the most common presenting symptom in patients with skeletal metastases and is directly proportional to its impact on the cancer patient's quality of life.[1] Two main types of cancer pain exist: ongoing pain and incident or breakthrough pain. Ongoing pain is typically described as a dull and aching pain that is constant in nature and progresses in accordance with the disease process. Incident or breakthrough pain is most commonly associated with bone metastases and is characterized by sharp pains that are intermittent in nature and exacerbated by both volitional and nonvolitional movements. The latter type of pain is difficult to treat because of its intermittence and intensity and can be found in as many as 80% of patients with advanced disease.[2]

MECHANISMS OF BONE CANCER PAIN

Animal Models of Cancer Pain

Rodent and canine models of osteolytic and osteoblastic bone cancer pain have been described in the literature in the past decade. Each model differs in the route of inoculation of tumor cells,[3–5] type of tumor cells,[6,7] immunocompetency of the host,[7] and species of host.[8,9] Despite these differences, a wealth of information has been gleaned on the pathophysiologic mechanisms that drive bone cancer pain. Bone cancer pain is a multifactorial process that arises from different cells within the cancerous bone and whose signals extend toward the central nervous system for sensory processing.

Peripheral Mechanisms of Cancer Pain

Pain generally occurs during tissue ischemia and/or damage and is a result of release of neurotransmitters, cytokines, and factors from damaged cells, adjacent blood vessels, and nerve terminals. Pain is transduced at the level of the primary afferent nerve fiber that innervates peripheral tissues including bone. Previous studies have shown that bone is densely innervated by both sensory and sympathetic nerve fibers within bone marrow, mineralized bone, and periosteum (Fig. 1).[10] Sensory and sympathetic neurons are present within all three anatomic locations and are influenced by fractures, ischemia, or the presence of tumor cells and may play a unique but coordinated role in the generation of bone cancer pain.

The majority of metastatic skeletal malignancies are destructive in nature and produce regions of osteolysis (bone destruction). This primarily occurs through activation, recruitment, and proliferation of osteoclasts and is characterized by an increased number and size of osteoclasts found in tumor-bearing sites.[11,12] The activation and proliferation of osteoclasts is mediated by the interaction between RANK/RANKL/osteoprotegerin (OPG) signaling between osteoblast lineage and pre-osteoclastic mononuclear cells. This complex interaction is further described elsewhere in chapters in this publication. An increased expression of both RANK and RANKL has been found in tumor-bearing sites.[13] Osteoclasts have now been shown to play a pivotal role in the development and progression of bone cancer pain. Selective destruction of osteoclasts using the soluble decoy receptor for RANKL, OPG, results in inhibition of cancer-induced osteolysis, cancer pain behaviors, and neurochemical markers of peripheral and central sensitization.[14,15]

Tumor-derived cytokines, growth factors, and peptides have been shown to activate primary afferent nerve fibers that innervate bone. Prostaglandins, interleukins, protons, bradykinin, chemokines, TNF-α, and endothelins are all examples of chemical mediators released from tumor cells and have been shown to sensitize nerve terminals[16–21] resulting in cancer pain. Each mediator has a specialized cognate channel or receptor to which it binds resulting in conversion from a chemical to electrical signal (Fig. 2). In the bone cancer pain state, chemical mediators are released and then bind to their respective receptor or channel, and pain transduction occurs and resolves over time. Peripheral sensitization occurs where the peripheral nerve becomes sensitized because of constant stimulation and activation. This may be in the form of decreased excitation thresholds, upregulation of receptors and/or channels in both nerve terminals and sensory neurons, or recruitment of previously silent pain receptors or nociceptors and have been shown to occur in bone cancer pain models.[5,22–25]

Destruction or direct damage to peripheral nerves innervating bone has been postulated as another mechanism by which bone cancer exists. This neuropathic pain condition is supported by clinical findings of resistance to traditional opioid therapy known to be ineffective in managing neuropathic pain. Animal models of bone cancer pain have also shown damaged peripheral nerve terminals within cancerous bone and increased expression of neurochemical markers of injury in sensory neurons that innervate bone. Gabapentin (Neurontin; Pfizer, New York, NY, USA), a common drug used to manage neuropathic pain, was efficacious in reducing neuropathic pain behaviors but did not affect tumor-induced bone destruction or tumor proliferation.[26]

Central Sensitization

Central sensitization refers to the heightened reactivity of central nervous system neurons in the face of sustained peripheral neural input. Whereas this may occur within the thalamus and cortex, research has been primarily focused on the dorsal horn of the spinal cord. Electrophysiologic and anatomic studies have shown a change in the activity and responsiveness of dorsal horn neurons in response to persistent painful stimulation.

The authors state that they have no conflicts of interest.

Key words: hyperalgesia, osteolysis, sensitization, osteoprotegerin, bisphosphonates, growth factors, analgesia, allodynia

© 2008 American Society for Bone and Mineral Research

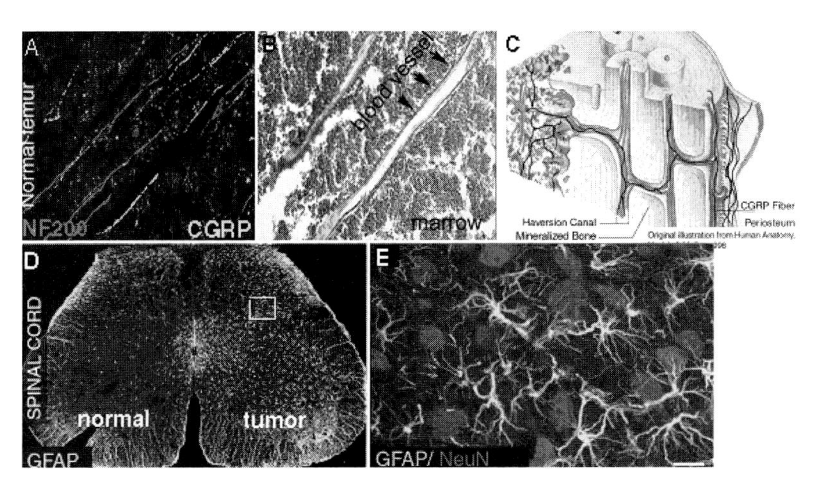

FIG. 1. Peripheral and central mechanisms of bone cancer. Histophotomicrographs of confocal (A) and histologic (B) serial images of normal bone and confocal images of spinal cord of tumor-bearing mice (D and E). Note the extensive myelinated (gray, NF 200) and unmyelinated (white, CGRP) nerve fibers within bone marrow that appear to course along blood vessels (arrowheads, B). (C) Schematic diagram showing the innervation within periosteum, mineralized bone, and bone marrow. All three tissues may be sensitized during the various stages of bone cancer pain. (D) Confocal imaging of GFAP expressed by astrocytes in a spinal cord of a tumor-bearing mouse. Note increased expression only on side ipsilateral to tumorous limb. (E) High-power magnification of tumor-bearing side of spinal cord showing hypertrophy of astrocytes (white) without changes in neuronal numbers (gray, stained with neuronal marker, NeuN). NF200, neurofilament 200; CGRP, calcitonin gene-related peptide; GFAP, glial fibrillary acidic protein; NeuN, neuronal marker.

Persistent stimulation of unmyelinated C fibers results in the increased activity and responsiveness of spinal neurons that receive input from the stimulated unmyelinated C fiber. This increased responsiveness is of short duration and is known as wind up. Sensitization can also occur when persistent stimulation results in phenotypic changes in neurons that do not receive, but are adjacent to, neurons that receive the persistent painful stimulation. Typically these neurons receive input from A-β fibers that normally do not transmit painful stimuli. However, once sensitized, these neurons are capable of transmitting both nonpainful and painful information. Central sensitization is mediated in part by glutamate, substance P, prostaglandins, and growth factors. These receptors/ion channels are known as *N*-methyl-D-aspartate (NMDA), neurokinin-1 (NK-1), prostaglandin E receptor (EP), and tyrosine kinase B (trkB), respectively. Upregulation of transient receptor potential (TRPV1) and sodium channels have also been reported in central sensitization. Allodynia is a condition where normally non-noxious stimulation is painful and a condition that can result from central sensitization.

Reorganization of Peripheral and Central Nervous System in Response to Cancer Pain

There are several examples of nociceptor peripheral sensitization in experimental cancer models.[3,8,16] In normal mice, the neurotransmitter substance P is synthesized by nociceptors and released in the spinal cord when noxious—but not non-noxious—mechanical stress is applied to the femur. In mice with bone cancer, what would normally be a nonpainful level of mechanical stress can induce the release of substance P from primary afferent fibers that terminate in the spinal cord. Substance P, in turn, binds to and activates the neurokinin-1 receptor that is expressed by a subset of spinal cord neurons.

Similar to phenotypic alterations and sensitization of peripheral nerves, studies involving a murine model of bone cancer pain also showed extensive neurochemical reorganization in the spinal cord segments that receive input from primary afferent neurons that innervate the tumor-bearing bone. This includes an astrocyte hypertrophy, which may be accompanied by a decreased expression of glutamate reuptake transporters resulting in increased extracellular levels of the excitatory neurotransmitter glutamate leading to excitotoxicity within the central nervous system. This and other spinal cord changes could be attenuated by blocking the tumor-induced tissue destruction and pain.[14,21,24,27,28] These findings indicate that cancer pain induces, and is at least partially maintained, by a state of central sensitization, in which neurochemical changes in the spinal cord and forebrain promote an increased transmission of nociceptive information, so that normally non-noxious input is amplified and perceived as a noxious stimulus.

The constant barrage of neurochemicals that are implicated in pain transmission results in changes in the activity and phenotype of nerve fibers and neurons within the peripheral and central nervous system. These changes result in a chronic and debilitating condition that is recalcitrant to many conventional modes of acute pain therapy. The chronic bone cancer pain patient has a "pain portrait" that is distinct from the acute pain patient and its management reflects this difference (Fig. 1).

THERAPEUTIC STRATEGIES: PAST, PRESENT, AND FUTURE

Pain research has significantly improved our understanding of acute and chronic pain mechanisms. By highlighting key molecular mechanisms involved in pain transmission, new drugs are currently being studied as potential novel and selec-

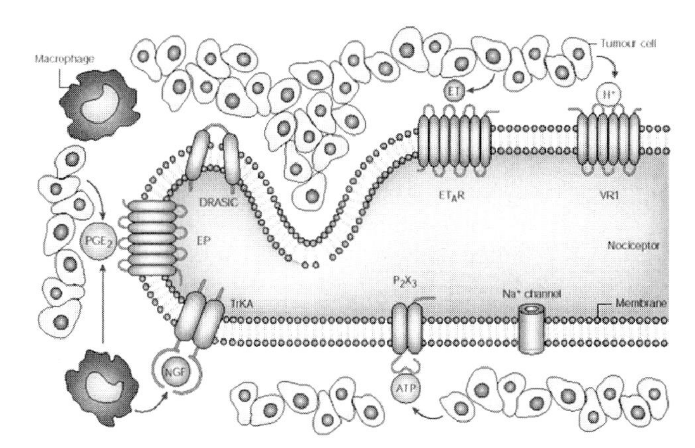

FIG. 2. Interaction between chemical mediators and receptors. Schematic diagram of a peripheral pain fiber expressing receptors and ion channels. Interaction between neurotransmitters and chemical mediators and their cognate receptor results in pain transduction and signaling (Reprinted with permission from Macmillan Publishers Ltd. from Mantyh PW, Clohisy DR, Koltzenburg M, Hunt SP 2002 Molecular mechanisms of cancer pain. Nat Rev Cancer **2**:201–209.) H+, protons; ET, endothelin; VR1, vanilloid receptor-1; ETAR, endothelin A receptor; DRASIC, dorsal root acid sensing ion channel; EP, prostaglandin E receptor; PGE2, prostaglandin E2; TrkA, high-affinity nerve growth factor tyrosine kinase receptor A; NGF, nerve growth factor; ATP, adenosine triphosphate; P2X3, purinergic ion-gated receptor; Na^+, sodium.

© 2008 American Society for Bone and Mineral Research

tive therapies. Currently available medications such as the opioids are certainly fraught with side effect profiles that may limit their clinical efficacy and patient quality of life. Research is now focused on specific receptor or channel targets within the nervous system that limits systemic complications.

Therapeutic Targets: Ion Channels

The TRPV1 family of channels is located on unmyelinated C fibers and spinal nociceptive neurons that mediate pain transmission. TRPV1 channels can be activated by noxious heat, capsaicin, and acid. Mice that lack the channel are unable to develop chronic pain states, and when antagonists to TRPV1 are administered orally or into the intrathecal space, chronic pain is markedly reduced.[19,29] Interestingly, intrathecal administration of resinifera toxin, a potent capsaicin analog, in a canine bone cancer model resulted in reductions in pain behaviors and selective destruction of small-, but not medium- or large-diameter, sensory neurons.[9] Long-term efficacy or safety has not been established in animal models, but TRPV1 is only expressed on nociceptive peripheral terminals and spinal neurons; therefore, its selective blockade may limit its side effect profile.

Therapeutic Targets: Growth Factors

Growth factors are important in the differentiation, maturation, and plasticity of cells within the nervous system. Key growth factors that have been implicated in pain include nerve growth factor (NGF), glial-derived growth factor (GDNF), and brain-derived growth factor (BDNF).[30] Receptors for such growth factors are expressed on C fibers and Aδ fibers that terminate within the spinal cord. In animal models of chronic pain, NGF levels are elevated in damaged peripheral tissues. Neutralizing antibodies against NGF are effective, at least in animal models, in reducing and, in some cases, preventing chronic pain. Interestingly, administration of NGF has also been shown to reduce neuropathic pain. Phase III clinical trials are currently underway assessing the efficacy of recombinant NGF to treat diabetic neuropathy. BDNF has been implicated in the modulation of central sensitization because its expression is increased in nociceptive neurons in models of peripheral neuropathy. BNDF sensitizes C fiber activity resulting in hyperalgesia and allodynia. Inhibition of BNDF and its cognate receptor, trkB, results in decreased C fiber firing and a reduction in pain behaviors. Finally, GDNF is important in the survival of sensory neurons and supporting cells. Neuropathic pain behaviors commonly observed in animal models of chronic pain are prevented or reversed after GDNF administration. The timing of administration determined whether treatment was protective or therapeutic nature.

Therapeutic Targets: Cytokines

Endothelin antagonists are another type of pharmacological agent that may show significant promise in the management of cancer pain.[20,31] Endothelins are a family of vasoactive peptides that are expressed by several tumors, especially prostate cancer, with levels that seem to correlate with pain severity.[32] Endothelins could contribute to cancer pain by directly sensitizing or exciting nociceptors, because a subset of small unmyelinated primary afferent neurons express endothelin A receptors.[33] Direct application of endothelin to peripheral nerves induces activation of primary afferent fibers and an induction of pain behaviors,[34] whereas selective blockade of endothelin A receptors blocked bone cancer pain–related behaviors and spinal changes indicative of peripheral and central sensitization.[20] Like prostaglandins, endothelins that are produced by cancer cells are also thought be involved in regulating angiogenesis[35] and tumor growth.[36] These findings indicate endothelin antagonists may be useful not only in inhibiting cancer pain but in reducing tumor growth and metastasis.

Therapeutic Targets: Osteoclast

Experiments in a murine model of bone cancer pain report that osteoclasts play an essential role in cancer-induced bone loss and that osteoclasts contribute to the etiology of bone cancer pain. OPG is a soluble target that holds significant promise for alleviating bone cancer pain.[13,37] Whereas OPG has been shown to decrease pain behaviors in an animal model of bone cancer,[14,27] it is still being developed for use in cancer patients. The reader is directed to other chapters within this primer for further descriptions of the OPG system and its clinical utility.

Bisphosphonates are pyrophosphate analogs that are commonly used to manage osteolytic and osteoblastic processes such as primary bone tumors (i.e., multiple myeloma), metastatic bone cancers, humoral hypercalcemia of malignancy, and moderate to severe osteoporosis and osteopenia. Nitrogen-containing bisphosphonates are available in both intravenous (zoledronic acid/Zometa, Novartis Pharmaceuticals; pamidronate/Aredia, Novartis Pharmaceuticals) and oral forms (alendronate/Fosamax, Merck and Co.; ibandronate/Boniva, Roche Laboratories; risedronate/Actonel, Proctor and Gamble) with differing potencies. Bisphosphonates act by disrupting the activity of osteoclasts, one of the principle bone cells responsible for bone metabolism.[38]

Administration of bisphosphonates have a clear and positive impact on skeletal health and quality of life in people who suffer from destructive bone diseases and have also been reported to reduce pain in patients with osteoclast-induced skeletal metastases.[39,40] The long-term beneficial effects of bisphosphonate treatment on reductions in bone pain and skeletal related events (e.g., spinal compression, fractures) and an improvement in overall quality of life are clearly obvious from clinical trials in multiple myeloma, breast cancer,[39,41] and prostate cancer[42] patients. Because of the potentially beneficial and skeletal protective effects of bisphosphonates, recommendations may evolve to include prophylactic treatment of prostate cancer patients without evidence of skeletal metastases with bisphosphonates.[42]

REFERENCES

1. Coleman RE 2006 Clinical features of metastatic bone disease and risk of skeletal morbidity. Clin Cancer Res **12:**6243–6249.
2. Mercadante S, Fulfaro F 2007 Management of painful bone metastases. Curr Opin Oncol **19:**308–314.
3. Schwei MJ, Honore P, Rogers SD, Salak-Johnson JL, Finke MP, Ramnaraine ML, Clohisy DR, Mantyh PW 1999 Neurochemical and cellular reorganization of the spinal cord in a murine model of bone cancer pain. J Neurosci **19:**10886–10897.
4. Zhang R-X, Liu B, Wang L, Ren K, Qiao J-T, Berman BM, Lao L 2005 Spinal glial activation in a new rat model of bone cancer pain produced by prostate cancer cell inoculation of the tibia. Pain **118:**125–136.
5. Cain DM, Wacnik PW, Turner M, Wendelschafer-Crabb G, Kennedy WR, Wilcox GL, Simone DA 2001 Functional interactions between tumor and peripheral nerve: Changes in excitability and morphology of primary afferent fibers in a murine model of cancer pain. J Neurosci **21:**9367–9376.
6. Halvorson KGBA, Sevcik MABA, Ghilardi JRBA, Rosol TJD-VMP, Mantyh PWPJD 2006 Similarities and differences in tumor growth, skeletal remodeling and pain in an osteolytic and osteoblastic model of bone cancer. Clin J Pain **22:**587–600.
7. Sabino MA, Luger NM, Mach DB, Rogers SD, Schwei MJ, Mantyh PW 2003 Different tumors in bone each give rise to a distinct pattern of skeletal destruction, bone cancer-related pain

© 2008 American Society for Bone and Mineral Research

behaviors and neurochemical changes in the central nervous system. Int J Cancer **104:**550–558.
8. Walker K, Medhurst SJ, Kidd BL, Glatt M, Bowes M, Patel S, McNair K, Kesingland A, Green J, Chan O, Fox AJ, Urban LA 2002 Disease modifying and anti-nociceptive effects of the bisphosphonate, zoledronic acid in a model of bone cancer pain. Pain **100:**219–229.
9. Brown DC, Iadarola MJ, Perkowski SZ, Erin H, Shofer F, Laszlo KJ, Olah Z, Mannes AJ 2005 Physiologic and antinociceptive effects of intrathecal resinifera toxin in a canine bone cancer model. Anesthesiology **103:**1052–1059.
10. Mach DB, Rogers SD, Sabino MC, Luger NM, Schwei MJ, Pomonis JD, Keyser CP, Clohisy DR, Adams DJ, O'Leary P, Mantyh PW 2002 Origins of skeletal pain: Sensory and sympathetic innervation of the mouse femur. Neuroscience **113:**155–166.
11. Taube T, Elomaa I, Blomqvist C, Beneton MN, Kanis JA 1994 Histomorphometric evidence for osteoclast-mediated bone resorption in metastatic breast cancer. Bone **15:**161–166.
12. Clohisy DR, Perkins SL, Ramnaraine ML 2000 Review of cellular mechanisms of tumor osteolysis. Clin Orthop Relat Res **373:**104-114.
13. Clohisy DR, Mantyh PW 2004 Bone cancer pain and the role of RANKL/OPG. J Musculoskelet Neuron Interact **4:**293–300.
14. Honore P, Luger NM, Sabino MA, Schwei MJ, Rogers SD, Mach DB, O'Keefe PF, Ramnaraine ML, Clohisy DR, Mantyh PW 2000 Osteoprotegerin blocks bone cancer-induced skeletal destruction, skeletal pain and pain-related neurochemical reorganization of the spinal cord. Nat Med **6:**521–528.
15. Roudier M, Bain S, Dougall W 2006 Effects of the RANKL inhibitor, osteoprotegerin, on the pain and histopathology of bone cancer in rats. Clin Exp Metastasis **23:**167–175.
16. Wacnik PW, Eikmeier LJ, Ruggles TR, Ramnaraine ML, Walcheck BK, Beitz AJ, Wilcox GL 2001 Functional interactions between tumor and peripheral nerve: Morphology, algogen identification, and behavioral characterization of a new murine model of cancer pain. J Neurosci **21:**9355–9366.
17. Wacnik PW, Eikmeier LJ, Simone DA, Wilcox GL, Beitz AJ 2005 Nociceptive characteristics of tumor necrosis factor-[alpha] in naive and tumor-bearing mice. Neuroscience **132:**479–491.
18. Khasabova IA, Stucky CL, Harding-Rose C, Eikmeier L, Beitz AJ, Coicou LG, Hanson AE, Simone DA, Seybold VS 2007 Chemical interactions between fibrosarcoma cancer cells and sensory neurons contribute to cancer pain. J Neurosci **27:**10289–10298.
19. Ghilardi JR, Rohrich H, Lindsay TH, Sevcik MA, Schwei MJ, Kubota K, Halvorson KG, Poblete J, Chaplan SR, Dubin AE, Carruthers NI, Swanson D, Kuskowski M, Flores CM, Julius D, Mantyh PW 2005 Selective blockade of the capsaicin receptor TRPV1 attenuates bone cancer pain. J Neurosci **25:**3126–3131.
20. Peters CM, Lindsay TH, Pomonis JD, Luger NM, Ghilardi JR, Sevcik MA, Mantyh PW 2004 Endothelin and the tumorigenic component of bone cancer pain. Neuroscience **126:**1043–1052.
21. Sabino MA, Ghilardi JR, Jongen JL, Keyser CP, Luger NM, Mach DB, Peters CM, Rogers SD, Schwei MJ, de Felipe C, Mantyh PW 2002 Simultaneous reduction in cancer pain, bone destruction, and tumor growth by selective inhibition of cyclooxygenase-2. Cancer Res **62:**7343–7349.
22. Hunt SP, Mantyh PW 2001 The molecular dynamics of pain control. Nat Rev Neurosci **2:**83–91.
23. Gilchrist LS, Cain DM, Harding-Rose C, Kov AN, Wendelschafer-Crabb G, Kennedy WR, Simone DA 2005 Re-organization of P2X3 receptor localization on epidermal nerve fibers in a murine model of cancer pain. Brain Res **1044:**197–205.
24. Halvorson KG, Kubota K, Sevcik MA, Lindsay TH, Sotillo JE, Ghilardi JR, Rosol TJ, Boustany L, Shelton DL, Mantyh PW 2005 A blocking antibody to nerve growth factor attenuates skeletal pain induced by prostate tumor cells growing in bone. Cancer Res **65:**9426–9435.
25. Honore P, Schwei J, Rogers SD, Salak-Johnson JL, Finke MP, Ramnaraine ML, Clohisy DR, Mantyh PW 2000 Cellular and neurochemical remodeling of the spinal cord in bone cancer pain. Prog Brain Res **129:**389–397.
26. Peters CM, Ghilardi JR, Keyser CP, Kubota K, Lindsay TH, Luger NM, Mach DB, Schwei MJ, Sevcik MA, Mantyh PW 2005 Tumor-induced injury of primary afferent sensory nerve fibers in bone cancer pain. Exp Neurol **193:**85–100.
27. Luger NM, Honore P, Sabino MA, Schwei MJ, Rogers SD, Mach DB, Clohisy DR, Mantyh PW 2001 Osteoprotegerin diminishes advanced bone cancer pain. Cancer Res **61:**4038–4047.
28. Sevcik MA, Ghilardi JR, Peters CM, Lindsay TH, Halvorson KG, Jonas BM, Kubota K, Kuskowski MA, Boustany L, Shelton DL, Mantyh PW 2005 Anti-NGF therapy profoundly reduces bone cancer pain and the accompanying increase in markers of peripheral and central sensitization. Pain **115:**128–141.
29. Cui M, Honore P, Zhong C, Gauvin D, Mikusa J, Hernandez G, Chandran P, Gomtsyan A, Brown B, Bayburt EK, Marsh K, Bianchi B, McDonald H, Niforatos W, Neelands TR, Moreland RB, Decker MW, Lee CH, Sullivan JP, Faltynek CR 2006 TRPV1 receptors in the CNS play a key role in broad-spectrum analgesia of TRPV1 antagonists. J Neurosci **26:**9385–9393.
30. Sah DW, Ossipo MH, Porreca F 2003 Neurotrophic factors as novel therapeutics for neuropathic pain. Nat Rev Drug Discov **2:**460–472.
31. Carducci MA, Jimeno A 2006 Targeting bone metastasis in prostate cancer with endothelin receptor antagonists. Clin Cancer Res **12:**6296–6300.
32. Nelson JB, Chan-Tack K, Hedican SP, Magnuson SR, Opgenorth TJ, Bova GS, Simons JW 1996 Endothelin-1 production and decreased endothelin B receptor expression in advanced prostate cancer. Cancer Res **56:**663–668.
33. Pomonis JD, Rogers SD, Peters CM, Ghilardi JR, Mantyh PW 2001 Expression and localization of endothelin receptors: Implications for the involvement of peripheral glia in nociception. J Neurosci **21:**999–1006.
34. Davar G 2001 Endothelin-1 and metastatic cancer pain. Pain Med **2:**24–27.
35. Dawas K, Loizidou M, Shankar A, Ali H, Taylor I 1999 Angiogenesis in cancer: The role of endothelin-1. Ann R Coll Surg Engl **81:**306–310.
36. Asham EH, Loizidou M, Taylor I 1998 Endothelin-1 and tumour development. Eur J Surg Oncol **24:**57–60.
37. Mantyh PW, Clohisy DR, Koltzenburg M, Hunt SP 2002 Molecular mechanisms of cancer pain. Nat Rev Cancer **2:**201–209.
38. Green J 2004 Bisphosphonates: Preclinical review. Oncologist **9**(Suppl 4)**:**3–13.
39. Lipton ASE, Saad F, Gleason D, Gordon D, Smith M, Rosen L, Kowalski MO, Reitsma D, Seaman J 2002 The new bisphosphonate, Zometa (zoledronic acid), decreases skeletal complications in both osteolytic and osteoblastic lesions: A comparison to pamidronate. Cancer Invest **20**(Suppl 2)**:**45–54.
40. Berenson JR, Rosen LS, Howell A, Porter L, Coleman RE, Morley W, Dreicer R, Kuross SA, Lipton A, Seaman JJ 2001 Zoledronic acid reduces skeletal-related events in patients with osteolytic metastases. Cancer **91:**1191–1200.
41. Carteni GBR, Giotta F, Lorusso V, Scalone S, Vinaccia V, Rondena R, Amadori D 2006 Efficacy and safety of zoledronic acid in patients with breast cancer metastatic to bone: A multicenter clinical trial. Oncologist **11:**841–848.
42. Saad FMJ, Eastham J 2006 Rationale for zoledronic acid therapy in men with hormone-sensitive prostate cancer with or without bone metastasis. Urol Oncol **24:**4–12.

© 2008 American Society for Bone and Mineral Research

Chapter 84. Skeletal Complications of Childhood Cancer

Ingrid A. Holm

Harvard Medical School Children's Hospital, Boston, Massachusetts

INTRODUCTION

With improvements in therapy, more and more children are surviving childhood cancer. The 5-yr survival rate for children with cancer is 80%, and the rate for acute lymphocytic leukemia (ALL), the most common childhood cancer, is 87%.[1] It has been estimated that 1 in 900 individuals 15–45 yr of age will be a childhood cancer survivor,[2] and by 2010, the number will be 1 in 250.[3] As more and more children with cancer are surviving, the number of adults with complications from cancer is increasing, and the cumulative incidence of a chronic health condition years after cancer diagnosis is 73%.[4] This chapter focuses on one of these long-term sequelae, deficits in BMD.

During adolescence, ~40% of peak bone mass is accumulated. Lack of adequate bone mineral acquisition during this period may compromise peak bone mass, leading to life-long deficits in bone mineral and predisposing to fractures. In children with cancer, nutritional deficiencies and prolonged decreases in activity may compromise bone accrual. Cranial irradiation may lead to hormonal deficiencies that further compromise bone accrual.[5] Finally, chemotherapeutic agents, such as glucocorticoids and methotrexate, interfere with bone mineral accretion and skeletal development.[5]

ACUTE LYMPHOCYTIC LEUKEMIA

ALL is the most common childhood malignancy, and much of the literature on the skeletal effects of cancer focuses on this population.

At Diagnosis of ALL

The ALL disease process may affect bone acquisition, because osteopenia (BMD Z-score < –1.0 SD) at diagnosis is reported in 10–46% of children with ALL[6–10]; a few found no decrease in BMD.[11,12] Osteopenia and/or fractures are evident on radiographs in 13% and 10% of patients, respectively, and musculo-skeletal pain is common.[6] Bone formation is impaired, and markers of bone formation are low,[13] including osteocalcin, type I collagen carboxy-terminal propeptide (PICP), and bone-specific alkaline phosphatase (BSALP).[6,7,9–11,14–16] Bone resorption may also be impaired.[13] Urinary N-telopeptide (NTX) is normal,[7] and type I collagen carboxyl-terminal telopeptide (ICTP) is normal[9,11] or reduced[10,14,15] in most, but not all,[12] studies. The reduction in bone formation and resorption markers suggests a low bone turnover state is present at diagnosis.[14,17] 1,25-dihydroxyvitamin D is reduced in at least some patients in most,[6,9,11,16] but not all,[8] studies.

During Treatment of ALL

Treatment of ALL is associated with additional loss of bone in some patients.[7–9,11,12,16] There is not a strong relationship between changes in growth velocity and BMD during therapy. A decreased height SDS during therapy has been reported to either be associated with,[8] or not associated with,[9] changes in BMD or BMD Z-scores. Overall, growth is not significantly compromised in most children undergoing cancer treatment, which may explain why the findings of studies using areal BMD do not differ significantly from those of studies using BMAD (apparent volumetric BMD) or QCT to correct for size differences.[9]

Chemotherapy is associated with a decrease in BMD, because BMD is decreased in patients treated with chemotherapy only.[9] Glucocorticoids and methotrexate are known to predispose to osteopenia and fractures,[18] and higher cumulative doses of methotrexate[11] and higher doses of glucocorticoids[7] are associated with a lower BMD. Glucocorticoids decrease bone formation, decrease intestinal absorption of calcium, and increase renal calcium excretion. Methotrexate may also inhibit new bone formation and fracture healing.[18] Cranial irradiation is associated with a drop in BMD Z-score during treatment.[8] Lifestyle factors, such as calcium intake and physical activity, are not correlated with BMD during therapy.[9]

Accumulation of normal bone mass may be impaired during therapy. Although induction therapy with prednisolone and methotrexate is associated with a further decrease in bone formation markers, with discontinuation of prednisolone, bone formation markers increase to normal[7,9–11,13,14]; the increase is less in children who received high-dose methotrexate.[14] During induction therapy with prednisolone and methotrexate, bone resorption markers (pyridinoline, ICTP) also decrease, and then increase after prednisolone is discontinued[14] to normal or elevated levels,[9,11,13,15] suggesting resorption is increased.[17] High doses of methotrexate are associated with an even greater increase in bone resorption markers.[12,14] These findings suggest that osteoblast and osteoclast activity are suppressed secondary to the actions of glucocorticoids on bone and that high doses of methotrexate continue to inhibit osteoblast proliferation but increase osteoclast activity.[14] Higher levels of bone formation markers at diagnosis (osteocalcin) and at 1 yr (PICP) have been correlated with a higher BMD at 1 yr.[11] 1,25-dihydroxyvitamin D levels have been reported to be reduced[7,11,19] or normal.[8,9] In one study,[9] although 1,25-dihydroxyvitamin D levels were within the normal range, the levels increased significantly in the first 32 wk of treatment, and the increase was positively correlated with the change in lumbar spine BMD over that period of time.[9] 25-Hydroxyvitamin D has been reported to be reduced[11,19] or normal.[7]

Fractures affect up to 40% of patients,[7,12,16] have been reported to be 6-fold increased above normal,[9] and tend to occur during or shortly after discontinuation of chemotherapy.[20] Fractures are associated with the osteopenia that develops during therapy[13]; children who fracture have a greater decrease in BMD Z-score than children who do not fracture,[9,11] and a decrease in BMC or BMD Z-score during treatment is a positive predictor,[6,16] and an increase in BMC is a negative predictive,[6] of subsequent fracture. In one study, the increase in NTX was correlated with the occurrence of a fracture.[7]

There have been a few pilot studies of oral alendronate during therapy in children with ALL.[21,22] Alendronate therapy results in an increase in the BMD Z-score in most patients treated for 6–24 mo. No placebo-controlled trials have

The author states that she has no conflicts of interest.

Key words: clinical/pediatrics, cancer, bone mineralization, drug effects, osteopenia

© 2008 American Society for Bone and Mineral Research

been carried out, and many issues remain, most notably what the criteria for treatment should be.

After Treatment of ALL

In most studies of ALL survivors, low BMD persists,[9,23–30] although a few studies found that BMD normalizes,[31,32] and bone formation and resorption markers normalize.[33] Most studies found some "catch-up" in BMD after discontinuation of therapy, and increasing length of time since cessation of therapy is associated with a higher BMD or BMC.[28,30] In the short term (0–3 yr after discontinuation of ALL therapy), a greater increase in total body BMD is seen in patients compared with controls.[33] In a longitudinal QCT study of ALL survivors at an average of 11 yr after ALL diagnosis, the mean lumbar spine BMD Z-score increased with time, although the percentage of individuals with a BMD Z-score < –1.0 SD increased over time,[34] suggesting that in a subgroup of patients, catch-up growth in BMD does not occur. Thus, although there may be catch-up in the BMD after discontinuation of therapy, peak bone mass may not be achieved.

Risk factors associated with a lower BMD in follow-up include white race,[23,29,34] older age at diagnosis of ALL,[18,34] poor nutritional status,[34] alcohol consumption,[34] male sex,[23,27,29,34] reduced exercise capacity (as measured by peak oxygen consumption),[28] and low activity levels.[27,28] This latter finding is particularly notable, because there is evidence that long-term ALL survivors are less likely to meet the Centers for Disease Control and Prevention physical activity recommendations and are more likely to report no leisure-time physical activity than the general population.[35]

The relative contribution of chemotherapy versus cranial irradiation, and the subsequent hormonal deficiencies on BMD is not clear. Patients who received chemotherapy in the absence of cranial irradiation have a low BMD in some,[10,20,27] but not all,[20] studies. Previous exposure to methotrexate is associated with a reduction in BMC at the spine,[28] and higher doses of methotrexate have been associated with a lower BMD.[27,34] A history of cranial irradiation has been associated with low BMD in some studies,[23,29,34,36] and BMD is inversely correlated with cranial irradiation dose.[29] Others have found that the use of cranial irradiation status did not predict BMD[30] and that the decrease in BMD seen in adults after treatment for ALL is not dependent on severity of growth hormone deficiency that is caused by cranial irradiation.[24] These findings taken together suggest that, although chemotherapy and cranial irradiation likely play a role in low BMD in follow-up, additional factors likely adversely affect BMD.

Fracture rates are increased in long-term survivors of ALL. From the time of diagnosis of ALL through the subsequent 5 yr, fractures rates are reported to be double that of controls.[33] In one study, the 5-yr cumulative incidence of fractures was 28%.[37] Factors associated with a higher risk of fractures include the diagnosis of ALL during adolescence,[18,37] male sex,[37] and treatment with dexamethasone versus prednisone.[37] In one study, a history of cranial irradiation was not associated with fracture rate.[38]

Children undergoing stem cell transplant (SCT) may be at particularly high risk for deficits in bone mass. A low BMD after transplant has been reported in 36–47% of survivors.[39–41] Risk factors for osteopenia include female sex[39,40] and total body irradiation (TBI),[39] and osteopenia is more common in patients with delayed pubertal growth, growth hormone deficiency, hypogonadism, and chronic renal insufficiency, all secondary effects of TBI.[39,40] The incidence of osteopenia may be lower in children transplanted at <3 yr of age,[42] and BMD Z-scores are lower in pubertal and postpubertal children post-transplant.[41]

OTHER CANCERS

Children with brain tumors may be particularly susceptible to osteopenia because of cranial irradiation and subsequent growth hormone deficiency and hypogonadism. Children surviving with a brain tumor have decreased BMD,[43–48] and cranial irradiation may be associated with an even high prevalence of a low BMD.[49] The degree of osteopenia is correlated with a reduction in health-related quality of life,[45,49] and children with osteopenia reported more pain that significantly limited physical activity.[45] Osteopenia has been described in survivors of bone tumors, including Ewing's sarcoma and osteosarcomas.[50,51] Low intake of vitamin D and low physical activity, as well as treatment with methotrexate, ifosfamide, bleomycin, and cisplatin, may contribute.[50] In addition, a large minority of survivors of malignant lymphomas have decreased lumbar spine BMD and trabecular bone at the wrist (pQCT), correlated with cumulative dose of corticosteroid administered.[52]

CONCLUSIONS

Bone mass accrual is impaired in children with cancer. There are a number of contributing factors, some of which are modifiable, such as optimizing nutrition (particularly vitamin D and calcium intake), improving physical activity, and early identification of growth hormone deficiency and hypogonadism. In recognition of the potential long-term morbidity associations with loss of bone mineral in children with cancer, the Children's Oncology Group has produced guidelines for the management of bone morbidity that occurs after treatment of childhood cancer.[13,53]

REFERENCES

1. National Cancer Institute Surveillance, Epidemiology and End Result, Section XXVIII, Childhood Cancer by Site: Incidence, Survival and Mortality.
2. Meadows AT, Black B, Nesbit ME Jr, Strong LC, Nicholson HS, Green DM, Hays DM, Lozowski SL 1993 Long-term survival. Clinical care, research, and education. Cancer **71**(10 Suppl):3213–3215.
3. Bleyer WA 1990 The impact of childhood cancer on the United States and the world. CA Cancer J Clin **40:**355–367.
4. Oeffinger KC, Mertens AC, Sklar CA, Kawashima T, Hudson MM, Meadows AT, Friedman DL, Marina N, Hobbie W, Kadan-Lottick NS, Schwartz CL, Leisenring W, Robison LL 2006 Chronic health conditions in adult survivors of childhood cancer. N Engl J Med **355:**1572–1582.
5. van Leeuwen BL, Kamps WA, Jansen HW, Hoekstra HJ 2000 The effect of chemotherapy on the growing skeleton. Cancer Treat Rev **26:**363–376.
6. Halton JM, Atkinson SA, Fraher L, Webber CE, Cockshott WP, Tam C, Barr RD 1995 Mineral homeostasis and bone mass at diagnosis in children with acute lymphoblastic leukemia. J Pediatr **126:**557–564.
7. Halton JM, Atkinson SA, Fraher L, Webber C, Gill GJ, Dawson S, Barr RD 1996 Altered mineral metabolism and bone mass in children during treatment for acute lymphoblastic leukemia. J Bone Miner Res **11:**1774–1783.
8. Henderson RC, Madsen CD, Davis C, Gold SH 1998 Longitudinal evaluation of bone mineral density in children receiving chemotherapy. J Pediatr Hematol Oncol **20:**322–326.
9. van der Sluis IM, van den Heuvel-Eibrink MM, Hahlen K, Krenning EP, de Muinck Keizer-Schrama SM 2002 Altered bone mineral density and body composition, and increased fracture risk in childhood acute lymphoblastic leukemia. J Pediatr **141:**204–210.
10. Boot AM, van den Heuvel-Eibrink MM, Hahlen K, Krenning EP, de Muinck Keizer-Schrama SM 1999 Bone mineral density in chil-

© 2008 American Society for Bone and Mineral Research

dren with acute lymphoblastic leukaemia. Eur J Cancer **35:**1693–1697.
11. Arikoski P, Komulainen J, Riikonen P, Voutilainen R, Knip M, Kroger H 1999 Alterations in bone turnover and impaired development of bone mineral density in newly diagnosed children with cancer: A 1-year prospective study. J Clin Endocrinol Metab **84:**3174–3181.
12. Arikoski P, Komulainen J, Riikonen P, Parviainen M, Jurvelin JS, Voutilainen R, Kroger H 1999 Impaired development of bone mineral density during chemotherapy: A prospective analysis of 46 children newly diagnosed with cancer. J Bone Miner Res **14:**2002–2009.
13. Sala A, Barr RD 2007 Osteopenia and cancer in children and adolescents: The fragility of success. Cancer **109:**1420–1431.
14. Crofton PM, Ahmed SF, Wade JC, Stephen R, Elmlinger MW, Ranke MB, Kelnar CJ, Wallace WH 1998 Effects of intensive chemotherapy on bone and collagen turnover and the growth hormone axis in children with acute lymphoblastic leukemia. J Clin Endocrinol Metab **83:**3121–3129.
15. Sorva R, Kivivuori SM, Turpeinen M, Marttinen E, Risteli J, Risteli L, Sorva A, Siimes MA 1997 Very low rate of type I collagen synthesis and degradation in newly diagnosed children with acute lymphoblastic leukemia. Bone **20:**139–143.
16. Atkinson SA, Halton JM, Bradley C, Wu B, Barr RD 1998 Bone and mineral abnormalities in childhood acute lymphoblastic leukemia: Influence of disease, drugs and nutrition. Int J Cancer Suppl **11:**35–39.
17. Mulder JE, Bilezikian JP 2004 Bone density in survivors of childhood cancer. J Clin Densitom **7:**432–442.
18. Davies JH, Evans BA, Jenney ME, Gregory JW 2005 Skeletal morbidity in childhood acute lymphoblastic leukaemia. Clin Endocrinol (Oxf) **63:**1–9.
19. Arikoski P, Kroger H, Riikonen P, Parviainen M, Voutilainen R, Komulainen J 1999 Disturbance in bone turnover in children with a malignancy at completion of chemotherapy. Med Pediatr Oncol **33:**455–461.
20. van der Sluis IM, van den Heuvel-Eibrink MM, Hahlen K, Krenning EP, de Muinck Keizer-Schrama SM 2000 Bone mineral density, body composition, and height in long-term survivors of acute lymphoblastic leukemia in childhood. Med Pediatr Oncol **35:**415–420.
21. Lethaby C, Wiernikowski J, Sala A, Naronha M, Webber C, Barr RD 2007 Bisphosphonate therapy for reduced bone mineral density during treatment of acute lymphoblastic leukemia in childhood and adolescence: A report of preliminary experience. J Pediatr Hematol Oncol **29:**613–616.
22. Wiernikowski JT, Barr RD, Webber C, Guo CY, Wright M, Atkinson SA 2005 Alendronate for steroid-induced osteopenia in children with acute lymphoblastic leukaemia or non-Hodgkin's lymphoma: Results of a pilot study. J Oncol Pharm Pract **11:**51–56.
23. Arikoski P, Komulainen J, Voutilainen R, Riikonen P, Parviainen M, Tapanainen P, Knip M, Kroger H 1998 Reduced bone mineral density in long-term survivors of childhood acute lymphoblastic leukemia. J Pediatr Hematol Oncol **20:**234–240.
24. Brennan BM, Rahim A, Adams JA, Eden OB, Shalet SM 1999 Reduced bone mineral density in young adults following cure of acute lymphoblastic leukaemia in childhood. Br J Cancer **79:**1859–1863.
25. Hoorweg-Nijman JJ, Kardos G, Roos JC, van Dijk HJ, Netelenbos C, Popp-Snijders C, de Ridder CM, Delemarre-van de Waal HA 1999 Bone mineral density and markers of bone turnover in young adult survivors of childhood lymphoblastic leukaemia. Clin Endocrinol (Oxf) **50:**237–244.
26. Nysom K, Holm K, Michaelsen K, Hertz H, Müller J, Mølgaard C 1998 Bone mass after treatment for acute lymphoblastic leukemia in childhood. J Clin Oncol **16:**3752–3760.
27. Tillmann V, Darlington AS, Eiser C, Bishop NJ, Davies HA 2002 Male sex and low physical activity are associated with reduced spine bone mineral density in survivors of childhood acute lymphoblastic leukemia. J Bone Miner Res **17:**1073–1080.
28. Warner J, Evans W, Webb D, Bell W, Gregory J 1999 Relative osteopenia after treatment for acute lymphoblastic leukemia. Pediatr Res **45:**544–551.
29. Kaste SC, Jones-Wallace D, Rose SR, Boyett JM, Lustig RH, Rivera GK, Pui CH, Hudson MM 2001 Bone mineral decrements in survivors of childhood acute lymphoblastic leukemia: Frequency of occurrence and risk factors for their development. Leukemia **15:**728–734.
30. Alikasifoglu A, Yetgin S, Cetin M, Tuncer M, Gumruk F, Gurgey A, Yordam N 2005 Bone mineral density and serum bone turnover markers in survivors of childhood acute lymphoblastic leukemia: Comparison of megadose methylprednisolone and conventional-dose prednisolone treatments. Am J Hematol **80:**113–118.
31. Kadan-Lottick N, Marshall JA, Baron AE, Krebs NF, Hambidge KM, Albano E 2001 Normal bone mineral density after treatment for childhood acute lymphoblastic leukemia diagnosed between 1991 and 1998. J Pediatr **138:**898–904.
32. Mandel K, Atkinson S, Barr RD, Pencharz P 2004 Skeletal morbidity in childhood acute lymphoblastic leukemia. J Clin Oncol **22:**1215–1221.
33. Marinovic D, Dorgeret S, Lescoeur B, Alberti C, Noel M, Czernichow P, Sebag G, Vilmer E, Leger J 2005 Improvement in bone mineral density and body composition in survivors of childhood acute lymphoblastic leukemia: A 1-year prospective study. Pediatrics **116:**e102–e108.
34. Kaste SC, Rai SN, Fleming K, McCammon EA, Tylavsky FA, Danish RK, Rose SR, Sitter CD, Pui CH, Hudson MM 2006 Changes in bone mineral density in survivors of childhood acute lymphoblastic leukemia. Pediatr Blood Cancer **46:**77–87.
35. Florin TA, Fryer GE, Miyoshi T, Weitzman M, Mertens AC, Hudson MM, Sklar CA, Emmons K, Hinkle A, Whitton J, Stovall M, Robison LL, Oeffinger KC 2007 Physical inactivity in adult survivors of childhood acute lymphoblastic leukemia: A report from the childhood cancer survivor study. Cancer Epidemiol Biomarkers Prev **16:**1356–1363.
36. Gilsanz V, Carlson ME, Roe TF, Ortega JA 1990 Osteoporosis after cranial irradiation for acute lymphoblastic leukemia. J Pediatr **117:**238–244.
37. Strauss AJ, Su JT, Dalton VM, Gelber RD, Sallan SE, Silverman LB 2001 Bony morbidity in children treated for acute lymphoblastic leukemia. J Clin Oncol **19:**3066–3072.
38. Barr RD, Halton J, Willan A, Cockshott WP, Gill G, Atkinson S 1998 Impact of age and cranial irradiation on radiographic skeletal pathology in children with acute lymphoblastic leukemia. Med Pediatr Oncol **30:**347–350.
39. Leung W, Ahn H, Rose SR, Phipps S, Smith T, Gan K, O'Connor M, Hale GA, Kasow KA, Barfield RC, Madden RM, Pui CH 2007 A prospective cohort study of late sequelae of pediatric allogeneic hematopoietic stem cell transplantation. Medicine (Baltimore) **86:**215–224.
40. Taskinen M, Kananen K, Valimaki M, Loyttyniemi E, Hovi L, Saarinen-Pihkala U, Lipsanen-Nyman M 2006 Risk factors for reduced areal bone mineral density in young adults with stem cell transplantation in childhood. Pediatr Transplant **10:**90–97.
41. Taskinen M, Saarinen-Pihkala UM, Hovi L, Vettenranta K, Makitie O 2007 Bone health in children and adolescents after allogeneic stem cell transplantation: High prevalence of vertebral compression fractures. Cancer **110:**442–451.
42. Perkins JL, Kunin-Batson AS, Youngren NM, Ness KK, Ulrich KJ, Hansen MJ, Petryk A, Steinberger J, Anderson FS, Baker KS 2007 Long-term follow-up of children who underwent hematopoeitic cell transplant (HCT) for AML or ALL at less than 3 years of age. Pediatr Blood Cancer **49:**958–963.
43. Petraroli M, D'Alessio E, Ausili E, Barini A, Caradonna P, Riccardi R, Caldarelli M, Rossodivita A 2007 Bone mineral density in survivors of childhood brain tumours. Childs Nerv Syst **23:**59–65.
44. Pietila S, Sievanen H, Ala-Houhala M, Koivisto AM, Liisa Lenko H, Makipernaa A 2006 Bone mineral density is reduced in brain tumour patients treated in childhood. Acta Paediatr **95:**1291–1297.
45. Barr RD, Simpson T, Webber CE, Gill GJ, Hay J, Eves M, Whitton AC 1998 Osteopenia in children surviving brain tumours. Eur J Cancer **34:**873–877.
46. Gurney JG, Kadan-Lottick NS, Packer RJ, Neglia JP, Sklar CA, Punyko JA, Stovall M, Yasui Y, Nicholson HS, Wolden S, McNeil DE, Mertens AC, Robison LL 2003 Endocrine and cardiovascular late effects among adult survivors of childhood brain tumors: Childhood Cancer Survivor Study. Cancer **97:**663–673.
47. Hesseling PB, Hough SF, Nel ED, van Riet FA, Beneke T, Wessels G 1998 Bone mineral density in long-term survivors of childhood cancer. Int J Cancer Suppl **11:**44–47.
48. Krishnamoorthy P, Freeman C, Bernstein ML, Lawrence S, Rodd C 2004 Osteopenia in children who have undergone posterior fossa or craniospinal irradiation for brain tumors. Arch Pediatr Adolesc Med **158:**491–496.
49. Odame I, Duckworth J, Talsma D, Beaumont L, Furlong W, Web-

© 2008 American Society for Bone and Mineral Research

ber C, Barr R 2006 Osteopenia, physical activity and health-related quality of life in survivors of brain tumors treated in childhood. Pediatr Blood Cancer **46**:357–362.

50. Azcona C, Burghard E, Ruza E, Gimeno J, Sierrasesumaga L 2003 Reduced bone mineralization in adolescent survivors of malignant bone tumors: Comparison of quantitative ultrasound and dual-energy x-ray absorptiometry. J Pediatr Hematol Oncol **25**:297–302.
51. Ruza E, Sierrasesumaga L, Azcona C, Patino-Garcia A 2006 Bone mineral density and bone metabolism in children treated for bone sarcomas. Pediatr Res **59**:866–871.
52. Sala A, Talsma D, Webber C, Posgate S, Atkinson S, Barr R 2007 Bone mineral status after treatment of malignant lymphoma in childhood and adolescence. Eur J Cancer Care (Engl) **16**:373–379.
53. Landier W, Bhatia S, Eshelman DA, Forte KJ, Sweeney T, Hester AL, Darling J, Armstrong FD, Blatt J, Constine LS, Freeman CR, Friedman DL, Green DM, Marina N, Meadows AT, Neglia JP, Oeffinger KC, Robison LL, Ruccione KS, Sklar CA, Hudson MM 2004 Development of risk-based guidelines for pediatric cancer survivors: The Children's Oncology Group Long-Term Follow-Up Guidelines from the Children's Oncology Group Late Effects Committee and Nursing Discipline. J Clin Oncol **22**:4979–4990.

Chapter 85. Treatment and Prevention of Bone Metastases and Myeloma Bone Disease

Jean-Jacques Body

Department of Internal Medicine, CHU Brugmann, Université Libre de Bruxelles (U.L.B.), Brussels, Belgium

CLINICAL ASPECTS

According to various large series, up to 90% of patients with advanced cancer will develop bone metastases. The skeleton is the most common site of metastatic disease. It is also the most frequent site of first distant relapse in breast and prostate cancers.

Breast Cancer

Metastatic bone disease causes considerable distress to breast cancer patients. Because of the long clinical course breast cancer may follow, morbidity caused by tumor bone disease also makes major demands on resources for health care provision. The term skeletal-related events (SREs) refers to the major complications of tumor bone disease, namely pathological fractures, need for radiotherapy, need for bone surgery, spinal cord compression, and hypercalcemia.[1] Such major complications will be observed in up to one third of the patients whose first relapse is in bone. Bone pain can be the source of great suffering causing most patient concern and physician visits.[2] A recent prospective study suggested that patients with osteolytic lesions, as evaluated by a CT scan of the most painful lesion, have the highest mean pain score and analgesic consumption and the least mean scores for quality of life compared with patients who have mixed or sclerotic lesions.[3] Hypercalcemia classically occurs in 10–15% of the cases, spinal cord compression in ~10% and, when long bones are invaded, fractures will occur in 10–20% of the cases.[4,5] Pathological fractures are a dramatic consequence of tumor bone disease and they occur with a median onset of 11 mo from the initial diagnosis of bone involvement.

Across all tumor types, patients with breast cancer have the highest incidence of skeletal complications. Taken from data in placebo groups of randomized bisphosphonates trials, the mean skeletal morbidity rate (i.e., the mean number of SREs per year) varies between 2.2 and 4.0.[4–8] Patients who have metastases only in the skeleton have a higher rate of SREs than patients who also have bone and visceral metastases (e.g., a 2- to 3-fold increase in pathological fractures). Survival from diagnosis of bone metastases is longest for patients with only bone metastases (median survival, 24 mo) and least for patients with bone and liver metastases (5.5 mo).[8]

Multiple Myeloma

Bone pain is a presenting feature in three fourths of the patients with multiple myeloma. Back pain correlates with the presence of vertebral fractures that are present in more than one half of the patients at diagnosis. Extensive osteolytic lesions are frequent in this aggressive bone disease and, typically, they do not heal despite successful antineoplastic treatment. Diffuse osteoporosis can also be a presenting and misleading feature. The increased fracture rate seems to be especially high around the time of diagnosis. In a large retrospective cohort study, fracture risk was increased 16-fold in the year before diagnosis and 9-fold thereafter. Fractures of the vertebrae and ribs were the most frequent.[9]

Prostate Cancer

Surprisingly, there are only a few studies documenting the frequency and the nature of bone metastatic complications in patients with hormone-refractory prostate cancer.[10–13] The incidence of SREs can be best estimated by analyzing the placebo group of the zoledronic acid–controlled trial.[12,13] Inclusion criteria required that patients had at least one bone metastasis and an augmentation of prostate-specific antigen (PSA) levels while on hormonal therapy. During a follow-up period of 2 yr, nearly one half of the patients developed one or more SREs, which, in this study, also included a change in antineoplastic therapy to treat bone pain. The two most frequent complications were the need for radiation therapy and the occurrence of pathological fractures. These fractures appeared more frequently at peripheral than at vertebral sites. The median time to the first SRE was 10.5 mo, whereas the mean skeletal morbidity rate per year was nearly 1.5. The median survival was 9.5 mo.[13]

The following sections relate to the use of bisphosphonates in tumor bone disease. The reader is referred to other sources for the other therapeutic modalities of bone metastases, namely analgesics,[14] radiotherapy,[15] radioisotopes,[16] surgery,[17] vertebroplasty[18] or kyphoplasty. This recently introduced technique consists in the introduction of inflatable bone tamps into compression fractures. Impressive results have

Dr. Body has consulted for Amgen, Novartis, and Roche.

© 2008 American Society for Bone and Mineral Research

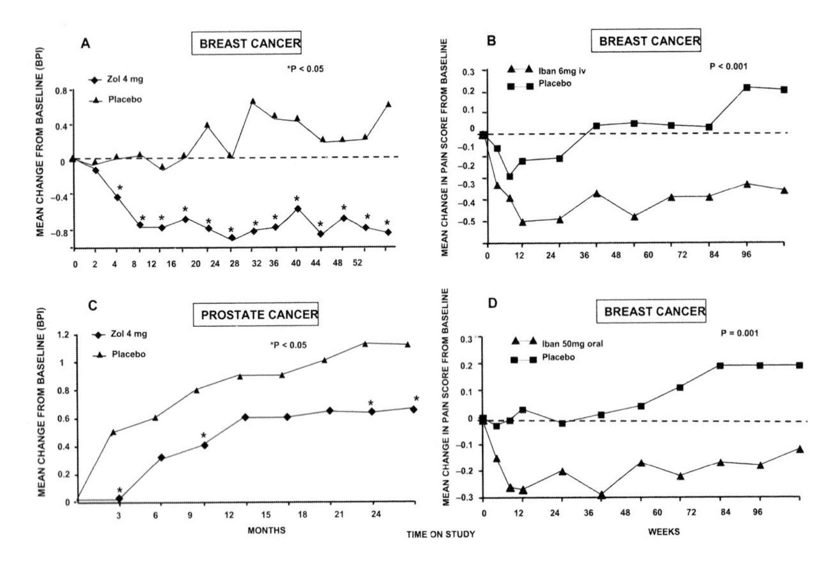

FIG. 1. Analgesic activity of long-term therapy with bisphosphonates in placebo-controlled trials. Effects of zoledronic acid (Zole) are shown in left panels (pain assessed by the Brief Pain Inventory score, quantitative 10-point scale): top, breast cancer[27]; bottom, prostate cancer.[12,13] Effects of ibandronate (Iban) in breast cancer are shown in the right panels (pain assessed on a qualitative five-point scale): top, intravenous route[6]; bottom, oral route.[28]

been reported in a series of 18 consecutive myeloma patients with osteolytic vertebral compression fractures. The technique was without major complications and, on average, 34% of height lost at the time of fracture was restored. Pain improved, as well as physical and social functioning.[19]

CURRENT USE OF BISPHOSPHONATES IN TUMOR BONE DISEASE

Cancer Hypercalcemia

Cancer hypercalcemia is reviewed in another chapter of the *Primer*.

Metastatic Bone Pain

Bisphosphonates can relieve metastatic bone pain and improve patient functioning and quality of life even if the mechanism of this analgesic effect remains largely unknown. The relative inability of first-generation oral bisphosphonates to reduce metastatic bone pain has been confirmed in a placebo-controlled study of oral clodronate in patients with progressing bone metastases after a median time on study of almost 2 mo.[20] In another short-term study, oral clodronate was inferior to intravenous pamidronate in relieving metastatic bone pain ($p < 0.01$) after 3 mo.[21] However, the opinion that the intravenous route has to be selected to obtain optimal analgesic effects is now challenged by the long-term analgesic effects of oral ibandronate, which have been shown in a placebo-controlled study.[22] Randomized placebo-controlled trials have shown that intravenous pamidronate, clodronate, ibandronate, and zoledronic acid exert useful pain relief.[5] A clinically meaningful relief of bone pain seems to occur in one half to two thirds of the patients treated with pamidronate and most of the effect is obtained after only one or two infusions. More recently, an open phase II trial has shown that intensive ibandronate dosing seems to provide rapid and effective relief from severe metastatic bone pain.[23] However, bisphosphonates have to be considered as co-analgesics and they cannot replace opioids for the treatment of cancer pain.

Over the long term, randomized placebo-controlled trials have also shown that clodronate, pamidronate, ibandronate, and zoledronic acid exert useful pain control. The American Society of Clinical Oncology (ASCO) panel considered "reasonable" to start intravenous bisphosphonates in women with abnormal bone scan with localized pain and normal plain radiographs but not if the abnormal bone scan is asymptomatic.[24] In two randomized, placebo-controlled trials, mean pain scores and use of analgesics in patients treated with pamidronate 90 mg monthly for 2 yr increased significantly less than in the placebo group.[25] In a phase III trial of patients with breast cancer, pain and analgesic scores were reduced to a similar extent at 13 mo with zoledronic acid 4 mg and pamidronate 90 mg.[26] Bone pain levels were assessed in a randomized, placebo-controlled study of zoledronic acid in Japanese patients with breast cancer and bone metastases.[27] Patients receiving zoledronic acid 4 mg for 12 mo experienced a significant decrease from baseline ($p < 0.001$) in their mean composite Brief Pain Inventory (BPI) score (Fig. 1, top left panel). In phase III trials of intravenous and oral ibandronate, bone pain was reduced and similarly maintained below baseline for 2 yr (Fig. 1, top right panel[6]; Fig. 1, bottom right panel[28]). There were also significant improvements in global quality of life and physical functioning with intravenous or oral ibandronate versus placebo.[29] Given the different patient populations and endpoints used between studies, direct comparative trials are needed before concluding that one bisphosphonate exerts more potent analgesic effects than another one.

In prostate cancer, uncontrolled trials have often been positive, whereas placebo-controlled studies were usually negative, whether for clodronate or for pamidronate. Initial pamidronate uncontrolled trials reported impressive results, but a more recent controlled trial suggests that pamidronate is no more effective than placebo in reducing bone pain or SREs over 6 mo.[11] In a placebo-controlled study of zoledronic acid including 643 men with bone metastases and progressive prostate cancer after androgen deprivation therapy, the effect on bone pain appeared to be less impressive[12,13] (Fig. 1, bottom left panel) than in breast cancer. A favorable pain response, defined as a 2-point difference on the 11-point scale of the BPI assessment, was observed in 33% of the zoledronic acid–treated patients compared with 25% for patients receiving placebo.[30] In the placebo-controlled trial in patients with bone metastases from other tumors, mainly non–small cell lung cancer, zoledronic acid had no significant effect on bone pain or quality of life.[31] The analgesic effect of bisphosphonates is thus well established in patients with breast cancer or myeloma but could be less in patients with other tumor types. The role of bisphosphonates as an alternative or an adjunct to radiotherapy remains unclear. However, preliminary data suggest

© 2008 American Society for Bone and Mineral Research

that a combination of the radioisotope Sr89 and zoledronic acid could have a greater analgesic effect that each therapeutic modality given alone.[32]

Prevention of Skeletal-Related Events

Breast Cancer. Bisphosphonates constitute a highly effective therapeutic means for the prevention of skeletal complications secondary to bone metastases in breast cancer. Several placebo-controlled trials and fewer comparative trials against another bisphosphonate have been performed. Assessment of treatment effects has often used the first-event analyses, such as the proportion of patients with at least one SRE or time to the first event. These are quite objective and conservative endpoints, but they do not take into account all subsequent events that occur in any given patient. From a clinical perspective, an aggregate score of symptomatic SREs is more relevant. Skeletal morbidity rate (SMR) or skeletal morbidity period rate (SMPR, number of periods with at least one SRE) take into account the occurrence of multiple SREs. SMPR is probably more adequate than SMR because events often occur in clusters and SMPR thus reduces the risk of multiple counting of the same event. More recently, more sophisticated analyses have emerged. Multiple-event analyses are able to model all events and the time between events. They allow the calculation of a hazard ratio that indicates the relative risk of skeletal events between two treatment groups.

Clinical trials of the bisphosphonates clodronate, in Europe, and pamidronate, in the United States and Europe, have established their effectiveness in breast cancer patients with bone metastases.[7,26,33–35] In these trials, clodronate has been shown to increase the time to the first event and to reduce the incidence of hypercalcemia and of vertebral fractures. However, clodronate is considered to be less effective than pamidronate for the prevention of SREs.[36] This has been shown in a limited comparative trial against pamidronate.[21] Two double-blind randomized placebo-controlled trials comparing 90 mg pamidronate infusions every 4 wk to placebo infusions for up to 2 yr in addition to chemo- or hormonal therapy in large series of breast cancer patients with at least one lytic bone metastasis showed that bisphosphonates can reduce SMR by more than one third, increase the median time to the occurrence of the first SRE by almost 50%, and reduce the proportion of patients having any SRE.[7,35]

In the past few years, more convenient and somewhat more effective aminobisphosphonates have emerged. Zoledronic acid is widely used for patients with bone metastases from various tumors,[13,31,37–39] and ibandronate has since been approved in many countries, but not the United States, for the prevention of skeletal events in patients with breast cancer and bone metastases.

Three randomized double-blind multicenter trials assessed the efficacy of zoledronic acid in patients with breast cancer and multiple myeloma, in prostate cancer, and in lung or other solid tumors. The primary efficacy endpoint was the proportion of patients with at least one SRE, defined as pathological fracture, spinal cord compression, radiation therapy to bone, and surgery to bone. Secondary endpoints included time to first SRE, SMR, and Andersen-Gill multiple-event analysis. Patients with breast cancer or multiple myeloma (n = 1648) were randomized to a 15-min infusion of zoledronic acid 4 or 8 mg or a 2-h infusion of pamidronate 90 mg every 3–4 wk.[31,37] The proportion of patients with at least one SRE was similar in all treatment groups and the preestablished criterion for noninferiority of zoledronic acid to pamidronate was thus met. Zoledronic acid 8 mg was not more effective than the 4-mg dose but was associated with an increased frequency of renal adverse events, explaining why all patients in that treatment arm were switched to the lower dose of zoledronic acid during all zoledronic acid trials. Median time to first SRE was ~1 yr in all three treatment groups, and SMRs were also not significantly different. A preplanned multiple-event analysis, according to the Andersen-Gill model, showed that zoledronic acid 4 mg reduced the risk of developing a skeletal complication by an additional 20% over that achieved by pamidronate 90 mg in the breast cancer subgroup ($p < 0.05$; Fig. 2, top graph).[37] In a subsequent subgroup analysis according to the radiological aspect of bone metastases, zoledronic acid was shown to be even more effective than pamidronate in patients with at least one osteolytic lesion. The short infusion time (15 min compared with 1 or 2 h for pamidronate) offers a quite convenient therapy.[39] On the other hand, limited data obtained in 31 patients on either pamidronate or clodronate for bone metastases from breast cancer suggest that switching to second-line zoledronic acid after an SRE or progression of bone metastases can significantly improve pain control and reduce bone turnover markers.[39]

The efficacy of intravenous and oral ibandronate has been assessed in randomized double-blind, placebo-controlled studies.[6,22] Breast cancer patients were randomized to ibandronate 6 mg or placebo infused over 1–2 h every 3–4 wk in the intravenous trial, whereas oral ibandronate 50 mg was given once daily 1 h before breakfast. The primary efficacy endpoint was the skeletal morbidity period rate (SMPR), defined as the number of 12-wk periods with skeletal complications (vertebral fractures, nonvertebral fractures, radiotherapy to bone, and surgery to bone) divided by the total observation time. Secondary endpoints also included a multiple-event analysis. Intravenous and oral ibandronate significantly reduced the SMPR compared with placebo ($p < 0.005$ for both). Multiple-event Poisson regression analysis showed that intravenous ibandronate led to a statistically significant 40% reduction in the risk of SREs compared with placebo (Fig. 2, bottom panel; $p < 0.005$). The effect of oral ibandronate 50 mg on the risk of SREs was similar (Fig. 2, bottom panel; 38% reduction versus placebo, $p < 0.0001$). In a 3-mo head-to-head trial between zoledronic acid and oral ibandronate, it was shown that the effects of both bisphosphonates on markers of bone turnover were similar.[40]

Myeloma. A recent systematic review of the various therapeutic options for the management of multiple myeloma has considered the introduction of bisphosphonates as one of the two most important therapeutic advances for this disease, the other one being the use of high-dose chemotherapy.[41] Updated ASCO guidelines recommend to start bisphosphonates in patients with lytic disease on plain X-rays or imaging studies or with spine compression fracture(s) from osteopenia. The panel still considers it reasonable to start bisphosphonates in patients with osteopenia based on plain radiographs or BMD measurement, but they are not recommended in patients with solitary plasmacytoma or smoldering or indolent myeloma.[42]

The Cochrane Myeloma Review Group has reported a meta-analysis based on 11 trials and involving 2183 assessable patients. This review concluded that both pamidronate and clodronate reduce the incidence of hypercalcemia, the pain index, and the number of vertebral fractures in myeloma patients.[43] In the largest placebo-controlled trial, the proportion of patients developing any SRE was significantly ($p < 0.001$) smaller in the pamidronate than in the placebo group, and, at the end of a second year extension of the trial, the mean number of skeletal events per year was 1.3 in the pamidronate group versus 2.2 in the placebo group.[44] The newer more potent bisphosphonate zoledronic acid has been shown to have

© 2008 American Society for Bone and Mineral Research

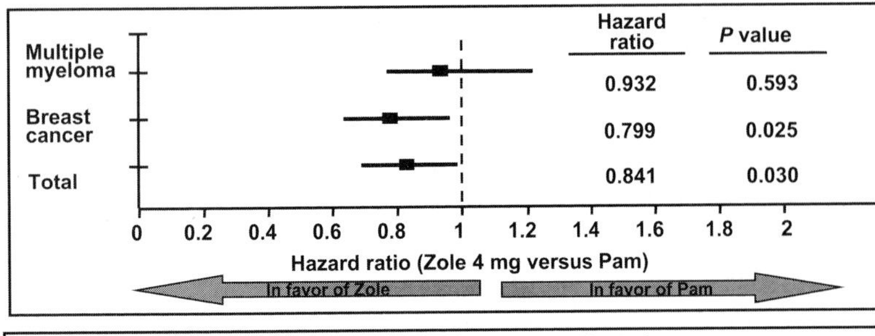

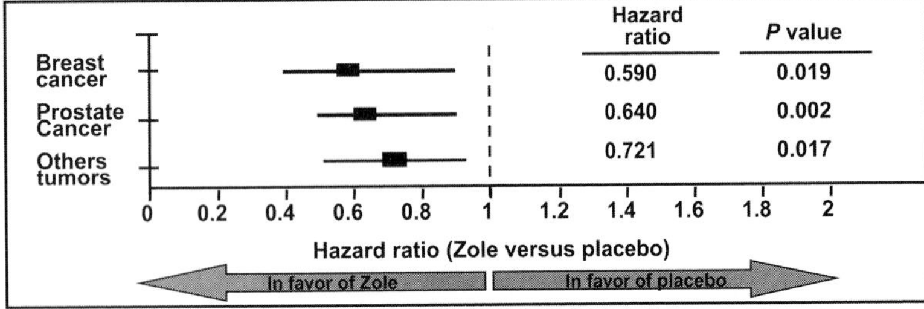

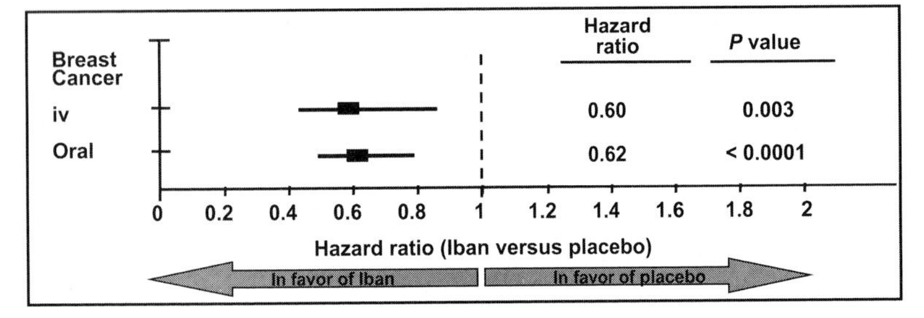

FIG. 2. Effects of long-term therapy with bisphosphonates on the risk of developing a skeletal complication in patients with bone metastases. Data are summarized by a multiple-event analysis (Andersen-Gill model for zoledronic acid and Poisson regression analysis for ibandronate). Hazard ratios (and 95% CI) are shown in the left part of each graph with corresponding p values indicated in the right parts. The top graph is taken from Ref. 37; data in the middle graph are from Refs. 27, 13, and 31; data in the bottom graph relate to ibandronate and are taken from Refs. 6 and 22.

a comparable efficacy to pamidronate in a randomized phase III trial including myeloma patients (Fig. 2, top panel).[37] Although there is no direct comparative trial between clodronate and pamidronate or zoledronic acid, the ASCO Panel recommends only intravenous pamidronate or zoledronic acid in light of the use of the time to the first event as the primary endpoint and a more complete assessment of bony complications.[42]

Prostate Cancer and Other Solid Tumors. Skeletal metastases from prostate cancer are typically osteoblastic. Therefore, it was traditionally not felt that this form of bone metastasis might respond to antiresorptive therapy. Meanwhile, histomorphometric analyses of bone biopsies and studies on biochemical markers of bone turnover showed that enhanced bone formation in osteoblastic lesions is accompanied by severe bone resorption.[45] There is only one study whose primary objective was to show a reduction in the frequency of objective SREs after bisphosphonate therapy. Hormone-refractory prostate cancer patients with bone metastases (n = 643) were randomized to intravenous zoledronic acid 8 or 4 mg or placebo every 3–4 wk.[12,13] As mentioned above, the group receiving 8 mg was switched during the trial to 4 mg because of renal toxicity. At the end of the core trial, there was a relative reduction of 25% in the number of patients presenting an objective bone complication. In a multiple-event analysis, zoledronic acid 4 mg significantly decreased the risk of developing skeletal complications by 36% compared with placebo ($p <$ 0.005; Fig. 2, middle panel). Other secondary endpoints, including the time to the first SRE or the percentage of patients who presented a fracture, were also significantly reduced in the 4 mg zoledronic acid group. One can speculate that part of the favorable effects of zoledronic acid, especially on the fracture rate, could be caused by effective therapy of castration-induced osteoporosis.

In a similar phase III study conducted in patients with lung and other solid tumors, the results were less impressive than in other cancers, partly because of the short survival of lung cancer patients.[46] At 9 mo, the primary endpoint (percentage of patients with an SRE) was not significantly lower with zoledronic acid 4 mg than with placebo, but a multiple-event analysis indicated a favorable effect (Fig. 2, middle panel). Retrospective subset analysis of patients with kidney cancer suggests a marked efficacy in that particular tumor.

Bisphosphonates as Adjuvant Therapy

Another potential major role for bisphosphonates is the prevention or at least a delay in the development of bone metastases. Bisphosphonates have the potential to reduce tumor burden in bone, whether indirectly by decreasing bone turnover or directly by one or several antitumor effects.[47] Published trials have only used clodronate thus far. The only double-blind placebo-controlled trial involving >1000 unselected breast cancer patients after surgery treated for 2 yr with 1600 mg clodronate or placebo indicated that clodronate could reduce the incidence of bone metastases (by 31% at 5 yr, p = 0.043) and may prolong survival (p = 0.048).[48] An ongoing trial is comparing clodronate with placebo in the adjuvant setting (The National Surgical Adjuvant Breast Project [NSABP] B-34 trial; n = 3200). The accrual is completed, and this study will prove to be the confirmatory trial for the effects of adjuvant clodronate. Newer aminobisphosphonates such as ibandronate and zoledronic acid are expected to inhibit bone metastases more effectively, and they are currently tested in the adjuvant setting. In patients with prostate cancer, there is no evidence that bisphosphonates are useful in the adjuvant setting. In a recent placebo-controlled trial, oral clodronate did not improve bone metastasis-free survival.[49]

© 2008 American Society for Bone and Mineral Research

Practical Recommendations for Possible Individualized Use, Safety Aspects, and Perspectives

Breast Cancer. The indications of bisphosphonate therapy in breast cancer patients nowadays go from the correction of tumor-induced hypercalcemia to the prevention of cancer treatment-induced bone loss.[50,51] Their main use is currently the prevention of SREs in patients with tumor bone disease. The fact that the efficacy of monthly 8 mg zoledronic acid infusions is not superior to the 4-mg dose[37,46] suggests that we have reached some form of a ceiling effect. Zoledronic acid is more efficient than pamidronate but the efficacy of new generation bisphosphonates is probably quite close, although comparative trials with clinical endpoints are needed to prove this statement. Increasing the dose in patients who seem to respond poorly to standard doses of zoledronic acid cannot be recommended because of possible renal toxicity, and caution is probably mandatory with all bisphosphonates on a long-term basis. Relying on the concept of metronomic chemotherapy that exerts potent anti-angiogenic effects, repeated low doses of zoledronic acid led to a prolonged decrease of vascular endothelial growth factor (VEGF) levels, suggesting a prolonged anti-angiogenic effect, but the clinical implications of these findings are unknown.[52]

There are few evidence-based medicine criteria for when bisphosphonate treatment should be started and stopped. Current guidelines from the ASCO recommend the routine use of intravenous pamidronate or zoledronic acid in patients with breast cancer and radiographic evidence of bone destruction with additional consideration for patients who have an abnormal bone scan and localized bone pain.[24] Oral bisphosphonates could be a preferred alternative for many patients on endocrine therapy to avoid the inconvenience of monthly infusions. Long-term compliance can, however, be a problem with oral bisphosphonates. Based on the available data, it is recommended to start bisphosphonates when metastatic bone disease is shown in weight-bearing bones or in vertebrae, when painful site(s) correspond to areas of bone destruction as shown by imaging techniques, or when bone disease is multifocal on presentation. However, when starting therapy, clinicians should now consider the risk of an excessive duration of bisphosphonate therapy in an asymptomatic patient with minimal bone disease, who has a good chance to respond to a first regimen of anti-neoplastic therapy and the likelihood of a prolonged survival.[53] Animal data indicate that high-dose bisphosphonates for 1 yr significantly increase microdamage accumulation and reduce bone toughness (i.e., its ability to absorb energy or sustain deformation without breaking). Both factors are significantly related to the suppression of bone turnover.[54] More importantly, bisphosphonate-associated osteonecrosis of the jaw is linked to the duration of bisphosphonate therapy[55] and might, at least in part, be a consequence of excessive bone turnover inhibition and suppression of angiogenesis (see Safety Aspects).

ASCO guidelines recommend that, once initiated, intravenous bisphosphonates should be continued until there is a substantial decline in the patient's general performance status.[24] However, criteria are lacking to determine if and how long an individual patient benefits from their administration. Promoting lifelong therapy is in contradiction with the paucity of data regarding the usefulness and the safety of treatment durations beyond 2 yr. Stopping bisphosphonate therapy, at least temporarily, or reducing the frequency of the infusions (e.g., an infusion every 3 mo), could be considered in patients whose bone disease is not "aggressive" and is well controlled by the antineoplastic treatment. Such recommendations would go along the same line as the ones recently updated for myeloma (see below), but no data or specific guidelines are available yet to support these recommendations.[53] Biochemical markers of bone resorption might help identify patients who continue to benefit, particularly because a high rate of bone resorption is a predictive factor of future SREs and a poor prognostic factor.[56] A recent retrospective study suggests that early normalization of elevated baseline N-telopeptide of collagen type 1 (NTX), a bone resorption marker, is associated with longer event-free and overall survival times compared with patients who still have elevated NTX after 3 mo of zoledronic acid therapy.[57] Using bone markers to guide clinical decisions is not currently recommended for the individual patient. However, when bone turnover markers are suppressed in a patient whose bone disease is well controlled by antineoplastic therapy, it might help in the decision to consider a temporary arrest or a decrease in the frequency of bisphosphonate infusions after a prolonged administration. The benefits of such a marker-directed therapy are currently being tested.

Other Neoplasms. Updated ASCO guidelines on multiple myeloma advise to seriously consider discontinuing bisphosphonates in patients with responsive or stable disease after 2 yr of therapy. Bisphosphonates should be resumed on relapse with new-onset SREs.[42] A 2-yr treatment period was also recommended by the Mayo Clinic, but the International Myeloma working Group recommends to stop therapy after 1 yr in patients who have achieved a complete response or a very good partial response with transplantation without active bone disease.

Although skeletal metastases from prostate cancer are typically osteoblastic, prolonged administration of zoledronic acid can significantly reduce the incidence of skeletal complications. Bisphosphonates should probably be recommended for all patients with hormone-refractory prostate cancer and bone metastases, especially when they are symptomatic. In other tumor types, it is reasonable to start bisphosphonates if the skeleton is one of (or the) predominant symptomatic metastatic site(s) and expected survival time is at least 4–6 mo.

Safety Aspects

Although generally well tolerated, bisphosphonates are occasionally associated with adverse events. Hypocalcemia is a side effect that may occur with all bisphosphonates but especially in vitamin D–deficient patients. It is advisable to administer calcium and vitamin D to all patients on prolonged bisphosphonate therapy to avoid hypocalcemia and the deleterious consequences of chronic secondary hyperparathyroidism. Characteristic adverse events with oral bisphosphonates are gastrointestinal, such as epigastric pain and esophagitis, even through oral clodronate and especially ibandronate have been shown to be well tolerated in controlled phase III trials.[22] The reported incidence of renal function deterioration in clinical trials of zoledronic acid was 10.7% in patients with multiple myeloma or breast cancer, which is not significantly different than the pamidronate figures in that trial.[37] Rare cases of renal failure with zoledronic acid have been subsequently reported, and renal monitoring is now recommended before each infusion of zoledronic acid.[58] Recent updates to the product label advocate stepwise dose reductions when baseline creatinine clearance is 30–60 ml/min. Zoledronic acid is not recommended in patients with severe renal deterioration or those taking nephrotoxic medications. Prolonged use of intravenous ibandronate in patients with breast cancer has shown a low incidence of renal adverse events that is comparable to placebo.[6]

Osteonecrosis of the jaw (ONJ) has been more recently

© 2008 American Society for Bone and Mineral Research

reported, especially after prolonged bisphosphonate therapy.[55,59] Although often devastating, ONJ can also present as an asymptomatic bony exposure. Its definition, early diagnosis, and follow-up have been recently reviewed by an ASBMR task force.[59] According to the series, its prevalence varies between 1% and 10% in patients on prolonged bisphosphonate therapy, with higher rates in myeloma than in solid tumors. It typically occurs after dental extraction but can also occur spontaneously. The prolonged inhibition of bone remodeling, the decreased intraosseous blood flow, the lack of repair of physiologic microfractures caused by constant stress from masticatory forces, and infections in the maxilla and/or mandible, all probably play a key role in the development of ONJ. The risk seems to be higher after zoledronic acid than after pamidronate but is essentially linked to the duration of therapy. Recent data suggest that the frequency of ONJ could be much lowered by appropriate preventive and regular dental care.

Inhibition of RANKL

Inhibition of the RANK–RANKL system is extremely effective in animal models to prevent tumor-induced osteolysis, confirming that activation of this system is the predominant mechanism of bone destruction, whatever the nature of secreted tumor factors. Initial data in humans indicate that blocking the RANK/RANKL system by osteoprotegerin (OPG) or denosumab, a "fully human" antiserum against RANKL, can indeed potently inhibit bone resorption in patients with bone metastases or myeloma bone disease.[60] In a 6-mo phase II study in bisphosphonate-naïve patients including six treatment cohorts (five different denosumab treatment regimens and one open-label intravenous bisphosphonate cohort), it was shown that denosumab treatment induced rapid and sustained reduction of bone turnover and, under the conditions of this particular study, seemed to be similar to intravenous bisphosphonate treatment at reducing the risk of SREs.[61] Large phase III trials comparing denosumab and zoledronic acid are now ongoing in patients with various tumors and bone metastases. This antibody against RANKL seems to be particularly promising because of its potency, the ease of its administration (subcutaneous route) and, thus far at least, the apparent lack of toxicity.

REFERENCES

1. Body JJ, Bartl R, Burckhardt P, Delmas PD, Diel IJ, Fleisch H, Kanis JA, Kyle RA, Mundy GR, Paterson AHG, Rubens RD, for the International Bone and Cancer Study Group 1998 Current use of bisphosphonates in oncology. J Clin Oncol **16:**3890–3899.
2. Cleeland CS, Janjan NA, Scott CB, Seiferheld WF, Curran WJ 2000 Cancer pain management by radiotherapists: A survey of radiation therapy oncology group physicians. Int J Radiat Oncol Biol Phys **47:**203–208.
3. Vassiliou V, Kalogeropoulou C, Giannopoulou E, Leotsinidis M, Tsota I, Kardamakis D 2007 A novel study investigating the therapeutic outcome of patients with lytic, mixed and sclerotic bone metastases treated with combined radiotherapy and ibandronate. Clin Exp Metastasis **24:**169–178.
4. Coleman R, Rubens R 1987 The clinical course of bone metastases from breast cancer. Br J Cancer **55:**61–66.
5. Body JJ 2006 Breast Cancer: Bisphosphonate Therapy for Metastatic Bone Disease. Clin Cancer Res **12:**6258s–6263s.
6. Body JJ, Diel IJ, Lichinitser MR, Kreuser ED, Dornoff W, Gorbunova VA, Budde M, Bergstrom B, MF 4265 Study Group 2003 Intravenous ibandronate reduces the incidence of skeletal complications in patients with breast cancer and bone metastases. Ann Oncol **14:**1399–1405.
7. Hortobagyi GN, Theriault RL, Lipton A, Porter L, Blayney D, Sinoff C, Wheeler H, Simeone JF, Seaman JJ, Knight RD, Heffernan M, Mellars K, Reitsma DJ 1998 Long-term prevention of skeletal complications of metastatic breast cancer with pamidronate: Protocol 19 Aredia Breast Cancer Study Group. J Clin Oncol **16:**2038–2044.
8. Plunkett TA, Smith P, Rubens RD 2000 Risk of complications from bone metastases in breast cancer. implications for management. Eur J Cancer **36:**476–482.
9. Melton LJ III, Kyle RA, Achenbach SJ, Oberg AL, Rajkumar SV 2005 Fracture risk with multiple myeloma: A population-based study. J Bone Miner Res **20:**487–493.
10. Berruti A, Dogliotti L, Bitossi R, Fasolis G, Gorzegno G, Bellina M, Torta M, Porpiglia F, Fontana D, Angeli A 2000 Incidence of skeletal complications in patients with bone metastatic prostate cancer and hormone refractory disease: Predictive role of bone resorption and formation markers evaluated at baseline. J Urol **164:**1248–1253.
11. Small EJ, Smith MR, Seaman JJ, Petrone S, Kowalski MO 2003 Combined analysis of two multicenter, randomized, placebo-controlled studies of pamidronate disodium for palliation of bone pain in men with metastatic prostate cancer. J Clin Oncol **21:**4277–4284.
12. Saad F, Gleason DM, Murray R, Tchekmedyian S, Venner P, Lacombe L, Chin JL, Vinholes JJ, Goas JA, Chen B 2002 Zoledronic Acid Prostate Cancer Study Group. Zoledronic Acid Prostate Cancer Study Group: A randomized, placebo-controlled trial of zoledronic acid in patients with hormone-refractory metastatic prostate carcinoma. J Natl Cancer Inst **94:**1458–1468.
13. Saad F, Gleason DM, Murray R, Tchekmedyian S, Venner P, Lacombe L, Chin JL, Vinholes JJ, Goas JA, Zheng M, Zoledronic Acid Prostate Cancer Study Group 2004 Long-term efficacy of zoledronic acid for the prevention of skeletal complications in patients with metastatic hormone-refractory prostate cancer. J Natl Cancer Inst **96:**879–882.
14. Ripamonti C, Fulfaro F 2000 Malignant bone pain: Pathophysiology and treatments. Curr Rev Pain **4:**187–196.
15. Janjan NA 1997 Radiation for bone metastases: Conventional techniques and the role of systemic radiopharmaceuticals. Cancer **80**(8 Suppl)**:**1628–1645.
16. Robinson RG, Preston DF, Schiefelbein M, Baxter KG 1995 Strontium 89 therapy for the palliation of pain due to osseous metastases. JAMA **274:**420–424.
17. Ecker RD, Endo T, Wetjen NM, Krauss WE 2005 Diagnosis and treatment of vertebral column metastases. Mayo Clin Proc **80:**1177–1186.
18. Peh WC, Gilula LA 2003 Percutaneous vertebroplasty: Indications, contraindications, and technique. Br J Radiol **76:**69–75.
19. Dudeney S, Lieberman IH, Reinhardt MK, Hussein M 2002 Kyphoplasty in the treatment of osteolytic vertebral compression fractures as a result of multiple myeloma. J Clin Oncol **20:**2382–2387.
20. Robertson A, Reed N, Ralston S 1995 Effect of oral clodronate on metastatic bone pain: A double-blind, placebo-controlled study. J Clin Oncol **13:**2427–2430.
21. Jagdev SP, Purohito P, Heatley S, Herling C, Coleman RE 2001 Comparison of the effect of intravenous pamidronate and oral clodronate on symptoms and bone resorption in patients with metastatic bone disease. Ann Oncol **12:**1433–1438.
22. Body JJ, Diel IJ, Lichinitzer M, Lazarev A, Pecherstorfer M, Bell R, Tripathy D, Bergstrom B 2004 Oral ibandronate reduces the risk of skeletal complications in breast cancer patients with metastatic bone disease: Results from two randomised, placebo-controlled phase III studies. Br J Cancer **90:**1133–1137.
23. Mancini I, Dumon JC, Body JJ 2004 Efficacy and safety of ibandronate in the treatment of opioid-resistant bone pain associated with metastatic bone disease: A pilot study. J Clin Oncol **22:**3587–3592.
24. Hillner BE, Ingle JN, Chlebowski RT, Gralow J, Yee GC, Janjan NA, Cauley JA, Blumenstein BA, Albain KS, Lipton A, Brown S, American Society of Clinical Oncology 2003 American Society of Clinical Oncology 2003 update on the role of bisphosphonates and bone health issues in women with breast cancer. J Clin Oncol **21:**4042–4057.
25. Lipton A, Theriault RL, Hortobagyi GN, Simeone J, Knight RD, Mellars K, Reitsma DJ, Heffernan M, Seaman JJ 2000 Pamidronate prevents skeletal complications and is effective palliative treatment in women with breast carcinoma and osteolytic bone metastases. Cancer **88:**1082–1090.
26. Rosen LS, Gordon D, Kaminski M, Howell A, Belch A, Mackey J, Apffelstaedt J, Hussein M, Coleman RE, Reitsma DJ, Seaman

© 2008 American Society for Bone and Mineral Research

JJ, Chen BL, Ambros Y 2001 Zoledronic acid versus pamidronate in the treatment of skeletal metastases in patients with breast cancer or osteolytic lesions of multiple myeloma: A phase III, double-blind, comparative trial. Cancer J **7**:377–387.

27. Kohno N, Aogi K, Minami H, Nakamura S, Asaga T, Iino Y, Watanabe T, Goessl C, Ohashi Y, Takashima S 2005 Zoledronic acid significantly reduces skeletal complications compared with placebo in Japanese women with bone metastases from breast cancer: A randomized, placebo-controlled trial. J Clin Oncol **23**:3314–3321.
28. Body JJ, Diel IJ, Bell R, Pecherstorfer M, Lichinitser MR, Lazarev AF, Tripathy D, Bergstrom B 2004 Oral ibandronate improves bone pain and preserves quality of life in patients with skeletal metastases due to breast cancer. Pain **111**:306–312.
29. Diel IJ, Body JJ, Lichinitser MR, Kreuser ED, Dornoff W, Gorbunova VA, Budde M, Bergstrom B, MF 4265 Study Group 2004 Improved quality of life after long-term treatment with the bisphosphonate ibandronate in patients with metastatic bone disease due to breast cancer. Eur J Cancer **40**:1704–1712.
30. Weinfurt KP, Anstrom KJ, Castel LD, Schulman KA, Saad F 2006 Effect of zoledronic acid on pain associated with bone metastasis in patients with prostate cancer. Ann Oncol **17**:986–989.
31. Rosen LS, Gordon D, Tchekmedyian S, Yanagihara R, Hirsh V, Krzakowski M, Pawlicki M, de Souza P, Zheng M, Urbanowitz G, Reitsma D, Seaman JJ 2003 Zoledronic acid versus placebo in the treatment of skeletal metastases in patients with lung cancer and other solid tumors: A phase III, double-blind, randomized trial–the Zoledronic Acid Lung Cancer and Other Solid Tumors Study Group. J Clin Oncol **21**:3150–3157.
32. Storto G, Klain M, Paone G, Liuzzi R, Molino L, Marinelli A, Soricelli A, Pace L, Salvatore M 2006 Combined therapy of Sr-89 and zoledronic acid in patients with painful bone metastases. Bone **39**:35–41.
33. Paterson AH, Powles TJ, Kanis JA, McCloskey E, Hanson J, Ashley S 1993 Double-blind controlled trial of oral clodronate in patients with bone metastases from breast cancer. J Clin Oncol **11**:59–65.
34. Body JJ, Dumon JC, Piccart M, Ford J 1995 Intravenous pamidronate in patients with tumor-induced osteolysis: A biochemical dose-response study. J Bone Miner Res **10**:1191–1196.
35. Theriault RL, Lipton A, Hortobagyi GN, Leff R, Gluck S, Stewart JF, Costello S, Kennedy I, Simeone J, Seaman JJ, Knight RD, Mellars K, Heffernan M, Reitsma DJ 1999 Pamidronate reduces skeletal morbidity in women with advanced breast cancer and lytic bone lesions: A randomized, placebo-controlled trial: Protocol 18 Aredia Breast Cancer Study Group. J Clin Oncol **17**:846–854.
36. Lipton A 2003 Bisphosphonates and metastatic breast carcinoma. Cancer **97**(Suppl.):848–853.
37. Rosen LS, Gordon D, Kaminski M, Howell A, Belch A, Mackey J, Apffelstaedt J, Hussein MA, Coleman RE, Reitsma DJ, Chen BL, Seaman JJ 2003 Long-term efficacy and safety of zoledronic acid compared with pamidronate disodium in the treatment of skeletal complications in patients with advanced multiple myeloma or breast carcinoma: A randomized, double-blind, multicenter, comparative trial. Cancer **98**:1735–1744.
38. Body JJ 2003 Zoledronic acid: An advance in tumour bone disease and a new hope for osteoporosis. Expert Opin Pharmacother **4**:567–580.
39. Clemons MJ, Dranitsaris G, Ooi WS, Yogendran G, Sukovic T, Wong BY, Verma S, Pritchard KI, Trudeau M, Cole DE 2006 Phase II trial evaluating the palliative benefit of second-line zoledronic acid in breast cancer patients with either a skeletal-related event or progressive bone metastases despite first-line bisphosphonate therapy. J Clin Oncol **24**:4895–4900.
40. Body JJ, Lichinitser M, Tjulandin S, Garnero P, Bergström B 2007 Oral ibandronate is as active as intravenous zoledronic acid for reducing bone turnover markers in women with breast cancer and bone metastases. Ann Oncol **18**:1165–1171.
41. Kumar A, Loughran T, Alsina M, Durie BG, Djulbegovic B 2003 Management of multiple myeloma: A systematic review and critical appraisal of published studies. Lancet Oncol **4**:293–304.
42. Kyle RA, Yee GC, Somerfield MR, Flynn PJ, Halabi S, Jagannath S, Orlowski RZ, Roodman DG, Twilde P, Anderson K, American Society of Clinical Oncology 2007 American Society of Clinical Oncology 2007 clinical practice guideline update on the role of bisphosphonates in multiple myeloma. J Clin Oncol **25**:2464–2472.
43. Djulbegovic B, Wheatley K, Ross J, Clark O, Bos G, Goldschmidt H, Cremer F, Alsina M, Glasmacher A 2002 Bisphosphonates in multiple myeloma. Cochrane Database Syst Rev **3**:CD003188.
44. Berenson JR, Lichtenstein A, Porter L, Dimopoulos MA, Bordoni R, George S, Lipton A, Keller A, Ballester O, Kovacs M, Blacklock H, Bell R, Simeone JF, Reitsma DJ, Heffernan M, Seaman J, Knight RD 1998 Long-term treatment of advanced multiple myeloma patients reduces skeletal events. J Clin Oncol **16**:593–602.
45. Garnero P, Buchs N, Zekri J, Rizzoli R, Coleman RE, Delmas PD 2000 Markers of bone turnover for the management of patients with bone metastases from prostate cancer. Br J Cancer **82**:858–864.
46. Rosen LS, Gordon D, Tchekmedyan NS, Yanagihara R, Hirsh V, Krzakowski M, Pawlicki M, De Souza P, Zheng M, Urbanowitz G, Reitsma D, Seaman J 2004 Long-term efficacy and safety of zoledronic acid in the treatment of skeletal metastases in patients with nonsmall cell lung carcinoma and other solid tumors: A randomized, phase III, double-blind, placebo-controlled trial. Cancer **100**:2613–2621.
47. Fromigue O, Kheddoumi N, Body JJ 2003 Bisphosphonates antagonize bone growth factors effects on human breast cancer cells survival. Br J Cancer **89**:178–184.
48. Powles T, Paterson S, Kanis JA, McCloskey E, Ashley S, Tidy A, Rosenqvist K, Smith I, Ottestad L, Legault S, Pajunen M, Nevantaus A, Mannisto E, Suovuori A, Atula S, Nevalainen J, Pylkkanen L 2002 Randomized, placebo-controlled trial of clodronate in patients with primary operable breast cancer. J Clin Oncol **20**:3219–3224.
49. Mason MD, Sydes MR, Glaholm J, Langley RE, Huddart RA, Sokal M, Stott M, Robinson AC, James ND, Parmar MK, Dearnaley DP, Medical Research Council PR04 Collaborators 2007 Oral sodium clodronate for nonmetastatic prostate cancer–results of a randomized double-blind placebo-controlled trial: Medical Research Council PR04 (ISRCTN61384873). J Natl Cancer Inst **99**:765–776.
50. Body JJ 2004 Hypercalcemia of malignancy. Semin Nephrol **24**:48–54.
51. Body JJ, Bergmann P, Boonen S, Boutsen Y, Devogelaer JP, Goemaere S, Reginster JY, Rozenberg S, Kaufman JM 2007 Management of cancer treatment-induced bone loss in early breast and prostatic cancer – a consensus paper of the Belgian Bone Club. Osteoporos Int **18**:1165–1171.
52. Santini D, Vincenzi B, Tonini G 2007 Zoledronic Acid and angiogenesis. Clin Cancer Res **13**:6850–6851.
53. Body JJ 2007 Individualization of bisphosphonate therapy. In: Piccart M, Wood WC, Hung MC, Solin LJ, Cardoso F (eds.) Breast Cancer Management and Molecular Medicine: Towards Tailored Approaches. Springer, Berlin, Germany, pp. 545–564.
54. Mashiba T, Hirano T, Turner CH, Forwood MR, Johnston CC, Burr DB 2000 Suppressed bone turnover by bisphosphonates increases microdamage accumulation and reduces some biomechanical properties in dog rib. J Bone Miner Res **15**:613–620.
55. Woo SB, Hellstein JW, Kalmar JR 2006 Narrative [corrected] review: Bisphosphonates and osteonecrosis of the jaws. Ann Intern Med **144**:753–761.
56. Coleman RE, Major P, Lipton A, Brown JE, Lee KA, Smith M, Saad F, Zheng M, Hei YJ, Seaman J, Cook R 2005 Predictive value of bone resorption and formation markers in cancer patients with bone metastases receiving the bisphosphonate zoledronic acid. J Clin Oncol **23**:4925–4935.
57. Lipton A, Cook RJ, Major P, Smith MR, Coleman RE 2007 Zoledronic acid and survival in breast cancer patients with bone metastases and elevated markers of osteoclast activity. Oncologist **12**:1035–1043.
58. Chang JT, Green L, Beitz J 2003 Renal failure with the use of zoledronic acid. N Engl J Med **349**:1676–1679.
59. Khosla S, Burr D, Cauley J, Dempster DW, Ebeling PR, Felsenberg D, Gagel RF, Gilsanz V, Guise T, Koka S, McCauley LK, McGowan J, McKee MD, Mohla S, Pendrys DG, Raisz LG, Rug-

giero SL, Shafer DM, Shum L, Silverman SL, Van Poznak CH, Watts N, Woo SB, Shane E, American Society for Bone and Mineral Research 2007 Bisphosphonate-associated osteonecrosis of the jaw: Report of a task force of the American Society for Bone and Mineral Research. J Bone Miner Res **22:**1479–1491.
60. Body JJ, Facon T, Coleman RE, Lipton A, Geurs F, Fan M, Holloway D, Peterson MC, Bekker PJ 2006 A study of the biologic receptor activator of nuclear factor-kB ligand inhibitor, Denosumab, in patients with multiple myeloma or bone metastases from breast cancer. Clin Cancer Res **12:**1221–1228.
61. Lipton A, Steger GG, Figueroa J, Alvarado C, Solal-Celigny P, Body JJ, de Boer R, Berardi R, Gascon P, Tonkin KS, Coleman R, Paterson AH, Peterson MC, Fan M, Kinsey A, Jun S 2007 Randomized active-controlled phase II study of denosumab efficacy and safety in patients with breast cancer-related bone metastases. J Clin Oncol **25:**4431–4437.

Chapter 86. Radiotherapy of Skeletal Metastases

Øyvind S. Bruland,[1] Amanda Hird,[2] and Edward Chow[2]

[1]*Department of Oncology, The Norwegian Radium Hospital, University of Oslo, Oslo, Norway;* [2]*Department of Radiation, University of Toronto, Oncology Odette Cancer Centre, Sunnybrook Health Sciences Centre, Toronto, Ontario, Canada*

INTRODUCTION

Bone is the most common site of symptomatic cancer metastasis. Two thirds to three quarters of patients with advanced disease from breast and prostate carcinomas have skeletal metastases, and lung, thyroid, and renal carcinoma metastasize to bone in ~30–40% of cases.[1] Pain is the most common symptom.[1,2] Additionally, clinical implications of skeletal metastases include pathological fracture, nerve entrapment/spinal cord compression (SCC), bone marrow insufficiency, and hypercalcemia. Hence, bone metastases have a devastating impact on a patient's quality of life.[1,3,4] SCC is of particular concern to cancer patients with a long expected survival[1] (e.g., those with the diagnosis of skeletal metastasis as the first and sole metastatic event).

Optimal management combines medical treatment, radiation therapy, surgery, bone-targeted radiopharmaceuticals, and bisphosphonates depending on the biology of the disease, extent of the skeletal involvement, and the life expectancy of the patient.

EXTERNAL BEAM RADIOTHERAPY

Skeletal metastases are the single most frequent indication for palliative radiotherapy. External beam radiotherapy (EBRT) effectively relieves pain from localized sites of skeletal metastases.[5] However, the lack of tumor-only selectivity limits its clinical use. Furthermore, because skeletal metastases usually are multiple and distributed throughout the axial skeleton,[2,4] larger or multiple fields of irradiation are often necessary. Table 1 outlines factors to be considered when prescribing palliative radiotherapy for bone metastases.

Pain Palliation

Solid empirical evidence has clearly documented that single-fraction (SF) EBRT provides equivalent pain relief compared with multi-fraction (MF) EBRT for uncomplicated bone metastases documented by >25 randomized clinical trials (RCTs) and 3 recent meta-analyses.[6–8]

One of the first RCTs was conducted by the Radiation Therapy Oncology Group.[9] Ninety percent of patients experienced some degree of pain relief, and 54% achieved complete pain palliation. The trial initially concluded that the low-dose, short-course schedules were as effective as the high-dose protracted programs. However, this study was criticized for using physician-based pain assessment. A reanalysis of the same set of data grouped solitary and multiple bone metastases and used the endpoint of pain relief, taking analgesia intake into account, as well as the need for retreatment. The authors concluded that the number of radiation fractions was significantly related to complete combined relief (absence of pain and use of narcotics) and that protracted dose-fractionation schedules were the most effective.[10] This was contrary to the initial report and highlights that the choice of endpoints will influence the outcome.[11]

More recently, results from several large-scale prospective RCTs have been published. The UK Bone Pain Trial Working Party randomized 765 patients with bone metastases to either an SF or MF regimen.[12] There were no significant differences in the time to first improvement in pain, time to complete pain relief, and time to first increase in pain at any time up to 12 mo after randomization, and no differences in the incidence of nausea, vomiting, SCC, or pathological fracture between the two groups. Retreatment was, however, twice as common after SF than after MF radiotherapy. The study concluded that an SF of 8 Gy is as safe and effective for the palliation of metastatic bone pain for at least 12 mo with greater convenience and lower cost than MF treatment.

Table 1. Factors to be Considered When Prescribing Palliative Radiotherapy for Bone Metastases

EBRT Single-fraction	*EBRT Multi-fraction*
Indication: pain relief	Indication: local tumor control
Short life expectancy	Expected long-term survival
Concomitant visceral metastases	Predominantly bone or bone only metastasis
Poor performance status	Good performance status
Inflammatory pain	Neuropathic pain
Aspects of cost and inconvenience	Spinal cord compression
	Postoperative EBRT after an orthopedic procedure in selected cases
	Impending fractures where surgery is not indicated

Dr. Bruland serves as Chairman of the Scientific Advisory Board for Algeta, ASA, Oslo, Norway, and is co-founder and minor stockholder in Algeta, ASA, Oslo, Norway, holding the patents of Alpharadin Radium-223. This bone-seeking Radiopharm is currently in phase II/III clinical trials. All other authors state that they have no conflicts of interest.

© 2008 American Society for Bone and Mineral Research

The large Dutch Bone Metastases Study included 1171 patients and confirmed the results mentioned above.[13] In this trial, the retreatment rates were 25% in the single 8-Gy arm, 7% in the MF arm, and more pathological fractures were observed in the SF group, but the absolute percentage was low. In a cost-utility analysis of this RCT, there was no difference in life expectancy or quality-adjusted life expectancy. The estimated cost of radiotherapy, including retreatments and nonmedical costs, was significantly lower for the SF than for the MF schedule.[14]

A Scandinavian RCT planned to recruit 1000 patients with painful bone metastases randomized to single 8 Gy or 30 Gy (3 Gy in 10 fractions).[15] The data monitoring committee recommended closure of the study after 376 patients had been recruited because interim analyses indicated that the treatment groups had similar outcomes. Equivalent pain relief within the first 4 mo was experienced, and no differences were found for fatigue, global quality of life, and survival between the groups.[15]

Two meta-analyses published in 2003 each showed no significant difference in complete and overall pain relief between SF and MF EBRT for bone metastases.[6,7] Results were remarkably similar, with the paper of Wu et al.[7] reporting a complete response rate (absence of pain) of 33% and 32% after SF and MF EBRT, respectively, compared with 34% and 32% for Sze et al.[6] Overall response rates from the two meta-analyses were 62% and 59%, compared with 60% and 59%,[6] for SF and MF, respectively. When restricted to evaluable patients, overall response rates became 73% for each arm.[7] Most patients experienced pain relief in the first 2–4 wk after EBRT.[7] Side effects were similar and generally consisted of nausea and vomiting.

An updated meta-analysis reviewed 16 RCTs that compared SF and MF schedules[8] involving a total of 2513 randomizations to SF arms and 2487 to MF arms. The overall response rate to SF EBRT was 58%, and complete response rate was 23%, which was not significantly different from the 59% and 24% experienced by patients randomized to MF EBRT. No differences in acute toxicity, pathological fracture (3.2% of patients fractured after SF versus 2.8% after MF), or SCC incidence were found, thus confirming the conclusions of the 2003 systematic reviews.

Neuropathic Pain and Spinal Cord Compression

There is some evidence that certain groups of patients would benefit from a protracted schedule. In a comparison of a single 8 Gy versus 20 Gy (4 Gy in five fractions) for 272 patients with a neuropathic pain component,[16] it was found that SF was not as effective as MF; however, it was also not significantly worse. The authors recommended MF as standard radiotherapy for patients with neuropathic pain. However, in patients with short survival or poor performance status, as well as when cost/inconvenience of MF is relevant, SF could be used instead.[16]

In the treatment of neoplastic SCC, the most appropriate radiation therapy schedule is still undefined. Patients with pending or complete SCC have been excluded and/or not addressed in the eligibility criteria in most RCTs. Only one RCT has studied the outcome in patients with SCC and an estimated outcome of 6 mo or less with no indication for primary surgery.[17] However, two EBRT schedules not commonly used were compared; 16 Gy in two fractions over 1 wk or a split course of 15 Gy in three fractions, followed by 4 days rest, and the additional 15 Gy in five fractions. No significant differences were reported between the two arms.[17]

Until the results from a recently published RCT comparing surgery and postoperative EBRT and EBRT alone[18] were presented in favor of primary surgery, the common view was that the outcome did not differ between EBRT and surgery for patients with vertebral metastases and SCC.[19]

Impending Fracture and Risk Prediction

An impending fracture has a significant likelihood of fracture under normal physiological stresses. Although some physicians believe that all patients with proximally located femoral metastases should undergo preventive surgery, this would result in a large number of unnecessary surgical procedures.[20] Furthermore, a proportion of patients will not be candidates for an operative procedure or will refuse surgical intervention. Often a minimum life expectancy (6–12 wk), a reasonable performance status, manageable co-morbidities, and adequate remaining bone to support the implanted hardware are required to justify the morbidity and mortality risk.[21]

If an orthopedic intervention is not appropriate, patients may receive EBRT alone. Although EBRT can provide pain relief and tumor control, it does not restore bone stability, and remineralization will take weeks to months.[22] Patients should be warned of the increased risk of fracture in the peri-radiation period because of an induced hyperemic response at the periphery of the tumor that temporarily weakens the adjacent bone. Pain relief may allow the patient to be more mobile and, hence, at greater risk for fracture. As such, measures to reduce anatomic forces across the lesion (crutches, a sling, or a walker) are routinely introduced during this time.

Although there is no consensus on appropriate dose fractionation, most authors recommend a MF course of EBRT in a patient with an impending or established fracture.[20] One retrospective series analyzed 27 pathologic fractures in various sites treated with doses of 40–50 Gy over 4–5 wk. Healing with remineralization was seen in 33%, with pain relief in 67%.[23] In practice, 20–40 Gy for established pathologic fracture is generally given over 1–3 wk. In patients with an apparently solitary, histologically confirmed metastasis, especially after a long disease-free interval, some clinicians may wish to give even a higher dose, 40–50 Gy, under the assumption that this will provide long-term control.

Reirradiation

Subsets of patients with metastatic disease have longer life expectancies than in the past because of advances in systemic therapy and may therefore outlive the duration of benefit provided by their initial palliative EBRT. This may require consideration of reirradiation of previously treated sites at a later date.[24]

The clinical indications, optimal dose and fractionation, and techniques for retreatment are controversial[25] because of lack of precise quantitative data on the time course, magnitude, and tissue specificity of long-term occult radiation injury recovery.[26]

Retreatment rates after SF EBRT varied from 18% to 25% compared with 7% to 9% after MF EBRT.[12,13,15,27] The Dutch Bone Metastases Study Group recently reanalyzed their data to specifically report the efficacy of reirradiation.[28] Of patients not responding to initial radiation, 66% who initially received a single 8 Gy responded to retreatment compared with 33% of patients who initially received a MF course. Retreatment in patients after pain progression was successful in 70% of those who received SF initially compared with 57% of those who received more than one fraction. Overall, reirradiation was effective in 63% of all such treated patients.

Hence, it is important to consider reirradiation of sites of metastatic bone pain after initial EBRT, particularly when this

© 2008 American Society for Bone and Mineral Research

follows an initial period of response. There is also evidence that a proportion of initial nonresponders will respond. The preferred dose schedule, however, is at present unknown, but a large, prospective, randomized intergroup study using common reirradiation schedules has been launched.[29]

BONE-SEEKING RADIOPHARMACEUTICALS

Treatment with intravenously injected bone-seeking radiopharmaceuticals (BSRs) is an intriguing alternative that selectively delivers ionizing radiation to targeted areas of amplified osteoblastic activity and targets multiple (symptomatic and asymptomatic) metastases simultaneously. The target is Ca-OH-apatite, which is particularly abundant in sclerotic metastases from prostate cancer but is also present, although more heterogeneously distributed, in mixed sclerotic/osteolytic metastases from breast cancer. This is evident from the biodistribution image common to all BSRs—exemplified as "hotspots" visualized on a routine diagnostic bone scan (by ^{99m}Tc-MDP; a radiolabeled bisphosphonate). BSRs effectively relieve pain and have been thoroughly reviewed.[30–34] In the commercially available formulations, the radioisotopes involved are β-emitters: strontium-89 dichloride (Metastron; GE Healthcare, Chalfont St. Giles, UK) and ^{153}Sm-EDTMP (Quadramet; Schering AG, Berlin, Germany, and Cytogen Co., Princeton, NJ, USA).

Because of the millimeter range of the emitted electrons, the cross-irradiation of the bone marrow represents an ever-present concern. After intravenous injection of a β-emitting BSR, bone marrow is an innocent bystander and the dose-limiting organ. Furthermore, disease-associated bone marrow suppression already present in these patients often results in delayed and unpredictable recovery. This severely limits the usefulness of β-emitting BSRs, especially when dosages are increased to deliver potential antitumor radiation levels and/or repeated treatments are attempted. Few clinical studies to date have reported on the feasibility of combining BSRs and chemotherapy.[35–38]

Because of short particle track-length and potent cell-killing, an α-emitting BSR could be an intriguing alternative.[39] In contrast to the β-emitters, the α-particle emitters deliver a much more energetic and localized radiation that produce densely ionizing tracks and predominantly nonreparable double DNA strand breaks. In a phase 1 study of single-dosage administration of escalating amounts of the natural bone-seeker ^{223}Ra in 25 patients with bone metastases from breast and prostate cancer,[40] dose-limiting hematological toxicity was not observed. Mild and reversible myelosuppression occurred, with only grade one toxicity for thrombocytes at the two highest doses. Results from a phase 2 RCT of external beam radiation plus either saline or ^{223}Ra injections (given four times at 4-wk intervals) have recently been published.[41] Radium-223 treatment resulted in a statistically significant decrease from baseline compared with placebo both in bone alkaline phosphatase and prostate-specific antigen. A favorable adverse event profile was observed, with minimal bone marrow toxicity for patients who received ^{223}Ra. Importantly, survival analyzes from this phase 2 trial showed a significant overall survival benefit for ^{223}Ra.[41]

REFERENCES

1. Coleman RE 2006 Clinical features of metastatic bone disease and risk of skeletal morbidity. Clin Cancer Res **12**(20 suppl):6243–6249.
2. Hage WD 2000 Incidence, location, and diagnostic evaluation of metastatic bone disease. Orthop Clin North Am **31**:515–528.
3. BASO: British Association of Surgical Oncology 1999 The management of metastatic bone disease in the United Kingdom. Eur J Surg Oncol **25**:3–23.
4. Hoegler D 1997 Radiotherapy for palliation of symptoms in incurable cancer. Curr Probl Cancer **21**:129–183.
5. Chow E, Wong R, Hruby G, Connolly R, Franssen E, Fung KW, Andersson L, Schueller T, Stefaniuk K, Szumacher E, Hayter C, Pope J, Holden L, Loblaw A, Finkelstein J, Danjoux C 2001 Prospective Patient-Based Assessment of Effectiveness of Palliative Radiotherapy for Bone Metastases in an Outpatient Radiotherapy Clinic. Radiother Oncol **61**:77–82.
6. Sze WM, Shelley M, Held I, Mason M 2003 Palliation of metastatic bone pain: Single fraction versus multifraction radiotherapy—a systemic review of randomized trials. Clin Oncol **15**:345–352.
7. Wu JSY, Wong R, Johnston M, Bezjak A, Whelan T 2003 Meta-analysis of dose-fractionation radiotherapy trials for the palliation of painful bone metastases. Int J Radiother Oncol Biol Phys **55**:594–605.
8. Chow E, Harris K, Fan G, Tsao M, Sze WM 2007 Palliative radiotherapy trials for bone metastases: A systemic review. J Clin Oncol **25**:1423–1436.
9. Tong D, Gillick L, Hendrickson F 1982 The palliation of symptomatic osseous metastases: Final results of the study by the Radiation Therapy Oncology Group. Cancer **50**:893–899.
10. Blitzer P 1985 Reanalysis of the RTOG study of the palliation of symptomatic osseous metastases. Cancer **55**:1468–1472.
11. Chow E, Wu JS, Hoskin P, Coia LR, Bentzen SM, Blitzer PH 2002 International consensus on palliative radiotherapy endpoints for future clinical trials in bone metastases. Radiother Oncol **64**:275–280.
12. Bone Pain Trial Working Party 1999 8 Gy single fraction radiotherapy for the treatment of metastatic skeletal pain: Randomized comparison with multi-fraction schedule over 12 months of patient follow-up. Radiother Oncol **52**:111–121.
13. Steenland E, Leer J, van Houwelingen H, Post WJ, van den Hout WB, Kievit J, de Haes H, Oei B, Vonk E, van der Steen-Banasik E, Wiggenraad RGJ, Hoogenhout J, Wárlám-Rodenhuis C, van Tienhoven G, Wanders R, Pomp J, van Reijn M, van Mierlo T, Rutten E 1999 The effect of a single fraction compared to multiple fractions on painful bone metastases: A global analysis of the Dutch Bone Metastasis Study. Radiother Oncol **52**:101–109.
14. Van den Hout WB, van der Linden YM, Steenland E, Wiggenraad RGJ, Kievit J, de Haes H, Leer JWH 2003 Single- versus multiple-fraction radiotherapy in patients with painful bone metastases: Cost-utility analysis based on a randomized trial. J Natl Cancer Inst **95**:222–229.
15. Kaasa S, Brenne E, Lund J, Fayers P, Falkmer U, Holmberg M, Lagerlund M, Bruland O 2006 Prospective randomized multicentre trial on single fraction radiotherapy (8 Gy X 1) versus multiple fractions (3 Gy X 10) in the treatment of painful bone metastases: Phase III randomized trial. Radiother Oncol **79**:278–284.
16. Roos DE, Turner SL, O'Brien PC, Smith JG, Spry NA, Burmeister BH, Hoskin PJ, Ball DL 2005 Randomized trial of 8 Gy in 1 versus 20 Gy in 5 fractions of radiotherapy for neuropathic pain due to bone metastases (Trans-Tasman Radiation Oncology Group, TROG 96.05). Radiother Oncol **75**:54–63.
17. Maranzano E, Bellavita R, Rossi R 2005 Radiotherapy alone or surgery in spinal cord compression? The choice depends on accurate patient selection. J Clin Oncol **23**:8270–8272.
18. Patchell RA, Tibbs PA, Regine WF, Payne R, Saris S, Kryscio RJ, Mohiuddin M, Young B 2005 Direct decompressive surgical resection in the treatment of spinal cord compression caused by metastatic cancer: A randomised trial. Lancet **366**:643–648.
19. Byrne TN 1992 Spinal cord compression from epidural metastases. N Engl J Med **32**:614–619.
20. van der Linden YM, Kroon HM, Dijkstra SP, Lok JL, Noordijk EM, Leer JWH Marijnen CAM and for the Dutch Bone Metastasis Study Group 2003 Simple radiographic parameter predicts fracturing in metastatic femoral bone lesions: Results from a randomised trial. Radiother Oncol **69**:21–31.
21. Healey JH, Brown HK 2000 Complications of bone metastases: Surgical management. Cancer **88**:2940–2951.
22. Agarawal JP, Swangsilpa T, van der Linden Y, Rades D, Jeremic B, Hoskin PJ 2006 The role of external beam radiotherapy in the management of bone metastases. Clin Oncol **18**:747–760.
23. Rieden K, Kober B, Mende U 1986 Radiotherapy of pathological fractures and skeletal lesions in danger of fractures. Strahlenther Onkol **162**:742–749.

© 2008 American Society for Bone and Mineral Research

24. Morris DE 2000 Clinical experience with retreatment for palliation. Semin Radiat Oncol **10**:210–221.
25. Jones B, Blake PR 1999 Retreatment of cancer after radical radiotherapy. Br J Radiol **72**:1037–1039.
26. Nieder C, Milas L, Ang KK 2000 Tissue tolerance to reirradiation. Semin Radiat Oncol **10**:200–209.
27. Hartsell WF, Scott CB, Bruner DW, Scarantino CW, Ivker RA, Roach M III, Suh JH, Demas WF, Movsas B, Petersen IA, Konski AA, Cleeland CS, Janjan NA, DeSilvio M 2005 Randomized trial of short- versus long-course radiotherapy for palliation of painful bone metastases. J Natl Cancer Inst **97**:798–804.
28. van der Linden YM, Lok JJ, Steenland E, Martijn H, van Houwelingen H, Marijnen CAM, Leer JWH 2004 Single fraction radiotherapy is efficacious: A further analysis of the Dutch Bone Metastasis Study controlling for the influence of retreatment. Dutch Bone Metastases Study Group. Int J Radiat Oncol Biol Phys **59**:528–537.
29. Chow E, Hoskin PJ, Wu J, Roos D, van der Linden Y, Hartsell W, Vieth R, Wilson C, Pater J 2006 A phase III international randomised trial comparing single with multiple fractions for re-irradiation of painful bone metastases: National Cancer Institute of Canada Clinical Trials Group (NCIC CTG) SC 20. Clin Oncol **18**:125–128.
30. Lewington VJ 2005 Bone-seeking radionuclides for therapy. J Nucl Med **46**:38s–47s.
31. Silberstein EB 2000 Systemic radiopharmaceutical therapy of painful osteoblastic metastases. Semin Radiat Oncol **10**:240–249.
32. Finlay IG, Mason MD, Shelley M 2005 Radioisotopes for the palliation of metastatic bone cancer: A systematic review. Lancet Oncol **6**:392–400.
33. Bauman G, Charette M, Reid R, Sathya J 2005 Radiopharmaceuticals for the palliation of painful bone metastases—a systematic review. Radiother Oncol **75**:258e1-258.e13.
34. Reisfield GM, Silberstein EB, Wilson GR 2005 Radiopharmaceuticals for the palliation of painful bone metastases. Am J Hosp Palliat Care **22**:41–46.
35. Tu SM, Kim J, Pagliaro LC, Vakar-Lopez F, Wong FC, Wen S, General R, Podoloff DA, Lin SH, Logothetis CJ 2005 Therapy tolerance in selected patients with androgen-independent prostate cancer following strontium-89 combined with chemotherapy. J Clin Oncol **23**:7904–7910.
36. Pagliaro LC, Delpassand ES, Williams D, Millikan RE, Tu SM, Logothetis CJ 2003 A Phase I/II study of strontium-89 combined with gemcitabine in the treatment of patients with androgen independent prostate carcinoma and bone metastases. Cancer **97**:2988–2994.
37. Sciuto R, Festa A, Rea S, Pasqualoni R, Bergomi S, Petrilli G, Maini CL 2002 Effects of low-dose cisplatin on 89Sr therapy for painful bone metastases from prostate cancer: A randomized clinical trial. J Nucl Med **43**:79–86.
38. Akerley W, Butera J, Wehbe T, Noto R, Stein B, Safran H, Cummings F, Sambandam S, Maynard J, Di Rienzo G, Leone L 2002 A multiinstitutional, concurrent chemoradiation trial of strontium-89, estramustine, and vinblastine for hormone refractory prostate carcinoma involving bone. Cancer **15**:1654–1660.
39. Bruland ØS, Nilsson S, Fisher DR, Larsen RH 2006 High-linear energy transfer irradiation targeted to skeletal metastases by the alpha-emitter 223Ra: Adjuvant or alternative to conventional modalities? Clin Cancer Res **12**:6250s–6257s.
40. Nilsson S, Balteskard L, Fosså SD, Westlin JE, Borch KW, Salberg G, Larsen RH, Bruland ØS 2005 First clinical experiences with alpha emitter radium-223 in the treatment of skeletal metastases from breast and prostate cancer. Clin Cancer Res **11**:4451–4459.
41. Nilsson S, Franzén L, Parker C, Tyrrell C, Blom R, Tennvall J, Lennernäs B, Petersson U, Johannessen DC, Sokal M, Pigott K, Yachnin J, Garkavij M, Strang P, Harmenberg J, Bolstad B, Bruland ØS 2007 Bone-targeted radium-223 in symptomatic, hormone refractory prostate cancer: A randomized, placebo-controlled, phase 2 study. Lancet Oncol **8**:587–594.

Chapter 87. Orthopedic Treatment of Metastatic Bone Disease

Kristy Weber and Scott Kominsky

Departments of Orthopaedic Surgery and Oncology, Johns Hopkins University, Baltimore, Maryland

INTRODUCTION

More than 1.4 million people are diagnosed with cancer each year,[1] and ~50% of those will develop bone metastasis. As treatments improve for primary and metastatic disease, patients are living longer with their disease. This often causes them to experience the morbidities of related bone disease. Although the most worrisome clinical problem is progressive disease in the skeleton, patients can also experience treatment-related osteoporosis. Additional physiologic disruptions in patients with bone metastasis include anemia and hypercalcemia. The bone lesions themselves can cause extreme pain and put the patient at risk for pathologic fractures. Patients become less mobile and may function at a lower level. Prolonged immobilization caused by pain or risk of fracture creates potential problems with thromboembolic disease or decubitus ulcers. Lesions in the vertebral region can cause progressive neurologic deficits. Overall quality of life is often markedly diminished.

Comprehensive treatment of bone metastasis is beyond the scope of this chapter, but advances in chemotherapy, targeted biologic therapy, and vaccines have been variably effective. Different forms of radiation are used to target cancer cells within the bone to provide palliative pain relief and potentially abrogate the need for surgical intervention. External beam radiation, cyberknife, and radiopharmaceuticals such as samarium are used depending on the location of disease. This chapter focuses on treatment that affects the neoplastic process and the bone microenvironment. A brief review of the molecular events related to metastatic bone disease will be discussed. The use of bisphosphonate therapy as well as surgical stabilization will be summarized.

BIOLOGY OF METASTATIC BONE LESIONS

Tumor–Bone Interface

The tumor cells interact within the bone microenvironment so that tumor growth is stimulated[2–7] (Fig. 1). The majority of work has been done in the area of breast cancer bone metas-

The authors state that they have no conflicts of interest.

Key words: bone, metastasis, orthopedic

© 2008 American Society for Bone and Mineral Research

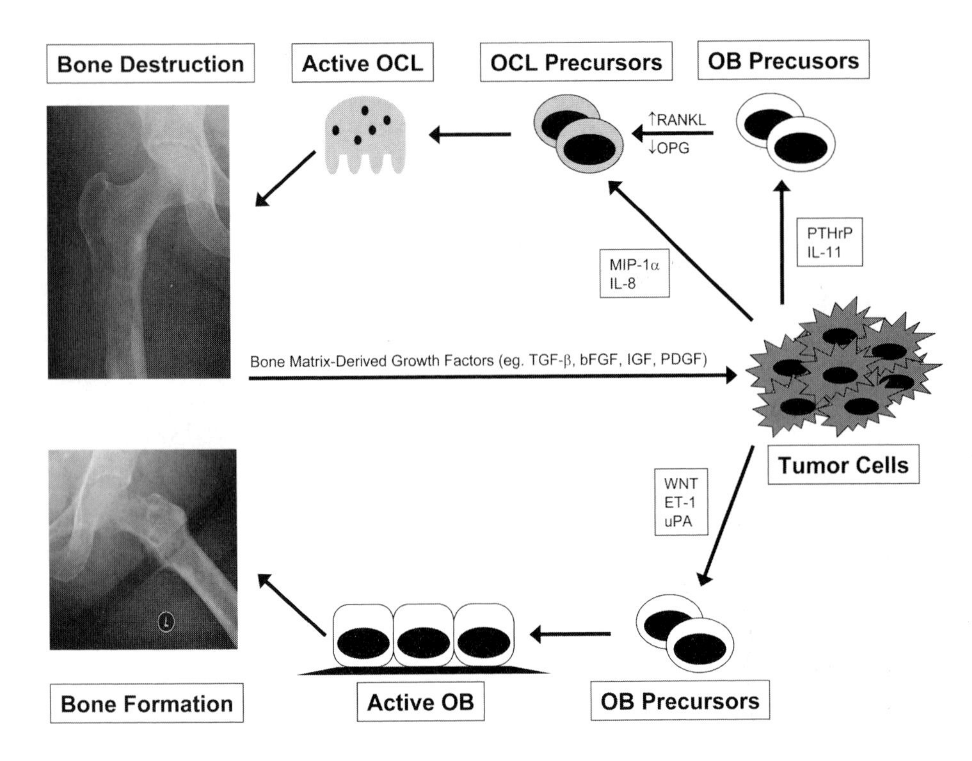

FIG. 1. This schematic shows the general cycle of bone destruction (osteolytic metastasis–lung cancer) and bone formation (osteoblastic metastasis–breast cancer).

tasis using osteolytic mouse models of disease.[8] In the course of normal bone remodeling, TGF-β is released, which stimulates breast cancer cells to secrete PTH-related protein (PTHrP). PTHrP stimulates osteoblast precursors to increase RANKL, which increases osteoclast differentiation. An increased number of active osteoclasts then destroy more bone, thereby restarting the cycle. Factors other than PTHrP can affect this cycle such as interleukin (IL)-8, IL-11, and macrophage inflammatory protein (MIP)-1α.[6] In addition, the bone microenvironment is rich in growth factors other than TGF-β including basic fibroblast growth factor (bFGF), insulin-like growth factor (IGF), and platelet-derived growth factor (PDGF).

To study osteoblastic metastasis, prostate cancer models have been used to identify important factors that stimulate osteoblasts to form new bone. Endothelin-1 (ET-1) is one of the factors secreted by prostate and breast cancer cells that stimulates osteoblast precursors.[9] Increased numbers of active osteoblasts are responsible for depositing new bone. A specific downstream target of ET-1, Dickkopf homolog 1 (Dkk1), plays a role in the bone destruction of patients with multiple myeloma and in patients with osteoblastic metastasis. Based on multiple studies, it is now known that both osteoblasts and osteoclasts play an important role in osteoblastic metastasis.[10]

Both PTHrP and the ET-1 receptor have been targeted in clinical trials to treat patients with breast and prostate cancer bone metastasis, respectively.

Use of Bisphosphonates

The use of bisphosphonates to treat patients with metastatic disease has significantly decreased the incidence of skeletally related events such as pathologic fractures.[11] These compounds bind preferentially to bone matrix and are known to inhibit osteoclastic bone resorption. Bisphosphonates are separated into two classes, nitrogen-containing and non–nitrogen-containing, each having a different mechanism of action. Non–nitrogen-containing bisphosphonates cause osteoclast apoptosis after breakdown into metabolites that compete with ATP during energy metabolism. Nitrogen-containing bisphosphonates target the mevalonate pathway, specifically the enzyme farnesyl diphosphate synthase (FPPS), and cause osteoclast inactivation by interfering with geranylgeranylation. Some studies have also shown direct antitumor effects by bisphosphonates.[12]

Bisphosphonates have been used with published success to treat bone pain and hypercalcemia in breast cancer and are most efficacious when used as an adjunct to systemic cancer therapies. They are becoming a routine part of the care for almost any patient with metastatic bone disease or multiple myeloma. The most commonly used drugs are zoledronic acid (Zometa; Novartis, Basel, Switzerland) and pamidronate (Aredia; Novartis) given as intravenous injections.[13,14] Based on a review of 30 randomized controlled trials of patients treated with oral or intravenous bisphosphonates for metastatic disease, these drugs are associated with a significant reduction in all skeletal morbidity endpoints with the exception of spinal cord compression.[11] They should be given when bone metastasis are first diagnosed, because they significantly increase the time to the first skeletal-related event.

SURGICAL TREATMENT OF METASTATIC BONE DISEASE

Impending Fractures

Because patients with bone metastasis are unlikely to be surgically cured, the primary focus of orthopedic oncologists is to improve quality of life. If a bone lesion is discovered at an early stage, radiation or systemic medical treatment may prevent further destruction and avoid surgery. However, if a lesion progresses despite nonsurgical treatment or is discovered only after there has been extensive cortical destruction and pain, surgical stabilization must be considered.[15] Patients who have prophylactic fixation of their extremity have a shorter hospitalization, quicker return to premorbid function, and less hardware complications.[16] Elective stabilization also

© 2008 American Society for Bone and Mineral Research

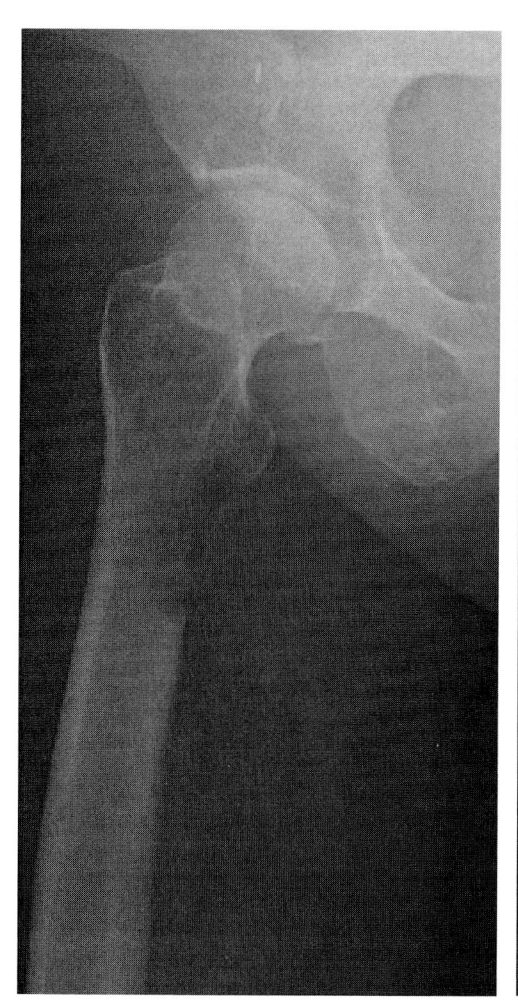

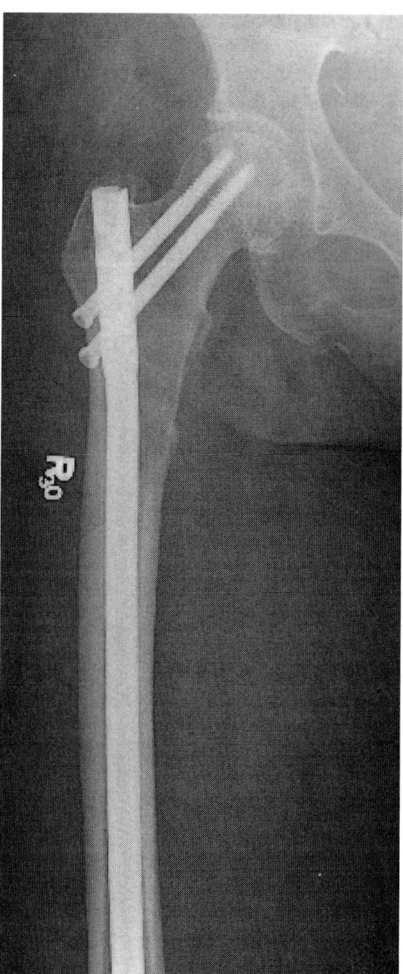

FIG. 2. (A) A radiograph of the right proximal femur in an 82-yr-old woman with metastatic breast cancer shows an osteolytic lesion in the subtrochanteric region at high risk for pathologic fracture. (B) Postoperative radiograph showing the stabilization of the femur with an intramedullary reconstruction nail.

allows the medical oncologist and surgeon to coordinate operative treatment and systemic chemotherapy. The difficulty lies in reliably determining which bone lesions will fracture. Several classifications have been proposed that involve determination of pain, cortical destruction, and/or size of the bone lesion.[17,18] Most widely quoted is the Mirels system, which recommends stabilization of lesions based on (1) radiographic appearance, (2) location, (3) presence of pain, and (4) size related to bone diameter.[18] Efforts to biomechanically assess the relative fracture risk of osteolytic lesions using quantitative CT scan analysis have been successful for the spine and extremities.[19]

Surgical Treatment

The goals of surgical treatment of patients with bone metastasis are to improve function and decrease pain. Treatment of impending or actual fractures secondary to metastatic bone disease uses different principles than those used for routine traumatic fractures. The underlying bone quality is often poor, and the patient may have progressive destruction despite treatment.

Upper extremity bone metastases are less common than those in the lower extremity and can often be treated nonoperatively. However, if patients require their upper extremities for weight bearing (i.e., have lower extremity lesions that require protected weight bearing with assistive devices), surgical treatment should be considered to improve function. Lesions in the scapula and clavicle are generally treated nonoperatively with radiation because most surgical options do not improve function in these areas. Bone destruction in the proximal humerus is treated by proximal humeral prosthetic replacement or an intramedullary device if secure fixation can be achieved. Diaphyseal humeral lesions are treated with intramedullary devices or occasionally intercalary metal spacers.[20,21] Distal humeral lesions are less common and can be stabilized with crossed intramedullary pins, dual plating, or segmental distal humeral prosthetic reconstruction. Bone metastases distal to the elbow are extremely rare and are treated on an individual basis.

Lower extremity metastasis are more common than those found in the upper extremities and have a large impact on quality of life because of the need for weight bearing. Pelvic lesions are usually treated nonoperatively or with minimally invasive techniques if the acetabulum is not affected. Acetabular lesions are treated according to specific classification schemes depending on the extent and location of bone loss.[22] Patients with severe bone loss should have a long enough predicted life span and a good performance status to make the procedure and recovery worthwhile. Metastases to the femoral neck are common, and patients often present with hip fractures.[23] The treatment is either bipolar or total hip reconstruction depending on the status of the acetabulum.[24] Inter-

© 2008 American Society for Bone and Mineral Research

nal fixation with plates and screws is not indicated, because there is a high risk of hardware failure with disease progression. In the intertrochanteric and subtrochanteric regions, options for prosthetic reconstruction or intramedullary fixation are available depending on the extent and location of bone loss as well as the tumor histology. Tumors that are less responsive to systemic treatment or radiation (i.e., renal cell carcinoma) are often treated more aggressively with surgical resection in this area. Femoral diaphyseal lesions are treated with intramedullary fixation (Fig. 21).[(24)] It is important that the intramedullary device includes femoral neck fixation to avoid future hip fractures. Distal femoral lesions are treated by intramedullary fixation, plate fixation or prosthetic reconstruction. Lesions distal to the knee are uncommon and treated on an individual basis.

The most common site of bone metastasis is the spine (thoracic). If patients are neurologically intact and there are no fracture fragments impinging on the spinal cord, radiation is often the first treatment option. If patients have intractable pain, significant or progressive neurologic deficits, or deformity progression, they should be stabilized surgically.[(25,26)]

Minimally Invasive Options

In selected patients, minimally invasive procedures provide an alternative to surgery and can produce long-lasting pain relief. Kyphoplasty and vertebroplasty are now commonly used techniques for patients who have osteolytic spine metastasis without neurologic compromise. Both techniques can be performed safely, stabilize the collapsed vertebral body, and yield quick pain relief.[(27)] Radiofrequency ablation (RFA) is used for metastasis in multiple bony sites with nearly all patients achieving some measure of pain relief.[(28)] Cyberknife treatment is a type of minimally invasive radiosurgery often used for spine metastasis that is performed on an outpatient basis with few side effects.[(29)] The goal is to totally destroy the tissue within the target volume with an effort to achieve local control. This technique is not appropriate for patients with cord compression and neurologic deficit, spinal instability, or radiosensitive tumors not previously treated with radiation.

REFERENCES

1. Jemal A, Siegel R, Ward E, Murray T, Xu J, Thun MJ 2007 Cancer statistics, 2007. CA Cancer J Clin **57:**43–66.
2. Chirgwin JM, Mohammad KS, Guise TA 2004 Tumor-bone cellular interactions in skeletal metastases. J Musculoskelet Neuronal Interact **4:**308–318.
3. Kominsky S, Doucet M, Brady K, Weber KL 2007 TGF-β Influences the Development of Renal Cell Carcinoma Bone Metastasis. J Bone Miner Res **22:**37–44.
4. Mundy GR 2002 Metastasis to bone: Causes, consequences and therapeutic opportunities. Nat Rev Cancer **2:**584–593.
5. Park JI, Lee MG, Cho K, Park BJ, Chae KS, Byun DS, Ryu BK, Park YK, Chi SG 2003 Transforming growth factor-beta1 activates interleukin-6 expression in prostate cancer cells through the synergistic collaboration of the Smad2, p38-NF-kappaB, JNK, and Ras signaling pathways. Oncogene **22:**4314–4332.
6. Roodman GD 1993 Role of cytokines in the regulation of bone resorption. Calcif Tissue Int **53**(Suppl 1):S94–S98.
7. Steeve KTMPST, Dominique H, Yannick F 2004 IL-6, RANKL, TNF-alpha/IL-1: Interrelations in bone resorption pathophysiology. Cytokine Growth Factor Rev **15:**49–60.
8. Kakonen SM, Selander KS, Chirgwin JM, Yin JJ, Burns S, Rankin WA, Grubbs BG, Dallas M, Cui Y, Guise TA 2002 Transforming growth factor-beta stimulates parathyroid hormone-related protein and osteolytic metastases via Smad and mitogen-activated protein kinase signaling pathways. J Biol Chem **277:**24571–24578.
9. Guise TA, Yin JJ, Mohammad KS 2003 Role of endothelin-1 in osteoblastic bone metastases. Cancer **97:**779–784.
10. Zhang J, Dai J, Yao Z, Lu Y, Dougall W, Keller ET 2003 Soluble receptor activator of nuclear factor κB-Fc diminishes prostate cancer progression in bone. Cancer Res **63:**7883–7890.
11. Ross JR, Saunders Y, Edmonds PM, Patel S, Broadley KE, Johnston SRD 2003 Systematic review of role of bisphosphonates on skeletal morbidity in metastatic cancer. BMJ **327:**469–475.
12. Senaratne SG, Mansi JL, Colston KW 2002 The bisphosphonate zoledronic acid impairs Ras membrane localisation and induces cytochrome c release in breast cancer cells. Br J Cancer **86:**1479–1486.
13. Berenson JR 2005 Recommendations for zoledronic acid treatment of patients with bone metastases. Oncologist **10:**52–62.
14. Conte P, Latreille J, Mauriac L, Calabresi F, Santos R, Campos D, Bonneterre J, Franchi G, Ford JM 1996 Delay in progression of bone metastases treated with intravenous pamidronate; results from a multicentre randomised controlled trial. J Clin Oncol **14:**2552.
15. Weber KL, Lewis VO, Randall RL, Lee AK, Springfield D 2004 An approach to the management of the patient with metastatic bone disease. Instr Course Lect **53:**663–676.
16. Katzer A, Meenen NM, Grabbe F, Rueger JM 2002 Surgery of skeletal metastases. Arch Orthop Trauma Surg **122:**251–258.
17. Beals RK, Lawton GD, Snell WE 1971 Prophylatic internal fixation of the femur in metastatic breast cancer. Cancer **28:**1350–1354.
18. Mirels H 1989 Metastatic disease in long bones: A proposed scoring system for diagnosing impending pathological fractures. Clin Orthop **249:**256–265.
19. Hong J, Cabe GD, Tedrow JR, Hipp JA, Snyder BD 2004 Failure of trabecular bone with simulated lytic defects can be predicted non-invasively by structural analysis. J Orthop Res **22:**479–486.
20. Damron TA, Sim FH, Shives TC, An KN, Rock MG, Pritchard DJ 1996 Intercalary spacers in the treatment of segmentally destructive diaphyseal humeral lesions in disseminated malignancies. Clin Orthop **324:**233–243.
21. Redmond BJ, Biermann JS, Blasier RB 1996 Interlocking intramedullary nailing of pathological fractures of the shaft of the humerus. J Bone Joint Surg Am **78:**891–896.
22. Marco RA, Sheth DS, Boland PJ, Wunder JS, Siegel JA, Healey JH 2000 Functional and oncological outcome of acetabular reconstruction for the treatment of metastatic disease. J Bone Joint Surg Am **82:**642–651.
23. Schneiderbauer MM, Von Knoch M, Schleck CD, Harmsen WS, Sim FH, Scully SP 2004 Patient survival after hip arthroplasty for metastatic disease of the hip. J Bone Joint Surg Am **86:**1684–1689.
24. O'Connor M, Weber K 2003 Indications and Operative Treatment for Long Bone Metastasis with a Focus on the Femur. Clin Orthop **415S:**276–278.
25. Bohm P, Huber J 2002 The surgical treatment of bony metastasis of the spine and limbs. J Bone Joint Surg Br **84:**521–529.
26. Holman PJ, Suki D, McCutcheon I, Wolinsky JP, Rhines LD, Gokaslan ZL 2005 Surgical management of metastatic disease of the lumbar spine: Experience with 139 patients. J Neurosurg Spine **2:**550–563.
27. Hentschel SJ, Burton AW, Fourney DR, Rhines LD, Mendel E 2005 Percutaneous vertebroplasty and kyphoplasty performed at a cancer center: Refuting proposed contraindications. J Neurosurg Spine **2:**436–440.
28. Goetz MP, Callstrom MR, Charboneau JW, Farrell MA, Maus TP, Welch TJ, Wong GY, Sloan JA, Novotny PJ, Petersen IA, Beres RA, Regge D, Capanna R, Saker MB, Gronemeyer DH, Gevargez A, Ahrar K, Choti MA, de Baere TJ, Rubin J 2004 Percutaneous image-guided radiofrequency ablation of painful metastases involving bone: A multicenter study. J Clin Oncol **22:**300–306.
29. Gerszten PC, Welch WC 2004 Cyberknife radiosurgery for metastatic spine tumors. Neurosurg Clin N Am **15:**491–501.

© 2008 American Society for Bone and Mineral Research

SECTION VIII

Sclerosing and Dysplastic Bone Diseases
(Section Editor: Richard Keen)

Please also see the following website on Chondrodysplasias: http://www.csmc.edu/10784.html

© 2008 American Society for Bone and Mineral Research

Chapter 88. Sclerosing Bone Disorders

Michael P. Whyte

Division of Bone and Mineral Diseases, Washington University School of Medicine at Barnes-Jewish Hospital and Center for Metabolic Bone Disease and Molecular Research, Shriners Hospitals for Children, St. Louis, Missouri

INTRODUCTION

Osteosclerosis and hyperostosis refer to trabecular and cortical bone thickening, respectively. Increased skeletal mass is caused by many rare (often hereditary) osteochondrodysplasias,[1,2] as well as by a variety of dietary, metabolic, endocrine, hematologic, infectious, or neoplastic disorders (Table 1).

OSTEOPETROSIS

Osteopetrosis (OMIM: 166600, 259700, 259710, 259720, 259730, 607634, 611490, 611497), sometimes called "marble bone disease," was first described in 1904 by Albers-Schönberg.[3] Traditionally, two major clinical forms are discussed[4]: the autosomal dominant, adult (benign) type that is associated with relatively few symptoms,[5] and the autosomal recessive, infantile (malignant) type that is typically fatal (if untreated) in early childhood.[6] Additional rarer types have included an "intermediate" form that presents during childhood where the impact on life expectancy is poorly understood.[7] Osteopetrosis with renal tubular acidosis and cerebral calcification is the inborn error of metabolism, carbonic anhydrase II deficiency.[4] Neuronal storage disease with malignant osteopetrosis has been considered a distinct entity.[8] Osteopetrosis, lymphedema, anhydrotic ectodermal dysplasia, and immunodeficiency (OL-EDA-ID) is an X-linked condition that affects boys.[9] Other unusual forms of osteopetrosis have been called "lethal," "transient infantile," and "postinfectious."[4] Drug-induced osteopetrosis was first described in 2003 in a boy who received high doses of pamidronate.[10] Recent revelation of the genetic defects (see below) responsible for most cases of osteopetrosis has greatly clarified this nosology while further illuminating osteoclast biology.[11]

Although defects in several genes cause osteopetrosis,[11] all true forms are caused by failure of osteoclast-mediated resorption of the skeleton.[4] Consequently, primary spongiosa (calcified cartilage deposited during endochondral bone formation) persists and is shown as a histopathological hallmark. Understandably, "osteopetrosis" has been used generically to describe radiodense skeletons yet lacking this finding. Now, it is crucial to appreciate that therapeutic approaches for genuine osteopetroses, for which this pathogenesis has been elucidated, may be inappropriate for other sclerosing bone disorders.[4]

Clinical Presentation

Infantile osteopetrosis manifests during the first year of life.[6] Nasal stuffiness caused by underdevelopment of the mastoid and paranasal sinuses is an early symptom. Cranial foramina do not widen, and optic, oculomotor, and facial nerves may become paralyzed. Hearing loss is common. Blindness can also be caused by retinal degeneration or raised intracranial pressure.[12] Some patients develop hydrocephalus or sleep apnea. Eruption of the dentition is delayed, and there is failure to thrive. Bones are dense, but fragile. Recurrent infection and spontaneous bruising and bleeding follow myelophthisis caused by the excessive bone, abundant osteoclasts, and fibrous tissue that crowds marrow spaces. Hypersplenism and hemolysis may exacerbate the anemia. Physical findings include short stature, macrocephaly, frontal bossing, "adenoid" appearance, nystagmus, hepatosplenomegaly, and *genu valgum.* Untreated patients usually die in the first decade of life from hemorrhage, pneumonia, severe anemia, or sepsis.[6]

Intermediate osteopetrosis leads to short stature, cranial nerve deficits, ankylosed teeth that predispose to osteomyelitis of the jaw, recurrent fractures, and mild or occasionally moderately severe anemia.[7]

Adult osteopetrosis features radiographic abnormalities that appear during childhood. In some kindreds, "carriers" show no disturbances.[5,13] The long bones are brittle and may fracture. Facial palsy, compromised vision or hearing, psychomotor delay, osteomyelitis of the mandible,[13] carpal tunnel syndrome, slipped capital femoral epiphysis, and osteoarthritis are potential complications. Two principal types of adult osteopetrosis have been proposed,[14] but so-called autosomal dominant osteopetrosis, type 1 (ADO 1) proved to be the high bone mass phenotype caused by *LRP5* gene activation (see later), whereas ADO 2 is a genuine osteopetrosis better called Albers-Schönberg disease.[4]

Neuronal storage disease with osteopetrosis features especially severe skeletal manifestations with epilepsy and neurodegeneration.[8] Lethal osteopetrosis manifests in utero and causes stillbirth.[4] Transient infantile osteopetrosis inexplicably resolves during the first months of life.[4]

Radiological Features

Generalized, symmetrical increase in bone mass is the major radiographic finding.[15] Trabecular and cortical bone appear thickened. In the severe forms, all three components of skeletal development are disrupted: growth, modeling, and remodeling. Increased density is typically uniform, but alternating sclerotic and lucent bands may appear in the iliac wings and metaphyses. Metaphyses become widened and can develop a club shape or "Erlenmeyer flask" deformity (Fig. 1). Rarely, distal phalanges in the hands are eroded (common in pycnodysostosis). Pathological fracture of long bones is not rare. Rachitic-like changes in growth plates may occur[16] because of hypocalcemia with secondary hyperparathyroidism. The skull is usually thickened and dense, especially at the base, and the paranasal and mastoid sinuses are underpneumatized. Vertebrae can show, on lateral view, a "bone-in-bone" (endobone) configuration. Albers-Schönberg disease selectively thickens the skull base with a "rugger jersey" appearance of the spine.[14] Skeletal scintigraphy helps show fractures and osteomyelitis. MRI can assess bone marrow transplantation, because successful engraftment will enlarge medullary spaces.[17] Cranial CT and MRI findings have been detailed for pediatric patients.[18]

Laboratory Findings

In infantile osteopetrosis, failure of bone resorption can lead to hypocalcemia because circulating calcium levels are increasingly dependent on dietary intake.[19] Secondary hyperparathyroidism with elevated serum levels of calcitriol is common. In Albers-Schönberg disease, this disturbance is mild.[14] Increased serum acid phosphatase and the brain isoenzyme of creatine kinase (BB-CK) are biomarkers for osteopetrosis.[20] Both enzymes seem to originate from the excessive or defective osteoclasts.[20,21]

The author states that he has no conflicts of interest.

© 2008 American Society for Bone and Mineral Research

TABLE 1. DISORDERS THAT CAUSE HIGH BONE MASS

Dysplasias and dysostoses
Autosomal dominant osteosclerosis
Central osteosclerosis with ectodermal dysplasia
Craniodiaphyseal dysplasia
Craniometaphyseal dysplasia
Dysosteosclerosis
Endosteal hyperostosis (van Buchem disease and sclerosteosis)
Frontometaphyseal dysplasia
Infantile cortical hyperostosis (Caffey disease)
Juvenile Paget's disease (osteoectasia with hyperphosphatasia or hyperostosis corticalis)
Melorheostosis
Metaphyseal dysplasia (Pyle disease)
Mixed sclerosing bone dystrophy
Oculodento-osseous dysplasia
Osteodysplasia of Melnick and Needles
Osteopathia striata
Osteopetrosis
Osteopoikilosis
Progressive diaphyseal dysplasia (Engelmann disease)
Pycnodysostosis
Tubular stenosis (Kenny-Caffey syndrome)
Metabolic
Carbonic anhydrase II deficiency
Fluorosis
Heavy metal poisoning
Hepatitis C–associated osteosclerosis
Hypervitaminosis A,D
Hyper-, hypo-, and pseudohypoparathyroidism
Hypophosphatemic osteomalacia
LRP5 activation (high bone mass phenotype)
Milk-alkali syndrome
Renal osteodystrophy
X-linked hypophosphatemia
Other
Axial osteomalacia
Diffuse idiopathic skeletal hyperostosis (DISH)
Erdheim-Chester disease
Fibrogenesis imperfecta ossium
Hypertrophic osteoarthropathy
Ionizing radiation
Leukemia
Lymphomas
Mastocytosis
Multiple myeloma
Myelofibrosis
Osteomyelitis
Osteonecrosis
Paget's disease
Sarcoidosis
Sickle cell disease
Skeletal metastases
Tuberous sclerosis

Histopathological Findings

The radiographic features of the osteopetroses[15] can be diagnostic; however, osteoclast failure during endochondral bone formation provides a pathognomonic histological finding because remnants of primary spongiosa persist as "islands" or "bars" of calcified cartilage within trabecular bone (Fig. 2).

Osteoclasts may be increased, normal, or rarely decreased in number.[22] In infantile osteopetrosis, they are usually abundant on bone surfaces. Nuclei are especially numerous, and ruffled borders or clear zones are absent.[23] Fibrous tissue often crowds marrow spaces.[23] Adult osteopetrosis may show increased osteoid and few osteoclasts also lacking ruffled borders, or osteoclasts can be especially numerous and large.[24] Immature "woven" bone is common. Rounded, hypermultinucleated osteoclasts are off of bone surfaces in bisphosphonate-induced osteopetrosis.[10]

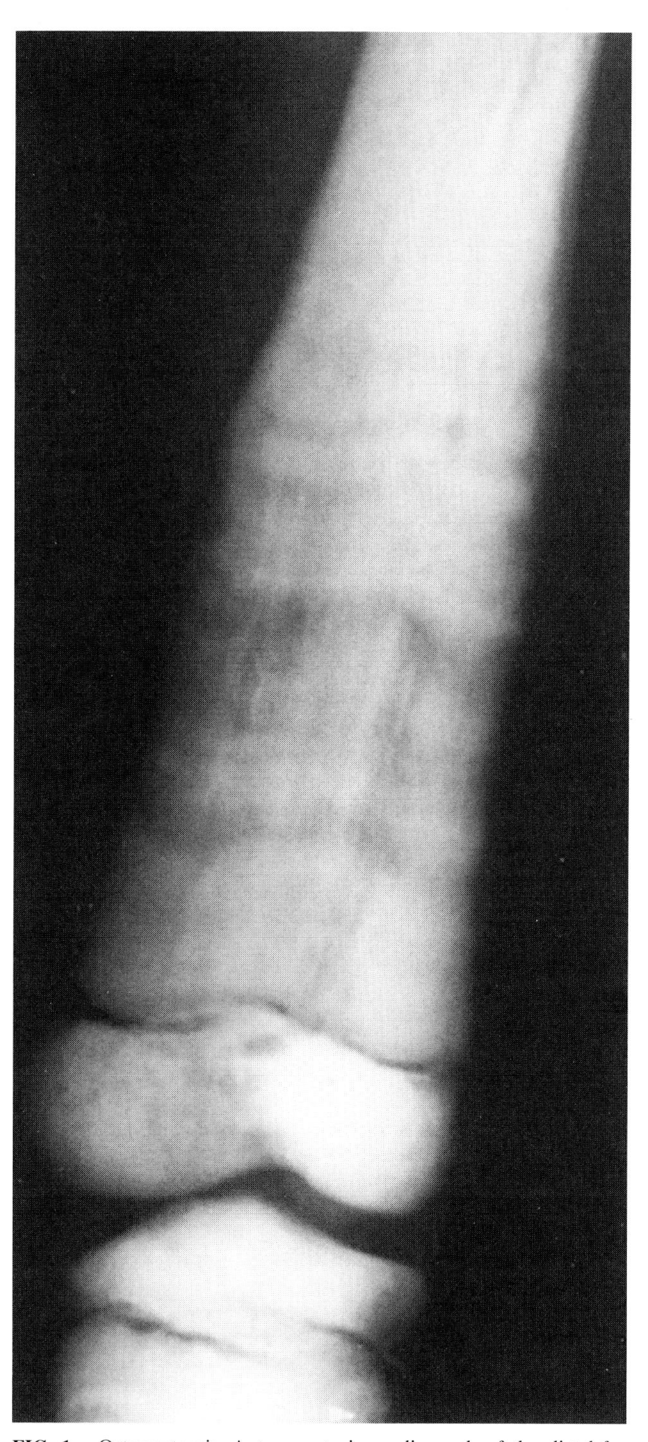

FIG. 1. Osteopetrosis. Anteroposterior radiograph of the distal femur of a 10-yr-old boy shows a widened metadiaphysis with characteristic alternating dense and lucent bands. (Reprinted with permission from Whyte MP, Murphy WA 1990 Osteopetrosis and other sclerosing bone disorders. In: Avioli LV, Krane SM (eds.) Metabolic Bone Disease, 2nd ed. Saunders, Philadelphia, PA, USA, p. 618.)

© 2008 American Society for Bone and Mineral Research

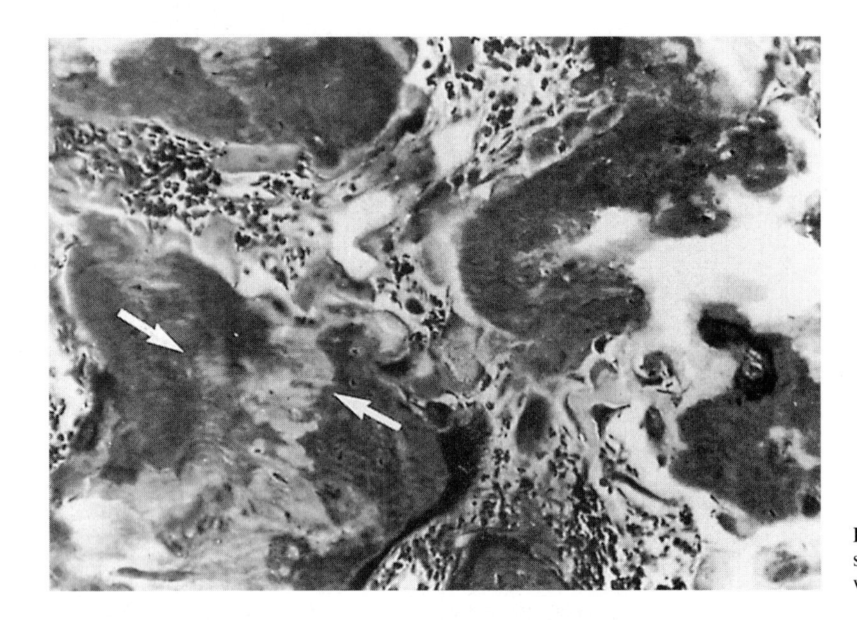

FIG. 2. Osteopetrosis. A characteristic area of lightly stained calcified primary spongiosa (arrows) is found within darkly stained mineralized bone.

Etiology and Pathogenesis

The potential causes of osteopetrosis are many and complex.[4] Defects could involve primarily the stem cell for osteoclastogenesis or its microenvironment, mononuclear precursor cell, or mature heterokaryon. Furthermore, an osteoblast defect has been reported.[25] In theory, the bone matrix could resist resorption.[4] In osteopetrosis with neuronal storage disease (featuring accumulation of ceroid lipofuscin), lysosomes could be defective.[8] Virus-like inclusions of uncertain significance have been found in the osteoclasts of a few cases of mild osteopetrosis.[26] Synthesis of an abnormal PTH or defective production of interleukin (IL)-2 or superoxide have been considered.[4] In fact, leukocyte function in infantile osteopetrosis may be abnormal.[27] Ultimately, impaired skeletal resorption causes bone fragility because collagen fibers do not interconnect osteons, and woven bone remodels poorly to compact bone.

The molecular basis for osteopetrosis is now known for the majority of patients.[11] Haploinsufficiency for chloride channel 7 activity caused by deactivating mutations in *CLCN7* causes Albers-Schönberg disease.[28] Autosomal recessive infantile osteopetrosis most often involves mutations in *TCIRG1* (*ATP6I*) encoding the α3 subunit of the vacuolar proton pump.[29] *CLCN7* defects can also cause autosomal recessive malignant or intermediate osteopetrosis.[30] Deactivation of *CA II* lends to carbonic anhydrase II deficiency.[4] Accordingly, mutation of three genes that regulate acidification by osteoclasts explains osteopetrosis in most patients. Loss-of-function of the GL (grey-lethal) gene encoding "osteopetrosis associated transmembrane protein 1" (*OSMT1*) causes especially severe osteopetrosis.[31] OL-EDA-ID is caused by inactivation of a key modulator of NF-κB.[9] Especially rare infants with osteopetrosis featuring few osteoclasts (who fail bone marrow transplantation) suffer from deactivation of the gene encoding RANKL.[22] Also, RANK may be deactivated.

Treatment

Because the causes and outcomes differ among the osteopetroses, a precise diagnosis is crucial before therapy is attempted. Diagnosis has depended on careful evaluation of the disease complications and progression as well as investigation of the family, but can now be improved by mutation analysis offered by commercial laboratories.[11]

Bone Marrow Transplantation. Bone marrow transplantation (BMT) from HLA-identical donors has remarkably improved some patients with infantile osteopetrosis.[32] Nevertheless, BMT may not be beneficial for all patients,[4] because the causal defect is sometimes extrinsic to the osteoclast (e.g., RANKL deficiency).[22] Hypercalcemia can occur as osteoclast function begins.[33] Severe, acute, pulmonary hypertension is a frequent complication of stem cell transplantation.[34] Patients with severely crowded medullary spaces seem less likely to engraft. Histomorphometric study of bone helps to predict the outcome of BMT, and this procedure early on seems best.[32] BMT from HLA-nonidentical donors warrants further study. Administration of progenitor cells in blood from HLA-haploidentical parents has been effective.[35]

Hormonal and Dietary Therapy. Some success has been reported with a calcium-deficient diet. Conversely, calcium supplementation may be necessary for hypocalcemia accompanying severe osteopetrosis.[16] High doses of calcitriol to stimulate quiescent osteoclasts, while dietary calcium is limited to prevent absorptive hypercalciuria and hypercalcemia, may improve infantile osteopetrosis.[36] Nevertheless, some patients seem to become resistant to this treatment. The observation that leukocytes produce less superoxide led to recombinant human interferon γ-1b treatment for malignant osteopetrosis.

High-dose glucocorticoid treatment stabilizes pancytopenia and hepatomegaly. Prednisone and a low-calcium/high-phosphate diet has been discussed as an alternative to BMT.[37] One case report describes reversal of malignant osteopetrosis after prednisone treatment.[38]

Supportive. Hyperbaric oxygenation can be helpful for osteomyelitis of the jaw. Surgical decompression of the optic and facial nerves and auditory canal[39] may benefit some patients. Joint replacement is challenging but possible[40]; internal fixation may be necessary for femoral fractures.

Radiographic studies occasionally detect malignant osteopetrosis late in pregnancy. Early prenatal diagnosis by sonography has generally been unsuccessful. Mutation analysis is increasingly feasible, with most severe cases caused by *TCIRG1* and *CLCN7* mutations.

© 2008 American Society for Bone and Mineral Research

CARBONIC ANHYDRASE II DEFICIENCY

In 1983, the autosomal recessive syndrome of osteopetrosis with renal tubular acidosis (RTA) and cerebral calcification was identified as carbonic anhydrase II (CA II) deficiency (OMIM: 611492).[41]

Clinical Presentation

There is considerable clinical variability.[42] In infancy or early childhood, patients can suffer fractures, failure to thrive, developmental delay, short stature, optic nerve compression with blindness, and dental malocclusion. Mental subnormality is common, but not invariable. RTA may explain the hypotonia, apathy, and muscle weakness that troubles some patients. Periodic hypokalemic paralysis can occur. Recurrent long bone fractures, although unusual, can cause significant morbidity.[42] Life expectancy does not seem threatened, but the oldest published cases have been young adults.[43] Autopsy studies have not been reported.[43]

Radiological Features

CA II deficiency resembles other osteopetroses radiographically, except cerebral calcification appears at age 2–5 yr, and the osteosclerosis and modeling defects diminish over years. Skeletal radiographs are typically abnormal at diagnosis, although findings can be subtle at birth. The cerebral calcification resembles this finding in idiopathic hypoparathyroidism or pseudohypoparathyroidism, increases during childhood, and affects cortical and basal ganglia gray matter.

Laboratory Findings

Metabolic acidosis manifests as early as the neonatal period. Both proximal and distal RTA have been described[43]; distal (type I) RTA seems better documented. Any anemia is generally mild.

Etiology and Pathogenesis

CA accelerates the first step in the reaction $CO_2 + H_2O \leftrightarrow H_2CO_3 \leftrightarrow H^+ + HCO_3^-$. CA II is present in many tissues, including brain, kidney, erythrocytes, cartilage, lung, and gastric mucosa.[44] Deactivating mutations in the gene encoding CA II cause this disorder and reveal significance for CA II in bone, kidney, and perhaps brain.[45] In heterozygous carriers, CA II levels in erythrocytes are ~50% of normal.[41,43] There is a CA II knockout mouse model.[45]

Treatment

Transfusion of CA II-replete erythrocytes in one patient did not improve the systemic acidosis.[46] The RTA has been treated with HCO_3^-, but the long-term impact is unknown. BMT has corrected the osteopetrosis and slowed the cerebral calcification, but not altered the RTA.[47]

PYCNODYSOSTOSIS

Pycnodysostosis is the autosomal recessive disorder that perhaps affected Henri de Toulouse-Lautrec (1864–1901).[48] More than 100 patients have been described since 1962.[49] Parental consanguinity is recorded in ~30% of cases. Most reports are from Europe or the United States, but some come from Israel, Indonesia, India, and Africa. Pycnodysostosis seems to be especially prevalent in the Japanese.[50]

Clinical Presentation

Pycnodysostosis is typically diagnosed during infancy or early childhood because of disproportionate short stature and a relatively large cranium, fronto-occipital prominence, small facies and chin, obtuse mandibular angle, high-arched palate, dental malocclusion with retained deciduous teeth, proptosis, and a beaked and pointed nose.[51] The anterior fontanel and other cranial sutures are usually open. Fingers are short and clubbed from acro-osteolysis or aplasia of terminal phalanges, and the hands are small and square. The thorax is narrow and there may be *pectus excavatum*, kyphoscoliosis, and increased lumbar lordosis. Sclerae can be blue. Recurrent fractures typically involve the lower limbs and cause *genu valgum*. Rickets has been described. Adult height ranges between 4 ft 3 in and 4 ft 11 in. Mental retardation affects ~10% of cases.[51] Recurrent respiratory infections and right heart failure may occur from chronic upper airway obstruction caused by micrognathia.

Radiographic Features

Pycnodysostosis resembles osteopetrosis because uniform osteosclerosis becomes apparent in childhood and increases with age and there are recurrent fractures. The calvarium and base of the skull are sclerotic, and the orbital ridges are radiodense. However, the marked modeling defects of osteopetrosis do not occur, although long bones have narrow medullary canals. Additional findings include delayed closure of cranial sutures and fontanels (prominently the anterior), obtuse mandibular angle, wormian bones, gracile clavicles with hypoplastic ends, partial absence of the hyoid bone, and hypoplasia of the distal phalanges and ribs.[52] Endobones and radiodense striations are absent.[15]

Laboratory Findings

Serum levels of calcium, inorganic phosphate, and alkaline phosphatase are usually unremarkable. There is no anemia. Electron microscopy has suggested that degradation of bone collagen might be defective.[53] In chondrocytes, inclusions have been described. Virus-like inclusions have been reported in the osteoclasts of brothers.[54] Diminished growth hormone secretion and low serum insulin-like growth factor 1 levels have been reported in five of six affected children.[55]

Etiology and Pathogenesis

In 1996, loss-of-function mutation within the gene encoding cathepsin K was discovered to cause pycnodysostosis.[50] Cathepsin K, a lysosomal cysteine protease, is highly expressed in osteoclasts.[56] Impaired collagen degradation seems to be a fundamental defect[57] and compromises bone quality.[58]

Treatment

There is no established medical therapy. BMT has not been reported. Fractures of long bones are typically transverse and heal at a satisfactory rate, although there can be delayed union and massive callus formation. Internal fixation of long bones or extraction of teeth is difficult because of skeletal hardness. Jaw fracture has occurred. Osteomyelitis of the mandible may require antibiotics and surgery.

PROGRESSIVE DIAPHYSEAL DYSPLASIA (CAMURATI-ENGELMANN DISEASE)

Progressive diaphyseal dysplasia (PDD; OMIM: 131300) was characterized by Cockayne in 1920.[59] Camurati recog-

© 2008 American Society for Bone and Mineral Research

nized the autosomal dominant inheritance. Engelmann described the severe form in 1929. In 2001, mutations were identified within a specific region of the gene that encodes TGFβ1.[60]

All races are affected. Clinical severity is quite variable.[61] Hyperostosis occurs gradually on both the periosteal and endosteal surfaces of long bones. In severe cases, the axial skeleton and skull are also involved. Some carriers have no radiographic changes, but bone scintigraphy is abnormal.

Clinical Presentation

PDD typically presents during childhood with limping or a broad-based and waddling gait, leg pain, muscle wasting, and decreased subcutaneous fat in the extremities mimicking a muscular dystrophy.[62] Severely affected individuals also have a characteristic body habitus that includes a large head with prominent forehead, proptosis, and thin limbs exhibiting thickened, painful bones and little muscle mass. Cranial nerve palsies may develop when the skull is involved. Puberty is sometimes delayed. Raised intracranial pressure can occur. Physical findings include palpable widened bones and skeletal tenderness. Some patients have hepatosplenomegaly, Raynaud's phenomenon, and other findings suggestive of vasculitis.[63] Although radiological studies typically show progressive skeletal disease, the clinical course is variable, and spontaneous improvement sometimes occurs during adult life.[64]

Radiological Features

Hyperostosis of major long bone diaphyses involves both the periosteal and endosteal surfaces.[15] The thickening is fairly symmetrical and gradually spreads to include metaphyses, yet spares epiphyses (Fig. 3). Diaphyses gradually widen and develop irregular surfaces. The tibias and femurs are most commonly affected; less frequently, the radii, ulna, humeri, scapulae, clavicles, and pelvis and, occasionally, short tubular bones are involved. Age-of-onset, rate of progression, and degree of bony change are highly variable. With relatively mild PDD, especially in adolescents or young adults, radiographic and scintigraphic abnormalities may be confined to the lower limbs. In severely affected children, regional osteopenia is possible.

Clinical, radiographic, and scintigraphic findings are generally concordant. Bone scanning typically shows focally increased radionuclide accumulation.[65] In some patients, however, advanced and metabolically quiescent disease features unremarkable bone scintigraphy.[65] Conversely, markedly increased radioisotope accumulation with minimal radiographic findings can represent early skeletal disease.[65] MRI and CT have delineated the cranial findings.[66]

Laboratory Findings

Serum alkaline phosphatase and urine hydroxyproline are elevated in some PDD patients. Modest hypocalcemia and significant hypocalciuria sometimes occur with severe disease, probably because of markedly positive calcium balance.[64] Other biochemical parameters of bone and mineral metabolism are typically normal. Mild anemia and leukopenia and elevated erythrocyte sedimentation rate may have been reported.[63]

Histopathology shows new bone formation along diaphyses with nascent woven bone undergoing centripetal maturation and then incorporation into the cortex. Electron microscopy of muscle has shown myopathic and vascular changes.[62]

Etiology and Pathogenesis

PDD is caused by mutation within one region of the gene

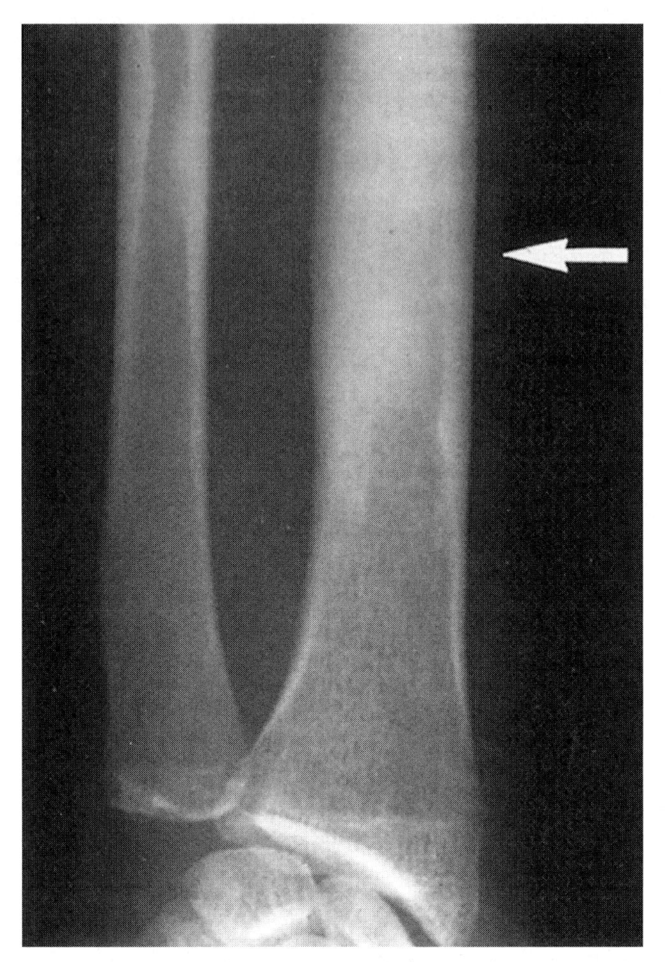

FIG. 3. Progressive diaphyseal dysplasia (Camurati-Engelmann disease). The distal radius of this 20-yr-old woman has characteristic patchy thickening (arrow) of the periosteal and endosteal surfaces of the diaphysis.

that encodes TGFβ1 (*TGFβ1*). A latency-associated peptide remains bound to TGFβ1, keeping it active in skeletal matrix.[67] Mild PDD can reflect variable penetrance. PDD has been described as more severe in ensuing generations ("anticipation").[68] Furthermore, there does seem to some locus heterogeneity.[69]

The clinical and laboratory features of severe PDD and its responsiveness to glucocorticoid treatment have suggested it is a systemic disorder (i.e., an inflammatory connective tissue disease).[63]

Treatment

PDD is somewhat unpredictable. Symptoms may remit during adolescence or adult life. Prednisone given in small doses on alternate days is effective for bone pain and can correct histological abnormalities of bone.[70] Resection of a "cortical window," has relieved localized bone pain. Bisphosphonate therapy may be helpful but has transiently increased symptoms.[71]

ENDOSTEAL HYPEROSTOSIS

In 1955, van Buchem and colleagues described hyperostosis corticalis generalisata.[72] Subsequently, this and further disorders were characterized as types of endosteal hyperostosis. The

© 2008 American Society for Bone and Mineral Research

autosomal dominant, relatively mild form is called Worth disease,[73] and an autosomal recessive, severe form is called sclerosteosis.[74]

Van Buchem Disease

Van Buchem disease (OMIM: 239100) is a severe, autosomal recessive disorder.[72]

Clinical Presentation. Progressive, asymmetrical enlargement of the jaw occurs during puberty causing marked thickening and a wide angle, but without prognathism. Dental malocclusion is uncommon. Carriers of the gene defect may be symptom free; however, recurrent facial nerve palsy, deafness, and optic atrophy from narrowing of cranial foramina are common and can begin as early as infancy. Long bones may be painful with applied pressure but are not fragile, and joint range-of-motion is unaffected. Sclerosteosis (see below) differs because of excessive height and syndactyly.[74]

Radiological Features. Endosteal thickening produces a dense diaphyseal cortex with narrow medullary canal.[15] The hyperostosis is selectively endosteal; long bones are properly modeled. However, osteosclerosis also affects the skull base, facial bones, vertebrae, pelvis, and ribs. The mandible enlarges.

Laboratory Findings. Serum alkaline phosphatase activity may be increased because of high levels of the skeletal isoform, whereas calcium and phosphate concentrations are unremarkable. Van Buchem and colleagues suggested that the excessive bone was essentially of normal quality.[72]

Etiology and Pathogenesis. Van Buchem disease and sclerosteosis were predicated to be allelic disorders, with their differences reflecting modifying genes.[74] Actually, loss-of-function mutations in *SOST*, the gene encoding sclerostin, cause sclerosteosis,[75] whereas van Buchem disease involves a 52-kb deletion that impairs downstream enhancement of *SOST*.[76] Sclerostin binds to LRP5/6, antagonizes canonical Wnt signaling,[77] and promotes osteoblast apoptosis.[78]

Treatment. There is no specific medical therapy. Decompression of narrowed foramina may help cranial nerve palsies.[79] Surgery has been used to recontour the mandible.[80]

Sclerosteosis

Sclerosteosis (cortical hyperostosis with syndactyly; OMIM: 269500), like van Buchem disease, is an autosomal recessive endosteal hyperostosis that affects primarily Afrikaners or others of Dutch ancestry.[74] Initially, sclerosteosis was distinguished from van Buchem disease by excessive height and syndactyly. In fact, the genetic defects differ.[75,76]

Clinical Presentation. At birth, only syndactyly may be noted. During early childhood, affected individuals become tall and heavy with skeletal overgrowth involving especially the skull causing facial disfigurement. Deafness and facial palsy are prominent problems. The mandible has a square configuration. Raised intracranial pressure and headache may result from a small cranial cavity. The brainstem can be compressed. Syndactyly from either cutaneous or bony fusion of the middle and index fingers is typical, but of variable severity. Patients are resistant to fracture. Life expectancy may be shortened.[81]

Radiological Features. Except for syndactyly, the skeleton appears normal in early childhood. Then, progressive bone acquisition widens the skull and mandible.[82] Long bones develop thickened cortices. Vertebral pedicles, ribs, tubular bones, and pelvis may also appear dense. Auditory ossicles may fuse and the internal canals and cochlear aqueducts become narrow.[83]

Histopathological Findings. Dynamic histomorphometry of one affected skull showed an increased rate of bone formation with thickened trabeculae and osteoidosis, whereas resorption appeared quiescent.[84]

Etiology and Pathogenesis. Enhanced osteoblast activity with failure of osteoclasts to compensate causes the dense bone of sclerosteosis.[84] No abnormality of calcium homeostasis or of pituitary gland function has been documented.[85] The pathogenesis of the neurological defects has been described.[84]

Loss-of-function mutations in *SOST* cause sclerosteosis.[75]

Treatment. There is no established medical treatment. Surgery for syndactyly is difficult if there is bony fusion. Management of the neurological complications has been reviewed.[84]

OSTEOPOIKILOSIS

Osteopoikilosis ("spotted bones") is an autosomal dominant, radiographic curiosity. With accompanying connective tissue nevi, dermatofibrosis lenticularis disseminata, the disorder is the Buschke-Ollendorff syndrome.[86] In 2004, deactivating mutations in the *LEMD3* gene were identified.[87]

Clinical Presentation

Osteopoikilosis (OMIM: 166700) is usually an incidental finding. The bony lesions are asymptomatic, but if not understood can precipitate investigation for metastatic disease to the skeleton.[88] Family members at risk should be screened with a radiograph of a wrist and knee after childhood. Joint contractions and limb length inequality can occur, especially in individuals with accompanying changes of melorheostosis. The nevi usually involve the lower trunk or extremities and are small asymptomatic papules. Sometimes they are yellow or white discs or plaques, deep nodules, or streaks.[86]

Radiological Features

There are numerous, small, usually round or oval, foci of osteosclerosis.[15] Commonly affected sites are the ends of the short tubular bones, metaepiphyses of long bones, and tarsal, carpal, and pelvic bones (Fig. 4). Lesions are unchanged for decades. Bone scanning is not abnormal.[88]

Histopathological Studies

Dermatofibrosis lenticularis disseminata consists of unusually broad, markedly branched, interlacing elastin fibers in the dermis; the epidermis is normal.[86] Foci of osteosclerosis are thickened trabeculae that merge with surrounding normal bone or are islands of cortical bone that include Haversian systems. Mature lesions appear to be remodeling slowly.

OSTEOPATHIA STRIATA

Osteopathia striata (OMIM: 166500) features linear striations at the ends of long bones and in the ileum.[15] Like os-

© 2008 American Society for Bone and Mineral Research

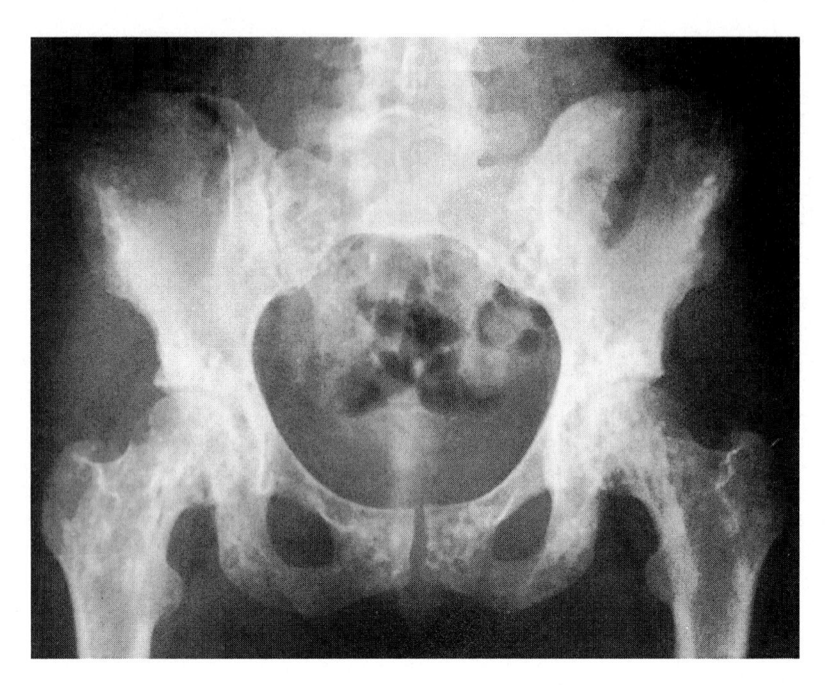

FIG. 4. Osteopoikilosis. The characteristic feature is the spotted appearance shown here in the pelvis and metaepiphyseal regions of the femora. [Reproduced with permission from Whyte MP 1995 Rare disorders of skeletal formation and homeostasis. In: Becker KN (ed.) Principles and Practice of Endocrinology and Metabolism, 2nd ed. Lippincott-Raven Publishers, Philadelphia, PA, USA, p. 598.)

teopoikilosis, it is usually a radiographic curiosity, but it can also occur in a variety of important disorders including osteopathia striata with cranial sclerosis[89] and osteopathia striata with focal dermal hypoplasia (OMIM: 300373).[90]

Clinical Presentation

Osteopathia striata alone is an autosomal dominant trait. Symptoms that may have led to the diagnosis are probably unrelated. With sclerosis of the skull, however, cranial nerve palsies are common,[89] and this is perhaps an X-linked dominant disorder.[91] Osteopathia striata with focal dermal hypoplasia (Goltz syndrome) is a serious condition of males, transmitted as an X-linked recessive trait, featuring widespread linear areas of dermal hypoplasia through which adipose tissue can herniate and a variety of bony defects in the limbs.[90]

Radiological Features

Gracile linear striations are found in cancellous bone, particularly within metaepiphyses of major long bones and the periphery of the iliac bones.[15] Carpal, tarsal, and tubular bones of the hands and feet are less often and more subtly affected. The striations appear unchanged for years. Radionuclide accumulation is not increased during bone scanning.[88]

Treatment

Histopathological studies of bone have not been described. Although unlikely to be misdiagnosed, radiographic screening after childhood of family members at risk seems prudent. Osteopathia striata with cranial sclerosis has been detected prenatally by sonography.

MELORHEOSTOSIS

Melorheostosis (OMIM: 155950), from the Greek, refers to "flowing hyperostosis." The radiographic appearance resembles wax that has dripped down a candle. About 200 cases have been published[92] since the first description in 1922.[93] Melorheostosis occurs sporadically, including when it accompanies osteopoikilosis.

Clinical Presentation

Melorheostosis typically presents during childhood. Monomelic involvement is usual; bilateral disease is characteristically asymmetrical. Cutaneous changes may overlie the skeletal lesions and include linear scleroderma-like patches and hypertrichosis. Fibromas, fibrolipomas, capillary hemangiomas, lymphangiectasia, and arterial aneurysms can also occur.[94] Soft tissue abnormalities are often noted before the hyperostosis. Pain and stiffness are the major symptoms. Affected joints may contract. Leg length inequality can follow premature fusion of epiphyses. Bone lesions seem to advance most rapidly during childhood. In adult life, melorheostosis may or may not progress.[95] Nevertheless, pain is more frequent when there is continuing subperiosteal bone formation.

Radiological Features

Dense, irregular, and eccentric hyperostosis of both periosteal and endosteal surfaces of a single bone, or several adjacent bones, is the hallmark of melorheostosis (Fig. 5).[15,92] Any bone may be affected, but the lower extremities are most commonly involved. Bone can also develop in soft tissues near skeletal lesions, particularly near joints. Melorheostotic bone is hyperemic and "hot" during bone scanning.

Laboratory Findings

Serum calcium, inorganic phosphate, and alkaline phosphatase levels are normal.

Histopathological Findings

Melorheostosis features endosteal thickening during growth and periosteal new bone formation during adult life.[92] Affected bones are sclerotic with thickened, irregular lamellae. Marrow fibrosis may be present.[92] In the skin, unlike in true scleroderma, the collagen of the scleroderma-like lesions appears normal and has therefore been called linear melorheostotic scleroderma.[96]

Etiology and Pathogenesis

The distribution of the bone and soft tissue lesions in sclero-

© 2008 American Society for Bone and Mineral Research

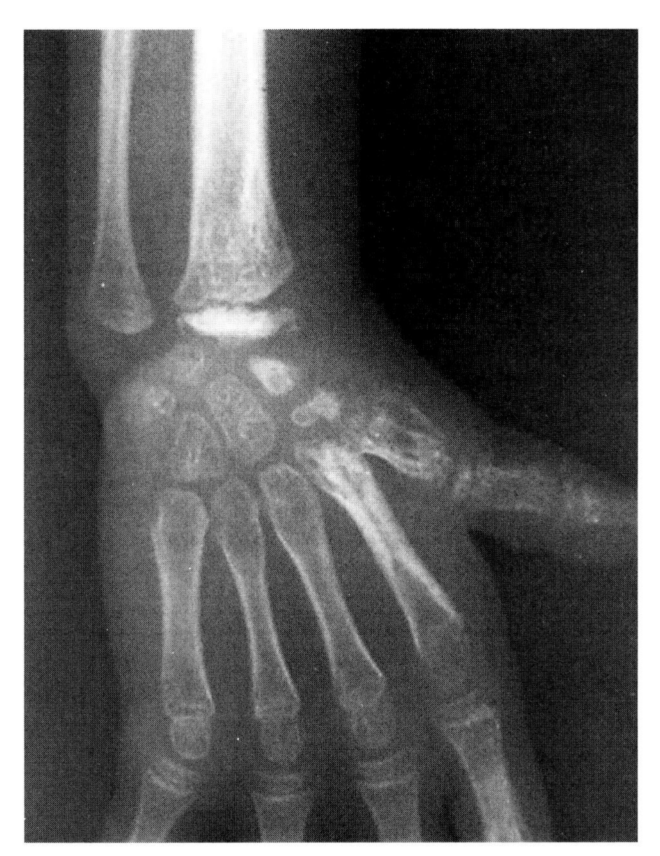

FIG. 5. Melorheostosis. Characteristic patchy osteosclerosis is most apparent in the radius and second metacarpal of this 8-yr-old girl.

tomes, myotomes, and dermatomes suggests a segmentary, embryonic defect.[96] Linear scleroderma may represent the primary abnormality that extends into the skeleton. In affected skin, there may be altered expression of several adhesion proteins.[97] The germline, loss-of-function mutations in *LEMD3* that cause osteopoikilosis and the Buschke-Ollendorff syndrome do not explain sporadic melorheostosis.[98]

Treatment

Surgical correction of contractures can be difficult; recurrent deformity is common. Distraction techniques seem promising.[99]

AXIAL OSTEOMALACIA

Axial osteomalacia (OMIM: 109130) features coarsening of trabecular bone in the axial, but not appendicular, skeleton.[100] Fewer than 20 patients have been described. Most have been sporadic cases, but an affected mother and son have been reported.[101]

Clinical Presentation

Most patients have been middle-age or elderly men. Radiographic manifestations are probably detectable earlier.[102] Dull, vague, chronic, axial bone pain (often in the cervical spine) usually prompts the radiographic discovery.

Radiological Features

Abnormalities are confined essentially to the spine and pelvis where the coarsened trabecular pattern resembles osteomalacia.[103] The appendicular skeleton is unremarkable. However, Looser zones (characteristic of osteomalacia) are not reported. The cervical spine and ribs seem most severely affected. Several patients have had features of ankylosing spondylitis.[102]

Laboratory Studies

Serum alkaline phosphatase (bone isoenzyme) may be increased. In a few patients, inorganic phosphate levels tended to be low.[103] For others, osteomalacia occurred despite normal levels of calcium, phosphate, 25-hydroxyvitamin D, and 1,25-dihydroxyvitamin D.

Histopathological Findings

Iliac crest specimens have distinct corticomedullary junctions, but the cortices can be especially wide and porous. Total bone volume may be increased. There is excess osteoid, but collagen has a normal lamellar pattern. Tetracycline labeling shows defective skeletal mineralization.[102] Osteoblasts are flat "lining" cells, but stain intensely for alkaline phosphatase. Changes of secondary hyperparathyroidism are absent.[102]

Etiology and Pathogenesis

Axial osteomalacia may be caused by an osteoblast defect.[104]

Treatment

Effective medical therapy has not been reported, but the natural history seems relatively benign. Methyltestosterone and stilbestrol or vitamin D_2 (as much as 20,000 U/d for 3 yr) have not been helpful.[104] Slight improvement in skeletal histology, but not in symptoms, was described after calcium and vitamin D_2. Long-term follow-up of one patient showed that symptoms and radiographic findings did not change.[104]

FIBROGENESIS IMPERFECTA OSSIUM

Fibrogenesis imperfecta ossium was identified in 1950. Approximately 10 cases have been reported.[105] Radiographic studies suggest generalized osteopenia, but coarse and dense appearing trabecular bone explains its designation as an osteosclerotic disorder. The clinical, biochemical, radiological, and histopathological features have been carefully contrasted with axial osteomalacia.[100]

Clinical Presentation

Presentation occurs during middle age or later. Both sexes are affected. Characteristically, gradual onset of intractable skeletal pain is followed by rapid deterioration. The course is debilitating with progressive immobility. Spontaneous fractures are prominent. Physical examination shows marked bony tenderness.

Radiological Features

Skeletal changes affect all but the skull. Initially, there may be osteopenia only and a slightly abnormal appearance of trabecular bone.[105] Subsequently, the findings are more consistent with osteomalacia with further alterations of the trabecular bone pattern, heterogeneous bone density, and cortical thinning. Corticomedullary junctions become indistinct. Areas of the skeleton may have a mixed lytic and sclerotic appearance.[105] Remaining trabeculae appear coarse and dense in a "fish-net" pattern. Pseudofractures may develop. Some pa-

© 2008 American Society for Bone and Mineral Research

tients have a "rugger jersey" spine. Diaphyses may show periosteal reaction. The radiographic features can distinguish fibrogenesis imperfecta ossium from axial osteomalacia (generalized versus axial, respectively). The histopathological findings are also different.[100]

Laboratory Findings

Serum calcium and inorganic phosphate levels are normal, but alkaline phosphatase is increased. Urine hydroxyproline may be normal or elevated.[105] Typically, there is no renal tubular dysfunction. Acute agranulocytosis and macroglobulinemia have been reported.

Histopathological Findings

The osseous lesion is an osteomalacia.[105] Collagen lacks birefringence where there is defective mineralization. Electron microscopy reveals thin and randomly organized collagen fibrils in a "tangled" pattern. Cortical bone in the femurs and tibias may show the least abnormality. Osteoblasts and osteoclasts can be abundant. In some regions, matrix structures of 300–500 nm diameter have been observed.[106] Unless bone is viewed with polarized-light or electron microscopy, fibrogenesis imperfecta ossium can be mistaken for osteoporosis or other forms of osteomalacia.[106]

Etiology and Pathogenesis

This is an acquired disorder of collagen synthesis in lamellar bone. The etiology is unknown. Genetic factors have not been implicated.

Treatment

There is no recognized therapy. Temporary improvement can occur. Treatment with vitamin D (or an active metabolite), calcium, salmon calcitonin, and sodium fluoride has not helped.[105] In fact, ectopic calcification has complicated high-dose vitamin D_2 therapy. Treatment with melphalan and prednisolone seemed to benefit one patient.

PACHYDERMOPERIOSTOSIS

Pachydermoperiostosis (hypertrophic osteoarthropathy: primary or idiopathic; OMIM: 167100) causes clubbing of the digits, hyperhidrosis and thickening of the skin especially on the face and forehead (*cutis verticis gyrata*), and periosteal new bone formation particularly in the distal extremities. Autosomal dominant inheritance with variable expression is established,[107] but there also seems to be autosomal recessive transmission.[108]

Clinical Presentation

Men seem to be more severely affected than women, and blacks more commonly than whites. Age at presentation is variable, but usually it is during adolescence.[107,108] All principal features (clubbing, periostitis, and pachydermia) trouble some patients; others have just one or two. Clinical manifestations emerge over a decade, and can then abate.[108] Progressive enlargement of the hands and feet may cause a "paw-like" appearance, and there may be excessive perspiration. Acro-osteolysis can occur. Fatigue and arthralgias of the elbows, wrists, knees, and ankles are common. Stiffness and limited mobility of both the appendicular and the axial skeleton may develop. Compression of cranial or spinal nerves has been described. Cutaneous changes include coarsening, thickening, furrowing, pitting, and oiliness of especially the scalp and face. Myelophthisic anemia with extramedullary hematopoiesis may occur. Life expectancy is not compromised.

Radiological Features

Severe periostitis thickens tubular bones distally; typically the radius, ulna, tibia, and fibula, and sometimes the metacarpals, tarsals/metatarsals, clavicles, pelvis, skull base, and phalanges. Clubbing is obvious, and acro-osteolysis can occur. The spine is rarely involved. Ankylosis of joints, especially in the hands and feet, may trouble older patients.[15]

The major challenge in differential diagnosis is secondary hypertrophic osteoarthropathy (pulmonary or otherwise). Here, however, the radiographic features are somewhat different, featuring periosteal reaction that is typically smooth and undulating.[109] In pachydermoperiostosis, periosteal proliferation is exuberant, irregular, and often involves epiphyses. Bone scanning in either condition reveals symmetrical, diffuse, regular uptake along the cortical margins of long bones, especially in the legs, causing a "double stripe" sign.

Laboratory Findings

Nascent periosteal bone roughens cortical bone surfaces and undergoes cancellous compaction so that centrally it can be difficult to distinguish histopathologically from original cortex. There may also be osteopenia of trabecular bone from quiescent formation.[15] Mild cellular hyperplasia and thickening of blood vessels is found near synovial membranes, but synovial fluid is unremarkable.[110] Electron microscopy shows layered basement membranes.

Etiology and Pathogenesis

Pachydermoperiostosis has not been mapped within the genome. A controversial hypothesis suggests that some circulating factor acts on vasculature to cause hyperemia, thereby altering soft tissues; later, blood flow is reduced.[111]

Treatment

There is no established treatment. Painful synovial effusions may respond to nonsteroidal anti-inflammatory drugs. Colchicine reportedly helped one patient. Contractures or neurovascular compression by osteosclerotic lesions may require surgical intervention.

HEPATITIS C–ASSOCIATED OSTEOSCLEROSIS

In 1992, a new disorder featured severe, generalized osteosclerosis and hyperostosis in former intravenous drug abusers infected with hepatitis C virus.[112] Approximately 20 cases have been reported. Hepatitis C virus infection proved common to all patients.

Periosteal, endosteal, and trabecular bone thickening occurs throughout the skeleton, except the cranium. During active disease, the forearms and legs are painful. DXA shows values 200–300% above control means. Remodeling of good quality excessive bone seems accelerated during active disease and may respond to antiresorptive therapy. Gradual, spontaneous remission with decreases in DXA values can occur. The IGF system features distinctive increases in circulating levels of IGF binding protein 2 and "big" IGF II.[113]

HIGH BONE MASS PHENOTYPE

Certain activating mutations of the *LRP5* gene (OMIM: 607636) encoding low-density lipoprotein receptor-related protein 5, inherited as an autosomal dominant trait, increase

© 2008 American Society for Bone and Mineral Research

skeletal mass with good quality bone.[114] Enhanced *Wnt* signaling stimulates osteoblasts.[114] Some patients have *torus palatinus*,[114] cranial nerve palsies, and oropharyngeal exostoses.[115]

OTHER SCLEROSING BONE DISORDERS

Table 1 lists the many conditions that cause focal or generalized increases in skeletal mass. Sarcoidosis characteristically causes cysts within coarsely reticulated bone. However, sclerotic areas occasionally appear in the axial skeleton or long bones. Although multiple myeloma typically features generalized osteopenia and discrete osteolytic lesions, in indolent forms widespread osteosclerosis can occur. Lymphoma, myelofibrosis, and mastocytosis may also increase bone mass. Metastatic carcinoma, especially from the prostate, can cause dense bones. Diffuse osteosclerosis is also frequent in secondary, but not primary, hyperparathyroidism (e.g., renal disease).

REFERENCES

1. Online Mendelian Inheritance in Man 2000 OMIM. Available online at http://www.ncbi.nlm.nih.gov/omim. Accessed September 2, 2008.
2. Castriota-Scanderbeg A, Dallapiccola B 2005 Abnormal Skeletal Phenotypes: From Simple Signs To Complex Diagnoses. Springer, New York, NY, USA.
3. Albers-Schönberg H 1904 Rontgenbilder einer seltenen, Knoch!-!enerkrankung. Meunch Med Wochenschr **51**:365.
4. Whyte MP 2002 Osteopetrosis. In: Royce PM, Steinmann B (eds.) Connective Tissue and Its Heritable Disorders, 2nd ed. Wiley-Liss, New York, NY, USA, pp. 789–807.
5. Johnston CC Jr, Lavy N, Lord T, Vellios F, Merritt AD, Deiss WP Jr 1968 Osteopetrosis: A clinical, genetic, metabolic, and morphologic study of the dominantly inherited, benign form. Medicine (Baltimore) **47**:149–167.
6. Loria-Cortes R, Quesada-Calvo E, Cordero-Chaverri E 1977 Osteopetrosis in children: A report of 26 cases. J Pediatr **91**:43–47.
7. Kahler SG, Burns JA, Aylsworth AS 1984 A mild autosomal recessive form of osteopetrosis. Am J Med Genet **17**:451–464.
8. Jagadha V, Halliday WC, Becker LE, Hinton D 1988 The association of infantile osteopetrosis and neuronal storage disease in two brothers. Acta Neuropathol (Berl) **75**:233–240.
9. Dupuis-Girod S, Corradini N, Hadj-Rabia S, Fournet JC, Faivre L, Le Deist F, Durand P, Doffinger R, Smahi A, Israel A, Courtois G, Brousse N, Blanche S, Munnich A, Fischer A, Casanova JL, Bodemer C 2002 Osteopetrosis, lymphedema, anhidrotic ectodermal dysplasia, and immunodeficiency in a boy and incontinentia pigmenti in his mother. Pediatrics **109**:1–6.
10. Whyte MP, Wenkert D, Clements KL, McAlister WH, Mumm S 2003 Bisphosphonate-induced osteopetrosis. N Engl J Med **394**:455–461.
11. Balemans W, Van Wesenbeeck L, Van Hul W 2005 A clinical and molecular overview of the human osteopetroses. Calcif Tissue Int **77**:263–274.
12. Vanier V, Miller R, Carson BS 2000 Bilateral visual improvement after unilateral optic canal decompression and cranial vault expansion in a patient with osteopetrosis, narrowed optic canals, and increased intracranial pressure. J Neurol Neurosurg Psychiatry **69**:405–406.
13. Waguespack SG, Hui SL, DiMeglio LA, Econs MJ 2007 Autosomal dominant osteopetrosis: Clinical severity and natural history of 94 subjects with a chloride channel 7 gene mutation. J Clin Endocrinol Metab **92**:771–778.
14. Bollerslev J 1989 Autosomal dominant osteopetrosis: Bone metabolism and epidemiological, clinical and hormonal aspects. Endocr Rev **10**:45–67.
15. Resnick D, Niwayama G 2002 Diagnosis of Bone and Joint Disorders, 4th ed. Saunders, Philadelphia, PA, USA.
16. Di Rocco M, Buoncompagni A, Loy A, Dellacqua A 2000 Osteopetrorickets: Case report. Eur J Paediatr Neurol **159**:579–581.
17. Rao VM, Dalinka MK, Mitchell DG, Spritzer CE, Kaplan F, August CS, Axel L, Kressel HY 1986 Osteopetrosis: MR characteristics at 1.5 T. Radiology **161**:217–220.
18. Elster AD, Theros EG, Key LL, Chen MYM 1992 Cranial imaging in autosomal recessive osteopetrosis (parts I & II). Radiology **183**:129–144.
19. Key LL, Carnes D, Cole S, Holtrop M, Bar-Shavit Z, Shapiro F, Arceci R, Steinberg J, Gundberg C, Kahn A, Teitelbaum S, Anast C 1984 Treatment of congenital osteopetrosis with high dose calcitriol. N Engl J Med **310**:409–415.
20. Whyte MP, Chines A, Silva DP Jr, Landt Y, Ladenson JH 1996 Creatine kinase brain isoenzyme (BB-CK) presence in serum distinguishes osteopetrosis among the sclerosing bone disorders. J Bone Miner Res **11**:1438–1443.
21. Alatalo SL, Ivaska KK, Waguespack SG, Econs MJ, Vaananen HK, Halleen JM 2004 Osteoclast-derived serum tartrate-resistant acid phosphatase 5b in AlbersSchönberg disease (type II autosomal dominant osteopetrosis). Clin Chem **50**:883–890.
22. Sobacchi C, Frattini A, Guerrini MM, Abinun M, Pangrazio A, Susani L, Bredius R, Mancini G, Cant A, Bishop N, Grabowski P, Del Fattore A, Messina C, Errigo G, Coxon FP, Scott DI, Teti A, Rogers MJ, Vezzoni P, Villa A, Helfrich MH 2007 Osteoclast-poor human osteopetrosis due to mutations in the gene encoding RANKL. Nat Genet **39**:960–962.
23. Helfrich MH, Aronson DC, Everts V, Mieremet RHP, Gerritsen EJA, Eckhardt PG, Groot CG, Scherft JP 1991 Morphologic features of bone in human osteopetrosis. Bone **12**:411–419.
24. Bollerslev J, Steiniche T, Melsen F, Mosekilde L 1986 Structural and histomorphometric studies of iliac crest trabecular and cortical bone in auto somal dominant osteopetrosis: A study of two radiological types. Bone **10**:19–24.
25. Lajeunesse D, Busque L, Ménard P, Brunette MG, Bonny Y 1996 Demonstration of an osteoblast defect in two cases of human malignant osteopetrosis. Correction of the phenotype after bone marrow transplant. Bone **98**:1835–1842.
26. Mills BG, Yabe H, Singer FR 1988 Osteoclasts in human osteopetrosis contain viralnucleocapsid-like nuclear inclusions. J Bone Miner Res **3**:101–106.
27. Beard CJ, Key L, Newburger PE, Ezekowitz RA, Arceci R, Miller B, Proto P, Ryan T, Anast C, Simons ER 1986 Neutrophil defect associated with malignant infantile osteopetrosis. J Lab Clin Med **108**:498–505.
28. Cleiren E, Benichou O, Van Hul E, Gram J, Bollerslev J, Singer FR, Beaverson K, Aledo A, Whyte MP, Yoneyama T, deVernejoul MC, Van Hul W 2001 Albers-Schönberg disease (autosomal dominant osteopetrosis, type II) results from mutations in the *ClCN7* chloride channel gene. Hum Mol Genet **10**:2861–2867.
29. Susani L, Pangrazio A, Sobacchi C, Taranta A, Mortier G, Savarirayan R, Villa A, Orchard P, Vezzoni P, Albertini A, Frattini A, Pagani F 2004 TCIRG1-dependent recessive osteopetrosis: Mutation analysis, functional identification of the splicing defects, and in vitro rescue by U1 snRNA. Hum Mutat **24**:225–235.
30. Campos-Xavier AB, Saraiva JM, Ribeiro LM, Munnich A, Cormier-Daire V 2003 Chloride channel 7 (CLCN7) gene mutations in intermediate autosomal recessive osteopetrosis. Hum Genet **112**:186–189.
31. Chalhoub N, Benachenhou N, Rajapurohitam V, Pata M, Ferron M, Frattini A, Villa A, Vacher J 2003 Grey-lethal mutation induces severe malignant autosomal recessive osteopetrosis in mouse and human. Nat Med **9**:399–406.
32. Dries sen GJ, Gerritsen EJ, Fischer A, Fasth A, Hop WC, Veys P, Porta F, Cant A, Steward CG, Vossen JM, Uckan D, Friedrich W 2003 Long-term outcome of haematopoietic stem cell transplantation in autosomal recessive osteopetrosis: An EBMT report. Bone Marrow Transplant **32**:657–663.
33. Rawlinson PS, Green RH, Coggins AM, Boyle IT, Gibson BE 1991 Malignant osteopetrosis: Hypercalcaemia after bone marrow transplantation. Arch Dis Child **66**:638–639.
34. Steward CG, Pellier I, Mahajan A, Ashworth MT, Stuart AG, Fasth A, Lang D, Fischer A, Friedrich W, Schulz AS 2004 The Working Party on Inborn Errors of the European Blood and Marrow Transplantation Group. Severe pulmonary hypertension: A frequent complication of stem cell transplantation for malignant infantile osteopetrosis. Br J Haematol **124**:63–71.
35. Tsuji Y, Ito S, Isoda T, Kajiwara M, Nagasawa M, Morio T, Mizutani S 2005 Successful nonmyeloablative cord blood transplantation for an infant with malignant infantile osteopetrosis. J Pediatr Hematol Oncol **27**:495–498.
36. Key LL Jr 1987 Osteopetrosis: A genetic window into osteoclast function. Cases Metab Bone Dis. A CPC Series. Triclinica Communications, New York, NY, USA, **2**:1–12.

© 2008 American Society for Bone and Mineral Research

37. Dorantes LM, Mejia AM, Dorantes S 1986 Juvenile osteopetrosis: Effects of blood and bone of prednisone and low calcium, high phosphate diet. Arch Dis Child **61**:666–670.
38. Iacobini M, Migliaccio S, Roggini M, Taranta A, Werner B, Panero A, Teti A 2001 Case Report: Apparent cure of a newborn with malignant osteopetrosis using prednisone therapy. J Bone Miner Res **16**:2356–2360.
39. Dozier TS, Duncan IM, Klein AJ, Lambert PR, Key LL Jr 2005 Otologic manifestations of malignant osteopetrosis. Otol Neurotol **26**:762–766.
40. Strickland JP, Berry DJ 2005 Total joint arthroplasty in patients with osteopetrosis: A report of 5 cases and review of the literature. J Arthoplasty **20**:815–820.
41. Sly WS, Hewett-Emmett D, Whyte MP, Yu YS, Tashian RE 1983 Carbonic anhydrase II deficiency identified as the primary defect in the autosomal recessive syndrome of osteopetrosis with renal tubular acidosis and cerebral calcification. Proc Natl Acad Sci USA **80**:2752–2756.
42. Whyte MP 1993 Carbonic anhydrase II deficiency. Clin Orthop **294**:52–63.
43. Sly WS, Shah GN 2001 The carbonic anhydrase II deficiency syndrome: Osteopetrosis with renal tubular acidosis and cerebral calcification. In: Scriver CR, Beaudet AL, Sly WS, Valle D, Child B, Vogelstein B (eds.) The Metabolic and Molecular Bases of Inherited Disease, 8th ed. McGraw-Hill Book Company, New York, NY, USA, pp. 5331–5343.
44. Roth DE, Venta PJ, Tashian RE, Sly WS 1992 Molecular basis of human carbonic anhydrase II deficiency. Proc Natl Acad Sci USA **89**:1804–1808.
45. Shah GN, Bonapace G, Hu PY, Strisciuglio P, Sly WS 2004 Carbonic anhydrase II deficiency syndrome (osteopetrosis with renal tubular acidosis and brain calcification): Novel mutations in CA2 identified by direct sequencing expand the opportunity for genotype-phenotype correlation. Hum Mutat **24**:272.
46. Whyte MP, Hamm LL III, Sly WS 1988 Transfusion of carbonic anhydrase-replete erythrocytes fails to correct the acidification defect in the syndrome of osteopetrosis, renal tubular acidosis, and cerebral calcification (carbonic anhydrase II deficiency). J Bone Miner Res **3**:385–388.
47. McMahon C, Will A, Hu P, Shah GN, Sly WS, Smith OP 2001 Bone marrow transplantation corrects osteopetrosis in the carbonic anhydrase II deficiency syndrome. Blood **97**:1947–1950.
48. Maroteaux P, Lamy M 1965 The malady of Toulouse-Lautrec. JAMA **191**:715–717.
49. Maroteaux P, Lamy M 1962 La pycnodysostose. Presse Med **70**:999–1002.
50. Gelb BD, Brömme D, Desnick RJ 2001 Pycnodysostosis: Cathepsin K deficiency. In: Scriver CR, Beaudet AL, Sly WS, Valle D, Child B, Vogelstein B (eds.) The Metabolic and Molecular Bases of Inherited Disease, 8th ed. McGraw-Hill Book Company, New York, NY, USA, pp. 3453–3468.
51. Elmore SM 1967 Pycnodysostosis: A review. J Bone Joint Surg Am **49**:153–162.
52. Soto TJ, Mautalen CA, Hojman D, Codevilla A, Piqué J, Pangaro JA 1969 Pycnodysostosis, metabolic and histologic studies. Birth Defects **5**:109–115.
53. Everts V, Aronson DC, Beertsen W 1985 Phagocytosis of bone collagen by osteoclasts in two cases of pycnodysostosis. Calcif Tissue Int **37**:25–31.
54. Beneton MNC, Harris S, Kanis JA 1987 Paramyxovirus-like inclusions in two cases of pycnodysostosis. Bone **8**:211–217.
55. Soliman AT, Rajab A, AlSalmi I, Darwish A, Asfour M 1996 Defective growth hormone secretion in children with pycnodysostosis and improved linear growth after growth hormone treatment. Arch Dis Child **75**:242–244.
56. Fratzl-Zelman N, Valenta A, Roschger P, Nader A, Gelb BD, Fratzl P, Klaushofer K 2004 Decreased bone turnover and deterioration of bone structure in two cases of pycnodysostosis. J Clin Endocrinol Metab **89**:1538–1547.
57. Everts V, Hou WS, Rialland X, Tigchelaar W, Saftig P, Bromme D, Gelb BD, Beertsen W 2003 Cathepsin K deficiency in pycnodysostosis results in accumulation of non-digested phagocytosed collagen in fibroblasts. Calcif Tissue Int **73**:380–386.
58. Edelson JG, Obad S, Geiger R, On A, Artul HJ 1992 Pycnodysostosis: Orthopedic aspects, with a description of 14 new cases. Clin Orthop **280**:263–276.
59. Engelmann G 1929 Ein fall von osteopathia hyperostotica (sclerotisans) multiplex infantilis. Fortschr Geb Roentgen **39**:1101–1106.
60. Saito T, Kinoshita A, Yoshiura KI, Makita Y, Wakui K, Honke K, Niikawa N, Taniguchi N 2001 Domain-specific mutations of a transforming growth factor (TGF)-f31 latency-associated peptide cause Camurati-Engelmann disease because of the formation of a constitutively active form of TGF-f31. J Biol Chem **276**:11469–11472.
61. Wallace SE, Lachman RS, Mekikian PB, Bui KK, Wilcox WR 2004 Marked phenotypic variability in progressive diaphyseal dysplasia (Camurati-Engelmann disease): Report of a four-generation pedigree, identification of a mutation in TGFB1, and review. Am J Med Genet A **129**:235–247.
62. Naveh Y, Ludatshcer R, Alon U, Sharf B 1985 Muscle involvement in progressive diaphyseal dysplasia. Pediatrics **76**:944–949.
63. Crisp AJ, Brenton DP 1982 Engelmann's disease of bone: A systemic disorder? Ann Rheum Dis **41**:183–188.
64. Smith R, Walton RJ, Corner BD, Gordon IR 1977 Clinical and biochemical studies in Engelmann's disease (progressive diaphyseal dysplasia). Q J Med **46**:273–294.
65. Kumar B, Murphy WA, Whyte MP 1981 Progressive diaphyseal dysplasia (Englemann's disease): Scintigraphic-radiologic-clinical correlations. Radiology **140**:87–92.
66. Applegate LJ, Applegate GR, Kemp SS 1991 MR of multiple cranial neuropathies in a patient with Camurati-Engelmann disease: Case report. Am Soc Neuroradiol **12**:557–559.
67. Janssens K, ten Dijke P, Ralston SH, Bergmann C, Van Hul W 2003 Transforming growth factor-beta 1 mutations in Camurati-Engelmann disease lead to increased signaling by altering either activation or secretion of the mutant protein. J Biol Chem **278**:7718–7724.
68. Saraiva JM 2000 Anticipation in progressive diaphyseal dysplasia. J Med Genet **37**:394–395.
69. Hecht JT, Blanton SH, Broussard S, Scott A, Hall CR, Mlunsky JM 2001 Evidence for locus heterogeneity in the Camurati-Engelmann (DPD1) Syndrome. Clin Genet **59**:198–200.
70. Naveh Y, Alon U, Kaftori JK, Berant M 1985 Progressive diaphyseal dysplasia: Evaluation of corticosteroid therapy. Pediatrics **75**:321–323.
71. Inaoka T, Shuke N, Sato J, Ishikawa Y, Takahashi K, Aburano T, Makita Y 2001 Scintigraphic evaluation of pamidronate and corticosteroid therapy in a patient with progressive diaphyseal dysplasia (Camurati-Engelmann disease). Clin Nucl Med **26**:680–682.
72. Van Buchem FSP, Prick JJG, Jaspar HHJ 1976 Hyperostosis Corticalis Generalisata Familiaris (Van Buchem's Disease). Excerpta, Amsterdam, The Netherlands.
73. Perez-Vicente JA, Rodriguez de Castro E, Lafuente J, Mateo MM, Gimenez-Roldan S 1987 Autosomal dominant endosteal hyperostosis. Report of a Spanish family with neurological involvement. Clin Genet **31**:161–169.
74. Beighton P, Barnard A, Hamersma H, van der Wouden A 1984 The syndromic status of sclerosteosis and van Buchem disease. Clin Genet **25**:175–181.
75. Brunkow ME, Gardner JC, Van Ness J, Paeper BW, Kovacevich BR, Proll S, Skonier JE, Zhao L, Sabo PJ, Fu Y, Alisch RS, Gillett L, Colbert T, Tacconi P, Galas D, Hamersma H, Beighton P, Mulligan J 2001 Bone dysplasia sclerosteosis results from loss of the SOST gene product, a novel cystine knot-containing protein. Am J Hum Genet **68**:577–589.
76. Loots GG, Kneissel M, Keller H, Baptist M, Chang J, Collette NM, Ovcharenko D, Plajzer-Frick I, Rubin EM 2005 Genomic deletion of a long-range bone enhancer misregulates sclerostin in Van Buchem disease. Genome Res **15**:928–935.
77. Li X, Zhang Y, Kang H, Liu W, Liu P, Zhang J, Harris SE, Wu D 2005 Sclerostin binds to LRP5/6 and antagonizes canonical Wnt signaling. J Biol Chem **280**:19883–19887.
78. Sutherland MK, Geoghegan JC, Yu C, Turcott E, Skonier JE, Winkler DG, Latham JA 2004 Sclerostin promotes the apoptosis of human osteoblastic cells: a novel regulation of bone formation. Bone **35**:828–835.
79. Ruckert EW, Caudill RJ, McCready PJ 1985 Surgical treatment of van Buchem disease. J Oral Maxillofac Surg **43**:801–805.
80. Schendel SA 1988 van Buchem disease: Surgical treatment of the mandible. Ann Plast Surg **20**:462–467.
81. Hamersma H, Gardner J, Beighton P 2003 The natural history of sclerosteosis. Clin Genet **63**:192–197.

© 2008 American Society for Bone and Mineral Research

82. Beighton P, Cremin BJ, Hamersma H 1976 The radiology of sclerosteosis. Br J Radiol **49:**934–939.
83. Hill SC, Stein SA, Dwyer A, Altman J, Dorwart R, Doppman J 1986 Cranial CT findings in sclerosteosis. AJNR Am J Neuroradiol **7:**505–511.
84. Stein SA, Witkop C, Hill S, Fallon MD, Viernstein L, Gucer G, McKeever P, Long D, Altman J, Miller NR, Teitelbaum SL, Schlesinger S 1983 Sclerosteosis, neurogenetic and pathophysiologic analysis of an American kinship. Neurology **33:**267–277.
85. Epstein S, Hamersma H, Beighton P 1979 Endocrine function in sclerosteosis. S Afr Med J **55:**1105–1110.
86. Uitto J, Santa Cruz DJ, Starcher BC, Whyte MP, Murphy WA 1981 Biochemical and ultrastructural demonstration of elastin accumulation in the skin of the BuschkeOllendorff syndrome. J Invest Dermatol **76:**284–287.
87. Hellemans J, Preobrazhenska O, Willaert A, Debeer P, Verdonk PC, Costa T, Janssens K, Menten B, Van Roy N, Vermeulen SJ, Savarirayan R, Van Hul W, Vanhoenacker F, Huylebroeck D, De Paepe A, Naeyaert JM, Vandesompele J, Speleman F, Verschueren K, Coucke PJ, Mortier GR 2004 Loss-of-function mutations in LEMD3 result in osteopoikilosis, Buschke-Ollendorff syndrome and melorheostosis. Nat Genet **36:**1213–1218.
88. Whyte MP, Murphy WA, Seigel BA 1978 99m Tc-pyrophosphate bone imaging in osteopoikilosis, osteopathia striata, and melorheostosis. Radiology **127:**439–443.
89. Rabinow M, Unger F 1984 Syndrome of osteopathia striata, macrocephaly, and cranial sclerosis. Am J Dis Child **138:**821–823.
90. Happle R, Lenz W 1977 Striation of bones in focal dermal hypoplasia: Manifestation of functional mosaicism? Br J Dermatol **96:**133–138.
91. Viot G, Lacombe D, David A, Mathieu M, de Broca A, Faivre L, Gigarel N, Munnich A, Lyonnet S, Le Merrer M, Cormier-Daire V 2002 Osteopathia striata cranial sclerosis: Non-random X-inactivation suggestive of X-linked dominant inheritance. Am J Med Genet **107:**1–4.
92. Campbell CJ, Papademetriou T, Bonfiglio M 1968 Melorheostosis: A report of the clinical, roentgenographic, and pathological findings in fourteen cases. J Bone Joint Surg Am **50:**1281–1304.
93. Leri A, Joanny J 1922 Une affection non decrite des os. Hyperostose "en coulee" sur toute la longueur d'un membre ou "melorheostose." Bull Mem Soc Med Hop Paris **46:**1141–1145.
94. Applebaum RE, Caniano DA, Sun CC, Azizkhan RA, Queral LA 1986 Synchronous left subclavian and axillary artery aneurysms associated with melorheostosis. Surgery **99:**249–253.
95. Colavita N, Nicolais S, Orazi C, Falappa PG 1987 Melorheostosis: Presentation of a case followed up for 24 years. Arch Orthop Trauma Surg **106:**123–125.
96. Wagers LT, Young AW Jr, Ryan SF 1972 Linear melorheostotic scleroderma. Br J Dermatol **86:**297–301.
97. Kim JE, Kim EH, Han EH, Park RW, Park IH, Jun SH, Kim JC, Young MF, Kim IS 2000 A TGF-3-inducible cell adhesion moleculae, Big-h3, is downregulated in melorheostosis and involved in oseogeneis. J Cell Biochem **77:**169–178.
98. Mumm S, Zhang X, McAlister WH, Wenkert D, Whyte MP 2005 Deactivating germline mutations in LEMD3 cause osteopoikilosis and Buschke-Ollendorff syndrome, but not melorheostosis. J Bone Miner Res **20:**S1;S418.
99. Atar D, Lehman WB, Grant AD, Strongwater AM 1992 The Ilizarov apparatus for treatment of melorheostosis: Case report and review of the literature. Clin Orthop **281:**163–167.
100. Christmann D, Wenger JJ, Dosch JC, Schraub M, Wackenheim A 1981 L'osteomalacie axiale: Analyse compare avec la fibrogenese imparfaite. J Radiol **62:**37–41.
101. Whyte MP, Fallon MD, Murphy WA, Teitelbaum SL 1981 Axial osteomalacia: Clinical, laboratory and genetic investigation of an affected mother and son. Am J Med **71:**1041–1049.
102. Nelson AM, Riggs BL, Jowsey JO 1978 Atypical axial osteomalacia: Report of four cases with two having features of ankylosing spondylitis. Arthritis Rheum **21:**715–722.
103. Cortet B, Berniere L, Solau-Gervais E, Hacene A, Cotton A, Delcambre B 2000 Axial osteomalacia with sacroiliitis and moderate phosphate diabetes: Report of a case. Clin Exp Rheumatol **18:**625–628.
104. Condon JR, Nassim JR 1971 Axial osteomalacia. Postgrad Med **47:**817–820.
105. Lang R, Vignery AM, Jenson PS 1986 Fibrogenesis imperfecta ossium with early onset: Observations after 20 years of illness. Bone **7:**237–246.
106. Ralphs JR, Stamp TCB, Dopping-Hepenstal PJC, Ali SY 1989 Ultrastructural features of the osteoid of patients with fibrogenesis imperfecta ossium. Bone **10:**243–249.
107. Rimoin DL 1965 Pachydermoperiostosis (idiopathic clubbing and periostosis). Genetic and physiologic considerations. N Engl J Med **272:**923–931.
108. Matucci-Cerinic M, Lott T, Jajic IVO, Pignone A, Bussani C, Cagnoni M 1991 The clinical spectrum of pachydermoperiostosis (primary hypertrophic osteoarthropathy). Medicine **79:**208–214.
109. Ali A, Tetalman M, Fordham EW 1980 Distribution of hypertrophic pulmonary osteoarthropathy. AJR Am J Roentgenol **134:**771–780.
110. Lauter SA, Vasey FB, Huttner I, Osterland CK 1978 Pachydermoperiostosis: Studies on the synovium. J Rheumatol **5:**85–95.
111. Cooper RG, Freemont AJ, Riley M, Holt PJL, Anderson DC, Jayson MIV 1992 Bone abnormalities and severe arthritis in pachydermoperiostosis. Ann Rheum Dis **51:**416–419.
112. Whyte MP, Teitelbaum SL, Reinus WR 1996 Doubling skeletal mass during adult life: The syndrome of diffuse osteosclerosis after intravenous drug abuse. J Bone Miner Res **11:**554–558.
113. Khosla S, Ballard FJ, Conover CA 2002 Use of site-specific antibodies to characterize the circulating form of big insulin-like growth factor II in patients with hepatitis C-associated osteosclerosis. J Clin Endocrinol Metab **87:**3867–3870.
114. Boyden LM, Mao J, Belsky J, Mitzner L, Farhi A, Mitnick MA, Wu D, Insogna K, Lifton RP 2002 High bone density due to a mutation in LDL-receptor-related protein 5. N Engl J Med **345:**1513–1521.
115. Rickels MR, Zhang X, Mumm S, Whyte MP 2005 Oropharyngeal skeletal disease accompanying high bone mass and novel LRP5 mutation. J Bone Miner Res **20:**878–885.

Chapter 89. Fibrous Dysplasia

Michael T. Collins,[1] Mara Riminucci,[2] and Paolo Bianco[2]

[1]Skeletal Clinical Studies Unit, Craniofacial and Skeletal Diseases Branch, National Institute of Dental and Craniofacial Research, National Institutes of Health, Department of Health and Human Services, Bethesda, Maryland; [2]Dipartimento di Medicina Sperimentale, Universita' La Sapienza, Rome, Italy

INTRODUCTION

Fibrous dysplasia of bone (FD; OMIM 174800) is an uncommon skeletal disorder with a broad spectrum of clinical expressions, ranging from an incidentally discovered asymptomatic radiographic finding, involving a single skeletal site, to a severe disabling disease. The disease may involve one bone (monostotic), multiple bones (polyostotic FD), or even the entire skeleton (panostotic FD). In polyostotic disease, lesions of dif-

The authors state that they have no conflicts of interest.

Key words: bone, Gs α, GNAS, fibroblast growth factor 23, osteomalacia, McCune-Albright syndrome, fibro-osseous

© 2008 American Society for Bone and Mineral Research

ferent limb bones are often (but not necessarily) ipsilateral.[1,2] FD may be associated with extraskeletal lesions or dysfunction, most commonly cutaneous hyperpigmentation (Figs. 1A and 1B), and hyperfunctioning endocrinopathies, including precocious puberty, hyperthyroidism, growth hormone (GH) excess, and Cushing syndrome. FD in combination with one or more of the extraskeletal manifestations is known as McCune-Albright syndrome (MAS).[3] A renal tubulopathy, which includes renal phosphate wasting, is one of the most common extraskeletal dysfunctions associated with polyostotic disease.[4] More rarely, FD may be associated with myxomas of skeletal muscle (Mazabraud's syndrome)[5] or dysfunction of heart, liver, pancreas, or other organs within the context of the MAS.[6]

ETIOLOGY AND PATHOGENESIS

All forms of FD are caused by dominant, gain-of-function (activating), missense mutations of the *GNAS* gene, encoding the α subunit of the stimulatory G protein, $G_s\alpha$.[7,8] Mutations occur postzygotically, are never inherited, and result in a somatic mosaic state. Single base transitions lead to replacement of arginine at position 201 with histidine or cysteine (most commonly), or rarely with other amino acids,[9] and recently, substitutions at Q227 have been reported.[10] The two most common mutations arise in early development as a consequence of methylation and deamination of cytosines within the CpG dinucleotide in the Arg 201 codon, presumably during de novo methylation of cells in the inner cell mass. The multipotency of inner cell mass cells may explain how the original mutation is transmitted to derivatives of all three germ layers, accounting for the broad organ distribution of severe forms of the disease.[11] Size and viability of the mutated clone arising from the single, original mutated cell may determine the variable distribution and frequency of the mutated cells in the postnatal organism, and the extent and severity of disease.[1] As a consequence of the mutation, the catalysis of GTP to GDP by $G_s\alpha$ is significantly decreased. Constitutive activation of adenylyl cyclase by the mutated $G_s\alpha$ ensues, and the resulting excess cAMP is thought to mediate a number of pathological effects in mutated cells.[1] In bone, mutations impact on cells of the osteogenic lineage, with adverse effects both on osteoprogenitor cells and differentiated osteoblasts.[12,13] Whereas the $G_s\alpha$ gene is not imprinted and is biallelically expressed in bone, asymmetric expression of the $G_s\alpha$ alleles in osteoprogenitors may account for the extent and severity of lesions.[14] Expansion of the osteoprogenitor cell pool leads to their accumulation in marrow spaces, resulting in local loss of hematopoietic tissue and marrow fibrosis. Osteogenic cells derived from mutated skeletal progenitors are functionally and morphologically abnormal and deposit abnormal matrix. Bone trabeculae are abnormal in shape (so-called Chinese writing, alphabet soup patterns), collagen orientation, and biochemical composition,[12] and in many cases, are severely undermineralized and abnormally compliant (Fig. 2D).[15,16] Elevated serum levels of fibroblast growth factor 23 (FGF23), a recently identified phosphate-regulating hormone produced by highly activated osteoblastic cells in FD tissue (Fig. 2E), have been shown to be the etiology of the renal phosphate wasting commonly seen in association with FD.[17,18] The histological pattern may be significantly different at different skeletal sites, and peculiar patterns are seen in craniofacial bones (Figs. 2A and 2B).[19] Specific microscopic features, such as Sharpey fibers (Fig. 3C) and retracted osteoblasts, may, however, be recognized at all skeletal sites as recurrent histological hallmarks of the disease.[13] The hormonal climate influences FD lesions[20] and may significantly alter the local rate of bone remodeling.[21] FD tissue is highly vascularized and therefore prone to bleeding, leading to posthemorrhagic cysts.[22]

CLINICAL FEATURES

The sites of skeletal involvement (the "map" of affected tissues) are established early in patients with FD. Ninety percent of the craniofacial lesions are established before 5 yr of age, and 75% of all sites of FD are evident by 15 yr of age.[23] Pathological effects of $G_s\alpha$ mutations in osteogenic cells are most pronounced and evident during the phase of rapid bone growth and account for that fact that childhood and adolescence are the periods during which the disease most commonly presents and the period of peak rate of fractures.[24,25] Presentation in infancy is rare and usually heralds severe, widespread

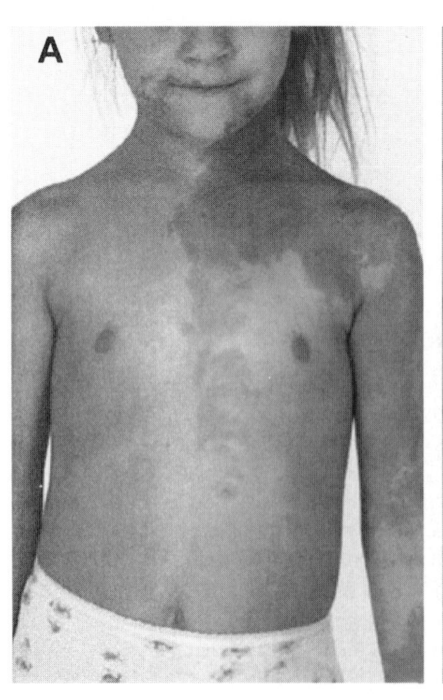

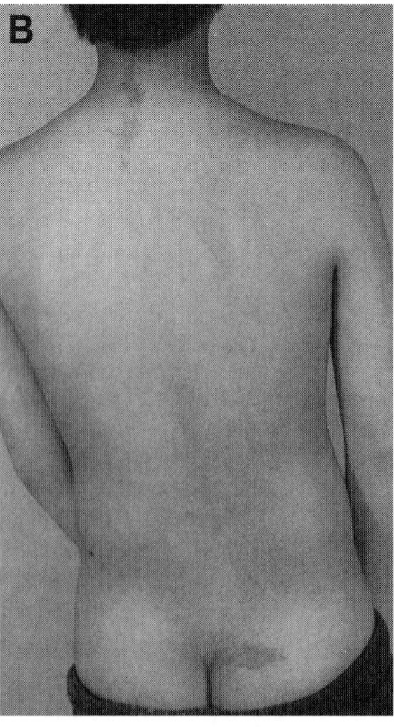

FIG. 1. Café-au-lait skin pigmentation. (A) A typical lesion on the face, chest, and arm of a 5-yr-old girl with MAS that shows jagged "coast of Maine" borders and the tendency for the lesions to both respect the midline and follow the developmental lines of Blashko. (B) Typical lesions that are often found on the nape of the neck and crease of the buttocks are shown.

© 2008 American Society for Bone and Mineral Research

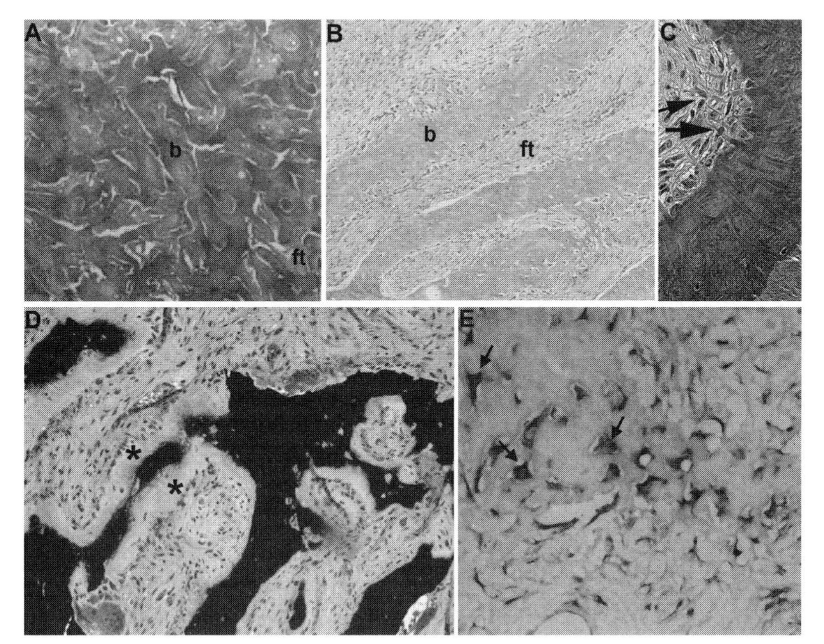

FIG. 2. Representative histological images of craniofacial FD. (A) Calvarial FD lesions are characterized by uninterrupted networks of bone trabeculae (b) embedded in the fibrous tissue (ft). (B) In FD lesions from gnathic bones, newly formed bone trabeculae (b) are deposited within the fibrous tissue (ft) in a typical discontinuous and parallel pattern. (C) Collagen fibers perpendicularly oriented to forming bone surfaces (Sharpey fibers, arrows) represent a recurrent histological feature of FD at all skeletal sites. (D and E) Osteomalacic changes and FGF23 production in FD. (D) In many cases of FD, processing for undecalcified embedding reveals excess osteoid (asterisks) and severe undermineralization of the fibrous dysplastic bone. (E) The mineralization defect of the FD tissue is related to elevated levels of FGF23 produced by activated FD osteogenic cells (arrows), as shown by in situ hybridization.

disease with multiorgan involvement. Pain, fracture, and deformity are the most common presenting features. In general, children are less likely to have pain and/or complain of pain, per se, and may instead report stiffness or tiredness.[26] In adults the complaint of pain is common, especially in the ribs, long bones, and craniofacial bones. It is often severe and may require narcotic analgesics.[26] Lesions in the spine and pelvis are usually less painful. Pathological fractures or stress fracture of weight-bearing limb bones is a prime cause of morbidity. Deformity of limb bones is caused by expansion and abnormal compliance of lesional FD, fracture treatment failure, and local complications such as cyst formation.[1] Deformity of craniofacial bone is solely the result of the overgrowth of lesional bone.

Although any bone may be affected, the proximal metaphysis of the femora and the skull base are the two sites most commonly involved.[26] Femoral disease usually presents in childhood with limp, fracture, pain, and deformity, ranging from coxa vara to the classical shepherd's crook deformity (Fig. 3A). Radiographically, the lesion may be limited to the metaphysis or extend along the diaphysis for variable length.[22] The picture most commonly observed in children and adolescents consists of an expansile, deforming, medullary lesion, with cortical thinning and an overall "ground glass" density (Fig. 3A). The radiographic picture is significantly affected by the evolution of the lesion over time and by the appearance of superimposed changes, such as aneurysmal bone cysts. Hence, lesions observed in adults tend to appear more sclerotic and less homogeneous (Fig. 3B). Sclerosis in FD lesions of the femur and other limb bones may signify less active disease.

In the skull, FD mostly involves the skull base and facial bones. The typical presentation is in childhood with facial asymmetry or a "bump" that persists, but symmetric expansion of the malar prominences and/or frontal bosses may also be seen. The disease can progress into adulthood and disfiguration may be marked when FD is accompanied by GH excess.[20] Abnormal growth and deformity of craniofacial bones may result in encroachment on cranial nerves. However, severe adverse consequences are rare, but much more common in patients with associated growth hormone excess.[20,27,28] FD tissue in craniofacial bones is especially prone to bleeding, herniation through cranial foramina and vascular passages, as well as formation of posthemorrhagic cysts. These events may precipitate blindness when they occur in the vicinity of the optic nerves.[29] Radiographically, craniofacial FD typically has a homogenous "ground glass" appearance in children (Fig. 3C), but in adults, lesions with a more sclerotic, "pagetoid" appearance are typical (Fig. 3D), which correlate with site-specific osteosclerotic histological changes.[19] Lesions in the spine, ribs, and pelvis are common, may be elusive on plain radiographs, but are easily detected by bone scintigraphy, the most sensitive imaging technique for the detection of FD lesions (Figs. 3E–3G). Disease in the spine is common and is frequently associated with scoliosis, may require surgery, and can be progressive into adulthood.[30]

Malignancy in FD is rare (<1%).[31] Whereas there is an association with the development of cancer with prior treatment with high-dose external beam radiation,[32,33] it may also occur independent of prior exposure to ionizing radiation. Rapid lesion expansion and disruption of the cortex on radiographs should alert the clinician to the possibility of sarcomatous change. Osteogenic sarcoma is the most common, but not the only type of bone tumor that may complicate FD. The clinical course is usually aggressive, surgery is the primary treatment, and chemotherapeutic regimens do not seem to improve prognosis significantly.

MANAGEMENT AND TREATMENT

Diagnosis of FD must be established based on expert assessment of clinical, radiographic, and histopathological features. Markers of bone turnover are usually elevated.[4] The extent of the skeletal disease is best determined with total body bone scintigraphy, which can be used to assess the skeletal disease burden and predict functional outcome.[34] The metabolic derangements associated with FD, especially hypophosphatemia and growth hormone excess, are associated with a significantly worse clinical outcome and therefore must be screened for and treated.[16,20,25,28]

Mutation analysis may be helpful in distinguishing FD from unrelated fibro-osseous lesions of the skeleton, which may

© 2008 American Society for Bone and Mineral Research

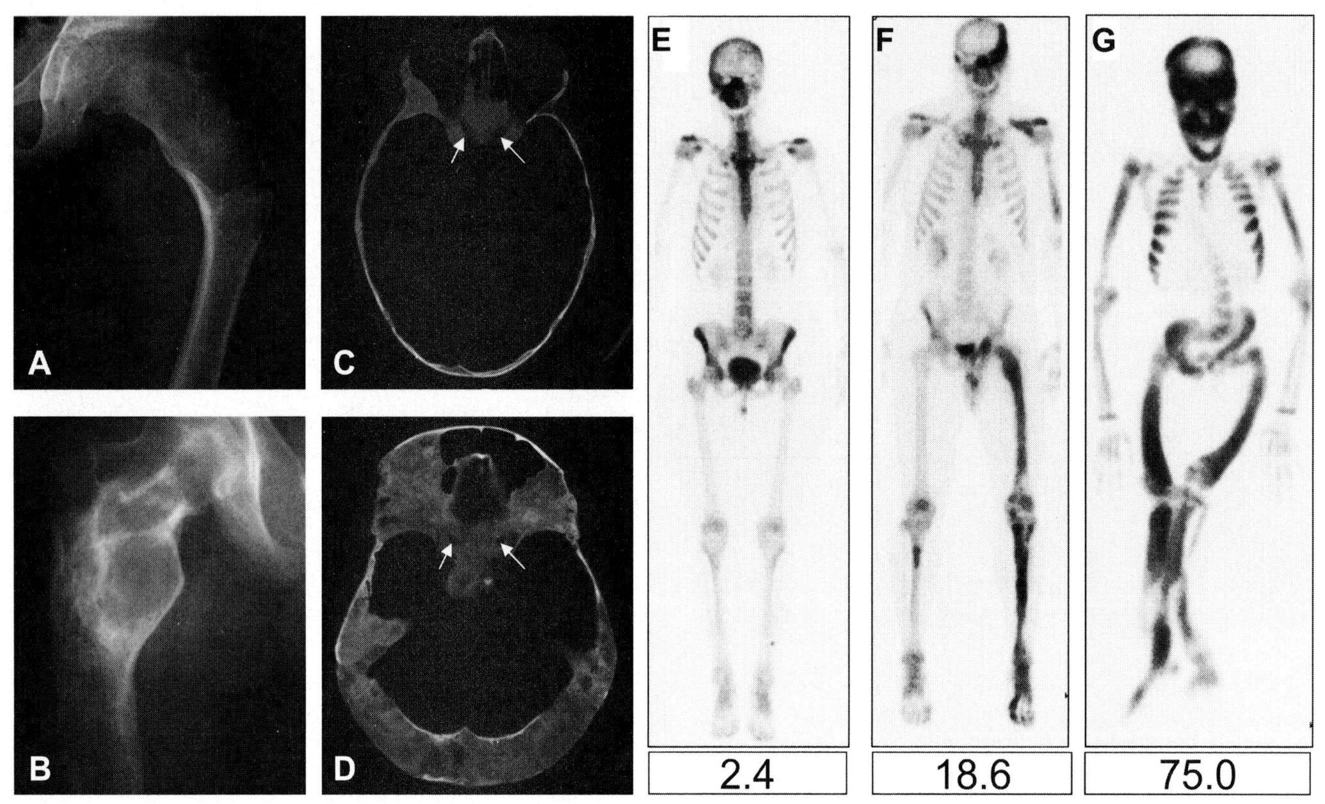

FIG. 3. Radiographic appearance of fibrous dysplasia. (A) A proximal femur with typical ground glass appearance and shepherd's crook deformity in a 10-yr-old child is shown. (B) The appearance of FD in the femur of an untreated 40-yr-old man shows the tendency for FD to appear more sclerotic with time. (C) The typical ground glass appearance of FD in the craniofacial region on a CT image of a 10-yr-old child is shown. The white arrows indicate the optic nerves, which are typically encased with FD. (D) A CT image in a 40-yr-old woman shows the typical appearance of craniofacial FD in an older person, with mixed solid and "cystic" lesions. The Houndsfield unit measurements of "cystic" lesions are quite useful in distinguishing soft tissue "cystic" lesions from true fluid-filled cysts, which are much more uncommon and tend to behave aggressively with rapid expansion and compression of vital structures. (E–G) Bone scintigraphy in FD. Representative ^{99}Tc-MDP bone scans that show tracer uptake at affected skeletal sites and the associated skeletal disease burden score[31] are shown. (E) A 50-yr-old woman with monostotic FD confined to a single focus involving contiguous bones in the craniofacial region. (F) A 42-yr-old man with polyostotic FD shows the tendency for FD to be predominantly (but not exclusively) unilateral and to involve the skull base and proximal femur. (G) A 16-yr-old boy with MAS and involvement of virtually all skeletal sites (panostotic) is shown.

mimic FD both clinically and radiographically (osteofibrous dysplasia, ossifying fibromas of jawbones[1]). Isolated lesions of the proximal femur in adults may be improperly diagnosed and classified as distinct fibro-osseous lesions. For example, all cases of so-called "liposclerosing myxofibrous tumor" in which GNAS mutations were sought were found to represent monostotic fibrous dysplasia.[35] Multiple nonossifying fibromas, skeletal angiomatosis, and Ollier's disease may sometimes enter the differential diagnosis, which again relies on histology and mutation analysis.

Disease of the proximal femur, in which there is fracture or impending fracture, is often best treated by insertion of intramedullary nails, in an effort to prevent serious deformity and limb length discrepancy.[22,36,37] Design of specific types of nails is felt to be necessary, and development of such devices is underway.[14] Surgery is not advocated for craniofacial disease unless hearing or vision loss are documented, and prophylactic optic nerve decompression seems to be contraindicated.[28] Treatment with bisphosphonates (pamidronate, etc.) has been advocated based on observational studies with claims of reduced pain, decreased serum and urine markers of bone metabolism, and improvement in the radiographic appearance of the disease.[38,39] However, a recently completed open label, prospective study with appropriate histological, radiographic, and clinical endpoints showed pain relief but showed no benefit radiographically or histologically.[40] Ongoing placebo-controlled studies in the United States and Europe may help to better define the role of bisphosphonates in treating FD.

REFERENCES

1. Bianco P, Gehron Robey P, Wientroub S 2003 Fibrous dysplasia. In: Glorieux F, Pettifor J, Juppner H (eds.) Pediatric Bone: Biology and Disease. Academic Press/Elsevier, New York, NY, USA, pp. 509–539.
2. Collins MT 2006 Spectrum and natural history of fibrous dysplasia of bone. J Bone Miner Res **21:**S2;P99–P104.
3. Danon M, Crawford JD 1987 The McCune-Albright syndrome. Ergeb Inn Med Kinderheilkd **55:**81–115.
4. Collins MT, Chebli C, Jones J, Kushner H, Consugar M, Rinaldo P, Wientroub S, Bianco P, Robey PG 2001 Renal phosphate wasting in fibrous dysplasia of bone is part of a generalized renal tubular dysfunction similar to that seen in tumor- induced osteomalacia. J Bone Miner Res **16:**806–813.
5. Cabral CE, Guedes P, Fonseca T, Rezende JF, Cruz Junior LC, Smith J 1998 Polyostotic fibrous dysplasia associated with intramuscular myxomas: Mazabraud's syndrome. Skeletal Radiol **27:**278–282.
6. Shenker A, Weinstein LS, Moran A, Pescovitz OH, Charest NJ, Boney CM, Van Wyk JJ, Merino MJ, Feuillan PP, Spiegel AM 1993 Severe endocrine and nonendocrine manifestations of the McCune-Albright syndrome associated with activating mutations of stimulatory G protein GS. J Pediatr **123:**509–518.

© 2008 American Society for Bone and Mineral Research

7. Weinstein LS, Shenker A, Gejman PV, Merino MJ, Friedman E, Spiegel AM 1991 Activating mutations of the stimulatory G protein in the McCune- Albright syndrome. N Engl J Med **325:**1688–1695.
8. Bianco P, Riminucci M, Majolagbe A, Kuznetsov SA, Collins MT, Mankani MH, Corsi A, Bone HG, Wientroub S, Spiegel AM, Fisher LW, Robey PG 2000 Mutations of the GNAS1 gene, stromal cell dysfunction, and osteomalacic changes in non-McCune-Albright fibrous dysplasia of bone. J Bone Miner Res **15:**120–128.
9. Riminucci M, Fisher LW, Majolagbe A, Corsi A, Lala R, De Sanctis C, Robey PG, Bianco P 1999 A novel GNAS1 mutation, R201G, in McCune-Albright syndrome. J Bone Miner Res **14:**1987–1989.
10. Idowu BD, Al-Adnani M, O'Donnell P, Yu L, Odell E, Diss T, Gale RE, Flanagan AM 2007 A sensitive mutation-specific screening technique for GNAS1 mutations in cases of fibrous dysplasia: The first report of a codon 227 mutation in bone. Histopathology **50:**691–704.
11. Riminucci M, Saggio I, Robey PG, Bianco P 2006 Fibrous dysplasia as a stem cell disease. J Bone Miner Res **21**(Suppl 2):125–131.
12. Riminucci M, Fisher LW, Shenker A, Spiegel AM, Bianco P, Gehron Robey P 1997 Fibrous dysplasia of bone in the McCune-Albright syndrome: Abnormalities in bone formation. Am J Pathol **151:**1587–1600.
13. Bianco P, Kuznetsov S, Riminucci M, Fisher LW, Spiegel AM, Gehron Robey P 1998 Reproduction of human fibrous dysplasia of bone in immunocompromised mice by transplanted mosaics of normal and Gs-alpha mutated skeletal progenitor cells. J Clin Invest **101:**1737–1744.
14. Michienzi S, Cherman N, Holmbeck K, Funari A, Collins MT, Bianco P, Robey PG, Riminucci M 2007 GNAS transcripts in skeletal progenitors: Evidence for random asymmetric allelic expression of Gs{alpha}. Hum Mol Genet (in press).
15. Riminucci M, Fisher LW, Shenker A, Spiegel AM, Bianco P, Gehron Robey P 1997 Fibrous dysplasia of bone in the McCune-Albright syndrome: Abnormalities in bone formation. Am J Pathol **151:**1587–1600.
16. Corsi A, Collins MT, Riminucci M, Howell PG, Boyde A, Robey PG, Bianco P 2003 Osteomalacic and hyperparathyroid changes in fibrous dysplasia of bone: Core biopsy studies and clinical correlations. J Bone Miner Res **18:**1235–1246.
17. Collins MT, Chebli C, Jones J, Kushner H, Consugar M, Rinaldo P, Wientroub S, Bianco P, Robey PG 2001 Renal phosphate wasting in fibrous dysplasia of bone is part of a generalized renal tubular dysfunction similar to that seen in tumor-induced osteomalacia. J Bone Miner Res **16:**806–813.
18. Riminucci M, Collins MT, Fedarko NS, Cherman N, Corsi A, White KE, Waguespack S, Gupta A, Hannon T, Econs MJ, Bianco P, Gehron Robey P 2003 FGF-23 in fibrous dysplasia of bone and its relationship to renal phosphate wasting. J Clin Invest **112:**683–692.
19. Riminucci M, Liu B, Corsi A, Shenker A, Spiegel AM, Gehron Robey P, Bianco P 1999 The histopathology of fibrous dysplasia of bone in patients with activating mutations of the Gs al[ha gene: Site-specific patterns and recurrent histological hallmarks. J Pathol **187:**249–258.
20. Akintoye SO, Chebli C, Booher S, Feuillan P, Kushner H, Leroith D, Cherman N, Bianco P, Wientroub S, Robey PG, Collins MT 2002 Characterization of gsp-mediated growth hormone excess in the context of McCune-Albright syndrome. J Clin Endocrinol Metab **87:**5104–5112.
21. Corsi A, Collins MT, Riminucci M, Howell PGT, Boyde A, Robey PG, Bianco P 2003 Osteomalacic and hyperparathyroid changes in fibrous dysplasia of bone:core biopsy studies and clinical correlations. J Bone Miner Res **18:**1235–1246.
22. Ippolito E, Bray EW, Corsi A, De Maio F, Exner UG, Robey PG, Grill F, Lala R, Massobrio M, Pinggera O, Riminucci M, Snela S, Zambakidis C, Bianco P 2003 Natural history and treatment of fibrous dysplasia of bone: A multicenter clinicopathologic study promoted by the European Pediatric Orthopaedic Society. J Pediatr Orthop B **12:**155–177.
23. Hart ES, Kelly MH, Brillante B, Chen CC, Ziran N, Lee JS, Feuillan P, Leet AI, Kushner H, Robey PG, Collins MT 2007 Onset, progression, and plateau of skeletal lesions in fibrous dysplasia, and the relationship to functional outcome. J Bone Miner Res **22:**1468–1474.
24. Harris WH, Dudley HR, Barry RJ 1962 The natural history of fibrous dysplasia. An orthopedic, pathological, and roentgenographic study. J Bone Joint Surg Am **44:**207–233.
25. Leet AI, Chebli C, Kushner H, Chen CC, Kelly MH, Brillante BA, Robey PG, Bianco P, Wientroub S, Collins MT 2004 Fracture incidence in polyostotic fibrous dysplasia and the McCune-Albright syndrome. J Bone Miner Res **19:**571–577.
26. Kelly MH, Brillante B, Collins MT 2008 Pain in fibrous dysplasia of bone: Age-related changes and the anatomical distribution of skeletal lesions. Osteoporos Int **19:**57–63.
27. Cutler CM, Lee JS, Butman JA, FitzGibbon EJ, Kelly MH, Brillante BA, Feuillan P, Robey PG, DuFresne CR, Collins MT 2006 Long-term outcome of optic nerve encasement and optic nerve decompression in patients with fibrous dysplasia: Risk factors for blindness and safety of observation. Neurosurgery **59:**1011–1017.
28. Lee JS, FitzGibbon E, Butman JA, Dufresne CR, Kushner H, Wientroub S, Robey PG, Collins MT 2002 Normal vision despite narrowing of the optic canal in fibrous dysplasia. N Engl J Med **347:**1670–1676.
29. Diah E, Morris DE, Lo LJ, Chen YR 2007 Cyst degeneration in craniofacial fibrous dysplasia: Clinical presentation and management. J Neurosurg **107:**504–508.
30. Leet AI, Magur E, Lee JS, Wientroub S, Robey PG, Collins MT 2004 Fibrous dysplasia in the spine: Prevalence of lesions and association with scoliosis. J Bone Joint Surg Am **86:**531–537.
31. Ruggieri P, Sim FH, Bond JR, Unni KK 1994 Malignancies in fibrous dysplasia. Cancer **73:**1411–1424.
32. Saglik Y, Atalar H, Yildiz Y, Basarir K, Erekul S 2007 Management of fibrous dysplasia. A report on 36 cases. Acta Orthop Belg **73:**96–101.
33. Hansen MR, Moffat JC 2003 Osteosarcoma of the skull base after radiation therapy in a patient with McCune-Albright syndrome: Case report. Skull Base **13:**79–83.
34. Collins MT, Kushner H, Reynolds JC, Chebli C, Kelly MH, Gupta A, Brillante B, Leet AI, Riminucci M, Robey PG, Bianco P, Wientroub S, Chen CC 2005 An instrument to measure skeletal burden and predict functional outcome in fibrous dysplasia of bone. J Bone Miner Res **20:**219–226.
35. Corsi A, De Maio F, Ippolito E, Cherman N, Gehron Robey P, Riminucci M, Bianco P 2006 Monostotic fibrous dysplasia of the proximal femur and liposclerosing myxofibrous tumor: Which one is which? J Bone Miner Res **21:**1955–1958.
36. Stanton RP 2006 Surgery for fibrous dysplasia. J Bone Miner Res **21:**S2;105–109.
37. Keijser LC, Van Tienen TG, Schreuder HW, Lemmens JA, Pruszczynski M, Veth RP 2001 Fibrous dysplasia of bone: Management and outcome of 20 cases. J Surg Oncol **76:**157–166.
38. Liens D, Delmas PD, Meunier PJ 1994 Long-term effects of intravenous pamidronate in fibrous dysplasia of bone. Lancet **343:**953–954.
39. Chapurlat RD, Delmas PD, Liens D, Meunier PJ 1997 Long-term effects of intravenous pamidronate in fibrous dysplasia of bone. J Bone Miner Res **12:**1746–1752.
40. Plotkin H, Rauch F, Zeitlin L, Munns C, Travers R, Glorieux FH 2003 Effect of pamidronate treatment in children with polyostotic fibrous dysplasia of bone. J Clin Endocrinol Metab **88:**4569–4575.

© 2008 American Society for Bone and Mineral Research

Chapter 90. Chondrodysplasias

Yasemin Alanay[1] and David L. Rimoin[2]

[1]*Clinical Genetics Unit, Department of Pediatrics, Hacettepe University School of Medicine, Ankara, Turkey;* [2]*Medical Genetics Institute, Cedars-Sinai Medical Center, Los Angeles, California*

INTRODUCTION

The skeletal dysplasias are a clinically and genetically heterogeneous group of disorders of bone and/or cartilage characterized by abnormalities in growth, development, and/or differentiation of the human skeleton.[1] They may be divided into three main subgroups: osteodysplasias, chondrodysplasias, and dysostoses. The osteodysplasias are related to abnormalities in BMD and mineralization, such as osteopenia or osteosclerosis, the chondrodysplasias are characterized by genetic abnormalities affecting cartilage, leading to short stature caused by defective linear growth, whereas the dysostoses have been defined as malformations of single bones, alone or in combination. However, it is now apparent that mutations in a single gene can result in disorders that can be classified in each of these categories.

Chondrodysplasias are usually generalized and progressive abnormalities, leading to changes in the size and shape of the limbs, trunk, and/or skull, frequently resulting in disproportionate short stature.[2] Whereas they have been considered to be generalized disorders of endochondral and/or membranous ossification, their pathogenesis is now beginning to be understood, but the extent of their clinical and molecular heterogeneity is still being elucidated. In the past, most disproportionate dwarfs were referred to as having either achondroplasia (MIM 100800; i.e., those with short limbs) or Morquio disease (MIM 253010; i.e., those with short trunks). It is now apparent that there are >370 distinct skeletal dysplasias that must be differentiated one from another for specific genetic counseling, prognosis, and treatment. The classification of the skeletal dysplasias was first limited to clinical and radiographic criteria. Indeed, many cases observed today still can only be classified on these grounds, and numerous cases still defy classification. A need to develop a uniform and consistent nomenclature and classification system for these conditions led to the "International Nomenclatures of Constitutional Diseases of Bone." These were initially formulated in 1972 in Paris and have since been regularly revised (International Working Group on Constitutional Diseases of Bone). The initial categorizations were purely descriptive and consisted of a mixture of the key clinical, radiographic, and pathologic features of each condition. The concept of "families" of disorders evolved, where conditions with similar genetic backgrounds are grouped together.[3] The latest classification is a hybrid that incorporates clinical (i.e., mesomelic dysplasia group), radiographic (i.e., metaphyseal dysplasia group), and molecular descriptors (i.e., type II collagenopathies group), as well as using various Greek terms (e.g., atelosteogenesis-omodysplasia group) to classify conditions (see online Appendix Table 1).

In the latest 2006 revision of the International Nosology and Classification of Genetic Skeletal Disorders, 372 different conditions have been listed in 37 groups defined by molecular, biochemical, and/or radiographic criteria.[4] It is now apparent that no single classification of these disorders will be adequate, necessitating the development of a multidimensional electronic classification tool, incorporating clinical, radiographic, morphologic, biochemical, molecular, and pathway data.

The authors state that they have no conflicts of interest.

MULTIDISCIPLINARY APPROACH

The evaluation of patients with chondrodysplasias mandates a multidisciplinary approach involving clinical geneticists, radiologists, molecular biologists, and biochemical geneticists for diagnosis and a host of surgical specialists for management of their many complications.[5] A careful clinical examination and detailed radiographic evaluation of the skeleton are the first steps toward an accurate diagnosis. A detailed history (length at birth, growth curves, etc.) and pedigree analysis, followed by anthropometric measurements with special emphasis on body proportions, are important. Serial radiographic evaluations are often necessary because some skeletal abnormalities may become evident at a later age.[3,6] (Fig. 1). In an adult patient, availability of a prepubertal skeletal survey, allowing the evaluation of the epiphyses and metaphyses before epiphyseal closure, can be essential for diagnostic purposes. Comparison with normal radiographs at the particular age can be a key for diagnosis. Ossified epiphyses, which are small and/or irregular for age, suggest an epiphyseal dysplasia. Widened, flared, and/or irregular metaphyses suggest a metaphyseal dysplasia, whereas a diaphyseal abnormality (widening, cortical thickening, or marrow space expansion) suggests an osteodysplasia (Fig. 1). A combination of the aforementioned findings with platyspondyly or vertebral irregularities is helpful in recognizing chondrodysplasias coined as spondylo-epiphyseal dysplasia (SED), spondylo-epi-meta-physeal dysplasia (SEMD), etc.[6] Recognition of other well-known, often pathognomonic skeletal changes in the skeletal survey is helpful in narrowing the list of entities considered in the differential diagnosis.[1,6]

Morphologic examination of cartilage and bone should also be a part of the diagnostic work-up when possible.[2] Patients undergoing surgery should be considered for bone and cartilage specimen collection. Details on the collection and analysis of such specimens can be found at www.csmc.edu/skeletaldysplasia.

FUTURE REFLECTIONS OF CURRENT RESEARCH

During the past decade, enormous progress has been made in our knowledge concerning the biochemistry and molecular genetic basis of these disorders.[7] In the 1980s, elucidation of defects in the type I collagen genes in osteogenesis imperfecta paved the way to further assessment of matrix protein defects in the skeletal dysplasias.[8] A large number of matrix protein defects have now been described in the chondrodysplasias, including types I, II, IX, X, and XI collagen, matrillin 3, COMP, and perlecan. In addition to these defects in extracellular (matrix) structural proteins, the chondrodysplasias can also be pathogenetically grouped into disorders with defects in metabolic pathways (including enzymes, ion channels, and transporters); defects in folding, processing, transport, and degradation of macromolecules; defects in hormones, growth factors, receptors, and signal transduction; defects in nuclear

Key words: chondrodysplasia, skeletal dysplasia, bone dysplasia, disproportionate short stature, dwarfism

Electronic Databases: 1. International Skeletal Dysplasia Registry—www.csmc.edu/skeletaldysplasia. 2. International Skeletal Dysplasia Society—www.isds.ch/. 3. Genetests—www.genetests.com.

© 2008 American Society for Bone and Mineral Research

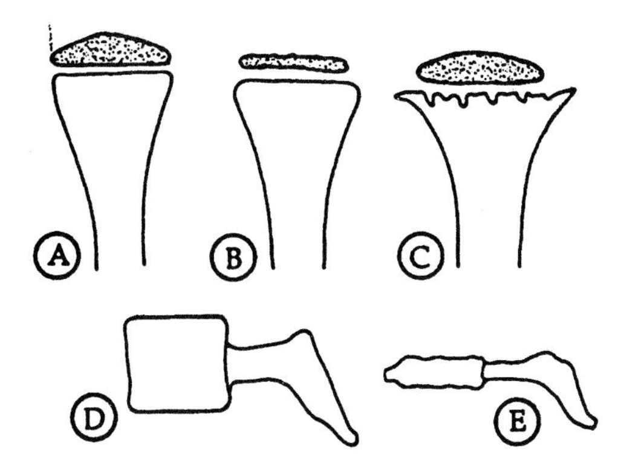

Involvement	Disease Category
A+D	Normal
B+D	Epiphyseal dysplasia
C+D	Metaphyseal dysplasia
B+E	Spondyloepiphyseal dysplasia
C+E	Spondylometaphyseal dysplasia
B+C+E	Spondyloepimetaphyseal dysplasia

FIG. 1. Chondrodysplasias. Classification based on radiographic involvement of long bones and vertebrae.[2] [Reprinted from Unger S, Lachman RS, Rimoin DL 2007 Chondrodysplasias. In: Rimoin DL, Connor JM, Pyeritz RE, Korf B (eds.) Emery and Rimoin's Principles and Practice of Medical Genetics, 5th ed., vol. 3. Elsevier, Philadelphia, PA, USA, pp. 3709–3753.]

proteins (transcription factors, homeobox genes); defects in RNA processing and metabolism; and defects in cytoskeletal proteins. There are also a number of disorders that have been mapped but the gene has not yet been identified; and many more disorders in which the locus has not yet been identified.[9]

Identification of the molecular basis of many skeletal dysplasias has shown that mutations in the same gene can produce quite distinct phenotypes, such as achondrogenesis II (MIM 200610) and Stickler syndrome (MIM 108300) with mutations in *COL2A1*[10] and Larsen syndrome (MIM 150250), atelosteogenesis I/III (MIM 108720/108721), and spondylo-carpal-tarsal syndrome (MIM 272460) with mutations in filamin B.[11] In contrast, similar phenotypes can be produced by mutations in different genes. For example, multiple epiphyseal dysplasia (MED) can result from mutations in six different genes, with the gene involved in one half of the MED cases still waiting to be discovered.[12] In most cases, it is difficult to correlate the precise mutation in a disease gene with the phenotype, making it often impossible to predict the specific phenotype from the molecular data alone (e.g., *COMP* mutations in pseudoachondroplasia or MED), or the converse, to predict the molecular defect from the clinical features seen in an individual (e.g., MED).

Research in this group of relatively rare monogenic disorders may well provide genetic models for more common multifactorial complex diseases of the bones and joints, such as osteoarthritis, osteoporosis, scoliosis, and disc herniation.

REFERENCES

1. Spranger J, Brill P, Poznanski A 2002 Bone Dysplasias. An Atlas of Genetic Disorders of Skeletal Development, 2nd ed. Oxford University Press, New York, NY, USA.
2. Unger S, Lachman RS, Rimoin DL 2007 Chondrodysplasias. In: Rimoin DL, Connor JM, Pyeritz RE, Korf B (eds.) Emery and Rimoin's Principles and Practice of Medical Genetics, 5th ed., vol. 3. Elsevier, Philadelphia, PA, USA, pp. 3709–3753.
3. Spranger J 1989 Radiologic nosology of bone dysplasias. Am J Med Genet **34:**96–104.
4. Superti-Furga A, Unger S 2007 Nosology and classification of genetic skeletal disorders: 2006 revision. Am J Med Genet **143:**1–18.
5. Mortier GR 2001 The diagnosis of skeletal dysplasias: A multidisciplinary approach. Eur J Radiol **40:**161–167.
6. Lachman RS 2007 Taybi and Lachman's Radiology of Syndromes, Metabolic Disorders and Skeletal Dysplasias, 5th ed. Elsevier, Philadelphia, PA, USA.
7. Ikegawa S 2006 Genetic analysis of skeletal dysplasia: Recent advances and perspectives in the post-genome-sequence era. J Hum Genet **51:**581–586.
8. Williams CJ, Prockop DJ 1983 Synthesis and processing of a type I procollagen containing shortened pro-alpha 1(I) chains by fibroblasts from a patient with osteogenesis imperfecta. J Biol Chem **258:**5915–5921.
9. Superti-Furga A, Bonafe L, Rimoin DL 2001 Molecular-pathogenetic classification of genetic disorders of the skeleton. Am J Med Genet **106:**282–293.
10. Rimoin D, Cohn D, Eyre D 2004 Clinical–molecular correlations in the skeletal dysplasias. Pediatr Radiol **24:**425–426.
11. Krakow D, Robertson SP, King LM, Morgan T, Sebald ET, Bertolotto C, Wachsmann-Hogiu S, Acuna D, Shapiro SS, Takafuta T, Aftimos S, Kim CA, Firth H, Steiner CE, Cormier-Daire V, Superti-Furga A, Bonafe L, Graham JM Jr, Grix A, Bacino CA, Allanson J, Bialer MG, Lachman RS, Rimoin DL, Cohn DH 2004 Mutations in the gene encoding filamin B disrupt vertebral segmentation, joint formation and skeletogenesis. Nat Genet **36:**405–410.
12. Jakkula E, Mäkitie O, Czarny-Ratajczak M, Jackson GC, Damignani R, Susic M, Briggs MD, Cole WG, Ala-Kokko L 2005 Mutations in the known genes are not the major cause of MED; distinctive phenotypic entities among patients with no identified mutations. Eur J Hum Genet **13:**292–301.

© 2008 American Society for Bone and Mineral Research

Chapter 91. Ischemic and Infiltrative Disorders

Richard W. Keen

Institute of Orthopaedics and Musculoskeletal Sciences, Royal National Orthopaedic Hospital, Middx, United Kingdom

INTRODUCTION

Regional interruption of blood flow to the skeleton can cause ischemic (aseptic or avascular) necrosis. Ischemia, if sufficiently severe and prolonged, will kill osteoblasts and chondrocytes. Clinical problems arise if subsequent resorption of necrotic tissue during skeletal repair compromises bone strength enough to cause fracture with deformity of bone and secondary damage to cartilage.

Epidemiology

There is little accurate data on the incidence of osteonecrosis, although it is estimated there are ~15,000 new cases per year in the United States.[1] The disease seems to occur more frequently in males than in females, with the overall male to female ratio being 8:1. The age of onset is variable, although the majority of cases are <50 yr of age. The average age of female cases is on average 10 yr older than male cases.

Pathogenesis

Osteonecrosis is often seen in association with a number of different conditions (Table 1). Trauma with fracture of the femoral neck interrupts the major part of the blood supply to the head and may lead to osteonecrosis. Glucocorticoids and alcoholism are two of the main iatrogenic factors known to predispose to osteonecrosis.

In the past 2 yr, an increasing number of reports have appeared in the medical literature that suggest that the use of bisphosphonates, especially intravenous bisphosphonates, is associated with osteonecrosis of the jaw (ONJ).[2] There is no universally accepted definition of ONJ, and it may represent a distinct form of osteonecrosis, different to the other forms. Clinically it typically appears as an area of exposed alveolar bone that can occur in the mandible or the maxilla. The majority of these reports are in patients with multiple myeloma, breast cancer, or other malignancies. In connection with these malignancies, the patients were receiving or did receive intravenous nitrogen-containing bisphosphonates in much higher doses than those used for osteoporosis and Paget's disease. However, a much smaller number of cases of ONJ have been associated with oral bisphosphonates used at lower doses to treat osteoporosis and Paget's disease. The current evidence and recommendations for patients who are on bisphosphonates and are about to have dental work undertaken have been reviewed.[2]

A mechanical interruption to the bone's blood supply is common to most of the conditions associated with osteonecrosis. For many types of nontraumatic ischemic necrosis, the predisposed sites seem to reflect the physiological conversion of red marrow to fatty marrow with aging.[3] This process occurs from distal to proximal in the appendicular skeleton. As the transition occurs, marrow blood flow decreases. Accordingly, disorders that increase the size and/or number of adipocytes within critical areas of medullary space (e.g., alcohol abuse, Cushing's syndrome) may ultimately compress sinusoids, thereby leading to infarction of bone. Other factors potentially involved in the pathogenesis of osteonecrosis include fat embolization, hemorrhage, and abnormalities in the quality of susceptible bone tissue. Infection and dental trauma seem to be important factors in the etiology of ONJ.

A genetic basis to idiopathic osteonecrosis of the femoral head is suggested by the occurrence of disease in twins and a clustering of cases in families.[4,5] Increased incidence of osteonecrosis in specific animal models also provides further evidence of the existence of susceptibility genes.[6] In sporadic cases of osteonecrosis of the femoral neck, a number of genetic association studies have been conducted, linking specific genes to the pathogenesis of disease. The majority of the studies have, to date, focused on gene polymorphisms affecting the coagulation and fibrinolytic system.

The factor V Leiden mutation (G1691A, Arg506Gln) is a common risk factor for thrombophilia. Three of four studies investigating the role of the mutation in osteonecrosis have reported a positive correlation of factor V Leiden with primary osteonecrosis.[7–10] Plasminogen-activating inhibitor-1 (PAI-1) was also studied in patients with osteonecrosis. Homozygosity for the *4G* allele (4G/4G) has been reported to significantly increase the plasma PAI-1 level, and in two studies by the same group of investigators, the 4G/4G genotype was found to be a risk factor of osteonecrosis.[10,11] Some studies have also investigated the role of the 5,10-methylenetetrahydrofolate reductase (*MTHFR*) gene polymorphism. The *MTHFR* C677T variant was over-represented in some groups of primary studies of osteonecrosis, but not in all.[10–12] In another study, the first that did not focus on the coagulation system, the role of endothelial NO synthase (eNOS) was investigated.[13] NO synthesized by eNOS has vasodilatory effects on vascular tone, inhibits platelet aggregation, and modulates smooth muscle proliferation. Allele *4a* of a VNTR polymorphism in intron 4 of the *eNOS* gene was found to be a risk factor for idiopathic osteonecrosis at the hip.

Genetic mutations were recently identified in three families with osteonecrosis and dominant inheritance. Mutations in the type II collagen (*COL2A2*) gene (mapped on chromosome 12q13) proved to be the genetic cause of the disease. Type II collagen is the major structural protein in the extracellular matrix of cartilage.[14]

Clinical Features

The clinical presentation will be dependent on many factors, including the age of the patient, anatomical site of involvement, and the extent and severity of this involvement. The femoral head is the most common location for the development of osteonecrosis, although it may also occur at other sites including distal femur, humeral head, wrist, and foot. Patients may develop pain that can persist for weeks to months before radiographs show any change, although patients can be asymptomatic. Avascular necrosis at the hip classically will present with pain in the groin, although this can be referred to the buttock, thigh, or knee. The pain is exacerbated by weight bearing but can also be present at rest. Gait may be affected, and patients can present with a limp. Once the femoral head has begun to collapse, range of hip movement will be reduced, and leg shortening may develop.

In osteonecrosis of the hip, involvement of the contralateral hip is present in 30–70% of cases at the time of first examina-

The author states that he has no conflicts of interest.

Key words: osteonecrosis, avascular necrosis, ischemia, mastocytosis, histiocytosis

© 2008 American Society for Bone and Mineral Research

TABLE 1. CAUSATIVE FACTORS ASSOCIATED WITH OSTEONECROSIS

Traumatic
Fracture of the femoral neck
Dislocation or fracture-dislocation of the hip
Minor fracture
Nontraumatic
Alcohol
Arteriosclerosis and other occlusive vascular disorders
Bisphosphonates
Carbon tetrachloride poisoning
Connective tissue diseases
Cushing's disease (OMIM 219090; OMIM 219080)
Diabetes mellitus (OMIM 222100, OMIM 125853)
Disordered lipid metabolism
Dysplasia
Fatty liver
Gaucher's disease (OMIM 231000)
Glucocorticoid treatment
Human immunodeficiency virus
Dysbaric conditions
Hyperuricemia and gout
Legg Calve Perthe disease (OMIM 150600)
Osteomalacia
Pancreatitis
Pregnancy
Radiotherapy
Sickle cell anemia (OMIM 603903)
Systemic lupus erythematosus (OMIM 152700)
Solid organ transplantation
Thrombophlebitis
Tumors

tion. Within 3 yr of diagnosis, >50% of cases will have progressed in the contralateral hip to such a stage where surgical intervention is required.[15,16]

Radiological Features

In the earliest stages of the disease, plain radiographs will be normal. MRI is useful to detect early pathological changes. The most characteristic image is a margin of low signal on T1- and T2-weighted images and is observed in 60–80% of cases. Radionucleotide isotope bone scans and CT can also be used in cases where MRI is either contraindicated or has been inconclusive. Osteonecrosis can be staged according to the sequence of the radiological changes. These are detailed in Table 2.

Laboratory Findings

In idiopathic osteonecrosis, laboratory investigations will generally be normal. Investigations may show potential contributory factors such as connective tissue disease, diabetes, hyperlipidemia, and gout.

Histopathological examination of tissue from affected bone is consistent with the pathogenesis that is suggested from the radiographic examinations. It shows that these various processes of skeletal death and repair are focal and may be occurring simultaneously.[17]

Differential Diagnosis

In stages 3 and 4 of the disease, the radiological features of the disease are specific. In the later stages 5 and 6, a differential diagnosis is not necessary, because by this stage, the bone and joint have been irreversibly damaged and the only treatment option would be joint replacement. In the earlier stages of the disease (1 and 2), other diseases of bone, cartilage, and synovial tissue should be considered in the differential diagnosis.

Treatment

Medical treatment includes non–weight bearing for osteonecrosis affecting load-bearing bones. This may be for between 4 and 8 wk. Vasoactive drugs such as prostacyclin may play a role in early stages of osteonecrosis.[18] Bisphosphonates have also been shown to be effective in the treatment of osteonecrosis. Data have been observed for alendronate[19] and more recently for zoledronate.[20]

Surgical treatment for osteonecrosis involves core decompression. This reduces the intramedullary pressure within the ischemic bone and has been postulated to improve circulation. The outcome from core decompression at the femoral head, with regards to resolution of radiographic changes and improvement in symptoms, varies from 34% to 95% in the early stages of the disease.[21] These results seem better than continuing with conservative measures such as non–weight bearing.

INFILTRATIVE DISORDERS

Systemic Mastocytosis

In systemic mastocytosis (OMIM 154800), there is wide-

TABLE 2. STAGING OF OSTEONECROSIS

Stage	*Findings*	*Techniques*
0	All techniques normal or nondiagnostic Necrosis on biopsy	Biopsy and histology
1	Radiographs and CT normal Positive result from at least one of the additional investigations listed opposite	Radionucleotide scan MRI Biopsy and histology
2	Radiographic abnormalities without collapse (sclerosis, cysts, osteopenia)	Radionucleotide scan MRI Biopsy and histology
3	Crescent sign	Radiographs CT
4	Flattening or evident collapse	Radiographs CT
5	As for stage 4, with narrowing of joint space	Radiographs
6	As for stage 5, with destruction of joint	Radiographs

© 2008 American Society for Bone and Mineral Research

spread infiltration of tissues with mast cells. These cells can be widely distributed in nearly every organ and originate from bone marrow progenitor cells. An activating mutation in the *C-KIT* gene coding for c-kit (the receptor for stem cell factor) that controls mast cell development has been identified.[22] The relationship, however, between these mutations and the clinical phenotype is not fully clarified.

The clinical features of mastocytosis are produced by liberation of mast cell products. Urticaria pigmentosa has been described in between 14% and 100% of patients with systemic mastocytosis. Radiographic appearances can be variable, and the diagnosis is often confirmed on bone biopsy and histological examination. The diagnosis can also be made by the measurement of urinary excretion of mast cell mediators such as *N*-methyl histamine.

Treatment of systemic mastocytosis must be "tailored" in individual patients.[23,24] Severe bone pain from advanced bone disease has been reported to respond to radiotherapy.[25] Bisphosphonates have controlled pain and improved BMD in early trials.[26]

Histiocytosis X

Histiocytosis-X is the term to unify what had been regarded as three distinct entities: Letterer-Siwe disease (OMIM 246400), Hand-Schüller-Christian disease (OMIM 267700), and eosinophilic granuloma.[27,28] An immature, clonal Langerhans cell is considered the pathognomonic and linking feature, and the condition is now called Langerhans cell histiocytosis (OMIM 604856).

Many tissues and organs can be involved, including brain, lung, oropharynx, gastrointestinal tract, skin, and bone marrow. Prognosis is age related; infants and the elderly have poorer outcomes. The signs and symptoms of the three principal clinical forms also differ.

Letterer-Siwe disease presents between several weeks and 2 yr of age with hepatosplenomegaly, lymphadenopathy, anemia, hemorrhagic tendency, fever, failure to grow, and skeletal lesions. Hand-Schüller-Christian disease is a chronic condition that begins in early childhood, although symptoms may not manifest until the third decade. The classic triad of findings consists of exophthalmos, diabetes insipidus, and bony lesions, although this is only seen in 10% of cases. Eosinophilic granuloma occurs most frequently in children between 3 and 10 yr of age, and it is rare after the age of 15 yr. A solitary and painful lesion in a flat bone is the most common finding.

Histiocytosis X tends to be benign and self-limiting when there is no systemic involvement. Treatment for severe disease includes chemotherapy, radiation therapy, and immunotherapy.[29,30] Allogeneic bone marrow transplantation has also been used in a patient with severe disease and poor prognosis.[31]

REFERENCES

1. Steinberg ME, Steinberg DR 1991 Avascular necrosis of the femoral head. In: Steinberg ME (ed.) The Hip and Its Disorders. WB Saunders, Philadelphia, PA, USA, pp. 623–647.
2. Shane E, Goldring S, Christakos S, Drezner M, Eisman J, Silverman S, Pendrys D 2006 Osteonecrosis of the jaw: More research needed. J Bone Miner Res **21:**1503–1505.
3. Edeiken J, Dalinka M, Karasick D 1990 Edeiken's Roentgen Diagnosis of Diseases of Bone, 4th ed. Williams and Wilkins, Baltimore, MD, USA.
4. Glueck CJ, Glueck HI, Welch M, Freiberg TT, Hamer T, Stroop D 1994 Familial idiopathic osteonecrosis mediated by familial hypofibrinolysis with high levels of plasminogen activator inhibitor. Thromb Haemost **71:**195–198.
5. Nobillot R, Le Parc JM, Benoit J, Paolaggi JB 1994 Idiopathic osteonecrosis of the hip in twins. Ann Rheum Dis **53:**702.
6. Boss JH, Misselevich I 2003 Osteonecrosis of the femoral head of laboratory animals: The lessons learned from a comparative study of osteonecrosis in man and experimental animals. Vet Pathol **40:**345–354.
7. Zalavras CG, Vartholomatos G, Dokou E, Malizos KN 2004 Genetic background of osteonecrosis: Associated with thrombophilic mutations? Clin Orthop Relat Res **422:**251–255.
8. Bjorkman A, Svensson PJ, Hillarp A, Burtscher IM, Runow A, Benoni G 2004 Factor V Leiden and prothrombin gene mutation: Risk factors for osteonecrosis of the femoral head in adults. Clin Orthop Relat Res **425:**168–172.
9. Bjorkman A, Burtscher IM, Svensson PJ, Hillarp A, Besjakov J, Benoni G 2005 Factor V Leiden and the prothrombin 20210A gene mutation and osteonecrosis of the knee. Arch Orthop Trauma Surg **125:**51–55.
10. Glueck CJ, Fontaine RN, Gruppo R, Stroop D, Sieve-Smith L, Tracy T, Wang P 1999 The plasminogen activator inhibitor-1 gene, hypofibrinolysis, and osteonecrosis. Clin Orthop Relat Res **366:**133–146.
11. Glueck CJ, Freiberg RA, Fontaine RN, Tracy T, Wang P 2001 Hypofibrinolysis, thrombophilia, osteonecrosis. Clin Orthop Relat Res **386:**19–33.
12. Zalavras CG, Malizos KN, Dokou E, Vartholomatos G 2002 The 677C→T mutation of the methylene-tetrahydrofolate reductase gene in the pathogenesis of osteonecrosis of the femoral head. Haematologica **87:**111–112.
13. Koo KH, Lee JS, Lee YJ, Kim KJ, Yoo JJ, Kim HJ 2006 Endothelial nitric oxide synthase gene polymorphisms in patients with nontraumatic femoral head osteonecrosis. J Orthop Res **24:**1722–1728.
14. Liu YF, Chen WM, Lin YF, Yang RC, Lin MW, Li LH, Chang YH, Jou YS, Lin PY, Su JS, Huang SF, Hsiao KJ, Fann CS, Hwang HW, Chen YT, Tsai SF 2005 Type II collagen gene variants and inherited osteonecrosis of the femoral head. N Engl J Med **352:**2294–2301.
15. Jacobs B 1978 Epidemiology of traumatic and nontraumatic osteonecrosis. Clin Orthop Rel Res **130:**51–67.
16. Bradway JK, Morrey BF 1993 The natural history of the silent hip in bilateral atraumatic necrosis of the femoral head. J Arthroplasty **8:**383–387.
17. Plenk H Jr, Hofmann S, Eschberger J, Gstettner M, Kramer J, Schneider W, Engel A 1997 Histomorphology and bone morphometry of the bone marrow edema syndrome of the hip. Clin Orthop **334:**73–84.
18. Disch AC, Matziolis G, Perka C 2005 The management of necrosis-associated and idiopathic bone-marrow oedema of the proximal femur by intravenous iloprost. J Bone Joint Surg Br **87:**560–564.
19. Lai KA, Shen WJ, Yang CY, Shao CJ, Hsu JT, Lin RM 2005 The use of alendronate to prevent early collapse of the femoral head in patients with nontraumatic osteonecrosis. A randomized clinical study. J Bone Joint Surg Am **87:**2155–2159.
20. Ramachandran M, Ward K, Brown RR, Munns CF, Cowell CT, Little DG 2007 Intravenous bisphosphonate therapy for traumatic osteonecrosis of the femoral head in adolescents. J Bone Joint Surg Am **89:**1727–1734.
21. Stulberg BN, Bauer TW, Belhobek GH 1990 Making core decompression work. Clin Orthop **261:**186–195.
22. Metcalfe DD, Atkin C 2001 Matsocytosis: Molecular mechanisms and clinical disease heterogeneity. Leuk Res **25:**577–582.
23. Valent P, Ghannadan M, Akin C, Krauth MT, Selzer E, Mayerhofer M, Sperr WR, Arock M, Samorapoompichit P, Horny HP, Metcalfe DD 2004 On the way to targeted therapy of mast cell neoplasms: Identification of molecular targets in neoplastic mast cells and evaluation of arising treatment concepts. Eur J Clin Invest **34**(Suppl 2)**:**41–52.
24. Krokowski M, Sotlar K, Krauth MT, Fodinger M, Valent P, Horny HP 2005 Delineation of patterns of bone marrow mast cell infiltration in systemic mastocytosis: Value of CD25, correlation with subvariants of the disease, and separation from mast cell hyperplasia. Am J Clin Pathol **124:**560–568.
25. Johnstone PA, Mican JM, Metcalfe DD, DeLaney TF 1994 Radiotherapy of refractory bone pain due to systemic mast cell disease. Am J Clin Oncol **17:**328–330.
26. Brumsen C, Hamady NAT, Papapoulos SE 2002 Osteoporosis and bone marrow mastocytosis: Dissociation of skeletal responses and

© 2008 American Society for Bone and Mineral Research

mast cell activity during long-term bisphosphonate therapy. J Bone Miner Res **17**:567–569.
27. Lam KY 1997 Langerhans cell histocytosis (histiocytosis X). Postgrad Med J **73**:391–394.
28. Coppes-Zantinga A, Egeler RM 2002 The Langerhans cell histiocytosis X files revealed. Br J Haematol **116**:3–9.
29. Bollini G, Jouve JL, Gentet JC, Jacquemier M, Bouyala JM 1991 Bone lesions in histiocytosis X. J Pediatr Orthop **11**:469–477.
30. Greenberger JS, Crocker AC, Vawter G, Jaffe N, Cassady JR 1981 Results of treatment of 127 patients with systemic histiocytosis (Letterer-Siwe syndrome, Schüller-Christian syndrome and multifocal eosinophilic granuloma). Medicine (Baltimore) **60**:311–388.
31. Ringdën O, Aohström L, Lönnqvist B, Boaryd I, Svedmyr E, Gahrton G 1987 Allogeneic bone marrow transplantation in a patient with chemotherapy-resistant progressive histiocytosis X. N Engl J Med **316**:733–735.

Chapter 92. Tumoral Calcinosis—Dermatomyositis

Nicholas Shaw

Department of Endocrinology, Birmingham Children's Hospital, Birmingham, United Kingdom

TUMORAL CALCINOSIS

Tumoral calcinosis (OMIM #211900) is a rare autosomal recessive metabolic disorder characterized by the progressive deposition of calcium phosphate crystals in periarticular spaces and soft tissues. The biochemical hallmark of this condition is hyperphosphatemia caused by increased renal tubular reabsorption of phosphate. However, tumoral calcinosis is also described in the absence of elevated phosphate. Although the first description of this condition was in 1898, the term tumoral calcinosis was not used until 1943.[1]

Clinical Features

Mineral deposition manifests as soft tissue masses around major joints. In one report, the order of frequency for the site of first lesions are hips, elbows, shoulders, and scapulae.[2] Onset of the lesions can vary from 22 mo of age to adulthood, with the majority manifesting by 20 yr of age. Many cases described in the Anglo-American literature have been of black ancestry, with a large number of cases reported from Africa. In many reports, familial cases are described with what seems to be either an autosomal recessive or dominant pattern of inheritance. The soft tissue masses are usually painless and can progressively grow in size to that of an orange or grapefruit. Although they occur around joints, they do not usually impair range of movement because they are extracapsular. They can compress adjacent neural structures such as the sciatic nerve and may also cause ulceration of overlying skin causing a sinus tract that leaks a chalky fluid and may become infected. Some affected subjects have been reported to have features of pseudoxanthoma elasticum (i.e., skin changes, vascular calcification, and angioid streaks of the retina). A specific dental abnormality may be seen with hypoplastic teeth containing short bulbous roots and almost complete obliteration of pulp cavities with pulp stones.

Radiographic Findings

On plain X-ray, early and small lesions are located in regions known to be occupied by bursae and are often distributed in a paraarticular fashion along the extensor surfaces of large joints.[3] These soft tissue lesions comprise multiple globular amorphous calcific components separated by radiolucent fibrous septae. Occasionally fluid levels can be seen indicating a cystic component. An inflammatory process "diaphysitis" may also be seen on plain X-ray, CT, or MRI, in some cases, usually occurring in the middle third of long bones. Vascular calcification has also been reported on plain X-ray or CT. An isotope bone scan is the most reliable and simplest method for detection, localization, and assessment of extension of the calcific masses. Periarticular masses that are radiologically indistinguishable from those described in tumoral calcinosis may be seen in patients with chronic renal failure.

Biochemical Findings

Many subjects with tumoral calcinosis have been shown to have elevated levels of plasma phosphate and serum 1,25-dihydroxyvitamin D_3.[4] The tubular maximum of tubular phosphate reabsorption in relation to glomerular filtration rate ($TmPO_4$/GFR) is elevated, but renal function is otherwise normal. However, the condition is also described with normal levels of plasma phosphate. Plasma calcium, alkaline phosphatase, and serum PTH are usually normal. Metabolic balance studies have shown positive calcium and phosphorous balances caused by increased gastrointestinal absorption and reduced renal excretion.

Histopathology

It is suggested that the early lesions are triggered by bleeding followed by aggregation of foamy histiocytes, which become transformed into cystic cavities lined by osteoclast-like giant cells and histiocytes. Movement and friction caused by the periarticular location of the lesions seem to be key to the transformation. In a review of 111 cases from Zaire collected over a 30-yr period, histology identified exuberant cellular proliferative changes adjacent to the classical cystic form.[5] These consisted either of ill-defined reactive-like perivascular solid cell nests admixed with mononuclear and iron-loaded macrophages or well-organized fibrohistiocytic nodules of variable size embedded in a dense collagenous stroma. Mature lesions are filled with calcareous material in a viscous milky fluid.

Etiology and Pathogenesis

The first identification of a genetic basis for tumoral calcinosis was in 2004 when, in a study of large Druze and black kindreds, the gene was mapped to 2q24-q31 with identification of biallelic mutations in the *GALNT3* gene in affected individuals.[6] The *GALNT3* gene encodes a glycosyltransferase enzyme responsible for initiating mucin-type O-glycosylation. Haplotype analysis in families with normophosphatemic familial tumoral calcinosis has excluded linkage to 2q24-q31. In a family previously felt to have autosomal dominant inheritance, individuals expressing the full phenotype were shown to have

The author states that he has no conflicts of interest.

© 2008 American Society for Bone and Mineral Research

biallelic mutations in *GALNT3*, whereas those who were heterozygous for the mutations showed incomplete expression of the condition, with increased plasma phosphate or 1,25 dihydroxyvitamin D_3 levels but no calcified deposits.[7] Thus, autosomal recessive inheritance of the condition was confirmed. Subsequently mutations in the gene for fibroblast growth factor (FGF)23 were identified in affected individuals who were negative for mutations in *GALNT3*.[8,9] In addition, elevated plasma levels of C-terminal FGF23 have been shown in these individuals with low plasma levels of intact FGF23, suggesting failure of secretion of the intact protein from cells.

GALNT3 is now known to produce an enzyme that selectively O-glycosylates a furin-like convertase recognition sequence in FGF23, thus preventing proteolytic processing of FGF23 and allowing secretion of intact FGF23. Thus, mutations in *GALNT3* result in defective secretion of intact FGF23, resulting in hyperphosphatemia and increased synthesis of 1,25-dihydroxyvitamin D_3. Recently, a mutation in the *KLOTHO* gene was reported in a 13-yr-old girl with tumoral calcinosis who, in addition to elevated plasma phosphate and 1,25-dihydroxyvitamin D_3 levels, also had hypercalcemia and high serum PTH.[10] She had evidence of elevated levels of both C-terminal and intact FGF23 but reduced FGF23 bioactivity. KLOTHO is a co-factor required by FGF23 to enable it to bind and signal through its FGF receptors.[11] Thus, mutations in three genes, *GALNT3, FGF23*, and *KLOTHO*, have been shown to cause the clinical and biochemical features of tumoral calcinosis.

Treatment

Surgical removal of the calcified masses may be required if they are painful, affect function, or for cosmetic reasons. Several different medical approaches to treatment have been reported, although they are usually individual case reports. Aluminium hydroxide in combination with dietary phosphate and calcium restriction has been reported to be successful.[12] Calcitonin has been used to induce phosphaturia,[13] and the combination of acetazolamide with aluminium hydroxide used for a 14-yr period in one patient was reported to be effective in improving the lesions.[14] Bisphosphonate therapy with alendronate was reported to alleviate symptoms within 12 wk in one patient.[15]

The recent identification of the role of FGF23 in phosphate metabolism and tumoral calcinosis may hopefully lead to new medical approaches to treatment.

DERMATOMYOSITIS IN CHILDREN

Juvenile dermatomyositis is an idiopathic inflammatory disorder of the skin and muscle. It is characterized by progressive weakness predominantly of the proximal muscles and a rash that particularly affects the face and the extremities. It differs from adult-onset dermatomyositis in that it is frequently associated with small vessel vasculitis in the skin, muscle, and gastrointestinal tract, and there is no association with malignancy. Dystrophic soft tissue calcification or "calcinosis" occurs in damaged or devitalized tissues in the presence of normal calcium/phosphorous metabolism (Fig. 1).

Clinical Presentation

It is a rare disorder, with an estimated incidence of 1.9–2.5 per million children <16 yr of age and is more common in girls than boys with a ratio of ~2:1. In a UK survey, the median age of onset was 6.8 yr, with two peak ages of onset in girls of 6 and 11 yr of age.[16] Eighty-eight percent of the reported cases were white. Calcinosis is not a feature at initial presentation, is usually noted 1–3 yr after the disease onset, and is reported to occur in 20–40% of affected individuals.[17] The duration of untreated dermatomyositis is associated with pathological calcifications, thus showing a clear link with chronic inflammation.[18] The dystrophic calcification can cause pain, skin ulceration, limited joint mobility, contractures, and predispose to abscess formation. The calcification, once present, typically remains stable, but rarely, some spontaneous resolution is reported. The clinical course of dermatomyositis in children is variable, with some having long-term relapsing or persistent disease, whereas others recover.

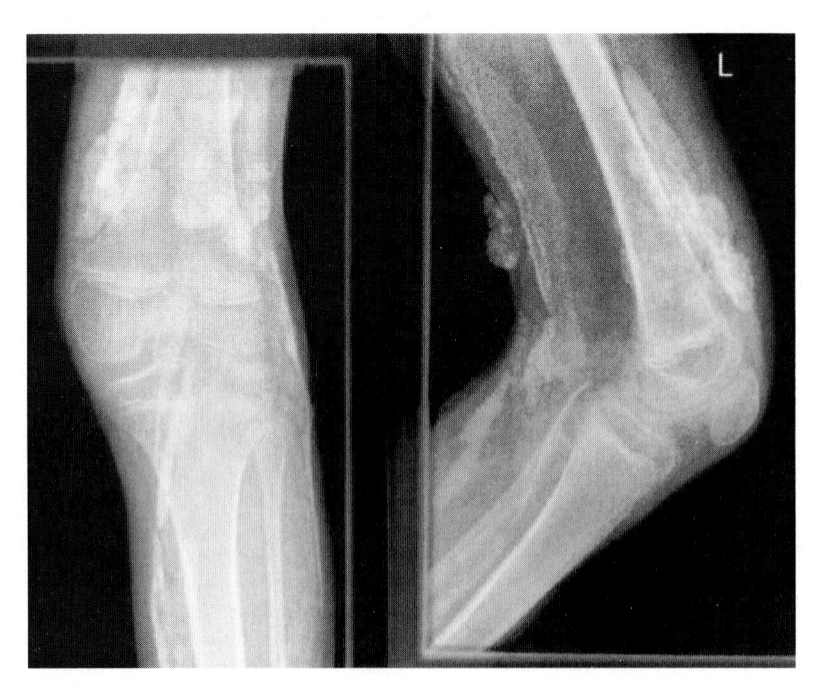

FIG. 1. Subcutaneous calcification in a child with dermatomyositis.

© 2008 American Society for Bone and Mineral Research

Biochemical and Histological Features

Levels of plasma calcium, phosphate, and alkaline phosphatase are usually normal. Urinary levels of γ-carboxyglutamic acid have been reported to be elevated, particularly if there is calcinosis. The mineral present in the calcified deposits has been shown to be a poorly crystallized hydroxyapatite.[17] They contain relatively more mineral than matrix compared with bone. Bone matrix proteins such as osteopontin, sialoprotein, and osteonectin are present within the calcifications, with more osteonectin than is found in human bone.

Radiographic Features

Four types of dystrophic calcification can occur: superficial masses within the skin; deep, discrete, subcutaneous nodular masses near joints that can impair movement (calcinosis circumscripta); deep, linear, sheet-like deposits within intramuscular fascial planes (calcinosis universalis); and lacy, reticular subcutaneous deposits that encase the trunk to form a generalized "exoskeleton."

"Milk of calcium" fluid collections are a rare complication of calcinosis.[19] Although established calcification can be readily seen on plain X-rays, MRI seems to be a sensitive method for detection and localization of muscle inflammation and edema. It is also an excellent modality for monitoring progression or remission of the disease.[20]

Treatment

High-dose corticosteroids soon after the onset of symptoms remains the mainstay of current treatment, reducing the potential risk of calcinosis by suppression of the inflammatory process. Additional agents that are used include methotrexate and infliximab. Several different therapies including bisphosphonates, diltiazem, and surgical extirpation have been used to treat calcinosis, with individual case reports suggesting benefit.[21,22] However, a review of the published literature over a 32-yr period concluded that no treatment has convincingly prevented or reduced calcinosis with a lack of systematic study and clinical therapeutic trials.[23]

REFERENCES

1. Inclan A, Leon P, Camejo MG 1943 Tumoral calcinosis. JAMA **121**:490–495.
2. Slavin RE, Wen J, Kumar D, Evans EB 1993 Familial tumoral calcinosis: A clinical, histopathologic, and ultrastructural study with an analysis of its calcifying process and pathogenesis. Am J Surg Pathol **17**:788–802.
3. Martinez S, Vogler JB, Harrelson JM, Lyles KW 1990 Imaging of tumoral calcinosis: New observations. Radiology **174**:215–222.
4. Lyles KW, Halsey DL, Friedman NE, Lobaugh B 1988 Correlations of serum concentrations of 1,25-dihydroxyvitamin D, phosphorous and parathyroid hormone in tumoral calcinosis. J Clin Endocrinol Metab **67**:88–92.
5. Pakasa NM, Kalengayi RM 1997 Tumoral calcinosis: A clinicopathological study of 111 cases with emphasis on the earliest changes. Histopathology **31**:18–24.
6. Topaz O, Shurman DL, Bergman R, Indelman M, Ratajczak P, Mizrachi M, Khamaysi Z, Behar D, Petronius D, Friedman V, Zelikovic I, Raimer S, Metzker A, Richard G, Sprecher E 2004 Mutations in GALNT3, encoding a protein involved in o-linked glycosylation, cause familial tumoral calcinosis. Nat Genet **36**:579–581.
7. Ichikawa S, Lyles KW, Econs MJ 2005 A novel GALNT3 mutation in a pseudoautosomal dominant form of tumoral calcinosis: Evidence that the disorder is autosomal recessive. J Clin Endocrinol Metab **90**:2420–2423.
8. Benet-Pages A, Orlik P, Strom TM, Lorenz-Depiereux B 2005 An FGF23 missense mutation causes familial tumoral calcinosis with hyperphosphataemia. Hum Mol Genet **14**:385–390.
9. Larsson T, Yu X, Davis SI, Draman MS, Mooney SD, Cullen MJ, White KE 2005 A novel recessive mutation in fibroblast growth factor-23 causes familial tumoral calcinosis. J Clin Endocrinol Metab **90**:2424–2427.
10. Ichikawa S, Imel EA, Kreiter ML, Yu X, Mackenzie DS, Sorenson AH, Goetz R, Mohammed M, White KE, Econs MJ 2007 A homozygous missence mutation in human KLOTHO causes severe tumoral calcinosis. J Clin Invest **117**:2684–2691.
11. Urakawa I, Yamazaki Y, Shimada T, Iijima K, Hasegawa H, Okawa K, Fujita T, Fukumoto S, Yamashita T 2006 Klotho converts canonical FGF receptor into a specific receptor for FGF23. Nature **444**:770–774.
12. Gregosiewicz A, Warda E 1989 Tumoral calcinosis: Successful medical treatment. J Bone Joint Surg Am **71**:1244–1249.
13. Salvi A, Cerudelli B, Cimino A, Zuccato F, Giustina G 1983 Phosphaturic action of calcitonin in pseudotumoral calcinosis. Horm Metab Res **15**:260.
14. Yamaguchi T, Sugimoto T, Imai Y, Fukase M, Fujita T, Chihara K 1995 Successful treatment of hyperphosphatemic tumoral calcinosis with long term acetazolimide. Bone **16**:247S–250S.
15. Jacob JJ, Mathew K, Thomas N 2007 Idiopathic sporadic tumoral calcinosis of the hip: Successful oral bisphosphonate therapy. Endocr Pract **13**:182–186.
16. Symmons DPM, Sills JA, Davis SM 1995 The incidence of juvenile dermatomyositis: Results from a nation-wide study. Br J Rheumatol **34**:732–736.
17. Pachman LM, Veis A, Stock S, Abbott K, Vicari F, Patel P, Giczewski D, Webb C, Spevak L, Boskey A 2006 Composition of calcifications in children with juvenile dermatomyositis. Arthritis Rheum **54**:3345–3350.
18. Pachman LM, Abbott K, Sinacore JM, Amoruso L, Dyer A, Lipton R, Ilowite N, Hom C, Cawkwell G, White A, Rivas-Chacon R, Kimura Y, Ray L, Ramsey-Goldman R 2006 Duration of illness is an important variable for untreated children with untreated dermatomyositis. J Pediatr **148**:247–253.
19. Samson C, Soulen RL, Gursel E 2000 Milk of calcium fluid collections in juvenile dermatomyositis: MR characteristics. Pediatr Radiol **30**:28–29.
20. Park JH, Vital TL, Ryder NM, Hernanz-Schulman M, Leon Partain C, Price RR, Olsen NJ 1994 Magnetic resonance imaging and P-31 magnetic spectroscopy provide unique quantitative data useful in the longitudinal management of patients with dermatomyositis. Arthritis Rheum **37**:736–746.
21. Mukamel M, Horev G, Mimouni M 2001 New insights into calcinosis of juvenile dermatomyositis: A study of composition and treatment. J Pediatr **138**:763–766.
22. Oliveri MB, Palermo R, Mautalen C, Hubscher O 1996 Regression of calcinosis during diltiazem treatment in juvenile dermatomyositis. J Rheumatol **23**:2152–2155.
23. Boulman N, Slobodin G, Rozenbaum M, Rosner I 2005 Calcinosis in rheumatic diseases. Semin Arthritis Rheum **34**:805–812.

© 2008 American Society for Bone and Mineral Research

Chapter 93. Vascular Calcification

Dwight A. Towler[1] and Linda L. Demer[2]

[1]*Department of Medicine, Division of Bone and Mineral Diseases, Washington University School of Medicine, St. Louis, Missouri;* [2]*Department of Medicine, Division of Cardiology and Vascular Medicine, David Geffen School of Medicine, UCLA, Los Angeles, California*

INTRODUCTION

The detailed molecular and cellular framework that controls skeletal calcium metabolism is merging. Interactions between a functional triumvirate of endothelial, mesenchymal, and hematopoietic cell lineages control bone formation and bone resorption—entrained to morphogenetic, metabolic, inflammatory, neuroendocrine, and mechanical demands placed on the skeleton. With advancing age, vascular inflammation, hypertension, and in certain dysmetabolic states (diabetes, dyslipidemia, uremia, hyperphosphatemia), calcium accumulates to a substantial extent in another venue—the arterial vasculature.[1] Our understanding of vascular calcified matrix metabolism significantly lags behind that of skeletal metabolism. However, as in bone, cells of endothelial, mesenchymal, and hematopoietic cell lineages control vascular mineral metabolism, entrained to morphogenetic, metabolic, inflammatory, neuroendocrine, and mechanical demands experienced by the particular vascular segment.[1] This chapter provides a very brief overview of vascular calcification, organized into histoanatomical categories that highlight known or probable differences in pathobiology.

ATHEROSCLEROTIC CALCIFICATION

The most common form of vascular calcification is atherosclerotic calcification (Table 1), in which hydroxyapatite mineral forms within an intimal plaque in association with lipid deposits and monocyte/macrophage infiltration. Mineralization is not only limited to end-stage disease; calcium deposition is in fact found in most atherosclerotic lesions.[2] With implementation of cardiac gated electron beam CT scanning (EBCT), calcium deposits have been identified in ~20% of young adults, 60% of middle-aged adults, and 90% of the elderly.[3]

Pathobiology of Atherosclerotic Calcification

Inflammatory Milieu of Atherosclerotic Lesions. Atherosclerosis is an inflammatory disease, better termed an "atherosclerítis." The process is initiated by a breach in the endothelial barrier and/or high serum lipid levels, both resulting in accumulation of lipoprotein particles in the subendothelial space between the endothelial monolayer and the underlying internal elastic lamina (Fig. 1).[1] When the retained lipoproteins undergo chemical oxidation, they generate bioactive phospholipids and oxysterols that induce influx of vascular smooth muscle cells (VSMCs), monocytes, and T lymphocytes.[1] Along with immunocyte-derived inflammatory cytokines, inflammatory oxylipids induce osteogenic differentiation and mineralization of VSMCs.[4] The earliest sites of atherosclerotic calcification are lipid- nucleated, located along the internal elastic lamina between the endothelial and medial layers. Proteases are activated in atherosclerosis, showing cryptic sites of lipid and calcium binding in matrix elastin. Normally the extracellular matrix of arteries includes collagen type III and collagen type I; elastin degradation with upregulation of type I collagen may accelerate osteogenic differentiation of vascular myofibroblasts by oxylipids.[1]

Ectopic Bone in Atherosclerosis. Although it is well known that calcium deposits in the artery wall are generally composed of hydroxyapatite, it is underappreciated that ~15% of calcified plaques contain lamellar bone.[5] Virchow[6] first noted that some cases of vascular calcification represent vascular "ossification with real plates of bone." A prerequisite for bone formation within arteries is angiogenic invasion, driven by vascular endothelial growth factor, as in endochondral bone formation. It is important to re-emphasize that arterial calcification does not always lead to histological ossification. Sites of atherosclerotic ossification arise from previously calcified vascular segments and frequently include marrow spaces with fat tissue and hematopoietic cells.[7] Angiogenesis and vasa vasorum biology play important roles in the formation of advanced atherosclerotic calcification, because mature bone tissue does not appear until vascular invasion of the deposit. When angiogenic potential is limiting, vascular calcification does not achieve this final ossification stage—even when osteochondrogenic mechanisms are clearly controlling mineral deposition at both histological stages.

Molecular Mechanisms, Vascular Stem Cells, and Relationships to Bone Formation. As with skeletal bone formation and remodeling, vascular mineralization is subject to both positive and negative regulation. Matrix vesicles are found in atherosclerotic calcification,[8] along with bone matrix proteins, such as osteopontin, osteocalcin, osteonectin, matrix GLA protein, and bone sialoprotein.[1] Calcified atherosclerotic lesions also contain osteogenic regulatory factors such as bone morphogenetic protein (BMP)-2, Msx2, Runx2, Osterix, and Wnts.[1,9] Current evidence indicates that, in vascular cells, BMP-2 signaling initiates ectopic osteoblast differentiation.[1]

In addition to bone, advanced atherosclerotic lesions often contain cartilage, marrow-like tissue, and adipose. The origin of all these tissues is not clear. One important consideration is the existence of resident mesenchymal stem cells in the artery wall.[10] The VSMCs of the tunica media contain heterogeneous subpopulations. Some—including aortic myofibroblasts, bovine aortic calcifying vascular cells (CVCs), and microvascular pericytes—exhibit multilineage potential in vitro; osteogenic, chondrogenic, leiomyogenic, marrow stromal, and adipogenic lineage allocation is controlled by BMP-2, TGF-β, and canonical Wnt ligands. Some multipotent cells in the artery wall arise from the adventitia[9,11] and may originate from bone marrow, immigrating to the atherosclerotic plaque through the circulation and the vasa vasorum. The relative contributions of regional vascular mesenchymal progenitors versus circulating marrow-derived mesenchymal progenitors have yet to be determined.

In Vitro and In Vivo Models of Vascular Calcification. A variety of cells harvested from the artery wall, with the exception of endothelial cells, produces hydroxyapatite mineral in vitro.[12,13] Unselected medial VSMCs, like osteoblasts, mineralize the film of extracellular matrix overlying cellular mono-

Dr. Towler has consulted for Wyeth, Lilly, and GlaxoSmithKline. He has received research funding from Barnes-Jewish Hospital. Dr. Demer states that she has no conflicts of interest.

© 2008 American Society for Bone and Mineral Research

TABLE 1. HISTOANATOMIC TYPES OF VASCULAR CALCIFICATION

Atherosclerotic intimal calcification
Medial artery calcification of diabetes and chronic kidney disease
Elastocalcinotic medial artery calcification (Marfan's syndrome, pseudoxanthoma elasticum)
Cardiac valve calcification (native and bioprosthetic)
Calcific uremic arteriolopathy ("cutaneous calciphylaxis")
Cardiac annular calcification
Post-infarct myocardial calcification
Pericardial calcification
Soft tissue calciphylaxis including vessels (acute hyperphosphatemia and renal failure)
Calcifying primary cardiac tumors
Portal vein calcification
Pelvic vein phleboliths

layers in the presence of exogenous phosphate donors. Occasionally, these cultures will produce a 3D nodular aggregate containing mineral. Bovine retinal pericytes—microvascular smooth muscle cells—require several weeks to produce calcified nodules. Cloned bovine CVCs and human VSMCs require ~10 days. The rate is affected by exogenous ascorbic acid, presumably because of the changes in extracellular type I collagen production. Mineralizing CVC nodules range widely in size—from ~100 to 1500 μm in diameter—and contain irregularly shaped hydroxyapatite mineral deposits within their core. The spacing of the nodules seems to be mediated by a reaction-diffusion process involving the morphogens BMP-2 and its inhibitor matrix GLA protein (MGP).[14]

Mice deficient in apolipoprotein E develop spontaneous vascular calcification, primarily in the form of cartilaginous metaplasia.[15] Mice deficient in the low-density lipoprotein receptor (LDLR) develop hyperlipidemia and vascular calcification when exposed to high-fat diabetogenic diets. Aortic osteogenic cascades in LDLR-null mice are entrained to diet-induced inflammatory signals provided by RANKL[16] and TNF-α.[17] Mice expressing the human Lp(a) gene also develop calcified aortic lesions.[18] Atherosclerotic calcification can be induced by vitamin D and calcium supplements in hyperlipidemic rabbits.[19] However, vitamin D intoxication also enhances warfarin-induced vascular calcification in rats, depositing apatite and whitlockite without atherosclerotic injury.[20]

Role of Hyperlipidemia and Inflammatory Lipids in Atherosclerotic Vascular Calcification. Atherosclerosis and atherosclerotic vascular calcification are associated with hyperlipidemia.[1] Progression of coronary calcification correlates with the severity of hyperlipidemia.[21] Conversely, when patients lower their cholesterol levels with lipid-lowering agents, the rate of progression of coronary calcification is reduced.[22] This epidemiological evidence—together with the close physical relationship between vascular calcification and atherosclerotic lesions—strongly suggests a mechanistic relationship. Inflammatory cytokines and oxidized lipoprotein lipids induce osteoblast differentiation of CVCs.[4,23] However, lipids physically nucleate hydroxyapatite mineral deposition as well. At the ultrastructural level, hydroxyapatite mineral crystals from atherosclerotic lesions are physically associated with microcrystals of cholesterol.[24] The phospholipids in matrix vesicles are crucial in vesicular nucleation.[25] Calcification can also nucleate in the elastin matrix, where lipids incorporate into the molecular structure.[26] In vivo, mice with hyperlipidemia develop both atherosclerosis and vascular calcification.[15] Within atherosclerotic plaques, lipidaceous apoptotic bodies accumulate that nucleate mineral deposition through mechanisms similar to those used by physically smaller matrix vesicle.[27] Vascular calcium hydroxyapatite crystals trigger inflammation,[28] and this response may fuel vascular disease progression through a procalcific "feedforward" vicious cycle.

Clinical Issues in Atherosclerotic Calcification

Clinical Significance. The correlation between the degree of calcification and atherosclerosis is strong enough that the "calcium score" is a reliable clinical marker for coronary artery disease, predicting cardiovascular events independently and more accurately than some conventional risk factors.[1] Clinical consequences of vascular calcification primarily stem from perturbed endothelial anti-thrombotic function and mechanical rigidity of the aortic arch and cardiac valves. These structures develop abnormally high flow impedance once calcified, resulting in systolic hypertension, left ventricular hypertrophy, heart failure, aortic stenosis, coronary ischemia, acute coronary syndrome, and myocardial infarction. Aortic elastic recoil is required for maintaining diastolic aortic pressure, which in turn, is required for coronary perfusion. Calcified aortas lack recoil, resulting in a low diastolic pressure, high pulse pressure, and coronary insufficiency.[1] Most plaques that rupture are calcified, and numerous calcium deposits increase the risk of rupture[29]; ~80% of significantly narrowed plaques are calcified.[30] Real-time imaging during human angioplasty and engineering considerations suggest that solid mechanical failure stresses are concentrated at the edges of calcium deposits.[1]

Inhibition of Atherosclerotic Calcification. Patients who successfully lower their cholesterol levels significantly reduce progression of their coronary calcification.[22] Atherogenic effects of oxidized lipids are also blocked by high-density lipo-

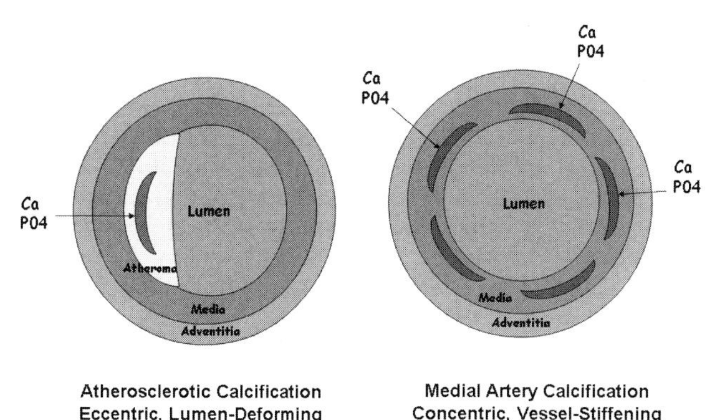

FIG. 1. Arterial calcification. In arterial cross-section, three layers—intima, media, and adventitia—are present from the lumen outward. In atherosclerosis, eccentric, subintimal atheroma formation, cholesterol deposition with inflammation, fibrosis, fragmented internal elastic lamina, apoptotic body formation, and calcium deposition demarcate the calcified atherosclerotic plaque.[2] Atherosclerosis deforms the lumen and potentially provides a focus for thrombosis and acute arterial occlusion. In medial artery calcification, calcium deposition is concentric, thus compromising vascular compliance without lumen deformation. Low-grade adventitial inflammation, elastinolysis, and vascular smooth muscle cell matrix vesicle formation drive concentric disease processes. Myofibroblast proliferation, vascular monocyte-macrophage infiltration, and microvessel formation (angiogenesis) are key components of osteogenic vascular calcification responses.[1,5] The vasa vasorum arises from the adventitia. $CaPO_4$, apatitic calcium phosphate deposition.

© 2008 American Society for Bone and Mineral Research

protein (HDL) and in vitro vascular calcification. Osteoclast-like cells have been identified in calcified atherosclerotic lesions.[7] The net effects of augmenting "vascular osteoclast" activity on vascular integrity and function are as yet unknown.

CARDIAC VALVE CALCIFICATION

Cardiac valve leaflets are remarkably thin and pliable, yet strong and inelastic, consisting of two layers of interstitial cell myofibroblasts surrounded on either side by endothelial monolayers.[31] Valvular sclerosis (fibrosis) is a common occurrence during hypertension, inflammation, diabetes, dyslipidemia, and advanced age and can occur in the absence of narrowing of the valvular opening. Once the scarring becomes sufficiently advanced to narrow the orifice (i.e., stenotic), it is usually calcified. Hence, a common disorder is calcific cardiac valve stenosis. Stenosis is most clinically apparent in calcific aortic sclerosis because the aortic valve is in a high pressure system, supplies coronary and systemic circulation, and hence can rapidly threaten hemodynamic stability when abnormal. Calcification is primarily on the aortic face of the valve, the layer known as the fibrosa. The two other layers, the interstitial spongiosa and the ventricularis (side facing the ventricle), are generally spared.[31] The primary cell type, known as the valvular interstitial cell, is a macrovascular myofibroblast akin to wound myofibroblasts. Valve interstitial cell myofibroblasts can function as osteoprogenitors and resemble the mural myofibroblasts that contribute to medial artery calcification of type II diabetes.[1]

Pathobiology of Cardiac Valve Calcification

Native Cardiac Valve Calcification. For years, cardiac valvular stenosis had been attributed solely to mechanical "wear and tear." However, recent evidence indicates that a majority of aortic calcific valvular stenosis is also promoted by the metabolic insults of dyslipidemia and diabetes.[32] As with atherosclerotic calcification, most risk factors can be categorized as inflammatory or oxidative stressors. In early lesions, increases in subendothelial thicknesses on the aortic face, with myofibroblasts intracellular lipid accumulation, expansion of the valve interstitial valve fibrosus, diffuse stippled calcium deposition, and monocyte-macrophage infiltration are histologically evident.[31] As in atherosclerotic calcification, mature lamellar or endochondral bone tissue is found in ~15% of stenotic cardiac valves.[5] Calcified aortic valves express many of the same osteogenic processes as described for atherosclerotic and medial artery calcification.[1]

Bioprosthetic Cardiac Valve Calcification. Cardiac valves can be replaced by mechanical or biological tissue prostheses. The biological prostheses are fashioned from devitalized, glutaraldehyde-treated, allograft, or xenograft valves. Bioprosthetic valves have the advantage of not requiring long-term anti-coagulation; however, these valves ultimately fail because of mineralization, usually within 10 yr. Cell-free bioprosthetic valves can mineralize in vitro with inorganic phosphate supplementation. Interestingly, the valve fixation procedure does not usually remove lipids; however, when lipids are removed experimentally, ex vivo mineralization is reduced.[33] Thus, cellular regulation occurs at the level of biogenesis and removal of pyrophosphate-like mineralization inhibitors and generation of matrix:lipid complexes that nucleate mineral deposition. As such, bioprosthetic valve calcification depends on cell-regulated mechanisms similar to those directing skeletal mineralization.[1]

Models of Valve Calcification. Rajamannan et al.[34] developed a rabbit model for studying calcific valvulopathy elicited by diet-induced hyperlipidemia. As with mural CVCs, valve myofibroblasts incorporate calcium and deposit hydroxyapatite mineral in their matrix. Osteogenic calcification is enhanced by oxysterols, RANKL, and canonical Wnt signaling.[35] Recent studies indicate that the LDL receptor–deficient murine[17,36] and Watanabe rabbit[35] models faithfully recapitulate the molecular pathobiology of human cardiac valve calcification.[37]

Clinical Issues in Cardiac Valve Calcification

Clinical Significance of Calcific Valvular Stenosis and Bioprosthetic Valve Calcification. Calcific aortic stenosis is the most frequent cardiac valve disorder in developed countries, afflicting 2–3% of those >65 yr of age. It confers high morbidity and mortality.[38] Valvular calcification can be diagnosed by echocardiography, and narrowing of the orifice is quantified by Doppler techniques.

Inhibition of Cardiac Valve Calcification. Etidronate, once used to inhibit heterotopic bone formation after hip surgery, seems to inhibit progression of aortic calcification in patients with end-stage renal disease (ESRD).[39] Some evidence suggests that lipid lowering may reduce valvular calcification. In the hyperlipidemic rabbit model, treatment to lower serum lipids levels reduced the severity of calcification[34] through effects on canonical Wnt signaling.[35] Strategies focused on routine LDL cholesterol reduction seem insufficient to reduce the progression of valvular calcification once disease initiated[40]; aggressive early intervention with potent statins may inhibit disease progression.[41] Because parameters on the metabolic syndrome–type II diabetes continuum incrementally increase aortic valve calcification risk,[32] multimodality strategies that correct these metabolic-inflammatory risk factors may be required for consistent clinical responses.

MEDIAL ARTERY CALCIFICATION

Medial artery calcification is a characteristic feature of diabetes and ESRD.[1] Although diabetes is the leading cause of ESRD, diabetes is an independent risk factor for vascular calcification.[42] Even in ESRD, the extent of medial artery calcification increases with worsening glycemic control.[43] Uremic and diabetic medial artery calcification proceed through matrix vesicle–dependent mineralization processes.[27] Whereas both types of medial calcification are sometimes denoted as Monckeberg's sclerosis, emerging differences in pathobiology are highlighted by individual consideration.

Pathobiology of Medial Artery Calcification

Medial Artery Calcification of Diabetes and Uremia. Medial artery calcification is characterized by the deposition of apatitic calcium phosphate in the tunica media of large vessels—with the notable absence of neointima formation. Medial artery calcium deposition is nucleated by lipidaceous extracellular vesicles that arise from a minimum of two sources: (1) the apoptotic bodies of dying VSMCs and (2) the regulated extrusion of mineralizing matrix vesicles from viable VSMCs.[27] Importantly, Reynolds et al.[44] recently showed that VSMC matrix vesicles can either promote or inhibit matrix calcification, dependent on whether serum-derived fetuin is recruited into MGP-containing complexes. Serum fetuin is taken up by VSMCs and packaged into matrix vesicles that serve to inhibit nucleation. Moreover, fetuin promotes VSMC "phagocytosis"

© 2008 American Society for Bone and Mineral Research

of pro-osteogenic matrix vesicles. This highlights the complexity of the VSMC-regulated matrix vesicle metabolism that controls initiation and propagation of vascular mineralization.

Production of osteogenic matrix vesicles entails the upregulation of bone alkaline phosphatase (ALP), an ectoenzyme that promotes deposition of calcified matrix by degrading inorganic pyrophosphate (PPi). PPi is a cell-generated organic anion that inhibits mineralization and is a physiologically relevant substrate for ALP hydrolysis. Johnson and Terkeltaub[45] elegantly showed that loss of extracellular PPi derived from either the extracellular enzyme NPP1 (ectonucleotide pyrophosphatase/phosphodiesterase I) or the cellular PPi exporter *ank* predisposes to massive arterial calcification.[45] In the setting of ESRD, circulating PPi levels are reduced.[46] Thus, along with the prevalent glucose intolerance, hyperphosphatemia, hypercalcemia, and fetuin deficiency, reduction in PPi synergistically promotes the profound calcific vasculopathy that assails patients with ESRD.[1] Strategies that restore serum PPi "tone" using nonhydrolyzable bisphosphonate PPi analogs may inhibit progression of vascular calcification.[39] Alternatively, if vascular (versus bone) tissue selectivity can be achieved, strategies that inhibit ALP degradation of arterial PPi[47]may impact vascular calcium load.

Molecular Mechanisms, Vascular Stem Cells, and Relationships to Bone Formation. The molecular mechanisms that regulate vascular calcification in diabetes are beginning to be understood from studies of murine models.[16,17,36] High-fat diets that induce obesity, insulin-resistant diabetes, and dyslipidemia promote vascular calcification in male low-density lipoprotein receptor (LDLR)-deficient mice.[36] In this physiologically relevant model of type II diabetes, the high-fat Western diet—a stimulus for obesity and vascular matrix vesicle formation[48]—promotes medial artery and aortic valve calcification.[36] A working model has emerged in which a paracrine BMP2-Msx2-Wnt signaling cascade, initiated by the adventitial oxidative stressors of type II diabetes, controls the osteogenic differentiation and mineralization of vascular progenitors through nonendochondral processes resembling craniofacial membranous mineralization.[49] The concentric medial calcification of diabetes arises in part from the anatomic relationship between the Msx2 expressing cells of the periaortic adventitia that elaborate a Wnt-laden osteogenic milieu[9] and CVCs in the tunica media that undergo osteogenic differentiation in response to signals or cells conveyed through the vasa vasorum.[10] The precise origins of the Msx2-expressing cells and CVCs are not known; however, MSC populations reside within the aorta.[10] Recent data showed that low-grade inflammatory signals provided by TNF-α mediate induction of aortic BMP2-Msx2-Wnt signaling.[36]

In uremia, a "perfect storm" of calcific vasculopathy occurs. Phosphate retention, secondary hyperparathyroidism, inflammation, deficiency of calcification inhibitors, and the accumulation of PTH fragments that perturb normal calcium phosphate homeostasis drive tremendous vascular calcium loads. Hyperphosphatemia stimulates VSMC elaboration of procalcific matrix vesicles.[27] In vitro, inhibition of the cellular phosphate transporter, PiT1, inhibits VSMC calcification and subsequent osteochondrogenic differentiation.[50] Phosphate retention of chronic kidney disease (CKD) presents a critically important pathogenic feature, but the disastrous milieu of ESRD confounds simple extrapolation of phosphate normalization to restoration of vascular health. Maintenance of normal rates of bone formation seems to be required to "buffer" the vasculature from transient elevations in phosphate and calcium that induce VSMC vesicle biogenesis. Low-turnover bone disease is associated with more severe arterial calcification in ESRD.[51] Excessive calcium intake through oval calcium carbonate phosphate binders suppresses PTH levels, reduces bone mass, and increases vascular calcium load in patients with ESRD.[52] Moreover, a proteolytic fragment of PTH, PTH(7-84), accumulates in ESRD and functions to induce resistance to PTH. PTH(7-84) binds the PTH1R and induces receptor internalization without eliciting downstream signaling cascades.[53] If PTH1R signaling maintains skeletal mineralization while simultaneously limiting vascular calcium accumulation, the accumulation of such antagonistic PTH fragments may contribute to the vasculopathy of ESRD.

Elastocalcinotic Vascular Calcification. During the progression of any form of vascular calcification, secondary perturbations in elastin metabolism likely contribute to mineralization. However, vascular calcification arising from primary alterations in elastin metabolism may represent a unique entry point in a feedforward cycle of medial artery calcification.[54] Aberrant arterial elastin organization and metabolism is characterized by aortic root dilatation, aneurysm formation, and medial degeneration with calcification. This is evident in Marfan's syndrome, where deficiencies in fibrillin 1 cause homeostatic failure in the microfibrillar array of the tunica adventitia that withstands physiological hemodynamic stress; metalloproteinase induction, disruption of the tunica media elastin network, and calcification subsequently occur.[54] This calcification mechanism is not initially associated with matrix vesicle formation; instead, calcium phosphate deposition occurs in association with degenerating elastin fibrils of the tunica media. Whereas molecular mechanisms are not understood, it is apparent that proteolysis of tropoelastin and elastin enhances vascular matrix calcium deposition. Calcification also occurs in the setting of pseudoxanthoma elasticum, arising from mutations in the *ABCC6* gene.[55] Electron microscopy confirms primary fragmentation of elastic lamina and deposition of calcium along thickened elastin fibers in the absence of matrix vesicles.

In Vivo Models of Medial Artery Calcification. The best known models are those of vitamin D intoxication with warfarin or nicotine.[56,57] These treatments result in an elastocalcinotic medial artery calcification, with apatite and whitlockite deposition. Another model is high-fat diabetogenic diet administration to male $LDLR^{-/-}$ mice[36]; this represents a more clinically relevant disease model that develops both medial and atherosclerotic calcium deposition.[36] Induction of chronic renal insufficiency profoundly accelerates calcification in $LDLR^{-/-}$ mice.[58] Osteoprotegerin (OPG)-deficient mice and Msx2 transgenic mice[9] represent two genetically engineered medial artery calcification models. With OPG deficiency, intimal and medial artery calcification arises,[59] potentially related to unopposed actions of vascular RANKL. Side-by-side comparisons have not been performed with these models to clarify mechanistic similarities and differences.

Clinical Issues in Medial Artery Calcification

Clinical Significance of Medial Artery Calcification. Epidemiological studies have shown that medial artery calcification increases cardiovascular morbidity and mortality in patients with diabetes[60]and uremia.[61] Compromised elastic artery compliance is likely a major contributor to the increased risk for lower extremity amputation of patients with type II diabetes.[60] In addition to patient morbidity, the economic costs of managing lower extremity amputation in type II diabetes are staggering—equivalent to the combined costs of treating fatal myocardial infarction (MI), nonfatal MI, and angina.[62]

© 2008 American Society for Bone and Mineral Research

Inhibition or Regression of Medial Artery Calcification. Few studies have explored whether medial artery calcification is preventable or reversible. Vitamin D plus warfarin-induced vascular calcification in the rat is inhibited by treatment with OPG.[57] Inhibition of the phosphate transporter, Pit1, inhibits VSMC calcification and osteochondrogenic differentiation.[50] The phosphate binding resin sevelamer inhibits the endochondral vascular calcification of apolipoprotein E–null mice.[63] Aggressive lipid-lowering therapy with statins suppresses cardiovascular calcification and vascular Wnt signaling in dyslipidemic rabbits.[35] Administration of the endothelin receptor antagonist darusentan induces regression of elastocalcinotic medial calcification in rodents by upregulation of carbonic anhydrase.[56] In humans, only sevelamer,[64] etidronate,[39] and rosuvastatin[41] have decreased progression of vascular calcification, but medial artery calcification has not been specifically assessed.

CALCIFIC UREMIC ARTERIOLOPATHY

A particularly severe form of vascular calcification observed in ESRD is calcific uremic arteriolopathy (CUA). Unlike the highly common macrovascular medial artery and atherosclerotic calcification of ESRD, CUA afflicts much smaller arteries, most notably the arterioles of the dermis.[65] Clinically, it presents as a vasculitis, with livido reticularis followed by cord-like dermal thickening and subsequent "dry" cutaneous necrosis. The histopathology is arteriolar (100–600 μm diameter) medial calcification with concomitant endovascular, fibroproliferative neointimal constriction; frequent small vessel thrombosis; and fat necrosis with panniculitis and acute inflammatory changes. Similar histopathology can occur in mesenteric arterioles.[65] The pathobiology of CUA is poorly understood. Antecedent hyperphosphatemia and elevated calcium–phosphate product is prevalent but insufficient to explain the disease process. However, treatment with warfarin before onset of CUA is observed in more than one half of the afflicted patients.[65] MGP is a highly important modulator of BMP signaling and inhibitor of osteochondrogenic vascular calcification. Zebboudj et al.[66] showed that MGP forms a complex with BMPs that precludes osteogenic signaling; bioactivity is dependent on modification of MGP by Gla residues. Moreover, MGP-fetuin complexes assembled by VSMCs form vesicles that can actually inhibit vascular calcium deposition.[44] Thus, MGP-fetuin deficiencies associated with uremia and warfarin treatment likely contribute to CUA pathogenesis. However, until a robust animal model of CUA is developed, these notions remain speculative. As previously noted,[65] the original calciphylaxis model of Selye that causes skin necrosis in experimental animals does not recapitulate the histopathology of CUA. Infusion of sodium thiosulfate with or without vitamin K supplementation has been empirically used to treat CUA[67] (DA Towler, unpublished data), but no randomized controlled trial of thiosulfate therapy has yet been reported.

MYOCARDIAL, PERICARDIAL, AND ANNULAR CALCIFICATION

Myocardial dystrophic calcification is visible by chest X-ray in ~5–10% of patients who have survived ≥5 yr after a left ventricular infarct.[1] Granulomatous diseases (tuberculosis, histoplasmosis, sarcoidosis) were historically the common causes of pericardial calcification; with the incidence of tuberculosis in decline, granulomatous pericardial calcification has also declined.[68] Uremia, systemic lupus, and postviral, post-irradiation, or post-hemopericardium pericardial inflammation are the more common settings in which pericardial calcification is seen today. Pericardial stiffening produces clinically significant hemodynamic abnormalities, leading to restrictive heart failure. The mitral annulus, a fibrous ring embedded in the myocardium surrounding each valve, often undergoes calcification through endochondral metaplasia. Rarely, primary cardiac tumors such as rhabdomyomas, endotheliomas, and myxomas can also calcify.

VENOUS VASCULAR CALCIFICATION

Calcification does occur in the venous vasculature; calcified pelvic venous thromboliths are commonly noted by radiography, with little if any clinical consequence. However, venous vasculature exposed to elevated transmural pressures become subject to "arterialization." Calcification of the saphenous vein grafts used for coronary bypass represents one clinically relevant example. Orthotopic venous calcification occurs in portal hypertension, with ~11% of patients with cirrhosis having portal and mesenteric venous calcification.[69]

SUMMARY

The pathogenetic mechanisms outlined above offer multiple strategies for potential therapeutic intervention. However, in patients, only sevelamer has been shown to consistently decrease progression of vascular calcification in CKD[64]; clinical studies of bisphosphonates and statins have been either too small, or yielded conflicting outcomes Novel noninvasive imaging methods—such as those being developed in murine aortic disease models[70]—are sorely needed to better evaluate efficacy of potential therapeutic strategies on arterial calcified matrix metabolism. The biological complexity of vascular calcification indicates that additional, thoughtfully crafted translational research will be needed to adequately address the clinical needs arising from this heterogeneous disorder of mineral metabolism.

ACKNOWLEDGMENTS

The authors are supported by grants from the National Institutes of Health.

REFERENCES

1. Abedin M, Tintut Y, Demer LL 2004 Vascular calcification: Mechanisms and clinical ramifications. Arterioscler Thromb Vasc Biol **24:**1161–1170.
2. Stary HC, Chandler AB, Dinsmore RE, Fuster V, Glagov S, Insull W Jr, Rosenfeld ME, Schwartz CJ, Wagner WD, Wissler RW 1995 A definition of advanced types of atherosclerotic lesions and a histological classification of atherosclerosis. A report from the Committee on Vascular Lesions of the Council on Arteriosclerosis, American Heart Association. Arterioscler Thromb Vasc Biol **15:**1512–1531.
3. Newman AB, Naydeck BL, Sutton-Tyrrell K, Feldman A, Edmundowicz D, Kuller LH 2001 Coronary artery calcification in older adults to age 99: Prevalence and risk factors. Circulation **104:**2679–2684.
4. Parhami F, Morrow AD, Balucan J, Leitinger N, Watson AD, Tintut Y, Berliner JA, Demer LL 1997 Lipid oxidation products have opposite effects on calcifying vascular cell and bone cell differentiation. A possible explanation for the paradox of arterial calcification in osteoporotic patients. Arterioscler Thromb Vasc Biol **17:**680–687.
5. Mohler ER III, Gannon F, Reynolds C, Zimmerman R, Keane MG, Kaplan FS 2001 Bone formation and inflammation in cardiac valves. Circulation **103:**1522–1528.
6. Virchow R 1863. Cellular Pathology as Based Upon Physiological and Pathological Histology. Dover Publications, New York, NY, USA.
7. Hunt JL, Fairman R, Mitchell ME, Carpenter JP, Golden M,

© 2008 American Society for Bone and Mineral Research

Khalapyan T, Wolfe M, Neschis D, Milner R, Scoll B, Cusack A, Mohler ER III 2002 Bone formation in carotid plaques: A clinicopathological study. Stroke **33:**1214–1219.
8. Tanimura A, McGregor DH, Anderson HC 1983 Matrix vesicles in atherosclerotic calcification. Proc Soc Exp Biol Med **172:**173–177.
9. Shao JS, Cheng SL, Pingsterhaus JM, Charlton-Kachigian N, Loewy AP, Towler DA 2005 Msx2 promotes cardiovascular calcification by activating paracrine Wnt signals. J Clin Invest **115:**1210–1220.
10. Tintut Y, Alfonso Z, Saini T, Radcliff K, Watson K, Bostrom K, Demer LL 2003 Multilineage potential of cells from the artery wall. Circulation **108:**2505–2510.
11. Hu Y, Zhang Z, Torsney E, Afzal AR, Davison F, Metzler B, Xu Q 2004 Abundant progenitor cells in the adventitia contribute to atherosclerosis of vein grafts in ApoE-deficient mice. J Clin Invest **113:**1258–1265.
12. Schor AM, Allen TD, Canfield AE, Sloan P, Schor SL 1990 Pericytes derived from the retinal microvasculature undergo calcification in vitro. J Cell Sci **97:**449–461.
13. Bostrom K, Watson KE, Horn S, Wortham C, Herman IM, Demer LL 1993 Bone morphogenetic protein expression in human atherosclerotic lesions. J Clin Invest **91:**1800–1809.
14. Garfinkel A, Tintut Y, Petrasek D, Bostrom K, Demer LL 2004 Pattern formation by vascular mesenchymal cells. Proc Natl Acad Sci USA **101:**9247–9250.
15. Rattazzi M, Bennett BJ, Bea F, Kirk EA, Ricks JL, Speer M, Schwartz SM, Giachelli CM, Rosenfeld ME 2005 Calcification of advanced atherosclerotic lesions in the innominate arteries of ApoE-deficient mice: Potential role of chondrocyte-like cells. Arterioscler Thromb Vasc Biol **25:**1420–1425.
16. Morony S, Tintut Y, Zhang Z, Cattley RC, Van G, Dwyer D, Stolina M, Kostenuik PJ, Demer LL 2008. Osteoprotegerin Inhibits Vascular Calcification Without Affecting Atherosclerosis in ldlr(−/−) Mice. Circulation **117:**411–420.
17. Weiss RM, Ohashi M, Miller JD, Young SG, Heistad DD 2006 Calcific aortic valve stenosis in old hypercholesterolemic mice. Circulation **114:**2065–2069.
18. Teivainen PA, Eliassen KA, Berg K, Torsdalen K, Svindland A 2004 Atherogenesis and vascular calcification in mice expressing the human LPA gene. Pathophysiology **11:**113–120.
19. Demer LL 1991 Effect of calcification on in vivo mechanical response of rabbit arteries to balloon dilation. Circulation **83:**2083–2093.
20. Price PA, Faus SA, Williamson MK 2000 Warfarin-induced artery calcification is accelerated by growth and vitamin D. Arterioscler Thromb Vasc Biol **20:**317–327.
21. Pohle K, Maffert R, Ropers D, Moshage W, Stilianakis N, Daniel WG, Achenbach S 2001 Progression of aortic valve calcification: Association with coronary atherosclerosis and cardiovascular risk factors. Circulation **104:**1927–1932.
22. Callister TQ, Raggi P, Cooil B, Lippolis NJ, Russo DJ 1998 Effect of HMG-CoA reductase inhibitors on coronary artery disease as assessed by electron-beam computed tomography. N Engl J Med **339:**1972–1978.
23. Tintut Y, Patel J, Parhami F, Demer LL 2000 Tumor necrosis factor-alpha promotes in vitro calcification of vascular cells via the cAMP pathway. Circulation **102:**2636–2642.
24. Hirsch D, Azoury R, Sarig S, Kruth HS 1993 Colocalization of cholesterol and hydroxyapatite in human atherosclerotic lesions. Calcif Tissue Int **52:**94–98.
25. Boskey AL, Posner AS 1977 The role of synthetic and bone extracted Ca-phospholipid-PO4 complexes in hydroxyapatite formation. Calcif Tissue Res **23:**251–258.
26. Noma A, Takahashi T, Wada T 1981 Elastin-lipid interaction in the arterial wall. Part 2. In vitro binding of lipoprotein-lipids to arterial elastin and the inhibitory effect of high density lipoproteins on the process. Atherosclerosis **38:**373–382.
27. Reynolds JL, Joannides AJ, Skepper JN, McNair R, Schurgers LJ, Proudfoot D, Jahnen-Dechent W, Weissberg PL, Shanahan CM 2004 Human vascular smooth muscle cells undergo vesicle-mediated calcification in response to changes in extracellular calcium and phosphate concentrations: A potential mechanism for accelerated vascular calcification in ESRD. J Am Soc Nephrol **15:**2857–2867.
28. Nadra I, Mason JC, Philippidis P, Florey O, Smythe CD, McCarthy GM, Landis RC, Haskard DO 2005 Proinflammatory activation of macrophages by basic calcium phosphate crystals via protein kinase C and MAP kinase pathways: A vicious cycle of inflammation and arterial calcification? Circ Res **96:**1248–1256.
29. Ehara S, Kobayashi Y, Yoshiyama M, Shimada K, Shimada Y, Fukuda D, Nakamura Y, Yamashita H, Yamagishi H, Takeuchi K, Naruko T, Haze K, Becker AE, Yoshikawa J, Ueda M 2004 Spotty calcification typifies the culprit plaque in patients with acute myocardial infarction: An intravascular ultrasound study. Circulation **110:**3424–3429.
30. Honye J, Mahon DJ, Jain A, White CJ, Ramee SR, Wallis JB, al-Zarka A, Tobis JM 1992 Morphological effects of coronary balloon angioplasty in vivo assessed by intravascular ultrasound imaging. Circulation **85:**1012–1025.
31. Otto CM, Kuusisto J, Reichenbach DD, Gown AM, O'Brien KD 1994 Characterization of the early lesion of 'degenerative' valvular aortic stenosis. Histological and immunohistochemical studies. Circulation **90:**844–853.
32. Katz R, Wong ND, Kronmal R, Takasu J, Shavelle DM, Probstfield JL, Bertoni AG, Budoff MJ, O'Brien KD 2006 Features of the metabolic syndrome and diabetes mellitus as predictors of aortic valve calcification in the Multi-Ethnic Study of Atherosclerosis. Circulation **113:**2113–2119.
33. Vyavahare NR, Jones PL, Hirsch D, Schoen FJ, Levy RJ 2000 Prevention of glutaraldehyde-fixed bioprosthetic heart valve calcification by alcohol pretreatment: Further mechanistic studies. J Heart Valve Dis **9:**561–566.
34. Rajamannan NM, Subramaniam M, Springett M, Sebo TC, Niekrasz M, McConnell JP, Singh RJ, Stone NJ, Bonow RO, Spelsberg TC 2002 Atorvastatin inhibits hypercholesterolemia-induced cellular proliferation and bone matrix production in the rabbit aortic valve. Circulation **105:**2660–2665.
35. Rajamannan NM, Subramaniam M, Caira F, Stock SR, Spelsberg TC 2005 Atorvastatin inhibits hypercholesterolemia-induced calcification in the aortic valves via the Lrp5 receptor pathway. Circulation **112:**I229–I234.
36. Al-Aly Z, Shao JS, Lai CF, Huang E, Cai J, Behrmann A, Cheng SL, Towler DA 2007 Aortic Msx2-Wnt calcification cascade is regulated by TNF-alpha-dependent signals in diabetic Ldlr−/− mice. Arterioscler Thromb Vasc Biol **27:**2589–2596.
37. Caira FC, Stock SR, Gleason TG, McGee EC, Huang J, Bonow RO, Spelsberg TC, McCarthy PM, Rahimtoola SH, Rajamannan NM 2006 Human degenerative valve disease is associated with up-regulation of low-density lipoprotein receptor-related protein 5 receptor-mediated bone formation. J Am Coll Cardiol **47:**1707–1712.
38. Rosenhek R, Binder T, Porenta G, Lang I, Christ G, Schemper M, Maurer G, Baumgartner H 2000 Predictors of outcome in severe, asymptomatic aortic stenosis. N Engl J Med **343:**611–617.
39. Nitta K, Akiba T, Suzuki K, Uchida K, Watanabe R, Majima K, Aoki T, Nihei H 2004 Effects of cyclic intermittent etidronate therapy on coronary artery calcification in patients receiving long-term hemodialysis. Am J Kidney Dis **44:**680–688.
40. Cowell SJ, Newby DE, Prescott RJ, Bloomfield P, Reid J, Northridge DB, Boon NA 2005 A randomized trial of intensive lipid-lowering therapy in calcific aortic stenosis. N Engl J Med **352:**2389–2397.
41. Moura LM, Ramos SF, Zamorano JL, Barros IM, Azevedo LF, Rocha-Goncalves F, Rajamannan NM 2007 Rosuvastatin affecting aortic valve endothelium to slow the progression of aortic stenosis. J Am Coll Cardiol **49:**554–561.
42. Reaven PD, Sacks J 2005 Coronary artery and abdominal aortic calcification are associated with cardiovascular disease in type 2 diabetes. Diabetologia **48:**379–385.
43. Ishimura E, Okuno S, Kitatani K, Kim M, Shoji T, Nakatani T, Inaba M, Nishizawa Y 2002 Different risk factors for peripheral vascular calcification between diabetic and non-diabetic haemodialysis patients–importance of glycaemic control. Diabetologia **45:**1446–1448.
44. Reynolds JL, Skepper JN, McNair R, Kasama T, Gupta K, Weissberg PL, Jahnen-Dechent W, Shanahan CM 2005 Multifunctional roles for serum protein fetuin-a in inhibition of human vascular smooth muscle cell calcification. J Am Soc Nephrol **16:**2920–2930.
45. Johnson K, Terkeltaub R 2005 Inorganic pyrophosphate (PPI) in pathologic calcification of articular cartilage. Front Biosci **10:**988–997.
46. Lomashvili KA, Khawandi W, O'Neill WC 2005 Reduced plasma pyrophosphate levels in hemodialysis patients. J Am Soc Nephrol **16:**2495–2500.
47. Narisawa, S, Harmey, D, Yadav, MC, O'Neill WC, Hoylaerts MF,

© 2008 American Society for Bone and Mineral Research

Millan JL 2007 Novel inhibitors of alkaline phosphatase suppress vascular smooth muscle cell calcification. J Bone Miner Res **22:**1700–1710.
48. Hsu HH, Camacho NP, Sun F, Tawfik O, Aono H 2000 Isolation of calcifiable vesicles from aortas of rabbits fed with high cholesterol diets. Atherosclerosis **153:**337–348.
49. Shao JS, Aly ZA, Lai CF, Cheng SL, Cai J, Huang E, Behrmann A, Towler DA 2007 Vascular Bmp Msx2 Wnt signaling and oxidative stress in arterial calcification. Ann N Y Acad Sci **1117:**40–50.
50. Giachelli CM 2003 Vascular calcification: In vitro evidence for the role of inorganic phosphate. J Am Soc Nephrol **14:**S300–S304.
51. London GM, Marty C, Marchais SJ, Guerin AP, Metivier F, de Vernejoul MC 2004 Arterial calcifications and bone histomorphometry in end-stage renal disease. J Am Soc Nephrol **15:**1943–1951.
52. Raggi P, James G, Burke SK, Bommer J, Chasan-Taber S, Holzer H, Braun J, Chertow GM 2005 Decrease in thoracic vertebral bone attenuation with calcium-based phosphate binders in hemodialysis. J Bone Miner Res **20:**764–772.
53. Sneddon WB, Magyar CE, Willick GE, Syme CA, Galbiati F, Bisello A, Friedman PA 2004 Ligand-selective dissociation of activation and internalization of the parathyroid hormone (PTH) receptor: Conditional efficacy of PTH peptide fragments. Endocrinology **145:**2815–2823.
54. Bunton TE, Biery NJ, Myers L, Gayraud B, Ramirez F, Dietz HC 2001 Phenotypic alteration of vascular smooth muscle cells precedes elastolysis in a mouse model of Marfan syndrome. Circ Res **88:**37–43.
55. Klement JF, Matsuzaki Y, Jiang QJ, Terlizzi J, Choi HY, Fujimoto N, Li K, Pulkkinen L, Birk DE, Sundberg JP, Uitto J 2005 Targeted ablation of the abcc6 gene results in ectopic mineralization of connective tissues. Mol Cell Biol **25:**8299–8310.
56. Essalihi R, Dao HH, Gilbert LA, Bouvet C, Semerjian Y, McKee MD, Moreau P 2005 Regression of medial elastocalcinosis in rat aorta: A new vascular function for carbonic anhydrase. Circulation **112:**1628–1635.
57. Price PA, June HH, Buckley JR, Williamson MK 2001 Osteoprotegerin inhibits artery calcification induced by warfarin and by vitamin D. Arterioscler Thromb Vasc Biol **21:**1610–1616.
58. Davies MR, Lund RJ, Hruska KA 2003 BMP-7 is an efficacious treatment of vascular calcification in a murine model of atherosclerosis and chronic renal failure. J Am Soc Nephrol **14:**1559–1567.
59. Bucay N, Sarosi I, Dunstan CR, Morony S, Tarpley J, Capparelli C, Scully S, Tan HL, Xu W, Lacey DL, Boyle WJ, Simonet WS 1998 Osteoprotegerin-deficient mice develop early onset osteoporosis and arterial calcification. Genes Dev **12:**1260–1268.
60. Lehto S, Niskanen L, Suhonen M, Ronnemaa T, Laakso M 1996 Medial artery calcification. A neglected harbinger of cardiovascular complications in non-insulin-dependent diabetes mellitus. Arterioscler Thromb Vasc Biol **16:**978–983.
61. London GM, Marchais SJ, Guerin AP, Metivier F 2005 Arteriosclerosis, vascular calcifications and cardiovascular disease in uremia. Curr Opin Nephrol Hypertens **14:**525–531.
62. Clarke P, Gray A, Legood R, Briggs A, Holman R 2003 The impact of diabetes-related complications on healthcare costs: Results from the United Kingdom Prospective Diabetes Study (UKPDS Study No. 65). Diabet Med **20:**442–450.
63. Phan O, Ivanovski O, Nguyen-Khoa T, Mothu N, Angulo J, Westenfeld R, Ketteler M, Meert N, Maizel J, Nikolov IG, Vanholder R, Lacour B, Drueke TB, Massy ZA 2005 Sevelamer prevents uremia-enhanced atherosclerosis progression in apolipoprotein E-deficient mice. Circulation **112:**2875–2882.
64. Chertow GM, Burke SK, Raggi P 2002 Sevelamer attenuates the progression of coronary and aortic calcification in hemodialysis patients. Kidney Int **62:**245–252.
65. Coates T, Kirkland GS, Dymock RB, Murphy BF, Brealey JK, Mathew TH, Disney AP 1998 Cutaneous necrosis from calcific uremic arteriolopathy. Am J Kidney Dis **32:**384–391.
66. Zebboudj AF, Imura M, Bostrom K 2002 Matrix GLA protein, a regulatory protein for bone morphogenetic protein-2. J Biol Chem **277:**4388–4394.
67. Cicone JS, Petronis JB, Embert CD, Spector DA 2004 Successful treatment of calciphylaxis with intravenous sodium thiosulfate. Am J Kidney Dis **43:**1104–1108.
68. Ling LH, Oh JK, Breen JF, Schaff HV, Danielson GK, Mahoney DW, Seward JB, Tajik AJ 2000 Calcific constrictive pericarditis: Is it still with us? Ann Intern Med **132:**444–450.
69. Verma V, Cronin DC II, Dachman AH 2001 Portal and mesenteric venous calcification in patients with advanced cirrhosis. AJR Am J Roentgenol **176:**489–492.
70. Aikawa E, Nahrendorf M, Sosnovik D, Lok VM, Jaffer FA, Aikawa M, Weissleder R 2007 Multimodality molecular imaging identifies proteolytic and osteogenic activities in early aortic valve disease. Circulation **115:**377–386.

Chapter 94. Fibrodysplasia (Myositis) Ossificans Progressiva

Frederick S. Kaplan and Eileen M. Shore

Division of Molecular Orthopaedics, Department of Orthopaedic Surgery, The University of Pennsylvania School of Medicine, Philadelphia, Pennsylvania

INTRODUCTION

Fibrodysplasia ossificans progressiva (FOP: MIM 135100) is a rare heritable disorder of connective tissue characterized by (1) congenital malformations of the great toes and (2) recurrent episodes of painful soft tissue swelling that lead to progressive heterotopic ossification.[1,2]

Post-traumatic myositis ossificans also features heterotopic bone and cartilage formation and may follow hip replacement, spinal cord injury, brain injury, athletic injury, burns, war wounds, and valvular heart disease.[3,4]

FOP, first described in 1692, has had >700 cases reported and is among the rarest of human afflictions, with an estimated incidence of one per two million individuals.[1,2] All races are affected.[2] Autosomal dominant transmission with variable expressivity is established[5]; however, reproductive fitness is low, and most cases are sporadic. Gonadal mosaicism has been described.[6]

CLINICAL PRESENTATION

Malformations of the great toes can be noted at birth (Fig. 1). In some cases, the thumbs are strikingly short.[1,2] FOP is usually diagnosed when soft tissue swellings and radiographic

The authors state that they have no conflicts of interest.

Key words: heterotopic ossification, fibrodysplasia ossificans progressiva, bone morphogenetic protein, bone morphogenetic protein receptors, ACVR1, ALK2

© 2008 American Society for Bone and Mineral Research

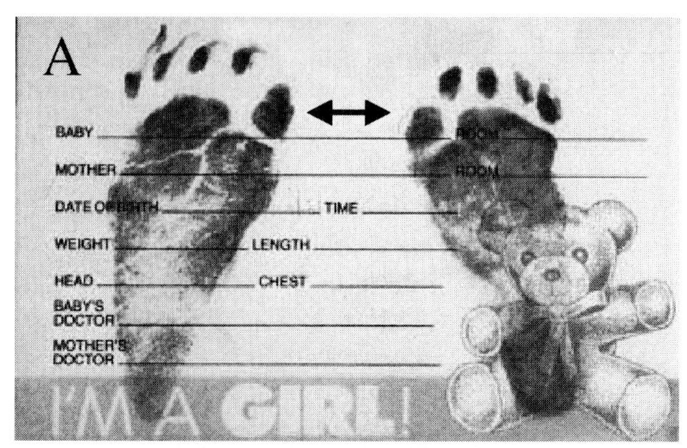

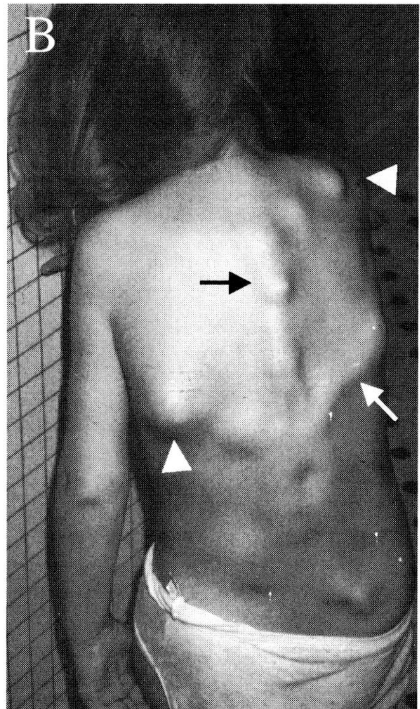

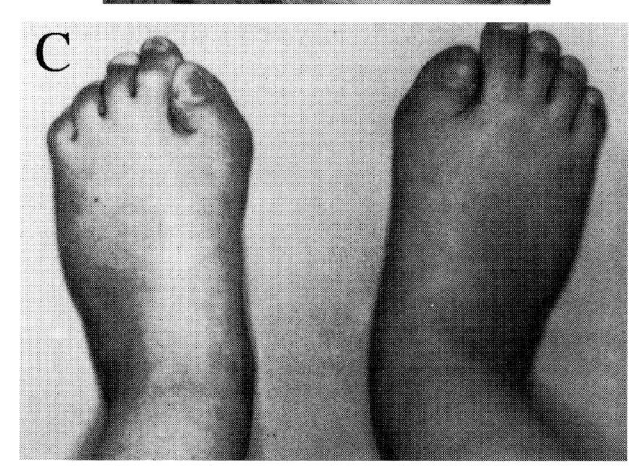

FIG. 1. Fibrodysplasia (myositis) ossificans progressiva. Characteristic features of FOP are seen in early childhood. The presence of short malformed great toes at birth (A, arrows) heralds the later spontaneous appearance of the preosseous soft tissue lesions on the neck and back (B, arrowheads) and should provoke suspicion of FOP even before the transformation to heterotopic bone (arrows). An inspection of the toes (C) and/or genetic DNA sequence analysis of ACVR1 will confirm the diagnosis and may alleviate the need for a lesional biopsy (trauma) that could exacerbate the condition. (Reprinted with permission from the American Society for Bone and Mineral Research from Kaplan FS, Smith RM 1997 Clinical vignette—Fibrodysplasia ossificans progressiva (FOP). J Bone Miner Res **12**:855.)

evidence of heterotopic ossification are noted; however, misdiagnosis is common and leads to unnecessary biopsies and invasive procedures that result in permanent harm.[1,2,7]

The severity of FOP differs greatly among patients.[5,8] Most become immobilized and confined to a wheelchair by the third decade of life.[1,2,9] Typically, episodes of soft tissue swelling (flare-ups) begin during the first decade of life (Fig. 1).[10] Wide variability in the rate of disease progression, even among identical twins, attests to the important postnatal influence of environmental factors.[11]

Flare-ups appear spontaneously or may be precipitated by minor trauma including intramuscular injections and influenza-like viral illnesses.[2,12,13] Swellings develop rapidly during the course of several days. Aponeuroses, fascia, tendons, ligaments, and connective tissue of voluntary muscles may be affected. Although some swellings may regress spontaneously, most mature by an endochondral pathway to form heterotopic bone.[14] Flare-ups recur with unpredictable frequency. Once ossification develops, it is permanent. Disability is cumulative.

Bony masses immobilize joints and cause contractures and deformity. Ossification around the hips, typically present by the third decade of life, often prevents ambulation.[9] Involvement of the muscles of mastication (frequently after injection of local anesthetic or overstretching of the jaw during dental procedures) can severely limit jaw movement and impair nutrition.[15,16] Ankylosis of the spine and ribs further restricts mobility and may imperil cardiopulmonary function (Fig. 1).[1,2,9,17] Scoliosis is common and associated with heterotopic bone that asymmetrically connects the ribs to the pelvis.[18] Restrictive lung disease may follow.[19] Vocal muscles, extraocular muscles, smooth muscles, diaphragm, and heart are spared.[1] Hearing impairment is common.[20]

RADIOLOGIC FEATURES

Skeletal anomalies and soft tissue ossification are characteristic radiologic features of FOP.[21] Principal malformations involve the great toe, although other skeletal anomalies may occur. Osteochondromas are frequent.[2,14] Progressive fusion of cervical vertebrae may be confused with Klippel-Feil syndrome.[1,22] The femoral necks may be broad yet short.[2] However, the remainder of the skeleton is generally unremarkable.[21]

Radiographs and bone scans suggest normal modeling and remodeling of heterotopic bone.[23] Fractures are not increased and respond similarly in either the heterotopic or normotopic skeleton.[24]

Bone scans detect abnormalities in soft tissue before heterotopic ossification can be shown by conventional radiographs.[23] CT and MRI of early lesions have been described.[25]

LABORATORY FINDINGS

Routine biochemical studies of mineral metabolism are usually normal, although serum alkaline phosphatase activity may be increased, especially during disease flare-ups.[1,2,26] Urinary basic fibroblast growth factor levels may be elevated during flare-ups and coincide with the preosseous angiogenic fibroproliferative stage.[27]

HISTOPATHOLOGY

Early FOP lesions consist of an intense aggregation of B and T lymphocytes in the perivascular spaces of otherwise normal-appearing skeletal muscle.[28] A nearly pure T-cell infiltration occurs between edematous muscle fibers at the leading edge of a fibroproliferative lesion and is indistinguishable from aggres-

© 2008 American Society for Bone and Mineral Research

sive juvenile fibromatosis.[28,29] Immunostaining with a monoclonal antibody against bone morphogenetic protein (BMP)2/4 is intense in FOP lesions but not in aggressive fibromatosis.[29] Mast cell infiltration is seen at all stages of FOP flare-ups.[30] Lesional evolution proceeds through six well-described stages.[28] Endochondral ossification is the major pathway for heterotopic bone formation.[14] Mature osseous lesions have haversian systems and may contain hematopoietic marrow.

ETIOLOGY AND PATHOGENESIS

Similarities between FOP and the effects of *Drosophila* decapentaplegic gene (*BMP4* homolog) mutations suggested involvement of the BMP signaling pathway in the pathogenesis of FOP.[31] In fact, the BMP signaling pathway is highly dysregulated in FOP cells.[32–37] FOP cells overexpress BMP4 and are unable to appropriately upregulate the expression of multiple BMP antagonists, including Noggin and Gremlin, in response to a BMP challenge.[32,34,35] Additionally, FOP cells exhibit a defect in BMP receptor internalization and increased activation of downstream signaling, suggesting that altered BMP receptor trafficking participates in ectopic bone formation in this disease.[36,37] BMP4 transgenic mice that develop an FOP-like phenotype have been described.[38]

An initial genome-wide linkage analysis mapped FOP to 4q27–31; however, subsequent DNA sequence analysis of candidate genes did not identify any mutations.[39,40] With the discovery of additional pedigrees, a more conservative genome-wide linkage analysis excluded 4q27–31 and identified linkage of FOP to 2q23–24, a locus that includes the activin A type I receptor gene, *ACVR1*, encoding a BMP type I receptor.[41] A recurrent heterozygous missense mutation (c.617G>A; R206H) in the glycine-serine (GS) activation domain of ACVR1 was identified in all affected individuals with classic features of either sporadic or inherited FOP.[41] Protein modeling predicted destabilization of the GS domain, consistent with constitutive activation of ACVR1 as the underlying cause of the ectopic chondrogenesis, osteogenesis, and joint fusions seen in FOP.[41] Upregulation of the mutant receptor by a putative pH-sensitive salt bridge has been suggested.[42] Basal and ligand-stimulated dysregulation of BMP signaling are found in connective tissues progenitor cells from FOP patients.[43]

TREATMENT

There is no established medical treatment for FOP.[1,2,44] The disorder's rarity, variable severity, and fluctuating clinical course pose substantial uncertainties when evaluating experimental therapies. Bone marrow transplantation is ineffective, because even a normal immune system may trigger FOP flare-ups in a genetically susceptible host.[45] Binders of dietary calcium, radiotherapy, and warfarin are ineffective.[1,2,46] Limited benefits have been reported using corticosteroids and disodium etidronate in combination during flare-ups or by using isotretinoin to prevent disease activation.[47,48] However, these impressions reflect uncontrolled studies. Accordingly, medical intervention is currently supportive.[44]

Removal of FOP lesions is often followed by significant recurrence. Surgical release of joint contractures is unsuccessful and risks new, trauma-induced heterotopic ossification.[1,2] Osteotomy of ectopic bone to mobilize a joint is generally counterproductive because additional heterotopic ossification develops at the operative site. Spinal bracing is ineffective, and surgical intervention is associated with numerous complications.[18] Dental therapy should preclude injection of local anesthetics and stretching of the jaw.[1,2,15,16] In fact, newer dental techniques for focused administration of anesthetic are available. Guidelines for general anesthesia have been reported.[15] Whereas physical therapy to maintain joint mobility may be harmful by provoking or exacerbating lesions, occupational therapy and mobility evaluations are often helpful.[49]

Intramuscular injections should be avoided.[8] Prevention of falls is crucial.[50] Measures against recurrent pulmonary infections and onset of cardiopulmonary complications of restrictive lung disease are important.[19] Focused efforts based on inhibition of BMP signaling offers hope for the future.[51,52]

PROGNOSIS

Despite widespread heterotopic ossification and severe disability, some patients live productive lives into the seventh decade. Most, however, die earlier from pulmonary complications secondary to restrictive chest wall involvement.[1,2,17]

PROGRESSIVE OSSEOUS HETEROPLASIA

Research on FOP led to the discovery of progressive osseous heteroplasia (POH: MIM 166350), a distinct developmental disorder of heterotopic ossification.[53–55] Like FOP, POH is an autosomal dominant genetic disorder of heterotopic ossification. However, unlike in FOP, heterotopic ossification in POH commonly begins in the dermis and progresses to deeper tissues by an intramembraneous rather than an endochondral pathway.[55] Identification of two patients with POH-like features who also had Albright hereditary osteodystrophy suggested the possibility of a genetic link between the two conditions,[55,56] which was confirmed in a third patient with pure POH.[57] These discoveries led to the rapid identification of paternally inherited inactivating mutations of the *GNAS* gene as the genetic cause of POH.[58] Reduced expression of Gs-α, one of several proteins encoded by *GNAS*, can induce an osteoblast-like phenotype in human mesenchymal stem cells.[59]

REFERENCES

1. Connor JM, Evans DAP 1982 Fibrodysplasia ossificans progressiva: The clinical features and natural history of 34 patients. J Bone Joint Surg Br **64:**76–83.
2. Kaplan FS, Glaser DL, Shore EM, Deirmengian GK, Gupta R, Delai P, Morhart P, Smith R, Le Merrer M, Rogers JG, Connor JM, Kitterman JA 2005 The phenotype of fibrodysplasia ossificans progressiva. Clin Rev Bone Miner Metab **3:**183–188.
3. Kaplan FS, Glaser DL, Hebela N, Shore EM 2004 Heterotopic ossification. J Am Acad Orthop Surg **12:**116–125.
4. Mohler ER III, Gannon F, Reynolds C, Zimmerman R, Keane MG, Kaplan FS 2001 Bone formation and inflammation in cardiac valves. Circulation **20:**1522–1528.
5. Shore EM, Feldman GJ, Xu M, Kaplan FS 2005 The genetics of fibrodysplasia ossificans progressiva. Clin Rev Bone Miner Metab **3:**201–204.
6. Janoff HB, Muenke M, Johnson LO, Rosenberg A, Shore EM, Okereke E, Zasloff M, Kaplan FS 1996 Fibrodysplasia ossificans progressiva in two half-sisters. Evidence for maternal mosaicism. Am J Med Genet **61:**320–324.
7. Kitterman JA, Kantanie S, Rocke DM, Kaplan FS 2005 Iatrogenic harm caused by diagnostic errors in fibrodysplasia ossificans progressiva. Pediatrics **116:**654–661.
8. Janoff HB, Tabas JA, Shore EM, Muenke M, Dalinka MK, Schlesinger S, Zasloff MA, Kaplan FS 1995 Mild expression of fibrodysplasia ossificans progressiva: A report of 3 cases. J Rheumatol **22:**976–978.
9. Rocke DM, Zasloff M, Peeper J, Cohen RB, Kaplan FS 1994 Age and joint-specific risk of initial heterotopic ossification in patients who have fibrodysplasia ossificans progressiva. Clin Orthop **301:**243–248.
10. Cohen RB, Hahn GV, Tabas JA, Peeper J, Levitz CL, Sando A, Sando N, Zasloff M, Kaplan FS 1993 The natural history of het-

© 2008 American Society for Bone and Mineral Research

erotopic ossification in patients who have fibrodysplasia ossificans progressiva. A study of 44 patients. J Bone Joint Surg Am **75:**215–219.

11. Hebela N, Shore EM, Kaplan FS 2005 Three pairs of monozygotic twins with fibrodysplasia ossificans progressiva: The role of environment in the progression of heterotopic ossification. Clin Rev Bone Miner Metab **3:**205–208.
12. Lanchoney TF, Cohen RB, Rocke DM, Zasloff MA, Kaplan FS 1995 Permanent heterotopic ossification at the injection site after diphtheria-tetanus-pertussis immunizations in children who have fibrodysplasia ossificans progressiva. J Pediatr **126:**762–764.
13. Scarlett RF, Rocke DM, Kantanie S, Patel JB, Shore EM, Kaplan FS 2004 Influenza-like viral illnesses and flare-ups of fibrodysplasia ossificans progressiva (FOP). Clin Orthop **423:**275–279.
14. Kaplan FS, Tabas JA, Gannon FH, Finkel G, Hahn GV, Zasloff MA 1993 The histopathology of fibrodysplasia ossificans progressiva: An endochondral process. J Bone Joint Surg Am **75:**220–230.
15. Luchetti W, Cohen RB, Hahn GV, Rocke DM, Helpin M, Zasloff M, Kaplan FS 1996 Severe restriction in jaw movement after routine injection of local anesthetic in patients who have progressiva. Oral Surg Oral Med Oral Pathol Oral Radiol Endod **81:**21–25.
16. Janoff HB, Zasloff M, Kaplan FS 1996 Submandibular swelling in patients with fibrodysplasia ossificans progressiva. Otolaryngol Head Neck Surg **114:**599–604.
17. Kussmaul WG, Esmail AN, Sagar Y, Ross J, Gregory S, Kaplan FS 1998 Pulmonary and cardiac function in advanced fibrodysplasia ossificans progressiva. Clin Orthop **346:**104–109.
18. Shah PB, Zasloff MA, Drummond D, Kaplan FS 1994 Spinal deformity in patients who have fibrodysplasia ossificans progressiva. J Bone Joint Surg Am **76:**1442–1450.
19. Kaplan FS, Glaser DL 2005 Thoracic insufficiency syndrome in patients with fibrodysplasia ossificans progressiva. Clin Rev Bone Miner Metab **3:**213–216.
20. Levy CE, Lash AT, Janoff HB, Kaplan FS 1999 Conductive hearing loss in individuals with fibrodysplasia ossificans progressiva. Am J Audiol **8:**29–33.
21. Mahboubi S, Glaser DL, Shore EM, Kaplan FS 2001 Fibrodysplasia ossificans progressiva (FOP). Pediatr Radiol **31:**307–314.
22. Schaffer AA, Kaplan FS, Tracy MR, O'Brien ML, Dormans JP, Shore EM, Harland RM, Kusumi K 2005 Developmental anomalies of the cervical spin in patients with fibrodysplasia ossificans progressiva are distinctly different from those in patients with Klippel-Feil syndrome. Spine **30:**1379–1385.
23. Kaplan FS, Strear CM, Zasloff MA 1994 Radiographic and scintigraphic features of modeling and remodeling in the heterotopic skeleton of patients who have fibrodysplasia ossificans progressiva. Clin Orthop **304:**238–247.
24. Einhorn TA, Kaplan FS 1994 Traumatic fractures of heterotopic bone in patients who have fibrodysplasia ossificans progressiva. Clin Orthop **308:**173–177.
25. Shirkhoda A, Armin A-R, Bis KG, Makris J, Irwin RB, Shetty AN 1995 MR imaging of myositis ossificans: Variable patterns at different stages. J Magn Reson Imaging **65:**287–292.
26. Lutwak L 1964 Myositis ossificans progressiva: Mineral, metabolic, and radioactive calcium studies of the effects of hormones. Am J Med **37:**269–293.
27. Kaplan F, Sawyer J, Connors S, Keough K, Shore E, Gannon F, Glaser D, Rocke D, Zasloff M, Folkman J 1998 Urinary basic fibroblast growth factor: A biochemical marker for preosseous fibroproliferative lesions in patients with FOP. Clin Orthop **346:**59–65.
28. Gannon FH, Valentine BA, Shore EM, Zasloff MA, Kaplan FS 1998 Acute lymphocytic infiltration in an extremely early lesion of fibrodysplasia ossificans progressiva. Clin Orthop **346:**19–25.
29. Gannon F, Kaplan FS, Olmsted E, Finkel G, Zasloff M, Shore EM 1997 Differential immunostaining with bone morphogenetic protein (BMP) 2/4 in early fibromatous lesions of fibrodysplasia ossificans progressiva and aggressive juvenile fibromatosis. Hum Pathol **28:**339–343.
30. Gannon FH, Glaser D, Caron R, Thompson LD, Shore EM, Kaplan FS 2001 Mast cell involvement in fibrodysplasia ossificans progressiva. Hum Pathol **32:**842–848.
31. Kaplan F, Tabas JA, Zasloff MA 1990 Fibrodysplasia ossificans progressiva: A clue from the fly? Calcif Tissue Int **47:**117–125.
32. Shafritz AB, Shore EM, Gannon FH, Zasloff MA, Taub R, Muenke M, Kaplan FS 1996 Dysregulation of bone morphogenetic protein 4 (BMP4) gene expression in fibrodysplasia ossificans progressiva. N Engl J Med **335:**555–561.
33. Lanchoney TF, Olmsted EA, Shore EM, Gannon FA, Rosen V, Zasloff MA, Kaplan FS 1998 Characterization of bone morphogenetic protein 4 receptors in fibrodysplasia ossificans progressiva. Clin Orthop **346:**38–45.
34. Olmsted EA, Kaplan FS, Shore EM 2003 Bone morphogenetic protein-4 regulation in fibrodysplasia ossificans progressiva. Clin Orthop **408:**331–343.
35. Ahn J, Serrano de La Peña L, Shore EM, Kaplan FS 2003 Paresis of a bone morphogenetic protein antagonist response in a genetic disorder of heterotopic skeletogenesis. J Bone Joint Surg Am **85:**667–674.
36. Serrano de la Peña L, Billings PC, Fiori JL, Ahn J, Kaplan FS, Shore EM 2005 Fibrodysplasia ossificans progressiva (FOP), a disorder of ectopic osteogenesis, misregulates cell surface expression and trafficking of BMPRIA. J Bone Miner Res **20:**1168–1176.
37. Fiori JL, Billings PC, Serrano de la Peña L, Kaplan FS, Shore EM 2006 Dysregulation of the BMP-p38 MAPK signaling pathway in cells from patients with fibrodysplasia ossificans progressiva (FOP). J Bone Miner Res **21:**902–909.
38. Kan L, Hu M, Gomes WA, Kessler JA 2004 Transgenic mice overexpressing BMP4 develop a fibrodysplasia ossificans progressiva (FOP)-like phenotype. Am J Pathol **165:**1107–1115.
39. Feldman G, Li M, Martin S, Urbanek M, Urtizberea JA, Fardeau M, LeMerrer M, Connor JM, Triffitt J, Smith R, Muenke M, Kaplan FS, Shore EM 2000 Fibrodysplasia ossificans progressiva (FOP), a heritable disorder of severe heterotopic ossification, maps to human chromosome 4q27-31. Am J Hum Genet **66:**128–135.
40. Xu MQ, Feldman G, Le Merrer M, Shugart YY, Glaser DL, Urtizberea JA, Fardeau M, Connor JM, Triffitt J, Smith R, Shore EM, Kaplan FS 2000 Linkage exclusion and mutational analysis of the noggin gene in patients with fibrodysplasia ossificans progressiva. Clin Genet **58:**291–298.
41. Shore EM, Xu M, Feldman GJ, Fenstermacher DA, Cho T-J, Choi IH, Connor JM, Delai P, Glaser DL, Le Merrer M, Morhart R, Rogers JG, Smith R, Triffitt JT, Urtizberea JA, Zasloff M, Brown MA, Kaplan FS 2006 A recurrent mutation in the BMP type I receptor ACVR1 causes inherited and sporadic fibrodysplasia ossificans progressiva. Nat Genet **38:**525–527.
42. Groppe JC, Shore EM, Kaplan FS 2007 Functional modeling of the ACVR1 (R206H) mutation in FOP. Clin Orthop **462:**87–92.
43. Billings PC, Fiori JL, Bentwood JL, O'Connell MP, Jiao X, Nussbaum B, Caron RJ, Shore EM, Kaplan FS 2008 Dysregulated BMP signaling and enhanced osteogenic differentiation of connective tissue progenitor cells from patients with fibrodysplasia ossificans progressiva (FOP). J Bone Miner Res **23:**305–313.
44. Glaser DL, Kaplan FS 2005 Treatment considerations for the management of fibrodysplasia ossificans progressiva. Clin Rev Bone Miner Metab **3:**243–250.
45. Kaplan FS, Glaser DL, Shore EM, Pignolo RJ, Xu M, Zhang Y, Senitzer D, Forman SJ, Emerson SG 2007 Hematopoietic stem-cell contribution to ectopic skeletogenesis. J Bone Joint Surg Am **89:**347–357.
46. Moore SE, Jump AA, Smiley JD 1986 Effect of warfarin sodium therapy on excretion of 4-carboxy-L-glutamic acid in scleroderma, dermatomyositis, and myositis ossificans progressiva. Arthritis Rheum **29:**344–351.
47. Brantus J-F, Meunier PJ 1998 Effects of intravenous etidronate and oral corticosteroids in fibrodysplasia ossificans progressiva. Clin Orthop **346:**117–120.
48. Zasloff MA, Rocke DM, Crofford LJ, Hahn GV, Kaplan FS 1998 Treatment of patients who have fibrodysplasia ossificans progressiva with isotretinoin. Clin Orthop **346:**121–129.
49. Levy CE, Berner TF, Bendixen R 2005 Rehabilitation for individuals with fibrodysplasia ossificans progressiva. Clin Rev Bone Miner Metab. **3:**251–256.
50. Glaser DM, Rocke DM, Kaplan FS 1998 Catastrophic falls in patients who have fibrodysplasia ossificans progressiva. Clin Orthop **346:**110–116.
51. Glaser DL, Economides AN, Wang L, Liu X, Kimble RD, Fandl JP, Wilson JM, Stahl N, Kaplan FS, Shore EM 2003 In vivo somatic cell gene transfer or an engineered noggin mutein prevents BMP4-induced heterotopic ossification. J Bone Joint Surg Am **85:**2332–2342.
52. Kaplan FS, Glaser DL, Pignolo RJ, Shore EM 2007 A new era of fibrodysplasia ossificans progressiva (FOP): A druggable target for the second skeleton. Expert Opin Biol Ther **7:**705–712.
53. Kaplan FS, Craver R, MacEwen GD, Gannon FH, Finkel G, Hahn

© 2008 American Society for Bone and Mineral Research

G, Tabas J, Gardner RJ, Zasloff MA 1994 Progressive osseous heteroplasia: A distinct developmental disorder of heterotopic ossification. J Bone Joint Surg Am **76:**425–436.
54. Rosenfeld SR, Kaplan FS 1995 Progressive osseous heteroplasia in male patients. Clin Orthop **317:**243–245.
55. Kaplan FS, Shore EM 2000 Progressive osseous heteroplasia. J Bone Miner Res **15:**2084–2094.
56. Eddy MC, Jan De Beur SM, Yandow SM, McAlister WH, Shore EM, Kaplan FS, Whyte MP, Levine MA 2000 Deficiency of the alpha-subunit of the stimulatory G protein and severe extraskeletal ossification. J Bone Miner Res **15:**2074–2083.
57. Yeh GL, Mathur S, Wivel A, Li M, Gannon FH, Ulied A, Audi L, Olmstead EA, Kaplan FS, Shore EM 2000 GNAS1 mutation and Cbfa1 misexpression in a child with severe congenital platelike osteoma cutis. J Bone Miner Res **15:**2063–2073.
58. Shore EM, Ahn J, Jan de Beur S, Li M, Xu M, Gardner RJ, Zasloff MA, Whyte MP, Levine MA, Kaplan FS 2002 Paternally-inherited inactivating mutations of the GNAS1 gene in progressive osseous heteroplasia. N Engl J Med **346:**99–106.
59. Leitman SA, Ding C, Cooke DW, Levine MA 2005 Reduction in Gs-alpha induces osteogenic differentiation in human mesenchymal stem cells. Clin Orthop **434:**231–238.

Chapter 95. Osteogenesis Imperfecta

Joan C. Marini

NICHD, Bone and Extracellular Matrix Branch, National Institutes of Health, Bethesda, Maryland

INTRODUCTION

Osteogenesis imperfecta (OI), also known as brittle bone disease, is a genetic disorder of connective tissue characterized by fragile bones and a susceptibility to fracture from mild trauma and even acts of daily living.[1–3] The clinical range of this condition is extremely broad, ranging from cases that are lethal in the perinatal period to cases that may be difficult to detect and can present as early osteoporosis. Individuals with OI may have varying combinations of growth deficiency, defective tooth formation (dentinogenesis imperfecta), hearing loss, macrocephaly, blue coloration of sclerae, scoliosis, barrel chest, and ligamentous laxity. Classical OI is an autosomal dominant condition caused by defects in type I collagen, the major structural component of the extracellular matrix of bone, skin, and tendon. Classical OI is generally described using the Sillence classification,[4] a nomenclature based on clinical and radiographic features, which was first proposed in 1979. Subsequent biochemical and molecular studies have shown that the mild Sillence type I OI is caused by quantitative defects in type I collagen,[5] whereas the moderate and severe types are caused by structural defects in either of the two chains that form the type I collagen heterotrimer.[2] Recurrence of classical OI types in the children of unaffected parents is caused by parental mosaicism.[6] Recent exciting developments have shown the genetic cause of recessive OI, a relatively rare condition in which the clinical range overlaps with the lethal and severe Sillence types.[7] Autosomal recessive OI is caused by defects in two of the components of the prolyl 3-hydroxylation complex, CRTAP[8,9] and P3H1[10]; this complex modifies the α1(I) chain of collagen in the endoplasmic reticulum.[11] About 5% of OI cases are not caused by defects in type I collagen or the P3H1 hydroxylation complex; their etiology is presently unknown.

CLINICAL PRESENTATION

Because the eight types of OI vary widely in symptoms and in the timing of their onset, the diagnosis and its differential varies with the age of the individual in question. A positive family history is usually not present, because most mutations occur de novo. Prenatally, severe types II, III, VII, or VIII OI may be difficult to distinguish from thanatophoric dysplasia, camptomelic dysplasia, and achondrogenesis type I.[12] Neonatally, type III or VIII OI and infantile hypophosphatasia may have an overlapping presentation, but infantile hypophosphatasia has the radiographic distinction of spurs extending from sides of knee and elbow joints and the biochemical distinction of a low alkaline phosphatase level. In childhood diagnoses of the milder forms of OI, the major distinctions are with juvenile and idiopathic osteoporosis and child abuse. The key diagnostic element for OI is the generalized nature of the connective tissue defect, with facial features (flat midface, triangular shape, bluish sclerae, yellowish or opalescent teeth), relative macrocephaly, thoracic configuration (barrel chest or pectus excavatum), joint laxity, vertebral compressions, and growth deficiency present in variable combinations in each case. The recessive forms of OI caused by defects in the P3H1 complex overlap with types II and III OI but have the distinction of white sclerae.[8,10] When a diagnosis is still in doubt, collagen biochemical studies and DNA sequencing of type I collagen, *CRTAP* and *LEPRE1* provide helpful information on the presence of a mutation.

CLINICAL TYPES

The classification proposed by Sillence in 1979[4] (Table 1) is based on clinical and radiographic criteria that distinguished four types. Although both clinical and laboratory practice have subsequently evolved, the classification has continued to be useful and is still in general use in a modified form. The Sillence types have autosomal dominant inheritance; recurrence of a collagen mutation in the children of unaffected parents is almost always caused by parental mosaicism. More recently the types of OI have been extended to include V through VIII, although they are defined by different criteria than types I–IV (OMIM 166200, 166210, 259420, 166220). Types V (OMIM %610967) and VI (OMIM %610968) are defined by histology and clinical/radiographic signs and have an unknown etiology. Types VII (OMIM 610682) and VIII (OMIM 610915) have autosomal recessive inheritance and white sclerae. Type VII was first described histologically and clinically; it was later shown to be caused by mutations in *CRTAP*. Type VIII was first defined biochemically and molecularly as deficiency of P3H1.

The author states that she has no conflicts of interest.

Key words: brittle bone disease, recessive osteogenesis imperfecta, collagen, CRTAP, P3H1

Electronic Databases: 1. OI Mutation Consortium—http://www.oiprogram.nichd.nih.gov.

© 2008 American Society for Bone and Mineral Research

TABLE 1. OI NOSOLOGY

	Type	*Inheritance*	*Phenotype*	*Defect*
Sillence types	I	AD	Mild	Null α1(I) allele Some glycine substitutions
	II	AD	Lethal	Structural defects in Type I collagen
	III	AD	Progressive deforming	Structural defects in Type I collagen
	IV	AD	Moderate	Structural defects in Type I collagen
New types	V	AD	Hypertrophic callus Dense metaph. band Distinctive histology	Unknown
	VI	?	Mineralization defect "Fish-scale" Lamellae	Unknown
	VII	AR	Severe to lethal	Mutations in *CRTAP*
	VIII	AR	Severe to lethal	Mutations in *LEPRE1*

AD, autosomal dominant; AR, autosomal recessive.

Type I OI is the mildest form of the disorder. There is postnatal onset of fractures, usually after ambulation is attained, and even beginning in early middle age when type I OI can present as early onset osteoporosis. Fractures decrease markedly after puberty. Individuals with type I OI usually have blue sclerae and often have easy bruising. They may have hearing loss (onset as early as late childhood, but usually in the 20s) or joint hyperextensibility. Growth deficiency and long bone deformity are generally mild. Type I has been divided into A and B subtypes, based on the absence or presence of dentinogenesis imperfecta.

Type II OI is usually lethal in the perinatal period, although survival for months is not uncommon, and survival to a year or more has been noted. These individuals are often born prematurely and are small for gestational age. Legs are usually held in the frog leg position with hips abducted and knees flexed. Radiographically, long bones are extremely osteoporotic, with in utero fractures and abnormal modeling (often a crumpled cylindrical shape). The skull is severely undermineralized with wide-open anterior and posterior fontanels. Scleral hue is blue-gray. The bones of these infants are composed predominately of woven bone without haversian canals or organized lamellae. Demise is generally of pulmonary origin, especially respiratory insufficiency and pneumonias.

Type III OI is known as the progressive deforming type. Most individuals with type III OI survive childhood with severe bone dysplasia. The presentation at birth may be similar to the mild end of the type II OI spectrum. They have extremely fragile bones and, over a lifetime, will have dozens to hundreds of fractures. The long bones are soft and deform from normal muscle tension and subsequent to fractures. These individuals have extreme growth deficiency; final stature is in the range of a prepubertal child. Almost all type III cases develop scoliosis. Radiographically, metaphyseal flaring and "popcorn" formation at growth plates are seen in addition to osteoporosis. They require intensive physical rehabilitation and orthopedic care to attain assisted ambulation in childhood; many will require wheelchairs for mobility. This form is compatible with a full lifespan, although many individuals have respiratory insufficiency and cor pulmonale in middle age, and some die in infancy and childhood from respiratory causes.

Type IV is the moderately severe Sillence form. The diagnosis may be made at birth or delayed until the toddler or school ages. Scleral hue is variable. These children often have several fractures a year and bowing of their long bones. Fractures decrease after puberty. Essentially all type IV individuals have short final stature, often in the range of pubertal children; many of these children are responsive to growth hormone for significant additional height. Radiographically, they have osteoporosis and mild modeling abnormalities. They may have platybasia. Many develop vertebral compressions and scoliosis. With consistent rehabilitation intervention and orthopedic management, these individuals should be able to attain independent mobility. This form is compatible with a full lifespan.

OI/EDS is a discrete subgroup of patients who have an overlap of the skeletal symptoms of OI (type IV, usually, or III) and the joint laxity of Ehlers-Danlos syndrome. Hip dysplasia occurs in some patients and early progressive scoliosis in others. Tissue is friable, requiring extra intervention for spinal fixation. They have a mutation in the amino terminal region of the type I collagen chains that interferes with collagen N-propeptide processing.[13]

Recently, types V–VIII OI have been classified. Although these types have continued the Sillence numeration, they are based on different criteria than the Sillence types. Types V and VI were defined using bone histology distinctions and generally have a phenotype that would be included in Sillence type IV. These individuals do not have defects in type I collagen.

Types VII and VIII are recessive forms, whose phenotypes overlaps with Sillence types II and III. These two recessive forms have deficiency of components of a collagen modification complex in the endoplasmic reticulum.

Type V OI is associated with a triad of findings.[14] First, there is a radiographically dense band adjacent to the growth plate of long bones. Second, they develop hypertrophic calluses at the sites of fractures or surgical procedures. Finally, there is calcification of the membrane between the radius and ulna, leading to restricted rotation. They have normal teeth and white sclerae. On microscopy, the bony lamellae are mesh-like.

Type VI OI can be distinguished only on bone biopsy.[15] The lamellae have a "fish-scale"-like appearance under the microscope. These individuals have moderate to severe skeletal disease, with normal teeth and sclerae. Alkaline phosphatase is slightly elevated.

Type VII OI is an autosomal recessive form caused by defects in CRTAP, cartilage-associated protein.[8,9] The index pedigree occurs in an isolated First Nations community in northern Quebec.[16] These individuals have rhizomelia and moderate bone disease, associated with a hypomorphic mutation in *CRTAP*. Null mutations in *CRTAP* have been shown to cause a lethal form of OI, with white sclerae, rhizomelia, and a small to normal cranium.

Type VIII OI is an autosomal recessive form caused by

© 2008 American Society for Bone and Mineral Research

defects in prolyl 3-hydroxylase 1 (P3H1, encoded by *LEPRE1*). P3H1 forms a complex in the ER with CRTAP, and there is considerable overlap in the phenotypes of OI VII and VIII. Null mutations in *LEPRE1* result in a phenotype that overlaps types II and III OI, but has distinct features, including white sclerae, extreme growth deficiency, and undermineralization.[10] For example, a 15-yr-old child is the length of a 3 yr old, and a 3 yr old is the length of a 3 mo old. There is a recurring *LEPRE1* mutation that occurs in West Africans and blacks and is lethal in homozygous individuals.

RADIOGRAPHIC FEATURES

The skeletal survey in classical OI shows generalized osteopenia. Long bones have thin cortices and a gracile appearance. In moderately to severely affected patients, long bones have bowing and modeling deformities, including cylindrical configuration from an apparent lack of modeling, metaphyseal flaring, and "popcorn" appearance at the metaphyses.[17] Long bones of the upper extremity often seem milder than those of the lower extremity, even without weight bearing. Vertebrae often have central compressions even in mild type I OI; these often appear first at the T_{12}–L_1 level, consistent with weight-bearing stress. In moderate to severe OI, vertebrae will have central and anterior compressions and may appear compressed throughout. The compressions are generally consistent with the patient's L_1–L_4 DXA Z-score but do not correlate in a straightforward manner with scoliosis. In the lateral plane film of the spine, it is not easy to assess the asymmetry of vertebral collapse, which, along with paraspinal ligamentous laxity, is generally the cause of OI scoliosis. The skull of OI patients with a wide range of severity has wormian bones, although this is not unique to OI. Patients with type III and IV OI may also have platybasia, which should be followed with periodic CT studies for basilar impression and invagination.[18]

The skeletal radiographic appearance of only a few infants and children with types VII and VIII OI have been described.[8–10] Both groups have extreme osteoporosis and abnormal long bone modeling, leading to a cylindrical appearance. The bone material appears cystic and disorganized. In surviving children with type VIII, there is flaring of the metaphyses. These children also have gracile hands that appear relatively long but have shortened metacarpals.

Bone densitometry by DXA (L_1–L_4) is useful over a wide age and severity range of OI. It aids diagnosis in milder cases and facilitates longitudinal follow-up in moderate to severe forms. There is a general correlation of Z-score and severity of OI. Type I individuals are generally in the –1 to –2 range, type IV Z-scores cluster in the –2 to –4 range, whereas type III spans –3 to –6. Children with type VIII OI have –6 to –7 Z-scores. It is important to remember that the Z-score compares the mineral quantity of the bone being studied to bone with a normal matrix structure and crystal alignment. In OI, many mutations result in irregular crystal alignment on the abnormal matrix, in addition to reduced mineral quantity. DXA does not measure bone quality, which includes bone geometry, histomorphometry, and mechanical properties.

LABORATORY FINDINGS

Serum chemistries related to bone and mineral metabolism are generally normal. Alkaline phosphatase may be elevated after a fracture and is slightly elevated in type VI.[15] Acid phosphatase is elevated in type VIII OI and can logically be expected to be elevated in type VII. Hormones of the growth axis have normal levels.[19]

Bone histomorphometry shows defects in bone modeling and in production and thickening of trabeculae.[20] Cortical width and cancellous bone volume are decreased in all types; trabecular number and width are decreased. Bone remodeling is increased, as are osteoblast and osteoclasts surfaces. When viewed under polarized light, the lamellae of OI bone are thinner and less smooth than in controls. Mineral apposition rate is normal; crystal disorganization may contribute to bone weakness.

ETIOLOGY AND PATHOGENESIS

About 85–90% of patients who have clinical OI have abnormalities of type I collagen, the major structural protein of the bone extracellular matrix. The recessive types VII and VIII OI have a defect in a component of a complex that interacts with collagen post-translationally and 3-hydroxylates α1(I)Pro986 in the endoplasmic reticulum.[7] These forms together can be estimated to make up 5–7% of OI cases. Patients with types V and VI OI and a small group of unclassified patients with clinical OI do not have mutations in type I collagen; type V OI has been estimated to make up 5% of OI cases.[14]

Cultured dermal fibroblasts are convenient cells in which to examine the collagen biochemistry of probands using gel electrophoresis. Probands with type I OI, who synthesize a reduced amount of structurally normal type I collagen because of a null *COL1A1* allele, display a relative increase in the COL3/COL1 ratio.[5] Probands with the clinically significant types II, III, and IV OI synthesize a mixture of normal collagen and collagen with a structural defects. With rare exceptions, the structural defects are either substitutions for one of the glycine residues that occur at every third position along the chain and are essential for proper helix folding (80%) or alternative splicing of an exon (20%), resulting in an in-frame deletion of a section of the chain. Structural abnormalities delay helix folding, expose the constituent chains to modifying enzymes for a longer time, and result in overmodification detectable as slower electrophoretic migration. The biochemical test does not accurately detect abnormalities in the amino one third of the α1(I) or amino one half of the α2(I) chain.[21] Cultured fibroblasts from patients with types VII and VIII OI also have overmodification of the helical regions of the collagen chains, suggesting that deficiency of the P3H1 complex delays helix folding.[8,10]

Mutation detection by direct sequencing is more sensitive than the biochemical test, although it does not provide functional information. Collagen sequencing is available either as exon-by-exon sequencing of DNA or transcript sequencing as cDNA. Each technique will miss a small percent of unusual mutations, such as large deletions or rearrangements or low-percentage splicing defects.

Genotype–phenotype modeling of the >800 mutations currently available has yielded different patterns for the two chains, supporting distinct roles in maintaining matrix integrity.[22] About one third of the substitutions in α1(I) are lethal, especially those to residues with a branched or charged side chain. Two exclusively lethal regions coincide with the proposed major ligand-binding regions for the collagen monomer with integrins, matrix metalloproteinases (MMPs), fibronectin, and cartilage oligomeric matrix protein (COMP). For the α2(I) chain, only one fifth of substitutions are lethal; these substitutions are clustered in eight regularly spaced regions along the chain, coinciding with the proteoglycan binding regions on the collagen fibril.

TREATMENT

Early and consistent rehabilitation intervention is the basis for maximizing the physical potential of individuals with OI.[23,24] Physical therapy should begin in infancy for the se-

© 2008 American Society for Bone and Mineral Research

verest types, promoting muscle strengthening, aerobic conditioning, and if possible, protected ambulation. Programs to assure that children have muscle strength to lift a limb against gravity should continue between orthopedic interventions using isotonic and aerobic conditioning. Swimming should be encouraged.

Orthopedic care should be in the hands of a surgeon with experience in OI. Fractures should not be allowed to heal without reduction to prevent loss of function. The goals of orthopedic surgery are to correct deformity for ambulation and to interrupt a cycle of fracture and refracture. The classic osteotomy procedure requires fixation with an intramedullary rod. The hardware currently in use includes telescoping rods (Bailey-Dubow[25] or Fassier-Duval[26] rods) and nonelongating rods (Rush rods). Important considerations include selection of a rod with the smallest diameter suited to the situation to avoid cortical atrophy. Children who are anticipated to have significant growth may require fewer rod revisions with either of the extensible rods.

The secondary features of OI, including abnormal pulmonary functions, hearing loss, and basilar invagination, are best managed in a specialized coordinated care program. The severe growth deficiency of OI is responsive to exogenous growth hormone administration in about one half of cases of type IV OI[27] and most type I OI[28]; some treated children can attain heights within the normal growth curves. Responders to recombinant growth hormone (rGH) also experience increased L_1–L_4 DXA, bone volume per total volume (BV/TV), and bone formation rate (BFR). Growth hormone remains under study for its effects on OI skeletal integrity.

Four controlled trials have shown the benefits and limitations of bisphosphonate treatment for OI.[29–32] The trabecular bone of vertebral bodies has the most positive response. BMD is increased, although the functional meaning of this measurement is difficult to assess because it also includes retained mineralized cartilage; the increase in Z-scores tapers after 1- to 2-yr treatment. More importantly, the vertebral ability to resist compressive forces is shown as increased vertebral area and decreased central vertebral compressions. The effect of bisphosphonate treatment on predominately cortical long bone is more equivocal. There is a combination of increased stiffness and load bearing that is balanced by weakened bone quality.[33] There is, at best, a trend toward reduced fracture incidence or a reduced relative risk rather than a clear statistical benefit. The functional changes in ambulation, muscle strength, and bone pain reported in the uncontrolled trails have been shown to be placebo effects. The prolonged half-life and recirculation of pamidronate in children up to 8 yr after treatment cessation may pose pediatric specific skeletal and reproductive risks.[34] Prolonged or high-dose administration to children can induce defective bone remodeling[35] and may lead to accumulation of bone microdamage. Delayed osteotomy healing was noted at conventional doses.[36] Our current management of bisphosphonates for OI is to treat for 2–3 yr and then discontinue the drug but continue to follow the patient.

REFERENCES

1. Marini JC 2004 Osteogenesis imperfecta. In: Behrman RE, Kliegman RM, Jenson HB (eds.) Nelson Textbook of Pediatrics, 17th ed. Saunders, Philadelphia, PA, USA, pp. 2336–2338.
2. Byers PH, Cole WG 2002 Osteogenesis imperfecta. In: Royce PM, Steinman B (eds.) Connective Tissue and Its Heritable Disorders. Wiley-Liss, New York, NY, USA, pp. 385–430.
3. Kuivaniemi H, Tromp G, Prockop DJ 1997 Mutations in fibrillar collagens (types I, II, III, and XI), fibril-associated collagen (type IX), and network-forming collagen (type X) cause a spectrum of diseases of bone, cartilage, and blood vessels. Hum Mutat **9:**300–315.
4. Sillence DO, Senn A, Danks DM 1979 Genetic heterogeneity in osteogenesis imperfecta. J Med Genet **16:**101–116.
5. Willing MC, Pruchno CJ, Byers PH 1993 Molecular heterogeneity in osteogenesis imperfecta type I. Am J Med Genet **45:**223–227.
6. Cohn DH, Starman BJ, Blumberg B, Byers PH 1990 Recurrence of lethal osteogenesis imperfecta due to parental mosaicism for a dominant mutation in a human type I collagen gene (COL1A1). Am J Hum Genet **46:**591–601.
7. Marini JC, Cabral WA, Barnes AM, Chang W 2007 Components of the collagen prolyl 3-hydroxylation complex are crucial for normal bone development. Cell Cycle **6:**1675–1681.
8. Barnes AM, Chang W, Morello R, Cabral WA, Weis M, Eyre DR, Leikin S, Makareeva E, Kuznetsova N, Uveges TE, Ashok A, Flor AW, Mulvihill JJ, Wilson PL, Sundaram UT, Lee B, Marini JC 2006 Deficiency of cartilage-associated protein in recessive lethal osteogenesis imperfecta. N Engl J Med **355:**2757–2764.
9. Morello R, Bertin TK, Chen Y, Hicks J, Tonachini L, Monticone M, Castagnola P, Rauch F, Glorieux FH, Vranka J, Bachinger HP, Pace JM, Schwarze U, Byers PH, Weis M, Fernandes RJ, Eyre DR, Yao Z, Boyce BF, Lee B 2006 CRTAP is required for prolyl 3- hydroxylation and mutations cause recessive osteogenesis imperfecta. Cell **127:**291–304.
10. Cabral WA, Chang W, Barnes AM, Weis M, Scott MA, Leikin S, Makareeva E, Kuznetsova NV, Rosenbaum KN, Tifft CJ, Bulas DI, Kozma C, Smith PA, Eyre DR, Marini JC 2007 Prolyl 3-hydroxylase 1 deficiency causes a recessive metabolic bone disorder resembling lethal/severe osteogenesis imperfecta. Nat Genet **39:**359–365.
11. Vranka JA, Sakai LY, Bachinger HP 2004 Prolyl 3-hydroxylase 1, enzyme characterization and identification of a novel family of enzymes. J Biol Chem **279:**23615–23621.
12. Marini JC, Chernoff EJ 2001 Osteogenesis imperfecta. In: Cassidy SB, Allanson JE (eds.) Management of Genetic Syndromes. Wiley-Liss, New York, NY, USA, pp. 281–300.
13. Cabral WA, Makareeva E, Colige A, Letocha AD, Ty JM, Yeowell HN, Pals G, Leikin S, Marini JC 2005 Mutations near amino end of alpha1(I) collagen cause combined osteogenesis imperfecta/Ehlers-Danlos syndrome by interference with N-propeptide processing. J Biol Chem **280:**19259–19269.
14. Glorieux FH, Rauch F, Plotkin H, Ward L, Travers R, Roughley P, Lalic L, Glorieux DF, Fassier F, Bishop NJ 2000 Type V osteogenesis imperfecta: A new form of brittle bone disease. J Bone Miner Res **15:**1650–1658.
15. Glorieux FH, Ward LM, Rauch F, Lalic L, Roughley PJ, Travers R 2002 Osteogenesis imperfecta type VI: A form of brittle bone disease with a mineralization defect. J Bone Miner Res **17:**30–38.
16. Ward LM, Rauch F, Travers R, Chabot G, Azouz EM, Lalic L, Roughley PJ, Glorieux FH 2002 Osteogenesis imperfecta type VII: An autosomal recessive form of brittle bone disease. Bone **31:**12–18.
17. Goldman AB, Davidson D, Pavlov H, Bullough PG 1980 "Popcorn" calcifications: A prognostic sign in osteogenesis imperfecta. Radiology **136:**351–358.
18. Charnas LR, Marini JC 1993 Communicating hydrocephalus, basilar invagination, and other neurologic features in osteogenesis imperfecta. Neurology **43:**2603–2608.
19. Marini JC, Bordenick S, Heavner G, Rose S, Hintz R, Rosenfeld R, Chrousos GP 1993 The growth hormone and somatomedin axis in short children with osteogenesis imperfecta. J Clin Endocrinol Metab **76:**251–256.
20. Rauch F, Travers R, Parfitt AM, Glorieux FH 2000 Static and dynamic bone histomorphometry in children with osteogenesis imperfecta. Bone **26:**581–589.
21. Cabral WA, Milgrom S, Letocha AD, Moriarty E, Marini JC 2006 Biochemical screening of type I collagen in osteogenesis imperfecta: Detection of glycine substitutions in the amino end of the alpha chains requires supplementation by molecular analysis. J Med Genet **43:**685–690.
22. Marini JC, Forlino A, Cabral WA, Barnes AM, San Antonio JD, Milgrom S, Hyland JC, Korkko J, Prockop DJ, De Paepe A, Coucke P, Symoens S, Glorieux FH, Roughley PJ, Lund AM, Kuurila-Svahn K, Hartikka H, Cohn DH, Krakow D, Mottes M, Schwarze U, Chen D, Yang K, Kuslich C, Troendle J, Dalgleish R, Byers PH 2007 Consortium for osteogenesis imperfecta mutations in the helical domain of type I collagen: Regions rich in lethal

© 2008 American Society for Bone and Mineral Research

mutations align with collagen binding sites for integrins and proteoglycans. Hum Mutat **28:**209–221.
23. Binder H, Conway A, Hason S, Gerber LH, Marini J, Berry R, Weintrob J 1993 Comprehensive rehabilitation of the child with osteogenesis imperfecta. Am J Med Genet **45:**265–269.
24. Gerber LH, Binder H, Weintrob J, Grange DK, Shapiro J, Fromherz W, Berry R, Conway A, Nason S, Marini J 1990 Rehabilitation of children and infants with osteogenesis imperfecta. A program for ambulation. Clin Orthop **251:**254–262.
25. Zionts LE, Ebramzadeh E, Stott NS 1998 Complications in the use of the Bailey-Dubow extensible nail. Clin Orthop **348:**186–195.
26. Fassier F 2005 Experience with the Fassier-Duval rod: Effectiveness and complications. 9th International Conference on Osteogenesis Imperfecta, June 13–16, 2005, Annapolis, MD, USA.
27. Marini JC, Hopkins E, Glorieux FH, Chrousos GP, Reynolds JC, Gundberg CM, Reing CM 2003 Positive linear growth and bone responses to growth hormone treatment in children with types III and IV osteogenesis imperfecta: High predictive value of the carboxyterminal propeptide of type I procollagen. J Bone Miner Res **18:**237–243.
28. Antoniazzi F, Bertoldo F, Mottes M, Valli M, Sirpresi S, Zamboni G, Valentini R, Tato L 1996 Growth hormone treatment in osteogenesis imperfecta with quantitative defect of type I collagen synthesis. J Pediatr **129:**432–439.
29. Gatti D, Antoniazzi F, Prizzi R, Braga V, Rossini M, Tato L, Viapiana O, Adami S 2005 Intravenous neridronate in children with osteogenesis imperfecta: A randomized controlled study. J Bone Miner Res **20:**758–763.
30. Glorieux FH, Rauch F, Ward LM, Smith PA, Verbruggen N, Heyden N, Lombardi A 2004 Alendronate in the treatment of pediatric osteogenesis imperfecta. J Bone Miner Res **19:**1043.
31. Letocha AD, Cintas HL, Troendle JF, Reynolds JC, Cann CE, Chernoff EJ, Hill SC, Gerber LH, Marini JC 2005 Controlled trial of pamidronate in children with types III and IV osteogenesis imperfecta confirms vertebral gains but not short-term functional improvement. J Bone Miner Res **20:**977–986.
32. Sakkers R, Kok D, Engelbert R, van Dongen A, Jansen M, Pruijs H, Verbout A, Schweitzer D, Uiterwaal C 2004 Skeletal effects and functional outcome with olpadronate in children with osteogenesis imperfecta: A 2-year randomised placebo-controlled study. Lancet **363:**1427–1431.
33. Uveges TE, Kozloff KM, Ty JM, Gronowicz G, Ledgard F, Goldstein SA, Marini JC 2004 Alendronate treatment of the Brtl mouse model for osteogenesis imperfecta improved bone geometry and loading before fracture but decreased bone mineral quality and alters osteoblast morphology. Eur J Hum Genet **12:**1208.
34. Papapoulos SE, Cremers SC 2007 Prolonged bisphosphonate release after treatment in children. N Engl J Med **356:**1075–1076.
35. Whyte MP, Wenkert D, Clements KL, McAlister WH, Mumm S 2003 Bisphosphonate-induced osteopetrosis. N Engl J Med **349:**457–463.
36. Munns CF, Rauch F, Zeitlin L, Fassier F, Glorieux FH 2004 Delayed osteotomy but not fracture healing in pediatric osteogenesis imperfecta patients receiving pamidronate. J Bone Miner Res **19:**1779–1786.

Chapter 96. Marfan Syndrome and Related Disorders of the Connective Tissue

Emilio Arteaga-Solis[1] and Francesco Ramirez[2]

[1]Pulmonary Medicine Division, Department of Pediatrics, Columbia University College of Physician and Surgeons, New York, New York; [2]Department of Pharmacology and Systems Therapeutics at the Mount Sinai School of Medicine, New York, New York

INTRODUCTION

Marfan syndrome (MFS; OMIM 1547000) is a dominantly inherited disorder of the connective tissue with cardinal manifestations in the cardiovascular, ocular, and skeletal systems.[1] MFS has an estimated incidence of 2–3 per 10,000 individuals and shows high penetrance and marked clinical variability. Diagnosis is based on the Ghent nosology that differentiates MFS from conditions with distinct genetic cause, repertoire of manifestations, natural history, and response to treatment.[2] MFS is caused by mutations in the gene coding for fibrillin-1 (FBN1), the major structural component of extracellular microfibrils.[3,4] Fibrillin-rich microfibrils (either associated with elastin in the elastic fibers or as elastin-free assemblies) impart physical properties to connective tissues, in addition to participating in the regulation of local TGF-β bioavailability.[3,4] Decreased tissue integrity and perturbed cell behavior are therefore believed to be pathogenic components of the MFS phenotype.[3] In this view, MFS is part of a larger group of connective tissue disorders that include Shprintzen-Goldberg syndrome (SGS; OMIM 182212) and Weill-Marchesani syndrome (WMS; OMIM 608328), which do not fulfill the Ghent nosology but are occasionally caused by *FBN1* mutations; congenital contractural arachnodactyly (CCA; OMIM 121050), which is caused by mutations in the gene coding for the structurally and functionally related fibrillin-2 protein (FBN2); and Loeys-Dietz syndrome (LDS)-I and -II (OMIM 609192 and 130050, respectively), which are both associated with mutations in TGF-β receptors (TGFBR1 and TGFBR2).[3,4]

MARFAN SYNDROME

Diagnosis of MFS requires a combination of manifestations in different organ systems that are defined as major and minor clinical criteria based on how frequently they appear in the general population.[2] In the case of the skeletal system, many of the MFS abnormalities are either commonly found in the general population or present in other connective tissue disorders (Table 1).[5–7] Hence, four of eight major skeletal manifestations need to be present in a proband to fulfill the high diagnostic specificity of a major criterion of MFS.[2] Craniofacial anomalies are the least specific and include high-arched palate, long narrow skull (dolichocephaly), recessed lower mandible (retrognathia), malar hypoplasia, sunken eye (enophthalmos), and downward-slanting palpebral fissures. Major manifestations in the axial skeleton include longitudinal rib overgrowth causing anterior or posterior displacement of the sternum (pectus carinatum or pectus excavatum, respectively), scoliosis >20° (generally thoracic and convex to the right), vertebral displacement (spondylolisthesis), and enlargement of the outer layer of the meningeal sac in the lumbosacral spine

The authors state that they have no conflicts of interest.

Key words: animal models—rodents, matrix proteins, noncollagenous proteins, gene/genetic research, knockout/in, growth and development, patterning of the limb, modeling and remodeling, molecular pathways, orthopedics

© 2008 American Society for Bone and Mineral Research

TABLE 1. SKELETAL FEATURES OF MFS AND RELATED DISORDERS

	MFS	*LDS-I*	*LDS-II*	*SGS*	*WMS*	*CCA*
Axial skeletal abnormalities						
Dural ectasia	+++	++	++	–	–	–
Pectus deformity	+++	+++	+++	+++	–	+++
Scoliosis	+++	+++	+++	++	–	+++
Appendicular skeletal abnormalities						
Arachnodactyly	+++	++	++	+++	–	+++
Brachydactyly	–	–	–	–	+++	–
Camptodactyly	+	++	++	++	++	++
Club foot deformity	–	++	++	+	–	–
Dolichostenomelia	+++	+/–	+/–	+++	–	+++
Joint laxity	++	+++	+++	++	–	+/–
Pes planus	++	++	++	–	–	++
Protusio acetabuli	+++	–	–	–	–	+
Short stature	–	–	–	–	+++	–
Osteoporosis	+/–	NR	NR	+	+/–	++
Craniofacial abnormalities						
Bifid uvula	–	+++	+/–	+	–	–
Cleft palate	–	++	–	+	–	–
Craniosynostotis	–	++	–	++	–	+/–
Dolichocephaly	+++	+/–	+/–	–	–	–
Down-slant of palpebral fissures	++	++	++	+++	–	–
Early and sever myopia	+++	–	–	+	+++	–
Ectopia lentis	+++	–	–	–	++	–
Enophthalmos	++	+	+	–	–	–
Exophthalmos	–	–	–	+++	–	–
High arched palate	+++	+++	+++	+++	–	+
Hypertelorism	–	+++	–	+++	–	+/–
Malar hypoplasia	+++	+++	+++	++	+	–
Micrognatia	++	+++	+++	+++	–	+

–, not associated; **+/–**, rare and or subtle; **+**, occasionally observed; **++**, commonly observed; **+++**, generally observed; **NR**, not reported.

(dural ectasia). The requirement of corrective surgery for pectus excavatum defines it as a major manifestation. Disproportionate long limbs (dolichostenomelia) and fingers (arachnodactyly) are often the most striking and evident manifestations in MFS. Dolichostenomelia is a major manifestation of the appendicular skeleton that is defined as an upper body to lower extremity segment ratio <0.85 or as an arm span to height ratio >1.05. Arachnodactyly, on the other hand, is considered a less specific finding except in combination with joint hypermobility, when it gives rise to the highly specific signs of the wrist (overlap between thumb and fifth finger when wrapped around the contralateral wrist) and thumb (projection of the thumb well beyond the ulnar border of the hand when folded across the palm). Other major findings in the appendicular skeleton include medial displacement of the medial malleolus associated with flat feet (pes planus), medial protrusion of the femoral head (protusio acetabuli), and reduced elbow extension.

Albeit relatively common, vertebral abnormalities, such as decreased pedicles width and laminar thicknesses, increased interpediculate distance (IPD), and vertebral scalloping, are not included in the Ghent nosology.[2,8,9] However, IPD might provide a new screening tool or diagnostic criterion for MFS because it has been reported to exhibit 75% sensitivity and 95% specificity in measuring L_4 widening.[10] Osteopenia is a controversial finding in MFS, especially in pediatric patients.[11–16] Whereas some studies using DXA have identified reduced BMD in pre- and postmenopausal women and adolescents with MFS, other studies have been more equivocal.[11–16] One of the main reasons for this discrepancy is that the interpretation of DXA data from MFS patients is more challenging than in other individuals because of the significantly longer bones. As a result, the etiology and significance of low BMD in relationship to a higher long-term risk for fractures remain uncertain, and application of current therapies of bone mineral replacement in MFS patients is generally viewed as premature.

Dural ectasia (usually at the L_5–S_2 level) is a common manifestation that affects 63–92% of adults and 40% of children with MFS.[17–21] The abnormality is often associated with herniation of the nerve root and promotion of secondary changes in the osseous structures that result in thinner pedicles and laminae and increased risk of dural tear and failure of corrective surgery. In many instances, dural ectasia is also associated with moderate to severe low back pain, headache, and proximal leg pain, as well as weakness and numbness.[22,23] Surgical repair of severe pectus excavatum is recommended to ameliorate cardiac and pulmonary function.[24–26] Corrective surgery is commonly performed after patients reach skeletal maturity to avoid recurrence of chest abnormality caused by continued rib overgrowth.[27] Spine deformities in MFS include progressive scoliosis, thoracic lordosis, and more rarely, thoracolumbar kyphosis, spondylolistethis, and cervical problems.[6] Scoliosis worsens in MFS patients with >30° curve at a higher rate than in individuals with idiopathic scoliosis.[7] In contrast to idiopathic scoliosis, spinal bracing in MFS is usually inadequate to manage severely progressive scoliosis, and surgical repair carries a greater risk of complications.[6,28–30] Infantile scoliosis, mainly associated with the severe form of MFS, is relatively rare (2–3%), and neither bracing nor surgery is indicated before 4 yr of age.[31] Similarly unnecessary is surgical correction of protusio acetabuli because hip function is usually unaffected.[32]

© 2008 American Society for Bone and Mineral Research

MFS PATHOGENESIS

Genotype–phenotype correlations in MFS have largely failed to identify fibrillin-1 domains with tissue-specific functions (UMD-FBN1 mutation database; http://www.umd.necker.fr/).[33] Moreover, similar presentations of the disease are associated with mutations that affect either the structure or the amount of fibrillin-1.[34] Studies of mice harboring missense or hypomorphic mutations of *Fbn1* have correlated aortic aneurysm, myxomatous cardiac valve disease, developmental emphysema, and muscle hypoplasia with elevated TGF-β activity in the respective tissues.[35–38] They have also documented the efficacy of TGF-β antagonism to rescue these abnormalities in mice, thus paving the way to a new drug-based strategy against the life-threatening manifestations of MFS.[37] The discovery that perturbed TGF-β signaling is a critical determinant of clinical severity in MFS reflects the emerging role of fibrillin-rich microfibrils in modulating TGF-β bioavailability through the sequestration of the latent complex into the matrix.[4] Furthermore, this instructive role of microfibril assemblies is also consistent with the notion that selected MFS manifestations, such as bone overgrowth, are more likely to originate from cellular rather than structural deficits.[1,4] However, information regarding the possible involvement of perturbed TGF-β signaling in the MFS skeleton is still limited.

Fibrillin gene expression in the forming skeleton begins before mesenchyme differentiation, and thereafter fibrillins accumulate widely and abundantly at many skeletal sites where they form distinct macromolecular assemblies, such as uninterrupted elastic fibers running along the entire length of the periosteal matrix of long bones, circumferential bundles of microfibrils wrapped around the Ranvier's groove, and compact fiber-like microfibrils deposited immediately around chondrocytes, osteocytes, and osteons, at the endochondral surface, and within the trabecular matrix.[39–42] Whereas some investigators have argued that elastic fibers in the perichondrial/periosteal matrix may constrain bone growth at both ends, others have assigned a comparable function to the microfibrils tightly packed around growth plate chondrocytes.[40,41] A recent study of young MFS patients has indicated that skeletal overgrowth is in part the result of an earlier and longer-duration peak velocity during adolescence.[43] Although mouse models of MFS display only a subset of skeletal manifestations,[44] insights into relevant mechanisms of pathogenesis are beginning to emerge. Dural attenuation in fibrillin-1 underexpressing (mgR/mgR) mice has in fact been associated with elastic fiber disorganization and increased matrix metalloproteinase (MMP-2) and TGF-β activity, whereas osteopenia in these mice seems to be accounted for by increased bone resorption.[7,45]

MARFAN-RELATED DISORDERS

In rare instances, *FBN1* mutations can also cause SGS or WMS.[46–48] The former is a rare autosomal dominant disorder in which cranial malformations and neurodevelopmental deficits associate with MFS-like manifestations; additionally, there is also significant clinical overlap between SGS and LDS (Table 1). In point of fact, common phenotypic features amongst MFS, SGS, and LDS have generated a few controversies, including the now refuted claim of genetic heterogeneity in MFS.[48–53] Skeletal manifestations and craniofacial abnormalities in SGS include craniosynostosis, micrognathia, arched palate, bifid uvula, hypertelorism, exophthalmos, bone overgrowth, arachnodactyly, joint hypermobility, pectus deformity, and scoliosis (Table 1). Disease-causing *FBN1* mutations have only been identified in two unrelated SGS patients, and neither of them maps to a region of the protein devoid of MFS-causing defects.[46,48] WMS is a clinically homogeneous but genetically heterogeneous disorder of the connective tissue characterized by short stature, short and stubby hands and feet (brachydactyly), joint stiffness, and characteristic eye anomalies (Table 1). An in-frame i deletion has been identified in a patient with autosomal dominant WMS, whereas nonsense and frame-shift mutations in the ADAMTS10 protease have been reported in two individuals with autosomal recessive WMS.[47,54] The pathophysiological significance of these findings remains to be elucidated.

LDS-I and LDS-II are autosomal dominant disorders caused by heterozygous mutations in the *TGFBR1* or *TGFBR2* genes.[52,55] They display both MFS-like and unique features, and differ from each other for the presence (LDS I) or absence (LDS II) of some craniofacial anomalies. LDS-II overlaps substantially with the vascular form of Ehlers-Danlos syndrome (EDS IV; OMIM 130050), a condition typically caused by mutations in type III collagen (COL3A1). Craniofacial and skeletal manifestations in LDS include hypertelorism, uvula malformations, malar hypoplasia, arched palate, retrognathia, pectus deformity, scoliosis, joint laxity, and dural ectasia; craniosynostosis, cleft palate, and clubfoot deformity can also be present, whereas overgrowth of long bones is mild or absent (Table 1). *TGFBR* mutations in LDS are predicted to impair receptor function and thus blunt TGF-β signaling. Paradoxically, they are instead associated with increased TGF-β signaling in mutant tissues and cells.[52,55] It has been argued that TGFBR dysfunctions in LDS either activate an unproductive compensatory response or have gain-of-function properties.[4]

CCA (Beals syndrome) is a dominantly inherited disorder akin to, but far more rare than, MFS. CCA is characterized by multiple joint contractures, arachnodactyly, dolichostenomelia, scoliosis, and osteopenia (Table 1).[56,57] Calf muscle hypoplasia and crumpled ears are other distinguishing features of CCA.[56,57] Differential diagnosis between MFS and CCA is sometimes difficult because of the presence of additional overlapping features. Cardiovascular abnormalities and ocular complications of different severity have been reported in CCA, and crumpled ears and finger contractures (camplodactyly) have been observed in patients with the most severe (neonatal) form of MFS.[57–60] Although generally milder than MFS, severe and neonatal forms of CCA have been described.[59,61] Elbow, knee, and finger contractures are found at birth in all CCA patients and generally improve with time. Kyphoscoliosis is often present at birth or early in childhood and contributes to morbidity later in life. Management of spinal deformity in CCA patients is symptomatic. *FBN2* mutations in CCA cluster to a region of fibrillin-2 that corresponds to the fibrillin-1 segment where mutations causing neonatal MFS are located.[62] The functional significance of this observation is still obscure.

Fibrillins 1 and 2 exhibit superimposable primary structures predominantly made of multiple cysteine-rich domains.[63] During development, fibrillin-2 is generally synthesized before and in lower amounts than fibrillin-1, and both molecules can be part of the same microfibril assemblies.[39] Studies of mice with graded deficiencies of fibrillins 1 and 2 have documented both unique and overlapping roles of the two proteins.[44,64,65] Preliminary data from *Fbn2*-null mice have implicated promiscuous TGF-β activation in the osteopenic phenotype of this animal model of CCA.[3] The same mutant mice have also linked fibrillin-2 microfibrils to the regulation of a bone morphogenetic protein (BMP)-driven program of digit formation that does not involve participation of fibrillin-1 molecules.[65] These and other data thus strongly suggest that the composi-

© 2008 American Society for Bone and Mineral Research

tional diversification of microfibril assemblies may provide a structural context to the activity of TGF-β/BMP signaling molecules.[4]

ACKNOWLEDGMENTS

The authors thank Karen Johnson for organizing the manuscript. Studies from the authors' laboratory that are described in the review were supported by grants from the National Institutes of Health (AR-42044, AR-049698).

REFERENCES

1. Judge DP, Dietz HC 2005 Marfan's syndrome. Lancet **366:**1965–1976.
2. De Paepe A, Devereux RB, Dietz HC, Hennekam RC, Pyeritz RE 1996 Revised diagnostic criteria for the Marfan syndrome. Am J Med Genet **62:**417–426.
3. Robinson PN, Arteaga-Solis E, Baldock C, Collod-Béroud G, De Paepe A, Dietz HC, Guo G, Handford PA, Judge DP, Kielty CM, Loeys B, Milewicz DM, Ney A, Ramirez F, Reinhardt DP, Tiedemann K, Whiteman P, Godfrey M 2006 The molecular genetics of Marfan syndrome and related disorders. J Med Genet **43:**769–787.
4. Ramirez F, Dietz HC 2007 Marfan syndrome: From molecular pathogenesis to clinical treatment. Curr Opin Genet Dev **17:**252–258.
5. Giampietro PF, Raggio C, Davis JG 2002 Marfan syndrome: Orthopedic and genetic review. Curr Opin Pediatr **14:**35–41.
6. Sponseller P, Shindle M 2004 Orthopedic problems in Marfan syndrome. In: Robinson PN, Godfrey M (eds.) Marfan Syndrome: A Primer for Clinicians and Scientists. Kluwer Academics/Plenum Publishers, New York, New York, USA, pp. 24–34.
7. Jones K, Sponseller PD, Erkula G, Sakai L, Ramirez F, Dietz HC, Kost-Byerly S, Bridwell KH, Sandell L 2007 Symposium on the musculoskeletal aspects of Marfan syndrome: Meeting report and state of the science. J Orthop Res **25:**413–422.
8. Sponseller PD, Hobbs W, Riley LH III, Pyeritz RE 1995 The thoracolumbar spine in Marfan syndrome. J Bone Joint Surg Am **77:**867–876.
9. Sponseller PD, Ahn NU, Ahn UM, Nallamshetty L, Rose PS, Kuszyk BS, Fishman EK 2000 Osseous anatomy of the lumbosacral spine in Marfan syndrome. Spine **25:**2797–2802.
10. Ahn NU, Ahn UM, Nallamshetty L, Rose PS, Buchowski JM, Garrett ES, Kebaish KM, Sponseller PD 2001 The lumbar interpediculate distance is widened in adults with the Marfan syndrome: Data from 32 cases. Acta Orthop Scand **72:**67–71.
11. Kohlmeier L, Gasner C, Bachrack LK, Marcus R 1995 The bone mineral status of patients with Marfan syndrome. J Bone Miner Res **10:**1550–1555.
12. Tobias JH, Dalzell N, Child AH 1995 Assessment of bone mineral density in women with Marfan syndrome. Br J Rheumatol **34:**516–519.
13. Le Parc JM, Plantin P, Jondeau G, Goldschild M, Albert M, Boileau C 1999 Bone mineral density in sixty adult patients with Marfan syndrome. Osteoporos Int **10:**475–479.
14. Carter N, Duncan E, Wordsworth P 2000 Bone mineral density in adults with Marfan syndrome. Rheumatol **39:**307–309.
15. Giampietro PF, Peterson M, Schneider R, Davis JG, Raggio C, Myers E, Burke SW, Boachie-Adjei O, Mueller CM 2003 Assessment of bone mineral density in adults and children with Marfan syndrome. Osteoporos Int **14:**559–563.
16. Moura B, Tubach F, Sulpice M, Boileau C, Jondeau G, Muti C, Chevallier B, Ounnoughene Y, Le Parc JM, Multidisciplinary Marfan Syndrome Clinic Group 2006 Bone mineral density in Marfan syndrome: A large case-control study. Joint Bone Spine **73:**733–735.
17. Pyeritz RE, Fishman EK, Bernhardt BA, Siegelman SS 1988 Dural ectasia is a common feature of the Marfan syndrome. Am J Hum Genet **43:**726–732.
18. Villeirs GM, Van Tongerloo AJ, Verstraete KL, Kunnen MF, De Paepe AM 1999 Widening of the spinal canal and dural ectasia in Marfan's syndrome: Assessment by CT. Neuroradiology **41:**850–854.
19. Fattori R, Nienaber CA, Descovich B, Ambrosetto P, Reggiani LB, Pepe G, Kaufmann U, Negrini E, von Kodolitsch Y, Gensini GF 1999 Importance of dural ectasia in phenotypic assessment of Marfan's syndrome. Lancet **354:**910–913.
20. Ahn NU, Sponseller PD, Ahn UM, Nallamshetty L, Rose PS, Buchowski JM, Garrett ES, Kuszyk BS, Fishman EK, Zinreich SJ 2000 Dural ectasia in the Marfan syndrome: MR and CT findings and criteria. Genet Med **2:**173–179.
21. Knirsch W, Kurtz C, Haffner N, Binz G, Heim P, Winkler P, Baumgartner D, Freund-Unsinn K, Stern H, Kaemmerer H, Molinari L, Kececioglu D, Uhlemann F 2006 Dural ectasia in children with Marfan syndrome: A prospective, multicenter, patient-control study. Am J Med Genet A **140:**775–781.
22. Ahn NU, Sponseller PD, Ahn UM, Nallamshetty L, Kuszyk BS, Zinreich SJ 2000 Dural ectasia is associated with back pain in Marfan syndrome. Spine **25:**1562–1568.
23. Foran JR, Pyeritz RE, Dietz HC, Sponseller PD 2005 Characterization of the symptoms associated with dural ectasia in the Marfan patient. Am J Med Genet A **134:**58–65.
24. Lawson ML, Mellins RB, Tabangin M, Kelly RE Jr, Croitoru DP, Goretsky MJ, Nuss D 2005 Impact of pectus excavatum on pulmonary function before and after repair with the Nuss procedure. J Pediatr Surg **40:**174–180.
25. Coln E, Carrasco J, Coln D 2006 Demonstrating relief of cardiac compression with the Nuss minimally invasive repair for pectus excavatum. J Pediatr Surg **41:**683–686.
26. Sigalet DL, Montgomery M, Harder J, Wong V, Kravarusic D, Alassiri A 2007 Long term cardiopulmonary effects of closed repair of pectus excavatum. Pediatr Surg Int **5:**493–497.
27. Arn PH, Scherer LR, Haller JA, Pyeritz RE 1989 Outcome of pectus excavatum in patients with Marfan syndrome and in the general population. J Pediatr **115:**954–958.
28. Sponseller PD, Bhimani M, Solacoff D, Dormans JP 2000 Results of brace treatment of scoliosis in Marfan syndrome. Spine **25:**2350–2354.
29. Jones KB, Erkula G, Sponseller PD, Dormans JP 2002 Spine deformity correction in Marfan syndrome. Spine **18:**2003–2012.
30. Di Silvestre M, Greggi T, Giacomini S, Cioni A, Bakaloudis G, Lolli F, Parisini P 2005 Surgical treatment for scoliosis in Marfan syndrome. Spine **30:**E597–E604.
31. Sponseller P, Sethi N, Cameron DE, Pyeritz RE 1997 Infantile scoliosis in Marfan syndrome. Spine **22:**509–516.
32. Sponseller P, Jones KB, Ahn NU, Erkula G, Foran JR, Dietz HC 2006 Protrusio acetabuli in Marfan syndrome: Age-related prevalence and associated hip function. J Bone Joint Surg Am **88:**486–495.
33. Faivre L, Collod-Beroud G, Loeys BL, Child A, Binquet C, Gautier E, Callewaert B, Arbustini E, Mayer K, Arslan-Kirchner M, Kiotsekoglou A, Comeglio P, Marzilliano N, Dietz HC, Halliday D, Beroud C, Bonithon-Kopp C, Claustres M, Muti C, Plauchu H, Robinson PN, Ades LC, Biggin A, Benetts B, Brett M, Holman KJ, DeBacker J, Coucke P, Francke U, DePaepe A, Jondeau G, Boileau C 2007 Effect of mutation type and location on clinical outcome in 1,013 probands with Marfan syndrome or related phenotypes and FBN1 mutations: An international study. Am J Hum Genet **81:**454–466.
34. de Vries BB, Pals G, Odink R, Hamel BC 2007 Homozygosity for a FBN1 missense mutation: Clinical and molecular evidence for recessive Marfan syndrome. Eur J Hum Genet **15:**930–935.
35. Neptune ER, Frischmeyer PA, Arking E, Myers L, Bunton TE, Gayraud B, Ramirez F, Sakai LY, Dietz HC 2003 Dysregulation of TGF-β activation contributes to pathogenesis in Marfan syndrome. Nat Genet **33:**407–411.
36. Ng CM, Cheng A, Myers LA, Martinez-Murillo F, Jie C, Bedja D, Gabrielson KL, Hausladen JM, Mecham RP, Judge DP, Dietz HC 2004 TGFβ-dependent pathogenesis of mitral valve prolapse in a mouse model of Marfan syndrome. J Clin Invest **114:**1586–1592.
37. Habashi JP, Judge DP, Holm TM, Cohn RD, Loeys B, Cooper TK, Myers L, Klein EC, Liu G, Calvi C, Podowski M, Neptune ER, Halushka MK, Bedja D, Gabrielson K, Rifkin DB, Carta L, Ramirez F, Huso DL, Dietz HC 2006 Losartan, an AT1 antagonist, prevents aortic aneurysm in a mouse model of Marfan syndrome. Science **312:**117–121.
38. Cohn RD, van Erp C, Habashi JP, Soleimani AA, Klein EC, Lisi MT, Gamradt M, ap Rhys CM, Holm TM, Loeys BL, Ramirez F, Judge DP, Ward CW, Dietz HC 2007 Angiotensin II type 1 receptor blockade attenuates TGF-β-induced failure of muscle regeneration in multiple myopathic states. Nat Med **13:**204–210.
39. Zhang H, Hu W, Ramirez F 1995 Developmental expression of fibrillin genes suggests heterogeneity of extracellular microfibrils. J Cell Biol **129:**1165–1176.
40. Gigante A, Specchia N, Nori S, Greco F 1996 Distribution of elastic fiber types in the epiphyseal region. J Orthop Res **14:**810–817.

© 2008 American Society for Bone and Mineral Research

41. Keene DR, Jordan CD, Reinhardt DP, Ridgway CC, Ono RN, Corson GM, Fairhurst M, Sussman MD, Memoli VA, Sakai LY 1997 Fibrillin-1 in human cartilage: Developmental expression and formation of special banded fibers. J Histochem Cytochem **45:**1069–1082.
42. Kitahama S, Gibson MA, Hatzinikolas G, Hay S, Kuliwaba JL, Evdokiou A, Atkins GJ, Findlay DM 2000 Expression of fibrillins and other microfibril-associated proteins in human bone and osteoblast-like cells. Bone **27:**61–67.
43. Erkula G, Jones KB, Sponseller PD, Dietz HC, Pyeritz RE 2002 Growth and maturation in Marfan syndrome. Am J Med Genet **109:**100–115.
44. Pereira L, Lee SY, Gayraud B, Andrikopoulos K, Shapiro SD, Bunton T, Biery NJ, Dietz HC, Sakai LY, Ramirez F 1999 Pathogenetic sequence for aneurysm revealed in mice underexpressing fibrillin-1. Proc Natl Acad Sci USA **96:**3819–3823.
45. Jones KB, Myers L, Judge DP, Kirby PA, Dietz HC, Sponseller PD 2005 Toward an understanding of dural ectasia: A light microscopy study in a murine model of Marfan syndrome. Spine **30:**291–293.
46. Sood S, Eldadah ZA, Krause WL, McIntosh I, Dietz HC 1996 Mutation in fibrillin-1 and the Marfanoid-craniosynostosis (Shprintzen-Goldberg) syndrome. Nat Genet **12:**209–211.
47. Faivre L, Gorlin RJ, Wirtz MK, Godfrey M, Dagoneau N, Samples JR, Le Merrer M, Collod-Beroud G, Boileau C, Munnich A, Cormier-Daire V 2003 In frame fibrillin-1 gene deletion in autosomal dominant Weill-Marchesani syndrome. J Med Genet **40:**34–36.
48. Kosaki K, Takahashi D, Udaka T, Kosaki R, Matsumoto M, Ibe S, Isobe T, Tanaka Y, Takahashi T 2006 Molecular pathology of Shprintzen-Goldberg syndrome. Am J Hum Genet **140A:**104–108.
49. Collod G, Babron MC, Jondeau G, Coulon M, Weissenbach J, Dubourg O, Bourdarias JP, Bonaiti-Pellie C, Boileau C 1994 A second locus for Marfan syndrome maps to chromosome 3p24.2-p25. Nat Genet **8:**264–268.
50. Dietz H, Francke U, Furthmayr H, Francomano C, DePaepe A, Devereux R, Ramirez F, Pyeritz R 1995 The question of heterogeneity in Marfan syndrome. Nat Genet **9:**228–231.
51. Mizuguchi T, Collod-Beroud G, Akiyama T, Abifadel M, Harada N, Morisaki T, Allard D, Varret M, Claustres M, Morisaki H, Ihara M, Kinoshita A, Yoshiura K, Junien C, Kajii T, Jondeau G, Ohta T, Kishino T, Furukawa Y, Nakamura Y, Niikawa N, Boileau C, Matsumoto N 2004 Heterozygous TGFBR2 mutations in Marfan syndrome. Nat Genet **36:**855–860.
52. Loeys BL, Chen J, Neptune ER, Judge DP, Podowski M, Holm T, Meyers J, Leitch CC, Katsanis N, Sharifi N, Xu FL, Myers LA, Spevak PJ, Cameron DE, De Backer J, Hellemans J, Chen Y, Davis EC, Webb CL, Kress W, Coucke P, Rifkin DB, De Paepe AM, Dietz HC 2005 A syndrome of altered cardiovascular, craniofacial, neurocognitive and skeletal development caused by mutations in TGFBR1 or TGFBR2. Nat Genet **37:**275–281.
53. Robinson PN, Neumann LM, Demuth S, Enders H, Jung U, Konig R, Mitulla B, Muller D, Muschke P, Pfeiffer L, Prager B, Somer M, Tinschert S 2005 Shprintzen-Goldberg syndrome: Fourteen new patients and a clinical analysis. Am J Med Genet A **135:**251–262.
54. Dagoneau N, Benoist-Lasselin C, Huber C, Faivre L, Megarbane A, Alswaid A, Dollfus H, Alembik Y, Munnich A, Legeai-Mallet L, Cormier-Daire V 2004 ADAMTS10 mutations in autosomal recessive Weill-Marchesani syndrome. Am J Hum Genet **75:**801–806.
55. Loeys BL, Schwarze U, Holm T, Callewaert BL, Thomas GH, Pannu H, De Backer JF, Oswald GL, Symoens S, Manouvrier S, Roberts AE, Faravelli F, Greco MA, Pyeritz RE, Milewicz DM, Coucke PJ, Cameron DE, Braverman AC, Byers PH, De Paepe AM, Dietz HC 2006 Aneurysm syndromes caused by mutations in the TGF-β receptor. N Engl J Med **355:**788–798.
56. Viljoen D 1994 Congenital contractural arachnoactyly (Beals syndrome). J Med Genet **31:**640–643.
57. Tuncbilek E, Alanay Y 2006 Congenital contractural arachnodactyly (Beals syndrome). Orphanet J Rare Dis. **1:**20–22.
58. Bawle E, Quigg MH 1992 Ectopia lentis and aortic root dilatation in congenital contractural arachnodactyly. Am J Med Genet **42:**19–21.
59. Wang M, Clericuzio CL, Godfrey M 1996 Familial occurrence of typical and severe lethal congenital contractural arachnodactyly caused by missplicing of exon 34 of fibrillin-2. Am J Hum Genet **59:**1027–1034.
60. Gupta PA, Wallis DD, Chin TO, Northrup H, Tran-Fadulu VT, Towbin JA, Milewicz DM 2006 FBN2 mutation associated with manifestations of Marfan syndrome and congenital contractural arachnodactyly. J Med Genet **41:**e56.
61. Currarino G, Friedman JM 1986 A severe form of congenital contractural arachnodactyly in two newborn infants. Am J Med Genet **25:**763–773.
62. Gupta PA, Putnam EA, Carmical SG, Kaitila I, Steinmann B, Child A, Danesino C, Metcalfe K, Berry SA, Chen E, Delorme CV, Thong MK, Ades LC, Milewicz DM 2002 Ten novel FBN2 mutations in congenital contractural arachnodactyly: Delineation of the molecular pathogenesis and clinical phenotype. Hum Mut **19:**39–48.
63. Ramirez F 1996 Fibrillin mutations in Marfan syndrome and related phenotypes. Curr Opin Genet Dev **6:**309–315.
64. Carta L, Pereira L, Emilio Arteaga-Solis E, Lee-Arteaga SY, Lenart B, Starcher B, Merkel CA, Sukoyan M, Kerkis A, Hazeki N, Keene DR, Sakai LY, Ramirez F 2006 Fibrillins 1 and 2 perform partially overlapping functions during aortic development. J Biol Chem **281:**8016–8023.
65. Arteaga-Solis E, Gayraud B, Lee SY, Shum L, Sakai L, Ramirez F 2001 Regulation of limb patterning by extracellular microfibrils. J Cell Biol **154:**275–281.

Chapter 97. Enzyme Defects and the Skeleton

Michael P. Whyte

Division of Bone and Mineral Diseases, Washington University School of Medicine at Barnes-Jewish Hospital and Center for Metabolic Bone Disease and Molecular Research, Shriners Hospitals for Children, St. Louis, Missouri

INTRODUCTION

Inborn errors of metabolism from enzyme deficiencies can importantly affect the skeleton. Five types are reviewed here.

HYPOPHOSPHATASIA

Hypophosphatasia (HPP) is a rare, heritable rickets or osteomalacia (OMIM 146300, 241500, 241510)[1] characterized biochemically by subnormal activity of the tissue-nonspecific isoenzyme of alkaline phosphatase (TNSALP).[2,3] Although TNSALP is normally present in all tissues, HPP disturbs predominantly the skeleton and teeth. Approximately 350 cases have been reported, showing a remarkable range of severity with four overlapping clinical forms described according to patient age when skeletal disease is discovered: perinatal, infantile, childhood, and adult. Odonto-HPP features dental manifestations only.[4] Generally, the earlier the skeletal problems, the more severe the clinical course.[2,3]

The author states that he has no conflicts of interest.

© 2008 American Society for Bone and Mineral Research

Perinatal HPP manifests in utero.[5] At birth, extreme skeletal hypomineralization causes caput membranaceum and short, deformed limbs.[3] Most affected newborns live briefly while suffering increasing respiratory compromise. Prolonged survival is very rare. The radiographic features are pathognomonic.[2] Sometimes the skeleton is so poorly calcified that only the skull base is seen.[3] Alternatively, the calvarium can be ossified only centrally, segments of the spinal column may appear missing, and severe rachitic changes manifest in the limbs.[3]

Infantile HPP presents before age 6 mo.[3,6] Development appears normal until poor feeding, inadequate weight gain, hypotonia, and wide fontanels are noted. Rachitic deformities are then recognized. Vitamin B_6-dependent seizures may appear.[7] Hypercalcemia and hypercalciuria can cause recurrent vomiting, nephrocalcinosis, and sometimes renal compromise.[6,8] With the illusion of widely open fontanels (hypomineralized calvarium), functional craniosynostosis may occur. A flail chest predisposes to pneumonia.[6] There can be spontaneous improvement or progressive skeletal deterioration,[3,6] with ~50% of patients dying in infancy.[2,3,7] Radiographic changes are characteristic but less striking than in perinatal HPP.[2] Abrupt transition from relatively normal-appearing diaphyses to hypomineralized metaphyses can suggest sudden metabolic deterioration. Progressive skeletal demineralization with fractures and thoracic deformity herald a lethal outcome.[6]

Childhood HPP causes premature loss of deciduous teeth (age, <5 yr), without root resorption, from hypoplasia of dental cementum.[4] Lower incisors are typically shed first, but the entire dentition can be lost. Permanent teeth fare better. Delayed walking with a waddling gait, short stature, and dolichocephaly are common.[3] Static myopathy is a poorly understood complication.[2] After puberty, patients seem improved, but skeletal symptoms will likely recur in middle age.[2,3,9] Radiographs show characteristic "tongues" of lucency projecting from growth plates into metaphyses (Fig. 1).[2] True, premature fusion of cranial sutures with craniosynostosis can cause a "beaten-copper" appearance of the skull.[2]

Adult HPP usually presents during middle age, often with poorly healing, recurrent, metatarsal stress fractures.[9] Discomfort localized in the thighs or hips can reflect femoral pseu-

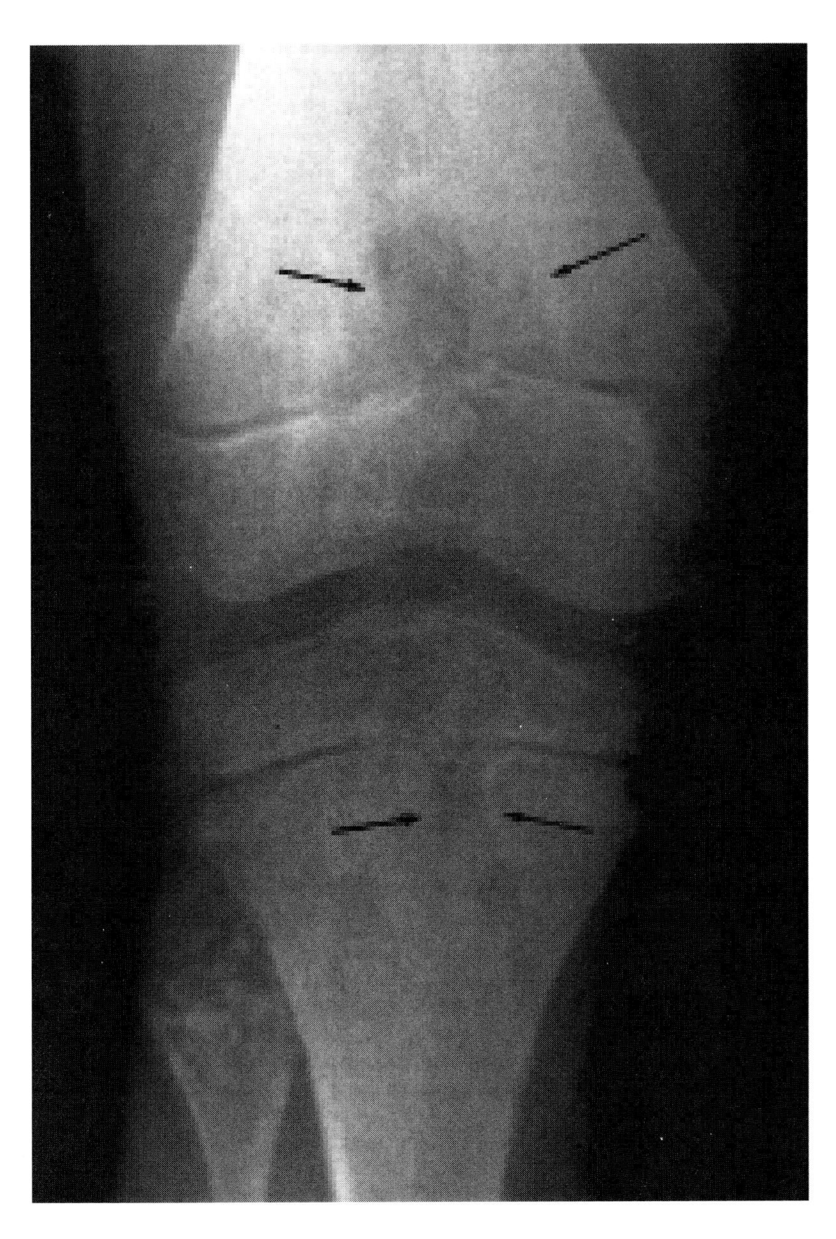

FIG. 1. The metaphyses at the knees of this 10-yr-old boy with childhood hypophosphatasia show characteristic "tongues" of radiolucency (arrows). Note, however, that his mineralization defect (rickets) does not manifest with widening of the growth plates.

© 2008 American Society for Bone and Mineral Research

dofractures.[9–11] Patients may recall rickets and/or premature loss of deciduous teeth during childhood.[9] Chondrocalcinosis from calcium pyrophosphate dihydrate crystal deposition may occur.[2] Radiographs can show osteopenia, metatarsal stress fractures, chondrocalcinosis, and proximal femoral pseudofractures.[2,9–11]

HPP rickets/osteomalacia is remarkable because serum levels of calcium and inorganic phosphate (Pi) are not reduced, and ALP activity is low, not high.[3] In fact, hypercalcemia occurs frequently in perinatal and infantile HPP,[6,8] apparently from dysynergy between gut absorption of calcium and the defective skeletal growth and mineralization.[2,3] In childhood and adult HPP, ~50% of patients are hyperphosphatemic because of enhanced renal reclamation of Pi (increased TmP/GFR).[2,3] Serum PTH and $1,25(OH)_2D$ concentrations are suppressed if there is hypercalcemia. Nondecalcified sections of HPP bone show rickets or osteomalacia without secondary hyperparathyroidism.[12]

Three phosphocompounds accumulate endogenously in HPP[2,3]: phosphoethanolamine (PEA), inorganic pyrophosphate (PPi), and pyridoxal 5′-phosphate (PLP). PPi assay is a research technique. If vitamin B_6 is not supplemented, elevated plasma PLP is an especially good marker for HPP.[2,3] The worse the hypophosphatasemia (low serum ALP activity), the greater the plasma PLP, and the more severe the clinical manifestations.[2,3]

Perinatal and infantile HPP are autosomal recessive traits[2,5,6]; carrier parents and siblings often have low or low-normal serum ALP activity, and sometimes mildly elevated plasma PLP levels. Pyridoxine given orally causes exaggerated increments in plasma PLP levels in all patients and in some carriers.[2,3] Milder forms of HPP (odonto-HPP, childhood HPP, and adult HPP) represent either autosomal dominant or autosomal recessive traits.[13,14]

HPP is diagnosed from a consistent clinical history and physical findings, radiographic or histopathological evidence of rickets or osteomalacia, and hypophosphatasemia together with TNSALP substrate accumulation.[2,3] Mutation analysis of *TNSALP* is available commercially. Approximately 190 mutations (~79% missense) have been identified.[14]

Disturbances in vitamin B_6 metabolism in HPP indicate that TNSALP functions as an ectoenzyme.[2,3] Extracellular accumulation of PPi, an inhibitor of hydroxyapatite crystal growth, impairs skeletal mineralization.[2,3,15] *TNSALP* knockout mice manifest infantile HPP[16] and have helped elucidate the role of TNSALP.[15]

There is no established treatment for HPP. Marrow cell transplantation seemed to rescue two severely affected infants.[6] Teriparatide appeared to stimulate TNSALP biosynthesis by osteoblasts and heal fractures in a woman with HPP.[10] Enzyme replacement therapy using a bone-targeted, recombinant form of TNSALP may be nearing clinical trials.[17] In childhood HPP, dietary restriction and binders to correct hyperphosphatemia (to thereby reduce inhibition of TNSALP by Pi) require study.[10]

Unless deficiencies are documented, avoiding traditional treatments for rickets or osteomalacia seems best in HPP, because circulating levels of calcium, Pi, and 25(OH)D are usually not reduced.[2,3] In fact, supplementation could provoke or exacerbate hypercalcemia or hypercalciuria.[6] Hypercalcemia in perinatal or infantile HPP responds to restriction of dietary calcium and perhaps to salmon calcitonin and/or glucocorticoid therapy.[6,8] Fractures can mend spontaneously, but healing may be delayed, including after osteotomy. Load-sharing intramedullary rods, rather than load-sparing plates, seem best for fractures and pseudofractures in adults.[11] Expert dental care is important. Soft foods and dentures may be necessary, even for pediatric patients.

Sonography and radiographs of fetuses have detected perinatal HPP in the second trimester.[5] First trimester diagnosis requires TNSALP mutation identification.[18] Importantly, some cases of HPP manifest worrisome bowing in utero that then corrects postnatally and resembles childhood HPP.[5]

MUCOPOLYSACCHARIDOSES

Mucopolysaccharidoses (e.g., Hunter, Hurler, Morquio disease) are a family of disorders caused by diminished activity of the lysosomal enzymes that degrade glycosaminoglycans (acid mucopolysaccharidoses).[19,20] Accumulation of these complex carbohydrates within marrow cells somehow alters the skeleton in a pattern referred to by radiologists as "dysostosis multiplex,"[21,22] featuring macrocephaly, dyscephaly, a J-shaped sella turcica, osteoporosis with coarsened trabeculae, oar-shaped ribs, widened clavicles, oval or hook-shaped vertebral bodies, dysplasia of the capital femoral epiphyses, coxa valga, epiphyseal and metaphyseal dysplasia, and proximal tapering of the second and fifth metacarpals. It may be that bone morphogenetic protein (BMP) signaling is disrupted.[23] Joint contractures are also common.[19,20] However, the severity and precise manifestations vary according to the specific enzymopathy and the underlying gene mutations.[24] Each disorder manifests with a broad range of severity.[19,20] Enzyme assays or genetic testing are available.[1,19,20]

Mucopolysaccharidoses are increasingly treated by marrow cell transplantation or enzyme replacement therapy.[25,26]

HOMOCYSTINURIA

Homocystinuria in its classic form is a rare (1:60,000–300,000) autosomal recessive disorder (OMIM 236200) caused by cystathionine β-synthase deficiency.[27] Consequently, homocysteine, an intermediate in methionine metabolism, accumulates endogenously, predisposing to thrombosis and embolism and modification of connective tissue proteins including fibrillin within periosteum and perichondrium. Total homocysteine is increased in plasma. Major clinical problems involve the eyes, CNS, vasculature, and skeleton.[27] Dislocation of the ocular lens can be the initial manifestation. Mental subnormality and thrombotic events are important complications. Patient appearance suggests Marfan syndrome; however, joint mobility is limited. Bones are elongated and overtubulated.[21,22] There may be pectus excavatum or carinatum, arachnodactyly, and genu valgum.[27] Generalized osteoporosis occurs with "codfish" vertebrae and kyphoscoliosis.[21,22] Mild manifestations are said to predict responsiveness (including skeletal disease) to pyridoxine (vitamin B_6) therapy, but this is controversial.[27,28] Treatment may instead include a low methionine-cystine diet and betaine.[27]

There are other causes of hyperhomocysteinemia being studied in relationship to common forms of osteoporosis.[28–31]

ALKAPTONURIA

Alkaptonuria is a rare (<1:250,000), autosomal recessive disorder (OMIM 203500)[1] caused by deficiency of homogentisic acid oxidase from loss-of-function mutation within the *AKU* gene.[32] Consequently, phenylalanine and tyrosine degradation is blocked, leading to tissue accumulation and urine excretion of homogentisic acid. Oxidation and polymerization of homogentisic acid explains the characteristic black appearance of urine exposed to air and discoloration of connective tissues.[32] Ochronosis refers to the pigmentation in the sclera,

© 2008 American Society for Bone and Mineral Research

skin, teeth, nails, bucchal mucosa, endocardium, intima of large vessels, hyaline cartilage of major joints, and intervertebral disks.[32] In elderly patients, pigmentation is also striking in costal, laryngeal, and tracheal cartilage and in fibrocartilage, tendons, and ligaments. Although its pathogenesis is not well understood, severe degenerative disease from tissue fragility occurs in the spinal column where disks calcify and vertebrae fuse, and in major peripheral joints, especially the hips and knees.[33] Perhaps homogentisic acid inhibits collagen synthesis by inhibiting lysyl hydroxylase.[32] Radiographic changes in the spine are almost pathognomonic, showing dense calcification of remaining disk material. Calcification of ear cartilage can also occur. The shoulders and hips are most likely involved by osteoarthritis.

There is no established medical treatment, but a low-protein or other special diets seem worthwhile. Ascorbic acid may block homogentisic acid polymerization.[32]

DISORDERS OF COPPER TRANSPORT

Wilson disease (OMIM 277900)[1] and Menkes disease (OMIM 309400)[1] are genetic disorders of copper (Cu^{2+}) caused by deficiencies of Cu^{2+}-transporting ATPases in the trans-Golgi of different tissues.[34]

Wilson disease affects ~1:55,000 individuals in the United States, causes impaired biliary excretion of Cu^{2+}, and leads to hepatic injury and then Cu^{2+} storage in additional tissues with marked variability in the severity of clinical sequelae. Kayser-Fleischer rings in the eyes, hepatitis and cirrhosis, renal tubular dysfunction and calculi, neurological disease, and hypoparathyroidism are potential complications.[34] Skeletal disease includes osteoporosis, osteomalacia, and chondrocalcinosis with osteoarthritis and joint hypermobility. There can be hyperphosphaturia and hypercalciuria. Loss-of-function mutation disrupts the ATPase, Cu^{2+} transporting, β polypeptide gene, *ATP7B*.[1] Cu^{2+} chelation using penicillamine is effective for most cases.[34]

In Menkes disease, inherited as an X-linked recessive trait,[1] boys develop Cu^{2+} deficiency that leads to kinky, sparse hair and central nervous system (CNS) disease including mental retardation, seizures, and intracranial hemorrhage.[34] The skeletal sequelae include short stature, microcephaly, brachycephaly, wormian bones, metaphyseal dysplasia featuring widening and spurs, osteoporosis, and joint laxity.[21,22] Death usually occurs by age 3 yr. A mild form is called "occipital horn syndrome."[34] Serum Cu^{2+} and ceruloplasmin levels are low because of mutation within the ATPase, Cu^{2+} transporting, α polypeptide gene, *ATP7A*.[1]

REFERENCES

1. McKusick-Nathans Institute of Genetic Medicine, Johns Hopkins University 2008 Online Mendelian Inheritance in Man Available online at http://www.ncbi.nlm.nih.gov/omim/. Accessed June 27, 2008.
2. Whyte MP 2002 Hypophosphatasia: Nature's window on alkaline phosphatase function in humans. In: Bilezikian J, Raisz L (eds.) Principles of Bone Biology, 3rd ed. Academic Press, San Diego, CA, USA. (in press)
3. Whyte MP 2001 Hypophosphatasia. In: Scriver CR, Beaudet AL, Sly WS, Valle D, Childs B, Vogelstein B (eds.) The Metabolic and Molecular Bases of Inherited Disease, 8th ed. McGraw-Hill, New York, NY, USA, pp. 5313–5329.
4. Van den Bos T, Handoko G, Niehof A, Ryan LM, Coburn SP, Whyte MP, Beertsen W 2005 Cementum and dentin in hypophosphatasia. J Dent Res **84:**1021–1025.
5. Pauli RM, Modaff P, Sipes SL, Whyte MP 1999 Mild hypophosphatasia mimicking severe osteogenesis imperfecta in utero: Bent but not broken. Am J Med Genet **86:**434–438.
6. Cahill RA, Wenkert D, Perlman SA, Steele A, Coburn SP, McAlister WH, Mumm S, Whyte MP 2007 Infantile hypophosphatasia: Transplantation therapy trial using bone fragments and cultured osteoblasts. J Clin Endocrinol Metab **92:**2923–2930.
7. Baumgartner-Sigl SB, Haberlandt E, Mumm S, Sergi C, Ryan L, Ericson KL, Whyte MP, Högler W 2007 Pyridoxine-responsive seizures as the first symptom of infantile hypophosphatasia caused by two novel missense mutations (c.677T>C, p.M226T; c.1112C>T, p.T371I) of the tissue-nonspecific alkaline phosphatase gene. Bone **40:**1655–1661.
8. Barcia JP, Strife CF, Langman CB 1997 Infantile hypophosphatasia: Treatment options to control hypercalcemia, hypercalciuria, and chronic bone demineralization. J Pediatr **130:**825–828.
9. Khandwala HM, Mumm S, Whyte MP 2006 Low serum alkaline phosphatase activity with pathologic fracture: Case report and brief review of adult hypophosphatasia. Endocr Pract **12:**676–680.
10. Whyte MP, Mumm S, Deal C 2007 Adult hypophosphatasia treated with teriparatide. J Clin Endocrinol Metab **92:**1203–1208.
11. Coe JD, Murphy WA, Whyte MP 1986 Management of femoral fractures and pseudofractures in adult hypophosphatasia. J Bone Joint Surg Am **68:**981–990.
12. Fallon MD, Weinstein RS, Goldfischer S, Brown DS, Whyte MP 1984 Hypophosphatasia: Clinicopathologic comparison of the infantile, childhood, and adult forms. Medicine (Baltimore) **63:**12–24.
13. Henthorn PS, Raducha M, Fedde KN, Lafferty MA, Whyte MP 1992 Different missense mutations at the tissue-nonspecific alkaline phosphatase gene locus in autosomal recessively inherited forms of mild and severe hypophosphatasia. Proc Natl Acad Sci USA **89:**9924–9928.
14. Mornet E 2005 Tissue nonspecific alkaline phosphatase gene mutations database. Available online at http://www.sesep.uvsq.fr/Database.html. Accessed May 13, 2005.
15. Harmey D, Hessle L, Narisawa S, Johnson KA, Terkeltaub R, Millan JL 2004 Concerted regulation of inorganic pyrophosphate and osteopontin by akp2, enpp1, and ank: An integrated model of the pathogenesis of mineralization disorders. Am J Pathol **164:**1199–1209.
16. Fedde KN, Blair L, Silverstein J, Coburn SP, Ryan LM, Weinstein RS, Waymire K, Narisawa S, Millan JL, MacGregor GR, Whyte MP 1999 Alkaline phosphatase knock-out mice recapitulate the metabolic and skeletal defects of infantile hypophosphatasia. J Bone Miner Res **14:**2015–2026.
17. Millán JL, Narisawa S, Lemire I, Loisel TP, Boileau G, Leonard P, Gramatikova S, Terkeltaub R, Pleshko Camacho N, McKee MD, Crine P, Whyte MP 2008 Enzyme replacement therapy for murine hypophosphatasia. J Bone Miner Res **23:**777–787.
18. Henthorn PS, Whyte MP 1995 Infantile hypophosphatasia: Successful prenatal assessment by testing for tissue-nonspecific alkaline phosphatase gene mutations. Prenat Diagn **15:**1001–1006.
19. Neufeld EF, Muenzer J 2001 The mucopolysaccharidoses. In: Scriver CR, Beaudet AL, Sly WS, Valle D, Childs B, Vogelstein B (eds.) The Metabolic and Molecular Bases of Inherited Disease, 8th ed. McGraw-Hill, New York, NY, USA, pp. 3421–3452.
20. Leroy JG, Wiesmann U 2001 Disorders of lysosomal enzymes. In: Royce PM, Steinmann B (eds.) Connective Tissue and Its Heritable Disorders. Wiley-Liss, New York, NY, USA, pp. 8494–8499.
21. Taybi H, Lachman RS 2006 Radiology of Syndromes, Metabolic Disorders, and Skeletal Dysplasias, 5th ed. Mosby, St. Louis, MO, USA.
22. Resnick D, Niwayama G 2002 Diagnosis of Bone and Joint Disorders, 4th ed. WB Saunders, Philadelphia, PA, USA.
23. Khan SA, Nelson MS, Pan C, Gaffney PM, Gupta P 2008 Endogenous heparan sulfate and heparin modulate bone morphogenetic protein-4 signaling and activity. Am J Physiol Cell Physiol **294:**C1387–C1397.
24. Muenzer J 2004 The mucopolysaccharidoses: A heterogeneous group of disorders with variable pediatric presentations. J Pediatr **144:**S27–S34.
25. Schiffmann R, Brady RO 2002 New prospects for the treatment of lysosomal storage diseases. Drugs **62:**733–742.
26. Braunlin EA, Stauffer NR, Peters CH, Bass JL, Berry JM, Hopwood JJ, Krivit W 2003 Usefulness of bone marrow transplantation in the Hurler syndrome. Am J Cardiol **92:**882–886.
27. Mudd SH, Levy HL, Kraus JP 2001 Disorders of transsulfuration. In: Scriver CR, Beaudet AL, Sly WS, Valle D, Childs B, Vogelstein B (eds.) The Metabolic and Molecular Bases of Inherited Disease, 8th ed. McGraw-Hill, New York, NY, USA, pp. 2007–2056.
28. Green TJ, McMahon JA, Skeaff CM, Williams SM, Whiting SJ

© 2008 American Society for Bone and Mineral Research

2007 Lowering homocysteine with B vitamins has no effect on biomarkers of bone turnover in older persons: A 2-y randomized controlled trial 1–3. Am J Clin Nutr **85:**460–464.
29. Cagnacci A, Bagni B, Zini A, Cannoletta M, Generali M, Volpe A 2008 Relation of folates, vitamin B12 and homocysteine to vertebral bone mineral density change in postmenopausal women. A five-year longitudinal evaluation. Bone **42:**314–320.
30. Herrmann W, Herrmann M 2008 Is hyperhomocysteinemia a risk factor for osteoporosis? Expert Rev Endocrinol Metab **3:**309–313.
31. Salari P, Larijani B, Abdollahi M 2008 Association of hyperhomocysteinemia with osteoporosis: A systematic review. Therapy **5:**215–222.
32. La Du BN 2001 Alkaptonuria. In: Scriver CR, Beaudet AL, Sly WS, Valle D, Childs B, Vogelstein B (eds.) The Metabolic and Molecular Bases of Inherited Disease, 8th ed. McGraw-Hill, New York, NY, USA, pp. 2109–2123.
33. Mannoni A, Selvi E, Lorenzini S, Giorgi M, Airó P, Cammelli D, Andreotti L, Marcolongo R, Porfirio B 2004 Alkaptonuria, ochronosis, and ochronotic arthropathy. Sem Arthrit Rheum **33:**239–248.
34. Culotta VC, Gitlin JD 2001 Disorders of copper transport. In: Scriver CR, Beaudet AL, Sly WS, Valle D, Childs B, Vogelstein B (eds.) The Metabolic and Molecular Bases of Inherited Disease, 8th ed. McGraw-Hill, New York, NY, USA, pp. 3105–3126.

© 2008 American Society for Bone and Mineral Research

SECTION IX

Approach to Nephrolithiasis
(Section Editor: Rajesh Thakker)

© 2008 American Society for Bone and Mineral Research

Chapter 98. Calcium Nephrolithiasis

David A. Bushinsky[1] and Peter A. Friedman[2]

[1]*University of Rochester School of Medicine, Rochester, New York;* [2]*University of Pittsburgh School of Medicine, Pittsburgh, Pennsylvania*

INTRODUCTION

Nephrolithiasis refers to the formation of a crystalline solid phase within the kidney, which generally leads to considerable morbidity but rare mortality.[1–3] Stones tend to form in the renal papilla and collecting system. Stones are composed of calcium oxalate (~26%), calcium phosphate (7%), mixed calcium oxalate and calcium phosphate (37%), uric acid (5%), magnesium ammonium phosphate (struvite, 22%), and/or cystine (2%).[1,2,4] Patients with nephrolithiasis often present with renal colic but may also commonly present with hematuria or urinary tract infection.[5] The severity of the clinical presentation is dependent on the location, size, and type of stone and may lead to hospitalizations, shock-wave lithotripsy, and/or invasive surgical procedures.

Men are affected more than women, and the peak age of onset is in the third to fourth decade.[6] The incidence of nephrolithiasis seems to be increasing over the last few decades, and the prevalence is between 4% and 9% in men and 2% and 4% in women. The lifetime risk of kidney stones in men is ~10–12% in men and somewhat lower in women.

RENAL CALCIUM TRANSPORT

An understanding of hypercalciuria and the pathophysiological formation of renal calculi is predicated on an appreciation of renal calcium absorption.

Nearly one half of the total serum calcium is bound to plasma proteins, which are not filtered. Only free calcium and calcium complexed to small anions are filtered by the glomerulus and enter the proximal nephron. Proximal tubules absorb some 60–70% of the filtered calcium. Here, calcium absorption proceeds primarily through the paracellular pathway between adjacent cells (Fig. 1).[7,8] A small fraction of proximal tubule calcium absorption is active and occurs across the cell.[9] Proximal tubule calcium absorption is governed by various physical forces affecting passive cation movement. Specific hormonal regulation is not known to occur.

Medullary and cortical thick ascending limbs reabsorb ~20% of the filtered calcium. In these segments, calcium is transported by a combination of cellular and paracellular routes. Basal calcium absorption proceeds through the paracellular pathway, where its rate of movement is determined by the prevailing electrochemical driving forces, which, in turn, are established by the extent of sodium absorption. The cellular component of calcium absorption in thick limbs is regulated by PTH in cortical thick limbs and by calcitonin in medullary thick limbs (Fig. 1).[10,11] Thus, in contrast to proximal tubules, where virtually all calcium absorption proceeds through the paracellular pathway, in medullary and cortical thick ascending limbs, transport follows parallel absorption routes, a passive mechanism through the paracellular path, and an active mechanism that is transcellular. A third permutation of this scheme occurs in distal tubules.

Distal tubules, consisting of distal convoluted tubules and connecting tubules, are responsible for reabsorbing 8–10% of the filtered load. In these segments, the calcium permeability of the tight junctions is low and passive calcium absorption is absent; calcium absorption, stimulated by PTH, is active and follows a transcellular route (Fig. 1).

PATHOGENESIS

A kidney stone can only form in urine that is supersaturated with respect to the specific components of the stone.[1–4] Supersaturation is dependent on the product of the free ion activities of the stone components. Whereas increases in concentration will tend to increase free ion activity, factors such as the presence of inhibitors of crystallization will decrease it.

Stones will not form in urine that is undersaturated with respect to the components of the stone.[1–4] As the ion activity product is increased, a concentration is reached where a previously formed stone will neither shrink nor grow. This is the equilibrium solubility product; however, a new stone will not yet form. The ion activity product must increase further in this now supersaturated, metastable solution to the so-called formation product (or the upper limit of metastability), which will permit a new stone to form. The ions will come together, nucleate, to form a more stable, solid phase. The spontaneous formation of a new solid phase in supersaturated urine is unusual. Generally a new crystal will form by heterogeneous nucleation when a solid phase forms on a preexisting surface such as cellular debris or another crystal. For crystals to become a clinically significant stone they must grow to sufficient size to irritate the urothelium or obstruct a lumen. Crystals generally increase in size by aggregating with other similar crystals.

The glomerular ultrafiltrate passes through the renal tubule so rapidly that it would be impossible for a clinically significant kidney stone to spontaneously form.[12] In patients with calcium oxalate nephrolithiasis, the initial crystal phase forms in the interstitium around the thin limbs of the loop of Henle.[13] This solid phase, which contains only calcium phosphate, increases in size and finally erodes into the urinary space forming a so called Randals' plaque. Calcium oxalate crystals may adhere to the Randals' plaque increasing in size and finally breaking off, leading to clinically significant stone disease.

CLINICAL PRESENTATION

Patients with nephrolithiasis often describe severe flank pain of increasing intensity that is abrupt in onset and is caused by the passage of a stone from the renal pelvis into the ureter.[5] The pain may migrate from the anterior abdomen to the testicle or labia majora. The patient may complain of nausea and may vomit. The mere presence of a stone on a radiograph does not exclude that the patient has another acute abdominal or pelvic emergency. The pain may be associated with hematuria; however, hematuria has many causes and again must not be attributed to a kidney stone until other causes are investigated. The pain and hematuria often do not resolve until the stone is passed into the bladder. Stones <5 mm in diameter generally will pass spontaneously. Although the acute presentation is common, some kidney stones are asymptomatic and only found serendipitously.

Radiographic studies are integral in determining if a patient has a kidney stone.[1] A traditional KUB (kidneys, ureter, and

Dr. Bushinsky has consulted for Amgen, Genzyme, Shire, Cytochroma, and Relypsa, owns stock in Amgen, and has received grants from the Renal Research Institute. Dr. Friedman states that he has no conflicts of interest.

© 2008 American Society for Bone and Mineral Research

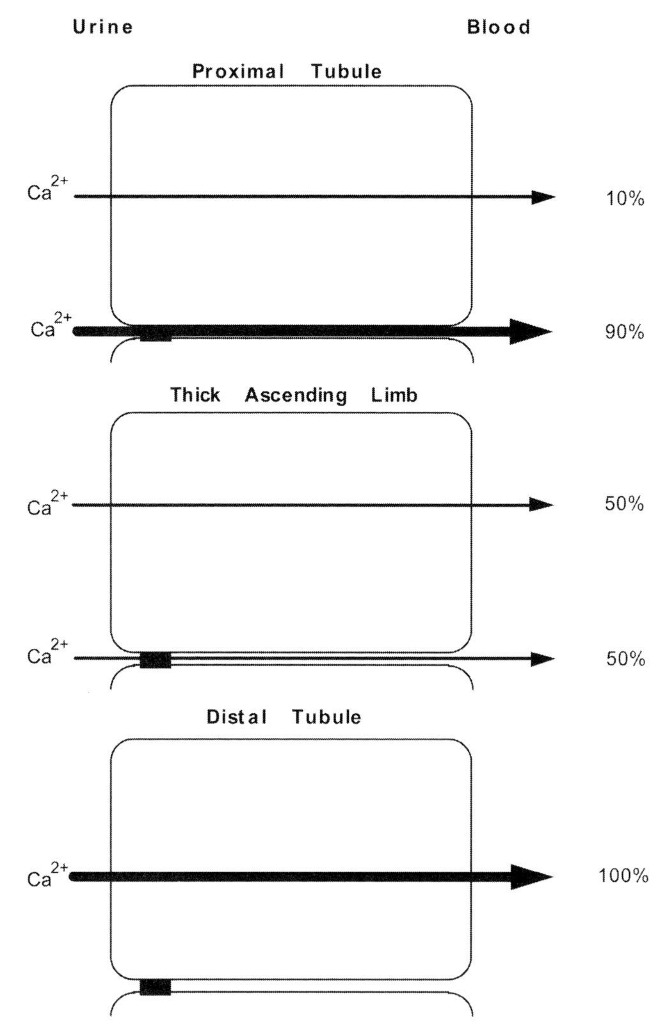

FIG 1. Relative magnitude of cellular and paracellular calcium absorption in proximal tubules, thick ascending limbs, and distal tubules.

bladder) may detect calcium-containing stones but will often miss those composed of uric acid. A renal ultrasound avoids radiation and is useful for finding stones in the renal parenchyma and for detecting obstruction. An intravenous pyelogram (IVP) can detect obstructing stones regardless of the composition but requires contrast. A helical (spiral) CT is highly sensitive and specific for detecting small stones, including those composed of uric acid and those in the ureters and does not require contrast. A helical CT has become the radiographic procedure of choice in evaluating a patient with nephrolithiasis, although it does expose the patient to more radiation that either a KUB or IVP. The addition of contrast will allow the visualization of non–radio-opaque stones and obstruction but should only be done if a noncontrast study fails to provide adequate information.

EVALUATION AFTER A SINGLE STONE

The intensity of the evaluation after a single stone is dependent on the likelihood of recurrence. Recurrence rates of untreated patients vary with different studies, but it is generally felt that 40–50% of patients will have a recurrent stone in 5 yr.[2] Once a patient has a recurrent stone, the next recurrence will occur in a shorter interval than the first.

Many, including an NIH consensus conference on the Prevention and Treatment of Kidney Stones, suggest a basic evaluation after an adult forms a single stone.[2] Children, patients with growing stones, and non–calcium stone formers merit a more extensive evaluation as detailed below. Often easily corrected factors are found in the basic evaluation, which can lead to a marked reduction in stone formation.

The basic evaluation consists of a medical history, family history, stone history, and an extensive review of medications, fluid intake, occupation, and diet.[1–3]

Medical History

A number of systemic diseases clearly increase the predisposition to form kidney stones. For example, granulomatous diseases such as sarcoid can lead to a $1,25(OH)_2D_3$-mediated increase in intestinal calcium absorption and urine calcium excretion. Intestinal malabsorption caused by inflammatory bowel disease or a short bowel can increase oxalate absorption and excretion.

Family History

Idiopathic hypercalciuria is an inherited polygenic disorder. Certain autosomal recessive disorders such as primary hyperoxaluria and cystinuria lead to nephrolithiasis at a young age.

Stone History

A chronology of stone events including age of the first stone, stone type and size, and the frequency of subsequent stones will guide the subsequent evaluation and therapy. Early first stone formation suggests a genetic disorder, and formation of repeated unilateral stones suggests an anatomic disorder.

Medications

A large number of medications, by increasing excretion of the ions that make up the solid phase, forming a solid phase themselves or altering the urinary environment, can promote stone formation. Calcium-containing antacids increase the total amount of absorbed and subsequently excreted calcium, and loop diuretics increase calcium excretion. Indinavir, acyclovir, and triamterene themselves can precipitate and form stones. The carbonic anhydrase inhibitors increase urine pH and promote calcium phosphate and calcium oxalate precipitation.

Occupation

In many occupations, the need to urinate may be disruptive. Outdoor activity in a hot, dry climate may increase insensible fluid losses that are not compensated for by increased oral hydration, resulting in a concentrated urine that is more prone to crystal formation.

Diet

The diet of many individuals leads directly to stone formation. A dietary history is often the most revealing aspect of the initial evaluation. Sodium-induced hypercalciuria can often be traced to consumption of processed foods. Hyperoxaluria can often be traced to excessive spinach, chocolate, or nut consumption and occasionally to rhubarb. A paucity of fluid intake may reflect habit more than a conscious desire not to be inconvenienced by urination. A surfeit of protein resulting in

© 2008 American Society for Bone and Mineral Research

metabolism to metabolic acids and hypercalciuria and hypocitraturia is a fixture of many diets.

LABORATORY EXAMINATION

In general, all patients should have a serum creatinine level to determine the degree of renal function, bicarbonate and potassium levels to exclude renal tubular acidosis, calcium and phosphorus levels to exclude hyperparathyroidism, and a uric acid level to exclude hyperuricemia. A urine analysis will give an indication of hematuria and increased concentration, often associated with stones, and pH, which may be high in the presence of urea splitting organisms and struvite and low in the presence of excessive protein intake.

All available stones should be analyzed for ion content and crystal phases optimally by X-ray diffraction crystallography or infrared spectroscopy. This inexpensive, risk-free analysis is often the single most important guide to therapy.

A 24-h urine collection for ion excretion and supersaturation is generally only performed when a patient has formed more than one stone, if stones are found to be enlarging or increasing in number, or in children. The collection should at a minimum be analyzed for volume, calcium, oxalate, sodium, uric acid, phosphorus, citrate, and creatinine. There are several commercial laboratories that will calculate supersaturation with respect to the mineral phases calcium oxalate, calcium phosphate, and uric acid. The goal of all therapy directed at preventing kidney stones is to lower the prevailing urinary supersaturation. Except in specific research settings, there is no utility in attempting to separate those patients who absorb excessive amounts of dietary calcium from those whose kidneys can not adequately reabsorb filtered calcium.

GENERAL ADVICE TO AVOID STONES

If a patient has only formed a single stone, most physicians opt not to collect a 24-h urine for ion excretion and supersaturation.

Patients are told to increase fluid intake, reduce dietary sodium and protein, and consume an age- and sex-appropriate amount of calcium.[14,15] The fluid intake should be increased so that the patient excretes at least 2–2.5 liters of urine a day, which will lead to a reduction in urinary supersaturation. Urine calcium excretion is directly correlated with dietary and thus urinary sodium excretion. Patients are instructed to limit sodium intake to <3 g/d. Animal protein causes an increase in metabolic acid production, which leads to increased urine calcium excretion and decreased urine citrate excretion, both of which promote calcium stone formation. Patients who consume inadequate amounts of dietary calcium have less intestinal calcium oxalate complex formation, allowing more oxalate absorption. Retrospective analysis of large databases has repeatedly shown that a low calcium diet is a significant risk for stone formation.[16] This general advice can reduce recurrence of the next stone by ~60% at 5 yr.[2]

MECHANISMS AND DIRECTED THERAPY FOR CALCIUM NEPHROLITHIASIS

Approximately 70% of all kidney stones contain calcium, usually with oxalate or with phosphate or urate. The principal mechanisms by which calcium containing kidney stones form are inadequate urinary volume, hypercalciuria, hyperoxaluria, hyperuricosuria, hypocitraturia, and renal tubular acidosis (Table 1).[1,3] Stone formation increases with increasing supersaturation in a continuous manner. Although normal ranges are given, a lower urinary calcium, oxalate, and uric acid and a higher urinary volume and citrate will all help prevent stone formation.

TABLE 1. CAUSES OF CALCIUM STONE FORMATION

- Inadequate urinary volume
- Hypercalciuria
  - Idiopathic hypercalciuria
  - Hypercalcemic disorders
    - Primary hyperparathyroidism
    - Malignancy
    - Granulomatous diseases
      - Sarcoid
  - Immobilization
  - Thyrotoxicosis
  - Milk-alkali syndrome
- Medications that promote calcium stone formation
  - Loop diuretics
  - Antacids (calcium and noncalcium antacids)
  - Vitamin D
  - Glucocorticoids
  - Theophylline
  - Vitamin C
  - Acetazolamide
  - Amphotericin B
- Medications that may precipitate into stones or crystals
  - Indinavir
  - Acyclovir (when infused rapidly intravenously)
  - Triamterene
- Hyperoxaluria
  - Dietary hyperoxaluria (oxalate 40–60 mg/d)
  - Enteric oxaluria (oxalate 60–100 mg/d)
    - Malabsorptive disorders
      - Crohn's disease
      - Sprue (celiac disease)
      - Jejuno-ilial bypass
      - Chronic pancreatitis
      - Biliary obstruction
  - Primary hyperoxaluria types 1 and 2 (oxalate 80–300 mg/d)
- Hyperuricosuria
- Hypocitraturia
  - Dietary protein excess
  - Metabolic acidosis
  - Hypokalemia
  - Exercise
  - Infection
  - Starvation
  - Hypomagnesemia
  - Androgens
- Renal tubular acidosis (distal, type1)
- Anatomic abnormalities of the genitourinary tract
  - Medullary sponge kidney
  - Tubular ectasia
  - Congenital megacalyx

Modified from References 1 and 3.

INADEQUATE URINARY VOLUME

The normal excretion of calcium and oxalate in a paucity of urine will insure stone formation because supersaturation must increase. Database analysis has shown that reduced urine volume is associated with increased stone formation.[16] A 5-yr intervention study confirmed that an increase in urine volume significantly reduces stone recurrence.[17] A urine volume of at least 2 liters will require a fluid intake of at least 3 liters because of normal insensible losses, which increase in hot and/or dry climates.

Hypercalciuria

Hypercalciuria (>250 mg/24 h in women and >300 mg/24 h in

© 2008 American Society for Bone and Mineral Research

men) is the most consistent metabolic abnormality found in patients with calcium nephrolithiasis. Idiopathic hypercalciuria, which is defined as excess calcium excretion with no identifiable metabolic cause, is found in up to 40% of stone-formers but has an incidence of <10% in the overall population.[18] Excess urinary calcium leads to increased supersaturation with respect to a solid phase, generally calcium oxalate or calcium phosphate.

Idiopathic hypercalciuria is an inherited metabolic abnormality as up to 75% of children with kidney stones have a family history of nephrolithiasis.[2,19] Of patients with hypercalciuria, the prevalence of nephrolithiasis in the family history was 69%.[18] Supporting a genetic basis for hypercalciuria is our ability to breed a strain of rats with this disorder.[19]

Idiopathic hypercalciuria is thought to be caused by dysregulation of mineral ion transport at sites where large ion fluxes must be precisely controlled, which are the intestine, kidney, and bone.[18] Excessive absorption of calcium by the gastrointestinal tract leads to an increase in serum calcium that would suppress PTH secretion and $1,25(OH)_2D_3$, which, along with the increased filtered load of calcium to the kidneys, would result in hypercalciuria. A primary defect in renal calcium reabsorption would lead directly to hypercalciuria and result in a fall in serum calcium that would stimulate production of PTH and $1,25(OH)_2D_3$, resulting in enhanced intestinal absorption. Hypercalciuria could also develop as a result of a renal defect in phosphorus reabsorption because the resultant hypophosphatemia would lead to increased $1,25(OH)_2D_3$ and stimulation of intestinal calcium and phosphorus absorption. The increased filtered load of calcium in the setting of suppressed PTH, because of the increased calcium, would also lead to hypercalciuria. Enhanced bone demineralization would increase the serum calcium concentration, which in turn would suppress PTH production. The increase in the filtered load of calcium would result in hypercalciuria.

Analysis of human data indicates that, whereas some patients have a single, specific genetic defect resulting in hypercalciuria,[20,21] most do not have a specific site of mineral ion transport dysregulation; rather, they have a systemic disorder of mineral ion homeostasis resulting in hypercalciuria.[1–3] Attempts to categorize patients into a specific site of dysregulation may lead to inconsistent results even in the best research laboratories.[4]

An animal model of idiopathic hypercalciuria has been developed to aid in the understanding of this disorder. Through >70 generations of inbreeding of the most hypercalciuric progeny of hypercalciuric rats, we have established a strain of rats, each of which excrete 8–10 times as much urine calcium as the wildtype rats.[1,19] The principal mechanism for the hypercalciuria in these rats is increased intestinal calcium absorption. However, when these hypercalciuric rats are deprived of dietary calcium, their urine calcium excretion remains elevated, indicating a defect in renal calcium reabsorption and/or an increase in bone resorption. Bone from these rats releases more calcium, compared with bone of control rats, when exposed to increasing amounts of $1,25(OH)_2D_3$. There is also a primary defect in renal calcium reabsorption. The intestine, bone, and kidney of these hypercalciuric rats have been shown to have an increased number of vitamin D receptors.[22] Thus, these hypercalciuric rats have a systemic abnormality in calcium homeostasis similar to observations in humans. As each of these hypercalciuric rats forms renal stones, we have termed the rats genetic hypercalciuric stone-forming (GHS) rats. An increased number of vitamin D receptors may be the underlying mechanism for hypercalciuria in these rats.[22] Initial studies have shown an increased number of vitamin D receptors in the monocytes of humans with idiopathic hypercalciuria.[23]

Patients with persistent hypercalciuria despite a diet low in sodium often benefit from the hypocalciuric action of a thiazide diuretic. Chlorthalidone is generally used because it requires only daily dosing at an initial dose of 25 mg. Thiazides increase serum lipid levels. In patients for whom this is a concern, indapamide at an initial dose of 1.25 mg can be used. Thiazides also lower serum potassium. If an increase in potassium-rich foods does not lead to an increase in serum potassium, patients are often started on potassium citrate. If hypokalemia persists, a potassium-sparing diuretic may be used. Amiloride, at a starting dose of 5 mg, is usually used because triamterine itself can precipitate into a solid phase.

Hyperoxaluria

Oxalate is produced through endogenous metabolism of glyoxylate and ascorbic acid or derived from dietary sources, such as rhubarb, cocoa, nuts, tea, and certain leafy green vegetables.[1,3,4] Oxalate is excreted in the urine and raises urinary supersaturation with respect to calcium oxalate. Hyperoxaluria (>45 mg/24 h) accounts for the formation of ~5% of all calcium stones. Hyperoxaluria results from excessive oxalate ingestion (dietary oxaluria), malabsorptive gastrointestinal disorders (enteric oxaluria), or excessive endogenous metabolism of oxalate caused by a hepatic enzyme deficiency (primary hyperoxaluria). The antifreeze ethylene glycol is metabolized to oxalate and ingestion of this nephrotoxic compound leads to nephrolithiasis.

Dietary oxaluria results in moderate increases in levels of urinary oxalate (40–60 mg/d). Treatment consists of avoiding foods that contain large amounts of oxalate. Intestinal calcium binds oxalate and mealtime calcium supplementation is often used to decrease oxalate absorption.

Enteric oxaluria often causes higher urinary oxalate levels (60–100 mg/d). Gastrointestinal malabsorptive disorders associated with normal colonic function often lead to enteric oxaluria. Here the malabsorbed fatty acids bind intestinal calcium, freeing oxalate for absorption. The malabsorbed bile acids increase colonic permeability. When possible, treatment consists of correcting the underlying disorder, or if that is not possible, treating the steatorrhea. As for dietary oxaluria, an oxalate restriction coupled to calcium supplementation is used. These patients are also at risk for low urinary volumes and hypocitraturia, increasing the risk of nephrolithiasis. The acidic, concentrated urine increases the risk of uric acid stones.

Primary hyperoxaluria (PH) is caused by hepatic enzyme deficiencies that lead to massive endogenous oxalate production, tissue deposition, and severe hyperoxaluria (80–300 mg/d).[24] PH type 1 is caused by a deficiency of hepatic alanine:glyoxylate aminotransferase (AGT). The more uncommon disorder PH type 2 is caused by a lack of D-glycerate reductase (DGDH) and glyoxylate reductase (GR). PH 1 patients may benefit from pyridoxine (vitamin B_6) therapy, because it reduces oxalate production. PH patients should consume large amount of fluids and be treated with potassium citrate, magnesium, and orthophosphate. The latter inhibits calcium-oxalate crystallization but should be avoided with chronic kidney disease. In PH 1 patients, liver transplantation will be curative. The development of renal failure, from oxalate deposition, often necessitates renal transplantation because dialysis is not as effective as a functioning kidney in oxalate removal.

© 2008 American Society for Bone and Mineral Research

HYPERURICOSURIA

Uric acid crystals may form the nidus of calcium oxalate stones. Patients often have elevated urinary uric acid levels (>750 mg/24 h) with relatively normal urinary calcium and oxalate levels. As opposed to the typical patient with uric acid stones who has a very acid urinary pH, these patients tend to have a higher urinary pH (>5.5).[25] They are generally treated with dietary purine restriction, increased fluid intake, and, if necessary, allopurinol, 100–300 mg/d.

HYPOCITRATURIA

Citrate binds calcium and forms a soluble complex reducing supersaturation with respect to both calcium oxalate and calcium phosphate. Consumption of large amounts of dietary protein, metabolic acidosis, exercise, infection, starvation, androgens, and acetazolamide are leading causes of hypocitraturia (<400 mg/24 h).[2] Treatment consists of addressing the underlying cause, which often means moderating dietary protein intake. If this is not successful, potassium citrate (30–75 mEq/d) is prescribed as a wax matrix formulation. Potassium and bicarbonate levels should be closely monitored, especially in patients with chronic kidney disease.

RENAL TUBULAR ACIDOSIS

In distal renal tubular acidosis (type 1 RTA), distal tubular hydrogen ion excretion is impaired, which results in a nonanion gap metabolic acidosis and a persistently alkaline urine. The acidemia increases calcium and phosphate efflux from bone and enhanced proximal tubular reabsorption of citrate. Excretion of calcium and phosphate increases; there is severe hypocitraturia, and an alkaline urine pH, all of which promote calcium-phosphate precipitation. Patients often present with nephrocalcinosis (renal parenchymal calcification). Therapy consists of potassium citrate or potassium bicarbonate supplementation, often 1–2 mEq/kg/d, to treat both the metabolic acidosis and hypocitraturia.

REFERENCES

1. Monk RD, Bushinsky DA 2008 Kidney stones. In: Larsen PR, Kronenberg HM, Melmed S, Polonsky KS (eds.) Williams Textbook of Endocrinology, 11th ed. Saunders, Philadelphia, PA, USA, pp. 1311–1326.
2. Bushinsky DA, Coe FL, Moe OW 2008 Nephrolithiasis. In: Brenner BM (ed.) The Kidney, 8th ed. Saunders, Philadelphia, PA, USA, pp. 1299–1349.
3. Monk RD, Bushinsky DA 2007 Nephrolithiasis and nephrocalcinosis. In: Johnson R, Frehally J (eds.) Comprehensive Clinical Nephrology, 3rd ed. Mosby, London, UK, pp. 641–669.
4. Coe FL, Evan A, Worcester E 2005 Kidney stone disease. J Clin Invest **115:**2598–2608.
5. Teichman JMH 2004 Acute renal colic from ureteral calculus. N Engl J Med **350:**684–693.
6. Stamatelou KK, Francis ME, Jones CA, Nyberg LM, Curhan GC 2003 Time trends in reported prevalence of kidney stones in the United States: 1976-1994. Kidney Int **63:**1817–1823.
7. Ng RC, Rouse D, Suki WN 1984 Calcium transport in the rabbit superdivial proximal convoluted tubule. J Clin Invest **74:**834–842.
8. Bomsztyk K, George JP, Wright FS 1984 Effects of luminal fluid anions on calcium transport by proximal tubule. Am J Physiol Renal Physiol **246:**F600–F608.
9. Ullrich KJ, Rumrich G, Kloss S 1976 Active Ca2+ reabsorption in the proximal tubule of the rat kidney. Dependence on sodium- and buffer transport. Pflugers Arch **364:**223–228.
10. Bourdeau JE, Burg MB 1980 Effect of PTH on calcium transport across the cortical thick ascending limb of Henle's loop. Am J Physiol **239:**F121–F126.
11. Suki WN, Rouse D 1981 Hormonal regulation of calcium transport in thick ascending limb renal tubules. Am J Physiol Renal Physiol **241:**F171–F174.
12. Bushinsky DA 2003 Nephrolithiasis: Site of the initial solid phase. J Clin Invest **111:**602–605.
13. Evan AP, Lingeman JE, Coe FL, Parks JH, Bledsoe SB, Shao Y, Sommer AJ, Paterson RF, Kuo RL, Grynpas M 2003 Randall plaque of patients with nephrolithiasis begins in basement membranes of thin loops of Henle. J Clin Invest **111:**607–616.
14. Borghi L, Schianchi T, Meschi T, Guerra A, Allegri F, Maggiore U, Novarini A 2002 Comparison of two diets for the prevention of recurrent stones in idiopathic hypercalciuria. N Engl J Med **346:**77–84.
15. Bushinsky DA 2002 Recurrent hypercalciuric nephrolithiasis—does diet help? N Engl J Med **346:**124–125.
16. Curhan GC, Willett WC, Rimm EB, Stampfer MJ 1993 A prospective study of dietary calcium and other nutrients and the risk of symptomatic kidney stones. N Engl J Med **328:**833–838.
17. Borghi L, Meschi T, Amato F, Briganti A, Novarini A, Giannini A 1996 Urinary volume, water, and recurrences in idiopathic calcium nephrolithiasis: A 5-year randomized prospective study. J Urol **155:**839–843.
18. Frick KK, Bushinsky DA 2003 Molecular mechanisms of primary hypercalciuria. J Am Soc Nephrol **14:**1082–1095.
19. Bushinsky DA, Frick KK, Nehrke K 2006 Genetic hypercalciuric stone-forming rats. Curr Opin Nephrol Hypertens **15:**403–418.
20. Moe OW, Bonny O 2005 Genetic hypercalciuria. J Am Soc Nephrol **16:**729–745.
21. Gambaro G, Vezzoli G, Casari G, Rampoldi L, D'Angelo A, Borghi L 2004 Genetics of hypercalciuria and calcium nephrolithiasis: From the rare monogenic to the common polygenic forms. Am J Kidnew Dis **44:**963–986.
22. Karnauskas AJ, van Leeuwen JP, van den Bemd GJ, Kathpalia PP, DeLuca HF, Bushinsky DA, Favus MJ 2005 Mechanism and function of high vitamin D receptor levels in genetic hypercalciuric stone-forming rats. J Bone Miner Res **20:**447–454.
23. Favus MJ, Karnauskas AJ, Parks JH, Coe FL 2004 Peripheral blood monocyte vitamin D receptor levels are elevated in patients with idiopathic hypercalciuria. J Clin Endocrinol Metab **89:**4937–4943.
24. Milliner DS 2005 The primary hyperoxalurias: An algorithm for diagnosis. Am J Nephrol **25:**154–160.
25. Maalouf NM, Cameron MA, Moe OW, Sakhaee K 2004 Novel insights into the pathogenesis of uric acid nephrolithiasis. Curr Opin Nephrol Hypertens **13:**181–189.

© 2008 American Society for Bone and Mineral Research

Chapter 99. Diagnosis and Evaluation of Nephrolithiasis

Stephen J. Knohl and Steven J. Scheinman

Division of Nephrology, Department of Medicine, SUNY Upstate Medical University, Syracuse, New York

INTRODUCTION

The physician evaluating a patient with nephrolithiasis needs to keep in mind the relative prevalence of various stone types. Calcium salts (oxalate, phosphate, or mixed) are the most common stone types, together accounting for 70–75% of nephrolithiasis; low urinary volume, hypercalciuria, hyperoxaluria, hypocitraturia, and hyperuricosuria are identified risk factors.[1] Struvite stones (magnesium ammonium phosphate) represent ~10–15%[1] and form in alkaline urine with high ammonium concentration that results from infection with urease-producing bacteria.[2] Uric acid stones, accounting for 5–10%, are associated primarily with an acidic urine; hyperuricosuria and low urinary volume are additional risk factors.[1,3] Cystine stones account for ~1%. They are the only clinical consequence of hereditary cystinuria, which is the only cause of cystine stones. Stones composed of crystallized drugs represent <1% of all stones worldwide.

BASIC EVALUATION OF THE STONE PATIENT

Essential to every evaluation of renal colic, whether an initial or repeat event, is a detailed history, basic metabolic profile, serum uric acid, urinalysis with urine culture, noncontrast helical CT of the abdomen (i.e., stone-protocol CT of the kidneys), and, if a stone has been retrieved, stone analysis.[4] The history should include a complete medical history in an effort to identify disease states associated with nephrolithiasis. Stone-formers have a higher incidence of obesity, dyslipidemia, hypertension, physical inactivity, and diabetes mellitus.[5] A family history of stone disease is associated with an increased risk of stone occurrence and recurrence.[6] A careful dietary history is useful to identify foods that contain excesses of oxalate (such as nuts, chocolate, legumes, soy, green leafy vegetables, berries, and ripe rhubarb), uric acid (examples include shellfish, organ meats, and yeast), animal protein, and sodium. Medications should be reviewed to identify drugs associated with stone disease, including those that precipitate to form stones (triamterene, indinavir, acyclovir, and sulfonamide antibiotics) and those that promote the formation of calcium stones (acetazolamide, loop diuretics, and topiramate).[7]

The basic metabolic profile is useful for clues to the etiology of stone disease. Hypokalemia promotes intracellular acidosis, which increases citrate reabsorption in the proximal tubule; hypocitraturia is a known risk factor for calcium stone formation. Hypobicarbonatemia with a normal anion gap may indicate the presence of a distal renal tubular acidosis (RTA), a risk factor for calcium phosphate crystallization. Hypercalcemia should alert one to the possibility of primary hyperparathyroidism, sarcoidosis, other granulomatous diseases, or malignancy as an underlying etiology of calcium stones. Blood urea nitrogen (BUN) and creatinine measurements (and the associated four-model "modification-of-diet-in-renal diseases" glomerular filtration rate (MDRD GFR) estimates that now often accompany basic metabolic profiles) mark the level of renal function, with acute changes indicating the need for more urgent medical attention as obstruction to urine flow may be present.

A serum uric acid is useful because hyperuricemia, when the result of overproduction or excessive intake, is associated with hyperuricosuria. The presence of hyperuricosuria has been linked with an increased risk of both uric acid and calcium oxalate stones, although the association with calcium oxalate has recently been called into question.[8]

The urinalysis can provide an abundance of useful information. A high specific gravity indicates a concentrated urine and is useful to follow to monitor compliance with fluid intake, as a low urine volume is a risk factor for stones of all types. Dipstick urine pH, although not as accurate as a metered urine pH, is useful if measured on fresh urine. An alkaline pH on a first-morning void is consistent with distal renal tubular acidosis, a risk factor for calcium phosphate stones. A pH of 8 or higher, however, is not typical of normal physiology and indicates infection with urease-producing bacteria, a risk factor for magnesium ammonium phosphate (struvite) stones, particularly when associated with positive tests for nitrite and leukocyte esterase. Hematuria could indicate current stone disease or, in children, the presence of hypercalciuria.[9] Crystals, if present, can identify the culprit stone type (Fig. 1).

RADIOGRAPHIC EVALUATION

In a patient with renal colic, radiography is essential in assessing stone burden, stone location(s), and the presence of obstruction. Studies commonly used include the plain abdominal film (KUB); ultrasound; intravenous pyelogram (IVP); and helical, noncontrast CT scanning. Historically, stones were classified as radio-opaque or radiolucent based on their appearance on KUB (Fig. 2). Struvite, cystine, and calcium-containing stones are radio-opaque. Cystine stones are radiodense because of the disulfide bonds. Indinavir stones and pure uric acid stones are radiolucent on KUB; uric acid stones, but not indinavir stones, are visible on helical CT scanning. A KUB, however, has an unfavorable sensitivity and specificity profile (45–59% and 71–77%, respectively) and is not usually the best option in an the initial stone evaluation.[10] Ultrasonography has high specificity (97%), provides excellent visualization of hydronephrosis, and avoids exposure to radiation, but has a lower sensitivity than KUB because of difficulty identifying calculi that are small or that are located in the ureter[10] (Fig. 3). For this reason, ultrasound is preferred only for patients in whom it is important to limit exposure to ionizing radiation (e.g., pregnancy). IVP, previously the gold standard, has better sensitivity (64–87%) than KUB or ultrasound and excellent specificity similar to that of ultrasonography.[10] It may also play a therapeutic role by promoting an osmotic diuresis. However, compared with noncontrast helical CT of the kidneys, IVP has a lower sensitivity, takes more time, and exposes the patient to iodinated contrast and higher doses of radiation.[11] Indications for IVP include evaluation of a patient in whom indinavir stones or medullary sponge kidney are suspected.

In most circumstances, the best option for imaging is noncontrast helical CT of the kidneys. Sensitivity and specificity for this modality are both at or above 95%[10] (Fig. 4). It is more expensive than other imaging modalities but the higher cost of helical CT is offset by speed of diagnosis, ease of use, and better sensitivity.[12] It has the added advantage of detecting other causes of abdominal pain that may present with symptoms resembling renal colic.[13]

The authors state that they have no conflicts of interest.

© 2008 American Society for Bone and Mineral Research

STONE ANALYSIS

Stone analysis should be performed on all patients in whom a specimen is available, because it is the most direct way to identify stone type and allows for stone-specific therapy. Currently accepted methods of analysis include polarization microscopy, infrared spectroscopy, and X-ray crystallography/diffractometry.[14] Polarization microscopy, a technique that identifies crystals based on their interaction with polarized light, is inexpensive, quick, and requires very little stone material; the disadvantages include less reliability in identifying calcium phosphate, uric acid, and mixed stones.[14] Infrared spectroscopy is based on the interaction of infrared light and the molecules making up a stone (Fig. 5). Advantages include ease of use, small sample requirement, and ability to identify noncrystalline material (which often makes up 5% of an individual stone), whereas the major disadvantage is the amount of time required to perform the test.[14] X-ray diffractometry is based on the diffraction of X-rays by the crystalline structure (Fig. 6). The advantage of this technique is its ability to identify any crystal type; the main drawback, however, is its inability to identify noncrystalline material.[14]

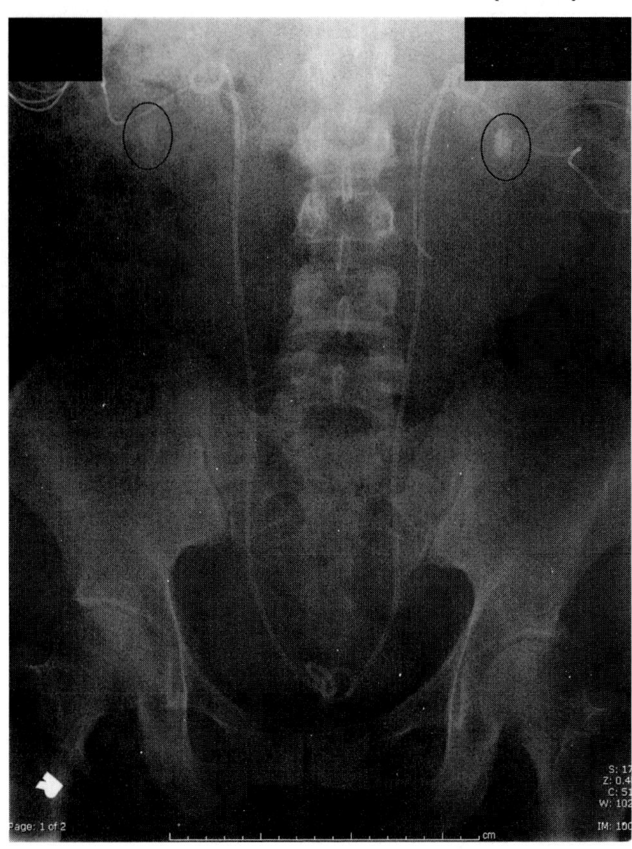

FIG. 2. The black ovals mark the bilateral radioopaque calculi identified on this plain film. Also noted are bilateral double J-stents in this 36-yr-old womane with calcium phosphate stones in the face of distal renal tubular acidosis secondary to Sjogren syndrome.

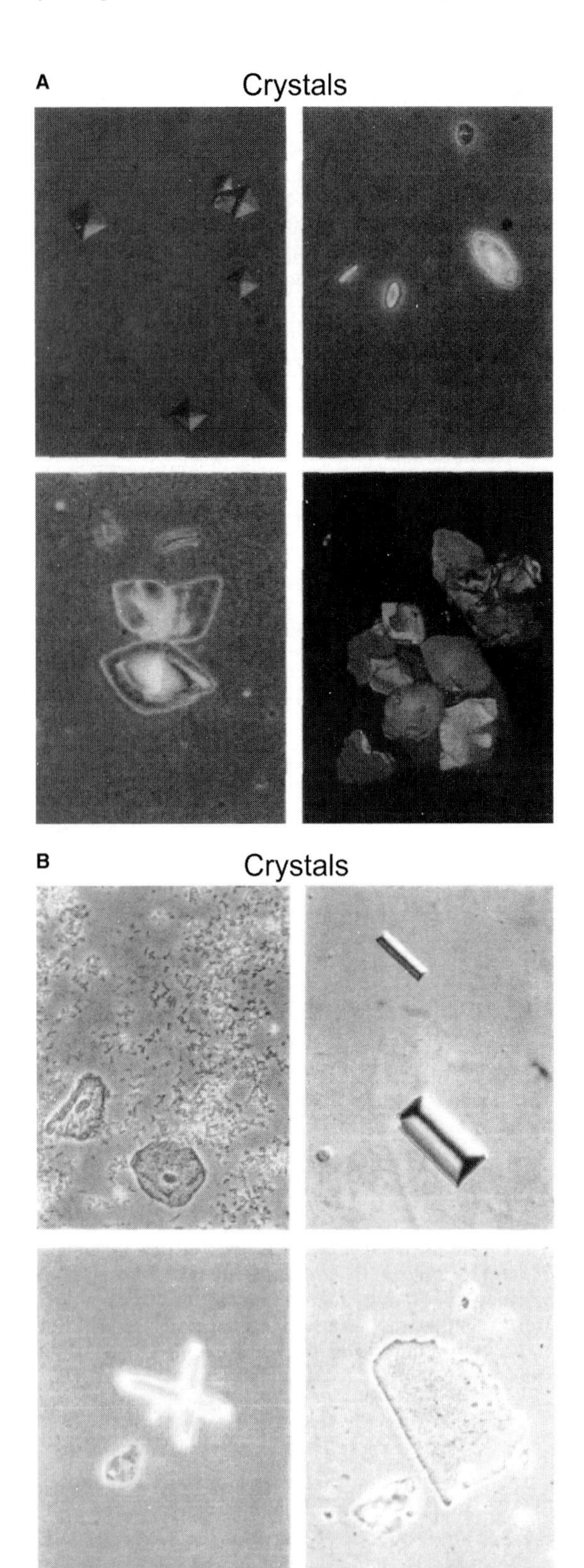

FIG. 1. (A) Clockwise from top left. Typical bipyramidal calcium oxalate dihydrate crystals (interference-contrast, ×640). Ovoid monohydrate calcium oxalate crystals (phase-contrast, ×640). Rhomboid uric acid crystals (phase-contrast, ×400). Uric acid crystals under polarized light (×250). (B) Clockwise from top left. Amorphous phosphates (phase contrast, ×400). Triple phosphate crystals (interference-contrast, ×400). A star-like calcium phosphate crystal (phase-contrast, ×400). Calcium phosphate plate (phase-contrast, ×400). (Reprinted with permission from Oxford University Press from Fogazzi GB 1996 Crystalluria: A neglected aspect of urinary sediment analysis. Nephrol Dial Transplant **11**:379–387.)

© 2008 American Society for Bone and Mineral Research

Renal Ultrasound

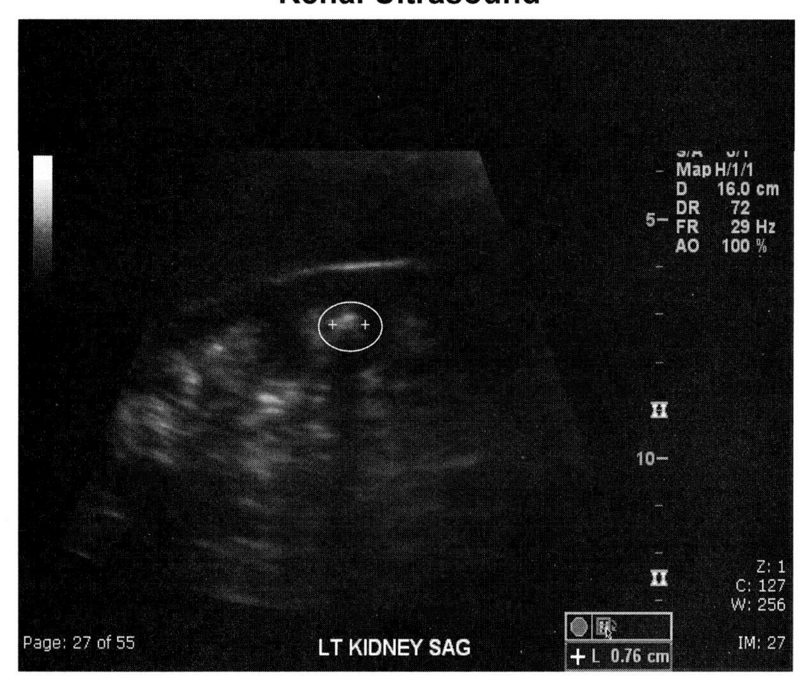

FIG. 3. The white oval highlights a calculus with posterior shadowing identified in the left kidney in the same patient described in Fig. 2.

WHICH PATIENTS SHOULD HAVE FULL METABOLIC PROFILING?

Beyond this evaluation of patients with acute renal colic, the cost-effectiveness of additional testing is not established for all adult first-time stone formers. It is clear that a child with stones, even on the first event, deserves a thorough metabolic evaluation given the increased likelihood of finding an inherited cause. Furthermore, metabolic evaluation of adults with multiple or recurrent stones has been shown to reduce overall

Non-contrast, helical CT of the Kidneys

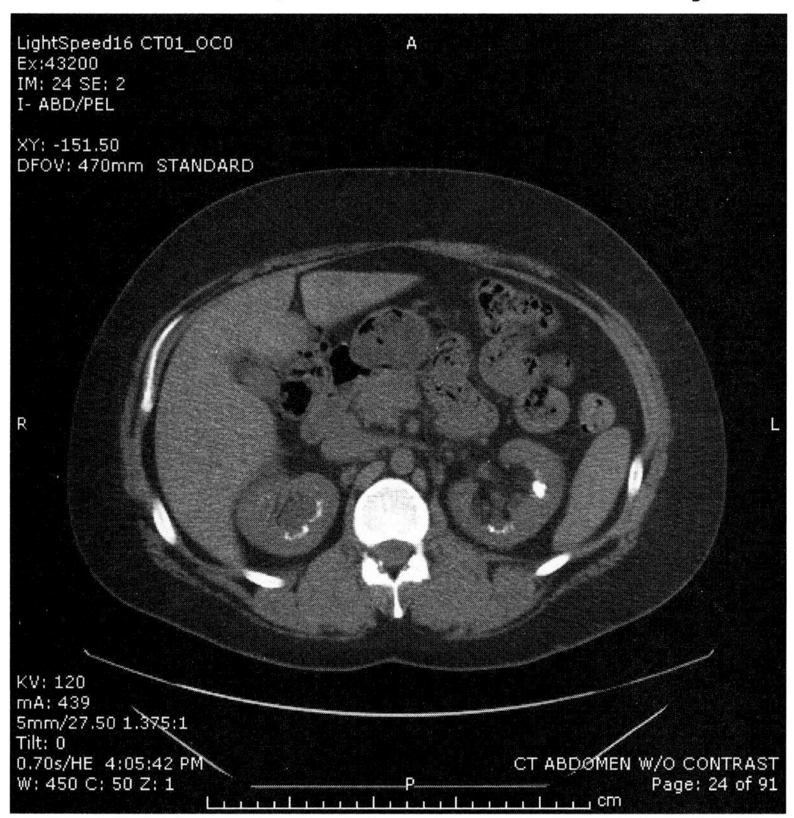

FIG. 4. A calculus is identified in the left kidney, and nephrocalcinosis is present bilaterally in the same patient described above in Figs. 2 and 3.

© 2008 American Society for Bone and Mineral Research

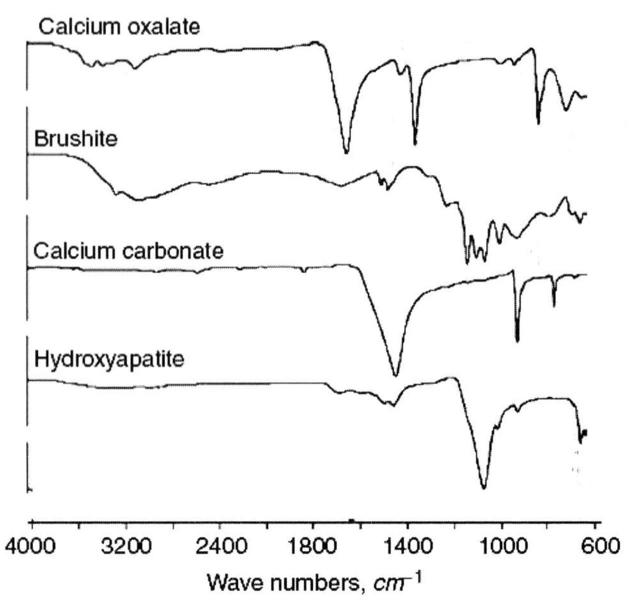

FIG. 5. Fourier transform infrared microspectroscopy (l-FTIR) spectra. (Reprinted with permission from Macmillan Publishers from Evan AP, Lingeman JE, Coe FL, Shao Y, Parks JH, Bledsoe SB, Phillips CL, Bonsib S, Worcester EM, Sommer AJ, Kim SC, Tinmouth WW, Grynpas M 2005 Crystal-associated nephropathy in patients with brushite nephrolithiasis. Kidney Int **67**:576–591, copyright 2005.)

costs.[15,16] Other patients in whom metabolic evaluation should be considered include patients with noncalcareous stones, reduced renal function, single functioning kidney, a strong family history of kidney stones, calcium oxalate stone-formers with an increased risk of enteric oxaluria (those with a history of bariatric surgery, short bowel syndrome, chronic diarrhea, or malabsorption), anatomic urinary tract abnormalities, and recurrent urinary tract infections.[4] The core of this additional testing is 24-h urine profiling, which should be performed at least 2–4 wk after an acute episode of renal colic has resolved. The components of a 24-h urine study include measurement of volume, creatinine, calcium, oxalate, citrate, uric acid, sodium, potassium, chloride, magnesium, phosphorus, ammonium, sulfate, urine urea nitrogen, and pH. Specialized laboratories will use these data to derive the protein catabolic rate and supersaturation measures for calcium oxalate, calcium phosphate, and uric acid. If the stone type is not known and cystine is in the differential diagnosis, qualitative cystine screening should be performed; the screen involves a colorimetric test using cyanide-nitroprusside on a 24-h urine collection. If this is positive, or if the stone is known to be cystine, quantitative testing should be performed using solid-phase assay.[17] Given the day-to-day variability in urine chemical composition, the initial metabolic evaluation should include two 24-h urine collections.[18] Adequacy of the urine collection can only be evaluated by reviewing the creatinine excretion of every collection, which should be 20–25 mg/kg lean body weight in men and 15–20 mg/kg in women. Additional laboratory studies depend on the results of the 24-h testing.

SUPERSATURATION

Unique to every salt is a characteristic termed the ion activity product (AP). The solubility product (SP) is defined as the AP below which crystallization does not occur. The formation product (FP) is defined as the AP at which crystallization does occur. Supersaturation is a value obtained by the ratio AP/SP: a value <1 indicates that crystallization will not occur; a value ≥1 indicates that crystallization can (but, in contradistinction to FP, does not have to) occur.[19] FP can be classified further into homogeneous or heterogeneous. Homogenous FP is the AP at which crystallization will occur independent of other urine components. Heterogeneous FP, however, is an AP dependent on the promoters and inhibitors present in the urine. The interval between SP and $FP_{homogeneous}$ is termed the metastable range of supersaturation; crystallization of a salt with an AP in this range depends on the balance between promoters (examples being calcium, oxalate, sodium, urine pH, and low urine volume) and inhibitors (examples being citrate, pyrophosphate, potassium, magnesium, and high urine volume) of stone formation.

In human urine, calcium oxalate AP normally exists in the metastable range of supersaturation.[20] A major goal of stone prevention is to reduce supersaturation, particularly for the salt of the stone type the patient has experienced, if that is known. Increasing urinary volume will reduce the supersaturation of all salts proportionately. Therapeutic measures for altering specific solutes are discussed in a separate chapter.

HYPERCALCIURIA

Hypercalciuria, the most commonly identified metabolic abnormality on 24-h urine profiling, is defined as excretion of >200 mg/d of calcium in women and >250 mg/d in men.[21] The differential diagnosis of hypercalciuria is discussed in another chapter. If hypercalciuria is documented, laboratory evaluation should include measurement of intact PTH, phosphorus, 1,25-dihydroxyvitamin D_3, magnesium, and thyroid-stimulating hormone (TSH). Given the growing evidence of bone demineralization in patients with hypercalciuria, a DXA scan to evaluate for osteoporosis may be considered as well.[22] Because dietary calcium restriction below a recommended daily intake of 800–1000 mg is not of therapeutic value in patients with stones, categorization of patients into subsets of hypercalciuria based on intestinal absorption, bone resorption, or renal reabsorption is not clinically useful.

HYPEROXALURIA

Hyperoxaluria is typically defined as an oxalate excretion of >45 mg/d.[21] Most cases of hyperoxaluria are diet-mediated. Ethylene glycol, an antifreeze, is metabolized to oxalate and when ingested can cause stones.

Oxalate levels >100 mg/d suggest that enteric oxaluria is likely contributing as well. Enteric oxaluria can result from a malabsorptive process (such as in pancreatic insufficiency, cystic fibrosis, or Crohn's disease, or after gastric bypass or bariatric surgery) in which undigested fatty acids in the colon saponify calcium, allowing more free oxalate to remain in solution and not complexed with calcium, resulting in enhanced oxalate absorption. Nonabsorbed bile salts also increase colonic permeability to oxalate. At levels >500 mg/d, primary hyperoxaluria, a rare inherited disease that results in enhanced conversion of glyoxylate to oxalate, should be considered as an etiology. This topic is also discussed further in another chapter.

HYPERURICOSURIA

Hyperuricosuria is defined as urinary uric acid excretion exceeding 750 and 800 mg/d in women and men, respectively.[21] It is a risk factor for both calcium oxalate and uric acid nephrolithiasis.[3] A purine-rich diet is often a factor, but it is possible that some patients may have mild metabolic overproduc-

© 2008 American Society for Bone and Mineral Research

X-ray Diffractometry

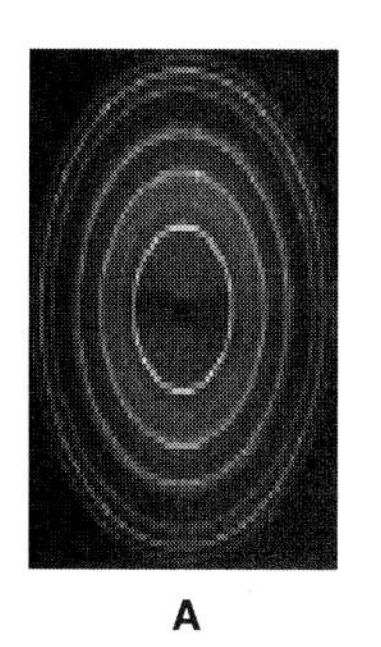
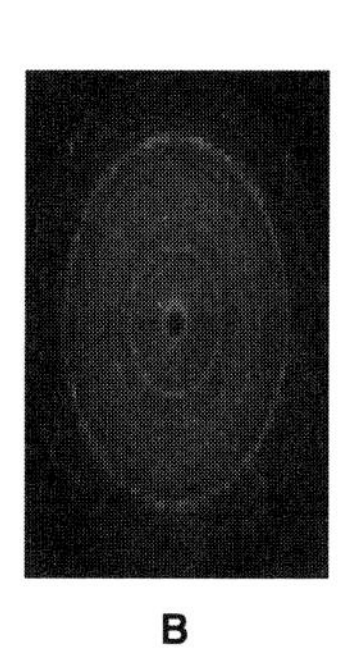
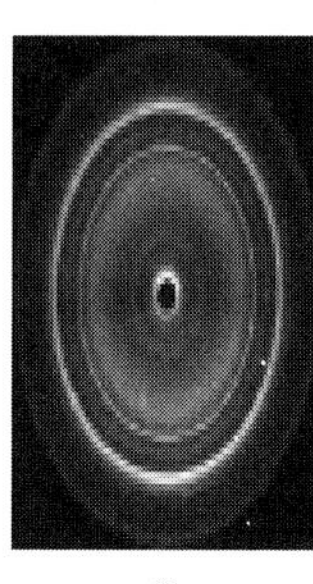
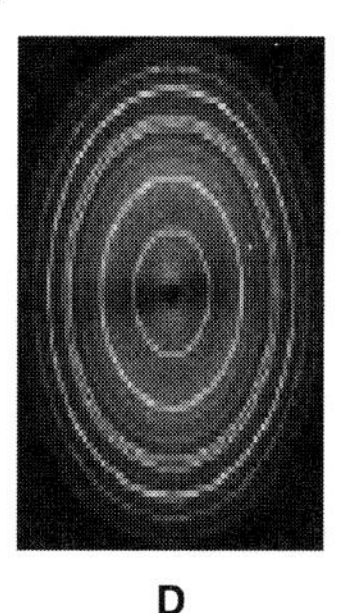
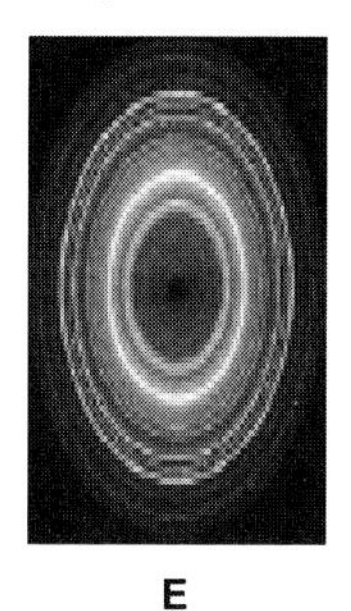
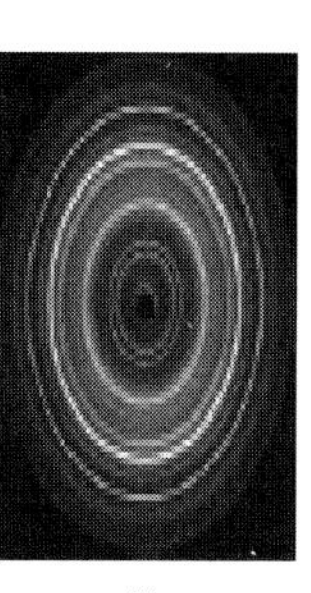
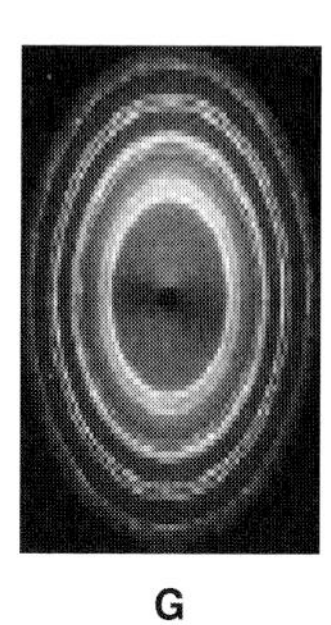

A B C D E F G

FIG. 6. Scatter patterns from X-ray diffractometry for the seven most common stone components: (A) calcium oxalate monohydrate, (B) calcium oxalate dihydrate, (C) calcium phosphate, (D) calcium phosphate dihydrate, (E) struvite, (F) uric acid, and (G) cystine. (Reprinted with permission from IOP Publishing from Davidson MT, Batchelar DL, Velupillai S, Denstedt JD, Cunningham IA 2005 Analysis of urinary stone components by X-ray coherent scatter: characterizing composition beyond laboratory X-ray diffractometry. Phys Med Biol **50:**3773–3786.)

tion of uric acid. Other less common causes include gout, leukemia, tumor lysis syndrome, Lesch-Nyhan syndrome, enhanced phosphoribosylpyrophosphate synthetase activity, xanthinuria, and 2,8-dihydroxyadeninuira.

HYPOCITRATURIA

Hypocitraturia is defined as a citrate excretion of <500 mg/d in women and <350 mg/d in men.[21] Citrate, by forming a soluble salt with calcium, inhibits calcium stone formation. Acidemia reduces citrate excretion, whereas alkalemia augments it. Most causes are related to the protein or acid content of the diet, but patients with hypocitraturia should be assessed for hypokalemia, diarrhea, infection, distal renal tubular acidosis, and for the use of acetazolamide.

URINE pH

Uric acid and cystine crystallization is favored by an acidic urine. Calcium phosphate and struvite stones form in an alkaline urine. A persistently alkaline urine combined with hypocitraturia is highly suggestive of a distal renal tubular acidosis.

OTHER FACTORS

Calcium excretion parallels sodium excretion; reducing sodium intake (and thus excretion) may reduce calciuria. It has also been suggested that excessive sodium intake is predictive of stone risk independently of its effects to increase the excretion of calcium, uric acid, and oxalate.[8] Hypokalemia, by causing intracellular acidosis, can result in hypocitraturia, thereby increasing stone risk. Increased potassium excretion, on the other hand, decreases the risk of stone formation.[23] Magnesium in the urine has been shown to inhibit calcium oxalate crystallization, and enteral magnesium may reduce intestinal oxalate absorption.[24] Phosphate measurements are required to evaluate urinary supersaturation and usually are reflective of dietary intake. When evaluated in the context of serum PTH, phosphate, and calcium levels, urinary phosphate excretion can yield information regarding disorders of the calcium–vitamin D–PTH axis. Ammonium levels are increased in urine that is infected with urease-producing bacteria, promoting struvite formation; ammonium excretion is physiologically increased in acid-loading states.[25,26] Under conditions of impaired acid excretion (i.e., a distal renal tubular acidosis), ammonium excretion may be low.

Markers of protein excretion, such as urinary sulfate, urine urea nitrogen, and protein catabolic rate, serve as an indirect measure of acid load; the acid, which requires buffering from bone, can result in hypercalciuria.[27,28]

EFFECT OF SEX ON NORMAL RANGES OF URINARY CONSTITUENTS

Women typically excrete less calcium and uric acid and more citrate than men, and the distribution curves that define the "normal" ranges in men and women reflect this. The normally higher levels of two important stone constituents and normally lower levels of citrate explain much, although not all, of the increased risk of stones in men. However, it should be remembered that the distribution of these solutes is continuous and that a given level of solute has the same relevance to stone risk in a man or a woman. In this regard, the supersaturation ratios for each stone salt that are provided by specialty laboratories are particularly helpful.

ACKNOWLEDGMENTS

The authors thank George S. Muthalakuzhy, MD, for valuable assistance in the preparation of this chapter.

REFERENCES

1. Moe OW 2006 Kidney stones: Pathophysiology and medical management. Lancet **367:**333–344.
2. Healy KA, Ogan K 2007 Pathophysiology and management of infectious staghorn calculi. Urol Clin North Am **34:**363–374.
3. Shekarriz B, Stoller ML 2002 Uric acid nephrolithiasis: Current concepts and controversies. J Urol **168:**1307–1314.
4. Miller NL, Lingeman JE 2007 Management of kidney stones. BMJ **334:**468–472.
5. Ramey SL, Franke WD, Shelley MC II 2004 Relationship among risk factors for nephrolithiasis, cardiovascular disease, and ethnicity: Focus on a law enforcement cohort. AAOHN J **52:**116–121.
6. Ljunghall S, Danielson BG, Fellstrom B, Holmgren K, Johansson G, Wikstrom B 1985 Family history of renal stones in recurrent stone patients. Br J Urol **57:**370–374.
7. Parmar MS 2004 Kidney stones. BMJ **328:**1420–1424.
8. Curhan GC, Taylor EN 2007 24-h uric acid excretion and the risk of kidney stones. Kidney Int **73:**489–496.
9. Stapleton FB, Roy S III, Noe HN, Jerkins G 1984 Hypercalciuria in children with hematuria. N Engl J Med **310:**1345–1348.
10. Portis AJ, Sundaram CP 2001 Diagnosis and initial management of kidney stones. Am Fam Physician **63:**1329–1338.
11. Catalano O, Nunziata A, Altei F, Siani A 2002 Suspected ureteral colic: Primary helical CT versus selective helical CT after unenhanced radiography and sonography. AJR Am J Roentgenol **178:**379–387.
12. Chen MY, Zagoria RJ 1999 Can noncontrast helical computed

© 2008 American Society for Bone and Mineral Research

tomography replace intravenous urography for evaluation of patients with acute urinary tract colic? J Emerg Med **17:**299–303.
13. Ha M, MacDonald RD 2004 Impact of CT scan in patients with first episode of suspected nephrolithiasis. J Emerg Med **27:**225–231.
14. Schubert G 2006 Stone analysis. Urol Res **34:**146–150.
15. Parks JH, Coe FL 1996 The financial effects of kidney stone prevention. Kidney Int **50:**1706–1712.
16. Robertson WG 2006 Is prevention of stone recurrence financially worthwhile? Urol Res **34:**157–161.
17. Coe FL, Clark C, Parks JH, Asplin JR 2001 Solid phase assay of urine cystine supersaturation in the presence of cystine binding drugs. J Urol **166:**688–693.
18. Parks JH, Goldfisher E, Asplin JR, Coe FL 2002 A single 24-hour urine collection is inadequate for the medical evaluation of nephrolithiasis. J Urol **167:**1607–1612.
19. Tiselius H 2005 Aetiological factors in stone formation. In: Ponticelli C (ed.) Oxford Textbook of Clinical Nephrology, 3rd ed. Oxford University Press, New York, NY, USA.
20. Mandel N 1996 Mechanism of stone formation. Semin Nephrol **16:**364–374.
21. Coe FL, Evan A, Worcester E 2005 Kidney stone disease. J Clin Invest **115:**2598–2608.
22. Asplin JR, Donahue S, Kinder J, Coe FL 2006 Urine calcium excretion predicts bone loss in idiopathic hypercalciuria. Kidney Int **70:**1463–1467.
23. Curhan GC, Willett WC, Rimm EB, Stampfer MJ 1993 A prospective study of dietary calcium and other nutrients and the risk of symptomatic kidney stones. N Engl J Med **328:**833–838.
24. Massey L 2005 Magnesium therapy for nephrolithiasis. Magnes Res **18:**123–126.
25. Griffith D 1983 Infection induced urinary stones. In: Roth RA, Finlayson B (eds.) Stones—Clinical Management of Urolithiasis—International Perspectives in Urology, vol. 6. Williams & Wilkins, Baltimore, MD, USA, pp. 210–227.
26. Parivar F, Low RK, Stoller ML 1996 The influence of diet on urinary stone disease. J Urol **155:**432–440.
27. Martini LA, Wood RJ 2000 Should dietary calcium and protein be restricted in patients with nephrolithiasis? Nutr Rev **58:**111–117.
28. Bingham SA 2003 Urine nitrogen as a biomarker for the validation of dietary protein intake. J Nutr **133**(Suppl 3)**:**921S–924S.
29. Finocchiaro R, D'Eufemia P, Celli M, Zaccagnini M, Viozzi L, Troiani P, Mannarino O, Giardini O 1998 Usefulness of cyanide-nitroprusside test in detecting incomplete recessive heterozygotes for cystinuria: A standardized dilution procedure. Urol Res **26:**401–405.

Chapter 100. Treatment of Renal Stones

John R. Asplin

Litholink Corp., Chicago, Illinois, Department of Medicine, University of Chicago Pritzker School of Medicine, Chicago, Illinois

INTRODUCTION

Renal stones form in urine that is supersaturated with respect to the salt of which the stone is composed. Crystallization can be induced at low levels of supersaturation if there are promoters of crystallization in the urine, whereas urinary inhibitors oppose crystallization. Medical treatment of nephrolithiasis works by lowering urine saturation of the stone forming salt, reducing promoters of crystallization, and/or increasing inhibitors of crystallization. Optimal treatment is based on the type of stone and the lithogenic risk factors identified in 24-h urine chemistries. Treatment may consist of diet, medication, or a combination of both.

Before a discussion of medical therapy, consideration must be given to the goals of treatment. The aim of treatment is to prevent new stone formation and reduce the rate of growth of any existing stones. Radiologic evaluation is a critical component of any treatment regimen. A knowledge of the number and location of all stones at the time treatment is initiated is required to monitor the success of therapy. If a patient develops renal colic and the stone passing is clearly a new stone, medical therapy has not been successful; the cause of the failure should be studied, and the treatment regimen should be re-evaluated. Not only is the knowledge of old versus new stone important for the physician, but also for the patient. Patients who pass a stone despite following the prescribed therapy will become discouraged and stop therapy unless they understand that the stone that passed was present all along. Serial radiologic evaluation (every 1–2 yr) is also needed to insure that new asymptomatic stones are not forming during treatment.

High fluid intake is the foundation of all medical regimens for renal stone prevention. By keeping urine dilute, urine saturation of all stone-forming salts is lowered. Although high fluid intake has long been considered the standard of care, Borghi et al.[1] have performed the only prospective randomized study of high fluid intake in the prevention stones. Over the 5 yr of follow-up, a 55% reduction of stone recurrence in the group treated with high fluid intake compared with the control group was observed. The high fluid group had an average 24-h urine volume of 2.5 liters, which provides a reasonable therapeutic goal for renal stone patients.

CALCIUM STONES

There are four abnormalities of urine chemistries known to cause calcium oxalate kidney stones: hypercalciuria, hyperoxaluria, hypocitraturia, and hyperuricosuria. Therapy is directed against whichever abnormality(s) the patient has.

Hypercalciuria

In the past, low calcium diets had often been prescribed for calcium stone formers. Although dietary calcium restriction clearly lowers urine calcium, it has never been shown to reduce stone formation in a prospective controlled trial. In addition, the long-term safety of a low calcium diet has been questioned because hypercalciuria has been associated with reduced BMD and increased risk of fractures.[2,3] However, other dietary interventions can be used to lower urine calcium. Both high sodium and high protein diets increase urine calcium excretion, so restriction of these dietary components seems a reasonable intervention. Borghi et al.[4] tested the efficacy of a normal calcium (1200 mg/d), low sodium, low protein intake diet com-

Dr. Asplin is an employee of Litholink Corp. and has served as a consultant for Altus Pharmaceuticals.

Key words: nephrolithiasis, hypercalciuria, cystinuria, struvite, thiazide, citrate, allopurinol, acetohydroxamic acid, tiopronin, dietary therapy

© 2008 American Society for Bone and Mineral Research

pared with the classic low calcium diet (400 mg/d) in 120 men with recurrent kidney stones. Stone recurrence was reduced by 48% in the group treated with the low sodium, low protein diet compared with the low calcium diet. Although this study does not answer the question of whether low calcium diets reduce stone formation, it does show there is an alternative diet therapy that is more efficacious and likely better for general health.

For patients with frequent stones or whose stone disease persists despite dietary intervention, thiazide diuretics should be used to lower urine calcium. There are three prospective, placebo-controlled, 3-yr trials of thiazide type diuretics, all of which have shown a significant reduction in stone formation rates in patients with recurrent kidney stones.[5–7] There has not been a comparison between the various thiazides to determine whether one has a greater degree of anti-hypercalciuric effect than the others. Diet sodium restriction should be continued when thiazides are used, because excess salt in the diet will reduce the effectiveness of a thiazide. Because thiazides cause potassium wasting, serum potassium should be monitored because hypokalemia can lead to a reduction in urine citrate excretion, creating a new risk factor for stone formation.[8] Thiazides also have an additional benefit in this population because they improve calcium balance and prevent osteopenia in hypercalciuric stone formers.[9,10]

Hyperoxaluria

Oxalate excreted in the urine is derived from both intestinal absorption of oxalate in the diet and from endogenous production of oxalate.[11] Oxalate is an endproduct of human metabolism: whatever oxalate is absorbed or produced by the body must be excreted by the kidneys. Because a significant portion of urine oxalate is derived from diet, low oxalate diets are recommended for all hyperoxaluric stone patients, although the efficacy of low oxalate diet in preventing stone formation has not been studied. General guidelines for low oxalate diets are available from many sources.[12,13] Other dietary advice focuses on foods that have a secondary effect on oxalate excretion. Low calcium diets can increase urine oxalate excretion because low calcium concentration in the intestinal lumen leaves more oxalate in the unbound state to be absorbed.[11] The magnitude of this effect in clinical practice is uncertain, but it seems prudent to have patients on a normal calcium intake of ~1200 mg/d. Endogenous production of oxalate may be increased by an increased intake of oxalate precursors. A number of amino acids can be metabolized to oxalate, so it is not surprising that high protein diets have been shown to increase urine oxalate excretion in some patients with hyperoxaluria.[14] Vitamin C can also be metabolized to oxalate; as a general recommendation, stone patients should avoid high doses of vitamin C supplements.[15]

There are no FDA-approved medications for hyperoxaluria at this time. Pyridoxine has been reported to lower urine oxalate in some patients with idiopathic hyperoxaluria, but there are no controlled trials showing that pyridoxine reduces stone formation in such patients.[16] Magnesium has been proposed as a therapy for stone disease, because it may lower urine oxalate by complexing oxalate in the intestine and act as a crystal inhibitor in the urine. There has been only one controlled trial of magnesium supplements in calcium stone disease, and it did not show a benefit to therapy.[7] Currently under study is the use of oxalate-degrading bacteria as probiotics.[17,18] These agents would act by lowering intestinal oxalate concentration, limiting absorption, and perhaps even increasing intestinal secretion of oxalate. However, further studies need to be done to prove their efficacy in humans with hyperoxaluric stone disease.[19]

Enteric hyperoxaluria and primary hyperoxaluria can lead to severe hyperoxaluria and cause kidney failure in addition to stone disease. Enteric hyperoxaluria can be seen in patients with extensive small bowel disease or bowel resection. Recently, it has been shown that modern bariatric surgery can also cause enteric hyperoxaluria.[20] To lower urine oxalate, low oxalate diets are prescribed, and calcium supplements are provided with meals to bind oxalate in the diet.[21] Calcium citrate is preferred over calcium carbonate because many bariatric procedures alter stomach anatomy, and there may not be sufficient acid exposure to dissolve calcium carbonate. If fat malabsorption is present, a low fat diet should be recommended; cholestyramine may also lower oxalate absorption in patients with steatorrhea. Primary hyperoxaluria (PH) is a rare genetic form of hyperoxaluria. In PH type 1 (OMIM 259900), ~30% of patients will have with a significant reduction in oxalate excretion with pyridoxine therapy.[22] If oxalate excretion cannot be lowered with pyridoxine, patients should be treated with neutral phosphate salts and/or potassium citrate, as well as high fluid intake.[23,24] PH type 2 (OMIM 260000) does not respond to pyridoxine therapy but otherwise is treated the same as PH type 1.

Hypocitraturia

Citrate is a normal constituent of human urine that prevents calcium stone formation by forming a soluble complex with calcium, thereby reducing the free calcium ion concentration in the urine. Low urine citrate may be seen with acidemic states, such as renal tubular acidosis and hypokalemia or it may be idiopathic in origin.[8] Urine citrate excretion is highly dependent on systemic acid-base status, increasing with alkali loading. Reducing dietary animal protein and increasing consumption of fruits and vegetables will increase urine citrate excretion, but dietary intervention is usually not sufficient to correct significant hypocitraturia.[25] Some authorities have proposed using lemonade as a treatment for hypocitraturia because of the high citric acid content of lemon juice.[26] However, significant increases in urine citrate have not been found in all studies.[27]

The standard treatment for hypocitraturia is alkali supplementation. Alkali may be provided as either citrate or bicarbonate, but citrate is most commonly used. Citrate is usually provided as potassium citrate, because sodium salts can increase urine calcium excretion. If the patient cannot tolerate a potassium load, sodium alkali can be used, although dietary sodium restriction should be encouraged to minimize the increase in urine calcium. There have been two randomized placebo-controlled trials of citrate salts in the treatment of calcium stone disease: one with potassium citrate and another using potassium-magnesium citrate.[28,29] Both showed significant reductions in stone recurrence rates in the group receiving active drug. In addition to the increase in citrate, alkali salts can also cause a mild reduction in the urine calcium excretion by buffering daily metabolic acid production. In fact, long-term treatment of calcium stone formers with citrate salts has been shown to increase BMD as well, presumably from the favorable effects on urine calcium excretion and calcium balance.[30] One potential adverse consequence of alkali therapy is an increase in urine pH, which, in patients with elevated urine calcium, can increase the risk of calcium phosphate stone formation. Such patients should have urine calcium controlled by the addition of a thiazide diuretic.

© 2008 American Society for Bone and Mineral Research

Hyperuricosuria

Uric acid is an endproduct of purine metabolism, produced by the oxidation of xanthine through the enzyme xanthine oxidase. Hyperuricosuria promotes calcium oxalate crystallization in the urine by a salting out phenomenon.[31] Hyperuricosuria is most commonly caused by diets high in purines, although some patients may have a metabolic defect that leads to overproduction of uric acid.[32] A low purine diet is a reasonable therapy for patients with hyperuricosuria, although its effectiveness in preventing stone formation has not been tested. Allopurinol, a xanthine oxidase inhibitor, has been shown to reduce stone formation in calcium oxalate stone formers with hyperuricosuria in a randomized prospective trial.[33] It has not been shown to be effective in treatment of patients without hyperuricosuria.

URIC ACID STONES

Three factors contribute to uric acid stone formation: urine flow rate, uric acid excretion rate, and urine pH.[34] Low urine pH (<5.8 in a 24-h urine collection) is the most common abnormality found in uric acid stone formers.[35] Uric acid stones seldom form without a low urine pH, unless the patient has marked overproduction of uric acid from a metabolic defect. Overly acidic urine may be seen in patients with a high intake of animal protein, chronic diarrheal states, or as an idiopathic state. It has recently been recognized that patients with the metabolic syndrome have an increased rate of uric acid stone formation caused by overly acidic urine.[36] There are no prospective placebo controlled trials for prevention of uric acid stones, nor are there likely to be, because increasing urine pH with alkali therapy is universally regarded as effective. Alkali at an initial dose of 40–60 meq/d in two or three divided doses will be sufficient for most patients. The desired endpoint is an average 24-h urine pH of 6.0–6.5. Excessive alkalinization does not offer additional protection from uric acid stone formation, but it will increase the risk of calcium phosphate stone formation. Low purine diets will help lower uric acid excretion, and lowering animal protein intake will decrease metabolic acid production. Allopurinol can be used for uric acid stones if the patient also has gout or very high levels of uric acid excretion or if the patient cannot tolerate sufficient alkali to raise urine pH >6.0. Generally, allopurinol is considered a second-line therapy for uric acid stones.

CYSTINE STONES

Cystine is an amino acid formed by the linkage of two cysteine molecules through a disulfide bond. It is poorly soluble in urine, and patients with cystinuria (OMIM 220100) can develop severe, recurrent stone disease. Therapy to reduce stone formation is focused on lowering cystine concentration in the urine or increasing urine cystine solubility. Patients with cystinuria are instructed to increase fluid intake and maintain a high urine flow rate to keep urine cystine concentration <250 mg/liter; most patients need to produce at least 3 liters of urine per day. The patient should awaken at least once per night to void and drink additional water. Dietary sodium and protein restriction lower urine cystine excretion moderately and should be encouraged in all patients.[37,38]

There are two treatment options to improve cystine solubility. Solubility increases as urine pH increases, although the effect is not significant until urine pH is >7.[39] Because the desired pH is higher than for other forms of stone disease, it may require larger doses of alkali, although the individual response is highly variable. Potassium alkali is preferred because of the undesirable effect of high sodium intake on urine cystine excretion, although sodium salts should be used if the patient is not able to excrete a high potassium load. The other way to increase solubility of cystine is to form a new disulfide complex that is more soluble than cystine itself. Tiopronin and D-penicillamine contain thiol groups and will undergo a disulfide exchange, leading to the formation of a drug–cysteine complex that is much more soluble than cystine. These medications have been shown to reduce cystine saturation in cystine stone formers.[40] Because of the high frequency of side effects,[41] thiol-containing drugs are usually reserved for patients with a very high level of cystine excretion or those who continue to form stones despite dietary and alkali therapy. Captopril, an angiotensin converting enzyme inhibitor, contains a thiol group and has been suggested as therapy for cystinuria. However, at the doses used, it is unlikely to complex sufficient cystine to alter stone risk and studies conflict regarding the effectiveness of captopril.[42,43]

STRUVITE STONES

Struvite (magnesium-ammonium-phosphate) stones require elevated urine pH and high ammonium concentration to form. In normal human physiology, urine ammonium excretion falls when urine pH rises above 7; the only time human urine displays high pH and high ammonium is when there is infection with a urease-containing organism such as *Proteus mirabilis*. Struvite forms large staghorn stones that can lead to loss of renal function. Treatment requires a combined medical and surgical approach. The best outcomes require complete removal of all stone material because the stone itself often harbors the offending bacteria, and if the stone is left in the urinary system, infection will recur, as will the stones. Extracorporeal shock wave lithotripsy and percutaneous nephrolithotomy may be used alone or in combination to remove all stone material.[44] Stones should be cultured to guide choice of antibiotics. Antibiotics should be used to sterilize the urinary system. In patients who are not surgical candidates or in whom stones continue to recur, the urease inhibitor acetohydroxamic acid (AHA) may be considered. Although AHA will not eradicate infection, it does prevent the breakdown of urea and the change in the urinary environment that leads to struvite stone formation. There are three placebo-controlled trials that have showed the effectiveness of AHA, but the significant number of side effects such as headache, tremulousness, and possibly deep venous thrombosis limits its use.[45–47]

REFERENCES

1. Borghi L, Meschi T, Amato F, Briganti A, Novarini A, Giannini A 1996 Urinary volume, water and recurrences in idiopathic calcium nephrolithiasis: A 5-year randomized prospective study. J Urol **155:**839–843.
2. Melton LJ III, Crowson CS, Khosla S, Wilson DM, O'Fallon WM 1998 Fracture risk among patients with urolithiasis: A population-based cohort study. Kidney Int **53:**459–464.
3. Lauderdale DS, Thisted RA, Wen M, Favus MJ 2001 Bone mineral density and fracture among prevalent kidney stone cases in the Third National Health and Nutrition Examination Survey. J Bone Miner Res **16:**1893–1898.
4. Borghi L, Schianchi T, Meschi T, Guerra A, Allegri F, Maggiore U, Novarini A 2002 Comparison of two diets for the prevention of recurrent stones in idiopathic hypercalciuria. N Engl J Med **346:**77–84.
5. Borghi L, Meschi T, Guerra A, Novarini A 1993 Randomized prospective study of a nonthiazide diuretic, indapamide, in preventing calcium stone recurrences. J Cardiovasc Pharmacol **22**(Suppl 6)**:**S78–S86.
6. Laerum E, Larsen S 1984 Thiazide prophylaxis of urolithiasis: A double- blind study in general practice. Acta Med Scand **215:**383–389.
7. Ettinger B, Citron JT, Livermore B, Dolman LI 1988 Chlorthali-

© 2008 American Society for Bone and Mineral Research

done reduces calcium oxalate calculous recurrence but magnesium hydroxide does not. J Urol **139:**679–684.
8. Hamm LL 1990 Renal handling of citrate. Kidney Int **38:**728–735.
9. Adams J, Song C, Kantorovich V 1999 Rapid recovery of bone mass in hypercalcuric, osteoporotic men treated with hydrochlorothiazide. Ann Intern Med **130:**658–660.
10. Coe F, Parks J, Bushinsky D, Langman C, Favus M 1988 Chlorthalidone promotes mineral retention in patients with idiopathic hypercalciuria. Kidney Int **33:**1140–1146.
11. Holmes RP, Goodman HO, Assimos DG 2001 Contribution of dietary oxalate to urinary oxalate excretion. Kidney Int **59:**270–276.
12. Harvard School of Public Health Nutrition Department's File Download Site 2007 Oxalate Table of Foods. Available online at https://regepi.bwh.harvard.edu/health/Oxalate/files/. Accessed October 23, 2007.
13. Oxalosis and Hyperoxaluria Foundation 2004 The Oxalate Content of Food. Available online at http://www.ohf.org/diet.html/. Accessed October 23, 2007.
14. Nguyen QV, Kalin A, Drouve U, Casez JP, Jaeger P 2001 Sensitivity to meat protein intake and hyperoxaluria in idiopathic calcium stone formers. Kidney Int **59:**2273–2281.
15. Traxer O, Huet B, Poindexter J, Pak CY, Pearle MS 2003 Effect of ascorbic acid consumption on urinary stone risk factors. J Urol **170:**397–401.
16. Edwards P, Nemat S, Rose GA 1990 Effects of oral pyridoxine upon plasma and 24-hour urinary oxalate levels in normal subjects and stone formers with idiopathic hypercalciuria. Urol Res **18:**393–396.
17. Hatch M, Cornelius J, Allison M, Sidhu H, Peck A, Freel RW 2006 Oxalobacter sp. reduces urinary oxalate excretion by promoting enteric oxalate secretion. Kidney Int **69:**691–698.
18. Campieri C, Campieri M, Bertuzzi V, Swennen E, Matteuzzi D, Stefoni S, Pirovano F, Centi C, Ulisse S, Famularo G, De Simone C 2001 Reduction of oxaluria after an oral course of lactic acid bacteria at high concentration. Kidney Int **60:**1097–1105.
19. Goldfarb DS, Modersitzki F, Asplin JR 2007 A randomized, controlled trial of lactic acid bacteria for idiopathic hyperoxaluria. Clin J Am Soc Nephrol. **2:**745–749.
20. Asplin JR, Coe FL 2007 Hyperoxaluria in kidney stone formers treated with modern bariatric surgery. J Urol **177:**565–569.
21. Worcester EM 2002 Stones from bowel disease. Endocrinol Metab Clin North Am **31:**979–999.
22. Monico CG, Rossetti S, Olson JB, Milliner DS 2005 Pyridoxine effect in type I primary hyperoxaluria is associated with the most common mutant allele. Kidney Int **67:**1704–1709.
23. Milliner DS, Eickholt JT, Bergstralh EJ, Wilson DM, Smith LH 1994 Results of long-term treatment with orthophosphate and pyridoxine in patients with primary hyperoxaluria. N Engl J Med **331:**1553–1558.
24. Leumann E, Hoppe B, Neuhaus T 1993 Management of primary hyperoxaluria: Efficacy of oral citrate administration. Pediatr Nephrol **7:**207–211.
25. Breslau NA, Brinkley L, Hill KD, Pak CY 1988 Relationship of animal protein-rich diet to kidney stone formation and calcium metabolism. J Clin Endocrinol Metab **66:**140–146.
26. Seltzer MA, Low RK, McDonald M, Shami GS, Stoller ML 1996 Dietary manipulation with lemonade to treat hypocitraturic calcium nephrolithiasis. J Urol **156:**907–909.
27. Odvina CV 2006 Comparative value of orange juice versus lemonade in reducing stone-forming risk. Clin J Am Soc Nephrol. **1:**1269–1274.
28. Ettinger B, Pak CYC, Citron JT, Thomas C, Adams-Huet B, Vangessel A 1997 Potassium-magnesium citrate is an effective prophylaxis against recurrent calcium oxalate nephrolithiasis. J Urol **158:**2069–2073.
29. Barcelo P, Wuhl O, Servitge E, Roussaud A, Pak C 1993 Randomized double- blind study of potassium citrate in idiopathic hypocitraturic calcium nephrolithiasis. J Urol **150:**1761–1764.
30. Pak CY, Peterson RD, Poindexter J 2002 Prevention of spinal bone loss by potassium citrate in cases of calcium urolithiasis. J Urol **168:**31–34.
31. Grover P, Ryall R, Marshall V 1993 Dissolved urate promotes calcium oxalate crystallization: Epitaxy is not the cause. Clin Sci **85:**303–307.
32. Coe FL, Moran E, Kavalich AG 1976 The contribution of dietary purine over-consumption to hyperpuricosuria in calcium oxalate stone formers. J Chronic Dis **29:**793–800.
33. Ettinger B, Tang A, Citron JT, Livermore B, Williams T 1986 Randomized trial of allopurinol in the prevention of calcium oxalate calculi. N Engl J Med **315:**1386–1389.
34. Asplin J 1996 Uric acid stones. Semin Nephrol **16:**412–424.
35. Sakhaee K, Adams-Huet B, Moe OW, Pak CY 2002 Pathophysiologic basis for normouricosuric uric acid nephrolithiasis. Kidney Int **62:**971–979.
36. Abate N, Chandalia M, Cabo-Chan AV Jr, Moe OW, Sakhaee K 2004 The metabolic syndrome and uric acid nephrolithiasis: Novel features of renal manifestation of insulin resistance. Kidney Int **65:**386–392.
37. Rodman JS, Blackburn P, Williams JJ, Brown A, Pospischil MA, Peterson CM 1984 The effect of dietary protein on cystine excretion in patients with cystinuria. Clin Nephrol **22:**273–278.
38. Lindell A, Denneberg T, Edholm E, Jeppsson JO 1995 The effect of sodium intake on cystinuria with and without tiopronin treatment. Nephron **71:**407–415.
39. Nakagawa Y, Asplin JR, Goldfarb D, Parks JH, Coe FL 2000 Clinical use of cystine supersaturation measurements. J Urol **164:**1481–1485.
40. Dolin DJ, Asplin JR, Flagel L, Grasso M, Goldfarb DS 2005 Effect of cystine-binding thiol drugs on urinary cystine capacity in patients with cystinuria. J Endourol **19:**429–432.
41. Pak CY, Fuller C, Sakhaee K, Zerwekh JE, Adams BV 1986 Management of cystine nephrolithiasis with alpha-mercaptopropionylglycine. J Urol **136:**1003–1008.
42. Cohen TD, Streem SB, Hall P 1995 Clinical effect of captopril on the formation and growth of cystine calculi. J Urol **154:**164–166.
43. Michelakakis H, Delis D, Anastasiadou V, Bartsocas C 1993 Ineffectiveness of captopril in reducing cystine excretion in cystinuric children. J Inherit Metab Dis **16:**1042–1043.
44. Preminger GM, Assimos DG, Lingeman JE, Nakada SY, Pearle MS, Wolf JS 2005 AUA guidelines on management of staghorn calculi:diagnosis and treatment recommendations. J Urol **173:**1991–2000.
45. Griffith DP, Gleeson MJ, Lee H, Longuet R, Deman E, Earle N 1991 Randomized, double-blind trial of Lithostat (acetohydroxamic acid) in the palliative treatment of infection-induced urinary calculi. Eur Urol **20:**243–247.
46. Griffith DP, Khonsari F, Skurnick JH, James KE 1988 A randomized trial of acetohydroxamic acid for the treatment and prevention of infection-induced urinary stones in spinal cord injury patients. J Urol **140:**318–324.
47. Williams JJ, Rodman JS, Peterson CM 1984 A randomized double-blind study of acetohydroxamic acid in struvite nephrolithiasis. N Engl J Med **311:**760–764.

© 2008 American Society for Bone and Mineral Research

Chapter 101. Genetic Basis of Renal Stones

Rajesh V. Thakker

Academic Endocrine Unit, Nuffield Department of Clinical Medicine, Oxford Centre for Diabetes, Endocrinology and Metabolism (OCDEM), Churchill Hospital, Headington, Oxford, United Kingdom

INTRODUCTION

Renal stones affect ~8% of the population by the seventh decade and are usually associated with a metabolic abnormality that may include hypercalciuria, hyperphosphaturia, hyperoxaluria, hypocitraturia, hyperuricosuria, cystinuria, a low urinary volume, and a defect of urinary acidification. The etiology of these metabolic abnormalities and of renal stones is multifactorial and involves interactions between environmental and genetic determinants. The environmental determinants include dietary intake of salt, protein, calcium, and other nutrients, fluid intake, urinary tract infections, socioeconomic status of the individual, lifestyle, and climate. This chapter will focus on the genetics of renal stones (Table 1), and in particular, those associated with hypercalciuric renal stone disease (nephrolithiasis) in man.

GENETICS

The greatest risk factor for nephrolithiasis, after controlling for known dietary determinants, is having an affected family member. Thus, between 35% and 65% of renal stone formers will have relatives with nephrolithiasis, whereas only 5–20% of nonrenal stone formers will have relatives with nephrolithiasis. The first-degree relative risk (λ_R) among recurrent stone formers has been estimated to be in the range of 2–16. The wide range of these estimates is largely because of differences in the study designs and the methods used to ascertain the occurrence of renal stones in relatives. Moreover, genetic contributions to nephrolithiasis and hypercalciuria have been confirmed by two studies that have investigated the occurrence of these conditions in twins and shown that the heritability (h^2) of kidney stones and urinary calcium excretion are 56% and 52%, respectively. Both of these studies support the notion that there is a strong genetic contribution to renal stone disease and in the regulation of renal calcium excretion. However, the responsible genes remain to be identified, because linkage studies using affected sibling pairs or small families in genome-wide searches have to date not been undertaken. This may be partly caused by the lack of availability of appropriate patients and their relatives and also because of difficulties in correctly ascertaining the phenotype, which may require radiological studies and 24-h urine collections. However, study of some families with monogenic forms of hypercalciuric nephrolithiasis have yielded insights into the rental tubular mechanisms regulating calcium excretion, and these will be briefly reviewed.

MONOGENIC FORMS OF HYPERCALCIURIC NEPHROLITHIASIS

Idiopathic Hypercalciuria

Families with idiopathic hypercalciuria (IH) and recurrent calcium oxalate stones usually reveal an autosomal dominant mode of inheritance. Studies of such families have established linkage between hypercalciuric nephrolithiasis and loci on chromosome 1q23.3-q24, which contains the human soluble adenylyl cyclase (*SAC*) gene; chromosome 12q12-q14, which contains the vitamin D receptor (*VDR*) gene; and chromosome 9q33.2-q34.2, from which an appropriate candidate gene remains to be identified.

Autosomal Dominant Hypocalcemic Hypercalciuria Caused by Calcium-Sensing Receptor Mutations

The human calcium-sensing receptor (CaSR) is a 1078 amino acid cell surface protein that is predominantly expressed in the parathyroids and kidney and is a member of the family of G protein–coupled receptors. The CaSR allows regulation of PTH secretion and renal tubular calcium reabsorption in response to alterations in extracellular calcium concentrations. The human *CaSR* gene is located on chromosome 3q21.1, and loss-of-function *CaSR* mutations have been reported in the hypercalcemic disorders of familial benign (hypocalciuric) hypercalcemia (FBHH), neonatal severe primary hyperparathyroidism (NSHPT), and familial isolated hyperparathyroidism (FIHP). However, gain-of-function *CaSR* mutations result in autosomal dominant hypocalcemia with hypercalciuria (ADHH) and Bartter's syndrome type V (see below).

Patients with ADHH usually have mild hypocalcemia that is generally asymptomatic, but may in some patients be associated with carpo-pedal spasm and seizures. The serum phosphate concentrations in patients with ADHH are either elevated or in the upper-normal range, and the serum magnesium concentrations are either low or in the low-normal range. These biochemical features of hypocalcemia, hyperphosphatasemia, and hypomagnesemia are consistent with hypoparathyroidism and pseudohypoparathyroidism. However, these patients have serum PTH concentrations that are in the low-normal range. Thus, they are not hypoparathyroid, which would be associated with undetectable serum PTH concentrations, or pseudohypoparathyroid, which would be associated with elevated serum PTH concentrations. These patients were therefore classified as having autosomal dominant hypocalcemia (ADH), and the association of hypercalciuria with this condition led to it being referred to as autosomal dominant hypocalcemia with hypercalciuria (ADHH). Treatment with active metabolites of vitamin D to correct the hypocalcemia has been reported to result in marked hypercalciuria, nephrocalcinosis, nephrolithiasis, and renal impairment, which was partially reversible after cessation of the vitamin D treatment. Thus, it is important to identify and avoid vitamin D treatment in such ADHH patients and their families whose hypocalcemia is caused by a gain-of-function *CaSR* mutation and not hypoparathyroidism. More than 40 different *CaSR* mutations have been identified in ADHH patients, and >50% of these are in the extracellular domain. Almost every ADHH family has its

The author states that he has no conflicts of interest.

Key words: idiopathic hypercalciuria, nephrolithiasis, calcium, magnesium, phosphate, inheritance, hereditary, renal tubular disorders

Electronic Databases: 1. OMIM—http://www.ncbi.nlm.nih.gov/sites/entrez?db=omim Use OMIM numbers in Table 1 to search for details of diseases and mutations. 2. EMSEMBL—http://www/emsembl.org/index.html Use gene symbols in Table 1 to obtain DNA sequences, exon-intron structure, evolutionary conservation and chromosomal locations. 3. PUBMED—http://www.ncbi.n/m.nih.gov/sites/entrez Use disease names and gene symbols in Table 1 to obtain published articles.

© 2008 American Society for Bone and Mineral Research

TABLE 1. GENETICS OF MONOGENIC FORMS OF NEPHROLITHIASIS

Renal stone disease	*Mode of inheritance*	*Gene*	*Chromosomal location*	*OMIM number**
Associated with hypercalciuria				
IH	A-d	*SAC*	1q23.2-q24	143870
IH	A-d	*VDR*	12q12-q14	601769
IH	A-d	?	9q33.2-q34.2	?
ADHH	A-d	*CASR*	3q21.1	601199
Hypercalcemia with hypercalciuria	A-d	*CASR*	3q21.1	601199
Bartter syndrome				
Type I	A-r	*SLC12A1/ NKCC2*	15q15-q21.1	601678
Type II	A-r	*KCNJ1 /ROMK*	11q24	241200
Type III	A-r	*CLCNKB*	1q36	607364
Type IV	A-r	*BSND*	1q31	602522
Type V	A-d	*CASR*	3q21.1	601199
Type VI	X-r	*CLCN5*	Xp11.22	300009
Dent's disease	X-r	*CLCN5*	Xp11.22	300009
Lowe syndrome	X-r	*OCRL1*	Xq25	309000
HHRH	A-r	*NPT2c/SLC34A3*	9q34	241530
Nephrolithiasis, osteoporosis and hypophosphatasemia	A-d	*NPT2a/SLC34A1*	5q35	182309
Familial hypomagnesemia with hypercalciuria and nephrocalcinosis	A-r	*PCLN1/CLDN16*	3q28	248250
Familial hypomagnesemia with hypercalciuria and nephrocalcinosis with ocular abnormalities	A-r	*CLDN19*	1p34.2	248190
dRTA	A-d	*SLC4A1/kAE1*	17q21.31	179800
dRTA with sensorineural deafness	A-r	*ATP6B1/ ATP6V1B1*	2p13	267300
dRTA with preserved hearing	A-r	*ATP6N1B/ ATP6V0A4*	7q34	602722
Not associated with hypercalciuria				
Primary hyperoxaluria type 1	A-r	*AGXT*	2q37.3	259900
Primary hyperoxaluria type 2	A-r	*GRHPR*	9p13.2	260000
APRT deficiency	A-r	*APRT*	16q24.3	102600
Cystinuria type A	A-r	*SLC3A1*	2p16.3	220100
Cystinuria type B	A-r	*SLC7A9*	19q13.1	604144
Wilson's disease	A-r	*ATP7B*	13q14.3	277900

* Online Mendelian Inheritance in Man (OMIM) reference number.

A-d, autosomal dominant; A-r, autosomal recessive; X-r, X-linked recessive; *SAC*, human soluble adenylyl cyclase; *VDR*, vitamin D receptor; *CASR*, calcium sensing receptor; *SLC12A1*, solute carrier family 12, member 1; *NKCC2*, sodium-potassium-chloride co transporter 2; *ROMK*, renal outer-medullary potassium channel; *CLCNKB*, chloride channel Kb; *BSND*, Barttin; *CLCN5*, chloride channel 5; *OCRL1*, oculocerebrorenal syndrome of Lowe; *NPT2c/a*, sodium-phosphate co transporter type 2c/a; *SLC34A1/3*, solute carrier family 34, member 1/3; *PCLN1*, paracellin; *CLDN16/19*, claudin 16/19; kAE1, kidney anion exchanger 1; *ATP6B1*, ATPase, H^+ transporting (vacuolar proton pump), V1 subunit B1; *ATP6N1B*, ATPase, H^+ transporting, lysosomal V0 subunit a4; *AGXT*, alanine glyoxylate aminotransferase; *GRHPR*, glyoxalate reductase/hydroxypyruvate reductase; *APRT*, adenine phosphoribosyl transferase; *SLC3A1*, solute carrier family 3, member 1; *SLC7A9*, solute carrier family 7, member 9; *ATP7B*, ATPase, Cu^{++} transporting, β polypeptide; ?, unknown.

own unique missense heterozygous *CaSR* mutation. Expression studies of the ADHH associated *CaSR* mutations have shown a gain-of-function, whereby there is a leftward shift in the dose–response curve, such that the extracellular calcium concentration needed to produce a half-maximal (EC_{50}) increase in the total intracellular calcium ions (or inositol trisphosphate [IP_3]), is significantly lower than that required for the wildtype receptor.

Bartter Syndrome

Bartter syndrome is a heterogeneous group of autosomal hereditary disorders of electrolyte homeostasis characterized by hypokalemic alkalosis, renal salt wasting that may lead to hypotension, hyper-reninemic hyperaldosteronism, increased urinary prostaglandin excretion, and hypercalciuria with nephrocalcinosis. Mutations of several ion transporters and channels have been associated with Bartter syndrome, and six types (Table 1) are now recognized. Thus, type I is caused by mutations involving the bumetanide-sensitive sodium–potassium chloride co-transporter (NKCC2 or SLC12A2); type II is caused by mutations of the renal outer-medullary potassium channel (ROMK); type III is caused by mutations of the voltage-gated chloride channel (CLC-Kb); type IV is caused by mutations of Barttin, which is a β subunit that is required for trafficking of CLC-Kb and CLC-Ka, and this form is also associated with deafness because Barttin, CLC-Ka, and CLC-Kb are also expressed in the marginal cells of the scala media of the inner ear that secrete potassium ion-rich endolymph; and type V is caused by activating mutations of the CaSR. Patients with Bartter syndrome type V have the classical features of the syndrome (i.e., hypokalemic metabolic alkalosis, hyperreninemia, and hyperaldosteronism). In addition, they develop hypocalcemia, which may be symptomatic and lead to carpopedal spasm, and an elevated fractional excretion of calcium, which may be associated with nephrocalcinosis. Such patients have been reported to have heterozygous gain-of-function *CaSR* mutations, and in vitro functional expression of these mutations not only revealed a leftward shift in the dose–response curve for the receptor but also showed them to have a much lower EC_{50} than that found in patients with ADHH. This suggests that the additional features that occur in Bartter syndrome type V compared with ADHH are caused by severe gain-of-function mutations of the CaSR. Bartter syndrome type VI has been reported in one child from Turkey and was

© 2008 American Society for Bone and Mineral Research

associated with a *CLCN5* mutation; mutations in this gene are usually observed in Dent's disease (see below).

Dent's Disease

Dent's disease is an X-linked recessive renal tubular disorder characterized by a low molecular weight proteinuria, hypercalciuria, nephrocalcinosis, nephrolithiasis, and eventual renal failure. Dent's disease is also associated with the other multiple proximal tubular defects of the renal Fanconi syndrome, which include aminoaciduria, phosphaturia, glycosuria, kaliuresis, uricosuria, and impaired urinary acidification. With the exception of rickets, which occurs in a minority of patients, there seem to be no extrarenal manifestations in Dent's disease. The gene causing Dent's disease, *CLCN5*, encodes the chloride/proton antiporter, CLC-5. CLC family members, which are usually voltage-gated chloride channels, have important diverse functions that include the control of membrane excitability, transepithelial transport, and regulation of cell volume. CLC-5, which is predominantly expressed in the kidney and in particular the proximal tubule, thick ascending limb of Henle, and the α intercalated cells of the collecting duct, has been reported to be critical for acidification in the endosomes that participate in solute reabsorption and membrane recycling in the proximal tubule. CLC-5 is also known to alter membrane trafficking through the receptor-mediated endocytic pathway that involves megalin and cubulin. *CLC-5* mutations associated with Dent's disease impair chloride flow and likely lead to impaired acidification of the endosomal lumen, and thereby, also disrupt trafficking of endosomes back to the apical surface. This will result in impairment of solute reabsorption by the renal tubule and in the defects observed in Dent's disease. Mice that are deficient for CLC-5 develop the phenotypic abnormalities associated with Dent's disease. Some patients with Dent's disease, who do not have CLC-5 mutations, have been reported to have mutations of the gene encoding an inositol polyphosphate 5-phosphatase, which also results in the Lowe syndrome (see below).

Occulo-Cerebro-Renal syndrome of Lowe

The oculo-cerebro-renal syndrome of Lowe (OCRL) is an X-linked recessive disorder that is characterized by congenital cataracts, mental retardation, muscular hypotonia, rickets, and defective proximal tubular reabsorption of bicarbonate, phosphate, and amino acids. Some patients may also develop hypercalciuria and renal calculi. The *OCRL1* gene is located on Xq25 and encodes a member of the type II family of inositol polyphosphate 5-phosphatases. These enzymes hydrolyze the 5-phosphate of inositol 1,4,5-trisphosphate and of inositol 1,3,4,5-tetrakisphosphate, phosphatidylinositol 4,5-bisphosphate, and phosphatidylinositol 3,4,5-trisphosphate, thereby presumably inactivating them as second messengers in the phosphatidylinositol signaling pathway. The preferred substrate of OCRL1 is phosphatidylinositol 4,5-bisphosphate, and this lipid accumulates in renal proximal tubular cells in patients with Lowe syndrome. OCRL1 has been localized to lysosomes in renal proximal tubular cells and to the trans-Golgi network in fibroblasts. OCRL1 has also been shown to interact with clathrin and indeed co-localizes with clathrin on endosomal membranes that contain transferrin and mannose 6-phosphate receptors. Thus, it seems likely the *OCRL1* mutations in Lowe syndrome patients result in OCRL1 protein deficiency, which leads to disruptions in lysosomal trafficking and endosomal sorting.

Hereditary Hypophosphatemic Rickets With Hypercalciuria

Two different heterozygous mutations (Ala48Phe and Val147Met) in NPT2a (also referred to as *SLC34A1*), the gene encoding a sodium-dependent phosphate transporter, have been reported in patients with urolithiasis or osteoporosis and persistent idiopathic hypophosphatemia caused by decreased renal tubular phosphate reabsorption. When expressed in *Xenopus laevis* oocytes, the mutant NPT2a showed impaired function. However, these in vitro findings were not confirmed in another study using oocytes and OK cells, raising the concern that the identified NPT2a mutation could not explain the findings in the described patients. However, homozygous ablation of *Npt2a* in mice (*Npt2a*$^{-/-}$) results in increased urinary phosphate excretion, hypophosphatemia, an appropriate elevation in the serum levels of 1,25-dihydroxyvitamin D, hypercalcemia, decreased serum PTH levels, increased serum alkaline phosphatase activity, and hypercalciuria. Some of these biochemical features are observed in patients with hereditary hypophosphatemic rickets with hypercalciuria (HHRH), but there are important differences. Thus, HHRH patients develop rickets, short stature, increased renal phosphate clearance, and hypercalciuria, but have normal serum calcium levels, an increased gastrointestinal absorption of calcium and phosphate caused by an elevated serum concentration of 1, 25-dihydroxyvitamin D, suppressed parathyroid function, and normal urinary cyclic AMP excretion. However, HHRH patients do not have NPT2a mutations, and studies have shown that HHRH patients harbor homozygous or compound heterozygous mutations of *SLC34A3*, the gene encoding the sodium–phosphate co-transporter NPT2c. These findings indicate that NPT2c has a more important role in phosphate homeostasis than previously thought.

Familial Hypomagnesemia With Hypercalciuria and Nephrocalcinosis Caused by Paracellin-1 (Claudin 16) Mutations

Familial hypomagnesemia with hypercalciuria and nephrocalcinosis (FHHNC) is an autosomal recessive renal tubular disorder that is frequently associated with progressive kidney failure. FHHNC often presents in childhood with seizures or tetany caused by hypocalcemia and hypomagnesemia. Other recurrent clinical manifestations include urinary tract infections, polyuria, polydipsia, and failure to thrive. Studies have revealed hypomagnesemia, hypocalcemia, hyperuricemia, hypermagnesuria, hypercalciuria, incomplete distal renal tubular acidosis, hypocitraturia, and renal calcification. Treatment consists of high-dose enteral magnesium to restore normomagnesemia. Children with FHHNC who receive such treatment early develop normally. Linkage studies in 12 FHHNC kindreds localized the disease locus to chromosome 3q27, and positional cloning studies identified mutations in the gene encoding paracellin-1 (PCLN-1), which is also referred to as claudin 16 (*CLDN16*). FHHNC patients were either homozygotes or compound heterozygotes for *PCLN-1* mutations, consistent with the autosomal recessive inheritance of the disorder. The *PCLN-1* mutations consisted of premature termination codons, splice-site mutations, and missense mutations. The PCLN-1 protein, which consists of 305 amino acids, has sequence and structural similarity to the members of the claudin family and is therefore also referred to as CLDN16. Claudins are membrane-bound proteins that form the intercellular tight junction barrier in a variety of epithelia. Claudins have four transmembrane domains and intracellular amino and carboxy termini. The two luminal loops mediate cell–cell adhesion

© 2008 American Society for Bone and Mineral Research

through homo- and heterotypic interactions with claudins on a neighboring cell. In addition, claudins form paracellular ion channels, which facilitate renal tubular paracellular transport of solutes. CLDN16 is exclusively expressed in the thick ascending limb of Henle's loop, where it forms the paracellular channels that are driven by an electrochemical gradient and allow reabsorption of calcium and magnesium. Hence, loss of function of CLDN16 that would arise from FHHNC mutations would result in urinary calcium and magnesium loss and lead to hypocalcemia and hypomagnesemia, respectively. A CLDN16 missense mutation (Thr233Arg) has also been identified in two families with self-limiting childhood hypercalciuria. The hypercalciuria decreased with age and was not associated with progressive renal failure. The Thr233Arg mutation resulted in inactivation of a PDZ-domain binding motif, and this disrupted the association with the tight junction scaffolding protein, ZO-1, with accumulation of the mutant CLDN16 protein in lysosomes and no localization to the tight junctions. Thus, CLDN16 mutations may result in different abnormalities of renal tubular cell function and hence lead to differences in the clinical phenotype. A form of FHHNC with severe ocular involvement reported in one Swiss and eight Spanish/Hispanic families was recently mapped to chromosome 1p34.2. This region contains *CLDN19*, the gene that encodes claudin 19, a tight-junction protein expressed in kidney and eye. A Gly20Asp mutation located in the first transmembrane domain of CLDN19 was identified in all but one of the Spanish/Hispanic families, and a Gln57Glu mutation in the first extracellular loop of CLDN19 was found in the Swiss family. In addition, a Leu90Pro mutation in CLDN19 was identified in a consanguineous family of Turkish origin with FHHNC and severe ocular involvement.

Distal Renal Tubular Acidosis

In distal renal tubular acidosis (dRTA), the tubular secretion of hydrogen ions in the distal nephron is impaired, and this results in a metabolic acidosis that is often associated with hypokalemia because of renal potassium wasting, hypercalciuria with nephrocalcinosis, and metabolic bone disease. dRTA may be familial, with autosomal dominant or recessive inheritance.

Autosomal Dominant dRTA Caused by Erythrocyte Anion Exchanger (Band 3, AE1) Mutations. The family of anion exchangers (AEs) are widely distributed and involved in the regulation of transcellular transport of acid and base across epithelial cells, cell volume, and intracellular pH. For example, AE1, which is a major glycoprotein of the erythrocyte membrane, mediates exchange of chloride and bicarbonate. AE1 is also found in the basolateral membrane of the α-intercalated cells of renal collecting ducts, which are involved in acid secretion. Patients with autosomal dominant dRTA, the majority of whom had hypercalciuria, renal stones, and nephrocalcinosis, and a few of them who had erythrocytosis, were found to have AE1 mutations. These AE1 mutations resulted in several functional abnormalities that included reductions in chloride transport and trafficking defects that lead to a cellular retention of AE1 or mistargeting of AE1 to the apical membrane. AE1 mutations may also be associated with autosomal recessive dRTA in Southeast Asian kindreds that have ovalocytosis.

Autosomal Recessive Distal Renal Tubular Acidosis Caused by Proton Pump (H^+-ATPase) Mutations. Proton pumps are ubiquitously expressed, and one such multiunit H^+-ATPase is found in abundance on the apical (luminal) surface of the α-intercalated cells of the cortical collecting duct, which regulates urinary acidification. Failure of vectorial proton transport by these α-intercalated cells results in an inability of urinary acidification and in disorders of dRTA. The molecular basis of two types of autosomal recessive dRTA caused by proton pump abnormalities have been characterized. The gene causing one type of autosomal recessive dRTA that was associated with sensorineural hearing loss was mapped to chromosome 2p13, which contained the *ATP6B1* gene that encodes the B1 subunit of the apical proton pump (H^+-ATPase). Mutations, which would likely result in a functional loss of ATP6B1, were identified in >30% of families with this form of autosomal recessive dRTA that occurred, with deafness in >85% of families. The association of dRTA and deafness is consistent with the renal and cochlear expression of ATP6B1. ATP6B1 plays a critical role in regulating the pH of the inner ear endolymph, and dysfunction of this would lead to an alkaline microenvironment in the inner ear, which has been proposed to impair hair cell function and result in progressive deafness. The gene causing autosomal recessive dRTA with normal hearing was localized to chromosome 7q33-q34, which contained the *ATP6N1B* gene that encodes the noncatalytic accessory subunit of the proton pump of the α-intercalated cells of the collecting duct. *ATP6N1B* mutations, which are predicted to result in a functional loss, were identified in >85% of kindreds with autosomal recessive dRTA associated with normal hearing, and this is consistent with the expression of ATP6N1B in the kidney and not other organs. Approximately 15% of families with autosomal recessive dRTA were not found to have mutations in ATP6B1 or ATP6N1B, and this indicates mutations in other genes are likely to be involved in the etiology of autosomal recessive dRTA.

CONCLUSIONS

Renal stone disease (nephrolithiasis) affects 5% of adults and is often associated with hypercalciuria. Hypercalciuric nephrolithiasis is a familial disorder in >35% of patients and may occur as a monogenic disorder or as a polygenic trait. Studies of monogenic forms of hypercalciuric nephrolithiasis (e.g., Bartter syndrome, Dent's disease, ADHH, hypercalciuric nephrolithiasis with hypophosphatasemia, and familial hypomagnesemia with hypercalciuria) have helped to identify a number of transporters, channels, and receptors that are involved in regulating the renal tubular reabsorption of calcium. These studies have provided valuable insights into the renal tubular pathways that regulate calcium reabsorption and predispose to kidney stones and bone disease.

SUGGESTED READING

1. Frick KK, Bushinsky DA 2003 Molecular mechanisms of primary hypercalciuria. J Am Soc Nephrol **14:**1082–1095.
2. Curhan GC, Willett WC, Rimm EB, Stampfer MJ 1997 Family history and risk of kidney stones. J Am Soc Nephrol **8:**1568–1573.
3. Trinchieri A, Mandressi A, Luongo P, Coppi F, Pisani E 1988 Familial aggregation of renal calcium stone disease. J Urol **139:**478–481.
4. Goldfarb DS, Fischer ME, Keich Y, Goldberg J 2005 A twin study of genetic and dietary influences on nephrolithiasis: A report from the Vietnam Era Twin (VET) Registry. Kidney Int **67:**1053–1061.
5. Hunter DJ, Lange M, Snieder H, MacGregor AJ, Swaminathan R, Thakker RV, Spector TD 2002 Genetic contribution to renal function and electrolyte balance: A twin study. Clin Sci (Lond) **103:**259–265.
6. Stechman MJ, Loh NY, Thakker RV 2007 Genetics of hypercalciuric nephrolithiasis (renal stone disease). Ann NY Acad Sci **1116:**461–484.
7. Pearce SH, Williamson C, Kifor O, Bai M, Coulthard MG, Davies M, Lewis-Barned N, McCredie D, Powell H, Kendall-Taylor P, Brown EM, Thakker RV 1996 A familial syndrome of hypocalce-

© 2008 American Society for Bone and Mineral Research

mia with hypercalciuria due to mutations in the calcium-sensing receptor. N Engl J Med **335:**1115–1122.
8. Thakker RV 2004 Diseases associated with the extracellular calcium-sensing receptor. Cell Calcium **35:**275–282.
9. Lloyd SE, Pearce SH, Fisher SE, Steinmeyer K, Schwappach B, Scheinman SJ, Harding B, Bolino A, Devoto M, Goodyer P, Rigden SP, Wrong O, Jentsch TJ, Craig IW, Thakker RV 1996 A common molecular basis for three inherited kidney stone diseases. Nature **379:**445–449.
10. Jentsch TJ, Neagoe I, Scheel O 2005 CLC chloride channels and transporters. Curr Opin Neurobiol **15:**319–325.
11. Hoopes RR Jr, Shrimpton AE, Knohl SJ, Hueber P, Hoppe B, Matyus J, Simckes A, Tasic V, Toenshoff B, Suchy SF, Nussbaum RL, Scheinman SJ 2005 Dent disease with mutations in OCRL1. Am J Hum Genet **76:**260–267.
12. Bergwitz C, Roslin NM, Tieder M, Loredo-Osti JC, Bastepe M, Abu-Zahra H, Frappier D, Burkett K, Carpenter TO, Anderson D, Garabedian M, Sermet I, Fujiwara TM, Morgan K, Tenenhouse HS, Juppner H 2006 SLC34A3 mutations in patients with hereditary hypophosphatemic rickets with hypercalciuria predict a key role for the sodium-phosphate cotransporter NaPi-IIc in maintaining phosphate homeostasis. Am J Hum Genet **78:**179–192.
13. Lorenz-Depiereux B, Benet-Pages A, Eckstein G, Tenenbaum-Rakover Y, Wagenstaller J, Tiosano D, Gershoni-Baruch R, Albers N, Lichtner P, Schnabel D, Hochberg Z, Strom TM 2006 Hereditary hypophosphatemic rickets with hypercalciuria is caused by mutations in the sodium-phosphate cotransporter gene SLC34A3. Am J Hum Genet **78:**193–201.
14. Simon DB, Lu Y, Choate KA, Velazquez H, Al-Sabban E, Praga M, Casari G, Bettinelli A, Colussi G, Rodriguez-Soriano J, McCredie D, Milford D, Sanjad S, Lifton RP 1999 Paracellin-1, a renal tight junction protein required for paracellular Mg2+ resorption. Science **285:**103–106.
15. Konrad M, Schlingmann KP, Gudermann T 2004 Insights into the molecular nature of magnesium homeostasis. Am J Physiol Renal Physiol **286:**F599–F605.
16. Bruce LJ, Cope DL, Jones GK, Schofield AE, Burley M, Povey S, Unwin RJ, Wrong O, Tanner MJ 1997 Familial distal renal tubular acidosis is associated with mutations in the red cell anion exchanger (Band 3, AE1) gene. J Clin Invest **100:**1693–1707.
17. Karet FE, Finberg KE, Nelson RD, Nayir A, Mocan H, Sanjad SA, Rodriguez-Soriano J, Santos F, Cremers CW, Di Pietro A, Hoffbrand BI, Winiarski J, Bakkaloglu A, Ozen S, Dusunsel R, Goodyer P, Hulton SA, Wu DK, Skvorak AB, Morton CC, Cunningham MJ, Jha V, Lifton RP 1999 Mutations in the gene encoding B1 subunit of H+-ATPase cause renal tubular acidosis with sensorineural deafness. Nat Genet **21:**84–90.
18. Smith AN, Skaug J, Choate KA, Nayir A, Bakkaloglu A, Ozen S, Hulton SA, Sanjad SA, Al-Sabban EA, Lifton RP, Scherer SW, Karet FE 2000 Mutations in ATP6N1B, encoding a new kidney vacuolar proton pump 116-kD subunit, cause recessive distal renal tubular acidosis with preserved hearing. Nat Genet **26:**71–75.
19. Simon DB, Karet FE, Hamdan JM, DiPietro A, Sanjad SA, Lifton RP 1996 Bartter's syndrome, hypokalemic alkalosis with hypercalciuria, is caused by mutations in the Na-K-2Cl cotransporter NKCC2. Nat Genet **13:**183–188.
20. Simon DB, Karet FE, Rodriguez-Soriano J, Hamdan JH, DiPietro A, Trachtman H, Sanjad SA, Lifton RP 1996 Genetic heterogeneity of Bartter's syndrome revealed by mutations in the K+ channel, ROMK. Nat Genet **14:**152–156.

Chapter 102. Kidney Stones in the Pediatric Patient

Amy E. Bobrowski and Craig B. Langman

Feinberg School of Medicine, Northwestern University Division of Kidney Diseases, Department of Pediatrics, Chicago, Illinois

INTRODUCTION

Kidney stone disease, or nephrolithiasis, results from urinary crystal aggregation in a protein matrix within the urinary tract. This condition reflects a deviation from the balance between stone promoters and inhibitors, in favor of promoters. In general, higher urinary pH (with the exception of struvite stones, which form in alkaline urine), higher urine volume and dilution, higher urinary citrate, and free flow of urine serve as natural stone inhibitors. Kidney stones occur about one tenth less frequently in children than in adults, representing from 1 in 1000 to 1 in 7600 pediatric hospital admissions.[1] They are most common in white children, and overall, tend to affect boys and girls equally. This differs from the adult disease, in which there is a male preponderance (3:1). Within subtypes of stone disease, however, male children do have a slightly higher incidence of stones related to hypercalciuria and urinary tract abnormalities.[2]

Predisposing factors for nephrolithiasis can be determined in the majority of children affected. These include metabolic abnormalities in 48–86%, urinary tract infection in 14–75%, and co-existing structural urinary tract abnormalities in 10–40%.[2–4] The recurrence rate of kidney stones in children has been reported as anywhere from 6.5% to 54%,[2,3,5–7] and children with metabolic disorders are nearly five times more likely to have a recurrence.[5] Calcium oxalate stones are the most common found in children, with a frequency of 45–65%. These are followed by calcium phosphate (14–30%), struvite (13%), cystine (5%), uric acid (4%), and mixed (4%) stones.[2,8]

The authors state that they have no conflicts of interest.

PREDISPOSING CONDITIONS

Abnormalities of the Urinary Tract

Anatomic abnormalities of the genito-urinary tract may produce urinary stasis, allowing for crystal aggregation by heterotopic nucleation and/or infection. Most stones in North American children are located within the kidney or ureter. When recurrent bladder stones are seen, they are most often in the face of complex urologic abnormalities.[9] However, because most patients with urologic anomalies do not form stones, a full metabolic evaluation should still be performed in such children presenting with nephrolithiasis.

Infection/Struvite Stones

Urinary tract infections may co-exist with kidney stones that arise from a metabolic origin, making it necessary to rule out metabolic etiologies in such cases. Infection-related stones are more common in males, and more than one half of children with these stones have genito-urinary abnormalities.[2] Patients with surgically augmented bladders, particularly those augmented with intestinal segments, are at especially high risk for developing struvite stones in the bladder.[10] Formation of struvite stones ($Mg\text{-}NH_4\text{-}PO_4$) and calcium phosphate apatite-based stones is favored by an alkaline pH, resulting from NH_4 production by urea-splitting bacteria. *Proteus* species are the

© 2008 American Society for Bone and Mineral Research

TABLE 1. CAUSES OF HYPERCALCIURIA

Associated with hypercalcemia	*Associated with normal serum calcium*
Primary hyperparathyroidism	Familial idiopathic hypercalciuria
Sarcoid	Dent's disease
Idiopathic infantile hypercalcemia	Bartter syndrome
Immobilization	Familial hypomagnesemia-hypercalciuria
Bartter syndrome	Immobilization
Thyrotoxicosis	Prematurity, often associated with furosemide
Bone metastases	Distal renal tubular acidosis
Hypervitaminosis D	Ketogenic diet
Williams syndrome	Activating mutation of the extracellular calcium-sensing gene (generally with hypocalcemia)
	Medullary sponge disease
	Inflammatory diseases (e.g., JRA)
	Corticosteroid therapy

most common culprit, but *Pseudomonas*, *Klebsiella*, *Steptococcus*, *Serratia*, *Staphyloococcus*, *Candida*, and *Mycoplasma* species can also produce urease. The stones produced in this setting tend to grow rapidly, often forming staghorn calculi, a term signifying complete filling of the urinary infundibulum with stone.

Metabolic Abnormalities

Hypercalciuria. Calcium oxalate and calcium phosphate stones in children are most frequently caused by hypercalciuria, defined as a urinary calcium excretion of >4 mg/kg/d. Patients with hypercalciuria may present with microscopic or gross hematuria, dysuria, or urgency, even in the absence of any stones. Such children often have a positive family history of kidney stones and may have up to a 17% chance of subsequently developing urolithiasis.[11,12]

Familial idiopathic hypercalciuria (FIH) is the most common subset of hypercalciuria. Although the genetic basis of this condition is unknown, it seems to be inherited in an autosomal dominant pattern with incomplete penetrance. The pathophysiology of FIH is not yet well defined but may include any combination of the following, to varying degrees: a primary kidney tubular reduction in calcium resorption, increased dietary calcium absorption in the gastrointestinal tract secondary to excessive 1,25-dihydroxy-vitamin D action, and increased bone resorption.[13] The contribution of bone resorption to this disorder has important clinical implications, because restricting calcium intake in these patients may worsen their propensity toward significant osteoporosis.[14]

Causes of hypercalciuria are listed in Table 1. Dent's disease is an X-linked recessive condition of nephrolithiasis and subsequent kidney failure, linked to mutations in the *CLCN5* gene located on chromosome Xp11.22.[15] This gene is responsible for the transduction of a voltage-gated chloride channel in the kidney, the lack of which leads to hypercalciuria, low molecular weight proteinuria, nephrolithiasis, nephrocalcinosis, and varying degrees of glycosuria, aminoaciduria, and phosphaturia.[16] Bartter syndrome occurs with one of a series of mutations of genes coding for transporters in the thick ascending limb of the Loop of Henle. These genes include *NKCC2*, which transduces the Na-K-2Cl transporter (type I Bartter syndrome); *ROMK*, which transduces the potassium channel (type II); and *CLCNKB*, which transduces the chloride channel (type III). There is also a type IV Bartter syndrome, or Bartter syndrome with sensorineural deafness, which is caused by a mutation in the gene for barttin, a β-subunit of the chloride channel. Type V has a similar phenotype as type IV but is caused by defects in one or both of the chloride channels that co-localize with barttin: ClC-Ka and ClC-Kb. Distal renal tubular acidosis (dRTA) is a condition of metabolic acidosis, growth retardation, hypercalciuria, and nephrocalcinosis. When associated with a mutation in the *ATP6B1* gene responsible for a vacuolar H+-ATPase, it is associated with deafness and an autosomal dominant inheritance. Familial hypomagnesemia-hypercalciuria is associated with a mutation in the *PLCN-1* gene for the tight junction protein paracellin-1. Pseudohypoaldosteronism type II is seen with mutations in WNK kinases expressed in the distal nephron and presents with hypertension, hyperkalemia, and metabolic acidosis in addition to hypercalciuria.[17]

Other causes of hypercalciuria with normocalcemia include medullary sponge kidney, systemic inflammatory diseases, and iatrogenic, resulting from medications such as loop diuretics and corticosteroids. If hypercalcemia is detected, primary hyperparathyroidism, sarcoidosis, immobilization, thyroid disease, osteolytic metastases, hypervitaminosis D, and Williams syndrome should be considered on the differential.[18]

Hypocitraturia. Hypocitraturia is a contributory cause of nephrolithiasis, because citrate is necessary for the formation of a soluble calcium salt to prevent calcium stone crystallization. Most commonly it is seen in renal tubular acidosis, but it is also present in a subset of patients with FIH. Hypocitraturia can also occur in concert with other forms of hypercalciuria, hyperuricosuria, or hyperoxaluria. Chronic diarrhea, a high-protein diet, and hypokalemia can also induce low urinary citrate levels and a predisposition to stone formation,[19] because citrate absorption in the proximal tubule is stimulated by intracellular acidosis and potassium depletion.[20]

Hyperoxaluria. Oxalate is a human metabolic product made in the liver and excreted by the kidney, but can also be ingested and absorbed from dietary sources. Type 1 primary hyperoxaluria (PH1) is an autosomal recessive reduction in or absence of alanine glyoxyalate aminotransferase (AGT) activity, leading to increased conversion of glyoxylate to oxalate. Excessive urinary oxalate excretion may lead to crystallization and deposition in the urinary tract and kidney parenchyma. This in turn can result in kidney failure and systemic oxalosis, a clinical situation in which calcium oxalate precipitates in multiple organs and joints. Disease severity in PH1 varies widely. The course may be mild and fully responsive to medical therapy such as vitamin B_6 (pyridoxine) or may present aggressively in infancy with rapid kidney failure and severe systemic manifestations.

Because AGT is predominantly expressed in the liver, diagnosis has in the past relied solely on liver biopsy to assess AGT presence and activity. The gene encoding AGT (*AGXT*), located on chromosome 2q37.3, to date has at least 83 mutations that have been described that either eliminate, or decrease substantially, enzyme activity.[21,22] The relative ease in modern laboratory medicine at performance of sequence analysis, and the delineation of the molecular basis of many of the mutations behind PH1, has led to the proposal of molecular diagnostic algorithms that may obviate the need for invasive biopsy procedures. Monico et al.[21] have recently reported comprehensive mutation screening across the entire *AGXT* coding region in 55 probands with PH1, showing a 96–98% sensitivity in this population. When limited to sequencing of

© 2008 American Society for Bone and Mineral Research

exons 1, 4, and 7, the sensitivity was 77%. Given the relatively small size of the gene, complete molecular analysis should not involve prohibitive expense. An algorithm beginning with limited sequencing of exons 1, 4, and 7, followed by direct sequencing of the entire gene if inconclusive, would make intuitive sense and would eliminate the need for liver biopsy in most patients.

Type 2 primary hyperoxaluria (PH2) results from a deficiency of activity in the enzyme glyoxylate reductase/hydroxypyruvate reductase (GRHPR), which is more widely distributed in the human than AGT1, with a predominance in muscle, liver, and kidney. As a group, patients with PH2 seem to have less morbidity and mortality than those with PHI, with a lower incidence of end-stage kidney disease (ESKD) and an older age at onset of symptoms.[23,24]

Unfortunately, up to one third of patients with PH in some case series present at end stage, when uremia develops.[25,26] For this reason, PH should be considered in patients with recurrent calcium oxalate nephrolithiasis, unexplained nephrocalcinosis, or unexplained ESKD in which the kidneys are echodense with calcium.

Secondary (enteric) hyperoxaluria can result from increased oxalate absorption in the colon caused by small bowel malabsorption of fatty and bile acids. These substances increase colonic permeability to oxalate by binding luminal calcium, freeing unbound oxalate to be absorbed. Epithelial damage in these states also increases colonic absorption, and low dietary calcium intake can exacerbate the condition. Depletion of *Oxalobacter formigenes*, an enteric oxalate-degrading bacterium, can also contribute to enteric hyperoxaluria. Other rare secondary causes are pyridoxine deficiency (a co-factor for AGT activity) and excessive intake of oxalate-containing foods (rhubarb gluttony) or oxalate precursors (ascorbic acid, ethylene glycol).

Hyperuricosuria. Uric acid is the end product of purine metabolism. Hyperuricosuria may occur either in the face of uric acid overproduction or with normal serum uric acid concentrations and can predispose to both uric acid stones and calcium oxalate nephrolithiasis, acting as a heterotopic nucleation factor.[27] Lesch-Nyhan syndrome (complete deficiency of hypoxanthine-guanine phosphoribosyltransferase) and type 1 glycogen storage disease (glucose-6-phosphatase deficiency) are both inborn errors of metabolism that may present with hyperuricemia and hyperuricosuria/urolithiasis.[28] Gout caused by a partial hypoxanthine-guanine phosphoribosyltransferase deficiency can also cause uric acid nephrolithiasis in older children. Myeloproliferative disorders and other causes of cell breakdown are other secondary causes of uric acid stones. Ketogenic diets, excessive protein intake, and uricosuric drugs such as high-dose aspirin, probenecid, and ascorbic acid can also cause hyperuricosuria. Normal or low serum uric acid levels may be associated with uricosuria secondary to proximal renal tubular defects. These may be caused by a single defect in the renal urate exchanger URAT1[29] or disorders of generalized proximal tubule dysfunction. Insulin resistance, as seen in type 2 diabetes mellitus and metabolic syndrome, may also predispose to uric acid stones and an overly acidic urine by decreasing renal ammonia excretion and impairing hydrogen ion buffering.[30] In another perturbation of uric acid metabolism, xanthine stones are formed in an autosomal recessive disorder of the gene for xanthine dehydrogenase, whereby uric acid cannot be formed from xanthine precursors.

Cystinuria. Cystinuria is an autosomal recessive disease of disordered dibasic amino acid transport in kidney and may occasionally be diagnosed by the discovery of flat hexagonally shaped cystine crystals in the urine. Children with this condition have elevated urinary cystine, ornithine, arginine, and lysine levels, because all of these amino acids share transporters. Mutations of the *SLC3A1* gene on chromosome 2 and the *SLC7A9* gene on chromosome 19 have been identified,[31,32] and patients may be either homozygous or compound or obligate heterozygotes.[33] Affected homozygous children usually excrete >1000 μmol/g creatinine of cystine by the age of 1 yr, with a mean excretion of 4500 μmol/g creatinine, exceeding its solubility and leading to lifelong recurrent nephrolithiasis.[9]

Other Causes of Kidney Stones

Additional clinical situations in which patients are predisposed to forming kidney stones include patients with cystic fibrosis, who may have an absence of the oxalate-degrading bacterium, *O. formigenes*, hyperoxaluria, hypercalciuria, and/or hypocitraturia. Patients taking protease inhibitors, especially the poorly soluble indinavir, may have urinary excretion of crystallized drug product. Patients on a ketogenic diet for seizure control are predisposed to hypercalciuria and/or hypocitraturia.

CLINICAL EVALUATION

The clinical presentation of urinary tract stones in children may differ from that in adults. Urinary tract stones may present with abdominal, flank, or pelvic pain in only ~50% of children with nephrolithiasis. Gross or microscopic hematuria (occurring in 33–90% of affected children), dysuria, frequency, emesis, and urinary tract infections are additional common presenting signs in younger patients. A detailed history and physical should guide evaluation of kidney stones. Family history should focus on members with kidney stones (positive in greater than one third of affected children), gout, arthritis, or chronic kidney disease.[2] The presence of a concomitant urinary tract infection must be sought but should not be accepted as the cause of the stone. Patients should also be advised to submit any passed stones or stone fragments for analysis by polarization microscopy or X-ray diffraction but not by simple chemical analysis.

Useful imaging studies may include plain abdominal radiology, ultrasonography, and helical CT. Conventional abdominal radiographs may show only radiopaque but not radiolucent stones, whereas ultrasound of the urinary tract may show both radiolucent and -opaque stones, in addition to the presence of urinary obstruction or nephrocalcinosis. Ultrasound has largely taken the place of intravenous pyelography (IVP) as an initial study for stone presence, secondary to concerns about radiation and contrast exposure with the latter procedure. Noncontrast helical CT has been found to have high sensitivity and specificity in identifying even small stones without requiring intravenous contrast administration. It may precisely localize stones, detect obstruction and hydronephrosis, and is much more sensitive than the previously mentioned imaging modalities.[34]

Because the majority of children with stones may have a metabolic problem that is discoverable and generally amenable to therapy, diagnostic urinary and blood tests for stone evaluation should be obtained while the patient is on their routine activity schedule and diet. At least two 24-h urine collections should be performed, waiting at least 2 wk after any acute stone event. This time frame allows for the resumption of the child's normal intake of food and fluids after recovery from pain and/or surgical intervention, which is critical for correct assignment of metabolic disturbances. These collections can

© 2008 American Society for Bone and Mineral Research

TABLE 2. NORMATIVE DATA FOR URINARY SOLUTE EXCRETION[13,54]

Substance	*Reference range*
Calcium	≤4 mg/kg/d
Citrate	>400 mg/g creatinine (spot citrate/Cr ratio > 0.51 g/g)
Oxalate	≤0.5 mmol/1.73 m^2/d (<40 mg/1.73 m^2/d)
Uric acid	Varies with age, up to 815 mg/1.73 m^2/d
Cystine	<60 mg/1.73 m^2/d

assess urinary volume as a reflection of fluid intake and creatinine excretion for completeness of the 24-h collection (at least 10–15 mg/kg/d in children >2 yr of age; 6–9 mg/kg/d <2 yr of age) and measurement of levels of lithogenic substances such as calcium, oxalate, uric acid, and cystine. The collections can evaluate for decreased stone inhibitor levels as well, such as citrate and magnesium. Normal values of these substances are shown in Table 2. Serum levels of uric acid, potassium, calcium, phosphorus, creatinine, bicarbonate (total CO_2), and biointact PTH (if hypercalcemia is present) should be obtained as well at the end of the urinary collections.

Consultation with an expert in pediatric stone disorders is encouraged if questions arise about the results of these diagnostic studies.

SURGICAL MANAGEMENT

The goals of the management of patients with kidney stones are to remove existing stones and prevent stone recurrence, with preservation of kidney function. Pediatric patients usually pass ureteral stones up to 5 mm in size.[5] In the absence of infection or persistent pain, such stones can be safely observed for up to 6 wk. Larger stones and kidney-located stones, however, require the consideration of surgical intervention, with a goal of achieving and maintaining a stone-free state. Choice of surgical modality depends on stone composition, size, and location along the urinary tract. Shock wave lithotripsy (SWL) uses the generation and focusing of shock wave energy toward the stone. Pulverized fragments are subsequently passed, and multiple treatment sessions are sometimes required. One large pediatric series (n = 344) showed a 92% stone-free rate for renal pelvis stones <1 cm, a 68% rate for stones 1–2 cm, and a 50% rate for stones >2 cm. Calyceal stone clearance rates were lower.[35] Overall, stone-free rates in children treated using this procedure have ranged from 67% to 99% in various studies,[36] with the highest success rates appearing to be in the youngest children.[37] This procedure seems to be safe in young children and infants, with no evidence of long-term changes in glomerular filtration rate or in functional renal parenchymal scarring before and after treatment in the affected kidney.[38,39] Minor complications such as bruising, renal colic, and hematuria may occur with SWL treatment. Small children may require the use of lung shielding to prevent pulmonary contusion, as well as reduced power settings to avoid injury. Ureteral stenting may also be required for larger stone burdens. In general, large stone burden (>2 cm) and anatomic abnormalities are risk factors for unsuccessful SWL, and alternative urological approaches should be considered in these cases.[36] Struvite, calcium oxalate dehydrate, and uric acid stones are especially amenable to fragmentation with SWL, whereas cystine, brushite, and calcium oxalate monohydrate stones are all resistant to SWL treatment.[40]

Percutaneous nephrolithotomy (PNL) is an alternative procedure that may be used alone or in conjunction with SWL in patients with large stone burden, significant renal obstruction, and/or staghorn calculi. It is also commonly used to remove lower pole calculi >1 cm in size. Percutaneous access to the collecting system of the kidney is achieved, and a wire is advanced to dilate the tract to accommodate a nephroscope. Stones may be removed or pulverized under direct visualization, making this approach ideal for complex upper tract stones. A nephrostomy tube is often placed postoperatively, although a small series in adults did show a decrease in pain and recovery time, with no increase in complications, with the use of smaller or no nephrostomy tubes.[41] The development of smaller nephroscopes has made PNL available for children, with stone-free rates ranging from 83% to 98%.[36]

Ureteroscopy is most ideally used for the removal and/or fragmentation of distal ureteral stones. Whereas SWL has good efficacy for some smaller ureteral stones, stone-free rates in those with stones >10 mm in size have been found to be markedly higher with ureteroscopy (93%) than with SWL (50%).[42] Smaller rigid and flexible ureteroscopes have made this procedure an option for pediatric patients and have made the need for concomitant ureteral balloon dilation (and possible risks of stricture and vesicoureteral reflux) less frequent. Once the ureteroscope is passed, laser energy is used to fragment any visualized stones, and flexible wire baskets can be used to remove fragments. Postoperative stenting may be used to facilitate passage of residual fragments or to prevent ureteral obstruction in the face of edema caused by trauma to the ureteral wall. Stenting is not usually done in uncomplicated procedures, with easy passage of the scope.[43] A summary of surgical treatment options by stone size and location is shown in Table 3.

MEDICAL MANAGEMENT

Nonspecific management of urolithiasis includes an increase in fluid intake to increase urinary volume, urinary dilution, and induce stone particle motion through the urinary tract. Other specific measures depend on the underlying predisposing diagnosis.

Hypercalciuria may be treated with a low sodium diet, thiazide diuretics, and adequate potassium intake. Thiazide therapy (e.g., hydrochlorothiazide 1 mg/kg/d, maximum of 25 mg/d) in FIH significantly decreases urinary calcium excretion and rate of stone formation in adults.[44] A decrease in urinary calcium excretion with thiazide treatment in children with FIH has also been shown.[45] In a population with hypercalciuria from immobilization, 18 of 42 children were found to be hypercalciuric, with a higher rate of fractures.[46] A 3-wk course of hydrochlorothiazide and amiloride reduced the mean urinary calcium to creatinine ratio by 57.7%. Dietary calcium

TABLE 3. SURGICAL TREATMENT OPTIONS BY STONE SIZE AND LOCATION

Location/size	*Shock wave lithotripsy*	*Ureteroscopy*	*Percutaneous nephrolithotomy*
Renal			
<1 cm	Most common	Optional	Optional
1–2 cm	Most common	Optional	Optional
>2 cm	Optional	Rare	Most common
Lower pole			
<1 cm	Most common	Optional	Optional
>1 cm	Optional	Optional	Most common
Ureteral			
Proximal	Most common	Optional	Occasional
Distal	Optional	Most common	Rare

Reprinted with permission from Elsevier from Durkee CT, Balcom A 2006 Surgical management of urolithiasis. Pediatr Clin North Am **53:**13.

© 2008 American Society for Bone and Mineral Research

intake should not be limited. Citrate therapy (e.g., potassium citrate 2 mmol/kg once daily) is also appropriate in cases of documented hypocitraturia.

Another important issue for some patients is that of BMD in FIH. One study of 40 girls with FIH and their premenopausal mothers showed that BMD lumbar spine Z-scores were significantly lower in these patients compared with controls.(47) Others have shown that thiazide treatment, in addition to decreasing urinary calcium excretion, can also improve BMD scores in children. Average Z-score improved from –1.3 to –0.22 over 1 yr of treatment with hydrochlorothiazide and potassium citrate in one study of 18 children.(45)

The treatment of struvite stones rests on the eradication of stones, correction of any urinary obstruction, and treatment/prevention of urinary tract infections. Urinary acidification could theoretically be used to prevent crystallization, but evidence for such an approach is thus far lacking. The urease inhibitor acetohydroxaminic acid (AHA) may have some clinical use,(48) but its use is limited by a high incidence of neurologic and gastrointestinal side effects.

Hyperuricosuria may be treated with dietary sodium limitation, oral bicarbonate or citrate supplementation, or addition of allopurinol if increased uric acid production and hyperuricemia are present.

Patients with suspected primary hyperoxaluria should be given a therapeutic course of vitamin B_6 (pyridoxine), and urinary oxalate levels should be used to monitor success or to suggest the need for dose escalation. For patients with reduced kidney function, intensive dialysis followed by liver/kidney transplant can be curative, because a new liver replaces the enzymatic defect in PH-1.(49,50) Pretransplant hemodialysis for five to six times per week, and perhaps with additional nightly peritoneal dialysis, is needed to lessen the systemic oxalate burden and prevent recurrence of disease in the transplant kidney. Prompt referral to a pediatric center with expertise in this disorder is suggested.

Secondary hyperoxaluria that results from enteric hyperoxaluria may be treated with a low sodium/low fat diet, high fluid intake, and a dietary calcium intake at the upper end of the daily recommended intake. Limitation of oxalate-containing foods such as chocolate, rhubarb, nuts, and spinach should be advised, as well as possible supplementation with magnesium, phosphorus, and citrate salts.

Cystinuria is treated with fluids (minimum of 3 liters/1.73m^2/d) and provision of alkali salts, such as citrate. Low sodium intake can also decrease urinary excretion of cystine. Chelating agents such as D-penicillamine, or more recently, α-mercaptopropionyl glycine (Thiola; Mission Pharmacal, San Antonio, TX, USA) may also be prescribed by someone skilled in pediatric stone disease. D-penicillamine may cause a severe serum sickness-like reaction, but side effects are less severe with Thiola.(19) Angiotensin converting enzyme (ACE) inhibition with captopril therapy has been found beneficial in some patients with cystinuria (captopril-cystine complexes are 200 times more soluble than cystine alone) resistant to alkalinization and fluid therapy alone, and perhaps with less bothersome side effects.(51) Captopril, however, is not as effective as thiol-compound therapy.

PROGNOSIS

Estimates of the recurrence rate of urolithiasis in children have ranged from 16% to 67%.(2,28) Patients with residual fragments after treatment are at risk for symptomatic growth of those fragments,(52) and the existence of metabolic disorders is a strong predictor of this growth in children.(53) Additionally, prognosis for nephrolithiasis depends on type of stones and adherence to therapy. Cystine stones have a high recurrence rate, and obstruction may impair kidney function. Primary hyperoxaluria type I is a progressive disease, often leading to progressive loss of kidney function even with optimal compliance to medical therapy, unless pyridoxine responsiveness is established. Kidney stones from hyperuricosuria may continue to occur with or without symptoms despite treatment. Therefore, pediatric patients with stone disease should be referred to a subspecialist for appropriate diagnosis, treatment, and long-term nephrology follow-up.

REFERENCES

1. Stapleton FB 1989 Nephrolithiasis in children. Pediatr Rev **11**:21–30.
2. Milliner DS, Murphy ME 1993 Urolithiasis in pediatric patients. Mayo Clin Proc **68**:241–248.
3. Diamond DA, Rickwood AM, Lee PH, Johnston JH 1994 Infection stones in children: A twenty-seven-year review. Urology **43**:525–527.
4. Coward RJ, Peters CJ, Duffy PG, Corry D, Kellett MJ, Choong S, van't Hoff WG 2003 Epidemiology of pediatric stone disease in the UK. Arch Dis Child **88**:962–965.
5. Pietrow PK, Pope JC, Adams MC, Shyr Y, Brock JW III 2002 Clinical outcome of pediatric stone disease. J Urol **167**:670–673.
6. Choi H, Snyder HM III, Duckett JW 1987 Urolithiasis in childhood: Current management. J Pediatr Surg **72**:158–164.
7. Gearhart JR, Herzberg GZ, Jeffs RD 1991 Childhood urolithiasis: Experiences and advances. Pediatrics **87**:445–450.
8. Stapleton FB, McKay CP, Noe HN 1987 Urolithiasis in Children: The role of hypercalciuria. Pediatr Ann **16**:980–992.
9. Milliner DS 2004 Urolithiasis. In: Avner ED, Harmon WE, Niaudet P (eds.) Pediatric Nephrology. Lippincott Williams & Wilkins, Philadelphia, PA, USA, pp. 1091–1111.
10. Gillespie RS, Stapleton FB 2004 Nephrolithiasis in children. Pediatr Rev **25**:131–138.
11. Stapleton FB 1990 Idiopathic hypercalciuria: Association with isolated hematuria and risk for urolithiasis in children: The Southwest Pediatric Nephrology Study Group. Kidney Int **37**:807–811.
12. Garcia CD, Miller LA, Stapleton FB 1991 Natural history of hematuria associated with hypercalciuria in children. Am J Dis Child **145**:1204–1207.
13. Stapleton FB 2002 Childhood stones. Endocrinol Metab Clin North Am **31**:1001–1015.
14. Langman CB, Schmeissing KJ, Sailer DM 1994 Children with genetic hypercalciuria exhibit thiazide-response to osteopenia. Pediatr Res **35**:368A.
15. Lloyd SE, Pearce SHS, Fisher JE, Steinmeyer K, Schwappach B, Scheinman SJ, Harding B, Bolino A, Devoto M, Goodyer P, Rigden SP, Wrong O, Jentsch TJ, Craig IW, Thakker RV 1996 A common molecular basis for three inherited molecular kidney stone diseases. Nature **379**:445–449.
16. Dent CE, Friedman M 1964 Hypercalciuric rickets associated with renal tubular damage. Arch Dis Child **39**:240–249.
17. Thomas SE, Stapleton FB 2000 Leave no "stone" unturned: Understanding the genetic basis of calcium-containing urinary stones in childhood. Adv Pediatr **47**:199–221.
18. Nicoletta JA, Lande MB 2006 Medical evaluation and treatment of urolithiasis. Pedatr Clin North Am **53**:479–491.
19. Bartosh SM 2004 Medical management of pediatric stone disease. Urol Clin North Am **31**:575–587.
20. Reddy ST, Wang CY, Sakhaee K, Brinkley L, Pak CYC 2002 Effect of low-carbohydrate high-protein diets on acid-base balance, stone-forming propensity, and calcium metabolism. Am J Kidney Dis **40**:265–274.
21. Monico CG, Rossetti S, Schwanz HA, Olson JB, Lundquist PA, Dawson DB, Harris PC, Milliner DS 2007 Comprehensive mutation screening in 55 probands with type 1 primary hyperoxaluria shows feasibility of a gene-based diagnosis. J Am Soc Nephrol **18**:1905–1914.
22. Williams E, Rumsby G 2007 Selected exonic sequencing of the AGXT gene provides a genetic diagnosis in 50% of patients with primary hyperoxaluria type 1. Clin Chem **53**:1216–1221.
23. Milliner D, Wilson D, Smith L 2001 Phenotypic expression of primary hyperoxaluria: Comparative features of types I and II. Kidney Int **59**:31–36.

© 2008 American Society for Bone and Mineral Research

24. Milliner D, Wilson D, Smith L 1998 Clinical expression and long-term outcomes of primary hyperoxaluria types 1 and 2. J Nephrol **11**(Suppl):56–59.
25. Hoppe B, Langman C 2003 A United States survey on diagnosis, treatment, and outcomes of primary hyperoxaluria. Pediatr Nephrol **18:**986–991.
26. Jamieson N 2007 The European PH1 Transplant Registry report 1984-2007: Twenty-three years of combined liver and kidney transplantation for primary hyperoxaluria PH1. Presented at the 8th International Primary Hyperoxaluria Workshop, University College London, London, UK, June 29–30, 2007.
27. Pak CY, Waters O, Arnold L, Holt K, Cox C, Barilla D 1977 Mechanism for calcium urolithiasis among patients with hyperuricosuria. J Clin Invest **59:**426–431.
28. Polinsky MS, Kaiser BA, Baluarte HJ 1987 Urolithiasis in childhood. Pediatr Clin North Am **34:**683–710.
29. Enomoto A, Kimura H, Chairoungdua A, Shigeta Y, Jutabha P, Cha SH, Hosoyamada M, Takeda M, Sekine T, Igarashi T, Matsuo H, Kikuchi Y, Oda T, Ichida K, Hosoya T, Shimokata K, Niwa T, Kanai Y, Endou H 2002 Molecular identification of a renal urate anion exchanger that regulates blood urate levels. Nature **417:**447–452.
30. Maalouf NM, Sahaee K, Parks JH, Coe FL, Adams-Huet B, Pak CY 2004 Association of urinary pH with body weight in nephrolithiasis. Kidney Int **65:**1422–1425.
31. Chesney RW 1998 Mutational analysis of patients with cystinuria detected by a genetic screening network: Powerful tools in understanding the several forms of the disorder. Kidney Int **54:**279–280.
32. Feliubadalo L, Font M, Purroy J, Rousaud F, Estivill X, Nunes V, Golomb E, Centola M, Aksentijevich I, Kreiss Y, Goldman B, Pras M, Kastner DL, Pras E, Gasparini P, Bisceglia L, Beccia E, Gallucci M, de Sanctis L, Ponzone A, Rizzoni GF, Zelante L, Bassi MT, George AL Jr, Manzoni M, De Grandi A, Riboni M, Endsley JK, Ballabio A, Borsani G, Reig N, Fernández E, Estévez R, Pineda M, Torrents D, Camps M, Lloberas J, Zorzano A, Palacín M International Cystinuria Consortium 1999 Non-type I cystinuria caused by mutations in SLC7A9, encoding a subunit of rBAT. Nat Genet **23:**52–57.
33. Goodyer P, Saadi I, Ong P, Elkas G, Rozen R 1998 Cystinuria subtype and the risk of nephrolithiasis. Kidney Int **54:**56–61.
34. Jackman SV, Potter SR, Regan F, Jarrett TW 2000 Plain abdominal x-rays versus computerized tomography screening: Sensitivity for stone localization after nonenhanced spiral computerized tomography. J Urol **164:**308–310.
35. Muslumanoglu AY, Tefekli A, Sarilar O, Binbay M, Altunrende F, Ozkuvanci U 2003 Extracorporeal shock wave lithotripsy as first line treatment alternative for urinary tract stones in children: A large scale retrospective analysis. J Urol **170:**2405–2408.
36. Desai M 2005 Endoscopic management of stones in children. Curr Opin Urol **15:**107–112.
37. Aksoy Y, Ozbey I, Atmaca AF, Polat O 2004 Extracorporeal shock wave lithotripsy in children: Experience using a mpi-9000 lithotriptor. World J Urol **22:**115–119.
38. Goel MC, Baserge NS, Babu RV, Sinha S, Kapoor R 1996 Pediatric kidney: Functional outcome after extracorporeal shock wave lithotripsy. J Urol **155:**2044–2046.
39. Lottmann HB, Archambaud F, Hallal B, Pageyral BM, Cendron M 1998 99mTechnetium-dimercapto-succinic acid renal scan in the evaluation of potential long-term renal parenchymal damage associated with extracorporeal shock wave lithotripsy in children. J Urol **159:**521–524.
40. Saw KC, Lingeman JE 1999 Management of calyceal stones: Lesson 20. AUA Update Series **20:**154–159.
41. Desai MR, Kukreja RA, Desai MM, Mhaskar SS, Wani KA, Patel SH, Bapat SD 2004 A prospective randomized comparison of type of nephrostomy drainage following percutaneous nephrostolithotomy: Large bore versus small bore versus tubeless. J Urol **172:**565–567.
42. Lam JS, Greene TD, Gupta M 2002 Treatment of proximal ureteric calculi: Holmium:YAG laser ureterolithotripsy versus ESWL. J Urol **167:**1972–1976.
43. Durkee CT, Balcom A 2006 Surgical management of urolithiasis. Pediatr Clin North Am **53:**465–477.
44. Ohkawa M, Tokunaga S, Nakashima T, Orito M, Hisazumi H 1992 Thiazide treatment for calcium urolithiasis in patients with idiopathic hypercalciuria. Br J Urol **69:**571–576.
45. Reusz GS, Dobos M, Vásárhelyi B, Sallay P, Szabó A, Horváth C, Szabó A, Byrd DJ, Thole HH, Tulassay T 1998 Sodium transport and bone mineral density in hypercalciuria with thiazide treatment. Pediatr Nephrol **12:**30–34.
46. Bentur L, Alon U, Berant M 1987 Hypercalciuria in chronically institutionalized bedridden children: Frequency, predictive factors and response to treatment with thiazides. Int J Pediatr Nephrol **8:**29–34.
47. Garcia-Nieto V, Navarro JF, Monge M, Garcia-Rodriguez VE 2003 Bone mineral density in girls and their mothers with idiopathic hypercalciuria. Nephron Clin Pract **94:**C81–C82.
48. Griffith DP, Gleeson MJ, Lee H, Longreit R, Deman E, Earle N 1991 Randomized, double-blind trial of lithostat (acetohydroxaminic acid) in the palliative treatment of infection-induced urinary calculi. Eur Urol **20:**243–247.
49. Jamieson N 2005 A 20-year experience of combined liver/kidney transplantation for primary hyperoxaluria (PH1): The European PH1 Transplant Registry Experience 1984-2004. Am J Nephrol **25:**282–289.
50. Cibrik D, Kaplan B, Arndorfer J, Mier-Kriesche H 2002 Renal allograft survival in patients with oxalosis. Transplantation **74:**707–710.
51. Coulthard MG, Richardson J, Fleetwood A 1995 The treatment of cystinuria with captopril. Am J Kidney Dis **25:**661–662.
52. Streem SB, Yost A, Mascha E 1996 Clinical implications of clinically insignificant stone fragments after extracorporeal shock wave lithotripsy. J Urol **155:**1186–1190.
53. Afshar K, McLorie G, Papanikolaou F, Malek R, Harvey E, Pippi-Salle JL, Bagli DJ, Khoury AE, Farhat W 2004 Outcome of small residual stone fragments following shock wave lithotripsy in children. J Urol **172:**1600–1603.
54. Santos-Victoriano M, Brouhard BH, Cunningham RJ III 1998 Renal stone disease in children. Clin Pediatr **37:**583–599.

© 2008 American Society for Bone and Mineral Research

SECTION X

Oral and Maxillofacial Biology and Pathology (Section Editor: Laurie McCauley)

Please also see the ASBMR ONJ Task Force Report at: http://www.jbmronline.org/doi/full/10.1359/jbmr.0707onj?prevSearch=authorsfield%3A%28shane%29 and The Jackson Laboratory website at www.jax.org

© 2008 American Society for Bone and Mineral Research

Chapter 103. Craniofacial Growth and Development

Andrew Ravanelli and John Klingensmith

Department of Cell Biology, Duke University Medical Center, Durham, North Carolina

OVERVIEW

In this chapter, we provide a brief overview of human craniofacial development. The head is one of the first structures to arise in the embryo, with morphologically distinct primordia apparent by the end of the third week of pregnancy. Here we describe key events in the development of the head and its major compartments, focusing on skeletal structures. We also consider the major congenital malformations arising from defects in head formation.

EARLY EMBRYOGENESIS

The embryo develops from conception into a trilayered disc of cells. These germ layers are called the ectoderm, mesoderm, and endoderm. Organ systems begin to form as these layers fold to create two tubes. The neural ectoderm folds dorsally to become the neural tube. The anterior end of the neural tube expands to form the brain rudiments: the prosencephalon, mesencephalon, and rhombencephalon (Fig. 1A). The endoderm folds ventrally to create the gut tube. The anterior end is the foregut, which gives rise to the majority of the oral cavity and pharynx. The brain and foregut of the early embryo are the centers around which the head is formed, and each is a source of signals that direct development of craniofacial tissues. Many tissues of the head are formed from neural crest cells (NCCs), a pluripotential migratory cell type in the midgestation embryo. During neurulation, NCCs arise from the converging crests of the neural folds as they fuse dorsally to form the neural tube. The NCCs undergo an epithelial-to-mesenchymal transition, delaminate, and migrate through the mesoderm to populate various structures of the embryo (Fig. 1A). Cranial NCCs migrate to the head to eventually form the bulk of the facial tissues, many bones of the skull, the dental papillae of the teeth, and much of the muscular, nervous, and vascular tissue of the head and neck. The mesoderm also contributes substantially to the early head. Paraxial mesoderm segments and condenses into somitomeres, just lateral to the neural folds. As the trunk somitomeres condense further, they form a series of segmented units (somites). The first seven cranial somitomeres never fully condense to form definitive somites. In all cases, these units each consist of three compartments: dermatome, myotome, and sclerotome. The dermatome contributes to the dermis of the skin. Myotomes give rise to the skeletal muscles of the body, limbs, and head. Sclerotome derivatives form the bones of the axial skeleton, including the base of the skull.[1]

PHARYNGEAL ARCHES

A large portion of the head and neck derives from the pharyngeal arches (Fig. 1B). These are five, bilaterally paired swellings formed along the pharyngeal foregut in the fourth week, as the rostral neuropore is closing. The arches are numbered 1, 2, 3, 4, and 6 by homology to the primitive gill arches of lower vertebrates. An early rostrocaudal subdivision of the first arch yields the maxillary and mandibular prominences. Each arch is composed of a mesodermal core lined with endoderm and ectoderm. Endodermal pouches and ectodermal clefts separate the arches. NCCs migrate into the arches, filling them with mesenchyme. Each pharyngeal arch gives rise to a variety of specific structures of the head and neck (Table 1). In general, the mesodermal cores yield skeletal structures of the neck and lower head. The clefts and pouches form the epithelial linings of the many of the organs, ducts, and mucosa of the mouth and neck. The cranial neural crest differentiates into bones, cartilages, nerves, tooth rudiments, and many other craniofacial tissues.[1] Within each arch, a cranial nerve (CN), an aortic arch artery, and an arch cartilage develops. Their derivatives are listed in Table 1. The cranial nerves innervate the structures derived from the arches in which they form. The left and right arteries of the arches undergo differential development and regression in a very complex morphogenetic progression, ultimately forming components of the aortic system. The arch cartilages form a variety of small bones and cartilages, primarily in the middle ear and neck.

THE NEUROCRANIUM

The bones and muscles of the skull that encase the brain comprise the neurocranium (Fig. 1C), consisting largely of the membranous bones that form the cranial vault, and the cartilaginous neurocranium, forming the floor of the skull. The thin, broad bones of the membranous neurocranium derive from cranial NCCs that migrate from brain regions to form the mesenchyme covering the sides and top of the brain. This mesenchyme condenses and differentiates directly into osteoblasts that secrete osteoid and form bone (intramembranous ossification). The resulting calvarial bones are separated by dense, connective tissue seams (sutures), leaving open spaces (the fontanelles). These flexible junctures accommodate deformation during birth and brain enlargement during infancy. The cartilaginous neurocranium, or neural chondrocranium, is formed by the fusion of cartilages, composed primarily of cells from the sclerotome of cranial somitomeres and occipital somites. The cartilage models undergo endochondral ossification. Some undergo regression as membranous bone.[2] The resulting bones comprise some of the facial bones, including the sphenoid, ethmoid, temporal, and the caudal portion of the occipital bone. These bones have complex origins: The ethmoid and sphenoid cartilages are formed from somitomeric mesoderm and NCC derivatives, whereas the caudal portion of the occipital bone is formed from the somatic sclerotomes of the first three occipital somites and the cranial half of the first cervical somite. The sensory capsules, also part of the cartilaginous neurocranium, encase the nasal passages, the eyes, and the inner ear.[1]

THE VISCEROCRANIUM

The viscerocranium is the skeleton of the face. These cartilages and bones form the skeletal structures of the mouth and the supporting structures of the orphopharynx and trachea. The bones of the viscerocranium are also formed by either intramembranous or endochondral ossification (Fig. 1C). The membranous viscerocranium consists of the maxilla, the palatine bones, the zygoma, the squamous temporal bones, and the mandible. The latter is formed by ossification of NCC-derived mesenchyme that condenses around the mesodermal core of the mandibular prominence (Meckel's cartilage). The endo-

The authors state that they have no conflicts of interest.

Key words: neurocranium, viscerocranium, birth defect, pharyngeal arch, neural crest

© 2008 American Society for Bone and Mineral Research

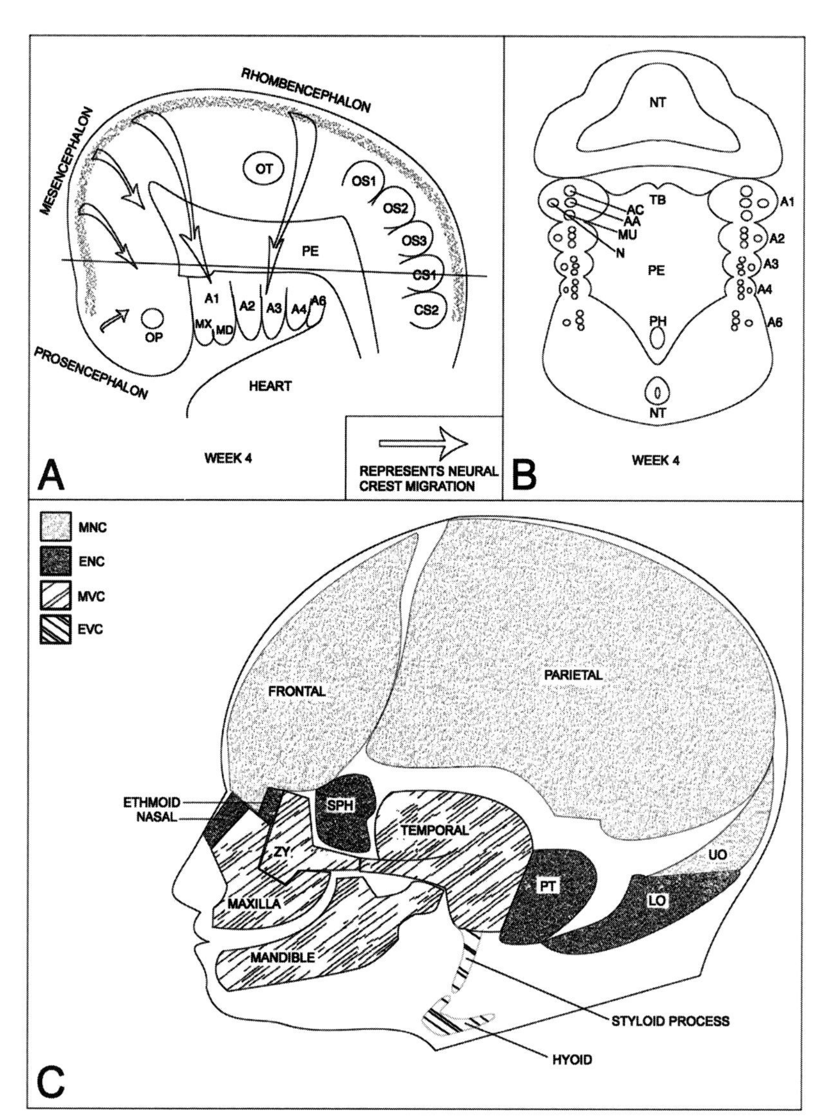

FIG. 1. Head precursors and skull derivatives. (A) The human embryo depicting formed pharyngeal arches (A1–A6), regions of the brain, optic vesicle (OP), otic vesicle (OT), occipital somites (OS1–OS3), the first two cervical somites (CS1–CS2), pharyngeal endoderm tube (PE), heart, and a schematic generalization of neural crest migration. MX and MD represent the maxillary and mandibular divisions of arch 1. Bisecting line represents plane of section for B. (B) Cross-section of embryo showing the pharyngeal endoderm (PE) and arch core components. NT, neural tube; TB, tongue bud; AC, arch cartilage; AA, arch artery; MU, arch muscle; N, arch nerve; PH, pharynx. (C) Bones of the skull. Shadings represent membranous neurocranium (MNC), neural chondrocranium (ENC), membranous viscerocranium (MVC), and endochondral viscerocranium (EVC). ZY, zygoma; SPH, sphenoid; PT, petrous temporal; UO and LO, upper and lower occipital.

chondral (cartilaginous) viscerocranium arises from the first two arches and consists of the middle ear bones, the styloid process of the temporalis, and the hyoid bone. The laryngeal cartilages are derived mainly from arches 4 and 6.

FACIAL MORPHOGENESIS

Mesenchymal masses of NCC derivatives yield the features and structures of the face from five primary swellings: two maxillary and two mandibular prominences derived from the first arch and a single frontonasal mass (or process) derived from head mesenchyme (Fig. 2A). These are organized around the stomodeum, the future opening to the mouth. The frontonasal mass forms the bulk of tissues rostral to the stomodeum, forming structures such as the forehead. The maxillary prominences form lateral tissues, ultimately producing much of the upper face. The mandibular prominences primarily form the tissues just caudal to the stomodeum, including the chin. The buccopharyngeal membrane, separating the stomodeum from the anterior foregut tube, ruptures at day 24 to create a broad, slit-like, embryonic mouth (the orphopharynx). This early mouth is reduced laterally by the fusion of the mandibular and maxillary prominences to form the cheeks. Nasal placodes thicken on the frontonasal mass and develop nasal pits that divide each side of the process into the medial and lateral nasal processes (Fig. 2B). The medial nasal processes grow down into the stomodeum and fuse with the maxillary processes (Fig. 2C). These medial nasal processes expand and fuse to form the intermaxillary process, thus forming the primitive upper lip and philtrum (Fig. 2D). Proliferating mesenchyme fills in the mandibular fusion to form the lower lip.[1,3] Rapid facial growth occurs from the fourth to eighth week.

THE ORONASAL CAVITY AND PALATOGENESIS

The oronasal cavity is created as the frontonasal mass enlarges and the first arches grow together to form the stomodeum. As the nasal pits invaginate, they create the nasal passage that grows inward toward the pharyngeal endoderm. The oronasal membrane, a layer of tissue separating the oral and nasal compartments, breaks down to form openings between the oral and nasal cavities (the primitive choana). The oral and nasal cavities become separated by the palate, which has several components. The primary palate is a small anterior domain contributed by the medial nasal processes, whereas the secondary palate comprises most of the soft and hard tissues of the roof of the mouth (Fig. 3C). The hard (bony) portion of the palate arises as palatine shelves grow together from the max-

© 2008 American Society for Bone and Mineral Research

TABLE 1. STRUCTURES DERIVED FROM PHARYNGEAL ARCH TISSUES

Pharyngeal tissue	*Major structures*	*Cranial nerve*	*Cartilages and bone*	*Skeletal muscles*	*Arch artery*
Arch 1	Mandibular arch, maxillary prominence	Trigeminal (V)	Palatopterygoid, maxilla, palatine, zygoma, squamous, temporal, incus	All muscles of mastication: masseter, temporalis, pterygoids	Terminal branch of maxillary artery
	Mandibular prominence		Meckel's, mandible, malleus	Mylohyoid, anterior digastric, tensor veli palatini, tensor tempani	
Pouch and cleft 1	Linings of auditory tube and external auditory meatus and tympanic membrane				
Arch 2	Hyoid arch	Facial (VII)	Reichert, styloid, hyoid, stapes, stylohyoid ligament	All muscles of facial expression, Posterior digastric, stylohyoid, stapedius	Stapedial artery, corticotympanic artery
Pouch 2	Lining of palatine tonsils				
Arch 3		Glossopharyngeal (IX)	Hyoid	Stylopharyngeus	Common carotid artery, root of internal carotid
Pouch and Cleft 3					
Dorsal tissues	Cells of inferior parathyroid gland				
Ventral tissues	Components of thymus gland				
Arch 4		Vagus: superior, laryngeal branch (X)	All laryngeal cartilages: thyroid, cricoid, arytenoids, corniculate, cuneiform, epiglottis (4 and 6)	Pharyngeal constrictors	Arch of aorta, right subclavian artery, base of pulmonary arteries
		Pharyngeal Branch (X)		All soft palate muscles except tensor veli palatini All intrinsic laryngeal muscles (4 and 6)	
Pouch 4					
Dorsal tissues	Cells of superior parathyroid gland				
Ventral tissues	Parafollicular cells of thymus				
Arch 6		Recurrent laryngeal branch (X)			Ductus arteriosus, roots of definitive pulmonary arteries

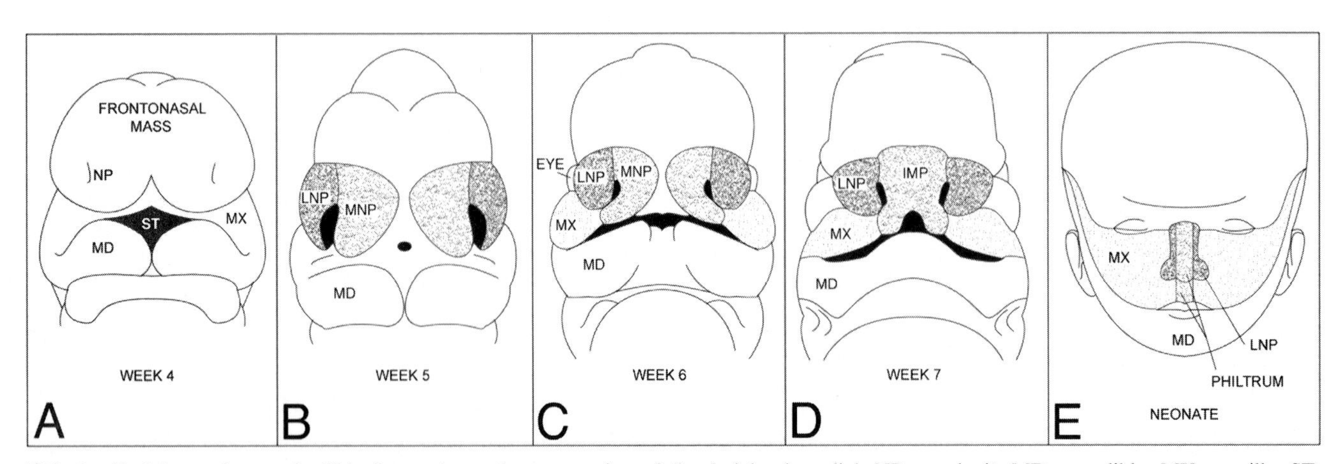

FIG. 2. Facial morphogenesis. This figure shows the progression of the facial primordial. NP, nasal pit; MD, mandible; MX, maxilla; ST, stomodeum; LNP and MNP, lateral and medial nasal processes; IMP, intermaxillary process.

© 2008 American Society for Bone and Mineral Research

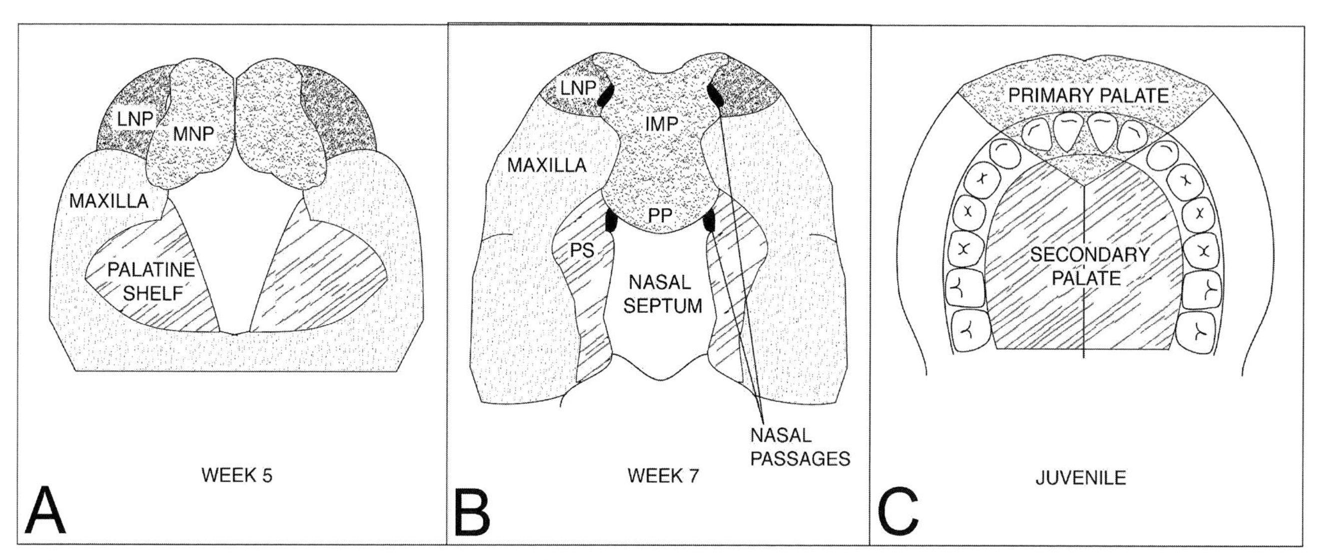

FIG. 3. Palatogenesis. This figure shows the movement of the primary palate (PP) and secondary palatine shelves (PS). LNP and MNP, lateral and medial nasal processes.

illary processes (Fig. 3A). As the secondary palatine shelves grow together, they become positioned above the tongue to allow for fusion in the midline. They also fuse anteriorly to the primary palate (Fig. 3B). Fusion of the palatine shelves with each other and with the nasal septum separates the nasal cavities from the oral cavity. The posterior parts of the palatine shelves do not ossify but fuse to form the soft palate. This is the fleshy portion of the mouth's roof, extending posteriorly from the hard palate. The tongue arises as a swelling from the floor of the pharynx in the fourth week. One medial and two lateral lingual buds swell and fuse to form the tongue primordium. The lateral buds outgrow the medial bud to form the oral portion of the tongue. The base of the tongue is formed with contributions from the second, third, and fourth arches. The teeth are formed from ectodermal (oral epithelium) and mesenchymal (neural crest derived) tissues of the mandibular and maxillary prominences. Each tooth bud is composed of an ectodermal dental lamina and a basal, mesenchymal dental papilla.[4,5] Their development into mature teeth is described in another chapter of this volume.

CRANIOFACIAL BIRTH DEFECTS

The head is most sensitive to developmental perturbations from 3–8 wk of gestation, the period when most of the cranial tissues and precursors of head structures are forming. The complexity of cell types contributing to the head and their dynamic reorganizations create many opportunities for error. Depending on when and how severe the insult is, a wide range of defects can occur—ranging from minor cosmetic concerns to serious medical disasters. Significant birth defects occur in ~1 in 50 live births, a third involving craniofacial malformations. Collectively, birth defects are the leading cause of infant mortality in the Western world.[6] In 2004, congenital malformations, deformations, and chromosomal abnormalities accounted for one fifth of all infant deaths in the United States.[6] Head malformations account for many of these deaths and often lead to significant disability in those who survive infancy.

Structural anomalies of craniofacial development have a varied etiology. They can result from genetic mutation(s) or from environmental disruption of developmental pathways. Such an environmental disruption could be chemical exposure to the mother, physical stress to the pregnancy, or anything else exogenous to the embryo. Increasingly, it seems likely that many malformations might result from an unfortunate interaction between a compromising allele of some key gene and an environmental trigger, neither of which would necessarily cause a birth defect on its own. A birth defect can occur as an isolated malformation or as part of a syndrome, in which multiple defects occur in a group of organs and structures. The latter case, a syndromic association of birth defects, is likely to reflect a disruption in a key developmental process or population of precursor cells. For example, a mutation causing a defect in the migration of NCCs could cause major malformations of the face, skull, heart, and many other structures. In contrast, a mutation in a gene required for palatal fusion might result in cleft palate but no other defects. Here we describe several representative classes of craniofacial birth defects and their origins (Table 2).

Holoprosencephaly

This major class of birth defects occurs because of insufficient tissue along the midline of the ventral prosencephalon and/or facial precursors. This deficiency of midline structures can show varying levels of severity. Defects can range from a single central upper incisor (Fig. 4A), to close-set eyes (hypotelorism), to either no nose (arhinia) or a single nostril (cebocephaly) (Fig. 4B), or even to cyclopia with a nose-like proboscis above the eye field (Fig. 4C). Alcohol consumption by pregnant women can result in midline facial deficiencies and mental retardation in the developing progeny. Unfortunately, pregnant women are normally unaware of their pregnancy during the critical period for forebrain formation, during weeks 3 and 4 of gestation. Holoprosencephalies (OMIM 236100) occur in as many as 1 in 250 conceptuses and 1 in 5,000–16,000 live births.[3,7]

Pharyngeal Arch Defects

Arch defects often involve the tissues of the viscerocranium and are frequently caused by improper development of NCC derivatives. These defects include micrognathia (Fig. 4E), agnathia, and palatal or mandibular clefting (Fig. 4D). Mandibulofacial dysostosis (pharyngeal arch defects) are included in a

© 2008 American Society for Bone and Mineral Research

TABLE 2. REPRESENTATIVE CONGENITAL MALFORMATIONS OF THE HEAD

Structure	Birth defect	Major origin	References: Embryological	References: Molecular	References: Clinical
Skull vault	Craniosynostosis	Premature fusion of the cranial sutures	3	7, 9	8
	Apert syndrome				
	Pfeiffer syndrome				
	Crouzon syndrome				
	Microcephaly	Severe underdevelopment of the skull and cerebrum			
Jaw	Micrognathia	Dysgenesis of the mandible	13, 14	4, 9, 14	8, 10
	Agnathia				
	DiGeorge syndrome				
Palate/mouth	Cleft lip/palate	Failure of fusion of a combination of the nasal and maxillary processes	12, 14	4, 9, 11, 14	8
	Pierre Robin sequence	U-shaped cleft palate, micrognathia, retracted tongue			
	DiGeorge syndrome	Velocardiofacial defects			
Face	Microsomia	Underdevelopment of various facial features	13, 14	6, 9, 14	8
	Apert syndrome	Abnormal fusion of facial bones			
	Pfeiffer syndrome				
	Crouzon syndrome				
	DiGeorge syndrome	Velocardiofacial defects			
	Holoprosencephaly	Midline deletions			
Teeth	Ectodermal dysplasia	Two or more ectodermal anomalies	3	4, 5, 9	8

large number of syndromes, including Treacher-Collins (OMIM 154500), Hallerman-Streiff, and Franceschetti's syndromes (OMIM 154500).[3,4]

Tissue Fusion Defects

Facial fusion defects occur when the epithelia of the facial primordia fail to fuse properly, causing facial clefting. The facial cleft sequence can involve a combination of complete or incomplete, bilateral or unilateral clefting of the lip, palate, or nostril (Fig. 4D). It can also include failure of fusion of the mandibular processes, or the mandible to the maxilla, causing improper lateral restriction of the mouth. These defects often cause problems in eating and breathing and require corrective plastic surgery.[3]

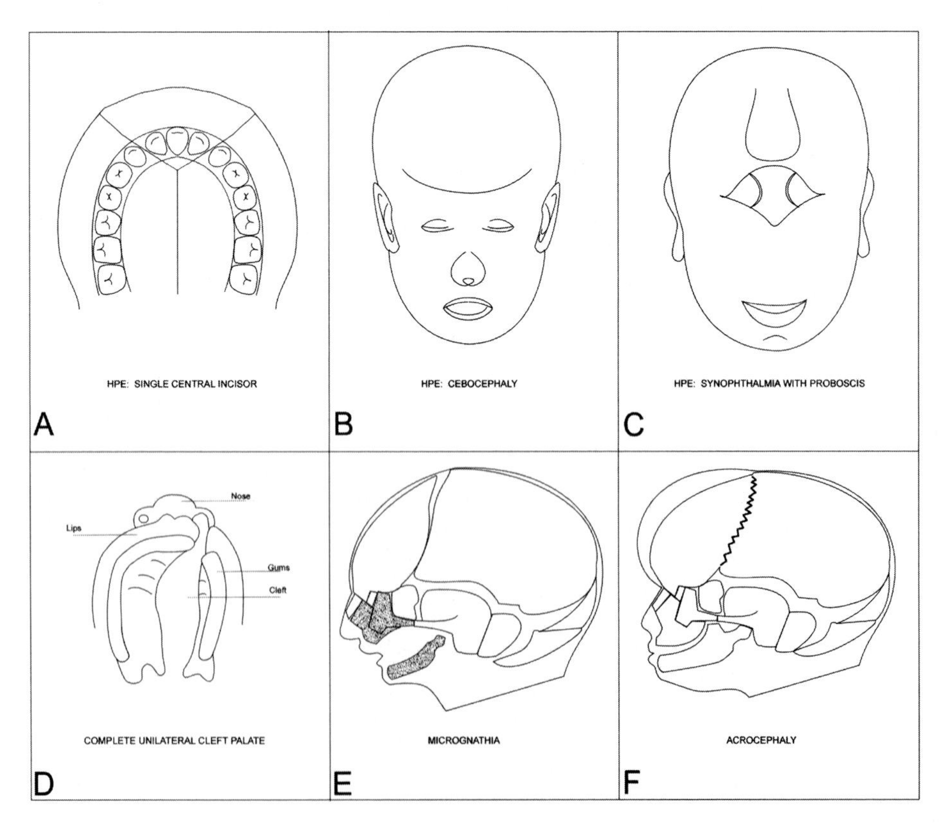

FIG. 4. Examples of craniofacial birth defects. This figure shows schematic diagrams of some craniofacial defects, including examples of holoprosencephaly (HPE) midline defects (A–C), facial clefting (D), pharyngeal arch defect (E), and cranial synostosis (F).

© 2008 American Society for Bone and Mineral Research

Vascular Defects

A defect in vascular development or incorrect blood flow can cause a variety of craniofacial defects involving hypoplasia of head tissues. For example, hemifacial microsomia (underdevelopment of one side of the face) is caused by a unilateral insufficiency of blood supply during facial development. This can cause clefting defects caused by underdevelopment of tissues that grow together and fuse. Vascular defects are also a cause of Goldenhar syndrome (OMIM 164210).[3]

Skeletal Dysplasias

This class of defects is caused by improper bone growth. Dysplasias that affect the head may involve cranial synostoses or premature fusion of the cranial sutures. This inhibits bone growth and puts pressure on the brain, which is forced to grow improperly and protrude where it can. Such defects include tower skull, or acrocephaly (Fig. 4F), and meningohydroencephaloceles. Skeletal dysplasias often involve other areas of the body including limbs, digits, and vertebrae. Some of the more common are Crouzon (OMIM 123500), Pfeiffer (OMIM 101600), Saethre-Chotzen (OMIM 101400), and Apert's syndromes (OMIM 101200).[3,8]

Experimental manipulations and genetic studies in animal models are identifying many of the genes and pathways that control both normal craniofacial development and its anomalies.[9] Most of the genes to date have been intercellular signaling factors, their transducers, or their targets, highlighting the prime importance of cell interactions to normal head formation. Genetic association and expression studies in humans can test whether these candidate genes are mutated or misexpressed in human congenital malformations. Rapidly advancing genomic, embryological, and medical technologies promise to bring soon a better understanding of the basis of craniofacial development and approaches toward minimizing its defects.

REFERENCES

1. Larsen WJ 1997 Human Embryology, 2nd ed. Churchill Livingstone, New York, NY, USA.
2. Holmbeck K, Bianco P, Chrysovergis K, Yamada S, Birkedal-Hansen H 2003 MT1-MMP-dependent, apoptotic remodeling of unmineralized cartilage: A critical process in skeletal growth. J Cell Biol **163:**661–671.
3. Thorogood P 1997 The head and face. In: Thorogood P (ed.) Embryos, Genes and Defects. John Wiley and Sons, Chichester, West Sussex, UK, pp. 197–229.
4. Cobourne MT, Sharpe PT 2003 Tooth and jaw: Molecular mechanisms of patterning in the first branchial arch. Arch Oral Biol **48:**1–14.
5. Laurikkala J, Mikkola M, Mustonen T, Aberg T, Koppinen P, Pispa J, Nieminen P, Galceran J, Grosschedl R, Thesleff I 2001 TNF signaling via the ligand-receptor pair ectodysplasin and edar controls the function of epithelial signaling centers and is regulated by Wnt and activin during tooth organogenesis. Dev Biol **229:**443–455.
6. Matthews TJ, MacDorman MF 2007 Infant Mortality Statistics From the 2004 Period Linked Birth/Infant Death Data Set. Center for Disease Control, National Vital Statistics Report, Atlanta, GA, USA.
7. Wallis D, Muenke M 2000 Mutations in holoprosencephaly. Hum Mutat **16:**99–108.
8. Yu HM, Jerchow B, Sheu TJ, Liu B, Costantini F, Puzas JE, Birchmeier W, Hsu W 2005 The role of Axin2 in calvarial morphogenesis and craniosynostosis. Development **132:**1995–2005.
9. Chai Y, Maxson RE Jr 2006 Recent advances in craniofacial morphogenesis. Dev Dyn **235:**2353–2375.
10. Jones KL 1988 Smith's Recognizable Patterns of Human Malformation, 4th ed. W.B. Saunders, Philadelphia, PA, USA.
11. Singh DJ, Bartlett SP 2005 Congenital mandibular hypoplasia: Analysis and classification. J Craniofac Surg **16:**291–300.
12. Houdayer C, Portnoi M, Vialard F, Soupre V, Crumiere C, Taillemite J, Couderc R, Vasquez M, Bahuau M 2001 Pierre Robin sequence and interstitial deletion 2q32.3-q33.2. Am J Med Genet **102:**219–226.
13. Schubert J, Jahn H, Berginski M 2005 Experimental aspects of the pathogenesis of Robin sequence. Cleft Palate Craniofac J **42:**372–376.
14. Hunt JA, Hobar PC 2003 Common craniofacial abnormalities: The facial dysostoses. Plast Reconstr Surg **112:**606–615.
15. Goodman FR 2003 Congenital abnormalities of body patterning: Embryology revisited. Lancet **362:**651–662.

Chapter 104. Development and Structure of Teeth and Periodontal Tissues

Alan Boyde and Sheila J. Jones

Centre for Oral Growth and Development, Queen Mary, University of London, London, United Kingdom

NORMAL DENTAL DEVELOPMENT

Three of the five distinct types of mineralized tissues found in the human body, enamel, dentine, and cementum, only occur in teeth. Because turnover in these tissues is nonexistent or minimal, they form a valuable, permanent record of conditions prevailing at their time of formation; this extends throughout fetal life and up to adulthood. Moreover, the enamel and dentine of the crowns of the deciduous teeth are available for analysis without surgical intervention when the teeth are shed naturally. Enamel is a surface tissue of epithelial origin, whereas dentine and cementum are avascular connective tissues of mesenchymal origin. Teeth[4] form at special locations within the jaws mapped out by the overlapping of molecular signals common to many developmental processes.[1] Tooth development is rigorously controlled by regulatory genes determining tooth type (incisor, canine, premolar, or molar) and shape.[2] Sequential local interactions at the interface between epithelium over the facial processes and mesenchyme derived from the cranial neural crest play a crucial role in tooth morphogenesis,[3] the main signaling molecules being members of the Hedgehog, bone morphogenetic protein, fibroblast growth factor, Wnt (wingless), and TNF families.[4–7]

The authors state that they have no conflicts of interest.

Electronic Databases: 1. Gene Expression in Tooth Database—http://bite-it.helsinki.fi/.

© 2008 American Society for Bone and Mineral Research

The embryonic tooth germ passes through three morphological stages, described as bud, cap, and bell, and has three main components, the enamel organ, the dental papilla, and the dental follicle. The epithelial enamel organ differentiates into a four-layered structure, within which the enamel knot is the signaling center that regulates tooth shape and size.[6] A complex sequence of epithelial—mesenchymal interactions results in waves of differentiation that start at the eventual enamel—dentine junction underlying the cusp tips—determined by the spatio-temporal induction of secondary enamel knots—and incisal central mammelons (rounded prominences on biting edges when incisors first erupt) and spread away, eventually delineating the whole junction between the tissues as the tooth germ grows. The expression of secretory signaling molecules varies continuously in the different cell types during tooth initiation and construction.[8] Odontoblasts, which make dentine, are postmitotic cells that differentiate from mesenchymal cells of the dental papilla at the interface with the inner enamel epithelial cells of the enamel organ, which themselves differentiate into pre-ameloblasts. Dentine formation triggers the pre-ameloblasts to differentiate into ameloblasts, the cells that produce enamel.[9] A bilayer of epithelial cells, the epithelial root sheath, extends from the enamel organ at the base of the developing crown to map out the dentine—cementum junction and initiate the differentiation of the odontoblasts of the root. The third tissue type, cementum, is the product of both fibroblasts and cementoblasts, which differentiate from mesenchymal cells of the dental follicle adjacent to the dentine once epithelial cells of the root sheath have moved away from the interface.[10] In human teeth, a small amount of afibrillar cementum may form on the enamel surface close to the junction between the crown and the root if there are interruptions in the covering layer of epithelial cells once enamel formation has been completed. Within the developing tooth, a core of loose connective tissue remains and eventually forms the dental pulp.

The dental follicle, also derived from cells of the cranial neural crest, gives rise to three components of the periodontium: cementum, alveolar bone, and the intervening periodontal ligament. The tooth germs are partially enclosed by the developing alveolar bone—this is initially typical woven bone, formed by osteoblasts, with enclosed osteocytes, and is remodeled to accommodate the growing teeth by osteoclasts of hematopoietic origin. The follicle, a sac of loose connective tissue that separates the developing tooth from its bony crypt, is essential for eruption and will become the periodontal ligament on tooth eruption.[11] This tissue contributes extrinsic collagen fibers to the cementum and alveolar bone, and its main cell type is the fibroblast (see Fig. 1 for a diagram of a mature tooth and its components).

NORMAL DENTAL STRUCTURE

Enamel

Enamel matrix is delicate when first secreted, at which time it is protected by the soft enamel organ. The mature, erupted enamel—the hardest of the hard tissues—is acellular and may contain 98% by weight or 93% by volume of an apatitic calcium phosphate of variable composition.[12,13] The final strength of enamel partly derives from the dentine mold on which it grows. The junction between these tissues is ill-defined and irregular on a microscopic scale, with tongues of dentine projecting into the enamel, crystals of indeterminate provenance at the common boundary, and many fine, short enamel tubules marking where ameloblast processes once contacted odontoblasts. Spindles, expanded continuations of dentine tubules within enamel, most likely result from the envelopment

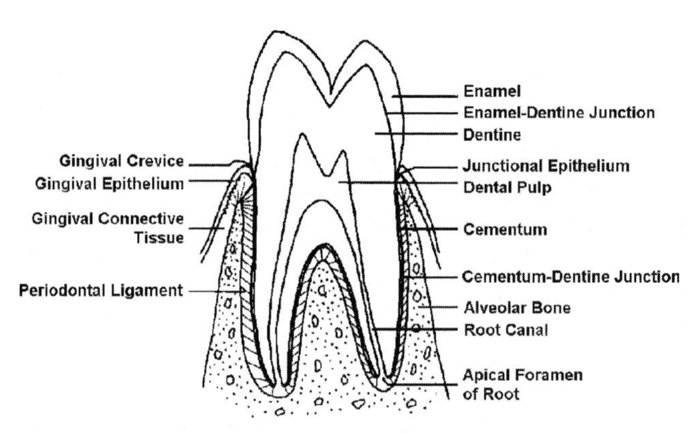

FIG. 1. Organization of dental and periodontal tissues in the erupted tooth.

of individual ameloblasts that died as amelogenesis commenced. The extracellular proteinaceous matrix of developing enamel is secreted by ameloblasts, which are highly polarized, tall cells. Its main component is amelogenin, a tissue-specific protein rich in proline, leucine, histidine, and glutamyl residues. Other, nonacidic proteins include enamelin, tuftelin, and ameloblastin (amelin, sheathlin): this 3D protein array is thought to control crystal growth.[14] To achieve enamel's high degree of mineralization, much of its organic matrix is degraded by neutral metalloproteinases and serine proteases and removed, even while ameloblasts are still secretory.[15] Enamel crystals are, even initially, very long and slender, with centers richer in carbonate; however, the net carbonate content falls as they thicken. In humans, relatively large amounts of mineral accumulate at early stages of development, and the enamel has a long postsecretory maturation period during which it becomes hard and the ameloblasts remain active. The maturation phase may last 5 years or more in human third permanent molars. In species with rapid enamel development, cyclical changes in morphology of the maturation ameloblasts are seen to coincide with episodic matrix removal. Enamel's final composition and mechanical properties are not uniform.

The most notable feature of enamel is the organization of the crystals into enamel "prisms" about 6 μm across and up to the enamel thickness in length, demarcated by a sharp change in crystal orientation (Figs. 2 and 3). Enamel crystals grow mainly with their long c-axes nearly parallel to each other and the larger sides of their flattened hexagonal cross-sections parallel within groups. Where the rate of formation is low, as in the superficial enamel, the secretory interface is nearly flat, and there is little variation in the underlying crystal orientation. However, during most of enamel formation, the secretory (Tomes') process of each ameloblast is lodged in a pit at the interface. Enamel matrix is released below a continuous belt of intercellular attachments so as to maintain the relatively constant shape of the interface between cells and matrix (Fig. 2).[16,17] The interpit phase is continuous and the crystals have their long axes perpendicular to the general plane of the developing enamel surface. In human enamel, the dividing lines of the prism junctions are generally incomplete, and the interlocking prisms are described as keyhole-shaped. The concentration of the cleavage products of the enamel proteins at the discontinuities in crystal orientation increases relatively during enamel maturation. Tufts and lamellae are other regions that finally contain less mineral and higher concentrations of proteins.[17]

As ameloblasts move away from the dentine, they travel in groups across the surface that they make. This results in decussation (crossing in an X fashion) of the enamel prisms, with

© 2008 American Society for Bone and Mineral Research

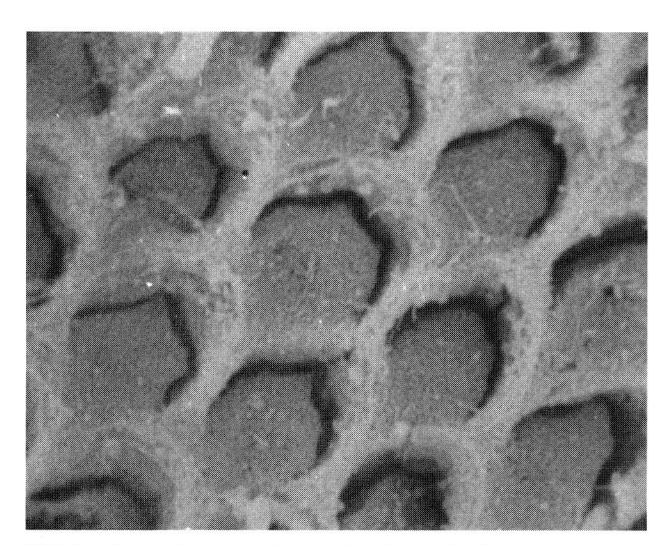

FIG. 2. External surface of cervical region of developing human permanent molar tooth, showing the morphology of the interface between ameloblasts and their calcified secretory product, the enamel. Scanning electron micrograph (SEM), field width 25 μm.

zones of prisms with contrasting 3D courses forming the Hunter-Schreger bands. The sides of the prisms show varicosities (Fig. 3) with the same period as cross-striations in the prisms, which are thought to be caused by circadian changes in the composition of the mineral component.[17] A prominence of the cross-striations occurs at 7- to 10-day intervals (the regular striae of Retzius), and major life events, such as birth (the neonatal line) or severe illness during enamel formation, may be recorded as conspicuous incremental lines. At the finished enamel surface, perikymata or imbrication lines are outcrops of the internal growth layers. They grade from horizontal bands displaying pits alternating with smoother regions at more incisal or occlusal levels, to near the neck of the tooth, small steps at the sharp boundary between the imbricating layers.

The unerupted crown is protected from resorption by a layer of cells termed the reduced enamel epithelium, comprising remnants of mature ameloblasts. These are lost once the tooth erupts. As the tooth wears during function, the surface features of the enamel become abraded, microcracks develop particularly along developmental faults, and the chemistry of the mineral exposed to the oral environment changes.

FIG. 3. Human enamel fractured to show the form of the prisms that are ~6 μm across. SEM, field width 82 μm.

Dentine

Dentine forms the bulk of the tooth and extends within both crown and root. It is a pale creamy yellow color, in contrast to the much whiter, harder enamel. Dentine is tough and elastic, and its prime feature is its penetration by odontoblast tubules that radiate out from the dental pulp to the periphery (Fig. 4). These, with their many side branches that remain in the tubules within the dentine, are analogous to the canaliculi that house osteocyte processes in bone. The peripheral, first formed dentine is termed mantle dentine, and the inner layer is termed circumferential dentine. After differentiating from cells of the dental papilla, the odontoblasts retreat centripetally as a cone-shaped monolayer sheet, depositing a collagenous predentine matrix and leaving lengthening cell processes.[18] The curved paths that the cell bodies take are therefore recorded in the extracellular matrix. This is similar to that of bone, comprising mainly type I collagen, acidic proteins, and proteoglycans. The predominant noncollagenous protein in dentine is the highly phosphorylated dentine phosphoprotein (phosphophoryn). This and dentine sialoprotein[19] are cleavage products of dentine sialophosphoprotein and are formed during the maturation of predentine into dentine. Dentine matrix protein 1 and other sialic acid-rich phosphoproteins common to dentine and bone are also present. Decorin, biglycan, lumican, and fibromodulin are the main proteoglycans in predentine.[20] The predentine matrix matures progressively, and the collagen fibrils thicken and compact and mineralize after a lag time of ~4 days.[21]

Dentine contains ~70% mineral (wet weight). Carbonate-rich calcium phosphate (hydroxyapatite) crystals initially form

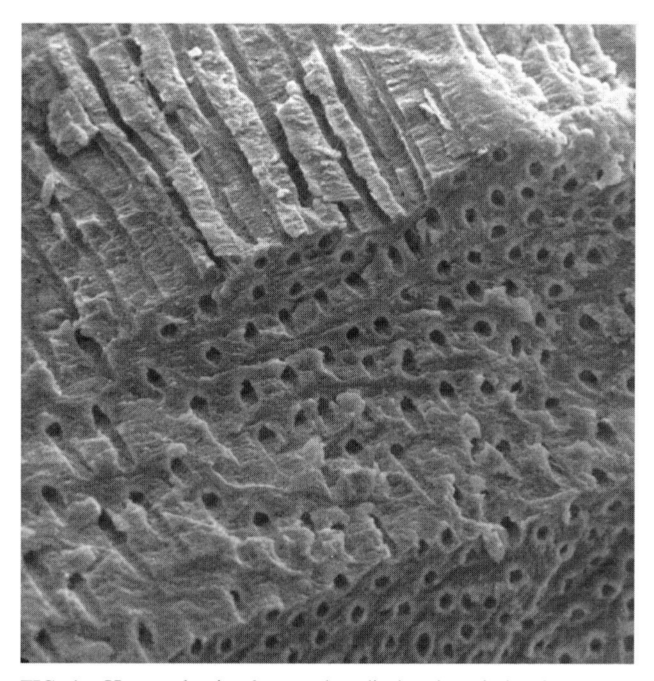

FIG. 4. Human dentine fractured to display the tubules that are ~2 μm across. SEM, field width 88 μm.

© 2008 American Society for Bone and Mineral Research

in relation to submicroscopic vesicles shed by the odontoblasts in the mantle layer or at sites on collagen fibrils rich in non-collagenous proteins. Mineralization extends radially from initial nucleation sites in the matrix, possibly by a process of secondary nucleation, forming regions of dentine known as calcospherites. These may fail to fuse, leaving unmineralized interglobular dentine between them. In a second, concurrent pattern of mineralization, crystals extend along the fine type I collagen fibrils that lie in a feltwork parallel to the incremental surface. Peritubular dentine is deposited within the tubules, partially or sometimes completely occluding them. It contains a negligible amount of collagen and mineralizes to a higher degree than the surrounding bulk intertubular dentine. Because it is harder and more wear-resistant than intertubular dentine, it stands proud on teeth worn through to dentine.

Like enamel, dentine is deposited rhythmically, leaving lines marking daily and approximately weekly increments.[22] Major life events, such as birth (the neonatal line) and illness, or dietary deficiencies are recorded as disturbances in the structure of the tissue forming at the time. Once eruption has occurred and root formation is complete, further dentine formation occurs as slowly deposited, regular secondary dentine or, irregularly, as a response of the pulp-dentine complex to attrition or disease. Nerves pass from the dental pulp between odontoblasts and extend into the dentine tubules for variable distances. Dentine is acquisitively painful if touched or subjected to large temperature or osmotic changes.

Like any other loose connective tissue, the dental pulp shows signs of aging, which may include diffuse or local calcifications and the formation of dental stones. In the roots of human teeth, occlusion of the tubules with peritubular dentine extends coronally from the root apex; the resulting transparent dentine can be used as a guide to the age of the tooth.

Cementum

Cementum is a calcified connective tissue that is deposited initially on the newly mineralized dentine matrix of the root by cells derived from the dental follicle.[10] Secretory proteins from the cells of the epithelial root sheath may be included in the first-formed matrix. Cementum is laid down centrifugally from the cement-dentine junction and is marked by incremental lines that are close together, continuous, and evenly spaced where apposition was slow and patchy and irregular otherwise. The tissue is similar to bundle (Sharpey fiber) bone in that it incorporates extrinsic collagen fibers formed by fibroblasts.[23] These fibers may be very closely packed, comprising the whole tissue in slowly forming acellular cementum (Fig. 5), or be separated from each other by intervening intrinsic collagen fibers, of cementoblast origin, which lie in the plane of the developing root surface (Fig. 3). Where cementum is deposited very rapidly, it is cellular, containing cementocytes that resemble the osteocytes of bundle bone (Fig. 6). In heavily remodeled root apices, there may be patches of cellular cementum without extrinsic fibers. Only in cementum-containing intrinsic fibers may a well-defined region of unmineralized precementum, equivalent to osteoid, be present at the surface of the tissue. The collagen of both the extrinsic and intrinsic fibers is type I. The main non-collagenous proteins of cementum identified so far (bone sialoprotein, osteopontin, osteocalcin, and α-2-HS-glycoprotein) vary in amount and distribution in the types of cementum and do not distinguish it from other calcified connective tissues[24,25] (Table 1).

Cementum mineralization reflects the rate of formation and the composition of the matrix. In afibrillar coronal cementum, the layer of noncollagenous proteins adsorbed on to the enamel surface mineralizes fully. At the cementum–dentine junction, collagen fibrils and non-collagenous constituents of the two tissues mingle without a regular, distinct border or osteopontin-rich hypermineralized cement line. The extrinsic fibers of slowly forming cementum mineralize completely, the advancing mineralized front across the fibers being relatively

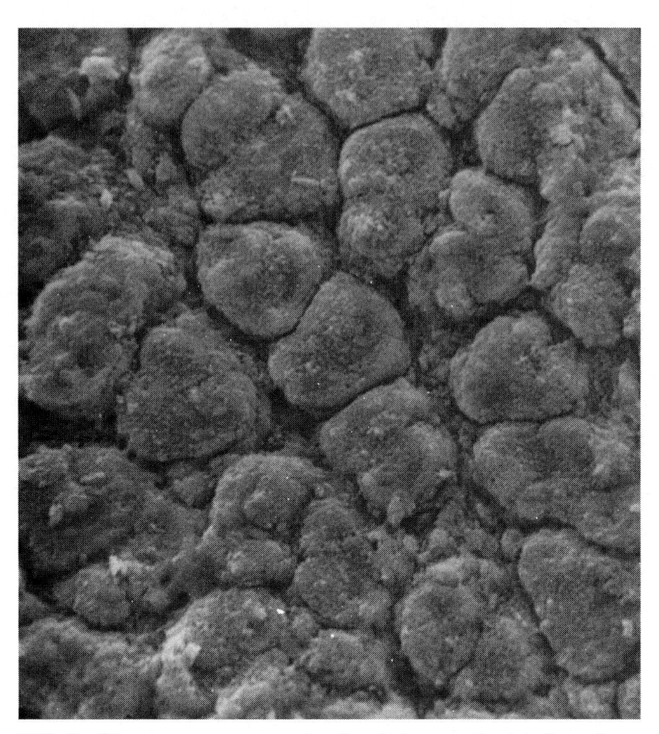

FIG. 5. Human cementum surface, made anorganic, showing mineralized ends of extrinsic fibers, ~6 μm diameter, separated by intrinsic fibers. SEM, field width 30 μm.

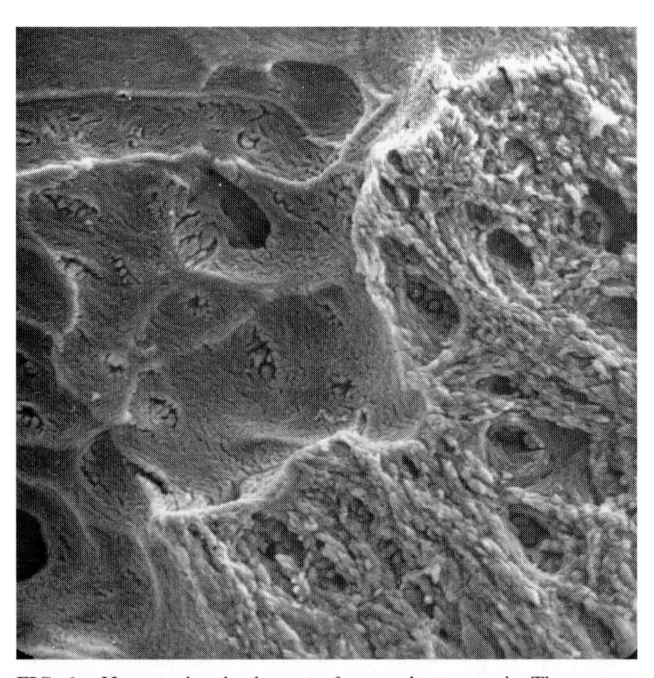

FIG. 6. Human alveolar bone surface made anorganic. The resorption lacunae reveal that the extrinsic (Sharpey's) fibers were only partly mineralized. The remainder of the surface was forming, as evidenced by incomplete mineralization of intrinsic and extrinsic fibers. SEM, field width 110 μm.

© 2008 American Society for Bone and Mineral Research

TABLE 1. MAIN CONSTITUENTS OF DENTAL TISSUES

	Enamel	*Dentine*	*Cementum*
Proteins	Amelogenin: major protein in immature enamel, secreted by ameloblasts, then degraded and removed. Non-amelogenin proteins (enamelins): includes Enamelin and Tuftelin; secreted by ameloblasts. Ameloblastin: also known as amelin or sheathlin, secreted by ameloblasts and odontoblasts	Collagen type 1 Dentine sialophosphoprotein (DSPP): in predentine, then degraded Dentine sialoprotein (DSP): in dentine, processed from DSPP Dentine phosphoprotein (DPP): in dentine, processed from DSPP Dentine matrix protein 1 (DMP1)	Collagen type 1 Bone sialoprotein Osteopontin Osteocalcin α2HS-glycoprotein
Proteoglycans		Decorin: chondroitin-sulphate-rich Biglycan: chondroitin-sulphate-rich Lumican: keratan-sulphate-rich Fibromodulin: keratan-sulphate-rich	Decorin: in cellular cementum only Biglycan: in cellular cementum only, in incremental lines Lumican: in cellular cementum only Versican: in cementocyte lacunae only
Proteinases	Enamelysin (MMP-20): metalloprotease in immature enamel, processes enamel proteins in secretory phase of amelogenesis Kallikrein 4 (KLK4): also known as Enamel matrix serine protease-1 (EMSP1); in immature enamel, also secreted by odontoblasts, clears enamel proteins during enamel maturation phase	Matrix metalloproteinases (MMP2): in predentine, degrade and process dentine proteins secreted by odontoblasts	

flat and defining a border between cementum and the dental sac or periodontal ligament. This type of cementum is more highly mineralized, more translucent, and paler than dentine. Where only a small proportion of intrinsic fibers exists in acellular cementum, the extrinsic fibers lead the mineralization front. As the rate of deposition of cementum increases and proportionately more intrinsic fibers are deposited, the likelihood that the extrinsic fibers will retain unmineralized cores increases. During periods of fast cellular cementogenesis, even the intrinsic fibers may retain unmineralized centers, and the mineralization front becomes irregular, with the extrinsic component lagging behind the intrinsic. This cementum type is the softest and least well mineralized of the calcified dental tissues. The mineralization front can be read to estimate the current rate of formation and the degree of mineralization of the fibers within the tissue indicates past rates.[23] The carbonate-rich apatite phase is similar to bone salt.

INTERRELATIONSHIPS OF TEETH, PERIODONTAL TISSUES, AND ALVEOLAR BONE

Teeth are a highly specialized part of an integrated functional unit (see Fig. 1), the primary (but not sole) purpose of which is the mastication of food. Unique among the human calcified tissues, enamel is destined to be exposed to an external environment. As the tooth erupts, the alveolar bone is resorbed to allow its passage, its root develops, and the crown pierces the oral mucosa that finally contributes to a tight ring seal of epithelial cells on the enamel close to the junction of crown and root. The complex molecular signaling cascades in the dental follicle controlling eruption and root growth are unclear.[11] At emergence, the root of the tooth is not yet fully formed, and the pulpal aspect of the root end (apex) resembles a large closing cone in bone. Root completion takes ~18 months more in the deciduous teeth and up to 3 years in the permanent teeth. During root development, the follicle becomes organized into the periodontal ligament that supports the tooth, provides nutrition and mechanosensation, and allows physiological tooth movement. Through the groups of fibers of the periodontal ligament, comprising types I and III collagen, functioning teeth are linked to each other, the gingiva, and the alveolar bone. On either side of the ligament, its principal fibers are incorporated within cementum and bundle bone: within the ligament, there is constant adaptive remodeling of the soft tissue.

Cementum in permanent teeth sees little remodeling, but the surface of alveolar bone (Fig. 6) is continually resorbing and forming to allow the tooth to move in response to eruption, growth drift, or changing functional forces. Resorption of deciduous tooth roots begins shortly after their completion, appearing first and most extensively on the aspect adjacent to the successional tooth. Interspersed between resorptive bursts are occasional short periods of repair by cemento(osteo)blasts. Odontoclasts—typical osteoclasts—resorb both cementum and dentine and, in deciduous molars, a small mount of enamel.

REFERENCES

1. Thesleff I 2000 Genetic basis of tooth development and dental defects. Acta Odontol Scand **58:**191–194.
2. Jernvall J, Thesleff I 2000 Reiterative signaling and patterning during mammalian tooth morphogenesis. Mech Dev **92:**19–29.
3. Miletich I, Sharpe PT 2004 Neural crest contribution to mammalian tooth formation. Birth Defects Res C Embryo Today **72:**200–212.
4. James CT, Ohazama A, Tucker AS, Sharpe PT 2002 Tooth development is independent of a Hox patterning programme. Dev Dyn **225:**332–335.
5. Thesleff I, Mikkola M 2002 The role of growth factors in tooth development. Int Rev Cytol **217:**93–135.
6. Mustonen T, Tummers M, Mikami T, Itoh N, Zhang N, Grindley T, Thesleff I 2002 Lunatic fringe, FGF, and BMP regulate the Notch pathway during epithelial morphogenesis of teeth. Dev Biol **248:**281–293.
7. Ohazama A, Sharpe PT 2004 TNF signalling in tooth development. Curr Opin Genet Dev **14:**513–519.
8. Smith AJ, Lesot H 2001 Induction and regulation of crown den-

© 2008 American Society for Bone and Mineral Research

tinogenesis: Embryonic events as a template for dental tissue repair? Crit Rev Oral Biol Med **12:**425–437.
9. Thesleff I, Åberg T 1997 Tooth morphogenesis and the differentiation of ameloblasts. In: Chadwick D, Cardew G (eds.) Dental Enamel. Wiley, Chichester, UK, pp. 1–17.
10. Diekwisch TG 2001 The developmental biology of cementum. Int J Dev Biol **45:**695–706.
11. Wise GE, Frazier-Bowers S, D'Souza RN 2002 Cellular, molecular, and genetic determinants of tooth eruption. Crit Rev Oral Biol Med **13:**323–334.
12. Elliott JC 1997 Structure, crystal chemistry and density of enamel apatites. In: Chadwick D, Cardew G (eds.) Dental Enamel. Wiley, Chichester, UK, pp. 54–67.
13. Elliott JC, Wong FS, Anderson P, Davis GR, Dowker SE 1998 Determination of mineral concentration in dental enamel from X-ray attenuation measurements. Connect Tiss Res **38:**61–79.
14. Diekwisch TGH, Berman BJ, Anderton X, Gurinsky B, Ortega AJ, Satchell PG, Williams M, Arumugham C, Luan X, McIntosh JE, Yamane A, Carlson DS, Sire J-Y, Shuler CF 2002 Membranes, minerals, and proteins of developing vertebrate enamel. Microsc Res Tech **59:**373–395.
15. Simmer JP, Hu JC 2002 Expression, structure, and function of enamel proteinases. Connect Tiss Res **43:**441–449.
16. Boyde A 1997 Microstructure of enamel. In: Chadwick D, Cardew G (eds.) Dental Enamel. Wiley, Chichester, UK. pp. 18–31.
17. Boyde A 1989 Enamel. In: Oksche A, Vollrath L. (eds.) Handbook of Microscopic Anatomy. Springer Verlag, Berlin, Germany, pp. 309–473.
18. Sasaki T, Garant PR 1996 Structure and organization of odontoblasts. Anat Rec **245:**235–249.
19. Butler WT, Brunn JC, Qin C, McKee MD 2002 Extracellular matrix proteins and the dynamics of dentin formation. Connect Tiss Res **43:**301–307.
20. Emberry G, Hall R, Waddington R, Septier D, Goldberg M 2001 Proteoglycans in dentinogenesis. Crit Rev Oral Biol Med **12:**331–349.
21. Linde A. Goldberg M 1993 Dentinogenesis. Crit Rev Oral Biol Med **4:**679–728.
22. Dean MC, Scandrett AE 1996 The relation between long-period incremental markings in dentine and daily cross-striations in enamel in human teeth. Arch Oral Biol **41:**233–241.
23. Jones SJ 1989 Cement. In: Osborn JW (ed.) Dental Anatomy and Embryology. Blackwell Scientific, Boston, MA, USA, pp. 193–205 and 286–294.
24. McKee MD, Zalzal S, Nanci A 1996 Extracellular matrix in tooth cementum and mantle dentin: Localization of osteopontin and other noncollagenous proteins, plasma proteins and glycoconjugates by electron microscopy. Anat Rec **245:**293–312.
25. Sasano Y, Maruya Y, Sato H, Zhu JX, Takahashi I, Mizoguchi I, Kagayama M 2001 Distinctive expression of extracellular matrix molecules at mRNA and protein levels during formation of cellular and acellular cementum in the rat. Histochem J **33:**91–99.

Chapter 105. Craniofacial Disorders Affecting the Dentition: Genetic

Yong-Hee Chun,[1] Paul H. Krebsbach,[2] and James P. Simmer[1]

[1]Department of Biological and Materials Sciences, University of Michigan Dental Research Laboratory, Ann Arbor, Michigan;
[2]Department of Biologic and Materials Sciences, School of Dentistry, University of Michigan, Ann Arbor, Michigan

GENETIC DISORDERS AFFECTING THE DENTITION

The skeleton contains two of the five mineralized tissues in the body: bone and calcified cartilage. The other three mineralized tissues, dentin, enamel, and cementum, are found in teeth. The mineral in each of these hard tissues is a biological apatite resembling calcium hydroxyapatite $Ca_{10}(PO_4)_6(OH)_2$ in structure, with the most common substitutions being carbonate (CO_3^{2-}) for phosphate (PO_4^{3-}) and fluoride (F^-) for hydroxyl (OH^-). Therefore, disorders involving the regulation of calcium and phosphate metabolism potentially affect multiple hard tissues. In every mineralizing tissue, biomineralization occurs in a defined extracellular space. Establishing these extracellular mineralizing environments involves the synthesis and secretion of extracellular matrix proteins, the transport of ions, and matrix turnover. Although each mineralizing tissue is in many ways unique, there are common elements that, when defective, lead to pathologies in multiple hard tissues. Changes in the dentition and its supporting oral structures may occur in response to disorders of mineral metabolism. The clinical presentations in these disorders may vary from mild asymptomatic changes to alterations that severely alter the form and function of craniofacial structures. In some cases, the oral phenotype may be the earliest or most obvious sign of a broader syndrome involving bone and mineral metabolism and lead to the original diagnosis. This chapter provides a concise overview of the dental manifestations of selected disorders of bone and mineral metabolism.

Genetic Diseases Affecting the Number of Teeth: Familial Tooth Agenesis and Supernumerary Teeth

The initiation of tooth development depends on interactions between oral epithelium and mesenchyme derived from cranial neural crest cells.[1] The initiation step in tooth development is formation of a dental placode, or local thickening of the oral epithelium. Similar ectodermal placodes initiate the development of hairs and nails, as well as the mammary, salivary, sweat, and sebaceous glands.[2] The principle molecular participants in the epithelial–mesenchymal interactions that initiate formation of ectodermal organs are signaling molecules, their receptors, and transcription factors. Some of the specific regulatory molecules involved contribute to the formation of multiple ectodermal organs. Genetic diseases affecting early events in tooth formation typically are manifested as misshapen teeth and by alterations in tooth number, as in supernumerary teeth or familial tooth agenesis. These dental

Key words: amelogenesis imperfecta, dentinogenesis imperfecta, dentin dysplasia, cleft lip and palate, familial tooth agenesis, supernumerary teeth, cherubism, mucopolysaccharidoses, vitamin D, hypophosphatemia
Electronic Databases: 1. Online Mendelian Inheritance in Man (OMIM)—http://www.ncbi.nlm.nih.gov/Omim/.

The authors state that they have no conflicts of interest.

© 2008 American Society for Bone and Mineral Research

phenotypes may occur in isolation or be associated with other developmental anomalies.

Hypohidrotic ectodermal dysplasia (HED) is an inherited condition featuring missing teeth, thin and sparse hair, missing sweat glands, and defective nails and salivary glands. HED is caused by mutations in ectodysplasin (*EDA*, Xq12-q13.1). Several related syndromes are caused by mutations in *FGF10* (5p13) and its receptors (i.e., FGFR3, 4p16.3), such as aplasia of the lacrimal and salivary glands (ALSG)[3] and lacrimo-auriculo-dento-digital syndrome (LADD).[4] LADD features aplasia or hypoplasia of the lacrimal and salivary systems associated with a variety of dental phenotypes: agenesis of the lateral maxillary incisors (hypodontia), or small (microdontia) and peg-shaped laterals, mild enamel dysplasia, and delayed tooth eruption.

Especially noteworthy among the genetic disturbances affecting early tooth formation are those that disrupt the Wnt signaling system. Wnt signaling is processed by an cytosolic complex of proteins consisting of axis inhibition protein 2 (Axin2), glycogen synthase kinase-3β (GSK-3β), and adenomatous polyposis coli (APC), which dephosphorylate and stabilize β-catenin. β-catenin translocates into the nucleus and regulates the activity of key transcription factors. Mutations in *APC* (5q21-q22) cause adenomatous polyposis of the colon, a syndrome featuring radiopaque lesions in the jaw comprised of clumped toothlets (odontomas) and gastrointestinal polyps that usually undergo malignant change by the fourth decade.[5] *AXIN2* (17q24) mutations cause a severe form of familial tooth agenesis also associated with gastrointestinal polyps that turn malignant by the fourth decade.[6] *APC* and *AXIN2* phenotypes are inherited in an autosomal dominant pattern. The discovery of odontomas or familial tooth agenesis on Panorex radiographs should raise concerns about gastrointestinal polyps, especially when the oral phenotype seems to have arisen spontaneously (not observed in either of the parents), because the family will not show a history of intestinal cancer.

Familial tooth agenesis occurs in the absence of other phenotypic features. Msh homeobox 1 (*MSX1*, 4p16.1) and paired box 9 (*PAX9*, 14q12-q13) express interacting transcription factors that are critical for the progression of tooth development beyond the bud stage.[7,8] Mutations in *MSX1* and *PAX9* cause similar patterns of familial tooth agenesis, with *PAX9* mutations being more likely to include second molars, whereas *MSX1* mutations are more likely to include maxillary first bicuspids among the teeth missing.[9]

Cleidocranial dysplasia is an autosomal dominant condition featuring skeletal and dental anomalies caused by mutations in the transcription factor gene *RUNX2* (6p21). The most notable feature is hypoplasia or aplasia of the clavicles permitting abnormal apposing of the shoulders. Exfoliation of the primary dentition is delayed or fails, and multiple supernumerary teeth, which typically fail to erupt, are observed on dental radiographs.[10]

Genetic Diseases Affecting Dentin: Osteogenesis Imperfecta, Dentinogenesis Imperfecta, and Dentin Dysplasia

Defects in either the α1 chains or α2 chain of type I collagen can cause osteogenesis imperfecta (OI). OI is associated with assorted dentin defects that are collectively designated as dentinogenesis imperfecta (DGI).[11] In rare cases, the dentin defects are the only prominent phenotype.[12] It has been reported that 10–50% of patients afflicted with OI also have DGI. This assessment, however, may underestimate the true prevalence, because mild forms of DGI may require microscopic analysis for diagnosis.[13] Dental abnormalities have been described in several subtypes of OI, but are most prevalent in OI types IB, IC, and IVB.[14] The range of dental defects observed in osteogenesis imperfecta are similar to those observed in kindreds with DGI and dentin dysplasia (DD).

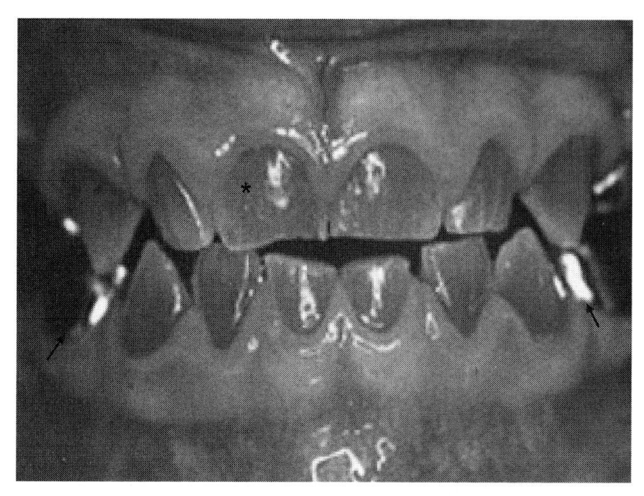

FIG. 1. Dental manifestations of teeth from a patient with dentinogenesis imperfecta. The permanent teeth of this patient exhibit the characteristic blue-gray or opalescent appearance associated with dentinogenesis imperfecta (asterisks). The enamel of the posterior teeth has fractured and the underlying dentin has undergone severe attrition. Crowns have been made to control further destruction (arrows).

The most abundant noncollagenous proteins in dentin are proteolytic cleavage products of a large chimeric protein known as dentin sialophosphoprotein (DSPP). In the past few years, nine different *DSPP* (4q21.3) mutations have been linked to inherited dentin defects in kindreds with DD type II, DGI type II, and DGI type III phenotypes.[15–21] No mutations in candidate genes encoding other extracellular matrix proteins, such as osteopontin, bone sialoprotein, and dentin matrix protein-1, have been identified. Like type I collagen, the *DSPP* gene is expressed in bone as well as dentin. Despite this, no bony defects have been reported in any of the kindreds with dentin defects linked to *DSPP* mutations.

The clinical classification system most often used to categorize inherited defects of dentin was established >30 yr ago, before recent discoveries concerning their genetic etiologies divided the phenotypes into two disease groups with five subtypes: DGI (types I–III) and DD (types I and II), with all forms showing an autosomal dominant pattern of inheritance.[22] Type I DGI is a collective designation for OI with DGI and has largely been abandoned in deference to the current OI classification system. An alternative designation for isolated inherited dentin defects is hereditary opalescent dentin.[23] Type II DGI is the most prevalent inherited dentin phenotype. Clinically, the teeth of individuals with DGI are characterized by an amber-like appearance (Fig. 1). The teeth are narrower at the cervical margins and thus exhibit a bulbous or bell-shaped crown. Microscopic anomalies of affected dentin include fewer and irregular dentin tubules containing vesicles and abnormally thick collagen fibers.[24] The mineral content of DGI teeth is reduced, being ~30% less than normal dentin and intrafibrillar collagen mineralization is absent. The structurally abnormal dentin may not provide adequate support for the overlying enamel. Although enamel is chemically and structurally normal in individuals with DGI, the lack of support from dentin leads to fracturing and severe attrition of the teeth, the distinguishing clinical feature of type II DGI. Type III DGI is a rare form that is also known as the "Brandywine isolate," after the prototype kindred identified in Bran-

© 2008 American Society for Bone and Mineral Research

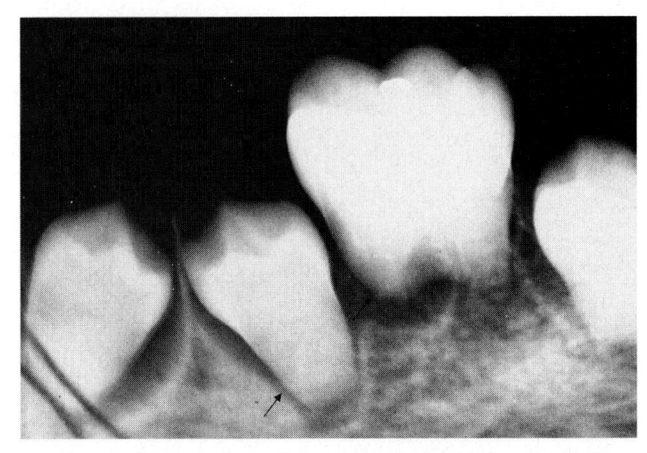

FIG. 2. Radiograph of teeth from a patient with dentin dysplasia. The roots are abnormally short or absent (arrows), and the pulp chamber is obliterated (courtesy of Dr. Sharon Brooks).

dywine, MD. This form features multiple pulp exposures in the deciduous teeth, which show considerable variation radiographically, ranging from shell teeth, to normal pulp chambers, to pulpal obliteration. The permanent teeth are the same as in type II DGI. In type I DD, both the permanent and deciduous teeth appear to have a normal shape and color clinically. Dental radiographs, however, show the teeth have short roots with periapical radiolucencies in noncarious teeth (Fig. 2). The primary teeth show total obliteration of the pulp. Type II DD seems to be a mild form of type II DGI, featuring amber tooth coloration with total pulpal obliteration in the primary teeth and a thistle-tube pulp configuration with ubiquitous pulp stones and normal to near normal coloration in the permanent teeth.[22,25,26]

Genetic Diseases Affecting Enamel: Amelogenesis Imperfecta

Amelogenesis imperfecta (AI) is a heterogeneous group of isolated inherited defects in the enamel layer of teeth. The enamel may be thin, soft, rough, and/or pigmented. When the various enamel phenotypes and the pattern of inheritance are considered, 14 subtypes are recognized.[27] Enamel formation is regulated by a small toolbox of specialized extracellular matrix molecules. Amelogenin (*AMELX*, Xp22.3) mutations cause X-linked amelogenesis imperfecta, with females showing vertical bands of normal and defective enamel, whereas males usually have little or no enamel at all. Enamelin (*ENAM*, 4q13.3) mutations cause autosomal dominant amelogenesis imperfecta, with distinctive horizontal bands often evident in the cervical third of the crown. The proteolytic enzymes enamelysin (*MMP20*, 11q22.3) and kallikrein 4 (*KLK4*, 19q13.41) cause autosomal recessive amelogenesis imperfecta, featuring relatively soft, pigmented enamel, usually of normal thickness, that may tend to chip off during function. Mutations in these genes only account for ~25% of all cases.[28] AI has also been used to describe enamel defects associated with inherited syndromes. There are >70 such conditions.

Mucopolysaccharidoses

Lysosomal storage disorders are made up of >40 inherited diseases that are caused primarily by defects in genes encoding lysosomal enzymes.[29,30] Among the lysosomal storage disorders are the mucopolysaccharidoses (MPSs), which are characterized by the accumulation of partially degraded glycosaminoglycans (previously called mucopolysaccharides) within lysosomes, as well as in the urine. There are 10 enzymes involved in the stepwise degradation of glycosaminoglycans, and deficiencies in these activities give rise to the MPSs.[31] There are seven MPS types: I, II, III, IV, VI, VII, and IX (types V and VIII have been retired, and type IX is extremely rare). The MPSs are distinguished from each other based on genetic, biochemical, and clinical analyses.[14] Although heterogeneous, several craniofacial characteristics are similar between the different types. The oral manifestations may include a short and broad mandible with abnormal condylar development and limited temporomandibular joint function. The teeth are often peg-shaped and exhibit increased interdental spacing, perhaps because of the frequently observed gingival hyperplasia and macroglossia. Some forms of MPS have abnormally thin enamel covering the clinical crowns or radiographic evidence of cystic lesions surrounding the molar teeth that contain excessive dermatan sulfate and collagen.[32–35]

MPS type IVA (Morquio A syndrome, MPS IVA) is an autosomal recessive disorder caused by deficiency of the lysosomal hydrolase, *N*-acetylgalactosamine 6-sulfatase (GALNS), encoded by a gene on human chromosome 16q24.3.[36,37] MPS type IVA is the only MPS associated with dental enamel malformations,[38] although mucopolysaccharides accumulate in the developing teeth in other MPS syndromes, such as Hurler (MPS I),[39] Hunter (MPS II), and Maroteaux-Lamy (MPS VI) syndromes. In MPS IVA, enamel malformations are a consistent feature. The enamel is dull gray in color, thin, pitted, and tends to flake off from the underlying dentin. The thin enamel layer is of normal hardness and radiodensity. MPS IVA patients often show severe bone dysplasia and dwarfism.

Cleft Lip and Palate

Cleft lip and/or palate are relatively common craniofacial malformations (1 in 700 births) that can have profound impact on nutritional, speech, dental, and psychological development. The causes of these disorders are complex and are known to involve both genetic and environmental factors. Most of these facial clefting birth defects are multifactorial and nonsyndromic. Although genetics may play a role in the cause of cleft lip and/or palate, they are not associated with well-defined syndromes. Between 15% and 50% of cleft lip and/or palate are associated with defined syndromes. In fact, there are nearly 300 recognized syndromes that may include a facial cleft as a manifestation. However, only ~10 genes have been identified as associated with these syndromes. Common syndromes with cleft palates include Apert's, Stickler's and Treacher Collins. Van der Woude's and Waardenberg's syndromes are associated with cleft lip with or without cleft palate. In at least one condition, Van der Woude's syndrome, a haplotype gene test may be used to identify variants in the *interferon regulatory factor 6* gene and provide a correlation with an increased risk of facial clefting.[40]

Cherubism

Cherubism is an autosomal dominant condition caused by mutations in *SH3BP2* (4p16.3).[41] *SH3BP2* expresses a protein that binds to c-Abl, a tyrosine kinase involved in diverse cell signaling cascades. In patients with cherubism, multiple cystic giant cell lesions of the jaw appear typically between 2 and 5 yr of age. The cysts replace bone and cause enlargement of the maxilla and mandible, which stabilizes or remits after puberty. Teeth are displaced during development, with root resorption, tooth agenesis, retention of deciduous teeth, ectopic eruption, and malocclusion observed as sequela.[42]

© 2008 American Society for Bone and Mineral Research

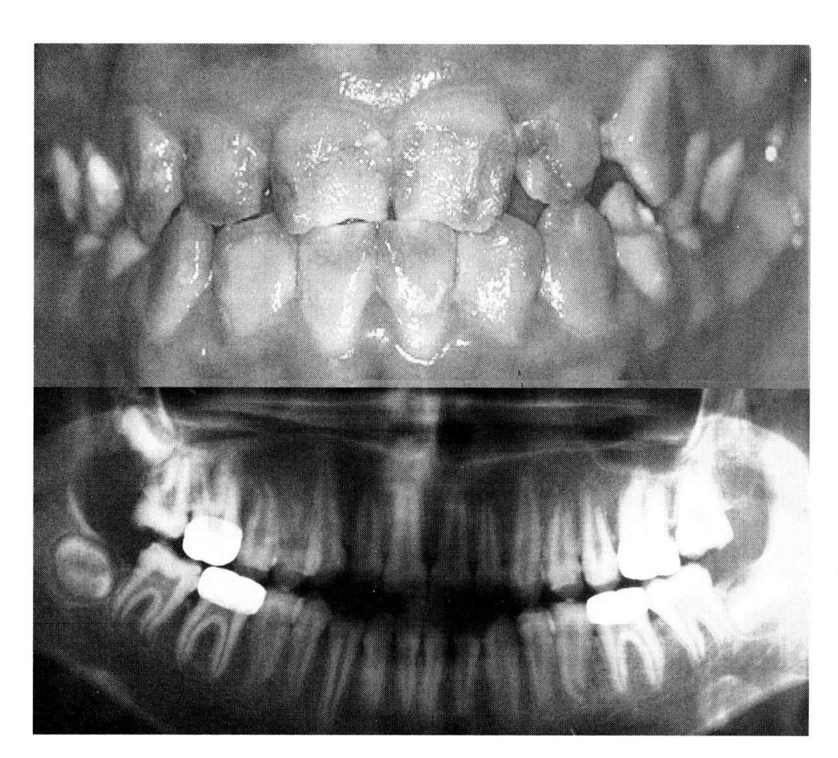

FIG. 3. Dental manifestations of vitamin D–dependent rickets type I. The oral photograph shows the dental presentation of a Hispanic patient having short stature for her age.[10] Both the enamel and the dentin are affected throughout the dentition. The oral radiograph (panorex) shows that the teeth generally had short roots with large pulp chambers. The radiodensity of the enamel crown is similar to that of the underlying dentin. Chronic periodontal disease is often associated with this condition (contributed by Dr. Jan C.-C. Hu).

ORAL MANIFESTATIONS OF METABOLIC BONE DISEASES OF GENETIC ORIGIN

Metabolic diseases of bone are disorders of bone remodeling that characteristically involve the entire skeleton and are often manifest in the oral cavity, which can lead to the diagnosis of the underlying systemic disease. Numerous studies suggest that subclinical derangements in calcium homeostasis and bone metabolism may also contribute to a variety of dental abnormalities including alveolar ridge resorption and periodontal bone loss in predisposed individuals. The significance of this spectrum of diseases and their overall impact on oral health and dental management are likely to increase as the elderly segment of the population increases in the coming decades.[43]

Vitamin D Deficiency

In vitamin D–resistant rickets, the primary oral abnormality is similar to dentin dysplasia. Enamel is usually reported to be normal, but in some instances, may be hypoplastic. Patients also suffer from delayed tooth eruption, and radiographically, teeth often display enlarged pulp chambers. Other salient radiographic findings include decreased alveolar bone density, thinning of bone trabeculae, loss of lamina dura, and retarded tooth calcification.[44] In familial hypophosphatemia, dental findings are often the first clinically noticeable signs of the disease and resemble those seen in rickets and osteomalacia. Patients may present with abscessed primary or permanent teeth that have no signs of dental caries.[45] Although the enamel is reported to be normal, microbial infection of the pulp is thought to occur though invasion of dentinal tubules exposed by attrition of enamel or through enamel microfractures.[46]

Vitamin D–dependent rickets type I is an autosomal recessive defect in vitamin D metabolism caused by mutations in the *CYP27B1* gene (12q13.3-q14) encoding 25-hydroxyvitamin D-1α-hydroxylase. Decreased $1,25(OH)_2$ vitamin D result in teeth with yellow-brown color, pitted enamel, short roots, and a tendency to develop chronic periodontal disease (Fig. 3).[47]

Hypophosphatemia

In familial hypophosphatemia, dental findings are often the first clinically noticeable signs of the disease and resemble those seen in rickets and osteomalacia. Patients may present with abscessed primary or permanent teeth that have no signs of dental caries.[45] Although the enamel is reported to be normal, microbial infection of the pulp is thought to occur though invasion of dentinal tubules exposed by attrition of enamel or through excessively extended pulp horns and enamel microfractures.[46]

Hypophosphatasia

Hypophosphatasia is an inherited disorder caused by a defect in the alkaline phosphatase (*ALPL,* 1p36.1-p34) gene. Osteoblasts show the highest level of ALPL expression, and profound skeletal hypomineralization occurs in the severest forms of hypophosphatasia. The dental hard tissue that appears to be most sensitive to an *ALPL* defect is cementum.[48] The classic oral presentation of childhood hypophosphatasia is the premature loss of fully rooted primary teeth. Histological examination indicates that these teeth lack cementum on their root surface, so that the attachment apparatus fails to develop properly. The periodontal ligament fibers do not connect the alveolar bone to the root, and the teeth exfoliate prematurely. In the permanent teeth, large pulp spaces, late eruption, and delayed apical closure are often observed. Bone loss is primarily horizontal, and in the adult form of the disease, there may be widespread dental caries.

REFERENCES

1. Chai Y, Jiang X, Ito Y, Bringas P Jr, Han J, Rowitch DH, Soriano P, McMahon AP, Sucov HM 2000 Fate of the mammalian cranial neural crest during tooth and mandibular morphogenesis. Development **127:**1671–1679.
2. Thesleff I 2006 The genetic basis of tooth development and dental defects. Am J Med Genet A **140:**2530–2535.
3. Entesarian M, Matsson H, Klar J, Bergendal B, Olson L, Arakaki R, Hayashi Y, Ohuchi H, Falahat B, Bolstad AI, Jonsson R,

© 2008 American Society for Bone and Mineral Research

Wahren-Herlenius M, Dahl N 2005 Mutations in the gene encoding fibroblast growth factor 10 are associated with aplasia of lacrimal and salivary glands. Nat Genet **37:**125–127.

4. Milunsky JM, Zhao G, Maher TA, Colby R, Everman DB 2006 LADD syndrome is caused by FGF10 mutations. Clin Genet **69:**349–354.
5. Oner AY, Pocan S 2006 Gardner's syndrome: A case report. Br Dent J **200:**666–667.
6. Lammi L, Arte S, Somer M, Jarvinen H, Lahermo P, Thesleff I, Pirinen S, Nieminen P 2004 Mutations in AXIN2 cause familial tooth agenesis and predispose to colorectal cancer. Am J Hum Genet **74:**1043–1050.
7. Stockton DW, Das P, Goldenberg M, D'Souza RN, Patel PI 2000 Mutation of PAX9 is associated with oligodontia. Nat Genet **24:**18–19.
8. Vastardis H, Karimbux N, Guthua SW, Seidman JG, Seidman CE 1996 A human MSX1 homeodomain missense mutation causes selective tooth agenesis. Nat Genet **13:**417–421.
9. Kim JW, Simmer JP, Lin BP, Hu JC 2006 Novel MSX1 frameshift causes autosomal-dominant oligodontia. J Dent Res **85:**267–271.
10. Cooper SC, Flaitz CM, Johnston DA, Lee B, Hecht JT 2001 A natural history of cleidocranial dysplasia. Am J Med Genet **104:** 1–6.
11. O'Connell AC, Marini JC 1999 Evaluation of oral problems in an osteogenesis imperfecta population. Oral Surg Oral Med Oral Pathol Oral Radiol Endod **87:**189–196.
12. Pallos D, Hart PS, Cortelli JR, Vian S, Wright JT, Korkko J, Brunoni D, Hart TC 2001 Novel COL1A1 mutation (G559C) [correction of G599C] associated with mild osteogenesis imperfecta and dentinogenesis imperfecta. Arch Oral Biol **46:**459–470.
13. Waltimo J, Ojanotko-Harri A, Lukinmaa PL 1996 Mild forms of dentinogenesis imperfecta in association with osteogenesis imperfecta as characterized by light and transmission electron microscopy. J Oral Pathol Med **25:**256–264.
14. Gorlin RJ, Cohen MMJ, Levin LS 1990 Syndromes of the head and neck. In: Motulsky AG, Harper PS, Bobrow M, Scriver C (eds.) Oxford Monographs on Medical Genetics, 3rd ed. Oxford University Press, New York, NY, USA.
15. Dong J, Gu T, Jeffords L, MacDougall M 2005 Dentin phosphoprotein compound mutation in dentin sialophosphoprotein causes dentinogenesis imperfecta type III. Am J Med Genet A **132:**305–309.
16. Kim JW, Hu JC, Lee JI, Moon SK, Kim YJ, Jang KT, Lee SH, Kim CC, Hahn SH, Simmer JP 2005 Mutational hot spot in the DSPP gene causing dentinogenesis imperfecta type II. Hum Genet **116:**186–191.
17. Kim JW, Nam SH, Jang KT, Lee SH, Kim CC, Hahn SH, Hu JC, Simmer JP 2004 A novel splice acceptor mutation in the DSPP gene causing dentinogenesis imperfecta type II. Hum Genet **115:**248–254.
18. Malmgren B, Lindskog S, Elgadi A, Norgren S 2004 Clinical, histopathologic, and genetic investigation in two large families with dentinogenesis imperfecta type II. Hum Genet **114:**491–498.
19. Rajpar MH, Koch MJ, Davies RM, Mellody KT, Kielty CM, Dixon MJ 2002 Mutation of the signal peptide region of the bicistronic gene DSPP affects translocation to the endoplasmic reticulum and results in defective dentine biomineralization. Hum Mol Genet **11:**2559–2565.
20. Xiao S, Yu C, Chou X, Yuan W, Wang Y, Bu L, Fu G, Qian M, Yang J, Shi Y, Hu L, Han B, Wang Z, Huang W, Liu J, Chen Z, Zhao G, Kong X 2001 Dentinogenesis imperfecta 1 with or without progressive hearing loss is associated with distinct mutations in DSPP. Nat Genet **27:**201–204.
21. Zhang X, Zhao J, Li C, Gao S, Qiu C, Liu P, Wu G, Qiang B, Lo WH, Shen Y 2001 DSPP mutation in dentinogenesis imperfecta Shields type II. Nat Genet **27:**151–152.
22. Shields ED, Bixler D, el-Kafrawy AM 1973 A proposed classification for heritable human dentine defects with a description of a new entity. Arch Oral Biol **18:**543–553.
23. Witkop CJ Jr 1971 Manifestations of genetic diseases in the human pulp. Oral Surg Oral Med Oral Pathol **32:**278–316.
24. Waltimo J 1994 Hyperfibers and vesicles in dentin matrix in dentinogenesis imperfecta (DI) associated with osteogenesis imperfecta (OI). J Oral Pathol Med **23:**389–393.
25. Giansanti JS, Allen JD 1974 Dentin dysplasia, type II, or dentin dysplasia, coronal type. Oral Surg Oral Med Oral Pathol **38:**911–917.
26. Lukinmaa PL, Ranta H, Ranta K, Kaitila I, Hietanen J 1987 Dental findings in osteogenesis imperfecta: II. Dysplastic and other developmental defects. J Craniofac Genet Dev Biol **7:**127–135.
27. Witkop CJ Jr 1988 Amelogenesis imperfecta, dentinogenesis imperfecta and dentin dysplasia revisited: Problems in classification. J Oral Pathol **17:**547–553.
28. Kim JW, Simmer JP, Lin BP, Seymen F, Bartlett JD, Hu JC 2006 Mutational analysis of candidate genes in 24 amelogenesis imperfecta families. Eur J Oral Sci **114**(Suppl 1)**:**3–12.
29. Meikle PJ, Hopwood JJ, Clague AE, Carey WF 1999 Prevalence of lysosomal storage disorders. JAMA **281:**249–254.
30. Winchester B, Vellodi A, Young E 2000 The molecular basis of lysosomal storage diseases and their treatment. Biochem Soc Trans **28:**150–154.
31. Neufeld E, Muenzer J 1995 The mucopolysaccharidoses. In: Scriver C, Beaudet A, Sly W, Valle D (eds.) The Metabolic and Molecular Bases of Inherited Disease. McGraw-Hill, New York, NY, USA, pp. 2465–2494.
32. Downs AT, Crisp T, Ferretti G 1995 Hunter's syndrome and oral manifestations: A review. Pediatr Dent **17:**98–100.
33. Keith O, Scully C, Weidmann GM 1990 Orofacial features of Scheie (Hurler-Scheie) syndrome (alpha-L- iduronidase deficiency). Oral Surg Oral Med Oral Pathol **70:**70–74.
34. Kinirons MJ, Nelson J 1990 Dental findings in mucopolysaccharidosis type IV A (Morquio's disease type A). Oral Surg Oral Med Oral Pathol **70:**176–179.
35. Smith KS, Hallett KB, Hall RK, Wardrop RW, Firth N 1995 Mucopolysaccharidosis: MPS VI and associated delayed tooth eruption. Int J Oral Maxillofac Surg **24:**176–180.
36. Baker E, Guo XH, Orsborn AM, Sutherland GR, Callen DF, Hopwood JJ, Morris CP 1993 The morquio A syndrome (mucopolysaccharidosis IVA) gene maps to 16q24.3. Am J Hum Genet **52:**96–98.
37. Masuno M, Tomatsu S, Nakashima Y, Hori T, Fukuda S, Masue M, Sukegawa K, Orii T 1993 Mucopolysaccharidosis IV A: Assignment of the human N-acetylgalactosamine-6-sulfate sulfatase (GALNS) gene to chromosome 16q24. Genomics **16:**777–778.
38. Witkop C, Sauk JJ 1976 Heretable defects in enamel. In: Steward RE, Presscott GH (eds.) Oral Facial Genetics. Mosby, St. Louis, MO, USA, pp. 151–226.
39. Gardner DG 1971 The oral manifestations of Hurler's syndrome. Oral Surg Oral Med Oral Pathol **32:**46–57.
40. Zucchero TM, Cooper ME, Maher BS, Daack-Hirsch S, Nepomuceno B, Ribeiro L, Caprau D, Christensen K, Suzuki Y, Machida J, Natsume N, Yoshiura K, Vieira AR, Orioli IM, Castilla EE, Moreno L, Arcos-Burgos M, Lidral AC, Field LL, Liu YE, Ray A, Goldstein TH, Schultz RE, Shi M, Johnson MK, Kondo S, Schutte BC, Marazita ML, Murray JC 2004 Interferon regulatory factor 6 (IRF6) gene variants and the risk of isolated cleft lip or palate. N Engl J Med **351:**769–780.
41. Lo B, Faiyaz-Ul-Haque M, Kennedy S, Aviv R, Tsui LC, Teebi AS 2003 Novel mutation in the gene encoding c-Abl-binding protein SH3BP2 causes cherubism. Am J Med Genet A **121:**37–40.
42. Pontes FS, Ferreira AC, Kato AM, Pontes HA, Almeida DS, Rodini CO, Pinto DS Jr 2007 Aggressive case of cherubism: 17-year follow-up. Int J Pediatr Otorhinolaryngol **71:**831–835.
43. Solt DB 1991 The pathogenesis, oral manifestations, and implications for dentistry of metabolic bone disease. Curr Opin Dent **1:**783–791.
44. Neville BW, Damm DD, Allen CM, Bouquot JE 1995 Oral and Maxillofacial Pathology, 1st ed. Saunders, New York, NY, USA.
45. Goodman JR, Gelbier MJ, Bennett JH, Winter GB 1998 Dental problems associated with hypophosphataemic vitamin D resistant rickets. Int J Paediatr Dent **8:**19–28.
46. Hillmann G, Geurtsen W 1996 Pathohistology of undecalcified primary teeth in vitamin D-resistant rickets: Review and report of two cases. Oral Surg Oral Med Oral Pathol Oral Radiol Endod **82:**218–224.
47. Zambrano M, Nikitakis NG, Sanchez-Quevedo MC, Sauk JJ, Sedano H, Rivera H 2003 Oral and dental manifestations of vitamin D-dependent rickets type I: Report of a pediatric case. Oral Surg Oral Med Oral Pathol Oral Radiol Endod **95:**705–709.
48. Chapple IL 1993 Hypophosphatasia: Dental aspects and mode of inheritance. J Clin Periodontol **20:**615–622.

© 2008 American Society for Bone and Mineral Research

Chapter 106. Pathology of the Hard Tissues of the Jaws

Paul C. Edwards

Department of Periodontics and Oral Medicine, School of Dentistry, University of Michigan, Ann Arbor, Michigan

INTRODUCTION

The mandible and maxilla are unique among the bones in that they contain an overlying osseous structure, the alveolar process, that functions to support 32 highly specialized hard tissue organs, the teeth. The teeth, composed of unique hard tissues that are not found anywhere else in the body (enamel, dentin, and cementum), develop through a process involving sequential and reciprocal interactions between oral epithelium and ectomesenchyme. By necessity, because growth of the jaws requires two sets of dentition, this process occurs over a period extending from the fetal period to the late teens.

As a result of two phenomena, namely the presence of specialized mineralized tissues that are exposed to the harsh oral environment and the unique developmental processes involved in the formation of teeth, the mandible and maxilla are home to a distinctive set of pathologic entities. This chapter provides a concise review of the more common and interesting of these entities: tooth demineralization/dental caries and odontogenic cysts/tumors.

TOOTH DEMINERALIZATION AND CARIES

Dental caries (tooth decay) is one of the most prevalent chronic diseases affecting modern society. Once viewed as a disease primarily of children, it has become evident that as a result of the reduction in tooth loss in the adult population over the last 30 yr, adults today are equally as likely to develop new carious lesions.[1]

The caries process represents the end result of a complex interaction between transmissible cariogenic oral microflora, primarily *Streptococcus mutans* and *Lactobacillus* spp., and fermentable dietary carbohydrates.[2] Oral microflora, vertically transmitted to the child through the mother, colonize the teeth through a process involving adhesion of bacterial surface proteins to salivary products adsorbed on the enamel surface.[3] Through the action of *S. mutans*–derived glucosyl transferase, an adherent extracellular polysaccharide matrix composed of water-insoluble glucan is produced.[2] Within this dental biofilm, refined carbohydrates, primarily sucrose, are metabolized, resulting in lactic acid production. The resultant drop in pH at the biofilm–tooth interface results in dissolution of the mineral component of the tooth. This process is a dynamic one, in which demineralization is countered by remineralization as the local pH level returns to normal through the buffering capacity of the saliva. During the remineralization phase, diffusion of phosphate and calcium ions back into the hydroxyapatite mineral component of the tooth predominates. Caries progression results when the rate of demineralization exceeds the remineralization rate (Fig. 1).

In addition to the frequency and duration of exposure to fermentable dietary carbohydrate, other factors involved in an individual's risk of developing caries include the virulence of the specific pathogenic genotypes of colonizing *S. mutans*, host immune response, and the buffering capacity and protein/mineral composition of saliva.[4] By disrupting the tooth surface biofilm, oral hygiene practices such as brushing and flossing can reduce the caries risk.

On exposure to trace quantities (~1 ppm) of fluoride ion, fluoridated hydroxyapatite and fluorapatite are formed. These are significantly more resistant to acid dissolution than hydroxyapatite. Studies[1] have suggested that the overall magnitude of caries reduction afforded by fluoride averages 25% in both children and adults, whether delivered professionally in the form of a fluoride gel or foam, self-administered in toothpaste, or by community water fluoridation.

Although the caries process in the heavily mineralized enamel outer layer of the tooth is primarily a physico-chemical process, the underlying dentin contains ~20% organic matrix, primarily type I collagen and a smaller component of noncollagenous proteins. Continued acid dissolution of the mineralized component of dentin results in exposure of the organic matrix to enzymatic degradation by both bacterial collagenases and host-derived matrix metalloproteinases.[5] The presence of tubules within the dentin through which odontoblast cell processes extend from the dental pulp results in the formation of a tightly integrated "dentin-pulp complex." Inflammation within the confined space of the pulp cavity, either as a result of direct extension of bacteria into the pulp tissue or by secondary strangulation of venous blood flow, leads to pulpal necrosis. Degradation products from bacteria and necrotic pulp tissue subsequently percolate to the apex of the necrotic tooth, leading, if untreated, to bone destruction and the formation of a mass of chronically inflamed granulation tissue (a periapical granuloma).

CYSTS AND TUMORS OF THE JAWS

Origin

Jaw lesions are categorized as being of odontogenic origin, referring to those that are related to structures involved in the development of the teeth, or nonodontogenic origin. By definition, odontogenic cysts and tumors are unique to the oral and maxillofacial region.

During tooth development, a thin epithelial structure from which the individual teeth ultimately form, termed the dental lamina, arises from an ingrowth of surface epithelium into the underlying connective tissue.[6] Remnants of these structures are believed to be the source of the epithelium involved in the formation of many developmental odontogenic cysts. The events involved in tumorigenesis of the odontogenic neoplasms remains poorly understood.[7]

Clinical and Radiographic Presentation

Most odontogenic cysts and tumors of the jaws originate in the tooth-bearing areas of the jaws above the inferior alveolar nerve canal and are characterized by replacement of bone by soft tissue or, less commonly, a mixture of soft and hard tissue. In the absence of secondary infection or significant expansion, odontogenic cysts and tumors typically cause no symptoms and are usually identified after routine radiographic examination of the jaws (Figs. 2A and 2B).

The radiographic and clinical presentations of these lesions, although often characteristic, are not pathognomonic. As with extragnathic bone lesions, assessment of radiographic features may be a necessary adjunct to the histology when attempting to arrive at a definitive diagnosis.

The author states that he has no conflicts of interest.

© 2008 American Society for Bone and Mineral Research

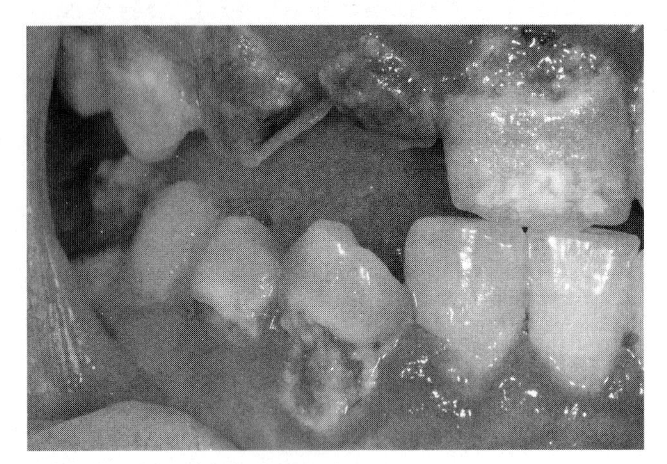

FIG. 1. Patient with rampant caries.

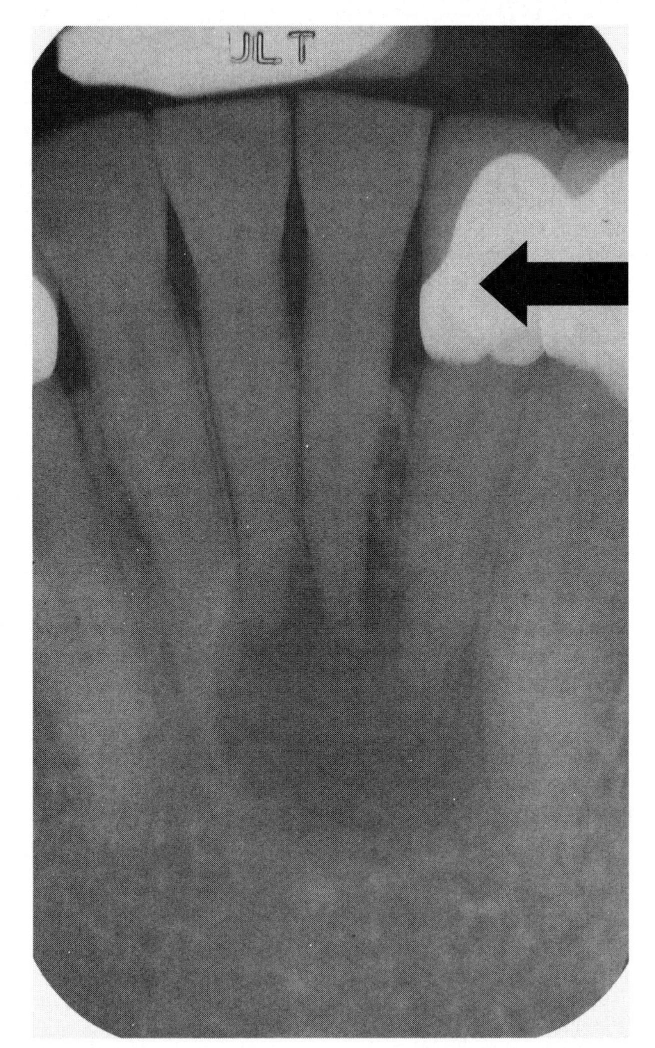

(A)

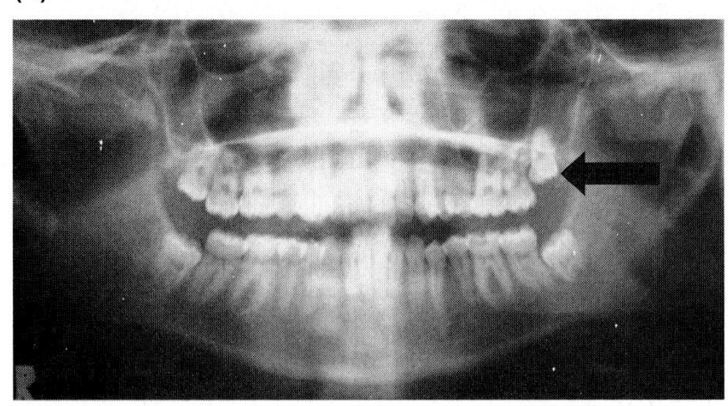

(B)

FIG. 2. The dentist's armamentarium of radiographic techniques includes intraoral dental radiographs supplemented with extraoral panoramic imaging. (A) Intraoral radiograph (periapical film) showing the outline of the mandibular anterior teeth and surrounding alveolar bone. A unilocular radiolucent lesion with a well-defined periphery is evident in the center of the radiograph. The left lateral incisor (arrow) was nonvital. The differential diagnosis was periapical cyst vs. granuloma. (B) Extraoral panoramic radiography provides a complete view of the maxillary and mandibular bones and neighboring structures. This radiographic technique is widely used as a screening tool. This panoramic radiograph, taken on an 18-yr-old patient, shows the presence of unerupted third molars (wisdom teeth). A small well-defined radiolucent area is evident around the crown of the maxillary left third molar (arrow). Because the pericoronal width is <4 mm, this likely represents a normal dental follicle surrounding the developing tooth.

ODONTOGENIC CYSTS

Odontogenic cysts are subclassified into inflammatory cysts and those of developmental origin.

Inflammatory Odontogenic Lesions: Periapical Granuloma and Periapical Cyst

Both the periapical granuloma and the periapical (radicular) cyst are common, slow-growing lesions that develop at the apex or midroot area of teeth exhibiting pulpal necrosis, usually representing the end result of caries extending into the pulp or previous trauma to the dental pulp. Continued bone destruction can lead to cortical bone perforation (Fig. 3). The periapical granuloma, composed of a localized collection of chronically inflamed granulation tissue, is the precursor to the periapical cyst. In the presence of degradation products from the necrotic pulp, epithelial cell rests are stimulated to proliferate under the influence of inflammatory mediators, cytokines, and growth factors released from mononuclear inflammatory cells and neighboring stromal cells.[8] Although the mechanism is poorly understood, the end result is the formation of a periapical cyst.

Nonvital teeth with periapical radiolucencies suggestive of a periapical granuloma or cyst are definitively treated by extraction of the causative tooth with conservative curettage of any lining from the cyst cavity. Failure to excise the cyst lining can lead to the development of a "residual cyst." When sufficient tooth structure remains to allow for restoration of the tooth, nonsurgical endodontic treatment, in which the degradation products are mechanically removed from the pulp chamber and canals, can be attempted. In this case, the patient should be followed radiographically to confirm bone regeneration. In extremely rare cases, the epithelial lining of a periapical cyst or residual cyst can undergo malignant transformation.[9]

Developmental Odontogenic Cysts

Dentigerous Cyst. The dentigerous (follicular) cyst, a slow-growing lesion capable of causing significant destruction of bone, is seen in association with the crown of an unerupted tooth. The third molar and maxillary canine teeth are most commonly involved, because these teeth are the last to erupt in the normal eruption sequence.[10]

Treatment involves enucleation of the cyst lining, usually with extraction of the associated impacted tooth. Rarely, the cyst lining can undergo transformation into an odontogenic tumor, most commonly the ameloblastoma.[11]

© 2008 American Society for Bone and Mineral Research

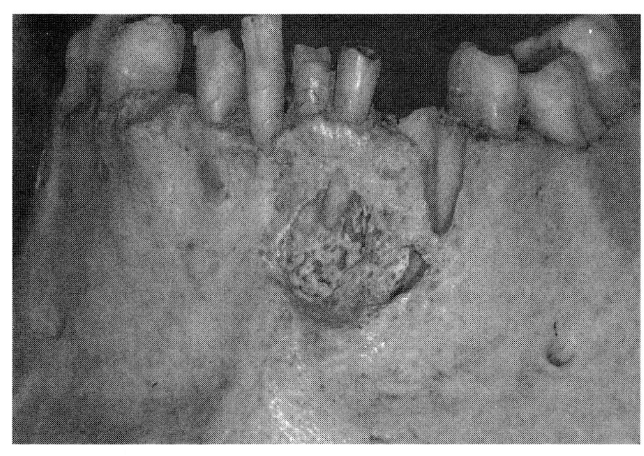

FIG. 3. Cadaver mandible with loss of buccal cortical plate. The crowns of the mandibular incisors are fractured, resulting in pulpal necrosis and the subsequent development of a periapical granuloma or cyst.

Odontogenic Keratocyst (Keratocystic Odontogenic Tumor). The odontogenic keratocyst (OKC) shows a preference for the posterior mandible and is characterized by a high recurrence rate (Fig. 4).[12] OKCs range in size from small unilocular radiolucent lesions, sometimes associated with an impacted tooth, to destructive multilocular lesions involving the entire mandible or maxilla.

Approximately 5–10% of OKCs are associated with the nevoid basal cell carcinoma syndrome (NBCCS, Gorlin syndrome), an autosomal dominantly inherited condition. Additional stigmata include the development of basal cell carcinomas at an early stage, the presence of small pit-like developmental defects in the palms of the hands and an increased incidence of neoplasms including medulloblastoma and meningioma.[13] NBCCS has been associated with germline loss of function mutations in the *patched-1* (*PTCH1*) gene, a tumor suppressor gene that is a component of the sonic hedgehog pathway.[14]

Similar *PTCH1* mutations have been documented in sporadic, non–syndrome-associated OKCs.[15] Recently, the World Health Organization has recommended that this entity be renamed "keratocystic odontogenic tumor" to emphasize its aggressive, neoplastic-like behavior.[16] Additional findings supportive of a neoplastic nature include the demonstration of increased proliferative activity in the epithelial lining[17] and loss of heterozygosity at loci associated with the *p16* and *p53* tumor suppressor genes.[18] However, this proposed change in terminology has yet to be universally adopted. Opponents to classifying the OKC as a "benign cystic neoplasm" point to similar *PTCH1* mutations in some dentigerous cysts.[19]

Treatment options vary from marsupialization and decompression before definitive treatment, to surgical curettage with adjuvant cryotherapy or chemical fixation, to surgical resection.[20] Some surgeons choose to treat maxillary lesions more aggressively because of the risk of recurrent lesions spreading into neighboring vital structures. Recurrences have been documented up to 10 yr after treatment.[21]

Lateral Periodontal Cyst. The lateral periodontal cyst (LPC) is an uncommon lesion with limited growth potential that is often overlooked in the clinical differential diagnosis of a radiolucency occurring along the lateral root surface of an anterior tooth.[22] In contrast to the periapical cyst, there is no causal relationship to pulpal necrosis. Treatment involves conservative debridement of the lesion with preservation of the associated teeth. To avoid unnecessary endodontic therapy or tooth extraction, assessment of pulpal vitality of all teeth with associated radiolucent lesions is recommended.

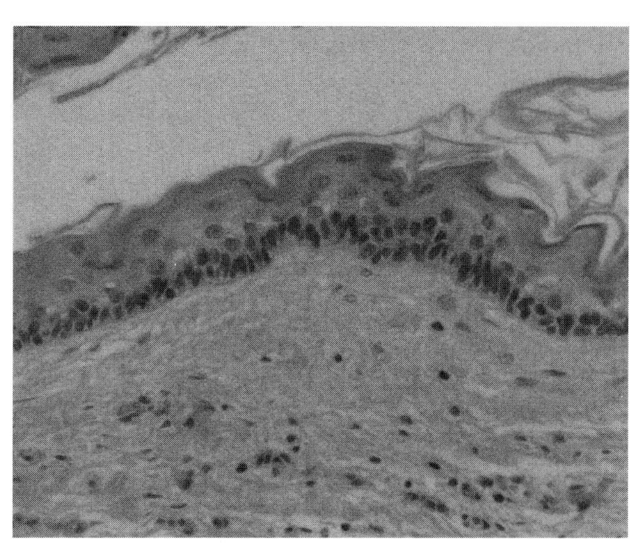

FIG. 4. H&E-stained section from an odontogenic keratocyst (keratocystic odontogenic tumor) shows a cyst lining composed of epithelial cells maturing to a parakeratotic surface. Keratin is evident in the lumen. The basal epithelial layer (adjacent to the basement membrane that separates the epithelium from the underlying connective tissue) is palisaded.

Glandular Odontogenic Cyst. The recently described[23] glandular odontogenic cyst (GOC) is a rare lesion (one study[24] documented only 11 cases among 55,000+ oral cavity biopsies) with histologic features reminiscent of both the lateral periodontal cyst and mucoepidermoid carcinoma, a malignant salivary gland neoplasm rarely identified in the jaws. The majority of GOCs occur in the anterior mandible, often crossing the midline.[25] The GOC exhibits a wide spectrum of clinical behavior, ranging from a benign process to a destructive lesion with features more suggestive of a malignant process.

ODONTOGENIC TUMORS

Benign Odontogenic Tumors of the Jaws

Ameloblastoma. The ameloblastoma is a locally destructive tumor with a propensity to cause significant cortical expansion, a marked predilection for the posterior mandible, and a high rate of recurrence.[26] Its incidence has been estimated at 0.3–2.3 new cases per million persons per year.[27]

The unicystic ameloblastoma, seen predominantly in teenagers at an average of 20 yr earlier than the conventional ameloblastoma, represents a cystic version of the conventional ameloblastoma, possibly associated with a lower risk of recurrence.[28] Often associated with an impacted mandibular third molar, it is frequently mistaken radiographically for a dentigerous cyst.

Malignant Odontogenic Tumors

Malignant counterparts to the benign odontogenic tumors are extremely rare.[29] They commonly present as destructive lesions with irregular, poorly defined radiographic margins. Pain, paresthesia, and a tendency for early lymph node metastasis are characteristic.

NON-ODONTOGENIC TUMORS OF THE JAWS: THE CENTRAL GIANT CELL GRANULOMA.

A large number of nonodontogenic cysts, pseudocysts, and tumors also occur in the jaws. Among these is a lesion unique

© 2008 American Society for Bone and Mineral Research

to the jaws: the central giant cell granuloma (CGCG). Interestingly, histologically identical lesions are seen in a number of other conditions including cherubism and hyperparathyroidism. Syndromes associated with an increased incidence of CGCG-like lesions include Noonan syndrome[30,31] and neurofibromatosis type-1.[32,33]

The classic CGCG is a variably aggressive nonneoplastic reactive lesion with an estimated incidence of 1.1 per million persons per year in the general population.[34] The CGCG is characterized histologically by the presence of multinucleated giant cells (MGCs) in a background of spindle-shaped mesenchymal cells. The MGCs are usually concentrated in areas of hemorrhage and are believed to develop from the fusion of mononuclear phagocytes.[35] Although they share similarities with the osteoclast,[36] phenotypic differences exist.[37]

Proposed etiologies[38] include intraosseous hemorrhage and abnormal repair of bone. CGCGs are also noted in association with other preexisting bone lesions, including fibrous dysplasia and ossifying fibroma.[39]

Isolated lesions are commonly treated by surgical curettage. Treatment options for large or multiple lesions include intralesional corticosteroid injection,[40] subcutaneous or intranasal administration of calcitonin,[41] and interferon α-2a.[42]

REFERENCES

1. Griffin SO, Regnier E, Griffin PM, Huntley V 2007 Effectiveness of fluoride in preventing caries in Adults. J Dent Res **86:**410–415.
2. Loesche WJ 1986 Role of *Streptococcus mutans* in human dental decay. Microbiol Rev **50:**353–380.
3. Napimoga MH, Hofling JF, Klein MI, Kamiya RU, Goncalves RB 2005 Transmission, diversity and virulence factors of *Streptococcus mutans* genotypes. J Oral Sci **47:**59–64.
4. Selwitz RH, Ismail AI, Pitts NB 2007 Dental caries. Lancet **369:**51–59.
5. Chaussain-Miller C, Fioretti F, Goldberg M, Menashi S 2006 The role of matrix metalloproteinases (MMPs) in human caries. J Dent Res **85:**22–32.
6. Cobourne MT, Sharpe PT 2003 Tooth and jaw: Molecular mechanisms of patterning in the first branchial arch. Arch Oral Biol **48:**1–14.
7. Kumamoto H 2006 Molecular pathology of odontogenic tumors. J Oral Pathol Med **35:**65–74.
8. Lin LM, Huang GT, Rosenberg PA 2007 Proliferation of epithelial cell rests, formation of apical cysts, and regression of apical cysts after periapical wound healing. J Endod **33:**908–916.
9. Whitlock RI, Jones JH 1967 Squamous cell carcinoma of the jaw arising in a simple cyst. Oral Surg Oral Med Oral Pathol **24:**530–536.
10. Daley TD, Wysocki GP 1995 The small dentigerous cyst: A diagnostic dilemma. Oral Surg Oral Med Oral Pathol Oral Radiol Endod **79:**77–81.
11. Holmlund HA, Anneroth G, Lundquist G, Nordenram A 1991 Ameloblastoma originating from odontogenic cysts. J Oral Pathol Med **20:**318–321.
12. Myoung H, Hong SP, Hong SD, Lee JI, Lim CY, Choung PH, Lee JH, Choi YJ, Seo BM, Kim MJ 2001 Odontogenic keratocyst: Review of 256 cases for recurrence and clinicopathologic parameters. Oral Surg Oral Med Oral Pathol Oral Radiol Endod **91:**328–333.
13. Kimonis VE, Goldstein AM, Pastakia B, Yang ML, Kase R, DiGiovanna JJ, Bale AE, Bale SJ 1997 Clinical Manifestations in 105 persons with nevoid basal cell carcinoma syndrome. Am J Med Genet **69:**299–308.
14. Hahn H, Wicking C, Zaphiropoulous PG, Gailani MR, Shanley S, Chidambaram A, Vorechovsky I, Holmberg E, Unden AB, Gillies S, Negus K, Smyth I, Pressman C, Leffell DJ, Gerrard B, Goldstein AM, Dean M, Toftgard R, Chenevix-Trench G, Wainwright B, Bale AE 1996 Mutations of the human homolog of Drosophila *patched* in the nevoid basal cell carcinoma syndrome. Cell **85:**841–851.
15. Gu XM, Zhao HS, Sun LS, Li TJ 2006 *PTCH* mutations in sporadic and Gorlin-syndrome-related odontogenic keratocysts. J Dent Res **85:**859–863.
16. Philipsen HP 2005 Keratocystic odontogenic tumor. In: Barnes L, Eveson JW, Reichart P, Sidransky D (eds.) Pathology and Genetics of Head and Neck Tumors. IARC Press, Lyons, France, pp. 306–307.
17. Slootweg PJ 1995 p53 protein and Ki-67 reactivity in epithelial odontogenic lesions. An immunohistochemical study. J Oral Pathol Med **24:**393–397.
18. Henley J, Summerlin DJ, Tomich C, Zhang S, Cheng L 2005 Molecular evidence supporting the neoplastic nature of odontogenic keratocyst: A laser capture microdissection study of 15 cases. Histopathology **47:**582–586.
19. Pavelic B, Levanat S, Crnic I, Kobler P, Anic I, Manojlovic S, Sutalo J 2001 *PTCH* gene altered in dentigerous cysts. J Oral Pathol Med **30:**569–576.
20. Blanas N, Freund B, Schwartz M, Furst IM 2000 Systematic review of the treatment and prognosis of the odontogenic keratocyst. Oral Surg Oral Med Oral Pathol Oral Radiol Endod **90:**553–558.
21. Kolokythas A, Fernandes RP, Pazoki A, Ord RA 2007 Odontogenic keratocyst: To decompress or not to decompress? A comparative study of decompression and enucleation versus resection/peripheral ostectomy. J Oral Maxillofac Surg **65:**640–644.
22. Fantasia JE 1979 Lateral periodontal cyst. An analysis of forty-six cases. Oral Surg Oral Med Oral Pathol Oral Radiol Endod **48:**237–243.
23. Gardner DG, Kessler HP, Morency R, Schaffner DL 1988 The glandular odontogenic cyst: An apparent entity. J Oral Pathol **17:**359–366.
24. Jones AV, Craig GT, Franklin CD 2006 Range and demographics of odontogenic cysts diagnosed in a UK population over a 30-year period. J Oral Pathol Med **35:**500–507.
25. Hussain K, Edmondson HD, Browne RM 1995 Glandular odontogenic cysts: Diagnosis and treatment. Oral Surg Oral Med Oral Pathol Oral Radiol Endod **79:**593–602.
26. Reichart PA, Philipsen HP, Sonner S 1995 Ameloblastoma: Biological profile of 3677 cases. Eur J Cancer B Oral Oncol **31B:**86–99.
27. Shear M, Singh S 1978 Age-standardized incidence rates of ameloblastoma and dentigerous cyst on the Witwatersrand, South Africa. Community Dent Oral Epidemiol **6:**195–199.
28. Philipsen HP, Reichart PA 1998 Unicystic ameloblastoma. A review of 193 cases from the literature. Oral Oncol **34:**317–325.
29. Slootweg PJ 2002 Malignant odontogenic tumors: An overview. Mund Kiefer Gesichtschir **6:**295–302.
30. Cohen MM Jr, Gorlin RJ 1991 Noonan-like/multiple giant cell lesion syndrome. Am J Med Genet **40:**159–166.
31. Edwards PC, Fox J, Fantasia JE, Goldberg J, Kelsch RD 2005 Bilateral central giant cell granulomas of the mandible in an eight year-old girl with Noonan syndrome (Noonan-like/multiple giant cell lesions syndrome). Oral Surg Oral Med Oral Pathol Oral Radiol Endod **99:**334–340.
32. Ruggieri M, Pavone V, Polizzi A, Albanase S, Magro G, Merino M, Duray P 1999 Unusual form of recurrent giant cell granuloma of the mandible and lower extremities in a patient with neurofibromatosis type 1. Oral Surg Oral Med Oral Pathol Oral Radiol Endod **87:**67–72.
33. Edwards PC, Fantasia JE, Saini T, Rosenberg T, Ruggiero S 2006 Clinically aggressive central giant cell granulomas in two patients with neurofibromatosis 1. Oral Surg Oral Med Oral Pathol Oral Radiol Endod **102:**765–772.
34. de Lange J, van den Akker HP 2005 Clinical and radiological features of central giant cell lesions of the jaw. Oral Surg Oral Med Oral Pathol Oral Radiol Endod **99:**464–470.
35. Abe E, Mocharla H, Yamate T, Taguchi Y, Manolagas SC 1999 Meltrin-alpha, a fusion protein involved in multinucleated giant cell and osteoclast formation. Calcif Tissue Int **64:**508–515.
36. Liu B, Yu SF, Li TJ 2003 Multinucleated giant cells in various forms of giant cell containing lesions of the jaws express features of osteoclasts. J Oral Pathol Med **32:**367–375.
37. Tobon-Arroyave SI, Franco-Gonzalez LM, Isaza-Guzman DM, Florez-Moreno GA, Bravo-Vasquez T, Castaneda-Pelaez DA, Vieco-Duran B 2005 Immunohistochemical expression of RANK, GR-alpha and CTR in central giant cell granulomas of the jaws. Oral Oncol **41:**480–488.
38. Dorfman HD, Czerniak B 1998 Giant cell lesions. In: Dorfman HD, Czerniak B (eds.) Bone Tumors. Mosby, St Louis, MO, USA, pp. 559–606.
39. Penfold CN, McCullagh P, Eveson JW, Ramsay A 1993 Giant cell

© 2008 American Society for Bone and Mineral Research

lesions complicating fibro-osseous conditions of the jaws. Int J Oral Maxillofac Surg **22:**158–162.
40. Terry BC, Jacoway JR 1994 Management of central giant cell lesion: An alternative to surgical therapy. Oral Maxillofac Surg Clin North Am **6:**579–600.
41. de Lange J, Rosenberg AJ, Van den Akker HP, Koole R, Wirds JJ, VandenBerg H 1999 Treatment of central giant cell granuloma of the jaw with calcitonin. Int J Oral Maxillofac Surg **28:**372–376.
42. Kaban LB, Mulliken JB, Ezekowitz RA, Ebb D, Smith PS, Folkman J 1999 Antiangiogenic therapy of a recurrent giant cell tumor of the mandible with interferon alfa-2a. Pediatrics **103:**1145–1149.

Chapter 107. Bisphosphonate-Associated Osteonecrosis of the Jaws

Nathaniel S. Treister[1] and Sook-Bin Woo[2]

[1]*Division of Oral Medicine and Dentistry, Brigham and Women's Hospital, Boston, Massachusetts;* [2]*Harvard School of Dental Medicine, Boston, Massachusetts*

DEFINITION OF BISPHOSPHONATE-ASSOCIATED OSTEONECROSIS OF THE JAWS

Bisphosphonate-associated osteonecrosis of the jaws (BONJ) is a recently described clinical entity that was first reported in 2003.[1–3] Of the >500 cases that have been reported in the literature, the vast majority are in patients with multiple myeloma and metastatic cancers (from the breast, prostate, and lung) to the skeletal system who were treated with intravenous bisphosphonates; however, this condition has been identified (albeit far less frequently) in patients with osteoporosis treated only with oral formulations.[3–5] All cases have affected either the maxilla or the mandible, except for a single report involving the external auditory canal after surgery.[6] The strong association between bisphosphonate use and osteonecrosis of the jaws is now well accepted, and appropriately designed prospective research is ongoing.

The American Association of Oral and Maxillofacial Surgeons (AAOMS) has published a working definition for BONJ in which the following three characteristics must be fulfilled: (1) current or previous treatment with a bisphosphonate; (2) exposed, necrotic bone in the maxillofacial region that has been present for at least 8 wk; and (3) no history of radiation therapy to the jaws.[7] The American Society for Bone and Mineral Research (ASBMR) has the same definition for confirmed cases; suspected cases fulfill all the criteria of a confirmed case except that the bone has been present for <8 wk.[8] The AAOMS definition excludes such conditions as poorly healing dental extraction sites (e.g., alveolar osteitis, or "dry socket"), benign sequestration of the lingual plate, necrotizing periodontitis, noma, and osteoradionecrosis.[7,9–13] These definitions are a critical initial step toward promoting uniform reporting for research and epidemiological reporting purposes and may need to be revised as new diagnostic parameters come to light. Application of such criteria in well-designed prospective studies will help to move research forward.[14] Whereas radiographic and other imaging studies may be of some diagnostic value, at the present time, we rely on a good patient history and clinical examination for the diagnosis of BONJ.

ETIOLOGY

The etiopathogenesis has been attributed to suppression of bone turnover (caused by bisphosphonate-induced osteoclast inhibition) coupled with conditions that are unique to the mandible and maxilla. First, the jaw bones are separated from the oral cavity and commensal microflora by only a very thin mucosal barrier that is readily breached during normal physiologic functions (e.g., chewing). Second, the jaws are involved frequently by infection either through the periodontium (i.e., periodontal disease) or endodontically, through the dental pulp (i.e., caries extending into the pulp chamber, through the roots, and into the bone). Third, dentoalveolar surgical procedures are common (e.g., tooth extractions and periodontal surgeries) during which bone is damaged and exposed to a bacteria-rich environment. Fourth, the rate of turnover of the jawbones is higher than for the long bones, which may result in greater uptake and higher local concentrations of bisphosphonates.[15] With profound osteoclast inhibition, the hypodynamic bone may be unable to respond to repair processes associated with physiologic trauma or infection, resulting in bone necrosis. There is also evidence that bisphosphonates may have anti-angiogenic properties that may also contribute to poor wound healing.[16,17] The fact that tooth sockets in patients who develop BONJ may be evident years after extraction is further evidence that bone turnover and remodeling in these areas are severely compromised.[18]

RISK FACTORS AND PREVALENCE

Several risk factors for BONJ have been identified. One of the most consistently observed is the use of nitrogen-containing bisphosphonates, and in particular zoledronic acid, which is the drug of choice for multiple myeloma and metastatic cancers in the United States because of its shorter infusion time and equivalent clinical efficacy compared with pamidronate; its potency is the highest among the bisphosphonates. Patients on zoledronic acid alone have a 9.5 times risk of developing BONJ compared with pamidronate alone and 4.5 times risk compared with pamidronate and zoledronic acid together.[19] The cumulative dose of bisphosphonates also has been identified as a key factor, with the risk increasing over time with a cumulative hazard of 1% in the first year of treatment with zoledronic acid and 15% at 4 yr.[20] The median time of treatment with zoledronic acid to development of BONJ ranges from 9 to 30 mo (although cases have been reported after only 3 mo) and seems to be significantly shorter compared with other agents.[20–23] The route of administration is also important, but this is likely a surrogate factor for the use of pamidronate and zoledronic acid, both of which are admin-

The authors state that they have no conflicts of interest.

Key words: bisphosphonates, osteonecrosis, oral lesions, oral medicine

© 2008 American Society for Bone and Mineral Research

istered intravenously, have high potency, and are used more frequently than in the nononcologic population.

Approximately 60% of cases are associated with either tooth extraction or other oral surgical procedures; however, it is unclear to what extent preexisting infection (e.g., periodontal and dental disease), the most common indication for dentoalveolar surgery, may be a contributing factor.[21,24,25] Additionally, nonsurgical trauma seems also to be a risk factor, because many cases occur on the mylohyoid ridge of the lingual mandible (where the mucosa is especially thin and susceptible to masticatory forces), on tori, and in patients wearing dentures, with no associated dental infection or oral surgical procedure.

Prevalence estimates range from 1% to 12% in the oncology population; however, some of these are based primarily on retrospective chart reviews or telephone interviews.[19,23,26–28] No true incidence studies have been reported; therefore, it is unclear how many patients exposed to bisphosphonates never develop BONJ after oral surgical procedures. Prevalence of BONJ in those taking oral bisphosphonates for osteoporosis has been reported to be between 1 in 2260–8470.[23]

CLINICAL PRESENTATION

BONJ presents with exposed, necrotic bone, typically yellowish-white in appearance, that varies in size from a few millimeters to several centimeters. The exposed bone may have rough and sharp edges or may be smooth. The mandible is more frequently affected than the maxilla (2:1 ratio), and lesions are more common in areas with very thin mucosa overlying bony prominences such as mandibular and maxillary tori, bony exostoses, and the mylohyoid ridge.[3,29] Lesions often develop in sites of previous dental extractions (Fig. 1A); however, in many cases, there is no antecedent surgical procedure (Fig. 1B). In some cases, the bone may project outward, causing trauma to the adjacent soft tissue. Loose bone fragments in the process of sequestration may also cause pain because of soft tissue irritation. Teeth that are located in the affected bone (often periodontally involved) may become progressively mobile and symptomatic. Pathologic fractures of the mandible may occur in severe cases with extensive involvement and lytic changes; however, this is an infrequent complication.[3,21] Approximately 60% of cases report some pain at initial presentation; however, many cases may remain asymptomatic for weeks, months, or years.[3,21,24]

The necrotic bone may become colonized by normal oral flora or there may be true infection of the marrow spaces (osteomyelitis). The surrounding soft tissue may become erythematous and edematous sometimes with an associated purulent discharge with intraoral or even extraoral sinus tract/fistula formation. Secondary maxillary sinusitis, with or without an oral-antral fistula, may be the initial presenting symptom in those with maxillary involvement[3,29] (Figs. 2A and 2B). Symptoms of paresthesia and anesthesia may develop as the local neurovascular bundle (e.g., inferior alveolar nerve) becomes affected by inflammation or infection around the necrotic bone.

RADIOGRAPHIC AND HISTOLOGICAL FINDINGS

A number of radiographic signs have been identified in BONJ from both plain films and more advanced imaging studies (e.g., CT and MRI).[30–33] These include osteosclerosis, osteolysis, mottling, thickening or loss of the lamina dura, widening of the periodontal ligament space, sequestrum formation, and persistent extraction sockets[18,32–34] (Figs. 3A–3C). Radiographic changes may mimic periapical pathology,

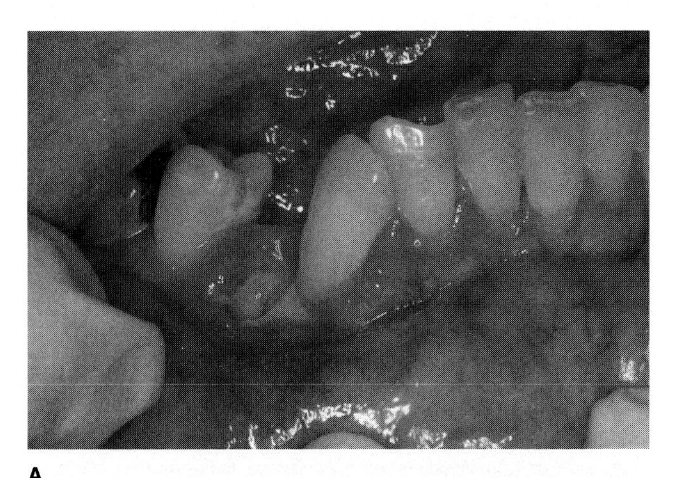

A

B

FIG. 1. (A) BONJ in the area of a nonhealing mandibular premolar extraction site. The surrounding soft tissue is healthy-appearing. (B) BONJ of the left mandibular torus with healthy-appearing surrounding soft tissue.

banal osteomyelitis, or in cancer patients, raise the suspicion of metastatic bone disease.[35] Because a bone biopsy may induce BONJ, this is only indicated when there is a strong clinical suspicion of metastatic disease and/or if the diagnosis will alter clinical management.

Plain films (e.g., periapical radiographs) and panoramic radiographs provide high resolution and show clear evidence of BONJ especially when disease is advanced[32]; however, they fail to detect many early changes or mild disease and are limited in their ability to evaluate extent of involvement because of their 2D nature. Whereas CT scans provide more accurate 3D information regarding the extent of changes, these also have not proved useful for early detection in asymptomatic individuals.[32] MRI may be of some use in evaluating the marrow and associated soft tissue changes but is not useful for interpreting bony pathology.[32] Nuclear medicine techniques have shown that Tc-99m is not taken up in clinical areas of BONJ (although these may be positive in early or subclinical BONJ); however, ^{18}F-fluorodeoxyglucose positron emission tomography (FDG-PET) integrated with CT may show focal uptake.[32,33] The precise role for these studies in risk assess-

© 2008 American Society for Bone and Mineral Research

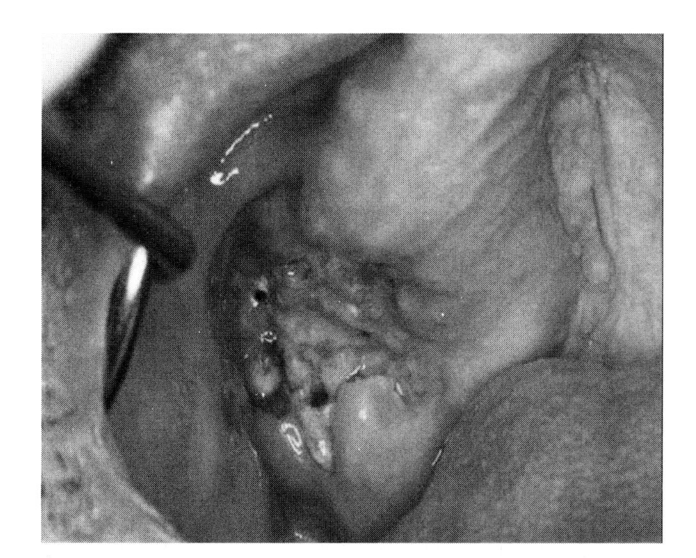

A

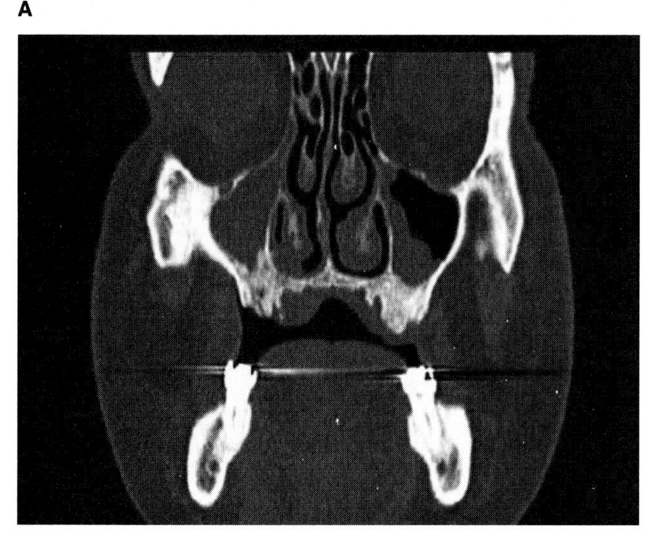

B

FIG. 2. (A) Extensive BONJ of the left posterior maxilla with associated soft tissue infection and purulence. (B) CT scan of patient from A with mottled right maxillary alveolar process and complete opacification of the right and partial opacification of the left maxillary sinuses.

ment, diagnosis, and clinical management remains to be determined.

Histopathological examination of débrided bone fragments shows necrotic bone often with extensive surface bacterial colonization, and granulation tissue, all fairly nonspecific findings.[36–38] Despite inhibition of osteoclastic activity, osteoclasts are present in the vicinity of the necrotic bone.[38] Bacterial cultures from the exposed surface or purulent discharge typically show normal oral flora and are therefore of limited clinical use.[37] Actinomycetes are a normal component of the oral flora, and their identification, either histopathologically or microbiologically, must be carefully interpreted. A diagnosis of actinomycosis should only be made if the culture was obtained from a sterile location (i.e., a nonexposed surface), if pain and suppuration/sinus tracts are present, and/or if sulfur granules are noted either clinically or histologically.[39]

MANAGEMENT

Management of BONJ is focused primarily on preserving quality of life by controlling pain and discomfort, managing infection, and avoiding procedures that may lead to the development of new areas of necrosis. Published treatment algorithms/guidelines are principally based on expert opinion consensus and not prospective studies.[7,29,34,40] For those exposed to bisphosphonates but without BONJ, treatment is directed

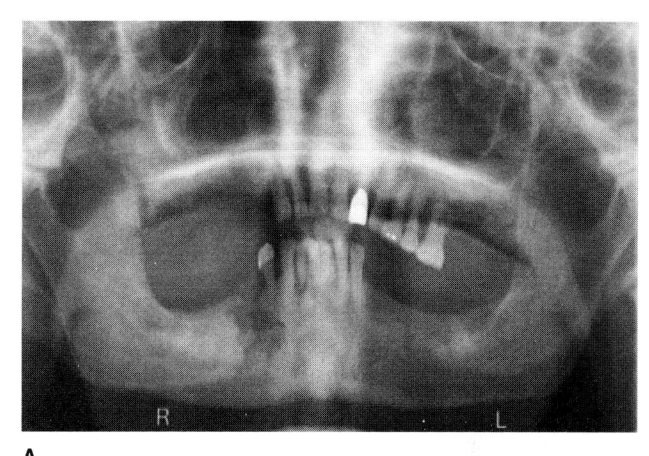

A

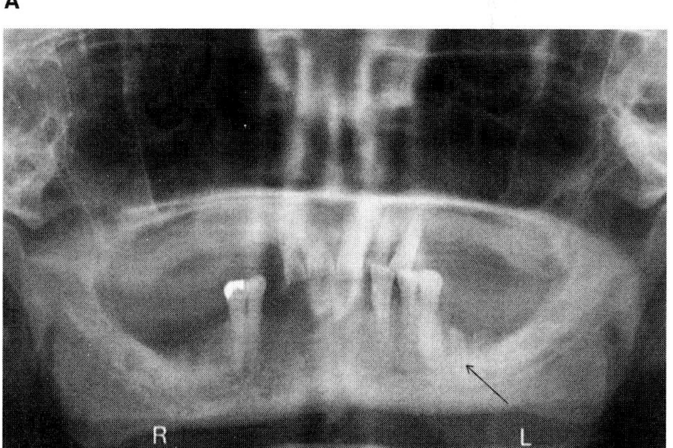

B

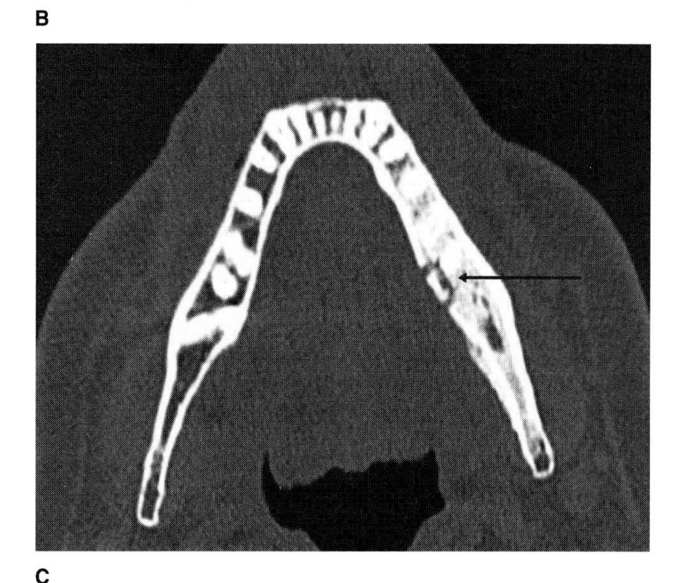

C

FIG. 3. (A) Panoramic radiograph showing a mixed radiolucent and radiopaque lesion of the right mandible with a larger area of diffuse osteosclerosis. (B) Panoramic radiograph showing a persistent extraction socket of the left mandible years after extraction. (C) Axial CT scan of the mandible showing sclerosis with lingual sequestrum formation (arrow).

© 2008 American Society for Bone and Mineral Research

TABLE 1. BONJ CLINICAL STAGING SYSTEM

Stage 1	Stage 2	Stage 3
Exposed, necrotic bone that is asymptomatic	Exposed, necrotic bone associated with pain and infection	Exposed, necrotic bone with pain, infection, and one or more of the following: • pathologic fracture • extraoral fistula • osteolysis extending to the inferior border

toward optimizing dental health by educating patients regarding their risk for developing BONJ, evaluating for odontogenic infections (through clinical examination and full mouth intraoral and panoramic radiographs), and treating active areas of infection. When a nonsurgical approach (such as endodontic therapy) is not feasible in a patient who has been exposed to bisphosphonates, there must be a discussion of the risks and benefits of surgical procedures, and patients should be followed closely postoperatively for any signs of infection or poor healing until the area is completely healed. The AAOMS guidelines suggest no alteration in planned surgery for patients who have had <3 yr of exposure to oral bisphosphonates, unless there has also been concomitant corticosteroid use.(7) A more recent publication showed that the mean duration of exposure to oral bisphosphonates before developing BONJ is 5.6 yr (range, 3.3–10.2 yr).(4)

A clinical staging system has been developed to categorize patients, guide treatment, and collect data(7) (Table 1). Complete resolution of BONJ does occur in 27–65% of cases without intervention in those taking oral bisphosphonates.(4,5) Resolution in those taking intravenous bisphosphonates, either of a spontaneous nature or after surgery, is not clear, and there has been little consistency in either surgical or nonsurgical treatment protocols. For the most part, the goal of therapy is to eliminate and prevent recurrence of secondary infection and symptoms. When sharp edges of necrotic bone traumatize the adjacent soft tissue, or when fragments are mobile, conservative removal and/or smoothing of the involved bone provides immediate symptomatic relief. When signs and symptoms of infection are present (i.e., inflamed surrounding soft tissue, purulence, and sinus tracts), management with both systemic antibiotics (e.g., amoxicillin/clavulanate, clindamycin) and topical antimicrobial agents (chlorhexidine gluconate) are effective in reducing or eliminating signs and symptoms. However, long-term daily therapy may be necessary to prevent recurrence. In the majority of symptomatic patients with stage 2 involvement, conservative measures alone are effective in at least stabilizing, if not improving, their BONJ (e.g., from stage 2 to stage 1).(34)

Only in severe cases with significant morbidity caused by extensive bone involvement or pathologic fracture is surgical intervention considered, because this treatment otherwise is generally ineffective and may result in disease progression.(3) In one small case series, hyperbaric oxygen therapy together with surgery and discontinuation of bisphosphonates for 6 mo resulted in 100% resolution of lesions(41); another study showed 50% of patients experiencing complete resolution on hyperbaric oxygen alone.(42) Additional case reports and small series have reported on the benefits of topically applied autologous platelet-derived growth factor with surgery, low-level laser therapy, ozone therapy, and teriparatide therapy.(43–46) Prospective randomized studies, however, are critical for evaluating the efficacy of any proposed treatments.

Long-term discontinuation of intravenous bisphosphonates may be beneficial in both stabilizing established lesions and preventing the development of new lesions.(7) However, given the long half-life of these agents, it is unclear whether this measure actually reduces the risk of developing BONJ.(47) Recently, a study showed that serum C-telopeptide cross-linked (CTX) levels rose after discontinuation of bisphosphonates, a not unexpected finding.(4) Another showed that patients who discontinued bisphosphonate therapy overall showed stabilization or resolution of disease compared with those who did not.(41)

Regardless, because of the growing recognition of BONJ as a potentially serious clinical complication of bisphosphonate therapy, a number of institutions have modified treatment protocols and reduced the dose and/or frequency of infusions according to disease signs and symptoms.(48–50)

At this time, monitoring of patients consists primarily of history and periodic examination. Radiographic evaluation may be of some use in assessing progression over time. No controlled studies of serologic markers for bone turnover such as bone-specific alkaline phosphatase, N-telopeptide cross-linked, and CTX have been conducted. Extrapolation of data from serum or urine markers to the risk of development of BONJ is also difficult because such measurements reflect remodeling activity in the entire skeletal system and not the jaws per se. Similar to treatments for BONJ, prospective controlled studies showing the usefulness and cost-effectiveness of these different modalities are necessary before their clinical application can be justified.

REFERENCES

1. Migliorati CA 2003 Bisphosphanates and oral cavity avascular bone necrosis. J Clin Oncol **21:**4253–4254.
2. Marx RE 2003 Pamidronate (Aredia) and zoledronate (Zometa) induced avascular necrosis of the jaws: A growing epidemic. J Oral Maxillofac Surg **61:**1115–1117.
3. Ruggiero SL, Mehrotra B, Rosenberg TJ, Engroff SL 2004 Osteonecrosis of the jaws associated with the use of bisphosphonates: A review of 63 cases. J Oral Maxillofac Surg **62:**527–534.
4. Marx RE, Cillo JE, Ulloa JJ 2007 Oral bisphosphonate-induced osteonecrosis: Risk factors, prediction of risk using serum CTX testing, prevention, and treatment. J Oral Maxillofac Surg **65:**2397–2410.
5. Yarom N, Yahalom R, Shoshani Y, Hamed W, Regev E, Elad S 2007 Osteonecrosis of the jaw induced by orally administered bisphosphonates: Incidence, clinical features, predisposing factors and treatment outcome. Osteoporos Int **18:**1363–1370.
6. Polizzotto MN, Cousins V, Schwarer AP 2006 Bisphosphonate-associated osteonecrosis of the auditory canal. Br J Haematol **132:**114.
7. American Association of Oral and Maxillofacial Surgeons 2007 American Association of Oral and Maxillofacial Surgeons position paper on bisphosphonate-related osteonecrosis of the jaws. J Oral Maxillofac Surg **65:**369–376.
8. Khosla S, Burr D, Cauley J, Dempster DW, Ebeling PR, Felsenberg D, Gagel RF, Gilsanz V, Guise T, Koka S, McCauley LK, McGowan J, McKee MD, Mohla S, Pendrys DG, Raisz LG, Ruggiero SL, Shafer DM, Shum L, Silverman SL, Van Poznak CH, Watts N, Woo SB, Shane E 2007 Bisphosphonate-associated osteonecrosis of the jaw: Report of a task force of the American Society for Bone and Mineral Research. J Bone Miner Res **22:**1479–1491.
9. Marx RE 1983 Osteoradionecrosis: A new concept of its pathophysiology. J Oral Maxillofac Surg **41:**283–288.

© 2008 American Society for Bone and Mineral Research

10. Peters E, Lovas GL, Wysocki GP 1993 Lingual mandibular sequestration and ulceration. Oral Surg Oral Med Oral Pathol **75:**739–743.
11. Novak MJ 1999 Necrotizing ulcerative periodontitis. Ann Periodontol **4:**74–78.
12. Houston JP, McCollum J, Pietz D, Schneck D 2002 Alveolar osteitis: A review of its etiology, prevention, and treatment modalities. Gen Dent **50:**457–463.
13. Enwonwu CO, Falkler WA Jr, Phillips RS 2006 Noma (cancrum oris). Lancet **368:**147–156.
14. Shane E, Goldring S, Christakos S, Drezner M, Eisman J, Silverman S, Pendrys D 2006 Osteonecrosis of the jaw: More research needed. J Bone Miner Res **21:**1503–1505.
15. Huja SS, Fernandez SA, Hill KJ, Li Y 2006 Remodeling dynamics in the alveolar process in skeletally mature dogs. Anat Rec A Discov Mol Cell Evol Biol **288:**1243–1249.
16. Wood J, Bonjean K, Ruetz S, Bellahcene A, Devy L, Foidart JM, Castronovo V, Green JR 2002 Novel antiangiogenic effects of the bisphosphonate compound zoledronic acid. J Pharmacol Exp Ther **302:**1055–1061.
17. Santini D, Vincenzi B, Dicuonzo G, Avvisati G, Massacesi C, Battistoni F, Gavasci M, Rocci L, Tirindelli MC, Altomare V, Tocchini M, Bonsignori M, Tonini G 2003 Zoledronic acid induces significant and long-lasting modifications of circulating angiogenic factors in cancer patients. Clin Cancer Res **9:**2893–2897.
18. Groetz KA, Al-Nawas B 2006 Persisting alveolar sockets-a radiologic symptom of BP-ONJ? J Oral Maxillofac Surg **64:**1571–1572.
19. Zervas K, Verrou E, Teleioudis Z, Vahtsevanos K, Banti A, Mihou D, Krikelis D, Terpos E 2006 Incidence, risk factors and management of osteonecrosis of the jaw in patients with multiple myeloma: A single-centre experience in 303 patients. Br J Haematol **134:**620–623.
20. Dimopoulos MA, Kastritis E, Anagnostopoulos A, Melakopoulos I, Gika D, Moulopoulos LA, Bamia C, Terpos E, Tsionos K, Bamias A 2006 Osteonecrosis of the jaw in patients with multiple myeloma treated with bisphosphonates: Evidence of increased risk after treatment with zoledronic acid. Haematologica **91:**968–971.
21. Marx RE, Sawatari Y, Fortin M, Broumand V 2005 Bisphosphonate-induced exposed bone (osteonecrosis/osteopetrosis) of the jaws: Risk factors, recognition, prevention, and treatment. J Oral Maxillofac Surg **63:**1567–1575.
22. Pozzi S, Marcheselli R, Sacchi S, Baldini L, Angrilli F, Pennese E, Quarta G, Stelitano C, Caparotti G, Luminari S, Musto P, Natale D, Broglia C, Cuoghi A, Dini D, Di Tonno P, Leonardi G, Pianezze G, Pitini V, Polimeno G, Ponchio L, Masini L, Musso M, Spriano M, Pollastri G 2007 Bisphosphonate-associated osteonecrosis of the jaw: A review of 35 cases and an evaluation of its frequency in multiple myeloma patients. Leuk Lymphoma **48:**56–64.
23. Mavrokokki T, Cheng A, Stein B, Goss A 2007 Nature and frequency of bisphosphonate-associated osteonecrosis of the jaws in Australia. J Oral Maxillofac Surg **65:**415–423.
24. Woo SB, Hande K, Richardson PG 2005 Osteonecrosis of the jaw and bisphosphonates. N Engl J Med **353:**99–102.
25. Maerevoet M, Martin C, Duck L 2005 Osteonecrosis of the jaw and bisphosphonates. N Engl J Med **353:**99–102.
26. Bamias A, Kastritis E, Bamia C, Moulopoulos LA, Melakopoulos I, Bozas G, Koutsoukou V, Gika D, Anagnostopoulos A, Papadimitriou C, Terpos E, Dimopoulos MA 2005 Osteonecrosis of the jaw in cancer after treatment with bisphosphonates: Incidence and risk factors. J Clin Oncol **23:**8580–8587.
27. Durie BG, Katz M, Crowley J 2005 Osteonecrosis of the jaw and bisphosphonates. N Engl J Med **353:**99–102.
28. Murad OM, Arora S, Farag AF, Guber HA 2007 Bisphosphonates and osteonecrosis of the jaw: A retrospective study. Endocr Pract **13:**232–238.
29. Woo SB, Hellstein JW, Kalmar JR 2006 Narrative [corrected] review: Bisphosphonates and osteonecrosis of the jaws. Ann Intern Med **144:**753–761.
30. 2005 Alendronate (Fosamax) and risedronate (Actonel) revisited. Med Lett Drugs Ther **47:**33–35.
31. Agarwala S, Jain D, Joshi VR, Sule A 2005 Efficacy of alendronate, a bisphosphonate, in the treatment of AVN of the hip. A prospective open-label study. Rheumatology (Oxford) **44:**352–359.
32. Chiandussi S, Biasotto M, Dore F, Cavalli F, Cova MA, Di Lenarda R 2006 Clinical and diagnostic imaging of bisphosphonate-associated osteonecrosis of the jaws. Dentomaxillofac Radiol **35:**236–243.
33. Catalano L, Del Vecchio S, Petruzziello F, Fonti R, Salvatore B, Martorelli C, Califano C, Caparrotti G, Segreto S, Pace L, Rotoli B 2007 Sestamibi and FDG-PET scans to support diagnosis of jaw osteonecrosis. Ann Hematol **86:**415–423.
34. Ruggiero SL, Fantasia J, Carlson E 2006 Bisphosphonate-related osteonecrosis of the jaw: Background and guidelines for diagnosis, staging and management. Oral Surg Oral Med Oral Pathol Oral Radiol Endod **102:**433–441.
35. Bedogni A, Saia G, Ragazzo M, Bettini G, Capelli P, D'Alessandro E, Nocini PF, Russo LL, Lo Muzio L, Blandamura S 2007 Bisphosphonate-associated osteonecrosis can hide jaw metastases. Bone **41:**942–945.
36. Merigo E, Manfredi M, Meleti M, Corradi D, Vescovi P 2005 Jaw bone necrosis without previous dental extractions associated with the use of bisphosphonates (pamidronate and zoledronate): A four-case report. J Oral Pathol Med **34:**613–617.
37. Badros A, Weikel D, Salama A, Goloubeva O, Schneider A, Rapoport A, Fenton R, Gahres N, Sausville E, Ord R, Meiller T 2006 Osteonecrosis of the jaw in multiple myeloma patients: Clinical features and risk factors. J Clin Oncol **24:**945–952.
38. Hansen T, Kunkel M, Weber A, James Kirkpatrick C 2006 Osteonecrosis of the jaws in patients treated with bisphosphonates—histomorphologic analysis in comparison with infected osteoradionecrosis. J Oral Pathol Med **35:**155–160.
39. Russo T 2005 Agents of actinomycosis. In: Mandell G, Bennett J, Dolin R (eds.) Principles and Practice of Infectious Diseases, 6th ed. Elsevier Churchill Livingston, Philadelphia, PA, USA, pp. 2924–2934.
40. Migliorati CA, Casiglia J, Epstein J, Jacobsen PL, Siegel MA, Woo SB 2005 Managing the care of patients with bisphosphonate-associated osteonecrosis: An American Academy of Oral Medicine position paper. J Am Dent Assoc **136:**1658–1668.
41. Magopoulos C, Karakinaris G, Telioudis Z, Vahtsevanos K, Dimitrakopoulos I, Antoniadis K, Delaroudis S 2007 Osteonecrosis of the jaws due to bisphosphonate use. A review of 60 cases and treatment proposals. Am J Otolaryngol **28:**158–163.
42. Freiberger JJ, Padilla-Burgos R, Chhoeu AH, Kraft KH, Boneta O, Moon RE, Piantadosi CA 2007 Hyperbaric oxygen treatment and bisphosphonate-induced osteonecrosis of the jaw: A case series. J Oral Maxillofac Surg **65:**1321–1327.
43. Vescovi P, Merigo E, Meleti M, Manfredi M 2006 Bisphosphonate-associated osteonecrosis (BON) of the jaws: A possible treatment? J Oral Maxillofac Surg **64:**1460–1462.
44. Adornato MC, Morcos I, Rozanski J 2007 The treatment of bisphosphonate-associated osteonecrosis of the jaws with bone resection and autologous platelet-derived growth factors. J Am Dent Assoc **138:**971–977.
45. Agrillo A, Ungari C, Filiaci F, Priore P, Iannetti G 2007 Ozone therapy in the treatment of avascular bisphosphonate-related jaw osteonecrosis. J Craniofac Surg **18:**1071–1075.
46. Harper RP, Fung E 2007 Resolution of bisphosphonate-associated osteonecrosis of the mandible: Possible application for intermittent low-dose parathyroid hormone [rhPTH(1-34)]. J Oral Maxillofac Surg **65:**573–580.
47. Ruggiero SL, Gralow J, Marx RE, Hoff AO, Schubert MM, Huryn JM, Toth B, Damato K, Valero V 2006 Practical guidelines for the prevention, diagnosis, and treatment of osteonecrosis of the jaw in patinets with cancer. J Oncol Pract **2:**7–14.
48. Lacy MQ, Dispenzieri A, Gertz MA, Greipp PR, Gollbach KL, Hayman SR, Kumar S, Lust JA, Rajkumar SV, Russell SJ, Witzig TE, Zeldenrust SR, Dingli D, Bergsagel PL, Fonseca R, Reeder CB, Stewart AK, Roy V, Dalton RJ, Carr AB, Kademani D, Keller EE, Viozzi CF, Kyle RA 2006 Mayo clinic consensus statement for the use of bisphosphonates in multiple myeloma. Mayo Clin Proc **81:**1047–1053.
49. Kyle RA, Yee GC, Somerfield MR, Flynn PJ, Halabi S, Jagannath S, Orlowski RZ, Roodman DG, Twilde P, Anderson K 2007 American Society of Clinical Oncology 2007 clinical practice guideline update on the role of bisphosphonates in multiple myeloma. J Clin Oncol **25:**2464–2472.
50. Weitzman R, Sauter N, Eriksen EF, Tarassoff PG, Lacerna LV, Dias R, Altmeyer A, Csermak-Renner K, McGrath L, Lantwicki L, Hohneker JA 2007 Critical review: Updated recommendations for the prevention, diagnosis, and treatment of osteonecrosis of the jaw in cancer patients–May 2006. Crit Rev Oncol Hematol **62:**148–152.

© 2008 American Society for Bone and Mineral Research

Chapter 108. Periodontal Diseases and Oral Bone Loss

Keith L. Kirkwood

Department of Periodontics and Oral Medicine, University of Michigan, Ann Arbor, Michigan

INTRODUCTION

Periodontal diseases constitute a variety of inflammatory conditions affecting the health of the periodontium. The primary etiological component is bacterial-derived plaque accumulation around teeth. The two major categories of periodontal diseases are gingivitis and periodontitis that are distinguished from one another based on the extent of tissue loss that directly supports the teeth. Gingivitis is limited to the soft tissues surrounding the teeth where the extension has not caused any bone loss, whereas periodontitis is characterized by soft tissue and alveolar bone loss, resulting in decreased supporting structures for the teeth.

ETIOLOGY

Although >500 bacterial species have been identified within the oral cavity, plaque-associated periodontal diseases are associated with a relatively narrow subset of periodontal pathogens.[1] Susceptibility to these mixed infections is often a highly variable host immune response to these pathogens. Both genetic and environmental factors can influence the immune response and modify susceptibility to infection (Fig. 1). Periodontal pathogens produce harmful products and enzymes (e.g., hyaluronidases, collagenases, proteases) that breakdown extracellular matrices such as collagen and host cell membranes to produce nutrients for their growth[2] and subsequent tissue invasion (Fig. 2). Many of the microbial surface proteins and lipopolysaccharide (LPS) molecules are responsible for eliciting an immune response in the host resulting in local tissue inflammation.[3] *Porphymonas gingivalis*, *Aggregalibacter actinomycetemcomitans,* and other periodontal pathogens possess multiple virulence factors such as cytoplasmic membranes, peptidoglycans, outer membrane proteins, LPS, capsules, and cell surface fimbriae (with *P. gingivalis*).[4] Once immune and inflammatory processes are initiated, various inflammatory molecules such as proteases, matrix metalloproteinases (MMPs), cytokines, prostaglandins, and host enzymes are released from leukocytes, gingival fibroblasts, osteoblasts, or other tissue-derived cells.[5,6] Proteases can degrade the collagen permitting leukocyte infiltration.[7] Although the production of collagenase from infiltrating neutrophils and other resident periodontal tissues is part of the natural host response to infection, in periodontal disease, there is an imbalance between the level of activated tissue-destroying MMPs and their endogenous inhibitors.[8,9]

The inflammatory infiltrate from the gingival tissue can initiate tissue and alveolar bone destruction through the activation of several pro-inflammatory cytokines including interleukin (IL)-1β, TNF-α, and IL-6. Within the diseased periodontal tissues, activated osteoclasts are an integral component of bone destruction.[10,11] Multiple inflammatory signals have been shown to modulate RANKL, RANK, or osteoprotegerin (OPG) within periodontal tissues.[12] In the presence of periodontal pathogens (e.g., *A. actinomycetemcomitans*), $CD4^+$ T cells display increased expression of RANKL, triggering the activation of osteoclasts that leads to bone loss.[13] As long as the subgingival plaque persists, and this increase in microbial density propagates, the destructive periodontal lesion will remain. As the pocket deepens, the flora becomes more anaerobic, and the host response becomes more destructive and chronic. Periodontal tissue destruction can result in tooth loss if left unabated.[14,15]

DIAGNOSIS AND CLASSIFICATION OF PERIODONTAL DISEASES

Despite our current appreciation regarding the etiology and progression of periodontal diseases, the diagnosis and classification are still made from clinical assessments.[16,17] Information routinely collected includes medical history, history of previous/current periodontal problems, and a thorough clinical exam. Clinical parameters of inflammation (e.g., bleeding on probing into the periodontal sulcus), pocket depth measurements, clinical attachment levels, and evidence of plaque/calculus are used to arrive at a clinical diagnosis. Radiographic assessments of the extent of bone loss are primary criteria to determine whether bone loss has occurred.

Plaque-induced gingival diseases have traditionally been divided into two general categories: gingivitis and periodontitis.[16] Gingivitis is characterized by inflammation without any loss of periodontal supporting structures (i.e., connective tissue and bone), whereas the hallmark of periodontitis is pathological detachment of connective tissue, leading to the resorption of alveolar bone support around teeth. In 1999, the International Workshop of Periodontal Diseases and Conditions recognized a new classification system of seven different forms of plaque induced periodontal diseases.[17] Table 1 shows the classification and differences between the various types of periodontal diseases. Table 2 highlights diagnostic assessment criteria used to make diagnoses.

TREATMENT

In general, periodontal therapy is directed toward the reduction and elimination of inflammation, thereby allowing the gingival tissue to heal. Both personal and professional maintenance is critical in preventing the recurrence of inflammation. Because gingivitis is defined as a reversible condition where no bone loss has been documented, therapy for gingivitis is usually limited to plaque removal and personal plaque control instructions.

Therapy for periodontitis varies considerably depending on the severity of attachment loss, anatomical location variations, type of periodontitis, and therapeutic objectives.[18] Because periodontitis destroys the supporting alveolar bone and surrounding connective tissues, therapy is directed toward arresting the progression of ongoing disease and resolution of inflammation along with repair or regeneration of destroyed tissues. Therapeutic approaches for periodontitis fall into two main categories: (1) nonsurgical therapy and (2) surgical therapy (Tables 3 and 4).[18]

The most effective nonsurgical intervention is scaling and root planing combined with personal plaque control. Clear

The author states that he has no conflicts of interest.

Key words: teeth and dental applications, dentistry, periodontics, periodontal diseases

Electronic Databases: 1. American Dental Association—http://www.ada.org/prof/resources/topics/gum.asp. 2. American Academy of Periodontology—http://www.perio.org/resources-products/posppr2.html.

© 2008 American Society for Bone and Mineral Research

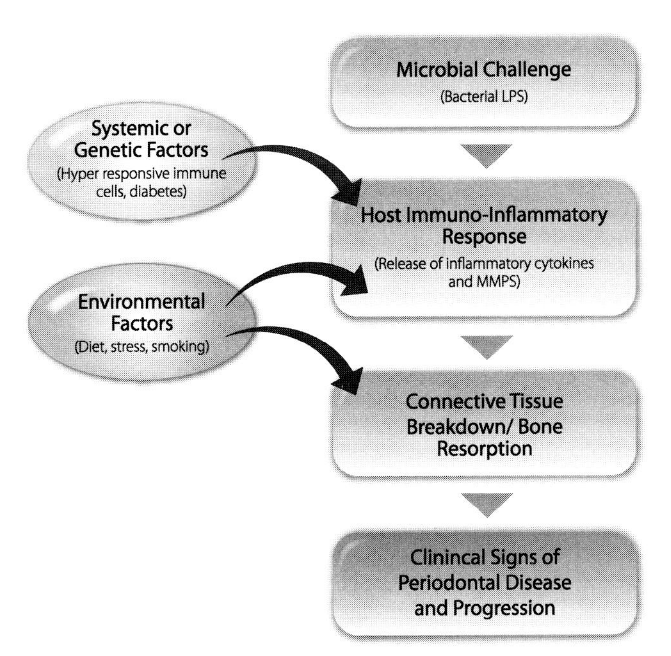

FIG. 1. Etiology of periodontal disease progression and factors that impact the disease progression. Periodontal diseases can occur when periodontal pathogenic bacteria is present in a susceptible host. Genetic and environmental factors modify the host immune response toward bacteria initiating tissue and bone destruction manifested as periodontitis.

benefits have been validated from scaling and root planing including reduction of inflammation, decreased pocket depths, gain of clinical attachment, and decreased progression of disease.[19]

Adjunctive pharmacological agents have been used to manage both gingivitis periodontitis. The benefit of topical antibacterial agents to help reduce bacterial plaque and prevent gingivitis has been documented in clinical trials. There are three agents currently approved by the American Dental Association (ADA) as anti-gingivitis agents (either mouth rinses or dentifrices).[19] One product lists essential oils (thymol, menthol, eucalyptol) and methyl salicylate as the active ingredients. The other two contain either chlorhexidine gluconate or triclosan. To carry the ADA seal, these agents must show plaque and inflammation reduction in 6-mo trials without any adverse effects. For periodontitis, both systemic and locally delivered chemotherapeutics have been used successfully for management of periodontal diseases. Systemically administered antibiotics have been used when mechanical therapy has not been sufficient (Table 3).[19] However, judicious use of antibiotics is recommended to treat acute infections or management of periodontitis in immunocompromised patients to avoid emergence of drug-resistant organisms.

Other systemic therapies focus on modulation of host–immune responses. Considerable efforts have focused on nonsteroidal anti-inflammatory (NSAIDS) and the collagenase inhibitor sub-antimicrobial dose doxycycline. Both have been shown to have beneficial effects, although long-term administration with any systemic agent may have potential negative side effects (Table 3). Controlled local delivery within periodontal pockets can change the pathogenic microflora and improve the clinical parameters of periodontitis. The Food and Drug Administration (FDA) has approved several systems including vinyl acetate fibers containing tetracycline, biodegradable chips containing chlorhexidine, and a minocycline-containing polymer for use as adjuncts to scaling and root planing.[19] In addition, a doxycycline bioabsorbable polymer gel has been approved as a monotherapy for reduction of periodontal disease clinical parameters.

Surgical therapy is primarily needed for access to facilitate mechanical instrumentation of the root surface debridement. The main objectives of periodontal surgery are to decrease bacterial etiological factors (e.g., subgingival calculus), reduce

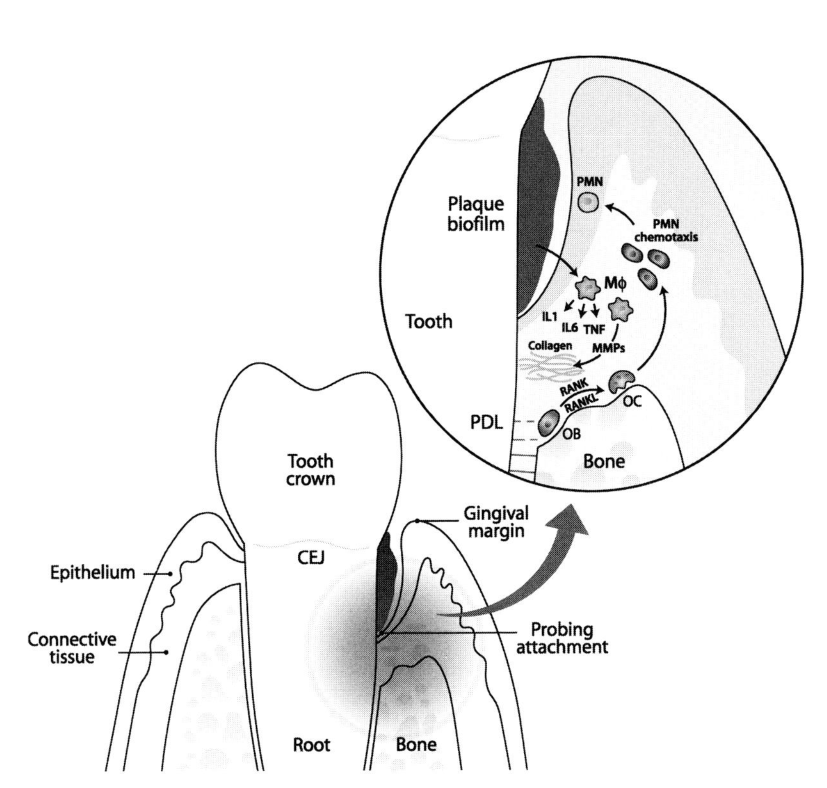

FIG. 2. Anatomy of periodontal tissues and cellular biology of inflammatory bone loss. The periodontal tissues surrounding the tooth include epithelium, connective tissue, periodontal ligament (PDL), and alveolar bone. Inset depicts cell responses to bacterial plaque biofilm on the root surface. Polymorphonuclear lymphocytes (PMNs) and macrophages (Mϕ) secrete inflammatory cytokines and matrix metalloproteinases (MMPs) to increase the immune response and degrade connective tissue matrix. Osteoclastogenesis is induced to these stimuli to increase bone resorption through RANKL/RANK system.

© 2008 American Society for Bone and Mineral Research

TABLE 1. GENERAL CLASSIFICATION OF PERIODONTAL DISEASES AND CONDITIONS*

Disease classification	*Comments*
Gingivitis	Generally plaque-induced but can be modified by endocrine factors, medications, and malnutrition. No alveolar bone loss.
Chronic periodontitis	Alveolar bone loss can be localized or generalized. Not based on age. Slow onset/progression.
Aggressive periodontitis	Alveolar bone loss can be localized or generalized. Not based on age. Rapid onset/progression.
Periodontitis as a manifestation of systemic diseases	Can be associated with hematological (e.g., netropenia), or genetic disorders (e.g., Papillon-Lefèvre syndrome)
Necrotizing periodontal diseases	Could be gingivitis or periodontitis forms.
Abscesses of the periodontium	Could be gingival or periodontal abscess.
Periodontitis associated with endodontic lesions	Associated with infections within the tooth.

*Adapted with permission from the American Academy of Periodontology from Armitage GC 1999 Development of a classification system for periodontal diseases and conditions. Ann Periodontol **4**:1–6.

TABLE 2. DIAGNOSIS OF PERIODONTAL DISEASES

Diagnostic assessments	*Specific parameter*	*Comments*
Clinical	Probing depth	Shallow pocket depths associated with less disease progression
	Bleeding on probing	Lack of bleeding associated with less inflammation
	Radiographs	Absence of bone loss associated with less disease progression
Microbiological	Culture for periodontal pathogens	May be useful to detect exact pathogens in patients refractory to conventional treatment
	DNA/biochemical probes	Identify noncultivatable organisms
Biomarkers	Inflammatory cytokines	IL-1 genotype may identify susceptible patients
	Pyridinoline cross-linked carboxyterminal telopeptide of type I collagen	Correlates well with clinical parameters and presence of periodontal pathogens

TABLE 3. NONSURGICAL TREATMENT OF PERIODONTITIS

	Treatment category	*Treatment or agent*	*Comments*
Mechanical	Mechanical debridement	Scaling and root planing—manual instrumentation and ultrasonics	Dec. inflammation, Dec. probing depth, clinical attachment levels
Chemotherapy	Mouth rinses dentifrices (toothpastes)	Chlorhexidine triclosan, essential oils	ADA approved for reduction of gingivitis
	Sustained release	Tetracyclines (doxycycline and minocycline), chlorhexidine	Can be used as adjunctive therapies with scaling and root planing
	Systemic antibiotics	Tetracyclines, clindamycin, metronidazole/amoxicillin	May be used in more aggressive periodontitis cases
Host response	MMP inhibitors	Low-dose doxycycline	May decrease progression of attachment loss
	NSAIDs	Flurbiprophen	May decrease progression of attachment loss

TABLE 4. SURGICAL TREATMENT OF PERIODONTITIS

	Treatment category	*Clinical procedure*	*Comments*
Pocket reduction	Pocket reduction surgery	Mucoperiosteal flap for access to debride root surface	Dec. pocket depthpatient access for oral hygiene. Can enhance access for restorative dentistry.
Regeneration	Guided tissue regeneration	Mucoperiosteal flap to gain access and a barrier membrane placed to facilitate new periodontal ligament, cementum and bone on previously diseased roots	Can use resorbable or nonresorbable membranes with or without bone grafts. Dec. pocket depths. Clinical attachment. Fill osseous defects.
	Bone grafts (autogenous, allogenic, xenografts)		Can use resorbable or nonresorbable membranes with or without bone grafts. Dec. pocket depths. Clinical attachment. Fill osseous defects. Autogenous graft is gold standard.
	Biologicals (enamel-derived matrix proteins, BMPs, platelet-derived growth factors [PDGFs])		Fill osseous defects. Improved clinical attachment and Dec. pocket depth. BMPs not FDA approved for intraoral use. PDGF with absorbable carrier is FDA approved.

© 2008 American Society for Bone and Mineral Research

the pocket depth, and regenerate lost periodontal supporting tissues. Flap access surgeries increase the surgeon's ability to instrument deep pocket depth areas and areas between roots (furcations) for more complete calculus removal. Osseous resection has traditionally been performed to more effectively reduce pocket depths. However, the optimal goal is to restore lost periodontal tissues. Several different regenerative surgical strategies have been used including bone grafting (autogenous and allografts), along with guided tissue regeneration (GTR) with or without bone grafts.[20] In the last 10 yr, several biological agents have been used to stimulate periodontal regeneration, including growth factors (e.g., platelet-derived growth factor), other proteins (e.g., enamel-derived growth factor), or synthetic peptides (e.g., a 15 amino acid type I collagen fragment; Table 4).[18,20]

FUTURE PERSPECTIVES

Current basic research has focused on a newer field termed osteoimmunology centered on molecular mechanisms of the immune system that impact on inflammatory bone loss and the skeletal system. Clinical diagnostics addressing novel salivary diagnostics measuring biomarkers that may be predictive of periodontal severity, susceptibility, or perhaps predict future disease are actively being pursued. In the clinical setting, tissue engineering and regeneration remain active areas of study. New biologicals are being evaluated in clinical trials to enhance predictability of grafting materials used in periodontal regeneration as well as dental implant therapy to replace missing teeth.

REFERENCES

1. Periodontology AAo 1999 Informational paper. The pathogenesis of periodontal diseases. J Periodontol **70:**457–470.
2. Bartold PM, Page RC 1986 The effect of chronic inflammation on gingival connective tissue proteoglycans and hyaluronic acid. J Oral Pathol **15:**367–374.
3. Darveau RP, Tanner A, Page RC 1997 The microbial challenge in periodontitis. Periodontol 2000 **14:**12–32.
4. Offenbacher S 1996 Periodontal diseases: Pathogenesis. Ann Periodontol **1:**821–878.
5. Graves DT, Jiang Y, Valente AJ 1999 The expression of monocyte chemoattractant protein-1 and other chemokines by osteoblasts. Front Biosci **4:**D571–D580.
6. Graves DT, Oskoui M, Volejnikova S, Naguib G, Cai S, Desta T, Kakouras A, Jiang Y 2001 Tumor necrosis factor modulates fibroblast apoptosis, PMN recruitment, and osteoclast formation in response to P. gingivalis infection. J Dent Res **80:**1875–1879.
7. Andrian E, Grenier D, Rouabhia M 2004 In vitro models of tissue penetration and destruction by Porphyromonas gingivalis. Infect Immun **72:**4689–4698.
8. Uchida M, Shima M, Shimoaka T, Fujieda A, Obara K, Suzuki H, Nagai Y, Ikeda T, Yamato H, Kawaguchi H 2000 Regulation of matrix metalloproteinases (MMPs) and tissue inhibitors of metalloproteinases (TIMPs) by bone resorptive factors in osteoblastic cells. J Cell Physiol **185:**207–214.
9. Golub LM, Lee HM, Greenwald RA, Ryan ME, Sorsa T, Salo T, Giannobile WV 1997 A matrix metalloproteinase inhibitor reduces bone-type collagen degradation fragments and specific collagenases in gingival crevicular fluid during adult periodontitis. Inflamm Res **46:**310–319.
10. Assuma R, Oates T, Cochran D, Amar S, Graves DT 1998 IL-1 and TNF antagonists inhibit the inflammatory response and bone loss in experimental periodontitis. J Immunol **160:**403–409.
11. Crotti T, Smith MD, Hirsch R, Soukoulis S, Weedon H, Capone M, Ahern MJ, Haynes D 2003 Receptor activator NF kappaB ligand (RANKL) and osteoprotegerin (OPG) protein expression in periodontitis. J Periodontal Res **38:**380–387.
12. Lerner UH 2004 New molecules in the tumor necrosis factor ligand and receptor superfamilies with importance for physiological and pathological bone resorption. Crit Rev Oral Biol Med **15:**64–81.
13. Teng YT, Nguyen H, Gao X, Kong YY, Gorczynski RM, Singh B, Ellen RP, Penninger JM 2000 Functional human T-cell immunity and osteoprotegerin ligand control alveolar bone destruction in periodontal infection. J Clin Invest **106:**R59–R67.
14. Kinane DF, Lappin DF 2001 Clinical, pathological and immunological aspects of periodontal disease. Acta Odontol Scand **59:**154–160.
15. Lappin DF, MacLeod CP, Kerr A, Mitchell T, Kinane DF 2001 Anti-inflammatory cytokine IL-10 and T cell cytokine profile in periodontitis granulation tissue. Clin Exp Immunol **123:**294–300.
16. Periodontology AAo 2003 Position paper. Diagnosis of peridontal diseases. J Periodontol **74:**1237–1247.
17. Armitage GC 1999 Development of a classification system for periodontal diseases and conditions. Ann Periodontol **4:**1–6.
18. Periodontology AAo 2001 Academy report. Treatment of plaque-induced gingivitis, chronic periodontitis, and other clinical conditions. J Periodontol **72:**1790–1800.
19. Periodontology AAo 2002 Informational paper. Modulation of host response in periodontal therapy. J Periodontol **73:**460–470.
20. Periodontology AAo 2005 Position paper. Periodontal regeneration. J Periodontol **76:**1601–1622.

Chapter 109. Oral Manifestations of Metabolic Bone Disease

Flavia Pirih,[1] Laurie K. McCauley,[1] and Jean Wactawski-Wende[2]

[1]Department of Periodontics and Oral Medicine, University of Michigan School of Dentistry, Ann Arbor, Michigan; [2]Department of Social and Preventive Medicine, University of Buffalo, Buffalo, New York

OSTEOPOROSIS AND ORAL BONE

Osteoporosis is a skeletal disorder characterized by compromised bone strength, predisposing to an increased risk of fracture. Bone strength is determined by BMD and bone quality. Periodontitis is an infection-mediated process characterized by resorption of the alveolar bone and loss of soft tissue attachment to the tooth. It is a major cause of tooth loss and edentulousness in adults. The interaction of host infection and host susceptibility (i.e., skeletal responsivity) in the incidence and progression of oral bone loss and periodontal disease is an area of intensive investigation, with most published studies support-

Dr. McCauley owns stock in Lilly and Amgen. All other authors state that they have no conflicts of interest.

Key words: periodontal diseases, osteoporosis, BMD, tooth loss, alveolar crestal height, oral bone loss, oral bone density, residual ridge resorption, Paget's, hyperparathyroidism, renal osteodystrophy

© 2008 American Society for Bone and Mineral Research

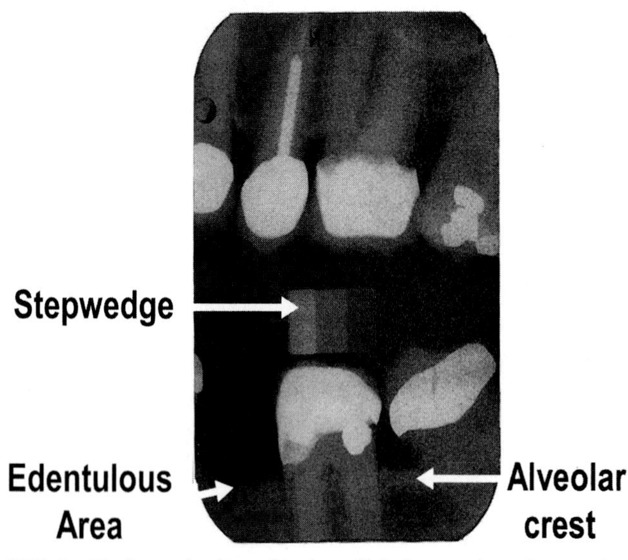

FIG. 1. Radiograph of a patient in a clinical research study to evaluate oral bone over time. The alveolar crest is cortical bone and typically radiopaque. With disease progression, density of the crest is lost as is horizontal and vertical bone height. The stepwedge provides a reference to compare osseous density with a standard. The edentulous area (after tooth loss) typically loses bone with time.

ing a positive association. However, because many studies are cross-sectional, include small sample sizes, and have inadequate control of potential confounding factors, our understanding of the interplay between these common conditions is limited.

METHODS TO ASSESS ORAL BONE

Techniques used to assess oral bone loss typically involve the use of radiographic measures. These include measures of alveolar crestal height (ACH), measures of resorption of the residual ridge after tooth loss (RRR), and assessment of oral BMD. ACH is assessed using bitewing oral radiographs (Fig. 1). Bitewings are digitized and distance measurements made from fixed points on the teeth (the cemento-enamel junction [CEJ]) to the top of the alveolar bone (crest) adjacent to the tooth. ACH is usually measured at two sites per tooth and is reported as the average loss of bone height in all teeth measured in the mouth (mean ACH). The larger the mean, the worse the bone loss surrounding the teeth. After a tooth is lost, there is loss of the bone in that region. This is called residual ridge resorption (RRR). Most often, the extent of RRR is described in edentulous subjects, but RRR can be described in dentate subjects that have lost one or more teeth (partially edentulous).

Measurement of oral BMD has been assessed using a variety of techniques that include measurement of absolute BMD, such as DXA, dual photon absorptiometry (DPA), QCT, and radiographic absorptiometry (RA), and studies that approximate change in oral density over time, such as computer-assisted densitometric image analysis (CADIA). All techniques used are limited by cost or precision. QCT provides a valuable assessment of oral density because it allows for assessment of density in regions of the oral bones not obstructed by teeth. However, it is expensive, and there is high exposure to radiation. DPA and DXA can be used to assess the mandible; however, positioning and reproducibility for oral density measurements and obstruction of teeth in dentate subjects makes the measurement more difficult. RA using bite-wing radiographs is a very reproducible method when positioning aids are used. However, this method is restricted to regions accessible to bitewing radiographs.

Density assessments in human cadavers vary by regions of the mandible, dentate status, age, and sex.[1] Often measurements are made in the mandible because of easier access. Each region (molar, premolar, or incisor) has different cortical thickness that may affect density, especially for techniques that are 2D. The ability to measure oral BMD is restricted in part by whether teeth are present, with edentulous subjects being easier to measure. However, edentulous subjects may have marked resorption of the residual ridge that, in turn, affects the area of comparison and ultimately density. Panoramic films have been used to predict skeletal density and have shown promise in some studies, primarily for measures of cortical bone at the inferior cortex of the mandible.[2,3]

STUDIES OF OSTEOPOROSIS, ALVEOLAR CRESTAL HEIGHT LOSS, AND RESIDUAL RIDGE RESORPTION

A relationship between osteoporosis or low skeletal BMD and ACH and RRR has been shown in most studies.[4,5] Loss of ACH and RRR is more predominant in women than men, particularly among older subjects. Women ≥70 yr of age were more than three times more likely to have moderate or severe ACH levels if they were osteoporotic.[6] Age is an important factor in loss of height of the residual ridge in edentulous adult mandibles in women but not in men.[7] In addition, older female subjects are more likely to be edentulous than older male and both male and female younger subjects.[8] Skeletal osteoporosis strongly affects RRR in edentulous patients, as does female sex and increasing age.[4]

Prospective studies of the association of ACH and skeletal BMD are limited. Evaluation of other parameters of periodontal bone loss have shown mixed correlation.[9,10] Brennan et al.[9] reported strong correlations between systemic BMD and clinical attachment loss in women. On the other hand, a study by Weyant et al.[10] reported a weak association. The main difficulty in coming to a consensus in this area is that studies are limited in number and are largely cross-sectional. Longitudinal studies composed of large populations are necessary to delineate the relationship between these two complex disease conditions.

OSTEOPOROSIS AND ORAL BMD

Studies have found that oral BMD correlates with systemic BMD and osteoporosis.[11] Osteoporotic women show significantly lower mandibular and forearm BMC values than control women with a history of osteoporotic fractures.[12] A significant correlation between forearm density and oral BMD has also been reported.[13] In addition, women with vertebral fractures had significantly lower mandibular bone mass and density.[13–15]

Postmortem studies in edentulous women and men found specific gravity of the mandible and radius decreased with increasing age.[16] Women had lower densities than men, and mandibular and radius measures were highly correlated. BMC in edentulous mandibles taken from cadavers by DPA was found to increase in male subjects with advancing age, whereas mandibular BMC of female subjects tended to decrease with advancing age.[17] Mandibular density measured by QCT differed between partially and totally edentulous postmenopausal women who had been edentulous >12 yr, suggesting years edentulous may be important when assessing the relationship between skeletal and mandibular density.[18]

© 2008 American Society for Bone and Mineral Research

Hormone therapy and endogenous hormone levels have been assessed for their role in BMD of oral and skeletal regions. Positive effects of hormone therapy on BMD have been reported.[19] A highly significant net gain in alveolar density was found in one study of 17-β-estradiol, (E2)-sufficient women compared with those E2-deficient using CADIA.[20] In a 2-yr study of the oral BMD and bone height changes in nonsmoking postmenopausal women with periodontal disease, women with low BMD had higher rates of density loss and ACH loss than those with higher BMD.[21] Estrogen deficiency was also found to be associated with greater loss of oral BMD and ACH in osteopenic women. In addition, oral and postcranial ACH and BMD were increased in postmenopausal women taking oral estrogen therapy compared with placebo.[22] It is important to note that the positive response to hormone replacement therapy in the femur was greater than in the alveolar bone, suggesting that such systemic bone-sparing agents have a greater effect in the appendicular skeleton than the craniofacial skeleton in dentate individuals.

In comparison with the dentate state, the edentulous mandible may be much more sensitive to metabolic changes. A study comparing the mandibular BMD (DXA), hip BMD (DXA), and ultrasound assessment of the calcaneus and hand in postmenopausal edentulous women found the largest change in density occurring in the mandible, followed by the femoral neck and Ward's triangle.[23]

OSTEOPOROSIS AND TOOTH LOSS

Most studies reported a positive relationship between osteoporosis and tooth loss. A cross-sectional study of mandibular BMD in osteoporotic women found tooth loss and edentulism were significantly more common in the osteoporotic group.[14] On average, the osteoporotic women lost more mandibular teeth compared with women with normal BMD. On the other hand, a second study observed that more osteoporotic subjects were edentulous, although the difference was not statistically significant.[15]

Taguchi et al.[24] studied the relation between tooth loss and oral BMD in both sexes 3–88 yr of age. In female subjects, a decrease in mandibular cortical bone width was positively correlated with tooth loss. In women >70 yr of age, the association was most apparent. Tooth loss was highly correlated with prevalence of spinal fracture.[25] In addition, there was a positive relationship between loss of the posterior teeth and alveolar and spinal BMD.[26]

Skeletal BMD in male subjects correlated with self-reported tooth loss, after controlling for age, BMI, and smoking.[27] However, the association of tooth loss with BMD was insignificant in women after controlling for the effects of age, BMI, and smoking. Overall, 24% of the men and 27% of the women were edentulous, which was less than expected from previous estimates from this region.

A study of estrogen replacement use after menopause and tooth retention in 488 women found estrogen users had more teeth than nonusers, and duration of estrogen use independently predicted number of remaining teeth.[28] Long-term users of estrogen had more teeth than never users. Estrogen use was shown to be protective for tooth loss, regardless of type of tooth or location in the mouth. Users of estrogen (1–4 yr) had 1.1 more teeth than nonusers.

A 3-yr prospective Swedish study of 14,375 older men and women found that women with the fewest teeth at baseline (lowest tertile) had a risk of hip fracture that was twice that of the women in the highest two tertiles.[29] The association between tooth loss and hip fracture was stronger in the men studied. The risk of fracture was 3-fold higher in men with the fewest teeth at baseline, although the absolute number of hip fractures were greater in women. A 10-yr prospective study found alveolar bone loss (ACH) at baseline was the strongest independent predictor of incident tooth loss in postmenopausal women.[30]

Not all studies have found an association between BMD and tooth loss. Klemetti and Vainio[31] did not find an association with tooth loss and BMD in a group of Finnish women; however, dental practices in Finland may have led to extractions for preventative purposes rather than as a result of underlying disease. One half of all women studied had all their maxillary teeth, and 25% of women had all their mandibular teeth extracted before the age of 30, suggesting the reason for tooth loss was likely not periodontal in nature. Civitelli et al.[22] did not find a relationship between patients on hormone replacement therapy and placebo controls; however, their patient population generally had good oral health. This suggests that the local factors operative to predispose to tooth loss are critical and, when combined with a systemic insult such as osteoporosis, the result is likely more deleterious.

Several studies of younger women have not shown an association between BMD and periodontal disease. Hildebolt et al.[32] found no association between spine or femur BMD and number of remaining teeth among subjects enrolled in a hormone replacement study trial. Subjects were relatively young postmenopausal women who had 10 or more teeth present and were generally periodontally healthy, limiting the ability to detect an association between BMD and tooth loss. Earnshaw et al.[33] found no relationship between BMD and tooth number in white women between the ages of 45 and 59, who were within 12 yr of menopause. Analysis included adjustment for age, years since menopause, hormone replacement, and center. A recent prospective study of a subgroup of participants from the Study of Osteoporotic Fractures found absolute BMD and percentage change in BMD were similar in women dentate and edentulous at baseline examination.[34] Additional studies are needed to further define this relationship in larger cohorts of women, especially prospective cohorts where temporality can be established.

OSTEOPOROSIS AND BONE REGENERATION IN THE ORAL CAVITY

A common concern in the dental treatment of patients with osteoporosis is whether they may present with compromised osseointegration of dental implants and osseous regeneration. Recent reviews suggest that this may be true with decreased success rates, especially in the maxilla[35,36]; however, the diagnosis of osteoporosis is not a contraindication for implant therapies. Furthermore, studies in this area are complicated by the frequent lack of information and standardization of therapies that patients are using for the treatment of their osteoporosis such as bisphosphonates. There is a clear need for studies evaluating the impact of bisphosphonates on osseous healing in the oral cavity.

POTENTIAL MECHANISMS AND THE BIOLOGICAL BASIS

Based on our knowledge of osteopenia and periodontal disease and the risk factors that affect both, it is reasonable to propose the following mechanisms.[5] First, low BMD in the oral bones may be associated with low systemic bone. This low BMD or loss of BMD may lead to more rapid resorption of alveolar bone after insult by periodontal bacteria. With less dense oral bone to start with or faster loss of bone, loss of bone surrounding the teeth may occur more rapidly. Second, sys-

© 2008 American Society for Bone and Mineral Research

temic factors affecting bone remodeling such as resorptive cytokines also modify local tissue response to periodontal infection. Third, genetic factors that predispose a person to systemic bone loss also influence or predispose an individual to periodontal destruction. Last, certain lifestyle factors, such as cigarette smoking and suboptimal calcium intake, among systemic factors like diabetes, may put individuals at risk for development of both systemic osteopenia and oral bone loss.

PAGET'S DISEASE

Paget's disease is the most common metabolic bone disease after osteoporosis and affects ~1.3% of the population over the age of 55.[37] Seventeen percent of Paget's disease patients have manifestations in the oral cavity.[38] The lesions in the oral cavity are more commonly found in the maxilla than in the mandible.[39] Alveolar ridge enlargement is observed and can lead to spreading of the teeth and an abnormal occlusal pattern. Because of alveolar ridge enlargement, edentulous patients with Paget's disease may require new dentures more frequently to compensate for the enlargement. Enlargement of the middle third of the face can also be observed.[40,41] Another complication of Paget's disease is hypercementosis (excessive deposition of the mineralized cementum structure of the tooth root), which may result in tooth ankylosis. Conversely, Paget's disease may lead to loosening of the teeth during its osteolytic phase.[39] Because the majority of affected patients have no symptoms, the diagnosis is primarily attained by incidental biochemical testing or radiologic examination.[42] Radiographically, in the beginning stage, the bone exhibits reduced radiodensity and may resemble cemento-osseous dysplasia. At a more advanced stage, numerous irregular radiopaque areas become evident, taking on a "cotton wool" appearance.[38] Histologically, there are increased numbers of osteoclasts; however, osteoblastic activity is also observed with evidence of a continuous process of abnormal deposition and resorption of bone.[43]

PRIMARY HYPERPARATHYROIDISM

Primary hyperparathyroidism (PHPT) is excessive secretion of PTH and is most commonly caused by a benign parathyroid adenoma. PHPT is the most common cause of hypercalcemia, usually diagnosed biochemically during a routine blood test.[44] Several manifestations of hyperparathyroidism have been reported to be observed in the oral cavity including partial or complete loss of lamina dura, increased periodontal ligament width, decreased alveolar bone density, and at more advanced stages, brown tumor formation.[45,46] Loss of lamina dura is not a pathognomonic sign of the disease because it is also seen in other diseases such as fibrous Cushing's syndrome and osteomalacia. Brown tumor formation is a chronic osseous lesion, with abundant hemorrhage. It is normally well demarcated and can be unilocular or multilocular. The Brown tumor of hyperparathyroidism is rarely observed because the disease is typically diagnosed at early stages.[45] Although several manifestations of hyperparathyroidism can be observed in the oral cavity, these lesions are usually seen later in the course of the disease. A recent study found reduced parameters of cortical density in the mandible but in non–tooth-associated bone.[47] Parameters of periodontal health such as attachment loss, probing depth, and bleeding on probing were unchanged in PHPT patients, but there was a correlation between increased widening of the periodontal ligament and serum PTH levels. Interestingly, increased presence of tori was found that may reflect an anabolic-like action of PHPT in the oral cavity. The incidence of oral tori and exostoses has been correlated with systemic BMD in postmenopausal white women and in patients with a mutation in the low-density lipoprotein (LDL) receptor–related protein-5.[48,49]

RENAL OSTEODYSTROPHY

Renal osteodystrophy is a term used to describe the skeletal complications of end-stage renal disease. The major contributors to the disease are secondary hyperparathyroidism and 1,25-dihydroxycalciferol deficiency. In the oral cavity, renal osteodystrophy shares many features of primary hyperparathyroidism including loss of lamina dura, "ground glass" appearance of the bone, loss of trabeculation, and brown tumor formation.[50] In addition, renal osteodystrophy may lead to an enlargement of the jaws and cementum resorption.[51] Although the manifestations of renal osteodystrophy in the oral cavity share many similarities with primary hyperparathyroidism, renal osteodystrophy does not lead to widening of the periodontal ligament. Furthermore, indices of periodontal disease are unchanged in patients with secondary hyperparathyroidism from chronic renal failure.[52]

SUMMARY

Several important questions remain regarding the correlation between systemic and oral BMD including determination of normal ranges of mandibular density by age and sex, further comparison of mandibular density in normal and metabolic bone disease patients, assessment of longitudinal progression of mandibular bone loss, comparison of the rate of bone loss in the mandible compared with other skeletal regions, and the effects of different therapies on mandibular density compared with other skeletal sites. Prospective study of the association between metabolic bone disease and oral bone status is needed in large cohorts where temporal sequence can be established and where adequate assessment and control of confounding variables can be done. Further study of measurement error in various techniques to assess oral density is also needed. Metabolic bone diseases and periodontal disease are major health concerns in the United States, especially in older populations. Studies that improve our understanding of the mechanisms by which metabolic bone diseases are associated with oral bone are needed and will be increasingly important in issues of quality of life related to these very prevalent disorders in older Americans.

REFERENCES

1. D'Amelio P, Panattoni GL, Di Stefano M, Nassisi R, Violino D, Isaia GC 2002 Densitometric study of dry human mandible. J Clin Densitom **5**:363–367.
2. Taguchi A, Ohtsuka M, Nakamoto T, Naito K, Tsuda M, Kudo Y, Motoyama E, Suei Y, Tanimoto K 2007 Identification of postmenopausal women at risk of osteoporosis by trained general dental practitioners using panoramic radiographs. Dentomaxillofac Radiol **36**:149–154.
3. Devlin H, Karayianni K, Mitsea A, Jacobs R, Lindh C, van der Stelt P, Marjanovic E, Adams J, Pavitt SKH 2007 Diagnosing osteoporosis by using dental panoramic radiographs:The OSTEODENT project. Oral Surg Oral Med Oral Pathol Oral Radiol Endod **104**:821–828.
4. Hirai T, Ishijima T, Hashikawa Y, Yajima T 1993 Osteoporosis and reduction of residual ridge in edentulous patients. J Prosthet Dent **69**:49–56.
5. Tezal M, Wactawski-Wende J, Grossi SG, Ho AW, Dunford R, Genco RJ 2000 The relationship between bone mineral density and periodontitis in postmenopausal women. J Periodontol **71**:1492–1498.
6. Wactawski-Wende J, Hausmann E, Hovey K, Trevisan M, Grossi

© 2008 American Society for Bone and Mineral Research

S, Genco RJ 2005 The association between osteoporosis and alveolar crestal height in postmenopausal women. J Periodontol **76**(11 Suppl):2116–2124.
7. Humphries S, Devlin H, Worthington H 1989 A radiographic investigation into bone resorption of mandibular alveolar bone in elderly edentulous adults. J Dent **17**:94–96.
8. Ortman LF, Hausmann E, Dunford RG 1989 Skeletal osteopenia and residual ridge resorption. J Prosthet Dent **61**:321–325.
9. Brennan RM, Genco RJ, Hovey KM, Trevisan M, Wactawski-Wende J 2007 Clinical Attachment Loss, Systemic Bone Density, and Subgingival Calculus in Postmenopausal Women. J Periodontol **78**:2104–2111.
10. Weyant RJ, Pearlstein ME, Churak AP, Forrest K, Famili P, Cauley JA 1999 The association between osteopenia and periodontal attachment loss in older women. J Periodontol **70**:982–991.
11. von Wowern N 1988 Bone mineral content of mandibles: Normal reference values–rate of age-related bone loss. Calcif Tissue Int **43**:193–198.
12. von Wowern N, Klausen B, Kollerup G 1994 Osteoporosis: A risk factor in periodontal disease. J Periodontol **65**:1134–1138.
13. Kribbs PJ, Smith DE, Chesnut CH III 1983 Oral findings in osteoporosis. Part II: Relationship between residual ridge and alveolar bone resorption and generalized skeletal osteopenia. J Prosthet Dent **50**:719–724.
14. Kribbs PJ, Chesnut CH III, Ott SM, Kilcoyne RF 1990 Relationships between mandibular and skeletal bone in a population of normal women. J Prosthet Dent **63**:86–89.
15. Kribbs PJ 1990 Comparison of mandibular bone in normal and osteoporotic women. J Prosthet Dent **63**:218–222.
16. Henrikson PA, Wallenius K 1974 The mandible and osteoporosis (1). A qualitative comparison between the mandible and the radius. J Oral Rehabil **1**:67–74.
17. Solar P, Ulm CW, Thornton B, Matejka M 1994 Sex-related differences in the bone mineral density of atrophic mandibles. J Prosthet Dent **71**:345–349.
18. Klemetti E, Vainio P, Lassila V 1994 Mineral density in the mandibles of partially and totally edentate postmenopausal women. Scand J Dent Res **102**:64–67.
19. Jacobs R, Ghyselen J, Koninckx P, van Steenberghe D 1996 Long-term bone mass evaluation of mandible and lumbar spine in a group of women receiving hormone replacement therapy. Eur J Oral Sci **104**:10–16.
20. Payne JB, Zachs NR, Reinhardt RA, Nummikoski PV, Patil K 1997 The association between estrogen status and alveolar bone density changes in postmenopausal women with a history of periodontitis. J Periodontol **68**:24–31.
21. Payne JB, Reinhardt RA, Nummikoski PV, Patil KD 1999 Longitudinal alveolar bone loss in postmenopausal osteoporotic/osteopenic women. Osteoporos Int **10**:34–40.
22. Civitelli R, Pilgram TK, Dotson M, Muckerman J, Lewandowski N, Armamento-Villareal R, Yokoyama-Crothers N, Kardaris EE, Hauser J, Cohen S, Hildebolt CF 2002 Alveolar and postcranial bone density in postmenopausal women receiving hormone/estrogen replacement therapy: A randomized, double-blind, placebo-controlled trial. Arch Intern Med **162**:1409–1415.
23. Drozdzowska B, Pluskiewicz W 2002 Longitudinal changes in mandibular bone mineral density compared with hip bone mineral density and quantitative ultrasound at calcaneus and hand phalanges. Br J Radiol **75**:743–747.
24. Taguchi A, Tanimoto K, Suei Y, Wada T 1995 Tooth loss and mandibular osteopenia. Oral Surg Oral Med Oral Pathol Oral Radiol Endod **79**:127–132.
25. Taguchi A, Tanimoto K, Suei Y, Otani K, Wada T 1995 Oral signs as indicators of possible osteoporosis in elderly women. Oral Surg Oral Med Oral Pathol Oral Radiol Endod **80**:612–616.
26. Taguchi A, Suei Y, Ohtsuka M, Otani K, Tanimoto K, Hollender LG 1999 Relationship between bone mineral density and tooth loss in elderly Japanese women. Dentomaxillofac Radiol **28**:219–223.
27. May H, Reader R, Murphy S, Khaw KT 1995 Self-reported tooth loss and bone mineral density in older men and women. Age Ageing **24**:217–221.
28. Krall EA, Dawson-Hughes B, Hannan MT, Wilson PW, Kiel DP 1997 Postmenopausal estrogen replacement and tooth retention. Am J Med **102**:536–542.
29. Astrom J, Backstrom C, Thidevall G 1990 Tooth loss and hip fractures in the elderly. J Bone Joint Surg Br **72**:324–325.
30. Tezal M, Wactawski-Wende J, Grossi SG, Dmochowski J, Genco RJ 2005 Periodontal disease and the incidence of tooth loss in postmenopausal women. J Periodontol **76**:1123–1128.
31. Klemetti E, Vainio P 1993 Effect of bone mineral density in skeleton and mandible on extraction of teeth and clinical alveolar height. J Prosthet Dent **70**:21–25.
32. Hildebolt CF, Pilgram TK, Dotson M, Yokoyama-Crothers N, Muckerman J, Hauser J, Cohen S, Kardaris E, Vannier MW, Hanes P, Shrout MK, Civitelli R 1997 Attachment loss with postmenopausal age and smoking. J Periodontal Res **32**:619–625.
33. Earnshaw SA, Keating N, Hosking DJ, Chilvers CE, Ravn P, McClung M, Wasnich RD 1998 Tooth counts do not predict bone mineral density in early postmenopausal Caucasian women. EPIC study group. Int J Epidemiol **27**:479–483.
34. Famili P, Cauley J, Suzuki JB, Weyant R 2005 Longitudinal study of periodontal disease and edentulism with rates of bone loss in older women. J Periodontol **76**:11–15.
35. Erdoğan O, Shafer D, Taxel P, Freilich MA 2007 A review of the association between osteoporosis and alveolar ridge augmentation. Oral Surg Oral Med Oral Pathol Oral Radiol Endod **106**:738e1–e13.
36. Alsaadi G, Quirynen M, Komarek A, van Steenberghe D 2007 Impact of local and systemic factors on the incidence of oral implant failures, up to abutment connection. J Clin Periodontol **34**:610–617.
37. Tiegs RD, Lohse CM, Wollan PC, Melton LJ 2000 Long-term trends in the incidence of Paget's disease of bone. Bone **27**:423–427.
38. Neville B, Damm D 2002 Bone Pathology in Oral and Maxillofacial Pathology. 2nd ed. Saunders, Philadelphia, PA, USA.
39. Smith BJ, Eveson JW 1981 Paget's disease of bone with particular reference to dentistry. J Oral Pathol **10**:233–247.
40. Akin RK, Barton K, Walters PJ 1975 Paget's disease of bone. Report of a case. Oral Surg Oral Med Oral Pathol **39**:707–712.
41. Carrillo R, Morales A, Rodriguez-Peralto JL, Lizama J, Eslava JM 1991 Benign fibro-osseous lesions in Paget's disease of the jaws. Oral Surg Oral Med Oral Pathol **71**:588–592.
42. Delmas PD, Meunier PJ 1997 The management of Paget's disease of bone. N Engl J Med **336**:558–566.
43. Gherardi G, Lo Cascio V, Bonucci E 1980 Fine structure of nuclei and cytoplasm of osteoclasts in Paget's disease of bone. Histopathology **4**:63–74.
44. Silverberg SJ, Bilezikian J 2001 Clinical presentation of primary hyperparathyroidism in the U.S. In: Bilezikian J (ed.) The Parathyroids. Academic Press, New York, NY, USA, pp. 349–360.
45. Silverman S Jr, Gordan G, Grant T, Steinbach H, Eisenberg E, Manson R 1962 The dental structures in primary hyperparathyroidism. Studies in forty-two consecutive patients. Oral Surg Oral Med Oral Pathol **15**:426–436.
46. Daniels JS 2004 Primary hyperparathyroidism presenting as a palatal brown tumor. Oral Surg Oral Med Oral Pathol Oral Radiol Endod **98**:409–413.
47. Padbury AD Jr, Tozum TF, Taba M Jr, Ealba EL, West BT, Burney RE, Gauger PG, Giannobile WV, McCauley LK 2006 The impact of primary hyperparathyroidism on the oral cavity. J Clin Endocrinol Metab **91**:3439–3445.
48. Belsky JL, Hamer JS, Hubert JE, Insogna K, Johns W 2003 Torus palatinus: A new anatomical correlation with bone density in postmenopausal women. J Clin Endocrinol Metab **88**:2081–2086.
49. Boyden LM, Mao J, Belsky J, Mitzner L, Farhi A, Mitnick MA, Wu D, Insogna K, Lifton RP 2002 High bone density due to a mutation in LDL-receptor-related protein 5. N Engl J Med **346**:1513–1521.
50. Silverman S Jr, Ware WH, Gillooly C Jr 1968 Dental aspects of hyperparathyroidism. Oral Surg Oral Med Oral Pathol **26**:184–189.
51. Goultschin J, Eliezer K 1982 Resorption of cementum in renal osteodystrophy. J Oral Med **37**:84–86.
52. Frankenthal S, Nakhoul F, Machtei EE, Green J, Ardekian L, Laufer D, Peled M 2002 The effect of secondary hyperparathyroidism and hemodialysis therapy on alveolar bone and periodontium. J Clin Periodontol **29**:479–483.

© 2008 American Society for Bone and Mineral Research

SECTION XI

Appendix
(Section Editor: Roger Bouillon)

Please visit the Seventh Edition *Primer* Website at www.asbmrprimer.org for the expanded and updated Appendix.

© 2008 American Society for Bone and Mineral Research

Subject Index

Note: Page numbers followed by t indicate tables.

B

D

P